D1126266

CHAPMAN & HALL/CRC COMPUTER and INFORMATION SCIENCE SERIES

The Practical Handbook of INTERNET COMPUTING

CHAPMAN & HALL/CRC COMPUTER and INFORMATION SCIENCE SERIES

Series Editor: Sartaj Sahni

PUBLISHED TITLES

HANDBOOK OF SCHEDULING: ALGORITHMS, MODELS, AND PERFORMANCE ANALYSIS
Joseph Y-T. Leung

THE PRACTICAL HANDBOOK OF INTERNET COMPUTING
Munindar P. Singh

FORTHCOMING TITLES

HANDBOOK OF COMPUTATIONAL MOLECULAR BIOLOGY
Srinivas Aluru

HANDBOOK OF ALGORITHMS FOR WIRELESS AND MOBILE NETWORKS AND COMPUTING
Azzedine Boukerche

DISTRIBUTED SENSOR NETWORKS
S. Sitharama Iyengar and Richard R. Brooks

SPECULATIVE EXECUTION IN HIGH PERFORMANCE COMPUTER ARCHITECTURES
David Kaeli and Pen-Chung Yew

HANDBOOK OF DATA STRUCTURES AND APPLICATIONS
Dinesh P. Mehta and Sartaj Sahni

HANDBOOK OF BIOINSPIRED ALGORITHMS AND APPLICATIONS
Stephan Olariu and Albert Y. Zomaya

HANDBOOK OF DATA MINING
Sanjay Ranka

SCALABLE AND SECURE INTERNET SERVICE AND ARCHITECTURE
Cheng Zhong Xu

CHAPMAN & HALL/CRC COMPUTER and INFORMATION SCIENCE SERIES

The Practical Handbook of INTERNET COMPUTING

WAGGONER LIBRARY
Trevecca Nazarene Univ
DISCARD

Edited by
Munindar P. Singh

WAGGONER LIBRARY
TREVECCA NAZARENE UNIVERSITY

CHAPMAN & HALL/CRC

A CRC Press Company
Boca Raton London New York Washington, D.C.

Library of Congress Cataloging-in-Publication Data

Singh, Munindar P. (Munindar Paul), 1964-
The practical handbook of Internet computing / Munindar P. Singh.
p. cm. -- (Chapman & Hall/CRC computer and information science series)
Includes bibliographical references and index.
ISBN 0-58488-381-2 (alk. paper)
1. Internet programming--Handbooks, manuals, etc. 2. Electronic data processing--Distributed processing--Handbooks, manuals, etc. I. Title. II. Series

QA76.625.S555 2004
006.7'6—dc22

2004049256

This book contains information obtained from authentic and highly regarded sources. Reprinted material is quoted with permission, and sources are indicated. A wide variety of references are listed. Reasonable efforts have been made to publish reliable data and information, but the author and the publisher cannot assume responsibility for the validity of all materials or for the consequences of their use.

Neither this book nor any part may be reproduced or transmitted in any form or by any means, electronic or mechanical, including photocopying, microfilming, and recording, or by any information storage or retrieval system, without prior permission in writing from the publisher.

All rights reserved. Authorization to photocopy items for internal or personal use, or the personal or internal use of specific clients, may be granted by CRC Press LLC, provided that $1.50 per page photocopied is paid directly to Copyright Clearance Center, 222 Rosewood Drive, Danvers, MA 01923 USA. The fee code for users of the Transactional Reporting Service is ISBN 1-58488-381-2/05/$0.00+$1.50. The fee is subject to change without notice. For organizations that have been granted a photocopy license by the CCC, a separate system of payment has been arranged.

The consent of CRC Press LLC does not extend to copying for general distribution, for promotion, for creating new works, or for resale. Specific permission must be obtained in writing from CRC Press LLC for such copying.

Direct all inquiries to CRC Press LLC, 2000 N.W. Corporate Blvd., Boca Raton, Florida 33431.

Trademark Notice: Product or corporate names may be trademarks or registered trademarks, and are used only for identification and explanation, without intent to infringe.

Visit the CRC Press Web site at www.crcpress.com

© 2005 by CRC Press LLC

No claim to original U.S. Government works

International Standard Book Number 1-58488-381-2
Library of Congress Card Number 2004049256
Printed in the United States of America 1 2 3 4 5 6 7 8 9 0
Printed on acid-free paper

To my mother and to the memory of my father
and to their grandchildren:
my nephews, Kavindar and Avi
my son, Amitoj
my daughter, Amika

Foreword

The time is absolutely ripe for this handbook. The title uses both "Internet" and "practical" in a broad sense, as is appropriate for a rapidly moving and complex field. The Internet itself is alive and well, both as infrastructure and as a major part of modern economic, scientific, and social life. Although the public perception is less overheated, the tremendous significance of the technology and what it supports is difficult to overstate.

This volume addresses a broad swath of technologies and concerns relating to the Internet. In addition to discussing the underlying network, it covers relevant hardware and software technologies useful for doing distributed computing. Furthermore, this handbook reviews important applications, especially those related to the World Wide Web and the widening world of Web-based services.

A great deal has been learned about how to use the Internet, and this handbook provides practical advice based on experience as well as standards and theory. *Internet* here means not only IP-based networking, but also the information services that support the Web and services as well as exciting applications that utilize all of these. There is no single source of information that has this breadth today.

The Handbook addresses several major categories of problems and readers.

- The architecture articles look at the components that make the Internet work, including storage, servers, and networking. A deeper, realistic understanding of such topics is important for many who seek to utilize the Internet, and should provide reference information for other articles.
- The technology articles address specific problems and solution approaches that can lead to excellent results. Some articles address relatively general topics (such as usability, multiagent systems, and data compression). Many others are aimed broadly at technologies that utilize or support the Internet (such as directory services, agents, policies, and software engineering). Another group focuses on the World Wide Web, the most important application of the Internet (such as performance, caching, search, and security). Other articles address newer topics (such as the semantic Web and Web services).
- It is not straightforward to implement good applications on the Internet: there are complicated problems in getting appropriate reach and performance, and each usage domain has its own peculiar expectations and requirements. Therefore, the applications articles address a variety of cross-cutting capabilities (such as mobility, collaboration, and adaptive hypermedia), others address vertical applications (government, e-learning, supply-chain management, etc.).
- Finally, there is an examination of the Internet as a holistic system. Topics of privacy, trust, policy, and law are addressed as well as the structure of the Internet and of the Web.

In summary, this volume contains fresh and relevant articles by recognized experts about the key problems facing an Internet user, designer, implementer, or policymaker.

Stuart Feldman
Vice President, Internet Technology, IBM

Preface

I define the discipline of *Internet computing* in the broad sense. This discipline involves considerations that apply not only to the public Internet but also to Internet-based technologies that are used within a single physical organization, over a virtual private network within an organization, or over an extranet involving a closed group of organizations. Internet computing includes not only a study of novel applications but also of the technologies that make such applications possible. For this reason, this area brings together a large number of topics previously studied in diverse parts of computer science and engineering. The integration of these techniques is one reason that Internet computing is an exciting area both for practitioners and for researchers.

However, the sheer diversity of the topics that Internet computing integrates has also led to a situation where most of us have only a narrow understanding of the subject matter as a whole. Clearly, we cannot be specialists in everything, but the Internet calls for a breed of super-generalists who not only specialize in their own area, but also have a strong understanding of the rest of the picture.

Like most readers, I suspect, I have tended to work in my own areas of specialty. I frequently have been curious about a lot of the other aspects of Internet computing and always wanted to learn about them at a certain depth. For this reason, I jumped at the opportunity when Professor Sartaj Sahni (the series editor) invited me to edit this volume. In the past year or so, I have seen a fantastic collection of chapters take shape. Reading this book has been a highly educational experience for me, and I am sure it will be for you too.

Audience and Needs

Internet computing has already developed into a vast area that no single person can hope to understand fully. However, because of its obvious practical importance, many people need to understand enough of Internet computing to be able to function effectively in their daily work.

This is the primary motivation behind this book: a volume containing 56 contributions by a total of 107 authors from 51 organizations (in academia, industry, or government). The special features of this handbook are as follows:

- An exhaustive coverage of the key topics in Internet computing.
- Accessible, self-contained, yet definitive presentations on each topic, emphasizing the concepts behind the jargon.
- Chapters that are authored by the world's leading experts.

The intended readers of this book are people who need to obtain in-depth, authoritative introductions to the major Internet computing topics. These fall into the following main groups:

Practitioners who need to learn the key concepts involved in developing and deploying Internet computing applications and systems. This happens often when a project calls for some unfamiliar techniques.

Technical managers who need quick, high-level, but definitive descriptions of a large number of applications and technologies, so as to be able to conceive applications and architectures for their own special business needs and to evaluate technical alternatives.

Students who need accurate introductions to important topics that would otherwise fall between the cracks in their course work, and that might be needed for projects, research, or general interest.

Researchers who need a definitive guide to an unfamiliar area, e.g., to see if the area addresses some of their problems or even to review a scientific paper or proposal that impinges on an area outside their own specialty.

The above needs are not addressed by any existing source. Typical books and articles concentrate on narrow topics. Existing sources have the following limitations for our intended audience and its needs.

- Those targeted at practitioners tend to discuss specific tools or protocols but lack concepts and how they relate to the subject broadly.
- Those targeted at managers are frequently superficial or concentrated on vendor jargon.
- Those targeted at students cover distinct disciplines corresponding to college courses, but sidestep much of current practice. There is no overarching vision that extends across multiple books.
- Those targeted at researchers are of necessity deep in their specialties, but provide only a limited coverage of real-world applications and of other topics of Internet computing.

For this reason, this handbook was designed to collect definitive knowledge about all major aspects of Internet computing in one place. The topics covered will range from important components of current practice to key concepts to major trends. The handbook is an ideal comprehensive reference for each of the above types of reader.

The Contents

The handbook is organized into the following parts.

Applications includes 11 chapters dealing with some of the most important and exciting applications of Internet computing. These include established ones such as manufacturing and knowledge management, others such as telephony and messaging that are moving into the realm of the Internet, and still others such as entertainment that are practically brand new to computing at large.

Enabling Technologies deals with technologies many of which were originally developed in areas other than Internet computing, but have since crossed disciplinary boundaries and are very much a component of Internet computing. These technologies, described in ten chapters, enable a rich variety of applications. It is fair to state that, in general, these technologies would have little reason to exist were it not for the expansion of Internet computing.

Information Management brings together a wealth of technical and conceptual material dealing with the management of information in networked settings. Of course, all of the Internet is about managing information. This part includes 12 chapters that study the various aspects of representing and reasoning with information — in the sense of enterprise information, such as is needed in the functioning of practical enterprises and how they do business. Some chapters introduce key representational systems; others introduce process abstractions; and still others deal with architectural matters.

Systems and Utilities assembles eight chapters that describe how Internet-based systems function and some of their key components or utilities for supporting advanced applications. These include the peer-to-peer, mobile, and Grid computing models, directory services, as well as distributed and network systems technologies to deliver the needed Web system performance over existing networks.

Engineering and Management deals with how practical systems can be engineered and managed. The ten chapters range from considerations of engineering of usable applications to specifying and executing policies for system management to monitoring and managing networks, to building overlays such as virtual private networks to managing networks in general.

Systemic Matters presents five eclectic chapters dealing with the broader topics surrounding the Internet. As technologists, we are often unaware of how the Internet functions at a policy level, what impacts such technologies have on humanity, how such technologies might diffuse into the real world, and the legal questions they bring up.

Munindar P. Singh
Raleigh, North Carolina

Acknowledgments

This book would not have been possible but for the efforts of several people. I would like to thank Bob Stern, CRC editor, for shepherding this book, and Sylvia Wood for managing the production remarkably effectively.

I would like to thank Wayne Clark, John Waclawsky, and Erik Wilde for helpful discussions about the content of the book. Tony Rutkowski has prepared a comprehensive list of the standards bodies for the Internet. I would like to thank the following for their assistance with reviewing the chapters: Daniel Ariely, Vivek Bhargava, Mahmoud S. Elhaddad, Paul E. Jones, Mark O. Riedl, Ranjiv Sharma, Mona Singh, Yathiraj B. Udupi, Francois Vernadat, Jie Xing, Pınar Yolum, and Bin Yu.

As usual, I am deeply indebted to my family for their patience and accommodation of my schedule during the long hours spent putting this book together.

Editor

Munindar P. Singh is a professor of computer science at North Carolina State University. From 1989 through 1995, he was with the Microelectronics and Computer Technology Corporation (better known as MCC). Dr. Singh's research interests include multiagent systems and Web services. He focuses on applications in e-commerce and personal technologies. His 1994 book, *Multiagent Systems*, was published by Springer-Verlag. He authored several technical articles and co-edited *Readings in Agents*, published by Morgan Kaufmann in 1998, as well as several other books. His research has been recognized with awards and sponsorship from the National Science Foundation, IBM, Cisco Systems, and Ericsson.

Dr. Singh was the editor-in-chief of *IEEE Internet Computing* from 1999 to 2002 and continues to serve on its editorial board. He is a member of the editorial board of the *Journal of Autonomous Agents and Multiagent Systems* and of the *Journal of Web Semantics*. He serves on the steering committee for the *IEEE Transactions on Mobile Computing*.

Dr. Singh received a B.Tech. in computer science and engineering from the Indian Institute of Technology, New Delhi, in 1986. He obtained an M.S.C.S. from the University of Texas at Austin in 1988 and a Ph.D. in computer science from the same university in 1993.

Contributors

Aberer, Karl
EPFL
Lausanne, Switzerland

Agha, Gul
Department of Computer Science
University of Illinois–Urbana-Champaign
Urbana, IL

Aparico, Manuel
Saffron Technology
Morrisville, NC

Arroyo, Sinuhe
Institute of Computer Science
Next Web Generation, Research Group
Leopold Franzens University
Innsbruck, Austria

Atallah, Mikhail J.
Computer Science Department
Purdue University
West LaFayette, IN

Avancha, Sasikanth V.R.
Department of CSEE
University of Maryland–Baltimore
Baltimore, MD

Bertino, Elisa
Purdue University
Lafayette, IN

Berka, David
Digital Enterprise Research Institute
Innsbruck, Austria

Bhatia, Pooja
Purdue University
West Lafayette, IN

Bigus, Jennifer
IBM Corporation
Rochester, MN

Bigus, Joseph P.
IBM Corporation
Rochester, MN

Black, Carrie
Purdue University
West Lafayette, IN

Bouguettaya, Athman
Department of Computer Science
Virginia Polytechnic Institute
Falls Church, VA

Branting, Karl
LiveWire Logic, Inc.
Morrisville, NC

Brownlee, Nevil
University of California–San Diego
La Jolla, CA

Brusilovsky, Peter
Department of Information Science and Telecommunications
University of Pittsburgh
Pittsburgh, PA

claffy, kc
University of California–San Diego
La Jolla, CA

Camp, L. Jean
Kennedy School of Government
Harvard University
Cambridge, MA

Casati, Fabio
Hewlett-Packard Company
Palo Alto, CA

Cassel, Lillian N.
Department of Computing Sciences
Villanova University
Villanova, PA

Chandra, Surendar
University of Notre Dame
Notre Dame, IN

Chakraborty, Dipanjan
Department of Computer Science and Electrical Engineering
University of Maryland
Baltimore, MD

Chawathe, Sudarshan
Department of Computer Science
University of Maryland
College Park, MD

Clark, Wayne C.
Cary, NC

Curbera, Francisco
IBM T.J. Watson Research Center
Hawthorne, NY

Darrell, Woelk
Elastic Knowledge
Austin, TX

Dellarocas, Chrysanthos
Massachusetts Institute of Technology
Cambridge, MA

Dewan, Prasun
Department of Computer Sciences
University of North Carolina
Chapel Hill, NC

Ding, Ying
Digital Enterprise Research Institute
Innsbruck, Austria

Duftler, Matthew
IBM T.J. Watson Research Center
Hawthorne, NY

Fensel, Dieter
Digital Enterprise Research Institute
Innsbruck, Austria

Ferrari, Elena
University of Insubria
Como, Italy

Fisher, Mark
Semagix, Inc.
Athens, GA

Fox, Edward A.
Virginia Polytechnic Institute
Blacksburg, VA

Frikken, Keith
Purdue University
West Lafayette, IN

Gomez, Juan Miguel
Digital Enterprise Research Institute
Galway, Ireland

Greenstein, Shane
Kellogg School of Management
Northwestern University
Evanston, IL

Grosky, William I.
Department of Computer and Information Science
University of Michigan
Dearborn, MI

Gudivada, Venkat
Engineering and Computer Science
Marshall University
Huntington, WV

Hauswirth, Manfred
EPFL
Lausanne, Switzerland

Helal, Sumi
Computer Science and Engineering
University of Florida
Gainesville, FL

Huhns, Michael N.
CSE Department
University of South Carolina
Columbia, SC

Ioannidis, John
Columbia University
New York, NY

Ivezic, Nenad
National Institute of Standards and Technology
Gaithersburg, MD

Ivory, Melody Y.
The Information School
University of Washington
Seattle, WA

Iyengar, Arun
IBM Corporation
T.J. Watson Research Center
Hawthorne, NY

Jones, Albert
National Institute of Standards and Technology
Gaithersburg, MD

Joshi, Anupam
Department of Computer Science and Electrical Engineering,
University of Maryland
Baltimore, MD

Kashyap, Vipul
National Library of Medicine
Gaithersburg, MD

Keromytis, Angelos
Computer Science
Columbia University
New York, NY

Khalaf, Rania
IBM Corporation
T.J. Watson Research Center
Hawthorne, NY

Kulvatunyou, Boonserm
National Institute of Standards and Technology
Gaithersburg, MD

Lara, Ruben
Digital Enterprise Research Institute
Innsbruck, Austria

Lavender, Greg
Sun Microsystems, Inc.
Austin, TX

Lee, Choonhwa
College of Information and Communications
Hanyang University
Seoul, South Korea

Lee, Wenke
Georgia Institute of Technology
Atlanta, GA

Lester, James C.
LiveWire Logic
Morrisville, NC

Liu, Ling
College of Computing
Georgia Institute of Technology
Atlanta, GA

Lupu, Emil
Department of Computing
Imperial College London
London, U.K.

Madalli, Devika
Documentation Research and Training Centre
Indian Statistical Institute
Karnataka, IN

Medjahed, Brahim
Department of Computer Science
Virginia Polytechnic Institute
Falls Church, VA

Miller, Todd
College of Computing
Georgia Institute of Technology
Atlanta, GA

Mobasher, Bamshad
School of Computer Science, Telecommunications and Information Systems
Depaul University
Chicago, IL

Mott, Bradford
LiveWire Logic, Inc.
Morrisville, NC

Mukhi, Nirmal
IBM Corporation
T.J. Watson Research Center
Hawthorne, NY

Nagy, William
IBM Corporation
T.J. Watson Research Center
Hawthorne, NY

Nahum, Erich
IBM Corporation
T.J. Watson Research Center
Hawthorne, NY

Nejdl, Wolfgang
L3S and University of Hannover
Hannover, Germany

Ouzzani, Mourad
Department of Computer Science
Virginia Polytechnic Institute
Falls Church, VA

Overstreet, Susan
Purdue University
West Lafayette, IN

Perich, Filip
Computer Science and Electrical Engineering
University of Maryland
Baltimore, MD

Prince, Jeff
Northwestern University
Evanston, IL

Raghavan, Vijay V.
University of Louisiana
Lafayette, LA

Rezgui, Abdelmounaam
Department of Computer Science
Virginia Polytechnic Institute
Falls Church, VA

Risch, Tore J. M.
Department of Information Technology
University of Uppsala
Uppsala, Sweden

Rutkowski, Anthony M.
VeriSign, Inc.
Dulles, VA

Sahai, Akhil
Hewlett-Packard Company
Palo Alto, CA

Sahni, Sartaj
University of Florida
Gainesville, FL

Schulzrinne, Henning
Department of Computer Science
Columbia University
New York, NY

Schwerzmann, Jacqueline
St. Gallen, Switzerland

Shaikh, Anees
IBM Corporation
T. J. Watson Research Center
Hawthorne, NY

Sheth, Amit
Department of Computer Science
University of Georgia
Athens, GA

Singh, Munindar P.
Department of Computer Science
North Carolina State University
Raleigh, NC

Sloman, Morris
Imperial College London
Department of Computing
London, U.K.

Sobti, Sumeet
Department of Computer Science
Princeton University
Princeton Junction, NJ

Steen, Maarten van
Vrije University
Amsterdam, Netherlands

Stephens, Larry M.
Swearingen Engineering Center
CSE Department
University of South Carolina
Columbia, SC

Subramanian, Mani
College of Computing
Georgia Institute of Technology
Atlanta, GA

Suleman, Hussein
University of Cape Town
Cape Town, South Africa

Sunderam, Vaidy
Department of Math and Computer Science
Emory University
Atlanta, GA

Tai, Stefan
IBM Corporation
Hawthorne, NY

Tewari, Renu
IBM Corporation
T.J. Watson Research Center
Hawthorne, NY

Touch, Joseph Dean
USC/ISI
Marina del Rey CA

Varela, Carlos
Department of Computer Science
Rensselaer Polytechnic Institute
Troy, NY

Vernadat, Francois B.
Thionville, France

Wahl, Mark
Sun Microsystems, Inc.
Austin, TX

Wams, Jan Mark S.
Vrije University
Amsterdam, Netherlands

Wellman, Michael P.
University of Michigan
Ann Arbor, MI

Wilde, Erik
Zurich, Switzerland

Williams, Laurie
Department of Computer Science
North Carolina State University
Raleigh, NC

Witten, Ian H.
Department of Computer Science
University of Waikato
Hamilton, New Zealand

Woelk, Darrell
Telcordia Technologies
Austin, TX

Wu, Zonghuan
University of Lousiana
Lafayette, LA

Yee, Ka-Ping
Computer Science
University of California–Berkeley
Berkeley, CA

Yianilos, Peter N.
Department of Computer Science
Princeton University
Princeton Junction, NJ

Yolum, Pınar
Vrije Universiteit
Amsterdam, The Netherlands

Young, Michael R.
Department of Computer Science
North Carolina State University
Raleigh, NC

Yu, Eric Siu-Kwong
Department of Computer Science
University of Toronto
Toronto, ON

Contents

PART 1 Applications

1 Adaptive Hypermedia and Adaptive Web 1-1
Peter Brusilovsky and Wolfgang Nejdl

2 Internet Computing Support for Digital Government 2-1
Athman Bouguettaya, Abdelmounaam Rezgui, Brahim Medjahed, and Mourad Ouzzani

3 E-Learning Technology for Improving Business Performance and Lifelong Learning 3-1
Darrell Woelk

4 Digital Libraries 4-1
Edward A. Fox, Hussein Suleman, Devika Madalli, and Lillian Cassel

5 Collaborative Applications 5-1
Prasun Dewan

6 Internet Telephony 6-1
Henning Schulzrinne

7 Internet Messaging 7-1
Jan Mark S. Wams and Maarten van Steen

8 Internet-Based Solutions for Manufacturing Enterprise Systems Interoperability — A Standards Perspective 8-1
Nenad Ivezic, Boonserm Kulvatunyou, and Albert Jones

9 Semantic Enterprise Content Management 9-1
Mark Fisher and Amit Sheth

10 Conversational Agents 10-1
James Lester, Karl Branting, and Bradford Mott

11 Internet-Based Games 11-1
R. Michael Young

PART 2 Enabling Technologies

12 Information Retrieval 12-1
Vijay V. Raghavan, Venkat N. Gudivada, Zonghuan Wu, and William I. Grosky

13 Web Crawling and Search 13-1
Todd Miller and Ling Liu

14 Text Mining 14-1
Ian H. Witten

15 Web Usage Mining and Personalization 15-1
Bamshad Mobasher

16 Agents 16-1
Joseph P. Bigus and Jennifer Bigus

17 Multiagent Systems for Internet Applications 17-1
Michael N. Huhns and Larry M. Stephens

18 Concepts and Practice of Personalization 18-1
Manuel Aparicio IV and Munindar P. Singh

19 Online Marketplaces 19-1
Michael P. Wellman

20 Online Reputation Mechanisms 20-1
Chrysanthos Dellarocas

21 Digital Rights Management 21-1
Mikhail Atallah, Keith Frikken, Carrie Black, Susan Overstreet, and Pooja Bhatia

PART 3 Information Management

22 Internet-Based Enterprise Architectures 22-1
Francois B. Vernadat

23 XML Core Technologies 23-1
Erik Wilde

24 Advanced XML Technologies 24-1
Erik Wilde

25 Semistructured Data in Relational Databases 25-1
Sudarshan Chawathe

26 Information Security 26-1
Elisa Bertino and Elena Ferrari

27 Understanding Web Services 27-1
Rania Khalaf, Francisco Curbera, William A. Nagy, Stefan Tai, Nirmal Mukhi, and Matthew Duftler

28 Mediators for Querying Heterogeneous Data 28-1
Tore Risch

29 Introduction to Web Semantics 29-1
Munindar P. Singh

30 Information Modeling on the Web 30-1
Vipul Kashyap

31 Semantic Aspects of Web Services 31-1
Sinuhe Arroyo, Ruben Lara, Juan Miguel Gomez, David Berka, Ying Ding, and Dieter Fensel

32 Business Process: Concepts, Systems, and Protocols 32-1
Fabio Casati and Akhil Sahai

33 Information Systems 33-1
Eric Yu

PART 4 Systems and Utilities

34 Internet Directories Using the Lightweight Directory Access Protocol 34-1
Greg Lavender and Mark Wahl

35 Peer-to-Peer Systems 35-1
Karl Aberer and Manfred Hauswirth

36 Data and Services for Mobile Computing 36-1
Sasikanth Avancha, Dipanjan Chakraborty, Filip Perich, and Anupam Joshi

37 Pervasive Computing 37-1
Sumi Helal and Choonhwa Lee

38 Worldwide Computing Middleware 38-1
Gul A. Agha and Carlos A. Varela

39 Metacomputing and Grid Frameworks 39-1
Vaidy Sunderam

40 Improving Web Site Performance 40-1
Arun Iyengar, Erich Nahum, Anees Shaikh, and Renu Tewari

41 Web Caching, Consistency, and Content Distribution 41-1
Arun Iyengar, Erich Nahum, Anees Shaikh, and Renu Tewari

42 Content Adaptation and Transcoding 42-1
Surendar Chandra

PART 5 Engineering and Management

43 Software Engineering for Internet Applications 43-1
Laurie Williams

44 Web Site Usability Engineering 44-1
Melody Y. Ivory

45 Distributed Storage 45-1
Sumeet Sobti and Peter N. Yianilos

46 System Management and Security Policy Specification 46-1
Morris Sloman and Emil Lupu

47 Distributed Trust 47-1
John Ioannidis and Angelos D. Keromytis

48 An Overview of Intrusion Detection Techniques 48-1
Wenke Lee

49 Measuring the Internet 49-1
Nevil Brownlee and kc claffy

50 What is Architecture? 50-1
Wayne Clark and John Waclawsky

51 Overlay Networks 51-1
Joseph D. Touch

52 Network and Service Management 52-1
Mani Subramanian

PART 6 Systemic Matters

53 Web Structure 53-1
Pınar Yolum

54 The Internet Policy and Governance Ecosystem 54-1
Anthony M. Rutkowski

55 Human Implications of Internet Technologies 55-1
L. Jean Camp and Ka-Ping Yee

56 The Geographical Diffusion of the Internet in the U.S. 56-1
Shane M. Greenstein and Jeff Prince

57 Intellectual Property, Liability, and Contract 57-1
Jacqueline Schwerzmann

Index I-1

PART 1

Applications

1

Adaptive Hypermedia and Adaptive Web

Peter Brusilovsky

Wolfgang Nejdl

CONTENTS

Abstract 1-1
1.1 Introduction 1-1
1.2 Adaptive Hypermedia 1-2
1.2.1 What Can Be Adapted in Adaptive Web and Adaptive Hypermedia 1-3
1.2.2 Adaptive Navigation Support 1-3
1.3 Adaptive Web 1-7
1.3.1 Adaptive Hypermedia and Mobile Web 1-7
1.3.2 Open Corpus Adaptive Hypermedia 1-8
1.3.3 Adaptive Hypermedia and the Semantic Web 1-8
1.4 Conclusion 1-12
References 1-12

Abstract

Adaptive Systems use explicit user models representing user knowledge, goals, interests, etc., that enable them to tailor interaction to different users. Adaptive hypermedia and Adaptive Web have used this paradigm to allow personalization in hypertext systems and the WWW, with diverse applications ranging from museum guides to Web-based education. The goal of this chapter is to present the history of adaptive hypermedia, introduce a number of classic but popular techniques, and discuss emerging research directions in the context of the Adaptive and Semantic Web that challenge adaptive hypermedia researchers in the new Millennium.

1.1 Introduction

Web systems suffer from an inability to satisfy the heterogeneous needs of many users. For example, Web courses present the same static learning material to students with widely differing knowledge of the subject. Web stores offer the same selection of "featured items" to customers with different needs and preferences. Virtual museums on the Web offer the same "guided tour" to visitors with different goals and interests. Health information sites present the same information to readers with different health problems. Adaptive hypermedia offers an alternative to the traditional "one-size-fits-all" approach. The use of adaptive hypermedia techniques allows Web-based systems to adapt their behavior to the goals, tasks, interests, and other features of individual users.

Adaptive hypermedia systems belong to the class of user-adaptive software systems [Schneider-Hufschmidt et al., 1993]. A distinctive feature of an adaptive system is an explicit user model that represents user knowledge, goals, interests, and other features that enable the system to distinguish among different

1-58488-381-2/05/$0.00+$1.50
© 2005 by CRC Press LLC

users. An adaptive system collects data for the user model from various sources that can include implicitly observing user interaction and explicitly requesting direct input from the user. The user model is employed to provide an adaptation effect, i.e., tailor interaction to different users in the same context. Adaptive systems often use intelligent technologies for user modeling and adaptation.

Adaptive hypermedia is a relatively young research area. Starting with a few pioneering works on adaptive hypertext in early 1990, it now attracts many researchers from different communities such as hypertext, user modeling, machine learning, natural language generation, information retrieval, intelligent tutoring systems, cognitive science, and Web-based education. Today, adaptive hypermedia techniques are used almost exclusively for developing various adaptive Web-based systems. The goal of this chapter is to present the history of adaptive hypermedia, introduce a number of classic but popular techniques, and discuss emerging research directions in the context of the Adaptive Web that challenge adaptive hypermedia researchers in the new Millennium.

1.2 Adaptive Hypermedia

Adaptive hypermedia research can be traced back to the early 1990s. At that time, a number of research teams had begun to explore various ways to adapt the output and behavior of hypertext systems to individual users. By the year 1996, several innovative adaptive hypermedia techniques had been developed, and several research-level adaptive hypermedia systems had been built and evaluated. A collection of papers presenting early adaptive hypermedia systems is available in [Brusilovsky et al., 1998b]. A review of early adaptive hypermedia systems, methods, and techniques is provided by [Brusilovsky, 1996].

The year of 1996 can be considered a turning point in adaptive hypermedia research. Before this time, research in this area was performed by a few isolated teams. However, since 1996, adaptive hypermedia has gone through a period of rapid growth. In 2000 a series of international conferences on adaptive hypermedia and adaptive Web-based systems was established. The most recent event in this series, AH'2002, has assembled more than 200 researchers. Two major factors account for this growth of research activity: The maturity of adaptive hypermedia as a research field and the maturity of the Word Wide Web as an application platform.

The early researchers were generally not aware of each other's work. In contrast, many papers published since 1996 cite earlier work and usually suggest an elaboration or an extension of techniques suggested earlier. Almost all adaptive hypermedia systems reported by 1996 were "classic hypertext" laboratory systems developed to demonstrate and explore innovative ideas. In contrast, almost all systems developed since 1996 are Web-based adaptive hypermedia systems, with many of them being either practical systems or research systems developed for real-world settings.

The change of the platform from classic hypertext and hypermedia to the Web has also gradually caused a change both in used techniques and typical application areas. The first "pre-Web" generation of adaptive hypermedia systems explored mainly adaptive presentation and adaptive navigation support, and concentrated on modeling user knowledge and goals [Brusilovsky et al., 1998b]. Empirical studies have shown that adaptive navigation support can increase the speed of navigation [Kaplan et al., 1993] and learning [Brusilovsky and Pesin, 1998], whereas adaptive presentation can improve content understanding [Boyle and Encarnacion, 1994]. The second "Web" generation brought classic technologies to the Web and explored a number of new technologies based on modeling user interests such as adaptive content selection and adaptive recommendation [Brusilovsky et al., 2000]. The first empirical studies report the benefits of using these technologies [Billsus et al., 2002]. The third New Adaptive Web generation strives to move adaptive hypermedia beyond traditional borders of closed corpus desktop hypermedia systems, embracing such modern Web trends as mobile Web, open Web, and Semantic Web.

Early adaptive hypermedia systems were focusing almost exclusively on such academic areas as education or information retrieval. Although these are still popular application areas for adaptive hypermedia techniques, most recent systems are exploring new, promising application areas such as kiosk-style information systems, e-commerce, medicine, and tourism. A few successful industrial systems [Billsus et al., 2002; Fink et al., 2002; Weber et al., 2001] show the commercial potential of the field.

1.2.1 What Can Be Adapted in Adaptive Web and Adaptive Hypermedia

In different kinds of adaptive systems, adaptation effects could be different. Adaptive Web systems are essentially Webs of connected information items that allow users to navigate from one item to another and search for relevant ones. The adaptation effect in this reasonably rigid context is limited to three major adaptation technologies: adaptive content selection, adaptive navigation support, and adaptive presentation. The first of these three technologies comes from the field of adaptive information retrieval (IR) and is associated with a search-based access to information. When the user searches for relevant information, the system can adaptively select and prioritize the most relevant items. The second technology was introduced by adaptive hypermedia systems [Brusilovsky, 1996] and is associated with a browsing-based access to information. When the user navigates from one item to another, the system can manipulate the links (e.g., hide, sort, annotate) to guide the user adaptively to most relevant information items. The third technology has some deep roots in the research on adaptive explanation and adaptive presentation in intelligent systems [Moore and Swartout, 1989; Paris, 1988]. It deals with presentation, not access to information. When the user gets to a particular page, the system can present its content adaptively.

Both adaptive presentation (content-level adaptation) and adaptive navigation support (link-level adaptation) have been extensively explored in a number of adaptive hypermedia projects. Early works on adaptive hypermedia were focused more on adaptive text presentation [Beaumont, 1994; Boyle and Encarnacion, 1994]. Later, the gradual growth of the number of nodes managed by a typical adaptive hypermedia system (especially, Web hypermedia) has shifted the focus of research to adaptive navigation support techniques. Since adaptive navigation support is the kind of adaptation that is most specific to hypertext context, we provide a detailed review of several major techniques in the next subsection.

1.2.2 Adaptive Navigation Support

The idea of adaptive navigation support techniques is to help users find their paths in hyperspace by adapting link presentation to the goals, knowledge, and other characteristics of an individual user. These techniques can be classified in several groups according to the way they adapt presentation of links. These groups of techniques are traditionally considered as different technologies for adapting link presentation. The most popular technologies are direct guidance, ordering, hiding, annotation, and generation.

Direct guidance is the simplest technology of adaptive navigation support. It can be applied in any system that can suggest the "next best" node for the user to visit according to the user's goals, knowledge, and other parameters represented in the user model. To provide direct guidance, the system can outline visually the link to the "best" node as done in Web Watcher [Armstrong et al., 1995], or present an additional dynamic link (usually called "next") that is connected to the next best node as done in the InterBook [Brusilovsky et al., 1998a] or ELM-ART [Weber and Brusilovsky, 2001] systems. The former way is clearer; the latter is more flexible because it can be used to recommend the node that is not connected directly to the current one (and not represented on the current page). A problem of direct guidance is that it provides no support to users who would not like to follow the system's suggestions. Direct guidance is useful, but it should be employed together with one of the more "supportive" technologies that are described below. An example of an InterBook page with direct guidance is shown on Figure 1.1.

The idea of *adaptive ordering* technology is to order all the links of a particular page according to the user model and some user-valuable criteria: the closer to the top, the more relevant the link is. Adaptive ordering has a more limited applicability: It can be used with noncontextual (freestanding) links, but it can hardly be used for indexes and content pages (which usually have a stable order of links), and can never be used with contextual links (hot words in text) and maps. Another problem with adaptive ordering is that this technology makes the order of links unstable: It may change each time the user enters the page. For both reasons this technology is most often used now for showing new links to the user in conjunction with link generation. Experimental research [Kaplan et al., 1993] showed that adaptive ordering can significantly reduce navigation time in IR hypermedia applications.

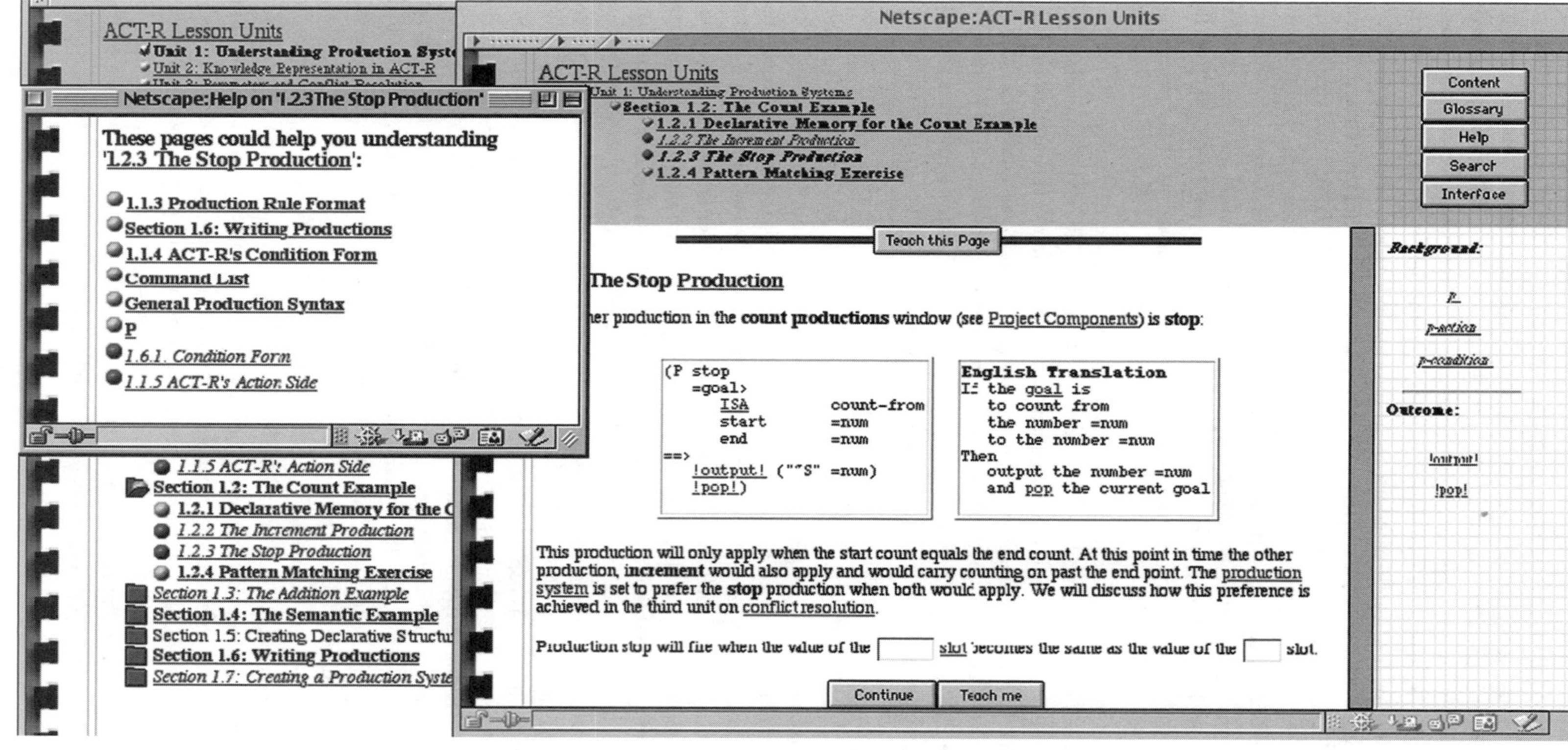

FIGURE 1.1 Adaptive guidance with "Teach Me" button and adaptive annotation with colored bullets in InterBook system.

The idea of navigation support by *hiding* is to restrict the navigation space by hiding, removing, or disabling links to irrelevant pages. A page can be considered as not relevant for several reasons, for example, if it is not related to the user's current goal or if it presents materials which the user is not yet prepared to understand. Hiding protects users from the complexity of the unrestricted hyperspace and reduces their cognitive overload. Early adaptive hypermedia systems have used a simple way of hiding: essentially removing the link together with the anchor from a page. De Bra and Calvi [1998] called this link removal and have suggested and implemented several other variants for link hiding. A number of studies of link hiding demonstrated that users are unhappy when previously available links become invisible or disabled. Today, link hiding is mostly used in reverse order — as gradual link enabling when more and more links become visible to the user.

The idea of *adaptive annotation* technology is to augment links with some form of comments that can tell the user more about the current state of the nodes behind the annotated links. These annotations can be provided in textual form or in the form of visual cues using, for example, different font colors [De Bra and Calvi, 1998], font sizes [Hohl et al., 1996], and font types [Brusilovsky et al., 1998a] for the link anchor or different icons next to the anchor [Brusilovsky et al., 1998a; Henze and Nejdl, 2001; Weber and Brusilovsky, 2001]. Several studies have shown that adaptive link annotation is an effective way of navigation support. For example, Brusilovsky and Pesin [1998] have compared the performance of students who were attempting to achieve the same educational goal using ISIS-Tutor with and without adaptive annotation. The groups working with enabled adaptive navigation support were able to achieve this educational goal almost twice as fast and with significantly smaller navigation overhead. Another study [Weber and Brusilovsky, 2001] reported that advanced users of a Web-based educational system have stayed with the system significantly longer if provided with annotation-based adaptive navigation support.

Annotation can be naturally used with all possible forms of links. This technology supports stable order of links and avoids problems with incorrect mental maps. For all the above reasons, adaptive annotation has gradually grown into the most often used adaptive annotation technology.

One of the most popular methods of adaptive link annotation is the traffic light metaphor that is used primarily in educational hypermedia systems. A green bullet in front of a link indicates recommended readings, whereas a red bullet indicates that the student might not have enough knowledge to understand the information behind the link yet. Other colors like yellow or white may indicate other educational states. This approach was pioneered in 1996 in ELM-ART and InterBook systems [Brusilovsky et al., 1998a; Weber and Brusilovsky, 2001] and used later in numerous other adaptive educational hypermedia systems. Figure 1.1 shows adaptive annotation in InterBook [Brusilovsky et al., 1998a] and Figure 1.2 in KBS-HyperBook system [Henze and Nejdl, 2001].

The last of the major adaptive navigation support technologies is *link generation.* It became popular in Web hypermedia in the context of recommender systems. Unlike pure annotation, sorting, and hiding technologies that adapt the way to present preauthored links, link generation creates new, nonauthored links for a page. There are three popular kinds of link generation: (1) discovering new useful links between documents and adding them permanently to the set of existing links; (2) generating links for similarity-based navigation between items; and (3) dynamic recommendation of relevant links. The first two kinds have been present in the neighboring research area of intelligent hypertext for years. The third kind is relatively new but already well-explored in the areas of IR hypermedia, online information systems, and even educational hypermedia.

Direct guidance, ordering, hiding, annotation, and generation are the primary technologies for adaptive navigation support. While most existing systems use exactly one of these ways to provide adaptive navigation support, these technologies are not mutually exclusive and can be used in combinations. For example, InterBook [Brusilovsky et al., 1998a] uses direct guidance, generation, and annotation. Hypadapter [Hohl et al., 1996] uses ordering, hiding, and annotation. Link generation is used almost exclusively with link ordering. Direct guidance technology can be naturally used in combination with any of the other technologies.

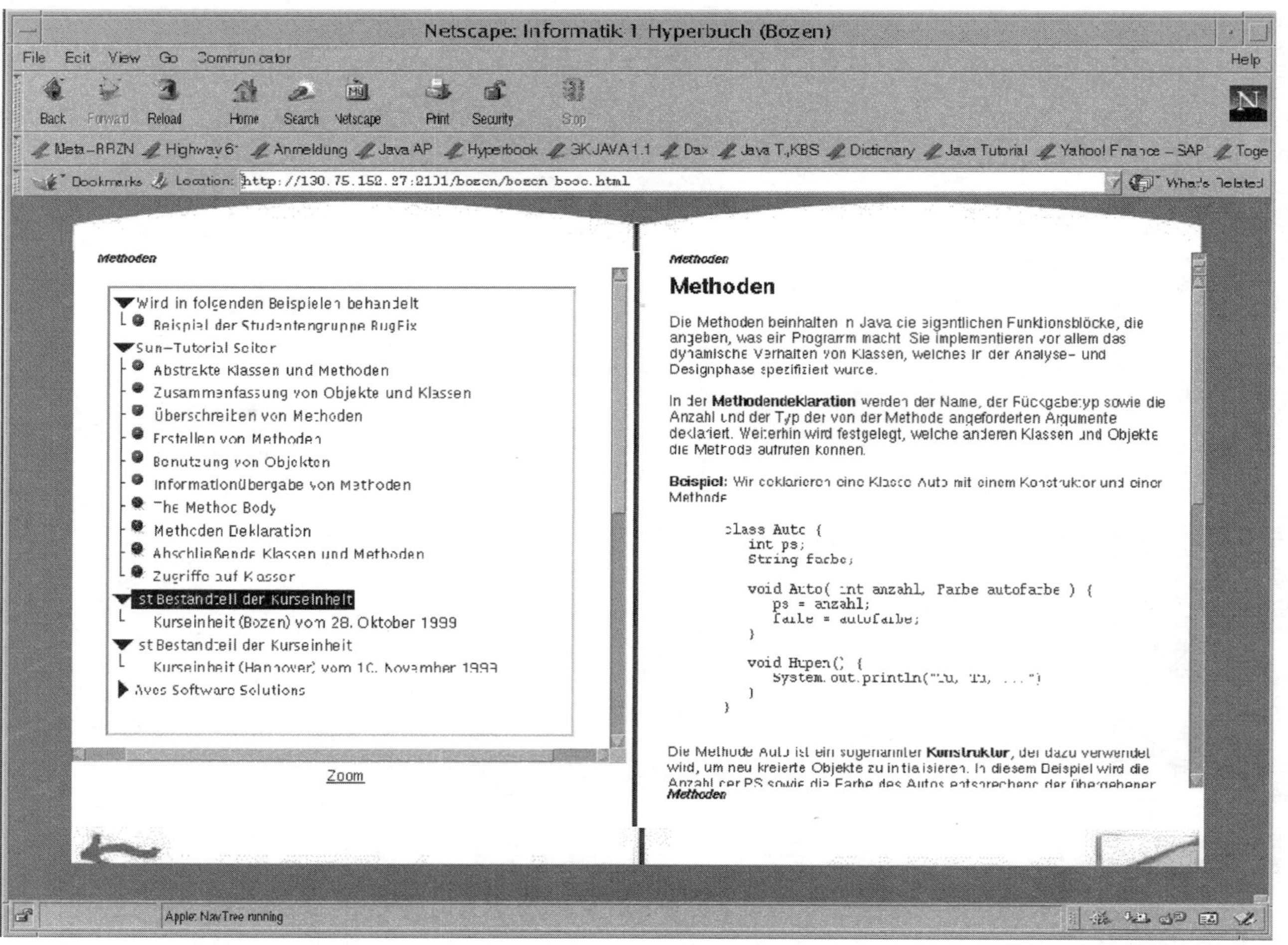

FIGURE 1.2 Adaptive annotation in KBS-HyperBook system.

1.3 Adaptive Web

The traditional direction of adaptive hypermedia research — bringing adaptivity to classic hypermedia systems — is being quite well explored. The recent 2 to 3 years have added few new methods, or techniques, or ideas. Most of the work being performed in this direction is now centered on developing specific variations of known methods and techniques and in developing practical systems. Although such work is important, some researchers may be more interested in expanding adaptive hypermedia beyond its traditional borders. We are now witnessing at least three exciting new directions of work towards an Adaptive Web, focused on mobile devices, open hypermedia, and semantic Web technologies. In the following three subsections we provide a brief overview of research currently being performed in these emerging areas. Most space is allocated to the Semantic Web direction, which is the most recent and probably the most challenging of the three.

1.3.1 Adaptive Hypermedia and Mobile Web

The work on adaptive hypermedia on handheld and mobile devices that was not originally connected to Web hypermedia is quickly moving towards an Adaptive Mobile Web. From one side, various handheld and mobile devices such as portable computers or personal information managers (PIM) provide an attractive platform to run a number of hypermedia applications such as aircraft maintenance support systems [Brusilovsky and Cooper, 1999], museum guides [Not et al., 1998], or news delivery systems [Billsus et al., 2002]. From another side, the need for adaptation is especially evident in Mobile Web applications. Technologies such as adaptive presentation, adaptive content selection, and adaptive navigation support that were an attractive luxury for desktop platform with large screens, high bandwidth, and rich interface become a necessity for mobile handheld devices [Billsus et al., 2002].

Mobile Web has brought two major research challenges to the adaptive hypermedia community. First, most of mobile devices have relatively small screens. Advanced adaptive presentation and adaptive navigation support techniques have to be developed to make a "small-screen interface" more useable. Second, user location and movement in a real space becomes an important and easy-to-get (with the help of such devices as GPS) part of a user model. A meaningful adaptation to user position in space (and time) is a new opportunity that has to be explored.

More generally, mobile devices have introduced a clear need to extend the borders of adaptation. In addition to adaptation to the personal characteristics of users, they demanded adaptation to the user's environment. Because users of the same server-side Web application can reside virtually everywhere and use different equipment, adaptation to the user's environment (location, time, computing platform, bandwidth) has become an important issue. A number of current adaptive hypermedia systems suggested some techniques to adapt to both the user location and the user platform. Simple adaptation to the platform (hardware, software, network bandwidth) usually involves selecting the type of material and media (i.e., still picture vs. movie) to present the content [Joerding, 1999]. More advanced technologies can provide considerably different interface to the users with different platforms and even use platform limitation to the benefits of user modeling. For example, a Palm Pilot version of AIS [Billsus and Pazzani, 2000] requires the user to explicitly request the following pages of a news story, thus sending a message to a system that the story is of interest. This direction of adaptation will certainly remain important and will likely provoke new interesting techniques. Adaptation to user location may be successfully used by many online information systems: SWAN [Garlatti and Iksal, 2000] demonstrates a successful use of user location for information filtering in a marine information system.

The currently most exciting kind of adaptive mobile Web applications are mobile handheld guides. Mobile adaptive guides were pioneered by the HYPERAUDIO project [Not et al., 1998] well before the emergence of the mobile Web and are now becoming very popular. Various recent projects explore a number of interesting adaptation techniques that take into account user location, direction of sight, and movements in both museum guide [Oppermann and Specht, 1999] and city guide [Cheverst et al., 2002] contexts.

1.3.2 Open Corpus Adaptive Hypermedia

Currently, almost all adaptive hypermedia systems work with a closed corpus set of documents assembled together at design time. This closed corpus is known to the system: knowledge about documents and links are traditionally obtained by manual indexing of documents and fragments with the user's possible knowledge, goals, background, etc. This approach cannot be applied to an open corpus such as open Web. To deal with the open Web, an adaptive hypermedia system should be able to extend its set of documents with minimal efforts from the human side. A simple approach to doing it is manually "extendable" hypermedia, which allows an adaptive system to take into account documents that have not been indexed at the design time. The real challenge is to develop systems that are able to extract some meaning from an open corpus of documents and work with the open Web without the help of a human indexer. The research on open corpus adaptive hypermedia has benefited from the existing streams of work on open hypermedia and ontologies.

Open hypermedia research, for a long time, evolved parallel to adaptive hypermedia research as well as to the World Wide Web, and have focused on hypermedia architectures that separate links from documents and allow the processing of navigational structures independent of the content objects served by the hypermedia system (for example, in the Microcosm system [Fountain et al., 1990], Chimera [Anderson et al., 1994], and Hyper-G [Andrews et al., 1995]).

Recently, the focus on research has led to approaches incorporating adaptive functionalities in such an open hypermedia environment. Bailey et al. [2002] for example, build on the Auld Linky system [Michaelides et al., 2001], a contextual link server that stores and serves the appropriate data structures for expressing information about content (data objects, together with context and behavior objects) and navigational structures (link structures, together with association and reference objects), among others. This makes it possible to provide basic adaptive functionalities (including link annotation or link hiding) and serve hypermedia content based on distributed content. Some pieces, however, still remain centralized in this architecture, e.g., the main piece of the hypermedia engine — the link server.

Once we want to integrate materials from different authors or heterogeneous sources, it becomes important to use commonly agreed sets of topics to index and characterize the content of the hypermedia pages integrated in the system [Henze and Nejdl, 2001, 2002]. This is addressed through the use of *ontologies*, which are "formal explicit specifications of shared conceptualizations" [Gruber, 1993]. In the process of ontology construction, communities of users and authors agree on a topic hierarchy, possibly with additional constraints expressed in first-order logic, which enables interoperability and exchangeability between different sources.

Furthermore, for really open adaptive hypermedia systems, which "operate on an open corpus of documents" [Henze and Nejdl, 2002], the data structures and metadata should be compatible with those defined by current Web standards. Therefore, the next step is to investigate which metadata standards and representation languages should be used in the context of the World Wide Web, and whether centralized link servers can be substituted by decentralized solutions.

1.3.3 Adaptive Hypermedia and the Semantic Web

The basic idea of the hypermedial/hypertext paradigm is that information is interconnected by links, and different information items can be accessed by navigating through this link structure. The World Wide Web, by implementing this basic paradigm in a simple and efficient manner, has made this model the standard way for information access on the Internet. Obviously, in an open environment like the World Wide Web, adaptive functionalities like navigational hints and other personalization features would arguably be even more useful. To do this, however, we have to extend adaptation functionalities from the closed architectures of conventional systems to an open environment, and we have to investigate the possibilities of providing additional metadata based on Semantic Web formalisms in this open environment as input to make these adaptation functionalities possible.

In the previous section we discussed how hypermedia system architectures can be extended into an open hypermedia environment. This allows us to accommodate distributed content, but still relies on a central server and central data structures to integrate and serve this content. Peer-to-Peer infrastructures go a step further and allow the provision of distributed services and content by a set of distributed peers, based on decentralized algorithms and data structures. An example of such a peer-to-peer infrastructure is the Edutella network (see, e.g., Nejdl et al., 2002a,b), which implements a peer-to-peer infrastructure based on Semantic Web technologies. In this network, information is provided by independent peers who can interchange information with others. Both data and metadata can be distributed in an arbitrary manner; data can be arbitrary digital resources, including educational content.

The crucial questions in such an environment are how to use standardized metadata to describe and classify information and to describe knowledge, preferences, and experiences of users accessing this information. Last but not least, adaptive functionalities as described in the previous sections have now to be implemented as queries in this open environment. Though many questions still remain to be answered in this area, we will sketch some possible starting points in the following text (see also Dolog et al., 2003).

1.3.3.1 Describing Educational Resources

Describing our digital resources is the first step in providing a (distributed) hypermedia system. One of the most common metadata schemas on the Web today is the Dublin Core Schema (DC) by the Dublin Core Metadata Initiative (DCMI). DCMI is an organization dedicated to promoting the widespread adoption of interoperable metadata standards and developing specialized metadata vocabularies for describing resources that enable more intelligent information discovery for digital resources.

Each Dublin Core element is defined using a set of 15 attributes for the description of data elements, including Title, Identifier, Language, and Comment. To annotate the author of a learning resource, DC suggests using the element creator, and thus we write, for example, *dc:creator(Resource) = nejdl.* Whereas Simple Dublin Core uses only the elements from the Dublin Core metadata set as attribute-value pairs, Qualified Dublin Core (DCQ) employs additional qualifiers to further refine the meaning of a resource. Since Dublin Core is designed for metadata describing any kind of (digital) resource, it pays no heed to the specific needs we encounter in describing learning resources. Therefore, the Learning Objects Metadata Standard (LOM) [IEEE-LTSC] by the IEEE Learning Technology Standards Committee (LTSC) was established as an extension of Dublin Core.

These metadata can be encoded in RDF [Lassila and Swick, 1999; Brickley and Guha, 2003], which makes distributed annotation of resources possible. Using RDF Schema, we can represent the schemas as discussed above, i.e., as the vocabulary to describe our resources. Specific properties are then represented as RDF triples *<subject, property, value>*, where *subject* identifies the resource we want to describe (using a URI), *property* specifies what property we use (e.g., *dc:creator),* and *value* the specific value, expressed as a string (e.g., "Nejdl") or another URI. We can then describe resources on the Web as shown in the following example:

```
<rdf:Description rdf:about="http://www.xyz.org/ai-2.html">
      <dc:title>Artificial Intelligence, Part 2</dc:title>
      <dc:author>Wolfgang Nejdl</dc:author>
      <dcq:requires resource="http://www.xyz.org/ai-1.html"/>
      <dcq:hasPart resource="http://www.xyz.org/ai-22.html"/>
</rdf:Description>
```

We can use any properties defined in the schemas we use, possibly mix different schemas without any problem, and also relate different resources to each other, for example, when we want to express interdependencies between these resources, hierarchical relationships, or others.

1.3.3.2 Topic Ontologies for Content Classification

Personalized access means that resources are tailored according to some relevant aspects of the user. Which aspects of the user are important or not depends on the personalization domain. For educational scenarios, it is important to take into account aspects such as whether the user is student or a teacher, whether he wants to obtain a certain qualification, has specific preferences, and, of course, what his knowledge level is for the topics covered in the course.

Preferences about learning materials can be easily exploited, especially if they coincide directly with the metadata and metadata values used. For users preferring Powerpoint presentations, for example, we can add the constraint *dc:format(Resource) = powerpoint* to queries searching appropriate learning materials.

Taking user knowledge about topics covered in the course into account is more tricky. The general idea is that we annotate each document by the topics covered in this document. Topics can be covered by sets of documents, and we will assume that a user fully knows a topic if he understands all documents annotated with this topic. However, though the standards we have just explored only provide one attribute *(dc:subject)* for annotating resources with topics, in reality we might want to have different kinds of annotations to distinguish between just mentioning a topic, introducing a topic, and covering a topic. In the following, we will simply assume that *dc:subject* is used for "covered" topics, but additional properties for these annotations might be useful in other contexts. Furthermore, we have to define which sets of documents for a given subject are necessary to "fully cover" a topic.

Additionally, it is obvious that self-defined keywords cannot be used, and we have to use an ontology for annotating documents and describing user knowledge (see also Henze and Nejdl, 2002). Defining a private ontology for a specific field works only in the closed microworld of a single university, so we have to use shared ontologies. One such ontology is the ACM Computer Classification system ([ACM, 2002]) that has been used by the Association for Computing Machinery since several decades to classify scientific publications in the field of computer science. This ontology can be described in RDF such that each entry in the ontology can be referenced by a URI and can be used with the *dc:subject* property as follows:

```
<rdf:Description rdf:about- "http://www.xyz.org/ai-2.html">
    <dc:subject resource=
    "http://www.xyz.org/acm/ccs.rdf#I.I.2.4_Semantic_Networks"/>
</rdf:Description>
```

1.3.3.3 Describing Users

Though user profile standardization is not yet as advanced as learning object metadata standards, there are two main ongoing efforts to standardize metadata for user profiles: the IEEE Personal and Private Information (PAPI) [IEEE] project and the IMS Learner Information Package (LIP) [IMS]. If we compare these standards, we realize that they have been developed from different points of view.

IMS LIP provides us with richer structures and aspects. Categories are rather independent and the relationships between different records which instantiate different categories can be accomplished via the instances of the relationships category of the LIP standard. The structure of the IMS LIP standard was derived from best practices in writing resumes. The IMS standard does not explicitly consider relations to other people, though these can be represented by relationships between different records of the identification category. Accessibility policies to the data about different learners are not defined.

PAPI, on the other hand, has been developed from the perspective of a learner's performance during his or her study. The main categories are thus performance, portfolio, certificates, and relations to other people (classmate, teacher, and so on). This overlaps with the IMS activity category. However, IMS LIP defines activity category as a slot for any activity somehow related to a learner. To reflect this, IMS activity involves fields that are related more to information required from management perspectives than from personalization based on level of knowledge. This can be solved in PAPI by introducing extensions and type of performance or by considering activity at the portfolio level because any portfolio item is the result of some activity related to learning. PAPI do not cover the goal category at all, which can be used for recommendation and filtering techniques, and does not deal with transcript category explicitly. IMS LIP defines transcript as a record that is used to provide an institutionally based summary of academic achievements. In PAPI, portfolio can be used, which will refer to an external document where the transcript is stored.

Using RDF's ability to mix features from more than one schema, we can use schema elements of both standards and also elements of other schemas. These RDF models can be accessible by different peers, and different and overlapping models are possible. Such distributed learner models were already discussed in Vassileva et al. [2003] in the context of distributed learner modeling, though not in the context of RDF-based environments.

1.3.3.4 Adaptive Functionalities as Queries in a Peer-to-Peer Network

Based on the assumption that all resources managed within the network are described by RDF metadata, the Edutella peer-to-peer network [Nejdl et al., 2002a] provides a standardized query exchange mechanism for RDF metadata stored in distributed RDF repositories using arbitrary RDFS schemata.

To enable different repositories to participate in the Edutella network, Edutella wrappers are used to translate queries and results from a common Edutella query and result exchange format to the local format of the peer and vice versa, and to connect the peer to the Edutella network by a JXTA-based P2P library [Gong, 2001]. For communication with the Edutella network, the wrapper translates the local data model into the Edutella Common Data Model (ECDM) and vice versa, and connects to the Edutella Network using the JXTA P2P primitives, transmitting the queries based on ECDM in RDF/XML form. The ECDM is based on Datalog (see, e.g., Garcia-Molina et al., 2002), which is a well-known nonprocedural query language based on Horn clauses without function symbols. Datalog queries, which are a subset of Prolog programs and of predicate logic, easily snap to relations and relational query languages like relational algebra or SQL, or to logic programming languages like Prolog. In terms of relational algebra, Datalog is capable of expressing selection, union, join, and projection and hence is a relationally complete query language. Additional features include transitive closure and other recursive definitions.

Based on the RDF metadata managed within the Edutella network, we can now cast adaptive functionalities as Datalog queries over these resources, which are then distributed through the network to retrieve the appropriate learning resources. Personalization queries are then sent not only to the local repository but to the entire Edutella network. In the following we use Prolog and first-order predicate logic notation to express these queries and use binary predicates to represent RDF statements.

In this way we can start to implement different adaptive hypermedia techniques, as described in the first part of this chapter. *Link annotation,* for example, can be implemented by an annotate (+Page, +User, -Color) redicate. We use the traffic-light metaphor to express the suitability of the resources for the user, taking into account the user profile. A green icon represents a document that is recommended for reading, for example. We can formalize that a document is recommended for the user if it has not been understood yet and if all its prerequisites have already been understood:

```
forall Page, User, Prereq:
annotate(Page, User, green) < --
not_understood_page(Page, User),
prerequisites(Page, Prereq),
forall P in Prereq understood_page(P,User).
```

In Prolog, the criterion above and a query asking for recommended pages for Nedl then looks as follows:

```
annotated(Page, User, green) :-
     not_understood_page(Page, User),
     prerequisites(Page, Prereq),
     not (member(P, Prereq),
     not_understood_page(P, User) ).

?- recommended (Page, nejdl, green)
```

Similar logic programs and queries have to be written for other adaptive functionalities as well. This not only leads to increased flexibility and openness of the adaptive hypermedia system but also allows us to logically characterize adaptive hypermedia systems without restricting the means for their actual implementation [Henze and Nejdl, 2003].

1.4 Conclusion

Adaptive Hypermedia systems have progressed a lot since their early days; we now have a large range of possibilities available for implementing them. Special purpose and educational adaptive hypermedia systems can be implemented on top of adaptive hypermedia engines or link servers with a large array of adaptive functionalities. In the World Wide Web context, adaptive functionalities and personalization features are gaining ground as well and will extend the current Web to a more advanced Adaptive Web.

References

ACM. 2002. The ACM computing classification system, http://www.acm.org/class/1.998/.

Anderson, Kenneth M., Richard N. Taylor, and E. James Whitehead. 1994. Chimera: Hypertext for heterogeneous software environments. In *Proceedings of the ACM European Conference on Hypermedia Technology,* September. ACM Press, New York.

Andrews, Keith, Frank Kappe, and Hermann Maurer. 1995. Serving information to the Web with Hyper-G. In *Proceedings of the 3rd International World Wide Web Conference,* Darmstadt, Germany, April. Elsevier Science, Amsterdam.

Armstrong, Robert, Dayne Freitag, Thorsten Joachims, and Tom Mitchell. 1995. WebWatcher: A learning apprentice for the World Wide Web. In C. Knoblock and A. Levy, Eds., *AAAI Spring Symposium on Information Gathering from Distributed, Heterogeneous Environments.* AAAI Press, Menlo Park, CA, pp. 6–12.

Bailey, Christopher, Wendy Hall, David Millard, and Mark Weal. 2002. Towards open adaptive hypermedia. In P. De Bra, P. Brusilovsky, and R. Conejo, Eds., *Proceedings of the 2nd International Conference on Adaptive Hypermedia and Adaptive Web-Based Systems (AH 2002),* Malaga, Spain, May. Springer-Verlag, London.

Beaumont, Ian. 1994. User modeling in the interactive anatomy tutoring system ANATOM-TUTOR. *User Modeling and User-Adapted Interaction,* 4(1): 21–45.

Billsus, Daniel, Clifford A. Brunk, Craig Evans, Brian Gladish, and Michael Pazzani. 2002. Adaptive interfaces for ubiquitous web access. *Communications of the ACM,* 45(5): 34–38.

Billsus, Daniel and Michael J. Pazzani. 2000. A learning agent for wireless news access. In Henry Lieberman, Ed., *Proceedings of the 2000 International Conference on Intelligent User Interfaces,* New Orleans, LA. ACM Press, New York, pp. 94–97.

Boyle, Craig and Antonio O. Encarnacion. 1994. MetaDoc: an adaptive hypertext reading system. *User Modeling and User-Adapted Interaction,* 4(l): 1–19.

Brickley, Dan and Ramanathan Guha. 2003. RDF Vocabulary Description Language 1.0: RDF Schema, January. http://www.w3.org/TR/rdf-schema/.

Brusilovsky, Peter. 1996. Methods and techniques of adaptive hypermedia. *User Modeling and User-Adapted Interaction,* 6(2–3): 87–129.

Brusilovsky, Peter and David W. Cooper. 1999. ADAPTS: Adaptive hypermedia for a Web-based performance support system. In P. Brusilovsky and P. De Bra, Eds., *2nd Workshop on Adaptive Systems and User Modeling on World Wide Web at 8th International World Wide Web Conference and 7th International Conference on User Modeling,* Toronto and Banff, Canada. Eindhoven University of Technology, Eindhoven, Netherlands, pp. 41–47.

Brusilovsky, Peter, John Eklund, and Elmar Schwarz. 1998a. Web-based education for all: A tool for developing adaptive courseware. *Computer Networks and ISDN Systems,* 30(1–7): 291–300.

Brusilovsky, Peter, Alfred Kobsa, and Julita Vassileva, Eds. 1998b. *Adaptive Hypertext and Hypermedia.* Kluwer Academic, Dordrecht, Netherlands.

Brusilovsky, Peter and Leonid Pesin. 1998. Adaptive navigation support in educational hypermedia: An evaluation of the ISIS-Tutor. *Journal of Computing and Information Technology,* 6(1): 27–38.

Brusilovsky, Peter, Olivero Stock, and Carlo Strapparava, Eds. 2000. *Adaptive Hypermedia and Adaptive Web-based Systems, AH2000,* Vol. 1892 of *Lecture Notes in Computer Science.* Springer-Verlag, Berlin.

Cheverst, Keith, Keith Mitchell, and Nigel Davies. 2002. The role of adaptive hypermedia in a context-aware tourist guide. *Communications of the ACM,* 45(5): 47–51.

De Bra, Paul and Licia Calvi. 1998. AHA! an open adaptive hypermedia architecture. *The New Review of Hypermedia and Multimedia,* 4: 115–139.

Dolog, Peter, Rita Gavriloaie, and Wolfgang Nejdl. 2003. Integrating adaptive hypermedia techniques and open rdf-based environments. In *Proceedings of the 12th World Wide Web Conference,* Budapest, Hungary, May, pp. 88–98.

Fink, Josef, Jürgen Koenemann, Stephan Noller, and Ingo Schwab. 2002. Putting personalization into practice. *Communications of the ACM,* 45(5): 41–42.

Fountain, Andrew, Wendy Hall, Ian Heath, and Hugh Davis. 1990. MICROCOSM: An open model for hypermedia with dynamic linking. In *Proceedings of the ACM Conference on Hypertext.* ACM Press, New York.

Garcia-Molina, Hector, Jeffrey Ullman, and Jennifer Widom. 2002. *Database Systems — The Complete* Book. Prentice Hall, Upper Saddle River, NJ.

Garlatti, Serge and Sébastien Iksal. 2000. Context filtering and spacial filtering in an adaptive information system. In P. Brusilovsky, O. Stock, and C. Strapparava, Eds., *International Conference on Adaptive Hypermedia and Adaptive Web-based systems,* Berlin. Springer-Verlag, London, pp. 315–318.

Gong, Li. April 2001. Project JXTA: A Technology Overview. Technical report, SUN Microsystems, http://www.jxta.org/project/www/docs/TechOverview.pdf.

Gruber, Tom. 1993. A translation approach to portable ontology specifications. *Knowledge Acquisition,* 5: 199–220.

Henze, Nicola and Wolfgang Nejdl. 2001. Adaptation in open corpus hypermedia. *International Journal of Artificial Intelligence in Education,* 12(4): 325–350.

Henze, Nicola and Wolfgang Nejdl. 2002. Knowledge modeling for open adaptive hypermedia. In P. De Bra, P. Brusilovsky, and R. Conejo, Eds., *Proceedings of the 2nd International Conference on Adaptive Hypermedia and Adaptive Web-Based Systems (AH 2002),* Malaga, Spain, May. Springer-Verlag, London, pp. 174–183.

Henze, Nicola and Wolfgang Nejdl. May 2003. Logically characterizing adaptive educational hypermedia systems. In *Proceedings of AH2003 Workshop, 13th World Wide Web Conference,* Budapest, Hungary.

Hohl, Hubertus, Heinz-Dieter Böcker, and Rul Gunzenhäuser. 1996. Hypadapter: An adaptive hypertext system for exploratory learning and programming, *User Modeling and User-Adapted Interaction,* 6(2–3): 131–156.

IEEE. IEEE P1484.2/D7, 2000-11-28. Draft standard for learning technology. Public and private information (PAPI) for learners. Available at: http:/ltse.ieee.org/wg2/.

IEEE-LTSC. IEEE LOM working draft 6.1. Available at: http://ltse.ieee.org/wgl2/index.html.

IMS. IMS learner information package specification. Available at: http://www.imsproject.org/profiles/index.cfm.

Joerding, Tanja. 1999. A temporary user modeling approach for adaptive shopping on the web. In P. Brusilovsky and P. De Bra, Eds., *2nd Workshop on Adaptive Systems and User Modeling on the World Wide Web at 8th International World Wide Web Conference and 7th International Conference on User Modeling,* Toronto and Banff, Canada. Eindhoven University of Technology, Eindhoven, Netherlands, pp. 75–79.

Kaplan, Craig, Justine Fenwick, and James Chen. 1993. Adaptive hypertext navigation based on user goals and context. *User Modeling and User-Adapted Interaction,* 3(3): 193–220.

Lassila, Ora and Ralph R. Swick. 1999. W3C Resource Description Framework model and syntax specification, February. http://www.w3.org/TR/REC-rdf-syntax/.

Michaelides, Danius T., David E. Millard, Mark J. Weal, and David De Roure. 2001. Auld Leaky: A contextual open hypermedia link server. In *Proceedings of the 7th Workshop on Open Hypermedia Systems, ACM Hypertext 2001 Conference,* Aarhus, Denmark. ACM Press, New York.

Moore, Johanna D. and William R. Swartout. 1989. Pointing: A way toward explanation dialogue. In *8th National Conference on Artificial Intelligence,* pp. 457–464.

Nejdl, Wolfgang, Boris Wolf, Changtao Qu, Stefan Decker, Michael Sintek, Ambjoern Naeve, Mikhael Nilsson, Matthias Palmr, and Tore Risch. 2002a. EDUTELLA: A P2P networking infrastructure based on RDF. In *Proceedings of the 11th International World Wide Web Conference (WWW 2002)*, Honolulu, Hawaii, June. ACM Press, New York.

Nejdl, Wolfgang, Boris Wolf, Steffen Staab, and Julien Tane. 2002b. EDUTELLA: Searching and annotating resources within an RDF-based P2P network. In M. Frank, N. Noy, and S. Staab, Eds., *Proceedings of the Semantic Web Workshop*, Honolulu, Hawaii, May.

Not, Elena, Daniela Petrelli, Marcello Sarini, Olivero Stock, Carlo Strapparava, and Massimo Zancanaro. 1998. Hypernavigation in the physical space: adapting presentation to the user and to the situational context. *New Review of Multimedia and Hypermedia,* 4: 33–45.

Oppermann, Reinhard and Marcus Specht. 1999. Adaptive information for nomadic activities: A process oriented approach. In *Software Ergonomie '99,* Walldorf, Germany. Teubner, Stuttgart, Germany, pp. 255–264.

Paris, Cécile. 1988. Tailoring object description to a user's level of expertise. *Computational Linguistics,* 14(3): 64–78.

Schneider-Hufschmidt, Matthias, Thomas Kühme, and Uwe Malinowski, Eds. 1993. *Adaptive User Interfaces: Principles and Practice.* Human Factors in Information Technology, No. 10. North-Holland, Amsterdam.

Vassileva, Julita, Gordon McCalla, and Jim Greer. February 2003. Multi-agent multi-user modelling in I-Help. *User Modeling and User-Adapted Interaction,* 13(l): 179–210.

Weber, Gerhard and Peter Brusilovsky. 2001. ELM-ART: An adaptive versatile system for Web-based instruction. *International Journal of Artificial Intelligence in Education,* 12(4): 351–384.

Weber, Gerhard, Hans-Christian Kuhl, and Stephan Weibelzahl. 2001. Developing adaptive internet based courses with the authoring system NetCoach. In P. De Bra, P. Brusilovsky, and A. Kobsa, Eds., *3rd Workshop on Adaptive Hypertext and Hypermedia,* Sonthofen, Germany. Technical University Eindhoven, Eindhoven, Netherlands, pp. 35–48.

2

Internet Computing Support for Digital Government

CONTENTS

Abstract 2-1
2.1 Introduction 2-2
2.1.1 A Brief History of Digital Government 2-3
2.2 Digital Government Applications: An Overview 2-4
2.2.1 Electronic Voting 2-4
2.2.2 Tax Filing 2-4
2.2.3 Government Portals 2-5
2.2.4 Geographic Information Systems (GISs) 2-5
2.2.5 Social and Welfare Services 2-5
2.3 Issues in Building E-Government Infrastructures 2-6
2.3.1 Data Integration 2-6
2.3.2 Scalability 2-6
2.3.3 Interoperability of Government Services 2-6
2.3.4 Security 2-7
2.3.5 Privacy 2-8
2.3.6 Trust 2-9
2.3.7 Accessibility and User Interface 2-9
2.4 A Case Study: The WebDG System 2-9
2.4.1 Ontological Organization of Government Databases 2-10
2.4.2 Web Services Support for Digital Government 2-10
2.4.3 Preserving Privacy in WebDG 2-13
2.4.4 Implementation 2-14
2.4.5 A WebDG Scenario Tour 2-14
2.5 Conclusion 2-15
Acknowledgment 2-15
References 2-15

Athman Bouguettaya
Abdelmounaam Rezgui
Brahim Medjahed
Mourad Ouzzani

Abstract

The Web has introduced new paradigms in the way data and services are accessed. The recent burst of *Web technologies* has enabled a novel computing paradigm: *Internet computing.* This new computing paradigm has, in turn, enabled a new range of applications built around Web technologies. These Web-enabled applications, or simply Web applications, cover almost every aspect of our everyday life (e.g., e-mail, e-shopping, and e-learning). *Digital Government* (DG) is a major class of *Web applications.* This chapter has a twofold objective. It first provides an overview of DG and the key issues and challenges in

1-58488-381-2/05/$0.00+$1.50
© 2005 by CRC Press LLC

building DG infrastructures. The second part of the chapter is a description of *WebDG,* an experimental DG infrastructure built around *distributed ontologies* and *Web services.*

2.1 Introduction

The Web has changed many aspects of our everyday life. The *e-revolution* has had an unparalleled impact on how people live, communicate, and interact with businesses and government agencies. As a result, many well-established functions of modern society are being rethought and redeployed. Among all of these functions, the *government* function is one where the *Web impact* is the most tangible.

Governments are the most complex organizations in a society. They provide the legal, political, and economic infrastructure to support the daily needs of citizens and businesses [Bouguettaya et al., 2002]. A government generally consists of large and complex networks of institutions and agencies. The Web is progressively, but radically, changing the traditional mechanisms in which these institutions and agencies operate and interoperate. More importantly, the Web is redefining the government–citizen relationship. Citizens worldwide are increasingly experiencing a new, Web-based paradigm in their relationship with their governments. Traditional, paper- and clerk-based functions such as voting, filing of tax returns, or renewing of driver licenses are swiftly being replaced by more efficient Web-based applications. People may value this development differently, but almost all appear to be accepting this new, promising form of government called *Digital Government* (DG).

DG or *E-Government* may be defined as the process of using information and communication technologies to enable the civil and political conduct of government [Elmagarmid and McIver, 2002]. In a DG environment (Figure 2.1), a complex set of interactions among government (local, state, and federal) agencies, businesses, and citizens may take place. These interactions typically involve an extensive transfer of information in the form of electronic documents. The objective of e-government is, in particular, to improve government–citizen interactions through the deployment of an infrastructure built around the "life experience" of citizens. DG is expected to drastically simplify the information flow among different government agencies and with citizens. Online DG services are expected to result in a significant reduction in the use of paper, mailing, and shipping activities, and, consequently, improving the services provided to citizens [Dawes et al., 1999].

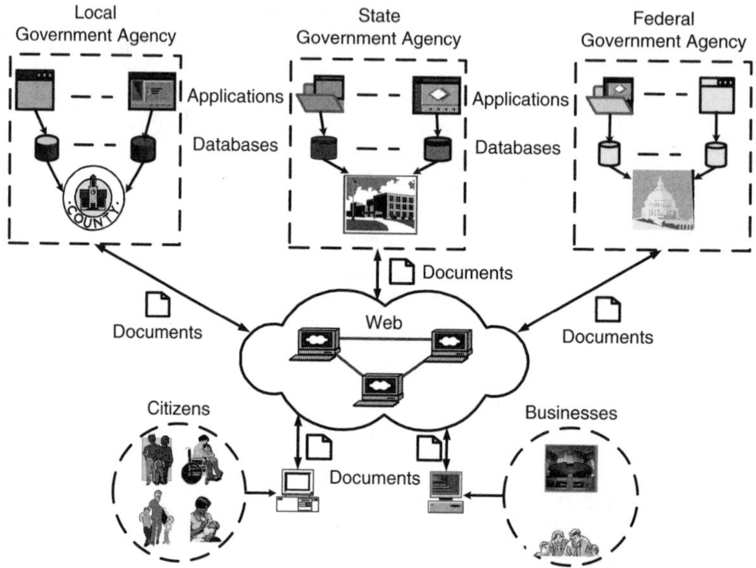

FIGURE 2.1 A Digital Government environment.

From a technical perspective, DG may be viewed as a particular *class* among other classes of Internet-based applications (e.g., e-commerce, e-learning, e-banking). Typically, a DG application is supported by a number of distributed hosts that interoperate to achieve a given government function. The Internet is the medium of choice for the interaction between these hosts. *Internet computing* is therefore the basis for the development of almost all DG applications. Indeed, Internet technologies are at the core of all DG applications. These technologies may be summarized in five major categories: (1) markup languages (e.g., SGML, HTML, XML), (2) scripting languages (e.g., CGI, ASP, Perl, PHP), (3) Internet communication protocols (e.g., TCP/IP, HTTP, ATM), (4) distributed computing technologies (e.g., CORBA, Java RMI, J2EE, EJB), and (5) security protocols (e.g., SSL, S-HTTP TSL). However, despite the unprecedented technological flurry that the Internet has elicited, a number of DG-related challenges still remain to be addressed. These include the interoperability of DG infrastructures, scalability of DG applications, and privacy of the users of these applications. An emerging technology that is particularly promising in developing the next generation of DG applications is *Web services.* A Web service is a functionality that can be programmatically accessible via the Web [Tsur et al., 2001]. A fundamental objective of Web services is to enable interoperability among different software applications running on a variety of platforms [Medjahed et al., 2003; Vinoski, 2002a; FEA Working Group, 2002]. This development grew against the backdrop of the *Semantic Web.* The Semantic Web is not a separate Web but an extension thereof, in which information is given well-defined meaning [W3C, 2001a]. This would enable machines to "understand" and automatically process the data that they merely display at present.

2.1.1 A Brief History of Digital Government

Governments started using computers to improve the efficiency of their processes as early as in the 1950s [Elmagarmid and McIver, 2002]. However, the real history of DG may be traced back to the second half of the 1960s. During the period from 1965 until the early 1970s, many technologies that would enable the vision of citizen-centered digital applications were developed. A landmark step was the development of packet switching that in turn led to the development of the ARPANET [Elmagarmid and McIver, 2002]. The ARPANET was not initially meant to be used by average citizens. However, being the ancestor of the Internet, its development was undoubtedly a milestone in the history of DG.

Another enabling technology for DG is *EDI (Electronic Data Interchange)* [Adam et al., 1998]. EDI can be broadly defined as the computer-to-computer exchange of information from one organization to another. Although EDI mainly focuses on business-to-business applications, it has also been adopted in DG. For example, the U.S. Customs Service initially used EDT in the mid- to late 1970s to process import paperwork more accurately and more quickly.

One of the pioneering efforts that contributed to boosting DG was the 1978 report [Nora and Minc, 1978] aimed at restructuring the society by extensively introducing telecommunication and computing technologies. It triggered the development, in 1979, of the French Télétel/Minitel videotext system. By 1995, Minitel provided over 26000 online services, many of which were government services [Kessler, 1995].

The PC revolution in the 1980s coupled with significant advances in networking technologies and dial-up online services had the effect of bringing increasing numbers of users to a computer-based lifestyle. This period also witnessed the emergence of a number of online government services worldwide. One of the earliest of these services was the Cleveland FREENET developed in 1986. The service was initially developed to be a forum for citizens to communicate with public health officials.

The early 1990s have witnessed three other key milestones in the DG saga. These were: (1) the introduction in 1990 of the first commercial dial-up access to the Internet, (2) the release in 1992 of the World Wide Web to the public, and (3) the availability in 1993 of the first general purpose Web browser Mosaic. The early 1990s were also the years when DG was established as a distinct research area. A new Internet-based form of DG had finally come to life. The deployment of DG systems also started in the 1990s. Many governments worldwide launched large-scale DG projects. A project that had a seminal effect was Amsterdam's Digital City project. It was first developed in 1994 as a local social information infrastructure. Since then, over 100 digital city projects have started across the world [Elmagarmid and McIver, 2002].

This chapter presents some key concepts behind the development of DG applications. In particular, we elaborate on the important challenges and research issues. As a case study, we describe our ongoing project named *WebDG* [Bouguettaya et al., 2001b,a; Rezgui et al., 2002]. The WebDG system uses distributed ontologies to organize and efficiently access government data sources. Government functions are *Web-enabled* by wrapping them with Web services.

In Section 2.2, we describe a number of widely used DG applications. Section 2.3 discusses some of the most important issues in building DG applications. In Section 2.4, we describe the major components and features of the WebDG system. We provide some concluding remarks in Section 2.5.

2.2 Digital Government Applications: An Overview

DG spans a large spectrum of applications. In this section we present a few of these applications and discuss some of the issues inherent to each.

2.2.1 Electronic Voting

E-voting is a DG application where the impact of technology on the society is one of the most straightforward. The basic idea is simply to enable citizens to express their opinions regarding local or national issues by accessing a government Web-based voting system. Examples of e-voting applications include electronic polls, political votes, and online surveys.

E-voting systems are particularly important. For example, a major political race may depend on an e-voting system. The "reliability" of e-voting systems must therefore be carefully considered. Other characteristics of a good e-voting system include (1) Accuracy (a vote cannot be altered after it is cast, and only all valid votes are counted), (2) Democracy (only eligible voters may vote and only once), (3) Privacy (a ballot cannot be linked to the voter who cast it and a voter cannot prove that he or she voted in a certain way), (4) Verifiability (it must be verifiable that all votes have been correctly counted), (5) Convenience (voters must be able to cast their votes quickly in one session and with minimal special skills), (6) Flexibility (the system must allow a variety of ballot question formats), and (7) Mobility (no restrictions must exist on the location from which a voter can cast a vote) [Cranor, 1996].

Many e-voting systems have been developed on a small and medium scale. Examples include the Federal Voting Assistance Program (instituted to allow U.S. citizens who happen to be abroad during an election to cast their votes electronically) [Rubin, 2002] and the *e-petitioner* system used by the Scottish Parliament [Macintosh et al., 2002]. Deploying large scale e-voting systems (e.g., at a country scale), however, is not yet a common practice. It is widely admitted that "the technology does not exist to enable remote electronic voting in public elections" [Rubin, 2002].

2.2.2 Tax Filing

A major challenge being faced by government financial agencies is the improvement of revenue collection and development of infrastructures to better manage fiscal resources. An emerging effort towards dealing with this challenge is the electronic filing (*e-filing*) of tax returns. An increasing number of citizens, businesses, and tax professionals have adopted e-filing as their preferred method of submitting tax returns. According to *Forrester Research,* federal, state, and local governments in the U.S. will collect 15% of fees and taxes online by 2006, which corresponds to $602 billion. The objective set by the U.S. Congress is to have 80% of tax returns filed electronically by 2007 [Golubchik, 2002].

One of the driving forces for the adoption of e-filing is the reduction of costs and time of doing business with the Tax Authority [Baltimore Technologies, 2001]. Each tax return generates several printable pages of data to be manually processed. Efficiency is improved by reducing employees' costs for manual processing of information and minimizing reliance upon traditional paper-based storage systems. For example, the U.S. Internal Revenue Service saves $1.20 on each electronic tax return it processes. Another demonstrable benefit of e-filing is the improvement of the quality of data collected. By reducing manual transactions, e-filing minimizes the potential for error. Estimations indicate that 25% of tax returns filed via traditional paper-based procedures are miscalculated either by the party submitting the return or by the internal revenue auditor [Baltimore Technologies, 2001].

2.2.3 Government Portals

Continuously improving public service is a critical government mission. Citizens and businesses are requiring on-demand access to basic government information and services in various domains such as finance, healthcare, transportation, telecommunications, and energy. The challenge is to deliver higher-quality services faster, more efficiently, and at lower costs. To face these challenges, many governments are introducing e-government portals. These Web-accessible interfaces aim at providing consolidated views and navigation for different government constituents. They simplify information access through a single sign-on to government applications. They also provide common look-and-feel user interfaces and prebuilt templates users can customize. Finally, e-government portals offer anytime–anywhere communication by making government services and information instantly available via the Web.

There are generally four types of e-government portals: Government-to-Citizen (G2C), Government-to-Employee (G2E), Government-to-Government (G2G), and Government-to-Business (G2B). G2C portals provide improved citizen services. Such services include transactional systems such as tax payment and vehicle registration. G2E portals streamline internal government processes. They allow the sharing of data and applications within a government agency to support a specific mission. G2G portals share data and transactions with other government organizations to increase operational efficiencies. G2B portals enable interactions with companies to reduce administrative expenses associated with commercial transactions and foster economic development.

2.2.4 Geographic Information Systems (GISs)

To conduct many of their civil and military roles, governments need to collect, store, and analyze huge amounts of data represented in graphic formats (e.g., road maps, aerial images of agricultural fields, satellite images of mineral resources). The emergence of Geographic Information Systems (GISs) had a revolutionary impact on how governments conduct activities that require capturing and processing images. A GIS is a computer system for capturing, managing, integrating, manipulating, analyzing, and displaying data that is spatially referenced to the Earth [McDonnell and Kemp, 1996].

The use of GISs in public-related activities is not a recent development. For example, the water and electricity [Fetch, 1993] supply industries were using GISs during the early 1990s. With the emergence of Digital Government, GISs have proven to be effective tools in solving many of the problems that governments face in public management. Indeed, many government branches and agencies need powerful GISs to properly conduct their functions. Examples of applications of GISs include mapmaking, site selection, simulating environmental effects, and designing emergency routes [U.S. Geological Survey, 2002].

2.2.5 Social and Welfare Services

One of the traditional roles of government is to provide social services to citizens. Traditionally, citizens obtain social benefits through an excessively effortful and time-consuming process. To assist citizens, case officers may have to manually locate and interrogate a myriad of government databases and/or services before the citizen's request can be satisfied.

Research aiming at improving government social and welfare services has shown that two important challenges must be overcome: (1) the distribution of service providers across several, distant locations, and (2) the heterogeneity of the underlying processes and mechanisms implementing the individual government social services. In Bouguettaya et al. [2001a], we proposed the *one-stop shop* model as a means to simplify the process of collecting social benefits for needy citizens. This approach was implemented and evaluated in our WebDG system described in detail later in this chapter.

2.3 Issues in Building E-Government Infrastructures

Building and deploying an e-government infrastructure entail a number of policy and technical challenges. In this section, we briefly mention some of the major issues that must be addressed for a successful deployment of most DG applications and infrastructures.

2.3.1 Data Integration

Government agencies collect, produce, and manage massive amounts of data. This information is typically distributed over a large number of autonomous and heterogeneous databases [Ambite et al., 2002]. Several challenges must be addressed to enable an efficient integrated access to this information. These include ontological integration, middleware support, and query processing [Bouguettaya et al., 2002].

2.3.2 Scalability

A DG infrastructure must be able to scale to support growing numbers of underlying systems and users. It also must easily accommodate new information systems and support a large spectrum of heterogeneity and high volumes of information [Bouguettaya et al., 2002]. For these, the following two important facets of the scalability problem in DG applications must be addressed.

2.3.2.1 Scalability of Information Collection

Government agencies continuously collect huge amounts of data. A significant challenge is to address the problem of the scalability of data collection, i.e., build DG infrastructures that scale to handle these huge amounts of data and effectively interact with autonomous and heterogeneous data sources [Golubchik, 2002; Wunnava and Reddy, 2000]. In particular, an important and challenging feature of many DG applications is their intensive use of data uploading. For example, consider a tax filing application through which millions of citizens file (i.e., upload) their income tax forms. Contrary to the problem of scalability of data downloading (where a large number of users *download* data from the same server), the problem of scalability of data uploading has not yet found its effective solutions. The *bistros* approach was recently proposed to solve this problem [Golubchik, 2002]. The basic idea is to first route all uploads to a set of intermediary Internet hosts (called bistros) and then forward the data from one or more bistros to the server.

2.3.2.2 Scalability of Information Processing

DG applications are typically destined to be used by large numbers of users. More importantly, these users may (or, sometimes, must) *all* use these applications in a short period of time. The most eloquent example for such a situation is certainly that of an e-voting system. On a vote day, an e-voting application must, within a period of only a few hours, process, i.e., collect, validate, and count the votes of tens of millions of voters.

2.3.3 Interoperability of Government Services

In many situations, citizens' needs cannot be fulfilled through one single e-government service. Different services (provided by different agencies) would have to interact with each other to fully service a citizen's request. A simple example is a child support service that may need to send an inquiry to a federal taxation service (e.g., IRS) to check revenues of a deadbeat parent. A more complex example is a government procurement service that would need to interact with various other e-government and business services.

The recent introduction of Web services was a significant advance in addressing the interoperability problem among government services. A Web service can easily and seamlessly discover and use any other Web service irrespective of programming languages, operating systems, etc. Standards to describe, locate, and invoke Web services are at the core of intensive efforts to support interoperability. Although standards like SOAP and WSDL are becoming commonplace, there is still some incoherence in the way that they are implemented by different vendors. For example, SOAP:Lite for Pert and .NET implement SOAP 1.1 differently. In addition, not all aspects of those standards are being adopted [Sabbouh et al., 2001].

A particular and interesting type of interoperability relates to *semantics.* Indeed, semantic mismatches between different Web services are major impediments to achieve full interoperability. In that respect, work in the Semantic Web is crucial in addressing related issues [Berners-Lee, 2001]. In particular, Web services would need mainly to be linked to ontologies that would make them meaningful [Trastour et al., 2001; Ankolekar et al., 2001]. Description, discovery, and invocation could then be made in a *semantics aware* way.

Web services *composition* is another issue related to Web services interoperability. Composition creates new value-added services with functionalities outsourced from other Web services [Medjahed et al., 2003]. Thus, composition involves interaction with different Web services. Enabling service composition requires addressing several issues including composition description, service composability, composition plan generation, and service execution.

2.3.4 Security

Digital government applications inherently collect and store huge amounts of sensitive information about citizens. Security is therefore a vital issue in these applications. In fact, several surveys and polls report that security is the main impediment citizens cite as the reason for their reluctance to use online government services. Applications such as e-voting, tax filing, or social e-services may not be usable if they are not sufficiently secured. Developing and deploying secure and reliable DG infrastructures require securing:

- The interaction between citizens and digital government infrastructures
- Government agencies' databases containing sensitive information about citizens and about the government itself
- The interaction among government agencies

Technically, securing DG infrastructures poses challenges similar to those encountered in any distributed information system that supports workflow-based applications across several domains [Joshi et al., 2001]. Advances in cryptography and protocols for secure Internet communication (e.g., SSL, S-HTTP) significantly contributed in securing information transfer within DG infrastructures. Securing DG infrastructures, however, involve many other aspects. For example, a service provider (e.g., a government agency) must be able to specify *who* may access the service, *how* and *when* accesses are made, as well as any other condition for accessing the service. In other words, *access control* models and architectures must be developed for online government services.

Part of the security problem in DG applications is also to secure the Web services that are increasingly used in deploying DG services. In particular, the issue of securing the interoperability of Web services is one that has been the focus of many standardization bodies. Many standards for securing Web services have been proposed or are under development. Examples include:

- **XML Encryption,** which is a W3C proposal for capturing the results of an encryption operation performed on arbitrary (but most likely XML) data [W3C, 2001c].
- **XML Signature,** which also is a W3C proposal that aims at defining XML schema for capturing the results of a digital signature operation applied to arbitrary data [W3C, 2001e].
- **SOAP Digital Signature,** which is a standard to use the XML digital signature syntax to sign SOAP messages [SOAP; W3C, 2001b].

- **XKMS** (XML Key Management Specification), which specifies protocols for distributing and registering public keys, suitable for use in conjunction with the proposed standard for XML Signature, and an anticipated companion standard for XML Encryption [W3C, 2001d].
- **XACML** (eXtensible Access Control Markup Language), which is an XML specification for expressing policies for information access over the Internet [OASIS, 2001].
- **SAML** (Security Assertion Markup Language), which is an XML-based security standard for exchanging authentication and authorization information [OASIS, 2002].
- **WS Security,** which aims at adding security metadata to SOAP messages [IBM et al., 2002].

Despite this intense standardization activity, securing e-government infrastructures still poses several challenges. In particular, two important aspects remain to be addressed:

- Developing *holistic* architectures that would implement the set of security standards and specify how to deploy these standards in real applications involving Web databases [IBM and Microsoft, 2002].
- Developing security models for Web databases that would consider Web services as (human) users are considered in conventional databases.

2.3.5 Privacy

Government agencies collect, store, process, and share information about millions of individuals who have different preferences regarding their privacy. This naturally poses a number of legal issues and technical challenges that must be addressed to control the information flow among government databases and between these and third-party entities (e.g., private businesses). The common approach in addressing this issue consists of enforcing privacy by law or by self-regulation. Few technology-based solutions have been proposed.

One of the legal efforts addressing the privacy problem was HIPAA, the Health Insurance Portability and Accountability Act passed by the U.S. Congress in 1996. This act essentially includes regulations to reduce the administrative costs of healthcare. In particular, it requires all health plans that transmit health information in an electronic transaction to use a standard format [U.S. Congress, 1996]. HIPAA is expected to play a crucial role in preserving individuals' rights to the privacy of their health information. Also, as it aims at establishing national standards for electronic healthcare transactions, HIPAA is expected to have a major impact on how Web-based healthcare providers and health insurance companies operate and interoperate.

Technical solutions to the privacy problem in DG have been *ad hoc.* For example, a number of protocols have been developed to preserve the privacy of e-voters using the Internet (e.g., Ray et al., 2001) or any arbitrary networks (e.g., Mu and Varadharajan, 1998). Another example of *ad hoc* technical solutions is the prototype system developed for the U.S. National Agricultural Statistics Service (NASS) [Karr et al., 2002]. The system disseminates survey data related to on-farm usage of chemicals (fertilizers, fungicides, herbicides, and pesticides). It uses *geographical aggregation* as a means to protect the identities of individual farms.

An increasingly growing number of DG applications are being developed using Web services. This reformulates the problem of privacy in DG as one of enforcing privacy in environments of interoperating Web services. This problem is likely to become more challenging with the envisioned semantic Web. The Semantic Web is viewed as an extension of the current Web in which machines become "much better able to process and understand the data that they merely display at present" [Berners-Lee, 2001]. To enable this vision, "intelligent" software agents will carry out sophisticated tasks on behalf of their users. In that process, these agents will manipulate and exchange extensive amounts of personal information and replace human users in making decisions regarding their personal data. The challenge is then to develop smart agents that autonomously enforce the privacy of their respective users, i.e., autonomously determine, according to the current context, what information is private.

2.3.6 Trust

Most of the research related to trust in electronic environments has focused on e-commerce applications. Trust, however, is a key requirement in almost all Web-based applications. It is particularly important in DG infrastructures. As many other social and psychological concepts, no single definition of the concept of trust is generally adopted. A sample definition states that trust is "the willingness to rely on a specific other, based on confidence that one's trust will lead to positive outcomes" [Chopra and Wallace, 2003]. The literature on trust in the DG context considers two aspects: (1) the trust of citizens in e-government infrastructures and (2) the impact of using e-government applications on people's trust of governments.

Citizens' trust in e-government is only one facet of the more general problem of users' trust in electronic environments. In many of these environments, users interact with parties that are either unknown or whose trustworthiness may not be easily determined. This poses the problem of building trust in environments where uncertainty is an inherent characteristic. A recent effort to address this problem has focused on the concept of *reputation*. The idea is to build reliable reputation management mechanisms that provide an objective evaluation of the trust that may be put in any given Web-based entity (e.g., a government agency's Web site). Users would then be able to use such a mechanism to assess the trust in these entities.

Results of studies addressing the second aspects (i.e., impact of e-government on citizens' trust) are still not conclusive. Some studies, however, assert that a relationship between citizens' access and use of e-government applications and their trust in governments may exist. A survey study conducted in 2002 concluded that "the stronger citizens believe that government Web sites provide reliable information, the greater their trust in government" [Wench and Hinnant, 2003]. This may be explained by factors that include: increased opportunities for participation, increased ease of communication with government, greater transparency, and perception of improved efficiency [Tolbeert and Mossberger, 2003].

2.3.7 Accessibility and User Interface

In a report authored by the US PITAC (President's Information Technology Advisory Committee) [U.S. PITAC, 2000], the committee enumerates, as one of its main findings, that "major technological barriers prevent citizens from easily accessing government information resources that are vital to their well being." The committee further adds that "government information is often unavailable, inadequate, out of date, and needlessly complicated."

E-government applications are typically built to be used by average citizens with, *a priori*, no special computer skills. Therefore, the user interfaces (UIs) used to access these applications must be easy to use and accessible to citizens with different aptitudes. In particular, some DG applications are targeted to specific segments of the society (e.g., citizens at an elderly age or with special mental and physical ability). These applications must provide a user interface that suits their respective users' abilities and skills. Recent efforts aim at building *smart UIs* that progressively "get acquainted" to their users' abilities and dynamically adapt to those (typically, decaying) abilities. A recent study [West, 2002] reports that 82% of U.S. government Web sites have some form of disability access, which is up from 11% in 2001.

2.4 A Case Study: The WebDG System

In this section, we describe our research in designing and implementing a comprehensive infrastructure for e-government services called WebDG (Web Digital Government). WebDG's major objective is to develop techniques to efficiently access government databases and services. Our partner in the WebDG project is the Family and Social Services Administration (FSSA). The FSSA's mission is to help needy citizens collect social benefits. The FSSA serves families and individuals facing hardships associated with low income, disability, aging, and children at risk for healthy development. To expeditiously respond to citizens' needs, the FSSA must be able to seamlessly integrate geographically distant, heterogeneous, and autonomously run information systems. In addition, FSSA applications and data need to be accessed through one single interface: the Web. In such a framework, case officers and citizens would transparently access data and applications as homogeneous resources. This section and the next discuss WebDG's major concepts and describe the essential components of its architecture.

2.4.1 Ontological Organization of Government Databases

The FSSA is composed of dozens of autonomous departments located in different cities and counties statewide. Each department's information system consists of a myriad of databases. To access government information, case officers first need to locate the databases of interest. This process is often complex and tedious due to the heterogeneity, distribution, and large number of FSSA databases. To tackle this problem, we segmented FSSA databases into *distributed ontologies.* An *ontology* defines a taxonomy based on the semantic proximity of information interest [Ouzzani et al., 2000]. Each ontology focuses on a single common information type (e.g., disability). It dynamically groups databases into a single collection, generating a conceptual space with a specific content and scope. The use of distributed ontologies elicits the filtering and reduction of the overhead of discovering FSSA databases.

Ontologies describe coherent slices of the information space. Databases that store information about the same topic are grouped together in the same ontology. For example, all databases that may be of interest to disabled people (e.g., *Medicaid* and *Independent Living*) are members of the ontology *Disability* (Figure 2.2). For the purpose of this project, we have identified eight ontologies within FSSA, namely *family, visually impaired, disability, low income, at risk children, mental illness and addiction, health and human services,* and *insurance.* Figure 2.2 represents some of those ontologies; each database is linked to the ontologies that it is member of. In this framework, individual databases join and leave ontologies at their own discretion. An overlap of two ontologies depicts the situation where a database stores information that is of interest to both of them. For example, the *Medicaid* database simultaneously belongs to three ontologies: *family, visually impaired,* and *disability.*

The FSSA ontologies are not isolated entities. They are related by *interontology relationships.* These relationships are dynamically established based on users' needs. They allow a query to be resolved by member databases of remote ontologies when it cannot be resolved locally. The interontology relationships are initially determined statically by the ontology administrator. They essentially depict a *functional* relationship that would dynamically change over time.

Locating databases that fit users' queries requires detailed information about the content of each database. For that purpose, we associate with each FSSA database a *co-database* (Figure 2.3). A co-database is an object-oriented database that stores information about its associated database, ontologies, and interontology relationships. A set of databases exporting a certain type of information (e.g., disability) is represented by a class in the co-database schema. This class inherits from a predefined class, *OntologyRoot,* that contains generic attributes. Examples of such attributes include *information-type* (e.g., "disability" for all instances of the class *disability*) and *synonyms* (e.g., "handicap" is a synonym of "disability"). In addition to these attributes, every subclass of the *OntologyRoot* class has some specific attributes that describe the domain model of the underlying databases.

2.4.2 Web Services Support for Digital Government

Several rehabilitation programs are provided within the FSSA to help disadvantaged citizens. Our analysis of the FSSA operational mechanisms revealed that the process of collecting social benefits is excessively time-consuming and frustrating. Currently, FSSA case officers must deal with different situations that depend on the particular needs of each citizen (disability, children's health, housing, employment, etc.) For each situation, they must typically delve into a potentially large number of applications and determine those that best meet the citizens' needs. For each situation, they must manually (1) determine applications that appropriately satisfy citizens' needs, (2) determine how to access each application, and (3) combine the results returned by those applications.

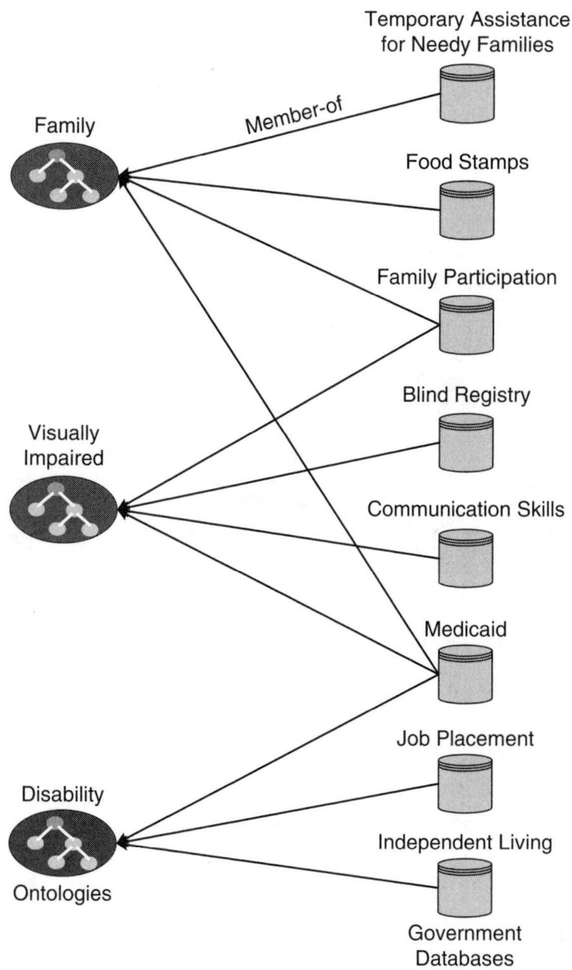

FIGURE 2.2 Sample FSSA ontologies.

To facilitate the process of collecting benefits, we wrapped each FSSA application with a Web service. Web Services are emerging as a promising middleware to facilitate application-to-application integration on the Web [Vinoski, 2002b]. They are defined as modular applications offering sets of related functions that can be programmatically accessed through the Web. Adopting Web services in e-government enables (1) *standardized* description, discovery, and invocation of welfare applications, (2) *composition* of preexisting services to provide *value-added* services, and (3) *uniform* handling of privacy.

The *providers* of WebDG services are bureaus within FSSA (e.g., Bureau of Family Resources) or external agencies (e.g., U.S. Department of Health and Human Services). They define descriptions of their services (e.g., operations) and publish them in the *registry*. *Consumers* (citizens, case officers, and other e-government services) access the registry to locate services of interest. The registry returns the description of each relevant service. Consumers use this description to "understand" how to use the corresponding Web service.

2.4.2.1 Composing WebDG Services

The incentive behind composing e-government services is to further simplify the process of searching and accessing these services. We propose a new approach for the *(semi)automatic composition* of Web services. Automatic composition is expected to play a major role in enabling the envisioned Semantic Web [Berners-Lee, 2001]. It is particularly suitable for e-government applications. Case officers and

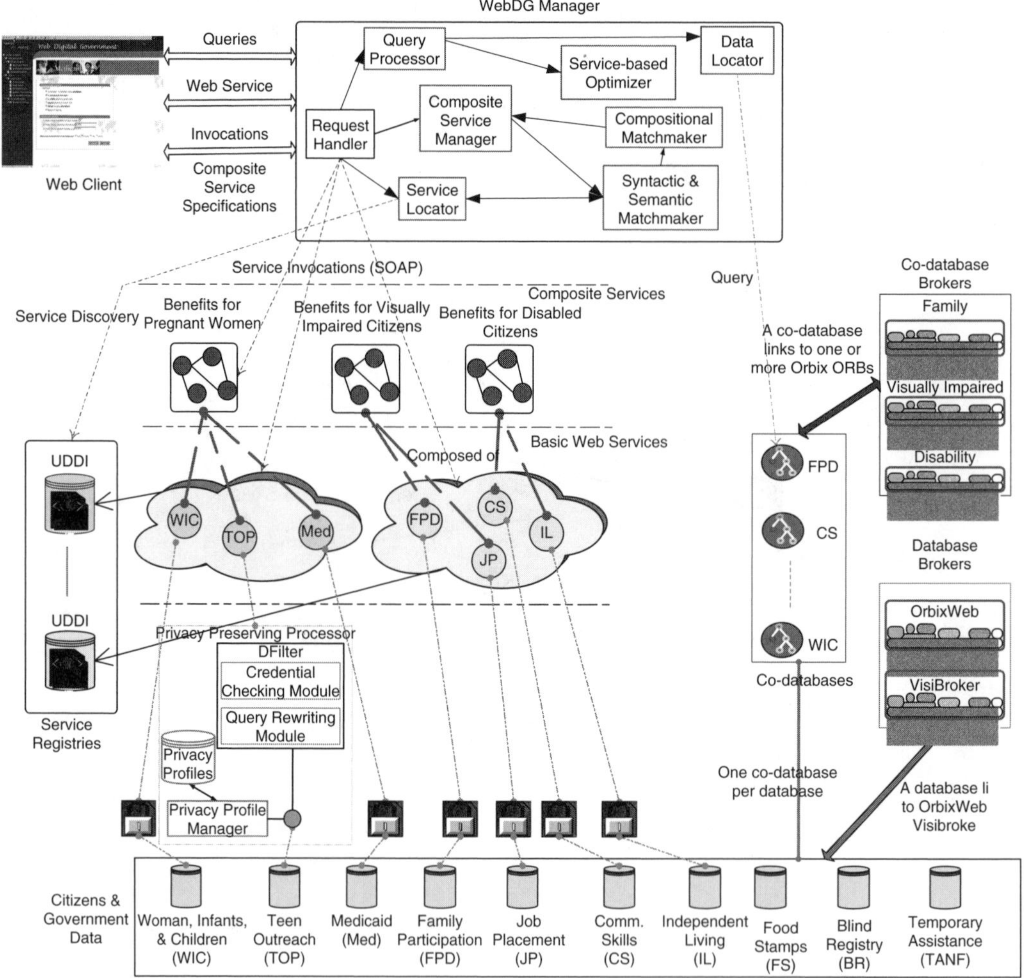

FIGURE 2.3 WebDG architecture.

citizens need no longer to search for services which might be otherwise a time-consuming process. Additionally, they are not required to be aware of the full technical details of the outsourced services.

WebDG's approach for service composition includes four phases: *specification, matchmaking, selection,* and *generation.*

Specification: Users define high-level descriptions of the desired composition via an XML-based language called CSSL (Composite Service Specification Language). CSSL uses a subset of WSDL service interface elements and extends it to allow the (1) description of semantic features of Web services and (2) specification of the control flow between composite services operations. Defining a WSDL-like language has two advantages. First, it makes the definition of composite services as simple as the definition of simple (i.e., noncomposite) services. Second, it allows the support of recursive composition.

Matchmaking: Based on users' specification, the matchmaking phase automatically generates composition plans that conform to that specification. A *composition plan* refers to the list of outsourced services and the way they interact with each other (plugging operations, mapping messages, etc). A major issue addressed by WebDG's matchmaking algorithm is *composability* of the outsourced services [Berners-Lee, 2001]. We propose a set of rules to check composability of e-government services. These include *operation semantics* composability and *composition soundness.* Operation semantics composability compares the categories or domains of interest (e.g., "healthcare," "adoption") of each pair of interacting operations. It also compares their types or functionalities (e.g., "eligibility," "counseling"). For that purpose, we define two ontologies: *category* and *type*. Our assumption is that both ontologies are predefined and agreed upon by government social agencies. Each operation includes two elements from the *category* and *type* ontologies, respectively. Composition soundness checks whether combining Web services in a specific way provides an added value. For that purpose, we introduce the notion of *composition template*. A composition template is built for each composition plan generated by WebDG. It gives the general structure of that plan. We also define a subclass of templates called *stored templates*. These are defined *a priori* by government agencies. Because stored templates inherently provide added values, they are used to test the soundness of composition plans.

Selection: At the end of the matchmaking phase, several composition plans may have been generated. To facilitate the *selection* of relevant plans, we propose to define *Quality of Composition* (*QoC*) parameters. Examples of such parameters include time, cost, and relevance of the plan with respect to the user's specification (based on ranking, for example). Composers define (as part of their profiles) thresholds corresponding to QoC parameters. Composition plans are returned only if the values of their QoC parameters are greater than their respective thresholds.

Generation: This phase aims at *generating* a detailed description of a composite service given a selected plan. This description includes the list of outsourced services, mappings between composite service and component service operations, mappings between messages and parameters, and flow of control and data between component services. Composite services are generated either in WSFL [WSFL] or XLANG [XLANG], two standardization efforts for composing services.

2.4.3 Preserving Privacy in WebDG

Preserving privacy is one of the most challenging tasks in deploying e-government infrastructures. The privacy problem is particularly complex due to the different perceptions that different users of e-government services may have with regard to their privacy. Moreover, the same user may have different privacy preferences associated with different types of information. For example, a user may have tighter privacy requirements regarding medical records than employment history. The user's perception of privacy also depends on the *information receiver*, i.e., who receives the information, and the *information usage*, i.e., the purposes for which the information is used.

Our approach to solving the privacy problem is based on three concepts: *privacy profiles*, *privacy credentials*, and *privacy scopes* [Rezgui et al., 2002]. The set of privacy preferences applicable to a user's information is called privacy profile. We also define privacy credentials that determines the privacy scope for the corresponding user. A privacy scope for a given user defines the information that an e-government service can disclose to that user. Before accessing an e-government service, users are granted privacy credentials. When a service receives a request, it first checks that the request has the necessary credentials to access the requested operation according to its privacy policy. If the request can be answered, the service translates it into an equivalent data query that is submitted to the appropriate government DBMS.

When the query is received by the DBMS, it is first processed by a privacy preserving data filter (*DFilter*). The DFilter is composed of two modules: the *Credential Checking Module* (CCM) and the *Query Rewriting Module* (QRM). The CCM determines whether the service requester is authorized to access the requested information based on credentials. For example, Medicaid may state that a case officer in a given state may not access information of citizens from another state. If the credential authorizes access to only part of the requested information, the QRM *redacts* the query (by removing unauthorized attributes) so that *all* the privacy constraints are enforced.

The *Privacy Profile Manager* (PPM) is responsible for enforcing privacy at a finer granularity than the CCM. For example, the local CCM may decide that a given organization can have access to local information regarding a group of citizens' health records. However, a subset of that group of citizens may explicitly request that parts of their records should not be made available to third-party entities. In this case, the local PPM will discard those parts from the generated result. The PPM is a translation of the consent-based privacy model in that it implements the privacy preferences *of individual* citizens. It maintains a repository of privacy profiles that stores individual privacy preferences.

2.4.4 Implementation

The WebDG system is implemented across a network of Solaris workstations. Citizens and case officers access WebDG via a *Graphical User Interface* (GUI) implemented using HTML/Servlet (Figure 2.3). Two types of requests are supported by WebDG: querying databases and invoking FSSA applications. All requests are received by the WebDG manager. The *Request Handler* is responsible for routing requests to the *Data Locator* (DL) or the *Service Locator* (SL). Queries are forwarded to the DL. Its role is to educate users about the information space and locate relevant databases. All information necessary to locate FSSA databases is stored in co-databases *(ObjectStore)*. The co-databases are linked to three different Orbix ORB (one ORB per ontology). Users can learn about the content of each database by displaying its corresponding documentation in HTML or text, audio, or video formats. Once users have located the database of interest, they can then submit SQL queries. The *Query Processor* handles these queries by accessing the appropriate database via JDBC gateways. Databases are linked to OrbixWeb or VisiBroker ORBS.

WebDG currently includes 10 databases and 7 FSSA applications implemented in Java (JDK 1.3). These applications are wrapped by WSDL descriptions. We use the *Axis's Java2WSDL* utility in *IBM's Web Services Toolkit* to automatically generate WSDL descriptions from Java class files. WSDL service descriptions are published into a UDDI registry. We adopt *Systinet's WASP UDDI Standard 3.1* as our UDDL toolkit. *Cloudscape* (4.0) database is used as a UDDI registry.

WebDG services are deployed using *Apache SOAP* (2.2). *Apache SOAP* provides not only server-side infrastructure for deploying and managing services but also client-side API for invoking those services. Each service has a *deployment descriptor*. The descriptor includes the unique identifier of the Java class to be invoked, session scope of the class, and operations in the class available for the clients. Each service is deployed using the *service management client* by providing its descriptor and the URL of the *Apache SOAP servlet rpcrouter*.

The SL allows the discovery of WSDL descriptions by accessing the UDDI registry. The SL implements *UDDI Inquiry Client* using WASP UDDI API. Once a service is discovered, its operations are invoked through *SOAP Binding Stub*, which is implemented using Apache SOAP API. Service operations are executed by accessing FSSA databases (*Oracle 8.0.5* and *Informix 7.0*). For example, TOP database contains sensitive information about foster families (e.g., household income). To preserve privacy of such information, operation invocations are intercepted by a *Privacy Preserving Processor*. The Privacy Preserving Processor is based on privacy credentials, privacy profiles, and data filters (Section 2.4.3). Sensitive information is returned *only* to authorized users.

2.4.5 A WebDG Scenario Tour

We present a scenario that illustrates the main features of WebDG. A demo of WebDG is available online at http://www.nvc.cs.vt.edu/~dgov. We consider the case of a pregnant teen *Mary* visiting case officer *John* to collect social benefits to which she is entitled. Mary would like to apply for a government-funded health insurance program. She also needs to consult a nutritionist to maintain an appropriate diet during her pregnancy. As Mary will not able to take care of the future newborn, she is interested in finding a foster family. The fulfillment of Mary's needs requires accessing different services scattered in and outside the local agency. For that purpose, John may either look for simple (noncomposite) Web services that fit Mary's specific needs or specify all those needs through one single composite service called Pregnancy Benefits (PB):

- **Step 1: Web Service Discovery** — To locate a specific Web service, John could provide either the service name, if known, or properties. This is achieved by selecting the "By Program Name" or "By Program Properties" nodes, respectively. WebDG currently supports two properties: *category* and *agency*. Assume John is interested in a service that provides help in finding foster families. He would select the *adoption* and *pregnancy* categories and the *Division of Family and Children* agency. WebDG would return the Teen Outreach Pregnancy (TOP) service. TOP offers childbirth and postpartum educational support for pregnant teens.

- **Step 2: Privacy-Preserving Invocation** — Assume that case worker John wants to use TOP service. For that purpose, he clicks on the service name. WebDG would return the list of operations offered by TOP service. As Mary is looking for a foster family, John would select the Search Family Adoption operation. This operation returns information about foster families in a given state (Virginia, for example). The value "No right" (for the attribute "Race") means that Mary does not have the right to access information about the race of family F1. The value "Not Accessible" (for the attribute "Household Income") means that family F1 does not want to disclose information about its income.
- **Step 3: Composing Web Services** — John would select the "Advanced Programs" node to specify the PB composite service. He would give the list of operations to be outsourced by PB without referring to any preexisting service. Examples of such operations include Find Available Nutritionist, Find PCP Providers (which looks for primary care providers), and Find Pregnancy Mentors. After checking composability rules, WebDG would return *composition plans* that conform to BP specification. Each plan has an ID (number), a graphical description, and a ranking. The ranking gives an approximation about the relevance of the corresponding plan. John would click on the plan's ID to display the list of outsourced services. In our scenario, WIC (a federally funded food program for Women, Infants, and Children), Medicaid (a healthcare program for low-income citizens and families), and TOP services would be outsourced by PB.

2.5 Conclusion

In this chapter, we presented our experience in developing DG infrastructures. We first gave a brief history of DG. We then presented some major DG applications. This is followed by a discussion of some key issues and technical challenges in developing DG applications. DG has the potential to significantly transform citizens' conceptions of civil and political interactions with their governments. It facilitates two-way interactions between citizens and government. For example, several U.S. government agencies (e.g., U.S. Department of Agriculture) now enable citizens to file comments online about proposed regulations. Citizens in Scotland can now create and file online petitions with their parliament.

The second part of the chapter is a description of our experimental DG infrastructure called WebDG. WebDG mainly addresses the development of customized digital services that aid citizens receiving services that require interactions with multiple agencies. During the development of WebDG, we implemented and evaluated a number of novel ideas in deploying DG infrastructures. The system is built around two key concepts: *distributed ontologies* and *Web services*. The ontological approach was used to organize government databases. Web services were used as *wrappers* that enable access to and interoperability among government services. The system uses emerging standards for the description (WSDL), discovery (UDDI), and invocation (SOAP) of e-government services. The system also provides a mechanism that enforces the privacy of citizens when interacting with DG applications.

Acknowledgment

This research is supported by the National Science Foundation under grant 9983249-EIA and by a grant from the Commonwealth Information Security Center (CISC).

References

Adam, Nabil, Oktay Dogramaci, Aryya Gangopadhyay, and Yelena Yesha. 1998. *Electronic Commerce: Technical, Business, and Legal Issues.* Prentice Hall, Upper Saddle River, NJ.

Ambite, José L., Yigal Arens, Luis Gravano, Vasileios Hatzivassiloglou, Eduard H. Hovy, Judith L. Klavans, Andrew Philpot, Usha Ramachandran, Kenneth A. Ross, Jay Sandhaus, Deniz Sarioz, Anurag Singla, and Brian Whitman. 2002. Data integration and access. In Ahmed K. Elmagarmid and William J. McIver, Eds., *Advances in Digital Government: Technology, Human Factors, and Policy,* Kluwer Academic, Dordrecht, Netherlands, pp. 85–106.

Ankolekar, Anupriya, Mark Burstein, Jerry R. Hobbs, Ora Lassila, David L. Martin, Sheila A. McIlraith, Srini Narayanan, Massimo Paolucci, Terry Payne, Katia Sycara, and Honglei Zeng. 2001. DAML-S: Semantic markup for Web services. In *Proceedings of the International Semantic Web Working Symposium (SWWS),* July 30–August 1.

Baltimore Technologies. 2001. Baltimore E-Government Solutions: E-Tax Framework. *White Paper,* http://www.baltimore.com/government/.

Berners-Lee, Tim. 2001. *Services and Semantics: Web Architecture.* W3C http://www.w3.org/2001/04/30-tbl.

Bouguettaya, Athman, Ahmed K. Elmagarmid, Brahim Medjabed, and Mourad Ouzzani. 2001a. Ontology-based support for Digital Government. In *Proceedings of the 27th International Conference on Very Large Databases (VLDB 2001),* Roma, Italy, September, Morgan Kaufmann, San Francisco.

Bouguettaya, Athman, Mourad Ouzzani, Brahim Medjahed, and J. Cameron. 2001b. Managing government databases. *Computer,* 34(2), February.

Bouguettaya, Athman, Mourad Ouzzani, Brahim Medjahed, and Ahmed K. Elmagarmid. 2002. Supporting data and services access in digital government environments. In Ahmed K. Elmagarmid and William J. McIver, Eds., *Advances in Digital Government: Technology, Human Factors, and Policy,* Kluwer Academic, pp. 37–52.

Chopra, Kari and William A. Wallace. 2003. Trust in electronic environments. In *Proceedings of the 36th Annual Hawaii International Conference on System Sciences (HICSS '03),* ACM Digital Library, ACM.

Cranor, Lorrie F. 1996. Electronic voting. *ACM Crossroads Student Magazine,* January.

Dawes, Sharon S., Peter A. Blouiarz, Kristine L. Kelly, and Patricia D. Fletcher. 1999. Some assembly required: Building a digital government for the 21st century. *ACM Crossroads Student Magazine,* March.

Elmagarmid, Ahmed K. and William J. McIver, Eds. 2002. *Advances in Digital Government: Technology, Human Factors, and Policy.* Kluwer Academic, Dordrecht, Netherlands.

FEA Working Group. 2002. E-Gov Enterprise Architecture Guidance (Common Reference Model). July.

Fetch, James. 1993. GIS in the electricity supply industry: An overview. In *Proceedings of the IEE Colloquium on Experience in the Use of Geographic Information Systems in the Electricity Supply Industry,* Seminar Digest No. 129, IEE, Herts, U.K.

Golubchik, Leana. 2002. Scalable data applications for Internet-based digital government applications. In Ahmed K. Elmagarmid and William J. McIver, Eds., *Advances in Digital Government: Technology, Human Factors, and Policy,* Kluwer Academic, Dordrecht, Netherlands, pp. 107–119.

IBM and Microsoft. 2002. Security in a web services world: A proposed architecture and roadmap. *White Paper.*

IBM, Microsoft, and Verisign. April 2002. *Web Services Security (WS-Security),* http://www106.ibm.com/developerworks/webservices/library/ws-secure, April.

Joshi, James, Arif Ghafoor, Walid G. Aref, and Eugene H. Spafford. 2001. Digital government security infrastructure design challenges. *Computer,* 34(2): 66–72, February.

Karr, Alan F., Jaeyong Lee, Ashish P. Sanil, Joel Hernandez, Sousan Karimi, and Karen Litwin. 2002. Web-based systems that disseminate information from databases but protect confidentiality. In Ahmed K. Elmagarmid and William J. McIver, Eds., *Advances in Digital Government: Technology, Human Factors, and Policy,* Kluwer Academic, Dordrecht, Netherlands, pp. 181–196.

Kessler, Jack. 1995. The French minitel: Is there digital life outside of the "US ASCII" Internet? A challenge or convergence? *D-Lib Magazine,* December.

Macintosh, Ann, Anna Malina, and Steve Farrell. 2002. Digital democracy through electronic petitioning. In Ahmed K. Elmagarmid and William J. McIver, Eds., *Advances in Digital Government: Technology, Human Factors, and Policy*, Kluwer Academic, Dordrecht, Netherlands, pp. 137–148.

McDonnell, Rachael and Karen K. Kemp. 1996. *International GIS Dictionary.* John Wiley & Sons, New York.

Medjahed, Brahim, Boualem Benatallah, Athman Bouguettaya, A. H. H. Ngu, and Ahmed K. Elmagarmid. 2003. Business-to-business interactions: Issues and enabling technologies. *The VLDB Journal,* 12(1), May.

Mu, Yi and Vijay Varadharajan. 1998. Anonymous secure e-voting over a network. In *Proceedings of the 14th Annual Computer Security Applications Conference (ACSAC '98),* IEEE Computer Society, Los Alamitos, CA.

Nora, Simon and Alain Minc. 1978. L'informatisation de la Société. *A Report to the President of France.*

OASIS. 2001. *eXtensible Access Control Markup Language,* http://www.oasis-open.org/committees/xacm/.

OASIS. 2002. *Security Assertion Markup Language,* http://www.oasis-open.org/committees/security/.

Ouzzani, Mourad, Boualem Benatallah, and Athman Bouguettaya. 2000. Ontological approach for information discovery in Internet databases. *Distributed and Parallel Databases,* 8(3), July.

Ray, Indrajit, Indrakshi Ray, and Natarajan Narasimhamurthi. 2001. An anonymous electronic voting protocol for voting over the Internet. In *Proceedings of the 3rd International Workshop on Advanced Issues of E-Commerce and Web-Based Information Systems (WECWIS '01),* June 21–22, ACM Digital Library, ACM, New York.

Rezgui, Abdelmounaam, Mourad Ouzzani, Athman Bouguettaya, and Brahim Medjahed. 2002. Preserving privacy in Web services. In *Proceedings of the 4th International Workshop on Web Information and Data Management (WIDM '02),* ACM Press, New York, pp. 56–62.

Rubin, Avi D. 2002. Security considerations for remote electronic voting. *Communications of the ACM,* 45(12), December.

Sabbouh, Marwan, Stu Jolly, Dock Allen, Paul Silvey, and Paul Denning. 2001. Interoperability. *W3C Web Services Workshop,* April 11–12, San Jose, CA.

SOAP. *Simple Object Access Protocol,* http://www.w3.org/TR/soap.

Tolbeert, Caroline and Karen Mossberger. 2003. The effects of e-government on trust and confidence in government. *DG.O 2003 Conference,* Boston, May.

Trastour, David, Claudio Bartolini, and Javier Gonzalez-Castillo. 2001. A Semantic Web approach to service description for matchmaking of services. In *Proceedings of the International Semantic Web Working Symposium (SWWS),* July 30–August 1.

Tsur, Shalom, Serge Abiteboul, Rakesh Agrawal, Umeshwar Dayal, Johannes Klein, and Gerhard Weikum. 2001. Are Web services the next revolution in e-commerce? (Panel). In *Proceedings of the 27th International Conference on Very Large Databases (VLDB),* Morgan Kaufmann, San Francisco.

U.S. Congress. 1996. *Health Insurance Portability and Accountability Act.*

U.S. Geological Survey. 2002. Geographic Information Systems.

U.S. PITAC. 2000. Transforming Access to Government Through Information Technology. *Report to the President,* September.

Vinoski, Steve. 2002a. Web services interaction models, Part 1: Current practice. *IEEE Internet Computing,* 6(3): 89–91, February.

Vinoski, Steve. 2002b. Where is middleware? *IEEE Internet Computing,* 6(2), March.

W3C. 2001a. *Semantic Web,* http://www.w3c.org/2001/sw/.

W3C. 2001b. *SOAP Security Extensions: Digital Signature,* http://www.w3.org/TR/SOAP-dsig.

W3C. 2001c. *XML Encryption, http://www.w3.org/Encryption.*

W3C. 2001d. *XML Key Management Specification (XKMS),* http://www.w3.org/TR/xkms/.

W3C. 2001e. *XML Signature,* http://www.w3.org/Signature/.

Welch, Eric W. and Charles C. Hinnant. 2003. Internet use, transparency, and interactivity effects on trust in government. In *Proceedings of the 36th Annual Hawaii International Conference on System Sciences (HICSS '03),* ACM Digital Library, ACM.

West, Darrell, M. 2002. Urban E-Government. Center for Public Policy, Brown University, Providence, RI, September.

WSFL. *Web Services Flow Language,* http://xml.coverpages.org/wsfl.html.

Wunnava, Subbarao V. and Madhusudhan V. Reddy. 2000. Adaptive and dynamic service composition in eFlow. In *Proceedings of the IEEE Southeastcon 2000,* IEEE, pp. 205–208.

XLANG. http://wwwcoverpages.org/xlang.html.

3

E-Learning Technology for Improving Business Performance and Lifelong Learning

CONTENTS

Abstract 3-1
Key Words 3-2
3.1 E-Learning and Business Performance 3-2
3.2 Evolution of Learning Technologies 3-3
3.3 Web-Based E-Learning Environments 3-4
3.3.1 Learning Theories and Instructional Design 3-4
3.3.2 Types of E-Learning Environments 3-5
3.3.3 Creation and Delivery of E-Learning 3-6
3.4 E-Learning Standards 3-7
3.4.1 Standards Organizations 3-7
3.4.2 SCORM Specification 3-8
3.5 Improving Business Performance Using E-Learning Technology 3-10
3.5.1 E-Learning Technology for Delivering Business Knowledge 3-10
3.5.2 E-Learning Technology for Improving Business Processes 3-12
3.5.3 E-Learning Technology for Lifelong Learning 3-13
3.5.4 Advancements in Infrastructure Technology To Support E-Learning 3-14
3.6 Conclusions 3-17
References 3-17

Darrell Woelk

Abstract

This chapter describes the impact that Internet-based e-learning can have and is having on improving the business performance for all types of organizations. The Internet is changing how, where, when, and what a student learns, how the progress of that learning is tracked, and what the impact of that learning is on the performance of the business. This chapter establishes a model for understanding how e-learning can impact business performance. It describes the history of e-learning technology and the present state of Internet-based e-learning architectures and standards. Finally, it lays out a vision for the future of e-learning that builds on advancements in semantic web and intelligent tutoring technology.

1-58488-381-2/05/$0.00+$1.50

© 2005 by CRC Press LLC

Key Words

E-Learning, knowledge management, business performance, lifelong learning, learning theory, standards, LMS, LCMS, SCORM, semantic web, ontology, web services

3.1 E-Learning and Business Performance

E-Learning is a critical component of the overall knowledge management strategy for an organization. Figure 3.1 is a representation of the phases of knowledge management and the role of e-learning in those phases [Woelk, 2002b] [Nonaka, 1995]. The *knowledge holder* on the left of Figure 3.1 has *tacit knowledge* that is valuable to the *knowledge seeker* on the right who is making business decisions and performing business tasks. This tacit knowledge can be transferred to the knowledge seeker either directly through a social exchange or indirectly by translating the tacit knowledge to *explicit knowledge* and storing it in the *knowledge repository* at the center of the figure. The knowledge seeker then translates the explicit knowledge back to tacit knowledge through a learning process and applies the tacit knowledge to business decisions and business tasks. Those decisions and tasks generate *operational data* that describe the performance of the business and the role of the knowledge seeker in that performance. Operational data can be analyzed to help determine if skills have been learned and to suggest additional learning experiences for the knowledge seeker.

The *knowledge organizer* in Figure 3.1 is a person (or software program) who relates new explicit knowledge created by the knowledge holder to other knowledge in the repository or further refines the created knowledge. The *instructional designer* is a person (or software program) who organizes the learning of the knowledge by adding such features as preassessments, additional learning aids, and postassessments. Web-based e-learning software improves the capability of an organization to transfer tacit knowledge from knowledge holders to knowledge seekers and assists knowledge seekers to learn explicit knowledge.

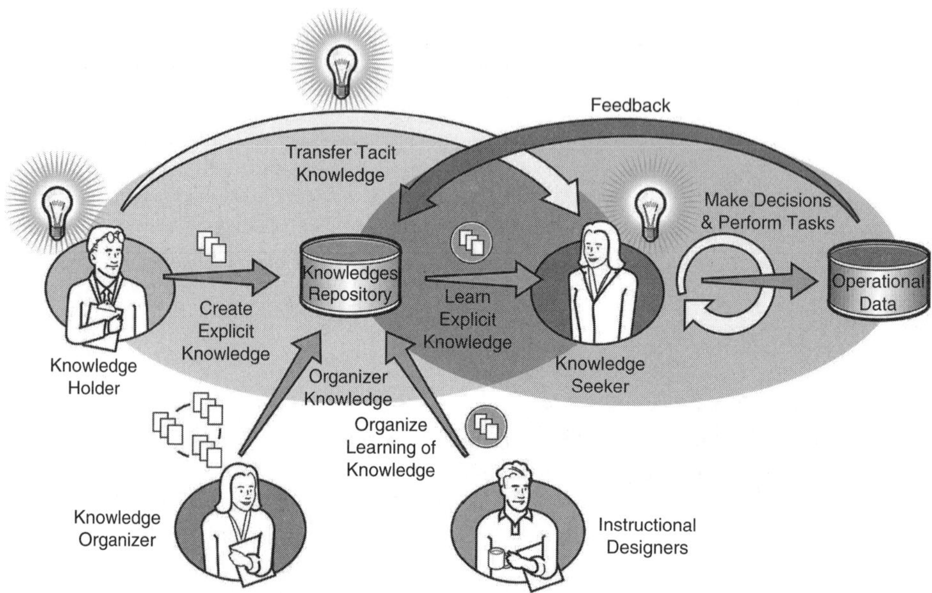

FIGURE 3.1 Knowledge management phases with e-learning enhancements.

3.2 Evolution of Learning Technologies

Learning technologies traditionally have focused on two areas shown in Figure 3.1: (1) software tools to help the instructional designer create learning experiences for the transfer of explicit knowledge to the knowledge seeker and (2) the runtime environment provided to the knowledge seeker for those learning experiences. Figure 3.2 (from [SCORM, 2002] and [Gibbons, 2000]) illustrates the evolution of learning technology in these areas. Investigation of techniques and algorithms for computer-based instruction began in the 1950s and 1960s with the focus on automating relatively simple notions of learning and instruction. This led to the development of procedural instructional languages that utilized instructional vocabulary understandable to training content developers.

Beginning in the late 1960s, the development of Computer Based Instruction (CBI) was split into two factions. One group (top of Figure 3.1) continued to follow an evolutionary path of improving the procedural instructional languages by taking advantage of general improvements in software technology. This led to commercial authoring systems that provided templates to simplify the creation of courses, thus lowering the cost and improving the effectiveness of authors. These systems were mostly client based with instructional content and procedural logic tightly bound together.

The second group (bottom of Figure 3.1), however, took a different approach to CBI. In the late 1960s, advanced researchers in this group began to apply the results of early artificial intelligence research to the study of how people learn. This led to development of a different approach called intelligent tutoring systems (ITS). This approach focuses on generating instruction in real time and on demand as required by individual learners and supporting mixed initiative dialogue that allows free-form discussion between technology and the learner.

The advent of the Internet and the World Wide Web in the early 1990s had an impact on both of these groups. The CBI systems developed by the first group began to change to take advantage of the Internet as a widely accessible communications infrastructure. The Internet provided neutral media formats and standard communications protocols that gradually replaced the proprietary formats and protocols of the commercially available systems. The ITS researchers also began to adapt their architectures to take

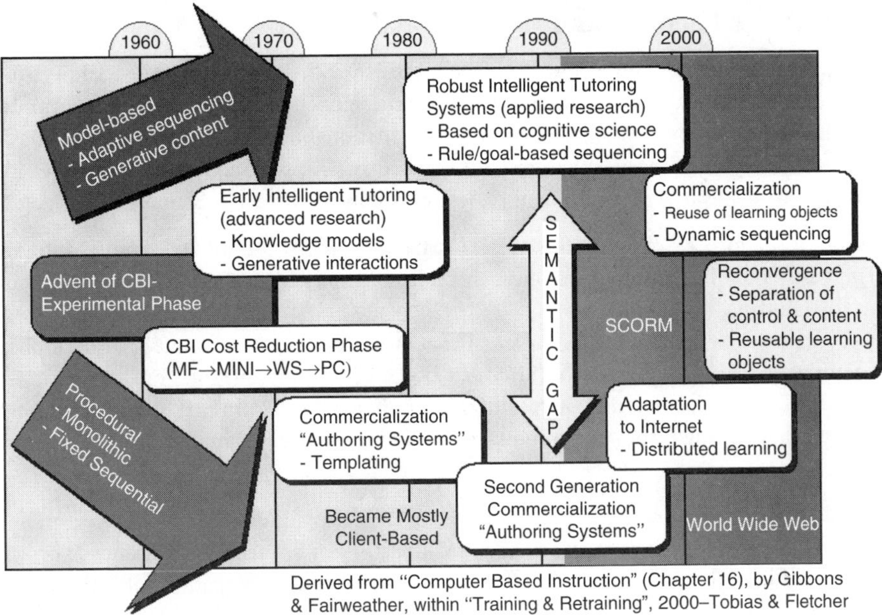

FIGURE 3.2 Evolution of learning technologies and intelligent tutoring.

advantage of the Internet. The generation of instruction in real time based on the needs of the individual learner became more feasible as more content became available dynamically online and real-time communication with the learner was improved.

This has set the stage for the reconvergence of commercial CBI products and advanced ITS research. The notion of dynamically assembling reusable learning objects [Longmire, 2000] into learning experiences has become a goal of both groups. Reusable learning objects are pieces of learning content that have the following attributes:

- Modular, free standing, and transportable among applications and environments
- Nonsequential
- Able to satisfy a single learning objective
- Accessible to broad audiences (such that it can be adapted to audiences beyond the original target audience)

The complexity of the algorithms for the dynamic selection of learning objects and the dynamic determination of the sequencing of the presentation of these learning objects to the user differs between the two groups. CBI products focus on providing authors with effective tools for specifying sequencing while ITS researchers focus on algorithms for dynamically adapting content and sequencing to individual learning styles and providing constant feedback on learning progress. Both groups, however, are moving towards a common architecture and infrastructure that will accommodate both existing products and the results of future research.

3.3 Web-Based E-Learning Environments

There are a variety of technologies and products that fall under the category of web based e-learning. This section will first discuss the learning theories and instructional design techniques that serve as the basis for e-learning. It will then describe the different categories of e-learning and describe the Web-based tools for creating and delivering an e-learning experience.

3.3.1 Learning Theories and Instructional Design

The development of theories about how people learn began with Aristotle, Socrates, and Plato. In more recent years, instructional design techniques have been developed for those learning theories that assist in the development of learning experiences. These instructional design techniques are the basis for the e-learning authoring software tools that will be discussed later in this section.

Although there are a variety of learning theories, the following three are the most significant:

- *Behaviorism* focuses on repeating a new behavioral pattern until it becomes automatic. The emphasis is on the response to stimulus with little emphasis on the thought processes occurring in the mind.
- *Cognitivism* is similar to behaviorism in that it stresses repetition, but it also emphasizes the cognitive structures through which humans process and store information.
- *Constructivism* takes a completely different approach to learning. It states that knowledge is constructed through an active process of personal experience guided by the learner himself/herself.

Present-day instructional design techniques have been highly influenced by behaviorism and cognitivism. A popular technique is to state a behavioral objective for a learning experience [Kizlik, 2002]. A well-constructed behavioral objective describes an intended learning outcome and contains three parts that communicate the *conditions* under which the behavior is performed, a *verb* that defines the behavior itself, and the degree (*criteria*) to which a student must perform the behavior. An example is "Given a stethoscope and normal clinical environment, the medical student will be able to diagnose a heart arrhythmia in 90% of affected patients."

A behaviorist–cognitivist approach to instructional design is prescriptive and requires analyzing the material to be learned, setting a goal, and then breaking this into smaller tasks. Learning (behavioral) objectives are then set for individual tasks. The designer decides what is important for the learner to know and attempts to transfer the information to the learner. The designer controls the learning experience although the learner may be allowed some flexibility in navigating the material. Evaluation consists of tests to determine if the learning objectives have been met.

A constructionist approach to instructional design is more facilitative than prescriptive. The designer creates an environment in which the learner can attempt to solve problems and build a personal model for understanding the material and skills to be learned. The learner controls the direction of the learning experience. Evaluation of success is not based on direct assessments of individual learning objectives but on overall success in problem solving. An example of a constructionist approach is a simulation that enables a student to experiment with using a stethoscope to diagnose various medical problems.

3.3.2 Types of E-Learning Environments

Figure 3.3 is a representation of a Web-based e-learning environment that illustrates two general types of Web-based learning environments: *synchronous* and *asynchronous*.

- *Synchronous:* A synchronous learning environment is one in which an instructor teaches a somewhat traditional class but the instructor and students are online simultaneously and communicate directly with each other. Software tools for synchronous e-learning include audio conferencing, video conferencing, and virtual whiteboards that enable both instructors and students to share knowledge.
- *Asynchronous:* In an asynchronous learning environment, the instructor only interacts with the student intermittently and not in real time. Asynchronous learning is supported by such technologies as online discussion groups, email, and online courses.

There are four general types of asynchronous e-learning [Kindley, 2002]:

- *Traditional Asynchronous E-Learning:* Traditional asynchronous e-learning courses focus on achieving explicit and limited learning objectives by presenting information to the learner and assessing the retention of that information by the student through tests. This type of e-learning is based on behaviorist–cognitivist learning theories and is sometimes called "page-turner" because of its cut-and-dried approach.
- *Scenario-Based E-Learning:* Scenario-based e-learning focuses more on assisting the learner to learn the proper responses to specifically defined behaviors. An example might be training a salesperson about how to react to a customer with a complaint. A scenario is presented and the student is asked to select from a limited set of optional responses.
- *Simulation-Based E-Learning:* Simulation-based e-learning differs from scenario-based e-learning in that a broader reality is created in which the student is immersed. The student interacts with this environment to solve problems. There are usually many different behavioral paths that can be successful. [Kindley, 2002] [Jackson, 2003]. This type of e-learning is based on constructivist learning theories.
- *Game-Based E-Learning:* Games are similar to simulations except that the reality is artificial and not meant to be an exact representation of the real world. Most game-based e-learning today is limited to simple games that teach particular skills. However, the use of more complex, massively multiplayer game techniques and technology for game-based e-learning is being investigated [DMC, 2003].

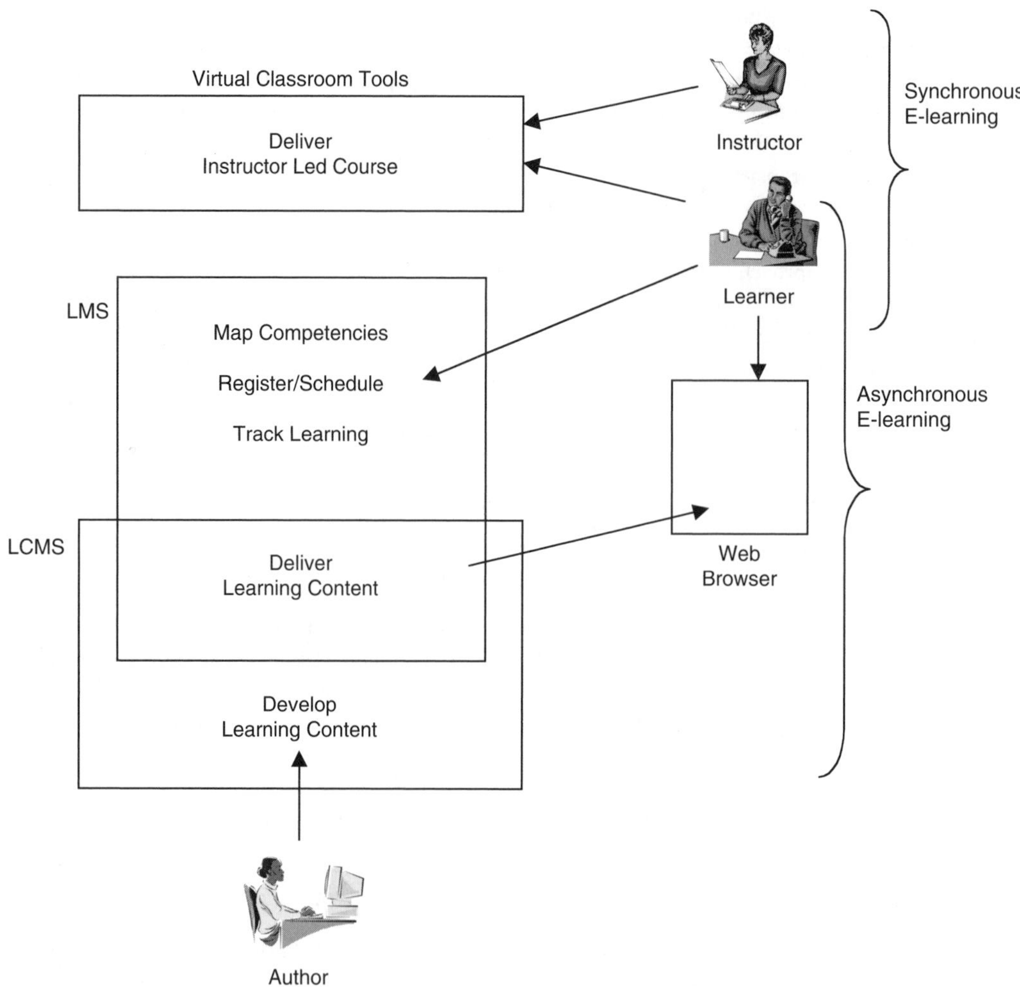

FIGURE 3.3 Typical Web-based e-learning environment.

3.3.3 Creation and Delivery of E-Learning

E-learning environments typically provide the following capabilities for creation and delivery of e-learning as shown in Figure 3.3.

- *Map Competencies to Courses:* An administrator can describe the competencies (skills) necessary for selected jobs within an organization and describe the learning content (courses) that will teach that skill.
- *Schedule Classes/Register Students:* An administrator can schedule synchronous classes or post links to courses for asynchronous classes. Students can register for synchronous and asynchronous classes.
- *Track Learning:* The system can track which classes a student takes and how the student scores on the assessments in the class.
- *Develop Learning Content:* Authors are provided with software tools for creating asynchronous courses made up of reusable learning objects.
- *Deliver Learning Content:* Asynchronous courses or individual learning objects that have been stored on the server are delivered to students via a Web browser client.

The capabilities described above are provided by three categories of commercial software products:

- *Virtual Classroom Tools* provide such tools as audio conferencing, video conferencing, and virtual whiteboards for synchronous e-learning.
- *Learning Management Systems (LMS)* provide mapping of competencies, scheduling of classes, registering of students, and tracking of students for both synchronous and asynchronous e-learning [Brandon-Hall, 2003]. The interface for most of this functionality is a Web browser.
- *Learning Content Management Systems (LCMS)* provide tools for authoring courses that include templates for commonly used course formats and scripting languages for describing sequencing of the presentation of learning content to the student [LCMS, 2003]. The interface for this functionality is usually a combination of a web browser for administrative functions and a Microsoft Windows based environment for more complex authoring functions.

3.4 E-Learning Standards

Standardization efforts for learning technology began as early as 1988 in the form of specifications for CBI hardware and software platforms. The Internet and the World Wide Web shifted the focus of these standards efforts to specifications for Internet protocols and data formats. This section will first review the organizations active in the creation of e-learning standards and then review the most popular standard in more detail.

3.4.1 Standards Organizations

There are two types of standards: *de facto* and *de jure*. *De facto* standards are created when members of an industry come together to create and agree to adopt a specification. *De jure* standards are created when an accredited organization such as IEEE designates a specification to be an official, or *de jure*, standard. The remainder of this chapter will refer to *de facto* standards as just specifications in order to avoid confusion.

The most popular *de facto* standards (specifications) are being developed by the following industry organizations:

- Aviation Industry CBT Committee (AICC) [AICC, 2003] is an international association of technology-based training professionals that develops guidelines for aviation industry in the development, delivery, and evaluation of Computer-Based Training (CBT) and related training technologies. AICC was a pioneer in the development of standards beginning with standards for CD-based courses in 1988.
- IMS (Instructional Management System) Global Learning Consortium [IMS, 2003] is a global consortium with members from educational, commercial, and government organizations that develops and promotes open specifications for facilitating online distributed learning activities such as locating and using educational content, tracking learner progress, reporting learner performance, and exchanging student records between administrative systems.
- Advanced Distributed Learning (ADL) Initiative [ADL, 2003], sponsored by the Office of the Secretary of Defense (OSD), is a collaborative effort between government, industry, and academia to establish a new distributed learning environment that permits the interoperability of learning tools and course content on a global scale. ADL's vision is to provide access to the highest quality education and training, tailored to individual needs and delivered cost-effectively anywhere and anytime.

In the last few years, these three organizations have begun to harmonize their specifications. The Sharable Content Object Reference Model (SCORM) specification from ADL, discussed in the following section, is the result of this harmonization.

The following official standards organizations are working on promoting some of the *de facto* e-learning specifications to *de jure* standards:

- The Institute of Electrical and Electronics Engineers (IEEE) Learning Technology Standards Committee (LTSC) [IEEELTSC, 2003] has formed working groups to begin moving the SCORM specification towards adoption as an IEEE standard.
- The International Organization for Standards (ISO) JTC1 SC36 subcommittee [ISO, 2003] has also recently created an *ad hoc* committee to study the IEEE proposal and proposals from other countries.

3.4.2 SCORM Specification

SCORM [SCORM 2002] is the most popular e-learning specification today. It assumes a Web-based infrastructure as a basis for its technical implementation. SCORM provides a specification for construction and exchange of learning objects, which are called Sharable Content Objects (SCOs) in the SCORM specification. The term SCO will be used in the remainder of this section instead of learning object. As the SCORM specification has evolved, it has integrated specifications from AAIC and IMS. This harmonization of competing specifications is critical to the acceptance of SCORM by the industry.

Figure 3.4 illustrates the types of interoperability addressed by the SCORM specification. SCORM does not address interoperability for the delivery of synchronous e-learning. It only addresses interoperability for asynchronous e-learning. In particular, it addresses the structure of online courses, the interface to a repository for accessing online courses, and the protocol for launching online courses and tracking student progress and scores.

The high-level requirements [SCORM 2002] that guide the scope and purpose of the SCORM specification are:

- The ability of a Web-based e-learning system to launch content that is authored by using tools from different vendors and to exchange data with that content during execution

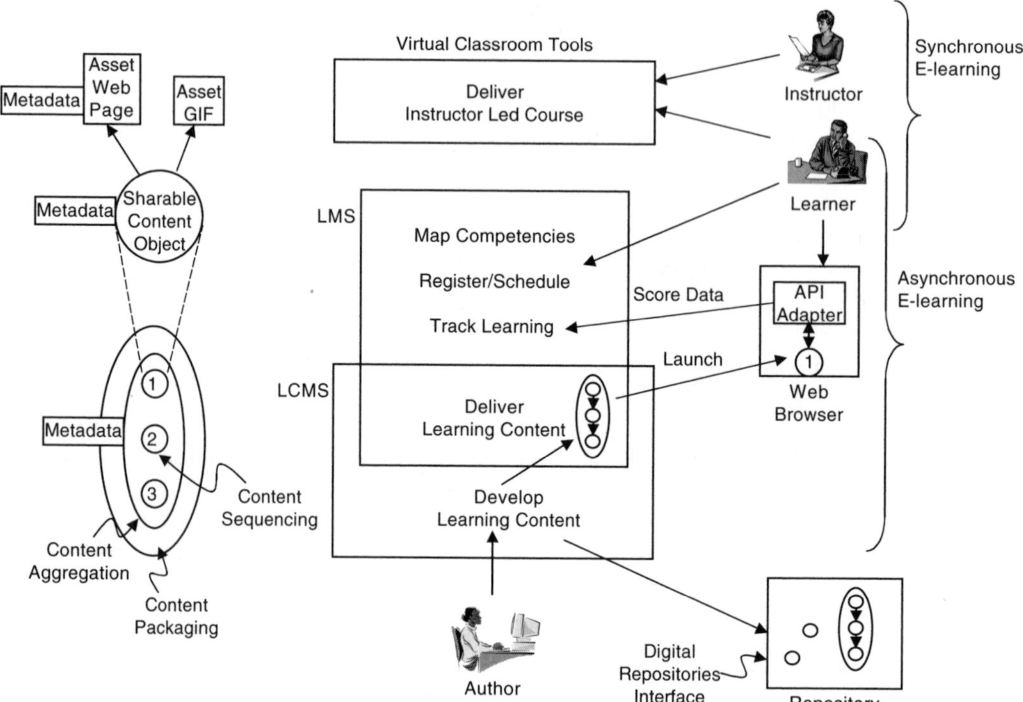

FIGURE 3.4 SCORM interoperability using SCOs (learning objects).

- The ability of Web-based e-learning systems from different vendors to launch the same content and exchange data with that content during execution
- The ability of multiple Web-based e-learning systems to access a common repository of executable content and to launch such content

The SCORM specification has two major components: the *content aggregation model* and the *run-time environment.*

3.4.2.1 SCORM Aggregation Model

The SCORM content aggregation model represents a pedagogical and learning-theory-neutral means for designers and implementers of learning content to aggregate learning resources. The most basic form of learning content is an *asset* that is an electronic representation of media, text, images, sound, web pages, assessment objects, or other pieces of data delivered to a Web client. The upper left corner of Figure 3.4 shows two assets: a Web page and a GIF file. An SCO is the SCORM implementation of a "learning object." It is a collection of one or more assets. The SCO in Figure 3.4 contains two assets: a Web page and a GIF file. An SCO represents the lowest granularity of learning content that can be tracked by an LMS using the SCORM run-time environment (see next section). An SCO should be independent of learning context so that it can be reusable. A content aggregation is a map that describes an aggregation of learning resources into a cohesive unit of instruction such as a course, chapter, and module.

Metadata can be associated with assets, SCOs, and content aggregations as shown in Figure 3.4. Metadata help make the learning content searchable within a repository and provide description information about the learning content. The metadata types for SCORM are drawn from the approximately 64 metadata elements defined in the IEEE LTSC Learning Object Metadata specification [IEEELOM, 2002]. There are both required and optional metadata. Examples of required metadata include title and language. An example of an optional metadata field is taxonpath, which describes a path in an external taxonomic classification. The metadata is represented as XML as defined in the IMS Learning Resources Metadata XML Binding Specification [IMS, 2003].

3.4.2.2 SCORM Content Packaging Model

The content aggregation specification only describes a map of how SCOs and assets are *aggregated* to form larger learning units. It does not describe how the smaller units are *actually* packaged together. The *content packaging* specification describes this packaging that enables exchange of learning content between two systems. The SCORM content packaging is based on the IMS Content Packaging Specification. A content package contains two parts: (1) a (required) special XML document describing the content organization and resources of the package. The special file is called the Manifest file because package content and organization is described in the context of manifests, and (2) the physical files referenced in the Manifest. Metadata may also be associated with a content package.

IMS recently released a specification for *simple sequencing* that is expected to become a part of the SCORM specification in the near future. The simple sequencing specification describes a format and an XML binding for specifying the sequencing of learning content to be delivered to the student. This enables an author to declare the relative order in which SCOs are to be presented to the student and the conditions under which an SCO is selected and delivered or skipped during a presentation. The specification incorporates rules that describe branching or flow of learning activities through content according to outcomes of a student's interaction with the content.

IMS has also recently released a Digital Repositories Interoperability (DRI) that defines a specific set of functions and protocols that enable access to a heterogeneous set of repositories as shown in the lower right of Figure 3.4. Building on specifications for metadata and content packaging, the DRI recommends XQuery for XML Metadata Search and simple messaging using SOAP with Attachments over HTTP for interoperability between repositories and LMS and LCMS systems. The specification also includes existing search technologies (Z39.50) that have successfully served the library community for many years.

3.4.2.3 SCORM Run-Time Environment

The SCORM run-time environment provides a means for interoperability between an SCO and an LMS or LCMS. This requires a common way to start learning resources, a common mechanism for learning resources to communicate with an LMS, and a predefined language or vocabulary forming the basis of the communication. There are three aspects of the Run-Time Environment:

- *Launch:* The launch mechanism defines a common way for an LMS or LCMS to start Web-based learning resources. This mechanism defines the procedures and responsibilities for the establishment of communication between the delivered learning resource and the LMS or LCMS.
- *Application Program Interface (API):* The API is the communication mechanism for informing the LMS or LCMS of the state of the learning resource (e.g., initialized, finished, or in an error condition), and is used for getting and setting data (e.g., score, time limits).
- *Data Model:* The data model defines elements that both the LMS or LCMS and SCO are expected to "know" about. The LMS or LCMS must maintain the state of required data elements across sessions, and the learning content must utilize only these predefined data elements if reuse across multiple systems is to occur. Examples of these data elements include:
  - Student id and student name
  - Bookmark indicating student progress in the SCO
  - Student score on tests in the SCO
  - Maximum score that student could attain
  - Elapsed time that student has spent interacting with the SCO and why the student exited the SCO (timeout, suspend, logout)

3.5 Improving Business Performance Using E-Learning Technology

Section 3.1 described how e-learning can be integrated with the knowledge management strategy of an organization to enhance the performance of the business. Sections 3.2, 3.3, and 3.4 described the evolution of e-learning technologies, the typical e-learning environment, and the present state of e-learning standards. This section will investigate further where e-learning and Internet technology can be used today and in the future to improve the delivery of business knowledge and improve business processes. It will then take a broader look at how e-learning and Internet technology can be used to enable a process of lifelong learning that spans multiple corporations and educational institutions. Finally, it will describe the advancements in infrastructure technology needed to support e-learning.

3.5.1 E-Learning Technology for Delivering Business Knowledge

The development and standardization of the learning object technology described in the previous sections makes it possible to more effectively integrate e-learning with delivery of business knowledge to employees of an organization. Figure 3.5 is a modification of Figure 3.1 that illustrates an example of the types of knowledge that must be delivered in an organization and the types of systems that hold that knowledge [Woelk, 2002b]. The knowledge repository that was shown in the center of Figure 3.1 is actually implemented in a typical organization by three distinct systems: a learning object repository, a content management system, and a knowledge management system.

The *learning object repository* holds reusable learning objects that have been created using an LCMS and will be delivered as part of a learning experience by an LMS. The *content management system* is used to develop and deliver enterprise content such as product descriptions that are typically delivered via the World Wide Web. Example content management systems are Vignette and Interwoven. The *knowledge management system* is used to capture, organize, and deliver more complex and unstructured knowledge such as email messages and text documents.

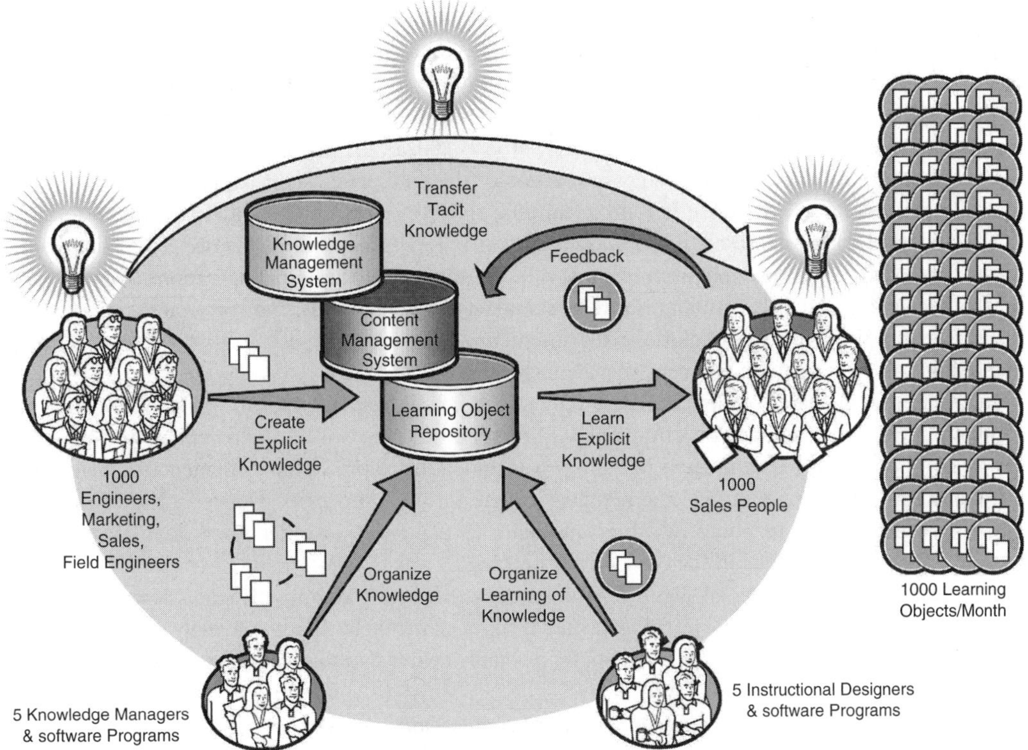

FIGURE 3.5 Delivering business knowledge from heterogeneous sources of learning content.

In most organizations today, these three types of systems are not interconnected with each other. If a learning object repository is being used to store learning objects for delivery to learners, the content in the repository is either developed from scratch or copied from some other file system with no tracking of where the content originated. This means that the instructional designers in the lower right corner of Figure 3.5 are typically talking directly to the engineers, marketing, sales, and field engineers on the left side of the figure to find existing online content or to learn enough to create learning objects themselves. The result is that there is content that the instructional designers either do not know about or that they know about but cannot find. This causes development resources to be wasted and inferior learning objects to be created.

The solution to this problem is to provide learning object repositories that can dynamically retrieve content from a content management system, thus providing learning objects that seamlessly include content from content management systems. E-learning vendors are beginning to support this feature. However, there is other content stored in knowledge management systems that should be also being included in learning objects. Therefore, learning object repositories must also be able to dynamically retrieve knowledge from knowledge management systems. This enables information such as email messages, memos, sales call notes, and audio messages that are related to a topic to be delivered with a learning experience.

Internet-based infrastructure such as Web services and enterprise integration applications will enable the integration of learning object repositories, content management systems, and knowledge management systems. However, while this integration will improve the caliber of content that is delivered by the e-learning system and decrease the cost of developing that content, it does not help ensure that the right people are receiving the right content at the time that they need it to be effective. The next section will discuss the integration of e-learning with the business processes of the organization.

3.5.2 E-Learning Technology for Improving Business Processes

A goal of corporate e-learning is to increase efficiency by identifying precisely the training that an employee needs to do his or her job and provide that training in the context of day-to-day job activities of the employee. Figure 3.6 illustrates the flow of activity to implement competency-based just-in-time learning services in an enterprise environment [Woelk, 2002a]. In the lower right hand corner is a representation of a couple of the business processes of the enterprise. The example shown here is the interaction between an Enterprise Resource Planning (ERP) process implemented using SAP software [SAP, 2003] and a Customer Relationship Management (CRM) process implemented using Siebel software [Siebel, 2003]. ERP software from SAP provides solutions for the internal operations of a company, such as managing financials, human resources, manufacturing operations, and corporate services. CRM software from Siebel provides solutions for interacting with a company's customers, such as sales, marketing, call centers, and customer order management.

In the upper left-hand corner of Figure 3.6 are the ontologies that capture knowledge about the company such as the products the company sells, the organizations within the company, and the competitors of the company. A competency ontology is included that captures the competencies an employee must possess to participate in specific activities of the business processes. Nodes in the competency ontology will be linked to nodes in other ontologies in Figure 3.6 in order to precisely describe the meaning of skill descriptions in the competency ontology.

The lower left-hand corner of Figure 3.6 illustrates the learning resources such as learning objects, courses, and e-mail. Each of these learning resources has been manually or automatically linked to various parts of one or more of the enterprise ontologies to enable people and software agents to more efficiently

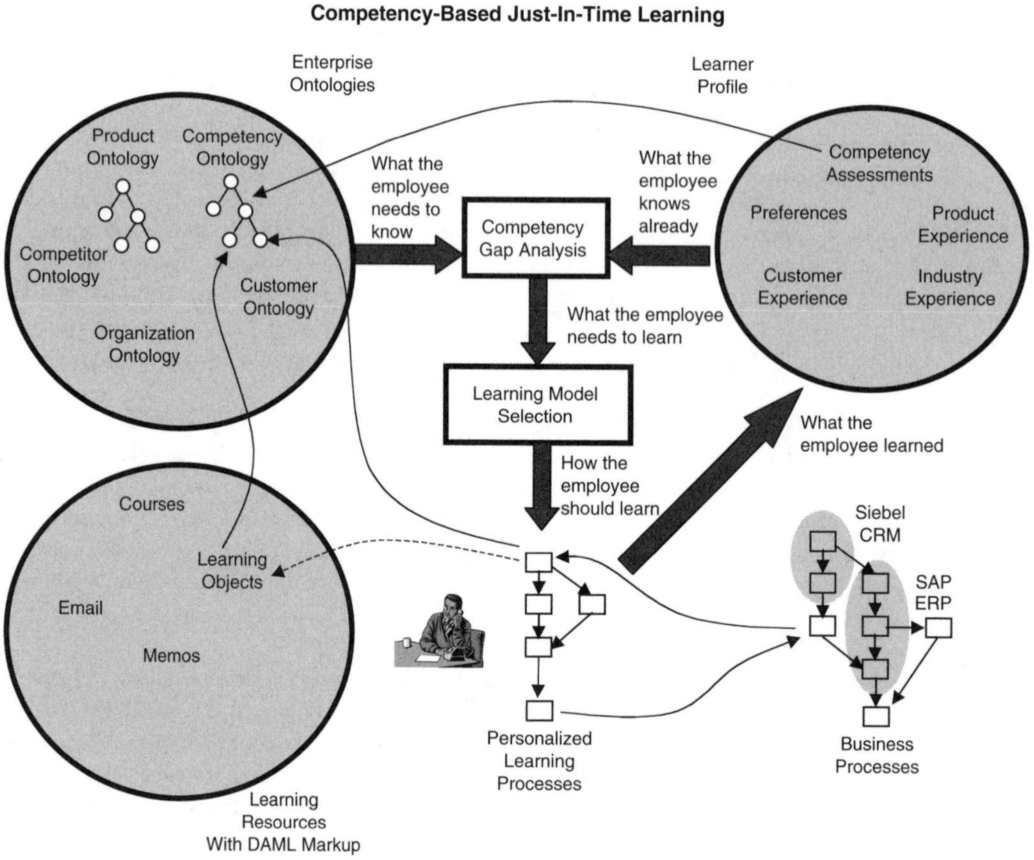

FIGURE 3.6 Competency-based just-in-time learning in a corporation.

discover the correct learning resources. The upper right hand corner illustrates the learner profile for an employee that contains preferences, experiences, and assessments of the employee's competencies.

The box in the upper middle of Figure 3.6 is a competency gap analysis that calculates what competencies the employee lacks to effectively carry out his or her job responsibilities. This calculation is based on what the employee needs to know and what he or she already knows. Once the competency gap has been identified, a learning model is selected either manually or automatically. This establishes the best way for the employee to attain the competency and enables the system to create the personalized learning process at the lower middle of Figure 3.6. The personalized learning process may be created and stored for later use or it may be created whenever it is needed, thus enabling dynamic access to learning objects based on the most recent information about the learner and the environment. The results of the personalized learning process are then returned to the learner's profile.

This system will enable the integration of personalized learning processes with the other business processes of the enterprise, thus enabling continuous learning to become an integral part of the processes of the corporation. Section 3.5.4 will describe the technologies necessary to implement the system described here and the state of each of the technologies.

3.5.3 E-Learning Technology for Lifelong Learning

Technology for e-learning and knowledge management described in the previous two sections have set the stage for the fulfillment of the vision for lifelong learning put forth by Wayne Hodgins for the Commission on Technology and Adult Learning in February, 2000 [Hodgins, 2000]. According to Hodgins, a key aspect of this vision is *performance-based learning.* Performance-based learning is the result of a transition from "teaching by telling" to "learning by doing," assisted by technological and human coaches providing low-level and high-level support. Furthermore, his key to the execution of performance-based learning is successful information management. Successful information management makes it possible "to deliver just the *right* information, in just the *right* amount, to just the *right* person in just the *right* context, at just the *right* time, and in a form that matches the way *that* person learns. When this happens, the recipient can act — immediately and effectively."

While the stage has been set for the fulfillment of a vision of performance-based lifelong learning, there are numerous social and technology obstacles that may yet stand in the way. These obstacles must first be understood and then proactive steps must be taken to overcome the obstacles or minimize their impact on attaining the vision.

There are separate obstacles to implementing a system to support performance-based learning and implementing a system to support *lifelong learning.* However, there are even greater obstacles to implementing an integrated system to support *performance-based lifelong learning* [Woelk, 2002c]:

- Performance-based learning requires that the system have *deep, exact* knowledge of what the persons are doing and what they already know about that task to determine what they should learn now.
- Lifelong learning requires that the system have *broad, general* knowledge over a number of years of what the persons have learned to determine what they should learn now.

Today, it is difficult to provide a system with deep, exact knowledge of what a person is doing and what he or she needs to learn. Therefore, this capability is restricted to handcrafted systems in large organizations. Furthermore, such a handcrafted system is limited in its access to broad, general knowledge about what the person has learned in the past because this knowledge is in databases that are not accessible to it.

The ultimate success of performance-based lifelong learning will be dependent on the sharing of knowledge and processes among the various organizations that make up the lifelong learning environment. Figure 3.7 describes some of these organizations along with the knowledge and the processes they must share: Learner Profile, Competency Ontology, Enterprise Ontologies, Learning Objects, Business

Processes, and Personalized Learning Processes. These types of knowledge were described in more detail in Section 5.2.

For performance-based lifelong learning to be successful, Figure 3.7 illustrates that the learner profile must be able to make reference to competency ontologies, learning objects, and business processes in multiple organizations. This requires standards for representations of the various types of knowledge and standards for the protocol to access the various types of knowledge. Furthermore, it is likely that multiple organizations may create knowledge such as competency ontologies separately and mappings among the ontologies will be required.

Technology advancements alone will not be sufficient to ensure the success of performance-based lifelong learning. Much of the success will depend on a commitment by the various organizations in Figure 3.7 to share their knowledge. This commitment includes not only an organizational commitment to sharing system knowledge, but just as important, a commitment by individuals in the organizations to interact with individuals in other organizations.

The following section will list some potential technology issues related to overcoming the obstacles to performance-based lifelong learning.

3.5.4 Advancements in Infrastructure Technology To Support E-Learning

This system will enable the integration of personalized learning processes with the other business processes of the enterprise, thus enabling continuous learning to become an integral part of the processes of the corporation. The following sections will describe the technologies necessary to implement the system described here and the state of each of the technologies.

3.5.4.1 Web Services

Each box in the business processes and the personalized learning processes in Figure 3.6 is an activity. These activities might be implemented as existing legacy applications or new applications on a variety of hardware and software systems. In the past, it would have been difficult to integrate applications executing on such heterogeneous systems. But the development of a set of technologies referred to as *Web services* [W3C, 2001a] [Glass, 2001] has simplified this integration. The most important of these technologies are Service Oriented Architecture Protocol (SOAP); Universal Description, Discovery, and Integration (UDDI); and Web Services Description Language (WSDL). IBM and Microsoft have both agreed on the specifications for these technologies, making it possible to integrate UNIX and Microsoft systems.

SOAP is a technology for sending messages between two systems. SOAP uses XML for representing the messages, and HTTP is the most common transport layer for the messages. UDDI is a specification for an online registry that enables publishing and dynamic discovery of Web services. WSDL is an XML representation that is used for describing the services that are registered with a UDDI registry. Many vendors of enterprise applications such as ERP and CRM are now providing Web service interfaces to their products. Once developers of learning objects and learning management systems begin to provide Web service interfaces to those objects, it will be possible to dynamically discover and launch learning objects as services on heterogeneous systems.

3.5.4.2 Semantic Web Services

UDDI and WSDL have limited capability for representing semantic descriptions of Web services. The discovery of services is limited to using restricted searches of keywords associated with the service. This is insufficient for the discovery of learning objects; but there a number of efforts underway to improve this situation.

The World Wide Web consortium has initiated an effort to develop specifications for a *semantic Web* [W3C, 2001b] where Web pages will include semantic descriptions of their content. One result of this effort has been the Resource Description Framework (RDF) model for describing the contents of a Web page [W3C, 2001c]. The U.S. Defense Advanced Research Projects Agency (DARPA) has also been sponsoring research as part of the DARPA Agent Markup Language (DAML) program [DAML, 2002]

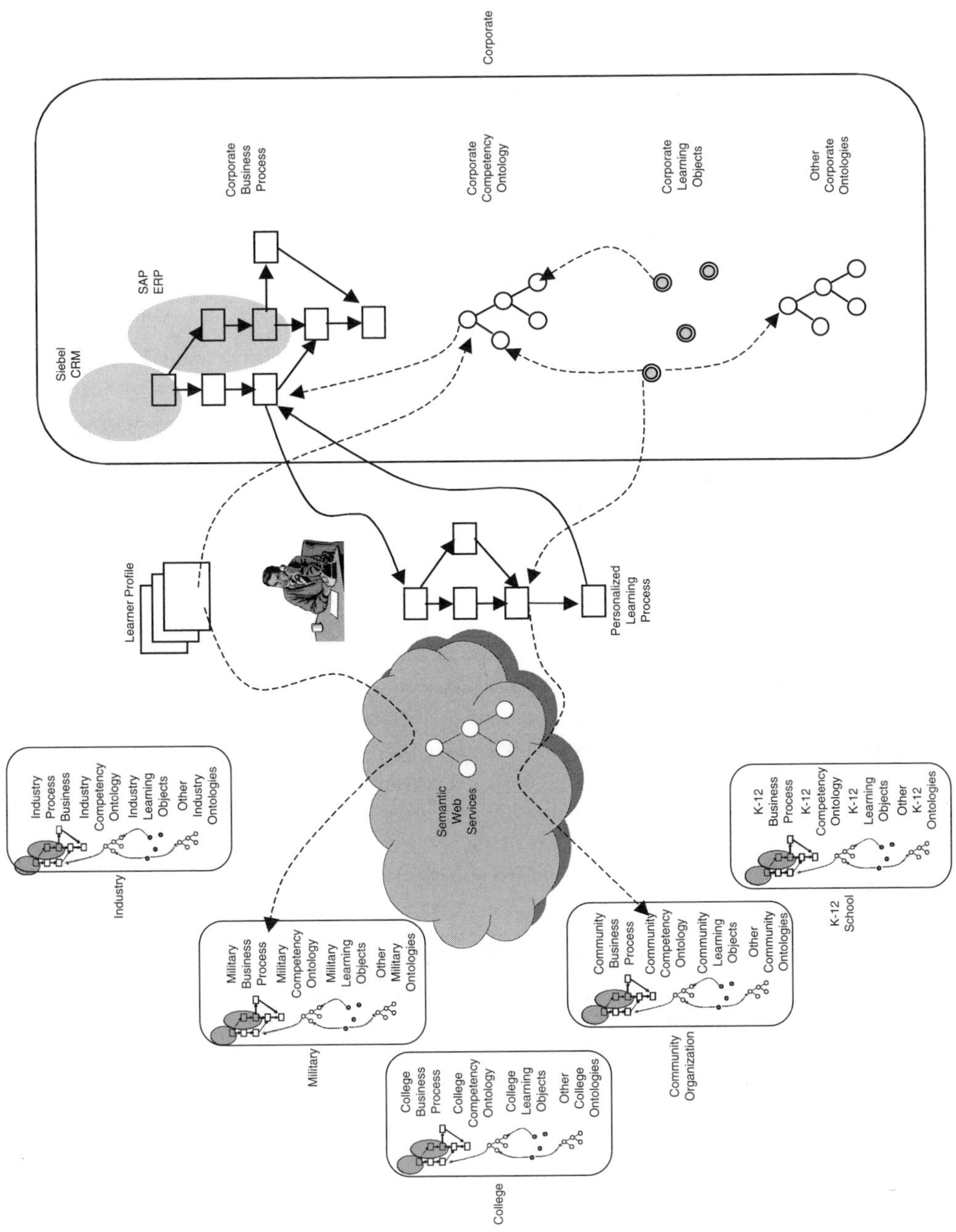

FIGURE 3.7 Sharing of knowledge and processes in a lifelong learning environment.

[McIlraith, 2001]. This program has focused on the development of a semantic markup language DAML+OIL based on RDF for Web pages that will enable software agents to understand and reason about the content of Web pages. The program has also developed a markup language for Web services called DAML-S [Ankolekar, 2001] that enables an improved semantic description of a Web service.

As described in Section 3.8, the e-learning industry is developing metadata standards for describing the semantics of learning objects. There is also an RDF representation of the learning object metadata [Dhraief, 2001] that should enable a DAML markup for learning objects and a DAML-S markup for learning object services.

3.5.4.3 Competency Ontologies

The concepts, relationships, and processes of an enterprise can be captured in a set of enterprise ontologies. Ontology representation and reasoning systems are available [Lenat, 1995], and the DARPA DAML program has also done significant research on the representation of ontologies in RDF. They have developed a large number of ontologies that can be referenced by the DAML markup language associated with a Web page in order to clarify the semantics of the Web page.

A few commercially available e-learning products use competency hierarchies to capture the skills necessary for various job types. The competencies in these hierarchies are then mapped to courses that can improve an employee's competency in a certain area. There is no industry standard representation for these competency hierarchies although there have been some efforts to create such a standard [HRXML, 2001]. These existing competency representations do not capture the rich semantics that could be captured using an ontology representation. A competency ontology can capture the relationships among various competencies and relationships with other ontologies, such as the product ontology for a corporation. A competency ontology will also allow reasoning about the competencies.

3.5.4.4 Representation of Business Processes and Learning Processes

A key requirement for the success of the system in Figure 3.6 is the ability to explicitly represent the processes in the enterprise. There must be a representation for the business processes so that competencies can be mapped to a specific activity in a business process. There must be a representation of personalized learning processes to enable integration of these learning processes with business processes.

There have been various attempts to standardize the representation of business processes ([Cichocki, 1998] [WFMC, 2001] [BMPL, 2002]). There are now numerous efforts underway to standardize a process representation for Web services. IBM and Microsoft have had competing proposals for a process representation [WSFL, 2001] [XLANG, 2001] but these have now been combined into a single specification [BPEL, 2003]. There has also been an effort within the DARPA DAML program and the DARPA CoABS [CoABS, 2002] program to standardize a more semantically expressive representation of processes.

3.5.4.5 Software Agents

There is a huge potential for the effective deployment of autonomous software agents in a system such as the one described in Figure 3.6 and Figure 3.7. In the past, there has been extensive research into the use of software agents for discovery of information [Woelk, 1994], collaboration planning and automation of processes [Tate, 1996], and numerous other applications [Bradshaw, 1997]. This research is now focusing on the use of software agents with the World Wide Web [Hendler, 2001]. Once the semantics of the services and processes in Figure 3.6 and Figure 3.7 have been adequately defined, autonomous software agents can be much more effective. These agents can proactively search for learning objects both inside and outside the enterprise that are needed to meet dynamically changing learning requirements. Furthermore, the role of simulation as a technique for training will be increased [Schank, 1997]. Developing a simulation of a business process using software agents will be simplified and the simulations can be integrated more directly with the business processes.

3.6 Conclusions

This chapter has reviewed the technology behind Internet-based e-learning and how e-learning can improve the business performance of an organization. It has described how e-learning, content-management, and knowledge-management technologies can be integrated to solve a broad variety of business problems. E-learning products and e-learning standards are still evolving, but many organizations today are benefiting from the capabilities of existing products. Before deploying an e-learning solution, however, it is important to specify what specific business problem is being solved. This chapter has described a range of business problems from delivering formal training classes online to providing knowledge workers with up-to-date information that they need to make effective decisions. Although a common technology infrastructure for solving these business problems is evolving, it does not exist yet. The best resource for information on e-learning products is www.brandonhall.com. Most of these products focus on creation and delivery of formal courses, but some e-learning companies are beginning to provide knowledge management capability. The best resource for information on knowledge management products is www.kmworld.com. However, it is important to remember that the successful deployment of e-learning technology is heavily dependent on first establishing and tracking the educational objectives. A good resource for information on how to create and execute a learning strategy for an organization is www.astd.org.

References

ADL. 2003. Advanced Distributed Learning (ADL) Initiative. www.adlnet.org.

AICC. 2003. Aviation Industry CBT Committee. www.aicc.org.

Ankolekar, Anupriya, Mark Burstein, Jerry R. Hobbs, Ora Lassila, David L. Martin, Sheila A. McIlraith, Srini Narayanan, Massimo Paolucci, Terry Payne, et al. DAML-S: Semantic Markup for Web Services, 2001. http://www.daml.org/services/SWWS.pdf.

Brandon-Hall. 2003. Learning Management Systems and Learning Content Management Systems Demystified, brandon-hall.com, http://www.brandohall.com/public/resources/lms_lcms/.

BPML. 2002. Business Process Modeling Language, Business Process Management Initiative, www.bpmi.org.

Bradshaw, Jeffrey. 1997. *Software Agents,* MIT Press, Cambridge, MA.

Brennan, Michael, Susan Funke, and Cushing Anderson. 2001 The Learning Content Management System, IDC, 2001. www.e-learningsite.com/download/white/lcms-idc.pdf.

Cichocki, Andrzej, Abdelsalam A. Helal, Marek Rusinkiewicz, and Darrell Woelk. 1998. *Workflow and Process Automation: Concepts and Technology,* Kluwer Academic, Dordrecht, Netherlands.

CoABS. 2002. DARPA Control of Agent Based Systems Program, http://www.darpa.mil/ito/research/coabs/.

BPEL. 2003. OASIS Web Services Business Process Execution Language Technical Committee. http://www.oasis-open.org/committees.

[DAML 2002] DARPA Agent Markup Language Program www.daml.org.

[Dhraief 2001] Hadhami Dhraief, Wolfgang Nejdl, Boris Wolf, and Martin Wolpers. Open Learning Repositories and Metadata Modeling, Semantic Web Working Symposium, July 2001.

DMC. 2003. IC^2 Institute Digital Media Collaboratory, University of Texas at Austin. http://dmc.ic2.org.

Gibbons, Andrew and Peter Fairweather. 2000. Computer-based Instruction. In S. Tobias and J.D. Fletcher (Eds.), *Training and Retraining: A Handbook for Business, Industry, Government, and the Military,* Macmillan, New York.

Glass, Graham. 2001. *Web Services: Building Blocks for Distributed Systems,* Prentice Hall PTR, www.phptr.com.

Hendler, James. 2001. Agents and the Semantic Web, *IEEE Intelligent Systems,* March/April.

Hodgins, Wayne. 2000. Into the Future: A Vision Paper, prepared for Commission on Technology and Adult Learning, www.learnitivity.com/download/MP7.pdf, February.

HRXML 2001. HR-XML Consortium Competencies Schema, http://www.hr-xml.org/subchannels/Competencies/index.htm.

IEEELTSC. 2003. Institute of Electrical and Electronics Engineers (IEEE) Learning Technology Standards Committee, http://grouper.ieee.org/groups/ltsc.

IEEELOM. 2002. Institute of Electrical and Electronics Engineers (IEEE) Learning Object Metadata (LOM) Standard. ltsc.ieee.org/wg12.

IMS. 2003. IMS Global Learning Consortium, www.imsproject.org.

ISO. 2003. International Organization for Standards (ISO) JTC1 SC36 Subcommittee, www.jtc1sc36.org.

Jackson, Melinda. 2003. Simulating Work: What Works, *eLearn Magazine,* October 2002.

Kindley, Randall. 2002. The power of Simulation-based e-Learning (SIMBEL), *The eLearning Developers' Journal,* September 17, www.learningguild.com.

Kizlik, Bob. How to Write Effective Behavioral Objectives. Adprima. www.adprima.com/objectives.htm.

LCMS. 2003. LCMS Council. www.lcmscouncil.org.

Lenat, Doug. November 1995. CYC: A large-scale investment in knowledge infrastructure, *Communications of the ACM,* Vol. 38 No. 11.

Longmire, Warren. 2000. A Primer on Learning Objects. *Learning Circuits: ASTD's Online Magazine All About E-Learning,* American Society for Training and Development (ASTD). www.learningcircuits.org/mar2000/primer.html.

McIlraith, Sheila A., Tran Cao Son, and Honglei Zeng. 2001. Semantic Web Services, IEEE Intelligent Mergel Systems, March/April.

Nonaka, Ikujiro and Hirotaka Takeuchi. 1995. *The Knowledge-Creating Company,* Oxford University Press, New York.

SAP. 2003. www.sap.com.

SCORM. 2002. *Sharable Content Object Reference Model, Version 1.2.* Advanced Distributed Learning Initiative, October 1, 2001. (www.adlnet.org).

Schank, Roger. 1997. *Virtual Learning: A Revolutionary Approach to Building a Highly Skilled Workforce,* McGraw-Hill, New York.

Siebel. 2003. www.siebel.com.

Tate, Austin. Representing plans as a set of constraints *N* the <I-N-OVA> model, *Proceedings of the 3rd International Conference on Artificial Intelligence Planning Systems (AIPS-96),* AAAI Press, Menlo Park, CA, pp. 221–228.

WFMC. 2001. Workflow Management Coalition, www.wfmc.org.

Woelk, Darrell and Christine Tomlinson. 1994. The infoSleuth project: Intelligent search management via semantic agents, *Proceedings of the 2nd International World Wide Web Conference,* October. NCSA.

Woelk, Darrell. 2002. E-Learning, semantic Web services and competency ontologies, *Proceedings of ED-MEDIA 2002 Conference,* June, ADEC.

Woelk, Darrell and Shailesh Agarwal. 2002b. Integration of e-learning and knowledge management, *Proceedings of E-Learn 2002 Conference,* Montreal, Canada, October.

Woelk, Darrell and Paul Lefrere. 2002. Technology for performance-based lifelong learning, *Proceedings of the 2002 International Conference on Computers in Education (ICCE 2002),* Auckland, New Zealand, December, IEEE Computer Society.

W3C. 2001a. World Wide Web Consortium Web Services Activity. www.w3c.org/2001/ws.

W3C. 2001b. World Wide Web Consortium Semantic Web Activity. www.w3c.org/2001/sw.

WSFL. 2001. Web Services Flow Language, IBM, http://www-4.ibm.com/software/solutions/webservices/pdf/WSFL.pdf.

XLANG. 2001. XLANG: Web Services for Process Design, Microsoft, http://www.gotdotnet.com/team/xml_wsspecs/xlang-c/default.htm.

4

Digital Libraries

Edward A. Fox

Hussein Suleman

Devika Madalli

Lillian Cassel

CONTENTS

Abstract 4-1
4.1 Introduction 4-1
4.2 Theoretical Foundation 4-3
4.2.1 Scenarios 4-4
4.3 Interfaces 4-5
4.4 Architecture 4-6
4.5 Inception 4-8
4.5.1 Digital Library Initiative 4-8
4.5.2 Networked Digital Libraries 4-8
4.5.3 Global DL Trends 4-9
4.6 Personalization and Privacy 4-10
4.7 Conclusions 4-10
References 4-11

Abstract

The growing popularity of the Internet has resulted in massive quantities of uncontrolled information becoming available to users with no notions of stability, quality, consistency, or accountability. In recent years, various information systems and policies have been devised to improve on the manageability of electronic information resources under the umbrella of "digital libraries." This article presents the issues that need to be addressed when building carefully managed information systems or digital libraries, including theoretical foundations, standards, digital object types, architectures, and user interfaces. Specific case studies are presented as exemplars of the scope of this discipline. Finally, current pertinent issues, such as personalization and privacy, are discussed from the perspective of digital libraries.

4.1 Introduction

Definitions of digital library (DL) abound (Fox and Urs, 2002), but a consistent characteristic across all definitions is an integration of technology and policy. This integration provides a framework for modern digital library systems to manage and provide mechanisms for access to information resources. This involves a degree of complexity that is evident whether considering: the collection of materials presented through a digital library; the services needed to address requirements of the user community; or the underlying systems needed to store and access the materials, provide the services, and meet the needs of patrons. Technologies that bolster digital library creation and maintenance have appeared over the last decade, yielding increased computational speed and capability, even with modest computing platforms. Thus, nearly any organization, and indeed many individuals, may consider establishing and presenting a digital library. The processing power of an average computer allows simultaneous service for multiple users, permits encryption and decryption of restricted materials, and supports complex processes for

1-58488-381-2/05/$0.00+$1.50
© 2005 by CRC Press LLC

user identification and enforcement of access rights. Increased availability of high-speed network access allows presentation of digital library contents to a worldwide audience. Reduced cost of storage media removes barriers to putting even large collections online. Commonly available tools for creating and presenting information in many media forms make content widely accessible without expensive special-purpose tools. Important among these tools and technologies are coding schemes such as JPEG, MPEG, PDF, and RDF, as well as descriptive languages such as SGML, XML, and HTML.

Standards related to representation, description, and display are critical for widespread availability of DL content (Fox and Sornil, 1999); other standards are less visible to the end user but just as critical to DL operation and availability. HTTP opened the world to information sharing at a new level by allowing any WWW browser to communicate with any information server, and to request and obtain information. The emerging standard for metadata tags is the Dublin Core (Dublin-Core-Community, 1999), with a set of 15 elements that can be associated with a resource: Title, Creator, Subject, Description, Publisher, Contributor, Date, Type, Format, Identifier, Source, Language, Relation, Coverage, and Rights. Each of the 15 elements is defined using 10 attributes specified in ISO/IEC 11179, a standard for the description of data elements. The 10 attributes are Name, Identifier, Version, Registration Authority, Language, Definition, Obligation, Datatype, Maximum Occurrence, and Comment. The Dublin Core provides a common set of labels for information to be exchanged between data and service providers.

The technologies are there. The standards are there. The resources are there. What further is needed for the creation of digital libraries? Though the pieces are all available, assembling them into functioning systems remains a complex task requiring expertise unrelated to the subject matter intended for the repository. The field is still in need of comprehensive work on analysis and synthesis, leading to a well-defined science of digital libraries to support the construction of specific libraries for specific purposes. Important issues remain as obstacles to making the creation of digital libraries routine. These have less to do with technology and presentation than with societal concerns and philosophy, and deserve attention from a wider community than the people who provide the technical expertise (Borgman, 1996). Among the most critical of these issues are Intellectual Property Rights, privacy, and preservation.

Concerns related to Intellectual Property Rights (IPR) are not new; nor did they originate with work to afford electronic access to information. Like many other issues, though, they are made more evident and the scale of the need for attention increases in an environment of easy widespread access. The role of IPR in the well-being and economic advance of developing countries is the subject of a report commissioned by the government of the U.K. (IPR-Commission, 2002). IPR serves both to boost development by providing incentives for discovery and invention, and to impede progress by denying access to new developments to those who could build on the early results and explore new avenues, or could apply the results in new situations. Further issues arise when the author of a work chooses to self-archive, i.e., to place the work online in a repository containing or referring to copies of his or her own work, or in other publicly accessible repositories. Key questions include the rights retained by the author and the meaning of those rights in an open environment. By placing the work online, the author makes it visible. The traditional role of copyright to protect the economic interest of the author (the ability to sell copies) does not then apply. However, questions remain about the rights to the material that have been assigned to others. How is the assignment of rights communicated to someone who sees the material? Does self-archiving interfere with possible publication of the material in scholarly journals? Does a data provider have responsibility to check and protect the rights of the submitter (ProjectRoMEO, 2002)?

Digital libraries provide opportunities for widespread dissemination of information in a timely fashion. Consequently, the openness of the information in the DL is affected by policy decisions for the developers of the information and those who maintain control of its representations. International laws such as TRIPS (Trade Related aspects of Intellectual Property Standards) determine rights to access information (TRIPS, 2003). Digital library enforcement of such laws requires careful control of access rights. Encryption can be a part of the control mechanism, as it provides a concrete barrier to information availability but adds complexity to digital library implementation.

Privacy issues related to digital libraries involve a tradeoff between competing goals: to provide personalized service (Gonçalves, Zafer, Ramakrishnan, and Fox, 2001) on the one hand and to serve users who are hesitant to provide information about themselves on the other. When considering these conflicting goals, it is important also to consider that information about users is useful in determining how well the DL is serving its users, and thus relates to both the practice and evaluation of how well the DL is meeting its goals.

Though the field of digital libraries is evolving into a science, with a body of knowledge, theories, definitions, and models, there remains a need for adequate evaluation of the success of a digital library within a particular context. Evaluation of a digital library requires a clear understanding of the purpose the DL is intended to serve. Who are the target users? What is the extent of the collection to be presented? Are there to be connections to other DLs with related information? Evaluation consists of monitoring the size and characteristics of the collection, the number of users who visit the DL, the number of users who return to the DL after the initial visit, the number of resources that a user accesses on a typical visit, the number of steps a user needs in order to obtain the resource that satisfies an information need, and how often the user goes away (frustrated) without finding something useful. Evaluation of the DL includes matching the properties of the resources to the characteristics of the users. Is the DL attracting users who were not anticipated when the DL was established? Is the DL failing to attract the users who would most benefit from the content and services?

4.2 Theoretical Foundation

To address these many important concerns and to provide a foundation to help the field advance forward vigorously, there is need for a firm theoretical base. While such a base exists in related fields, e.g., the relational database model, the digital library community has relied heretofore only on a diverse set of models for the subdisciplines that relate. To simplify this situation, we encourage consideration of the unifying theory described in the 5S model (Gonçalves, Fox, Watson, and Kipp, 2003a).

We argue that digital libraries can be understood by considering five distinct aspects: Societies, Scenarios, Spaces, Structures, and Streams. In the next section we focus on Scenarios related to services because that is a key concern and distinguishing characteristic in the library world. In this section we summarize issues related to other parts of the model.

With a good theory, we can give librarians interactive, graphical tools to describe the digital libraries they want to develop (Zhu, 2002). This can yield a declarative specification that is fed into a software system that generates a tailored digital library (Gonçalves and Fox, 2002). From a different perspective, digital library use can be logged in a principled fashion, oriented toward semantic analysis (Gonçalves et al., 2003b). 5S aims to support Societies' needs for information. Rather than consider only a user, or even collaborating users or sets of patrons, digital libraries must be designed with broad social needs in mind. These involve not only humans but also agents and software managers.

In order to address the needs of Societies and to support a wide variety of Scenarios, digital libraries must address issues regarding Spaces, Structures, and Streams. Spaces cover not only the external world (of 2 or 3 dimensions plus time, or even virtual environments — all connected with interfaces) but also internal representations using feature vectors and other schemes. Work on geographic information systems, probabilistic retrieval, and content-based image retrieval falls within the ambit of Spaces.

Because digital libraries deal with organization, Structures are crucial. The success of the Web builds upon its use of graph structures. Many descriptions depend on hierarchies (tree structures). Databases work with relations, and there are myriad tools developed as part of the computing field called "Data Structures." In libraries, thesauri, taxonomies, ontologies, and many other aids are built upon notions of structure.

The final "S," Streams, addresses the content layer. Thus, digital libraries are content management systems. They can support multimedia streams (text, audio, video, and arbitrary bit sequences) that afford an open-ended extensibility. Streams connect computers that send bits over network connections. Storage, compression and decompression, transmission, preservation, and synchronization are all key aspects of working with Streams. This leads us naturally to consider the myriad Scenarios that relate to Streams and the other parts of digital libraries.

4.2.1 Scenarios

Scenarios "consist of sequences of events or actions that modify the states of a computation in order to accomplish a functional requirement" (Gonçalves et al., 2003a). Scenarios represent services as well as the internal operation of the system. Overall, scenarios tell us what goes on in a digital library. Scenarios relate to societies by capturing the type of activity that a user group requires, plus the way in which the system responds to user needs.

An example scenario for a particular digital library might be access by a young student who wishes to learn the basics of a subject area. In addition to searching and matching the content to the search terms, the DL should use information in the user profile and metadata tags in the content to identify material compatible with the user's level of understanding in this topic area. A fifth grader seeking information on animal phyla for a general science report should be treated somewhat differently than a mature researcher investigating arthropoda subphyla. While both want to know about butterflies, the content should be suited to the need. In addition to providing a user interface more suited to a child's understanding of information organization, the DL should present materials with appropriate vocabulary before other materials. Depending on choices made in the profile setup, the responses could be restricted to those with a suitable reading level, or all materials could be presented but with higher ranking given to age-appropriate resources.

Scenarios are not limited to recognizing and serving user requests. Another scenario of interest to the designer of a digital library concerns keeping the collection current. A process for submission of new material, validation, description, indexing, and incorporation into the collection is needed. A digital library may provide links to resources stored in other digital libraries that treat the same topics. The contents of those libraries change and the DL provider must harvest updated metadata in order to have accurate search results. Here the activity is behind the scenes, not directly visible to the user, but important to the quality of service provided.

Other scenarios include purging the digital library of materials that have become obsolete and no longer serve the user community. While it is theoretically possible to retain all content forever, this is not consistent with good library operation. Determining which old materials have value and should be retained is important. If all material is to be kept forever, then there may be a need to move some materials to a different status so that their presence does not interfere with efficient processing of requests for current materials. If an old document is superseded by a new version, the DL must indicate that clearly to a user who accesses the older version.

Services to users go beyond search, retrieval, and presentation of requested information. A user may wish to see what resources he or she viewed on a previous visit to the library. A user may wish to retain some materials in a collection to refer to on later visits, or may simply want to be able to review his or her history and recreate previous result lists. In addition, the library may provide support for the user to do something productive with the results of a search. In the case of the National Science Digital Library, focused on education in the STEM (science, technology, engineering, and mathematics) areas, the aim is to support teaching and learning (http://www.nsdl.org). For example, in the NSDL collection project called CITIDEL (Computing and Information Technology Interactive Digital Educational Library) (CITIDEL, 2002), a service called VIADUCT allows a user to gather materials on a topic and develop a syllabus for use in a class. The syllabus includes educational goals for the activity, information about the time expected for the activity, primary resources and additional reference materials, preactivity, activity, and postactivity procedures and directions, and assessment notes. The resulting entity can be presented to students, saved for use in future instances of the class, and shared with other faculty with similar interests.

GetSmart, another project within the NSF NSDL program, provides tools for students to use in finding and organizing useful resources and for better learning the material they read (Marshall et al., 2003). This project provides support for concept maps both for the individual student to use and for teams of students who work together to develop a mutual understanding.

Scenarios also may refer to problem situations that develop within the system. A scenario that would need immediate attention is deterioration of the time to respond to user requests beyond an acceptable threshold (Fox and Mather, 2002). A scenario that presents the specter of a disk crash and loss of data needs to be considered in the design and implementation of the system. An important design scenario is the behavior by a community of users to achieve the level of traffic expected at the site and the possibility of DL usage exceeding that expectation.

4.3 Interfaces

User interfaces for digital libraries span the spectrum of interface technologies used in computer systems. The ubiquitous nature of the hyperlinked World Wide Web has made that the *de facto* standard in user interfaces. However, many systems have adopted approaches that either use the WWW in nontraditional ways or use interfaces not reliant on the WWW.

The classical user interface in many systems takes the form of a dynamically generated Web site. Emerging standards, such as W3C's XSLT transformation language, are used to separate the logic and workflow of the system from the user interface. See Figure 4.1 for a typical system using such an approach. Such techniques make it easier to perform system-wide customization and user-specific personalization of the user interface.

Portal technology offers the added benefit of a component model for WWW-based user interfaces. The uPortal project (JA-SIG, 2002) defines "channels" to correspond to rectangular portions of the user interface. Each of these channels has functionality that is tied to a particular service on a remote server. This greatly aids development, maintenance, and personalization of the interface.

Some collections of digital objects require interfaces that are specific to the subject domain and nature of the data. Geospatial data, in particular, has the characteristic that users browse by physical proximity in a 2-dimensional space. The Alexandria Digital Earth Prototype (Smith, Janee, Frew, and Coleman, 2001) allows users to select a geographical region to use as a search constraint when locating digital

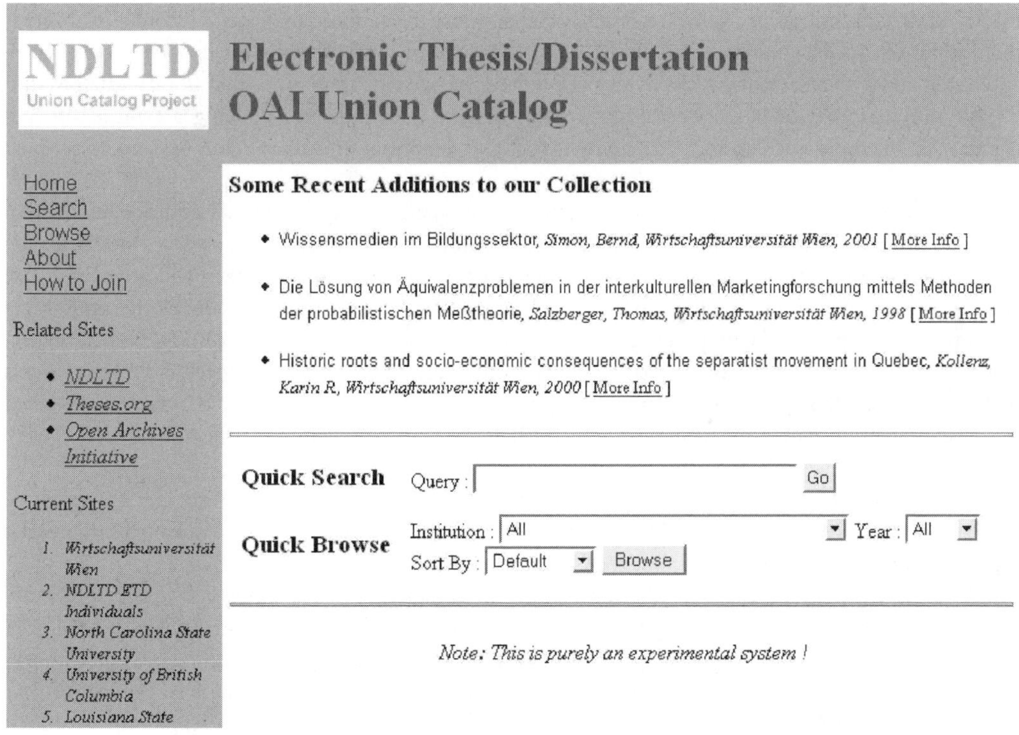

FIGURE 4.1 An example of a WWW-based system using a component-based service architecture and XSLT transformations to render metadata in HTML.

objects related to that region. Terraserver offers a similar interface to locate and navigate through aerial photographs that are stitched together to give users the impression of a continuous snapshot of the terrain (Microsoft, 2002). Both systems offer users the ability to switch between keyword searching and map browsing, where the former can be used for gross estimation and the latter to locate an exact area or feature.

In a different context, multifaceted data can be visualized using 2- and 3-dimensional discovery interfaces where different facets are mapped to dimensions of the user interface. As a simple example, the horizontal axis is frequently used to indicate year. The Envision interface expands on this notion by mapping different aspects of a data collection or subcollection to shape, size, and color, in addition to X and Y dimensions (Heath et al., 1995). Thus, multiple aspects of the data may be seen simultaneously. The SPIRE project analyzes and transforms a data collection so that similar concepts are physically near each other, thus creating an abstract but easily understandable model of the data (Thomas et al., 1998). Virtual reality devices can be used to add a third dimension to the visualization. In addition to representing data, collaborative workspaces in virtual worlds can support shared discovery of information in complex spaces (Börner and Chen, 2002).

In order to locate audio data such as music, it is sometimes desirable to search by specifying the tune rather than its metadata. Hu and Dannenberg (2002) provide an overview of techniques involving such sung queries. Typically, a user hums a tune into the microphone and the digitized version of that tune then is used as input to a search engine. The results of the search can be either the original audio rendering of the tune or other associated information. In this as well as the other cases mentioned above, it is essential that user needs are met, and that usability is assured (Kengeri, Seals, Harley, Reddy, and Fox, 1999) along with efficiency.

4.4 Architecture

Pivotal to digital libraries are software systems that support them; these manage the storage and access to information. To-date, many digital library systems have been constructed, some by loosely connecting applicable and available tools, some by extending existing systems that supported library catalogs and library automation (Gonçalves et al., 2002). Most systems are built by following a typical software engineering life cycle, with an increasing emphasis on architectural models and components to support the process.

Kahn and Wilensky (1995) specified a framework for naming digital objects and accessing them through a machine interface. This Repository Access Protocol (RAP) provides an abstract model for the services needed in order to add, modify, or delete records stored in a digital library. Dienst (Lagoze and Davis, 1995) is a distributed digital library based on the RAP model, used initially as the underlying software for the Networked Computer Science Technical Reference Library (NCSTRL) (Davis and Lagoze, 2000). Multiple services are provided as separate modules, communicating using well-defined protocols both within a single system and among remote systems. RAP, along with similar efforts, has informed the development of many modern repositories, such as the DSpace software platform developed at MIT (MIT, 2003).

Other notable prepackaged systems are E-Prints (http://www.eprints.org/) from the University of Southampton and Greenstone (http://www.greenstone.org) from the University of Waikato. Both provide the ability for users to manage and access collections of digital objects.

Software agents and mobile agents have been applied to digital libraries to mediate with one or more systems on behalf of a user, resulting in an analog to a distributed digital library. In the University of Michigan Digital Library Project (Birmingham, 1995), DLs were designed as collections of autonomous agents that used protocol-level negotiation to perform collaborative tasks. The Stanford InfoBus project (Baldonado, Chang, Gravano, and Paepcke, 1997) not only worked on standards for searching distributed collections (Gravano, Chang, Garca-Molina, and Paepcke, 1997; Paepcke et al., 2000), but also developed an approach for interconnecting systems using distinct protocols for each purpose, with CORBA as the transport layer. Subsequently, CORBA was used as a common layer in the FEDORA project (http://www.fedora.info/), which defined abstract interfaces to structured digital objects.

The myriad of different systems and system architectures has historically been a stumbling block for interoperability attempts (Paepcke, Chang, Garcia-Molina, and Winograd, 1998). The Open Archives Initiative (OAI, http://www.openarchives.org), which emerged in 1999, addressed this problem by developing the Protocol for Metadata Harvesting (PMH) (Lagoze, Van de Sompel, Nelson, and Warner, 2002), a standard mechanism for digital libraries to exchange metadata on a periodic basis. This allows providers of services to obtain all, or a subset, of the metadata from an archive ("data provider") with a facility for future requests to be satisfied with only incremental additions, deletions, and changes to records in the collection. Because of its efficient transfer of metadata over time, this protocol is widely supported by many current digital library systems.

The Open Digital Library (ODL) framework (Suleman and Fox, 2001) attempts to unify architecture with interoperability in order to support the construction of componentized digital libraries. ODL builds on the work of the OAI by requiring that every component support an extended version of the PMH. This standardizes the basic communications mechanism by building on the well-understood semantics of the OAI–PMH. Both use HTTP GET to encode the parameters of a typical request, with purpose-built XML structures used to specify the results and encapsulate metadata records where appropriate.

The model for a typical ODL-based digital library is illustrated in Figure 4.2. In this system, data is collected from numerous sources using the OAI–PMH, merged together into a single collection, and subsequently fed into components that support specific interactive service requests (hence the bidirectional arrows), such as searching. Other efforts have arisen that take up a similar theme, often viewing DLs from a services perspective (Castelli and Pagano, 2002).

Preservation of data is being addressed in the Lots of Copies Keeps Stuff Safe (LOCKSS, http://lockss.stanford.edu/) project, which uses transparent mirroring of popular content to localize access and enhance confidence in the availability of the resources. The Internet2 Distributed Storage Initiative (Beck, 2000) had somewhat similar goals and uses network-level redirection to distribute the request load to mirrors.

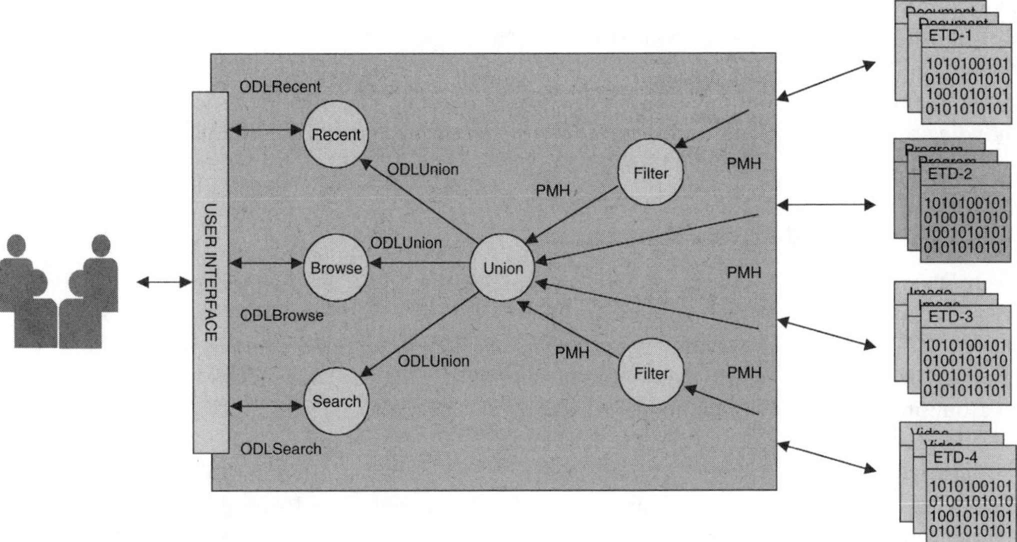

FIGURE 4.2 Architecture of digital library based on OAI and ODL components.

4.5 Inception

The concept behind digital libraries has its roots in libraries disseminating "knowledge for all" (Wells, 1938). Digital libraries break the barrier of physical boundaries and strive to give access to information across varied domains and communities. Though the terms Digital Library and Web both were initially popularized in the early 1990s, they trace back to projects dealing with linking among distributed systems (Englebart, 1963), automated storage and retrieval of information (Salton and McGill, 1983), library networks, and online resource sharing efforts. Though similar and mutually supportive in concept and practice, Digital Library and Web differ in emphasis, with the former more focused on quality and organization, and packaged to suit particular sets of users desiring specialized content and services akin to organized library services rendered by information professionals. Accordingly, many digital library projects have helped clarify theory and practice, and must be considered as case studies that illustrate key ideas and developments.

4.5.1 Digital Library Initiative

The core projects of the U.S. Digital Library Initiative (DLI) Phase I (http://www.dli2.nsf.gov/dlione), started in 1994 as a joint initiative of the National Science Foundation (NSF), Department of Defense Advanced Research Projects Agency (DARPA), and the National Aeronautics and Space Administration (NASA). Phase I involved a total funding of $24 million for a period of 4 years from 1994 to 1998. The intent in the first phase was to concentrate on the *investigation and development of underlying technologies for digital libraries*. The Initiative targeted research on information storage, searching, and access. The goals were set as developing technologies related to:

- Capturing, categorizing, and organizing information
- Searching, browsing, filtering, summarizing, and visualization
- Networking protocols and standards

DLI brought focus and direction to developments in the digital libraries arena. Various architectures, models, and practices emerged and precipitated further research. The NSF announced Phase II in February 1998. In addition to the NSF, the Library of Congress, DARPA, the National Library of Medicine (NLM), the National Aeronautics and Space Administration (NASA), and the National Endowment for the Humanities (NEH) served as sponsors. The second phase (1999 to 2004) went past an emphasis on technologies to focus on applying those technologies and others in real-life library situations.

The second phase aims at intensive study of the architecture and usability issues of digital libraries including research on (1) human-centered DL architecture, (2) content and collections-based DL architecture, and (3) systems-centered DL architecture.

4.5.2 Networked Digital Libraries

Many DL projects have emerged in the Web environment where content and users are distributed; some are significant in terms of their collections, techniques, and architecture. For example, NSF partially funded NCSTRL (http://www.ncstrl.org), a digital repository of technical reports and related works. By 2001, however, the Dienst services and software used by NCSTRL no longer fit with needs and practices, so a transition began toward the model advocated by OAI. OAI ushered in a simple and distributed model for exchange of records.

Colleges and universities, along with diverse partners interested in education, also are working on a distributed infrastructure for courseware. NSDL, already involving over 100 project teams, is projected to have a great impact on education, with the objective of facilitating enhanced communication between and among educators and learners. The basic objective of NSDL is to "catalyze and support continual improvements in the quality of Science, Mathematics, Engineering, and Technology education" (Manduca, McMartin, and Mogk, 2001).

4.5.3 Global DL Trends

Since the early 1990s, work on digital libraries has unfolded all around the globe (Borgman, 2000), with many heads of state interested in deploying them to preserve and disseminate the cultural and historic record (Fox, Moore, Larsen, Myaeng, and Kim, 2002). There has been some support for research, but more support for development and application, often as extensions to traditional library and publishing efforts.

In Europe there is an annual digital library conference (ECDL), and there have been projects at regional, national, and local levels. The Telematics for Libraries program of the European Commission (EC) aims to facilitate access to knowledge held in libraries throughout the European Union while reducing disparities between national systems and practices. Though not exclusively devoted to digital libraries — the program covers topics such as networking (OSI, Web), cataloging, imaging, multimedia, and copyright — many of the more than 100 projects do cover issues and activities related to digital libraries. In addition there have emerged national digital library initiatives in Denmark, France, Germany, Russia, Spain, and Sweden, among others.

In the U.K., noteworthy efforts in digital libraries include the ELINOR and the eLib projects. The Electronic Libraries Programme (eLib, http://www.jisc.ac.uk/elib/projects.html), funded by the Joint Information Systems Committee (JISC), aims to provide exemplars of good practice and models for well-organized, accessible hybrid libraries. The Ariadne magazine (http://www.ariadne.ac.uk/) reports on progress and developments within the eLib Programme and beyond.

The Canadian National Library hosts the Canadian Inventory of Digital Initiatives that provides descriptions of Canadian information resources created for the Web, including general digital collections, resources centered around a particular theme, and reference sources and databases. In Australia, libraries (at the federal, state, and university levels) together with commercial and research organizations are supporting a diverse set of digital library projects that take on many technical and related issues. The projects deal both with collection building and with services and research, especially related to metadata. Related to this, and focused on retrieval, are the subject gateway projects (http://www.nla.gov.au/initiatives/sg/), which were precursors to the formal DL initiatives.

In Asia the International Conference of Asian Digital Libraries (ICADL, http://www.icadl.org) provides a forum to publish and discuss issues regarding research and developments in the area of digital libraries. In India, awareness of the importance of digital libraries and electronic information services has led to conferences and seminars hosted on these topics. Several digital library teams are collaborating with the Carnegie Mellon University Universal Digital Library project. The collaboration has resulted in the Indian National Digital Library Initiative (www.dli.gov.in). The University of Mysore and University of Hyderabad are among those participating as members in the Networked Digital Library of Theses and Dissertations. In the area of digital library research, Documentation Research and Training (DRTC, www.drtc.isibang.ac.in), at the Indian Statistical Institute, researches and implements the technology and methodologies in digital library architecture, multilingual digital information retrieval, and related tools and techniques. Other digital library initiatives in Asia are taking shape through national initiatives such as the Indonesian Digital Library Network (http://idln.lib.itb.ac.id), the Malaysian National Digital library (myLib, http://www.mylib.com.my), and the National Digital Library of Korea (http://www.dlibrary.go.kr).

In general, the U.S. projects emphasized research and techniques for digital library architectures, storage, access, and retrieval. In the U.K., the initial focus was on electronic information services and digitization. Major projects under the Australian Digital Library initiative concentrated on storage and retrieval of images (and other media), and also on building subject gateways. In the Asian and European efforts, work in multilingual and cross-lingual information figures prominently because of diverse user communities seeking information in languages other than English.

4.6 Personalization and Privacy

Digital libraries allow us to move from the global to the personal. Personalization (Gonçalves et al., 2001) allows the DL to recognize a returning user and to restore the state of the user relationship with the library to where it was at the time of the last visit. It saves the user time in reconstructing prior work and allows saving the state of the user–DL interaction. It also allows the system to know user preferences and to tailor services to special needs or simple choices. Personalization depends on user information, generally in the form of a user profile and history of prior use. The user profile to some extent identifies an individual and allows the system to recognize when a user returns. In addition to making it possible for the library to provide services, the identification of a user allows evaluation of how well the library is serving that user. If a given user returns frequently, seems to find what was wanted, uses available services, keeps a supply of materials available for later use, and participates in user options such as annotation and discussions, it is reasonable to assume that the user is well served. Thus, an analysis of user characteristics and activities can help determine if the library is serving its intended audience adequately.

There also is a negative side to personalization. Many people are increasingly conscious of diminished privacy, and anxious about sharing data about their personal preferences and contact information. The concerns are real and reasonable and must be addressed in the design of the DL. Privacy statements and a clear commitment to use the information only in the service of the user and for evaluation of the DL can alleviate some of these concerns. Confidence can be enhanced if the information requested is limited to what is actually needed to provide services and if the role of the requested information is clearly explained. For example, asking for an e-mail address is understandable if the user is signing up for a notification service. Similarly, a unique identifier, not necessarily traceable to any particular individual, is necessary to retain state from one visit to another.

With the increasing numbers of digital libraries, repeated entry of user profile data becomes cumbersome. We argue for one way to address these issues — have the users' private profile information kept on their own systems. The user will be recognized at the library because of a unique identifier, but no other information is retained at the library site. In this way, the library can track returns and successes in meeting user needs, and could even accumulate resources that belong to this user. All personal details, however, remain on the user system and under user control. This can include search histories, resource collections, project results such as concept maps, and syllabi. With the growing size of disk storage on personal computers, storing these on the user's system is not a problem. The challenge is to allow the DL to restore state when the user returns.

4.7 Conclusions

Digital libraries afford many advantages in today's information infrastructure. Technology has enabled diverse distributed collections of content to become integrated at the metadata and content levels, for widespread use through powerful interfaces that will become increasingly personalized. Standards, advanced technology, and powerful systems can support a wide variety of types of users, providing a broad range of tailored services for communities around the globe. Varied architectures have been explored, but approaches like those developed in OAI, or its extension into Open Digital Libraries, show particular promise. The recent emergence of sophisticated but extensible toolkits supporting open architectures — such as EPrints, Greenstone, and DSpace — provide would-be digital archivists with configurable and reasonably complete software tools for popular applications, while minimizing the risk associated with custom development. While many challenges remain — such as integration with traditional library collections, handling the needs for multilingual access, and long-term preservation — a large research establishment is well connected with development efforts, which should ensure that digital libraries will help carry the traditional library world forward to expand its scope and impact, supporting research, education, and associated endeavors.

References

Baldonado, M., Chang, C.-C. K., Gravano, L., and Paepcke, A. (1997). The Stanford Digital Library metadata architecture. *International Journal on Digital Libraries, 1*(2), 108–121. http://www-diglib.stanford.edu/cgi-bin/-WP/get/SIDL-WP-1996-0051.

Beck, M. (2000). Internet2 Distributed Storage Infrastructure (I2-DSI) home page: UTK, UNCCH, and Internet2. http://dsi.internet2.edu.

Birmingham, W. P. (1995). An Agent-Based Architecture for Digital Libraries. *D-Lib Magazine, 1*(1). http://www.dlib.org/dlib/July95/07birmingham.html

Borgman, C. (1996). Social Aspects of Digital Libraries (NSF Workshop Report). Los Angeles: UCLA. Feb. 16–17. http://is.gseis.ucla.edu/research/dl/index.html.

Borgman, C. L. (2000). *From Gutenberg to the Global Information Infrastructure: Access to Information in the Networked World.* Cambridge, MA: MIT Press.

Börner, K., and Chen, C. (Eds.). (2002). *Visual Interfaces to Digital Libraries* (JCDL 2002 Workshop). New York: Springer-Verlag.

Castelli, D., and Pagano, P. (2002). OpenDLib: A digital library service system. In M. Agosti and C. Thanos (Eds.), *Research and Advanced Technology for Digital Libraries, Proceedings of the 6th European Conference, ECDL 2002,* Rome, September 2002. *Lecture Notes in Computer Science 2548* (pp. 292–308). Springer-Verlag, Berlin.

CITIDEL. (2002). CITIDEL: Computing and Information Technology Interactive Digital Educational Library. Blacksburg, VA: Virginia Tech. http://www.citidel.org.

Davis, J. R., and Lagoze, C. (2000). NCSTRL: Design and deployment of a globally distributed digital library. *J. American Society for Information Science, 51*(3), 273–280.

Dublin-Core-Community. (1999). Dublin Core Metadata Initiative. The Dublin Core: A Simple Content Description Model for Electronic Resources. WWW site. Dublin, Ohio: OCLC. http://purl.org/dc/.

Englebart, D. C. (1963). A conceptual framework for the augmentation of man's intellect. In P. W. Howerton and D. C. Weeks (Eds.), *Vistas in Information Handling* (pp. 1–20). Washington, D.C: Spartan Books.

Fox, E., Moore, R., Larsen, R., Myaeng, S., and Kim, S. (2002). Toward a Global Digital Library: Generalizing US–Korea Collaboration on Digital Libraries. *D-Lib Magazine, 8*(10). http://www.dlib.org/dlib/october02/fox/10fox.html.

Fox, E. A., and Mather, P. (2002). Scalable storage for digital libraries. In D. Feng, W. C. Siu, and H. Zhang (Eds.), *Multimedia Information Retrieval and Management* (Chapter 13). Springer-Verlag. http://www.springer.de/cgi/svcat/search_book.pl?isbn=3-540-00244-8

Fox, E. A., and Sornil, O. (1999). Digital libraries. In R. Baeza-Yates and B. Ribeiro-Neto (Eds.), *Modern Information Retrieval* (Ch. 15, pp. 415–432). Harlow, England: ACM Press/Addison-Wesley-Longman.

Fox, E. A., and Urs, S. (2002). Digital libraries. In B. Cronin (Ed.), *Annual Review of Information Science and Technology* (Vol. 36, Ch. 12, pp. 503–589), American Society for Information Science and Technology.

Gonçalves, M., and Fox, E., A. (2002). 5SL — A Language for declarative specification and generation of digital libraries, *Proceedings of JCDL 2002, 2nd ACM/IEEE-CS Joint Conference on Digital Libraries,* July 14–18, Portland, Oregon. (pp. 263–272), ACM Press, New York.

Gonçalves, M. A., Fox, E. A., Watson, L. T., and Kipp, N. A. (2003a). *Streams, Structures, Spaces, Scenarios, Societies (5S): A Formal Model for Digital Libraries* (Technical Report TR-03-04, preprint of paper accepted for ACM TOIS: 22(2), April 2004). Blacksburg, VA: Computer Science, Virginia Tech. http://eprints.cs.vt.edu:8000/archive/00000646/.

Gonçalves, M. A., Mather, P., Wang, J., Zhou, Y., Luo, M., Richardson, R., Shen, R., Liang, X., and Fox, E. A. (2002). Java MARIAN: From an OPAC to a modern Digital Library system, In A. H. F Laender and A.L. Oliveira (Eds.), *Proceedings of the 9th International Symposium on String Processing and Information Retrieval (SPIRE 2002),* September, Lisbon, Portugal. Springer-Verlag, London.

Gonçalves, M. A., Panchanathan, G., Ravindranathan, U., Krowne, A., Fox, E. A., Jagodzinski, F., and Cassel., L. (2003b). The XML log standard for digital libraries: Analysis, evolution, and deployment, *Proceedings of JCDL 2003, 3rd ACM/IEEE-CS Joint Conference on digital libraries,* May 27–31, Houston. ACM Press, New York, 312–314.

Gonçalves, M. A., Zafer, A. A., Ramakrishnan, N., and Fox, E. A. (2001). Modeling and building personalized digital libraries with PIPE and 5SL, *Proceedings of the 2nd DELOS-NSF Network of Excellence Workshop on Personalisation and Recommender Systems in Digital Libraries,* sponsored by NSF, June18–20, 2001. Dublin, Ireland. ERCIM Workshop Proceedings No. 01/W03, European Research Consortium for Information and Mathematics. http://www.ercim.org/publication/ws-proceedings/DelNoe02/Goncalves.pdf.

Gravano, L., Chang, C.-C. K., Garca-Molina, H., and Paepcke, A. (1997). STARTS: Stanford proposal for Internet meta-searching, In Joan Peckham (Ed.), *Proceedings of the 1997 ACM SIGMOD Conference on Management Data,* Tucson, AZ. (pp. 207–218), ACM Press, New York.

Heath, L., Hix, D., Nowell, L., Wake, W., Averboch, G., and Fox, E. A. (1995). Envision: A user-centered database from the computer science literature. *Communications of the ACM, 38*(4), 52–53.

Hu, N., and Dannenberg, R. B. (2002). A Comparison of melodic database retrieval techniques using sung queries, *Proceedings of the 2nd ACM/IEEE-CS Joint Conference on Digital Libraries,* 14–18 July, Portland, OR. (pp. 301–307), ACM Press, New York.

IPR-Commission. (2002). IPR Report 2002. Online report. http://www.iprcommission.org/papers/text/final_report/reportwebfinal.htm.

JA-SIG. (2002). uPortal architecture overview. Web site. JA-SIG (The Java in Administration Special Interest Group). http://mis105.mis.udel.edu/ja-sig/uportal/architecture/uPortal_architecture_overview.pdf.

Kahn, R., and Wilensky, R. (1995). A Framework for Distributed Digital Object Services. Technical report. Reston, VA: CNRI. http://www.cnri.reston.va.us/k-w.html.

Kengeri, R., Seals, C. D., Harley, H. D., Reddy, H. P., and Fox, E. A. (1999). Usability study of digital libraries: ACM, IEEE-CS, NCSTRL, NDLTD. *International Journal on Digital Libraries, 2*(2/3), 157–169. http://link.springer.de/link/service/journals/00799/bibs/9002002/90020157.htm.

Lagoze, C., and Davis, J. R. (1995). Dienst: An architecture for distributed document libraries. *Communications of the ACM, 38*(4), 47.

Lagoze, C., Van de Sompel, H., Nelson, M., and Warner, S. (2002). The Open Archives Initiative Protocol for Metadata Harvesting–Version 2.0, Open Archives Initiative. Technical report. Ithaca, NY: Cornell University. http://www.openarchives.org/OAI/2.0/openarchivesprotocol.htm.

Manduca, C. A., McMartin, F. P., and Mogk, D. W. (2001). Pathways to Progress: Vision and Plans for Developing the NSDL. NSDL, March 20, 2001. http://doclib.comm.nsdlib.org/PathwaysToProgress.pdf (retrieved on 11/16/2002).

Marshall, B., Zhang, Y., Chen, H., Lally, A., Shen, R., Fox, E. A., and Cassel, L. N. (2003). Convergence of knowledge management and e-Learning: the GetSmart experience, *Proceedings of JCDL 2003, 3rd ACM/IEEE-CS Joint Conference on Digital Libraries,* May 27–31, Houston. ACM Press, New York, 135–146.

Microsoft. (2002). TerraServer. Web site. Microsoft Corporation. http://terraserver.microsoft.com.

MIT. (2003). DSpace: Durable Digital Depository. Web site. Cambridge, MA: MIT. http://dspace.org.

Paepcke, A., Brandriff, R., Janee, G., Larson, R., Ludaescher, B., Melnik, S., and Raghavan, S. (2000). Search Middleware and the Simple Digital Library Interoperability Protocol. *D-Lib Magazine, 6*(3). http://www.dlib.org/dlib/march00/paepcke/03paepcke.html.

Paepcke, A., Chang, C.-C. K., Garcia-Molina, H., and Winograd, T. (1998). Interoperability for digital libraries worldwide. *Communications of the ACM, 41*(4), 33–43.

ProjectRoMEO. (2002). Project RoMEO, JISC project 2002-2003: Rights MEtadata for Open Archiving. Web site. UK: Loughborough University. http://www.lboro.ac.uk/departments/ls/disresearch/romeo/index.html.

Salton, G., and McGill, M. J. (1983). *Introduction to Modern Information Retrieval.* New York: McGraw-Hill.

Smith, T. R., Janee, G., Frew, J., and Coleman, A. (2001). The Alexandria digital earth prototype, *Proceedings of the 1st ACM/IEEE-CS Joint Conference on Digital Libraries, JCDL 2001,* 24-28 June. Roanoke, VA. (pp. 118–199), ACM Press, New York.

Suleman, H., and Fox, E. A. (2001). A Framework for Building Open Digital Libraries. *D-Lib Magazine, 7*(12). http://www.dlib.org/dlib/december01/suleman/12suleman.html.

Thomas, J., Cook, K., Crow, V., Hetzler, B., May, R., McQuerry, D., McVeety, R., Miller, N., Nakamura, G., Nowell, L., Whitney, P., and Chung Wong, P. (1998). Human Computer Interaction with Global Information Spaces — Beyond Data Mining. Pacific Northwest National Laboratory, Richland, WA. http://www.pnl.gov/infoviz/papers.html.

TRIPS. (2003). TRIPS: Agreement on Trade-Related Aspects of Intellectual Property Rights. Web pages. Geneva, Switzerland: World Trade Organization. http://www.wto.org/english/tratop_e/trips_e/t_agm1_e.htm

Wells, H. G. (1938). *World brain.* Garden City, New York: Doubleday.

Zhu, Q. (2002). 5SGraph: A Modeling Tool for Digital Libraries. Unpublished Master's Thesis, Virginia Tech, Computer Science, Blacksburg, VA. http://scholar.lib.vt.edu/theses/available/etd-11272002-210531/.

5

Collaborative Applications

CONTENTS

Abstract 5-1
5.1 Introduction 5-2
5.2 Dual Goals of Collaborative Applications 5-2
5.3 Toward Being There: Mimicking Natural Collaboration ... 5-2
5.3.1 Single Audio/Video Stream Transmission 5-2
5.3.2 Overview + Speaker 5-3
5.3.3 Multipoint Lecture 5-4
5.3.4 Video-Production-Based Lecture 5-4
5.3.5 Slides Video vs. Application Sharing 5-5
5.3.6 State-of-the-Art Chat 5-6
5.3.7 Horizontal Time Line 5-6
5.3.8 Vertical Time Line 5-8
5.3.9 Supporting Large Number of Users 5-8
5.3.10 Graphical Chat 5-8
5.4 Beyond Being There: Augmenting Natural Collaboration .. 5-9
5.4.1 Anonymity 5-9
5.4.2 Multitasking 5-10
5.4.3 Control of Presence Information 5-10
5.4.4 Meeting Browsing 5-10
5.4.5 Divergent Views and Concurrent Input 5-12
5.4.6 Chat History 5-12
5.4.7 Scripted Collaboration 5-13
5.4.8 Threaded Chat 5-14
5.4.9 Threaded E-Mail 5-15
5.4.10 Threaded Articles Discussions and Annotations 5-16
5.4.11 Variable-Granularity Annotations to Changing Documents .. 5-16
5.4.12 Robust Annotations 5-18
5.4.13 Notifications 5-18
5.4.14 Disruptions Caused by Messages 5-19
5.4.15 Prioritizing Messages 5-20
5.4.16 Automatic Redirection of Message and Per-Device Presence and Availability Forecasting 5-21
5.5 Conclusions 5-23
Acknowledgments 5-24
References 5-24

Prasun Dewan

Abstract

Several useful collaboration applications have been developed recently that go beyond the email, bulletin board, videoconferencing, and instant messaging systems in use today. They provide novel support for threading in mail, chat, and bulletin boards; temporal ordering of chat conversations; graphical, mediated,

1-58488-381-2/05/$0.00+$1.50
© 2005 by CRC Press LLC

and synchronous chat; variable-grained, multimedia annotations; document-based notifications; automatic presence, location, and availability identification; automatic camera placement and video construction in lecture presentations and discussion groups; and compression and browsing of stored video.

5.1 Introduction

This chapter surveys some of the recent papers on successful collaborative applications. There are three related reasons for doing this survey. First, it provides a concise description of the surveyed work. Second, in order to condense the information, it abstracts out several aspects of the work. In addition to reducing detail, the abstraction can be used to identify related areas where some research result may be applied. For example, discussions of lecture videos and research articles are abstracted to discussion of computer-stored artifacts, which is used to present flexible document browsing as a possible extension of flexible video browsing. Finally, it integrates the various research efforts, showing important relationships among them. It does so both by abstracting and by making explicit links between the surveyed papers so that they together tell a cohesive story. The integration is a first step towards a single, general platform for supporting collaboration.

This chapter is targeted at beginners to the field of collaboration who would like to get a flavor of the work in this area; practitioners interested in design, implementation and evaluation ideas; and researchers interested in unexplored avenues. It focuses on the semantics and benefits of collaborative applications without looking at their architecture or implementation, which are discussed elsewhere (Dewan, 1993; Dewan, 1998).

5.2 Dual Goals of Collaborative Applications

There are two main reasons for building collaborative applications. The popular reason is that it can allow geographically dispersed users to collaborate with each other in much the same way colocated ones do by trying to mimic, over the network, natural modes of collaboration, thereby giving the collaborators the illusion of "being there" in one location. For instance, it can support videoconferencing. However, (Hollan and Stornetta, May 1992) have argued that for collaboration technology to be really successful, it must go "beyond being there" by supporting modes of collaboration that cannot be supported in face-to-face collaboration. A simple example of this is allowing users in a meeting to have private channels of communication. We first discuss technology (studied or developed by the surveyed work) for mimicking natural collaboration, and then technology for augmenting or replacing natural collaboration. Sometimes the same technology supports both goals; in that case we first discuss those aspects that support the first goal and then those that support the second goal. Unless otherwise stated, each of the discussed technologies was found, in experiments, to be useful. Thus, failed efforts are not discussed here, though some of them are presented in the surveyed papers.

5.3 Toward Being There: Mimicking Natural Collaboration

Perhaps the simplest way to transport people to the worlds of their collaborators is to provide audio-based collaboration through regular phones. The most complex is to support telepresence through a "sea of cameras" that creates, in real time, a 3-D virtual environment for remote participants. The surveyed work shows several intermediate points between these two extremes.

5.3.1 Single Audio/Video Stream Transmission

The video and audio of a site is transmitted to one or more remote sites, allowing a meeting among multiple sites (Jancke, Venolia et al., 2001). This technology can be used to support a meeting between remote individuals or groups. An example of this is video walls, an implementation of which has been recently evaluated to connect three kitchens (Figure 5.1). In addition to the two remote kitchens, the screen also shows the image captured by the local camera and an image (such as a CNN program) that

FIGURE 5.1 Connected kitchens. (From Jancke, G., G.D. Venolia et al. [2001]. Linking public spaces: Technical and social issues. *Proceedings of CHI 2002,* ACM Digital Library.)

attracts the attention of the local visitors to the kitchen. This technology was found to be moderately useful, enabling a few of the possible spontaneous collaborations.

The possible collaborations could increase if the kitchen videos were also broadcast to desktops. On the other hand, this would increase privacy concerns as it results in asymmetric viewing; the kitchen users would not be able to see the desktop users and thus would not know who was watching them.

This technology was developed to support social interaction, but it (together with the next two technologies that improve on it) could just as well support distributed meetings, as many of them involve groups of people at different sites collaborating with each other (Mark, Grudin et al., 99).

5.3.2 Overview + Speaker

When the meeting involves a remote group, the above technique does not allow a remote speaker to be distinguished from the others. Moreover, if a single conventional camera is used, only members of the group in front of the camera will be captured. In Rui, Gupta et al., 2001, an omnidirectional camera (consisting of multiple cameras) sends an overview image to the remote site. In addition, a shot captured by the camera, whose position can be determined automatically by a speaker detection system or manually by the user, sends the image of the current speaker to the remote site (Figure 5.2). A button is created at the bottom for each participant that can be pressed to select the person whose image is displayed as the speaker.

A simple approach to detecting the speaker is to have multiple microphones placed at different locations in the room and use the differences between the times a speaker's voice reaches the different microphones.

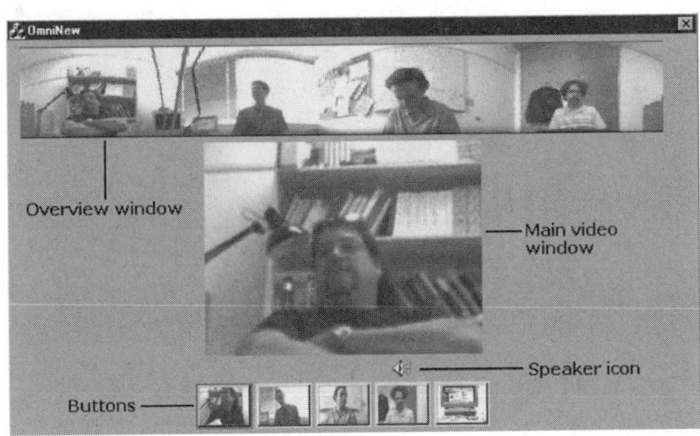

FIGURE 5.2 Overview, speaker, and persons selection buttons. (From Rui, Y., A. Gupta et al. [2001]. Viewing meetings captured by an omni-directional camera. *Proceedings of CHI 2001,* ACM Press, New York.)

For example, if each person has his or her own microphone, that microphone would receive the audio first and thus would indicate the location of the speaker. More complex triangulation techniques would be necessary if there is not a one-to-one mapping between speakers and microphones — for example, if there was a fixed-size microphone array in the room.

5.3.3 Multipoint Lecture

Neither of the techniques above can accommodate multiple remote sites. In the special case of a lecture to multiple remote sites, the following configuration has been tested (Jancke, Grudin et al., 2000). In the lecture room, a large screen shows a concatenation of representations (videos, images, text descriptions) of remote participants. Remote participants can ask questions and vote. Information about their vote and whether they are waiting for questions is attached to their representations. The lecturer uses audio to communicate with the remote attendees, while the latter use text to communicate with the former. The current question is also shown on the large screen.

Each of the desktops of the remote sites shows images of the speaker and the slides (Figure 5.3, right window). Similarly, representations of the remote audience members are shown, together with their vote status in a scrollable view at the lecture site (Figure 5.3, left window). So far, experience has shown that questioners never queue up; in fact, questions are seldom asked from remote sites.

5.3.4 Video-Production-Based Lecture

Unlike the previous scheme, Liu, Rui et al. (2001) and Rui, Gupta et al. (2003) show the local audience to the remote participants. The same screen region is used to show both the audience and the lecturer. This region is fed images captured by a lecturer-tracking camera; an audience-tracking camera, which can track an audience member currently speaking; and an overview camera, which shows both the audience and the speaker.

The following rules, based on practices of professional video producers, are used to determine how the cameras are placed, how their images are multiplexed into the shared remote window, and how they are framed:

- Switch to speaking audience members as soon as they are reliably tracked.
- If neither an audience member nor the lecturer is currently being reliably tracked, show the overview image.
- If the lecturer is being reliably tracked and no audience member is speaking, occasionally show a random audience member shot.
- Frame the lecturer so that there is half a headroom above him in the picture.

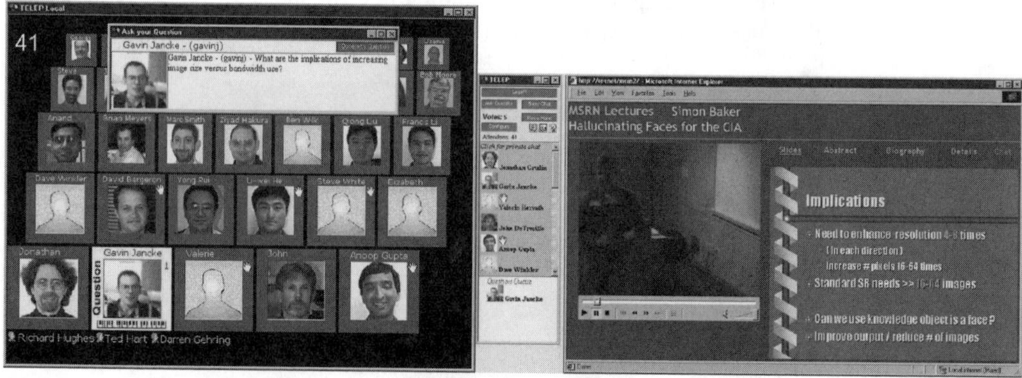

FIGURE 5.3 Display at lecture (left) and remote site (right). (From Jancke, G., J. Grudin et al. [2000]. Presenting to local and remote audiences: Design and use of the TELEP system. *Proceedings of the SIGCHI Conference on Human Factors in Computing Systems [CHI 2000]*, ACM Press, New York.)

- The time when a particular shot (which depends on the current camera and its position) is being displayed should have minimum and maximum limits that depend on the camera. The above rule helps in satisfying this rule.
- Two consecutive shots should be very different; otherwise a jerky effect is created.
- Start with a shot of the overview camera.
- Place the lecturer-tracking camera so that any line along which a speaker moves is not crossed by the camera tracking the speaker; that is, always show a moving lecturer from the same side (Figure 5.4).
- Similarly, place the audience-tracking camera so that a line connecting the lecturer and a speaking audience member is never crossed (Figure 5.4).

5.3.5 Slides Video vs. Application Sharing

There are two ways of displaying slides remotely: one is to transmit a video of the slides, while the other is to allow the remote site to share the application displaying the slides. The projects described above have taken the first approach, which has the disadvantage that it consumes more communication bandwidth, and is thus more costly. Researchers have also investigated sharing of PowerPoint slides using NetMeeting (Mark, Grudin et al., 1999). Their experience identifies some problems with this system or approach:

- Homogeneous computing environment: The groups studied used different computers. NetMeeting does not work well with screens with different resolutions and the time of the study did not work on Unix platforms. Thus, special PCs had to be bought at many sites, especially for conference rooms, to create a uniform computing environment, and managers were reluctant to incur this cost.
- Firewalls: Special conference servers had to be created that bridged the intranet behind the firewall and the internet.
- Heavyweight: The extra step required by this approach of creating and maintaining a shared session was a serious problem for the users, which was solved in some sites by having special technical staff responsible for this task.
- Delay: It often took up to 15 min to set up a shared session and sometimes as much as 30 min, a large fraction of the meeting time.

Despite these problems, people preferred to use NetMeeting from their buildings rather than travel. Moreover, application sharing provides several features such as remote input not supported by video-conferencing. Nonetheless, the above findings show several directions along which application sharing can be improved.

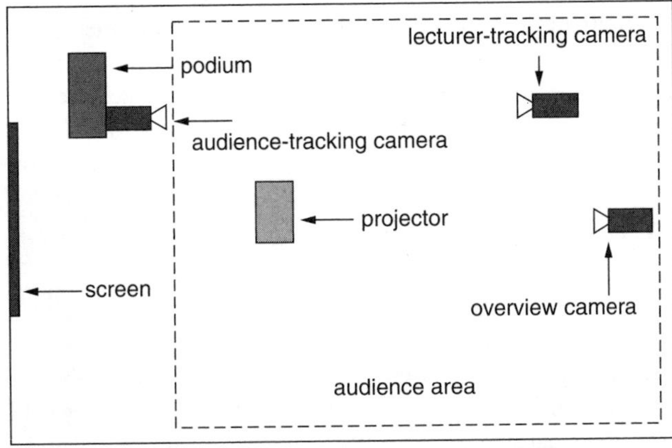

FIGURE 5.4 Cameras and their placement. (From Liu, Q., Y. Rui et al. [2001]. Automating camera management for lecture room environments. *Proceedings of CHI 2001,* ACM Press, New York.)

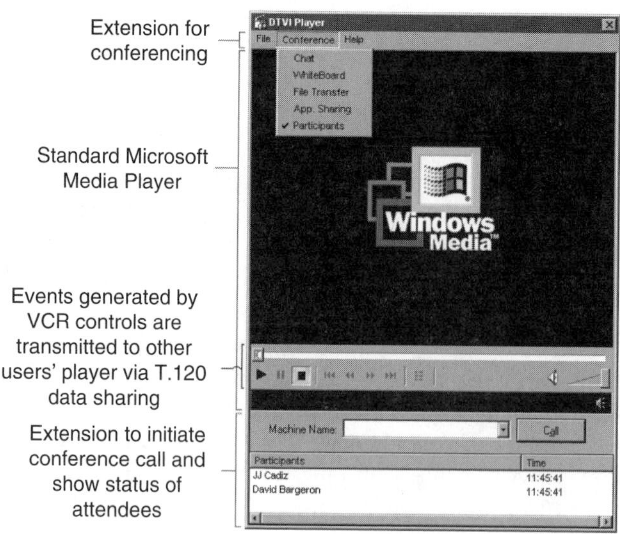

FIGURE 5.5 Collaborative video viewing. (From Cadiz, J.J., A. Balachandran et al. [2000]. Distance learning through distributed collaborative video viewing. *Proceedings of the ACM 2000 Conference on Computer Supported Cooperative Work [CSCW 2000]*, December 1–6, 1999, Philadelphia, PA. ACM Press, New York.)

One example where remote input is useful is collaborative viewing (and discussion) of a video, which has been studied in Cadiz, Balachandran et al. (2000). Figure 5.5 shows the interface created to support this activity.

The video windows of all users are synchronized, and each user can execute VCR controls such as stop, play, rewind, and fast forward.

5.3.6 State-of-the-Art Chat

So far, we have looked primarily at audio and video interfaces for communication among collaborators. Chat can also serve this purpose, as illustrated by the questioner's text message displayed in Figure 5.3. It is possible to have more elaborate chat interfaces that show the whole history of the conversation. Figure 5.6 shows an actual use of such an interface between t-pdewan and krishnag for scheduling lunch. It also shows a serious problem with such interfaces — the conversing parties can concurrently type messages without being aware of what the other is typing, leading to misinterpretation. In this example, it seems that t-pdewan's message asking krishnag to come down to his office was a response to the question the latter asked him regarding where they meet. In fact, these messages were sent concurrently, and had t-pdewan seen krishnag's concurrent message, he would indeed have gone up — a different outcome from what actually happened.

5.3.7 Horizontal Time Line

This is a well-recognized problem. Flow Chat (Vronay, Smith et al., 1999), shown in Figure 5.7, addresses it using two techniques.

1. To allow users typing concurrently to view each other's input, it displays a user's message to others, not when it is complete, as in the previous picture, but as it is typed. The color of the text changes in response to the user's pausing or committing the text.
2. To accurately reflect the sequencing of user's input in the history, it shows the conversation on a traditional time line. Each message in the history is put in a box whose left edge and right edge corresponds to the times when the message was started and completed, respectively. The top and bottom edges fit in a row devoted to the user who wrote the message. Next to the row is text box that can be used by the user to enter a new message. The contents of the message are shown in

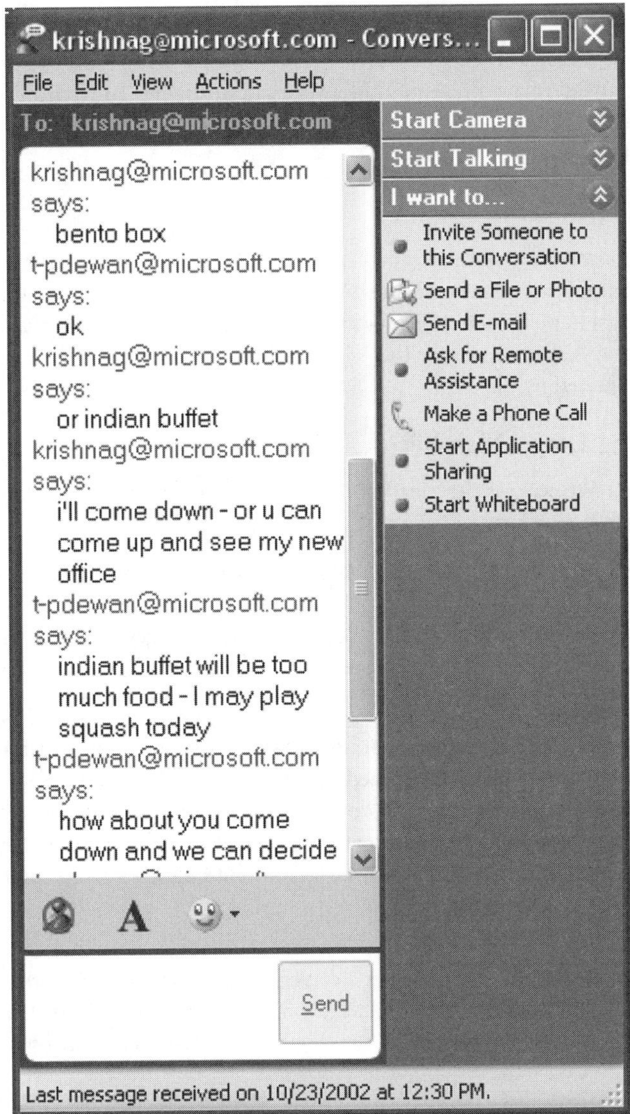

FIGURE 5.6 Misleading concurrent input in chat.

FIGURE 5.7 Horizontal time line in Flow Chat. (From Vronay, D., M. Smith et al. [1999]. Alternative interfaces for chat. *Proceedings of UIST 1999*.)

the time line after it is finished. However, a box with zero width is created in the time line when the message is started, whose width is increased as time passes.

The first technique would have prevented the two users of Figure 5.6 from misunderstanding each other, and the second one would have prevented a third party observing the conversation history from misunderstanding what happened.

5.3.8 Vertical Time Line

A scrolling horizontal time line uses space effectively when the number of users is large and the conversations are short. When this is not the case, the dual approach of creating vertical time lines can be used (Vronay, 2002), which is implemented in Freeway (Figure 5.8). Each user now is given a column, which consists of messages that "balloon" over the user's head as in cartoons. The balloons scroll or flow upwards as new messages are entered by the user.

5.3.9 Supporting Large Number of Users

Neither Flow Chat nor Freeway, as described above, is really suitable for a very large number of users (greater than 10), for two reasons:

1. It becomes distracting to see other users' input incrementally when a large number of users are typing concurrently, possibly in multiple threads of conversation.
2. There is too much vertical (horizontal) space between messages in the same thread if the rows (columns) of the users participating in the thread are far way from each other in the horizontal (vertical) time line. The likelihood of this happening increases when there are a large number of users.

Freeway addresses the first problem by not showing incremental input. Instead, it only shows a placeholder balloon with stars, which get replaced with actual text when the message is committed. Both Flow Chat and Freeway address the second problem by allowing users to move near each other by adjusting their row and column numbers, respectively. However, users did not use this feature much because of the effort required, which motivates research in automatic row and column management.

5.3.10 Graphical Chat

Making one's row or column near another user's corresponds to, in the real world, moving closer to a person to communicate more effectively. V-Chat (Smith, Farnham et al., 2000) better supports this concept by supporting avatars in a chat window (Figure 5.9). Users can move their avatars in a 3-D

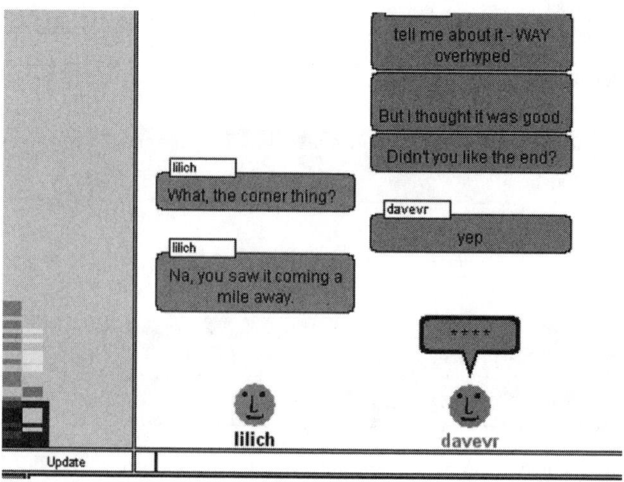

FIGURE 5.8 Vertical time line in Freeway. (From Vronay, D. [2002]. UI for Social Chat: Experimental Results.)

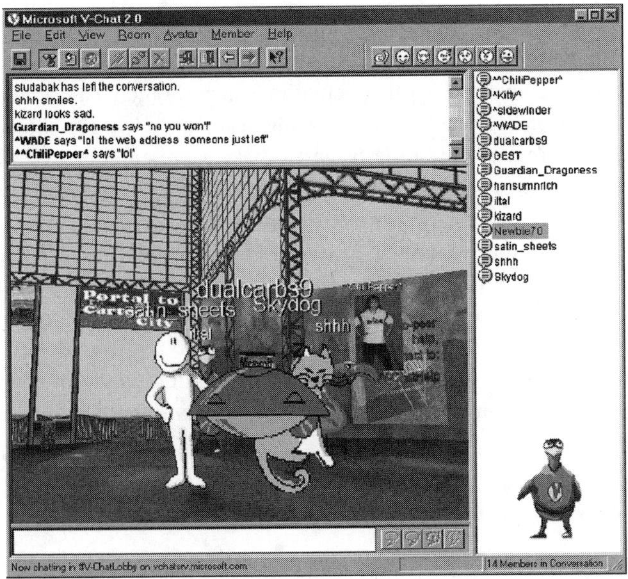

FIGURE 5.9 Avatars in V-Chat. (From Smith, M.A., S.D. Farnham et al. [2000]. The social life of small graphical chat spaces. *Proceedings of CHI 2000*, ACM Press, New York.)

space. Users can communicate with users whose avatars are within the lines of site of their avatars. They can also make their avatars perform gestures to express anger, shrugs, flirting, sadness, smiles, silliness, and waves. This concept has not been integrated with time lines, but could be, conceivably. Users close to each other in the graphical space could be placed in nearby rows or columns. Studies indicate that users conversing with each other automatically move their avatars close to each other.

The discussion in this section started with video-based communication and ended with text-based communication. This is just one way in which the range of technologies supporting "being there" can be organized in a logical progression. There are other effective ways of doing so. For example, in Chidambaram and Jones (1993), these technologies are placed on a continuum from lean media such as text to rich media such as face-to-face interaction.

5.4 Beyond Being There: Augmenting Natural Collaboration

So far, we have looked at how natural collaboration can be approximated by collaboration technology. As mentioned earlier, a dual goal of such technology is to support modes not found in face-to-face collaboration. As Grudin (2002) points out, we must be careful in the ways we try to change natural collaboration, which has remained constant over many years and may well be tied to our fundamental psychology.

5.4.1 Anonymity

It is possible for collaborators to perform anonymous actions. This can be useful even in face-to-face collaboration. An example is studied in Davis, Zaner et al. (2002), where PDAs are used to propose ideas anonymously to the group, thereby allowing these ideas to be judged independently of the perceived status of the persons making them. In an experiment that studied the above example scenario, it was not clear if perceived status did actual harm. The example, however, demonstrates an important example of use of small wireless computers in collaboration.

5.4.2 Multitasking

Another example of augmenting natural collaboration is multitasking. For instance, a user viewing a presentation remotely can be involved in other activities, which should be a useful feature, given that, as we see below in the discussion of asynchronous meeting browsing, a live presentation is not an efficient mechanism to convey information; the time taken to make a presentation is far more than the time required to understand it. Thus, viewing with multitasking can be considered an intermediate point between focused viewing and asynchronous meeting browsing.

A study has found that remote viewers used multitasking frequently but felt that it reduced their commitment to the discussion and they were less engaged (Mark, Grudin et al., 1999). The lack of engagement of some of the participants may not be a problem when they come with different skills. Experience with a multipoint lecture system (White, Gupta et al., 2000) shows that this lack of engagement helped in corporate training as the more experienced students could tune out of some discussions, and knowing this was possible, the lecturers felt more comfortable talking about issues that were not of general interest.

5.4.3 Control of Presence Information

A related feature not found in face to face collaboration is the absence of presence information about remote students. (Presence information about a person normally refers to data regarding the location, in-use computers, and activities of the person.) This is a well-liked feature. As shown in the TELEP figure (Figure 5.3), several remote students preferred to transmit static images or text rather than live video. On the other hand, lack of presence information was found to be a problem in other situations, as remote collaborators were constantly polled to determine if they were still there (Mark, Grudin et al., 1999). This justifies a TELEP-like feature that allows users to determine if presence information is shown at remote sites.

A more indirect and automatic way to determine presence information, which works in a discussion-based collaboration, is to show the recent activity of the participants.

5.4.4 Meeting Browsing

Another way to augment natural collaboration is to relax the constraint that everyone has to meet at the same time. A meeting can be recorded and then replayed by people who did not attend. The idea of asynchronously[1] replaying meetings is not new: videotaping meetings achieves this purpose. (Li, Gupta et al., 2000) show it is possible to improve on this idea by providing a user interface that is more sophisticated than current video or media players.

Figure 5.10 shows the user interface. It provides several features missing in current systems:

- Pause removal control: All pauses in speech and associated video are filtered out.
- Time compression control: The playback speed is increased without changing the audio pitch. The speeded-up video can be stored at a server or the client; the choice involves trading off flexibility in choosing playback speed for reduced network traffic (Omoigui, He et al., 1999).
- Table of contents: This is manually generated by an editor.
- User bookmarks: A user viewing the video can annotate portions of it, which serve as bookmarks for later revisiting the video.
- Shot boundaries: These are automatically generated by detecting shot transitions.
- Flexible jump-back/next: It is possible to jump to a next or previous boundary, bookmark, or slide transition, or to jump by a fixed time interval, using overloaded next and previous commands.

[1] A comparison of synchronous and asynchronous collaboration is beyond the scope of this paper. Dewan (1991) shows that there are multiple degrees of synchrony in collaboration and presents scenarios where the different degrees may be appropriate.

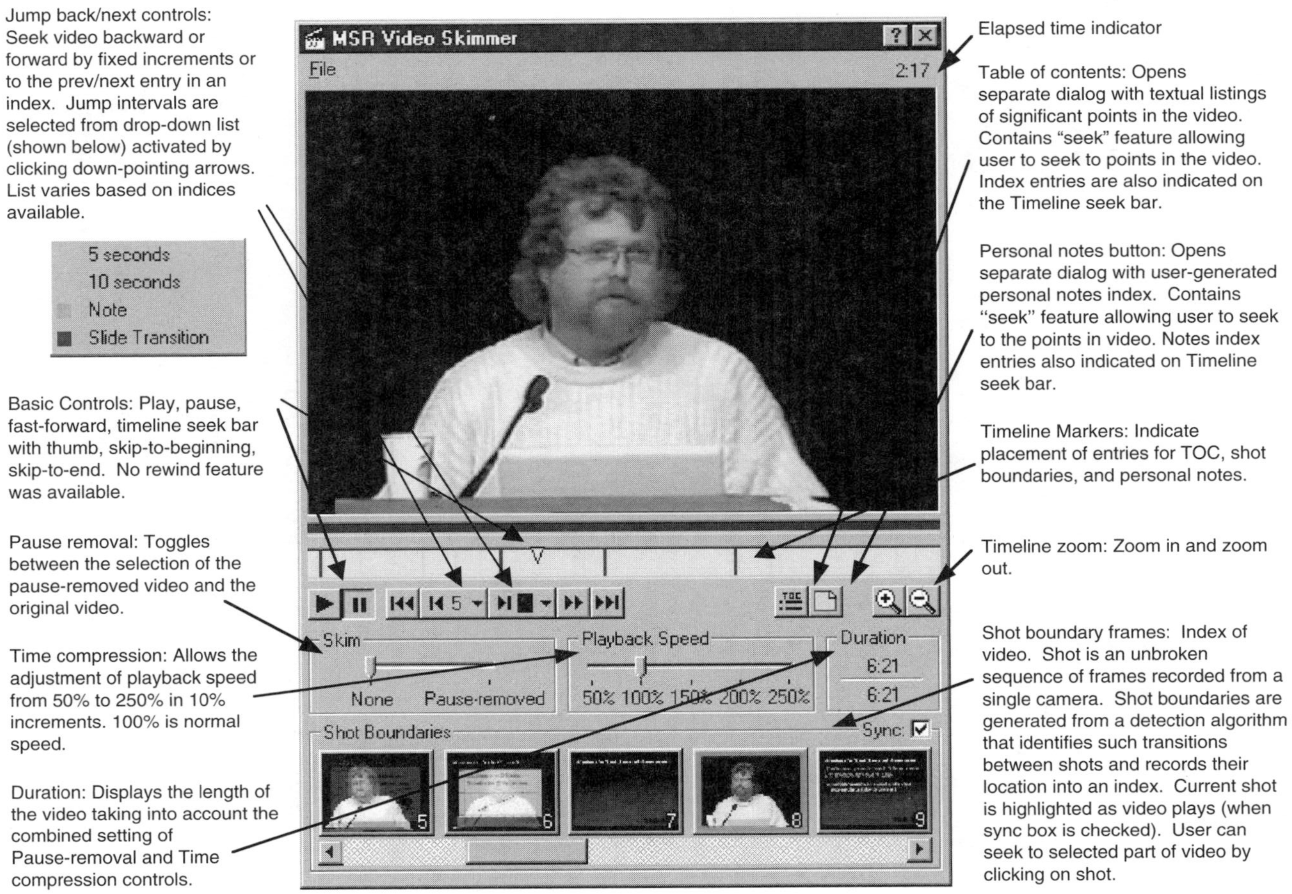

Jump back/next controls: Seek video backward or forward by fixed increments or to the prev/next entry in an index. Jump intervals are selected from drop-down list (shown below) activated by clicking down-pointing arrows. List varies based on indices available.

5 seconds
10 seconds
Note
Slide Transition

MSR Video Skimmer
File
2:17
Skim
None
Pause-removed
Playback Speed
50% 100% 150% 200% 250%
Duration
6:21
6:21
Shot Boundaries
Sync:
TOC

Elapsed time indicator

Table of contents: Opens separate dialog with textual listings of significant points in the video. Contains "seek" feature allowing user to seek to points in the video. Index entries are also indicated on the Timeline seek bar.

Personal notes button: Opens separate dialog with user-generated personal notes index. Contains "seek" feature allowing user to seek to the points in video. Notes index entries also indicated on Timeline seek bar.

Basic Controls: Play, pause, fast-forward, timeline seek bar with thumb, skip-to-beginning, skip-to-end. No rewind feature was available.

Timeline Markers: Indicate placement of entries for TOC, shot boundaries, and personal notes.

Pause removal: Toggles between the selection of the pause-removed video and the original video.

Timeline zoom: Zoom in and zoom out.

Time compression: Allows the adjustment of playback speed from 50% to 250% in 10% increments. 100% is normal speed.

Shot boundary frames: Index of video. Shot is an unbroken sequence of frames recorded from a single camera. Shot boundaries are generated from a detection algorithm that identifies such transitions between shots and records their location into an index. Current shot is highlighted as video plays (when sync box is checked). User can seek to selected part of video by clicking on shot.

Duration: Displays the length of the video taking into account the combined setting of Pause-removal and Time compression controls.

FIGURE 5.10 Browsing video. (From Li, F.C., A. Gupta et al. [2000]. Browsing digital video, *Proceedings of CHI 2000,* ACM Press, New York.)

Pause removal and time compression were found to be useful in sports videos and lectures, where there is a clear separation of interesting and noninteresting portions, but not in carefully crafted TV dramas. Up to 147% playback speed was attained in the studies. Similarly, shot boundaries were found to be particularly useful in sports programs, which have high variations in video contents as opposed to lectures, which have low variations in video contents. The table of contents was found to be particularly useful in lecture presentations.

He, Sanocki et al. (1999) explore two additional alternatives for summarizing audio and video presentations of slides:

1. Assume that the time spent on a slide is proportional to the importance of the slide. Assume also that important things about a slide are explained in the beginning of the presentation of the slide. Summarize a slide by allocating to a slide a time at the beginning portion of the slide discussion whose length is proportional to the total time given to the slide by the speaker.
2. Pitch-based summary: It has been observed that the pitch of a user's voice increases when explaining a more important topic. Summarize a presentation by including portions associated with high-pitch speech.

Both techniques were found to be acceptable and about equally good though not as good as summaries generated by authors of the presentations.

Cutler, Rui et al. (2002) explore two additional ways for browsing an archived presentation, which requires special instrumentation while the meeting is being carried out:

1. Whiteboard content-based browsing: Users can view and hear the recording from the point a particular whiteboard stroke was created. This feature was found be moderately useful.
2. Speaker-based filtering: Users can filter out portions of a video in which a particular person was speaking.

He, Sanocki et al. (2000) propose additional text-based summarization schemes applicable to slide presentations:

- Slides only: The audio or video of the speaker or the audience is not presented.
- Slides + text transcript: Same as above except that a text transcript of speaker audio is also seen.
- Slides + highlighted text transcript: Same as above except that the key points of the text transcript are highlighted.

For slides with high information density, all three methods were found to be as effective as author-generated audio or video summaries. For slides with low information density, the highlighted text transcript was found to be as effective as the audio or video summaries.

5.4.5 Divergent Views and Concurrent Input

Yet another related feature is allowing collaborators to see different views of shared state and concurrently edit it. It allows users to create their preferred views and to work on different parts of the shared state, thereby increasing the concurrency of collaboration. This idea has been explored earlier in several works (Stefik, Bobrow et al., April 1987; Dewan and Choudhary, April 1991). It requires special support for concurrency control (Munson and Dewan, November 1996), access control (Dewan and Shen, November 1998; Shen and Dewan, November 1992), merging (Munson and Dewan, June 1997), and undo (Choudhary and Dewan, October 1995), which are beyond the scope of this chapter.

5.4.6 Chat History

We have seen above what seems to be another example of augmenting natural collaboration; chat programs show the history of messages exchanged, which would not happen in the alternative of a face-to-face audio-based conversation. On the other hand, this would happen in a face-to-face conversation carried out by exchanging notes. Thus, whether the notion of chat history augments natural collaboration depends on what we consider is the alternative. The user interfaces for supporting it, however, can be far more sophisticated than that supported by note exchanges, as we see below.

Chat history can inform a newcomer to the conversation about the current context. However, it is not used much because looking at it distracts users from the current context. Freeway addresses these problems in two ways:

1. Snap back scrolling: It is possible to press a button and drag the scrollbar to any part of the history. When the button is released, the view scrolls back to the previous scroll position.
2. Overview pane: Only a portion of the history is shown in the chat window. A miniature of the entire history is shown in a separate window (on the left in Figure 5.8). A rectangle marks the portion of the miniature that is displayed in the scroll window, which can be dragged to change the contents of the chat window. As before, at the completion of the drag operation, the chat window snaps back to its original view.

While users liked and used these features, newcomers still asked other participants about previous discussions rather than looking at the history. Therefore, more work needs to be done to make it effective. An important disadvantage of keeping a history that latecomers can look at is that people might be careful in what they say because they do not know who may later join the conversation (Grudin, 2002). Thus, there should be a way to enter messages that are not displayed to late joiners, which brings up new user interface issues.

5.4.7 Scripted Collaboration

Farnham, Chesley et al. (2000) show that another way to augment natural chat is to have the computer provide a script for the session that suggests the discussion topics and how long each topic should be discussed. Figure 5.11 shows the use of a script to discuss interview candidates. The script automates the role of a meeting facilitator (Mark, Grudin et al., 1999), which has been found to be useful. After using it automatically in a scripted discussion, users manually enforced its rules in a subsequent nonscripted interview discussion.

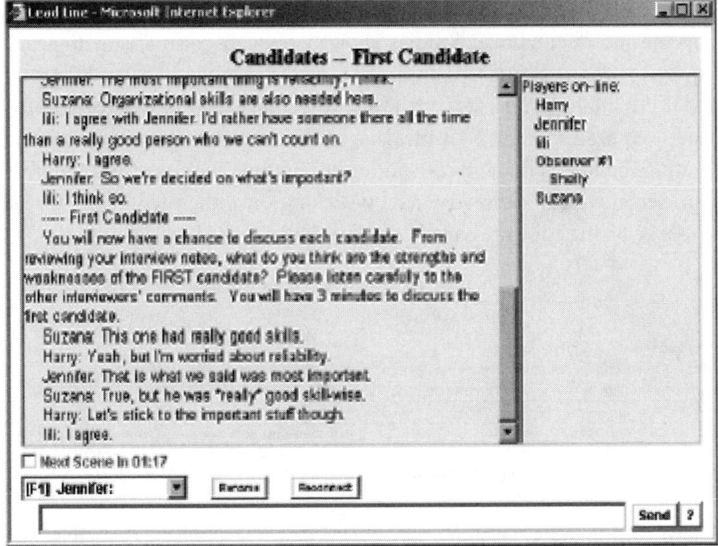

FIGURE 5.11 Scripted collaboration. (From Farnham, S.D., H. Chesley et al. [2000]. Structured on-line interactions: Improving the decision-making of small discussion groups. CSCW.)

5.4.8 Threaded Chat

A computer-provided script is one way of structuring a chat conversation. Threading the discussion is another method. As we saw in the chat discussion, in a large chat room, it is important to separate the various threads of discussion. Moving representations of communicating users close to each other is one way to achieve this effect, but does not work when a user is in more that one thread concurrently or the threads are hierarchical. Therefore, one can imagine a chat interface that supports bulletin-board-like threaded messages. Smith, Cadiz et al. (2000) have developed such an interface, shown in Figure 5.12. A new chat message can be entered as a response to a previous message and is shown below it after indenting it with respect to the latter. Independently composed messages are shown at the same indentation level and are sorted by message arrival times. Messages are thus arranged into a tree in which other messages at the same indentation level as a message are its siblings and those immediately following it at the next indentation level are its children. A user responds to a message by clicking on it and typing new text. As soon as the user starts typing, a new entry is added at the appropriate location in the window. However, this entry does not show incremental user input. It simply displays a message indicating that new text is being entered, which is replaced with the actual text when the user commits the input. Studies have shown that users pay too much attention to typing correctness when their incremental input is broadcast. In nonincremental input, they simply went back and corrected spelling errors before sending their message. It seems useful to give users the option to determine if incremental input is transmitted or received in the manner described in Dewan and Choudhary (1991).

Grudin (2002) wonders whether users should care about spelling errors in chat; it is meant to be an informal, lightweight alternative to e-mail, which at one time was meant to be an informal alternative to postal mail. He points out that one of the reasons for using chat today is to escape from the formality of e-mail; thus, focusing on spelling issues may be counterproductive.

In comparison with a traditional chat interface, the interface as described above makes it difficult for the newcomer to determine what the latest messages are. This problem is addressed by fading the font of new items to grey.

Messages can be further characterized as questions, answers, and comments, and can be used in recording and displaying statistics about the kinds of chat messages, which are shown in a separate pane. The system automatically classifies the message depending on the presence of a question mark in it or its parent message (the message to which it is a response).

The basic idea of threads has been useful in bulletin boards, but is it useful in the more synchronous chats supported by the interface above? Studies showed that, in comparison to traditional chat, the interface above required fewer messages to complete a task and resulted in balanced participation, though task performance did not change and users felt less comfortable with it. One reason for the discomfort may be the extra step required to click on a message before responding to it. Perhaps this problem can be solved by integrating threaded chat with the notion of moving one's computer representation (avatar, column, or row); a user moves to a thread once by clicking on a message in the thread. Subsequently, the most recent message or the root message is automatically clicked by default.

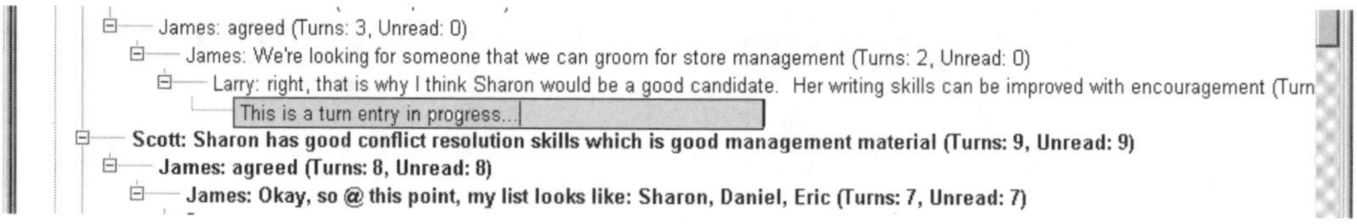

FIGURE 5.12 Threaded chat. (From Smith, M., J. Cadiz et al. [2000]. Conversation trees and threaded chats. *Proceedings of CSCW 2000.*)

5.4.9 Threaded E-Mail

If threads can be useful in organizing bulletin boards and chat messages, what about e-mail? It seems intuitive, at least in retrospect, that concepts from bulletin boards transfer to chat, because both contain messages broadcast to multiple users, only some of which may be of interest to a particular user. Moreover, in both cases, there is no notion of deleting messages. In contrast, all messages in a mailbox are directed at its owner, who can delete messages. Are threads still useful?

This question is addressed by Venolia, Dabbish et al (2001). They give four reasons for supporting threaded e-mail. Threads keep a message with those related to it, thereby giving better local context. They also give better global context as the contents of the mailbox can be decomposed into a small number of threads as opposed to large number of individual messages. This is particularly important when a user encounters a large number of unread messages. Moreover, one can perform a single operation such as delete or forward on the root of a thread that applies to all of its children. Finally, one can define thread-specific operations such as "delete all messages in the thread and unsubscribe future messages in it," and "forward all messages in the thread and subscription to future messages in it." The first two reasons apply also to chat and bulletin boards. It would be useful to investigate how the above thread-based operations can be applied to bulletin boards and chat interfaces.

Venolia, Dabbish et al. designed a new kind of user interface, shown in Figure 5.13, to test threaded mail. It differs from conventional thread-based user interfaces in three related ways. First, it uses explicit lines rather than indentation to indicate child–parent relationship, thereby saving on scarce display space. Second, because it shows this relationship explicitly, it does not keep all children of a node together. Instead, it intermixes messages from different threads, ordering them by their arrival time, thereby allowing the user to easily identify the most recent messages. Third, it groups threads by day. Finally, for each thread, it provides summary information such as the users participating in it. Users have liked this user interface, specially the local context it provides. One can imagine porting some of its features, such as grouping by day or summary information, to chat and bulletin boards.

Yet another interface for threads has been developed by Smith and Fiore (2001), which is shown in Figure 5.14. Like the previous interface, it shows explicit lines between parent and children nodes. However, the nodes in the tree display do not show the text of the messages. Instead, they are rendered

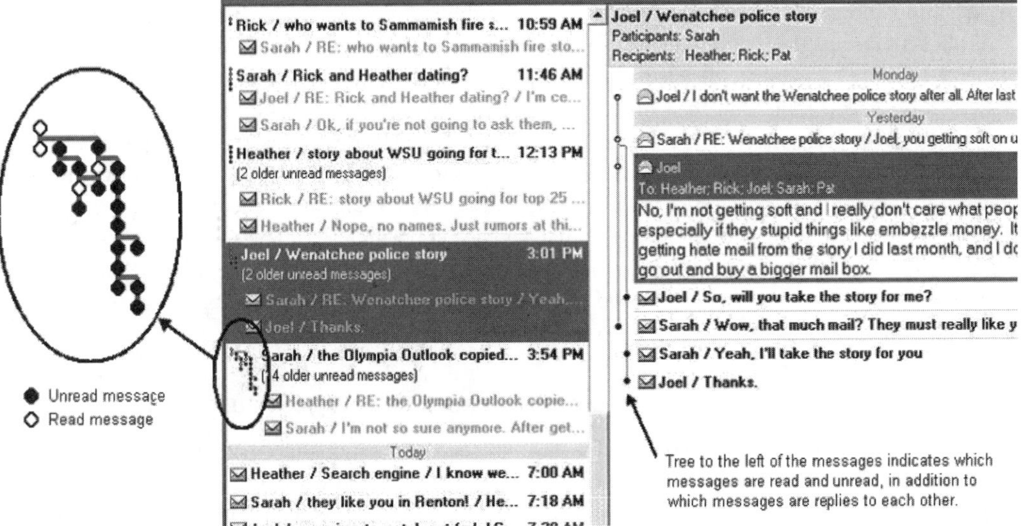

FIGURE 5.13 Threaded e-mail. (From Venolia, G.D., L. Dabbish et al. Supporting E-mail Workflow. Microsoft Research Technical Report MSR-TR-2001-88.)

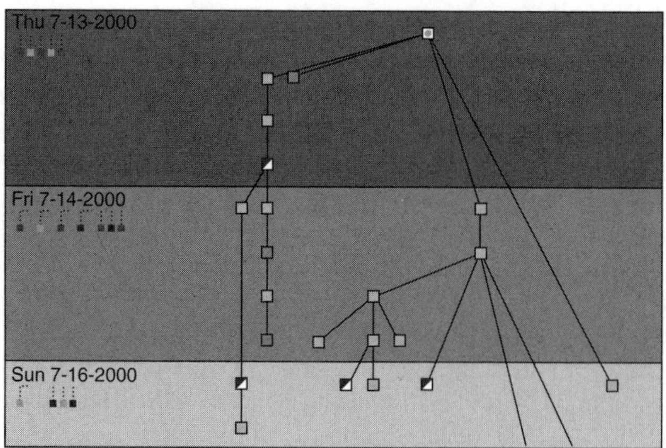

FIGURE 5.14 Graphical overview of threads. (From Smith, M.A. and A.T. Fiore [2001]. Visualization components for persistent conversations. *Proceedings of CHI 2001,* ACM Press, New York.)

as rectangular boxes giving summary information about the message. For example, a dotted box is used for a message from the author of the root post, and a half-shaded box for a message from the most prolific contributor. Clicking over a box displays the contents of the message in a different pane of the user interface. This interface allows the viewer to get a quick summary of the discussion and the people involved in it. It was developed for bulletin boards but could be applicable to chat and mail also.

5.4.10 Threaded Articles Discussions and Annotations

Chat, bulletin boards, and e-mail provide support for general discussions. Some of these discussions are about documents. It is possible to build a specialized user interface for such discussions. Two examples of such an interface are described and compared by Brush, Bargeron et al. (2002). The first example links a document to the discussion threads about it. The second example provides finer-granularity linking, associating fragments of a document with discussion threads, which are essentially threaded annotations. The annotation-based system also provides mechanisms to summarize the whole document and make private annotations. The summaries, however, are not discussions and thus not threaded.

Studies comparing the user interfaces, not surprisingly, found that the finer-granularity linkage allowed students to more easily make detailed points about the article because they did not have to reproduce the target of their comment in their discussions and thus created more comments. On the other hand, they had a slight preference for the coarser granularity. One reason was that they read paper copies of the article, often at home, where they did not have access to the tool. As a result, they had to redo work when commenting. Second, and more interesting, the coarser granularity encouraged them to make high-level comments about the whole article that were generally preferred. The annotation-based system did not provide an easy and well-known way to associate a discussion with the whole document. To create such an association, people attached the discussion to the document title or to a section header, which was not elegant or natural to everyone.

5.4.11 Variable-Granularity Annotations to Changing Documents

This problem is fixed by Office 2000 by allowing threaded annotations to be associated both with the whole document and a particular fragment (Figure 5.15). Like Brush, Bargeron et al. (2002), Cadiz, Gupta et al. (2000) studied the use of these annotations, focusing not on completed (and, in fact, published) research articles, about which one is more interested in general comments indicating what was learnt, but on specification drafts, where more comments about fragments can be expected. Since specification drafts can (and are expected to) change, the following issue is raised: When a fragment changes, what should happen to its annotations, which are essentially now "orphans?" As indicated in

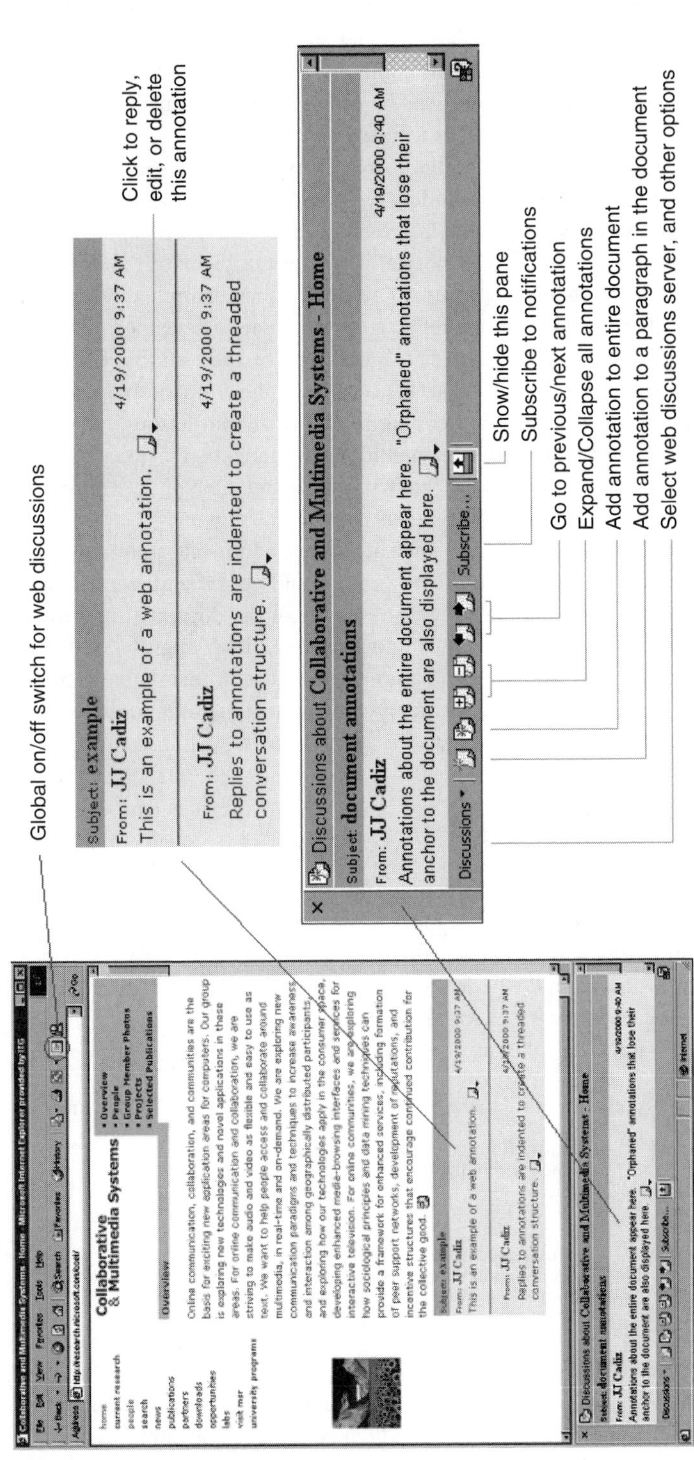

FIGURE 5.15 Variable-grained annotations with orphaning. (From Cadiz, J.J., A. Gupta et al. [2000]. Using Web annotations for asynchronous collaboration around documents. *Proceedings of the ACM 2000 Conference on Computer Supported Cooperative Work (CSCW 2000)*, December 1–6, 1999, Philadelphia, PA. ACM Press, New York.)

Figure 5.15 (right, bottom window), orphan annotations are displayed with annotations about the whole document.

Annotations are an alternative to the more traditional channels of commenting such as e-mail, telephone, or face-to-face meetings. However, the latter provide not only a way to send comments but also a mechanism to notify concerned parties about the comment. To make annotations more useful, automatic notifications, shown in Figure 5.16, were sent by e-mail when documents were changed. Users could decide on the frequency of notification sent to them.

To what extent would people really use annotations over the more traditional commenting channels? In a large field study carried out over 10 months, Cadiz, Gupta et al. (2000) found that there was significant use with an average of 20 annotations per person. Interestingly, users did not make either very high-level comments (because the author would probably not get it) or very nitpicky comments such as spelling mistakes (because most would not be interested in it.) Moreover, they continued to use other channels when they needed immediate response because delivery of notifications depended on subscription frequency and thus was not guaranteed to be immediate. Furthermore, they felt that the notifications did not give them enough information — in particular, the content of the annotation. In addition, a person making a comment does not know who is subscribing to automatic notification that is automatically generated, and often ends up manually sending e-mail to the subscribers. Finally, a significant fraction of people stopped using the system after making the first annotation. One of the reasons given for this is that they did not like orphan annotations losing their context.

A fix to the lack of nitpicky comments may be to create special editor-like annotations for document changes, which could be simply applied by the authors, who would not have to retype the correction. In fact, these could be generated by the "annotator" editing a copy of the document or a fragment copied from the document in the spirit of live text (Fraser and Krishnamurthy, August 1990). Users may still not be willing to put the effort into making such comments because in this shared activity the person making the effort is not the one who reaps its fruits, a problem observed in organizations (Grudin, 2001). We see in the following text fixes to other problems mentioned above.

5.4.12 Robust Annotations

A simple way to address the orphan annotation problem seems to be to attach them not to the whole document but to the smallest document unit containing the fragment to which they are originally attached. Brush, Bargeron et al. (2001) discuss a more sophisticated algorithm that did not orphan an annotation if the fragment to which it was attached changed in minor ways. More specifically, it saved a deleted fragment and cut words from the back and front of it until it was partially matched with some fragment in the changed document or it was less than 15 characters long. In case of match, it attached annotations of the deleted fragment to the matched fragment. In lab studies, users liked this algorithm

The following change(s) happened to the document http://product/overview/index.htm:

Event:	Discussion items were inserted or modified in the document
By:	rsmith
Time:	7/28/99 11:01:04 AM
Event:	Discussion items were inserted or modified in the document
By:	ajones
Time:	7/28/99 12:09:27 PM

Click here to stop receiving this notification.

FIGURE 5.16 Automatically generated notification. (From Cadiz, J.J., A. Gupta et al. [2000]. Using Web annotations for asynchronous collaboration around documents. *Proceedings of the ACM 2000 Conference on Computer Supported Cooperative Work (CSCW 2000),* December 1–6, 1999, Philadelphia, PA. ACM Press, New York.)

when the difference between the original and matched fragment was small and not when it was large. The authors of PREP have had to also wrestle with the problem of finding corresponding pieces of text in documents, and have developed a sophisticated and flexible diffing (Neuwirth, Chandok et al., October 1992) scheme to address this problem. It seems useful and straightforward to apply their algorithm to the orphan annotation problem. The users of the annotation system suggested a more intriguing approach — identify and use keywords to determine corresponding text fragments.

It is useful to provide threaded annotations for not only documents but also other objects such as lecture presentations, as shown in Bargeron, Gupta et al. (1999) and Bargeron, Grudin et al. (2002). A discussion is associated not with a fragment of a document, but with a point in the video stream and the associated slide, as shown in Figure 5.17. Similarly, it may be useful to create annotations for spreadsheets, programs, and PowerPoint slides.

However, as we see here, separate kinds of annotation- or thread-based mechanisms exist for different kinds of objects. For example, as we saw in Section 5.3 on video browsing, it is useful to create a flexible overloaded Next commands for navigating to the next annotation, next section, and, in general, the next unit, where the unit can change. Such a command could be useful for navigating through documents also. Thus, it would be useful to create a single, unified annotation- or thread-based mechanism.

5.4.13 Notifications

Consider now the issue of notifications about document comments. As mentioned earlier, the Web Discussions notification scheme already described was found to have several problems. Brush, Bargeron et al. made some modifications to address these problems. They allowed an annotation maker to determine who will receive notifications about it, thereby saving on duplicate mail messages. They also generated notifications that were more descriptive, giving the comment, identifying the kind of annotation (reply or comment) and, in case of a reply, giving a link to the actual annotation that can be followed

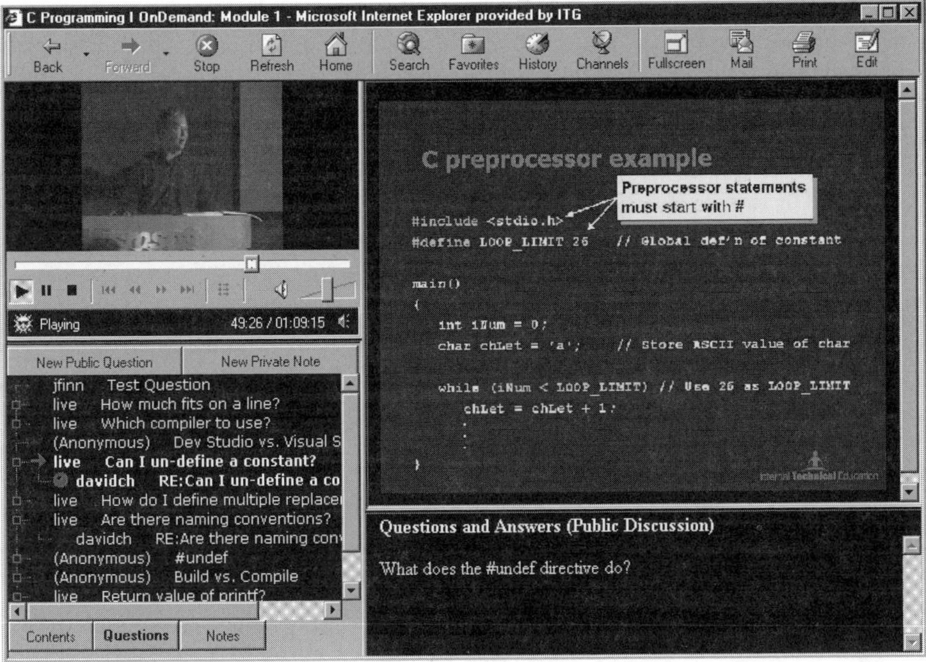

FIGURE 5.17 Annotating a presentation. (From Brush, A.J.B., D. Bargeron et al. Notification for shared annotation of digital documents. *Proceedings of CHI 2002,* Minneapolis, MN, April 20–25.)

This is an automatic notification. More information...
Click here to update your notification settings.
The changes that just occurred are:
On http://server/Notify.htm

colinb added a reply to a comment by duncanbb on 9/12/2001 3:20 PM
RE: test annotation
This is the text of an example annotation.

Click to update your notification settings.

FIGURE 5.18 A descriptive notification. (From Brush, A.J.B., D. Bargeron et al. Notification for shared annotation of digital documents. *Proceedings of CHI 2002,* Minneapolis, MN, April 20–25.)

to look at its context in the containing thread (Figure 5.18). As in Web discussions, these were sent as e-mail messages.

Sometimes, a user wished to continuously poll for information rather than receive a notification for each kind of change. Brush, Bargeron et al. supported this information awareness through a separate window, called a Slideshow (Cadiz, Venolia et al., 2002), created for viewing all information about which the user was expected to have only peripheral awareness. The source of each piece of information was associated with an icon called a ticket that appeared in the display of the source. A user subscribed to the source by dragging the ticket to the Slideshow (Figure 5.19a). When contained in the Slideshow, the ticket shows summary information about changes to the source. In the case of an annotated document, it shows the number of annotations and the number created on the current day (Figure 5.19b, right window). When the mouse is moved over the ticket, a new window called the tool tip window is displayed, which contains more detailed information, as shown in Figure 5.19b, left window. Studies found that users liked annotation awareness provided by the automatically e-mailed notifications and the Slideshow window. However, using them over Web Discussions did not seem to improve task performance.

It may be useful to integrate the two awareness mechanisms by inserting links or copies of the notifications that are currently sent by e-mail into the tooltip window, thereby reducing the clutter in mailboxes. Another way to integrate the two is to not send notifications when it is known that the user is polling the tooltip window or the document itself. Grudin (1994) observed that managers and executives who with the aid of their staff constantly polled the calendar found meeting notifications a nuisance. It is quite likely that spurious change notifications are as annoying. Perhaps the application-logging techniques developed by Horvitz et al. (2002), discussed later, can be adapted to provide this capability.

5.4.14 Disruptions Caused by Messages

While a message (such as e-mail, instant message, or document comment notification) about some activity improves the performance of that activity, it potentially decreases the performance of the foreground task of the person to whom it is sent. Czerwinski, Cutrell et al. (2000) studied the effect of instant messages on the performance of two kinds of tasks: a mundane search task requiring no thinking, and a more complex search task requiring abstract thinking. They found that the performance of the straightforward task decreased significantly because of the instant messages, but the performance of the complex task did not change. As society gets more sophisticated, the tasks they perform will also get more abstract, and thus if one can generalize the above results, messages will not have a deleterious effect. Nonetheless, it may be useful to build a mechanism that suppresses messages of low priority — especially if the foreground task is a mundane one.

5.4.15 Prioritizing Messages

Horvitz, Koch et al. (2002) have built such a system for e-mail. It prioritizes unread e-mail messages and lists them by priority, as shown in Figure 5.20. The priority of a message is a function of the cost of

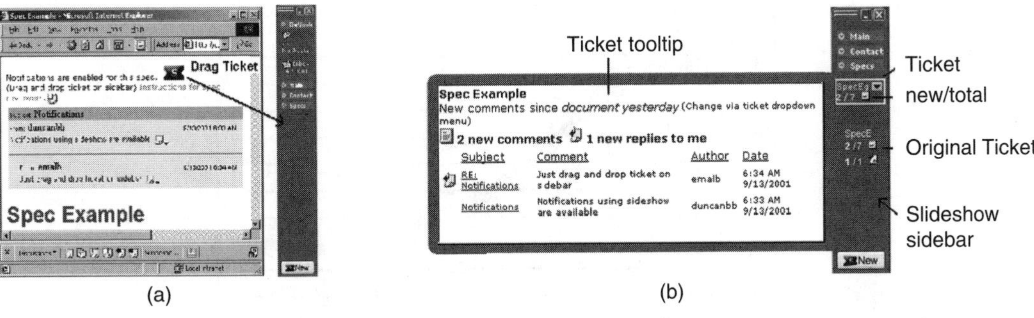

FIGURE 5.19 Slideshow continuous awareness. (From Brush, A.J.B., D. Bargeron et al. Notification for shared annotation of digital documents. *Proceedings of CHI 2002,* Minneapolis, MN, April 20–25.)

P...	Source	Title
90	External Caller ...	Voice Mail Message (6 seconds)
87	Naresh Sundar...	RE: Cross-group discussion on notific
80	Paul Bennett	RE: talk today?
77	Richard Draves	govt expenses
69	timpaek	RE: Phone Call

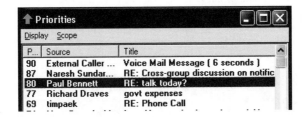

FIGURE 5.20 Automatically prioritizing messages. (From Horvitz, E., P. Koch et al. [2002]. Coordinate: Probabilistic forecasting of presence and availability. *Proceedings of the Eighteenth Conference on Uncertainty and Artificial Intelligence [UAI-2002],* August 2–4, Edmonton, Alberta, Canada. Morgan Kaufman, San Francisco.)

delayed review, which is calculated based on several criteria, including the organizational relationship with the sender, how near the sending time is to key times mentioned in messages scanned so far, the presence of questions, and predefined phrases in the messages, tenses, and capitalization.

5.4.16 Automatic Redirection of Message and Per-Device Presence and Availability Forecasting

This prioritization is also used for determining which messages should be sent to a user's mobile device. The goal is to cause disruption to the mobile user only if necessary. If the user has not been active at the desktop for a time greater than some parameter, and if the message has priority greater than some threshold, then the message can be sent to the mobile device. (These parameters can be set dynamically, based on whether the user is in a meeting or not.) It would be even better if this is done only when the person is likely to be away for some time, *a*. A person's presence in a location is forecast using calendar information for the period, if it exists. If it does not, then it can be calculated based on how long the user has been away from the office; log of the user's activities for various days of the week; and phases within the day such as morning, lunch, afternoon, evening, and night. This information is used to calculate the probability of users returning within some time, *r*, given that they have been away for some time, *a*, during a particular phase of a particular day, as shown in Figure 5.21. It was found that this estimate was fairly reliable.

This (continuously updated) estimate can be used to automatically fill unmarked portions of a user's calendar, as shown in Figure 5.22, which can be viewed by those who have access to it. In addition, it can be sent as "out-of-office e-mail" response to urgent messages (Figure 5.24). One can imagine providing this information in response to other incoming messages such as invitation to join a NetMeeting conference.

So far, we have assumed that users have two devices: their office desktop and mobile device. In general, they may have access to multiple kinds of devices. Horvitz, Koch et al. generate logs of activities for all

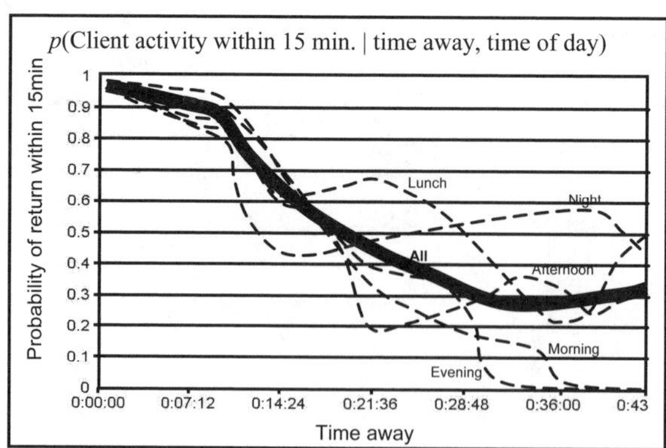

FIGURE 5.21 Probability of returning within 15 min. (From Horvitz, E., P. Koch et al. [2002]. Coordinate: Probabilistic forecasting of presence and availability. *Proceedings of the Eighteenth Conference on Uncertainty and Artificial Intelligence [UAI-2002],* August 2–4, Edmonton, Alberta, Canada. Morgan Kaufman, San Francisco.)

FIGURE 5.22 Presence prediction in shared calendar. (From Horvitz, E., P. Koch et al. [2002]. Coordinate: Probabilistic forecasting of presence and availability. *Proceedings of the Eighteenth Conference on Uncertainty and Artificial Intelligence [UAI-2002],* August 2–4, Edmonton, Alberta, Canada. Morgan Kaufman, San Francisco.)

of their devices, which are used to provide fine-grained presence information by device. With each device, its capabilities are also recorded. This information can be used, for instance, to determine how long it will be before a user has access to a device allowing teleconferencing.

Presence, of course, is not the same as availability. For example, a user may have access to a teleconferencing device but not be available for the conference. Similarly, a user may be at the desktop but not be ready to read new e-mail. Moreover, current availability is not enough to carry out some collaboration for an extended period of time. For example, a user who has returned to his office may not stay long enough to carry out the collaboration. Horvitz, Koch et al. address these problems also, that is, they try to forecast continuous presence (for some period of time) and availability. For instance, they can forecast the likelihood that a person will return to his office for at least 15 min given that he has been away for 25 min (Figure 5.23).

Predictions about continuous presence and availability of a person are made by reading calendars; monitoring attendance of scheduled meetings based on meeting kind; tracing application start, focus, and interaction times; and allowing users to set interruptability levels. For example, monitoring application usage can be used to predict when a person will next read e-mail. Similarly, the probability that a person will actually attend a scheduled meeting depends on whether attendance is optional or required, the number of attendees and, if it is a recurrent meeting, the person's history of attending the meeting.

One potential extension to this work is to use information about deadlines to determine availability. For example, I do not wish to be interrupted an hour before class time or the day before a paper or proposal deadlines. Information about deadlines could be determined from:

- The calendar — the beginning of a meeting is a deadline to prepare items for it.
- To-do lists, if they list the time by which a task has to be done.
- Project tracking software.
- Documents created by the user, which may have pointers to dates by which they are due. For example, an NSF electronic proposal contains the name of the program to which it is being submitted, which can be used to find on the web the date by which it is due.

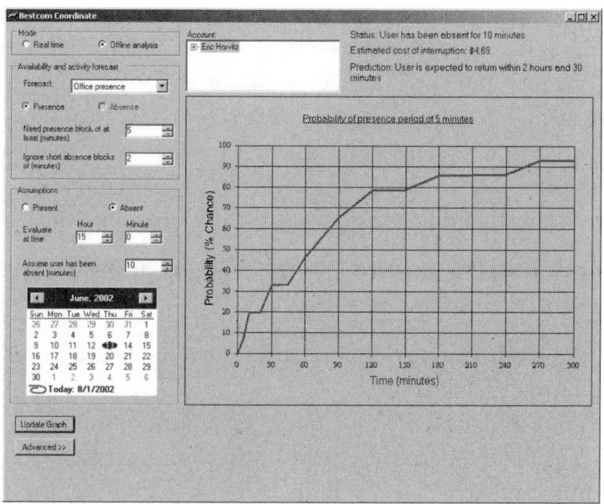

FIGURE 5.23 Automated presence response. (From Horvitz, E., P. Koch et al. [2002]. Coordinate: Probabilistic forecasting of presence and availability. *Proceedings of the Eighteenth Conference on Uncertainty and Artificial Intelligence [UAI-2002]*, August 2–4, Edmonton, Alberta, Canada. Morgan Kaufman, San Francisco.)

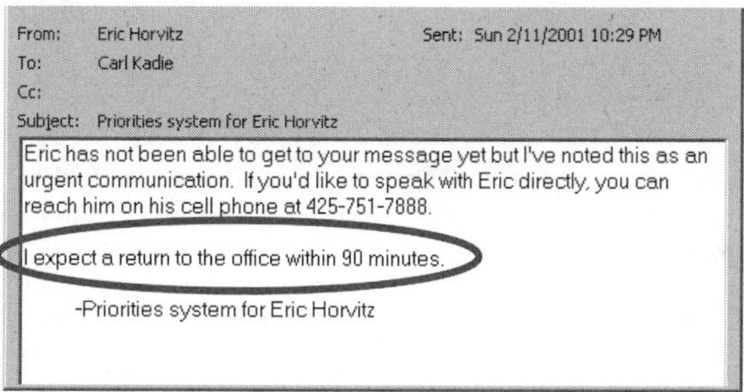

FIGURE 5.24 Forecasting continuous presence. (From Horvitz, E., P. Koch et al. [2002]. Coordinate: probabilistic forecasting of presence and availability. *Proceedings of the Eighteenth Conference on Uncertainty and Artificial Intelligence [UAI-2002]*, August 2–4, Edmonton, Alberta, Canada. Morgan Kaufman, San Francisco.)

Another possible extension is to use application logging to determine if a notification should actually be sent or not; if a user has been polling some data, then there is no need to send him a notification when it is changed. As mentioned earlier, Grudin observed that managers and executives, who with the aid of their staff constantly polled the calendar, found meeting notifications a nuisance.

Application logging could also be used to automatically convert a series of e-mail message exchanges in real time to an instant message conversation.

5.5 Conclusions

This paper has several lessons for practitioners looking to learn from existing research. Today, collaboration products are divided into systems supporting mail, instant messaging, presence, application-sharing infrastructures, and custom extensions to popular applications such as word processors, spreadsheets, and program development environments. The surveyed work presents opportunities for extending these products with new features:

- Messaging (Instant or Mail): Those working with messaging may wish to determine if the benefits of snapback scrolling, time flow, scripted collaboration, threads, and graphical chat apply to their target audience.

- Presence: Those looking at presence can learn from the user interfaces shown here that allow a collaborator to be aware of remote users — in particular, the TELEP user interface for showing a large number of remote students. Related to this is the work on notifications and automatic forecasting of presence.
- Custom extensions: Most of the custom extensions to single-user tools support annotations. The surveyed work evaluates the usefulness of existing annotation support and proposes several new techniques such as robust annotations to overcome its weakness.

By comparing and contrasting the surveyed efforts, this paper also identifies some specific new areas of research that extend existing directions of research:

- An integrated thread-based annotation mechanism that applies to instant messaging, mail, news, and commenting of multimedia objects
- Automatically creating annotations that request rephrasing by editing a copy of the document or fragment, which can then be simply accepted by the author to create the rephrase
- Creating robust annotations by flexibly diffing a revision with the original
- Extensions to application logging that use deadlines to detect interruptability, convert real-time mail messages into instant messaging conversations, and suppress a notification if the user is constantly polling the source of the notification
- Control over when a user's input is transmitted to other users in a chat window, whiteboard, or some other application.

Acknowledgments

Jonathan Grudin and the anonymous reviewer provided numerous comments for improving this chapter. This work was supported in part by Microsoft Corporation and NSF grants IIS 9977362, ANI 0229998, IIS 0312328, and EIA 0303590.

References

Bargeron, D., J. Grudin et al. (2002). Asynchronous collaboration around multimedia applied to on-demand education. *Journal of MIS* 18(4).

Bargeron, D., A. Gupta et al. (1999). Annotations for Streaming Video on the Web: System Design and Usage Studies. WWW 8.

Brush, A. J. B., D. Bargeron et al. (2002). Supporting interaction outside of class: Anchored discussions vs. discussion boards. *Proceedings of CSCL 2002,* January 7–11, Boulder, CO.

Brush, A. J. B., D. Bargeron et al. Notification for shared annotation of digital documents. *Proceedings of CHI 2002,* Minneapolis, MN, April 20–25.

Brush, A. J. B., D. Bargeron et al. (2001). Robust Annotation Positioning in Digital Documents. Proceedings ACM S16CH1 Conference on Human Factors in Computing Systems (?) 285–292.

Cadiz, J. J., A. Balachandran et al. (2000). Distance learning through distributed collaborative video viewing. *Proceedings of the ACM 2000 Conference on Computer Supported Cooperative Work (CSCW 2000),* December 1–6, 1999, Philadelphia, PA. ACM Press, New York.

Cadiz, J. J., A. Gupta et al. (2000). Using Web annotations for asynchronous collaboration around documents. *Proceedings of the ACM 2000 Conference on Computer Supported Cooperative Work (CSCW 2000),* December 1–6, 1999, Philadelphia, PA. ACM Press, New York.

Cadiz, J. J., G. D. Venolia et al. (2002). Designing and deploying an information awareness interface. *Proceedings of the ACM 2002 Conference on Computer Supported Cooperative Work (CSCW 2002),* November 16–20, 1999, New Orelans, LA. ACM Press, New York.

Chidambaram, L. and B. Jones (1993). Impact of communication and computer support on group perceptions and performance. *MIS Quarterly* 17(4): 465–491.

Choudhary, R. and P. Dewan (October 1995). A general multi-user Undo/Redo model. H. Marmolin, Y. Sundblad, and K. Schmidt (Eds.), *Proceedings of the Fourth European Conference on Computer-Supported Cooperative Work,* Kluwer, Dordrecht, Netherlands.

Cutler, R., Y. Rui, et al. (2002). Distributed Meetings: A Meeting Capture and Broadcast System. ACM Multimedia.

Czerwinski, M., E. Cutrell et al. (2000). Instant messaging and interruption: Influence of task type on performance. *Proceedings of OZCHI 2000,* December 4–8, Sydney, Australia.

Davis, J. P., M. Zaner et al. (2002). Wireless Brainstorming: Overcoming Status Effects in Small Group Decisions. Social Computing Group, Microsoft.

Dewan, P. (1993). Tools for implementing multiuser user interfaces. *Trends in Software: Special Issue on User Interface Software* 1: 149–172.

Dewan, P. (1998). Architectures for collaborative applications. *Trends in Software: Computer Supported Co-operative Work* 7: 165–194.

Dewan, P. and R. Choudhary (April 1991). Flexible user interface coupling in collaborative systems. *Proceedings of the ACM Conference on Human Factors in Computing Systems, CHI '91,* ACM Digital Library, ACM, New York.

Dewan, P. and H. Shen (November 1998). Flexible meta access-control for collaborative applications. *Proceedings of ACM 1998 Conference on Computer Supported Cooperative Work (CSCW 1998),* November 14–18, Seattle, WA. ACM Press, New York.

Farnham, S. D., H. Chesley et al. (2000). Structured on-line interactions: Improving the decision-making of small discussion groups. CSCW.

Fraser, C. W. and B. Krishnamurthy (August 1990). Live text. *Software Practice and Experience* 20(8).

Grudin, J. (1994). Groupware and Social Dynamics: Eight Challenges for Developers. *Communications of the ACM,* 37(1): 92–105.

Grudin, J. (2001). Emerging Norms: Feature Constellations Based on Activity Patterns and Incentive Differences. Microsoft.

Grudin, J. (2002). Group Dynamics and Ubiquitous Computing. *Communications of the ACM,* 45(12): 74–78.

He, L., E. Sanocki et al. (1999). Auto-Summarization of Audio-Video Presentations. ACM Multimedia.

He, L., E. Sanocki et al. (2000). Comparing Presentation Summaries: Slides vs. Reading vs. Listening. CHI.

Hollan, J. and S. Stornetta (May 1992). Beyond being there. *Proceedings of the ACM Conference on Human Factors in Computing Systems, CHI '92,* ACM Digital Library, ACM, New York.

Horvitz, E., P. Koch et al. (2002). Coordinate: Probabilistic forecasting of presence and availability. *Proceedings of the Eighteenth Conference on Uncertainty and Artificial Intelligence (UAI-2002),* August 2–4, Edmonton, Alberta, Canada. Morgan Kaufman, San Francisco.

Jancke, G., J. Grudin et al. (2000). Presenting to local and remote audiences: Design and use of the TELEP system. *Proceedings of the SIGCHI Conference on Human Factors in Computing Systems (CHI 2000),* ACM Press, New York.

Jancke, G., G. D. Venolia et al. (2001). Linking public spaces: Technical and social issues. *Proceedings of CHI 2002,* ACM Digital Library.

Li, F. C., A. Gupta et al. (2000). Browsing digital video, *Proceedings of CHI 2000,* ACM Press, New York.

Liu, Q., Y. Rui et al. (2001). Automating camera management for lecture room environments. *Proceedings of CHI 2001,* ACM Press, New York.

Mark, G., J. Grudin et al. (1999). Meeting at the Desktop: An Empirical Study of Virtually Collocated Teams. ECSCW.

Munson, J. and P. Dewan (June 1997). Sync: A Java framework for mobile collaborative applications. *IEEE Computer* 30(6): 59–66.

Munson, J. and P. Dewan (November 1996). A concurrency control framework for collaborative systems. *Proceedings of the ACM 1996 Conference on Computer Supported Cooperative Work,* ACM Press, New York.

Neuwirth, C. M., R. Chandok et al. (October 1992). Flexible diff-ing in a collaborative writing system. *Proceedings of ACM 1992 Conference on Computer Supported Cooperative Work,* ACM Press, New York.

Omoigui, N., L. He et al. (1999). Time-compression: Systems concerns, usage, and benefits. *Proceedings of CHI 1999,* ACM Digital Library.

Rui, Y., A. Gupta et al. (2001). Viewing meetings captured by an omni-directional camera. *Proceedings of CHI 2001,* ACM Press, New York.

Rui, Y., A. Gupta et al. (2003). Videography for telepresentations. *Proceedings of CHI 2003.*

Shen, H. and P. Dewan (November 1992). Access control for collaborative environments. *Proceedings of the ACM Conference on Computer Supported Cooperative Work.*

Smith, M., J. Cadiz et al. (2000). Conversation trees and threaded chats. *Proceedings of CSCW 2000.*

Smith, M. A., S. D. Farnham et al. (2000). The social life of small graphical chat spaces. *Proceedings of CHI 2000,* ACM Press, New York.

Smith, M. A. and A. T. Fiore (2001). Visualization components for persistent conversations. *Proceedings of CHI 2001,* ACM Press, New York.

Stefik, M., D. G. Bobrow et al. (April 1987). WYSIWIS Revised: Early Experiences with Multiuser Interfaces. *ACM Transactions on Office Information Systems* 5(2): 147–167.

Venolia, G. D., L. Dabbish et al. (2001). Supporting Email Workflow. Microsoft Research Technical Report MSR-TR-2001-88.

Vronay, D. (2002). UI for Social Chat: Experimental Results.

Vronay, D., M. Smith et al. (1999). Alternative interfaces for chat. *Proceedings of UIST 1999.*

White, S. A., A. Gupta et al. (2000). Evolving use of a system for education at a distance. *Proceedings of HICSS-33* (Short version in CHI 99 Extended Abstracts, pp. 274–275.).

6

Internet Telephony

Henning Schulzrinne

CONTENTS

Abstract 6-1
6.1 Introduction 6-2
6.2 Motivation 6-3
6.2.1 Efficiency 6-3
6.2.2 Functionality 6-3
6.2.3 Integration 6-4
6.3 Standardization 6-4
6.4 Architecture 6-5
6.5 Overview of Components 6-7
6.5.1 Common Hardware and Software Components 6-7
6.6 Media Encoding 6-8
6.6.1 Audio 6-8
6.6.2 Video 6-10
6.7 Core Protocols 6-10
6.7.1 Media Transport 6-10
6.7.2 Device Control 6-11
6.7.3 Call Setup and Control: Signaling 6-12
6.7.4 Telephone Number Mapping 6-17
6.7.5 Call Routing 6-18
6.8 Brief History 6-18
6.9 Service Creation 6-19
6.10 Conclusion 6-19
6.11 Glossary 6-20
References 6-21

Abstract

Internet telephony, also known as voice-over-IP, replaces and complements the existing circuit-switched public telephone network with a packet-based infrastructure. While the emphasis for IP telephony is currently on the transmission of voice, adding video and collaboration functionality requires no fundamental changes.

Because the circuit-switched telephone system functions as a complex web of interrelated technologies that have evolved over more than a century, replacing it requires more than just replacing the transmission technology. Core components include speech coding that is resilient to packet losses, real-time transmission protocols, call signaling, and number translation. Call signaling can employ both centralized control architectures as well as peer-to-peer architectures, often in combination.

Internet telephony can replace traditional telephony in both enterprise (as IP PBXs) and carrier deployments. It offers the opportunity for reduced capital and operational costs, as well as simplified introduction of new services, created using tools similar to those that have emerged for creating Web services.

1-58488-381-2/05/$0.00+$1.50
© 2005 by CRC Press LLC

6.1 Introduction

The International Engineering Consortium (IEC) describes Internet Telephony as follows:

> Internet telephony refers to communications services — voice, facsimile, and/or voice-messaging applications — that are transported via the Internet, rather than the public switched telephone network (PSTN). The basic steps involved in originating an Internet telephone call are conversion of the analog voice signal to digital format and compression/translation of the signal into Internet protocol (IP) packets for transmission over the Internet; the process is reversed at the receiving end.

More technically, Internet telephony is the real-time delivery of voice and possibly other multimedia data types between two or more parties, across networks using the Internet protocols, and the exchange of information required to control this delivery.

The terms Internet telephony, IP telephony, and voice-over-IP (VoIP) are often used interchangeably. Some people consider IP telephony a superset of Internet telephony, as it refers to all telephony services over IP, rather than just those carried across the the Internet. Similarly, IP telephony is sometimes taken to be a more generic term than VoIP, as it de-emphasizes the voice component. While some consider telephony to be restricted to voice services, common usage today includes all services that have been using the telephone network in the recent past, such as modems, TTY, facsimile, application sharing, whiteboards, and text messaging. This usage is particularly appropriate for IP telephony because one of the strengths of Internet telephony is the ability to *be media-neutral,* that is, almost all of the infrastructure does not need to change if a conversation includes video, shared applications, or text chat.

Voice services can also be carried over other packet networks without a mediating IP layer; for example, voice-over-DSL (VoDSL) [Ploumen and de Clercq, 2000] for consumer and business DSL subscribers, and voice-over-ATM (VoATM) for carrying voice over ATM [Wright, 1996, 2002], typically as a replacement for interswitch trunks. Many consider these as transition technologies until VoIP reaches maturity. They are usually designed for single-carrier deployments and aim to provide basic voice transport services, rather than competing on offering multimedia or other advanced capabilities. For brevity, we will not discuss these other voice-over-packet technologies (VoP) further in this chapter.

A related technology, multimedia streaming, shares the point-to-point or multipoint delivery of multimedia information with IP telephony. However, unlike IP telephony, the source is generally a server, not a human being, and, more importantly, there is no bidirectional real-time media interaction between the parties. Rather, data flows in one direction, from media server to clients. Like IP telephony, streaming media requires synchronous data delivery where the short-term average delivery rate is equal to the native media rate, but streaming media can often be buffered for significant amounts of time, up to several seconds, without interfering with the service. Streaming and IP telephony share a number of protocols and codecs that will be discussed in this chapter, such as RTP and G.711. Media streaming can be used to deliver the equivalent of voice mail services. However, it is beyond the scope of this chapter.

In the discussion below, we will occasionally use the term *legacy telephony* to distinguish plain old telephone service (POTS) provided by today's time-division multiplexing (TDM) and analog circuits from packet-based delivery of telephone-related services, the Next-Generation Network (NGN). Apologies are extended to the equipment and networks thus deprec(i)ated. The term public switched telephone network (PSTN) is commonly taken as a synonym for "the phone system," although pedants sometimes prefer the postmonopoly term GSTN (General Switched Telephone Network).

IP telephony is one of the core motivations for deploying quality-of-service into the Internet, since packet voice requires one-way network latencies well below 100 msec and modest packet drop rates of no more than about 10% to yield usable service quality [Jiang and Schulzrinne, 2003; Jiang et al., 2003]. Most attempts at improving network-related QoS have focused on the very limited use of packet prioritization in access routers. Because QoS has been widely covered and is not VoIP specific, this chapter will not go into greater detail. Similarly, authentication, authorization, and accounting (AAA) are core telephony services, but not specific to VoIP.

6.2 Motivation

The transition from circuit-switched to packet switched telephone services is motivated by cost savings, functionality, and integration, with different emphasis on each depending on where the technology is being used.

6.2.1 Efficiency

Traditional telephone switches are not very cost effective as traffic routers; each 64 kb/sec circuit in a traditional local office switch costs roughly between $150 and $500, primarily because of the line interface costs. Large-scale PBXs have similar per-port costs. A commodity Ethernet switch, on the other hand, costs only between $5 and $25 per 100 Mb/sec port, so switching packets has become significantly cheaper than switching narrowband circuits even if one discounts the much larger capacity of the packet switch and only considers per-port costs [Weiss and Hwang, 1998].

Free long-distance phone calls were the traditional motivation for consumer IP telephony even if they were only free incrementally, given that the modem or DSL connection had already been paid for. In the early 1990s, US long-distance carriers had to pay about $0.07/min to the local exchange carriers, an expense that gatewayed IP telephony systems could bypass. This allowed Internet telephony carriers to offer long-distance calls terminating at PSTN phones at significant savings. This charge has now been reduced to less than $0.01/min, decreasing the incentive [McKnight, 2000].

In many developing countries, carriers competing with the monopoly incumbent have found IP telephony a way to offer voice service without stringing wires to each phone, using DSL, or satellite uplinks. Also, leased lines were often cheaper, on a per-bit basis, than paying international toll charges, opening another opportunity for arbitrage [Vinall, 1998].

In the long run, the cost differential in features such as caller ID, three-way calling, and call waiting may well be more convincing than lower per-minute charges.

For enterprises, the current cost of a traditional circuit-switched PBX and a VoIP system are roughly similar, at about $500 a seat, due to the larger cost of IP phones. However, enterprises with branch offices can reuse their VPN or leased lines for intracompany voice communications and can avoid having to lease small numbers of phone circuits at each branch office. It is well known that a single large trunk for a large user population is more efficient than dividing the user population among smaller trunks, due to the higher statistical multiplexing gain. Enterprises can realize operational savings because moves, adds, and changes for IP phones are much simpler, only requiring that the phone be plugged in at its new location.

As described in Section 6.2.3, having a single wiring plant rather than maintaining separate wiring and patch panels for Ethernet and twisted-pair phone wiring is attractive for new construction.

For certain cases, the higher voice compression and silence suppression found in IP telephony (see Section 6.5.1) may significantly reduce bandwidth costs. There is no inherent reason that VolP has better compression, but end system intelligence makes it easier and more affordable to routinely compress all voice calls end-to-end. As noted, silence suppression is not well supported in circuit switched networks outside high-cost point-to-point links. (Indeed, in general, packetization overhead can eat up much of this advantage.)

6.2.2 Functionality

In the long run, increased functionality is likely to be a prime motivator for transitioning to IP telephony, even though current deployment is largely limited to replicating traditional PSTN features and functionality. PSTN functionality, beyond mobility, has effectively stagnated since the mid-1980 introduction of CLASS features [Moulton and Moulton, 1996] such as caller ID. Attempts at integrating multimedia, for example, have never succeeded beyond a few corporate teleconferencing centers.

Additional functionality is likely to arise from services tailored to user needs and vertical markets (Section 6.7.5), created by or close to users, integration with presence, and other Internet services, such as Web and e-mail. Since Internet telephony completes the evolution from in-band signaling found in analog telephony to complete separation of signaling and media flows, services can be offered equally well by businesses and specialized non-facility-based companies as by Internet service providers or telephone carriers.

Because telephone numbers and other identifiers are not bound to a physical telephone jack, it is fairly easy to set up virtual companies where employee home phones are temporarily made part of the enterprise call center, for example.[1]

It is much easier to secure VoIP services via signaling and media encryption, although legal constraints may never make this feature legally available.

6.2.3 Integration

Integration has been a leitmotif for packet-based communications from the beginning, with integration occurring at the physical layer (same fiber, different wavelengths), link layer (SONET), and, most recently, at the network layer (everything-over-IP). Besides the obvious savings in transmission facilities and the ability to allocate capacity more flexibly, managing a single network promises to be significantly simpler and to reduce operational expenditures.

6.3 Standardization

While proprietary protocols are still commonly found in the applications for consumer VoIP services and indeed dominate today for enterprise IP telephony services (Cisco Call Manager protocol), there is a general tendency towards standardizing most components needed to implement VoIP services.

Note that standardization does not imply that there is only one way to approach a particular problem. Indeed, in IP telephony, there are multiple competing standards in areas such as signaling, while in others different architectural approaches are advocated by different communities. Unlike telephony standards, which exhibited significant technical differences across different countries, IP telephony standards so far diverge mostly for reasons of emphasis on different strengths of particular approaches, such as integration with legacy phone systems vs. new services or maturity vs. flexibility.

A number of organizations write standards and recommendations for telephone service, telecommunications, and the Internet. Standards organizations used to be divided into official and industry standards organizations, where the former were established by international treaty or law, while the latter were voluntary organizations founded by companies or individuals. Examples of such treaty-based organizations include the International Telecommunications Union (ITU, www.itu.int) that in 1993 replaced the former International Telephone and Telegraph Consultative Committee (CCITT). The CCITT's origins are over 100 years old. National organizations include the American National Standards Institute (www.ansi.org) for the U.S, and the European Telecommunications Standards Institute (ETSI) for Europe. Because telecommunications is becoming less regional, standards promulgated by these traditionally regional organizations are finding use outside those regions.

In the area of IP telephony, 3GPP, the 3rd Generation Partnership Project, has been driving the standardization for third-generation wireless networks using technology "based on evolved GSM core networks and the radio access technologies that they support." It consists of a number of organizational partners, including ETSI. A similar organization, 3GPP2, deals with radio access technologies derived

[1] Such an arrangement requires that the residential broadband access provider offer sufficiently predictable quality-of-service (QoS), either by appropriate provisioning or explicit QoS controls. It remains to be seen whether Internet service providers will offer such guaranteed QoS unbundled from IP telephony services. Initial deployments of consumer VoIP services indicate that QoS is sufficient in many cases without additional QoS mechanisms.

from the North American CDMA (ANSI/TIA/EIA-41) system; it inherits most higher-layer technologies, such as those relevant for IP telephony, from 3GPP.

When telecommunications were largely a government monopoly, the ITU was roughly the "parliament of monopoly telecommunications carriers," with a rough one-country, one-vote rule. Now, membership appears in the ITU to be open to just about any manufacturer or research organization willing to pay its dues. Thus, today there is no substantial practical difference between these different major standardization organizations. Standards are not laws or government regulations and obtain their force if customers require that vendors deliver products based on standards.

The Internet Engineering Task Force (IETF) is "a large open international community of network designers, operators, vendors, and researchers"[2] that specifies standards for the Internet Protocol, its applications such as SMTP, IMAP, and HTTP, and related infrastructure services such as DNS, DHCP, and routing protocols. Many of the current IP telephony protocols described in this chapter were developed within the IETF.

In a rough sense, one can distinguish primary from secondary standardization functions. In the primary function, an organization develops core technology and protocols for new functionality, while the emphasis in secondary standardization is on adapting technology developed elsewhere to new uses or describing it more fully for particular scenarios. As an example, 3GPP has adopted and adapted SIP and RTP, developed within the IETF, for the Internet multimedia subsystem in 3G networks. 3GPP also develops radio access technology, which is then in turn used by other organizations.

In addition, some organizations, such as the International Multimedia Telecommunications Consortium (IMTC) and the SIP Forum, provide interoperability testing, deployment scenarios, protocol interworking descriptions, and educational services.

6.4 Architecture

IP telephony, unlike other Internet applications, is still dominated by concerns about interworking with older technology, here, the PSTN. Thus, we can define three classes [Clark, 1997] of IP telephony operation (Figure 6.1), depending on the number of IP and traditional telephone end systems.

In the first architecture, sometimes called trunk replacement, both caller and callee use circuit-switched telephone services. The caller dials into a gateway, which then connects via either the public Internet or a private IP-based network or some combination to a gateway close to the callee. This model requires no changes in the end systems and dialing behavior and is often used, without the participants being aware of it, to offer cheap international prepaid calling card calls. However, it can also be used to connect two PBXs within a corporation with branch offices. Many PBX vendors now offer IP trunk interfaces that simply replace a T-1 trunk by a packet-switched connection.

Another hybrid architecture, sometimes called hop-on or hop-off depending on the direction, places calls from a PSTN phone to an IP-based phone or vice versa. In both cases, the phone is addressed by a regular telephone number, although the phone may not necessarily be located in the geographic area typically associated with that area code. A number of companies have started to offer IP phones for residential and small-business subscribers that follow this pattern. A closely related architecture is called an *IP PBX*, where phones within the enterprise connect to a gateway that provides PSTN dial tone.

If the IP PBX is shared among several organizations and operated by a service provider, it is referred to as *IP Centrex* or *hosted IP PBX*, as the economic model is somewhat similar to the centrex service offered by traditional local exchange carriers. Like classical centrex, IP centrex service reduces the initial capital investment for the enterprise and makes system maintenance the responsibility of the service provider. Unlike PSTN centrex, where each phone has its own access circuit, IP centrex only needs a fraction of the corporate Internet connectivity to the provider and is generally more cost-efficient. If the enterprise uses standards-compliant IP phones, it is relatively straightforward to migrate between IP centrex and IP PBX architectures, without changing the wiring plant or the end systems.

[2] IETF web site.

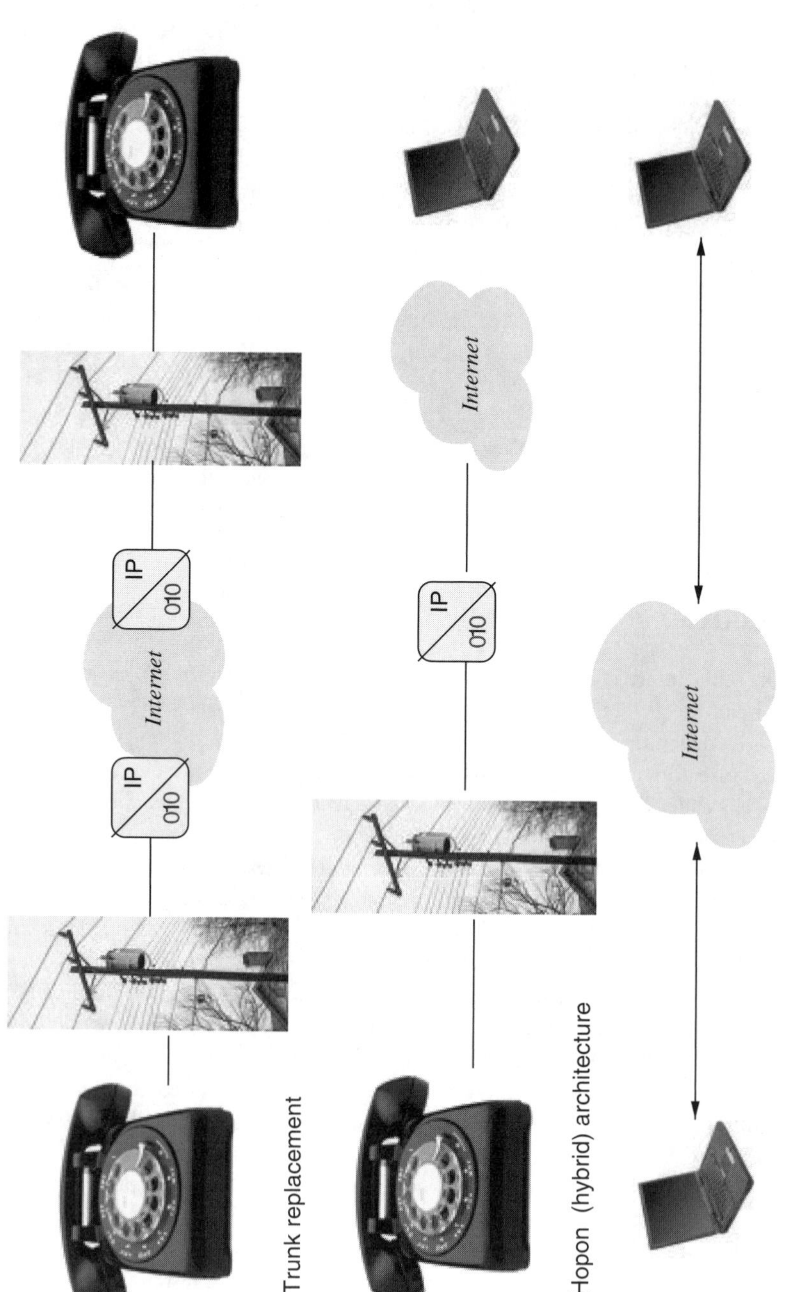

FIGURE 6.1 Internet telephony architectures.

This architecture is also found in some cable systems where phone service is provided by the cable TV operator (known as a multisystem operator, MSO) [Miller et al., 2001; Wocjik, 2000]. Note, however, that not all current cable TV–phone arrangements use packet voice; some early experiments simply provide a circuit switched channel over coax and fiber.

The third architecture dispenses with gateways and uses direct IP-based communications end-to-end between caller and callee. This arrangement dominated early PC-based IP telephony, but only works well if all participants are permanently connected to the Internet.

The most likely medium-term architecture is a combination of the hybrid and end-to-end model, where calls to other IP phones travel direct, whereas others use gateways and the PSTN. If third-generation mobile networks succeed, the number of IP-reachable devices may quickly exceed those using the traditional legacy interface. If devices are identified by telephone numbers, there needs to be a way for the caller to determine if a telephone number is reachable directly. The ENUM directory mechanism described in Section 6.7.4 offers one such mapping.

6.5 Overview of Components

At the lower protocol layers, Internet components are easily divided into a small number of devices and functions that rarely cause confusion. For example, hosts, routers, and DNS servers have clearly defined functionality and are usually placed in separate hardware. Usually, servers are distinguished by the protocols they speak: a web server primarily deals with HTTP, for example. Things are not nearly as simple for IP telephony, where an evolving understanding, the interaction with the legacy telephony world and marketing have created an abundance of names that sometimes reflect function and sometimes common bundlings into a single piece of hardware.

In particular, the term “softswitch” is often used to describe a set of functions that roughly replicate the control functionality of a traditional telephone switch. However, this term is sufficiently vague that it should be avoided in technical discussions.

The International Packet Communications Consortium [International Packet Communications Consortium] has attempted to define these functional entities and common physical embodiments.

6.5.1 Common Hardware and Software Components

The most common hardware components in IP telephony are IP phones, access gateways, and integrated access devices (IADs).

IP phones are end systems and endpoints for both call setup (signaling) and media, usually audio. There are both hardware phones that operate stand-alone, and softphones, software applications that run on common operating system platforms on personal computers. Hardware phones typically consist of a digital signal processor with analog-to-digital (A/D) and digital-to-analog (D/A) conversion, general-purpose CPU, and network interface. The CPU often runs an embedded operating system and usually supports standard network protocols such as DNS for name resolution, DHCP for network autoconfiguration, NTP for time synchronization, and tftp and HTTP for application configuration. Modern IP phones offer the same range of functionality as analog and digital business telephones, including speakerphones, caller ID displays, and programmable keys. Some IP phones have limited display programmability or have a built-in Java environment for service creation. (See Figure 6.2)

Access gateways connect the packet and circuit-switched world, both in the control and media planes. They packetize bit streams or analog signals coming from the PSTN into IP packets and deliver them to their IP destination. In the opposite direction, they convert sequences of IP packets containing segments of audio into a stream of voice bits and “dial” the appropriate number in the legacy phone system. Small (residential or branch-office) gateways may support only one or two analog lines, while carrier-class gateways may have a capacity of a T1 (24 phone circuits) or even a T3 (720 circuits). Large-scale gateways may be divided into a media component that encodes and decodes voice and a control component, often a general-purpose computer, that handles signaling.

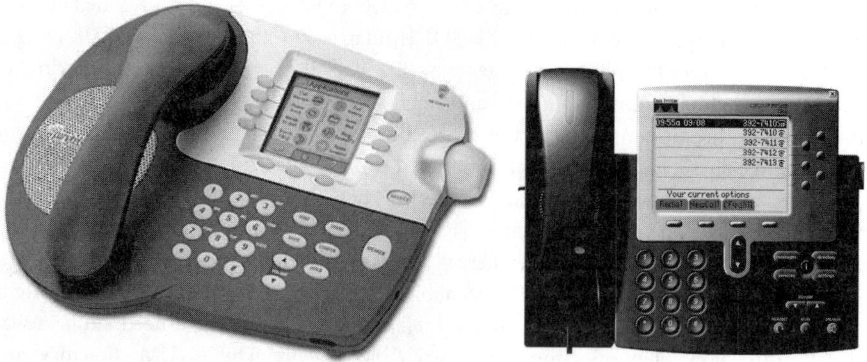

FIGURE 6.2 Some examples of IP phones.

An integrated access device (IAD) typically features a packet network interface, such as an Ethernet port, and one or more analog phone (so-called FXS, i.e., station) interfaces. They allow commercial and residential users to reuse their large existing investment in analog and digital phones, answering machines, and fax machines on an IP-based phone network. Sometimes the IAD is combined in the same enclosure with a DSL or cable modem and then, to ensure confusion, labeled a residential gateway (RG).

In addition to these specialized hardware components, there are a number of software functions that can be combined into servers. In some cases, all such functions reside in one server component (or a tightly coupled group of server processes), while in other cases they can be servers each running on its own hardware platform. The principal components are:

Signaling conversion: Signaling conversion servers transform and translate call setup requests. They may translate names and addresses, or translate between different signaling protocols. Later on, we will encounter them as gatekeepers in H.323 networks (Section 6.7.3.1), proxy servers in Session Initiation Protocol (SIP) (Section 6.7.3.2) networks, and protocol translators in hybrid networks [Liu and Mouchtaris, 2000; Singh and Schulzrinne, 2000].

Application server: An application server implements service logic for various common or custom features, typically through an API such as JAIN, SIP servlets, CPL, or proprietary versions, as discussed in Section 6.9. Often, they provide components of the operational support system (OSS), such as accounting, billing, or provisioning. Examples include voice mail servers, conference servers, and calling card services.

Media server: A media server manipulates media streams, e.g., by recording, playback, codec translation, or text-to-speech conversion. It may be treated like an end system, i.e., it terminates both media and signaling sessions.

6.6 Media Encoding

6.6.1 Audio

In both legacy and packet telephony, the most common way of representing voice signals is as a logarithmically companded[3] byte stream, with a rate of 8000 samples of 8 bits each per second. This telephone-quality audio codec is known as G.711 [International Telecommunication Union, 1998b], with two regional variations known as μ-law or A-law audio, which can reproduce the typical telephone frequency range of about 300 to 3400 Hz. Typically, 20 to 50 msec worth of audio samples are transmitted in one audio packet. G.711 is the only sample-based codec in wide use.

[3] Smaller audio loudness values receive relatively more bits of resolution than larger ones.

As noted earlier, one of the benefits of IP telephony is the ability to compress telephone-quality voice below the customary rate of 64 kb/sec found in TDM networks. All of commonly used codecs operate at a sampling rate of 8000 Hz and encode audio into frames of between 10 and 30 msec duration. Each audio frame consists of speech parameters, rather than audio samples. Only a few audio codecs are commonly used in IP telephony, in particular G.723.1 [International Telecommunication Union, 1996c] operating at 5.3 or 6.3 kb/sec and modest speech quality, G.729 [International Telecommunication Union, 1996a] at 8 kb/sec, and the GSM full-rate (FR) codec at 13 kb/sec.

More recently, two new royalty-free low-bitrate codecs have been published: iLBC [Andersen et al., 2003] operating at 13.33 or 15.2 kb/sec, with a speech quality equivalent to G.729 but higher loss tolerance, and Speex [Herlein et al., 2003], operating at a variable bit rate ranging between 2.15 and 24.6 kb/sec.

All codecs can operate in conjunction with silence suppression, also known as *voice activity detection* (VAD). VAD measures speech volume to detect when a speaker is pausing between sentences or letting the other party talk. Most modern codecs incorporate silence detection, although it is a separate speech processing function in codecs like G.711. Silence suppression can reduce the bit rate by 50 to 60%, depending on whether short silences between words and sentences are removed or not [Jiang and Schulzrinne, 2000a]. The savings can be much larger in multiparty conferences; there, silence suppression is required also to avoid the summed background noise of the listeners interfering with audio perception.

During pauses, no packets are transmitted, but well-designed receivers will play *comfort noise* [Gierlich and Kettler, 2001] that avoids the impression to the listener that the line is dead. The sender occasionally updates [Zopf, 2002] the loudness and spectral characteristics, so that there is no unnatural transition when the speaker breaks his or her silence.

Silence suppression not only reduces the average bit rate but also simplifies playout delay adaptation, which is used by the receiver to compensate for the variable queueing delays incurred in the network.

DTMF ("touchtone") and other voiceband data signals such as fax tones pose special challenges to high-compression codecs and may not be rendered sufficiently well to be recognizable by the receiver. Also, it is rather wasteful to have an IP phone generate a waveform for DTMF signals just to have the gateway spend DSP cycles recognizing it as a digit. Thus, many modern IP phones generate tones as a special encoding [Schulzrinne and Petrack, 2000].

While the bit rate and speech quality are generally the most important figures of merit for speech codecs, codec complexity, resilience to packet loss, and algorithmic delay are other important considerations. The algorithmic delay is the delay imposed by the compression operation, as the compression operation needs to have access to a certain amount of audio data (block size) and may need to look ahead to estimate parameters.

Music codecs such as MPEG 2 Layer 3, commonly known as MP3, or MPEG-2 AAC can also compress voice, but because they are optimized for general audio signals rather than speech, they typically produce much lower audio quality for the same bit rate. The typical MP3 encoding rates, for example, range from 32 kb/sec for "better than AM radio" quality to 96 and 128 kb/sec for "near CD quality." (Conversely, many low-bit-rate speech codecs sound poor with music because their acoustic model is tuned towards producing speech sounds, not music.)

Generally, the algorithmic delay of these codecs is too long for interactive conversations, for example, about 260 msec for AAC at 32 kb/sec. However, the new AAC MPEG-4 low-delay codec reduces algorithmic delays to 20 msec.

In the future, it is likely that "better-than-phone-quality" codecs will become more prevalent, as more calls are placed between IP telephones rather than from or into the PSTN. So-called conference-quality or wideband codecs typically have an analog frequency range of 7 kHz and a sampling rate of 16 kHz, with a quality somewhat better than static-free AM radio. Examples of such codecs include G.722.1 [International Telecommunication Union, 1999a; Luthi, 2001] at 24 or 32 kb/sec, Speex [Herlein et al., 2003] at 4 to 44.2 kb/sec, AMR WB [Sjoberg et al., 2002; International Telecommunication Union, 2002 3GPP], a,b at 6.6 to 23.85 kb/sec.

The quality of audio encoding with packet loss can be improved by using forward error correction (EEC) and packet loss concealment (PLC) [Jiang et al., 2003; Jiang and Schulzrinne, 2002b; Rosenberg and Schulzrinne, 1999; Jiang and Schulzrinne, 2002c,a, 2000b; Schuster et al., 1999; Bolot et al., 1995; Toutireddy and Padhye, 1995; Carle and Biersack, 1997; Stock and Adanez, 1996; Boutremans and Boudec, 2001; Jeffay et al., 1994].

6.6.2 Video

For video streams, the most commonly used codecs are H.261 [International Telecommunication Union, 1993b], which is being replaced by more modern codecs such as H.263 [International Telecommunication Union, 1998c], H.263+ and H.264. Like MPEG-1 and MPEG-2, H.261 and H.263 make use of interframe correlation and motion prediction to reduce the video bit rate. The most recent standardized video codec is H.264, also known as MPEG-4 AVC or MPEG-4 Part 10. Like MPEG-2, H.264/AVC is based on block transforms and motion-compensated predictive coding. H.264 features improved coding techniques, including multiple reference frames and several block sizes for motion compensation, intra-frame prediction, a new 4×4 integer transform, a 1/4 pixel precision motion compensation, an in-the-loop deblocking filter, and improved entropy coding, roughly halving the bitrate compared to earlier standards for the same fidelity.

Sometimes, motion JPEG is used for high-quality video, which consists simply of sending a sequence of JPEG images. Compared to motion-compensated codecs, its quality is lower, but it also requires much less encoding effort and is more tolerant of packet loss.

6.7 Core Protocols

Internet telephony relies on five types of application-specific protocols to offer services: media transport (Section 6.7.1), device control (Section 6.7.2), call setup and signaling (Section 6.7.3), address mapping (Section 6.7.4), and call routing (Section 6.7.5). These protocols are not found in all Internet telephony implementations.

6.7.1 Media Transport

As described in Section 6.6.1, audio is transmitted in frames representing between 10 and 50 msec of speech content. Video, similarly, is divided into frames, at a rate of between 5 and 30 frames a second. However, these frames cannot simply be placed into UDP or TCP packets, as the receiver would not be able to tell what kind of encoding is being used, what time period the frame represents, and whether a packet is the beginning of a talkspurt.

The Real-Time Transport Protocol (RTP [Schulzrinne et al., 1996]) offers this common functionality. It adds a 12-byte header between the UDP packet header and the media content.[4] The packet header labels the media encoding so that a single stream can alternate between different codecs [Schulzrinne, 1996], e.g., for DTMF [Schulzrinne and Petrack, 2000] or different network conditions. It has a timestamp increasing at the sampling rate that makes it easy for the receiver to correctly place packets in a playout buffer, even if some packets are lost or packets are skipped due to silence suppression. A sequence number provides an indication of packet loss. A secure profile of RTP [Baugher et al., 2003] can provide confidentiality, message authentication, and replay protection. Finally, a synchronization source identifier (SSRC) provides a unique 32-bit identifier for multiple streams that share the same network identity.

Just as IP has a companion control protocol, ICMP [Postel, 1981], RTP uses RTCP for control and diagnostics. RTCP is usually sent on an adjacent UDP port number to the main RTP stream and is paced to consume no more than a set fraction of the main media stream, typically 5%. RTCP has three main functions: (1) it identifies the source by a globally unique user@host-style identifier and adds labels such as the speaker's name; (2) it reports on sender characteristics such as the number of bytes

[4] TCP is rarely used because its retransmission-based loss recovery mechanism may not recover packets in the 100 msec or so required and congestion control may introduce long pauses into the media stream.

and packets transmitted in an interval; (3) receivers report on the quality of the stream received, indicating packet loss and jitter. More extensive audio-specific metrics have been proposed recently [Friedman et al., 2003).

Although RTP streams are usually exchanged unmodified between end systems, it is occasionally useful to introduce processing elements into these streams. RTP *mixers* take several RTP streams and combine them, e.g., by summing their audio content in a conference bridge. RTP *translators* take individual packets and manipulate the content, e.g., by converting one codec to another. For mixers, the RTP packet header is augmented by a list of contributing sources that identify the speakers that were mixed into the packet.

6.7.2 Device Control

Some large-scale gateways are divided into two parts, a media-processing part that translates between circuit-switched and packet-switched audio and a media gateway controller (MGC) or call agent (CA) that directs its actions. The MGC is typically a general-purpose computer and terminates and originates signaling, such as the Session Initiation Protocol (SIP) (see Section 6.7.3.2), but does not process media.

In an enterprise PBX or cable modem context (called network-based call signaling there [CableLabs, 2003]), some have proposed that a central control agent provides low-level instructions to user end systems, such as IADs and IP phones, and receives back events such as numbers dialed or on/off hook status. There are currently two major protocols that allow such device control, namely the older MGCP [Arango et al., 1999] and the successor Megaco/H.248 [Groves et al., 2003]. Currently, MGCP is probably the more widely used protocol. MGCP is text-based, while Megaco/H.248 has a text and binary format, with the latter apparently rarely implemented due to its awkward design.

Figure 6.3 gives a flavor of the MGCP protocol operation, drawn from CableLabs [2003]. First, the CA sends a NotificationRequest (RQNT) to the client, i.e., the user's phone. The N parameter identifies the call agent, the X parameter identifies the request, and the R parameter enumerates the events, where hd stands for off hook. The 200 response by the client indicates that the request was received. When the user picks up the phone, a Notify (NTFY) message is sent to the CA, including the O parameter that describes the event that was observed. The CA then instructs the devices with a combined *CreateConnection* (*CRCX*) and NotificationRequest command to create a connection, labeled with a call ID C, provides dial tone (dl in the S parameter) and collects digits according to digit map D. The digitmap spells out the combinations of digits and time-outs (T) that indicate that the complete number has been dialed. The client responds with a 200 message indicating receipt of the CRCX request and includes a session description so that the CA knows where it should direct dialtone to. The session description uses the Session Description Protocol (SDP) [Handley and Jacobson, 1998]; we have omitted some of the details for the sake of brevity. The C line indicates the network address, the m line the media type, port, the RTP profile (here, the standard audio/video profile), and the RTP payload identifier (0, which stands for G.711 audio). To allow later modifications, the connection gets its own label (I). The remainder of the call setup proceeds apace, with a notification when the digits have been collected. The CA then tells the calling client to stop collecting digits. It also creates a connection on the callee side and instructs that client to ring. Additional messages are exchanged when the callee picks up and when either side hangs up. For this typical scenario, the caller generates and receives a total of 20 messages, while the callee side sees an additional 15 messages.

As in the example illustrated, MGCP and Megaco/H.248 instruct the device in detailed operations and behavior and the device simply follows these instructions. The device exports low-level events such as hook switch actions and digits pressed, rather than, say, calls. This makes it easy to deploy new services without upgrades on the client side, but also keeps all service intelligence in the network, i.e., the CA. Since there is a central CA, device control systems are limited to single administrative domains. Between domains, CAs use a peer-to-peer signaling protocol, such as SIP or H.323, described in Section 6.7.3.1, to set up the call.

```
RQNT 1201 aaln/1@ec-1.whatever.net MGCP 1.0 NCS 1.0
N: ca@ca1.whatever.net:5678
X: 0123456789AB
R: hd
------------------------------
200 1201 OK
------------------------------
NTFY 2001 aaln/1@ec-1.whatever.net MGCP 1.0 NCS 1.0
N: ca@ca1.whatever.net:5678
X: 0123456789AB
O: hd

------------------------------
CRCX 1202 aaln/1@ec-1.whatever.net MGCP 1.0 NCS 1.0
C: A3C47F21456789F0
L: p:10, a:PCMU
M: recvonly
N: ca@ca1.whatever.net:5678
X: 0123456789AC
R: hu, [0-9#*T](D)
D: (0T | 00T | [2-9]xxxxxx | 1[2-9]xxxxxxxxx | 011xx.T)
S: dl

--------------------------------
200 1202 OK
I: FDE234C8

c=IN P4 128.96.41.1
m=audio 3456 RTP/AVP 0
```

FIGURE 6.3 Sample call flow [CableLabs, 2003]

6.7.3 Call Setup and Control: Signaling

One of the core functions of Internet telephony that distinguishes it from, say, streaming media is the notion of call setup. Call setup allows a caller to notify the callee of a pending call, to negotiate call parameters such as media types and codecs that both sides can understand, to modify these parameter in mid-call, and to terminate the call.

In addition, an important function of call signaling is "rendezvous," the ability to locate an end system by something other than just an IP address. Particularly with dynamically assigned network addresses, it would be rather inconvenient if callers had to know and provide the IP address or host name of the destination. Thus, the two most prevalent call signaling protocols both offer a binding (or registration) mechanism where clients register their current network address with a server for a domain. The caller then contacts the server and obtains the current whereabouts of the client.

The protocols providing these functions are referred to as signaling protocols; sometimes, they are also further described as peer-to-peer signaling protocols since both sides in the signaling transactions have equivalent functionality. This distinguishes them from device control protocols such as MGCP and Megaco/H.248, where the client reacts to commands and supplies event notifications.

Two signaling protocols are in common commercial use at this time, namely H.323 (Section 6.7.3.1) and SIP (Section 6.7.3.2). Their philosophies differ, although the evolution of H.323 has brought it closer to SIP.

6.7.3.1 H.323

The first widely used standardized signaling protocol was provided by the ITU in 1996 as the H.323 family of protocols. H.323 has its origins in extending ISDN multimedia conferencing, in Recommendation H.320 [International Telecommunication Union, 1999b], to LANs and inherits aspects of ISDN circuit-switched signaling. Also, H.323 has evolved considerably, through four versions, since its original design. This makes it somewhat difficult to describe its operation definitively in a modest amount of

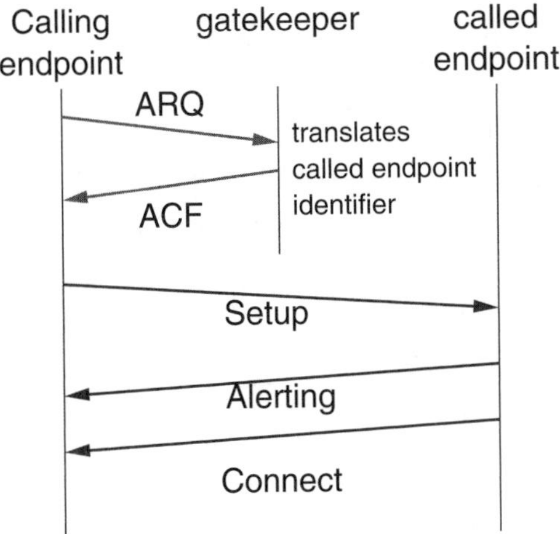

FIGURE 6.4 Example H.323 call flow, fast-connect.

space. In addition, many common implementations, such as Microsoft NetMeeting, only support earlier versions, typically version 2, of the protocol. Most of the trunking gateway deployments are using H.323 versions 2, 3, and 4, while version 2 still predominates in the LAN market. Version 5 was published in July 2003. (Later versions are supposed to support all earlier versions and fall back to the less-functional version if necessary.)

H.323 is an umbrella term for a whole suite of protocol specifications. The basic architecture is described in H.323 [International Telecommunication Union, 2003], registration and call setup signaling ("ringing the phone") is described in H.225.0 [International Telecommunication Union, 1996d], and media negotiation and session setup in H.245 [International Telecommunication Union, 1998a]. The ISDN signaling messages that are carried in H.225.0 are described in Q.931 [International Telecommunication Union, 1993a]. The two sub-protocols for call and media setup, Q.931 and H.245, use different encodings. Q.931 is a simple binary protocol with mostly fixed-length fields, while H.245, H.225.0 call setup, and H.450 service invocations are encoded as ASN.1 and are carried as user-to-user (UU) information elements in Q.931 messages.

H.225.0, H.245, H.450, and other parts of H.323 use the packet ASN.1 encoding rules (PER). [International Telecommunication Union, 1997a]. Generally, H.323 applications developers rely on libraries or ASN.1 code generators.

The protocols listed so far are sufficient for basic call functionality and are those most commonly implemented in endpoints. Classical telephony services such as call forwarding, call completion, or caller identification are described in the H.450.x series of recommendations. Security mechanisms are discussed in H.235. Functionality for application sharing and shared whiteboards, with its own call setup mechanism, is described in the T.120 series of recommendations [International Telecommunication Union, 1996b].

H.323 uses similar component labels as we have seen earlier, namely terminals (that is, end systems) and gateways. It also introduces gatekeepers, which route signaling messages between domains and registered users, provide authorization and authentication of terminals and gateways, manage bandwidth, and provide accounting, billing, and charging functions. Finally, from its origin in multimedia conferencing, H.323 describes multipoint control units (MCUs), the packet equivalent to a conference bridge.

Each gatekeeper is responsible for one *zone*, which can consist of any number of terminals, gateways, and MCUs.

Figure 6.4 shows a typical fast-connect call setup between two terminals within the same zone. The gatekeeper translates the H.323 identifier, such as a user name, to the current terminal network address, which is then contacted directly. (Inter-gatekeeper communications is specified in H.323v3). Figure 6.5 shows the original non-fast-connect call setup, where the H.245 messages are exchanged separately, rather than being bundled into the H.225.0 messages.

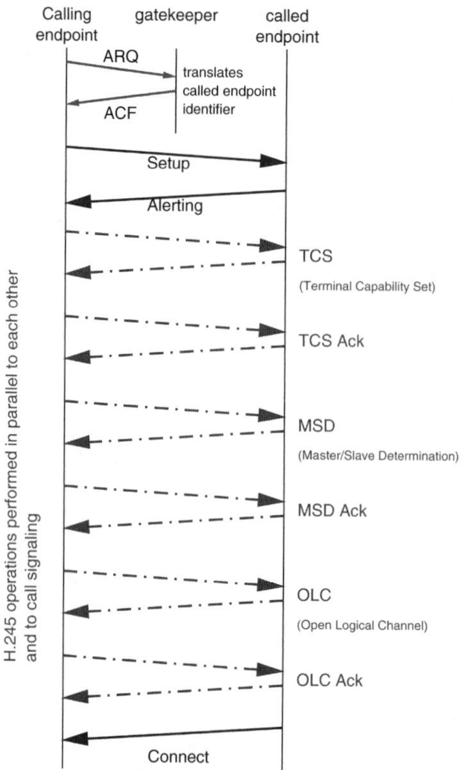

FIGURE 6.5 Example H23 call-flow without fast-connect.

6.7.3.2 Session Initiation Protocol (SIP)

The Session Initiation Protocol (SIP) is a protocol framework originally designed for establishing, modifying, and terminating multimedia sessions such as VoIP calls. Beyond the session setup functionality, it also provides event notification for telephony services such as supervised call transfer and message waiting indication and more modern services such as presence.

SIP does not describe the audio and media components of a session; instead, it relies on a separate session description carried in the body of INVITE and ACK messages. Currently, only the Session Description Protocol (SDP) [Handley and Jacobson, 1998] is being used, but an XML-based replacement [Kutscher et al., 2003] is being discussed. The example in Figure 6.6 [Johnston, 2003] shows a simple audio session originated by user *alice* to be received by IP address 192.0.2.101 and port 49172 using RTP and payload type 0 (μ-law audio).

Besides carrying session descriptions, the core function of SIP is to locate the called party, mapping a user name such as sip:alice@atlanta.example.com to the networkaddresses used by devices owned by Alice. Users can reuse their e-mail address as a SIP URI or choose a different one. As for e-mail addresses, users can have any number of SIP URIs with different providers that all reach the same device.

```
v=0
o=alice 2890844526 2890844526 IN IP4 client.atlanta.example.com
s=-
c=IN IP4 192.0.2.101
t=0 0
m=audio 49172 RTP/AVP 0
a=rtpmap:0 PCMU/8000
```

FIGURE 6.6 Example session description

User devices such as IP phones and conferencing software run SIP *user agents*; unlike for most protocols, such user agents usually can act as both clients and servers, i.e., they both originate and terminate SIP requests.

Instead of SIP URIs, users can be identified also by telephone numbers, expressed as "tel" URIs [Schulzrinne and Vaha-Sipila, 2003] such as tel: +1-212-555-1234. Calls with these numbers are then either routed to an Internet telephony gateway or translated back into SIP URIs via the ENUM mechanism described in Section 6.7.4.

A user provides a fixed contact point, a so-called SIP proxy , that maps incoming requests to network devices registered by the user. The caller does not need to know the current IP addresses of these devices. This decoupling between the globally unique user-level identifier and device network addresses supports *personal mobility*, the ability of a single user to use multiple devices, and deals with the practical issue that many devices acquire their IP address temporarily via DHCP. The proxy typically also performs call routing functions, for example, directing unanswered calls to voice mail or an auto-attendant. The SIP proxy plays a role somewhat similar to an SMTP Mail Transfer Agent (MTA) [rfc, 2001], but naturally does not store messages. Proxies are not required for SIP; user agents can contact each other directly.

A request can traverse any number of proxies, but typically at least two, namely, one *outbound* proxy in the caller's domain and the *inbound* proxy in the caller's domain. For reliability and load balancing, a domain can use any number of proxies. A client identifies a proxy by looking up the DNS SRV [Gulbrandsen et al., 2000] record enumerating primary and fallback proxies for the domain in the SIP URI.

Session setup messages and media generally traverse independent paths, that is, they only join at the originating and terminating client. Media then flows directly on the shortest network path between the two terminals. In particular, SIP proxies do not process media packets. This makes it possible to route call setup requests through any number of proxies without worrying about audio latency or network efficiency. This *path-decoupled* signaling completes the evolution of telephony signaling from in-band audio signaling to out-of-band, disassociated channel signaling introduced by Signaling System No. 7 (SS7). Because telephony signaling needs to configure switch paths, it generally meets up with the media stream in telephone switches; there is no such need in IP telephony.

Just as a single phone line can ring multiple phones within the same household, a single SIP address can contact any number of SIP devices with one call, albeit potentially distributed across the network. This capability is called *forking* and is performed by proxies. These forking proxies gather responses from the entities registered under the SIP URI and return the best response, typically the first one to pick up. This feature makes it easy to develop distributed voicemail services and simple automatic call distribution (ACD) systems.

Figure 6.7 shows a simple SIP message and its components. SIP is a textual protocol, similar to SMTP and HTTP [Fielding et al., 1999]. A SIP request consists of a request line containing the request method and the SIP URI identifying the destination, followed by a number of header fields that help proxies and user agents to route and identify the message content. There are a large number of SIP request methods, summarized in Table 6.1.

SIP messages can be requests or responses that only differ syntactically in their first lines. Almost all SIP requests generate a final response indicating whether the request succeeded or why it failed, with some requests producing a number of responses that update the requestor on the progress of the request via provisional responses.

Unlike other application-layer protocols, SIP is designed to run over both reliable and unreliable transport protocols. Currently, UDP is the most common transport mechanism, but TCP and SCTP, as well as secure transport using TLS [Dierks and Allen, 1999] are also supported. To achieve reliability, a request is retransmitted until it is acknowledged by a provisional or final response. The INVITE transaction, used to set up sessions, behaves a bit differently since considerable time may elapse between the call arrival and the time that the called party picks up the phone. An INVITE transaction is shown in Figure 6.8.

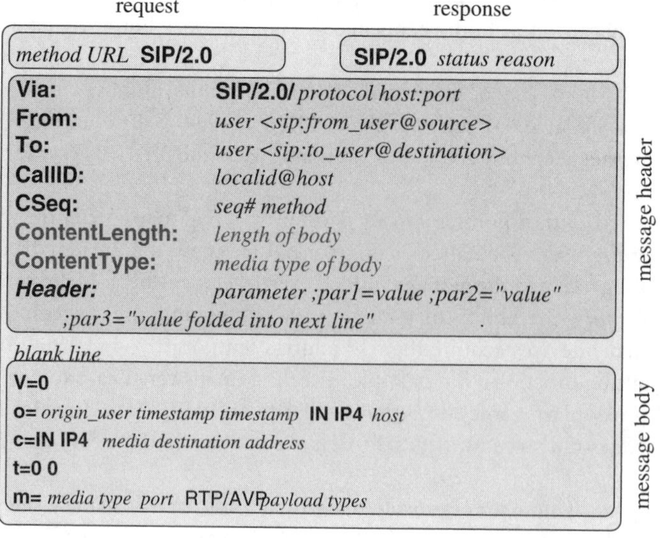

FIGURE 6.7 Example SIP INVITE message.

TABLE 6.1 SIP Request Methods

Transaction	Reference	Functions
ACK	[Rosenberg et al., 2002b]	Acknowledges final INVITE response
BYE	[Rosenberg et al., 2002b]	Terminates session
CANCEL	[Rosenberg et al., 2002b]	Cancels INVITE
INFO	[Donovan, 2000]	Mid-session information transfer
INVITE	[Rosenberg et al., 2002b]	Establishes session
NOTIFY	[Roach, 2002]	Event notification
OPTIONS	[Rosenberg et al., 2002b]	Determine capabilities
PRACK	[Rosenberg and Schulzrinne, 2002]	Acknowledge provisional response
REGISTER	[Rosenberg et al., 2002b]	Register name–address mapping
SUBSCRIBE	[Roach, 2002]	Subscribe to event
UPDATE	[Rosenberg, 2002]	Update session description
MESSAGE	[rfc, 2002]	User-to-user messaging
REFER	[Sparks, 2003]	Transfer call

Once a request has reached the right destination, the two parties negotiate the media streams using an offer–answer model, where the caller typically offers a capability and the callee makes a counterproposal. Sessions can be changed in the middle of one in progress, e.g., to add or remove a media stream.

SIP can be extended by adding new methods, message body types, or header fields. Generally, receivers and proxies are free to ignore header fields that they do not understand, but a requestor can require that the receiver understand a particular feature by including a Require header field. If the receiver does not implement that feature, it must reject the request.

SIP user agents can initiate sessions between two other entitles, acting as third-party call controllers or back-to-back user agents (B2BUAs) [Rosenberg et al., 2003b].

While the basic protocol mechanisms are stable, components of the SIP infrastructure are currently still under active development within the IETF and, for third-generation mobile networks, in 3GPP. The features include support for legacy telephone characteristics such as overlap dialing, as well as advanced call routing features such as caller preferences [Rosenberg et al., 2003a; Rosenberg and Kyzivat, 2003].

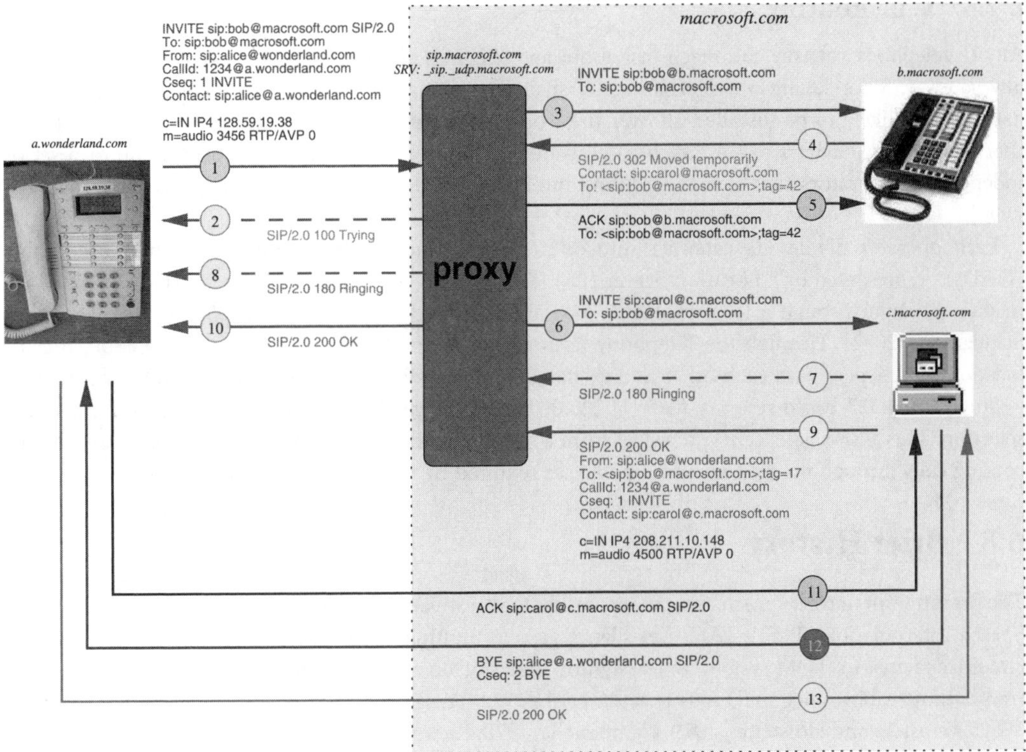

FIGURE 6.8 Example SIP call flow.

6.7.4 Telephone Number Mapping

In the long run, VoIP destinations may well be identified by textual SIP URIs, probably derived automatically from a person's e-mail address. However, familiarity, deployed infrastructure, and end system user interface limitations dictate the need to support telephone numbers [International Telecommunication Union, 1997b] for the foreseeable future. To facilitate the transition to an all-IP infrastructure, it is helpful if telephone numbers can be mapped to SIP and other URIs. This avoids, for example, a VoIP terminal needing to go through a gateway to reach a terminal identified by a telephone number, even though that terminal also has VoIP capability.

The ENUM service [Faltstrom, 2000; Faltstrom and Mealling, 2003] offers a standardized mapping service from global telephone numbers to one or more URIs. It uses the Dynamic Delegation Discovery System (DDDS) system [Mealling, 2002] and a relatively new DNS record type, NAPTR. NAPTR records allow for mapping of the name via a regular expression, as shown in Figure 6.9 for the telephone number +46-89761234. Because the most significant digit for telephone numbers is on the left, while the most significant component of DNS names is on the right, the telephone number is reversed and converted into the DNS name "4.3.2.1.6.7.9.8.6.4.e164.arpa" in this example.

```
$ORIGIN 4.3.2.1.6.7.9.8.6.4.e164.arpa.
   IN NAPTR 10 100 "u" "E2U+sip" "!^.*$!sip:info@example.com!"
   IN NAPTR 10 101 "u" "E2U+h323" "!^.*$!h323:info@example.com!"
   IN NAPTR 10 102 "u" "E2U+msg:mailto" "!^.*$!mailto:info@example.com!"
```

FIGURE 6.9 ENUM example. [From Faltstrom, P. and M. Mealling. The E.164 to URI DDDS application (ENUM). Internet draft, Internet Engineering Task Force, May 2003. URL http://www.ietf.org/internet-drafts/draft-ietf-enum-rfc2916his-06.txt. Work in progress.].

6.7.5 Call Routing

Any IP telephony gateway can reach just about any telephone number, and any VoIP device can reach any gateway. Since saving on international transit is a major motivation for deploying IP telephony, gateways are likely to be installed all over the world, with gateways in each country handling calls for that country or maybe a region. Such gateways may be operated by one large corporation or a set of independent operators that exchange billing information via a clearinghouse [Hoffman and Yergeau, 2000].

Each operator divides its gateways into one or more Internet Telephony administrative domains (ITADs) , represented by a Location Server (LS). The location servers learn about the status of gateways in their domain through a local protocol, such as TGREP [Bangalore et al., 2003] or SLP [Zhao and Schulzrinne, 2002]. Through the Telephony Routing over IP protocol (TRIP) [Rosenberg et al., 2002a], location servers peer with each other and exchange information about other ITADs and their gateways.

Today, in H.323-based systems, RAS (H.225.0) LRQ messages and H.501 are widely used for gateway selection. This allows gatekeepers to select from a number of known destination devices quickly without routing calls through interior signaling nodes, as required by the TRIP approach.

6.8 Brief History

The first attempt to treat speech as segments rather than a stream of samples was probably Time-Assigned Speech Interpolation (TASI). TASI uses silence gaps to multiplex more audio streams than the nominal circuit capacity of a TDM system by reassigning time slots to active speech channels. It has been used in transoceanic cables since the 1960s [Easton et al., 1982; Fraser et al., 1962; Miedema and Schachtman, 1962; Weinstein and Hofstetter, 1979; Campanella, 1978; Rieser et al., 1981]. Although TASI is not packet switching, many of the analysis techniques to estimate the statistical multiplexing gains apply to packet voice as well.

Attempts to transmit voice across IP-based packet networks date back to the earliest days of ARPAnet, with the first publication in 1973, only 2 years after the first e-mail. [Magill, 1973; Cohen, 1976a,b, 1977b, 1978; Anonymous, 1983]. In August 1974, real-time packet voice was demonstrated between USC/IISI and MIT Lincoln Laboratories, using CVSD (Continuous Variable Slope Delta Modulation) and Network Voice Protocol (NVP) [Cohen, 1977a]. In 1976, live packet voice conferencing was demonstrated between USC/ISI, MIT Lincon Laboratories, Chicago, and SRI, using linear predictive audio coding (LPC) and the Network Voice Control Protocol (NVCP). These initial experiments, run on 56 kb/sec links, demonstrated the feasibility of voice transmission, but required dedicated signal processing hardware and thus did not lend themselves to large-scale deployments. Development appears to have been largely dormant since those early experiments.

In 1989, the Sun SPARCstation 1 introduced a small form-factor Unix workstation with a low-latency audio interface. This also happened to be the workstation of choice for DARTnet, an experimental T-1 packet network funded by DARPA (Defense Advanced Research Projects Agency). In the early 1990s, a number of audio tools such as vt, vat [Jacobson, 1994; Jacobson and McCanne, 1992] and nevot [Schulzrinne, 1992], were developed that explored many of the core issues of packet transmission, such as playout delay compensation [Montgomery, 1983; Ramjee et al., 1994; Rosenberg et al., 2000; Moon et al., 1998], packet encapsulation, QOS, and audio interfaces. However, outside of the multicast backbone overlay network (Mbone) [Eriksson, 1993; Chuang et al., 1993] that reached primarily research institutions and was used for transmitting IETF meetings [Casner and Deering, 1992] and NASA space launches, the general public was largely unaware of these tools. More popular was Cu-SeeMe, developed in 1992/1993 [Cogger, 1992].

The ITU standardized the first audio protocol for general packet networks in 1990 [International Telecommunication Union, 1990] , but this was used only for niche applications, as there was no signaling protocol to set up calls.

In about 1996, VocalTec Communications commercialized the first PC-based packet voice applications, primarily used initially to place free long-distance calls between PCs. Since then, standardization of signaling protocols like RTP and H.323 in 1996 [Thom, 1996] have started the transition from experimental research to production services.

6.9 Service Creation

Beyond basic call setup and teardown, the legacy telephone has developed a number of services or *features*, including such common ones as call forwarding on busy or three-way calling and more specialized ones such as distributed call center functionalities. Almost all such services were designed to be developed on PSTN or PBX switches and deployed as a general service, with modest user parameterization.

Both SIP and 11.323 can support most SS7 features [Lennox et al., 1999] through protocol machinery, although the philosophy and functionality differs between protocols [Glasmann et al., 2001]. Unlike legacy telephones, both end systems and network servers can provide services [Wu and Schulzrinne, 2003, 2000], often in combination. End system services scale better and can provide a more customized user interface, but may be less reliable and harder to upgrade.

However, basic services are only a small part of the service universe. One of the promises of IP telephony is the ability for users or programmers working closely with small user groups to create new services or customize existing ones. Similar to how dynamic, data-driven web pages are created, a number of approaches have emerged for creating IP telephony services. Java APIs such as JAIN and SIP servlets are meant for programmers and expose almost all signaling functionality to the service creator. They are, however, ill-suited for casual service creation and require significant programming expertise.

Just like common gateway interface (cgi) services on Web servers, SIP-cgi [Lennox et al., 2001] allows programmers to create user-oriented scripts in languages such as Perl and Python. A higher-level representation of call routing services is exposed through the Call Processing Language (CPL) [Lennox and Schulzrinne, 2000a; Lennox et al., 2003].

With distributed features, the problem of feature interaction [Cameron et al., 1994] arises. IP telephony removes some of the common causes of feature interaction such as ambiguity in user input, but adds others [Lennox and Schulzrinne, 2000b] that are just beginning to be explored.

6.10 Conclusion

IP telephony promises the first major fundamental rearchitecting of conversational voice services since the transition to digital transmission in the 1970s. Like the Web, it does not consist of a single breakthrough technology, but the combination of pieces that are now becoming sufficiently powerful to build large-scale operational systems, not just laboratory experiments.

Recent announcements indicate that major telecommunications carriers will be replacing their class-5 telephone switches by IP technology in the next 5 years or so. Thus, even though the majority of residential and commercial telephones will likely remain analog for decades, the core of the network will transition to a packet infrastructure in the foreseeable future. Initially, just as for the transition to digital transmission technology, these changes will largely be invisible to end users.

For enterprises, there are now sufficiently mature commercial systems available from all major PBX vendors, as well as a number of startups, that offer equivalent functionality to existing systems. Specialty deployments, such as in large call centers, hotels, or banking environments, remain somewhat more difficult, as end systems (at appropriate price points) and operations and management systems are still lacking. While standards are available and reaching maturity, many vendors are still transitioning from their own proprietary signaling and transmission protocols to IETF or ITU standards. Configuration and management of very large, multivendor deployments pose severe challenges at this point, so that most installations still tend to be from a single vendor, despite the promise of open and interoperable architectures offered by IP telephony.

In some cases, hybrid deployments make the most technical and economic sense in an enterprise, where older buildings and traditional users continue to be connected to analog or digital PBXs, while new buildings or telecommuting workers transition to IP telephony and benefit from reduced infrastructure costs and the ability to easily extend the local dialing plan to offsite premises.

Widespread residential use hinges on the availability of broadband connections to the home. In addition, the large deployed infrastructure of inexpensive wired and cordless phones, and answering and fax machines currently have no plausible replacement, except by limited-functionality integrated access devices (IADs). Network address translators (NATs) and limited upstream bandwidth further complicate widespread rollouts, so that it appears likely that Internet telephony in the home will be popular mostly with early adopters, typically heavy users of long-distance and international calls that are comfortable with new technology.

Deployment of IP telephony systems in enterprises is only feasible if the local area network is sufficiently robust and reliable to offer acceptable voice quality. In some circumstances, Ethernet-powered end systems are needed if phone service needs to continue to work even during power outages; in most environments, a limited number of analog emergency phones will be sufficient to address these needs.

Internet telephony challenges the whole regulatory approach that has imposed numerous rules and regulations on voice service but left data services and the Internet largely unregulated. Emergency calling, cross-subsidization of local calls by long-distance calls and interconnect arrangements all remain to be addressed. For example, in the U.S., billions of dollars in universal service fund (USF) fees are at stake, as the traditional notion of a telephony company becomes outdated and may become as quaint as an e-mail company would be today. In the long run, this may lead to a split between network connectivity providers and service providers, with some users relying on third parties for e-mail, Web, and phone services, while others operate their own in-house services.

The transition from circuit-switched to packet-switched telephony will take place slowly in the wireline portion of the infrastructure, but once third-generation mobile networks take off, the majority of voice calls could quickly become packet-based.

This transition offers an opportunity to address many of the limitations of traditional telephone systems, empowering end users to customize their own services just like Web services have enabled myriads of new services far beyond those imagined by the early Web technologists. Thus, instead of waiting for a single Internet telephony "killer application," the Web model of many small but vital applications appears more productive. This evolution can only take shape if technology goes beyond re-creating circuit-switched transmission over packets.

6.11 Glossary

The following glossary lists common abbreviations found in IP telephony. It is partially extracted from International Packet Communications Consortium.

3G — Third Generation (wireless)
3GPP — 3G Partnership Project (UMTS)
3GPP2 — 3G Partnership Project 2 (UMTS)
AAA — Authentication, Authorization, and Accounting (IETF)
AG — Access Gateway
AIN — Advanced Intelligent Network
AS — Application Server
BICC — Bearer Independent Call Control (ITU Q.1901)
CPL — Call Processing Language
CSCF — Call State Control Function (3GPP)
DTMF — Dual Tone/Multiple Frequency
ENUM — E.164 Numbering (IETF RFC 2916)
GK — Gatekeeper
GPRS — General Packet Radio Service
GSM — Global System for Mobility
IAD — Integrated Access Device

IETF — Internet Engineering Task Force
IN — Intelligent Network
INAP — Intelligent Network Application Protocol
ISDN — Integrated Services Digital Network
ISUP — Integrated Services Digital Network User Part (SS7)
ITU — International Telecommunications Union
IUA — ISDN User Adaptation
IVR — Interactive Voice Response
JAIN — Java Application Interface Network
LDAP — Lightweight Directory Access Protocol (IETF)
M3UA — MTP3 User Adaptation (IETF SIGTRAN)
MEGACO — Media Gateway Control (IETF RFC 3015 or ITU H.248)
MG — Media Gateway
MGC — Media Gateway Controller
MGC-F — Media Gateway Controller Function (IPCC)
MGCP — Media Gateway Control Protocol (IETF, ITU-T J.162)
MPLS — Multi-Protocol Label Switching
MS — Media Server
MSC — Mobile Services Switching Center (GSM, 3GPP)
MSO — Multi-System Operator
MTA — Multimedia Terminal Adaptor (PacketCable)
NCS — Network Call/Control Signaling (PacketCable MGCP)
NGN — Next Generation Network
OSS — Operational Support System
PBX — Private Branch Exchange
POTS — Plain Old Telephone Service
PSE — Personal Service Environment (3GPP)
PSTN — Public Switched Telephone Network
QoS — Quality of Service
RAN — Radio Access Network
RFC — Request For Comment (IETF)
RG — Residential Gateway
RSVP — Resource Reservation Protocol (IETF)
RTCP — Real Time Transport Control Protocol (IETF)
RTP — Real Time Transport Protocol (IETF RFC 1889)
SCP — Service Control Point
SCTP — Stream Control Transmission Protocol
SDP — Session Description Protocol (IETF RFC 2327)
SG — Signaling Gateway
SIGTRAN — Signaling Transport (IETF)
SIP — Session Initiation Protocol (IETF)
SIP-T — SIP For Telephony (IETF)
SS7 — Signaling System 7 (ITU)
TDM — Time Division Multiplexing
TRIP — Telephony Routing over IP (IETF RFC 2871)
UMTS — Universal Mobile Telecommunications System
VAD — Voice Activity Detection
VLR — Visitor Location Register (GSM, 3GPP)
VoDSL — Voice over DSL
VoIP — Voice over IP
VoP — Voice over Packet

References

Internet message format. RFC 2822, Internet Engineering Task Force, April 2001. URL http://www.rfc-editor.org/rfc/rfc2822.txt.

Session initiation protocol (SIP) extension for instant messaging. RFC 3428, Internet Engineering Task Force, December 2002. URL http://www.rfc-editor.org/rfc/rfc3428.txt.

3GPP. AMR speech codec, wideband; Frame structure. TS 26.201, 3rd Generation Partnership Project (3GPP), a. URL http://www.3gpp.org/ftp/Specs/archive/26_series/26.201/.

3GPP. Mandatory Speech Codec speech processing functions AMR Wideband speech codec; Transcoding functions. TS 26.190, 3rd Generation Partnership Project (3GPP), b. URL http://www.3gpp.org/ftp/Specs/archive/26_series/26.190/.

Andersen, S. C. et al. Internet low bit rate codec. Internet draft, Internet Engineering Task Force, July 2003. URL http://www.ietf.org/internet-drafts/draft-ietf-avt-ilbc-codec-02.txt. Work in progress.

Anonymous. Special issue on packet switched voice and data communication. *IEEE Journal on Selected Areas in Communications,* SAC-1(6), December 1983.

Arango, M., A. Dugan, I. Elliott, C. Huitema, and S. Pickett. Media gateway control protocol (MGCP) version 1.0. RFC 2705, Internet Engineering Task Force, October 1999. URL http://www.rfc-editor.org/rfc/rfc2705.txt.

Bangalore, M. et al. A telephony gateway REgistration protocol (TGREP). Internet draft, Internet Engineering Task Force, July 2003. URL http://www.ietf.org/internet-drafts/draft-ietf-iptel.-tgrep-02.txt. Work in progress.

Baugher, Mark et al. The secure real-time transport protocol. Internet draft, Internet Engineering Task Force, July 2003. URL http://www.ietf.org/internet-drafts/draft-ietf-avt-srtp-09.txt. Work in progress.

Bolot, J. C., H. Crepin, and Anilton Garcia. Analysis of audio packet loss in the Internet. In *Proceedings of the International Workshop on Network and Operating System Support for Digital Audio and Video (NOSSDAV),* Lecture Notes in Computer Science, pages 163–174, Durham, New Hampshire, April 1995. Springer. URL http://www.nossdav.org/1995/papers/bolot.ps.

Boutremans, Catherine and Jean-Yves Le Boudec. Adaptive delay aware error control for Internet telephony. In *Internet Telephony Workshop,* New York, April 2001, URL http://www.cs.columbia.edu/hgs/papers/iptel2001/34.ps.

CableLabs. Packetcable network-based call signaling protocol specification. Specification PKT SP-EC MGCP-107–0, Cable Television Laboratories, April. 2003. URL http://www.packetcable.com/downloads/specs/PKT-SP-MGCP-I07-030415.pdf.

Cameron, E. J., N. Griffeth, Y. Lin, Margaret E. Nilson, William K. Schure, and Hugo Velthuijsen. A feature interaction benchmark for IN and beyond. In *Feature Interactions in Telecommunications Systems,* pages 1–23, Elsevier, Amsterdam, Netherlands, 1994.

Campanella, S. J., Digital speech interpolation techniques. In *Conference record of the IEEE National Telecommunications Conference,* volume 1, pages 14.1.1–14.1.5, Birmingham, Alabama, December 1978. IEEE.

Carle, G. and Ernst Biersack. Survey of error recovery techniques for IP-Based audio-visual multicast applications *IEEE Network,* 11(6):24–36, November 1997. URL http://207.127.135.8/ni/private/1997/nov/Carle.html, http://www.eurecom.fr/btroup.

Casner, Stephen and S. E. Deering. First IETF Internet audiocast. *ACM Computer Communication Review,* 22(3): 92–97, July 1992. URL http://www.acm.org/sigcomm/ccr/archive/1992/jul92/casner.ps.

Chuang, S., Jon Crowcroft, S. Hailes, Mark Handley, N. Ismail, D. Lewis, and Ian Wakeman. Multimedia application requirements for multicast communications services. In *International Networking Conference (INET),* pages BFB-1–BFB-9, San Francisco, CA, August 1993. Internet Society.

Clark, David D. A taxonomy of Internet telephony applications. In *25th Telecommunications Policy Research Conference,* Washington, D.C. September 1997.URL http://itc.mit.edu/itel/pubs/ddc.tprc97.pdf.

Cogger, R. CU-SeeMe Cornell desktop video, December 1992.

Cohen, Danny. Specifications for the network voice protocol (NVP). RFC 741, Internet Engineering Task Force, November 1977a. URL http://www.rfc-editor.org/rfc/rfc741.txt.

Cohen, Danny. The network voice conference protocol (NVCP). NSC Note 113, February 1976a.

Cohen, Danny. Specifications for the network voice protocol. Technical Report ISI/RR-75-39 (AD A02, USC/Information Sciences Institute, Marina del Rey, CA, March 1976b. Available from DTIC).

Cohen, Danny. Issues in transnet packetized voice communications. In *5th Data Communications Symposium,* pages 6–10–6–I3, Snowbird, UT, September 1977b. ACM, IEEE.

Cohen, Danny. A protocol for packet-switching voice communication. *Computer Networks,* 2(4/5): 320–331, September/October 1978.

Dierks, T. and C. Allen. The TLS protocol version 1.0. RFC 2246, Internet Engineering Task Force, January 1999. URL http://www.rfc-editor.org/rfc/rfc2246.txt.

S. Donovan. The SIP INFO method. RFC 2976, Internet Engineering Task Force, October 2000. URL http://www.rfc-editor.org/rfc/rfc2976.txt.

Easton, Robert E., P. T. Hutchison, Richard W. Kolor, Richard C, Mondello, and Richard W. Muise. TASI-E communications system. *IEEE Transactions on Communications,* COM-30(4): 803–807, April 1982.

Eriksson, Hans. MBone — the multicast backbone. In *International Networking Conference (INET),* pages CCC-1-CCC-5, San Francisco, CA, August 1993. Internet Society.

Faltstrom, P., E.164 number and DNS. RFC 2916, Internet Engineering Task Force, September 2000. URL http://www.rfc-editor.org/rfc/rfc2916.txt.

Faltstrom, P. and M. Mealling. The E.164 to URI DDDS application (ENUM). Internet draft, Internet Engineering Task Force, May 2003. URL http://www.ietf.org/internet-drafts/draft-ietf-enum-rfc2916his-06.txt. Work in progress.

Fielding, R., J. Gettys, J. C. Mogul, H. Frystyk, L. Masinter, P. J. Leach, and T. Berners-Lee. Hypertext transfer protocol — HTTP/1.1. RFC 2616, Internet Engineering Task Force, June 1999. URL http:/ /www.rfc-editor.org/rfc/rfc2616.txt.

Fraser, Keir et al. Over-all characteristics of a TASI-system. *Bell System Technical Journal,* 41: 1439–1473, 1962.

Friedman, Timur et al. RTP control protocol extended reports (RTCP XR). Internet draft, Internet Engineering Task Force, May 2003. URL draft-ietf-avt-rtcp-report-extns-06.txt,.pdf. Work in progress.

Gierlich, Hans and Frank Kettler. Conversational speech quality — the dominating parameters in VoIP systems. In *Internet Telephony Workshop,* New York, April 2001. URL http://www.cs.columbia.edu/hgs/papers/iptel2001/9.ps.

Glasmann, Josef, Wolfgang Kellerer, and Harald Mller. Service development and deployment in H.323 and SIP. In *IEEE Symposium on Computers and Communications,* pages 378–385, Hammamet, Tunisia, July 2001. IEEE.

Groves, C., M. Pantaleo, Thomas Anderson, and Tracy M. Taylor. Editors. Gateway control protocol version 1. RFC 3525, Internet Engineering Task Force, June 2003. URL http://www.rfc-editor.org/rfc/rfc3525.txt.

Gulbrandsen, A., P. Vixie, and L. Esibov. A DNS RR for specifying the location of services (DNS SRV). RFC 2782, Internet Engineering Task Force, February 2000, URL http://www.rfc-editor.org/rtc/rfc2782.txt.

Handley, M. and V. Jacobson. SDP: session description protocol. RFC 2327, Internet Engineering Task Force, April 1998. URL http://www.rfc-editor.org/rfc/rfc2327.txt.

Herlein, G. et al. RTP payload format for the speex codec. Internet draft, Internet Engineering Task Force, July 2003. URL http://www.ietf.org/internet-drafts/draft-herlein-speex-rtp-profile-01.txt. Work in progress.

Hoffman, P. and F. Yergeau. UTF-16, an encoding of ISO 10646. RFC 2781, Internet Engineering Task Force, February 2000. URL http://www.rfc-editor.org/rfc/rfc2781.txt.

International Packet Communications Consortium. http://www.softswitch.org/.

International Telecommunication Union. Voice packetization — packetized voice protocols. Recommendation G.764, Telecommunication Standardization Sector of ITU, Geneva, Switzerland, 1990. URL http://wwvl.itu.int/itudoc/itu-t/rec/g/g700--799/g/64.html.

International Telecommunication Union. Digital subscriber signalling system no. 1 (DSS 1) — ISDN user-network interface layer 3 specification for basic call control. Recommendation Q.931, ITU, Geneva, Switzerland, March 1993a. URL http://www.itu.int/itudocs/itu-t/rec/q/q500-999/q931.24961.html.

International Telecommunication Union. Video codec for audiovisual services at px64 kbit/s. Recommendation H.261, Telecommunication Standardization Sector of ITU, Geneva, Switzerland, March 1993b.

International Telecommunication Union. Coding of speech at 8 kbit/s using conjugate-structure algebraic-code-excited linear-prediction. Recommendation G.729, Telecommunication Standardization Sector of ITU, Geneva, Switzerland, March 1996a. URL http://www.itu.int/itudoc/itu-t/rec/g/g700-799/g7293_2350.html.

International Telecommunication Union. Data protocols for multimedia conferencing. Recommendation T.120, Telecommunication Standardization Sector of ITU, Geneva, Switzerland, July 1996b. URL http://www.itu.int.

International Telecommunication Union. Dual rate speech coder for multimedia communications transmitting at 5.3 and 6.3 kbit/s. Recommendation G,723.1, Telecommunication Standardization Sector of ITU, Geneva, Switzerland, March 1996c. URL http://www.itu.int/itudoc/itu-t/rec/g/g700-799/g723-1.html.

International Telecommunication Union. Media stream packetization and synchronization on non-guaranteed quality of service LANs. Recommendation H.225.0, Telecommunication Standardization Sector of ITU, Geneva, Switzerland, November 1996d. URL http://www.itu.int.

International Telecommunication Union. ASN.1 encoding rules — specification of packed encoding rules (PER). Recommendation X.691, Telecommunication Standardization Sector of ITU Geneva, Switzerland, December 1997a. URL http://www.itu.int.

International Telecommunication Union. The international public telecommunication numbering plan. Recommendation E.164, Telecommunication Standardization Sector of ITU, Geneva, Switzerland, May 1997b. URL http://www.itu.int.

International Telecommunication Union. Control protocol for multimedia communication. Recommendation H.245, Telecommunication Standardization Sector of ITU, Geneva, Switzerland, February 1998a. URL http://www.itu.int.

International Telecommunication Union. Pulse code modulation (PCM) of voice frequencies. Recommendation G.711, Telecommunication Standardization Sector of ITU, Geneva, Switzerland, November 1998b.

International Telecommunication Union. Video coding for low bit rate communication. Recommendation H.263, Telecommunication Standardization Sector of ITU, Geneva, Switzerland, February 1998c.

International Telecommunication Union. Coding at 24 and 32 kbit/s for hands-free operation in systems with low frame loss. Recommendation & 722,1, International Telecommunication Union, September 1999a. URL http://www.itu.int/rec/recommendation.asp?type=folders&lang=e&parent=T-REC-G.722

International Telecommunication Union. Narrow-band visual telephone systems and terminal equipment. Recommendation H.320, Telecommunication Standardization Sector of ITU, Geneva, Switzerland, May 1999b. URL http://www.itu.int/itu-t/rec/h/h320.html.

International Telecommunication Union. Wideband coding of speech at around 16 kbit/s using adaptive multi-rate wideband (AMR-WB). Recommendation, International Telecommunication Union, January 2002. URL http://www.itu.int/rec/recommendation.asp?type=folders&long=e&parent=T-REC-G.722

International Telecommunication Union. Packet based multimedia communication systems. Recommendation H.323, Telecommunication Standardization Sector of ITU, Geneva, Switzerland, July 2003. URL http://www.itu.int/.

Jacobson, V. Multimedia conferencing on the Internet. In *SIGCOMM Symposium on Communications Architectures and Protocols,* London, August 1994. URL ftp://cs.ucl.ac.uk/darpa/vjtut.ps.Z. Tutorial slides.

Jacobson, V. and Steve McCanne. vat — LBNL audio conferencing tool, July 1992. URL http://www-nrg.ee.lbl.gov/vat/. Available at http://www-nrg.ee.lbl.gov/vat/.

Jeffay, K., D. Stone, and F. Smith. Transport and display mechanisms for multimedia conferencing across packet-switched networks. *Computer Networks and ISDN Systems,* 26(10):1281-1304, July 1994. URL http://www.elsevier.com/locate/comnet.

Jiang, Wenyu, Kazuumi Koguchi, and Henning Schulzrinne. QoS evaluation of VoIP end-points. In *Conference Record of the International Conference on Communications (ICC),* May 2003.

Jiang, Wenyu and Henning Schulzrinne. Analysis of on-off patterns in VolP and their effect on voice traffic aggregation. In *International Conference on Computer Communication and Network,* Las Vegas, Nevada, October 2000a. URL http://www.cs.columbia.edu/IRT/papers/Jian0010Analysis.pdf.

Jiang, Wenyu and Henning Schulzrinne. Modeling of packet loss and delay and their effect on real-time multimedia service quality. In *Proceedings of the International Workshop on Network and Operating System Support for Digital Audio and Video (NOSSDAV),* June 2000b. URL http://www.nossdav.org/2000/papers/27.pdf.

Jiang, Wenyu and Henning Schulzrinne. Comparison and optimization of packet loss repair methods on VoIP perceived quality under bursty loss. In *Proc. International Workshop on Network and Operating System Support for Digital Audio and Video (NOSSDAV),* Miami Beach, Florida, May 2002a. URL http://www.cs.columbia.edu/IRT/papers/Jian0205_Comparison.pdf.

Jiang, Wenyu and Henning Schulzrinne. Comparisons of FEC and cadet robustness on VoIP quality and bandwidth efficiency. In *ICN,* Atlanta, Georgia, August 2002b. URL http://www.cs.columhia.edu/IRT/papers/Jian0208_Comparisons.pdf.

Jiang, Wenyu and Henning Schulzrinne. Speech recognition performance as an effective perceived quality predictor. In *IWQoS,* Miami Beach, May 2002c. URL http://www.cs.columbia.edu/hgs/papers/Jian0205:Speech.pdf.

Jiang, Wenyu and Henning Schulzrinne. Assessment of VoIP service availability in the current Internet. In *Passive and Active Measurement Workshop, San* Diego, CA, April 2003. URL http://www.cs.columbia.edu/IRT/papers/Jian03041_Assessment.pdf.

Johnston, A. R. Session initiation protocol basic call flow examples. Internet draft, Internet Engineering Task Force, April 2003. URL http://www.ietf.org/internet-drafts/draft-ietf-sipping-basic-call-flows-02.txt. Work in progress.

Kutscher, D., Juerg Ott, and Carsten Bormann. Session description and capability negotiation. Internet draft, Internet Engineering Task Force, March 2003. URL http://www.ietf.org/internet-drafts/draft--ietf-mmusic-sdpng-06.txt. Work in progress.

Lennox, J. and Henning Schulzrinne. Call processing language framework and requirements. RFC 2824, Internet Engineering Task Force, May 2000x. URL http://www.rfc-editor.org/rfc/rfc2824.txt.

Lennox, J. Henning Schulzrinne, and Thomas F. La Porta. Implementing intelligent network services with the session initiation protocol. Technical Report CUCS-002-99, Columbia University, New York, January 1999. URL ftp://ftp.cs.columbia.edu/reports/reports-1999/cucs-002-99.ps.gz.

Lennox, J. Henning Schulzrinne, and J. Rosenberg. Common gateway interface for SIP. RFC 3050, Internet Engineering Task Force, January 2001. URL http://www.rfc-editor.org/rfc/rfc3050.txt.

Lennox, Jonathan and Henning Schulzrinne. Feature interaction in Internet telephony. In *Feature Interaction in Telecommunications and Software Systems VI,* Glasgow, U.K., May 2000b. URL http://www.cs.columbia.edu/IRT/papers/Lenn0005_Feature.pdf.

Lennox, Jonathan, Xinzhou Wu, and Henning Schulzrinne. CPL: a language for user control of Internet telephony services. Internet draft, Internet Engineering Task Force, August 2003. URL draft-ietf-iptel-cpl-07.txtps. Work in progress.

Liu Hong, and Petros N. Mouchtaris. Voice over IP signaling: H.323 and beyond. *IEEE Communications Magazine,* 38(10). October 2000. URL http://www.comsoc.org/livepubs/ci1/public/2000/oct/index.html.

Luthi, P. RTP payload format for ITU-T recommendation G.722.1. RFC 3047, Internet Engineering Task Force, January 2001. URL http://www.rfc-editor.org/rfc/rfc3047.txt.

Magill, D. T., Adaptive speech compression for packet communication systems. In *Conference record of the IEEE National Telecommunications Conference,* pages 29D-1–29D-5, 1973.

McKnight, Lee. Internet telephony markets: 2000–3001. In *Carrier Class IP Telephony,* San Diego, CA, January 2000. URL http://www.cs.columbia.edu/IRT/papers/others /McKn0001_Internet.ppt.gz.

Mealling, M. Dynamic delegation discovery system (DDDS) part one: The comprehensive DDDS. RFC 3401, Internet Engineering Task Force, October 2002. URL http://www.rfc-editor.org/rfc/rfc3401.txt.

Miedema, H. and M. G. Schachtman. TASI quality — effect of speech detectors and interpolation. *Bell System Technical Journal,* 41(4):1455–1473, July 1962.

Miller, Ed, Flemming Andreasen, and Glenn Russell. The PacketCable architecture. *IEEE Communications Magazine,* 39(6), June 2001. URL http://www.comsoc.org/livepubs/ci1/public/2001/jun/index.html.

Montgomery, Warren A. Techniques for packet voice synchronization. *IEEE Journal on Selected Areas in Communications,* SAC-1(6):1022–1028, December 1983.

Moon, Sue, James F. Kurose, and Donald F. Towsley. Packet audio playout delay adjustment: performance bounds and algorithms. *Multimedia Systems,* 5(1):17–28, January 1998. URL ftp://gaia.cs.umass.edu/pub/Moon95_Packet.ps.gz.

Moulton, Pete and Jeremy Moulton. Telecommunications technical fundamentals. Technical handout, The Moulton Company, Columbia, MD, 1996. URL http://www.moultonco.com/semnotes/telecomm/teladd.htm. see http://www.moultonco.com/semnotes/telecomm/teladd.htm.

Ploumen, Frank M. and Luc de Clercq. The all-digital loop: benefits of an integrated voice-data access network. In *Communication Technology (ICCT),* Beijing, China, August 2000. IEEE.

Postel, John. Internet control message protocol. RFC 792, Internet Engineering Task Force, September 1981. URL http://www.rfc-editor. org/rfc/rfc792.txt.

Ramjee, R., James F. Kurose, Donald F. Towsley, and Henning Schulzrinne. Adaptive playout mechanisms for packetized audio applications in wide-area networks. In *Proceedings of the Conference on Computer Communications (IEEE Infocom),* pages 680–688, Toronto, Canada, June 1994. IEEE Computer Society Press, Los Alamitos, California. URL ftp://gaia.cs.umass.edu/pub/Ramj94:Adaptive.ps.Z.

Rieser, J. H., H. G. Suyderhood, and Y. Yatsuzuka. Design considerations for digital speech interpolation. In *Conference Record of the International Conference on Communications (ICC),* pages 49.4.1–49.4.7, Denver, Colorado, June 1981. IEEE.

Roach, A. B. Session initiation protocol (SIP)-specific event notification. RFC 3265, Internet Engineering Task Force, June 2002. URL http://www.rfc-editor.org/rfc/rfc3265.txt.

Rosenberg, J. The session initiation protocol (SIP) UPDATE method. RFC 3311, Internet Engineering Task Force, October 2002. URL http://www.rfc-editor.org/rfc/rfc3311.txt.

Rosenberg, J., H. F. Salama, and M. Squire. Telephony routing over IP (TRIP). RFC 3219, Internet Engineering Task Force, January 2002a. URL http://www.rfc-editor.org/rfc/rfc3219.txt.

Rosenberg, J. and Henning Schulzrinne. An RTP payload format for generic forward error correction. RFC 2733, Internet Engineering Task Force, December 1999. URL http://www.rfc-editor.org/rfc/rfc2733.txt.

Rosenberg, J. and Henning Schulzrinne. Reliability of provisional responses in session initiation protocol (SIP). RFC 3262, Internet Engineering Task Force, June 2002. URL http://www.rfc-editor.org/rfc/rfc3262.txt.

Rosenberg, J., Henning Schulzrinne, G. Camarillo, A. R. Johnston, J. Peterson, R. Sparks, M. Handley, and E. Schooler. SIP: session initiation protocol. RFC 3261, Internet Engineering Task Force, June 2002b. URL http://www.rfc-editor.org/rfc/rfc3261.txt.

Rosenberg, J. et al. Indicating user agent capabilities in the session initiation protocol (SIP). Internet draft, Internet Engineering Task Force, June 2003a. URL http://www.ietf.org/internet-drafts/draft-ietf-sip-callee-caps-00.txt. Work in progress.

Rosenberg, J. and P Kyzivat. Guidelines for usage of the session initiation protocol (SIP) caller preferences extension. Internet draft, Internet Engineering Task Force, July 2003. URL http://www.ietf.org/internet-drafts/draft-ietf-sipping-callerprefs-usecases-00.t> Work in progress.

Rosenberg, J., James L. Peterson, Henning Schulzrinne, and Gonzalo Camarillo. Best current practices for third party call control in the session initiation protocol. Internet draft, Internet Engineering Task Force, July 2003b. URL http://www.ietf.org/internet-drafts/draft-ietf-sipping-3pcc-04.txt. Work in progress.

Rosenberg, J., Lili Qiu, and Henning Schulzrinne. Integrating packet FEC into adaptive voice playout buffer algorithms on the Internet. In *Proceedings of the Conference on Computer Communications (IEEE Infocom),* Tel Aviv, Israel, March 2000. URL http://www.cs.columbia.edu/hgs/papers/Rose0003_Integrating.pdf.

Schulzrinne, Henning. Voice communication across the internet: A network voice terminal. Technical Report TR 92–50, Dept. of Computer Science, University of Massachusetts, Amherst, MA, July 1992. URL http://www.cs.columbia.edu/hgs/papers/Schu9207_Voice.ps.gz.

Schulzrinne, Henning. RTP profile for audio and video conferences with minimal control. RFC 1890, Internet Engineering Task Force, January 1996. URL http://www.rfc-editor.org/rfc/rfc1890.txt.

Schulzrinne, Henning, S. Casner, R. Frederick, and V. Jacobson. RTP: a transport protocol for real-time applications. RFC 1889, Internet Engineering Task Force, January 1996. URL http://www.rfc-editor.org/rfc/rfc1889.txt.

Schulzrinne, Henning and S. Petrack. RTP payload for DTMF digits, telephony tones and telephony signals. RFC 2833, Internet Engineering Task Force, May 2000. URL http://www.rfc-editor.org/rfc/rfc2833.txt.

Schulzrinne, Henning and A. Vaha-Sipila. The tel URI for telephone calls. Internet draft, Internet Engineering Task Force, July 2003. URL http://www.ietf.org/internet-drafts/draft-ietf-iptel-rfc2806bis-02.txt. Work in progress.

Schuster, Guido, Jerry Mahler, Ikhlaq Sidhu, and Michael S. Borella. Forward error correction system for packet based real time media. U.S. Patent US5870412, 3Com, Chicago, Illinois, February 1999. URL http://www.patents.ibm.com/patlist?icnt=US&patent_number=5870412.

Singh, Kundan and Henning Schulzrinne. Interworking between SIP/SDP and H.323. In *IP-Telephony Workshop (IPtel),* Berlin, Germany, April 2000. URL http://www.cs.columbia.edu/hgs/papers/Sing0004_Interworking.pdf.

Sjoberg, J., M. Westerlund, A. Lakaniemi, and Q. Xie. Real-time transport protocol (RTP) payload format and file storage format for the adaptive multi-rate (AMR) and adaptive multi-rate wide-band (AMR-WB) audio codecs. RFC 3267, Internet Engineering Task Force, June 2002. URL http://www.rfc-editor.org/rfc/rfc3267.txt.

Sparks, R. The Session Initiation Protocol (SIP) refer method. RFC 3515, Internet Engineering Task Force, April 2003. URL http://www.rfc-editor.org/rfc/rfc3515.txt.

Stock, T. and Xavier Garcia Adanez. On the potentials of forward error correction mechanisms applied to real-time services carried over B-ISDN. In Bernhard Plattner (Ed.,) *International Zurich seminar on Digital Communications, IZS (Broadband Communications Networks, Services, Applications, Future Directions*), Lecture Notes in Computer Science, pages 107–118, Zurich, Switzerland, February 1996. Springer-Verlag. URL http://tcomwww.epfl.ch/garcia/publications/publications.html/izs96.ps.

Thom, Gary A. H.323: the multimedia communications standard for local area networks. *IEEE Communications Magazine,* 34(12), December 1996. URL http://www.comsoc.org/pubs/ci/comsoc/private/1996/dec/Thom.html.

Toutireddy, Kiran and J. Padhye. Design and simulation of a zero redundancy forward error correction technique for packetized audio transmission. Project report, Univ. of Massachusetts, Amherst, Massachusetts, December 1995. URL http://www-ccs.cs.umass.edu/jitu/fec/.

Vinall, George. Economics of Internet telephony. In *Voice on the Net,* San Jose, California, March/April 1998. URL http://www.pulver.com/oldslides.

Weinstein, Clifford J. and Edward M. Hofstetter. The tradeoff between delay and TASI advantage in a packetized speech multiplexer. *IEEE Transactions on Communications,* COM-27(11):1716–1720, November 1979.

Weiss, Mark Allen and Jenq-Neng Hwang. Internet telephony or circuit switched telephony: Which is cheaper? In *Telecommunications Policy Research Conference,* Washington, D.C., October 1998. URL http://www2.sis.pitt.edu/mweiss/papers/itel.pdf.

Wocjik, Ronald J. Packetcable network architecture. In *Carrier Class IP Telephony,* San Diego, California, January 2000.

Wright, David J. Voice over ATM: an evaluation of network architecture alternatives. *IEEE Network,* 10(5):22–27, September 1996. URL http://207.127.135.8/ni./private/1996/sep/Wright.html.

Wright, David J. Voice over MPLS compared to voice over other packet transport technologies. *IEEE Communications Magazine,* 40(11):124–132, November 2002.

Wu, Xiaotao and Henning Schulzrinne. Where should services reside in Internet telephony systems? In *IP Telecom Services Workshop,* Atlanta, Georgia, September 2000. URL http://www.cs.columbia.edu/hgs/papers /Wu0009_Where.pdf.

Wu, Xiaotao and Henning Schulzrinne. Programmable end system services using SIP. In *Conference Record of the International Conference on Communications (ICC),* May 2003.

Zhao, W. and Henning Schulzrinne. Locating IP-to-Public switched telephone network (PSTN) telephony gateways via SLP. Internet draft, Internet Engineering Task Force, August 2002. URL http://www.ietf.org/internet-drafts/draft-zhao-iptel-gwloc-slp-05.txt. Work in progress.

Zopf, R. Real-time transport protocol (RTP) payload for comfort noise (CN). RFC 3389, Internet Engineering Task Force, September 2002. URL http://www.rfc-editor.org/rfc/rfc3389.txt.

7

Internet Messaging

CONTENTS

7.1 Introduction 7-1
7.2 Current Internet Solution 7-2
 7.2.1 Electronic Mail 7-2
 7.2.2 Network News 7-5
 7.2.3 Instant Messaging 7-7
 7.2.4 Web Logging 7-10
7.3 Telecom Messaging 7-11
 7.3.1 Principal Operation 7-11
 7.3.2 Naming 7-11
 7.3.3 Short Message Service 7-11
7.4 A Comparison 7-12
7.5 A Note on Unsolicited Messaging 7-14
 7.5.1 Spreading Viruses 7-14
 7.5.2 Spam 7-15
 7.5.3 Protection Mechanisms 7-15
7.6 Toward Unified Messaging 7-16
7.7 Outlook 7-17
References 7-18

Jan Mark S. Wams

Maarten van Steen

7.1 Introduction

It might be argued that messaging is the *raison d'etre* of the Internet. For example, as of 2002, the number of active electronic mailboxes is estimated to be close to 1 billion and no less than 30 billion e-mail messages are sent daily with an estimated 60 billion by 2006. Independent of Internet messaging, one can also observe an explosion in the number of messages sent through the Short-Message Services (SMS); by the end of 2002, the number of these messages has been estimated to exceed 2 billion per day. Messaging has established itself as an important aspect of our daily lives, and one can expect that its role will only increase.

As messaging continues to grow, it becomes important to understand the underlying technology. Although e-mail is perhaps still the most widely applied instrument for messaging, other systems are rapidly gaining popularity, notably instant messaging. No doubt there will come a point at which users require that the various messaging systems be integrated, allowing communication to take place independent of specific protocols or devices. We can already observe such an integration of, for example, e-mail and short-messaging services through special gateways.

In this chapter, we describe how current Internet messaging systems work, but also pay attention to telephony-based messaging services (which we refer to as telecom messaging) as we expect these to be widely supported across the Internet in the near future. An important goal is to identify the key shortcomings of current systems and to outline potential improvements. To this end, we introduce a taxonomy

1-58488-381-2/05/$0.00+$1.50

© 2005 by CRC Press LLC

by which we classify and compare current systems and from which we can derive the requirements for a unified messaging system.

7.2 Current Internet Solutions

We start by considering the most dominant messaging system on the Internet: electronic mail. We will discuss e-mail extensively and take it as a reference point for the other messaging systems. These include network news (a bulletin-board service), and the increasingly popular instant messaging services. Our last example is Web logging, which, considering its functionality, can also be thought of as an Internet messaging service.

7.2.1 Electronic Mail

Electronic mail (referred to as e-mail) is without doubt the most popular Internet messaging application, although its popularity is rivaled by applications such as instant messaging and the telecom messaging systems that we discuss in Section 7.3. The basic model for e-mail is simple: a user sends an electronic message to one or more explicitly addressed recipients, where it is subsequently stored in the recipient's mailbox for further processing.[1] One of the main advantages of this model is the asynchronous nature of communication: the recipient need not be online when a message is delivered to his or her mailbox, but instead, can read it at any convenient time.

7.2.1.1 Principal Operation

The basic organization of e-mail is shown in Figure 7.1(a) and consists of several components. From a user's perspective, a mailbox is conceptually the central component. A mailbox is simply a storage area that is used to hold messages that have been sent to a specific user. Each user generally has one or more mailboxes from which messages can be read and removed. The mailbox is accessed by means of a mail user agent (MUA), which is a management program that allows a user to, for example, edit, send, and receive messages.

Messages are composed by means of a user agent. To send a message, the user agent generally contacts a local message submit agent (MSA), which temporarily queues outgoing messages. A crucial component in e-mail systems is mail servers, also referred to as message transfer agents (MTAs). The MTA at the sender's site is responsible for removing messages that have been queued by the MSA and transferring them to their destinations, possibly routing them across several other MTAs. At the receiving side, the MTA spools incoming messages, making them available for the message delivery agent (MDA). The latter is responsible for moving spooled messages into the proper mailboxes.

Assume that Alice at site *A* has sent a message *m* to Bob at site *B*. Initially, this message will be stored by the MSA at site *A*. When the message is eventually to be transferred, the MTA at site *A* will set up a connection to the MTA at site *B* and pass it message *m*. Upon its receipt, this MTA will store the message for the MDA at *B*. The MDA will look up the mailbox for Bob to subsequently store *m*. In the Internet, mail servers are generally contacted by means of the Simple Mail Transfer Protocol (SMTP), which is specified in RFC 2821 [Klensin, 2001].

Note that this organization has a number of desirable properties. In the first place, if the mail server at the destination's site is currently unreachable, the MTA at the sender's site will simply keep the message queued as long as necessary. As a consequence, the actual burden of delivering a message in the presence of unreachable or unavailable mail servers is hidden from the e-mail users.

Another property is that a separate mail spooler allows easy forwarding of incoming messages. For example, several organizations provide a service that allows users to register a long-lived e-mail address. What is needed, however, is that a user also provides an actual e-mail address to which incoming messages

[1] It should be noted that many users actually have multiple mailboxes. For simplicity, we will often speak in terms of a single mailbox per recipient.

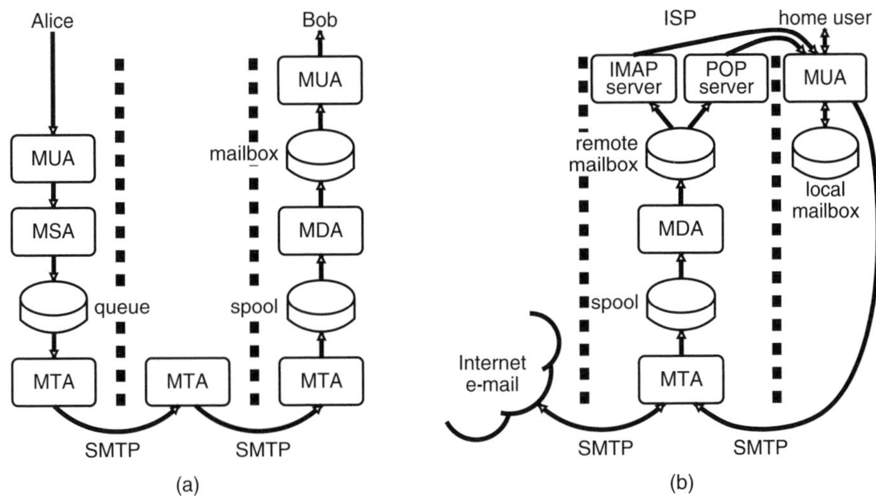

FIGURE 7.1 (a) The general organization of e-mail. (b) How e-mail is supported by an ISP.

can be forwarded. In the case of forwarding, a mail server simply passes an incoming message to the MSA, but this time directed to the actual address.

7.2.1.2 Remote Access

The organization as sketched in Figure 7.1 assumes that the user agent has continuous (local) access to the mailbox. In many cases, this assumption does not hold. For example, many users have e-mail accounts at an Internet Service Provider (ISP). In such cases, mail sent to a user is initially stored in the mailbox located at his or her ISP. To allow a user to access his or her mailbox, a special server is needed, as shown in Figure 7.1(b).

The remote access server essentially operates as a proxy for the user agent. There are two models for its operation. In the first model, which has been adopted in the Post Office Protocol (POP3) described in RFC 2449 [Gellens et al., 1998], the remote access server transfers a newly arrived message to the user, who is then responsible for storing it locally. Although POPS does allow users to keep a transferred message stored at the ISP, it is customary to configure user agents to instruct the server to delete any message just after the agent has fetched it. This setup is often necessary due to the limited storage space that an ISP provides to each mailbox. However, even when storage space is not a problem, POP3 provides only minimal mailbox search facilities, making the model not very popular for managing messages.

As an alternative, there is also a model in which the access server does not normally delete messages after they have been transferred to the user. Instead, it is the ISP that takes responsibility for mailbox management. This model is supported by the Internet Message Access Protocol (IMAP), which is specified in RFC 2060 [Crispin, 1996]. In this case, the access server provides an interface that allows a user to browse, read, search, and maintain his or her mailbox. IMAP is particularly convenient for mobile users, and in principle can support even handheld wireless access devices such as GSM cell phones (although special gateways are needed).

7.2.1.3 Naming

To enable the transfer of messages, a scheme for addressing the source and destination is necessary. For Internet e-mail, an address consists of two parts: the name of the site to which a message needs to be sent, which, in turn, is prefixed by the name of the user for whom it is intended. These two parts are separated by an at-sign ("@"). Given a name, the e-mail system should be able to set up a connection between the sending and receiving MTA to transfer a message, after which it can be stored in the addressed user's mailbox. In other words, what is required is that an e-mail name can be resolved to the network address of the destination mail server.

```
;<<>> DIG 9.1.0<<>> mx cs.vu.nl
;; global options: printcmd
;; Got answer:
;;->>HEADER<<- opcode: QUERRY, status: NOERROR, id: 43753
:: flags: qr rd ra; QUERRY: 1, ANSWER: 2, AUTHORITY: 4, ADDITIONAL: 3

;; QUESTION SECTION:
;cs.vu.nl.                         IN      MX

;;ANSWER SECTION:
cs.vu.nl.            86069         IN      MX      1 tornado.cs.vu.nl.
cs.vu.nl.            86069         IN      MX      2 zephyr.cs.vu/nl.

;;ADDITIONAL SECTION:
tornado.cs.vu.nl.    86069         IN      A       192.31.231.152
zephyr.cs.vu.nl.     86069         IN      A       192.31.231.66
```

FIGURE 7.2 Response to a DNS query using the dig tool (edited).

Resolving an e-mail name requires support from the Internet Domain Name System (DNS) [Mockapetris, 1987; Albitz and Liu, 1998]. Consider sending an e-mail to an address johndoe@cs.vu.nl. In this example, johndoe identifies the user at site cs.vu.nl. To send a message, it is necessary to identify a mail server that can handle incoming e-mail traffic. For Internet e-mail, DNS stores such information in what are known as mail exchange records or simply MX records. For example, using a program called domain information groper (dig), a DNS query requesting an MX record for cs.vu.nl returns the answer shown in Figure 7.2.

The most important part of the response is the answer section (shown in boldface in Figure 7.2), which states that there are two mail servers for cs.vu.nl. The preferred mail server is named tornado.cs.vu.nl, while a secondary server named zephyr.cs.vu.nl is also available. To initiate an SMTP session, the sender's MTA usually sets up a TCP connection to the preferred MTA at the destination site, for which it needs the server's IP address. In our example, this would require resolving the name tornado.cs.vu.nl, which is, in principle, done by means of another DNS query. DNS anticipates such additional queries when asked for an MX record, and includes an additional section containing the IP addresses of the returned snail servers, thus avoiding another query.

Once the message has been transferred to the destination MTA, it is the task of the latter to resolve the user name that is part of the e-mail address to the appropriate mailbox. How this user-name resolution is done is not prescribed by SMTP.

7.2.1.4 Message Formats: MIME

An important issue in any messaging system is that sender and receiver agree on the format of the message content. Such an agreement is possible by including the description of that format as part of the message header. This is the principle underlying Multipurpose Internet Mail Extensions (MIME), which we briefly discuss next.

An e-mail message is a string of values that is mapped into readable text by a character set. The best known character set is the North American ASCII set, which has 96 characters, but lacks European characters such as "ß," "å," "ç" and "" Moreover, in Asian, Russian, Arabic, and other languages, totally different word and character sets are in use. MIME is a standard defined to accommodate these different sets, as well as graphics, sound, and encodings like HTML (see, e.g., RFC 2231 [Freed and Moore, 1997]). The MIME standard is not exclusively used for e-mail. Other Internet messaging systems, such as netnews, use MIME too.

An e-mail message is formed after a memo: it has a body with the actual message and a header containing information about the author, date of creation, subject, and so on. MIME specifies a number of header fields that are column-separated keyword-value pairs, to define the structure of the message body. For example, the header fields in Figure 7.3 describe that the message body is composed — and should be displayed — using ASCII-text mapping. Figure 7.4 indicates that the body is to be interpreted as a Chinese HTML message that is (base64) encoded for transfer.

Another useful MIME content type, which indicates that a message has multiple body parts separated by a unique string, is called multipart. There are four multipart subtypes. The most common subtype is mixed to indicate a series of generic body parts that carry their own header-fields. There are other subtypes-like alternatives for representing the same data in different formats, such as plain text and HTML, and parallel for parts that need to be simultaneously processed, such as sound and graphics. Body parts are bracketed by lines with only the unique string and usually have their own MIME header fields, as shown in Figure 7.5.

7.2.2 Network News

Network news, also abbreviated to netnews, gained its popularity as part of USENET, a logical network mainly consisting of many computers that used simple dialup phone lines for message transfer. The

```
MIME-VERSION: 1.0
Content-Type: text/plain; charset=us-ascii
Content-Transfer-Encoding: 8bit
```

FIGURE 7.3 MIME header example: plain text message.

```
MIME-VERSION: 1.0
Content-Type: text/html; charset=big5
Content-Transfer-Encoding: base64
```

FIGURE 7.4 MIME header example: An HTML/Chinese (Big5) message.

```
MIME-VERSION: 1.0
Content-Type: multipart/mixed; boundary=unique-135711

--unique-135711
Content-type: text/plain; charset=US-ASCII

Look at this picture.
--unique-135711
Content-Type: image/jpeg; name=picture.jpg
Content-Transfer-Encoding: base64
Content-Disposition: inline

P9jJ4AAQsk ... DBkSEw8UHRof
    ⋮
UKJgkjhUUOTaOGZs5wP712Q==
--unique-135711
Content-type: text/enriched

<bold><italic>BYE!</italic></bold>
--unique-135711--
```

FIGURE 7.5 Multipart message example.

netnews model is that of an electronic bulletin board: messages are put up on the board to be read and reacted to by others. In netnews, messages are referred to as articles that are posted in a specific newsgroup. A newsgroup is thus a collection of logically related articles and forms the electronic representation of a bulletin board. A nontechnical overview of network news is given by Comer [2000].

A user provides the netnews system with the name of a newsgroup in order to read articles. The header of any new article that has not yet been read by the user is then transferred to the user. If the user wants to read the entire article, he or she will request the transfer of the article's body. After reading an article, a user can respond by posting a reaction in that same newsgroup (which again appears as just another article). Cross postings by which an article refers to an article in a different newsgroup is also possible.

Note that users do not actively delete articles. However, to prevent articles from consuming storage indefinitely, system administrators generally remove articles after some time. Unlike e-mail where messages are permanently stored until explicitly deleted by a recipient, this policy makes news articles impermanent unless special measures are taken to store them permanently.

7.2.2.1 Principal Operation

The core of the network news system is formed by a huge collection of news servers spread across thousands of different sites. A news server, also referred to as News Transfer Agent (NTA), is capable of receiving, sending, and storing articles. The basic organization of the network news system is shown in Figure 7.6. A client, called a News User Agent (NUA), connects to a news server to read and post articles for one or several newsgroups. Likewise, servers connect to each other to exchange articles, as we will discuss in more detail. Although the figure suggests a principal difference between clients and servers, no such difference actually exists. In fact, the protocol that is used between a client and server and the one used between two servers is the same. All information exchange follows the Network News Transfer Protocol (NNTP), specified in RFC 977 [Kantor and Lapsley, 1986].

A news server should connect to one or more existing news feeds, which are just other news servers that are willing to exchange articles. In many cases, a news feed is operated by a separate organization such as an ISP. When a news server contacts a news feed, it requests the transfer of new articles. Several operations are available to establish such a transfer, of which some important ones are listed in Figure 7.7.

The transfer protocol is relatively simple and has not been changed since its specification in 1986. However, practice has shown that extensions and deviations from the original specification were needed. In particular, the communication between servers, and that between a client and a server, are different enough to warrant further refinements, effectively leading to two very similar yet different protocols. These refinements are described in RFC 2980 [Barber, 2000].

7.2.2.2 Naming

An important difference between netnews and e-mail is that there is no need to explicitly name and look up news servers. Instead, the address of a news feed is assumed to be known at the time a news client

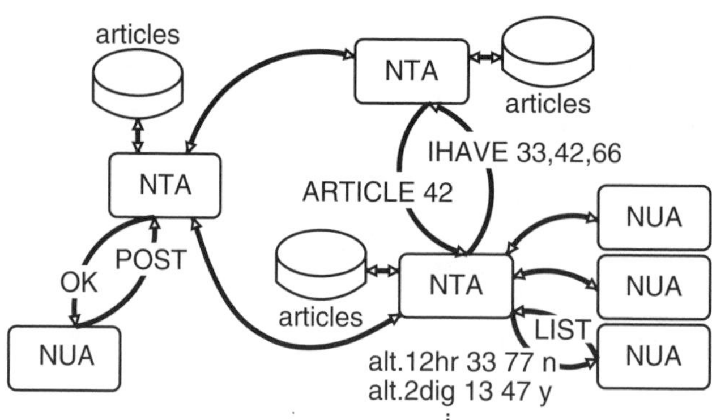

FIGURE 7.6 The general organization of network news.

Operation	Description
LIST	Returns a list of newgroups available at the callee with each entry identifying the first and last article in that group.
GROUP	Makes a specified group "current," and returns and estimate of the number of articles at the callee in that group.
ARTICLE	Transfers (to the caller) a specified article in the current group.
POST	Tells the callee that an article has been posted at the caller.
IHAVE	Tells the callee that specific article is available to be sent.
NEWNEWS	Returns a list of articles that have been posted at the callee in specific news groups.
NEWGROUPS	Returns a list of newsgroups that have been created at the callee.

FIGURE 7.7 Commonly used operations to establish the transfer of articles between two news programs.

or server is configured so that its address can be readily used to set up an NNTP session. Jointly, these sessions ensure that articles are flooded through the network consisting of netnews servers. In contrast, for e-mail it is necessary to devise a naming scheme by which users and mail servers can be looked up at runtime. This naming scheme is needed to support the point-to-point communication in e-mail systems.

Naming in news therefore restricts itself to newsgroups and implicitly also to articles. In particular, it is important to have a suitable naming scheme for the tens of thousands of newsgroups that currently exist. To this end, a hierarchical naming scheme has been devised that is simple yet flexible enough to support a large number of newsgroups. A newsgroup name is a series of strings separated by a dot, such as comp.os.research. In this example, comp identifies the broad category of newsgroups related to computer science, which is further divided into newsgroups dealing with operating systems (os) and, in particular, the one containing articles on research in this area (research).

Each article has a unique identifier consisting of two parts separated by the at-sign ("@"). An example of such an identifier is 3e1ed38c$1@news.cs.vu.nl (see also RFC 1036 [Horton and Adams, 1987]). The second part identifies the host where the article was first entered into the news system, in this example news.cs.vu.nl. The first part is a unique identifier normally generated by the host named in the second part (and hidden for the user). In principle, an article's identifier is globally unique and is never reused: it is a so-called true identifier [Wieringa and de Jonge, 1995].

7.2.3 Instant Messaging

One of the upcoming means of usercentric communication across the Internet is instant messaging. The model underlying instant messaging is that of synchronous communication; a message can be successfully transferred only if the destination is willing to receive it at the time it is sent. In many other respects, instant messaging strongly resembles e-mail, and the two forms are sometimes integrated into a single system. One of the first one-to-one instant messaging system was called "term-talk" and ran on the Plato system as early as 1973 [Woolley, 1994]. One of the earliest full-blown instant messaging systems appeared in M.I.T.'s Athena system [Belville, 1990]. Instant messaging on the Internet originated as Internet Relay Chat, (IRC, described in RFC 1459 [Oikarinen and Reed, 1993] and updated in RFCs 2810-2813 [Kalt, 2000]), but became really popular with the introduction of ICQ (pronounced as "I seek you"). Currently, there are many instant messaging clients, whereas instant messaging services are provided by large organizations such as AOL and Microsoft.

An instant messaging system generally has a separate component, called a presence information service, which is used to inform users of each other's presence (see also RFC 2778 [Day et al., 2000b]). Such a service allows a user to see whether it is possible to send a message to someone else. When a user logs in, his or her presence is published and forwarded to subscribers of that information. Likewise, a user can indicate that he or she is temporarily not reachable or has logged out. Managing presence information is increasingly becoming an important issue as it strongly affects the privacy of publishers. We return to presence information in more detail in the following text.

7.2.3.1 Principal Operation

The principal operation of an instant messaging service is quite simple. In all cases, we need to first set up a channel between the communicating parties. Let us assume that Alice wants to communicate with Bob. If Alice has Bob's address, then, in principle, she can set up a channel directly to Bob's user agent (UA) as shown in Figure 7.8(a).

The main drawback of this approach is that Bob's contact address must be fixed (i.e., the address where Alice can reach Bob), and also that Alice can set up unsolicited channels to Bob. To alleviate these problems, many instant messaging services adopt the scheme shown in Figure 7.8(b). In this case, a central server keeps track of online clients (whose contact address may be different each time they come online). Alice sends a setup request to the server, which subsequently returns Bob's address, possibly after checking whether Alice is authorized to set up a channel to Bob. Note that the central server may also be used as an intermediary for all communication between Alice and Bob, including the instant messages sent between them, as shown in Figure 7.8(c).

The obvious drawback of the central server is that it forms a potential bottleneck. This centralized approach, even when multiple servers are used, is recognized as one of the main scalability problems in IRC. To circumvent these scalability problems, several instant messaging servers can be used, as shown in Figure 7.8(d). In this solution, Alice contacts a local instant-messaging server and requests a communication channel to Bob. Her local server then contacts a server that controls connections to Bob and requests a communication endpoint to Bob's client. After the proper security checks have been made and Alice has indeed been found to be authorized to contact Bob, Bob's address is returned to allow the set up of a connection.

This use of a distributed instant messaging service scales well as only the servers that are local to Alice and Bob need to assist in setting up a connection. However, it also introduces a lookup problem because the server local to Alice needs to locate Bob's local instant messaging server. A simple solution, but one that has not been widely deployed yet, is to follow the same approach as in e-mail. In principle, every site makes use of a single instant messaging server and users simply identify themselves by their e-mail address. Let us assume Bob's e-mail address is bob@cs.vu.nl. When Alice wants to contact Bob, her (well-known) local instant messaging server queries the Domain Name System (DNS) for the instant mes-

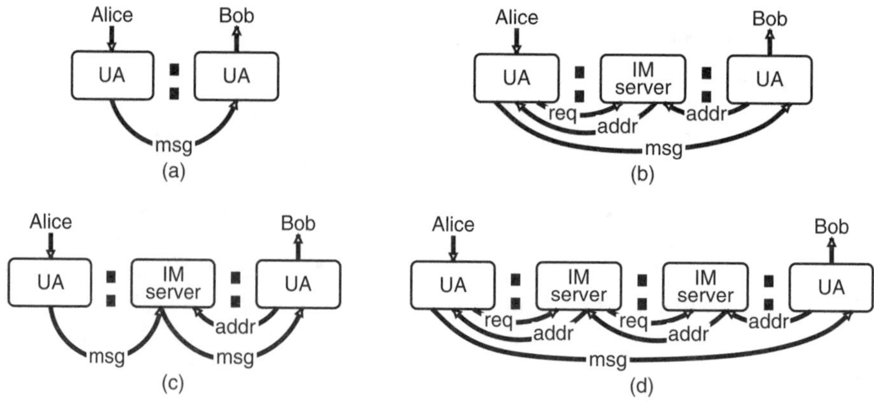

FIGURE 7.8 Setting up an instant-messaging connection: (a) directly, (b) through a central server, (c) centralized, including messaging, and (d) through different servers.

saging server at cs.vu.nl, analogous to asking DNS for the name of the mail server at that site. Once its name is returned, its IP address can then be looked up using DNS again.

So far, we have assumed that instant messaging takes place only between pairs of individuals. In general, this need not be the case. Two different forms of multiparty instant messaging exist. First, setting up a connection between two parties can easily be extended by inviting another party, leading to an *ad hoc* group or chat session. In this case, each message is sent to all members participating in the session. An invited party can join the session, and any joined party can later leave again. A session dissolves when the last member leaves.

The second type of multiparty instant messaging is through so-called chat rooms, which are effectively permanent sessions. To enter a chat room, a user needs to set up a connection to a well-known server that handles all communication for that chat room, effectively leading to the communication scheme shown in Figure 7.8(c). Each message sent to the server is multicast to every other client that has entered the chat room. Unlike *ad hoc* groups, chat rooms continue to exist even after the last member has left. By nature, a chat room is useful for online discussions on a very specific subject, and this is indeed the way in which they are generally organized.

7.2.3.2 Presence Information Service

As mentioned before, an important component of an instant messaging service is a service that provides presence information. In a minimalist approach, such a service merely reports whether a user is online or offline, allowing an initiator to see whether it makes sense to even try to set up an instant messaging connection. However, presence information can, and often is, extended so as to include other possible states, as shown in Figure 7.9.

Many variations on these states exist. For example, some presence services automatically switch a recipient from online to away when there has been no interaction with the instant messaging client for some time. Likewise, when an invitation is sent out to a recipient who is currently busy, the inviting client may receive a message saying that the other party does not want to be disturbed.

It is not difficult to see that, although instant messaging by itself is relatively simple, a presence information service can easily grow into a sophisticated and complex part of an instant messaging service. Following the general architecture as described in RFC 2778 [Day et al., 2000b], a presence information service may also provide the means to send notifications when a client's status changes. Such a notification may be useful, for example, when Alice wants to contact Bob as soon as he can accept invitations again.

Despite its attractiveness, the real problem with this functionality emerges when one starts thinking about security. In effect, Alice subscribes to notifications concerning state changes of Bob. Although it may seem obvious that Bob should be in full control of permissible subscriptions, practice shows that this is not always the case. However, as also laid down in RFC 2779 [Day et al., 2000a], a client should always be in full control concerning who is allowed to send instant messages and who is allowed to subscribe to presence-state changes. This model has been adopted by the IETF working group on Extensible Messaging and Presence Protocol (XMPP).

In essence, before Alice can subscribe to presence information concerning Bob, XMPP requires that she send Bob a request for subscription. If this request is granted, Bob can pass her the appropriate

Status	Alert	Description
OFFLINE	No	No instant-messaging client is currently running at the recipient.
ONLINE	Yes	The recipient's instant-messaging client is currently running.
AWAY	Yes	The recipient's client is running, but invitations cannot be accepted.
BUSY	No	The recipient's client is running, but invitations are not notified.

FIGURE 7.9 Examples of different states maintained by a presences information service. The column alert indicates whether the recipient is notified when a setup request arrives.

credentials by which she can obtain a subscription at the presence information service. Bob, in turn, can always request the presence service to unsubscribe Alice.

No matter how simple this model may seem, it has serious implications for the design and implementation of a presence information service. The simplest situation is when the presence service is implemented as a single (trusted) centralized server. In that case, managing subscriptions boils down to checking lists of subscribers and sending notifications as needed. However, when dealing with a distributed presence information service, we are essentially dealing with the same problems that general publish/subscribe systems have. It is very hard to manage many users who may be geographically widely dispersed, scalability problems suddenly become paramount, and obvious solutions do not exist (see, e.g., [Carzaniga et al., 2001]). Relatively little is known about the development of large-scale wide-area notification systems, especially systems that can face up to various security attacks.

7.2.3.3 Naming

Naming is generally straightforward in instant-messaging systems and mainly concerns identifying users. For this reason, systems are gradually adopting the e-mail naming scheme. In the case of chat rooms, instant-messaging service providers generally offer a list of topics for which a chat room is hosted. By selecting a topic, a user is then allowed to join a chat session. Naming in such cases is therefore implicit and of less importance than it is with core instant messaging.

However, naming in many popular instant-messaging systems is still much of a nuisance. In particular, several systems, such as ICQ, simply provide a unique (long) number that is to be used as ID. The drawback of using these numbers is similar to that of using network addresses instead of host names: they are difficult to remember by humans. To circumvent problems, users simply build local lists of aliases for those people who are regularly contacted.

7.2.4 Web Logging

As a final example of Internet messaging, we briefly consider an increasingly popular form known as Web logging or simply "blogging" [Blood, 2002]. A Web log (or "blog") can be viewed as a unidirectional form of messaging: a user simply maintains a log of messages that others can generally only read. However, the number of Web logs that make it possible for readers to react to messages is growing rapidly. Web logging can be considered analogous to columns and commentaries in newspapers. As the popularity of this form of messaging continues to increase, communities of similar logs are starting to grow in which loggers (or "bloggers" as they are normally called) are referring and reacting to each other's work. In a sense, this utilization resembles the use of network news.

There is a lot of Web logging going on, but because most Web logs are self-published, precise numbers are difficult to obtain. Experts, however, agree that Web logging is growing ever more popular, probably due to the existence of tools and sites that make it easy to start blogging. Many companies have an in-house Web log, millions of private Web logs exist on the WWW, and there are some high-volume Web logs that serve a huge number of readers.

One of the bigger sites that allows everybody with a Web browser to easily maintain a Web log is Blogger (blogger.com). According to Blogger, over a million people have used their service to start a Web log, and subscriptions show exponential growth. There are a few big and influential Web logs. Over the course of 2002, the Web log Drudge (drudgereport.com) served around a billion pages. The Web log Slashdot (slashdot.org), for example, also serves millions of pages per day with "news for nerds." Needless to say that Slashdot has multiple editors and that sophisticated distributed moderating takes place to keep this huge volume usable.

7.2.4.1 Principal Operation

The principal operation of Web logging is extremely simple: a user simply publishes material on a single site that can be read by anyone accessing that site. Many tools are available that ease the process of updating and managing published material, effectively hiding the technical intricacies related to Web servers.

An important difference with all messaging systems discussed so far is that there are, in principle, no recipients. All material related to Web logging is conceptually published at a single site that needs to be polled regularly if a reader wants to keep track of changes. Alternatively, some systems already offer subscription facilities by which readers are automatically notified when updates occur. In practice, update notifications are simply sent by e-mail.

7.3 Telecom Messaging

Internet messaging is rivaled in popularity by telecom messaging. The two are becoming increasingly intertwined and there is no doubt that they will be fully integrated at some point in the future. Traditionally, telecom messaging was implemented in-band, that is, making direct use of voice channels. Examples of in-band telecom messaging are the fax system, bulletin board systems, and the French Minitel. A notable exception to in-band telecom messaging is paging, which usually has a dedicated radio frequency channel. The introduction of the global standard for telecommunications, Signaling System number Seven (SS7), allowed for out-of-band handling of data.

7.3.1 Principal Operation

Both in-band and out-of-band telecom messaging rely on the SS7 protocol stack. ITU defined SS7 as a packet-switching four-layer stack resembling the seven layer ISO/OSI network stack with the top four layers integrated into one (for an overview of SS7, see Dreher and Harte [2002]). Data packets are sent over the SS7 network to set up (and tear down) voice connections. Routing information is stored in databases called Home Location Registers (HLRs).

From the perspective of messaging, it is interesting to note that (limited-size) data packets, too, can be sent over the SS7 network, just like IP-packets can be sent over the Internet. This enables the implementation of efficient messaging schemes like the Short Message Service (SMS), which we briefly discuss below.

7.3.2 Naming

For telecom messaging, names take the form of telephone numbers. Traditionally, these numbers have been directly used for routing. Dialing 1234 would get a phone connected to exit four of exit three of exit two of exit one of the exchange office that the telephone was connected to. Nowadays, most telephone numbers have three basic parts: a country code, an area code, and a subscriber number. For local calls, the first two parts need not be explicitly provided.

Partly due to the introduction of cellular phones, routing schemes needed to be adjusted, and thus affected the way naming was deployed. To facilitate a fixed number for these roaming devices, the area code in a telephone number was used to also designate a particular cell-phone operator, effectively diverging from the geographical interpretation initially tied to area codes. This approach has nowadays been taken a step further, as the original area code is now also used to designate different types of services, such as toll-free calls, premium-rate calls, normal cell phones, paging, and so on.

A simple aliasing scheme is used to help users to remember telephone numbers by associating several letters with a single digit. This may lead to telephone numbers such as 555-shoe-shine which actually stands for the number 555-7463-74463.

7.3.3 Short Message Service

One particularly popular telecom messaging system in Europe and Asia is the Short Message Service (SMS). As mentioned earlier, by the end of 2002, over 2 billion SMS messages were sent daily, and this number is still rapidly increasing. As a side note, in Japan the more advanced i-mode messaging is similarly popular.

Besides SMS, there is the Enhanced Messaging Service (EMS), which combines multiple SMSs into one EMS, and the (technically unrelated) Multimedia Messaging Service (MMS), which can handle much larger messages. Both EMS and MMS currently (early 2003) cannot rival SMS in volume. There is one important drawback to SMS that is related to its use of SS7: it can carry only very short messages — around 160 characters long, in Europe. There are more characters in an SMS message, but some are used to store the callers' telephone number and other data. On the other hand, SMS messages can be sent independent of voice traffic. Even if the network is loaded up to the point where it becomes impossible to make voice calls, SMS messages can still get through. The SMS system is a store-and-forward system, allowing SMS messages to be delayed if necessary. This makes SMS a very robust service.

As an interesting side note, SMS messages originally were never intended for subscriber-to-subscriber usage. They were designed for voice-mail notification.

7.3.3.1 Principal Operation

SMS can be seen as a service that allows the transfer of a set of characters implemented in the Mobile Application Part layer (MAP) on top of SS7. There are two similar standards, the American IS-41 (or ANSI-41) and the international GSM-MAP.

SMS messages are routed by what are known as SMSC (SMSC). An SMSC sends an SMS request to the addressed HLR to find the usually roaming recipient. The HLR has two possible responses. In the case of an inactive response, the recipient is currently offline. The HLR will send an active response when the subscriber becomes available.

The SMSC tries to forward the SMS message to active, that is, online subscribers and receives a reply message stating whether or not the SMS message was successfully delivered. An SMSC keeps trying to send an SMS message for a limited time. In the end it always sends back a report to the original sender of the SMS message, stating success or failure.

Nowadays, SMSCs are connected to many systems and networks like fax, e-mail, voice mail, WWW, IP networks, and so on. It is feasible to send SMSs from — and often to — a wired phone, Web page, PDAs, satellite phone, and so on.

7.3.3.2 Naming

Telecom messaging, by definition, uses telephone numbers to name the recipient, and SMS is no exception. However, since the SMS centers are interconnected to many nontelecom messaging systems, several extensions have been proposed to allow cross-system sending.

7.3.3.3 SMS Cross Messaging

Two ways to address a nontelecom messaging system are currently popular: address prefixing and keywording. With address prefixing, the first part of the SMS message is reserved for the address of the recipient's naming scheme and is sent to a gateway that has a telephone number. For example, an SMS message such as "johndoe@cs.vu.nl Dinner at 8?" may be addressed to telephone recipient 8008, which acts as an e-mail gateway, delivering the message "Dinner at 8?" to the e-mail box of johndoe@cs.vu.nl.

With keywording, the telephone number of a service is used in combination with a keyword-like message. For example, the message "weather adam" would signal to a service provider that it should send an SMS message with the weather forecast for Amsterdam back to the sender.

Often a special short (four-digit) telephone number is used in conjunction with address prefixing or keywording. Given the shortage of four-digit telephone numbers, it is common for gateways and service providers to share a single short telephone number at which they can offer several services. It could be argued that prefixes and keywords actually become part of the naming and addressing scheme used in SMS cross messaging.

7.4 A Comparison

To compare the various messaging systems, we have developed a simple taxonomy. This taxonomy is organized along the four most important aspects from the perspective of a user, as opposed to a technical

Dimension	Values
Time	immediate, impermanent, permanent
Direction	simples, duplex
Audience	group, world
Address	single, list, all

FIGURE 7.10 The four dimensions of a taxonomy for comparing messaging systems.

or design perspective. With this taxonomy, any messaging system can be scaled with respect to four independent dimensions, which are shown in Figure 7.10.

A messaging system can have one of three values in the time dimension: (1) immediate, meaning that all messages are short-lived or available only once during a short period, (2) impermanent, meaning that all messages are available pending their expiration or revocation by some set of rules, and (3) permanent, meaning that all messages are available indefinitely unless a message is explicitly revoked by an authorized user.

With respect to direction, we distinguish two values. The value simplex means that a write-only storage or channel is used for sending a message. The recipient cannot use the same storage or channel to reply. A reply has to be directed towards another storage or channel. The value duplex means that one store or channel is used for both reading and writing.

The audience describes the set of potential recipients of messages. We distinguish the following two values: (1) world, which stands for every user who has the hardware, software, and connectivity to use the system, and (2) group, which stands for a true subset of all potential recipients. In a grouped messaging system, users cannot send messages to a user outside their audience even if this outsider is ready for any message and uses the same system. Restriction of audience (grouping) can be the result of restrictions related to the infrastructure or implementation. The system can also limit the audience as a service, security measure, or due to politics.

In the address dimension, a messaging system can have three values: (1) single — if the system allows only one recipient per message, (2) list — if the system allows for addressing more than one explicitly addressed recipient (i.e., a list of singles), and (3) all — if the system allows for some form of broadcasting. Note that when all members of an audience are addressed, this does not mean that every addressed recipient will necessarily receive each message. As we have seen for Web logging, a system may require an addressee to explicitly fetch a message from storage or a channel.

The four dimensions are truly independent, although not all of the 36 combinations are equally useful. Note that it is easy to confuse audience and address: both are subsets of recipients. The audience comes with the system, and users have no direct influence on it. The address, on the other hand, is something the user determines and the system cannot influence. The intersection of audience and address is the set of recipients that will receive the message.

Using this taxonomy, we can easily compare the messaging systems discussed so far. In Figure 7.11 we show how e-mail, network news, instant messaging, web logging, and short messaging can be classified according to this taxonomy.

The easiest classification is that for e-mail: messages are kept in the system until explicitly destroyed. Also, it should be clear that e-mail employs unidirectional communication, whereas, in principle, there are no limitations concerning to whom a message can be sent. E-mail supports both single and multiple-addressed messages.

In the network news systems, articles are normally removed after some time. Special archives are used to permanently store messages, but these do not form an intrinsic part of news. Communication can be considered bidirectional as messages can be written to and read from the same channel. The targeted audience is always a group, namely the subscribers to a specific newsgroup. An article is always posted to all subscribers, which effectively implies broadcasting.

Dimension	E-mail	News	Web log	IM/1-1	IM/group	SMS
Time	permanent	impermanent	permanent	immediate	immediate	permanent
Direction	simplex	duplex	duplex	simplex	duplex	simplex
Audience	world	group	world	group	group	world
Address	list	all	all	list	all	single

FIGURE 7.11 Classification of current messaging systems.

Web logging messages generally have an impermanent status as they are regularly updated. However, many sites keep old messages available, giving them a permanent status. Initially, Web logging made use of unidirectional storage. However, modern systems provide an interface to allow readers to post comments at the logger's site. This facility essentially turns Web logging into a duplex system. Clearly, there are no restrictions on who can read the logs, meaning that the targeted audience is the entire world. That the system also employs broadcasting is because everyone from the targeted audience can actually access the logs. Note that broadcasting is implemented through polling. As a result, if a reader does not access the logger's site, the addressed recipient will not be able to read the message.

We need to divide instant messaging into two groups. When one considers the one-to-one way of messaging, it is clear that messages have an immediate character: they are never stored. Likewise, communication is unidirectional, with a targeted group that is, or rather should be, restricted by the service provider. As in e-mail, the addressed recipients are always a single user. When one looks at chat sessions, one can see two major differences. First, communication is now essentially duplex: the same channel is used to send and receive messages. Also, messages are always addressed to the entire group of members who participate in the chat session.

Finally, the short messaging systems as supported by telephony-based infrastructures provide permanent messages (that can have an expire date). Messaging is, as in e-mail, unidirectional, just as the targeted audience can be anyone. Finally, messages in these systems are addressed to a single recipient.

7.5 A Note on Unsolicited Messaging

Control concerning the receipt of messages has been briefly touched upon in the discussion so far. Although it is not our goal to go into these matters in great detail, there is at least one issue that needs to be addressed in this chapter: unsolicited messaging. The e-mail system is especially known for its great "potential" to send unsolicited messages to its users. In the following text, let us take a closer look at two forms of such messaging: the spreading of viruses and spamming.

7.5.1 Spreading Viruses

The e-mail system has become the primary habitat of viruses, as concluded by a survey conducted in 2001 by ICSA Laboratories [Bridwell and Tippett, 2001] (see Figure 7.12). This stands to reason because

Medium	1996	1997	1998	1999	2000	2001
Diskette	74%	88%	67%	39%	7%	1%
internet e-mail	9%	26%	32%	56%	87%	83%
Other Internet	12%	24%	14%	16%	2%	20%
Other	15%	7%	5%	9%	2%	1%

A virus can have more than one medium, so totals can exceed 100%

FIGURE 7.12 Changes in virus distribution media over time.

e-mail offers the largest homogeneous audience, but there is more that makes e-mail the number one target.

The primary success factor of most viruses is impersonation. With impersonation, a virus uses the victim's identity and messaging address list to spread itself to the next collection of gullible users taken in by the apparent trustworthy origin of the message. This works especially well with e-mail, where users generally trust the apparent origin.

Furthermore, there are two strategies that work particularly well in combination with e-mail. In the case of self-replication, a program is sent as part of the message and this program handles its own replication after being delivered to the attacked messaging client. The other strategy is coercion, by which a message (often called a hoax) is voluntarily forwarded by the attacked user, simply because it solicits proliferation.

The success of the self replicating e-mail virus is largely due to the predominance of Microsoft's e-mail client "Outlook." The mere popularity of this product makes it attractive to exploit its security vulnerabilities so that by targeting Outlook, a virus maximizes the number of potential victims. There have been two successful variants of self-replicating viruses. Worms exploit bugs in the attacked messaging client. Worms start to replicate as soon as the victim reads the message. Trojans form malicious software (often called malware) that poses as a picture or other harmless data that starts replicating when the victim opens the attachment.

E-mail facilitates multirecipient messages, making it easy, cheap, and effortless to send a single message to many users. Therefore e-mail is an attractive attack medium for hoaxes. Two main indicators that an e-mail is a hoax are the forwarding request and the lack of a date. Basically, a hoax is an e-mail version of the traditional chain letter, rumor, pyramid game, or Ponzi scheme. A hoax traditionally travels without a malware payload and is usually platform independent. A hoax can be tenacious because well-intentioned users keep refurbishing it and the most successful variants thrive on.

7.5.2 Spam

E-mail seems very prone to spams, also called junk-mail or unsolicited commercial e-mail (UCE). Other messaging systems, like the ones discussed, can expect a similar fate. This situation will not change any time soon because most spam is actually very effective. There is a simple reason for this effectiveness. Even though the vast majority of spam receivers simply discard the incoming spam messages, a small fraction actually does react, either willingly or accidentally. Considering the size of the targeted group of recipients, and the small amount of money involved in reaching a group through e-mail, any small response is already enough to warrant success for the sender.

Another related cause of the effectiveness of spam is that its senders actually hope for complaints by recipients. Each time a complaint comes in that identifies the recipient, the spammer will be left with a messaging address that is known to be read (a so-called active account). This information can be sold for a much higher price than addresses of unverified accounts.

At the moment of writing, spam is largely distributed through e-mail, but spam has long been a problem for the USENET News system. It is now also becoming an issue for telecom and instant messaging services.

7.5.3 Protection Mechanisms

In several countries, antispam laws are active or under consideration. However, it is doubtful whether any legislation will help. First of all, it will not help against acquaintance seam, that is, spam following a solicited message. Second, many messaging systems can be fooled easily, making it hard to track down the perpetrator.

Getting rid of unwanted messages like viruses and spam is generally difficult. In principle, unwanted messages can be filtered out at any hop a message makes. For e-mail, there are three logical moments to delete unwanted messages: when it is sent, during message transfer, and when it is to be delivered. Removing (or marking) of unwanted messages is referred to as filtering.

In principle, filtering is simple. A message is checked against a list of signatures, which are known (nearly unique) combinations of bytes, and subsequently deleted or marked, if the signature is found in the message. For example; if a message contains a word referring to a specific commercial product, chances are it is spam so that it should be deleted.

The problem with filters, however, is that there is always a fraction of false positives, that is, messages that are deleted as spam, while, in fact, they are not. Likewise, filters will also lead to a fraction of false negatives, being spam messages that are not recognized as such. Practice shows that keeping both fractions small is difficult. Effective filtering will delete most spam messages (and likewise almost all viruses), while occasionally mistaking a message for spam, but almost never wrongly identify a virus. Note that filters need to be updated regularly for recognizing new viruses.

Filtering on signatures requires careful construction of the signatures to minimize false positives. Most messaging systems allow filtering rules to be handcrafted, but this approach is often not very effective. As an alternative, signature-delete-rules can also be constructed by specialists and subsequently downloaded (often by means of a paid subscription). This approach has traditionally been used for virus filtering, mainly because it is the only way a virus' signature can be spread faster than the virus itself. More sophisticated rules would still be needed to catch what is known as a polymorphic virus, that is, a virus that changes itself to avoid detection. The blacklist filter, which is also usually downloaded, is a signature-delete rule that filters on the sender's address. This type of rule has been successfully used against spam in the past. However, spam has evolved to evade this type of filtering, rendering it increasingly less useful.

A more sophisticated approach is to generate rules based on statistical comparison of the content of previous unwanted messages. Generated rules, because of their adaptive property, have been proven to be very successful against various new types of spam that could not be detected using blacklists or other static signatures.

7.6 Toward Unified Messaging

Given that so many different messaging systems exist, it is not surprising that several attempts are being made toward their integration into a unifying system. Unification is generally interpreted as integration of the existing systems such that users can send and retrieve any type of message using only a single interface. Proposals range from relatively simple integration, such as that described by Yeo et al. [2000], to advanced architectures that actually integrate many Internet and telecom services [Wang et al., 2000].

What many unified messaging systems do is actually concentrate on the "integration of technology rather than the integration of messaging models." The result is often that a single messaging technology is used as the nexus for all other systems. E-mail often plays such a role. However, rather than placing technology as the key integrator, it can be argued that integration of messaging models is the key issue. This approach has essentially been adopted in, for example, the Mobile People Architecture [Maniatis et al., 1999].

Continuing to follow the user's perspective, unified messaging is more about models than about technology. A relatively simple model that can cover all four dimensions from our taxonomy is the Unified Messaging (UM) model. In this model, each message is said to be targeted to a specific user or a group of users. For simplicity, we assume that each message is immutable (i.e., it cannot be changed after being sent), and that it is usually short. These properties lead us to use the term Targeted Immutable Short Message or TISM for short. We use the name target to denote the destination of a TISM.

Any UM model should address the following requirements:

1. Large-scale messaging. Any model should be able to handle hundreds of billions of messages a day between billions of users.

2. Independence of trusted sites. A UM model can be implemented using a client-server system with a trusted server, but also a peer-to-peer communication model [Milojicic et al., 2002], or even a combination of both.
3. Orthogonality of the four dimensions. Every UM model should allow any combination of time, direction, audience, and address.
4. Prevention of spam. Each model must offer maximum control to prevent unsolicited messages without restricting the freedom of speech.

The prevention of unsolicited messages should be adequately dealt with. The following UM model, described by Wams and Van Steen [2003], does exactly that. A target protects each TISM with public key encryption and a Message Authentication Code (MAC) [Schneier, 1996]. In this model, each target is associated with a unique post-key/read-key pair. To post a TISM, the proper post-key is needed. Likewise, to read a TISM, the proper read-key is needed. Without a read-key, it is sufficiently hard to reconstruct a TISM, even if a post-key and a copy of the encrypted TISM are available. Without a post-key, it is very hard to spoof a TISM even if the read-key is available. The UMS implementing this UM model will generate a post-key/read-key pair for every new target.

A target is identified by a system-wide unique binary string. We define a target-ID as this unique binary string. A (post-key, read-key, target-ID) tuple is denoted as post-read-tuple. Likewise, we use read-tuple and post-tuple. When the UMS creates a target for a user, a post-read-tuple is returned to the user, from which a separate post-tuple and read-tuple can be created. Typically, a user might create a target and distribute its read-tuple to others, enabling them to get the encoded TISMs from the target (using the target-ID) and to decode those TISMs (with the read-key). This setup resembles a Web-log messaging system. Had the user distributed the past-tuple, an e-mail life system would have resulted.

The UMS user has a number of ways to distribute (key, target-ID) tuples. A user could pass on a tuple wrapped in a TISM. Alternatively, he or she could distribute a tuple through the World Wide Web or some other generic distribution system, or could store a tuple in a (local) name space system specially designed for the UMS. Other lookup models are also feasible.

To utilize the fine-grained control the UMS offers, the user needs a separate target for each different communication partner or group. This may sound complex, especially to users that manage all their Internet e-mail from one mailbox. However, most e-mail users already have many sub-mailboxes. Likewise, most instant-messaging systems allow users to create any channel/room they want to. As another example, every netnews user can create a new alt.* group at will (like the actually existing alt.swedish.chef.bork.bork. - bork). Creating a new box or channel in one of these legacy messaging systems is limited by the ability to create a new entry in the accompanying name space. For example, finding a meaningful name for a new alt.* newsgroup that does not already exist is hard, as is the case for instant-messaging channels/rooms. Moreover, the e-mail sub-mailboxes are not usually publicly addressable.

7.7 Outlook

Considering the popularity of e-mail and other messaging systems, and the convergence of Internet and telecommunications, it is beyond doubt that we can expect to see exponential growth in network-based messaging in the near future. Unified messaging, in any form, will play a crucial role. Users will demand a simple messaging model that is independent of the devices they use for composing, sending, and receiving messages.

Another important observation is that messaging will further integrate with the many Web-based information systems. Message-based ordering is already common practice, but seamless integration of messaging, electronic commerce, and information services can be expected. Again, unified messaging will prove to be crucial.

An interesting development in this context is moving user agents toward Web servers, effectively offering end users no more than just a simple interface to messaging operations that are carried out at a remote server. This approach is already being taken in Web-based e-mail, which allows users to access their mailbox from any place provided they have access to the Web. Using Web clients as the universal means to access messaging systems implies that unification and integration must take place at the server side. It is as yet unclear whether this approach will succeed if we simply integrate technologies instead of unifying messaging models.

References

Albitz, P. and C. Liu. *DNS and BIND,* 3rd edition. O'Reilly & Associates, Sebastopol, CA, 1998.

Barber, S. Common NNTP Extensions. RFC 2980, October 2000.

Belville, S. Zephyr on Athena. MIT, Cambridge, MA, February 1990.

Blood, R. *The Weblog Handbook.* Perseus Publishing, Cambridge, MA, 2002.

Bridwell, L. and P. Tippett. ICSA Labs 7th Annual Computer Virus Prevalence Survey, 2001.

Carzaniga, A., D. S. Rosenblum, and A. L. Wolf. Design and Evaluation of a Wide-Area Event Notification Service. ACM Transactions on Computer Systems, 19(3): 332–383, August 2001.

Comer, D. *The Internet Book,* 3rd edition. Prentice Hall, Upper Saddle River, NJ, 2000.

Crispin, M. Internet Message Access Protocol — Version 4rev1. RFC 2060, December 1996.

Day, M., S. Aggarwal, G. Mohr, and J. Vincent. Instant Messaging/Presence Protocol Requirements. RFC 2779, February 2000a.

Day, M., J. Rosenberg, and H. Sugano. A Model for Presence and Instant Messaging. RFC 2778, February 2000b.

Dreher, R. and L. Harte. *Signaling System 7 Basics,* 2nd edition, APDG Publishing, Fuquay-Varina, NC, 2002.

Freed, N. and K. Moore. MIME Parameter Value and Encoded Word Extensions: Character Sets, Languages, and Continuations. RFC 2231, November 1997.

Gellens, R., C. Newman, and L. Lundblade. POP3 Extension Mechanism. RFC 2449, November 1998.

Horton, M. and R. Adams. Standard for Interchange of USENET Messages. RFC 1036, December 1987.

Kalt, C. Internet Relay Chat. RFCs 2810–2813, April 2000.

Kantor, B. and P. Lapsley. Network News Transfer Protocol: A Proposed Standard for the Stream-Based Transmission of News. RFC 977, February 1986.

Klensin, J., Simple Mail Transfer Protocol. RFC 2821, April 2001.

Maniatis, P., M. Roussopoulos, E. Swierk, K. Lai, G. Appenzeller, X. Zhao, and M. Baker. The Mobile People Architecture. *ACM Mobile Comput. Commun. Rev.,* 3(3): 36–42, July 1999.

Milojicic, D. S., V. Kalogeraki, R. Lukose. K. Nagaraja, J. Pruyne, B. Richard, S. Rollins, and Z. Xu. Peer-to-Peer Computing. Technical Report HPL-2002-57, Hewlett Packard Laboratories, Palo Alto, CA, March 2002.

Mockapetris, P. Domain Names — Concepts and Facilities. RFC 1034, November 1987.

Oikarinen, J. and D. Reed. Internet Relay Chat Protocol. RFC 1459, May 1993.

Schneier, B. *Applied Cryptography,* 2nd edition, John Wiley & Sons, New York, 1996.

Wams, J.M.S. and M. van Steen. Pervasive Messaging. In *Proceedings of the First International Conference on Pervasive Computing and Communications (PerCom),* Los Alamitos, CA, March 2003. IEEE, IEEE Computer Society Press.

Wang, H. J., B. Raman, C. Chuah, R. Biswas, R. Gummadi, B. Hohlt, X. Hong, E. Kiciman, Z. Mao, J. S. Shih, L. Subramanian, B. Y. Zhao, A. D. Joseph, , and R. H. Katz. ICEBERG: An Internet-core Network Architecture for Integrated Communications. *IEEE Pers. Commun.,* 7(4): 10–19, August 2000.

Wieringa, R. and W. de Jonge. Object identifiers, keys, and surrogates — Object identifiers revisited. *Theory and Practice of Object Systems,* 1(2): 101–114, 1995.

Woolley, D. PLATO: The Emergence of Online Community. *Matrix News,* January, 1994.

Yeo, C. K., S. C. Hui, I. Y. Soon, and G. Manik. Unified Messaging: A System for the Internet. *Int. J. Comp. Internet Manage.,* 10(3), September 2000.

8

Internet-Based Solutions for Manufacturing Enterprise Systems Interoperability — A Standards Perspective

CONTENTS

8.1 Introduction 8-1
8.2 A General Overview of Approaches for Interoperable Manufacturing Enterprise Systems 8-2
8.2.1 General Concepts 8-2
8.2.2 Selected Approaches 8-3
8.3 Interoperable Information Systems within the Manufacturing Enterprise 8-6
8.3.1 An Intra-enterprise Manufacturing Interoperability Description Framework 8-6
8.3.2 Manufacturing Enterprise Information Systems 8-7
8.3.3 OAG Semantic Integration Standards in the Manufacturing Sector 8-8
8.4 Interoperable Information Systems Outside of Manufacturing Enterprises 8-8
8.4.1 An Inter-enterprise Manufacturing Interoperability Classification Framework 8-10
8.4.2 Example Inter-enterprise Scenarios of Integration 8-11
8.5 Advanced Developments in Support of Interoperable Manufacturing Enterprise Systems 8-12
8.5.1 OAG 8-15
8.5.2 ebXML 8-15
8.5.3 RosettaNet 8-16
8.5.4 National Institute of Standards and Technology 8-16
8.5.5 Semantic Web Activity 8-17
8.5.6 Disclaim 8-17
References 8-17

Nenad Ivezic

Boonserm Kulvatunyou

Albert Jones

8.1 Introduction

This chapter reviews efforts of selected standards consortia to develop Internet-based approaches for interoperable manufacturing enterprise information systems. The focus of the chapter is on the efforts

1-58488-381-2/05/$0.00+$1.50

© 2005 by CRC Press LLC

to capture common meaning of data exchanged among interoperable information systems inside and outside a manufacturing enterprise.

We start this chapter by giving a general overview of the key concepts in standards approaches to enable interoperable manufacturing enterprise systems. These approaches are compared on the basis of several characteristics found in standards frameworks such as horizontal or vertical focus of the standard, the standard message content definitions, the standard process definitions, and dependence on specific standard messaging solutions.

After this initial overview, we establish one basis for reasoning about interoperable information systems by recognizing key manufacturing enterprise objects managed and exchanged both inside and outside the enterprise. Such conceptual objects are coarse in granularity and are meant to drive semantic definitions of data interchanges by providing a shared context for data dictionaries detailing the semantics of these objects and interactions or processes involved in data exchange.

In the case of intra-enterprise interoperability, we recognize enterprise information processing activities, responsibilities, and those high-level conceptual objects exchanged in interactions among systems to fulfill the assigned responsibilities. Here, we show a mapping of one content standard onto the identified conceptual objects.

In the case of inter-enterprise interoperability, we recognize key business processes areas and enumerate high-level conceptual objects that need to be exchanged among supply-chain or trading partners. Here, we also show example mappings of representative content standards onto the identified conceptual objects.

We complete this chapter by providing an account of some advanced work to enhance interoperability of manufacturing enterprise information systems in the context of the enterprise standards development.

8.2 A General Overview of Approaches for Interoperable Manufacturing Enterprise Systems

Here, we provide a general overview of the key concepts and selected standards approaches to enable interoperable manufacturing enterprise systems. We compare these approaches with respect to several key characteristics found in interoperable solutions.

8.2.1 General Concepts

To understand the focus of this chapter, its place within the general interoperability architecture, and to characterize selected approaches, we identify three key characteristics of the approaches: interoperability focus, industry focus, and integration objective. Other important aspects of interoperability solutions related to security, network protocols, trading partner agreements, and registry and repository solutions are issues that transcend all enterprises and are not considered in this discussion.

8.2.1.1 Interoperability Focus

This is the first key characteristic of an interoperable standards solution and identifies the scope of that solution within an interoperability stack. Figure 8.1 shows one such abstract interoperability stack, based in part on previous studies [Business Internet Consortium, 2002], with four layers typically taken into

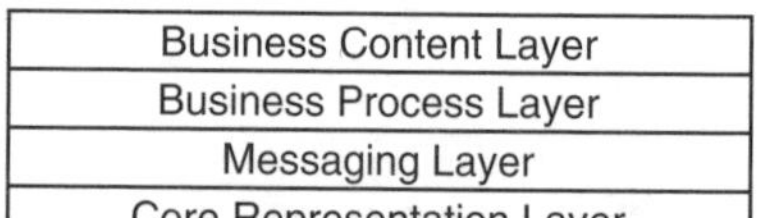

Business Content Layer
Business Process Layer
Messaging Layer
Core Representation Layer

FIGURE 8.1 An abstract interoperability stack defining scope of Internet-based interoperable solutions.

account when developing current interoperable solutions. A standards approach may develop interoperability specifications in one or more of these layers:

- *Core Representation Layer* defines the syntax of messages, usually as sequence of data fields. The syntax supports specifications in the layers above for defining messages, process, and content. For the Internet-based approaches in this chapter, we assume the W3C standards, such as XML DTD, XSLT, and XML Schema, that define message structure, document types, and data access within the documents [W3C, 2003].
- *Messaging Layer* includes standardized message and envelope structure definitions. Within this layer, session recording and communication setup for message transport are addressed so that coordination between interacting parties is assured. The issues of reliable and secure messaging are dealt with here. This layer is the foundation of communications among the other layers as it provides support for the message exchange and content packaging.
- *Business Process Layer* defines the way business processes are encoded so that the semantics of these processes may be shared and executed in a repeatable manner. Within this layer, business processes are defined that may be either broadly applicable or specific to an industry. The processes comprise simple interactions such as request/response or complex interactions such as collaborative product development or supply-chain planning.
- *Business Content Layer* includes business definitions, data dictionary entries, business documents, and attachments that may constitute the meaning of a business message. Within this layer, one may specify composition of a valid business content — data structures, data types, constraints, and code lists. Also, this layer includes definitions of business terminology and accepted values that may be used in messages in support of many industries. The content covers many application domains such as product development, logistics, finance, and quality.

8.2.1.2 Industry Focus

This is the second key characteristic of an interoperability approach and may be either horizontal or vertical. Many of the interoperability approaches originate and are fixed on a specific industry sector; we call them vertical industry standards. Others, to a lesser or greater extent, are focused on tying enterprises across industry sectors; we call them horizontal standards. For example, virtually all organization types deal with sales, procurement, and human resources in a generic sense. In addition, manufacturing companies need to exchange data within their respective cross-industry supply chains. This, becomes a significant issue when content standards developed by different organizations within one sector need to be translated and "understood" by information processing systems in organizations from another sector.

8.2.1.3 Integration Objective

This is the third key characteristic of an interoperability approach and may be either architecture/application integration, supply-chain integration, or trading network integration. In the architecture/application integration case, the interoperability approach enables interoperability of applications and information processing systems that coexist within some enterprise architecture. In the supply-chain integration case, the interoperability approach supports interactions that take place in an industry specific or cross-industry supply chains. In the trading network integration case, the interoperability approach addresses the needs of advertising the manufacturing or trading capabilities, identifying partners, establishing partnerships, and negotiating terms of trade among involved parties that may take on a wide range of roles such as customers, suppliers, logistics, retailer, broker, and warehouse.

8.2.2 Selected Approaches

Most of the developments to enhance enterprise interoperability are taking place within voluntary consortia that develop standards for business processes, business content, enabling technologies, and the overall business architectures. We focus on three prominent standardization efforts, summarized in Table

TABLE 8.1 A Comparison of Three Standard Approaches: OAGIS, RosettaNet, and ebXML

	OAG	RosettaNet	ebXML
Date formed	1995	1998	1999
Founders	Enterprise software vendors	Information Technology, Electronic Components, and Semiconductor Manufacturing companies	UN/CEFACT and OASIS
Objectives	To build specifications that define the business object interoperability between enterprise business applications	To create, implement, and promote open e-business process standards in support of supply-chain integration	To provide an open XML-based infrastructure enabling the global use of electronic business information in an interoperable, secure manner by all parties
Interoperability focus	Business Content Layer	Messaging Layer, Business Process Layer, Content Layer	Messaging Layer, Business Process Layer, Content Layer
Deliverables	Business Object Document (BOD) specifications; Integration scenarios (non-normative)	RosettaNet Implementation Framework (RNIF), Business Dictionary, Technical Dictionary, Partner Interface Processes (PIPs)	Business Process (BPSS), Core Components (CC), Messaging (ebMS), Collaborative Protocol Profile and Agreement (CPP/A), Registry, and others
Industry focus	Horizontal (automotive, aerospace, logistics, telecommunications)	Vertical to horizontal: High-technologies to also include retail industries	Horizontal
Integration driver	Architecture/Application Interoperablity	Supply-chain integration	Trading Partner Network Integration
Interactions	RNIF, ebXML (current)	ebXML (future)	OAG, xCBL, SWIFT (future)
Technology	W3C XML Schema	W3C XML DTD	UML and W3C XML

8.1, that influence manufacturing enterprise interoperability: Open Applications Group, RosettaNet, and ebXML.

8.2.2.1 Open Applications Group

The Open Applications Group (OAG) is building specifications that define the business object interoperability between enterprise business applications [OAG, 2003]. The OAG Integration Specification (OAGIS) is the common content model needed to represent information objects that enable communication between business applications [Rowell, 2002]. Such a content model provides a common basis of understanding among developers who specify intent of the messages to be processed by enterprise information systems.

OAGIS includes a large set of Business Object Documents (BODs) and integration scenarios that can be used in different business environments, such as application-to-application (A2A) and business-to-business (B2B). BODs are message content definitions that can be used broadly across many different industries (for example, telecommunications and automotive) and aspects of Supply Chain Automation (for example, Ordering, Catalog Exchange, Quotes).

OAGIS implies an architecture/application integration approach enabling interoperability of applications and systems that need to coexist within some inter- or intra-enterprise architecture.

OAGIS does not specify an implementation architecture and can be utilized over different messaging and transport solutions such as RosettaNet Implementation Framework (RNIF) and ebXML Messaging [RNIF, 2003; EbMS, 2003].

OAGIS have been adopted and used in aerospace, automotive, and telecommunications manufacturing industries. As shown later in this chapter, the OAGIS content standards support interaction among information systems typically found in a manufacturing enterprise Product Data Management system, Enterprise Resource Planning system, and Factory Planning System. The BOD structures can be extended to accommodate alternative integration scenarios. In situations where existing BOD structures are not available for customization, new BODs can be developed. An example is the Standards for Technology in Automotive Retail (STAR) consortium that defines standard XML message for dealer-to-OEM business transactions (i.e., Parts Order, Sales Lead, Credit Application) within the STAR/XML project [STAR, 2003].

OAGIS specifications are currently represented using W3C XML Schema and make use of advanced features such as XSLT, overlays, constraints specification, and validation using Schematron [Schematron, 2003].

8.2.2.2 RosettaNet

RosettaNet is a consortium of Electronic Components (EC), Information Technology (IT), and Semiconductor Manufacturing (SM) companies working to create, implement, and promote open e-business process standards [RosettaNet, 2003]. The RosettaNet standards propose solutions for the three layers of the interoperability stack shown in Figure 8.1: messaging, business processes, and business content. In this manner, the standard is self-sufficient and complete; it can be implemented independent of other standards with the exception of core representation standards.

The RosettaNet specifications include RNIF that provides for data exchange protocols; Business Dictionary that defines the properties used in basic business activities; Technical Dictionary that provides common language for defining products and services; and Partner Interface Processes (PIPs) that are system-to-system, XML-based dialogs that define business processes between trading partners.

The Rosettanet specifications can be applied to a variety of supply-chain integration scenarios and trading-partner data exchanges. The industry focus is mostly vertical (i.e., high-technology) but with plans to be extended horizontally: RosettaNet recently joined the Uniform Code Council (UCC) and indications are that additional retail-oriented industries will be included within the scope of the standard [UCC, 2003]. The RosettaNet specification, although complete and independent of other standards, is planning for future interfacing with other messaging and transport solutions such as ebXML Messaging. Also, some advanced business process standards such as ebXML BPSS are planned for use in development of new PIPs [ebBPSS, 2003].

The specification is a supply-chain-driven interoperability approach. The standard has been adopted and implemented in a variety of software products in support of high-technology supply chains throughout IT, EC, and SM sectors. The primary deliverables, PIPs, provide building blocks for inter-enterprise manufacturing integration. The PIPs are categorized in a number of cluster groups, such as Product Information, Order Management, Inventory Management, Marketing Information Management, Service and Support, and Manufacturing. As shown later in this chapter, these PIPs contain guidelines that prescribe the content of the messages exchanged using the prescribed PIP choreography. Different from BOD structures that are intentionally left to be extensible by the implementers, the PIP structures can only be changed through a RosettaNet-sanctioned, and formally managed, change-submission process. When existing PIP structures are not available for customization, new projects are started and PIPs are developed [iHUB, 2002]. RosettaNet specifications are represented using W3C XML DTD specifications.

8.2.2.3 ebXML

ebXML (Electronic Business using eXtensible Markup Language) is an effort cosponsored by UN/CEFACT and OASIS to develop a modular suite of specifications that enables enterprises of any size and in any geographical location to conduct business over the Internet [ebXML.org]. ebXML is developing a series of standards specifications to exchange business messages, establish trading relationships, communicate data in common terms, and define and register business processes.

ebXML produces a wide range of specifications including Business Process Specification Schema (BPSS), Core Components, Collaboration Protocol Profile and Agreement (i.e., a mechanism for declaring a trading partner's capabilities and agreement), standardized messaging service, and others [ebBPSS, 2003; ebCC, 2003; ebCPPA, 2003]. In addition, ebXML is working on a standardized UML-based modeling methodology for modeling business processes and translating those models into XML documents.

Different ebXML specifications are at different levels of maturity. For example, the messaging specification is well advanced and has been adopted by software vendors. In contrast, due to its sheer complexity, the core component specifications that provide for a methodology leading to a unified, well-defined semantics of message content are only now being validated.

The ebXML specifications could be applied to a variety of supply-chain integration scenarios and trading-partner data exchanges. The industry focus is definitely horizontal, cutting through virtually all industry sectors. The ebXML effort is a trading-partner-driven interoperability approach supporting the needs of publishing, discovering, and establishing trading agreements for a general trading-partner context irrespective of the industry.

The ebXML specification is complete and independent of other standards, in principle. However, with respect to its content-standard development process, the adopted ebXML development process is to recognize a number of existing content standards (such as OAGIS) and, over time, drive its own content standards process based on these existing standards.

The mature parts of the standard (e.g., messaging and business processes) have been adopted by a variety of software product vendors. Pilot efforts, reference implementations, and initial adoptions of ebXML exist in automotive industry with more such efforts advertised for the future [STAR, 2003]. ebXML specifications are represented using UML modeling and XML representations.

8.3 Interoperable Information Systems within the Manufacturing Enterprise

We now focus on the semantic issues for interoperable information systems *within* the manufacturing enterprise. We adopt one proposed *interoperability framework* for describing processing activities, responsibilities, and high-level interface objects in the manufacturing enterprise information systems. Then, we give an objective basis for a possible mapping for the OAGIS manufacturing content standard onto the adopted high-level interface objects. Also, we illustrate how these mapped content standards would be supportive of one interoperable manufacturing enterprise architecture.

8.3.1 An Intra-enterprise Manufacturing Interoperability Description Framework

An interoperability framework was proposed to describe features of interoperable information systems within a manufacturing enterprise [OMG, 2003]. This framework outlines high-level processing activities, information-processing responsibilities, and high-level interface objects.

Manufacturing information processing activities include Product Development, Process Design, Process Prototyping, Requirements Planning, Production Planning, Resource Scheduling, Preparation and Setup, Process Operations, Work-in-Process Reporting, and Cost and Usage Reporting.

Information processing responsibilities that are supported by these manufacturing activities include Capture/maintain product specification, Capture/maintain item descriptions, and Capture/maintain Manufacturing Bill of Materials, to name only a few.

Based on the established activities and responsibilities, we can identify conceptual interface objects. An example interface object may be identified from the Manufacturing Process Definition activity. Within this activity, one responsibility is Capture and Maintain Manufacturing Bill of Materials. On the other hand, another information processing activity is Resource Scheduling with a responsibility to provide Effective Manufacturing Bill of Materials. With the potential to assign the two responsibilities to two different systems (e.g., ERP and PDM systems), there is a need to exchange Bill of Materials. These conceptual objects form the context for identification of data dictionaries detailing the semantics of these objects and processes involved in data exchange.

In the previous example, the Bill of Materials (BOM) is a conceptual object necessary for manufacturing systems integration. Such an object provides a context to identify elements of data dictionary such

TABLE 8.2 High-Level Interface Objects Supporting Intra-enterprise Manufacturing Systems Interoperability

Bill of Material	List of the material items needed to create a particular configuration of a final product in a particular manufacturing facility in a certain time frame
Cost and Usage Report	Reports from the manufacturing facility to the enterprise management systems on actual costs of operating the production facility and materials and resources used, including relationships to specific orders, products and yields, and work-in-process inventory
Inventory	Body of business information that tracks the available supply of parts, tools and materials, and possibly the warehousing of finished goods
Item	Individual instances of the materials — the pieces or packaged units — that include final products, component parts/products, raw/stock materials, "work-in-process"
Item Description	Specifications for a kind of item that is used in, or results from, the manufacturing processes. The information describes the properties that are common to all instances, such as its name, part number, its mechanical, electrical, or chemical properties, etc.
Labor	Information about employee and contractor time and effort expended on fulfillment of a particular order, when the enterprise uses the information to define cost of manufacture
Labor Resources	Information about individual employees and contract personnel directly related to their use as manufacturing resources
Lot	A unit of product that is in work in the manufacturing facility. As such it is a collection of product items, possibly accompanied by other materials, in some state of manufacture
Manufacturing Order	Vehicle by which the manufacturing facility is directed to produce a quantity of product items
Master Schedule	List of all manufacturing orders to be fulfilled in a given factory over a certain time period, with target volumes and completion dates
Plant and Equipment Resources	Information about plant facilities and individual machines and other equipment that is specifically allocated for use, or specifically accounted for, in fulfilling manufacturing orders
Process Specifications	List of operations and steps required in order to manufacture a finished good
Product Specification	Engineering descriptions of a finished good that results from the manufacturing processes
Purchase Request	Request from the manufacturing systems to the purchasing system to purchase (additional) materials, based on a demand discovered in the factory
Tooling	Items needed for the manufacturing process but not part of the finished goods

as Batch Size Quantity (i.e., the number of items that can be produced in each run of the BOM) and Effective Period (i.e., the time period during which the BOM is effective). The related interactions that may take place among manufacturing systems for this conceptual object may include requests to get and synchronize a BOM and interactions requesting and showing BOM detail.

Using a similar analysis, a collection of high-level interface objects may be identified to assure interoperability among systems implementing different manufacturing information processing. Table 8.2 identifies these objects and gives a summary semantics definition for the objects based on [OMG, 2003].

8.3.2 Manufacturing Enterprise Information Systems

The information systems that are found in a manufacturing enterprise include Product Data Management systems, Enterprise Resource Planning systems, Factory Scheduler/Dispatchers, Factory Planning Systems, Process Control Systems, Human Resources Management Systems, and Manufacturing Execution Systems.

An example architecture supporting manufacturing enterprise information processing activities is shown in Figure 8.2, based on OMG [2003]. The architecture shows information systems communicating using the interface objects identified in Table 8.2 that are the subjects of the information flows identified on the wires linking the software component boxes. (The unlabeled dashed edges indicate an additional control interface that exists between the Manufacturing Execution and each individual tool Control system.)

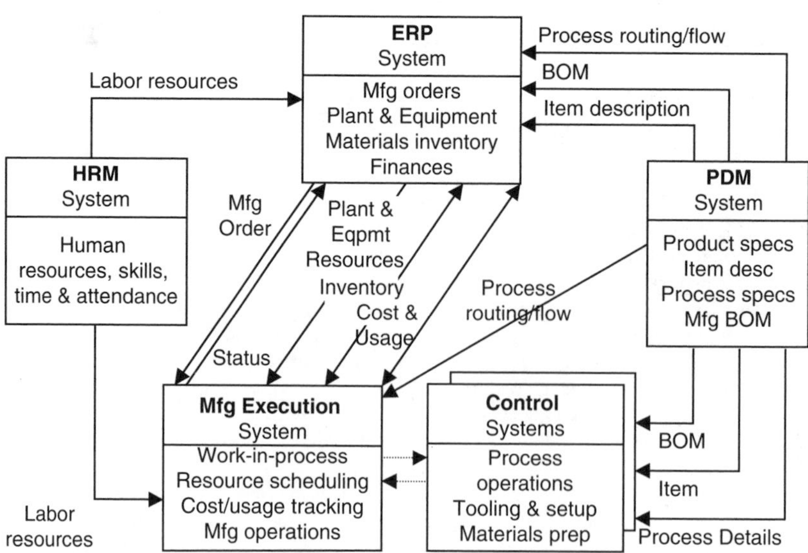

FIGURE 8.2 An example architecture of a manufacturing enterprise information system.

8.3.3 OAG Semantic Integration Standards in the Manufacturing Sector

Table 8.3 gives one possible mapping of the OAGIS content standards onto the identified conceptual objects. The OAGIS BODs consist of verb and noun parts (e.g., ProcessPurchaseOrder — verb Process and noun PurchaseOrder). For the purposes of this mapping, we have identified the nouns that constitute the BODs. As can be seen from the table, the mapping is not always one-to-one because OAGIS may have multiple nouns that express semantics of the corresponding conceptual object; for example, Cost and Usage Report is mapped onto Consumption, Costing Activity, and variants of WIP nouns. In addition, a number of OAGIS nouns are included that do not have a mapping onto the proposed high-level objects.

To supplement the semantic definitions of its BODs, OAGIS includes (nonnormatively) possible communications between various software modules as starting points for the OAGIS users to find and adopt to their own needs. (Note that the actual scenarios of integration that need to include the actual flow control can be derived from these proposed communications.) Figure 8.3 shows possible communications using OAG BODs that are similar to the example manufacturing enterprise system architecture shown in Figure 8.2.

For example, Manufacturing Order in Figure 8.2 corresponds to Production Order (as indicated in Table 8.3) in the suggested CreateProductionOrder BOD used in Figure 8.3. Similarly, BOM and Process Routing/Flow in Figure 8.2 correspond to BOM and Routing exchanged between BOM/Configuration/PDM component and Production/WIP component in Figure 8.3.

8.4 Interoperable Information Systems Outside of Manufacturing Enterprises

In this section, we look at the approaches that support interoperable systems across multiple manufacturing enterprises. Here, without the luxury of an existing high-level manufacturing inter-enterprise interoperability framework, we adapt a proposed classification of enterprise business processes as a starting point for constructing such a framework. We show how two alternative approaches to enable interoperable inter-enterprise manufacturing systems, OAGIS and RosettaNet, may be mapped onto the classification structure. We illustrate two inter-enterprise integration scenarios by using the OAGIS and RosettaNet concept objects.

TABLE 8.3 A Possible Mapping of the OAG Content Standard Nouns onto the Identified Conceptual Objects

High-Level Concept Object	OAG Noun	OAG Meaning
Bill of Material	Bill of Material	List of items to be produced in a specified time period. The Bill of Material structure is broken down into three ways to represent the Item. An Item may be included by itself or may be represented as part of a set of options or as an option within a class of options
Cost and Usage Report	Consumption	Process whereby a certain amount or quantity of inventory, resources, or product is utilized that likely lead to the need for some form of replenishment
	Costing Activity	Details of the activities in the Manufacturing Application that caused the entries in the Journal
	WIP Confirm	Work-in-Progress confirmation represents confirmation of the movement of WIP materials. The noun refers to general information about the entire WIP transaction, as well as line item detail about the specific WIP operation or routing step. This may apply to the movement of raw materials or finished products
	WIP Merge	WIP Merge is used to notify a Manufacturing Application of the creation of a single production lot from multiple production lots of a product being made on a production order
	WIP Move	WIP Move is used to communicate which processing step the product is coming from and which step it is being moved to, along with the quantity moving and the time this event occurred
	WIP Recover	WIP Recover is used to notify a Manufacturing Application of the creation of usable production materials from material previously considered unsuitable for production use. This is most often likely to represent a return to production of scrap material
	WIP Split	WIP Split is used to notify a Manufacturing Application of the creation of multiple production lots from a single production lot of a product being made on a production order
	WIP Status	WIP Status is used to notify a Manufacturing Application of the progress of a production order at a point in time
Inventory	Inventory Balance	Stocked items and the quantities of each item by location. Other item-by-location information, such as serial numbers or lot numbers, can also be included
	Inventory Count	Results of a physical inventory or cycle count of the actual on-hand quantities of each item in each location. Compare to the noun InventoryBalance, which represents system-maintained on-hand quantities
	Inventory Issue	Request to process an issue or request information about an issue
	Inventory Movement	Identify items being moved, source, and destination of movement
	Inventory Receipt	Intended for use in Unplanned Receipt Scenarios
Item	Item Cross Reference	Item Cross Reference describes both alternate and related items. Alternate items could specify items that have alternative universal identifiers such as EAN, UPC, or party-specific identifiers such as supplier part number or customer part number. Related items could be spares, accessories, or substitutes
	Item Master	Represents any unique purchased part or manufactured product. Item, as used here, refers to the basic information about an item, including its attributes, cost, and locations. It does not include item quantities. Item is used as the Item Master
Item Description	Item Master Value Class	Grouping to determine the General Ledger accounting effect. These are user-defined values, with the exception of the values TOTAL, MATERIAL, LABOR, BURDEN, OVERHEAD, SUBCONTRACT
Labor	Employee Time	Time sheet information for an employee
	Employee Work Schedule	Planned work hours for an employee

TABLE 8.3 A Possible Mapping of the OAG Content Standard Nouns onto the Identified Conceptual Objects (continued)

High-Level Concept Object	OAG Noun	OAG Meaning
Labor Resources	Personnel	Human resource information maintained for each employee. It includes such data as job code, employee status, department or place in the organization, and job-related skills. Although generally maintained in a Human Resource Management System (HRMS), this information may also be needed and updated by manufacturing applications (workforce scheduling) or project management
Lot	Lot	Manufacturing lot
Manufacturing Order	Production Order	Document requesting the manufacture of a specified product and quantity
Master Production Schedule	Sequence Schedule	A Sequence Schedule is used to indicate sequential scheduling of ordered items in the manufacturing process. Commonly, the sequence schedule is generated by a work-in-process application and transmitted to an order or material planning application
Plant and Equipment Resources	Resource	An abstract type describing the allocation of persons, equipment, or materials likely in a manufacturing environment
	Resource Allocation	Identifies the resources that are need for a production order and indicates where they are to be assigned
Process Specifications	Routing	Description of the resources, steps, and activities associated with a path or routing connected with a manufacturing process. Typically, a routing contains people, machines, tooling, operations, and steps
Product Specification	Project	A set of tasks with the following attributes: a singular purpose, a start and end date, those that accumulate costs, and those that may have materials and overhead. SYNONYMS: Job, Process Model, WBS
Purchase Request	Requisition	Request for the purchase of goods or services. Typically, a requisition leads to the creation of a purchase order to a specific supplier
Tooling	Tool	A tool needed for a given task
	Dispatch List	A prioritized detail status of orders and operations scheduled or in-process at a specific work center
	Engineering Change Document	A request for a change to a manufactured item. This document allows the change to progress through the different states from being a request and going through the review process to becoming an approved Engineering Change Order
	Engineering Work Document	Carries product structure information and information on what is to be changed in it as the result of a project design activity
	Maintenance Order	Order for a machine, building, tooling, or fixed asset to be repaired or for preventive maintenance to be performed
	Pick List	List of materials to be retrieved from various locations in a warehouse in order to fill a production order, sales order, or shipping order
	Product Requirement	Request to reserve or allocate a specified quantity of a specified item Typically, this requirement would be received by an inventory or production system
	Planning Schedule	Indicates a demand forecast sent from a customer to a supplier, or a supply schedule sent from a supplier to a customer

8.4.1 An Inter-Enterprise Manufacturing Interoperability Classification Framework

A classification of enterprise business processes has been proposed [ebXML, 2003]. This classification is based on general inter-enterprise activities and includes business areas such as procurement/sales, design, recruitment and training, logistics, and others. The classification is not complete and was meant to be a basis for constructing an evolving framework of a common business processes catalog. Nevertheless, we take this classification as a basis for identifying some high-level interface objects commonly identified in an inter-enterprise activity.

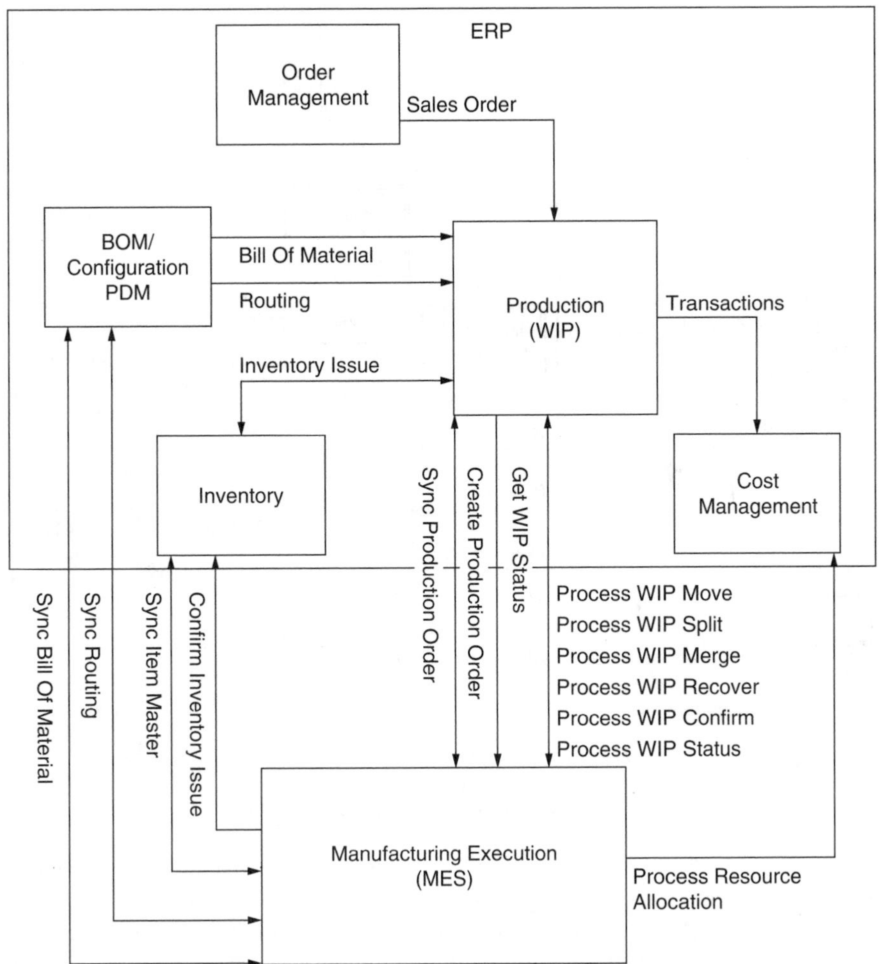

FIGURE 8.3 An example intra-enterprise integration scenario using OAG BODs.

We focus only on a subset of the inter-enterprise information processing activities that fall into the following business areas: procurement and sales, logistics, manufacturing, and financial services. We map the OAG and RosettaNet higher-level concepts onto the classification categories and arrive at the following table with semantics for the higher-level interface objects.

8.4.2 Example Inter-Enterprise Scenarios of Integration

Figure 8.4 gives a family of supply-chain integration scenarios by identifying communication links (and associated BODs or simpler communication patterns) among software modules. As stated before, OAG does not provide control of flow to define integration scenarios in these nonnormative specifications.

Software modules are indicated with single-line boxes. The software modules outside the enterprise are indicated with the dashed-line boxes. The arrows indicate possible communication channels and associated BODs. The double-lined boxes indicate simpler communication patterns that may be found among the other nonnormative OAG integration scenarios. In this way, OAG integration scenarios are defined recursively from simpler to more complex scenarios.

Definitions of the OAG BODs used in this figure can be found in Table 8.4. In the figure, the inter-enterprise conceptual objects (i.e., OAG nouns) identified in Table 8.4 are used to form BODs such as ProcessPurchaseOrder, AddPurchaseOrder, and others to support the supply-chain integration scenario.

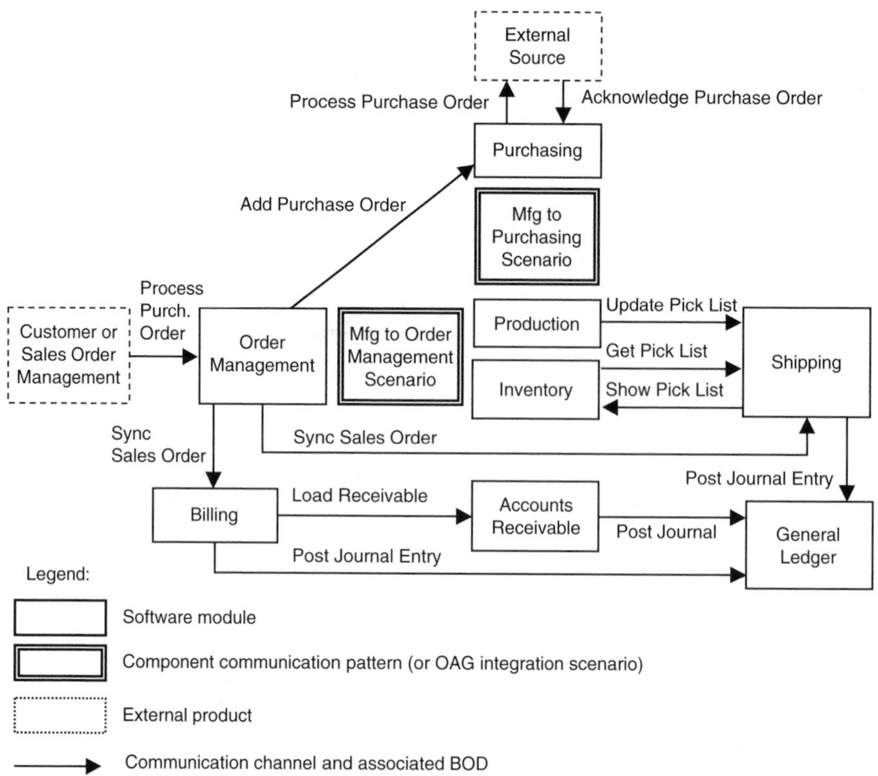

FIGURE 8.4 A possible supply-chain integration scenario supported by OAG BODs.

One possible scenario of integration may start with the ProcessPurchaseOrder BOD from an external customer order management system into the supplier's order management system. The local order management system may determine whether the items are to come from the inventory and, in that case, the SyncSalesOrder is sent to the shipping module. The shipping module may communicate with the inventory system using GetPickList and ShowPickList pair of BODs. Alternatively, if the items are to be obtained from an external source, the order management system may use AddPurchaseOrder BOD to communicate this decision to the purchasing system that, in turn, can send ProcessPurchaseOrder to the external supplier.

Although in case of RosettaNet, supply-chain integration scenarios are supported by identifying PIPs, one can also identify, in early phases of an interoperability effort, support of these integration scenarios using high-level RosettaNet business data entities such as those recognized in Table 8.3. For example, the RosettaNet iHUB project was put together to use and further develop PIPs in support of supply and demand planning within a collaborative, dynamic trading network [iHUB, 2003]. One of the high-level conceptual integration diagrams identified the roles in such a trading network and the conceptual objects exchanged between these roles and the network hub. In that way, identification of conceptual objects such as those in Table 8.4 represents a basis for data exchange among the partners. Table 8.5, summarizes the roles and objects of interest or objects exchanged by these roles.

8.5 Advanced Developments in Support of Interoperable Manufacturing Enterprise Systems

We conclude this chapter by giving an account of advanced developments in support of interoperable manufacturing enterprise information systems that are taking place in the standards organizations including ebXML, OAG, RosettaNet, NIST, and W3C.

TABLE 8.4 A Mapping between the OAG Content Standard Nouns and RosettaNet Business Data Entities

OAG Noun and Definition	RosettaNet Business Data Entity/PIP and Definition
Procurement/Sales	
Consumption Process whereby a certain amount or quantity of inventory, resources, or product is utilized that likely lead to the need for some form of replenishment	**Consumption Notice** Business document to trading partner who owns consigned product that communicates material, material quantities, and dates the material quantities were consumed
Delivery Receipt Transaction for the receiving of goods or services. It may be used to indicate receipt of goods in conjunction with a purchase order system	**Receipt Information** The collection of business properties that describes the receipt of a quantity of a product delivered
Electronic Catalog A list of items or commodities. Each item can be classified into one or more categories, and the specifications of each item can be identified. A catalog has at least one publisher and one or many suppliers for the items in the catalog	**Sales Catalog** The collection of business properties that describe a seller's catalog of products
Inspection Report on the inspection of items identified in the source document	**Inspection Results** The collection of business properties that describe the results of a product inspection
Invoice Invoice document to the customer	**Invoice** An itemized list of goods or services specifying the price and the terms of sale
Party Information use by business applications to reference parties that may play different roles within an integration (e.g., Supplier, Customer, Carrier)	**Partner Role Description** The collection of business properties that describe a business partners' role in a partner interface process
Price List List of items with their base price, price breaks, discounts, and qualifiers	**Price List** The collection of business properties that describe product pricing in a price-list document
Product Availability Information on the availability of a specified item at a specified inventory location for a specified date	**Product Availability** The collection of properties that describe a product's time frame for being available
Purchase Order An order to purchase goods from a buyer to a supplier	**Purchase Order** The collection of business properties that describe a buyer's offer to purchase a quantity of products at an agreed price and schedule
Quote Document describing the prices of goods or services provided by a vendor. The quote includes the terms of the purchase, delivery proposals, identification of goods or services ordered, and their quantities	**Quote** The collection of business properties that describe an offer to supply a quantity of products at an agreed price and schedule
Request for Quote Document describing goods or services desired from a vendor. The RFQ includes the terms of the purchase, delivery requirements, identification of goods or services ordered, as well as their quantities	**Request Quote (PIP)**
Sales Order A customer order, a step beyond a Purchase Order in that the receiving entity of the order also communicates Sales Information about the Order along with the Order itself	**Sales Information** The collection of business properties that describe the sale of a product
Cart List of items selected for purchase	**Shopping Cart** The collection of product descriptions, quantities, and prices that comprise a buyer's intent to purchase
Manufacturing	
Bill of Materials List of items to be produced in a specified time period. (Same as in inter-enterprise)	**Bill of Material** The collection of business properties that describes a bill of material for a product

TABLE 8.4 A Mapping between the OAG Content Standard Nouns and RosettaNet Business Data Entities (continued)

OAG Noun and Definition	RosettaNet Business Data Entity/PIP and Definition
	Procurement/Sales
Engineering Change Document A request for a change to a manufactured item. (Same as in inter-enterprise)	**Engineering Change Request** The collection of business properties that enables a party proposing an engineering change to send an engineering change request to a change review forum
Inventory Balance Stocked items and the quantities of each item by location. Other item-by-location information, such as serial numbers or lot numbers, can also be included. (Same as inter-enterprise)	**Inventory Report** The collection of business properties that describe a product in inventory at a specific point in time
Maintenance Order Order for a machine, building, tooling, or fixed asset to be repaired or for preventive maintenance to be performed. (Same as inter-enterprise)	**Service Event Information** The collection of business properties that describe the data elements and entities associated with performing repair or maintenance service on a part or unit of product
Planning Schedule Indicates a demand forecast sent from a customer to a supplier, or a supply schedule sent from a supplier to a customer. (Same as inter-enterprise)	**Forecast** The collection of business properties that describe the advance indication of the opportunity for selling or demand
Sequence Schedule Sequential scheduling of ordered items in the manufacturing process. (Same as inter-enterprise)	**Product Release Schedule** The collection of business properties that identifies the dates(s), quantity(s), times(s) and release number for Material Release
Engineering Work Document Carries product structure information and information on what is to be changed in it as the result of a project design activity. (Same as inter-enterprise)	**Engineering Information** The information for engineering purpose, i.e., technical data necessary for process of the device
Work In Process Movement of material or finished products; also, production order status. (Same as inter-enterprise)	**Work In Process** The collection of business properties that describe the manufacturing steps that must be performed
	Financial Services
Credit Customer credit information used in the context of credit checking new sales orders	**Credit Reference** The collection of business properties that describe the current credit status of an account of party
Credit Status Credit approval status of a customer or a specific customer order	**Credit Reference** The collection of business properties that describe the current credit status of an account of party
Exchange Rate Information that applies to the currency exchange rate ratio	**Currency Conversion** The collection of business properties that describe the exchange of money in circulation
	Logistics
Shipment A document that identifies and describes a collection of goods to be transported by a carrier and delivered to one or more destinations	**Shipment** The collection of business properties that describe a consignment tendered for transportation from one point to another
Shipment Schedule Commonly, a shipment schedule is generated by a material planning application and transmitted to an order or material planning application	**Shipping Information** The collection of business properties that describe information regarding shipments tendered for transportation
Carrier Route Describes a scheduled journey that a transportation service provider (freight carrier) is requested to perform for a shipper, customer, or coordinator	**Routing Information** The collection of business properties that describes a leg used in the routing of a shipment

TABLE 8.5 The Roles and Objects Exchanged by These Roles in RosettaNet iHUB Project

	Component Manufacturer	Original Equipment Manufacturer	Component Supplier	Distributor
Forecast	+	+	+	+
Inventory report	+	+	+	+
Replenishment/ Consumption	+			
Purchase order	+	+		+
Sales order	+		+	+
Shipment	+	+	+	+
BOM	+			
Product master	+	+	+	
Work order	+			
Product catalog				+

8.5.1 OAG

OAG has recently enhanced its OAGIS specifications by adopting W3C XML Schema recommendations to represent its object concepts along with other advanced related technologies such as XSL, XSLT, and Schematron [Schematron]. OAG has also taken steps to ensure that adoption of XML Schema as its representation approach continues to reflect its extensible design and support of other vertical and horizontal industry standards [OAGIS, 2003]. One important practical development is that OAGIS keeps separate the content structure definitions from the content validation specifications to reflect the fact that validation procedures are typically specific to the users of the interoperability standards. Although this practice encourages a common content model within and across different industries, it leaves much room to different interpretations of standards specification meaning and, consequently, may cause interoperability issues. For that reason, researchers are investigating complementary technologies such as RDF and Semantic Web to provide additional rigor and to formally define the meaning of a content model [Kulvatunyou, 2002].

OAG is also embarking on collaboration with other standards specification efforts that are focused on either a specific vertical industry sector or addresses a portion of horizontal efforts. An example of the former is a recent inclusion of TranXML specifications within the OAGIS [TranXML, 2003]. TranXML provides specifications for the transportation and logistics industries. An example of the latter is the collaborative effort with HR-XML that provides a cross-industry standard specification for procurement of human resources [HR-XML,2003].

8.5.2 ebXML

The ebXML effort at UN/CEFACT is developing an advanced approach for content and structure definition named Core Components (CC) to execute business collaborations in complex application contexts [ebCC, 2003]. Presently, the CC specification is at the verification stage under UN/CEFACT Open Development Process (ODP). The CC approach starts with a syntax-independent construct of information (using UML class diagrams). The class diagrams of the CC specification together with adopted naming conventions and rules enable serialization of the object class diagrams into data vocabularies. The CC approach starts with a basic vocabulary (called core components) and employs a context mechanism to apply to that vocabulary. The context mechanism defines eight context categories (shown in Table 8.6), which when applied to the core components results in an application context-specific business object called business information entity (BIE) that is used for actual business data exchange. Different business-specific semantic constraints and restrictions may be applied at the time of creation of BIEs. The context mechanism introduced in this way provides means to narrow down the intended meaning of business terms through a flexible mechanism while allowing an important generality at the

TABLE 8.6 The Core Components Context Mechanism Defines Eight Context Categories

Context Category	Description	Example
Business Process	The type of process	Ordering, Delivery
Product Classification	The type of products that the collaboration is about	Parts, Consumer goods
Industry Classification	The sector in which the collaboration takes place	Aerospace, Electronic Components
Geopolitical	The location of the partners	International, Europe
Official constraints	The legislation that applies	US law, EU law
Business Process Role	The roles the partners play in the process	Buyer, Seller
Supporting Role	Roles of relevant parties outside the collaboration	Shipping Agent
System Capabilities	Specific system requirements	SAP, Intuit

basic dictionary level. It is expected that the users of standardized core components will be able to agree on the semantics at the basic dictionary level and, then, arrive at a common meaning of the business information entities (BIEs) by applying the context mechanism to the dictionary (i.e., core components).

Early adoptions of the CC approach have begun by several organizations including OAG, UCC, and ebXML UBL Technical Committee [OAG, 2003; UCC, 2003; UBL, 2003]. A number of issues have been revealed in these initial steps which broadly fall into one of the two categories.

The first category of issues deals with the *usage of the CC approach* where one may proceed with a top-down approach or a bottom-up approach to develop core components. The users (such as UCC EAN and SWIFT) that employ a top-down approach attempt to derive core components from the business requirements (as recommended by the CC specifications) without looking back at the existing content models. On the other hand, the users (such as OAG and UBL) who use the bottom-up approach attempt to derive core components from existing content models. A tradeoff between the two approaches is obvious.

In the case of top-down approaches, any harmonization of the results among such efforts will be supported well as the business requirements provide a common basis for defining context and common semantics without a regard for backward compatibility. However, breaking the backward compatibility will cause issues for existing users of these standards specifications. On the other hand, in the case of bottom-up approaches, a harmonization of the results among different efforts will not have a common basis and the differences in the starting models may be reflected in the identified core components. However, the backward compatibility may be preserved allowing continuity in the standards adoption.

The second category of issues deals with the *interpretation of the specification and information modeling.* These issues are always present where there is an attempt to model information and information types need to be determined. Questions such as "Whether an entity should be an object or a simple type?" or "Whether an object characteristic should be a contextual property or simple property?" abound in any standards adoption process. A core component primer [CCSD, 2003] has been developed to assist the adopting organizations in the process of addressing these issues.

8.5.3 RosettaNet

RosettaNet has made a significant investment to enhance interoperability among products implementing RosettaNet specification by developing its own conformance certification program. The RosettaNet Ready program provides tools and services required to measure compliance of a product implementation with RNIF and PIP specifications. In addition, RosettaNet has put in place the RosettaNet Interoperability program to improve implementation interoperability through education and testing activities (RosettaNet, 2003). The program's objective is to drive down the cost of connecting trading partners and especially to enable small and medium enterprises to get involved in trading networks. The initial problems being addressed by the program include the new trading partner transport, routing, and packaging concerns and security issues.

8.5.4 National Institute of Standards and Technology

The National Institute of Standards and Technology (NIST) has been developing a Manufacturing B2B Interoperability Testbed in collaboration with the OAG. The main objective for the testbed is to advance available technology for on-demand, highly available, and efficient interoperability demonstration, piloting, and testing [NIST]. The testbed project has utilized technologies such as Semantic Web and W3C XML technologies, and standards specifications such as OAGIS and ebXML to develop tools such as content checking, business process monitoring, and virtual trading partners in support of interoperability testing and demonstration. The NIST Testbed also collaborates with industry partners and consortia such as OASIS ebXML Interoperability, Implementation, and Conformance (IIC) to advance the state of the art of automated testing facilities that is accessible in a distributed fashion [ebXML IIC, 2003].

Another important activity at NIST addresses the issue of convergence and reuse in developing B2B and other eBusiness standards. The eBusiness Standards Convergence (eBSC) Forum has been initiated to provide a forum for advancing collaboration among different eBusiness initiatives and achieving cross-industry interoperability and convergence [NIST-eBSC, 2003]. The participating organizations include industry organizations and initiatives (e.g., Aerospace Industries Association, Automotive Industry Action Group), standards development organizations (e.g., OAG, OASIS, ebXML), eBusiness software testing organizations (e.g., Drake Certivo, Drummond Group), and various NIST organization units. The forum has established a work plan that includes a number of deliverables to improve convergence of eBusiness standards including:

- Recommendations on what is needed for the paradigm shift to cross-industry standards convergence
- Agreement on eBusiness architecture framework and opportunities for convergence
- A common conceptual model for eBusiness capabilities stack
- Recommendations on Generic Industry Roadmap for industry adaptation

8.5.5 Semantic Web Activity

Semantic Web technologies have been put forward by W3C to develop new methods of data encoding on the Web to give well-defined meaning to information. To achieve this, existing formal logic systems are adopted for Web-based representation [SW]. There are two basic ways to employ Semantic Web technologies to enhance enterprise systems interoperability. In the first approach, Semantic Web technologies are used to annotate information and provide semantic formalism to the information exchanged between applications or enterprises. Such annotation enhances clarity of the information at design as well as run times and allows more efficient information integration processes [Peng, Kulvatunyou, 2003].

In the other approach, Semantic Web technologies are used to provide a well-defined meaning for the whole integration task and create ontology service and software modules that can be dynamically composed to achieve certain functionality. An example effort of this kind is Semantic Web for Web Services [DAML-S, 2003]. Obviously, the latter approach is significantly harder but carries a promise of potentially significantly changing the enterprise integration industry.

8.5.6 Disclaimer

Certain commercial software products are identified in this paper. These products were used only for demonstration purposes. This use does not imply approval or endorsement by NIST, nor does it imply that these products are necessarily the best available for the purpose.

References

Business Internet Consortium. XML.org Web Site, accessed May 2003. *High-Level Conceptual Model for B2B Integration*, version 2.0. Available online via < http://www.xml.org/xml/b2b_conceptual_model_2-0.pdf>.

CCSD. Core Components Supplemental Documentation (CCSD) Project Web Site, accessed May 2003. *CCSD Home Page.* Available online via http://webster.disa.org/cefact-groups/tmg/ccsd.html.

DAML-S. DAML Services Web Site, accessed May 2003. DAML-S Home Page. Available online via < http://www.daml.org/services/>.

ebBPSS. UN/CEFACT Web Site, accessed May 2003. *EbXML Business Process Specification Schema version 1.01*. Available online at < http://www.ebxml.org/specs/ebBPSS.pdf>.

ebCC. DISA UN/CEFACT Web Site, accessed May 2003. *EbXML Core Component Specification version 1.9*. Available online at < http://webster.disa.org/cefact-groups/tmg/downloads/CCWG/for_review/CCTS_V_1pt90.zip>.

ebCPPA. UN/CEFACT Web Site, accessed May 2003. *EbXML Business Process Specification Schema version 1.01*. Available online at < http://www.ebxml.org/specs/ebcpp-2.0.pdf>.

ebMS. OASIS ebXML Messaging Services Specification Technical Committee Web Site, accessed May 2003. *EbXML Messaging Service Specification version 2.0* . Available online via < http://www.oasis-open.org/committees/documents.php?wg_abbrev=ebxml-msg>.

ebXML. ebXML Web Site, accessed May 2003. *ebXML Home Page*. Available online via < http://www.ebxml.org/>.

ebXML IIC. OASIS ebXML Implementation, Interoperability, and Conformance Technical Committee Web Site, accessed May 2003. *EbXML IIC Test Framework version 1.0.* Available online via < http://www.oasis-open.org/committees/documents.php?wg_abbrev=ebxml-iic>.

HR-XML. HR-XML Consortium Web Site, accessed May 2003. *HR-XML Home Page* . Available online via <http://www.hr-xml.org/channels/home.htm>.

iHUB. RosettaNet Web Site, accessed May 2003. *IHub Program Home Page* . Available online via <http://www.rosettanet.org/ihub>.

Kulvatunyou, Boonserm and Nenad Ivezic. Semantic Web for Manufacturing Web Services. In *Electronic Proceedings of International Symposium on Manufacturing and Applications (ISOMA),* June 9–13, 2002.

Kulvatunyou, Boonserm and Nenad Ivezic, Rick Wysk, and Albert Jones. Integrated product and process data for B2B collaboration, to be published in the *Journal of Artificial Intelligence in Engineering, Design, Analysis, and Manufacturing, special issue in New AI Paradigm for Manufacturing* ,September 2003.

NIST. NIST B2B Interoperability Testbed Web Site, accessed May 2003. *OAG/NIST Testbed Home Page.* Available online via < http://www.mel.nist.gov/msid/oagnisttestbed>.

NIST-eBSC. The eBusiness Standards Convergence Forum, accessed August 2003. *The eBSC Homepage.* Available online via < http://www.mel.nist.gov/div826/msid/sima/ebsc/index.htm>.

OAG. The Open Applications Web Site, accessed May 2003. *OAG Homepage.* Available online via < http://www.openapplications.org/>.

OAGIS. Open Application Groups Web Site, accessed May 2003. *Open Application Group Integration Specification version 8.0* . Available online via <http://www.openapplications.org/downloads/oagi-downloads.htm>.

OMG. OMG Web Site, accessed May 2003. *Request for Proposals: Release for Production* . An Object Management Group Request for Proposals. Available online via < http://www.omg.org/cgi-bin/doc?mfg/98-07-05>.

Peng, Yun, Youyong Zou, Xiocheng Luan, Nenad Ivezic, Michael Gruninger, and Albert Jones. Semantic resolution for e-commerce. In *Proceedings of The First International Joint Conference on Autonomous Agents and Multiagent Systems, AAMAS 2002,* July 15–19, 2002, pages 1037–1038, ACM Digital Library, ACM, New York.

RNIF. The RosettaNet Web Site, accessed May 2003. *RosettaNet Implementation Framework (RNIF) 2.0 version 2.0*. Available online via < http://www.rosettanet.org/RosettaNet/Doc/0/TAO5O8VV3E7KLCRIDD1BMU6N38/RNIF2.1.pdf>.

RosettaNet. The RosettaNet Web Site, accessed May 2003. *RosettaNet Homepage* . Available online via <http://www.rosettanet.org/RosettaNet/Rooms/DisplayPages/LayoutInitial>

Rowell, Michael. Using OAGIS for Integration. *XML Journal*, 3(11), 2002.

Schematron. The Schematron Web Site, accessed May 2003. *Schematron Homepage.* Available online via <http://www.ascc.net/xml/resource/schematron/schematron.html>.

STAR. Standards for Technology in Automotive Retail Web Site, accessed May 2003. *Making the Case for IT Standards in Retail Automotive* . STAR publication. Available online via <http://www.starstandard.org/site_services/articles/Case%20Study%201-27-03%20FINAL.pdf>.

SW. W3C Semantic Web Activity Web Site, accessed May 2003. Semantic Web Activity Home Page. Available online via < http://www.w3.org/2001/sw/>.

TranXML. TranXML Web Site, accessed May 2003. *TranXML Home Page* . Available online via <http://www.transentric.com/products/commerce/tranxml.asp>.

UBL. Universal Business Language Technical Committee Web Site, accessed May 2003. *UBL Home Page* . Available online via <http://www.oasis-open.org/committees/tc_home.php?wg_abbrev=ubl>.

UCC. The UC-Council Web Site, accessed May 2003. *RosettaNet Merges With the Uniform Code Council. Available online via* < http://www.uc-council.org/documents/doc/UCC_-_RSNT_Final.doc>

W3C. The World Wide Web Consortium Web Site, accessed May 2003. *W3C Homepage.* Available online via < http://www.w3c.org/>.

9

Semantic Enterprise Content Management

Mark Fisher

Amit Sheth

CONTENTS

Abstract ... 9-1
9.1 Introduction ... 9-1
9.2 Primary Challenges for Content Management Systems ... 9-3
9.2.1 Heterogeneous Data Sources ... 9-3
9.2.2 Distribution of Data Sources ... 9-3
9.2.3 Data Size and the Relevance Factor ... 9-4
9.3 Facing the Challenges: The Rise of Semantics ... 9-4
9.3.1 Enabling Interoperability ... 9-4
9.3.2 The Semantic Web ... 9-5
9.4 Core Components of Semantic Technology ... 9-6
9.4.1 Classification ... 9-6
9.4.2 Metadata ... 9-6
9.4.3 Ontologies ... 9-9
9.5 Applying Semantics in ECM ... 9-13
9.5.1 Toolkits ... 9-13
9.5.2 Semantic Metadata Extraction ... 9-14
9.5.3 Semantic Metadata Annotation ... 9-15
9.5.4 Semantic Querying ... 9-15
9.5.5 Knowledge Discovery ... 9-16
9.6 Conclusion ... 9-18
References ... 9-19

Abstract

The emergence and growth of the Internet and vast corporate intranets as information sources has resulted in new challenges with regard to scale, heterogeneity, and distribution of content. Semantics is emerging as the critical tool for enabling more scalable and automated approaches to achieve interoperability and analysis of such content. This chapter discusses how a Semantic Enterprise Content Management system employs metadata and ontologies to effectively overcome these challenges.

9.1 Introduction

Systems for high-volume and distributed data management were once confined to the domain of highly technical and data-intensive industries. However, the general trend in corporate institutions over the past three decades has led to the near obsolescence of the physical file cabinet in favor of computerized data storage. With this increased breadth of data-rich industries, there is a parallel increase in the demand for handling a much wider range of data source formats with regard to syntax, structure, accessibility, and

1-58488-381-2/05/$0.00+$1.50
© 2005 by CRC Press LLC

physical storage properties. Unlike the data-rich industries of the past that typically preferred to store their data within the highest possible degree of structure, many industries today require the same management capabilities across a multitude of data sources of vastly different degrees of structure. In a typical company, employee payroll information is stored in a database, accounting records are stored in spreadsheets, internal company policy reports exist in word-processor documents, marketing presentations exist alongside white papers and Web-accessible slideshows, and company financial briefing and technical seminars are available online as a/v files and streaming media.

Thus, the "Information Age" has given rise to the ubiquity of Content Management Systems (CMS) for encompassing a wide array of business needs from Human Resource Management to Customer Resource Management, invoices to expense reports, and presentations to e-mails. This trend has affected nearly every type of enterprise — financial institutions, governmental departments, media and entertainment management — to name but a few. Moreover, the growth rate of data repositories has accelerated to the point that traditional CMS no longer provides the necessary power to organize and utilize that data in an efficient manner. Furthermore, CMS are often the backbone of more dynamic internal processes within an enterprise, such as content analytics, and the more public face of an enterprise as seen through its enterprise portal. The result of not having a good CMS would mean lost or misplaced files, inadequate security for highly sensitive information, nonviable human resource requirements for tedious organizational tasks, and, in the worst case, it may even lead to unrecognized corruption or fraud perpetrated by malevolent individuals who have discovered and exploited loopholes that will undoubtedly exist within a mismanaged information system.

Current demands for business intelligence require information analysis that acts upon massive and disparate sources of data in an extremely timely manner, and the results of such analysis must provide actionable information that is highly relevant for the task at hand. For such endeavors, machine processing is an indisputable requirement due to the size and dispersal of data repositories in the typical corporate setting. Nonetheless, the difficulty in accessing highly relevant information necessitates an incredibly versatile system that is capable of traversing and "understanding" the meaning of content regardless of its syntactic form or its degree of structure. Humans searching for information can determine with relative ease the meaning of a given document, and during the analytical process will be unconcerned, if not unaware, of differences in the format of that document (e.g., Web page, word processor document, e-mail). Enabling this same degree of versatility and impartiality for a machine requires overcoming significant obstacles, yet, as mentioned above, the size and distribution of data leave no choice but to confront these issues with machine-processing. A human cannot possibly locate relevant information within a collection of data that exceeds millions or even billions of records, and even in a small set of data, there may be subtle and elusive connections between items that are not immediately apparent within the limits of manual analysis. By applying advanced techniques of semantic technology, software engineers are able to develop robust content management applications with the combined capabilities of intelligent reasoning and computational performance.

"Content," as used throughout this chapter, refers to any form of data that is stored, retrieved, organized, and analyzed within a given enterprise. For example, a particular financial institution's content could include continuously updated account records stored in a Relational Database Management System (RDBMS), customer profiles stored in a shared file system in the form of spreadsheets, employee policies stored as Web pages on an intranet, and an archive of e-mail correspondence among the company employees. In this scenario, several of the challenges of CMS are apparent.

This chapter will focus on three such challenges, and for each of these, we will discuss the benefits of applying *semantics* to create an enhanced CMS. Throughout we will emphasize that the goal of any such system should be to increase overall efficiency by maximizing return on investment (ROI) for employees who manage data, while minimizing the technical skill level required of such workers, even as the complexity of information systems grows inevitably in proportion to the amount of data. The trends that have developed in response to these challenges have propelled traditional CMS into the realm of semantics where quality supersedes quantity in the sense that a small set of highly relevant information offers much more utility than a large set of irrelevant information. Three critical enablers of semantic technology — *classification, metadata,* and *ontologies* — are explored in this chapter. Finally we show how the combined application of these three core components may aid in overcoming the challenges as traditional content management evolves into semantic content management.

9.2 Primary Challenges for Content Management Systems

9.2.1 Heterogeneous Data Sources

First, there is the subtle yet highly complicated issue that most large-scale information systems comprise heterogeneous data sources. These sources differ structurally and syntactically [Sheth, 1998]. Retrieving data from an RDBMS, for instance, involves programmatic access (such as ODBC) or, minimally, the use of a query language (SQL). Likewise, the HTML pages that account for a significant portion of documents on the Internet and many intranets are actually composed of marked-up, or *tagged*, text (tags provide stylistic and structural information) that is interpreted by a browser to produce a more human-readable presentation. One of the more challenging environments is when the transactional data needs to be integrated with documents or primarily textual data. Finally, a document created within a word-processing application is stored as binary data and is converted into text using a proprietary interpreter built into the application itself (or an associated "viewer"). Some of the applications, such as Acrobat, provide increasing support for embedding manually entered metadata in RDF and based on the Dublin Core metadata standard (to be discussed later). A system that integrates these diverse forms of data in a way that allows for their *interoperability* must create some normalized representation of that data in order to provide equal accessibility for human and machine alike. In other words, while the act of reading an e-mail, a Web page, and a word-processor document is not altogether different for a human, a machine is "reading" drastically different material in regard to structure, syntax, and internal representation. Add to this equation the need to manage content that is stored in rich media formats (audio and video files), and the difficulty of such a task is compounded immensely. Thus, for any system that enables automation for managing such diverse content, this challenge of interoperability must be overcome.

9.2.2 Distribution of Data Sources

Inevitably, a corporation's content is not only stored in heterogeneous formats, but its data storage systems will likely be distributed among various machines on a network, including desktops, servers, network file-systems, and databases. Accessing such data will typically involve the use of various protocols (HTTP, HTTPS, FTP, SCP, etc.). Security measures, such as firewalls and user-authentication mechanisms, may further complicate the process of communication among intranets, the Internet, and the World Wide Web. Often, an enterprise's business depends not only on proprietary and internally generated content, but also subscribed syndicated content, or open source and publicly available content. In response to these complexities, an information management system must be extremely adaptable in its traversal methods, highly configurable for a wide variety of environments, and noncompromising in regard to security.

Increasingly, institutions are forming partnerships based upon the common advantage of sharing data resources. This compounds the already problematic nature of data distribution. For example, a single corporation will likely restrict itself to a single database vendor in order to minimize the cost, infrastructure, and human resources required for maintenance and administration of the information system. However, a corporation should not face limitations regarding its decisions for such resource sharing partnerships simply based on the fact that a potential partner employs a different database management system. Even after issues of compatibility are settled, the owner of a valuable data resource will nevertheless want to preserve a certain degree of autonomy for their information system in order to retain control of its contents [Sheth and Larson, 1990]. This is a necessary precaution regardless of the willingness to share the resource. Understandably, a corporation may want to limit the shared access to certain views or subsets of its data, and even more importantly, it must protect itself given that the partnership may expire. Technologies within the growing field of Enterprise Application Integration are overcoming such barriers with key developments in generic transport methods (XML, IIOP, SOAP, and Web Services). These technologies are proving to be valuable tools for the construction of secure and reliable interface mechanisms in the emerging field of Semantic Enterprise Information Integration (SEII) systems.

9.2.3 Data Size and the Relevance Factor

The third, and perhaps most demanding, challenge arises from the necessity to find the most relevant information within a massive set of data. Information systems must deal with content that is not only heterogeneous and distributed but also exceptionally large. This is a common feature of networked repositories (most notably the World Wide Web). A system for managing, processing, and analyzing such data must incorporate filtering algorithms for eliminating the excessive "noise" in order for users to drill down to subsets of relevant information. Such challenges make the requirements for speed and automation critical. Ideally, a CMS should provide increased quality of data management as the quantity of data grows. In the example of a search engine, increasing the amount of data available to the search's indexing mechanism should enable an end user to find not only more but *better* results. Unfortunately it is all too often the case that an increased amount of data leads to exactly the opposite situation where the user's results are distorted due to a high number of false positives. Such distortion results from the system's combined inabilities to determine the contextual meaning of its own contents or the intentions of the end user.

9.3 Facing the Challenges: The Rise of Semantics

The growing demands for integrating content, coupled with the unfeasibility of actually storing the content within a single data management system, have given rise to the field of Enterprise Content Management (ECM). Built upon many of the technical achievements of the Document Management (DM) and CMS communities, the applications of ECM must be more generic with regard to the particularities of various data sources, more versatile in its ability to process and aggregate content, more powerful in handling massive and dynamic sources in a timely manner, more scalable in response to the inevitable rise of new forms of data, and more helpful in providing the most relevant information to its front-end users. While encompassing each of these features, an ECM system must overcome the pervasive challenge of reducing the requirements of manual interaction to a minimum. In designing the functional specifications for an ECM application, system architects and developers focus upon any management task that has traditionally been a human responsibility and investigate the possibilities of devising an automated counterpart. Typically the most challenging of these tasks involves text analytics and decision-making processes. Therefore, many developments within ECM have occurred in parallel with advances in the Artificial Intelligence (AI), lexical and natural language processing, and data management and information retrieval communities. The intersection of these domains has occurred within the realm of semantics.

9.3.1 Enabling Interoperability

The first two challenges presented in the previous section, heterogeneity and distribution, are closely related with regard to their resulting technical obstacles. In both cases, the need for interoperability among a wide variety of applications and interfaces to data sources presents a challenge for machine processibility of the content within these sources. The input can vary widely, yet the output of the data processing must create a normalized view of the content so that it is equally usable (i.e., *machine-readable*) in an application regardless of source. Certain features of data storage systems are indispensable for the necessary administrative requirements of their users (e.g., automated backup, version-tracking, referential integrity), and no single ECM system could possibly incorporate all such features. Therefore, an ECM system must provide this "normalized view" as a portal layer, which does not infringe upon the operational procedures of the existing data infrastructure, yet provides equal access to its contents via an enhanced interface for the organization and retrieval of its contents.

While this portal layer exists for the front-end users, there is a significant degree of processing required for the back-end operations of data aggregation. As the primary goal of the system is to extract the most relevant information from each piece of content, the data integration mechanism must not simply duplicate the data in a normalized format. Clearly, such a procedure would not only lead to excessive storage capacity requirements (again this is especially true in dealing with data from the World Wide Web) but would also accomplish nothing for the relevance factor. One solution to this predicament is an indexing mechanism that analyses the content and determines a subset of the most relevant information, which may be stored within the content's *metadata* (to be discussed in detail later). Because a computer typically exploits structural and syntactic regularities, the complexity of analysis grows more than linearly in relation to the inconsistencies within these content sources. This is the primary reason that many corporations have devoted vast human resources to tasks such as the organization and analysis of data. On the other hand, corporations for which data management is a critical part of the operations typically store as much data as possible in highly structured systems such as RDBMS, or for smaller sets of data, use spreadsheet files. Still other corporations have vast amounts of legacy data that are dispersed in unstructured systems and formats, such as the individual file-systems of desktop computers in a Local Area Network or e-mail archives or even in a legacy CMS that no longer supports the needs of the corporation.

9.3.2 The Semantic Web

Ironically, the single largest and rapidly growing source of data — the World Wide Web — is a collection of resources that is extremely nonrestrictive in terms of structural consistency. This is a result of a majority of these existing as HTML documents, which are inherently flexible with regard to structure. In hindsight, this issue may be puzzling and even frustrating to computer scientists who, in nearly every contemporary academic or commercial environment, will at some point be confronted with such inconsistencies while handling data from the World Wide Web. Nevertheless the very existence of this vast resource is owed largely to the flexibility provided by HTML, as this is the primary enabling factor for nonspecialists who have added countless resources to this global data repository. The guidelines for the HTML standard are so loosely defined that two documents which appear identical within a browser could differ drastically in the actual HTML syntax. Although this presents no problem for a human reading the Web page, it can be a significant problem for a computer processing the HTML for any purpose beyond the mere display in the browser. With XML (eXtensible Markup Language), well-formed structure is enforced, and the result is increased consistency and vastly more reliability in terms of machine readability. Additionally, XML is customizable (extensible) for any domain-specific representation of content. When designing an XML Schema or DTD, a developer or content provider outlines the elements and attributes that will be used and their hierarchical structure. The developer may specify which elements or attributes are required, which are optional, their cardinality, and basic constraints on the values. XML, therefore, aids considerably in guaranteeing that the content is machine readable because it provides a template describing what may be expected in the document. XML also has considerably more semantic value because the elements and attributes will typically be named in a way that provides meaning as opposed to simple directives for formatting the display.

For these reasons, the proponents of the Semantic Web have stressed the benefits of XML for Web-based content storage as opposed to the currently predominant HTML. XML has been further extended by the Resource Description Framework (RDF, described in the section on ontologies in the following text), which enables XML tags to be labeled in conjunction with a referential knowledge representation. This in turn allows for machine-based "inferencing agents" to operate upon the contents of the Web. Developed for information retrieval within particular domains of knowledge, these specialized agents might effectively replace the Web's current "search engines." These are the concepts that may transform the state of the current World Wide Web into a much more powerful and seemingly intelligent resource, and researchers who are optimistic about this direction for the Web propose that it will not require a heightened technical level for the creators or consumers of its contents [Berners-Lee et al., 2001]. It is true that many who upload information to today's Web use editors that may completely preclude the need to learn HTML syntax. For the Semantic Web to emerge pervasively, analogous editors would need to provide this same ease of use while infusing semantic information into the content.

9.4 Core Components of Semantic Technology

9.4.1 Classification

Classification is, in a sense, a coarse-level method of increasing the relevancy factor for a CMS. For example, imagine a news content provider who publishes 1000 stories a day. If these stories were indexed en masse by a search engine with general keyword searching, it could often lead to many irrelevant results. This would be especially true in cases where the search terms are ambiguous in regard to context. For example, the word "bear" could be interpreted as a sports team's mascot or as a term to describe the current state of the stock market. Likewise, names of famous athletes, entertainers, business executives, and politicians may overlap — especially when one is searching only by last name. However, these ambiguities can be reduced if an automatic classification system is applied. A simple case would be a system that is able to divide the set of stories into groups of roughly a couple of hundred stories each within five general categories, such as World News, Politics, Sports, Entertainment, and Business. If the same keyword searches mentioned above were now applied within a given category, the results would be much more relevant, and the term "bear" will likely have different usage and meaning among the stories segregated by the categories.

Such a system is increasingly beneficial as the search domain becomes more focused. If a set of 1000 documents were all within the domain of Finance and the end users were analysts with finely tuned expectations, the search parameters might lead to unacceptable results due to a high degree of overlap within the documents. While the layman may not recognize the poor quality of these results, the analyst, who may be particularly interested in a merger of two companies, would only be distracted by general industry reports that happen to mention these same two companies. In this case the information retrieval may be extremely time-critical (even more critical cases exist, such as national security and law enforcement). A highly specialized classification system could divide this particular set of documents into categories such as "Earnings," "Mergers," "Market Analysis," etc. Obviously, such a fine-grained classification system is much more difficult to implement than the earlier and far more generalized example. Nevertheless, with a massive amount of data becoming available each second, such classification may be indispensable. Several techniques of classification may be used to address such needs, including statistical analysis and pattern matching [Joachims, 1998], rule-based methods [Ipeirotis et al., 2000], linguistic analysis [Losee, 1995], probabilistic methods employing Bayesian theory [Cheeseman and Stutz, 1996], and machine-learning methods [Sebastiani, 2002], including those based on Hidden Markov Models [Frasconi et al., 2002]. In addition, ontology-driven techniques, such as named-entity and domain-phrase recognition, can vastly improve the results of classification [Hammond et al., 2002]. Studies have revealed that a committee-based approach will produce the best results because it maximizes the contributions of the various classification techniques [Sheth et al., 2002]. Furthermore, studies have also shown that classification results are significantly more precise when the documents to be classified are tagged with metadata resources (represented in XML) and conform to a predetermined schema [Lim and Liu, 2002].

9.4.2 Metadata

Metadata can be loosely defined as "data about data." For a discussion of enterprise applications and their metadata-related methodologies for infusing Content Management Systems with semantic capabilities, and to reveal the advantages offered by metadata in semantic content management, we will outline our description of metadata as progressive levels from the perspective of increasing utility. These levels of metadata are not mutually exclusive; on the contrary, the accumulative combination of each type of metadata provides a multifaceted representation of the data including information about its syntax, structure, and semantic context. For this discussion, we use the term "document" to refer to a piece of textual content — the data itself. Given the definition above, each form of metadata discussed here may be viewed in some sense as data *about* the data within this hypothetical document. The goal of incorporating metadata into a CMS is to enable the end user to find actionable and contextually relevant information. Therefore, the utility of these types of metadata is judged against this requirement of contextual relevance.

9.4.2.1 Syntactic Metadata

The simplest form of metadata is *syntactic* metadata, which provides very general information, such as the document's size, location, or date of creation. Although this information may undoubtedly be useful in certain applications, it provides very little in the way of context determination. However, the assessment of a document's relevance may be partially aided by such information. The date of creation or date of modification for a document would be particularly helpful in an application where highly time-critical information is required and only the most recent information is desired. For example, a news agency competing to have the first release of breaking news headlines may constantly monitor a network of reports where the initial filtering mechanism is based upon scanning only information from the past hour. Similarly, a brokerage firm may initially divide all documents based on date and time before submitting to separate processing modules for long-term market analysis and short-term index change reports. These attributes, which describe the document's creators, modifiers, and times of their activity, may also be exploited for the inclusion of version-tracking and user-level access policies into the ECM system. Most document types will have some degree of syntactic metadata. E-mail header information provides author, date, and subject. Documents in a file-system are tagged with this information as well.

9.4.2.2 Structural Metadata

The next level of metadata is that which provides information regarding the *structure* of content. The amount and type of such metadata will vary widely with the type of document. For example, an HTML document may have many tags, but as these exist primarily for purposes of formatting, they will not be very helpful in providing contextual information for the enclosed content. XML, on the other hand, offers exceptional capabilities in this regard. Although it is the responsibility of the document creator to take full advantage of this feature, structural metadata is generally available from XML. In fact, the ability to enclose content within meaningful tags is usually the fundamental reason one would choose to create a document in XML. Many "description languages" that are used for the representation of knowledge are XML-based (some will be discussed in the section on *ontologies* in the following text). For determining contextual relevance and making associations between content from multiple documents, structural metadata is more beneficial than merely syntactic metadata, because it provides information about the topic of a document's *content* and the items of interest within that content. This is clearly more useful in determining context and relevance when compared to the limitations of syntactic metadata for providing information about the document itself.

9.4.2.3 Semantic Metadata

In contrast to the initial definition of metadata above, we may now construct a much more pertinent definition for *semantic* metadata as "data that may be associated explicitly or implicitly with a given piece of content (i.e., a document) and whose relevance for that content is determined by its ontological position (its context) within one or more domains of knowledge." In this sense, metadata is the building block of semantics. It offers an invaluable aid in classification techniques, it provides a means for high-precision searching, and, perhaps most important, it enables interoperability among heterogeneous data sources.

How does semantic metadata empower a Content Management System to better accomplish each of these tasks? In the discussion that follows, we will provide an in-depth look at how metadata can be leveraged against an ontology to provide fine-grained contextual relevancy for information within a given domain or domains. As we briefly mentioned in the discussion of classification techniques in the previous section, the precision of classification results may be drastically augmented by the use of domain knowledge. In this case, the method is *named entity recognition.*

Named entity recognition involves finding items of potential interest within a piece of text. A named entity may be a person, place, thing, or event. If these entities are stored within an ontology, then a vast amount of information may be available. It is precisely this *semantic* metadata that allows for interoperability across a wide array of data storage systems because the metadata that is extracted from any document may be stored as a "snapshot" of that document's relevant information. The metadata contained within this snapshot simply references the instances of named entities that are stored in the ontology. Therefore, there is a rich resource of information available for each named entity including synonyms, attributes, and other related entities. This enables further "linking" to other documents on three levels: those containing the same explicit metadata (mention the exact same entities), those containing the same metadata implicitly (such as synonyms or hierarchically related named entities), and those related by ontological associations between named entities (one document mentions a company's name while another simply mentions its ticker symbol). This process in effect normalizes the vastly different data sources by referencing the back-end ontology, and while this exists "behind the scenes," it allows for browsing and searching within the front-end portal layer.

9.4.2.4 Metadata Standards

The use of metadata for integrating heterogeneous data [Bornhövd, 1999; Snijder, 2001] and managing heterogeneous media [Sheth and Klas, 1998; Kashyap et al., 1995] has been extensively discussed, and an increasing number of metadata standards are being proposed and developed throughout the information management community to serve the needs of various applications and industries. One such standard that has been well accepted is the Dublin Core Metadata Initiative (DCMI). Figure 9.1 shows the 15 elements defined by this metadata standard. It is a very generic element set flexible enough to be used in content management regardless of the domain of knowledge. Nevertheless, for this same reason, it is primarily a set of syntactic metadata as described above; it offers information about the document but offers very little with regard to the structure or content of the document. The semantic information is limited to the "Resource Type" element, which may be helpful for classification of documents, and the inclusion of a "Relation" element, which allows for related resources to be explicitly associated. In order to provide more semantic associations through metadata, this element set could be extended with domain-specific metadata tags. In other words, the Dublin Core metadata standard may provide a useful parent class for domain-specific document categories.

The Learning Technology Standards Committee (LTSC), a division of the IEEE, is developing a similar metadata standard, known as Learning Object Metadata (LOM). LOM provides slightly more information regarding the structure of the object being described, yet it is slightly more specialized with a metadata element set that focuses primarily upon technology-aided educational information [LTSC, 2000].

The National Library of Medicine has created a database for medical publications known as MEDLINE, and the search mechanism requires that the publications be submitted according to the PubMed XML specification (the DTD is located at: http://www.ncbi.nlm.nih.gov/entrez/query/static/PubMed.dtd). Once more, the information is primarily focused upon authorship and creation date, but it does include an element for uniquely identifying each article, which is helpful for indexing the set of documents. This would be particularly helpful if some third-party mechanism were used to traverse, classify, and create associations between documents within this repository.

The next section will demonstrate how *ontologies* provide a valuable method for finding implicit semantic metadata in addition to the explicitly mentioned domain-specific metadata within a document (see Figure 9.2). This ability to discover implicit metadata enables the annotation process to proceed to the next level of *semantic enhancement,* which in turn allows the end user of a semantic CMS to locate contextually relevant content. The enhancement of content with nonexplicit semantic metadata will also enable analysis tools to discover nonobvious relationships between content.

Element	Description
Title	A name given to the resource.
Contributor	An entity responsible for making contributions to the content of the resource.
Creator	An entity primarily responsible for making the content of the resource.
Publisher	An entity responsible for making the resource available.
Subject and Keywords	The topic of the content of the resource.
Description	An account of the content of the resource.
Date	A date associated with an event in the life cycle of the resource.
Resource Type	The nature or genre of the content of the resource.
Format	The physical or digital manifestation of the resource.
Resource Identifier	An unambiguous reference to the resource within a given context.
Language	A language of the intellectual content of the resource.
Relation	A reference to a related resource.
Source	A Reference to a resource from which the present resource is aderived.
Coverage	The extent or scope of the content of the resource.
Rights Management	Information about rights held in and over the resource.

FIGURE 9.1 Dublin Core Metadata Initiative as described in the Element Set Schema. (From DCMI. Dublin Core Metadata Initiative, 2002. URL: http://dublincore.org/2002/08/13/dces.)

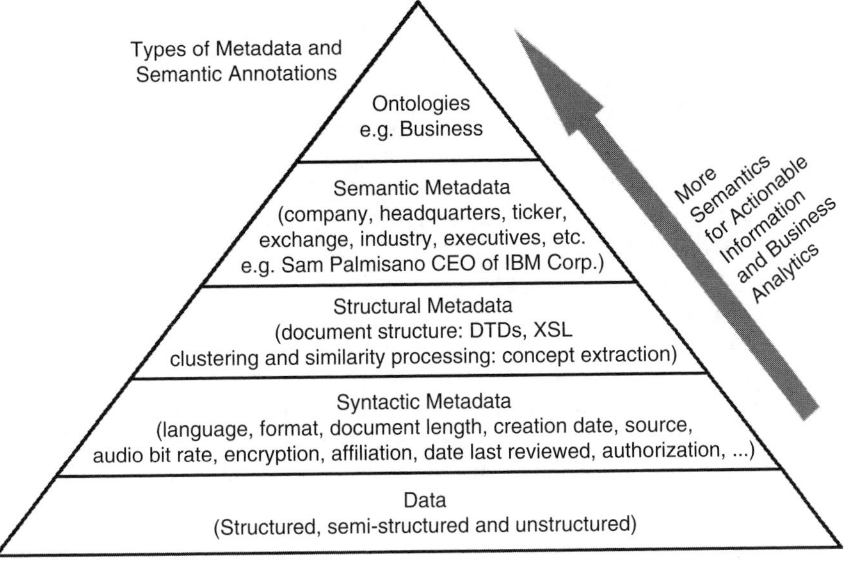

FIGURE 9.2 Filtering to highly relevant information is achieved as the type of semantic annotations and metadata progress toward domain-modeling through the use of ontologies.

9.4.3 Ontologies

Although the term ontology originated in philosophy where it means the "study of existence" (*ontos* is the Greek word for "being"), there is a related yet more pragmatic and concrete meaning for this term in computer science; an ontology is a representation of a domain of knowledge. To appreciate the benefits offered by an ontological model within a content management system, we will convey the intricacies and the features of such a system in comparison with other, more basic forms of knowledge representation. In this way, the advantages of using an ontological model will be presented as a successive accumulation of its forebears. The use of ontologies to provide underpinning for information sharing, heterogeneous database integration, and semantic interoperability has been long realized [Gruber, 1991; Kashyap and Sheth, 1994; Sheth, 1998; Wache et al., 2001].

9.4.3.1 Forms of Knowledge Representation

The simplest format for knowledge representation is a *dictionary.* In a sense, a dictionary may be viewed as nothing more than a table where the "terms" are the keys and their "definitions" are the values. In the most basic dictionary — disregarding etymological information, example sentences, synonyms, and antonyms — there are no links between the individual pieces of knowledge (the "terms"). Many more advanced forms of knowledge organization exist, yet the differences are sometimes subtle and thus terminology is often misused (see http://www.kmconnection.com/C_and_R_definitions.htm). From a theoretical viewpoint, when antonyms and synonyms are included, one is dealing with a *thesaurus* as opposed to a dictionary. The key difference is a critical one and one that has massive implications for network technologies: the pieces of knowledge are *linked.* Once the etymological information is added (derivation) and the synonyms are organized hierarchically (inheritance), the thesaurus progresses to the next level, *taxonomy.* The addition of hierarchical information to a thesaurus means, for instance, that no longer is "plant" simply synonymous with "flower," but a flower *is a type of* (or *subclass of*) plant. Additionally, we know that a tulip *is a type of* flower. In this way, the relations between the pieces of knowledge, or *entities,* take the form of a tree structure as the representation progresses from thesaurus to taxonomy. Now, with the tree structure, one may derive other forms of association besides "is a subclass/ is a superclass"; for example, the tulip family and the rose family are both subclasses of flower, and therefore they are related to each other as siblings. Despite this, a basic taxonomy limits the forms of associativity to these degrees of relatedness, and although such relationships can create a complex network and may prove quite useful for certain types of data analysis, there are many other ways in which entities may be related. In an all-inclusive knowledge representation, a rose may be *related to* love in general or Valentine's Day in particular. Similarly, the crocus may be associated with spring, and so on. In other words, these associations may be emotional, cultural, or temporal. The fundamental idea here is that some of the most interesting and relevant associations may be those that are discovered or traversed by a data-analysis system utilizing a reference knowledge base whose structure of entity relationships is much deeper than that of a basic taxonomy; rather than a simple tree, such a knowledge structure must be visually represented as a Web. Finally, in adding one last piece to this series of knowledge representations, we arrive at the level of *ontology,* which is most beneficial for *semantic* content management. This addition is the *labeling* of relationships; the associations are provided with contextual information. From the example above, we could express that "a rose-symbolizes-love" or "a crocus-blooms-in-Spring." Now, these entities are not *merely* associated but are associated *in a meaningful way.* Labeled relationships provide the greatest benefit in cases where two types of entities may be associated in more than one way. For example, we may know that Company A is associated with Company B, but this alone will not tell us if Company A is a competitor of Company B, or if Company A is a subsidiary of Company B, or vice versa.

9.4.3.2 Ambiguity Resolution

Returning to the flower example above, we will present an even greater challenge. Assuming an application uses a reference knowledge base which is in the form of a general but comprehensive ontology (such as the lexical database, WordNet, described below), determining the meaning of a given entity, such as "plant" or "rose," may be quite difficult. It is true that there is a well-defined instance of each word in our ontology with the meanings and associations as intended in the examples outlined previously. Still, the application also may find an instance of "plant," which is synonymous with "factory," or a color known as "rose." To resolve such ambiguities, the system must analyze associated data from the context of the extracted entity within its original source. If several other known flowers whose names are not used for describing colors were mentioned in the same document, then the likelihood of that meaning would become evident. More complex techniques may be used, such as linguistic analysis, which could determine that the word was used as a noun, while the color, "rose," would most likely have been used as an adjective. Another technique would rely upon the reference ontology where recognition of associated concepts or terms would increase the likelihood of one meaning over the other. If the document also mentioned "Valentine's Day," which we had related to "roses" in our ontology, this would also increase the likelihood of that meaning. Programmatically, the degree of likelihood may be represented as a "score" with various parameters contributing weighted components to the calculation. For such forms of analysis, factors such as proximity of the terms and structure of the document would also contribute to the algorithms for context determination.

9.4.3.3 Ontology Description Languages

With steadily growing interest in areas such as the Semantic Web, current research trends have exposed a need for standardization in ontology representation. For semantic content management, such standardization would clearly be advantageous. The potential applications for knowledge sharing are innumerable, and the cost benefit of minimizing redundancy in the construction of comprehensive domain ontologies is indisputable. Nevertheless, there are two key obstacles for such endeavors. First, the construction of a knowledge model for a given domain is a highly subjective undertaking. Decisions regarding the granularity of detail, hierarchical construction, and determination of relevant associations each offer an infinite range of options. Second, there is the inevitable need for combining independently developed ontologies via intersections and unions, or analyzing subsets and supersets. This integration of disparate ontologies into a normalized view requires intensive heuristics. If one ontology asserts that a politician *is affiliated with* a political party while another labels the same relationship as "politician *belongs to* party," the integration algorithm would need to decide if these are two distinct forms of association or if they should be merged. In the latter case, it must also decide which label to retain. Although the human ability to interpret such inconsistencies is practically instinctual, to express these same structures in a machine-readable form is another matter altogether.

Among the prerequisites of the Semantic Web that are common with those of semantic content management is this ability to deal with multiple ontologies. In one sense, the Semantic Web may be viewed as a *global* ontology that reconciles the differences among *local* ontologies and supports query processing in this environment [Calvanese et al., 2001]. Such query processing should enable the translation of the query terms into their appropriate meanings across different ontologies in order to provide the benefits of semantic search as compared to keyword-based search [Mena et al., 1996]. The challenges associated with ontology integration vary with regard to the particularities of the task. Some examples are the reuse of an existing ontological representation as a resource for the construction of a new ontology, the unification or merging of multiple ontologies to create a deeper or broader representation of knowledge, and the incorporation of ontologies into applications that may benefit from their structured data [Pinto et al., 1999].

Recently there have been many key developments in response to these challenges of ontology assimilation. XML lies at the foundation of these *ontology description languages* because the enforcement of consistent structure is a prerequisite to any form of knowledge model representation that aspires to standardization. To evolve from the structural representations afforded by XML to an infrastructure suitable for representing a semantic network of information requires the inclusion of capabilities for the representation of associations. One of the most accepted candidates in this growing field of research is RDF [W3C, 1999] and its outgrowth RDF-Schema (RDF-S) [W3C, 2003].

RDF-S provides a specification that moves beyond ontological representation capabilities to those of ontological modeling. The addition of a "schema" brings object-oriented design aspects into the semantic framework of RDF. In other words, hierarchically structured data models may be constructed with a separation between class-level definitions and instance-level data. This representation at the "class level" is the actual schema, also known as the *definitional* component, while the instances constitute the factual, or *assertional*, component. When a property's class is defined, constraints may be applied with regard to possible values, as well as which types of resource a particular instance of that property may describe.

The DARPA Agent Markup Language (DAML), in its latest manifestation as DAML+OIL (Ontology Inference Layer), expands upon RDF-S with extensions in the capabilities for constructing an ontology model based on constraints. In addition to specifying hierarchical relations, a class may be related to other classes in disjunction, union, or equality. DAML+OIL provides a description framework for restrictions in the mapping of property values to data types and objects. These restriction definitions outline such constraints as the required values for a given class or its cardinality limitations (maximum and minimum occurrences of value instances for a given property). The W3C Web Ontology Working Group (WebOnt) has created a Web Ontology Language, known as OWL (http://www.w3.org/TR/owl-ref/), which is derived from DAML+OIL and likewise follows the RDF specification.

The *F-Logic* language has also been used in ontology building applications. F-Logic, which stands for "Frame Logic," is well suited to ontology description although it was originally designed for representing any object-oriented data model. It provides a comprehensive mechanism for the description of object-oriented class definitions `including` "object identity, complex objects, inheritance, polymorphic types, query methods, encapsulation, and others" [Kifer et al., 1990]. The OntoEdit tool, developed at the AIFB of the University of Karlsruhe, is a graphical environment for building ontologies. It is built upon the framework of F-Logic, and with the OntoBroker "inference engine" and associated API, it allows for the importing and exporting of RDF-Schema representations. F-Logic also lies at the foundation of other systems that have been developed for the integration of knowledge representation models through the transformation of RDF. The first of these "inference engines" was SiLRI (Simple Logic-based RDF Interpreter [Decker et al., 1998]), which has given way to the open source transformation language, TRIPLE [Sintek and Decker, 2002], and a commercial counterpart offered by Ontoprise GmbH (http://www.ontoprise.com).

9.4.3.4 Sample Knowledge Bases

Several academic and industry-specific projects have led to the development of shareable knowledge bases as well as tools for accessing and adding content. One such knowledge base is the lexical database, WordNet, whose development began in the mid-1980s at Princeton University. WordNet is structured as a networked thesaurus in the form of a "lexical matrix," which maps *word forms* to *word meanings* with the possibility of many-to-many relationships [Miller et al., 1993]. The full range of a thesaurus' *semantic relations* may be represented in WordNet. The set of all word meanings for a given word form a *synset*. The synset may represent any of the following lexical relations: synonymy (same or similar meaning), antonymy (opposite meaning), hyponymy/hypernymy (hierarchical *is a/has a* relation), and meronymy/holonymy (*has a part/is a part of* relation). Additionally, WordNet allows for correlations between morphologically inflected forms of the same word, such as plurality, possessive forms, gerunds, participles, different verb tenses, etc. Although WordNet has been a popular and useful resource as a comprehensive thesaurus with a machine-readable syntax, it is not a formal ontology because it only represents the lexical relations listed above and does not provide contextual associations. It is capable of representing that a "branch" is synonymous with a "twig" or a "department" within an institution. If the first meaning is intended, then it will reveal that a branch is part of a tree. However, for the second meaning, it will not discover the fact that an administrative division typically has a chairman or vice president overseeing its operations. This is an example of the labeled relationships required for the representation of "real-world" information. Such associations are lacking in a thesaurus but may be stored in an ontology. WordNet has still been useful as a machine-readable lexical resource and, as such, is a candidate for assimilation into ontologies. In fact, there have been efforts to transform WordNet into an ontology with a greater ability to represent the world as opposed to merely representing language [Oltramari et al., 2002].

In the spirit of cooperation that will be required for the Semantic Web to succeed, the Open Directory Project is a free and open resource, and on the Website (http://www.dmoz.org/), it claims to be the "largest, most comprehensive human-edited directory of the Web." The directory structure is designed with browsing in mind as opposed to searching and is primarily a hierarchical categorization of Web resources, which allows multiple classifications for any given resource. Therefore, it is not an ontology; rather, it may be loosely referred to as a taxonomy of Web resources that have been manually, and therefore subjectively, classified. It is maintained by volunteers who each agree to supervise a category. While this undoubtedly raises questions with regard to the authority and consistency of the resource, its success and growth are promising signs for the future of Semantic Web.

The National Library of Medicine has developed an ontology-driven system, known as the Unified Medical Language System (UMLS), for the assimilation, organization, and retrieval of medical information. Intended for integration of data sources ranging from biology, anatomy, and organic chemistry to pharmacology, pathology, and epidemiology, it provides an invaluable resource for medical researchers and practitioners alike (http://www.nlm.nih.gov/research/umls/).

Because many of the researchers and institutions involved in the creation of these and other large knowledge bases are constantly striving for increased shareability, it is feasible that the level of standardization will soon enable the construction of a single high-level Reference Ontology that integrates these various domains of knowledge [Hovy, 1997].

9.5 Applying Semantics in ECM

9.5.1 Toolkits

Any semantic CMS must be designed in a generic way that provides flexibility, extensibility, and scalability for customized applications in any number of potential domains of knowledge. Because of these requirements, such a system should include a toolkit containing several modules for completing these necessary customization tasks. The user of such a toolkit should be able to manage the components outlined in the previous section. For each task the overall goal should be to achieve the optimum balance between configurability and automation. Ideally, these tasks are minimally interactive beyond the initialization phase. In other words, certain components of the system, such as content extraction agents and classifiers, should be fully automated after the configurable parameters are established, but a user may want to tweak the settings in response to certain undesired results. The highest degree of efficiency and quality would be achieved from a system that is able to apply heuristics upon its own results in order to maximize its precision and minimize its margin of error. For example, if in the early stages, a user makes adjustments to a particular data extractor or classification module after manually correcting a document's metadata associations and category specification, the system could recognize a pattern in the adjustments so that future occurrences of such a pattern would trigger automatic modifications of the corresponding configuration parameters.

The classification procedure should be configurable in terms of domain specification and granularity within each domain. Additionally, if such a feature is available, the user should be able to fine-tune the scoring mechanisms involved in the interaction of multiple classifier methods. If the classification module requires training sets, the method for accumulating data to be included in these sets should be straightforward. Creation of content extraction agents should also be handled within a user-friendly, graphical environment. Because tweaking parameters for the crawling and extraction agents may be necessary, the toolkit should include a straightforward testing module for these agents that produces feedback to guide the user in constructing these rules for gathering metadata as precisely as possible. The same approach should be taken when designing an ontology-modeling component. Due to the inherent complexity of an ontological knowledge representation and the importance of establishing this central component of the system, it is critical that this component provide an easily navigable, visual environment. A general description of the requirements for ontology editing and summary of several tools that address these needs may be found in Denny [2002]. Finally, an important feature for any content management system is some form of auditing mechanism. In many industries, there is a need for determining the reliability of content sources. Keeping track of this information aids in the determination of the "reliability" of content. Once again, the World Wide Web is the extreme case where it is very difficult to determine if a source is authoritative. Likewise, tracking the date and time of content entering the system is important, especially for the institutions where timeliness has critical implications — news content providers, law enforcement, financial services, etc.

The Karlsruhe Ontology and SemanticWeb Tool Suite (KAON) is an example of a semantic content management environment. It consists of a multilayered architecture of services and management utilities [Bozsak et al., 2002]. This is the same set of tools that contains the OntoEdit GUI environment for ontology modeling, which has been described in the discussion of ontology description languages above. KAON has been developed with a particular focus on the Semantic Web. Another suite of tools is offered by the ROADS project, which has been developed by the Access to Networked Resources section of eLib (the Electronic Libraries Programme). ROADS provides tools for creating and managing information portals, which they refer to as "subject gateways" [http://www.ilrt.bristol.ac.uk/roads/].

9.5.2 Semantic Metadata Extraction

Traditionally, when dealing with heterogeneous, dispersed, massive and dynamic data repositories, the overall quality or relevance of search results within that data may be inversely proportional to the number of documents to be searched. As can be seen from any major keyword-based search engine, as the size of data to be processed grows, the number of false-positives and irrelevant results grows accordingly. When these sources are dynamic (the World Wide Web again being an extreme case), the resulting "links" may point to nothing at all, or — what is often even worse for machine processing applications — they may point to different content than that which was indexed. Therefore, two major abilities are favorable in any system that crawls and indexes content from massive data repositories: the extraction of the semantic metadata from each document for increased relevance, and an automated mechanism, so that this extraction will maintain reliable and timely information. The complexity of metadata extraction among documents of varying degrees of structure presents an enormous challenge for the goal of automation.

A semantic ECM toolkit should provide a module for creating extractor agents, which act as *wrappers* to content sources (e.g., a Website or file-system). The agent will follow certain rules for locating and extracting the relevant metadata. Obviously, this is not a trivial task when dealing with variance in the source structure. While the World Wide Web offers the greatest challenges in this regard, it is understandably the most popular resource for extraction. Several extraction wrapper technologies have focused upon crawling and retrieving data from Web pages such as the WysiWyg Web Wrapper Factory (W4F), which provides a graphical environment and a proprietary language for formulating retrieval and extraction rules [Sahuget and Azavant, 1999]. ANDES is a similar wrapper technology that incorporates regular expressions and XPath rules for exploiting structure within a document [Myllymaki, 2001]. Semi-automatic wrapper-generation is possible with the XML-based XWRAP toolkit, which enables interactive rule formulation in its test environment. Using an example input document, the user selects "semantic tokens," and the application attempts to create extraction rules for these items, but because the structure of input documents may vary considerably, the user must enter new URLs for testing and adjust the rules as necessary [Liu et al., 2000]. Likewise, S-CREAM (Semi-automatic CREAtion of Metadata) allows the user to manually annotate documents and later applies these annotations within a training mechanism to enable automated annotation based on the manual results. The process is aided by the existence of an ontology as a reference knowledge base for associating the "relational metadata" with a given document [Handschuh et al., 2002]. A fully automatic method for extracting and enhancing metadata is not only the preferred method for the obvious reason that it minimizes manual supervision of the system, but such a method is also most flexible in that it will be equally integrated into a push-or-pull data-aggregation environment [Hammond et al., 2002]. Although the majority of research in crawling and extraction technologies has been undertaken in academic institutions, commercial metadata extraction products have been developed by corporations such as Semagix [Sheth et al., 2002] and Ontoprise (http://www.ontoprise.com).

9.5.3 Semantic Metadata Annotation

It has been stressed that achieving interoperability among heterogeneous and autonomous data sources in a networked environment requires some ability to create a normalized view. Minimally, this could be a "metadata snapshot" generated by semantic annotation. If an ontology that is comprehensive for the domain at hand exists in the back-end and the interfacing mechanisms for handling distributed data of various formats reside on the front-end, then after filtering the input through a classifier to determine its contextual domain, the system will be able to apply "tagging" or "markup" for the recognized entities. Because of the inclusion of the classification component, the tagged entities would be contextually relevant. An advanced system would also have the ability to *enhance* the content by analyzing known relationships between the recognized entities and those that should be associated with the entity due to implied reference. For example, a story about a famous sports personality may or may not mention that player's team, but the metadata enhancement process would be able to include this information. Similarly, a business article may not include a company's ticker symbol, but a stock analyst searching for documents by ticker symbol may be interested in the article. If the metadata enhancement had added the ticker symbol, which it determined from its relationship with the company name, then the analyst would be able to find this article when searching with the ticker symbol parameter alone. No keyword-based search engine would have returned such an article in its result set because the ticker symbol's value is simply not present in the article. Implied entities such as these may be taken for granted by a human reading a document, but when a machine is responsible for the analysis of content, an ontology-driven classifier coupled with a domain-specific metadata annotator, will enable the user to find highly relevant information in a timely manner. Figure 9.3 shows an example of semantic annotation of a document. Note that the entities are not only highlighted but the types are also labeled.

9.5.4 Semantic Querying

Two broad categories account for a majority of human information gathering: searching and browsing. Searching implies a greater sense of focus and direction whereas the connotations of "browsing" are that of aimless wandering with no predefined criteria to satisfy. Nevertheless, it is increasingly the case that browsing technologies are employed to locate highly precise information. For example, in law enforcement, a collection of initial evidence may not provide any conclusive facts, yet this same evidence may reveal nonobvious relationships when taken as a starting point within an ontology-driven semantic browsing application. Ironically, searching for information with most keyword-based search engines typically leads the user into the process of browsing before finding the intended information if indeed it is found at all.

The term *query* is more accurate for discussing the highly configurable type of search that may be performed in a semantic content management application. A query consists of not only the search term or terms, but also a set of optional parameter values. For example, if these parameters correspond to the same categories that drive the classification mechanism, then the search term or terms may be mapped into the corresponding entities within the domain-specific ontology. The results of the query thus consist of documents whose metadata had been extracted and which contained references to these same entities. In this manner, semantic querying provides much higher precision than keyword-based search owing to its ability to retrieve contextually relevant results. Clearly, semantic querying is enabled by the semantic ECM system that we have outlined in this chapter. It requires the presence of a domain-specific ontology and the processes that utilize this ontology — the ontology-driven classification of content, and the extraction of domain-specific and semantically enhanced metadata for that content.

To fully enable custom applications of this semantic querying in a given enterprise, a semantic ECM system should also include flexible and extensible APIs. In most cases, the users of such a system will require a custom front-end application for accessing information that is of particular interest within their organization. For example, if an API allows for the simple creation of a dynamic Web-based interface to

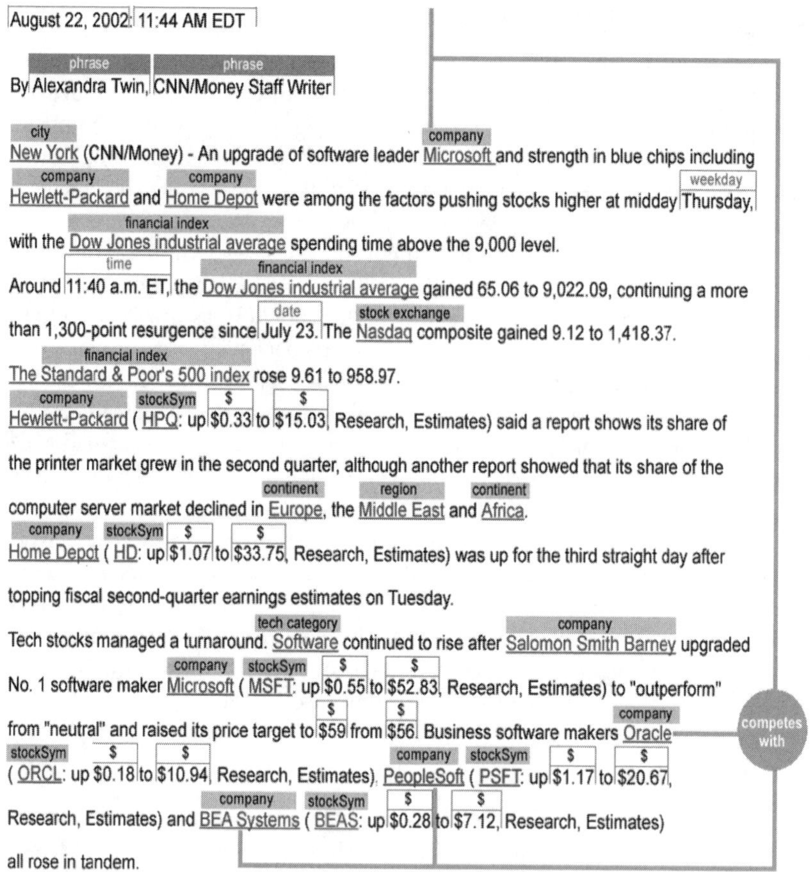

FIGURE 9.3 Example of semantic metadata annotation. *Note that named entities, currency values, and dates and times are highlighted and labeled according to their classification. Also relationships between entities are labeled. This information comes from a reference ontology (Semagix, Inc.).*

the underlying system, then the application will appeal to a wide audience without compromising its capabilities. While APIs enable easier creation and extension of the ontology, visualization tools offer a complimentary advantage for the browsing and viewing of the ontology on a schema or instance level. Figure 9.4 shows one such tool, the Semagix Visualizer (http://www.semagix.com).

9.5.5 Knowledge Discovery

It has been stressed that machine processing is indispensable when dealing with massive data sources within humanly insurmountable time constraints. Another major benefit of machine processing related to semantic content management is the ability to discover nonobvious associations within that content. For example, while manually sifting through documents or browsing files, it is highly unlikely that one would happen to discover a relationship between two persons that consisted of a chain of three or more associations. For example, in a law enforcement scenario where two suspects, "Person A" and "Person B," are under investigation, it may be important to know that Person A lived in the same apartment complex as the brother of a man who was a coworker of a woman who shared a bank account with Person B. Similarly complex associations may be pertinent for a financial institution processing credit reports or a federal agency doing a background check for job applicants.

Obviously, the exact definition of such scenarios will differ considerably dependent upon the application. Therefore, any semantic content management system that aims to support automated knowledge

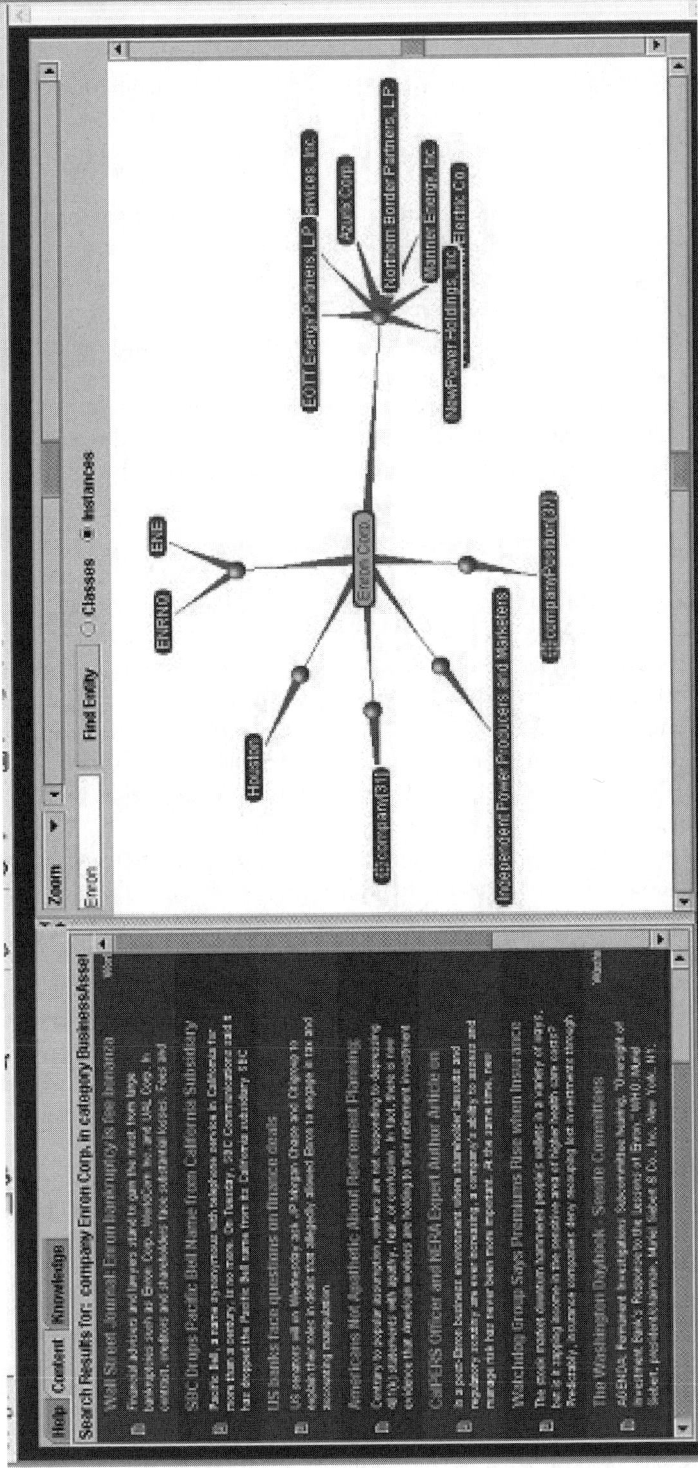

FIGURE 9.4 An example of an ontology visualization tool. The Semagix Visualizer provides a navigable view of the ontology on the right-hand side while the left-hand panel displays either associated documents or more detailed knowledge.

discovery should have a highly configurable module for designing templates for such procedures. It would be necessary for a user to determine which types of entities may be meaningfully associated with each other, which relationships are important for traversal between entities, and possibly even a weighted scoring mechanism for calculating the relative level of association for a given relationship. For example, two people working for the same company would most likely receive a higher "weight" than two people living in the same city. Nevertheless, the procedure could be programmed to handle even more advanced analytics such as factoring in the size of the company and the size of the city so that two people living in New York City would receive very little "associativity" compared with two people in Brunswick, Nebraska.

Other less-directed analysis may be employed with very similar processing. For example, when dealing with massive data repositories, knowledge discovery techniques may find associations between entities that were mentioned together in documents more than 10 times (or some predetermined threshold) and flag these as "related" entities to be manually confirmed. This type of application may be applied for finding nonobvious patterns in large data sets relatively quickly. As a filtering mechanism, such a procedure could significantly amplify timeliness and relevance, and in many cases these results would have been impossible to obtain from manual analysis regardless of the time constraints. A framework of complex semantic relationships is presented in Sheth et al. [2003], and a formal representation of one type of complex relationships called *semantic associations* is presented in Anyanwu and Sheth [2003].

9.6 Conclusion

Enterprise information systems comprise heterogeneous, distributed, and massive data sources. Content from these sources differs systemically, structurally, and syntactically, and accessing that content may require using multiple protocols. Despite these challenges, timeliness and relevance are absolutely required when searching for information, and therefore the amount of manual interaction must be minimized. To overcome these challenges, a system for managing this content must achieve interoperability, and the key to this is semantics. However, enabling a machine to read in documents of varying degrees of structure from heterogeneous data sources and "understand" the meaning of each document in order to find associations among those documents is not a trivial task.

Advanced classification techniques may be employed for filtering data into precise categories or domains. The domains should be defined as metadata schemas, which basically outline the items of interest that may occur within a document in a given category (such as "team" in the sports domain or "ticker symbol" in the business domain). Therefore, each piece of content may be annotated (or "tagged") with the instances of these metadata classes. As a collection of semantic metadata, a document can become significantly more machine-readable than in its original format. Moreover, the excess has been removed so that only the contextually relevant information remains.

This notion of tagging documents with the associated metadata according to a predefined schema is fundamental for the proponents of the Semantic Web. Metadata schemas also lie at the foundation of most languages used for describing ontologies. An ontology provides a valuable resource for any semantic content management system because the metadata within a document may be more or less relevant depending upon its location within the referential knowledge base. Furthermore, an ontology may be used to actually enrich the metadata associated with a document by including implicit entities that are closely related to the explicitly mentioned entities in the given context.

Applications that make use of ontology-driven metadata extraction and annotation are becoming increasingly popular within both the academic and commercial environments. Because of their versatility and extensibility, such applications are suitable candidates for a wide range of content management systems, including Document Management, Web Content Management, Digital Asset Management, and Enterprise Application Integration. The leading vendors have developed refined toolkits for managing and automating the necessary tasks of semantic content management. As the visibility of these products increases, traditional content management systems will be superseded by systems that enable heightened relevance in information retrieval by employing ontology-driven classification and metadata extraction. These semantic-based systems will permeate the enterprise market.

References

Anyanwu, Kemafor and Amit Sheth. The ρ Operator: Discovering and Ranking Associations on the Semantic Web, Proceedings of the Twelfth International World Wide Web Conference, Budapest, Hungary, May 2003.

Berners-Lee, Tim, James Hendler, and Ora Lassila. The Semantic Web: A new form of Web content that is meaningful to computers will unleash a revolution of new possibilities, *Scientific American,* May 2001.

Bornhövd, Christof. Semantic Metadata for the Integration of Web-based Data for Electronic Commerce, International Workshop on Advance Issues of E-Commerce and Web-Based Information Systems, 1999.

Bozsak, A. E., M. Ehrig, S. Handschuh, Hotho et al. KAON — Towards a Large Scale Semantic Web. In: K. Bauknecht, A. Min Tjoa, G. Quirchmayr (Eds.): Proceedings of the 3rd International Conference on E-Commerce and Web Technologies (EC-Web 2002), pp. 304–313, 2002.

Calvanese, Diego, Giuseppe De Giacomo, and Maurizio Lenzerini. A Framework for Ontology Integration, In Proceedings of the First Semantic Web Working Symposium, pp. 303–316, 2001.

Cheeseman, Peter and John Stutz. Bayesian classification (AutoClass): Theory and results, in *Advances in Knowledge Discovery and Data Mining,* Usama M. Fayyad, Gregory Piatetsky-Shapiro, Padhraic Smyth, and Ramasamy Uthurusamy, Eds., pp. 153–180, 1996.

DCMI. Dublin Core Metadata Initiative, 2002. URL: http://dublincore.org/2002/08/13/dces.

Decker, Stefan, Dan Brickley, Janne Saarela, and Jurgen Angele. A Query and Inference Service for RDF, in Proceedings of the W3C Query Languages Workshop (QL-98), Boston, MA, December 3–4, 1998.

Denny, M. Ontology Building: A Survey of Editing Tools, 2002. available at: http://www.xml.com/pub/a/2002/11/06/ontologies.html.

Frasconi, Paolo, Giovanni Soda, and Alessandro Vullo. Hidden Markov models for text categorization in multi-page documents, *Journal of Intelligent Information Systems,* 18(2–3), pp. 195–217, 2002.

Gruber, Thomas. The role of common ontology in achieving sharable, reusable knowledge bases, in *Principles of Knowledge Representation and Reasoning,* James Allen, Richard Fikes, and Erik Sandewall, Eds., Morgan Kaufman, San Mateo, CA, pp. 601–602, 1991.

Hammond, Brian, Amit Sheth, and Krzysztof Kochut. Semantic enhancement engine: A modular document enhancement platform for semantic applications over heterogeneous content," in *Real World Semantic Web Applications,* V. Kashyap and L. Shklar, Eds., IOS Press, 2002.

Handschuh, Siegfried, Steffen Staab, and Fabio Ciravegna. "S-CREAM: Semi-automatic Creation of Metadata," in 13th International Conference on Knowledge Engineering and Knowledge Management, October 2002.

Hovy, Eduard. A Standard for Large Ontologies, Workshop on Research and Development Opportunities in Federal Information Services, Arlington, VA, May 1997. Available at: http://www.isi.edu/nsf/papers/hovy2.htm.

Ipeirotis, Panagiotis, Luis Gravano, and Mehran Sahami. Automatic Classification of Text Databases through Query Probing, in Proceedings of the ACM SIGMOD Workshop on the Web and Databases, May 2000.

Joachims, Thorsten. Text Categorization with Support Vector Machines: Learning with Many Relevant Features, in *Proceedings of the Tenth European Conference on Machine Learning,* pp. 137–142, 1998.

Kashyap, Vipul and Amit Sheth. Semantics-Based Information Brokering, in Proceedings of the Third International Conference on Information and Knowledge Management (CIKM), pp. 363–370, November 1994.

Kashyap, Vipul, Kshitij Shah, and Amit Sheth. Metadata for building the MultiMedia Patch Quilt, in *Multimedia Database Systems: Issues and Research Directions,* S. Jajodia and V. S. Subrahmaniun, Eds., Springer-Verlag, pp. 297–323, 1995.

Kifer, Michael, Georg Lausen, and James Wu. Logical Foundations of Object-Oriented and Frame-Based Languages. Technical Report 90/14, Department of Computer Science, State University of New York at Stony Brook (SUNY), June 1990.

Lim, Ee-Peng, Zehua Liu, and Dion Hoe-Lian Goh. A Flexible Classification Scheme for Metadata Resources, in Proceedings of Digital Library — IT Opportunities and Challenges in the New Millennium, Beijing, China, July 8–12, 2002.

Liu, Ling, Calton Pu, and Wei Han. XWRAP: An XML-Enabled Wrapper Construction System for Web Information Sources, in Proceedings of the International Conference on Data Engineering, pp. 611–621, 2000.

Losee, Robert M. and Stephanie W. Haas, Sublanguage terms: dictionaries, usage, and automatic classification, in *Journal of the American Society for Information Science,* 46(7), pp. 519–529, 1995.

LTSC. Draft Standard for Learning Object Metadata, 2000. IEEE Standards Department. URL: http://ltsc.ieee.org/doc/wg12/LOM_WD6-1_1.doc.

Mena, Eduardo, Arantza Illarramendi, Vipul Kashyap, and Amit Sheth. OBSERVER: An Approach for Query Processing in Global Information Systems based on Interoperation across Pre-existing Ontologies, in Conference on Cooperative Information Systems. pp. 14–25, 1996.

Miller, George, Richard Beckwith, Christiane Fellbaum, Derek Gross, and Katherine Miller. Introduction to WordNet: An On-line Lexical Database. Revised August 1993.

Myllymaki, Jussi. Effective Web Data Extraction with Standard XML Technologies, in Proceedings of the 10th International Conference on World Wide Web, pp. 689–696, 2001.

Oltramari, Alessandro, Aldo Gangemi, Nicola Guarino, and Claudio Masolo. Restructuring WordNet's Top-Level: The OntoClean approach, Proceedings of LREC 2002 (OntoLex Workshop) 2002.

Pinto, H. Sofia. Asuncion Gomez-Perez, and Joao P. Martins. Some Issues on Ontology Integration, 1999.

Sahuget, Arnaud and Fabien Azavant. Building Lightweight Wrappers for Legacy Web Data-Sources Using W4F. Proceedings of the International Conference on Very Large Data Bases, pp. 738–741, 1999.

Sebastiani, Fabrizio. Machine learning in automated text categorization, in *ACM Computing Surveys,* 34(1), March 2002.

Sheth, Amit. Changing Focus on interoperability in information systems: From system, syntax, structure to semantics, in *Interoperating Geographic Information Systems,* M. Goodchild, M. Egenhofer, R. Fegeas, and C. Kottman, Eds., Kluwer, Dordrecht, Netherlands, 1998.

Sheth, Amit and Wolfgang Klas, Eds. *Multimedia Data Management: Using Metadata to Integrate and Apply Digital Data,* McGraw Hill, New York, 1998.

Sheth, Amit, Clemens Bertram, David Avant, Brian Hammond, Krzysztof Kochut, and Yash Warke. Semantic content management for enterprises and the Web, *IEEE Internet Computing,* July/August 2002.

Sheth, Amit, I. Budak Arpinar, and Vipul Kashyap. Relationships at the heart of Semantic Web: Modeling, discovering, and exploiting complex semantic relationships, *Enhancing the Power of the Internet: Studies in Fuzziness and Soft Computing,* M. Nikravesh, B. Azvin, R. Yager, and L. Zadeh, Eds., Springer-Verlag, 2003.

Sheth, Amit and James Larson. Federated database systems for managing distributed, heterogeneous, and autonomous databases, *ACM Computing Surveys,* 22(3), pp. 183–236, September 1990.

Sintek, Michael and Stefan Decker. TRIPLE — A Query, Inference, and Transformation Language for the Semantic Web, International Semantic Web Conference, Sardinia, June 2002.

Snijder, Ronald. Metadata Standards and Information Analysis: A Survey of Current Metadata Standards and the Underlying Models, Electronic resource, available at http://www.geocities.com/ronald-snijder/, 2001.

Sure, York, Juergen Angele, and Steffen Staab. OntoEdit: Guiding Ontology Development by Methodology and Inferencing, Proceedings of the International Conference on Ontologies, Databases and Applications of Semantics ODBASE 2002.

Wache, Holger, Thomas Vögele, Ubbo Visser, Heiner Stuckenschmidt, Gerhard Schuster, Holger Neumann, and Sebastian Hübner. Ontology-based integration of information: A survey of existing approaches, in *IJCAI-01 Workshop: Ontologies and Information Sharing,* H. Stuckenschmidt, Ed., pp. 108–117, 2001.

W3C. Resource Description Framework (RDF) Model and Syntax Specification, 1999. URL: http://www.w3.org/TR/REC-rdf-syntax/.

W3C. RDF Vocabulary Description Language 1.0: RDF Schema, 2003. URL: http://www.w3.org/TR/rdf-schema/.

10

Conversational Agents

James Lester

Karl Branting

Bradford Mott

CONTENTS

Abstract 10-1
10.1 Introduction 10-1
10.2 Applications 10-2
10.3 Technical Challenges 10-3
10.3.1 Natural Language Requirements 10-3
10.3.2 Enterprise Delivery Requirements 10-7
10.4 Enabling Technologies 10-8
10.4.1 Natural Language Processing Technologies 10-8
10.4.2 Enterprise Integration Technologies 10-12
10.5 Conclusion 10-14
References 10-15

Abstract

Conversational agents integrate computational linguistics techniques with the communication channel of the Web to interpret and respond to statements made by users in ordinary natural language. Web-based conversational agents deliver high volumes of interactive text-based dialogs. Recent years have seen significant activity in enterprise-class conversational agents. This chapter describes the principal applications of conversational agents in the enterprise, and the technical challenges posed by their design and large-scale deployments. These technical challenges fall into two categories: accurate and efficient natural-language processing; and the scalability, performance, reliability, integration, and maintenance requirements posed by enterprise deployments.

10.1 Introduction

The Internet has introduced sweeping changes in every facet of contemporary life. Business is conducted fundamentally differently than in the pre-Web era. We educate our students in new ways, and we are seeing paradigm shifts in government, healthcare, and entertainment. At the heart of these changes are new technologies for communication, and one of the most promising communication technologies is the conversational agent, which marries agent capabilities with computational linguistics.

Conversational agents exploit natural-language technologies to engage users in text-based information-seeking and task-oriented dialogs for a broad range of applications. Deployed on retail Websites, they respond to customers' inquiries about products and services. Conversational agents associated with financial services' Websites answer questions about account balances and provide portfolio information. Pedagogical conversational agents assist students by providing problem-solving advice as they learn. Conversational agents for entertainment are deployed in games to engage players in situated dialogs about the game-world events. In coming years, conversational agents will support a broad range of applications in business enterprises, education, government, healthcare, and entertainment.

1-58488-381-2/05/$0.00+$1.50
© 2005 by CRC Press LLC

Recent growth in conversational agents has been propelled by the convergence of two enabling technologies. First, the Web emerged as a universal communications channel. Web-based conversational agents are scalable enterprise systems that leverage the Internet to simultaneously deliver dialog services to large populations of users. Second, computational linguistics, the field of artificial intelligence that focuses on natural-language software, has seen major improvements. Dramatic advances in parsing technologies, for example, have significantly increased natural-language understanding capabilities.

Conversational agents are beginning to play a particularly prominent role in one specific family of applications: enterprise software. In recent years, the demand for cost-effective solutions to the customer-service problem has increased dramatically. Deploying automated solutions can significantly reduce the high proportion of customer service budgets devoted to training and labor costs. By exploiting the enabling technologies of the Web and computational linguistics noted above, conversational agents offer companies the ability to provide customer service much more economically than with traditional models. In *customer-facing* deployments, conversational agents interact directly with customers to help them obtain answers to their questions. In *internal-facing* deployments, they converse with customer service representatives to train them and help them assist customers.

In this chapter we will discuss Web-based conversational agents, focusing on their role in the enterprise. We first describe the principal applications of conversational agents in the business environment. We then turn to the technical challenges posed by their development and large-scale deployments. Finally, we review the foundational natural-language technologies of interpretation, dialog management, and response execution, as well as an enterprise architecture that addresses the requirements of conversational scalability, performance, reliability, "authoring," and maintenance in the enterprise.

10.2 Applications

Effective communication is paramount for a broad range of tasks in the enterprise. An enterprise must communicate clearly with its suppliers and partners, and engaging clients in an ongoing dialog — not merely metaphorically but also literally — is essential for maintaining an ongoing relationship. Communication characterized by information-seeking and task-oriented dialogs is central to five major families of business applications:

- *Customer service:* Responding to customers' general questions about products and services, e.g., answering questions about applying for an automobile loan or home mortgage.
- *Help desk:* Responding to internal employee questions, e.g., responding to HR questions.
- *Website navigation:* Guiding customers to relevant portions of complex Websites. A "Website concierge" is invaluable in helping people determine where information or services reside on a company's Website.
- *Guided selling:* Providing answers and guidance in the sales process, particularly for complex products being sold to novice customers.
- *Technical support:* Responding to technical problems, such as diagnosing a problem with a device.

In commerce, clear communication is critical for acquiring, serving, and retaining customers. Companies must educate their potential customers about their products and services. They must also increase customer satisfaction and, therefore, customer retention, by developing a clear understanding of their customers' needs. Customers seek answers to their inquiries that are correct and timely. They are frustrated by fruitless searches through Websites, long waits in call queues to speak with customer service representatives, and delays of several days for email responses.

Improving customer service and support is essential to many companies because the cost of failure is high: loss of customers and loss of revenue. The costs of providing service and support are high and the quality is low, even as customer expectations are greater than ever. Achieving consistent and accurate customer responses is challenging and response times are often too long. Effectiveness is, in many cases, further reduced as companies transition increasing levels of activity to Web-based self-service applications, which belong to the customer relationship management software sector.

Over the past decade, customer relationship management (CRM) has emerged as a major class of enterprise software. CRM consists of three major types of applications: sales-force automation, marketing, and customer service and support. Sales-force automation focuses on solutions for lead tracking, account and contact management, and partner relationship management. Marketing automation addresses campaign management and email marketing needs, as well as customer segmentation and analytics. Customer-service applications provide solutions for call-center systems, knowledge management, and e-service applications for Web collaboration, email automation, and live chat. It is to this third category of customer service systems that conversational agent technologies belong.

Companies struggle with the challenges of increasing the availability and quality of customer service while controlling their costs. Hiring trained personnel for call centers, live chat, and email response centers is expensive. The problem is exacerbated by the fact that service quality must be delivered at a level where customers are comfortable with the accuracy and responsiveness.

Companies typically employ multiple channels through which customers may contact them. These include expensive support channels such as phone and interactive voice response systems. Increasingly, they also include Web-based approaches because companies have tried to address increased demands for service while controlling the high cost of human-assisted support. E-service channels include live chat and email, as well as search and automated email response.

The tradeoff between cost and effectiveness in customer support presents companies with a dilemma. Although quality human-assisted support is the most effective, it is also the most expensive. Companies typically suffer from high turnover rates which, together with the costs of training, further diminish the appeal of human-assisted support. Moreover, high turnover rates increase the likelihood that customers will interact with inexperienced customer service representatives who provide incorrect and inconsistent responses to questions.

Conversational agents offer a solution to the cost vs. effectiveness tradeoff for customer service and support. By engaging in automated dialog to assist customers with their problems, conversational agents effectively address sales and support inquiries at a much lower cost than human-assisted support. Of course, conversational agents cannot enter into conversations about all subjects — because of the limitations of natural-language technologies they can only operate in circumscribed domains — but they can nevertheless provide a cost-effective solution in applications where question-answering requirements are bounded. Fortunately, the applications noted above (customer service, help desk, Website navigation, guided selling, and technical support) are often characterized by subject-matter areas restricted to specific products or services. Consequently, companies can meet their business objectives by deploying conversational agents that carry on dialogs about a particular set of products or services.

10.3 Technical Challenges

Conversational agents must satisfy two sets of requirements. First, they must provide sufficient language processing capabilities that they can engage in productive conversations with users. They must be able to understand users' questions and statements, employ effective dialog management techniques, and accurately respond at each "conversational turn." Second, they must operate effectively in the enterprise. They must be scalable and reliable, and they must integrate cleanly into existing business processes and enterprise infrastructure. We discuss each of these requirements in turn.

10.3.1 Natural Language Requirements

Accurate and efficient natural language processing is essential for an effective conversational agent. To respond appropriately to a user's utterance,[1] a conversational agent must (1) interpret the utterance, (2) determine the actions that should be taken in response to the utterance, and (3) perform the actions,

[1] An *utterance* is a question, imperative, or statement issued by a user.

which may include replying with text, presenting Web pages or other information, and performing system actions such as writing information to a database. For example, if the user's utterance were:

(1) I would like to buy it now

the agent must first determine the literal meaning of the utterance: the user wants to purchase something, probably something mentioned earlier in the conversation. In addition, the agent must infer the goals that the user sought to accomplish by making an utterance with that meaning. Although the user's utterance is in the form of an assertion, it was probably intended to express a request to complete a purchase.

Once the agent has interpreted the statement, it must determine how to act. The appropriate actions depend on the current goal of the agent (e.g., selling products or handling complaints), the dialog history (the previous statements made by the agent and user), and information in databases accessible to the agent, such as data about particular customers or products. For example, if the agent has the goal of selling products, the previous discussion identified a particular consumer item for sale at the agent's Website, and the product catalog shows the item to be in stock, the appropriate action might be to present an order form and ask the user to complete it. If instead the previous discussion had not clearly identified an item, the appropriate action might be to elicit a description of a specific item from the user. Similarly, if the item were unavailable, the appropriate action might be to offer the user a different choice.

Finally, the agent must respond with appropriate actions. The appropriate actions might include making a statement, presenting information in other modalities, such as product photographs, and taking other actions, such as logging information to a database. For example, if the appropriate action were to present an order form to the user and ask the user to complete it, the agent would need to retrieve or create a statement such as "Great! Please fill out the form below to complete your purchase," create or retrieve a suitable Web page, display the text and Web page on the user's browser, and log the information. Figure 10.1 depicts the data flow in a conversational agent system.

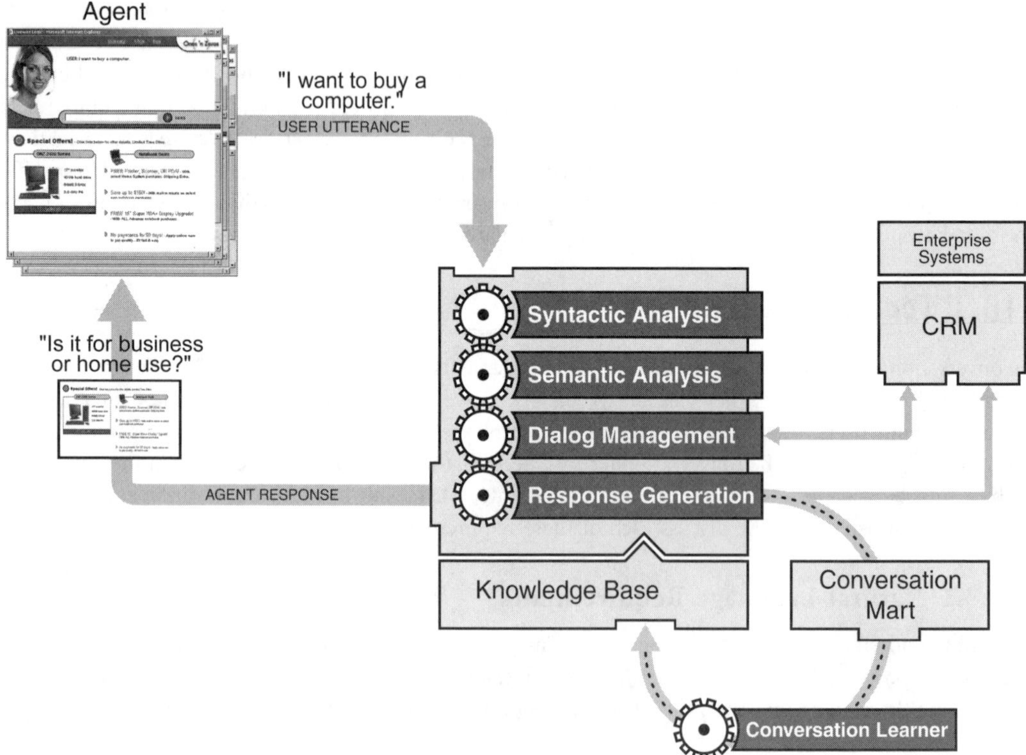

FIGURE 10.1 Data flow in a conversational agent.

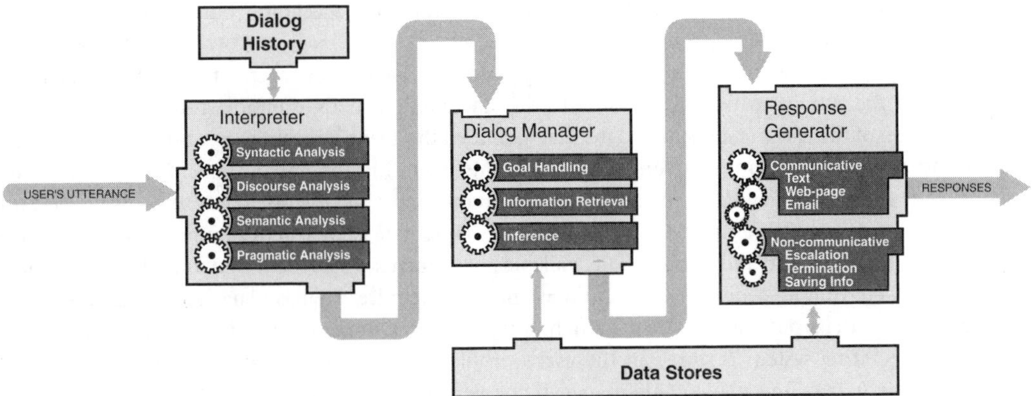

FIGURE 10.2 The primary natural-language components of a conversational agent.

The three primary components in the processing of each utterance are shown in Figure 10.2. The first component in this architecture, the Interpreter, performs four types of analysis of the user's statement: syntactic, discourse, semantic, and pragmatic. Syntactic analysis consists of determining the grammatical relationships among the words in the user's statement. For example, in the sentence:

(2) I would like a fast computer

syntactic analysis would produce a parse of the sentence showing that "would like" is the main verb, "I" is the subject, and "a fast computer" is the object. Although many conversational agents (including the earliest) rely on pattern matching without any syntactic analysis [Weizenbaum, 1966], this approach cannot scale. As the number of statements that the agent must distinguish among increases, the number of patterns required to distinguish among the statements grows rapidly in number and complexity.[2*] Discourse analysis consists of determining the relationships among multiple sentences. An important component of discourse analysis is *reference resolution,* the task of determining the entity denoted by a referring expression, such as the "it" in "I would like to buy it now." A related problem is interpretation of *ellipsis,* that is, material omitted from a statement but implicit in the conversational context. For example, "Wireless" means "I would like the wireless network," in response to the question, "Are you interested in a standard or wireless network?", but the same utterance means "I want the wireless PDA," in response to the question, "What kind of PDA would you like?"

Semantic analysis consists of determining the meaning of the sentence. Typically, this consists of representing the statement in a canonical formalism that maps statements with similar meaning to a single representation and that facilitates the inferences that can be drawn from the representation. Approaches to semantic analysis include the following:

- Replace each noun and verb in a parse with a word sense that corresponds to a set of synonymous words, such as WordNet synsets [Fellbaum, 1999].
- Represent the statement as a case frame [Fillmore, 1968], dependency tree [Harabagiu et al., 2000], or logical representation, such as first-order predicate calculus.

Finally, the Interpreter must perform pragmatic analysis, determining the pragmatic effect of the utterance, that is, the speech (or communication) act [Searle, 1979] that the utterance performs. For example, "Can you show me the digital cameras on sale?" is in the form of a question, but its pragmatic effect is a request to display cameras on sale. "I would like to buy it now" is in the form of a declaration, but its pragmatic effect is also a request. Similarly, the pragmatic effect of "I don't have enough money,"

[2*] In fact, conversational agents must address two forms of scalability: *domain scalability,* as discussed here, and *computational scalability,* which refers to the ability to handle large volumes of conversations and is discussed in Section 10.3.2.

is a refusal in response to the question "Would you like to proceed to checkout?" but a request in response to "Is there anything you need from me?"

The interpretation of the user's statement is passed to a Dialog Manager, which is responsible for determining the actions to take in response to the statement. The appropriate actions depend on the interpretation of the user's statement and the dialog state of the agent, which represents the agent's current conversation goal. In the simplest conversational agents, there may be only a single dialog state, corresponding to the goal of answering the next question. In more complex agents, a user utterance may cause a transition from one dialog state to another. The new dialog state is, in general, a function of the current state, the user's statement, and information available about the user and the products and services under discussion. Determining a new dialog state may therefore require database queries and inference. For example, if the user's statement is, "What patch do I need for my operating system?" and the version of the user's operating system is stored in the user's profile, the next dialog state may reflect the goal of informing the user of the name of the patch. If the version of the operating system is unknown, the transition may be to a dialog state reflecting the goal of eliciting the operating system version.

The Dialog Manager is responsible for detecting and responding to changes in topic. For example, if a user's question cannot be answered without additional information from the user, the dialog state must be revised to reflect the goal of eliciting the additional information. Similarly, if a user fails to understand a question and asks for a clarification, the dialog state must be changed to a state corresponding to the goal of providing the clarification. When the goal of obtaining additional information or clarification is completed, the Dialog Manager must gracefully return to the dialog state at which the interruption occurred.

The final component is the Response Generator. Responses fall into two categories: communications to the user, such as text, Web pages, email, or other communication modalities; and noncommunication responses, such as updating user profiles (e.g., if the user's statement is a declaration of information that should be remembered, such as "My OS is Win-XP"), escalating from a conversational agent to a customer service representative (e.g., if the agent is unable to handle the conversation), and terminating the dialog when it is completed. The responses made by the agent depend on the dialog state resulting from the user's statement (which represents the agent's current goals) and the information available to the agent through its dialog history, inference, or other data sources. For example, if the current dialog state corresponds to the goal of informing the user of the cost of an item for sale at a Website and the price depends on whether the user is a repeat customer, the response might depend on the information in the dialog history concerning the status of the user and the result of queries to product catalogs concerning alternative prices.

Responses typically include references to existing content that has been created throughout the enterprise. Repurposing content is particularly important when the products and services that a response addresses change continually. Centralized authoring, validation, and maintenance of responses facilitate consistency and drastically reduce maintenance costs.

Enterprise applications of conversational agents impose several constraints not generally present in other forms of conversational agents. First, high accuracy and graceful degradation of performance are very important for customer satisfaction. Misunderstandings in which the agent responds as though the user had stated something other than what the user intended to say (false positives) can be very frustrating and difficult for the agent to recover from, particularly in dialog settings. Once the agent has started down the wrong conversational path, sophisticated dialog management techniques are necessary to detect and recover from the error. Uncertainty by the agent about the meaning of a statement (false negatives) can also be frustrating to the user if the agent repeatedly asks users to restate their questions. It is often preferable for the agent to present a set of candidate interpretations and ask the users to choose the interpretation they intended.

Second, it is essential that authoring be easy enough to be performed by nontechnical personnel. Knowledge bases are typically authored by subject matter experts in marketing, sales, and customer care departments who have little or no technical training. They cannot be expected to create scripts or programs; they certainly cannot be expected to create or modify grammars consisting of thousands of productions (grammar rules). Authoring tools must therefore be usable by personnel who are nontechnical but who can nonetheless provide examples of questions and answers. State-of-the-art authoring suites exploit machine learning and other corpus-based and example-based techniques. They induce linguistic knowledge from examples, so authors are typically not even aware of the existence of the grammar. Hiding the details of linguistic knowledge and processing from authors is essential for conversational agents delivered in the enterprise.

10.3.2 Enterprise Delivery Requirements

In addition to the natural language capabilities outlined above, conversational agents can be introduced into the enterprise only if they meet the needs of a large organization. To do so, they must provide a "conversational QoS" that enables agents to enter into dialogs with thousands of customers on a large scale. They must be scalable, provide high throughput, and guarantee reliability. They must also offer levels of security commensurate with the conversational subject matter, integrate well with the existing enterprise infrastructure, provide a suite of content creation and maintenance tools that enable the enterprise to efficiently author and maintain the domain knowledge, and support a broad range of analytics with third-party business intelligence and reporting tools.

10.3.2.1 Scalability

Scalability is key to conversational agents. Because the typical enterprise that deploys a conversational agent does so to cope with extraordinarily high volumes of inbound contacts, conversational agents must scale well. To offer a viable solution to the contemporary enterprise, conversational agents must support on the order of tens of thousands of conversations each day. Careful capacity planning must be undertaken prior to deployment. Conversational agents must be architected to handle ongoing expanded rollouts to address increased user capacity. Moreover, because volumes can increase to very high levels during crisis periods, conversational agents must support rapid expansions of conversations on short notice. Because volume is difficult to predict, conversational agents must be able to dynamically increase all resources needed to handle unexpected additional dialog demand.

10.3.2.2 Performance

Conversational agents must satisfy rigorous performance requirements, which are measured in two ways. First, agents must supply a conversational throughput that addresses the volumes seen in practice. Although the loads vary from one application to another, agents must be able to handle on the order of hundreds of utterances per minute, with peak rates in the thousands. Second, agents must also provide guarantees on the number of simultaneous conversations as well as the number of simultaneous utterances that they can support. In peak times, a large enterprise's conversational agent can receive a very large volume of questions from thousands of concurrent users that must be processed as received in a timely manner to ensure adequate response times. As a rough guideline, agents must provide response tunes in a few milliseconds so that the total response time (including network latency) is within the range of one or two seconds.[3]

10.3.2.3 Reliability

For all serious enterprise deployments, conversational reliability and availability are critical. Conversational agents must be able to reliably address users' questions in the face of hardware and software failures. Failover mechanisms specific to conversational agents must be in place. For example, if a conversational agent server goes down, then ongoing and new conversations must be processed by the remaining active servers, and conversational transcript logging must be continued uninterrupted. For some mission-critical conversational applications, agents may need to be geographically distributed to ensure availability, and both conversational knowledge bases and transcript logs may need to be replicated.

[3] In well-engineered conversational agents, response times are nearly independent of the size of the subject matter covered by the agent.

10.3.2.4 Security

The security requirements of the enterprise as a whole, as well as those of the particular application for which a conversational agent is deployed, determine its security requirements. In general, conversational agents must provide at least the same level of security as the site on which it resides. However, because conversations can cover highly sensitive topics and reveal critical personal information, the security levels at which conversational agents must operate are sometimes higher than the environment they inhabit. Agents therefore must be able to conduct conversations over secure channels and support standard authentication and authorization mechanisms. Furthermore, conversational content creation tools (see the following text) must support secure editing and promotion of content.

10.3.2.5 Integration

Conversational agents must integrate cleanly with existing enterprise infrastructure. In the presentation layer, they must integrate with content management systems and personalization engines. Moreover, the agent's responses must be properly synchronized with other presentation elements, and if there is a visual manifestation of an agent in a deployment (e.g., as an avatar), all media must also be coordinated. In the application layer, they must easily integrate with all relevant business logic. Conversational agents must be able to access business rules that are used to implement escalation policies and other domain-specific business rules that affect dialog management strategies. For example, agents must be able to integrate with CRM systems to open trouble tickets and populate them with customer-specific information that provides details of complex technical support problems. In the data storage layer, conversational agents must be able to easily integrate with back-office data such as product catalogs, knowledge management systems, and databases housing information about customer profiles. Finally, conversational agents must provide comprehensive (and secure) administrative tools and services for day-to-day management of agent resources.

To facilitate analysis of the wealth of data provided by hundreds of thousands of conversations, agents must integrate well with third-party business intelligence and reporting systems. At runtime, this requirement means that transcripts must be logged efficiently to databases. At analysis time, it means that the data in "conversation marts" must be easily accessible for reporting on and for running exploratory analyses. Typically, the resulting information and its accompanying statistics provide valuable data that are used for two purposes: improving the behavior of the agent and tracking users' interests and concerns.

10.4 Enabling Technologies

The key enabling technologies for Web-based conversational agents are empirical, corpus-based computational linguistics techniques that permit development of agents by subject-matter experts who are not expert in computer technology, and techniques for robustly delivering conversations on a large scale.

10.4.1 Natural Language Processing Technologies

Natural language processing (NLP) is one of the oldest areas of Artificial intelligence research, with significant research efforts dating back to the 1960s. However, progress in NLP research was relatively slow during its first decades because manual construction of NLP systems was time consuming, difficult, and error-prone. In the 1990s, however, three factors led to an acceleration of progress in NLP. The first was development of large corpora of tagged texts, such as the Brown Corpus, the Penn Treebank [LDC, 2003], and the British National Corpus [Bri, 2003]. The second factor was development of statistical, machine learning, and other empirical techniques for extracting grammars, ontologies, and other information from tagged corpora. Competitions, such as MUC and TREC [Text Retrieval, 2003], in which alternative systems were compared head-to-head on common tasks, were a third driving force. The combination of these factors has led to rapid improvements in techniques for automating the construction of NLP systems.

The first stage in the interpretation of a user's statement, syntactic analysis, starts with tokenization of the user's statement, that is, division of the input in a series of distinct lexical entities. Tokenization can be surprisingly complex. One source of tokenization complexity is contraction ambiguity, which can require significant contextual information to resolve, e.g., "John's going to school" vs. "John's going to school makes him happy." Other sources of tokenization complexity include acronyms (e.g., "arm" can mean "adjustable rate mortgage" as well as a body part), technical expressions (e.g., "10 BaseT" can be written with hyphens, or spaces as in "10 Base T"), multiword phrases (e.g., "I like diet coke" vs. "when I diet coke is one thing I avoid"), and misspellings.

The greatest advances in automated construction of NLP components have been in syntactic analysis. There are two distinct steps in most implementations of syntactic analysis: part-of-speech (POS) tagging, and parsing. POS tagging consists of assigning to each token a part of speech indicating its grammatical function, such as singular noun or comparative adjective. There are a number of learning algorithms capable of learning highly accurate POS tagging rules from tagged corpora, including transformation-based and maximum entropy-based approaches [Brill, 1995; Ratnaparkhi, 1996].

Two distinct approaches to parsing are appropriate for conversational agents. Chunking, or robust parsing, consists of using finite-state methods to parse text into chunks, that is, constituent phrases with no posthead modifiers. There are very fast and accurate learning methods for chunk grammars [Cardie et al., 1999; Abney, 1995]. The disadvantage of chunking is that finite-state methods cannot recognize structures with unlimited recursion, such as embedded clauses (e.g., "I thought that you said that I could tell you that ..."). Context-free grammars can express unlimited recursion at the cost of significantly more complex and time-consuming parsing algorithms. A number of techniques have been developed for learning context-free grammars from tree banks [Statistical, 2003]. The performance of the most accurate of these techniques, such as lexicalized probabilistic context-free grammars [Collins, 1997], can be quite high, but the parse time is often quite high as well. Web-based conversational agents may be required to handle a large number of user statements per second, so parsing time can become a significant factor in choosing between alternative approaches to parsing. Moreover, the majority of statements directed to conversational agents are short, without complex embedded structures.

The reference-resolution task of discourse analysis in general is the subject of active research [Proceedings, 2003], but a circumscribed collection of rules is sufficient to handle many of the most common cases. For example, recency is a good heuristic for the simplest cases of anaphora resolution, e.g., in the sentence (3), "one" is more likely to refer to "stereo" than to "computer."

(3) I want a computer and a stereo if one is on sale.

Far fewer resources are currently available for semantic and pragmatic analysis than for syntactic analysis, but several ongoing projects provide useful materials. WordNet, a lexical database, has been used to provide lexical semantics for the words occurring in parsed sentences [Fellbaum, 1999]. In the simplest case, pairs of words can be treated as synonymous if they are members of a common WordNet synonym set.[4] FrameNet is a project that seeks to determine the conceptual structures, or frames, associated with words [Baker et al., 1998]. For example, the word "sell" is associated, in the context of commerce, with a seller, a buyer, a price, and a thing that is sold. The "sell" frame can be used to analyze the relationships among the entities in a sentence having "sell" as the main verb. FrameNet is based on a more generic case frame representation that organizes sentences around the main verb phrase, assigning other phrases to a small set of roles, such as agent, patient, and recipient [Filmore, 1967]. Most approaches to pragmatic analysis have relied on context to disambiguate among a small number of distinguishable communicative acts or have used *ad hoc,* manually constructed rules for communicative-act classification.

Figure 10.3 displays the steps in the interpretation of sentence (1) in Section 10.3.1. The first step is POS tagging, using the Penn Treebank POS tags. Next, the tagged text is parsed with a simple context-free grammar. The pronouns, "I" and "it," are replaced in the discourse analysis step, based on the rules that "I" refers to the user, in this case customer 0237, and that "it" refers to the most recently mentioned

[4] For example, "tail" and "tag" belong to a WordNet synset that also includes "chase," "chase after," "trail," and "dog."

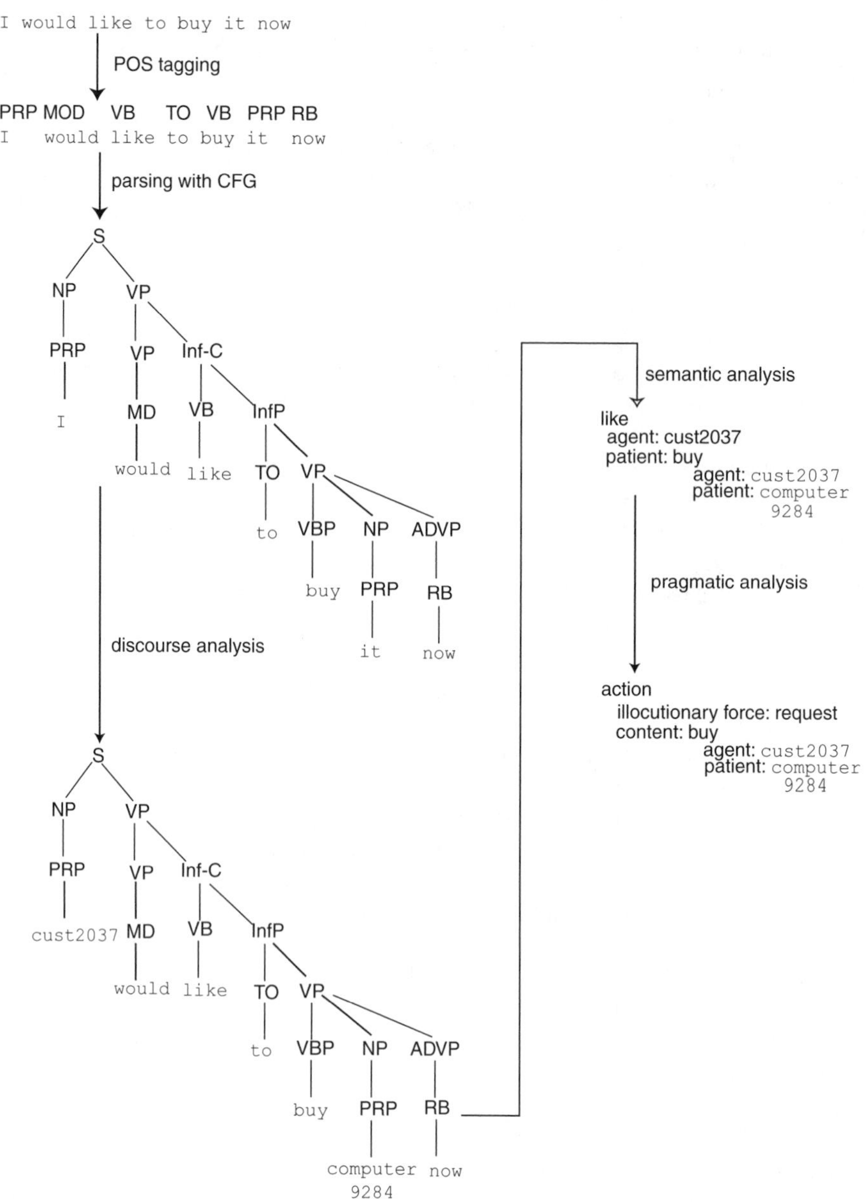

FIGURE 10.3 Steps in the processing of sentence (1).

noun phrase. The resulting parse tree is converted to a case-frame representation in which "like" is the main verb, customer 0237 is the agent, and a case frame involving "buy" is the patient. In the pragmatic analysis step, the verb "like" with a case frame as its patient is interpreted as the illocutionary action "request" with the embedded case frame as the thing requested. The final result of the interpretation process is then passed to the Dialog Manager.

Three approaches have commonly been applied to dialog management. The first approach uses a finite-state automaton (FSA) as a dialog model. Transitions in the FSA correspond to individual statements, and states correspond to agent goals. The FSA approach is suitable for simple dialogs in which all possible sequences of dialog states can be anticipated. However, it is ill-suited for dialogs in which the order of statements is unpredictable, the user may supply several pieces of information in a single utterance, or in the case of mixed initiative dialogs, i.e., dialogs in which conversational control can move between interlocutors. Mixed initiative occurs when one person answers a question with another question, as when additional information is needed to answer a question. Consider the following dialog:

```
User: Do I qualify for a loan?
Agent: What was your income last year?
User: Does money I can get from my parents count as income?
Agent: Do you have to pay the money back?
User: Yes.
Agent: Money you have to pay back is not income.
User: Then my income was about $45,000
Agent: Do you own anything that could be used as collateral?
```

The dialog starts with a question from the user about qualifying for a loan, but to answer the question the agent needs information about the user's income. The agent therefore changes the focus to the user's income. However, the user needs additional information about what qualifies as income to answer the agent's question, so the user takes the initiative again. Once again, the agent can only answer the question by asking an additional question about whether a transfer of money was income. After the user provides the information needed by the agent, the agent can answer the previous question by the user concerning what counts as income, allowing the user to answer the previous question about what his income was. The agent then returns to the goal of eliciting the information needed to answer the original question.

A second approach to dialog management, suited for information elicitation systems, uses templates or frames with slots corresponding to the information to be elicited. This handles unpredictable statement order and compound statements more effectively than the FSA approach, but provides little support for mixed-initiative dialog.

The third approach uses a goal stack or an agenda mechanism to manage dialog goals. This approach can change topics by pushing a goal state corresponding to a new topic onto the stack, then popping the stack when the topic is concluded. The goal-stack approach is more complex to design than the FSA or template approaches, but is able to handle mixed-initiative dialogs.

Continuing the example of sentence (1) because the Dialog Manager has received a "request" communicative act from the Interpreter with content

```
Buy
Agent: cust0237
Patient: computer9284
```

the Dialog Manager should change state, either by following a transition in an FSA corresponding to a request to buy or by pushing onto a goal stack a goal to complete a requested sale. If the patient of the buy request had been unspecified, the transition would have been to a dialog state corresponding to the goal to determine the thing that the user wishes to buy.

A change in dialog state by the Dialog Manager gives rise to call to the Response Generator to take one or more appropriate actions, including communications to the user and noncommunication responses. Typically, only canned text is used, but sometimes template instantiation [Reiter, 1995] is used. In the current example, a dialog state corresponding to the goal completing a requested purchase of a computer might cause the Response Generator to instantiate a template with slots for the computer model and price. For example, the template

```
Great! <computer model> is on sale this week for just <price>!
```

might be instantiated as

```
Great! Power server 1000 is on sale this week for just $1,000.00!
```

Similarly, other communication modalities, such as Web pages and email messages, can be implemented as templates instantiated with context-specific data.

Over the course of a deployment, the accuracy of a well-engineered conversational agent improves. Both false positives and false negatives diminish over time as the agent learns from its mistakes. Learning begins before the go-live in "pretraining" sessions and continues after the agent is in high-volume use. Even after accuracy rates have climbed to very high levels, learning is nevertheless conducted on an ongoing basis to ensure that the agent's content knowledge is updated as the products and services offered by the company change.

Typically, three mechanisms have been put in place for quality improvement. First, transcripts of conversations are logged for offline analysis. This "conversation mining" is performed automatically and augmented with a subject matter expert's input. Second, enterprise-class conversational agent systems include authoring suites that support semiautomated assessment of the agent's performance. These suites exploit linguistic knowledge to summarize a very large number of questions posed by users since the most recent review period (i.e., frequently on the order of several thousand conversations) into a form that is amenable to human inspection. Third, the conversational agent performs a continuous self-assessment to evaluate the quality of its behavior. For example, well-engineered conversational agents generate confidence ratings for each response which they then use both to improve their performance and to shape the presentation of the summarized logs for review.

10.4.2 Enterprise Integration Technologies

Conversational. agents satisfy the scalability, performance, reliability, security, and integration requirements by employing the deployment scheme depicted in Figure 10.4. They should be deployed in an n-tier architecture in which clustered conversational components are housed in application-appropriate security zones. When a user's utterance is submitted via a browser, it is transported using HTTP or HTTPS. Upon reaching the enterprise's outermost firewall, the utterance is sent to the appropriate Web server, either directly or using a dedicated hardware load balancer. When a Web server receives the utterance, it is submitted to a conversation server for processing. In large deployments, submission to conversation servers must themselves be load balanced.

When the conversation server receives the utterance, it determines whether a new conversation is being initiated or whether the utterance belongs to an ongoing conversation. For new conversations, the conversation server creates a conversation instance. For ongoing conversations, it retrieves the corresponding conversation instance, which contains the state of the conversation, including the dialog history.[5] Next, the conversation server selects an available dialog engine and passes the utterance and conversation instance to it for interpretation, dialog management, and response generation. In some cases, the conversation server will invoke business logic and access external data to select the appropriate response or take the appropriate action. Some business rules and data sources will be housed behind a second firewall for further protection. For example, a conversation server may use the CRM system to inspect the user's profile or to open a trouble ticket and populate it with data from the current conversation. In the course of creating a response, the conversation agent may invoke a third-party content management system and personalization engines to retrieve (or generate) the appropriate response content.

Once language processing is complete, the conversation instance is updated and relevant data is logged into the conversation mart, which is used by the enterprise for analytics, report generation, and continued improvement of the agent's performance. The response is then passed back to the conversation server and relayed to the Web server, where an updated view of the agent presentation is created with the new response. Finally, the resulting HTML is transmitted back to the user's browser.

10.4.2.1 Scalability

This deployment scheme achieves the scalability objectives in three ways. First, each conversation server contains a pool of dialog engines. The number of dialog engines per server can be scaled according to

[5] For deployments where conversations need to be persisted across sessions (*durable conversations*), the conversation server retrieves the relevant *dormant* conversation by indexing on the user's identification and then reinitiating it.

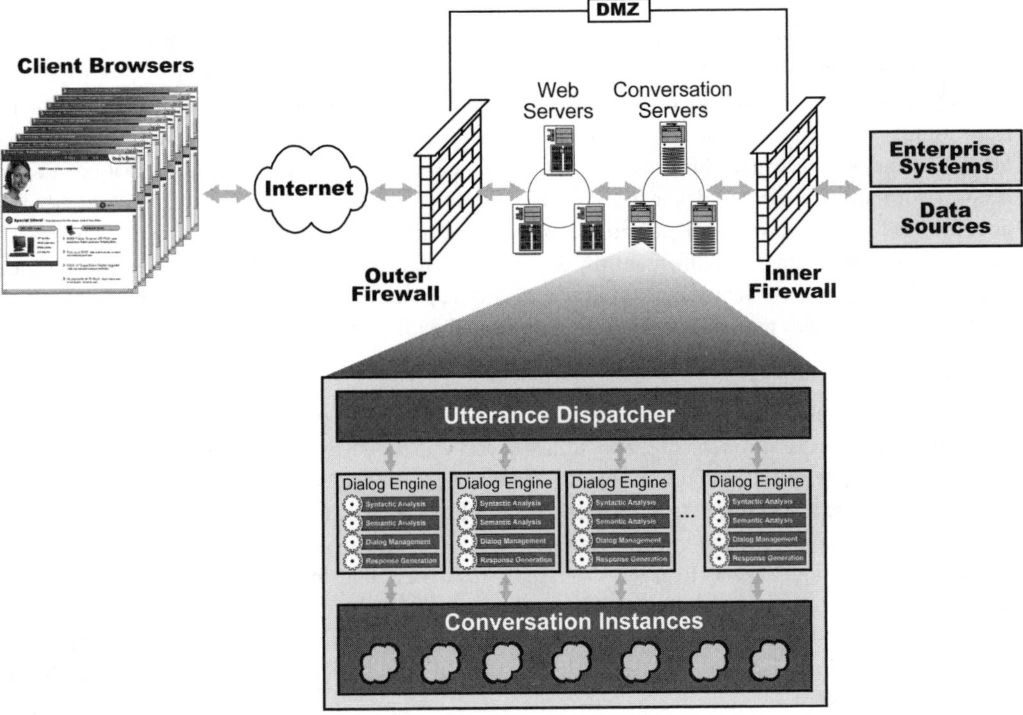

FIGURE 10.4 An enterprise deployment scheme for conversational agents.

the capabilities of the deployment hardware. Second, conversation servers themselves can be clustered, thereby enabling requests from the Web servers to be distributed across the cluster. Conversation instances can be assigned to any available conversation server. Third, storage of the knowledge base and conversation mart utilize industry-standard database scaling techniques to ensure that there is adequate capacity for requests and updates.

10.4.2.2 Performance

Conversational agents satisfy the performance requirements by providing a pool of dialog engines for each conversation server and clustering conversation servers as needed. Guarantees on throughputs are achieved by ensuring that adequate capacity is deployed within each conversation server and its dialog engine pool. Guarantees on the number of simultaneous conversations that can be held are achieved with the same mechanisms; if a large number of utterances are submitted simultaneously, they are allocated across conversation servers and dialog engines. Well-engineered conversational agents are deployed on standard enterprise-class servers. Typical deployments designed to comfortably handle up to hundreds of thousands of questions per hour consist of one to four dual-processor servers.

10.4.2.3 Reliability

A given enterprise can satisfy the reliability and availability requirements by properly replicating conversation resources across a sufficient number of conversation servers, Web servers, and databases, as well as by taking advantage of the fault tolerance mechanisms employed by enterprise servers. Because maintaining conversation contexts, including dialog histories, is critical for interpreting utterances, in some deployments it is particularly important that dialog engines be able to access the relevant context information but nevertheless be decoupled from it for purposes of reliability. This requirement is achieved by disassociating conversation instances from individual dialog engines.

10.4.2.4 Security

The deployment framework achieves the security requirements through four mechanisms. First, conversational traffic over the Internet can be secured via HTTPS. Second, conversation servers should be deployed within a DMZ to provide access by Web servers but to limit access from external systems. Depending on the level of security required, conversation servers are sometimes placed behind internal firewall to increase security. Third, using industry standard authentication and authorization mechanisms, information in the knowledge base, as well as data in the conversation mart, can be secured from unauthorized access within the organization. For example, the content associated with particular knowledge-base entries should be modified only by designated subject-matter experts within a specific business unit. Finally, for some conversational applications, end users may need to be authenticated so that only content associated with particular roles is communicated with them.

10.4.2.5 Integration

Conversational agents in the framework integrate cleanly with the existing IT infrastructure by exposing agent integration APIs and accessing and utilizing APIs provided by other enterprise software. They typically integrate with J2EE- and NET-based Web services to invoke enterprise-specific business logic, content management systems, personalization engines, knowledge management applications, and CRM modules for customer segmentation and contact center management. In smaller environments it is also useful for conversational agents to access third-party databases (housing, for example, product catalogs and customer records) via mechanisms such as JDBC and ODBC.

In summary, well-engineered conversational agents utilizing the deployment scheme described above satisfy the high-volume conversation demands experienced in the enterprise. By housing dialog engines in a secure distributed architecture, the enterprise can deliver a high throughput of simultaneous conversations reliably, integrate effortlessly with the existing environment, and scale as needed.

10.5 Conclusion

With advances in computational linguistics, well-engineered conversational agents have begun to play an increasingly important role in the enterprise. By taking advantage of highly effective parsing, semantic analysis, and dialog management technologies, conversational agents clearly communicate with users to provide timely information that helps them solve their problems. While a given agent cannot hold conversations about arbitrary subjects, it can nevertheless engage in productive dialogs about a specific company's products and services. With large-scale deployments that deliver high volumes of simultaneous conversations, an enterprise can employ conversational agents to create a cost-effective solution to its increasing demands for customer service, guided selling, Website navigation, and technical support. Unlike the monolithic CRM systems of the 1990s, which were very expensive to implement and whose tangible benefits were questionable, self-service solutions such as conversational agents are predicted by analysts to become increasingly common over the next few years. Because well-engineered conversational agents operating in high-volume environments offer a strong return on investment and a low total cost of ownership, we can expect to see them deployed in increasing numbers. They are currently in use in large-scale applications by many Global 2000 companies. Some employ external-facing agents on retail sites for consumer products, whereas others utilize internal-facing agents to assist customer service representatives with support problems.

To be effective, conversational agents must satisfy the linguistic and enterprise architecture requirements outlined above. Without a robust language-processing facility, agents cannot achieve accuracy rates necessary to meet the business objectives of an organization. Conversational agents that are not scalable, secure, reliable, and interoperable with the IT infrastructure cannot be used in large deployments.

In addition to these two fundamental requirements, there are three additional practical considerations for deploying conversational agents. First, content reuse is critical. Because of the significant investment in the content that resides in knowledge management systems and on Websites, it is essential for conversational agents to have the ability to leverage content that has already been authored. For example, conversational agents for HR applications must be able to provide access to relevant personnel policies and benefits information. Second, all authoring activities must be simple enough to be performed by nontechnical personnel. Some early conversational agents required authors to perform scripting or programming. These requirements are infeasible for the technically untrained personnel typical of the divisions in which agents are usually deployed, such as customer care and product management. Finally, to ensure a low level of maintenance effort, conversational agents must provide advanced learning tools that automatically induce correct dialog behaviors. Without a sophisticated learning facility, maintenance must be provided by individuals with technical skills or by professional service organizations, both of which are prohibitively expensive for large-scale deployments.

With advances in the state-of-the-art of their foundational technologies, as well as changes in functionality requirements within the enterprise, conversational agents are becoming increasingly central to a broad range of applications, As parsing, semantic analysis, and dialog management capabilities continue to improve, we are seeing corresponding increases in both the accuracy and fluidity of conversations. We are also seeing a gradual movement towards multilingual deployments. With globalization activities and increased internationalization efforts, companies have begun to explore multilingual content delivery. Over time, it is expected that conversational agents will provide conversations in multiple languages for language-specific Website deployments. As text-mining and question-answering capabilities improve, we will see an expansion of agents' conversational abilities to include an increasingly broad range of "source" materials. Coupled with advances in machine learning, these developments are further reducing the level of human involvement required in authoring and maintenance. Finally, as speech recognition capabilities improve, we will begin to see a convergence of text-based conversational agents with voice-driven help systems and IVR. While today's speech-based conversational agents must cope with much smaller grammars and limited vocabularies — conversations with speech-based agents are much more restricted than those with text-based agents — tomorrow's speech-based agents will bring the same degree of linguistic proficiency that we see in today's text-based agents. In short, because conversational agents provide significant value, they are becoming an integral component of business processes throughout the enterprise.

References

Abney, Steven. Partial parsing via finite-state cascades. *Natural Language Engineering,* 2(4): 337–344, 1995.

Baker, Collin F., Charles J. Fillmore, and John B. Lowe. The Berkeley FrameNet project. In Christian Boitet and Pete Whitelock, Eds., *Proceedings of the 36th Annual Meeting of the Association for Computational Linguistics and 17th International Conference on Computational Linguistics,* pages 86–90. Morgan Kaufmann, San Francisco, CA, 1998.

Brill, Eric. Transformation-based error-driven learning and natural language processing: a case study in part-of-speech tagging. *Computational Linguistics,* 21(4): 543–565, 1995.

British national corpus, 2003. http://www.natcorp.ox.ac.uk/.

Cardie, Claire, Scott Mardis, and David Pierce. Combining error-driven pruning and classification for partial parsing. In *Proceedings of the 16th International Conference on Machine Learning,* pages 87–96. Morgan Kaufmann, San Francisco, CA, 1999.

Collins, Michael. Three generative, lexicalised models for statistical parsing. In *Proceedings of the 35th Annual Meeting of the Association for Computational Linguistics,* Madrid, 1997.

Fellbaum, Christiane, Ed. *Wordnet: An Electronic Lexical Database.* MIT Press, Cambridge, MA 1999.

Fillmore, Charles. The case for case. In *Universals in Linguistic Theory,* pages 1-90. Holt, Rinehart & Winston, New York, 1968.

Harabagiu, Sanda, Marius Pasca, and Steven Maiorano. Experiments with open-domain textual question answering. In *Proceedings of COLING-2000,* Saarbrüken, Germany, August 2000.

LDC catalog, 2003. http://www.ldc.upenn.edu/Catalog/, University of Pennsylvania.

Text retrieval competition, 2003. National Institute of Standards and Technology, http://trec.nist.gov/.

Proceedings of the 2003 International Symposium on Reference Resolution and Its Applications to Question Answering and Summarization, Venice, Italy, June 23–24, 2003.

Ratnaparkhi, Adwait. A maximum entropy model for part-of-speech tagging. In Eric Brill and Kenneth Church, Eds., *Proceedings.of the Conference on Empirical Methods in Natural Language Processing,* pages 133–142. Association for Computational Linguistics, Somerset, NJ, 1996.

Reiter, Ehud. NLG vs. templates. In *Proceedings of the 5th European Workshop on Natural-Language Generation,* Leiden, Netherlands, 1995.

Searle, John. *Expression and Meaning: Studies in the Theory of Speech Acts.* Cambridge University Press, New York, 1979.

Statistical natural language processing and corpus-based computational linguistics: An annotated list of resources, 2003. Stanford University, http://www-nlp.stanford.edu/links/statnlp.html.

Weizenbaum, Joseph. ELIZA — a computer program for the study of natural language communication between man and machine. *Communications of the Association for Computing Machinery,* 9(1): 36–45, 1966.

11

Internet-Based Games

CONTENTS

Abstract........ 11-1
11.1 Introduction........ 11-1
11.2 Background and History........ 11-2
11.2.1 Genre........ 11-2
11.2.2 A Short History of Online Games........ 11-3
11.3 Games, Gameplay, and the Internet........ 11-5
11.4 Implementation Issues........ 11-7
11.4.1 System Architecture........ 11-10
11.4.2 Consistency........ 11-10
11.5 Future Directions: Games and Mobile Devices........ 11-11
11.6 Summary........ 11-13
11.7 Further Information........ 11-13
Acknowledgements........ 11-13
References........ 11-14

R. Michael Young

Abstract

Networked computer games — currently played by close to 100 million people in the U.S. — represent a significant portion of the $10 billion interactive entertainment market. Networked game implementations build on a range of existing technology elements, from broadband network connectivity to distributed database management. This chapter provides a brief introduction to networked computer games, their characteristics and history, and a discussion of some of the key issues in the implementation of current and future game systems.

11.1 Introduction

Computer game history began in 1961 when researchers developed Spacewar, a game that drew small lines and circles on a monitor in order to demonstrate the capabilities of the first PDP-1 computer installed at Massachusetts Institute of Technology (MIT). Data from Jupiter Research [2002] indicates that, in 2002, 105 million people in the U.S. played some form of computer game. The computer game industry now generates over $10 billion annually, having exceeded Hollywood domestic box office revenues for the last 5 years. Current research is extending the technology of computer games into applications that include both training and education. For example, University of Southern California's Institute for Creative Technologies, a collaboration between artificial intelligence researchers, game developers, and Hollywood movie studios, is using game technology to create advanced game-like training simulations for the U.S. Army [Hill et al., accepted]. Similarly, the Games-to-Teach project, a collaboration between MIT, Carnegie Mellon University, and Microsoft Research is exploring the effectiveness of games specifically designed and built as teaching tools [Squire, accepted].

1-58488-381-2/05/$0.00+$1.50
© 2005 by CRC Press LLC

While not all computer games use the Internet, network computer games account for a considerable portion of the sales figures reported above. This chapter provides a brief introduction to networked computer games, their characteristics and history, and a discussion of some of the key issues in the implementation of current and future game systems.[1]

11.2 Background and History

11.2.1 Genre

The structure of a network-based game is determined, to a large extent, by the type of gameplay that it must support. Much like a film or other conventional entertainment media, a game's style of play can be categorized by genre. Given below is a brief description of the most popular genres for network-based games. This list is not meant to be definitive, however. Just as a film may cross genre boundaries, many successful games have elements of gameplay from more than one genre.

Action games are real-time games, that is, games that require the user to continuously respond to changes in the game's environment. The action game genre is dominated by *first-person shooter* (or FPS) titles. In an FPS, the player views the game world through the eyes of an individual character, and the main purpose of gameplay is to use weapons to shoot opponents. Because these combat-oriented games tend to be quick-paced and demand rapid and accurate player responses, they place high demands on both the effectiveness of the graphics rendering capabilities as well as the network throughput of the player's computer. Figure 11.1 shows a screenshot from one of the more popular first-person shooters, Epic Games' Unreal Championship.

In *adventure games*, players typically control characters whose main tasks are to solve puzzles. Adventure games embed these puzzle-solving activities within a storyline in which the player takes on a role. Historically, adventure games have been turn-based, where a player and the computer take turns snaking changes to effect the game world's state. Recently, hybrid games that cross the boundaries between action and adventure have become popular. These hybrids use a strong storyline to engage the player in puzzle solving, and use combat situations to increase the tension and energy levels throughout the game's story.

In *role-playing games* (RPGs), the player directs one or more characters on a series of quests. Each character in an RPG has a unique set of attributes and abilities, and, unlike most action and adventure games, RPG gameplay revolves around the player increasing her characters' skill levels by the repeated accomplishment of more and more challenging goals. Role-playing games are typically set in worlds rich with detail; this detail serves to increase the immersion experienced by the player as well as to provide sufficient opportunities for the game designer to create exciting and enjoyable challenges for the player to pursue.

Strategy games require players to manage a large set of resources to achieve a predetermined goal. Player tasks typically involve decisions about the quantity and type of raw material to acquire, the quantity and type of refined materials to produce, and the way to allocate those refined materials in the defense or expansion of one's territory. Historically, strategy games have been turn-based; with the advent of network-based games, a multiplayer version of strategy games, called the *real-time strategy* (RTS) game, sets a player against opponents controlled by other players, removing the turn-based restrictions. In real-time strategy games, all players react to the dynamics of the game environment asynchronously.

The principal goal of a *simulation game* is to recreate some aspect of a real-world environment. Typical applications include simulations of complex military machinery (e.g., combat aircraft flight control, strategic level theater-of-operations command) or social organizations (e.g., city planning and management). Simulations are often highly detailed, requiring the player to learn the specifics of the simulated context in order to master the game. Alternatively, *arcade simulations* provide less-complicated interfaces

[1] Note that this article focuses on *online games*, computer games played on the Internet, not *computer gaming*, a term typically used to describe gambling-based network applications.

FIGURE 11.1 Epic Games' Unreal Championship is one of the most popular first-person action titles.

for similar applications, appealing to more casual game players that want to enjoy participating in a simulation without needing to master the required skills of the real-world model.

Sports games are quite similar in definition to simulation games but focus on the simulation of sports, providing the player with the experience of participating in a sporting event either as a player or as a team coach. *Fighting games* typically involve a small number of opposing characters engaged in a physical combat with each other. Gameplay in fighting games is built around the carefully timed use of a wide set of input combinations (e.g., multiple keystrokes and mouse clicks) that define an array of offensive and defensive character moves.

Casual games are already-familiar games from contexts beyond the computer, such as board games, card games, or television game shows. While the interfaces to casual games are typically not as strikingly visual as games from other genres, they represent a substantial percentage of the games played online. In part, their popularity is due to their pacing, which affords the opportunity for players to reflect more on their gameplay and strategy and to interact more with their opponents or teammates in a social context through chat.

11.2.2 A Short History of Online Games

The first online games were written for use in PLATO, the first online computer-aided instructional system, developed by Don Bitzer at the University of Illinois [Bitzer and Easley, 1965]. Whereas the intent behind PLATO's design was to create a suite of educational software packages that were available via a network to a wide range of educational institutions, the PLATO system was a model for many different kinds of network-based applications that followed it. In particular, PLATO supported the first online community, setting the stage for today's massively multiplayer game worlds. Early games on PLATO included the recreation of MIT's Spacewar, computer versions of conventional board games like checkers and backgammon, and some of the first text-based adventure and role-playing games.

In contrast to the formal educational context in which PLATO was developed and used, MUDs (Multi-User Dungeons), another form of multiuser online environment, were developed and put to use in more informal contexts. MUDs are text-based virtual reality systems in which a central server maintains a database representing the gameworld's state and runs all functions for updating and changing that state. Clients connect to the server using the Telnet protocol; players type commands that are transmitted to the server, the server translates text commands into program calls, executes those programs to update the gameworld, and then sends the textual output from the executed programs back to the client for display.

The first MUDs, developed in the 1970s, were originally designed as online multiuser game worlds where users played text-based role-playing games much like the pen-and-paper game Dungeons and Dragons. Many of the early MUD systems were models for later commercial (but nonnetworked) text-based games such as Zork and Adventure, and their design and use further extended the notions of online community experienced by PLATO users.

With the advent of commercial online service providers in the early 1980s, network games began the move toward the mainstream. Providers such as Compuserve, Delphi, and Prodigy began offering single-player games that ran on remote servers. Soon they were offering multiuser games similar to MUDs, with more elaborate interfaces and world design. Among the most influential of these games was Lucas-Film's Habitat, developed by Morningstar and Farmer [1990]. Habitat was a graphical virtual world in which users could modify the presentation of their own characters and create new objects in the game. This level of customization created a highly individualized experience for Habitat's users. A sample screen from the Habitat interface is shown in Figure 11.2. Habitat administrators regularly organized events with the Habitat world to engage its users in role-playing; Habitat's successes and failures were widely studied by game developers and academics, and the principles pioneered by Habitat's developers served as a model for the design of many later online role-playing games.

As Internet access increased in the early 1990's, game developers began to incorporate design approaches that had proved successful for nonnetworked PC games in the development of a new type of network game, the *persistent world.* Unlike many network games up to that point, persistent worlds maintained or extended their game state once a player logged off. A player could return many times to the same game world, assuming the same identity and building upon previous game successes, inter-character relationships, and personal knowledge of the game world. The first persistent world game was Ultima Online, a role-playing game that was designed to offer an expansive, complete world in which

FIGURE 11.2 In 1980, LucasFilm's Habitat was one of the first graphical online persistent worlds.

players interacted. Subsequently, successful persistent worlds (e.g., Microsoft's Asheron's Call, Sony's Everquest, and Mythic Entertainment's Dark Age of Camelot) have been among the most financially viable online games. NCSoft's Lineage, played exclusively in South Korea, is currently the persistent world with the largest subscriber base, with reports of more than 4 million subscribers (played in a country of 47 million). Architectures for persistent world games are described in more detail below.

11.3 Games, Gameplay, and the Internet

The Internet serves two important roles with respect to current games and game technology: game *delivery* and game *play.* Those games that use the Internet only as a means for distribution typically execute focally on the user's PC, are designed for a single player, and run their code within a Web browser. Most often, these games are developed in Flash, Shockwave, or Java, and run in the corresponding browser plug-ins or virtual machines. In addition to these development environments, which have a wide applicability outside Web-based game development, a number of programming tools are available that are targeted specifically at Web-based game development and distribution. These products, from vendors such as Groove Alliance and Wild Tangent, provide custom three-dimensional rendering engines, scripting languages, and facilities targeted at Internet-based product distribution and sales.

Web browser games often have simpler interfaces and gameplay design than those games that are stand-alone applications. Aspects of their design can be divided into three categories, depending on the revenue model that is being used by their provider. *Informal games* are downloaded from a Website to augment the site's main content, for instance, when a children's tic-tac-toe game is available from a Website focusing on parental education.

Advertising games are used to advertise products, most often through product placement (e.g., a game on the lego.com Website that might feature worlds built out of lego bricks, a game on disney.com that might include characters and situations from a newly released film). *Teaser games* are used to engage players in a restricted version of a larger commercial game and serve to encourage players to download or purchase the full version of the game for a fee.

While browser-based games are popular and provide a common environment for execution, most game development is targeted at games running on PCs as stand-alone applications. This emphasis is the result of several factors. For one, many games can be designed to play both as network games and as single-player games. In the latter case, a single-player version does not require a PC with network access. Further, the environments in which Web browser plug-ins execute are often restricted for security reasons, limiting the resources of the host PC accessible to the game developer when writing the code. Finally, users must often download and install special-purpose plug-ins in order to execute the games targeted at Web browsers (Figure 11.3) toward. In contrast, most of the APIs for stand-alone game development are available as part of the operating system or as components that can be included with a stand-alone games installer.

A growing number of games are now written for *game consoles,* consumer-market systems specially designed for home gameplay. Console systems have all aspects of their operating system and operating environment built into hardware, and run games distributed on CD-ROM. They connect directly to home televisions though NTSC and PAL output, and take input from special-purpose game controllers. One advantage of the game console over the PC for both the game player and the game developer is that the console presents a known and fixed system architecture. No customization, installation, or configuration is needed, making gameplay more reliable for the end user and software development more straightfoward for the game developer.

Game consoles have been a major element of the electronic entertainment industry since 1975, when Atari released the Tele-Game system, the first home game console (dedicated to playing a single game, Pong). The first console system with network capability — an integral 56k modem — was Sega's Dreamcast, released in 1999. Currently, there are three main competitors in the market for consoles, all of them with broadband capability: Sony's Playstation 2, Microsoft's X-Box, and Nintendo's Gamecube.

FIGURE 11.3 Lego's Bionicle game is played in a Web browser using a Flash plug-in.

The structure of network access for games developed for these platforms varies across the manufacturers, who impose different restrictions through the licenses granted to game developers. Microsoft requires that all network games developed for its X-Box console use its existing online games service, X-Box Live. X-Box Live provides support for network infrastructure to game developers and handles all aspects of the user interface for customers, from billing to in-game "lobbies" where players gather to organize opponents and teammates prior to the start of a game. In contrast, both Sony and Nintendo adopt models that give developers substantial freedom in creating and managing the online interfaces, communities, and payment options for their games.

The majority of games that make use of the Internet, whether PC- or console-based, do so for multiplayer connectivity. Single-player games that use the Internet use the network primarily for downloading only, as mentioned earlier. Recently, several game architectures have been developed in academic research laboratories that use the Internet to distribute processing load for single-player games. Because these systems use complicated artificial intelligence (AI) elements to create novel gameplay, their computational requirements may exceed the capabilities of current PC processors. These systems exploit Internet-based approaches in order to balance computational demand across client/server architectures [Laird, 2001; Young and Riedl, 2003; Cavazza et al., 2002].

Multiplayer games are oriented either toward a single session or toward a persistent world model. In single-session games (the most prevalent form of multiplayer games), players connect to a server in order to join other players for a single gameplay session. The server may be a dedicated server hosted by the game publisher or it may be running on one of the players PCs. Most single-session games involve between two and ten players. In this style of game, players connect either through the private advertisement of IP numbers (for instance, by personal communication between friends) or through *lobby services* provided by third-party hosts. Lobby services allow game clients to advertise their players' identities, form teams, and chat before and after games. Some lobby services also act to gather and post statistics, winnings, and other public competitive data.

When using a typical lobby service, users that are running their PCs or game consoles as *servers* (that is, those users that are hosting the main processing of a game on their own machines) register their availability as a server with the lobby service. Users that are seeking to connect their PC as *clients* to an existing game server (that is, those users wanting to use their computers to interact with the game world but not providing the computation supporting the main game logic) register themselves with the lobby service as well. They then may request that the lobby service automatically connect them to a server after matching their interests (e.g., type of game, number of human players, level of difficulty) with those servers currently looking for additional players. Alternatively, client users may request a list of available servers, along with statistics regarding the server's load, network latency, particulars about the game that will be played when connected to it and so on. Users can then choose from the list in hopes of connecting to a server that better suits their interests.

Some lobby services also provide the ability to create "friend lists" similar to instant messaging buddy lists. As players connect to lobby services that use friend lists, their friends are notified of their connection status. A player can send an invitation message to a friend that is connected at the same time, requesting that the friend join the player's server, or asking permission to join the server currently hosting the friend. Players can typically configure their game servers to allow teams composed of (1) human users playing against human users, (2) humans against computer-controlled opponents (called "bats") or (3) a mix of users and bats against a similarly composed force.

Aside from player and team statistics, little data is kept between play sessions in single-session games. In contrast, persistent world games maintain all game state across all player sessions. Gameplay in persistent worlds involves creating a character or persona, building up knowledge, skill, and relationships within the gameworld, and using the resulting skills and abilities to participate in joint tasks, typically quests, missions, or other fantasy or play activities. Because persistent worlds typically have several orders of magnitude more players at one time than do single-session games, they are referred to as *massively multiplayer* (MMP) games. Persistent fantasy worlds are called massively multiplayer online role-playing games, or MMORPGs. Connections between players in MMP games are usually made through in-game mechanisms such as social organizations, guilds, noble houses, political parties, and so on.

While players pay once at the time of purchase for most single-session games, persistent world games are typically subscription based. In general, the greater the number of subscribers to a persistent world, the more appealing the gameplay, since much of the diversity of activity in a persistent world emerges from the individual activities of its subscribers. Everquest. Sony's MMP game, has close to a half a million subscribers, and it is estimated that over 40,000 players are online on Everquest at any one time. Figure 11.4 shows a sample Everquest screenshot. The following section provides a short discussion of the design of a typical MMP network architecture.

11.4 Implementation Issues

In this section, we focus on implementation issues faced by the developers of massively multiplayer persistent worlds, the online game genre requiring the most complex network architecture. MMP game designers face many of the same problems faced by the developers of real-time high-performance distributed virtual reality (VR) systems [Singhal and Zyda, 2000]. MMP games differ from virtual reality systems in one important factor. VR systems often have as a design goal the creation of a virtual environment indistinguishable from its real-world correlate [Rosenblum, 1999]. In contrast, MMP games seek to create the illusion of reality only to the extent that it contributes to overall gameplay. Due both to the emphasis in MMP game design on gameplay over simulation and to the games' inherent dependency on networked connectivity, MMPS have unique design goals, as we discuss in the following text.

Although most games that are played across a network are implemented using custom (i.e., proprietary) network architectures, Figure 11.5 shows an example of a typical design. As shown in this figure, a massively multiplayer architecture separates game presentation from game simulation. Players run a client program on their local game hardware (typically a PC or a game console) that acts only as an input/output device. The local machine accepts command input from a remote game server and renders the player's view of the game world, sending keystroke, mouse, or controller commands across the network

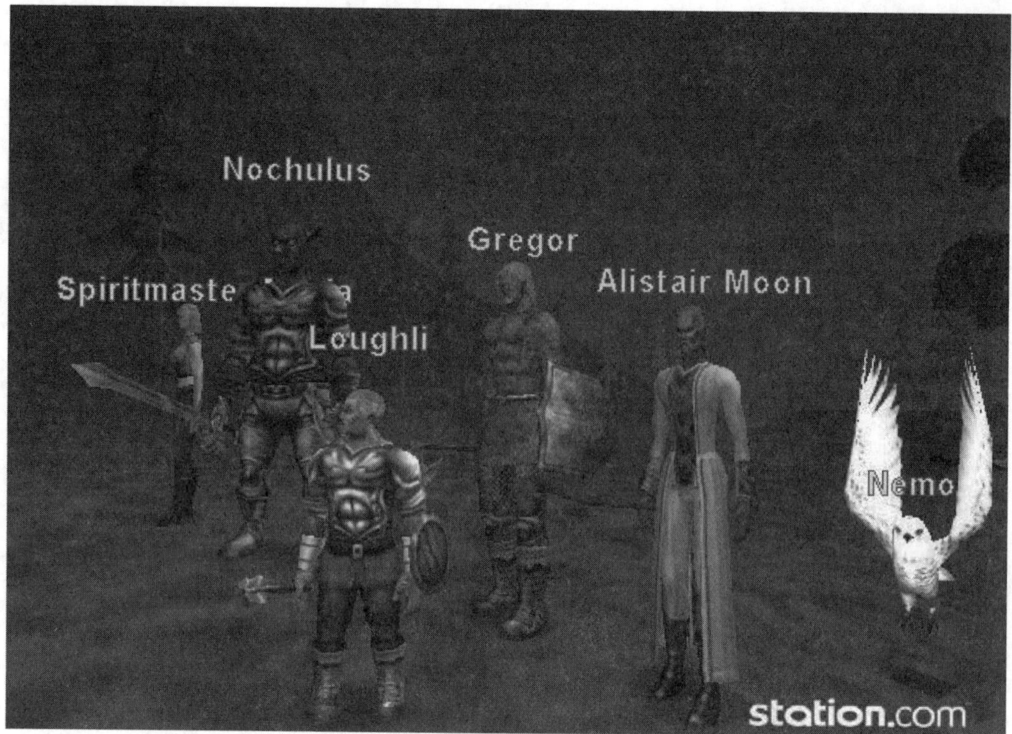

FIGURE 11.4 Sony Online Entertainment's Everquest is one of the most popular massively multiplayer games.

to signal the player's moves within the game. Most (or all) of the processing required to manage the game world's state is handled by the game logic executing on a remote server.

Communication between client and server can be either via UDP or TCP, depending on the demands of the game. Because the UDP protocol does not guarantee packet delivery, overhead for packet transmission is low and transfer rates are consequently higher. As a result, UDP packets are typically used in games when high data rates are required, for instance, in action games where game state changes rapidly. TCP is used when high reliability is the focus, for instance, in turn-based games where single packets may carry important state change information. Within a packet, whether UDP or TCP, XML is commonly used to encode message content. Some developers, however, find the added message length imposed by XML structure to critically impact their overall message throughput. In those cases, proprietary formats are used, requiring the development of special-purpose tokenizers and parsers. The content of each message sent from client to server is designed to minimize the amount of network traffic; for example, typical message content may include only position and orientation data for objects, and then only for those objects whose data has changed since the previous time frame. Efficiency is achieved by preloading the data describing three-dimensional models and their animations, the world's terrain, and other invariant features of the environment on client machines before a game begins (either by downloading the data in compressed format via the network or, for larger data files, by distributing the data on the game's install CD).

The game logic server typically performs all computation needed to manage the game world state. In large-scale MMPs, the simulation of the game world happens on distinct *shards*. A shard is a logical partition of the game environment made in such a way as to ensure little or no interaction within the game between players and objects in separate shards. Limits are placed on the number of players and objects within each shard, and all shards are then mapped onto distinct server clusters. The most common means for defining a shard is to partition a game world by dividing distinct geographical regions within the world into *zones*, assigning one zone to a single server or server cluster. MMP worlds that are zoned

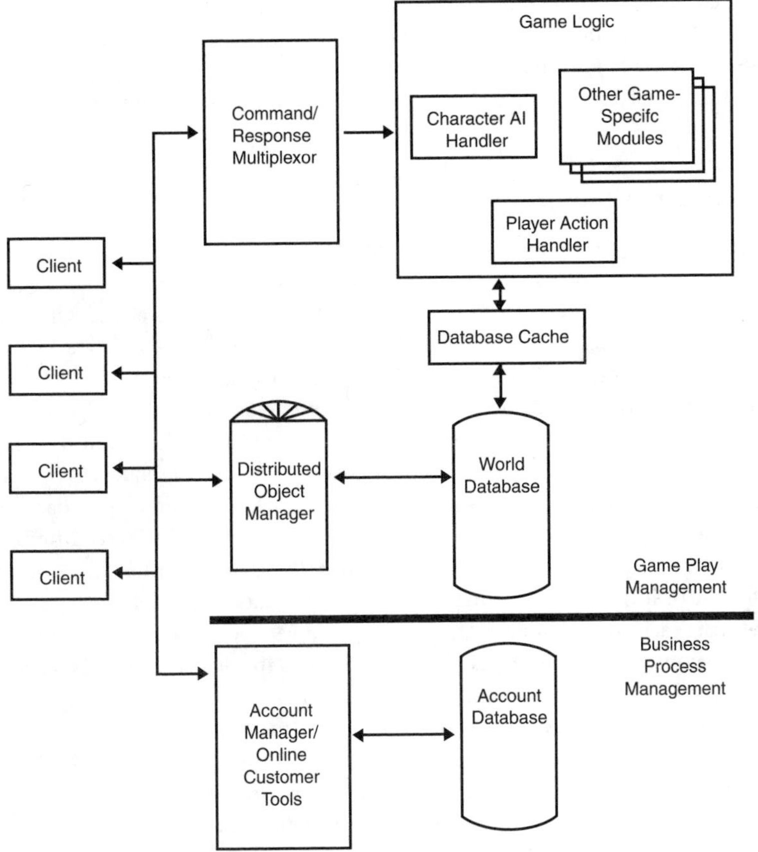

FIGURE 11.5 A typical massively multiplayer system architecture.

either prohibit player movement from one zone to another or have special-purpose portals or border-crossing points within the game that correspond to zone entry and exit points. Zone portals use functions that transfer processes and data from one shard to the next as players move between zones.

In contrast to zoned worlds, MMPs can be designed to create *seamless* worlds in which a player may interact with objects, computer-controlled characters, or other players that are themselves executing on servers other than the player's. In seamless worlds, objects that are located spatially near the boundary between server *a* and another server, server *b*, can be viewed by players from both servers. In order to ensure that all players see consistent worldviews when considering such objects, proxy objects are created on server *b* for each object that resides on server *a* that is visible from server *b*. Objects on server *a* are responsible for communicating important state changes (e.g, orientation, location, animation state) to their proxies on server *b*. This process is complicated when characters move objects (or themselves) into or out of border regions or completely across the geographical boundaries between servers, requiring the dynamic creation and destruction of proxy objects.

Despite these complications, there are a number of benefits to the use of seamless world design for MMP games. First, players are not presented with what can appear to be arbitrary geographical partitions of their world in those locations where distinct shard boundaries occur. Second, seamless worlds can have larger contiguous geographical areas for gameplay; space need not be divided based on the processor capabilities of individual server clusters hosting a particular zone. Finally, seamless worlds are more scalable than zoned ones because boundaries between server clusters in a seamless world can be adjusted after the release of a game. In fact, some seamless MMP games adjust the boundaries between servers dynamically during gameplay based on player migration within the game world.

Some MMP designs disassociate the physical location of a player from the set of servers that maintain that player's state. In these approaches, the system may dynamically allocate players to shards in order to balance processor load or anticipated memory and disk access. Partitioning of MMP servers may also be done along functional lines, with distinct clusters handling physics simulation, game AI, or other factors.

11.4.1 System Architecture

To manage clients' connections to an MMP, a multiplexor sits between the clients and the shard servers. The multiplexor acts as a login management system, routes players to the correct shards and, as mentioned earlier, may dynamically shift clients from one server to another, acting as a load balancer.

The game world state consists of the physical layout of the game's world (e.g., its geography and architecture), the properties of each character in the world, and the properties of each inanimate object that appears in it. In many MMP games, there may be hundreds of thousands of players, an equal or greater number of computer-controlled characters, and millions of objects. All of their properties must be readily available to the game logic server in order to determine the consequences of character actions. To facilitate this computation, the game-world state may be held in memory, in a database, or in some combination. Typical approaches use transaction-based database control to record important or slowly changing in-game transactions (e.g., the death of players, the achievement of in-game goal states) but hold more dynamic or less critical information (e.g., player location) in memory.

A high-speed data cache often connects the game logic server and the world database. Use of this cache increases server response time because many of the updates to the database made by clients are relatively local (e.g., the location of a player may change many times in rapid succession as the player's character runs from one location to the next). The cache also serves to prevent denial of service attacks on the database from collections of malicious clients making high-volume demands for database updates.

In order to maintain a consistent worldview across all clients, the MMP system must communicate changes in world state from the world database to the clients. Often, MMPs use a distributed object update model, sending updates about the world just to those clients that currently depend upon that data.

In addition to their responsibility for maintaining the game's world state, MMP systems typically also provide a distinct set of servers to handle business process management. These servers provide user authentication, log billing information, maintain usage statistics, and perform other accounting and administrative functions. They may also provide Web-based access to customers' account information and customer service facilities. Often, an MMP system will provide limited server access and disk space for community support services such as player-modifiable Web pages and in- and out-of-game player-to-player chat.

Most current MMP servers run on Linux, with the exception of those games published by Microsoft. Given the market penetration of the Windows desktop, client platforms are, by and large, developed for Windows. The process of developing an MMP server has become easier recently due to a growing number of middleware vendors whose products provide graphics, physics, artificial intelligence, and networking solutions to game developers. Third-party network solutions from vendors such as Zona, Butterfly.net, and Turbine Games provide a range of middleware functionalities, from simple network programming APIs to full-scale development and deployment support for both gameplay and business applications.

11.4.2 Consistency

Because much of the gameplay in MMP games revolves around groups of players interacting with each other in a shared space, care must be taken to ensure as consistent a shared state as possible across the all clients. Problems arise due to the effects of network latency; as players act on the game state asynchronously, updates resulting from their actions must propagate from the server to all clients that share some aspect of the affected state. When those updates are not processed by all clients simultaneously, anomalies in gameplay can arise.

Consider a prototypical example: aiming and firing a laser weapon at a computer-controlled opponent that is moving across a battlefield. Two teammates, *a* and *b*, both observe the computer-controlled opponent *o* in motion across the field. Player *b*'s client has very low latency, and so *b*'s view of the world is identical to that of the server's. Network latency, however, has delayed packets to *a*'s client; as a result, *a* sees *o* in a position that lags behind its actual position on the server (and on *b*'s client). Player *a* targets *o* using her client's out-of-date position data and fires her laser. From *a*'s laser target data that arrives at the server, the game logic determines that *a*'s shot trails behind *o*'s actual position and sends a message to the clients indicating that the laser blast has missed its target.

Player *a*, knowing that she had targeted the opponent accurately, assumes that the server's game logic is faulty. As a result, Player *a* may become discouraged with the game and stop playing it. Player *b*, having seen *a* target *o* well behind *o*'s actual position, assumes that *a* is a poor shot and may decide not to team with *a* due to the erroneous evaluation of her skill level.

A wide range of techniques are used in MMP games to deal with the effects of network latency. Perhaps the most widely used technique is *dead reckoning*. In dead reckoning, a client maintains a partial state of the visible game world. In particular, those objects that are in motion in the client's current field of view are tracked. As latency increases and packets from the server are slow to arrive, the client makes estimates about those objects' new positions, drawing the objects in positions extrapolated from their velocity, acceleration, and location specified in the last packet from the server.

While dead reckoning keeps objects from suspending their movement in times of high latency, it can also result in the need for sudden corrections in object location when packets from the server do arrive and the extrapolation of the object's position has been in error. Some games choose simply to move the incorrectly placed objects into the position specified by the server, resulting in a sudden, observable jump in location that can disrupt gameplay. Other MMPs will use a smoothing approach, in which the client and server negotiate over a window of time, adjusting the object's position gradually until its location is in line with both client and server representations.

Dead reckoning for computer-controlled characters is made easier when those characters navigate using *pathnodes*. Pathnodes are data structures placed in an MMP world, unseen by the players but used by the server as way-points for path navigation. Computer-controlled characters moving between two locations construct a path for themselves that runs along a series of pathnodes; these paths can be described to clients so that, when position information from the server is delayed, position prediction made by dead reckoning can be more accurate.

Dead reckoning is a client-side technique that simulates the position computations being made on the server. Another interesting approach to deal with latency is for the server to simulate the state of the world represented on each client [Bernier, 2001; Olsen, 2000]. In this approach, all commands between client and server are time stamped, and the server polls each client at a fixed high rate to keep an accurate history of the client's latency. From this history and the list of server-to-client messages that the client has acknowledged, the server can make an estimate of the state of the world on the side of the client at any given time. As messages from the client arrive at the server, the server determines what the state of the world was *on the client* at the time that the command was issued. So, when Player *a* fires her laser, the server can determine that her target, as seen on her client, was in a position directly in front of her weapon, and so can signal to the clients that Player *a*'s action has succeeeded.

11.5 Future Directions: Games and Mobile Devices

At the 2003 Electronic Entertainment Expo, a major trade show for the computer and console-gaming industry, Microsoft introduced their notion of the *digital entertainment lifestyle*, a market direction that seeks to integrate the X-Box Live gaming service with their other online personal/lifestyle products and services, including messaging, music, video, email, and the Web. As Microsoft's leaders suggest by this emphasis, the future for networked games lies in their integration into the broader context of everyday life. As game designers adapt their games to appeal to a mass market, the technology to support pervasive forms of gameplay will also need extension.

One of the principal Internet technologies that will see rapid expansion in games development over the next 5 years lies in the area of mobile gaming and the use of Internet technologies on cell phones and other small portable devices. Though not all mobile gaming applications involve gameplay that is network-based, there is a substantial population of gamers that currently play games on mobile computing platforms. IDC Research estimates that there are currently over 7 million mobile game players in the U.S. alone (the per capita statistics for Europe and Japan are much higher). This number is expected to grow to over 70 million by 2007. The handsets currently used by mobile game players in the U.S. are equipped with 32-bit RISC ARM9 processors, Bluetooth wireless network capability, and a data transfer rate via GSM or CDMA signal of 40 to 100 kbps. These devices have 128 ∞ 128 pixel color displays, directional joysticks, and multikey press keypads, allowing a restricted but still effective input and output interface for game developers to utilize.

Efforts are underway by collections of cellphone manufacturers and software companies to create well-defined specifications for programming standards for mobile devices; these standards include elements central to mobile game design, such as advanced graphics capability, sound and audio, a range of keypad and joystick input features, and network connectivity. Sun Microsystem's Java is emerging as one such standard programming language for wireless devices. Java 2 Micro Edition (J2ME) specifies a subset of the Java language targeting the execution capabilities of smaller devices such as PDAs, mobile phones, and pagers. Many cellphone manufacturers have joined with Sun to create the Mobile Information Device Profile (MIDP 2.0 [Sun Microsystems, 2003]), a specification for a restricted version of the Java virtual machine that can be implemented across a range of mobile devices.

Qualcomm has also created a virtual machine and language, called the Binary Run-time Environment for Wireless (BREW [Qualcomm, 2003]). BREW has been embedded in many of the current handsets that use Qualcomm's chipsets. The specification for BREW is based on C++, with support for other languages (e.g., Java). Included in BREW's definition is the BREW distribution system (BDS), a specification for the means by which BREW program developers and publishers make their applications accessible to consumers, and the methods by which end users can purchase BREW applications online and download them directly to their mobile devices,

Programmers using either standard have access to a wide range of features useful for game development, including graphical sprites, multiple layers of graphical display, and sound and audio playback. With current versions of both development environments, full TCP, UDP, HTTP, and HTTPS network protocols are also supported.

The computing capacity of current and soon-to-market handsets, while sufficient for many types of games, still lags behind the capacity of PCs and consoles to act as clients for games that require complex input and high network bandwidth. In order to connect a player to a fast-paced multiplayer action/adventure game, a game client must be able to present a range of choices for action to the player and quickly return complicated input sequences as commands for the player's character to act upon. Handsets are limited in terms of the view into the game world that their displays can support. They are further limited by their processors in their ability to do client-side latency compensation (e.g., the dead-reckoning techniques mentioned above). Finally, they are restricted in the size and usability of their keypads in their ability to generate high-volume or complex input sequences.

One means of addressing these limitations might lie in the use of client *proxies* [Fox, 2002], machines that would sit between a game server and a handset (or a collection of handsets) and emulate much of the functionality of a full-scale game client. These proxies could perform any client-side game logic in place of the handset, communicating the results to the handset for rendering. Further, by using a proxy scheme, the handset client could present a simplified view of the player's options to her, allowing her to use the handset's restricted input to select her next moves. Artificial intelligence routines running on the client proxy could then be used to translate the simplified player input into commands for more complex behavior. The more complex command sequence would then be sent on to the game server as if it had come from a high-end game client. The function of such a proxy service could be expanded in both directions (toward both client and server), augmenting the information sent to the game server and enhancing the filtering and summarization process used to relay information from the server back to the client.

11.6 Summary

Networked game implementations build on a range of existing technology elements, from broadband network connectivity to distributed database management. As high-speed Internet access continues to grow, the design of games will shift to encompass a wider market. Jupiter Research estimates that the number of computer and console gamers in the U.S. alone in 2002 is close to 180 million, and anticipates these numbers rising to 230 million within the next 5 years. Just as interestingly, annual revenues from games, PC and console combined, are expected to rise from $9 billion to $15 billion during that same time frame.

The combined increases in market size, potential revenue, network access, and processor capability are certain to transform current notions of network games as niche applications into mainstream entertainment integrated with other aspects of ubiquitous social and personal computing. It is likely that future versions of computer games will not just be played between activities (e.g., playing Tetris while waiting for a bus) or during activities (e.g., playing Minesweeper while listening to a college lecture) but will become an integral part of those activities.

Adoption of networked games by the mainstream is likely to prompt a corresponding shift toward the development of new types of games and of new definitions of gameplay. Appealing to a mass market, future multiuser game development may merge with Hollywood film production, creating a single, inseperable product. The convergence of pervasive network access and new models of interactivity within games will result in games that are integrated with the physical spaces in which their users find themselves. For example, location-aware network technology will allow game developers to create new educational titles that engage students at school, at home, and in many of the informal contexts found in day-to-day life.

11.7 Further Information

There are a growing number of scientific journals that publish research on Internet-based computer games, including the *International Journal of Games Research, The Journal of Game Development,* and *the International Journal of Intelligent Games and Simulation.* Work on Internet games is published also in the proceedings of a range of computer science conferences, such as the ACM Conference on Computer Graphics and Interactive Techniques (SIGRAPH), the CHI Conference on Human Factors in Computing Systems, and the National Conference for the American Association for Artificial Intelligence. The Game Developers' Conference is the primary conference reporting on current industry state-of-the-art. More specialized conferences devoted exclusively to the design and development of computer games, encompassing Internet-based systems, include the NetGames Conference, the Digital Games Research Conference, and the International Conference on Entertainment Computing.

Wolf and Perron [2003] provide a useful overview of the emerging field of game studies, much of which concerns itself with the technical, sociological, and artistic aspects of Internet games and gaming. An in-depth discussion of the issues involved in implementing massively multiplayer games is given in Alexander [2002]. An effective guide for the design of the online games themselves can be found in Friedl [2002].

Acknowledgements

The author wishes to thank Dave Weinstein of Red Storm Entertainment for discussions about the current state of network gaming. Further, many of the concepts reported here are discussed in more depth in the International Game Developers Association's *White Paper on Online Games* [Jarret et al., March 2003].

The work of the author was supported by the National Science Foundation through CAREER Award #0092586.

References

Alexander, Thor *Massively Multi-Player Game Development (Game Development Series).* Charles River Media, New York, 2002.

Bernier, Yahn Latency compensation methods in client/server in-game protocol design and optimization. In *Proceedings of the 2001 Game Developers Conference,* pages 73–85, 2001.

Bitzer, D.L. and J.A. Easley. PLATO: A computer-controlled teaching system. In M.A. Sass and W.D. Wilkinson, Eds., *Computer Augmentation of Human Reasoning.* Spartan Books, Washington, D.C., 1965.

Cavazza, Mare, Fred Charles, and Steven Mead. Agents' interaction in virtual storytelling. In *Proceedings of the 3rd International Workshop on Intelligent Virtual Agents,* Madrid, Spain, 2002.

Duchaineau, Mark, Murray Wolinsky, David Sigeti, Mark C. Miller, Charles Aldrich, and Mark Mineev. ROAMing terrain: Real-time optimally adapting meshes. In *Proceedings of IEEE Visualization,* 1997.

Fox, David Small portals: Tapping into MMP worlds via wireless devices. In Thor Alexander, Ed., *Massively Multiplayer Game Development.* Charles River Media, New York, 2002.

Friedl, Markus *Online Game interactivity Theory (Advances in Graphics and Game Development Series).* Charles River Media, New York, 2002.

Hill, R.W., J. Gratch, S. Marsella, J. Rickel, W. Swartout, and D. Traum. Virtual humans in the mission rehearsal exercise system. *Künstliche Intelligenz,* Special Issue on Embodied Conversational Agents, accepted for publication.

Jarret, Alex, Jon Stansiaslao, Elinka Dunin, Jannifer MacLean, Brian Roberts, David Rohrl, John Welch, and Jeferson Valadares. IGDA online games white paper, second edition. Technical report, International Game Developers Association, March 2003.

Jupiter Research. Jupiter games model, 2002. Jupiter Research, a Division of Jupitermedia Corporation.

Laird, John. Using a computer game to develop advanced artificial intelligence. *IEEE Computer,* 34(7): 70–75, 2001.

Lindstrom, P.D., W. Koller, L.F. Ribarsky, N. Hodges, Faust, and G.A. Turner. Real-time, continuous level of detail rendering of height fields. In *ACM SIGGRAPH 96,* pages 109–118, 1996.

Morningstar, C. and F. Farmer. The lessons of LucasFilm's Habitat. In Michael L. Benedickt, Ed., *Cyberspace: First Steps.* MIT Press, Cambridge, MA, 1990.

Olsen, John Interpolation methods, In *Game Programming Gems 3.* Charles River Media, New York, 2000.

Qualcomm. BREW White Paper [On-Line], 2003. Available via http://www.qualcomm.com/brew/about/whitepaper10.html.

Rosenblum, Andrew Toward an image indistinguishable from reality. *Communications of the ACM,* 42(68): 28–30,1999.

Singhal, Sandeep and Michael Zyda. *Networked Virtual Environments.* Addison-Wesley, New York, 2000.

Squire, Kurt Video games in education. *International Journal of Simulations and Gaming,* accepted for publication.

Sun Microsystems. JSR-000118 Mobile Information Device Profile 2.0 Specification [On-Line], 2003. Available via http://jcp.org/aboutJava/communityproress/final/jsr118/.

Wolf, Mark J.P. and Bernard Perron. *The Video Game Theory Reader.* Routledge, New York, 2003.

Young, R. Michael and Mark O. Riedl. toward an architecture for intelligent control of narrative in interactive virtual worlds. In *International Conference on Intelligent User Interfaces,* January 2003.

PART 2

Enabling Technologies

12

Information Retrieval

CONTENTS

Abstract 12-1
12.1 Introduction 12-2
12.2 Indexing Documents 12-2
12.2.1 Single-Term Indexing 12-3
12.2.2 Multiterm or Phrase Indexing 12-5
12.3 Retrieval Models 12-6
12.3.1 Retrieval Models Without Ranking of Output 12-7
12.3.2 Retrieval Models With Ranking of Output 12-7
12.4 Language Modeling Approach 12-8
12.5 Query Expansion and Relevance Feedback Techniques... 12-9
12.5.1 Automated Query Expansion and Concept-Based Retrieval Models 12-9
12.5.2 Relevance Feedback Techniques 12-11
12.6 Retrieval Models for Web Documents 12-12
12.6.1 Web Graph 12-13
12.6.2 Link Analysis Based Page Ranking Algorithm 12-14
12.6.3 HITS Algorithm 12-14
12.6.4 Topic-Sensitive PageRank 12-15
12.7 Multimedia and Markup Documents 12-15
12.7.1 MPEG-7 12-15
12.7.2 XML 12-15
12.8 Metasearch Engines 12-16
12.8.1 Software Component Architecture 12-16
12.8.2 Component Techniques For Metasearch Engines 12-16
12.9 IR Products and Resources 12-18
12.10 Conclusions and Research Direction 12-19
Acknowledgments 12-20
References 12-20

Vijay V. Raghavan
Venkat N. Gudivada
Zonghuan Wu
William I. Grosky

Abstract

This chapter provides a succinct yet comprehensive introduction to Information Retrieval (IR) by tracing the evolution of the field from classical retrieval models to the ones employed by the Web search engines. Various approaches to document indexing are presented followed by a discussion of retrieval models. The models are categorized based on whether or not they rank output, use relevance feedback to modify initial query to improve retrieval effectiveness, and consider links between the Web documents in assessing their importance to a query. IR in the context of multimedia and XML data is discussed. The chapter also provides a brief description of metasearch engines that provide unified access to multiple Web search engines. A terse discussion of IR products and resources is provided. The chapter is concluded by indicating research directions in IR. The intended audience for this chapter are graduate students desiring to pursue research in the IR area and those who want to get an overview of the field.

1-58488-381-2/05/$0.00+$1.50
© 2005 by CRC Press LLC

12.1 Introduction

An information retrieval (IR) problem is characterized by a collection of documents, possibly distributed and hyperlinked, and a set of users who perform queries on the collection to find a right subset of the documents. In this chapter, we trace the evolution of IR models and discuss their strengths and weaknesses in the contexts of both unstructured text collections and the Web.

An IR system is typically comprised of four components:(1) document indexing — representing the information content of the documents, (2) query indexing — representing user queries, (3) similarity computation — assessing the relevance of documents in the collection to an user query request, and (4) query output ranking — ranking the retrieved documents in the order of their relevance to the user query.

Each of sections 12.2 to 12.4 discusses important issues associated with one of the above components. Various approaches to document indexing are discussed in Section 12.2. In respect of the query output ranking component, Section 12.3 describes retrieval models that are based on exact representations for documents and user queries, employ similarity computation based on exact match that results in a binary value (i.e., a document is either relevant or nonrelevant), and typically do not rank query output. This section also introduces the next generation retrieval models that are more general in their query and document representations and rank the query output. In Section 12.4, recent work on a class of IR models based on the so-called *Language modeling approach* is overviewed. Instead of separating the modeling of retrieval and indexing, these models unify approaches for indexing of documents and queries with similarity computation and output ranking.

Another class of retrieval models recognize the fact that document representations are inherently subjective, imprecise, and incomplete. To overcome these issues, they employ *learning* techniques, which involve user feedback and *automated query expansion.* Relevance feedback techniques elicit user assessment on the set of documents initially retrieved, use this assessment to modify the query or document representations, and reexecute the query. This process is iteratively carried out until the user is satisfied with the retrieved documents. IR systems that employ query expansion either involve statistical analysis of documents and user actions or perform modifications to the user query based on *rules.* The rules are either handcrafted or automatically generated using a dictionary or thesaurus. In both relevance feedback and query expansion based approaches, typically weights are associated with the query and document terms to indicate their relative importance in manifesting the query and document information content. The query expansion-based approaches recognize that the terms in the user query are not just literal strings but denote domain concepts. These models are referred to as Concept-based or Semantic retrieval models. These issues are discussed in Section 12.5.

In Sections 12.6 and 12.8 we discuss Web-based IR systems. Section 12.6 examines recent retrieval models introduced specifically for information retrieval on the Web. These models employ information content in the documents as well as citations or links to other documents in determining their relevance to a query. IR in the context of multimedia and XML data is discussed in Section 12.7. There exists a multitude of engines for searching documents on the Web including AltaVista, InfoSeek, Google, and Inktomi. They differ in terms of the scope of the Web they cover and the retrieval models employed. Therefore, a user may want to employ multiple search engines to reap their collective capability and effectiveness. Steps typically employed by a *metasearch* engine are described in Section 12.8. Section 12.9 provides a brief overview of a few IR products and resources. Finally, conclusions and research directions are indicated in Section 12.10.

12.2 Indexing Documents

Indexing is the process of developing a document representation by assigning content descriptors or terms to the document. These terms are used in assessing the relevance of a document to a user query and directly contribute to the retrieval effectiveness of an IR system. Terms are of two types: objective and nonobjective. *Objective terms* apply integrally to the document, and in general there is no disagreement about how to assign them. Examples of objective terms include author name, document URL, and date of publication. In contrast, there is no agreement about the choice or the degree of applicability of *nonobjective terms* to the document. These are intended to relate to the information content manifested in the document. Optionally, a weight may be assigned to a nonobjective term to indicate the extent to which it represents or reflects the information content manifested in the document.

The effectiveness of an indexing system is controlled by two main parameters: indexing exhaustivity and term specificity [Salton, 1989]. *Indexing exhaustivity* reflects the degree to which all the subject matter or domain concepts manifested in a document are actually recognized by the indexing system. When indexing is exhaustive, it results in a large number of terms assigned to reflect all aspects of the subject matter present in the document. In contrast, when the indexing is *nonexhaustive,* the indexing system assigns fewer terms that correspond to the major subject aspects that the document embodies. *Term specificity* refers to the degree of breadth or narrowness of the terms. The use of broad terms for indexing entails retrieving many useful documents along with a significant number of nonrelevant ones. Narrow terms, on the other hand, retrieve relatively fewer documents, and many relevant items may be missed.

The effect of indexing exhaustivity and term specificity on retrieval effectiveness is explained in terms of recall and precision — two parameters of retrieval effectiveness used over the years in the IR area. *Recall* (R) is defined as the ratio of the number of relevant documents retrieved to the total number of relevant documents in the collection. The ratio of the number of relevant documents retrieved to the total number of documents retrieved is referred to as *precision* (P). Ideally, one would like to achieve both high recall and high precision. However, in reality, it is not possible to simultaneously maximize both recall and precision. Therefore, a compromise should be made between the conflicting requirements. Indexing terms that are narrow and specific (i.e., high term specificity) result in higher precision at the expense of recall. In contrast, indexing terms that are broad and nonspecific result in higher recall at the cost of precision. For this reason, an IR system's effectiveness is measured by the precision parameter at various recall levels.

Indexing can be carried out either manually or automatically. *Manual indexing* is performed by trained indexers or human experts in the subject area of the document by using a *controlled vocabulary* made available in the form of terminology lists and *scope notes* along with instructions for the use of the terms. Because of the sheer size of many realistic document collections (e.g., the Web) and the diversity of subject material present in these collections, manual indexing is not practical. Automatic indexing relies on a less tightly controlled vocabulary and entails representing many more aspects of a document than is possible under manual indexing. This helps to retrieve a document with respect to a great diversity of user queries.

In these methods, a document is first scanned to obtain a set of terms and their frequency of occurrence. We refer to this set of terms as *term set of the document.* Grammatical function words such as *and, or,* and *not* occur with high frequency in all the documents and are not useful in representing their information content. A precompiled list of such words is referred to as *stopword list.* Words in the stopword list are removed from the term set of the document. Further, stemming may be performed on the terms. *Stemming is* the process of removing the suffix or tail end of a word to broaden its scope. For example, the word *effectiveness* is first reduced to *effective* by removing *ness,* and then to *effect* by dropping *ive.*

12.2.1 Single-Term Indexing

Indexing, in general, is concerned with assigning nonobjective terms to documents. It can be based on single or multiple terms (or words). In this section, we consider indexing based on single terms and describe three approaches to it: statistical, information-theoretic, and probabilistic.

12.2.1.1 Statistical Methods

Assume that we have N documents in a collection. Let tf_{ij} denote the frequency of the term T_j in document D_i. The term frequency information can be used to assign weights to the terms to indicate their degree of applicability or importance as index terms.

Indexing based on term frequency measure fulfills only one of the indexing aims — recall. Terms that occur rarely in individual documents of a collection are not captured as index terms by the term frequency measure. However, such terms are highly useful in distinguishing documents in which they occur from those in which they do not occur, and help to improve precision. We define the document frequency of the term T_j, denoted by df_{jl}, as the number of times T_j occurs in a collection of N documents. Then, the inverse document frequency (*idf*), given by log N/df_j, is an appropriate indicator of T_j as a document discriminator.

Both the term frequency and the inverse document frequency measures can be combined into a single frequency-based indexing model. Such a model should help to realize both the recall and precision aims of indexing because it generates indexing terms that occur frequently in individual documents and rarely in the remainder of the collection. To reflect this reasoning in the indexing process, we assign an importance or *weight* to a term based on both term frequency (*tf*) and inverse document frequency (*idf*). The weight of a term T_j in document D_i, denoted w_{ij}, is given by $w_{ij} = tf_{ij}$, $\log N/df_j$. The available experimental evidence indicates that the use of combined term frequency and document frequency factors (i.e., *tf-idf*) provides a high level of retrieval effectiveness [Salton, 1989].

Some important variations of the above weighting scheme have been reported and evaluated. In particular, highly effective versions of $tf \cdot idf$ weighting approaches can be found in Croft [1983]; Salton and Buckley [1988]. More recently, a weighting scheme known as Okapi has been demonstrated to work very well with some very large test collections [Jones et al., 1995].

Another statistical approach to index is based on the notion of *term discrimination value.* Given that we have a collection of N documents and each document is characterized by a set of terms, we can think of each document as a point in the document space. Then the distance between two points in the document space is inversely proportional to the similarity between the documents corresponding to the points. When two documents are assigned very similar term sets, the corresponding points in the document space will be closer (that is, the density of the document space is increased); and the points are farther apart if their term sets are different (that is, the density of the document space is decreased).

Under this scheme, we can approximate the value of a term as a document discriminator based on the type of change that occurs in the document space when a term is assigned to the documents of the collection. This change can be quantified based on the increase or decrease in the average distance between the documents in the collection. A term has a good discrimination value if it increases the average distance between the documents. In other words, terms with good discrimination value decrease the density of the document space. Typically, high document frequency terms increase the density, medium document frequency terms decrease the density, and low document frequency terms produce no change in the document density. The term discrimination value of a term T_j, denoted dv_j, is then computed as the difference of the document space densities before and after the assignment of term T_j to the documents in the collection. Methods for computing document space densities are discussed in Salton [1989].

Medium-frequency terms that appear neither too infrequently nor too frequently will have positive discrimination values; high-frequency terms, on the other hand, will have negative discrimination values. Finally, very low-frequency terms tend to have discrimination values closer to zero. A term weighting scheme such as $w_{ij} = tf_{ij} \cdot dv_j$ which combines terra frequency and discrimination value, produces a somewhat different ranking of term usefulness than the $tf \cdot idf$ scheme.

12.2.1.2 Information-Theoretic Method

In information theory, the least predictable terms carry the greatest information value [Shannon, 1951]. Least predictable terms are those that occur with the smallest probabilities. Information value of a term with occurrence probability p is given as $-\log_2 p$. The average information value per term for t distinct terms occurring with probabilities $p_1, p_2, \ldots, p_t$, respectively, is given by:

$$\vec{H} = -\sum_{i=1}^{t} p_i \log_2 p_i \tag{12.1}$$

The average information value given by equation 12.1 has been used to derive a measure of term usefulness for indexing — *signal-noise ratio.* The signal–noise ratio favors terms that are concentrated in particular documents (i.e., low document frequency terms). Therefore, its properties are similar to those of the inverse document frequency. The available data shows that substituting signal-noise ratio measure for inverse document frequency (*idf*) in $tf \cdot idf$ scheme or for discrimination value in $tf \cdot dv$ scheme did not produce any significant change or improvement in the retrieval effectiveness [Salton, 1989].

12.2.1.3 Probabilistic Method

Term weighting based on the probabilistic approach assumes that relevance judgments are available with respect to the user query for a training set of documents. The training set might result from the top-ranked documents by processing the user query using a retrieval model such as the vector space model. The relevance judgments are provided by the user.

An initial query is specified as a collection of terms. A certain number of top-ranking documents, with respect to the initial query, are used to form a training set. To compute the term weight, the following conditional probabilities are estimated using the training set: *document relevant to the query, given that the term appears in the document,* and *document nonrelevant to the query, given that the term appears in the document* are estimated using the training set [Yu and Salton, 1976; Robertson and Sparck-Jones, 1976].

Assume that we have a collection of N documents of which R are relevant to the user query; that R_t of the relevant documents contain term t; that t occurs in f_t documents. Various conditional probabilities are estimated as follows:

Pr [t is present in the document | document is relevant] = R_t/R
Pr [t is present in the document | document is nonrelevant] = $(f_t\ NR_t)/N\ NR$
Pr [t is absent in the document | document is relevant] = – $R\text{-}R_t/R$
Pr [t is absent in the document | document is nonrelevant] = $((N\ R)-(f_t\ NR_t))/(N\ NR)$

From these estimates, the weight of term t, denoted w_t, is derived using Bayes's theorem as:

$$w_t = \log \frac{R_t/(R-R_t)}{(f_t-R_t)/(N-f_t-(R-R_t))} \tag{12.2}$$

The numerator (denominator) expresses the odds of term t occurring in a relevant (nonrelevant) document. Term weights greater than 0 indicate that the term's occurrence in the document provides evidence that the document is relevant to the query; values less than 0 indicate to the contrary. While the discussion of the above weight may be considered inappropriate in the context of determining term weights in the absence of relevant information, it is useful to consider how methods of computing a term's importance can be derived by proposing reasonable approximations of the above weight under such a situation [Croft and Harper, 1979].

12.2.2 Multiterm or Phrase Indexing

The indexing schemes described above are based on assigning single-term elements to documents. Assigning single terms to documents is not ideal for two reasons. First, single terms used out of context often carry ambiguous meaning. Second, many single terms are either too specific or too broad to be useful in indexing. Term phrases, on the other hand, carry more specific meaning and thus have more discriminating power than the individual terms. For example, the terms *joint* and *venture* do not carry much indexing value in financial and trade document collections. However, the phrase *joint venture* is a highly useful index term. For this reason, when indexing is performed manually, indexing units are composed of groups of terms such as noun phrases that permit unambiguous interpretation. To generate complex index terms or term phrases automatically, three methods are used: statistical, probabilistic, and linguistic.

12.2.2.1 Statistical Methods

These methods employ *term grouping* or *term clustering* methods that generate groups of related words by observing word cooccurrence patterns in the documents of a collection. The term-document matrix is a two-dimensional array consisting of n rows and t columns. The rows are labeled D_1, D_2, ..., D_n and correspond to the documents in the collection; columns are labeled T_1, T_2, ..., T_t and correspond to the term set of the document collection. The matrix element corresponding to row D_i and column T_j represents the importance or weight of the term T_j assigned to document D_i. Using this matrix, term groupings or classes are generated in two ways. In the first method, columns of the matrix are compared to each other to assess whether the terms are jointly assigned to many documents in the collection. If so, the terms are assumed to be related and are grouped into the same class. In the second method, the term–document matrix is processed row-wise. Two documents are grouped into the same class if they have similar term assignments. The terms that cooccur frequently in the various document classes form a term class.

12.2.2.2 Probabilistic Methods

Probabilistic methods generate complex index terms based on term-dependence information. This requires considering an exponential number of term combinations, and for each combination an estimate of joint cooccurrence probabilities in relevant and nonrelevant documents. However, in reality, it is extremely difficult to obtain information about occurrences of term groups in the documents of a collection. Therefore, only certain dependent term pairs are considered in deriving term classes [Van Rijsbergen, 1977; Yu et al., 1983]. In both the statistical and probabilistic approaches, cooccurring terms are not necessarily related semantically. Therefore, these approaches are not likely to lead to high-quality indexing units.

12.2.2.3 Linguistic Methods

There are two approaches to determine term relationships using *linguistic methods:* term-phrase formation and thesaurus-group generation. A *term phrase* consists of the phrase head, which is the principal phrase component, and other components. A term with document frequency exceeding a stated threshold (e.g., $df > 2$) is designated as phrase head. Other components of the phrase should be medium- or low-frequency terms with stated cooccurrence relationships with the phrase head. Cooccurrence relationships are those such as the phrase components should cooccur in the same sentence with the phrase head within a stated number of words of each other. Words in the stopword list are not used in the phrase formation process. The use of only word cooccurrences and document frequencies do not produce high-quality phrases, however. In addition to the above steps, the following two syntactic considerations can also be used. Syntactic class indicators (e.g., adjective, noun, verb) are assigned to terms, and phrase formation is then limited to sequences of specified syntactic indicators (e.g., noun–noun, adjective–noun). A simple syntactic analysis process can be used to identify syntactic units such as subject, noun, and verb phrases. The phrase elements may then be chosen from within the same syntactic unit.

While phrase generation is intended to improve precision, thesaurus-group generation is expected to improve recall. A thesaurus assembles groups of related specific terms under more general, higher-level class indicators. The thesaurus transformation process is used to broaden index terms whose scope is too narrow to be useful in retrieval. It takes low-frequency, overly specific terms and replaces them with thesaurus class indicators that are less specific, medium-frequency terms. Manual thesaurus construction is possible by human experts, provided that the subject domain is narrow. Though various automatic methods for thesaurus construction have been proposed, their effectiveness is questionable outside of the special environments in which they are generated.

Others considerations in index generation include case sensitivity of terms (especially for recognizing proper nouns), and transforming dates expressed in various diverse forms into a canonical form.

12.3 Retrieval Models

In this section we first present retrieval models that do not rank output, followed by those that rank the output.

12.3.1 Retrieval Models Without Ranking of Output

12.3.1.1 Boolean Retrieval Model

Boolean retrieval model is a representative of this category. Under this model, documents are represented by a set of index terms. Each index term is viewed as a Boolean variable and has the value *true* if the term is present in the document. No term weighting is allowed and all the terms are considered to be equally important in representing the document content. Queries are specified as arbitrary Boolean expressions formed by linking the terms using the standard Boolean logical operators *and, or,* and *not.* The retrieval status value (*RSV*) is a measure of the query-document similarity. The *RSV* is 1 if the query expression evaluates to *true;* otherwise the RSV is 0.

Documents whose RSV evaluates to 1 are considered relevant to the query. The Boolean model is simple to implement, and many commercial systems are based on this model. User queries can be quite expressive since they can be arbitrarily complex Boolean expressions. Boolean model-based IR systems tend to have poor retrieval performance. It is not possible to rank the output since all retrieved documents have the same RSV. The model does not allow assigning weights to query terms to indicate their relative importance. The results produced by this model are often counterintuitive. As an example, if the user query specifies ten terms linked by the logical connective *and,* a document that has nine of these terms is not retrieved. User relevance feedback is often used in IR systems to improve retrieval effectiveness. Typically, a user is asked to indicate the relevance or nonrelevance of a few documents placed at the top of the output. Because the output is not ranked, the selection of documents for relevance feedback elicitation is difficult.

12.3.2 Retrieval Models With Ranking of Output

Retrieval models under this category include Fuzzy Set, Vector Space, Probabilistic, and Extended Boolean or *p-norm.*

12.3.2.1 Fuzzy Set Retrieval Model

Fuzzy set retrieval model is based on fuzzy set theory [Radecki, 1979]. In conventional set theory, a member either belongs to or does not belong to a set. In contrast, fuzzy sets allow partial membership. We define a membership function F that measures the degree of importance of a term T_j in document D_i by $F(D_i, T_j) = k$, for $0 \leq k \leq 1$. Term weights w_{ij} computed using the $tf \cdot idf$ scheme can be used for the value of k. Logical operators *and, or,* and *not* are appropriately redefined to include partial set membership. User queries are expressed as in the case of the Boolean model and are also processed in a similar manner using the redefined Boolean logical operators. The query output is ranked using the *RSVs.* It has been found that fuzzy-set-based IR systems suffer from lack of discrimination among the retrieved output nearly to the same extent as systems based on the Boolean model. This leads to difficulties in the selection of output documents for elicitation of relevance feedback. The query output is often counterintuitive. The model does not allow assigning weights to user query terms.

12.3.2.2 Vector Space Retieval Model

The *vector space retrieval model* is based on the premise that documents in a collection can be represented by a set of vectors in a space spanned by a set of normalized term vectors [Raghavan and Wong, 1986]. If the set of normalized term vectors is linearly independent, then each document will be represented by an n-dimensional vector. The value of the first component in this vector reflects the weight of the term in the document corresponding to the first dimension of the vector space, and so forth. A user query is similarly represented by an n-dimensional vector. The *RSV* of a query-document is given by the scalar product of the query and the document vectors. The higher the *RSV,* the greater is the document's relevance to the query.

The strength of the model lies in its simplicity. Relevance feedback can be easily incorporated into this model. However, the rich expressiveness of query specification inherent in the Boolean model is sacrificed in the vector space model. The vector space model is based on the assumption that the term vectors spanning the space are orthogonal and existing term relationships need not be taken into account. Furthermore, the query-document similarity measure is not specified by the model and must be chosen somewhat arbitrarily.

12.3.2.3 Probabilistic Retrieval Model

Probabilistic retrieval models take into account the term dependencies and relationships, and major parameters such as the weights of the query terms and the form of the query-document similarity are specified by the model itself. The model is based on two main parameters, *Pr*(*rel*) and *Pr*(*nonrel*), which are probabilities of relevance and nonrelevance of a document to a user query. These are computed by using the probabilistic term weights (Section 12.2.1) and the actual terms present in the document. Relevance is assumed to be a binary property so that *Pr*(*rel*) = I *Pr*(*nonrel*). In addition, the model uses two cost parameters, a_1 and a_2, to represent the loss associated with the retrieval of a nonrelevant document and nonretrieval of a relevant document, respectively.

As noted in Section 12.2.1, the model requires term-occurrence probabilities in the relevant and nonrelevant parts of the document collection, which are difficult to estimate. However, the probabilistic retrieval model serves an important function for characterizing retrieval processes and provides a theoretical justification for practices previously used on an empirical basis (e.g., introduction of certain term-weighting systems).

12.3.2.4 Extended Boolean Retrieval Model

In the extended Boolean model, as in the case of the vector space model, a document is represented as a vector in a space spanned by a set of orthonormal term vectors. However, the query-document similarity is measured in the *extended Boolean* (or *p-norm*) model by using a generalized scalar product between the corresponding vectors in the document space [Salton et al., 1983]. This generalization uses the well-known L_p norm defined for an *n*-dimensional vector, $\vec{d}$, where the length of $\vec{d}$ is given by $\|d\| = \|(w_1, w_2 \quad, w_n)\| = \left(\Sigma_{j=1}^{n} w_j^p\right)^{1/p}$, where $1 \leq p \leq \infty$, and $w_1, w_2, \ldots, w_n$ are the components of the vector $\vec{d}$.

Generalized Boolean *or* and *and* operators are defined for the *p*-norm model. The interpretation of a query can be altered by using different values for *p* in computing query-document similarity. When p = 1, the distinction between the Boolean operators *and* and *or* disappears as in the case of the vector space model. When the query terms are all equally weighted and $p = \infty$, the interpretation of the query is the same as that in the fuzzy set model. On the other hand, when the query terms are not weighted and $p = \infty$, the *p*-norm model behaves like the strict Boolean model. By varying the value of *p* from 1 to ∞, we obtain a retrieval model whose behavior corresponds to a point on the continuum spanning from the vector space model to the fuzzy and strict Boolean models. The best value for *p* is determined empirically for a collection, but is generally in the range $2 \leq p \leq 5$.

12.4 Language Modeling Approach

Unlike the classical probabilistic model, which explicitly models user relevance, a language model views documents themselves as the source for modeling the processes of querying and ranking documents in a collection. In such models, the rank of a document is determined by the probability that a query Q would be generated by repeated random sampling from the document model M_D : $P(Q|M_D)$ [Ponte and Croft, 1998; Lavrenko and Croft, 2001]. As a new altemative paradigm to the traditional IR approach, it integrates document indexing and document retrieval into a single model. In order to estimate the conditional probability $P(Q|M_D)$, explicitly or implicitly, a two-stage process is needed: the indexing stage estimates the language model for each document and the retrieval stage computes the query likelihood based on the estimated document model.

In the simplest case, for the first stage, the maximum likelihood estimate of the probability of each term *t* under the term distribution for each document *D* is calculated as:

$$\hat{p}_{ml}(t|M_D) = \frac{tf_{(t,D)}}{dl_D}$$

where $tf_{(t,D)}$ is the raw term frequency of term t in document D, and dl_D is the total number of tokens in D [Ponte and Croft, 1998]. For the retrieval stage, given the assumption of independence of query terms, the ranking formula can simply be expressed as:

$$\prod_{t \in D} p_{ml}(t|M_D)$$

for each document. However, the above ranking formula will assign zero probability to a document that is missing one or more query terms. To avoid this, various smoothing techniques are proposed with the aim of adjusting the maximum likelihood estimator of a language model so that the unseen terms can be assigned proper non-zero probabilities. A typical smoothing method called *linear interpolation smoothing* [Berger and Lafferty, 1999], which adjusts maximum likelihood model with the collection model $p(t|C)$ whose influence is controlled by a coefficient parameter λ, can be expressed as:

$$p(Q|M_D) = \prod_{t \in Q} \left(\lambda p(t|M_D) + (1-\lambda)p(t|C)\right)$$

The effects of different smoothing methods and different settings of smoothing parameter on retrieval performance can be referred to Zhai and Lafferty [1998].

Obviously, the language model provides a well-interpreted estimation technique to utilize collection statistics. However, the lack of explicit models of relevance makes it conceptually difficult to incorporate the language model with many popular techniques in Information Retrieval, such as relevance feedback, pseudo-relevance feedback, and automatic query expansion [Lavrenko and Croft, 2001]. In order to overcome this obstacle, more sophisticated frameworks are proposed recently that employ explicit models of relevance and incorporate the language model as a natural component, such as risk minimization retrieval framework [Lafferty and Zhai, 2001] and relevance-based language models [Lavrenko and Croft, 2001].

12.5 Query Expansion and Relevance Feedback Techniques

Unlike the database environment, ideal and precise representations for user queries and documents are difficult to generate in an information retrieval environment. It is typical to start with an imprecise and incomplete query and iteratively and incrementally improve the query specification and, consequently, retrieval effectiveness [Aalbersberg, 1992; Efthimiadis, 1995; Haines and Croft, 1993]. There are two major approaches to improve retrieval effectiveness: automated query expansion and relevance feedback techniques. The following section discusses automated query expansion techniques and Section 12.5.2 presents relevance feedback techniques.

12.5.1 Automated Query Expansion and Concept-Based Retrieval Models

Automated query expansion methods are based on term co-occurrences [Baeza-Yates and Ribeiro-Neto, 1999], Pseudo-Relevance Feedback (PRF) [Baeza-Yates and Ribeiro-Neto, 1999], concept-based retrieval [Qiu and Frei, 1993], and language analysis [Bodner and Song, 1996; Bookman and Woods, 2003; Mitra et al., 1998; Sparck-Jones and Tait, 1984]. Language analysis based query expansion methods are not discussed in this chapter.

12.5.1.1 Term Cooccurrences Based Query Expansion

Term cooccurrences based methods involve identifying terms related to the terms in the user query. Such terms might be synonyms, stemming variations, or terms that are physically close to the query terms in the document text. There are two basic approaches to term cooccurrence identification: global and local analysis. In the global analysis, a similarity thesaurus based on term–term relationships is generated. This approach does not work well in general because the term relationships captured in the similarity thesaurus are often invalid in the local context of the user query [Baeza-Yates and Ribeiro-Neto, 1999]. Automatic local analysis employs clustering techniques. Term cooccurrences based clustering is performed on top-ranked documents retrieved in response to the user's initial query. Local analysis is not suitable in the Web context because it requires accessing the actual documents from a Web server. The idea of applying global analysis techniques to a local set of documents retrieved is referred to as *local context analysis.* A study reported in Xu and Croft [1996] demonstrate the advantages of combining local and global analysis.

12.5.1.2 Pseudo-Relevance Feedback Based Query Expansion

In the PRF method, multiple top-ranked documents retrieved in response to the user's initial query are assumed to be relevant. This method has been found to be effective in cases where the initial user query is relatively comprehensive but precise [Baeza-Yates and Ribeiro-Neto, 1999]. However, it has been noted that the method often results in adding unrelated terms, which has a detrimental effect on retrieval effectiveness.

12.5.1.3 Concept-Based Retrieval Model

Compared to term phrases, which capture more conceptual abstraction than the individual terms, concepts are intended to capture even higher levels of domain concepts. Concept-based retrieval treats the terms in the user query as representing domain concepts and not as literal strings of letters. Therefore, it can fetch documents even if they do not contain the specific words in the user query.

There have been several investigations into concept-based retrieval [Belew, 1989; Bollacker et al., 1998; Croft, 1987; Croft et al., 1989; McCune et al., 1989; Resnik, 1995]. RUBRIC (Rule-Based Information Retrieval by Computer) is a pioneer system in this direction. It uses *production rules* to capture user query concepts (or topics). Production rules define a hierarchy of retrieval subtopics. A set of related production rules is represented as an AND/OR tree referred to as *rule-based tree.* RUBRIC facilitates users to define detailed queries starting at a conceptual level.

Only a few concept-based information retrieval systems have been used in the real domains for the following reasons. These systems focus on representing concept relationships without addressing the acquisition of the knowledge. The latter itself is challenging in its own right. Users would prefer to retrieve documents of interest without having to define the rules for their queries. If the system features predefined rules, users can then simply make use of the relevant rules to express concepts in their queries.

The work reported in [Kim, 2000] provides a logical semantics for RUBRIC rules, defines a framework for defining rules to manifest user query concepts, and demonstrate a method for automatically constructing rule-based trees from typical thesauri. The latter has the following fields: USE, BT (Broad Term), NT (Narrow Term), and RT (Related Term). The USE field represents the terms to be used instead of the given term with almost the same meaning. For example, Plant associations and Vegetation types can be used instead of the term Habitat types. As the names imply, BT, NT, and RT fields list more general terms, more specific terms, and related terms of the thesaurus entry. Typically, NT, BT, and RT fields contain numerous terms. Indiscriminately using all the terms results in an explosion of rules. Kim [2000] has suggested a method to select a subset of terms in NT, BT, and RT fields. Experiments conducted on a small corpus with a domain-specific thesaurus show that concept-based retrieval based on automatically constructed rules is more effective than handmade rules in terms of precision.

An approach to constructing query concepts using document features is discussed in Chang et al. [2002]. The approach involves first extracting features from the documents, deriving primitive concepts by clustering the document features, and using the primitive concepts to represent user queries.

The notion of *Concept Index* is introduced in Nakata et al. [1998]. Important concepts in the document collection are indexed, and concepts are cross-referenced to enable concept-oriented navigation of the document space. An incremental approach to cluster document features to extract domain concepts in the Web context is discussed in Wong and Fu [2000]. The approach to concept-based retrieval in Qiu and Frei [1993] is based on language analysis. Their study reveals that language analysis approaches require a deep understanding of queries and documents, which entails a higher computational cost. Furthermore, deep understanding of language still stands as an open problem in the Artificial Intelligence field.

12.5.2 Relevance Feedback Techniques

The user is asked to provide evaluations or relevance feedback on the documents retrieved in response to the initial query. This feedback is used subsequently in improving retrieval effectiveness. Issues include methods for relevance feedback elicitation and means to utilize the feedback to enhance retrieval effectiveness. Relevance feedback is elicited in the form of either *two-level* or *multilevel* relevance relations. In the former, the user simply labels a retrieved document as *relevant* or *nonrelevant,* whereas in the latter, a document is labeled as *relevant, somewhat relevant,* or *nonrelevant.* Multilevel relevance can also be specified in terms of relationships. For example, for three retrieved documents d_1, d_2, and d_3, we may specify that d_1 is more relevant than d_2 and that d_2 is more relevant than d_3. For the rest of this section, we assume two-level relevance and the vector space model. The set of documents deemed relevant by the user comprise *positive feedback,* and the nonrelevant ones comprise *negative feedback.*

As shown in Figure 12.1, two major approaches to utilizing relevance feedback are based on modifying the query and document representations. Methods based on modifying the query representation affect only the current user query session and have no effect on other user queries. In contrast, methods based on modifying the representation of documents in a collection can affect the retrieval effectiveness of future queries. The basic assumption for relevance feedback is that documents relevant to a particular query resemble each other in the sense that the corresponding vectors are similar.

12.5.2.1 Modifying Query Representation

There are three ways to improve retrieval effectiveness by modifying the query representation.

12.5.2.1.1 Modification of Term Weights

The first approach involves adjusting the query term weights by adding document vectors in the positive feedback set to the query vector. Optionally, negative feedback can also be made use of by subtracting the document vectors in the negative feedback set from the query vector. The reformulated query is expected to retrieve additional relevant documents that are similar to the documents in the positive feedback set. This process can be carried out iteratively until the user is satisfied with the quality and number of relevant documents in the query output [Rocchio and Salton, 1965].

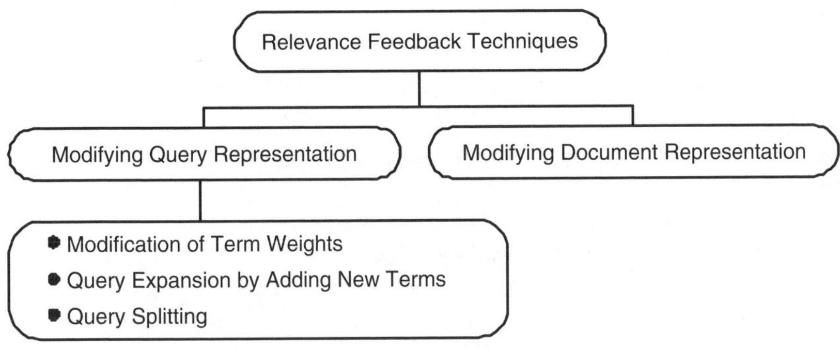

FIGURE 12.1 A taxonomy for relevance feedback techniques.

Modification of query term weights can be based on the positive feedback set, the negative feedback set, or a combination of both. Experimental results indicate that positive feedback is more consistently effective. This is because documents in the positive feedback set are generally more homogeneous than the documents in the negative feedback set. However, an effective feedback technique, termed *dec hi*, uses all the documents in the positive feedback set and subtracts from the query only the vectors of the highest ranked nonrelevant documents in the negative feedback set [Harman, 1992].

The above approaches only require a weak condition to be met with respect to ensuring that the derived query is optimal. A stronger condition, referred to as *acceptable ranking*, was introduced, and an algorithm that can iteratively learn an optimal query has been introduced in Wong and Yao [1990].

More recent advances relating to deterministic strategies for deriving query weights optimally are reported in Herbrich et al. [1998]; Tadayon and Raghavan [1999]. Probabilistic strategies to obtain weights optimally have already been mentioned in Sections 12.2.1 and 12.2.2.

12.5.2.1.2 Query Expansion by Adding New Terms

The second method involves modifying the original query by adding new terms to it. The new terms are selected from the positive feedback set and are sorted using measures such as noise (a global term distribution measure similar to *idf*), postings (the number of retrieved relevant documents containing the term), noise within postings (where frequency is the $\log_2$ of the total frequency of the term in the retrieved relevant set), *noise* × *frequency* × *postings*, and *noise* × *frequency*. A predefined number of top terms from the sorted list are added to the query. Experimental results show that the last three sort methods produced the best results, and adding only selected terms is superior to adding all terms. There is no performance improvement by adding terms beyond 20 [Harman, 1992]. Probabilistic methods that take term dependencies into account may also be included under this category, and they have been mentioned in Section 12.2. There have also been proposals for generating term relationships based on user feedback [Yu, 1975; Wong and Yao, 1993; Jung and Raghavan, 1990].

12.5.2.1.3 Query Splitting

In some cases, the above two techniques do not produce satisfactory results because the documents in the positive feedback set are not homogeneous (i.e., they do not form a tight cluster in the document space) or because the nonrelevant documents are scattered among certain relevant ones. One way to detect this situation is to cluster the documents in the positive feedback set to see if more than one homogeneous cluster exists. If so, the query is split into a number of sub-queries such that each sub-query is representative of one of the clusters in the positive feedback set. The weight of terms in the sub-query can then be adjusted or expanded as in the previous two methods.

12.5.2.2 Modifying Document Representation

Modifying the document representation involves adjusting the document vector based on relevance feedback, and is also referred to as *user-oriented clustering* [Deogun et al., 1989; Bhuyan et al., 1997]. This is implemented by adjusting the weights of retrieved and relevant document vectors to move them closer to the query vector. The weights of retrieved nonrelevant documents vectors are adjusted to move them farther from the query vector. Care must be taken to insure that individual document movement is small because user relevance assessments are necessarily subjective. In all the methods, it has been noted that more than two or three iterations may result in minimal improvements.

12.6 Retrieval Models for Web Documents

The IR field is enjoying a renaissance and widespread interest as the Web is getting entrenched more deeply into all walks of life. The Web is perhaps the largest, dynamic library of our times and sports a large collection of textual documents, graphics, still images, audio, and video collections [Yu and Meng, 2003]. Web search engines debuted in mid 1990s to help locate relevant documents on the Web. IR techniques need suitable modifications to work in the Web context for various reasons. The Web documents are highly distributed — spread over hundreds of thousands of Web Servers. The size of the Web is growing exponentially. There is no quality control on, and hence no authenticity or editorial process in, Web document creation. The documents are highly volatile; they appear and disappear at the will of the document creators.

These issues create unique problems for retrieving Web documents. The first issue is what portion of the document to index. Choices are document title, author names, abstract, and full text. Though this problem is not necessarily unique to the Web context, it is accentuated given the absence of editorial process and the diversity of document types. Because of the high volatility, any such index can get outdated very quickly and needs to be rebuilt quite frequently. It is an established goodwill protocol that the Web servers not be accessed for the full text of the documents in determining its relevance to user queries. Otherwise, the Web servers will get overloaded very quickly. The full text of the documents is retrieved once it has been determined that the document is relevant to a user query. Typically, document relevance to a user query is determined using the index structure built *a priori*. Users of Web search engines, on average, use only two or three terms to specify their queries. About 25% of the search engine queries were found to contain only one term [Baeza-Yates and Ribeiro-Neto, 1999]. Furthermore, the *ploysemy problem* — having multiple meanings for a word — is more pronounced in the Web context due to the diversity of documents.

Primarily, there are three basic approaches to searching the Web documents: Hierarchical Directories, Search Engines, and Metasearch Engines. Hierarchical Directories, such as the ones featured by Yahoo (www.yahoo.com) and Open Directory Project (dmoz.org), feature a manually created hierarchical directory. At the top level of the directory are categories such as Arts, Business, Computers, and Health. At the next (lower) level, these categories are further refined into more specialized categories. For example, the Business category has Accounting, Business and Society, and Cooperatives (among others) at the next lower level. This refinement of categories can go to several levels. For instance, Business/Investing/Retirement Planning is a category in Open Directory Project (ODP). At this level, the ODP provides hyperlinks to various Websites that are relevant to Retirement Planning. This level also lists other related categories such as Business/Financial Services/Investment Services and Society/People/Senior/Retirement. Directories are very effective in providing guided navigation and reaching the relevant documents quite quickly. However, the Web space covered by directories is rather small. Therefore, this approach entails high precision but low recall. Yahoo pioneered the hierarchical directory concept for searching the Web. ODP is a collaborative effort in manually constructing hierarchical directories for Web search.

Early search engines used Boolean and Vector Space retrieval models. In the case of the latter, document terms were weighted. Subsequently, HTML (Hypertext Markup Language) introduced meta-tags, using which Web page authors can indicate suitable keywords to help the search engines in the indexing task. Some of the search engines even incorporated relevance feedback techniques (Section 12.5) to improve retrieval effectiveness. Current generation search engines (e.g., Google) consider the (hyper)link structure of Web documents in determining the relevance of a Web page to a query. Link-based ranking is of paramount importance given that Web page authors often introduce spurious words using HTML meta-tags to alter their page ranking to potential queries. The primary intent of rank altering is to improve Web page hits to promote a business, for example. Link-based ranking helps to diminish the effect of spurious words.

12.6.1 Web Graph

Web graph is a structure obtained by considering Web pages as nodes and the hyperlinks (simply, links) between the pages as directed edges. It has been found that the average distance between connected Web pages is only 19 clicks [Efe et al., 2000]. Furthermore, the Web graph contains densely connected regions that are in turn only a few clicks away from each other. Though an individual link from page p_1 to page p_2 is weak evidence that the latter is related to the former (because the link may be there just for navigation), an aggregation of links is a robust indicator of importance. When the link information is supplemented with text-based information on the page (or the page text around the anchor), even better search results that are both important and relevant have been obtained [Efe et al., 2000].

When only two links are considered in the Web graph, we obtain a number of possible basic patterns: endorsement, cocitation, mutual reinforcement, social choice, and transitive endorsement. Two pages pointing to each other — *endorsement* — is a testimony to our intuition about their mutual relevance. *Cocitation* occurs when a page points to two other pages. Bibliometric studies reveal that relevant papers are often cited together [White and McCain, 1989]. A page that cites the home page of the *New York Times* is most likely to cite the home page of the *Washington Post* also — *mutual reinforcement. Social choice* refers to two documents linking to the same page. This pattern implies that the two pages are related to each other because they point to the same document. Lastly, *transitive endorsement* occurs when a page p_1 points to another page p_2, and p_2 in turn points to p_3. Transitive endorsement is a weak measure of p_3 being relevant to p_1.

Blending these basic patterns gives rise to more complex patterns of the Web graph: *complete bipartite graph, clan graph, in-tree,* and *out-tree.* If many different pages link (directly or transitively) to a page — that is, the page has high in-degree — it is likely that the (heavily linked) page is an authority on some topic. If a page links to many authoritative pages (e.g., a survey paper) — that is, the page has high out-degree — then the page is considered to be a good source (i.e., hub) for searching relevant information. In the following section we briefly discuss two algorithms for Web page ranking based on Web graph.

12.6.2 Link Analysis Based Page Ranking Algorithm

Google is a search engine that ranks Web pages by importance based on link analysis of a Web graph. This algorithm is referred to as *PageRank*[1] [Brin and Page, 1998]. The rank of a page depends on the number of pages pointing to it as well as the rank of those pointing pages. Let r_p be the rank of a page p and x_p be the number of outgoing links on a page. The rank of p is recursively computed as:

$$r_p = (1-d) + d \sum_{\forall p;q \rightarrow p} \frac{r_q}{x_q} \qquad (12.3)$$

where d is a damping factor whose value is selected to be between 0 and 1. It assigns higher importance to pages with high in-degrees or pages that are linked to by highly ranked pages.

12.6.3 HITS Algorithm

Unlike the PageRank algorithm (which computes page ranks offline, independent of user query), the HITS (Hyperlink-Induced Topic Search) algorithm relies on deducing authorities and hubs in a subgraph comprising results of a user query and the local neighborhood of the query result [Kleinberg, 1998]. *Authorities* are those pages to which many other pages in the neighborhood point. *Hubs,* on the other hand, point to many good authorities in the neighborhood. They have mutually reinforcing relationships: Authoritative pages on a search topic are likely to be found near good hubs, which in turn link to many good sources of information on the topic. The kind of relationships of interest are modeled by special subgraph structures such as bipartite and clan. One challenging problem that arises in this context is called *topic drift,* which refers to the tendency of the HITS algorithm to converge to a strongly connected region that represents just a single topic.

The algorithm has two major steps: sampling and weight propagation. In the first step, using one of the commercially available search engines, about 200 pages are selected using keyword-based search. This set of pages is referred to as the *root set.* The set is expanded into a *base set* by adding any page on the Web that has a link to/from a page in the root set. The second step computes a weight for each page in the base set. This weight is used to rank the relevance of the page to a query. The output of the algorithm is a short list of pages with the largest hub weights and a separate list of pages with the largest authority weights.

[1] Our reference is to the original PageRank algorithm.

12.6.4 Topic-Sensitive PageRank

The PageRank algorithm computes the rank of a page statically — page-rank computation is independent of user queries. There have been extensions to PageRank in which page ranks are computed for each topic in a predetermined set offline [Haweliwala, 2002]. This is intended to capture more accurately the notion of importance of a page with respect to several topics. The page-rank value corresponding to a topic that closely corresponds to the query terms is selected in ranking the pages.

12.7 Multimedia and Markup Documents

Though the current Web search engines primarily focus on textual documents, ubiquity of multimedia data (graphics, images, audio, and video) and markup text (e.g., XML documents) on the Web mandate future search engines be capable of indexing and searching multimedia data. Multimedia information retrieval area addresses these issues, and the results have culminated in MPEG-7 [International, 2002] — a standard for describing multimedia data to facilitate efficient browse, search, and retrieval. The standard is developed under the auspices of the Moving Pictures Expert Group (MPEG).

12.7.1 MPEG-7

MPEG-7 is called *multimedia content description interface* and is designed to address the requirements of diverse applications — Internet, Medical Imaging, Remote Sensing, Digital Libraries, E-commerce, to name a few. The standard specifies a set of descriptors (i.e., syntax and semantics of features/index terms) and description schemes (i.e., semantics and structure of relationships between descriptions and description schemes), an XML-based language to specify description schemes, and techniques for organizing descriptions to facilitate effective indexing, efficient storage, and transmission. However, it does not encompass the automatic extraction of descriptors and features. Furthermore, it does not specify how search engines can make use of the descriptors.

Multimedia feature extraction/indexing is a manual and subjective process especially for semantic-level features. Low-level features such as color histograms are extracted automatically. However, they have limited value for content-based multimedia information retrieval. Because of the semantic richness of audiovisual content, difficulties in speech recognition, natural language understanding, and image interpretation, fully automated feature extraction tools are unlikely to appear in the foreseeable future. Robust semiautomated tools for feature extraction and annotation are yet to emerge in the marketplace.

12.7.2 XML

The eXtensible Markup Language (XML) is a W3C standard for representing and exchanging information on the Internet. In recent years, documents in widely varied areas are increasingly represented in XML, and by 2006 about 25% of LAN traffic will be in XML [EETimes, 2003]. Unlike HTML, XML tags are not predefined and are used to mark up the document *content.*

An XML document collection, *D*, contains a number of XML documents (*d*). Each such *d* contains XML elements (*p*) and associated with elements are words (*w*). An element *p* can have zero or more words (*w*) associated with it, a sub-element *p*, or zero or more attributes (*a*) with values (*w*) bound to them. From the information-content point of view, *a* is similar to *p*, except that *p* has more flexibility in information expression and access. An XML document, *d*, therefore is a hierarchical structure.

D can be represented in (*d, p, w*) format. This representation has one more component than a typical full-text collection, which is represented as (d, w). Having *p* in XML document collection *D* entails benefits including the following: *D* can be accessed by content-based retrieval; *D* can be displayed in different formats; and *D* can be evolved by regenerating *p* with *w*. Document *d* can be parsed to construct a *document tree* (by DOM parser) or to identify events for corresponding event-handlers (by SAX parser). Information content extracted via parsing is used to build an index file and to convert into a database format.

Indexing d encompasses building occurrence frequency table freq(p, w) — the number of times w occurs in p. Frequency of occurrence of w in d, freq(d, w), is defined as freq(d, w) = $\Sigma_p$freq(p, w). Based on the value of freq(p, w), d is placed in a fast-search data structure.

12.8 Metasearch Engines

A metasearch engine (or metasearcher) is a Web-based distributed IR system that supports unified access to multiple existing search engines. Metasearch Engines emerged in the early 1990s and provided simple common interfaces to a few major Web search engines. With the fast growth of Web technologies, the largest metasearch engines become complex portal systems that can now search on around 1,000 search engines. Some early and current famous metasearch engines are WAIS [Kahle and Medlar, 1991], STARTS [Gravano et al., 1997], MetaCrawler [Selberg and Etzioni, 1997], SavvySearch [Howe and Dreilinger, 1997] and Profusion [Gauch et al., 1996].

In addition to rendering convenience to users, metasearch engines increase the search coverage of the Web by combining the coverage of multiple search engines. A metasearch engine does not maintain its own collection of Web pages, but it may maintain information about its underlying search engines in order to achieve higher efficiency and effectiveness.

12.8.1 Software Component Architecture

A generic metasearch engine architecture is shown in Figure 12.2. When a user submits a query, the metasearch engine selects a few underlying search engines to dispatch the query; when it receives the Web pages from its underlying search engines, it merges the results into a single ranked list and displays them to the user.

12.8.2 Component Techniques For Metasearch Engines

Several techniques are applied to build efficient and effective metasearch engines. In particular, we introduce two important component technologies used in query processing: database selection and result merging.

12.8.2.1 Database Selection

When a metasearch engine receives a query from a user, the database selection mechanism is invoked (by the database selector in Figure 12.2) to select local search engines that are likely to contain useful Web pages for the query. To enable database selection, the database representative, which is some characteristic information representing the contents of the document database of each search engine, needs to be collected at the global representative database and made available to the selector. From all the underlying search engines, by comparing database representatives, a metasearch engine can decide to select a few search engines that are most useful to the user query. Selection is especially important when the number of underlying search engines is large because it is unnecessary, expensive, and unrealistic to send query to many search engines for one user query.

Database selection techniques can be classified into three categories: Rough Representative, Statistical Representative, and Learning-based approaches [Meng et al., 2002]. In the first approach, the representative of a database contains only a few selected key words or paragraphs that are relatively easy to obtain, require little storage, but are usually inadequate. These approaches are applied in WAIS [Kahle and Medlar, 1991] and other early systems. In the second approach, database representatives have detailed statistical information (such as document frequency of each term) about the document databases, so they can represent databases more precisely than rough representatives. Meng et al. [2002] provide a survey of several of these types of approaches. In the learning-based approach, the representative is the historical knowledge indicating the past performance of the search engine with respect to different queries; it is then used to determine the usefulness of the search engine for new queries. Profusion

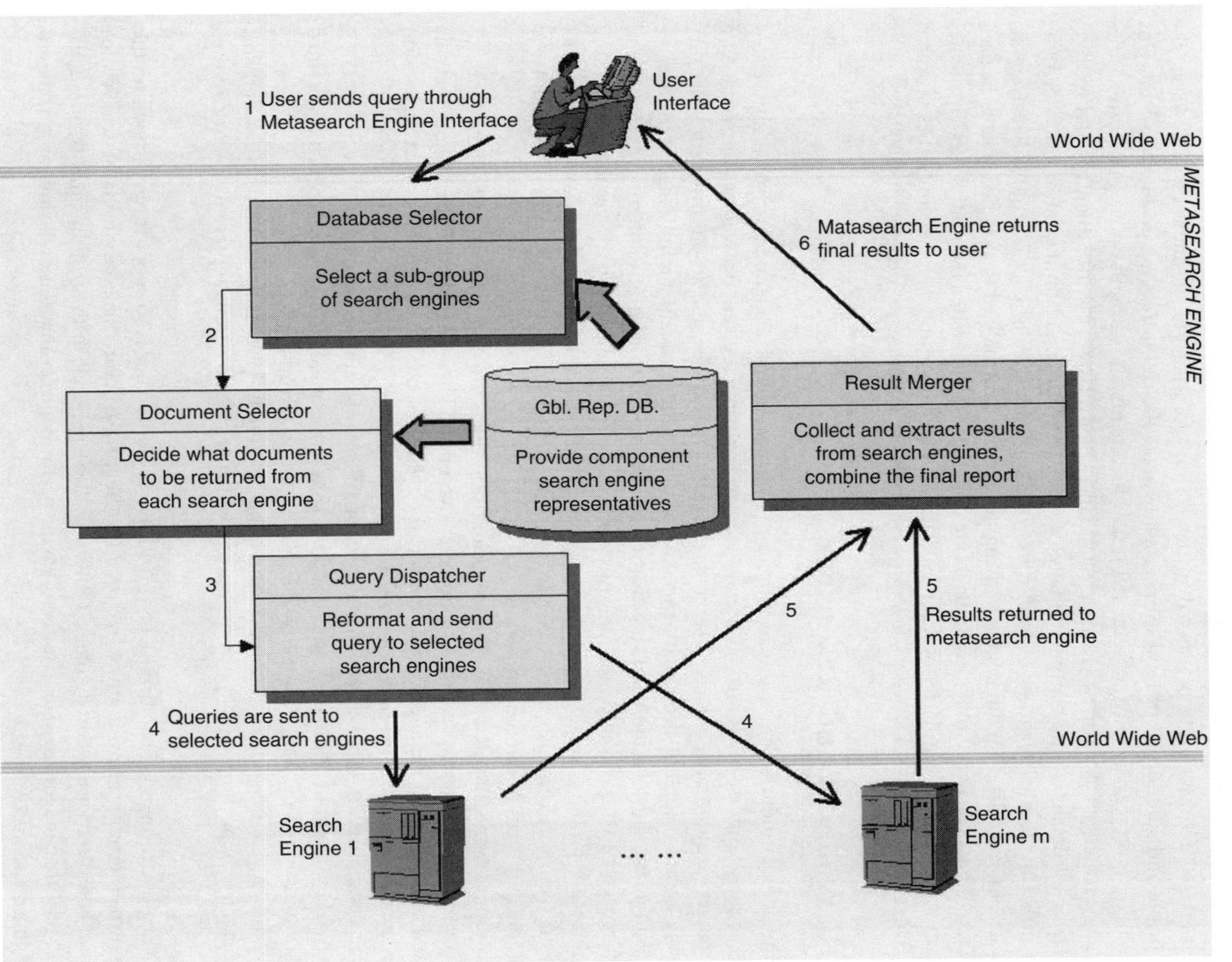

FIGURE 12.2 Metasearch engine reference software component architecture with the flow of query processing. (Numbers associated with the arrows indicate the sequence of steps for processing a query).

(www.profusion.com) and SavvySearch (www.search.com), which are current leading metasearch engines, both fall into this category.

12.8.2.2 Result Merging

Result merging is the process in which, after dispatching a user query to multiple search engines and receiving results back from those search engines, a metasearch engine arranges results from different sources into a single ranked list to provide to users.

Ideally, merged results should be ranked in descending order of global similarities. However, the heterogeneities and autonomies of local search engines make the result merging problem difficult. One simple solution is to actually fetch all returned result documents and compute their global similarities in the metasearch engine (Inquirus) [Lawrence and Giles, 1998]. However, since the process of fetching and analyzing all documents is computationally expensive and time consuming, most result merging methods utilize the local similarities or local ranks of returned results to effect merging. For example, local similarities of results from different search engines can be renormalized to a unified scale to be used as global ranking scores. For another example, if one document d is returned by multiple search engines in a certain way, the global similarity of d can be calculated by combining its local similarities in search engines. These approaches, along with a few others, have been discussed in Meng et al. [2002].

12.9 IR Products and Resources

We use the phrase products and resources to refer to commercial and academic IR systems and related resources. A good number of IR systems are available today; some are generic, whereas others target a specific market — automotive, financial services, government agencies, and so on. In recent years they are evolving toward being full-fledged, off-the-shelf product suites providing a full range of services — indexing, automatic categorization and classification of documents, collaborative filtering, graphical and natural language based query specification, query processing, relevance feedback, results ranking and presentation, user profiling and automated alerts, and support for multimedia data. Not every product provides all these services. Also, they differ in indexing models employed, algorithms for similarity computation, and types of queries supported.

Due to rapid advances in IR in the Web scenario, IR products are also evolving fast. These products primarily work with textual media in a distributed environment. Those that claim to handle other media such as audio, images, and video essentially convert the media to text. For example, broadcast television programs content is represented by the text of closed-captions. Video scene titles and captions are used as content descriptors of the former. Digital text of titles and captions is obtained by using OCR technology. The content of audio clips and sound tracks is represented by digital text, which is obtained by textual transcription of the media using speech recognition technology.

A survey of 23 vendors located in the U.S. and Canada, done in 1996, is presented in Kuhns [1996]. We also list a few important, well-known resources, which are by no means representative or comprehensive:

- TREC (trec.nist.gov): The purpose of Text REtrieval Conference is to support research within the information retrieval community by providing the infrastructure necessary for large-scale evaluation of text retrieval methodologies. TRBC provides large-scale test sets of documents, questions, and relevance judgments. These testbeds enable performance evaluation of various approaches to IR in a standardized way.
- Lemur (www-2.cs.cmu.edu/~lemur/): Lemur is a toolkit for Language Modeling and Information Retrieval.
- SearchTools.com (www.searchtools.com): Provides a list of tools that you can use to construct your own search engines.

12.10 Conclusions and Research Direction

In this chapter, we trace the evolution of theories, models, and practice relevant to the development of IR systems in the contexts of both unstructured text collections and documents on the Web that are semistructured and hyperlinked. A retreival model usually refers to the techniques employed for similarity computation and ranking the query output. Often, multiple retrieval models are based on the same indexing techniques and differ mainly in the approaches to similarity computation and output ranking. Given this context, we have described and discussed strengths and weaknesses of various retrieval models.

The following considerations apply when selecting a retrieval model for Web documents: computational requirements, retrieval effectiveness, and ease of incorporating relevance feedback. Computational requirements refer to both the disk space required for storing document representations as well as the time complexity of crawling, indexing, and computing query-document similarities. Specifically, strict Boolean and fuzzy-set models are preferred over vector space and p-norm models on the basis of lower computational requirements. However, from a retrieval effectiveness viewpoint, vector space and p-norm models are preferred over Boolean and fuzzy-set models. Though the probabilistic model is based on a rigorous mathematical formulation, in typical situations where only a limited amount of relevance information is available, it is difficult to accurately estimate the needed model parameters. All models facilitate incorporating relevance feedback, though learning algorithms available in the context of Boolean models is too slow to be practical for use in real-time adaptive retrieval. Consequently, deterministic approaches for optimally deriving query weights, of the kind mentioned in Section 12.5.2, offer the best promise for achieving effective and efficient adaptive retrieval in real-time.

More recently, a number of efforts are focusing on unified retrieval models that incorporate not only similarity computation and ranking aspects, but also document and query indexing issues. The investigations along these lines, which fall in the category of the language modeling approach, are highlighted in Section 12.4. Interest in methods for incorporating relevance in the context of the language modeling approach is growing rapidly. It is important to keep in mind that several interesting investigations have already been made in the past, even as early as two decodes ago, that can offer useful insight for future work on language modeling [Robertson et al., 1982; Jung and Raghavan, 1990; Wong and Yao. 1993; Yang and Chute, 1994].

Another promising direction of future research is to consider the use of the language modeling approach at other Ievels of document granularity. In other words, the earlier practice has been to apply indexing methods like *tf* idf* and Okapi not only at the granularity of a collection for document retrieval, but also at the levels of a single document or mutiple search engines (i.e., multiple collections) for passage retrieval or search engine selection, respectively. Following through with this analogy suggests that the language modeling approach ought to be investigated with the goals of passage retrieval and search engine selection (the latter, of course, in the context of improving the effectiveness of metasearch engines).

In addition to (and, in some ways, as an alternative to) the use of relevance feedback for enhancing the effectiveness of retrieval systems, there have been several important advances in the direction of automated query expansion and concept-based retrieval. While some effective techniques have emerged, much room still exists for additional performance enhancements, and more future research on how rule bases can be automatically generated is warranted.

Among the most exciting advances with respect to retrieval models for Web documents are the development of methods for ranking Web pages on the basis of analyzing the hyperlink structure of the Web. While early work ranked pages independently of a particular query, more recent research emphasizes techniques that derive topic-specific page ranking. Results in this area, while promising, still need to be more rigorously evaluated. It is also important to explore ways to enhance the efficiency of methods available for topic-specific page ranking.

Metasearch engine technologies are still far away from being mature. Scalability is still a big issue. It is an expensive and labor-intensive task to build and maintain a metasearch engine that searches on a few hundred search engines. Researches is being conducted to solve problems such as automatically connecting to search engines, categorizing search engines, automatically and effectively extracting search engine representatives and so on. It is predictable that in the near future, metasearch engines will be built on hundreds of thousands of search engines. Web-searchable databases will become unique and effective tools to retrieve the Deep Web contents. The latter is estimated to be hundreds of times larger than the Surface Web contents [Bergman, 2002].

Other active IR research directions include Question Answering (QA), Text Categorization, Human Interaction, Topic Detection and Tracking (TDT), multimedia IR, Cross-lingual Retrieval. The Website of ACM Special Interest Group on Information Retrieval (www.sigir.org) is a good place to visit to know more about current research activities in the IR field.

Acknowledgments

The authors would like to thank Kemal Efe, Jong Yoon, Ying Xie, and anonymous referees for their insight, constructive comments, and feedback. This research is supported by a grant from the Louisiana State Governor's Information Technology Initiative (GITI).

References

Aalbersberg, I.J. Incremental relevance feedback. *Proceedings of the 15th Annual International ACM SIGIR Conference,* ACM Press, New York, pp. 11–22, June 1992.

Baeza-Yates, R. and B. Ribeiro-Neto. *Modern Information Retrieval.* Addison-Wesley, Reading, MA, 1999.

Belew, R. Adaptive information retrieval: Using a connectionist representation to retrieve and learn about documents. *Proceedings of the 12th Annual International ACM SIGIR Conference,* ACM Press, New York, pp. 11–20, 1989.

Berger, A. and J. Lafferty. Information retrieval as statistical translation. *ACM SIGIR Conference on Research and Development in Information Retrieval,* pp. 222–229, 1999.

Bergman, M. The Deep Web: Surfacing the hidden value. *BrightPlanet,* available at www.completeplanet.com/Tutorials/DeepWeb/index.asp. (Date of access: April 25, 2002).

Bhuyan, J.N., J.S. Deogun, and V.V. Raghavan. Algorithms for the boundary selection problem. *Algorithmica,* 17: 133–161, 1997.

Bodner, R. and F. Song. In *Lecture Notes in Computer Science,* Vol. 1081, pp. 146–158, available at http://citeseer.nj.nec.com/bodner96knowledgebased.html, 1996.

Bookman, L. and W. Woods. Linguistic Knowledge Can Improve Information Retrieval. http://acl.ldc.upenn.edu/A/A00/A00-1036.pdf (Date of access: January 30th, 2003).

Bollacker, K.D., S. Lawrence, and C.L. Giles. CiteSeer: An autonomous Web agent for automatic retrieval and identification of interesting publications. *Proceedings of the 2nd International Conference on Autonomous Agents,* ACM Press, New York, pp. 116–123, May 1998.

Bookstein, A. and D.R. Swanson. Probabilistic model for automatic indexing. *Journal of the American Society for Information Science,* 25(5): 312–318, 1974.

Brin, S. and L. Page. The Anatomy of a Large Scale Hypertextual Web Search Engine. In *Proceedings of the WWW7/Computer Networks,* 30(1–7): 107–117, April 1998.

Chang, Y., I. Choi, J. Choi, M. Kim, and V.V. Raghavan. Conceptual retrieval based on feature clustering of documents. In *Proceedings of the ACM SIGIR Workshop on Mathematical/Formal Methods in Information Retrieval,* Tampere, Finland, August 2002.

Croft, W.B. Experiments with representation in a document retrieval system. *Information Technology,* 2: 1–21, 1983.

Croft, W.B. Approaches to intelligent information retrieval. *Information Processing and Management,* 23(4): 249–254, 1987.

Croft, W.B. and D.J. Harper. Using probabilistic models of document retrieval without relevance information. *Journal of Documentation,* 35: 285–295, 1979.

Croft, W.B., T.J. Lucia, J. Cringean, and P Willett. Retrieving documents by plausible study: an experimental study. *Information Processing and Management,* 25(6): 599–614, 1989.

Deogun, J.S., V.V. Raghavan, and P. Rhee. Formulation of the term refinement problem for user-oriented information retrieval. In *The Annual AI Systems in Government Conference,* pp. 72–78, Washington, D.C., March 1989.

EETimes. URL: www.eetimes.com, July 2003.

Efe, K., V.V. Raghavan, C.H. Chu, A.L. Broadwater, L. Bolelli, and S. Ertekin. The shape of the web and its implications for searching the web. In *International Conference on Advances in Infrastructure for Electronic Business, Science, and Education on the Internet,* Proceedings at http://www.ssgrr.it/en/ssgrr2000/proccedings.htm. Rome, Italy, July–August, 2000.

Efthimiadis, E. User choices: a new yardstick for the evaluation of ranking algorithms for interactive query expansion. *Information Processing and Management,* 31(4): 605–620, 1995.

Fellbaum, C. *WordNet: An Electronic Lexical Database.* The MIT Press, Cambridge, MA, 1998.

Gauch, S., G. Wang, and M. Gomez. ProFusion: Intelligent fusion from multiple, distributed search engines. *Journal of Universal Computer Science,* 2(9): 637–649, 1996.

Gravano, L., C. Chang, H. Garcia-Molina, and A. Paepcke. Starts: Stanford proposal for Internet mesa-searching. *ACM SIGMOD Conference,* Tucson, AZ, ACM Press, New York, pp. 207–219, 1997.

Haines, D. and W. Bruce Croft. Relevance feedback and inference networks. *Proceedings of the 16th Annual International ACM SIGIR Conference,* ACM Press, New York, pp. 2–11, June 1993.

Harman, D. Relevance feedback revisited. *Proceedings of the 15th Annual International ACM SIGIR Conference,* ACM Press, New York, pp. 1–10, June 1992.

Haweliwala, T. Topic-Sensitive PageRank. *Proceedings of WWW2002,* May 2002.

Herbrich, R., T. Graepel, P. Bollmann-Sdorra, and K. Obermayer. Learning preference relations in IR. In *Proceedings of the Workshop Text Categorization and Machine Learning, International Conference on Machine Learning-98,* pp. 80–84, March 1998.

Howe, A. and D. Dreilinger. SavvySearch: A MetaSearch Engine that Learns Which Search Engines to Query. *Al Magazine,* 18(2): 19–25, 1997.

Jones, S., M.M. Hancock-Beaulieu, S.E. Robertson, S. Walker, and M. Gatford. Okapi at TREC-3. In *The Third Text Retrieval Conference (TREC-3),* Gaithersburg, MD, pp. 109–126, April 1995.

Jung G.S. and V.V. Raghavan. Connectionist learning in constructing thesaurus-like knowledge structure. In *Working Notes of AAAI Symposium on Text-based Intelligent Systems,* pp. 123–127, Palo Alto, CA, March 1990.

Kahle, B. and A. Medlar. An information system for corporate users: Wide area information servers. *Technical Report TMC1991,* Thinking Machines Corporation, 1991.

Kim, M., F. Lu, and V. Raghavan. Automatic Construction of Rule-based Trees for Conceptual Retrieval. *SPIRE-2000,* pp. 153–161, 2000.

Kleinberg, J. Authoritatives sources in a hyperlinked environment. In *Proceedings of ACM-SIAM Symposium on Discrete Algorithms,* pp. 668–677, January 1998.

Kuhns, R. A Survey of Information Retrieval Vendors. *Technical Report: TR-96-56,* Sun Microsystems, Santa Clara, CA, October 1996.

Lafferty, J. and C. Zhai. Document Language Models, Query Models, and Risk Minimization for Information Retrieval, *ACM SIGIR Conference on Research and Development in Information Retrieval,* pp. 111–119, 2001.

Lavrenko, V. and W. Croft. Relevance-Based Language Models, *ACM SIGIR Conference on Research and Development in Information Retrieval,* pp. 120–127, 2001.

Lawrence, S. and C.L. Giles. Inquirus, the NECI meta search engine. *7th International World Wide Web Conference,* Brisbane, Australia, pp. 95–105, 1998.

McCune, B.P., R.M. Tong, J.S. Dean, and D.G. Shapiro. RUBRIC: A system for Rule-Based Information Retrieval. *IEEE Transactions on Software Engineering,* 11(9): 939–945, September, 1985.

Meng, W., C. Yu, and K. Liu. Building efficient and effective metasearch engines. *ACM Computing Surveys,* 34(1): 48–84, 2002.

Mitra, M., A. Singhal, and C. Buckley. Improving automatic query expansion. In *Proceedings of the 21st ACM SIGIR conference,* pp. 206–214, Melbourne, Australia, August 1998.

International Standards Organization (ISO). MPEG-7 Overview (version 8). ISO/IEC JTC1/SC29/WG11 N4980, July 2002. URLs: mpeg.tilab.com, www.mpeg-industry.com.

Nakata, K., A. Voss, M. Juhnke, and T. Kreifelts. Collaborative concept extraction from documents. *Proceedings of the 2nd International Conference on Practical Aspects of Knowledge Management (PAKM 98),* Basel, Switzerland, pp. 29–30,1999.

Ponte, J. and W. Croft. A language modeling approach to Information Retrieval, *ACM SIGIR Conference on Research and Development in Information Retrieval,* pp. 275–281, 1998.

Qiu, Y. and H. Frei. Concept based query expansion. *Proceedings of the 16th Annual International ACM SIGIR Conference,* ACM Press, New York, pp. 160–170, June 1993.

Radecki, T. Fuzzy set theoretical approach to document retrieval. *Information Processing and Management,* 15: 247–259, 1979.

Raghavan, V. and S.K.M. Wong. A critical analysis of vector space model for information retrieval. *Journal of the American Society for Information Science,* 37(5): 279–287, 1986.

Resnik, P. Using information content to evaluate semantic similarity in a taxonomy. *Proceedings of the 14th International Joint Conference on Artificial Intelligence,* pp. 448–453, 1995.

Robertson, S.E., M.E. Maron, and W.S. Cooper. Probability of relevance: A unification of two competing models for document retrieval. *Information Technology, Research, and Development,* 1: 1–21,1982.

Robertson, S.E. Okapi. http://citeseer.nj.nec.com/correct/390640.

Robertson, S.E. and K. Sparck-Jones. Relevance weighting of search terms. *Journal of American Society of Information Sciences,* pp. 129–146, 1976.

Rocchio, J.J. and G. Salton. Information optimization and interactive retrieval techniques. In *Proceedings of the AFIPS-Fall Joint Computer Conference 27 (Part 1),* pp. 293–305,1965,

Salton, G. *Automatic Text Processing.* Addison-Wesley, Reading, MA, 1989.

Salton, G. and C. Buckley. Term-weighting approaches in automatic text retrieval. *Information Processing and Management,* 24: 513–523, 1988.

Salton, G., E.A. Fox, and H. Wu. Extended boolean information retrieval. *Communications of the ACM,* 36: 1022–1036, 1983.

Selberg, E. and O. Etzioni. The MetaCrawler architecture for resource aggregation on the Web. *IEEE Expert,* 12(1): 8–14, 1997.

Sparck-Jones, K. and J.I. Tait. Automatic search term variant generation. *Journal of Documentation,* 40: 50–66,1984.

Shannon, C.E. Prediction and entropy in printed English. *Bell Systems Journal,* 30(1): 50–65, 1951.

Tadayon, N. and V.V. Raghavan. Improving perceptron convergence algorithm for retrieval systems. *Journal of the ACM,* 20(11–13): 1331–1336, 1999.

Van Rijsbergen, C.J. A theoretical basis for the use of co-occurrence data in information retrieval. *Journal of Documentation,* 33: 106–119, June 1977.

White, H. and K. McCain. Bibliometrics. In *Annual review of Information Science and Technology,* Elsevier, Amsterdam, pp. 119–186, 1989.

Wong, W. and A. Fu. Incremental document clustering for web page classification. *IEEE 2000 International Conference on Information Society in the 21st Century: Emerging Technologies and New Challenges* (IS 2000), pp. 5–8, 2000.

Wong, S.K.M. and Y.Y. Yao. Query formulation in linear retrieval models. *Journal of the American Society for Information Science,* 41: 334–341, 1990.

Wong, S.K.M. and Y.Y. Yao. A probabilistic method for computing term-by-term relationships. *Journal of the American Society for Information Science,* 44(8): 431–439, 1993.

Xu, J. and W. Croft. Query exapnsion using local and global document analysis. *Proceedings of 19th ACM SIGIR Conference on Research and Development in Information Retrieval,* pp. 4–11, 1996.

Yang, Y. and C.G. Chute. An example-based mapping method for text categorization and retrieval. *ACM Transactions on Information Systems,* 12: 252–277, 1994.

Yu, C.T. A formal construction of term classes. *Journal of the ACM,* 22: 17–37, 1975.

Yu, C.T., C. Buckley, K. Lam, and G. Salton. A generalized term dependence model in information retrieval. *Information Technology, Research, and Development,* 2: 129–154, 1983.

Yu, C. and W. Meng. Web search technology. In *The Internet Encyclopedia,* H. Bidgoli, Ed., John Wiley and Sons, New York, (to appear), 2003.

Yu, C.T. and G. Salton. Precision weighing — an effective automatic indexing method. *Journal of the ACM,* pp. 76–88, 1976.

Zhai, C. and J. Lafferty. A study of smoothing methods for language models applied to *Ad Hoc* information retrieval, *ACM SIGIR Conference on Research and Development in Information Retrieval,* pp. 334–342, 2001.

13

Web Crawling and Search

CONTENTS

Abstract 13-1
13.1 Introduction 13-1
13.2 Essential Concepts and Well-Known Approaches 13-2
13.2.1 Crawler 13-2
13.2.2 Indexer 13-7
13.2.3 Relevance Ranking 13-9
13.2.4 Databases 13-12
13.2.5 Retrieval Engine 13-12
13.2.6 Improving Search Engines 13-13
13.3 Research Activities and Future Directions 13-14
13.3.1 Searching Dynamic Pages — The Deep Web 13-14
13.3.2 Utilizing Peer-to-Peer Networks 13-15
13.3.3 Semantic Web 13-16
13.3.4 Detecting Duplicated Pages 13-16
13.3.5 Clustering and Categorization of Pages 13-17
13.3.6 Spam Deterrence 13-17
13.4 Conclusion 13-18
References 13-18

Todd Miller

Ling Liu

Abstract

Search engines make finding information on the Web possible. Without them, users would spend countless hours looking through directories or blindly surfing from page to page. This chapter delves into the details of how a search engine does its job by exploring Web crawlers, indexers, and retrieval engines. It also examines the ranking algorithms used to determine what order results should be shown in and how they have evolved in response to the spam techniques of malicious Webmasters. Lastly, current research in search engines is presented and used to speculate on the search of the future.

13.1 Introduction

The amount of information on the World Wide Web (Web) is growing at an astonishing speed. Search engines, directories, and browsers have become ubiquitous tools for accessing and finding information on the Web. Not surprisingly, the explosive growth of the Web has made Web search a harder problem than ever.

Search engines and directories are the most widely used services for finding information on the Web. Both techniques share the same goal of helping users quickly locate Web pages of interest. Internet directories, however, are manually constructed. Only those pages that have been reviewed and categorized

1-58488-381-2/05/$0.00+$1.50
© 2005 by CRC Press LLC

are listed. Search engines, on the other hand, automatically scour the Web, building a massive index of all the pages that they find. Today, popular Internet portals (such as Yahoo!) utilize both directories and a search engine, giving users the choice of browsing or searching the Web.

Search engines first started to appear shortly after the Web was born. The need for a tool to search the Web for information was quickly realized. The earliest known search engine was called the World Wide Web Worm (WWWW). It was called a worm for its ability to automatically traverse the Web, going page by page just as an inch worm goes inch by inch. Early search engines ran on only one or two computers and indexed only a few hundred thousand pages. However, as the Web quickly grew, search engines had to become more efficient, distributed, and learn to battle attempts by some Webmasters to mislead the automated facilities in an effort to get more traffic to their sites.

Search engines today index billions of pages and use tens of thousands of machines to answer hundreds of millions of queries daily. At the time of writing, the most popular search engine is Google, which quickly rose to its position of prominence after developing a new methodology for ranking pages and combining it with a clean and simple user interface. Google has done so well that Yahoo!, AOL, and Netscape all use Google to power their own search engines. Other top search engines include AltaVista, AskJeeves, and MSN. Other companies, such as Overture and Inktomi, are in the business providing search engines with advertising and indices of the Web.

Even with all of these different companies working to index and provide a search of the Web, the Web is so large that each individual search engine is estimated to cover less than 30% of the entire Web. However, the indices of the search engines have a large amount of overlap, as they all strive to index the most popular sites. This means that even when combined, search engines index less than two thirds of the total Web [Lawrence & Giles, 1998]. What makes matters even worse is the fact that when compared to earlier studies, search engines are losing ground every day, as they cannot keep up with the growth rate of the Web [Bharat and Broder, 1998].

Even though search engines only cover a portion of the Web and can produce thousands of results for the user to sift through for an individual query, they have made an immeasurable contribution to the Web. Search engines are the primary method people use for finding information on the Web, as over 74% use search engines to find new Websites. In addition, search engines generate approximately 7% of all Website traffic. These and other [Berry & Browne, 1999; Glossbrenner & Glossbrenner, 1999] statistics and more illustrate the important role that Web search engines play in the Internet world.

The rest of this chapter will focus on how search engines perform their jobs and present some of the future directions that this technology might take. We first describe the essential concepts, the well-known approaches, and the key tradeoffs of Web crawling and search. Then we discuss the technical challenges faced by current search engines, the upcoming approaches, and the research endeavors that look to address these challenges.

13.2 Essential Concepts and Well-Known Approaches

A modern Web search engine consists of four primary components: a crawler, an indexer, a ranker, and a retrieval engine, connected through a set of databases. Figure 13.1 shows a sketch of the general architecture of a search engine. The crawler's job is to effectively wander the Web retrieving pages that are then indexed by the indexer. Once the crawler and indexer finish their job, the ranker will precompute numerical scores for each page indexed, determining its potential importance. Lastly, the retrieval engine acts as the mediator between the user and the index, performing lookups and presenting results. We will examine each of these components in detail in the following sections.

13.2.1 Crawler

Web crawlers are also known as robots, spiders, worms, walkers, and wanderers. Before any search engine can provide services, it must first discover information on the Web. This is accomplished through the use of a crawler. The crawler starts to crawl the Web from a list of seed URLs. It retrieves a Web page

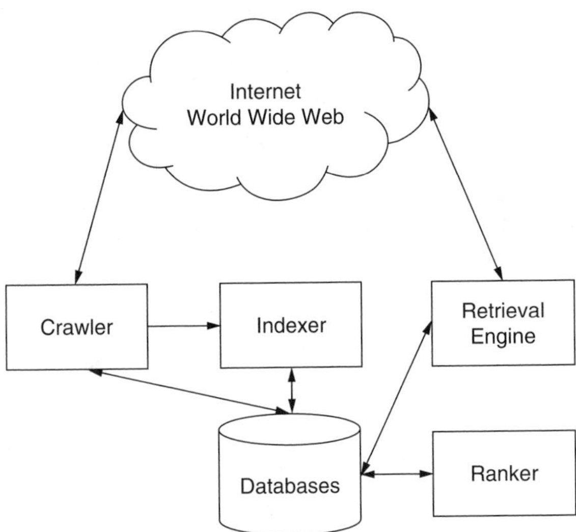

FIGURE 13.1 General architecture of a modern search engine.

using one seed URL, finds links to other pages contained in the page retrieved, follows those links to retrieve more pages, and thus discovers more links. This is how it has become known as a crawler, as it crawls the Web page by page, following links.

Web crawling is the primary means of discovering information on the Internet and is the only effective method to date for retrieving billions of documents with minimal human intervention [Pinkerton, 1994]. In fact, the goal of a good crawler is to find as many pages as possible within a given time. This algorithm is dependant upon links to other pages in order to find resources. If a page exists on the Web but has no links pointing to it, this method of information discovery will never find the page. Crawling also lends itself to a potentially endless process, as the Web is a constantly changing place. This means that a crawl will never be complete and that a crawl must be stopped at some point. Thus some pages will inevitably not be visited. Since a crawl is not complete, it is desirable to retrieve the more useful pages before retrieving less useful ones. However, what determines the usefulness of a page is a point of great contention.

Every crawler consists of four primary components: lists of URLs, a picker, retriever, and link extractor. Figure 13.2 illustrates the general architecture of a Web crawler. The following sections will present the details of each component.

13.2.1.1 Lists of URLs

The URL lists are the crawler's memory. A URL is a Universal Resource Location, and in our context a URL can be viewed as the address of a Web page. The crawler maintains two lists of URLs, one for pages that it has yet to visit and one for pages that have already been crawled. Generally, the "to be crawled" list is prepopulated with a set of "seed URLs." The seed URLs will act as the starting point for the crawler when it first begins retrieving pages. The pages used for seeding are picked manually and should be very popular pages with a large number of outgoing links.

At the first glance, these lists of URLs seem quite simplistic. However, when crawling the whole Web, the lists will encompass billions of URLs. If we assume that a URL can be uniquely represented by 16 bytes, a billion URLs would require 16 GB of disk space. Lists so large will not fit in memory, so the majority is stored on disk. However, a large cache of the most frequently occurring links is kept in memory to assist with the overall crawler performance, avoiding the need to access the disk for every operation.

The lists must provide three functions in order for the crawler to operate effectively. First, there must be a method for retrieving uncrawled URLs so that the crawler can decide where to go next. The second function is to add extracted links that have not been seen by the crawler before to the uncrawled URL

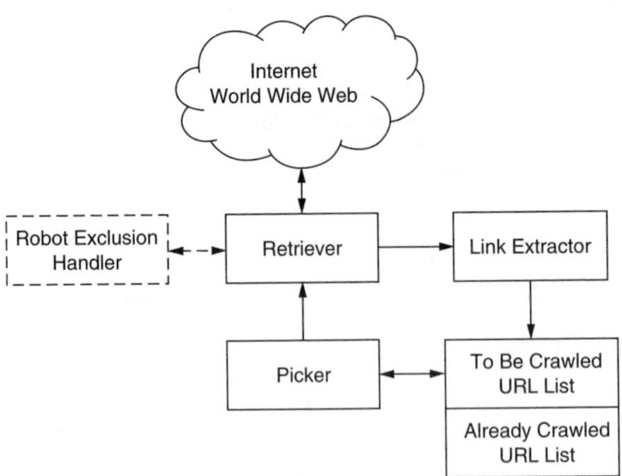

FIGURE 13.2 General web crawler architecture.

list. Both the crawled list and to-be-crawled list must be checked for each link to be added as to whether or not it has previously been extracted or crawled. This is essential to avoid crawling the same page multiple times, which is frowned upon by Webmasters as a crawler can overstrain a site or even bring it down if it is not careful. Lastly, the lists have to be updated when a page has been successfully crawled such that it can be moved from the to-be-crawled to the crawled list.

13.2.1.2 Picker

An efficient crawler must predict which page would be the best page to crawl next, out of the pages yet uncrawled. Because the amount of time which a crawler can spend wandering the Web is limited, and the Web is virtually infinite in size, the path that a crawler takes will have a tremendous impact on the quality and coverage of the search engine. Thus the picker mechanism, the algorithm for deciding which page to visit next, is arguably the most important part of a crawler.

Breadth-First Search (BFS) and Depth-First Search (DFS) are the two frequently used graph traversal mechanisms. The BFS advocate would debate that a good crawl would cover small portions of a large number of sites, thus giving it great breadth [Najork and Wiener, 2001; Pinkerton, 1994], whereas the DFS promoter would argue that a good crawl covers a small number of sites but each in great depth [Chakrabarti et al. 1999]. With BFS, rather than simply picking the next URL on the list of unvisited pages, the crawler could pick one from a site that has not been visited yet at all. This would increase the breadth of the crawl by forcing the crawler to visit at least one page from all sites it knows of before exploring any one site in depth. Similarly, with DFS, the crawler could pick one URL that is from the same Website as the previously crawled page. This technique will give the crawler great in-depth knowledge of each site indexed, but may lack in its coverage of the Web as a whole. More recently, crawlers have started to strike a middle ground and crawl pages in order of reputation [Cho et al., 1998; Aggarwal, 2001]. Reputation in this case is based upon a mathematical algorithm that is heavily based upon the links between Web pages, where a link represents a vote of confidence in another page. In the reputation-based scheme, the unvisited URL with the highest reputation score would be selected. One drawback to the reputation-based selection is that reputations must be recomputed as more pages are discovered by the crawler. This means an increasing amount of computing power must be dedicated to calculating the reputations of the remaining pages as the unvisited URL list grows.

13.2.1.3 Retriever

Once the picker has decided which page to visit next, the retriever will request the page from the remote origin Web server. Because the sending and receiving data over a network takes time, the retriever is usually the slowest component of the crawler. For performance reasons, modern crawlers will have multiple retrievers working in parallel, minimizing the effect of network delay. However, introducing multiple retrievers also introduces the possibility of a crawler overloading a Web server with too many requests simultaneously. It is suggested that the retrievers be coordinated in such a fashion that no more than two simultaneously access the same Web server. The retriever must also decide how to handle URLs that have either gone bad (because the page no longer exists), been moved, or are temporarily out of service (because the Web server is down), as all commonly occur.

13.2.1.4 Link Extractor

The final essential component of a crawler is the link extractor. The extractor takes a page from the retriever and extracts all the (outgoing) links contained in the page. To effectively perform this job, the extractor must be able to recognize the type of document it is dealing with and use a proper means for parsing it. For example, HTML (the standard format for Web pages) must be handled differently than a Word document, as both can contain links but are encoded differesntly. Link extractors use a set of parsers, algorithms for extracting information from a document, with one parser per type of document. Simple crawlers handle only HTML, but they still must be capable of detecting what is not an HTML document. In addition, HTML is a rather loosely interpreted standard that has greatly evolved as the Web has grown. Thus, most Web pages have errors or inconsistencies that make it difficult for a parser to interpret the data. Since a Web crawler does not need to display the Web page, it does not need to use a full parser such as that used in Web browsers. Link extractors commonly employ a parser that is "just good enough" to locate the links in a document. This gives the crawler yet another performance boost, in that it only does the minimal processing on each document retrieved.

13.2.1.5 Additional Considerations

There are a number of other issues with Web crawlers that make them more complex than they first appear. In this section, we will discuss the ways that crawlers can be kept out of Websites, or maliciously trapped within a Website, methods for measuring the efficiency of a crawler, and scalability and customizability issues of a crawler design.

13.2.1.5.1 Robot Exclusion

The Web is a continuously evolving place where anyone can publish information at will. Some people use the Web as a means of privacy and anonymity, and wish to keep their Website out of the search engines' indices. Other Webmasters push the limits of the Web, creating new means of interactivity and relationships between pages. Sometimes a crawler will stumble across a set of these pages and cause unintended results as it explores the Web without much supervision. To address the need for privacy and control of which pages a crawler can crawl without permission, a method of communication between Webmasters and crawlers was developed called robot exclusion. A standard for this is provided by Martijn Koster at http://www.robotstxt.org/wc/norobots.html.

This communication is called the robot exclusion standard. It is a mutual agreement between crawler programmers and Webmasters that allows Webmasters to specify which parts of their site (if any) are acceptable for crawlers. A Webmaster who wishes to guide a crawler through his site creates a file named "robots.txt" and places it in his root Web directory. Many major Websites use this file, such as CNN (http://cnn.com/robots.txt).

The concept is that a good crawler will request this file before requesting any other page from the Web server. If the file does not exist, it is implied that the crawler is free to crawl the entire site without restriction. If the file is present, then the crawler is expected to learn the rules defined in it and to obey them. There is nothing to prevent a crawler from ignoring the file completely, but rather there is an implicit trust placed upon the crawler programmers to adhere to the robot exclusion standard.

13.2.1.5.2 Measuring Efficiency

Because the Web is growing without limit and crawlers need to crawl as many pages as they can within a limited time, efficiency and speed are primary concerns. Thus it is important to understand the popular means for measuring the efficiency of a crawler. Until now the only agreed-upon measure for a crawler is its speed, which is measured in pages per second. The fastest published speeds of crawlers are over 112 pages/sec using two machines [Heydon & Najork, 1999]. Commercial crawlers have most likely pushed this speed higher as they seek to crawl more of the Web in a shorter time. Coverage can be measured by examining how many different Web servers were hit during the crawl, the total number of pages visited, and the average depth per site. While speed and coverage can be measured, there is no agreed-upon means for measuring the quality of crawl, as factors like speed and coverage do not endorse the quality or usefulness of the pages retrieved [Henzinger et al., 1999].

13.2.1.5.3 Scalability and Customizability

Scalability and customizability are another desirable property of the modern crawler. Designing a scalable and extensible Web crawler comparable to the ones used by the major search engines is a complex endeavor. By scalable, we mean the design of the crawler should scale up to the growth of the Web. By extensible, we mean that the crawler should be designed in a modular way to allow new functionality to be incorporated easily and seamlessly, allowing the crawler to be adapted quickly to changes in the Web.

Web crawlers are almost as old as the Web itself. The first crawler, Matthew Gray's Wanderer, was written in the spring of 1993, roughly coinciding with the first release of NCSA Mosaic [Gray, 1996a]. Several papers about Web crawling appeared between 1994 and 1998 [Eichmann, 1994; McBryan, 1994; Pinkerton, 1994]. However, at the time, the Web was several orders of magnitude smaller than it is today. So the earlier systems did not address the scaling problems inherent in a crawl of today's Web.

Mercator, published in 1999 [Heydon and Najork, 1999], is a research effort from the HP SRC Classic group, aiming at providing a scalable and extensible Web crawler. A key technique for scaling in Mercator is to use a bounded amount of memory, regardless of the size of the crawl; thus the vast majority of the data structures are stored on disk. One of the initial motivations of the Mercator was to collect a variety of statistics about the Web, which can be done by a random walker program to perform a series of random walks of the Web [Henzinger et al. 1999]. Thus, the Mercator crawler was designed to be extensible. For example, one can reconfigure Mercator as a random walker without modifying the crawler's core. As reported by the Mercator author [Heydon and Najork, 1999], a random walker was reconfigured by plugging in modules totaling 360 lines of Java source code.

13.2.1.6 Attacks on Crawlers

Although robot exclusion is used to keep crawlers out, some people create programs designed to keep crawlers trapped in a Website. These programs, known as crawler traps, produce an endless stream of pages and links that all stay in the same place but look like unique pages to an unsuspecting crawler. Some crawler traps are simply malicious in intent. However, most are designed to ensnare a crawler long enough to make it think that a site is very large and important and thus raise the ranking of the overall site. This is a form of "crawler spam" that most search engines have now wised up to. However, detecting and avoiding crawler traps automatically is still a difficult task. Traps are often the one place where human monitoring and intervention is required [Heydon and Najork, 1999].

Malicious Webmasters may also try to fool crawlers though a variety of other methods, such as keyword stuffing or ghost sites. Because the Web is virtually infinite in size, all crawlers have to make an attempt to prioritize the sites that they will visit next so that they are sure to see the most important and most authoritative sites before looking at lesser ones. Keyword stuffing involves putting hundreds, or even thousands, of keywords into a Web page purely for the purpose of trying to trick the crawler into thinking that the page is highly relevant to topics related to the keywords stuffed in. Sometimes this is done in an attempt to make a page appear to be about a very popular topic, but in actuality is about something completely different — usually in an attempt to attract additional Web surfers in the hopes of making more sales of a product or service. Keyword stuffing is now quickly detected by modern crawlers and can be largely avoided.

As another tactic, malicious Webmasters took to making a myriad of one-page sites (sometimes referred to as ghost sites) that only serve to direct traffic to a primary site. This makes their primary site look deceptively important because each link is treated as an endorsement of the site's reputation or authoritativeness by a crawler. Ghost sites are much more difficult for a crawler to detect, especially if the sites are hosted on many different machines. While progress in detection has been made, a constant battle rages on between malicious Webmasters out to do anything to make money versus the search engines, which aim to provide their users with highly accurate and spam-free search results.

13.2.2 Indexer

In terms of what pages are actually retrieved by a query, indexing can be even more critical than the crawling process. The index (also called the catalog) contains the content extracted from every page that the robot finds. If a Web page changes, this catalog is updated with the new information. An indexer is the program that actually performs the process of building the index. The goal is to extract words from the documents (Web pages) that will allow the retrieval engine to efficiently find a set of documents matching a given query. The indexing process usually takes place in parallel to the crawler and is performed on separate machines because it is very computation intensive. The process is actually done in two steps: producing two indices that the search engine will use to find documents and comparing them to each other. Efficient use of disk space becomes an enormous concern in indexing, as the typical index is about 30% the size of the corpus indexed [Brin and Page, 1998]. For example, if you indexed 1 billion pages, with the average page being 10k in size, the index would be approximately 3 Terabytes (TB) in size. Thus, every bit that can be eliminated saves hundreds of megabytes (e.g., for an index of 1 billion documents, a single bit adds 122 MB to the size of the index). Estimates on the size of the Web vary greatly, but at the time of publication, Google claimed that over 3 billion pages were in its index. This would mean that they most likely have an index over 9 TB in size — not counting all the space needed for the lexicon, URL lists, robot exclusion information, page cache, and so on.

Once the indices are complete, the retrieval engine will take each word in a user's search query, find all the documents that contain that word using an index, and combine all of the documents found from each of the keywords into one set. Each document in the combined set is then compared to each other to produce a ranking. After ranking is finished, the final result set is shown to the user.

To accomplish the creation of the necessary indices, most indexers break the work into three components: a document preprocessor, a forward index builder, and an inverted index builder [Berry & Browne, 1999].

13.2.2.1 Document Preprocessor

Before a page can be indexed, it first must be preprocessed (parsed) so that the content can be extracted from the page. The document preprocessor analyzes each document to determine which parts are the best indicators of the document's topic. This information is used for future search and ranking. Document preprocessing is also referred to as term extraction and normalization.

Everything the crawler finds goes into the second part of a search engine, the indexer. An obvious question is how to select or choose which words to use in the index. Distinct terms have varying relevance when used to describe a document's contents. Deciding on the importance of an index term for summarizing the contents of a document is not a trivial issue.

A limit is usually placed on the number of words or the number of characters or lines that are used to build an index for any one document, so as to place a maximum limit on the amount of space needed to represent any document. Additionally, common words that are nondescriptive are removed. These words, such as "the," "and," and "I," are called stop words. Stop words are not likely to assist in the search process and could even slow it down by creating a document set that is too large to reasonably handle. Thus they can be safely removed without compromising the quality of the index and saving precious disk space at the same time.

There are several techniques for term extraction and normalization. The goal of term extraction and normalization is to extract right items for indexing and normalize the selected terms into a standard format by, for example, taking the smallest unit of the document (in most cases, this is individual words) and constructing a searchable data structure.

There are three main steps: identification of processing tokens (e.g., words); characterizations of tokens, such as removing stop words from the collection of processing tokens; and stemming of the tokens, i.e., the removing of suffixes and sometimes prefixes to reduce a word to its root form. Stemming has a long tradition in the IR index-building process. For example, reform, reformative, reformulation, reformatory, reformed, and reformism can all be stemmed to the root word reform. Thus all six words would map to the word reform in the index, leading to a space savings of five words in the index.

Some search engines claim to index all the words from every page. The real catch is what the engines choose to regard as a "word." Some have a list of stop words (small, common words that are considered insignificant enough to be ignored) that they do not include in the index. Some leave out obvious candidates such as articles and conjunctions. Others leave out other high-frequency, but potentially valuable, words such as "Web" and "Internet." Sometimes numerals are left out, making it difficult, for example, to search for "Troop 13." Most search engines index the "high-value" fields, areas of the page that are near the top of the document, such as the title, major headings, and sometimes even the URL itself. Metatags are usually indexed, but not always. Metatags are words, phrases, or sentences that are placed in a special section of the HTML code as a way of describing the content of the page. Metatags are not displayed when you view a page, though you can view them if you wish by viewing the Web page's source. Some search engines choose not to index information contained in metatags because they can be abused by Web-page developers in order to get their page a higher placement in the search engines' ranking algorithms. Most engines today have automatic, reasonably effective ways of dealing with such abuses. Some search engines also index the words in hypertext anchors and links, names of Java "applets," links within image maps, etc. Understanding that there are these variations in indexing policy goes a long way towards explaining why relevant pages, even when in the search engines' database, may not be retrieved by some searches.

The output of the document preprocessor is usually a list of words for each document, in the same order as they appear in the document, with each word having some associated metadata. This metadata is used to indicate the context and the location of the word in the document, such as it was in a title, heading, or appeared in bold or italic face, and so forth. This additional information will be used by the index builders to determine how much emphasis to give each word as it processes the page.

Most indexers use Inverted File Structures to organize the pair of document ID and the list of words that summarize the document. The Inverted File Structure provides a critical shortcut in the search process. It has three components:

1. The forward index or the so-called Document Index, where each document is given a unique number identifier and all the index terms (processing tokens) within the document are identified.
2. The Dictionary, a sorted list of all the index terms in the collection along with pointers to the Inversion List. For each term extracted in the Document Index, the dictionary builder extracts the stem word and counts its occurrence in the document. A record of the dictionary consists of the term and the number of its occurrences in the document.
3. The Inverted Index, which contains a pointer from the term to all the documents that contain this term. Each record of the inverted index consists of a term and a list of pairs (document number and position) to show which documents contain that term, and where in the document it occurs. For example, the word "Adam" might appear in document number 5 as the 10th word in the page. "Adam" might also appear in document 43 as the 92nd word. Thus, the entry in the inverted index for the word "Adam" would look like this: (5,10), (43,92). In the next two subsections we discuss the key issues in building the forward index and the inverted index.

13.2.2.2 Forward Index Builder

Once a page has been successfully parsed, a forward index must be built for the page. The forward index is also called the document index. Each entry in it consists of a key and associated data. The key is a unique identifier for the page (such as a URL). The data associated with a particular key is a list of words contained in the document. Each word has a weight associated with it, indicating how descriptive that particular term is of the whole document. In the end, the list of words and weights is used in comparing documents against each other for similarity and relevance to a query. During the forward index building process, the dictionary of a sorted list of all the unique terms can be generated.

Often to save additional disk space, the URL is not used as the identifier for the page. Rather, a hash code, fingerprint, or an assigned document id is used. This allows as few as 4 bytes to uniquely identify over 4 billion pages in place of potentially hundreds of bytes per page using the full URL. It is also common to represent words using unique identifiers rather than the full word as well, as often a couple of bytes can be saved by an alternate representation. When utilizing an encoding scheme such as this, a translation table must also be created for converting between the identifier and the original URL or word.

13.2.2.3 Inverted Index Builder

After the forward index has been built, it is possible to build an inverted index. The inverted index is the same type of index as one would find in the back of a book. Each entry in the inverted index contains a word (or its unique identifier) and the unique IDs of all pages that contained that word. This index is of primary importance because it will be the first step in locating documents that match a user's query.

Generating the inverted index is done using the forward index and the dictionary (which is also called the lexicon). The simplest procedure is to step through the forward index one entry at a time. For each entry, the vector detailing a single document's contents is retrieved. For each word referenced in the vector, the document's ID is added to the entry for that word in the inverted index. The process is repeated until eventually all forward index entries have been processed. Generally, due to the size of the inverted index, only a small portion of the index can be built on a single machine. These portions are then combined later (either programmatically or logically) to form a complete index.

It is possible to build search engines that do not use a forward index or that throw away the forward index after generating the inverted index. However, this makes it more difficult to respond to a search query. The way that indices are stored on disk make it easy to get the data when given a key, but difficult to get the key when given the data. Thus a search engine with only an inverted index would be able to find documents that contain the words in the user's query, but would then be unable to readily compare the documents it finds to decide which are most relevant. This is because it would have to search the entire inverted index to reconstruct the forward vectors for each document in the result set. Due to the sheer size of the inverted index, this could take minutes or even hours to complete for a single document [Berry & Browne, 1999; Glossbrenner & Glossbrenner, 1999].

13.2.3 Relevance Ranking

Without ranking the results of a query, users would be left to sort through potentially hundreds of thousands of results, manually searching for the document that contains the information they seek. Obviously, the search engine needs to make a first pass on behalf of the user to order the list of matched Web pages that are most likely to be relevant appear at the top. This means that users should have to explore only a few sites, assuming that their query was well formed and that the data they sought was indexed.

Ranking is a required component for relevancy searching. The basic premise of relevancy searching is that results are sorted, or ranked, according to certain criteria. Most of the criteria are classified into connectivity-based criteria or content-based criteria. Connectivity-based ranking resembles citation-based ranking in classic IR. The ranking criteria consider factors such as the number of links made to a page or the number of times a page is accessed from a results list. Content-based criteria can include the number of terms matched, proximity of terms, location of terms within the document, frequency of terms (both within the document and within the entire database, document length, and other factors.

- Term Frequency: Documents with more occurrences of the search term receive a higher weight. Also the number of occurrences relative to the document length is considered, and shorter documents are ranked higher than a longer document with the same number of occurrences.
- Term Location: Terms in the title, headings, or metatags are weighted higher than terms only within the text. In addition, the number of occurrences relative to the document length is considered, and shorter documents are ranked higher than a longer document with the same number of occurrences.

- Proximity: For documents that contain all keywords in a search, the documents that contain search terms as a contiguous phrase are ranked higher than those that do not.

In addition to retrieving documents that contain the search terms, some search engines such as Excite analyze the content of the documents for related phrases in a process, called Intelligent Concept Extraction (ICE). Thus, a search on "elderly people" may also retrieve documents on "senior citizens."

The exact "formula" for how these criteria are combined with the "ranking algorithm" varies among search engines.

Most search engine companies give a general description of criteria they consider in computing a page's ranking "score" and its placement in the results list. However, concrete ranking algorithms are closely guarded company secrets in the highly competitive search engine industry. There are good reasons for such secrecy: releasing the details of the ranking mechanism to the public would make it easy for malicious Webmasters to figure out how to defeat the algorithm to make their site appear higher than it should for a query.

In general, there are two potential stages in the ranking process. The first stage is a precomputed global ranking for each page. This method is usually based upon links between pages and uses mathematical algorithms to determine the overall importance of a page on the Web. The second stage is an on-the-fly ranking that is performed for each individual query over the set of documents relevant to the query. This is used to measure how relevant a document is to the original query. Modern search engines use a combination of both techniques to develop the final rankings of results presented to a user for a query.

13.2.3.1 Query-Dependent (Local) Ranking

In early search engines, traditional information retrieval techniques were employed. The basic concept comes from vector algebra and is referred to as vector ranking [Salton and McGill, 1983]. We call it a local ranking or query-dependent ranking because it ranks only those documents that are relevant to the user's query. This type of ranking is computed on-the-fly and cannot be done in advance. Once a user has submitted a query, the first step in the process is to create a vector that represents the user's query. This vector is similar to that of the forward index, in that it contains words and weights.

Once the query vector is created, the inverted index is used to find all the documents that contain at least one of the words in the query. The search engine retrieves the forward index for each of the relevant documents and creates a vector for each, representing its contents. This vector is called a forward vector. Then, each of the forward vectors of the relevant documents is compared to the query vector by measuring the angle between the two. A small (or zero angle) means a very close match, signifying that the document should be highly relevant to the query. A large angle means that the match is not very good and the relevance is probably low. Documents are then ranked by their angular difference to the query, presenting those with the smallest angles at the top of the result set, followed by those of increasing distance.

A representative connectivity-based ranking algorithm for the query-dependent approach is the HITS [Kleinberg, 1999]. It ranks the returned documents by analyzing the (incoming) links to and the (outgoing) links from the result pages based on the concept of hubs and authorities.

13.2.3.2 Hubs and Authorities

In a system using hubs and authorities, a Web page is either a hub or an authority [Kleinberg, 1999]. Authoritative pages are those that are considered to be a primary source of information on a particular topic. For example, a news Website such as CNN could be considered an authority. Hubs are pages that link to authoritative pages in a fashion similar to that of a directory such as Yahoo!. The basic idea is that a good authoritative page will be pointed to lots of good hubs and that a good hub will point to lots of good authorities.

Hubs and authorities are identified by an analysis of links between pages in a small collection. In practice, the collection is specific to a user's query. The collection starts off as the most relevant pages in relation to the query. The set is then expanded by including every page that links to one of the relevant pages, along with every page that is linked from a relevant page. This creates a base set that is of sufficient size to properly analyze the connections between pages to identify the hubs and authorities.

The idea of hubs and authorities has been shown to work well; however it does suffer from some drawbacks. First, it is quite computationally expensive and requires more computing resources to perform its task on each individual query. Because little can be precomputed, this method has difficulty in scaling to the volume of traffic received by today's top search-engines. Hubs and authorities is also susceptible to search engine spam, as malicious Webmasters can easily create sites that mimic good hubs and authorities in an attempt to receive higher placement in search results.

13.2.3.3 Query-Independent (Global) Ranking

While local (query-dependent) rankings can provide results that are highly tailored to a particular query, the amount of computation that can be performed for each query is limited. Thus concepts of global, query-independent rankings were developed. A global ranking is one that can be performed in advance and then used across all queries as a basis for ranking. Several methods of global ranking have developed around the concept of citation counting.

Citation counting, in its simplest form, is counting the number of pages that cite a given page. The thinking is that a page that is linked to by many other sites must be more reputable and thus important than a page with fewer links to it. The citation count can be combined with localized rankings, putting those documents with high citation counts and minimal angular difference on the top of the result set. As mentioned in the section about attacks on crawlers, malicious Webmasters can create pages designed to make a Website appear more important by exploiting citation-based ranking systems.

The Google PageRank algorithm is a representative connectivity-based ranking for the query-independent approach.

13.2.3.4 PageRank

One of Google's co-creators invented a system known as PageRank. PageRank is based upon the concept of a "random surfer," and can be considered a variation on the link citation count algorithm [Page et al., 1998]. In PageRank, we can think of a Web surfer who just blindly follows links from page to page. This means not all links are given equal weight, as each link on a page has an equal probability of being followed. If the surfer continues long enough, this would mathematically reduce to the fact that a page with lots of links will contribute less of citation to the pages that it links to than a page with only a few links.

The basic concept behind PageRank is illustrated in Figure 13.3. This shows four pages, each with links coming into and going out of it. The PageRank of each page is shown at the top of the page. The page with a rank of 60 and three outgoing links will contribute a rank of 20 to each page that it links to. The page with a rank of 10 and two outgoing links will contribute a rank of 5 to each page it links to. Thus the page with a rank of 25 received that rank from the two other pages shown. The end result is that pages will contribute a portion of their reputation to each site that they link to. Having one link from a large site such as Yahoo! will increase a page's rank by more than several links for smaller, less reputable sites. This helps to create an even playing field in which defrauding the system is not as simple a task as it is with simple citation counts.

The process is iterative, in that cycles are eventually formed that will cause a previously ranked page to have a different rank than before, which in turn will affect all of the pages in the cycle. To eliminate these cycles and ensure that the overall algorithm will mathematically converge, a bit of randomness has been added in. As the random surfer surfs, with each link followed, there is a possibility that the surfer will randomly jump to a completely new site not linked to by the current page (hence the "random surfer"). This prevents the surfer from getting caught in an endless ring of pages that link only to each other for the purpose of building up an unusually high rank.

However, even this system is not perfect. For example, newly added pages, regardless of how useful or important they might be, will not be ranked highly until sufficiently linked to by other pages. This can be self-defeating as a page with a low PageRank will be buried in the search results, which limits the number of people who will know about it. If no one knows about it, then no one will link to it, and thus the PageRank of the page will never increase.

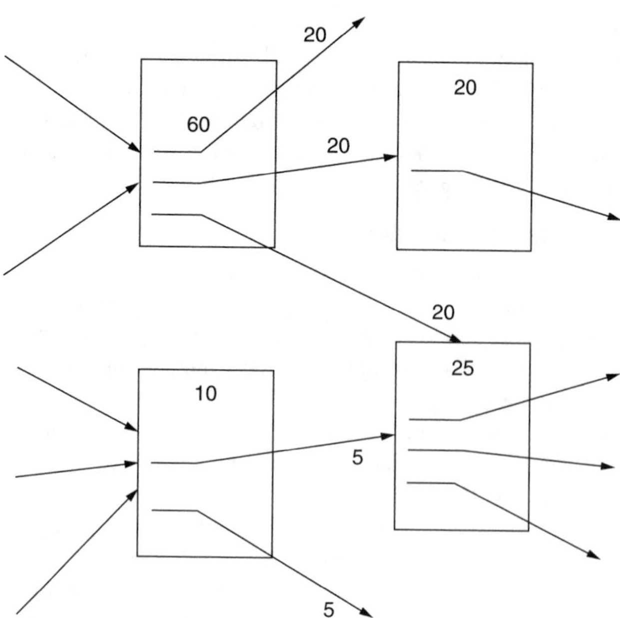

FIGURE 13.3 Simplified view of PageRank.

13.2.4 Databases

Another major part of any search engine is its databases. All of the data generated by the crawling and indexing processes must be stored in large, distributed databases. The total size of the indices and other data will always exceed the data storage capacity of any one machine. While improvements in storage technology may make it cheaper, smaller, and higher in capacity, the growth of the Web will always outpace it. For this, as well as redundancy and efficiency, search engine databases are highly distributed in a manner that provides high-speed, parallel access.

Most search engines will divide their data into a set of tables designed to be accessed through a single primary key. This type of structure lends itself to high-efficiency retrieval, which is vital in responding to searches as quickly as possible. The minimal tables required for a search engine are a URL look-up, lexicon, and forward and inverted indices. The URL lookup table provides the means for translating from a unique identifier to the actual URL. Additional data about a particular page may also be stored in this table, such as the size of the page found at that location and the date it was last crawled. The lexicon serves as a translation table between a keyword and its unique ID. These unique IDs are used in the indices as a way to greatly reduce storage space requirements. Lastly, the inverted index provides the means of locating all documents that contain a particular keyword.

With these tables, the search engine can translate a user's query into keyword IDs (using the lexicon), identify the URL IDs that contain one or all of the keywords (using the inverted index), rank the documents relevant to a query (using the forward index), and then translate the URL IDs to actual URLs for presentation to the user (using the URL table). Detailed information on process of retrieval information from the databases is presented in the following section, Retrieval Engine.

13.2.5 Retrieval Engine

The final component of a search engine is the retrieval engine. The retrieval engine is responsible for parsing a user's query, finding relevant documents, ranking them, and presenting the results to the user. This is the culmination of the work of all the other parts, and the results it produces largely depends on how well the other parts performed. The process is fairly straightforward. First, a query is received from a user. The retrieval engine will parse the query, throwing out overly common words (just as the indexer did for the Web pages) and eliminating duplicates. Advanced search engines that support Boolean expressions or phrases will determine the conditions and possibly break the query down into multiple smaller queries.

Once the query has been parsed, the retrieval engine uses the inverted index to find all documents that contain at least one of the words in the query. Some search engines, such as Google, require that a document contain all words from the query in order to be considered relevant. After the documents have been identified, their forward index and ranking entries are retrieved. This information is used, along with a local ranking algorithm, to produce a ranking over the set of documents. The set is then sorted according to rank, with the most relevant to the query at the top. Once sorted, the results can be shown to the user.

13.2.6 Improving Search Engines

Search engines are often more than just crawlers, indexers, rankers, and retrieval engines. They also consist of user interface, which allows users to access and utilize the search engine. There are other nuances that have been added along the way in an effort to improve the efficiency with which a search engine can do its job. This section will present more information on these other aspects of search engines and how researchers continue to develop ways to further improve search.

13.2.6.1 User Interface

The user interface, the means by which a user interacts with a search engine, has gone largely unchanged over the course of Internet evolution. All major search engines have a simple text entry box for the user's query and present results in the form of a textual listing (usually 10 pages at a time). However, a few improvements have been made.

Google introduced the concept of dynamic clippings in the search results. This means that each Web page in the results has one or two lines that have been excerpted from the page with search terms bolded. The idea is to provide users with a glimpse into the contents of each page listed, allowing them to make a decision as to whether or not the page is truly relevant to their search. This frees users from having to visit each Web page in the search results until they find the desired information.

Even with innovations such as dynamic clipping, finding the desired information from search results can be as frustrating as trying to find a needle in a haystack. Some research is now concentrating on finding new ways to interact with search engines. Projects like Kartoo (www.kartoo.com), VisIT (www.visit.uiuc.edu), and Grokker (www.groxis.com) aim to visualize search results using graphical maps of the relevant portions of the Web. The idea behind these projects is that a more graphical and interactive interface will allow users to see more easily patterns and relationships among pages and determine on their own which ones are the most likely to be useful.

13.2.6.2 Metadata

Another complaint with the World Wide Web is that documents found when searching may have the keywords specified in the search, but the document is of little relevance to the user. This is partly due to the fact that many words have multiple meanings, such as mouse can refer to both a rodent and a computer peripheral. Users typically only use two or three words in their query to find information out of billions of documents, so search engines are left to do a lot of guessing about what a user's true intentions are with their search. One solution that has been proposed is to have Web pages incorporate more data about what their contents represent.

The idea of metadata in Web pages is to provide search engines with more contextual information about what information is truly contained on a page. Because crawlers and indexers cannot understand language, they cannot understand the real content of a page. However, if special hypertext tags were developed, along with a system of categories and classifications, search engines would be able to read and understand the tags and make more informed decisions about what data is on the page. The initial obstacle that this idea faces is the massive amount of standardization that has to be done on deciding what type of metadata is appropriate for Web pages, useful for search engines, and easy for Webmasters to incorporate into their pages. The largest initiative to develop such a standard is Dublin Core Metadata Initiative (www.dublincore.org), which is already gaining some acceptance.

After a standard is developed, it will take a long time for Webmasters to adopt the concept and incorporate metadata into their pages. Metadata will not directly assist, or even be presented to, the user browsing the Webpage. Thus, Webmasters will be spending many hours adding metadata to pages only for the sake of search engines so that their pages may be better represented in the index. It remains to be seen if this is incentive enough for Webmasters to spend the time required to update their sites. The biggest problem facing metadata is honesty. There is nothing to prevent Webmasters from misclassifying their site or creating ghost sites that are classified under different categories but seek to direct traffic to a singular main site. This is one of the reasons why most search engines ignore the existing metadata tags, as malicious Webmasters use them as a way to deceive the search engine into thinking that a page is about something that it is not.

13.2.6.3 Metasearch

Another approach to searching the Web is to create a search engine of search engines. This technique, known as metasearch, does not do any crawling or indexing of its own. Rather, users' queries are submitted in parallel to multiple search engines. Each result set is collected and combined into one giant result set. In theory, this will provide the user with a more complete search, as each search engine is likely to have covered some part of the Web that the others have not.

The combination of result sets is the tricky part though, as a number of different issues arise. First, because search engines do not make their ranking data available, there is no easy method of deciding which of the first entries in the result sets should be the first entry in the combined set. Also, conflicting rankings may occur if two search engines return the same document, but with very a large difference in rank. Lastly, the speed of the overall metasearch is limited by the slowest search engine it consults, thus making metasearch slower than a direct search.

13.3 Research Activities and Future Directions

Even though no search engine has taken the Web by storm since Google's introduction in 1998, there is still a large amount of ongoing research in the area of search engine technology. This section presents some of the research activities and possible future directions that Web crawling and search might take in the years to come. It should be noted that this is presented from the authors' perspective and is not all-inclusive of the research work being done. The omission of other research does not signify that it is less viable than the ideas presented here.

13.3.1 Searching Dynamic Pages — The Deep Web

Dynamic Web pages refer to the Web pages behind the forms. They are generated in part or whole by a computer program upon a search request. The number of dynamic pages has been growing exponentially. The huge and rapidly growing number of dynamic pages forms a hidden Web, out of the reach of search engines. Consequently, current search engines and their crawlers are mostly limited to accessing what is called the static or indexable Web. The static Web consists of pages that physically reside on Web servers' local disks and are not generated on-the-fly by a computer program. The dynamic pages today make up the vast majority of the Web, but are largely hidden from search engines because a form must be used to gain access. This dynamic, infinitely sized part of the Web is called the deep Web or the hidden Web, and it poses a number of problems to search engines as the dynamic content on the Web continues to grow at an astonishing speed.

13.3.1.1 Types of Dynamic Pages

For our purposes, we will divide dynamic pages into two simple categories: database access pages and session-specific pages. Database access pages are the ones that contains information retrieved from a database upon a search request. Examples of this type of page include product information at an online store, a newspaper article archive, and stock quotes. This information can even be more computational in nature, such as getting driving directions between two locations or search results for a query of the Web. It is obvious that a large amount of this information could be of potential interest to search engines, as people often use them to search for things such as products or articles.

Session-specific pages contain information that is specific to a particular user session on a Website. An example of this would be a user's shopping cart inventory while browsing an online store. Something so specific to the user is useless to a general search engine, and we need not worry about indexing such pages. However, some session-specific pages are merely augmented with personalized information, whereas the majority of the page is common across users. One example of this would be an online store where personalized recommendations are given in a column on the right side of the page, and the rest of the page is not specific to the given user. This means that a deep-Web crawler would have to potentially determine what is session specific and what is not on any given page.

13.3.1.2 Accessing Dynamic Pages

The other large problem facing deep-Web crawling is figuring out how to access the information behind the forms. In order to achieve this, the crawler must be able to understand the form either through its own analysis or with the help of metadata to guide it. Another potential approach is to develop server-side programs that Webmasters can use to open up their databases to search engines in a controlled manner. While no one has found a solution to this problem yet, it is receiving great attention as it is the first barricade to the deep Web [Raghavan and Garcia-Molina, 2001]. Most of the existing approaches to providing access to dynamic pages from multiple Websites are built through the use of wrapper programs. A wrapper is a Web source-specific computer program that transforms a Website search and result presentation into a more structured format. Wrappers can be generated semiautomatically by wrapper generation programs [Liu et al., 2001]. However, most of wrapper-based technology for dynamic Web access have been restricted to finding information about products for sale online. This is primarily because wrapper technology can only apply to such focused services, and it cannot scale up to manage the vast diversity of domains in the entire dynamic Web.

13.3.1.3 Content Freshness and Validity

With these two types of dynamic pages in mind, it starts to become obvious that the time for which data is valid will vary widely. A dynamic page that presents the current weather conditions is of no use in an index a week from now, whereas an old news story may be valuable indefinitely. One large obstacle for deep-Web crawlers to overcome will be determining the freshness of the page contents and what is worth putting into the index without requiring human invention.

13.3.1.4 Yellow Pages of the Web

Some have conjectured that searching the dynamic Web will be more like searching the yellow pages. Rather than having one generic search engine that covers the deep Web, there would be a directory of smaller, topic-, or site-specific search engines. Users would find the desired specialized search engine by navigating a classification hierarchy, and only then would perform a highly focused search using a search engine designed for their particular task.

13.3.2 Utilizing Peer-to-Peer Networks

In an attempt to solve the current problems with search engines, researchers are exploring new ways of performing search and its related tasks. Peer-to-peer search has recently received a great deal of attention by the search community because of the fact that a large, volunteer network of computing power and bandwidth could be established virtually overnight with no cost other than the development of the software. A traditional search engine requires millions of dollars worth of hardware and bandwidth to operate effectively, making a virtually free infrastructure very appealing. In a peer-to-peer search engine, each peer would be responsible for crawling, indexing, ranking, and providing search for a small portion of the Web. When connected to a large number of other peers, all the small portions can be tied together, allowing the Web as a whole to be searched. However, this method is also fraught with a number of issues, the biggest of which is speed.

Peer-to-peer networks are slower for accessing data than a centralized server architecture (like that of today's commercial search engines). In a peer-to-peer network, data has to flow through a number of peers in response to a request. Each peer is geographically separated and thus messages must pass through several internet routers before arriving at their destination. The more distributed the data in a peer-to-peer network is, the more peers that must be contacted in order to process a request. Thus, achieving the tenths-of-a-second search speeds that traditional search engines are capable of is simply not possible in a peer-to-peer environment.

In addition to issues with the speed comes the issue of work coordination. In a true peer-to-peer environment, there is no central server through which activities such as crawling and indexing can be coordinated [Singh et al., 2003]. Instead, what Web pages each individual peer is responsible for processing must be decided upon in a collective manner. This requires a higher level of collaboration between peers than what has been developed previously for large-scale file-sharing systems. Lastly, data security and spam resistance becomes more difficult in a peer-to-peer network. Because anyone can be a peer in the network, malicious Webmasters could attempt to hack their peer's software or data to alter its behavior so as to present their Website more favorably in search results. A means for measuring and developing trust among peers is needed in order to defend against such attacks on the network.

13.3.3 Semantic Web

The semantic Web has the ability to revolutionize the accuracy with which search engines would be able to assist people in locating information of interest. The semantic Web is based upon the idea of having data on the Web defined and linked in a way that it can be more easily understood by machines not just for display purposes. This will help crawlers of the future to comprehend the actual contents of a page based upon an advanced set of markup languages. Rather than trying to blindly derive the main topic of a page through word frequency and pattern analysis, the topic could be specified directly by the author of the page in a manner that would make it immediately apparent to the crawler. Of course, this will also open up new ways for malicious Webmasters to spam search engines, allowing them to provide false data and mislead the crawler. Although the idea of a semantic Web is not new, it has not yet become a reality. There is still a large amount of ongoing research in this area, most of which is devoted to developing the means by which semantics can be given to a Web page [Decker et al. 2000; Broekstra et al. 2001].

13.3.4 Detecting Duplicated Pages

It is a common practice for portions of a Website, manual, or other document to be duplicated across multiple Websites. This is done often to make it easier for people to find and to overcome the problems of a global network, such as slow transfers and down servers. However, this duplication is problematic for search engines because users can receive the same document hundreds of times in response to a query. This can make it more difficult to find the proper document if the one so highly duplicated is not it. Ideally, search engines would detect and group all replicated content under a single listing but still allow users to explore the individual replications.

Although it seems that detecting this duplicated content should be easy, there are a number of nuances that make it quite difficult for search engines to achieve. One example is that individually duplicated pages may be the same in appearance but different in their actual HTML code because of different URLs for the document's links. Some replicated documents add a header or footer so that readers will know where the original came from. Another problem with content-based detection is that documents often have multiple versions but are largely the same. This leads to the question of how versions should be handled and detected.

One way to overcome these problems is to look for collections of documents that are highly similar, both in appearance and structure, but not necessarily perfect copies. This makes it easier for a search engine to detect and handle the duplication. However, it is common to only duplicate a portion of a collection, which makes it difficult for a search engine to rely upon set analysis to detect replication. Lastly, even with duplicated pages being detected, there is still the question of automatically determining which one is the original copy. A large amount of research continues to be put into finding better ways of detecting and handling these duplicated document collections [Bharat and Broder, 1999; Cho et al. 1999].

13.3.5 Clustering and Categorization of Pages

In a problem similar to detecting duplicated pages, it would be of great benefit to searchers if they could ascertain what type of content is on a page or to what group of pages it belongs. This falls into the areas of clustering and categorizing pages.

Clustering focuses on finding ways to group pages into sets, allowing a person to more easily identify patterns and relationships between pages [Broder et al., 1997]. For example, when searching a newspaper Website, it might be beneficial to see the results clustered into groups of highly related articles. Clustering is commonly done based upon links between pages or similarity of content. Another approach is categorization. Categorization is similar to clustering, but involves taking a page and automatically finding a place inside of a hierarchy. This is done based upon the contents of the page, who links to it, and other attributes [Chakrabarti et al., 1998]. Currently, the best categorization is done by hand because computers cannot understand the content of a page. This lack of understanding makes it difficult for a program to automatically categorize pages with a high degree of accuracy.

13.3.6 Spam Deterrence

As long as humans are the driving force of the online economy, Webmasters will seek out new ways to get their site listed higher in search results in an effort to get more traffic and, hopefully, sales. Not all such efforts are malicious, however. Search optimization is a common practice among Webmasters. The goal of the optimization is to find good combinations of keywords and links to help their site appear higher in the search results, while not purposefully trying to deceive the search engine or user.

However, some Webmasters take it a step further with keyword spamming. This practice involves adding lots of unrelated keywords to a page in an attempt to make their page appear in more search results, even when the user's query has nothing to do with what their Web page is actually about. Search engine operators quickly wised up to this practice, as finding relevant sites became more difficult for their users. This led some Webmasters to get more creative in their attempts, and they began to "spoof" the Web crawlers. Crawler spoofing is the process of detecting when a particular Web crawler is accessing a Website and returning a different page than what the surfer would actually see. This way, the Web crawler will see a perfectly legitimate page, but one that has absolutely nothing to do with what the user will see when they visit the site.

Lastly, with the popularity of citation-based ranking systems such as PageRank, Web masters have begun to concentrate on link optimization techniques. A malicious Webmaster will create a large network of fake sites, all with links to their main site in an effort to get a higher citation rating. This practice is known as a link farm, in which, in some cases, Webmasters will pool their resources to create a larger network that is harder for the search engines to detect.

Search engines have a variety of ways of dealing with spoofing and link farms, but it continues to be a battle between the two. Each time the search engines are able to find a way to block spam, the Webmasters find a new way in. Web crawlers that have greater intelligence, along with other collaborative filtering techniques, will help keep the search engines ahead of the spam for the time being.

13.4 Conclusion

This chapter has covered essential concepts, techniques, and key tradeoffs from early Web search engines to the newest technologies currently being researched. However, it is recognized that Internet search is still in its infancy, with a large room for growth and development ahead. One of the largest challenges that will always plague search engines is the growth of the Web. As the Web increasingly becomes more of an information repository, marketplace, and social space, it also continues to grow at an amazing pace that can only be estimated. A statistics done in 1997 projects that the Web is estimated to double in size every six months [Gray, 1996b] .

In addition to its exponential growth, millions of existing pages are added, updated, deleted, or moved every day. This makes Web crawling and search a problem that is harder than ever, because crawling the Web once a month is not good enough in such a dynamic environment. Instead, a search engine needs to be able to crawl and recrawl a large portion of the Web on a high-frequency basis. The dynamics of the Web will be a grand challenge to any search-engine designer. The Web and its usage will continue to evolve, and so will the way in which we use search engines in our daily lives.

References

Aggarwal, Charu, Fatima Al-Garawi, and Phillip Yu. *Intelligent Crawling on the World Wide Web with Arbitrary Predicates.* The 10th International World Wide Web Conference, Hong Kong, May 2001.

Berry, Michael and Murray Browne. *Understanding Search Engines: Mathematical Modeling and Text Retrieval.* Society for Industrial and Applied Mathematics, 1999.

Bharat, Krishna and Andrei Broder. *A Technique for Measuring the Relative Size and Overlap of Public Web Search Engines.* Proceedings of the 7th International World Wide Web Conference, Brisbane, Australia, April 1998.

Bharat, Krishna and Andrei Broder. *A study of host pairs with replicated content.* Proceedings of the 8th International World Wide Web Conference, Toronto, Canada, May 1999.

Brin, Sergey and Lawrence Page. *The Anatomy of a Large-Scale Hypertextual Web Search Engine.* Proceedings of the 7th International World Wide Web Conference, pp. 107–117, Brisbane, Australia, April 1998.

Broder, Andrei, Steven Glassman, and Mark Manasse. *Syntactic Clustering of the Web.* Proceedings of the 6th International World Wide Web Conference, pp. 391–404, Santa Clara, California, April 1997.

Broekstra, Jeen, Michel C. A. Klein, Stefan Decker, Dieter Fensel, Frank van Harmelen, and Ian Horrocks. *Enabling Knowledge Representation on the Web by Extending RDF Schema.* Proc of the tenth World Wide Webb Conference (www 2001), Hong Kong, pp. 467–478, 2001.

Chakrabarti, Soumen, Martin van den Berg, and Byron Dom. *Focused Crawling: A New Approach to Topic-Specific Web Resource Discovery.* Proceedings of the 8th International World Wide Web Conference, Toronto, Canada, May 1999.

Chakrabarti, Soumen, Byron Dom, and Piotr Indyk. *Enhanced Hypertext Categorization Using Hyperlinks.* Proceedings of SIGMOD-98, Seattle, Washington, pp. 307–318, 1998.

Cho, Junghoo, Hector Garcia-Molina, and Lawrence Page. *Efficient Crawling through URL Ordering.* Proceedings of the 7th International World Wide Web Conference, pp. 161–172, Brisbane, Australia, April 1998.

Cho, Junghoo, Narayana Shivakumar, and Hector Garcia-Molina. *Finding Replicated Web Collections.* Technical Report (http://www-db.stanford.edu/pub/papers/cho-mirror.ps), Department of Computer Science, Stanford University, 1999.

Decker, Stefan, Sergey Melnik, Frank van Harmelen, Dieter Fensel, Michel C. A. Klein, Jeen Broekstra, Michael Erdmann, and Ian Horrocks. *The Semantic Web: The Roles of XML and RDF. IEEE Internet Computing,* Vol. 4, No. 5, pp. 63–74, 2000.

Eichmann, David. *The RBSE Spider — Balancing Effective Search Against Web Load.* Proceedings of the 1st International World Wide Web Conference, pp. 113–120, CERN, Geneva, 1994.

Glossbrenner, Alfred and Emily Glossbrenner. *Search Engines for the Word Wide Web.* 2nd Edition, Peachpit Press, 1999.

Gray, Matthew. *Web Growth Summary.* On the World Wide Web, http://www.mit.edu/people/mkgray/net/web-growth-summary.html, 1996a.

Gray, Matthew. *Internet Growth and Statistics: Credits and Background.* On the World Wide Web, http://www.mit.edu/people/mkgray/net/background.html, 1996b.

Henzinger, Monkia, Allan Heydon, Michael Mitzenmacher, and Marc A. Najork. *Measuring Index Quality Using Random Walks on the Web.* Proceedings of the 8th International World Wide Web Conference, pp. 213–225, Toronto, Canada, May 1999.

Heydon, Allan and Marc Najork. *Mercator: A Scalable, Extensible Web Crawler.* World Wide Web, December 1999, pp. 219–229.

Kleinberg, Jon M. *Authoritative Sources in a Hyperlinked Environment. Journal of the ACM,* Vol. 46, No. 5, pp. 604–632, 1999.

Lawrence, Steve and C. Lee Giles. *How Big Is the Web? How Much of the Web Do the Search Engines Index? How up to Date Are the Search Engines?* On the World Wide Web, http://www.neci.nec.com/~lawrence/websize.html, 1998.

Liu, Ling, Carlton Pu, and Wei Han. *An XML-Enabled Data Extraction Tool for Web Sources. International Journal of Information Systems, Special Issue on Data Extraction, Cleaning, and Reconciliation.* (Mokrane Bouzeghoub and Maurizio Lenzerini, Eds.), 2001.

McBryan, Oliver. *GENVL and WWWW: Tools for Taming the Web.* Proceedings of the 1st International World Wide Web Conference, CERN, Geneva, May 1994.

Najork, Marc and Janet L. Wiener. Breadth-First Crawling Yields High-Quality Pages. Proceedings of the 10th International World Wide Web Conference, Hong Kong, pp. 114–118, May 2001.

Page, Lawrence, Serget Brin, Rajeev Motwani, and Terry Winograd. *The PageRank Citation Ranking: Bringing Order to the Web.* Stanford Digital Libraries working paper, 1997.

Pinkerton, Brian. *Finding What People Want: Experiences with the WebCrawler.* Proceedings of the 1st International World Wide Web Conference, CERN, Geneva, May 1994.

Raghavan, Sriram and Hector Garcia-Molina. *Crawling the Hidden Web.* Proceedings of the 27th International Conference on Very Large Databases, Rome, September 2001.

Salton, Gerard and Michael J. McGill. *Introduction to Modern Information Retrieval,* 1st ed. McGraw-Hill, New York, 1983.

Singh, Aameek, Mudhakar Srivatsa, Ling Liu, and Todd Miller. *Apoidea: A Decentralized Peer-to-Peer Architecture for Crawling the World Wide Web.* Proccedings of the ACM SIGIR workshop on Distributed IR, Springer-Verlag, New York, 2003.

14

Text Mining

CONTENTS

14.1 Introduction 14-1
14.1.1 Text Mining and Data Mining 14-2
14.1.2 Text Mining and Natural Language Processing 14-3
14.2 Mining Plain Text 14-4
14.2.1 Extracting Information for Human Consumption 14-4
14.2.2 Assessing Document Similarity 14-6
14.2.3 Language Identification 14-8
14.2.4 Extracting Structured Information 14-9
14.3 Mining Structured Text 14-14
14.3.1 Wrapper Induction 14-14
14.4 Human Text Mining 14-16
14.5 Techniques and Tools 14-17
14.5.1 High-Level Issues: Training vs. Knowledge Engineering... 14-17
14.5.2 Low-Level Issues: Token Identification 14-18
14.6 Conclusion 14-19
References 14-20

Ian H. Witten

14.1 Introduction

Text mining is a burgeoning new field that attempts to glean meaningful information from natural language text. It may be loosely characterized as the process of analyzing text to extract information that is useful for particular purposes. Compared with the kind of data stored in databases, text is unstructured, amorphous, and difficult to deal with algorithmically. Nevertheless, in modern culture, text is the most common vehicle for the formal exchange of information. The field of text mining usually deals with texts whose function is the communication of factual information or opinions, and the motivation for trying to extract information from such text automatically is compelling, even if success is only partial.

Four years ago, Hearst [Hearst, 1999] wrote that the nascent field of "text data mining" had "a name and a fair amount of hype, but as yet almost no practitioners." It seems that even the name is unclear: the phrase "text mining" appears 17 times as often as "text data mining" on the Web, according to a popular search engine (and "data mining" occurs 500 times as often). Moreover, the meaning of either phrase is by no means clear: Hearst defines data mining, information access, and corpus-based computational linguistics and discusses the relationship of these to text data mining — but does not define that term. The literature on data mining is far more extensive, and also more focused; there are numerous textbooks and critical reviews that trace its development from roots in machine learning and statistics. Text mining emerged at an unfortunate time in history. Data mining was able to ride the back of the high technology extravaganza throughout the 1990s and became firmly established as a widely-used practical technology — though the dot com crash may have hit it harder than other areas [Franklin, 2002]. Text mining, in contrast, emerged just before the market crash — the first workshops were held

1-58488-381-2/05/$0.00+$1.50

© 2005 by CRC Press LLC

at the *International Machine Learning Conference* in July 1999 and the *International Joint Conference on Artificial Intelligence* in August 1999 — and missed the opportunity to gain a solid foothold during the boom years.

The phrase text mining is generally used to denote any system that analyzes large quantities of natural language text and detects lexical or linguistic usage patterns in an attempt to extract probably useful (although only probably correct) information [Sebastiani, 2002]. In discussing a topic that lacks a generally accepted definition in a practical handbook such as this, I have chosen to cast the net widely and take a liberal viewpoint of what should be included, rather than attempt a clear-cut characterization that will inevitably restrict the scope of what is covered.

The remainder of this section discusses the relationship between text mining and data mining, and between text mining and natural language processing, to air important issues concerning the meaning of the term. The chapter's major section follows: an introduction to the great variety of tasks that involve mining plain text. We then examine the additional leverage that can be obtained when mining semistructured text such as pages of the World Wide Web, which opens up a range of new techniques that do not apply to plain text. Following that we indicate, by example, what automatic text mining techniques may aspire to in the future by briefly describing how human "text miners," who are information researchers rather than subject-matter experts, may be able to discover new scientific hypotheses solely by analyzing the literature. Finally, we review some basic techniques that underpin text-mining systems and look at software tools that are available to help with the work.

14.1.1 Text Mining and Data Mining

Just as data mining can be loosely described as looking for patterns in data, text mining is about looking for patterns in text. However, the superficial similarity between the two conceals real differences. Data mining can be more fully characterized as the extraction of implicit, previously unknown, and potentially useful information from data [Witten and Frank, 2000]. The information is implicit in the input data: It is hidden, unknown, and could hardly be extracted without recourse to automatic techniques of data mining. With text mining, however, the information to be extracted is clearly and explicitly stated in the text. It is not hidden at all — most authors go to great pains to make sure that they express themselves clearly and unambiguously — and, from a human point of view, the only sense in which it is "previously unknown" is that human resource restrictions make it infeasible for people to read the text themselves. The problem, of course, is that the information is not couched in a manner that is amenable to automatic processing. Text mining strives to bring it out of the text in a form that is suitable for consumption by computers directly, with no need for a human intermediary.

Though there is a clear difference philosophically, from the computer's point of view the problems are quite similar. Text is just as opaque as raw data when it comes to extracting information — probably more so.

Another requirement that is common to both data and text mining is that the information extracted should be "potentially useful." In one sense, this means *actionable* — capable of providing a basis for actions to be taken automatically. In the case of data mining, this notion can be expressed in a relatively domain-independent way: Actionable patterns are ones that allow nontrivial predictions to be made on new data from the same source. Performance can be measured by counting successes and failures, statistical techniques can be applied to compare different data mining methods on the same problem, and so on. However, in many text-mining situations it is far harder to characterize what "actionable" means in a way that is independent of the particular domain at hand. This makes it difficult to find fair and objective measures of success.

It is interesting that data mining also evolved out of a history of difficult relations between disciplines, in this case machine learning and statistics: the former rooted in experimental computer science, with ad hoc evaluation methodologies; the latter well-grounded theoretically, but based on a tradition of testing explicitly-stated hypotheses rather than seeking new information.

This is necessary whenever the result is intended for human consumption rather than (or as well as) a basis for automatic action. This criterion is less applicable to text mining because, unlike data mining, the input itself is comprehensible. Text mining with comprehensible output is tantamount to summarizing salient features from a large body of text, which is a subfield in its own right — text summarization.

14.1.2 Text Mining and Natural Language Processing

Text mining appears to embrace the whole of automatic natural language processing and, arguably, far more besides — for example, analysis of linkage structures such as citations in the academic literature and hyperlinks in the Web literature, both useful sources of information that lie outside the traditional domain of natural language processing. But, in fact, most text-mining efforts consciously shun the deeper, cognitive aspects of classic natural language processing in favor of shallower techniques more akin to those used in practical information retrieval.

The reason is best understood in the context of the historical development of the subject of natural language processing. The field's roots lie in automatic translation projects in the late 1940s and early 1950s, whose aficionados assumed that strategies based on word-for-word translation would provide decent and useful rough translations that could easily be honed into something more accurate using techniques based on elementary syntactic analysis. But the sole outcome of these high-profile, heavily-funded projects was the sobering realization that natural language, even at an illiterate child's level, is an astonishingly sophisticated medium that does not succumb to simplistic techniques. It depends crucially on what we regard as "common-sense" knowledge, which despite — or, more likely, because of — its everyday nature is exceptionally hard to encode and utilize in algorithmic form [Lenat, 1995].

As a result of these embarrassing and much-publicized failures, researchers withdrew into "toy worlds" — notably the "blocks world" of geometric objects, shapes, colors, and stacking operations — whose semantics are clear and possible to encode explicitly. But it gradually became apparent that success in toy worlds, though initially impressive, does not translate into success on realistic pieces of text. Toy-world techniques deal well with artificially-constructed sentences of what one might call the "Dick and Jane" variety after the well-known series of eponymous children's stories. But they fail dismally when confronted with real text, whether painstakingly constructed and edited (like this article) or produced under real-time constraints (like informal conversation).

Meanwhile, researchers in other areas simply had to deal with real text, with all its vagaries, idiosyncrasies, and errors. Compression schemes, for example, must work well with all documents, whatever their contents, and avoid catastrophic failure even when processing outrageously deviant files (such as binary files, or completely random input). Information retrieval systems must index documents of all types and allow them to be located effectively whatever their subject matter or linguistic correctness. Keyphrase extraction and text summarization algorithms have to do a decent job on any text file. Practical, working systems in these areas are topic-independent, and most are language-independent. They operate by treating the input as though it were data, not language.

Text mining is an outgrowth of this "real text" mindset. Accepting that it is probably not much, what can be done with unrestricted input? Can the ability to process huge amounts of text compensate for relatively simple techniques? Natural language processing, dominated in its infancy by unrealistic ambitions and swinging in childhood to the other extreme of unrealistically artificial worlds and trivial amounts of text, has matured and now embraces both viewpoints: relatively shallow processing of unrestricted text and relatively deep processing of domain-specific material.

It is interesting that data mining also evolved out of a history of difficult relations between disciplines, in this case machine learning — rooted in experimental computer science, with *ad hoc* evaluation methodologies — and statistics — well-grounded theoretically, but based on a tradition of testing explicitly-stated hypotheses rather than seeking new information. Early machine-learning researchers knew or cared little of statistics; early researchers on structured statistical hypotheses remained ignorant of parallel work in machine learning. The result was that similar techniques (for example, decision-tree building and nearest-neighbor learners) arose in parallel from the two disciplines, and only later did a balanced rapprochement emerge.

14.2 Mining Plain Text

This section describes the major ways in which text is mined when the input is plain natural language, rather than partially structured Web documents. In each case we provide a concrete example. We begin with problems that involve extracting information for human consumption — text summarization and document retrieval. We then examine the task of assessing document similarity, either to categorize documents into predefined classes or to cluster them in "natural" ways. We also mention techniques that have proven useful in two specific categorization problems — language identification and authorship ascription — and a third — identifying keyphrases — that can be tackled by categorization techniques but also by other means. The next subsection discusses the extraction of structured information, both individual units or "entities" and structured relations or "templates." Finally, we review work on extracting rules that characterize the relationships between entities.

14.2.1 Extracting Information for Human Consumption

We begin with situations in which information mined from text is expressed in a form that is intended for consumption by people rather than computers. The result is not "actionable" in the sense discussed above, and therefore lies on the boundary of what is normally meant by text mining.

14.2.1.1 Text Summarization

A text summarizer strives to produce a condensed representation of its input, intended for human consumption [Mani, 2001]. It may condense individual documents or groups of documents. Text compression, a related area [Bell et al., 1990], also condenses documents, but summarization differs in that its output is intended to be human-readable. The output of text compression algorithms is certainly not human-readable, but neither is it actionable; the only operation it supports is decompression, that is, automatic reconstruction of the original text. As a field, summarization differs from many other forms of text mining in that there are people, namely professional abstractors, who are skilled in the art of producing summaries and carry out the task as part of their professional life. Studies of these people and the way they work provide valuable insights for automatic summarization.

Useful distinctions can be made between different kinds of summaries; some are exemplified in Figure 14.1 (from Mani [2001]). An *extract* consists entirely of material copied from the input — for example, one might simply take the opening sentences of a document (Figure 14.1a) or pick certain key sentences scattered throughout it (Figure 14.1b). In contrast, an *abstract* contains material that is not present in the input, or at least expresses it in a different way — this is what human abstractors would normally produce (Figure 14.1c). An *indicative* abstract is intended to provide a basis for selecting documents for closer study of the full text, whereas an *informative* one covers all the salient information in the source at some level of detail [Borko and Bernier, 1975]. A further category is the *critical abstract* [Lancaster, 1991], which evaluates the subject matter of the source document, expressing the abstractor's views on the quality of the author's work (Figure 14.1d). Another distinction is between a *generic* summary, aimed at a broad readership, and a *topic-focused* one, tailored to the requirements of a particular group of users.

While they are in a sense the archetypal form of text miners, summarizers do not satisfy the condition that their output be actionable.

14.2.1.2 Document Retrieval

Given a corpus of documents and a user's information need expressed as some sort of query, document retrieval is the task of identifying and returning the most relevant documents. Traditional libraries provide catalogues (whether physical card catalogues or computerized information systems) that allow users to identify documents based on surrogates consisting of *metadata* — salient features of the document such as author, title, subject classification, subject headings, and keywords. Metadata is a kind of highly structured (and therefore actionable) document summary, and successful methodologies have been developed for manually extracting metadata and for identifying relevant documents based on it, methodologies that are widely taught in library school (e.g., Mann [1993]).

If	(*wheat* & *farm*)
	or (*wheat* & *commodity*)
	or (*bushels* & *export*)
	or (*wheat* & *tonnes*)
	or (*wheat* & *winter* & *soft*)
then	WHEAT

FIGURE 14.3 Rule for assigning a document to the category WHEAT.

lection of preclassified news articles, which is widely used for document classification research (e.g., Hayes et al. [1990]). WHEAT is the name of one of the categories.

Rules like this can be produced automatically using standard techniques of machine learning [Mitchell, 1997; Witten and Frank, 2000]. The training data comprises a substantial number of sample documents for each category. Each document is used as a positive instance for the category labels that are associated with it and a negative instance for all other categories. Typical approaches extract "features" from each document and use the feature vectors as input to a scheme that learns how to classify documents. Using words as features — perhaps a small number of well-chosen words, or perhaps all words that appear in the document except stop words — and word occurrence counts as feature values, a model is built for each category. The documents in that category are positive examples and the remaining documents negative ones. The model predicts whether or not that category is assigned to a new document based on the words in it, and their occurrence counts. Given a new document, each model is applied to determine which categories need to be assigned. Alternatively, the learning method may produce a likelihood of the category being assigned, and if, say, five categories were sought for the new document, those with the highest likelihoods could be chosen.

If the features are words, documents are represented using the "bag of words" model described earlier under document retrieval. Sometimes word counts are discarded and the "bag" is treated merely as a set (Figure 14.3, for example, only uses the presence of words, not their counts). Bag (or set) of words models neglect word order and contextual effects. Experiments have shown that more sophisticated representations — for example, ones that detect common phrases and treat them as single units — do not yield significant improvement in categorization ability (e.g., Lewis [1992]; Apte et al. [1994]; Dumais et al. [1998]), although it seems likely that better ways of identifying and selecting salient phrases will eventually pay off. Each word is a "feature." Because there are so many of them, problems arise with some machine-learning methods, and a selection process is often used that identifies only a few salient features. A large number of feature selection and machine-learning techniques have been applied to text categorization [Sebastiani, 2002].

14.2.2.2 Document Clustering

Text categorization is a kind of "supervised" learning where the categories are known beforehand and determined in advance for each training document. In contrast, document clustering is "unsupervised" learning in which there is no predefined category or "class," but groups of documents that belong together are sought. For example, document clustering assists in retrieval by creating links between similar documents, which in turn allows related documents to be retrieved once one of the documents has been deemed relevant to a query [Martin, 1995].

Clustering schemes have seen relatively little application in text-mining applications. While attractive in that they do not require training data to be preclassified, the algorithms themselves are generally far more computation-intensive than supervised schemes (Willett [1988] surveys classical document clustering methods). Processing time is particularly significant in domains like text classification, in which instances may be described by hundreds or thousands of attributes. Trials of unsupervised schemes include Aone et al. [1996], who use the conceptual clustering scheme COBWEB [Fisher, 1987] to induce natural groupings of close-captioned text associated with video newsfeeds; Liere and Tadepalli [1996], who explore the effectiveness of AutoClass; Cheeseman et al. [1988] in producing a classification model for a portion of the Reuters corpus; and Green and Edwards [1996], who use AutoClass to cluster news items gathered from several sources into "stories," which are groupings of documents covering similar topics.

14.2.3 Language Identification

Language identification is a particular application of text categorization. A relatively simple categorization task, it provides an important piece of metadata for documents in international collections. A simple representation for document categorization is to characterize each document by a profile that consists of the "n-grams," or sequences of n consecutive letters, that appear in it. This works particularly well for language identification. Words can be considered in isolation; the effect of word sequences can safely be neglected. Documents are preprocessed by splitting them into word tokens containing letters and apostrophes (the usage of digits and punctuation is not especially language-dependent), padding each token with spaces, and generating all possible n-grams of length 1 to 5 for each word in the document. These n-grams are counted and sorted into frequency order to yield the document profile.

The most frequent 300 or so n-grams are highly correlated with the language. The highest ranking ones are mostly unigrams consisting of one character only, and simply reflect the distribution of letters of the alphabet in the document's language. Starting around rank 300 or so, the frequency profile begins to be more specific to the document's topic. Using a simple metric for comparing a document profile with a category profile, each document's language can be identified with high accuracy [Cavnar and Trenkle, 1994].

An alternative approach is to use words instead of n-grams, and compare occurrence probabilities of the common words in the language samples with the most frequent words of the test data. This method works as well as the n-gram scheme for sentences longer than about 15 words, but is less effective for short sentences such as titles of articles and news headlines [Grefenstette, 1995].

14.2.3.1 Ascribing Authorship

Author metadata is one of the primary attributes of most documents. It is usually known and need not be mined, but in some cases authorship is uncertain and must be guessed from the document text. Authorship ascription is often treated as a text categorization problem. However, there are sensitive statistical tests that can be used instead, based on the fact that each author has a characteristic vocabulary whose size can be estimated statistically from a corpus of their work.

For example, *The Complete Works of Shakespeare* (885,000 words) contains 31,500 different words, of which 14,400 appear only once, 4,300 twice, and so on. If another large body of work by Shakespeare were discovered, equal in size to his known writings, one would expect to find many repetitions of these 31,500 words along with some new words that he had not used before. According to a simple statistical model, the number of new words should be about 11,400 [Efron and Thisted, 1976]. Furthermore, one can estimate the total number of words known by Shakespeare from the same model: The result is 66,500 words. (For the derivation of these estimates, see Efron and Thisted [1976].) This statistical model was unexpectedly put to the test 10 years after it was developed [Kolata, 1986]. A previously unknown poem, suspected to have been penned by Shakespeare, was discovered in a library in Oxford, England. Of its 430 words, statistical analysis predicted that 6.97 would be new, with a standard deviation of ±2.64. In fact, nine of them were (*admiration, besots, exiles, inflection, joying, scanty, speck, tormentor,* and *twined*). It was predicted that there would be 4.21 ± 2.05 words that Shakespeare had used only once; the poem contained seven — only just outside the range. 3.33 ± 1.83 should have been used exactly twice before; in fact five were. Although this does not prove authorship, it does suggest it — particularly since comparative analyses of the vocabulary of Shakespeare's contemporaries indicate substantial mismatches.

Text categorization methods would almost certainly be far less accurate than these statistical tests, and this serves as a warning not to apply generic text mining techniques indiscriminately. However, the tests are useful only when a huge sample of preclassified text is available — in this case, the life's work of a major author.

14.2.3.2 Identifying Keyphrases

In the scientific and technical literature, keywords and keyphrases are attached to documents to give a brief indication of what they are about. (Henceforth we use the term "keyphrase" to subsume keywords, that is, one-word keyphrases.) Keyphrases are a useful form of metadata because they condense documents into a few pithy phrases that can be interpreted individually and independently of each other.

Given a large set of training documents with keyphrases assigned to each, text categorization techniques can be applied to assign appropriate keyphrases to new documents. The training documents provide a predefined set of keyphrases from which all keyphrases for new documents are chosen — a controlled vocabulary. For each keyphrase, the training data define a set of documents that are associated with it, and standard machine-learning techniques are used to create a "classifier" from the training documents, using those associated with the keyphrase as positive examples and the remainder as negative examples. Given a new document, it is processed by each keyphrase's classifier. Some classify the new document positively — in other words, it belongs to the set of documents associated with that keyphrase — while others classify it negatively — in other words, it does not. Keyphrases are assigned to the new document accordingly. The process is called keyphrase *assignment* because phrases from an existing set are assigned to documents.

There is an entirely different method for inferring keyphrase metadata called keyphrase *extraction*. Here, all the phrases that occur in the document are listed and information retrieval heuristics are used to select those that seem to characterize it best. Most keyphrases are noun phrases, and syntactic techniques may be used to identify these and ensure that the set of candidates contains only noun phrases. The heuristics used for selection range from simple ones, such as the position of the phrase's first occurrence in the document, to more complex ones, such as the occurrence frequency of the phrase in the document vs. its occurrence frequency in a corpus of other documents in the subject area. The training set is used to tune the parameters that balance these different factors.

With keyphrase assignment, the only keyphrases that can be assigned are ones that have already been used for training documents. This has the advantage that all keyphrases are well-formed, but has the disadvantage that novel topics cannot be accommodated. The training set of documents must therefore be large and comprehensive. In contrast, keyphrase *extraction* is open-ended: phrases are selected from the document text itself. There is no particular problem with novel topics, but idiosyncratic or malformed keyphrases may be chosen. A large training set is not needed because it is only used to set parameters for the algorithm.

Keyphrase extraction works as follows. Given a document, rudimentary lexical techniques based on punctuation and common words are used to extract a set of candidate phrases. Then, features are computed for each phrase, such as how often it appears in the document (normalized by how often that phrase appears in other documents in the corpus); how often it has been used as a keyphrase in other training documents; whether it occurs in the title, abstract, or section headings; whether it occurs in the title of papers cited in the reference list, and so on. The training data is used to form a model that takes these features and predicts whether or not a candidate phrase will actually appear as a keyphrase — this information is known for the training documents. Then the model is applied to extract likely keyphrases from new documents. Such models have been built and used to assign keyphrases to technical papers; simple machine-learning schemes (e.g., Naïve Bayes) seem adequate for this task.

To give an indication of the success of machine learning on this problem, Figure 14.4 shows the titles of three research articles and two sets of keyphrases for each one [Frank et al., 1999]. One set contains the keyphrases assigned by the article's author; the other was determined automatically from its full text. Phrases in common between the two sets are italicized. In each case, the author's keyphrases and the automatically extracted keyphrases overlap, but it is not too difficult to guess which are the author's. The giveaway is that the machine-learning scheme, in addition to choosing several good keyphrases, also chooses some that authors are unlikely to use — for example, *gauge*, *smooth*, and especially *garbage*! Despite the anomalies, the automatically extracted lists give a reasonable characterization of the papers. If no author-specified keyphrases were available, they could prove useful for someone scanning quickly for relevant information.

14.2.4 Extracting Structured Information

An important form of text mining takes the form of a search for structured data inside documents. Ordinary documents are full of structured information: phone numbers, fax numbers, street addresses,

Protocols for secure, atomic transaction execution in electronic commerce	
anonymity	*atomicity*
atomicity	*auction*
auction	customer
electronic commerce	*electronic commerce*
privacy	intruder
real-time	merchant
security	protocol
transaction	*security*
	third party
	transaction

Neural multigrid for gauge theories and other disordered systems	
disordered systems	Disordered
gauge fields	gauge
multigrid	*gauge fields*
neural multigrid	interpolation kernels
neural networks	length scale
	multigrid
	smooth

Proof nets, garbage, and computations	
cut-elimination	cut
linear logic	*cut elimination*
proof nets	garbage
sharing graphs	proof net
typed lambda-calculus	weakening

FIGURE 14.4 Titles and keyphrases, author- and machine-assigned, for three papers.

e-mail addresses, e-mail signatures, abstracts, tables of contents, lists of references, tables, figures, captions, meeting announcements, Web addresses, and more. In addition, there are countless domain-specific structures, such as ISBN numbers, stock symbols, chemical structures, and mathematical equations. Many short documents describe a particular kind of object or event, and in this case elementary structures are combined into a higher-level composite that represent the document's entire content. In constrained situations, the composite structure can be represented as a "template" with slots that are filled by individual pieces of structured information. From a large set of documents describing similar objects or events, it may even be possible to infer rules that represent particular patterns of slot-fillers.

Applications for schemes that identify structured information in text are legion. Indeed, in general interactive computing, users commonly complain that they cannot easily take action on the structured information found in everyday documents [Nardi et al., 1998].

14.2.4.1 Entity Extraction

Many practical tasks involve identifying linguistic constructions that stand for objects or "entities" in the world. Often consisting of more than one word, these terms act as single vocabulary items, and many document processing tasks can be significantly improved if they are identified as such. They can aid searching, interlinking, and cross-referencing between documents, the construction of browsing indexes, and can comprise machine-processable metadata which, for certain operations, act as a surrogate for the document contents.

Examples of such entities are:

- Names of people, places, organizations, and products
- E-mail addresses, URLs
- Dates, numbers, and sums of money
- Abbreviations
- Acronyms and their definition
- Multiword terms

Some of these items can be spotted by a dictionary-based approach, using lists of personal names and organizations, information about locations from gazetteers, abbreviation and acronym dictionaries, and so on. Here the lookup operation should recognize legitimate variants. This is harder than it sounds — for example (admittedly an extreme one), the name of the Libyan leader *Muammar Qaddafi* is represented in 47 different ways on documents that have been received by the Library of Congress [Mann, 1993]! A central area of library science is devoted to the creation and use of standard names for authors and other bibliographic entities (called "authority control").

In most applications, novel names appear. Sometimes these are composed of parts that have been encountered before, say *John* and *Smith*, but not in that particular combination. Others are recognizable by their capitalization and punctuation pattern (e.g., *Randall B. Caldwell*). Still others, particularly certain foreign names, will be recognizable because of peculiar language statistics (e.g., *Kung-Kui Lau*). Others will not be recognizable except by capitalization, which is an unreliable guide, particularly when only one name is present. Names that begin a sentence cannot be distinguished on this basis from other words. It is not always completely clear what to "begin a sentence" means; in some typographic conventions, itemized points have initial capitals but no terminating punctuation. Of course, words that are not names are sometimes capitalized (e.g., important words in titles; and, in German, all nouns), and a small minority of names are conventionally written unpunctuated and in lower case (e.g., some English names starting with *ff*, the poet *e e cummings*, the singer *k d lang*). Full personal name-recognition conventions are surprisingly complex, involving baronial prefixes in different languages (e.g., *von*, *van*, *de*), suffixes (*Snr*, *Jnr*), and titles (*Mr.*, *Ms.*, *Rep.*, *Prof.*, *General*).

It is generally impossible to distinguish personal names from other kinds of names in the absence of context or domain knowledge. Consider places like *Berkeley*, *Lincoln*, *Washington*; companies like *du Pont*, *Ford*, even *General Motors*; product names like *Mr. Whippy* and *Dr. Pepper*; book titles like *David Copperfield* or *Moby Dick*. Names of organizations present special difficulties because they can contain linguistic constructs, as in *the Food and Drug Administration* (contrast *Lincoln and Washington*, which conjoins two separate names) or the *League of Nations* (contrast *General Motors of Detroit*, which qualifies one name with a different one).

Some artificial entities like e-mail addresses and URLs are easy to recognize because they are specially designed for machine processing. They can be unambiguously detected by a simple grammar, usually encoded in a regular expression, for the appropriate pattern. Of course, this is exceptional: these items are not part of "natural" language.

Other entities can be recognized by explicit grammars; indeed, one might define structured information as "data recognizable by a grammar." Dates, numbers, and sums of money are good examples that can be captured by simple lexical grammars. However, in practice, things are often not so easy as they might appear. There may be a proliferation of different patterns, and novel ones may occur. The first step in processing is usually to divide the input into lexical tokens or "words" (e.g., split at white space or punctuation). While words delimited by nonalphanumeric characters provide a natural tokenization for many examples, such a decision will turn out to be restrictive in particular cases, for it precludes patterns that adopt a nonstandard tokenization, such as *30Jul98*. In general, any prior division into tokens runs the risk of obscuring information.

To illustrate the degree of variation in these items, Figure 14.5 shows examples of items that are recognized by IBM's "Intelligent Miner for Text" software [Tkach, 1998]. Dates include standard textual forms for absolute and relative dates. Numbers include both absolute numbers and percentages, and can be written in numerals or spelled out as words. Sums of money can be expressed in various currencies.

Most abbreviations can only be identified using dictionaries. Many acronyms, however, can be detected automatically, and technical, commercial, and political documents make extensive use of them. Identi-

Dates	Numbers	Sums of money
"March twenty-seventh, nineteen ninety-seven"	"One thousand three hundred and twenty-seven"	"Twenty-seven dollars"
"March 27, 1997"	"Thirteen twenty-seven"	"DM 27"
"Next March 27th"	"1327"	"27,000 dollars USA"
"Tomorrow"	"Twenty-seven percent"	"27,000 marks Germany"
"A year ago"	27%	

FIGURE 14.5 Sample information items.

fying acronyms and their definitions in documents is a good example of a text-mining problem that can usefully be tackled using simple heuristics.

The dictionary definition of "acronym" is: A word formed from the first (or first few) letters of a series of words, as *radar*, from *ra*dio *d*etection *a*nd *r*anging. Acronyms are often defined by following (or preceding) their first use with a textual explanation, as in this example. Heuristics can be developed to detect situations where a word is spelled out by the initial letters of an accompanying phrase. Three simplifying assumptions that vastly reduce the computational complexity of the task while sacrificing the ability to detect just a few acronyms are to consider (1) only acronyms made up of three or more letters; (2) only the first letter of each word for inclusion in the acronym, and (3) acronyms that are written either fully capitalized or mostly capitalized. In fact, the acronym *radar* breaks both (2) and (3); it involves the first *two* letters of the word *radio*, and, like most acronyms that have fallen into general use, it is rarely capitalized. However, the vast majority of acronyms that pervade today's technical, business, and political literature satisfy these assumptions and are relatively easy to detect. Once detected, acronyms can be added to a dictionary so that they are recognized elsewhere as abbreviations. Of course, many acronyms are ambiguous: the Acronym Finder Web site (at www.mtnds.com/af/) has 27 definitions for *CIA*, ranging from *Central Intelligence Agency* and *Canadian Institute of Actuaries* to *Chemiluminescence Immunoassay*. In ordinary text this ambiguity rarely poses a problem, but in large document collections, context and domain knowledge will be necessary for disambiguation.

14.2.4.2 Information Extraction

"Information extraction" is used to refer to the task of filling templates from natural language input [Appelt, 1999], one of the principal subfields of text mining. A commonly cited domain is that of terrorist events, where the template may include slots for the perpetrator, the victim, type of event, and where and when it occurred, etc. In the late 1980s, the Defense Advanced Research Projects Agency (DARPA) instituted a series of "Message understanding conferences" (MUC) to focus efforts on information extraction on particular domains and to compare emerging technologies on a level basis. MUC-1 (1987) and MUC-2 (1989) focused on messages about naval operations; MUC-3 (1991) and MUC-4 (1992) studied news articles about terrorist activity; MUC-5 (1993) and MUC-6 (1995) looked at news articles about joint ventures and management changes, respectively; and MUC-7 (1997) examined news articles about space vehicle and missile launches. Figure 14.6 shows an example of a MUC-7 query. The outcome of information extraction would be to identify relevant news articles and, for each one, fill out a template like the one shown.

Unlike text summarization and document retrieval, information extraction in this sense is not a task commonly undertaken by people because the extracted information must come from each individual article taken in isolation — the use of background knowledge and domain-specific inference are specifically forbidden. It turns out to be a difficult task for people, and inter-annotator agreement is said to lie in the 60 to 80% range [Appelt, 1999].

Query	"A relevant article refers to a vehicle launch that is scheduled, in progress, or has actually occurred and must minimally identify the payload, the date of the launch, whether the launch is civilian or military, the function of the mission, and its status."
Template	Vehicle: Payload: Mission date: Mission site: Mission type (military, civilian): Mission function (test, deploy, retrieve): Mission status (succeeded, failed, in progress, scheduled):

FIGURE 14.6 Sample query and template from MUC-7.

The first job of an information extraction system is entity extraction, discussed earlier. Once this has been done, it is necessary to determine the relationship between the entities extracted, which involves syntactic parsing of the text. Typical extraction problems address simple relationships among entities, such as finding the predicate structure of a small set of predetermined propositions. These are usually simple enough to be captured by shallow parsing techniques such as small finite-state grammars, a far easier proposition than a full linguistic parse. It may be necessary to determine the attachment of prepositional phrases and other modifiers, which may be restricted by type constraints that apply in the domain under consideration. Another problem, which is not so easy to resolve, is pronoun reference ambiguity. This arises in more general form as "coreference ambiguity": whether one noun phrase refers to the same real-world entity as another. Again, Appelt [1999] describes these and other problems of information extraction.

Machine learning has been applied to the information extraction task by seeking pattern-match rules that extract fillers for slots in the template (e.g., Soderland et al. [1995]; Huffman [1996]; Freitag [2000]). As an example, we describe a scheme investigated by Califf and Mooney [1999], in which pairs comprising a document and a template manually extracted from it are presented to the system as training data. A bottom-up learning algorithm is employed to acquire rules that detect the slot fillers from their surrounding text. The rules are expressed in pattern-action form, and the patterns comprise constraints on words in the surrounding context and the slot-filler itself. These constraints involve the words included, their part-of-speech tags, and their semantic classes.

Califf and Mooney investigated the problem of extracting information from job ads such as those posted on Internet newsgroups. Figure 14.7 shows a sample message and filled template of the kind that might be supplied to the program as training data.

This input provides support for several rules. One example: is "a noun phrase of 1 or 2 words, preceded by the word *in* and followed by a comma and a noun phrase with semantic tag *State*, should be placed in the template's *City* slot." The strategy for determining rules is to form maximally specific rules based on each example and then generalize the rules produced for different examples. For instance, from the phrase *offices in Kansas City, Missouri,* in the newsgroup posting in Figure 14.7, a maximally specific rule can be derived that assigns the phrase *Kansas City* to the *City* slot in a context where it is preceded by *offices in* and followed by "*Missouri*," with the appropriate parts of speech and semantic tags. A second newsgroup posting that included the phrase *located in Atlanta, Georgia*, with *Atlanta* occupying the filled template's *City* slot, would produce a similar maximally specific rule. The rule generalization process takes these two specific rules, notes the commonalities, and determines the general rule for filling the *City* slot cited above.

Newsgroup posting	Telecommunications. SOLARIS Systems Administrator. 38-44K. Immediate need Leading telecommunications firm in need of an energetic individual to fill the following position in our offices in Kansas City, Missouri: SOLARIS SYSTEMS ADMINISTRATOR Salary: 38-44K with full benefits Location: Kansas City, Missouri
Filled template	Computer_science_job Title: SOLARIS Systems Administrator Salary: 38-44K State: Missouri City: Kansas City Platform: SOLARIS Area: telecommunication

FIGURE 14.7 Sample message and filled template.

If *Language* contains both *HTML* and *DHTML*, then *Language* also contains *XML*
If *Application* contains *Dreamweaver 4* and *Area* is *Web design*, then *Application* also contains *Photoshop 6*
If *Application* is *ODBC*, then *Language* is *JSP*
If *Language* contains both *Perl* and *HTML*, then *Platform* is Linux

FIGURE 14.8 Sample rules induced from job postings database.

14.2.4.3 Learning Rules from Text

Taking information extraction a step further, the extracted information can be used in a subsequent step to learn rules — not rules about how to extract information, but rules that characterize the content of the text itself. Following on from the project described above to extract templates from job postings in internet newsgroups, a database was formed from several hundred postings, and rules were induced from the database for the fillers of the slots for *Language*, *Platform*, *Application*, and *Area* [Nahm and Mooney, 2000]. Figure 14.8 shows some of the rules that were found.

In order to create the database, templates were constructed manually from a few newsgroup postings. From these, information extraction rules were learned as described above. These rules were then used to extract information automatically from the other newsgroup postings. Finally, the whole database so extracted was input to standard data mining algorithms to infer rules for filling the four chosen slots. Both prediction rules — that is, rules predicting the filler for a predetermined slot — and association rules — that is, rules predicting the value of *any* slot — were sought. Standard techniques were employed: C4.5Rules [Quinlan, 1993] and Ripper [Cohen, 1995] for prediction rules and Apriori [Agarwal and Srikant, 1994] for association rules.

Nahm and Mooney [2000] concluded that information extraction based on a few manually constructed training examples could compete with an entire manually constructed database in terms of the quality of the rules that were inferred. However, this success probably hinged on the highly structured domain of job postings in a tightly constrained area of employment. Subsequent work on inferring rules from book descriptions on the Amazon.com Website [Nahm and Mooney, 2000] produced rules that, though interesting, seem rather less useful in practice.

14.3 Mining Structured Text

Much of the text that we deal with today, especially on the Internet, contains explicit structural markup and thus differs from traditional plain text. Some markup is internal and indicates document structure or format; some is external and gives explicit hypertext links between documents. These information sources give additional leverage for mining Web documents. Both sources of information are generally extremely noisy: they involve arbitrary and unpredictable choices by individual page designers. However, these disadvantages are offset by the overwhelming amount of data that is available, which is relatively unbiased because it is aggregated over many different information providers. Thus Web mining is emerging as a new subfield, similar to text mining, but taking advantage of the extra information available in Web documents, particularly hyperlinks, and even capitalizing on the existence of topic directories in the Web itself to improve results [Chakrabarti, 2003].

We briefly review three techniques for mining structured text. The first, wrapper induction, uses internal markup information to increase the effectiveness of text mining in marked-up documents. The remaining two, document clustering and determining the "authority" of Web documents, capitalize on the external markup information that is present in hypertext in the form of explicit links to other documents.

14.3.1 Wrapper Induction

Internet resources that contain relational data, such as telephone directories, product catalogs, etc., use formatting markup to clearly present the information they contain to users. However, with standard

HTML, it is quite difficult to extract data from such resources in an automatic way. The XML markup language is designed to overcome these problems by encouraging page authors to mark their content in a way that reflects document structure at a detailed level; but it is not clear to what extent users will be prepared to share the structure of their documents fully in XML, and even if they do, huge numbers of legacy pages abound.

Many software systems use external online resources by hand-coding simple parsing modules, commonly called "wrappers," to analyze the page structure and extract the requisite information. This is a kind of text mining, but one that depends on the input having a fixed, predetermined structure from which information can be extracted algorithmically. Given that this assumption is satisfied, the information extraction problem is relatively trivial. But this is rarely the case. Page structures vary; errors that are insignificant to human readers throw automatic extraction procedures off completely; Websites evolve. There is a strong case for automatic induction of wrappers to reduce these problems when small changes occur and to make it easier to produce new sets of extraction rules when structures change completely.

Figure 14.9 shows an example taken from Kushmerick et al. [1997], in which a small Web page is used to present some relational information. Below is the HTML code from which the page was generated; below that is a wrapper, written in an informal pseudo-code, that extracts relevant information from the HTML. Many different wrappers could be written; in this case the algorithm is based on the formatting information present in the HTML — the fact that countries are surrounded by <B> … </B> and country codes by <I> … </I>. Used in isolation, this information fails because other parts of the page are rendered in boldface too. Consequently the wrapper in Figure 14.9c uses additional information — the <P> that precedes the relational information in Figure 14.9b and the <HR> that follows it — to constrain the search. This wrapper is a specific example of a generalized structure that parses a page into a *head*, followed by a sequence of relational items, followed by a *tail*; where specific delimiters are used to signal the end of the head, the items themselves, and the beginning of the tail.

It is possible to infer such wrappers by induction from examples that comprise a set of pages and tuples representing the information derived from each page. This can be done by iterating over all choices of delimiters, stopping when a consistent wrapper is encountered. One advantage of automatic wrapper

(a)

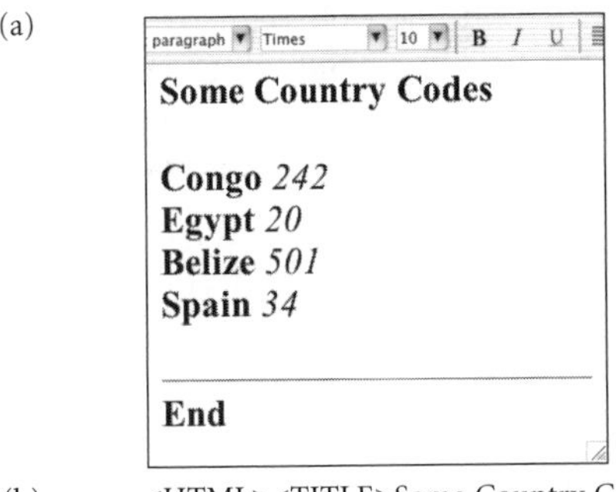

(b)

```
<HTML><TITLE>Some Country Codes</TITLE>
<BODY><B>Some Country Codes</B><P>
<B>Congo</B> <I>242</I><BR>
<B>Egypt</B> <I>20</I><BR>
<B>Belize</B> <I>501</I><BR>
<B>Spain</B> <I>34</I><BR>
<HR><B>End</B></BODY></HTML>
```

(c)

```
ExtractCountryCodes(page P)
Skip past first occurrence of <P> in P
While next <B> is before next <HR> in P
For each [s, t] ∈ {[<B>,</B>], [<I>,</I>]}
Skip past next occurrence of s in P
Extract attribute from P to next occurrence of t
Return extracted tuples
```

FIGURE 14.9 Web page, underlying HTML, and wrapper extracting relational information.

induction is that recognition then depends on a minimal set of cues, providing some defense against extraneous text and markers in the input. Another is that when errors are caused by stylistic variants, it is a simple matter to add these to the training data and reinduce a new wrapper that takes them into account.

14.3.1.1 Document Clustering with Links

Document clustering techniques are normally based on the documents' textual similarity. However, the hyperlink structure of Web documents, encapsulated in the "link graph" in which nodes are Web pages and links are hyperlinks between them, can be used as a different basis for clustering. Many standard graph clustering and partitioning techniques are applicable (e.g., Hendrickson and Leland [1995]). Link-based clustering schemes typically use factors such as these:

- The number of hyperlinks that must be followed to travel in the Web from one document to the other
- The number of common ancestors of the two documents, weighted by their ancestry distance
- The number of common descendents of the documents, similarly weighted

These can be combined into an overall similarity measure between documents. In practice, a textual similarity measure is usually incorporated as well, to yield a hybrid clustering scheme that takes account of both the documents' content and their linkage structure. The overall similarity may then be determined as the weighted sum of four factors (e.g., Weiss et al. [1996]). Clearly, such a measure will be sensitive to the stylistic characteristics of the documents and their linkage structure, and given the number of parameters involved, there is considerable scope for tuning to maximize performance on particular data sets.

14.3.1.2 Determining "Authority" of Web Documents

The Web's linkage structure is a valuable source of information that reflects the popularity, sometimes interpreted as "importance," "authority," or "status" of Web pages. For each page, a numeric rank is computed. The basic premise is that highly ranked pages are ones that are cited, or pointed to, by many other pages. Consideration is also given to (1) the rank of the citing page, to reflect the fact that a citation by a highly ranked page is a better indication of quality than one from a lesser page, and (2) the number of outlinks from the citing page, to prevent a highly ranked page from artificially magnifying its influence simply by containing a large number of pointers. This leads to a simple algebraic equation to determine the rank of each member of a set of hyperlinked pages [Brin and Page, 1998]. Complications arise because some links are "broken" in that they lead to nonexistent pages, and because the Web is not fully connected, but these are easily overcome.

Such techniques are widely used by search engines (e.g., Google) to determine how to sort the hits associated with any given query. They provide a social measure of status that relates to standard techniques developed by social scientists for measuring and analyzing social networks [Wasserman and Faust, 1994].

14.4 Human Text Mining

All scientific researchers are expected to use the literature as a major source of information during the course of their work to provide new ideas and supplement their laboratory studies. However, some feel that this can be taken further: that new information, or at least new hypotheses, can be derived directly from the literature by researchers who are expert in information seeking but not necessarily in the subject matter itself. Subject-matter experts can only read a small part of what is published in their fields and are often unaware of developments in related fields. Information researchers can seek useful linkages between related literatures that may be previously unknown, particularly if there is little explicit cross-reference between the literatures.

We briefly sketch an example to indicate what automatic text mining may eventually aspire to but is nowhere near achieving yet. By analyzing chains of causal implication within the medical literature, new hypotheses for causes of rare diseases have been discovered, some of which have received supporting experimental evidence [Swanson, 1987; Swanson and Smalheiser, 1997]. While investigating causes of migraine headaches, Swanson extracted information from titles of articles in the biomedical literature, leading to clues like these:

Stress is associated with migraines
Stress can lead to loss of magnesium
Calcium channel blockers prevent some migraines
Magnesium is a natural calcium channel blocker
Spreading cortical depression is implicated in some migraines
High levels of magnesium inhibit spreading cortical depression
Migraine patients have high platelet aggregability
Magnesium can suppress platelet aggregability

These clues suggest that magnesium deficiency may play a role in some kinds of migraine headache, a hypothesis that did not exist in the literature at the time. Swanson found these links. Thus a new and plausible medical hypothesis was derived from a combination of text fragments and the information researcher's background knowledge. Of course, the hypothesis still had to be tested via nontextual means.

14.5 Techniques and Tools

Text mining systems use a broad spectrum of different approaches and techniques, partly because of the great scope of text mining and consequent diversity of systems that perform it, and partly because the field is so young that dominant methodologies have not yet emerged.

14.5.1 High-Level Issues: Training vs. Knowledge Engineering

There is an important distinction between systems that use an automatic training approach to spot patterns in data and ones that are based on a knowledge engineering approach and use rules formulated by human experts. This distinction recurs throughout the field but is particularly stark in the areas of entity extraction and information extraction. For example, systems that extract personal names can use handcrafted rules derived from everyday experience. Simple and obvious rules involve capitalization, punctuation, single-letter initials, and titles; more complex ones take account of baronial prefixes and foreign forms. Alternatively, names could be manually marked up in a set of training documents and machine-learning techniques used to infer rules that apply to test documents.

In general, the knowledge-engineering approach requires a relatively high level of human expertise — a human expert who knows the domain and the information extraction system well enough to formulate high-quality rules. Formulating good rules is a demanding and time-consuming task for human experts and involves many cycles of formulating, testing, and adjusting the rules so that they perform well on new data.

Markup for automatic training is clerical work that requires only the ability to recognize the entities in question when they occur. However, it is a demanding task because large volumes are needed for good performance. Some learning systems can leverage unmarked training data to improve the results obtained from a relatively small training set. For example, an experiment in document categorization used a small number of labeled documents to produce an initial model, which was then used to assign probabilistically weighted class labels to unlabeled documents [Nigam et al., 1998]. Then a new classifier was produced using all the documents as training data. The procedure was iterated until the classifier remained unchanged. Another possibility is to bootstrap learning based on two different and mutually reinforcing perspectives on the data, an idea called "co-training" [Blum and Mitchell, 1998].

14.5.2 Low-Level Issues: Token Identification

Dealing with natural language involves some rather mundane decisions that nevertheless strongly affect the success of the outcome. Tokenization, or splitting the input into words, is an important first step that seems easy but is fraught with small decisions: how to deal with apostrophes and hyphens, capitalization, punctuation, numbers, alphanumeric strings, whether the amount of white space is significant, whether to impose a maximum length on tokens, what to do with nonprinting characters, and so on. It may be beneficial to perform some rudimentary morphological analysis on the tokens — removing suffixes [Porter, 1980] or representing them as words separate from the stem — which can be quite complex and is strongly language-dependent. Tokens may be standardized by using a dictionary to map different, but equivalent, variants of a term into a single canonical form. Some text-mining applications (e.g., text summarization) split the input into sentences and even paragraphs, which again involves mundane decisions about delimiters, capitalization, and nonstandard characters.

Once the input is tokenized, some level of syntactic processing is usually required. The simplest operation is to remove stop words, which are words that perform well-defined syntactic roles but from a nonlinguistic point of view do not carry information. Another is to identify common phrases and map them into single features. The resulting representation of the text as a sequence of word features is commonly used in many text-mining systems (e.g., for information extraction).

14.5.2.1 Basic Techniques

Tokenizing a document and discarding all sequential information yield the "bag of words" representation mentioned above under document retrieval. Great effort has been invested over the years in a quest for document similarity measures based on this representation. One is to count the number of terms in common between the documents: this is called *coordinate matching*. This representation, in conjunction with standard classification systems from machine learning (e.g., Naïve Bayes and Support Vector Machines; see Witten and Frank [2000]), underlies most text categorization systems.

It is often more effective to weight words in two ways: first by the number of documents in the entire collection in which they appear ("document frequency") on the basis that frequent words carry less information than rare ones; second by the number of times they appear in the particular documents in question ("term frequency"). These effects can be combined by multiplying the term frequency by the inverse document frequency, leading to a standard family of document similarity measures (often called "tf × idf"). These form the basis of standard text categorization and information retrieval systems.

A further step is to perform a syntactic analysis and tag each word with its part of speech. This helps to disambiguate different senses of a word and to eliminate incorrect analyses caused by rare word senses. Some part-of-speech taggers are rule based, while others are statistically based [Garside et al., 1987] — this reflects the "training" vs. "knowledge engineering" referred to earlier. In either case, results are correct about 95% of the time, which may not be enough to resolve the ambiguity problems.

Another basic technique for dealing with sequences of words or other items is to use Hidden Markov Models (HMMs). These are probabilistic finite-state models that "parse" an input sequence by tracking its flow through the model. This is done in a probabilistic sense so that the model's current state is represented not by a particular unique state but by a probability distribution over all states. Frequently, the initial state is unknown or "hidden," and must itself be represented by a probability distribution. Each new token in the input affects this distribution in a way that depends on the structure and parameters of the model. Eventually, the overwhelming majority of the probability may be concentrated on one particular state, which serves to disambiguate the initial state and indeed the entire trajectory of state transitions corresponding to the input sequence. Trainable part-of-speech taggers are based on this idea: the states correspond to parts of speech (e.g., Brill [1992]).

HMMs can easily be built from training sequences in which each token is pre-tagged with its state. However, the manual effort involved in tagging training sequences is often prohibitive. There exists a "relaxation" algorithm that takes untagged training sequences and produces a corresponding HMM [Rabiner, 1989]. Such techniques have been used in text mining, for example, to extract references from plain text [McCallum et al., 1999].

If the source documents are hypertext, there are various basic techniques for analyzing the linkage structure. One, evaluating page rank to determine a numeric "importance" for each page, was described above. Another is to decompose pages into "hubs" and "authorities" [Kleinberg, 1999]. These are recursively defined as follows: A good hub is a page that points to many good authorities, while a good authority is a page pointed to by many good hubs. This mutually reinforcing relationship can be evaluated using an iterative relaxation procedure. The result can be used to select documents that contain authoritative content to use as a basis for text mining, discarding all those Web pages that simply contain lists of pointers to other pages.

14.5.2.2 Tools

There is a plethora of software tools to help with the basic processes of text mining. A comprehensive and useful resource at nlp.stanford.edu/lionks/statnlp.html lists taggers, parsers, language models, and concordances; several different corpora (large collections, particular languages, etc.); dictionaries, lexical, and morphological resources; software modules for handling XML and SGML documents; and other relevant resources such as courses, mailing lists, people, and societies. It classifies software as freely downloadable and commercially available, with several intermediate categories.

One particular framework and development environment for text mining, called General Architecture for Text Engineering or GATE [Cunningham, 2002], aims to help users develop, evaluate, and deploy systems for what the authors term "language engineering." It provides support not just for standard text-mining applications such as information extraction but also for tasks such as building and annotating corpora and evaluating the applications.

At the lowest level, GATE supports a variety of formats including XML, RTF, HTML, SGML, email, and plain text, converting them into a single unified model that also supports annotation. There are three storage mechanisms: a relational database, a serialized Java object, and an XML-based internal format; documents can be reexported into their original format with or without annotations. Text encoding is based on Unicode to provide support for multilingual data processing, so that systems developed with GATE can be ported to new languages with no additional overhead apart from the development of the resources needed for the specific language.

GATE includes a tokenizer and a sentence splitter. It incorporates a part-of-speech tagger and a gazetteer that includes lists of cities, organizations, days of the week, etc. It has a semantic tagger that applies handcrafted rules written in a language in which patterns can be described and annotations created as a result. Patterns can be specified by giving a particular text string, or annotations that have previously been created by modules such as the tokenizer, gazetteer, or document format analysis. It also includes semantic modules that recognize relations between entities and detect coreference. It contains tools for creating new language resources and for evaluating the performance of text-mining systems developed with GATE.

One application of GATE is a system for entity extraction of names that is capable of processing texts from widely different domains and genres. This has been used to perform recognition and tracking tasks of named, nominal, and pronominal entities in several types of text. GATE has also been used to produce formal annotations about important events in a text commentary that accompanies football video program material.

14.6 Conclusion

Text mining is a burgeoning technology that is still, because of its newness and intrinsic difficulty, in a fluid state — akin, perhaps, to the state of machine learning in the mid-1980s. Generally accepted characterizations of what it covers do not yet exist. When the term is broadly interpreted, many different problems and techniques come under its ambit. In most cases, it is difficult to provide general and meaningful evaluations because the task is highly sensitive to the particular text under consideration. Document classification, entity extraction, and filling templates that correspond to given relationships between entities are all central text-mining operations that have been extensively studied. Using structured data such as Web pages rather than plain text as the input opens up new possibilities for extracting information from individual pages and large networks of pages. Automatic text-mining techniques have a long way to go before they rival the ability of people, even without any special domain knowledge, to glean information from large document collections.

References

Agarwal, R. and Srikant, R. (1994) Fast algorithms for mining association rules. *Proceedings of the International Conference on Very Large Databases VLDB-94.* Santiago, Chile, pp. 487–499.

Aone, C., Bennett, S.W., and Gorlinsky, J. (1996) Multi-media fusion through application of machine learning and NLP. *Proceedings of the AAAI Symposium on Machine Learning in Information Access.* Stanford, CA.

Appelt, D.E. (1999) Introduction to information extraction technology. *Tutorial, International Joint Conference on Artificial Intelligence IJCAI'99.* Morgan Kaufmann, San Francisco. Tutorial notes available at www.ai.sri.com/~appelt/ie-tutorial.

Apte, C., Damerau, F.J., and Weiss, S.M. (1994) Automated learning of decision rules for text categorization. *ACM Trans Information Systems,* Vol. 12, No. 3, pp. 233–251.

Baeza-Yates, R. and Ribiero-Neto, B. (1999) *Modern information retrieval.* Addison-Wesley Longman, Essex, U.K.

Bell, T.C., Cleary, J.G. and Witten, I.H. (1990) *Text Compression.* Prentice Hall, Englewood Cliffs, NJ.

Blum, A. and Mitchell, T. (1998) Combining labeled and unlabeled data with co-training. *Proceedings of the Conference on Computational Learning Theory COLT-98.* Madison, WI, pp. 92–100.

Borko, H. and Bernier, C.L. (1975) *Abstracting concepts and methods.* Academic Press, San Diego, CA.

Brill, E. (1992) A simple rule-based part of speech tagger. *Proceedings of the Conference on Applied Natural Language Processing ANLP-92.* Trento, Italy, pp. 152–155.

Brin, S. and Page, L. (1998) The anatomy of a large-scale hypertextual Web search engine. *Proceedings of the World Wide Web Conference WWW-7.* In *Computer Networks and ISDN Systems,* Vol. 30, No. 1–7, pp. 107–117.

Califf, M.E. and Mooney, R.J. (1999) Relational learning of pattern-match rules for information extraction. *Proceedings of the National Conference on Artificial Intelligence AAAI-99.* Orlando, FL, pp. 328–334.

Cavnar, W.B. and Trenkle, J.M. (1994) N-Gram-based text categorization. *Proceedings of the Symposium on Document Analysis and Information Retrieval.* Las Vegas, NV, pp. 161–175.

Cheeseman, P., Kelly, J., Self, M., Stutz., J., Taylor, W., and Freeman, D. (1988) AUTOCLASS: A Bayesian classification system. *Proceedings of the International Conference on Machine Learning ICML-88.* San Mateo, CA, pp. 54–64.

Cohen, W.W. (1995) Fast effective rule induction. *Proceedings of the International Conference on Machine Learning ICML-95.* Tarragona, Catalonia, Spain, pp. 115–123.

Cunningham, H. (2002) GATE, a General Architecture for Text Engineering. *Computing and the Humanities,* Vol. 36, pp. 223–254.

Dumais, S.T., Platt, J., Heckerman, D., and Sahami, M. (1998) Inductive learning algorithms and representations for text categorization. *Proceedings of the International Conference on Information and Knowledge Management CIKM-98.* Bethesda, MD, pp. 148–155.

Efron, B. and Thisted, R. (1976) Estimating the number of unseen species: how many words did Shakespeare know? *Biometrika,* Vol. 63, No. 3, pp. 435–447.

Fisher, D. (1987) Knowledge acquisition via incremental conceptual clustering. *Machine Learning,* Vol. 2, pp. 139–172.

Frank, E., Paynter, G., Witten, I.H., Gutwin, C., and Nevill-Manning, C. (1999) Domain-specific keyphrase extraction. *Proceedings of the International Joint Conference on Artificial Intelligence IJCAI-99.* Stockholm, Sweden, pp. 668–673.

Franklin, D. (2002) New software instantly connects key bits of data that once eluded teams of researchers. *Time,* December 23.

Freitag, D. (2000) Machine learning for information extraction in informal domains. *Machine Learning,* Vol. 39, No. 2/3, pp. 169–202.

Garside, R., Leech, G., and Sampson, G. (1987) *The Computational Analysis of English: A Corpus-Based Approach*. Longman, London.

Green, C.L., and Edwards, P. (1996) Using machine learning to enhance software tools for Internet information management. *Proceedings of the AAAI Workshop on Internet Based Information Systems.* Portland, OR, pp. 48–56.

Grefenstette, G. (1995) Comparing two language identification schemes. *Proceedings of the International Conference on Statistical Analysis of Textual Data JADT-95*. Rome, Italy.

Harman, D.K. (1995) Overview of the third text retrieval conference. In *Proceedings of the Text Retrieval Conference TREC-3*. National Institute of Standards, Gaithersburg, MD, pp. 1–19.

Hayes, P.J., Andersen, P.M., Nirenburg, I.B., and Schmandt, L.M. (1990) Tcs: a shell for content-based text categorization. *Proceedings of the IEEE Conference on Artificial Intelligence Applications CAIA-90*. Santa Barbara, CA, pp. 320–326.

Hearst, M.A. (1999) Untangling text mining. *Proceedings of the Annual Meeting of the Association for Computational Linguistics ACL99*. University of Maryland, College Park, MD, June.

Hendrickson, B. and Leland, R.W. (1995) A multi-level algorithm for partitioning graphs. *Proceedings of the ACM/IEEE Conference on Supercomputing*. San Diego, CA.

Huffman, S.B. (1996) Learning information extraction patterns from examples. In S. Wertmer, E. Riloff, and G. Scheler, Eds. *Connectionist, Statistical, and Symbolic Approaches to Learning for Natural Language Processing*, Springer-Verlag, Berlin, pp. 246–260.

Kleinberg, J.M. (1999) Authoritative sources in a hyperlinked environment. *Journal of the ACM*, Vol. 46, No. 5, pp. 604–632.

Kolata, G. (1986) Shakespeare's new poem: an ode to statistics. *Science*, No. 231, pp. 335–336, January 24.

Kushmerick, N., Weld, D.S., and Doorenbos, R. (1997) Wrapper induction for information extraction. *Proceedings of the International Joint Conference on Artificial Intelligence IJCAI-97*. Nayoya, Japan, pp. 729–735.

Lancaster, F.W. (1991) *Indexing and abstracting in theory and practice*. University of Illinois Graduate School of Library and Information Science, Champaign, IL.

Lenat, D.B. (1995) CYC: a large-scale investment in knowledge infrastructure. *Communications of the ACM*, Vol. 38, No. 11, pp. 32–38.

Lewis, D.D. (1992) An evaluation of phrasal and clustered representations on a text categorization task. *Proceedings of the International Conference on Research and Development in Information Retrieval SIGIR-92*. pp. 37–50. Copenhagen, Denmark.

Liere, R. and Tadepalli, P. (1996) The use of active learning in text categorization. *Proceedings of the AAAI Symposium on Machine Learning in Information Access*. Stanford, CA.

Mani, I. (2001) *Automatic summarization*. John Benjamins, Amsterdam.

Mann, T. (1993) *Library research models*. Oxford University Press, New York.

Martin, J.D. (1995) Clustering full text documents. *Proceedings of the IJCAI Workshop on Data Engineering for Inductive Learning at IJCAI-95*. Montreal, Canada.

McCallum, A., Nigam, K., Rennie, J., and Seymore, K. (1999) Building domain-specific search engines with machine learning techniques. *Proceedings of the AAAI Spring Symposium*. Stanford, CA.

Mitchell, T.M. (1997) *Machine Learning*. McGraw Hill, New York.

Nahm, U.Y. and Mooney, R.J. (2000) Using information extraction to aid the discovery of prediction rules from texts. *Proceedings of the Workshop on Text Mining, International Conference on Knowledge Discovery and Data Mining KDD-2000*. Boston, pp. 51–58.

Nahm, U.Y. and Mooney, R.J. (2002) Text mining with information extraction. *Proceedings of the AAAI-2002 Spring Symposium on Mining Answers from Texts and Knowledge Bases*. Stanford, CA.

Nardi, B.A., Miller, J.R. and Wright, D.J. (1998) Collaborative, programmable intelligent agents. Communications of the ACM, Vol. 41, No. 3, pp. 96-104.

Nigam, K., McCallum, A., Thrun, S., and Mitchell, T. (1998) Learning to classify text from labeled and unlabeled documents. *Proceedings of the National Conference on Artificial Intelligence AAAI-98*. Madison, WI, pp. 792–799.

Porter, M.F. (1980) An algorithm for suffix stripping. *Program*, Vol. 13, No. 3, pp. 130–137.

Quinlan, R. (1993) *C4.5: Programs for Machine Learning*. Morgan Kaufmann, San Mateo, CA.

Rabiner, L.R. (1989) A tutorial on hidden Markov models and selected applications in speech recognition. *Proceedings of the IEEE* Vol. 77, No. 2, pp. 257–286.

Salton, G. and McGill, M.J. (1983) *Introduction to Modern Information Retrieval*. McGraw Hill, New York.

Sebastiani, F. (2002) Machine learning in automated text categorization. *ACM Computing Surveys*, Vol. 34, No. 1, pp. 1–47.

Soderland, S., Fisher, D., Aseltine, J., and Lehnert, W. (1995) Crystal: inducing a conceptual dictionary. *Proceedings of the International Conference on Machine Learning ICML-95*. Tarragona, Catalonia, Spain, pp. 343–351.

Swanson, D.R. (1987) Two medical literatures that are logically but not bibliographically connected. *Journal of the American Society for Information Science*, Vol. 38, No. 4, pp. 228–233.

Swanson, D.R. and Smalheiser, N.R. (1997) An interactive system for finding complementary literatures: a stimulus to scientific discovery. *Artificial Intelligence*, Vol. 91, pp. 183–203.

Tkach, D. (Ed.). (1998) Text Mining Technology: Turning Information into Knowledge. IBM White Paper, February 17, 1998.

Wasserman, S. and Faust, K. (1994) *Social Network Analysis: Methods and Applications*. Cambridge University Press, Cambridge, U.K.

Weiss, R., Velez, B., Nemprempre, C., Szilagyi, P., Duda, A., and Gifford, D.K. (1996) HyPursuit: A hierarchical network search engine that exploits content-link hypertext clustering. *Proceedings of the ACM Conference on Hypertext*. Washington, D.C., March, pp. 180–193.

Willett, P. (1988) Recent trends in hierarchical document clustering: a critical review. *Information Processing and Management*, Vol. 24, No. 5, pp. 577–597.

Witten, I.H., Moffat, A., and Bell, T.C. (1999) *Managing Gigabytes: Compressing and Indexing Documents and Images*. Morgan Kaufmann, San Francisco, CA.

Witten, I.H. and Frank, E. (2000) *Data Mining: Practical Machine Learning Tools and Techniques with Java Implementations*. Morgan Kaufmann, San Francisco, CA.

Witten, I.H. and Bainbridge, D. (2003) *How to Build a Digital Library*. Morgan Kaufmann, San Francisco, CA.

15

Web Usage Mining and Personalization

CONTENTS

Abstract 15-1
15.1 Introduction and Background 15-1
15.2 Data Preparation and Modeling 15-3
15.2.1 Sources and Types of Data 15-5
15.2.2 Usage Data Preparation 15-6
15.2.3 Postprocessing of User Transactions Data 15-8
15.2.4 Data Integration from Multiple Sources 15-10
15.3 Pattern Discovery from Web Usage Data 15-12
15.3.1 Levels and Types of Analysis 15-12
15.3.2 Data-Mining Tasks for Web Usage Data 15-13
15.4 Using the Discovered Patterns for Personalization 15-20
15.4.1 The *k*NN-Based Approach 15-20
15.4.2 Using Clustering for Personalization 15-21
15.4.3 Using Association Rules for Personalization 15-22
15.4.4 Using Sequential Patterns for Personalization 15-23
15.5 Conclusions and Outlook 15-26
15.5.1 Which Approach? 15-26
15.5.2 The Future: Personalization Based on Semantic Web Mining 15-27
References 15-28

Bamshad Mobasher

Abstract

In this chapter we present a comprehensive overview of the personalization process based on Web usage mining. In this context we discuss a host of Web usage mining activities required for this process, including the preprocessing and integration of data from multiple sources, and common pattern discovery techniques that are applied to the integrated usage data. We also presented a number of specific recommendation algorithms for combining the discovered knowledge with the current status of a user's activity in a Website to provide personalized content. The goal of this chapter is to show how pattern discovery techniques such as clustering, association rule mining, and sequential pattern discovery, performed on Web usage data, can be leveraged effectively as an integrated part of a Web personalization system.

15.1 Introduction and Background

The tremendous growth in the number and the complexity of information resources and services on the Web has made Web personalization an indispensable tool for both Web-based organizations and end users. The ability of a site to engage visitors at a deeper level and to successfully guide them to useful and pertinent information is now viewed as one of the key factors in the site's ultimate success. Web

1-58488-381-2/05/$0.00+$1.50
© 2005 by CRC Press LLC

personalization can be described as any action that makes the Web experience of a user customized to the user's taste or preferences. Principal elements of Web personalization include modeling of Web objects (such as pages or products) and subjects (such as users or customers), categorization of objects and subjects, snatching between and across objects and/or subjects, and determination of the set of actions to be recommended for personalization.

To date, the approaches and techniques used in Web personalization can be categorized into three general groups: manual decision rule systems, content-based filtering agents, and collaborative filtering systems. Manual decision rule systems, such as Broadvision (www.broadvision.com), allow Website administrators to specify rules based on user demographics or static profiles (collected through a registration process). The rules are used to affect the content served to a particular user. Collaborative filtering systems such as Net Perceptions (www.netperceptions.com) typically take explicit information in the form of user ratings or preferences and, through a correlation engine, return information that is predicted to closely match the user's preferences. Content-based filtering systems such as those used by WebWatcher [Joachims et al., 1997] and client-side agent Letizia [Lieberman, 1995] generally rely on personal profiles and the content similarity of Web documents to these profiles for generating recommendations.

There are several well-known drawbacks to content-based or rule-based filtering techniques for personalization. The type of input is often a subjective description of the users by the users themselves, and thus is prone to biases. The profiles are often static, obtained through user registration, and thus the system performance degrades over time as the profiles age. Furthermore, using content similarity alone may result in missing important "pragmatic" relationships among Web objects based on how they are accessed by users. Collaborative filtering [Herlocker et al., 1999; Konstan et al., 1997; Shardanand and Maes, 1995] has tried to address some of these issues and, in fact, has become the predominant commercial approach in most successful e-commerce systems. These techniques generally involve matching the ratings of a current user for objects (e.g., movies or products) with those of similar users (nearest neighbors) in order to produce recommendations for objects not yet rated by the user. The primary technique used to accomplish this task is the *k*-Nearest-Neighbor (*k*NN) classification approach that compares a target user's record with the historical records of other users in order to find the top *k* users who have similar tastes or interests.

However, collaborative filtering techniques have their own potentially serious limitations. The most important of these limitations is their lack of scalability. Essentially, *k*NN requires that the neighborhood formation phase be performed as an online process, and for very large data sets this may lead to unacceptable latency for providing recommendations. Another limitation of *k*NN-based techniques emanates from the sparse nature of the data set. As the number of items in the database increases, the density of each user record with respect to these items will decrease. This, in turn, will decrease the likelihood of a significant overlap of visited or rated items among pairs of users, resulting in less reliable computed correlations, Furthermore, collaborative filtering usually performs best when explicit nonbinary user ratings for similar objects are available. In many Websites, however, it may be desirable to integrate the personalization actions throughout the site involving different types of objects, including navigational and content pages, as well as implicit product-oriented user events such as shopping cart changes or product information requests.

A number of optimization strategies have been proposed and employed to remedy these shortcomings [Aggarwal et al., 1999; O'Conner and Herlocker, 1999; Sarwar et al., 2000a; Ungar and Foster, 1998; Yu, 1999]. These strategies include similarity indexing and dimensionality reduction to reduce real-time search costs, as well as offline clustering of user records, allowing the online component of the system to search only within a matching cluster. There has also been a growing body of work in enhancing collaborative filtering by integrating data from other sources such as content and user demographics [Claypool et al., 1999; Pazzani, 1999].

More recently, Web usage mining [Srivastava et al., 2000], has been proposed as an underlying approach for Web personalization [Mobasher et al., 2000a]. The goal of Web usage mining is to capture and model the behavioral patterns and profiles of users interacting with a Website. The discovered patterns are usually represented as collections of pages or items that are frequently accessed by groups of users with common needs or interests. Such patterns can be used to better understand behavioral characteristics of visitors or user segments, to improve the organization and structure of the site, and to create a personalized experience for visitors by providing dynamic recommendations. The flexibility provided by Web usage mining can help enhance many of the approaches discussed in the preceding text and remedy many of their shortcomings. In particular, Web usage mining techniques, such as clustering, association rule mining, and navigational pattern mining, that rely on offline pattern discovery from user transactions can be used to improve the scalability of collaborative filtering when dealing with clickstream and e-commerce data.

The goal of personalization based on Web usage mining is to recommend a set of objects to the current (active) user, possibly consisting of links, ads, text, products, or services tailored to the user's perceived preferences as determined by the matching usage patterns. This task is accomplished by snatching the active user session (possibly in conjunction with previously stored profiles for that user) with the usage patterns discovered through Web usage mining. We call the usage patterns used in this context *aggregate usage profiles* because they provide an aggregate representation of the common activities or interests of groups of users. This process is performed by the recommendation engine which is the online component of the personalization system. If the data collection procedures in the system include the capability to track users across visits, then the recommendations can represent a longer-term view of user's potential interests based on the user's activity history within the site. If, on the other hand, aggregate profiles are derived only from user sessions (single visits) contained in log files, then the recommendations provide a "short-term" view of user's navigational interests. These recommended objects are added to the last page in the active session accessed by the user before that page is sent to the browser.

The overall process of Web personalization based on Web usage mining consists of three phases: data preparation and transformation, pattern discovery, and recommendation. Of these, only the latter phase is performed in real time. The data preparation phase transforms raw Web log files into transaction data that can be processed by data-mining tasks. This phase also includes data integration from multiple sources, such as backend databases, application servers, and site content. A variety of data-mining techniques can be applied to this transaction data in the pattern discovery phase, such as clustering, association rule mining, and sequential pattern discovery. The results of the mining phase are transformed into aggregate usage profiles, suitable for use in the recommendation phase. The recommendation engine considers the active user session in conjunction with the discovered patterns to provide personalized content.

In this chapter we present a comprehensive view of the personalization process based on Web usage mining. A generalized framework for this process is depicted in Figure 15.1 and Figure 15.2. We use this framework as our guide in the remainder of this chapter. We provide a detailed discussion of a host of Web usage mining activities necessary for this process, including the preprocessing and integration of data from multiple sources (Section 15. 1) and common pattern discovery techniques that are applied to the integrated usage data (Section 15.2.4). We then present a number of specific recommendation algorithms for combining the discovered knowledge with the current status of a user's activity in a Website to provide personalized content to a user. This discussion shows how pattern discovery techniques such as clustering, association rule mining, and sequential pattern discovery, performed on Web usage data, can be leveraged effectively as an integrated part of a Web personalization system (Section 15.3.2.3).

15.2 Data Preparation and Modeling

An important task in any data mining application is the creation of a suitable target data set to which data-mining algorithms are applied. This process may involve preprocessing the original data, integrating data from multiple sources, and transforming the integrated data into a form suitable for input into specific data-mining operations. Collectively, we refer to this process as data preparation.

The data preparation process is often the most time-consuming and computationally intensive step in the knowledge discovery process. Web usage mining is no exception: in fact, the data preparation process in Web usage mining often requires the use of especial algorithms and heuristics not commonly

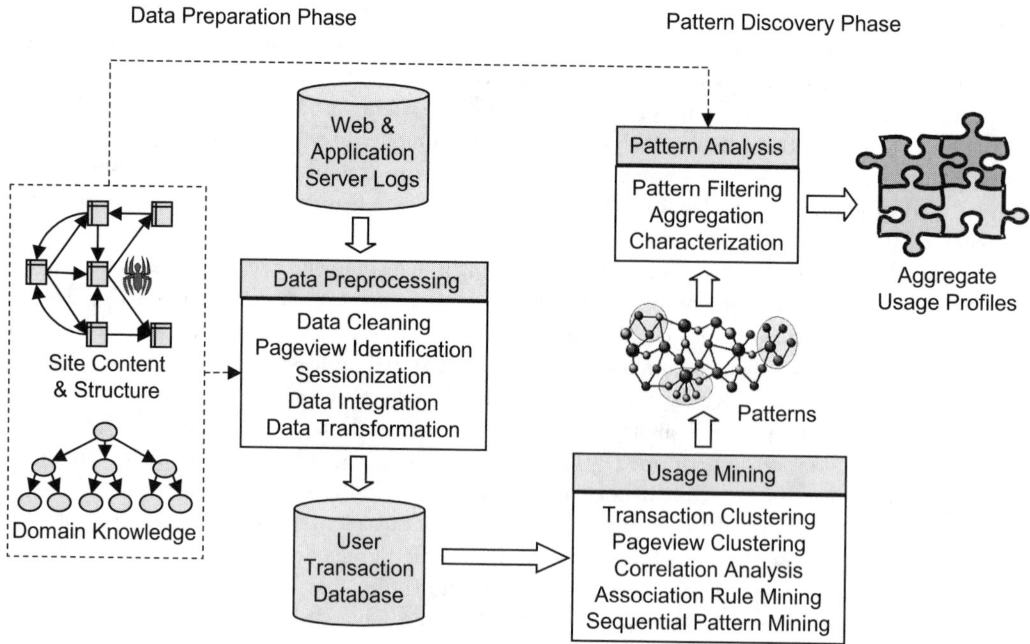

FIGURE 15.1 The offline data preparation and pattern discovery components.

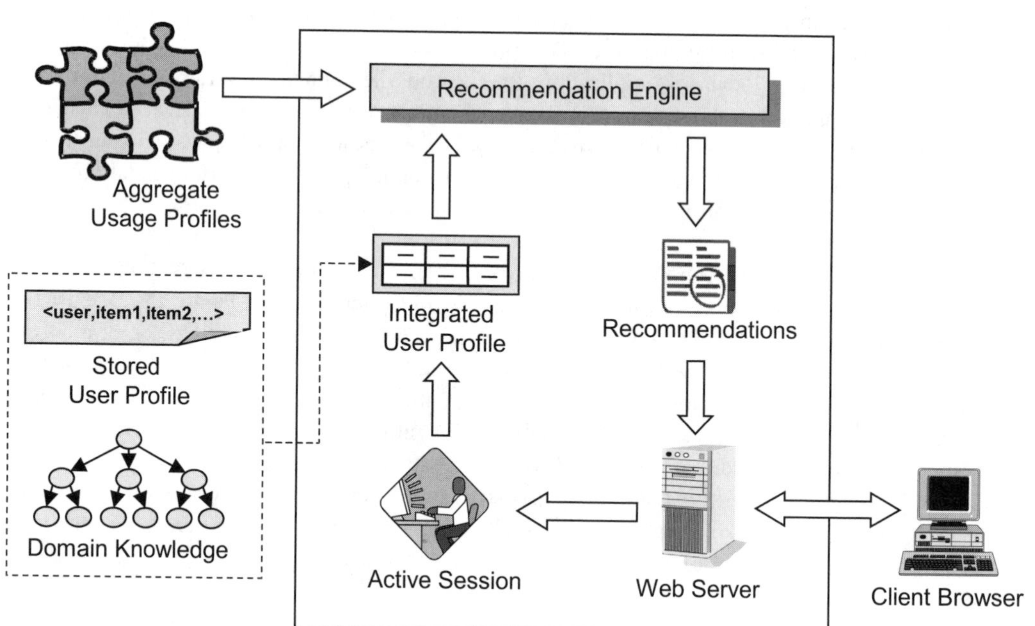

FIGURE 15.2 The online personalization component.

employed in other domains. This process is critical to the successful extraction of useful patterns from the data. In this section we discuss some of the issues and concepts related to data modeling and preparation in Web usage mining. Although this discussion is in the general context of Web usage analysis, we are focused especially on the factors that have been shown to greatly affect the quality and usability of the discovered usage patterns for their application in Web personalization.

15.2.1 Sources and Types of Data

The primary data sources used in Web usage mining are the server log files, which include Web server access logs and application server logs. Additional data sources that are also essential for both data preparation and pattern discovery include the site files and metadata, operational databases, application templates, and domain knowledge. Generally speaking, the data obtained through these sources can be categorized into four groups [Cooley et al., 1999; Srivastava et al., 2000]:

15.2.1.1 Usage Data

The log data collected automatically by the Web and application servers represents the fine-grained navigational behavior of visitors. Depending on the goals of the analysis, this data needs to be transformed and aggregated at different levels of abstraction.

In Web usage mining, the most basic level of data abstraction is that of a pageview. Physically, a pageview is an aggregate representation of a collection of Web objects contributing to the display on a user's browser resulting from a single user action (such as a clickthrough). These Web objects may include multiple pages (such as in a frame-based site), images, embedded components, or script and database queries that populate portions of the displayed page (in dynamically generated sates). Conceptually, each pageview represents a specific "type" of user activity on the site, e.g., reading a news article, browsing the results of a search query, viewing a product page, adding a product to the shopping cart, and so on.

On the other hand, at the user level, the most basic level of behavioral abstraction is that of a server session (or simply a session). A session (also commonly referred to as a "visit") is a sequence of pageviews by a single user during a single visit. The notion of a session can be further abstracted by selecting a subset of pageviews in the session that are significant or relevant for the analysis tasks at hand. We shall refer to such a semantically meaningful subset of pageviews as a transaction (also referred to as an episode according to the W3C Web Characterization Activity [W3C]). It is important to note that a transaction does not refer simply to product purchases, but it can include a variety of types of user actions as captured by different pageviews in a session.

15.2.1.2 Content Data

The content data in a site is the collection of objects and relationships that are conveyed to the user. For the most part, this data is comprised of combinations of textual material and images. The data sources used to deliver or generate this data include static HTML/XML pages, images, video clips, sound files, dynamically generated page segments from scripts or other applications, and collections of records from the operational databases. The site content data also includes semantic or structural metadata embedded within the site or individual pages, such as descriptive keywords, document attributes, semantic tags, or HTTP variables.

Finally, the underlying domain ontology for the site is also considered part of the content data. The domain ontology may be captured implicitly within the site or it may exist in some explicit form. The explicit representations of domain ontologies may include conceptual hierarchies over page contents, such as product categories, structural hierarchies represented by the underlying file and directory structure in which the site content is stored, explicit representations of semantic content and relationships via an ontology language such as RDF, or a database schema over the data contained in the operational databases.

15.2.1.3 Structure Data

The structure data represents the designer's view of the content organization within the site. This organization is captured via the interpage linkage structure among pages, as reflected through hyperlinks. The structure data also includes the intrapage structure of the content represented in the arrangement of HTML or XML tags within a page. For example, both HTML and XML documents can be represented as tree structures over the space of tags in the page.

The structure data for a site is normally captured by an automatically generated "site map" that represents the hyperlink structure of the site. A site mapping tool must have the capability to capture and represent the inter- and intra-pageview relationships. This necessity becomes most evident in a frame-based site where portions of distinct pageviews may represent the same physical page. For dynamically generated pages, the site mapping tools must either incorporate intrinsic knowledge of the underlying applications and scripts, or must have the ability to generate content segments using a sampling of parameters passed to such applications or scripts.

15.2.1.4 User Data

The operational databases for the site may include additional user profile information. Such data may include demographic or other identifying information on registered users, user ratings on various objects such as pages, products, or movies, past purchase or visit histories of users, as well as other explicit or implicit representations of a user's interests.

Obviously, capturing such data would require explicit interactions with the users of the site. Some of this data can be captured anonymously, without any identifying user information, so long as there is the ability to distinguish among different users. For example, anonymous information contained in client-side cookies can be considered a part of the users' profile information and can be used to identify repeat visitors to a site. Many personalization applications require the storage of prior user profile information. For example, collaborative filtering applications usually store prior ratings of objects by users, though such information can be obtained anonymously as well.

15.2.2 Usage Data Preparation

The required high-level tasks in usage data preprocessing include data cleaning, pageview identification, user identification, session identification (or sessionization), the inference of missing references due to caching, and transaction (episode) identification. We provide a brief discussion of some of these tasks below; for a more detailed discussion see Cooley [2000] and Cooley et al. [1999].

Data cleaning is usually site-specific and involves tasks such as removing extraneous references to embedded objects, graphics, or sound files, and removing references due to spider navigations. The latter task can be performed by maintaining a list of known spiders and through heuristic identification of spiders and Web robots [Tan and Kumar, 2002]. It may also be necessary to merge log files from several Web and application servers. This may require global synchronization across these servers. In the absence of shared embedded session IDs, heuristic methods based on the "referrer" field in server logs along with various sessionization and user identification methods (see the following text) can be used to perform the merging.

Client- or proxy-side caching can often result in missing access references to those pages or objects that have been cached. Missing references due to caching can be heuristically inferred through path completion, which relies on the knowledge of site structure and referrer information from server logs [Cooley et al., 1999]. In the case of dynamically generated pages, form-based applications using the HTTP POST method result in all or part of the user input parameter not being appended to the URL accessed by the user (though, in the latter case, it is possible to recapture the user input through packet sniffers on the server side).

Identification of pageviews is heavily dependent on the intrapage structure of the site, as well as on the page contents and the underlying site domain knowledge. For a single frame site, each HTML file has a *one-to-one* correlation with a pageview. However, for multiframed sites, several files make up a given pageview. Without detailed site structure information, it is very difficult to infer pageviews from a Web server log. In addition, it may be desirable to consider pageviews at a higher level of aggregation, where each pageview represents a collection of pages or objects — for example, pages related to the same concept category.

Not all pageviews are relevant for specific raining tasks, and among the relevant pageviews some may be more significant than others. The significance of a pageview may depend on usage, content and structural characteristics of the site, as well as on prior domain knowledge (possibly specified by the site designer and the data analyst). For example, in an e-commerce site, pageviews corresponding to product-oriented events (e.g., shopping cart changes or product information views) may be considered more significant than others. Similarly, in a site designed to provide content, content pages may be weighted higher than navigational pages. In order to provide a flexible framework for a variety of data-mining activities, a number of attributes must be recorded with each pageview. These attributes include the pageview ID (normally a URL uniquely representing the pageview), duration, static pageview type (e.g., information page, product view, or index page), and other metadata, such as content attributes.

The analysis of Web usage does not require knowledge about a user's identity. However, it is necessary to distinguish among different users. In the absence of registration and authentication mechanisms, the most widespread approach to distinguishing among users is with client-side cookies. Not all sites, however, employ cookies, and due to abuse by some organizations and because of privacy concerns on the part of many users, client-side cookies are sometimes disabled. IP addresses alone are not generally sufficient for mapping log entries onto the set of unique users. This is mainly due the proliferation of ISP proxy servers that assign rotating IP addresses to clients as they browse the Web. It is not uncommon, for instance, to find a substantial percentage of IP addresses recorded in server logs of a high-traffic site as belonging to America Online proxy server or other major ISPs. In such cases, it is possible to more accurately identify unique users through combinations IP addresses and other information such as user agents, operating systems, and referrers [Cooley et al., 1999].

Since a user may visit a site more than once, the server logs record multiple sessions for each user. We use the phrase user activity log to refer to the sequence of logged activities belonging to the same user. Thus, sessionization is the process of segmenting the user activity log of each user into sessions. Websites without the benefit of additional authentication information from users and without mechanisms such as embedded session IDs must rely on heuristics methods for sessionization. A sessionization heuristic is a method for performing such a segmentation on the basis of assumptions about users' behavior or the site characteristics.

The goal of a heuristic is the reconstruction of the real sessions, where a real session is the actual sequence of activities performed by one user during one visit to the site. We denote the "conceptual" set of real sessions by $\mathfrak{R}$. A sessionization heuristic h attempts to map $\mathfrak{R}$ into a set of constructed sessions, which we denote as $C \equiv C_h$. For the ideal heuristic, h^*, we have $C \equiv C_{h^*} = \mathfrak{R}$. Generally, sessionization heuristics fall into two basic categories: time-oriented or structure-oriented. Time-oriented heuristics apply either global or local time-out estimates to distinguish between consecutive sessions, while structure-oriented heuristics use either the static site structure or the implicit linkage structure captured in the referrer fields of the server logs. Various heuristics for sessionization have been identified and studied [Cooley et al., 1999]. More recently, a formal framework for measuring the effectiveness of such heuristics has been proposed [Spiliopoulou et al., 2003], and the impact of different heuristics on various Web usage mining tasks has been analyzed [Berendt et al., 2002b].

Finally, transaction (episode) identification can be performed as a final preprocessing step prior to pattern discovery in order to focus on the relevant subsets of pageviews in each user session. As noted earlier, this task may require the automatic or semiautomatic classification of pageviews into different functional types or into concept classes according to a domain ontology. In highly dynamic sites, it may also be necessary to map pageviews within each session into "service-base" classes according to a concept hierarchy over the space of possible parameters passed to script or database queries [Berendt and Spiliopoulou, 2000]. For example, the analysis may ignore the quantity and attributes of an items added to the shopping cart and focus only on the action of adding the item to the cart.

The above preprocessing tasks ultimately result in a set of n pageviews, $P = \{p_1, p_2, \cdots, p_n\}$, and a set of m user transactions, $T = \{t_1, t_2, \cdots, t_m\}$, where each $t_i \in T$ is a subset of P. Conceptually, we can view each transaction t as an l-length sequence of ordered pairs:

$$t = \langle (p_1^t, w(p_1^t)), (p_2^t, w(p_2^t)), \cdots, (p_l^t, w(p_l^t)) \rangle$$

where each $p_i^t = p_j$ for some $j \in \{1, \cdots, n\}$, and $w\langle p_i^t \rangle$ is the weight associated with pageview p_i^t in the transaction t.

The weights can be determined in a number of ways, in part based on the type of analysis or the intended personalization framework. For example, in collaborative filtering applications, such weights may be determined based on user ratings of items. In most Web usage mining tasks, the focus is generally on anonymous user navigational activity where the primary sources of data are server logs. This allows us to choose two types of weights for pageviews: weights can be binary, representing the existence or nonexistence of a pageview in the transaction, or they can be a function of the duration of the pageview in the user's session. In the case of time durations, it should be noted that usually the time spent by a user on the last pageview in the session is not available. One commonly used option is to set the weight for the last pageview to be the mean time duration for the page taken across all sessions in which the pageview does not occur as the last one.

Whether or not the user transactions are viewed as sequences or as sets (without taking ordering information into account) is also dependent on the goal of the analysis and the intended applications. For sequence analysis and the discovery of frequent navigational patterns, one must preserve the ordering information in the underlying transaction. On the other hand, for clustering tasks as well as for collaborative filtering based on *k*NN and association rule discovery, we can represent each user transaction as a vector over the n-dimensional space of pageviews, where dimension values are the weights of these pageviews in the corresponding transaction. Thus given the transaction t above, the n-dimensional transaction vector $\vec{t}$ is given by:

$$\vec{t} = \langle w^t_{p_1}, w^t_{p_2}, \cdots, w^t_{p_n} \rangle$$

where each $w^t_{p_j} = w(p^t_i)$, for some $i \in \{1, \cdots, n\}$, in case p_j appears in the transaction t, and $w^t_{p_j} = 0$, otherwise. For example, consider a site with 6 pageviews A, B, C, D, E, and F. Assuming that the pageview weights associated with a user transaction are determined by the number of seconds spent on them, a typical transaction vector may look like: ⟨11, 0, 22, 5,127, 0⟩. In this case, the vector indicates that the user spent 11 sec on page A, 22 sec on page C, 5 sec on page D, and 127 sec on page E. The vector also indicates that the user did not visit pages B and F during this transaction.

Given this representation, the set of all m user transactions can be conceptually viewed as an $m \times n$ transaction–pageview matrix that we shall denote by TP. This transaction–pageview matrix can then be used to perform various data-mining tasks. For example, similarity computations can be performed among the transaction vectors (rows) for clustering and *k*NN neighborhood formation tasks, or an association rule discovery algorithm, such as Apriory, can be applied (with pageviews as items) to find frequent itemsets of pageviews.

15.2.3 Postprocessing of User Transactions Data

In addition to the aforementioned preprocessing steps leading to user transaction matrix, there are a variety of transformation tasks that can be performed on the transaction data. Here, we highlight some of data transformation tasks that are likely to have an impact on the quality and actionability of the discovered patterns resulting from raining algorithms. Indeed, such postprocessing transformations on session or transaction data have been shown to result in improvements in the accuracy of recommendations produced by personalization systems based on Web usage mining [Mobasher et al., 2001b].

15.2.3.1 Significance Filtering

Using binary weights in the representation of user transactions is often desirable due to efficiency requirements in terms of storage and computation of similarity coefficients among transactions. However, in this context, it becomes more important to determine the significance of each pageview or item access. For example, a user may access an item p only to find that he or she is not interested in that item, subsequently backtracking to another section of the site. We would like to capture this behavior by discounting the access to p as an insignificant access. We refer to the processing of removing page or item requests that are deemed insignificant as *significance filtering*.

The significance of a page within a transaction can be determined manually or automatically. In the manual approach, the site owner or the analyst is responsible for assigning significance weights to various pages or items. This is usually performed as a global mapping from items to weights, and thus the significance of the pageview is not dependent on a specific user or transaction. More commonly, a function of pageview duration is used to automatically assign signficance weights. In general, though, it is not sufficient to filter out pageviews with small durations because the amount of time spent by users on a page is not merely based on the user's interest on the page. The page duration may also be dependent on the characteristics and the content of the page. For example, we would expect that users spend far less time on navigational pages than they do on content or product-oriented pages.

Statistical significance testing can help capture some of the semantics illustrated above. The goal of significance filtering is to eliminate irrelevant items with time duration significantly below a certain threshold in transactions. Typically, statistical measures such as mean and variance can be used to systematically define the threshold for significance filtering.

In general, it can be observed that the distribution of access frequencies as a function of the amount of time spent on a given pageview is characterized by a log-normal distribution. For example, Figure 15.3 (left) shows the distribution of the number of transactions with respect to time duration for a particular pageview in a typical Website. Figure 15.3 (right) shows the distribution plotted as a function of time in a log scale. The log normalization can be observed to produce a Gaussian distribution. After this transformation, we can proceed with standard significance testing: the weight associated with an item in a transaction will be considered to be 0 if the amount of time spent on that item is significantly below the mean time duration of the item across all user transactions. The significance of variation from the mean is usually measured in terms of multiples of standard deviation. For example, in a given transaction t, if the amount of time spent on a pageview p is 1.5 to 2 standard deviations lower than the mean duration for p across all transactions, then the weight of p in transaction t might be set to 0. In such a case, it is likely that the user was either not interested in the contents of p, or mistakenly navigated to p and quickly left the page.

15.2.3.2 Normalization

There are also some advantages in rising the fully weighted representation of transaction vectors (based on time durations). One advantage is that for many distance- or similarity-based clustering algorithms, more granularity in feature weights usually leads to more accurate results. Another advantage is that, because relative time durations are taken into account, the need for performing other types of transformations, such as significance filtering, is greatly reduced.

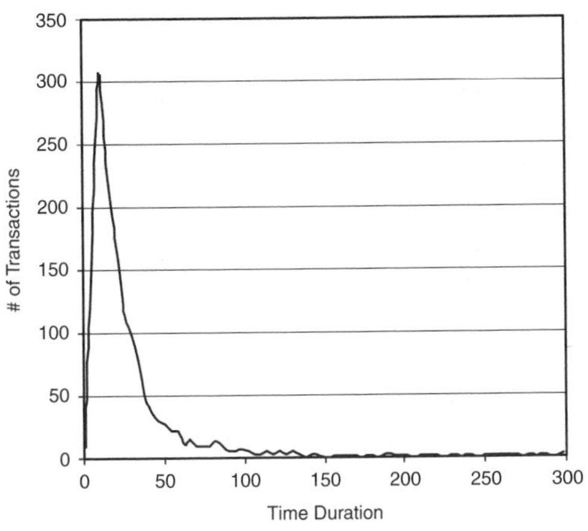

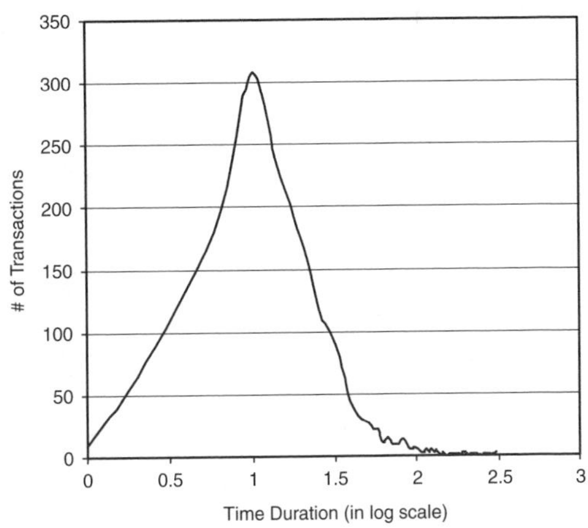

FIGURE 15.3 Distribution of pageview durations: raw-time scale (left), log-time scale (right).

However, raw time durations may not be an appropriate measure for the significance of a pageview. This is because a variety of factors, such as structure, length, and the type of pageview, as well as the user's interests in a particular item, may affect the amount of time spent on that item. Appropriate weight normalization can play an essential role in correcting for these factors.

Generally, two types of weight normalization are applied to user transaction vectors: normalization across pageviews in a single transaction and normalization of a pageview weights across all transactions. We call these transformations transaction normalization and pageview normalization, respectively. Pageview normalization is useful in capturing the relative weight of a pageview for a user with respect to the weights of the same pageview for all other users. On the other hand, transaction normalization captures the importance of a pageview to a particular user relative to the other items visited by that user in the same transaction. The latter is particularly useful in focusing on the "target" pages in the context of short user histories.

15.2.4 Data Integration from Multiple Sources

In order to provide the most effective framework for pattern discovery and analysis, data from a variety of sources must be integrated. Our earlier discussion already alluded to the necessity of considering the content and structure data in a variety of preprocessing tasks such as pageview identification, sessionization, and the inference of missing data. The integration of content, structure, and user data in other phases of the Web usage mining and personalization processes may also be essential in providing the ability to further analyze and reason about the discovered patterns, derive more actionable knowledge, and create more effective personalization tools.

For example, in e-commerce applications, the integration of both user data (e.g., demographics, ratings, purchase histories) and product attributes from operational databases is critical. Such data, used in conjunction with usage data in the mining process, can allow for the discovery of important business intelligence metrics such as customer conversion ratios and lifetime values. On the other hand, the integration of semantic knowledge from the site content or domain ontologies can be used by personalization systems to provide more useful recommendations. For instance, consider a hypothetical site containing information about movies that employs collaborative filtering on movie ratings or pageview transactions to give recommendations. The integration of semantic knowledge about movies (possibly extracted from site content) can allow the system to recommend movies, not just based on similar ratings or navigation patterns but also perhaps based on similarities in attributes such as movie genres or commonalities in casts or directors.

One direct source of semantic knowledge that can be integrated into the mining and personalization processes is the collection of content features associated with items or pageviews on a Website. These features include keywords, phrases, category names, or other textual content embedded as meta information. Content preprocessing involves the extraction of relevant features from text and metadata. Metadata extraction becomes particularly important when dealing with product-oriented pageviews or those involving nontextual content. In order to use features in similarity computations, appropriate weights must be associated with them. Generally, for features extracted from text, we can use standard techniques from information retrieval and filtering to determine feature weights [Frakes and Baeza-Yates, 1992]. For instance, a commonly used feature-weighting scheme is tf.idf, which is a function of the term frequency and inverse document frequency.

More formally, each pageview p can be represented as a k-dimensional feature vector, where k is the total number of extracted features from the site in a global dictionary. Each dimension in a feature vector represents the corresponding feature weight within the pageview. Thus, the feature vector for a pageview p is given by:

$$p = \langle fw(p, f_1), fw(p, f_2), \cdots, fw(p, f_k) \rangle$$

where $fw(p, f_j)$, is the weight of the *j*th feature in pageview $p \in P$, for $1 \leq j \leq k$. For features extracted from textual content of pages, the feature weight is usually the normalized tf.idf value for the term. In order to combine feature weights from metadata (specified externally) and feature weights from the text content, proper normalization of those weights must be performed as part of preprocessing.

Conceptually, the collection of these vectors can be viewed as a $n \times k$ pageview-feature matrix in which each row is a feature vector corresponding to one of the *n* pageviews in *P*. We shall call this matrix PF. The feature vectors obtained in this way are usually organized into an inverted file structure containing a dictionary of all extracted features and posting files for each feature specifying the pageviews in which the feature occurs along with its weight. This inverted file structure corresponds to the transpose of the matrix PF.

Further preprocessing on content features can be performed by applying text-mining techniques. This would provide the ability to filter the input to, or the output from, usage-mining algorithms. For example, classification of content features based on a concept hierarchy can be used to limit the discovered usage patterns to those containing pageviews about a certain subject or class of products. Similarly, performing clustering or association rule mining on the feature space can lead to composite features representing concept categories. The mapping of features onto a set of concept labels allows for the transformation of the feature vectors representing pageviews into concept vectors. The concept vectors represent the semantic concept categories to which a pageview belongs, and they can be viewed at different levels of abstraction according to a concept hierarchy (either preexisting or learned through machine-learning techniques). This transformation can be useful both in the semantic analysis on the data and as a method for dimensionality reduction in some data-raining tasks, such as clustering.

A direct approach for the integration of content and usage data for Web usage mining tasks is to transform user transactions, as described earlier, into "content-enhanced" transactions containing the semantic features of the underlying pageviews. This process, performed as part of data preparation, involves mapping each pageview in a transaction to one or more content features. The range of this mapping can be the full feature space or the concept space obtained as described above. Conceptually, the transformation can be viewed as the multiplication of the transaction–pageview matrix TP (described in Section 15.2.2) with the pageview–feature matrix PF. The result is a new matrix $\text{TF} = \{t_1', t_2', \cdots, t_m'\}$, where each t_i' is a *k*-dimensional vector over the feature space. Thus, a user transaction can be represented as a content feature vector, reflecting that user's interests in particular concepts or topics. A variety of data-raining algorithms can then be applied to this transformed transaction data.

The above discussion focused primarily on the integration of content and usage data for Web usage mining. However, as noted earlier, data from other sources must also be considered as part of an integrated framework. Figure 15.4 shows the basic elements of such a framework.

The content analysis module in this framework is responsible for extracting and processing linkage and semantic information from pages. The processing of semantic information includes the steps described above for feature extraction and concept mapping. Analysis of dynamic pages may involve (partial) generation of pages based on templates, specified parameters, or database queries based on the information captured from log records. The outputs from this module may include the site map capturing the site topology as well as the site dictionary and the inverted file structure used for content analysis and integration.

The site map is used primarily in data preparation (e.g., in pageview identification and path completion). It may be constructed through content analysis or the analysis of usage data (using the referrer information in log records). Site dictionary provides a mapping between pageview identifiers (for example, URLs) and content or structural information on pages; it is used primarily for "content labeling" both in sessionized usage data as well as the integrated e-commerce data. Content labels may represent conceptual categories based on sets of features associated with pageviews.

The data integration module is used to integrate sessionized usage data, e-commerce data (from application servers), and product or user data from databases. User data may include user profiles, demographic information, and individual purchase activity. E-commerce data includes various product-oriented events, including shopping cart changes, purchase information, impressions, clickthroughs, and

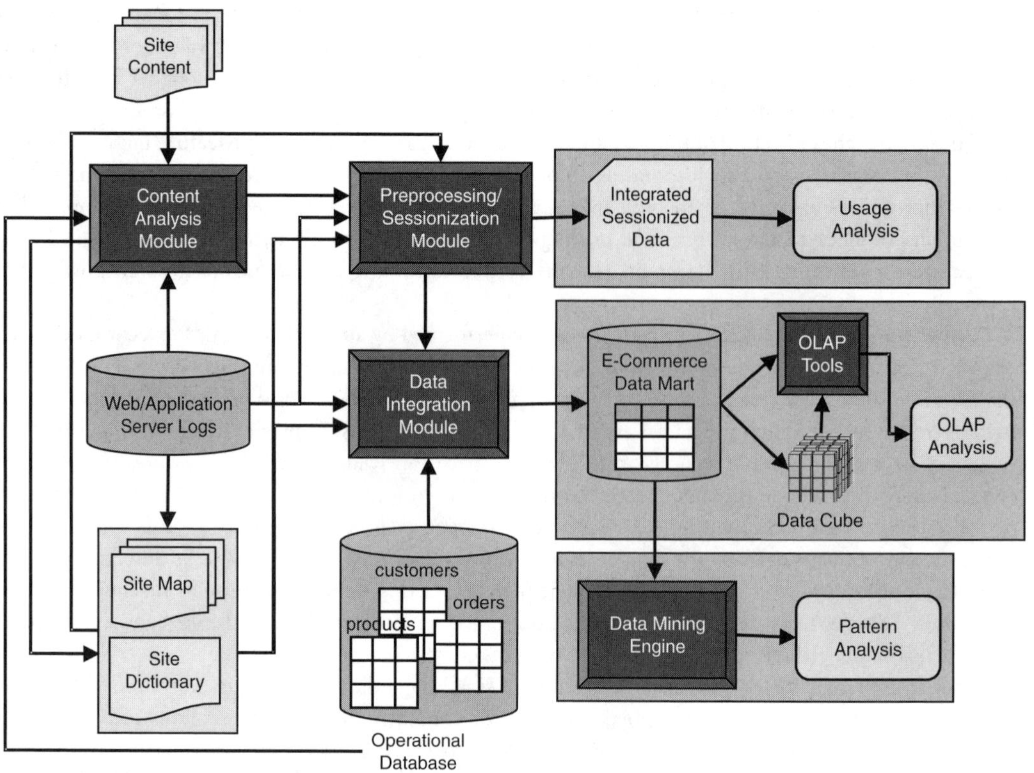

FIGURE 15.4 An integrated framework for Web usage analysis.

other basic metrics primarily used for data transformation and loading mechanism of the Data Mart. The successful integration of this type of e-commerce data requires the creation of a site-specific "event model" based on which subsets of a user's clickstream are aggregated and mapped to specific events such as the addition of a product to the shopping cart. Product attributes and product categories, stored in operational databases, can also be used to enhance or expand content features extracted from site files.

The e-commerce data mart is a multidimensional database integrating data from a variety of sources, and at different levels of aggregation. It can provide precomputed e-metrics along multiple dimensions, and is used as the primary data source in OLAP analysis, as well as in data selection for a variety of data-mining tasks (performed by the data-mining engine).

We discuss different types and levels of analysis that can be performed based on this basic framework in the next section.

15.3 Pattern Discovery from Web Usage Data

15.3.1 Levels and Types of Analysis

As shown in Figure 15.4, different kinds of analysis can be performed on the integrated usage data at different levels of aggregation or abstraction. The types and levels of analysis, naturally, depend on the ultimate goals of the analyst and the desired outcomes.

For instance, even without the benefit of an integrated e-commerce data mart, statistical analysis can be performed on the preprocessed session or transaction data. Indeed, static aggregation (reports) constitutes the most common form of analysis. In this case, data is aggregated by predetermined units such as days, sessions, visitors, or domains. Standard statistical techniques can be used on this data to gain knowledge about visitor behavior. This is the approach taken by most commercial tools available for Web log analysis (however, most such tools do not perform all of the necessary preprocessing tasks described earlier, thus resulting in erroneous or misleading outcomes). Reports based on this type of analysis may include information about most frequently accessed pages, average view time of a page, average length of a path through a site, common entry and exit points, and other aggregate measure.

The drawback of this type of analysis is the inability to "dig deeper" into the data or find hidden patterns and relationships. Despite a lack of depth in the analysis, the resulting knowledge can be potentially useful for improving the system performance and providing support for marketing decisions. The reports give quick overviews of how a site is being used and require minimal disk space or processing power. Furthermore, in the past few years, many commercial products for log analysis have incorporated a variety of data-mining tools to discover deeper relationships and hidden patterns in the usage data.

Another form of analysis on integrated usage data is Online Analytical Processing (OLAP). OLAP provides a more integrated framework for analysis with a higher degree of flexibility. As indicated in Figure 15.4, the data source for OLAP analysis is a multidimensional data warehouse which integrates usage, content, and e-commerce data at different levels of aggregation for each dimension. OLAP tools allow changes in aggregation levels along each dimension during the analysis.

Indeed, the server log data itself can be stored in a multidimensional data structure for OLAP analysis [Zaiane et al., 1998]. Analysis dimensions in such a structure can be based on various fields available in the log files, and may include time duration, domain, requested resource, user agent, referrers, and so on. This allows the analysis to be performed, for example, on portions of the log related to a specific time interval, or at a higher level of abstraction with respect to the URL path structure. The integration of e-commerce data in the data warehouse can enhance the ability of OLAP tools to derive important business intelligence metrics. For example, in Buchner and Mulvenna [1999], an integrated Web log data cube was proposed that incorporates customer and product data, as well as domain knowledge such as navigational templates and site topology.

OLAP tools, by themselves, do not automatically discover usage patterns in the data. In fact, the ability to find patterns or relationships in the data depends solely on the effectiveness of the OLAP queries performed against the data warehouse. However, the output from this process can be used as the input for a variety of data-mining algorithms. In the following sections we focus specifically on various data-mining and pattern discovery techniques that are commonly performed on Web usage data, and we will discuss some approaches for using the discovered patterns for Web personalization.

15.3.2 Data-Mining Tasks for Web Usage Data

We now focus on specific data-mining and pattern discovery tasks that are often employed when dealing with Web usage data. Our goal is not to give detailed descriptions of all applicable data-mining techniques but to provide some relevant background information and to illustrate how some of these techniques can be applied to Web usage data. In the next section, we present several approaches to leverage the discovered patterns for predictive Web usage running applications such as personalization.

As noted earlier, preprocessing and data transformation tasks ultimately result in a set of n pageviews, $P = \{p_1, p_2, \cdots, p_n\}$ and a set of m user transactions, $T = \{t_1, t_2, \cdots, t_m\}$, where each $t_i \in T$ is a subset of P. Each transaction t is an l-length sequence of ordered pairs: $t = \langle (p_1^t, w(p_1^t)), (p_2^t, w(p_2^t)), \cdots (p_l^t, w(p_l^t)) \rangle$, where each $p_i^t = p_j$ for some $j \in \{1, \cdots, n\}$, and $w(p_i^t)$ is the weight associated with pageview p_i^t in the transaction t.

Given a set of transactions as described above, a variety of unsupervised knowledge discovery techniques can be applied to obtain patterns. Techniques such as clustering of transactions (or sessions) can lead to the discovery of important user or visitor segments. Other techniques such as item (e.g., pageview) clustering, association rule mining [Agarwal et al., 1999; Agrawal and Srikant, 1994], or sequential pattern discovery [Agrawal and Srikant, 1995] can be used to find important relationships among items based on the navigational patterns of users in the site. In the cases of clustering and association rule discovery, generally, the ordering relation among the pageviews is not taken into account; thus a transaction is viewed as a set (or, more generally, as a bag) of pageviews $s_t = \left\{p_i^t \middle| 1 \le i \le l \text{ and } w\left(p_i^t\right) = 1\right\}$ In the case of sequential patterns, however, we need to preserve the ordering relationship among the pageviews within transactions in order to effectively model users' navigational patterns.

15.3.2.1 Association Rules

Association rules capture the relationships among items based on their patterns of cooccurrence across transactions (without considering the ordering of items). In the case of Web transactions, association rules capture relationships among pageviews based on the navigational patterns of users. Most common approaches to association discovery are based on the Apriori algorithm (Agrawal and Srikant, 1994, 1995] that follows a generate-and-test methodology. This algorithm finds groups of items (pageviews appearing in the preprocessed log) occurring frequently together in many transactions (i.e., satisfying a user-specified minimum support threshold). Such groups of items are referred to as frequent itemsets.

Given a transaction T and a set $I = \{I_1, I_2, \cdots, I_k\}$ of frequent itemsets over T, the *support* of an itemset $I_i \in I$ is defined as

$$\sigma(I_i) = \frac{\left|\{t \in T : I_i \subseteq t\}\right|}{T}.$$

An important property of support in the Apriori algorithm is its downward closure: if an itemset does not satisfy the minimum support criteria, then neither do any of its supersets. This property is essential for pruning the search space during each iteration of the Apriori algorithm.

Association rules that satisfy a minimum confidence threshold are then generated from the frequent itemsets. An association rule r is an expression of the form $X \Rightarrow Y(\sigma_r, \alpha_r)$, where X and Y are itemsets, $\sigma_r = \sigma(X \cup Y)$ is the support of $X \cup Y$ representing the probability that X and Y occur together in a transaction. The confidence for the rule r, α_r is given by $\sigma(X \cup Y) / \sigma(X)$ and represents the conditional probability that Y occurs in a transaction given that X has occured in that transaction.

The discovery of association rules in Web transaction data has many advantages. For example, a high-confidence rule such as {special-offers/ , /products/software/} ⇒ {shopping-cart/} might provide some indication that a promotional campaign on software products is positively affecting online sales. Such rules can also be used to optimize the structure of the site. For example, if a site does not provide direct linkage between two pages A and B, the discovery of a rule {A} ⇒ {B} would indicate that providing a direct hyperlink might aid users in finding the intended information.

The result of association rule mining can be used in order to produce a model for recommendation or personalization systems [Fu et al., 2000; Lin et al., 2002; Mobasher et al., 2001a; Sarwar et al., 2000b]. The top-*N* recommender systems proposed in [Sarwar et al., 2000b] uses the association rules for making recommendations. First, all association rules are discovered on the purchase information. Customer's historical purchase information then is matched against the left-hand side of the rule in order to find all rules supported by a customer. All right-hand-side items from the supported rules are sorted by confidence and the first *N* highest-ranked items are selected as recommendation set. One problem for association rule recommendation systems is that a system cannot give any recommendations when the data set is sparse. In Fu et al. [2000], two potential solutions to this problem were proposed. The first solution is to rank all discovered rules calculated by the degree of intersection between the left-hand side of the rule and a user's active session and then to generate the top *k* recommendations. The second solution is to utilize collaborative filtering technique: the system finds "close neighbors" who have similar interest to a target user and makes recommendations based on the close neighbor's history. In Lin et al. [2002], a collaborative recommendation system was presented using association rules. The proposed mining algorithm finds an appropriate number of rules for each target user by automatically selecting the minimum support. The recommendation engine generates association rules for each user among both users and items. Then it gives recommendations based on user association if a user minimum support is greater than a threshold. Otherwise, it uses article association.

In Mobasher et al. [2001a], a scalable framework for recommender systems using association rule mining was proposed. The recommendation algorithm uses an efficient data structure for storing frequent itemsets and produces recommendations in real time without the need to generate all association rules from frequent itemsets. We discuss this recommendation algorithm based on association rule mining in more detail in Section 15.3.2.3.

A problem with using a global minimum support threshold in association rule mining is that the discovered patterns will not include "rare" but important items that may not occur frequently in the transaction data. This is particularly important when dealing with Web usage data; it is often the case that references to deeper content or product-oriented pages occur far less frequently than those of top-level navigation-oriented pages. Yet, for effective Web personalization, it is important to capture patterns and generate recommendations that contain these items. Liu et al. [1999] proposed a mining method with multiple minimum supports that allows users to specify different support values for different items. In this method, the support of an itemset is defined as the minimum support of all items contained in the itemset. The specification of multiple minimum supports allows frequent itemsets to potentially contain rare items that are nevertheless deemed important. It has been shown that the use of multiple support association rules in the context of Web personalization can be useful in dramatically increasing the coverage (recall) of recommendations while maintaining a reasonable precision [Mobasher et al., 2001a].

15.3.2.2 Sequential and Navigational Patterns

Sequential patterns (SPs) in Web usage data capture the Web page trails that are often visited by users in the order that they were visited. Sequential patterns are those sequences of items that frequently occur in a sufficiently large proportion of transactions. A sequence $\langle s_1, s_2, \cdots, s_n \rangle$ occurs in a transaction $t = \langle p_1, p_2, \cdots, p_m \rangle$ (where $n \le m$) if there exist n positive integers $1 \le a_1 < a_2 < \cdots < a_n \le m$ and $s_i = p_{ai}$ for all i. We say that $\langle cs_1, cs_2, \cdots, cs_n \rangle$ is a contiguous sequence in t if there exists an integer $0 \le b \le m - n$, and $cs_i = p_{b+i}$ for all $i = 1$ to n. In a contiguous sequential pattern (CSP), each pair of adjacent elements, s_i and s_{i+1}, must appear consecutively in a transaction t which supports the pattern, while a sequential pattern can represent noncontiguous frequent sequences in the underlying set of transactions.

Given a transaction set T and a set $S = \{S_1, S_2, \cdots, S_n\}$ of frequent sequential (respectively, contiguous sequential) pattern over T, the support of each S_i is defined as follows:

$$\sigma(S_i) = \frac{|\{t \in T : S_i \text{ is (contiguous) subsequence of } t\}|}{|T|}$$

The confidence of the rule $X \Rightarrow Y$, where X and Y are (contiguous) sequential patterns, is defined as

$$\alpha(X \Rightarrow Y) = \frac{\sigma(X o Y)}{\sigma(X)},$$

where o denotes the concatenation operator. Note that the support thresholds for SPs and CSPs also satisfy downward closure property, i.e., if a (contiguous) sequence of items, S, has any subsequence that does not satisfy the minimum support criteria, then S does not have minimum support. The Apriori algorithm used in association rule mining can also be adopted to discover sequential and contiguous sequential patterns. This is normally accomplished by changing the definition of support to be based on the frequency of occurrences of subsequences of items rather than subsets of items [Agrawal and Srikant, 1995].

In the context of Web usage data, CSPs can be used to capture frequent navigational paths among user trails [Spiliopoulou and Faulstich, 1999; Schechter et al., 1998]. In contrast, items appearing in SPs, while preserving the underlying ordering, need not be adjacent, and thus they represent more general navigational patterns within the site. Frequent item sets, discovered as part of association rule mining, represent the least restrictive type of navigational patterns because they focus on the presence of items rather than the order in which they occur within the user session.

The view of Web transactions as sequences of pageviews allows us to employ a number of useful and well-studied models that can be used to discover or analyze user navigation patterns. On such approach is to model the navigational activity in the Website as a Markov chain. In general, a Markov model is characterized by a set of states $\{s_1, s_2, \cdots, s_n\}$ and a transition probability matrix

$$\{p_{1,1}, \ldots, p_{1,n}, \ldots, p_{2,1}, \ldots, p_{2,n}, \ldots, p_{n,1}, \ldots, p_{n,n}\}$$

where $p_{i,j}$ represents the probability of a transition from state s_i to state s_j.

Markov models are especially suited for predictive modeling based on contiguous sequences of events. Each state represents a contiguous subsequence of prior events. The order of the Markov model corresponds to the number of prior events used in predicting a future event. So, a *k*th-order Markov model predicts the probability of next event by looking the past *k* events. Given a set of all paths *R*, the probability of reaching a state s_j from a state s_i via a (noncyclic) path $r \in R$ is given by $p(r) = \prod pk, k+1$, where *k* ranges from *i* to $j - 1$. The probability of reaching s_j from s_i is the sum over all paths: $p(j \mid i) = \sum_{r \in R} p(r)$.

In the context of Web transactions, Markov chains can be used to model transition probabilities between pageviews. In Web usage analysis, they have been proposed as the underlying modeling machinery for Web prefetching applications or for minimizing system latencies [Deshpande and Karypis, 2001; Palpanas and Mendelzon, 1999; Pitkow and Pirolli, 1999; Sarukkai, 2000]. Such systems are designed to predict the *next* user action based on a user's previous surfing behavior. In the case of first-order Markov models, only the user's current action is considered in predicting the next action, and thus each state represents a single pageview in the user's transaction. Markov models can also be used to discover high-probability user-navigational trails in a Website. For example, in Borges and Levene [1999], the user sessions are modeled as a hypertext probabilistic grammar (or alternatively, an absorbing Markov chain) whose higher probability paths correspond to the user's preferred trails. An algorithm is provided to efficiently mine such trails from the model.

As an example of how Web transactions can be modeled as a Markov model, consider the set of Web transaction given in Figure 15.5 (left). The Web transactions involve pageviews A, B, C, D, and E. For each transaction the frequency of occurrences of that transaction in the data is given in table's second column (thus there are a total of 50 transactions in the data set). The (absorbing) Markov model for this data is also given in Figure 15.5 (right). The transitions from the "start" state represent the prior probabilities for transactions starting with pageviews A and B. The transitions into the "final" state represent the probabilities that the paths end with the specified originating pageviews, For example, the transition probability from the state A to B is 16/28 = 0.57 because, out of the 28 occurences of A in transactions, B occurs immediately after A in 16 cases.

Higher-order Markov models generally provide a higher prediction accuracy. However, this is usually at the cost of lower coverage and much higher model complexity due to the larger number of states. In order to remedy the coverage and space complexity problems, Pitkow and Pirolli [1999] proposed all-*k*th-order Markov models (for coverage improvement) and a new state reduction technique called longest repeating subsequences (LRS) (for reducing model size). The use of all-*k*th-order Markov models generally requires the generation of separate models for each of the *k* orders; if the model cannot make a prediction using the *k*th order, it will attempt to make a prediction by incrementally decreasing the model order. This scheme can easily lead to even higher space complexity because it requires the representation of all possible states for each *k*. Deshpande and Karypis [2001] propose selective Markov models, introducing several schemes in order to tackle the model complexity problems with all-*k*th-order Markov models. The proposed schemes involve pruning the model based on criteria such as support, confidence, and error rate. In particular, the support-pruned Markov models eliminate all states with low support determined by a minimum frequency threshold.

Transactions	Frequency
A, B, E	10
B, D, B, C	4
B, C, E	10
A, B, E, F	6
A, D, B	12
B, D, B, E	8

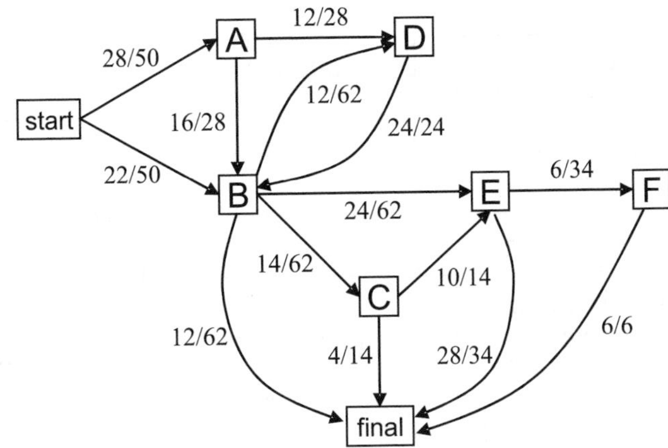

FIGURE 15.5 An example of modeling navigational trails as a Markov chain.

Another way of efficiently representing navigational trails is by inserting each trail into a trie structure [Spiliopoulou and Faulstich, 1999]. It is also possible to insert frequent sequences (after or during sequential pattern raining) into a trie structure [Pei et al., 2000]. A well-known example of this approach is the notion of aggregate tree introduced as part of the WUM (Web Utilization Miner) system [Spiliopoulou and Faulstich, 1999]. The aggregation service of WUM extracts the transactions from a collection of Web lags, transforms them into sequences, and merges those sequences with the same prefix into the aggregate tree (a trie structure). Each node in the tree represents a navigational subsequence from the root (an empty node) to a page and is annotated by the frequency of occurrences of that subsequence in the transaction data (and possibly other information such as markers to distinguish among repeat occurrences of the corresponding page in the subsequence). WUM uses a powerful mining query language, called MINT, to discover generalized navigational patterns from this trie structure. MINT includes mechanism to specify sophisticated constraints on pattern templates such as wildcards with user-specified boundaries, as well as other statistical thresholds such as support and confidence.

As an example, again consider the set of Web transaction given in the previous example. Figure 15.6 shows a simplified version of WUM's aggregate tree structure derived from these transactions. The advantage of this approach is that the search for navigational patterns can be performed very efficiently and the confidence and support for the sequential patterns can be readily obtained from the node annotations in the tree. For example, consider the navigational sequence ⟨A, B, E, F⟩. The support for this sequence can be computed as the support of F divided by the support of first pageview in the sequence, A, which is $6/28 = 0.21$, and the confidence of the sequence is the support of F divided by support of its parent, E, or $6/16 = 0.375$. The disadvantage of this approach is the possibly high space complexity, especially in a site with many dynamically generated pages.

15.3.2.3 Clustering Approaches

In general, there are two types of clustering that can be performed on usage transaction data: clustering the transactions (or users) themselves, or clustering pageviews. Each of these approaches is useful in different applications and, in particular, both approaches can be used for Web personalization, There has been a significant amount of work on the applications of clustering in Web usage mining, e-marketing, personalization, and collaborative filtering.

For example, an algorithm called PageGather has been used to discover significant groups of pages based on user access patterns [Perkowitz and Etzioni, 1998]. This algorithm uses, as its basis, clustering of pages based on the Clique (complete link) clustering technique. The resulting clusters are used to automatically synthesize alternative static index pages for a site, each reflecting possible interests of one user segment. Clustering of user rating records has also been used as a prior step to collaborative filtering in order to remedy the scalability problems of the k-nearest-neighbor algorithm [O'Conner and Her-

Transactions	Frequency
A, B, E	10
B, D, B, C	4
B, C, E	10
A, B, E, F	6
A, D, B	12
B, D, B, E	8

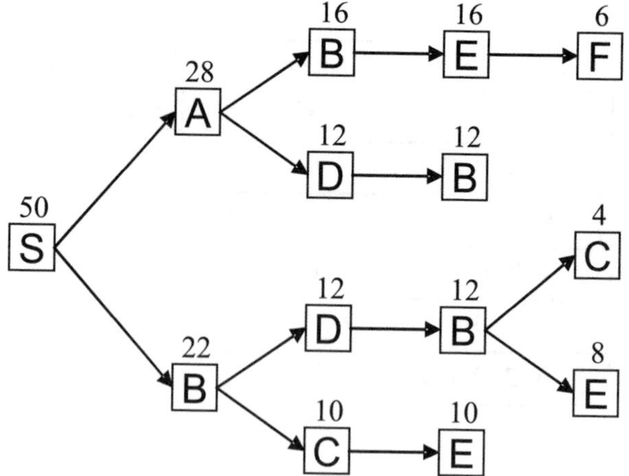

FIGURE 15.6 An example of modeling navigational trails in an aggregate tree.

locker, 1999]. Both transaction clustering and pageview clustering have been used as an integrated part of a Web personalization framework based on Web usage raining [Mobasher et al., 2002b].

Given the mapping of user transactions into a multidimensional space as vectors of pageviews (i.e., the matrix TP in Section 15.2.2), standard clustering algorithms, such as k-means, generally partition this space into groups of transactions that are close to each other based on a measure of distance or similarity among the vectors. Transaction clusters obtained in this way can represent user or visitor segments based on their navigational behavior or other attributes that have been captured in the transaction file. However, transaction clusters by themselves are not an effective means of capturing an aggregated view of common user patterns. Each transaction cluster may potentially contain thousands of user transactions involving hundreds of pageview references. The ultimate goal in clustering user transactions is to provide the ability to analyze each segment for deriving business intelligence, or to use them for tasks such as personalization.

One straightforward approach in creating an aggregate view of each cluster is to compute the centroid (or the mean vector) of each cluster. The dimension value for each pageview in the mean vector is computed by finding the ratio of the sum of the pageview weights across transactions to the total number of transactions in the cluster. If pageview weights in the original transactions are binary, then the dimension value of a pageview p in a cluster centroid represents the percentage of transactions in the cluster in which p occurs. Thus, the centroid dimension value of p provides a measure of its significance in the cluster. Pageviews in the centroid can be sorted according to these weights and lower-weight pageviews can be filtered out. The resulting set of pageview-weight pairs can be viewed as an "aggregate usage profile" representing the interests or behavior of a significant group of users. We discuss how such aggregate profiles can be used for personalization in the next section.

As an example, consider the transaction data depicted in Figure 15.7 (left). In this case, the feature (pageview) weights in each transaction vector is binary. We assume that the data has already been clustered using a standard clustering algorithm such as k-means, resulting in three clusters of user transactions. The table in the right portion of Figure 15.7 shows the aggregate profile corresponding to cluster 1. As indicated by the pageview weights, pageviews B and F are the most significant pages characterizing common interests of users in this segment. Pageview C, however, only appears in one transaction and might be removed given a filtering threshold greater than 0.25.

Note that it is possible to apply a similar procedure to the transpose of the matrix TP, resulting a collection of pageview clusters. However, traditional clustering techniques, such as distance-based methods, generally cannot handle this type clustering. The reason is that instead of using pageviews as dimensions, the transactions must be used as dimensions, whose number is in tens to hundreds of thousands in a typical application. Furthermore, dimensionality reduction in this context may not be

		A	B	C	D	E	F
Cluster 0	user 1	0	0	1	1	0	0
	user 4	0	0	1	1	0	0
	user 7	0	0	1	1	0	0
Cluster 1	user 0	1	1	0	0	0	1
	user 3	1	1	0	0	0	1
	user 6	1	1	0	0	0	1
	user 9	0	1	1	0	0	1
Cluster 2	user 2	1	0	0	1	1	0
	user 5	1	0	0	1	1	0
	user 8	1	0	1	1	1	0

Aggregate Profile for Cluster 1	
Weight	Pageview
1.00	B
1.00	F
0.75	A
0.25	C

FIGURE 15.7 An example of deriving aggregate usage profiles from transaction clusters.

appropriate, as removing a significant number of transactions may result in losing too much information. Similarly, the clique-based clustering approach of PageGather algorithm [Perkowitz and Etzioni, 1998] discussed above can be problematic because finding all maximal cliques in very large graphs is not, in general, computationally feasible.

One approach that has been shown to be effective in this type (i.e., item-based) clustering is Association Rule Hypergraph partitioning (ARHP) [Han et al., 1998]. ARHP can efficiently cluster high-dimensional data sets and provides automatic filtering capabilities. In the ARHP, first-association rule mining is used to discover a set I of frequent itemsets among the pageviews in P. These itemsets are used as hyperedges to form a hypergraph $H = \langle V, E\rangle$, where $V \subseteq P$ and $E \subseteq I$. A hypergraph is an extension of a graph in the sense that each hyperedge can connect more than two vertices. The weights associated with each hyperedge can be computed based on a variety of criteria such as the confidence of the association rules involving the items in the frequent itemset, the support of the itemset, or the "interest" of the itemset.

The hypergraph H is recursively partitioned until a stopping criterion for each partition is reached resulting in a set of clusters C. Each partition is examined to filter out vertices that are not highly connected to the rest of the vertices of the partition. The connectivity of vertex v (a pageview appearing in the frequent itemset) with respect to a cluster c is defined as:

$$conn(v, c) = \frac{\Sigma_{e\subseteq c, v\in e} weight(e)}{\Sigma_{e\subseteq c} weight(e)}$$

A high connectivity value suggests that the vertex has strong edges connecting it to other vertices in the partition. The vertices with connectivity measure that are greater than a given threshold value are considered to belong to the partition, and the remaining vertices are dropped from the partition. The connectivity value of an item (pageviews) defined above is important also because it is used as the primary factor in determining the weight associated with that item within the resulting aggregate profile. This approach has also been used in the context of Web personalization [Mobasher et al., 2002b], and its performance in terms of recommendation effectiveness has been compared to the transaction clustering approach discussed above.

Clustering can also be applied to Web transactions viewed as sequences rather than as vectors. For example in Banerjee and Ghosh [2001], a graph-based algorithm was introduced to cluster Web transactions based on a function of longest common subsequences. The novel similarity metric used for clustering takes into account both the time spent on pages as well as a significance weight assigned to pages.

Finally, we also observe that the clustering approaches such as those discussed in this section can also be applied to content data or to the integrated content-enhanced transactions described in Section 15.2.2. For example, the results of clustering user transactions can be combined with "content profiles" derived from the clustering of text features (terms or concepts) in pages [Mobasher et al., 2000b]. The feature clustering is accomplished by applying a clustering algorithm to the transpose of the pageview-feature matrix PF, defined earlier. This approach treats each feature as a vector over the space of pageviews. Thus the centroid of a feature cluster can be viewed as a set (or vector) of pageviews with associated weights. This representation is similar to that of usage profiles discussed above, however; in this case the weight of a pageview in a profile represents the prominence of the features in that pageview that are associated with the corresponding cluster. The combined set of content and usage profiles can then be used seamlessly for more effective Web personalization. One advantage of this approach is that it solves the "new item" problem that often plagues purely usage-based or collaborative approaches; when a new item (e.g., page or product) is recently added to the site, it is not likely to appear in usage profiles due to the lack of user ratings or access to that page, but it may still be recommended according to its semantic attributes captured by the content profiles.

15.4 Using the Discovered Patterns for Personalization

As noted in the Introduction section, the goal of the recommendation engine is to match the active user session with the aggregate profiles discovered through Web usage raining and to recommend a set of objects to the user. We refer to the set of recommended object (represented by pageviews) as the recommendation set. In this section we explore the recommendation procedures to perform the matching between the discovered aggregate profiles and an active user's session. Specifically, we present several effective recommendation algorithms based on clustering (which can be seen as an extension of standard *k*NN-based collaborative filtering), association rule mining (AR), and sequential pattern (SP) or contiguous sequential pattern (CSP) discovery. In the cases of AR, SP, and CSP, we consider efficient and scalable data structures for storing frequent itemset and sequential patterns, as well as recommendation generation algorithms that use these data structures to directly produce real-time recommendations (without the apriori generation of rule).

Generally, only a portion of the current user's activity is used in the recommendation process. Maintaining a history depth is necessary because most users navigate several paths leading to independent pieces of information within a session. In many cases these sub-sessions have a length of no more than three or four references. In such a situation, it may not be appropriate to use references a user made in a previous sub-session to make recommendations during the current sub-session. We can capture the user history depth within a sliding window over the current session. The sliding window of size n over the active session allows only the last n visited pages to influence the recommendation value of items in the recommendation set. For example, if the current session (with a window size of 3) is $\langle A, B, C\rangle$, and the user accesses the pageview D, then the new active session becomes $\langle B, C, D\rangle$. We call this sliding window the user's active session window.

Structural characteristics of the site or prior domain knowledge can also be used to associate an additional measure of significance with each pageview in the user's active session. For instance, the site owner or the site designer may wish to consider certain page types (e.g., content vs. navigational) or product categories as having more significance in terms of their recommendation value. In this case, significance weights can be specified as part of the domain knowledge.

15.4.1 The *k*NN-Based Approach

Collaborative filtering based on the *k*NN approach involves comparing the activity record for a target user with the historical records of other users in order to find the top k users who have similar tastes or interests. The mapping of a visitor record to its neighborhood could be based on similarity in ratings of items, access to similar content or pages, or purchase of similar items. The identified neighborhood is then used to recommend items not already accessed or purchased by the active user. Thus, there are two primary phases in collaborative filtering: the neighborhood formation phase and the recommendation phase.

In the context of personalization based on Web usage mining, *k*NN involves measuring the similarity or correlation between the active session $\vec{s}$ and each transaction vector $\vec{t}$ (where $t \in T$). The top *k*-most similar transactions to $\vec{s}$ are considered to be the neighborhood for the session *s*, which we denote by $NB(s)$ (taking the size *k* of the neighborhood to be implicit):

$$NB(s) = \{\vec{t}_{s_1}, \vec{t}_{s_2}, \cdots, \vec{t}_{s_k}\}$$

A variety of similarity measures can be used to find the nearest neighbors. In traditional collaborative filtering domains (where feature weights are item ratings on a discrete scale), the Pearson *r* correlation coefficient is commonly used. This measure is based on the deviations of users' ratings on various items from their mean ratings on all rated items. However, this measure may not be appropriate when the primary data source is clickstream data (particularly in the case of binary weights). Instead we use the cosine coefficient, commonly used in information retrieval, which measures the cosine of the angle between two vectors. The cosine coefficient can be computed by normalizing the dot product of two vectors with respect to their vector norms. Given the active session $\vec{s}$ and a transaction $\vec{t}$, the similarity between them is obtained by:

$$sim(\vec{t}, \vec{s}) = \frac{\vec{t} \cdot \vec{s}}{|\vec{t}| \times |\vec{s}|}.$$

In order to determine which items (not already visited by the user in the active session) are to be recommended, a recommendation score is computed for each pageview $p_i \in P$ based on the neighborhood for the active session. Two factors are used in determining this recommendation score: the overall similarity of the active session to the neighborhood as a whole, and the average weight of each item in the neighborhood.

First we compute the mean vector (centroid) of $NB(s)$. Recall that the dimension value for each pageview in the mean vector is computed by finding the ratio of the sum of the pageview's weights across transactions to the total number of transactions in the neighborhood. We denote this vector by $cent(NB(s))$. For each pageview *p* in the neighborhood centroid, we can now obtain a recommendation score as a function of the similarity of the active session to the centroid vector and the weight of that item in this centroid. Here we have chosen to use the following function, denoted by $rec(\vec{s}, p)$:

$$rec(\vec{s}, p) = \sqrt{weight(p, NB(s)) \times sim(\vec{s}, cent(NB(s)))}$$

where $weight(p, NB(s))$ is the mean weight for pageview *p* in the neighborhood as expressed in the centroid vector. If the pageview *p* is in the current active session, then its recommendation value is set to zero.

If a fixed number *N* of recommendations are desired, then the top *N* items with the highest recommendation scores are considered to be part of the recommendation set. In our implementation, we normalize the recommendation scores for all pageviews in the neighborhood (so that the maximum recommendation score is 1), and return only those that satisfy a threshold test. In this way, we can compare the performance of *k*NN across different recommendation thresholds.

15.4.2 Using Clustering for Personalization

The transaction clustering approach discussed in Section 15.3.2 will result in a set $TC = \{c_1, c_2, \cdots, c_k\}$ of transaction clusters, where each c_i is a subset of the set of transactions *T*. As noted in that section, from each transaction cluster we can derive and aggregate usage profile by computing the centroid vectors for that cluster. We call this method PACT (Profile Aggregation Based on Clustering Transactions) [Mobasher et al., 2002b].

In general, PACT can consider a number of other factors in determining the item weights within each profile and in determining the recommendation scores. These additional factors may include the link distance of pageviews to the current user location within the site or the rank of the profile in terms of its significance. However, to be able to consistently compare the performance of the clustering-based approach to that of *k*NN, we restrict the item weights to be the mean feature values of the transaction cluster centroids. In this context, the only difference between PACT and the *k*NN-based approach is that we discover transaction clusters offline and independent of a particular target user session.

To summarize the PACT method, given a transaction cluster *c*, we construct an aggregate usage profile pr_c as a set of pageview-weight pairs:

$$pr_c = \{\langle p, weight(p, pr_c)\rangle \mid p \in P, weight(p, pr_c) \geq \mu\}$$

where the significance weight, $weight(p, pr_c)$, of the pageview *p* within the usage profile pr_c is:

$$weight(p, pr_c) = \frac{1}{|c|} \cdot \sum_{t \in c} w_p^t$$

and w_p^t is the weight of pageview *p* in transaction $t \in c$. The threshold parameter μ is used to prune out very low support pageviews in the profile. An example of deriving aggregate profiles from transaction clusters was given in the previous section (see Figure 15.7).

This process results in a number of aggregate profiles, each of which can, in turn, be represented as a vector in the original *n*-dimensional space of pageviews. The recommendation engine can compute the similarity of an active session $\vec{s}$ with each of the discovered aggregate profiles. The top matching profile is used to produce a recommendation set in a manner similar to that for the *k*NN approach discussed in the preceding text. If $\vec{pr}$ is the vector representation of the top matching profile *pr*, we compute the recommendation score for the pageview *p* by

$$rec(\vec{s}, p) = \sqrt{weight(p, pr) \times sim(\vec{s}, \vec{pr})},$$

where $weight(p, pr)$ is the weight for pageview *p* in the profile *pr*. As in the case of *k*NN, if the pageview *p* is in the current active session, then its recommendation value is set to zero.

Clearly, PACT will result in dramatic improvement in scalability and computational performance because most of the computational cost is incurred during the offline clustering phase. We would expect, however, that this decrease in computational costs be accompanied also by a decrease in recommendation effectiveness. Experimental results [Mobasher et al., 2001b] have shown that through proper data preprocessing and using some of the data transformation steps discussed earlier, we can dramatically improve the recommendation effectiveness when compared to *k*NN.

It should be noted that the pageview clustering approach discussed in Section 15.3.2 can also be used with the recommendation procedure detailed above. In that case, also, the aggregate profiles are represented as collections of pageview-weight pairs and thus can be viewed as vectors over the space of pageviews in the data.

15.4.3 Using Association Rules for Personalization

The recommendation engine based on association rules matches the current user session window with frequent itemsets to find candidate pageviews for giving recommendations. Given an active session window *w* and a group of frequent itemsets, we only consider all the frequent itemsets of size $|w|+1$

containing the current session window. The recommendation value of each candidate pageview is based on the confidence of the corresponding association rule whose consequent is the singleton containing the pageview to be recommended.

In order to facilitate the search for itemsets (of size $|w|+1$) containing the current session window w, the frequent itemsets are stored in a directed acyclic graph, here called a Frequent Itemset Graph. The Frequent Itemset Graph is an extension of the lexicographic tree used in the "tree projection algorithm" [Agarwal et al., 1999]. The graph is organized into levels from 0 to k, where k is the maximum size among all frequent itemsets. Each node at depth d in the graph corresponds to an itemset I, of size d, and is linked to itemsets of size $d + 1$ that contain I at level $d + 1$. The single root node at level 0 corresponds to the empty itemset. To be able to match different orderings of an active session with frequent itemsets, all itemsets are sorted in lexicographic order before being inserted into the graph. The user's active session is also sorted in the same manner before matching with patterns.

Given an active user session window w, sorted in lexicographic order, a depth-first search of the Frequent Itemset Graph is performed to level $|w|$. If a match is found, then the children of the matching node n containing w are used to generate candidate recommendations. Each child node of n corresponds to a frequent itemset $w \cup \{p\}$. In each case, the pageview p is added to the recommendation set if the support ratio $\sigma(w \cup \{p\}) / \sigma(w)$ is greater than or equal to α, where α is a minimum confidence threshold. Note that $\sigma(w \cup \{p\}) / \sigma(w)$ is the confidence of the association rule $w \Rightarrow \{p\}$. The confidence of this rule is also used as the recommendation score for pageview p. It is easy to observe that in this algorithm the search process requires only $O(|w|)$ time given active session window w.

To illustrate the process, consider the example transaction set given in Figure 15.8. Using these transactions, the Apriori algorithm with a frequency threshold of 4 (minimum support of 0.8) generates the itemsets given in Figure 15.9. Figure 15.10 shows the Frequent Itemsets Graph constructed based on the frequent itemsets in Figure 15.9. Now, given user active session window $\langle B, E \rangle$, the recommendation generation algorithm finds items A and C as candidate recommendations. The recommendation scores of item A and C are 1 and 4/5, corresponding to the confidences of the rules $\{B, E\} \rightarrow \{A\}$ and $\{B, E\} \rightarrow \{C\}$, respectively.

15.4.4 Using Sequential Patterns for Personalization

The recommendation algorithm based on association rules can be adopted to work also with sequential or contiguous sequential patterns. In this case, we focus on frequent (contiguous) sequences of size $|w|+1$ whose prefix contains an active user session w. The candidate pageviews to be recommended are the last items in all such sequences. The recommendation values are based on the confidence of the patterns. If

T1: {*ABDE*}
T2: {*ABECD*}
T3: {*ABEC*}
T4: {*BEBAC*}
T5: {*DABEC*}

FIGURE 15.8 Sample Web Transactions involving pageviews A; B, C, D, and E.

Size 1	Size 2	Size 3	Size 4
{A}(5)	{A,B}(5)	{A,B,C}(4)	{A,B,C,E}(4)
{B}(6)	{A,C}(4)	{A,B,E}(5)	
{C}(4)	{A,E}(5)	{A,C,E}(5)	
{E}(5)	{B,C}(4)	{B,C,E}(4)	
	{B,E}(5)		
	{C,E}(4)		

FIGURE 15.9 Example of discovered frequent itemsets.

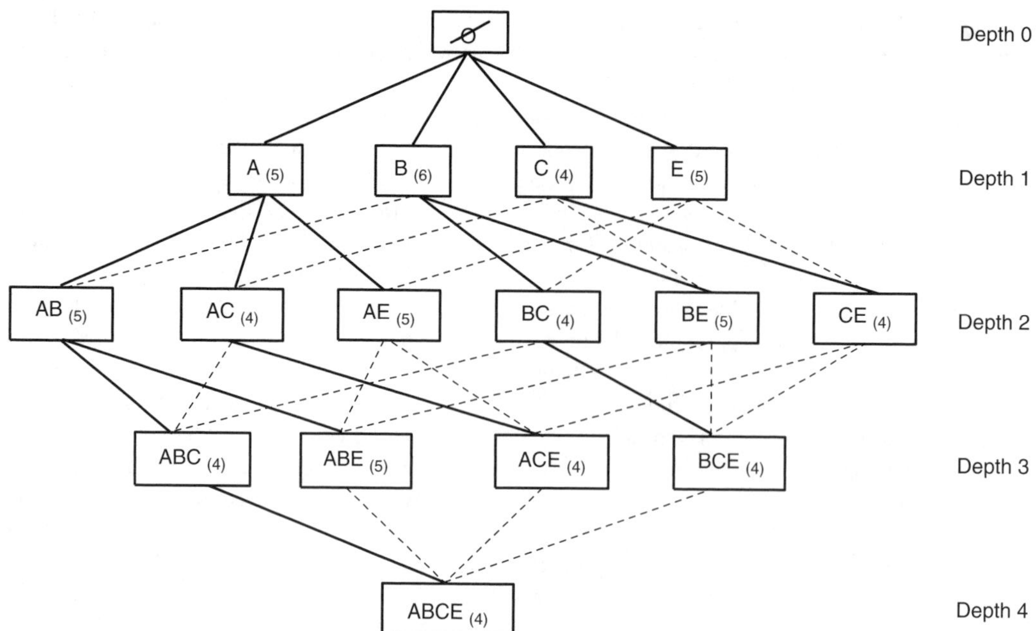

FIGURE 15.10 An example of a Frequent Itemsets Graph.

the confidence satisfies a threshold requirement, then the candidate pageviews are added to the recommendation set.

A simple trie structure, which we call Frequent Sequence Trie (FST), can be used to store both the sequential and contiguous sequential patterns discovered during the pattern discovery phase. The FST is organized into levels from 0 to k, where k is the maximal size among all sequential or contiguous sequential patterns. There is the single root node at depth 0 containing the empty sequence. Each nonroot node N at depth d contains an item s_d and represents a frequent sequence $\langle s_1, s_2, \cdots, s_{d-1}, s_d \rangle$ whose prefix $\langle s_1, s_2, \cdots, s_{d-1} \rangle$ is the pattern represented by the parent node of N at depth $d - 1$. Furthermore, along with each node we store the support (or frequency) value of the corresponding pattern. The confidence of each pattern (represented by a nonroot node in the FST) is obtained by dividing the support of the current node by the support of its parent node.

The recommendation algorithm based on sequential and contiguous sequential patterns has a similar structure as the algorithm based on association rules. For each active session window $w = \langle w_1, w_2, \cdots, w_n \rangle$, we perform a depth-first search of the FST to level n. If a match is found, then the children of the matching node N are used to generate candidate recommendations. Given a sequence $S = \langle w_1, w_2, \cdots, w_n, p \rangle$ represented by a child node of N, the item p is then added to the recommendation set as long as the confidence of S is greater than or equal to the confidence threshold. As in the case of the frequent itemset graph, the search process requires $O(|w|)$ time given active session window size $|w|$.

To continue our example, Figure 15.11 and Figure 15.12 show the frequent sequential patterns and frequent contiguous sequential patterns with a frequency threshold of 4 over the example transaction set

Size 1	Size 2	Size 3
⟨*A*⟩(5)	⟨*A,B*⟩(4)	⟨*A,B,E*⟩(4)
⟨*B*⟩(6)	⟨*A,C*⟩(4)	⟨*A,E,C*⟩(4)
⟨*C*⟩(4)	⟨*A,E*⟩(4)	
⟨*E*⟩(5)	⟨*B,C*⟩(4)	
	⟨*B,E*⟩(5)	
	⟨*C,E*⟩(4)	

FIGURE 15.11. Example of discovered sequential patterns.

Size 1	Size 2
⟨A⟩(5)	⟨A,B⟩(
⟨B⟩(6)	4)
⟨C⟩(4)	⟨B,E⟩(
⟨E⟩(5)	4)

FIGURE 15.12. Example of discovered contiguous sequential patterns.

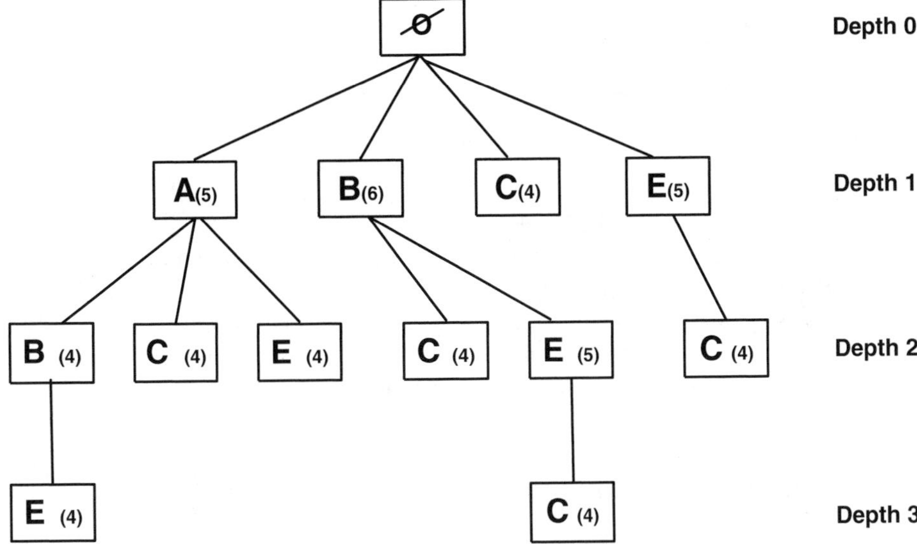

FIGURE 15.13 Example of a Frequent Sequence Trie (FST).

given in Figure 15.8. Figure 15.13 and Figure 15.14 show the trie representation of the sequential and contiguous sequential patterns listed in the Figure 15.11 and Figure 15.12, respectively. The sequential pattern ⟨A, B, E⟩ appears in the Figure 15.13 because it is the subsequence of 4 transactions T_1, T_2, T_3, and T_5. However, ⟨A, B, E⟩ is not a frequent contiguous sequential pattern because only three transactions (T_2, T_3, and T_5) contain the contiguous sequence ⟨A, B, E⟩. Given a user's active session window ⟨A, B⟩, the recommendation engine using sequential patterns finds item E as a candidate recommendation. The recommendation score of item E is 1, corresponding to the rule ⟨A, B⟩ ⇒ ⟨E⟩. On the other hand, the recommendation engine using contiguous sequential patterns will, in this case, fail to give any recommendations.

It should be noted that, depending on the specified support threshold, it might be difficult to find large enough itemsets or sequential patterns that could be used for providing recommendations, leading to reduced coverage. This is particularly true for sites with very small average session sizes. An alternative to reducing the support threshold in such cases would be to reduce the session window size. This latter choice may itself lead to some undesired effects since we may not be taking enough of the user's activity history into account. Generally, in the context of recommendation systems, using a larger window size over the active session can achieve better prediction accuracy. But, as in the case of higher support threshold, larger window sizes also lead to lower recommendation coverage.

In order to overcome this problem, we can use the all-kth-order approach discussed in the previous section in the context of Markov chain models. The above recommendation framework for contiguous sequential patterns is essentially equivalent to kth-order Markov models; however, rather than storing all navigational sequences, only frequent sequences resulting from the sequential pattern raining process are stored. In this sense, the above method is similar to support pruned models described in the previous section [Deshpande and Karypis, 2001], except that the support pruning is performed by the Apriori algorithm in the mining phase. Furthermore, in contrast to standard all-kth-order Markov models, this

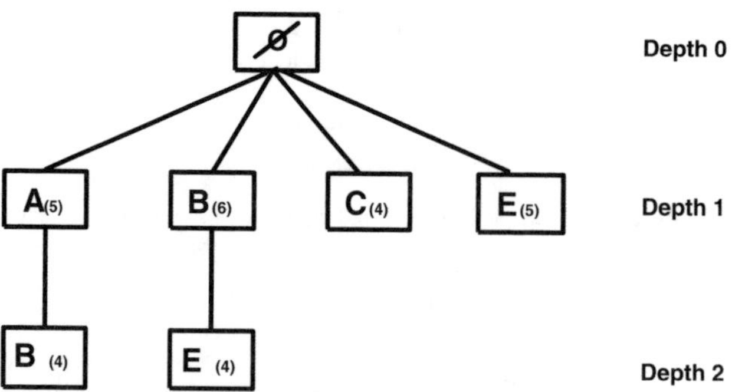

FIGURE 15.14 Example of an FST for contiguous sequences.

framework does not require additional storage because all the necessary information (for all values of k) is captured by the FST structure described above.

The notion of all-kth-order models can also he easily extended to the context of general sequential patterns and association rule. We extend these recommendation algorithms to generate all-kth-order recommendations as follows. First, the recommendation engine uses the largest possible active session window as an input for the recommendation engine. If the engine cannot generate any recommendations, the size of active session window is iteratively decreased until a recommendation is generated or the window size becomes 0.

15.5 Conclusions and Outlook

In this chapter we have attempted to present a comprehensive view of the personalization process based on Web usage mining. The overall framework for this process was depicted in Figure 15.1 and Figure 15.2. In the context of this framework, we have discussed a host of Web usage mining activities necessary for this process, including the preprocessing and integration of data from multiple sources, and pattern discovery techniques that are applied to the integrated usage data. We have also presented a number of specific recommendation algorithms for combining the discovered knowledge with the current status of a user's activity in a Website to provide personalized content to a user. The approaches we have detailed show how pattern discovery techniques such as clustering, association rule mining, and sequential pattern discovery, performed on Web usage data, can be leveraged effectively as an integrated part of a Web personalization system.

In this concluding section, we provide a brief discussion of the circumstances under which some of the approaches discussed might provide a more effective alternative to the others. We also identify the primary problems, the solutions of which may lead to the creation of the next generation of more effective and useful Web-personalization and Web-mining tools.

15.5.1 Which Approach?

Personalization systems are often evaluated based on two statistical measures, namely precision and coverage (also known as recall). These measures are adaptations of similarly named measures often used in evaluating the effectiveness of information retrieval systems. In the context of personalization, precision measures the degree to which the recommendation engine produces accurate recommendations (i.e., the proportion of relevant recommendations to the total number of recommendations), while coverage (or recall) measures the ability of the recommendation engine to produce all of the pageviews that are likely to be visited by the user (i.e., proportion of relevant recommendations to all pageviews that will be visited, according to some evaluation data set). Neither of these measures individually is sufficient to evaluate the performance of the recommendation engine; however, they are both critical. A low precision in this context will likely result in angry customers or visitors who are not interested in the recommended items, whereas low coverage will result in the inability of the site to produce relevant cross-sell recommendations at critical points in the user's interaction with the site. In previous work [Mobasher et al., 2001a, 2002b, a], many of the approaches presented in this chapter have been evaluated based on these measures using real usage data. Here we present a summary of the findings.

In the case of clustering approaches, we have compared the performance of transaction clustering method, PACT, with the pageview clustering approach based on hypergraph partitioning, ARHP [Mobasher et al., 2002a]. In general, the ARHP approach performs better when the data set is filtered to focus on more "interesting" objects (e.g., content-oriented pages that are situated more deeply within the site). It seems to produce a smaller set of high quality, and more specialized, recommendations even when a small portion of the user's clickstream is used by the recommendation engine. On the other hand, PACT provides a clear performance advantage when dealing with all the relevant pageviews in the site, particularly as the session window size is increased. Thus, if the goal is to provide a smaller number of highly focused recommendations, then the ARHP approach may be a more appropriate method. This is particularly the case if only specific portions of the site (such as product-related or content pages) are to be personalized. On the other hand, if the goal is to provide a more generalized personalization solution integrating both content and navigational pages throughout the whole site, then using PACT as the underlying aggregate profile generation method seems to provide clear advantages.

More generally, clustering, in contrast to association rule or sequential pattern mining, provides a more flexible mechanism for personalization even though it does not always lead to the highest recommendation accuracy. The flexibility comes from the fact that many inherent attributes of pageviews can be taken into account in the mining process, such as time durations and possibly relational attributes of the underlying objects.

The association rule (AR) models also performs well in the context of personalization. In general, the precision of AR models are lower than the models based on sequential patterns (SP) and contiguous sequential patterns (CSP), but they often provide much better coverage. Comparison to *k*NN have shown that all of these techniques outperform *k*NN in terms of precision. In general, *k*NN provides better coverage (usually on par with the AR model), but the difference in coverage is diminished if we insist on higher recommendation thresholds (and thus more accurate recommendations).

In general, the SP and the AR models provide the best choices for personalization applications. The CSP model can do better in terms of precision, but the coverage levels are often too low when the goal is to generate as many good recommendations as possible. This last observation about the CSP models, however, does not extend to other predictive applications such as prefetching, where the goal is to predict the immediate next action of the user (rather than providing a broader set of recommendations). In this case, the goal is not usually to maximize coverage, and the high precision of CSP makes it an ideal choice for this type of application.

The structure and the dynamic nature of a Website can also have an impact on the choice between sequential and nonsequential models. For example, in a highly connected site, reliance on fine-grained sequential information in user trails is less meaningful. On the other hand, a site with many dynamically generated pages, where often a contiguous navigational path represents a semantically meaningful sequence of user actions, each depending on the previous actions, the sequential models are better suited in providing useful recommendations.

15.5.2 The Future: Personalization Based on Semantic Web Mining

Usage patterns discovered through Web usage mining are effective in capturing item-to-item and user-to-user relationships and similarities at the level of user sessions. However, without the benefit of deeper domain knowledge, such patterns provide little insight into the underlying reasons for which such items or users are grouped together. It is possible to capture some of the site semantics by integrating keyword-based content-filtering approaches with collaborative filtering and usage-mining techniques. These approaches, however, are incapable of capturing more complex relationships at a deeper semantic level based on the attributes associated with structured objects.

Indeed, with the growing interest in the notion of semantic Web, an increasing number of sites use structured semantics and domain ontologies as part of the site design, creation, and content delivery. The primary challenge for the next generation of personalization systems is to effectively integrate semantic knowledge from domain ontologies into the various parts of the process, including the data preparation, pattern discovery, and recommendation phases. Such a process must involve some or all of the following tasks and activities:

1. Ontology learning, extraction, and preprocessing: Given a page in the Web site, we must be able to extract domain-level structured objects as semantic entities contained within this page. This task may involve the automatic extraction and classification of objects of different types into classes based on the underlying domain ontologies. The domain ontologies, themselves, may be prespecified or may be learned automatically from available training data [Craven et al., 2000]. Given this capability, the transaction data can be transformed into a representation that incorporates complex semantic entities accessed by users during a visit to the site.
2. Semantic data mining: In the pattern discovery phase, data-mining algorithms must be able to deal with complex semantic objects. A substantial body of work in this area already exists. These include extensions of data-mining algorithms (such as association rule mining and clustering) to take into account a concept hierarchy over the space of items. Techniques developed in the context of "relational" data mining are also relevant in this context. Indeed, domain ontologies are often expressed as relational schema consisting of multiple relations. Relational data mining techniques have focused on precisely this type of data.
3. Domain-level aggregation and representation: Given a set of structured objects representing a discovered pattern, we must then be able to create an aggregated representation as a set of pseudo objects each characterizing objects of different types occurring commonly across the user sessions. Let us call such a set of aggregate pseudo objects a Domain-level Aggregate Profile. Thus, a domain-level aggregate profile characterizes the activity of a group of users based on the common properties of objects as expressed in the domain ontology. This process will require both general and domain-specific techniques for comparison and aggregation of complex objects, including ontology-based semantic similarity measures.
4. Ontology-based recommendations: Finally, the recommendation process must also incorporate semantic knowledge from tide domain ontologies. This requires further processing of the user's activity record according to the ontological structure of the objects accessed and the comparison of the transformed "semantic transactions" to the discovered domain-level aggregate profiles. To produce useful recommendations for users, the results of this process must be instantiated to a set of real objects or pages that exist in the site.

The notion of "Semantic Web Mining" was introduced in Berendt et al. [2002a]. Furthermore, a general framework was proposed for the extraction of a concept hierarchy from the site content and the application of data-mining techniques to find frequently occurring combinations of concepts. An approach to integrate domain ontologies into the personalization process based on Web usage raining was proposed in Dai and Mobasher [2002], including an algorithm to construct domain-level aggregate profiles from a collection of semantic objects extracted from user transactions.

Efforts in this direction are likely to be the most fruitful in the creation of much more effective Web usage raining and personalization systems that are consistent with the emergence and proliferation of the semantic Web.

References

Agarwal, R., C. Aggarwal, and V. Prasad. A Tree Projection Algorithm for Generation of Frequent Itemsets. In *Proceedings of the High Performance Data Mining Workshop,* Puerto Rico, April 1999.

Aggarwal, C. C., J. L. Wolf, and P. S. Yu. A New Method for Similarity Indexing for Market Data. In *Proceedings of the 1999 ACM SIGMOD Conference,* Philadelphia, PA, June 1999.

Agrawal, R. and R. Srikant. Fast Algorithms for Mining Association Rules. In *Proceedings of the 20th International Conference on Very Large Data Bases (VLDB'94),* Santiago, Chile, September 1994.

Agrawal, R. and R. Srikant. Mining Sequential Patterns. In *Proceedings of the International Conference on Data Engineering (ICDE'95),* Taipei, Taiwan, March 1995.

Banerjee, A. and J. Ghosh. Clickstream Clustering Using Weighted Longest Common Subsequences. In *Proceedings of the Web Mining Workshop at the 1st SIAM Conference on Data Mining,* Chicago, IL, April 2001.

Berendt, B., A. Hotho, and G. Stumme. Towards Semantic Web Mining. In *Proceedings of the First International Semantic Web Conference (ISWC'02),* Sardinia, Italy, June 2002a.

Berendt, B., B. Mobasher, M. Nakagawa, and M. Spiliopoulou. The Impact of Site Structure and User Environment on Session Reconstruction in Web Usage Analysis. In *Proceedings of the 4th WebKDD 2002 Workshop, at the ACM-SIGKDD Conference on Knowledge Discovery in Databases (KDD'2000),* Edmonton, Alberta, Canada, July 2002b.

Berendt, B. and M. Spiliopoulou. Analysing navigation behaviour in web sites integrating multiple information systems. *VLDB Journal, Special Issue on Databases and the Web,* 9(1): 56–75, 2000.

Borges, J. and M. Levene. Data Mining of User Navigation Patterns. In B. Masand and M. Spiliopoulou, Eds., *Web Usage Analysis and User Profiling: Proceedings of the WEBKDD'99 Workshop,* LNAI 1836, pp. 92–111. Springer-Verlag, New York, 1999.

Buchner, A. and M. D. Mulvenna. Discovering internet marketing intelligence through online analytical Web usage mining. *SIGMOD Record,* 4(27), 1999.

Claypool, M., A. Gokhale. T. Miranda, P. Murnikov, D. Netes, and M. Sartin. Combining Content-based and Collaborative Filters in an Online Newspaper. In *Proceedings of the ACM SIGIR'99 Workshop on Recommender Systems: Algorithms and Evaluation,* Berkeley, CA, August 1999.

Cooley, R. Web Usage Mining: Discovery and Application of Interesting Patterns from Web Data. Ph.D. dissertation, Department of Computer Science, University of Minnesota, Minneapolis, MN, 2000.

Cooley, R., B. Mobasher, and J. Srivastava. Data preparation for mining World Wide Web browsing patterns. *Journal of Knowledge and Information Systems,* 1(1), 1999.

Craven, M., D. DiPasquo, D. Freitag, A. McCallum, T. Mitchell, K. Nigam, and S. Slattery. Learning to construct knowledge bases from the World Wide Web. *Artificial Intelligence,* 118(1–2): 69–113, 2000.

Dai, H. and B. Mobasher. Using Ontologies to Discover Domain-Level Web Usage Profiles. In *Proceedings of the 2nd Semantic Web Mining Workshop at ECML/PKDD 2002,* Helsinki, Finland, August 2002.

Deshpande, M. and G. Karypis. Selective Markov Models for Predicting Web-Page Accesses. In *Proceedings of the First International SIAM Conference on Data Mining,* Chicago, April 2001.

Frakes, W. B. and R. Baeza-Yates. *Information Retrieval: Data Structures and Algorithms.* Prentice Hall, Englewood Cliffs, NJ, 1992.

Fu, X., J. Budzik, and K. J. Hammond. Mining Navigation History for Recommendation. In *Proceedings of the 2000 International Conference on Intelligent User Interfaces,* New Orleans, LA, ACM Press, New York, January 2000.

Han, E., G. Karypis, V. Kumar, and B. Mobasher. Hypergraph based clustering in high-dimensional data sets: A summary of results. *IEEE Data Engineering Bulletin,* 21(1): 15–22, March 1998.

Herlocker, J., J. Konstan, A. Borchers, and J. Riedl. An Algorithmic Framework for Performing Collaborative Filtering. In *Proceedings of the 22nd ACM Conference on Research and Development in Information Retrieval (SIGIR'99),* Berkeley, CA, August 1999.

Joachims, T., D. Freitag, and T. Mitchell. WebWatcher: A Tour Guide for the World Wide Web. In *Proceedings of the International Joint Conference in Al (IJCAI97),* Los Angeles, CA, August 1997.

Konstan, J., B. Miller, D. Maltz, J. Herlocker, L. Gordon, and J. Riedl. Grouplens: Applying collaborative filtering to Usenet news. *Communications of the ACM,* 40(3), 1997.

Lieberman, H. Letizia: An Agent that Assists Web Browsing. In *Proceedings of the 1995 International Joint Conference on Artificial Intelligence, IJCAI'95,* Montreal, Canada, August 1995.

Lin, W., S. A. Alvarez, and C. Ruiz. Efficient adaptive-support association rule mining for recommender systems. *Data Mining and Knowledge Discovery,* 6: 83–105, 2002.

Liu, B., W. Hsu, and Y. Ma. Association Rules with Multiple Minimum Supports. In *Proceedings of the ACM SIGKDD International Conference on Knowledge Discovery and Data Mining (KDD'99, poster),* San Diego, CA, August 1999.

Mobasher, B., R. Cooley, and J. Srivastava, Automatic personalization based on Web usage mining. *Communications of the ACM,* 43(8): 142–151, 2000a.

Mobasher, B., H. Dai, T. Luo, and M. Nakagawa. Effective Personalization Based on Association Rule Discovery from Web Usage Data. In *Proceedings of the 3rd ACM Workshop on Web Information and Data Management, (WIDM'01),* Atlanta, GA, November 2001a.

Mobasher, B., H. Dai, T. Luo, and M. Nakagawa. Improving the Effectiveness of Collaborative Filtering on Anonymous Web Usage Data. In *Proceedings of the IJCAI 2001 Workshop on Intelligent Techniques for Web Personalization (ITWP'01),* Seattle, WA, August 2001b.

Mobasher, B., H. Dai, T. Luo, and M. Nakagawa. Using Sequential and Non-Sequential Patterns in Predictive Web Usage Mining Tasks. In *Proceedings of the 2002 IEEE International Conference on Data Mining (ICDM'02),* Maebashi City, Japan, December 2002a.

Mobasher, B., H. Dai, T. Luo, Y. Sun, and J. Zhu. Integrating Web Usage and Content Mining for More Effective Personalization. In *E-Commerce and Web Technologies: Proceedings of the EC-WEB 2000 Conference, Lecture Notes in Computer Science (LNCS)* 1875, pp. 165–176. Springer-Verlag, New York, September 2000b.

Mobasher, B., H. Dai, M., Nakagawa, and T. Luo. Discovery and evaluation of aggregate usage profiles for Web personalization. *Data Mining and Knowledge Discovery,* 6: 61–82, 2002b.

O'Conner, M. and J. Herlocker. Clustering Items for Collaborative Filtering. In *Proceedings of the ACM SIGIR Workshop on Recommender Systems,* Berkeley, CA, August 1999.

Palpanas, T. and A. Mendelzon. Web Prefetching Using Partial Match Prediction. In *Proceedings of the 4th International Web Caching Workshop (WCW99),* San Diego, CA, March 1999.

Pazzani, M. A Framework for Collaborative, Content-Based, and Demographic Filtering. *Artificial Intelligence Review,* 13(5–6): 393–408. 1999.

Pei, J., J. Han, B. Mortazavi-Asl, and H. Zhu. Mining Access Patterns Efficiently from Web Logs. In *Proceedings of the 4th Pacific-Asia Conference on Knowledge Discovery and Data Mining (PAKDD'00),* Kyoto, Japan, April 2000.

Perkowitz, M. and O. Etzioni. Adaptive Web Sites: Automatically Synthesizing Web Pages. In *Proceedings of the 15th National Conference on Artificial Intelligence,* Madison, WI, July 1998.

Pitkow, J. and P. Pirolli. Mining Longest Repeating Subsequences to Predict WWW Surfing. In *Proceedings of the 2nd USENIX Symposium on Internet Technologies and Systems,* Boulder, CO, October 1999.

Sarukkai, R. R. Link Prediction and Path Analysis Using Markov Chains. In *Proceedings of the 9th International World Wide Web Conference,* Amsterdam, May 2000.

Sarwar, B., G. Karypis, J. Konstan, and J. Riedl. Application of Dimensionality Reduction in Recommender Systems — A Case Study. In *Proceedings of the WebKDD 2000 Workshop at the ACM-SIGKDD Conference on Knowledge Discovery in Databases (KDD'2000),* Boston, MA, August 2000a.

Sarwar, B. M., G. Karypis, J. Konstan, and J. Riedl. Analysis of Recommender Algorithms for E-Commerce. In *Proceedings of the 2nd ACM E-Commerce Conference (EC'00),* Minneapolis, MN, October 2000b.

Schechter, S., M. Krishnan, and M. D. Smith. Using Path Profiles to Predict HTTP Requests. In *Proceedings of the 7th International World Wide Web Conference,* Brisbane, Australia, April 1998.

Shardanand, U. and P. Maes. Social Information Filtering: Algorithms for Automating "Word of Mouth." In *Proceedings of the Computer-Human Interaction Conference (CHI'95),* Denver, CO, May 1995.

Spiliopoulou, M. and H. Faulstich. WUM: A Tool for Web Utilization Analysis. In *Proceedings of EDBT Workshop at WebDB'98,* LNCS 1590, pp. 184–203. Springer-Verlag, New York, 1999.

Spiliopoulou, M., B. Mobasher, B. Berendt, and M. Nakagawa. A framework for the evaluation of session reconstruction heuristics in Web usage Analysis. *INFORMS Journal of Computing — Special Issue on Mining Web-Based Data for E-Business Applications,* 15(2), 2003.

Srivastava, J., R. Cooley, M. Deshpande, and P. Tan. Web Usage Mining: Discovery and Applications of Usage Patterns from Web Data. *SIGKDD Explorations,* 1(2): 12–23, 2000.

Tan, P. and V. Kumar. Discovery of Web robot sessions based on their navigational patterns. *Data Mining and Knowledge Discovery,* 6: 9–35, 2002.

Ungar, L. H. and D. P. Foster. Clustering Methods for Collaborative Filtering. In *Proceedings of the Workshop on Recommendation Systems at the 15th National Conference on Artificial Intelligence,* Madison, WI, July 1998.

W3C, World Wide Web Committee. Web Usage Characterization Activity. http://www.w3.org/WCA.

Yu, P. S. Data Mining and Personalization Technologies. In *Proceedings of the International Conference on Database Systems for Advanced Applications (DASFAA'99),* Hsinchu, Taiwan, April 1999.

Zaiane, O., M. Xin, and J. Han. Discovering Web Access Patterns and Trends by Applying OLAP and Data Mining Technology on Web Logs. In *Proceedings of the IEEE Conference on Advances in Digital Libraries (ADL'98).* Santa Barbara, CA, April 1998.

16

Agents

CONTENTS

Abstract 16-1
16.1 Introduction 16-2
16.2 What Is an Intelligent Agent? 16-2
16.3 Anatomy of an Agent 16-2
16.3.1 An Agent Architecture 16-3
16.3.2 Sensors: Gathering Input 16-3
16.3.3 Perception 16-4
16.3.4 Decision-Making Behavior 16-4
16.3.5 Communication 16-6
16.3.6 Effectors: Taking Action 16-7
16.3.7 Mobility 16-7
16.3.8 An Example Agent 16-7
16.4 Multiagent Teams 16-8
16.5 Intelligent Agents on the Internet 16-8
16.5.1 Bots 16-8
16.5.2 Agents behind Websites 16-9
16.6 Research Issues Related to Agents and the Internet ... 16-10
16.6.1 Agents and Human–Computer Interfaces 16-10
16.6.2 Agents and Privacy 16-10
16.6.3 Agents and Security 16-11
16.6.4 Autonomic Computing and Agents 16-11
16.7 Summary 16-12
16.8 Further Information 16-12
16.9 Glossary 16-12
Acknowledgments 16-13
References 16-13

Joseph P. Bigus

Jennifer Bigus

Abstract

The rise of distributed networked computing and of the Internet has spurred the development of autonomous intelligent agents (also called software robots or bots). Software agents are used to advise or assist users in performing tasks on the Internet, to help automate business processes, and to manage the network infrastructure. In this chapter, we explore the essential attributes of intelligent agents, describe an abstract agent architecture, and discuss the functional components required to implement the architecture. We examine the various types of intelligent agents and application-specific bots that are available to help users perform Internet-related tasks and describe how agents are used behind commercial Websites. Finally, we discuss some of the issues surrounding the widespread adoption of agent technology, including the human–computer interface, security, privacy, trust, delegation, and management of the Internet computing infrastructure.

1-58488-381-2/05/$0.00+$1.50
© 2005 by CRC Press LLC

16.1 Introduction

The rise of distributed networked computing and of the Internet has spurred the development of autonomous intelligent agents (also called software robots or bots). Software agents are used to advise or assist users in performing tasks on the Internet, to help automate business processes, and to manage the network infrastructure. While the last decade of Internet growth has provided a fertile and highly visible ground for agent applications, autonomous software agents can trace their heritage back to research in artificial intelligence spanning the past half century.

In this chapter, we explore the essential attributes of intelligent agents, describe a prototypical agent architecture, and discuss the functional components required to implement the architecture. We examine the various types of intelligent agents and application-specific bots that are available to help users perform Internet-related tasks and describe how agents are used behind commercial Websites. Finally, we discuss some of the issues surrounding the widespread adoption of agent technology, including the human–computer interface, security, privacy, trust, delegation, and management of the Internet computing infrastructure.

16.2 What Is an Intelligent Agent?

So, what is an intelligent agent and how does it differ from other software that is used on the desktop or on the Internet? There is no generally agreed-upon definition of intelligent agent, but there are several generally agreed-upon attributes of an agent [Franklin and Graesser, 1996]. First and foremost, agents are autonomous, meaning they can take actions on their own initiative. The user delegates authority to them, so they can make decisions and act on the user's behalf. Second, agents typically run for days or weeks at a time. This means that agents can monitor and collect data over a substantial time interval. Because they are long running, agents can "get to know" the user by watching the user's behavior while performing repetitive tasks [Maes, 1994] and by detecting historical trends in Web data sources that are of interest. While most agents stay in one place, some are mobile, meaning they can move between computer systems on the network. Finally, all agents must be able to communicate, either directly with people using a human–computer interface or with other agents using an agent communication language [Bradshaw, 1997].

16.3 Anatomy of an Agent

In this section we explore the basic architectural elements of an autonomous intelligent software agent. We start with a formal definition to motivate our discussion of the essential technical attributes of an agent.

> An intelligent agent is an active, persistent software component that can perceive, reason, act, and communicate [Huhns and Singh, 1998].

Let us examine this definition in more detail. Whereas intelligence, especially artificial intelligence, is somewhat in the eye of the beholder, a commonly accepted notion of intelligence is that of rationality. People are considered rational if they make decisions that help them achieve their goals. Likewise, a rational software agent is one that behaves with human-like decision-making abilities [Russell and Norvig, 1995]. An intelligent agent is a persistent software component, meaning it is not a transient software program but a long-running program whose current state, past experiences, and actions are maintained in persistent memory. For an agent to perceive, it must be able to sense its environment and process external data either through events or by polling. For an agent to reason, it must contain some form of domain knowledge and associated reasoning logic. To take action, an agent must have a decision-making component with the ability to change its environment through direct actions on hardware devices or by

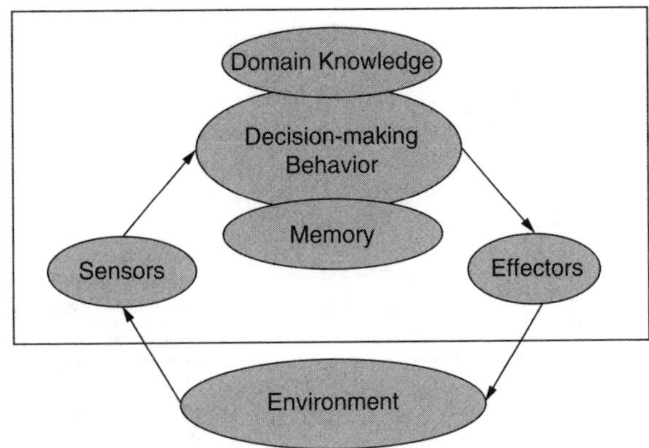

FIGURE 16.1 A basic intelligent agent architecture.

invoking actions on other software components. Finally, an agent must be able to communicate with humans and with other agents.

16.3.1 An Agent Architecture

In Figure 16.1 we show a diagram of the major components of an autonomous intelligent agent. The input and output modules, called sensors and effectors, are the interfaces between the agent and its world. The sensory input is preprocessed by an optional perceptual component and passed into a decision-making or behavioral component. This component must include decision-making logic, some domain knowledge that allows the agent to perform the prescribed task, and working memory that stores both short- and long-term memories. Once behavioral decisions are made by the decision-making component, any actions are taken through the effector component.

There are several major variations on the basic architecture described in Figure 16.1. Most of these variations deal with the structure of the behavioral or decision-making component. These include reactive agents that are typically simple stimulus–response agents [Brooks, 1991], deliberative agents that have some reasoning and planning components, and Belief–Desire–Intention (BDI) agents that contain complex internal models representing their beliefs about the state of the world, their current desires, and their committed intentions of what goals are to be achieved [Rao and Georgeff, 1991]. These architectures can be combined and layered, with lower levels being reactive and higher levels being more deliberative [Sloman, 1998].

While simple agents may have their goals hard coded into the decision-making logic, most agents have a control interface that allows a human user to specify goals and to provide feedback to the agent as it tries to meet those goals. Social agents also include the ability to communicate and cooperate with other agents. Many agents also include a learning component that allows the agent to adapt based on experience and user feedback.

Figure 16.2 shows a more complete and flexible agent architecture. In the following sections we describe the purpose and technical requirements of the various components used in the agent architectures, with a special focus on the elements of the behavioral subsystem.

16.3.2 Sensors: Gathering Input

Any software program needs a way to get input data. For agents, the inputs are provided by sensors, which allow the agent to receive information from its environment. Sensors can be connectors to a web server to read HTML pages, to an NNTP server to read articles in newsgroups, to an FTP connection to download files, or to messaging components that receive external events containing requests from users or other agents. A sensor could also be implemented to actively poll or request input data from an external

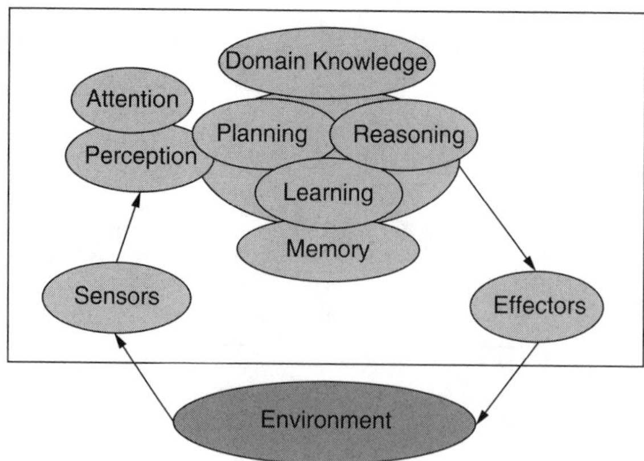

FIGURE 16.2 An expanded intelligent agent architecture.

hardware device, from a software component, or from another agent in the environment. In this architecture, sensors provide a unified input channel for low-level environmental inputs as well as high-level communications. An alternative would provide a separate channel for human–agent and agent–agent communications.

16.3.3 Perception

Although having sensors to gather data from the environment is necessary, the real trick is to turn that data into useful information. Like animals that have very specialized preprocessing systems for touch, smell, taste, hearing, and sight, software agents need similar preprocessing capabilities. The sensors gather raw data, and the perceptual subsystem converts the data into a format that can be easily digested by the decision-making component. For example, as events are streaming into an agent, it may need some way to detect patterns in those events. This job can be performed by an event correlation engine that takes events and turns them into higher-level percepts or situations. Doing this as a preprocessing step greatly lessens the burden on the agent's decision-making components.

Another important job of the perceptual component is to filter out noise. The agent must be able to differentiate between the usual, normal inputs and the unusual, abnormal ones. People have a natural ability to tune out noise and focus their attention on novel or interesting inputs. Agents need this same capability. In order to detect what is abnormal, it is often necessary to build internal models of the world. When data comes in, it is checked against the model of what is expected. If the data matches, then it is a normal occurrence, but if there is a mismatch, it is a new situation that has been detected. This function can also be performed by a separate attentional subsystem that works closely with the perceptual system to not only transform the raw sensor data but also to indicate the relative importance of that data. When exceptional conditions are detected, some agent designs provide special mechanisms for distribution of alarm signals that are processed at a higher priority than other input signals.

16.3.4 Decision-Making Behavior

The behavioral or decision-making component of an agent can range from a simple Tcl/Tk or Perl script to C++ or Java code, to a complex reasoning and inferencing engine. There are three major pieces of this subsystem: domain knowledge, working memory, and decision-making logic. When the behavior is defined by procedural code, the domain-knowledge and decision-making logic can be one and the same. When the behavior is defined by rules or semantic networks and processed by inferencing engines, then the domain knowledge is separate from the decision-making logic. The working memory can be as simple as local data or variables stored as part of the agent, or as sophisticated as an associative or content-addressable memory component. In the next sections we explore details of the decision-making component, including domain knowledge, reasoning, planning, and learning.

16.3.4.1 Domain Knowledge

How do we represent the information that an agent needs to perform its assigned task? The answer depends on the type of knowledge that must be represented in the agent. Perhaps the most common type of knowledge is procedural knowledge. Procedural knowledge is used to encode processes — step-by-step instructions for what to do and in what order. Procedural knowledge can be directly represented by computer programs.

A second type of knowledge is relational knowledge. A common format for relational knowledge is relational databases, where groups of related information are stored in rows or tuples, and the set of related attributes are stored in the columns or fields of the database table. Although this is relational, it does not explicitly allow for the definition of the relationships between the fields. For the latter, graph-based representations such as semantic networks can be used to represent the entities (nodes) and the relationships (links).

Another type of knowledge representation is hierarchical or inheritable knowledge. This type of knowledge representation allows "kind-of," "is-a," and "has-a" relationships between objects and allows reasoning about classes using a graph data structure.

Perhaps the most popular knowledge representation is simple if–then rules. Rules are easily understood by nontechnical users and support a declarative knowledge representation. Because each rule stands alone, the knowledge is declared and explicitly defined by the antecedent conditions on the left-hand side of the rule and the consequent actions on the right-hand side of the rule. Note that whereas individual rules are clear and concise, large sets of rules or rulesets are often required to cover any nontrivial domain, introducing complex issues related to rule management, maintenance, priorities, and conflict resolution.

16.3.4.2 Reasoning

Machine reasoning is the use of algorithms to process knowledge in order to infer new facts or to prove that a goal condition is true. If the most common knowledge representation is if–then rules, then the most common machine reasoning algorithms are inferencing using forward or backward chaining. With rule-based reasoning, there are three major components: data, rules, and a control algorithm. In forward chaining, the initial set of facts is expanded through the firing of rules whose antecedent conditions are true, until no more rules can fire [Forgy, 1982]. This process uses domain knowledge to enhance the understanding of a situation using a potentially small set of initial data. For example, given a customer's age and income, forward chaining can be used to infer if the customer is a senior citizen or is entitled to silver, gold, or platinum level discounts. The backward-chaining algorithm can use the same set of rules as forward chaining, but works back from the goal condition through the antecedent clauses of the rules to find a set of bindings of data to variables such that the goal condition is proved true or false [Bratko, 1986]. For example, an expert system whose goal is to offer product selection advice can backward chain through the rules to guide the customer through the selection process.

A popular alternative to inferencing using Boolean logic is the use of fuzzy logic [Zadeh, 1994]. An advantage of fuzzy rule systems is that linguistic variables and hedges provide an almost natural language-like knowledge representation, allowing expressions such as "almost normal" and "very high" in the rules.

16.3.4.3 Planning or Goal-Directed Behavior

When a task-oriented agent is given a goal, the agent must first determine the sequence of actions that must be performed to reach the goal. Planning algorithms are used to go from the initial state of the world to the desired end state or goal by applying a sequence of operators to transform the initial state into intermediate states until the goal state is reached. The sequence of operators defines the plan. Once the planning component determines the sequence of actions, the plan is carried out by the agent to accomplish the task. The main algorithms in AI planning are the operator-based STRIPS approach and the hierarchical task network (HTN) approach [IEEE Expert, 1996].

Although planning looks easy at first glance, planning algorithms become very complex due to constraints on the order of operations and changes in the world that occur after the plan was calculated, but before it is completed. Taking into account uncertainties and choosing between alternative plans also complicate things. Because multiple solutions may exist, the number of possible combinations of actions can overwhelm a planning agent due to a combinatorial explosion.

16.3.4.4 Learning

A key differentiator for an agent is the ability to adapt and learn from experience. Despite our best efforts, there is no way to anticipate, *a priori*, all of the situations that an agent will encounter. Therefore, being able to adapt to changes in the environment and to improve task performance over time is a big advantage that adaptive agents have over agents that cannot learn.

There are several common forms of learning, such as rote learning or memorization; induction or learning by example where the important parameters of a problem are extracted in order to generalize to novel but similar situations; and chunking, where similar concepts are clustered into classes. There are a wide variety of machine-learning algorithms that can discern patterns in data. These include decision trees, Bayesian networks, and neural networks. Neural networks have found a large niche in applications for classification and predication using data sets and are a mainstay of business data-mining applications. [Bigus, 1996]

There are three major paradigms for learning: supervised, unsupervised, and reinforcement learning. In supervised learning, explicit examples of inputs and corresponding outputs are presented to the learning agent. These could be attributes of an object and its classification, or elements of a function and its output value. Common supervised learning algorithms include decision trees, back propagation neural networks, and Bayesian classifiers. As an example, data-mining tools can use these algorithms to classify customers as good or bad credit risks based on past experience with similar customers and to predict future profitability based on a customer's purchase history. In unsupervised learning, the data is presented to the learning agent and common features are used to group or cluster the data using a similarity or distance metric. Examples of unsupervised learning algorithms include Kohonen map neural networks and K-nearest-neighbor classifiers. A common use of unsupervised learning algorithms is to segment customers into affinity groups and to target specific products or services to members of each group. Reinforcement learning is similar to supervised learning in that explicit examples of inputs are presented to the agent, but instead of including the corresponding output value, a nonspecific reinforcement signal is given after a sequence of inputs is presented. Examples of reinforcement learning algorithms include temporal difference learning and Q-learning.

In addition to the general learning paradigm is the issue of whether the learning agent is trained using all the data at once (batch mode or offline) or if it can learn from one example at a time (incremental mode or online). In general, incremental learning and the ability to learn from a small number of examples are essential attributes of agent learning.

16.3.5 Communication

The communication component must be able to interact with two very different types of partners. First is the human user who tells the agent what to do. The interface can range from a simple command line or form-based user interface to a sophisticated natural language text or speech interface. Either way, the human must be able to tell the agent what needs to be done, how it should be done, and when the task must be completed. Once this is accomplished, the agent can autonomously perform the task. However, there may be times when the agent cannot complete its assigned task and has to come back to the user to ask for guidance or permission to take some action. There are complex issues related to how humans interact and relate to intelligent agents. Some of the issues are discussed in detail in the final section of this chapter.

The second type of communication is interaction with other agents. There are two major aspects to any communication medium: the protocol for the communication and the content of the communication.

The protocol determines how two agents find each other, and how messages are formatted so that the receiving agent can read the content. This metadata, or data about the data, is handled by an agent communication language such as KQML (knowledge query and manipulation language) [Labrou and Finin, 1997]. The second aspect of communication is the semantic content or meaning of the message. This is dependent on a shared ontology, so that both agents know what is meant when a certain term is used. The Semantic Web project is an attempt to define ontologies for many specialized domains [Berners-Lee et al., 2001]. Agents use these ontologies to collaborate on tasks or problems.

16.3.6 Effectors: Taking Action

Effectors are the way that agents take actions in their world. An action could be sending a motion command to a robotic arm, sending an FTP command to a file server, displaying an HTML Web page, posting an article to an NNTP newsgroup, sending a request to another agent, or sending an e-mail notification to a user. Note that effectors may be closely tied to the communication component of the agent architecture.

16.3.7 Mobility

Agents do not have to stay in one place. Mobility allows an agent to go where the data is, which can be a big advantage if there is a large amount of data to examine. A disadvantage is that there must be a mobile agent infrastructure in place for agents to move around a network. Long-running processes must be available on each computer system where the agent may wish to reside. Security becomes a major issue with mobile agents because it is sometimes difficult to differentiate legitimate agents acting on behalf of authorized users from illegitimate agents seeking to steal data or cause other havoc with the computing systems.

16.3.8 An Example Agent

In this section, we describe the design and development of an intelligent agent for filtering information, specifically articles in Internet newsgroups. The goal is to build an intelligent assistant to help filter out spam and uninteresting articles posted to one or more newsgroups and to rank and present the interesting articles to the user.

The mechanics of how we do this is straightforward. The agent interacts with the newsgroup server using the NNTP protocol. Its sensors and effectors are sockets through which NNTP commands and NNTP responses are sent and received. The human–computer interface is a standard newsgroup reader interface that allows the user to select the newsgroup to monitor, download the articles from the selected newsgroups, and display their subject line in a list so the user can select them for viewing.

What we have described so far is a standard newsgroup reader. Simple bots could be used to automate this process and automatically download any unread articles for the user when requested. A more powerful agent could allow the user to specify specific keywords that are of interest. The agent could score the articles posted to the newsgroup based on those keywords. The articles could be presented to the user, ordered by their score.

The next level of functionality would be to allow the user to provide feedback to the agent so that the agent could adapt and tune its scoring mechanism. Instead of a preset list of keywords, the agent could use the feedback to add weighting to certain keywords and refine the scoring mechanism. Neural networks could be used to build a model of articles and keyword counts mapped to the expected interest level of the user. Using feedback over time, the scoring mechanism would be tuned to reflect the user's weightings of the various keywords. An agent of this type was developed in Bigus and Bigus [2001] using a Java agent framework.

16.4 Multiagent Teams

While individual agents can be useful for simple tasks, most large-scale applications involve a large number of agents. Each agent plays specific roles in the application and contains task- or domain-specific knowledge and capabilities. When agents collaborate to solve a problem, they often must reason not only about their individual goals but also about other agents' intentions, beliefs, and goals. They must make commitments to other agents about what goals and actions they intend to pursue and, in turn, depend on other agents to fulfill their commitments [Cohen and Levesque, 1990; Sycara, 1998].

A community of agents requires a set of common services to operate efficiently, much like a city needs basic services and infrastructure to work well. These include yellow pages or directory services where agents can register themselves and their capabilities and interests so that other agents can find them. The agents need a communication infrastructure so they can send messages, ask questions, give answers, and plan and coordinate group operations.

16.5 Intelligent Agents on the Internet

In this section, we discuss common uses of agents on the Internet today. This includes a review of software robots or bots commonly available on the Web and of agents used as part of Websites and e-business applications.

16.5.1 Bots

A whole cottage industry has been spawned for highly specialized agents that are useful for Web-oriented tasks. These cover the gamut from Web searching, information tracking, downloading software, and surfing automation agents to Internet auctions, monitoring stocks, and Internet games [Williams, 1996]. These agents are available for download and personal use, and range from $5 to $50 or more depending on their sophistication and power.

One of the first and most useful applications of agents on the Internet was their use in solving the problem of finding information. Commercial web search sites such as Alta Vista, Yahoo, and Google utilize the combination of an information taxonomy constructed by hand and the content information gleaned from hundreds of thousands of Websites using specialized agents called spiders. These agents scour the Web for information and bring it back to the search site for inclusion in the search database.

Search bots allow you to enter queries and then submit the queries to multiple Internet search sites. They collect and interpret the results from the search and provide a unified set of results.

Tracking bots allows you to keep an eye on Websites and Web pages of interest. They can notify you when site content has been updated and even provide snapshots of the old and new content with changes highlighted. These bots specialize in news and stock information tracking and can also be used to monitor the health of your own Website.

File-sharing bots such as KaZaA, WinMX, and Morpheus allow you to share data on your computer's hard drive with hundreds or thousands of other Internet users. These bots enable peer-to-peer computing because your computer can share data directly with other file-sharing bot users without going through a central server computer.

Download bots help Internet users automate the process of downloading programs, music, and videos from Websites. They add functionality to browsers by improving the download speed by using multiple threads and by being able to recover and pick up where they left off when the network connection was lost.

Personal assistant bots read Web-page and e-mail content aloud to users (for the visually impaired). Some automatically translate text from one national language to another.

Surf bots can help users avoid those annoying pop-up ads that have proliferated at some commercial Websites by immediately closing the pop-up ads. Browsers store URLs and cache Web pages, and sites leave cookies of information on a user's computer. An intrusive or malicious user could see where you have been and what you have been up to on the Internet. Privacy bots can be used to remove all traces of your activities while Web surfing.

Shopping bots help consumers find the potential sellers of products as well as comparison shop by price. The economic benefit of these agents to the buyer is obvious. The potential impact on sellers is less obvious. The widespread use of shopping bots on the Internet could possibly lead to price wars and loss of margins for most commodities.

Auction bots have been developed to assist Internet users who make use of Internet auction sites such as eBay or Yahoo to buy or sell items. These agents monitor the bidding on specific auctions and inform their owner of significant changes in the status of the auction.

Stock bots perform a similar function by notifying the user regarding the movement of stock prices over the course of a trading day, based on user-specified conditions.

Chatterbots are agents that can perform a natural language dialogue or discussion with a human. The earliest chatterbot was the AI program eLiza, which was a simple pattern-matching program intended to simulate a psychoanalyst. Modern chatterbots use a combination of natural language understanding and modeling in an attempt to educate and entertain.

Game bots were developed to act as computer players in Internet games. They allow a single user to interact with multiple agent players to try alternative strategies for practice or just to play the game.

16.5.2 Agents behind Websites

Whereas most users of the Internet think of the World Wide Web as a collection of HTML and dynamic HTML pages, commercial Websites often have intelligent agents working behind the scenes to provide personalization, customization, and automation. For example, Website reference bots allow site owners to check their current standings on the most popular search engines. They also provide automated submission of their site for inclusion in the search engine's index.

One of the promises of the Internet from a customer-relationship perspective is to allow businesses to have a personalized relationship with each customer. This one-to-one marketing can be seen in the personalization of a user's Web experience. When a consumer logs onto a Website, different content is displayed based on past purchases or interests, whether explicitly expressed via profiles or by past browsing patterns. Companies can offer specials based on perceived interests and the likelihood that the consumer would buy those items. Personalization on Websites is done using a variety of intelligent technologies, ranging from statistical clustering to neural networks.

Large e-commerce Websites can get hundreds or thousands of e-mail inquiries a day. Most of these are of rather routine requests for information or simple product requests that can be handled by automated e-mail response agents. The e-mail is first analyzed and classified and then an appropriate response is generated using a knowledge base of customer and product information. Agents are also used behind customer self-service applications, such as product configurators and product advisors.

The future of applications on the Internet seems to center on the development of electronic commerce and Web Services. Web Services depend on standards including WSDL (Web Services Definition Language) and SOAP (Simple Object Access Protocol), which allow companies to specify the services they can provide as well as the methods and bindings to invoke those services. Intelligent agents will certainly play a role in the definition and provisioning of these Internet services.

Another application area where intelligent agents play a role is in business-to-business purchasing and supply-chain replenishment. A typical company has hundreds of suppliers, and a manufacturing company may have thousands of part suppliers. Managing the ordering and shipment of parts to manufacturing plants to insure the speedy output and delivery of finished products is a complex problem. Aspects of this application handled by intelligent agents include the solicitation of part providers, negotiation of terms and conditions of purchase agreements, and tracking of delivery dates and inventory levels so that stocks can be kept at appropriate levels to maximize manufacturing output while minimizing the carrying costs of the parts inventory.

16.6 Research Issues Related to Agents and the Internet

In this section we discuss issues related to the adoption and widespread use of agents on the Internet, including human–computer interfaces, privacy, security, and autonomic computing.

16.6.1 Agents and Human–Computer Interfaces

The interface between humans and autonomous agents has some elements that are similar to any human–computer interface, but there are some additional issues to consider when designing the user interface for an intelligent agent. Unlike computer software tools that are used under the direct control and guidance of the user, intelligent agents are autonomous, acting on the authority of the user, but outside the user's direct control. This model, in which the user delegates tasks to the agent, has implications and effects that must be considered when designing the human–computer interface.

Studies conducted in the organizational and management sciences have shown that delegation is an important skill for successful managers. However, they often fail to delegate for a number of reasons including the amount of time it takes to explain what needs to be done, the loss of control over the task while still being held accountable for the results, and fear that the task will not get done or will not be done well. Trust is an issue in human-to-agent delegation as well.

Intelligent agents are more suitable for some tasks than for others. The user interface must be designed in such a way that the user can opt out of delegating to the agent if the cost of using the agent outweighs the benefits. Part of the cost of using an intelligent agent is the time it takes to communicate with the agent. The interface design should allow the user to convey intentions and goals to the agent using natural language interfaces or by demonstrating what needs to be done. The agent must be capable of conveying its understanding of the task to the user, often through the anthropomorphic use of gestures, facial expressions, and vocalization [Milewski and Lewis, 1997].

Anthropomorphic agents are becoming a more common user interface paradigm, especially for agents that act as advisors, assistants, or tutors. An anthropomorphic interface makes an agent more personable, helping to establish trust and a comfort level beyond what is normally experienced with a traditional human–computer interface. Agents represented as three-dimensional characters are judged by users to have both a high degree of agency and of intelligence [King and Ohya, 1996]. Care must be taken, however, to ensure that the social interaction capabilities of the agent are sophisticated enough to meet the user's expectations. Failure to measure up to those expectations can negatively impact the user's perception of the agent [Johnson, 2003].

The anthropomorphic characteristics of an agent are often used to convey emotion. A character that expresses emotion increases the credibility and believability of an agent [Bates, 1994]. Emotion also plays an important role in motivation of students when interacting with a pedagogical agent. Having an agent that "cares" about how a student is doing can encourage the student to also care about his or her progress. In addition, enthusiasm for the subject matter, when conveyed by the agent, may foster enthusiasm in the student. Agents that have personality make learning more fun for the student, which, in turn, can make the student more receptive to learning the subject matter [Johnson et al. 2000].

Overall, the human–computer interface of an intelligent agent provides an additional dimension to the agent that can further add to the effectiveness of the agent as a personal assistant. The affective dimension of autonomous agent behavior can give the interface an emotional component to which the user reacts and relates. The challenge lies in building agents that are believable and lifelike, enhancing the interaction between the agent and the user without being condescending, intrusive, or annoying.

16.6.2 Agents and Privacy

One of the biggest impediments to wider use of the Internet for electronic commerce is the concern consumers have about privacy. This includes the ability to perform transactions on the Internet without leaving behind a trail of cookies and Web logs. Another equally important aspect is the privacy and security of their personal, financial, medical, and health data maintained by companies with whom they do business. The ease-of-use and vast array of information are enticing, but the thought of unauthorized or criminal access to personal information has a chilling effect for many people.

What role can agents play in this space? One could argue that sending out an autonomous (and anonymous) agent to make a purchase could assuage those concerns. Of course, this assumes that the privacy of any information contained in the agent is maintained. A secure network connection protocol, such as HTTPS, could be used by the agent to communicate with the Websites.

16.6.3 Agents and Security

Another major issue for the use of autonomous agents on the Internet is related to security. Although agents can help legitimate users perform tasks such as searching for information on Websites and bidding in online auctions, they can also be used by unscrupulous users to flood Websites with requests or send e-mail spam. How can we leverage intelligent agent technology to enhance the usefulness of the Internet without providing new opportunities for misuse?

One approach is to ensure that each and every agent can be traced back to a human user. The agent would be able to adopt the authorization of the user and perform transactions on the user's behalf. Each agent would require a digital certificate used to prove the agent's identity and to encrypt any sensitive messages. The same security mechanisms that make electronic commerce safe for human users on Websites could be used to secure agent-based applications.

16.6.4 Autonomic Computing and Agents

A major computing initiative and grand challenge in computer science is in the area of self-configuring, self-healing, self-optimizing, and self-protecting systems known as autonomic computing. Researchers in industry and academia have identified complexity in the distributed computing infrastructure as a major impediment to realizing sustainable growth and reliability in the Internet [Kephart and Chess, 2003].

An autonomic computing architecture has been developed to help manage this complexity. The architecture, shown is Figure 16.3 is made up of four major elements, known as the MAPE loop, where M stands for monitoring, A represents data analysis, P is for planning, and E is for execution. A central component of the architecture is the domain knowledge. The autonomic manager components can be implemented as intelligent agents using sensors and effectors to interact with other autonomic elements and the resources they are managing [Bigus et al., 2002].

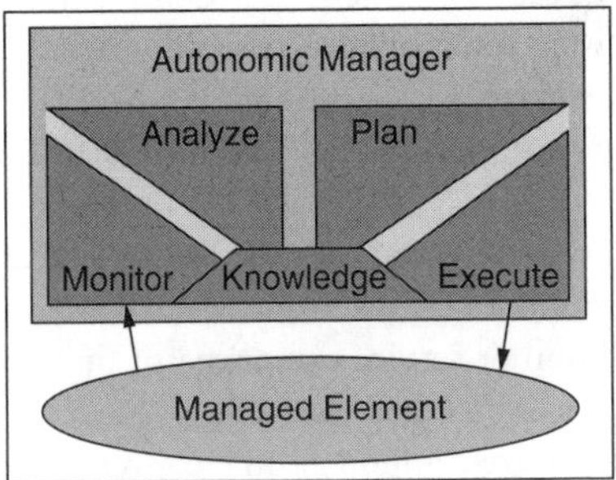

FIGURE 16.3 An autonomic computing architecture.

16.7 Summary

In this chapter, we defined intelligent agents and described their essential attributes, such as autonomy, persistence, reasoning, and communication. Basic and expanded intelligent agent architectures were presented, and the various functional components including sensors, effectors, perception, reasoning, planning, and learning were described. We identified many types of bots used on the Internet today and introduced several research issues related to the successful application of intelligent agent technology.

16.8 Further Information

The research, development and deployment, of intelligent agent applications on the Internet are very active areas in computer science today. Information on the latest research can be found at the major agent conferences including the International Joint Conference on Autonomous Agents and Multiagent Systems (AAMAS), the National Conference on Artificial Intelligence (AAAI), and the International Joint Conference on Artificial Intelligence (IJCAI). Major magazines and journals include *IEEE Internet Computing, IEEE Intelligent Systems, Artificial Intelligence Magazine (AAAI), Artificial Intelligence Journal* (Elsevier), *Journal of Artificial Intelligence Research (JAIR)*, and *The Journal of Experimental and Theoretical Artificial Intelligence (JETAI)*. Popular Web resources include the BotSpot site found at www.botspot.com, the PC AI magazine site at http://www.pcai.com/ and the University of Maryland Baltimore County (UMBC) Agent Web site at http://agents.umbc.edu.

16.9 Glossary

Agent communication language: A formal language used by agents to talk with one another.

Agent platform: A set of distributed services including agent lifecycle, communication and message transport, directory, and logging services on which multiple agents can run.

Anthropomorphic agent: An agent whose representation has virtual human characteristics such as gestures, facial expressions, and vocalization used to convey emotion and personality.

Artificial intelligence: Refers to the ability of computer software to perform activities normally thought to require human intelligence.

Attentional subsystem: The software component that determines the relative importance of input received by an agent.

Autonomous agent: An agent that can take actions on its own initiative.

Behavioral subsystem: The decision-making component of an agent that includes domain knowledge, working memory, and decision-making logic.

Belief–Desire–Intention agent: A sophisticated type of agent that holds complex internal states representing its beliefs, desires, and intended actions.

Bot: A shorthand term for a software robot or agent, usually applied to agents working on the Internet.

Effector: The means by which an agent takes action in its world.

Intelligent Agent: An active, persistent software component that can perceive, reason, act, and communicate.

Learning: The ability to adapt behavior based on experience or feedback.

Mobility: The ability to move around a network from system to system.

Multiagent system: An application or service comprised of multiple intelligent agents that communicate and use the services of an agent platform.

Perceptual subsystem: The software component that converts raw data into the format used by the decision-making component of an agent.

Planning: The ability to reason from an initial world state to a desired final state producing a partially ordered set of operators.

Reasoning: The ability to use inferencing algorithms with a knowledge representation.

Sensor: The means by which an agent received input from its environment.

Social agent: An agent with the ability to interact and communicate with other software agents and to join in cooperative and competitive activities.

Acknowledgments

The authors would like to acknowledge the support of the IBM T.J. Watson Research Center and the IBM Rochester eServer Custom Technology Center.

References

Bates, Joseph. The role of emotion in believable agents. *Communications of the ACM*, 37(7): 122–125, July 1994.

Berners-Lee, Tim, James Hendler, and Ora Lassila. The Semantic Web. *Scientific American*, May 2001.

Bigus, Joseph P. *Data Mining with Neural Networks*. McGraw Hill, New York, 1996.

Bigus, Joseph P. and Jennifer Bigus. *Constructing Intelligent Agents using Java*, 2nd ed., John Wiley & Sons, New York, 2001.

Bigus, Joseph P., Donald A. Schlosnagle, Jeff R. Pilgrim, W. Nathanial Mills, and Yixin Diao. ABLE: A toolkit for building multiagent autonomic systems. *IBM Systems Journal*, 41(3): 350–370, 2002.

Bradshaw, Jeffrey M. Ed. *Software Agents*. MIT Press, Cambridge, MA, 1997.

Bratko, Irving. *Prolog Programming for Artificial Intelligence*. Addison-Wesley, Reading, MA, 1986.

Brooks, Rodney A. Intelligence without representation. *Artificial Intelligence Journal*, 47: 139–159, 1991.

Cohen, Philip R. and Hector J. Levesque. Intention is choice with commitment. *Artificial Intelligence Journal*, 42(2–3): 213–261, 1990.

Forgy, Charles L. Rete: A fast algorithm for the many pattern/many object pattern match problem. *Artificial Intelligence*, 19: 17–37, 1982.

Franklin, Stan and Art Graesser. Is it an agent, or just a program? A taxonomy for autonomous agents. *Proceedings of the 3rd International Workshop on Agent Theories, Architectures, and Languages*. Springer-Verlag, New York, 1996.

Huhns, Michael N. and Munindar P. Singh, Eds. *Readings in Agents*. Morgan Kaufmann, San Francisco, 1998.

IEEE Expert. AI planning systems in the real world. *IEEE Expert*, December, 4–12, 1996.

Johnson, W. Lewis. Interaction tactics for socially intelligent pedagogical agents. *International Conference on Intelligent User Interfaces*, 251–253, 2003.

Johnson, W. Lewis, Jeff W. Rickel, and James C. Lester. Animated pedagogical agents: face-to-face interaction in interactive learning environments. *International Journal of Artificial Intelligence in Education*, 11: 47–78, 2000.

Kephart, Jeffrey O. and David M. Chess. The vision of autonomic computing. *IEEE Computer*, January 2003.

King, William J. and Jun Ohya. The representation of agents: anthropomorphism, agency, and intelligence. *Proceedings CHI'96 Conference Companion*, 289–290, 1996.

Labrou, Yannis and Tim Finin. Semantics and conversations for an agent communication language. *Proceedings of the 15th International Conference on Artificial Intelligence*, 584–491. International Joint Conferences on Artificial Intelligence, 1997.

Maes, Patti. Agents that reduce work and information overload. *Communications of the ACM*, 7: 31–40, 1994.

Milewski, Allen E. and Steven H. Lewis. Delegating to software agents. *International Journal of Human–Computer Studies*, 46: 485–500, 1997.

Rao, Anand S. and Michael P. Georgeff. Modeling rational agents within a BDI-architecture. *Proceedings of the International Conference on Principles of Knowledge Representation and Reasoning*, 473–484, 1991.

Sloman, Aaron. Damasio, Descartes, alarms, and meta-management. *Proceedings of the IEEE International Conference on Systems, Man, and Cybernetics*, 2652–2657, San Diego, CA, 1998.

Sycara, Katia. Multiagent Systems. *AI Magazine* 19(2): 79–92, 1998.

Russell, Stewart and Peter Norvig. *Artificial Intelligence: A Modern Approach*. Prentice Hall, Englewood Cliffs, NJ, 1995.

Williams, Joseph, Ed. *Bots and other Internet Beasties*. Sams.net, Indianapolis, IN, 1996.

Zadeh, Lotfi A. Fuzzy logic, neural networks, and soft computing. *Communications of the ACM*, 3: 78–84, 1994.

17

Multiagent Systems for Internet Applications

Michael N. Huhns

Larry M. Stephens

CONTENTS

Abstract ... 17-1
17.1 Introduction ... 17-1
17.1.1 Benefits of an Approach Based on Multiagent Systems ... 17-3
17.1.2 Brief History of Multiagent Systems ... 17-3
17.2 Infrastructure and Context for Web-Based Agents ... 17-4
17.2.1 The Semantic Web ... 17-4
17.2.2 Standards and Protocols for Web Services ... 17-4
17.2.3 Directory Services ... 17-5
17.3 Agent Implementations of Web Services ... 17-6
17.4 Building Web-Service Agents ... 17-6
17.4.1 Agent Types ... 17-6
17.4.2 Agent Communication Languages ... 17-14
17.4.3 Knowledge and Ontologies for Agent ... 17-15
17.4.4 Reasoning Systems ... 17-16
17.4.5 Cooperation ... 17-17
17.5 Composing Cooperative Web Services ... 17-17
17.6 Conclusion ... 17-17
Acknowledgment 17
References ... 17-17

Abstract

The World Wide Web is evolving from an environment for *people* to obtain information to an environment for *computers* to accomplish tasks on behalf of people. The resultant Semantic Web will be computer friendly through the introduction of standardized Web services. This chapter describes how Web services will become more agent-like, and how the envisioned capabilities and uses for the "Semantic Web" will require implementations in the form of multiagent systems. It also describes how to construct multiagent systems that implement Web-based software applications.

17.1 Introduction

Web services are the most important Internet technology since the browser. They embody computational functionality that corporations and organizations are making available to clients over the Internet. Web services have many of the same characteristics and requirements as simple software agents, and because of the kinds of demands and expectations that people have for their future uses, it seems apparent that Web services will soon have to be more like complete software agents. Hence, the best way to construct Web services will be in terms of multiagent systems. In this chapter, we describe the essential character-

1-58488-381-2/05/$0.00+$1.50
© 2005 by CRC Press LLC

istics and features of agents and multiagent systems, how to build them, and then how to apply them to Web services.

The environment for Web services, and computing in general, is fast becoming ubiquitous and pervasive. It is ubiquitous because computing power and access to the Internet is being made available everywhere; it is pervasive because computing is being embedded in the very fabric of our environment. Xerox Corporation has coined the phrase "smart matter" to capture the idea of computations occurring within formerly passive objects and substances. For example, our houses, our furniture, and our clothes will contain computers that will enable our surroundings to adapt to our preferences and needs. New visions of interactivity portend that scientific, commercial, educational, and industrial enterprises will be linked, and human spheres previously untouched by computing and information technology, such as our personal, recreational, and community life, will be affected.

This chapter suggests a multiagent-based architecture for all of the different devices, components, and computers to understand each other, so that they will be able to work together effectively and efficiently. The architecture that we describe is becoming canonical. Agents are used to represent users, resources, middleware, security, execution engines, ontologies, and brokering, as depicted in Figure 17.1. As the technology advances, we can expect such specialized agents to be used as standardized building blocks for information systems and Web services.

Multiagent systems are applicable not only to the diverse information soon to be available locally over household, automobile, and environment networks but also to the huge amount of information available globally over the World Wide Web being made available as Web services.

Organizations are beginning to represent their attributes, capabilities, and products on the Internet as services that can be invoked by potential clients. By invoking each other's functionalities, the Web services from different organizations can be combined in novel and unplanned ways to yield larger, more comprehensive functionalities with much greater value than the individual component services could provide.

Web services are XML-based, work through firewalls, are lightweight, and are supported by all software companies. They are a key component of Microsoft's .NET initiative, and are deemed essential to the business directions being taken by IBM, Sun, and SAP. Web services are also central to the envisioned *Semantic Web* [Berners-Lee et al., 2001], which is what the World Wide Web is evolving into. But the Semantic Web is also seen as a friendly environment for software agents, which will add capabilities and functionality to the Web. What will be the relationship between multiagent systems and Web services?

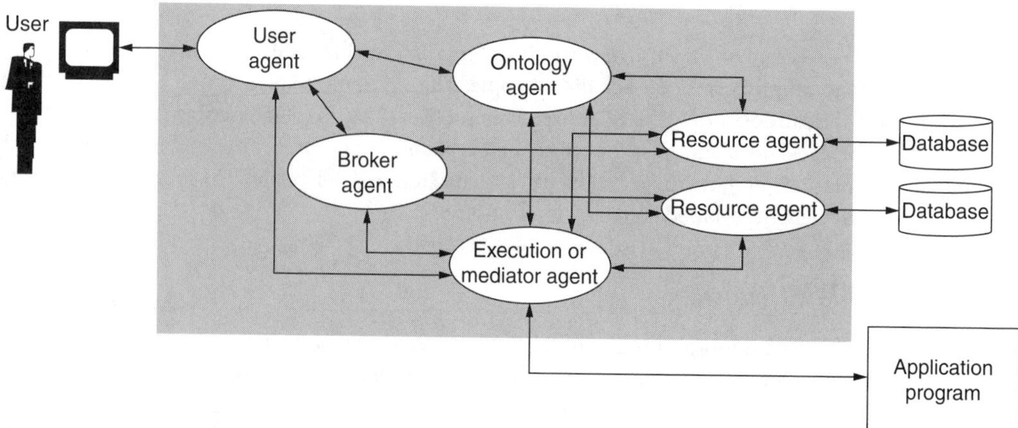

FIGURE 17.1 The agents in a multiagent Internet application first determine a user's request (the responsibility of the user agent) and then satisfy it by managing its processing. Under the control of the execution agent, the request might be sent to one or more databases or Websites, which are managed by resource agents.

17.1.1 Benefits of an Approach Based on Multiagent Systems

Multiagent systems can form the fundamental building blocks for not only Web services but also software systems in general, even if the software systems do not themselves require any agent-like behaviors [Jennings, 2000]. When a conventional software system is constructed with agents as its modules, it can exhibit the following characteristics:

- Agent-based modules, because they are active, more closely represent real-world things, that are the subjects of many applications.
- Modules can hold beliefs about the world, especially about themselves and others; if their behavior is consistent with their beliefs, then it will be more predictable and reliable.
- Modules can negotiate with each other, enter into social commitments to collaborate, and can change their mind about their results.

The benefits of building software out of agents are [Coelho et al., 1994; Huhns, 2001]:

1. Agents enable dynamic composibility, where the components of a system can be unknown until runtime.
2. Agents allow interaction abstractions, where interactions can be unknown until runtime.
3. Because agents can be added to a system one-at-a-time, software can continue to be customized over its lifetime, even potentially by end users.
4. Because agents can represent multiple viewpoints and can use different decision procedures, they can produce more robust systems. The essence of multiple viewpoints and multiple decision procedures is redundancy, which is the basis for error detection and correction.

An agent-based system can cope with a growing application domain by increasing the number of agents, each agent's capability, the computational resources available to each agent, or the infrastructure services needed by the agents to make them more productive. That is, either the agents or their interactions can be enhanced.

As described in Section 17.2.3, agents share many functional characteristics with Web services. We show how to build agents that implement Web services and achieve the benefits listed above. In addition, we describe how personal agents can aid users in finding information on the Web, keeping data current (such as trends in stocks and bonds), and alerts to problems and opportunities (such as bargains on eBay). We next survey how agent technology has progressed since its inception, and indicate where it is heading.

17.1.2 Brief History of Multiagent Systems

Agents and agency have been the object of study for centuries. They were first considered in the philosophy of action and ethics. In this century, with the rise of psychology as a discipline, human agency has been studied intensively.

Within the five decades of artificial intelligence (AI), computational agents have been an active topic of exploration, The AI work in its earliest stages investigated agents explicitly, albeit with simple models. Motivated by results from psychology, advances in mathematical logic, and concepts such as the Turing Test, researchers concentrated on building individual intelligent systems or one of their components, such as reasoning mechanisms or learning techniques. This characterized the first 25 years of AI. However, the fact that some problems, such as sensing a domain, are inherently distributed, coupled with advances in distributed computing, led several researchers to investigate distributed problem solving and distributed artificial intelligence (DAI). Progress and directions in these areas became informed more by sociology and economics than by psychology.

From the late seventies onward, the resultant DAI research community [Huhns, 1987; Bond and Gasser, 1988; Gasser and Huhns, 1989] concerned itself with agents as computational entities that interacted with each other to solve various kinds of distributed problems. To this end, whereas AI at large borrowed abstractions such as beliefs and intentions from psychology, the DAI community borrowed abstractions and insights from sociology, organizational theory, economics, and the philosophies of language and linguistics. These abstractions complement rather than oppose the psychological abstractions, but — being about groups of agents — are fundamentally better suited to large distributed applications.

With the expansion of the Internet and the Web in the 1990s, we witnessed the emergence of software agents geared to open information environments. These agents perform tasks on behalf of a user or serve as nodes — brokers or information sources — in the global information system. Although software agents of this variety do not involve specially innovative techniques, it is their synthesis of existing techniques and their suitability for their application that makes them powerful and popular. Thus, much of the attention they have received is well deserved.

17.2 Infrastructure and Context for Web-Based Agents

17.2.1 The Semantic Web

The World Wide Web was designed for humans. It is based on a simple concept: information consists of pages of text and graphics that contain links, and each link leads to another page of information, with all of the pages meant to be viewed by a person. The constructs used to describe and encode a page, the Hypertext Markup Language (HTML), describe the appearance of the page but not its contents. Software agents do not care about appearance, but rather the contents. The Semantic Web will add Web services that are envisioned to be

- Understandable to computers
- Adaptable and personalized to clients
- Dynamically composable by clients
- Suitable for robust transaction processing by virtual enterprises.

There are, however, some agents that make use of the Web as it is now. A typical kind of such agent is a *shopbot*, an agent that visits the online catalogs of retailers and returns the prices being charged for an item that a user might want to buy. The shopbots operate by a form of "screen-scraping," in which they download catalog pages and search for the name of the item of interest, and then the nearest set of characters that has a dollar-sign, which presumably is the item's price. The shopbots also might submit the same forms that a human might submit and then parse the returned pages that merchants expect are being viewed by humans. The Semantic Web will make the Web more accessible to agents by making use of semantic constructs, such as ontologies represented in OWL, RDF, and XML, so that agents can *understand* what is on a page.

17.2.2 Standards and Protocols for Web Services

A Web service is a functionality that can be engaged over the Web. Web services are currently based on the triad of functionalities depicted in Figure 17.2. The architecture for Web services is founded on principles and standards for connection, communication, description, and discovery. For providers and requestors of services to be connected and to exchange information, there must be a common language. This is provided by the eXtensible Markup Language (XML). Short descriptions of the current protocols for Web service connection, description, and discovery are in the following paragraphs; more complete descriptions are found elsewhere in this book.

A common protocol is required for systems to communicate with each other, so that they can request services, such as to schedule appointments, order parts, and deliver information. This is provided by the Simple Object Access Protocol (SOAP) [Box et al., 2000].

The services must be described in a machine-readable form, where the names of functions, their required parameters, and their results can be specified. This is provided by the Web Services Description Language (WSDL).

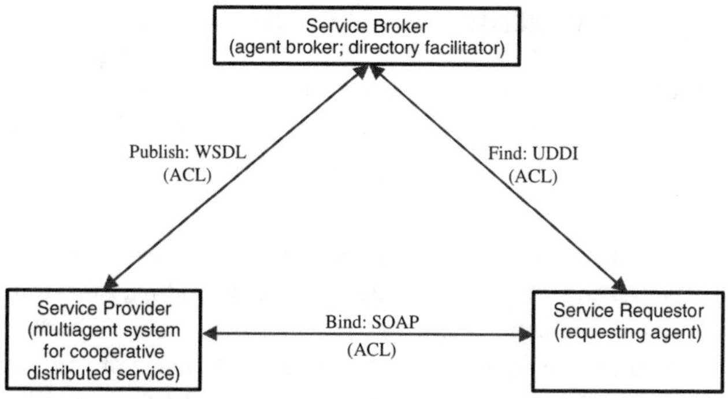

FIGURE 17.2 The general architectural model for Web services, which rely on the functionalities of publish, find, and bind. The equivalent agent-based functionalities are shown in parentheses, where all interactions among the agents are via an agent-communication language (ACL). Any agent might serve as a broker. Also, the service provider's capabilities might be found without using a broker.

Finally, clients — users and businesses — need a way to find the services they need. This is provided by Universal Description, Discovery, and Integration (UDDI), which specifies a registry or "yellow pages" of services.

Besides standards for XML, SOAP, WSDL, and UDDI, there is a need for broad agreement on the semantics of specific domains. This is provided by the Resource Description Framework (RDF) [Decker et al., 2000a,b], the OWL Web Ontology Language [Smith et al., 2003], and, more generally, ontologies [Heflin and Hendler, 2000].

17.2.3 Directory Services

The purpose of a directory service is for components and participants to locate each other, where the components and participants might be applications, agents, Web service providers, Web service requestors, people, objects, and procedures. There are two general types of directories, determined by how entries are found in the directory: (1) name servers or *white pages*, where entries are found by their name, and (2) *yellow pages*, where entries are found by their characteristics and capabilities.

The implementation of a basic directory is a simple database-like mechanism that allows participants to insert descriptions of the services they offer and query for services offered by other participants. A more advanced directory might be more active than others, in that it might provide not only a search service but also a brokering or facilitating service. For example, a participant might request a brokerage service to recruit one or more agents that can answer a query. The brokerage service would use knowledge about the requirements and capabilities of registered service providers to determine the appropriate providers to which to forward a query. It would then send the query to those providers, relay their answers back to the original requestor, and learn about the properties of the responses it passes on (e.g., the brokerage service might determine that advertised results from provider X are incomplete and so seek out a substitute for provider X).

UDDI is itself a Web service that is based on XML and SOAP. It provides both a white-pages and a yellow-pages service, but not a brokering or facilitating service.

The DARPA (Defense Advanced Research Projects Agency) DAML (DARPA Agent Markup Language) effort has also specified a syntax and semantics for describing services, known as DAML-S (now migrating to OWL-S http://www.daml.org/services). This service description provides

- Declarative ads for properties and capabilities, used for discovery
- Declarative APIs, used for execution
- Declarative prerequisites and consequences, used for composition and interoperation

17.3 Agent Implementations of Web Services

Typical agent architectures have many of the same features as Web services. Agent architectures provide yellow-page and white-page directories where agents advertise their distinct functionalities and where other agents search to locate the agents in order to request those functionalities. However, agents extend Web services in several important ways:

- A Web service knows only about itself but not about its users/clients/customers. Agents are often self-aware at a metalevel and, through learning and model building, gain awareness of other agents and their capabilities as interactions among the agents occur. This is important because without such awareness a Web service would be unable to take advantage of new capabilities in its environment and could not customize its service to a client, such as by providing improved services to repeat customers.
- Web services, unlike agents, are not designed to use and reconcile ontologies. If the client and provider of the service happen to use different ontologies, then the result of invoking the Web service would be incomprehensible to the client.
- Agents are inherently communicative, whereas Web services are passive until invoked. Agents can provide alerts and updates when new information becomes available. Current standards and protocols make no provision for even subscribing to a service to receive periodic updates.
- A Web service, as currently defined and used, is not autonomous. Autonomy is a characteristic of agents, and it is also a characteristic of many envisioned Internet-based applications. Among agents, autonomy generally refers to social autonomy, where an agent is aware of its colleagues and is sociable, but nevertheless exercises its independence in certain circumstances. Autonomy is in natural tension with coordination or with the higher-level notion of a commitment. To be coordinated with other agents or to keep its commitments, an agent must relinquish some of its autonomy. However, an agent that is sociable and responsible can still be autonomous. It would attempt to coordinate with others where appropriate and to keep its commitments as much as possible, but it would exercise its autonomy in entering into those commitments in the first place.
- Agents are cooperative and, by forming teams and coalitions, can provide higher-level and more comprehensive services. Current standards for Web services do not provide for composing functionalities.

17.4 Building Web-Service Agents

17.4.1 Agent Types

To better communicate some of the most popular agent architectures, this chapter uses UML diagrams to guide an implementer's design. However, before we describe these diagrams, we need to review some of the basic features of agents. Consider the architecture in Figure 17.3 for a simple agent interacting with an information environment, which might be the Internet, an intranet, or a virtual private network (VPN). The agent senses its environment, uses what it senses to decide upon an action, and then performs the action through its effectors. Sensory input can include received messages, and the action can be the sending of messages.

To construct an agent, we need a more detailed understanding of how it functions. In particular, if we are to construct one using conventional object-oriented design techniques, we should know in what ways an agent is more than just a simple object. Agent features relevant to implementation are unique identity, proactivity, persistence, autonomy, and sociability [Weiß, 1999].

An agent inherits its *unique identity* simply by being an object. To be *proactive*, an agent must be an object with an internal event loop, such as any object in a derivation of the Java thread class would have. Here is simple pseudocode for a typical event loop, where events result from sensing an environment:

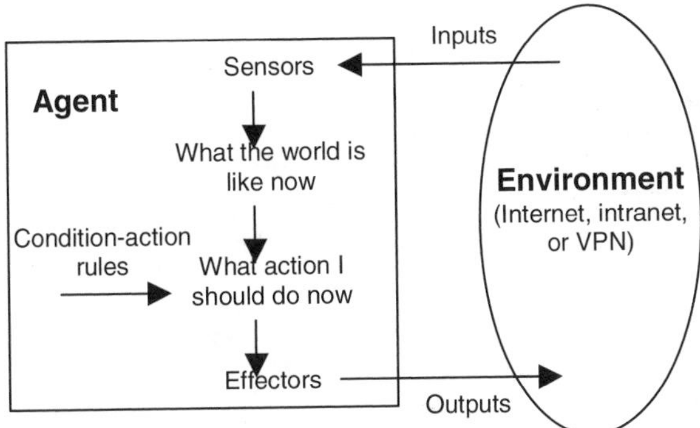

FIGURE 17.3 A simple interaction between an agent and its information environment (Adapted from Russell, Stuart J. and Peter Norvig. *Artificial Intelligence: A Modern Approach, 2nd Ed.* Prentice-Hall, Upper Saddle River, NJ, 2003.). "What action I should do now" depends on the agent's goals and perhaps ethical considerations, as noted in Figure 17.6.

```
Environment e;
RuleSet r;
while (true) {
      state = senseEnvironment(e);
      a = chooseAction(state, r);
      e.applyAction(a);
}
```

This is an infinite loop, which also provides the agent with *persistence.* Ephemeral agents would find it difficult to converse, making them, by necessity, asocial. Additionally, persistence makes it worthwhile for agents to learn about and model each other. To benefit from such modeling, they must be able to distinguish one agent from another; thus, agents need unique identities.

Agent *autonomy* is akin to human free will and enables an agent to choose its own actions. For an agent constructed as an object with methods, autonomy can be implemented by declaring all the methods private. With this restriction, only the agent can invoke its own methods, under its own control, and no external object can force the agent to do anything it does not intend to do. Other objects can communicate with the agent by creating events or artifacts, especially messages, in the environment that the agent can perceive and react to.

Enabling an agent to converse with other agents achieves *sociability.* The conversations, normally conducted by sending and receiving messages, provide opportunities for agents to coordinate their activities and cooperate, if so inclined. Further sociability can be achieved by generalizing the input class of objects an agent might perceive to include a piece of sensory information and an event defined by the agent. Events serving as inputs are simply "reminders" that the agent sets for itself. For example, an agent that wants to wait 5 min for a reply would set an event to fire after 5 min. If the reply arrives before the event, the agent can disable the event. If it receives the event, then it knows it did not receive the reply in time and can proceed accordingly.

The UML diagrams in Figure 17.4 and Figure 17.5 can help in understanding or constructing a software agent. These diagrams do not address every functional aspect of an agent's architecture. Instead, they provide a general framework for implementing traditional agent architectures [Weiß, 1999).

17.4.1.1 Reactive Agents

A reactive agent is the simplest kind to build because it does not maintain information about the state of its environment but simply reacts to current perceptions. Our design for such an agent, shown in Figure 17.4, is fairly intuitive, encapsulating a collection of behaviors, sometimes known as plans, and the means for selecting an appropriate one. A collection of objects, in the object-oriented sense, lets a developer add and remove behaviors without having to modify the action selection code, since an iterator

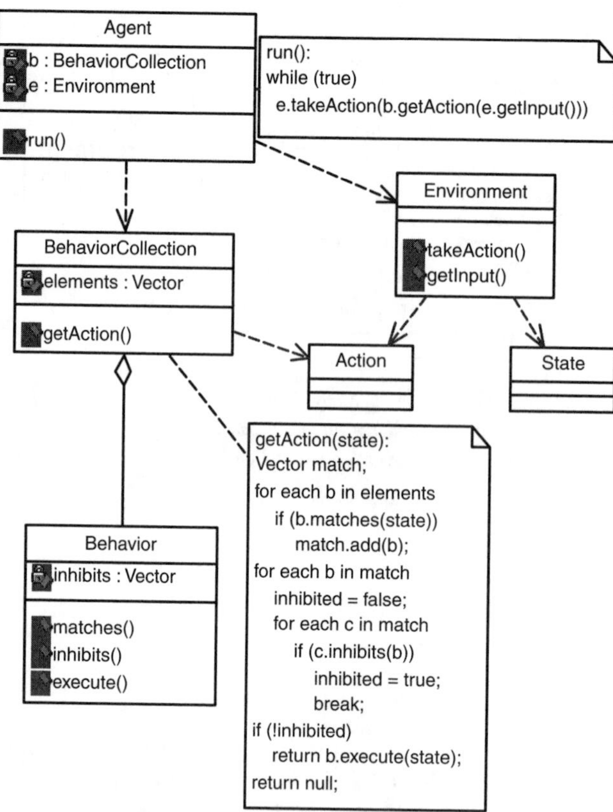

FIGURE 17.4 Diagram of a simple reactive architecture for an agent. The agent's run () method executes the action specified by the current behavior and state.

can be used to traverse the list of behaviors. Each behavior fires when it matches the environment, and each can inhibit other behaviors. Our action-selection loop is not as efficient as it could be, because getAction operates in $O(n)$ time (where n is the number of behaviors). A better implementation could lower the computation time to $O(logn)$ using decision trees, or $O(1)$ using hardware or parallel processing. The developer is responsible for ensuring that at least one behavior will match for every environment. This can be achieved by defining a default behavior that matches all inputs but is inhibited by all other behaviors that match.

17.4.1.2 BDI Agents

A belief–desire–intention (BDI) architecture includes and uses an explicit representation for an agent's beliefs (state), desires (goals), and intentions (plans). The beliefs include self-understanding ("I believe I can perform Task-A"), beliefs about the capabilities of other agents ("I believe Agent-B can perform Task-B"), and beliefs about the environment ("Based on my sensors, I believe I am 3 ft from the wall"). The intentions persist until accomplished or are determined to be unachievable. Representative BDI systems — the Procedural Reasoning System (PRS) and JAM — all define a new programming language and implement an interpreter for it. The advantage of this approach is that the interpreter can stop the program at any time, save its state, and execute some other intention if it needs to. The disadvantage is that the interpreter — not an intention — runs the system; the current intention may no longer be applicable if the environment changes.

The BDI architecture shown in Figure 17.5 eliminates this problem. It uses a voluntary multitasking method instead, whereby the environment thread constantly checks to make sure the current intention is applicable. If not, the agent executes stopCurrentIntention ( ), which will call the intention's stopEx-

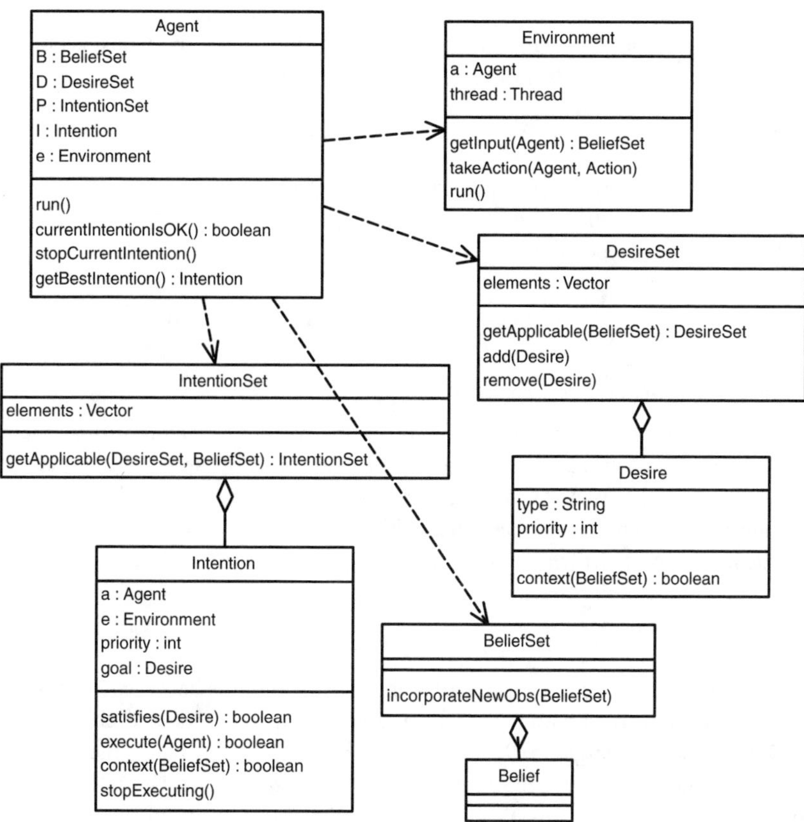

FIGURE 17.5 Diagram of a belief–desire–intention architecture for an agent.

ecuting ( ) method. Thus, the intention is responsible for stopping itself and cleaning up. By giving each intention this capability, we eliminate the possibility of a deadlock resulting from the intentions having some resource reserved when it was stopped. The following pseudocode illustrates the two main loops, one for each thread, of the BDI architecture. The variables a, B, D, and I represent the agent and its beliefs, desires, and intentions.

The agent's run method consists of finding the best applicable intention and executing it to completion. If the result of the execution is true, the meaning is that the desire was achieved, so the desire is removed from the desire set. If the environment thread finds that an executing plan is no longer applicable and calls for a stop, the intention will promptly return from its execute ( ) call with a false. Notice that the environment thread modifies the agent's set of beliefs. The belief set needs to synchronize these changes with any changes that the intentions make to the set of beliefs.

```
Agent::run () {
      Environment e;
      e.run (); //start environment in its own thread
      while (true) {
            I = a.getBestIntention();
            If (I.execute(a)) // true if intention was achieved
                  a.D.remove(I.goal); // I.goal is a desire
      }
}
```

Finally, the environment thread's sleep time can be modified, depending on the systems real-time requirements. If we do not need the agent to change intentions rapidly when the environment changes, the thread can sleep longer. Otherwise, a short sleep will make the agent check the environment more frequently, using more computational resources. A more efficient callback mechanism could easily replace the current run method if the agents input mechanism supported it.

```
Environment::run() {
      while (true) {
            a.B.incorporateNewObservations (e.getInput (a)) ;
            if (! a.currentIntentionIsOK() )
                  a.stopCurrentIntention();
            sleep(someShortTime);
      }
}
```

17.4.1.3 Layered Architectures

Other common architectures for software agents consist of layers of capabilities where the higher layers perform higher levels of reasoning. For example, Figure 17.6 shows the architecture of an agent that has a philosophical and ethical basis for choosing its actions. The layers typically interact in one of three ways: (case 1) inputs are given to all of the layers at the same time; (case 2) inputs are given to the highest layer first for deliberation, and then its guidance is propagated downward through each of the lower layers until the lowest layer performs the ultimate action; and (case 3) inputs are given to the lowest layer first, which provides a list of possible actions that are successively filtered by each of the higher layers until a final action remains.

The lowest level of the architecture enables an agent to react to immediate events [Müller et al., 1994]. The middle layers are concerned with an agent's interactions with others [Castelfranchi, 1998; Castelfranchi et al., 2000; Rao and Georgeff, 1991; Cohen and Levesque, 1990], whereas the highest level enables the agent to consider the long-term effects of its behavior on the rest of its society [Mohamed and Huhns, 2001]. Agents are typically constructed starting at the bottom of this architecture, with increasingly more abstract reasoning abilities layered on top.

Awareness of other agents and of one's own role in a society, which are implicit at the social commitment level and above, can enable agents to behave coherently [Gasser, 1991]. Tambe et al. [2000] have shown how a team of agents flying helicopters will continue to function as a coherent team after their

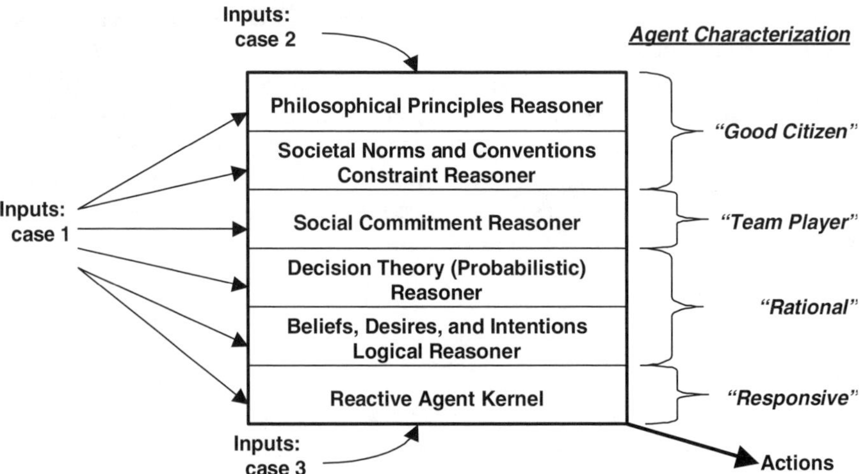

FIGURE 17.6 Architecture for a philosophical agent. The architecture defines layers of deliberation for enabling an agent to behave appropriately in a society of agents.

leader has crashed because another agent will assume the leadership role. More precisely, the agents will adjust their individual intentions in order to fulfill the commitments made by the team.

17.4.1.4 Behaviors and Activity Management

Most popular agent architectures, including the three we diagrammed, include a set of behaviors and a method for scheduling them. A behavior is distinguished from an action in that an action is an atomic event, whereas a behavior can span a longer period of time. In multiagent systems, we can also distinguish between physical behaviors that generate actions, and conversations between agents. We can consider behaviors and conversations to be classes inheriting from an abstract activity class. We can then define an activity manager responsible for scheduling activities.

This general activity manager design lends itself to the implementation of many popular agent architectures while maintaining the proper encapsulation and decomposability required in good object-oriented programming. Specifically, activity is an abstract class that defines the interface to be implemented by all behaviors and conversations. The behavior class can implement any helper functions needed in the particular domain (for example, subroutines for triangulating the agents position). The conversation class can implement a finite-state machine for use by the particular conversations. For example, by simply filling in the appropriate states and adding functions to handle the transitions, an agent can define a contracting protocol as a class that inherits from conversation. Details of how this is done depend on how the conversation class implements a finite-state machine, which varies depending on the system's real-time requirements.

Defining each activity as its own independent object and implementing a separate activity manager has several advantages. The most important is the separation between domain and control knowledge. The activities will embody all the knowledge about the particular domain the agent inhabits, while the activity manager embodies knowledge about the deadlines and other scheduling constraints the agent faces. When the implementation of each activity is restricted to a separate class, the programmer must separate the agent's abilities into encapsulated objects that other activities can then reuse. The activity hierarchy forces all activities to implement a minimal interface that also facilitates reuse. Finally, placing the activities within the hierarchy provides many opportunities for reuse through inheritance. For example, the conversation class can implement a general lost-message error-handling procedure that all conversations can use.

17.4.1.5 Architectural Support

Figure 17.4 and Figure 17.5 provide general guidelines for implementing agent architectures using an object-oriented language. As agents become more complex, developers will likely have to expand upon our techniques. We believe these guidelines are general enough that it will not be necessary to rewrite the entire agent from scratch when adding new functionality.

Of course, a complete agent-based system requires an infrastructure to provide for message transport, directory services, and event notification and delivery. These are usually provided as operating system services or, increasingly, in an agent-friendly form by higher-level distributed protocols such as Jini (http://www.sun.com/jini/), Bluetooth (http://www.bluetooth.com), and FIPA's (the Foundation of Intelligent Physical Agents, at http://www.fipa.org/) emerging standards. See http://www.multiagent.com/, a site maintained by José Vidal, for additional information about agent tools and architectures.

The canonical multiagent architecture, shown in Figure 17.1, is suitable for many applications. The architecture incorporates a variety of resource agents that represent databases, Websites, sensors, and file systems. Specifications for the behaviors of two types of resource agents are shown in Figure 17.7 and Figure 17.8. Figure 17.7 contains a procedural specification for the behavior of a resource (wrapper) agent that makes a database system active and accessible to other agents. In a supply-chain management scenario, the database agent might represent a supplier's order-processing system; customers could send orders to the agent (inform) and then issue queries for status and billing information. A procedural specification for the behavior of an Internet agent that actively monitors a Website for new or updated

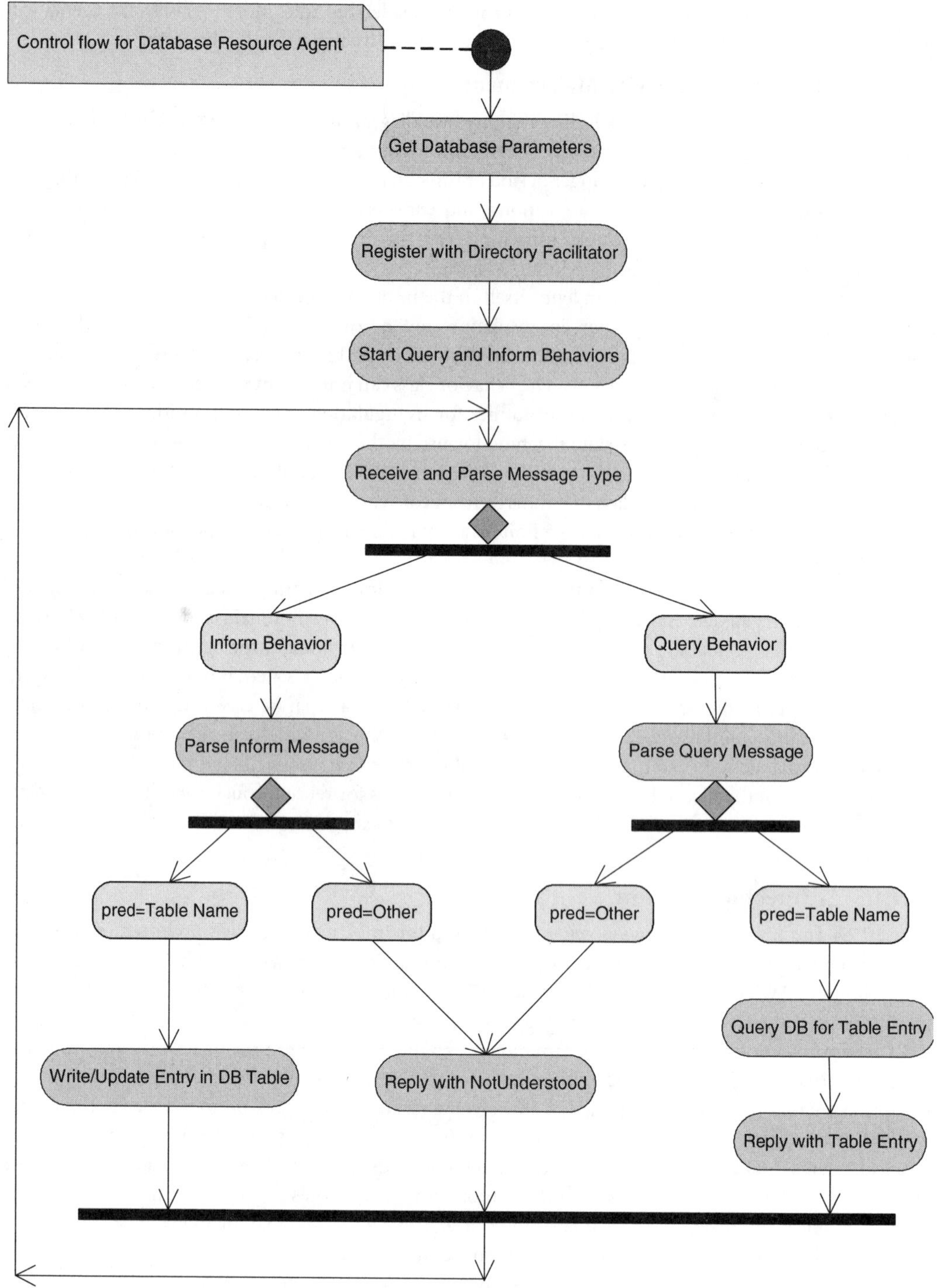

FIGURE 17.7 Activity diagram showing the procedural behavior of a resource agent that makes a database system active and accessible to other agents.

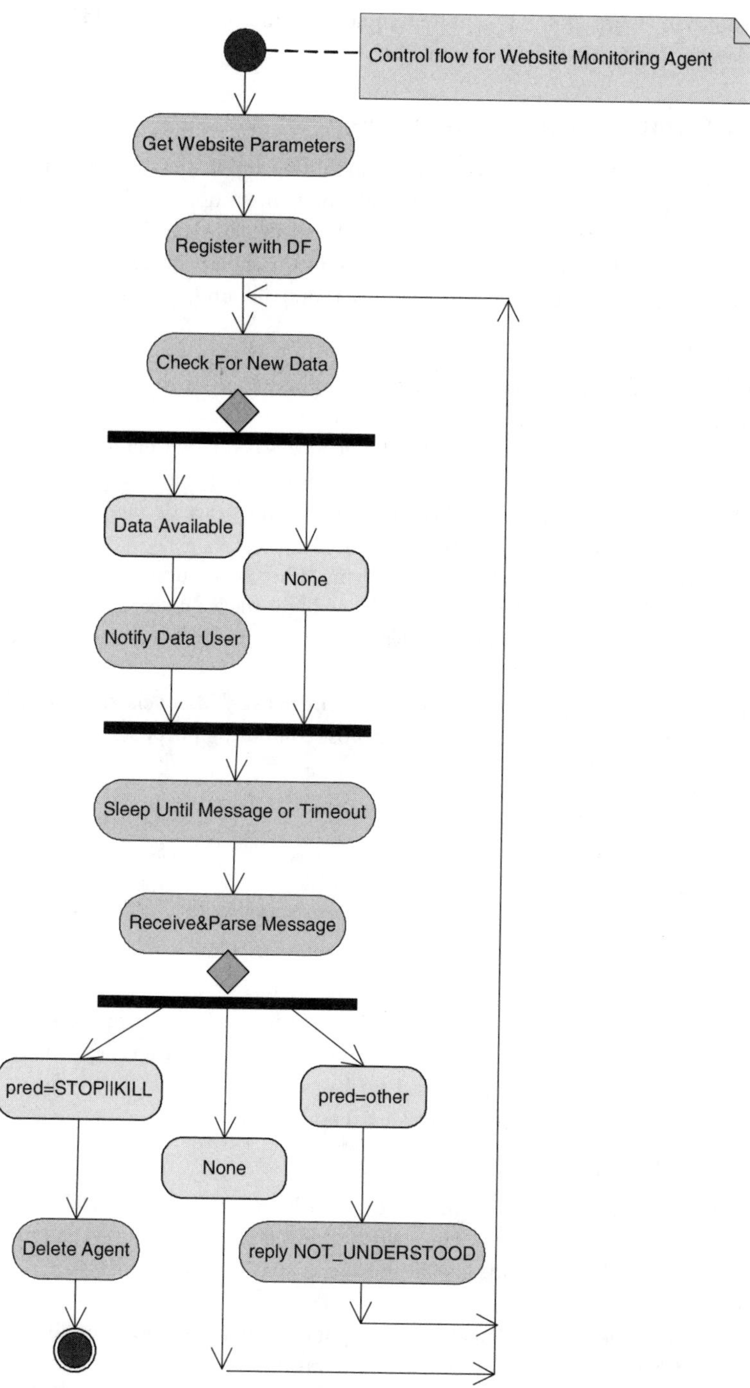

FIGURE 17.8 Activity diagram showing the procedural behavior of an Internet agent that actively monitors a Website for new or updated information.

information is shown in Figure 17.8. An agent that implements this procedure could be used by customers looking for updated pricing or product information.

17.4.2 Agent Communication Languages

Agents representing different users might collaborate in finding and fusing information, but compete for goods and resources. Similarly, service agents may collaborate or compete with user, resource, and other service agents. Whether they are collaborators or competitors, the agents must interact purposefully with each other. Most purposeful interactions — whether to inform, query, or deceive — require the agents to talk to one another, and talking intelligibly requires a mutually understood language.

17.4.2.1 Speech Acts

Speech acts have to do with communication; they have nothing to do with speaking as such, except that human communication often involves speech. Speech act theory was invented in the fifties and sixties to help understand human language [Austin, 1962]. The idea was that with language you not only make statements but also perform actions. For example, when you request something, you do not just report on a request; you actually cause the request. When a justice of the peace declares a couple man and wife, she is not reporting on their marital status but changing it.

The stylized syntactic form for speech acts that begins "I hereby request ..." or "I hereby declare ..." is called a *performative*. With a performative, literally, saying it makes it so. Verbs that cannot be put in this form are not speech acts. For example, "solve" is not a performative because "I hereby solve this problem" is not sufficient.

Several thousand verbs in English correspond to performatives. Many classifications have been suggested for these, but the following are sufficient for most computing purposes:

- Assertives (informing)
- Directives (requesting or querying)
- Commissives (promising)
- Prohibitives
- Declaratives (causing events in themselves as, for example, the justice of the peace does in a marriage ceremony)
- Expressives (expressing emotions)

In natural language, it is not easy to determine what speech act is being performed. In artificial languages, we do not have this problem. However, the meanings of speech acts depend on what the agents believe, intend, and know how to perform, and on the society in which they reside. It is difficult to characterize meaning because all of these things are themselves difficult.

17.4.2.2 Common Language

Agent projects investigated languages for many years. Early on, agents were local to each project, and their languages were mostly idiosyncratic. The challenge now is to have any agent talk to any other agent, which suggests a common language; ideally, all the agents that implement the (same) language will be mutually intelligible.

Such a common language needs an unambiguous syntax so that the agents can all parse sentences the same way. It should have a well-defined semantics or meaning so that the agents can all understand sentences the same way. It should be well known so that different designers can implement it and so it has a chance of encountering another agent who knows the same language. Further, it should have the expressive power to communicate the kinds of things agents may need to say to one another.

So, what language should you give or teach your agent so that it will understand and be understood? The current popular choice is being administered by the Foundation for Intelligent Agents (FIPA), at http://www.fipa.org. The FIPA ACL separates the domain-dependent part of a communication — the content — from the domain-independent part — the packaging — and then provides a standard for the domain-independent part.

FIPA specifies just six performatives, but they can be composed to enable agents to express more complex beliefs and expectations. For example, an agent can request to be informed about one of several alternatives. The performatives deal explicitly with actions, so requests are for communicative actions to be done by the message recipient.

The FIPA specification comes with a formal semantics, and it guarantees that there is only one way to interpret an agent's communications. Without this guarantee, agents (and their designers) would have to choose among several alternatives, leading to potential misunderstandings and unnecessary work.

17.4.3 Knowledge and Ontologies for Agents

An ontology is a computational model of some portion of the world. It is often captured in some form of a semantic network, a graph whose nodes are concepts or individual objects and whose arcs represent relationships or associations among the concepts. This network is augmented by properties and attributes, constraints, functions, and rules that govern the behavior of the concepts.

Formally, an ontology is an agreement about a shared conceptualization, which includes frameworks for modeling domain knowledge and agreements about the representation of particular domain theories. Definitions associate the names of entities in a universe of discourse (for example, classes, relations, functions, or other objects) with human-readable text describing what the names mean, and formal axioms that constrain the interpretation and well-formed use of these names.

For information systems, or for the Internet, ontologies can be used to organize keywords and database concepts by capturing the semantic relationships among the keywords or among the tables and fields in a database. The semantic relationships give users an abstract view of an information space for their domain of interest.

17.4.3.1 A Shared Virtual World

How can such an ontology help our software agents? It can provide a shared virtual world in which each agent can ground its beliefs and actions. When we talk with our travel agent, we rely on the fact that we all live in the same physical world containing planes, trains, and automobiles. We know, for example, that a 777 is a type of airliner that can carry us to our destination.

When our agents talk, the only world they share is one consisting of bits and bytes — which does not allow for a very interesting discussion! An ontology gives the agents a richer and more useful domain of discourse.

The previous section described FIPA, which specifies the syntax but not the semantics of the messages that agents can exchange. It also allows the agents to state which ontology they are presuming as the basis for their messages.

Suppose two agents have access to an ontology for travel, with concepts such as airplanes and destinations, and suppose the first agent tells the second about a flight on a 777. Suppose further that the concept "777" is not part of the travel ontology. How could the second agent understand? The first agent could explain that a 777 is a kind of airplane, which is a concept in the travel ontology. The second agent would then know the general characteristics of a 777. This communication is illustrated in Figure 17.9.

17.4.3.2 Relationships Represented

Most ontologies represent and support relationships among classes of meaning. Among the most important of these relationships are:

- *Generalization and inheritance,* which are abstractions for sharing similarities among classes while preserving their differences. Generalization is the relationship between a class and one or more refined versions of it. Each subclass inherits the features of its superclass, adding other features of its own. Generalization and inheritance are transitive across an arbitrary number of levels. They are also antisymmetric.
- *Aggregation,* the part–whole or part-of relationship, in which classes representing the components of something are associated with the class representing the entire assembly. Aggregation is also

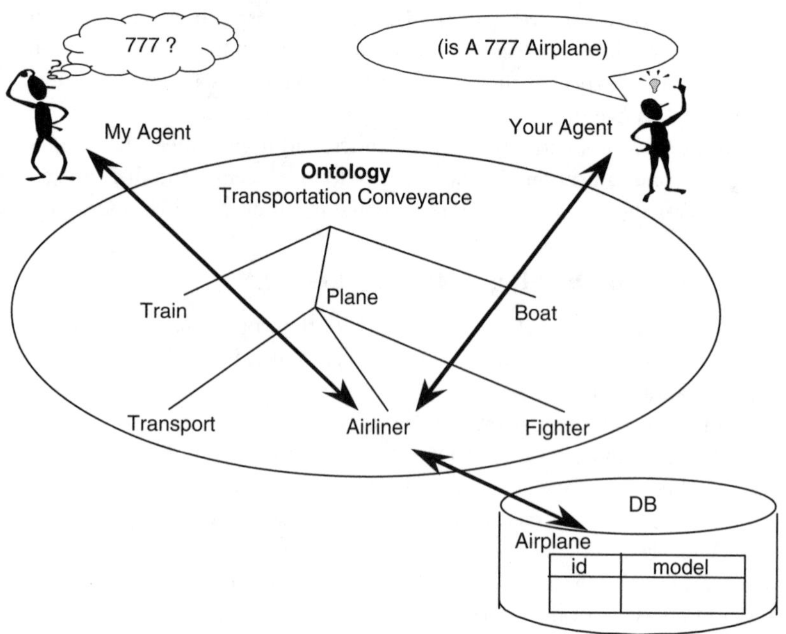

FIGURE 17.9 Communication between agents sharing a travel ontology.

transitive, as well as antisymmetric. Some of the properties of the assembly class propagate to the component classes.

- *Instantiation*, which is the relationship between a class and each of the individuals that constitute it.

Some of the other relationships that occur frequently in ontologies are *owns*, *causes*, and *contains*. Causes and contains are transitive and antisymmetric; owns propagates over aggregation because when you own something, you also own all of its parts.

17.4.4 Reasoning Systems

A simple and convenient means to incorporate a reasoning capability into an Internet software agent is via a rule-execution engine, such as JESS [Friedman-Hill, 2003]. With JESS, knowledge is supplied in the form of declarative rules. There can be many or only a few rules, and Jess will continually apply them to data in the form of a knowledge base. Typically, the rules represent the heuristic knowledge of a human expert in some domain, and the knowledge base represents the state of an evolving situation.

An example rule in JESS is

```
(defrule recognize-airliner
     "If an object ?X is a plane and carries passengers,
     then assert that ?X is an airliner."
     (isA ?X plane)
     (carries ?X passengers)
     =>
     (assert (isA ?X airliner)))
```

The associated knowledge base might contain facts about an airline company concerning their equipment and its characteriztics, such as

```
(assert (isA 777-N9682 plane))
(assert (carries 777-N9682 passengers))
```

Many of the common agent development environments, such as JADE [Bellifemine and Trucco, 2003], ZEUS [Nwana et al., 1999], and FIPA-OS [Nortel Networks, 2003], include facilities for incorporating JESS into the agents a developer is constructing.

17.4.5 Cooperation

The most widely used means by which agents arrange to cooperate is the contract-net protocol. This interaction protocol allows an initiating agent to solicit proposals from other agents by sending a *Call for Proposals*, evaluating their proposals, and then accepting the preferred one (or even rejecting all of them). Any agent can initiate the protocol, so it can be applied recursively.

The initiator sends a message with a CFP speech act that specifies the action to be performed and, if needed, conditions upon its execution. The responders can reply by sending a PROPOSE message that includes any preconditions for their action, such as their cost or schedule. Alternatively, responders may send a REFUSE message to indicate their disinterest or a NOT-UNDERSTOOD message to indicate a communication problem. The initiator then evaluates the received proposals and sends an ACCEPT-PROPOSAL message to the agents whose proposal will be accepted and a REJECT-PROPOSAL message to the others. Once the chosen responders have completed their task, they respond with an INFORM of the result of the action or with a FAILURE if anything went wrong.

17.5 Composing Cooperative Web Services

Imagine that a merchant would like to enable a customer to be able to track the shipping of a sold item. Currently, the best the merchant can do is to point the customer to the shipper's Website, and the customer can then go there to check on delivery status. If the merchant could compose its own production notification system with the shipper's Web services, the result would be a customized delivery notification service by which the customer — or the customer's agents — could find the status of a purchase in real time.

As Web uses (and thus Web interactions) become more complex, it will be increasingly difficult for one server to provide a total solution and increasingly difficult for one client to integrate solutions from many servers. Web services currently involve a single client accessing a single server, but soon applications will demand federated servers with multiple clients sharing results. Cooperative peer-to-peer solutions will have to be managed, and this is an area where agents have excelled. In doing so, agents can balance cooperation with the interests of their owner.

17.6 Conclusion

Web services are extremely flexible, and a major advantage is that a developer of Web services does not have to know who or what will be using the services being provided. They can be used to tie together the internal information systems of a single company or the interoperational systems of virtual enterprises. But how Web services tie the systems together will be based on technologies being developed for multi-agent systems. The result will be a Semantic Web that enables work to get done and better decisions to be made.

Acknowledgment

The US National Science Foundation supported this work under grant number IIS-0083362.

References

Austin, John L. *How to Do Things with Words*. Clarendon Press, Oxford, 1962.

Bellifemine, Fabio and Tiziana Trucco. Java agent development framework, 2003. http://sharon.cselt.it/projects/jade/.

Berners-Lee, Tim, James Hendler, and Ora Lassila. The semantic web. *Scientific American,* 284(5): 34–43, 2001.

Bond, Alan and Les Gasser, Eds. *Readings in Distributed Artificial Intelligence.* Morgan Kaufmann, San Francisco, 1988.

Box, Don, David Ehnebuske, Gopal Kakivaya, Andrew Layman, Noah Mendelsohn, Henrik Frystyk Nielsen, Satish Thatte, and Dave Winer. Simple object access protocol (SOAP) 1.1, 2000. www.w3.org/TR/SOAP.

Castelfranchi, Cristiano. Modelling social action for AI agents. *Artificial Intelligence,* 103: 157–182, 1998.

Castelfranchi, Cristiano, Frank Dignum, Catholyn M. Jonker, and Jan Treur. Deliberate normative agents: Principles and architecture. In Nicholas R. Jennings and Yves Lesperance, Eds. *Intelligent Agents VI: Agent Theories, Architectures, and Languages (ATAL-99),* volume 1757, pp. 364–378, Springer-Verlag, Berlin, 2000.

Coelho, Helder, Luis Antunes, and Luis Moniz. On agent design rationale. In *Proceedings of the XI Simposio Brasileiro de Inteligencia Artificial (SBIA),* pp. 43–58, Fortaleza, Brazil, 1994.

Cohen, Philip R. and Hector J. Levesque. Persistence, intention, and commitment. In Philip Cohen, Jerry Morgan, and Martha Pollack, Eds., *Intentions in Communication.* MIT Press, Cambridge, MA, 1990.

Decker, Stefan, Sergey Melnik, Frank van Harmelen, Dieter Fensel, Michel Klein, Jeen Broekstra, Michael Erdmann, and Ian Horrocks. The semantic web: the roles of XML and RDF. *IEEE Internet Computing,* 4(5): 63–74, September 2000a.

Decker, Stefan, Prasenjit Mitra, and Sergey Melnik. Framework for the semantic web: an RDF tutorial. *IEEE Internet Computing,* 4(6): 68–73, November 2000b.

Friedman-Hill, Ernest J. Jess, the Java expert system shell, 2003, http://herzberg.ca.sandia.gov/jess.

Gasser, Les. Social conceptions of knowledge and action: DAI foundations and open systems semantics, *Artificial Intelligence,* 47: 107–138,1991.

Gasser, Les and Michael N. Huhns, Eds. *Distributed Artificial Intelligence,* Vol. 2. Morgan Kaufmann, London, 1989.

Heflin, Jeff and James A. Hendler. Dynamic ontologies on the Web. In *Proceedings of American Association for Artificial Intelligence Conference (AAAI),* pp. 443–449, Menlo Park, CA, 2000. AAAI Press.

Huhns, Michael N., Ed. *Distributed Artificial Intelligence.* Morgan Kaufmann, London, 1987.

Huhns, Michael N. Interaction-oriented programming. In Paulo Ciancarini and Michael Wooldridge, Eds., *Agent-Oriented Software Engineering,* Vol. 1957 of *Lecture Notes in Artificial Intelligence,* pp. 29–44, Springer-Verlag, Berlin, 2001.

Huhns, Michael N. and Munindar P. Singh, Eds. *Readings in Agents.* Morgan Kaufmann, San Francisco, 1998.

Jennings, Nicholas R. On agent-based software engineering. *Artificial Intelligence,* 117(2): 277–296, 2000.

Mohamed, Abdulla M. and Michael N. Huhns. Multiagent benevolence as a societal norm. In Rosaria Conte and Chrysanthos Dellarocas, Eds., *Social Order in Multiagent Systems,* pp. 65–84, Kluwer, Boston, 2001.

Müller, Jörg P. Markus Pischel, and Michael Thiel. Modeling reactive behavior in vertically layered agent architectures. In Michael J. Wooldridge and Nicholas R. Jennings, Eds., *Intelligent Agents,* Vol. 890 of *Lecture Notes in Artificial Intelligence,* pp. 261–276, Springer-Verlag, Berlin, 1994.

Nortel Networks. FIPA-OS, 2003. http://fipa-os.sourceforge.net/.

Nwana, Hyacinth, Divine Ndumu, Lyndon Lee, and Jaron Collis. ZEUS: A tool-kit for building distributed multi-agent systems. *Applied Artificial Intelligence,* 13(1),1999.

Rao, Anand S. and Michael P Georgeff. Modeling rational agents within a BDI-architecture. In *Proceedings of the International Conference on Principles of Knowledge Representation and Reasoning,* pp. 473–484,1991. Reprinted in Huhns and Singh [1998].

Russell, Stuart J. and Peter Norvig. *Artificial Intelligence: A Modern Approach, 2nd ed.* Prentice-Hall, Upper Saddle River, NJ, 2003.

Smith, Michael K., Chris Welty, and Deborah McGuiness. Web ontology language (OWL) guide version 1.0, 2003. http://www.w3.org/TR/2003/WD-owl-guide-20030210/.

Tambe, Milind, David V. Pynadath, and Nicolas Chauvat. Building dynamic agent organizations in cyberspace. *IEEE Internet Computing,* 4(2): 65–73, February 2000.

Weiß, Gerhard, Ed. *Multiagent Systems: A Modern Approach to Distributed Artificial Intelligence.* MIT Press, Cambridge, MA, 1999.

18

Concepts and Practice of Personalization

Manuel Aparicio IV

Munindar P. Singh

CONTENTS

Abstract 18-1
18.1 Motivation 18-1
18.2 Key Applications and Historical Development 18-3
18.2.1 Desktop Applications such as E-mail 18-3
18.2.2 Web Applications such as E-commerce 18-4
18.2.3 Knowledge Management 18-6
18.2.4 Mobile Applications 18-7
18.3 Key Concepts 18-8
18.3.1 Individual vs. Collaborative 18-9
18.3.2 Representation and Reasoning 18-9
18.4 Discussion 18-11
18.4.1 Advice 18-11
18.4.2 Metrics 18-12
18.4.3 Futures 18-13
References 18-14

Abstract

Personalization is how a technical artifact adapts its behavior to suit the needs of an individual user. Information technology provides both a heavy need for personalization because of its inherent complexity and the means to accomplish it through its ability to gather information about user preferences and to compute user's needs. Although personalization is important in all information technology, it is particularly important in networked applications. This chapter reviews the main applications of personalization, the major approaches, the tradeoffs among them, the challenges facing practical deployments, and the themes for ongoing research.

18.1 Motivation

Personalization is one of the key attributes of an intelligent system that is responsive to the user's needs. Personalization emerged with the early desktop applications. The emergence of the Internet as a substrate for networked computing has enabled sharing and collaborative applications. More generally, the Internet enables a dynamic form of trading, where users can carry on individualized monetary or information trades. Further, because Internet environments involve parties engaging each other in interactions from a distance where there is a greater risk that users' needs would be misunderstood, there is concomitantly an increased need for personalization. This led to a corresponding evolution in the techniques of personalization.

1-58488-381-2/05/$0.00+$1.50
© 2005 by CRC Press LLC

It is helpful to make a distinction between personalization and customization. Although sometimes these terms are used interchangeably, the distinction between them is key from the standpoints of both users and technology. Customization involves selecting from among some preset options or setting some parameter values that are predefined to be settable. For example, if you choose the exterior color and upholstery for your vehicle, you are customizing it from a menu of options that the manufacturer made available to you. Likewise, if you choose the ring-tones with which your mobile telephone rings, you are also customizing this functionality of the telephone. In the same vein, when you select the elements to display on your personal page at a leading portal, such as Yahoo!, you are customizing your page based on the menu of options made available by the portal. A desktop example is the reordering of menus and restriction of visible items on a menu to those used recently. Such approaches do not attempt to model what a user might care about, but simply remember the recent actions of users with little or no regard for the context in which they were performed.

A variation is when an application helps a user or a system administrator construct a user profile by hand. Such profiles are inherently limited in not being able to anticipate situations that have not been explicitly included in the profile. They can represent only a few dimensions of interests and preferences, often missing out on subtle relationships. Moreover, they require user input for all the profile data, which is a major reason why they are limited to a few dimensions. Such profiles also suffer from the problem of not being able to adapt to any changes in users' interests. The profile offers degrading performance as the user's preferences drift and must be reconstructed by hand whenever the user's interests have changed significantly. Such systems work to a small extent even if they do not work very well. Consequently, users are reluctant to change anything for fear of inadvertently breaking something that does work.

By contrast, your experience with a leading e-commerce site, such as Amazon.com, is personalized. This is because a site such as Amazon.com will behave in a manner (chiefly by presenting materials and recommendations) that responds to your needs without the need for any explicit selection of parameters on your part.

Therefore, the main difference between personalization and customization lies in the fact that customization involves a more restrictive and explicit choice from a small, predetermined set of alternatives, whereas personalization involves a more flexible, often at least in parts implicit, choice from a larger, possibly changing set of alternatives.

A simple approach for personalization involves the development of a domain model or taxonomy. For example, in the world of news portals, we may have a category called business and another category called sports. Business and sports are siblings under news. Under sports, we would have further classifications such as football and hockey. This approach is called *category matching*. For customization, the user would simply drill down this hierarchy and select the news categories of interest. For personalization, we first try to infer the right categories for the user and then supply the materials that the match the user's needs.

The basic philosophy underlying the main classes of applications of personalization can be understood in terms of *decision support*. The key point of decision support is to help a human user decide, efficiently and effectively. Specifically, this concept is to be contrasted with automation, conventionally the goal of information technology deployments. The key point of automation is that an agent would *replace* the human. That is, the agent would be expected to take important decisions on its own accord. It has been shown time and again that such unfettered uses of computational intelligence lead to problems. By keeping the human in the loop and supporting the human's decision making, we can improve the quality of the typical decision and reduce the delay in obtaining decisions while ensuring that a human is involved, so that the resulting behaviors are trustworthy. We consider decision support as the essence of workflow and logistics applications where personalization can apply. The idea of decision-support extends naturally into other kinds of collaborative applications where peers exchange information about their experiences and expertise. And it even accommodates e-commerce, where the decisions involved are about what to purchase.

Along the lines of decision-support, we formulate a user's actions as choices made by the user. Personalization is about learning the kinds of choices that a user would find relevant in his specific circumstances. The term *choice* indicates the more general extent of this topic than items or products to be reviewed or purchased, which are simply one family of applications.

Personalization is inherently tied to the vision of an intelligent agent acting as the user's *personal assistant*. In simple, if somewhat idealized terms, an agent is something that watches its user's environment and the user's decisions and actions, learns its user's preferences, and helps its user in making further decisions. From the perspective of personalization, an agent needs the ability to perceive the user's environment. For desktop applications, this would involve perceiving the user interface to the level of graphical widgets that are displayed and selected. For Internet applications, a personal assistant still can benefit from perceiving the elements of desktop interaction. However, often personalization in the Internet setting is based on Website servers, where the perceptions are limited to browser actions that are transmitted to the Website, the so-called click-stream data. Applications that support human-to-human collaboration involve a combination of the two.

Personal assistants can be implemented via a variety of techniques. The chapter on intelligent agents discusses techniques for building agents in general. Of significance in practice are rule-based approaches as well as machine-learning techniques.

This chapter concentrates on the concepts of personalization. Some other relevant topics, especially the specific techniques involved, are discussed in the chapters on Web mining, business processes, policies, mobile services, and pervasive computing. Section 18.2 provides an overview of the main traditional and emerging applications for personalization in networked applications. Section 18.3 reviews the major approaches for personalization. Section 18.4 discusses some practical considerations and the remaining challenges that are guiding current developments in personalization, and summarizes the main themes that drive this large area.

18.2 Key Applications and Historical Development

Personalization is, or rather should be, ubiquitous in computing. It has significant usefulness in a variety of applications. The arrival of the Internet in the commercial sense has created a number of applications where personalization is expected. There are three main reasons for this shift in expectations.

- When the direct human touch is lost, as in many Internet applications, there is a greater need for intelligence and personalization in supporting effective interactions.
- Modern computers make it possible to carry out the computations necessary to produce effective personalization.
- The networked nature of the applications means that often many users will be aggregated at a single site, which will then have the data available to produce effective personalization based on collaborative filtering. When little data was available, the best we could do was to offer the user a fixed set of choices about some aspect of the interaction. But across many users, choices can be collected and recommendations made based on shared, common experiences.

18.2.1 Desktop Applications such as E-mail

Desktop computing is synonymous with personal computing and provided the first environment for personalization. Two technical approaches, rule-based and machine-learning systems, were part of many early experiments, and both approaches continue to find their way into all current and future areas of personalization.

Machine learning has been slower to develop because early algorithms were largely intended for batch-oriented, highly parametric data mining. Data mining has its place in personalization as will be discussed, but the individual modeling of desktop end users implies the use of methods that can learn on-the-fly, watching and predicting the user's situation and actions automatically, without any knowledge engineering. Early desktop operating systems also inhibited the development of adaptive personalization because most of the operating systems were too "opaque" for learning systems to observe the user's situations and actions. The MAC OS developed a semantic layer for system and application events, beyond keystroke and mouse events, and this allowed some adaptive personalization systems such as Open Sesame to use some simple but incremental machine-learning techniques.

Without access to each application's events and support of third-party observation of such events, the only other early attempts were by operating systems themselves to model users for more adaptive help systems. For instance, both OS2 and Windows used various combinations of rule-based and adaptive systems to infer the user's level of expertise and intention in order to provide appropriate support. As Microsoft has also developed desktop applications that rely on elements of the operating system, a Bayesian form of machine learning is increasingly being made available to all applications for modeling and helping end users. In fact, ever since the earliest experiments with such personalization (as in the failed "Bob" interface), Microsoft has persisted in its research and development and has made clear statements about machine learning becoming part of the operating system. Clearly, the value of the system learning about the user rather than the user having to learn about the system remains the future of more intelligent computers.

Early desktop applications lacked the machine-learning algorithms and operating system support that would have helped transfer many research ideas for personalization into actual product, but e-mail is an interesting story in the development of rule-based personalization, which continues to survive and has relevance to current issues with spam. Such rule-based systems are not adaptive. Users are typically assumed to be the authors of the rules to suit themselves, and as such can be seen as an advanced form of customization rather than personalization. Still, e-mail filtering was an early implementation of the core idea behind personalization: the computer as an intelligent agent that can be instructed (by one method or another) to assist the user's individual needs and desires.

Rule-based e-mail filtering itself has had limited success although it has been and still is a staple feature of all e-mail products. Rule authoring is notoriously difficult, but early experiences with this problem have lead to some improvements. For instance, rule types are typically preauthored and provided as templates, such as "If from:X and subject keyword:Y, then delete." Users simply fill in the slots to build such rules for managing their inbox. As well, many corporations discovered that end users tend not to write rules, but that IT can develop them and successfully support their use across an organization.

All of this experience is now finding a place in the war against spam mail. Both rule-based and machine-learning systems are available. E-mail applications and desktop operating systems (dominantly Windows and NT, of course) provide much more transparency of APIs and events, allowing third parties to try many approaches and algorithms. For instance, Bayesian learning is a popular method for building individual models — what each individual user decides is spam or not. Spam filtering is largely a case for personalization. On the other hand, "blacklist filtering" and corporate/ISP policies are more or less like rule-based filtering. Users tend not to write such rules, but a central IT function can successfully support a group of users, at least to the extent that much of spam is commonly and clearly decided as spam.

E-mail and spam filtering are also good examples of how the Internet has exploded over the last many years, blurring the distinction between desktop and Web-based applications. Web applications are largely based on transparent, machine-readable, standards-based content, which enables agents and algorithms to better attach to the content the user sees and actions the user performs. Most critically, however, the Internet and the Web have enabled the networking of users beyond their individual desktops. Approaches to spam also include ideas for adaptive collaboration, but this kind of personalization was best developed through the rise of the Web and e-commerce personalization.

18.2.2 Web Applications such as E-commerce

The Internet explosion of technical and financial interests provided fertile ground for personalization. The Web grew the scope of computer users to include a mass of human browsers and consumers. E-commerce in particular combined general ideas of computer personalization with marketing ideas of personalization. Ideal marketing would entail one-to-one marketing in which each customer is known and treated as a unique individual, and it seemed that Internet technologies would allow electronic capture and use of consumer behaviors, all during the browsing process and at point-of-sale.

E-commerce, especially of the business-to-consumer variety, provides the most well-known examples of personalization. A classical scenario is where a customer is recommended some items to purchase based on previous purchases by the given customer as well as on purchases by other customers. At the larger e-commerce sites, with as many as millions of items in their catalogs, such recommendations are essential to help customers become aware of the available products. Customers would often not be able to search effectively through such large catalogs and would not be able to evaluate which of the items were of best use to them. A recommendation can effectively narrow down the search space for the customers. Because of the obvious payoff and prospective returns on investment, e-commerce has been a prime historical motivator for research into personalization.

The ideal of personalized commerce requires individual consumer models that are richly detailed with the customer's browsing and purchase behavior. In particular, the model should understand an individual's preferences in terms of product features and how they map to the situational desires. In the same way that desktop systems would like to include a personal secretary for every user, e-commerce systems would like to include a personal shopper for every user — an intelligent agent that would learn all the details of each person's needs and preferences. If not a personal shopper for every consumer, the system should at least be like a well-known salesperson who can understand particular needs, know who they mapped to product features, and remember the consumer's past preferences when making new recommendations in repeat visits.

However, the earliest online catalogues and available other technology left much to be desired. First, understanding the intention of a consumer remains a hard research problem. Search engines must still advance to include individual meaning of user queries. Systems like AskJeeves continue to work on natural language input rather than just keywords, but personalized meaning is a problem even for sentences or keywords. For instance, consider a consumer looking for a "phone." Assuming that the search engine will retrieve actual phone items (not all the products, such as accessories, that might contain the keyword "phone" for one reason or another), the meaning of phone differs between the corporate user needing a two-line speakerphone and the home consumer wanting a wall phone for the kitchen. Keyword conjunction, navigation of taxonomic trees, and disambiguation dialogs are all also appropriate answers to such problems, but user modeling, knowing what the user probably means and probably wants based on past experience, is largely the method used by great personal shoppers and sales agents who individually know their customers.

Second but most historically significant, early online catalogs were very impoverished, backed by databases of only product SKUs without any schematized product features. In other words, the transparency of HTML of the Web made it possible to observe purchase behaviors, but there was little to record and understand of the behavior other than to record it. Necessity is the mother of invention, and given this situation, "collaborative filtering" was invented as a new inference for still being able to make product recommendations from such impoverished systems. The basic idea was this: based only on consumer IDs and product SKUs: given a purchase by one consumer, look for other consumers who have made the same purchase and see what else they have bought as well. Recommend these other products to the consumer.

Collaborative filtering enjoyed the hype of the Internet boom and many companies and products such as Firefly, NetPerceptions, and Likeminds and others provided more or less this kind of personalization technology. However, the inference of collaborative filtering is actually a form of stereotyping, akin to market segmentation of individuals into groups. While the personalization industry continued to sell the idea of one-to-one individualized marketing, the technology did not live up to this promise. As well, many mistakes were made in various integrations of the technology. Even if this example is apocryphal, poor consumer experiences were often reported such as when buying a child's book as a gift and then being persistently stereotyped and faced with children's' books recommendations on return visits.

Given the general effects of the bursting Internet bubble and general reductions in IT spending over the last few years, personalization companies have also suffered. Even at the time of this writing, industry leaders such as NetPerceptions are planning to liquidate. However, Amazon.com continues to be a strong example of taking lessons learned and using personalization to great effect. For instance, the problem of persistently clustering a customer as purchaser of children's books is handled by making the collaborative inference only within the context of the current book purchase. In-house experience and persistence in applying personalization have been successful here. Also note that Amazon.com has a general strategy toward personalization, including customization and even basic human factors. For instance, one-click purchasing is low tech but perhaps the most profound design for knowing the customer (in simple profiles) and increasing ease-of-use and purchase.

Collaborative filtering also suffered from low accuracy rates and thus low return on investment (ROI). Like marketing campaigns based on market segmentation (again, based on group statistics rather than each individual), hit rates tend to be very low. The "learning" rate of the system is also very slow. For example, consider when a new and potentially "hot" book is first introduced to a catalogue. The collaborative recommendation system is unlikely to recommend it until a large-enough group of consumers has already purchased it because the inference is based on looking at past purchases of product SKUs. In contrast, feature-based inferencing is more direct and faster to inject products as recommendations. If a consumer seems to like books with particular features (authors, subjects, etc.), then a new book with such desirable features should be immediately matched to users, even before a single purchase.

However, whereas the business of personalization has suffered, understanding of the technical issues continues and will serve to improve accuracy as personalization reemerges for desktop and Web applications. For instance, the argument between supporters of collaborative inferencing and individual modeling has been *the* argument in personalization. One side argues about the power of collective knowledge. The other side argues about the power of focused knowledge. Recent research (Ariely, Lynch, and Aparicio, in press) found that both are right and both are wrong. Individual modeling is best when the retailer knows a lot of about the user. Individual modeling allows the feature-based inferencing just mentioned above, which is faster to recommend products and more likely to make accurate recommendations. Such modeling is possible with today's feature-rich catalogues. However, whenever the retailer is faced with a new consumer (or the consumer expresses interests that the retailer has not yet observed in the consumer) and there is no individual model, then collaborative filtering provides additional information to help.

Future personalization systems will combine all this industry experience. Just as the current war on spam is already leveraging experience with early e-mail and collaborative systems, stronger future requirements for personalization will also include the hybridization of individual and collaborative models as well as the hybridization of rule-based and adaptive systems. Before the burst of the Internet bubble, the personalization industry was heading toward richer, feature-based, more dynamic modeling. As such interests recover, these directions will reemerge.

Beyond just e-commerce, such requirements for easier, faster, and more accurate search of products will become a requirement for general search. The major search engines are still clearly not ideal. As the Internet continues to grow, personalization will be an inevitable requirement to understand the user's needs and wants.

18.2.3 Knowledge Management

The topic of Knowledge Management deserves a chapter of its own, and in fact many books have been written on this very broad topic. In regard to personalization, it should be noted that many of the personalization companies and technologies crossed into this application as well. The main justification is that user models of desires and preferences as in consumer modeling can be transferred to user models of situations and practices in expertise modeling.

All the same issues apply to knowledge management as raised for other applications such as e-mail and e-commerce. Rule-based systems and Knowledge Engineering are likely to be included in a hybrid system, but here too, knowledge engineering is a secondary task that is difficult and takes users away from their primary task. People are less inclined to read and write best practices than they are to simply practice. From a corporate perspective, primary tasks are more related to making money. As such, tacit or implicit knowledge management is preferred. Tacit knowledge can be captured by machine-learning techniques that, as in the case of personalization, can watch user situations, actions, and outcomes to build individual and collaborative models of user practice.

Beyond the recommendation of books and CDs, knowledge management applications tend to be more serious, in the sense that the costs and consequences of decisions extend beyond \$10 to \$20 consumer purchases. In particular, advanced techniques of personalization will be required for "heavy lifting" sorts of knowledge management and decision support, such as for individualized medical practice. The importance of individual differences in pharmacology is increasingly outweighing population-level statistics about a drug's safety and efficacy. Such individual differences can be fundamentally one-to-one between a patient and a drug, not predicted by group memberships based on race, age, geography, or behavior.

Workflow enactment is another important application area where personalization finds use. Workflows involve executing a set of tasks with a variety of control and data-flow dependencies between them. Importantly, some of the tasks are performed by humans and some by machines. Thus, workflows involve repeated human involvement. Often, the human is faced with multiple choices for each decision. Personalization of some of the decisions might be appropriate and would reduce the cognitive overload on the human. Workflows can be enacted over business intranets or even over the public Internet. Often, they are wrapped through e-mail.

Workflow is a good example of an application area that underpins many heavy lifting applications and that can be improved by individual user modeling and personalization. Whereas workflows tend to be rigid and developed through hard knowledge-engineering of the "correct" procedures, real organizational systems also operate by informal *ad hoc* workflows between people. These workflows are defined by user preferences and experiences. For instance, if a user routes an activity to a particular person with particular other orders, the user is likely to use the same routing for a future activity that is similar to the first. As the system builds an individual model for each user to learn various situations and actions, each model becomes a personal assistant for helping to manage the workflow inbox, much like for e-mail but within the context of more structured activities and processes.

18.2.4 Mobile Applications

The recent expansion of mobile applications has opened up another major arena for personalization. Mobile applications are like other networked applications in some respects. However, they are more intense than applications that are executed in wireline environments because they typically involve devices with smaller displays, limited input capabilities, reduced computing power, and low bandwidth. Further, these devices are typically employed when the user is on the run, where there is a reduced opportunity for careful interactions. For these reasons, personalization is more important in mobile than in other environments. Further, mobile devices tend to be more intensely personal by their very nature. For example, a phone or PDA is typically with their user at all times and are used for purposes such as communication and personal information management (e.g., calendar and address book), which inherently tie them to the user. Thus they provide a natural locus for personalization.

Mobile environments can and often support an ability to estimate, in real time, the position of a mobile device. A device's position corresponds to its X–Y coordinates. Position must be mapped to a geographical location, which is a meaningful abstraction over position. That is, location corresponds to something that would relate to a user's viewpoint, such as work or home, or maybe even the 3rd aisle in the supermarket. Although the precision of position determination varies, technology is becoming quite accurate and can support a variety of applications. This has led to an increasing interest in location-based services in wide-area settings. Currently proposed location-based services are limited and may well not become widely popular. Examples include pushing coupons to a user based on what stores are close to his or her current location. We believe that more conventional applications with a location component would perhaps be more valuable. For example, a user's participation in a logistics workflow may be guided based on his or her proximity to a site where a particular step of the workflow is being performed. Both kinds of applications involve an adaptation of a system's actions to suit the (perceived) needs of a user and thus are a form of personalization. Further, any application-level heuristics would be applied in conjunction with user-specific preferences, e.g., whether a user wishes to have coupons pushed to him or her and whether he or she truly wishes to participate in the workflow when he or she is in a particular location.

A more general way to think of personalization in mobile environments is to consider two concepts that apply not merely to mobile environments but to all networked environments. Networked environments involve an ability to discern the *presence* of users. Presence deals with whether a given user is present on a given network. Network here could be a physical subnet or an application-level construct such as a chat room. Presence and location together lead to the richer concept of *availability,* which deals with more than just whether a user is in a certain place or on a certain network, but whether the user is available for a certain task or activity. Ultimately, knowing a user's availability is what we care about. However, a user's availability for a task depends on whether the task is relevant for him or her right now. This cannot easily be captured through simplistic rules, but requires more extensive personalization. Indeed, personalization in many applications can be framed as inferring a user's availability for various tasks. For example, the decision whether to send a user a coupon or an alert can be based on an estimation of the user's availability for the task or tasks with which the given coupon or alert is associated.

Mobile computing is still in its early days of development as with desktop and Web-based applications. The ideas for personalization are emerging from research just as they were in the early days of desktop and Web applications. For instance, Remembrance Agent (MIT) is an intelligent agent that remembers what a mobile user does within the context of locale (and other aspects of the entire physical situation) [Rhodes and Maes, 2000]. If the user returns to some place and begins working on a document that was opened in that place in the past, Remembrance Agent could recall other documents associated with the given document and place. Again, the basic vision is of a personal assistant that can watch and learn from the user, making relevant recommendations that are highly accurate (without annoying false alarms).

One early example of adaptivity for handheld devices is found in the handwriting recognition approach invented by Jeff Hawkins for the Palm Pilot and Handspring. Rather than have the user learn how to completely conform to the letter templates, Graffiti can adapt to example provided by the user. This particular example is on the fringe of personalization interests and is more akin to user-dependent input recognition systems (also including voice recognition), but it represents the philosophy and technology of how adaptive systems will be established one point at a time and then grow to be pervasive. Jeff Hawkins used an associative memory for such machine learning and suggests that, one day, more silicon will be devoted to associative memories than for any other purpose [Hawkins, 1999].

As with Microsoft's interest in Bayesian learning as part of the operating system, personalization will find a fundamental place in all operating systems, including handheld and other new computing devices. For instance, the proliferation of television channels and content makes browsing of a viewer's guide increasingly difficult. TV viewing is usually a matter of individual mood, interest, and preference, and personalization technology is already being introduced to set-top devices. Aside from CD purchasing, which was one of the original applications of collaborative filtering, music listening will also benefit from personalization technology to learn the moods and preferences of each individual.

18.3 Key Concepts

After discussing the applications of personalization, we are now ready to take a closer look at the key concepts, especially as these are instantiated in the major technical approaches. Even as personalization has expanded its role into both desktop and Internet settings, and especially into applications such as e-mail that combine elements of desktop and Internet applications, it has had some challenges.

One set of challenges arises from the hype associated with some of the early work on personalization. Because people expected personalization to work near perfectly with little or no user input, they were disappointed with the actual outcome. The trade press was harsh in its criticisms of personalization. The main problem identified was that personalization was too weak to be cost effective in practical settings where there are many users and the users are not technically savvy. In such cases, simplistic techniques could only yield limited personalization, which was not adequate for effectively guiding operations. The techniques had an extremely low learning rate and yielded high errors. For example, in e-commerce settings, it was difficult to introduce new products because the approach would require a number of customers to have bought a new product before being able to incorporate it into a recommendation. The systems were not personal enough and yielded at best coarse-grained market segmentation, not mass customization or one-to-one marketing as promised.

The criticisms of the IT press are in many ways valid. There is clearly a need for richer, true individualization, which is being widely recognized in the industry. This realization has led to an increased focus on richer modeling for higher accuracy and resulting value. In the newer approaches, personalization applies to both individual taste and individual practice.

The evolution of personalization as a discipline can be understood in terms of a fundamental debate, which attracted a lot interest in the 1990s. This debate is between two main doctrines about personalization: should it be based on individual models or on collaborative representations. This debate has petered out to some extent because of the recognition that some sort of a hybrid or mixin approach is preferable in practice. However, the concepts involved remain essential even today.

18.3.1 Individual vs. Collaborative

Individual modeling involves creating a model for each user. Modulo the difficulties of constructing such models, they can be effective because they seek to capture what exactly a user needs. Collaborative modeling, of which the best known variant is collaborative filtering, involves relating a user to other users and using the choices of other users to filter the choices to be made available to the given user.

In its canonical form, each of the above approaches has the serious deficiency of narcissism.

- With individual modeling, it is difficult to introduce any novelty into the model. Specifically, unless the user has experienced some choices or features, it is not possible to predict what the relevance might be. In this sense, the risk of narcissism is obvious.
- With collaborative modeling too, it is difficult to introduce novelty, although the scope is expanded in that exploration by one or a few users can, in principle, help other users. The risk of narcissism still remains at a system level because, if the users chosen to collaborate with the given user have not yet experienced some choices or features, those choices or features will not be recommended to the given user.

The above debate has a simple resolution. Hybridization of the individual and collaborative approaches proves more effective in practice. In a setting such as collaborative filtering, where a recommendation is based on an aggregation by the system of the choices made by relevant users, hybridization can be understood based on the following conceptual equation. Here user_weight corresponds to the relative importance assigned to the individual and the collaborative aspects. The equation could be applied to make overall judgments about relevance or at the level of specific attributes or features.

Relevance = (user_experience * user_relevance) + ((1 – user_experience) * collaborative_relevance)

Hybridization does not necessarily have to apply in an aggregated manner. An alternative family of approaches would reveal to the user identities of the other users and their choices and let the given user make an informed choice.

18.3.2 Representation and Reasoning

Early collaborative filtering inference was too weak to be cost effective. A general weakness of collaborative approaches is that they are not individualized — they essentially stereotype users. If a user is stereotyped incorrectly, then suboptimal recommendations result. An example of such an error is in the approach known as k-means clustering. This approach involves constructing clusters of users based on an appropriate notion of proximity. A user is then stereotyped according to the cluster in which the user falls, that is, the user is treated as if he or she were the centroid of the cluster to which he or she is the closest. Thus a user who is an outlier in his chosen cluster would be likely to be misunderstood. By contrast, the k-nearest-neighbors approach constructs a cluster for the given user where the user is close to being the centroid. The recommendation is then based on the choices of the user's nearest neighbors. Figure 18.1 illustrates this point schematically.

Some basic concepts of standard machine learning are worth reviewing. Machine-learning approaches can be supervised or unsupervised and online or offline. Supervised approaches require inputs from users in order to distinguish between good and bad cases, the user explicitly or implicitly marking the relevant and irrelevant cases. Unsupervised approaches seek to learn a user's preferences without explicit inputs from the user but based instead on observed actions or other coincidences that are somehow reinforced or self-organized. In our setting, such reinforcement may be inferred based on heuristics such as that a user who visits a Web page repeatedly likes the content presented therein. An example of self-organizing coincidences would be in learning to associate all the items of a shopping cart. Given another shopping cart, recalling such associations to additional items provides a method of cross-selling.

The evaluation of machine-learning approaches is based on estimations of true vs. false positives and negatives usually called "hits," "misses," "false alarms," and "missed opportunities." Sometimes the utility of a correct decision and the cost of an error must be balanced to settle on an optimal risk–reward payoff. For example, if the potential payoff from an unexamined good alternative is high, and the cost of examining an undesirable alternative is low, it might be safer to err on the side of making additional suggestions.

Finally, personalization can be applied in two main kinds of business settings: back room and front room. Back-room settings are those where the data pertaining to personalization are gathered up and processed offline, for example, to plan marketing campaigns and to understand whether a Website ought to be reorganized. By contrast, front-room settings seek to apply the results of personalization in a customer-facing activity. Examples of these include carrying on dialogs with users as for customer relationship management (CRM), and making real-time recommendations for products. These call for rapid adaptation to adjust quickly to what the user needs. On the other hand, back-room applications, such as sales campaigns, can be slower and more static. Here, rule-based systems are appropriate to

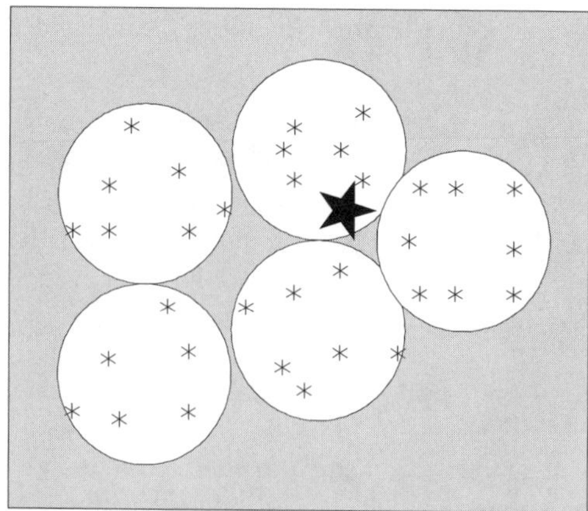
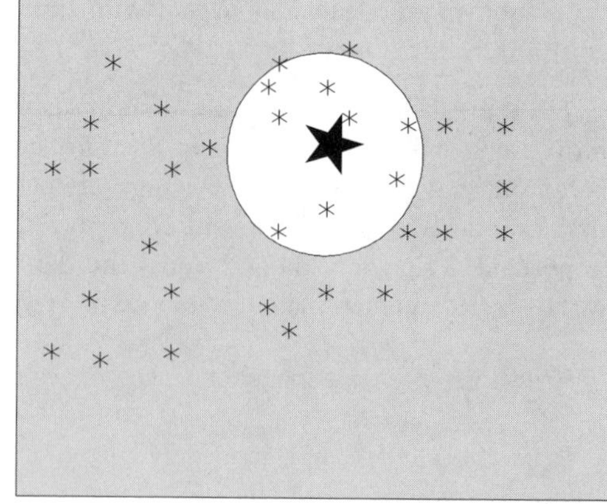

FIGURE 18.1 The clusters approach distributes the data points (shown by small stars) among a number of clusters. The given user (shown by a large star) falls into one of the clusters and is treated like the center of that cluster. The nearest-neighbors approach associates the given user with other users that are the closest to it. In effect, the given user is the center of its own cluster. The nearest-neighbors approach can thus make more relevant predictions.

implement sales policies and explicit sales promotions. Online machine-learning approaches function incrementally as the data comes in, and are appropriate for front-room settings, whereas offline approaches operate on data that is gathered up through several interactions, perhaps across a large data warehouse and are appropriate for back-room settings.

18.4 Discussion

Aside from the past ups and downs of the Internet revolution, personalization is a fundamental requirement and will continue to gather lessons learned for building effective total solutions. So what is one now to do? This section provides a review of past experiences, how to integrate and measure, and what is yet to be discovered. Difficulties remain for search in both e-commerce and in general across the Internet. Search engines in general seem to have waves of progress from one sort of technology to another, and personalization will eventually be required for more reasonable accuracy.

18.4.1 Advice

The following pieces of advice include some practical matters of interest to customers of personalization as well as industry consultants' beliefs about personalization vendors.

18.4.1.1 Hybridization

First, consider all the methods that have been described. As evidenced by the breath of topics in one of the last major reviews (Riecken, 2000), personalization means different things to different people. This is true because personalization solutions need to include issues of human factors as well as underlying technology. Think low tech as well as high tech. Think of customization as well as personalization. For instance, Amazon.com includes "one-click" as a simple or low-tech but powerful method as well as its higher-tech collaborative filtering methods.

As described above, also consider the hybridization of individual and collective approaches. The question is no longer about which is better, but about which is better when, and how the combination can provide best accuracy overall.

Also consider the use of rule-based systems, but not for end-user authoring. Whereas adaptive technologies are preferred for automatically tuning in to customer behaviors, rules still have a strong place if centrally administered, especially for back-end functions such as sales policy and campaign management.

18.4.1.2 Richer Detail

Just as SKU-based online catalogues matured into feature-based descriptions and this allowed better forms of personalization, expect increasing richness in data source descriptions and in models themselves.

On the data side, industry standard schemas will help better identify the meaning of items and features. For instance, taxonomies will help organize product categories, and more complete ontologies (more than just taxonomies) will help describe the structural meaning of items and features. As these improved descriptions are fed into personalization systems, more intelligent algorithms can be included such as for analogical reasoning, partially transferring what is learned from trucks to cars and vice versa, for example.

On the model side, algorithms must become more dynamic and internally complex. Many current implementations suffer from their linearity. In other words, many approaches keep only independent variable accounts of each item or feature. Such linear modeling is inexpensive. Naïve Bayesian modeling, for instance, shows some very good results. However, as decisions become more complex and better accuracy is required, individual models must be based on nonlinear techniques. In other words, the models must assume context-dependency, in that features can be dependent on one another. The desirability of one feature is often largely dependent on the other features. In heavy-lifting decision support, the interaction of elements within a given situation will become very critical in representation and modeling.

Of course, nonlinear modeling in more difficult and computationally expensive. For one problem, many nonlinear methods, such as most neural networks, are highly parametric (require much backroom knob-tuning) and are trained in the batch mode (cannot be dynamically updated case by case on-the-fly). In contrast, a class of modeling called "lazy learning" includes case-based, memory-based, and other similar techniques that could provide nonparametric, incremental, nonlinear modeling as required.

Newer algorithms must emerge to handle vast numbers of individual models, while also becoming internally richer and more powerful in prediction.

18.4.1.3 Grounding

Without the intelligence and common sense of real sales people and experts to assist the user, the grounding problem — recording situations and actions and ensuring what they mean — has been difficult. For instance, the apparent time a user spends on a page can be totally misleading if the user were to walk away for a meeting or just to get coffee. This will continue to be a matter of research for online question and answer systems, but practical progress is also possible.

New ideas in information retrieval will be presented next under Metrics, but before situations and actions can be measured, they must at least be available. In the design of the user interface, it is critically important to gather and display the context of the user. What is the task? What is already included — whether it be a shopping basket or a research paper? What did the user do? What did the user really intend to do? How can we measure the final outcome? For instance, is the user browsing or intending to buy? But even a purchase is not the final outcome. A fully integrated system must also know whether an item is returned or not, to ground a better sense of customer satisfaction, not merely the first act of purchasing.

18.4.2 Metrics

Personalization is a specialized enhancement of information retrieval. In general, a user submits a query and the system returns a set of answers. Depending on the task, these answers might be documents, products, or both. Therefore, information retrieval (IR) measures are often quoted as measures for personalization. For instance, precisions and recall are classic. Precision measures how well the system recommends "hits" rather than "false alarms." Recall, in compliment, measures how well the system recommends *all* the hits. The *relevance* score of a system is often measured by marking the "right" answers in the catalog that the system should recommend for any given query, then reporting the precision and recall for the query.

However, recent IR research is moving to better definitions of relevance. For one, relevance is understood as more personal. Rather than assuming some set of right answers, a philosophy of personalized relevance is coming to dominate. What the user intends by even any single query word is often different from person to person. Therefore, the precision and recall of a recommendation system can be measured only as a matter for each person's measure of success.

IR is also moving to notions of effective relevance. This again is part of the grounding problem, but we can measure the effectiveness of an answer in several ways. First, if a user thinks it is worth the effort to at least explore the recommendation, we can consider this as some measure of potential relevance. IR is also moving to allow partial relevance. Rather than marking an item as a hit or miss, as being absolutely relevant or irrelevant, many cases of browsing and querying are more complex and fuzzy, and therefore, recommendations can only by more or less relevant. To the degree that the user makes a cost-benefit tradeoff in deciding to investigate, we can attempt to measure some degree of partial relevance.

Of course, business ROI must be grounded in business revenue as much as possible. Therefore, we can consider the purchase action as marking more effective relevance of the item; as mentioned, nonreturn of an item marks even greater effectiveness. However, even this should not be the ultimate measure.

Instead, customer satisfaction should be measured by repeat visits by the customer, which is typically the real bottom line for customer relationship management.

18.4.3 Futures

Effective personalization systems can be deployed and measured as described, but several remaining issues should also be kept in mind. These topics are a matter of research, but near-term awareness and action are also advised.

18.4.3.1 Novelty Injection

Repeat visits by the customer should be the ultimate measure of personalized recommendation systems, but ongoing research indicates the requirement for more advanced technologies that will be able to achieve it. There are two problems to overcome.

The first problem stems from the assumptions of traditional information retrieval: that precision and recall to a given query can be measured by return of the right answers, which are absolutely relevant, avoiding return of false positives that are absolutely not relevant. The underlying assumption here is also that there is only one query — one shot to get it right or wrong.

The second problem stems from the nature of personalization technology itself. All adaptive recommendation systems will tend to be narcissistic. Whether individual or collaborative, recommendation systems observe user behaviors and make new recommendations based on such past experience. The danger is that the recommendation system focuses the user on only such past behaviors, which creates myopic users and limited recommenders.

Instead, personalization technologies must assume partial relevance and the need to explore more uncertain items to determine relevance or not. Personalization technologies must also explicitly inject novelty and variety into recommendation sets. For e-commerce systems, this is the definition of real choice. If a recommendations system suggests all very relevant but similar items, then it is not really giving the user a choice. In decision support, such novelty and variety is a matter of thinking out of the box, considering alternative hypotheses, and learning about new sources.

Indeed, the injection of novelty is suited to building a long-term customer relationship (Dan Ariely and Paulo Oliveira), but there is a delicate tradeoff. Early sessions with the customer need to optimize relevance for each single query in order to build trust and to optimize purchase "hits" for the single occasion. However, long-term relationships are built by some cost to pure relevance in favor of novelty. By blending items most likely to be purchased based on past experience with items that are partially relevant but somewhat unfamiliar, the user optimizes his or her breadth of knowledge about the product catalog while the system optimizes the breadth of knowledge about the user. Technically, this requires user models that can report levels of user experience expressed as novelty in a given product, and ensure variety, the dissimilarity between a set of relevant suggestions.

18.4.3.2 Context Switching

The role of context is a growing issue within artificial intelligence. Even the most principled rules are not thought of as true or false, but true or false under a set of other specific conditions (other rules). Machine learning is also grappling with contextual issues of different perspectives and situational dependencies. Personalization can be seen as resonant with such thought. Rather than think of absolute right answers, individual modeling aims to capture the perspectives of different users as at least one dimension of context.

Yet, the representation, capture, and use of context will remain a difficult set of issues. As mentioned above, one problem is in how to extract and hold context to understand the intentions and current situation of the user. For instance, in searching a topic for a research paper, it would be helpful for the system to also "look" at any already selected references and perhaps even the status of the paper as it is being written. For applications such as e-commerce and entertainment, it would be helpful to understand the location and mood of the recipients before making recommendations. Carrying on a dialog with the user without being annoying and intrusive is a major user-interface and application-design problem.

Context switching is a deeper and harder problem. Situations change. Consumer tastes change. The fundamental problem of context for adaptive systems is this: when is a model wrong or merely inappropriate? Ideally, a truly intelligent sales person remembers the customer's habits for appropriate repetition, injects reasonable variety into recommendations, and moves quickly to learn new tastes as the customer changes. However, even when tastes change, the old knowledge is not exactly wrong and no longer needed: When the consumer again changes taste, perhaps back to the first model, the salesperson should easily remember (not relearn) the earlier habits of such a well-known customer.

Knowing when to switch and how to switch across radical changes, either in e-commerce customers or decision-support situations, remains a challenge. There is currently no elegant technical answer. Some absolute context switching, such as a model of "on diet" and a separate model for "off diet" can be explicitly made and controlled. However, human brains do this more elegantly and with ease, and we should expect intelligent personalization agents of the future to be richly detailed and dynamic enough even to this extreme. Neuroscience and psychology have known how associative memories are formed, inhibited (but not forgotten), and then re-recalled. Some computational modeling of these phenomena has been advanced [Aparicio and Strong, 1992] but not yet fully researched and applied. Once fully developed, such advances will provide long-term knowledge and responsiveness to each customer, which is the holy grail of personalization.

References

Aparicio, Manuel, Donald Gilbert, B. Atkinson, S. Brady, and D. Osisek, "The role of intelligent agents in the information infrastructure." *1st International Conference on the Practical Application of Intelligent Agents and Multi-Agent Technology (PAAM),* London, 1995.

Aparicio, Manuel and P.N. Strong, Propagation controls for true Pavlovian conditioning. In *Motivation, Emotion, and Goal Direction in Neural Networks.* D.S. Levine and S.J. Level (Eds.). Lawrence Erlbaum, Hillsdale, NJ, 1992.

Ariely, Dan, John G. Lynch Jr., and Manuel Aparicio, Learning by collaborative and individual-based recommendation agents. *Journal of Consumer Psychology.* In press.

Breese, John S. David Heckerman, and Carl Kadie, Empirical analysis of predictive algorithms for collaborative filtering. *Proceedings of the 14th Conference on Uncertainty in Artificial Intelligence*, 1998, pp. 43–52.

Good, Nathaniel, J. Ben Schafer, Joseph A. Konstan, Al Berchers, Badrul Sarwar, Jon Herlocker, and John Riedl, Combining collaborative filtering with personal agents for better recommendations. *Proceedings of the National Conference on in Artificial Intelligence*, 1999, pp. 439–446.

Hawkins, Jeff, "That's not how my brain works — Q&A with Charles C. Mann." *MIT Technology Review*, July/August 1999, pp. 76–79.

Horvitz, Eric Principles of mixed-initiative user interfaces. *Proceedings of the SIGCHI Conference on Human Factors in Computing Systems (CHI)*, ACM Press, New York, 1999, pp. 159–166.

Riecken, D. (Guest Ed.). Special issue on personalization. *Communications of the ACM*, 43(9), Sept. 2000.

Resnick, Paul, Neophytos Iacovou, Mitesh Suchak, Peter Bergstorm, and John Riedl, *GroupLens: an open architecture for collaborative filtering of netnews*, *Proceedings of the ACM Conference on Computer Supported Cooperative Work*, ACM Press, New York, 1994, 175–186.

Rhodes, Bradley J. and Pattie Maes, Just-in-time information retrieval agents. *IBM Systems Journal* (special issue on the MIT Media Laboratory), Vol. 29, Nos. 3 and 4, pp. 685–704, 2000.

Sarwar, Badrul M., George Karypis, Joseph A. Konstan, and John Riedl, Analysis of recommendation algorithms for e-commerce. *ACM Conference on Electronic Commerce*, 2000, pp. 158–167.

Shardanand, Upendra and Pattie Maes, Social Information Filtering: Algorithms for Automating "Word of Mouth." *Proceedings of the ACM SIGCHI Conference on Human Factors in Computing Systems (CHI),* Vol. 1, ACM Press, 1995, pp. 210–217.

19

Online Marketplaces

CONTENTS

19.1 What Is an Online Marketplace? 19-1
19.2 Market Services 19-2
 19.2.1 Discovery Services 19-3
 19.2.2 Transaction Services 19-4
19.3 Auctions 19-4
 19.3.1 Auction Types 19-5
 19.3.2 Auction Configuration and Market Design 19-6
 19.3.3 Complex Auctions 19-8
19.4 Establishing a Marketplace 19-9
 19.4.1 Technical Issues 19-9
 19.4.2 Achieving Critical Mass 19-10
19.5 The Future of Online Marketplaces 19-11
References 19-12

Michael P. Wellman

Even before the advent of the World Wide Web, it was widely recognized that emerging global communication networks offered the potential to revolutionize trading and commerce [Schmid, 1993]. The Web explosion of the late 1990s was thus accompanied immediately by a frenzy of effort attempting to translate existing markets and introduce new ones to the Internet medium. Although many of these early marketplaces did not survive, quite a few important ones did, and there are many examples where the Internet has enabled fundamental change in the conduct of trade. Although we are still in early days, automating commerce via online markets has in many sectors already led to dramatic efficiency gains through reduction of transaction costs, improved matching of buyers and sellers, and broadening the scope of trading relationships.

Of course, we could not hope to cover in this space the full range of interesting ways in which the Internet contributes to the automation of market activities. Instead, this chapter addresses a particular slice of electronic commerce in which the Internet provides a new medium for marketplaces. Since the population of online marketplaces is in great flux, we focus on general concepts and organizing principles, illustrated by a few examples rather than attempting an exhaustive survey.

19.1 What Is an Online Marketplace?

Marketplace is not a technical term, so unfortunately, there exists no precise and well-established definition clearly distinguishing what is and is not an online marketplace. However, we can attempt to delimit its meaning with respect to this chapter. To begin, what do we mean by a "market"? This term, too, lacks a technical definition, but for present purposes, we consider a market to be an interaction mechanism where the participants establish deals (trades) to exchange goods and services for monetary payments (i.e., quantities of standard currency).

Scoping the "place" in "marketplace" can be difficult, especially given the online context. Some would say that the Web itself is a marketplace (or many marketplaces), as it provides a medium for buyers and

1-58488-381-2/05/$0.00+$1.50
© 2005 by CRC Press LLC

sellers to find each other and transact in a variety of ways and circumstances. However, for this chapter we adopt a narrower conception, limiting attention to sites and services attempting to provide a well-scoped environment for a particular class of (potential) exchanges.

Many preexisting marketplaces are now online simply because the Internet has provided an additional interface to existing protocols. For example, online brokerages have enabled any trader to route orders (with some indirection) to financial exchanges and electronic crossing networks (e.g., Island or REDI-Book). Although such examples certainly qualify as online marketplaces, the plethora of different interfaces, and usually nontransparent indirections, do make them less pure instances of online marketplaces. For high-liquidity marketplaces like equity exchanges, these impurities may not substantially impede vibrant trade. For newer and more completely online marketplaces, directness and transparency are hallmarks of the value they provide in facilitating exchange.

Perhaps the most well-known and popularly used online marketplace is eBay [Cohen, 2002], an auction site with over 12 million items (in hundreds of categories and subcategories) available for bid every day. The canonical "person-to-person" marketplace, eBay has upwards of 69 million registered users.[1] Whereas many eBay sellers (and some buyers) earn their livelihood trading on the site (which is why the "consumer-to-consumer" label would be inaccurate), participation requires only a lightweight registration process, and most aspects of the transaction (e.g., shipping, payment) are the ultimate responsibility of the respective parties to arrange. Note the contrast with the brokered trading model employed in financial markets, where securities are generally exchanged between broker–dealers on behalf of clients.

Many online marketplaces define commerce domains specific to an industry or trading group. One of the most prominent of these is Covisint, formed in 2000 by a consortium of major automobile manufacturers (Ford, General Motors, DaimlerChrysler, and Renault–Nissan, later joined by Peugeot Citroen) to coordinate trading processes with a large universe of suppliers.[2] Covisint provides electronic catalog tools, operates online procurement auctions, and supports a variety of document management and information services for its trading community.

Although many of the online marketplaces launched by industry consortia in the late 1990s have since failed, as of 2002 there were still dozens of such exchanges, with projections for renewed (albeit slower) growth [Woods, 2002]. Similarly, the number of person-to-person sites had reached into the hundreds during the speculative Internet boom. Clearly, eBay dominates the field, but many niche auctions remain as well.

The examples of person-to-person auctions (eBay), industry-specific supplier networks (Covisint), and online brokerages illustrate the diversity of online marketplaces that have emerged on the Internet over the past decade. Another category of major new markets are the exchanges in electric power and other commodities corresponding to recently (partially) deregulated industries. Many of these are hidden from view, running over private (or virtually private) networks, but these, too, constitute online marketplaces, and play an increasingly significant role in the overall economy.

19.2 Market Services

What does a marketplace do? In order to facilitate conduct of trade, a marketplace may support any or all phases in the lifecycle of a transaction. It can be useful to organize commerce activities into three stages, representing the fundamental steps that parties must go through in order to conduct a transaction.

1. The **Connection**: searching for and *discovering* the opportunity to engage in a commercial interaction
2. The **Deal**: *negotiating* and agreeing to terms
3. The **Exchange**: *executing* a transaction

[1] Source: http://pages.ebay.com/community/aboutebay and internetnews.com, May 2003.

[2] 76,000 members as of January 2003. Source: http://www.covisint.com/about/history.

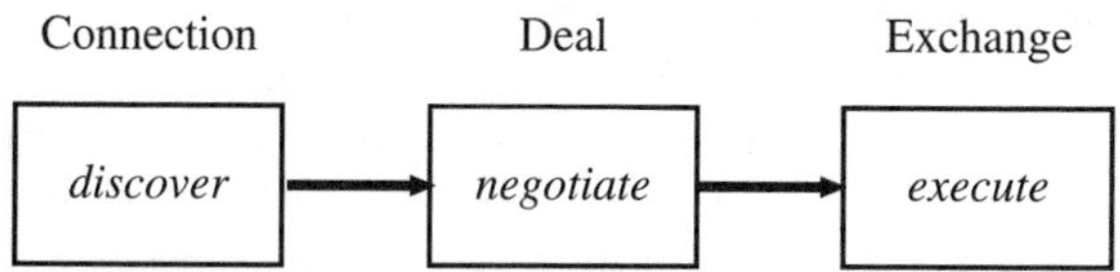

FIGURE 19.1 The fundamental steps of a commerce interaction.

These steps are illustrated in Figure 19.1. Of course, the boundaries between steps are not sharp, and these activities may be repeated, partially completed, retracted, or interleaved along the way to a complete commercial transaction. Nevertheless, keeping in mind the three steps is useful as a way to categorize particular marketplace services, which tend to focus on one or the other.

In this chapter, we focus on the negotiation phase, not because it is necessarily the most important, but because it often represents the core functionality of an online marketplace. Discovery and exchange are relatively open-ended problems, with services often provided by third parties outside the scope of a particular marketplace, as well as within the marketplace itself. Moreover, several aspects of these services are covered by other chapters of this handbook. Nevertheless, a brief overview of some discovery and transaction facilities is helpful to illustrate some of the opportunities provided by the online medium, as well as the requirements for operating a successful marketplace.

19.2.1 Discovery Services

At a bare minimum, marketplaces must support discovery to the extent of enabling users to navigate the opportunities available at a site. More powerful discovery services might include electronic catalogs, keyword-based or hierarchical search facilities, and so forth. The World Wide Web has precipitated a resurgence in the application of information retrieval techniques [Belew, 2000], especially those based on keyword queries over large textual corpora.

Going beyond generic search, a plethora of standards have been proposed for describing and accessing goods and services across organizations (UDDI [Ariba Inc. et al., 2000], SOAP, a variety of XML extensions), all of which support discovering connections between parties to a potential deal. For the most part these are designed to support search using standard query-processing techniques. Some recent proposals have suggested using *semantic Web* [Berners-Lee et al., 2001] techniques to provide matchmaking services based on inference over richer representations of goods and services offered and demanded [Di Noia et al., 2003, Li and Horrocks, 2003].

The task of discovering commerce opportunities has inspired several innovative approaches that go beyond matching of descriptions to gather and disseminate information relevant to comparing and evaluating commerce opportunities. Here we merely enumerate some of the important service categories:

- *Recommendation* [Resnick and Varian, 1997; Schafer et al., 2001]. Automatic recommender systems suggest commerce opportunities (typically products and services to consumers) based on prior user actions and a model of user preferences. Often this model is derived from cross-similarities among activity profiles across a collection of users, in which case it is termed *collaborative filtering* [Riedl and Konstan, 2002]. A familiar example of collaborative filtering is Amazon.com's "customers who bought" feature.
- *Reputation.* When unfamiliar parties consider a transaction with each other, third-party information bearing on their reliability can be instrumental in establishing sufficient trust to proceed. In particular, for person-to-person marketplaces, the majority of exchanges represent one-time interactions between a particular buyer and seller. *Reputation systems* [Dellarocas, 2003; Resnick et al., 2002] fill this need by aggregating and disseminating subjective reports on transaction results across a trading community. One of the most prominent examples of a reputation system is eBay's "Feedback Forum" [Cohen, 2002; Resnick and Zeckhauser, 2002], which some credit significantly for eBay's ability to achieve a critical-mass network of traders.

- *Comparison shopping.* The ability to obtain deal information from a particular marketplace suggests an opportunity to collect and compare offerings across multiple marketplaces. The emergence on the Web of *price comparison services* followed soon on the heels of the proliferation of searchable retail Web sites. One early example was BargainFinder [Krulwich, 1996], which compared prices for music CDs available across nine retail Web sites. The University of Washington ShopBot [Doorenbos et al., 1997] demonstrated the ability to automatically learn how to search various sites, exploiting known information about products and regularity of retail site organization. Techniques for rapidly adding sites and product information have continued to improve, and are employed in the many comparison-shopping services active on the Web today.
- *Auction aggregation.* The usefulness of comparison shopping for fixed-price offerings suggested that similar techniques might be applicable to auction sites. Such information services might be even more valuable in a dynamically priced setting, as there is typically greater inherent uncertainty about the prevailing terms. The problem is also more challenging, however, as auction listings are often idiosyncratic, thus making it difficult to recognize all correspondences. Nevertheless, several auction aggregation services (BidFind, AuctionRover, and others) were launched in the late 1990s. Concentration in the online auction industry combined with the difficulty of delivering reliable information has limited the usefulness of such services, however, and relatively few are operating today.

19.2.2 Transaction Services

Once a deal is negotiated, it remains for the parties to execute the agreed-upon exchange. Many online marketplaces support transaction services to some extent, recognizing that integrating "back-end" functions — such as logistics, fulfillment, and settlement — can reduce overall transaction costs and enhance the overall value of a marketplace [Woods, 2002].

A critical component of market-based exchange, of course, is *payment,* the actual transfer of money as part of an overall transaction. The online medium enables the automation of payment in new ways, and, indeed, the 1990s saw the introduction of many novel *electronic payment mechanisms* [O'Mabony et al., 1997], offering a variety of interesting features [MacKie-Mason and White, 1997), including many not available in conventional financial clearing systems. For example, some of the schemes supported anonymity [Chaum, 1992], micropayments [Manasse, 1995], or atomic exchange of digital goods with payment [Sirbu and Tygar, 1995].

As it turned out, none of the innovative electronic payment mechanisms really caught on. There are several plausible explanations [Crocker, 1999], including inconvenience of special-purpose software, network effects (i.e., the need to achieve a critical mass of buyers and sellers), the rise of advertising-supported Internet content, and decreases in credit-card processing fees. Nevertheless, some new payment services have proved complementary with marketplace functions, and have thrived. The most well-known example is PayPal, which became extremely popular among buyers and sellers in person-to-person auctions, who benefited greatly from simple third-party payment services. PayPal's rapid ascension was in large part due to an effective "viral marketing" launch strategy, in which one could send money to any individual, who would then be enticed to open an account.

19.3 Auctions

Until a few years ago, if one said the word "auction," most bearers would conjure up images of hushed rooms with well-dressed art buyers bidding silently while a distinguished-looking individual leads the proceeding from a podium with a gavel. Or, they might have envisioned a more rowdy crowd watching livestock while yelling out their bids to the slick auctioneer speaking with unintelligible rapidity. Another common picture may have been the auctioneer at the fishing dock lowering the price until somebody agrees to haul away that day's catch. Today, one is just as likely to suggest a vision (based on direct experience) of an auction happening online. Thus is the extent to which online auctions have emerged as a familiar mode of commercial interaction.

Speculations abound regarding the source of the popularity of online auctions. For some, it is a marketing gimmick–enticing customers by making a game of the buying process. Indeed, participating in auctions can be fun, and this factor undoubtedly plays a significant role. More fundamentally; however, auctions support dynamic formation of prices, thereby enabling exchanges in situations where a fixed price — unless it happened to be set exactly right — would not support as many deals. Dynamic market pricing can improve the equality of trades to the extent there is significant value uncertainty, such as for sparsely traded goods, high demand variability, or rapid product obsolescence. Distribution of information is, of course, the rationale for auctions in offline contexts as well. The online environment is particularly conducive to auctions, due to at least two important properties of the electronic medium.

First, the network supports inexpensive, wide-area, dynamic communication, Although the primitive communication protocol is point-to-point, a mediating server (i.e., the auction) can easily manage a protocol involving thousands of participants. Moreover, the information revelation process can be carefully controlled. Unlike the human auctioneer orchestrating a room of shouting traders, a network auction mediator can dictate exactly which participants receive which information and when, according to auction rules.

Second, to the extent that auction-mediated negotiation is tedious, it can be automated. Not only the auctioneer, but also the participating traders, may be represented by computational processes. For example, many sellers employ listing software tools to post large collections of goods for sale over time. To date, trading automation appears to be only minimally exploited by buyers in popular Internet auctions, for example, via "sniping" services that submit bids automatically at designated times, thus freeing the bidder from the necessity of manual monitoring.

19.3.1 Auction Types

Despite the variety in imagery of the auction scenarios above, most people would recognize all of them as auctions, with items for sale, competing buyers and a progression of tentative prices, or bids, until the final price, or clearing price, is reached. How the initial price is chosen, whether the tentative prices are announced by the auctioneer or the traders (i.e., bidders) themselves, or even whether the prices go up or down toward the result, are defining details of the particular type of auction being executed. Although the specific rules may differ, what makes all of these auctions is that they are organized according to well-defined rules, and at the end of the process, these rules will dictate what deal, if any, is struck as a consequence of the bidding activity by the auction participants.

Many obvious variants on the above scenarios will clearly qualify as auctions as well. For example, there might be several items for sale instead of one, or the bidders might compete to sell, rather than buy, the good or goods in question.

Once we consider how auction rules can vary, we see that auctions naturally group themselves into *types*, where auctions of a given type share some distinctive feature. For example, the scenarios described above are all instances of *open outcry* auctions, which share the property that all status information (e.g., the tentative prices) is conveyed immediately and globally to all participants.

Another form of open outcry auction is the familiar "trading pit" of a commodities or securities exchange. Although this might not always be viewed as an auction in common parlance, it shares with the examples above some essential features. Even the seemingly chaotic trading pit operates according to rules governing who is allowed to shout what and when, and what the shouts entail in terms of offers of exchange.

The most immediately distinguishing feature of the trading pit is that it is *two-sided:* both buyers and sellers play bidders in this protocol. In contrast, the art, livestock, and fish auctions alluded to above are *one-sided:* a single seller offers an item to multiple bidding buyers.

The inverse one-sided auction — where a single buyer receives bids from multiple competing sellers — is sometimes called a *reverse auction*, and is often employed by businesses in procurement, where it is often called a *request for quotations* or RFQ.

Unlike open outcry events, in *sealed-bid auctions* the participants do not learn the status of the auction until the end, if then, or until some other explicit action by the auctioneer. Familiar examples of sealed-bid auctions include government sales of leases for offshore oil drilling and procedures by which real-estate developers let construction contracts. Note that the latter is an example of a procurement auction, as it is the sellers of construction services doing the bidding.

Sealed-bid auctions may be one-shot, or may involve complex iterations over multiple rounds, as in the prominent U.S. FCC spectrum auctions held in recent years [McAfee and McMillan, 1996]. Like open outcry auctions, sealed-bid auctions may also come in one-sided or two-sided varieties.

Internet auctions, like their offline counterparts, also come in a range of types and variations. Today, almost all consumer-oriented auctions are one-sided (i.e., they allow buy bids only), run by the sellers themselves or by third parties. The prevailing format can be viewed as an attempt to mimic the familiar open outcry auctions, posting the current high bid and the bidders identity. The very familiarity of these mechanisms is an advantage, and may be the reason they have proliferated on the Internet.

19.3.2 Auction Configuration and Market Design

Auctions operated in business-to-business marketplaces are also predominantly one-sided (typically procurement or reverse auctions), though some two-sided auctions (often called *exchanges*) persist. Familiarity is also a factor in designing business-oriented auctions, though we should expect less of a tendency for a one-size-fits-all approach for several reasons.

- Industry trade groups may have preexisting prevailing conventions and practices, which will be important to accommodate in market designs.
- Participants earn their livelihood by trading, and so are more willing to invest in learning market-specific trading rules. Adopting more complex rules may be worthwhile if they enable more efficient trading over time.
- Transactions will involve higher stakes on average, and so even small proportional gains to customization may be justified.

The Michigan Internet AuctionBot [Wurman et al., 1998] was an early attempt to support general configurability in auction operation. Although the ability to customize auction rules proved not to be very useful for consumer-oriented marketplaces, for the reasons stated above, this capability provides potentially greater value for specialized commercial trade. The AuctionBot provided a model for the auction platform now distributed as Ariba Sourcing™, which underlies several business-to-business marketplaces.

19.3.2.1 Dimensions of Market Design

Flexible market infrastructure supports a variety of market rules, covering all aspects of operating a market. The infrastructure is *configurable* if the designer can mix and match operating rules across the various categories of market activity. We have found it particularly useful to organize market design around three fundamental dimensions (Figure 19.2), which correspond to three core activities performed by the market.

1. **Bidding rules.** Traders express offers to the market in messages called *bids*, describing deals in which they are willing to engage. The market's bidding policy defines the form of bids, and dictates what bids are admissible, when, and by whom, as a function of bids already received.
2. **Clearing policy.** The object of the market is to determine exchanges, or deals, by identifying compatible bids and *clearing* them as trades. The clearing policy dictates how and when bids are matched and formed into trades, including determining the terms of the deals in cases where there exist many consistent possibilities.
3. **Information revelation policy.** Markets typically post intermediate information about the status of bidding prior to the determination of final trades. Determining what information is available to whom, when, and in what form, is the subject of information revelation policy.

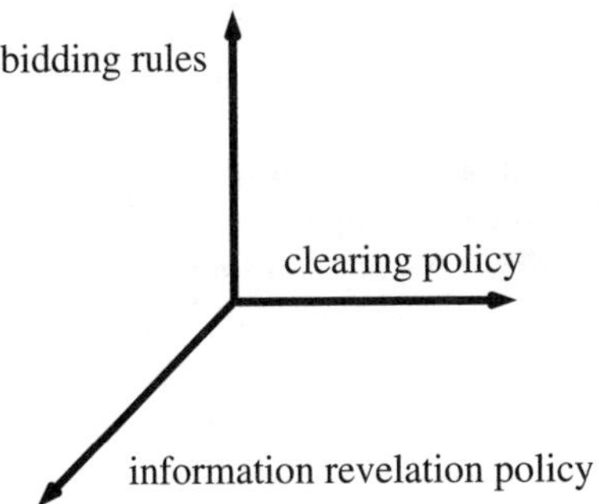

FIGURE 19.2 Three dimensions of market design.

19.3.2.2 Bidding Rules

To illustrate the specification of market rules across these dimensions, let us consider some of the possible range of bidding rules an auction can impose. The outline below is far from exhaustive — even for the bidding dimension alone. I include it merely to illustrate the great variety of separately definable auction features. A more comprehensive and technically precise exposition of auction design space is presented by Wurman et al. [2001].

We generally assume that any trader can always submit a bid to an auction. The bidding rules determine whether it will be *admitted* to the auction. Admitted bids are entered into the auction's *order book,* which stores the bids considered currently active.

Most bidding rules can be defined to hold for everyone or specialized to hold for particular classes of traders. In general, a bidding rule may consider the current order book, previous bids by this trader or, for that matter, any aspect of the auction's history. However, it is helpful to focus the examples on forms of bidding rules corresponding to particularly useful categories.

- **Allowable bid modifications.** These rules regulate when a bid revision is permitted, as a function of the previous pattern of bids by this trader or others.
  - *Withdrawal/replace allowed:* Whether or when a new bid may be submitted to supersede a previous one.
  - *Bid frequency restrictions:* Set over the entire course of the auction or for designated periods. For example, an auction might define a notion of stage, or round, and allow each trader to bid once per round.
- **Static restrictions on bid content.** Content rules define what bids are admissible, based on the specifics of the offer. A content rule is static if it can be defined independently of other bids that have been submitted by this trader or others.
  - *One- vs. two-sided:* Competitive bidding on both the buy and sell sides, or just one or the other As discussed above, in a one-sided auction, only one distinguished trader is allowed to sell (buy); all others can submit only buy (sell) bids.
  - *Bid quantities:* Offers can be for single or multiple units, and if multiunit, the allowable offer patterns. For example, a multiunit offer may be limited to a single price point, or arbitrary *price-quantity schedules* may be allowed. Similarly, quantity bidding rules control such issues as whether or not indivisible ("all-or-none") bids are allowed.
- **Dynamic restrictions on bid content.** A content rule is dynamic if it depends on previous bids by this trader or the current order book.
  - *Beat-the-quote:* A new bid must be better than some designated benchmark, such as the best offer received so far. These rules can be used to implement an ascending (or descending) auction, where prices progress in a given direction until the final price is reached.

- *Bid dominance:* In a manner analogous to beat-the-quote, we can require that a new bid improves the trader's own bid. There are various versions of this rule, based on different criteria for comparing bids.
- *Eligibility:* Defines the conditions under which a trader is eligible to submit bids, or the prices or quantities allowed in those bids. Eligibility is typically based on trader qualifications (e.g., credit ratings) or prior bidding history. For example, *activity rules* define eligibility based on the extent of current bids — the stronger the current bids, the greater the trader's eligibility for subsequent bidding.

- **Payments.** Sometimes, restrictions such as those above can be waived on agreement to pay a fixed or variable fee. For example, an initial bid may require an entry fee (refundable or not) or withdrawals may be allowed on payment of a decommitment penalty.

19.3.2.3 Criteria for Auction Design

Given the wide range of possible ways to run an auction, how is the designer to choose the policies for a particular market? The first step is to define one's objectives. There are many characteristics of a market we may care about. These may be categorized roughly into *process*- and *outcome*-oriented features.

Process-oriented features bear on the operation of the market and the participation effort required of traders or other interested parties. For example, we generally prefer that market rules be as simple and familiar as possible, all else being equal, as this promotes ease of learning and participation. Markets may also differ on how much time they impose for bid preparation and monitoring or how much information they require the traders to reveal. Some market structures might be considered more transparent than others or otherwise present perceived differences in fairness. All of these may be important issues for marketplace designers.

Outcome-oriented features represent properties of the results that would be reasonably expected form the market. Natural measures include expected revenue from a seller-run auction or expected expenditures in a procurement auction. Often we care most directly about overall *efficiency,* that is, how well the market allocates resources to their most valuable uses. A natural index of efficiency is *total surplus,* the aggregate gain (measured in currency units) from trade summed over all participants. Other considerations include the resistance of the mechanism to market manipulation, collusion, or various forms of cheating.

To take such issues into account, the designer, of course, needs some way to relate the market rules to these desired characteristics. Fortunately, there exists a substantial body of theory surrounding auctions [Klemperer, 1999; Krishna, 2002], starting from the seminal (Nobel Prize-winning) work of Vickrey [1961]. Auction theory tends to focus on outcome-oriented features, analyzing markets as games of incomplete information [Fudenberg and Tirole, 1991]. One of the key results of the field of *mechanism design* is the impossibility of guaranteeing efficiency through a mechanism where rational agents are free to participate or not, without providing some subsidy [Myerson and Satterthwaite, 1983]. It follows that auction design inevitably requires tradeoffs among desirable features, In recent years, the field has accumulated much experience from designing markets for privatization [Milgrom, 2003], yielding many lessons about market process as well as performance characteristics.

19.3.3 Complex Auctions

The discussion of market types above focused attention on "simple" auctions, where a single type of good (one or more units) is to be exchanged, and the negotiation addresses only price and quantity. In a multidimensional auction, bids may refer to multiple goods or features of a good. Although such complex auctions are not yet prevalent, automation has only recently made them feasible, and they are likely to grow in importance in online marketplaces.

A *combinatorial* auction [de Vries and Vohra, 2003] allows indivisible bids for bundles of goods. This enables the bidder to indicate a willingness to obtain goods if and only if the combination is available. Such a capability is particularly important when the goods are *complementary,* that is, the value of obtaining some is increased when the others are obtained as well. For example, a bicycle assembler needs both wheels and frames; neither part constitutes a bicycle without the other. Bidding rules for combinatorial auctions dictate what bundles are expressible, and clearing policy defines the method for calculating overall allocations and payments.

A *multiattribute* auction [Bichler, 2001] allows bids that refer to multiple features of a single good. For example, a shipment of automobile tires might be defined by wheel diameter, tread life, warranty, delivery date, and performance characteristics (antiskid, puncture resistance), as well as the usual price and quantity. Multiattribute bids may specify the value of particular feature vectors or express a correspondence of values over extended regions of the attribute space. The form of such bids are defined by the bidding rules, and the clearing policy dictates the method of matching such multiattribute offers.

Multidimensional negotiation constitutes an area of great potential for online marketplaces — enabling a form of trading automation not previously possible. Ultimately, combinatorial and multiattribute negotiation could, in principle, support the negotiation of general contracts [Reeves et al., 2002]. Before that vision becomes reality, however, numerous technical issues in multidimensional negotiation must be addressed, including such problems as:

- What are the best forms for expressing combinatorial and multiattribute bids?
- What intermediate information should be revealed as these auctions proceed?
- How can we reduce the complexity of participating in multidimensional negotiations?
- What strategies should we expect from combinatorial and multiattribute bidders?
- What is the appropriate scope of a market? Combining related goods in a single negotiation avoids market coordination failures, but imposes synchronization delays and other potential costs in computation, communication, and organization.

Although several existing proposals and models address these questions in part, multidimensional negotiation remains an active research topic in market design.

19.4 Establishing a Marketplace

To build an effective online marketplace, one needs to identify unfulfilled trading opportunities, design a suitable negotiation mechanists, and provide (directly or through ancillary parties) well-integrated discovery and transaction services. This is, of course, quite a tall order, and the specifics are dauntingly open ended. Nevertheless, assembling all these functions still is not sufficient to ensure marketplace success.

Unfortunately, despite several useful sources of advice on establishing an online marketplace [Kambil and van Heck, 2002; Woods, 2002], much of the prevailing wisdom is based on anecdotal experience — accrued within a dynamic technological and economic environment — and continues to evolve rapidly. In this section, I briefly note some of the additional technical and organizational issues that can prove instrumental in making an online marketplace really work. As the field matures, we can expect that some of these will become routinely addressed by common infrastructure, and others will become more precisely understood through accumulation and analysis of experience.

19.4.1 Technical Issues

The section on auctions above discusses economic as well as technical issues in the design and deployment of negotiation mechanisms, focusing on the logic of market procedures. To underpin the market logic, we require a robust computational infrastructure to ensure its proper operation under a range of conditions, loads, and extraordinary events. By their very nature, online marketplaces operate over distributed networks, typically accessed by a heterogeneous collection of traders and observer nodes. For example, one user might submit a simple bid through a Web page accessed via telephone modem, whereas another might automatically submit large arrays of trades through a programmatic interface from a fast workstation connected through a high-bandwidth network. Access is generally asynchronous and conducted over public networks.

In many respects, the processing issues faced by a marketplace are identical to those in other transaction-processing applications. We naturally care to a great extent about general system reliability and availability, and transparency of operation. It is important that transactions be atomic (i.e., an operation either completes or has no effect), and that state is recoverable in case of an outage, system crash, or other fault event.

There may also be some additional issues particularly salient for market applications. For example, maintaining *temporal integrity* can be critically important for correct and fair implementation of market rules. In the market context, temporal integrity means that the outcome of a negotiation is a function of the sequence of communications received from traders, independent of delays in computation and communication internal to the market. One simple consequence of temporal integrity is that bids be processed in order received, despite any backlog that may exist. (Bid processing may in general require a complex computation on the order book, and it would be most undesirable to block incoming messages while this computation takes place.) Another example involves synchronization with market events. If the market is scheduled to clear at time t, then this clear should reflect all bids received before t, even if they are not all completely processed by this time. One way to enforce this kind of temporal integrity is to maintain a *market logical time*, which may differ (i.e., lag somewhat behind) the actual clock time [Wellman and Wurman, 1998]. Given this approach, any information revealed by the market can be associated with a logical time, thus indicating the correct state based on bids actually received by this logical time.

Despite its apparent importance, strikingly few online markets provide any meaningful guarantees of temporal integrity or even indications relating information revealed to the times of the states they reflect. For example, in a typical online brokerage, one is never sure about the exact market time corresponding to the posted price quotes. This makes it difficult for a trader (or even a regulator!) to audit the stream of bids to ensure that all deals were properly determined. The likely explanation is that these systems evolved on top of semi-manual legacy systems for which such fine-grained accounting was not feasible. As a result, to detect improper behavior it is often necessary to resort to pattern-matching and other statistical techniques [Kirkland et al., 1999]. With increased automation, a much higher standard of temporal integrity and accountability should be possible to achieve normally, and this should be the goal of new market designers.

Finally, one cannot deploy a marketplace without serious attention to issues of privacy and security. We cannot do justice to such concerns here, and hence we will just note that simply by virtue of their financial nature, markets represent an obvious security risk. In consequence, the system must carefully authenticate and authorize all market interactions (e.g., both bidding and access of revealed market information). Moreover, online marketplaces are often quite vulnerable to denial-of-service and other resource-oriented attacks. Because negotiation necessarily discloses sensitive information (as do other market activities, such as search and evaluation), it is an essential matter of privacy to ensure that the market reveals no information beyond that dictated by the stated revelation policy.

19.4.2 Achieving Critical Mass

If we build an electronic marketplace for a compelling domain with rich supporting services, sound economic design, and technically solid in all respects — will the traders come? Alas, it (still) depends …

Trading in markets is a *network* activity, in the sense that the benefit of participating depends on the participation of others. Naturally, it is a waste of effort searching for deals in a market where the attractive counterparties are scarce. To overcome these *network effects* [Shapiro and Varian, 1998; Shy, 2001], it is often necessary to invest up front to develop a critical mass of traders that can sustain itself and attract additional traders. In effect, the marketplace may need to subsidize the early entrants, helping them overcome the initial fixed cost of entry until there are sufficient participants such that gains from trading itself outweigh the costs. Note that enticing entrants by promising or suggesting some advantage in the market itself is generally counterproductive, as it inhibits the traders on the "other side" who will ultimately render the market profitable overall.

It is commonplace to observe in this context that the key to a successful marketplace is achieving sufficient *liquidity.* A market is liquid to the extent it is readily possible to make a trade at the "prevailing" price at any time. In a *thin market,* in contrast, it is often the case that one can execute a transaction only at a disadvantageous price due to frequent temporary imbalances caused by the sparseness of traders. For example, markets in equities listed on major stock exchanges are famously liquid, due to large volume as well as the active participation of *market makers* or specialists with express obligations (incurred in return for their privileged status in the market) to facilitate liquidity by trading on their own account when necessary.

It is perhaps unfortunate that the financial markets have provided the most salient example of a functionally liquid market. It appears that many of the first generation of online marketplaces have attempted to achieve liquidity by emulation of these markets, quite often hiring key personnel with primary experience as traders in organized equity or commodity exchanges. For example, traditional financial securities markets employ variants of the *continuous double auction* [Friedman and Rust, 1993], which matches buy and sell orders instantaneously whenever compatible offers appear on the market. However, many eminently tradeable goods inherently lack the volume potential of financial securities, and for such markets instantaneous matching might be reasonably sacrificed for more designs likely to produce more robust and stable prices. In principle, new marketplaces provide an opportunity for introducing customized market designs. In practice, however, familiarity and other factors introduce a bias toward "legacy" trading processes.

19.5 The Future of Online Marketplaces

Anyone contemplating a prediction of the course of online marketplaces will be cautioned by the memory of prevailing late-1990s forecasts that proved to be wildly optimistic. Though many online marketplaces came and went during "the Bubble," the persistence of some through the pessimistic "Aftermath" is surely evidence that online marketplaces can provide real value. Even the failed attempts have left us with cautionary tales and other learning experiences [Woods, 2002], and in some instances, useful technologies. So without offering any specific prognostications with exponential growth curves, this chapter ends with a generally positive outlook plus a few suggestions about what we might see in the next generation of online marketplaces.

First, while specific marketplaces will come and go, the practice of online trading will remain, and likely stabilize over time through recognition of successful models and standardization of interfaces. Decisions about joining marketplaces or starting new ones should perhaps be driven less by strategic concerns (e.g., the "land grab" mentality that fueled the Bubble), and more by the objective of supporting trading activities that improve industry efficiency and productivity.

Second, as discussed above, there is currently a large amount of research attention, as well as some commercial development, devoted to the area of multidimensional negotiation. Combinatorial and multiattribute auctions support richer expressions of offers, accounting for multiple facets of a deal, and interactions between parts of a deal. Whereas multidimensional negotation is not a panacea (presenting additional costs and complications, and unresolved issues), it does offer the potential to get beyond some of the rigidities inhibiting trade in online marketplaces.

Finally, trading is a labor-intensive activity. Whereas online marketplaces can provide services to assist discovery and monitoring of trading opportunities, it may nevertheless present too many plausible options for a person to reasonably attend. Ultimately, therefore, it is reasonable to expect the trading function itself to be automated, and for online marketplaces to become primarily the province of programmed traders. Software agents can potentially monitor and engage in many more simultaneous market activities than could any human. A recently inaugurated annual trading agent competition [Wellman et al., 2003] presents one vision of a future of online markets driven by autonomous trading agents.

References

Ariba Inc., IBM Corp., and Microsoft Corp. Universal description, discovery, and integration (UDDI). Technical white paper, UDDI.org, 2000.

Belew, Richard K. *Finding Out About.* Cambridge University Press, 2000.

Berners-Lee, Tim, James Hendler, and Ora Lassila. The semantic web. *Scientific American,* 284(5): 34–43, 2001.

Bichler, Martin. *The Future of e-Markets: Multidimensional Market Mechanisms.* Cambridge University Press, Cambridge, U.K., 2001.

Chaum, David. Achieving electronic privacy. *Scientific American,* 267(2): 96–101, 1992.

Cohen, Adam. *The Perfect Store: Inside eBay.* Little, Brown, and Company, New York, 2002.

Crocker, Steve. The siren song of Internet micropayments. *iMP: The Magazine on Information Impacts,* 1999.

de Vries, Sven and Rakesh Vohra. Combinatorial auctions: a survey. *INFORMS Journal on Computing,* 15, 284–309, 2003.

Dellarocas, Chrysanthos. The digitization of word-of-mouth: promise and challenges of online reputation mechanisms. *Management Science,* 1407–1424, 2003.

Di Noia, Tommaso, Eugenio Di Sciascio, Francesco M. Donini, and Marina Mongiello. A system for principled matchmaking in an electronic marketplace. In *Twelfth International World Wide Web Conference,* Budapest, 2003, in press *Int. J. Elec. Comm.* 2004.

Doorenbos, Robert B., Oren Etzioni, and Daniel S. Weld. A scalable comparison-shopping agent for the world-wide web. In *First International Conference on Autonomous Agents,* pages 39–48, 1997.

Friedman, Daniel and John Rust, Eds. *The Double Auction. Market.* Addison-Wesley, Reading, MA, 1993.

Fudenberg, Drew and Jean Tirole. *Game Theory.* MIT Press, Cambridge, MA, 1991.

Kambil, Ajit and Eric van Heck. *Making Markets.* Harvard Business School Press, Boston, 2002.

Kirkland, J. Dale, Ted E. Senator, James J. Hayden, Tomasz Dybala, Henry G. Goldberg, and Ping Shyr. The NASD Regulation advanced-detection system (ADS). *AI Magazine,* 20(1): 55–67, 1999.

Klemperer, Paul D. Auction theory: A guide to the literature. *Journal of Economic Surveys,* 13: 227–286,1999.

Krishna, Vijay. *Auction Theory.* Academic Press, San Diego, CA, 2002.

Krulwich, Bruce T. The BargainFinder agent: Comparison price shopping on the Internet. In Joseph Williams, Ed., *Bots and Other Internet Beasties,* chapter 13, pages 257–263. Sams Publishing, Indianapolis, IN, 1996.

Li, Lei and Ian Horrocks. A software framework for matchmaking based on semantic web technology. In *Twelfth International World Wide Web Conference,* Budapest, 2003, in press *Int J Elec. Comm.* 2004.

MacKie-Mason, Jeffrey K. and Kimberly White. Evaluating and selecting digital payment mechanisms. In Gregory L. Rosston and David Waterman, Eds., *Interconnection and the Internet: Selected Papers from the 24th Annual Telecommunications Policy Research Conference.* Lawrence Erlbaum, Hillsdale, NJ, 1997.

Manasse, Mark S. The Millicent protocols for electronic commerce. In *First USENIX Workshop on Electronic Commerce,* pages 117–123, New York, 1995.

McAfee, R. Preston and John McMillan. Analyzing the airwaves auction. *Journal of Economic Perspectives,* 10(1): 159–175, 1996.

Milgrom, Paul. *Putting Auction Theory to* Work. Cambridge University Press, Cambridge, U.K., 2003.

Myerson, Roger B. and Mark A. Satterthwaite. Efficient mechanisms for bilateral trading. *Journal of Economic Theory,* 29: 265–281, 1983.

O'Mahony, Donal, Michael Pierce, and Hitesh Tewari. *Electronic Payment Systems.* Artech House, Norwood, MA, 1997.

Reeves, Daniel M., Michael P. Wellman, and Benjamin N. Grosof. Automated negotiation from declarative contract descriptions. *Computational Intelligence,* 18: 482–500, 2002.

Resnick, Paul and Hal R. Varian. Recommender systems. *Communications of the ACM,* 40(3): 56–58, 1997.

Resnick, Paul and Richard Zeckhauser. Trust among strangers in Internet transactions: Empirical analysis of eBay's reputation system. In Michael R. Baye, editor, *The Economics of the Internet and E-Commerce,* volume 11 of *Advances in Applied Microeconomics.* Elsevier Science, Amsterdam, 2002.

Resnick, Paul, Richard Zeckhauser, Eric Friedman, and Ko Kuwabara. Reputation systems. Com*munications of the ACM,* 43(12): 45–48, 2002.

Riedl, John and Joseph A. Konstan. *Word of Mouse: The Marketing Power of Collaborative Filtering.* Warner Books, New York, 2002.

Schafer, J. Ben, Joseph A. Konstan, and John Riedl. E-commerce recommendation applications. *Data Mining and Knowledge Discovery,* 5: 115–153, 2001.

Schmid, Beat F. Electronic markets. *Electronic Markets,* 3(3), 1993.

Shapiro, Carl and Hal R. Varian. *Information Rules: A Strategic Guide to the Network Economy.* Harvard Business School Press, Boston, 1998.

Shy, Oz. *The Economics of Network Industries.* Cambridge University Press, Cambridge, MA, 2001.

Sirbu, Marvin and J. D. Tygar. NetBill: an Internet commerce system optimized for network delivered services. *IEEE Personal Communications,* 2(4): 34–39, 1995.

Vickrey, William. Counterspeculation, auctions, and competitive sealed tenders. *Journal of Finance,* 16: 8–37,1961.

Wellman, Michael P., Amy Greenwald, Peter Stone, and Peter R. Wurman. The 2001 trading agent competition. *Electronic Markets,* 13: 4–12, 2003.

Wellman, Michael P. and Peter R. Wurman. Real time issues for Internet auctions. In *IEEE Workshop on Dependable and Real-Time E-Commerce Systems,* Denver, CO, 1998.

Woods, W. William A. *B2B Exchanges 2.0.* ISI Publications, Hong Kong, 2002.

Wurman, Peter R., Michael P. Wellman, and William E. Walsh. The Michigan Internet AuctionBot: A configurable auction server for human and software agents. In *Second International Conference on Autonomous Agents,* pages 301–308, Minneapolis, 1998.

Wurman, Peter R., Michael P. Wellman, and William E. Walsh. A parameterization of the auction design space. *Games and Economic Behavior,* 35: 304–338, 2001.

20

Online Reputation Mechanisms

CONTENTS

20.1 Introduction 20-1
20.2 An Ancient Concept In a New Setting 20-3
20.3 A Concrete Example: eBay's Feedback Mechanism.......... 20-5
20.4 Reputation in Game Theory and Economics 20-8
20.4.1 Basic Concepts 20-8
20.4.2 Reputation Dynamics 20-9
20.5 New Opportunities and Challenges of Online Mechanisms 20-12
20.5.1 Understanding the Impact of Scalability 20-12
20.5.2 Eliciting Sufficient and Honest Feedback 20-12
20.5.3 Exploiting the Information Processing Capabilities of Feedback Mediators 20-13
20.5.4 Coping with Easy Name Changes 20-14
20.5.5 Exploring Alternative Architectures 20-14
20.6 Conclusions 20-15
References 20-16

Chrysanthos Dellarocas

Online reputation mechanisms harness the bidirectional communication capabilities of the Internet in order to engineer large-scale, word-of-mouth networks. They are emerging as a promising alternative to more established assurance mechanisms such as branding and formal contracting in a variety of settings ranging from online marketplaces to peer-to-peer networks. This chapter surveys our progress in understanding the new possibilities and challenges that these mechanisms represent. It discusses some important dimensions in which Internet-based reputation mechanisms differ from traditional word-of-mouth networks and surveys the most important issues related to designing, evaluating, and using them. It provides an overview of relevant work in game theory and economics on the topic or reputation. It further discusses how this body of work is being extended and combined with insights from computer science, information systems, management science, and psychology in order to take into consideration the special properties of online mechanisms such as their unprecedented scalability, the ability to precisely design the type of feedback information they solicit and distribute, and challenges associated with the relative anonymity of online environments.

20.1 Introduction

A fundamental aspect in which the Internet differs from previous technologies for mass communication is its bidirectional nature: Not only has it bestowed upon organizations a low-cost channel through which to reach audiences of unprecedented scale but also, for the first time in human history, it has enabled

1-58488-381-2/05/$0.00+$1.50

© 2005 by CRC Press LLC

individuals to almost costlessly make their personal thoughts and opinions accessible to the global community of Internet users.

An intriguing family of electronic intermediaries are beginning to harness this unique property, redefining and adding new significance to one of the most ancient mechanisms in the history of human society. *Online reputation mechanisms,* also known as *reputation systems* [Resnick, Zeckhauser, Friedman, and Kubwara, 2000] and *feedback mechanisms* [Dellarocas, 2003b] are using the Internet's bidirectional communication capabilities in order to artificially engineer large-scale word-of-mouth networks in online environments.

Online reputation mechanisms allow members of a community to submit their opinions regarding other members of that community. Submitted feedback is analyzed, aggregated with feedback posted by other members, and made publicly available to the community in the form of member *feedback profiles.* Several examples of such mechanisms can already be found in a number of diverse online communities (Figure 20.1).

Perhaps the best-known application of online reputation mechanisms to date has been as a technology for building trust in electronic markets. This has been motivated by the fact that many traditional trust-building mechanisms, such as state-enforced contractual guarantees and repeated interaction, tend to be less effective in large-scale online environments [Kollock, 1999]. Successful online marketplaces such as eBay are characterized by large numbers of small players, physically located around the world and often known to each other only via easily changeable pseudonyms. Contractual guarantees are usually difficult or too costly to enforce due to the global scope of the market and the volatility of identities. Furthermore, the huge number of players makes repeated interaction between the same set of players less probable, thus reducing the incentives for players to cooperate on the basis of hoping to develop a profitable relationship.

Online reputation mechanisms have emerged as a viable mechanism for inducing cooperation among strangers in such settings by ensuring that the behavior of a player towards any other player becomes

Web Site	Category	Summary of reputation mechanism	Format of solicited feedback	Format of feedback profiles
eBay	Online auction house	Buyers and sellers rate one another following transactions	Positive, negative or neutral rating plus short comment; ratee may post a response	Sums of positive, negative and neutral ratings received during past 6 months
eLance	Professional services marketplace	Contractors rate their satisfaction with subcontractors	Numerical rating from 1–5 plus comment; ratee may post a response	Average of ratings received during past 6 months
Epinions	Online opinions forum	Users write reviews about products/service; other members rate the usefulness of reviews	Users rate multiple aspects of reviewed items from 1–5; readers rate reviews as "useful", "not useful", etc.	Averages of item ratings; % of readers who found a review "useful"
Google	Search engine	Search results are rank ordered based on how many sites contain links that point to them [Brin and Page, 1998]	How many links point to a page, how many links point to the pointing page, etc.	Rank ordering acts as an implicit indicator of reputation
Slashdot	Online discussion board	Postings are prioritized or filtered according to the rating they receive from readers	Readers rate posted comments	Rank ordering acts as an implicit indicator of reputation

FIGURE 20.1 Some examples of online reputation mechanisms used in commercial Websites.

publicly known and may therefore affect the behavior of the entire community towards that player in the future. Knowing this, players have an incentive to behave well towards each other, even if their relationship is a one-time deal. A growing body of empirical evidence seems to demonstrate that these systems have managed to provide remarkable stability in otherwise very risky trading environments (see, for example, Bajari and Hortacsu [2003]; Dewan and Hsu [2002]; Houser and Wonders [2000]; Lucking-Reiley et al. [2000]; Resnick and Zeckhauser [2002]).

The application of reputation mechanisms in online marketplaces is particularly interesting because many of these marketplaces would probably not have come into existence without them. It is, however, by no means the only possible application domain of such systems. Internet-based feedback mechanisms are appearing in a surprising variety of settings: For example, Epinions.com encourages Internet users to rate practically any kind of brick-and-mortar business, such as airlines, telephone companies, resorts, etc. Moviefone.com solicits and displays user feedback on new movies alongside professional reviews and Citysearch.com does the same for restaurants, bars, and performances. Even news sites, perhaps the best embodiment of the unidirectional truss media of the previous century, are now encouraging readers to provide feedback on world events alongside professionally written news articles.

The proliferation of online reputation mechanisms is already changing people's behavior in subtle but important ways. Anecdotal evidence suggests that people now increasingly rely on opinions posted on such systems in order to make a variety of decisions ranging from what movie to watch to what stocks to invest on. Only 5 years ago the same people would primarily base those decisions on advertisements or professional advice. It might well be that the ability to solicit, aggregate, and publish mass feedback will influence the social dynamics of the 21st century in a similarly powerful way in which the ability to mass broadcast affected our societies in the 20th century.

The rising importance of online reputation systems not only invites but also necessitates rigorous research on their functioning and consequences. How do such mechanisms affect the behavior of participants in the communities where they are introduced? Do they induce socially beneficial outcomes? To what extent can their operators and participants manipulate them? How can communities protect themselves from such potential abuse? What mechanism designs work best in what settings? Under what circumstances can these mechanisms such as contracts, legal guarantees, and professional reviews become viable substitutes (or complements) of more established institutions? This is just a small subset of questions that invite exciting and valuable research.

This chapter surveys our progress so far in understanding the new possibilities and challenges that these mechanisms represent. Section 20.2 discusses some important dimensions in which Internet-based reputation mechanisms differ from traditional word-of-mouth networks. Section 20.3 provides a case study of eBay's feedback mechanism, perhaps the best known reputation system at the time of this chapter's writing. The following two sections survey our progress in developing a systematic discipline that can help answer those questions. First, Section 20.4 provides an overview of relevant past work in game theory and economics. Section 20.5 then discusses how this body of work is being extended in order to take into consideration the special properties of online mechanisms. Finally, Section 20.6 summarizes the main points and lists opportunities for future research.

20.2 An Ancient Concept In a New Setting

Word-of-mouth networks constitute an ancient solution to a timeless problem of social organization: the elicitation of good conduct in communities of self-interested individuals who have short-term incentives to cheat one another. The power of such networks to induce cooperation without the need for costly and inefficient enforcement institutions has historically been the basis of their appeal. Before the establishment of formal law and centralized systems of contract enforcement backed by the sovereign power of a state, most ancient and medieval communities relied on word-of-mouth as the primary enabler of economic and social activity [Benson, 1989; Greif, 1993; Milgrom, North, and Weingast, 1990]. Many aspects of social and economic life still do so today [Klein, 1997].

What makes online reputation mechanisms different from word-of-mouth networks of the past is the combination of (1) their unprecedented scale, achieved through the exploitation of the Internet's low-cost, bidirectional communication capabilities, (2) the ability of their designers to precisely control and monitor their operation through the introduction of automated feedback mediators, and (3) new challenges introduced by the unique properties of online interaction, such as the volatile nature of online identities and the almost complete absence of contextual cues that would facilitate the interpretation of what is, essentially, subjective information.

- *Scale enables new applications.* Scale is essential to the effectiveness of word-of-mouth networks. In an online marketplace, for example, sellers care about buyer feedback primarily to the extent that they believe that it might affect their future profits; this can only happen if feedback is provided by a sufficient number of current customers and communicated to a significant portion of future prospects. Theory predicts that a minimum scale is required before reputation mechanisms have any effect on the behavior of rational agents [Bakos and Dellarocas, 2002]. Whereas traditional word-of-mouth networks tend to deteriorate with scale, Internet-based reputation mechanisms can accumulate, store, and flawlessly summarize unlimited amounts of information at very low cost. The vastly increased scale of Internet-based reputation mechanisms might therefore make such mechanisms effective social control institutions in settings where word-of-mouth previously had a very weak effect. The social, economic, and perhaps even political consequences of such a trend deserve careful study.
- *Information technology enables systematic design.* Online word-of-mouth networks are artificially induced through explicitly designed information systems (feedback mediators). Feedback mediators specify who can participate, what type of information is solicited from participants, how it is aggregated, and what type of information is made available to them about other community members. They enable mechanism designers to exercise precise control over a number of parameters that are very difficult or impossible to influence in brick-and-mortar settings. For example, feedback mediators can replace detailed feedback histories with a wide variety of summary statistics, they can apply filtering algorithms to eliminate outlier or suspect ratings, they can weight ratings according to some measure of the rater's trustworthiness, etc. Such degree of control can impact the resulting social outcomes in nontrivial ways (see Section 20.5.2, Section 20.5.3, and Section 20.5.4). Understanding the full space of design possibilities and the consequences of specific design choices introduced by these new systems is an important research challenge that requires collaboration between traditionally distinct disciplines such as computer science, economics, and psychology, in order to be properly addressed.
- *Online interaction introduces new challenges.* The disembodied nature of online environments introduces several challenges related to the interpretation and use of online feedback. Some of these challenges have their roots at the subjective nature of feedback information. Brick-and-mortar settings usually provide a wealth of contextual cues that assist in the proper interpretation of opinions and gossip (such as the fact that we know the person who acts as the source of that information or can infer something about her through her clothes, facial expression, etc.). Most of these cues are absent from online settings. Readers of online feedback are thus faced with the task of making sense out of opinions of complete strangers. Other challenges have their root at the ease with which online identities can be changed. This opens the door to various forms of strategic manipulation. For example, community members can build a good reputation, milk it by cheating other members, and then disappear and reappear under a new online identity and a clean record [Friedman and Resnick, 2001]. They can use fake online identities to post dishonest feedback for the purpose of inflating their reputation or tarnishing that of their competitors [Dellarocas, 2000; Mayzlin, 2003]. Finally, the mediated nature of online reputation mechanisms raises questions related to the trustworthiness of their operators. An important prerequisite for the widespread acceptance of online reputation mechanisms as legitimate trust-building institu-

tions is, therefore, a better understanding of how such systems can be compromised as well as the development of adequate defenses.

20.3 A Concrete Example: eBay's Feedback Mechanism

The feedback mechanism of eBay is, arguably, the best known online reputation mechanism at the time of this writing. Founded in September 1995, eBay is the leading online marketplace for the sale of goods and services by a diverse global community of individuals and businesses. Today, the eBay community includes more than 100 million registered users.

One of the most remarkable aspects of eBay is that the transactions performed through it are not backed up by formal legal guarantees. Instead, cooperation and trust are primarily based on the existence of a simple feedback mechanism. This mechanism allows eBay buyers and sellers to rate one another following transactions and makes the history of a trader's past ratings public to the entire community. On eBay, users are known to each other through online pseudonyms (eBay IDs). When a new user registers to the system, eBay requests that she provide a valid email address and a valid credit card number (for verification purposes only).

Most items on eBay are sold through English auctions. A typical eBay transaction begins with the seller listing an item he has on sale, providing an item description (including text and optionally photos), a starting bid, an optional reserve price, and an auction closing date/time. Buyers then place bids for the item up until the auction closing time. The highest bidder wins the auction. The winning bidder sends payment to the seller. Finally, the seller sends the item to the winning bidder. It is easy to see that the above mechanism incurs significant risks for the buyer. Sellers can exploit the underlying information asymmetries to their advantage by misrepresenting an item's attributes (adverse selection) or by failing to complete the transaction (moral hazard), i.e., keeping the buyer's money without sending anything back.

It is clear that without an adequate solution to these adverse selection and moral hazard problems, sellers have an incentive to always cheat and/or misrepresent, and therefore expecting this, buyers would either not use eBay at all or place very low bids. To address these problems, eBay uses online feedback as its primary trust building mechanism. More specifically, following completion of a transaction, both the seller and the buyer are encouraged to rate one another. A rating designates a transaction as positive, negative, or neutral, together with a short text comment. eBay aggregates all ratings posted for a member into that member's feedback profile. An eBay feedback profile consists of four components (Figure 20.2 and Figure 20.3):

1. A member's overall profile makeup: a listing of the sum of positive, neutral, and negative ratings received during that member's entire participation history with eBay
2. A member's summary feedback score equal to the sum of positive ratings received by unique users minus the number of negative ratings received by unique users during that member's entire participation history with eBay
3. A prominently displayed eBay ID card, which displays the sum of positive, negative, and neutral ratings received during the most recent 6-month period (further subdivided into ratings received during the past week, month, and past 6 months)
4. The complete ratings history, listing each individual rating and associated comment posted for a member in reverse chronological order

Seller feedback profiles are easily accessible from within the description page of any item for sale. More specifically, all item descriptions prominently display the seller's eBay ID, followed by his summary feedback score (component B in Figure 20.2). By clicking on the summary feedback score, prospective buyers can access the seller's full feedback profile (components A, B, and C) and can then scroll through the seller's detailed ratings history (component D).

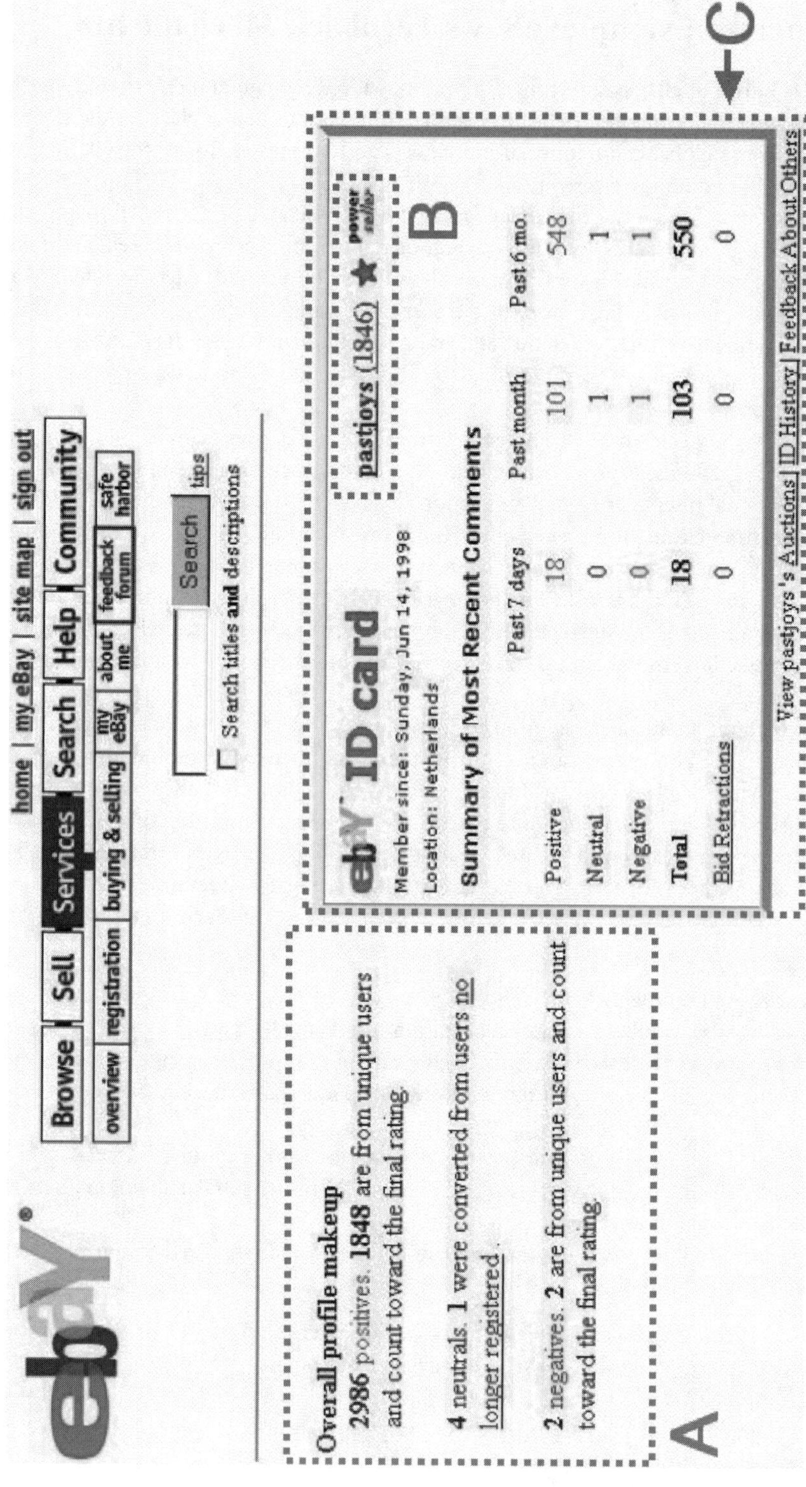

FIGURE 20.2 Profile summary of eBay member.

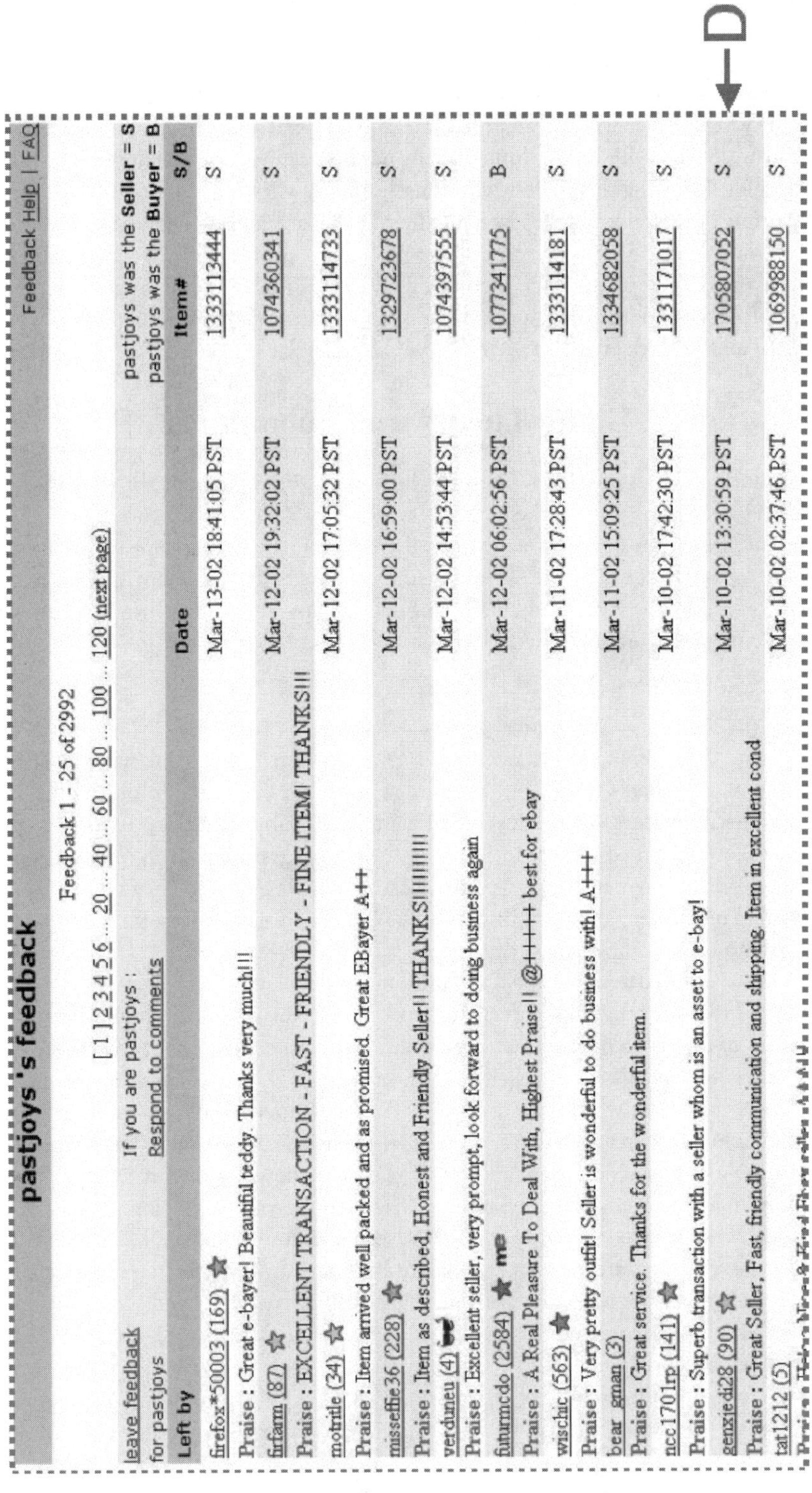

pastjoys 's feedback Feedback Help | FAQ

Feedback 1 - 25 of 2992

[1] 2 3 4 5 6 ... 20 ... 40 ... 60 ... 80 ... 100 ... 120 (next page)

leave feedback for pastjoys | If you are pastjoys : Respond to comments | pastjoys was the **Seller** = **S** pastjoys was the **Buyer** = **B**

Left by	Date	Item#	S/B
firefox*50003 (169)	Mar-13-02 18:41:05 PST	1333113444	S
Praise : Great e-bayer! Beautiful teddy. Thanks very much!!!			
firfarm (87)	Mar-12-02 19:32:02 PST	1074360341	S
Praise : EXCELLENT TRANSACTION - FAST - FRIENDLY - FINE ITEM! THANKS!!!			
motritle (34)	Mar-12-02 17:05:32 PST	1333114733	S
Praise : Item arrived well packed and as promised. Great EBayer A++			
misseffie36 (228)	Mar-12-02 16:59:00 PST	1329723678	S
Praise : Item as described, Honest and Friendly Seller!! THANKS!!!!!!!!!!			
verduneu (4)	Mar-12-02 14:53:44 PST	1074397555	S
Praise : Excellent seller, very prompt, look forward to doing business again			
futurmcdo (2584) me	Mar-12-02 06:02:56 PST	1077341775	B
Praise : A Real Pleasure To Deal With, Highest Praise!! @+++++ best for ebay			
wischic (563)	Mar-11-02 17:28:43 PST	1333114181	S
Praise : Very pretty outfit! Seller is wonderful to do business with! A+++			
bear_gman (3)	Mar-11-02 15:09:25 PST	1334682058	S
Praise : Great service. Thanks for the wonderful item.			
ncc1701rp (141)	Mar-10-02 17:42:30 PST	1331171017	S
Praise : Superb transaction with a seller whom is an asset to e-bay!			
genxjedi28 (90)	Mar-10-02 13:30:59 PST	1705807052	S
Praise : Great Seller, Fast, friendly communication and shipping. Item in excellent cond			
tat1212 (5)	Mar-10-02 02:37:46 PST	1069988150	S

D

FIGURE 20.3 Detailed feedback history.

The feedback mechanism for eBay is based on two important assumptions: first, that members will indeed leave feedback for each other (feedback is currently voluntary and there are no concrete rewards or penalties for providing it or for failing to do so); second, that in addition to an item's description, buyers will consult a seller's feedback profile before deciding whether to bid on a seller's auction. Based on the feedback profile information, buyers will form an assessment of the seller's likelihood to be honest in completing the transaction, as well as to accurately describe the item's attributes. This assessment will help buyers decide whether they will indeed proceed with bidding. It will, further, influence the amounts they are willing to bid. Sellers with "bad" profiles (many negative ratings) are therefore expected to receive lower bids or no bids. Knowing this, sellers with long horizons will find it optimal to behave honestly even towards onetime buyers to avoid jeopardizing their future earnings on eBay. At equilibrium, therefore, the expectation is that buyers will trust sellers with "good" profiles to behave honestly and sellers will indeed honor the buyers' trust. Initial theoretical and empirical evidence suggests that, despite its simplicity, eBay's feedback mechanism succeeds to a large extent in achieving these objectives (see Dellarocas [2003b] and Resnick et al. [2002] for surveys of relevant studies).

20.4 Reputation in Game Theory and Economics

Given the importance of word-of-mouth networks in human society, reputation formation has been extensively studied by economists using the tools of game theory. This body of work is perhaps the most promising foundation for developing an analytical discipline of online reputation mechanism design. This section surveys past work in this area, emphasizing the results that are most relevant to the design of online reputation mechanisms. Section 20.5 then discusses how this body of work is being extended to address the unique properties of online systems.

20.4.1 Basic Concepts

According to Wilson [1985] reputation is a concept that arises in repeated game settings when there is uncertainty about some property (the "type") of one or more players in the mind of other players. If "uninformed" players have access to the history of past stage game (iteration) outcomes, reputation effects then often allow informed players to improve their long-term payoffs by gradually convincing uninformed players that they belong to the type that best suits their interests. They do this by repeatedly choosing actions that make them appear to uninformed players as if they were of the intended type (thus "acquiring a reputation" for being of that type).

The existence of some initial doubt in the mind of uninformed players regarding the type of informed players is crucial in order for reputation effects to occur. To see this, consider an eBay seller who faces an sequence of sets of one-time buyers in a marketplace where there are only two kinds of products: high-quality products that cost 0 to the seller and are worth 1 to the buyers and low-quality products that cost 1 to the seller and are worth 3 to the buyers. Buyers compete with one another on a Vickrey (second-price) auction and therefore bid amounts equal to their expected valuation of the transaction outcome. The winning bidder sends payment to the seller and the seller then has the choice of either "cooperating" (producing a high quality good) or "cheating" (producing a low quality good). The resulting payoff matrix is depicted in Figure 20.4. If the seller cannot credibly precommit to cooperation and buyers are certain that they are facing a rational, utility-maximizing seller, the expected outcome of all transactions will be the static Nash equilibrium: sellers will always cheat and, expecting this, buyers always place low bids. This outcome is socially inefficient in that the payoffs of both parties are equal to or lower to those they could achieve if they cooperated.

The concept of reputation allows the long-run player to improve his payoffs in such settings. Intuitively, a long-run player who has a track record of playing a given action (e.g., cooperate) often enough in the past acquires a reputation for doing so and is "trusted" by subsequent short-run players to do so in the future as well. However, why would a profit-maximizing long-term player be willing to behave in such a way and why would rational short-term players use past history as an indication of future behavior?

	Cooperate	Cheat
Bid high	0,2	-2,3
Bid low	2,0	0.1

FIGURE 20.4 Payoff matrix of a simplified "eBay" bilateral exchange stage game (first number in each cell represents buyer payoff, second number represents seller payoff).

To explain such phenomena, Kreps, Milgrom, Roberts, and Wilson [1982], Kreps and Wilson [1982], and Milgrom and Roberts [1982] introduced the notion of "commitment" types. Commitment types are long-run players who are locked into playing the same action.[1] An important subclass of commitment types are Stackelberg types: long-run players who are locked into playing the so-called Stackelberg action. The Stackelberg action is the action to which the long-run player would credibly commit if he could. In the above eBay-type example, the Stackelberg action would be to cooperate; cooperation is the action that maximizes the seller's lifetime payoffs if the seller could credibly commit to an action for the entire duration of the game; therefore, the Stackelberg type in this example corresponds to an "honest" seller who never cheats. In contrast, an "ordinary" or "strategic" type corresponds to a profit-maximizing seller who cheats whenever it is advantageous for him to do so.

Reputation models assume that short-run players know that commitment types exist, but are ignorant of the type of the player they face. An additional assumption is that short-run players have access to the entire history of past stage game outcomes. The traditional justification for this assumption is that past outcomes are either publicly observable or explicitly communicated among short-run players. The emergence of online feedback mechanisms provides, of course, yet another justification (however, the *private* observability of outcomes in online systems introduces a number of complications; see Section 20.5.2). A player's reputation at any given time, then, consists of the conditional posterior probabilities over that player's type, given a short-run player's prior probabilities over types and the repeated application of Bayes' rule on the history of past stage game (iteration) outcomes.

In such a setting, when selecting his next move, the informed player must take into account not only his short-term payoff, but also the long-term consequences of his action based on what that action reveals about his type to other players. As long as the promised future gains due to the increased (or sustained) reputation that comes from playing the Stackelberg action offset whatever short-term incentives he might have to play otherwise, the equilibrium strategy for an "ordinary" informed player will be to try to "acquire a reputation" by masquerading as a Stackelberg type (i.e., repeatedly play the Stackelberg action with high probability.)

In the eBay-type example, if the promised future gains of reputation effects are high enough, rational sellers are induced to overcome their short-term temptation to cheat and to try to acquire a reputation for honesty by repeatedly producing high quality. Expecting this, buyers will then place high bids, thus increasing the seller's long-term payoffs.

20.4.2 Reputation Dynamics

The derivation of equilibrium strategies in repeated games with reputation effects is, in general, quite complicated. Nevertheless, a small number of specific cases have been extensively studied. They provide interesting insight into the complex behavioral dynamics introduced by reputational considerations.

[1] Commitment types are sometimes also referred to as "irrational" types because they follow fixed, "hard-wired" strategies as opposed to "rational" profit-maximizing strategies. An alternative way to justify such players is to consider them as players with non-standard payoff structures such that that the "commitment" action is their dominant strategy given their payoffs.

20.4.2.1 Initial Phase

In most cases, reputation effects begin to work immediately and, in fact, are strongest during the initial phase, when players must work hard to establish a reputation. Holmstrom [1999] discusses an interesting model of reputational considerations in the context of an agent's "career" concerns: Suppose that wages are a function of an employee's innate ability for a task. Employers cannot directly observe an employee's ability; however, they can keep track of the average value of her past task outputs. Outputs depend both on ability and labor. The employee's objective is to maximize her lifetime wages while minimizing the labor she has to put in. At equilibrium, this provides incentives to the employee to work hard right from the beginning of her career in order to build a reputation for competence. In fact, these incentives are strongest at the very beginning of her career when observations are most informative.

During the initial phase of a repeated game it is common that some players realize lower, or even negative, profits while the community "learns" their type. In those cases players will only attempt to build a reputation if the losses from masquerading as a Stackelberg type in the current round are offset by the present value of the gains from their improved reputation in the later part of the game. In trading environments, this condition usually translates to the need of sufficiently high profit margins for "good quality" products in order for reputation effects to work. This was first pointed out in [Klein and Leffler, 1981] and explored more formally in [Shapiro, 1983].

Another case where reputation effects may fail to work is when short-run players are "too cautious" *vis-à-vis* the long-run player and therefore update their beliefs too slowly in order for the long-run player to find it profitable to try to build a reputation. Such cases may occur when, in addition to Stackelberg ("good") types, the set of commitment types also includes "bad" or "inept" types: players who always play the action that the short-run players like least. In the eBay-type example, a "bad" type corresponds to a player who always cheats. If short-run players have a substantial prior belief that the long-run player may be a "bad," type then the structure of the game may not allow them to update their beliefs fast enough to make it worthwhile for the long-run player to try to acquire a reputation.

Diamond's [1989] analysis of reputation formation in debt markets presents an example of such a setting, In Diamond's model there are three types of borrowers: safe borrowers, who always select safe projects (i.e., projects with zero probability of default); risky borrowers, who always select risky projects (i.e., projects with higher returns if successful but with nonzero probability of default); and strategic borrowers who will select the type of project that maximizes their long term expected payoff. The objective of lenders is to maximize their long term return by offering competitive interest rates, while at the same time being able to distinguish profitable from unprofitable borrowers. Lenders do not observe a borrower's choice of projects, but they do have access to her history of defaults. In Diamond's model, if lenders believe that the initial fraction of risky borrowers is significant, then despite the reputation mechanism, at the beginning of the game interest rates will be so high that strategic players have an incentive to select risky projects. Some of them will default and will exit the game. Others will prove lucky and will begin to be considered safe players. It is only after lucky strategic players have already acquired some initial reputation (and therefore begin to receive lower interest rates) that it becomes optimal for them to begin "masquerading" as safe players by consciously choosing safe projects in order to maintain their good reputation.

20.4.2.2 Steady State (or Lack Thereof)

Reputation games are ideally characterized by an equilibrium in which the long-run player repeatedly plays the Stackelberg action with high probability and the player's reputation converges to the Stackelberg type.

The existence of such equilibria crucially depends on the ability to perfectly monitor the outcomes of individual stage games. In games with perfect public monitoring of stage game outcomes, such a steady state almost always exists. For example, consider the "eBay game" that serves as an example throughout this section, with the added assumption that buyers perfectly and truthfully observe and report the seller's action. In such cases, the presence of even a single negative rating on a seller's feedback history reveals the fact that the seller is not honest. From then on, buyers will always choose the low bid in perpetuity. Since such an outcome is not advantageous for the seller, reputation considerations will induce the seller to cooperate forever.

The situation changes radically if monitoring of outcomes is imperfect. In the eBay example, imperfect monitoring means that even when the seller produces high quality there is a possibility that an eBay buyer will post a negative rating, and, conversely, even when the seller produces low quality, the buyer may post a positive rating. A striking result is that in such "noisy" environments, reputations cannot be sustained indefinitely. If a strategic player stays in the game long enough, short-run players will eventually learn his true type and the game will inevitably revert to one of the static Nash equilibria [Cripps, Mailath, and Samuelson, 2002].

To see the intuition behind this result, note that reputations under perfect monitoring are typically supported by a trigger strategy. Deviations from the equilibrium strategy reveal the type of the deviator and are punished by a switch to an undesirable equilibrium of the resulting complete-information continuation game. In contrast, when monitoring is imperfect, individual deviations neither completely reveal the deviator's type nor trigger punishments. Instead, the long-run convergence of beliefs ensures that eventually any current signal of play has an arbitrarily small effect on the uniformed player's beliefs. As a result, a player trying to maintain a reputation ultimately incurs virtually no cost (in terms of altered beliefs) from indulging in a single small deviation from Stackelberg play. But the long-run effect of many such small deviations from the commitment strategy is to drive the equilibrium to full revelation.

Holmstrom's "career concerns" paper provides an early special case of this striking result: the longer an employee has been on the market, the more "solid" the track record she has acquired and the less important her current actions in influencing the market's future assessment of her ability. This provides diminishing incentives for her to keep working hard. Cripps, Mailath, and Samuelson's result, then, states that if the employee stays on the market for a really long time, these dynamics will lead to an eventual loss of her reputation.

These dynamics have important repercussions for systems like eBay. If eBay makes the entire feedback history of a seller available to buyers (as it does today) and if an eBay seller stays on the system long enough, the above result predicts that once he establishes an initial reputation for honesty, he will be tempted to occasionally cheat buyers. In the long run, this behavior will lead to an eventual collapse of his reputation and therefore of cooperative behavior. The conclusion is that, if buyers pay attention to a seller's entire feedback history, eBay's current mechanism fails to sustain long-term cooperation.

20.4.2.3 Endgame Considerations

Since reputation relies on a tradeoff between current "restraint" and the promise of future gains, in finitely-repeated games incentives to maintain a reputation diminish and eventually disappear as the end of the game comes close.

A possible solution is to assign some postmortem value to reputation, so that players find it optimal to maintain it throughout the game. For example, reputations can be viewed as assets that can be bought and sold in a market for reputations. Tadelis [1999] shows that a market for reputations is indeed sustainable. Furthermore, the existence of such a market provides "old" agents and "young" agents with equal incentives to exert effort [Tadelis, 2002]. However, the long-run effects of introducing such a market can be quite complicated since good reputations are then likely to be purchased by "inept" agents for the purpose of depleting them [Mailath and Samuelson, 2001; Tadelis, 2002]. Further research is needed in order to fully understand the long-term consequences of introducing markets for reputations as well as for transferring these promising concepts to the online domain.

20.5 New Opportunities and Challenges of Online Mechanisms

In Section 20.2, a number of differences between online reputation mechanisms and traditional word-of-mouth networks were discussed. This section surveys our progress in understanding the opportunities and challenges that these special properties imply.

20.5.1 Understanding the Impact of Scalability

Bakos and Dellarocas [2002] model the impact of information technology on online feedback mechanisms in the context of a comparison of the social efficiency of litigation and online feedback. They observe that online feedback mechanisms provide linkages between otherwise disconnected smaller markets (each having its own informal word-of-mouth networks) in which a firm operates. This, in turn, is equivalent to increasing the discount factor of the firm when considering the future impacts of its behavior on any given transaction. In trading relationships, a minimum discount factor is necessary to make reputation effects productive at all in inducing cooperative behavior. Once this threshold is reached, however, the power of reputation springs to life in a discontinuous fashion and high levels of cooperation can be supported. Thus, the vastly increased potential scale of Internet-based feedback mechanisms and the resulting ability to cover a substantial fraction of economic transactions are likely to render these mechanisms into powerful quality assurance institutions in environments where the effectiveness of traditional word-of-mouth networks has heretofore been limited. The social, economic, and perhaps even political consequences of such a trend deserve careful study.

20.5.2 Eliciting Sufficient and Honest Feedback

Most game theoretic models of reputation formation assume that stage game outcomes (or imperfect signals thereof) are *publicly* observed. Online reputation mechanisms, in contrast, rely on *private* monitoring of stage game outcomes and voluntary feedback submission. This introduces two important new considerations (1) ensuring that sufficient feedback is, indeed, provided and (2) inducing truthful reporting.

Economic theory predicts that voluntary feedback will be underprovided. There are two main reasons for this. First, feedback constitutes a public good: once available, everyone can costlessly benefit from it. Voluntary provision of feedback leads to suboptimal supply, since no individual takes account of the benefits that her provision gives to others. Second, provision of feedback presupposes that the rater will assume the risks of transacting. Such risks are highest for new products: prospective consumers may be tempted to wait until more information is available. However, unless somebody decides to take the risk of becoming an early evaluator, no feedback will ever be provided.

Avery, Resnick, and Zeckhauser [1999] analyze mechanisms whereby early evaluators are paid to provide information and later evaluators pay so as to balance the budget. They conclude that any two of three desirable properties for such a mechanism can be achieved, but not all three, the three properties being voluntary participation, no price discrimination, and budget balance.

Since monitoring is private and assessments usually subjective, an additional consideration is whether feedback is honest. Miller, Resnick, and Zeckhauser [2002] propose a mechanism for eliciting honest feedback based on the technique of proper scoring rules. A scoring rule is a method for inducing decision makers to reveal their true beliefs about the distribution of a random variable by rewarding them based on the actual realization of the random variable and their announced distribution [Cooke, 1991]. A proper scoring rule has the property that the decision maker maximizes the expected score when he truthfully announces his belief about the distribution.

Their mechanism works as long as raters are assumed to act independently. Collusive behavior can defeat proper scoring rules. Unfortunately, online environments are particularly vulnerable to collusion. The development of effective mechanisms for dealing with collusive efforts to manipulate online ratings is currently an active area of research. Dellarocas [2000, 2004] explores the use of robust statistics in aggregating individual ratings as a mechanism for seducing the effects of coordinated efforts to bias ratings. To this date, however, there is no effective solution that completely eliminates the problem.

20.5.3 Exploiting the Information Processing Capabilities of Feedback Mediators

Most game theoretic models of reputation assume that short-run players have access to the entire past history of stage game outcomes and update their prior beliefs by repeated application of Bayes' rule on that information.

Online feedback mediators completely control the amount and type of information that is made available to short-run players. This opens an entire range of new possibilities: For example, feedback mediators can hide the detailed history of past feedback from short-term players and replace it with a summary statistic (such as the sum, mean, or median of past ratings) or with any other function of the feedback history. They can filter outlying or otherwise suspect ratings. They can offer *personalized* feedback profiles; that is, present different information about the same long-run player to different short-run players.

Such information transformations can have nontrivial effects in the resulting equilibria and can allow online reputation mechanisms to induce outcomes that are difficult or impossible to attain in standard settings. The following are two examples of what can be achieved:

As discussed in Section 20.4.2.2, in environments with imperfect monitoring, traditional reputation models predict that reputations are not sustainable; once firms build a reputation they are tempted to "rest on their laurels"; this behavior, ultimately, leads to a loss of reputation. Economists have used a variety of devices to construct models that do not exhibit this undesirable behavior. For instance, Mailath and Samuelson [1998] assume that in every period there is a fixed, exogenous probability that the type of the firm might change. Horner [2002] proposes a model in which competition among firms induces them to exert sustained effort.

Online feedback mediators provide yet another, perhaps much more tangible, approach to eliminating such problems: By designing the mediator to publish only recent feedback, firms are given incentives to constantly exert high effort. In the context of eBay, this result argues for the elimination of the detailed feedback history from feedback profile and the use of summaries of recent ratings as the primary focal point of decision-making. Dellarocas [2003a] studied the equilibria induced by a variation of eBay's feedback mechanism in which the only information available to buyers is the sum of positive and negative ratings posted on a seller during the most recent N transactions. He found that, in trading environments with opportunistic sellers, imperfect monitoring of a seller's effort level, and two possible transaction outcomes (corresponding to "high" and "low" quality respectively), such a mechanism induces high levels of cooperation that remain stable over time. Furthermore, the long-run payoffs are independent of the size of the window N. A mechanism that only publishes the single most recent rating is just as efficient as a mechanism that summarizes larger numbers of ratings.

A second example of improving efficiency through proper mediator design can be found in Dellarocas [2002], which studied settings in which a monopolist sells products of various qualities and announces the quality of each product. The objective of a feedback mechanism in such settings is to induce truthful announcements. Once again, Cripps, Mailath, and Samuelson's [2002] result predicts that, in noisy environments, a mechanism that simply publishes the entire history of feedback will not lead to sustainable truth-telling. Dellarocas proposes a mechanism that acts as an intermediary between the seller and the buyers. The mechanism does not publish the history of past ratings. Instead, it keeps track of discrepancies between past seller quality announcements and corresponding buyer feedback and then punishes or rewards the seller by "distorting" the seller's subsequent quality announcements so as to compensate for whatever "unfair" gains or losses he has realized by misrepresenting the quality of his items. If consumers are risk-averse, at equilibrium this induces the seller to truthfully announce quality throughout the infinite version of the game.

20.5.4 Coping with Easy Name Changes

In online communities it is usually easy for members to disappear and reregister under a completely different online identity with zero or very low cost. Friedman and Resnick [2001] refer to this property as "cheap pseudonyms." This property hinders the effectiveness of reputation mechanisms: community members can build a reputation, milk it by cheating other members, and then vanish and reenter the community with a new identity and a clean record.

Friedman and Resnick [2001] discuss two classes of approaches to this issue: Either make it more difficult to change online identities or structure the community in such a way so that exit and reentry with a new identity becomes unprofitable. The first approach makes use of cryptographic authentication technologies and is outside the scope of this chapter. The second approach is based on imposing an upfront cost to each new entrant, such that the benefits of "milking" one's reputation are exceeded by the cost of subsequent reentry. This cost can be an explicit entrance fee or an implicit cost of having to go through an initial reputation-building (or "dues paying") phase with low or negative profits. Friedman and Resnick [2001] show that, although dues paying approaches incur efficiency losses, such losses constitute an inevitable consequence of easy name changes.

Dellarocas [2003a] shows how such a dues-paying approach can be implemented in an eBay-like environment where feedback mediators only publish the sum of recent ratings. He proves that, in the presence of easy name changes, the design that results in optimal social efficiency is one where the mechanism sets the initial profile of new members to correspond to the "worst" possible reputation.[2] He further shows that, although this design incurs efficiency losses relative to the case where identity changes are not possible, its efficiency is the highest possible attainable by any mechanism if players can costlessly change their identities.

20.5.5 Exploring Alternative Architectures

The preceding discussion has assumed a centralized architecture in which feedback is explicitly provided and a single trusted mediator controls feedback aggregation and distribution. Though the design possibilities of even that simple architecture are not yet fully understood, centralized reputation mechanisms do not begin to exhaust the new possibilities offered by information technology.

In recent years the field of multiagent systems [Jennings, Sycara, and Wooldridge, 1998] has been actively researching online reputation systems as a technology for building trust and inducing good behavior in artificial societies of software agents. Two lines of investigation stand out as particularly novel and promising:

20.5.5.1 Reputation Formation Based on Analysis of "Implicit Feedback"

In our networked society, several traces of an agent's activities can be found on publicly accessible databases. Instead of (or in addition to) relying on explicitly provided feedback, automated reputation mechanisms can then potentially infer aspects of an agent's attributes, social standing, and past behavior through collection and analysis of such "implicit feedback" information.

Perhaps the most successful application of this approach to date is exemplified by the Google search engine. Google assigns a measure of reputation to each Web page that matches the keywords of a search request. It then uses that measure in order to rank order search hits. Google's page reputation measure is based on the number of links that point to a page, the number of links that point to the pointing page, and so on [Brin and Page, 1998]. The underlying assumption is that if enough people consider a page to be important enough in order to place links to that page from their pages, and if the pointing pages

[2] For example, if the mechanism summarizes the 10 most recent ratings, newcomers would begin the game with a profile that indicates that all 10 recent ratings were negative. An additional assumption is that buyers cannot tell how long a given seller has been on the market and therefore cannot distinguish between newcomers with "artificially tarnished" profiles and dishonest players who have genuinely accumulated many negative ratings.

are "reputable" themselves, then the information contained on the target page is likely to be valuable. Google's success in returning relevant results is testimony to the promise of that approach.

Pujol, Sangüesa, and Delgado [2002] propose a generalization of the above algorithm that "extracts" the reputation of nodes in a general class of social networks. Sabater and Sierra [2002] describe how direct experience, explicit, and implicit feedback can be combined into a single reputation mechanism.

Basing reputation formation on implicit information is a promising solution to problems of eliciting sufficient and truthful feedback. Careful modeling of the benefits and limitations of this approach is needed in order to determine in what settings it might be a viable substitute or complement of voluntary feedback provision.

20.5.5.2 Decentralized Reputation Architectures

Our discussion of reputation mechanisms has so far implicitly assumed the honesty of feedback mediators. Alas, mediators are also designed and operated by parties whose interests may sometimes diverge from those of community participants.

Decentralizing the sources of reputation is a promising approach for achieving robustness in the presence of potentially dishonest mediators and privacy concerns. A number of decentralized reputation mechanisms have recently been proposed (Zacharia, Moukas, and Maes [2000]; Mui, Szolovits, and Ang [2001], Sen and Sajja [2002], Yu and Singh [2002]).

The emergence of peer-to-peer networks provides a further motivation for developing decentralized reputation systems. In such networks, reputation system represent a promising mechanism for gauging the reliability and honesty of the various network nodes. Initial attempts to develop reputation mechanisms for peer-to-peer networks are reported in Aberer and Despotovic [2001], Kamvar, Schlosser, and Garcia-Molina [2003], and Xiong and Liu [2003].

Though novel and intriguing, none of these works provides a rigorous analysis of the behavior induced by the proposed mechanisms or an explicit discussion of their advantages relative to other alternatives. More collaboration is needed in this promising direction between computer scientists, who better understand the new possibilities offered by technology, and social scientists, who better understand the tools for evaluating the potential impact of these new systems.

20.6 Conclusions

Online reputation mechanisms harness the remarkable ability of the Internet not only to disseminate but also collect and aggregate information from large communities at very low cost in order to artificially construct large-scale word-of-mouth networks. Such networks have historically proven to be effective social control mechanisms in settings where information asymmetries can adversely impact the functioning of a community and where formal contracting is unavailable, unenforceable, or prohibitively expensive. They are fast emerging as a promising alternative to more established trust building mechanisms in the digital economy.

The design of such mechanisms can greatly benefit from the insights produced by more than 20 years of economics and game theory research on the topic of reputation. These results need to be extended to take into account the unique new properties of online mechanisms such as their unprecedented scalability, the ability to precisely design the type of feedback information that is solicited and distributed, the volatility of online identities, and the relative lack of contextual cues to assist interpretation of what is, essentially, subjective information.

The most important conclusion drawn from this survey is that reputation is a powerful but subtle and complicated concept. Its power to induce cooperation without the need for costly and inefficient enforcement institutions is the basis of its appeal. On the other hand, its effectiveness is often ephemeral and depends on a number of additional tangible and intangible environmental parameters.

In order to translate these initial results into concrete guidance for implementing and participating in effective reputation mechanisms further advances are needed in a number of important areas. The following list contains what the author considers to be the imperatives in the most important open areas of research in reputation mechanism design:

- Scope and explore the design space and limitations of mediated reputation mechanisms. Understand what set of design parameters work best in what settings. Develop formal models of those systems in both monopolistic and competitive settings.
- Develop effective solutions to the problems of sufficient participation, easy identity changes, and strategic manipulation of online feedback.
- Conduct theory-driven experimental and empirical research that sheds more light into buyer and seller behavior *vis-à-vis* such mechanisms.
- Compare the relative efficiency of reputation mechanisms to that of more established mechanisms for dealing with information asymmetries (such as state-backed contractual guarantees and brand-name building) and develop theory-driven guidelines for deciding which set of mechanisms to use and when.
- Identify new domains where reputation mechanisms can be usefully applied.

Online reputation mechanisms attempt to artificially engineer heretofore naturally emerging social phenomena. Through the use of information technology, what had traditionally fallen within the realm of the social sciences is, to a large extent, being transformed into an engineering design problem. The potential to engineer social outcomes through the introduction of carefully crafted information systems is opening a new chapter on the frontiers of information technology. It introduces new methodological challenges that require collaboration between several traditionally distinct disciplines, such as economics, computer science, management science, sociology, and psychology, in order to be properly addressed. Our networked societies will benefit from further research in this exciting area.

References

Aberer, K., Z. Despotovic, 2001. Managing trust in a peer-2-peer information system. *Proceedings of the tenth international conference on Information and Knowledge Management.* Association for Computing Machinery, Atlanta, GA. 310–317.

Avery, C., P. Resnick, R. Zeckhauser. 1999. The Market for Evaluations. *American Economics Review* 89(3): 564–584.

Bajari, P., A. Hortacsu. 2003. Winner's Curse, Reserve prices and Endogenous Entry: Empirical Insights from eBay Auctions. *Rand Journal of Economics* 34(2).

Bakos, Y., C. Dellarocas. 2002. Cooperation without Enforcement? A Comparative Analysis of Litigation and Online Reputation as Quality Assurance Mechanisms. L. Applegate, R. Galliers, J. I. DeGross, Eds. *Proceedings of the 23rd International Conference on Information Systems (ICIS 2002).* Association for Information Systems, Barcelona, Spain, 127–142.

Benson, B. 1989. The Spontaneous Evolution of Commercial Law. *Southern Economic Journal.* 55(January) 644–661. Reprinted in D. B. Klein, Ed. 1997. *Reputation: Studies in the Voluntary Elicitation of Good Conduct.* University of Michigan Press, Ann Arbor, MI, 165–189.

Brin S., L. Page. 1998. The Anatomy of a Large-Scale Hypertextual Web Search Engine. *Computer Networks and ISDN Systems* 30(1–7): 107–117.

Cripps, M., G. Mailath, L. Samuelson. 2002. Imperfect Monitoring and Impermanent Reputations. Penn Institute for Economic Research Working Paper 02-021, University of Pennsylvania, Philadelphia, PA. (available at: http://www.econ.upenn.edu/Centers/pier/Archive/02-021.pdf)

Dellarocas, C. 2000. Immunizing online reputation reporting systems against unfair ratings and discriminatory behavior. *Proceedings of the 2nd ACM Conference on Electronic Commerce.* Association for Computing Machinery, Minneapolis, MN, 150–157.

Dellarocas, C. 2002. Goodwill Hunting: An economically efficient online feedback mechanism in environments with variable product quality. J. Padget, O. Shehory, D. Parkes, N. Sadeh, W. Walsh, Eds. *Agent-Mediated Electronic Commerce IV. Designing Mechanisms and Systems. Lecture Notes in Computer Science* 2351, Springer-Verlag, Berlin, 238–252.

Dellarocas C. 2003a. Efficiency and Robustness of Binary Feedback Mechanisms in Trading Environments with Moral Hazard. MIT Sloan Working Paper No. 4297-03, Massachusetts Institute of Technology. Cambridge, MA (available at: http://ssrn.com/abstract=393043)

Dellarocas C. 2003b. The Digitization of Word-of-Mouth: Promise and Challenges of Online Feedback Mechanisms. *Management Science* (October 2003).

Dellarocas, C. 2004. Building Trust On-Line: The Design of Robust Reputation Mechanisms for Online Trading Communities. G. Doukidis, N. Mylonopoulos, N. Pouloudi, Eds. *Social and Economic Transformation in the Digital Era.* Idea Group Publishing, Hershey, PA.

Dewan, S., V. Hsu, 2002. Adverse Selection in Reputations-Based Electronic Markets: Evidence from Online Stamp Auctions. Working Paper, Graduate School of Management, University of California, Irvine, CA (available at: http://web.gsm.uci.edu/~sdewan/Home%20Page/Adverse%20Selection%20in%20Reputations.pdf)

Diamond, D. 1989. Reputation Acquisition in Debt Markets. *Journal of Political Economy* 97(4): 828–862.

Friedman, E., P. Resnick. 2001. The Social Cost of Cheap Pseudonyms. *Journal of Economics and Management. Strategy* 10(1): 173–199.

Fudenberg, D., D. Levine. 1992. Maintaining a Reputation When Strategies are Imperfectly Observed, *Review of Economic Studies* 59(3): 561–579.

Greif, A. 1993. Contract Enforceability and Economic Institutions in Early Trade: The Maghribi Traders' Coalition. *American Economic Review* 83(June) 525–548.

Holmstrom, B. 1999. Managerial Incentive Problems: A Dynamic Perspective. *Review of Economic Studies* 66(1): 169–182.

Horner, J. 2002. Reputation and Competition, *American Economic Review* 92(3): 644–663.

Houser, D., J. Wonders. 2000. Reputation in Auctions: Theory and Evidence from eBay. Department of Economics Working Paper 00-01, The University of Arizona, Tucson, AZ (available at: http://info-center.ccit.arizona.edu/~econ/working-papers/Internet_Auctions.pdf)

Jennings, N., K. Sycara, M. Wooldridge, 1998. A Roadmap of Agent Research and Development. *Autonomous Agents and Multi-Agent Systems* 1(1): 275–306.

Kamvar, S. D., M. T. Schlosser, H. Garcia-Molina, 2003. The Eigentrust algorithm for reputation management in P2P networks. *Proceedings of the 12th international conference on World Wide Web.* Association for Computing Machinery, Budapest, Hungary, 640–651.

Klein, D., Ed. 1997. *Reputation: Studies in the Voluntary Elicitation of Good Conduct.* University of Michigan Press, Ann Arbor, MI.

Klein, B., K. Leffler. 1981. The Role of Market Forces in Assuring Contractual Performance. *Journal of Political Economy* 89(4): 615–641.

Kollock, P. 1999. The Production of Trust in Online Markets. E. J. Lawler, M. Macy, S. Thyne, and H. A. Walker, Eds. *Advances in Group Processes* (Vol. 16). JAI Press, Greenwich, CT.

Kreps, D., Milgrom, P., Roberts, J., Wilson, R. 1982. Rational Cooperation in the Finitely Repeated Prisoners' Dilemma. *Journal of Economic Theory* 27(2): 245–252.

Kreps, D., R. Wilson. 1982. Reputation and Imperfect Information. *Journal of Economic Theory* 27(2): 253–279.

Lucking-Reiley, D., D. Bryan, et al. 2000. Pennies from eBay: the Determinants of Price in Online Auctions, Working Paper, University of Arizona, Tucson, AZ (available at: http://eller.arizona.edu/~reiley/papers/PenniesFromEBay.html.)

Mailath, G. J., L. Samuelson. 1998. Your Reputation Is Who You're Not, Not Who You'd Like to Be. Center for Analytic Research in Economics and the Social Sciences (CARESS) Working Paper 98–11, University of Pennsylvania, Philadelphia, PA (available at: http://www.ssc.upenn.edu/~gmailath/wpapers/rep-is-sep.html)

Mailath, G. J., L. Samuelson. 2001. Who Wants a Good Reputation? *Review of Economic Studies* 68(2): 415–441.

Mayzlin, D. 2003. Promotional Chat on the Internet. Working Paper #MK-14, Yale School of Management, New Haven, CT.

Milgrom, P. R., D. North, B. R. Weingast. 1990. The Role of Institutions in the Revival of Trade: The Law Merchant, Private Judges, and the Champagne Fairs. *Economics and Politics* 2(1): 1–23. Reprinted in D. B. Klein, Ed. 1997. *Reputation: Studies in the Voluntary Elicitation of Good Conduct.* University of Michigan Press, Ann Arbor, MI, 243–266.

Milgrom, P., J. Roberts. 1982. Predation, Reputation and Entry Deterrence. *Journal of Economic Theory* 27(2): 280–312.

Miller, N., P. Resnick, R. Zeckhauser. 2002. Eliciting Honest Feedback in Electronic Markets, Research Working Paper RWP02-039, Harvard Kennedy School, Cambridge, MA (available at: http://www.si.umich.edu/~presnick/papers/elicit/index.html)

Mui, L., P. Szolovits, C. Ang. 2001. Collaborative sanctioning: applications in restaurant recommendations based on reputation. *Proceedings of the 5th International Conference on Autonomous Agents,* Association for Computing Machinery, Montreal, Quebec, Canada, 118–119.

Pujol, J. M., R. Sanguesa, J. Delgado. 2002. Extracting reputation in multi agent systems by means of social network topology. *Proceedings of the 1st International Joint Conference on Autonomous Agents and Multiagent Systems,* Association for Computing Machinery, Bologna, Italy, 467–474.

Resnick, P., R. Zeckhauser, E. Friedman, K. Kuwabara. 2000. Reputation Systems. *Communications of the ACM* 43(12): 45–48.

Resnick, P., R. Zeckhauser. 2002. Trust Among Strangers in Internet Transactions: Empirical Analysis of eBay's Reputation System. Michael R. Baye, Ed. *The Economics of the Internet and E-Commerce (Advances in Applied Microeconomics, Vol. II).* JAI Press, Greenwich, CT.

Resnick, P, R. Zeckhauser, J. Swanson, K. Lockwood. 2002. The Value of Reputation on eBay: A Controlled Experiment. Working Paper, University of Michigan, Ann Arbor, MI (available at: http://www.si.umich.edu/~presnick/papers/postcards/index.html.)

Sabater, J., C. Sierra. 2002, Reputation and social network analysis in multi-agent systems. *Proceedings of the First International Joint Conference on Autonomous Agents and Multiagent Systems,* Association for Computing Machinery, Bologna, Italy, 475–482.

Sen, S., N. Sajja. 2002. Robustness of reputation-based trust: boolean case. *Proceedings of the First International Joint Conference on Autonomous Agents and Multiagent Systems,* Association for Computing Machinery, Bologna, Italy, 288–293.

Shapiro, C. 1983. Premiums for High Quality Products as Returns to Reputations. *The Quarterly Journal of Economics* 98(4): 659–680.

Tadelis, S. 1999. What's in a Name? Reputation as a Tradeable Asset. *The American Economic Review* 89(3): 548–563.

Tadelis, S. 2002. The Market for Reputations as an Incentive Mechanism. *Journal of Political Economy* 92(2): 854–882.

Wilson, R. 1985. Reputations in Games and Markets. A. Roth, Ed. *Game-Theoretic Models of Bargaining,* Cambridge University Press, Cambridge, U.K., 27–62.

Xiong, L., L. Liu, 2003. A Reputation-Based Trust Model for Peer-to-Peer ecommerce Communities. *IEEE Conference on E-Commerce (CEC'03),* Newport Beach, KY, June, 2003.

Yu, B., M. Singh. 2002. An evidential model of distributed reputation management. *Proceedings of the First International Joint Conference on Autonomous Agents and Multiagent Systems,* Association for Computing Machinery, Bologna, Italy, 294–301.

Zacharia, G., A. Moukas, P. Maes. 2000. Collaborative Reputation Mechanisms in Electronic Marketplaces. *Decision Support Systems* 29(4): 371–388.

21

Digital Rights Management

Mikhail Atallah

Keith Frikken

Carrie Black

Susan Overstreet

Pooja Bhatia

CONTENTS

Abstract 21-1
21.1 Introduction 21-1
21.2 Overview 21-2
21.3 Digital Rights Management Tools 21-4
21.3.1 Software Cracking Techniques and Tools 21-5
21.3.2 Protection Mechanisms 21-5
21.3.3 Further Remarks About Protection Mechanisms 21-12
21.4 Legal Issues 21-13
Acknowledgment 21-15
References 21-15

Abstract

This chapter surveys Digital Rights Management (DRM). The primary objective of DRM is to protect the rights of copyright owners for digital media, while protecting the privacy and the usage rights of the users. We review the various approaches to DRM, their weaknesses and strengths, advantages and disadvantages. We also survey the tools that are used to circumvent DRM protections, and techniques that are commonly used for making DRM protections more resilient against attack. We also take a brief look at the legal issues of DRM. The chapter gives an overview and a brief glimpse of these techniques and issues, without going into intricate technical details. While the reader cannot expect to find all DRM topics in this chapter, nor will the chapter's coverage of its topics be complete, the reader should be able to obtain sufficient information for initial inquiries and references to more in-depth literature.

21.1 Introduction

The ability to create exact replicas of digital data reduces the ability to collect payment for an item, and since there is no degradation of quality for these easily made copies, this "piracy" is damaging to the owners of digital content. The fear of piracy has kept many content owners from embracing the Internet as a distribution channel, and has kept them selling their wares on physical media and in physical stores. Compared to the digital distribution of data which requires no physical media and no middlemen, this is not an economically efficient way of doing business, but it has been lucrative for content owners; they are in no rush to make it easier for pirates to steal their movies. However, this "physical delivery" model's profitability for the content owners has decreased with the online availability of pirated versions of their content, which is more convenient to download than taking a trip to the store (and this pirated version is also "free" to the unscrupulous). Some content owners have recently turned to online delivery, using a mix of DRM technology and legal means to protect their rights, with mixed results. The protections are typically

1-58488-381-2/05/$0.00+$1.50
© 2005 by CRC Press LLC

defeated by determined attackers, sometimes after considerable effort. Some of the deployed and proposed techniques for protecting digital data have potential for serious damage to consumer privacy.

Digital Rights Management, more commonly known as DRM, aims at making possible the distribution of digital content while protecting the rights of the parties involved (mainly the content owners and the consumers). The content owners run risks such as the unauthorized duplication and redistribution of their digital content (piracy), or the unauthorized violations of use policies by a user, whereas the consumers run risks that include potential encroachments on their privacy as a result of DRM technology. Any computer user also runs the risk that, because of a combination of DRM-related legislation (that mandates one particular approach to DRM) and corporate decisions from within the IT and entertainment industries, future computers may become too constrained in their architectures and too inflexible to continue being the wonderful tools for creation and tinkering that they have been in the past. While it may not be possible to eliminate all these risks using purely technological means (because legal, ethical, business, and other societal issues play important roles), sound DRM technology will play a crucial role in mitigating these risks. DRM technology also has many other "side uses," such as ensuring that only authorized applications run (e.g., in a taximeter, odometer, or any situation where tampering is feared) and that only valid data is used. This chapter is a brief introduction to DRM.

Before we delve further into DRM-related issues, we need to dispel several misconceptions about DRM including that DRM is primarily about antipiracy, that DRM and access control are the same problem, and that DRM is a purely technical problem. DRM is about more than antipiracy. It is just as much about new business models, new business and revenue opportunities for content owners and distributors, and wide-ranging policy enforcement (preventing uncontrolled copying and distribution is but one kind of policy). As we shall see later in this chapter, DRM technologies can also have profound implications for antivirus protection, integrity checking and preservation, and the security of computer systems and networks in general. The line between DRM and access control is blurred by many DRM proponents. Access control is a different problem than DRM; the existing and proposed solutions for access control are simpler, less restrictive, and with less wide-ranging implications for system architectures, than those for DRM. How does DRM differ from access controls? The latter is about control of clients' access to server-side content, whereas the former is about control of clients' access to client-side content. For example, if the server wants to make sure that only authorized personnel access a medical records database through the network, this is access control, but if the software on the client's machine wants to allow a user to view a movie but only in a specified manner (for example the user can only view the movie three times, or within the next 30 days, or the user cannot copy the movie, etc.) then this is DRM. A remarkable number of otherwise knowledgeable corporate spokespersons misrepresent DRM technology or proposed DRM legislation as "protecting the user," when the truth is that the user is the adversary from whom they seek protection and whom they wish to constrain. DRM is largely about constraining and limiting the user's ability to use computers and networks. It may be socially desirable to so constrain the user, and thus to protect the revenue stream of digital content owners and distributors (the creators of digital content have grocery bills to pay, too). But it should be stated for what it is: protection from the user, not of the user. The interested user may consent to such constraints as a condition for using a product or system, viewing a movie, etc. However, as we will see below, there are attempts to impose an exorbitant dose of DRM-motivated costs on *all* users, even legally constraining what digital devices (including computers) can and cannot do.

The rest of this chapter proceeds as follows: In Section 2, we give an overview of the general techniques used in DRM; section 3 discusses the tools to protect DRM (encryption wrappers, watermarking, code obfuscation, trusted hardware, etc.) and the tools the attackers use to circumvent DRM technologies; and Section 4 discusses legal (also social, political) issues related to DRM.

21.2 Overview

The purpose of this section is to provide an overview of how DRM systems operate. There are various design questions that need to be addressed, and in many situations there is no clear, strictly superior solution.

Content owners will have differing requirements that they desire to place upon their content (e.g., how many times it can be viewed, whether it can be copied, modified, etc.). Hence, to have a general DRM system there must be some mechanism for describing this range of policies. Metadata is used for this purpose; metadata describes the content and policies relating to data (ownership, usage, distribution, etc.). Standards and techniques exist for expressing, in a way that is computer-interpretable yet convenient for humans to use, the policies concerning the access and use of digital content. As elsewhere, open standards are desirable, as they ensure interoperability. Examples include XrML (originated from XEROX PARC) and ODRL (Open Digital Rights Language). XrML is becoming the dominant industry standard. The rights to be managed are not limited to duplication, but also to modification, printing, expiration, release times, etc. They are described in a license, whose terms and conditions are verified and enforced by the reader (e.g., a movie viewer) that mediates access to the digital content.

With a standard for expressing metadata, there is a design issue that needs to be addressed. Should the metadata be embedded into the content or should it be stored separately? An advantage in keeping content and the Metadata separate is that the policies can be updated after purchase (e.g., if the customer wants to buy three additional viewings of the movie then it can be done without having to redownload the movie). When the metadata is embedded, on the other hand, the policies for the file are self-contained and any attack against the file either changes the metadata or prevents applications from looking for metadata. These attacks can also be done when the metadata is stored separately, but there is another avenue of attack that tries to modify the linking of the content to its metadata. Thus, embedding the data is likely to be more secure, but it is more convenient to keep the metadata separate.

In some commercial products the policy is stored in a remote secure server, allowing more central control, and rapid modification and enforcement of policies. This has been popular with some corporations. Corporate emails and internal documents are therefore always sent in encrypted form. Viewing an email then means that the mail application has to obtain the decryption key from the server that can enforce the policy related to that email (which could be "view once only," "not printable," "not forwardable," etc.). Such a corporation can now remotely "shred" emails and documents by having the server delete the key that corresponds to them (so that they become impossible to read, even though they remain on various employee computers). A common policy like "all emails should be deleted after three months" that is next to impossible to enforce otherwise (employees simply do not comply) now becomes easily enforceable. This may be socially undesirable because it facilitates the automatic deletion of evidence of wrongdoing, but it appeals to many corporations who are worried about ancient emails causing them huge liability (as so often happens with lawsuits alleging harassment, hostility, discrimination, etc.).

What are some of the approaches one can take to prevent the unrestricted replication and dissemination of digital content? We begin with the case of media (audio, video, text, and structured aggregates thereof) because the question of protecting media ultimately reduces to the question of protecting software (as will become clear in what follows). One way is to force access to the media to take place through approved applications ("players") that enforce the DRM policy (embedded in the media or separate). One policy could take the form of "tying" the media to the hardware, so that copying it to another machine would make it no longer viewable. (Of course, the DRM data attached to the media would now specify the approved hardware's identifying parameters.) This is problematic if the customer repairs hardware or upgrades it and would seem to require the customer obtaining a certification from some approved dealer that she did indeed perform the upgrade (or repair) of her hardware and thus deserves a cost-free copy of the media (that would now be tied to the new hardware). The policy could alternatively take the form of tying the media to a particular customer, who can view it on all kinds of different hardwares after undergoing some form of credentials-checking (possibly involving communication with the media owner's server). Finally, the policy could be that the customer never actually acquires the media as a file to be played at the customer's convenience. The player plays a file that resides on a remote server under the control of the media owner. However, requiring that media files reside on a remote server will add a large communication overhead and will disallow the playing of media files on systems that are not connected to a network.

Given a choice between being able to prevent the violation of policy and being able to detect the violation and take legal action, it is obvious that the prevention of such violations is preferred. However, it is often easier to detect violations than to prevent them. This is evident if one contrasts the mechanisms that need to be in place for detection versus protection. In either case some data needs to be bound to the object (in the case of detection this would be the ownership information, while in the prevention case this would be the policies allowed for the file). However, for prevention there is an additional step of requiring "viewers" of the data to enforce the rules, and steps need to be taken to prevent renegade players. Thus, the mechanisms that need to be in place for prevention are a superset of those for detection. Relying on the legal system for DRM protection is problematic since many violators are in countries outside the reach of our legal system. Even with domestic violators, legal means can be difficult. For example, if there is piracy by a large number of parties it is tedious and expensive to prosecute each case. There are advantages and disadvantages to both approaches, and it is often best to use a hybrid of the two. To create a DRM system that is resilient to attack, one must understand how an attacker would compromise such a system. In all such approaches that rely on an approved player for enforcing DRM policy, the point of attack for digital bandits becomes the player. "Cracking" the player produces a version of it that is stripped of the functionality that the digital pirates dislike (the one that does credentials-checking and enforces other DRM policies), while preserving the player's ability to view the digital media. Many "cracked" versions of such player software can be found on the Internet (e.g., on "warez" sites). Thus the question of preventing the cracking (unauthorized modification) of software is central to the protection of media. If the player proves unbreakable then another avenue of attack would be to modify the policy files (i.e., the metadata) in some way as to allow more liberal access. This could be done by either removing the policy altogether or by replacing it with a more "liberal" policy. A design decision for a player is how to handle files without policies. There are two options: (1) treat all such files as files where users have complete control, or (2) treat all such files as pirated files and refuse to "play" the file for the user. If the second approach is used, then one does not need to worry about attacks that remove policies, but it is dangerous since it obsoletes many current systems (including ones where users have legitimate access), which may cause users to boycott the new technology.

A final issue that is addressed in this section is why is DRM difficult? DRM is difficult because it is not enough to just prevent an average user from breaking the DRM system. It is not difficult to prevent the average person from being able to "crack" a DRM scheme, but if an expert creates a "cracked" player or a tool for removing copyright protections, then this expert could make it available to the public. If this happens, then even a novice user can use it and effectively be as powerful as the expert. Thus DRM is difficult because a single breaking of the system possibly allows many users, even those with limited technical knowledge, to circumvent the system. DRM is easier in some situations where it is enough to postpone the defeat of a protection "long enough" rather than stop it altogether; what "long enough" means varies according to the business in question. For example, computer game publishers make most of the revenue from a product in the first few weeks after its initial release.

21.3 Digital Rights Management Tools

In this section we take a look at the various tools used for software cracking and for DRM protection. The attacker's toolkit consists of decompilers, debuggers, performance profilers, and static and dynamic program analysis tools. The defender's toolkit can be divided into two groups: hardware and software. Software techniques include encryption wrappers, watermarking, finger-printing, code obfuscation, and software aging. Hardware techniques include dongles and trusted hardware. This section should be viewed as a survey of these techniques, since detailed discussion of these tools in depth would require much more space than available for this chapter.

21.3.1 Software Cracking Techniques and Tools

The following is a brief survey of the tools that exist or that are likely to be developed to attack software protection mechanisms (some of these are not normally thought of as attack tools).

1. Disassemblers, decompilers, and debuggers. These allow attackers to:
   - Locate events in the code. The places where a variable (say password) is used, a routine is called (say to open a information box), or a jump is made (say to "exit message") can be identified.
   - Trace variable use in the code. One can follow the password and variables derived from it. This is tedious to do unless the use is simple, but it is faster than reading the code.
   - Create more understandable version of the code. This is the principal use of these tools when code protections are complex.
2. Performance Profilers: These can be used to identify short, computation intensive code segments. Some protection mechanisms (like encryption wrappers) have this property.
3. Static and dynamic program analysis tools: These, and their natural extensions, provide valuable information to the attacker. Program slicing identifies which instructions actually affect the value of a variable at any point in the program. Value tracing identifies and displays the ancestors and descendents of a particular variable value. Distant dependencies are also displayed — dependencies that are widely separated. This tool can help locate protection mechanisms and show how values propagate through code. Constant tracing identifies all variables that always acquire the same value at run-time in a program, that is, values that are independent of program input. These are important in protection, as they are keys to authorizations (passwords, etc.), to protection (checksums), and to implicit protections (constants in identities), so that knowing these values gives the attacker important clues on how a code is protected.
4. Pattern Matching: A protected program may have many protection-related code fragments and identities inserted into it. These are normally short pieces of code so one can look for repeated patterns of such codes. More sophisticated versions of these tools look for approximate matches in the patterns.

21.3.2 Protection Mechanisms

There are a plethora of protection mechanisms for DRM. These include encryption wrappers, watermarking, fingerprinting, code obfuscation, software aging, and various hardware techniques. The following section presents an overview of these tools; their technical details are omitted. None of these tools is a "magic bullet" that fixes all DRM problems, and in most cases a combination is used.

21.3.2.1 Encryption Wrappers

In this protection technique the software file (or portions thereof) is encrypted, and self-decrypts at run-time. This, of course, prevents a static attack and forces the attacker to run the program to obtain a decrypted image of it. The protection scheme usually tries to make this task harder in various ways, some of which we briefly review next.

One is to include in the program antidebug, antimemory-dump, and other defensive mechanisms that aim to deprive the attacker of the use of various attack tools. For example, to prevent the attacker from running the program in a synthetic (virtual machine) environment in which automated analysis and attack tools are used, the software can contain instructions that work fine in a real machine but otherwise cause a crash. This is sometimes done by having one instruction x corrupt an instruction y in memory at a time when it is certain that y would be in the cache on a real machine; so it is the uncorrupted version of y, the one in the cache, that executes on a real machine, but otherwise it is the corrupted one that tries to execute and often causes a crash. These protections can usually be defeated by a determined adversary (virtual machines exist that emulate a PC very faithfully, including the cache behavior).

Another way to make the attacker's task harder is to take care to not expose the whole program in unencrypted form at any one time — code decrypts as it needs to for execution, leaving other parts of the program still encrypted. This way a snapshot of memory does not expose the whole decrypted program, only parts of it. This forces the attacker to take many such snapshots and try to piece them together to obtain the overall unencrypted program. But there is often a much less tedious route for the attacker: to figure out the decryption key which, after all, must be present in the software. (Without the key the software could not self-decrypt.)

Encryption wrappers usually use symmetric (rather than public key) encryption for performance considerations, as public key encryption is much slower than symmetric encryption. The encryption is often combined with compression, resulting in less space taken by the code and also making the encryption harder to defeat by cryptanalysis (because the outcome of the compression looks random). To reap these benefits the compression has to be done prior to the encryption, otherwise there is no point in doing any compression. (The compression ratio would be poor because of compressing random-looking data, and the cryptanalyst would have structured rather than random cleartext to figure out from the ciphertext.)

21.3.2.2 Watermarking and Fingerprinting

The goal of watermarking is to embed information into an object such that the information becomes part of the object and is hard to remove by an adversary without considerably damaging the object. There are many applications of watermarking, including inserting ownership information, inserting purchaser information, detecting modification, placing caption information, etc. If the watermarking scheme degrades the quality of an item significantly, then the watermarking scheme is useless since users will not want the reduced-quality object. There are various types of watermarks, and the intended purpose of the watermark dictates which type is used. Various decisions need to be made about the type of watermark for a given application. The following is a list of such decisions.

1. Should the watermark be visible or indiscernible? Copyright marks do not always need to be hidden, as some watermarking systems use visible digital watermarks which act as a deterrent to an attacker. Most of the literature has focused on indiscernible (e.g., invisible, inaudible) digital watermarks which have wider applications, hence we will focus on invisible watermarks in this section.
2. Should the watermark be different for each copy of the media item or should it be the same? A specific type of watermarking is fingerprinting, which embeds a unique message in each instance of digital media so that a pirated version can be traced back to the culprit. This has consequences for the adversary's ability to attack the watermark, as two differently marked copies often make possible a "diff" attack that compares the two differently marked copies and often allows the adversary to create a usable copy that has neither one *of* the two marks.
3. Should a watermark be robust or fragile? A fragile watermark is destroyed if even a small alteration is done to the digital media, while a robust one *is* designed to withstand a wide range of attacks. These attacks include anything that attempts to remove or modify the watermark without destroying the object.

In steganography, the very existence of the mark must not be detectable. The standard way of securing communication is by encrypting the traffic, but in some situations the very fact that an encrypted message has been sent between two parties has to be concealed. Steganography hides the very existence of a secret message, as it provides a covert communication channel between two parties, a channel whose existence is unknown to a possible attacker. A successful attack now consists of detecting the existence of this communication (e.g., using statistical analysis of images with and without hidden information), so the attacker need not actually learn the secret message to be considered successful.

Since invisible watermarking schemes often rely on inserting information into the redundant parts of data, inserting watermarks into compressed or encrypted data is problematic since data in this form either appears random (encryption) or does not contain redundancy (compression). Although watermarks may be embedded in any digital medium, by far most of the published research on watermarking has dealt with images. Robust image watermarking commonly takes two forms: spatial domain and frequency domain. The former inserts the mark in the pixels of the image, whereas the latter uses the transform (Discrete Cosine, Fourier, etc.) of the image. One spatial domain watermarking technique mentioned in the literature slightly modifies the pixels in one of two randomly-selected subsets of an image. Modification might include, for example, flipping the low-order bit of each pixel representation. The easiest way to watermark an image is to change directly the values of the pixels in the spatial domain. A more advanced way to do it is to insert the watermark in the frequency domain, using one of the well known transforms: FFT, DCT, or DWT. There are other techniques for watermarking images; for example, there are techniques that use fractals to watermark an image.

Software is also watermarkable, usually at design time by the software publisher. Software watermarks can be static, readable without running the software, or could appear only at run-time, possibly in an evanescent form that does not linger for long. In either case, reading the watermark can require knowing a secret key, without which the watermark remains invisible.

Watermarks may be used for a variety of applications, including proof of authorship or ownership of a multimedia creation, fingerprinting for identifying the source of illegal copies of multimedia objects, authentication of multimedia objects, tamper-resistant copyright protection of multimedia objects, and captioning of multimedia objects to provide information about the object. We now give a more detailed explanation of each of these applications.

1. *Proof of ownership:* The creators or owners of the digital object need to prove that it is theirs and insert a watermark to that effect. For example, a photographer may want to be able to prove that a specific photograph was taken by him in order to prevent a magazine from using the photo without paying the royalties. By inserting a robust watermark into the item stating ownership information, the owner would be able to prove ownership of the item.
2. *Culprit tracing:* Inserting a robust watermark with information about the copyright owner, as well as the entity who is purchasing (downloading) the item, will allow traceback to the entity if the item were to be illegally disseminated to others by this entity. This is a form of fingerprinting. Breaking the security of the scheme can have nasty consequences, allowing the attacker to "frame" an innocent victim. One problem with this model is that a user could claim that another party stole his copy and pirated it, and such claims could be true and are difficult to disprove.
3. *Credentials checking:* If a trusted player wants to verify that the user has a license for an item, then a robust watermark inserted into the item can be used for this purpose.
4. *Integrity checking:* There are many cases where modification unauthorized of an object is to be detected. In this case placing a fragile watermark into an item (one that would be destroyed upon modification), and requiring the presence of the watermark at viewing-time, would prevent such modifications. If requirements are such that both a robust and fragile watermark need to be inserted into an object, then obviously the robust watermark should be inserted first.
5. *Restricting use:* Another use of robust watermarks is to embed captions into the information that contain the metadata for an object. For example, if an object is read-only then a watermark specifying this would allow a player to enforce this policy.

The goal of an attack depends on the type of watermark and the attacker's goal. For a robust watermark, the goal is often to remove or modify the watermark; for a fragile watermark, the goal is to make changes without destroying the watermark. The attacker may not care about removing an existing watermark. In this case, a second watermark may be inserted that is just as likely to be authentic as the first and prevent the effective use of the first in court. Any attack that removes a watermark must not destroy too much of the data being protected (e.g., the image must not become too blurry as a result), otherwise doing so will defeat the purpose of the attacker. In the case where the object being protected is a fingerprint, an attack must be resilient against multiple entities colluding together to remove watermarks by comparing their copies of the item (a "diff attack"). In summary, watermarks are a powerful DRM tool. However, their use can require trusted players to check for the existence of the watermarks, which implies that watermarking requires other DRM technologies to be effective.

21.3.2.3 Code Obfuscation

Code obfuscation is a process which takes code and "mangles" it so that it is harder to reverse engineer. As pointed out in [Collberg and Thomborson, 2002], with code being distributed in higher level formats such as Java bytecode, protecting code from reverse engineering is even more important. Code obfuscation has many applications in DRM. For example, a company that produces software with algorithms and data structures that give them a competitive edge over their competition has to prevent their competition from reverse engineering their code and then using the proprietary algorithms and data structures in their own software. Code obfuscation also has applications besides protecting the rights of software companies; it is also useful when protecting players from being "cracked." A trusted player has specific areas where policy checks are made, and this will likely be a target of a person trying to remove these checks. Obfuscating the code makes it more difficult to find the code fragments where such checks are made.

Many transformations can be used to obfuscate code. In Collberg and Thomborson [2002], there are a set of requirements for these transformations, which are summarized here. It is important that these transformations do not change what the program does. To make the obfuscation resilient to attack, it is desirable to maximize the obscurity of the code, and these transformations need to be resilient against tools designed to automatically undo them. These transformations should not be easily detectable by statistical means, which will help prevent automatic tools from finding the locations of transformations. A limiting factor on these transformations is that the performance of the code should not be effected too much. For more information on code obfuscation see Collberg et al. [1997], Collberg and Thomborson [2002], and Wroblewski [2002]. An interesting note about obfuscation is that if the transformations hide information by adding crude, inefficient ways of doing simple tasks, then the code optimizer in the compiler may remove much of the obscurity and "undo" much of the obfuscation. If, on the other hand, the obfuscation "fools" the optimizer and prevents it from properly doing its job, then experience has shown that the performance hit due to obfuscation is considerable; so either way one loses something. This seems to speak in favor of low-level obfuscation (assembly level) because it does not prevent the code optimizer from doing its job, yet most existing automatic obfuscators are essentially source-to-source translators.

Below are some of the types of obfuscation transformations that have been proposed:

1. Layout obfuscation involves manipulation of the code in a "physical appearance" aspect. Examples of this includes replacing important variables with random strings, removing all formatting (making nested conditional statements harder to read), and the removing of comments. These are the easiest transformations to make and they are not as effective as the other means; yet, when combined with the other techniques, they do contribute confusion to the overall picture.
2. Data obfuscation focuses on the data structures used within a program. This includes manipulation of the representation and the methods of using that data, merging of data (arrays or inheritance relationships) that is independent, or splitting up data that is dependent, and allows the reordering of the layout of the objects used. This is a helpful tactic, since the data structures are the elements that contain important information that any attacker needs to comprehend.
3. Control obfuscation attempts to manipulate the control flow of a program in such a way that a person is not able to discern its "true" (pre-obfuscation) structure. This is achieved through merging or splitting various fragments of code, reordering any expressions, loops or blocks, etc. The overall process is quite similar to creating a spurious program that is embedded into the original program and "tangled" with it, which aids in obfuscating the important features of a program.
4. Preventive transformations are manipulations that are made to stop any possible deobfuscation tool from extracting the "true" program from within the obfuscation. This can be done through the use of what Collberg et al. [1997] call *opaque predicates.* These are the main method that an obfuscation has to prevent any portions of the inner spurious program from being identified by the attacker. An opaque predicate is basically a conditional statement that always evaluates to true, but in a manner that is extremely hard to recognize, thereby confusing the attacker.

Like most protections, obfuscation only delays a determined attacker; in other words it "raises the bar" against reverse engineering, but does not prevent it. The paper by Barak et al. [2001] gives a family of functions that are probably impossible to completely obfuscate. Thus code obfuscation is good for postponing reverse engineering, and if the time by which it can be postponed is acceptable to an organization, then obfuscation can be a viable approach.

21.3.2.4 Software Aging

Software aging uses a combination of software updates and encryption to age the software and protect the intellectual property of the software owner. More explicitly, through the use of mandated updates, the software is able to maintain its functionality with respect to other copies of the same software. The focus of this technique is to deter pirating, where the user obtains a copy of the software and then (with or without modification) redistributes it at a discount cost. Unlike most other DRM techniques, this method can only be applied to specific types of software; it is applicable only for those that generate files or messages that will need to be interpreted or viewed by another user with the same software. Examples of this include such products as word processors or spreadsheets, since the users often create a document and then send it to other users to view or modify.

The way software aging works is that every piece of software sold has two unique identifiers, a registration number that is different for every copy, and a key that is the same for all copies of the software. The registration number is necessary for the software update process, while the key is used to encrypt every file or message that a user creates with the software. More importantly, the key is also used to decrypt every file or message that it reads, whether it is their own file or someone else's. With an out-of-date key, the user will not be able to read documents encrypted with a more recent key. In applications that have a continual exchange of information with other users, this forces users to continually update their software, since they cannot read objects created by other versions. The software is "aged" by making documents produced by newer versions of the software incompatible with older software; thus forcing a software pirate to create a new cracked version for each update.

One advantage of this scheme is that it allows for possible piracy detection. If any illegal purchaser would happen to try to download the updates, then this could lead to a correlation that this user has purchased or obtained an illegal copy of the software. So by the interaction of the illegal users with the pirate, where they would be attempting to receive the updates, linkage between the illegal users and the pirate could be obtained. The possible piracy detection could deter future piracy, which could thereby reduce the amount of pirated software.

There are several disadvantages to this scheme. First, this technique is only possible with certain types of software, thereby making this scheme not very useful for various types of media or other types of software. It is not as extensible as other types of DRM technologies, which can be used for several different types of media. Also, with this scheme, there could exist a substantial overhead cost with the encryption and decryption process. Furthermore this scheme will be a nuisance to many users whenever they get a document that they cannot use. This nuisance is amplified if there is an illegal user that receives an update for a valid user before that user. Within this scheme, there could also exist some privacy concerns. Because the registration number is being sent to the vendor's central server every time a piece of software receives a software update, this does allow for a form of ID linkage between the purchaser and a particular copy of the software. Also, because the user is continually downloading every time there is a new update available, the constant interaction between the purchaser and the vendor makes it necessary to be careful that there is no user-tracking software also in the download. Thus there needs to be a check in place so that there is no abuse of this constant interaction between the user's PC and the vendor. To summarize, software aging provides a useful mechanism to deter piracy for a class of software, but the disadvantages of this technique make its usage cumbersome.

21.3.2.5 Hardware Techniques

Any DRM protection mechanism that requires software players to make checks to determine if a user has the required license to use a file is vulnerable to an attacker modifying the player so that these checks are rendered ineffective (bypassed or their outcome ignored). Thus, this type of system does not protect against highly skilled attackers with large resources, and if such an attacker "cracks" a player and makes the cracked version publicly available then any user can circumvent the protection mechanisms. In Lipton et al. [2002] a proof is given that no fully secure all-software protection exists and that some form of trusted hardware is needed for really foolproof protection. This is strong motivation to use some form of tamper-resistant hardware in situations where a crack needs to be prevented rather than merely postponed "long enough." If a DRM solution requires the presence of a hardware mechanism that cannot be circumvented with software alone, then any circumvention scheme would require that the hardware be cracked as well, a much harder task. Even if the hardware is cracked, the attacker who cracked it may not have an easy way to make the attack available to others. Whether he does depends on the details of the system and the extent to which the hardware protection is defeated; a compromise of secret keys within the hardware could, in some cases, lead to massive consequences.

The presence of hardware can be used to prove authorization (e.g., to access a file), or the hardware can itself control the access and require proof of authorization before allowing access. In one type of protection the software checks for the presence of the hardware and exits if the hardware is absent, but that is open to attack — for example by modifying the software so it does it not perform the check (or still performs it but ignores its outcome). Another form of protection is when the tamperproof hardware performs some essential function without which the software cannot be used (such as computing a function not easily guessed from observing its inputs and outputs). Another type of hardware protection controls access and requires the user to prove adequate authority for using the file for the specific use. An example of this is for a sound card to not play a copyrighted sound file until proper authorization has been proven. Another example is to have the operating system and the hardware prevent the usage of unauthorized players. These mechanisms are covered in the section titled "other hardware approaches" below.

21.3.2.1.1 Dongles

Some software applications use a copy protection mechanism most typically referred to by end-users as a dongle. A dongle is a hardware device that connects to the printer port, serial port, or USB port on a PC. Dongles were originally proposed for software piracy protection, but they can be used to protect media files. The purpose of dongles is to require the presence of the dongle in order to play a media file, which ties a piece of hardware to the digital media. Hence, in order to pirate the media file one must either (1) work around the need for the hardware or (2) duplicate the hardware. The difficulty of reversing a dongle depends on the complexity of the dongle; there are various kinds of dongles including (1) a device that just outputs a serial number (vulnerable to a "replay attack"), (2) a device that engages in a challenge-response protocol (different data each time, which prevents replay attack), (3) a device that decrypts content, and (4) a device that provides some essential feature of a program or media file. Dongles are still used by many specialized applications, typically those with relatively higher pricing. Dongles have several disadvantages that have limited their usage. Users dislike them for a variety of reasons, but mainly because they can be troublesome to install and use since they often require a special driver and can interfere with the use of peripherals such as printers and scanners. Since no standard exists for dongles, each protected program requires an additional dongle, which causes an unwieldy number of connected dongles. Dongles are also not an option for many software companies, since they add an additional manufacturing expense to each copy of the program. Dongles also do not facilitate Internet-based distribution of software since a dongle must be shipped to each customer to allow operation of the software. In summary, dongles "raise the bar" for the attacker but are not a particularly good copy-protection mechanism for most software applications.

There has been some recent work to make dongles practical for DRM. Dongles are traditionally used for software piracy protection; however, in Budd [2001] a modification to the dongle was introduced in the form of a smartcard with a lithium battery that wears out in order to provide a viable method for media protection. The key idea of this system was that there was a universal key that would decrypt any digital object, and that the profit from battery sales would be distributed among the owners of such protected objects. To distribute the profit fairly, the smartcards would keep track of usage information that would be collected through battery returning programs for which economic incentives are given. It is pointed out in Budd [2001] that a weakness in such a scheme is that if one smartcard is broken then all digital objects protected with that key become unprotected.

Programs that use a dongle query the port at startup and at programmed intervals thereafter and terminate if it does not respond with the dongle's programmed validation response. Thus, users can make as many copies of the program as they want but must pay for each dongle. The idea was clever, but it was initially a failure, as users disliked tying up a serial port this way. Almost all dongles on the market today will pass data through the port and monitor for certain codes with minimal if any interference with devices further down the line. This innovation was necessary to allow daisy-chained dongles for multiple pieces of software. The devices are still not widely used, as the software industry has moved away from copy-protection schemes in general.

21.3.2.1.2 Trusted Hardware Processors

In any DRM mechanism, there are certain places in the code where important choices are made, e.g., whether or not to allow a user access to a movie. If the software's integrity is not preserved at these crucial checkpoints then the DRM mechanism can be compromised. If checks are made in software, then one must prevent the software from being modified. To ensure that every single component running on a machine is legally owned and properly authorized, including the operating system, and that the infringement of any copyright (software, images, videos, text, etc.) does not occur, one must work from the lowest level possible in order to verify the integrity of everything on a particular machine. This could be done by a tamper resistant chip or some other form of trusted hardware. The tamper-resistant hardware would check and verify every piece of hardware and software that exists or that requests to be run on a computer, starting at the boot-up process, as described in Arbaugh et al. [1997]. This tamper-resistant, trusted hardware could guarantee integrity through a chain-like process by checking one entity at a time when the machine boots up, and every entity that wants to be run or used on that machine after the machine is already booted. Trusted hardware would restrict certain activities of the software through those particular hardware devices.

We now give an example of how such trusted hardware could operate. The trusted hardware stores all of the the keys necessary to verify signatures, decrypt licenses and software before running it, and to encrypt messages in online handshakes and online protocols that it may need to run with a software vendor, or with another trusted server on the network. Software downloaded onto a machine would be stored in encrypted form on the hard drive and would be decrypted and executed by the trusted hardware, which would also encrypt and decrypt information it sends and receives from the random-access memory. The same software or media would be encrypted in a different way for each trusted processor that would execute it because each processor would have a distinctive decryption key. This would put quite a dent in the piracy problem, as disseminating one's software or media files to others would not do them much good (because it would not be "matched" to the keys in their own hardware). It would also be the ultimate antivirus protection: A virus cannot attach itself to a file that is encrypted with an unknown key and, even if it could somehow manage that feat (say, because the file is not encrypted), it would nevertheless be rendered harmless by the fact that files could also have a signed hash of their content attached to them as a witness to their integrity. The trusted hardware would verify their integrity by verifying the signature and comparing the result to a hash of the file that it computes. (If they don't match, then the trusted hardware knows that the file is corrupted).

Note the inconvenience to the user if, for example, an electric power surge "fries" the processor and the user has to get it replaced by an approved hardware repair shop. Even though the disk drive is intact, the software and movies on it are now useless and must be downloaded again (because they are "matched" to the old processor but not to the new one). To prevent abuse from claims of fake accidents, this redownloading may necessitate for the hapless user to prove that the power-surge accident did occur (presumably with the help of the repair shop).

The trusted hardware could enforce the rule that only "approved" sound cards, and video cards, output devices, etc., are a part of the computer system. It could enforce such rules as "No bad content is played on this machine" — the type of each content can be ascertained via digital signatures and allowed to play only if it is "approved" (i.e., signed by an authority whose signature verification key is known to the trusted hardware). The potential implications for censorship are chilling.

One of the disadvantages of these types of systems is the time spent encrypting and decrypting; these disadvantages can be made acceptable through special hardware, but for low-end machines the boot-up time could be an annoyance.

Trusted hardware also limits the use and functionality of many hardware devices that the public currently uses and has certain expectations about the continued usage of these devices. By limiting and hindering certain activities, this could cause a negative reaction from consumers; it is unlikely that consumers will appreciate the fact that they will be limited to purchase only certain "approved" hardware, software, and media.

There are also serious privacy concerns relating to these hardware-based protection schemes because they could enable more stealthy ways for software publishers to spy on users, to know what is on a user's computer, to control what die the user can and cannot execute, view, connect to the computer, print, etc. Censorship becomes much easier to enforce, whistleblowing by insiders at their organization's misdeeds becomes harder, and the danger increases of unwise legislation that would mandate the use of DRM-motivated hardware that coerces the sale of severely restricted computers only. This would have disastrous consequences to IT innovation, and to the exportability of our IT technology; foreigners will balk at buying machines running only encrypted software that may do all kinds of unadvertised things besides the advertised word processing.

Another drawback of hardware protections is their inflexibility. They are more awkward to modify and update than software-based ones. Markets and products change rapidly, consumers respond unpredictably, business and revenue models evolve, and software-based protections' flexibility may in some cases offset their lower level of security. The largest software producer and the largest hardware producer are nevertheless both planning the introduction of such hardware-based technology, and at the same time the lobbyists from the movie industry and other content-providers have attempted to convince Congress to pass laws making them mandatory.

21.3.3 Further Remarks About Protection Mechanisms

To summarize the protection mechanisms, we give a succinct summary of the protection mechanism outlined in this section.

1. **Watermarking:** A watermark allows information to be placed into an object. This information can be used to provide a trusted player with authorization or authentication information.
2. **Fingerprinting:** This is a special case of watermarking, in which the watermark is different for each user so that pirated copies can be traced to the original culprit.
3. **Code obfuscation:** This makes reverse engineering more difficult and can protect the trusted player from modification.
4. **Software aging:** This is an obscure protection mechanism that deters piracy by making pirates update their software frequently. However, this works only for certain kinds of software.
5. **Dongles:** A dongle is a hardware device that provides some information (or some capability) needed to use a file.
6. **Hardware-based integrity:** One of the difficulties with creating a DRM system is protecting the trusted players against unauthorized modification. An "integrity chip" can verify that a player has not been modified before allowing it to run and access copyrighted material.

In summary, the functions DRM protection mechanisms perform include, mainly, to: (1) protect the Trusted Player (column PROTECT in the table below), (2) provide information to the Trusted Player (INFO), (3) make piracy easier to trace (TRACE), (4) make piracy more difficult by confining the digital object in antidissemination ways (CONFINE) in some other ways than (1) to (3). The table below summarizes the above techniques in these regards.

Technique	(PROTECT)	(INFO)	(TRACE)	(CONFINE)
Watermarking		√		
Fingerprinting			√	
Code Obfuscation	√			
Software Aging				√
Dongles		√		√
Integrity Chip	√			√

Each protection method needs to be evaluated according to its impact on the following software cost and performance characteristics:

- Portability
- Ease of use
- Implementation cost
- Maintenance cost
- Compatibility with installed base
- Impact on running time
- Impact on the size of the program

Applying a particular protection mechanism improves a security feature but can simultaneously degrade one or more of the above traditional (nonsecurity) characteristics. In summary, the application of DRM techniques is situational, and no single technique stands out as best in all situations. A hybrid of many mechanisms is often desirable.

21.4 Legal Issues

DRM is a problem where issues from such areas as law, ethics, economics, business, etc., play a role at least as important (and probably more) than issues that are purely technical. A crucial role will be played by the insurance industry: An insurance rate is a valuable "price signal" from the marketplace about the effectiveness (or lack thereof) of a deployed DRM technology, just like the home insurance rebate one gets for a home burglar alarm system is a valuable indicator of how much the risk of burglary is decreased by such a system. It is now possible to buy insurance against such Internet-related risks as the inadvertent use of someone else's intellectual property.

Any DRM technology is bound to eventually fail in the face of a determined attacker who deploys large enough resources and can afford to use a long enough amount of time to achieve his goal. It is when technology fails that the law is most valuable. Section 1201(a) of the 1998 Digital Millennium Copyright Act (DMCA) says: "No person shall circumvent a technological measure that effectively controls access to a work protected under this title." This can detract from security in the following way. In cryptography, it is through repeated attempts by researchers to break certain cryptosystems that we have acquired the confidence we have today in their security (many others fell by the wayside after being found to not be resistant enough to cryptanalysis). If reputable researchers are prevented from probing the weaknesses of various protection schemes and reporting their findings, then we may never acquire such confidence in them: They may be weak, and we would never know it (to the ultimate benefit of evildoers who will be facing weaker protections than otherwise).

Also forbidden is the design and dissemination of tools whose primary use is to defeat the DRM protections. This should not apply to tools whose primary intended use is legal, but that can be misused as attack tools against DRM protections. For example, software debuggers are a major attack tool for software crackers, and yet creating and distributing debuggers is legal. However, there are huge potential problems for someone who creates a tool whose main purpose is legal but specialized and not "mainstream," a tool that is later primarily used to launch powerful attacks on DRM protections. This could generate much DMCA-related headache for its creator (the argument that would be made by the entities whose revenue-stream it threatens is that its main use is so esoteric that its creator must surely have known that it would be used primarily as an attack tool). The freedom to "creatively play" that was behind so much of the information technology revolution can be threatened.

Legislation often has unintended consequences, and the DMCA, enacted to provide legal weapons against digital pirates, has been no exception. This is especially so for its antitamper provisions that forbid circumventing (or developing tools to circumvent) the antipiracy protections embedded into digital objects. These banned both acts of circumvention and the dissemination of techniques that make it possible. We review some of these next.

First, it has resulted in lawsuits that had a chilling effect on free speech and scientific research. This is a short-sighted and truly Pyrrhic victory that will ultimately result in less protection for digital objects: Cryptographic tools (and our trust in them) would today be much lower had something like the DMCA existed at their inception and "protected" them by preventing the scientific investigations that have succeeded in cracking and weeding out the weaker schemes. We may end up knowing even less than we do today as to whether the security technologies we use every day are vulnerable or not. No one will have dared to probe them for vulnerabilities or publicize these vulnerabilities for fear of being sued under the DMCA.

Second, by banning even legitimate acts of circumvention, the DMCA has inadvertently protected anticustomer and anticompetitive behavior on the part of some manufacturers and publishers. What is a legitimate act of tampering with what one has bought? Here is an example: Suppose Alice buys a device that runs on a widely available type of battery, but later observes that the battery discharges faster when she uses any brand of battery other than the "recommended" (and particularly overpriced) one. Alice is a tinkerer. She digs deeper and discovers to her amazement that the device she bought performs a special challenge-response protocol with the overpriced battery to become convinced that this is the type of battery being used. If the protocol fails to convince the device, then the device concludes that another kind of battery is being used and then it misbehaves deliberately (e.g., by discharging the battery faster). This kind of anticustomer and anticompetitive behavior is well documented. (It occurs in both consumer and industrial products). There are all-software equivalents of the above "Alice's story" that the reader can easily imagine (or personally experience). The point is that such anticustomer and anticompetitive practices do not deserve the protection of a law that bans both tampering with them and developing/disseminating techniques for doing so. Who will argue that it is not Alice's right to "tamper" with what she bought, to overcome what looks like an unreasonable constraint on her use of it?

For a more in-depth treatment of this issue, and many specific examples of the unintended consequences of the DMCA, we refer the reader to the excellent Website of the Electronic Frontiers Foundation (www.eff.org).

When they cannot prevent infringement, DRM technologies report it, and publishers then use the legal system. The targets of such lawsuits are often businesses rather than individuals. Businesses cannot claim that they are mere conduits for information, the way phone companies or ISPs do to be exempt from liability; after all, businesses already exercise editorial and usage control by monitoring and restricting what their employees can do (they try to ferret out hate speech, sexual harassment, viruses and other malware, inappropriate Web usage, and the leakage of corporate secrets, etc.). Businesses will increase workplace monitoring of their employees as a result of DRM for fear that they will be liable for infringements by their employees. In fact, a failure to use DRM by an organization could one day be viewed as failure to exercise due care and by itself expose the buyer to legal risks.

But what is infringement on intellectual property in cyberspace? What is trespassing in cyberspace? As usual, there are some cases that are clear-cut, but many cases are at the "boundary" and the legal system is still sorting them out. Throughout history, the legal system has considerably lagged behind major technological advances, sometimes by decades. This has happened with the automobile, and is now happening with the Internet. Many Internet legal questions (of liability, intellectual property, etc.) have no clear answer today. The issue of what constitutes "trespass" in cyberspace is still not completely clear. Even linking to someone's Website can be legally hazardous. While it is generally accepted that linking to a site is allowed without having to ask for permission, and in fact doing so is generally viewed as "flattering" and advantageous to that site, there have nevertheless been lawsuits for copyright infringement based on the fact that the defendant had linked deep into the plaintiff's site (bypassing the intended path to those targeted Web pages that included the viewing of commercial banners). Generally speaking, although explicit permission to link is not required, one should not link if the target site explicitly forbids such linking. This must be explicitly stated, but it might be done as part of the material that one must "click to accept" before entering the site, and thus can easily be overlooked.

Acknowledgment

Portions of this work were supported by Grants ETA-9903545 and ISS-0219560 from the National Science Foundation, Contract N00014-02-1-0364 from the Office of Naval Research, by sponsors of the Center for Education and Research in Information Assurance and Security, and by Purdue Discovery Park's E-Enterprise Center.

References

Anderson, Ross. *Security Engineering: A Guide to Building Dependable Distributed Systems.* John & Wiley Sons, New York, February 2001, pp. 413–452.

Anderson, Ross. TCPA/Palladium Frequently Asked Questions *www.cl.cam.ac.uk/users/rja14/tcpa-faq.html,* 2003.

Anderson, Ross. Security in Open versus Closed Systems — Dance of Blotzmann, Coase, and Moore *www.ftp.cl.cam.ac.uk/ftp/users/rja14/toulouse.pdf,* 2002.

Arbaugh, William, David Farber, Jonathan Smith. A Secure and Reliable Bootstrap Architecture Proceedings of the IEEE Symposium on Security and Privacy, Oakland, CA 1997.

Barak, Boaz, Oded Goldreich, Russell Impagliazzo, Steven Rudich, Amit Sahai, Salil Vadhan, Ke Yang. *On the (Im)possibility of Obfuscating Program.* Electronic Colloquium on Computational Complexity, Report No. 57, 2001.

Budd, Timothy. Protecting and Managing Electronic Content with a Digital Battery. *Computer,* vol. 34, no. 8, pp. 2–8, Aug 2001.

Caldwell, Andrew E., Hyun-Jin Choi, Andrew B. Kahng, Stefanus Mantik, Miodrag Potkonjak, Gang Qu, Jennifer L. Wong. Effective Iterative Techniques for Fingerprinting Design IP. *Design Automation Conference,* New Orleans, LA, 1999, pp. 843–848.

Camenisch, Jan. Efficient Anonymous Fingerprinting with Group Signatures. In *ASIACRYPT,* Kyoto, Japan, 2000, LNCS 1976, pp. 415–428.

Collberg, Christian, Clark Thomborson, Douglas Low. *A taxonomy of obfuscating transformations.* Tech. Report # 148, Department of Computer Science, University of Auckland, 1997.

Collberg, Christian and Clark Thomborson. Watermarking, Tamper-Proofing, and Obfuscation Tools for Software Protection. *IEEE Transactions on Software Engineering,* vol. 28 no. 8 pages 735–746, 2002.

Jakobsson, Markus and Michael Reiter. Discouraging Software Piracy Using Software Aging. Digital Rights Management Workshop, Philadelphia, PA, 2001 pp. 1–12.

Kahng, A., J. Lach, W. Mangione-Smith, S. Mantik, I. Markov. Watermarking Techniques far Intellectual Property Protection. *Design Automation Conference* 1998, San Francisco, CA, pp. 776–781.

Maude, Tim and Derwent Maude. Hardware Protection Against Software Piracy *Communication of ACM,* vol. 27, no. 9, pp. 950–959, September 1984.

Pfitzmann, Birgit and Ahmad-Reza Sadeghi. Coin-Based Anonymous Fingerprinting. In *EURO-CRYPT,* Prague, 1999, LNCS 1592, pp. 150–164.

Pfitzmann, Birgit and Matthias Schunter. Asymmetric Fingerprinting. In *Advances in Cryptology EURO-CRYPT* Saragossa, Spain, 1996, vol. 1070 of LNCS, pp. 84–95.

Qu, Gang and Miodrag Potkonjak. Fingerprinting Intellectual Property Using Constraint-Addition. *Design Automation Conference,* Los Angeles, CA, 2000, pp. 587–592.

Qu, Gang, Jennifer L. Wong, and Miodrag Potkonjak. Optimization-Intensive Watermarking Techniques for Decision Problems In *Proceedings 36th ACM/IEEE Design Automation Conference* 1999, pp. 33–36.

Schneier, Bruce. *Secrets and Lies: Digital Security in a Networked World.* John Wiley & Sons, New York, 2000.

Seadle, M., J. Deller, and A. Gurijala. Why Watermark? The Copyright Need for an Engineering Solution. Joint Conference on Digital Libraries, Portland, OR, July 2002.

Wang, Chenxi, Jonathan Hill, John Knight, Jack Davidson. *Software Tamper Resistance: Obstructing Static Analysis of Programs.* Ph.D Dissertation, University of Virginia, Charlottesville, VA.

Wroblewski, Gregory. *General Method of Program Code Obfuscation.* Ph.D Dissertation, Wroclaw University of Technology, Institute of Engineering Cybernetics, Wrodaw, Poland, 2002.

Part 3

Information Management

22

Internet-Based Enterprise Architectures

CONTENTS

22.1 Introduction 22-2
22.2 From Client–Server to *n*-Tier Architectures 22-3
22.2.1 Client–Server Architecture 22-3
22.2.2 Remote Procedure Calls (RPC) 22-4
22.2.3 Messaging Systems 22-5
22.2.4 *n*-Tier Architectures 22-5
22.3 New Keys to Interoperability: XML, SOAP, Web Services, Meta Data Registries, and OMGs MDA 22-6
22.3.1 XML 22-6
22.3.2 SOAP 22-7
22.3.3 Web Services 22-7
22.3.4 Meta Data Registries 22-9
22.3.5 OMG Model-Driven Architecture (MDA) 22-10
22.4 The J2EE Architecture 22-12
22.4.1 The J2EE Layered Approach 22-12
22.4.2 J2EE Container Model 22-13
22.4.3 Web Container 22-14
22.4.4 EJB Container 22-14
22.4.5 Java Message Service (JMS) 22-15
22.4.6 Java Naming and Directory Interface (JNDI) 22-15
22.4.7 Java Database Connectivity (JDBC) 22-16
22.4.8 J2EE Application Architectures 22-16
22.5 The .NET Architecture 22-16
22.5.1 Basic Principles 22-17
22.5.2 .NET Application Architectures 22-18
22.5.3 Microsoft Transaction Server (MTS) and Language Runtime 22-18
22.5.4 Microsoft Message Queue (MSMQ) 22-20
22.5.5 Active Directory 22-20
22.6 Comparison of J2EE and .NET 22-20
22.7 Conclusion: The Global Architecture 22-21
References 22-22

François B. Vernadat

This chapter describes and evaluates common industry approaches such as J2EE and .NET to build Internet-based architectures for interoperable information systems within business or administrative organizations of any size. It discusses the directions in which enterprise architectures are moving, especially with respect to supporting business process operations and offering information and services both to internal and external users (staff, clients, or partners).

1-58488-381-2/05/$0.00+$1.50

© 2005 by CRC Press LLC

22.1 Introduction

Today's organizations, be they profit or non-profit organizations, are facing major challenges in terms of flexibility and reactivity — also called *agility* — to manage day-to-day internal or interorganizational business processes. Indeed, they have to cope with changing conditions of their environment, stronger connectivity with their trading partners (e.g., networked enterprise, integrated supply chains), interoperability of their information systems, and innovation management in their business area. Among the business needs frequently mentioned, we can list:

- Wider distribution support to data, applications, services, and people
- More intense Web-based operations with a variety of partners and clients
- Intense transaction across heterogeneous platforms
- Integration of different information systems (legacy and new ones)
- High service availability (preferably, 7 d a week, 24 h/d)
- Scalability and growth capabilities of application systems
- High security and performance standards
- Ability to reuse existing building blocks (also called enterprise components)
- Ability to quickly adapt the current architecture to business evolution needs

There is, therefore, a need to build open, flexible, and modular enterprise architectures on top of a sound IT architecture [Vernadat, 2000]. However, the IT architecture should by no means constrain the enterprise architecture, as it happened too often in the past in the form of large monolithic enterprise information systems (ERP systems still suffer from this deviation, especially in small and medium sized enterprises). The enterprise architecture, which deals with people, customers, suppliers, partners, business processes, application systems, and inter-organizational relationships (e.g., CRM, SCM), now looks like a constellation of nodes in a networked organization (Figure 22.1). The IT architecture plays the role of the enabling infrastructure that supports operations of the enterprise architecture. It must be as much as possible transparent to the enterprise architecture.

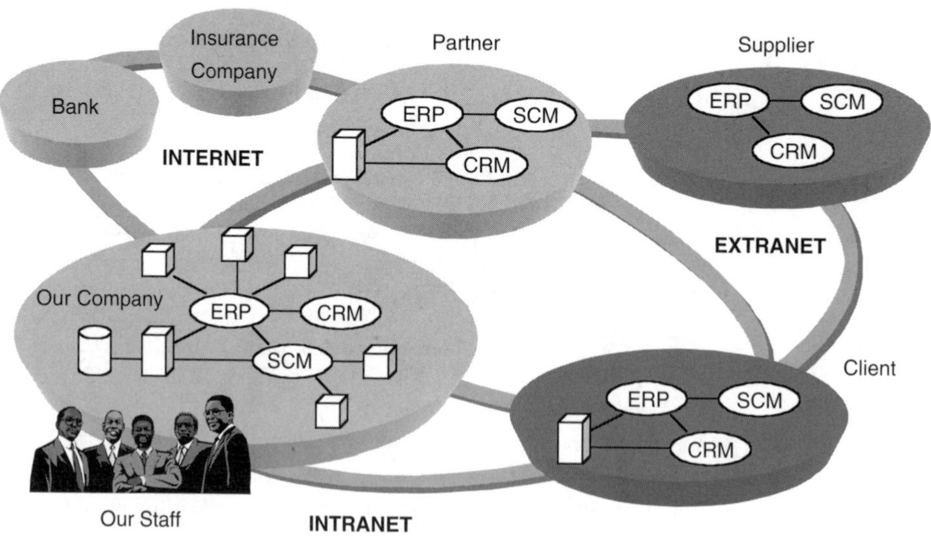

FIGURE 22.1 Principle of the networked organization.

Recent technological advances that contribute to make all this become a reality include:

- Greater computing power now widely available. (It is still doubling every 18 months and makes possible to have PC-based data centers for small and medium-sized organizations.)
- Increased connectivity by means of low cost, broad-reach Internet, wireless, or broadband access.
- Electronic device proliferation that can be easily connected to networks nearly anywhere (PC*s*, personal digital assistants or PDA*s*, cellular phones, etc.).
- Internet standards, and especially XML and ebXML, allowing XML-based integration.
- Enterprise Application Integration (EAI), i.e., methods and tools aimed at connecting and coordinating the computer applications of an enterprise. Typically, an enterprise has an existing base of installed legacy applications and databases, wants to continue to use them while adding new systems, and needs to have all systems communicating to one another.

22.2 From Client–Server to *n*-Tier Architectures

Interoperation over computer networks and Internet has become a must for agile enterprise architectures, especially to coordinate and synchronize business processes within a single enterprise or across networked enterprises [Sahai, 2003; Vernadat, 1996]. Interoperability is defined in *Webster's* dictionary as "the ability of a system to use the parts of another system."

Enterprise interoperability can be defined as the ability of an enterprise to use information or services provided by one or more other enterprises. This assumes on the part of the IT architecture the ability to send and receive messages made of requests or data streams (requests and responses). These kinds of exchange can be made in synchronous or asynchronous mode. Interoperability also assumes extensive use of standards in terms of data exchange in addition to computing paradigms. This section reviews some of the key paradigms for computer systems interoperability.

22.2.1 Client–Server Architecture

Wherever the information is stored, there comes a time at which it needs to be transferred within or outside of the enterprise. To support this, the client–server architecture, together with Remote Procedure Calls (RPC*s*), has been the systemwide exchange computing paradigm of the 1990s.

The basic idea of the client–server architecture is to have a powerful machine, called the *server*, providing centralized services, which can be accessed from remotely located, less powerful machines, called *clients*. This two-level architecture characterizes client–server systems, in which the client requests a service (i.e., function execution or access to data) and the server provides it.

The client–server architecture makes possible the connection of different computer environments on the basis of a simple synchronous message passing protocol. For instance, the server can be a Unix machine hosting a corporate relational database management system (RDBMS) while the clients could be PC*s* under Apple OS or MS-Windows running interface programs, spreadsheets, or local database applications, and used as front-ends by business users.

Figure 22.2 illustrates the basic principle of the client–server approach. A client issues a request (via a computer network) to a server. The server reads the request, processes it, and sends the result(s) back to the client. The server can provide one or more services to clients and is accessed via one of its interfaces, i.e., a set of callable functions. Callable functions (e.g., open, close, read, write, fetch, update, or execute) are defined by their signature (i.e., name and list of formal parameters). Their implementation is internal to the server and is transparent to the client.

On the server side, the server can itself become a client to one more other servers, which in turn can be clients to other servers, and so forth, making the architecture more complex. Clients and servers can be distributed over several machines connected by a computer network or may reside on the same node of the network.

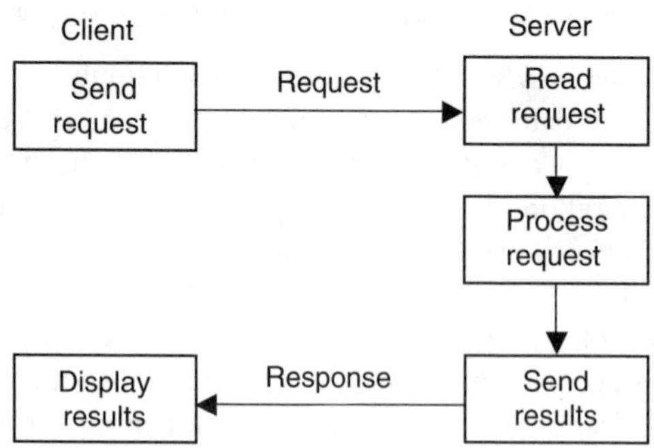

FIGURE 22.2 Principle of the client–server approach.

On the client side, the client can be a simple program (i.e., an interface) used to interact with the application hosted on the server. In this case, it implements no complex features and all heavy computing duties are performed by the server. It is then called a *thin client.* Conversely, due to the increasing power and disk capacity of PCs, the client can be a sophisticated program able to locally execute a significant part of the functionality of the server to reduce the burden of computer communication exchange and improve processing speed. It is then called a *thick client.*

Good examples of universal thin clients in the Internet computing world are Web browsers (e.g., Netscape, Internet Explorer, Opera). These are widely available tools on most operating systems. They can provide access to any Web site provided that one knows the site URL to be accessed. They can also give access to enterprise portals from anywhere in the world, thus making possible for employees on travel to remain connected with their enterprise.

22.2.2 Remote Procedure Calls (RPC)

RPC can be defined as an interprocess communication mechanism to start some processes on remotely located processors. In pure computer science terms, it is a way to start a process located in a separate address space from the calling code. It can, therefore, be used as one of the synchronous messaging protocols for implementing a client–server system, especially in non-Windows environments (Windows environments can, for instance, use Windows sockets).

The goal of the RPC mechanism is to ease the development of client–server applications in large-scale heterogeneous and distributed environments and to free system developers from the burden of dealing with the complexity of low level layers of networks. The RPC mode of operation is similar to a local procedure call but extended to a network environment. In other words, for the calling application, the procedure is executed as if it were executed locally.

The RPC distribution mechanism considers a *RPC client* and a *RPC server.* The RPC client is the part of an application that initiates a remote procedure call, while the RPC server is the part of the application that receives the remote call and takes in charge the procedure execution (Figure 22.3). Communication between the two applications (connection, transfer, and disconnection) is handled by the RPC mechanism installed as a software extension of both applications and interacting with the respective operating system hosting each application. It consists of a software layer made of a *RPC library* and a *client stub* on the client machine and a *server stub* on the server machine.

The client stub is an interface that prepares the parameters (or *arguments*) of the procedure called by the client application and puts them in a package that will be transmitted by the RPC routines (provided by the RPC Library). The client application can explicitly identify the server on which the procedure will be executed or let the RPC Library locate itself the corresponding server. Each application server is identified by its unique UUID name (Universal Unique IDentifier). This name is known to the client stub. A name server is available on the network. This name server maintains a table containing, among other things, the list of all servers running server applications (identified by their UUID). On the server side, the RPC Library

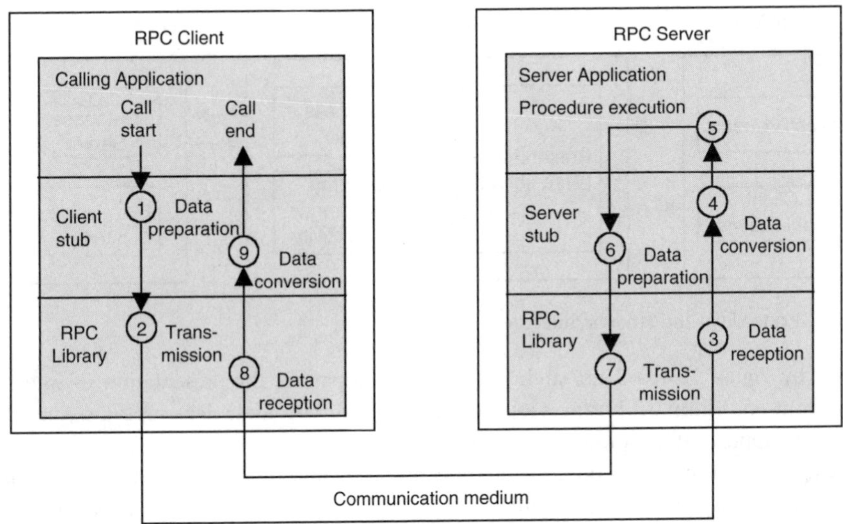

FIGURE 22.3 Principle of the Remote Procedure Call (RPC).

receives the package sent by the client and transmits it to the server stub (RPC interface between the RPC Library and the server application) for the application concerned by the call (there can be several server stubs on the same server). The server stub decodes the parameters received and passes them to the called procedure. Results are prepared as a new package and sent back by the server application to the calling application in a more symmetric way than the call was issued by the client application (Figure 22.3).

22.2.3 Messaging Systems

Messaging systems are middleware components used to handle the exchange of *messages*, i.e., specially-formatted data structures describing events, requests, and replies, among applications that need to communicate. They are usually associated with a message queuing system to serialize, prioritize, and temporarily store messages.

IBM MQSeries, Sun Microsystems Java Messaging System (JMS), and Microsoft MSMQ are common examples of messaging systems, also called *message-oriented middleware* (MOM).

There are two major messaging models: the *point-to-point model* (in which the addresses of the issuer and of the receiver need to be known) and the *publish/subscribe model* (in which the message is broadcasted by the issuer and intercepted by relevant recipients that may notify receipt of the message).

Messaging makes it easier for programs to communicate across different programming environments (languages, compilers, and operating systems) since the only thing that each environment needs to understand is the common messaging format and protocol.

22.2.4 *n*-Tier Architectures

Client–server architectures assume tight and direct peer-to-peer communication and exchange. They have paved the way to more general and flexible architectures able to serve concurrently many clients, namely 3-tier architectures and, by extension, *n*-tier architectures.

3-tier architectures are architectures organized in three levels as illustrated by Figure 22.4:

- The level of clients or service requestors (thin or thick clients)
- The level of the application server (also called middle-tier or business level) supposed to provide services required by clients
- The level of secondary servers, usually database or file servers but also specialized servers and mainframes, which provide back-end services to the middle tier

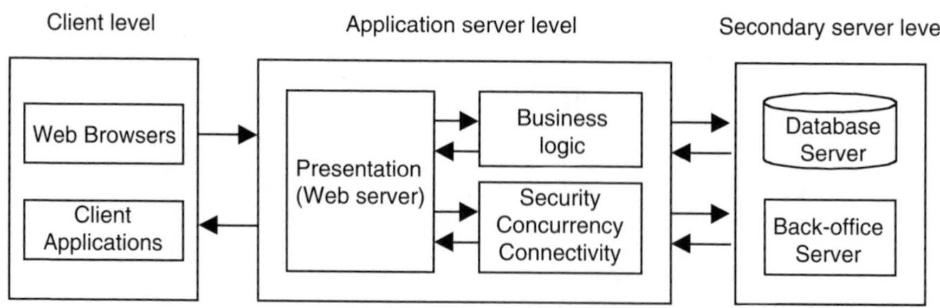

FIGURE 22.4 Principle of the 3-tier architecture.

As shown by Figure 22.4, a 3-tier architecture clearly separates the presentation of information to requesters (front-end) from the business logic of the application (middle tier), itself dissociated from the data servers or support information systems (back-end).

Software applications based on the 3-tier architecture principle are becoming very popular nowadays. Since the year 2000, it is estimated that more than half of new application developments use this approach. It is especially well-suited for the development of enterprise portals, data warehouses, business intelligence applications, or any other type of applications required to serve a large number of clients (users or systems), access, reformat, or even aggregate data issued from several sources; to support concurrent transactions; and to meet security and performance criteria.

The idea of making applications talk to one another is not new. Currently, distributed architectures such as OSF/DCE,[1] OMG/CORBA,[2*] or Microsoft DCOM,[3**] already support exchange among applications using the RPC mechanism. However, each of these architectures is using a proprietary exchange protocol, and therefore they are not easily interoperable. For instance, it is difficult to make a CORBA object communicate with a COM object. Furthermore, protocols used to transport these objects are often blocked by enterprise firewalls.

To go one step further towards enterprise interoperability, *n*-tier architectures based on open and standardized protocols are needed. This is the topic of the next section.

22.3 New Keys to Interoperability: XML, SOAP, Web Services, Meta Data Registries, and OMGs MDA

22.3.1 XML

XML (eXtended Mark-up Language) is a language derived from SGML (Standard Generalized Markup Language–ISO 8879) [Wilde, 2003; W3C, 2000a]. It is a standard of the World Wide Web Consortium (W3C) intended to be a universal format for structured content and data on the Web. Like HTML (Hypertext Markup Language), XML is a tagged language, i.e., a language which bounds information by tags. HTML has a finite set of predefined tags oriented to information presentation (e.g., title, paragraph, image, hypertext link, etc.). Conversely, XML has no predefined tags. The XML syntax uses matching start and end tags, such as <name> and </name>, to mark up information. It is a meta-language that allows the creation of an unlimited set of new tags to characterize all pieces or aggregates of pieces of basic information that a Web page can be made of. No XML tag conveys presentation information. This is the role of XSL and XSLT.

XSL (eXtensible Stylesheet Language) is another language recommended by the World Wide Web Consortium (W3C) to specify the representation of information contained in an XML document [W3C,

[1] OSF/DCE: Open Software Foundation/Distributed Computing Environment

[2*] OMG/CORBA: Object Management Group/Common Object Request Broker Architecture

[3**] DCOM: Distributed Component Object Model

2000b]. XSL is used to produce XSL pages that define the style sheets for the formatting of XML documents. XSL can be divided into two components:

- XSLT (eXtensible Stylesheet Transformation): allows transforming of the hierarchical structure of an XML document into a different structure by means of template rules
- The data layout language: allows defining the page layout of textual or graphical elements contained in the information flow issued from the XSLT transformation

Thanks to the clear separation between content and layout, XML provides a neutral support for data handling and exchange. It is totally independent of the physical platform that transports or publishes the information because XSL style sheets are used to manage the layout of the information in function of the support or destination, the content of the basic XML file containing data remaining the same. It is also totally independent of the software that will process it (nonproprietary format). The only dependence on a specific domain appears in the way the information structured in an XML file must be interpreted with respect to this domain. This can be specified by means of Data Type Documents (DTDs), which define the structure of the domain objects.

This so-called data neutralization concept is essential to achieve systems interoperability.

22.3.2 SOAP

SOAP stands for Simple Object Access Protocol. It is a communication protocol and a message layout specification that defines a uniform way of passing XML-encoded data between two interacting software entities. SOAP has been submitted to the World Wide Web Consortium to become a standard in the field of Internet computing [W3C, 2002a].

The original focus of SOAP is to provide:

- A framework for XML-based messaging systems using RPC principles for data exchange
- An envelope to encapsulate XML data for transfer in an interoperable manner that provides for distributed extensibility and evolvability, as well as intermediaries such as proxies, caches, and gateways
- An operating system neutral convention for the content of the envelope when messages are exchanged by RPC mechanisms (sometimes called "RPC over the Web")
- A mechanism to serialize data based on XML schema data types
- A non-exclusive mechanism layered on HTTP transport layer (to be able to interconnect applications — via RPC, for instance — across firewalls)

Although the SOAP standard does not yet specify any security mechanisms (security handling is delegated to the transport layer), SOAP is likely to become the standard infrastructure for distributed applications. First, because, as its name suggests, it keeps things simple. Second, because it nicely complements the data neutralization principle of XML (i.e., transfers ASCII data). Furthermore, it is independent of the use of any particular platform, language, library, or object technology (it builds on top of these). Finally, it does not dictate a transport mechanism. (SOAP specification defines how to send SOAP messages over HTTP but other transport protocols can be used such as SMTP, or raw TCP.)

22.3.3 Web Services

The aim of Web Services is to connect computers and devices with each other using the Internet to exchange data and to process data dynamically, i.e., on-the-fly. It is a relatively new concept, still rapidly evolving, but which has the power to drastically change the way we build computer-based applications and which can make computing over the Web a reality soon. No doubt that it will become a fundamental building block in implementing agile distributed architectures [W3C, 2002b].

Web Services can be defined as interoperable software objects that can be assembled over the Internet using standard protocols and exchange formats to perform functions or execute business processes [IST

Diffuse Project, 2002; Khalaf et al., 2003]. A concise definition is proposed by Desti*Corp* [IST Diffuse Project, 2002]. It states that "Web Services are encapsulated, loosely coupled, contracted software functions, offered via standard protocols."

In a sense, Web Services can be assimilated to autonomous software agents hosted on some servers connected to the Web. These services can be invoked by means of their URL by any calling entity via Internet and using XML and SOAP to send requests and receive responses. Their internal behavior can be implemented in whatever language (e.g., C, Java, PL/SQL) or even software system (e.g., SAS, Business Objects, Ilog Rules). Their granularity can be of any size. The great idea behind Web Services is that a functionality can be made available on the Web (or published) and accessed (or subscribed) by whoever needs it without having to know neither its location nor its implementation details, i.e., in a very transparent way.

Thanks to Web Services, direct exchanges are made possible among applications in the form of XML flows only. This makes possible on-the-fly software execution as well as business process execution through the use of loosely coupled, reusable software components. In other words, the Internet itself becomes a programming platform to support active links, or transactions, between business entities. Because businesses will no longer be tied to particular applications and to underlying technical infrastructures, they will become more agile in operation and adaptive to changes of their environment. Partnerships and alliances can be set up, tested, and dissolved much more rapidly. Figure 22.5 shows the evolution and capabilities of programming technology culminating in Web Services and clarifies this point.

Web Services rely on the following components, which are all standards used in industry:

- WSDL (Web Service Description Language): a contract language used to declare Web Service interfaces and access methods using specific description templates [W3C, 2001].
- UDDI (Universal Description, Discovery, and Integration): an XML-based registry or catalog of businesses and the Web Services they provide (described in WSDL). It is provided as a central facility on the Web to publish, find, and subscribe Web Services [OASIS, 2002].
- XML: the language used to structure and formulate messages (requests or responses).
- SOAP: used as the messaging protocol for sending requests and responses among Web Services.
- SMTP/HTTP(S) and TCP/IP: used as ubiquitous transport and Internet protocols.

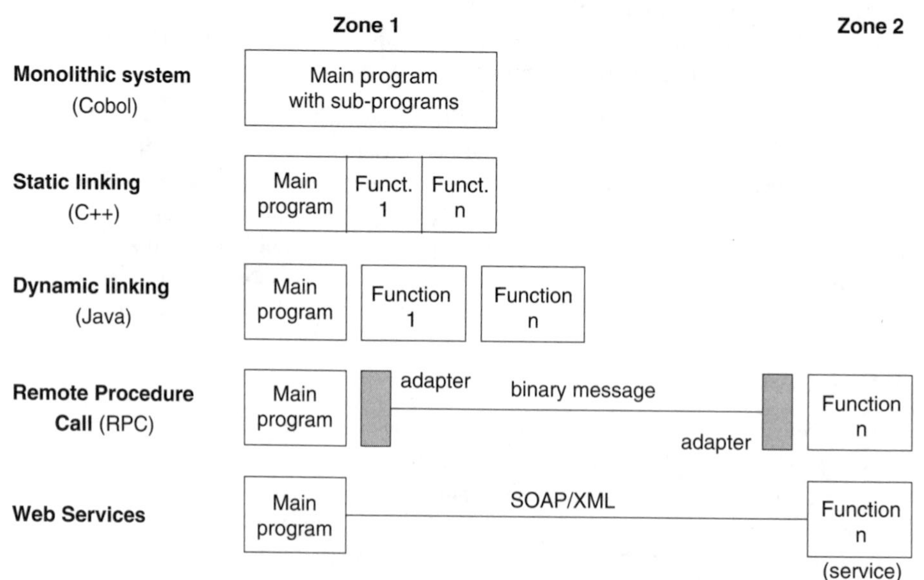

FIGURE 22.5 Evolution of programming technology.

Because all these are open and noninvasive standards and protocols, messages can go through enterprise firewalls. In business terms, this means that services can be delivered and paid for as fluid streams of messages as opposed to the economic models of packaged software products, which in most cases require license, operation, and maintenance costs. Using Web Services, a business can transform assets into licensable, traceable, and billable products running on its own or third-party computers. The subscribers can be other applications or end-users. Moreover, Web Services are technology-independent and rely on a loosely-coupled Web programming model. They can therefore be accessed by a variety of connecting devices able to interpret XML or HTML messages (PCs, PDAs, portable phones). Figure 22.6 gives an overview of what the next generation of applications is going to look like, i.e., "Plug and Play" Web Services over Internet.

22.3.4 Meta Data Registries

Meta Data Registries (MDRs) are database systems used to collect and manage descriptive and contextual information (i.e., metadata) about an organization's information assets [Lavender, 2003; Metadata Registries, 2003]. In layman's terms, these are databases for metadata (i.e., data about data) used to make sure that interacting systems talk about the same thing. For instance, examples of metadata in national statistical institutes concern the definition of statistical objects, nomenclatures, code-lists, methodological notes, etc., so that statistical information can be interpreted the same way by all institutes. Metadata Registries constitute another type of fundamental building blocks of enterprise architectures to achieve enterprise interoperability.

For instance, UDDI, as used for Web Services registry, is a kind of MDR. An LDAP database (i.e., a database for Lightweight Directory Access Protocol used to locate organizations, individuals, and other resources such as files and devices in a network, whether on the public Internet or on a corporate intranet) is another kind. A thesaurus of specialized terms and a database of code lists in a specific application domain are two other examples.

The main roles of Metadata Registries are:

- To enable publishing and discovery of information (e.g., user lists, code lists, XML files, DTDs, UML models, etc.) or services (e.g., Web Services)
- To allow organizations to locate relevant business process information
- To provide content management and cataloging services (Repository)
- To provide services for sharing information among systems

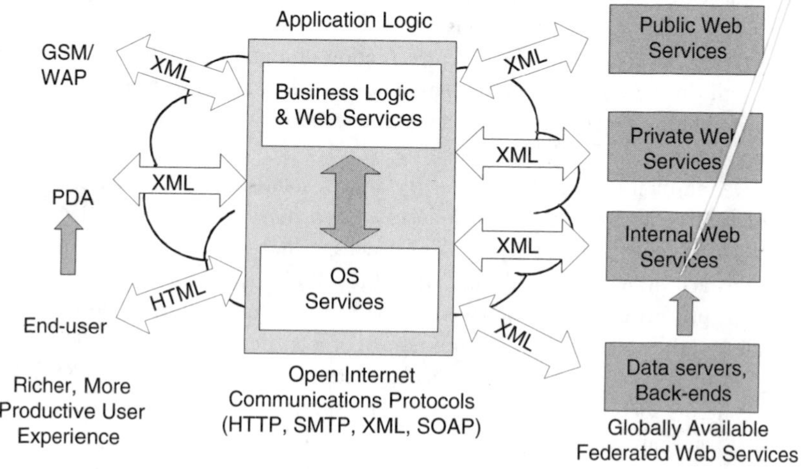

FIGURE 22.6 Web Services and business applications.

Depending on their scope, three types of registries can be defined:

- Private scope registries: Their access is strictly limited to a few applications or groups of users (for instance, business application integration catalogs).
- Semi-private scope registries: industry or corporation specific (for instance, an enterprise portal catalog of users).
- Public scope registries: unrestricted access.

Because of their ability to describe context, structure, and life cycle stages of operational data as well as of system components and their states (e.g., modules, services, applications), MDRs can play an essential role as semantic mediators at the unification level of systems interoperability. This role is enforced by a number of standards recently published concerning the creation and structure of registries (e.g., ISO/IEC 11179 and ISO/IEC 20944, OASIS/ebXML Registry Standard) or their contents (e.g., ISO 704, ISO 1087, ISO 16642, Dublin Core Registries) [Metadata Registries, 2003].

22.3.5 OMG Model-Driven Architecture (MDA)

The Object Management Group (OMG) has contributed a number of IT interoperability standards and specifications, including the Common Object Request Broker Architecture (CORBA), the Unified Modeling Language (UML), the Meta Object Facility (MOF), XML Metadata Interchange (XMI), and the Common Warehouse Metamodel (CWM). Details on each of these standards can be found on the Web at www.omg.org. However, due to recent IT trends, OMG has realized the need for two modifiers: First, there are limits to the interoperability level that can be achieved by creating a single set of standard programming interfaces. Second, the increasing need to incorporate Web-based front ends and to make links to business partners, who may be using proprietary interface sets, can force integrators back to low-productivity activities of writing glue code to hold multiple components together.

OMG has, therefore, produced a larger vision necessary to support increased interoperability with specifications that address integration through the entire system life cycle: from business modeling to system design, to component construction, to assembly, integration, deployment, management, and evolution. The vision is known as the OMG's Model Driven Architecture (MDA), which is described in [OMG, 2001, 2003] and from which this section is adapted.

MDA first of all defines the relationships among OMG standards and how they can be used today in a coordinated fashion. Broadly speaking, MDA defines an approach to IT system specification that separates the specification of system functionality from the specification of the implementation of the functionality on a specific platform. To this end, the MDA defines an architecture for models that provides a set of guidelines for structuring specifications expressed as models.

As defined in [OMG, 2001], "a model is a representation of a part of the function, structure, and/or behavior of a system (in the system-theoretic sense and not restricted to software systems). A platform is a software infrastructure implemented with a specific technology (e.g., Unix platform, CORBA platform, Windows platform)."

Within the MDA, an interface standard — for instance, a standard for the interface to an Enterprise Resource Planning (ERP) system — would include a *Platform-Independent Model* (PIM) and at least one *Platform-Specific Model* (PSM). The PIM provides formal specifications of the structure and function of the system that abstracts away technical details, while the PSM is expressed in terms of the specification model of the target platform. In other words, the PIM captures the conceptual design of the interface standard, regardless of the special features or limitations of a particular technology. The PSM deals with the "realization" of the PIM in terms of a particular technology — for instance, CORBA — wherein target platform concepts such as exception mechanisms, specific implementation languages, parameter formats and values, etc. have to be considered.

Figure 22.7 illustrates how OMG standards fit together in MDA. It provides an overall framework within which the roles of various OMG and other standards can be positioned.

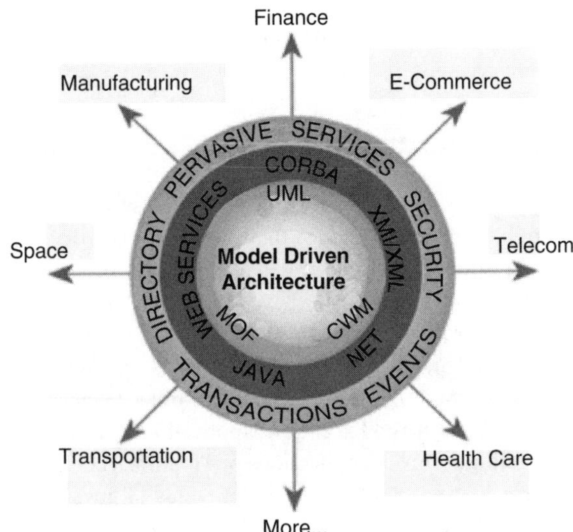

FIGURE 22.7 OMG's Model Driven Architecture.

The core of the architecture (inner layer) is based on OMG's modeling standards: UML, the MOF, and CWM. It comprises a number of UML profiles for different computing aspects. For example, one will deal with Enterprise Computing with its component structure and transactional interaction (business processes); another will deal with Real-Time Computing with its special needs for resource control. The total number of profiles in MDA is planned to remain small. Each UML profile is intended to represent common features to be found on all middleware platforms for its category of computing. Its UML definition must be independent of any specific platform.

The first step when developing an MDA-based application is to create a PIM of the application expressed in UML using the appropriate UML profile. Then, the application model (i.e., the PIM) must be transformed into a UML, platform-specific, model (i.e., a PSM) for the targeted implementation platform (e.g., CORBA, J2EE, COM+ or .NET). The PSM must faithfully represent both the business and technical run-time semantics in compliance with the run-time technology to be used. This is expressed by the second inner layer in Figure 22.7.

Finally, all applications, independent of their context (Manufacturing, Finance, e-Commerce, Telecom), rely on some essential services. The list varies somewhat depending on the application domain and IT context but typically includes Directory Services, Event Handling, Persistence, Transactions, and Security. This is depicted by the outer layer of Figure 22.7 that makes the link with application domains.

The MDA core, based on OMG technologies, is used to describe PIMs and PSMs. Both PIMs and PSMs can be refined n-times in the course of their development until the desired system description level and models consistency are obtained. These core technologies include:

- UML (Unified Modeling Language): It is used to model the architecture, objects, interactions between objects, data modeling aspects of the application life cycle, as well as the design aspects of component-based development including construction and assembly.
- XMI (XML Metadata Interchange): It is used as the standard interchange mechanism between various tools, repositories, and middleware. It plays a central role in MDA because it marries the world of modeling (UML), metadata (MOF and XML), and middleware (UML profile for Java, EJB, CORBA, .NET, etc.).
- MOF (Meta Object Facility): Its role is to provide the standard modeling and interchange constructs that are used in MDA. These constructs can be used in UML and CWM models. This common foundation provides the basis for model/metadata interchange and interoperability, and is the mechanism through which models are analyzed in XMI.

- CWM (Common Warehouse Metamodel): This is the OMG data warehouse standard. It covers the full life cycle of designing, building, and managing data warehouse applications and supports management of the life cycle.

22.4 The J2EE Architecture

The Java language was introduced by Sun Microsystems in the 1990s to cope with the growth of Internet-centric activities and was soon predicted to change the face of software development as well as the way industry builds software applications. This indeed has really happened, especially for Internet applications, as many key players in the software industry have joined the bandwagon, and there is now a large and growing Java community of programmers around the world with its specific Java jargon.

Although robust, performing, open, portable, based on object-oriented libraries, and dynamic linking, Java by itself cannot fulfill all integration, interoperation, and other IT needs of industry.

The J2EE (Java To Enterprise Edition and Execution) application server framework has been proposed as a *de facto* standard for building complex applications or IT infrastructures using Java.

A J2EE-compliant platform is supposed to preserve previous IT investments, be a highly portable component framework, deal with heterogeneity of applications, offer scalability and distributed transactions, and, as much as possible, ensure security and a high degree of availability.

This standard has been accepted by industry and most software and computer vendors are providing J2EE compliant application server environments such as WebLogic by BEA Systems, WebSphere by IBM, Internet Operating Environment (IOE) by HP, or iAS by Oracle, to name only four.

22.4.1 The J2EE Layered Approach

The J2EE framework [SUN Microsystems, 2001] adopts a three-layer approach to build *n*-tier applications as illustrated by Figure 22.8. It assumes execution of these applications on a Java Virtual Machine (JVM) to be in principle independent of any specific operating system intricacies. It is made of three layers:

- *Presentation Layer.* Deals with connection with the outside environment and the way information is presented to users and client applications.

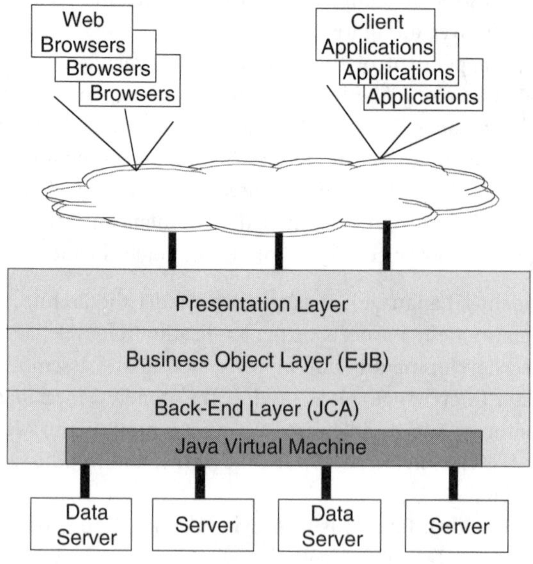

FIGURE 22.8 The J2EE layered approach.

- *Business Object Layer.* Performs the business processing (or application server part) offered by the Java application to be developed. The application logic is programmed by means of Enterprise Java Beans or EJBs. An EJB is a (reusable) platform independent software component written in Java with self-descriptive capabilities and invoked via its methods.
- *Back-End Layer* (or JCA for J2EE Connector Architecture). Deals with connection with back-end systems such as database servers, ERP systems, remote systems, or other J2EE or non-J2EE platforms.

This three-layered approach is implemented according to the architecture specified by the J2EE Container Model. This model defines a number of architectural parts, called Containers, and some satellite components (JMS, JNDI, JDBC).

22.4.2 J2EE Container Model

The J2EE Container Model provides a global view of the elements that need to be developed or to be used for developing the three layers of a J2EE platform (Figure 22.9). A container is an interface between a component and a specific platform functionality. Container settings customize the underlying support provided by the J2EE Server, including services such as transaction management, security, naming and directory services, or remote connectivity. The role of the container is also to manage other services such as component life cycle, database connection, resource pooling, and so forth. The model defines three types of containers: *Application Client Container, Web Container,* and *EJB Container.*

On the client side (Client Machine), access to the J2EE Server can be made by means of a Web browser via HTTP in the case of a Web-based application, for instance to present HTML pages to the client, or by an application program encapsulated in an Application Client Container. In the latter case, the application can access HTML pages (i.e., talk to the Web Container), or request functions written in Java (i.e., invoke the EJB Container), or both.

On the server side, the J2EE Server is the middleware that supports communication with a Web client by means of server-side objects called Web components (Servlets and Java Server Pages). Web components are activated in the Web Container environment. In addition to Web page creation and assemblage, the role of the Web Container is to provide access to built-in services of the J2EE platform such as request dispatching, security authorization, concurrency and transaction management, remote connectivity, or life cycle management. These services are provided by components of the EJB Container. These compo-

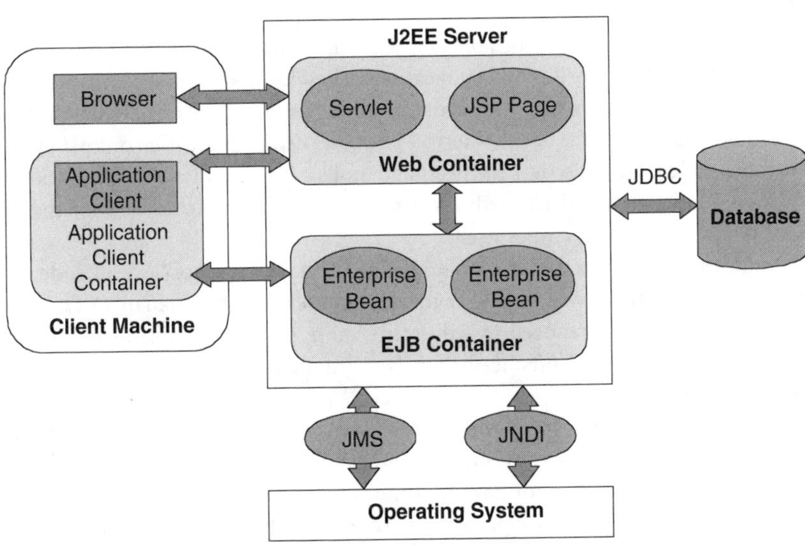

FIGURE 22.9 The J2EE Container Model and satellite components.

nents are called Enterprise Java Beans (EJBs). Because an EJB is a server-side Java component that encapsulates some piece of business logic, it can be accessed by a client application through one of its methods defined in the bean's interface.

22.4.3 Web Container

As mentioned earlier, the Web Container is an environment of the J2EE Server that supports Web-based communication (usually via the HTTP protocol) with a client (browser or Web application). There are two types of Web components in a Web Container: Java servlets and Java Server Pages (JSPs).

A *servlet* is a Java programming language class used to extend the capabilities of servers that host applications accessed via a request-response programming model. Although servlets can respond to any type of request, they are commonly used to extend the applications hosted by Web servers. For such applications, the Java Servlet technology defines HTTP-specific servlet classes.

Java Server Pages (JSPs) are easier to use than servlets. JSP*s* are simple HTML files containing embedded code in the form of scripts (HTML files with extension.jsp and containing clauses of the form <script> script code </script>). They allow the creation of Web contents that have both static and dynamic components. JSP*s* can provide all the dynamic capabilities of servlets but they provide a more natural approach for the creation of static content. The main features of JSP*s* are:

- A language for developing JSP pages, which are text-based documents that describe how to process a request and to construct a response.
- Constructs for accessing server-side objects.
- Mechanisms to define extensions to the JSP language in the form of custom tags. Custom tags are usually distributed in the form of tag libraries, each one defining a set of related custom tags and containing the objects that implement the tags.

In developing Web applications, servlets and JSP pages can be used interchangeably. Each one has its strengths:

- Servlets are best suited for the management of the control functions of an application, such as dispatching requests, and handling non-textual data.
- JSP pages are more appropriate for generating text-based markup languages.

22.4.4 EJB Container

Enterprise Java Beans (EJBs) are both portable and reusable components for creating scalable, transactional, multiuser, and secure enterprise-level applications. These components are Java classes dealing with the business logic of an application and its execution on the server side. They are invoked by clients by means of the methods declared in the bean's interface (*Remote Method Invocation* or RMI principle similar to the RPC method). Clients can discover the bean's methods by means of a self-description mechanism called *introspection*. These are standard methods that allow an automatic analysis of the bean, which, thanks to its syntactic structure, describes itself.

Because they are written in Java, programming with EJBs makes the application code very modular and platform-independent and, hence, simplifies the development and maintenance of distributed applications. For instance, EJBs can be moved to a different, more scalable platform should the need arise. This allows "plug and work" with off-the-shelf EJBs without having to develop them or have any knowledge of their inner workings.

The role of the EJB Container is to provide system-level services such as transaction management, life-cycle management, security authorization, and connectivity support to carry out application execution. Thus, the client developer does not have to write "plumbing" code, i.e., routines that implement either transactional behavior, access to databases, or access to other back-end resources because these are provided as built-in components of the EJB Container. The EJB Container can contain three types of EJB*s*: Session Beans, Entity Beans, and Message-Driven Beans.

Session Beans: A session bean is an EJB that acts as the counterpart of a single client inside the J2EE Server. It performs work for its client, shielding the client from server-side complexity. To access applications deployed on the server, the client invokes the session bean's methods. There are two types of session beans: stateful and stateless session beans.

- In a *stateful session bean*, the instance variables represent the state of a unique client-bean session for the whole duration of that client-bean session.
- A *stateless session bean* does not maintain any conversational state for a client other than for the duration of a method invocation. Except during method invocation, all instances of a stateless bean are equivalent, thereby allowing better scalability.

Entity Beans: An entity bean represents a business object in a persistent storage mechanism, e.g., RDBMS. Typically, entity beans correspond to an underlying table and each instance of the bean corresponds to a row in that table. Entity beans differ from session beans in several ways. Entity beans are persistent, allow shared access, have primary keys, and may participate in relationships with other beans. There are two types of persistence for entity beans: *bean-managed (BMP)* and *container-managed (CMP)*. Being a part of an entity bean's deployment descriptor (DD), the abstract schema defines the bean's persistent fields and relationships and is referenced by queries written in the *Enterprise Java Beans Query Language (EJB QL)*.

Message-Driven Beans (MDBs): An MDB is an EJB that allows J2EE applications to process messages asynchronously by acting as a Java Message Service (JMS) listener (see next section). Messages may be sent by any J2EE component or even a system that does not use J2EE technology. An MDB has only a bean class., i.e., the instance variables can contain state across the handling of client messages.

22.4.5 Java Message Service (JMS)

The Java Message Service (JMS) is an application program interface (JMS API) ensuring a reliable, flexible service for the asynchronous exchange of critical business data and events throughout the enterprise. It provides queuing system facilities that allow a J2EE application to create, send, receive, and read messages to/from another application (J2EE or not) using both a point-to-point and publish/subscribe policy (Figure 22.10). JMS, which has been part of J2EE platforms since version 1.3, has the following features:

- Application clients, EJBs, and Web components can send or synchronously receive a JMS message
- Application clients can in addition receive JMS messages asynchronously
- MDBs enable the asynchronous consumption of messages. A JMS provider may optionally implement concurrent processing of messages by MDBs
- Messages sent and received can participate in distributed transactions

22.4.6 Java Naming and Directory Interface (JNDI)

The Java Naming and Directory Interface (JNDI) is a J2EE resource location service accessible via its application programming interface (JNDI API). It provides naming and directory functionality to appli-

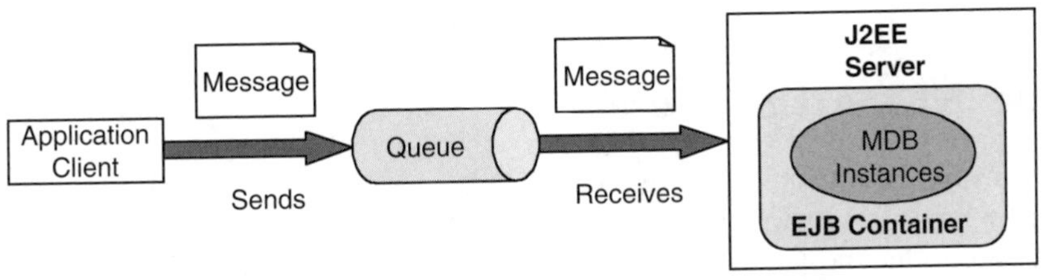

FIGURE 22.10 Java Message Service (JMS) principles.

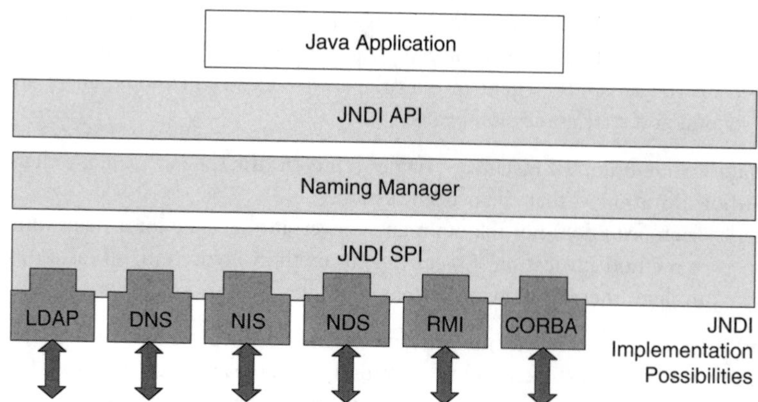

FIGURE 22.11 The JNDI layered approach.

cations. It has been designed to be independent of any specific directory service implementation (e.g., LDAP, DNS, UDDI). It uses a layered approach (Figure 22.11) in which the JNDI SPI is a specific presentation interface for the selected directory implementation possibility (e.g., LDAP, DNS, CORBA).

A JNDI name is a user-friendly name for an object. Each name is bound to its corresponding object by the naming directory service provided by the J2EE Server. This way, a J2EE component can locate objects by invoking the JNDI lookup method. However, the JNDI name of a resource and the name of the resource referenced are not the same. This allows clean separation between code and implementation.

22.4.7 Java Database Connectivity (JDBC)

The Java Database Connectivity (JDBC) is an other application programming interface (API) provided by the J2EE Server to provide standard access to any relational database management system (RDBMS). It has been designed to be independent of any specific database system. This application program interface is used to encode access request statements in standard Structured Query Language (SQL) that are then passed to the program that manages the database. It returns the results through a similar interface.

22.4.8 J2EE Application Architectures

Figure 22.12 presents what could be a typical J2EE application architecture. Many of the components mentioned in Figure 22.12 have been presented earlier. The Web browser talks to the servlets/JSP components exchanging XML or HTML-formatted messages via HTTP, possibly using HTTPS, a secured connection with SSL (Secured Socket Layer). Other applications (either internal or external to the enterprise) can access the application server via a RMI/IIOP connection, i.e., by Remote Method Invocations using an Internet Inter ORB (object request broker) Protocol. Legacy systems can be connected to the application server via a J2EE Connector Architecture (JCA) in which Java Beans will handle the communication with each legacy system. (This is made much easier when the legacy system has a properly defined API.) The role of the JCA is to ensure connection management, transaction management, and security services.

22.5 The .NET Architecture

The Microsoft's Dot Net Architecture (DNA or .NET) is the counterpart in the Windows world of the J2EE technology in the Java world [Microsoft, 2001]. The aim is to set a software technology for connecting information, people, systems, and various types of devices using various communication media (e.g., e-mail, faxes, telephones, PDAs) .NET will have the ability to make the entire range of

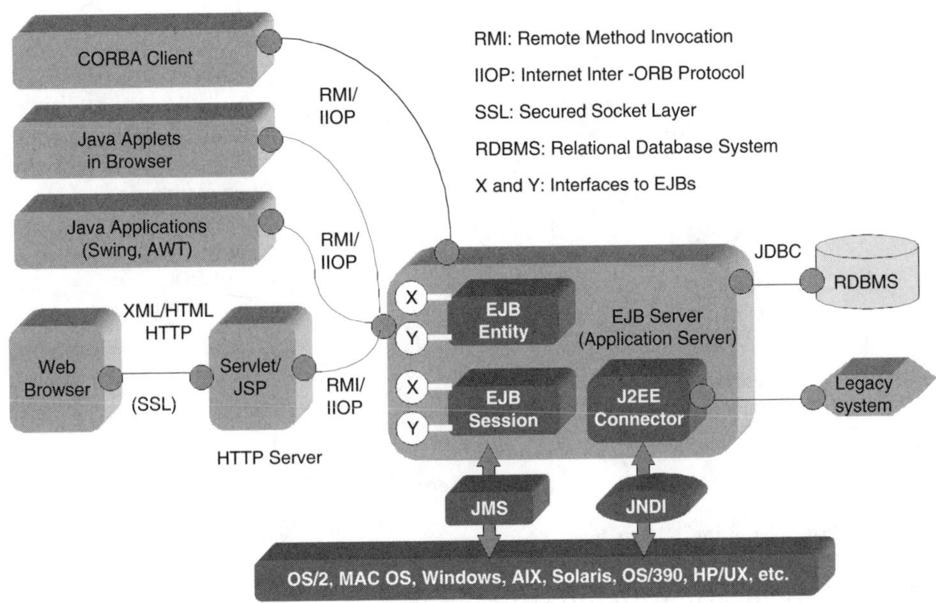

FIGURE 22.12 Typical J2EE *n*-tier application architecture.

computing devices work together and to have user information automatically updated and synchronized on all of them. It will provide centralized data storage (for wallet, agenda, contacts, documents, services, etc.), which will increase efficiency and ease of access to information, as well as synchronization among users and services (see Figure 22.6).

22.5.1 Basic Principles

According to Microsoft, .NET is either a vision, i.e., a new platform for the digital era, and a set of technologies and products to support Windows users in implementing their open enterprise architectures. This vision fully relies on a Web Service strategy and the adoption of major Internet computing standards (more specifically, HTTP, SMTP, XML, SOAP).

A Web Service is defined by Microsoft as a programmable entity that provides a particular element of functionality, such as application logic, and is accessible to a number of potentially disparate systems through the use of Internet standards.

The idea is to provide individual and business users with a seamlessly interoperable and Web-enabled interface for applications and computing devices to make computing activities increasingly Web browser-oriented. The .NET platform includes servers, building-block services (such as Web-based data storage or e-business facilities), developer tools, and device software for smart clients (PCs, lap-tops, personal digital assistants, WAP or UMTS phones, video-equipped devices, etc.) as illustrated by Figure 22.13.

As seen in Figure 22.13, smart clients (application software and operating systems) enable PC*s* and other smart computing devices to act on XML Web Services, allowing anywhere, anytime access to information. XML Web Services, either internal, private, or public, are small, reusable applications interacting in XML. They are being developed by Microsoft and others to form a core set of Web Services (such as authentication, contacts list, calendaring, wallet management, ticketing services, ebiz-shops, news providers, and weather forecast services) that can be combined with other services or used directly by smart clients. Developer tools include Microsoft Visual Studio and Microsoft .NET Framework as a complete solution for developers to build, deploy, and run XML Web Services. Servers can belong to the MS Windows 2000 server family, the .NET Enterprise Servers, and the upcoming Windows Server 2003 family to host the Web Services or other back-end applications.

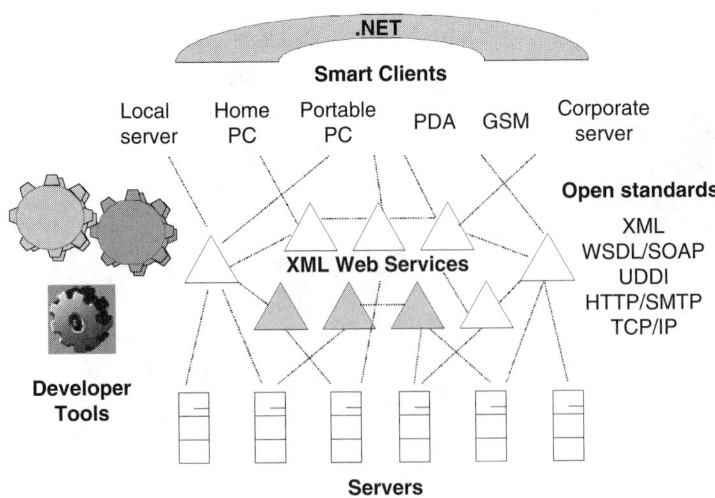

FIGURE 22.13 Principles of the .NET architecture.

22.5.2 .NET Application Architectures

.NET, like J2EE platforms, also clearly separates information presentation and layout to browsers from the business logic handled by the application server platform (Figure 22.14) .NET extends the classical Microsoft's OLE-COM-DCOM-COM+ development chain for application integration technology. The counterpart of EJB*s* in Microsoft environments are ActiveX and COM+ objects. COM+ components are activated in the Microsoft Transaction Server (MTS), the counterpart of the EJB Server. The connection with Web browsers is managed by the Microsoft Web server called Internet Information Server (IIS) and its Active Server Pages (ASP.NET, formerly called ASP).

ASP.NET allows a Web site builder to dynamically build Web pages on the fly, as required by clients, by inserting elements in Web pages which will let their contents be dynamically generated. These elements can be pure presentation elements or queries to a relational database but also result from business process execution carried out by ActiveX business components. ASP.NET supports code written in compiled languages such as Visual Basic, C++, C#, and Perl. ASP.NET files can be recognized by their.aspx extension.

The application can serve calling Web Services and call support Web Services to do its job.

Figure 22.14 presents what could be a typical .NET application architecture, to be compared to the J2EE one in Figure 22.12. Some similarities between the two architectures are obvious, such as JNDI which becomes an Active Directory, JMS which is replaced by MSMQ, the Microsoft message queuing system, and ODBC (Open Database Connectivity) which plays a similar role as JDBC. (Microsoft also offers ActiveX Data Objects [ADO] and OLE DB solutions to access data servers.) External applications can talk to the application server via DCOM, if they are developed with Microsoft technology or need a bridge with DCOM, otherwise. For instance, a CORBA client needs a CORBA/COM bridge to transform CORBA objects into COM objects.

Users accessing the system with a Web browser to interact through XML or HTML pages can be connected by a HTTP/HTTPS connection. In this case the session is managed by the IIS/ASP module of the architecture previously mentioned.

22.5.3 Microsoft Transaction Server (MTS) and Language Runtime

The Microsoft Transaction Server (MTS), based on the Component Object Model (COM) which is the middleware component model for Windows NT, is used for creating scalable, transactional, multiuser, and secure enterprise-level server side components. It is a program that manages application and database transaction requests or calls to Web Services on behalf of a client computer user. The Transaction Server

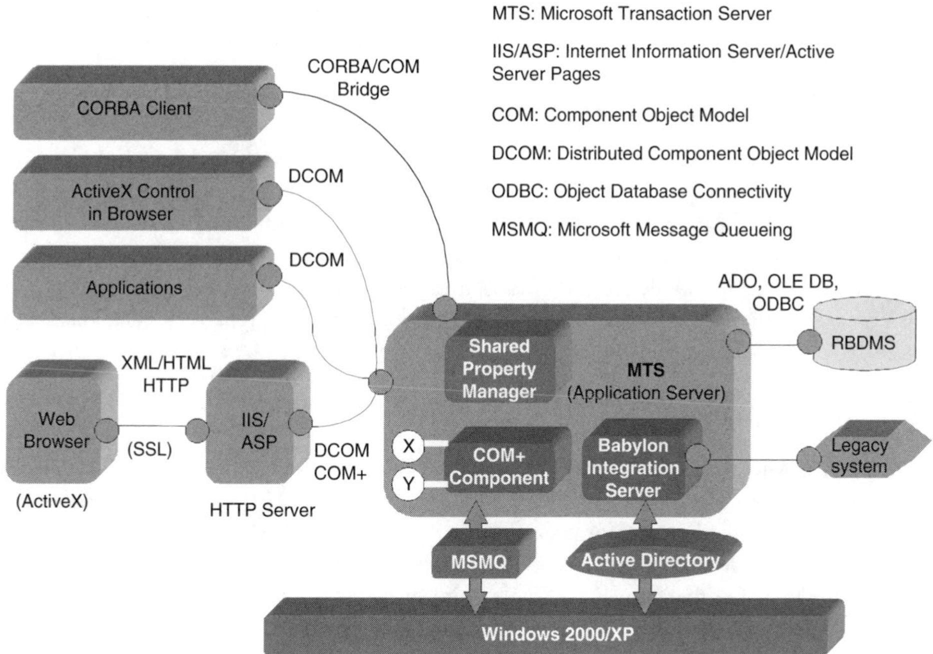

FIGURE 22.14 Typical Microsoft .NET *n*-tier application architecture.

screens the user and client computer to remove the need for having to formulate requests for unfamiliar databases and, if necessary, forwards the requests to database servers or to other types of servers. It also manages security, connection to servers, and transaction integrity.

MTS can also be defined as a component-based programming model (for managed components). An MTS component is a type of COM component that executes in the MTS run-time environment. MTS supports building enterprise applications using ready-made MTS components and allows "plug and work" with off-the-shelf MTS components developed by component developers.

It is important to realize that MTS is a stateless component model, whose components are always packaged as an in-proc DLL (Dynamic Link Library). Since they are made of COM objects, MTS components can be implemented in an amazing variety of different languages including C++, Java, Object Pascal (Delphi), Visual Basic and even COBOL. Therefore, business objects can be implemented in these various languages as long as they are compatible with the Common Language Runtime (CLR), a controlled environment that executes transactions of the Microsoft Transaction Server.

The difference with J2EE servers is that in Microsoft's .NET application servers the business objects (COM+ components) are compiled in an intermediary language called MSIL (Microsoft Intermediate Language). MSIL is a machine-independent language to be compiled into machine code by a just-in-time compiler and not interpreted as with Java. This peculiarity makes the approach much more flexible and much faster at run-time than Java-based systems.

Sitting on top of CLR, MTS allows the easy deployment of applications based on the 3-tier architecture principle using objects from COM/COM+ libraries. It also allows deployment of DCOM, the Distributed Component Object Model of Microsoft COM. It is enough to install MTS on a machine to be able to store and activate COM components, for instance as a DLL if the component is written in Visual Basic. These components can be accessed by a remote machine by registering them on this new server. Like EJBs, these components are able to describe all their interfaces when called by specific methods. The components can be clustered into groups, each group executing a business process. Such a process group is called a MTS package. Each package can be maintained separately, which ensures its independence and its isolation from the transactional computation point of view. MTS ensures the management of transactions and packages. For instance, packages frequently solicited will be automatically maintained in main memory while less frequently used packages will be moved to permanent storage.

22.5.4 Microsoft Message Queue (MSMQ)

MSMQ is the message-oriented middleware provided by Microsoft within .NET. It is the analog of JMS in J2EE. The Microsoft Message Queue Server (MSMQ) guarantees a simple, reliable, and scalable means of synchronous communication freeing up client applications to do other tasks without waiting for a response from the other end. It provides loosely-coupled and reliable network communications services based on a message queuing model. MSMQ makes it easy to integrate applications, implement a push-style business event delivery environment between applications, and build reliable applications that work over unreliable but cost-effective networks.

22.5.5 Active Directory

Active Directory is Microsoft's proprietary directory service, a component of the Windows 2000 architecture. Active Directory plays the role of J2EE's JNDI in .NET. Active Directory is a centralized and standardized object-oriented storage organization that automates network management of user data, security, and distributed resources, and enables interoperation with other directories (e.g., LDAP). It has a hierarchical organization that provides a single point of access for system administration (regarding management of user accounts, client servers, and applications, for example). It organizes domains into organizations, organizations into organization units, organizations units into elements.

22.6 Comparison of J2EE and .NET

The two technologies, J2EE and .NET, are very similar in their principles. Both are strongly based on new standards, a common one being XML. Figure 22.12 and Figure 22.14 have been intentionally drawn using the same layout to make the comparison more obvious. They, however, differ a lot at the system and programming levels, although both are object-oriented and service-based.

While the .NET technology is younger than J2EE and is vendor-dependent, J2EE remains a specification and .NET is a product. However, there are many J2EE compliant products on the market and, therefore, both technologies are operational and open to each other, despite some compatibility problems among J2EE platforms when the standard is not fully respected.

From a practical standpoint, .NET offers solutions that are faster to develop and that execute faster, while J2EE offers a stronger component abstraction. Especially, the library interfaces and functionality is well-defined and standardized in the J2EE development process. Hence, a key feature of J2EE-based application development is the ability to interchange plug-compatible components of different manufacturers without having to change a line of code. J2EE is recommended, even for Windows environments, when there are many complex applications to be integrated and there is a large base of existing applications to be preserved (legacy systems). However, performance tuning of J2EE platforms (JVM in particular) remains an issue for experts due to their inherent complexity. This is less critical for the CLR of .NET and its language interface, which recognizes 25 programming languages. However, load balancing is better managed in J2EE products because EJB*s* can dynamically replicate the state of a session (stateful session entities) while .NET allows session replication for ASP.NET pages but not for COM+ components. Finally, in terms of administration, a .NET platform, thanks to its better integration, is easier to manage than a J2EE platform.

Table 22.1 provides a synthetic comparison of the two types of platform. The main difference lies in the fact that with .NET Microsoft addresses the challenge of openness in terms of interoperability while Sun addresses it in terms of portability. Experience has proved that .NET is easier to integrate to the existing environment of the enterprise and that Web Service philosophy is not straightforward for J2EE platforms.

TABLE 22.1 J2EE and .NET Comparison

J2EE	.NET/DNA
J2EE = a specification **Middle-tier components: EJB** • Stateless and stateful service components plus data components **Naming and Directory: JNDI** **J2EE features** • Portability is the key • Supported by key players in middleware vendors (30+) • Rapidly evolving specifications (J2EE connector, EJB) • Strong competencies in object technology required • Many operating systems supported and strong portability of JVM **J2EE at work** • Dynamic Web pages: Servlets/JSP • Database access: JDBC, SQL/J • Interpreter: Java Runtime Environment • SOAP, WSDL, UDDI: Yes • *de facto* industry standard	**.NET = a product** **Middle-tier components: COM+** • Stateless service components (.NET managed components) **Naming and Directory: Active Directory** **DNA features** • Interoperability is the key • Integrated environment that simplifies its administration • Reduced impact of code granularity on system performances • Fully dependent on Windows 2000 • Open to other platforms **.NET at work** • Dynamic Web pages: ASP.NET • Database access: ADO.NET • Interpreter: Common Language Runtime • SOAP, WSDL, UDDI: Yes • Productivity and performances • Reduced time-to-market

22.7 Conclusion: The Global Architecture

Interoperability of enterprise systems is the key to forthcoming e-business and e-government organizations or X2X organizations, be they business-to-customer (B2C), business-to-business (B2B), business-to-employee (B2E), government-to-citizen (G2C), government-to-employee (G2E), government-to-business (G2B), or government-to-nongovernment (G2NG) relationships. In any case, Enterprise Interoperability (EI), i.e., the ability to make systems work better together, is, first of all, an organizational issue and next a technological issue [Vernadat, 1996]. The aim of this chapter was to review emerging technological issues to define and build a sound Corporate Information Systems Architecture on which an open, flexible, and scalable Enterprise Architecture, dealing with the management of business processes, information and material flows, as well as human, technical, and financial resources, can sit. Among these, the J2EE and .NET architectures are going to play a premier role because of their component-oriented approach, their Web Service orientation, and their ability to support execution of inter-organizational business processes in a totally distributed operational environment preserving legacy systems.

As illustrated by Figure 22.15, there are two other essential components in the big picture of Internet-based IT systems architecture to achieve enterprise interoperability that have not been discussed in this chapter. These are Enterprise Application Integration (EAI) and Enterprise Information Portals (EIP).

Enterprise Application Integration: Implementing business processes requires the integration of new or legacy applications of different nature (e.g., SQL databases, bills-of-materials processors, ERP systems, home-made analytical systems, etc.). EAI tools and standards facilitate communication, i.e., data and message exchange, among applications within a company or with partner companies. EAI techniques are strongly based on the data neutralization principle by means of common exchange formats (e.g., EDI, STEP, Gesmes, HTML, XML).

Enterprise Information Portals: These are front-end systems providing a personalized user interface and single-point-of-access based on Web browsing technology to provide access to different internal and external information sources, services or applications, and to integrate or substitute existing user interfaces for host access or client computing.

To complete the global picture of Figure 22.15, a number of features can be added to the architecture, especially at the level of the application server. One of these is the single sign-on (SSO) facility to prevent

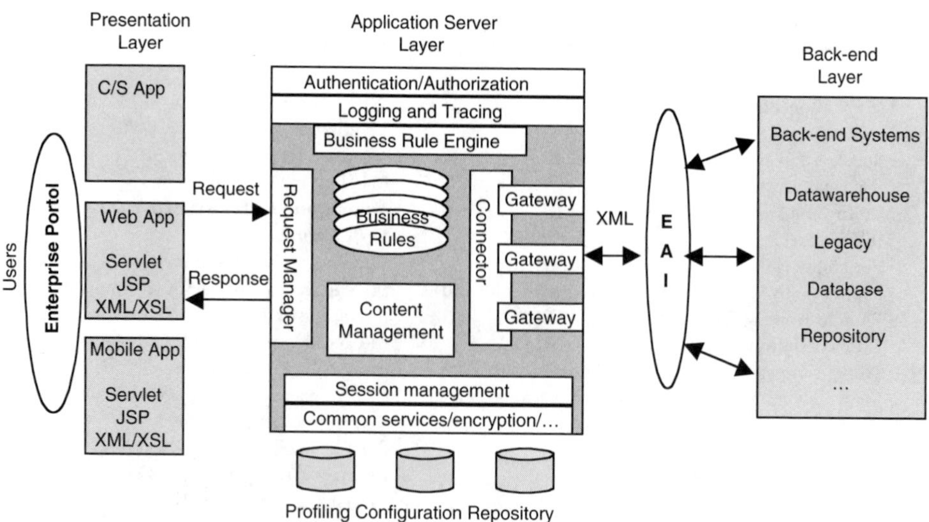

FIGURE 22.15 The global architecture of future IT systems.

users from logging on again each time they access another system in the architecture (authentication/authorization module). Another one concerns the content management module to deal with storage and management of any kind of electronic documents to be accessed to serve client applications or be presented within Web pages to users. Profiling is another very interesting feature that allows to define different groups of users and to associate with each group specific access rights and specific sets of functionality. In addition, encryption/decryption capabilities have proven to be very useful features to protect confidential or sensitive data. The business rule engine module is a component that allows to control execution of business processes and to enforce the respect of predefined declarative rules about enterprise operations.

With J2EE, .NET, Enterprise Portal, and Web Service technologies, the IT community is providing the business and administration communities with the necessary tools and techniques to build interoperable enterprise architectures, i.e., open, expandable, modular, reusable, and cooperative solutions to their integration needs.

References

IST Diffuse Project. Will Web services revolutionize e-Commerce? Proceedings of the Second Annual Diffuse Conference, Brussels, Belgium, February 6, 2002.

Khalaf, R., Curbera, F., Nagy, W., Mukhi, N., Tai, S., Duftler, M. Web services, In *Practical Handbook of Internet Computing* (M. Singh, Ed.), CRC Press, Boca Raton, FL, 2005.

Lavender, G. Directory services, In *Practical Handbook of Internet Computing* (M. Singh, Ed.), CRC Press, Boca Raton, FL, 2005.

Metadata Registries. Web site of Open Forum 2003 on Metadata Registries, http://metadata-stds.org/openforum2003/frmain.htm, 2003.

Microsoft. The Dot Net Architecture, http://www.microsoft.com/net, 2001.

OASIS. UDDI: Universal Description, Discovery, and Integration, http://www.uddi.org, 2002.

OMG. Model Driven Architecture (MDA), Document number ormsc/2001-07-01, Object Management Group, 2001 (http://www.omg/org/mda/).

OMG. MDA Guide Version 1.01, Document number omg/2003-06-01, Object Management Group, June 2003.

Sahai, A. Business processes integration, In *Practical Handbook of Internet Computing* (M. Singh, Ed.), CRC Press, Boca Raton, FL, 2005.

SUN Microsystems. J2EE: Java to Enterprise Edition and Execution, http://java.sun.com, 2001.

Vernadat, F.B. *Enterprise Modeling: Principles and Applications,* Chapman & Hall, London, 1996.

Vernadat, F.B. Enterprise Integration and Management in Agile Organizations, In *Agile Manufacturing* (A. Gunasekaran, Ed.), Springer-Verlag, Berlin, 2000.

Wilde, E. Advanced XML technologies, In *Practical Handbook of Internet Computing* (M. Singh, Ed.), CRC Press, Boca Raton, FL, 2005.

W3C. World Wide Web Consortium, XML: eXtensible Markup Language, http://www.w3.org/xml, 2000a.

W3C. World Wide Web Consortium, XSL: eXtensible Stylesheet Language, http://www.w3.org/Style/XSL, 2000b.

W3C. World Wide Web Consortium, WSDL: Web Service Description Language, http://www.w3.org/TR/wsdl, 2001.

W3C. World Wide Web Consortium, SOAP: Simple Object Access Protocol, http://www.w3.org/TR/SOAP, 2002a.

W3C. World Wide Web Consortium, Web Service, http://www.w3.org/2002/ws, 2002b.

23

XML Core Technologies

CONTENTS

23.1 Introduction 23-1
23.2 Core Standards 23-2
23.2.1 XML 23-2
23.2.2 XML Namespaces 23-5
23.3 XML Data Models 23-6
23.3.1 XML Information Set (XML Infoset) 23-6
23.3.2 XML Path Language (XPath) 23-7
23.3.3 XML Application Programming Interfaces 23-9
23.4 XML Schema Languages 23-11
23.4.1 XML Schema 23-12
23.4.2 RELAX NG 23-14
23.4.3 Document Schema Definition Languages (DSDL) 23-15
References 23-17

Erik Wilde

23.1 Introduction

The *Extensible Markup Language (XML)* has become the foundation of very diverse activities in the context of the Internet. Its uses range from an XML format for representing Internet Request For Comments (RFC) (created by Rose [1999]) to very low-level applications such as data representation in remote procedure calls.

Put simply, XML is a format for structured data that specifies a framework for structuring mechanisms and defines a syntax for encoding structured data. The application-specific constraints can be defined by XML users, and this openness of XML makes it adaptable to a large number of application areas.

Ironically, XML today resembles the *Abstract Syntax Notation One (ASN.1)* of the *Open Systems interconnection (OSI)* protocols, which at the time it was invented was rejected by the Internet community as being unnecessarily complex. However, XML is different from ASN.1 in that it has only one set of "encoding rules," the XML document format, which is character-based. There is an ongoing debate whether that is a boon (because XML documents are very bulky and handling binary data is a problem) or a bane (because applications only need to know one encoding format), but so far no alternative encodings have been successful on a large scale. Apart from this difference in encoding rules, XML can be viewed as the "presentation layer" (referring to the OSI reference model) of many of today's Internet-based applications. XML users can be divided into two camps, the "document-oriented" and the "data-oriented." Whereas the document-oriented users will probably reject XML as a data presentation format as being too simplistic, the overwhelming majority of XML users — the data-oriented users — will probably agree with this view.

In this chapter, XML and a number of accompanying specifications are examined. The topics covered are by no means exhaustive, as the number of XML-related or XML-based technologies is rising practically daily. This chapter discusses the most important technologies and standards; for a more exhaustive and up-to-date list, please refer to the online glossary of XML technologies at http://dret.net/glossary/.

1-58488-381-2/05/$0.00+$1.50

© 2005 by CRC Press LLC

This chapter discusses XML technologies as they are and as they depend on each other. Each of the technologies may be used in very different contexts, and sometimes the context may be very important to decide which technology to use and how to use it. To help make these decisions, a set of general guidelines for using XML in Internet protocols has been published as *RFC 3470* by Hollenbeck et al. [2003].

In Section 23.2, XML core technologies that are the foundation of almost all applications of XML today are described. XML's data model is discussed in Section 23.3, which describes XML on a more abstract level. In many application areas, it is necessary to constrain the usage of XML. This is the task of XML schema languages, which are described in Section 23.4.

In Chapter 24, more technologies built on top of XML are discussed, in particular style sheet languages for using XML as the source for presentation, and technologies for supporting XML processing in various environments and application scenarios.

23.2 Core Standards

XML itself is a rather simple standard. Its specification is easy to read and defines the syntax for structured documents, the "XML documents," and a schema language for constraining documents, the *Document Type Definition (DTD)*. The complexity of XML today is a result of the multitude of standards that in one way or the other build on top of XML. In this first section, another core standard is described, which in a way can be regarded as being part of "Core XML" today. This standard is *XML Namespaces,* a mechanism for making names in an XML environment globally unique. *XML Base* by Marsh [2001], a specification for the interpretation of relative URIs in XML documents, and *XML Inclusions* by Marsh and Orchard [2004], as a standard syntax for including documents or document fragments, are also considered a core part of XML, but are not covered in detail here.

23.2.1 XML

The XML specification was first released in February 1998 by Bray et al. [1998], followed by a "second edition" in October 2000 and then a "third edition" in February 2004 by Bray et al. The second and third editions of XML 1.0 did not make any substantial changes to the specification; they simply corrected known errors of the preceding editions. XML 1.1 by Cowan [2004] reached recommendation status in early 2004, but is mainly concerned with character set and encoding issues,[1] and thus also leaves most of XML as it is.

XML, when seen as an isolated specification, is rather simple. It is a subset of the much older *Standard Generalized Markup Language (SGML)* by the International Organization for Standardization [1986], which has been an ISO standard since 1986. XML, which started out under the title "SGML on the Web," has been developed by the World Wide Web Consortium (W3C). It was an effort to overcome the limited and fixed vocabulary of the *Hypertext Markup Language (HTML)* and enable users to define and use their own vocabularies for document structures. It turned out that the result of this effort, XML, also was extremely useful to people outside of the document realm, who started using XML for any kind of structured data, not just documents. Since its very beginning, XML has been used in application areas that were not part of the original design goals, which is the reason for some of the shortcomings of XML in today's application areas, such as the support of application-oriented datatypes, or some way to make XML documents more compact.

Figure 23.1 shows a simple example of an XML document. It starts with an (optional) XML declaration, which is followed by the document's content, in this case, a book represented by a number of elements. Element names are enclosed in angle brackets, and each element must have an opening and a closing

[1] Specifically, XML 1.1 is a reaction to Unicode moving from 2.0 past 3.0. Further, to facilitate XML processing on mainframe computers, the NEL character (U + 85) has been added to XML's list of end-of-line characters.

```
<?xml version="1.0"?>
<book editor="Munindex P. Singh">
  <title>Practical Handbook of Internet Computing</title>
  ...
  <chapter id="core" author="Erik Wilde" date="2003-09-12">
    <title>XML Core Technologies</title>
    <para>The <emph>Extensible Markup Language (XML)</emph> ... </para>
    <section id="xml">
      <title>XML</title>
      <para>The XML specification has been first released in February
       1998 by <ref id="xml10-spec"/>, followed by a <quote>second
       edition</quote> of the specification in October 2000 ... </para>
      ...
     </section>
   </chapter>
   ...
</book>
```

FIGURE 23.1 Example XML 1.0 document.

tag. Elements must be properly nested, so that the structure of elements of the document can be regarded as a tree, with the so-called *document element* acting as the tree's root. In addition to elements, XML knows a number of other structural concepts, such as attributes (for example, the `editor` attribute of the `book` element) and text (the `title` elements contain text instead of other elements). Elements may also be empty (such as the `ref` element), in which case they may use two equivalent forms of markup, full (`<ref id="xml10-spec"></ref>`) or abbreviated (`<ref id="xml10-spec"/>`). If a document conforms to the syntactic rules defined by the XML specification, it is said to be *well-formed.*

The XML specification is rather easy to read and, in fact, can serve as a reference when questions about syntactic details of XML arise. DuCharme [1998] has published an annotated version of the specification (the original version, not the later editions), which makes it even easier to read. Further, the W3C provides an increasing number of translated specifications,[2] which are intended to increase awareness and global ease of use.

Logically, the XML specification defines two rather separate things:

- *A syntax for structured data:* An *XML document* is a document using a *Markup Language* as a structuring mechanism. XML markup is a special case of the more generalized model of SGML markup usinga fixed set of markup characters. (The most visible ones are the "<" and ">" characters for delimiting tags.) For HTML users, XML markup is straightforward; XML simply disallows some markup abbreviations that HTML provides through its use of SGML markup minimization features. Specifically, XML always requires full markup (the only exception is an abbreviation for empty elements), so the HTML method of omitting markup (such as element end tags and attribute value delimiters) is not possible.

  The most important structural parts of XML documents are elements and attributes. Elements can be nested and thus make it possible to create arbitrarily complex tree structures (see Figure 23.1 for an example). XML also allows processing instructions, comments, and some other con-

[2] http://www.w3.org/Consortium/Translation/

structs, but these are less frequently used in XML applications. If an XML document conforms to the syntax defined in the XML specification, it is said to be *well-formed.*

- *A schema language for constraining XML documents:* As many applications want to restrict the structural possibilities of XML, for example, by defining a fixed set of elements and their allowed combinations, XML also defines a schema language, the *DTD.* A DTD defines element types, and the most important aspects of element types are their attributes and content models (see Figure 23.2 for an example). If an XML document conforms to all constraints defined in a DTD, it is said to be *valid* with regard to this DTD.

  DTDs are part of XML because they are also part of SGML, and it is this heritage from the document-oriented world that makes it easier to understand the limitations of DTDs. For example, DTDs provide only very weak datatyping support, and these types are applicable only to attributes. With XML's success and its application in very diverse areas, the need for alternative schema languages became apparent, and alternative schema languages were developed (see Section 23.4 for further information).

Even though XML documents and DTDs are not strictly separated (it is possible to embed a DTD or parts of it into an XML document), they should be regarded as logically separate. In a (purely hypothetical) new major version of XML, it is conceivable that DTDs would be left out of the core and become a separate specification.

Figure 23.2 shows an example of a DTD, in this case a DTD for the XML document shown in Figure 23.1. All elements and attributes must be declared. Elements are declared by defining the allowed content, which may be none (for example, the `ref` element is declared to be `EMPTY`), only elements (for example, the `book` element is declared to have as content a sequence of one `title` and any number of `chapter` elements), or text mixed with elements (for example, the `para` element may contain text or `emph`, `quote`, or `ref` elements). The element declarations, in effect, define a grammar for the legal usage of elements. In addition, a DTD defines attribute lists that declare which attributes may or must be used with which elements (for example, a `chapter` element must have `id` and `date` attributes and may have an `author` attribute). If a document conforms to the definitions contained in a DTD, it is said to be *valid* with regard to this DTD.

```
<!ELEMENT    book       (title, chapter*)>
<!ATTLIST    book
             editor     CDATA #REQUIRED >
<!ELEMENT    chapter    (title, para+, section*)>
<!ATTLIST    chapter
             id         ID     #REQUIRED
             data       CDATA #REQUIRED
             author     CDATA #IMPLIED >
<!ELEMENT    section    (title, para+)>
<!ATTLIST    section
             id         CDATA #REQUIRED >
<!ELEMENT    title      (#PCDATA)>
<!ELEMENT    para       (#PCDATA | emph | quote | ref)*>
<!ELEMENT    emph       (#PCDATA | quote | ref)*>
<!ELEMENT    quote      (#PCDATA | emph )*>
<1ELEMENT    ref        EMPTY>
<!ATTLIST    ref
             id         ID     #REQUIRED >
```

FIGURE 23.2 Example XML 1.0 DTD.

To summarize, XML itself is a rather simple specification for structured documents (using a markup language) and for a schema language to define constraints for these documents.

23.2.2 XML Namespaces

XML Namespaces, defined by Bray et al. [1999], are a mechanism for associating URLs and one or more names. A new version of the specification by Bray et al. [2004], reached recommendation status in early 2004. The URI of a namespace serves as a globally unique identifier, which makes it possible to refer to the names in the namespace in a globally unique way. Namespaces are important when XML documents may contain names (elements and attributes) from different schemas. In this case, the names must be associated with a namespace, and XML Namespaces use a model of declarations and references to these declarations.

Figure 23.3 shows an example of using namespaces. It uses the *XML Linking Language (XLink)* defined by DeRose et al. [2001], which has a vocabulary for embedding linking information into XML documents. XLink is based on namespaces; all information in a document that is relevant for XLink can be easily identified through belonging to the XLink namespace. In order to use a namespace, it must be declared in the XML document using attribute syntax (`xmlns:xlink="http://www.w3.org/1999/xlink"`). The namespace declaration binds a prefix to a *namespace name,* in this case the prefix `xlink` to the namespace name `http://www.w3.org/1999/xlink`. This namespace name is defined in the XLink specification, and thus serves as an identifier that all names referenced from this namespace have the semantics defined in the XLink specification. An element or attribute is identified as belonging to a namespace by prefixing its name with a namespace prefix, such as in `xlink:type="simple"`. In this case, the application interpreting the document knows that this attribute is the `type` attribute from the `http://www.w3.org/1999/xlink` namespace, and can take the appropriate action. Namespace declarations are recursively inherited by child elements and can be redeclared (in XML Namespaces 1.1, they can also be undeclared). There also is the concept of the *default namespace,* which is not associated with a prefix (the default namespace is the only namespace that can be undeclared in

```
<book editor="Munindar P. Singh"
      xmlns:xlink="http://www.w3.org/1999/xlink">
  <title>Practical Handbook of Internet Computing</title>
   ...
  <chapter id="core" author="Erik Wilde" date="2003-09-12">
    <title>XML Core Technologies</title>
    <para>The <emph>Extensible Markup Language (XML)</emph> ... </para>
    <section id="xml">
      <title>XML</title>
      <para>The <link xlink:type="simple" xlink:title="W3C XML spec"
      xlink:href="http://www.w3.org/TR/REC-xml">XML
      specification</link> has been first released in February 1998 by
     <ref id="xml10-spec"/>, followed by a <quote>second
     edition</quote> of the specification in October 2000 ... </para>
      ...
    </section>
  </chapter>
  ...
</book>
```

FIGURE 23.3 Namespace example.

XML Namespaces 1.0). It is important to note that namespace prefixes have only local significance; they are the local mechanism that associates a name with a namespace URL.

XML Namespaces impose some additional constraints on XML documents, the most important being that colons may not be used within names (they are required for namespace declarations and qualified names) and that the use of namespace declarations and qualified names must be consistent (e.g., no undefined prefixes are used). As namespaces are used in almost all XML application areas today, care should be taken that all XML documents are namespace-compliant.

23.3 XML Data Models

Although XML defines a syntax for representing structured data, from the application point of view, the goal in most cases is to access that structured data and not the syntax. In fact, much of XML's success is based on the fact that application programmers never have to deal with XML syntax directly, but can use existing tools to handle the syntax. Thus, from the application point of view, XML is often used in abstractions of various levels. For example, in most applications, whitespace in element tags (such as `<element attribute= "value" >`) is irrelevant, but for programmers building an XML editor, which should preserve the input without any changes, this whitespace is significant and must be accessible through the XML tools they are using.

According to the terminology specified by Pras and Schönwälder [2003], XML itself implements a *data model.* At a more abstract level, there is the question about XML's *information model* (to quote Pras and Schönwälder [2003], "independent of any specific implementations or protocols used to transport the data"). There is an ongoing debate about whether XML should have had an information model and not just a syntax, but there is no simple answer to this question, and XML's success demonstrates that a data model without an information model can indeed be very successful. However, from the developers' point of view, this answer is unsatisfactory. The question remains: What is relevant in an XML document and what is not? The relevance of attribute order or of whitespace in element tags is one example of this question, and another is whether character references (such as ` `) are preserved or resolved to characters.

As XML itself does not define an information model, and accentuated by the process of many other standards emerging in the XML world, it became apparent that the same questions about the information model come up again and again. In an effort to provide a solution to this problem, the W3C defined the *XML Information Set,* which is used by a number of W3C specifications as an information model. However, there still are several other data models in use, and it is unlikely that there will be a single and universally accepted XML information model soon. In the following sections, some of the more important information models are described.

23.3.1 XML Information Set (XML Infoset)

The *XML Information Set (XML Infoset)* defined by Cowan and Tobin [2001] was the first attempt to solve the problem of different views of XML's information model in different application areas. The Infoset is not intended to be a specification that is directly visible to XML users; it is intended to be used by specification writers as a common XML information model. The Infoset defines 11 types of information items (such as element and attribute items), which are each described by a number of properties. The Infoset is defined informally, and does not prescribe any particular data format or API.[3] The Infoset also explicitly states that users are free to omit parts of it or add parts to it, even though it is unclear how this should be done.

The Infoset omits a number of syntax-specific aspects of an XML document, such as whitespace in element tags and attribute order. The Infoset specification contains a list of omissions; some of them are

[3] Seen in this way, an XML document is just one way to represent the information specified by the Infoset.

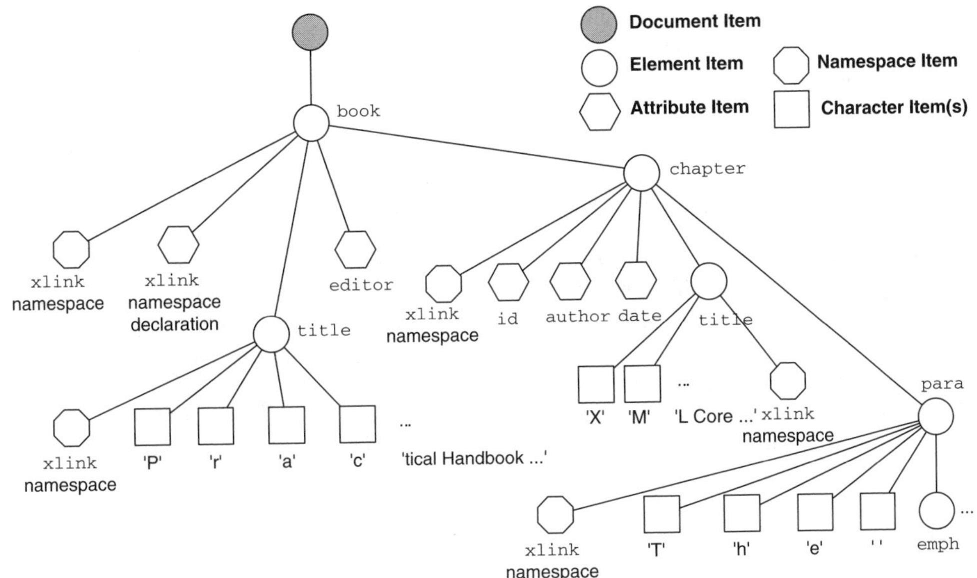

FIGURE 23.4 Example XML Infoset.

minor, whereas others are important and considered by many people to be problematic. Among these omissions are CDATA sections, the entity structure of the XML document, and character references.

Figure 23.4 shows an example of an XML Infoset. It depicts a part of the Infoset of the XML document shown in Figure 23.3. In particular, it shows how namespace declarations appear as two information items (an attribute and a namespace item) on the element where the namespace is declared, and as namespace items on all descendants of this element. This explains why in Figure 23.3, the `xlink` prefix for the XLink namespace can be used for the attributes of the `link` element even though the namespace is declared on another element (specifically, the `book` document element).

Even though it could be argued that the Infoset could be improved in terms of extensibility (Wilde [2002]), its current informal version makes it hard to extend in an interoperable way. However, the Infoset is used by a number of other specifications, and as such, is certainly the most successful information model of XML.

One interesting application of the Infoset is *Canonical XML*, defined by Boyer [2001]. It defines a normalization of XML documents, which makes it easier to compare documents (based on their Infoset content, so the goal is to compare Infosets rather than XML documents with all their possible syntactic variations). Canonical XML simply defines a serialization process of an Infoset into XML document syntax, and so could be specified rather easily by building on top of the abstraction provided by the Infoset.

23.3.2 XML Path Language (XPath)

One of the most interesting XML technologies is the *XML Path Language (XPath)*, defined by Clark and DeRose [1999]. It is a language for selecting parts of an XML document, based on a structural view of the document. XPath came into existence in an interesting way. When the *Extensible Stylesheet Language (XSL)* (described in Section 24.1.2) was created, it was discovered that the transformation part of the process would be useful in its own right, and so it was subsequently created as a stand-alone technology, the *XSL Transformations (XSLT)* (described in Section 24.2.2). Working on XSLT, it was discovered that the core of the language, the ability to select parts of an XML document, also had the potential to be useful in contexts other than XSLT. Consequently, XPath was created, so that currently XPath is referenced by XSLT, which in turn is referenced by XSL. In the meantime, this design has shown its usefulness, as XPath has already been used by a number of other XML technologies, most notably *XML Schema, Schematron,* the *Document Object Model (DOM),* and the *XML Query Language (XQuery).*

XPath is a language that can be most easily understood through the analogies of XML as a tree structure and a file system as a tree structure. In the same way that most operating systems provide tools or commands for navigating through the file system's structure, XPath provides a language for navigating through an XML document's structure. In comparison to the simple but powerful *cd* command of operating systems, XPath provides the following generalizations:

- *Node types:* While a file system tree only has nodes of one type (files), the nodes of an XML document tree can be of different types. The XML Infoset described in the previous section defines 11 types of information items. XPath reduces the complexity of the Infoset model to seven node types. The most important XPath node types are element and attribute nodes.
- *Axes:* Navigation in a file system tree implicitly uses children for navigation, and XPath generalizes this by introducing *axes,* which are used to select nodes with relationships different from that of direct children (to remain as intuitive as possible, the child axis is the default axis, which means that if no axis is specified, the child axis is chosen).

  To illustrate these concepts, the XPath `/html/head/title` selects the `title` child of the head child of the hurl document element (assuming a HTML document structure), and is equivalent to the more explicit `/child::html/child::head/child::title` XPath. On the other hand, the `/descendant::title` XPath selects all `title` descendants (children, children's children, and so on) from the document root, and thus uses a different axis.

  XPath defines a number of axes, most of them navigating along parent/child relationships, with the exception of a special axis for attributes and another for namespaces (both of these are XPath nodes).
- *Predicates:* Axes and node types make it possible to select certain nodes depending on their relationship with other nodes. To make selection more powerful, it is possible to apply predicates to a set of selected nodes. Each predicate is evaluated for every node that has been selected by the node type and axis, and only if it evaluates to `true` will this node be ultimately selected. Predicates are simply appended after the axis and node specifiers, as in the XPath `/descendant::meta[position()=2]`, which selects the second meta element in the document.

The three concepts described above are the three constituents of a *location step.* Many XPaths are so-called *location paths,* and a location path is simply a sequence of location steps separated by slashes. The examples given above are XPath location paths, concatenating several steps with an axis (if the axis is omitted, the default child axis is taken), a node test,[4] and optional predicates.

XPath's type system is rather simple; there are strings, numbers, booleans, and node sets. The first three types are well-known, for example, from programming languages. Node sets are specific to XPath and can contain any number of nodes from the XPath node tree.

XPath makes it easy to select parts of an XML document, based on complex structural criteria. Depending on the context where XPath is being used, this selection can be supported by additional concepts such as variables. XPath provides a function library of some basic string, numerical, and boolean functions. However, XPath's strength is the support for accessing XML structures, and the function library is sufficient for basic requirements, but lacks more sophisticated functionality (such as regular expression handling for strings). XPath applications are free to extend XPaths function library, and most XPath applications (such as XSLT) exploit this feature.

Figure 23.5 shows three examples of XPath expressions, which all refer to the XML example document shown in Figure 23.1. The first XPath selects the `title` "children" of the `book` element, and thus selects the `title` element with the "Practical Handbook of Internet Computing" content. The second XPath

[4] The node test can also be a *name test,* not only testing for node type but also testing for a specific name from among named nodes (i.e., elements and attributes).

```
1. /book/title
2. /book/chapter[@id='core']/@author
3. //ref[@id='xml10=spec']/ancestor::chapter
```

FIGURE 23.5 XPath examples.

selects the `book` element, then its `chapter` child where the `id` attribute has the value "core," and then the `author` attribute of this element. Consequently, this XPath selects the attribute with the value "Erik Wilde". The third XPath selects all `ref` elements within the document (regardless of their location in the element tree), and from these only the ones with the `id` attribute set to "xml10-spec." From these elements, it selects all ancestor element (i.e., all the elements that are hierarchically above them in the element tree) that are `chapter` elements. In effect, this XPath selects all chapters containing references to the XML 1.0 specification.

For a common data model to be used with the *XML Query Language (XQuery)* (described in Section 24.3.2) and for a new version of XSLT (XSLT 2.0), a heavily extended version (2.0) of XPath is currently under construction by Berglund et al. [2003]. Formally, the new version of XPath will be based on an explicit data model (Fernández et al. [2003]), which is in turn based on the XML Infoset as well as on the *Post Schema Validation Infoset (PSVI) contributions of* `XML Schema` (see Section 23.4.1 for PSVI details). XPath 2.0 will be much more powerful than XPath 1.0, and it is planned to design it as backward compatible as possible (probably not 100%).

23.3.3 XML Application Programming Interfaces

In many cases, XML users are not interested in handling XML markup directly, but want to have access to XML documents through some *Application Programming Interface (API)*. In fact, this is one of the biggest reasons for XML's success: A multitude of tools are available on very different platforms, and it is very rarely the case that users must deal with XML directly. To further underscore this point, every place in a system where XML is accessed directly (e.g., by reading markup from a file and inspecting it) should be reviewed with great care, because it usually means that code will be written that already exists in XML (in this case, in XML parsers) has to be rewritten. This is not only a waste of resources but also dangerous, because it usually results in inferior XML parsing code, which is not robust against the great variety of representations that XML allows.

There are two main APIs for XML; one is the *Document Object Model (DOM)* described in Section 23.3.3.1, and the other one is the *Simple API for XML (SAX)* described in Section 23.3.3.2. Many modern parsers provide a DOM as well as a SAX interface, and Figure 23.6 shows how this is usually implemented. While SAX is a rather simple interface, which simply reports events while parsing a stream of characters from an XML document, DOM provides a tree model, which can be used for rather complex tasks. Many parsers implement a thin SAX layer on top of their parsing engine and provide access to this layer. However, parsers also implement a DOM layer on top of this SAX layer, which consumes the SAX events and uses them to create an in-memory representation of the document tree.

Consequently, if an application only needs the sequence of events generated by a parser, then SAX is the right API. However, if a more sophisticated representation is necessary, and the application wants to have random access to the document structure, then a DOM parser should be chosen, which consumes more resources than a SAX parser (in particular, memory for the document tree structure), but on the other hand provides a more powerful abstraction of the XML document.

Apart from the DOM and SAX APIs described in the following two sections, there are a number of alternative APIs for XML. Two of the more popular among these are JDOM and *Java API for XML Parsing (JAXP)*. Whereas JDOM can be regarded as an essentially DOM-like interface optimized for Java, JAXP is not an XML API itself, but an abstraction layer (i.e., an API for other APIs). JAXP can be used to access XML-specific functionality from within Java programs. It supports parsing XML documents using the

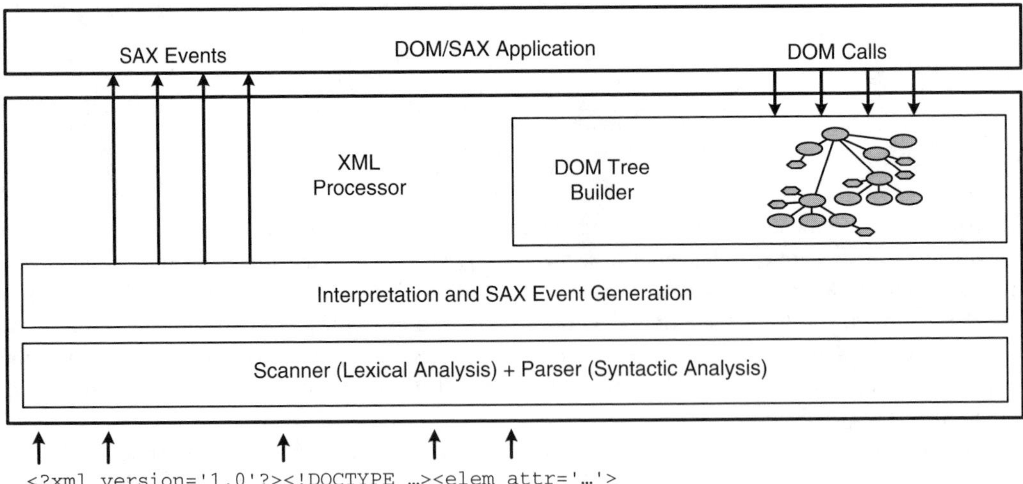

FIGURE 23.6 XML processor supporting SAX and DOM.

DOM or SAX APIs, and processing XML documents with XSLT using the TrAX API (more information about XSLT can be found in Section 24.2.2).

23.3.3.1 Document Object Model (DOM)

The *Document Object Model (DOM)* is the oldest API for XML. In fact, the DOM existed before XML was created, and started as an API that provided an interface between JavaScript code and HTML pages. This first version of the DOM is often referred to as *DOM Level 0.* When XML was created, it quickly became apparent that the DOM would be good starting point for an XML API, and a first version of the DOM was created that supported HTML as well as XML. This first version, defined by Wood et al. [2000], is called the *DOM Level 1 (DOM1).*

DOM quickly established itself as the most popular API for XML, one of the reasons being that it is not restricted to a particular programming language. The DOM is defined using the *Interface Definition Language (IDL),* which is a language for specifying interfaces independent of any particular programming language. The DOM specification also contains two *languages,* which are mappings from IDL to the peculiarities of a specific programming language. The two programming language bindings contained in the DOM specifications cover Java and ECMAScript (which is the standardized version of JavaScript), but DOM language bindings for a multitude of other languages are also available. The advantage of this independence from a programming language is that programmers can quickly transfer their knowledge about the DOM from one language to another, by simply looking at the new language binding. The disadvantage is that the DOM is not a very elegantly designed interface because it cannot benefit from the features of a particular programming language. This is the reason why JDOM was created, which has been inspired by the DOM but has been specifically designed for Java.

Since the DOM has been very successful, many programmers were using it and requested additional functionality. Consequently, the DOM has evolved into a module-based API. *DOM Level 2 (DOM2)* defined by Le Hors et al. [2000] and *DOM Level* 3 (*DOM3*), defined by Le Hors et al. [2003], are both markers along this road, with the DOM3 being the current state of the art. DOM3 even contains an *XPath Module,* which makes it possible to select nodes from the document by using XPath expressions (for more information about XPath refer to Section 23.3.2). Implementations of the DOM (usually XML processors) may support only a module subset, so before selecting a particular software that supports the DOM, it is important to check the version of the DOM and the modules it supports.

23.3.3.2 Simple API for XML (SAX)

Although the DOM is a very powerful API, it also is quite complicated and requires a lot of resources (because a DOM implementation needs to create an in-memory representation of the document tree, which can be an issue when working with large documents). In an effort to create a lightweight XML API, the *Simple API for XML* (*SAX*) was created. As shown in Figure 23.6, it works differently from DOM because it is event-based. This means that SAX applications register event handlers with the API, which are then called whenever the corresponding event is triggered during parsing. These events are closely related to markup structures, such as recognizing an element, an attribute, or a processing instruction while parsing the document.

This event-based model results in a fundamentally different way of designing an application. While DOM-based applications operate in a "pull-style" model, pulling the relevant information from an XML document by accessing the corresponding parts of the tree, SAX-based applications operate in a "push-style" model. In this model, the document is "pushed" through the parser, which simply calls the event-handlers that have been registered. Essentially, the program flow of a DOM-based application is controlled by the application writer, whereas the program flow of a SAX-based application is mostly determined by the document being processed.[5]

A SAX parser is a more lightweight piece of software because it does not have to build an internal tree model of the document. However, this also means that applications cannot access a document in the random access mode that DOM supports. Consequently, the question of whether a particular application should be built on top of a SAX or DOM interface depends on the application's requirements. For rather simple streaming applications, SAX is probably sufficient, whereas applications needing a more flexible way of accessing the document are probably better supported by a DOM interface.

23.4 XML Schema Languages

As described in Section 23.2.1, the XML specification defines a syntax for XML documents as well as a schema language for these documents, the *Document Type Definition (DTD)*. However, as XML became successful and was used in very diverse application areas, it quickly became apparent that the document-oriented and rather simple features of DTDs were not sufficient for a growing number of users.

In principle, everything that can be described in a schema language (such as a DTD) can also be checked programmatically, so that it would be possible to use XML without any schema language and implement constraint checking in the application. However, a schema has some very substantial advantages over code: it is declarative, it is easier to write, understand, modify, and extend than code, and it can be processed with a variety of software tools, as long as these tools support the schema language. Basically, in the same way that XML is useful because people can use existing tools to process XML documents, a schema language is useful because people can use existing tools to do constraint checking for XML documents.

Lee and Chu [2000] and van der Vlist [2002] have published comparisons of different schema languages that have been proposed by various groups. Since a schema language must reach critical mass to be useful (the language must be supported by a number of tools to have the benefit of being able to exchange schemas between different platforms), many of these proposals never had significant success.

The W3C developed a schema language of its own, which took a number of proposals and used them to create a new and powerful schema language. The result of this effort is *XML Schema* described in Section 23.4.1. However, XML Schema has received some criticism because it is very big and bulky, and because it tries to solve too many problems at once. As an alternative, a simpler and less complicated schema language has been developed outside of W3C, the *RELAX NG* language described in Section 23.4.2.

---

[5] This push vs. pull model is also an issue with the XSLT programming language (described in Section 24.2.2), which can be used in both ways.

It is questionable whether there ever will be one schema language to satisfy the requirements from all areas where XML is used, and the likely answer to this question is that it is impossible. If it is ever developed, this schema language will be so complex and powerful that it would essentially have become a programming language, departing from the advantages of a small and declarative language. Consequently, a reasonable approach to schema languages is to live with a number of them, provide features to combine them, and handle validation as a modular task. This is the approach taken by the *Document Schema Definition Languages (DSDL)* activity, which is described in Section 23.4.3.

23.4.1 XML Schema

XML Schema has been designed to meet user needs that go beyond the capabilities of DTDs. With one exception (the definition of *entities*, which are not supported by XML Schema), XML Schema is a superset of DTDs, which means that everything that can be done with DTDs can also be done with XML Schema. XML Schema is a two-part specification; the first part by Thompson et al. [2001] defines the structural capabilities, whereas the second part by Biron and Malhotra [2001] defines the datatypes that may be used in instances or for type derivation.

Figure 23.7 shows an example of an XML Schema. It is based on the DTD shown in Figure 23.2, but only parts are shown because the complete schema is much longer than the DTD. Obviously, the XML Schema differs from DTD in that it uses XML syntax, whereas DTDs use a special syntax. Also, the

```
<xs:schema xmlns:xs="http://www.w3.org/2001/XMLSchema">
  <xs:element name="book">
    <xs:complextype>
      <xs:sequence>
        <xs:element ref=title"/>
        <xs:element ref="chapter" minOccurs="0" maxOccurs="30"/>
      </xs:sequence>
      <xs:attribute name="editor" type="xs:string" use="required"/>
    </xs:complexType>
    <xs:unique name="chapter-section-id-unique">
      <xs:selector xpath="chapter | .//section"/>
      <xs:field xpath="@id"/>
    </xs:unique>
  </xs:element>
  <xs:element name="chapter">
    <xs:complexType>
      <xs:sequence>
        <xs:element ref="title"/>
        <xs:element ref="para" maxOccurs="unbounded"/>
        <xs:element ref="section" minOccurs="0" maxOccurs="20"/>
      </xs:sequence>
      <xs:attribute name="id" type="xs:ID" use="required"/>
      <xs:attribute name="data" type="xs:date" use="required"/>
      <xs:attribute name="author" type="xs:string"/>
    </xs:complexType>
  </xs:element>
  ...
</xs:schema>
```

FIGURE 23.7 XML Schema example.

occurrences of elements can be specified more precisely, in this case limiting the maximum number of `chapters` in a `book` to 30, and the maximum number of `sections` per `chapter` to 20. XML Schema supports a wide variety of datatypes; in this example, the `date` attribute is defined to use the `xs:date` type, and thus may only contain dates (in the DTD, the attribute was defined as `CDATA` and thus could contain any string of characters). Generally, when comparing XML Schema with DTDs, the following two aspects are the most important differences:

- *Types and type derivation:* XML Schema introduces a type system of *simple* and *complex types. Simple types* may be used for elements and attributes, defining types that do not include any XML structures. *Complex types* are types defining XML structures (similar to the content models and attribute lists of DTDs), and thus may only be used for elements.

  Types are derived from each other by a number of type derivation mechanisms: *restriction, list,* and *union* derivation for simple types, and *restriction* and *extension* for complex types. Consequently, all types are related to each other in a *type derivation hierarchy.* Elements and attributes use types from this type hierarchy.
- *Built-in datatypes:* Since many applications need the same simple datatypes, XML Schema defines a large number of built-in datatypes, which may either be used directly or as starting points for further type derivations (for example, by restricting the interval of number types). Figure 23.8 shows XML Schema's built-in datatypes, and can be regarded as the initial type hierarchy that users start with before defining their own types.

In an interesting generalization of the `ID/IDREF` attribute types of DTDs (basically an attribute with associated semantics, ensuring the integrity of intradocument references created with attributes), XML Schema defines *Integrity Constraints,* which can be used to model the same semantics as with `ID/IDREF` attributes in DTDs (see the `xs:unique` element in Figure 23.7 for an example of an identity constraint, specifying that `id` attributes must have unique values among all `chapter` children and `section` descendants of the `book` element). XML Schema integrity constraints use XPaths (see Section 23.3.2) to select keys and key references, and thus may be applied to element content as well as to attributes. Furthermore, keys and key references may contain combinations of elements and attributes, in effect defining multipart keys or key references. Finally, since integrity constraints are defined by selecting the respective parts using XPath(s), being part of an integrity constraint is in addition to having a type (as opposed to DTDs, where `ID/IDREF` are attribute types). Consequently, integrity constraints are evaluated using values (rather than lexical representations), so that `2` and `+02` are identical (if the type is numerical).

XML Schema validation is not defined in terms of documents, but in terms of Infosets (as described in Section 23.3.1). XML Schema validation starts with an Infoset, and augments the Infoset with additional information items and properties, which are called *Past Schema Validation Infoset (PSVI) contributions.* The advantage of this approach is that XML Schema does not require any additional mechanisms to define validation, and that validation is fine-grained (instead of the usual binary approach, where a document is either valid or not). The disadvantage of this approach is that there is no serialization for the PSVI contributions, so that the result of XML Schema validation cannot be expressed as an XML document. Another disadvantage is that there is no standard API (see Section 23.3.3) for the PSVI contributions, as the standard APIs only cover the standard Infoset and do not provide access to Infoset extensions.

Consequently, XML Schema has received some criticism for being too heavyweight and complicated, and one reaction to this increase in complexity is a smaller and lighter schema language, *RELAX NG,* described in the following section. However, since XML Schema (and in particular, its type system) will be the foundation of some influential XML technologies (XQuery 1.0 as described in Section 24.3.2, and XSLT 2.0 as described in Section 23.3.2), it is unlikely that any other schema language will entirely displace XML Schema in the near future. But the development of lightweight alternatives as well as of a modular framework for multiple schema languages (see Section 23.4.3) is an important development that will probably gain momentum in the light of the increasing variety of XML application areas.

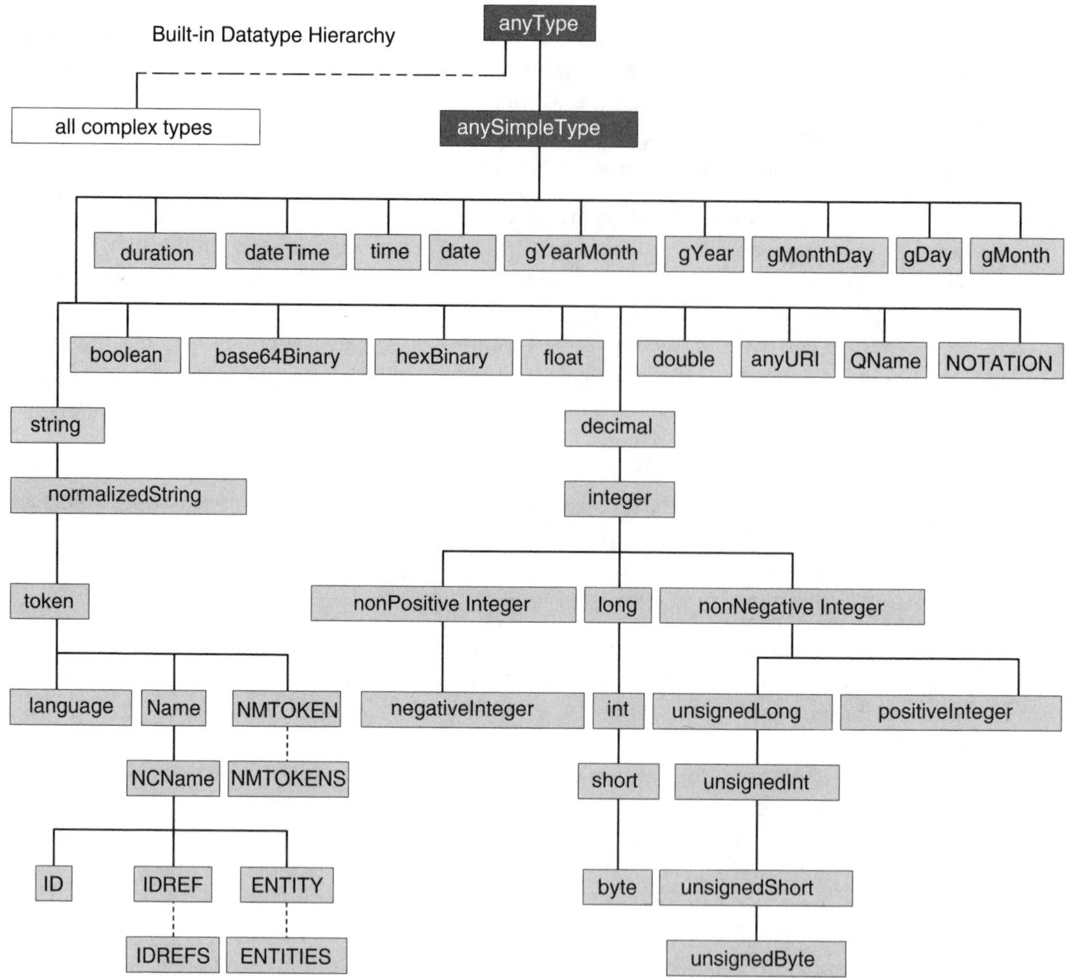

FIGURE 23.8 XML Schema built-in datatypes. Taken from Biron and Malhotra [2001] (http://www.w3.org/TR/xmlschema-2); Copyright ©2003 World Wide Web Consortium, (Massachusetts Institute of Technology, European Research Consortium for Informatics and Mathematics, Keio University). All Rights Reserved, http://www.w3.org/Consortium/legal/2002/copyright-documents-20021231.

23.4.2 RELAX NG

During the period when it was clear that DTDs were not powerful enough but no alternative schema language had had enough success to be the obvious successor to DTDs, two fairly small schema languages were designed, *Regular Language description for XML (RELAX)* and *Tree Regular Expressions for SML (TREX)*. Both were designed to create a schema language that was smaller than XML Schema, but still addressed the main weaknesses of DTDs. Since both approaches were similar in nature (providing a lightweight alternative to XML Schema), they were merged and the resulting schema language was named *RELAX NG* (defined by Clark [2001]). It is important to note that RELAX NG does not attempt to replace XML Schema. It is similar in certain ways (the most important being that it also is a grammar-based schema language), but also deliberately excludes some of the features of XML Schema, such as the type system, built-in datatypes, and identity constraints.

When comparing RELAX NG with DTDs, there is some added functionality that is not present in DTDs. Apart from the syntax issue (RELAX NG is based on an XML syntax[6]), RELAX NG supports datatyping, integrates attributes into content models, supports XML Namespaces, supports unordered

```
start   = book
book    = element book    { attribute editor { text }, title, chapter* }
chapter = element chapter { attribute id { xsd:ID },
                            attribute date { text },
                            attribute author { text }?,
                            title, para+, section* }
section = element section { attribute id { text }, title, para+ }
title   = element title   { text }
para    = element para    { (text | emph | quote | ref)* }
emph    = element emph    { (text | quote | ref)* }
quote   = element quote   { (text | emph)* }
ref     = element ref     { attribute id { xsd:ID }, empty }
```

FIGURE 23.9 Example RELAX NG Schema.

content, and supports context-sensitive content models. In trying to strictly separate document validation from document interpretation, RELAX NG does not support entity or notation declarations, and does also not support default values for attributes to be specified. Further, identity constraints (the `ID/IDREF` attribute types of DTDs) are not supported by RELAX NG, leaving the task of checking co-constraints (i.e., constraints within documents) to other mechanisms.

RELAX NG supports datatyping, but does not define a datatype library of its own. Instead, it provides a mechanism for referencing externally defined datatypes (using their Namespace URI). These datatypes can then be used for typing elements and attributes. One of the most popular datatype libraries used with RELAX NG is the XML Schema datatypes library defined in XML Schema's second part by Biron and Malhotra [2001].

Two features that are unique to RELAX NG and not supported by DTDs and XML Schema are the integration of attributes into the content model, and the possibility of specifying ambiguous content models. Integrating attributes into the content model makes it possible to specify content models that contain a choice of an element and an attribute, which in practice is very useful. Ambiguous (or non-deterministic) content models sometimes may be useful, but they should be used with care because they introduce the possibility that there are different interpretations of a document (which is not problematic in terms of validation, but may create problems in later stages of document processing).

Figure 23.9 shows an example of a RELAX NG schema. It uses the compact syntax, which is easier to read for human users, and is equivalent to the DTD shown in Figure 23.2. The most obvious difference apart from the syntax is the seamless integration of attributes into the content models, where they are used in conjunction with normal occurrence specifiers (such as `?`) to mark them as required or optional. Furthermore, RELAX NG uses an explicit `start` symbol to mark the document element of a schema.

23.4.3 Document Schema Definition Languages (DSDL)

Despite their differences, DTDs, XML Schema, and RELAX NG share many properties, the most common being that they all are grammar-based.[7] This means that they define a schema in terms of a grammar that can be used to construct valid instances according to the schema. The grammars have slightly different forms and features, but basically the DTD's ideas of element types and content models, and attribute

---

[6] RELAX NG uses XML as the syntax for schemas, but to make schema handling easier for human users, it also defines a *compact syntax* (defined by Clark [2002]), which uses a non-XML approach to represent RELAX NG schemas.

[7] It should be noted that the word "grammar" here and in the remainder of this section always refers to *context-free grammars*, excluding more complex (and powerful) classes of grammars.

lists associated with element types have not changed much in XML Schema or RELAX NG (with RELAX NG having the cleaner model). However, when thinking of schema languages in general, it quickly becomes obvious that grammar-based schemas are only one way to describe a schema. Other ways are easily imaginable, and this is where the *Document Schema Definition Languages (DSDL)* framework developed by the International Organization for Standardization comes into play.

DSDL is a framework for combining different schema languages and thus departs from the idea that there is one schema language that fits all needs and which should be used for all applications. DSDL's model is to provide a framework for modular validation, so that users can freely combine different schema languages, picking the ones that best fit their requirements. This obviously introduces the problem that a more complex infrastructure must be there to perform validation. However, the approach to combine schema languages follows an established tradition in computer science to break down a complex problem into smaller and simpler parts, and since validation may be a complex problem (depending on the application requirements), it may be advisable to apply this principle. DSDL is still evolving, and as it is going to be a multipart standard, some of the parts are briefly described in the following list:

- *Interoperability framework:* DSDL is a framework for validation, and consequently it is necessary to have a language for describing how individual schemas will be used for validation. This language can be thought of as a "metaschema" language, and because it needs to describe the combination of different processing steps for an XML document, it can be thought of as a *Processing Pipeline Language,* a concept discussed in more detail in Section 24.2.3.
- *Grammar-based validation:* This is the area XML users are familiar with because this is what DTDs are for. XML Schema and RELAX NG are part of this type of schema language, and DSDL will probably define RELAX NG as the preferred grammar-based schema language.
- *Rule-based validation:* A schema of a rule-based (schema) language simply defines a set of rules that must be satisfied by documents. These rules may be specific and detailed, or they may be rather loose and allow for a lot of variation. This is an interesting alternative to the generative nature of grammar-based schema languages and is described in more detail later in this section.
- *Datatypes:* One of the most required features not provided by DTDs was an application-oriented set of datatypes. One example of such a set is the XML Schema datatypes defined in XML Schema's second part by Biron and Malhotra [2001]. Since RELAX NG does not define a datatype library of its own, XML Schema's datatypes are the most likely candidate for this part.
- *Path-based integrity constraints:* DTDs have the `ID/IDREF(S)` mechanism, and XML Schema allows XPath-based integrity constraints; something along these lines is referred to in this part of DSDL. RELAX NG does not support integrity constraints of any kind, so if RELAX NG is used as the grammar-based schema language and integrity constraints are required, then some supplemental mechanism is needed.
- *Character repertoire validation:* XML is based on Unicode, and consequently allows a huge number of characters to appear as part of the markup or character data. However, many applications accepting XML-based input are not prepared to handle all possible Unicode characters. Consequently, there should be a mechanism for limiting the allowed character repertoire for different syntactic parts of XML documents.

This list of DSDL parts is not complete, and it is possible that it will be substantially expanded or that parts of it will be removed before DSDL proceeds to more mature standardization stages. However, the general idea of DSDL which is to look at validation as a combination of different schema languages that may be selected according to application requirements, will remain viable.

One interesting aspect in the context of DSDL is rule-based validation. *Schematron*[8] is a language for describing assertions that must be satisfied in certain contexts. Both the assertion as well as the contexts are specified as XPaths, which makes it easy to describe complex structural criteria that must be met by

[8] See `http://www.ascc.net/xml/resource/schematron/schematron.html`

a document. In particular, the structural possibilities go far beyond grammar-based approaches and identity constraints, so that Schematron may be used as an extension to grammar-based schema,[9] or as a complete replacement.[10]

References

Berglund, Anders, Scott Boag, Don Chamberlin, Mary F. Fernández, Michael Kay, Jonathan Robie, and Jérôme Siméon. *XML Path Language (XPath) 2.0.* World Wide Web Consortium, Working Draft WD-xpath20-20030822, August 2003.

Biron, Paul V. and Ashok Malhotra. *XML Schema Part 2: Datatypes.* World Wide Web Consortium, Recommendation REC-xmlschema-2-20010502, May 2001.

Boyer, John. *Canonical XML Version 1.0.* World Wide Web Consortium, Recommendation REC-xml-c14n-20010315, March 2001.

Bray, Tim, Dave Hollander, and Andrew Layman. *Namespaces in XML.* World Wide Web Consortium, Recommendation REC-xml-names-19990114, January 1999.

Bray, Tim, Dave Hollander, Andrew Layman, and Richard Tobin. *Namespaces in XML 1.1.* World Wide Web Consortium, Recommendation REC-xml-names11-20040204, February 2004.

Bray, Tim, Jean Paoli, and C. M. Sperberg-McQueen. *Extensible Markup Language (XML) 1.0.* World Wide Web Consortium, Recommendation REC-xml-19980210, February 1998.

Bray, Tim, Jean Paoli, C. M. Sperberg-McQueen, Eve Maler, and François Yergeau. *Extensible Markup Language (XML) 1.0 (Second Edition).* World Wide Web Consortium, Recommendation *REC-xml-20001006*, October 2004.

Clark, James. *RELAX NG Specification.* Organization for the Advancement of Structured Information Standards, Committee Specification, December 2001.

Clark, James. *RELAX NG Compact Syntax.* Organization for the Advancement of Structured Information Standards, Committee Specification, November 2002.

Clark, James and Steven J. DeRose. *XML Path Language (XPath) Version 1.0.* World Wide Web Consortium, Recommendation REC-xpath-19991116, November 1999.

Cowan, John. *XML 1.1.* World Wide Web Consortium, Recommendation REC-xml11-20040204, February 2004.

Cowan, John and Richard Tobin. *XML Information Set.* World Wide Web Consortium, Recommendation REC-xml-infoset-20011024, October 2001.

DeRose, Steven J., Eve Maler, and David Orchard. *XML Linking Language (XLink) Version 1.0.* World Wide Web Consortium, Recommendation REC-xlink-20010627, June 2001.

DuCharme, Bob. *XML: The Annotated Specification.* Prentice Hall, Upper Saddle River, NJ, December 1998.

Fernández, Mary F., Ashok Malhotra, Jonathan Marsh, Marton Nagy, and Norman Walsh, *XQuery 1.0 and XPath 2.0 Data Model.* World Wide Web Consortium, Working Draft WD-xpath-datamodel-20030502, May 2003.

Hollenbeck, Scott, Marshall T. Rose, and Larry Masinter. *Guidelines for the Use of Extensible Markup Language (XML) within IETF Protocols.* Internet best current practice RFC 3470, January 2003.

International Organization for Standardization. *Information Processing — Text and Office Systems — Standard Generalized Markup Language (SGML).* ISO 8879, 1986.

International Organization for Standardization. *Information Technology — Document Schema Definition Languages (DSDL)*, to be published as ISO/IEC 19757.

[9] Only modeling those assertions that may not be expressed in a grammar-based language.

[10] However, modeling grammar-like assertions in Schematron is less concise than in a grammar-based language, so this approach should be chosen only if there are no or only few grammar-like assertions.

Le Hors, Arnaud, Philippe Le Hégaret, Lauren Wood, Gavin Thomas Nicol, Jonathan Robie, Mike Champion, and Steven Byrne. *Document Object Model (DOM) Level 2 Core Specification.* World Wide Web Consortium, Recommendation REC-DOM-Level-2-Core-20001113, November 2000.

Le Hors, Arnaud, Philippe Le Hégaret, Lauren Wood, Gavin Thomas Nicol, Jonathan Robie, Mike Champion, and Steven Byrne. *Document Object Model (DOM) Level 3 Core Specification.* World Wide Web Consortium, Working Draft WD-DOM-Level-3-Core-20030609, June 2003.

Lee, Dongwon and Wesley W. Chu. *Comparative Analysis of Six XML Schema Languages. ACM SIGMOD Record,* 29(3):76–87, September 2000.

Marsh, Jonathan. *XML Base.* World Wide Web Consortium, Recommendation REC-xmlbase-20010627, June 2001.

Marsh, Jonathan and David Orchard. *XML Inclusions (XInclude) Version 1.0.* World Wide Web Consortium, Candidate Recommendation CR-xinclude-20040413, April 2004.

Pras, Aiko and Jürgen Schönwälder. *On the Difference between Information Models and Data Models.* Internet informational RFC 3444, January 2003.

Rose, Marshall T. *Writing I-Ds and RFCs using XML.* Internet informational RFC 2629, June 1999.

Thompson, Henry S., David Beech, Murray Maloney, and Noah Mendelsohn. *XML Schema Part 1: Structures.* World Wide Web Consortium, Recommendation REC-xmlschema-1-20010502, May 2001.

van der Vlist, Eric. *XML Schema Languages.* Proceedings of XML 2002, Baltimore, MD, December 2002.

Wilde, Erik. *Making the Infoset Extensible.* Proceedings of XML 2002, Baltimore, MD, December 2002.

Wood, Lauren, Arnaud Le Hors, Vidur Apparao, Steven Byrne, Mike Champion, Scott Isaacs, Ian Jacobs, Gavin Thomas Nicol, Jonathan Robie, Robert Sutor, and Chris Wilson. *Document Object Model (DOM) Level 1 Specification (Second Edition).* World Wide Web Consortium, Working Draft WD-DOM-Level-1-20000929, September 2000.

24

Advanced XML Technologies

CONTENTS

Introduction 24-1
24.1 Style Sheet Languages 24-1
24.1.1 Cascading Style Sheets (CSS) 24-2
24.1.2 Extensible Stylesheet Language (XSL) 24-3
24.2 XML Processing 24-4
24.2.1 Programming with XML 24-4
24.2.2 Transforming XML 24-4
24.2.3 Processing Pipelines 24-6
24.2.4 Distributed Programming 24-7
24.3 XML and Databases 24-7
24.3.1 XML and Relational Databases 24-8
24.3.2 Native XML Databases 24-8
References 24-9

Erik Wilde

Introduction

In Chapter 23, a number of XML core technologies are described. These are the technologies that most users of XML will probably encounter when working with XML. However, since XML is used in a wide range of applications scenarios, there are numerous other technologies that built on top of the core XML technologies. To cover some of these technologies, Section 24.1 briefly discusses XML style sheet languages for formatting XML documents. In many application scenarios the presentation of XML documents is not an issue, but the automated processing of XML data is very important. Thus, in Section 24.2 it is discussed how XML documents may be processed in different application scenarios. Finally, Section 24.3 discusses some of the questions surrounding the issue of how to handle XML in an infrastructure based on database management systems.

24.1 Style Sheet Languages

Even though XML is most successful as a format for exchanging structured data, its original purpose was to serve as document format on the Web. As such, there has to be a way to present an XML document in a browser or on any other kind of presentation device. For HTML, the elements and attributes and their (formatting) semantics are known because they are part of the HTML specification, and they are built into Web browsers. XML documents, on the other hand, use elements and attributes that are user-defined and consequently do not have any predefined formatting semantics. In order to format a document, a browser must have additional information, and this is where style sheet languages come into

1-58488-381-2/05/$0.00+$1.50

© 2005 by CRC Press LLC

play. The basic idea is that the presentation of an XML document requires the document itself and additional presentation information, and this additional information comes in the form of a style sheet.

The two most popular style sheet languages for XML are *Cascading Style Sheets (CSS)* and the *Extensible Stylesheet Language (XSL)*. CSS is the older of the two languages and was originally created for HTML. XSL, on the other hand, has been created specifically for XML and is the more powerful of the two languages. Figure 24.1 shows an overview of style sheet languages for XML; details of this figure are discussed in the following sections.

24.1.1 Cascading Style Sheets (CSS)

Cascading Style Sheets (CSS) were developed for HTML when it became apparent that HTML was increasingly burdened with formatting-specific information such as font and color. The idea of CSS was to separate the structure of an HTML page from the instructions of how to format it. With HTML, CSS is used to specify the formatting information that should differ from the formatting that is built into HTML browsers. With CSS for XML, the situation is different because there is no built-in (or default) formatting information for XML and, consequently, CSS for XML must fully specify how to format an XML document.

Apart from that, CSS for HTML and XML are similar. Basically, CSS can be split by half into the selectors (selecting the parts of a document that some formatting should be applied to) and associated properties (specifying the formatting that should be applied to the selected parts). The properties of CSS remain exactly the same for HTML and XML, and the selectors are simply specifying element and attribute names from the XML document, rather than HTML element and attribute names. Consequently, CSS for XML is written for a document class, with a given set of element and attribute names (in exactly the same way as CSS for HTML is written for the HTML document class, using HTML's element and attribute names).

Figure 24.2 shows an example of a CSS style sheet. It uses the element and attribute names from the document introduced in Figure 23.1. For example, `title` elements are formatted differently according to their position in the document tree, so that `book` and `chapter titles` use a larger font than `section titles`. As some more examples of CSS formatting, `para` elements are formatted as individual paragraphs (as a separate display block), and `quote` elements use special properties for inserting content before and after the element's occurrence.

The advantage of CSS is that it can be written very quickly, and that users with CSS know-how from HTML can take their knowledge and apply it without any modification to formatting XML documents.

The disadvantage of CSS is that it has certain limitations. The most important limitation is that CSS is not able to modify the structure of a document. The formatting of each part of a document may be specified, but it is impossible to apply any structural changes; for example, changing the sequence of a number of paragraphs, or even generating new content, such as creating and inserting a table of contents are not possible.

Another limitation of CSS is that it does not adapt very well to the flexibility of XML. For example, there is no way in CSS to specify that an element is a link. CSS supports the built-in link functionality of HTML's linking elements, but if an XML document uses different element names for links (maybe even a standardized link vocabulary such as XLink), it is impossible to create a CSS style sheet that will format this element as a link.

For the above reasons, the more powerful *Extensible Stylesheet Language (XSL)* has been created, which is described in detail in the following section.

CSS has been evolving since it has been invented. The first version is known as *Cascading Style Sheets, Level 1 (CSS1)* and has been defined by Lie and Bos [1999]. *CSS2* was published by Bos et al. [1998] as the next version of CSS, and currently CSS3 is under development by Meyer and Bas [2001]. CSS3 is the first version of CSS that has been modularized, defining modules for functionally different areas such as multicolumn layout and list formatting.

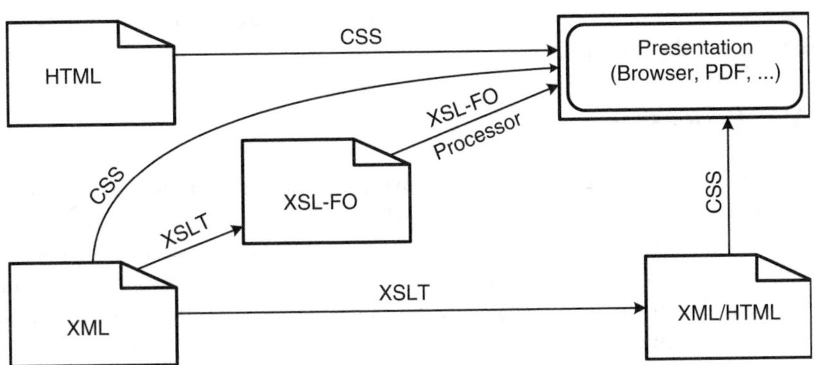

FIGURE 24.1 XML Style Sheet Languages.

```
book > title          { font-size: xx-large ; display : block }
chapter > title       { font-size: x-large ; display : block }
section > title       { font-size: large ; display : block }
para                  { display : block }
emph                  { font-style : italic }
quote::before         { content : open-quote }
quote::after          [ content : close-quote }
ref                   { content : attr(id) }
```

FIGURE 24.2 Example CSS Style Sheet.

24.1.2 Extensible Stylesheet Language (XSL)

The *Extensible Stylesheet Language (XSL)* defined by Adler et al. [2001] is a more powerful way to format XML documents than CSS. In XSL, formatting is a two-step process, the first step being a structural transformation and the second step being the actual formatting of the transformed XML document.

The transformation part of XSL is done using the *XSL Transformations (XSLT)* described in detail in Section 24.2.2. XSLT is a programming language which takes as an input an XML document and produces as output either a text document, HTML, or XML. In the scenario of XSL, XSLT is used to transform the XML document into an intermediary XML document, which uses *XSL Formatting Objects (XSL-FO)*. Figure 24.1 shows these possibilities, where the original XML document is transformed to either XSL-FO or something else (which then can be presented using non-XSL means).

XSL-PO is a formatting vocabulary that is used as a source to produce high-quality formatting results. It is technically aligned with CSS3, which means that the formatting functionality present in XSL-FO is similar to CSS3 (there are some minor differences). The difference between XSL-FO and CSS3 is that XSL-FO is an XML vocabulary using elements and attributes, which are produced from a source document using XSLT, while CSS3 formatting is applied to an XML document using CSS selectors. This difference is the reason for the greater flexibility of the XSL approach, which allows structural transformations to be applied before actual formatting takes place.

Once an XSL-FO document has been produced, it has to be presented. This is the task of an *XSL-FO processor,* which takes as input an XSL-FO, and produces as output a formatted result. This result can either be an on-screen presentation or a format intended for capturing formatted documents, such as the *Portable Document Format (PDF).*

While XSL was designed for the formatting of XML documents, its split into a transformation language (XSLT) and the formatting objects (XSL-FO) has made it possible to reuse these individual parts of the language in a variety of application contexts. The transformation part of XSL is by far the more popular part of the language, with transformations being applied in different application contexts, only a tiny fraction of them actually being formatting applications, and an even smaller fraction of these using XSL-FO as the target format of the transformation.

24.2 XML Processing

Even though XML has been designed as a new format for documents on the Web, its success has been based on its ability to represent structured data. XML's major application area today is as an exchange format between applications, and as such it needs to be processed in a variety of ways: to be generated from application data, to be transformed between different XML representations, to be consumed by applications, and to be stored in databases. In the following sections, these different areas of XML processing are described in greater detail.

24.2.1 Programming with XML

In many cases, XML needs to be accessed from programs. This access may either be read-only access, when a program gets an XML document as input, or access including modifications. Modifying XML may mean making changes to an existing XML document, such as changing textual content or attribute values or making structural changes, or it may mean creating an XML document. Creating XML documents often is necessary when programs need to communicate with other programs through the exchange of XML documents.

In all these cases, programs will likely use existing XML tools for supporting them, and use the tools through APIs such as DOM, SAX, JDOM, or JAXP as described in Section 23.3.3. Using standardized APIs not only makes it easier for programmers to reuse their knowledge of how these APIs work, it also makes programs independent of the underlying XML tools, which thus may be replaced with other tools if necessary.

It is therefore highly advisable to restrict program access to XML tools to standardized interfaces. This makes it possible to avoid dependencies from certain tools, which can be problematic if the tools turn out to have some untolerable bugs or if they are simply not available on another platform which the program has to be ported to.

24.2.2 Transforming XML

One special case of working with XML is the transformation of an XML document into another document. In many cases, this is best done using *XSL Transformations (XSLT)*, defined by Clark [1999]. XSLT is a specialized programming language for transforming XML, which has been developed as one part of the *Extensible Stylesheet Language (XSL)* as described in Section 24.1.2. Even though the name of XSLT seems to imply that it is for style sheet purposes only, it can be regarded as a general transformation language for XML. Some features of XSLT show its heritage from the style sheet area, but apart from these minor features, XSLT can be considered a general transformation language.

XSLT is a programming language which accepts as input an XML Infoset, and produces as output either XML, or HTML, or a text document. This makes it possible to chain multiple transformation processes in a pipeline, which is described in more detail in Section 24.2.3.

The most important part of XSLT is the *XML Path Language (XPath)* as described in Section 23.3.2. In XSLT, XPath is used to select parts of the input document, and in general as an expression language to calculate results and structures for processing and output. The remaining part of XSLT (i.e., the non-XPath part of XSLT) is a functional programming language based on a runtime system that processes the document and chooses parts (so-called *templates*) of the XSLT code to execute upon processing of certain nodes of the document tree.

Figure 24.3 shows an example of an XSLT program. This program is intended to transform documents that have been used as examples in the previous chapter, so it may process any document conforming to the document class defined by the DTD of Figure 23.2, and as a specific example the XML document

```
<xsl:stylesheet version="1.0"
                xmlns:xsl="http://www.w3.org/1999/XSL/Transform">
  <xsl:template match="book">
    <html>
      <head><title><xsl:value-of select="title"/></title></head>
      <body><xsl:apply-templates/></body>
    <html>
  </xsl:template>
  <xsl:template match="title">
    <xsl:choose>
      <xsl:when test="parent::book">
        <h1><xsl:value-of select="."/></h1>
      </xsl:when>
      <xsl:when test="parent::chapter">
        <h2><xsl:value-of select="."/></h2>
      </xsl:when>
      <xsl:otherwise>
        <xsl:element name="{concat('h',count(ancestor::section)+2)}">
          <xsl:value-of select="."/>
        </xsl:element>
      </xsl:otherwise>
    </xsl:choose>
  </xsl:template>
  <xsl:template match="para">
    <p><xsl:apply-templates/></p>
  </xsl:template>
  <xsl:template match="emph">
    <em><xsl:apply-templates/></em>
  </xsl:template>
  <xsl:template match="quote">
    "<xsl:apply-templates/>"
  </xsl:template>
  <xsl:template match="ref">
    <a href="{concat('references#',@id)}">
      <xsl:value-of select="@id"/>
    </a>
  </xsl:template>
</xsl:stylesheet>
```

FIGURE 24.3 Example XSLT program.

shown in Figure 23.1. XSLT works by defining *templates,* which are used to transform specific parts of the input document. In the above example, the elements and attributes of the example XML document are transformed into HTML element, so that the result of the transformation process (i.e., the XSLT program execution) is an HTML page that can be used by any Web browser.

Figure 24.4 shows the result of the transformation process. The whitespace has been cleaned up manually to make the example easier to read, but since Web browsers are ignoring whitespace anyway, this is not a problem. The code shown in Figure 24.3 is all that is required to perform this transformation between a proprietary XML document class and HTML, which demonstrates that XSLT programs can be very short and rather easy to write. Most of the work of programming in XSLT is constructing XPaths.

```
<html>
  <head>
    <title>Practical Handbook of Internet Computing</title>
  </head>
  <body>
    <h1>Practical Handbook of Internet Computing</h1>
    <h2>XML Core Technologies</h2>
    <p>The <em>Extensible Markup Language (XML)</em> ... </p>
    <h3>XML</h3>
    <p>the XML specification has been first released in February 1998
      by <a href="references#xml10-spec">xml10-spec</a>, followed by
      a "second edition" of the specification in October 2000 ... </p>
  </body>
</html>
```

FIGURE 24.4 Resulting HTML.

For example, the code part generating the titles for `section` elements generates an element (using the `xs:element` instruction) in the result document whose name is the concatenation of the character "h" and the sum of 2 and the number of `section` ancestors of the `title` element. For the `section` directly inside a `chapter`, this yields an HTML `h3` element, for a `section` inside this `section` an `h4` element and so forth.

XSLT is powerful for performing tasks based on the structure of the XML input because XPath makes it easy to construct elaborate expressions for selecting parts of an XML document. However, for tasks beyond structural transformations of an XML document (such as string processing or numerical calculations), XSLT is not the right tool. Since a structural transformation is often one part of processing an XML input, the *Transformation API for XML (TrAX)* can be used to access an XSLT processor from within application programs.

XSLT 1.0 has some limitations which were criticized by many developers (such as the inability to produce multiple output documents) and were addressed in an updated version, XSLT 1.1 defined by Clark [2001]. XSLT 1.1 never reached recommendation status because it became apparent that XSLT needed a major revision in cooperation with the activities surrounding the *XML Query Language (XQuery)* (described in the following section) and, consequently, XSLT 2.0 is under development by Kay [2003].

24.2.3 Processing Pipelines

One of the main reasons for XML's success is the availability of a large number of tools and components, which can be used and reused when working with XML and programming XML-based applications. Processing of XML data can often be thought of as a pipeline, for example, with the first stages being validations of various kinds and the later stages being application-dependant processing of parts of the XML document. Instead of rebuilding the infrastructure to support this kind of pipeline processing for every application, it makes sense to define a generic pipeline model and reuse it in different application contexts.

However, so far there is no established model for XML processing pipelines. There are proposals by Walsh and Maler [2002], the *XPipe* language by McGrath [2002], and the *XML Pipeline Definition Language (XPL)* by Bruchez [2002], but none of these so far has been accepted as a general way to build XML processing pipelines.

Furthermore, there are application-specific solutions such as Cocoon's pipeline concept described by Brogden et al. [2002] and DSDL's *Interoperability Framework* as described Section 23.4.3. These application-specific languages do not attempt to provide a general solution to the processing pipeline model, but could probably also benefit (or even be replaced by) a generic XML processing pipeline language.

Generally speaking, the idea of XML processing pipelines could bring the same flexibility and modularity to XML processing as the pipe model brought to the Unix operating system. Unix pipes connect processes on a more basic level, using a byte stream, while an XML processing pipeline should connect XML processing components with a mechanism to pass XML structures, based on some XML data model (see Section 23.3 for more information about XML data models).

24.2.4 Distributed Programming

XML as a data exchange format not only is used locally between processes or applications (as discussed in the previous section about processing pipelines) but often is used as an exchange format across networks. Basically, XML can be exchanged between applications by using any transport protocol supporting the exchange of octet strings, such as the *File Transfer Protocol (FTP)*, the *Hypertext Transport Protocol (HTTP)*, or the *Simple Mail Transfer Protocol (SMTP)*. However, this basic model of exchanging XML would require application designers to solve the same problems again and again, for example designing some kind of "envelope format" for communicating the application-level semantics of the XML being exchanged.

To avoid this, the idea of "Web Services" emerged, which basically is an infrastructure for loosely coupled applications using Web and Internet technologies (see Chapter 31 for more information about Web Services). The current Web Services technologies are the *Simple Object Access Protocol (SOAP)* defined by Mitra [2003] for the actual exchange of XML-based messages, and the *Web Services Description Language (WSDL)* defined by Chinnici et al. [2003] for defining the interfaces of a Web Service. Both technologies are XML-based and therefore fit well into an XML-based application scenario.

Web Services are a way of distributing programming, although at a basic level. Data representation for distributed systems has also been the target of *Abstract Syntax Notation One (ASN.1)*, a language designed in the context of the *Open Systems Interconnection (OSI)* architecture. ASN.1's information model differs from that of XML, but the structural differences are minor: both models are tree-based. Unlike XML, ASN.1 makes a clear distinction between the information model (the abstract syntax) and data models, which are defined by so-called *encoding rules*. The *XML Encoding Rules (XER)* defined by the International Telecommunication Union [2001] are a way to encode ASN.1's information model using an XML syntax.

The role of XML in distributed applications is only beginning, and slowly some standards appear which make it easier to build distributed applications. However, it will take considerable time until a universally accepted framework for XML-based distributed applications exists, and until then developers are forced to make their choices between competing technologies.

24.3 XML and Databases

XML is increasingly popular as a format for data exchange, and also as a format for storing data. The bulk of data today is stored in *Database Management Systems (DBMS)*. The majority of DBMS use the *relational data model* and are consequently called *Relational Database Management Systems (RDBMS)*. With the advent of XML, the structural differences between the relational data model and XML have led to RDBMS systems that support XML in a variety of ways, and *native XML databases*, which are DBMS using XML as their data model. Abiteboul et al. [1999] describe the fundamental differences between relations and XML's data model, which lead to the following options when dealing with XML and database systems:

- *Storing XML as character data:* In this case, the XML document is simply stored as character data. This is an efficient way to store the XML, but it makes it inefficient to query into the XML data.
- *Shredding XML into relational structures:* If the structure of the XML to be stored is known in advance, the database system can disassemble the XML document and store its contents in rela-

tional structures. This makes it possible to use relational queries to access the XML data. However, if the XML has to be retrieved, it has to be reconstructed from the relational structures.

- *Storing XML natively in a relational structure:* An increasing number of relational databases support XML as a datatype, making it possible to store the XML data as an object of type XML, which can then be queried using XML-specific queries. To execute these queries efficiently, the database must build indexes over the XML data.
- *Storing XML natively in an XML database:* In this case, the data model of the database is XML-based, so that the XML is not stored in a relational column of type XML, but in a database that uses XML as the underlying data model. As in the former case, the database must build indexes over the XML data for efficient querying.

The question whether users should rely on the XML support of RDBMS products or make the transition to native XML databases is hard to answer. It mainly depends on the application area. Many business data models are inherently relational (or at least reasonably easy to cast into a relational model), and thus are adequately supported by relational databases. However, other application areas such as document-centric businesses may have semistructured data as the center of their data model, and for these application areas it may make sense to move to XML as the data model.

24.3.1 XML and Relational Databases

Relational data basically is a collection of tables, with each table consisting of a fixed number of columns and an arbitrary number of rows. Tables can be linked through *keys* and *foreign keys,* and these links represent conceptual connections between the rows of different tables. Data is retrieved from an RDBMS by using a query language, the single most successful language being the *Structured Query Language (SQL).*

In many cases, an RDBMS exchanges data with business partners using XML-based technologies, so the question is how to extract data from an RDBMS and encode it in XML, and vice versa. The extraction of data from an RDBMS into some form of XML encoding has been supported by proprietary functions of database vendors for some time, but the *SQL/XML* extension of SQL now defines a standardized way to write SQL queries that result in XML documents. The other direction may be handled in different ways, for example by using XSLT to transform an XML document into an SQL statement that inserts or updates data.

In either case, the question is how to map the data models of the RDBMS and XML, or how to derive one data model from the other, if one is given and the other is still undefined. An increasing number of tools is available for generating data models automatically or semiautomatically, but for data models that will have a long life-span and need to be understood and used by humans, it is still recommended to treat data modeling as a very important step in a project, and thus carefully design the data model by hand. Since the relational model is structurally simpler than XML, it usually is not a problem to derive an XML schema from a relational schema. However, if a complex XML schema has to be mapped to a relational model (and in particular, if the XML schema uses a lot of mixed content), it may be rather complicated to create a relational schema that covers the full XML schema, is easy to understand, and reasonably efficient to query.

24.3.2 Native XML Databases

XML's data model is richer than the relational model, and for some application areas, particularly document-centric applications, it makes more sense to use XML as the data model than to map between XML for transfer and processing, and relational data for querying and storage. In this case, a *native XML database* is required, which is a database that uses XML as the underlying data model. Consequently, native XML databases can store and generate XML directly without the need to map between data models. Native XML databases are often optimized for different application scenarios than relational databases, focusing on large collections of small and irregularly structured XML documents, which are not handled so well by relational databases, even if they support XML as a datatype.

input.xml

```
<?xml version="1.0" encoding="ISO-8859-1"?>
<documentelement attribute2 = '&apos;' attribute1 = "&quot;">
  <emptylement></emptylement>
  <emptylement/>
  <element >Character References: &#97;&#x61;a</element >
</documentelement>
```

output.xml

```
<?xml version="1.0" encoding="UTF-8"?>
<documentelement attribute1='"' attribute2="'">
  <emptylement/>
  <emptylement/>
  <element>Character References: aaa<element>
</document/element>
```

FIGURE 24.5 Example for XML round-tripping.

If the database uses XML as the data model, it is not quite clear what that means. As discussed in Section 23.3, XML itself does not define a data model, and there is a variety of data models that have been defined for XML. However, the XML Infoset as described in Section 23.3.1 is the data model that is used for many applications and specifications building on top of XML. Depending on the requirements, an XML database implementing the XML Infoset as the internal data model can have unexpected results.

Storing and retrieving a document in such a database can yield the result shown in Figure 24.5. The input and the output document have the exact same XML Infoset, so from the database's point of view they are identical. Since different native XML databases behave differently with regard to this problem, XML database users should know their requirements before selecting a specific product. For most data-oriented applications, Infoset-based round-tripping is probably sufficient, while document-oriented applications often require support for XML structures (such as character references and a document's entity structure) that are not part of the Infoset.

Storing and retrieving XML documents only is part of what a native XML database has to do. Another very important aspect is efficient querying. Native XML databases often used their own, proprietary query language to access XML structures in the database. In an effort to establish a standard query language, the *XML Query Language (XQuery)* defined by Boag et al. [2003] currently is under development. It is based on *XPath 2.0* as described in Section 23.3.2, and most of the functionality of XQuery is provided by XPath 2.0. XQuery adds constructors which have a purpose similar to the `JOIN` statement of relational databases: They make it possible to recombine the results of queries and then query this combination.

One serious limitation of XQuery as it is currently standardized is that it does not provide inserts or updates, which will probably be included in future versions of the language. Until then, updating XML still requires functionality beyond that of the standardized XQuery language.

References

Abiteboul, Serge, Peter Buneman, and Dan Suciu. *Data on the Web: From Relations to Semistructured Data and XML.* Morgan Kaufmann, San Francisco, CA, October 1999. ISBN 0-55860-622-X.

Adler, Sharon, Anders Berglund, Jeff Caruso, Stephen Deach, Paul Grosso, Eduardo Gutentag, R. Alexander Milowski, Scott Parnell, Jeremy Richman, and Stephen Zilles. *Extensible Stylesheet Language (XSL) Version 1.0.* World Wide Web Consortium, Recommendation REC-xsl-20011015, October 2001.

Boag, Scott, Don Chamberlin, Mary F. Fernández, Daniela Florescu, Jonathan Robie, and Jérôme Siméon. *XQuery 1.0: An XML Query Language.* World Wide Web Consortium, Working Draft WD-xquery-20030822, August 2003.

Bos, Bert, Håkon Wium Lie, Chris Lilley, and Ian Jacobs. *CSS2 Specification.* World Wide Web Consortium, Recommendation REC-CSS2-19980512, May 1998.

Brogden, Bill, Conrad D'Cruz, and Mark Gaither. *Cocoon 2 Programming: Web Publishing with XML and Java.* Sybex, Berkeley, CA, October 2002. ISBN 0782141315.

Bruchez, Erik. *An Introduction to XML Pipelines.* Technical report, Orbeon, Inc., Mountain View, CA, October 2002.

Chinnici, Roberto, Martin Gudgin, Jean-Jacques Moreau, and Sanjiva Weerawarana. *Web Services Description Language (WSDL) Version 1.2 Part 1: Core Language.* World Wide Web Consortium, Working Draft WD-wsdl12-20030611, June 2003.

Clark, James. *XSL Transformations (XSLT) Version 1.0.* World Wide Web Consortium, Recommendation REC-xslt-19991116, November 1999.

Clark, James. *XSL Transformations (XSLT) Version 1.1.* World Wide Web Consortium, Working Draft WD-xslt11-20010824, August 2001.

International Telecommunication Union. *Information Technology — ASN.1 Encoding Rules — XML Encoding Rules (XER).* ITU-T Recommendation X.693, December 2001.

Kay, Michael. *XSL Transformations (XSLT) Version 2.0.* World Wide Web Consortium, Working Draft WD-xslt20-20030502, May 2003.

Lie, Håkon Wium and Bert Bos. Cascading Style Sheets, level 1. World Wide Web Consortium, Recommendation REC-CSSI-19990111, January 1999.

McGrath, Sean. *XPipe — A Pipeline Based Approach To XML Processing.* In *Proceedings of XML Europe 2002,* Barcelona, Spain, May 2002.

Meyer, Eric A. and Bert Bos. *CSS3 Introduction.* World Wide Web Consortium, Working Draft WD-css3-roadmap-20010523, May 2001.

Mitra, Nilo. *SOAP Version 1.2 Part 0: Primer.* World Wide Web Consortium, Proposed Recommendation PR-soap12-part0-20030507, May 2003.

Walsh, Norman and Eve Maler. *XML Pipeline Definition Language Version 1.0.* World Wide Web Consortium, Note NOTE-xml-pipeline-20020228, February 2002.

25

Semistructured Data in Relational Databases

CONTENTS

Abstract 25-1
25.1 Introduction 25-1
25.1.1 Sources of Semistructured Data 25-2
25.1.2 Running Example 25-3
25.2 Relational Schemas for Semistructured Data 25-4
25.2.1 Using Tuple-Generating Elements 25-4
25.2.2 Representing Deep Structure 25-4
25.2.3 Representing Ancestors of TGEs 25-5
25.2.4 Varying Components 25-8
25.2.5 Semistructured Components 25-9
25.2.6 Graph Representation 25-10
25.2.7 Storing Unparsed XML 25-12
25.3 Using XML Features of Relational Database Systems 25-13
25.3.1 IBM DB2 25-13
25.3.2 Microsoft SQL Server 25-16
25.3.3 Oracle XSU 25-17
25.3.4 Sybase 25-18
References 25-18

Sudarshan Chawathe

Abstract

Semistructured data is data whose structure is irregular, incomplete, and frequently changing. Traditional database methods are ill-suited to storing and querying such data because they rely on a well-defined structure (schema). While such data occurs in several formats, our focus is on the XML format. We present several alternatives for storing semistructured data in relational databases and discuss their merits and limitations. We also discuss the methods used in commercial database management systems from IBM, Oracle, Microsoft, and Sybase. Throughout, we use a concrete running example to illustrate the features of semistructured data, explain the methods, clarify their features, and motivate further improvements.

25.1 Introduction

A collection of Web pages from a typical online store illustrates the organization typical of semistructured data. A page describing a book for sale is structured to indicate information such as the title, author, price, editorial review, and ISBN. Although such structuring suggests a traditional schema, there are variations in this structure that hinder the application of standard database design methods. There may be no information on the ISBN or publisher of some books. Other books may be listed with multiple prices (perhaps from different sellers in a used-book marketplace). The editorial review may consist of

1-58488-381-2/05/$0.00+$1.50
© 2005 by CRC Press LLC

a numerical score, a text description, or both. The publisher may be listed in various forms (name only, name followed by address as text, name followed by address as a multifield record, etc.).

25.1.1 Sources of Semistructured Data

It is natural to inquire why some data is semistructured. After all, there are well developed methods for modeling data and for transforming abstract models into concrete database schemas (e.g., Entity Relationship modeling and relational design theory [Garcia-Molina et al., 2002]). One source of semistructured data is information that resides natively in a format that resembles a document more than a traditional database. For example, it is more natural to view system documentation (e.g., help files, Unix *man* pages, GNU *Info* files) as a collection of documents rather than as a collection of entities and relationships. Such *document data* is increasingly common on the Internet. There is a vast body of literature on information retrieval methods for managing documents (Baeza-Yates and Ribeiro-Neto, 1999]. The majority of this work has focused on finding documents that match a query in a Boolean or vector-space model. Our focus here is on methods for querying documents as structured data repositories, permitting, for example, matching parts of a document, joins on matches across documents, and other features typically found in database query languages. The structure of a document is irregular and dynamic because, in a document-centric environment, changing the structure of a document is often as natural and common as is changing document content.

Even when it is possible to accurately model data using traditional methods, the task may be too cumbersome relative to the importance of the data. For example, a detailed study of the standard forms and memos used in an organization is likely to yield a database schema for such data. However, this database design task, which requires discovering and formalizing the conventions and informal standards used within the organization, is likely to be too laborious for an implementation of an application for searching the forms and memos. We refer to such data as *incidental data:* it is too important to discard or ignore, but not important enough to justify a traditional data modeling effort. Incidental data is also marked by frequent changes in structure. If this structure is encoded in a database schema, every change in structure potentially triggers a laborious schema modification. Therefore, it may be prudent to not encode the structure of such data in database schemas.

Yet another source of semistructured data is *data integration.* When information from several sources is combined, the schema assumptions made by one source are likely to be invalid in another. For example, the schema of an employee database may be based on the assumption that an employee has at most one manager. Integrating this database with another that contains multiway employee–manager relationships requires adopting a more general schema. As the number of integrated sources rises, the schema for the integrated data becomes increasingly general. Since traditional database methods require all data to strictly adhere to a fixed schema, a single exception to a generally valid assumption necessitates the adoption of a more general schema that does not make the assumption. (For example, consider a collection of electronic catalogs, all but one of which list the full address of the source of each item in the form of street, city, state, and postal code, while one catalog lists the address only as unstructured text.)

When the amount of semistructured data is small or when it is accessed using only a few, well-known access patterns, it is possible to encapsulate such data in one or more applications using *ad hoc* methods. However, when a large amount of such data needs to be queried by several applications or when the access patterns cannot be predicated well, it is necessary to store such data in a database management system that supports *ad hoc* queries along with other database features such as concurrency control, access control, and durability. One option in such a situation is to use special database systems designed for semistructured data or object database systems with some extensions [Chawathe, 2003]. However, there are significant benefits to using standard relational database systems instead. For instance, applications typically need to access the semistructured data not in isolation, but in conjunction with a structured database. Storing semistructured data in the relational database system used for the structured data simplifies system design and permits query optimization across the boundary between structured and semistructured data. Relational storage of semistructured data is therefore the focus of this chapter.

25.1.2 Running Example

As a running example, we use the domain of datasets that describe photographs. Typical information stored for each photograph includes the date and time the photograph was taken, the location pictured, the camera settings, the photographer, and so on. Figure 25.1 depicts a small fragment of such a dataset, rendered as XML.

A complete description of XML and related standards is beyond the scope of this chapter. For our purposes, the following simplified definition of XML suffices: An XML document is a text serialization of a hierarchy of *elements.* An element is a unit of data that may contain text and other (nested) elements, called its subelements. Each element has a *name,* which may be thought of as the informal type of the element. Syntactically, an element named *foo* is represented by enclosing its contents between a begin tag, written as <foo>, and an end tag, written as </foo>. Each element may be adorned with zero or more *attribute-value* pairs, represented in the begin tag of the element using the syntax <foo attr1="val1" attr2="val2">. The XML fragment of Figure 25.1 has two *photometa* elements at the top level. The location element near the bottom of the figure is nested within the second photometa element and in turn contains <*line 1*>, city, state, and note as subelements. The time element on line 3 has a zone attribute with value PST.

At first glance, our sample data appears to possess a uniform structure. However, a closer examination reveals several variations: The second photometa element is missing a camera subelement as found in the first. The two time elements differ in their formats. The first includes a zone attribute indicating the time zone and uses a 24-hour format that includes seconds whereas the second has no time zone information and uses a 12-hour format without seconds. The location elements also exhibit differing structures. The first specifies the location using a brief label, while the second specifies the different components of a location separately using a set of subelements.

```
 1: <photometa>
 2:    <date>2002-12-21</date>
 3:    <time zone="PST">15:03:07</time>
 4:    <flen>10mm</flen>
 5:    <F>8.0</F>
 6:    <shutter>1/125</shutter>
 7:    <flash>n</flash>
 8:    <location>Death Valley</location>
 9:    <camera>Nikon 995</camera>
10: </photometa>
11: <photometa>
12:    <date>2002-12-23</date>
13:    <time>10:37 PM</time>
14:    <flen>28mm</flen>
15:    <F>2.2</F>
16:    <shutter>1/250</shutter>
17:    <flash>Y</flash>
18:    <location>
19:        <line1>2127 Firewood Ln</line1>
20:        <city>Springfield</city>
21:        <state>MA</state>
22:        <note>Living Room</note>
23:    </location>
24: <photometa>
```

FIGURE 25.1 Photo metadata in XML.

Photometa

ID int	Date date	Time time	Flen float	F float	Shutter float	Flash boolean	Location varchar(100)	Camera varchar(50)
1001	2002=12-21	23:03:07	10	8.0	0.008	0	Death Valley	Nikon 995
1002	2002-12-23	22:37:00	28	2.2	0.004	1	Springfield	*null*

FIGURE 25.2 A simple relational schema.

In Section 25.2, we explore several schemes for coping with such variations in structure. Our discussion is based on general methods for encoding semistructured data, specifically XML, into relational tables. In Section 25.3, we turn our attention to specific XML features found in commercial products from four database vendors. We compare the features with each other and relate there to the general methods introduced in Section 25.2. Throughout this chapter, we use our running example to tie together ideas from different methods and database products into a common framework.

25.2 Relational Schemas for Semistructured Data

In this section, we present some simple methods for representing semistructured data in relational form. For concreteness, we will focus on data in XML format, using the data of Figure 25.1 as a running example.

25.2.1 Using Tuple-Generating Elements

Perhaps the simplest method for storing the photo data is to use a relation with one attribute for each subelement of the photometa element, as suggested by Figure 25.2. In general, this method is based on choosing one XML element type as the *tuple-generating element (TGE) type.* We use a relation that has one attribute for each possible subelement (child) of the tuple-generating element, along with an ID attribute that serves as an artificial key of the relation. Each instance of a tuple-generating element is mapped to a tuple. The content of each subelement of this tuple-generating element is the value of the corresponding attribute in the tuple.

One drawback of this method is that instances of the tuple-generating element that do not have every possible subelement result in tuples with nulls. In the sample data, the missing camera subelement of the second photometa element results in a null in the second tuple of Figure 25.2. Another drawback is that subelements with subelements of their own (i.e., those with *element* or *mixed* content [Bray et al., 1998]) are not represented well. In the sample data, the location element of the second photometa element has several subelements; the relational representation stores only the city. This problem is more severe for data that has a deeply nested structure.

25.2.2 Representing Deep Structure

We may address the second drawback above by creating relational attributes for not only the immediate subelements of a tuple-generating element, but also the subelements at deeper levels. Essentially, this method flattens the nested structure that occurs within tuple-generating elements. Figure 25.3 illustrates this method for the sample data. In general, this method is based on selecting a tuple-generating element type and creating a relation with an ID attribute and an attribute for every possible subelement (direct or indirect) of such elements. For our running example, we have chosen photometa as the tuple-generating element to yield the representation of Figure 25.3. (Rows beginning with ... represent continuations of the preceding rows.) Subelements that have only *element* content are not mapped to attributes in this method. The location subelement of the second photometa element does not generate an attribute for this reason. Although this method addresses the second drawback of our earlier method, it exacerbates the problem of nulls. Note that we have chosen to represent the location information of

Photometa

<u>ID</u> int	Date date	Time time	Flen float	F float	Shutter float	Flash boolean	...
1001	2002-12-21	23:03:07	10	8.0	0.008	0	...
1002	2002-12-23	22:37:00	28	2.2	0.004	1	...

Photometa (contd.)

...	Addr varchar(40)	City varchar(40)	State char(2)	Camera varchar(50)
...	Death Valley	*null*	*null*	Nikon 995
...	2127 Firewood Ln	Springfield	MA	*null*

FIGURE 25.3 The schema of Figure 25.2 modified for detailed location information.

the first photometa element in the Addr attribute of the relation. The alternative — of including a location attribute in the relation — results in more nulls.

These simple representations have the advantage of being easy to query. For example, to locate the IDs of photos taken at a location with the string "field" somewhere in its name we may use the following query for the first scheme:

```
Select ID where Photometa like '%field%';
```

The query for the second scheme is only slightly longer:

```
select ID from Photometa
where Addr like '%field%' or city like '%field%'
    or State like '%field%';
```

25.2.3 Representing Ancestors of TGEs

Both the methods described above ignore XML content that lies outside the scope of the selected tuple-generating elements. For example, consider the data depicted in Figure 25.4. This data is similar to that of Figure 25.1, except that the photometa elements are now not the top-level elements, but are grouped within elements representing the trips on which photographs were taken, while trips may themselves be grouped into collections. Our earlier scheme based on photometa as a tuple-generating element ignores the trip and collection elements. Consequently, there would be no way to search for photographs from a given trip or collection.

We may address this shortcoming by adding to the relation schema one attribute for each possible ancestor of the tuple-generating element. The resulting relational representation of the data in Figure 25.4 is depicted in Figure 25.5. This representation allows us to search for tuple-generating elements based on the contents of their ancestors in addition to the contents of their descendants. For example, we may find photometa elements for photos from the Winter 2002 trip that have field in the address using the following query:

```
select P.ID
from Photometa P
where P.Trip_name = 'Winter 2002' and (P.Addr like '%field%'
    or P.City like '%field%' or P.State like '%field%');
```

We need not restrict our queries to returning only the tuple-generating elements. For example, the following query returns the names of trips on which at least one flash picture was taken:

```
select distinct P.Trip_name
from Photometa P
```

```
<collection name="c1">
  <trip>
    <name>California</name>
    <photometa>
      <data>2002-12-21</data>
      <time zone="PST">15:03:07</time>
      <flen>10mm</flen>
      <F>8.0</F>
      <shutter>1/125</shutter>
      <flash>n</flash>
      <location>Death Valley</location>
      <camera>Nikon 995</camera>
    </photometa>
  <trip>
</collection name>
<trip>
  <name>Winter 2002</name>
  <photometa>
    <date>2002-12-23</date>
    <time>10:37 PM</time>
    <flen>28mm</flen>
    <f>2.2</F>
    <shutter>1/250</shutter>
    <flash>y</flash>
    <location>
      <line1>2127 Firewood Ln</line1>
      <city>Springfield</city>
      <state>MA</state
      <note>Living Room</note>
    <location>
  </photometa>
  </trip>
```

FIGURE 25.4 Photo metadata with additional structure.

Photometa

<u>ID</u> int	Date date	Time time	Flen float	F float	Shutter float	Flash boolean	Addr varchar(40)	...
1001	2002-12-21	23:03:07	10	8.0	0.008	0	Death Valley	...
1002	2002-12-23	22:37:00	28	2.2	0.004	1	2127 Firewood Ln	...

Photometa (contd.)

...	City varchar(40)	State char(2)	Camera varchar(50)	Collection_Name varcar(40)	Trip_Name varchar(40)
...	*null*	*null*	Nikon 995	cl	California
...	Springfield	MA	*null*	*null*	Winter 2002

FIGURE 25.5 The schema of Figure 25.3 modified to store data about ancestors of tuple-generating elements.

```
where P.flash = 1;
```

Similarly, the following query finds trips on which no flash pictures were taken. (We do not need a distinct keyword in the above query because the minus operator in SQL has set semantics and automatically eliminates duplicates.)

```
      (select Trip_name from Photometa)
minus
      (select Trip_name from Photometa where P.flash = 1);
```

Let us augment our running example with an XML rendition of an address-book, as depicted in Figure 25.6. A relational representation, using the method of Section 25.2.3 is depicted in Figure 25.7.

```
<abentry>
  <name>Alice Angler</name>
  <addr loc="home">
    <line1>111 Avocado</line2>
    <line2>Apt 1A</line2>
    <city>Anchorage</city>
    <state>Alaska</state>
    <zip>40210</zip>
</abentry>
<abentry>
  <name>Bob Baker</name>
  <addr>
    <line1>2127 Firewood Ln</line1>
    <city>Springfield</city>
    <state>Massachusetts</state>
  </addr>
</abentry>
```

FIGURE 25.6 An XML address-book.

The following query returns, for each person in the address book, the number of photographs taken in that person's city, if that number is nonzero, For the purpose of matching photographs to persons, we interpret any city listed in a person's address book record to be that person's city.

AddressBook

ID int	Name varchar(50)	Loc varchar(50)	Line 1 vachar(50)	Line2 varchar(50)	... ...
1101	Alice Angler	home	111 Avocado	Apt 1A	...
1104	Bob Baker	null	2127 Firewood Ln	*null*	...

AddressBook (contd.)

... ...	City varchar(50)	State varchar(50)	ZIP varchar(10)
...	Anchorage	Alaska	40210
...	Springfield	Massachusetts	*null*

FIGURE 25.7 A relational representation of the address book of Figure 25.6.

```
select A.name as Person; count(P.ID) as Num_Pics
from Addresshook A, Photometa P
where A.city = P.city
group by A.name
order by A.name;
```

This query illustrates the advantages of a SQL-like query language as a tool for accessing semistructtured data. There is no way to express this query using simpler query models, such as the Boolean or vector-space models used by text search engines.

The methods described above require a judicious choice of the tuple-generating elements. More precisely, the kinds of queries that are easily expressible in the relational schema generated using these methods depends on the choice of the tuple-generating elements. For example, if we were to choose location as the tuple generating element in our running example, it would not be possible to express our "field in location" query using the methods of Sections 25.2.1 and 25.2.2 because the Photometa elements, being ancestors of the tuple-generating elements, would not be reflected in the relational schema.

25.2.4 Varying Components

The problem of differing structure for an element type (e.g., the location element in our sample data) may be addressed by storing different structures separately. Figure 25.8 illustrates a scheme that separates location elements with text content from those with structured content. In general, this method produces one relation for each variant structure. Our example has only two variants. However, in principle the number of variants may be larger.

Our "field in location" query for the scheme of Figure 25.8 may be expressed as follows:

```
(select P.ID
from Photometa P, TextLocations T
where P.LocID = T.LocID and T.LocID like '%field%')
union
(select P.ID
from Photometa P, StructLocations S
where P.LocID = s.LocID and
      (S.Addr like '%field%' or S.City like '%field%'));
```

It is easy to observe that the number of subqueries required to express our "field in location" query using this scheme is equal to the number of variants. The situation is exacerbated in queries that join

Photometa

ID int	Date date	Time time	Flen float	F float	Shutter float	Flash boolean	LocID int	Camera varchar(50)
1001	2002-12-21	23:03:07	10	8.0	0.008	0	2001	Nikon 995
1002	2002-12-23	22:37:00	28	2.2	0.004	1	3002	*null*

TextLocations

LocID int	Desc varchar(100)
2001	Death Valley

StructLocations

LocID int	Addr varchar(40)	City varchar(40)	State char (2)
3002	2127 Firewood Ln	Springfield	MA

FIGURE 25.8 The schema of Figure 25.2 modified to permit different location types.

two or more relations, each of which has several variants. Therefore, this method should be used only when domain constraints rule out all but a few variants.

25.2.5 Semistructured Components

In order to avoid the problems resulting from a large number of variant structures we may collapse several variants into a generic representation that encodes both the schema of a variant and its content. Figure 25.9 illustrates this idea for our running example. In this representation, the input data is conceptually divided into two parts. The first part, which we call the *specific part,* is represented using one of the methods we have discussed so far. The relational schemas for this part of the data depend on the domain and on the representation method used. For our running example, the Photometa relation of Figure 25.9 represents the Photometa elements and their subelements, excluding the location elements and their subelements. The second part, which we call the *generic part* is represented using a ternary relation of fixed schema. This relation has attributes denoting the identifiers, names, and values of elements in parts of data it encodes. The identifier attribute is a foreign key that references the ID attribute of the main relation. Since there are, in general, several XML elements in the generic part corresponding to a tuple-generating element in the specific part, this identifier is not a key.

In Figure 25.9, the Locations relation represents all variants of the Location element and its subelements. As in the method of Section 25.2.2, there is one tuple for each XML element, excluding elements that have element content (i.e., those that do not have any content besides subelements). In fact, we can think of this method as starting with the representation of Section 25.2.2 and separating the relational attributes corresponding to the XML elements that lie in the subtrees rooted at the Location subelements into the ternary relation described above.

The relation used to represent the generic part has three features worth noting: First, it flattens the nested structure of the XML elements it represents. In our running example, for instance, the nesting of the city and state elements within the location element for the second Photometa element of Figure 25.1 is lost. This loss of structure is not very obvious in Figure 25.9 because the location element that is the parent of the city and state elements (line 18 of Figure 25.1) has element content and does not generate a tuple in the Locations relation. Second, we have implicitly assumed that all element values in the generic part can be stored in a single relational attribute (implying they have the same type). When this assumption does not hold, we may need to add additional attributes to this relation. For XML, it is convenient to use the text serialization of data in this representation in order to avoid the need for explicitly

Photometa

ID int	Date date	Time time	Flen float	F float	Shutter float	Flash boolean	LocID int	Camera varchar(50)
1001	2002-12-21	23:03:07	10	8.0	0.008	0	2001	Nikon 995
1002	2002-12-23	22:37:00	28	2.2	0.004	1	3002	*null*

Locations

LocID int	Name varchar(20)	Val varchar(40)
2001	Desc	Death Valley
3002	Addr	2127 Firewood Ln
3002	City	Springfield
3002	State	MA

FIGURE 25.9 The schema of Figure 25.8 modified to permit different semistructured location descriptions.

representing other types. However, we must still reconcile differences in the size of such text types. In our running example, we have arbitrarily used varchar(40) as the type for the Val attribute. Another option is to use a datatype such as *CLOB* in order to permit large text values. (CLOB, an abbreviation of Character Large Object, is a datatype provided by database systems for storing text data that is too large for the character datatypes such as varchar. When the data occurs in a compressed format such as WBXML, a binary datatype such as *BLOB* — binary large object — may be used.) Third, as indicated by Figure 25.9, we have assumed that LocID and Attr form a key for the the Locations relation. This assumption implies that element names are unique within the subtrees nested at the location element of each Photometa element. In cases where such an assumption may not hold, we add an artificial key to the schema of this relation.

Our "field in locations" query is easily expressed in this representation:

```
select P.ID
from Photometa P, Locations L
where P.LocID = L.LocID and L.Val like '%field%';
```

Comparing this query with the analogous one from Section 25.2.4 may suggest that this method is always superior to that one. However, queries that access individual elements within the generic part of data are more difficult in this representation. For example, to find the identifiers of photographs taken at the address 2127 Firewood Ln, Springfield, MA, we need a multiway join that includes self joins on the Locations relation:

```
select P.ID
from Photometa P, Locations L1, Locations L2, Locations L3
where P.ID = L1.LocID and P.ID = L2.LocID and P.ID = L3.LocID
      and L1.Name = 'Addr' and L1.Val = '2127 Firewood. Ln'
      and L2.Name = 'City' and L2.Val = 'Springfield'
      and L3.Name = 'State' and L3.Val = 'MA';
```

In general, the number of joins in such a query equals the number of subelements in the generic part that are queried (three, in our example: Addr, City, and State). By creating an index on the LocID field, such joins can be rendered efficient.

25.2.6 Graph Representation

We may carry the idea of a generic representation further by using a simple relational encoding of the XML tree. In this method, the relational schema is fixed, and consists of four relations that encode a graph: Vertices, Contents, Attributes, and Edges. Figure 25.10 illustrates such a representation for our running example. This scheme uses artificial keys to identify nodes in the XML tree (or any graph in general). The Vertices relation has one tuple for each XML element and indicates the element type (name). The Contents relation indicates the text content of each element that has text (#PCDATA) content. (There are no tuples in this relation representing XML elements with empty or element content.) The Attributes relation indicates the set of attribute-value pairs associated with each XML element. Finally, the Edges relation stores the tree edges in the form (parent, child).

The flexibility of this scheme comes at the expense of increased complexity of typical queries. For example, in order to find photos containing the string field in the text of a location element or in the text of an element one level below a location element, we need a query such as the following:

```
(select V1.ID
from Vertices V1, Vertices V2, Contents C, Edges E
where V1.Name = 'Photometa' and V1.ID = E.SID and V2.ID = E.DID
      and V2.Name = 'location' and V2.ID = C.ID and
      C.Data like '%field%')
union
```

Vertices

ID int	Name varchar(40)
101	Photometa
102	date
103	time
104	flen
105	F
106	shutter
107	flash
108	location
109	camera
111	Photometa
112	data
113	time
114	flen
115	F
116	shutter
117	flash
118	location
119	line1
120	city
121	state
122	note

Contents

ID int	Data varchar(100)
102	2002-12-21
103	15:03:07
104	10mm
105	8.0
106	1/125
107	n
108	Death Valley
112	2002-12-23
113	10:37 PM
114	28mm
115	2.2.
116	1/250
117	y
119	2127 Firewood Ln
120	Springfield
121	MA
122	Living Room

Attrib utes

ID int	AttrName varchar(100)	AttrVal varchar(100)
103	zone	PST

Edges

SID int	DID int
101	102
101	103
101	104
101	105
101	106
101	107
101	108
101	109
111	112
111	113
111	114
111	115
111	116
111	117
111	118
118	119
118	120
118	121
118	122

FIGURE 25.10 The data of Figure 25.1 stored in a generic schema for graphs.

```
(select V1.ID
from Vertices V1, Vertices V2, Vertices V3, Contents C, Edges E1,
     Edges E2
where V1.Name = 'Photometa' and V1.ID = E1.SID and V2.ID = E1.DID
     and V2.Name = 'location' and E1.DID = E2.SID and V3.ID = E2.DID
     and V3.ID = C.ID and C.Data like '%field%');
```

In order to express the "field in location" query used in earlier sections (where the string field may occur in any subelement of a location element), we need to use recursion. Using the recursive-query syntax of SQL-99, we may express this query as follows:

```
with
     recursive LocationComp(PID, LID) as
     (select V1.ID, V2.ID
     from Vertices V1, Vertices V2, Edges E
     where V1.Name = 'Photometa' and V2.Name = 'location'
```

```
            and V1.ID = E.SID and V2.ID = E.DID)
      union
      (select L.PID, E.DID
      from LocationComp L, Edges E
      where L.LID = E.SID)
select L.PID
from LocationComp L
where L.LID = C.ID and C.Data like '%field%';
```

This query illustrates the trade-off between representations that flatten some of the XML tree structure (e.g., the representation of Section 25.2.5) and those that do not.

As described, our scheme permits encoding of graphs that are not trees. This feature may be exploited to encode an interpretation of XML other than the usual DOM-tree interpretation. In particular, IDREF attributes in the XML data can be interpreted as graph edges with source being the element anchoring the attribute and target being the element with a matching ID attribute. When such an encoding is used, it may be beneficial to distinguish between tree edges (from the XML DOM tree) and graph edges corresponding to IDREF and other constructs because some applications may require easy access to only the former. For example, producing a serialized version of the data encoded in relations is easier when only tree edges are traversed. The two kinds of edges can be distinguished by either adding a Boolean attribute to the Edges relation or storing them in separate relations.

25.2.7 Storing Unparsed XML

The methods discussed so far are based on parsing XML data to determine its tree structure and storing the parsed representation. This approach has the benefit of avoiding parsing during query processing. Since parsing is typically an expensive operation, it makes sense to parse the input data only once, when it is first loaded into a relational database. It is often necessary to produce an XML serialization of the relational representation of data, functionally identical to the XML data from which it was initially generated. Such a serialization query needs to perform two tasks: First, a tree-structured representation of data must be generated from the flat relational representation. Second, the tree-structured representation must be converted into text form using XML syntax. These two tasks, called structuring and tagging, have been studied in detail [Shanmugasundaram et al., 2001]. Although we do not discuss them in detail in this chapter, a key observation is that serialization can be time-consuming due to the large number of string operations involved in the process. In applications that require frequent XML serializations of data, it may be prudent to store some data in an unparsed, XML form.

Storing unparsed XML data also offers an easy solution to the problem of variant structures. Figure 25.11 illustrates this method for our running example. Location elements and their subelements are stored in the XMLText attribute of the LocationsXML relation. The LocID attribute of that relation is used to map Location elements to the corresponding Photometa elements in the Photometa relation. In this example, we have used a separate LocID attribute in the Photometa relation, permitting multiple locations in a photometa element. If it is known that photometa elements have at most one location element each, we may use the ID attribute for this purpose instead. We have used varchar(100) as the type of the XMLText attribute. However, when the unparsed data values are likely to be large, it may be more appropriate to use CLOB as the datatype. In fact, as we discuss in the next section, many database systems provide specialized datatypes for this purpose. Storing data in unparsed form makes it difficult to query. For example, although we may express our ongoing "field in location" query using a join on the LocID attribute, this query returns a superset of the desired results (because it matches locations in which the string field occurs, say, within an element or attribute name). However, we may remedy this situation by using user-defined functions (UDFs) in queries, a feature supported by many database systems. In particular, if we define a function getXMLContent that returns only the text content (minus markup for element tags and attributes) of its argument, we may express our query as follows:

Photometa

<u>ID</u> int	Date date	Time time	Flen float	F float	Shutter float	Flash boolean	LocID int	Camera varchar(50)
1001	2002-12-21	23:03:07	10	8.0	0.008	0	2001	Nikon 995
1002	2002-12-23	22:37:00	28	2.2	0.004	1	3002	*null*

LocationsXML

LocID int	XMLText varchar(100)
2001	<location>Death Valley</location>
3002	<location> <line>2127 Firewood Ln</line1> <line>Apt 311</line1> <city>Springfield</city> <state>MA</state> <note>Living Room</note> </location>

FIGURE 25.11 The data of Figure 25.1 stored using an XML-valued relational attribute.

```
select P.ID
from Photometa P, LocationsXML L
where P.LocID = L.LocID and getXMLContent(L.XMLText) like '%field%';
```

As we discuss below, many database systems provide special functions for extracting desired portions of unparsed XML attribute values, further simplifying such queries.

25.3 Using XML Features of Relational Database Systems

In this section, we discuss the XML features offered by some relational database products, focusing on features that aid in storing and querying XML in relational form. We do not describe a number of other XML features provided by these systems. In particular, most systems provide ways to efficiently produce a serialized XML form of data stored in relational tables. For details on such XML publishing of relational data, we refer the reader to Chawathe [2003].

25.3.1 IBM DB2

The DB2 XML Extender provides an interesting combination of parsed and unparsed options for storing XML data [Selinger, 2001; IBM, 2000]. It provides three datatypes for storing unparsed XML: XMLVARCHAR for small XML fragments, XMLCLOB for larger fragments, and XML-FILE for fragments that are stored in files outside the database system proper. We refer to attributes of these three types as XML-typed attributes. For our running example, Figure 25.12 suggests a Photometa table that uses the XMLCLOB datatype to store all location information. As in Section 25.2.7, the location attribute for a tuple representing a photometa element stores the unparsed XML text corresponding to its location element (and subelements).

For each XML-typed attribute, DB2 permits the creation of one or more supporting tables that contain parsed equivalents of parts of the XML-typed attribute. These tables, called *side tables,* may be viewed as user-level indexes on the XML-typed attributes. A database designer must decide on the number and type of side tables to use; such decisions are similar to those guiding index selection in conventional databases. The mapping between the XML-typed attribute and side tables is specified using *Data Access*

```
create table Photometa(
     ID int primary key no null
     -- ... attributes Date, Time, flen, F, Shutter, Flash, and Camera
     location XMLCLOB);

<dad>
     <dtdid>photometa,dtd</dtdid>
     <validation>yes</validation>
     <Xcolumn>
          <table name="line_side_table">
               <column name="name" type="varchar(40)"
               path="/location/line" multi_occurrence="yes"/>
          </table>
          <table name="city_side_table">
               <column name="city" type="varchar(40)"
               path="/location/city"
               multi_occurrence="no"/>
               <column name="state" type="varchar(40)"
               path="/location/state"
               multi_occurrence="no"/>
          </table>
     </Xcolumn>
</dad>
```

FIGURE 25.12 Storing the data of Figure 25.1 using the features of IBM DB2 XML Extender.

Definitions (DADs), which are XML-format specifications of the schemas of side tables along with the XPath expressions mapping table attributes to XML elements or attributes. Side tables may also be thought of as materialized views that are automatically maintained by the database system. Once defined, the user is freed of the burden of ensuring that the side tables remain consistent with the value of the corresponding XML-typed attribute in the main table.

A side-table DAD for our sample table appears in Figure 25.12. This DAD defines two side tables. The first, called line_side_table, has a column named line that is populated with the contents of line subelements of location elements (as indicated by the path /location/line). Each side table also has implicit columns corresponding to the primary keys of the main table. In our example, an ID column is implicitly added to all side tables of the Photometa table. The side table resulting from this specification is illustrated in Figure 25.13. The second side-table defined by the DAD, city_side_table, contains city and state columns in addition to the implicit ID column. The multi_occurrence attributes in the DAD indicate whether it is permissible for a single ID value to be associated with multiple values of the column defined by the corresponding column element. In our example, we have assumed that there is at most one line, city, and state associated with each tuple in Photometa. However, multiple locations can easily be supported by setting the relevant multi_occurrence attributes to yes.

line_side_table

ID int	Line varchar(40)
3002	2127 Firewood Ln
3002	Apt 311

city_side_table

ID int	city varchar(40)	state varchar(40)
3002	Sprinfield	MA

FIGURE 25.13 DB2 side tables for the specification in Figure 25.12.

```
<!ELEMENT location (#PCDATA|line|city|state|note)*>
<!ELEMENT line (#PCDATA)>
<!ELEMENT city (#PCDATA)>
<!ELEMENT state (#PCDATA)>
<!ELEMENT note (#PCDATA)>
<ATTLIST line num CDATA>
```

FIGURE 25.14 A candidate DTD for the structured location elements in the data of Figure 25.1.

The DAD associates a *DTD (document type definition)* with the XML-typed attribute. The DTD for our example, photometa.dtd is exhibited in Figure 25.14. For our purpose, the following simplified description of DTDs suffices: A DTD contains two main kinds of declarations. The first is an ELEMENT declaration, which defines the allowable contents of an element, identified by its name. For example, the third line in Figure 25.14 indicates that city elements have text content (The notation #PCDATA is used to represent parsed character data, informally, text.) There is at most one element declaration for each element type (name). The declaration for the location element type indicates that such elements contain zero or more occurrences of elements of type line, city, state, and note, along with text. The syntax, following common conventions, uses | to denote choice and * to denote zero or more occurrences. The second kind of declaration found in DTDs is an ATTLIST declaration, which defines the set of attribute-value pairs that may be associated with an element type. For example, the last line in Figure 25.14 indicates that line elements may include a num attribute of text type. (The notation CDATA indicates character data, or text. A technical point is that, unlike #PCDATA, this character data is not parsed and may thus include special characters.)

As indicated above, we may think of side tables as user-level indexes. Essentially, whenever a query on the main table accesses an XML element or attribute (within an XML-typed attribute) that is referenced by a column in a side table, the query may be made more efficient by using the side table to access the column. The increased efficiency is due to the side tables providing access to the required XML elements and attributes without query-time parsing. Although it is possible to use side tables in this manner, the task is made easier by views that are automatically defined by the system when side tables are initialized. Briefly, there is one view for each main table, consisting of a join with all its side tables on the key attribute. For our running example, the system creates the following view:

```
create view Photometa.view(ID, Date, Time, Flen, F, Shutter,
            Flash, Camera, Line, City, State) as
      select P.ID, P.Date, P.Time, P.Flen, P.F, P.Shutter, P.Flash,
            P.Camera, L.Line, C.City, C.State
      from Photometa P, line_side_table L, city_side_table C
      where P.ID = L.ID and P.ID = C.ID;
```

Our "field in location" query is easily expressed using this view:

```
select P.ID
from Photometa_view P
where City like '%field%';
```

It is not necessary to create side-tables on all parts of the XML-typed attribute that may be queried. Indeed, since there is a maintenance overhead associated with side tables, a database designer must carefully select the side tables to instantiate. Queries that need to access parts of an XML-typed attribute that are not part of a side table must use *extracting functions.* The general scheme of extracting functions is extractType (*Col, Path*), where *Type* is a SQL type such as Integer or Varchar, *Col* is the name of an XML-typed column, and *Path* is an XPath expression indicating the part of the XML attribute that is to be extracted. Such a function returns objects of the type in its name, assuming the XML element or attribute to which the path points can be appropriately coerced (indicating an error otherwise). For our example, we may use the following query to find photo IDs that have "field" in the text content of either a location element or a city element nested within a location element:

```
select P.ID
from Photometa P, LocationsXML L
where P.LOCID = L.LocID and
      (extractVarchar(L.Location,/location) like '%field%' or
      extractVarchar(L.location,/location//city) like '%field%');
```

The path expression in the argument of an extracting function must match no more than one value within the XML-typed attribute for each tuple of the main table. For example, the above query generates an error if there are multiple city elements within the location element corresponding to one photometa element (and tuple). Fortunately, DB2 provides *table extracting functions* for coping with this situation. The names of table extracting functions are obtained by pluralizing the names of the corresponding (scalar) extracting functions. These functions return a table that can be used in the from clause of a SQL query in the usual manner. For our running example, we may use the following query to find the IDs of photometa elements with "field" in the city, when multiple city elements (perhaps in multiple location elements) are possible.

```
select P.ID
from Photometa P, LocationsXML L
where P.LocID = L.LocID and exists
      (select 1
      from table(extractVarchars(L.location,/location//city)) as C
      where C.returnedVarchar like '%field%');
```

25.3.2 Microsoft SQL Server

Using Microsoft's SQL Server, we may separate the structured and semistructured parts of our data as follows: The structured part is stored in a Photometa table similar to the one used in Section 25.3.1, as suggested by Figure 25.12. However, instead of storing the XML fragment describing the location of each Photometa element in a column of the Photometa table, we store all the location data (for all Photometa elements) in a separate file outside the database system. Let us assume this file is called LocationsXML. In order to maintain the mapping between Location and Photometa elements, we shall further assume that location elements have a LocID attribute that is a intuitively a foreign key referencing the ID attribute of the appropriate Photometa tuple.

The main tool used to access external XML data is the *OpenXML* rowset function [Microsoft, 2003]. It parses XML files into a user-specified tabular format. Queries using the OpenXML function include a with clause that provides a sequence of triples consisting of a relational attribute name, the type of the attribute, and a path expression that indicates how the attribute gets its value from the XML data. For example, consider the following query, which finds the identifiers of Photometa elements that contain "field" in the city or description elements:

```
select P.ID
from Photometa P,
      openXML(@locationsFile, '/location', 2)
      with (LOCID int '@LocID,
            Line varchar(40) 'line',
            City varchar(40) 'city',
            State varchar(40) 'state',
            Desc varchar(100) 'text()') L
where P.LocID = L.LocID and
      (L.City like '%field%' or L.Desc like '%field%');
```

The first argument to the OpenXML function is a file handle (declared and initialized elsewhere) pointing to the LocationsXML file. The second argument is an XPath expression that matches elements that are to be extracted as tuples. These elements are essentially the tuple-generating elements discussed in Section 25.2.1. The third argument is a flag that indicates that the following with clause is to be used for mapping elements to tuples. The with clause declares the names and types of columns in the generated table in a manner similar to that used in a create table statement. Each attribute name is followed by a type and an XPath expression. This expression is evaluated starting at nodes matching the expression in the OpenXML function (/location in our example) as context nodes. Thus, the first triple in the "with" clause above indicates that the LocID attribute of a location element is used to populate the LocID attribute of the corresponding tuple.

Although our example uses rather simple forms of XPath, it is possible to use more complex forms. For example, the parent or sibling of a context node may be accessed using the appropriate XPath axes (e.g., . . /, . . /alternate). If the XPath expression for some attribute in the with clause does not match anything for some instance of a context node (matching the expression in the OpenXML function), that position in the generated tuple has a null. It is an error if an XPath expression in the with clause matches more than one item for a given context node. SQL Server also provides an alternate method for storing XML that uses an Edges table to store the graph representation of XML. This method is essentially the method of Section 25.2.6.

25.3.3 Oracle XSU

Oracle's *XML SQL Utility (XSU)* includes features for parsing XML and storing the result in a relational table [Higgins, 2001, Chapter 5]. However, such storage works only if there is a simple mapping from the structure of the input XML data to the relational schema of the database. By simple mapping, we mean one similar to the one illustrated in Section 25.2.1. Although this method may be useful for storing well-structured XML data, it does not fare well with semistructured data because it necessitates cumbersome restructuring and schema redesign whenever the input data changes form.

XSU also provides some features for storing unparsed XML text in a relational attribute. It provides the *XMLType* datatype, which is similar to the XMLVarchar, XMLCLOB, and XMLFile types of DB2. The extraction functions used to parse such attributes at query-execution time are also similar to those used by DB2, but use an object-oriented syntax. For example, the method *extract(/product/price)* may be used on a XMLType attribute to extract the price element; the numeric value of the result is extracted by a *getRealVal* method. (Similar methods exist for other types.) For our running example, we use a table similar to the one suggested by Figure 25.12, replacing XMLCLOB with XMLType. The following query may be used to locate the Photometa elements that contain "field" in their city elements:

```
select P.ID
from Photometa P
where P.location.extract(/location/city).getStringVal()
     like '%field%';
```

In addition to the extract function, XSU provides a Boolean function *existsNode* for checking the existence of specified nodes in an XMLType field. For example. the following query finds Photometa tuples whose locations have a city subelement:

```
select P.ID
from Photometa P
where P.location.existsNode(/location//city);
```

Unlike DB2, XSU does not use side tables as a tool for improving access to selected elements. However, query performance can be improved by using functional indexes based on the extract function. For our running example, we may speed up the execution of queries similar to the "field" query above by creating an index on the city elements as follows:

```
create index photometa_city_idx on Photometa(
      location.extract(/location/city).getStringVal());
```

One may also create a text index on an XML attribute in order to support efficient searches on that attribute. Such an index is implemented by Oracle as a function-based index on the getCLOBVal( ) method of XMLType objects [Kaminaga, 2002b, a]. For our running example, we may create a text index on the contents of the location attribute as follows:

```
create index photometa_location_txtidx on Photometa(location)
      indextype is ctxsys.context
      parameters ('SECTION GROUP ctxsys.path_section_group');
```

This index may now be used in an alternate version of the "field" query described above:

```
select P.ID
from photometa P
where contains(P.location 'field inpath(/location//city)');
```

25.3.4 Sybase

The XML features of the *Sybase Adaptive Server Enterprise (ASE)* use a tight coupling with Java classes and methods to aid storing XML in relational tables [Sybase, 2001, 1999]. A database designer creates a set of Java classes that are customized by the database designer, ASE provides three methods for storing XML: element, document, and hybrid.

In the element storage method, the XML elements of interest are stored separately in relational attributes. This method is similar to the method of Section 25.2.7. For our running example, we may use a Photometa table similar to that suggested by Figure 25.12. Our query to find photographs with "field" in the cities of their locations can then be expressed as follows:

```
select P.ID
from Photometa P
where P.location>>getLocationElement(1, "city") like '%field%';
```

In this query, the getLocationElement method is invoked for the location column (which is XML-valued) of each tuple in the Photometa table. The second argument to this method specifies the name of the element that is to be extracted, and the first argument is the ordinal number of the element (among those with the same name). In our example query, the method returns the string value of the contents of the first city element.

In the document storage method, XML elements of interest are stored together (coalesced) in a single document, which in turn is stored in a table of suitable schema. In our continuing example, consider a table LocationFiles that has a schema similar to the table LocationsXML described earlier. The difference is that now all location data is stored in a single attribute value. We will assume that location elements have an ID subelement that is a foreign key referencing the ID attribute of the corresponding Photometa tuple. This scheme has the advantage of permitting easy and efficient retrieval of all location data in XML form (perhaps for display or export to another application). However, querying the document-format XML data is difficult. For example, we may attempt to write our "field in city" query as follows:

```
          (select F.XMLText>>getLocationElement(1, "ID")
          from LocationFiles F
          where F.XMLText>>getLocationElement(1,"city")like'\%field%')
     union \\
          (select F.XMLText>>getLocationElement(2,"ID")
          from LocationFiles F
          where F.XMLText>>getLocationElement(2,"city")like'\%field\%');
```

However, this query checks only the cities of the first two locations in the XML-valued location attribute. We may extend it to check more cities; however, checking all cities requires the assistance of the host program.

The hybrid storage method is essentially a combination of the element and document storage methods. The relevant data is stored in both unparsed (document) and parsed (element) form, with the former providing efficient document-centric access and the latter providing efficient data-centric access. In this respect, the tables used for element storage in the hybrid method are user-level indexes analogous to side tables in DB2 (described in Section 25.3.1). However, unlike side tables, these user-level indexes are not automatically maintained by the database system; it is the responsibility of application programs to ensure that they remain consistent with the data in document storage.

Although the term semistructured data has not been in use for very long, the kind of data it describes is not new. Traditionally, such data has either been ignored or handled using non-database techniques such as text search engines and application-specific methods. As for structured data, using a database management system for such data provides benefits such as efficient evaluation of ad hoc queries, consistency, concurrency control, and durability. In addition, applications that use both structured and semistructured data benefit from the ability to use a single system for both kinds of data. We have presented several methods for relational storage of semistructured data. As noted, current database systems provide a variety of features for managing such data. Mapping these features to the methods discussed in this chapter provides a framework for determining the best approach for the domain under consideration.

Acknowledgment

This material is based upon work supported by the National Science Foundation under grants IIS-9984296 (CAREER) and IIS-0081860 (ITR). Any opinions, findings, and conclusions or recommendations expressed in this material are those of the author and do not necessarily reflect the views of the National Science Foundation.

References

Chawathe, Sudarshan S. *Managing Historical XML Data,* volume 57 of *Advances to Computers,* pages 109–169. Elsevier Science, 2003. To appear.

Garcia-Molina, Hector, Jeffrey D. Ullman, and Jennifer Widom. *Database Systems: The Complete Book.* Prentice-Hall, 2002.

Higgins, Shelley. Oracle9i application developer's guide — XML. Available at http://www.oracle.com/, June 2001. Release 1 (9.0.1) part number A88894–01.

IBM. XML Extender administration and programming, version 7. Product information. Available at http://www. fbm.com/, 2000.

Kaminaga, Garrett. Oracle Text 9.0.1 XML features overview. Oracle Technology Network. Available at http//otn.oracle.com/, November 2002a.

Kaminaga, Garrett. Oracle Text 9.2.0 technical overview. Oracle Technology Network. Available at http/ /otn.oracle.com/, June 2002b.

Microsoft. Using OPENXM. Microsoft SQL Server Documentation http://msdn. microsoft.com/, 2003.

Selinger, Pat. What you should know about DB2 support for XML: A starter kit. *The IDUG Solutions. Journal,* 8(1), May 2001. International DB2 Users Group. Available at http://www.idug.org/.

Shanmugasundaram, Jayavel, Eugene Shekita, Rimon Barr, Michael Carey, Bruce Lindsay, Hamid Pirahesh, and Berthold Reinwald. Efficiently publishing relational data as XML documents. *The VLDB Journal,* 10(2–3): 133–154, 2001.

Sybase. Using XML with the Sybase Adaptive Server SQL databases. Technical White Paper. Available at http://www. sybase.com/, 1999.

Sybase. XML technology in Sybase Adaptive Server Enterprise. Technical White Paper. Available at http:/ /www.sybase.com/, 2001.

26

Information Security

E. Bertino

E. Ferrari

CONTENTS

26.1 Introduction 26-1
26.2 Basic Concepts 26-2
26.2.1 Access Control Mechanisms: Foundations and Models 26-2
26.2.2 A Brief Introduction to XML 26-6
26.3 Access Control for Web Documents 26-7
26.3.1 Access Control: Requirements for Web Data 26-7
26.3.2 A Reference Access Control Model for the Protection of XML Documents 26-9
26.4 Authentication Techniques for XML Documents 26-12
26.4.1 An Introduction to XML Signature 26-12
26.4.2 Signature Policies 26-13
26.5 Data Completeness and Filtering 26-14
26.5.1 Data Completeness 26-14
26.5.2 Filtering 26-15
26.6 Conclusions and Future Trends 26-16
References 26-17

26.1 Introduction

As organizations increase their reliance on Web-based systems for both day-to-day operations and decision making, the security of information and knowledge available on the Web becomes crucial. Damage to and misuse of the data representing information and knowledge affect not only a single user or an application but they may also have disastrous consequences on the entire organization. Security breaches are typically categorized into *unauthorized data observation, incorrect data modification,* and *data unavailability.* Unauthorized data observation results in disclosure of information to users not entitled to gain access to such information. All organizations we may think of, ranging from commercial organizations to social organizations such as healthcare organizations, may suffer heavy losses from both financial and human points of view upon unauthorized data observation. Incorrect modifications of data, either intentional or unintentional, may result in inconsistent and erroneous data. Finally, when data are unavailable, information crucial for the proper functioning of the organization is not readily available.

A complete solution to the information security problem must thus meet the following three requirements:

1. *Secrecy* or *confidentiality* — which refers to the protection of data against unauthorized disclosure
2. *Integrity* — which means the prevention of unauthorized or improper data modification
3. *Availability* — which refers to the prevention of and recovery from software errors and from malicious denials making data not available to legitimate users

1-58488-381-2/05/$0.00+$1.50
© 2005 by CRC Press LLC

In particular, when data to be secured refer to information concerning individuals, the term *privacy* is used. Privacy is today becoming increasingly relevant, and enterprises have thus begun to actively manage and promote the level of privacy they provide to their customers. In addition to those traditional requirements, the development of Web-based networked information systems has introduced some new requirements. New requirements that are relevant in such contexts include *data completeness, self-protection,* and *filtering.* It is important to note that whereas the traditional requirements are mainly meant to protect data against illegal access and use, the new requirements are meant to protect subjects and users. In particular, by data completeness we mean that a subject receiving an answer to an access request must be able to verify the completeness of the response, that is, the subject must be able to verify that it has received all the data that it is entitled to access, according to the stated access control policies.

As an example, consider a Website publishing information about medical drugs. In such a case, a completeness policy may require that if a subject has access to information about a specific drug, information concerning side effects of the drug must not to be withheld from the subject. By self-protection and filtering, we mean that a subject must be able to specify what is unwanted information and therefore be guaranteed not to receive such information. Such a requirement is particularly relevant in push-based information systems, automatically sending information to users, and in protecting specific classes of users, such as children, from receiving improper material. Data security is ensured by various components in a computer system. In particular, the *access control mechanism* ensures data secrecy. Whenever a subject tries to access a data item, the access control mechanism checks the right of the subject against a set of *authorizations,* stated usually by some security administrators. An authorization states which user can perform which action on which data item. Data security is further enhanced by *cryptographic mechanisms* that protect the data when being transmitted across a network. Data integrity is jointly ensured by several mechanisms. Whenever a subject tries to modify some data item, the access control mechanism verifies that the subject has the right to modify the data, whereas the *semantic integrity mechanism* verifies that the updated data are semantically correct. In addition, content authentication techniques, such as the ones based on *digital signatures,* may be used by a subject to verify the authenticity of the received data contents with respect to the original data. Finally, the *error recovery mechanism* ensures that data are available and correct despite hardware and software failures. Data availability is further enhanced by intrusion detection techniques that are able to detect unusual access patterns and thus prevent attacks, such as query floods that may result in denial of service to legitimate users. The interaction among some of the components described above is shown in Figure 26.1.

In this chapter, we will first focus on access control mechanisms and related authorization models for Web-based information systems. In order to make the discussion concrete, we will discuss the case of data encoded according to XML [Extensible Markup Language, 2000]. However, the concepts and techniques we present can be easily extended to the cases of other data models or data representation languages. We then briefly discuss authentication techniques for XML data because this is the key issue for Internet computing, whereas we refer the reader to Bertino [1998] and to any database textbook for details on semantic integrity control, and to Stallings [2000] for cryptography techniques. We then outline recent work on data completeness and filtering, and finally present future research directions.

26.2 Basic Concepts

In this section, we introduce the relevant concepts for the discussion in the subsequent sections. In particular, we first introduce the basic notions of access control mechanisms and present a brief survey of the most relevant models. We then introduce the XML language.

26.2.1 Access Control Mechanisms: Foundations and Models

An access control mechanism can be defined as a system that regulates the operations that can be executed on data and resources to be protected. Its goal is thus to control operations executed by subjects in order

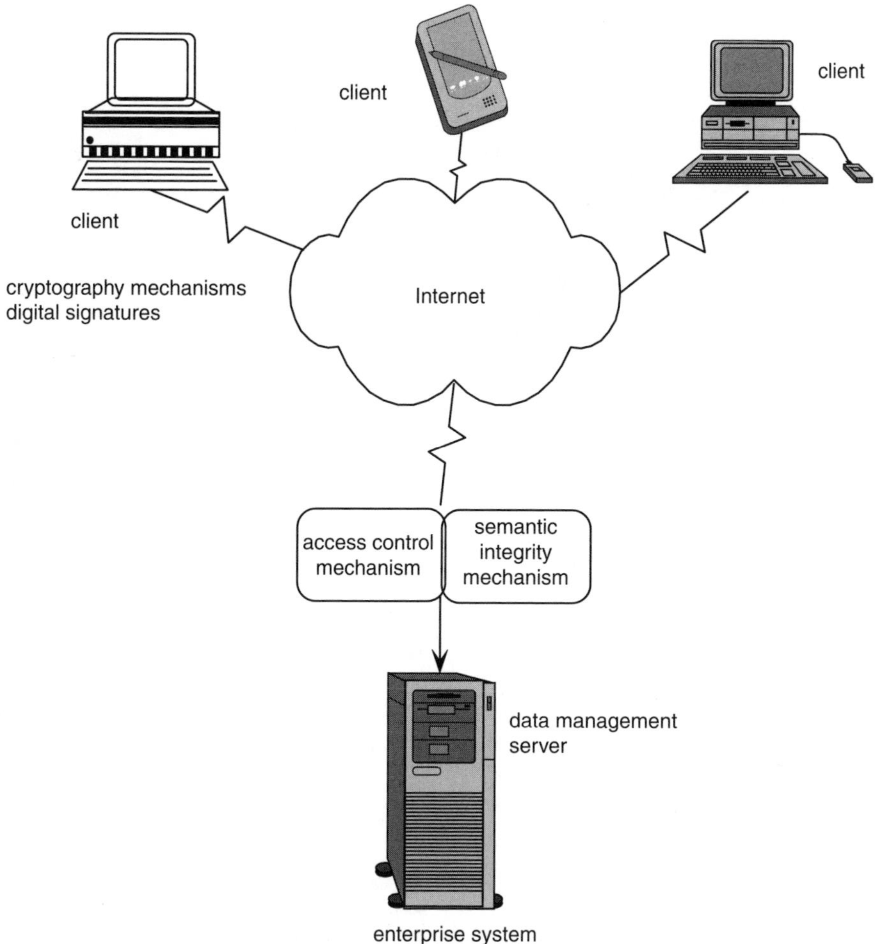

FIGURE 26.1 Main security modules in an enterprise scenario.

to prevent actions that could damage data and resources. The basic concepts underlying an access control system are summarized in Figure 26.2.

Access control policies specify what is authorized and can thus be used like requirements. They are the starting point in the development of any system that has security features. Adopted access control policies mainly depend on organizational, regulatory, and user requirements. They are implemented by mapping them into a set of *authorization rules*. Authorization rules, or simply authorizations, establish the operations and rights that subjects can exercise on the protected objects. The reference monitor is the control mechanism; it has the task of determining whether a subject is permitted to access the data. Any access control system is based on some *access control model*. An access control model essentially defines the various components of the authorizations and all the authorization-checking functions. Therefore, the access control model is the basis on which the authorization language is defined. Such a language allows one to enter and remove authorizations into the system and specifies the principles according to which access to objects is granted or denied; such principles are implemented by the reference monitor.

Most relevant access control models are formulated in terms of *objects, subjects*, and *privileges*. An object is anything that holds data, such as relational tables, documents, directories, inter-process messages, network packets, I/O devices, or physical media. A subject is an abstraction of any active entity that performs some computation in the system. Subjects can be classified into *users* — single individuals connecting to the

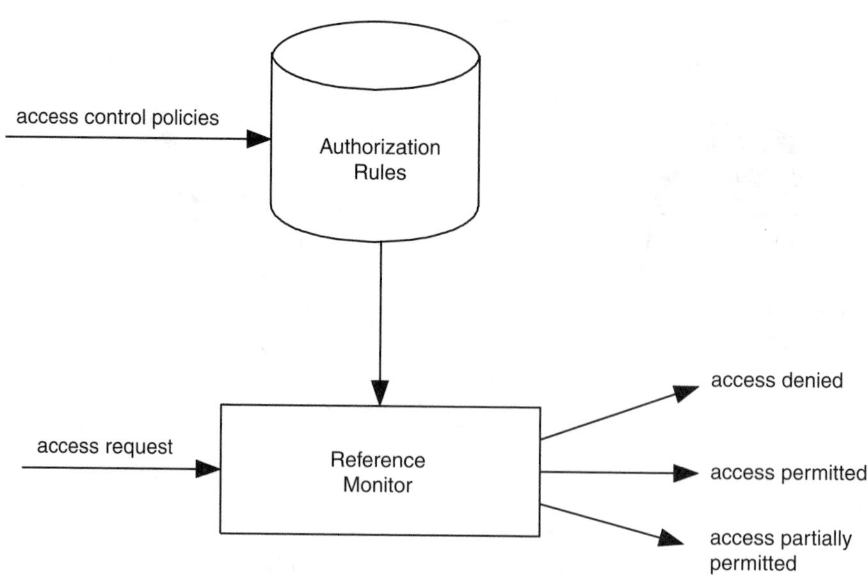

FIGURE 26.2 Main components of an access control system.

system, *groups* — sets of users, *roles* — named collections of privileges or functional entities within the organization, and *processes* — executing programs on behalf of users. Finally, privileges correspond to the operations that a subject can exercise on the objects in the system. The set of privileges thus depends on the resources to be protected; examples of privileges are read, write, and execute privileges for files in a file system, and select, insert, update, and delete privileges for relational tables in a relational DBMS.

Objects, subjects, and privileges can be organized into hierarchies. The semantics of a hierarchy depends on the specific domain considered. An example of hierarchy is the composite object hierarchy, typical of object-based DBMSs, that relates a given object to its component objects. Another relevant example is represented by the role hierarchy that relates a role in a given organization to its more specialized roles, referred to as junior roles. Hierarchies allow authorizations to be implicitly propagated, thus increasing the conciseness of authorization rules and reducing the authorization administration load. For example, authorizations given to a junior role are automatically propagated to its ancestor roles in the role inheritance hierarchy, and thus there is no need of explicitly granting these authorizations to the ancestors.

The most well-known types of access control models are the *discretionary model* and the *mandatory model.* Discretionary access control (DAC) models govern the access of subjects to objects on the basis of the subjects identity and on the authorization rules. Authorization rules state, for each subject, the privileges it can exercise on each object in the system. When an access request is submitted to the system, the access control mechanism verifies whether there is an authorization rule authorizing (partially or totally) the access. In this case, the access is authorized; otherwise, it is denied. Such models are called discretionary in that they allow subjects to grant, at their discretion, authorizations to other subjects to access the protected objects. Because of such flexibility, DAC models are adopted in most commercial DBMSs. An important aspect of discretionary access control is related to the authorization administration, that is, the function of granting and revoking authorizations. It is the function by which authorizations are entered into (or removed from) the access control system. Common administration approaches include *centralized* administration, by which only some privileged users may grant and revoke authorizations, and *ownership-based* administration, by which grant and revoke operations on a data object are issued by the creator of the object. The ownership-based administration is often extended with features for administration delegation. Administration delegation allows the owner of an object to assign other users the right to grant and revoke authorizations, thus enabling decentralized authorization administration. Most commercial DBMSs adopt the ownership-based administration with administration delegation. More sophisticated administration approaches have been devised such as the *joint-based* administration, by which several users are jointly responsible for authorization administration; these administration approaches are particularly relevant for cooperative, distributed applications such as workflow systems and computer-supported cooperative work (CSCW).

One of the first DAC models to be proposed for DBMSs is the model defined by Griffiths and Wade [1976] in the framework of the System RDBMS, which introduced the basic notions underlying the access control models of current commercial DBMSs. Such a model has been then widely extended with features such as negative authorizations, expressing explicit denials, and thus supporting the formulation of exceptions with respect to authorizations granted on sets of objects; more articulated revoke operations; and temporal authorizations, supporting the specification of validity intervals for authorizations. Even though DAC models have been adopted in a variety of systems because of their flexibility in expressing a variety of access control requirements, their main drawback is that they do not impose any control on how information is propagated and used once it has been accessed by subjects authorized to do so. This weakness makes DAC systems vulnerable to malicious attacks, such as attacks through Trojan Horses embedded in application programs or through *covert channels.* A covert channel [Bertino, 1998] is any component or feature of a system that is misused to encode or represent information for unauthorized transmission. A large variety of components or features can be exploited to establish covert channels, including the system clock, the operating system interprocess communication primitives, error messages, the concurrency control mechanism, and so on.

Mandatory access control (MAC) models address the shortcoming of DAC models by controlling the flow of information among the various objects in the system. The main principle underlying MAC models is that information should only flow from less protected objects to more protected objects. Therefore, any flow from more protected objects to less protected objects is illegal and it is thus forbidden by the reference monitor. In general, an MAC system specifies the accesses that subjects have to objects based on subjects–objects classification. The classification is based on a partially ordered set of *access classes,* also called *labels,* that are associated with every subject and object in the system. An access class generally consists of two components: a security level and a set of categories. The security level is an element of a hierarchically ordered set. A very well known example of such a set is the one including the levels Top Secret (TS), Secret (S), Confidential (C), and Unclassified (U), where TS > S > C > U. An example of the set of categories in an unordered set is NATO, Nuclear, Army. Access classes are partially ordered as follows: An access class c_i *dominates* (>) a class c_j iff the security level of c_i is greater than or equal to that of c_j and the categories of c_i include those of c_j. The security level of the access class associated with a data object reflects the sensitivity of the information contained in the object, whereas the security level of the access class associated with a user reflects the user's trustworthiness not to disclose sensitive information to users not cleared to see it. Categories are used to provide finer-grained security classification of subjects and objects than classifications provided by security levels alone and are the basis for enforcing *need-to-know* restrictions. Access control in a MAC system is based on the following two principles formulated by Bell and LaPadula [1976]:

- No read-up: a subject can read only those objects whose access class is dominated by the access class of the subject.
- No write-flown: a subject can write only those objects whose access class dominates the access class of the subject.

Verification of these principles prevents information in a sensitive object from flowing into objects at lower or incomparable levels. Because of such restrictions, this type of access control has also been referred to as *multilevel security.* Database systems that satisfy multilevel security properties are called multilevel secure database management systems (MLS/DBMSs).

The main drawback of MAC models is their lack of flexibility, and therefore, even though some commercial DBMSs exist that provide MAC, these systems are seldom used. In addition to DAC and MAC models, a third type, known as role-based access control (RBAC) model, has been more recently proposed. The basic notion underlying such a model is one of *role.* Roles are strictly related to the

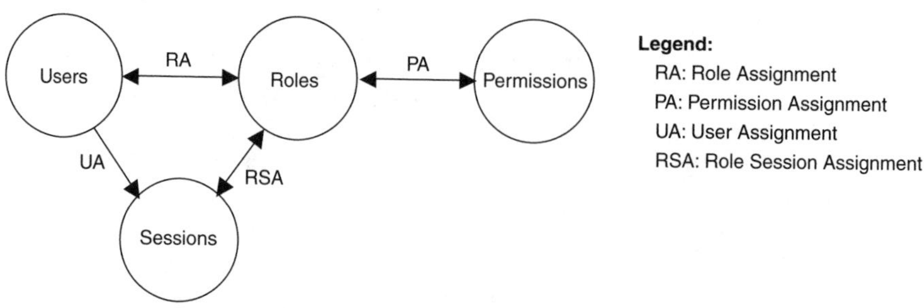

FIGURE 26.3 Building blocks of RBAC models.

organization and can be seen as a set of actions or responsibilities associated with a particular working activity. Under RBAC models, all authorizations needed to perform a certain activity are granted to the role associated with that activity, rather than being granted directly to users. Users are then made members of roles, thereby acquiring the roles' authorizations. User access to objects is mediated by roles; each user is authorized to play certain roles and, on the basis of the role, he or she can perform "accesses" on the objects. Because a role groups a (possibly large) number of related authorizations, authorization administration is greatly simplified. Whenever a user needs to perform a certain activity or is assigned a certain activity, the user only needs to be granted the authorization to play the proper role, rather than being directly granted the required authorizations. A session is a particular instance of a connection of a user to the system and defines the subset of activated roles. At each moment, different sessions for the same user can be active. When users log into the system, they establish a session and, during this session, can request activation of a subset of the roles they are authorized to play. Basic building blocks of RBAC models are presented in Figure 26.3.

The use of roles has several well-recognized advantages from an enterprise perspective. First, because roles represent organizational functions, an RBAC model can directly support security policies of the organization. Authorization administration is also greatly simplified. If a user moves to a new function within the organization, there is no need to revoke the authorizations he had in the previous function and grant the authorizations he needs in the new function. The security administrator simply needs to revoke and grant the appropriate role membership. Last, but not least, RBAC models have been shown to be policy-neutral [Sandhu, 1996]; in particular, by appropriately configuring a role system, one can support different policies, including the mandatory and discretionary ones. Because of this characteristic, RBAC models have also been termed *policy-neutral models.* This is extremely important because it increases the flexibility in supporting the organization's security policies. Although current DBMSs support RBAC concepts, they do not exploit the full potential of these models. In particular, several advanced RBAC models have been developed, supporting, among other features, role hierarchies by which roles can be defined as subroles of other roles, and constraints supporting separation of duty requirements [Sandhu, 1996]. An RBAC model supporting sophisticated separation of duty constraints has also been recently proposed for workflow systems [Bertino et al., 1999].

26.2.2 A Brief Introduction to XML

XML [Extensible Markup Language, 2000] is today becoming the standard for data representation and exchange over the web. Building blocks of any XML document are *elements* and *attributes.* Elements can be nested at any depth and can contain other elements *(subelements),* in turn originating a hierarchical structure. An element contains a portion of the document delimited by two *tags:* the *start tag,* at the beginning of the element, and the *end tag,* at the end of the element. Attributes can have different types allowing one to specify element identifiers (attributes of type ID), additional information about the element (e.g., attributes of type CDATA containing textual information), or links to other elements (attributes of type IDREF(s)/URI(s)). An example of XML document modeling an employee dossier is

```
<Employee dossier Emp ID="EID">
    <Resume Date="8/8/2003">
        <Personal_Data Name="John" ... >
            <Reserved>
                <Health> ... </Health>
                <Criminal> ... </Criminal>
            </Reserved>
        </Personal Data>
        <Education>
            <Qualification> ... </Qualification>
            ...
        </Education>
        <Activity>
            <Professional_Experience> ... </Professional Experience>
            ...
        </Activity>
        </Resume>
        <Evaluation>
            <Mngr Eval> ... </Mngr Eval>
            <Board Direct Eval> ... </Board Direct Eval>
            <HR Eval> ... </HR_Eval>
        </Evaluation>
        <Career>
            <Position Role="Secretary" Salary="1000" Date-"8/8/2001"/>
            ...
            <Promotion Role="..." Salary="1500" eval_result="yes"/>
        </Career>
</Employee_dossier>
```

FIGURE 26.4 An example of XML document.

presented in Figure 26.4. Another relevant feature of XML is the support for the definition of application-specific document types through the use of *Document Type Definitions* (DTDs) or *XMLSchemas.*

26.3 Access Control for Web Documents

In this section, we start by discussing the main access control requirements of Web data by highlighting how protecting Web data is considerably different from protecting data stored in a conventional information system. Then, to make the discussion more concrete, we present a reference access control model for the protection of XML documents [Bertino et al., 2001b, Bertino and Ferrari, 2002a]. Data exchange over the Web often takes the form of documents that are made available at Web servers or those that are actively broadcasted by Web servers to interested clients. Thus, in the following text we use the terms data and documents as synonyms.

26.3.1 Access Control: Requirements for Web Data

In conventional Data Management Systems, access control is usually performed against a set of authorization rules stated by security administrators or users according to some access control policies. An authorization rule, in general, is specified on the basis of three parameters: (s, o, p) and specifies that subject s is authorized to exercise privilege p on object o. This simple identity-based paradigm, however, does not fit very well in the Web environment. The main reason is that Web data are characterized by peculiar characteristics that must be taken into account when developing an access control model suitable for their protection. These requirements are mainly dictated by the need of flexibility, which is necessary to cope with a dynamic and evolving environment like the Web. In the following, we briefly discuss the main issues that must be taken into account.

26.3.1.1 Subject Qualification

The first aspect when dealing with the protection of Web data is how to qualify subjects to whom an access control policy applies. In conventional systems, subjects are usually referred to on the basis of an ID-based mechanism. Examples of identity information could be a user-ID or an IP address. Thus, each access control policy is bound to a subject (or a set of subjects) simply by including the corresponding IDs in the policy specification. This mechanism, although it has the advantage of being very simple, is no longer appropriate for the Web environment where the community of subjects is usually highly dynamic and heterogeneous. This means that more flexible ways of qualifying subjects should be devised that keep into account, in addition to the subject identities, other characteristics of the subject (either personal or deriving from relationships the subject has with other subjects).

26.3.1.2 Object Qualification

The second issue regards the number and typology of *protection objects* to which a policy may apply, where by protection object we mean a data portion over which a policy can be specified. An important security requirement of the Web environment is the possibility of specifying policies at a wide range of protection granularity levels. This is a crucial need because Web data source contents usually have varying protection requirements. In some cases, the same access control policy may apply to a set of documents in the source. In other cases, different policies must be applied to different components within the same document, and many other intermediate situations may also arise. As an example of this need, if you consider the XML document in Figure 26.4, while the name and the professional experiences of an employee may be available to everyone, access to the health record or the manager evaluation record must be restricted to a limited class of users. To support a differentiated and adequate protection of Web data, the access control model must thus be flexible enough to support a spectrum of protection granularity levels. For instance, in the case of XML data, examples of granularity levels can be the whole document, a document portion, a single document component (e.g., an element or attribute), or a set of documents. Additionally, this wide range of protection granularity levels must be complemented with the possibility of specifying policies that exploit both the structure and the content of a Web document. This is a relevant feature because there can be cases in which all the documents with the same structure have the same protection needs and thus can be protected by specifying a unique policy, as well as situations where documents with the same structure have contents with very different sensitivity degrees.

26.3.1.3 Exception Management and Propagation

Supporting fine-grained policies could lead to the specification of a possibly high number of access control policies. An important requirement is that the access control model has the capability of reducing as much as possible the number of policies that need to be specified. A means to limit the number of policies that need to be specified is the support for both positive and negative policies. Positive and negative policies provide a flexible and concise way of specifying exceptions in all the situations where a whole protection object has the same access control requirements, apart from one (or few) of its components. A second feature that can be exploited to limit the number of policies that need to be specified is *policy propagation*, according to which policies (either positive or negative) specified for a protection object at a given granularity level apply by default to all protection objects related to it according to a certain relationship. For instance, in the case of XML documents, the relationships that can be exploited derive from the hierarchical structure of an XML document. For example, one can specify that a policy specified on a given element propagates to its direct and indirect subelements.

26.3.1.4 Dissemination Strategies

The last fundamental issue when dealing with access control for Web data is what kind of strategies can be adopted to release data to subjects in accordance with the specified access control policies. In this respect, there are two main strategies that can be devised:

- **Pull mode.** This represents the simplest strategy and the one traditionally used in conventional information systems. Under this mode, a subject explicitly requests data to the Web source when needed. The access control mechanism checks which access control policies apply to the requesting subject and, on the basis of these policies, builds a view of the requested data that contains all and only those portions for which the subject has an authorization. Thus, the key issue in supporting information pull is how to efficiently build data views upon an access request on the basis of the specified access control policies.
- **Push mode.** Under this approach, the Web source periodically (or when the same relevant event arises) sends the data to its subjects, without the need for an explicit request. Even in this case, different subjects may have the right to access different views of the brodcasted documents. Because this mode is mainly conceived for data dissemination to a large community of subjects, the key issue is how to efficiently enforce information push by ensuring at the same time the confidentiality requirements specified by the defined policies.

In the following section, we illustrate an access control model for XML documents addressing the above requirements. The model has been developed in the framework of the Author-*X* project [Bertino et al., 2001b; Bertino and Ferrari, 2002a].

26.3.2 A Reference Access Control Model for the Protection of XML Documents

We start by illustrating the core components of the model, and then we deal with policy specification and implementation techniques for the access control model.

26.3.2.1 Credentials

Author-*X* provides a very flexible way for qualifying subjects, based on the notion of *credentials.* The idea is that a subject is associated with a set of information describing his characteristics (e.g., qualifications within an organization, name, age). Credentials are similar to roles used in conventional systems; the main difference is that a role can just be seen as a name associated with a set of privileges, denoting an organizational function that needs to have such privileges for performing its job, whereas credentials have an associated set of properties that can be exploited in the formulation of access control policies. Thus, credentials make easier the specification of policies that apply only to a subset of the users belonging to a role, which show some common characteristics (for instance, a policy that authorizes all the doctors with more than 2 years of experience). Because credentials may contain sensitive information about a subject, an important issue is how to protect this information. For instance, some credential properties (such as the subject name) may be made accessible to everyone, whereas other properties may be visible only to a restricted class of subjects. To facilitate credential protection, credentials in Author-*X* are encoded using an XML-based language called *X*-Sec [Bertino et al., 2001a]. This allows a uniform protection of XML documents and credentials in that credentials themselves are XML documents and thus can be protected using the same mechanisms developed for the protection of XML documents. To simplify the task of credential specification, *X*-Sec allows the specification of *credential-types,* that is, DTDS which are templates for the specification of credentials with a common structure. A credential-type models simple properties of a credential as empty elements and composite properties as elements with element content, whose subelements model composite property components. A credential is an instance of a credential type and specifies the set of property values characterizing a given subject against the credential type itself. *X*-Sec credentials are certified by the credential issuer (e.g., a certification authority) using the techniques proposed by the W3C XML Signature Working Group [XML Signature Syntax, 2002].

The credential issuer is also responsible for certifying properties asserted by credentials. An example of an *X*-Sec credential type that can be associated with a department head is shown in Figure 26.5(a), whereas Figure 26.5(b) shows one of its possible instances.

A subject can be associated with different credentials, possibly issued by different authorities, that describe the different roles played during the subject's everyday life. For instance, a subject can have, in addition to a dept_head credential, also a credential named IEEE_member, which qualifies him or her as a member of IEEE. To simplify the process of evaluating subject credentials against access control policies, all the credentials a subject possesses are collected into an XML document called *X-profile.*

26.3.2.2 Protection Objects

Author-*X* provides a spectrum of protection granularity levels, such as a whole document, a document portion, a single document component (i.e., an attribute/element or a link), and a collection of documents. These protection objects can be identified both on the basis of their structure and their content. Additionally, Author-*X* allows the specification of policies both at the instance and at the schema level (i.e., DTD/XMLSchema), where a policy specified at the schema level propagates by default to all the corresponding instances. To simplify the task of policy specification, Author-*X* supports different explicit propagation options that can be used to reduce the number of access control policies that need to be defined. Propagation options state how policies specified on a given protection object of a DTD/XML-Schema/document propagate (partially or totally) to lower-level protection objects. We use a natural number *n* or the special symbol '*' to denote the *depth* of the propagation, where symbol '*' denotes that the access control policy propagates to all the direct and indirect subelements of the elements specified in the access control policy specification, whereas symbol *n* denotes that the access control policy propagates to the subelements of the elements specified in the policy specification, which are, at most, *n* level down in the document/DTD/XML-Schema hierarchy.

26.3.2.3 Access Control Modes

Author-*X* supports two different kinds of access control policies: *browsing policies* that allow subjects to see the information in a document and/or to navigate through its links, and *authoring policies* that allow the modification of XML documents under different modes. More precisely, the range of access modes provided by Author-*X* is the following: {view,navigate,append,write,browse-all,auth_all}, where the view privilege authorizes a subject to view an element or some of its components, the navigate privilege authorizes a subject to see the existence of a specific link or of all the links in a given element, the append privilege allows a subject to write information in an element (or in some of its parts) or to include a link in an element, without deleting any preexisting information, whereas the write privilege allows a subject to modify the content of an element and to include links in the element. The set of access modes is complemented by two additional privileges that respectively subsume all the browsing and authoring privileges.

```
<!DOCTYPE dept_head[
<!ELEMENT dept_head(dept_name, name,
address,
phone_number*, email?, company)>
<!ELEMENT dept_name(#PCDATA)>
<!ELEMENT name(fname lname)>
<!ELEMENT address(#PCDATA)>
<!ELEMENT phone_number(#PCDATA)>
<!ELEMENT email(#PCDATA)>
<!ELEMENT company(#PCDATA)>
<!ELEMENT fname(#PCDATA)>
<!ELEMENT lname(#PCDATA)>
<!ATTLIST dept_head credID #REQUIRED>
]>
```

```
<dept_head credID= "154">
<dept_name>HR</dept_name>
<name>
<fname>John</fname>
<lname>Smith<lname>
</name>
<address>168 Bright Street, 1709 New
York City<address>
<company>Data2000</company>
</dept_head><phone
<phone_number>709 854345 </phone_number>
<email>john.smith@data 2000.com</email>
```

FIGURE 26.5 (a) An example of *X*-Sec credential type and (b) one of its instances.

```
<!DOCTYPE acc_policy_base
<!ELEMENT acc_policy_base(acc_policy_spec)*>
<!ELEMENT acc_policy_spec(objs_spec)>
<!ELEMENT objs_spec EMPTY>
<!ATTLIST acc_policy_spec cred_expr CDATA #REQUIRED>
<!ATTLIST objs_spec target CDATA #REQUIRED>
<!ATTLIST objs_spec path CDATA>
<!ATTLIST acc_policy_spec priv CDATA #REQUIRED>
<!ATTLIST acc_policy_spec type CDATA #REQUIRED>
<!ATTLIST acc_policy_spec prop CDATA #REQUIRED>
]>
```

FIGURE 26.6 *X*-Sec access control policy base template..

26.3.2.4 *X*-Sec Access Control Policies

Access control policies are encoded in *X*-Sec, according to the template reported in Figure 26.6.

The template is a DTD where each policy specification is modeled as an element (acc_policy-spec) having an attribute/element for each policy component. The meaning of each component is explained in Table 26.1.

The template allows the specification of both positive and negative credential-based access control policies. An access control policy base is therefore an XML document instance of the *X*-Sec access control policy base template. An example of access control policy base, referring to the portion of XML source illustrated in Figure 26.4, is reported in Figure 26.7.

26.3.2.5 Implementation Techniques for the Access Control Model

To be compliant with the requirements discussed at the beginning of this section, Author-*X* allows the release of XML data according to both a push and a pull mode. In the following, we focus on the techniques devised for information push because this constitutes the most innovative way of distributing data on the Web. We refer the interested reader to [Bertino et al., 2001b] for the details on information push enforcement. The idea exploited in Author-*X* is that of using encryption techniques for an efficient support of information push. More precisely, the idea is that of selectively encrypting the documents to be released under information push on the basis of the specified access control policies. All document portions to which the same access control policies apply are encrypted with the same key. Then, the same document encryption is sent to all the subjects, whereas each subject only receives the decryption keys corresponding to document portions he or she is allowed to access. This avoids the problem of generating and sending a different view for each subject (or group of subjects) that must receive a document according to a push mode. A relevant issue in this context is how keys can be efficiently and securely distributed to the interested subjects. Towards this end, Author-*X* provides a variety of ways for the delivering of decryption keys to subjects. The reason for providing different strategies is that the key delivery method may depend on many factors, such as for instance the number of keys that need to be delivered, the number of subjects, specific subject preferences, or the security requirements of the considered domain. Basically, the Security Administrator may select between *offline* and *online* key delivery

TABLE 26.1 Attribute Specification

Attr_name	Parent_node	Meaning
Cred_expr	acc_policy_spec	XPath expression on *X*-profiles denoting the subjects to which the policy applies
Target	objs_spec	Denotes the documents/DTDs/XMLSchemas to which the policy applies
Path	objs_spec	Denotes selected portions within the target (through *X*Path)
Priv	acc_policy_spec	Specifies the policy access mode
Type	acc_policy_spec	Specifies whether the access control policy is positive or negative
Prop	acc_policy_spec	Specifies the policy propagation option

```
<acc_policy_base>
<acc_policy_spec cred_expr="//Manager" priv="BROWSE_ALL"type="GRANT"
propt_opt="*"
<objs_spec target="Employee_dossier.dtd"/>
</acc_policy_spec>
<acc_policy_spec cred_expr="//Employee" priv="BROWSE_ALL"type="GRANT"
propt_opt="*"
<objs_spec target="Employee_dossier.xml" path="//Resume|//Position|//
      Promotion[@eval_result=yes]"/>
</acc_policy_spec>
<acc_policy_spec cred_expr="//
Board_directors_members"priv="VIEW"type="Grant"propt_opt="*"
<objs_spec target="Employee_dossier.xml" path="Board_direct_eval|//Resume|//
Career"/>
</acc_policy_spec>
<acc_policy_spec cred_expr="//
Board_directors_members"priv="VIEW"type="Deny"propt_opt="*"
<objs_spec target="Employee_dossier.xml" path="//Reserve|//@salary"/>
</acc_policy_spec>
</acc_policy_base>
```

FIGURE 26.7 An example of access control policy base.

modes [Bertino and Ferrari, 2002a]. According to an online mode, it is the source that delivers to subjects both the encrypted document and the keys (either into a unique message or into separate messages). By contrast, in the offline mode, the source locally stores the keys (for example in an LDAP directory) and the subjects query the directory for retrieving the necessary keys.

26.4 Authentication Techniques for XML Documents

Digital signature techniques are the means for ensuring the authenticity of the data transmitted over the Web. The most common approach to digital signatures is based on public-key cryptography [Stallings, 2000]. Such an approach requires the creation of a one-way hash value (called *message digest*) of the document, and then the encryption of the obtained digest value with the private key of the signer. The recipient receives this encrypted digest along with the original data. He then decrypts it with the signer public key and locally calculates the digest of the received data. If these two values match, the signature is validated. Although the problem of XML document confidentiality has been widely investigated and several access control models have been so far defined [Pollmann], no comparable amount of work has been carried out for providing a comprehensive solution to XML document authenticity. In particular, whereas several signature techniques are today available allowing the data receiver to verify the authenticity of the received XML data, few proposals exist encompassing all aspects of the signature process. A key requirement is the definition of a model supporting the specification of *signature policies*. Similar to access control policies, a signature policy has to specify which subjects must sign a document (or document portions).

In the following text, we first briefly review the standardization effort carried out by the W3C XML Signature Working Group [XML Signature Syntax, 2002]. We then discuss the support for signature policies provided by Author-*X*, since this is the unique model we are aware of that provides the support for signature policies.

26.4.1 An Introduction to XML Signature

The W3C XML Signature Working Group, in conjunction with IETF, is working on a standard called *XML Signature,* with the twofold goal of defining an XML representation for the digital signature of

```
<Signature>
     <SignedInfo>
     (CanonicalizationMethod)
       (SignatureMethod)
       (<Reference (URI=)?>
          (Transforms)?
          (DigestMethod)
          (DigestValue)
     (/SignedInfo)
     (SignatureValue)
     (KeyInfo)?
     (Object)*
</Signature>
```

FIGURE 26.8 Basic structure of an XML Signature.

arbitrary data contents (called data objects in what follows), not necessarily XML-encoded data, but which, at the same time, is particularly well suited for digitally signing XML documents. The idea is to take advantage of the semantically rich and structured nature of XML to provide a flexible framework for the representation of digital signatures of arbitrarily digital contents. A fundamental feature of the proposal is that a single XML Signature can sign more than one type of digital content; for instance, a single XML Signature can be used to sign an HTML file and all the JPEG files containing images linked to the HTML page. The overall idea is that data objects to be signed are digested, each digest value is placed into a distinct XML element with other additional information, and then all the resulting XML elements are collected in a parent element that is digested and cryptographically signed. The resulting XML structure is the one reported in Figure 26.8. Another key feature of XML Signature, which is not supported by older digital signature standards, is the possibility of signing only selected portions of an XML document. This is a relevant feature when documents flow among parties and each party wishes to sign only those portions of the documents for which it is responsible. Additionally, the standard proposal supports a variety of strategies for locating the objects being signed. These objects can be either external data or local data, that is, a portion of the XML document containing the signature itself. In particular, the standard proposal supports three different kinds of signature, which differ for the localization of the data objects being signed:

- **Enveloping Signature:** The signature is the parent of the signed data.
- **Enveloped Signature:** The data object embeds its signature, and thus it is the parent of its signature.
- **Detached Signature:** The data object is either an external data, or a local data object included as a sibling element of its signature.

26.4.2 Signature Policies

The main difference between an access control policy and a signature policy is that the first expresses the possibility of exercising a privilege on a given document, whereas the second expresses the *duty* of signing a document. In the text that follows, we describe signature policies in the framework of Author-*X* [Bertino et.al., 2003]. Author-*X* supports two different types of signature policies: simple and joint. Simple signature policies require the signature of a single subject on a document or document portions. A joint signature is by contrast required when the same document portion must be signed by several subjects. This could be the case, for instance, of a travel approval that must be signed by both the administrative manager and the technical manager. Similar to access control policies, *X*-Sec provides an XML template for the specification of signature policies, which is illustrated in Figure 26.9.

```
<!DOCTYPE   sign_policy_base[
<!ELEMENT   sign_policy_base(sign_policy_spec)*>
<!ELEMENT   sign_policy_spec(sbjs_spec, objs_sped)>
<!ELEMENT   sbjs_spec EMPTY>
<!ELEMENT   objs_spec EMPTY>
<!ATTLIST   sbjs_spec cred_expr CDATA #REQUIRED>
<!ATTLIST   objs_spec target CDATA #REQUIRED>
<!ATTLIST   objs_spec path CDATA>
<!ATTLIST   sign_policy_spec duty CDATA #REQUIRED>
<!ATTLIST   sign_policy_spec type CDATA #REQUIRED>
<!ATTLIST   sign_policy_spec prop CDATA #REQUIRED>
]>
```

FIGURE 26.9 *X*-Sec signature policy base template.

The structure of the template is very similar to that provided for access control policies; the only differences are attribute **duty** and element subjs_spec. Attribute **duty** specifies the kind of signature policy and can assume two distinct values: *sign* and *joint-sign*. The *sign duty* imposes that at least a subject, whose *X*-profile satisfies the credential expression in the signature policy, signs the protection objects to which the policy applies, whereas the *joint-sign* duty imposes that, for each credential expression specified in the signature policy, at least one subject, whose *X*-profile satisfies the specified credential expression, signs the protection objects to which the policy applies. To support both simple and joint signatures, the subjs_spec element contains one or more XPath compliant expressions on *X*-profiles. A signature policy base is an instance of the signature policy base template previously introduced. An example of signature policy base for the XML document in Figure 26.4 is reported in Figure 26.10. For instance, the first signature policy in Figure 26.10 imposes that each employee signs his resume, whereas according to the second policy, the manager must sign the employee's evaluation. The third signature policy requires a joint signature by two members of the board of directors on the employee evaluation made by the board of directors.

26.5 Data Completeness and Filtering

In this section, we discuss possible approaches for the enforcement of both data completeness and self-protection and filtering, two additional security requirements that are particularly crucial when dealing with Web-based information systems.

26.5.1 Data Completeness

By data completeness we mean that any subject requesting data from a Web data source must be able to verify that he or she has received all the data (or portions of data) that the subject is entitled to access, according to the stated access control policies. This requirement is particularly crucial when data are released according to a *third-party architecture.* Third-party architectures for data publishing over the Web are today receiving growing attention due to their scalability properties and to their ability to efficiently manage a large number of subjects and a great amount of data. In a third-party architecture, there is a distinction between the *Owner* and the *Publisher* of information. The Owner is the producer of the information, whereas Publishers are responsible for managing (a portion of) the Owner information and for answering subject queries. A relevant issue in this architecture is how the Owner can ensure a complete publishing of its data even if the data are managed by an untrusted third party. A possible approach to completeness verification [Bertino et al., 2002b] relies on the use of the *secure structure* of an XML document. In a third-party architecture, the secure structure is sent by the Owner to the Publisher

```
<sign_policy_base>
    <sign_policy_spec duty="sign" propt_opt="*">
        <subjs_spec cred_expr="Employee/>
        <objs_spec_target="Employee_dossier.dtd" path="//Resume/>
    <sign_policy_spec>
    <sign_policy_spec duty="sign"propt_opt="*">
        <subjs_spec cred_expr="//manager"/>
        <objs_spec target="Employee_dossier.dtd" path="//Evaluation"/>
    </sign_policy_spec>
    <sign_policy_spec duty="joint_sign" propt_opt="*">
        <subjs_spec cred_exp1="//Board_directors_member"
        cred_expr2="//Board_directors_member"/>
        <objs_spec target="Employee_dossier.xml" path="//
Board_directors_eval"/>
    </sign_policy_spec>
    </sign_policy_spec duty="sign "propt_opt="*">
        <subjs_spec_cred_expr="Human_resources_head"/>
        <objs_spec target="Employee_dossier.xml" path="//hr_eval"/>
    </sign_policy_spec>
    </sign_policy_spec duty="sign "propt_opt="*">
        <subjs_spec cred_expr="//Payroll_Department_head|
            //Human_resources_responsible">
        <objs_spec target="Employee_dossier.xml" path="//career"/>
    </sign_policy_spec>
<sign_policy_base>
```

FIGURE 26.10 An example of signature policy base.

and successively returned to subjects together with the answer to a query on the associated document. By contrast, if a traditional architecture is adopted, the secure structure is sent directly to subjects when they submit queries to the information owner. This additional document contains the tagname and attribute names and values of the original XML document, hashed with a standard hash function, so that the receiving subject cannot get information he is not allowed to access. To verify the completeness of the received answer, the subject performs the same query submitted to the Publisher on the secure structure. Clearly, the query is first transformed to substitute each tag and attribute name and each attribute value with the corresponding hash value. By comparing the result of the two queries, the subject is able to verify completeness for a wide range of XPath queries [Bertino et.al., 2002b], without accessing any confidential information.

26.5.2 Filtering

The Web has the undoubted benefit of making easily available to almost everyone a huge amount of information. However, the ease with which information of any kind can be accessed through the Web poses the urgent need of protecting specific classes of users, such as children, from receiving improper material. To this purpose, *Internet filtering systems* are now being developed that can be configured according to differing needs. The aim of a filtering system is to block accesses to improper Web contents. Filtering systems operate according to a set of *filtering policies* that state what can be accessed and what should be blocked. Thus, differently from what happens in access control mechanisms, filtering policies should not be specified by content producers (i.e., target-side), but by content consumers (i.e., client-side). Clearly, a key requirement for Internet filtering is the support for content-based filtering. This feature relies on the possibility of associating a computer-effective description of the semantic content (the meaning) of Web documents. The current scenario for implementing and using content-based filtering systems implies that content providers or independent organizations associate PICS-formatted labels with Web documents. However, such an approach has serious shortcomings. The first drawback is related to completeness, i.e., the possibility of representing the semantic content of Web documents to the necessary level of detail. A second important problem concerns the low level of neutrality of the current rating/filtering proposals. A third drawback is that current filtering systems do not have the ability of associating different policies to different classes of users on the basis of their credentials. This is a relevant feature when filtering systems must be used by institutional users (private users such as publishers or public users such as NGOs or parental associations) that must specify different filtering policies for different classes of users. Thus, a promising research direction is that of developing *multistrategy* filtering systems, able to support a variety of ways of describing the content of Web documents and a variety of different filtering policies. An example of this research trend is the system called MaX (E. Bertino. E. Ferrari, Perego, 2003), developed in the framework of the *IAP EUFORBIA* project. MaX provides the ability of specifying credential-based filtering policies, and it is almost independent from the techniques used to describe the content of a Web document. It can support either standard PICS-based rating systems, as well as keyword-based and concept-based descriptions. However, much more work needs to be done in the field of multistrategy filtering systems. A first research direction is related to the fact that most of the filtering systems so far developed are mainly conceived for textual data. An interesting issue is thus how to provide content-based filtering for other kinds of multimedia data such as, for instance, images, audio, and video. Another relevant issue is how to complement existing filtering systems with *supervision mechanisms* according to which accesses to a Website are allowed, provided that a subject or a group of subjects (e.g., parents) are preventively informed of these accesses.

26.6 Conclusions and Future Trends

Information security, and in particular, data protection from unauthorized accesses, remain important goals of any networked information system. In this chapter, we have outlined the main concepts underlying security, as well as research results and innovative approaches. Much more, however, needs to be done. As new applications emerge on the Web, new information security mechanisms are required. In this chapter, we could not discuss many emerging applications and research trends. In conclusion, however, we would like to briefly discuss some important research directions.

The first direction is related to the development of trust negotiation systems. The development of such systems is motivated by the fact that the traditional identity-based approach to establishing trust is not suitable for decentralized environments, where most of the interactions occur between strangers. In this context, the involved parties need to establish mutual trust before the release of the requested resource. A promising approach is represented by trust negotiation, according to which mutual trust is established through an exchange of property-based digital credentials. Disclosure of credentials, in turn, must be governed by policies that specify which credentials must be received before the requested credential can be disclosed. Several approaches to trust negotiation have been proposed (e.g., T. Yu, 2003). However, all such proposals focus on only some of the aspects of trust negotiation, such as the language to express policies and credentials, the strategies for performing trust negotiations, or the efficiency of the interactions, and none of them provide a comprehensive solution. We believe that key ingredients of such a comprehensive solution are first of all a standard language for expressing credentials and policies. In this respect, an XML-based language could be exploited for the purpose. The language should be flexible enough to express a variety of protection needs because the environments in which a negotiation could take place can be very heterogeneous. Another important issue is to devise different strategies to carry on the negotiation, on the basis of the sensitivity of the requested resource, the degree of trust previously established by the involved parties, and the requested efficiency. For instance, there can often be cases in which two parties negotiate the same or a similar resource. In such a case, instead of performing several times the same negotiation from scratch, the results of such previous negotiations can be exploited to speed up the current one.

Another research direction is related to providing strong privacy. Even though privacy-preserving data-releasing techniques have been widely investigated in the past in the area of statistical databases, the Web makes available a large number of information sources that can then be combined by other parties to infer private, sensitive information, perhaps also through the use of modern data-mining techniques. Therefore, individuals are increasingly concerned about releasing their data to other parties, as they do not know with what other information the released information could be combined, thus resulting in the disclosure of their private data. Current standards for privacy preferences are a first step towards addressing this problem. However, much more work needs to be done, in particular in the area of privacy-preserving data-mining techniques.

Another research direction deals with mechanisms supporting distributed secure computations on private data. Research proposals in this direction include mechanisms for secure multiparty computation — allowing two or more parties to jointly compute some computation of their inputs while hiding their inputs from each other — and protocols for private information retrieval — allowing a client to retrieve a selected data object from a database while hiding the identity of this data object from the server managing the database. Much work, however, needs to be carried out to apply such approaches to the large variety of information available on the Web and to the diverse application contexts.

Finally, interesting research issues regarding how protection mechanisms that have been developed and are being developed for Web documents can be easily incorporated into existing technology and Web-based enterprise information system architectures. The Web community regards XML as the most important standardization effort for information exchange and interoperability. We argue that compatibility with XML companion technologies is an essential requirement in the implementation of any access control and protection mechanism.

References

Bell, D.L., and L.J. LaPadula. Secure computer systems: Unified exposition and multics interpretation. Technical report, MITRE, March 1976.

Bertino, E., Data security. *Data and Knowledge Engineering,* 25(1/2): 199–216, March 1998.

Bertino, E., V Atluri and E. Ferrari. The specification and enforcement of authorization constraints in workflow managenent systems. *ACM Transactions on Information and System Security,* 2(1): 65–104, February 1999.

Bertino, E. and E. Ferrari. Secure and selective dissemination of xml documents. *ACM Transactions on Information and System Security,* 5(3): 290–331, August 2002a.

Bertino, E., E. Ferrari, and S. Castano. On specifying security policies for web documents with an xml-based language. Lecture Notes in Computer Science, pages 57–65. 1st ACM Symposium on Access Control Models and Technologies (SACMAT'01), ACM Press, New York May 2001a.

Bertino, E., E. Ferrari, and S. Castano. Securing xml documents with author-*X. IEEE Internet Computing,* 5(3): 21–31, May/June 2001b.

Bertino, E., E. Ferrari, and L. Parasiliti. Signature and access control policies for xml documents. 8th European Symposium on Research in Computer Security (ESORICS 2003), Gjovik, Norway, October 2003.

Bertino, E., E. Ferrari, B. Thuraisingam, A. Gupta, and B. Carminati. Selective and authentic third-party distribution of xml documents. Technical Report, *IEEE Transactions on Knowledge and Data Engineering,* in press.

Bertino, E., A. Vinai, and B. Catania. *Encyclopedia of Computer Science and Technology,* volume 38, in the chapter "Transaction Models and Architectures," pages 361–400. Marcel Dekker, New York, 1998.

Extensible Markup Language (XML) 1.0 (Second Edition). World Wide Web Consortium, W3C Recommendation 6 October 2000. URL http://www.w3.org/TR/REC-xml.

Griffiths, P.P. and B.W. Wade. An authorization mechanism for a relational database system. *ACM Transactions on Database* Systems, 1(3): 242–255, September 1976.

Pollmann, C. Geuer. The xml security page. URL http://www.nue.et-inf.uni-siegen.de/euer-pollmann/xml_security.html.

Sandhu, R. Role hierarchies and constraints for lattice-based access controls. *Lecture Notes in* Computer Science 1146, pages 65–79. 4th European Symposium on Research in Computer *Security (ESORICS 2003)*, Rome, September 1996.

Stallings, W. *Network Security Essentials: Applications and Standards,* Prentice Hall, Englewood Cliffs, NJ, 2000.

T. Yu, M. Winslett. A unified scheme for resource protection in automated trust negotiation. IEEE Symposium on Security and Privacy, Oakland, CA, 2003.

Yu, T., K. E. Seamons, and M. Winslett. Supporting structured credentials and sensitive policies through interoperable strategies for automated trust negotiation. *ACM Transactions on Information and System Security,* 6(1): 1–42, February 2003.

XML Path (XPATH) 1.0, 1999. World Wide Web Consortium. URL http://www.w3.org/TR/1999/REC-xpath-19991116.

XML Signature Syntax and Processing. World Wide Web Consortium, W3C Recommendation February 12, 2002. URL http://www.w3.org/TR/xmldsig-core/.

27

Understanding Web Services

CONTENTS

Abstract ... 27-1
27.1 Introduction ... 27-1
27.2 Service Oriented Computing ... 27-3
27.3 Understanding the Web Services Stack ... 27-4
27.4 Transport and Encoding ... 27-5
27.4.1 SOAP ... 27-5
27.5 Quality of Service ... 27-7
27.5.1 Security ... 27-7
27.5.2 Reliability ... 27-8
27.5.3 Coordination ... 27-9
27.6 Description ... 27-11
27.6.1 Functional Definition of a Web Services ... 27-11
27.6.2 A Framework for Defining Quality of Service ... 27-13
27.6.3 Service Discovery ... 27-13
27.7 Composition ... 27-15
27.7.1 Choreography ... 27-15
27.8 Summary ... 27-17
References ... 27-18

Rania Khalaf
Francisco Curbera
William Nagy
Stefan Tai
Nirmal Mukhi
Matthew Duftler

Abstract

Web services aim to support highly dynamic integration of applications in natively interorganizational environments. They pursue a platform independent integration model, based on XML standards and specifications, that should allow new and existing applications created on proprietary systems to seamlessly integrate with each other. This chapter provides an overview of the Web services paradigm, starting with the motivations and principles guiding the design of the Web services stack of specifications. Practical descriptions are provided of the key specifications that define how services may communicate, define, and adhere to quality of service requirements; provide machine readable descriptions to requesting applications; and be composed into business processes. An example is discussed throughout the chapter to illustrate the use of each specification.

27.1 Introduction

Web services is one of the most powerful and, at the same time, more controversial developments of the software industry in the last 5 years. Starting with the initial development of the Simple Object Access Protocol specification (SOAP 1.0) in 1999, followed by the Universal Description Discovery and Integra-

1-58488-381-2/05/$0.00+$1.50
© 2005 by CRC Press LLC

tion (UDDI 1.0) and the Web Services Description Language (WSDL 1.0) in 2000, the computer industry quickly bought into a vision of cross-vendor interoperability, seamless integration of computing systems, falling information technology (IT) costs, and the rapid creation of an Internet-wide service economy. The vision was soon backed by an array of products supporting the emerging pieces of the Web services framework and endorsed by most industry analysts.

It is easy to miss, among this unprecedented level of industry support, the flurry of new specifications and the skepticism of many (often from the academic community), the main motivating factors behind the Web services effort and its distinguishing features. There are three major motivating forces:

1. A fundamental shift in the way enterprises conduct their business, toward greater integration of their business processes with their partners. The increased importance of the business-to-business (B2B) integration market is a reflection of this trend [Yates et al., 2000].
2. The realization that the installed computing capacity is largely underutilized, and a renewed focus on efficient sharing. The development of scientific computing Grids are the most prominent example of this trend [Foster et al., 2001].
3. A recognition of the power of Internet standards to drive technical and business innovation.

Two important trends in the way business is conducted are particularly relevant to our first point. On the one hand, enterprises are moving to concentrate on their core competencies and outsource other activities to business partners. Businesses today assume that fundamental parts of their core processes may be carried out by trusted partners and will be effectively out of their direct control. On the other hand, a much more dynamic business environment and the need for more efficient processes have made "just-in-time" techniques key elements of today's business processes (in production, distribution, etc.). Just-in-time integration of goods and services provided by partners into core processes is now commonplace for many businesses.

A parallel trend has taken place in the scientific community. The idea that scientific teams are able to remotely utilize other laboratories' specialized applications and scientific data instead of producing and managing it all locally is the counterpart of the business trends we just described. In addition, the Grids initiative has helped focus the industry's attention on the possibility of achieving higher levels of resource utilization through resource sharing. A standard framework for remote application and data sharing will necessarily allow more efficient exploitation of installed computing capacity.

These changes in the way we think about acquiring services and goods have taken place simultaneously with the expansion of the Web into the first global computing platform. In fact, the wide availability of Internet networking technologies has been the technical underpinning for some of these developments. More importantly, however, the development of the Internet has shown the power of universal interoperability and the need to define standards to support it. HTTP and HTML were able to spark and support the development of the Web into a global human-centric computing and information sharing platform. Would it be possible to define a set of standards to support a global, interoperable, and application-centric computing platform? The Web services effort emerges in part as an attempt to answer this question.

Web services is thus an effort to define a distributed computing platform to support a new way of doing business and conducting scientific research in a way that ensures universal application to application interoperability. The computing model required to support this program necessarily has to differ from previous distributed computing frameworks if it is to support the requirements of dynamism, openness, and interoperability outlined here. Service Oriented Computing (SOC) provides the underlying conceptual framework on which the Web services architecture is being built. We review the main assumptions of this model in the next section.

In the rest of this chapter we will provide an overview of the key specifications of the Web services framework. Section 27.2 discusses the principal assumptions of the SOC model. Section 27.3 contains an introduction to the Web services specification stack. Section 27.4 covers the basic remote Web services interaction mechanisms, and Section 27.5 describes the quality of service protocols that support Web services interactions. In Section 27.6 we discuss how Web services are described and in Section 27.7 we explain how Web services can be combined to form service compositions. We conclude in Section 27.8 with a summary of the contents of this chapter.

27.2 Service Oriented Computing

Service Oriented Computing (SOC) tries to capture essential characteristics of a distributed computing system that natively supports the type of environment and the goals which we have described in the previous section. SOC differentiates the emerging platform from previous distributed computing architectures. Simplifying the discussion, we may summarize the requirements implied by our discussion above in three main ideas:

- The assumption that computing relies on networks of specialized applications owned and maintained by independent providers
- The concept of just-in-time, automated application integration, including pervasive support for a dynamic binding infrastructure
- The assumption that a set of basic protocols will provide universal interoperability between applications regardless of the platform in which they run

The exact nature of service oriented computing has not yet been clearly and formally articulated (though there is a growing interest in the space, see ICSOC, 2001; Papazoglou and Georgakopoulos [2003]), but we can state a set of key characteristics of SOC platforms that follow directly from these assumptions.

Platform independence: The set of protocols and specifications that support the SOC framework should avoid any specific assumption about the capabilities of the implementation platforms on which services run. The realization of the wide heterogeneity of platforms and programming models and the requirement for universal support and interoperability motivate this principle.

Explicit metadata: Applications ("services") must declaratively define their functional and nonfunctional requirements and capabilities ("service metadata") in an agreed, machine readable format. The aim is to reduce the amount of out-of band and implicit assumptions regarding the operation and behavior of the application by making all technical assumptions about the service explicit. Implicit assumptions about service properties limit its ability to operate in interorganizational environments.

Metadata driven dynamic binding: Based on machine readable declarative service descriptions, automated service discovery, selection, and binding become native capabilities of SOC middleware and applications, allowing just-in-time application integration. A direct consequence of the dynamic binding capability is a looser coupling model between applications. Applications express their dependencies on other services in terms of a set of behavioral characteristics and potentially discover the actual services they will utilize at a very late stage of their execution (late binding).

A componentized application model: In a SOC environment, services are basic building blocks out of which new applications are created. New applications are built out of existing ones by creating service compositions. A service composition combines services following a certain pattern to achieve a business goal, solve a scientific problem, or provide new service functions in general. Whether these services are found inside or outside the organization is essentially irrelevant from the application integration perspective, once we assume a SOC-enabled middleware platform. Service composition thus provides a mechanism for application integration which seamlessly supports cross-enterprise (business to business, B2B) and intra-enterprise application integration (EAI). Service-oriented computing is thus naturally a component oriented model.

A peer-to-peer interaction model: The interaction between services must be able to naturally model the way organizations interact, which does not necessarily follow the traditional client-server asymmetric model. Rather, they are much like business interactions: bidirectional and conversational. A typical interaction involves a series of messages exchanged between two parties over a possibly long-running conversation. Different modes of service coupling should be possible; however, loosely-coupled services combine traditional decoupled messaging-style interactions (data-excbange) with the explicit application contracts of tighter-coupled (interface-driven) object-oriented interaction styles. Note that with tightly-coupled services the notion of an application contract similar to those in object-oriented systems has a different flavor, since it relies on explicit specification of behavior and properties in metadata, as opposed to implicit assumptions about the state of an object.

27.3 Understanding the Web Services Stack

The Web services framework consists of a set of XML standards and specifications that provide an instantiation of the service-oriented computing paradigm discussed in the previous section. In this section we provide an overview of these different specifications. We organize them into a Web services stack, which is illustrated in Figure 27.1.

The Web services stack is extensible and modular. As the framework continues to mature and evolve, new specifications may be added in each of the layers to address additional requirements. The modularity enables a developer to use only those pieces of the stack deemed necessary for the application at hand. Using alternative transport and encoding protocols is an example of this extensibility. The framework is split into four main areas: transport and encoding, quality of service, descriptions, and business processes.

The transport and quality of service layers define basic "wire" protocols that every service will be assumed to support to ensure interoperability. A basic messaging protocol (SOAP, Gudgin et al. [2003]) and encoding format (XML) provide basic connectivity. The specifications in the quality of service layer (QoS) define protocols for handling QoS requirements such as exchanging messages reliably, security, and executing interactions with transactional semantics. The focus of these protocols is to provide "on the wire" interoperability by describing the normative requirements of the exchanged sequences of messages. Implementation and programming model implications are necessarily absent in order to guarantee full implementation and platform independence. It is important to observe that the Web services framework only requires following standard protocols for interoperability reasons. They may be replaced by platform specific ones when appropriate, while still complying with higher levels of the specification stack.

Description layer specifications deal with two problems. The first one is how to represent service behavior, capabilities, and requirements in a machine readable form. The second is how to enable potential service users to discover and dynamically access services. The Web Services Description Language, WSDL [Christensen et al., 2001] is used to define the functional capabilities of a service, such as the operations, service interfaces, and message types recognized by the service. WSDL also provides deployment information such as network address, transport protocol, and encoding format of the interaction. Quality of service requirements and capabilities are declared using the WS-Policy framework [Box et al., 2002b]. WS-Policy enables Web services quality of service "policies" to be attached to different parts of a WSDL definition. Different policy "dialects" are defined to represent specific types of QoS protocols, such reliable messaging, security, etc. Thus, the description layer enables the creation of

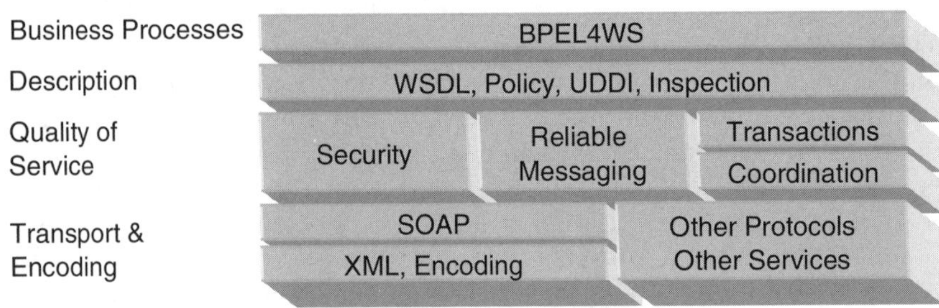

FIGURE 27.1 The Web services stack.

standard metadata associated with the protocols that form the other layers. A second set of specifications define how to publish, categorize, and search for services based on their service descriptions. The Universal Description, Discovery, and Integration (UDDI) specification [Bellwood et al., 2002] and WS-Inspection are the two protocols in this category.

Finally, the business process layer deals with the composition of services. Building on the service (component) descriptions defined using WSDL and WS-Policy, the specifications in this layer will define how services may be combined to create new applications or services. Currently only one specification, the Business Process Execution Language for Web Services (BPEL4WS) [Curbera et al., 2002b], provides this type of functionality. BPEL4WS defines a process or workflow oriented composition model particularly well suited to deal with business applications. Other composition mechanisms will likely emerge in the near future.

In order to understand how these layers fit together, consider a simple example of a loan request interaction. We will detail this example throughout this chapter as we cover the key specifications from the different parts of the stack.

In this example, Joe, a college student, wants to request a loan to pay for his college tuition. From his friends Joe learns about a company called LoanCo that processes such requests using Web services and decides to work with it. He looks at the functional definition of the service as contained in its WSDL document and sees that he can access its loan processing operation ("request") using SOAP over HTTP. Joe then sends a secure, reliable loan request to the service. On the other side of the wire, LoanCo used BPEL4WS to create its loan service. As part of its operation, the loan service interacts with two other Web services, one from a credit reporting agency and another from a highly-reputable financial institution to decide whether or not to provide the loan. LoanCo's process is not tied to specific agencies or banks, and it can decide at runtime which ones it would like to use for each customer, perhaps finding them through a query to a UDDI registry.

In the following sections we present the different layers of the stack in more detail, focusing on the specifications that are most relevant in each layer. The example which we just described will be used in each section to illustrate how the corresponding specifications are used.

27.4 Transport and Encoding

To achieve the goal of universal interoperability, a small set of application-to-application interaction protocols needs to be defined and adopted across the industry. The SOAP specification defines a messaging model which provides basic interoperability between applications, regardless of the platforms in which they are implemented. In this respect, SOAP is for Web services what the HTTP protocol is for the Web. As we mentioned before, the definition and widespread adoption of a common interaction protocol for Web services does not preclude the use of other platform or application specific protocols. For example, two applications may replace SOAP by a shared technology whose use provides additional benefits such as increased quality of service.

27.4.1 SOAP

SOAP is a lightweight, platform independent, XML-based messaging and RPC protocol. SOAP messages can be carried on any existing transport such as HTTP, SMTP, TCP, or proprietary messaging protocols like IBM's MQ Series. SOAP does not require the use of any of these, but it does define a standard way of carrying SOAP messages inside an HTTP request.

In its simplest form, a SOAP message is nothing but an envelope containing an optional header and a mandatory body. The envelope, header, and body are each expressed as XML elements. The header and the body sections can contain arbitrary XML elements. The following XML snippet shows an example of a (particularly simple) SOAP request that carries no header section.

```
<SOAP-ENV:Envelope
      xmlns:SOAP-ENV="http://schemas.xmlsoap.org/soap/envelope/"
      xmlns:xsi="http://www.w3.org/2001/XMLSchema-instance"
      xmlns:xsd="http://www.w3.org/2001/XMLSchema">
    <SOAP-ENV:Body>
        <ns1:request
            xmlns:ns1="http://loans.org/wsdl/loan-approval"
            SOAP-ENV:encodingStyle="http://schemas.xmlsoap.org/soap/encoding/">
        <name xsi:type="xsd:string">Joe</name>
        <amount xsi:type="xsd:int">10000</amount>
      </ns1:request>
      </SOAP-ENV:Body>
</SOAP-ENV:Envelope>
```

In addition SOAP introduces the notion of "actors," which can be used to indicate the intended recipient of various parts of a SOAP message. The actor construct is used to guide a SOAP message through a sequence of intermediaries, with each intermediary processing its portion of the message and forwarding the remainder.

27.4.1.1 Messaging using SOAP

The basic level of functionality provided by SOAP is that of a messaging protocol. The SOAP envelope wraps the application message in its body. The header section contains XML elements that support middleware level protocols such as reliable messaging, authentication, distributed transactions, etc. The envelop may be transmitted using any number of protocols, such as HTTP; it may even flow over different protocols as it reaches its destination. In a typical example, Joe's client may utilize LoanCo's service by sending it a SOAP envelope as the content of an HTTP POST request. When messaging semantics are used, the SOAP body contains arbitrary XML data; the service and client must come to an agreement as to how the data is to be encoded.

27.4.1.2 Remote Procedure Calls using SOAP

In addition to pure messaging semantics, the SOAP specification also defines a mechanism for performing remote procedure calls (RPC). This mechanism places constraints upon the messaging protocol defined above, such as how the root element in the body of the SOAP envelope is to be named. In order to carry the structured and typed information necessary for representing remote procedure calls, the SOAP specification provides suggestions as to how the data should be encoded in the messages which are exchanged.

XML Schema [Fallside, 2001], another W3C specification, provides a standard language for defining the structure of XML documents as well as the data type of XML structures. While SOAP allows one to use whatever encoding style or serialization rules one desires, SOAP does define an encoding style based on XML schema that may be used. This encoding style will allow for the generation of an XML representation for almost any type of application data. The requests and responses of RPC calls are represented using this XML encoding.

No matter what one's platform and transport protocol of choice are, there is most likely a SOAP client and server implementation available. There are literally dozens of implementations out there, many of which are capable of automatically generating and/or processing SOAP messages. So long as the generated messages conform to the SOAP specification, SOAP peers can exchange messages without regard to implementation language or platform.

That being said, it is important to once again point out that supporting SOAP is not a requirement for being considered a Web service; it simply provides a fall-back if no better suited communication mechanism exists.

27.5 Quality of Service

Just like any other distributed computing platform, the Web services framework must provide mechanisms to guarantee specific quality of service (QoS) properties such as performance, reliability, security, and others. Business interactions are just not feasible if they cannot be assumed to be secure and reliable. These requirements are often formalized through contracts and service agreements between organizations.

In a SOC environment, protocols are defined as strict "wire protocols," in that they only describe the external or "visible" behavior of each participant ("what goes on the wire"), avoiding any unnecessary specification of how the protocol should be implemented. This is a consequence of the platform and implementation independence design principle, and distinguishes Web service protocol specifications from their counterparts in other distributed computing models. In addition the SOC model requires that the QoS requirement and capabilities of services be declaratively expressed in machine readable form to enable open access and dynamic binding.

In this section we examine three key QoS protocols: security, reliability, and the ability to coordinate activities with multiple parties (for example, using distributed transaction capabilities). The policy framework described in Section 27.6.2 provides a generic mechanism to describe the QoS characteristics of a service.

27.5.1 Security

Security is an integral part of all but the most trivial of distributed computing interactions. The security mechanisms applied may range from a simple authentication mechanism, such as that which is provided by HTTP Basic Authentication, to message encryption and support for nonrepudiation. In most cases, the issues that must be addressed in Web services are almost identical to those faced by other computing systems, and so it is generally a matter of applying existing security concepts and techniques to the technologies being used. As always, securing anything is a very complex task and requires a delicate balance between providing security and maintaining the usability of a system.

27.5.1.1 Authentication/Authorization

Web services are used to expose computational resources to outside consumers. As such, it is important to be able to control access to them and to guarantee that they are used only in the prescribed manner. This guarding of resources is performed by the code responsible for authentication and authorization.

There are many existing programming models and specifications which provide authentication and authorization mechanisms that can be used in conjunction with Web services. As with communications protocols, the support for preexisting technologies, such as Kerberos [Steiner et al., 1988], allows Web services to be integrated into an existing environment, although such a choice does limit future integration possibilities and may only prove useful for single-hop interactions.

A set of new proposals, such as the Security Assertions Markup Language (SAML) [Hallam-Baker et al., 2002] and WS-Security [Atkinson et al., 2002], have been developed to provide authentication and authorization mechanisms which fit naturally into XML-based technologies. SAML provides a standard way to define and exchange user identification and authorization information in an XML document. The SAML specification also defines a profile for applying SAML to SOAP messages. WS-Security defines a set of SOAP extensions that can be used to construct a variety of security models, including authentication and authorization mechanisms. Both SAML and WS-Security allow authentication and authorization information to be propagated along a chain of intermediaries, allowing a "single sign-on" to be achieved.

In our example, if LoanCo requires that Joe be authenticated before he can access the service, then his initial SOAP request may contain a SOAP header like the following (assuming that LoanCo uses WS-Security).

```
<wsse:Securiry xmlns:wsse="...">
      <wsse:UsernameToken>
            <wsse:Username>Joe</wsse:Username>
            <wsse:Password>money</wsse:Password>
      </wsse:UsernameToken>
</wsse:Security>
```

27.5.1.2 Confidentiality

The data traveling between Web services and their partners are often confidential in nature, and so we need to be able to guarantee that the messages can only be read by the intended parties. As with other distributed systems, this functionality is typically implemented through some form of encryption. In scenarios where IP is being used for the communication transport, we can use SSL/TLS to gain confidentiality on a point-to-point level. If SOAP is being used, the WS-Security specification defines how XML–Encryption [Dillaway et al., 2002] may be applied to allow us to encrypt/decrypt part or all of the SOAP message, thereby allowing finer-grain access to the data for intermediaries and providing a flexible means for implementing end-to-end message level confidentiality. Again, the choice of technologies depends upon the environment in which we are deploying.

27.5.1.3 Integrity

In addition to making sure that prying eyes are unable to see the data, there is a need to be able to guarantee that the message which was received is the one that was actually sent. This is usually implemented through some form of a digital signature or through encryption. If we are using IP, we can again use SSL/TLS to gain integrity on a point-to-point level. WS-Security defines how XML-SIG (XML Signature Specification) [Bartel et al., 2002] may be used to represent and transmit the digital signature of a message and the procedures for computing and verifying such signatures to provide end-to-end message level integrity.

27.5.2 Reliability

Messages sent over the Internet using an unreliable transport like SOAP-over-HTTP may never reach their target, due to getting lost in transit, unreachable recipients, or other failures. Additionally, they may be reordered by the time they arrive, resulting in surprising and unintended behavior. Interactions generally depend on the reliable receipt and proper ordering of the messages that constitute them. Most high-level communication protocols themselves do not include reliability semantics. Reliable messaging systems, on the other hand, enable fire-and-forget semantics at the application level.

In order to address reliability in a uniform way, a Web services specification named WS-Reliable-Messaging [Bilorusets et al., 2003], WS-RM for short, has been proposed. WS-RM is not tied to a particular transport protocol or implementation strategy; however, a binding for its use in SOAP has been defined in the specification.

The basic idea is for a recipient to send an acknowledgment of the receipt of each message back to the sender, possibly including additional information to ensure certain requirements. A WS-RM enabled system provides an extensible set of delivery assurances, four of which are defined in the specification: at most once, at least once, exactly once, and in order (which may be combined with one of the other three). If an assurance is violated, a fault must be thrown.

In order to track and ensure the delivery of a message, the message must include a "sequence" element that contains a unique identifier for the sequence, a message number signaling where the message falls in the sequence, and optional elements containing an expiration time and/or indicating the last message in the sequence. An acknowledgment is sent back, using the "sequenceAcknowledgment" element containing the sequence identifier and a range of the numbers of the successfully received message. Using the message numbers in a sequence, a recipient is able to rearrange messages if they arrive out of order. A "SequenceFault" element is defined to signal erroneous situations. WS-RM defines a number of faults to signal occurrences such as invalid acknowledgments and exceeding the maximum message number in a sequence.

Continuing with our example, let's assume that both Joe's middleware and LoanCo implement the WS-RM specification. When Joe's client sends his request to his middleware, it may decide to chop the message into two smaller messages before transmitting it over SOAP. The SOAP header of each would contain a < wsrm : sequence > element, such as the one below for the second message:

```
<wsrm:Sequence xmlns:wsrm="..." xmlns:wsn="...">
      <wsu:Identifier>http://loanapp235.com/RM/xyz</wsu:Identifier>
      <wsrm:MessageNumber>2</wsrm:MessageNumber>
      <wsrm:LastMessage/>
</wsrm:Sequence>
```

Now assume that the LoanCo service acknowledges the first message but not the second:

```
<wsrm:SequenceAcknowledgment xmlns:wsrm="..." xmlns:wsu="...">
      <wsu:Identifier>http://loanapp235.com/RM/xyz</wsu:Identifier>
      <wsrm:AcknowledgmentRange Upper="1" Lower="1"/>
</wsrm:SequenceAcknowledgment>
```

After some time, Joe's system assumes that the second part has been lost and resends it. Upon receiving the resent message, LoanCo's middleware recombines the two messages and hands the appropriate input to the service. Later, LoanCo may actually receive the original second message, but knows from the sequence identifier and the message number that it is a duplicate it can safely ignore.

Note that WS-RM itself does not specify which part of a system assembles the pieces or how the assembly is accomplished. That is left to the implementation. A very basic system without reliability support might do it in the application itself, but ideally reliability would be handled in the middleware layer. The specification does, however, include the concept of message sources and targets (Joe and LoanCo's middleware) that may be different from the initial sender and ultimate receiver of the message (Joe and LoanCo).

So far, we have not mentioned how a service may declare the delivery assurances it offers but simply how reliable message delivery can be carried out in the Web services framework. In section 27.6.2, we illustrate the use of a pluggable framework for declaring such information on a service's public definition.

27.5.3 Coordination

Multiparty service interactions typically require some form of transactional coordination. For example, a set of distributed services that are invoked by an application may need to reach a well-defined, consistent agreement on the outcome of their actions. In Joe's case, he may wish to coordinate the loan approval with an application to college. Joe does not need a loan without being accepted into college, and he needs to decline the college acceptance should he not get a loan. The two activities form an atomic transaction, where either all or none of the activities succeed.

27.5.3.1 Coordination

WS-Coordination [Cabrera et al., 2002b] addresses the problem of coordinating multiple services. It is a general framework defining common mechanisms that can be used to implement different coordination models. The framework compares to distributed object frameworks for implementing extended transactions, such as the J2EE Activity service [Houston et al., 2001].

Using WS-Coordination, a specific coordination model is represented as a coordination type supporting a set of coordination protocols. A coordination protocol is the set of well-defined messages that are exchanged between the services that are the coordination participants. Coordination protocols include completion protocols, synchronization protocols, and outcome notification protocols. The WS-Transaction specification (described below) exemplifies the use of WS-Coordination by defining two coordination types for distributed transactions.

In order for a set of distributed services to be coordinated, a common execution context needs to exist. WS-Coordination defines such a coordination context, which can be extended for a specific coordination type. A middleware system implementing WS-Coordination can then be used to attach the context to application messages so that the context is propagated to the distributed coordination participants.

WS-Coordination further defines two generic coordination services and introduces the notion of a coordinator. The two generic services are the Activation service and the Registration service. A coordinator groups these two services as well as services that represent specific coordination protocols.

The Activation service can be used by applications wishing to create a coordination context. The context contains a global identifier, expiration data, and coordination type-specific information, including the endpoint reference for the Registration service. The endpoint reference is a WSDL definition type that is used to identify an individual port; it consists of the URI of the target port as well as other contextual information such as service-specific instance data.

A coordination participant can register with the Registration service for a coordination protocol (using the endpoint reference obtained from the context). The participant may also choose to use its own coordinator for this purpose. Figure 27.2 illustrates the sequence of a WS-Coordination activation, service invocation with context propagation, and protocol registration, using two coordinators.

27.5.3.2 Distributed Transactions

Transactions are program executions that transform the shared state of a system from one consistent state into another consistent state. Two principle kinds of transactions exist: short-running transactions where locks on data resources can be held for the duration of the transaction and long-running business transactions, where resources cannot be held.

WS-Transaction [Cabrera et al., 2002a] leverages WS-Coordination by defining two coordination types for Web services transactions. Atomic Transaction (AT) is the coordination type supporting short-running transactions, and Business Activity (BA) is the coordination type supporting long-running business transactions.

ATs render the well-known and widely used distributed transaction model of traditional middleware and databases for Web services. A set of coordination protocols supporting atomicity of Web services execution, including the two-phase commit protocol, are defined. ATs can be used to coordinate Web services within an enterprise, when resources can be held. ATs may also be used across enterprises if a tight service coupling for transactional coordination is desired.

BAs model potentially long-lived activities. They do not require resources to be held, but do require business logic to handle exceptions. Participants here are viewed as business tasks that are children to the BA for which they register. Compared to ATs, BAs suggest a more loosely-coupled coordination model in that, for example, participants may choose to leave a transaction or declare their processing outcome before being solicited to do so.

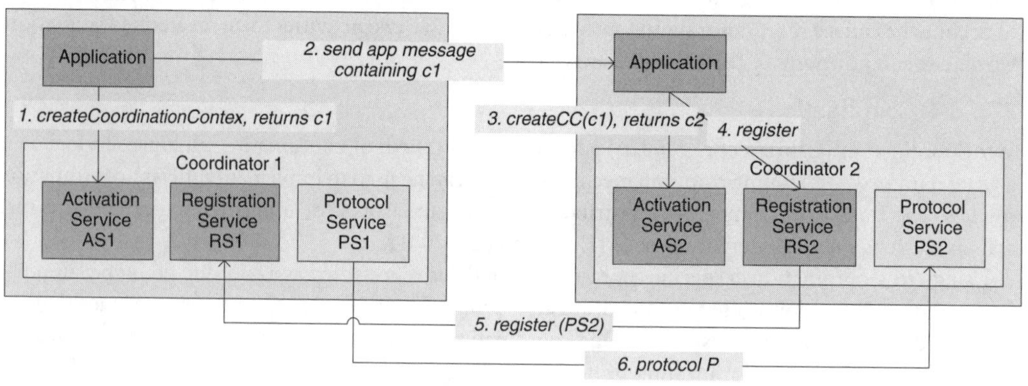

FIGURE 27.2 Coordinating services.

In our example, Joe would use middleware that supports the WS-Transaction Atomic Transaction coordination type to create a coordination context. He would also require the two services, the loan approval service and the college application service, to support the AT protocols of the WS-Transaction specification. Joe would invoke the two services and have the middleware propagate the coordination context with the messages; the two services being invoked will then register their resources as described. Should any one of the two services fail, the transaction will abort (none of the two services will commit). An abort can be triggered by a system or network failure, or by the application (Joe) interpreting the results of the invocations.

The WS-Transaction AT and BA coordination types are two models for Web services transactions, which can be implemented using the WS-Coordination framework. Other related specifications have been published in the area of Web services transactions. These include the Business Transaction Protocol [Ceponkus et al., 2002], and the Web services Composite Application Framework (WS-CAF) [Bunting et al., 2003] consisting of a context management framework, a coordination framework, and a set of transaction models. The BTP and the WS-CAF transaction protocols may be implemented using WS-Coordination, otherwise, the BTP and the WS-Transaction protocols may be implemented using the WS-CAF context and coordination framework. In general, there is significant overlap between these specifications defining models, protocols, and mechanisms for context-based transactional coordination.

27.6 Description

Service descriptions are central to two core aspects of the SOC model. First, the ability of applications to access services that are owned and managed by third party organizations relies not only on the availability of interoperability protocols, but also on the assumption that all relevant information needed to access the service is published in an explicit, machine readable format. In addition, the ability to perform automatic service discovery and binding at runtime relies on selecting services and adapting to their requirements based on those service descriptions. This section describes the Web services languages that can be used to encode service descriptions, and the discovery mechanisms that can be built on top of those descriptions.

27.6.1 Functional Definition of a Web Services

The functional description of a Web service is provided by the Web Services Description Language (WSDL) [Christensen et al., 2001]. A complete WSDL description provides two pieces of information: an application-level description of the service (which we will also call the "abstract interface") and the specific protocol-dependent details that need to be followed to access the service at concrete service endpoints. This separation of the abstract from the concrete enables the definition of a single abstract component that is implemented by multiple code artifacts and deployed using different communication protocols and programming models.

The abstract interface in WSDL consists of the operations supported by a service and the definition of their input/output messages. A WSDL message is a collection of named parts whose structure is formally described through the use of an abstract type system, usually XML Schema. An operation is simply a combination of messages labeled *input, output,* or *fault.* Once the messages and operations have been defined, a portType element is used to group operations supported by the endpoint.

For example, the "request" operation that processes the loan application in our example could be defined in the following snippet taken from the loan approval sample in [Curbera et al., 2002b]:

```
<wsdl:portType name="loanServicePT" xmlns:wsdl="...">
     <wsdl:operation name="request">
          <wsdl:input message="lns:creditInformationMessage"/>
          <wsdl:output message="lns:approvalMessage"/>
          <wsdl:fault name="unableToHandleRequest"
```

```
                    message="lns:errormessage"/>
        </wsdl:operation>
</wsdl:portType>
```

where, for example, the approval message is defined to contain one part that is of the type defined by XML Schema's *string:*

```
<wsdl:message name="approvalMessage" xmlns:wsdl="...">
        <wsdl:part name="accept" type="xsd:string"/>
</wsdl:message>
```

The abstract definition of a service, as defined by WSDL, provides all of the information necessary for a user of the service to program against (assuming that their middleware is capable of dealing with the transport/protocol details.)

The second portion of a WSDL description contains the concrete, implementation-specific aspects of a service: what protocols may be used to communicate with the service, how a user should interact with the service over the specified protocols, and where an artifact implementing the service's interface may be found.

This information is defined in the binding, port, and service elements. A binding element maps the operations and messages in a portType (abstract) to a specific protocol and data encoding format (concrete). Bindings are extensible, enabling one to define access to services over multiple protocols in addition to SOAP. A pluggable framework such as that defined in [Mukhi et al., 2002] may then be used for multiprotocol Web services invocations. A port element provides the location of a physical endpoint that implements a specific portType using a specific binding. A service is a collection of ports.

An example is shown below of LoanCo's offering of an implementation of the "loanServicePT" port type over a SOAP binding. Notice how the abstract "request" operation is mapped to a SOAP-encoded rpc style invocation using SOAP over HTTP:

```
<wsdl:binding name="SOAPBinding" type="tns:loanServicePT"
                    xmlns:wsdl="...":xmlns:soap="...">
            soap:binding style="rpc"
                    transport="http://schemas.xmlsoap.org/soap/http"/>
<wsdl:operation name="request">
        <soap:operation soapAction="" style="rpc"/>
        <wsdl:input>
            <soap:body use="encoded" namespace="lns:loanapproval"
                    encodingStyle="http://schemas.xmlsoap.org/soap/encoding/"/>
        </wsdl:input>
        <wsdl:output>
            <soap:body use="encoded" namespace="lns:loanapproval"
                    encodingStyle="http://schemas.xmlsoap.org/soap/encoding/"/>
            </wsdl:output>
        </wsdl:operation>
</wsdl:binding>
```

Finally, the LoanService service element contains a port at which an endpoint is located that can be communicated with using the information in the associated binding (and implementing the portType associated with that binding).

```
<wsdl:service name="LoanService">
        <wsdl:documentation>Loan Service</documentation>
        <wsdl:port name="SOAPPort" binding="tns:SOAPBinding">
            <soap:address location="http://www.loanco.com/loanservice"/>
        </wsdl:port>
</wsdl:service>
```

27.6.2 A Framework for Defining Quality of Service

As we have seen, WSDL provides application-specific descriptions of the abstract functionality and concrete bindings and ports of Web services. The quality-of-service aspects of a Web service, however, are not directly expressed in WSDL. QoS characteristics may comprise reliable messaging, security, transaction, and other capabilities, requirements, or usage preferences of a service. In our example, the loan approval service may wish to declare that authentication is required before using the service or that the service supports a specific transaction protocol and allows applications to coordinate the service according to the transaction protocol.

The Web Services Policy Framework (WS-Policy) [Box et al., 2002b] provides a general-purpose model to describe and communicate such quality-of-service information. WS-Policy is a domain-neutral framework which is used to express domain-specific policies such as those for reliable messaging, security, and transactions.

WS-Policy provides the grammar to express and compose both simple declarative assertions as well as conditional expressions. For example, the following security policy (taken from the WS-Policy specification) describes that two types of security authentication, Kerberos and X509, are supported by a service, and that Kerberos is preferred over X509 authentication.

```
<wsp:Policy xmlns:wsp="..." xmlns:wsse="...">
      <wsp:ExactlyOne>
            <wsse:SecurityToken TokenType="wsse:Kerberosv5TGT"
                        wsp:Usage="wsp:Required" wsp:Preference="100"/>
            <wsse:SecurityToken TokenType="wsse:X509v3"
                        wsp:Usage="wsp:Required" wsp:Preference="1"/>
      </wsp:ExactlyOne>
</wsp:policy>
```

The example illustrates the three basic components of the WS-Policy grammar: the top level < wsp : Policy > container element, policy operators (here: < wsp : ExactlyOne >) to group statements, and attributes to distinguish usage. Different policy operators and values for the usage attributes are defined in WS-Policy, allowing all kinds of policy statements to be made.

Policies can be associated with services in a flexible manner and may be used by both clients and the services themselves. Specific attachment mechanisms are defined in the WS-PolicyAttachment specification [Box et al., 2002a], including the association of policies with WSDL definitions and UDDI entities.

The Web services policy framework also provides common policy assertions that can be used within a policy specification: for example, assertions for encoding textual data and for defining supported versions of specifications. These common assertions are defined in the WS-PolicyAssertions Language (WS-PolicyAssertions) specification [Box et al., 2002c].

27.6.3 Service Discovery

To allow developers and applications to use a service, its description must be published in a way that enables easy discovery and retrieval, either manually at development time or automatically at runtime. Two of the specifications which facilitate the location of service information for potential users are the Universal Description, Discovery, and Integration (UDDI) [Bellwood et al., 2002] and WS-Inspection [Ballinger et al., 2001].

27.6.3.1 UDDI

UDDI is cited as the main Web service query and classification mechanism. From an architectural perspective, UDDI takes the form of a network of queryable business/service information registries, which may or may not share information with one another. The UDDI specification defines the structure of data that may be stored in UDDI repositories and the APIs which may be used to interact with a repository, as well as the guidelines under which UDDI nodes operate with each other.

The UDDI consortium, uddi.org, manages an instance of UDDI called the UDDI Business Registry (UBR) , which functions as a globally known repository at which businesses can register and discover Web services. Service seekers can use the UBR to discover service providers in a unified and systematic way, either directly through a browser or using UDDI's SOAP APIs for querying and updating registries. A variety of "private" UDDI registries have been created by companies and industry groups to provide an implementation of the functionality for their own internal services.

UDDI encodes three types of information about Web services: "white pages" information such as name and contact data; "yellow pages" or categorization information about business and services; and the so-called "green pages" information, which includes technical data about the services.

A service provider is represented in UDDI as a "businessEntity" element, uniquely identifiable by a business key, containing identifying information about the provider and a list of its services. Each such "businessService" contains one or more binding templates that contain the technical information needed for accessing different endpoints of that service that possibly have different technical characteristics.

The most interesting field in a binding template is "tModelInstanceDetails," which is where the technical description of the service is provided in a list of references to technical specifications, known as "tModels," that the service complies with. The tModels represent a technical specification, such as a WSDL document, that has already been registered in the directory and assigned a unique key. They enable service descriptions to contain arbitrary external information that is not defined by UDDI itself.

Taxonomical systems can be registered in UDDI as tModels to enable categorized searching. Three standard taxonomies have been preregistered in the UBR: an industry classification (NAICS), a classification of products and services (UNSPSC), and a geographical identification system (ISO 3166).

In our example, banks and credit agencies will have published information about their services in UDDI so that it may be easily retrieved by new or existing customers. When LoanCo needs their services, it may query a UDDI registry to discover the necessary description information. For example, it may submit a query to find a bank using an NAICS [Bureau, 2002] code as the search key and may submit another query to find a credit agency using an existing WSDL interface as the search key. UDDI may return any number of matching entries in the responses, allowing LoanCo to pick the one that best fulfills their requirements. Once a service record is located within UDDI, any registered information, such as WSDL interfaces or WS-Policies, may be retrieved and consumed.

27.6.3.2 WS-Inspection

The WS-Inspection specification provides a completely decentralized mechanism for locating service related information. WS-Inspection operates in much the same as the Web; XML-based WS-Inspection documents are published at well known locations and then retrieved through some common protocol such as HTTP.

WS-Inspection documents are used to aggregate service information, and contain "pointers" to service descriptions, such as WSDL documents or entries in UDDI repositories. In our example, LoanCo may publish a WS-Inspection document that looks like the following

```
<wsil:inspection xmlns:wsil="...">
      <wsil:service>
            <wsil:description referencedNamespace=
"http://schemas.xmlsoap.org/wsdl/"
                  location="http://loanapp235/loanservice.wsdl" />
      </wsil:service>
</wsil:inspection>
```

on its Website to advertise its services to any interested clients. If Joe knows that he is interested in working with LoanCo, he may search their Website for a WS-Inspection document which would tell him what services they have available. In fact, if Joe were using a WS-Inspection-aware Web browser, it might simply present him with an interface for directly interacting with the service.

27.7 Composition

In a service-oriented architecture, application development is closely tied to the ability to compose services. A collection of services can be composed into an aggregate service, which is amenable to further composition. Alternatively, a set of services may be orchestrated by specifying their interactions upfront. In the latter case the aggregate entity may not define a composable service, but can be defined using the same mechanisms as a service composition.

The two most often cited use cases for service composition are Enterprise Application Integration (EAI) in which it defines the interactions between applications residing within the same enterprise, and Business Process Integration (BPI) in which it does so for applications spread across enterprises. In both of these cases, the fundamental problem is that of integration of heterogeneous software systems. Service composition offers hope in attacking these issues since once each application is offered as a service described in a standard fashion, integration reduces to the orchestration of services or the creation of aggregate services through composition operations.

27.7.1 Choreography

The Business Process Execution Language for Web Services [Curbera et al., 2002b], or BPEL for short, is a language for specifying service compositions/business processes. BPEL lets process designers create two kinds of processes using the same language, barring a small set of language features whose use depends on the kind of process being defined:

1. *Abstract* processes, which are used to define protocols between interacting services usually controlled by different parties. Such processes cannot be interpreted directly as they do not specify a complete description of business logic and service interactions. Rather, each party involved may use the process description to verify that his/her own executable business logic matches the agreed upon protocol.
2. *Executable* processes, which specify a complete set of business rules governing service interactions and represent the actual behavior of each party; these can be interpreted, and such a process is similar to a program written in some high level programming language to define a composition.

A key feature of BPEL is that every process has a WSDL description, and this description can in fact be derived from the process definition itself. This is significant since the BPEL process then becomes nothing but the specification of an implementation of a Web service described using WSDL, which may consequently be composed recursively as the service defined through the process. may be used in another process definition.

BPEL derives its inspiration from the workflow model for defining business processes. The process definition can be viewed as a flowchart-like expression of an algorithm. Each step in the process is called an activity. There are two sets of activities in BPEL:

1. *Primitive* activities, such as the invocation of a service (specified using the *invoke* activity), receiving a message (the *receive* activity, responding to a *receive* the *reply* activity), signaling a fault (the *throw* activity), termination of the process (the *terminate* activity), etc.
2. *Structured* activities, which combine other activities into complex control structures. Some of these are the *sequence* activity, which specifies that all the contained activities must execute in order; the *flow* activity wherein contained activities run in parallel, with control dependencies between them specified using *links;* and the *scope* activity, which defines a unit of fault handling and compensation handling.

The services being composed through a BPEL process are referred to as *partners* of the process. The relationship between the partner and the process is defined using a *service link type*. This specifies the functionality (in terms of WSDL port types supported) that the process and the partner promise to provide.

Consider an example of a BPEL process definition that defines how a LoanCo's process may be implemented, illustrated in Figure 27.3. This service is in fact implemented through the composition of two other services, which may be offered by third parties: a credit agency that determines the suitability of an applicant for a loan, and a bank that is capable of making a decision for high risk applications such as those with a high loan amount or a requester with a bad credit history.

This process views the credit agency and bank as partners whose functionality it makes use of. Additionally, the user of the composite service itself is also a partner. Viewing each interacting entity as a partner in this way defines a model in which each service is a peer.

The process logic is straightforward, defined by grouping activities in a *flow* construct and specifying control flow through the use of links with transition conditions that determine whether a particular link is to be followed or not. The incoming loan application is processed by a *receive* activity. For loan requests less than $10,000, an *invoke* activity hands off the application to the credit agency service for a risk assessment. If that assessment comes back as high or if the amount had been large to start with, then the bank service is invoked. On the other hand, if the credit agency had determined that the applicant was low risk, an *assign* activity creates a message approving the applicant. Finally, a *reply* activity sends a message back to the applicant, which at this point contains either the bank's decision or the positive approval created in the *assign*. The full BPEL definition of this flow is available in [Curbera et al., 2002b] and with the samples distributed with the prototype runtime BPWS4J [Curbera et al., 2002a].

The user, of course, views the process merely through its WSDL interface. It sees that the service offers a request-response operation for loan approval, sends the loan application in the required format, and receives an answer in return. Each request-response operation on the WSDL is matched at runtime to a *receive* and *reply* activity pair.

BPEL processes may be stateless, but in most practical cases a separate stateful process needs to be created to manage the interaction with a particular set of partners. In our example, each loan applicant needs to use his own version of the loan approval process, otherwise we might have Joe and Bob making simultaneous loan applications and receiving counter-intuitive responses. BPEL does not have explicit lifecycle control, instead processes are created implicitly when messages are received. Data contained within the exchanged messages are used to identify the particular stateful interaction. In BPEL, sets of these key data fields are known as *correlation sets.*

A loan application generally contains a set of business data, such as the user's Social Security number, which serve as useful correlation fields for a process. A process definition allows specification of sets of such correlation fields and also specifies how they can be mapped to business data contained within application messages. Thus, correlation data is not an opaque middleware-generated token but instead consists of a set of fields contained in the application messages.

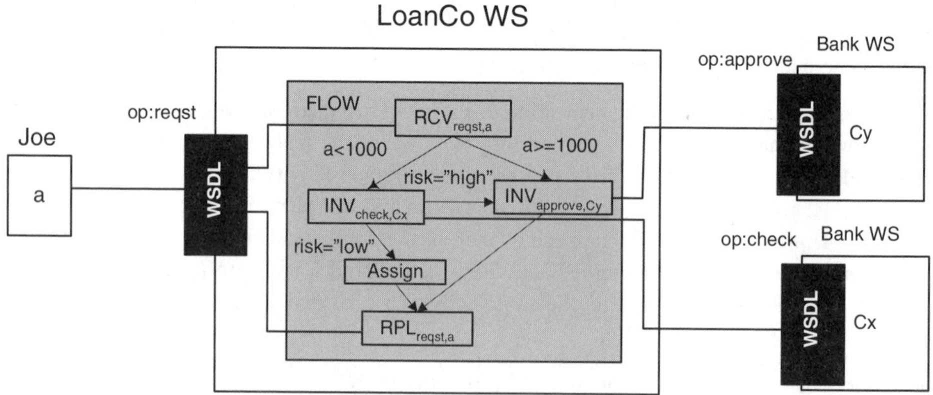

FIGURE 27.3 Sample loan approval flow using two services and exposed as a Web service.

BPEL also provides facilities for specifying fault handling, similar to the exception handling facility in Java, and allows for the triggering of compensation and the specification of compensation handlers. Compensation is used to undo the effects of earlier actions and is used to implement the rollback of a transaction or activities of a similar nature.

Now that we have some understanding of how compositions are created using BPEL, let's take a step back and see why BPEL is necessary. One might immediately conclude that it is possible to specify such compositions directly in one's favorite programming language such as Java. Although this is true, the key here is that BPEL allows standard specification of such a composition, portable across-Web service containers, and it is a recognition of the fact that programming at the service-level is a separate issue from programming at the object-level. It is not meant to replace someone's favorite language. The latter has its place in defining fine grained logic, and BPEL's features in this space do not match up. However, it is in wiring together services, each of which may implement fine grained logic, where BPEL's capabilities are meant to be harnessed.

Complex business applications typically involve interactions between multiple independent parties, and such interactions are usually transactional in nature. For this purpose, one may use WS Coordination and WS-Transaction in tandem. with BPEL4WS as described in [Curbera et al., 2003].

27.8 Summary

Web services aim to support highly dynamic integration of applications in natively inter-organizational environments. Web services also aims at achieving universal application to application interoperability based on open Internet standards such as XML. Recognizing the intrinsic heterogeneity of the existing computing infrastructure, Web services pursues a platform independent integration model that should allow new and existing applications created on proprietary systems to seamlessly integrate with each other.

In this chapter, we have reviewed the motivating factors behind the Web services effort, and the architectural principles that guide the design of the Web services stack of specifications. We have reviewed the main specifications of the stack and illustrated them by providing a practical description of key specifications for each area. These areas define how services may communicate, define, and adhere to quality of service requirements, provide machine readable descriptions to requesting applications, and be composed into business processes. An example is discussed throughout the chapter to show how each specification may be used in a simple scenario. Figure 27.4 illustrates this scenario and summarizes how the different pieces of the technology fit together.

A large number of resources are available, especially on-line, where further information may be found about Web services, including the specifications proposed, their status in the standardization process, implementation strategies, and available implementations. Dedicated Web sites with articles and developer information include IBM's "developeTWorks" (http://www.ibm.com/developer-works), Microsoft's "gotdotnet" (http://www.gotdotnet.com), and third-party pages like "webservices.org" (http://www,webservices.org). Additionally, a number of academic conferences have been specifically addressing this space, including the relation between Web services and related Internet technologies such as the Semantic Web and Grid Computing.

Service Oriented Computing is still in its youth. The specifications presented in this chapter are the building blocks leading to a complete standards-based framework to support service orientation. With its extensible, modular design, the Web services stack is having specifications filling in the remaining gaps and industry support consolidating behind a set of basic standards. Over the next few years, we will likely see the deployment and adoption of the full SOC model by business and scientific communities.

As SOC evolves, we believe the way to stay in step is to design projects with the SOC principles in mind. As the technology matures and develops, systems designed in such a manner will be more agile and able to adopt emerging specifications along the way.

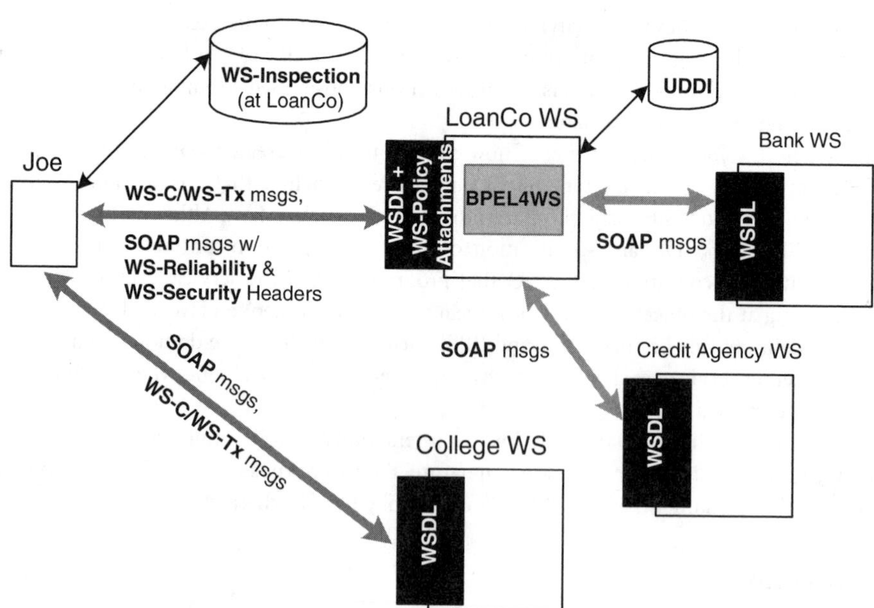

FIGURE 27.4 Example: services are exposed using WSDL (w/possible WS-Policy attachments defining QoS), communicated with using SOAPIHTTP, coordinated into an atomic transaction using WS-Coordination and WS-Transaction, and discovered using UDDI and WS-Inspection.

References

Atkinson, Bob, Giovanni Della-Libera, Satoshi Hada, Maryann Hondo, Phillip Hallam-Baker, Chris Kaler, Johannes Klein, Brian LaMacchia, Paul Leach, John Manferdelli, Hiroshi Maruyarna, Anthony Nadalin, Nataraj Nagaratnam, Hemma Prafullchandra, John Shewchuk, and Dan Simon. Web Services Security (WS-Security) 1.0. Published online by IBM, Microsoft, and VeriSign at http://www-106.ibm.com/developerworks/library/ws-secure, 2002.

Ballinger, Keith, Peter Brittenham, Ashok Malhotra, William A. Nagy, and Stefan Pharies. Web Services Inspection Language (WS-Inspection) 1.0. Published on the World Wide Web by IBM Corp. and Microsoft Corp. at http://www.ibm.com/developerworks/webservices/library/ws-wsilspec.html, November 2001.

Bartel, Mark, John Boyer, Donald Eastlake, Barb Fox, Brian LaMacchia, Joseph Reagle, Ed Simon, and David Solo. XML-Signature Syntax and Processing, W3C Recommendation, published online at http://www.w3.org/TR/xmldsig-core/, 2002.

Bellwood, Tom, Luc Clement, David Ehnebuske, Andrew Hately, Maryann Hondo, Yin Leng Husband, Karsten Januszewski, Sam Lee, Barbara McKee, Joel Munter, and Claus von Riegen. The Universal Description, Discovery and Integration (UDDI) protocol. Published on the World Wide Web at http:/Iwww.uddi.org, 2002.

Bilorusets, Rusian, Adam Bosworth, Don Box, Felipe Cabrera, Derek Collison, Jon Dart, Donald Ferguson, Christopher Ferris, Tom Freund, Mary Ann Hondo, John Ibbotson, Chris Kaler, David Langworthy, Amelia Lewis, Rodney Limprecht, Steve Lucco, Don Mullen, Anthony Nadalin, Mark Nottingham, David Orchard, John Shewchuk, and Tony Storey. Web Services Reliable Messaging Protocol (WS-ReliableMessaging). Published on the World Wide Web by IBM, Microsoft, BEA, and TIBCO at http://www-106.ibm.com/developerworks/webservices/library/ws-rm/, 2003.

Box, Don, Francisco Curbera, Maryann Hondo, Chris Kaler, Hiroshi Maruyama, Anthony Nadalin, David Orchard, Claus von Riegen, and John Shewchuk. Web Services Policy Attachment (WS-PolicyAttachment). Published online by IBM, BEA, Microsoft, and SAP at http://www.106.ibm.com/developerworks/webservices/library/ws-polatt, 2002a.

Box, Don, Francisco Curbera, Dave Langworthy, Anthony Nadalin, Nataraj Nagaratnam, Mark Nottingham, Claus von Riegen, and John Shewchuk. Web Services Policy Framework (WS-Policy Framework). Published online by IBM, BEA, and Microsoft at http://www-106.ibm.com/developerworks/webservices/library/ws-polfram, 2002b.

Box, Don, Maryann Hondo, Chris Kaler, Hiroshi Maruyama, Anthony Nadalin, Nataraj Nagaratnam, Paul Patrick, and Claus von Riegen. Web Services Policy Assertions (WS-PolicyAssertions). Published online by IBM, BEA, Microsoft, and SAP at http://www-106.ibm.com/developerworks/webservices/library/ws-polas, 2002c.

Bunting, Doug, Martin Chapman, Oisin Hurley, Mark Little, Jeff Mischkinsky, Eric Newcomer, Jim Webber, and Keith Swenson. Web Services Composite Application Framework (WS-CAF) version 1.0. Published online by Arjuna, Fujitsu, IONA, Oracle, and Sun at http://developers.sun.com/techtopics/webservices/wscaf/index.html, 2003.

Cabrera, Felipe, George Copeland, Bill Cox, Tom Freund, Johannes Klein, Tony Storey, and Satish Thatte. Web Services Transactions (WS-Transaction) 1.0. Published online by IBM, BEA, and Microsoft at http://www-106.ibm.com/developerworks/library/ws-transpec, 2002a.

Cabrera, Felipe, George Copeland, Tom Freund, Johannes Klein, David Langworthy, David Orchard, John Shewchuk, and Tony Storey. Web Services Coordination (WS-Coordination) 1.0. Published online by IBM, BEA, and Microsoft at http://www-106.ibm.com/developerworks/library/ws-coor, 2002b.

Ceponkus, Alex, Sanjay Dalal, Tony Fletcher, Peter Furniss, Alastair Green, and Bill Pope. Business Transaction Protocol. Published on the World Wide Web at http://www.oasis-open.org, 2002.

Christensen, Erik, Francisco Curbera, Greg Meredith, and Sanjiva Weerawarana. Web Services Description Language (WSDL) 1.1. Published on the World Wide Web by W3C at http://www.w3.org/TR/wsdl, March 2001.

Curbera, Francisco, Matthew Duftler, Rania Khalaf, Nirmai Mukhi, William Nagy, and Sanjiva Weerawarana. BPWS4J. Published on the World Wide Web by IBM at http://www.alphaworks.ibm.com/tech/bpws4j, August 2002a.

Curbera, Francisco, Yaron Goland, Johannes Klein, Frank Leymann, Dieter Roller, Satish Thatte, and Sanjiva Weerawarana. Business Process Execution Language for Web Service (BPEL4WS) 1.0. Published on the World Wide Web by BEA, IBM, and Microsoft at http://www.ibm.com/developerworks/library/ws-bpel, August 2002b.

Curbera, Francisco, Rania Khalaf, Nirmal Mukhi, Stefan Tai, and Sanjiva Weerawarana. Web Services, The next step: robust service composition. *Communications of the ACM: Service Oriented Computing,* 46(10), 2003.

Dillaway, Blair, Donald Eastlake, Takeshi Imamura, Joseph Reagle, and Ed Simon. XML Encryption Syntax and Processing. W3C Recommendation, published online at http://www.w3.org/TRJxmlenc-core/, 2002.

Fallside, D.C. XML Schema Part 0: primer. W3C Recommendation, published online at http://www.w3.org/TR/xmlschema-0/, 2001.

Foster, Ian, Carl Kesselman, and Steven Tuecke. The anatomy of the grid: enabling scalable virtual organizations. *International Journal of Supercomputing Applications,* 15(3), 2001.

Gudgin, Martin, Marc Hadley, Noah Mendelsohn Jean-Jacques Moreau, and Henrik Frystyk Nielsen. SOAP Version 1.2. W3C Proposed Recommendation, published online at http://www.w3c.org/2000/xp/Group/, 2003.

Hallam-Baker, Phillip, Eve Maler, Stephen Farrell, Irving Reid, David Orchard, Krishna Sankar, Simon Godik, Hal Lockhart, Carlisle Adams, Tim Moses, Nigel Edwards, Joe Pato, Marc Chanliau, Chris McLaren, Prateek Mishra, Charles Knouse, Scott Cantor, Darren Platt, Jeff Hodges, Bob Blakley, Marlena Erdos, and R.L. "Bob" Morgan. Assertions and Protocol for the OASIS Security Assertion Markup Language (SAML). Published on the World Wide Web at http://www.oasis-open.org, 2002.

Houston, Iain, Mark C. Little, Ian Robinson, Santosh K. Shrivastava, and Stuart M. Wheater. The CORBA Activity Service Framework for Supporting Extended Transactions. In *Proceedings of Middle-ware 2001,* number 2218 in LNCS, pages 197+. Springer-Verlag, 2001.

Orlowska, M., S. Weerawarana, M. Papazoglou, and J. Yang, Proceedings of the First International Conference on Service-Oriented Computing (ISOC 2003), Springer-Verlag, LNCS 2910, December 2003.

Mukhi, Nirmal, Rania Khalaf, and Paul Fremantle. Multi-protocol Web Services for Enterprises and the Grid. In *Proceedings of the EuroWeb Conference,* December 2002.

Papazoglou, Mike P. and Dimitri Georgakopoulos, Eds. *Communications of the ACM: Service-Oriented Computing,* 46(10) 2003.

Steiner, J., C. Neuman, and J. Schiller. Kerberos: An Authentication Service for Open Network Systems. In *Usenix Conference Proceedings,* Dallas, TX, February 1988.

U.S. Census Bureau, editor. *North American Industry Classification System (NAICS).* U.S. Government, 2002.

Yates, Simon, Charles Rutstein, and Christopher Voce. Demystifying b2b integration. The Forrester Report, September 2000.

28

Mediators for Querying Heterogeneous Data

CONTENTS

Abstract 28-1
28.1 Introduction 28-1
28.2 Mediator Architectures 28-2
28.2.1 Wrappers 28-3
28.2.2 Reconciliation 28-4
28.2.3 Composable Mediators 28-5
28.3 The Amos II Approach to Composable Mediation 28-6
28.3.1 The Functional Data Model of Amos II 28-7
28.3.2 Composed Functional Mediation 28-11
28.3.3 Implementing Wrappers 28-15
28.4 Conclusions 28-16
References 28-17

Tore Risch

Abstract

The mediator approach to integrating heterogeneous sources introduces a virtual middleware mediator database system between different kinds of wrapped data sources and applications. The mediator layer provides a view over the data from the underlying heterogeneous sources, which the applications can access using standard query-based database APIs. The sources can be not only conventional database servers available over the Internet but also web documents, search engines, or any data-producing system. The architecture of mediator systems is first overviewed. An example illustrates how to define mediators in an object-oriented setting. Finally, some general guidelines are outlined of how to define mediators.

28.1 Introduction

The mediator architecture was originally proposed by Wiederhold [31] as an architecture for integrating heterogeneous data sources. The general idea was that mediators are relatively simple distributed software modules that transparently encode domain-specific knowledge about data and share abstractions of that data with higher layers of mediators or applications. A mediator module thus contains rules for semantic integration of its sources, i.e., how to resolve semantic similarities and conflicts. Larger networks of mediators can then be defined through these primitive mediators by logically composing new mediators in terms of other mediators and data sources. It is an often overlooked fact that a mediator (Wiederhold in [31]) was a relatively simple knowledge module in the data integration and that mediators could be combined to integrate many sources. There also need to be a distinction between mediator modules and the system interpreting these modules, the *mediator engine.* Different mediator modules may actually be interpreted by different kinds of mediator engines.

1-58488-381-2/05/$0.00+$1.50
© 2005 by CRC Press LLC

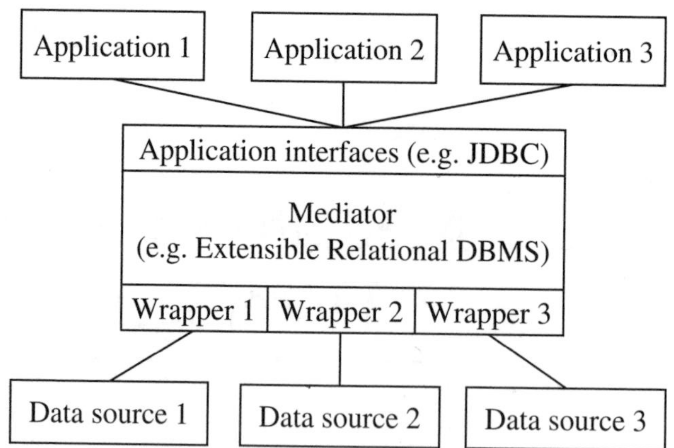

FIGURE 28.1 Central mediator architecture.

Many systems have since then been developed based on the mediator approach. [4, 1, 6, 8, 13, 29] Most of these systems regard the mediator as a central system with interfaces to different data sources called *wrappers.* We will call this a *central mediator.* Often the mediator engines are relational database systems extended with mechanisms for accessing other databases and sources. The central mediators provide (e.g., SQL) queries to a mediator schema that includes data from external sources. Important design aspects are performance and scalability over the amounts of data retrieved. Some problems with the central mediator approach are that a universal global schema is difficult to define, in particular when there are many sources.

Another important issue when integrating data from different sources is the choice of common data model[1] (CDM) for the mediator. The CDM provides the language in which the mediating views are expressed. The mediator engine interprets the CDM and queries are expressed in terms of it. If the CDM is less expressive than some of the sources, semantics will be lost. For example, if some sources have object-oriented (OO) abstractions, a mediator based on the relational model will result in many tables where the OO semantics is hidden behind some conventions of how to map OO abstractions to less expressive tabular representations.

We will describe in Section 28.1 the architecture of central mediators and wrappers, followed by a description of *composable mediators* where mediators may wrap other mediators, in Section 28.2.3. As an example of a composable mediator system, in Section 28.2.3 we make an overview of the Amos II mediator engine and show a simple example of how to mediate heterogeneous data using Amos II. Amos II is based on a distributed mediator architecture and a functional data model that permit simple and powerful mediation of both relational and object-oriented sources. Our example illustrates how data from a relational database can be mediated with data from an XML-based Web source using a functional and object-oriented common data model.

28.2 Mediator Architectures

The central mediator/wrapper architecture is illustrated in Figure 28.1. A central mediator engine is interfaced to a number of data sources through a number of *wrappers.* The engine is often a relational database manager. The central mediator contains a universal mediator schema that presents to users and applications a transparent view of the integrated data. The mediator schema must further contain meta-information of how to reconcile differences and similarities between the wrapped data sources.[2] SQL

[1] We use the term *data model to* mean the language used for defining a schema.

[2] Sometimes the term *ontology* is used for such semantically enriched schemas.

queries posed to the mediator in terms of the universal schema are translated to data access calls to the source data managers. Applications interact with the mediator manager using standard interfaces for the kind of database management system used (e.g., JDBC or ODBC for relational databases).

Most mediator systems so far are based on the relational database model, but other data models are conceivable too. Being a database manager of its own, the mediator will contain its own data tables. These tables are normally an operational large database. One can regard the central mediator architecture as being a conventional database server extended with possibilities to efficiently access external data sources.

Wrappers are interfaces between the mediator engine and various kinds of data sources. The wrappers implement functionality to translate SQL queries to the mediator into query fragments (subqueries) or interface calls to the sources. If a mediator has many sources, there will be many query execution strategies for calling the sources and combining their results. One important task of the mediator engine is to do such query decomposition to utilize, in an optimized way, the query capabilities of the sources.

The problem of updates in mediators has not gained much attention. One reason for this is that mediators normally wrap autonomous sources without having updates rights to these. Another reason is that mediator updates are problematic because mediators are essentially views of wrapped data. Therefore, updates in mediators are similar to the problem of updating views in relational databases, which is possible only in special cases. [2, 16] Vidal and Loscio [30] propose some view update rules for mediators.

Notice here that the mediator approach is different from the *data warehouse* approach of integrating data. The idea of integrating data in data warehouses is to import them to a central very large relational database for subsequent data analysis using SQL and OLAP (Online Analytical Processing) tools. The data importation is done as regular database applications that convert external data to tabular data inserted into the data warehouse. Such data importations are run *offline* regularly, e.g., once a day.

In contrast, the mediator approach retains the data in the sources. Queries to the central mediator schema are dynamically translated by the query processor of the mediator engine into queries or subroutine calls retrieving data from the sources at query time. The data integration in a mediator system is thus *online*. This makes data and decisions based on data current. From an implementation point of view the mediator approach is more challenging because it may be difficult to efficiently process dynamically provided data. In the data warehouse solution, efficiency can rather easily be obtained by careful physical design of the central relational database tables. A mediator engine must dynamically access external data in real time, which is more challenging.

28.2.1 Wrappers

Figure 28.2 illustrates the architecture of a general wrapper component of a mediator system. A wrapper may contain both physical interfaces to a source and rules or code for translating the data of each source to the schema of the mediator represented by its CDM. The different layers do the following tasks:

1. On the lowest layer there is a *physical interface* to the data source. For example, if relational databases are accessed, there need to be interfaces for connecting to the sources, sending SQL queries to the sources as strings, and iterating over the result tuples. For relational databases this can be implemented using the standardized JDBC/ODBC APIs that are based on sending SQL strings to the database server for evaluation. SQL Management of External Data, SQL/MED, [23] is an ISO standard that provides interface primitives for wrapping external data sources from an extensible relational database system. SQL/MED provides wrapper-specific interface primitives regarding external data as foreign tables. Using SQL/MED, the wrapper implementor provides tabular wrapper abstractions by internally calling data-source-specific physical interfaces to obtain the information required for the foreign table abstractions. The physical interface layer can therefore be seen as hidden inside the interface to a wrapped source. SQL/MED is supported by IBM's mediation product DB2 Information Integrator[3] having SQL/MED-based wrappers implemented for all major relational database systems, XML-documents, Web search engines, etc.

[3] http: *//www.ibm.com/software/data/integration/db2ii/*

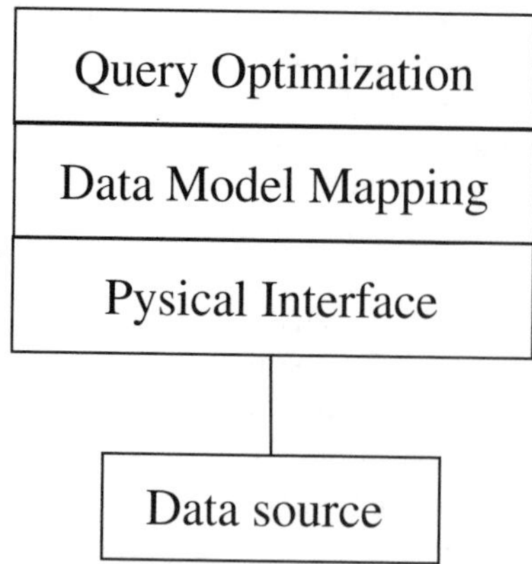

FIGURE 28.2 Wrapper layers.

In some systems the term wrapper is used to mean simply grammar rules for parsing, e.g., HTML-documents.[7] This would here mean a physical interface to HTML sources.

2. A data source may use a different data model than the CDM of the mediator engine, The wrapper must therefore contain code expressing how to map schema constructs between different data models, here called data *model mappings.* For example, if a source is an object-oriented database and the mediator engine uses a relational data model, the object-oriented concepts used by the source need to be translated to relational (tabular) data. In such a case the wrapper for that kind of source will contain general application-independent knowledge of how to map objects to relations. In the SQL/MED standard [23] this is achieved by delivering accessed data as foreign relational tables to the central mediator through external relations that are implemented as functions in, e.g., C, that deliver result table rows tuple-wise.
3. Some sources may require source-specific *wrapper query optimization methods.* Wrapper query optimization is needed, e.g., if one wants to access a special storage manager indexing data of a particular kind, such as free text indexing. The query optimizer of the mediator engine will then have to be extended with new query optimization rules and algorithms dealing with the kind of query operators the storage manager knows how to index. For example, for text retrieval there might be special optimization rules for phrase-matching query functions. The optimizations may have to be extended with costing information, which is code to estimate how expensive a query fragment to a source is to execute, how selective it is, and other properties. Furthermore, source-specific query transformation rules may be needed that generate optimized query fragments for the source. These query fragments are expressed in terms of the source's query language, e.g., some text retrieval language for an Internet search engine.

Once a wrapper is defined, it is possible to make queries to the wrapped data source in terms of queries of the mediator query language. For example, if the common data model is a relational database, SQL can be used for querying the wrapped data source as an external table.

28.2.2 Reconciliation

Different sources normally represent the same and similar information differently from the integrated schema. The schema of the mediator therefore must include view definitions describing how to map the schema of a particular source into the integrated mediator schema. Whereas the data model mapping rules are source and domain independent, these *schema mapping rules* contain knowledge of how the schema of a particular wrapped data source is mapped into the mediator's schema.

The most common method to specify the schema mapping rules is *global as view.* [6, 8, 11, 22, 29] With global as view, the mediator schema is defined in terms of a number of views that map wrapped source data (external relations) to the mediator's schema representations. Thus views in the integrated schema are defined by matching and transforming data from the source schemas. In defining these views, common keys and data transformations need to be part of the view in order to reconcile similarities and differences between source data. For relational mediators, SQL can be used for defining these views. However, data integration often involves rather sophisticated matching and reconciliation of data from the different sources. The same and similar information may be present in more than one source and the mediator needs to deal with how to handle cases when there are conflicting and overlapping data retrieved from the sources. Such reconciliation operations are often difficult or impossible to express in basic SQL. For example, rather complex string matching may be needed to identify equivalent text retrieved from the Web, and advanced use of outer joins may be needed for dealing with missing information. Therefore, in a mediator based on a relational database engine, such functionality will often have to be performed as user-defined functions (UDFs) plugged into the relational database engine. The schema mapping rules can then be expressed by view definitions containing calls to these UDFs.

With global as view, whenever a new source is to be integrated, the global mediator view definitions have to be extended accordingly. This can be problematic when there are many similar sources to integrate.

With *local as view* [19], there is a common fixed-mediator schema. Whenever a new source is to be integrated, one has to define how to map data from the global schema to the new source without altering the global schema. One thus, so to speak, includes new sources by defining an inverse top-down mapping from the mediator schema to the source schema. Local as view has the advantage that it is simple to add new sources. There are, however, problems of how to reconcile differences when there are conflicts and overlaps between sources. Usually, local as view provides some default reconciliation based on accessing the "best" source, e.g., by best covering the data needed for a user query. If one needs careful reconciliation management, local as view does not provide good mechanisms for that; local as view is more suited for "fuzzy" matching such as for retrieving documents.

There are also tools to semiautomatically generate schema mappings, e.g., the Clio system. [24] Clio uses some general heuristics to automatically generate schema mappings as view definitions, and these mappings can be overridden by the user if they are incorrect.

28.2.3 Composable Mediators

The central mediator architecture is well suited for accessing external data sources of different kinds from an extensible relational database server. This provides relational abstractions of all accessed data, and these abstractions can be made available to applications and users through regular SQL APIs such as JDBC/ODBC. One can predict that different information providers will set up many such information integration servers on the Internet. Each information provider provides transparent views over its mediated data. It may even be so that the same data source is transparently mediated by different mediator servers. When many such mediator servers are available, there will be need for mediating the mediators too, i.e., to define mediator servers that access other mediator servers along with other data sources. This we call *composable mediators* and is illustrated by Figure 28.3.

There is another reason that next generation mediators should be composable. That is to be able to scale the data integration process by modularizing the data integration of large numbers of data sources. Each single meditator should be a module that contains the knowledge of how to integrate just a small number of sources and provide view abstractions of these sources to higher-level mediators and applications. Rather than trying to integrate all sources through one mediator as in the central mediator approach, compositions are defined in terms of other mediator compositions.

Composable mediators reduce the complexity of defining mediators over many sources because the data integration is modularized and logically composed. This is important in particular when integrating the large numbers of different kinds of sources available on the Internet. It is possible to achieve similar effects by view compositions inside a central mediator server too, but it is not always realistic to have one mediator server integrating all data. Furthermore, in many cases it seems less natural to pass through

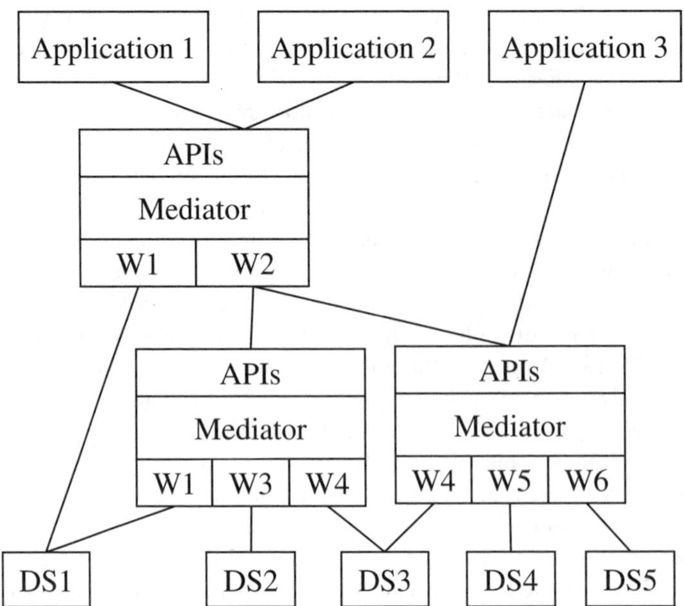

FIGURE 28.3 Composed mediator architecture.

the same large relational database server whenever data integration is required, but rather one would like to have compositions of many more mediator servers not always based on tabular data.

Even though composable mediators provide a possible solution to large-scale data integration design, it also poses some implementation challenges. For example, if each mediator is regarded as a database view, the regular multilevel view expansion used in relational DBMSs produce very large queries. Because the cost-based query optimization used in relational databases is NP-complete over the size of the query, it will be impossible to optimize large compositions of mediator views with cost-based query optimization. This has been investigated in [14] where incomplete view expansion for composable mediators is proposed. There it is shown that view expansion is favorable in particular when there are common sources hidden inside the views. Therefore, knowledge of what sources a mediator view is dependent on can be used in deciding whether or not to expand it. Another problem is the optimization of queries in a composable mediation framework that contains more or less autonomous mediator peers having their own query processors. In such an environment the different mediator query optimizers will have to cooperate to compile query fragments, as investigated in Josifovski et al. [9, 12]. In the DIOM project [22] there is a similar framework for integration of relational data sources with a centrally performed compilation process.

28.3 The Amos II Approach to Composable Mediation

As an example of a mediator system we give an overview of the Amos II system [25, 26] followed by a simple example of how to mediate using Amos II. The example illustrates some of the problems involved in defining views over distributed and heterogeneous data.

Amos II is based on a functional data model having its roots in the Daplex data model. [28] One of the original purposes of Daplex was data integration and similar functional data models have been used for data integration in several other systems, including Multibase, [3] Pegasus, [4] and PIFDM. [17] Most mediator systems use a relational data model. However, a functional model turns out to be a very well-suited data model for integrating data because it is more expressive and simple than both OO and relational models and has in Amos II been extended with mediator reconciliation primitives. [10, 11]

Amos II is a distributed system where several mediator peers communicate over the Internet. Each mediator peer appears as a virtual functional database layer. We say therefore that Amos II is a *peer mediator* system. [14] Functional views provide transparent access to data sources from clients and other mediator peers. Conflicts and overlaps between similar real-world entities being modeled differently in different data sources are reconciled through the mediation primitives [10, 11] of the multi-mediator query language AmosQL. The mediation services allow transparent access to similar data structures represented differently in different data sources. Applications access data from distributed data sources through queries to views in some mediator peer.

Logical composition of mediators is achieved when *multidatabase views* in mediators are defined as Amos II functions in terms of functions in other Amos II peers or data sources. The functional multi-database views make the mediator peers appear to the user as a single virtual database containing a set of function definitions that transparently combine underlying data. General queries can be specified over these mediator functions using the functional query language AmosQL. A distributed query optimizer will thereby translate query expressions into optimized query fragments over the sources of a mediator while respecting the autonomy of sources and mediator peers.

In order to access data from external data sources rather than other Amos II peer, one or several *wrappers* can be defined inside an Amos II mediator. An Amos II wrapper for a particular kind of data source consists of an interface to physically access such sources and a *translator* for wrapper query optimization. The wrappers are defined by a set of functions that map the data source's data into the schema of the mediator; i.e., global as view is used. Wrappers have been defined for ODBC-based access to relational databases, [5] access to XML documents, [20] CAD systems, [18] or Internet search engines. [15] External Amos II peers known to a mediator can also be regarded as external data sources, and there is a special wrapper for accessing other Amos II peers. However, among the Amos II peers, special query optimization methods are used that take into account the distribution, capabilities, costs, etc., of the different peers. [12, 14, 9]

28.3.1 The Functional Data Model of Amos II

The basic concepts of the data model are *objects, types,* and *functions.* Each mediator database schema is defined in terms of these basic concepts. In order to wrap a non-Amos Il data source inside an Amos II mediator database, the system has mechanisms for defining user-defined functions and types along with query processing rules over these. Furthermore, in order to compose views of data in other mediator peers, the basic concepts are orthogonally extended with *proxy* objects that are placeholders for corresponding objects in other mediators.

28.3.1.1 Objects

Objects model all entities in a mediator database. The system is reflective in the sense that everything in Amos II is represented as objects managed by the system, both system and user-defined objects. There are two main kinds of representations of objects: *literals* and *surrogates.* The surrogates have associated object identifiers (OIDs) that are explicitly created and deleted by the user or the system. Examples of surrogates are objects representing real-world entities such as persons, meta-objects such as functions, or even Amos II mediators as meta-mediator objects stored in some Amos Il mediator database.

The literal objects are self-described system-maintained objects that do not have explicit OIDs. Examples of literal objects are numbers and strings. Objects can also be *collections,* representing collections of other objects. The system-supported collections are bags (unordered sets with duplicates) and *vectors* (order-preserving collections). Literal objects and collections are automatically deleted by an incremental garbage collector when they are no longer referenced in the database.

Proxy objects in a mediator peer are local OIDs having associated descriptions of corresponding surrogate objects stored in other mediators or data sources. They provide a general mechanism to define references to remote surrogate objects in other mediators or in data sources wrapped by a mediator.

Proxy objects are created implicitly by the system whenever OIDs need to be exchanged with other mediators. They are garbage collected by the system when no longer needed.

28.3.1.2 Types

Objects are classified into *types*, making each object an *instance* of one or several types. The set of all instances of a type is called the *extent* of the type. The types are organized in a multiple-inheritance, supertype/subtype hierarchy. If an object is an instance of a type, then it is also an instance of all the supertypes of that type, and the extent of a type is a subset of all extents of the supertypes of that type (extent-subset semantics). For example, if the type Student is a subtype of type Person, the extent of type Student is also a subset of the extent of type Person. The extent of a type that is multiple inherited from other types is a subset of the intersection of the extents of its supertypes.

There are two kinds of types, *stored* and *derived* types:

- A *stored type* is a type whose extent is explicitly stored in the local store of a mediator. Stored types are used for representing metadata about data retrieved from other mediators. Amos II can also be used as a stand-alone database server and then the stored types represent stored database objects.
- In contrast, a *derived type* is a virtual type whose extent is defined through a query over the (virtual) database in a mediator. The derived types are used for combining and reconciling differences between data retrieved from heterogeneous schemas in different mediators, as will be explained later.

Stored types are defined and stored in an Amos II peer through the create type statement. For example, assume we have a database named Uppsala with these types:

```
create type Person;
create type Employee under Person;
create type Teacher under Employee;
create type Student under Person;
create type Course;
create type Attendance;
```

The above statements define in a mediator a schema with six new types; the extent of type Person is the union of all objects of types Person, Employee, Student, and Teacher.

The types themselves are represented as instances of a system type named *Type*. For defining types stored in other mediators and sources, the system internally uses *proxy types*, which are proxy objects for objects of type *Type* in other mediators. A proxy object is an instance of some proxy type or types, and the extent of a proxy type is a set of proxy objects. Proxy types are defined implicitly by the system when the user references types in other mediators. For example, the following statement defines in a mediator a derived type UppsalaStudent that inherits its contents from the type named Student in a mediator named Uppsala:

```
create derived type UppsalaStudent under Student@Uppsala;
```

The mediator where UppsalaStudent is defined will internally use a proxy-type object to reference the external type Student@Uppsala. The extent of type UppsalaStudents is a set of proxy objects for the extent of type Student@Uppsala. This is a primitive example of integration of data from other mediators. One may then ask for all objects of type UppsalaStudent by the query:

```
select s from UppsalaStudent s;
```

The query will return proxy objects for all objects of type Student in peer Uppsala.

28.3.1.3 Functions

Functions model properties of objects, computations over objects, and relationships between objects. Functions are basic primitives in AmosQL queries and views. A function consists of two parts, the *signature* and the *implementation:* The *signature* defines the types and optional names of the argument or arguments and the result of a function. For example, the signature of the function modeling the attribute name of type Person has the signature:

```
name(Person)->Charstring;
```

The *implementation* of a function specifies how to compute its result given a tuple of argument values. For example, the function name could obtain the name of a person by accessing a wrapped data source. The implementation of a function is normally nonprocedural, i.e., a function only computes result values for given arguments and does not have any side effects.

Furthermore, Amos II functions are often *multidirectional,* meaning that the system is able to inversely compute one or several argument values if the expected result value is known. Inverses of multidirectional functions can be used in database queries and are important for specifying general queries with function calls over the database. For example, in the following query that finds the age of the person named "Tore," if there is an index on the result of function name, the system will use the inverse of function name to avoid iterating over the entire extent of type Person:

```
select age(p) from Person p where name(p) = 'Tore';
```

Depending on their implementation, the basic functions can be classified into *stored, derived,* and *foreign* functions.

- *Stored functions* represent properties of objects (attributes) locally stored in an Amos II mediator. Stored functions correspond to attributes in object-oriented databases and tables in relational databases. In a mediator, stored functions are used for representing metadata about data in sources, private data, or data materialized in the mediator by the mediator engine.
- *Derived functions* are functions defined in terms of functional queries over other Amos II functions. Derived functions cannot have side effects, and the query optimizer is applied when they are defined. Derived functions correspond to side-effect-free methods in object-oriented models and views in relational databases. AmosQL uses an SQL-like *select* query statement for defining derived functions.
- *Foreign functions* (user-defined functions) provide the basic interfaces for wrapping external systems from Amos II. For example, data structures stored in external storage managers can be manipulated through foreign functions as well as basic interfaces to external query systems, e.g., JDBC and ODBC interfaces. Foreign functions can also be defined for updating external data structures, but foreign functions used in queries must be side-effect free.

Foreign functions correspond to methods in object-oriented databases. Amos II provides a possibility to associate several implementations of inverses of a given foreign function — *multidirectional foreign functions* — which declares to the query optimizer that there are several access paths implemented for the function. To help the query processor, each associated access path implementation may have associated cost and selectivity functions. The multidirectional foreign functions provide access to external storage structures similar to data "blades," "cartridges," or "extenders" in object-relational databases. [27] The basis for the multidirectional foreign function was developed in Litwin and Risch, [21] where the mechanisms are further described.

Amos II functions can furthermore be *overloaded,* meaning that they can have different implementations, called *resolvents,* depending on the type or types of their arguments. For example, the salary may be computed differently for types Student and Teacher. Resolvents can be any of the basic function types.

Example of functions in the previous Amos IT database schema are:

```
create function ssn(Person) -> Integer; /* Stored function */
create function name(Person) -> Character;
create function pay(Employee) -> Integer;
create function subject(Course) -> Character;
```

```
create function teacher(Course) -> Teacher;
create function student(Attendance) -> Student;
create function course(Attendance) -> Course;
create function score(Attendance) -> Integer;
create function courses(Student s) -> Bag of Course c
                                                   /* Derived function */
     as select c
          from Attendance a
          where student(a) = s and
               course (a) = c;
create function score(Student s, Course c) -> Integer sc
     as select score(a)
          from Attendance a
          where student(a) = s and
               course(a) = c;
create function teaches(Teacher t) -> Bag of Course c
                                                   /* Inverse of teacher */
     as select c
          where teacher(c) = t;
```

The function name is overloaded on types Person and Course. The functions courses and score are derived functions that use the inverse of function course. The function courses returns a set (bag) of values. If "Bag of" is declared for the value of a function, it means that the result of the function is a bag (multiset), e.g., function courses.

Functions (attributes) are inherited, so the above statement will make objects of type Teacher have the attributes name, ssn, dept, pay, and teaches.

As for types, function definitions are also system objects belonging to a system type named Function. Functions in one mediator can be referenced from other mediators by defining *proxy functions* whose OIDs are proxy objects for functions in other mediators. The creation of a proxy function is made implicitly when a mediator function is defined in terms of a function in another mediator. For example:

```
create function uppsala_students() -> Bag of Charstring
as select name(s) from Student@Uppsala s;
```

returns the names of all objects of type Student in peer Uppsala. The system will internally generate a proxy type for Student@Uppsala and a proxy function for the function name in Uppsala.

Proxy objects can be used in combination with local objects. This allows for general multidatabase queries over several mediator peers. The result of such queries may be literals (as in the example), proxy objects, or local objects. The system stores internally information about the origin of each proxy object so it can be identified properly. Each local OID has a locally unique OID number, and two proxy objects are considered equal if they represent objects created in the same mediator or source with equal OID numbers.

Proxy types can be used in function definitions as any other type. In the example, one can define a derived function returning the teacher proxy object of a named teacher peer `Uppsala`:

```
create function uppsala_teacher_named(Charstrirng nm)
                                        -> Teacher@Uppsala
     as select nm
          from Teachers@Uppsala p
          where name(t) = nm;
```

In this case the local function Uppsala_teacher_named will return a proxy object instance of the type Teacher in mediator named Uppsala for which it holds that the value of function name in Uppsala returns nm. The function can be used freely in local queries and function definitions and as proxy functions in multidatabase queries from other mediator peers. For example, this query returns proxy objects for the students of a course taught by a person named Carl in peer Uppsala:

```
select students(teaches(Uppsala_teacher_named('Carl')));
```

The so-called Daplex-semantics is used for function composition, meaning that bag-valued function calls are automatically unnested. This can also be seen as a form of extended path expressions through functional notation.[4]

28.3.2 Composed Functional Mediation

The multidatabase primitives of AmosQL provides the basis for defining derived types and functions in mediators that combine and reconcile data from several sources. Types, functions, and queries provide powerful primitives for mediating both relational and object-oriented data sources without losing data semantics.

As an example, assume we have two sources:

1. A relational database in Stockholm stores details about students and teachers at Stockholm university and the names and social security numbers of all Swedish residents. It has the relations

```
residents(ssn, firstname, lastname, address)
course(cid, name, teacherssn)
takes(ssn, cid, score)
```

2. An XML-based database in Uppsala stores details about students at Uppsala University. It is stored as an XML document on the Web wrapped by Amos II. The XML file is loaded into the system and there represented by the Uppsala schema given above. The wrapper imports and converts the data by reading the XML document and stores it in the mediator's database using database update statements. The details of how to translate XML documents to Amos II is not detailed here; basic XML primitives can be mapped to Amos II data elements automatically, [20] or some XML wrapping tool (e.g. Xerces2 [32]) can read the source based on an XML-Schema definition. The alternative to retaining the data in the XML-source is also possible if the source is managed by some XQuery tool.[5] In that case the wrapper will be more complex and needs to translate mediator queries into XQuery statements.

Now we define a mediator named StudMed to be used by students attending both Uppsala and Stockholm universities. It will access data wrapped by the two Amos II peers named Uppsala and Stockholm, both of which are assumed to be set up as autonomous mediator peers.

In our scenario we furthermore semantically enrich the wrapped relational data in Stockholm by providing an object-oriented view of some of the tabular data. This is done by defining a derived type Student along with object navigation functions that connect instances through object references rather than foreign key references.

We begin by showing how to define a mediator that wraps and semantically enrich the relational database in Stockholm, and then we show how to define the mediator StudMed fusing data from both wrapped sources.

28.3.2.1 Wrapping the Relational Database

Entities from external sources are linked to a mediator by calling the system function access (source, entities) that calls the wrapper to import to the data source schema. If the source is a wrapped relational

[4] With an extended path notation, the above query could have been written as select "Carl".uppsala-teacher-named.teaches.students.

[5] http. //www.w3.org/XMLQuery#implementations.

database, the specified source relations become proxy types in Amos II and the columns become proxy functions.

The relational database is accessed from the Stockholm mediator through these commands:

```
set :a = jdbc("ibds", "interbase.interclient.Driver");
connect(:a, "jdbc:interbase://localhost/stockholm.gdb",
                 "SYSDBA", "masterkey");
access(:a, {"resident", "course", "teaches", "takes"});
```

Here the Interbase Interclient JDBC driver[6*] is used to connect to the database stockholm.gdb running on the same host as the mediator peer Stockholm. After the above commands, the following types and functions are available in the mediator Stockholm:

```
type Resident
function ssn(Resident) -> Integer
function firstname(Resident) -> Charstring
function lastname(Resident) -> Charstring
function address(Resident) -> Charstring
type Course
function cid(Course) -> Integer
function name(Course) -> Charstring
type Teaches
function ssn(Teaches) -> Integer
function cid(Teaches) -> Course
type Takes
function ssn(Takes) -> Integer
function cid(Takes) -> Integer
function score(Takes) -> Integer
```

Because there is no explicit type Student in the wrapper, we define it as a derived type:

```
create function student?(Resident p) -> boolean
      as select true
            where some(select true
                  from takes t
                  where ssn(t)=ssn(p));
create derived type Student under Resident p where student?(p);
```

A resident is thus a student if he takes some course. This is an example of a schema mapping.

We also need functions to represent the relationship between a course enrollment and its student:

```
create function student(Takes t) -> Student s
      as select s
            where ssn(t) = ssn(s);
create function course(Takes t) -> Course c
      as select c
            where cid(c) = cid(t);
```

These two functions allow direct object references between objects of types Takes,Student, and Course.

At this point we can set up Stockholm as a mediator peer on the Internet. Other peers and applications can access it using AmosQL.

[6*] http://prdownloads. sourceforge.net/firebird.

28.3.2.2 Composing Mediators

The mediator StudMed provides a composed view of some combined data from mediators Stockholm and Uppsala. The mediator schema has the following types and function signatures:

```
type Student
function id(Person) -> Integer
function name(Person) -> Charstring
type Course
function subject(Course) -> Character
type Takes
function student(Takes) -> Student
function course(Takes) -> Course
function score(Takes) -> Integer
```

All these types and functions are proxies for data in one of or both the other peers.

We begin with accessing the desired types and functions of the other mediators by a system call:

```
access("Uppsala",
      {"Student", "Course","Teacher", "Attendance"});
access("Stockholm",
      ("Student", "Course", "Takes"});
```

Here the call to access defines a proxy type Student@Stockholm with proxy functions ssn, name, address (inherited). Analogously for the other types accessed from the mediator peers.

The data of the type Student in mediator StudMed is derived from the derived type Student in mediator Stockholm and the stored type Student in mediator Uppsala. In StudMed we wish to model the *union* of the students in Uppsala and Stockholm along with their properties. This is modeled in Amos II as a *derived supertype* of the proxy types Student@Stockholm and Student@Uppsala. In order to reconcile corresponding students, we need a *key* for Students in Uppsala and Stockholm, and the social security number (SSN) provides this. Since various properties of students are computed differently in different sources, we also need to define how to compute equivalent properties from different sources and how to reconcile differences, conflicts, and overlaps. A special derived-type syntax called IUT (Integration Union Type) [11] provides the primitives to do the reconciliation:

```
create derived type Student
     key Integer id  /* Common key = SSN */
     supertype of
          Student@Stockholm s: ssn(s),/* Key mapping */
          Student@Uppsala u: ssn(u)/* Key mapping */
     functions
          (name Charstring)/* Mediator view function */
     case s:
          name = firstname(s) + " "÷ lastname(s);
                                       /* Concatenated names */
     end functions;
```

Because function name is not directly available in Stockholm, it must be reconciled through string concatenation (case s) when a student attends only Stockholm classes. If a student attends classes from both cities, the name function from Uppsala is used.

The integrated type Course can similarly be defined using the following definition:

```
create derived type Course
     key Charstring subject
     supertype of
```

```
        Course@Stockholm st: name(st),
        Course@Uppsala u: subject(u);
```

Finally, we need to specify the derived-type Takes that links students to courses. Inheritance between objects from different mediators provides a convenient way to reference objects of types Student and Course.

First we need to define two utility functions that compute the composite key for the two different sources as a vector ({} notation):

```
create function takesKey(Takes@Stockholm s) -> Vector
    as select {ssn(s),name(c)}
        from Course@Stockholm c
        where cid(s) = cid(c);
create function takesKey(Attendance@Uppsala u) -> Vector
    as select (ssn(student(u)), title(course(u))};
```

Now we can define the IUT for Takes simply as:

```
create derived type Takes
    key vector
    supertype of
        Takes@stockholm s: takesKey(s),
        Attendance@uppsala u: takesKey(u)
    functions
    (student Student, course Course, score Integer)
    case s,u:
        student = student(u);
        course = course(u);
        score = min(score(u),score(s));
    end functions;
```

The definition becomes this simple because there are already corresponding functions student, course, and score in both mediator peers Uppsala and Stockholm. The functions student in Stockholm and Uppsala return objects of types Student@Stockhom and Student@Uppsala, respectively, which are inherited from type Student in mediator StudMed. Analogously for function course returning type Course also inherited between the mediators.

28.3.2.3 Querying the Mediator

From the user's point of view, the mediator looks like any other Amos II database. It can be freely queried using AmosQL. In our example, we can ask general queries such as the names and SSNs of all students with a score on some course larger than 5:

```
select distinct name(s), id(s)
from Student s, Takes t
where score(t)>5 and student(t)=s;
```

The distributed mediator query optimizer translates each query to interacting query optimization plans in the different involved mediators and sources using an set of distributed query optimization and transformation techniques. [9, 12, 14] Since types and functions are specified declaratively in AmosQL through functional queries, the system utilizes knowledge about type definitions to eliminate overlaps and simplify the queries before generating the distributed query execution plans. [10, 11] The plans are generated through interactions between the involved mediators. Global query decomposition strategies are used for obtaining an efficient global execution plan. [9, 12]

28.3.3 Implementing Wrappers

The physical wrapper interface is implemented as a set of foreign AmosQL functions, while the data model mapping and wrapper query optimization are implemented through a set of source specific rewrite rules that transforms general AmosQL query fragments into calls to the interface foreign functions. Finally, the schema mappings are defined through derived functions and types as in the example discussed earlier.

28.3.3.1 Foreign Functions

As a very simple example of how to wrap a data source using a multidirectional foreign function, assume we have an external disk-based hash table on strings to be accessed from Amos II. We can then implement it as follows:

```
create function get_string(Charstring x)-> Charstring r
      as foreign "JAVA:Foreign/get_hash";
```

Here the foreign function get_string is implemented as a Java method get_hash of the public Java class Foreign. The Java code is dynamically loaded when the function is defined or the mediator initialized. The Java Virtual Machine is interfaced with the Amos II kernel through the Java Native Interface to C.

Multidirectional foreign functions include declarations of inverse foreign function implementations. For example, our hash table can not only be accessed by keys but also scanned, allowing queries to find all the keys and values stored in the table. We can generalize it by defining:

```
create function get_string(Charstring x)->Charstring y
      as multidirectional
            ("bf" foreign "JAVA:Foreign/get_hash"
                  cost (100,1))
            ("ff" foreign "JAVA:Foreign/scan_hash"
                  cost "scan_cost");
```

Here, the Java method scan_hash implements scanning of the external hash table. Scanning will be used, e.g., in queries retrieving the hash key for a given hash value. The *binding patterns,* bf and ff, indicate whether the argument or result of the function must be bound (b) or free (f) when the external method is called.

The cost of accessing an external data source through an external method can vary heavily depending on, e.g., the binding pattern, and to help the query optimizer, a foreign function can have associated costing information defined as user functions. The cost specifications estimate both *execution costs* in internal cost units and *result sizes* (fanouts) for a given method invocation. In the example, the cost specifications are constant for get_hash and computed through the Amos II function scan_cost for scan_hash.

For relational sources there is a primitive foreign access function to send parameterized query strings to a relational database source:

```
create function sql Relational s,
                    Charstring query,
                    Vector params)
      -> bag of Vector as foreign "PrepareAndExecute";
```

Example of call:

```
sql(:s,
      "select c.cid from course c where c.name = ?",
      {"Programming"});
```

For a given relational data source s, the function executes the specified query with parameters marked ? substituted with the corresponding values in params. The function sql is implemented as an overloaded foreign function calling ODBC or JDBC, depending on the type of the source.

28.3.3.2 Defining Translator Rules

The wrapping of an external hash table requires no further data model mapping or query optimization. However, for more advanced sources, such as relational databases, the wrapper must also include rewrite rules that translate functional query expressions into optimized source queries. [5] It is the task of the translator part of a wrapper to transform AmosQL queries to the wrapper into calls to primitive functions like `sql`. In Fahl and Risch [5] it is explained what rewrites are needed for relational databases.

To provide for a convenient way to integrate new kinds of data sources, we have developed a general mechanism to define wrapper query optimization by translation rules for different kinds of sources. The system includes mechanisms to define different *capabilities* of different kinds of sources, where a capability basically specifies which kind of query expressions can be translated by a particular source. The capabilities are specified by a set of rewrite rules associated with the wrapped source. These rules identify which query expressions a particular source can handle and transform them into primitive calls to interface functions for the source.

For wrappers of relational databases, the transformation rules identify connected subqueries for a given source and then translate the graph into SQL query strings. Cartesian products and calls to functions not executable as SQL are processed as queries in the wrapping mediator. Thus the rewrite rules specify what parts of an AmosQL query to a wrapped source can be translated to source queries and what parts need to be treated by Amos II.

28.4 Conclusions

Mediators provide a general framework for specifying queries and views over combinations of heterogeneous data sources. Each kind of data source must have a wrapper which is a program module transforming source data into the common data model used by the mediator engine. A given wrapper provides the basic mechanisms for accessing any source of the wrapped kind. The mediator will contain views defining mappings between each source and the schema of the mediator. We discussed various overall architectures of mediator systems and wrappers. As an example of a mediator system we gave an overview of the Amos II mediator system and an example of how it integrates wrapped heterogeneous data.

In summary, the following is involved in setting up a mediator framework:

- *Classify kinds of data sources.* First one needs to investigate what kinds of sources are involved in the mediation. Are wrappers already defined for some of the sources? If not, the available APIs to the different kinds sources are investigated in order to design new wrappers.
- *Implement wrappers.* Wrappers needs to be designed and implemented that translate queries in the mediator query language into queries of the source. The results from the source queries are translated back from the source representation into the data abstractions of the mediator model. This is the most challenging task in the data integration process. For example, if an Internet search engine is to be wrapped and a relational data model is used in the mediator, SQL queries to a source need to be translated into query search strings of the particular search engine. The wrapper will pass the translated query strings to the search engine using some API of the source permitting this. In the same way, the results passed back from the search engine need to be passed to the relational mediator as rows in relations. [15] The wrappers need to translate API calls used by the mediator engine into API calls of the source, e.g., adhering to the SQL/MED standard if that is used by the mediator system.
- *Define source schemas.* The particular data sources to be mediated need to be identified and analyzed. The structure of the data to access need to be investigated and their schema defined. For example, if a relational database model is used, the sources are modeled as a number of external source relations.
- *Define mediator schema.* The schema of the mediator needs to be defined in terms of the source schemas. Views in the integrated schema are defined by matching data from the source schemas.

In our example, we illustrated the reconciliation using a functional data model. In the relational data model, reconciliation means defining views that join source relations. In defining these views, common keys and data transformations need to be defined in order to reconcile similarities and differences between source data. A problem here can be that SQL does not support reconciliation and therefore the mediating view may be complex, involving user defined functions (UDFs) to handle some matchings and transformations.

Most present mediator frameworks are central in that a central mediator schema is used for integrating a number of sources in a two-tier mediator framework. The *composable* mediator framework [9, 14] generalizes this by allowing transparent definitions of autonomous mediators in terms of other mediators without knowing internals of source mediators. Thus a multi-tier network of interconnected mediators can be created where higher-level mediators do not know that lower-level mediators in their turn access other mediators. Such a peer mediator architecture poses several challenges, e.g., for query optimization. [14] We have already illustrated how to compose mediators using the composable mediator system Amos II.

References

[1] O. Bukhres and A. Elmagarmid (Eds.): *Object-Oriented Multidatabase Systems.* Prentice Hall, Englewood Cliffs, NJ, 1996.

[2] U. Dayal and P.A. Bernstein: On the Correct Translation of Update Operations on Relational Views. *Transactions on Database Systems,* 7(3), 381–416, 1981.

[3] U. Dayal and H-Y Hwang: View Definition and Generalization for Database Integration in a Multidatabase System. *IEEE Transactions on Software Engineering,* 10(6), 628–645, 1984.

[4] W. Du and M. Shan: Query Processing in Pegasus. In O. Bukhres, A. Elmagarmid (Eds.): *Object-Oriented Multidatabase Systems.* Prentice Hall, Englewood Cliffs, NJ, 449–471, 1996.

[5] G. Fahl and T. Risch: Query Processing over Object Views of Relational Data. *The VLDB Journal,* 6(4), 261–281, 1997.

[6] H. Garcia-Molina, Y. Papakonstantinou, D. Quass, A. Rajaraman, Y.Sagiv, J. Ullman, V. Vassalos, and J. Widom: The TSIMMIS approach to mediation: Data models and languages. *Journal of Intelligent Information Systems (JIIS),* 8(2), 117–132, 1997.

[7] J-R. Gruser, L. Raschid, M.E. Vidal, and L. Bright: Wrapper Generation for Web Accessible Data Sources, *3rd Conference on Cooperative Information Systems (CoopIS'98),* 1998.

[8] L. Haas, D. Kossmann, E. Wimmers, and J. Yang: Optimizing Queries accross Diverse Data Sources. *Proceedings of the International Conference on Very Large Data Based (VLDB'97),* pp. 276–285, Athens, 1997.

[9] V. Josifovski, T. Katchaounov, and T. Risch: Optimizing Queries in Distributed and Composable Mediators. *4th Conference on Cooperative Information Systems (CoopIS'99),* pp. 291–302, 1999.

[10] V. Josifovski and T. Risch: Functional Query Optimization over Object-Oriented Views for Data Integration. *Intelligent Information Systems (JIIS),* 12(2–3), 165–190, 1999.

[11] V. Josifovski and T. Risch: Integrating Heterogeneous Overlapping Databases through Object-Oriented Transformations. *25th Conference on Very Large Databases (VLDB'99),* 435–446, 1999.

[12] V. Josifovski and T.Risch: Query Decomposition for a Distributed Object-Oriented Mediator System. *Distributed and Parallel Databases,* 11(3), 307–336, May 2001.

[13] V. Josifovski, P. Schwarz, L. Haas, and E. Lin: Garlic: A New Flavor of Federated Query Processing for DB2, *ACM SIGMOD Conference,* 2002.

[14] T. Katchaounov, V. Josifovski, and T. Risch: Scalable View Expansion in a Peer Mediator System, *Proceedings of the 8th International Conference on Database Systems for Advanced Applications (DASFAA 2003),* Kyoto, Japan, March 2003.

[15] T. Katchaounov, T. Risch, and S. Zürcher: Object-Oriented Mediator Queries to Internet Search Engines, *International Workshop on Efficient Web-based Information Systems (EWIS),* Montpellier, France, September 2, 2002.

[16] A.M. Keller: The role of semantics in translating view updates, *IEEE Computer,* 19(1), 63–73, 1986.

[17] G.J.L. Kemp, J.J. Iriarte, and P.M.D. Gray: Efficient Access to FDM Objects Stored in a Relational Database, Directions in Databases, *Proceedings of the 12th British National Conference on Databases (BNCOD 12),* pp. 170–186, Guildford, U.K., 1994.

[18] M. Koparanova and T. Risch: Completing CAD Data Queries for Visualization, *International Database Engineering and Applications Symposium (IDEAS'*2002), Edmonton, Alberta, Canada, July 17–19, 2002.

[19] A.Y. Levy, A. Rajaraman, and J.J. Ordille: Querying heterogeneous information sources using source descriptions. In *Proceedings of the International Conference on Very Large Databases (VLDB'96),* Mumbai, India, 1996.

[20] H. Lin, T. Risch, and T. Katchaounov: Adaptive data mediation over XML data. In special issue on "Web Information Systems Applications" of *Journal of Applied System Studies (JASS),* 3(2), 2002.

[21] W. Litwin and T. Risch: Main memory oriented optimization of OO queries using typed datalog with foreign predicates. *IEEE Transactions on Knowledge and Data Engineering,* 4(6), 517–528, 1992.

[22] L. Liu and C. Pu: An adaptive object-oriented approach to integration and access of heterogeneous information sources. *Distributed and Parallel Databases,* 5(2), 167–205, 1997.

[23] J. Melton, J. Michels, V. Josifovski, K. Kulkarni, P. Schwarz, and K. Zeidenstein: SQL and management of external data, *SIGMOD Record,* 30(1), 70–77, March 2001.

[24] L. Popa, Y. Velegrakis, M. Hernandez, R. J. Miller, and R. Fagin: Translating Web Data, *28th International Conference for Very Large Databases (VLDB 2002),* Hong Kong, August 2002.

[25] T. Risch and V. Josifovski: Distributed Data Integration by Object-Oriented Mediator Servers. *Concurrency and Computation: Practice and Experience,* 13(11), September, 2001.

[26] T. Risch, V. Josifovski, and T. Katchaounov: Functional data integration in a distributed mediator system. In P. Gray, L. Kerschberg, P. King, and A. Poulovassilis (Eds.): *Functional Approach to Computing with Data,* Springer-Verlag, New York, 2003.

[27] M. Stonebraker and P. Brown: *Object-Relational DBMSs: Tracking the Next Great Wave.* Morgan Kaufmann, San Francisco, CA, 1999.

[28] D. Shipman: The functional data model and the data language DAPLEX. *ACM Transactions on Database Systems,* 6(1), 140–173, 1981.

[29] A. Tomasic, L. Raschid, and P Valduriez: Scaling access to heterogeneous data sources with DISCO. *IEEE Transactions on Knowledge and Date Engineering,* 10(5), 808–823, 1998.

[30] V.M.P. Vidal and B.F. Loscio: Solving the Problem of Semantic Heterogeneity in Defining Mediator Update Translations, *Proceedings of ER '99, 18th International Conference on Conceptual Modeling, Lecture Notes in Computer Science 1728,* 1999.

[31] G. Wiederhold: Mediators in the architecture of future information systems. *IEEE Computer,* 25(3), 38–49,1991.

[32] Xerces2 Java Parser, http://Xml.apache.org/Xerces2-j/, 2002.

29

Introduction to Web Semantics

CONTENTS

Abstract 29-1
29.1 Introduction 29-1
29.2 Historical Remarks 29-2
29.3 Background and Rationale 29-3
29.3.1 Understanding Information on the Web 29-3
29.3.2 Creating Information on the Web 29-3
29.3.3 Sharing Information over the Web 29-4
29.4 Ontologies 29-4
29.5 Key Ontology Languages 29-6
29.5.1 RDF and RDF Schema 29-6
29.5.2 OWL 29-8
29.6 Discussion 29-10
29.6.1 Domain-Specific Ontologies 29-11
29.6.2 Semantic Web Services and Processes 29-11
29.6.3 Methodologies and Tools 29-11
29.7 Summary 29-12
References 29-13

Munindar P. Singh

Abstract

The Web as it exists currently is limited in that, although it captures content, it does so without any explicit representation of the meaning of that content. The current approach for the Web may be adequate as long as humans are intended to be the direct consumers of the information on the Web. However, involving humans as direct consumers restricts the scale of several applications. It is difficult for unassisted humans to keep up with the complexity of the information that is shared over the Web. The research program of Web semantics seeks to encode the meaning of the information on the Web explicitly so as to enable automation in the software tools that create and access the information. Such automation would enable a richer variety of powerful applications than have previously been possible.

29.1 Introduction

The Web we know and love today has evolved into a practically ubiquitous presence in modern life. The Web can be thought of as a set of abstractions over data communication (in the nature of the hypertext transport protocol, better known as HTTP) and information markup (in the nature of the hypertext markup language or HTML).

1-58488-381-2/05/$0.00+$1.50

© 2005 by CRC Press LLC

The Web has clearly been successful or we would not be talking about it here. However, its success has also exposed its limitations. For one, it is difficult to produce and consume the content on today's Web. This is because a human must be involved in order to assign meaning to the content because the data structures used for that content are weak and concentrate on the presentation details. Thus the meaning, if any, is confined to text or images, which must be read and interpreted by humans.

Researchers have realized almost since the inception of the Web that it is limited in terms of its representation of meaning. This has led to the vision of the Semantic Web in which the meaning of the content would be captured explicitly in a declarative manner and be reasoned about by appropriate software tools. The main advantage of such an encoding of Web semantics is that it would enable greater functionality to be shifted from humans to software. Software tools would be able to better exploit the information on the Web. They would be able to find and aggregate the right information to better serve the needs of the users and possibly to produce information for other tools to consume. Further, the availability of information with explicitly represented semantics would also enable negotiation among the various parties regarding the content.

This vision of the Semantic Web was first promulgated by Tim Berners-Lee. We use the term *Semantic Web* to refer to Berners-Lee's project and the term *Web semantics* to refer to semantics as dealing with the Web in general. Web semantics owes much of its intellectual basis to the study of knowledge modeling in artificial intelligence, databases, and software engineering. The arrival of the Web has given a major impetus to knowledge modeling simply because of the scale and complexity of it, which severely limits the effectiveness of *ad hoc* methods. Web semantics has expanded to become a leading subarea of Internet computing with a large number of active researchers and practitioners. Over the past few years, there has been much scientific activity in this area. Recently, a conference, *International Conference on the Semantic Web*; and an academic journal, *Journal on Web Semantics*, have been launched in this area.

This chapter provides a high-level introduction to Web semantics. It deals with the key concepts and some of the techniques for developing semantically rich representations. Some other chapters carry additional relevant topics. Specifically, Kashyap [this volume] discusses ontologies and metadata; Brusilovsky and Nejdl [this volume] introduce adaptive hypermedia; Fisher and Sheth [this volume] present enterprise portals as an application of semantic technologies; and, [Arroyo et al., this volume] describe semantic Web services, which in simple terms involve an application of semantic techniques to the modeling of Web services.

29.2 Historical Remarks

In simple terms, the history of the Web can be understood in terms of the increasing explicitness of what is represented. The earliest Web was cast in terms of HTML. HTML provides a predetermined set of tags, which are primarily focused on the presentation of content. An inability to express the structure of the content proves to be a serious limitation when the purpose is to mark up content in a general enough manner that it can be processed based on meaning.

The work on HTML was predated by several years by work on markup techniques for text. The information retrieval community had developed the standardized general markup language (SGML) as a powerful approach to create arbitrary markups. Unfortunately, SGML proved too powerful. It was arcane and cumbersome and notoriously difficult to work through. Consequently, robust tools for creating and parsing SGML did not come into existence. Thus whereas the expressiveness and flexibility of SGML were attractive, its complexity was discouraging. This combination led to the design of a new language that was simple yet sufficiently expressive and — most importantly — extensible to accommodate novel applications.

This language was formalized as the extensible markup language (XML) [Wilde, this volume]. The move to XML in conjunction with stylesheets separates the presentation from the intrinsic structure of the content. XML provides an ability to expand the set of tags based on what an application needs. Thus, it enables a richer variety of content structure to be specified in a manner that potentially respects the needs of the given applications. Additional stylesheets, also expressible in XML, enable the rendering of the content specified in XML in a manner that can be processed by existing browsers. Stylesheets also have another important function, which is to transform XML content from one form into another.

XML provides structure in terms of syntax, meaning that an XML document corresponds to a unique parse tree, which can be traversed in a suitable manner by the given application. However, XML does not nail down the structure that an application may give to the content that is specified. Consequently, there can be lots of nonstandard means to express the same information in XML.

This led to a series of languages that capture content at a higher level of abstraction than XML. This chapter introduces the most established of these languages, namely, Resource Description Framework (RDF), RDF Schema (RDFS), and Web Ontology Language (OWL). In principle, these higher layers do not have to be mapped into XML, but usually they are, so as to best exploit the tools that exist for XML. But occasionally, especially when the verbosity of XML is a consideration, other representations can be used. The above development can be understood as a series of layers where the lower layers provide the basic syntax and the upper layers provide increasing shades of meaning. Berners-Lee presented his vision for the layers as what is known as the *Layer Cake.*

29.3 Background and Rationale

Let's consider some major use cases for the Web, which motivate the need for Web semantics.

29.3.1 Understanding Information on the Web

The Web is designed for human consumption. The pages are marked up in a manner that is interpreted by the popular Web browsers to display a page so that a human viewing that page would be able to parse and understand the contents of the page. This is what HTML is about. HTML provides primitives to capture the visual or, more generally, the presentational aspects of a page. These pages do not directly reflect the structure of the content of the page. For example, whereas HTML captures the recommended type faces and type sizes of the text in a document, it does not capture the sections or subsections of the document.

An even more telling example is when you access a form over the Web. The form may ask for various inputs, giving slots where a user can enter some data. The only clue as to the information required is in the names of the fields that are given in the adjacent text. For example, a field that appears next to a label "first name" would be understood as asking for the user's first name. However, there are two potential shortcomings of such an approach. First, the user needs to assign meaning to words based on his or her tacit knowledge of what the given application may require. A software application can work with such a form only based on some rough and ready heuristics — for example, that the words "first name" indicate the first name. Second, when the meaning is subtle, it is unwieldy for both humans and software, because there is no easy way for the correct interpretation to be specified via *ad hoc* label names.

For example, if the form were really meant to request certain relevant information, but the information to be requested was not known in advance, there would be no way for an application to guess the correct meaning. More concretely, assume you are using a software application that tries to order some medical supplies for your hospital. Would this application be able to correctly fill out the forms at a relevant site that it visits? Only to the extent that you can hard code the forms. Say it knows about shipping addresses. Would it be able to reason that the given site needs a physical address rather a post office box? It could, but only if the knowledge were appropriately captured.

29.3.2 Creating Information on the Web

The Web is designed for the creation of information by humans. Information is gathered up on Web pages by humans and its markup is created, in essence, by hand. Tools can help in gathering the information and preparing the markup, but the key decisions must be made by humans.

It is true that Web pages can be generated by software applications, either from documents produced by hand or from databases based on further reasoning. However, the structure that is given even to such dynamically generated Web pages is determined in an *ad hoc* manner by the programmers of the applications that produce such pages.

It would be great if the schema for a Web page were designed or customized on the fly, based on the needs of the user for whom it was being prepared. The contents of the page could then automatically be generated based on the concepts needed to populate the page.

29.3.3 Sharing Information over the Web

More generally, consider the problem of two or more parties wishing to share information. For example, these parties could be independent enterprises that are engaged in e-business. Clearly, the information to be shared must be transmitted in some appropriate fashion from one to the other. The information must be parsed in an unambiguous manner. Next it must be interpreted in the same way by the parties involved.

Imagine the software applications that drive the interactions between the interacting parties. The applications interpret the information that is exchanged. The interpretation can be in the nature of the data structures (say, the object structures) that the applications serialize or stream into the information that they send over the wire, and the structures that they materialize or construct based on the information that they receive.

Web semantics as understood here is about declaratively specifying the meaning of the information that is exchanged. Naturally this meaning takes the form of describing the object structures to which the exchanged information corresponds.

An important special case of the above is configuring Web applications. Often information is not accessed directly but is used indirectly through specialized tools and applications, e.g., for capturing information and presenting it in a suitable manner to users. For example, when an enterprise resource planning (ERP) system is deployed in a particular enterprise, it must be instantiated in an appropriate manner for that specific enterprise. For example, a hospital billing system must deal with the hospital's human resources (payroll) systems, with insurance companies, and with local government agencies, Thus installing a new billing system can be cumbersome. Even maintaining an existing system in the light of external changes, say, to the government regulations, is difficult. Likewise for Web applications involving e-business. For example, to operate a supply chain may involve suitable software applications at the interacting companies to be configured in such a manner as to respect the common information model that the companies have agreed upon.

Traditionally, the knowledge models underlying such applications are implicit in sections of the procedural code for the given applications. In such cases, when an application is to be deployed in a new installation (e.g., at a new enterprise), it must be painstakingly tuned through a long process involving expensive consultants. However, if the application is modeled appropriately, it would be a simpler matter of refreshing its models for the particular enterprise or business context where it is being deployed.

29.4 Ontologies

Although the current Web has its strong points, it is notorious for its lack of meaning. Since the content of the Web is captured simply in terms of the natural language text that is embedded in HTML markup, there is only a little that we can do with it. For example, the best that current search engines can do is, upon crawling Web pages, to index the words that occur on those pages. Users can search based on the words. And, when a user conducts a search, the engine can produce pages that include the given words. However, such searches miss out on the meaning of interest to the user. Capturing the meaning of the pages is an example of what Web semantics is about. Given an explicit representation of the meaning of a given page, a search engine would index the pages based on the meanings captured by pages rather than just the words that happen to occur on a page. In that manner, the engine would be able to support meaningful searches for its users.

We would not be able to capture the meaning for each page — or, more generally, information resource — in a piecemeal manner. Instead, we must model the knowledge with which the meaning of the given resource can be captured. That is, the knowledge of the domain of interest (i.e., the *universe of discourse)* would provide a basis for the semantics underlying the information even if across disparate resources.

To put the above discussion of ontologies into a practical perspective, let us consider a series of examples from a practical domain involving a business transaction that might serve as one step of a supply chain. Let us consider a simple setting where some medical parts are ordered by one enterprise from another. First we consider a fragment of an order placed in XML.

```
<catheter>
    <type>central</type>
    <tunneled>yes</tunneled>
</catheter>
```

This could be alternatively expressed as follows:

```
<catheter type="central" tunneled="yes"/>
```

Or even as the following fragment:

```
<central_catheter tunneled="yes"/>
```

How can such varied forms be understood? One idea is to define transforms among these variations so that messages written in one form can be morphed into a second form and thus understood by software that is designed only to accommodate the latter form. Such transforms are introduced elsewhere [Wilde, this volume].

The question of interest here is how can we justify such syntactic transformations. Clearly, this must be through the meanings of the terms used; hence, the need for ontologies.

An *ontology* is a formal representation of the conceptualization of a domain of interest. From the computer science perspective, there are three main kinds of things that constitute an ontology.

- An ontology expresses classes or concepts. For example, a medical ontology may have concepts such as catheter, procedure, catheter insertion, and angioplasty. An ontology captures taxonomic relationships among its concepts. For example, we may have that angioplasty is a kind of catheter insertion, and catheter insertion is a kind of procedure. Similarly, we can define blood vessel, artery, vein, jugular vein, and carotid artery with the obvious taxonomic relationships among them. Such taxonomic relationships are treated specially because they are at the very core of the space of concepts expressed by an ontology.
- An ontology expresses relationships among the concepts. For example, we may have a relationship usedIn in our medical ontology, which relates catheters to catheter insertion. Cardinality constraints may be stated over these relationships. For example, we may require that at least one catheter be used in a catheter insertion.
- An ontology may express additional constraints.

Ontologies can be represented in various ways. Well-known approaches include the following.

- *Frame systems.* Each frame is quite like a class in an object-oriented programming language. The connection is not coincidental; frame systems predate object-oriented languages. Thus each frame corresponds to a concept. Relationships among concepts are captured as the slots of various frames. For example, we may have a frame called catheter insertion, which has a slot called cathetersUsed. This slot is filled with a set of catheters. And we may have another slot called intoVessel to capture the blood vessel into which the insertion takes place.
- *Description logics.* In these approaches, concepts are defined through formal expressions or descriptions that refer to other concepts. Based on the formal semantics for the language, it is possible to determine whether one description *subsumes* another, i.e., refers to a larger class than the other. The language chosen determines the complexity of this computation. For example, we may define

a new procedure called venal catheter insertion as a kind of catheter insertion whose intoVessel relationship must refer to a vein.

- *Rules,* which capture constraints on the taxonomic and other relationships that apply. In recent work, rules are used mainly to capture constraints on data values and taxonomic relationships are left to one of the above approaches. An example of a rule might be that for credit card payments by new customers or above a certain amount, the shipping address must be the same as the billing address of the credit card.

The subsumption hierarchy computed in description logics recalls the explicit class hierarchy captured by the frame system. However, in description logics, the hierarchy is derived from the definitions of the classes, whereas the hierarchy is simply given in frame systems. Frame systems have a certain convenience and naturalness, whereas description logics have a rigorous formal basis but can be unintuitive for untrained users.

The past several years have seen a convergence of the two approaches. This research, with support from the U.S. Defense Advanced Research Projects Agency (DARPA), led to the DARPA Agent Markup Language (DAML), and with support from the European Union to the Ontology Interchange Language (OIL). OIL was sometimes referred to as the Ontology Inference Language. DAML and OIL were combined into DAML+OIL, which has evolved into a W3C draft known as the Web Ontology Language (OWL) [McGuiness and van Harmelen, 2003]. This chapter considers only OWL, since that is the current direction.

29.5 Key Ontology Languages

This section introduces the main ontology languages used for Web semantics. Extensive literature is available on these languages describing their formal syntax and semantics. Such details are beyond the scope of this chapter. Instead, this chapter seeks only to introduce the languages at a conceptual level.

In principle, ontologies can be represented in a variety of ways. Consider the ontology dealing with medical terms that was informally described above. We can certainly encode this in XML. For example, we may come up with the following simple, if somewhat contrived, solution:

```
<?xml version="1.0"?>
<class name="Catheter"/>
<class name="Procedure"/>
<class name="CatheterInsertion" extends="Procedure"/>
<class name="Angioptasty" extends="CatheterInsertion"/>
<relationship name="usedIn" firstArg="what" secondArg="where"/>
<usedIn what="Catheterv where="Catheterlnsertion"/>
```

Before even considering the merits of the above approach, we can see that it has some immediate shortcomings along the lines of the shortcomings of XML that were described above. A multiplicity of such ontology representations is possible. Unless we nail down the representation, we cannot exchange the ontologies and we cannot create tools that would process such representations. For this reason, it became clear that a standardized representation was needed in which ontologies could be captured and which lay above the level of XML.

29.5.1 RDF and RDF Schema

The first such representation is the Resource Description Framework (RDF), which provides a standard approach to express graphs [Decker et al., 2000]. RDF is not exclusively tied to XML, but can be rendered into a variety of concrete syntaxes, including a standard syntax that is based on XML. An RDF specification describes a simple information model, and corresponds to a particular knowledge representation that the given model can have. RDF is general enough to capture any knowledge representations that correspond to graphs. For example, RDF versions of taxonomies such as conventionally used in software class diagrams can be readily built.

The basic concept of RDF is that it enables one to encode *statements.* Following simplified natural language, each statement has an object, a subject, and a predicate. For example, given the English sentence, "catheters are used in angioplasty," we would say that "catheters" is the subject, "angioplasty" is the object, and "used" is the predicate. This analysis is quite naive in that we do not consider any of the subtleties of natural language, such as tense or active vs. passive voice. However, this analysis provides an elegant starting point for computational representation.

RDF enables us to encode information in the form of *triples,* each of which consists of a *subject,* an *object,* and a *property* or predicate. Subjects of RDF statements must be *resources,* i.e., entities. These entities must have an identity given via a URI. As usual, the main point about URIs is that they are unique; there is no assumption that the URI corresponds to a physical network address or that the entity so named is a physical entity. The objects of RDF statements must be resources or literals, which are based on general data types such as integers. If they are resources, they can be described further by making them subjects of other statements; if they are literals, the description would bottom out with them. Properties must also be resources. By linking statements via the subjects or objects that are common to them we can construct general graphs easily. The vertices of a graph would correspond to resources that feature as subjects or objects. The edges of a graph would correspond to statements — with the origin of an edge being the subject, the target of the edge being the object, and the label of the edge being the property.

Because each RDF statement has exactly three parts — its subject, object, and property — it is essential to have some additional mechanisms by which more complex structures can be encoded in RDF. One of them is the use of certain *containers.* RDF defines Bag (unordered collection with duplicates allowed), Seq (ordered collection also with duplicates allowed), and Alt (disjoint union). Members of containers are asserted via rdf:li. RDF schematically defines properties to indicate membership in the containers. These are written _1, _2, and so on.

The second important mechanism in RDF is *reification.* That is, RDF statements can be reified, meaning that they can be referred to from other statements. In other words, statements can be treated as resources and can be subjects or objects of other statements. Thus complex graphs can be readily encoded in RDF. For example, we can encode a statement that asserts (as above) that angioplasty is a kind of catheter insertion, and another statement which asserts that the first statement is false, and yet another statement which asserts that the first statement is true except for neonatal patients. The essence of this example is that when statements are reified, we can assert further properties of them in a natural manner.

The following is a description of our example ontology in RDF. Using XML namespaces, RDF is associated with a standard namespace in which several general RDF terms are defined. By convention, the namespace prefix rdf is used to identify this namespace. The primitives in rdf include rdf:type, which is a property that states the class of a given resource. To support reification, RDF includes a type called rdf:Statement; all statements are resources and are of rdf:type rdf:Statement. Each rdf:Statement has three main properties, namely, rdf:subject, rdf:object and rdf:predicate, which can be thought of as acecssor functions for the three components of a statement. Other useful primitives are rdf:Description, and attributes rdf:resource, rdf:ID and rdf:about. The primitive rdf:Description creates the main element about which additional properties are stated.

However, RDF, too, leaves several key terms to be defined by those who use RDF. For example, one person may build an RDF model of a taxonomy using a term subCategory, and another person may use the term subset for the same purpose. When such arbitrary terms are selected by each modeler, the models cannot be related, compared, or merged without human intervention. To prevent this problem, the RDF schema language (RDFS) specifies a canonical set of terms using which simple taxonomies can be unambiguously defined.

In more general terms, RDFS is best understood as a system for defining application-specific RDF vocabularies. RDFS defines a standard set of predicates that enable simple semantic relationships to be captured by interpreting vertices as classes and edges as relationships. RDFS standardizes a namespace, which is conventionally abbreviated as rdfs.

In simple terms, RDFS defines primitives that build on rdf and impose standard interpretations. The main primitives include rdfs:Class (a set of instances, each of which has rdf:type equal to the given class); rdfs:subClassOf (a property indicating that instances of its subject class are also instances of its object class, thereby defining the taxonomy); and rdfs:Resource, which is the set of resources. Next, for properties, RDFS includes rdf:Property, the class of properties, which is defined as an instance of rdfs:Class and rdfs:subPropertyOf forms a taxonomy over properties. The properties rdfs:range and rdfs:domain apply to properties and take classes as objects. Multiple domains and ranges allowed for a single property and are interpreted as conjunctions of the given domains and ranges. Further, rdfs:Resource is an instance of rdfs:Class; rdfs:Literal is the class of literals, i.e., strings and integers; rdfs:Datatype is the class of data types; each data type is a subclass of rdfs:Literal.

The following listing gives an RDF rendition of the above example ontology. This formulation uses RDF Schema primitives introduced above, so it is a standard formulation in that respect.

```
<?xml version='1.0'?>
<rdf:RDF
      xmlns:rdf="http://www.w3.org/1999/02/22-rdf-syntax -ns#"
      xmlns:rdfs="http://www.w3.org/2000/01/rdf-schema#">
<rdfs:Class rdf:about="Catheter"/>
<rdfs:Class rdf:about="Procedure"/>
<rdfs:Class rdf:about="CatheterInsertion">
      <rdfs:subClassOf rdf:resource="#Procedure"/>
</rdfs:Class>
<rdfs:Class rdf:about="Angioplasty">
      <rdfs:subClassOf rdf:resource="#CatheterInsertion"/>
</rdfs:Class>
<rdfs:Property rdf:about="usedIn">
      <rdfs:domain rdf:resource="#Catheterv/>
      <rdfs:range rdf:resource="#CatheterInsertion"/>
</rdfs:Property>
```

29.5.2 OWL

Although RDFS provides the key primitives with which to capture taxonomies, there are often several other kinds of refinements of meaning that must be specified to enable the unambiguous capture of knowledge about information resources. These are captured through the more complete ontology languages, the leading one of which is the Web Ontology Language (OWL) [McGuiness and van Harmelen, 2003]. Specifications in OWL involve classes and properties that are defined in terms of various constraints and for which the taxonomic structure (e.g., the subclass relationship) can be inferred from the stated definitions. Such specifications include constraints on cardinality and participation that are lacking from RDFS. Of course, such constraints could be syntactically encoded in XML (as, indeed, they are). However, what OWL does, in addition, is give them a standard interpretation. Any compliant implementation of OWL is then required to process such specifications in the standard manner.

OWL is defined as a set of three dialects of increasing expressivity and named OWL Lite, OWL DL (for description logic), and OWL Full, respectively. In this introduction, we will simply describe the main features of OWL without regard to the dialect.

OWL includes the class and property primitives derived from RDF and RDF Schema. These are enhanced with a number of new primitives, For classes, we have equivalenceClass and disjointWith, among others. Booleans such as intersection, union, and complementation can also be asserted. For properties, we can declare properties as transitive and symmetric and whether they are functional, meaning that each domain element is mapped to at most range element. Cardinality restrictions can also be stated about the properties. The most interesting are the property type restrictions. The primitive allValuesFrom restricts a property as applied on a class (which should be a subclass of its domain) to take values only from a specified class (which should be a subclass of its range). The primitive some-ValuesFrom works analogously.

The following listing gives an OWL representation for our example ontology. For most of its entries, this listing resembles the RDF Schema version given above. However, it leads to more expressiveness when we consider the cardinality and the property restrictions.

```
<?xml version="1.0" ?>
<rdf:RDF
      xmlns:owl ="http://www.w3.org/2002/07/owl#"
      xmlns:rdf ="http://www.w3.org/1999/02/22-rdf--syntax-ns#"
      xmlns:rdfs="http://www.w3.org/2000/01/rdf-schema#">

<owl:Ontology rdf:about="Surgery">
      <owl:versionInfo>$ld: Surgery.owl,v1.0 2003/11/01$
      </owl:versionlnfo>
      <rdfs:comment>An Ontology for Surgery</rdfs:comment>
</owl:Ontology>

<owl:Class rdf:about="Catheter">
      <rdfs:label>Catheter</rdfs:label>
</owl:Class>

<owl:Class rdf:about="Procedure">
      <rdfs:label>Procedure</rdfs:label>
</owl:Class>

<owl:Class rdf:about="CatheterInsertion">
      <rdfs:label>Catheter Insertion</rdfs:label>
      <rdfs:subClassOf rdf:resource="#Procedure"/>
</owl:Class>

<owl:Class rdf:about="Angioplasty">
      <rdfs:label>Angioplasty</rdfs:label>
      <rdfs:subClassOf rdf:resource="#CatheterInsertion"/>
</owl:Class>

<owl:ObjectProperty rdf:about="usedIn">
      <rdfs:domain rdf:resource="#Catheter"/>
      <rdfs:range rdf:resource="#Catheterlnsertion"/>
</owl:ObjectProperty>

</owl:Ontology>
```

The above is largely self-explanatory. However, OWL enables further structure. The simplest enhancement is to assert constraints about cardinality, an example of which follows.

```
<owl:ObjectProperty rdf:about="cathetersUsed">
      <owl:inverseOf rdf:resource=#"usedIn"/>
      <owl:minCardinality>1</owl:minCardinality>
</owl:ObjectProperty>
```

```
</owl:Ontology>
```

The power of description logics is most apparent when we capture more subtle kinds of ontological constraints. Let us consider the example about venal catheter insertion, which is defined as catheter insertion that operates on a vein. To illustrate this, we also define arterial catheter insertion. Also note that veins and arteries are defined to be disjoint kinds of blood vessels. Now a description logic reasoner can infer that venal catheter insertion must be disjoint with arterial catheter insertion.

```
<owl:Class rdf:about="Bloodvessel">
      <rdfs:label>Blood Vessel</rdfs:label>
</owl:Class>

<owl:Class rdf:about="Arteryv>
      <rdfs:subClassOf rdf:resource="#BloodVessel"/>
</owl:Class>

<owl:Class rdf:about="Vein">
      <rdfs:subClassOf rdf:resource="#BloodVessel"/>
      <owl.disjointWith rdf:resource="#Artery"/>
</owl:Class>

<owl:ObjectProperty rdf:about="intoVessel">
      <rdfs:domain rdf:resource="#CatheterInsertion"/>
      <rdfs:range rdf:resource="#Vessel"/>
</owl:ObjectProperty>

<owl:Class rdf:about="VenalCatheterInsertion">
      <rdfs:subClassOf rdf:resource="#CatheterInsertion"/>
      <owl:Restriction>
            <owl:onProperty rdf:resourec="intoVessel"/>
            <owl:allValuesFrom rdf:resource="#Vein"/>
      </owl:Restriction>
</owl:Class>

<owl:Class rdf:about="ArterialCatheterInsertion">
      <rdfs:subClassOf rdf:resource="#CatheterInsertion"/>
      <owl:Restriction>
            <owl:onProperty rdf:resource="#intoVessel"/>
            <owl:allValuesFrom rdf:resource="#Artery"/>
      </owl:Restriction>
</owl:Class>
```

29.6 Discussion

The above development was about the representation of domain knowledge in standard representations. In some cases, an ontology that is agreed upon by interacting parties would be sufficient for them to work together. For example, an ontology of surgical components might form the basis of a surgical supplies catalog and the basis for billing for medical components and procedures. In many other cases, such an ontology would merely be the starting point for capturing additional application-specific knowledge. Such knowledge would be used to mark up information resources in a manner that can be comprehended and processed by others across the Web.

A number of tools now exist that offer a variety of key functionality for developing ontologies. This is, however, a fast-changing area. Jena is an open-source RDF and RDPS toolkit, which includes support for querying and inference over RDF knowledge bases [McBride, 2002]. Protégé [Protégé, 2000] and OilEd [Bechhofer et al., 2001] are well-known tools for building and maintaining ontologies. Ontology tools can require fairly sophisticated reasoning, which can easily prove intractable. Although the formal foundations are well-understood now, work is still ongoing to identify useful sublanguages that are tractable. The work of Horrocks and colleagues is an important contribution in this regard, e.g., [Horrocks et al., 1999].

29.6.1 Domain-Specific Ontologies

OWL gives us the essence of Web semantics, at least as far as the basic information is concerned. However, a lot of work based on OWL is required so as to make ontologies practical. The first tasks deal with specifying standardized ontologies for different application domains. These enable narrower knowledge models that are built on top of them to be comprehended by the various parties, For example, a medical (surgery) ontology would define concepts such as "incision" and "catheter" in a manner that is acceptable to surgeons given their current practice. Thus, someone could use that ontology to describe a particular cardiac procedure in which an incision is made and a catheter inserted into an artery. Then another surgery tool would be able to interpret the new cardiac procedure or at least to relate it to existing cardiac procedures.

Clearly, these are details that are specific to the given application domain. Equally clearly, if there were no agreement about these terms (say, among surgeons), it would be difficult for the given procedures to be understood the same way (even if everyone used OWL). Understanding the key terms is crucial so that their meanings can be combined — or, more precisely, so that if the meanings are automatically combined, then the result is comprehensible and sensible to the participants in the domain. A challenge that large-scale vocabularies open up is ensuring their quality [Schulz et al., 1998]. Ideally a vocabulary would have no more than one term for the same concept, but even such a simple constraint can be difficult to enforce.

29.6.2 Semantic Web Services and Processes

Another natural extension to the above deals with information processing. In other words, whereas the above deals with information as it is represented on the Web, we must supplement it with representations of how the information may be modified. Such representations would enable the processes to interoperate and thereby lead to superior distributed processes.

Semantic Web Services apply the techniques of Web semantics to the modeling and execution of Web services. In simple terms, semantic Web services approaches apply semantics in the following ways:

- Modeling the information that is input to or output from Web services
- Describing ontologies for the nonfunctional attributes of Web services, such as their performance, reliability, and the quality of their results, among others
- Describing ontologies that capture the process structure of Web services and conversations that they support

The above lead into formalizations of more general models of processes and protocols as well as of contract languages and policies, e.g., Grosof and Poon [2003]. The motivation for these formalization is similar to that for domain-specific ontologies in that they streamline the design of tools and applications that deal with information processing.

29.6.3 Methodologies and Tools

Ontology development is a major challenge. It is sometimes said to suffer from the "*two Ph.D.s*" problem meaning that a knowledge modeler must be a specialist both in the domain of interest and in the knowledge modeling profession. Consequently, a lot of the work on ontologies is about making this task simpler and scalable.

A conventional view is that knowledge modeling is an effort that is separate from and precedes knowledge use. Clearly, when an ontology exists in a given domain, it should be used. However, an existing ontology would often not be adequate. In such a case, it would need to be extended while it is being used. We suggest that such extensions would and should be attempted primarily on demand; otherwise, the modeling effort will be unmotivated and expensive and will only tend to be put off.

Ontology management has drawn a lot of attention in the literature. Ontology management refers to a number of functionalities related to creating, updating, maintaining, and versioning ontologies [Klein, 2001]. These challenges have been around since the earliest days of ontologies and have been especially well-studied in the context of clinical terminology development. For example, the Galapagos project considered the challenges in merging terminologies developed in a distributed manner [Campbell et al., 1996]. Galapagos encountered the challenges of version management through locking, as well as the semantic problems of resolving distinctions across terminologies.

The key aspects of ontology management studied in the recent literature include how mismatches occur among ontologies — language, concept, paradigm (e.g., how time is modeled by different people), and so on — and how ontology versions are created and maintained. This is especially important because distributed development over the Web is now the norm [Heflin and Hendler, 2000]. A recent evaluation of existing tools by Das et al. [2001] reveals that while the knowledge representation capabilities of these tools surpass the requirements, their versioning capabilities remain below par.

A number of heuristics have been defined to resolve ontology mismatches, e.g., see Noy and Musen [2002]. Potential matches are suggested to a user, who can decide if the matches are appropriate. Under some reasonable assumptions, the task can be made greatly tractable for tools for knowledge modeling.

- The various models are derived from a common model, which could have been proposed say by a standardization group in the domain of the given ontology. When the upper parts of the model are fixed, changes in the rest of it are easier to accommodate.
- Editing changes made to the models are available, so that several changes can be reapplied to the original model when a derived model is to be consolidated in it. If immutable internal identifiers are used for the terms, that facilitates applying the editing changes unambiguously.
- Simple granular locking can be applied whenever the tasks of developing a model are parceled out to members of a domain community. The people granted the locks have an advantage in that their suggested changes would propagate by default. Other users should restrict their changes in the components where there is a potential conflict. This is clearly pessimistic, but can still be effective in practice.

29.7 Summary

This chapter provided a brief introduction to Web semantics. The main take-away message is that Web semantics is here to stay. Further, Web semantics pervades Internet computing. It applies not only to traditional Web applications such as Web browsing but to all aspects of information management over the Web. Web semantics promises not only improved functionality to users but also enhanced productivity for programmers and others who manage information resources. However, important challenges remain, the handling of which will determine how quickly Web semantics propagates into applications of broad appeal, but already efforts are under way to address those challenges.

References

Arroyo, Sinuhe, Ruben Lara, Juan Miguel Gomez, David Berka, Ying Ding, and Dieter Fensel. Semantic aspects of Web services. In *The Practical Handbook of Internet Computing,* Munindar P. Singh, Ed., CRC Press, Boca Raton, FL, 2005.

Bechhofer, Sean, Ian Horrocks, Carole Goble, and Robert Stevens. OilEd: A reasonable ontology editor for the semantic Web. In *Proceedings of KI-2001, Joint German/Austrian Conference on Artificial Intelligence,* volume 2174 of *Lecture Notes in Computer Science,* pages 396–408. Springer-Verlag, Berlin, September 2001.

Brusilovsky, Peter and Wolfgang Nejdl. Adaptive hypermedia and adaptive Web. In *The Practical Handbook of Internet Computing,* Vol. 2 Munindar P. Singh, Ed., CRC Press, Boca Raton, FL, 2005.

Campbell, Keith E., Simon P. Cohn, Christopher G. Chute, Glenn D. Rennels, and Edward H. Shortliffe. Gálapagos: Computer-based support for evolution of a convergent medical terminology. In *Proceedings of the AMIA Annual Fall Symposium,* Washington, D.C., pages 269–273, 1996.

Das, Aseem, Wei Wu, and Deborah L. McGuinness. Industrial strength ontology management. In *Proceedings of the International Semantic Web Working Symposium (SWWS),* Palo Alto, CA, pages 17–37, 2001.

Decker, Stefan, Prasenjit Mitra, and Sergey Melnik. Framework for the semantic Web: An RDF tutorial. *IEEE Internet Computing,* 4(6): 68–73, November 2000.

Fisher, Mark and Amit Sheth. Semantic enterprise content management. In *The Practical Handbook of Internet Computing,* Vol. 2 Muninder P. Singh, Ed., CRC Press, Boca Raton, FL, in press.

Grosof, Benjamin N. and Terrence C. Poon. SweetDeal: Representing agent contracts with exceptions using XML rules, ontologies, and process descriptions. In *Proceedings of the 12th International Conference on the World Wide Web,* Budapest, Hungary, pages 340–349, 2003.

Heflin, Jeff and James A. Hendler. Dynamic ontologies on the Web. In *Proceedings of American Association for Artificial Intelligence Conference (AAAI),* Cape Cod, MA, pages 443–449, 2000.

Horrocks, Ian, Ulrike Sattler, and Stephan Tobies. Practical reasoning for expressive description logics. In *Proceedings of the 6th International Conference an Logic for Programming and Automated Reasoning (LPAR),* Tbilisi, pages 161–180, 1999.

Kashyap, Vipul. Information modeling on the Web. In *The Practical Handbook of Internet Computing,* Vol. 2 Munindar P. Singh, Ed., CRC Press, Boca Raton, FL, 2005.

Klein, Michel. Combining and relating ontologies: An analysis of problems and solutions. In *Proceedings of the IJCAI Workshop on Ontologies and Information Sharing,* Seattle, WA, 2001.

McBride, Brian. Jena: A semantic Web toolkit. *IEEE Internet Computing,* 6(6): 55–59, November 2002.

McGuiness, Deborah L. and Frank van Harmelen. Web Ontology Language (OWL): Overview. www.w3.org/TR/2003/WD-owl-features-20030210/, W3C working draft, February 2003.

Noy, Natalya Fridman and Mark A. Musen. PROMPTDIFF: A fixed-point algorithm for comparing ontology versions. In *Proceedings of the National Conference on Artificial Intelligence (AAAI),* Edmonton, Alberta, Canada, pages 744–750, 2002.

Protégé. The Protégé ontology editor and knowledge acquistion system. http://protege.stanford.edu, 2000.

Schulz, Erich B., James W. Barrett, and Colin Price. Read code quality assurance: From simple syntax to semantic stability. *Journal of the American Medical Informatics Association,* 5: 337–346,1998,

Wilde, Erik. XML core technologies. In *The Practical Handbook of Internet Computing,* Vol. 2 Munindar P. Singh, Ed., CRC Press, Boca Raton, FL, 2005.

30

Information Modeling on the Web: The Role of Metadata, Semantics, and Ontologies

CONTENTS

30.1 Introduction 30-2
30.2 What is Metadata? 30-3
30.2.1 Metadata Usage in Various Applications 30-3
30.2.2 Metadata: A Means for Modeling Information 30-4
30.3 Metadata Expressions: Modeling Information Content 30-6
30.3.1 The InfoHarness System: Metadata-Based Object Model for Digital Content 30-7
30.3.2 Metadata-Based Logical Semantic Webs 30-10
30.3.3 Modeling Languages and Markup Standards 30-13
30.4 Ontology: Vocabularies and Reference Terms for Metadata 30-14
30.4.1 Terminological Commitments: Constructing an Ontology 30-14
30.4.2 Controlled Vocabulary for Digital Media 30-16
30.4.3 Ontology-Guided Metadata Extraction 30-17
30.4.4 Medical Vocabularies and Terminologies: The UMLS Project 30-17
30.4.5 Expanding Terminological Commitments across Multiple Ontologies 30-20
30.5 Conclusions 30-20
References 30-21

Vipul Kashyap

The Web consists of huge amounts of data available in a variety of digital forms stored in thousands of repositories. Approaches that use semantics of information captured in metadata extracted from the data underlying are being viewed as an appealing approach, especially in the context of the *Semantic Web* effort.

We present in this chapter a discussion on approaches adopted for metadata-based information modeling on the web. Various types of metadata developed by researchers for different media are reviewed and classified with respect to the extent they model data or information content. The reference terms and ontology used in the metadata are classified with respect to their dependence on the application domain. We discuss approaches for using metadata to represent the context of the information request, the interrelationships between various pieces of data, and for their exploitation for search, browsing, and querying the information. Issues related to the use of terminologies and ontologies, such as establishing and maintaining terminological commitments, and their role in metadata design and extraction are also discussed.

1-58488-381-2/05/$0.00+$1.50
© 2005 by CRC Press LLC

Modeling languages and formats, including the most recent ones, such as the Resource Description Framework (RDF) and the DARPA Agent Markup Language (DAML+OIL) are also discussed in this context.

30.1 Introduction

The World Wide Web [Berners-Lee et al., 1992] consists of huge amounts of digital data in a variety of structured, unstructured (e.g., image) and sequential (e.g., audio, video) formats that are either stored as Web data directly manipulated by Web servers, or retrieved from underlying database and content management systems and served as dynamically generated Web content. Whereas content management systems support creation, storage, and access functions in the context of the content managed by them, there is a need to support correlation across different types of digital formats in *media-independent, content-based* manner.

Information relevant to a user or application need may be stored in multiple forms (e.g., structured data, text, image, audio and video) in different repositories and Websites. Responding to a user's information request typically requires correlation of such information across multiple forms and representations. There is a need for association of various pieces of data either by preanalysis by software programs or dynamic correlation of information in response to an information request. Common to both the approaches is the ability to describe the *semantics* of the information represented by the underlying data.

The use of semantic information to support correlation of heterogeneous representations of information is one of the aims of the current *Semantic Web* effort [Berners-Lee et al., 2001]. This capability of modeling information at a semantic level both across different types of structured data (e.g., in data warehouses) and across different types of multimedia content, is missing on the current Web and has been referred to as the "semantic bottleneck" [Jain, 1994]. Machine-understandable metadata and standardized representations thereof form the foundation of the Semantic Web. The *Resource Description Framework (RDF)* [Lassila and Swick] and *XML* [Bray et al.] based specifications are currently being developed in an effort to standardize the formats for representing metadata. It is proposed that the vocabulary terms used to create the metadata will be chosen from third-party ontologies available from the Web. Standardized specifications for representing ontologies include XML and RDF schemas, *DARPA Agent Markup Language* (DAML+OIL) [DAML+OIL] and the *Web Ontology Language (OWL)* [OWL].

In this chapter, we present issues related to the use of *metadata, semantics,* and *ontologies* for modeling information on the Web organized in a three-level framework (Figure 30.1):

- The middle level represents the **metadata** component involving the use of metadata descriptions to capture the information content of data stored in Websites and repositories. Intensional descriptions constructed from metadata are used to abstract from the structure and organization of data and specify relationships across pieces of interest.

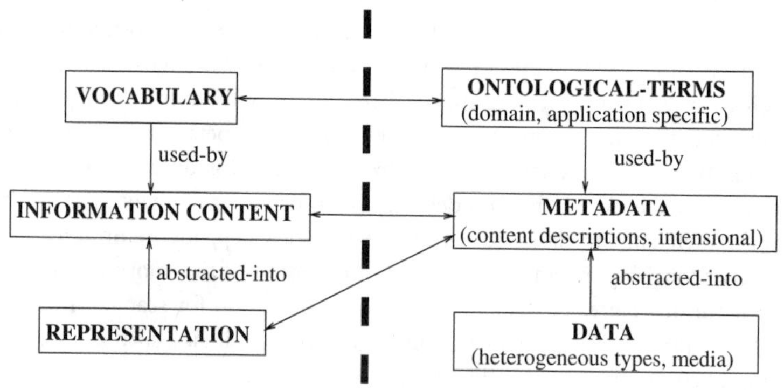

FIGURE 30.1 Key issues for information modeling.

- The top level represents the **ontology** component, involving terms (concepts, roles) in domain-specific ontologies used to characterize metadata descriptions. These terms capture pieces of domain knowledge that describe relationships between data items (via association with the terms) across multiple repositories, enabling semantic interoperability.

The organization of this chapter is as follows. In Section 30.2, we discuss a definition of metadata, with various examples. A classification of metadata based on the information content they capture is presented along with its role in modeling information. In Section 30.3, we discuss how metadata expressions can be used to model interrelationships between various pieces of information within a dataset and across multiple datasets. We also present an account of various modeling and markup languages that may be used to model the information represented in the data. Finally, in Section 30.4, we present issues related to the use of reference terms and ontological concepts for creating metadata descriptions. Section 30.5 presents the conclusions.

30.2 What is Metadata?

Metadata in its most general sense is defined as data or information about data. For structured databases, the most common example of metadata is the schema of the database. However, with the proliferation of various types of multimedia data on the Web, we shall refer to an expanded notion of metadata, of which the schema of structured databases is a (small) part. Metadata may be used to store derived properties of media useful in information access or retrieval. They may describe, or be a summary of the information content of the data described in an intensional manner. They may also be used to represent properties of, or relationships between individual objects of heterogeneous types and media. Figure 30.1 illustrates the components for modeling information on the Web. Metadata is the pivotal idea on which both the (ontology and metadata) components depend. The function of the metadata descriptions is twofold:

- To enable the abstraction of representational details such as the format and organization of data and capture the information content of the underlying data independent of representational details. These expressions may be used to represent useful relationships between various pieces of data within a repository or Website.
- To enable representation of domain knowledge describing the information domain to which the underlying data belongs. This knowledge may then be used to make inferences about the underlying data to determine the *relevance* and identify relationships across data stored in different repositories and Websites.

We now discuss issues related to metadata from two different perspectives identified in Boll et al. [1998], such as the usage of metadata in various applications and the information content captured by the metadata.

30.2.1 Metadata Usage in Various Applications

We discuss a set of application scenarios that require functionality for manipulation and retrieval of digital content that are relevant to the Web. The role of metadata, especially in the context of modeling information to support this functionality, is discussed.

30.2.1.1 Navigation, Browsing, and Retrieval from Image Collections

An increasing number of applications, such as those in healthcare, maintain large collections of images. There is a need for semantic content based navigation, browsing, and retrieval of images. An important issue is to associate a user's semantic impression with images, e.g., image of a brain tumor. This requires knowledge of spatial content of the image and the way it changes or evolves over time, which can be represented as metadata annotations.

30.2.1.2 Video

In many applications relevant to news agencies, there exist collections of video footage which need to be searched based on semantic content, e.g., videos containing field goals in a soccer game. This gives rise to the same set of issues as described above, such as the change in the spatial positions of various objects in the video images (spatial evolution). However, there is a temporal aspect to videos that was not captured above. Sophisticated time-stamp based schemes can be represented as a part of the metadata annotations.

30.2.1.3 Audio and Speech

Radio stations collect many, if not all, of their important and informative programs, such as radio news, in archives. Parts of such programs are often reused in other radio broadcasts. However, to efficiently retrieve parts of radio programs, it is necessary to have the right metadata generated from, and associated with, the audio recordings. An important issue here is capturing in text the essence of the audio, in which vocabulary plays a central role. Domain-specific vocabularies can drive the metadata extraction process, making it more efficient.

30.2.1.4 Structured Document Management

As the publishing paradigm is shifting from popular desktop publishing to database-driven Web-based publishing, processing of structured documents becomes more and more important. Particular document information models, such as SGML [SGML] and XML, introduce structure and content-based metadata. Efficient retrieval is achieved by exploiting document structure, as the metadata can be used for indexing, which is essential for quick response times. Thus, queries asking for documents with a title containing "Computer Science" can be easily optimized.

30.2.1.5 Geographic and Environmental Information Systems

These systems have a wide variety of users that have very specific information needs. Information integration is a key requirement, which is supported by provision of descriptive information to end users and information systems. This involves issues of capturing descriptions as metadata and reconciling the different vocabularies used by the different information systems in interpreting the descriptions.

30.2.1.6 Digital Libraries

Digital libraries offer a wide range of services and collections of digital documents, and constitute a challenging application area for the development and implementation of metadata frameworks. These frameworks are geared towards description of collections of digital materials such as text documents, spatially referenced datasets, audio, and video. Some frameworks follow the traditional library paradigm with metadata-like subject headings [Nelson et al., 2001] and thesauri [Lindbergh et al., 1993].

30.2.1.7 Mixed-Media Access

This is an approach that allows queries to be specified independent of the underlying media types. Data corresponding to the query may be retrieved from different media such as text and images and "fused" appropriately before being presented to the user. Symbolic metadata descriptions may be used to describe information from different media types in a uniform manner.

30.2.2 Metadata: A Means for Modeling Information

We now characterize different types of metadata based on the amount of information content they capture and present a classification of metadata types used by various researchers (Table 30.1).

30.2.2.1 Content Independent Metadata

This type of metadata captures information that does not depend on the content of the document with which it is associated. Examples of this type of metadata are location, modification date of a document, and type of sensor used to record a photographic image. There is no information content captured by these metadata, but these might still be useful for retrieval of documents from their actual physical

TABLE 30.1 Metadata for Digital Media

Metadata	Media/Metadata type
Q-Features	Image, video/Domain specific
R-Features	Image, video/Domain independent
Impression vector	Image/Content descriptive
NDVI, Spatial registration	Image/Domain specific
Speech feature index	Audio/Direct content-based
Topic change indices	Audio/Direct content-based
Document vectors	Text/Direct content-based
Inverted indices	Text/Direct content-based
Content classification metadata	Multimedia/Domain specific
Document composition metadata	Multimedia/Domain independent
Metadata templates	Media independent/Domain specific
Land-cover, relief	Media independent/Domain specific
Parent–child relationships	Text/Domain independent
Contexts	Structured databases/Domain specific
Concepts from Cyc	Structured databases/Domain specific
User's data attributes	Text, Structured databases/Domain specific
Medical subject headings	Text databases/Domain specific
Domain-specific ontologies	Media independent/Domain specific

locations and for checking whether the information is current or not. This type of metadata helps to encapsulate information into units of interest and organizes their representation within an object model.

30.2.2.2 Content Dependent Metadata

This type of metadata depends on the content of the document it is associated with. Examples of content-dependent metadata are size of a document, maxcolors, number of rows, and number of columns of an image. These type of metadata typically capture representational and structural information and provide support for browsing and navigation of the underlying data. Content-dependent metadata can be further subdivided as follows:

30.2.2.2.1 Direct Content-Based Metadata

This type of metadata is based directly on the contents of a document. A popular example of this is full-text indices based on the document text. Inverted tree and document vectors are examples of this type of metadata. *Media-specific metadata* such as color, shape, and texture are typically direct content-based metadata.

30.2.2.2.2 Content-Descriptive Metadata

This type of metadata describes information in a document without directly utilizing its contents. An example of this metadata is textual annotations describing the contents of an image. This metadata comes in two flavors:

30.2.2.2.2.1 Domain-Independent Metadata — These metadata capture information present in the document independent of the application or subject domain of the information, and are primarily structural in nature. They often form the basis of indexing the document collection to enable faster retrieval. Examples of these are C/C++ parse trees and HTML/SGML document type definitions. Indexing a document collection based on domain independent metadata may be used to improve retrieval efficiency.

30.2.2.2.2.2 Domain-Specific Metadata — Metadata of this type is described in a manner specific to the application or subject domain of the information. Issues of vocabulary become very important in this case, as the metadata terms have to be chosen in a domain-specific manner. This type of metadata, which helps abstract out representational details and capture information meaningful to a particular application or subject domain, is domain-specific metadata. Examples of such metadata are relief, land-cover from the geographical information domain, and medical subject headings (MeSH) from the medical domain.

In the case of structured data, the database schema is an example of domain-specific metadata, which can be further categorized as:

Intra-Domain-Specific Metadata These type of metadata capture relationships and associations between data within the context of the same information domain. For example, the relationship between the CEO and his corporation is captured within a common information domain, such as the business domain.

Inter-Domain-Specific Metadata These type of metadata capture relationships and associations between data across information domains. For example, the relationship between (medical) instrument and (legal) instrument spans across the medical and legal information domains.

30.2.2.2.2.3 Vocabulary for Information Content Characterization — Domain-specific metadata can be constructed from terms in a controlled vocabulary of terms and concepts, e.g., the biomedical vocabularies available in the Unified Medical Language System (UMLS) [Lindbergh et al., 1993], or a domain-specific ontology, describing information in an application or subject domain. Thus, we view ontologies as metadata, which themselves can be viewed as a vocabulary of terms for construction of more domain-specific metadata descriptions.

30.2.2.2.2.4 Crisp vs. Fuzzy Metadata — This is an orthogonal dimension for categorization. Some of the metadata referred to above are fuzzy in nature and are modeled using statistical methods, e.g., document vectors. On the other hand, other metadata annotations might be of a crisp nature, e.g., author name.

In Table 30.1 we have surveyed different types of metadata used by various researchers. Q-Features and R-Features were used for modeling image and video data [Jain and Hampapur, 1994]. Impression vectors were generated from text descriptions of images [Kiyoki et al., 1994]. NDVI and spatial registration metadata were used to model geospatial maps, primarily of different types of vegetation [Anderson and Stonebraker, 1994]. Interesting examples of mixed media access are the speech feature index [Glavitsch et al., 1994] and topic change indices [Chen et al., 1994]. Metadata capturing information about documents are document vectors [Deerwester et al., 1990], inverted indices [Kahle and Medlar, 1991], document classification and composition metadata [Bohm and Rakow, 1994], and parent–child relationships (based on document structure) [Shklar et al., 1995c]. Metadata Templates [Ordille and Miller, 1993] have been used for information resource discovery. Semantic metadata such as contexts [Sciore et al., 1992; Kashyap and Sheth, 1994], land-cover, relief [Sheth and Kashyap, 1996], Cyc concepts [Collet et al., 1991], concepts from domain ontologies [Mena et al., 1996] have been constructed from well-defined and standardized vocabularies and ontologies. Medical Subject headings [Nelson et al., 2001] are used to annotate biomedical research articles in MEDLINE [MEDLINE]. These are constructed from biomedical vocabularies available in the UMLS [Lindbergh et al., 1993]. An attempt at modeling user attributes is presented in Shoens et al. [1993]. The above discussion suggests that domain-specific metadata capture information which is more meaningful with respect to a specific application or a domain. The information captured by other types of metadata primarily reflect the format and organization of underlying data.

30.3 Metadata Expressions: Modeling Information Content

We presented in the previous section different types of metadata that capture information content to different extents. Metadata has been used by a wide variety of researchers in various contexts for different functionality relating to retrieval and manipulation of digital content. We now discuss approaches for combining metadata to create information models based on the underlying data. There are two broad approaches:

- Use of content- and domain-independent metadata to encapsulate digital content within an infrastructural object model.
- Use of domain-specific metadata to specify existing relationships within the same content collection or across collections.

We present both these approaches, followed by a brief survey of modeling and markup languages that have been used.

30.3.1 The InfoHarness System: Metadata-Based Object Model for Digital Content

We now discuss the InfoHarness [Shklar et al., 1994, 1995c, a, b; Sheth et al., 1995] system, which has been the basis of many successful research projects and commercial products. The main goal of InfoHarness is to provide uniform access to information independent of the formats, location, and organization of the information in the individual information sources. We discuss how content-independent metadata (e.g., type, location, access rights, owner, creation date, etc.) may be used to encapsulate the underlying data and media heterogeneity and represent information in an object model. We then discuss how the information spaces might be logically structured and discuss an approach for an interpreted modeling language.

30.3.1.1 Metadata for Encapsulation of Information

Representational details are abstracted out of the underlying data, and metadata is used to capture information content. This is achieved by encapsulation of the underlying data into units of interest called information units, and extraction of metadata describing information of interest. The object representation is illustrated in Figure 30.2 and is discussed below.

A metadata entity that is associated with the lowest level of granularity of information available to InfoHarness is called an *information unit* (IU). An IU may be associated with a file (e.g., a Unix man page or help file, a Usenet news item), a portion of a file (e.g,, a C function or a database table), a set of files (e.g., a collection of related bitmaps), or any request for the retrieval of data from an external source (e.g., a database query). An InfoHarness Object (IHO) may be one of the following:

1. A single information unit
2. A collection of InfoHarness objects (either indexed or nonindexed)
3. A single information unit and a nonindexed collection of InfoHarness objects

Each IHO has a unique object identifier that is recognized and maintained by the system. An IHO that encapsulates an IU contains information about the location of data, retrieval method, and any parameters needed by the method to extract the relevant portion of information. For example, an IHO associated with a C function will contain the path information for the .c file that contains the function, the name and location of the program that knows how to extract a function from a .c file, and the name of the function to be passed to this program as a parameter. In addition, each IHO may contain an arbitrary number of attribute-value pairs for attribute-based access to the information. An InfoHarness Repository (IHR) is a collection of IHOs. Each IHO (known as the *parent*) that encapsulates a collection of IHOs stores unique object identifiers of the members of the collection. We refer to these members as *children* of the IHO. IHOs that encapsulate indexed collections store information about the location of both the index and the query method.

30.3.1.2 Logical Structuring of the Information Space

We now discuss the various types of logical structure that can be imposed on the content in the context of the functionality enabled by such a structuring. This structuring is enabled by the extraction of the different kinds of metadata discussed above.

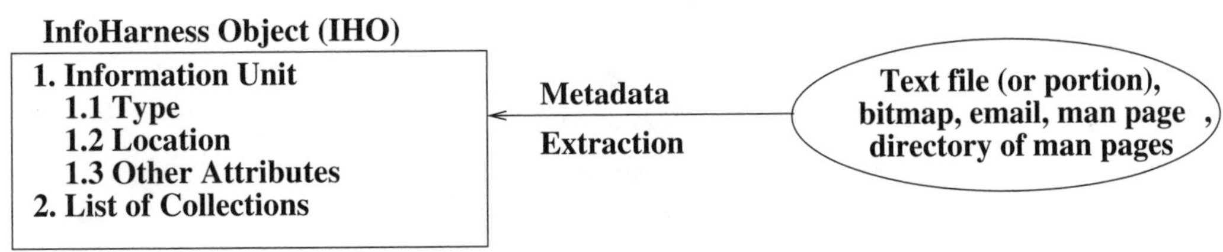

FIGURE 30.2 Metadata encapsulation in InfoHarness.

Consider the scenario illustrated in Figure 30.3. Case I depicts the actual physical distribution of the various types of documents required in a large software design project. The different documents are spread all over the file system as a result of different members of the project putting the files where they are deemed appropriate. Appropriate metadata extractors preprocess these documents and store important information like *type* and *location* and establish appropriate parent–child relationships. Case II illustrates the desired logical view seen by the user. Information can be browsed according to units of interest as opposed to browsing the information according to physical organization in the underlying data repositories.

A key capability enabled by the logical structuring is the ability to seamlessly plug in third-party indexing technologies to index document collections. This is illustrated in Figure 30.3, Case II, where the same set of documents is indexed using different third-party indexing technologies. Each of these document collections so indexed can be now queried using a *keyword-based query* without the user having to worry about the details of the underlying indexing technology.

Attribute-based access provides a powerful complementary or alternative search mechanism to traditional content-based search and access [Sheth et al., 1995]. While attribute-based access can provide better precision [Salton, 1989], it can be more complex as it requires that appropriate attributes have been identified and the corresponding metadata instantiated before accessing data.

In Figure 30.4 we illustrate an example of attribute-based access in InfoHarness. Attribute-based queries by the user result in SQL queries to the metadata repository and retrieval of the news items that satisfy the conditions specified. The advantages of attribute-based access are:

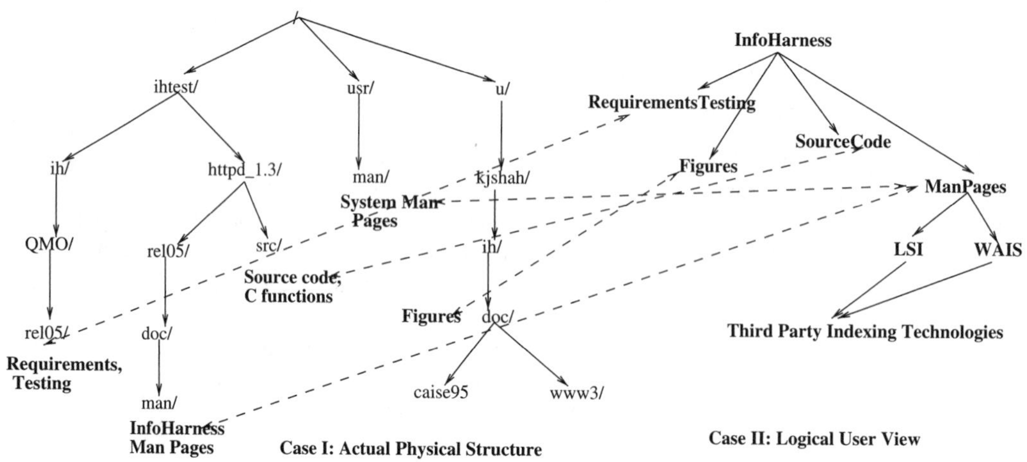

FIGURE 30.3 Logical structuring of the Information Space.

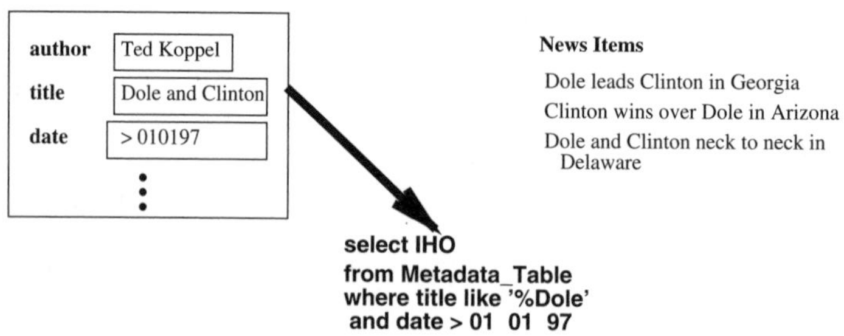

FIGURE 30.4 Attribute-based access in InfoHarness.

Enhance the semantics of the keywords. When a user presents a keyword (e.g., "Ted Koppel") as the value of an attribute (e.g., author), there are more constraints on the keyword as compared to when it appears by itself in a keyword-based query. This improves the precision of the answer.

Attributes can have associated types. The attribute submission date could have values of type date. Simple comparison operators (<, >, ,) can now be used for specifying constraints.

Querying content-independent information. One cannot query content-independent information such as modification date using keyword-based access as such information will never be available from the analysis of the content of the document.

30.3.1.3 IRDL: A Modeling Language for Generating the Object Model

The creation of an IHR amounts to the generation of metadata objects that represent IHOs and indexing of physical information encapsulated by members of indexed collections. The IHR can either be generated manually by writing metadata extractors or created automatically by interpreting IRDL (InfoHarness Repository Definition Language) statements. A detailed discussion of IRDL can be found in Shklar et al. [1995a], and its use in modeling heterogeneous information is discussed in Shklar et al. [1995b]. There are three main IRDL commands:

Encapsulate. This command takes as input information the *type* and *location* of physical data and returns a set of IHOs, each of which encapsulates a piece of data. Boundaries of these pieces are determined by the type. For example, in the case of e-mail, a set IHOs, each of which is associated with a separate mail message, is returned.

Group. This command generates an IHO associated with a collection and establishes parent–child relationships between the collection IHO and the member IHOs. In case a parameter indicating the indexing technology is specified, an index on the physical data associated with the member IHOs is created.

Merge. This command takes as input an IHO and associated references and creates a composite IHO.

We explain the model generation process by discussing an example for C programs (Figure 30.5). The steps that generate the model displayed in Figure 30.5, Case I, are as follows:

1. For each C file do the following:
   (a) Create simple IHOs that encapsulate individual functions that occur in this file.
   (b) Create a composite IHO that encapsulates the file and points to IHOs created in step 1.1.
2. Create an indexed collection of the composite IHOs created in step 1, using LSI for indexing physical data.

The IRDL statements that generate the model discussed above are:

```
BEGIN
     COLLTYPE LSI;
     DATATYPE TXT, C;
     VAR IHO: File_IHO, LSI_Collection;
     VAR SET IHO: File_IHO_SET, Function_IHO_SET;
     File_IHO_SET = ENCAP TXT "/usr/local/test/src";
     FORALL File_IHO IN File_IHO_SET
     {
          Function_IHO_SET = ENCAP C File_IHO;
          File_IHO = COMBINE IHO Function_IHO_SET;
          WRITE File_IHO, Function_IHO_SET;
     }
     LSI_Collection = INDEX LSI File_IHO_SET "/usr/local/db/c";
     WRITE LSI_Collection;
END
```

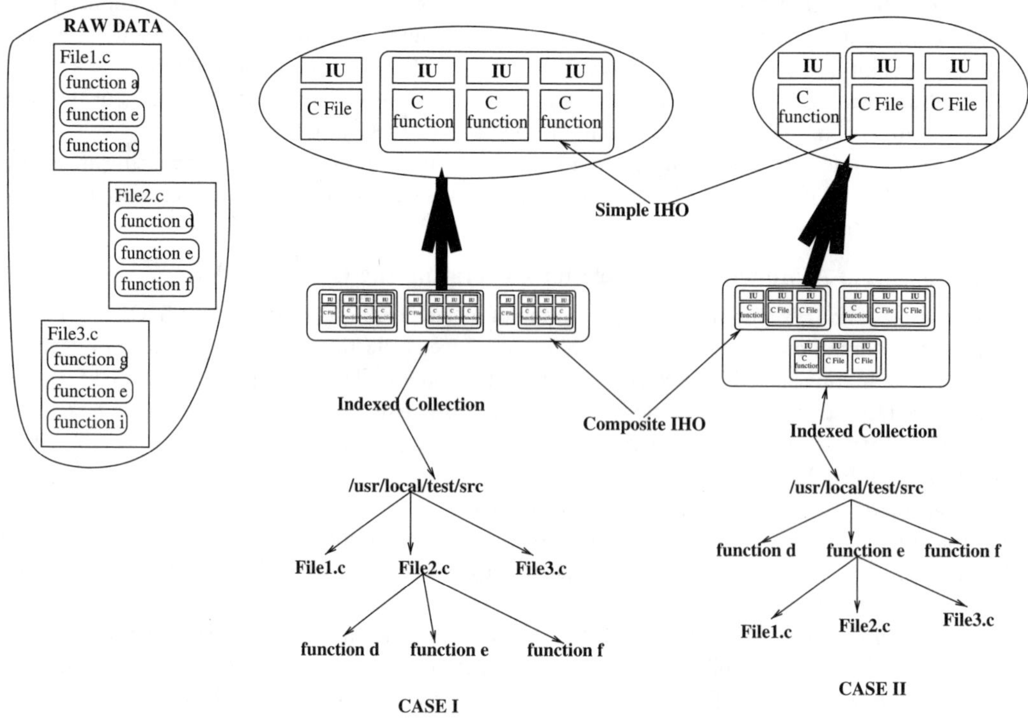

FIGURE 30.5 Object model generation for a C program.

This is another example of logical structuring using parent–child relationships to set up different logical views of the same underlying physical space. Case I (Figure 30.5) illustrates the case where a directory containing C code is viewed as a collection of C files, each of which is a collection of C functions. Case II (Figure 30.5), on the other hand, illustrates the case where the directory is viewed as a collection of C functions, each of which is a collection of the C files in which it appears.

30.3.2 Metadata-Based Logical Semantic Webs

The Web as it exists today is a graph of information artifacts and resources, where graph nodes are represented by embedded HREF tags. These tags enable linking of related (or unrelated) Web artifacts. This Web is very suitable for browsing but provides little or no direct help for searching, information gathering, or analysis. Web crawlers and search engines try to impose some sort of an order by building indices on top of Web artifacts, which are primarily textual. Thus a keyword query may be viewed as imposing a correlation (logical relationship) at a very basic (limited) level between the artifacts that make up the result set for that query. For example, let us say a search query "Bill Clinton" (Q) retrieves http://www.billclinton.com (Resource1) and http://www.whitehouse.gov/billclinton.html (Resource2). An interesting viewpoint would be that Resource1 and Resource2 are "correlated" with each other through the query Q. The above may be represented graphically with Resource1 and Resource2 represented as nodes linked by an edge labeled by the string corresponding to Q. Metadata is the key to this correlation. The keyword index (used to process keyword queries) may be conceptually viewed as content-dependent metadata, and the keywords in the query as specific resource descriptors for the index, the evaluation of which would result in a set of linked or correlated resources.

We discussed in the previous section the role played by metadata in encapsulating digital content into an object model. An approach using an interpreted modeling language for metadata extraction and generation of the object model was presented. We now present a discussion on how a metadata-based formalism, the metadata reference link (MREF) [Sheth and Kashyap, 1996; Shah and Sheth, 1998] can be used to enable semantic linking correlation, an important prerequisite for building logical Semantic Webs. MREF is a generalization of the <A HREF> construct used by the current Web to specify links and is defined as follows.

- <A MREF KEYWORDS=[keyword-list] THRESHOLD=[real] >Document Description< /A>
- <A MREF ATTRIBUTES([attr-value-pairdist])>Document Description< /A>

Different types of correlation are enabled based on the type of metadata that is used. We now present examples of correlation.

30.3.2.1 Content-Independent Correlation

This type of correlation arises when content-independent metadata (e.g., the location expressed as a URL) is used to establish the correlation. The correlation is typically media independent as content-independent metadata typically do not depend on media characteristics. In this case, the correlation is done by the designer of the document as illustrated in the following example:

```
<TITLE>A Scenic Sunset at Lake Tahoe</TITLE>
Lake Tahoe is a very popular tourist spot and
<A HREF="http://www1.server.edu/lake-tahoe.txt">some interesting
facts</A> are available here. The scenic beauty of Lake Tahoe can
be viewed in this photograph:
<center>
<IMG ALIGN=MIDDLE SRC="http://www2.server.edu/lake-tahoe.img">
</center>
```

The correlation is achieved by using physical links and without using any higher-level specification mechanism. This is predominantly the type of correlation found in the HTML documents on the World Wide Web [Berners-Lee et al., 1992].

30.3.2.2 Correlation Using Direct Content-Based Metadata

We present in the following text an example based on a query in Ogle and Stonebraker [1995] to demonstrate a correlation involving attribute-based metadata. One of the attributes is color, which is a *media-specific* attribute. Hence we view this interesting case of correlation as *media-specific* correlation.

```
<TITLE>Scenic waterfalls</TITLE>
Some interesting
<A MREF ATTRIBUTES (keyword="scenic waterfalls"; color="blue")>
information on scenic waterfalls</A> is available here.
```

30.3.2.3 Correlation Using Content-Descriptive Metadata

In Kiyoki et al. [1994], keywords are associated with images, and a full-text index is created on the keyword descriptions. Because the keywords describe the contents of an image, we consider these as *content-descriptive* metadata. Correlation can now be achieved by querying the collection of image documents and text documents using the same set of keywords as illustrated in this example:

```
<TITLE>Scenic Natural Sights</TITLE>
some interesting
<A MREF KEYWORDS="scenic waterfall mountain",THRESH=0.9>
information on Lake Tahoe</A> is available here.
```

This type of correlation is more meaningful than content-independent correlation. Also the user has more control over the correlation, as he or she may be allowed to change the thresholds and the keywords. The keywords used to describe the image are media independent and hence correlation is achieved in a media-independent manner.

30.3.2.4 Domain-Specific Correlation

To better handle the information overload on the fast-growing global information infrastructure, there needs to be support for correlation of information at a higher level of abstraction independent of the medium of representation of the information [Jain, 1994]. Domain-specific metadata, which is necessarily media independent, needs to be modeled. Let us consider the domain of a site location and planning application supported by a Geographic Information System and a correlation query illustrated in the following example:

```
<TITLE>Site Location and Planning</TITLE>
To identify potential locations for a future shopping mall, we present
below all regions having a population greater than 500 and area greater
than 50 sq feet having an urban land cover and moderate relief
<A MREF ATTRIBUTES(population &gt; 500; area &gt; 50; regiontype = block;
landcover = urban; relief = moderate)> can be viewed here</A>
```

The processing of the preceding query results in the structured information (area, population) and the snap of the regions satisfying the above constraints being included in the HTML document. The query processing system will have to map these attributes to image processing and other SQL-based routines to retrieve and present the results.

30.3.2.5 Example: RDF Representation of MREF

These notions of metadata-based modeling are fundamental to the notion of the emerging Semantic Web [Berners-Lee et al., 2001]. Semantic Web researchers have focused on markup languages for representing machine-understandable metadata. We now present a representation of the example listed above using the RDF markup language.

```
<HEAD>
<OBJECT declare id="mall-loc" type="application/x-mref"
          data-"<?namespace href="http://www.foo.com/SitePlanning"
as="SP"?>
          <?namespace href="http://www.w3.org/schemas/rdf-schema"
as="RDF"?>
               <RDF:serialization>
               <RDF:bag id="MREF:mall-loc>
<SP:attribute>
          <RDF:resource id="constraint_001">
     <SP:name>population</SP:color>
     <SP:type>number</SP:type>
<SP:operator>greater</SP:operator>
     <RDF:PropValue>500</RDF:PropValue>
          </RDF:resource>
          </SP:attribute>
               </RDF:bag>
                    </RDF:serialization>">
</OBJECT>
</HEAD>
<BODY>
To identify potential locations for a future shopping mall, all regions
having a population greater than 500 and area greater than 50 acres
having an urban land cover and moderate relief
<OBJECT classid="http://www.foo.con/sp.mref"
standby="Loading MREF..." data="#mall-loc">can be viewed here.</OBJECT>
</BODY>
```

30.3.3 Modeling Languages and Markup Standards

The concept of a simple, declarative language to support modeling is not new. Although modeling languages borrow from the classical hierarchical, relational, and network approaches, a number of them incorporate and extend the relational model. The languages examined in the following text may be categorized as:

Algebraic model formulation generators AMPL [Fourer et al.,1987], GAMS [Kendrick and Meeraus, 1987] and GEML [Neustadter, 1994] belong to this group.

Graphical model generators GOOD [Gyssens et al., 1994] and GYNGEN [Forster and Mevert, 1994] belong to this group.

Hybrid/compositional model generators These languages have an underlying representation based on mathematical and symbolic properties, e.g., CML [Falkenhainer et al., 1994] and SHSML [Taylor, 1993].

GOOD attempts to provide ease of high-level conceptualizing and manipulation of data. Sharing similarities with GOOD, GYNGEN focuses on process modeling by capturing the semantics underlying planning problems. CNFL and SHSML facilitate the modeling of dynamic processes. GEML is a language based on sets and has both primitive and derived data types. Primitive types may be defined by the user or built-in scalars. Derived types are recursive applications of operations such as the cartesian product or subtyping.

GOOD, GYNGEN, SHSML, and CML all employ graphs for defining structures. For the individual languages, variations arise when determining the role of nodes/edges as representations of the underlying concepts, and composing and interconnecting them to produce meaningful representations. GOOD relies on the operations of node addition/deletion, edge addition/deletion, and abstraction to build directed graphs. SHSML and CML are designed specifically to handle data dependencies arising from dynamic processes with time-varying properties.

Structure in CML is domain-theory dependent, defined by a set of top-level forms. Domain theories are composed from components, processes, interaction phenomena, logical relations, etc. The types in this language include symbols, lists, terms composed of lists, sequences, and sets of sequences. The language promotes reuse of existing domain theories to model processes under a variety of conditions.

A host of initiatives have been proposed by the W3C consortium that have a lot in common with the modeling languages listed above. The effort has been to standardize the various features across a wide variety of potential applications on the Web and specify markup formats for them. A list of such markup formats are:

XML. XML is a markup language for documents containing structured information. It is a meta-language for describing markup languages, i.e., it provides a facility to define tags and the structural relationships between them. Because there is no predefined tag set, there cannot be any preconceived semantics. All of the semantics of an XML document are defined by specialized instantiations, applications that process XML specifications, or by stylesheets. The vocabulary that makes up the tags and associated values may be obtained from ontologies and thesauri possibly available on the Web.

XSLT and XPath. The Extensible Stylesheet Language Transformations (XSLT) and the XML Path Language (XPath) are essentially languages that support transformation of XML specifications from one language to another.

XML Schema. The XML Schema definition language is a markup language that describes and constrains the content of an XML document. It is analogous to the database schema for relational databases and is a generalization of DTDs.

XQuery. The XQuery language is a powerful language for processing and querying XML data. It is analogous to the structured query language (SQL) used in the context of relational databases.

RDF. The Resource Description Framework (RDF) is a format for representing machine-understandable metadata on the Web. It has a graph-based data model with *resources* as nodes, *properties* as labeled edges, and *values* as nodes.

RDF Schema. Though RDF specifies a data model, it does not specify the vocabulary, e.g., what properties need to be represented, of the metadata description. These vocabularies (ontologies) are represented using RDF Schema expressions and can be used to constrain the underlying RDF statements.

DAML+OIL. The DARPA Agent Markup Language (DAML+OIL) is a more sophisticated specification (compared to RDF Schema) used to capture semantic constraints that might be available in an ontology/vocabulary.

Topic Maps. Topic Maps share with RDF the goal of representing relationships amongst data items of interest. A topic map is essentially a collection of topics used to describe key concepts in the underlying data repositories (text and relational databases). Relationships to these topics are represented using links also, called *associations*. Links that associate a given topic with the information sources in which it appears are called *occurrences*. Topics are related together independently of what is said about them in the information being indexed. A topic map defines a multidimensional topic space, a space in which the locations are topics, and in which the distances between topics are measurable in terms of the number of intervening topics that must be visited in order to get from one topic to another. It also includes the kinds of relationships that define the path from one topic to another, if any, through the intervening topics, if any.

Web Services. Web Services are computations available on the Web that can be invoked via standardized XML messages. Web Services Description Language (WSDL) describes these services in a repository — the Universal Description, Discovery, and Integration Service (UDDI) — which can be invoked using the Simple Object Access Protocol (SOAP) specification.

30.4 Ontology: Vocabularies and Reference Terms for Metadata

We discussed in the previous sections how metadata-based descriptions can be an important tool for modeling information on the Web. The degree of semantics depends on the nature of these descriptions, i.e., whether they are domain specific. A crucial aspect of creating metadata descriptions is the vocabulary used to create them. The key to utilizing the knowledge of an application domain is to identify the basic vocabulary consisting of terms or concepts of interest to a typical user in the application domain and the interrelationships among the concepts in the ontology.

In the course of collecting a vocabulary or constructing an ontology for information represented in a particular media type, some concepts or terms may be independent of the application domain. Some of them may be media specific while others may be media independent. There might be some application-specific concepts for which interrelationships may be represented. They are typically independent of the media of representation. Information represented using different media types can be modeled with application-specific concepts.

30.4.1 Terminological Commitments: Constructing an Ontology

An ontology may be defined as the specification of a representational vocabulary for a shared domain of discourse that may include definitions of classes relations functions and other objects [Gruber, 1993]. A crucial concept in creating an ontology is the notion of *terminological commitment*, which requires that subsribers to a given ontology agree on the semantics of any term in that ontology. This makes it incumbent upon content providers subscribing to a particular ontology to ensure that the information stored in their repositories is somehow *mapped* to the terms in the ontology. Content users, on the other hand, need to specify their information requests by using terms from the same ontology. A terminological commitment may be achieved via various means, such as alignment with a dominant standard or ontology, or via a negotiation process. Terminological commitments act as a bridge between various content providers and users. This is crucial as this terminological commitment then carries forward to the metadata descriptions constructed from these ontological concepts. However, in some cases, content providers and subscribers may subscribe to different ontologies, in which case terminological commitments need to be expanded to multiple ontologies, a situation we discuss later in this chapter. We view

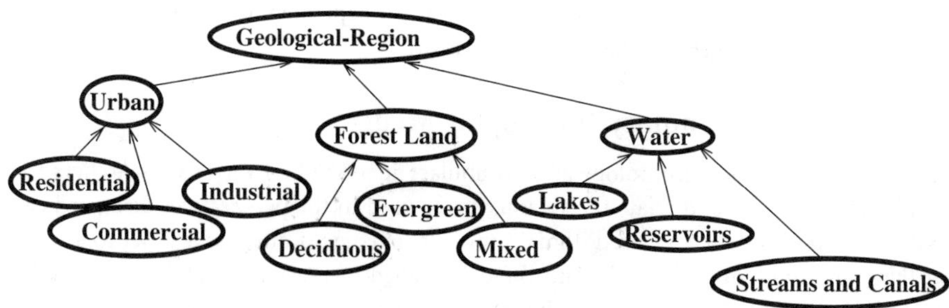

A classification using a generalization hierarchy

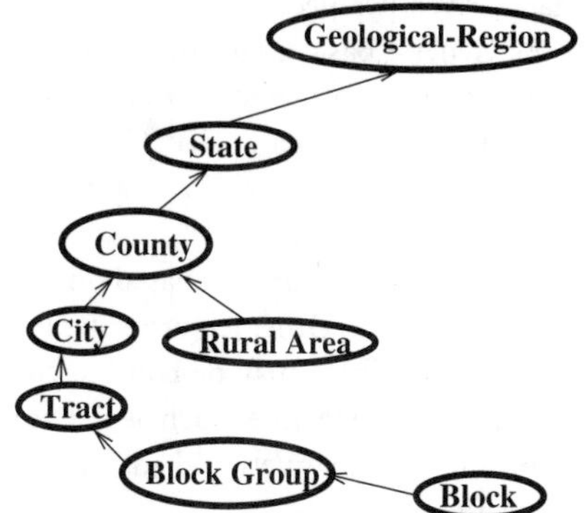

A classification using an aggregation hierarchy

FIGURE 30.6 Hierarchies describing a domain vocabulary.

terminological commitments as a very important requirement for capturing the semantics of domain-specific terms.

For the purposes of this chapter, we assume that media types presenting related information share the same domain of discourse. Typically, there may be other terms in the vocabulary that may not be dependent on the domain and may be media specific. Further, it may be necessary to translate between descriptive vocabularies that involve approximating, abstracting, or eliminating terms as a part of the negotiated agreement reached by various content managers. It may also be important to translate domain-specific terms to domain-independent media-specific terms by using techniques specialized to that media type. An example of a classification that can serve as a vocabulary for constructing metadata is illustrated in Figure 30.6.

In the process of construction, we view the ontology from the following two different dimensions:

1. *Data-Driven* vs. *Application-Driven* Dimension

   Data-driven perspective. This refers to concepts and relationships designed by interactive identification of objects in the digital content corresponding to different media types.

   Application-driven perspective. This refers to concepts and relationships inspired by the class of queries for which the related information in the various media types is processed. The concept *Rural Area* in Figure 30.6 is one such example.

2. *Domain-Dependent* vs. *Domain-Independent* Dimension

   Domain-dependent perspective. This represents the concepts that are closely tied to the application domain we wish to model. These are likely to be identified using the application-driven approach.

   Domain-independent perspective. This represents concepts required by various media types, e.g., color, shape, and texture for images, such as R features [Jain and Hampapur, 1994], to identify the domain-specific concepts. These are typically independent of the application domain and are generated by using a data-driven approach.

30.4.2 Controlled Vocabulary for Digital Media

In this section we survey the terminology and vocabulary identified by various researchers for characterizing multimedia content and relate the various terms used to the perspectives discussed earlier.

Jain and Hampapur [1994] have used domain models to assign a qualitative label to a feature (such as *pass, dribble,* and *dunk* in basketball), and these are called Q-Features. Features which rely on low-level domain-independent models such as object trajectories are called R-Features. Q-Features may be considered as an example of the domain-dependent application-driven perspective, whereas R-Features may be associated with the domain-independent data-driven perspective.

Kiyoki et al. [1994] have used basic words from the General Basic English Dictionary as features that are then associated with the images. These features may be considered as examples of the domain-dependent data-driven perspective. Color names defined by ISCC (Inter Society Color Council) and NBS (National Bureau of Standard) are used as features and may be considered as examples of the domain-independent data-driven perspective.

Anderson and Stonebraker [1994] model some features that are primarily based on the measurements of Advanced Very High Resolution Radiometer (AVHRR) channels. Other features refer to spatial latitude/longitude and temporal (begin date, end date) information. These may be considered as examples of domain-independent data-driven perspective. However, there are features such as the normalized difference vegetation index (NDVI) that are derived from different channels and may be considered as an example of the domain-dependent data-driven perspective.

Glavitsch et. al. [1994] have determined from experiments that good indexing features lay between phonemes and words. They have selected three special types of subword units VCV-, CV- and VC-. The letter V stands for a maximum sequence of vowels and C for a maximum sequence of consonants. They process a set of speech and text documents to determine a vocabulary for the domain. The same vocabulary is used for both speech and text media types, and may be considered as examples of the domain-dependent data-driven perspective.

Chen et. al. [1994] use the keywords identified in text and speech documents as their vocabulary. Issues of restricted vs. unrestricted vocabulary are very important. These may be considered as examples of the domain-dependent data and application-driven perspectives. A summary of the above discussion is presented in Table 30.2.

TABLE 30.2 Controlled Vocabulary for Digital Media

Vocabulary Feature	Media Type	Domain Dependent or Independent	Application or Data Driven
Q Features [Jain and Hampapur, 1994]	Video, Image	Domain Dependent	Application Driven
R Features [Jain and Hampapur, 1994]	Video, Image	Domain Independent	Data Driven
English Words [Kiyoki et. al., 1994]	Image	Domain Dependent	Data Driven
ISCC and NBS colors [Kiyoki et. al., 1994]	Image	Domain Independent	Data Driven
AVHRR features [Anderson and Stonebraker, 1994]	Image	Domain Independent	Data Driven
NDVI [Anderson and Stonebraker, 1994]	Image	Domain Dependent	Data Driven
Subword units [Glavitsch et. al., 1994]	Audio, Text	Domain Dependent	Data Driven
Keywords [Chen et. al., 1994]	Image, Audio Text	Domain Dependent	Application and Data Driven

30.4.3 Ontology-Guided Metadata Extraction

The extraction of metadata from the information in various media types can be primarily guided by the domain-specific ontology, though it may also involve terms in the domain-independent ontology.

Kiyoki et. al. [1994] describe the automatic extraction of impression vectors based on English words or ISCC and NBS colors. The users, when querying an image database, use English words to query the system. One way of guiding the users could be to display the list of English words used to construct the metadata in the first place. Glavitsch et. al. [1994] describe the construction of a speech feature index for both text and audio documents based on a common vocabulary consisting of subword units. Chen et al. [1994] describe the construction of keyword indices, topic change indices, and layout indices. These typically depend on the content of the documents, and the vocabulary is dependent on keywords present in the documents.

In the above cases, the vocabulary is not predefined and depends on the content of documents in the collection. Also, interrelationships between the terms in the ontology are not identified. A controlled vocabulary with terms and their interrelationships can be exploited to create metadata that model domain-dependent relationships as illustrated by the GIS example discussed in Kashyap and Sheth [1997].

Example: Consider a decision-support query across multiple data repositories possibly representing data in multiple media.

Get all regions having a population greater then 500, area greater than 50 acres, having an urban land-cover and moderate relief.

The metadata (referred to as m-context) can be represented as:

(AND region (population > 500) (area > 50) (= land-cover "urban") (= relief "moderate"))

Suppose the ontology from which the metadata description is constructed supports complex relationships. Furthermore, let:

CrowdedRegion @ (AND region (population > 200))

Inferences supported by the ontology enable determination that the regions required by the query metadata discussed earlier are instances of CrowdedRegion. Thus the metadata description (now referred to as c-context) can be rewritten as:

(AND CrowdedRegion (population > 500) (area > 50) (= land-cover "urban") (= relief "moderate"))

The above example illustrates how metadata expressions, when constructed using ontological concepts, can take advantage of ontological inferences to support metadata computation.

30.4.4 Medical Vocabularies and Terminologies: The UMLS Project

Metadata descriptions constructed from controlled vocabularies have been used extensively to index and search for information in medical research literature. In particular, articles in the MEDLINE (R) bibliographic database has used terms obtained from the MeSH vocabulary to annotate medical research articles. Besides this, there are a wide variety of controlled vocabularies in medicine used to capture information related to disesases, drugs, laboratory tests, etc. Efforts have been made to integrate various perspectives by creating a "Meta" Thesaurus or vocabulary that links these vocabularies together. This was the goal of the UMLS project, initiated in 1986 by the U.S. National Library of Medicine (NLM) [Lindbergh et al., 1993]. The UMLS consists of biomedical concepts and associated strings (Metathesaurus), a semantic network, and a collection of lexical tools, and has been used in a large variety of applications. The three main Knowledge Sources in the UMLS are:

1. The UMLS Metathesaurus provides a common structure for more than 95 source biomedical vocabularies, organized by concept or meaning. A concept is defined as a cluster of terms (one or more words representing a distinct concept) representing the same meaning (e.g., synonyms, lexical variants, translations). The 2002 version of the Metathesaurus contains 871,584 concepts named by 2.1 million terms. Interconcept relationships across multiple vocabularies, concept

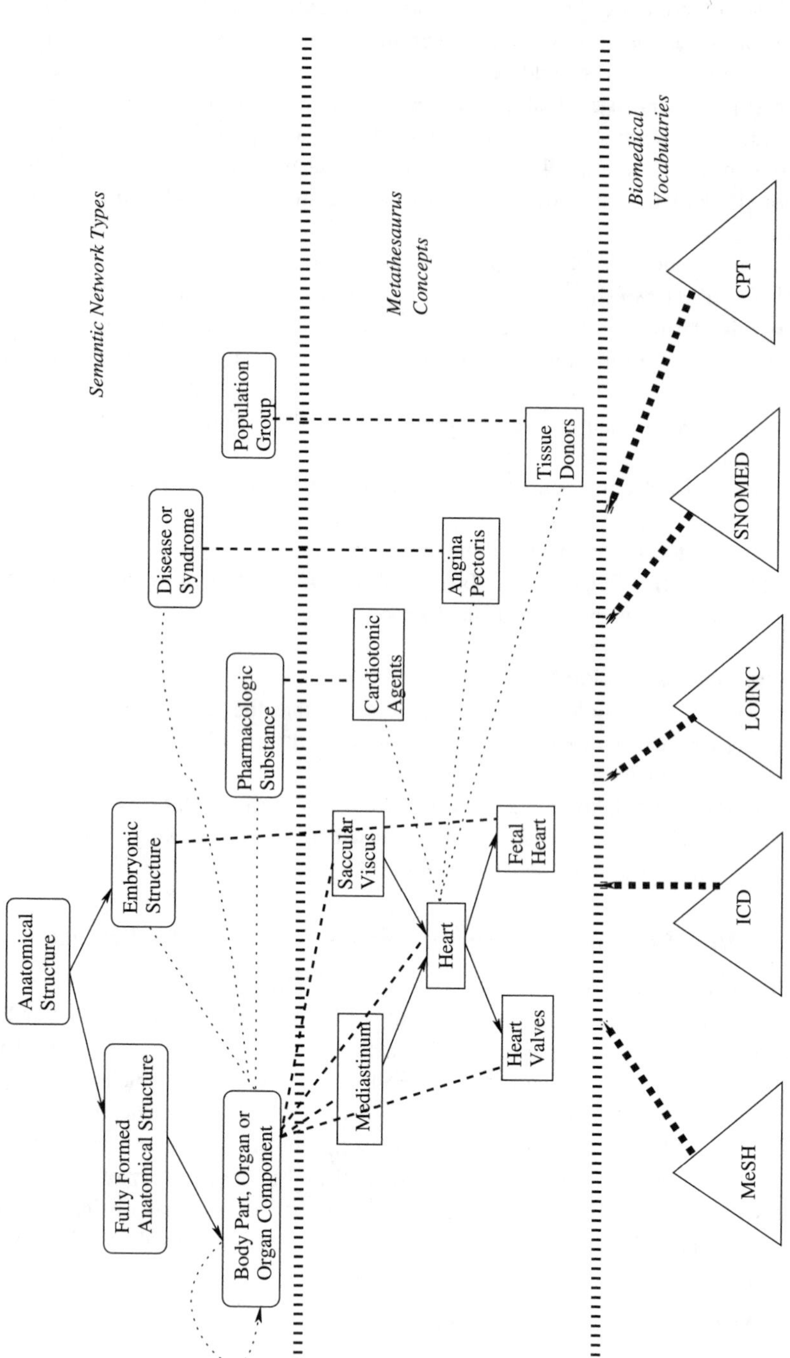

FIGURE 30.7 Biomedical vocabularies and the Unified Medical Language System.

categorization, and information on concept cooccurrence in MEDLINE are also included [McCray and Nelson, 1995].

2. The UMLS Semantic Network categorizes Metathesaurus concepts through semantic types and relationships [McCray and Nelson, 1995].
3. The SPECIALIST lexicon contains over 30,000 English words, including many biomedical terms. Information for each entry, including base form, spelling variants, syntactic category, inflectional variation of nouns, and conjugation of verbs are used by the lexical tools [McCray et al., 1994]. There are over 163,000 records in the 2002 SPECIALIST lexicon representing over 268,000 distinct strings.

Some of the prominent medical vocabularies are as follows:

Medical Subject Headings (MeSH). The Medical Subject Headings (MeSH) [Nelson et al., 2001] have been produced by the NLM since 1960. The MeSH thesaurus is NLM's controlled vocabulary for subject indexing and searching of journal articles in PubMed, and books, journal titles, and nonprint materials in NLM's catalog. Translated into many different languages, MeSH is widely used in indexing and cataloging by libraries and other institutions around the world. An example of the MeSH expression used to index and search for the concept "Mumps pancreatitis" is illustrated in Figure 30.8.

International Classification of Diseases (ICD). The World Health Organization's International Classification of Diseases, 9th Revision (ICD-9)[ICD] is designed for the classification of morbidity and mortality information for statistical purposes, for the indexing of hospital records by disease and operations, and for data storage and retrieval. ICD-9-CM is a clinical modification of ICD-9. The term "clinical" is used to emphasize the modification's intent: to serve as a useful tool in the area of classification of morbidity data for indexing of medical records, medical care review, and ambulatory and other medical care programs, as well as for basic health statistics. To describe the clinical picture of the patient, the codes must be more precise than those needed only for statistical groupings and trend analysis.

Systematized Nomenclature for Medicine (SNOMED). The SNOMED [Snomed] vocabulary was designed to address the need for a detailed and specific nomenclature to accurately reflect, in computer readable format, the complexity and diversity of information found in a patient record. The design ensures clarity of meaning, consistency in aggregation, and ease of messaging. The SNOMED is compositional in nature, i.e., new concepts can be created as compositions of existing ones, and has a systematized hierarchical structure. Its unique design allows for the full integration of electronic medical record information into a single data structure. Overall, SNOMED has contributed to the improvement in patient care, reduction of errors inherent in data coding, facilitation of research, and support of compatibility across software applications.

Current Procedural Terminology (CPT). The Current Procedural Terminology (CPT) codes [CPT] are used to describe services in electronic transactions. CPT was developed by the American Medical Association (AMA) in the 1960s, and soon became part of the standard code set for Medicare and

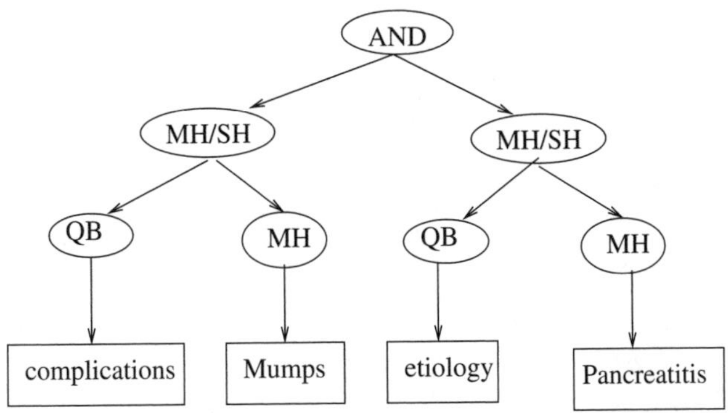

FIGURE 30.8 A MeSH descriptor for information retrieval.

Medicaid. In subsequent decades, it was also adopted by private insurance carriers and managed care companies, and has now become the *de facto* standard for reporting healthcare services.

Logical Observation Identifier Names and Codes (LOINC). The purpose of the Logical Observation Identifier Names and Codes (LOINC) database [LOINC] is to facilitate the exchange and pooling of results, such as blood hemoglobin, serum potassium, or vital signs, for clinical care, outcomes management, and research. Its purpose is to identify observations in electronic messages such as Health Level Seven (HL7) [HL7] observation messages, so that when hospitals, health maintenance organizations, pharmaceutical manufacturers, researchers, and public health departments receive such messages from multiple sources, they can automatically file the results in the right slots of their medical records, research, and/or public health systems.

30.4.5 Expanding Terminological Commitments across Multiple Ontologies

We discussed in the beginning of this section the desirability of expanding the process of achieving terminological commitments across multiple ontologies. The UMLS system described earlier may be viewed as an attempt to establish terminological commitments against a multitude of biomedical vocabularies. The UMLS Metathesaurus may be viewed as a repository of intervocabulary relationships. Establishing terminological commitments across users of the various biomedical vocabularies would require using the relationships represented in the UMLS Metathesaurus to provide translations from a term in a source vocabulary to a term or expression of terms in a target vocabulary. This requires the ability to integrate the two vocabularies in a common graph structure and navigation of the graph structure for suitable translation. This is illusrated in an abstract manner in Figure 30.9 and is being investigated in the context of the Semantic Vocabulary Interoperation Project at the NLM [SVIP].

30.5 Conclusions

The success of the World Wide Web has led to the availability of tremendous amounts of heterogeneous digital content. However, this has led to concerns about scalability and information loss (e.g., loss in precision/recall). Information modeling is viewed as an approach for enabling scalable development of the Web for access to information in an information-preserving manner. Creation and extraction of machine-understandable metadata is a critical component of the Semantic Web effort that aims at enhancing the current Web with the "semantics" of information.

In this chapter we presented a discussion on metadata, its use in various applications having relevance to the Web, and a classification of various metadata types, capturing different levels of information content. We discussed approaches that use metadata descriptions for creating information models and spaces, and various ways by which the semantics of information embedded in the data can be captured.

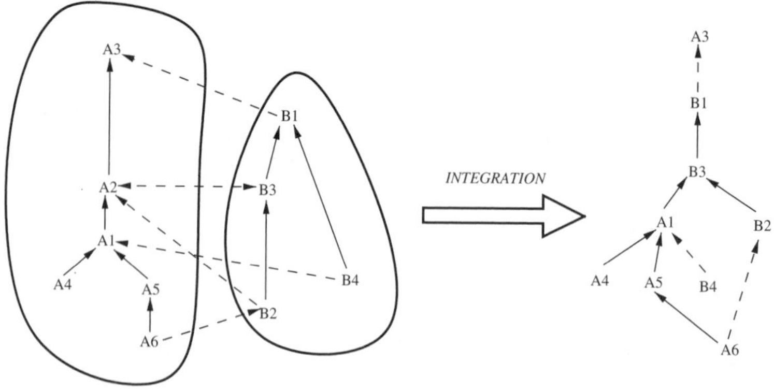

FIGURE 30.9 Expanding terminological commitments by integration of ontologies.

In this context, we also discussed the role played by controlled vocabularies and ontologies in providing reference terms and concepts for constructing metadata descriptions. Examples from the domain of biomedical information were presented and issues related to the establishment of terminological commitments across multiple user communities were also discussed. The role played by metadata and ontologies is crucial in modeling information and semantics, and this chapter provides an introduction to these technologies from that perspective.

References

Anderson, J. and M. Stonebraker. Sequoia 2000 Metadata schema for satellite images, in W. Klaus and A. Sheth, Eds., *SIGMOD Record, special issue on Metadata for Digital Media,* 1994.

Berners-Lee, T. et al. World Wide Web: the information universe. *Electronic Networking: Research, Applications, and Policy,* 1(2), 1992.

Berners-Lee, T., James Hendler, and Ora Lassila. The Semantic Web. *Scientific American,* May, 2001.

Bohm, K. and T. Rakow. Metadata for multimedia documents, in W. Klaus and A. Sheth, Eds., *SIGMOD Record, special issue on Metadata for Digital Media,* 1994.

Boll, S., W. Klas, and A. Sheth. Overview on using metadata to manage multimedia data. In A. Sheth and W. Klas, Eds., *Multimedia Data Management.* McGraw-Hill, New York, 1998.

Bray, T., J. Paoli, and C. M. Sperberg-McQueen. Extensible markup language (XML) 1.0. http://www.w3.orgfrR/REC-xml.

Chen, F., M. Hearst, J. Kupiec, J. Pederson, and L. Wilcox. Metadata for mixed-media access, in W. Klaus and A. Sheth, Eds., *SIGMOD Record, special issue on Metadata for Digital Media,* 1994.

Collet, C., M. Huhns, and W. Shen. Resource integration using a large knowledge base in Carnot. *IEEE Computer,* December 1991.

CPT. Current Procedural Terminology. http://www.ama-assn.org/ama/pub/category/3113.html.

DAML+OIL. The DARPA Agent Markup Language. http://www.daml.org/.

Deerwester, S., S. Dumais, G. Furnas, T. Landauer, and R. Hashman. Indexing by latent semantic indexing. *Journal of the American Society for Information Science,* 41(6), 1990.

Falkenhainer, B. et al. CML: A Compositional Modeling Language, 1994. Draft.

Forster, M. and P. Mevert. A tool for network modeling. *European Journal of Operational Research,* 72, 1994.

Fourer, R., D. M. Gay, and B. W. Kernighan. AMPL: A Mathematical Programming Language. Technical Report 87-03, Department of Industrial Engineering and Management Sciences, Northwestern University, Chicago, IL, 1987.

Glavitsch, U., P. Schauble, and M. Wechsler. Metadata for integrating speech documents in a text retrieval system, in W. Klaus and A. Sheth, Eds., *SIGMOD Record, special issue on Metadata for Digital Media,* 1994.

Gruber, T. A translation approach to portable ontology specifications. *Knowledge Acquisition: International Journal of Knowledge Acquisition for Knowledge-Based Systems,* 5(2), June 1993.

Gyssens, M. et al. A graph-oriented object database model. *IEEE Transactions on Knowledge and Data Engineering,* 6(4), 1994.

HL7. The Health Level Seven Standard. http://www.hl7.org.

ICD. The International Classification of Diseases, 9th Revision, Clinicial Modification. http://www.cdc.gov/nchs/about/otheract/icd9/abticd9.htm.

Jain, R. Semantics in multimedia systems. *IEEE Multimedia.* 1(2), 1994.

Jain, R. and A. Hampapur. Representations of video databases, in W. Klaus and A. Sheth, Eds., *SIGMOD Record, special issue on Metadata for Digital Media,* 1994.

Kahle, B. and A. Medlar. An information system for corporate users: wide area information servers. *Connexions — The Interoperability Report,* 5(11), November 1991.

Kashyap, V. and A. Sheth. Semantics-based Information Brokering. In *Proceedings of the Third International Conference on Information and Knowledge Management (CIKM),* November 1994.

Kashyap, V. and A. Sheth. Semantic heterogeneity: role of metadata, context, and ontologies. In M. Papazoglou and G. Schlageter, Eds., *Cooperative Information Systems: Current Trends and Directions.* Academic Press, San Diego, CA, 1997.

Kendrick, D. and A. Meeraus. GAMS: An Introduction. Technical Report, Development Research Department, The World Bank, 1987.

Kiyoki, Y., T. Kitagawa, and T. Hayama. A meta-database system for semantic image search by a mathematical model of meaning, in W. Klaus and A. Sheth, Eds., *SIGMOD Record, special issue on Metadata for Digital Media,* 1994.

Klaus, W. and A. Sheth. Metadata for digital media. *SIGMOD Record, special issue on Metadata for Digital Media,* W. Klaus and A. Sheth, Eds., 23(4), December 1994.

Lassila, O. and R. R. Swick. Resource description framework (RDF) model and syntax specification. http://www.w3.org/TR/REC-rdf-syntax/.

Lindbergh, D., B. Humphreys, and A. McCray. The unified medical language system: *Methods of Information in Medicine,* 32(4), 1993. http://umlsks.nlm.nih.gov.

LOINC. The logical observation identifiers names and codes database. http://www.loinc.org.

McCray, A. and S. Nelson. The representation of meaning in the UMLS. *Methods of Information in Medicine,* 34(1–2): 193–201, 1995.

McCray, A., S. Srinivasan, and A. Browne. Lexical Methods for Managing Variation in Biomedical Terminologies. In *Proceedings of the Annual Symposium on Computers in Applied Medical Care,* 1994.

MEDLINE. The PubMed MEDLINE system. http://www.ncbi.nlm.nih.gov/entrez/query.fcgi?db=PubMed.

Mena, E., V. Kashyap, A. Sheth, and A. Illarramendi. OBSERVER: An Approach for Query Processing in Global Information Systems Based on Interoperation Across Pre-existing Ontologies. In *Proceedings of the First IFCIS International Conference on Cooperative Information Systems* (CoopIS '96), June 1996.

Nelson, S.J., W.D. Johnston, and B.L. Humphreys. Relationships in medical subject headings (MeSH). In C.A. Bean and R.Green, Eds., *Relationships in the Organization of Knowledge.* Kluwer Academic, Dordrecht, The Netherlands, 2001.

Neustadter, L. A Formalization of Expression Semantics for an Executable Modeling Language. In *Proceedings of the Twenty Seventh Annual Hawaii International Conference on System Sciences,* 1994.

Ogle, V. and M. Stonebraker. Chabot: retrieval from a relational database of images. *IEEE Computer; special issue on Content-Based Image Retrieval Systems,* 28(9), 1995.

Ordille, J. and B. Miller. Distributed Active Catalogs and Meta-Data Caching in Descriptive Name *Services.* In *Proceedings of the 13th International Conference on Distributed Computing Systems,* May 1993.

OWL. The Web Ontology Language. http://www.w3.org/TR/owl_guide/.

Salton, G. *Automatic Text Processing.* Addison-Wesley, Reading, MA, 1989.

Sciore, E., M. Siegel, and A. Rosenthal. Context interchange using meta-attributes. In *Proceedings of the CIKM,* 1992.

SGML. The Standard Generalized Markup Language. http://www.w3.org/MarkUp/SGML/.

Shah, K. and A. Sheth. Logical information modeling of web-accessible heterogeneous digital assets. In *Proceedings of the IEEE Advances in Digital Libraries (ADL) Conference,* April 1998.

Sheth, A. and V. Kashyap. Media-independent correlation of information. What? How? In *Proceedings of the First IEEE Metadata Conference,* April 1996. http://www.computer.org/conferences/meta96/sheth/index.html.

Sheth, A., V. Kashyap, and W. LeBlanc. Attribute-based Access of Heterogeneous Digital Data. In *Proceedings of the Workshop on Web Access to Legacy Data, Fourth International WWW Conference,* December 1995.

Shklar, L., K. Shah, and C. Basu. The InfoHarness repository definition language. In *Proceedings of the Third International WWW Conference,* May 1995a.

Shklar, L., K. Shah, C. Basu, and V. Kashyap. Modelling Heterogeneous Information. In *Proceedings of the Second International Workshop on Next Generation Information Technologies (NGITS '95)*, June 1995b.

Shklar, L., A. Sheth, V. Kashyap, and K. Shah. InfoHarness: Use of Automatically Generated Metadata for Search and Retrieval of Heterogeneous Information. In *Proceedings of CAiSE '95, Lecture Notes in Computer Science* #932, June 1995c.

Shklar, L., S. Thatte, H, Marcus, and A. Sheth. The InfoHarness Information Integration Platform. In *Proceedings of the Second International WWW Conference,* October 1994.

Shoens, K., A. Luniewski, P. Schwartz, J. Stamos, and J. Thomas. The Rufus System: Information organization for semi-structured data. In *Proceedings of the 19th VLDB Conference,* September 1993.

Snomed. The systematized nomenclature of medicine. http://www.snomed.org.

SVIP. The Semantic Vocabulary Interoperation Project, http://cgsb2.nlm.nih.gov/kashyap/projects/SVIP/.

Taylor, J.H. Towards a Modeling Language Standard for Hybrid Dynamical Systems. In *Proceedings of the 32nd Conference on Decision and Control,* 1993.

31

Semantic Aspects of Web Services

Sinuhé Arroyo
Rubén Lara
Juan Miguel Gómez
David Berka
Ying Ding
Dieter Fensel

CONTENTS

Abstract 31-1
31.1 Introduction 31-1
31.2 Semantic Web 31-2
31.2.1 Related Projects 31-3
31.3 Semantic Web Services 31-3
31.4 Relevant Frameworks 31-5
31.4.1 WSMF 31-6
31.4.2 WS-CAF 31-8
31.4.3 Frameworks Comparison 31-10
31.5 Epistemological Ontologies for Describing Services .. 31-11
31.5.1 DAML-S and OWL-S 31-11
31.5.2 DAML-S Elements 31-11
31.5.3 Limitations 31-14
31.6 Summary 31-14
Acknowledgements 31-15
References 31-15

Abstract

Semantics promise to lift the Web to its full potential. The combination of machine-processable semantics provided by the Semantic Web with current Web Service technologies has coined the term Semantic Web Services. Semantic Web Services offer the means to achieve a higher order level of value-added services by automating the task-driven assembly of interorganization business logics, thus making the Internet a global, common platform where agents communicate with each other. This chapter provides an introduction to the Semantic Web and Semantic Web Services paying special attention to its automation support. It details current initiatives in the E.U. and U.S., presents the most relevant frameworks towards the realization of a common platform for the automatic task-driven composition of Web Services, sketches a comparison among them to point out their weaknesses and strengths, and finally introduces the most relevant technologies to describe services and their limitations.

31.1 Introduction

Current Web technology exploits very little of the capabilities of modern computers. Computers are used solely as information-rendering devices, which present content in a human-understandable format. Incorporating semantics is essential in order to exploit all of the computational capabilities of computers for information processing and information exchange. Ontologies enable the combination of data and information with semantics, representing the backbone of a new technology called Semantic Web. By

1-58488-381-2/05/$0.00+$1.50
© 2005 by CRC Press LLC

using ontologies, the Semantic Web will lift the current WWW to a new level of functionality where computers can query each other, respond appropriately, and manage semistructured information in order to perform a given task.

If ontologies are the backbone of the Semantic Web, Web Services are its arms and legs. Web Services are self-contained, self-describing, modular applications that can be published, located, and invoked over the Web [Tidwell, 2000]. In a nutshell, Web Services are nothing but distributed pieces of software that can be accessed via the Web, roughly just another implementation of RPC. The potential behind such a simple concept resides in its possibilities of being assembled *ad hoc to* perform tasks or execute business processes. Web Services will significantly further the development of the Web by enabling automated program communication. Basically, they will allow the deployment of new complex value-added services. Web Services constitute the right means to accomplish the objectives of the Semantic Web. They not only facilitate the resources to access semantically enriched information, but its assembly and combination possibilities, enhanced with semantic descriptions of their functionalities, also will provide higher order functionality that will lift the Web to its full potential.

The combination of the Semantic Web and the Web Service technology has been named Semantic Web Services. Semantic Web Services may be the killer application of this emerging Web. The Semantic Web will provide the means to automate the use of Web Services, namely *discovery, composition,* and *execution* [Mcllraith et al., 2001]. The automation of these processes will make the task-driven assembly of interorganization business logics a reality. Such automation will transform the Internet in a global, common platform where agents (organizations, individuals, and software) communicate with each other to carry out various commercial activities, providing a higher order level of value-added services. The effects of this new technological development will expand areas such as Knowledge Management, Enterprise Application Integration, and e-Business.

In a nutshell, the Semantic Web promises a revolution in human information access similar to the one caused by the telephone and comparable to the invention of the steam engine. Such a revolution will count with ontologies and Web Services as its most important champions. The way computers are seen and the WWW is understood will be changed completely, and effects will be felt in every aspect of our daily life.

This chapter provides an overview and detailed analysis of Semantic Web Services and related technologies. The contents are organized as follows: Section 31.2 presents an overview of the Semantic Web, a little bit of history, what is the Semantic Web, and a prospective of its possibilities; Section 31.3 introduces Semantic Web Services, fundamentals, actual state of development, and future trends and directions; 31.4 presents the most important initiatives towards the development of frameworks for Semantic Web Services, why frameworks are necessary, and what benefits they provide; Section 31.5 includes an overview of the most relevant upper ontologies developed to describe Semantic Web Services from a functional and nonfunctional point of view; and finally Section 31.6 provides a summary together with a view of the future direction the Semantic Web Services technology will take.

31.2 Semantic Web

The Semantic Web is the next generation of the WWW where information has machine-processable and machine-understandable semantics. This technology will bring structure to the meaningful content of Web pages, being not a separate Web but an augmentation of the current one, where information is given a well-defined meaning. The Semantic Web will include millions of small and specialized reasoning services that will provide support for the automated achievement of tasks based on accessible information.

One of the first to come up with the idea of the Semantic Web was the inventor of the current WWW, Tim Berners-Lee. He envisioned a Web where knowledge is stored on the meaning or content of Web resources through the use of machine-processable meta-data. The Semantic Web can be defined as "*an extension of the current Web in which information is given well-defined meaning, better enabling computers and people to work in co-operation*" [Berners et al., 2001].

The core concept behind the Semantic Web is the representation of data in a machine-interpretable way. Ontologies facilitate the means to realize such representation. Ontologies represent formal and consensual specifications of conceptualizations, which provide a shared and common understanding of a domain as data and information machine-processable semantics, which can be communicated among agents (organizations, individuals, and software) [Fensel, 2001]. Many definitions of *ontology* have been given during the last years. One that really fits its application in the computer science field is the one given by Gruber [1993]: "*An ontology is a formal, explicit specification of a shared conceptualization.*"

Ontologies bring together two essential aspects that help to push the Web to its full potential. On the one hand they provide (a) *machine processability* by defining formal semantics for information making computers able to process it; on the other, they allow (b) *machine-human understanding* due to their ability to specify real-world semantics that permit to link machine-processable content with human meaning using a consensual terminology as connecting element [Fensel, 2001].

Regarding Web Services, machine processability enabled by ontologies represent their most valuable contribution. They offer the necessary means to describe the capabilities of a concrete Web Service in a shared vocabulary that can be understood by every service requester. Such a shared functionality description is the key element towards task-driven automatic Web Service discovery, composition, and execution of interorganization business logics. It will allow to (1) locate different services that, solely or in combination, will provide the means to solve a given task, (2) combine services to achieve a goal, and (3) facilitate the replacement of such services by equivalent ones that, solely or in combination, can realize the same functionality, e.g., in case of failure during execution.

As ontologies are built (models of how things work), it will be possible to use them as common languages to describe Web Services and the payloads they contain in much more detail [Daconta et, al 2003].

31.2.1 Related Projects

Currently there are many initiatives, private and public, trying to bring the Berners-Lee vision of the Semantic Web to a plausible reality [Ding et al., 2003]. Among the most relevant ones financed by the European Comission are projects like Onto-Knowledge [Ontoknowledge], DIP [DIP], SEKT [SEKT], SWWS [SWWS], Esperonto [Esperonto], Knowledge Web [KowledgeWeb], and OntoWeb [OntoWeb]. In the U.S. one of the most relevant initiatives is the DARPA Agent Markup Language [DAML], supported by the research funding agency of the U.S. Department of Defense. Other relevant DARPA-funded initiatives are High Performance Knowledge Bases [HPKB], which is now completed, and its follow up, Rapid Knowledge Formation [RKF]. The National Science Foundation (NSF) has also sponsored some Semantic Web efforts. On its side, the W3C [W3C] has made an important effort towards the standardization of Semantic Web related technologies.

European and U.S. researchers have been collaborating actively on standardization efforts [Ding et al., 2003]. A result of this collaboration is the joint Web Ontology Language OWL.

Some of the initiatives presented in this section count among their main partners major software vendors such as BT or HP, providing a valuable business point of view that allows aligning the technology with market needs, and proving the interest that exists towards its development.

31.3 Semantic Web Services

In simple terms, Web Services are a piece of software accessible via the Web. By Service can be understood any type of functionality that software can deliver, ranging from mere information providers (such as stock quotes, weather forecasts, or news aggregation) to more elaborate ones that may have some impact in the real world (such as booksellers, plane-ticket sellers, or e-banking), basically any functionality offered by the current Web can be envisioned as a Web Service. The big issue about Web Services resides in their capabilities for changing the Web from a static collection of information to a dynamic place where different pieces of software can be assembled on the fly to accomplish users' goals (expressed, perhaps, in any natural language). This is a very ambitious goal, which at the moment cannot be achieved with the current state of the art technologies around Web Services. UDDI [Bellwood et al., 2002], WSDL [Christensen et al., 2001], and SOAP [Box et al., 2000] facilitate the means to advertise, describe, and invoke them, using semiformal natural language terms but do not say anything about what services can do, nor how they do it in a machine understandable and processable way. An alternative to UDDI, WSDL, and SOAP is ebXML (electronic business XML), which permits exchanging and transporting business documents over the Web using XML. It converges with the previously cited technologies, differing with them to the degree in which it recognizes business process modeling as a core feature [Newcomer, 2002]. All these initiatives lack proper support for semantics, and therefore human intervention is needed to actually discover, combine, and execute Web Services. The goal is to minimize any human intervention, so the assembly of Web Services can be done in a task-driven automatic way.

The Semantic Web in general and ontologies in particular are the right means to bridge the gap and actually realize a dynamic Web. They provide the machine-processable semantics that, added on top of current Web Services, actualize the potential of the Semantic Web Services. Semantic Web Services combine Semantic Web and Web Service Technology. *Semantic Web Services* are defined as *self-contained, self-describing, semantically marked-up software resources that can be published, discovered, composed and executed across the Web in a task driven automatic way.*

What really makes a difference with respect to traditional Web Services is that they are semantically marked-up. Such enhancement enables *Automatic Web Service Discovery*: *the location of services corresponding to the service requester specification for a concrete task.*

Ideally, a task will be expressed using any natural language, and then translated to the ontology vocabulary suitable for the concrete application domain. Due to the fact that there will most likely be thousands of different ontologies for a concrete domain, (different service providers will express their business logics using different terms and conventions, and different service requesters will express their requirements using different vocabularies) the appropriate support for ontology merging and alignment must be available. Semantically marked-up services will be published in semantically enhanced services' repositories where they can be easily located and their capabilities matched against the user's requirements. In a nutshell, these repositories are traditional UDDI registries augmented with a semantic layer on top that provides machine-processable semantics for the services registered.

As an example a service requester might say, "Locate all the services that can solve mathematical equations." An automatic service discovery engine will then surf all available repositories in search for services that fulfill the given task. The list of available services will probably be huge, so the user might impose some limitations (*functional* — how the service is provided — and *nonfunctional* — execution time, cost, and so on) in order to get an accurate and precise set of services.

Automatic Web Service composition is an assembly of services based on its functional specifications in order to achieve a given task and provide a higher order of functionality.

Once a list of available Web Services has been retrieved, it could happen that none of the available services completely fulfils the proposed task by itself or that other cheaper, faster, vendor-dependent combinations of services are preferred. In this case, some of the services would need to be assembled, using programmatic conventions to accomplish the desired task. During this stage, Web Services are organized in different possible ways based on functional requirements (*pre-conditions:* conditions that must hold before the service is executed; and *post-conditions:* conditions that hold after the service execution) and non functional (processing time and cost) requirements.

As an example, a service requester might say, "Compose available services to solve the following set of mathematical operations $[((a \times b) + c) - d]$." During the discovery phase different multiplication, addition and subtraction services might have been found, each one of them with its particular functional and nonfunctional attributes. Let us suppose that some multiplication services have been located, but they are all too slow and expensive, so we are not interested in using them. Instead we want to use additions to perform the multiplication. Such knowledge, the fact that multiple addition can realize multiplication, should be stated in some *domain knowledge* — characterization of relevant information for a specific area — in a way that the service composer can understand and is able to present all different possibilities to solve our set of operations. Different alternatives that lead towards the task accomplishments are then presented to the service requester who chooses among them the composition path that suits its needs the best.

Automatic Web Service execution is an invocation of a concrete set of services, arranged in a particular way following programmatic conventions that realizes a given task.

When the available services have been composed and the service requester has chosen the execution path that best suits its needs according to the nonfunctional requirements, the set of Web Services is to be executed. Each Web Service specifies one or more APIs that allows its execution. The semantic markup facilitates all the information regarding inputs required for service execution and facilitates outputs returned once this has finished.

As an example a service requester might say, "Solve the following mathematical operations using the execution path that contains no multiplication services [((3 × 3) + 4)–1]." So what the user is actually saying is solve: [((3 + 3 + 3) + 4)–1]. The addition service that realized the multiplication will be put into a loop, which will add 3 to itself three times, the result will be added to 4 and then 1 will be subtracted to obtain the result, as stated in the equation.

Prior to execution, the service may impose some limitations to the service requester, such restrictions are expressed in terms of *assumptions* — conditions about the state of the world. A service provider may state that the service requester has to have a certain amount of money in the bank prior to the execution of the service, or whatever other requirements are considered necessary as part of a concrete business logic. In case any of the execution's constituents fails to accomplish its goal (e.g., network breaks down, service provider server breaks down, etc.), a recovery procedure must be applied to replace the failed service with another service or set of services capable of finishing the work. Once the composed service has been successfully executed, it might be registered in any of the available semantically enhanced repositories. By doing so, a new level of functionality is achieved, making available for reuse already composed and successfully executed services.

Current technology does not fully realize any of the parts of the Web Services' automation process. Among the reasons are:

- Lack of fully developed markup languages
- Lack of marked up content and services
- Lack of semantically enhanced repositories
- Lack of frameworks that facilitate discovery, composition, and execution
- Lack of tools and platforms that allow semantic enrichment of current Web content

Essentially, the technology is not yet mature enough, and there is a long path that should be walked in order to make it a reality. Some academic and private initiatives are very interested in developing these technologies, both entities already making very strong, active efforts to grow them. Among them, Sycara et al., [1999] presented a Web Service discovery initiative using matchmaking based on a representation for annotating agent capabilities so that they can be located and brokered. Sirin et al. [2003] have developed a prototype that guides the user in the dynamic composition of Web Services. Web Service Modeling Ontology (WSMO) an initiative carried by the next Web generation research group at Innsbruck University aims at providing a conceptual model for describing the various aspects of Semantic Web Services [WSMO].

31.4 Relevant Frameworks

Various initiatives aim to provide a full-fledged modeling framework for Web Services that allows their automatic discovery, composition, and execution. Among them, the most relevant ones are WSMF and WS-CAF. The benefits that can be derived from these developments will constitute a significant evolution in the way the Web and e-business are understood, providing the appropriate conceptual model for developing and describing Web Services and their assembly.

31.4.1 WSMF

The Web Service Modeling Framework (WSMF) is an initiative towards the realization of a full-fledged modeling framework for Semantic Web Services that counts on ontologies as its most important constituents. WSMF aims at providing a comprehensive platform to achieve automatic Web Service discovery, selection, mediation, and composition of complex services; that is, to make Semantic Web services a reality and to exploit their capabilities. The WSMF description given in this section is based on Fensel and Bussler [2002]. WSMF specification is currently evolving, but none of the key elements presented here is likely to undergo a major change.

31.4.1.1 WSMF Objectives

The WSMF objectives are as follows. ***Automated discovery*** includes the means to mechanize the task of finding and comparing different vendors and their offers by using machine-processable semantics. ***Data mediation*** involves ontologies to facilitate better mappings among the enormous variety of data standards. ***Process mediation*** provides mechanized support to enable partners to cooperate despite differences in business logics, which are numerous and heterogeneous.

31.4.1.2 WSMF Principles

WSMF revolves around two complementary principles, namely (1) ***strong decoupling***, where the different components that realize an e-commerce application should be as disaggregated as possible, hiding internal business intelligence from public access, allowing the composition of processes based on their public available interfaces, and carrying communication among processes by means of public message exchange protocols; and (2) ***strong mediation***, which will enable scalable communications, allowing anybody to speak with everybody. To allow this m:m communication style, terminologies should be aligned, and interaction styles should be intervened.

In order to achieve such principles, a mapping among different business logics, together with the ability to establish the difference between public processes and private processes of complex Web Services, are key characteristics the framework should support. Mediators provide such mapping functionalities and allow expressing the difference among publicly visible workflows and internal business logics.

31.4.1.3 WSMF Elements

WSMF consists of four different main elements: (1) ontologies that provide the terminology used by other elements, (2) goal repositories that define the tasks to be solved by Web Services, (3) Web Services as descriptions of functional and nonfunctional characteristics, and (4) mediators that bypass interoperability problems. A more detailed explanation of each element is given below.

Ontologies. They interweave human understanding with machine processability. In WSMF ontologies provide a common vocabulary used by other elements in the framework. They enable reuse of terminology, as well as interoperability between components referring to the same or linked terminology.

Goal repositories. Specify possible objectives a service requester may have when consulting a repository of services. A goal specification consists of two elements:

- **Preconditions.** Conditions that must hold previous to the service execution and enable it to provide the service.
- **Postconditions.** Conditions that hold after the service execution and that describe what happens when a service is invoked.

Due to the fact that a Web Service can actually achieve different goals (i.e., Amazon can be used to buy books, and also as an information broker on bibliographic information about books), the Web Services descriptions, and the goals they are able to achieve are kept separately, allowing n:m mapping among services and goals. Conversely, a goal can be achieved but by different and eventually competing Web Services.

Keeping goal specifications separate from Web Service descriptions enhances the discovery phase, as it enables goal-based search of Web Services instead of functionality based search.

Web Services. In WSMF the complexity of a Web Service is measured in terms of its external visible description, contrary to the flow followed by most description languages, which differentiate them based on the complexity of the functionality and whether they can be broken into different pieces or not. Under traditional conventions, a complex piece of software such as an inference engine with a rather simplistic interface can be defined as elementary, whereas a much simpler software product such as an e-banking aggregator that can be broken down into several Web Services is considered complex. This reformulation of the definition that may look trivial has some relevant consequences:

- Web Services are not described themselves but rather their interfaces, which can be accessed via a network. By these means service providers can hide their business logic, which usually is reflected in the services they offer.
- The complexity of a Web Service description provides a scale of complexity, which begins with some basic description elements, and gradually increases the description density by adding further means to portray various aspects of the service.

As can be inferred from the previous paragraph, the framework describes services as black boxes, i.e., it does not model the internal aspects of how such a service is achieved, hiding all business logic aspects. The black box description used by WSMF consists of the following main elements:

- *Web Service name.* Unique service identifier used to refer to it
- *Goal reference* Description of the objective that can be stored in a goal repository
- *Input and output data.* Description of the data structures required to provide the service
- *Error data.* Indicates problems or error states
- *Message exchange protocol.* Provides the means to deal with different types of networks and their properties providing an abstraction layer
- *Nonfunctional parameters.* Parameters that describe the service such as execution time, price, location, or vendor

In addition to this basic Web Service description, WSMF considers other properties of the service such as:

1. *Failure* — When an error occurs affecting one of the invoked elements of a service and recovery is not possible, information about the reason of the failure must be provided.
2. *Concurrent execution* — If necessary it should allow the parallel execution of different Web Services as a realization of the functionality of a particular one.
3. *Concurrent data input and output* — In case input data is not available while a service is executing, it enables providing such input at a later stage and, if required, pass it from one invoker to another until reaching the actual requester.
4. *Dynamic service binding* — In case a service is required to invoke others to provide its service, a new proxy call is declared; the proxy allows referral to a service without knowing a definite time to which concrete service is bound.
5. *Data flow* — Refers to the concrete proxy ports where data has to be forwarded.
6. *Control flow* — Defines the correct execution sequence among two or more services.
7. *Exception handling* — Upon failure, services may return exception codes, if this is the case, the means to handle a concrete exception must be defined.
8. *Compensation* — Upon failure of an invoked Web Service, a compensation strategy that specifies what to do can be defined.

Mediators. Deal with the service requester and provider heterogeneity, performing the necessary operations to enable full interoperation among them. It uses a peer-to-peer approach by means of a third party, facilitating in this way a higher degree of transparency and better scalability for both requester and provider. Mediators facilitate coping with the inherit heterogeneity of Web-based computing environments, which are flexible and open by nature. This heterogeneity refers to:

- *Data mediation.* In In terms of data representation, data types and data structures.

- *Business logics mediation.* Compensate for the mismatches in business logics.
- *Message protocols mediation.* Deals with protocol's heterogeneity.
- *Dynamic service invocation mediation.* In terms of cascading Web Service invocation. It can be done in a hard-wired way, but also, it can be more flexible by referring to certain (sub)-goals.

These main elements of the framework together with message understanding and message exchange protocol layers are put together to provide automatic Web Service discovery, selection, mediation, and composition of complex services.

31.4.2 WS-CAF

The Web Service Composite Application Framework (WS-CAF) [Bunting et al., 2003] represents another initiative that tries to address the application composition problem with the development of a framework that will provide the means to coordinate long-running business processes in an architecture and transaction model independent way. WS-CAF has been designed with the aim of solving the problems that derive from the combined use of Web Services to support information sharing and transaction processing. The discussion of WS-CAF is based on a draft of the specification made available on July 28, 2003, by Arjuna Technologies, Fujitsu Software, IONA Technologies, Oracle, and Sun Microsystems. Due to the draft nature of the specification, it is very likely that changes will occur, even though the main elements of it may remain stable.

31.4.2.1 WS-CAF Objectives

The main objectives of WS-CAF are: (1) ***interoperability***, to support various transaction models across different architectures; (2) ***complementarity***, to accompany and support current state of the art of business process description languages standards such as BPEL [Andrews et al., 2002], WSCI [Arkin et al., 2002], or BPML [Arkin, 2002] to compose Web Services; (3) ***compatibility***, to make the framework capable of working with existing Web Service standards such as UDDI [Bellwood et al., 2002], WSDL [Christensen et al., 2001], or WS-Security [Atkinson et al., 2002]; and (4) ***flexibility*** based on a stack architecture, to support the specific level of service required by Web Services combination.

31.4.2.2 WS-CAF Principles and Terminology

The framework definition is based on two principles, namely: (1) ***interrelation*** or how participants share information and coordinate their efforts to achieve predictable results despite failure; and (2) ***cooperation*** to accomplish shared purposes. Web Services cooperation can range from performing operations over a shared resource to its own execution in a predefined sequence.

The WS-CAF terminology makes use of the concepts of *participant*, *context*, *outcome*, and *coordinator* throughout the specification:

- **Participants**. Cooperating Web Services that take part in the achievement of a shared purpose.
- **Context**. Allows storing and sharing relevant information to participants in a composite process to enable work correlation. Such data structure includes information like identification of shared resources, collection of results, and common security information.
- **Outcome**. Summary of results obtained by the execution of cooperating Web Services.
- **Coordinator**. Responsible for reporting participants about the outcome of the process, context management, and persisting participants outcome.

31.4.2.3 WS-CAF Elements

The framework consists of three main elements: (1) *Web Services Context (WS-CTX)* represents the basic processing unit of the framework, allowing multiple cooperating Web Services to share a common context; (2) *Web Services Coordination Framework (WS-*cf.*)* builds on top of WS-CTX and distributes organizational information relevant for the activity to participants, allowing them to customize the coordination protocol that better suits their needs; and (3) *Web Services Transaction Management (WS-TXM)* represents a layer on top of WS-CTX and WS-cf., defining transaction models for the different types of B2B interaction [Bunting et al., 2003].

31.4.2.4 WS-CTX

WS-CTX is a lightweight mechanism that allows multiple Web Services participating in an activity to share a common context. It defines the links among Web Services through their association with a Web Service Context Service, which manages the shared context for the group. WS-CTX defines the context, the scope of the context sharing, and basic rules for context management [Bunting et al., 2003a].

WS-CTX allows stating starting and ending points for activities, provides registry facilities to control which Web Services are taking part in a concrete activity, and disseminates context information. WS-CTX is composed of three main elements, namely:

- *Context Service:* Defines the scope of an activity and how information about the context can be referenced and propagated.
- *Context:* Defines basic information about the activity structure. It contains information on how to relate multiple Web Services with a particular activity. The maintenance of contexts and its association with execution environments is carried by the Context Service, which keeps a repository of contexts.
- *The Activity Lifecycle Service (ALS):* An extension of the Context Service, which facilitates the activity's enhancement with higher-level interfaces. Whenever a context is required for an activity and it does not exactly suit the necessities of the particular application domain, the Context Service issues a call to the registered ALS, which provides the required addition to the context.

Essentially WS-CTX allows the definition of activity in regard to Web Services, provides the means to relate Web Services to one another with respect to a particular activity, and defines Web Services mappings onto the environment.

31.4.2.5 WS-CF

WS-CF is a sharable mechanism that allows management of lifecycles, and context augmentation, and guarantees message delivery, together with coordination of the interactions of Web Services that participate in a particular transaction, by means of outcome messages [Bunting et al., 2003b].

WS-CF allows the definition of the starting and ending points for coordinated activities, the definition of points where coordination should take place, the registration of participants to a concrete activity, and propagation of coordination information to activity participants.

The WS-CF specification has three main architectural components, namely:

- *Coordinator:* Provides the means to register participants for a concrete activity
- *Participant:* Specifies the operation(s) performed as part of the coordination sequence processing
- *Coordination Service:* Determines a processing pattern used to define the behavior of a particular coordination model

In a nutshell, WS-CF defines the core infrastructure for Web Services Coordination Service, provides means to define Web Services mappings onto the environment, delineates infrastructure support, and finally allows concreting the responsibilities of the different WS-CF subcomponents.

31.4.2.6 WS-TXM

WS-TXM comprises protocols that facilitate support for various transaction processing models, providing interoperability across multiple transaction managers by means of different recovery protocols [Bunting et al., 2003c].

WS-TXM gives the core infrastructure for Web Services Transaction Service, facilitates means to define Web Services mappings onto the environment, defines an infrastructure to support an event communication mechanism, and finally establishes the roles and responsibilities of the WS-TXM components.

Roughly speaking, WS-CAF enables the sharing of Web Service's context viewed as an independent resource; provides a neutral and abstract transaction protocol to map to existing implementation; presents an application transaction level dependent upon application needs; includes a layered architecture that allows applications to the level of service needed, not imposing more functionalities than required; and includes a great degree of interoperability, thanks to the use of vendor neutral protocols.

31.4.3 Frameworks Comparison

Both of the initiatives introduced so far present a solution to compose applications out of multiple services using different approaches, and currently at different development stages. Whereas WSMF adopts a more formal philosophy that focuses on how Web Services should be described to achieve composition based on the paradigms of the Semantic Web (and providing an extensive covering of all the different aspects in the field), WS-CAF takes a more hands-on approach from a service execution point of view and its requirements to address the same problem.

WS-CAF puts special emphasis on dealing with failure, context management, transaction support, and effort coordination, whereas WSMF uses ontologies as a pivotal element to support its scalable discovery, selection, mediation and composition aim.

WS-CAF already counts with a layered architecture for services execution, which shows a more mature state of development, whereas WSMF efforts have focused on establishing the pillars of what a complete framework to model Semantic Web Services should look like, providing a broader coverage of the subject.

Both initiatives count on the support of major software vendors, which will result in the development of a solid technology with a clear business- and user-driven aim.

Table 31.1 summarizes the functionality provided by both frameworks regarding its automation support. As a comment, the mediation must account for process, protocol, data, and service invocation mediation, whereas execution must take care of failure, context management, transaction support, effort coordination, concurrent execution, concurrent input and output, exception handling, and compensation strategies in case of failure.

The Semantic Web is the future of the Web. It counts on ontologies as a key element to describe services and provide a common understanding of a domain. Semantic Web Services is its killer application. The development foreseen for this technology presents WSMF as a stronger candidate from an impact and visibility point of view, as well as a long-term runner, due to the use of the paradigms of the Semantic Web. WS-CAF does not use ontologies in its approach, but puts strong emphasis in the coordination of multiple and possibly incompatible transaction processing models across different architectures which, in the short term, will bridge a technology gap and will enable the cooperation of both frameworks in the near future.

TABLE 31.1 Summary of Intended Purpose of WSMF and WS-CAF

Framework	Automation Support
WSMF	— Discovery — Selection — Mediation — Composition — Execution
WS-CAF	— Mediation — Composition — Execution

31.5 Epistemological Ontologies for Describing Services

Upper level ontologies supply the means to describe the content of on-line information sources. In particular, and regarding Web Services, they provide the means to mark them up, describing their capabilities and properties in unambiguous, computer-interpretable form. In the coming sections a description of DAML-S and its actual relation with OWL-S is presented.

31.5.1 DAML-S and OWL-S

The Defense Advanced Research Projects Agency (DARPA) Agent Markup Language (DAML) for Services (DAML-S) [DAML-S 2003] is a collaborative effort by BBN Technologies, Carnegie Mellon University, Nokia, Stanford University, SRI International, and Yale University to define an ontology for semantic markup of Web Services.

DAML-S sits on top of WSDL at the application level, and allows the description of knowledge about a service in terms of what the service does — represented by messages exchanged across the wire between service participants as to *why* and *how* it functions [DAML-S 2003].

The aim of DAML-S is to make Web Services computer interpretable enabling their automated use. Current DAML-S releases (up to 0.9) have been built upon DAML+OIL, but to ensure a smooth transition to OWL (Web Ontology Language), 0.9 release and subsequent ones will be based also on OWL. Roughly speaking, DAML-S refers to the ontology built upon DAML+OIL, whereas OWL-S refers to that built upon OWL.

31.5.2 DAML-S Elements

DAML-S ontology allows the definition of knowledge that states what the service requires from agents, what it provides them with, and how. To answer such questions, it uses three different elements: (1) ***the Service Profile,*** which facilitates information about the service and its provider to enable its discovery, (2) ***the Service Model***, which makes information about how to use the service available, and (3) ***Service Grounding***, which specifies how communications among participants are to be carried on and how the service will be invoked.

Service Profile. The Service Profile plays a dual role; service providers use it to advertise the services they offer, whereas service requesters can use it to specify their needs. It presents a public high-level description of the service that states its intended purpose in terms of: (1) *the service description*, information presented to the user, browsing service registries, about the service and its provider that helps to clarify whether the service meets concrete needs and constraints such as security, quality requirements, and locality; (2) *functional behavior,* description of duties of the service; and (3) *functional attributes*, additional service information such as time response, accuracy, cost, or classification of the service.

Service Model. This allows a more detailed analysis of the matching among service functionalities and user needs, enabling service composition, activity coordination and execution monitoring. It permits the description of the functionalities of a service as a *process*, detailing control and data flow structures. The Service Model includes two main elements, namely: (1) *Process Ontology*, which describes the service in terms of inputs (information necessary for process execution), outputs (information that the process provides), preconditions (conditions that must hold prior the process execution), effects (changes in the world as a result of the execution of the service) and, if necessary, component subprocess; and (2) *Process Control Ontology* that describes process in terms of its state (activation, execution and completion). Processes can have any number of inputs, outputs, preconditions, and effects. Both outputs and effects can have associated conditions.

The process ontology allows the definition of *atomic*, *simple*, and *composite* processes:

- **Atomic processes.** They are directly invocable, have no subprocesses, and execute in a single step from the requester perspective.
- **Simple processes.** They are not directly invocable and represent a single-step of execution. They are intended as elements of abstraction that simplify the composite process representation or allow different views of an atomic process.
- **Composite processes.** They decompose into other processes either composite or noncomposite, by means of control constructs.

Service Grounding. Service Grounding specifies details regarding how to invoke the service (protocol, messages format, serialization, transport, and addressing). The grounding is defined as mapping from abstract to concrete realization of service's descriptions in terms of inputs and outputs of the atomic process, realized by means of WSDL and SOAP. The service grounding shows how inputs and outputs of an atomic process are realized as messages that carry inputs and outputs.

In brief, DAML-S is an upper ontology used to describe Web Services and includes elements intended to provide automated support for the Semantic Web Service's tasks. Table 31.2 summarizes the facilities covered by each upper level concept in the DAML-S ontology:

The following example (adapted from Ankolekar et al. [2002]) provides detailed information on how to use the DAML-S ontology to describe the pieces of software that can build a service and how to define the grounding of each one of these basic processes. The example presents a Web aggregation service. Given the user's login, his password, and the area of interest to which the aggregation is to be performed, it consolidates information disseminated over different Web sources, (i.e., banks in which the user has an account, telephone companies the user works with, and news services to which the user has signed), and presents it avoiding the process of login and browsing each one of the sources, in search for the desired piece of information.

First the service must be described in terms of the different constituents that comprise it, specifying the type of process (atomic, simple, composite). In this case a description of the news aggregation service, as an atomic process, is provided.

```
<daml:Class rdf:ID="NewsAggregation">
      <rdfs:subClassOf rdf:resource="&process;#AtomicProcess"/>
</daml:Class>
```

Then, the set of different properties associated with each one of the programs of the service must be defined. An input for the news aggregation service could be the language used to write the gathered news.

```
<rdf:Property rdf:ID="language">
      <rdfs:subPropertyOf rdf:resource="&process;#input"/>
      <rdfs:domain rdf:resource="#NewsAggregation"/>
      <rdfs:range rdf:resource="&xsd;#string"/>
</rdf:Property>
```

TABLE 31.2 Summary of Intended Purpose of DAML-S Upper Level Concepts

Upper level concept	Automation Support
Profile	— Discovery
Model	— Planning — Composition — Interoperation — Execution monitoring
Grounding	— Invocation

Next the grounding must be defined and related to the service constituent. For this purpose a restriction tag is used, establishing that the NewsAggregation program has *grounding* and that it is identified by the name NewsAggregationGrounding. Basically we are stating that every instance of the class has an instance of the hasGrounding property, with the value NewsAggregationGrounding.

```
<daml:Class rdf:about="NewsAggregation ">
      <daml:sameClassAs>
            <daml:Restriction daml:cardinality="1">
                  <daml:onProperty rdf:resource="#hasGrounding"/>
                  <daml:hasValue   rdf:resource="#
            NewsAggregationGrounding"/>
      </daml:Restriction>
      </daml:sameClassAs>
</daml:Class>
```

Finally, an example of a DAML-S grounding instance is presented. It is important to notice that URIs (#ConsolidateNews, #NewsAggregationInput, etc.) correspond to constructs in a WSDL document, which is not shown here.

```
<grounding:WsdlGrounding rdf:ID=" NewsAggregationGrounding ">
      <grounding:wsdlReference rdf:resource="http://www.w3.org/TR/2001/
NOTE-wsdl-20010315">
      <grounding:otherReferences rdf:parseType="daml:collection">
            "http://www.w3.org/TR/2001/NOTE-wsdl-20010315"
            "http://schemas.xmlsoap.org/wsdl/soap/"
            "http://schemas.xmlsoap.org/soap/http/"
      </grounding:otherReferences>
      <grounding:wsdlDocuments rdf:parseType="daml:collection">
            "http://service.com/aggregation/newsAggregation.wsdl"
      </grounding:wsdlDocuments>
      <grounding:wsdlOperation
            rdf:resource="http://service.com//
newsAggregation#ConsolidateNews"/>
      <grounding:wsdlInputMessage
      rdf:resource="http://service.com//
newsAggregation.wsdl#NewsAggregationInput"/>
      <grounding:wsdlInputMessageParts rdf:parseType="daml:collection">
            <grounding:wsdlMessageMap>
            <grounding:damlsParameter rdf:resource=#NewsLanguage>
                  <grounding:wsdlMessagePart
      rdf:resource="http://service.com//newsAggregation.wsdl
#aggregatedNews">
            </grounding:wsdlMessageMap>

      … other message map elements…

</grounding:wsdlInputMessageParts>
<grounding:wsdlOutputMessage
            rdf:resource="http://service.com//
newsAggregtion.wsdl#NewsAggregationOutput"/>
      <grounding:wsdlOutputMessageParts rdf:parseType="daml:collection">

      … similar to wsdlInputMessageParts…

</grounding:wsdlOutputMessageParts>
<grounding:WsdlGrounding>
```

31.5.3 Limitations

DAML-S and OWL-S early beta stage can be seen as a Gruyere cheese with lots of holes to fill [Payne, 2003]. It has numerous limitations that will be overcome in a near future:

- **One-to-one mapping between Profiles and Service Models.** It prevents the reuse of profiles and n:m mappings between profiles and concrete services.
- **Separation between public and private business logic.** It does not allow hiding the internals of the service, exposing its business logic to requesters.
- **Lack of conversation interface definition.** The service model could be used as the conversational interface if the service is published as a composite service, but this is not stated as the intended purpose of the service model.
- **Interface overloading.** The defined WSDL grounding doesn't support publishing one single service interface for a service accepting different inputs.
- **Preconditions and effects.** No means to describe pre-conditions and effects are given, and the definitions of these elements are not clear enough.
- **Constructs.** The constructs defined for the process model may not be sufficient.

The path followed by this standard seems to be the correct one. Much effort is being put to work to overcome many of the actual problems and limitations, which will undoubtedly be solved in the near future, providing a solid standard to consistently describe Semantic Web Services.

31.6 Summary

The Semantic Web is here to stay. It represents the next natural step in the evolution of the current Web. It will have a direct impact in areas such as e-Business, Enterprise Application Integration, and Knowledge Management, and an indirect impact on many other applications affecting our daily life. It will help create emerging fields where knowledge is the most precious value, and will help to further the development of existing ones.

The Semantic Web will create a complete new concept of the Web by extending the current one, alleviating the information overload problem, and bringing back computers to their intended use as computational devices, and not just information-rendering gear. Ontologies are the backbone of this revolution due to their potential for interweaving human understanding of symbols with machine processability [Fensel, 2002]. Ontologies will enable semantical enhancement of Web Services, providing the means that facilitate the task-driven automatic discovery, composition, and execution of interorganization business logics, thus making Semantic Web Services the killer application of the Semantic Web.

A shared functionality description is the key element towards task-driven automatic Web Service discovery, composition, and execution of interorganization business logics. It will permit the (1) location of different services that, solely or in combination, will provide the means to solve a given task, (2) combination of services to achieve a goal, and (3) facilitation of the replacement of such services by equivalent ones that, solely or in combination, can realize the same functionality, such as in case of failure during execution.

Many business areas are giving a warm welcome to the Semantic Web and Semantic Web Services. Early adopters in the fields of biotechnology or medicine are already aware of its potentials to organize knowledge and infer conclusions from available data. The juridical field has already realized its benefits to organize and manage large amounts of knowledge in a structured and coherent way. The interest of professionals from this sector towards the Semantic Web is rapidly gaining momentum, being forecasted as an area where the Semantic Web will make a difference. Engineers, scientists, and basically anyone dealing with information will soon realize the benefits of Semantic Web Services due to the new business paradigm it provides. Business services will be published on the Web and will be marked up with machine-processable semantics to allow their discovery and composition, providing a higher level of functionality, adding increasingly more complex layers of services, and relating business logics from different companies in a simple and effective way.

The hardest problem the Semantic Web Services must overcome is its early development stage. Nowadays, the technology is not mature enough to accomplish its promises. Initiatives in Europe and the U.S. are gaining momentum, and the appropriate infrastructure, technology, and frameworks are currently being developed. Roughly speaking there is a gap between the current Web and the Semantic one in terms of annotations. Pages and Web resources must be semantically enriched in order to allow automatic Web Services interoperation. Once this is solved, the Semantic Web and Semantic Web Services will become a plausible reality and the Web will be lifted to its full potential, causing a revolution in human information access. The way computers are perceived and the Web is understood will be completely changed, and the effects will be felt in every detail of our daily life.

Acknowledgements

The authors would like to thank SungKook Han, Holger Lausen, Michael Stolberg, Jos de Brujin, and Anna Zhdanova for the many hours spent discussing different issues that helped in great manner to make this work possible. Also thanks to Alexander Bielowski for his feedback regarding English writing.

References

[Andrews et al., 2002] Tony Andrews, Francisco Curbera, Hitesh Dholakia, Yaron Goland, Johannes Klein, Frank Leymann, Kevin Liu, Dieter Roller, Doug Smith, Ivana Trickovic, and Sanjiva Weerawarana. Business Process Execution Language for Web Service (BPEL4WS) 1.1. http://www.ibm.com/developerworks/library/ws-bpel, August 2002.

[Ankolekar et al., 2002] Anupriya Ankolekar, Mark Burstein, Jerry R. Hobbs, Ora Lassila, David Martin, Drew McDermott, Sheila A. McIlraith, Srini Narayanan, Massimo Paolucci, Terry Payne, and Katya Sycara. DAML-S: Web Service Description for the Semantic Web. International Semantic Web Conference, ISWC 2002.

[Arkin, 2002] Assaf Arkin. Business Process Modelling Language. http://www.bpmi.org/, 2002.

[Arkin et al., 2002] Assaf Arkin, Sid Askary, Scott Fordin, Wolfgang Jekeli, Kohsuke Kawaguchi, David Orchard, Stefano Pogliani, Karsten Riemer, Susan Struble, Pal Takacsi-Nagy, Ivana Trickovic, and Sinisa Zimek. Web Service Choreography Interface 1.0. http://wwws.sun.com/software/xml/developers/wsci/wsci-spec-10.pdf, 2002.

[Atkinson et al., 2002] Bob Atkinson, Giovanni Della-Libera, Satoshi Hada, Maryann Hondo, Phillip Hallam-Baker, Johannes Klein, Brian LaMacchia, Paul Leach, John Manferdelli, Hiroshi Maruyama, Anthony Nadalin, Nataraj Nagaratnam, Hemma Prafullchandra, John Shewchuk, and Dan Simon. Web Services Security (WS-Security) 1.0. http://www-106.ibm.com/developerworks/library/ws-secure, 2002.

[Bellwood et al., 2002] Tom Bellwood, Luc Clément, David Ehnebuske, Andrew Hately, Maryann Hondo, Yin Leng Husband, Karsten Januszewski, Sam Lee, Barbara McKee, Joel Munter, and Claus von Riegen. UDDI Version 3.0. Published Specification. http://uddi.org/pubs/uddi-v3.00-published-20020719.htm, 2002.

[Berners-Lee et al., 2001] Tim Berners-Lee, James Hendler, and Ora Lassila. The Semantic Web. *Scientific American*, 284(5): 34–43, 2001.

[Box et al., 2000] Don Box, David Ehnebuske, Gopal Kakivaya, Andrew Layman, Noah Mendel-sohn, Henrik F. Nielsen, and Satish Thatte, Dave Winer. Simple Object Access Protocol (SOAP) 1.1. http://www.w3.org/TR/SOAP/, 2000.

[Bunting et al., 2003] Doug Bunting, Martin Chapman, Oisin Hurley, Mark Little, Jeff Mischkinsky, Eric Newcomer, Jim Webber, and Keith Swenson. Web Service Composite Application Framework (WS-CAF). http://developers.sun.com/techtopics/Webservices/wscaf/primer.pdf, 2003.

[Bunting et al., 2003a] Doug Bunting, Martin Chapman, Oisin Hurley, Mark Little, Jeff Mischkinsky, Eric Newcomer, Jim Webber, and Keith Swenson. Web Service Context (WS-Context). http://developers.sun.com/techtopics/Webservices/wscaf/wsctx.pdf, 2003.

[Bunting et al., 2003b] Doug Bunting, Martin Chapman, Oisin Hurley, Mark Little, Jeff Mischkinsky, Eric Newcomer, Jim Webber, and Keith Swenson. Web Service Coordination Framework (WS-cf.). http://developers.sun.com/techtopics/Webservices/wscaf/wscf.pdf, 2003.

[Bunting et al., 2003c] Doug Bunting, Martin Chapman, Oisin Hurley, Mark Little, Jeff Mischkinsky, Eric Newcomer, Jim Webber, and Keith Swenson. Web Service Transaction management (WS-TXM). http://developers.sun.com/techtopics/Webservices/wscaf/wstxm.pdf, 2003.

[Christensen et al., 2001] Erik Christensen, Francisco Curbera, Greg Meredith, and Sanjiva Weerawarana. WSDL Web Services Description Language (WSDL) 1.1. http://www.w3.org/TR/wsdl, 2001.

[Daconta et al., 2003] Michael C. Daconta, Leo J. Obrst, and Kevin T. Smith. *The Semantic Web: A Guide to the Future of XML, Web Services, and Knowledge Management.* John Wiley and Sons, Indianapolis, IN, 2003.

[DAML-S, 2003] The DAML services coalition. DAML-S: Semantic Markup for Web Services (version 0.9). http://www.daml.org/services/daml-s/0.9/daml-s.pdf, 2003.

[DAML] DARPA Agent Markup Language (DAML). www.daml.org.

[Ding et al., 2003] Ying Ding, Dieter Fensel, and Hans-Georg Stork. The semantic web: from concept to percept. *Austrian Artificial Intelligence Journal* (OGAI), 2003, in press.

[DIP] Data, Information and Process Integration with Semantic Web Services. http://dip.semanticweb.org/.

[ebXML] electronic business XML (ebXML). www.ebxml.org/specs.

[Esperonto] Esperonto. esperonto.semanticWeb.org.

[Fensel, 2001] Dieter Fensel. *Ontologies: Silver Bullet for Knowledge Management and Electronic Commerce.* Springer-Verlag, Berlin, 2001.

[Fensel, 2002] Dieter. Fensel. Semantic Enabled Web Services XML-Web Services ONE Conference, June 7, 2002.

[Fensel and Bussler, 2002] Dieter Fensel and Christoph Bussler. The Web Service Modeling Framework WSMF. *Electronic Commerce Research and Applications*, 1(2), 2002.

[Gruber, 1993] Thomas R. Gruber. A Translation Approach to Portable Ontology Specifications. *Knowledge Acquisition*, 5: 199–220, 1993.

[HPKB] High Performance Knowledge Bases (HPKB). reliant.teknowledge.com/HPKB.

[KowledgeWeb] Knowledge Web. knowledgeWeb.semanticWeb.org

[Martin, 2001] James Martin. Web Services: The Next Big Thing. *XML Journal*, 2, 2001. http://www.sys-con.com/xml/archivesbad.cfm.

[McIlraith et al., 2001] Sheila A. McIlraith, Tran C. Son, and Honglei Zeng. Semantic Web Services. *IEEE Intelligent Systems*, 16(2), March/April, 2001.

[Newcomer, 2002] Eric Newcomer. *Understanding Web Services: XML, WSDL, SOAP UDDI.* Addison Wesley, Reading, MA, 2002.

[OntoKnowledge] Ontoknowledge. www.ontoknowledge.org.

[OntoWeb] OntoWeb. www.ontoWeb.org.

[Payne, 2003]. Terry Payne. The First European Summer School on Ontological Engineering and the Semantic Web. http://minsky.dia.fi.upm.es/sssw03, Cercedilla, Spain, July 21-26, 2003.

[Peer, 2002] Joachim Peer. Bringing Together Semantic Web and Web Services. *First International Semantic Web Conference*, Sardinia, Italy, June 2002.

[RKF] Rapid Knowledge Formation (RKF). reliant.teknowledge.com/RKF.

[SEKT] Semantic Knowledge Technologies. sekt.semanticWeb.org.

[Sirin et al., 2003] Evren Sirin, James Hendler, and Bijan Parsia. Semi-automatic Composition of Web Services using Semantic Descriptions. *Proceedings of the 1st Workshop on Web Services: Modeling, Architecture and Infrastructure (WSMAI-2003).* In conjunction with ICEIS 2003, Angers, France, pp. 17–24, April 2003.

[Sycara et al., 1999] Katya Sycara, Matthias Klusch, and Seth Widoff. Dynamic service matchmaking among agents in open information environments. *ACM SIGMOD Record*, 28(1): 47–53, March 1999.

[SWWS] Semantic Web Enabled Web Services. swws.semanticWeb.org.

[WSMO] Web Service Modeling Ontology, http://www.wsmo.org/.

[Tidwell, 2000] Doug Tidwell. Web Services: the Web's Next Revolution. http://www-106.ibm.com/developerworks/edu/ws-dw-wsbasics-i.html.

[W3C] World Wide Web Consortium (W3C). www.w3.org/2001/sw.

32

Business Process: Concepts, Systems, and Protocols

Fabio Casati

Akhil Sahai

CONTENTS

32.1 Introduction 32-1
- 32.1.1 The Need for Business Process Automation 32-2
- 32.1.2 Workflow Management Systems Overview 32-4
- 32.1.3 Scheduling 32-5
- 32.1.4 Resource Assignment 32-6
- 32.1.5 Data Management 32-7
- 32.1.6 Failure and Exception Handling 32-7
- 32.1.7 WfMS Architectures 32-9

32.2 Web Services and Business Processes 32-10
- 32.2.1 Web Services 32-10
- 32.2.2 Web Services and Business Process Orchestration 32-11

32.3 Conclusion 32-13

References 32-13

32.1 Introduction

Companies and organizations are constantly engaged in the effort of providing better products and services at lower costs, of reducing the time to market, of improving and customizing their relationships with customers and ultimately, of increasing customer satisfaction and the company's profits. These objectives push companies and organizations towards continuously improving the *business processes* that are performed in order to provide services or produce goods. The term *business process* denotes a set of tasks executed by human or automated resources according to some ordering criteria that collectively achieves a certain business goal. For example, the set of steps required to process a travel reimbursement request constitutes a business process. A new execution of the process begins when an employee files a reimbursement request. At that point, an approving manager is selected, and the request is routed to him. If the manager approves the request, then the funds are transferred to the employee bank account, otherwise a notification is sent to the employee informing him or her about the causes of the rejection.

This process is graphically depicted in Figure 1. Of course, the process shown here is much simplified with respect to the way expense reimbursement requests are actually handled, but this example suffices to introduce the main concepts. Many different business processes are executed within a company. Requesting quotes, procuring goods, processing payments, and hiring or firing employees are all examples of business processes. The term "business" denotes that these are processes that perform some business function, and this distinguishes them from operating system processes.

1-58488-381-2/05/$0.00+$1.50
© 2005 by CRC Press LLC

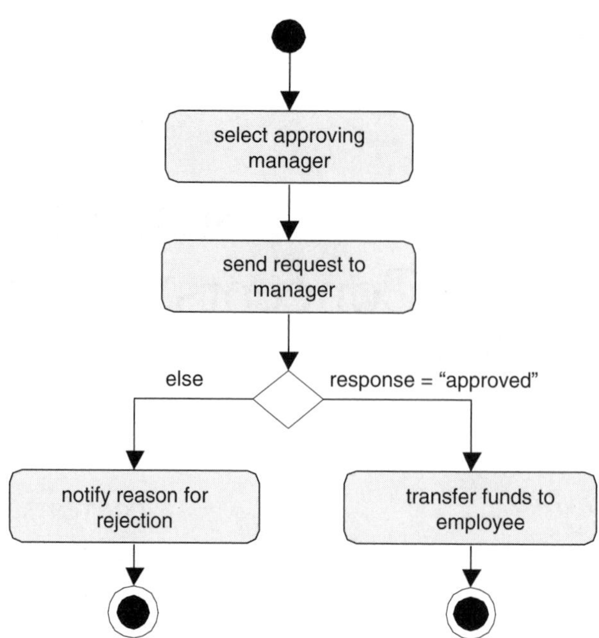

FIGURE 32.1 The travel expense reimbursement process.

32.1.1 The Need for Business Process Automation

Business processes are really at the heart of what a company does (Figure 2). Every activity of a company can actually be seen as a business process. This means that improving the quality and reducing the cost of the business processes is key to the success of a company. For example, if a company can have a better and more effective supply chain operations, then this will typically result in both higher revenues and higher profits.

From an information technology perspective, business process improvement and optimization are closely related to businesss *automation*. In fact, if the different steps of the process (and even its overall execution) can be enacted and supervised in an automated fashion, there is a reduced need for human involvement (i.e., reduced costs), and the execution is faster and more accurate.

Business process automation can be achieved at many different levels that also correspond to the historical evolution in the technology within this domain. At the basic level, the different enterprise functions can be automated. This corresponds to automating the individual steps in a business process. For example, in the expense reimbursement process, fund transfers can be performed through banking transactions (e.g., SWIFT wire transfer) and by accessing ERP systems for internal accounting purposes.

Automating the individual steps provides great benefit to companies because it generates significant improvement in terms of cost and quality: The execution of each individual step is now faster and more accurate. However, this in itself does not suffice to reach the goal of a streamlined process. Indeed, it leads to *islands of automation*, where the individual functions are performed very efficiently, but the overall process still requires much manual work for transferring data among the different applications and for synchronizing and controlling their execution. For example, in the expense-reimbursement process, even if the individual steps are automated, there is still lots of manual work involved in entering data into a system (e.g., the human resource database), getting the desired output (e.g., the name and email address of the approving manager), providing the data for the next step (sending an email to the manager), collecting once again the results (i.e., the approval or rejection) and, based on them,

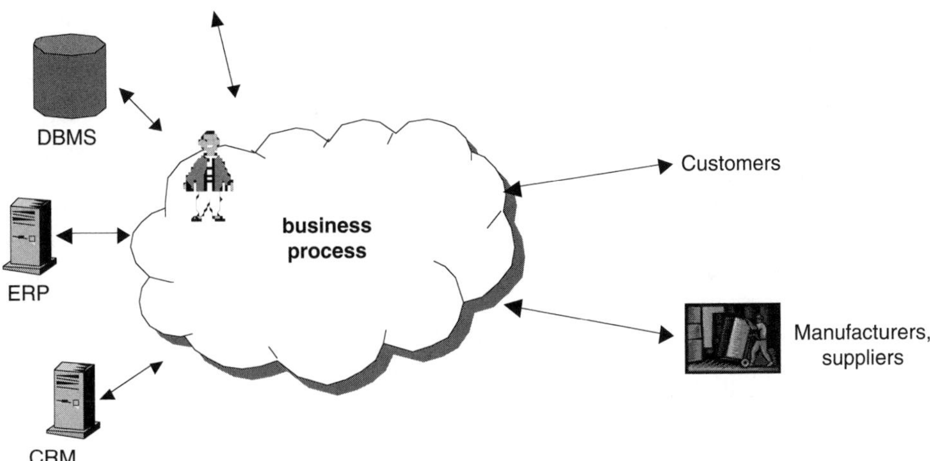

FIGURE 32.2 Business processes are at the heart of what a company does.

interacting with the funds-transfer system or sending an email to the employee notifying him or her of the rejection.

Even in such a simple process, someone must take care of interacting with many different systems, entering the same data many times, making sure that the steps are performed in the correct order and that nothing is skipped, etc. All this is clearly prone to errors. Furthermore, exception detection and handling is also left to the human. For example, the administrative employees in charge of executing the process need to continuously monitor each case to make sure that the approving manager sends a response in due time, otherwise a reminder should be sent or the request should be routed to a different manager.

Another problem is that of *process tracking*. It happens frequently that, once a request is made, the employee calls back at a later time to know the status of the request, or a customer enquires about the progress of an order. If only the individual process steps are automated, but the overall process logic is executed manually, it will be difficult to be able to assess the stage in which a request is.

The discussion above shows just a few of the reasons why automating the individual steps, and not the overall process, is not sufficient to achieve the quality and efficiency goals required by today's corporations.

The obvious next step, therefore, consists in automating the process logic by developing an application that goes through the different steps of the process, which automatically transfers data from one step to the next and manages exceptions. This is conceptually simple, but the implementation can prove to be very challenging. The reason is simple: the systems that support the execution of each individual step are very different from each other. They have different interfaces, different interaction paradigms, different security requirements, different transport protocols, different data formats, and more.

Therefore, if we take the simplistic approach of coding the process logic in some third generation language without any specific process automation support, then not only do we have to code the sequencing of the steps, the transfer of data between one step and the next, and the exception handling, but we also have to deal with the heterogeneity of the invoked applications.

The amount of coding needed is usually quite large, and it is one of the hardest tasks in process automation. The problem gets even worse if the invoked applications belong to a different company (e.g., a product database maintained by a vendor), as this requires crossing firewalls and trust domains.

A solution frequently adopted to deal with this problem consists in adding a layer that shields the programmers from the heterogeneity and makes all systems look alike from the perspective of the developer who has to code the process logic. The middleware tools that support this effort are typically called *Enterprise Application Integration* (EAI) platforms.

Essentially, their functionality consists in providing a single data model and interaction mechanism to the developers (e.g., Java objects and JMS messages), and in converting these homogeneous models and mechanisms into the ones supported by each of the invoked applications, by means of adapters that perform appropriate mappings.

Many commercial EAI platforms exist, supporting integration within and across enterprises. Some examples are IBM WebSphere MQ [1], Tibco Rendez-Vous and related products [2], WebMethods Enterprise Integrator [3], and Vitria BusinessWare [4].

EAI platforms therefore aim at solving one part of the problem: that of managing heterogeneity. To achieve process automation, the "only" remaining step consists in encoding the process logic. This is actually what most people refer to when speaking about process automation. It is also the most interesting problem from a conceptual and research perspective. In the following we, therefore, focus on this aspect of business process automation, examine the challenges in this domain, and show what has been done in the industry and in academia to address them.

32.1.2 Workflow Management Systems Overview

In order to support the definition and execution of business processes, several tools have been developed within the last decade. These tools are collectively known as *Workflow Management Systems* (WfMS). In WfMS terminology, a *workflow* is a formal representation of a business process, suitable for being executed by an automated tool. Specifications of a process done by means of a workflow language are called *workflow schemas.* Each execution of a workflow schema (e.g., each travel expense reimbursement) is called *workflow instance.*

A WfMS is then a platform that supports the specification of workflow schemas and the execution of workflow instances. In particular, a WfMS supports the following functionality:

- *Schedule* activities for execution, in the appropriate order, as defined by the workflow schema. For example, in the expense reimbursement process, it schedules the manager selection, then the notification of the request to the manager, and then either the employee notification or the fund transfer, depending on the manager's approval.
- *Assign* activities to *resources.* In fact, a different person or component can execute each task in a process. The WfMS identifies the appropriate resource, based on the workflow definition and on resource assignment criteria.
- *Transfer data* from one activity to the next, thereby avoiding the problem of repeated data entry, one of the most labor-intensive and error-prone aspects of manual process execution.
- *Manage exceptional situations* that can occur during process execution. In the travel expense reimbursement process, these can include situations in which the manager does not send the approval, the Human Resources database system is not available, or the employee cancels the reimbursement request.
- *Efficiently execute high process volumes.* In workflows such as payrolls or order management, the number of instances to be executed can be in the order of the tens of thousands per day. Therefore, it is essential that the WfMS be able to support volumes of this magnitude.

There are currently hundreds of tools in the market that support these functionalities, developed by small and large vendors. Some examples are HP Process Manager, IBM MQ Series Workflow [5] (now called WebSphere workflow), Tibco Process Management [6], and Microsoft BizTalk Orchestration [7]. Although these systems differ in the detail, they are essentially based on similar concepts, and all of them try to address the issues described above. We now examine each of these issues in more detail, also showing several possible alternatives to approach each issue. The reader interested in more details is referred to [8].

32.1.3 Scheduling

Scheduling in a WfMS is typically specified by means of a flow chart, just like the one shown in Figure 1. More specifically, a workflow is described by a directed graph that has four different kinds of nodes:

- *Work nodes* represent the invocation of activities (also called *services*), assigned for execution to a human or automated *resource*. They are represented by rounded boxes in Figure 1.
- *Route nodes* are decision points that route the execution flow among nodes based on an associated *routing rule*. The diamond of Figure 1. is an example of a routing node that defines conditional execution: one of the activities connected in output is selected, based on some runtime branching condition. In general, more than one output task can be started, thereby allowing for parallel execution.
- *Start nodes* denote the entry point to the processes.
- *Completion nodes* denote termination points.

Arcs in the graph denote execution dependencies among nodes: When a work node execution is completed, the output arc is "fired," and the node connected to this arc is activated. Arcs in output to route nodes are, instead, fired based on the evaluation of the routing rules.

The model described above is analogous in spirit to activity diagrams [9]. Indeed, many models follow a similar approach because this is the most natural way for developers to think of a process. It is also analogous to how programmers are used to coding applications, essentially involving the definition of a set of procedure calls, and of the order and conditions under which these procedures should be invoked.

A variation on the same theme consists in using Petri nets as a workflow-modeling paradigm, although modified to make them suitable for this purpose. Petri-net-based approaches have been discussed in detail in [10].

Other techniques have also been proposed although they are mostly used as internal representation of a workflow schema and, therefore, are not exposed to the designers. For example, one such modeling technique consists in specifying the workflow as a set of *Event-Condition-Action* (ECA) rules that define which step should be activated as soon as another step is completed. For example, the following rule can be defined to specify part of the process logic for the travel expense reimbursement process:

```
WHEN COMPLETED(SEND_REQUEST_TO_MANAGER)
IF RESPONSE="APPROVED"
THEN START(TRANSFER_FUNDS_TO_EMPLOYEE)
```

The entire process can then be specified by a set of such rules. Examples of rule-based approaches can be found in [11, 12, 13].

In the late 1990s, an industry-wide effort was started with the goal of standardizing scheduling languages and more generally, standardizing workflow languages. The consortium supervising the standardization is called *Workflow Management Coalition* [14]. Despite initial enthusiasm, the efforts of the coalition did not manage to achieve consensus among the major players, and therefore few vendors support the workflow language proposed by this consortium.

As a final remark on process modeling we observe that, until very recently, processes were designed to work in isolation, that is, each process was designed independently of others, and it was very hard to model interactions among them (e.g., points at which processes should synchronize or should exchange data). WfMSs provided no support for this, neither in the modeling language nor in the infrastructure, so that interoperability among processes had to be implemented in *ad hoc* ways by the developer.

The only form of interaction supported by the early system was that a process was able to start another process. More recently, approaches to support interoperability were proposed both in academia (see, e.g., [15, 16]) and in industry, especially in the context of service composition (discussed later in this chapter). These approaches were mainly based on extending workflow models with the capability of publishing and subscribing to messages and of specifying points in the flow where a message was to be sent or received. In this way, requests for messages acted both as a synchronization mechanism (because such requests are blocking) and as a means to receive data from other processes.

32.1.4 Resource Assignment

Once a task has been selected for execution, the next step that the WfMS must perform consists in determining the best resource that can execute it. For example, in the travel expense reimbursement process, once the WfMS has scheduled the task *send request to manager* for execution, it needs to determine the manager to whom the request should be routed.

The ability of dynamically assigning work nodes to different resources requires the possibility of defining *resource rules*, i.e., specifications that encode the logic necessary for identifying the appropriate resource to be assigned to each work node execution, based on instance-specific data. WfMSs typically adopt one of the two following approaches for resource selection:

1. In one approach, the workflow model includes a *resource model* that allows system administrators to specify resources and their properties. The resource rule will then select a resource based on the workflow instance data and the attributes of the resource. For example, a workflow model may allow the definition of resources characterized by a name and a set of *roles*, describing the capabilities or the authorizations of a resource. A role may be played by multiple resources, and the same resource can have several roles. For example, the resource *John Smith* may play the roles of *IT manager* and *Evaluator of supply chain projects*. Once resources and roles have been defined, then work node assignment is performed by stating the roles that the resource to which the work is assigned must have. For example, an assignment rule can be:

   ```
   if %EMPLOYEE_DEPT="IT" then role="IT manager"
   else role="manager"
   ```

   where EMPLOYEE_DEPT is the name of a workflow data item (discussed next). In this case, the rule states that the node with which the rule is associated should be assigned to a resource playing role *IT manager* if the requesting employee is in the IT department and to a (generic) *manager* otherwise. Roles can typically be organized into a specialization hierarchy, with the meaning that if role A is an ancestor of role B, then a resource playing the specialized role B can also play the more generic role A. For example, role *IT manager* may be a child of role *manager*. Role hierarchies simplify the specifications of resource rules, as it is possible to assign nodes to a super-role (such as *manager*), rather than explicitly listing all the subroles for which *manager* is the ancestor. This is particularly useful if roles change over time (e.g., a new subrole of *manager* is introduced), because the existing resource rules that assign work items to role *manager* do not need to be modified to take the new role into account.
2. In an alternative approach, the workflow model does not include a resource model. Instead, work nodes are assigned to resources by contacting a *resource broker* each time a work node needs to be executed. In this case, the resource rule language is determined by the resource broker, not by the WfMS. The rule is simply passed to the resource broker for execution. The broker is expected to return the names or identifiers of one or more resources to which the work node should be assigned.

These two approaches present both advantages and disadvantages. When an internal resource model is present, the assignment is typically faster, as it does not require the invocation of the resource broker. In addition, it is more practical if the resources are limited in number and change slowly with time as there is no need to install and configure an external broker, and it is possible to specify assignments with the simple, point-and-click interface provided by the workflow design tool.

However, the internal resource model presents several drawbacks. The main problem is that often the resources are actually controlled by a component that is external with respect to the WfMS. This is a frequent situation in large organizations. For example, work nodes may need to be executed by employees who have certain attributes (e.g., work in a specific location). The database describing listed active employees, along with their roles and permissions, is maintained externally, typically by the Human Resource (HR) department, and its contents may change daily. It is therefore impractical to keep changing the workflow resource definition to keep it synchronized with the HR database. In addition, resource rules may need to be based on a variety of attributes that go beyond the resource name or role (e.g., the location). Therefore, in such cases, WfMSs offering an external broker are more flexible because users can typically plug-in the broker they need to contact in an external resource directory and query the appropriate resource, based on a broker-specific language.

32.1.5 Data Management

Once the WfMS has identified the activity to be executed and the resource that will execute it, the next step consists in preparing the data necessary for the task execution. For example, invoking the *send request to manager* activity essentially involves sending an email to a manager, thereby passing the relevant information (e.g., the employee name, the travel data, the reimbursement amount, and other information). In general, this data is derived from workflow invocation parameters (provided as a new instance is created) or from the output of a previously executed activity.

An important aspect of a workflow execution therefore involves transferring data from one activity to the next, so that the proper data is made available to the resource. Data transfers between activities are typically specified as part of a workflow schema, by means of *workflow variables*. Basically, each workflow schema includes the declaration of a set of variables; just as is done in conventional programming languages. Variables are typed, and the data types can be the "usual" integer, real, or string, or they can be more complex types, ranging from vectors to XML schemas. The variables act as data containers that store the output of the executed activities as well as the workflow invocation data passed to the WfMS as a new instance is started. For example, data about the employee reimbursement request, provided as a new instance is started, can be stored in variables *employeeName*, *employeeID*, *travelDestination*, and *requestedAmount*. The value of workflow variables can then be used to determine the data to be passed to the resource once the activity is invoked. For example, the input for the activity *send request to manager* can be specified as being constituted by variables *employeeName*, *travelDestination*, and *requestedAmount*. Variables can also be used to describe how the execution flow should be routed among activities as they are typically used as part of branch conditions. For example, Figure 1 shows that the variable *response* is used in a branch condition to determine whether the workflow should proceed by refunding the employees or by notifying them of the rejection. This approach is the one followed by most workflow models, including for example BPEL4WS [17] or Tibco [6]. It is also analogous in spirit to how programming languages work.

An alternative approach consists in performing data transfer by directly linking the output of a node A with the input of a subsequent node B, with the meaning that the output of A (or a subset) is used as the input of B. This is called the *data flow* approach, and it is followed by a few models, in particular, those specified by IBM, including MQ Series Workflow [5] and WSFL [18].

32.1.6 Failure and Exception Handling

Although research in workflow management has been very active for several years, and the need for modeling exceptions in information systems has been widely recognized (see, e.g., [19, 20]), only recently has the workflow community tackled the problem of exception handling. One of the first contributions came from Eder and Liebhart [21], who analyzed the different types of exceptions that may occur during workflow execution and provided a classification of such exceptions. They divided exceptional situations into *basic failures*, corresponding to failures at the system level (e.g., DBMS, operating system, or network failure), *application failures*, corresponding to failures of the applications invoked by the WfMS in order to execute a given task, *expected exceptions*, corresponding to predictable deviations from the normal behavior of a process, and *unexpected exceptions*, corresponding to inconsistencies between the business process in the real world and its corresponding workflow description.

Basic failures are not specific to business processes and workflow management; approaches to failure handling have in fact been developed in several different contexts, particularly in the area of transaction processing. Therefore, WfMSs may (and in fact do) handle failures by relying on existing concepts and technology. For instance, basic failures are handled at the system level, by relying on the capability of the underlying DBMS to maintain a persistent and consistent state, thus supporting forward recovery.

A generic approach to handling application failures involves the integration of workflow models with advanced transaction models [22]. In fact, if the workflow model provides "traditional" transaction capabilities such as the partial and global rollback of a process, application failures can be handled by rolling back the process execution until a decision (split) point in process is reached, from which forward execution can be resumed along a different path. This model is supported by several systems; for instance, ConTracts provide an execution and failure model for workflow applications [23]; a ConTract is a transaction composed of *steps*. Isolation between steps is relaxed so that the results of completed steps are visible to other steps. In order to guarantee semantic atomicity, each step is associated with a *compensating step* that (semantically) undoes its effect. When a step is unable to fulfill its goal, backward recovery is performed by compensating completed steps, typically, in the reverse order of their forward execution, up to a point from which forward execution can be resumed along a different path. WAMO [21, 24] and Crew [25] extend this approach by providing more flexible and expressive models, more suitable for workflow applications. Recently, commercial systems (such as the above mentioned BizTalk) started to provide this kind of functionality.

Expected exceptions are *predictable deviations* from the normal behavior of the process. Examples of expected exceptions are:

- In a travel reservation process, the customer cancels the travel reservation request.
- In a proposal presentation process, the deadline for the presentation has expired.
- In a car rental process, an accident occurs to a rented car, making it unavailable for subsequent rentals.

Unlike basic and application failures, expected exceptions are strictly related to the workflow domain: They are part of the semantics of the process, and it should be possible to model them within the process, although they are not part of its "normal" behavior.

Different models offer different approaches to define how to handle expected exceptions:

- A modeling paradigm often proposed in the literature but rarely adopted in commercial systems consists in modeling the exception by means of an ECA rule, where the event describes the occurrence of a potentially exceptional situation (e.g., the customer cancels the order), the condition verifies that the occurred event actually corresponds to an exceptional situation that must be managed, whereas the action reacts to the exception (e.g., aborts the workflow instance) [11, 12, 26]. Although this approach is conceptually feasible, it did not gain popularity due to the complexity of rule-based modeling (viable only if there are very few rules) and due to the need of supporting two different languages (one for defining the normal flow and the other for defining the exceptional flow).
- Another viable paradigm consists in mimicking the exception-handling scheme followed by programming languages such as Java: A part of the schema is enclosed into a logical unit, and exception-handling code is attached to the unit to catch and handle the exception, much like a pair of *try/catch* statements [27].
- Another possible approach consists in defining a specialized type of node, called *event node*, that "listens" to an event explicitly raised by a user or by the WfMS, and if the event is detected, activates an exception handling portion of the flow, specified using the same techniques used for defining the normal schema. This approach is for example discussed in [28, 29].

The common aspect of all these techniques is that they allow capturing events that occur asynchronously with respect to the workflow execution, i.e., that can occur at any point in a workflow instance (or in a logical unit, for the Java-like approach) and not just in correspondence with the execution of a certain task. This is important as exceptions typically have this asynchronous nature.

Finally, unexpected exceptions correspond to errors in modeling the process. Therefore, the only (obvious) way to handle them consists in modifying the process definition, and hence we do not discuss them further.

32.1.7 WfMS Architectures

After having presented the different aspects of a workflow model, we now briefly discuss workflow architectures. A WfMS is typically characterized by the following components (Figure 3):

- The *workflow designer*, a lightweight client-side tool that allows users to define and deploy new workflows as well as modify existing workflows. It is typically characterized by a graphical user interface through which designers can quickly and easily draw a schema. Definitions are then translated and saved in a textual format (typically in some XML representation).
- The *workflow definition manager* receives process definitions from the designer and stores them into a file repository or a relational database. In addition, it deploys the workflows so that it can be instantiated, manages versions, and controls concurrent access, for example, preventing multiple users from modifying the same workflow at the same time.
- A *workflow engine* executes workflow instances, as discussed earlier in the section. The engine accesses both the workflow definition and the workflow instance databases, in order to determine the nodes to be executed.
- A *resource broker* determines the resources that are capable and authorized to execute a work item.
- The *worklist handler* receives work items from the engine and delivers them to users (by pushing work to the users or by delivering upon request). For human resources, access to the worklist handler is typically performed via a browser, and automated resources can access the worklist through an API, typically a Java or C++ API. Note that the worklist handler hides the heterogeneity

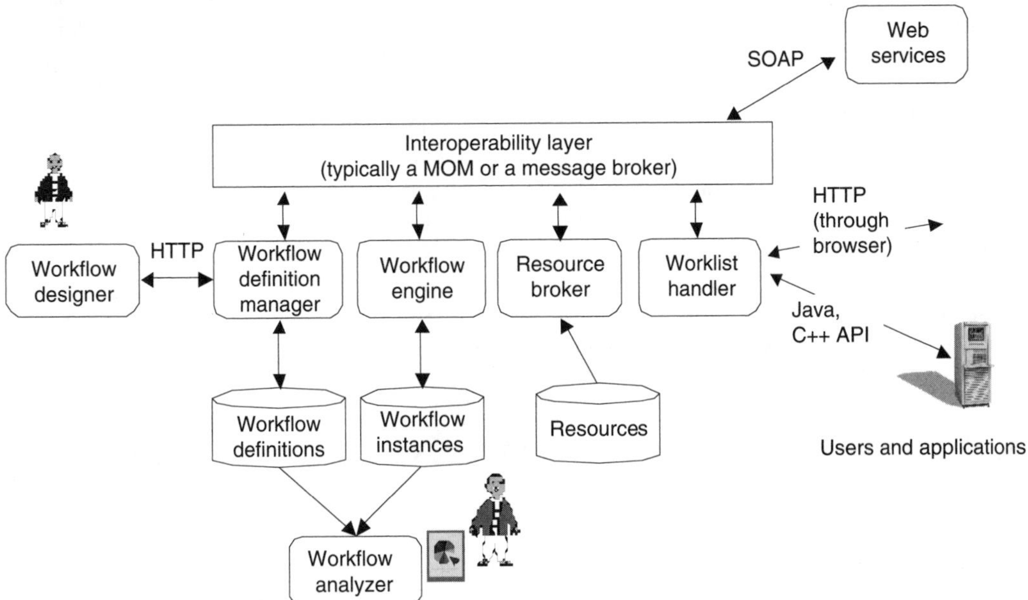

FIGURE 32.3 A typical WfMS architecture.

of the resources from the WfMS. In fact, a WfMS simply places tasks in the resources' *worklists.* It is up to the resource (or to a suitably defined adapter) to access the worklist and deliver the job to the resource. For example, if the resource is a SAP system, then the adapter will have to access the worklist corresponding to the SAP workflow resource and, based on the task, invoke one of the SAP interface operations.

- The *workflow analyzer* allows users (either through a browser or through a programmatic API) to access both status information about active processes and aggregate statistics on completed processes, such as the average duration of a process or of a node.
- The *interoperability layer* enables interaction among the different WfMS components. Typically, this layer is a CORBA broker or a message-oriented middleware (MOM) implementation.

This architecture allows the efficient execution of thousands of concurrent workflow instances. As with TP monitors, having a single operating system process (the workflow engine) manage all workflow instances avoids overloading the execution platform with a large number of operating system processes. In addition, when multiple machines are available, the WfMS components can be distributed across them, thereby allowing the workflow engine (which is often the bottleneck) to have more processing power at its disposal, in the form of a dedicated machine. WfMSs may also include a load balancing component to allow the deployment of multiple workflow engines over different machines within the same WfMS. Modern WfMSs can typically schedule hundreds of thousands of work nodes per hour, even when running on a single workstation.

We observe here that, recently, WfMSs based on a completely different architecture have appeared, especially in the context of Web service composition. One of the novelties that Web services brings is *standardization*, so that each component supports the same interface definition language and interaction protocol (or, at least, heterogeneity is considerably reduced with respect to traditional integration problems).

Uniformity in the components assembled by the workflow makes integration easier and in particular, makes it possible to directly push the work item to the resource (a Web service, in this case), without requiring the presence of a worklist handler. In fact, when a workflow composes Web services rather than generic applications, the steps in the workflow correspond to invocation of Web service operations. Because the underlying assumption is that the operation can be invoked by means of standard protocols (typically SOAP on top of HTTP), the engine can directly invoke the operation through the middleware (Figure 3). There is no need for an intermediate stage and intermediate format, where the engine deposits the work items and the resource picks up the work to do.

Web services bring another important differentiation: Up till now the business processes inside the enterprise could collaborate with other processes in a mutually agreed upon but *ad hoc* manner. Web services are providing standardized means through which business processes can compose with other processes. This enables dynamic composition (Figure 4). In fact, not only is it possible to develop complex applications by composing Web services through a workflow, but it is also possible to expose the composition (the workflow) as yet another Web service, more complex and at a higher level of abstraction. These newly created services (called *composite* services) can be dynamically composed (just like basic Web services are composed), thereby enabling the definition of complex interactions among business processes.

32.2 Web Services and Business Processes

32.2.1 Web Services

Web services are described as distributed services that are identified by URIs, whose interfaces and binding can be defined, described, and discovered by XML artifacts, and that support direct XML message-based interactions with other software applications via Internet-based protocols. Web Services have to be described in such a way that other Web services may use them.

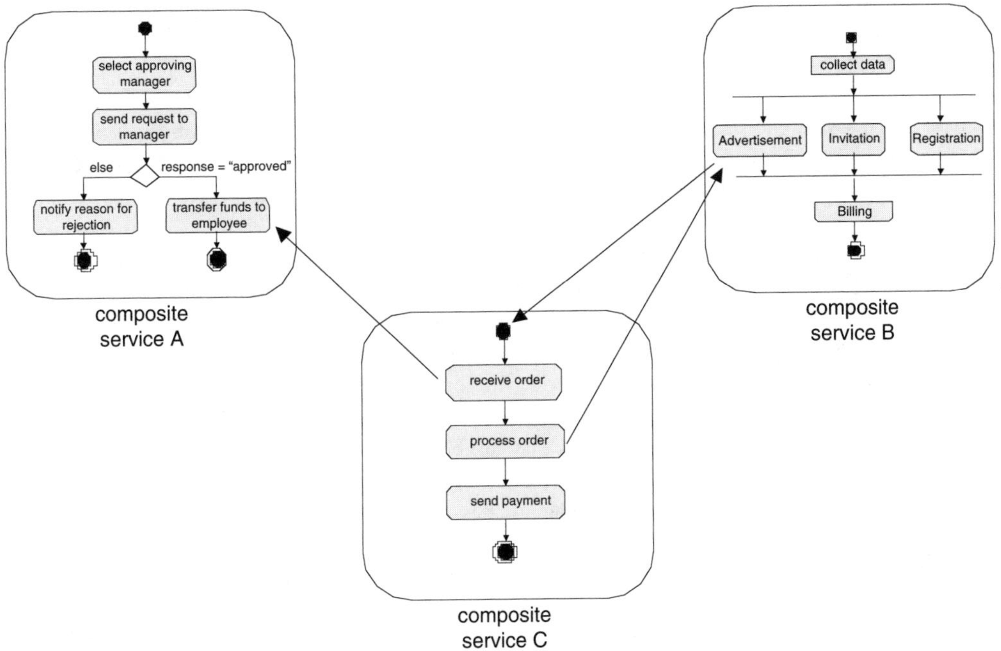

FIGURE 32.4 Services can be iteratively composed into more complex (composite) services.

Interface definition languages (IDL) have been used in traditional software environments. The IDL describes the interfaces that the object exposes. These interfaces are implemented by the object, and are useful for interoperation between objects. Web Services Description Language (WSDL) is an attempt to use similar concepts in Web services. Web Services Definition Language (WSDL) enables dynamic interaction and interoperation between Web services. WSDL not only describes the interfaces, but also the corresponding bindings involved. For example, in WSDL a service is described through a number of endpoints. An endpoint is composed of a set of operations. An operation is described in terms of messages received or sent out by the Web service:

- Message — an abstract definition of data being communicated, consisting of message parts.
- Operation — a unit of task supported by the service. Operations are of the following types, namely, one-way, request–response, solicit–response, and notification.
- Port type — an abstract collection of operations that may be supported by one or more end points.
- Binding — a concrete protocol and data format specification for a particular port type.
- Port — a single end point defined as a combination of a binding and a network address.
- Service — a collection of related end points.

32.2.2 Web Services and Business Process Orchestration

Web Services usually expose a subset of enterprise business processes and activities to the external world through a set of operations defined in their WSDL. The enterprise business processes have to be defined and some of their activities have to be linked to the WSDL operations. This requires modeling of Web service's back-end business processes. In addition, Web services need to interact with other Web services by exchanging a sequence of messages. A sequence of message exchange is termed a *conversation*. The conversations can be described independently of the internal flows of the Web services and could be simply described by sequencing the exposed operations and messages of the Web services (as mentioned in their WSDLs). However, such an inter-Web service interaction automatically leads to coupling of the internal business processes of the Web services to form what is called a *global process*. The participating Web services may or may not be aware of the whole global process depending on their understanding with each other and the internal information they are willing to expose to each other.

32.2.2.1 Business Process Languages

Business processes languages like XLANG/WSFL/BPEL4WS are used to model business processes. The Web Services Flow Language (WSFL) introduces the notion of flows and activities. XLANG is another technology from Microsoft that provides a mechanism for defining business processes and global flow coordination.

WSFL [18] models business processes as a set of activities and links. An activity is a unit of useful work. The links could be data links where data is fed into an activity from another, or control links where decisions are made to follow one activity or another. These activities are made available through one or more operations that are grouped through end points (as defined in WSDL). A service is made up of a set of end points. A service provider can provide one or more services. Just like internal flows, global flows can be defined. Global flow consists of plug links that link up operations of two service providers. This helps in the creation of complex services that can be recursively defined.

XLANG defines services by extending WSDL. The extension elements describe the behavioral aspects. A behavior spans multiple operations. A behavior has a header and a body. An action is an atomic component of a behavior. The action elements could be an *operation*, a *delay* element or a *raise* element. The delay elements, namely delayFor and delayUntil, introduce delays in the execution of the process to either wait for something to happen (for example a timeout) or wait till an absolute date–time has been reached, respectively. The raise construct is used to create exceptions. It handles the exceptions by calling the handlers registered with the raise definition. Processes put actions together in useful ways. A process form could be a sequence, switch, while, All, Pick, Context, Compensate, or Empty.

Business Process Execution Language for Web Services (BPEL4WS) [30] combines the WSFL and XLANG capabilities. It is an attempt to standardize business process language. A single BPEL4WS process describes the global process that links multiple Web services. Entry points are defined in the BPEL4WS specification of a global process. These entry points either consume WSDL operations' incoming messages from input-only or input–output operations. BPEL4WS only utilizes the input-only and input–output (request–response) operations of WSDL. It does not require or support the output-only (notification) and output–input (solicit–response) operations.

The BPEL4WS process itself is composed of activities. There are a collection of primitive activities: invoking an operation on a Web service (<invoke>), waiting to receive a message to operation of the service (<receive>), creating a response to an input/output operation (<reply>), waiting for some time without doing anything (<wait>), indicating an error (<throw>), copying data from one place to another (<assign>), closing the entire service instance down(<terminate>), or doing nothing through (<empty>). These basic primitives may be: combining through (sequence), branching through (switch), defining loops (while), executing one of the chosen paths (pick), or executing activities in parallel (flow). Within activities executing in parallel, one can indicate execution order constraints by using *links*. BPEL4WS provides the capability to combine activities into complex algorithms that may represent Web service to Web service interactions.

32.2.2.2 e-Business Languages

Web Service to Web Service interactions need to follow certain business protocols to actually undertake business on the Web. X-EDI, ebXML, BTP, TPA-ML, cXML, and CBL are some of the B2B technologies that have been proposed to enable this paradigm with Web services.

In ebXML [31], parties register their Collaboration Protocol Profiles (CPP) at ebXML registries. Each CPP is assigned a GUID by the ebXML registry. Once a party discovers another party's CPP and they decide on doing business, they form a Collaboration Protocol Agreement (CPA). CPAs are formed after negotiation between the parties. The intent of the CPA is not to expose internal business process of the parties but to expose the part of the process that is visible and that involves interactions between the parties. The messages exchanged between the involved parties or business partners may utilize ebXML Messaging Service (ebMS). A *conversation* between parties is defined through the CPA and the business process specification document it references. This conversation involves one or more *Business Transactions*. A business transaction may involve exchange of messages as requests and replies. The CPA may refer to multiple business process specification documents. Any one conversation will involve only a single process specification document, however. Conceptually, the B2B server at the parties is responsible for managing the CPAs and for keeping track of the conversations. It also interfaces the functions defined in the CPA with the internal business processes. The CPP contains the following:

- Process specification layer: This details the business transactions that form the collaboration. It also specifies the relative order of business transactions.
- Delivery channel: It describes party's characteristics for receiving and sending messages. A specification can contain more than one delivery channels.
- Document exchange layer: This describes the processing of the business documents like encryption, reliable delivery, and digital signatures.
- Transport layer: The transport layer identifies the transport protocols to be used in the end point addresses, along with other properties of the transport layer. The possible protocols being SMTP, HTTP, and FTP.

Web services can be used to enable business process composition. Web service to Web service interactions happen through message-based conversations. However, in order to undertake business on the Web, it is essential that enterprises provide guarantees to each other on performance, response times, reliability, and availability of these business processes and Web services. The guarantees are usually specified through service level agreements (SLA). The e-business operations have to be semantically analyzed [32], SLA violations have to be monitored [33], and the guarantees have to be assured.

32.3 Conclusion

Enterprises have used business processes and workflows for automating their tasks. Business processes have been traditionally managed by using workflow-management systems. Web services and the ensuing standardization process enable dynamic composition of business processes. Many issues remain unresolved regarding conversation definition, guaranteed negotiation, specification, and assurance.

References

1. IBM. WebSphere MQ Integrator Broker: Introduction and Planning. June 2002. Available from www.ibm.com.
2. TIBCO Software. TIBCO Enterprise Application Integration solutions. 2002. Available from www.tibco.com.
3. WebMethods. WebMethods Enterprise Integrator: User's Guide. 2002. Available from www.webmethods.com.
4. Vitria. BusinessWare: The Leading Integration Platform. 2002. Available from www.vitria.com.
5. IBM. MQ Series Workflow for Business Integration. 1999. Available from www.ibm.com.
6. Tibco. TIBCO Business Process Management solutions. 2002. Available from www.tibco.com.
7. Microsoft. Microsoft BizTalk Server 2002 Enterprise Edition. 2002. Available from www.microsoft.com.
8. F. Leymann and D. Roller. *Production Workflow: Concepts and Techniques.* 1999. Prentice-Hall, New York.
9. OMG. Unified Modeling Language Specifications. Version 1.3. 1999.

10. W. van der Aalst and Kees van Hee. *Workflow Management: Models, Methods, and Systems.* 2001. MIT Press, Cambridge, MA.
11. D. Chiu, Q. Li, Kamalakar Karlapalem. ADOME-WfMS: Towards cooperative handling of workflow exceptions. In A. Romanowsky, C. Dony, J. L. Knudesn, A. Tripathi, Eds. *Advances in Exception Handling Techniques.* 2000. Springer, New York.
12. G. Kappel, P. Lang, S. Rausch-Schott, W. Retschitzegger. Workflow Management Based on Objects, Rules, and Roles. IEEE Data Engineering Bulletin 18(1): 11–18, 1995.
13. F. Casati, S. Ceri, B. Pernici, G. Pozzi.Deriving active rules for workflow enactment. Proceedings of Database and Expert System Applications (DEXA'96). Zurich, Switzerland. 94–115, 1996.
14. The Workflow Management Coalition. Process Definition Interchange. Document number WfMC-TC-1025. 2002.
15. F. Casti and A. Discenza. Supporting Workflow Cooperation Within and Across Organizations. Symposium on Applied Computing (SAC'2000). Como, Italy. 196–202, June 2000.
16. W. van der Aalst and M. Weske. The P2P Approach to Interorganizational Workflows. Proceedings of the Conference on Advanced Information Systems Engineering (CAISE'01). Interlaken, Switzerland. 140–156, June 2001.
17. T. Andrews, F. Curbera, H. Dholakia, Y. Goland, J. Klein, F. Leymann, K. Liu, D. Roller, D. Smith, S. Thatte, I. Trickovic, S. Weerawarana. Business Process Execution Language for Web Services. Version 1.1. 2003.
18. Frank Leymann. Web Services Flow Language. Version 1.0. 2001.
19. A. Borgida. Language features for flexible handling of exceptions in information systems. *ACM Transactions on Database Systems*, 10(4): 565–603, 1985.
20. H. Saastamoinen. On the Handling of Exceptions in Information Systems. PhD thesis, University of Jyvaskyla, Finland, 1995.
21. J. Eder and W. Liebhart. The Workflow Activity Model WAMO. Proceedings of the 3rd International Conference on Cooperative Information Systems (CoopIs'95). Wien, Austria. 87–98, 1995.
22. D. Worah and A. Sheth. Transactions in transactional workflows. In S. Jajodia and L. Kerschberg, Eds., *Advanced Transaction Models and Architectures.* Kluwer Academic, New York, 1997.
23. A. Reuter, K. Schneider, and F. Schwenkreis. Contracts revisited. In S. Jajodia and L. Kerschberg, Eds., *Advanced Transaction Models and Architectures.* Kluwer Academic, New York, 1997.
24. J. Eder and W. Liebhart. Contributions to exception handling in workflow management. Proceedings of the EDBT Workshop on Workflow Management Systems, Valencia, Spain. 3–10, 1998.
25. M. Kamath and K. Ramamritham. Failure handling and coordinated execution of concurrent workflows. Proceedings of the 14th International Conference on Data Engineering (ICDE'98). Orlando, FL. 334–341, 1998.
26. Fabio Casati, Stefano Ceri, Stefano Paraboschi, Giuseppe Pozzi. Specification and Implementation of exceptions in workflow management systems. ACM TODS 24(3): 405–451, 1999.
27. Alexander Wise. Little-JIL 1.0: Language Reports. University of Massachussets, Amherst, MA. Document number UM-CS-1998-024. 1998.
28. C. Hagen and G. Alonso. Flexible exception handling in the OPERA process support system. In Proceedings of the 18th International Conference on Distributed Computing Systems (ICDCS'98), Amsterdam, The Netherlands. 526–533, May 1998.
29. Fabio Casati and Giuseppe Pozzi. Modeling and Managing Exceptional Behaviors in Workflow Management Systems. Proceedings of Cooperative Information Systems (CoopIS) 99. Edinburgh, Scotland. 127–138, 1999.
30. Business Process Execution Language for Web Services (BPEL4WS) http://www.ibm.com/developerworks/library/ws-bpel/.
31. EbXML: http://www.ebxml.org.
32. M. Sayal, A. Sahai, V. Machiraju, F. Casati. Semantic analysis of e-business operations. *Journal of Network and System Management*, March 2003 (special issue on E-Business Management).

33. Akhil Sahai, Vijay Machiraju, Mehmet Sayal, Aad van Moorsel, Fabio Casati, Li Jie Jin. Automated SLA Monitoring for Web Services. IEEE/IFIP Distributed Systems: Operations and Management (DSOM 2002). Montreal, Canada. 28–41, 2002.
34. Gustavo Alonso, Divyakant Agrawal, Amr El Abbadi, Mohan Kamath, Roger Günthör, C. Mohan. Advanced transaction model in workflow context. Proceedings of the 12th International Conference on Data Engineering (ICDE'96), New Orleans, LA, 574–581, February 1996.

33

Information Systems

CONTENTS

Abstract 33-1
33.1 Introduction 33-2
33.2 Information Systems Before the Advent of the Internet.... 33-2
33.2.1 Processes 33-2
33.2.2 Products 33-3
33.2.3 Nonfunctional Qualities 33-3
33.2.4 Social Structures 33-4
32.2.5 Automation 33-4
33.3 The World As Seen by Information Systems 33-5
33.3.1 Processes and Products 33-5
33.3.2 Nonfunctional Qualities 33-5
33.3.3 Social Structures 33-6
33.3.4 Automation 33-7
33.4 What Is New about Internet Computing? 33-7
33.5 Information Systems Challenges in the Internet Age... 33-8
33.5.1 Products and Processes 33-8
33.5.2 Nonfunctional Qualities 33-9
33.5.3 Social Structures 33-10
33.5.4 Automation 33-11
33.6 Conceptual Abstractions for Information Systems ... 33-12
33.6.1 Conceptualizing "What" and "When" 33-12
33.6.2 Conceptualizing "Where" 33-13
33.6.3 Conceptualizing "How" and "Why" 33-13
33.6.4 Conceptualizing "Who" 33-14
33.7 Summary and Conclusions 33-15
Acknowledgements 33-16
References 33-16

Eric Yu

Abstract

Internet computing is changing the nature and scope of information systems (IS). Most IS methods and techniques were invented before the advent of the Internet. What will the world of IS practice be like in the age of the Internet? What methods and techniques will be relevant? We review the world of information systems in terms of processes and products, qualities, social structures, and the role of automation. Given the rapid adoption of Internet thinking, not only among technical professionals but also by the public, we outline the prospects and challenges for information systems in the emerging landscape. In particular, we highlight the need for richer modeling abstractions to support the diversity of services and modes of operation that are required in the new age of worldwide, open network information systems.

1-58488-381-2/05/$0.00+$1.50
© 2005 by CRC Press LLC

33.1 Introduction

How will Internet computing change the world of information systems?

Following the widespread commercial availability of computing technologies, IS has been the dominant application area of computing. Organizations large and small, private and public, have come to rely on IS for their day-to-day operation, planning, and decision-making. Effective use of information technologies has become a critical success factor in modern society. Yet, success is not easily achieved. Many of the failures occur not in the technology, but in how technology is used in the context of the application domain and setting [Lyytinen, 1987; Standish, 1995]. Over the years, many methods and techniques have been developed to overcome the challenges to building effective information systems.

For many segments of society, the Internet has already changed how people work, communicate, or even socialize. Many of the changes can be attributed to information systems that now operate widely over the Internet. Internet computing is changing the scope and nature of information systems and of IS work.

What opportunities, problems, and challenges do Internet computing present to the IS practitioner? What makes the new environment different? Which existing techniques continue to be applicable and what adaptations are necessary? What new IS methods and techniques are needed in the Internet world?

IS is a multifaceted field, and requires multidisciplinary perspectives. In this chapter, we will only be able to explore some of the issues from a particular perspective — primarily that of IS engineering, with an emphasis on the interplay between the technical world of system developers and programmers on the one hand, and the application or problem-domain world of users, customers, and stakeholders on the other. This perspective highlights some of the key IS issues as the bridge between raw technology and the application domain.

The chapter is organized as follows. Section 33.2 considers the world of IS practice before the advent of the Internet. In Section 33.3, we ask how the users and applications are seen through the eyes of the IS practitioner, pre-Internet. Section 33.4 focuses on the new environment for information systems, brought about by Internet computing. Section 33.5 considers the implications and challenges of the Internet age for IS practice and research. As conceptual abstractions are at the heart of IS engineering, we focus in Section 33.6 on the kinds of abstractions that will be needed in the Internet age. We close in Section 33.7 with a summary and conclusions.

33.2 Information Systems Before the Advent of the Internet

Let us first consider the world of IS practice, focusing on methods and techniques used before the advent of the Internet.

What kinds of tasks and processes do IS professionals engage in? What products do the processes produce? What quality concerns drive their daily work and improvement initiatives? How is the division of work organized among professional specialties, and within and across organizations and industry sectors? Which areas of work can be automated, and which are to be retained as human tasks?

33.2.1 Processes

The overarching organizing concept in most IS curricula is that of the system development life cycle [Gorgone et al., 2002].

The overall process of creating and deploying an information system is broken down into a number of well-defined interdependent processes. These typically include planning, requirements elicitation, analysis, specification, design, implementation, operations and support, maintenance, and evolution. Verification and validation, including testing, is another set of activities that needs to be carried out in parallel with the main production processes. Some of the life-cycle activities involve participation of users and stakeholders. For example, technical feasibility, and business priorities and risks are reviewed at predefined checkpoints. When externally provided components or subsystems are involved, there are processes for procurement and integration. Processes are also needed to manage the information content

— during system development (e.g., defining the schemas) and during operation (e.g., ensuring information quality) [Vassiliadis et al., 2001].

A systematic process methodology is therefore a central concept in the field, imported initially from practices in large-scale engineering projects. The systematic approach is used to control budget, schedule, resources, and opportunities to change course, e.g., to reduce scope, or to realign priorities. Nevertheless, lack of a systematic process methodology continues to be a concern, as a contributing factor to poor quality or failure of software and information systems. Substantial efforts are used to institutionalize good practices in processes, through standards, assessment, and certification, and process improvement initiatives, e.g., Capability Maturity Model Integrated (CMMI) [Chrissi et al., 2003] and ISO 9000 [ISO, 1992].

Many IS projects adopt methodologies offered by vendors or consulting companies, which prescribe detailed processes supported by associated tools. Prescriptive processes provide guidance and structure to the tasks of system development. They may differ in the stages and steps defined, the products output at each step, and how the steps may overlap or iterate (e.g., the waterfall model [Royce, 1970], the spiral model [Boehm, 1988], and the Rational Unified Process [Kruchten, 2000]). Although prescriptive processes aim to create order out of chaos, they are sometimes felt to be overrestrictive or requiring too much effort and time. Alternative approaches that have developed over the years include rapid prototyping, Joint Application Development (JAD) [Wood and Silver, 1995], Rapid Application Development (RAD) [McConnell, 1996], and more recently agile development [Cockburn, 2001]. All of these make use of a higher degree of human interaction among developers, users, and stakeholders.

33.2.2 Products

Complementary to and intertwined with processes are the products that they output. These include products and artifacts that are visible to the end user such as executable code, documentation, and training material, as well as intermediate products that are internal to the system development organization. When more than one organization is involved in the creation and maintenance of a system, there are intermediate products that are shared among or flow across them.

Most of the products are informational — plans, requirements, specifications, test plans, designs, budgets and schedules, work breakdowns and allocations, architectural diagrams and descriptions, and so on. Some products are meant for long-term reference and record keeping, whereas others are more ephemeral and for short-term coordination and communication.

These informational products are encoded using a variety of modeling schemes, languages, and notations. Information-modeling techniques continue to be a central area of research [Brodie et al., 1984; Webster, 1988; Loucopoulos and Zicari, 1992; Boman et al., 1997; Mylopoulos, 1998]. Widely used techniques include Entity-Relationships (ER) modeling [Chen, 1976], Integrated Definition for Function Modelling (IDEF0) [NIST, 1993] (based on the Structured Analysis and Design Technique (SADT) [Ross and Shoman, 1977]), and the Unified Modelling Language (UML) [Rumbaugh et al., 1999].

Large system projects involve many kinds of processes, producing a great many types of information products related to each other in complex ways. Metamodeling and repository technologies [Brinkkemper and Joosten, 1996; Jarke et al., 1998; Bernstein et al., 1999; Bernstein, 2001] are often used to manage the large amounts and variety of information produced in a project. These technologies support retrieval, update, and coordination of project information among project team members. Metamodels define the types of processes and products and their interrelationships. Traceability from one project artifact or activity to another is one of the desired benefits of systematic project information management [Ramesh and Jarke, 2001].

33.2.3 Nonfunctional Qualities

Although processes and products constitute the most tangible aspects of IS work, less tangible issues of quality are nevertheless crucial for system success. Customers and users want systems that not only provide the desired functionalities, but also a whole host of nonfunctional requirements that are often conflicting — performance, costs, delivery schedules, reliability, safety, accuracy, usability, and so on. Meeting competing quality requirements has been and remains a formidable challenge for software and IS professionals [Boehm and In, 1996]. Not only are system developers not able to guarantee the correctness of large systems, they frequently fail to meet nonfunctional requirements as well. Many of the issues collectively identified as the software crisis years ago are still with us today [Gibbs, 1994].

Research subspecialties have arisen with specific techniques to address each of the many identified areas of quality or nonfunctional requirements — performance, reliability, and so forth. However, many quality attributes are hard to characterize, e.g., evolvability and reusability. When multiple requirements need to be traded off against one another, systematic techniques are needed to deal with the synergistic and conflicting interactions among them. Goal-oriented approaches [Chung et al., 2000] have recently been introduced to support the systematic refinement, interaction analysis, and operationalization of nonfunctional requirements. On the project management level, institutionalized software process improvement programs (such as CMMI) target overall project quality improvements. Quality improvements need to be measured, with results fed back into new initiatives [Basili and Caldiera, 1995].

33.2.4 Social Structures

Most information systems require teams of people to develop and maintain them. The structuring of projects into process steps and artifacts implies a social organization among the people performing the work, with significant degrees of task specialization. Some tasks require great familiarity with the application domain, whereas others require deep knowledge about specific technologies and platforms. Some require meticulous attention to detail, whereas others require insight and vision.

Interpersonal skills are as important as technical capabilities for project success [Weinberg, 1998; DeMarco and Lister, 1999]. Every product requires time and effort to create, so the quality depends on motivation, reward structures, and priorities, as well as on personnel capabilities. Yet the social organization is often implicit in how processes and products are structured, rather than explicitly designed, as there are few aids beyond generic project management tools.

Processes are judged to be too heavy (excessive regimentation) or too light (chaotic) based on the perceived need for human creativity, initiative, and flexibility for the task at hand. Factors influencing the determination of social structure include project and team size, familiarity with the application domain, and maturity of the technologies, as well as sociocultural and economic factors. Industry categories and structures (e.g., Enterprise Resource Planning [ERP] vendors vs. ERP implementers) and human resource categories (database designers vs. database administrators) are larger social structures that specific projects must operate within.

The social nature of IS work implies that its structure is a result of conflicting as well as complementary goals and interests. Individuals and groups come together and cooperate to achieve common objectives, but they also compete for resources, pursue private goals, and have different visions and values. Processes and products that appear to be objectively defined are in fact animated by actors with initiatives, aspirations, and skills.

The human intellectual capital perspective [Nonaka and Takeuchi, 1995] highlights the importance of human knowledge and ingenuity in systems development. Although considerable knowledge is manifested in the structure of processes and products, a great deal of knowledge remains implicit in human practices and expertise. There are limits on how much and what kinds of knowledge can be made explicit, encoded in some language or models, and systematically managed.

In reflecting on IS practices and software development as professional disciplines, authors acknowledge the human challenges of the field [Banville and Landry, 1992; Humphrey, 1995].

32.2.5 Automation

The quest for higher degrees of automation has been a constant theme in information systems and in software engineering. The large amounts of complex information content and the numerous, complicated relationships, the need for meticulous detail and accuracy, the difficulty of managing large teams, and the desire for ever quicker delivery and higher productivity — all call for more and better automated tools.

Numerous tools to support various stages and aspects of IS work have been offered — from Computer Aided Software Engineering (CASE) tools that support modelling and analysis, to code generators, test tools, simulation tools, repositories, and so on. They have met with varying degrees of success in adoption and acceptance among practitioners.

Automation relies on the formalization of processes and products. Those areas that are more amenable to mathematical models and semantic characterization have been more successful in achieving automated tool support. Thus, despite great efforts and many advances, IS work remains labour-intensive and requires social collaboration. Many issues are sociotechnical, e.g., requirements elicitation, reuse, agile development, and process improvement.

The difficulties encountered with automation in the developer's world may be contrasted with that in the user's world, where automation is the mandate and expectation of the IS practitioner.

33.3 The World As Seen by Information Systems

Information systems convey and manipulate information about the world. The kind of world (the application setting and the problem domain) that is perceived by the IS analyst is filtered through presuppositions of what the technology of the day can support. In the preceding section, we reflected upon the world of the IS practitioner in terms of processes and products, quality, social structures, and automation. Let us now use the same categories to consider how IS practitioners treat the world that they serve — the world that users and stakeholders inhabit.

33.3.1 Processes and Products

The predominant conceptualization of the world as seen by IS analysts is that of processes and products. The main benefit of computers was thought to be the ability to process and store large amounts of encoded information at high speeds and with great accuracy. In early applications, information systems were used to replace humans in routine, repetitive information processing tasks, e.g., census data processing and business transaction processing. The processes automate the steps that humans would otherwise perform. Processes produce information artifacts that are fed into other processes. The same concept can be applied to systems that deal with less routine work, e.g., management information systems, decision support systems, executive information systems, and strategic information systems.

Models and notations, usually graphical with boxes and arrows, were devised to help describe and understand what processes are used to transform what kinds of inputs into what kinds of outputs, and state transitions. Data Flow Diagrams (DFD) [DeMarco, 1979], SADT [Ross and Shoman, 1977], ER modelling [Chen, 1976], and UML [Rumbaugh et al., 1999] are in common use. These kinds of models shape and constrain how IS analysts perceive the application domain [Curtis et al., 1992].

We note that processes and products in the developer's world are treated somewhat differently than those in the user's world. In the latter, attention is focused on those that are potentially automatable. In the former, there is an understanding that a large part of the processes and products will be worked on by humans, with limited degrees of automation. We will return to this point in Section 33.4.

33.3.2 Nonfunctional Qualities

Most projects aim to achieve some improvement or change in qualitative aspects of the world — faster processing, fewer delays, information that is more accurate and up-to-date, lower costs, and so forth. In Section 32.3, we considered the pursuit of quality during a system development project. Here, we are concerned with the quality attributes of processes and products in the application domain in which the target system is to function. Many of the same considerations apply, except that now the IS professional is helping to achieve quality objectives in the client's world.

Quality issues may be prominent when making the business case for a project, and may be documented in the project charter or mandate. However, the connection of these high level objectives to the eventual definition of the system in terms of processes and products may be tenuous. Quality attributes are not easily expressible in the models that are used to define systems, as the latter are defined in terms of processes and products. Quality concerns may appear as annotations or comments accompanying the text (e.g., a bottleneck or missing information flow). Furthermore, a model typically describes only one situation at a time, e.g., the current system as it exists or a proposed design. Comparisons and alternatives are hard to express, as are pros and cons and justifications of decisions. These kinds of information, if recorded at all, are recorded outside of the modelling notations. Some quality attributes can be quantified, but many cannot. Specialized models can be used for certain quality areas (e.g., economic models and logistical models), but analyzing cross-impacts and making tradeoffs among them is difficult, as noted in Section 33.2.3. Design reasoning is therefore hard to maintain and keep up to date when changes occur.

33.3.3 Social Structures

Information systems change the social structures of the environment in which they operate. In performing some aspects of work that would otherwise be performed by people, they change how work is divided and coordinated. Bank tellers take on broader responsibilities as customer service representatives, phone inquiries are funnelled into centralized call centers, and data entry tasks are moved from clerical pools to end users and even to customers. Each time a system is introduced or modified, responsibilities and relationships are reallocated, and possibly contested and renegotiated. Reporting structures, and other channels of influence and control, are realigned. The nature of daily work and social interactions are altered. Reward structures and job evaluation criteria need to be readjusted.

The importance of social factors in information systems have long been recognized (e.g., [Kling, 1996; Lyytinen, 1987]). Many systems fail or fall into disuse not because of technical failure, but due to a failure in how the technology is matched to the social environment. Alternative methodologies have been proposed that pay attention to the broader context of information systems, e.g., Soft Systems Methodology [Checkland, 1981], ethnographic studies of work practices [Goguen and Jirotka, 1994], Participatory Design [Muller and Kuhn, 1993], Contextual Design [Holtzblatt and Beyer, 1995], and so on. Each has developed a following, and has produced success stories. Workplace democracy approaches have a long history in Scandinavian countries [Ehn, 1988].

Nevertheless, despite the availability of these alternative methods, social issues are not taken into account in-depth in most projects. When an information system operates within an organizational context, the corporate agenda of the target system dominates, e.g., to improve productivity and profitability. Users, who are employees, are expected to fit their work practices to the new system. Although users and other stakeholders may be given opportunities, in varying degrees, to participate and influence the direction of system development, their initiatives are typically limited.

Existing modelling techniques, most of which focus on process-and-product, are geared primarily to achieving the functionalities of the system, deferring or side-stepping quality or social concerns. For example, in the Structured Analysis paradigm, people and roles that appear in "physical" DFDs are abstracted away in going to the "logical" DFD, which is then used as the main analysis and design vehicle [DeMarco, 1979]. Actors in UML Use Case Diagrams [Rumbaugh et al., 1999] are modelled in terms of their interactions with the system, but not with each other. Given the lack of representational constructs for describing social relationships and analyzing their implications, IS practitioners are hard pressed to take people issues into account when considering technical alternatives. Conversely, stakeholders and users cannot participate effectively in decision making when the significance and implications of complex design alternatives are not accessible to them. It is hard for technical developers and application domain personnel to explore, analyze, and understand the space of possibilities together.

33.3.4 Automation

The responsibility of the IS professional is to produce automated information systems that meet the needs of the client. Although the success of the system depends a great deal on the environment, the mandate of the IS professional typically does not extend much beyond the automated system.

In the early 1990s, the concept of business process reengineering (BPR) overturned the narrow focus of traditional IS projects. Information systems are now seen as enablers for transforming work processes, not just to automate them in their existing forms [Hammer, 1990; Davenport and Short, 1990]. The transformation may involve radical and fundamental change. Process steps and intermediate products judged to be unnecessary are eliminated, together with the associated human roles, in order to achieve dramatic efficiency improvements and cost reductions. IS therefore has been given a more prominent role in the redesign of organizations and work processes. Yet, IS professionals do not have good techniques and tools for taking on this larger mandate. Many BPR efforts failed due to inadequate attention to social and human issues and concerns. A common problem was that implicit knowledge among experienced personnel is frequently responsible for sustaining work processes, even though they are not formally recognized. Existing IS modelling techniques, based primarily on a mechanistic view of work, are not helpful when one needs a sociotechnical perspective to determine what processes can be automated, eliminated or reconfigured.

33.4 What Is New about Internet Computing?

Why cannot IS practice carry on as before, i.e., as outlined in the preceding two sections? What parameters have changed as a result of Internet computing?

From a technology perspective, the Internet revolution can be viewed, simplistically, as one in connectivity, built upon a core set of protocols and languages: TCP/IP, HTTP, and HTML or XML. With their widespread adoption through open standards and successful business models (e.g., affordable connection fees and free browsers), the result, from the user's point of view, is a worldwide, borderless infrastructure for accessibility to information content and services — information of all types (as long as they are in digital format), regardless of what "system" or organization they originate from. Digital connectivity enabled all kinds of information services to coexist on a common, interoperable network infrastructure. Service providers have ready access to a critical mass of users, through the network effect of Metcalfe's Law [Gilder, 1993]. Automated services can access, invoke, and interact with each other.

Universal connectivity at the technology level makes feasible universal accessibility at the information content and services level. Internet computing is, therefore, triggering and stimulating the removal of technology-induced barriers in the flow and sharing of information. Previously compartmentalized information services and user communities are now reaching out to the rest of the world. Information systems, with Internet computing, find themselves broadening in scope with regard to content types, system capabilities, and organizational boundaries in the following ways:

1. Information systems have traditionally focused on structured data. The Internet, which gained momentum by offering information for the general public, unleashed an enormous appetite for unstructured information, especially text and images, but also multimedia in general. Corporations and other organizations quickly realized that their IS capabilities must address the full range of information content, to serve their public as well as streamline their internal workings. They can do this relatively easily, by embracing the same Internet technology for internal use as intranets.
2. Users working with information do not want to have to deal with many separate systems, each with its own technical idiosyncrasies. Internet computing, by offering higher level platforms for application building, makes it possible for diverse technical capabilities to appear to the user as a single system, as in the concept of portals. Thus, Internet computing vastly expands what a user may expect of a system.
3. Most information systems in the past had an internal focus and operated within the boundaries of an organization, typically using proprietary technologies from a small number of selected

vendors. Internet computing is inverting that, both from a technological viewpoint, and from an IS viewpoint. Technologically, the momentum and economics of Internet computing are such that corporate internal computing infrastructures are being converted to open Internet standards [IETF; W3C; OpenGroup]. At the information services level, organizations are realizing that much can be gained by opening up their information systems to the outside world — to customers and constituents, to suppliers, partners and collaborators, as in Business to Business (B2B) e-commerce and virtual enterprises [Mowshowitz, 1997]. The boundaries of organizations have become porous and increasingly fluid, defined by the shifting ownership and control of information and information flow, rather than by physical locations or assets.

33.5 Information Systems Challenges in the Internet Age

With the apparently simple premise of universal connectivity and accessibility, Internet computing is changing IS fundamentally. It is redrawing the map of information systems. As barriers to connectivity are removed, products and processes are being redefined. Quality criteria are shifting. New social structures are emerging around systems, both in the user's world and in the developer's world. People's conceptions of what computers can do, and what they can be trusted to do, are evolving.

33.5.1 Products and Processes

Let us first consider the impact of Internet computing on processes and products in the IS user's world. Over the years, a large organization would have deployed dozens or hundreds of information systems to meet their various business and organizational needs. Each system automates its own area of work processes and products, with databases, forms, reports, and screens for input and output. Soon, it was realized that these independently developed systems should be interacting with each other directly and automatically.

Thus, long before the Internet, numerous approaches emerged for extending the reach of information processes and products beyond the confines of a single system. For example, information in separate databases often represents different aspects of the same entity in the world. A customer, a purchase, an insurance policy, or a hospitalization — each of these has many aspects that may end up in many databases in the respective organizations. Database integration techniques were introduced to make use of data across multiple databases. Data warehousing provided powerful tools for understanding trends by enabling multidimensional analysis of data collected from the numerous operational databases in an organization. Data mining and knowledge discovery techniques enhanced these analyses.

Enterprise-wide information integration has also been motivated by the process perspective. BPR stimulated cross-functional linking of previously stand-alone "stove pipe" systems. Workflow management systems and document management systems were used to implement end-to-end business processes that cut across functional, departmental lines.

Different approaches were used to achieve integration or interoperability at various levels. Middleware technologies provided interprocess communication at a low level, requiring handcrafting of the interactions on an application-to-application basis. Enterprise application integration (EAI) products offered application-level interoperability. ERP systems offered integrated package solutions for many standard back office business processes. Integration is achieved at the business process level by adopting process blueprints from a single vendor [Curran and Keller, 1997]. When systems had disparate conceptual models of the world, metamodelling techniques were used to map across them.

Internet computing technologies come as a boon to the mishmash of technologies and approaches that have proliferated in the IS world. By offering a common network computing and information infrastructure that is readily accessible to everyone — regardless of organizational and other boundaries — the integration and interoperability challenges that organizations had been confronting individually at an enterprise level are now being addressed collectively on a worldwide scale [Yang and Papazoglou, 2000]. Organizations that had already been opening up their operations to the external world through

IS-enabled concepts such as supply chain management, customer relationships management (CRM), and virtual enterprise now have the momentum of the whole world behind them. Interorganizational interoperability initiatives (also known as B2B e-commerce) no longer need to begin from scratch between partner and partner, but are undertaken by entire industries and sectors through consortiums that set standards for business application level protocols, e.g., Rosettanet, ebXML, HL7, UN/CEFACT, OASIS, and BPMI.

Once the interaction protocols are set up, processes in one organization can invoke automated services in another without human intervention (WebServices). In an open world, anyone (individuals or organizations, and their information systems) has potential access to the full range of products and services offered on the open network, in contrast to the closed, proprietary nature of pre-Internet interactions. End-to-end process redesign can now be done, not only from one end of an organization to another, but across multiple organizations through to the customer and back.

To support flexible, open interactions, products and services increasingly need to be accompanied by rich metadata, e.g., by using XML and its semantic extensions [Berners-Lee et al., 2001]. Catalogs and directories are needed for locating desired products and services. Brokers, translators, and other intermediaries are also needed [Wiederhold and Genesereth, 1997].

Internet computing is stimulating coordinated use of multimedia and multichannel user interactions. The same user — a sales representative, a student, or a community services counsellor — may be drawing on material that combines text, images, voice, music, and video on a desktop, laptop, PDA, mobile phone, or other device. There will be increasing demands to enable higher level automated processing of digital information in all formats. The semantic web initiative, for example, aims to enhance semantic processing of web content through formal definitions of meanings (ontologies) for various subject domains and communities [Gomez-Peres and Corcho, 2002; SemanticWeb].

In terms of products and processes, the challenges brought about by Internet computing can be summarized as one of diversity. Standardization is one way to overcome the excessive proliferation of diversity. Yet, in an open world, the capacity to cope with diversity must recognize the inherent need to differentiate, and not inhibit innovation. So the great challenge is to have processes that can interoperate seamlessly, and products that are intelligible and useful to their intended users.

Given these recent transformations in the user's world, the character of work in the developer's world has seen rapid changes in the past decade or so. Development work that used to be organized vertically (from requirements to design to implementation) is now dealing increasingly with horizontal interactions, coordination, and negotiations. Each layer in the developer's world, from business process analysis to architectural design to implementation platforms, must address interaction with peers, coping with diversity and interoperability at that level [Bussler, 2002]. As a result, each level is working with new kinds of information artifacts (e.g., using new languages and metamodels [Mylopoulos et al., 1990; Yu et al. 1996]) and new development processes (e.g., understanding and negotiating peer-level protocols and interactions) [Isakowitz et al., 1998].

33.5.2 Nonfunctional Qualities

In the Internet world, when we are pursuing quality goals such as faster processing, greater accuracy and reliability, better usability, and so forth, we are dealing with processes that cut across many systems and organizations, and with information products from many sources. Unlike in the traditional world of closed systems, Internet computing implies that one may need to rely on many processes and products over which one has limited control or influence. Achieving quality in an open network environment requires new techniques not in common use in the traditional environment.

For example, if the product or service is commodity-like, one can switch to an alternate supplier when the supplier is unsatisfactory. This presupposes efficient market mechanisms, with low transaction costs. Accurate descriptions of functionalities as well as quality attributes are required, using metrics that allow meaningful comparison by automated search engines and shopbots. This may involve third-party assessors and certifiers of quality, and regulatory protection and legal recourse when obligations are not met.

The situation is complicated by the dynamic nature of Internet collaborations, where automated processes can come together for one transaction fleetingly, then in the next moment go their own way to participate in new associations. When market mechanisms fail, one would need to establish more stable associations among players based on past experiences of trust [Rosenbloom, 2000; Falcone et al., 2001].

As for developers, due to the open network environment, one can expect special emphasis on certain nonfunctional requirements such as scalability, reliability, usability, security (including availability, integrity, and confidentiality), time-to-market, costs, and performance. Design tradeoffs may be more challenging, as the designer attempts to cater to market-based, dynamically changing clientele, as well as to stakeholders in more stable long-term relationships.

Design techniques have traditionally been weak in dealing with quality or nonfunctional requirements. With Internet computing, there is the added need to support the more complex decision making involving competing demands from multiple dynamically configured stakeholders.

33.5.3 Social Structures

Traditional information systems that are function-specific and narrowly focused imply that there are well-defined human roles and responsibilities associated with each function, e.g., planning vs. execution, or product lines vs. geographic regions. Internet computing, by facilitating ready access to a wide range of information system capabilities, enables much greater flexibility in social organizational arrangements. When a common platform is used, learning curves are reduced, and movement across roles and positions is eased. For example, as more routine tasks are automated, the same personnel can monitor a wider range of activities, respond to problems and exceptions, and engage in process improvement and redesign. More fundamental changes are occurring at the boundaries of organizations. The Internet has made the online consumer/citizen a reality. Many transactions (e.g., catalogue browsing and ordering, banking and investments, tax filing, and proposal submissions) are now handled online, with the user directly interacting with automated information systems. The organization is effectively pushing some of its processes to the customer's side. Similar boundary renegotiations are taking place among suppliers and partners. These shifts in boundaries are changing internal organizational structures as well as broad industry structures. New business models are devised to take advantage of newly created opportunities [Timmers, 2000]. Disintermediation and reintermediation are occurring in various sectors of society and business. Organizations are experimenting with different kinds of decentralization, recentralization, and market orientation mechanisms, as well as internal coordination mechanisms. Citizens groups are organizing their activities differently, using chat rooms and other web-based media.

Many of the social organizational relationships are being shifted into the automated realm, as software agents act on behalf of their human counterparts, as referred to in the preceding sections. New partners may be found via automated directory services, e.g., Universal Description, Discovery, and Integration (UDDI).

Social dynamics is therefore becoming important in the analysis and design of information systems in the Internet age. Unfortunately, there are few techniques in the information system practitioner's toolbox that take social structures and dynamics into account [deMichelis et al., 1998].

The social organization of system development organizations is also rapidly changing, most directly resulting from changes in development processes and the types of artifacts they produce. New professional categories arise, as specialized knowledge and skills are sought. New dependencies and relationships among teams and team members need to be identified and negotiated. Education and training, upgrading, and obsolescence — these and other labour-market and human-resource issues are often critical for project success.

Larger changes, analogous to those happening in the user's world, are also happening in the developer's world. Industry structures are changing, and technology vendors are specializing or consolidating. Component creators and service providers are springing up to take advantage of the Internet computing platform. Outsourcing or insourcing, proprietary or open, commercial off-the-shelf (COTS) systems, and open source development — all these alter the dynamics of IS work. The adoption of particular system architectures has direct significance for the social structures around it.

As in the user's world, some processes in the developer's world will be carried out by software agents, with the social dynamics carried over into the automated realm. Again, there is little theoretical or practical support for the IS practitioner facing these issues.

33.5.4 Automation

With Internet computing, the broad range of IS capabilities is now accessible to the user on a single, consistent platform. The ability of these functions and capabilities to interoperate creates a powerful synergistic effect because they can make use of information that is already in digital form and that is machine processable. Automated functions can invoke each other at electronic speeds. For example, programmed trading of commodities and financial instruments has been operating for some time. It is feasible to have medical test results sent to and responded to by one's family physician, specialist, pharmacist, and insurer within seconds rather than days or weeks, if all the processing is automated. Governments can potentially collect electronic dossiers on the activities and movements of citizens for tax collection and law enforcement. Almost all knowledge work in organizations will be conducted through computers, as the technological support for searching, indexing, cross-referencing, multimedia presentation, and so forth, becomes a matter of routine expectation. More and more documents and other information content are "born digital" and will remain digital for most of their life cycle.

In the past, what was to become automated was decided for each system within a well-defined context of use. Significant investments and efforts were required for each application system because each system required its own underlying computing support (vertical technology stack) and operational procedures (including data entry and output). Cost–benefits analysis led to automation only in selected areas or processes, typically based on economic and efficiency criteria. This was typically done by system analysts at the early stages of system definition, with the application system as the focal point and unit of analysis.

The Internet has turned the tide in automation. We are witnessing that the concept of an isolated application system is dissolving. Information content — public or private — may pass through numerous systems on the network, invoking processing services from many operators and developed by many system vendors (e.g., via web services). Because of synergy and the network effect, it will be irresistible in economic or efficiency terms to automate [Smith and Fingar, 2002]. The investments have already been made, the technology infrastructure is there, and the content is already in machine-processable form. It will take a conscious effort to decide what not to automate.

The decision as to what to automate requires difficult analysis and decision making, but is crucial for the success and sustainability of systems. There will often be a clash of competing interests among stakeholders involving issues of trust, privacy, security, reliability, vulnerability, risks, and payoffs. Even economic and speed advantages that are the usual benefits are not necessarily realizable in the face of the potential downsides. One needs to understand broad implications and long-term consequences. Heavily interconnected networks imply many far-reaching effects that are not immediately discernable.

With the digital connectivity infrastructure in place, one has to take decisions on the degrees of automation. Information processing can range from the minimal (e.g., message transmission and representation at the destination, with no processing in between) to the sophisticated (extracting meaning and intent, and acting upon those interpretations). But even in messaging services, traffic patterns can be monitored and analyzed. So the analysis of what a system or service should do and should not do is much more complicated than in the pre-Internet world, and will involve complex human and machine processes as well as conflicting interests of many parties and conflicting perspectives.

The same factors apply to the developer's world. The increased demand from the great variety of IS capabilities will lead to pressure for more automation. When automation is raised to such a level that technical details can be hidden from the user, the entire development process can be pushed into the user's world.

Much human knowledge and experience cannot be made explicit and codified symbolically. Where and how implicit knowledge and human judgement interact and combine with automated machine processes remains a difficult design challenge.

IS practitioners have few tools that can support the analysis of these issues to help make these important decisions for users, service providers, and for society.

33.6 Conceptual Abstractions for Information Systems

IS practice is based on conceptual abstractions with well-defined properties. Abstractions focus attention on aspects of the world that are relevant for information systems development.

As we saw in Section 33.3, in order to develop information systems that can serve in some application setting, the complexity of that setting needs to be reduced through a set of modelling abstractions. The user's world needs to be expressed in terms of models that can be analyzed, leading to decisions about what aspects of that world will become the responsibility of the intended system. During systems development (Section 33.2), the models are translated stepwise through a different set of abstractions, from ones that describe the user's world (e.g., travel plans and bookings) to ones that describe the machine's world (data and operations in computers that store those plans and execute those bookings) [Jarke et al., 1992]. At each stage or level of translation, analyses are performed to understand the situation; decisions are made on how elements at the current level should translate into or correspond to elements at the next level.

Notations are important to help communication between the worlds of stakeholders and users on the one hand, and system designers on the other. They need to have sufficient expressiveness to convey the desired needs and requirements. Yet the notations need to be simple and concise enough for widespread adoption and standardization. Furthermore, they need to support analysis and inference, preferably automated, so as to be scalable.

A necessary consequence of using modelling is restriction in what can be said about the world. Aspects of the world that cannot be expressed tend to be left out, and will no longer be the focus of attention during system development. Therefore, the design of notations requires a difficult balance [Potts and Newstetter, 1997]. With too much detail, one can get bogged down; with too little, one can get a wrong system that does not do what is needed or intended. Whatever is chosen, it is the modelling techniques used that shape the analyst's perception of the world.

As we surveyed the user's and developer's worlds, we noted that not all the relevant aspects are equally well supported by existing modelling abstractions. Processes and products are the mainstay of most existing modelling techniques, but quality and social structure are not well captured. Therefore, those issues are not systematically dealt with in mainstream methodologies.

In the Internet age, it will be especially important to have conceptualizations and abstractions that relate concerns about social relationships and human interests to the technical alternatives in systems design, and vice versa.

The preceding sections revealed that, with Internet computing, information systems are now expected to deal with a much wider range of conceptualizations than before. We will consider the abstractions for expressing what, when, where, how and why, and who.

33.6.1 Conceptualizing "What" and "When"

The "what" refers to things that exist, events that occur, and properties and relationships that hold. These aspects of the world are most heavily addressed in existing modelling schemes. An online transaction needs to identify products bought and when payments take effect. Patient records need to distinguish different kinds of diseases and symptoms and document the nature and timing of treatments.

Product and process proliferation triggered by Internet computing will test the limits of these modelling techniques. Current work in ontologies [Guarino and Welty, 2002] is revealing subtleties and limitations in earlier work. Knowledge structuring mechanisms such as classification, generalization, and aggregation

[Greenspan et al., 1994] that have been used in object-oriented modelling will be utilized extensively. The global reach of Internet computing is likely to push each classification and specialization scheme to the limits of its applicability, for example, to organize the types and features of financial instruments that may be transacted electronically and that are becoming available in the global investment marketplace. Metamodelling techniques are especially relevant for working with the conceptual structures spanning multiple domains or contexts [Nissen and Jarke, 1999]; an example is the classification of medical conditions by physicians as opposed to insurance companies, and in the context of one country or culture compared to another.

The "what" and "when" cover the static and dynamic aspects of the world. Time is not always explicitly represented in dynamic models, but may appear as sequence or precedence relationships, e.g., coordinating multistep financial transactions that traverse many systems, countries, time zones, and organizations. Internet computing brings more complex temporal issues into play. Multiple systems cooperating on the network may operate on different time scales, interact synchronously or asynchronously at different periods. They will have different development and evolution lifecycles that require coordination. Conventional modelling techniques typically deal with only first-order dynamics, as in process execution and interaction. Second or higher order dynamics, such as change management, are usually not well integrated in the same modelling framework. When Internet computing is relied upon as a platform for long term continuity, there will be processes that have time horizons extending into years and decades (e.g., inter-organizational workflow, managing the impacts of legislative change). Over the long term, process execution and process change will have human and automated components, involving users and developers. Similarly, the long-term presentation and preservation of information content over generations of information systems will be significant issues (e.g., identification and referencing of objects and how to make objects interpretable by humans and machines in future generations) [GAO, 2002].

33.6.2 Conceptualizing "Where"

It should be no surprise that the Internet challenges conventional conceptions of geographic space. On the one hand, it enables users to transcend physical space, reaching out to others wherever they are. On the other hand, worldwide coverage means that users do come from many different geographic regions and locales, and that these differences are significant or can be taken advantage of in many applications. Peoples' preferences and interests, linguistic and cultural characteristics, legal frameworks and social values — all of these can be correlated to physical locations. Mobile and ubiquitous computing, silent commerce, and intelligent buildings can make use of fine-grained location awareness. Modelling techniques developed in geographical information systems (GIS) can be expected to find wider applications stimulated by Internet computing, e.g., to offer location sensitive services to users in vehicles, to help visitors navigate unfamiliar territory, or to track material goods in transit. Physical locations will often need to be mapped to jurisdictional territories, e.g., in enforcing building security.

33.6.3 Conceptualizing "How" and "Why"

The distinction between "what" and "how" is often made within software engineering and in systems design. Requirements are supposed to state the "what" without specifying the "how." Here the "what" refers to essential characteristics, whereas the "hows" used to achieve the "what" reflect incidental characteristics that may be peculiar to the implementation medium or mechanisms. This is one of the core principles of abstraction in dealing with large systems. The what and how distinction can be applied at multiple levels or layers, each time focusing on features and issues relevant to that level, while hiding "details" that can be deferred.

Structured analysis techniques (e.g., SADT, DFD) rely heavily on a layered, hierarchical structure for the gradual revealing of details. Although the vertical layering of processes or functions can be viewed as embodying the how (downwards) and why (upwards) for understanding the structure of a system, much of the reasoning leading to the structure is not captured. There is almost always more than one answer when considering how to accomplish a task. Yet most modelling techniques only admit one possible refinement in elaborating on the "how." Alternatives, their pros and cons, and why one of them is chosen typically cannot be described and analyzed within the notation and methodology. The lack or loss of this information in systems descriptions makes system evolution difficult and problematic.

Understanding how and why will be critical in the Internet environment, where systems can be much more dynamic and contingent. Systems are typically not conceived in terms of a single coherent system with a top-level overview that can then be decomposed into constituent elements to be designed and constructed by the same project team. Instead, systems could arise from network elements that come together in real time to participate in a cooperative venture, then dissolve and later participate in some other configuration. There will be many ways to assemble a system from a network of potential participants. Components can come together in real time for short periods to form a cooperative venture.

Designers, or the systems component themselves, must have ways of identifying possible solutions (the hows) and ways for judging which ones would work, and work well, according to quality criteria and goals (the whys).

Reasoning about how and why is needed at all levels in systems development, e.g., at the application service level (a navigation system recommends an alternate route based on traffic conditions and user preferences) and at the systems and networks level (a failure triggering diagnostics that lead to system recovery).

Representing how and why has been addressed in the areas of goal-oriented requirements engineering [Mylopoulos et al., 1999; van Lamsweerde, 2000], design rationales [Lee, 1997], the quality movement (e.g., [Hauser and Clausing, 1988], and partly in requirements traceability [Ramesh and Jarke, 2001].

33.6.4 Conceptualizing "Who"

The most underdeveloped aspect is the conceptualization of the notion of "who" to support systems analysis. The Internet environment will bring many actors into contact with each other. There will be individuals, groups, organizations, and units within organizations such as teams, task forces, and so forth. They will be acting in many different capacities and roles, with varying degrees of sustained identity. They will have capabilities, authorities, and responsibilities. Information system entities (e.g., software agents) may be acting on behalf of human actors, taking on some of their rights and obligations.

Traditional information systems tend to exist in closed environments, e.g., within the authority structure of a single organization. Social structures are more easily defined and instituted, e.g., as used in role-based access control techniques in computer security [SACMAT, 2003]. With Internet computing, there can be much greater numbers of participants and roles (both in types and instances), with dynamic and evolving configurations of relationships. Consider, for example, healthcare information systems that connect patients at home to hospitals and physicians, later expanding to community care centers, and eventually to insurers and government regulatory agencies and registries. New configurations can arise from time to time due to innovation (e.g., in business models) or regulatory change. There are complex issues of reliability, trust, privacy, and security, as well as operational responsibilities. There are difficult analysis and tradeoffs, arising from the complex social relationships. Virtuality of the Internet creates many new issues for notions of "who," e.g., identity and personae, influence and control, authority and power, ownership and sharing [Mulligan, 2003]. All of these are crucial in analyzing new organizational forms (centralize vs. decentralize, internal vs. external) and social relationships. Notions of community are important in knowledge management and in managing meaning in conceptual models.

These issues are not well addressed in traditional information systems techniques. Some of them are beginning to be studied in agent-oriented approaches to software systems and information systems [Papazoglou, 2001; Huhns and Singh, 1998]. However much of the work on agent-oriented software engineering is currently focused on the design of software agents [Giunchiglia et al., 2003]. For information systems, more attention needs to be paid to modelling and analyzing conceptions of "who" as applied to complex relationships involving human as well as software agents [Yu, 2002].

33.7 Summary and Conclusions

Internet computing is changing the world of information systems. Information systems started historically as computer applications designed specifically for a well defined usage setting. They implemented a narrow range of repetitive processes, producing predetermined types of information products. Most often, these are automated versions of manual processes. Systems development was primarily "vertically" oriented; the main activities or processes were to convert or translate a vision of a new system into functioning procedures (executable code) and populated databases. A system project involves significant investments and lead time because it typically requires its own technology infrastructure, including networking.

With Internet computing, information systems projects will become more and more "horizontal." The larger proportion of the effort will be to coordinate interactions with other system and information resources that already exist or may exist in the future. They will potentially interact with a much wider range of users, with different quality expectations and offerings, and evolving usage patterns. Development work can be more incremental, as new systems are built from ever higher-level platforms and components. These developments will enable information systems professionals to concentrate on the application level, helping users and stakeholders to formulate and understand their problems and aspirations in ways that can take advantage of information technology solutions.

Given the broad spectrum of technological capabilities that are now available on a common infrastructure with ever higher-level interoperability, information systems are coming into their own as embodying and realizing the wishes and visions of the user's world instead of reflecting the limitations and inherent structures of the underlying technologies.

The chief limitations in this regard are those imposed by the modelling techniques of the day. As we have reviewed in this chapter, traditional IS techniques have focused on those aspects that lead most directly to the computerization or automation of existing information processes and products, as these are perceived to be the most tangible results of the project investment. The compartmentalized, vertical system development perspective means that the perception and conceptualization of the world is filtered through preconceived notions of what can be automated, based on the technological implementation capabilities of the day. Hence traditional techniques have focused on the modelling and analysis of processes and products, activities and entities, objects and behaviors. Ontologies for analysis and design are well developed for dealing with the static and dynamic dimensions of the world. Much less attention has been paid to the quality and social aspects, even though these are known to be important success factors and to have contributed to many failures.

In the horizontal world brought about by massive networking, quality and social dimensions will come to the fore. Modelling techniques must cover the full range of expressiveness needed to reasoning about the *what, where, when, how, and why*, and especially the *who* of information systems in their usage and development contexts. Refined characterizations of the notions of *who* will be crucial for tackling the human and social issues that will increasingly dominate systems analysis and design. Privacy and security, trust and risks, ownership and access, rights and obligations, these issues will be contested, possibly down to level of transactions and data elements, by a complex array of stakeholders in the open networked world of Internet computing.

Development processes and organizations are benefiting from the same advances experienced in user organizations. Systems development work is taking advantage of support tools that are in effect specialized information systems for its own work domain. As the level of representation and analysis is raised closer to and becomes more reflective of the user's world and their language and conceptual models, the developer's world blends in with the user's world, providing faster and tighter change cycles achieving more effective information systems.

Despite connectivity and potential accessibility, the networked world will not be of uniform characteristics or without barriers. There will continue to be differentiation and heterogeneity in technical capabilities as well as great diversity in information content and services. Internet computing allows information systems to transcend many unwanted technological barriers, yet it must allow user communities to create, maintain, and manage boundaries and identities that reflect the needs for locality and autonomy. Techniques that support the management of homogeneity within a locality and heterogeneity across localities will be a crucial challenge in the Internet age.

Acknowledgements

The author is grateful to the editor and Julio Leite for many useful comments and suggestions.

References

Banville, Claude and Maurice Landry (1992). Can the Field of MIS be Disciplined? In: Galliers, R., Ed., *Information Systems Research: Issues Methods and Practical Guidelines.* London et al.: Blackwell, 61–88.

Basili, Victor R. and G. Caldiera (1995). Improve software quality by reusing knowledge and experience. *Sloan Management Review,* 37(1): 55–64, Fall.

Berners-Lee, Tim, Jim Hendler, and Ora Lassila (2001). The semantic web. *Scientific American,* May, 2001.

Bernstein, Philip A. (2001). Generic model management — a database infrastructure for schema management. Proceedings of the 9th International Conference on Cooperative Information Systems (CoopIS 01), Trento, Italy, LNCS 2172, Springer, New York, 1–6.

Bernstein, Philip A., Bergstraesser, T., Carlson, J., Pal, S., Sanders, P., Shutt, D. (1999). Microsoft repository version 2 and the open information model. *Information Systems,* 24(2): 71–98.

Boehm, Barry (1981). *Software Engineering Economics.* Prentice Hall, Englewood Cliffs, NJ.

Boehm, Barry and Hoh In (1996). Identifying Quality-Requirement Conflicts. *IEEE Software,* March, 25–35.

Boman, Magnus, Janis Bubenko, Paul Johannesson, and Benkt Wangler (1997). *Conceptual Modeling.* Prentice Hall, Englewood Cliffs, NJ.

BPMI. The Business Process Management Initiative. www.bpmi.org

Brinkkemper, Sjaak and S. Joosten (1996). Method engineering and meta-modelling: editorial. *Information and Software Technology,* 38(4): 259.

Bussler, Christoph (2002). P2P in B2BI. Proceedings of the 35th Hawaii International Conference on System Sciences. IEEE Computer Society. Vol. 9: 302–311.

Checkland, Peter B. (1981). *Systems Thinking, Systems Practice.* John Wiley and Sons, Chichester, U.K.

Chen, Peter (1976). The Entity-Relationship Model: towards a unified view of data. *ACM Transactions on Database Systems* 1(1): 9–36.

Chrissis, Marybeth, Mike Konrad, and Sandy Shrum (2003). *CMMI : Guidelines for Process Integration and Product Improvement.* Addison-Wesley, Reading, MA.

Chung, Lawrence, Brian Nixon, Eric Yu, and John Mylopoulos (2000). *Non-Functional Requirements in Software Engineering.* Kluwer Academic , Dordrecht, The Netherlands.

Cockburn, Alistair (2001). *Agile Software Development.* Addison-Wesley, Reading, MA.

Curran, Thomas and Gerhard Keller (1997). *SAP R/3 Business Blueprint: Understanding the Business Process Reference Model.* Pearson Education, Upper Saddle River, NJ.

Curtis, Bill, Mark Kellner, and James Over (1992). Process modelling. *Communications of the ACM,* 35(9): 75–90, September.

Davenport, Thomas H. and J.E. Short (1990). The new industrial engineering: information technology and business process redesign. *Sloan Management Review,* 1990, 31 (4): 11–27.

DeMarco, Thomas (1979). *Structured Analysis and System Specification.* Prentice Hall, Englewood Cliffs, NJ.

DeMarco, Thomas and T. Lister (1999). *Peopleware,* 2nd ed. Dorset House, New York.

deMichelis, Giorgio, Eric Dubois, Matthias Jarke, Florian Matthes, John Mylopoulos, Michael Papazoglou, Joachim W. Schmidt, Carson Woo, and Eric Yu (1998). A three-faceted view of information systems: the challenge of change. *Communications of the ACM,* 41(12): 64–70 ebXML. www.ebxml.org.

Ehn, Pelle (1988). *Work-Oriented Development of Software Artifacts.* Arbetslivscentrum, Stockholm.

Falcone, Rino, Munindar P. Singh, and Yao-Hua Tan (2001). Trust in Cyber-Societies: Integrating the Human and Artificial Perspectives. *Lecture Notes in Artificial Intelligence* 2246. Springer, New York.

General Accounting Office (GAO) (2002). Information Management: Challenges in Managing and Preserving Electronic Records. Report number GAO-02-586. GAO, Washington, D.C.

Gibbs, W. Wayt (1994). Software's chronic crisis. *Scientific American* (International edition) September 1994. 72–81.

Gilder, George (1993). Metcalfe's Law and Legacy. Forbes ASAP, September 13, 1993. http://www.gildertech.com/public/telecosm_series/metcalf.html

Giunchiglia, Fausto, James Odell, and Gerhard Weib, Eds. (2003). Agent-Oriented Software Engineering III, Third International Workshop, Bologna, Italy, July 15, 2002. *Lecture Notes in Computer Science*, 2585. Springer, New York.

Goguen, Joseph and Marina Jirotka, Eds. (1994). *Requirements Engineering: Social and Technical Issues.* Academic Press, London.

Gomez-Peres, A. and O. Corcho (2002). Ontology languages for the semantic web. *IEEE Intelligent Systems* 17(1): 54–60.

Gorgone, John T. et al. (2002). IS 2002 — Model Curriculum and Guidelines for Undergraduate Degree Programs in Information Systems. Association for Computing Machinery (ACM), Association for Information Systems (AIS), Association of Information Technology Professionals (AITP). http://www.acm.org/education/is2002.pdf.

Greenspan, Sol J., John Mylopoulos, and Alexander Borgida (1994). On Formal Requirements Modeling Languages: RML Revisited. International Conference on Software Engineering. ACM Press, New York. pp. 135–147.

Guarino, Nicola and Christopher A. Welty (2002) Evaluating ontological decisions with OntoClean. *Communications of the ACM,* 45(2): 61–65.

Hammer, Michael (1990) .Reengineering work: don't automate, obliterate. *Harvard Business Review,* July. 104–112.

Hauser, J.R. and D. Clausing (1988). The house of quality. *Harvard Business Review,* (3), May, 63–73.

HL7. www.hl7.org.

Holtzblatt, Karen and H.R. Beyer (1995). Requirements gathering: the human factor. *Communications of the ACM,* 38(5): 30–32.

Huhns, Michael and Munindar P. Singh (1998). *Readings in Agents.* Morgan Kaufmann, San Francisco, CA.

Humphrey, Watts (1995). *A Discipline for Software Engineering.* Addison-Wesley, Reading, MA.

IETF. The Internet Engineering Task Force. http://www.ietf.org/.

Isakowitz, Tomas, Michael Bieber, and Fabio Vitali (1998). Web information systems. *Communications of the ACM,* 41(7): 78–80.

ISO (1992) ISO9000 International Standards for Quality Management. Geneva, International Organization for Standardization.

Jarke, Matthias, John Mylopoulos, Joachim Schmidt, and Yannis Vassiliou (1992). DAIDA: an environment for evolving information systems. *ACM Transactions on Information Systems,* 10(1): 1–50, 1992.

Jarke, Matthias, K. Pohl, K. Weidenhaupt, K. Lyytinen, P. Martiin, J.-P. Tolvanen, and M. Papazoglou (1998) Meta modeling: a formal basis for interoperability and adaptability. In B. Krämer, M. Papazoglou, H.-W. Schmidt, Eds., *Information Systems Interoperability.* John Wiley and Sons, New York, pp. 229–263.

Kling, Rob (1996). *Computerization and Controversy: Value Conflicts and Social Choices,* 2nd ed. Academic Press, San Diego, CA.

Kruchten, Philippe (2000). *The Rational Unified Process: An Introduction,* 2nd ed. Addison-Wesley, Reading, MA.

Lamsweerde, Axel van (2000). Requirements Engineering in the Year 00: A Research Perspective. 22nd International Conference on Software Engineering, Limerick, Ireland, ACM Press, New York.

Lee, Jintae (1997). Design rationale systems: understanding the issues. *IEEE Exper.*, 12(3): 78–85.

Lyytinen, Kalle (1987). Different perspectives on information systems: problems and their solutions. *ACM Computing Surveys,* 19(1): 5–44.

Loucopoulos, Pericles and R. Zicari, Eds. (1992). *Conceptual Modeling, Databases and CASE: An Integrated View of Information System Development.* John Wiley and Sons, New York.

McConnell, Steve (1996). *Rapid Development: Taming Wild Software Schedules.* Microsoft Press.

Mowshowitz, Abbe (1997). Virtual Organization — introduction to the special section. *Communications of the ACM*, 40(9): 30–37

Muller, Michael J. and Sarah Kuhn (1993). Participatory design. *Communications of the ACM*, 36(6): 24–28, June.

Mulligan, Deirdre K. (2003). Digital rights management and fair use by design. Special issue. *Communications of the ACM*, 46(4): 30–33.

Mylopoulos, John (1998). Information modeling in the time of the revolution. Invited review. *Information Systems*, 23(3, 4): 127–155.

Mylopoulos, John, Alex Borgida, Matthias Jarke, and Manolis Koubarakis (1990). Telos: representing knowledge about information systems. *ACM Transactions on Information Systems*, 8(4): 325–362.

Mylopoulos, John, K. Lawrence Chung, and Eric Yu (1999). From object-oriented to goal-oriented analysis. *Communications of the ACM*, 42(1): 31–37, January 1999.

Nissen, Hans W. and Matthias Jarke (1999). Repository support for multi-perspective requirements engineering. *Information Systems*, 24(2): 131–158.

NIST (1993). Integrated Definition for Function Modeling (IDEF0). 1993, National Institute of Standards and Technology.

Nonaka, Ikujiro and Hirotaka Takeuchi (1995) *The Knowledge-Creating Company.* Oxford University Press, Oxford, U.K.

OASIS. Organization for the Advancement of Structured Information Standards. http://www.oasis-open.org/.

OpenGroup. The Open Group. http://www.opengroup.org/.

Papazoglou, Michael (2001). Agent-oriented technology in support of e-business. *Communications of the ACM* 44(4): 71–77.

Potts, Colin and Wendy Newstetter (1997). Naturalistic Inquiry and Requirements Engineering: Reconciling Their Theoretical Foundations. Proceedings of the 3rd IEEE International Symposium on Requirements Engineering, Annapolis, MD, January 1997. 118–127.

Ramesh, Bala and Matthias Jarke (2001). Towards reference models for requirements traceability. *IEEE Transactions on Software Engineering*, 27(1): 58–93.

Rosenbloom, Andrew (2000). Trusting technology: introduction to special issue. *Communications of the ACM*, 43(12), December.

Ross, Douglas T. and D. Schoman (1977). Structured analysis for requirements definition. *IEEE Transactions on Software Engineering* 3(1): 6–15. Special Issue on Requirements Analysis, January.

Royce, W.W. (1970). Managing the Development of Large Software Systems: Concepts and Techniques. Proceedings of WESCON, IEEE Computer Society Press, Los Alamitos, CA.

Rumbaugh, James, Ivar Jacobson, and Grady Booch (1999). *The Unified Modeling Language Reference Manual.* Addison-Wesley, Reading, MA.

SACMAT (2003). 8th ACM Symposium on Access Control Models and Technologies, June 2-3, 2003, Villa Gallia, Como, Italy, Proceedings.

SemanticWeb. The Semantic Web community portal. http://semanticweb.org/.

Smith, Howard and Peter Fingar (2002). *Business Process Management: The Third Wave.* Meghan-Kiffer Press, Tampa, FL.

Standish Group (1995). Software Chaos. http://www.standishgroup.com/chaos.html.

Timmers, Paul (2000). *Electronic Commerce: Strategies and Models for Business-to-Business Trading*. John Wiley and Sons, New York.

UDDI. www.uddi.org.

UN/CEFACT. United Nations Centre for Trade Facilitation and Electronic Business. http://www.unece.org/cefact/.

Vassiliadis, Panos, Christoph Quix, Yannis Vassiliou, and Matthias Jarke (2001). Data warehouse process management. *Information Systems*, 26(3): 205–236.

W3C. The World Wide Web Consortium. http://www.w3.org/.

WebServices. WebServices.org portal. http://www.webservices.org/.

Webster, Dallas E. (1988). Mapping the Design Information Representation Terrain. *IEEE Computer*, 21(12): 8–23.

Weinberg, Jerry (1998). *The Psychology of Computer Programming*, Silver Anniversary Edition. Dorset House, New York.

Wiederhold, Gio and Michael R. Genesereth (1997). The conceptual basis for mediation services. *IEEE Expert*, 12(5): 38–47.

Wood, Jane and Denise Silver (1995). *Joint Application Development*, 2nd ed. John Wiley and Sons, New York.

Yang, Jian and Mike P. Papazoglou (2000) Interoperation support for electronic business. *Communications of the ACM*, 43(6): 39–47.

Yu, Eric (January 1997). Towards Modeling and Reasoning Support for Early-Phase Requirements Engineering. Proceedings IEEE International Symposium on Requirements Engineering, Annapolis, MD, 226–235.

Yu, Eric and John Mylopoulos (1994).Understanding "Why" in Software Process Modeling, Analysis and Design. Proceedings 16th International Conference on Software Engineering, Sorrento, Italy.

Yu, Eric and John Mylopoulos (1994). From E–R to A–R — Modeling Strategic Actor Relationships for Business Process Reengineering. Proceedings 13th International Conference on the Entity-Relationship Approach, Manchester, U.K., December 1994; P. Loucopoulos, Ed., *Lecture Notes in Computer Science* 881, Springer-Verlag, New York, 548–565.

Yu, Eric, John Mylopoulos, and Yves Lespérance (1996). AI models for business process reengineering. *IEEE Expert*, 11(4).

Yu, Eric S. K. (2002). Agent-oriented modelling: software versus the world. In Michael Wooldridge, Gerhard Weib, and Paolo Ciancarini, Eds. *Agent-Oriented Software Engineering II, Lecture Notes in Computer Science* 2222 Springer, New York, 206–225.

Part 4

Systems and Utilities

34

Internet Directory Services Using the Lightweight Directory Access Protocol

CONTENTS

Abstract 34-2
34.1 Introduction 34-2
34.2 The Evolution of LDAP 34-4
34.2.1 The Past, Present, and Future Generations of LDAP Directories 34-4
34.2.2 First- and Second-Generation Directory Services 34-6
34.2.3 Next-Generation Directory Services 34-8
34.3 The LDAP Naming Model 34-9
34.3.1 The X.500 Naming Model 34-9
34.3.2 Limitations of the X.500 Naming Model 34-10
34.3.3 Early Alternatives to the X.500 Naming Model 34-10
34.3.4 Internet Domain-Based Naming 34-10
34.3.5 Naming Entries within an Organization 34-11
34.4 The LDAP Schema Model 34-12
34.4.1 Attribute-Type Definitions 34-12
34.4.2 Object-Class Definitions 34-12
34.4.3 Object Classes for Entries Representing People 34-13
34.4.4 Other Typical Object Classes 34-14
34.5 LDAP Directory Services 34-14
34.5.1 Basic Directory Services 34-15
34.5.2 High Availability Directory Services 34-18
34.5.3 Master–Slave Replication 34-18
34.5.4 LDAP Proxy Server 34-19
34.5.5 Multimaster Replication 34-21
34.5.6 Replication Standardization 34-22
34.6 LDAP Protocol and C Language Client API 34-23
34.6.1 LDAPv3 Protocol Exchange 34-23
34.6.2 General Result Handling 34-24
34.6.3 Bind 34-25
34.6.4 Unbind 34-26
34.6.5 Extended Request 34-26
34.6.6 Searching 34-26
34.6.7 Search Responses 34-28
34.6.8 Abandoning an Operation 34-28
34.6.9 Compare Request 34-29
34.6.10 Add, Delete, Modify, and ModifyDN Operations 34-29
34.11 Conclusion 34-30

Greg Lavender

Mark Wahl

1-58488-381-2/05/$0.00+$1.50
© 2005 by CRC Press LLC

Acknowledgments 34-30
References 34-31
Author Bios 34-32

Abstract

We survey the history, development, and usage of directory services based on the Lightweight Directory Access Protocol (LDAP). We present a summary of the naming model, the schema model, the principal service models, and the main protocol interactions in terms of a C language application programming interface.

34.1 Introduction

The landscape of network-based directory technology is fascinating because of the evolution of distributed systems ideas and Internet protocol technologies that have contributed to the success of the Internet as a collection of loosely coordinated, interoperable network-based systems. The success of open-systems directory technology based on the Lightweight Directory Access Protocol (LDAP) is attributed to the persistence of many people in academia, industry, and within international standards organizations. Today, LDAP-based technology is widely used within national and multinational intranets, wired and wireless service provider value-added networks, and the public Internet. This success is due to an Internet community process that worked to define and evolve practical X.500 and LDAP directory specifications and technologies towards a more ubiquitous Internet directory service.

There are many different types of directory services, each providing useful capabilities for users and applications in different network-based settings. We are primarily concerned with directories that have a structured data model upon which well-defined operations are performed, most notably search and update. Directories are used to service a much higher percentage of authentication operations (technically called "bind" operations) and search operations, rather than update operations, which requires them to optimize for reading rather than writing information. However, as directories become increasingly authoritative for much of the information that is used to enable a wider range of Web services as part of private intranet and public Internet infrastructures, they are increasingly required to provide the kind of update rates one expects of a transactional database management system.

Directories in general can be characterized by a hierarchical tree-based naming system that offers numerous distributed system advantages:

- Names are uniquely determined by concatenating hierarchical naming components starting at a distinguished root node (e.g., "com")
- Object-oriented schema and data model supporting very fast search operations
- Direct navigation or key-based querying using fully qualified or partially specified names
- Distribution and replication based on named sub-tree partitions
- Delegated or autonomous administration based on named sub-tree partitions

In addition, Internet directories add various authentication services and fine-grained access control mechanisms to ensure that access to information contained in the directory is only granted to authorized users. The type of information contained in a directory can be either general purpose and extensible, or specialized for a particular optimized directory service. In the following section, we briefly distinguish a general-purpose directory service based on LDAP from more specialized directory services, all of which typically coexist as part of an organizational information service, but each providing an important component of a multitiered networked information service.

The most widely known directory service on the Internet today is the Domain Name System (DNS). Others that have appeared and are either lightly used or no longer used include *whois*, *whois*$^{++}$, and *WAIS*. The primary purpose of DNS is to provide name to address lookups where a name is a uniquely determined hierarchical Internet domain name and an address is an IP or other network layer protocol address, resulting in a very specialized name-to-address mapping service. DNS has a hierarchical name structure, it is widely replicated, and it is autonomously administered. However, we distinguish DNS as a specialized directory service because its information model and query mechanism are specialized to the purpose of providing very specific host addressing information in response to direct or reverse lookup queries. There have been various extensions to DNS to extend its rather limited data model, namely in the area of service (SRV) records, but such records do not significantly enhance the main DNS hostname-to-address lookup service.

Another class of directories provides various specialized naming services for network operating systems, called NOS directories. Popular NOS directories include NIS for UNIX® systems, NDS for Novell Netware™, and Active Directory for Microsoft Windows™. NOS directories are typically based on proprietary protocols and services that are tightly integrated with the operating system, but may include some features and services adopted from X.500 and LDAP directory specifications in order to permit a basic level of interoperability with client applications that are written to such open-systems specifications. NOS directories are very well suited to local area network environments in support of workgroup applications (e.g., file, print, and LAN email address book services), but have historically failed to satisfy other business-critical applications requiring Internet scale levels of service and reliability.

Network-based file systems, such as NFS or AFS, may also be considered as specialized directory services in that they support an RPC-based query language that uses hierarchical file names as keys to access files for the purposes of reading and writing those files across a network. There have been attempts to create Internet-scale distributed file systems, but most file systems are highly specialized for efficient reading and writing of large amounts of file-oriented data on high-bandwidth local area networks or as part of a storage area network. Network file systems are not typically intended for use as general-purpose directory services distributed or replicated across wide-area networks, although there are various attempts underway to define Internet-scale file systems.

Various *ad hoc* directory services have also been constructed using custom database systems (e.g., using sequentially accessed flat files or keyed record access) or relational database management systems. For example, a common application of a directory service is a simple "white pages" address book, which allows various lookup queries based on search criteria ranging from people's names, email addresses, or other searchable identity attributes. Given that RDBMS products are widely deployed and provide robust information storage, retrieval, and transactional update services, it is straightforward to implement a basic white pages directory service on top of an RDBMS. However, there can be complications due to limitations imposed by the fact that most directory services are defined in terms of a hierarchical data model and the mapping of this hierarchical data model into a relational data model is often less than satisfactory for many network based directory-enabled applications that require ultrafast search access and cannot tolerate the inherent overhead of mapping between the two data models. In addition, network based authentication and access control models may require an extra abstraction layer on top of the RDBMS system. This mapping may result in less functionality in these areas as well as potential inefficiencies.

Finally, numerous "yellow pages" directories have arisen on the Internet that provide a way to find information. While called "directories" because of the way information is organized hierarchically and presented via a Web browser, such directories are more properly text-based information search and retrieval systems that utilize Web spiders and some form of semantic qualification to group textual information into categories that meet various search criteria. These Web directories are highly useful for searching and accessing very large information spaces that are otherwise difficult to navigate and search without the assistance of some form of browsable navigation directory (e.g., Yahoo! and Google). While powerful, we distinguish this type of directory from LDAP-based directories that rely on an extensible but consistent data model that facilitates a more highly structured search mechanism. Some yellow pages directories actually utilize LDAP directories to augment their generalized text search mechanisms by creating structured repositories of metainformation, which is used to guide future searches based on stored topological information and/or historical search results.

34.2 The Evolution of LDAP

The LDAPv3 protocol and modern directory servers are the result of many years of research, international standardization activities, commercial product developments, Internet pilot projects, and an international community of software engineers and network administrators operating a globally interconnected directory infrastructure. LDAP directories originated from projects and products that originally implemented directories based on the 1988 and 1993 International Telecommunications Union (ITU) X.500 series of international standards [ITU X.500, 1993; ITU X.501, 1993; ITU X.511, 1993; ITU X.518–X.521, 1993; ITU X.525, 1993] under the assumption that electronic messaging based on the ITU X.400 series of standards would eventually become the dominant e-mail system in the international business community. For several technical, economic, and sociopolitical reasons, X.400 has not become the dominant email transport system that was envisioned, but X.500 has survived, and many LDAP protocol features and the schema and data model are derived from the X.500 specifications, as well as the early work on deploying practical Internet directories by the IETF OSI-DS working group [Barker and Kille, 1991].

Before the emergence of the Internet as a ubiquitous worldwide public network, there was an assumption that the world's data networks, and the messaging and directory infrastructures deployed on them, would be operated much like the world's voice communications networks were managed in the 1980s. These networks were managed by large telecommunications companies that had international bilateral agreements for handling such international communication services. The emergence of the worldwide Internet as a viable commercial network undercut many of the underlying service model assumptions that had gone into designing these technologies by participants in the International Telecommunications Union standards organizations. As universities, industrial R&D organizations, and technology companies increasingly embraced the Internet as a model for doing business, they created technologies that effectively created a value-added network on top of the bandwidth leased from the Telcos. This new style of networking created new demand for innovations that would better fit the Internet style of networked computing, where there is a very high degree of local autonomy with regard to deploying and managing network services as part of an enterprise's core IT function, rather than leasing expensive application services from a telecommunications company, as opposed to simply leasing bandwidth.

34.2.1 The Past, Present, and Future Generations of LDAP Directories

Since the mid-1990s, directory servers based on LDAP have become a significant part of the network infrastructure of corporate intranets, business extranets, and service providers in support of many different kinds of mission-critical networked applications. The success of LDAP within the infrastructure is due to the gradual adoption of directory servers based on the LDAPv3 protocol. The use of LDAP as a client access protocol to X.500 servers via an LDAP-to-X.500 gateway has been replaced by pure LDAPv3 directory servers. In this section, we briefly review the technological evolution that led to the current adoption of LDAPv3 as part of the core network infrastructure.

During the 1980s, both X.400 and X.500 technologies were under active development by computer technology vendors, as well as by researchers working in networking and distributed systems. Experimental X.400/X.500 pilots were being run on the public research Internet based on open source technology called the ISO Development Environment (ISODE), which utilized a *convergence protocol* allowing OSI layer 7 applications to operate over TCP/IP networks through a clever mapping of the OSI class 0 transport protocol (TP0) onto TCP/IP. TP0 was originally designed for use with a reliable, connection-oriented *network* service, such as that provided by the X.25 protocol. However, the principal inventor of ISODE, Marshall Rose (who was to go on to develop SNMP), recognized that the reliable, connection-oriented *transport* service provided by TCP/IP could effectively (and surprisingly efficiently) masquerade as a reliable connection-oriented network service underneath TP0. By defining the simple (5-byte header) convergence protocol specified in RFC 1006 [Rose and Cass, 1987], and implementing an embedding of IP addresses into OSI presentation addresses, the OSI session and presentation layers could be directly mapped onto TCP. Hence any OSI layer 7 application requiring the upper-layer OSI protocols could instantly be made available on the public Internet. Given the lack of public OSI networks at the time, this pragmatic innovation enabled the rapid evolution of X.500-based directory technology through real-world deployments across the Internet at universities, and industrial and government R&D labs, and forward-looking companies working to commercially exploit emerging Internet technologies. The lack of any real competitive technology led to the rapid adoption of network-based directory services on the Internet (unlike X.400, which failed to displace the already well-established SMTP as the primary Internet protocol for email).

LDAP arose primarily in response to the need to enable a fast and simple way to write directory client applications for use on desktop computers with limited-memory capacity (< 16 MB) and processing power (< 100 MHz). The emergence of desktop workstations and the PC were driving demand for increasingly sophisticated client applications. In order for X.500 to succeed independently of X.400, client applications were needed that could run on these rather limited desktop machines by today's standards. A server is only as good as the service offered to client applications, and one of the major inhibitors to the adoption of X.500 as a server technology outside its use as an address routing service for X.400 was the lack of sophisticated client applications. The X.500 client protocol, DAP, was used principally by X.400 to access information required to route X.400 messages. Since X.400 was another server application based on the full OSI stack, the complexity of using the DAP protocol for access to X.500 was "in the noise" given the overall complexity and computing resources required by a commercial-grade X.400 system. Many early X.400/X.500 vendors failed to grasp that simply specified protocols and the "good enough" services they enable were a key driver in the growth of the Internet and the resulting market demand for server software. As a result, X.400 and X.500 infrastructures were not as widely deployed commercially as anticipated, except in selected markets (e.g., some military and governmental organizations) where a high degree of complexity is not necessarily an inherent disadvantage.

The simplest application for a directory is as a network-based white pages service, which requires a rather simple client to issue a search using a string corresponding to a person's name or a substring expression that will match a set of names, and return the list of entries in the directory that match the names. This type of simple white pages client application was the original motivation for defining LDAP as a "lightweight" version of DAP. Some people like to claim that the "L" in LDAP no longer stands for "lightweight" because LDAP is now used in servers to implement a full-blown distributed directory service, not just a simple client-access protocol. However, the original motivation for making LDAP a lightweight version of DAP was to eliminate the requirement for the OSI association control service element (ACSE), the remote operations service element (ROSE), the presentation service, and the rather complicated session protocol over an OSI transport service (e.g., TP4). Even with the convergence protocol defined in RFC 1006, the upper layer OSI protocols required a rather large in-memory code and data footprint to execute efficiently. LDAP was originally considered "lightweight" precisely because it operated directly over TCP (eliminating all of the OSI layers except for use of ASN.1 and BER [ITU X.681, 1994; ITU X.690, 1994]), had a much smaller binary footprint (which was critical for clients on small-memory desktop PCs of the time), and had a much simpler API than DAP. In this context, the use of the term lightweight meant a small memory footprint for client application, fewer bits on the wire, and a simpler programming model, not lightweight functionality.

The LDAPv2 specification [Yeong et al., 1995] was the first published version of the lightweight client directory access protocol. While supporting DAP-like search and update operations, the interface to LDAPv2 was greatly simplified in terms of the information required to establish an association with an X.500 server, via an LDAP-to-DAP gateway. The introduction of this application-level protocol gateway mapped the client operations to the full DAP and OSI protocol stack. So, in this sense, LDAPv2 was a proper subset of the services offered by DAP, and no changes were required to an X.500 server to support these lightweight client applications, such as address book client services, as part of a desktop email application. LDAPv2 enabled rapid development of client applications that could then take advantage of what was expected to be a global X.500-based directory system. As client applications began to be developed with LDAPv2, some operational shortcomings manifested themselves. The most notable was the lack of a strong authentication mechanism. LDAPv2 only supports anonymous and simple password-based authentication (note: Yeong et al. [1995] predated the emergence of SSL/TLS). Such security concerns, the mapping of the X.500 geopolitical naming model to the Internet domain names, the need for referrals to support distributed directory data, the need for an Internet standard schema (e.g., inetOrgPerson), and the desire for a mechanism for defining extensions led to the formation of IETF LDAPEXT working group that began defining a richer set of services based on LDAPv2, which became a series of specifications for LDAPv3 (RFCs 2251–2256 [Wahl, 1997; Wahl et al., 1997a; 1997b; 1997c; Howes, 1997; Howes and Smith, 1997).

This process of gradually realizing that a simpler solution will likely satisfy the majority of user and application requirements is a recurring theme in today's technology markets. In addition, the emergence of business-critical Web-based network services and the widespread adoption of LDAPv3 based technologies as part of the infrastructure enabling Web services has enabled a sustainable market for LDAP directories and ensured that someday in the future LDAP directories will be considered entrenched legacy systems to be coped with by some future infrastructure initiatives.

34.2.2 First- and Second-Generation Directory Services

As just discussed, LDAP originated as a simplification of the X.500 DAP protocol to facilitate client development on small machines. The first LDAPv2 client applications were used with an X.500 server called "Quipu," which was developed as a collaboration among various European and American universities, research organizations, and academic network providers and administrators.[1]

Quipu was based on the 1988 X.500 standards specifications and, as such, implemented DAP as its client access protocol, requiring either a full OSI protocol stack or, as discussed previously, the OSI upper layers over RFC 1006. Quipu was deployed on the research Internet in the late 1980s and gained substantial exposure as an early directory service on the Internet at relatively small scale (e.g., 100k directory entries was considered a large directory at the time). Quipu was deployed at a number of universities as part of the Paradise directory project, which was administered from University College London where Quipu was primarily developed. In cooperation with researchers at the University of Michigan (i.e., Tim Howes), individuals at the Internet service provider PSI (i.e., Marshall Rose and Wengyik Yeong), researchers at University College London (e.g., Steve Kille), and other individuals, LDAPv2 emerged, and an application layer gateway called the LDAP daemon (ldapd) was developed at the University of Michigan that mapped LDAPv2 operations to DAP operations that were then forwarded onto an X.500 server, such as Quipu. As a result of LDAPv2 and this LDAP-to-DAP gateway, lightweight client applications were rapidly developed that could run on Windows and Macintosh PCs.

The success of Quipu as an early prototype X.500 directory on the Internet and LDAPv2 as a client led to further innovation. One of the main advantages of Quipu was that it was extremely fast in responding to search operations. This was due to its internal task switching (co-routine) architecture, which predated POSIX threads on most UNIX® systems, and the fact that on startup, it cached all of the directory entries into memory. This feature also severely limited the scalability of Quipu because of the expense and limitations of physical memory on 32-bit server machines and the potentially long startup time required to build the in-memory cache. Work was begun in 1992, both at the University of Michigan and at the ISODE Consortium, to produce a more scalable and robust directory server. The ISODE Consortium was an early open source organization that was a spinout of University College London and the Microelectronics and Computer Technology Corporation (MCC) in Austin, Texas.

The University of Michigan team first developed a server that exploited X.500 chaining to create an alternative back-end server process for Quipu that utilized a disk-based database built from the UNIX® dbm package as the underlying data store. Client requests were first sent to the main Quipu server instance that maintained topology information in it cache that allowed it to chain the requests to the back-end server. Effectively, Quipu was turned into a "routing proxy" and scalability was achieved with one or

[1] A Quipu (pronounced *key-poo)* is a series of colored strings attached to a base rope and knotted so as to encode information. This device was used by peoples of the ancient Inca empire to store encoded information.

more back-end servers hosting the data in its disk-based database, using caching for fast access. This approach proved the viability of a disk-based database for Quipu, but without integrating the disk-based database into the core of the Quipu server. This approach was taken for simplicity, and also because POSIX threads were finally viable in UNIX® and the new back-end server was based on a POSIX threading model instead of a task-based co-routine model. However, it suffered from the drawback of now having two separate server processes between the LDAP client and the actual data. An LDAP client request had to first go through the LDAP-to-DAP gateway, then through the Quipu server, then to the back-end server over the X.500 DSP protocol [ITU X.525, 1993], then back to the client.

During this same period, the ISODE Consortium began work on a new X.500 server that was essentially a rewrite of much of Quipu. Based on the promising performance and scalability results from the University of Michigan's back-end server implementation, the goal was to integrate the disk-based backend and POSIX threading model into a new, single-process directory server that could scale well and deliver search performance comparable to that of Quipu. This result was achieved with a new directory server from ISODE in 1995, based on the 1993 X.500 ITU standards. The protocol front-end was also redesigned so that additional protocol planes could be added as needed to accommodate additional server protocols. At the time, there was serious consideration given to implementing a DNS protocol plane, so that the directory server could be used to provide a DNS service in addition to both the LDAP and X.500 directory services. However, this work was never done. Instead, work was done to provide an integrated LDAPv3 protocol plane alongside of the X.500 DAP and DSP protocols, resulting in the first dual-protocol X.500+LDAP directory server.

The work at the University of Michigan continued in parallel and it became obvious that one could do away with the LDAP-to-DAP protocol gateway and the routing Quipu server, and simply map LDAPv2 directly to the new disk-based back-end server. All that was required was to implement the LDAPv2 protocol as a part of the back-end server and eliminate the X.500 DSP protocol. This was the key observation that led to the first pure directory server based on LDAPv2 with some extensions, called the standalone LDAP daemon (slapd).

These two separate architectural efforts led to the definition of the LDAPv3 protocol within the IETF, which was jointly defined and authored by individuals at the University of Michigan (Tim Howes and Mark Smith) and at the ISODE Consortium (Steve Kille and Mark Wahl). Both the slapd and ISODE server were the first directory servers to implement LDAPv3 as a native protocol directly, and validated its utility for implementing a directory service, not just as a client access protocol to an X.500 directory. In addition, both servers adopted the Berkeley DB b-tree library package as the basis for the disk-based backend, which added to the scalability, performance, and eventual robustness of both servers. In 1996, Netscape hired the principal inventors of the University of Michigan slapd server, which became the Netscape Directory Server that was widely adopted as an enterprise-scale LDAPv3 directory server. In 1995, the ISODE Consortium converted from a not-for-profit open-source organization to a for-profit OEM technology licensing company and shipped an integrated X.500 and LDAPv3 server.

Also in 1996, Critical Angle was formed by former ISODE Consortium engineers and they developed a carrier-grade LDAPv3 directory server for telecommunications providers and ISPs. This was the first pure LDAPv3 server to implement chaining via LDAPv3, and also the first server to have multimaster replication using LDAPv3. In addition, Critical Angle developed the first LDAP Proxy Server, which provided automatic referral following for LDAPv2 and LDAPv3 clients, as well as LDAP firewall, load balancing, and failover capability, which are critical features for large-scale directory service deployments.

With the emerging market success of LDAPv3 as both a client and a server technology on the Internet and corporate intranets, vendors of previously proprietary LAN-based directory servers launched LDAPv3-based products, most notably Microsoft Active Directory and Novell eDirectory. Microsoft had originally committed to X.500 as the basis for its Windows™ directory server, but adopted LDAP as part of its Internet strategy.

IBM built its SecureWay LDAP directory product using the University of Michigan slapd open-source code and designed a mapping onto DB2 as the database backend. Like IBM, Oracle implemented an LDAP gateway onto its relational database product. However, it is generally the case that mapping of the hierarchical data model of LDAP and X.500 onto a relational data model has inherent limitations. For some directory service deployments, the overhead inherent in the mapping is a hindrance in terms of performance.

Most X.500 vendors continue to provide an LDAP-to-DAP gateway as part of their product offerings, but their marketing does not usually mention either the gateway or X.500, and instead calls the X.500 server an LDAP server.

The Critical Angle LDAP products were acquired in 1998 by Innosoft International. In 1999, AOL acquired Netscape. Sun Microsystems and AOL/Netscape entered into a joint marketing and technology alliance called iPlanet shortly thereafter. In March 2000, Sun acquired Innosoft and consolidated its directory server expertise by combining the Netscape technology, the Innosoft/Critical Angle technology, and its own Solaris LDAP-based directory technology initiatives into a single-directory product line under the Sun ONE™ brand.

34.2.3 Next-Generation Directory Services

Now that LDAPv3 directory servers are widely deployed, whether for native LDAP implementations or using LDAP gateways mapping operations onto X.500 servers or relational database systems, new types of directory-based services are being deployed. The most recent of these are identity management systems that provide authentication, authorization, and policy-based services using information stored in the directory. The most common type of identity management service is Web single sign-on. With the proliferation of network based systems, the need for a common authentication mechanism has dictated that a higher-level service be deployed that abstracts from the various login and authentication mechanisms of different Web-based services. Identity servers built on top of directory services are providing this functionality. At present, such services are primarily being deployed within an enterprise, but there are efforts underway to define standards for federating identity information across the Internet. It will take some time before these standards, activities, and the technologies that implement them are deployed, but the foundation on which most of them are being built is a directory service based on LDAPv3.

Another area where LDAP directories are gaining widespread usage is among wireless carriers and service providers. Next-generation wireless services are providing more sophisticated hand-held devices with the opportunity to interact with a more functional server infrastructure. Directories are being deployed to provide a network-based personal address book and calendars for mobile phones and PDAs that can be synchronized with the handheld devices, a laptop computer and a desktop computer. Directory services are being deployed as part of the ubiquitous network infrastructure that is supporting the management of personal contact and scheduling information for hundreds of millions of subscribers. Fortunately, LDAP directory technology has matured to the point at which it is capable of providing the performance, scalability, and reliability required to support this "always on" service in a globally connected world.

In order to simplify the integration of directory data access into Web service development environments, new directory access protocols building on LDAP are being defined. Instead of ASN.1 and TCP, these protocols use XML [Bray et al., 2000] as the encoding syntax and Simple Object Access Protocol (SOAP) [Gudgin et al., 2003] as a session protocol overlying HTTP or persistent message bus protocols based on Java Messaging Service (JMS) APIs. The standards body where these protocols are being developed is OASIS, the Organization for the Advancement of Structured Information Standards.

One group within OASIS in particular, the Directory Services working group, has published version 2 of the Directory Services Markup Language (DSMLv2), which leverages the directory semantics from LDAP — a hierarchical arrangement of entries consisting of attributes, but expressing the LDAP operations in SOAP. Just as X.500 servers added support for LDAP either natively or through an LDAP to X.500 gateway, there are implementations of DSMLv2 both as native protocol responders within an LDAP server and as DSMLv2 to LDAP gateway.

Other working groups have already or are in the process of defining protocols for more specialized directory access, such as Universal Description, Discovery, and Integration of Web Services (UDDI)

[Bellwood, 2002], ebXML Registry Services [OASIS, 2002], and Service Provisioning Markup Language (SPML) [Rolls, 2003]. In the future, market dynamics may favor the adoption of one or more of these XML-based protocols to augment and eventually supplant LDAP as the primary client access protocol for directory repositories in the Web services environment.

34.3 The LDAP Naming Model

The primary contents of most directory services are entries that represent people, but entries may also represent organizations, groups, facilities, devices, applications, access control rules, and any other information object. The directory service requires every entry have a unique name assigned when the entry is created, and most services have names for users that are based on attributes that do not change frequently and are human readable.

Entries in the directory are arranged in a single-rooted hierarchy. The Distinguished Name (DN) of an entry consists of the list of one or more distinguished attribute values chosen from the entry itself, followed by attributes from that entry's parent, and so on up the tree to the root. In most deployments, only a single-attribute value is chosen from each entry.

The most widely used naming attributes are defined in the following table.

dc	domainComponent: one element of a DNS domain name, e.g., dc=sun, dc=com
uid	userid: a person's account name, e.g., uid=jbloggs
cn	commonName: the full name of a person, group, device, etc. e.g., cn=Joe Bloggs
l	localityName: the name of a geographic region, e.g., l=Europe
st	stateOrProvinceName: used in the United States and Canada
o	organizationName: the name of an organization, e.g., o=Sun Microsystems
ou	organizationalUnitName: the name of a part of an organization: ou=Engineering
c	countryName: the two letter ISO 3166 code for a country, e.g., c=US

The hierarchy of entries allows for delegated naming models in which the organization managing the name space near the root of the tree agrees on the name for a particular entry with an organization to manage that entry, and delegates to that organization the ability to construct additional entries below that one. Several naming models have been proposed for LDAP.

34.3.1 The X.500 Naming Model

The original X.500 specifications assumed a single, global directory service, operating based on interconnections between national service providers. In the X.500 naming model, the top levels of the hierarchy were to have been structured along political and geographic boundaries. Immediately below the root would have been one entry for each country, and the entries below each country entry would have been managed by the telecommunications operator for that country. (In countries where there were multiple operators, the operators would have been required to coordinate the management of this level of the tree.)

As there is no one international telecommunications operator, an additional set of protocol definitions was necessary to define how the country entries at the very top of the tree were to be managed. The telecommunications operator for each country would be able to define the structure of entries below the country entry. The X.500 documents suggested that organizations that had a national registration could be represented by organization entries immediately below the country entry. All other organizations would be located below entries that represented that country's internal administrative regions, based on where that organization had been chartered or registered. In the U.S. and Canada, for example, there would have been intermediate entries for each state and province, as states and provinces operate as registrars for corporations. For example, a corporation that had been chartered as "A Inc." in the state of California might have been represented in the X.500 naming model as an entry with the name o=A Inc.,st=California,c=US, where "o" is the attribute for the organization name that was registered within the state, "st" for state or province name within the country, and "c" the attribute for the country code. It should be noted that some certificate authorities that register organizations in order to issue them X.509 public key certificates, e.g., for use with SSL or secure email, assume this model for naming organizations.

The entries for people who were categorized as part of an organization (e.g., that organization's employees) would be represented as entries below the organization's entry. However X.500 did not suggest a naming model for where entries representing residential subscribers would be represented.

34.3.2 Limitations of the X.500 Naming Model

The first limitation is that there is no well-defined place in the X.500 model to represent international and multinational organizations. Organizations such as NATO and agencies of the United Nations and the European Community, as well as multinational corporations were some of the first to attempt to pilot standards-based directory services, yet ran into difficulties as there was no obvious place for the organization's entry to be located in a name space that requires a root based on a national geopolitical naming structure.

A related problem is that a corporation typically operates in additional locales beyond the one where they are legally incorporated or registered. In the U.S., for example, many corporations are registered in Delaware for legal and tax reasons, but may have no operating business presence in that state beyond a proxy address. A naming structure that has the organization based in Delaware may hinder users searching the directory, who might not anticipate this location as the most likely place to find the directory information for the corporation.

In some cases, organizations preferred to have an entry created for them in a logically appropriate place in the X.500 hierarchy, yet the telecommunications operator implied by the naming model as being authoritative for that region of the directory tree may have had no plans to operate X.500. Conversely, some parts of the directory tree had conflicting registration authorities as a result of political turf wars and legal disputes, not unlike those that plagued the Internet Assigned Numbers Authority (IANA) and the administrators of the root DNS servers. For use within the U.S. and Canada, the North American Directory Forum (NADF) proposed a set of attribute and server extensions to address the problems of overlapping registration authorities creating entries for individual and business subscribers in entries; however, these extensions were not implemented, and no X.500 service saw significant deployment in these countries.

34.3.3 Early Alternatives to the X.500 Naming Model

Many LAN-based directory services predating LDAP suggested a simpler naming model. Unlike the complete interconnection in a single, global directory service, these models assumed that interconnection only occurred between pairs or small groups of cooperating organizations, and that relatively few organizations worldwide would interconnect. Instead of the geographic divisions of X.500, this naming model is based on a single registration authority that would assign names to organizations immediately below the root of the tree, e.g., o=Example, resulting in a flattened namespace. This model did not readily accommodate conflicts in names between organizations and relied on one registration authority to ensure uniqueness.

As the Internet became more widely used by organizations for electronic mail, a variant of the flat namespace model was to register the organization's Internet domain name as the value for the organizationName attribute, e.g., o=example.com. By relying on an external naming authority for managing the actual assignment of names to organizations, potential conflicts would be resolved before they reached the directory name registration authority, and the use of the hierarchical domain name space would allow for multiple organizations with the same name that were registered in different countries or regions, e.g., o=example.austin.tx.us and o=example.ca.

34.3.4 Internet Domain-Based Naming

The single-component organization naming model described above addresses the difficulty that organizations have when faced with getting started using the X.500 model, but this approach suffers from a serious limitation. While domain names themselves are hierarchical, placing the full domain name as a string into the organizationName attribute prevented the hierarchical structure of the directory from being used. In particular, it was not defined in that approach how an organization that was itself structured and represented internally with multiple domain names, e.g., east.example.com and west.example.com, would be able to manage these as part of a hierarchy below the organization entry.

These limitations were removed in the mapping defined in RFC 2247 [Kille, 1988] between Internet domain names and LDAP distinguished names. In this approach, each domain-name component is mapped into its own value of the "dc" attribute to form the distinguished name of the entry (typically an organization) to which the domain corresponds. For example, the domain name cs.ucl.ac.uk would be transformed into dc=cs,dc=ucl,dc=ac,dc=uk, and the domain name example.com into dc=example,dc=com.

Follow-on documents have proposed how DNS SRV records can be used to specify the public directory servers that an organization provides, in a similar manner to the MX records for specifying the mail servers for that organization. In combination with RFC 2247 naming, a directory client that knows only a domain name can use these techniques to locate the LDAP server to contact and construct the search base to use for the entry that corresponds to that domain.

RFC 2247 assumes that when performing operations on the directory entries for an organization, the organization's domain name is already known to the client so that it can be automatically translated into a distinguished name to be used as an LDAP search base. RFC 2247 does not address how to programmatically locate an organization when the organization's domain name is not known; this is currently an unsolved problem in the Internet.

34.3.5 Naming Entries within an Organization

There are currently no Internet standards that are widely adopted for naming entries representing people within an organization. Initial deployments of LDAP made extensive use of the organizationalUnit entries to construct a directory tree that mirrored the internal divisions of the organization and use the "cn" attribute as the distinguished attribute for the person's entry, as in the following:

```
cn=Joe Bloggs, ou=Western, ou=Sales, ou=People, dc=example, dc=com
```

That approach, however, resulted in organizations needing to frequently restructure their directory tree as the organization's internal structure changed, and even placing users within organizational units did not eliminate the potential for name conflicts between entries representing people with the same full name.

Currently, the most common approach for directory deployments, in particular those used to enable authentication services, is to minimize the use of organizationalUnit entries, and to name users by their login name in the uid attribute.

Many deployments now have only a single organizationalUnit entry: ou=People. Some multinational organizations use an organizationalUnit for each internal geographic or operating division, in particular when there are different provisioning systems in use for each division, or it is necessary to partition the directory along geographic lines in order to comply with privacy regulations. For example, an organization that has two operating subsidiaries X and Y might have entries in their directory named as follows:

```
uid=jbloggs, ou=X, ou=People, dc=example, dc=com
uid=jsmith, ou=France, ou=Y, ou=People, dc=example, dc=com
```

For service provider directories or directories that offer a hosted directory service for different organization entities, the DNS domain name component naming is most often used to organize information in the directory naming tree as follows:

```
uid=jbloggs, ou=X, dc=companyA, dc=com
uid=jsmith, ou=Y, dc=companyA, dc=com
uid=jwilliams, ou=A, dc=companyB, dc=com
```

uid=mjones, ou=B, dc=companyB, dc=com

In a typical service provider or hosted directory environment, the directory data for different organizations is physically partitioned and an LDAP proxy server is used to direct queries to the appropriate directory server that holds the data for the appropriate naming context.

34.4 The LDAP Schema Model

Directory servers based on LDAP implement an extensible object-oriented schema model that is derived from the X.500 schema model, but with a number of simplifications as specified in RFC 2256 [Wahl, 1997]. Two additional schema definition documents are RFC 2798, which defines the inetOrgPerson object class, and RFC 2247, which defines the dcObject and domain object classes. The schema model as implemented in most LDAP servers consists of two types of schema elements: attribute types and object classes. Object classes govern the number and type of attributes that an entry stored in the directory may contain, and the attribute types govern the type of values that an attribute may have. Unlike many database schema models, LDAP schema has the notion of multivalued attributes that allows a given attribute to have multiple values. The types of the values that may be associated with a given attribute are defined by the attribute type definition. The most common attribute type is a UTF-8 string, but many other types occur, such as integer, international telephone number, email address, URL, and a reference type that contains one or more distinguished names, representing a pointer to another entry in the directory.

A directory entry may have multiple object classes that define the attributes that are required to be present, or may optionally be present. Directory servers publish their internal schema as an entry in the directory. It can be retrieved by LDAP clients performing a baseObject search on the a special entry that is defined by the directory server to publish schema information (e.g., cn=schema), with the attributes attributeTypes and objectClasses specified as part of the search criteria. This schema entry maintains the schema definitions that are active for a given directory server instance. The format of these two attributes is defined in RFC 2252 [Wahl et al., 1997b].

34.4.1 Attribute-Type Definitions

An attribute-type definition specifies the syntax of values of the attribute and whether the attribute is restricted to having at most one value, and the rules that the server will use for comparing values. Most directory attributes have the Directory String syntax, allowing any UTF-8 encoded Unicode character, and use matching rules that ignore letter case and duplicate white space characters.

User attributes can have any legal value that the client provides in the add or modify request, but a few attributes are defined as operational, in which the attributes are managed or used by the directory server itself and may not be added or changed by most LDAP clients directly. Examples of operational attributes include createTimestamp and modifyTimestamp.

34.4.2 Object-Class Definitions

Each entry has one or more object classes, which specifies the real-world or information object that the entry represents, as well as the mandatory and permitted attributes defined in the entry. Object classes come in one of three kinds: *abstract*, *structural*, or *auxiliary*.

There are only two abstract classes. The top class is present in every entry, and it requires that the objectClass attribute be present. The other abstract class is alias, and it requires that the aliasedObjectName attribute be present.

Structural object classes define what the entry represents, and every entry must contain at least one structural object class; for example: organization, device, or person. A structural object class inherits either from top or from another structural object class, and all the structural object classes in an entry must form a single "chain" leading back to top. For example, the object class organizationalPerson inherits from person and permits additional attributes to be present in the user's entry that describe the person within an organization, such as title. It is permitted for an entry to be of object classes top, person and organizationalPerson, but an entry cannot be of object classes top, person and device, because device does not inherit from person, nor person from device.

Auxiliary classes allow additional attributes to be present in a user's entry, but do not imply a change in what the entry represents. For example, the object class strongAuthenticationUser allows the attribute userCertificate;binary to be present, but this class could be used in an entry with object class person, device, or some other structural class.

34.4.3 Object Classes for Entries Representing People

The person structural class requires the attributes "cn" (short for commonName) and "sn" (short for surname). This class is subclassed by the organizationalPerson class, and the organizationalPerson class is subclassed by the inetOrgPerson class. Most directory servers for enterprise and service provider applications use inetOrgPerson, or a private subclass of this class, as the structural class for representing users. In addition to the mandatory attributes cn, sn, and objectClass, the following attributes are typically used in entries of the inetOrgPerson object class:

departmentNumber	a numeric or alphanumeric code
description	a single line description of the person within the organization
displayName	name of the user as it should be displayed by applications
employeeNumber	unique employee number
employeeType	a descriptive text string, such as "Employee" or "Contractor"
facsimileTelephoneNumber	fax number, in international dialing format (e.g., +1 999 222 5555)
givenName	first or given name
homePhone	home phone number, in international dialing format
homePostalAddress	home mailing address, with "$" inserted between lines
jpegPhoto	photograph in JPEG format
labeledURI	the URI for a Web home page
mail	Internet email address
manager	distinguished name of the entry for a manager
mobile	mobile phone number in international dialing format (e.g., +1 999 222 4444)
ou	organizational Unit, if different from department
pager	pager phone number and codes, if any
postalAddress	mailing address, with "$" inserted between lines
roomNumber	office room number
secretary	distinguished name of the entry for a secretary
surname (sn)	Last or surname
telephoneNumber	telephone number, in international dialing format
title	Job title
uid	user id, typically part of the person's distinguished name
userPassword	user password compared against during LDAP authentication

For example, the following definition is an LDAP Interchange Format (LDIF) text representation of a typical user entry in an LDAP directory, as specified in RFC 2489 [Good, 2000]. The order of appearance of most attributes and value pairs does not imply any specific storage requirements, but it is convention to present the objectClass attribute first, after the distinguished name. Several additional attributes are permitted in entries of this object class, but are no longer widely used. For further details consult RFC 2256 [Wahl, 1997] and RFC 2798 [RFC 2798].

```
dn: uid=jbloggs,ou=people,dc=example,dc=com
objectClass: top
objectClass: person
objectClass: organizationalPerson
objectClass: inetOrgPerson
cn: Joe Bloggs
```

```
sn: Bloggs
departmentNumber: 15
description: an employee
displayName: Joe Bloggs
employeeNumber: 655321
employeeType: EMPLOYEE
facsimileTelephoneNumber: +1 408 555 1212
givenName: Joe
homePhone: +1 408 555 1212
homePostalAddress: Joe Bloggs $ 1 Mulberry Street $ Anytown AN 12345
labeledURI: http://eng.example.com/~jbloggs
mail: jbloggs@example.com
manager: uid=jsmith,ou=people,dc=example,dc=com
mobile: +1 408 555 1212
ou: Internet Server Engineering
pager: +1 408 555 1212
postalAddress: Joe Bloggs $ 1 Main Street $ Anytown AN 12345
roomNumber: 2114
telephoneNumber +1 408 555 1212 x12
title: Engineering Manager
uid: jbloggs
userPassword: secret
```

34.4.4 Other Typical Object Classes

The organization structural class requires the "o" attribute (short for organizationName) to be present in the entry, and permits many attributes to optionally also be present, such as telephoneNumber, facsimileTelephoneNumber, postalAddress, and description. This class is normally used to represent corporations, but could also represent other organizations that have employees, participants, or members and a registered name.

The organizationalUnit structural class requires the "ou" (short for organizationalUnitName) be present, and permits the same list of optional attributes as the organization class. This class is normally used to represent internal structures of an organization, such as departments, divisions, or major groupings of entries (e.g., ou=People).

The domain structural class requires the "dc" (domainComponent) attribute be present. This class is used to represent objects that have been created with domain component naming, but about which no other information is known. The dcObject auxiliary class is used to permit the dc attribute to be present in entries of the organization or organizationalUnit classes, typically so that the dc attribute can be used for naming the entry, although the o or ou attribute is still required to be present.

The groupOfNames structural class uses cn for naming the group, which is represented by the attribute member. This requires one or more attribute values each containing the distinguished name of another entry that is a member of the group. A similar and more widely used class, groupOfUniqueNames, uses the uniqueMember attribute.

34.5 LDAP Directory Services

There are several architectural applications of LDAP in today's Internet: email address book services, Web-based white pages lookup services, Web authentication/authorization services, email server routing and address list expansion services, and literally hundreds of uses that are generally categorized as a network-based repository for application-specific information (e.g., application configuration information, directory-enabled networking such as router tables, and policy-based user authentication and access authorization rules). In other words, basic LDAP directory services have become a critical part of the network infrastructure for many applications, just as DNS, FTP, SMTP, and HTTP are core infrastructure services. LDAP is very often there behind-the-scenes of many end-user applications and embedded in a number of other services that are not end-user visible.

One of the most common questions that arises in corporate directory services deployments is the following: "Why not just use a relational database system rather than a new kind of database?" The answer to this question is often as varied as the application services in support of which an LDAP directory is being considered. In some cases, such as a simple Web-based white pages service, there is no real compelling advantage over using an RDBMS that may already contain all the information about people and their contact details. Directories are a distinct type of distributed database and are best suited to a new generation of network-based applications whose data access and service availability requirements do not require a relational data model or a SQL-based transactional database.

An important distinction that networking system novices make is to distinguish a protocol and the service implemented in terms of the protocol. A protocol defines the communication pattern and the data exchanges between two end points in a distributed system. Typically, one end point is a client and the other a server, but both end points could be peers. The semantics offered by a server often extends beyond the information exchange rules that are specified by the protocol. In other words, the server may require additional features to implement a reliable, maintainable, highly available service that transcend the basic information exchange implied by a protocol. For example, LDAPv3 has the concept of extended operations and special controls, some of which are standardized and some of which are not. The result is that vendors have created extensions to the core protocol specifications to enable additional services such as configuration and management of the server over LDAP without ever having to shut down the server to ensure high availability. This is not necessarily a bad thing because extended operations and controls are useful from an administrative perspective, enabling network-based management of an LDAP service using the LDAP protocol, or special server-to-server communication enabling a distributed directory service. In a competitive market, where technology vendors compete with one another by enabling proprietary client visible features, complete interoperability between clients and servers may be broken. This situation is typically avoided by having periodic interoperability testing forums where competing vendors demonstrate interoperability. As long as the core protocol and basic service model is not violated, then client interoperability is maintained.

Another confusing aspect of standards-based network services is the difference between standards conformance vs. demonstrated interoperability. Within some standards organizations, the emphasis has been on demonstration of static conformance to a written specification. Static conformance is often achieved through the successful demonstration that a server passes some set of conformance tests. Conformance testing is useful to the vendor, but what is most useful to a user is interoperability testing, i.e., does the server from vendor A work with clients from vendors B, C, and D? Conformance is easier to achieve than interoperability. LDAPv3 has been shown to be a highly interoperable protocol and most clients work without complication with most servers. Vendors of LDAPv3 client and server products regularly meet to perform interoperability testing forums, sponsored by The Open Group, to ensure that products work together. In the remainder of this section we discuss the basic and advanced modes of operations for common LDAP-based directory services.

34.5.1 Basic Directory Services

The LDAP service model is a basic request/response client-server model. A client issues either a bind, search, update, unbind, or abort request to the server. All protocol operations may either be initiated directly over a TCP connection or encrypted via an SSL/TLS session.

The bind operation is used to pass authentication credentials to the server, if required. LDAP also supports anonymous search, subject to access control restrictions enforced by the server. Bind credentials consist of the user's distinguished name along with authentication credentials. A distinguished name typically identifies a user corresponding to a logical node in the directory information tree (DIT). For example: uid=jbloggs, ou=People, dc=sun, dc=com. Authentication credentials may be a simple clear text password (optionally over a SSL or TLS session), information obtained from a digital certificate required to strongly authenticate, or other encrypted or hashed password authentication mechanisms enabled by a particular server.

There are three principal models of distributed operations: a simple client-server model, a referral model, and a chaining model. The simple client-server interaction is not depicted, but operates as one would expect: a client performs a bind operation against the server (either anonymously or with bind credentials), issues an LDAP operation (e.g., search), obtains a result (possibly empty or an error), and then either issues other operations or unbinds from the server. In this mode of operation, an LDAP server will typically handle hundreds to thousands of operations per second on behalf of various types of LDAP-enabled clients and applications.

In some deployments, most notably those on the public Internet or in government, university and enterprise directory service environments where anonymous clients may connect and search a forest of directories, a referral model may be appropriate. The referral model assumes that either (1) all clients will bind anonymously, or (2) authentication information is replicated among the set of directory servers, or (3) there is some mechanism for proxy authentication at one server on behalf of a network of trusted directory servers that will accept authentication credentials via a proxy [Weltman, 2003].

Figure 34.1 depicts the most common referral model situation and assumes that a client is anonymous, such that it can bind and search any one of the two directory servers depicted. In the referral model, if a client requests an operation against one directory server (e.g., a search operation) and that directory

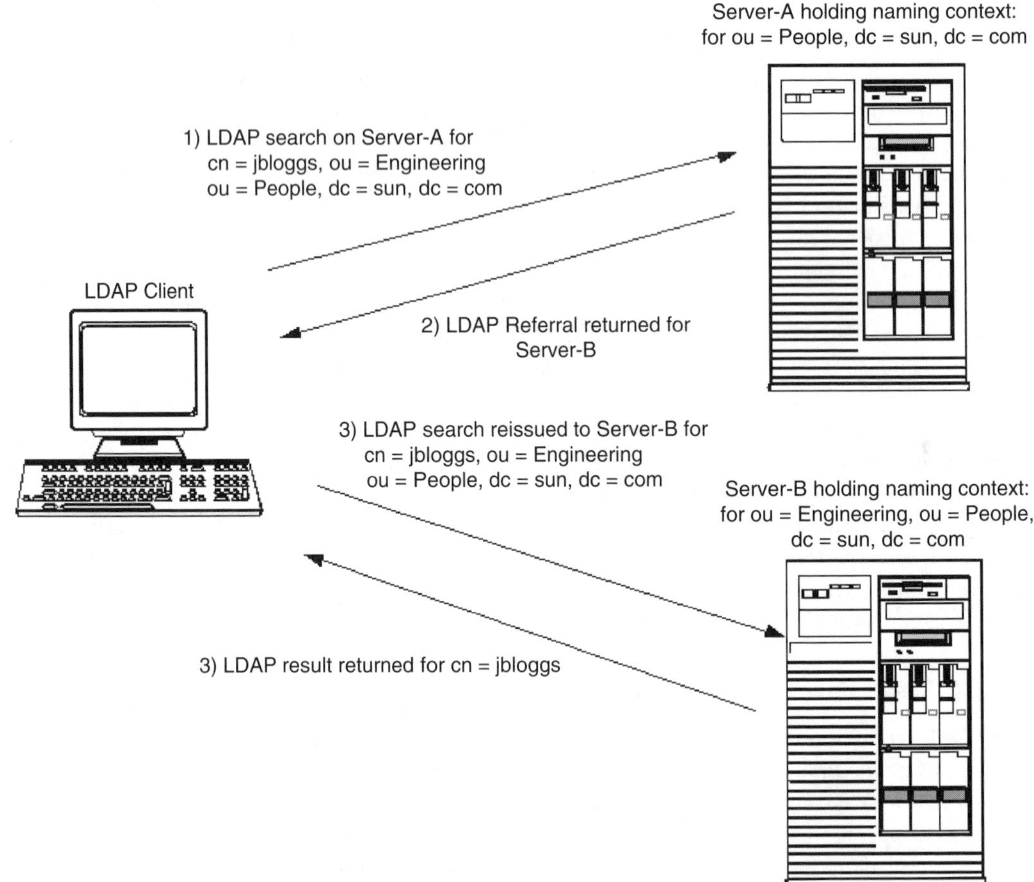

FIGURE 34.1 LDAP referral model.

server does not hold the entry or entries that satisfy that operation, then a referral may be returned to the client. A referral is a list of LDAP URLs that point to other directory servers that the original server is configured to refer queries to. A referring server may have out-of-date information and the referral may not succeed. Referral processing is the responsibility of the client application and is most often handled transparently as part of the LDAP API from which the client application is built. The referral model is most often appropriate in directory service deployments where there are no stringent requirements on authentication because servers may be configured to accept unauthenticated anonymous operations, such as searches. In fact, one of the major disadvantages of the referral model is that it facilitates trawling of a large distributed directory service and allows a snooping client application to probe and discover a directory service's topology. This may be undesirable even in a public Internet environment, such as a university network. For this reason, LDAP proxy servers were invented to provide an additional level of control in these more open network environments.

The chaining model is similar to the referral model, but provides a higher degree of security and administrative control. LDAP chaining is similar to chaining in the X.500 directory service model, but it is done without requiring an additional sever-to-server protocol as in the case of X.500. Figure 34.2 illustrates the chaining model. In this case, a client issues a search request to Server A, which uses its knowledge of which server holds the subordinate naming context, and it chains the request to Server B.

Chaining assumes an implied trust model between Server A and Server B because typically Server A will authenticate to Server B as itself, not as the client. For efficiency in the chaining model, it is typical

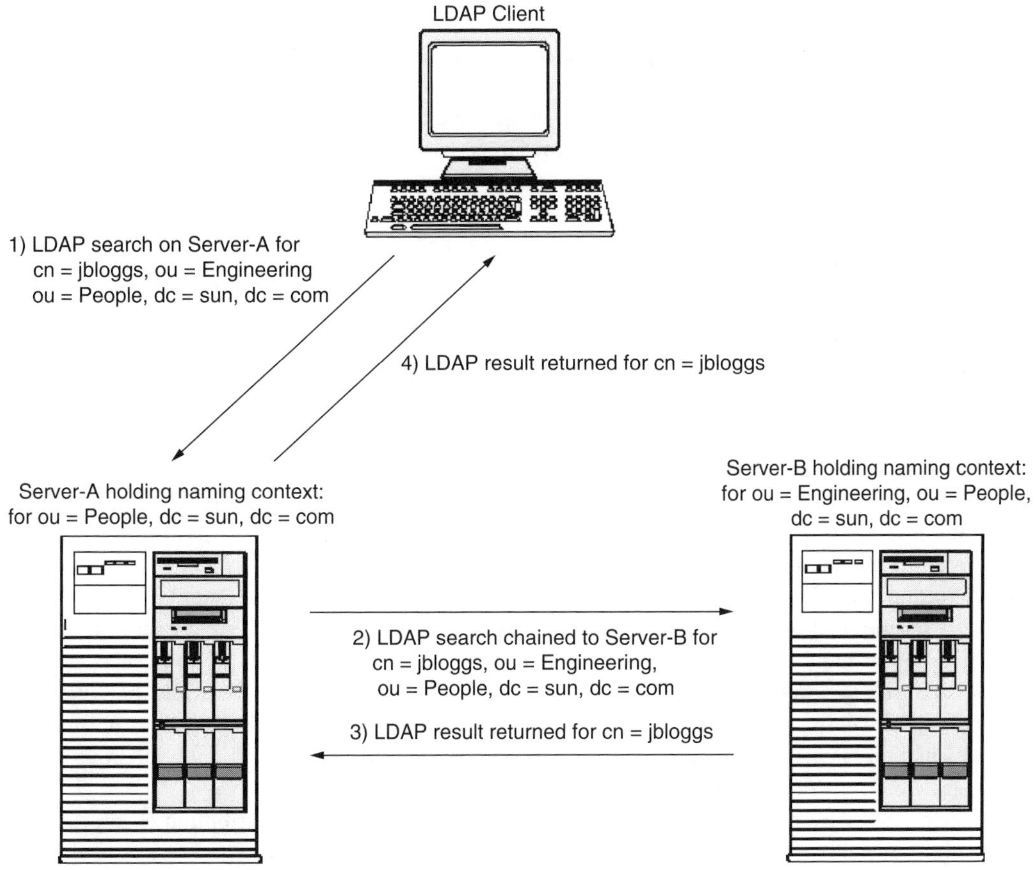

FIGURE 34.2 LDAP chaining model.

for Server A to maintain a persistent open network connection to Server B to eliminate the overhead of binding for each chained operation. In some cases, such as anonymous clients, there is no authentication information to proxy, so Server A will often maintain a separate, nonauthentication connection to Server B for such requests. A proxied authentication model could also be used in which case the client credentials are passed along as part of the chained operation, requiring both Server A and Server B to authenticate or proxy the authentication of the client.

34.5.2 High Availability Directory Services

First generation X.500 and LDAP directory servers focused primarily on implementing as much of the protocol specifications as possible, providing flexible and extensible schema mechanisms, and ensuring very fast search performance. During the early adoption phase of a new technology, these are the critical elements to get right and it is necessary for rapid feedback and technology evolution as the result of real world deployments. However, as LDAP server technology has become more central to the network infrastructure backing up mission-critical business operations (e.g., as a network user authentication service), security, reliability, scalability, performance, and high availability of the service are the dominant operational requirements. In the remainder of this section, we briefly discuss high availability features and issues related to deployment of LDAP directory services in support of business critical applications.

34.5.3 Master–Slave Replication

The X.500 specifications define a replication protocol called DISP (Directory Information Shadowing Protocol) that provides a simple master–slave replication service. Technically, directory replication is based on a *weakly consistent* supplier–consumer model, but the master–slave terminology has become dominant. In this weakly consistent replication model, one directory server acts a supplier (master) of data to another directory server that is the consumer (slave). At any given time a replica may be inconsistent with the authoritative master, and so a client accessing data at the replica might not see the latest modifications to the data. If this situation is likely to cause problems for the client applications, there are deployment techniques (e.g., using a proxy server) that ensure that a client application will be connected with an authoritative master server so that it may obtain up-to-date information.

Within the X.500 model it is possible for either the supplier to initiate a replication data flow or for the consumer to request a replication data flow from the supplier. Replication can occur as soon as a change is detected in a supplier, called on-change, or according to some periodic schedule. Supplier-initiated replication works best when network links between the supplier and the consumer are reliable, with low latency and reasonably high bandwidth. Consumer-initiated replication is most often used in situations where the consumer might frequently become disconnected from the network either due to the unscheduled network outages or high-latency, low-bandwidth networks that require potentially large replication data exchanges to be done during off-peak times as determined by the consumer (e.g., a consumer on a computer in a submarine).

Replication typically requires that not only directory entries be replicated from a supplier to a consumer but also schema and access control information. If not done automatically via a protocol, manual configuration is required for each consumer, and there could be thousands of consumers in large distributed directory system (e.g., a directory consumer in every airport in the world holding flight schedule information). Some directory servers do not implement schema and access control such that the information can be replicated, so manual configuration or some other out-of-band technique is used to replicate this type of operational information.

While master–slave replication provides high availability via geographic distribution of information, facilitating scalability for search operations, it does not provide high-availability for modify operations. In order to achieve write-failover, it is necessary to employ systems engineering techniques to enable either a cluster system running the single master or enable a hot standby mode so that in the event of the failure of the single master the hot standby server can be brought online without delay. Another technique is to allow a slave server to become a master server through a process of *promotion*, which involves special control logic to allow a slave to begin receiving updates and to notify other slaves that it is now the authoritative master. Many directory deployments only allow write operations to the master server and route all search operations, using DNS or an LDAP Proxy server, to one of the slaves so that the load on the master server is restricted to only modify operations.

A common scenario that is employed in single-master, multiple-slave, directory deployments is to deploy a small set of replica hubs, each being a read-only replica from which other slaves may obtain updates. In the event a master fails, a hub is easily promoted. In this model, depicted in Figure 34.3, the replica hubs are both consumers and suppliers because they consume their updates from an authoritative master server, but also supply updates to other slaves. This scenario is most useful when a large number of slave servers would put unnecessary load on a single master, and so a hierarchy of servers is established to distribute the replication update load from the master to the hubs; otherwise, the master might spend all of its time updating consumers.

34.5.4 LDAP Proxy Server

An LDAP proxy server is an OSI layer 7 application gateway that does LDAP protocol data unit forwarding, possibly requiring examination, and possibly on-the-fly modification, of the LDAP protocol message to determine application-layer semantic actions. The objective of the LDAP proxy server is to provide an administrative point of control in front of a set of deployed directories, but does not answer LDAP queries itself, instead chaining the queries when appropriate to the directory servers. The proxy allows a degree of transparency to the directory services, in conjunction with DNS maps pointing LDAP clients to one or more proxy services instead of at the actual directory servers. The LDAP proxy provides a number of useful functions that are best done outside of the core directory server. These functions include:

1. LDAP schema rewriting to map server schema to client schema in the cases where client schema is either hard-coded or for historical reasons does not match the extensible schema of the directory server. Once thousands of clients are deployed, it is difficult to correct the problem, and so server applications must often adapt for the sake of seamless interoperability.
2. Automatic LDAP referral following on behalf of both LDAPv2 and LDAPv3 clients. The LDAPv2 protocol did not define a referral mechanism, but a proxy server can map an LDAPv2 client request into an LDAPv3 client request so that referrals can be used with a mixed set of LDAPv2 and LDAPv3 clients.
3. An LDAP firewall that provides numerous control functions to detect malicious behavior on the part of clients, such as probing, trawling, and denial-of-service attacks. The firewall functions include rate limiting, host and TCP/IP-based filters similar to the TCP wrappers package, domain access control rules, and a number of LDAP-specific features, including operations blocking, size limits, time limits, and attribute filters. The rate-limiting feature allows a statistical back-off capability using TCP flow control so that clients attempting to overload the directory are quenched.
4. The proxy provides a control point for automatic load balancing and failover/failback capability. It may also be able to maintain state information about load on a set of directory servers and redirect LDAP traffic based on load criteria or other semantic criteria at the LDAP protocol level. In addition, it can detect the failure of a directory server and rebalance the incoming load to other directory servers, and detect when a failed directory server rejoins the group.
5. The proxy also provides the point-of-entry advertised to clients in the DNS to the directory service, thereby providing a level of indirection to client applications that facilitates maintenance, upgrades, migrations, and other server administrative tasks done on the back-end directory servers, in a manner that is transparent to clients so that a highly available directory service is delivered.

An LDAP proxy is unlike an IP firewall in that it does application-layer protocol security analysis. The proxy is also unlike an HTTP proxy in that it does not do caching of directory data because it is unable to apply directory access rules to clients. The LDAP proxy is also unlike an IP load balancer in that it is

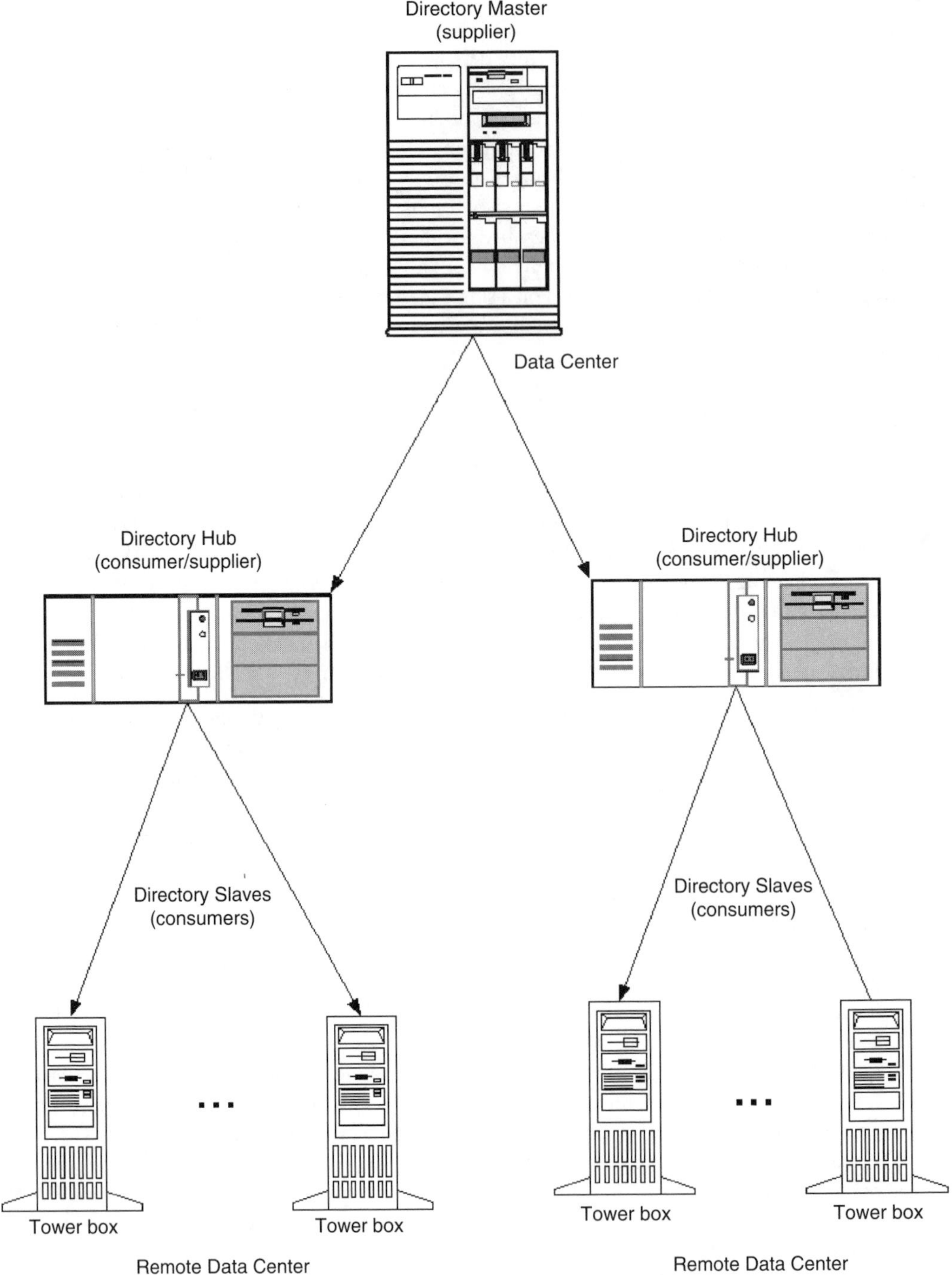

FIGURE 34.3 Master–slave replication.

able to make application-level load balancing decisions based on knowledge of directory server topology, query rates, query types, and load and availability metrics published by the directory servers.

Figure 34.4 is an illustration of a typical deployment of a pair of LDAP proxy servers that sit behind an IP firewall and accept LDAP connection requests and operations on port 389.

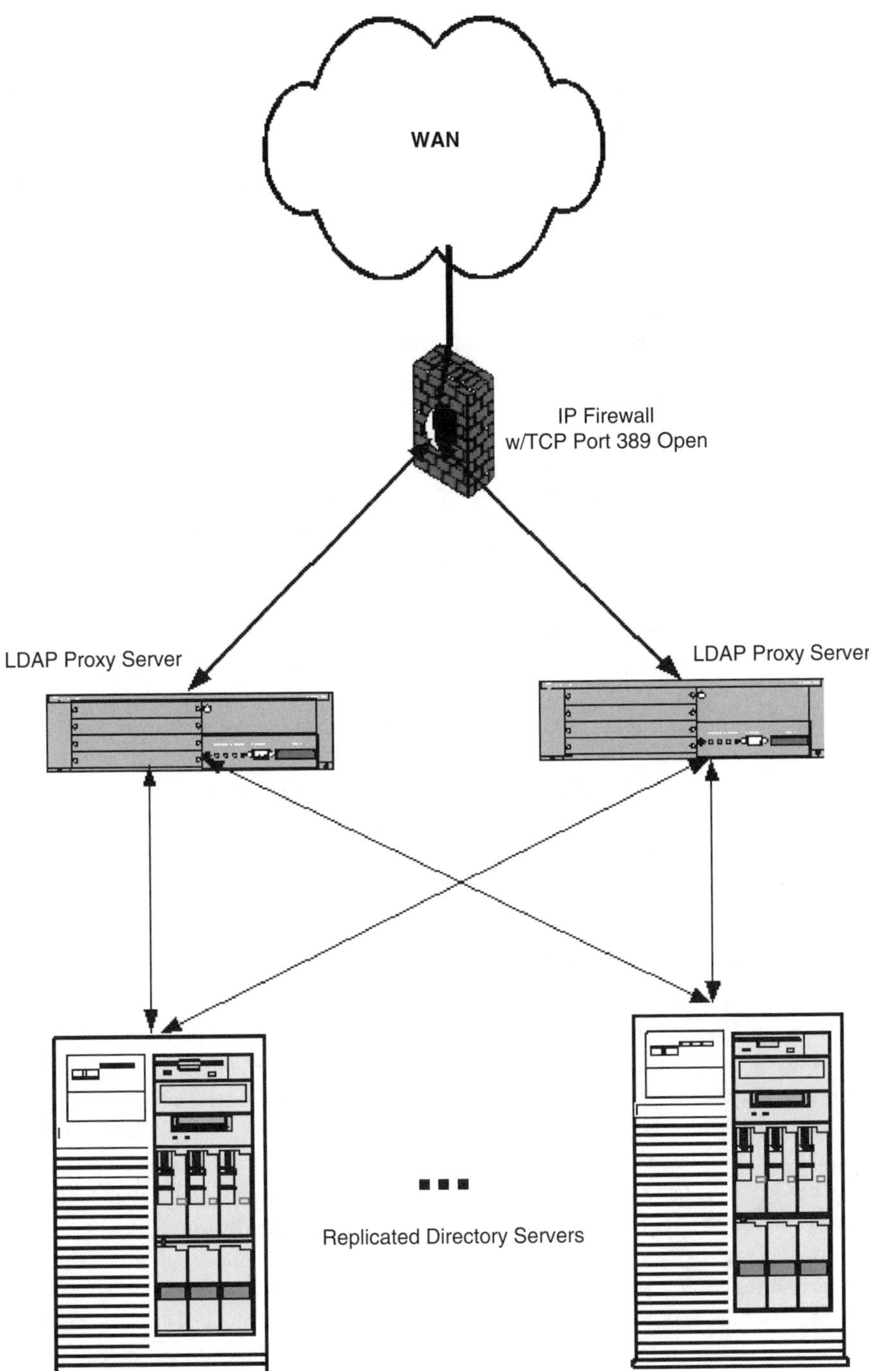

FIGURE 34.4 LDAP proxy servers.

34.5.5 Multimaster Replication

Multimaster replication with high-speed RAID storage, combined with multiple LDAP proxy servers and dynamic DNS provides a very high availability directory service. In some cases, a clustered operating system platform may provide additional availability guarantees. There are different techniques for implementing multimaster replication, but they all share in common the goal of maintaining write availability of a distributed directory service by ensuring that more than one master server is available and reachable for modifications that will *eventually* be synchronized with all of the other masters. As discussed previously, LDAP replication is based on a weakly-consistent distributed data model, and so any given master may be in a state of not having processed all updates seen at other master servers. This feature of an LDAP directory service is sometimes criticized as a weakness of the LDAP service model, but in well-designed directory service deployments with high bandwidth, low latency-LANs and WANs, it is possible to have weak consistency and still provide a very high service level for most application environments. In loosely coupled distributed systems over WANS, global consistency is very difficult to achieve economically. For those environments that require total consistency at any point in time, another network-based distributed data service is probably more appropriate.

Ideally, any directory server in a deployment of a directory service could be a master for some or all of the data that it holds, thereby providing n-way multimaster replication. In practice, however, the masters are typically placed in a controlled environment like a data center, where the server's authoritative database can be properly administered, regularly backed up, etc. In the case of geographically dispersed data centers, each data center may contain one or more master servers that are interconnected by a high-speed LAN and that stay in close synchronization, whereas masters in another data center are connected by a slower WAN and might often be out of synchronization. Different high availability goals will dictate how the masters are deployed, how the data is partitioned, and how proxy servers are deployed to help enable client applications to get at replicas or the masters themselves.

Whatever the choice of topology, multimaster replication combined with hub and slave replicas and proxy servers offers a highly available directory service. In this scenario, proxy servers provide a critical piece of functionality because replicas and hubs will often be configured to return referrals to one or more master servers for a client that requests to do an update on a replica. Alternatively, if the directory server used as a replica offers chaining, then it may be able to chain the operation to the master. For modifications, it is often desirable to have the client application authenticate to the master with its own credentials for auditing purposes, rather than have a replica proxy the modification on behalf of the client applications.

34.5.6 Replication Standardization

There is no official standard for how replication is to be done with LDAP. Each LDAP directory server vendor has implemented a specialized mechanism for replication. There are many reasons why no LDAP replication standard was reached within the IETF, but the principal reason was that no consensus could be reached on standard access control mechanism. A common access control mechanism, or consistent mapping, is required before LDAP replication interoperability can be achieved between servers from different vendors. In addition, market politics inhibited the successful definition, as the access control model and other semantic features implemented as part of a directory service, independent of the LDAP directory protocol, were viewed as competitive elements of various vendor products.

The question is often asked by the LDAP standards working group did not simply adopt the X.500 access control model and replication protocol, as they were already standardized. The reasons are complicated, but there is an important fact about X.500 replication that is often not often well understood. The X.500 replication protocol suffers from numerous flaws itself, and most X.500 products implement proprietary workarounds in their products to enable replication to work at a scale beyond a couple of hundred thousand directory entries. Briefly, the DISP requires a replication data unit to be sent as a single protocol data unit for both the total update during initialization, as well as during incremental updates for changes. Using the OSI upper layer protocols, the DISP protocol is defined in terms of a remote procedure call service and a session layer service that was not well designed to take procedure arguments that could be on the order of several megabytes to several gigabytes. Hence, implementations of DISP that use the remote operation service element (ROSE) of the OSI stack are most often severely limited in their ability to perform replication updates of any appreciable size. The alternative is to implement DISP in terms of the reliable transfer service element (RTSE), which is used by X.400, but no two X.500 vendors convincingly demonstrated interoperability at scale. As a result, X.500 total update replication in practice is neither scalable nor interoperable as a practical matter between any two X.500 serves from different vendors. The vendors had no choice but to make changes to the protocol as implemented in the product of each vendor to achieve practical replication between each vendor's own servers, and those changes usually make all but the most basic level of interoperability unachievable in practice. Fully interoperable and scalable replication between disparate directory servers, whether X.500 or LDAP, has not yet been achieved. The replication problem remains an active area of research, especially with respect to performance and scalability over WANs, and topology and replication agreement manageability of potentially hundreds-to-thousands of master–master and master–slave replication agreements.

34.6 LDAP Protocol and C Language Client API

The LDAPv3 protocol is defined in RFC 2251 [Wahl et al., 1997a]. Closely related specifications are RFC 2222 [Myers, 1997] and RFC 2252–2256 [Wahl, 1997; Wahl et al., 1997b; 1997c; Howes, 1997]. Many Internet application protocols, such as SMTP and HTTP, are defined as text-based request-response interactions; however, LDAPv3 (like SNMP) is defined using Abstract Syntax Notation One (ASN.1) so that application protocol data units (PDUs) are strongly typed in, that structural type information is sent along with user data in the form of encoded type tags. Clients and servers implementing LDAPv3 use the Basic Encoding Rules (BER to encode protocol data units as compact network byte order binary strings before transmission via TCP. This encoding/decoding process introduces slight computational overhead to protocol processing, but processing of LDAP operations is less computational intensive than other ASN.1-represented protocols such as DAP- or XML-encoded protocols, and even some text based protocols that require a lot of string parsing and end-of-message handling. This efficiency is due to the restricted use of only the basic and efficiently encoded ASN.1 data types in defining the LDAP protocol structure. Most data elements are represented as strings that are easily encoded using BER. This optimization allows very compact and efficient encoders/decoders to be written.

34.6.1 LDAPv3 Protocol Exchange

As with typical client-server protocols, the LDAPv3 protocol exchange is initiated by a client application requesting a TCP connection to a server, typically on reserved TCP port 389. Either endpoint of the connection can send an LDAPMessage PDU to the other endpoint at any time, although only request forms of the LDAPMessage are sent by the client and response forms sent by the server. Typically, a client will initiate a *bind*, *operation*$_1$, ..., *operation*$_N$, *unbind* sequence, where each operation is either a search, compare, or one of a family of update operations. A client may also choose to *abandon* an operation before unbinding and closing or aborting the TCP connection or the client may choose to rebind on an existing TCP connection with new credentials.

The LDAPMessage is defined in ASN.1 as follows:

An LDAPMessage is converted to bytes using a BER encoding, and the resulting series of bytes is sent on the TCP connection. In LDAP, only the BER definite length fields are used, so the receiver of a PDU knows how long the PDU will be as soon as the type tag and length of the outermost LDAPMessage SEQUENCE has been read from the network. The use of definite length encodings allows LDAP PDU processing on the server side to be done very efficiently because knowing the full length of the incoming PDU from reading the first few bytes leads to efficient memory management and minimization of data copying while reading bytes from the network.

The messageID field is an INTEGER. When the client sends a request, it chooses a value for the messageID, which is distinct from that of any other message the client has recently sent on that connection, typically by incrementing a counter for each message. Any LDAPMessage PDUs returned by the server in response to that request will use the same messageID field. This enables a client to send multiple

```
LDAPMessage ::= SEQUENCE {
        messageID        MessageID,
        protocolOp       CHOICE {
                bindRequest         BindRequest,
                bindResponse        BindResponse,
                unbindRequest       UnbindRequest,
                searchRequest       SearchRequest,
                searchResEntry      SearchResultEntry,
                searchResDone       SearchResultDone,
                searchResRef        SearchResultReference,
                modifyRequest       ModifyRequest,
                modifyResponse      ModifyResponse,
                addRequest          AddRequest,
                addResponse         AddResponse,
                delRequest          DelRequest,
                delResponse         DelResponse,
                modDNRequest        ModifyDNRequest,
                modDNResponse       ModifyDNResponse,
                compareRequest      CompareRequest,
                compareResponse     CompareResponse,
                abandonRequest      AbandonRequest,
                extendedReq         ExtendedRequest,
                extendedResp        ExtendedResponse },
        controls         [0] Controls OPTIONAL }
```

requests consecutively on the same connection, and servers that can process operations in parallel (for example, if they are multithreaded) will return the results to each operation as it is completed.

Normally, the server will not send any LDAPMessage to a client except in response to one of the above requests. The only exception is the unsolicited notification, which is represented by an extendedResp form of LDAPMessage with the messageID set to 0. The notice of disconnection allows the server to inform the client that it is about to abruptly close the connection. However, not all servers implement the notice of disconnection, and it is more typical that a connection is closed due to problems with the network or the server system becoming unavailable.

The controls field allows the client to attach additional information to the request, and the server to attach data to the response. Controls have been defined to describe server-side sorting, paging, and scrolling of results, and other features that are specific to particular server implementations.

In the C API, an application indicates that it wishes to establish a connection using the *ldap_init* call.

```
LDAP *ldap_init (const char *host,int port);
```

The *host* argument is either the host name of a particular server or a space-separated list of one or more host names. The default reserved TCP port for LDAP is 389. If a space-separated list of host names is provided, the TCP port for each host can be specified for each host, separated by a colon, as in "server-a:41389 server-b:42389." The TCP connection will be established when the client makes the first request call.

34.6.2 General Result Handling

The result PDU for most requests that have a result (all but Abandon and Unbind) is based on the following ASN.1 data type:

The resultCode will take the value zero for a successfully completed operation, except for the compare operation. Other values indicate that the operation could not be performed, or could only partially be performed. Only LDAP resultCode values between 0 and 79 are used in the protocol, and most indicate

```
LDAPResult ::= SEQUENCE {
          resultCode        ENUMERATED,
          matchedDN         LDAPDN,
          errorMessage      LDAPString,
          referral          [3] Referral OPTIONAL }
```

error conditions; for example, noSuchObject, indicating that the requested object does not exist in the directory. The LDAP C API uses resultCode values between 80 and 97 to indicate errors detected by the client library (e.g., out of memory).

In the remainder of this section, the C API will be described using the LDAP synchronous calls that block until a result, if required by the operation, is returned from the directory server. These API calls are defined by standard convention to have the suffix "_s" appended to the procedure names. Client applications that need to multiplex several operations on a single connection, or to obtain entries from a search result as they are returned asynchronously by the directory server, will use the corresponding asynchronous API calls. The synchronous and asynchronous calls generate identical LDAP messages and hence are indistinguishable to the server. It is up to the client application to define either a synchronous or asynchronous model of interaction with the directory server.

34.6.3 Bind

The first request that a client typically sends on a connection is the bind request to authenticate the client to the directory server. The bind request is represented within the LDAPMessage as follows:

```
BindRequest ::= [APPLICATION 0]   SEQUENCE {
          version                 INTEGER (1.. 127),
          name                    LDAPDN,
          authentication          AuthenticationChoice }
AuthenticationChoice ::=  CHOICE {
          simple                  [0] OCTET STRING,
          sasl                    [3] SaslCredentials }
```

There are two forms of authentication: simple password-based authentication and SASL (Simple Authentication and Security Layer). The SASL framework is defined in RFC 2222 [Myers, 1997]. A common SASL authentication mechanism is DIGEST-MD5 as defined in RFC 2831 [Leach and Newman, 2000].

Most LDAP clients use the simple authentication choice. The client provides the user's distinguished name in the name field, and the password in the simple field. The SaslCredentials field allows the client to specify a SASL security mechanism to authenticate the user to the server without revealing a password on the network, or by using a non-password-based authentication service, and optionally to authenticate the server as well.

```
SaslCredentials ::= SEQUENCE {
          mechanism               LDAPString,
          credentials             OCTET STRING OPTIONAL }
```

Some SASL mechanisms require multiple interactions between the client and the server on a connection to complete the authentication process. In these mechanisms the server will provide data back to the client in the serverSaslCreds field of the bind response.

```
BindResponse ::= [APPLICATION 1] SEQUENCE {
     COMPONENTS OF LDAPResult,
     serverSaslCreds             [7] OCTET STRING OPTIONAL }
```

The client will use the server's credential to compute the credentials to send to the server in a subsequent bind request.

In the C API, an application can perform a simple bind and block waiting for the result using the *ldap_simple_bind_s* call.

```
int ldap_simple_bind_s (LDAP *ld,const char *dn,const char *password);
```

34.6.4 Unbind

The client indicates to the server that it intends to close the connection by sending an unbind request. There is no response from the server.

```
UnbindRequest ::= [APPLICATION 2] NULL
```

In the C API, an application can send an unbind request and close the connection using the *ldap_unbind* call.

```
int ldap_unbind (LDAP *ld);
```

34.6.5 Extended Request

The extended request enables the client to refer operations that are not part of the LDAP core protocol definition. Most extended operations are specific to a particular server's implementation.

```
ExtendedRequest ::= [APPLICATION 23] SEQUENCE {
        requestName          [0] LDAPOID,
        requestValue         [1] OCTET STRING OPTIONAL }
ExtendedResponse ::= [APPLICATION 24] SEQUENCE {
        COMPONENTS OF LDAPResult,
        responseName         [10] LDAPOID OPTIONAL,
        response             [11] OCTET STRING OPTIONAL }
```

34.6.6 Searching

The search request is defined as follows:

```
SearchRequest ::= [APPLICATION 3] SEQUENCE {
        baseObject      LDAPDN,
        scope           ENUMERATED {
            baseObject                  (0),
            singleLevel                 (1),
            wholeSubtree                (2) },
        derefAliases    ENUMERATED,
        sizeLimit       INTEGER (0 .. maxInt),
        timeLimit       INTEGER (0 .. maxInt),
        typesOnly       BOOLEAN,
        filter Filter,
        attributes      AttributeDescriptionList }
```

The baseObject DN and scope determine which entries will be considered to locate a match. If the scope is baseObject, only the entry named by the distinguished name in baseObject field will be searched. If the scope is singleLevel, only the entries immediately below the baseObject entry will be searched. If the scope is wholeSubtree, then the entry named by baseObject and all entries in the tree below it are searched.

The derefAliases specifies whether the client requests special processing when an alias entry is encountered. Alias entries contain an attribute with a DN value that is the name of another entry, similar in concept to a symbolic link in a UNIX® file system. Alias entries are not supported by all directory servers and many deployments do not contain any alias entries.

The sizeLimit indicates the maximum number of entries to be returned in the search result, and the timeLimit the number of seconds that the server should spend processing the search. The client can provide the value 0 for either to specify "no limit."

The attributes field contains a list of the attribute types that the client requests be included from each of the entries in the search result. If this field contains an empty list, then the server will return all attributes of general interest from the entries. The client may also request that only types be returned, and not values.

The LDAP Filter is specified in the protocol encoding using ASN.1; however, most client APIs allow a simple text encoding of the filter to be used by applications. This textual encoding is defined in RFC 2254 [Howes, 1997].

In LDAP search processing, a filter, when tested against a particular entry, can evaluate to TRUE, FALSE, or Undefined. If the filter evaluates to FALSE or Undefined, then that entry is not returned in the search result set. Each filter is grouped by parenthesis, and the most common types of filter specify the "present," "equalityMatch," "substrings," "and," and "or" filter predicates. The "present" filter evaluates to TRUE if an attribute of a specified type is present in the entry. It is represented by following the type of the attribute with "=*," as in (telephoneNumber=*). The "equalityMatch" filter evaluates to TRUE if an attribute in the entry is of a matching type and value to that of the filter. It is represented as the type of the attribute, followed by a "=," then the value, as in (cn=John Smith). The "substring" filter evaluates to TRUE by comparing the values of a specified attribute in the entry to the pattern in the filter. It is represented as the type of the attribute, followed by a "=," and then any of following, separated by "*" characters:

- A substring that must occur at the beginning of the value
- Substrings that occur anywhere in the value
- A substring that must occur at the end of the value

For example, a filter (cn=John*) would match entries that have a commonName (cn) attribute beginning with the string "John." A filter (cn=*J*Smith) matches entries that have a value that contains the letter "J," and ends with "Smith." Many servers have restrictions on the substring searches that can be performed. It is typical for servers to restrict the minimum substring length.

An "and" filter consists of a set of included filter conditions, all of which must evaluate to TRUE if an entry is to match the "and" filter. This is represented using the character "&" followed by the set of included filters, as in (&(objectClass=person)(sn=smith)(cn=John*)).

An "or" filter consists of a set of included filter conditions, any of which must evaluate to TRUE if an entry is to match the "or" filter. This is represented using the character "|" followed by the set of included filters, as in (|(sn=smith)(sn=smythe)).

The "not" filter consists of a single included filter whose sense is inverted: TRUE becomes FALSE, FALSE becomes TRUE, and Undefined remains as Undefined. The negation filter is represented using the character "!" followed by the included filter, as in (!(objectClass=device)). Note that the negation filter applies to a single search filter component, which may be compound. Most LDAP servers cannot efficiently process the "not" filter, so it should be avoided where possible.

Other filters include the approxMatch filter, the greaterOrEqual, the lessOrEqual, and the extensible filter, which are not widely used. The approximate matching filter allows for string matching based on algorithms for determining phonetic matches, such as soundex, metaphone, and others implemented by the directory server.

34.6.7 Search Responses

The server will respond to the search with any number of LDAPMessage PDUs with the SearchResultEntry choice, one for each entry that matched the search, as well as any number of LDAPMessage PDUs with the SearchResultReference choice, followed by an LDAPMessage with the SearchResultDone choice.

```
SearchResultEntry ::= [APPLICATION 4] SEQUENCE {
          objectName      LDAPDN,
          attributes      PartialAttributeList }

SearchResultReference ::= [APPLICATION 19] SEQUENCE OF LDAPURL

SearchResultDone ::= [APPLICATION 5] LDAPResult
```

The SearchResultReference is returned by servers that do not perform chaining, to indicate to the client that it must progress the operation itself by contacting other servers. For example, if server S1 holds dc=example,dc=com, servers S2 and S3 hold ou=People,dc=example,dc=com and server S4 holds ou=Groups,dc=example,dc=com, then a wholeSubtree search sent to server S1 would result in the two LDAPMessage PDUs containing SearchResultReference being returned, one with the URLs:

```
ldap://S2/ou=People,dc=example,dc=com
ldap://S3/ou=People,dc=example,dc=com
```

and the other with the URL:

```
ldap://S4/ou=Groups,dc=example,dc=com
```

followed by a SearchResultDone.
Invoking a search request in the C API, and blocking for the results, uses the *ldap_search_s* call.

```
int ldap_search_s (
          LDAP              *ld,
          const char        *basedn,
          int               scope,
          const char        *filter,
          char              **attrs,
          int               attrsonly,
          LDAPControl       **serverctrls,
          LDAPControl       **clientctrls,
          struct timeval    *timeout,
          int               sizelimit,
          LDAPMessage        **res);
```

The scope parameter can be one of *LDAP_SCOPE_BASE, LDAP_SCOPE_ONELEVEL*, or *LDAP_SCOPE_SUBTREE.*

34.6.8 Abandoning an Operation

While the server is processing a search operation, the client can indicate that it is no longer interested in the results by sending an abandon request, containing the messageID of the original search request.

```
AbandonRequest ::= [APPLICATION 16] MessageID
```

The server does not reply to an abandon request, and no further results for the abandoned operation are sent.

In the C API, the client requests that an operation it invoked on that connection with an earlier asynchronous call be abandoned using the *ldap_abandon* call.

```
int ldap_abandon (LDAP *ld,int msgid);
```

34.6.9 Compare Request

The compare operation allows a client to determine whether an entry contains an attribute with a specific value. The typical server responses will be the compareFalse or compareTrue results codes, indicating that the comparison operation failed or succeeded. In practice, few client applications use the compare operation.

```
CompareRequest ::= [APPLICATION 14] SEQUENCE {
        entry               LDAPDN,
        ava                 AttributeValueAssertion}
AttributeValueAssertion ::= SEQUENCE {
        attributeDesc       AttributeDescription,
        assertionValue      OCTET STRING }
```

In the C API, the client requests a comparison on an attribute with a string syntax using the *ldap_compare_s* call. Note that a successful comparison is expressed with the result code *LDAP_COMPARE_TRUE* rather than *LDAP_SUCCESS*.

```
int ldap_compare_s (LDAP *ld,const char *dn,
                    const char *type, const char *value);
```

34.6.10 Add, Delete, Modify, and ModifyDN Operations

The Add, Delete, Modify, and ModifyDN operations operate on individual entries in the directory tree.

```
ModifyRequest ::= [APPLICATION 6] SEQUENCE {
        object              LDAPDN,
        modification        SEQUENCE OF SEQUENCE {
            operation               ENUMERATED {
                                            add         (0),
                                            delete      (1),
                                            replace     (2) },
            modification            AttributeTypeAndValues } }
```

In the C API, the client can invoke the modify operation using the ldap_modify_s call.

```
int ldap_modify_s (LDAP *ld,const char *dn,LDAPMod **mods);
```

```
typedef struct LDAPMod {
        int mod_op;
#define LDAP_MOD_ADD0x0
#define LDAP_MOD_DELETE0x1
#define LDAP_MOD_REPLACE0x2
#define LDAP_MOD_BVALUES0x80
        char *mod_type;
        union mod_vals_u {
                char **modv_strvals;
#define mod_values mod_vals.modv_strvals
                struct berval **modv_bvals;
#define mod_bvalues mod_vals.modv_bvals
        } mod_vals;
LDAPMod;
```

The mod_op field is based on one of the values LDAP_MOD_ADD, LDAP_MOD_DELETE or LDAP_MOD_REPLACE.

The Add operation creates a new entry in the directory tree.

```
AddRequest ::= [APPLICATION 8] SEQUENCE {
          entry             LDAPDN,
          attributes        SEQUENCE OF AttributeTypeAndValues }
```

In the C API, the client can invoke the add operation using the *ldap_add_s call.*

```
int ldap_add_s (LDAP *ld,const char *dn,LDAPMod **attrs);
```

The Delete operation removes a single entry from the directory.

```
DelRequest ::= [APPLICATION 10] LDAPDN
```

In the C API, the client can invoke the delete operation using the ldap_delete_s call.

```
int ldap_delete_s (LDAP *ld,const char *dn);
```

The ModifyDN operation can be used to rename or move an entry or an entire branch of the directory tree. The entry parameter specifies the DN of the entry at the base of the tree to be moved. The newrdn parameter specifies the new relative distinguished name (RDN) for that entry. The deleteoldrdn parameter controls whether the previous RDN should be removed from the entry or just be converted by the server into ordinary attribute values. The newSuperior field, if present, specifies the name of the entry that should become the parent of the entry to be moved.

```
ModifyDNRequest ::= [APPLICATION 12] SEQUENCE {
          entry             LDAPDN,
          newrdn            RelativeLDAPDN,
          deleteoldrdn      BOOLEAN,
          newSuperior       [0] LDAPDN OPTIONAL }
```

Many directory servers do not support the full range of capabilities implied by the ModifyDN operation (e.g., subtree rename), so this operation is not frequently used by clients.

34.7 Conclusion

This history of the evolution of LDAP technology is indeed a unique and fascinating case study in the evolution of a key Internet protocol and client-server technology used worldwide. There are many excellent client and server products based on LDAP available from several companies, each providing its various advantages and disadvantages. However, it is fair to say that it was through the diligent efforts of a few dedicated individuals that the emergence of the Internet as a commercially viable technology was accomplished, and it was the financial investment of several research organizations and corporations in this technology that has made directories based on the lightweight directory access protocol critical components of the worldwide public Internet, most corporate and organizational wired and wireless intranets, and the global wireless phone network.

Acknowledgments

The authors wish to thank K. C. Francis, Steve Kille, Scott Page, Stephen Shoaff, Kenneth Suter, Nicki Turman, and Neil Wilson for their insightful suggestions and editorial comments on earlier drafts of this article.

References

Barker, Paul and Steve Kille. *The COSINE and Internet X.500 Schema.* Internet RFC 1274, November 1991.

Bellwood, Tom (Ed.). *UDDI Version 2.04 API Specification.* July 2002.

Bray, Tim, Jean Paoli, C. M. Sperberg-McQueen, and Eve Maler, (Eds.). *Extensible markup language (XML) 1.0 (second edition).* W3C, October 2000.

Howes, Tim. *The String Representation of LDAP Search Filters,* Internet RFC 2254, December 1997.

Howes, Tim and Mark Smith. *The LDAP URL Format.* Internet RFC 2255, December 1997.

Good, Gordon. *The LDAP Data Interchange Format (LDIF) — Technical Specification,* Internet RFC 2849, June 2000.

Gudgin, Martin, Marc Hadley, Noah Mendelsohn, Jean-Jacques Moreau, and Henrick Nielsen. *SOAP version 1.2 part 1: Messaging framework.* W3C, June 2003.

Kille, Steve et al., *Using Domains in LDAP/X.500 Distinquishing Names,* Internet RFC 2247, January 1998.

International Telecommunications Union (ITU). *Information technology — Open Systems Interconnection Recommendation X.500 — The Directory: Overview of concepts, models, and services,* 1993.

International Telecommunications Union (ITU). *Information technology — Open Systems Interconnection Recommendation X.501 — The Directory: Models,* 1993.

International Telecommunications Union (ITU). *Information technology — Open Systems Interconnection Recommendation X.511 — The Directory: Abstract service definition,* 1993.

International Telecommunications Union (ITU). *Information technology — Open Systems Interconnection Recommendation X.518 — The Directory: Procedures for distributed operation,* 1993.

International Telecommunications Union (ITU). *Information technology — Open Systems Interconnection Recommendation X.519 — The Directory: Protocol specifications,* 1993.

International Telecommunications Union (ITU). *Information technology — Open Systems Interconnection Recommendation X.520 — The Directory: Selected attribute types,* 1993.

International Telecommunications Union (ITU). *Information technology — Open Systems Interconnection Recommendation X.521 — The Directory: Selected object classes,* 1993.

International Telecommunications Union (ITU). *Information technology — Open Systems Interconnection Recommendation X.525 — The Directory: Replication,* 1993.

International Telecommunications Union (ITU). *Information technology Recommendation X.681 — Abstract Syntax Notation One (ASN.1): Specification of basic notation,* 1994.

International Telecommunications Union (ITU). *Information technology Recommendation X.690 — ASN.1 encoding rules: Specification of Basic Encoding Rules (BER), Canonical Encoding Rules (CER) and Distinguished Encoding Rules (DER),* 1994.

Leach, Paul and Chris Newman. *Using Digest Authentication as a SASL Mechanism.* Internet RFC 2831, May 2000.

Myers, John. *Simple Authentication and Security Layer (SASL),* Internet RFC 2222, October 1997.

OASIS. Organization for the Advancement of Structured Information Standards. *Directory Services Markup Language v2.0,* December 2001.

OASIS. Organization for the Advancement of Structured Information Standards. *OASIS/ebXML Registry Services Specification v2.0.* April 2002.

Rolls, Darren (Ed.). *Service Provisioning Markup Language (SPML) Version 1.0.* OASIS Technical Committee Specification, June 2003.

Rose, Marshall T. and Dwight E. Cass. *ISO Transport Services on top of TCP: Version 3.* Internet RFC 1006, May 1987.

Smith, Mark, *Definition of the inet OrgPerson LDAP Object Class,* Internet RFC 2798, April 2000.

Smith, Mark, Andrew Herron, Tim Howes, Mark Wahl, and Anoop Anantha. *The C LDAP Application Program Interface.* Internet draft-ietf-ldapext-ldap-c-api-05.txt, November 2001.

Wahl, Mark. *A Summary of X.500(96) User Schema for use with LDAPv3,* Internet RFC 2256 December 1997.

Wahl, Mark, Tim Howes, and Steve Kille. *Lightweight Directory Access Protocol (v3)*. Internet RFC 2251, December 1997a.

Wahl, Mark, Andrew Coulbeck, Tim Howes, and Steve Kille. *Lightweight Directory Access Protocol (v3): Attribute Syntax Definitions*. Internet RFC 2252, December 1997b.

Wahl, Mark, Steve Kille, and Tim Howes. *Lightweight Directory Access Protocol (v3): UTF-8 String Representation of Distinguished Names*. Internet RFC 2253, December 1997c.

Weltman, Rob. *LDAP Proxied Authorization Control*. Internet draft-weltman-ldapv3-proxy-12.txt, April 2003.

Yeong, Wengyik, Tim Howes, and Steve Kille. *Lightweight Directory Access Protocol*. Internet RFC 1777, March 1995.

Yergeau, Frank. *UTF-8, a transformation format of Unicode and ISO 10646*. Internet RFC 2044, October 1996.

Authors

Greg Lavender is currently a director of software engineering and CTO for Internet directory, network identity, communications, and portal software products at Sun Microsystems. He was formerly Vice President of Technology at Innosoft International, co-founder and Chief Scientist of Critical Angle, an LDAP directory technology startup company, and co-founder and Chief Scientist of the ISODE Consortium, which was an open-source, not-for-profit, R&D consortium that pioneered both X.500 and LDAP directory technology. He is also an adjunct associate professor in the Department of Computer Sciences at the University of Texas at Austin.

Mark Wahl is currently a senior staff engineer and Principal Directory Architect at Sun Microsystems. He was previously Senior Directory Architect at Innosoft International, co-founder and President of Critical Angle, and lead directory server engineer at the ISODE Consortium. He was the co-chair of the IETF Working Group on LDAPv3 Extensions, and co-author and editor-in-chief of the primary LDAPv3 RFCs within the IETF.

35

Peer-to-Peer Systems

CONTENTS

Abstract 35-1
35.1 Introduction 35-1
35.2 Fundamental Concepts 35-3
35.2.1 Principles of P2P Architectures 35-3
35.2.2 Classification of P2P Systems 35-4
35.2.3 Emergent Phenomena in P2P Systems 35-5
35.3 Resource Location in P2P Systems 35-7
35.3.1 Properties and Categories of P2P Resource Location Systems 35-7
35.3.2 Unstructured P2P Systems 35-8
35.3.3 Hierarchical P2P Systems 35-10
35.3.4 Structured P2P Systems 35-10
35.4 Comparative Evaluation of P2P Systems 35-16
35.4.1 Performance 35-16
35.4.2 Functional and Qualitative Properties 35-17
35.5 Conclusions 35-20
References 35-20

Karl Aberer

Manfred Hauswirth

Abstract

Peer-to-peer (P2P) systems offer a new architectural alternative for global-scale distributed information systems and applications. By taking advantage of the principle of resource sharing, i.e., integrating the resources available at end-user computers into a larger system, it is possible to build applications that scale to a global size. Peer-to-peer systems are decentralized systems in which each participant can act as a client and as a server and can freely join and leave the system. This autonomy avoids single-point-of-failures and provides scalability but implies considerably higher complexity of algorithms and security policies. This chapter gives an overview of the current state in P2P research. We present the fundamental concepts that underly P2P systems and offer a short, but to-the-point, comparative evaluation of current systems to enable the reader to understand their performance implications and resource consumption issues.

35.1 Introduction

The Internet has enabled the provision of global-scale distributed applications and some of the most well-known Internet companies' success stories are based on providing such applications. Notable examples are eBay, Yahoo!, and Google. Although these applications rely on the Internet's networking infrastructure to reach a global user community, their architecture is strictly centralized, following the client-server paradigm. This has a significant impact on the computing resources required to provide the services. For example, Google centrally collects information available on the Web by crawling Web sites, indexing the retrieved data, and providing a user interface to query this index. Recent numbers reported

1-58488-381-2/05/$0.00+$1.50
© 2005 by CRC Press LLC

by Google (http://www.google.com/) show that their search engine service requires a workstation cluster of about 15,000 Linux servers.

From this observation one might be tempted to conclude that providing a global-scale application necessarily implies a major development, infrastructure, and administration investment. However, this conclusion has been shown to be inaccurate by a new class of applications, initially developed for the purpose of information sharing (e.g., music files, recipes, etc.). These systems are commonly denoted as *P2P file sharing systems.* The essential insight on which this new type of system builds is to take advantage of the principle of *resource sharing:* The Internet makes plenty of resources available at the end-user computers, or at its "edges," as is frequently expressed. By integrating these into a larger system it is possible to build applications at a global scale without experiencing the "investment bottleneck" mentioned above.

From a more architectural point of view, client-server systems are *asymmetric* whereas P2P systems are *symmetric.* In the client-server approach to system construction, only the clients request data or functionality from a server and consequently client-server systems are inherently *centralized.* To deal with the possibly very large number of clients, replication of servers and load balancing techniques are applied. Still, this centralized virtual server is a single-point of failure and a network bandwidth bottleneck because all network traffic uses the same Internet connection, i.e., if the server crashes or if it is not reachable, the system cannot operate at all. On the other hand, this approach offers a more stringent control of the system that facilitates the use of simple yet efficient algorithms and security policies.

In contrast, P2P systems are decentralized systems. There no longer exists a distinction between clients and servers but each peer can act both as a client and as a server, depending on which goal needs to be accomplished. The functionality of a P2P system comes into existence by the cooperation of the individual peers. This approach avoids the problems of client-server systems, i.e., single-point of failure, limited scalability, and "hot spots" of network traffic, and allows the participants to remain autonomous in many of their decisions. However, these advantages come at the expense of a considerably higher complexity of algorithms and more complicated security policies. Nevertheless, autonomy, scalability, load-sharing, and fault-tolerance are of such premiere importance in global-scale distributed systems that the P2P approach is of major interest.

Napster was the first and most famous proponent of this new class of systems. By the above definition, Napster is not a pure P2P system but a hybrid one: Coupling of resources was facilitated through a central directory server, where clients (Napster terminology!) interested in sharing music files logged in and registered their files. Each client could send search requests to the Napster server, which searched its database of currently registered files and returned a list of matches to the requesting client (client-server style). The requester then could choose from this list and download the files directly from an offering computer (P2P style).

The centralized database of Napster provided efficient search functionality and data consistency but limited the scalability of the system. Nevertheless, Napster took advantage of a number of resources made available by the community of cooperating peers by its P2P approach:

Storage and bandwidth: Because the most resource-intensive operations in music file sharing, i.e., the storage and the exchange of usually large data files, were not provided by a central server but by the participating peers, it was possible to build a global-scale application with approximately 100 servers running at the Napster site. This is lower than Google's requirement by orders of magnitude, even if the resource consumption of the system is of the same scale. At the time of its top popularity in February 2001, Napster had 1.57 million online users, who shared 220 files on average, and 2.79 billion files were downloaded during that month.

Knowledge: Users annotate their files (author, title, rating, etc.) before registering at the Napster server. In total this amounts to a substantial investment in terms of human resources, which facilitates making searches in Napster more precise. This effort can be compared to the one invested by Yahoo! for annotating Web sites.

Ownership: The key success factor for the fast adoption of the Napster system was the possibility to obtain music files for free. This can be viewed as "sharing of ownership." Naturally this large-scale copyright infringement provoked a heavy reaction from the music industry, which eventually led to the shutdown of Napster. The interpretation of the underlying economic processes is not the subject of this chapter, but they are quite interesting and a topic of vivid debates (see, e.g., http://www.dklevine.com/general/intellectual/napster.htm).

In parallel, another music file sharing system that was a pure P2P system entered the stage: Gnutella [Clip2, 2001]. Though Gnutella provides essentially the same functionality as Napster, no central directory server is required. Instead, in Gnutella each peer randomly chooses a small number of "neighbors" with whom it keeps permanent connections. This results in a connection graph in which each peer forwards its own or other peers' search requests. Each peer that receives a query and finds a matching data item in its store, sends an answer back to the requester. After a certain number of hops queries are no longer forwarded to avoid unlimited distribution, but still the probability of finding a peer that is able to satisfy it is high due to certain properties of the Gnutella graph, which will be discussed in detail later. A more technical and detailed discussion of Gnutella is provided in Section 35.3.2.1.

Gnutella's approach to search, which is commonly denoted as *constrained flooding* or *constrained broadcast,* involves no distinguished component that is required for its operation. It is a *fully decentralized system.* Strictly speaking, Gnutella is not a system, but an open protocol that is implemented by Gnutella software clients. As Gnutella can run without a central server, there is no single-point of failure, which makes attacks (legal, economic, or malicious ones) very difficult.

The downside of Gnutella's distributed search algorithm, which involves no coordination at all, is high network bandwidth consumption. This severely limits the throughput of the system and the network, which must support a plenitude of other services as well. So a key issue which is subject to current research is to find less bandwidth-intensive solutions: Do systems exist that offer similar functional properties as Gnutella or is decentralization necessarily paid by poor performance and high bandwidth consumption?

In addition, the performance of message flooding with respect to search latency depends critically on the global network structure that emerges from aggregate local behaviors of the Gnutella peers. In other words, the global network structure is the result of a *self organization* process. This leads to the interesting question of how it is possible to construct large self-organizing P2P systems such as Gnutella, yet with predictable behavior and quality-of-service guarantees.

In the following we will describe the current state-of-the-art in research and real-world systems to shed more light on these questions. We will present the concepts of P2P computing in general and discuss the existing P2P approaches.

35.2 Fundamental Concepts

35.2.1 Principles of P2P Architectures

The fundamental difference between a P2P architecture and a client–server architecture is that there no longer exists a clear distinction between the clients consuming a service and the servers providing a service. This implies partial or complete *symmetry of roles* regarding the system architecture and the interaction patterns. To clearly denote this, often the term *servent* is used for peers. The P2P architecture has two benefits: It enables resource sharing and avoids bottlenecks.

Pure P2P systems such as Gnutella, which feature completely symmetric roles, no longer require a coordinating entity or globally available knowledge to be used for orchestrating the evolution of the system. None of the peers in a pure P2P system maintains the global state of the system and all interactions of the peers are strictly *local.* The peers interact with only a very limited subset of other peers, which is frequently called "neighborhood" in the literature to emphasize its restricted character. Therefore, the control of the system is truly *decentralized.*

Decentralization is beneficial in many situations but its foremost advantage is that it enables *scalability*, which is a key requirement for designing large-scale, distributed applications: Because the individual components (peers) of a decentralized system perform local interactions only, the number of interactions remains low compared to the size of the system. As no component has a global view of the state, the resource consumption in terms of information storage and exchange compared to the size of the system also stays low. Therefore these systems scale well to very large system sizes. However, the development of efficient distributed algorithms for performing tasks in a decentralized system such as search or data placement is more complex than for the centralized case.

Even though no central entity exists, pure, decentralized P2P systems develop certain global structures that emerge from the collective behavior of the system's components (peers), and that are important for the proper operation of the system. For example, in the case of Gnutella, the global network structure emerges from the local pair-wise interactions that Gnutella is based on. Such global structures emerge without any external or central control. Systems with distributed control-evolving emergent structures are called *self-organizing systems* (decentralization implies *self organization*). A significant advantage of self-organizing systems is their *failure resilience*, which — in conjunction with the property of scalability — makes such systems particularly attractive for building large-scale, distributed applications.

Understanding the global behavior of P2P systems based on these principles is not trivial. To alleviate the comprehension of these phenomena and their interplay we will provide further details and discussions in Section 35.2.3.

35.2.2 Classification of P2P Systems

The P2P architectural paradigm is not new to computer science. In particular, the Internet's infrastructure and services exploit the P2P approach in many places at different levels of abstraction — for example, in routing, the domain name service (DNS), or in Usenet News. These mechanisms share a common goal: locating or disseminating resources in a network.

What is new with respect to the recent developments in P2P systems and more generally in Web computing is that the P2P paradigm also appears increasingly at other system layers. We can distinguish the following layers for which we observe this development:

Networking layer: basic services to route requests over the physical networks to a network address in an application-independent way

Data access layer: management of resource membership to specific applications; search and update of resources using application-specific identifiers in a distributed environment

Service layer: combination and enhancement of data access layer functionalities to provide higher-level abstractions and service ranging from simple data exchange, such as file sharing, to complex business processes

User layer: interactions of users belonging to user communities using system services for community management and information exchange

It is interesting to observe that the P2P paradigm can appear at each of these layers independently. The analysis to what degree and at which layers the P2P paradigm is implemented in a concrete system facilitates a more precise characterization and classification of the different types of P2P systems. Examples of how the P2P paradigm is used at the different layers are given in Figure 35.1.

Layer	Application domain	Service	Example system
Network	Internet	Routing	TCP/IP, DNS
Data access	Overlay networks	Resource location	Gnutell, Freenet
Service	P2P applications	Messaging, distibuted processing	Napster, Seti, Groove
User Layer	User communities	Collaboration	eBay, Ciao

FIGURE 35.1 The P2P paradigm at different layers.

For the networking layer of the Internet we have already pointed out the fact that it relies strongly on P2P principles, in particular, to achieve failure resilience and scalability. After all, the original design goal of Arpanet, the ancestor of today's Internet, was to build a scalable and failure-resilient networking infrastructure.

An important aspect in recent P2P systems, such as Freenet and Gnutella, is resource location. These systems construct so-called *overlay networks* over the physical network. In principle, applications could use the services provided by the networking layer to locate their resources of interest. However, having a separate, application-specific overlay network has the advantage of supporting application-specific identifiers and semantic routing and offers the possibility to provide additional services for supporting network maintenance, authentication, trust, etc., all of which would be hard to integrate into and support at the networking layer. The introduction of overlay networks is probably the essential innovation of P2P systems.

The P2P architecture can also be exploited at the service layer. In fact, with Napster we already discussed a system in which only this layer is organized in a P2P way (i.e., the download of files), whereas resource location is centralized. Also the state-of-the-art architectures for Web services such as J2EE and .NET build on a similar, directory-based architecture: Services are registered at a directory (e.g., a UDDI directory), where they can be looked up. The service invocation is then directly between the service provider and the requester. The main difference, though, between Napster and Web services is that Napster deals with only one type of service. Another well-known example for the P2P paradigm used at the service layer is SETI@Home, which exploits idle processing time on desktop computers to analyze radio signals for signs of extraterrestrial intelligence.

Because social and economic systems are typically organized in a P2P style and are also typical examples of self-organizing systems, it is natural that this is reflected in applications that support social and economic interactions. Examples are eBay or recommender systems such as Ciao (www.ciao.com). From a systems perspective these systems are centralized both at the resource location and the service layer, but the user interactions are P2P.

Another classification in P2P approaches can be given with respect to the level of generality at which they are applicable. Generally, research and development on P2P systems is carried out at three levels of generalization:

P2P applications: This includes the various file sharing systems, such as Gnutella, Napster, or Kazaa.

P2P platforms: Here, most notably SUN's JXTA platform [Gong, 2001] is an example of a generic architecture standardizing the functional architecture, the component interfaces, and the standard services of P2P systems.

P2P algorithms: This area deals with the development of scalable, distributed, and decentralized algorithms for efficient P2P systems. The current focus in research is on algorithms for efficient resource location.

35.2.3 Emergent Phenomena in P2P Systems

A fair share of the fascination of P2P systems but probably also their major challenges stem from the emergent phenomena that play a main role at all system layers. These phenomena arise from the fact that the systems operate without central control and thus it is not always predictable how they will evolve. Self-organizing systems are well-known in many scientific disciplines, in particular in physics and biology. Prominent examples are crystallization processes or insect colonies. Self-organization is basically a process of evolution of a complex system with only local interaction of system components, where the effect of the environment is minimal. Self-organization is driven by randomized variation processes — movements of molecules in the case of crystallization, movements of individual insects in insect colonies, or queries and data insertions in the case of P2P systems. These "fluctuations" or "noise," as they are also called, lead to a continuous perturbation of the system and allow the system to explore a global state space till it finds stable (dynamic) equilibrium states. These states correspond to the global, emergent structures [Heylighen, 1997].

In the area of computer science self-organization and the resulting phenomena have been studied in particular in the field of artificial intelligence for some time (agent systems, distributed decision making, etc.), but with today's P2P systems large-scale, self-organizing applications have become reality for the first time. Having P2P systems widely deployed offers an unforeseen opportunity for studying these phenomena in concrete large-scale systems. Theoretical work on self-organization can now be verified in the real world. Conversely, insights gained from self-organizing P2P systems may and do also contribute to the advancement of theory.

An implicit advantage of self-organizing systems that makes them so well applicable for constructing large-scale distributed applications is their inherent *failure resilience:* The global behavior of self-organizing systems is insensitive to local perturbations such as local component failures or local overloads. Other components can take over in these cases. Thus self-organizing systems tend to be extremely robust against failures. Put differently, in self-organizing systems random processes drive the exploration of the global state space of the system in order to "detect" the stable subspaces that correspond to the stable, emergent structures. Thus failures simply add to the randomization and thus to the system's evolution.

A number of experimental studies have been performed in order to elicit the emergent properties of P2P systems. In particular Gnutella has been subject to a number of exemplary studies, due to its openness and simple accessibility. We cite here some of these studies that demonstrate the types of phenomena that are likely to play a role in any P2P system and more generally on the Web.

Studies of Gnutella's network structure (that emerges from the local strategies used by peers to maintain their neighborhood) have revealed two characteristic properties of the resulting global Gnutella network [Ripeanu and Foster, 2002]:

Power-law distribution of node connectivity: Only few Gnutella peers have a large number of links to other peers whereas most peers have a very low number of links.

Small diameter: The Gnutella network graph has a relatively short diameter of approximately seven hops. This property ensures that a message flooding approach for search works with a relatively low time-to-live.

The power-law distribution of peer connectivity is explained as a result of a process of preferential attachment, i.e., nodes arriving at the network or changing their connectivity attach with higher probability to already well connected nodes [Barabási and Albert, 1999], whereas the low diameter is usually related to the small-world phenomenon [Kleinberg, 2000], which has first been observed for social networks. Small-world graphs have been identified as a class of graphs that combine the property of having a short diameter as found in random graphs with a high degree of local clustering typical for regular graphs [Watts and Strogatz, 1998]. In addition, they enable the efficient discovery of short connections between any nodes [Kleinberg, 2000]. Examples of small-world graphs are typically constructed from existing regular graphs by locally rewiring the graph structures, very much as they would result from a self-organization process. An example of such a process has been provided in the Freenet P2P system [Clarke et al., 2001, 2002].

Self-organization processes also play a role at the user layer in P2P systems. As P2P systems can also be viewed as social networks, they face similar problems. For example, "free riding" [Adar and Huberman, 2000] has become a serious problem. As in the real world, people prefer to consume resources without offering similar amounts of their own resources in exchange. This effect is also well-known from other types of economies. Most Gnutella users are free-riders, i.e., do not provide files to share, and if sharing happens, only a very limited number of files is of interest at all. Adar and Huberman show that 66% of Gnutella users share no files and nearly 47% of all responses are returned by the top 1% of the sharing hosts. This starts to transform Gnutella into a client–server-like system with a backbone structure, which soon may exhibit the same problems as centralized systems. Another economic issue that is becoming highly relevant for P2P systems is reputation-building [Aberer and Despotovic, 2001]. Simple reputation mechanisms are already in place in some P2P file sharing systems such as Kazaa (http://www.kazaa.com/). The full potential of reputation management can be seen best from online citation indices such as CiteSeer [Flake et al., 2002].

In the future we may see many other emergent phenomena. In particular, the possibility not only to have *emergent structures,* such as the network structure, but also *emergent behaviors,* e.g., behaviors resulting from evolutionary processes or a decentralized coordination mechanism, such as swarm intelligence [Bonabeau et al., 1999], appears to be a promising and exciting development.

A merit of P2P architectures is that they have introduced the principle of self-organization into the domain of distributed application architecture on a broad scale. We have seen that self-organization exhibits exactly the properties of scalability and failure resilience that mainly contributed to the success of P2P systems. This has become especially important recently, as the dramatic growth of the Internet has clearly shown the limits of the standard client–server approach and thus many new application domains inherently are P2P. For example, mobile ad-hoc networking, customer-to-customer e-commerce systems, or dynamic service discovery and workflow composition, are areas where self-organizing systems are about to become reality as the next generation of distributed systems.

However, it can be seen that the P2P architectural principle and self-organization can appear in many different ways. We explore this in the following sections in greater detail.

35.3 Resource Location in P2P Systems

The fundamental problem of P2P systems is resource location. In fact, P2P systems in the narrow sense, i.e., P2P information sharing systems such as Gnutella or Napster, are resource location systems. Therefore we provide an overview of the fundamental issues and approaches of P2P resource location in this section.

35.3.1 Properties and Categories of P2P Resource Location Systems

The problem of resource location can be stated as follows: A group G of peers with given addresses $p \in P$ holds resources $R(p) \subseteq R$. The resources can be, for example, media files, documents, or services. Each resource $r \in R(p)$ is identified by a resource key $k \in K$. The resource keys can be numbers, names, or metadata annotations. Each peer identified by address p is thus associated with a set of keys $K(p) \subseteq K$ identifying the resources it holds.

The problem of resource location is to find, given a resource identifier k, or more generally a predicate on a resource identifier k, a peer with address p that holds that resource, i.e., $k \in K(P)$. In other words, the task is managing and accessing the binary relation $I = \{(p, k)|k \in K(p)\} \subseteq K \times P$, the *index information.* Generally, each peer will hold only a subset $I(p) \subseteq I$ of the index information. Thus, in general it will not be able to answer all requests for locating resources itself. In that case, a peer p can contact another peer in its neighborhood $N(p) \subseteq G$.

For interacting with other peers two basic protocols need to be supported:

Network maintenance protocol: This protocol enables a node to join and leave a group G of peers. In order to join, a peer p needs to know a current member $p' \in G$ to which it can send a join message: $p' \rightarrow$ *join*(p'). This may cause a number of protocol-specific messages to reorganize the neighborhood and the index information kept at different peers of the group. Similarly, a peer can announce that it departs from a group by sending a message $p' \rightarrow$ *leave* (p'). A peer p joining a group G may already belong to another group G', and thus different networks may merge and split as a result of local join-and-leave interactions. Whereas joining usually is done explicitly in all protocols, leaving typically is implicit (network separation, peer failure, etc.) and the system has to account for that.

Data management protocol: This protocol allows nodes to search, insert, and delete resources, once they belong to a group. The corresponding messages are $p' \rightarrow$ *search*(k), which returns one or more peers holding the resource or the resource itself, $p' \rightarrow$ *insert*(k,r) for inserting a resource, and $p' \rightarrow$ *delete*(k) for deleting it. Peers receiving search, insert, or delete messages may forward them to other peers they know in case they cannot satisfy them. Updates as in database systems are usually not considered because P2P resource location systems usually support the publication of information rather than online data management, and strong consistency of data is not required.

These protocols must be supported by all kinds of P2P systems discussed below although the degree and strategies may vary according to the specific approach.

The distribution of index information, the selection of the neighborhood, the additional information kept at peers about the neighborhood, and the specific types of protocols supported defines the variants of resource location mechanisms. A direct comparison of resource location approaches is difficult for various reasons: (1) The problem is a very general one, with many functional and performance properties of the systems to be considered simultaneously, (2) the environments in which the systems operate are complex with many parameters defining their characteristics, and (3) the approaches are very heterogeneous because they are designed for rather different target applications. Nevertheless, it is possible to identify certain basic categories of approaches. We may distinguish them along the following three dimensions:

Unstructured vs. structured P2P systems: In unstructured P2P systems no information is kept about other nodes in terms of which resources they hold, i.e., the index information corresponding to a resource $k \in K(p)$ is kept only at peer p itself and no other information related to resources is kept for peers in its neighborhood $N(p)$. This is Gnutella's approach, for example. The main advantage of unstructured P2P systems is the high degree of independence among the nodes, which yields high flexibility and failure resilience. Structured P2P systems on the other hand can perform search operations more efficiently by exploiting additional information that is kept about other nodes, which enables directing search requests in a goal-oriented manner.

Flat vs. hierarchical P2P systems. In flat P2P systems all nodes are equivalent, i.e., there exists no distinction in the roles that the nodes play in the network. Again, Gnutella follows this approach. In hierarchical P2P systems distinctive roles exist, e.g., only specific nodes can support certain operations, such as *search*. An extreme example is Napster, where only a single node supports search. The main advantage of hierarchical P2P systems is improved performance, in particular for search, which comes at the expense of giving up some benefits of a "pure" P2P architecture, such as failure resilience.

Loosely coupled vs. tightly coupled P2P systems. In tightly coupled P2P systems there exists only one peer group at a time and only a single peer may join or leave the group at a time. Upon joining the group, the peer obtains a static, logical identifier that constrains the possible behavior of the peer with respect to its role in the group, e.g., the type of resources it keeps and the way it processes messages. Typically the key space for logical peer identifiers and for resources are identical. In loosely coupled P2P systems different peer groups can evolve, merge, or split. Though peers may play a specific role at a given time, this role and thus the logical peer address may change over time dynamically. Gnutella is an example of a loosely coupled P2P system, whereas Napster is an (extreme) case of a tightly coupled P2P system.

In the following we overview the main representatives of current P2P resource location systems in each of these categories and describe their functional properties. Following that, we will provide a comparative evaluation of important functional properties and performance criteria for these systems.

35.3.2 Unstructured P2P Systems

All current unstructured P2P systems are flat and loosely coupled. The canonical representative of this class of systems is Gnutella.

35.3.2.1 Gnutella

Gnutella is a decentralized file-sharing system that was originally developed in a 14-day "quick hack" by Nullsoft (Winamp) and was intended to support the exchange of cooking recipes. Its underlying protocol has never been published but was reverse-engineered from the original software.

The Gnutella protocol consists of 5 message types that support network maintenance (*Ping, Pong*) and data management (*Query, QueryHit, Push*). Messages are distributed using a simple constrained broadcast mechanism: Upon receipt of a message a peer decrements the message's time-to-live (TTL) field. If the TTL is greater than 0 and it has not seen the message's identifier before (loop detection), it resends the message to the other peers it has an open connection to. Additionally, the peer checks whether it should respond to the message, e.g., send a *QueryHit* in response to a *Query* message if it can satisfy the query. Responses are routed along the same path, i.e., the same peers, as the originating message.

To join a Gnutella network a peer must connect to a known Gnutella peer and send a *Ping* message that announces its availability and probes for other peers. To obtain peer addresses to start with, dedicated servers return lists of peers. This is outside the Gnutella protocol specification. Every peer receiving the *Ping* message can cache the new peer's address and can respond with a *Pong* message holding its IP address, port number and total size of the files it shares. This way a peer obtains many peer addresses that it caches (also *QueryHit* and *Push* messages contain IP address/port pairs, which can additionally be used to fill a peer's address cache). Out of the returned addresses it selects C neighbors (typically $C = 4$) and opens permanent connections to those. This defines its position in the Gnutella network graph. If one of these connections is dropped the peer can retry or choose another peer from its cache.

To locate a file a peer issues a *Query* message to all its permanently connected peers. The message defines the minimum speed and the search criteria. The search criteria can be any text string and its interpretation is up to the receivers of the message. Though this could be used as a container for arbitrary structured search requests, the standard is simple keyword search. If a peer can satisfy the search criteria it returns a *QueryHit* message listing all its matching entries. The originator of the query can then use this information to download the file via a simplified HTTP GET interaction.

In case the peer which sent the *QueryHit* message is behind a firewall, the requester may send a *Push* message (along the same way as it received the *QueryHit*) to the firewalled peer. The *Push* message specifies where the firewalled servent can contact the requesting peer to run a "passive" GET session. If both peers are behind firewalls, then the download is impossible.

Gnutella is a simple yet effective protocol: Hit rates for search queries are reasonably high, it is fault-tolerant toward failures of servents, and adopts well to dynamically changing peer populations. However, from a networking perspective, this comes at the price of very high bandwidth consumption: Search requests are broadcast over the network and each node receiving a search request scans its local database for possible hits.

For example, assuming a typical TTL of seven and an average of four connections C per peer (i.e., each peer forwards messages to three other peers) the maximum possible number of messages originating from a single Gnutella message is $2 * \Sigma_{i=0}^{TTL} C * (C-1)^i = 26{,}240$.

35.3.2.2 Improvements of Message Flooding

The high bandwidth consumption of Gnutella is a major drawback. Thus better ways to control the number of messages have been devised [Lv et al., 2002].

Expanding ring search starts a Gnutella-like search with small TTL (e.g., $TTL = 1$) and if there is no success it iteratively increases the TTL (e.g., $TTL = TTL + 2$) up to a certain limit. A more radical reduction of the message overhead is achieved by the *random walker* approach: To start a search, a peer sends out k random walkers. In contrast to Gnutella each peer receiving the request (a random walker) forwards it to only one neighbor, but search messages have a substantially higher TTL. The random walker periodically checks back with the originally requesting peer to stop the search in case another random walker found the result. Simulation studies and analytical results show that this model reduces message bandwidth substantially at the expense of increased search latency. To further improve the performance, replication schemes are applied.

A recent work shows that the random walker model can be further improved by using percolation-based search [Sarshar et al., 2003]. By using results from random graph theory, it is shown that searches based on probabilistic broadcast, i.e., multiple search messages, are forwarded to neighbors only with a probability $q < 1$ and will be successfully answered traversing qkN links where k is the average degree of nodes in the graph and q is fairly small (e.g., $q = 0.02$ in a network with $N = 20{,}000$ nodes).

35.3.3 Hierarchical P2P Systems

Hierarchical P2P Systems store index information at some dedicated peers or servers in order to improve search performance. Simple hierarchical P2P systems were already discussed in detail at the beginning of this chapter by the example of Napster as the best-known example of this class of systems. Thus, the following discussions focus on advanced hierarchical systems that are known under the term *super-peer architecture.*

35.3.3.1 Super-Peer Architectures

The super-peers approach was devised to combine some of the advantages of Napster with the robustness of Gnutella. In this class of systems three types of peers are distinguished:

1. A super-super-peer that serves as the entry point, provides lists of super-peers to peers at startup, and coordinates super-peers.
2. Super-peers that maintain the index information for a group of peers connected to them and interact with other super-peers through a Gnutella-like message forwarding protocol. Multiple super-peers can be associated with the same peer groups.
3. Ordinary peers that contact super-peers in order to register their resources and to obtain index information, just as in Napster. The file exchanges are then performed directly with the peers holding the resources.

A certain number of super-peers (a small number compared to the number of peers) is selected among the ordinary peers dynamically. An important criterion for selecting a super-peer is its available physical resources, such as bandwidth and storage space. Experimental studies [Yang and Garcia-Molina, 2002] show that the use of redundant super-peers is of advantage and that super-peers should have a substantially higher number of outgoing connections than in Gnutella (e.g., > 20) to minimize the *TTL*. In this way the performance of this approach can come close to Napster's with respect to search latency.

Most prominently, the super-peer approach has been implemented by FastTrack (http://www. fasttrack.nu/), which is used in media sharing applications such as Kazaa (http://www.kazaa.com/). This approach has also been implemented in the now defunct Clip2 Reflector and the JXTA search implementation.

35.3.4 Structured P2P Systems

Structured P2P systems distribute index information among all participating peers to improve search performance. In contrast to hierarchical P2P systems, this uniform distribution of index information avoids potential bottlenecks and asymmetries in the architecture. The difficult issue in this approach is that not only the index information but also the data access structure to this index information needs to be distributed as illustrated in Figure 35.2. These problems do not appear in the hierarchical approaches, where the data access structure (such as a hash table or a search tree) is only constructed locally.

All structured P2P systems share the property that each peer stores a routing table that contains part of the global data access structure, and searches are performed by forwarding messages selectively to peers using the routing table. The routing tables are constructed in a way that their sizes scale gracefully

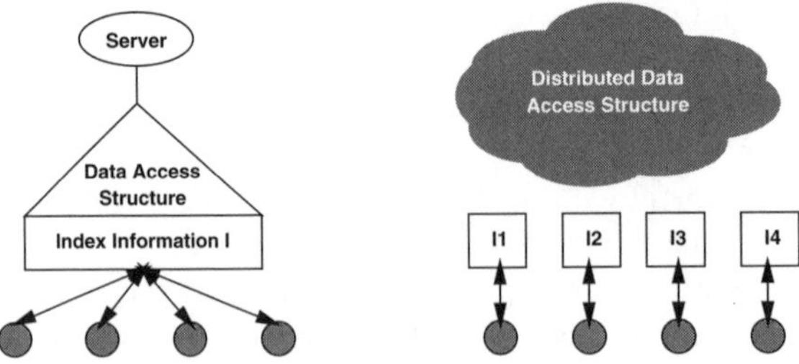

FIGURE 35.2 Centralized vs. distributed index.

and searches are fast (e.g., both table size and search latency are logarithmic in the total number of peers). Using routing tables to direct search messages reduces the total bandwidth consumed by searches and therefore these systems also scale well with respect to throughput. This advantage distinguishes them from unstructured P2P networks.

There are various strategies to organize the distributed data access structure, i.e., how the routing tables of the peers are being maintained:

- Ad hoc caching and clustering of index information along search paths (Freenet [Clarke et al., 2001, 2002])
- Prefix/suffix routing (P-Grid [Aberer et al., 2003], Chord [Dabek et al., 2001], Pastry [Rowstron and Druschel, 2001], and Tapestry [Zhao et al., 2004])
- Routing in a d-dimensional space (CAN [Ratnasamy et al., 2001])

Besides the applied routing approach, the main differences among the systems are found in their functional flexibility regarding issues such as replication, load balancing, and supported search predicates and in the properties of the network maintenance protocol. Loosely coupled structured P2P systems such as Freenet and P-Grid share the flexibility of network evolution with unstructured P2P networks, whereas tightly coupled structured P2P systems, such as CAN, Chord, Tapestry, and Pastry, impose a more strict control on the global properties and structure of the network for the sake of controllable performance.

In the following text we will discuss the corresponding systems and some of their technical details for each of the three strategies.

35.3.4.1 Freenet

Freenet [Clarke et al., 2001, 2002] is a P2P system for the publication, replication, and retrieval of data files. Its central goal is to provide an infrastructure that protects the anonymity of authors and readers of the data. It is designed in a way that makes it infeasible to determine the origin of files or the destination of requests. It is also difficult for a node to determine what it stores, as the files are encrypted when they are stored and sent over the network. Thus — following the reasoning of the designers of Freenet — nobody can be legally challenged, even if he or she stores and distributes illegal content, for example.

Freenet has an adaptive routing scheme for efficiently routing requests to the physical locations where matching resources are most likely to be stored. The routing tables are continually updated as searches and insertions of data occur. Additionally, Freenet uses dynamic replication of files to replicate them along search paths. Thus, search hits may occur earlier in the search process, which further improves search efficiency.

When a peer joins a Freenet network, it has to know some existing node in the network. By interacting with the network it will fill its routing table, which is initially empty, and the Freenet network structure will evolve. Figure 35.3 shows a sample Freenet routing table.

The routing tables in Freenet store the addresses of the neighboring peers and additionally the keys of the data files that this peer stores along with the corresponding data. When a search request arrives, it may be that the peer stores the data in its table and can immediately answer the request. Otherwise, it has to forward the request to another peer. This is done by selecting the peer that has the most similar key in terms of lexicographic distance. When an answer arrives the peer stores the answer in its data store. If the data store is already full this might require evicting other entries using a LRU (least recently used) strategy. As can be seen in Figure 35.3 the peer may also decide to evict only the data corresponding to a key before it evicts the address of the peer in order to save space while maintaining the routing

Key	Data	Address
8e4768isdd0932uje89	ZT38we01h02hdhgdzu	tcp/125.45.12.56:6474
456r5wero04d903iksd0	Rhweui12340jhd091230	tcp/67.12.4.65:4711
712345jb89b8nbopledh		tcp/40.56.123.234:1111

FIGURE 35.3 Sample routing table in Freenet.

information. Because peers route search requests only to the peer with the closest key, Freenet implements a depth-first search strategy rather than a breadth-first strategy as Gnutella. Therefore, the time-to-live of messages is also substantially longer, typically 500. As in Gnutella, Freenet messages carry identifiers in order to detect cycles.

Figure 35.4 shows Freenet's search mechanism and network reorganization strategy.

Peer A is sending a search request for file X.mp3 to B. As it does not have the requested data, B forwards the request to C, which has the closest key in its routing table. Because the TTL is 2 and C does not hold the data, this request fails. Therefore, B next forwards the request to D (next-similar key) where the data is found. While sending back the response containing X.mp3, the file is cached at all peers, i.e., at B and A. In addition A learns about a new node D and a new connection is created in the network.

Adding new information to Freenet, i.e., adding a file, is done in a sophisticated manner that tries to avoid key collisions: First, a key is calculated which is then sent out as a proposal in an insert message with a time-to-live value. The routing of insert messages uses the same key similarity measure as searching. Every receiving peer checks whether the proposed key is already present in its local store. If yes, it returns the stored file and the original requester must propose a new key. If no, it routes the insert message to the next peer for further checking. The message is forwarded until the time-to-live is 0 or a collision occurs. If the time-to-live is 0 and no collision was detected, the file is inserted along the path established by the initial insert message.

As the caching strategy for the routing tables is designed in a way that tends to cluster similar data (keys) at nodes in the network, the assumption is that nodes get more and more specialized over time. This assumption turns out to hold in simulations of Freenet, which show that the median path length converges to logarithmic length in the size of the network. An explanation why the search performance improves so dramatically is found in the properties of the graph structure of the network. Analyses of the resulting Freenet networks reveal that they have the characteristics of small-world graphs.

35.3.4.2 Prefix/Suffix Routing

The idea of prefix/suffix routing is usually credited to Plaxton [Plaxton et al., 1997] and in the meanwhile a number of variants of this approach have been introduced. We describe here only the underlying principle in a simplified form, to illustrate the strategy to construct a scalable, distributed data access structure.

Without constraining generality, assume that the keys K for identifying resources are binary. Then we may construct a binary tree, where each level of the tree corresponds to one bit of a key. The edges to the left are marked 0 whereas the edges to the right are marked 1. Following some path of such a tree starting at the root produces every possible key (more precisely the structure described is a binary *trie*).

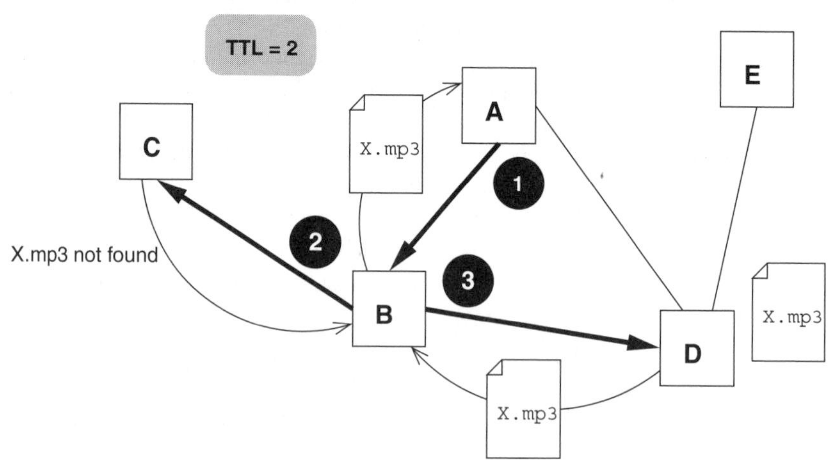

FIGURE 35.4 Searching in Freenet.

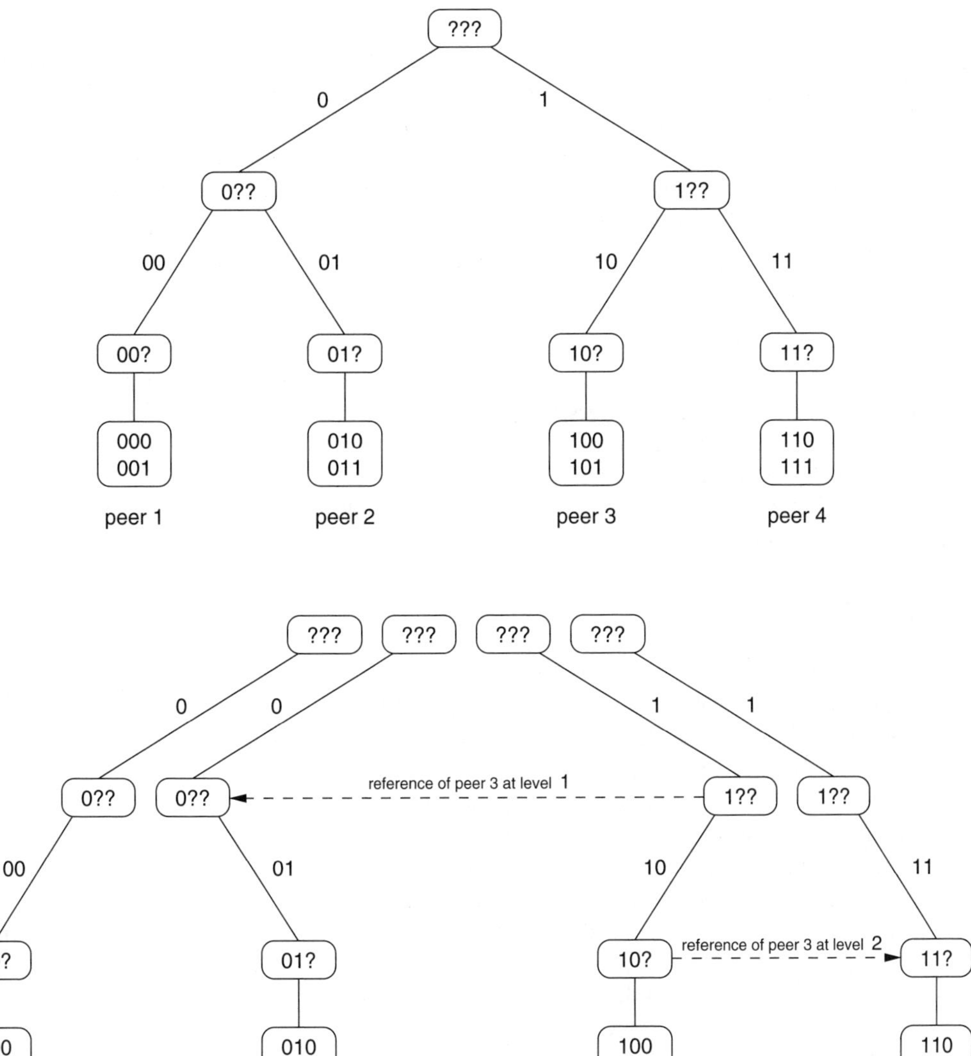

FIGURE 35.5 Prefix routing.

For simplicity we assume that the tree is balanced and has depth d. The upper part of Figure 35.5 shows such a tree for $d = 2$.

Now we associate each leaf of the tree with one peer and make the peer responsible for holding index information on those resources whose keys have the path leading from the root to the leaf (peer) as a prefix. As we cannot store the tree centrally, the key idea for distributing the tree (i.e., the data access structure) is that a peer associated with a specific leaf stores information on all nodes in the path from the root of the tree to its own leaf. More precisely, for each node along the path (each level) it stores the address(es) of some peer(s) that can be reached by following the alternative branch of the tree at the respective level. This information constitutes its routing information. The lower part of Figure 35.5 shows how the tree is decomposed into routing tables by this strategy and gives two sample routing table entries for peer 3.

By this decomposition we achieve the following:

- Searches for resource keys can be started from any node and by either successively continuing the search along the peer's own path in case the key matches at that level, or by forwarding the search to a peer whose address is in the routing table at the corresponding level.
- As the depth of the tree is logarithmic in the number of peers, the sizes of routing tables and the number of steps and thus messages for search are logarithmic. Therefore, storage, search latency, and message bandwidth scale well.

The most important criteria to distinguish the different variations of prefix/suffix routing used in practice are the following:

- The method used to assign peers to their position in the tree and to construct the routing tables
- Not all leaves are associated with peers and neighboring peers are taking over their responsibility
- *N*-ary trees instead of binary trees, which reduces the tree depth and search latency substantially
- Multiple routing entries at each level to increase failure resilience
- Using unbalanced trees for storage load balancing
- Replication of index information, either by assigning multiple peers to the same leaf or by sharing index information among neighboring leaves

We will go into some of the differences in the following when introducing several structured P2P systems.

35.3.4.3 Distributed Hash Table Approaches

Distributed hash table (DHT) approaches, such as Chord [Dabek et al., 2001], Pastry [Rowstron and Druschel, 2001], and Tapestry [Zhao et al., 2004], hash both peer addresses and resource keys into the same key space underlying the construction of the prefix/suffix routing scheme. This key space consists of binary or *n*-ary keys of fixed length. The key determines the position of the node in the tree and the index information it has to store. Often, it is assumed that a peer also stores the associated resources. Typically a peer stores all resources with keys numerically closest to it, e.g., in Pastry, or with keys in the interval starting from its own key to a neighbor's key, such as in Chord.

When a peer wants to join a network, it first has to obtain a globally unique key that is determined from the key space typically by applying a hash function to the node's physical (IP) address. Upon entry, it has to contact a peer already in the network, which it has to know by some out-of-band means, as in the other P2P approaches. After contacting this peer, the new peer starts a search for its key. At each step of this search, i.e., when traversing down the tree level by level, it will encounter peers that share a common prefix with its key. This allows the new peer to obtain entries from the routing tables of those peers that are necessary to fill its own routing table properly. Also it allows the current peers to learn about the new peer and enter its address into their routing tables. After the new peer has located the peer that is currently responsible for storing index information related to its own key, it can take over from this peer that part of the index information that it is responsible for. Similarly, the network is reorganized upon the departure of a peer. Given this approach for joining and leaving a peer network, different subnetworks cannot develop independently and be merged later (such as in Gnutella or Freenet), which explains why DHT approaches are tightly coupled.

35.3.4.4 P-Grid

P-Grid [Aberer et al., 2003] is a loosely coupled peer-to-peer resource location system combining the flexibility of unstructured P2P networks with the performance of a prefix routing scheme. It differs from DHT approaches by its approach for constructing the routing tables.

P-Grid constructs the routing tables by exploring the network through randomized interactions among peers. The random meetings are initiated either by the peers themselves, similar to how Gnutella uses *Ping* messages, or to how Freenet exploits queries and data insertions for this purpose. Initially each peer covers the complete key space (or in other words is associated with the root of the search tree). When

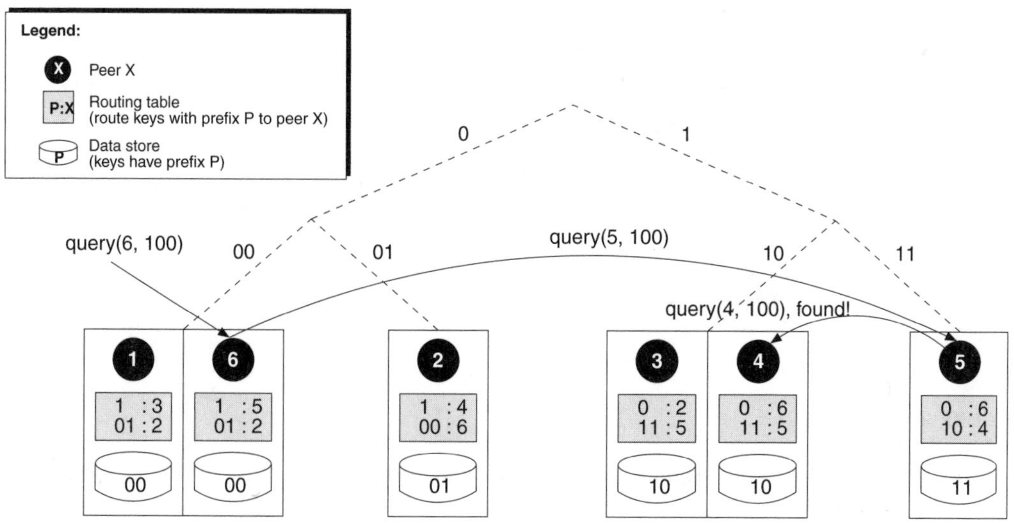

FIGURE 35.6 Example P-Grid.

two peers meet that cover the same key space, they can split the space into two parts and each peer associates itself with one of the two parts, i.e., one of the new paths in the search tree. The two peers reference each other in order to construct their routing tables at the newly created level of the search tree. Such splits are only performed if a sufficiently large number of data items is stored at the peers for the newly created subspaces. Otherwise, the peers do not split but replicate their data, i.e., they stay responsible for the same index information and become replicas of each other. Figure 35.6 shows an example P-grid.

If two peers meet that do not cover the same key space, they use their already existing routing structure to search for more candidates they could contact and then can use to exchange routing information to improve their routing tables. Through this adaptive method of path specialization, P-Grid's distributed tree adapts its shape to the data distribution, which makes it particularly suitable for skewed data distributions. This may result in unbalanced trees but the expected number of messages to perform a search remains logarithmic [Aberer, 2002]. By virtue of this construction process peers adopt their index responsibility incrementally within the P-Grid routing infrastructure (in terms of data keys for which they are responsible), and different networks can develop independently and be merged.

35.3.4.5 Topological Routing

Topological Routing, as introduced in CAN [Ratnasamy et al., 2001], uses a model different from prefix routing for organizing routing tables. As in DHT approaches, peer addresses and resource keys are hashed into a common key space. For topological routing these keys are taken from a d-dimensional space, i.e., the keys consist of a d-dimensional vector of simple keys, e.g., binary keys of fixed length. More precisely the d-dimensional space is a d-dimensional torus because all calculations on keys are performed modulo a fixed base. The dimension d is usually low, e.g., $d = 2 \ldots 10$. Each peer is responsible for a d-dimensional, rectangular sub-volume of this space. An example for dimension $d = 2$ is shown in Figure 35.7.

Peers are only aware of other peers that are responsible for neighboring subvolumes. Thus the routing tables contain the neighborhood information together with the direction of the neighbor. Searches are processed by routing the request to the peer that manages the subvolume with coordinates that are closer to the search key than the requester's coordinates. When a peer joins the network it selects (or obtains) randomly a coordinate of the d-dimensional space. Then it searches for this coordinate and will encounter

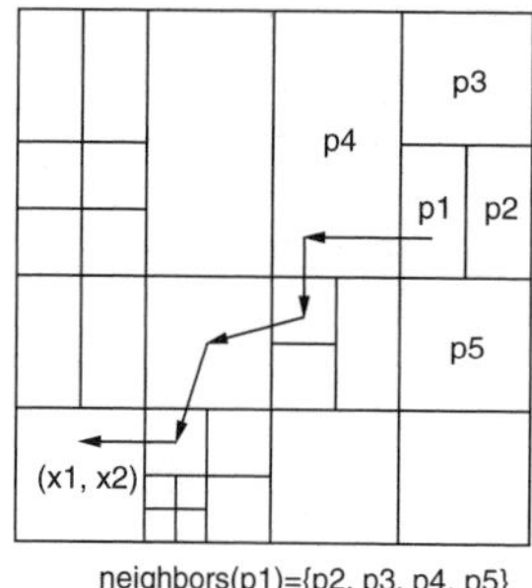

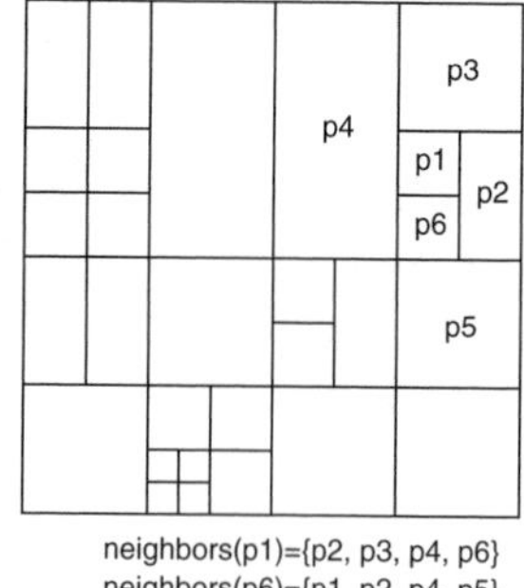

FIGURE 35.7 Example CAN.

the peer responsible for the subvolume holding the coordinate. The subvolume will be split in two along one dimension and the routing tables containing the neighbors will be reorganized.

The search-and-join costs in CAN do not only depend on the network size but in particular on the choice of the dimension d. Searches in CAN have an expected cost *of* $O(dn)^{1/d}$ and the insertion of a node incurs an additional constant cost of $O(d)$ for reorganizing the neighborhood. CAN is a strongly coupled P2P network as the identity of peers is fixed when peers join the network.

35.4 Comparative Evaluation of P2P Systems

This section compares the various flavors of P2P networks in respect to a set of criteria. Due to space limitations we discuss only functional and nonfunctional properties that we consider most important or which are not discussed in related work. For more detailed comparisons we refer the reader to Milojicic et al., [2002].

35.4.1 Performance

The key factors for comparing P2P systems are the search and maintenance costs. With respect to search cost, from a user's perspective, search latency should be low, whereas from a system's perspective the total communication costs, i.e., number and size of messages and number of connections (permanent and on-demand), are relevant. With respect to maintenance costs, which are only relevant from a system's perspective, we can distinguish communication costs, both for updates of data and for changes in the network structure, and the additional storage costs for supporting search, in terms of routing and indexing information required to establish the overlay network. As storage costs are likely to become less and less critical, it even seems to be possible that peers might become powerful enough to store indices for complete P2P networks. However, this simplistic argument does not apply, as keeping the index information consistent in such systems would incur unacceptably high costs. Thus distribution of routing and indexing information is not just a matter of reducing storage cost, but in particular of reducing the maintenance costs for this information.

Figure 35.8 informally summarizes the performance characteristics of the important classes of P2P systems that have been discussed earlier. We have excluded the random walker approach because no analytical results are available yet and have included the approach of fully replicating the index information for comparison, i.e., every peer would act as a Napster server.

The following notation is used: n denotes the number of peers in a network. n, $log(n)$, $dn^{1/d}$, 1, etc. are used as shorthand for the $O()$ notation, i.e., within a constant factor these functions provide an upper bound. We do not distinguish different bases of logarithms as these are specific to choices of parameters

Approach	Latency	Messages	Update cost	Storage
Gossiping (Gnutella)	$\log(n)$	n	1	1
Directory server (Napster)	1	1	1	n (max), 1 (avg)
Full replication	1	1	n	n
Super-peers	$\log(C)$	C	1	C (max), 1 (avg)
Prefix routing	$\log(n)$	$\log(n)$	$\log(n)$	$\log(n)$
Topological Routing (CAN)	$dn^{1/d}$	$dn^{1/d}$	$dn^{1/d}$	d

FIGURE 35.8 Performance comparison of P2P approaches.

in the systems. d and C are constants that are used to parameterize the systems. Where necessary we distinguish average from maximum bounds.

For Gnutella we can see that search latency is low due to the structure of its network graph. However, network bandwidth consumption is high and grows linearly in n. More precisely, if c is the number of outgoing links of a node, cn is an upper bound for the number of messages generated. On the other hand update and storage costs are constant because no data dependencies exist. On the other end of the spectrum we find Napster and full replication: They exhibit constant search costs but high update and storage costs. Super-peers trade a modest increase of search costs — assuming they use the same gossiping scheme as Gnutella — for a reduction of the storage load on the server peers. C is the number of super-peers in the system.

As we can see, structured networks balance all costs, namely search latency, bandwidth consumption, update costs, and storage costs. This balancing is the reason that makes these approaches so attractive as the foundation of the next generation of P2P systems. The schemes that are based on some variation of prefix routing incur logarithmic costs for all of these measures. Also Freenet exhibits similar behavior, but as analytical results are lacking (only simulations exist so far) we cannot include it in the comparison. Topological routing deviates from the cost distribution scheme of structured networks, but balanced cost distributions can be achieved by proper choices of the dimension parameter d.

A more detailed comparison of the approaches would have to include the costs for different replication schemes and the costs incurred for network maintenance, e.g., for joining the network or repairing the network after failures. Also other parameters such as failure rates for nodes or query and data distributions influence the relative performance. However, a complete comparison at this level of detail is beyond the scope of this overview and also still part of active research.

35.4.2 Functional and Qualitative Properties

35.4.2.1 Search Predicates

Besides performance, an important distinction among P2P approaches concerns the support of different types of search predicates. A substantial advantage of unstructured and hierarchical P2P systems is the potential support of arbitrary predicates, as their use is not constrained by the resource location infrastructure. In both of these classes of P2P systems, search predicates are evaluated only locally, which enables the use of classical database query or information retrieval processing techniques. In part, this explains the success of these systems despite their potentially less advantageous global performance characteristics.

As soon as the resource location infrastructure exploits properties of the keys in order to structure the search space and to decide on the forwarding of search requests if they cannot be answered locally, the support of search predicates is constrained. A number of approaches, including Chord, CAN, Freenet, and Pastry intentionally hash search keys, either for security or load balancing purposes, and thus lose all potential semantics contained in the search keys that could be used to express search predicates. Therefore, they support exclusively the search for keys identical to the search key, or in other words they support only the equality predicate for search.

More complex predicates, such as range or similarity, cannot be applied in a meaningful way. Nevertheless, these approaches have the potential to support more complex predicates. The prefix-routing-based approaches implement a distributed trie structure that can support prefix queries and thus range queries. For P-Grid prefix-preserving hashing of keys has been applied to exploit this property. Freenet uses routing based on lexicographic similarity. Thus, if the search keys were not hashed (which is done for security purposes to support anonymity in Freenet), queries could also find similar keys and not only the identical one. CAN uses a multidimensional space for keys, and thus data keys that, for example, bear spatial semantics might be mapped into this key space preserving spatial relationships. Thus spatial neighborhood searches would be possible.

Beyond the support of more complex atomic predicates also the ability of supporting value-based joins among data items, similar to relational databases, is of relevance and subject of current research [Harren et al., 2002].

35.4.2.2 Replication

Because the peers in a P2P network are assumed to be unreliable and frequently offline or unreachable, most resource location systems support replication mechanisms to increase failure resilience. Replication exists in two flavors in P2P systems: at the data level and possibly at the index level. Replication of data objects is applied to increase the availability of the data objects in the peer network. Hierarchical and structured P2P networks additionally replicate index information to enhance the probability of successfully routing search requests.

For *data replication* we can distinguish four different methods that are employed, depending on the mechanism to initiate the replication:

Owner replication: A data object is replicated to the peer that has successfully located it through a query. This form of replication occurs naturally in P2P file sharing systems such as Gnutella (unstructured), Napster (hierarchical), and Kazaa (super-peers) because peers implicitly make available to other users the data that they have found and downloaded (though this feature can be turned off by the user).

Path replication: A data object is replicated along the search path that is traversed as part of a search. This form of replication is used in Freenet, which routes results back to the requester along the search path in order to achieve a data clustering effect for accelerating future searches. This strategy would also be applicable to unstructured P2P networks in order to replicate data more aggressively.

Random replication: A data object is replicated as part of a randomized process. In P-Grid random replication is part of the construction of the P2P network. If peers do not find enough data to justify a further refinement of their routing tables, they replicate each others' data. For unstructured networks it has been shown that random replication, initiated by searches and implemented by selecting random nodes visited during the search process, is superior to owner and path replication [Lv et al., 2002].

Controlled replication: Here data objects are actively replicated a pre-specified number of times when they are inserted into the network. This approach is used in strongly coupled P2P networks such as Chord, CAN, Tapestry, and Pastry. We can distinguish two principal approaches: Either a fixed number of structured networks is constructed in parallel or multiple peers are associated with the same or overlapping parts of the data key space.

Index replication is applied in structured and in hierarchical P2P networks. For the super-peer approach it has been shown that having multiple replicated super-peers maintaining the same index information increases system performance [Yang and Garcia-Molina, 2002]. Structured P2P networks typically maintain multiple entries for the same routing path to have alternative routing paths at hand in case a referenced node fails or is offline.

35.4.2.3 Security

As in any distributed system, security plays a vital role in P2P systems. However, only a limited amount of work has been dedicated to this issue so far. At the moment, trust and reputation management, anonymity vs. identification, and denial-of-service (DOS) attacks seem to be the most relevant security aspects related to P2P systems:

Trust and reputation management: P2P systems depend on the cooperation of the participants of the system. Phrased differently, this means that each participant *trusts* the other participants that it interacts with in terms of proper routing, exchange of index information, and provision of proper (noncorrupted) data or, more generally, quality of service. For example, a peer could return false hits that hold advertisements instead of the content the requester originally was looking for, or a quality of service guarantee given by a peer could not be fulfilled. These are just two examples, but in fact trust and reputation management are crucial to make P2P systems a viable architectural alternative for systems beyond mere sharing of free files. The main requirements of trust and reputation management in a P2P setting are (full) decentralization of the trust information and robustness against positive/negative feedback and collusions. Several approaches have have already been proposed but have mostly been applied in experimental settings — for example for Gnutella [Damiani et al., 2003] and for P-Grid [Aberer and Despotovic, 2001], and have not found their way into generally available software distributions yet.

Anonymity vs. identification: Anonymity and identification in P2P systems serve conflicting purposes: Anonymity tries to protect "free speech" [Clarke et al., 2002] in the broadest sense whereas identification is mandatory in commercial systems to provide properties such as trust, non-repudiation, and accountability. So far anonymity and identification issues have only been addressed in some P2P systems. For example, Freenet uses an approach that makes it impossible to find out the origin of data or what data a peer stores (caches) [Clarke et al., 2002]. Thus nobody can be legally challenged even if illegal content is stored or distributed. For identification purposes, existing public key infrastructures (PKI) could be used. However, this would introduce a form of centralization and may harm scalability. P-Grid proposes a decentralized yet probabilistically secure identification approach [Aberer et al., 2004] that can also provide PKI functionality [Aberer et al., to be published]. Other P2P systems discussed in this chapter do not address anonymity or identification explicitly.

DOS attacks: At the moment, the proper functioning of P2P systems depends on the well-behaved cooperation of the participants. However, as P2P systems are large-scale distributed systems, the possibilities for DOS attacks are numerous. For example, in systems that use a distributed index such as Chord, Freenet, or P-Grid, the provision of false routing information would be disastrous or query flooding in unstructured systems such as Gnutella could easily overload the network. Currently only little work exists on the prevention of attacks in P2P systems (although more work exists on the analysis of attack scenarios [see Daswani and Garcia-Molina, 2002]): Pastry provides an approach to secure routing [Castro et al., 2002] which, however, is rather costly in terms of bandwidth consumption and Freenet [Clarke et al., 2002] can secure data so that it can only be changed by the owner. Otherwise the P2P systems discussed in this chapter do not address this issue.

35.4.2.4 Autonomy

A P2P system is composed of *autonomous peers* by definition, i.e., peers that belong to different users and organizations with no or only limited authority to influence their operation or behavior. In technical terms autonomy means that peers can decide independently on their role and behavior in the system, which, if done properly, is a key factor to provide scalability, robustness, and flexibility in P2P systems. Consequently, a higher degree of peer autonomy also implies that a higher degree of self-organization mechanisms are required to provide a system with meaningful behavior. Though all P2P systems claim their peers to be autonomous, a closer look at existing systems reveals that this statement is true only to a varying degree.

Hierarchical systems such as Napster and Kazaa limit the autonomy of peers to a considerable degree by their inherent centralization. This makes them less scalable, which must be compensated by considerable investments into their centralized infrastructures. Also robustness is harder to achieve because special point-of-failures (super-peers) exist. On the other hand, overall management is simpler than in systems with greater autonomy.

Unstructured systems like Gnutella offer the highest degree of autonomy. Such systems are very robust in scale but pay these advantages with considerable resource consumption.

Structured systems, as the third architectural alternative, balance the advantages of autonomy with resource consumption. Within this family many degrees of autonomy exist: Freenet offers a degree of autonomy that nearly reaches Gnutella's. However, the applied mechanisms have inhibited the development of an analytical model for Freenet so far and thus its properties have only been evaluated by simulations. P-Grid offers peer autonomy at a similar level to Freenet but additionally provides a mathematical model that enables quantitative statements on the system and its behavior. Freenet and P-Grid are loosely coupled and thus share the flexibility of network evolution with unstructured systems like Gnutella, i.e., peer communities can develop independently, merge, and split. Tightly coupled systems such as CAN, Chord, Tapestry, and Pastry, impose stricter control on the peers in terms of routing table entries, responsibilities, and global knowledge. This offers some advantages but limits the flexibility of the systems, for example, splitting and merging independent peer communities is impossible.

35.5 Conclusions

In this chapter we have tried to provide a concise overview of the current state of research in P2P systems. Our goal was to communicate the fundamental concepts that underly P2P systems, to provide further insights by a detailed presentation of the problem of resource location in P2P environments, and give a short, but to-the-point, comparative evaluation of state-of-the-art systems to enable the reader to understand the performance implications and resource consumption issues of the various systems. We have, however, omitted some areas from the discussion due to space limitations.

For example, we did not discuss advanced functionalities beyond resource location such as support for update functionality [Datta et al., 2003] or applications in information retrieval [Aberer and Wu, 2003], which would help to broaden the functionalities of P2P systems and in turn would increase the applicability of the P2P paradigm to domains beyond mere file sharing.

Some interesting fields of study are the application of economic principles to P2P systems [Golle et al., 2001], implications of the "social behavior" of the peers in a system [Adar and Huberman, 2000], and trust and reputation management in P2P systems [Aberer and Despotovic, 2001]. Interdisciplinary research among other disciplines studying complex systems such as economy, biology, and sociology will definitely be a direction attracting growing interest in the future.

There are a number of other interesting topics related to P2P that we could not address. However, we believe that we have provided the interested reader with sufficient basic know-how to conduct further studies.

References

Aberer, Karl. Scalable Data Access in P2P Systems Using Unbalanced Search Trees. In *Proceedings of Workshop on Distributed Data and Structures (WDAS-2002)*, Paris, 2002.

Aberer, Karl and Zoran Despotovic. Managing Trust in a Peer-2-Peer Information System. In *Proceedings of the 10th International Conference on Information and Knowledge Management (2001 ACM CIKM)*, pages 310–317. ACM Press, New York, 2001.

Aberer, Karl, Philippe Cudré-Mauroux, Anwitaman Datta, Zoran Despotovic, Manfred Hauswirth, Magdalena Punceva, and Roman Schmidt. P-Grid: A Self-organizing Structured P2P System. *SIGMOD Record*, 32(3), September 2003.

Aberer, Karl, Anwitaman Datta, and Manfred Hauswirth. Efficient, self-contained handling of identity in peer-to-peer systems. *IEEE Transactions on Knowledge and Data Engineering*, 16(7): 858–869, July 2004.

Aberer, Karl, Anwitaman Datta, and Manfred Hauswirth. A decentralized public key infrastrcuture for customer-to-customer e-commerce. *International Journal of Business Process Integration and Management.* In press.

Aberer, Karl and Jie Wu. A Framework for Decentralized Ranking in Web Information Retrieval. In *Proceedings of the Fifth Asia Pacific Web Conference (APWeb 2003)*, number 2642 in Lecture Notes in Computer Science, pages 213–226, 2003.

Adar, Eytan and Bernardo A. Huberman. Free Riding on Gnutella. *First Monday*, 5(10), 2000. *http://firstmonday.org/issues/issue5-10/adar/index.html.*

Barabási, Albert-László and Réka Albert. Emergence of scaling in random networks. *Science*, 286: 509–512, 1999.

Bonabeau, Eric, Marco Dorigo, and Guy Theraulaz. *Swarm intelligence: from natural to artificial systems.* Oxford University Press, Oxford, U.K., 1999.

Castro, Miguel, Peter Druschel, Ayalvadi Ganesh, Antony Rowstron, and Dan S. Wallach. Secure routing for structured peer-to-peer overlay networks. In *Proceedings of Operating Systems Design and Implementation (OSDI, 2002).*

Clarke, Ian, Scott G. Miller, Theodore W. Hong, Oskar Sandberg, and Brandon Wiley. Protecting free expression online with freenet. *IEEE Internet Computing*, 6(1): 40–49, January/February 2002.

Clarke, Ian, Oskar Sandberg, Brandon Wiley, and Theodore W. Hong. Freenet: A Distributed Anonymous Information Storage and Retrieval System. In *Designing Privacy Enhancing Technologies: International Workshop on Design Issues in Anonymity and Unobservability*, number 2009 in Lecture Notes in Computer Science, 2001.

Clip2. The Gnutella Protocol Specification v0.4 (Document Revision 1.2), June 2001. http://www9.limewire.com/developer/gnutella-protocol-0.4.pdf.

Dabek, Frank, Emma Brunskill, M. Frans Kaashoek, David Karger, Robert Morris, Ion Stoica, and Hari Balakrishnan. Building Peer-to-Peer Systems with Chord, a Distributed Lookup Service. In *Proceedings of the 8th Workshop on Hot Topics in Operating Systems (HotOS-VIII)*, pages 81–86, 2001.

Damiani, Ernesto, Sabrina De Capitani di Vimercati, Stefano Paraboschi, and Pierangela Samerati. Managing and sharing servents' reputations in P2P systems. *Transactions on Knowledge and Data Engineering*, 15(4): 840–854, July/August 2003.

Daswani, Neil and Hector Garcia-Molina. Query-flood DoS attacks in Gnutella. In *Proceedings of the 9th ACM conference on Computer and Communications Security (CCS)*, pages 181–192, 2002.

Datta, Anwitaman, Manfred Hauswirth, and Karl Aberer. Updates in Highly Unreliable, Replicated Peer-to-Peer Systems. In *Proceedings of the 23rd International Conference on Distributed Computing Systems (ICDCS'03)*, pages 76–87, 2003.

Flake, Gary William, Steve Lawrence, C. Lee Giles, and Frans Coetzee. Self-organization and identification of web communities. *IEEE Computer*, 35(3): 66–71, 2002.

Golle, Philippe, Kevin Leyton-Brown, and Ilya Mironov. Incentives for sharing in peer-to-peer networks. In *Proceedings of the Second International Workshop on Electronic Commerce (WELCOM 2001)*, number 2232 in Lecture Notes in Computer Science, pages 75–87, 2001.

Gong, Li. JXTA: A Network Programming Environment. *IEEE Internet Computing*, 5(3): 88–95, May/June 2001.

Harren, Matthew, Joseph M. Hellerstein, Ryan Huebsch, Boon Thau Loo, Scott Shenker, and Ion Stoica. Complex Queries in DHT-based Peer-to-Peer Networks. In *Proceedings of the 1st International Workshop on Peer-to-Peer Systems (IPTPS '02)*, volume 2429 of *Lecture Notes in Computer Science*, pages 242–259, 2002.

Heylighen, Francis. Self-organization. Principia Cybernetica Web, January 1997. http://pespmc1.vub.ac.be/SELFORG.html.

Kleinberg, Jon. The Small-World Phenomenon: An Algorithmic Perspective. In *Proceedings of the 32nd ACM Symposium on Theory of Computing*, pages 163–170, 2000.

Lv, Qin, Pei Cao, Edith Cohen, Kai Li, and Scott Shenker. Search and Replication in Unstructured Peer-to-Peer Networks. In *Proceedings of the 2002 International Conference on Supercomputing*, pages 84–95, 2002.

Milojicic, Dejan S., Vana Kalogeraki, Rajan Lukose, Kiran Nagaraja, Jim Pruyne, Bruno Richard, Sami Rollins, and Zhichen Xu. Peer-to-Peer Computing. Technical Report HPL-2002-57, HP Laboratories Palo Alto, CA, March 2002. http://www.hpl.hp.com/techreports/2002/HPL-2002-57.pdf.

Plaxton, C. Greg, Rajmohan Rajaraman, and Andréa W. Richa. Accessing Nearby Copies of Replicated Objects in a Distributed Environment. In *Proceedings of the 9th Annual Symposium on Parallel Algorithms and Architectures,* pages 311–320, 1997.

Ratnasamy, Sylvia, Paul Francis, Mark Handley, Richard Karp, and Scott Shenker. A Scalable Content-Addressable Network. In *Proceedings of the 2001 Conference on Applications, Technologies, Architectures, and Protocols for Computer Communications (SIGCOMM),* pages 161–172, 2001.

Ripeanu, Matei and Ian Foster. Mapping the Gnutella Network: Macroscopic Properties of Large-Scale Peer-to-Peer Systems. In *Proceedings of the 1st International Workshop on Peer-to-Peer Systems (IPTPS '02),* volume 2429 of *Lecture Notes in Computer Science,* pages 85–93, 2002.

Rowstron, Antony and Peter Druschel. Pastry: Scalable, Distributed Object Location and Routing for Large-Scale Peer-to-Peer Systems. In *IFIP/ACM International Conference on Distributed Systems Platforms (Middleware),* number 2218 in Lecture Notes in Computer Science, pages 329–350, 2001.

Sarshar, N., V. Roychowdury, and P. Oscar Boykin. Percolation-Based Search on Unstructured Peer-To-Peer Networks, 2003. http://www.ee.ucla.edu/~nima/Publications/search_ITPTS.pdf.

Watts, Duncan and Steven Strogatz. Collective dynamics of small-world networks. *Nature,* 393, 1998.

Yang, Beverly and Hector Garcia-Molina. Improving Search in Peer-to-Peer Networks. In *Proceedings of the 22nd International Conference on Distributed Computing Systems (ICDS '02),* pages 5–14, 2002.

Zhao, Ben Y., Ling Huang, Jeremy Stribling, Sean C. Rhea, Anthony D. Joseph, and John Kubiatowicz. Tapestry: A Resilient Global-scale Overlay for Service Deployment. *IEEE Journal on Selected Areas in Communications,* 22(1):41–53, January 2004.

36

Data and Services for Mobile Computing

CONTENTS

Abstract 36-1
36.1 Introduction 36-1
36.2 Mobile Computing vs. Wired-Network Computing 36-3
36.3 M-Services Application Architectures 36-4
36.4 Mobile Computing Application Framework 36-5
36.4.1 Communications Layer 36-6
36.4.2 Discovery Layer 36-7
36.4.3 Location Management Layer 36-8
36.4.4 Data Management Layer 36-9
36.4.5 Service Management Layer 36-11
36.4.6 Security Plane 36-12
36.4.7 System Management Plane 36-12
36.5 Conclusions 36-12
Acknowledgments 36-13
References 36-13

Sasikanth Avancha
Dipanjan Chakraborty
Filip Perich
Anupam Joshi

Abstract

The advent and phenomenal growth of low-cost, lightweight, portable computers concomitant with that of the Internet have led to the concept of mobile computing. Protocols and mechanisms used in Internet computing are being modified and enhanced to adapt to "mobile" computers. New protocols and standards are also being developed to enable mobile computers to connect to each other and to the Internet through both wired and wireless interfaces. The primary goal of the mobile computing paradigm is to enable mobile computers to accomplish tasks using all possible resources, i.e., data and services available in the network, anywhere, anytime. In this chapter we survey the state of the art of mobile computing and its progress toward its goals. We also present a comprehensive, flexible framework to develop applications for mobile computing. The framework consists of protocols, techniques, and mechanisms that enable applications to discover and manage data and services in wired, infrastructure-supported wireless and mobile *ad hoc* networks.

36.1 Introduction

The term *computing device* or *computer* usually evokes the image of a big, powerful machine, located in an office or home, that is always on and possibly connected to the Internet. The rapid growth of lightweight, easily and constantly available devices — available even when one is on the move — has dramatically altered this image. Coupled with the potential for easy network access, the growth of these

1-58488-381-2/05/$0.00+$1.50
© 2005 by CRC Press LLC

mobile devices has tremendously increased our capability to take computing services with us wherever we go. The combination of device mobility and computing power has resulted in the mobile computing paradigm. In this paradigm, computing power is constantly at hand irrespective of whether the mobile device is connected to the Internet or not. The smaller the devices, the greater their portability and mobility, but the lesser their computing capability. It is important to understand that the ultimate goal of the mobile computing paradigm is to enable people to accomplish tasks using computing devices, *anytime, anywhere.* To achieve this goal, network connectivity must become an essential part of mobile computing devices. The underlying network connectivity in mobile computing is, typically, wireless. *Portable Computing* is a variant of mobile computing that includes the use of wired interfaces (e.g., a telephone modem) of mobile devices. For instance, a laptop equipped with both a wireless and a wired interface connects via the former when the user is walking down a hallway (mobile computing), but switches to the latter when the user is in the office (portable computing).

The benefits of mobility afforded by computing devices are greatly reduced, if not completely eliminated, if devices can only depend on a wired interface for their network connectivity (e.g., telephone or network jack). It is more useful for a mobile computing device to use wireless interfaces for network connectivity when required. Additionally, networked sources of information may also become mobile. This leads to a related area of research called *ubiquitous computing.*

Let us now discuss the hardware characteristics of current-generation mobile computing devices. The emphasis in designing mobile devices is to conserve energy and storage space. These requirements are evident in the following characteristics, which are of particular interest in mobile computing:

Size, form factor, and *weight.* Mobile devices, with the exception of high-end laptops, are hand-helds (e.g., cell phones, PDAs, pen computers, tablet PCs). They are lightweight and portable. Mobility and portability of these devices are traded off for greater storage capacity and higher processing capability.

Microprocessor. Most current-generation mobile devices use low-power microprocessors, such as the family of ARM and XScale processors, in order to conserve energy. Thus, high performance is traded off for energy consumption because the former is not as crucial to mobile devices as the latter.

Memory size and type. Primary storage sizes in mobile devices range anywhere between 8 and 64 MB. Mobile devices may additionally employ flash ROMs for secondary storage. Higher-end mobile devices, such as pen computers, use hard drives with sizes of the order of gigabytes. In mid-range devices, such as the iPAQ or Palm, approximately half of the primary memory is used for the kernel and operating system leaving the remaining memory for applications. This limited capacity is again used to trade off better performance for lower energy consumption.

Screen size and type. The use of LCD technology and viewable screen diagonal lengths between 2 and 10 in. are common characteristics of mobile devices. The CRT technology used in desktop monitors typically consumes approximately 120 W, whereas the LCD technology used in PDAs consumes only between 300 and 500 mW. As with the other characteristics, higher screen resolution is traded off for lower power consumption; however, future improvements in LCD technology may provide better resolution with little or no increase in power consumption.

Input mechanisms. The most common input mechanisms for mobile devices are built-in key-pads, pens, and touch-screen interfaces. Usually, PDAs contain software keyboards; newer PDAs may also support external keyboards. Some devices also use voice as an input mechanism. Mobility and portability of devices are primary factors in the design of these traditional interfaces for cell phones, PDAs, and pen computers. Human–computer interaction (HCI) is a topic of considerable research and impacts the marketability of a mobile device. For example, a cell phone that could also be used as a PDA should not require user input via keys or buttons in the PDA mode; rather, it should accept voice input.

Communication interfaces. As discussed above, mobile devices can support both wired and wireless communication interfaces, depending on their capabilities. We shall concentrate on wireless interfaces in this context. As far as mobile devices are concerned, wireless communication is either short range or long range. Short-range wireless technologies include infrared (IR), Global System for Mobile Communications (GSM), IEEE 802.11a/b/g, and Bluetooth. IR, which is part of the optical spectrum, requires line-of-sight communication whereas the other three, which are part of the radio spectrum, can function as long as the two devices are in radio range and do not require line of sight. Long-range wireless technologies include satellite communications, which are also part of the radio spectrum. Although wireless interfaces provide network connectivity to mobile devices, they pose some serious challenges when compared to wired interfaces. Frequent disconnections, low and variable bandwidth — and most importantly, increased security risks — are some of these challenges.

The discussion thus far clearly suggests that mobile computing is not limited to the technical challenges of reducing the size of the computer and adding a wireless interface to it. It encompasses the problems and solutions associated with enabling people to use the computing power of their devices anytime, anywhere, possibly with network connectivity.

36.2 Mobile Computing vs. Wired-Network Computing

We now compare mobile computing and wired-network computing from the network perspective. For the purposes of this discussion, we consider only the wireless networking aspect of mobile computing. We shall also use the terms *wired-network computing* and *wired computing* interchangeably. We shall compare mobile computing and wired computing based on layers 1 through 4 of the standard 7-layer Open Standards Interconnection (OSI) stack. Figure 36.1 shows the Physical, Data Link (comprising the Link Management and Medium Access Control sub-layers), Network, and Transport layers of the two stacks.

The Physical layer: In the network stack for mobile computing, the physical layer consists of two primary media — the radio spectrum and the optical spectrum. The radio spectrum is divided into licensed and unlicensed frequency bands. Cellular phone technologies use the licensed bands whereas technologies such as Bluetooth and IEEE 802.11b use the unlicensed band. The optical spectrum is mainly used by infrared devices. The network stack for wired computing consists of cable technologies such as coaxial cable and optical fiber.

The Medium Access Control (MAC) sub-layer: The most frequently adopted MAC mechanism in the wired computing network stack is the well-known Carrier Sense Multiple Access with Collision Detection (CSMA/CD). It is also well known that CSMA/CD cannot be directly applied to the mobile computing stack because it would cause collisions to occur at the receiver as opposed to the sender. To prevent this situation, researchers and practitioners have designed different mechanisms based on collision avoidance and synchronizing transmissions. CSMA with Collision Avoidance (CSMA/CA) helps transmitters determine if other devices around them are also preparing to transmit, and, if so, to avoid collisions by deferring transmission. Time Division Multiple Access (TDMA), Frequency Division Multiple Access (FDMA), Code Division Multiple Access (CDMA), and Digital Sense Multiple Access with Collision Detection (DSMA/CD) are other popular MAC protocols used by mobile device network stacks to coordinate transmissions.

The Link Management sub-layer: This layer is present in only a few network stacks of mobile devices. For example, the IEEE 802.11b standard describes only the Physical and MAC layers as part of the specification. Some of the link management protocols on mobile device network stacks are required to handle voice connections (usually connection-oriented links), in addition to primarily connection-less data links. The Logical Link Control and Adaptation Protocol (L2CAP) in Bluetooth is an example of such a protocol. GSM uses a variant of the well-known Link Access Protocol D-channel (LAPD) called $LAPD_m$. High-level Data Link Control (HDLC), Point-to-Point Protocol (PPP), and Asynchronous Transfer Mode (ATM) are the most popular data link protocols used in wired networks.

The Network layer: Mobility of devices introduces a new dimension to routing protocols, which reside in the network layer of the OSI stack. Routing protocols for mobile networks, both *ad hoc* and infrastructure supported, need to be aware of mobile device characteristics such as mobility and energy consumption. Unlike static devices, mobile devices cannot always depend on a static address, such as an IP address. This is because they need to be able to attach to different points in the network, public or private. Routing protocols such as Mobile IP [Perkins, 1997] enable devices to dynamically obtain an IP address and connect to any IP-based network while they are on the move. This solution requires the

Mobile	Layer	Wired
TCP & its variants, Wireless Transaction Protocol	TRANSPORT	TCP, TP4
IP, Mobile IP, Routing Protocols for MANETs	NETWORK	IP (with IPSec), CLNP
L2CAP (Bluetooth), $LAPD_m$ (GSM)	LINK	HDLC, PPP (IP), ATM
CSMA/CA (802.11b), TDMA (Bluetooth), TD-FDMA (GSM), CDMA	MAC	CSMA/CD (Ethernet), Token Ring, FDDI
Radio Transceiver (802.11b, Bluetooth, GSM) Optical Transceiver (IR, Laser)	PHYSICAL	Co-axial Cable, Optical Fiber

FIGURE 36.1 Network stack comparison of mobile and wired-network computing.

existence of a central network (i.e., home network) that tracks the mobile device and knows its current destination network. Routers are the linchpins of the Internet. They decide how to route incoming traffic based on addresses carried by the data packets. In *ad hoc* networks, no static routers exist. Many nodes in the network may have to perform the routing function because the routers may be mobile, and thus move in and out of range of senders and receivers. All of these considerations have focused research on developing efficient routing protocols in mobile *ad hoc* networks (MANET).

The Transport layer: TCP has been the protocol of choice for the Internet. TCP performs very well on wired networks, which have high bandwidth and low delay. However, research on TCP performance over wireless networks has shown that it typically fails if non-congestion losses (losses due to wireless channel errors or client mobility) occur on the wireless link. This is because TCP implicitly assumes that all losses are due to congestion and reduces the window on the sender. If the losses are not due to congestion, TCP unnecessarily reduces throughput, leading to poor performance. Solutions to this problem include designing new transport protocols, such as CentaurusComm [Avancha et al., 2002b], that are more mobile-aware and modifying TCP to make it more mobile-aware. Modified versions of TCP [Barke and Badrinath, 1995; Brown and Singh, 1997; Goff et al., 2000] are well known in the research community. The Wireless Transaction Protocol (WTP) is part of the well-known Wireless Application Protocol (WAP) stack and provides reliable data transmission using retransmission, segmentation, and reassembly, as required.

Both academia and industry have contributed significantly to mobile-computing research aimed at designing the best possible network stack that takes into account the challenges of reduced computer size and computing power, energy conservation, and low-bandwidth, high-delay wireless interfaces. The discussion in this section provides a glimpse of the solutions applied to the most significant layers of the wired-network stack in order to address these challenges.

36.3 M-Services Application Architectures

Mobile computing applications can be classified into three categories — client-server, client-proxy-server, and peer-to-peer — depending on the interaction model. Evolution of mobile applications started from common distributed object-oriented systems such as CORBA and DCOM [Sessions, 1997], which primarily follow client-server architecture. The emergence of heterogeneous mobile devices with varying capabilities has subsequently popularized the client-proxy-server architecture. Increasing computational capabilities of mobile devices and the emergence of *ad hoc* networks is leading to a rapid growth of peer-to-peer architectures, similar to Gnutella in the Internet.

In the client-server architecture, a large number of mobile devices can connect to a small number of servers residing on the wired network, organized as a cluster. The servers are powerful machines with high bandwidth, wired-network connectivity and the capability to connect to wireless devices. Primary data and services reside on and are managed by the server, whereas clients locate servers and issue requests. Servers are also responsible for handling lower level networking details such as disconnection and retransmission. The advantages of this architecture are simplicity of the client design and straightforward cooperation among cluster servers. The main drawback of this architecture is the prohibitively large overhead on servers in handling each mobile client separately in terms of transcoding and connection handling, thus severely affecting system scalability.

In the client-proxy-server architecture, a proxy is introduced between the client and the server, typically on the edge of the wired network. The logical end-to-end connection between each server and client is split into two physical connections, server-to-proxy and proxy-to-client. This architecture increases overall system scalability because servers only interact with a fixed number of proxies, which handle transcoding and wireless connections to the clients. There has been substantial research and industry effort [Brooks et al., 1995; Zenel, 1995; Bharadvaj et al., 1998; Joshi et al., 1996] in developing client-proxy-server architectures. Additionally, intelligent proxies [Pullela et al., 2000] may act as computational platforms for processing queries on behalf of resource-limited mobile clients.

Transcoding, i.e., conversion of data and image formats to suit target systems, is an important problem introduced by client-server and client-proxy-server architectures. Servers and proxies are powerful machines that, unlike mobile devices, can handle data formats of any type and image formats of high resolution. Therefore, data on the wired network must be transcoded to suit different mobile devices. It is therefore important for the server or proxy to recognize the characteristics of a client device. Standard techniques of transcoding, such as those included in the WAP stack, include XSLT [Muench and Scardina, 2001] and Fourier transformation. The W3C CC/PP standard [Klyne et al., 2001] enables clients to specify their characteristics when connecting to HTTP servers using profiles.

In the peer-to-peer architecture, all devices, mobile and static, are peers. Mobile devices may act servers and clients. *Ad hoc* network technologies such as Bluetooth allow mobile devices to utilize peer resources in their vicinity in addition to accessing servers on the wired network. Server mobility may be an issue in this architecture, and so the set of services available to a client is not fixed. This may require mobile devices to implement service discovery [Rekesh,1999; Chakraborty et al., 2002a], collaboration, and composition [Chakraborty et al., 2002b; Mao et al., 2001]. The advantage of this architecture is that each device may have access to more up-to-date, location-dependent information and may interact with peers without infrastructure support. The disadvantage of this architecture is the burden on the mobile devices in terms of energy consumption and network-traffic handling.

Client-server and client-proxy-server architectures remain the most popular models of practical use from both commercial and non-commercial perspectives. Both these architectures provide users with certain guarantees, such as connectivity, fixed bandwidth, and security, because of the inherent power of the proxies and servers. From a commercial perspective, they guarantee increased revenues to infrastructure and service providers, as the number of wireless users increases. Peer-to-peer architectures, which truly reflect the goal of anytime, anywhere computing, are largely confined to academia, but possess the potential to revolutionize mobile computing in the decades to come.

36.4 Mobile Computing Application Framework

In this section, we describe a comprehensive framework for enabling the development of a mobile application using one of the three architectures described above. Figure 36.2 depicts the different components of the framework. Depending on the selected model, some of the components may not be required to build a complete mobile application. However, other components, such as the communications layer, form an intrinsic part of any mobile application. The design of this framework takes into consideration such issues. We describe the different layers and components of the framework in the next few subsections.

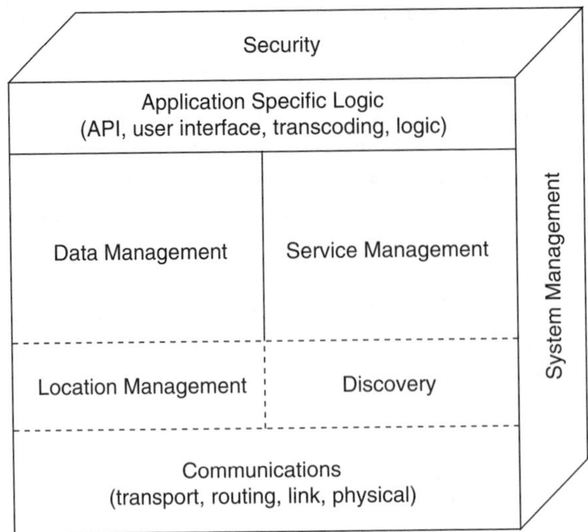

FIGURE 36.2 Mobile computing application framework.

36.4.1 Communications Layer

The communications layer in this framework encompasses the physical, MAC, link, network, and transport layers of the mobile computing stack illustrated in Figure 36.1. This layer is responsible for establishing and maintaining logical end-to-end connections between two devices, and for data transmission and reception.

The physical and MAC layers are primarily responsible for node discovery, and establishment and maintenance of physical connections between two or more wireless entities. These functions are implemented in different ways in different technologies. For example, in Bluetooth, node discovery is accomplished through the use of the *inquiry* command by the baseband (MAC) layer. In IEEE 802.11b, the MAC layer employs the RTS-CTS (i.e., Request-To-Send and Clear-To-Send) mechanism to enable nodes to discover each other when they are operating in the *ad hoc* mode. When IEEE 802.11b nodes are operating in *infrastructure* mode, the base station broadcasts beacons which the nodes use to discover the base station and establish physical connections with it. The establishment of physical connections is a process in which the nodes exchange operational parameters such as baud rate, connection mode (e.g., full-duplex or half-duplex), power mode (e.g., low-power or high-power), and timing information for synchronization, if required. In order to maintain the connection, some or all of these parameters are periodically refreshed by the nodes.

The link layer may not be part of the specifications of all wireless technologies. Some, such as IEEE 802.1.1b, use existing link layer protocols such as HDLC or PPP (for point-to-point connections) to establish data or voice links between the nodes. Bluetooth, on the other hand, uses a proprietary protocol, L2CAP, for establishing and maintaining links. This protocol is also responsible for other common link-layer functions such as framing, error correction, and quality of service. The task of the link layer is more difficult in wireless networks than in wired networks because of the high probability of errors either during or after transmission. Thus, error correction at the link layer must be robust enough to withstand the high bit-error rate of wireless transmissions.

The network layer in mobile computing stacks must deal with device mobility, which may cause existing routes to break or become invalid with no change in other network parameters. Device mobility may also be the cause of packet loss. For example, if the destination device, to which a packet is already *en route,* moves out of range of the network, then the packet must be dropped, Thus, both route establishment and route maintenance are important problems that the network layer must tackle. As the mobility of a network increases, so do route failures and packet losses. Thus, the routing protocol must be robust enough to either prevent route failures or recover from them as quickly as possible. In particular, routing protocols for mobile *ad hoc* networks have received considerable attention in the recent past. Many routing protocols for MANETs have been developed primarily for research purposes. These include *Ad hoc* On Demand Distance Vector (AODV) routing protocol, Dynamic Source Routing (DSR), and Destination-Sequenced Distance Vector (DSDV) routing protocol, However, for most applications that use some variant of the client-server model, the standard Internet Protocol (IP) is quite sufficient.

Mobile applications, unlike wired-network applications, tend to generate or require small amounts of data (of the order of hundreds or at the most thousands of bytes). Thus, protocols at the transport layer should be aware of the short message sizes, packet delays due to device mobility, and non-congestion packet losses. TCP is ill-suited for wireless networks. Numerous variations of TCP and transport protocols designed exclusively for wireless networks ensure that both ends of a connection agree that packet loss has occurred before the source retransmits the packet. Additionally, some of these protocols choose to defer packet transmission if they detect that current network conditions are unsuitable.

The functionality of the communications layer in this framework is usually provided by the operating system running on the mobile device. Therefore, the mobile application can directly invoke the lower level system functions via appropriate interfaces.

36.4.2 Discovery Layer

The discovery layer helps a mobile application discover data, services, and computation sources. These may reside in the vicinity of the mobile device or on the Internet. Due to resource constraints and mobility, mobile devices may not have complete information about all currently available sources. The discovery layer assumes that the underlying network layer can establish a logical end-to-end connection with other entities in the network. The discovery layer provides upper layers with the knowledge and context of available sources.

There has been considerable research and industry effort in service discovery in the context of wired and wireless networks. Two important aspects of service discovery are the *discovery architecture* and the *service matching mechanism.*

Discovery architectures are primarily of two types: *lookup-registry-based* and *peer-to-peer.* Lookup-registry-based discovery protocols work by registering information about the source at a registry. Clients query this registry to obtain knowledge about the source (such as its location, how to invoke it etc.). This type of architecture can be further subdivided into two categories: centralized registry-based, and federated or distributed registry-based architectures. A centralized registry-based architecture contains one monolithic centralized registry, whereas a federated registry-based architecture consists of multiple registries distributed across the network. Protocols such as Jini [Arnold et al., 1999], Salutation and Salutation-lite, UPnP [Rekesh, 1999], UDDI, and Service Location Protocol [Veizades et al., 1997] are examples of a lookup-registry-based architecture.

Peer-to-peer discovery protocols query each node in the network to discover available services on that node. These types of protocols treat each node in the environment equally in terms of functional characteristics. Broadcasting of requests and advertisements to peers is a simple, albeit inefficient, service-discovery technique in peer-to-peer environments. Chakraborty et al. [2002a] describe a distributed, peer-to-peer service discovery protocol using caching that significantly reduces the need to broadcast requests and advertisements. Bluetooth Service Discovery Protocol (SDP) is another example of a peer-to-peer service discovery protocol. In SDP, services are represented using 128-bit unique identifiers. SDP does not provide any information on how to invoke the service. It only provides information on the availability of the service on a specific device.

The service discovery protocols discussed in this section use simple interface, attribute, or unique identifier-based matching techniques to locate appropriate sources. Jini uses interface matching and SDP uses identifier matching, whereas the Service Location Protocol and Ninja Secure Service Discovery Systems discover services using attribute-based matching. The drawbacks of these techniques include lack of rich representation of services, inability to specify constraints on service descriptions, lack of inexact matching of service attributes, and lack of ontology support [Chakraborty et al., 2001]. Semantic matching is an alternative technique that addresses these issues. DReggie [Chakraborty et al., 2001] and Bluetooth Semantic Service Discovery Protocol (SeSDP) [Avancha et al., 2002a] both use a semantically rich language called DARPA Agent Markup Language (DAML) to describe and match both services and data. Semantic descriptions of services and data allow greater flexibility in obtaining a match between the query and the available information. Matching can now be inexact. This means that parameters such as functional characteristics, and hardware and device characteristics of the service provider may be used in addition to service or data attributes to determine whether a match can occur.

36.4.3 Location Management Layer

The location management layer deals with providing location information to a mobile device. Location information dynamically changes with mobility of the device and is one of the components of context awareness. It can be used by upper layers to filter location-sensitive information and obtain location-specific answers to queries, e.g., weather of a certain area and traffic condition on a road. The current location of a device relative to other devices in its vicinity can be determined using the discovery layer or the underlying communications layer. Common technologies use methods such as triangulation and signal-strength measurements for location determination. GPS [Hofmann-Wellenhof et al., 1997] is a well-known example of the use of triangulation based on data received from four different satellites. Cell phones use cell tower information to triangulate their position. On the other hand, systems such as RADAR [Bahl and Padmanabhan, 2000], used for indoor location tracking, work as follows. Using a set of fixed IEEE 802.11b base stations, the entire area is mapped. The map contains (x, y) coordinates and the corresponding signal strength of each base station at that coordinate. This map is loaded onto the mobile device. Now, as the user moves about the area, the signal strength from each base station is measured. The pattern of signal strengths from the stored map that most closely matches the pattern of measured signal strengths is chosen. The location of the user is that corresponding to the (x, y) coordinates associated with the stored pattern. Outdoor location management technologies have achieved technical maturity and have been deployed in vehicular and other industrial navigational systems. Location management, indoor and outdoor, remains a strong research field with the rising popularity of technologies such as IEEE 802.11b and Bluetooth.

The notion of location can be dealt with at multiple scales. Most "location determination" techniques actually deal with position determination, with respect to some global (latitude or longitude) or local (distances from the "corner" of a room) grid. Many applications are not interested in the absolute position as much as they are in higher-order location concepts (inside or outside a facility, inside or outside some jurisdictional boundary, distance from some known place, at a mountaintop, in a rain forest region, etc.) Absolute position determinations can be combined with GIS-type data to infer locations at other levels of granularity.

Expanding the notion of location further leads us to consider the notion of context. Context is any information that can be used to characterize the situation of a person or a computing entity [Dey and Abowd, 2000]. So, context covers data such as location, device type, connection speed, and direction of movement. Arguably, context even involves a user's mental state (beliefs, desires, intentions, etc.). This information can be used by the layers described next for data and service management. However, the privacy issues involved are quite complex. It is not clear who should be allowed to gather such information, under what circumstances should it be revealed, and to whom. For instance, a user may not want his or her GPS chip to reveal his or her current location, except to emergency response personnel. Some of these issues, specifically related to presence and availability, are being discussed in the PAM working group of Parlay. A more general formulation of such issues can be found in the recent work of Chen et al. [2003] who are developing OWL-based policies and a Decision-Logic-based reasoner to specify and reason about a user's privacy preferences as related to context information.

36.4.4 Data Management Layer

The data management layer deals with access, storage, monitoring, and data manipulation. Data may reside locally and also on remote devices. Similar to data management in traditional Internet computing, this layer is essential in enabling a device to interact and exchange data with other devices located in its vicinity and elsewhere on the network. The core difference is that this layer must also deal with mobile computing devices. Such devices have limited battery power and other resources in comparison to their desktop counterparts. The devices also communicate over wireless logical links that have limited bandwidth and are prone to frequent failures. Consequently, the data management layer often attempts to extend data management solutions for Internet computing by primarily addressing mobility and disconnection of a mobile computing device.

Work on data management can be classified along four orthogonal axes [Ozsu and Valduriez, 1999; Dunham and Helal, 1995]: autonomy, distribution, heterogeneity, and mobility. We can apply the classification to compare three architecture models adopted by existing data management solutions.

The client-server model is a two-level architecture with data distributed among servers. Servers are responsible for data replication, storage, and update. They are often fixed and reside on the wired infrastructure. Clients have no autonomy as they are fully dependent on servers, and may or may not be mobile and heterogeneous. This model was the earliest adopted approach for distributed file and database systems because it simplifies data management logic and supports rapid deployment [Satyanarayanan et al., 1990]. The model delegates all data management responsibility to only a small subset of devices, the servers. Additionally, the model addresses the mobility problem by simply not dealing with it or by using traditional time-out methods.

The client-proxy-server model extends the previous approach by introducing an additional level in the hierarchy. Data remains distributed on servers residing on a wired infrastructure. Clients still depend on servers, and may or may not be mobile and heterogeneous. However, a proxy, residing on the wired infrastructure, is placed between clients and servers. The proxy takes on a subset of server responsibilities, including disconnection management, caching, and transcoding. Consequently, servers no longer differentiate between mobile and fixed clients. They can treat all clients uniformly because they communicate with devices on the wired infrastructure only. Proxy devices are then responsible for delivering data to clients and for maintaining sessions when clients change locations [Dunham et al., 1997].

The peer-to-peer model takes a completely different approach from the other two models. This model is highly autonomous as each computing device must be able to operate independently. There is no distinction between servers and clients, and their responsibilities. The model also lies at the extreme of the other three axes because data may reside on any device, and each device can be heterogeneous and mobile. In this model, any two devices may interact with each other [Perich et al., 2002]. Additionally, unlike client-server-based approaches, the model is open in that there is no strict set of requirements that each device must follow. This may cause the data management layer to be implemented differently on each device. Consequently, each peer must address both local and global data-management issues; the latter are handled by servers or proxies in client-server-based models.

Local data management, logically operating at the end-user level, is responsible for managing degrees of disconnection and query processing. The least degree of disconnection encourages the device to constantly interact with other devices in the environment. The highest degree represents the state when the device only utilizes its local resources. The mobility of a device can affect both the type of queries as well as the optimization techniques that can be applied. Traditional query-processing approaches advocate location transparency. These techniques only consider aspects of data transfer and processing for query optimization. On the other hand, in the mobile computing environment, query-processing approaches promote *location awareness* [Kottkamp and Zukunft, 1998]. For example, a mobile device can ask for the location of the closest Greek restaurant, and the server should understand that the starting point of the search refers to the current position of the device [Perich et al., 2002; Ratsimor et al., 2001].

Global data management, logically operating at the architecture-level, deals with data addressing, caching, dissemination, replication, and transaction support. As devices move from one location to another or become disconnected, it is necessary to provide a naming strategy to locate a mobile station and its data. There have traditionally been three approaches for data addressing: *location-dependent, location-transparent,* and *location-independent* [Pinkerton et al., 1990; Sandberg et al., 1985]. To allow devices to operate disconnected, they must be able to cache data locally. This requirement introduces two challenges: *data selection* and *data update.* Data selection can be explicit [Satyanarayanan et al., 1990] or proactively inferred [Perich et al., 2002]. In the former approach, a user explicitly selects files or data that must be cached. The latter approach automatically predicts and proactively caches the required information. Data update of local replicas usually requires a weaker notion of consistency as the mobile device may have to operate on stale data without the knowledge that the primary copy was altered. This is especially the case when devices become disconnected from the network and cannot validate consistency of their data. Either subscription-based callbacks [Satyanarayanan et al., 1990] or latency- and recency-based refreshing [Laura Bright and Louiqa Raschid, 2002] can address this issue. In subscription-based approaches, a client requests the server to notify it (the client) when a particular datum is modified. In turn, when a server modifies its data, it attempts to inform all clients subscribed to that data. In the latter approaches, a client or proxy uses timestamp information to compare its local replicas with remote copies in order to determine when to refresh its copy.

Data dissemination models are concerned with read-only transactions where mobile clients can "pull" information from sources, or the sources can "push" data to them automatically [Acharya et al., 1995]. The latter is applicable when a group of clients share the same sources and they can benefit from accepting responses addressed to other peers.

To provide consistent and reliable computing support, the data management layer must support transaction and replica control. A transaction consists of a sequence of database operations executed as an atomic action [Ozsu and Valduriez, 1999]. This definition encompasses the four important properties of a transaction: atomicity, consistency, isolation, and durability (i.e., ACID properties). Another important property of a transaction is that it always terminates, either by committing the changes or by aborting all updates. The principal concurrency-control technique used in traditional transaction management relies on locking [Ozsu and Valduriez, 1999; Eswaran et al., 1976]. In this approach, all devices enter a state in which they wait for messages from one another. Because mobile devices may become involuntarily disconnected, this technique raises serious problems such as termination blocking and reduction in the availability of data. Current-generation solutions to the mobile transaction management problem often relax the ACID [Walborn and Chrysanthis, 1997] properties or propose completely different transaction processing techniques [Dunham et al., 1997].

Having relaxed the ACID properties, one can no longer guarantee that all replicas are synchronized. Consequently, the data management layer must address this issue. Traditional replica control protocols, based on voting or lock principles [Ellis and Floyd, 1983], assume that all replica holders are always reachable. This is often invalid in mobile environments and may limit the ability to synchronize the replica located on mobile devices. Approaches addressing this issue include data division into volume groups and the use of versions for pessimistic [Demers et al., 1994] or optimistic updates [Satyanarayanan et al., 1990; Guy et al., 1998]. Pessimistic approaches require epidemic or voting protocols that first modify the primary copy before other replicas can be updated and their holders can operate on them. On the other hand, optimistic replication allows devices to operate on their replicas immediately, which may result in a conflict that will require a reconciliation mechanism [Holliday et al., 2000]. Alternatively, the conflict must be avoided by calculating a voting quorum [Keleher and Cetintemel, 1999] for distributed data objects. Each replica can obtain a quorum by gathering weighted votes from other replicas in the system and by providing its vote to others. Once a replica obtains a voting quorum, it is assured that a majority of the replicas agree with the changes. Consequently, the replica can commit its proposed updates.

36.4.5 Service Management Layer

Service management forms another important component in the development of a mobile application. It consists of service discovery monitoring, service invocation, execution management, and service fault management. The service management layer performs different functions depending on the type of mobile application architecture. In the client-server architecture, most of the management (e.g., service execution state maintenance, computation distribution, etc.) is done by the server side of the application. Clients mostly manage the appropriate service invocation, notifications, alerts, and monitoring of local resources needed to execute a query. In the client-proxy-server architecture, most of the management (session maintenance, leasing, and registration) is done at the proxy or the lookup server. Disconnections are usually managed by tracking the state of execution of a service (mostly at the server side) and by retransmitting data once connection is established. One very important function of this layer is to manage the integration and execution of multiple services that might be required to satisfy a request from a client; this is referred to as *service composition*. Such requests usually require interaction of multiple services to provide a reply. Most of the existing service management platforms [Mao et al., 2001; Mennie and Pagurek, 2000] for composite queries are centralized and oriented toward services in the fixed wired infrastructure. Distributed broker-based architectures for service discovery, management, and composition in wireless *ad hoc* environments are current research topics [Chakraborty et al., 2002b]. Fault tolerance and scalability are other important components, especially in environments with many short-lived services. The management platform should degrade gracefully as more services become unavailable. Solutions for managing services have been incorporated into service discovery protocols designed for wired networks but not for mobile environments.

36.4.5.1 Service Transaction Management

This sub-layer deals with the management of transactions associated with m-services, i.e., services applicable to mobile computing environments. We discuss service transaction management as applied to client-server, client-proxy-server, and peer-to-peer architectures. Service transaction management in mobile computing environments is based on the same principles used by e-commerce transaction managers in the Internet. These principles are usually part of a transaction protocol such as the Contract Net Protocol [FIPA, 2001]. A Contract Net Protocol involves two entities, the buyer (also known as manager) and the seller (also known as contractor), who are interested in conducting a transaction. The two entities execute actions as specified in the protocol at each step of the transaction. Examples of these actions include *Call for Proposal (CFP), Refuse, Propose, Reject-Proposal, Accept-Proposal, Failure,* and *Inform-Done.* In the wired computing environment, the two entities execute all actions explicitly. In a mobile computing environment, complete execution of the protocol may be infeasible due to memory and computational constraints. For example, the *Refuse* action, performed by a seller who refuses the *CFP*, is implicit if the seller does not respond to the *CFP*. Thus, service transaction managers on mobile computing devices use simplified versions of transaction protocols designed for the Internet [Avancha et al., 2003].

In mobile computing environments using the client-server or client-proxy-server architecture, the service transaction manager would choose to use the services available on the Internet to successfully complete the transaction. For example, if a person is buying an airline ticket at the airport using her PDA, she could invoke the airline software's payment service and specify her bank account as the source of payment. On the other hand, in a peer-to-peer environment, there is no guarantee of a robust, online payment mechanism. In such situations, the transaction manager may choose other options, such as micropayments. For example, if a person were buying a music video clip from another person for $1, he may pay for it using digital cash. Both industry and academia have engaged in core research in the area of micropayments in past few years [Cox et al., 1995; Choi et al., 1997].

Three of the most important e-service transactional features that must be applied to m-services are: *Identification, Authentication,* and *Confidentiality.* Every entity in a mobile environment must be able to uniquely and clearly identify itself to other entities with whom it wishes to transact. Unlike devices on the Internet, a mobile device may not be able to use its IP address as a unique identifier. Every mobile device must be able to authenticate transaction messages it receives from others. In a mobile environment where air is the primary medium of communication, anybody can eavesdrop and mount man-in-the-middle attacks against others in radio range. Confidentiality in a mobile environment is achieved through encryption mechanisms. Messages containing payment and goods information must be encrypted to prevent theft of the data. However, mobile devices are constrained by computational and memory capacities to perform expensive computations involved in traditional encryption mechanisms. Technologies such as Smartcards [Hansmann et al., 2000] can help offload the computational burden from the mobile device at the cost of higher energy consumption.

36.4.6 Security Plane

Security has greater significance in a mobile environment than in a wired environment. The two main reasons for this are the lack of any security on the transmission medium, and the real possibility of theft of a user's mobile device.

Despite the increased need for security in mobile environments, the inherent constraints on mobile devices have prevented large-scale research and development of secure protocols. Lightweight versions of Internet security protocols are likely to fail because they ignore or minimize certain crucial aspects of the protocols in order to save computation and memory. The travails of the Wired Equivalent Privacy (WEP) protocol designed for the IEEE 802.11b are well known [Walker, 2000]. The IEEE 802.11b working group has now released WEP2 for the entire class of 802.1x protocols. Bluetooth also provides a link layer security protocol consisting of a *pairing procedure* that accepts a user-supplied passkey to generate an initialization key. The initialization key is used to calculate a link key, which is finally used in a challenge-response sequence, after being exchanged. The current Bluetooth security protocol uses procedures that have low computation complexity, which makes them susceptible to attacks. To secure data at the routing layer in client-server and client-proxy-server architectures, IPSec [Kent and Atkinson, 1998] is used in conjunction with Mobile IP. Research in securing routing protocols for networks using peer-to-peer architectures has resulted in interesting protocols such as Ariadne [Hu et al., 2002] and Security-Aware Ad hoc Routing [Yi et al., 2001]. The Wireless Transport Layer Security protocol is the only known protocol for securing transport layer data in mobile networks. This protocol is part of the WAP stack. WTLS is a close relative of the Secure Sockets Layer protocol that is *de jure* in securing data in the Internet. Transaction and application layer security implementations are also based on SSL.

36.4.7 System Management Plane

The system management plane provides interfaces so that any layer of the stack in Figure 36.2 can access system level information. System level information includes data such as current memory level, battery power, and the various device characteristics. For example, the routing layer might need to determine whether the current link layer in use is IEEE 802.11b or Bluetooth to decide packet sizes. Transaction managers will use memory information to decide whether to respond to incoming transaction requests or to prevent the user from sending out any more transaction requests. The application logic will acquire device characteristics from the system management plane to inform the other end (server, proxy, or peer) of the device's screen resolution, size, and other related information. The service discovery layer might use system-level information to decide whether to use semantic matching or simple matching in discovering services.

36.5 Conclusions

Mobile devices are becoming popular in each aspect of our everyday life. Users expect to use them for multiple purposes, including calendaring, scheduling, checking e-mail, and browsing the web. Current-generation mobile devices, such as iPAQs, are powerful enough to support more versatile applications that may already exist on the Internet. However, applications developed for the wired Internet cannot be directly ported onto mobile devices. This is because some of the common assumptions made in building Internet applications, such as the presence of high-bandwidth disconnection-free network connections, and resource-rich tethered machines and computation platforms, are not valid in mobile environments. Mobile applications must take these issues into consideration. In this chapter, we have discussed the modifications to each layer of the OSI stack, which are required to enable mobile devices to communicate with wired networks and other mobile devices. We have also discussed three popular application architectures, i.e., client-server, client-proxy-server, and peer-to-peer, that form an integral part of any mobile application. Finally, we have presented a general framework that mobile applications should use in order to be functionally complete, flexible, and robust in mobile environments. The framework consists of an abstracted network layer, discovery layer, location management, data management, service management, transaction management, and application-specific logic. Depending on the architecture requirements, each application may use only a subset of the described layers. Moreover, depending on the type of architecture, different solutions apply for the different layers. In conclusion, we have presented a sketch of the layered architecture and technologies that make up the state of the art of mobile computing and mobile applications. Most of these have seen significant academic research and, more recently, commercial deployment. Many other technologies are maturing as well, and will move from academic and research labs into products. We feel that the increasing use of wireless local and personal area networks (WLANs and WPANs), higher-bandwidth wireless telephony, and a continued performance/price improvement in handheld and wearable devices will lead to a significant increase in the deployment of mobile computing applications in the near future, even though not all of the underlying problems would have completely wrapped up solutions in the short term.

Acknowledgments

This work was supported in part by NSF awards IIS 9875433, IIS 0209001, and CCR 0070802, the DARPA DAML program, IBM, and Fujitsu Labs of America, Inc.

References

Acharya, Swarup, Rafael Alonso, Michael Franklin, and Stanley Zdonik. Broadcast Disks: Data Management for Asymmetric Communication Environments. In Michael J. Carey and Donovan A. Schneider, Eds., *ACM SIGMOD International Conference on Management of Data,* pp. 199–210, San Jose, CA, June 1995. ACM Press.

Arnold, Ken, Bryan O'Sullivan, Robert W. Scheifler, Jim Waldo and Ann Wollrath. *The Jini Specification (The Jini Technology).* Addison-Wesley, Reading, MA, June 1999.

Avancha, Sasikanth, Pravin D'Souza, Filip Perich, Anupam Joshi, and Yelena Yesba. P2P M-Commerce in pervasive environments. *ACM SIGecom Exchanges,* 3(4): 1–9, January 2003.

Avancha, Sasikanth, Anupam Joshi, and Tim Finin. Enhanced service discovery in Bluetooth. *IEEE Computer,* 35(6): 96–99, June 2002.

Avancha, Sasikanth, Vladimir Korolev, Anupam Joshi, Timothy Finin, and Y. Yesha. On experiments with a transport protocol for pervasive computing environments. *Computer Networks,* 40(4): 515–535, November 2002.

Bahl, Paramvir and Venkata N. Padmanabhan. RADAR: An in-building RF-based user location and tracking system. In *IEEE INFOCOM,* Vol. 2, pp. 775–784, Tel Aviv, Israel, March 2000.

Barke, Ajay V. and B. R. Badrinath. I-TCP: Indirect TCP for Mobile Hosts. In *15th International Conference on Distributed Computing Systems,* pp. 136–113, Vancouver, BC, Canada, June 1995. IEEE Computer Society Press.

Bharadvaj, Harini, Anupam Joshi, and Sansanee Auephanwiriyakyl. An Active Transcoding Proxy to Support Mobile Web Access. In *17th IEEE Symposium on Reliable Distributed Systems (SRDS),* pp. 118–123, West Lafayette, IN, October 1998.

Brooks, Charles, Murray S. Mazer, Scott Meeks, and Jim Miller. Application-Specific Proxy Servers as HTTP Stream Transducers. In *4th International World Wide Web Conference,* pp. 539–548, Boston, MA, December 1995.

Bright, Laura and Louiqa Raschid. Using Latency-Recency Profiles for Data Delivery on the Web. In *International Conference on Very Large Data Bases (VLDB),* pp. 550–561, Morgan Kaufmann, Kowloon Shangri-La Hotel, Hong Kong, China, August 2002.

Brown, Kevin and Suresh Singh. M-TCP: TCP for mobile cellular networks. *ACM Computer Communications Review,* 27(5): 19–43, October 1997.

Chakraborty, Dipanjan, Anupam Joshi, Tim Finin, and Yelena Yesha. GSD: A novel group-based service discovery protocol for MANETS. In *4th IEEE Conference on Mobile and Wireless Communications Networks (MWCN),* pp. 301–306. Stockholm, Sweden, September 2002a.

Chakraborty, Dipanjan, Filip Perich, Sasikanth Avancha, and Anupam Joshi. DReggie: A smart Service Discovery Technique for E-Commerce Applications. In *Workshop at 20th Symposium on Reliable Distributed Systems,* October 2001.

Chakraborty Dipanjan, Filip Perich, Anupam Joshi, Tim Finin, and Yelena Yesha. A Reactive Service Composition Architecture for Pervasive Computing Environments. In *7th Personal Wireless Communications Conference (PWC),* pp. 53–62, Singapore, October 2002b.

Chen, Harry, Tim Finin, and Anupam Joshi. Semantic Web in a Pervasive Context-Aware Architecture. In *Artificial Intelligence in Mobile System at UBICOMP,* Seattle, WA, October 2003.

Choi, Soon-Yong, Dale O. Stahl, and Andrew B. Whinston. Cyberpayments and the Future of Electronic Commerce. In *International Conference on Electronic Commerce, Cyberpayments Area,* 1997.

Cox, Benjamin, Doug Tygar, and Marvin Sirbu. NetBill Security and Transaction Protocol. In *1st USENIX Workshop of Electronic Commerce.,* pp. 77–88, New York. July 1995.

Demers, Alan, Karin Petersen, Mike Spreitzer, Douglas Terry, Marvin Theimer, and Brent Welch. The Bayou Architecture: Support for Data Sharing among Mobile Users. In *Proceedings IEEE Workshop on Mobile Computing Systems and Applications,* pp. 2–7, Santa Cruz, CA, December 8–9, 1994.

Dey, Anind K. and Gregory D. Abowd, Eds. Towards a Better Understanding of Context and Context-Awareness. In *Proceedings of the CHI 2000,* The Hague, Netherlands, April 2000. Also in GVU Technical Report GIT-99-22, College of Computing, Georgia Institute of Technology, Atlanta, GA.

Dunham, Margaret H. and Abdelsalam (Sumi) Helal. Mobile Computing and Databases: Anything New? In *ACM SIGMOD Record,* pp. 5–9. ACM Press, New York, December 1995.

Dunham, Margaret, Abdelsalam Helal, and Santosh Balakrishnan. A mobile transaction model that captures both the data movement and behavior. *ACM/Baltzer Journal of Mobile Networks and Applications,* 2(2): 149–162,1997.

Ellis, Carla Schlatter and Richard A. Floyd. The Roe File System. In *3rd Symposium on Reliability in Distributed Software and Database Systems,* pp. 175–181, Clearwater Beach, FL, October 1983. IEEE.

Eswaran, Kapali P., Jim Gray, Raymond A. Lorie, and Irving L. Traiger. The notion of consistency and predicate locks in a database system. *Communications of the ACM,* 19(11): 624–633, December 1976.

FIPA. FIPA Contract Net Interaction Protocol Specification. World Wide Web, http://www.fipa.org/specs/fipa00029/XC00029F.pdf,2001.

Goff, Tom, James Moronski, Dhananjay S. Phatak, and Vipul Gupta. Freeze-TCP: A True End-to-End TCP Enhancement Mechanism for Mobile Environments. In *INFOCOM,* Vol. 3, pp. 1537–1545, Tel Aviv, Israel, March 2000.

Guy, Richard, Peter Reiher, David Ratner, Michial Gunter, Wilkie Ma, and Gerald Popek. Rumor: Mobile Data Access through Optimistic Peer-to-Peer Replication. In *Workshop on Mobile Data Access in conjunction with 17th International Conference on Conceptual Modeling (ER),* pp. 254–265, Singapore, November 1998. World Scientific.

Hansmann, Uwe, Martin S. Nicklous, Thomas Schack, and Frank Seliger. *Smart Card Application Development using Java.* Springer-Verlag, New York, 2000.

Hofmann-Wellenhof, Bernhard, Herbert Lichtenegger, and James Collins. *Global Positioning System: Theory and Practice,* 4th ed., Springer-Verlag, New York, May 1997.

Holliday, JoAnne, Divyakant Agrawal, and Amr El Abbadi. Database Replication Using Epidemic Communication. In Arndt Bode, Thomas Ludwig II, Wolfgang Karl, and Ronal Wism, Eds., *6th Euro-Par-Conference,* Vol. 1900, pp. 427–434, Munich, Germany, September 2000. Springer.

Hu, Yih-Chun, Adrian Perrig, and David B. Johnson. Ariadne: A Secure On-Demand Routing Protocol for *Ad Hoc* Networks. In *8th ACM International Conference on Mobile Computing and Networking,* pp. 12–23, Atlanta, GA, September 2002. ACM Press.

Joshi, Anupam, Ranjeewa Weerasinghe, Sean P. McDermott, Bun K. Tan, Gregory Bernhardt, and Sanjiva Weerawarana. Mowser: Mobile Platforms and Web Browsers. *Bulletin of the Technical Committee on Operating Systems and Application Environments (TCOS),* 8(1), 1996.

Keleher, Peter J. and Ugur Cetintemel. Consistency management in Deno. *ACM Mobile Networks and Applications,* 5: 299–309, 1999.

Kent, Stephen and Randall Atkinson. IP Encapsulating Security Payload. World Wide Web, http://www.ietf.org/rfc/rfc2406.txt, November 1998.

Klyne, Graham, Franklin Raynolds, and Chris Woodrow. Composite Capabilities/Preference Profiles (CC/PP): Structure and Vocabularies. World Wide Web, http://www.w3.org/TR/CCPP-struct-vocab/, March 2001.

Kottkamp, Hans-Erich and Olaf Zukunft. Location-aware query processing in mobile database systems. In *ACM Symposium on Applied Computing,* pp. 416–423, Atlanta, GA, February 1998.

Mao, Zhuoqing Morley, Eric A. Brewer, and Randy H. Katz. Fault-tolerant, Scalable, Wide-Area Internet Service Composition. Technical report, CS Division, EECS Department, University of California, Berkeley, January 2001.

Mennie, David and Bernard Pagurek. An Architecture to Support Dynamic Composition of Service Components. In *5th International Worshop on Component-Oriented Programming,* June 2000.

Muench, Steve and Mark Scardina. XSLT Requirements. World Wide Web, http://www.w3.org/TR/xslt20req, February 2001.

Ozsu, M. Tamer and Patrick Valduriez. *Principles of Distributed Database Systems,* 2nd ed., Prentice Hall, Hillsdale, NJ, 1999.

Perich, Filip, Sasikanth Avancha, Dipanjan Chakraborty, Anupam Joshi, and Yelena Yesha. Profile Driven Data Management for Pervasive Environments. In *3rd International Conference on Database and Expert Systems Applications (DEXA),* pp. 361–370, Aix en Provence, France, September 2002.

Perkins, Charles E. *Mobile IP Design Principles and Practices.* Wireless Communication Series, Addison-Wesley, Reading, MA, 1997.

Pinkerton, C. Brian, Edward D. Lazowska, David Notkin, and John Zahorjan. A Heterogeneous Distributed File System. In *10th International Conference on Distributed Computing Systems,* pp. 424–431, May 1990.

Pullela, Chaitanya, Liang Xu, Dipanjan Chakraborty, and Anupam Joshi. A Component based Architecture for Mobile Information Access. In *Workshop in conjunction with International Conference on Parallel Processing,* pp. 65–72, August 2000.

Ratsimor, Olga, Vladimir Korolev, Anupam Joshi, and Timothy Finin. Agents2Go: An Infrastructure for Location-Dependent Service Discovery in the Mobile Electronic Commerce Environment. In *ACM Mobile Commerce Workshop in Conjunction with MobiCom,* pp. 31–37, Rome, Italy, July 2001.

Rekesh, John. UPnP, Jini and Salutation — A look at some popular coordination frameworks for future network devices. Technical report, California Software Labs, 1999. URL http://www.cswl.com/whitepaper/tech/upnp.html.

Sandberg, Russel, David Goldberg, Steve Kleiman, Dan Walsh, and Bob Lyon. Design and Implementation of the Sun Network Filesystem. In *Summer USENIX Conference,* pp. 119–130, Portland, OR, 1985.

Satyanarayanan, Mahadev, James J. Kistler, Puneet Kumar, Maria E. Okasaki, Ellen H. Siegel, and David C. Steere. Coda: A Highly Available File System for a Distributed Workstation Environment. *IEEE Transactions on Computers,* 39(4): 447–459, 1990.

Sessions, Roger. *COM and DOOM: Microsoft's Vision for Distributed Objects.* John Wiley & Sons, New York, October 1997.

Veizades, John, Erik Guttman, Charles E. Perkins, and Scott Kaplan. RFC 2165: Service location protocol, June 1997.

Walborn, Gary D. and Panos K. Chrysanthis. PRO-MOTION: Management of Mobile Transactions. In *ACM Annual Symposium on Applied Computing,* pp. 101–108, San Jose, CA, February 1997.

Walker, Jesse R. Unsafe at any key size; An analysis of the WEP encapsulation. IEEE Document 802.11–00/362, October 2000.

Yi, Seung, Prasad Naldurg and Robin Kravets. Security-aware *ad hoc* Routing for Wireless Networks. In *2nd RCM Symposium on Mobile Ad Hoc Networking and Computing,* pp. 299–302, Long Beach. California. USA., October 2001.

Zenel, Bruce. A Proxy Based Filtering Mechanism for The Mobile Environment. Ph.D. thesis, Department of Computer Science, Columbia University, New York, December 1995.

37

Pervasive Computing

Sumi Helal
Choonhwa Lee

CONTENTS

37.1 The Vision of Pervasive Computing 37-1
37.2 Pervasive Computing Technologies 37-2
37.2.1 Device Technology 37-3
37.2.2 Network Technology 37-4
37.2.3 Environment Technology 37-4
37.2.4 Software Technology 37-5
37.2.5 Information Access Technology 37-6
37.2.6 User Interface Technology 37-7
37.3 Ubiquitous Computing Systems 37-7
37.3.1 Active Bat 37-8
37.3.2 Classroom 2000 37-9
37.3.3 Lancaster Guide System 37-9
37.3.4 Matilda's Smart House 37-11
37.4 Conclusion 37-12
References 37-13

37.1 The Vision of Pervasive Computing

In the early 1990s, Mark Weiser called for a paradigm shift to *ubiquitous computing* in his seminal article [Weiser, 1991] opening with "The most profound technologies are those that disappear. They weave themselves into the fabric of everyday life until they are indistinguishable from it." This insight was influential enough to reshape computer research, and since then the research community has put forth enormous efforts to pursue his vision.

Writing and *electricity* are examples of such ubiquitous technologies "disappearing" into the background of our daily activities [Weiser and Brown, 1996]. Writing, perhaps our first information technology, is found in every corner of life in the "civilized" world: dashboards, road signs, toothbrushes, clothes, and even candy wrappers. As a part of physical objects to help us use them better, writing remains almost unnoticeable until for some specific reason we need to call them to the center of our consciousness. In other words, our everyday practices are not interfered with by their surrounding presence, but their meanings are readily available so that we can catch the conveyed message when necessary. For example, little attention is given to exit signs on a highway until we approach our destination. Also, dashboard speedometers on jammed downtown roads are rarely given much focus. Electricity permeates every aspect of our world, and has become so inseparable from modern life that we tend to forget about it and take it for granted.

Likewise, computation will be available everywhere, but unobtrusively, in the future ubiquitous computing world. It will be embedded in physical objects such as clothing, coffee cups, tabletops, walls, floors, doorknobs, roadways, and so forth. For example, while reading a morning newspaper, our coffee placed on a tabletop will be kept warm at our favorite temperature. A door will open automatically by sensing the current in our hand; another door, more intelligently, will detect our fingerprints or sense our

1-58488-381-2/05/$0.00+$1.50
© 2005 by CRC Press LLC

approach. This seamless computation will be invisibly integrated into physical artifacts within the environment, and will always be ready to serve us without distracting us from our daily practices. In a future world saturated with computation [Satyanarayanan, 2001], time-saving, nondistracting ease of use requiring minimum human attention will be a prime goal of ubiquitous computing and an inspiration to innovation as well. Eventually, computers will be everywhere and "nowhere."

The computer research community responded to Mark Weiser's call for a paradigm shift to ubiquitous computing by eagerly investing enormous resources to investigate enabling technologies. Although his view has been enthusiastically followed by numerous ensuing research projects, it often has been given different terms: *pervasive, invisible, calm, augmented, proactive, ambient*, and so forth to reflect their scope or emphasis on particular areas. Among them, the term *pervasive computing* gained popularity in the mid-1990s, and is now used with ubiquitous computing interchangeably.

Pervasive computing calls for interdisciplinary efforts, embracing nearly all sub-disciplines within computer science. Included are hardware, software, network, human–computer interface, and information technologies. Also, it subsumes distributed and mobile computing and builds on what has been achieved [Satyanarayanan, 2001]. Presented in the following section is an illustrative scenario depicting the future pervasive computing world to identify enabling technologies of pervasive computing. Next, we present a detailed look into each technology through a sampling of representative current developments and future challenges.

37.2 Pervasive Computing Technologies

First, a visionary scenario is presented that could commonly happen in the pervasive computing norm of the future. The scenario serves as a basis for identifying enabling technologies of pervasive computing and to support further discussions.

Bob from headquarters is visiting a branch office of his company for a meeting on the West Coast. At the reception desk in the lobby, he downloads a tag application into his cellular phone through a wireless connection, instead of wearing a visitor tag. This allows him access to the building without an escort. Moreover, he is constantly being tracked by the building location system, so his location may be pinpointed when necessary. Because this is Bob's first time in the branch, his secretary agent on the phone guides him to a reserved meeting room by contacting his scheduler on his office computer and referring him to a floor plan from the branch location system.

Alice had called for this meeting. The meeting room was automatically reserved when she marked her schedule one week previously. In addition, the time was consulted and confirmed among scheduling agents of all the intended attendees. Her scheduling agent notified the building receptionist of expected visitors, including Bob.

As Alice greets Bob in the meeting room, the room "knows" a meeting is about to begin and "notices" another attendee from the branch side, Charlie, is missing. The room sends a reminder to a display that Charlie has been staring at for hours while finalizing his departmental budget. With the deadline for the next year's budget approaching, Charlie has completely forgotten about the meeting. While rushing to get there, he tries to enter the wrong room where he is redirected by a display on the door that provides directions to lead him to the appropriate meeting room.

Using his phone, Bob instructs a projector to download his presentation slides from his office computer. The projector has been personalized to know what cues he has been using for his presentation; for example, he snaps his finger to signify the next slide and waives his left hand for the previous slide. The voice and video annotation are automatically indexed during the entire meeting and stored for later retrieval by both headquarters and branch computer systems.

This scenario involves six categories of various enabling technologies of pervasive computing that include device, network, environment, software, information access, and user interface technology, as shown in Table 37.1.

We will take a closer look at the enabling technologies of pervasive computing by discussing key subtechnologies and research issues of each category along with representative developments in the area.

TABLE 37.1 Pervasive Computing Technologies

Category	Constituent Technologies and Issues
Device	Processing capability, form factors, power efficiency, and universal information appliance
Network	Wireless communication, mobility support, automatic configuration, resource discovery, and spontaneous interaction
Environment	Location awareness, context sensing, sensor network, security, and privacy
Software	Context information handling, adaptation, dynamic service composition, and partitioning
Information access	Ubiquitous data access, agent, collaboration, and knowledge access
User interface	Perceptual technologies, biometrics, context-aware computing, multiple modality, implicit input, and output device coordination

It is followed by more futuristic trials and innovative approaches the ubiquitous computing research community is taking to cope with the challenges. This discussion will aid in understanding the direction ubiquitous computing is headed, as well as where it has come from.

37.2.1 Device Technology

The last decade has seen dramatic improvements of device technology. The technology available in the early 1990s fell short in meeting the ubiquitous computing vision; the Xerox ParcTab had just 128K of memory, 128 × 64 monochrome LCD display, and IR support [Weiser, 1993]. Today's typical PDAs are armed with powerful processing capability, 32M RAM, 320 × 240 transflective color TFT, and IEEE 802.11b support and/or Bluetooth.

The most remarkable advances are found in processing, storage, and display capabilities [Want et al., 2002]. For more than 35 years, microprocessor technology has been accelerating in accordance with the self-fulfilling Moore's law that states transistor density on a microprocessor doubles every 18 months. The smaller the processor, the higher the performance, as it can be driven by a faster clock. Also implied is less power consumption, which is another key issue for pervasive computing. Thanks to chip technology progress, today's PDAs are equipped with a processing capability equivalent to the early to mid-1990s desktop computer power [Estrin et al., 2002]. This advance is matched by storage capability improvement. Common today are high-end PDAs equipped with up to 64M RAM. Also, a matchbook-size Compact-Flash card (i.e., small removable mass storage) provides up to 1G storage capability. Yet another area of drastic improvement is display technology: TFT-LCD and PDP. Their capability crossed the 60" point recently, and prototype costs continue to drop.

Predictions were made that in the post-PC era we would carry multiple gadgets such as cellular phones, PDAs, and handhelds at all times. The need for multiple devices sparked the integration of specialized devices into a multifunctional device. For example, phone-enabled PDAs are already seen on the market, and many cellular phones have gained Internet browsing capability and personal organizer functionality traditionally performed by PDAs. Along with these developments, a more radical approach is being explored. Rather than building multiple hardware functionalities into a single device, software that runs on general-purpose hardware can transform it into a universal information appliance, with variety of gadgets including a cellular phone, a PDA, a digital camera, a bar-code scanner, and an AM/FM radio.

The MIT *SpectrumWare* project has already built a prototype of a software radio out of a general PC. All signal processing except for antenna and sampling parts is handled by software, enabling the performance of any device, once loaded with appropriate software modules in RAM [Guttag, 1999]. The feasibility of this multifunctional device is further backed by MIT Raw chip technology. By exposing raw hardware to software at the level of logic gates, a new opportunity opens up in which a chip itself can be dynamically customized to fit the particular needs of desired devices. More specifically, a software compiler can customize wirings on the chip to direct and store signals in a logic-wire-optimal way for target applications. This approach yields unprecedented performance, energy efficiency, and cost-effectiveness to perform the function of desired gadgets [Agarwal, 1999].

37.2.2 Network Technology

Wireless communication technology is another area with huge successes in technical advancement and wide deployment since the mid-1990s. According to their coverage, wireless communication networks can be classified into short range, local area, and wide area networks. Initially, Bluetooth was developed as a means of cable replacement, thereby providing coverage for short range (in-room coverage, typically 10 m). Communicating mobile devices form a small cell called a *piconet*, which is made up of one master and up to seven slave nodes, and supports about 1M data rate. Smaller coverage means lower power consumption, which is an indispensable feature for the tiny devices being targeted. Infrared is another local connectivity protocol that was developed well before Bluetooth, yet failed to gain popularity. The disadvantage of its line-of-sight requirement is ironically useful in some pervasive computing applications such as an orientation detection system or a location system that can detect physical proximity in terms of physical obstacles or containment. Next, as a midrange wireless protocol covering hundreds of meters, 11Mbps IEEE 802.11b was widely deployed in the past few years. It is no longer a surprise to see wireless-LAN-covered streets, airport lounges, and restaurants. More recently, 54Mbps IEEE 802.11a has started its deployment. Finally, we have seen explosive cellular market growth; as of early 2001, one out of 10 people in the world (680 million) own cellular phones [Parry, 2002]. Moreover, the next generation digital cellular networks such as 2.5G and 3G networks will bring communication capability closer to what ubiquitous computing vision mandates. For instance, the 3G networks support a data rate of 2Mbps for stationary users and 144Kbps for mobile users. The wireless network technologies with different coverage complement one another through vertical handoff by which a device switches to the best network available in a given environment [Stemm and Katz, 1998]. While moving around within a network or across several networks, seamless connectivity can be supported by IP mobility protocols [Perkins, 1996] [Campbell and Gomez-Castellanos, 2001].

Aside from the basic capability of communication, network technology must be able to foster spontaneous interaction, which encompasses mobility support, network resource discovery and automatic configuration, dynamic service discovery, and subsequent spontaneous interactions. Some of these issues are addressed by IETF Zero Configuration Working Group [Hattig, 2001]. A device discovers network resources and configures its network interface using stateful DHCP or stateless automatic configuration [Cheshire et al., 2002] [Thomson and Narten, 1996]. Once base communication protocol is configured and enabled, users may locate a service to meet their needs specified in a particular query language of service discovery protocols [Guttman et al., 1999] [Sun Microsystems, 1999] [UPnP Forum, 2000]. The service discovery problem brings about the ontology issue, which is about how the functionality of services we have never encountered before can be described.

37.2.3 Environment Technology

Smart environments instrumented with a variety of sensors and actuators enable sensing and activating a physical world, thereby allowing tight coupling of the physical and virtual computing worlds. Environmental technology is unique to and inherent in ubiquitous computing, bringing with it new challenges that do not have a place in distributed and mobile computing.

Early works completed in the beginning of the 1990s focused on office environments with location awareness [Want et al., 1992] [Want et al., 1995]. The first indoor location system seems to be the Active Badge project at the Olivetti Research Lab from 1989 to 1992. This system is able to locate people through networked sensors that pick up IR beacons emitted by wearable badges. Later, it evolved into Active Bat, the most fine-grained 3D location system [Addlesee et al., 2001], which can locate 3D position of objects within the accuracy of 3 cm for 95% of the readings. Users carry the Active Bats of 8.0 cm × 4.1 cm × 1.8 cm dimension, synchronized with a ceiling sensor grid using a wireless radio network. A particular Bat emits an ultrasonic pulse, when instructed by a RF beacon, and the system measures the time-of-flight of the ultrasonic pulse to the ceiling receivers. Measured arrival times are conveyed to a central computer, which calculates the position of the Bat. The fine-grained accuracy enables interesting location-aware applications such as location maps of people in an office complex, automatic call routing to the nearest phone, and Follow-me desktop application. The Follow-Me desktop follows a user in that the user's VNC desktop can be displayed on the nearest computer screen by clicking a button on the Bats.

Location systems can be classified into either positioning or tracking systems. With positioning systems, users receive location information from the environment and calculate their own position, whereas user locations are continuously kept track of by a central computer in tracking systems. Various techniques to detect user locations are used, including Infrared, ultrasonic, radio frequency, physical contacts, and even camera vision. Without requiring dedicated infrastructure to be deployed throughout the environments, some systems exploit cellular proximity, i.e., wireless communication signal strength and limited coverage of existing wireless communication infrastructure [Bahl and Padmanabhan, 2000]. For outdoor positioning systems, GPS can determine locations within the accuracy of 10 m in most areas, which is more accurate than most indoor location systems considering the large scale of outdoor features on the earth's surface. A comprehensive survey on the location systems was conducted by Hightower and Borriello [2001].

Until now, most smart space trials have focused on the instrumentation of spatially limited office or home environments, resulting in moderate systems scale. (The largest deployment reported is the Active Bat system consisting of 200 Bats and 750 receiver units covering three-floors of 10,000 ft^2 [Addlesee et al., 2001]). The environments were spatially limited and relatively static, so the infrastructure deployments were well planned via presurvey and offline calibration with emphasis on location determination. Because the environments may be unpredictable, unknown, and span heterogeneous, wide areas, they will be better sensed by taking various metrics collectively. To cope with this uncertainty, researchers are investigating relevant issues such as massive scale, autonomous, and distributed sensor networks using rich environmental information sensing rather than simply location. The various metrics may include light, temperature, acceleration, magnetic field, humidity, pressure, and acoustics [Estrin et al., 2002]. The possibility for new environmental technologies is being raised by recent advances in microelectromechanical systems (MEMS) and miniaturization technology. Researchers have been able to build a 1-in. scale system that involves processing, communication, and sensors, expecting to break the barrier of a 1 mm^3 computer within 10 years [Warneke et al., 2001]. The new wireless sensor network will find new applications and uses that were previously not possible, e.g., being sprinkled in an *ad hoc* manner to monitor the weather, forests, and wildfires, as well as civil infrastructures including buildings, roads, and bridges. The most challenging issues of sensor networks are scalability and autonomy [Estrin et al., 2002]; unlike the early systems in which sensed information was collected and processed by a central computer, information in sensor networks ought to be preprocessed by the source or intermediate node(s) because of energy concern, uncertainty, and dynamism of the environment. Because individual nodes have limited processing power and view of the environment, the intermediate data may be sent to another node for further processing.

37.2.4 Software Technology

Early mobile and ubiquitous computing applications focused primarily on service mobility (i.e., device independent service access), and context information processing. First, service mobility means services on any device, i.e., services must be able to be invoked on whatever devices happen to be with users at the moment. Services are negotiated and further adapted or transformed to the devices' capability [Klyne et al., 2002; Gimson et al., 2001]. In addition to the device capability context information, ubiquitous computing applications must deal with various context information by a variety of sensors. Even for the same type of context, it may have to switch to a different source. For example, an application will have to switch to an indoor location service from a GPS location service, when a user enters a building. Thus, an important issue becomes achieving context awareness while separating the context-sensing portion from the rest of applications. The context widget [Dey et al., 1999] provides abstraction components that pass logical information to applications, while hiding unnecessary specifics of raw information from physical sensors. This eliminates applications' dependency on a particular sensor type.

Pervasive computing vision requires more revolutionary changes to the notion of applications, i.e., a new model of applications. An application should not be designed with regard to a rigid decomposition of interactions with users. Rather a task (i.e., an application component) must be described in high-level logic in order to be instantiated later on according to the logic. More specifically, considering the application will run in a dynamic and unknown environment, the task (and constituent subtasks) logic must be described to abstract away any specific details relating to user device capability, the environments, and available services [Banavar et al., 2000]. The problem of how to describe abstracted task interfaces relates to the service description issue discussed in the network technology subsection. An abstract description of application components (i.e., services) facilitates dynamic service discovery and composition. Components compatible according to the application's task interface description and appropriate in the given context are integrated to synthesize an application. But the new model does not require all components to be loaded on a single device. The constituent modules spread through the ubiquitous environments, and an application is instantiated in just-in-time fashion (i.e., when needed). Thus the model of the distributed application components is characterized as "disappearing software," application functionality partitioning over multiple devices that facilitates recomposition, migration, and adaptation [Want et al., 2002].

The real power of a ubiquitous computing application does not come solely from the application itself. Rather, it comes from an orchestration of the application, supporting services, and environments as a cooperative whole. The new model has numerous advantages. First, it allows for adaptation to dynamically changing environments and fault-tolerance. Applications autonomously respond to changes and failures by migrating affected components only. Second, it enables gradual evolution of the environments and an application itself without disrupting the function as a whole. Introduction of new devices does not affect the entire application directly, and better services can be incorporated into the application as they become available. Therefore, no need exists for global upgrade or the associated downtime. Finally, this approach enhances the scalability of software infrastructure by allowing it to be shared among multiple application instances [Banavar and Bernstein, 2002].

37.2.5 Information Access Technology

In addition to the challenges brought forth by the ubiquity of computing, we face new requirements raised by the omnipresence of information as well. Ubiquitous information access means more than just ensuring that information is readily available under any condition; *Anytime, anywhere, and any device* computing has been more or less a motto of distributed and mobile computing research for the past 30 years. Built on predecessors' achievements, ubiquitous information research topics include remote information access, fault tolerance, replication, failure recovery, mobile information access [Kistler and Satyanarayanan, 1992], adapted access such as proxies and transcoding [Noble et al., 1997], CSCW (Computer Supported Cooperative Work), software agent, data mining, and knowledge representation [Berners-Lee et al., 2001].

However, ubiquitous information access requires a new level of sophistication atop distributed and mobile computing; personalized, context-aware information systems equipped with intelligence to infer users' implicit intentions by capturing relevant contextual cues. The whole process of information processing, i.e., information acquisition, indexing, and retrieval, should be augmented to handle implicit user needs.

Acquired information ought to be organized and indexed according to relevant contextual information. For example, a smart meeting room gathers and stores all information regarding on-going meetings from various sources, including meeting agendas and time, presentation slides, attendees, and relevant materials such as Web pages accessed during the meeting. To enable ubiquitous information access, the information access system first needs to establish useful contexts by sensing environments and monitoring users' activities. It may consider users' personal information and past access patterns, nearby people and objects, and inferred social activities they are presently engaging in. This context metadata becomes an inherent part of the information set, enabling personalized information access.

For information retrieval, a users' query needs to be refined according to relevant context, and its result as well. Given a search query of "pervasive computing," the system will return information relating to ubiquitous information access technology if able to infer that the user has recently been working for a mobile file system. It may, however, return information regarding wireless sensor nodes if the individual is a hardware engineer.

37.2.6 User Interface Technology

"Disappearing technology" implies the need for new modalities concerning input and output. Traditional WIMP (Windows, Icons, Menus, Pointing devices) based user interface requires constant attention and often turns out to be an annoying experience. It will not allow computing to disappear from our consciousness. One enlightening example of a new modality is the "dangling string" [Weiser and Brown, 1996]. This string hangs from a ceiling and is somehow electrically connected to an Ethernet cable. More specifically, a motor rotates the string at different speeds proportional to traffic volume on the network. When traffic is light, a gentle waving movement is noted; the string whirls fast under heavy traffic with a characteristic noise indicating the traffic volume. This aesthetically designed interface conveys certain information without moving into the foreground of our consciousness and overwhelming our attention, which may be opportunistic in certain situations. This interface can be compared to a fast scrolling, dizzy screen that quickly dumps the details of traffic monitoring and analysis.

The primary challenges of the ubiquitous computing user interface are *implicit input and distributed output,* which are based on diverse modalities as pointed out by Abowd et al. [2002]. Unobtrusive interfaces that employ appropriate modalities rather than plain keyboard and display allow for natural, comfortable interactions with computers in everyday practices. This makes forgetting their existence possible, hence invisible computing.

There is a trend of input technology advancing toward a more natural phenomena for humans, going from text input to the desktop metaphor of GUI-based windowing systems and perceptual technologies such as handwriting, speech, and gesture recognition. Further advances in recognition technologies include face, gait, and biometric recognition such as fingerprint and iris scanning. Along with the environmental sensing technology (including user identity, location, and other surrounding objects), the new development of various input modalities raises possibilities of natural interactions to catalyze the disappearance of computing. For example, when we say "Open it" in front of a door or while pointing to it, our intention is inferred by multiple input sources; primary user intention interpretation via speech recognition can be refined by physical proximity sensing or image recognition techniques.

Three form factor scales, i.e., the inch, foot, and yard scale, were prototyped by the early Xerox ParcTab project [Weiser, 1993], and were called ParcTab, MPad, and Liveboard, respectively. These form factors remain prevalent in today's computing devices; the ParcTab is similar to today's PDAs in size, whereas current laptops or tablet PCs are the MPad scale. Also, we can easily find high-resolution wall displays (the Liveboard class devices) in an office environment or public place. Aside from the classical display interface, there are several trials for innovative output modes like the "dangling string." A true challenge is to coordinate multiple output devices to enhance user experiences with minimal user attention and cognitive effort. For instance, a presentation system can orchestrate appropriate output devices among wall displays, projectors, stereo speaker systems, and microphones at the right time. While giving a presentation, if the presenter points to a slide containing a pie chart representing this year's sales, a big table of detailed numbers, categorized by product classes, will be displayed on a wall display. An example of output coordination across multiple devices can be found in WebSplitter [Han et al., 2000].

37.3 Ubiquitous Computing Systems

Despite all efforts in the last decade to reach for the compelling vision, there are very few ubiquitous computing systems that have undergone significant field trials. The examples are Active Bat [Addlesee et al., 2001], Classroom 2000 [Abowd, 1999], Lancaster Guide System [Davies et al., 2001], and Matilda's

Smart House [Helal et al., 2003]. The system descriptions below not only give insight into the essential and unique features of ubiquitous computing systems, but help identify gaps between state-of-the-art technologies and reality. As we will see in the descriptions, these systems focus on certain aspects of ubiquitous computing to dig into relevant technologies among the classification discussed in the previous section.

37.3.1 Active Bat

The first indoor location system, the Active Badge project (1989 to 1992), later evolved into the Active Bat at AT&T Laboratories Cambridge, which is the most fine-grained 3D location system. Users carry the Active Bat of 8.0 cm × 4.1 cm × 1.8 cm dimensions, which consists of a radio transceiver, controlling logic, and an ultrasonic transducer. Ultrasonic receiver units mounted on the ceiling are connected by a wired daisy-chain network. A particular Bat emits an ultrasonic pulse, when informed by an RF beacon from the cellular network base stations. Simultaneously, the covered ceiling-mounted receivers are reset via the wired network. The system then measures the time-of-flight of the ultrasonic pulse to the ceiling receivers. The measured arrival times are conveyed to a central computer, which calculates the 3D position of the Bat. In addition, the Bat has two buttons for input and a buzzer and two LEDs for output. The control button messages are sent to the system over the wireless cellular network.

The interaction with the environment is based on a model of the world constructed and updated by sensor data. Changes in the real world objects are immediately reflected by the corresponding objects of the model. Thus the model, i.e., a shared view of the environment by users and computers, allows application to control and query the environment. For example, Figure 37.1 is a visualization of an office model that mirrors a users' perception of the environment. It displays users, workstations, telephones with an extension number, and furniture in the office. The map allows users to easily locate a colleague. They can also locate the nearest phone not in use and place a call by simply clicking the phone.

Another interesting application enabled by the fine-grained 3D location sensing is a *virtual mouse panel*. An information model sets a mouse panel space around several 50 in. display objects in the office building. When a Bat enters the space, its positions are projected into the panel and translated into mouse pointer readings in the imaginary pad.

The project concentrated on the environment and device categories of the technology classification in Section 37.2, which include environmental instrumentation and accurate 3D location sensing, the model of the real world shared by users and applications, and the formalization of spatial relationships

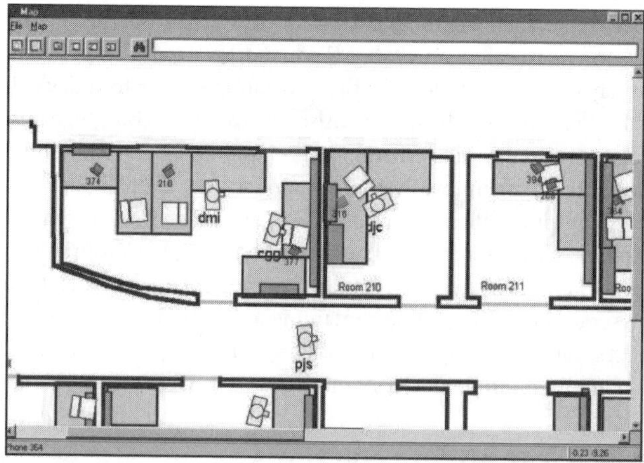

FIGURE 37.1 A model of an office environment. [From Addlesee, M., R. Curwen, S. Hodges, J. Newman, P. Steggles, A. Ward, and A. Hopper. Implementing a sentient computing system. *IEEE Computer*, Vol. 34, No. 8, August 2001, pp. 50–56.]

between the objects and 2D spaces around them. In addition, the project dealt with some issues of the software and user interface categories. For example, its programming support for location events based on the spatial relationship formalization enabled several interesting applications. Also, it was demonstrated that novel user interfaces such as the virtual mouse panel can be enabled when the Bat's limited processing and input/output capabilities are supplemented by environmental supports.

37.3.2 Classroom 2000

The Classroom 2000 project at Georgia Institute of Technology captures the traditional lecture experience (teaching and learning) in an instrumented classroom environment. The Classroom 2000 system supports the automated capture, integration, and access of a multimedia record of a lecture. For example, to facilitate student reviews after a lecture, class presentation slides are annotated with automatically captured multimedia information such as audio, video, and URLs visited during the class. A student's personal notes can also be incorporated. In addition to the whole class playback, the system supports advanced access capabilities, e.g., clicking on a particular handwritten note gives access to the lecture at the time it was written. This project was a long-term, large-scale experiment that started in 1995. Its more than three-year project lifetime includes extensive trials of over 60 courses at Georgia Institute of Technology, nine courses at Kennesaw State University, GA, and a few others.

The classroom is instrumented with microphones and video cameras mounted to the ceiling. An electronic whiteboard (a 72 in. diagonal upright pen-based computer used to capture the instructor's presentation slides and notes) and two ceiling-mounted LCD projectors are connected to the classroom network reaching the seats of 40 students. After class, the instructor makes the lectures available for review by the students using any networked computer. Also, students may use a hand-held tablet computer in the classroom to take their private notes, which will be automatically consolidated into the lecture during the integration phase.

Figure 37.2 is an example of class notes taken by the Classroom 200 system. The class progression is indicated on the left by a time line decorated with activities during the class. The left decoration frame includes covered slides and URLs visited in the lecture, providing an overview of the lecture to facilitate easy browsing. Clicking on a decoration causes either the slide to be displayed in the right frame or the URL page to be opened in a separate browser. The figure also shows a presentation slide annotated with the instructor's handwritten notes on the electronic whiteboard.

With great emphasis given to the information access and user interface technology categories, the project used existing technologies for the device, network, environment, and software categories. The classroom was instrumented with standard classroom equipments, an electronic whiteboard, and ordinary client and server machines connected to the classroom network. The primary concern was facilitation of the automatic capture and easy access later on, i.e., whether all significant activities were captured with time stamp information, and the captured multimedia were integrated and indexed based on the time stamps for later class reviews. For example, speech recognition software was used to generate a time-stamped transcript of the lecture, to be used for keyword search. The search result pointed to the moment of the lecture when the keyword was spoken.

37.3.3 Lancaster Guide System

The goal of the Lancaster Guide project, which started in 1997, was to develop and deploy a context-sensitive tourist guide for visitors to the city of Lancaster, U.K. In 1999, a field trial involving the general public was conducted to gain practical experiences for further understanding the requirements of ubiquitous computing applications. The Lancaster Guide system provides a customized tour guide for visitors considering their current location, preferences, and environmental context. For example, an architect may be more interested in old buildings and castles rather than a stroll along a river bank, whereas second-time visitors will probably be interested in the places that they missed during their last visit. The

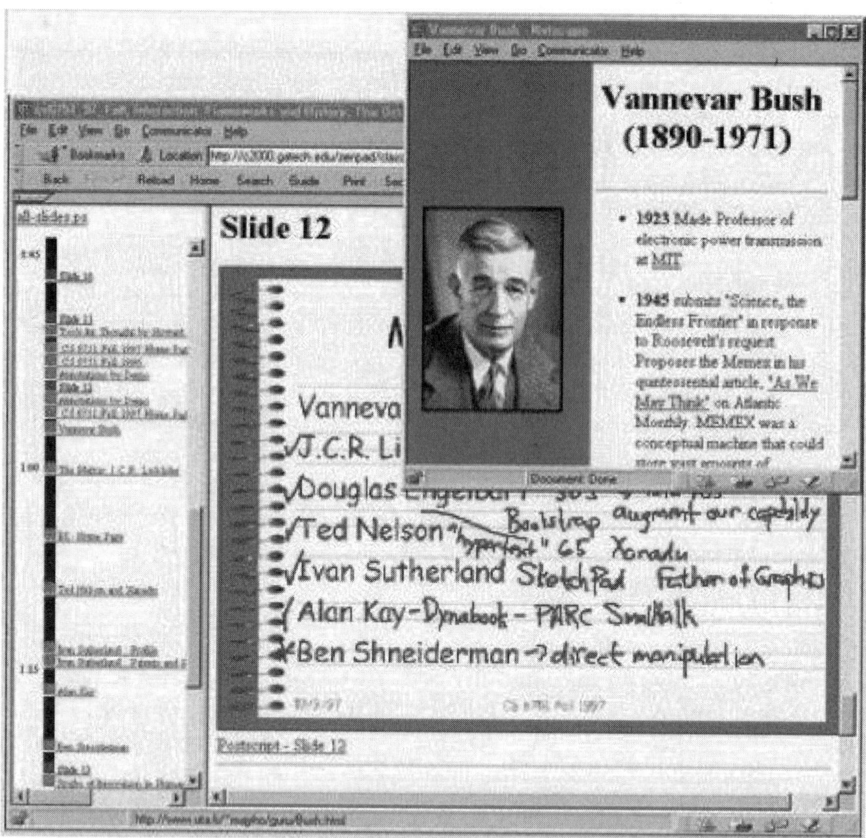

FIGURE 37.2 Access tool user interface. [From Abowd, G. D. Classroom 2000: an experiment with the instrumentation of a living educational environment. *IBM Systems Journal*, Vol. 38, No. 4, 1999, pp. 508–530.]

system also considers the visitors' languages, financial budgets, time constraints, and local weather conditions.

Approximate user location is determined by a cell of the 802.11 wireless network deployed over the city. A cell consists of a cell server machine and several base stations. As end-user systems, a tablet-based PC (Fujitsu TeamPad 7600 with a transflexive 800 x 600 pixel display) equipped with a 2M WaveLAN card communicates with the cell server via the base stations. The cell server periodically broadcasts location beacons containing a location identifier, and also caches Web pages frequently accessed by users in the cell. The cached pages are periodically broadcast to the cell, which in turn is cached by the end-user systems. When the cell server is informed of a cache miss on the client devices, it will add the missed page to its broadcast schedule for the next dissemination cycle. The information broadcast and caching is to reduce information access latency in the cell.

The Guide System's geographical information model associates a location with a set of Web pages. The Web pages presented to the visitors are dynamically created to reflect their context and preferences. Special Guide *tags* are used to indicate their context and preferences in the template HTML pages. In other words, the special tags (i.e., hooks between the HTML pages and information model) are expanded into personalized information pieces. Figure 37.3 shows a dynamically created tour guide page to list attractions based on proximity to a user's current location and the attractions' business hours.

To provide context-sensitive tour information to visitors, the project focused on the information access, context-aware user interface, and environment technology categories with regard to the technology classification. The 802.11 base stations over the city point to a location that is associated with a set of Web pages linked to the geographical information model. Based on the location information along with

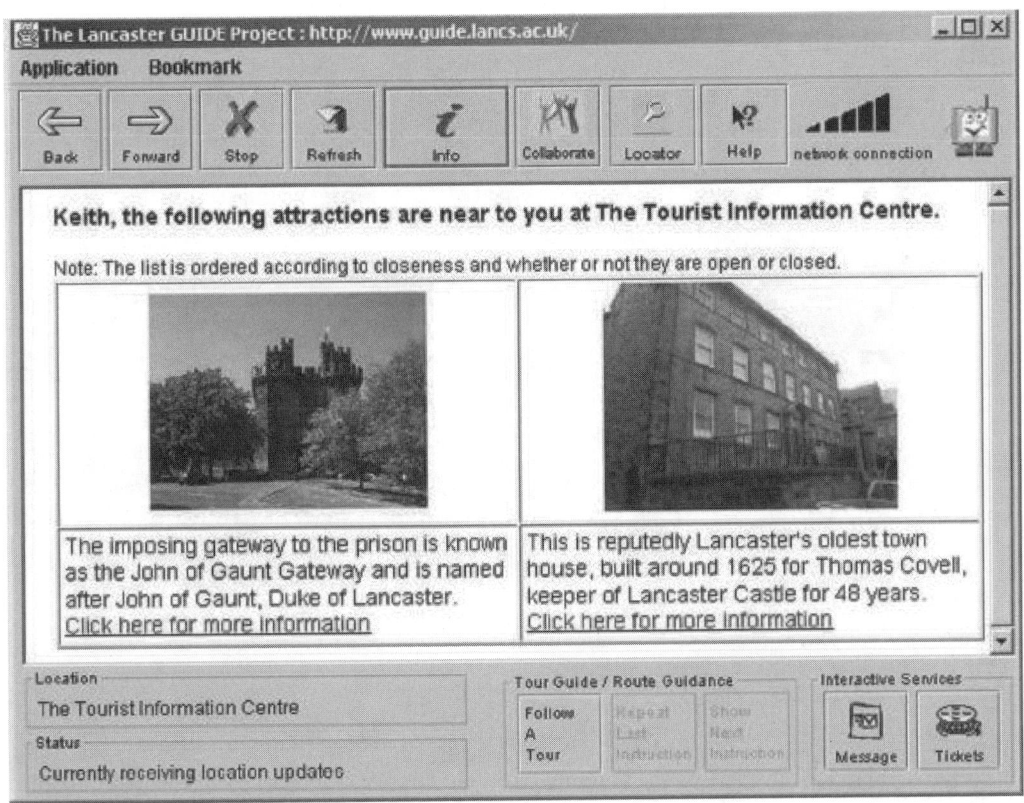

FIGURE 37.3 Lancaster Guide System. [From Davies, N., K. Cheverst, K. Mitchell, and A. Efrat. Using and determining location in a context-sensitive tour guide. *IEEE Computer*, Vol. 34, No. 8, August 2001, pp. 35–41.]

other context information, tailor-made HTML guide pages are dynamically created and presented to tourists. However, the system's ability to recommend a custom tour is dependent on the preference information a visitor inputs during the time of a user unit pickup at the tour information center.

37.3.4 Matilda's Smart House

Matilda's Smart House project is a relatively recent effort launched in 2002 to innovate pervasive applications and environments specifically designed to support the elderly, i.e., Matilda, in the RERC — Pervasive Computing Laboratory at the University of Florida. The project explores the use of emerging smart phone and other wireless technologies to create "magic wands", which enable elder persons with disabilities to interact with, monitor, and control their surroundings. By integrating the smart phone with smart environments, elders are able to turn appliances on and off, check and change the status of locks on doors and windows, aid in grocery shopping, and find other devices such as car keys or a TV remote. It also explores the use of smart phones as devices that can proactively provide advice such as reminders to take medications or call in prescription drug refills automatically.

Figure 37.4 is the top view of Matilda's Smart House consisting of a kitchen, living room, bathroom, and bedroom from left to right. Comparing the smart environment instrumentation with the early ubiquitous computing systems reveals computing technology advances. For the Xerox ParcTab project, every piece of the system needed to be hand-crafted from scratch, including computing devices, sensors, network, and user interface, because they were not available at that time. In contrast, many commercial off-the-shelf products readily available on the market are used for the Matilda's Smart House project. Included are J2ME smart phones as user devices, ultrasonic receivers in the four corners of the mock-up house, X-10 controlled devices (door, mailbox, curtain, lamp, and radio), and networked devices

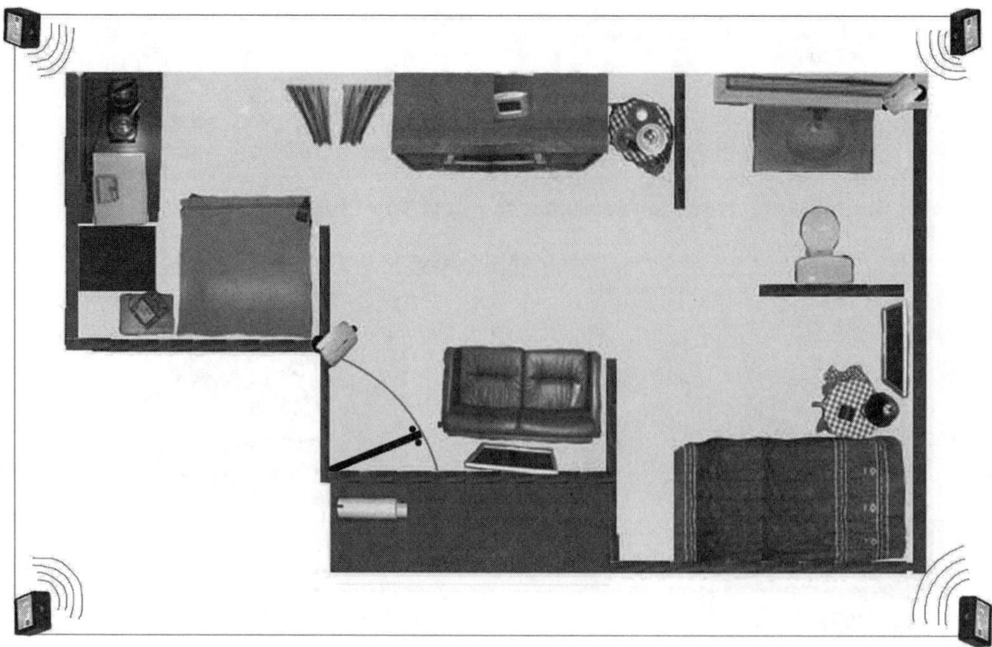

FIGURE 37.4 Matilda's Smart House.

(microwave, fridge, LCD displays on the wall, and cameras). The OSGi (Open Service Gateway initiative) is adopted as base software infrastructure facilitating the smart house resident's interactions with the smart environments. It also supports remote monitoring and administration by family members and caregivers. Building the smart space using COTS components freed the project team to focus more on the integration of the smart phone with the smart environments and various pervasive computing applications [Helal et al., 2003]. For example, medication reminders are provided on the LCD display Matilda is facing, with her orientation and location being sensed by the smart house. An audio warning is issued if she picks up a wrong medicine bottle, which is detected by a barcode scanner attached to her phone. Also, Matilda can use her phone to open the door for the delivery of automatically requested prescription drug refills.

The project spans the environment, device, network, and user interface technology categories. The smart environments are built using various networked appliances, X-10 devices, sensors, and an ultrasonic location system. J2ME smart phones are utilized as a user device to control and query the environment. Also, some user interface issues such as multiple modality and output device coordination are addressed. Several pervasive computing applications enabled through the project involve service discovery and interaction issues.

37.4 Conclusion

Five major university projects have been exploring pervasive computing issues, including the Oxygen project at MIT (http://oxygen.lcs.mit.edu); the Endeavor project at the University of California, Berkeley (http://endeavour.cs.berkeley.edu); the Aura Project at Carnegie Mellon University (http://www.cs.cmu.edu/~aura); the Portolano Project at the University of Washington (http://portolano.cs.washington.edu); and the Infosphere Project at Georgia Tech and Oregon Graduate Institute (http://www.cc.gatech.edu/projects/infosphere). These large-scale projects are investigating comprehensive issues of pervasive computing, so their scope is much broader than industry projects such as PIMA at

IBM Research (http://www.research.ibm.com/PIMA), Cooltown at HP (http://www.cooltown.hp.com), and the Easy Living Project at Microsoft Research (http://research.microsoft.com/easyliving).

Despite its compelling vision and enormous efforts from academia and industry during the past 10 years, the pervasive computing world does not seem close to us yet. Only a few ubiquitous computing systems were deployed and evaluated in the real world on a realistic scale. The changes and impacts pervasive computing systems will bring are not yet thoroughly understood. There were privacy and security controversies sparked by early deployment of pervasive computing systems, and user groups were reluctant to use those systems. The more information the system knows about an individual, the better he or she will be served. To what extent should personal information be allowed to enter the system and how can it prevent the information from being abused? These are questions that need to be answered for wide acceptance of the pervasive computing technologies. Besides, an introduction of new technologies may cause unexpected responses and confusions as seen in "The Pied Piper of Concourse C" behavior [Jessup and Robey, 2002]. People need time to develop notions of appropriate behaviors and new practices to adapt themselves for new technologies. In order to gain wide acceptance of pervasive computing systems, nontechnical factors such as social, economical, and legal issues should not be underestimated. A deeper understanding of these issues can be developed through real field trials out of laboratories, and practical experiences and feedback from the experiments will redefine pervasive computing systems.

References

Abowd, G. D. Classroom 2000: an experiment with the instrumentation of a living educational environment. *IBM Systems Journal*, Vol. 38, No. 4, 1999, pp. 508–530.

Abowd, G. D., E. D. Mynatt, and T. Rodden. The human experience. *IEEE Pervasive Computing*, Vol. 1, No. 1, January–March 2002, pp. 48–57.

Addlesee, M., R. Curwen, S. Hodges, J. Newman, P. Steggles, A. Ward, and A. Hopper. Implementing a sentient computing system. *IEEE Computer*, Vol. 34, No. 8, August 2001, pp. 50–56.

Agarwal, A. Raw computation. *Scientific American*, Vol. 281, No. 2, August 1999.

Bahl, P. and V. Padmanabhan. RADAR: An In-Building RF-based User Location and Tracking System. In Proceedings of the IEEE Infocom 2000, March 2000, pp. 775–784.

Banavar, G., J. Beck, E. Gluzberg, J. Munson, J. Sussman, and D. Zukowski. Challenges: An Application Model for Pervasive Computing. In Proceedings of the 6th ACM/IEEE International Conference on Mobile Computing and Networks (MobiCom'00), August 2000, pp. 266–274.

Banavar, G. and A. Bernstein. Software infrastructure and design challenges for ubiquitous computing applications. *Communications of the ACM*, Vol. 45, No. 12, December 2002, pp. 92–96.

Berners-Lee, T., J. Hendler, and O. Lassila. The semantic web. *Scientific American*, Vol. 284, No. 5, May 2001.

Campbell, A. T. and J. Gomez-Castellanos. IP micro-mobility protocols. *ACM SIGMOBILE Mobile Computer and Communication Review* (MC2R), Vol. 4, No. 4, October 2001, pp. 42–53.

Cheshire, S., B. Aboba, and E. Guttman. Dynamic Configuration of IPv4 Link-Local Addresses. IETF Internet Draft, August 2002.

Davies, N., K. Cheverst, K. Mitchell, and A. Efrat. Using and determining location in a context-sensitive tour guide. *IEEE Computer*, Vol. 34, No. 8, August 2001, pp. 35–41.

Dey, A. K., G. D. Abowd, and D. Salber. A Context-Based Infrastructure for Smart Environment. In Proceeding of the 1st International Workshop on Managing Interactions in Smart Environments (MANSE'99), December 1999, pp. 114–128.

Estrin, D., D. Culler, K. Pister, and G. Sukatme. Connecting the physical world with pervasive networks. *IEEE Pervasive Computing*, Vol. 1, No. 1, January–March 2002, pp. 59–69.

Gimson, R., S. R. Finkelstein, S. Maes, and L. Suryanarayana. Device Independence Principles. http://www.w3.org/TR/di-princ/, September 2001.

Guttag, J. V. Communication chameleons. *Scientific American*, Vol. 281, No. 2, August 1999.

Guttman, E., C. Perkins, J. Veizades, and M. Day. Service Location Protocol, Version 2. IETF RFC 2608, June 1999.

Han, R., V. Perret, and M. Naghshineh. WebSplitter: A Unified XML Framework for Multi-Device Collaborative Web Browsing. In Proceedings of the 2000 ACM Conference on Computer Supported Cooperative Work (CSCW 2000), December 2000, pp. 221–230.

Hattig, M. Zeroconf Requirements. IETF Internet Draft, March 2001.

Helal, S., B. Winkler, C. Lee, Y. Kaddoura, L. Ran, C. Giraldo, S. Kuchibhotla, and W. Mann. Enabling Location-aware Pervasive Computing Applications for the Elderly. In Proceedings of the 1st IEEE International Conference on Pervasive Computing and Communications (PerCom 2003), March 2003, pp. 531–536.

Hightower, J. and G. Borriello. Location systems for ubiquitous computing. *IEEE Computer*, Vol. 34, No. 8, August 2001, pp. 57–66.

Intanagonwiwat, C., R. Govindan, and D. Estrin. Directed Diffusion: a Scalable and Robust Communication Paradigm for Sensor Networks. In Proceedings of the 6th ACM/IEEE International Conference on Mobile Computing and Networks (MobiCom'00), August 2000, pp. 56–67.

Jessup, L. M. and D. Robey. The relevance of social issues in ubiquitous computing environments. *Communications of the ACM*, Vol. 45, No. 12, December 2002, pp. 88–91.

Kistler, J. J. and M. Satyanarayanan. Disconnected operation in the Coda File System. *ACM Transactions on Computer Systems*, Vol. 10, No. 1, February 1992, pp. 3–25.

Klyne, G., F. Reynolds, C. Woodrow, H. Ohto, and M. H. Butler. Composite Capabilities/Preference Profiles: Structure and Vocabularies. http://www.w3.org/TR/2002/WD-CCPP-struct-vocab-20021108, November 2002.

Noble, B. D., M. Satyanarayanan, D. Narayanan, J. E. Tilton, J. Flinn, and K. R. Walker. Agile Application-Aware Adaptation for Mobility. In Proceedings of the 16th ACM Symposium on Operating Systems Principles, October 1997, pp. 276–287.

Parry, R. Overlooking 3G. *IEEE Potentials*, Vol. 21, No. 4, October/November 2002, pp. 6–9.

Perkins, C. E. IP Mobility support. IETF RFC 2002, October 1996.

Satyanarayanan, M. Pervasive computing: vision and challenges. *IEEE Personal Communications*, Vol. 8, No. 4, August 2001, pp. 10–17.

Stemm, M. and R. H. Katz. Vertical handoffs in wireless overlay networks. *ACM Mobile Networking* (MONET), Vol. 3, No. 4, 1998, pp. 335–350.

Sun Microsystems. Jini Technology Architectural Overview http://www.sun.com/jini/whitepapers/architecture.html, January 1999.

Thomson, S. and T. Narten. IPv6 Stateless Address Autoconfiguration. IETF RFC 1971, August 1996.

UPnP Forum. Universal Plug and Play Device Architecture. http://www.upnp.org/download/UPnPDA10_20000613.htm, June 2000.

Want, R., A. Hopper, V. Falcao, and J. Gibbons. The active badge location system. *ACM Transaction on Information Systems*, Vol. 10, No. 1, January 1992, pp. 91–102.

Want, R., T. Pering, G. Borriello, and K. I. Farkas. Disappearing hardware. *IE,EE Pervasive Computing*, Vol. 1, No. 1, January/March 2002, pp. 36–47.

Want, R., B. N. Schilit, N. I. Adams, R. Gold, K. Petersen, D. Goldberg, J. R. Ellis, and M. Weiser. An overview of the ParcTab ubiquitous computing experiment. *IEEE Personal Communications*, Vol. 2, No. 6, December 1995, pp. 28–43.

Warneke, B., M. Last, B. Liebowitz, and K. S.J. Pister. Smart dust: communicating with a cubic-millimeter computer. *Computer*, Vol. 34, No. 1, Jan 2001, pp. 44–51.

Weiser, M. The computer for the 21st century. *Scientific American*, Vol. 256, No. 3, September 1991, pp. 94–104.

Weiser, M. Some computer science issues in ubiquitous computing. *Communications of the ACM*, Vol. 36, No. 7, July 1993, pp. 75–84.

Weiser, M. and J. S. Brown. The coming age of calm technology. *PowerGrid Journal*, Version 1.01, July 1996.

38

Worldwide Computing Middleware

CONTENTS

Abstract 38-1
38.1 Middleware 38-1
38.1.1 Asynchronous Communication 38-2
38.1.2 Higher-Level Services 38-3
38.1.3 Virtual Machines 38-4
38.1.4 Adaptability and Reflection 38-4
38.2 Worldwide Computing 38-4
38.2.1 Actor Model 38-4
38.2.2 Language and Middleware Infrastructure 38-5
38.2.3 Universal Actor Model and Implementation 38-6
38.2.4 Middleware Services 38-7
38.2.5 Universal Naming 38-8
38.2.6 Remote Communication and Mobility 38-9
38.2.7 Reflection 38-11
38.3 Related Work 38-13
38.3.1 Worldwide Computing 38-13
38.3.2 Languages for Distributed and Mobile Computation 38-15
38.3.3 Naming Middleware 38-16
38.3.4 Remote Communication and Migration Middleware 38-16
38.3.5 Adaptive and Reflective Middleware 38-17
38.4 Research Issues and Summary 38-17
38.5 Further Information 38-18
38.6 Glossary 38-18
Acknowledgments 38-19
References 38-19

Gul A. Agha

Carlos A. Varela

Abstract

The Internet provides the potential for utilizing enormous computational resources that are globally distributed. Such worldwide computing can be facilitated by *middleware* — software layers that deal with distribution and coordination, such as naming, mobility, security, load balancing, and fault tolerance. This chapter describes the *World-Wide Computer*, a worldwide computing infrastructure that enables distribution and coordination. We argue that the World-Wide Computer enables application developers to concentrate on their domain of expertise, reducing code and complexity by orders of magnitude.

38.1 Middleware

The wide variety of networks, devices, operating systems, and applications in today's computing environment create the need for abstraction layers to help developers manage the complexity of engineering

1-58488-381-2/05/$0.00+$1.50

© 2005 by CRC Press LLC

distributed software. A number of models, tools, and architectures have evolved to address the composition of objects into larger systems: Some of the widely used *middleware,* ranging in support from basic communication infrastructure to higher-level services, include CORBA [Object Management Group, 1997], DCOM [Brown and Kindel, 1996], Java RMI [Sun Microsystems and JavaSoft, 1996], and more recently Web Services [Curbera et al., 2002].

Middleware abstracts over operating systems, data representations, and distribution issues, enabling developers to program distributed heterogeneous systems largely as though they were programming a homogeneous environment. Many middleware systems accomplish this transparency by enabling heterogeneous objects to communicate with each other. Because the communication model for object-oriented systems is synchronous method invocation, middleware typically attempts to give programmers the illusion of local method invocation when they invoke remote objects. The middleware layers are in charge of low-level operations, such as marshaling and unmarshaling arguments to deal with heterogeneity, and managing separate threads for network communication.

Middleware toolkits provide compilers capable of creating code for the client (aka *stub*) and server (aka *skeleton)* components of objects providing remote application services, given their network-unaware implementation. Intermediate brokers help establish interobject communication and provide higher-level services, such as naming, event and life-cycle services.

38.1.1 Asynchronous Communication

In order to give the illusion of local method invocation when invoking remote objects, a calling object is blocked — waiting for a return value from a remote procedure or method call. When the value returns, the object resumes execution (see Figure 38.1). This style of communication is called *remote procedure call or RPC.* Users of middleware systems realized early on that network latencies make RPC much slower in nature than local communication and that the consequence of its extensive use can be prohibitive in overall application performance. Transparency of communication may thus be a misleading design principle [Waldo et al., 1997].

Asynchronous (aka *event-based)* communication services enable objects to communicate in much more flexible ways [Agha, 1986]. For example, the result of invoking a method may be redirected to a third party or *customer,* rather than going back to the original method caller. Moreover, the target object need not synchronize with the sender in order to receive the message, thus retaining greater scheduling flexibility and reducing the possibilities of deadlocks.

Because asynchronous communication creates the need for intermediate buffers, higher-level communication mechanisms can be defined without significant additional overhead. For example, one can define a communication mechanism that enables objects to communicate with peers without knowing in advance the specific target for a given message. One such model is a shared memory abstraction used

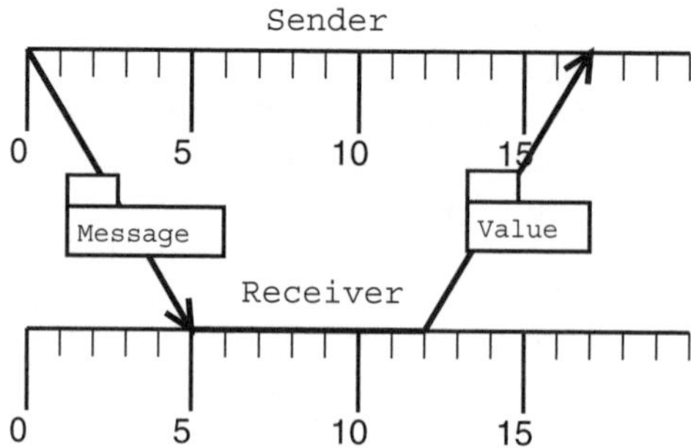

FIGURE 38.1 Synchronous communication semantics requires the sender of a message to block and wait until the receiver has processed the message and returned a value. In this example, the message processing takes only 7 time units, whereas the network communication takes 10 time units.

in Linda [Carriero and Gelernter, 1990]. Linda uses a shared tuple-space from which different processes (active objects) read and write. Another communication model is ActorSpaces [Callsen and Agha, 1994]; in ActorSpaces, actors use name patterns for directing messages to groups of objects (or a representative of a group). This enables secure communication that is transparent for applications. A more open but restrictive mechanism, called *publish-and-subscribe* [Banavar et al., 1999], has been used more recently. In publish-and-subscribe, set membership can be explicitly modified by application objects without pattern matching by an ActorSpace manager.

38.1.2 Higher-Level Services

Besides communication, middleware systems provide high-level services to application objects. Such high-level services include, for example, object naming, lifecycle, concurrency, persistence, transactional behavior, replication, querying, and grouping [Object Management Group, 1997]. We describe these services to illustrate what services middleware may be used to provide.

A *naming* service is in charge of providing object name uniqueness, allocation, resolution, and location transparency. Uniqueness is a critical condition for names so that objects can be uniquely found, given their name. This is often accomplished using a name context. Object names should be object location independent so that objects can move, preserving their name. A global naming context supports a universal naming space, in which context-free names are still unique. The implementation of a naming service can be centralized or distributed; distributed implementations are more fault-tolerant but create additional overhead.

A *life-cycle* service is in charge of creating new objects, activating them on demand, moving them, and disposing of them, based on request patterns. Objects consume resources and therefore cannot be kept on systems forever — in particular, memory is often a scarce shared resource. Life-cycle services can create objects when new resources become available, can deactivate an object — storing its state temporarily in secondary memory — when it is not being actively used and its resources are required by other objects or applications, and can also reactivate the object, migrate it, or can dispose of (garbage collect) it if there are no more references to it.

A *concurrency* service provides limited forms of protection against multiple threads sharing resources by means of lock management. The service may enable application threads to request exclusive access to an object's state, read-only access, access to a potentially dirty state, and so on, depending on concurrency policies. Programming models such as actors provide higher-level support for concurrency management, preventing common errors, such as corrupted state or deadlocks, which can result from the use of a concurrency service.

A *persistence* or *externalization* service enables applications to store an object's state in secondary memory for future use, e.g., to provide limited support for transient server failures. This service, even though high level, can be used by other services such as the life-cycle service described above.

A *transactional* service enables programming groups of operations with atomicity, consistency, isolation, and durability guarantees. Advanced transactional services may contain support for various forms of transactions such as nested transactions and long-lived transactions.

A replication service improves locality of access for objects by creating multiple copies at different locations. In the case of objects with mutable state, a master replica is often used to ensure consistency with secondary replicas. In the case of immutable objects, cloning in multiple servers is virtually unrestricted.

A *query* service enables manipulating databases with object interfaces using highly declarative languages such as SQL or OQL. Alternative query services may provide support for querying semistructured data such as XML repositories.

A *grouping* service supports the creation of interrelated collections of objects, with different ordering and uniqueness properties, such as sets and lists. Different object collections provided by programming language libraries have similar functionality, albeit restricted to a specific programming language.

38.1.3 Virtual Machines

Although CORBA's approach to heterogeneity is to specify interactions among object request brokers to deal with different data representations and object services, an alternative approach is to hide hardware and operating system heterogeneity under a uniform virtual machine layer (e.g., see Lindholm and Yellin [1997]). The virtual machine approach provides certain benefits but also has its limitations [Agha et al., 1998].

The main benefit of a virtual machine is platform independence, which enables safe remote code execution and dynamic program reconfiguration through bytecode verification and run-time object migration [Varela and Agha, 2001]. In principle, the virtual machine approach is programming language independent because it is possible to create bytecode from different high-level programming languages. In practice, however, bytecode verification and language safety features may prevent compiling arbitrary code in unsafe languages, such as C and C++, into Java bytecodes without using loopholes such as the Java native interface, which break the virtual machine abstraction.

The main limitations of the *pure* virtual machine approach are bytecode interpretation overhead in program execution and the inability to control heterogeneous resources as required in embedded and real-time systems. Research on just-in-time and dynamic compilation strategies has helped overcome the virtual machine bytecode execution performance limitations [Krall, 1998]. Open and extensible virtual machine specifications attempt to enable the development of portable real-time systems satisfying hard scheduling constraints and embedded systems with control loops for actuation [OVM Consortium, 2002; Schmidt et al., 1997; Bollela et al., 2000].

38.1.4 Adaptability and Reflection

Next-generation distributed systems will need to satisfy varying levels of quality of service, dynamically adapt to different execution environments, provide well-founded failure semantics, have stringent security requirements, and be assembled on-the-fly from heterogeneous components developed by multiple service providers.

Adaptive middleware [Agha, 2002] will likely prove to be a fundamental stepping stone to building next-generation distributed systems. Dynamic run-time customization can be supported by a reflective architecture. A reflective middleware provides a representation of its different components to the applications running on top of it. An application can *inspect* this representation and modify it. The modified services can be installed and immediately mirrored in further execution of the application (see Figure 38.2). We will describe the reflective model of actors in the next section.

38.2 Worldwide Computing

Worldwide computing research addresses problems in viewing dynamic networked distributed resources as a coordinated global computing infrastructure. We have developed a specific actor-based worldwide computing infrastructure, the *World-Wide Computer* (WWC), that provides naming, mobility, and coordination middleware, to facilitate building widely distributed computing systems over the Internet.

Worldwide computing applications view the Internet as an execution environment. Because Internet nodes can join and leave a computation at runtime, the middleware infrastructure needs to provide dynamic reconfiguration capabilities: In other words, an application needs to be able to decompose and recompose itself while running, potentially moving its subcomponents to different network locations.

38.2.1 Actor Model

In traditional object-oriented systems, the interrelationship between objects — as state containers, as threads, and as process abstractions — is highly intertwined. For example, in Java [Gosling et al., 1996], multiple threads may be concurrently accessing an object, creating the potential for state corruption. A class can declare all its member variables to be `private` and all its methods to be `synchronized` to prevent

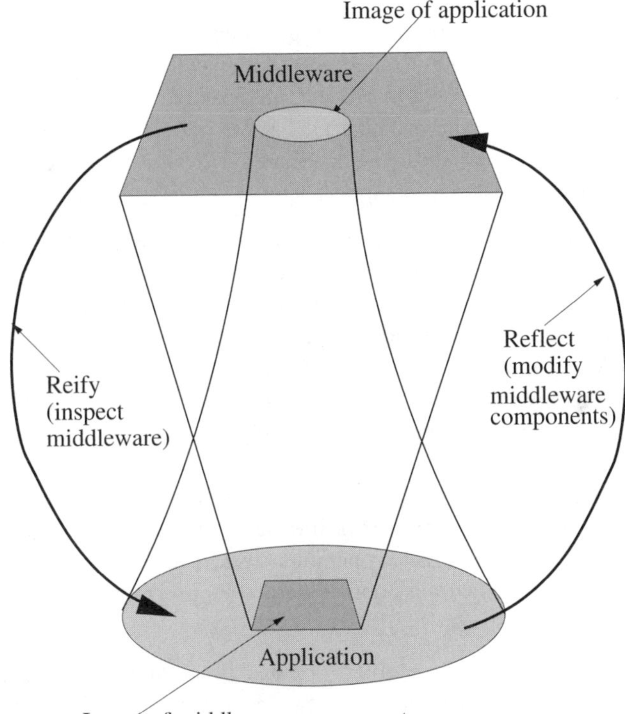

FIGURE 38.2 Using reflection, an application can inspect and modify middleware components.

state corruption because of multiple concurrent thread accesses. However, this practice is inefficient and creates potential for deadlocks (see e.g., citepvarela-agha-www7-98). Other languages, such as C++, do not even have a concurrency model built-in, requiring developers to use thread libraries.

Such passive object computation models severely limit application reconfigurability. Moving an object in a running application to a different computer requires guaranteeing that active threads within the object remain consistent after object migration and also requires very complex invocation stack migration, ensuring that references remain consistent and that any locks held by the thread are safely released.

The actor model of computation is a more natural approach to application reconfigurability because an actor is an autonomous unit abstracting over state encapsulation and state processing. Actors can only communicate through asynchronous message passing and do not share any memory. As a consequence, actors provide a very natural unit of mobility and application reconfigurability. Actors also provide a unit of concurrency by processing one message at a time. Migrating an actor is then as simple as migrating its encapsulated state along with any buffered unprocessed messages.

Reconfiguring an application composed of multiple actors is as simple as migrating a subset of the actors to another computer. Because communication is asynchronous and buffered, the application semantics remains the same as long as actor names can be guaranteed to be unique across the Internet. The universal actor model is an extension of the actor model, providing actors with a specific structure for universal names.

38.2.2 Language and Middleware Infrastructure

Several libraries that support the Actor model of computation have been implemented in different object-oriented languages. Three examples of these are the Actor Foundry [Open Systems Lab, 1998], Actalk [Briot, 1989], and Broadway [Sturman, 1996]. Such libraries essentially provide high-level middleware services, such as universal naming, communication, scheduling, and migration.

An alternative is to support distributed objects in a language rich enough to enable coordination across networks. Several actor languages have also been proposed and implemented to date, including ABCL, [Yonezawa, 1990], Concurrent Aggregates [Chien, 1993], Rosette [Tomlinson et al., 1989], and Thal [Kim, 1997]. An actor language can also be used to provide interoperability between different object systems; this is accomplished by wrapping traditional objects in actors and using the actor system to provide the necessary services. There are several advantages associated with directly using an actor programming language, as compared to using a library to support actors:

- **Semantic constraints:** Certain semantic properties can be guaranteed at the language level. For example, an important property is to provide complete encapsulation of data and processing within an actor. Ensuring that there is no shared memory or multiple active threads within an otherwise passive object is very important to guarantee safety and efficient actor migration.
- **API evolution:** Generating code from an actor language, it is possible to ensure that proper interfaces are always used to communicate with and create actors. In other words, programmers cannot incorrectly use the host language. Furthermore, evolutionary changes to an actor API need not affect actor code.
- **Programmability:** Using an actor language improves the readability of programs developed. Often, writing actor programs using a framework involves using language-level features (e.g., method invocation) to simulate primitive actor operations (e.g., actor creation or message sending). The need for a permanent semantic translation, unnatural for programmers, is a very common source of errors.

Our experience suggests that an active object-oriented programming language — one providing encapsulation of state and a thread manipulating that state — is more appropriate than a passive object-oriented programming language (even with an actor library) for implementing concurrent and distributed systems to be executed on the Internet.

38.2.3 Universal Actor Model and Implementation

The universal actor model extends the actor model [Agha, 1986] by providing actors with universal names, location awareness, remote communication, migration, and limited coordination capabilities [Varela, 2001].

We describe *Simple Actor Language System and Architecture* (SALSA), an actor language and system that has been developed to provide support for worldwide computing on the Internet. Associated with SALSA is a runtime system that provides the necessary services at the middleware level [Varela and Agha, 2001]. By using SALSA, developers can program at a higher level of abstraction.

SALSA programs are compiled into Java bytecode and can be executed stand-alone or on the World-Wide Computer infrastructure. SALSA programs are compiled into Java bytecode to take advantage of Java virtual machine implementations in most existing operating systems and hardware platforms. SALSA-generated Java programs use middleware libraries implementing protocols for universal actor naming, mobility, and coordination in the World-Wide Computer. Table 38.1 relates different concepts in the World Wide Web to analogous concepts in the World-Wide Computer.

TABLE 38.1 Comparison of WWW and WWC Concepts

	World Wide Web	World Wide Computer
Entities	Hypertext documents	Universal actors
Transport protocol	HTTP	RMSP/UANP
Language	HTML/MIME types	Java bytecode
Resource naming	URL	UAN/UAL
Linking	Hypertext anchors	Actor references
Run-time support	Web browsers/servers	Theaters/Naming servers

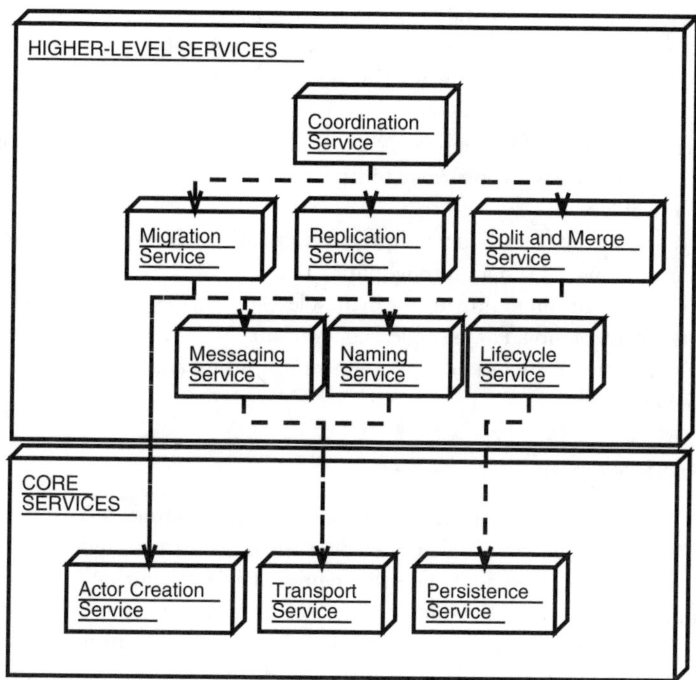

FIGURE 38.3 Core services include actor creation, transportation, and persistence. Higher-level services include actor communication, naming, migration, and coordination.

38.2.4 Middleware Services

Services implemented in middleware to support the execution of SALSA programs over the World-Wide Computer can be divided into two groups: *core services* and *higher-level services*, as depicted in Figure 38.3.

38.2.4.1 Core Services

An *actor creation* service supports the creation of new actors (which consists of an initial state, a thread of execution, and a mailbox) with specific behaviors. Every created actor is in a continuous loop, sequentially getting messages from its mailbox and processing them. Concurrency is a consequence of the fact that multiple actors in a given program may execute in parallel.

A *transport* service supports reliable delivery of data from one computer to another. The transport service is used by the higher-level remote communication, migration, and naming services.

A *persistence* service supports saving an actor's state and mailbox into secondary memory, either for checkpoints, fault tolerance or improved resource (memory, processing, power) consumption.

38.2.4.2 Higher-Level Services

A *messaging* service supports reliable asynchronous message delivery between peer actors. A message in SALSA is modelled as a potential Java method invocation. The message along with optional arguments is placed in the target actor's mailbox for future processing. A message-sending expression returns immediately after delivering the message — not after the message is processed as in traditional method invocation semantics. Section 38.2.6 discusses this service in more detail.

A *naming* service supports universal actor naming. A *universal naming* model enables developers to uniquely name resources worldwide in a location-independent manner. Location independence is important when resources are mobile. Section 38.2.5 describes the universal actor naming model and protocol used by this service.

A *life-cycle* service can deactivate an actor into persistent storage for improved resource utilization. It can also reactivate the actor on demand. Additionally, it performs distributed garbage collection.

A *migration* service enables actor mobility, preserving universal actor names and updating universal actor locations. Migration can be triggered by the programmer using SALSA messages, or it may be triggered by higher-level services such as load balancing and coordination. Section 38.2.6 provides more details on the actor migration service.

A *replication* service can be used to improve locality and access times for actors with immutable state. It can also be used for improving concurrency in parallel computations when additional processing resources become available.

A *split-and-merge* service can be used to fine-tune the granularity of homogeneous actors doing parallel computations to improve overall system throughput.

Coordination services are meant to provide the highest level of services to applications, including those requiring reflection and adaptation. For example, a load-balancing service can profile resource utilization and automatically trigger actor migration, replication, and splitting and merging behaviors for coordinated actors [Desell et al., 2004].

38.2.5 Universal Naming

Because universal actors are mobile (their location can change arbitrarily) it is critical to provide a universal naming system that guarantees that references remain consistent upon migration.

Universal Actor Names (UANs) are identifiers that represent an actor during its life time in a location-independent manner. An actor's UAN is mapped by a naming service into a *Universal Actor Locator* (UAL), which provides access to an actor in a specific location. When an actor migrates, its UAN remains the same, and the mapping to a new locator is updated in the naming system. As universal actors refer to their peers by their names, references remain consistent upon migration.

38.2.5.1 Universal Actor Names

A UAN refers to an actor during its life time in a location-independent manner. The main requirements on universal actor names are location independence, worldwide uniqueness, human readability, and scalability.

We use the Internet's Domain Name System (DNS) [Mockapetris, 1987] to hierarchically guarantee name uniqueness over the Internet in a scalable manner. More specifically, we use Uniform Resource Identifiers (URI) [Berners-Lee et al., 1998] to represent UANs. This approach does not require actor names to have a specific naming context because we build on unique Internet domain names.

The universal actor name for a sample address book actor is:

uan://wwc.yp.com/~smith/addressbook/

The protocol component in the name is `uan`. The DNS server name represents an actor's *home*. An optional port number represents the listening port of the naming service — by default 3030. The remaining name component, the *relative UAN*, is managed locally at the home name server to guarantee uniqueness.

38.2.5.2 Universal Actor Locators

An actor's UAN is mapped by a naming service into a UAL, which provides access to an actor in a specific location. For simplicity and consistency, we also use URIs to represent UALs. Two UALs for the address book actor above are:

rmsp://wwc.yp.com/~smith/addressbook/

and

rmsp://smith.pda.com:4040/addressbook/

The protocol component in the locator is rmsp, which stands for the *Remote Message Sending Protocol*. The optional port number represents the listening port of the actor's current *theater*, or single-node runtime system — by default 4040. The remaining locator component, the *relative UAL* is managed locally at the theater to guarantee uniqueness.

Although the address book actor can migrate from the user's laptop to his or her personal digital assistant (PDA) or cellular phone, the actor's UAN remains the same, and only the actor's locator changes. The naming service is in charge of keeping track of the actor's current locator.

38.2.5.3 Universal Actor Naming Protocol

When an actor migrates, its UAN remains the same, and the mapping to a new locator is updated in the naming system. The *Universal Actor Naming Protocol* (UANP) defines the communication between an actor's theater and an actor's home during its life time, which involves creation and initial binding, migration, and garbage collection.

UANP is a text-based protocol resembling HTTP with methods to create a UAN to UAL mapping, to retrieve a UAL given the UAN, to update a UAN's UAL, and to delete the mapping from the naming system. The following table shows the different UANP methods:

Method	Parameters	Action
PUT	Relative UAN, UAL	Creates a new entry in the database
GET	Relative UAN	Returns the UAL entry in the database
DELETE	Relative UAN	Deletes the entry in the database
UPDATE	Relative UAN, UAL	Updates the UAL entry in the database

A distributed naming service implementation can use consistent hashing to replicate UAN to UAL mappings in a ring of hosts and provide a scalable and reasonable level of fault tolerance. The logarithmic lookup time can further be reduced to a constant lookup time in most cases [Tolman, 2003].

38.2.5.4 Universal Naming in SALSA

The SALSA pseudo-code for a sample address book management program is shown in Figure 38.4. The program creates an address book manager and binds it to a UAN. After the program successfully terminates, the actor can be remotely accessed by its name.

38.2.6 Remote Communication and Mobility

The underlying middleware used by SALSA-generated Java code uses an extended version of Java-object serialization for both remote communication and actor migration.

```
behavior AddressBook {
    String getEmail(String name){...}
    void act(String[] args){
        try [
        AddressBook AddressBook=new AddressBook()at
            ("uan://wwc.yp.com/~smith/addressbook/";
            "rmsp://wwc.yp.com/~smith/addressbook/");
        } catch (Exception e){
            standardOutput<-println(e);
        }
    }
}
```

FIGURE 38.4 Universal actor name and locator binding in SALSA.

38.2.6.1 Remote Message Sending Protocol

Universal actors communicate with peers by passing messages asynchronously. When actors are executing in remote theaters, an Internet-based protocol is used for such communication — the *Remote Message Sending Protocol* (RMSP).

RMSP is a protocol implemented as an extension to Java object serialization. An actor's theater contains an RMSP server that listens for incoming messages from actors in remote theaters. Such messages are targeted to an actor with a locator local to the receiving theater. The theater keeps track of hosted actors and their locators so that incoming messages can be properly passed to the target actor.

Messages are represented as potential method invocations along with an optional continuation. Arguments are passed by value for primitive types and Java *serializable objects,* and by reference for universal actors.

In addition to passing information by using object serialization, RMSP updates universal actor references so that most efficient access can be performed for peer actors that are hosted in the target theater. Interested readers are referred to Varela [2001] for details.

38.2.6.2 Universal Actor Migration Protocol

Universal actors can move from one theater to another by processing a migration-request message. A universal actor implementation contains a thread of execution and encapsulated state. In response to a migration request, a universal actor's state (including buffered, unprocessed messages) is serialized, and a new thread of execution is started at the receiving theater.

We reuse RMSP for universal actor migration. The theater's RMSP server accepts incoming objects and acts upon them based on their type. Currently, there are two types of objects: Message for asynchronous message passing and UniversalActor for actor migration. An actor migration involves several steps: updating the naming service to reflect the actor's new locator; serializing the actor's state to the new theater; updating the actor's references to local resources, which we call *environment actors;* updating the theaters' metadata; and restarting the actor's thread in the new location.

To avoid potential race conditions that arise because of messages *en route* to migrating actors, we do not release the lock protecting the migrating actor's mailbox until the actor has completed the migration. Once the actor has acknowledged migration completion, the actor mailbox lock is released, and messages are rerouted appropriately by the runtime system, with the assurance that the actor is ready to receive them.

38.2.6.3 Remote Communication and Actor Migration in SALSA

The code for sending a getEmail () message to the address book manager created in the previous section, is shown in Figure 38.5. The code gets a reference to the address book manager using its UAN, sends a message requesting a user's email address, and prints it out in the console.

```
//
// Getting a remote actor reference and sending a message:
//

AddressBook addressBook = (AddressBook)
             AddressBook AddressBook.getReferenceByName
             ("uan://wwc.yp.com/~smith/addressbook/");
addressBook<-getEmail("David") @
     standardOutput<-println(token);
```

FIGURE 38.5 Remote communication in SALSA.

```
//
// Migrating an address book actor to a remote theater:
//
AddressBook addressBook=(AddressBook)
              AddressBook.getReferenceByName
              ("uan://wwc.yp.com/~smith/addressbook/");
a<-migrate ("rmsp://smith.pda.com/addressbook/");
```

FIGURE 38.6 Actor migration in SALSA.

SALSA also enables migrating an actor to a given theater. For example, the code for migrating the address book manager above is shown in Figure 38.6. In this case, the actor is migrated to a new location given by a UAL (rmsp://smith.pda.com/addressbook/).

38.2.7 Reflection

We use an explicit representation of the implementation of actors to provide a mechanism for customization of middleware [Astley and Agha, 1998]. Such a representation is called the *meta-level* system. Thus, in a reflective architecture, a system is composed of two kinds of actors: base-level (application) actors and meta-level actors or *meta-actors*. Meta-actors are part of the middleware that manages system resources and implements the base-actor's runtime semantics.

In implementation terms, actors do not directly interact with one another. Instead, actors make *system calls* to the middleware — these calls correspond to invocations of methods in meta-actors. A system call by an actor is always blocking and the actor waits till the call returns. A meta-actor executes the method that is invoked by another actor and returns control on completion of the execution.

A meta-actor is capable of customizing the behavior of another actor by executing the method invoked by it. Multiple customizations may be applied to a single actor by building a *meta-level stack* (see Figure 38.7). A meta-level stack consists of a single base-actor and a stack of meta-actors on top of it, where each meta-actor customizes the actor that is just below it in the stack. Messages received by an actor in a meta-level stack are always delegated to the top of the stack so that the meta-actor always controls the delivery of messages to its base-actor (See Figure 38.8). Similarly, messages sent by an actor pass through all the meta-actors in the stack.

Conceptually, we can translate actor operations into method calls to a meta-actor. These operations are:

- **transmit(msg):** This method is invoked when an actor sends a message (msg). If the actor has a meta-actor on its top, it calls the transmit method of the meta-actor and waits for its return. The method returns without any value. Otherwise, if the actor is not customized by a meta-actor, it passes the message to the system for sending.
- **create(beh):** This method is invoked when the actor wants to create another actor with a given behavior (beh). If there is a meta-actor on top of the actor, it calls the create method of the meta-actor and waits for it return. The method returns the address of the new actor. Otherwise, the actor passes the create request to the system.
- **ready:** The ready method is invoked when an actor has completed processing the current message and is waiting for another message. If the actor has a meta-actor on its top, it calls the ready method of the meta-actor and waits for its return. The method returns a message to the base-

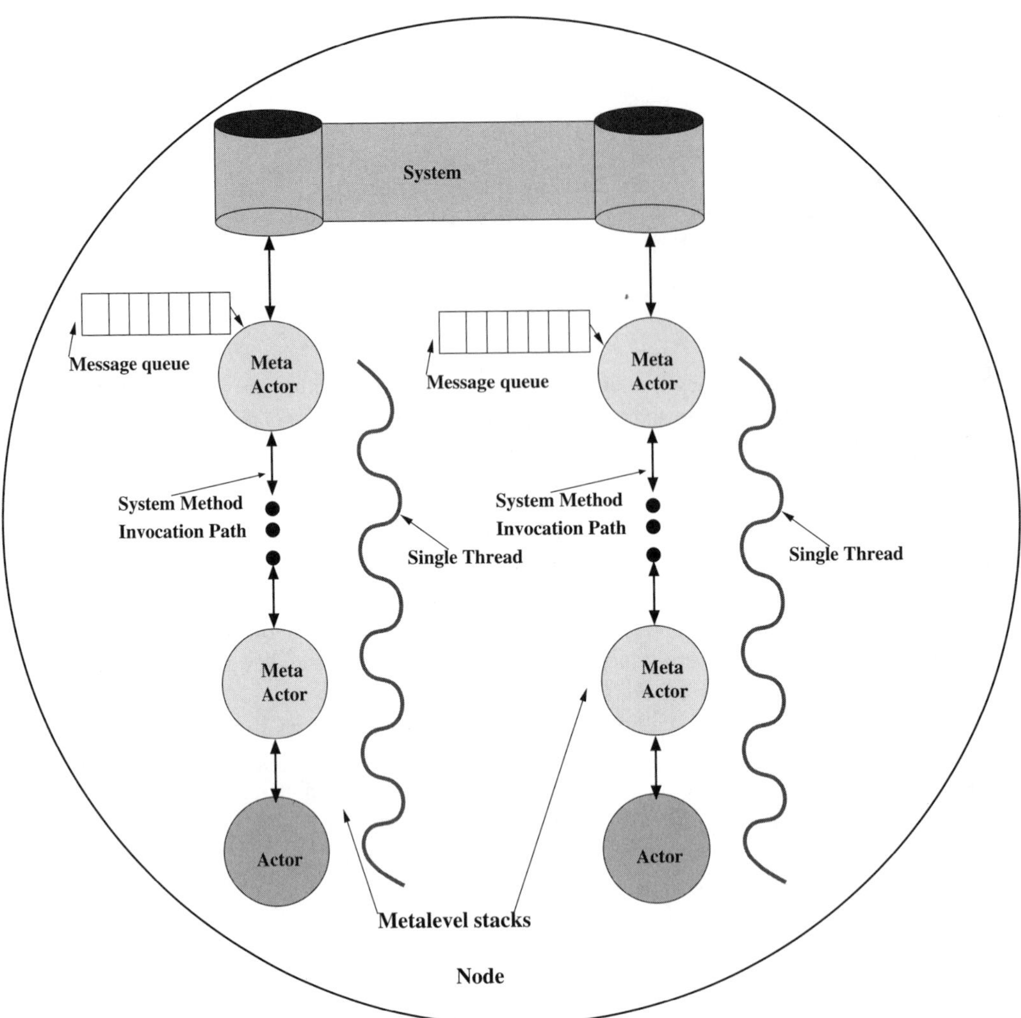

FIGURE 38.7 Meta-level stacks in an embedded node.

actor. Otherwise, the actor picks up a message from its mail queue and processes it. Note that there is a single mail-queue for a given meta-level stack.

Every meta-actor has a default implementation of the three-system method, which is given below:

- **transmit(msg):** If there is a meta-actor on its top, it calls the transmit(msg) method of that meta-actor and waits for it to return. Otherwise, it asks the system to send the message to the target and returns.
- **ready():** If there is a meta-actor on top of it, it calls the ready() method of that meta-actor and waits for it to return a message. Otherwise, the actor, by definition located at the top of the meta-level stack, dequeues a message from the mail queue. After getting the message, the actor returns the message to the base actor.
- **create(beh):** If the actor has a meta-actor at its top, it calls the create(beh) method of that meta-actor and waits for the actor to return with an actor address. Otherwise, the actor passes the create request to the system and waits till it gets an actor address from the system. After receiving new actor address, the actor returns it to the base actor.

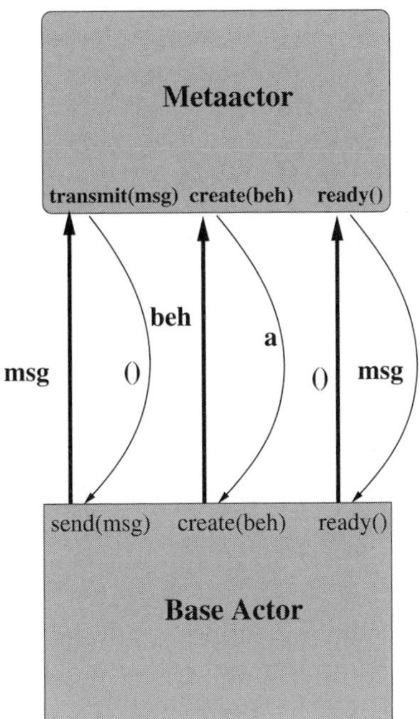

FIGURE 38.8 Interaction between meta-actor and base-actor.

As an example of how we may customize actors under this model, consider the encryption of messages between a pair of actors. Figure 38.9 gives pseudo-code for a pair of meta-actors, which may be installed on each endpoint. The Encrypt meta-actor implements the transmit method, which is called by the base-actor while sending a message. Within transmit, a message is encrypted before it is sent to its target. The Decrypt meta-actor implements the ready method, which is called when the base-actor is ready to process a message. The method ready decrypts the message before returning it to the base-actor.

Thus, the abstraction of the middleware in terms of meta-actors gives the power of dynamic customization. Meta-actors can be installed or pulled out dynamically (see Figure 38.10). The installation and removal of meta-actors by the application itself makes it capable of customizing the middleware. Applications can thus control adaptive middleware by reification and reflection. This means that although middleware can provide default policies for resource profiling, secure communication, load balancing, and coordination, applications can override default mechanisms and provide their own resource management policies.

38.3 Related Work

38.3.1 Worldwide Computing

Several research groups have been trying to achieve distributed computing on a large scale. Berkeley's NOW project has been effectively distributing computation on a "building-wide" scale [Anderson et al., 1995], and Berkeley's Millennium project is exploiting a hierarchical cluster structure to provide distributed computing on a "campus-wide" scale [Buonadonna et al., 1998]. The Globus project seeks to enable the construction of larger *computational grids* [Foster and Kesselman, 1998]. Caltech's Infospheres project has a vision of a worldwide pool of millions of objects (or agents), much like the pool of documents on the World Wide Web today [Chandy et al., 1996]. WebOS seeks to provide operating system services, such as client authentication, naming, and persistent storage, to wide area applications [Vahdat et al., 1998]. UIUC's 2K is an integrated operating system architecture addressing the problems of resource

```
actor Encrypt(){
  actor server;
  //Instantiated with name of server
  init(actor):={
    server:=S;
  }
  //Encrypt outgoing messages if they
  //are targeted to the server
  method transmit(Msg msg){
    actor target = msg.dest;
    if(target==server)
       target ← encrypt(msg);
    else
     target ← msg;
    continue();
  }
}
actor Decrypt(){
  //Decrypt incoming messages targeted for
  //base actor (if necessary)
  method rcv(Msg msg){
    if (encrypted(msg))
      deliver(decrypt(msg));
    else
      deliver(msg);
  }
}
```

FIGURE 38.9 The Encrypt policy actor intercepts transmit signals and encrypts outgoing messages. The Decrypt policy actor intercepts messages targeted for the server (i.e., the rcv method) and, if necessary, decrypts an incoming message before delivering it.

management in heterogeneous networks, dynamic adaptability, and configuration of component-based distributed applications [Kon et al., 1999].

Most approaches to worldwide computing are either operating system dependent or application dependent, or require a set of computers under the administrative control of its users.

In application-dependent worldwide computing, a domain-specific program is downloaded by users to provide their computing power to the data analysis task at hand. For example, the Search for Extra Terrestrial Intelligence (SETI) project at Berkeley uses idle computers from participants around the world to analyze space data in search of patterns and potential "signals" from intelligent life in outer space [Sullivan et al., 1997]. Interesting extensions include: a separation between office (SETI@Work) and home computers (SETI@Home) in the global computing task; and another project started at Stanford to understand how proteins fold (Folding@Home).

In other middleware approaches to worldwide computing, such as the Grid [Foster and Kesselman, 1998], it is required of users to have accounts (logins and passwords) in the systems that take part in the global computation. Although this approach may work well for certain groups of users with multiple supercomputer accounts, it does not properly scale to Internet computing. On the Internet, there is a wide variety of hardware architectures, operating systems, and domains of administrative control.

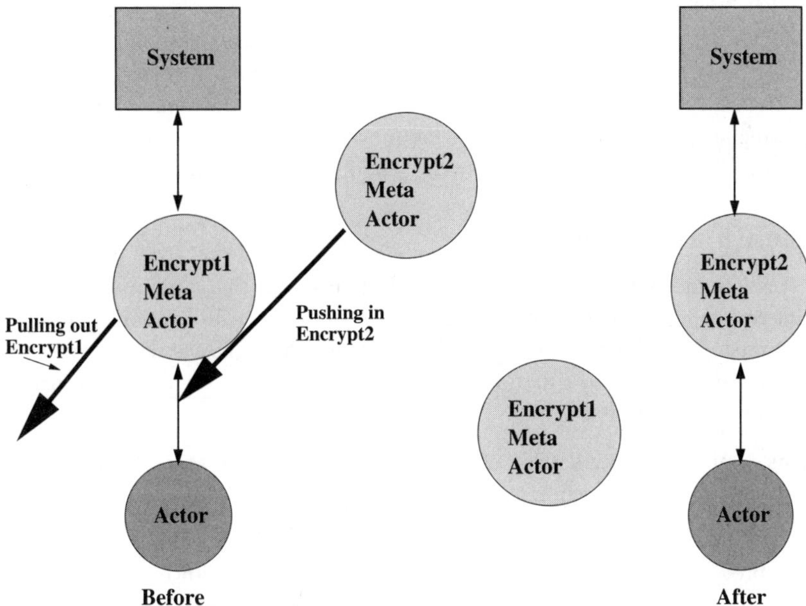

FIGURE 38.10 Dynamic customization is enabled by polling out and pushing in new meta-actors for the implementation of the encryption algorithm.

The WWC attempts to provide middleware for distributed system developers to use a heterogeneous network of Internet-connected computers, under multiple administrative domains, in an application-independent manner.

38.3.2 Languages for Distributed and Mobile Computation

The ABCL family of languages has been developed by Yonezawa's research group [Yonezawa, 1990] to explore an object-oriented concurrent model of computation, based on actors. ABCL has been developed in Common Lisp. One significant difference is that the order of messages from one object to another is preserved in their model. There are also three types of message-passing mechanisms: past, now, and future. The *past* type of message passing is nonblocking, as in actors. The *now* type is a blocking (RPC) message, with the sender waiting for a reply. The *future* type is a nonblocking message with a reply expected in the future. SALSA's more general continuation-passing style can be used to implement *now* and *future* message passing.

THAL, an extension to HAL (High-level Actor Language) [Houck and Agha, 1992], was developed by Kim [1997] to explore compiler optimizations and high performance actor systems. As a high-performance implementation, THAL has taken away features such as reflection and inheritance from HAL. THAL provides several communication abstractions including concurrent call/return communication, delegation, broadcast, and local synchronization constraints. THAL has shown that with proper compilation techniques, parallel actor programs can run as efficiently as their nonactor counterparts. Future research includes studying optimizations of SALSA actor programs, in particular, the actor model's data encapsulation enables eliminating most of Java's synchronization overhead.

Gray et al. [2000] present a very complete survey of mobile agent systems categorized by the programming languages they support. Agent systems supporting multiple programming languages include: Ara, D'Agents, and Tacoma. Java-based systems include Aglets [Lange and Oshima, 1998], Concordia, Jumping Beans, and Voyager. Other systems supporting a non-Java single programming language include: Messengers, Obliq, Telescript, and Nomadic Pict [Wojciechowski and Sewell, 1999].

Obliq [Cardelli, 1995] is a lexically scoped, untyped, interpreted language with an implementation relying on Modula 3's network objects. Obliq has higher-order functions, and static scope: closures transmitted over the network retain network links to sites that contain their free (global) variables.

Emerald [Jul et al., 1988], one of the first systems supporting fine-grained migration, includes different parameter-passing styles: call by reference, call by move, and call by visit. Instance variables can be declared *attached,* allowing arbitrary depth traversals in object serialization.

38.3.3 Naming Middleware

ActorSpaces [Callsen and Agha, 1994] is a communication model that compromises the efficiency of point-to-point communication in favor of an abstract pattern-based description of groups of message recipients. ActorSpaces are computationally passive containers of actors. Messages may be sent to one or all members of a group defined by a destination pattern. The model decouples actors in space and time and introduces three new concepts:

- **Patterns** — allow the specification of groups of message receivers according to their attributes
- **ActorSpaces** — provide a scoping mechanism for pattern matching
- **Capabilities** — give control over certain operations of an actor or actorspace

ActorSpaces provide the opportunity for actors to communicate with other actors by using their attributes. The model provides the equivalent of a Yellow Pages service, where actors may publish (in ActorSpace terminology, "make visible") their attributes to become accessible. Berners-Lee, in his original conception of Uniform Resource Citations [1998], intended to use this type of metadata to facilitate semiautomated access to resources. ActorSpaces bridge the gap between actors searching for a particular service and actors providing it.

Smart Names or *Active Names* (WebOS) provide scalability of read-only resources in the World Wide Web by enabling application-dependent name resolution in Web clients [Vahdat et al., 1998]. Using smart names, a client may find the most highly available resource that, depending on the application, may be the closest in distance or the one that provides the best quality of service.

The 2K distributed operating system [Ken et al., 1999] builds upon and enhances CORBA's naming system [Hydari, 1999]. CORBA objects have Interoperable Object References (IORs), which 2K objects can find through locally available clerks. *Junctions* enable the use of other resource naming spaces, such as DNS or a local file system. A special junction could be used to refer to WWC actors from 2K objects.

38.3.4 Remote Communication and Migration Middleware

The Common Object Request Broker Architecture (CORBA) [Object Management Group, 1997] has been designed with the purpose of handling heterogeneity in object-based distributed systems. Sun's JINI [Waldo, 1998] architecture has a goal similar to that of CORBA; the main difference between the two is that the former is Java-centric. One of the main components of JINI is the Remote Method Invocation (RMI) [Sun Microsystems and JavaSoft, 1996]. JavaSpaces [Sun Microsystems and JavaSoft, 1998] is another important component of JINI that uses a Linda-like [Carriero and Gelernter, 1990] model to share, coordinate, and communicate tasks in Jini-based distributed systems.

Java [Gosling et al., 1996] was the first programming language allowing Web-enabled secure execution of remote mobile code. Such mobile code, called *applets,* is downloaded, verified, and interpreted in a virtual sandbox, protecting the executing host from potentially insecure operations such as reading and writing from secondary memory and opening arbitrary network connections. IBM Aglets [Lange and Oshima, 1998] is a more recent framework for the development of Internet agents that can migrate, preserving their state.

38.3.5 Adaptive and Reflective Middleware

Adaptive middleware in distributed systems has been studied by several researchers. Gul Agha et al. [1993] have introduced meta-actors to implement different interaction services such as fault tolerance, security, and synchronization. Fabio Kon et al. [2002] have presented a model of reflective middleware that allows dynamic inspection and modification of the execution semantics of running applications as a response to changing resources in a distributed environment in order to improve performance. Research has also been done at the level of middleware security. Venkatasubramanian [2002] discussed the safe composibility of reflective middleware services to ensure the trustworthiness of systems. The Two-Level Actor Machine (TLAM) model [Venkatasubramanian and Talcott, 1995] is a reasoning framework for specifying and proving properties about interactions of middleware components.

Varela and Agha [1999] introduced a hierarchical model that groups actors into casts coordinated by *directors*. The cast directors are meta-level actors that filter incoming messages to group members. The hierarchical model is more general than the stack model presented in this chapter, in that a single meta-actor can coordinate more than one base actor. The hierarchical model does not restrict actor creation or message sending, only message reception. This restriction is valid in that actors are reactive entities, i.e., all computation proceeds in response to messages.

38.4 Research Issues and Summary

Worldwide computing is the coordinated use of large-scale, network-connected resources for global computations and human collaboration. In this chapter, we defined the universal actor model and linguistic abstractions for the coordination of globally distributed actors. We also described middleware services, such as naming and mobility, that are needed to implement actors on the World-Wide Computer.

There are still several open research problems that need to be solved before worldwide computing will become more common. These problems include:

- A security model that enables participants to safely volunteer their resources without risking loss of their information, and that enables worldwide computing users to trust the validity of global computations and the privacy of their data.
- A fine-grained resource management framework, which enables participants to control and be compensated for the computing resources they provide. For example, participants may wish to volunteer only a fixed percentage of their computing, communication, and storage capabilities.
- Higher-level coordination abstractions to facilitate the programming of worldwide computing systems in a way that is largely transparent to systems issues such as load balancing and fault tolerance.

A worldwide computing infrastructure including a universal actor programming language (SALSA) and several middleware services at different stages of development (WWC/IO) are freely available at http://www.cs.rpi.edu/wwc/salsa/. This infrastructure can be used as a modular starting point for implementing additional middleware services, or for developing distributed computing applications to be executed over the Internet or over Grid-like environments.

Worldwide computing will enable new classes of applications, where the dividing line between the physical world of communicating devices and the logical world of computing actors will get thinner and thinner. Physical devices are becoming more powerful and interconnected and logical actors are becoming more mobile and autonomous.

Mobility of devices and actors with different granularities in heterogeneous networks will induce *ad hoc* emergent coordination and interaction behavior patterns. These self-coordinated actor systems will enable efficient use of scarce resources and will clear the way for the creation of complex computing systems on very large scales. The availability of virtually unlimited storage, communication, and data-processing capabilities will open a new door for applications in many domains including science, education, business, government, and technology.

38.5 Further Information

Worldwide computing research is published in several computer science journals, including *ACM Transactions of Internet Technologies* and *IEEE Internet Computing.* There are several conferences devoted to special topics critical to worldwide computing, e.g., the International Conference on Coordination Models and Languages (COORDINATION), the ACM/IFIP/USENIX International Middleware Conference (MIDDLEWARE), the IEEE International Symposium on Cluster Computing and the Grid (CCGRID), and the International World Wide Web Conference (WWW) series. Online bibliographies for many of these conferences and journals can be found on the World Wide Web.

38.6 Glossary

Common Object Request Broker Architecture (CORBA): Suite of protocols, languages, and software systems for object-based distributed computing.

HyperText Markup Language (HTML): Language used to write Web content, including hypertext references or anchors.

HyperText Transfer Protocol (HTTP): Lightweight network protocol designed for information exchange between Web servers and clients.

Local Area Network (LAN): Computer network physically co-located, which is characterized by low latencies and high bandwidth.

Middleware: Software layers in between applications and operating systems dealing with distributed computing issues, such as naming, mobility, security, load balancing, and fault tolerance.

Remote Method Invocation (RMI): Java programming language API and framework for RPC-style synchronous interactions between objects.

Remote Message Sending Protocol (RMSP): Network protocol for asynchronous message exchange between WWC actors. Used by Java code generated from code written in the SALSA programming language.

Remote Procedure Call (RPC): Protocol for invoking remote procedures and marshaling and unmarshaling arguments across a network.

Simple Actor Language System and Applications (SALSA): Programming language facilitating the development of WWC applications using the universal actor model.

Uniform Resource Identifier (URI): Generic term uniformly denoting names, locators, or citations for worldwide resources.

Uniform Resource Locator (URL): URI specifying the location of a Web resource.

Universal Actor Locator (UAL): URI specifying the location of a universal actor.

Universal Actor Name (UAN): URI specifying the name of a universal actor, transparent to its location.

Universal Actor Naming Protocol (UANP): Network protocol to exchange information with a naming service regarding universal actor names and. locations.

Wide Area Network (WAN): Computer network spread over an area larger than a single building, potentially across continents. Characterized by higher latencies and lower bandwidths. AU: O

World-Wide Computer (WWC): Suite of protocols, languages, and software systems for world-wide computing using universal actors.

Worldwide computing: Area in computer science and engineering that studies all aspects related to using a wide area network as a computing and collaboration platform.

World Wide Web (WWW): Suite of protocols, languages, and software systems for worldwide information exchange.

Acknowledgments

Many ideas presented here are the result of countless discussions in the Open Systems Laboratory at the University of Illinois at Urbana-Champaign; in particular, we would like to express our gratitude to Mark Astley, Marcelo D' Amorim, Nadeem Jamali, Yusuke Tada, Prassanna Thati, Koushik Sen, James Waldby, and Reza Ziaei for helpful discussions about this work, and for some of the figures. We also thank members of the Worldwide Computing Laboratory at Rensselaer Polytechnic Institute for many fruitful discussions about worldwide computing; in particular, we would like to express our gratitude to Travis Desell, Kaoutar El Maghraoui, and Abe Stephens for continued development of the SALSA programming language and IOS middleware framework. The work described here has been supported in part by DARPA IXO NEST Program under contract F33615-01-C-1907, by the DARPA IPTO TASK Program under contract F30602-00-2-0586, and by ONR under MURI contract N00014-02-1-0715.

References

Agha, G. *Actors: A Model of Concurrent Computation in Distributed Systems.* MIT Press, Cambridge, MA, 1986.

Agha, Gul A. Introduction: Adaptive middleware. *Communications of the ACM,* 45(6): 30–32, June 2002.

Agha, G., M. Astley, J. Sheikh, and C. Varela. Modular heterogeneous system development: A critical analysis of Java. In J. Antonio, Ed., *Proceedings of the Seventh Heterogeneous Computing Workshop (HCW '98),* pp. 144–155. IEEE Computer Society, March 1998. http://osl.cs.uiuc.edu/Papers/HCW98.ps.

Agha, G., S. Frølund, R. Panwar, and D. Sturman. A linguistic framework for dynamic composition of dependability protocols. In *Dependable Computing for Critical Applications III,* pp. 345–363. International Federation of Information Processing Societies (IFIP), Elsevier, Amsterdam, 1993.

Anderson, Thomas E., David E. Culler, and David A. Patterson. A case for networks of workstations: NOW. *IEEE Micro,* February 1995.

Astley, M. and G. A. Agha. Customization and composition of distributed objects: Middleware abstractions for policy management. In *Sixth International Symposium on the Foundations of Software Engineering (FSE-6, SIGSOFT '98),* November 1998.

Banavar, G., T. Chandra, B. Mukherjee, J. Nagarajarao, R. Strom, and D. Sturman. An efficient multicast protocol for content-based publish-subscribe systems. In *19th International Conference on Distributed Computing Systems (19th ICDCS '99),* IEEE, Austin, TX, May 1999.

Berners-Lee, T., R. Fielding, and L. Masinter. Uniform Resource Identifiers (URI): Generic Syntax. IETF Internet Draft Standard RFC 2396, August 1998. http://www.ietf.org/rfc/rfc2396.txt.

Bollela, G., J. Gosling, B. Brosgoland, P. Dibble,. S. Furr, and M. Turnbull. *The Real-Time Specication for Java.* Addison-Wesley, Reading, MA, June 2000. Available from http://www.rtj.org/rtsj-V1.0.pdf.

Briot, J.-P. Actalk: A testbed for classifying and designing actor languages in the Smalltalk-80 environment. In *Proceedings of the European Conference on Object Oriented Programming* (ECOOP '89), pp. 109–129. Cambridge University Press, Cambridge, U.K., 1989.

Brown, N. and C. Kindel. Distributed component object model protocol — dcom/1.0. Technical report, Microsoft, May 1996. http://dsl.internic.net/internet-drafts/draft-brown-dcom-v1-spec-00.txt.

Buonadonna, Philip, Andrew Geweke, and David E. Culler. An implementation and analysis of the virtual interface architecture. In *Proceedings of Supercomputing* '98, Orlando, FL, November 1998.

Callsen, C. and G. Agha. Open heterogeneous computing in ActorSpace. *Journal of Parallel and Distributed Computing,* pp. 289–300, 1994.

Cardelli, L. A language with distributed scope. *Computing Systems,* 8(1): 27–59, January 1995. URL: http://research .microsoft.com/Users/luca/Papers/Obliq.A4.pdf.

Carriero, N. and D. Gelernter. *How to Write Parallel Programs.* MIT Press, Cambridge, MA, 1990.

Chandy, K. M., A. Rifkin, P. A. G. Sivilotti, J. Mandelson, M. Richardson, W. Tanaka, and L. Weisman. A worldwide distributed system using Java and the Internet. In *Proceedings of the 5th IEEE International Symposium on High Performance Distributed Computing*, New York, August 1996.

Chien, A. *Concurrent Aggregates: Supporting Modularity in Massively Parallel Programs.* MIT Press, Cambridge, MA, 1993.

Curbera, Francisco, Matthew Duftler, Rania Khalaf, William Nagy, Nirmal Mukhi, and Sanjiva Weerawarana. Unraveling the Web services web: An introduction to SOAP, WSDL, and UDDI. *IEEE Distributed Systems Online*, 3(4), 2002. URL: http://dsonline.computer.org/0204/features/wp2spot-htm.

Desell, Travis, Kaoutar El Maghraoui, and Carlos Varela. Load balancing of autonomous actors over dynamic networks. In *Proceedings of the Adaptive and Evolvable Software Systems: Techniques, Tools, and Applications Minitrack of the Software Technology Track of the Hawaii International Conference on System Sciences (HICSS '37)*, January 2004.

Foster, I. and C. Kesselman. The Globus project: A status report. In J. Antonio, Ed., *Proceedings of the 7th Heterogeneous Computing Workshop (HCW '98)*, pp. 4–18. IEEE Computer Society, March 1998.

Gosling, J., B. Joy, and G. Steele, *The Java Language Specification.* Addison-Wesley, Reading, MA, 1996.

Gray, R., D. Kotz, G. Cybenko, and D. Rus. Mobile agents: Motivations and state-of-the-art systems. Technical report, Darmouth College, April 2000. Available at ftp://ftp.cs.dartmouth.edu/TR/TR2000-365.ps.Z.

Houck, C. and G. Agha. HAL: A high-level actor language and its distributed implementation. In *Proceedings of the 21st International Conference on Parallel Processing (ICPP '92)*, Vol. 2, pp. 158–165, St. Charles, IL, August 1992.

Hydari, M. Design of the 2K Naming Service. M.S. thesis. Department of Computer Science. University of Illinois, Urbana-Champaign, IL, February 1999.

Jul, Eric, Henry M. Levy, Norman C. Hutchinson, and Andrew P. Black. Fine-grained mobility in the Emerald system. *TOGS*, 6(1): 109–133, 1988.

Kim, W. THAL: An Actor System for Efficient and Scalable Concurrent Computing. Ph.D. thesis, University of Illinois, Urbana-Champaign, IL, May 1997.

Kon, F., R. Campbell, M. Dennis Mickunas, and K. Nahrstedt. 2K: A Distributed Operating System for Dynamic Heterogeneous Environments. Technical report, Department of Computer Science, University of Illinois, Urbana-Champaign, IL, December 1999.

Kon, F., F. Costa, G. Blair, and Roy H. Campbell. The case for reflective middleware. *Communications of the ACM*, 45(6): 33–38, 2002.

Krall, Andreas. Efficient JavaVM just-in-time compilation. In Jean-Luc Gaudiot, Ed., *International Conference on Parallel Architectures and Compilation Techniques*, pp. 205–212, Paris, 1998. North-Holland. URL citeseer.nj.nec.com/krall98efficient.html.

Lange, D. and M. Oshima. *Programming and Deploying Mobile Agents with Aglets.* Addison-Wesley, Reading, MA, 1998.

Lindholm, T. and F. Yellin. *The Java Virtual Machine Specification.* Addison-Wesley, Reading, MA, 1997.

Mockapetris, P. Domain Names — Concepts and Facilities. IETF Internet Draft Standard RFC 1034, November 1987. http://www.ietforg/rfc/rfc1034.txt.

Object Management Group. CORBA services: Common object services specification version 2. Technical report, Object Management Group, June 1997. http://www.omg.org/corba/.

Open Systems Lab. The Actor Foundry: A Java-based Actor Programming Environment, 1998. Work in Progress. http://osl.cs.uiuc.edu/foundry/.

OVM Consortium. OVM An Open RTSJ Compliant JVM. http://www.ovmj.org/, 2002.

Schmidt, Douglas C., Aniruddha Gokhale, Timothy H. Harrison, and Guru Parulkar. A high-performance endsystem architecture for real-time CORBA. *IEEE Communications Magazine*, 14(2), 1997. URL: citeseer.nj.nec.com/schmidt97highperformance.html.

Sturman, D. *Modular Specification of Interaction Policies in Distributed Computing*, Ph.D. thesis, University of Illinois, Urbana-Champaign, IL, May 1996. TR UIUCDCS-R-96-1950.

Sullivan, W. T., D. Werthimer, S. Bowyer, J. Cobb, D. Gedye, and D. Anderson. A New Major SETI Project based on project SERENDIP data and 100,000 Personal Computers. In *Proceedings* of *the 5th International Conference on Bioastronomy.* Editrice Compositori, Bologna, Italy, 1997.

Sun Microsystems and JavaSoft. Remote Method Invocation Specification, 1996. http://www.javasoft.com/products/jdk/rmi/.

Sun Microsystems and JavaSoft. JavaSpaces, 1998. http://www.javasoft.com/products/Javaspaces/.

Tolman, Camron. A Fault-Tolerant Home-Based Naming Service for Mobile Agents. M.S. thesis, Rensselaer Polytechnic Institute, Troy, NY, April 2003. http://www.cs.rpi.edu/wwc/theses/fhns/cam_thesis_final.pdf.

Tomlinson, C., W. Kim, M. Schevel, V. Singh, B. Will, and G. Agha. Rosette: An object oriented concurrent system architecture. *Sigplan Notices,* 24(4): 91–93, 1989.

Vahdat, Amin, Thomas Anderson, Michael Dahlin, David Culler, Eshwar Belani, Paul Eastham, and Chad Yoshikawa. WebOS: Operating System Services For Wide Area Applications. In *Proceedings of the 7th IEEE Symposium on High Performance Distributed Computing,* July 1998.

Varela, C. Worldwide Computing with Universal Actors: Linguistic Abstractions for Naming, Migration, and Coordination. Ph.D. thesis, University of Illinois, Urbana-Champaign, IL, April 2001.

Varela, C. and G. Agha. A hierarchical model for coordination of concurrent activities. In P. Ciancarini and A. Wolf, Eds., *3rd International Conference on Coordination Languages and Models (COORDINATION '99),* Springer-Verlag, Berlin, LNCS 1594, pp. 166–182, April 1999. http://osl.cs.uiuc.edu/Papers/Coordination99.ps.

Varela, Carlos and Gul Agha. Programming dynamically reconfigurable open systems with SALSA. *ACM SIGPLAN Notices. OOPSLA '2001 Intriguing Technology Track Proceedings,* 36(12): 20–34, December 2001. http://www.cs.rpi.edu/~cvarela/oopsla2001.pdf.

Venkatasubramanian, N. Safe composibility of middleware services. *Communications of the ACM,* 45(6): 49–52, 2002.

Venkatasubramanian, Nalini and Carolyn Talcott. Meta-architectures for resource management in open distributed systems. In *Proceedings of the ACM Symposium on Principles of Distributed Computing,* ACM Press, New York, pp. 144–153, August 1995.

Waldo, J. JINI Architecture Overview. Work in progress. 1998, http://www.javasoft.com/products/jini/.

Waldo, Jim, Geoff Wyant, Ann Wollrath, and Sam Kendall. A note on distributed computing. In *Mobile Object Systems: Towards the Programmable Internet,* pp. 49–64. Springer-Verlag, Heidelberg, Germany, 1997. URL: citeseer.nj.nec.com/waldo94note.html.

Wojciechowski, Pawel and Peter Sewell. Nomadic Pict: Language and Infrastructure Design for Mobile Agents. In *1st International Symposium on Agent Systems and Applications (ASA '99)lThird International Symposium on Mobile Agents (MA '99),* Palm Springs, CA, 1999. URL: citeseer.nj.nec.com/article/wojciechowski99nomadic.html.

Yonezawa, A., Ed. ABCL *An Object-Oriented Concurrent System.* MIT Press, Cambridge, MA, 1990.

39

Metacomputing and Grid Frameworks

Vaidy Sunderam

CONTENTS

Abstract 39-1
39.1 Introduction 39-1
39.2 Historical Evolution of Network Computing 39-2
39.3 MPI and Network Computing 39-4
39.4 Computational Grids 39-4
39.4.1 Definitions 39-5
39.4.2 Example Grid Infrastructures 39-8
39.4.3 Programming Grids 39-10
39.5 Applicability Issues 39-10
39.5.1 Simplicity and Flexibility 39-10
39.5.2 Clusters and Standards 39-11
39.5.3 Performance Issues 39-11
39.5.4 Grid Environments 39-11
39.6 Current Trends 39-11
39.6.1 OGSA, OGSI, and GTK3 39-12
39.6.2 Components and Portals 39-14
39.7 Summary 39-14
References 39-15

Abstract

In this chapter, we discuss software infrastructures for resource aggregation and sharing across computer networks. Such technologies encompass clustering, network computing, metacomputing, and grids, which are all unified by their common goal of parallel distributed computing, typically for high performance applications. Until recently, these modes of computing were viewed as addressing a different set of concerns than were client–server platforms and distributed databases. However, convergence is possible if both domains adopt Web-service-based paradigms. This chapter traces the evolution of metacomputing frameworks over the past two decades, highlighting the strengths of and distinctions between different approaches. A summary discussion of each framework outlining the programming model supported, runtime software support, and generic usage scenarios is presented. A discussion of recent trends in metacomputing systems and directions for future research concludes the chapter.

39.1 Introduction

Network-based concurrent computing is attractive for a number of reasons, especially its ability to aggregate resources for high performance. This key ability distinguishes metacomputing frameworks and

1-58488-381-2/05/$0.00+$1.50
© 2005 by CRC Press LLC

grids from traditional distributed computing [1,2], in which the entities are inherently distributed or are replicated for fault tolerance, and from networking applications that focus on communications rather than computing. In network computing, resources, primarily compute cycles, from a number of *independent* computer systems are pooled to deliver parallel computing capabilities for compute-intensive applications. In various forms, therefore, network computing systems emulate parallel processors, albeit with generally slower interprocessor communication and synchronization capabilities. In exchange, they offer nearly infinite flexibility, permitting heterogeneous collections of machines of widely varying capabilities and running general purpose operating systems to be pooled via COTS networks. They can also be scaled incrementally, deployed on local or wide area networks across organizational boundaries, and are resistant to obsolescence.

Network computing is facilitated by software or "middleware" frameworks that essentially perform the logistics of aggregation and present a coherent concurrent programming environment to applications. Over the past decade, de-facto standard distributed computing environments such as Parallel Virtual Machine (PVM) [3] and certain implementations of the Message Passing Interface (MPI) [4], e.g., MPICH (or MPI-Chameleon), have been popular. These systems harness a collection of specified computers that have independent operating systems and network interfaces, and provide a message-passing API implemented as a library that application programs link against. Applications are typically SPMD, although other models are possible; individual processes communicate and synchronize with their peers via the provided interfaces and with assistance from the middleware substrate, cooperatively solve the problem at hand.

Recently however, this model has been expanded to suggest that high performance and unconventional applications are best served by sharing geographically distributed resources in a well-controlled, secure, and mutually fair way. Such coordinated worldwide resource sharing requires an infrastructure called a *grid.* Although grids are viewed as the successors of distributed computing environments in many respects, there are some substantial differences between the two that have not been clearly articulated, partly because there is no widely accepted definition for grids. There are common views held about grids; some define them as high-performance distributed environments, some take into consideration their geographically distributed, multidomain nature, and others define grids based on the number of resources they unify, and so on. In this chapter, we trace the evolution of network computing systems and their dovetailing into grids, highlight the fundamental characteristics and functionalities of grids, and outline current trends in metacomputing.

39.2 Historical Evolution of Network Computing

Distributed access to remote resources has existed for many years, and Remote Job Entry in the 1960s and Remote Procedure Call in the early 1980s may be considered precursors to network computing. However, it was only with the advent of high-performance RISC workstations in the late 1980s that network computing received serious consideration as a high-performance computing platform. Early software frameworks that supported workstation aggregation for HPC included P4, Linda, Express, and PVM. The P4 system, named after the book describing it (Portable Programs for Parallel Processors), is a library of macros and subroutines for programming a variety of parallel machines in C and Fortran. Although shared- and distributed-memory multiprocessors were the primary targets, a TCP/IP socket-based implementation was also available. Express, a commercial product, provided a similar library for explicitly parallel computing, but emphasized graphical tools for program development, debugging, and administration. Linda was also based on explicit parallelism but proposed the notion of a tuple-space, or abstract associative memory, for communication and synchronization between peer processes.

PVM (Parallel Virtual Machine) is a software system that permits the utilization of a heterogeneous network of parallel and serial computers as a unified, general, and flexible concurrent computational resource. The PVM system [3] initially supported message passing, shared memory, and hybrid paradigms; thus allowing applications to use the most appropriate computing model for the entire application or for individual subalgorithms. However, support for emulated shared-memory was omitted as the system evolved, as the message-passing paradigm was the model of choice for most scientific parallel processing applications. The PVM system is composed of a suite of user-interface primitives and supporting software that together enable concurrent computing on loosely coupled networks of processing elements. PVM may be implemented on a hardware base consisting of different machine architectures, including single-CPU systems, vector machines, and multiprocessors — permitting users to dynamically select the best-suited computing resource for each component of an application. These computing elements may be interconnected by one or more networks, which may themselves be different (e.g., one implementation of PVM operates on Ethernet, the Internet [TCP/UDP/TP protocols], and a fiber optic network), These computing elements are accessed by applications via a standard interface that supports common concurrent processing paradigms in the form of well-defined primitives embedded in procedural host languages. Application programs are composed of *components* that are subtasks at a moderately large level of granularity. During execution, multiple *instances of* each component may be initiated. Figure 39.1 depicts a simplified architectural overview of the PVM computing model as well as the system.

Application programs view the PVM system as a general and flexible parallel computing resource. Translucent layering permits flexibility while retaining the ability to exploit particular strengths of individual machines on the network. The PVM user interface is strongly typed; support for operating in a heterogeneous environment is provided in the form of special constructs that selectively perform machine-dependent data conversions where necessary. Interinstance communication constructs include those for the exchange of data structures as well as high-level primitives such as broadcast, barrier synchronization, mutual exclusion, and rendezvous. Application programs under PVM may possess arbitrary control and dependency structures. In other words, at any point in the execution of a concurrent application, the processes in existence may have arbitrary relationships between each other and, further, any process may communicate or synchronize with any other. The PVM system is implemented in two parts. The first part is a daemon, called *pvmd,* that executes on all the computers comprising the virtual machine. PVM is designed so that any user having normal access rights to each host in the pool may install and operate the system. To run a PVM application, the user executes the daemons on a selected host pool, and the set of daemons cooperate via distributed algorithms to initialize the virtual machine. PVM applications may then be started by executing a program on any of these machines. The usual method is for this manually started program to spawn other application processes, using PVM facilities. Multiple users may configure overlapping virtual machines, and each user can execute several PVM applications simultaneously. The second part of the system is a library of PVM interface routines that contains user-callable routines for message passing, spawning processes, coordinating tasks, and modi-

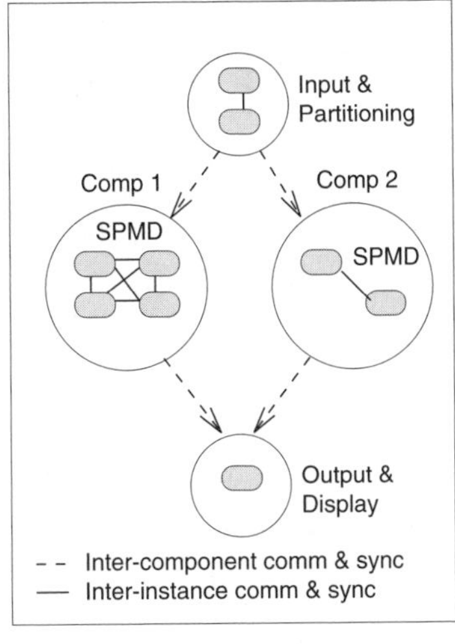

(a) PVM Computation Model

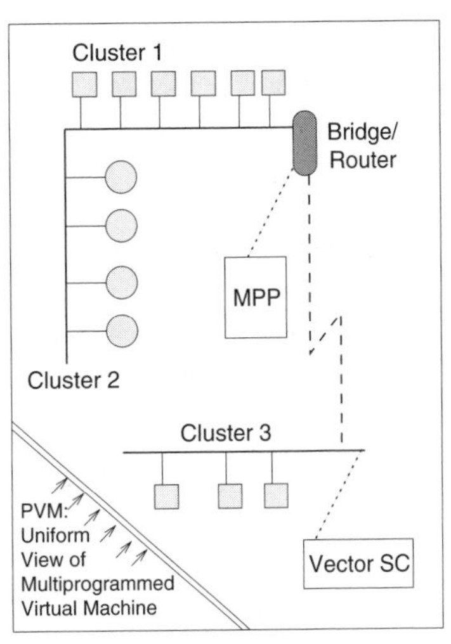

(b) PVM Architectural Overview

FIGURE 39.1 PVM programming and platform models

fying the virtual machine. PVM continues to be used widely, particularly in heterogeneous settings and for applications that exhibit dynamism in their resource and process needs.

39.3 MPI and Network Computing

Message Passing Interface (MPI) is a library specification for message-passing, proposed as a standard by a broad-based committee of vendors, implementers, and users. Although not ratified by conventional standards organizations, it is indeed a de-facto standard for message-passing parallel computing. However, MPI is only a specification of an application programming interface (API) for message-passing; the standard (version 1.1) defines the syntax and semantics of calls for message passing, collective communications, and synchronization among a group of tasks. The specification assumes that tasks have been instantiated and that they belong to a "communicator" or context within which they are identified through enumeration. The repertoire of MPI calls is very rich and comprehensive; in particular, very sophisticated facilities are available for specifying process topologies and for a variety of complex collective operations. Recent efforts (MPI versions 1.2 and 2.0) extend the standard to include process creation and management, one-sided communications, extended collective operations, external interfaces, I/O, and additional language bindings.

MPI implementations fall into several categories. Vendors of parallel machines and high-performance multiprocessors typically provide proprietary implementations of MPI that are tailored to match the hardware and interconnect architecture of the target platform. Efficiency is of paramount importance, and lowering overheads to attain communication performance within a few percent of that delivered by the raw hardware is not uncommon. Switch manufacturers and Beowulf cluster resellers contribute the second category of MPI implementations. Again, maximizing performance (as opposed to supporting heterogeneity or portability) is the overarching goal of these systems; typically, they are manifested as low-level software combined with device drivers appropriate to the switch (e.g., Myrinet or GigE) and to the computing nodes (typically Linux on x86 architectures). In straight parallel processing scenarios, these two types of MPI deployments are overwhelmingly the rule.

With respect to Internet computing, several implementations of MPI over TCP/IP protocols exist. Two well-known examples are MPICH and LAM MPI. The MPICH implementation (MPI/Chameleon) is available for both Unix and Windows platforms, and is designed to work over an abstract transport layer termed ADI (abstract device interface), thus permitting its deployment over different network interconnects. The TCP/IP implementation is most common, and is used in local clusters as well as across wide area networks. Given its historical evolution as a reference implementation that was developed concurrently with the standard, MPICH is very comprehensive and adheres to the semantics of the specification, both literally and implicitly. Currently, a newer version termed MPICH2 is being developed to incorporate both the MPI 1.1 and MPI 2 programming interfaces.

LAM/MPI is a high-quality open-source implementation of the Message Passing Interface specification, including all of MPI-1.2 and much of MPI-2. Intended for production as well as research use, LAM/MPI includes a rich set of features for system administrators, parallel programmers, application users, and parallel computing researchers. From its beginnings, LAM/MPI was designed to operate on heterogeneous clusters, and can also work with clusters of clusters. Several transport layers are supported by LAM/MPI and overheads are claimed to be low with TCP/IP, even on fast networks.

39.4 Computational Grids

During the late 1990s, the notion of computational grids evolved. Networking, driven by the Web, became ubiquitous at the low end but that phenomenon and other developments necessitated and otherwise motivated the increased deployment of high-speed wide area networks. Beginning with the I-Way experiment, such high-speed networks were viewed as facilitating large-scale distributed computing across countries and continents, and the notion of the "grid" began to take shape. The term itself is derived from the vision and eventual goal of delivering computation on demand, via the equivalent of electrical wall sockets, where providers and consumers are interconnected orthogonally to each other. However, the technical intent was essentially to provide a framework for large-scale metacomputing — "increasing delivered computation" is the principal motivation cited [5] as being the driving force behind grids.

39.4.1 Definitions

Definitions of computational grids vary widely and include:

"A flexible, secure, coordinated resource sharing among dynamic collections of individuals, institutions, and resources." [7]

"A single seamless computational environment in which cycles, communication, and data are shared, and in which the workstation across the continent is no less than one down the hall." [6]

"Wide area environment that transparently consists of workstations, personal computers, graphic rendering engines, supercomputers and nontraditional devices, e.g., TVs, toasters, etc." ?

"A collection of geographically separated resources (people, computers, instruments, and databases) connected by a high speed network [... distinguished by ...] a software layer, often called middleware, which transforms a collection of independent resources into a single, coherent, virtual machine." [9]

Upon careful inspection, however, it appears that these definitions (or visions) do not really distinguish grids from previous metacomputing systems. Many distributed systems have transparency as a goal, several frameworks have attempted to integrate a variety of devices, pervasiveness and dependability are standard objectives, and constructing a uniform abstraction is a common and well-established theme.

Two important aspects, however, may be considered as distinctions that differentiate computational grids from other types of distributed systems, namely, *resource virtualization* and *user virtualization*. Grids assume a virtual pool of *resources* rather than computational nodes. Although current systems mostly focus on computational resources (CPU cycles + memory) [10], grid systems are expected to operate on a wider range of resources, from storage, network, data, and software [6] to unusual resources such as graphical and audio input/output devices, manipulators, sensors and so on [8]. All these resources typically exist within nodes that are geographically distributed, and span multiple administrative domains. The virtual machine is made up of a set of resources taken from the pool. In grids, the virtual pool of resources is dynamic and diverse. Since computational grids are aimed at large-scale resource sharing, these resources can be added and withdrawn at any time at their owner's discretion, and their performance or load can change frequently over time. The typical number of resources in the pool is on the order of 1,000 or more [11]. Due to all these reasons, the user (and any agent acting on behalf of the user) has very little or no *a priori* knowledge about the actual type, state, and features of the resources constituting the pool.

Contrary to conventional systems that try to first find an appropriate node to map processes to, and then satisfy the resource needs locally, grids are based on the assumption of an abundant and common pool of resources. Thus, first the resources are selected and then the mapping is done according to the resource selection. The resource needs of a process are abstract in the sense that they are expressed in terms of resource types and attributes in general, e.g., 64 MB of memory or a processor of a given architecture or 200 MB of storage, etc. These needs are satisfied by certain physical resources, e.g., 64 MB memory on a given machine, an Intel PIII processor, and a file system mounted on the machine. Processes are mapped onto a node where these requirements can be satisfied. Since the virtual pool is large, dynamic, diverse, and the user has little or no knowledge about its current state, matching the abstract resources to physical ones cannot be solved at the user level or at the application level by selecting the right nodes, as is possible in the case of conventional environments. The virtual machine is constituted by the selected resources.

This virtualization of resources that are mapped to their physical counterparts within multiple administrative domains relates to the other major characteristic feature of grids, namely, user virtualization.

When resource access has to transcend administrative boundaries, a number of security issues, primarily dealing with access control, must become part of the distributed computing infrastructure, Access to the nodes hosting the needed resources cannot be controlled based on login access due to the large number of resources in the pool and the diversity of local security policies. It is unrealistic that a user have a login account on thousands of nodes simultaneously. Instead, higher level credentials are introduced at the virtual level that can identify the user to the resource owners, and based on this authentication they can authorize the execution of their processes as if they were local users.

39.4.1.1 Intrinsic Differences

The virtual machine of a conventional distributed application is constructed from the nodes available in the pool (Figure 39.2). Yet, this is just a different view of the physical layer and not really a different level of abstraction. Nodes appear on the virtual level exactly as they are at physical level, with the same names (e.g., *n1, n2* in Figure 39.2), capabilities, etc. There is an *implicit* mapping from the abstract resources to their physical counterparts because once the process has been mapped, resources local to the node can be allocated to it. Users have the same identity, authentication, and authorization procedure at both levels: they log in to the virtual machine as they would log in to any node of it.

On the contrary, in grid systems, both users and resources appear differently at virtual and physical layers. *Resources* appear as entities distinct from the physical node in the virtual pool. A process' resource needs can be satisfied by various nodes in various ways. There must be an *explicit* assignment provided by the system between abstract resource needs and physical resource objects. The actual mapping of processes to nodes is driven by the resource assignment.

Furthermore, in a grid, a user of the virtual machine is different from users (account owners) at the physical levels. Operating systems are based on the notion of processes; therefore, granting a resource involves starting or managing a local process on behalf of the user. Obviously, running a process is possible for local account holders. In a grid, a user has a valid access right to a given resource proven by some kind of credential (e.g., user Smith in Figure 39.2). However, the user is not authorized to log in and start processes on the node to which the resource belongs. A grid system must provide a functionality that finds a proper mapping between a user (a real person) who has the credentials to the resources and on whose behalf the processes work, and a local user ID (not necessarily a real person) that has a valid

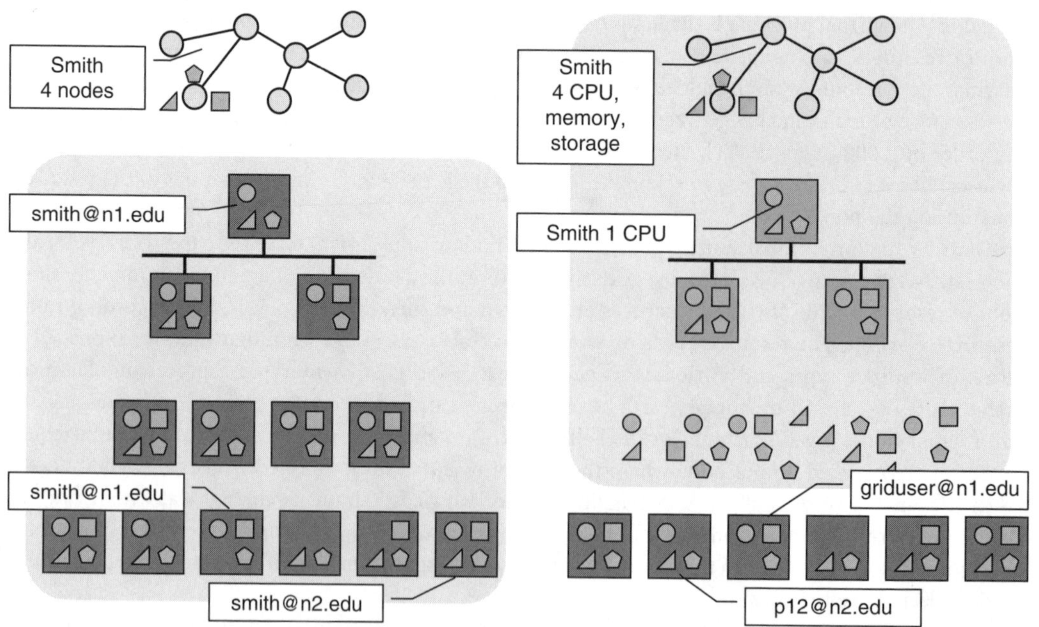

FIGURE 39.2 Conventional DCEs versus grids.

account and login rights on a node. The grid-authorized user temporarily has the rights of a local user for placing and running processes on the node.

Thus, in these respects, the physical and virtual levels in a grid are completely distinct, but there is a mapping between resources and users of the two layers. According to [12] these two fundamental features of grids are termed *user- and resource-abstraction*, and constitute the intrinsic difference between grids and other distributed systems.

39.4.1.2 Technical Differences

The fundamental differences introduced in the previous sections are at too high a level of abstraction to be patently evident in existing grid systems. In practice, the two fundamental functionalities of resource and user abstraction are realized on top of several services.

The key to resource abstraction is the selection of available physical resources based on their abstract appearance. First, there must be a notation provided in which the abstract resource needs can be expressed (e.g., Resource Specification Language (RSL) [13] of Globus and Collection Query Language [14] of Legion.) This specification must be matched to available physical resources. Since the user has no knowledge about the currently available resources and their specifics, resource abstraction in a real implementation must be supported at least by the following components that are independent of the application:

1. An *information system* that can provide information about resources upon a query and that can support both discovery and lookup. (Examples include the Grid Index Information Service (GIIS) [15] in Globus and Collection [14] in Legion).
2. A local *information provider* that is aware of the features of local resources, their current availability, load, and other parameters or, in general, a module that can update records of the information system either on its own or in response to a request. (Examples are the Grid Resource Information Service (GRIS) [15] in Globus and information reporting methods in Host and Vault objects in Legion [14].)

A user abstraction in a grid is a mapping of valid credential holders to local accounts. A valid (unexpired) credential is accepted through an authentication procedure, and the authenticated user is authorized to use the resource. Just as in the case of resource abstraction, this facility assumes other assisting services:

1. A *security mechanism* that accepts global user certificates and that authenticates users. In Globus this is the resource proxy process that is implemented as the gatekeeper as part of the GRAM [16].
2. Local *resource management* that authorizes authentic users to use certain resources. This is realised by the mapfile records in Globus that essentially control the mapping of users [16]. Authorization is then up to the operating system, based on the rights associated with the local account. (In Legion both authentication and authorization are delegated to the objects, i.e., there is no centralized mechanism, but every object is responsible for its own security [17].)

The subsystems listed above, with examples from the two currently popular grid frameworks viz. Globus and Legion, are the services that directly support user and resource abstraction. In addition to these basic functions, some grid environments and applications often insist on a greater level of assurance, both concerning resource availability from partner providers as well as security. In other words, provisions for "service level agreements" must be made. In the Globus toolkit, a protocol called SNAP[18] is supported for negotiating SLAs. Similarly, the nature and functionality of security mechanisms might be enhanced, for example, as in the "community authorization service" or CAS scheme[19]. These are examples of aspects relating to grids that are not strictly fundamental differences, and yet do not represent just logistical details. However, in practical grid implementations, other more mundane services are also necessary, e.g., staging, coallocation, etc., but these are more technical issues and are answers to the question of *how* grid mechanisms are realized, rather than to the question of *what* the grid model intrinsically contains.

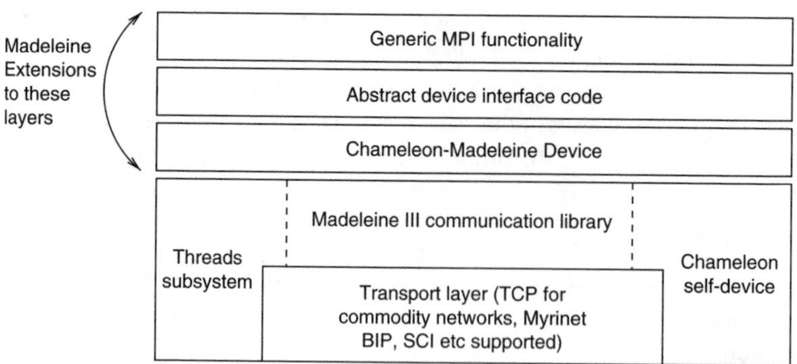

FIGURE 39.3 Madeleine Architecture.

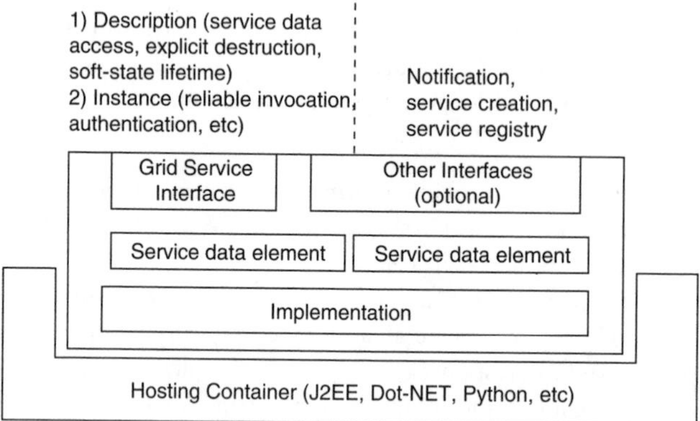

FIGURE 39.4 Grid Service Model.

39.4.2 Example Grid Infrastructures

The best-known software system that attempts to realize the vision of computational grids is Globus[10]. Currently in its third major incarnation, Globus is in fact a toolkit of related software subsystems that includes software services and libraries for resource monitoring, discovery, and management, besides security and file management. It is important to note that grids, in general, and Globus, in particular, have more to do with resource management than concurrent programming; indeed, most grid applications are developed and executed using (variants of) parallel programming environments such as MPI or CCA[20]. The major components of the production (release 2.4) version of Globus handle (1) information services; (2) resource management; (3) data management; and (4) security. The fourth component, Grid Security Infrastructure (GSI), augments the other three components with authentication and authorization facilities that they use to control distributed resource sharing.

The Globus Toolkit uses GSI for enabling secure authentication and communication over an open network. GSI provides a number of useful services for grids, including mutual authentication and single sign-on. GSI is based on public key encryption, X.509 certificates, and the Secure Sockets Layer (SSL) communication protocol. Extensions to these standards have been added for single sign-on and delegation. The Globus Toolkit's implementation of the GSI adheres to the Generic Security Service API (GSS–API), which is a standard API for security systems promoted by the Internet Engineering Task Force (IETF). The GSI uses public key cryptography as the basis of its functionality. A central concept in GSI authentication is the certificate. Every user and service on the grid is identified via a certificate, which contains information vital to identifying and authenticating the user or service. A GSI certificate includes four primary pieces of information: a subject name, the public key belonging to the subject, the identity of a Certificate Authority (CA) that has signed the certificate to certify that the public key and

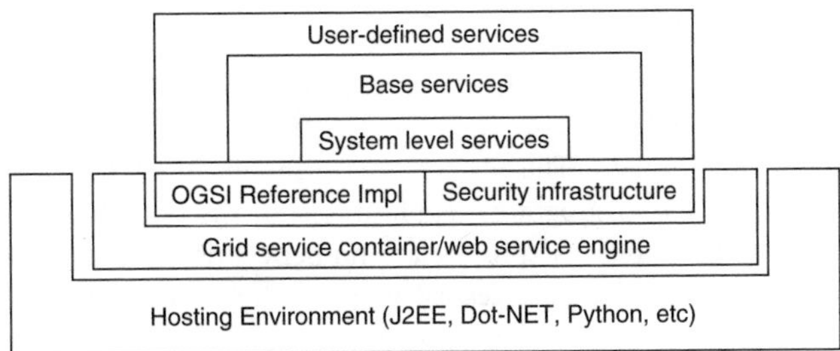

FIGURE 39.5 GT3 System Architecture.

the identity both belong to the subject, and the digital signature of the named CA. The GSI uses the Secure Sockets Layer (SSL) for its mutual authentication protocol, when both parties trust the CAs that signed each others certificates. By default, the GSI ensures data integrity and does not perform encrypted communication, but may be used to establish a shared key if needed. The major extensions to normal security mechanisms that the GSI supports are delegation and single-sign-on, particularly useful in grid computing given the large number of resources typically aggregated by users.

GRAM, the Globus resource management subsystem, layers high-level global resource management services on top of local resource allocation services. There are three main components to GRAM: an extensible resource specification language, the interface to local resource management tools, and a coallocator. This resource management service works in conjunction with the information service (MDS) and a broker (if one is available and is used). Requests specified in RSL are passed to the coallocator, which then attempts to allocate all the required resources by interfacing to multiple local allocation managers. The Globus Metacomputing Directory Service (MDS) serves as a clearinghouse for the whereabouts of such local resources. With assistance from the auxiliary Grid Resource Information Service (GRIS) and Grid Index Information Service (GIIS) subsystems, this service forms an LDAP-based information infrastructure for computational grids, and provides a uniform means of querying system information from a rich variety of system components, including computational nodes, data storage systems, scientific instruments, network links, and databases. The third major component of Globus, GridFTP, is a high-performance, secure, and reliable data transfer protocol optimized for high-bandwidth wide-area networks. Based on FTP, this service provides GSI security on control and data channels, multiple data channels for parallel transfers, partial file transfers, third-party (direct server-to-server) transfers, authenticated data channels, reusable data channels, and command pipelining.

A number of tools and portals related to Globus exist; representative examples will be mentioned later. In terms of middleware, alternatives include Legion, a metasystem software framework. Scalability, security and access control, and customizability are its main strengths. Although there are some similarities, Legion's is an integrated architecture as opposed to Globus' sum-of-services model. Legion's is a layered software system that comprises object management services (or core objects), a distributed file system, directory services resource management services, and security services, which are accessible via a unified, global address-space method invocation service. Legion is thus based on a single object-model architecture, and may be contrasted with Globe ?, which assumes a distributed object space that is shared and replicated locally for access. Globe objects use peer-to-peer communication: applications load (part of) the object implementation in their address space to participate in the distributed object. Users may contact any copy to have methods performed, but they know nothing about the internal structure and protocols used inside the object. This scheme allows different objects to use different algorithms for data partitioning, replication, consistency, and fault tolerance in a way transparent to the users. Objects have location-independent names and are constructed from a control subobject, a communications subobject, a replication subobject, a security subobject, and a semantics subobject that does the actual work.

39.4.3 Programming Grids

In the distributed BPC arena, there typically exists a three-way tension between support for legacy applications, support for quasi-portable code (e.g., that written in MPI), and the "native" programming model supported by the environment. The former class of applications tend to consist primarily of sequential legacy (e.g. 1960s Fortran programs) applications, or those applications that can be run in parameter-sweep mode and that often use proprietary numerical libraries that make them difficult to parallelize. Grid environments support this category of applications via interfaces to local scheduling or queueing systems such as PBS or NQS, enhancing such remote execution modalities with security and access control, staging, and convenient interfaces for resource requisitions and monitoring. Both Legion and Globus provide "portal" interfaces that are most useful for such applications. These portals may be Web-based or command-line-based, and provide a variety of functions that ease the housekeeping burden of organizing distributed resources on which to execute applications. At the other extreme, some grid computing environments (e.g., Legion and Globe) provide their own programming environments and models, in which framework-aware applications may be developed.

Substantial efforts, however, are devoted to executing portable parallel programs, typically written in MPI, on distributed metacomputing environments and grids. For example, MPICH-G2 [22] is a grid-enabled implementation of the MPI v1.1 standard, i.e., using services from the Globus Toolkit (e.g., job startup and security), MPICH-G2 allows users to couple multiple machines (potentially of different architectures) to run MPI applications. MPICH-G2 automatically converts data in messages sent between machines of different architectures and supports multiprotocol communication by automatically selecting TCP for intermachine messaging and (where available) vendor-supplied MPI for intramachine messaging. As another example, the library PACX-MPI (PArallel Computer eXtension) enables scientists and engineers to seamlessly run MPI-conforming parallel applications on a cluster or MPPs interconnected via the Internet. In a similar vein, MPICH–Madeleine is a free, MPICH-based implementation of the MPI standard that provides a multiprotocol implementation of MPI on top of a generic and multiprotocol communication layer called Madeleine III, the communication subsystem of the Parallel Multithreaded Machine (PM2) runtime environment. It was especially targeted at single clusters with several interconnection networks and clusters of clusters (possibly with several interconnection networks). Although these and similar systems combine the standardized MPI programming paradigm with the ability of grids to harness geographically distributed autonomic resources, they do not offer an elegant heterogeneous programming model that is naturally appropriate for grids. Several other limitations, including lack of support for operation across firewalls and performance variations caused by heterogeneous networks, also characterize these frameworks.

39.5 Applicability Issues

Given that numerous freely available middleware systems exist, it can often be confusing for end users to select the most appropriate framework for their needs. Unfortunately, this selection process is further complicated by the fact that *fundamental* differences between the various frameworks are few; with the exception of security-related issues, multiple systems might be usable in any given situation with appropriate costs and benefits. Therefore, we outline *one* possible characterization based on an orthogonal set of criteria that may be useful in selecting network computing middleware.

39.5.1 Simplicity and Flexibility

For beginning users with limited and varied resources, PVM is still considered the simplest and most flexible entry point. With a small resource footprint (for installation as well as use), PVM offers a complete, heterogeneous computing environment including dynamic resource and process management facilities. Yet, the programming interface, with its few tens of API calls, is intuitive and easy to learn. PVM is therefore almost universally acknowledged as a viable starting point for users who wish to be "up and running" in a short period. One price that PVM pays for this flexibility is performance. Since the most generic protocols (with multistage relay mechanisms) are used, communication efficiency is slightly suboptimal, which detracts from its viability for production runs. The API is also not an endorsed standard, although it is the only one that currently accomodates heterogeneous architectures as well as heterogeneous messages.

39.5.2 Clusters and Standards

For users willing to make medium-range investments in hardware, cluster solutions with an MPI implementation offer the best combination of performance and standardization. For a few thousand dollars per node, clusters of identical processors can be configured with high-speed network interconnects such as GigE or Myrinet, thereby providing some semblance of hardware multiprocessors. More importantly, several implementations of MPI that are specifically optimized for such networks exist. Therefore, applications based on the MPI standard are readily ported to such environments, which deliver near-optimal performance when managed appropriately.

39.5.3 Performance Issues

Appropriate management policies are as crucial to delivering high performance in networked systems as the components that comprise them. In other words, since components (both CPUs and networks) in such platforms are general purpose, often under control of a timesharing OS, resource allocation policies are extremely important. Normally, performance-critical applications are driven solely by time-to-completion criteria (as opposed, for example, to high-throughput or maximal utilization criteria). For such usage, *dedicated access* to the resources in question are critical. This is normally achieved, both in clusters and in grids, by (1) stripping the OS and daemon configurations to approximate a dedicated executive; and (2) running under control of a batch processing scheduler that serializes jobs. Unlike PVM, where the timesharing nature of each computer is reflected in a *timeshared virtual machine*, batch schedulers effectively provide dedicated access to resources, thus ensuring assured, if not guaranteed, performance.

39.5.4 Grid Environments

As discussed, the essence of grid middleware lies in its support for aggregating resources across multiple administrative domains. Given the (potential) differences in policies and authentication schemes in different parts of a virtual organization, only grid middleware among the various technologies provides an appropriate and integrated solution. On the other hand, such frameworks are large and complex, exhibit a fairly steep learning curve, and require substantial administrative support. Therefore, only a correspondingly small set of applications might consider the benefits worth the added effort. Indeed, Section D of the book *Grid Computing: Making the Global Infrastructure a Reality* [23] lists five or six applications, *all* of which are large-scale problems in extreme-high-end science and technology. Notwithstanding this, although much of the rationale and use case scenarios are based on high-end applications, grids may indeed be used on a smaller scale, if operation across multiple administrative domains is necessary. The second caveat with respect to grids concerns relatively less mature programming paradigms. Much of the effort to date has focused on resource management, and it is only recently that programming models, languages, and tools for grid computing have started receiving attention. These are discussed in the following section.

39.6 Current Trends

In the high-performance computing arena, grids continue to be the mainstay and primary focus of metasystems, and are viewed as complementary to compute engines such as Beowulf clusters and high-end supercomputers (e.g., the Earth Simulator and the Cray X1). As such, there has been a paradigm shift in metacomputing towards component- and service-based computing. In this section, we describe current examples of these approaches to Internet computing for HPC.

39.6.1 OGSA, OGSI, and GTK3

Given that grids transcend boundaries in terms of a multitude of aspects (machine architectures, programming languages, security mechanisms, and allocation policies, to name a few), it is not surprising that standardization, portability, and interoperability are crucial to any grid framework. As grid software evolved, proponents drew upon the Web services model that is rapidly gaining in popularity in commercial applications. Web services leverage XML and its derivative WSDL (Web services description language) to export services with assistance from standard Web components (the http protocol, hosting platforms, and backend database interfaces). Web services leverage the ubiquitous availability of these standard components, and add discovery mechanisms (e.g., UDDI), invocation frameworks (WSIF), and composability (BPEL4WS) to greatly enhance numerous commercial computing and interaction applications. Drawing upon these technologies, and motivated by the benefits of resource virtualization and standards, the grid community has developed the Open Grid Service Architecture (OGSA) as the basis for future grid efforts.

The central concept in OGSA is a grid service that is similar to a Web service in its use of interfaces/implementations, with access via multiple protocol bindings, transparency, and language interoperability. This model exemplifies the container abstraction found in many other environments, (e.g., J2EE) where common functions such as memory management, security, and transaction support are embedded in the container. However, in the case of grid services, extensions such as service semantics, lifetime management, statefulness, and certain other extensions are added. A typical grid service is described in WSDL and consists of syntax and semantics for clients to interact with service instances. A grid service instance embodies state and has one or more unique grid service handles (GSH) and one or more grid service references (GSR). Service instances are created by "factories" and are destroyed explicitly or via soft state. All grid services must implement several standard interfaces (e.g., FindServiceData and SetTerminationTime) and a factory that generates stubs, skeletons, and deployment mechanisms.

The Open Grid Services Infrastructure (OGSI) is a formal and technical specification of the concepts described in OGSA, including grid services. It defines a component model that extends WSDL and XML schema definitions to incorporate the concept of stateful Web services, asynchronous notifications, collections of and references to service instances, and extensions to describe service state data. There are two core requirements for describing Web services based on the OGSI: the ability to describe interface inheritance, which is a core concept with most distributed object systems; and the ability to describe additional information elements with the interface definitions, including service data. A service data declaration is a mechanism for publicly expressing the available state information of a service through a known schema, and can be static or dynamic.

An example of a software that implements the OGSI specification, which in turn provides specifics of the OGSA definition, is version 3 of the Globus toolkit (GTK3). This toolkit is Java based on its initial incarnation and involves the following steps for writing, deploying, and invoking a grid service:

- The grid service interface is written in GSDL, either by hand or by generating it automatically from a Java or IDL interface using a tool such as Java2WSDL.
- Both server and client side stubs are then generated from this decription using the GTK3-provided tool GSDL2Java.
- The implementation of the service is then written, derived from the "GridServiceImpl" base class and implementing the service interface. The methods specified in the interface are implemented and the superclass constructor is invoked.
- Having written or generated (1) a service interface, (2) a WSDL file, (3) server ad client side stub files, and (4) an implementation of the service, deployment is accomplished by creating a deployment descriptor that specifies publishing information for the service, compiling and creating "gar"

(grid archive) files containing the appropriate classes, and deploying it using "Globus-start-container."
- Combining the published information about the service with the location of the container constructs the grid service handle (GSH). Clients may obtain a reference to the service using the GSH and invoke methods on the service.

Figure 39.6 suggests that deploying OGSA grid services using GTK3 is similar to the procedures involved in deploying RMI, Web services, and certain types of enterprise applications, and follows the same interface-implementation-marshalling-container paradigm. The manual steps may be automated via tools such as Ant, but the process is essentially one of deploying services with standardized access techniques. GTK3 and OGSA, however, differ from simple RMI services or Web services in several major respects. First, as previously mentioned, grid service factories may be defined, thereby permitting multiple instances of services, generally one per client. Second, these services can be stateful. Finally, they can be transient, i.e., the service instance exists as long as needed, typically by clients.

To improve functionality and flexibility, grid services may also be implemented by delegation (in addition to implementation by inheritance from GridServiceImpl). Delegation permits the distribution of operations in a given port type into several classes, each of which may inherit from different base classes. This facility is supported by implementing an interface called "OperationProvider" that provides the base grid-service functionality; specific methods in this interface are used to indicate the functionality exported to the container. In conjunction with a modified deployment descriptor that uses a generic GridServiceImpl as its base class, and indicates that the services will be provided by an operation provider, substantial power and flexibility are gained in the service definition and deployment process. GT3 also provides tools to manage the lifecycle of grid services. One is the GridServiceCallback interface that includes methods invoked at various points in the lifetime of a grid service. A LifeCycleMonitor class is also provided and may be extended as appropriate to a specific grid service.

Grid services differ from typical Web services and other enterprise applications in that they are generally not instance-specific in terms of the service they provide. For example, numerous instances of a linear algebra service might be available, differentiated only by speed and capacity, not by the operations provided. OGSA and GT3 use the notion of "service data" to qualify grid service instances. Each instance of a grid service has an associated "service data set," containing zero or more "service data elements" (SDE). Finally, a "service data description" (SDD) describes a type of SDE; for a given service, the SDD

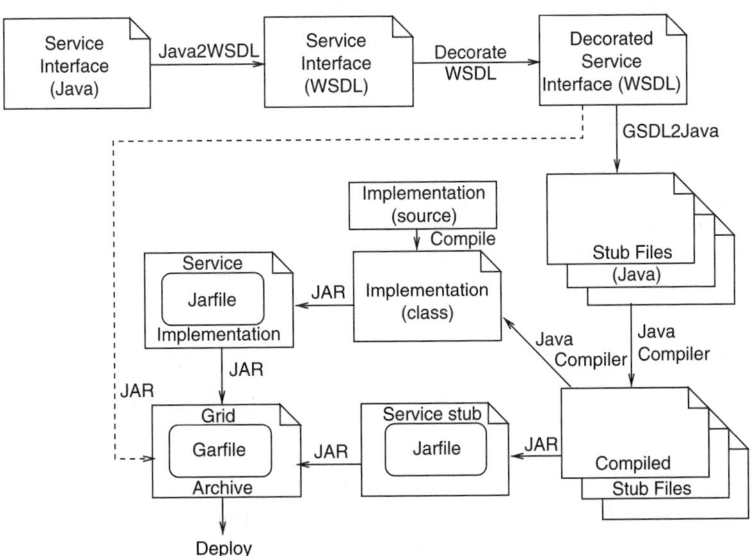

FIGURE 39.6 GT3 operational overview.

is expressed in XML and imported into the GWSDL description of the service. For a specific instance, a Java bean is generated containing methods to set the values of the SDE and may be invoked by the executing instance of the service. Clients may then use the GSH to query SDE values and to locate instances of the service satisfying specific characteristics.

Notifications complement service data in GT3. Observers may subscribe to particular service data elements that cause callbacks as appropriate. In order to use this facility, services are written to invoke a "notifyChange" method on a ServiceData object; pull- or push-mode notifications are possible, depending on whether any values are to be transmitted as part of the notification. Clients establish listeners that are invoked when values change and may or may not have to invoke additional methods on the service to obtain more information associated with the notification, depending on its type.

39.6.2 Components and Portals

Concurrent programming on an aggregated collection of resources using parallel computing models such as MPI is possible, but frameworks such as GT3 are more suited to service-oriented computations. For example, the CCA (Common Component Architecture) effort [20] defines a component architecture for large-scale scientific computing applications, and provides frameworks permitting components to be composed, deployed, and invoked. CCA defines the ports or public interfaces of components, event and dataflow facilities between components providing different services, interface description methods, component creation, and communications support. XCAT [24] is a component framework for distributed grid applications based on CCA that conforms to OGSI specifications. Conversion of a CCA-based XCAT component to a grid service consists of adding an OGSI GridService port and SDEs containing references to each of the other ports, creating a GSH corresponding to the ComponentID, and using a WSDL representation for the GSR. There are also some differences in messaging and factories, but overall, component architectures and grid service frameworks are well matched to each other. One reason for this suitability is that components tend to be large-grained modules possessing an interface that can be defined precisely — a definition that almost exactly fits a grid service. Second, different components in an application may be contributed by different organizations, again meshing well with the notion of resource sharing across multiple administrative domains. Finally, the nature of components makes them amenable to description using standardized markups (e.g., GSDL), deployment using the notion of factories, and appropriate entities for notifications.

In the context of Internet computing, there is also a trend toward a portal-based approach to accessing grids. In a typical scenario (e.g., Portal Expedition [25]) users log in to a (usually Web-based) portal, and within the resulting grid context, utilize a set of "portlets" to access grid resources. Portlets are configurable by users and can communicate with each other; with a GUI on the front end, they are each bound to a grid service to permit the user to interact with the service. This architecture greatly eases the housekeeping and logistics tasks associated with using grids, and in addition, can be used for related tasks such as collaboration, conferencing, and remote visualization. The main idea of portals is to provide an *application-oriented* Web-based gateway in which scientific-domain knowledge and tools are presented to users in terms of the application science and not in terms of complex distributed computing protocols[26].

39.7 Summary

Internet computing in the context of high-performance scientific applications is currently focused on two complementary technologies, namely, clusters and grids. Cluster platforms use a variety of network interconnects, including those that run standard Internet protocols, and are the mainstay of large number-crunching applications. Grids, often composed of clusters at multiple sites, deal primarily with resource access across multiple administrative domains, interoperability, and aggregation using a service-oriented paradigm.

References

1. H. Attiya and J. Welch. *Distributed Computing: Fundamentals, Simulations and Advanced Topics.* McGraw-Hill, 1998.
2. G. Coulouris, J. Dollimore and T. Kindberg. *Distributed Systems: Concepts and Design.* Addison-Wesley, Pearson Education 2001.
3. V.S. Sunderam, PVM: A framework for parallel distributed computing, *Concurrency: Practice and Experience*, Vol. 2, No. 4, pp. 315–339, December 1990.
4. W. Gropp et al., *MPI: The Complete Reference.* MIT Press, 1998.
5. I. Foster and C. Kesselman. *The Grid: Blueprint for a New Computing Infrastructure.* Morgan Kaufmann Publishers, 1999.
6. A.S. Grimshaw et al., Legion: The Next Logical Step Toward a Nationwide Virtual Computer. Technical report No. CS-94-21, University of Virginia, Charlottesville, June, 1994.
7. I. Foster, C. Kesselman and S. Tuecke. The anatomy of the grid, *International Journal of Supercomputer Applications*, Vol. 15(3), 2001.
8. A.S. Grimshaw and W.A. Wulf. *Legion — A View From 50,000 Feet. Proc. Fifth IEEE International Symposium on High Performance Distributed Computing*, IEEE Computer Society Press, Los Alamitos, California, August 1996.
9. G. Lindahl, A. Grimshaw, A. Ferrari and K. Holcomb. Metacomputing — What's in it for me? White paper tp://legion.virginia.edu/papers.html, University of Virginia, Charlottesville, 1998.
10. I. Foster, C. Kesselman. The Globus Toolkit. In [5]pp. 259–278.
11. S. Brunet et al. *Application Experiences with the Globus Toolkit, Proc. 7th IEEE Symp. on High Performance Distributed Computing*, IEEE Computer Society Press, Los Alamitos, California, 1998.
12. Z. Nemeth and V. Sunderam, Characterizing grids: Attributes and formalisms, *Journal of Grid Computing*, Vol. 1, No. 1, pp. 9–23, 2003.
13. K. Czajkowski et al., *A Resource Management Architecture for Metacomputing Systems, Proc. IPPS/SPDP 1998 Workshop on Job Scheduling Strategies for Parallel Processing*, IEEE Computer Society Press, Los Alamitos, California, 1998.
14. S.J. Chapin, D. Karmatos, J. Karpovich and A. Grimshaw, *The Legion Resource Management System, Proc. 5th Workshop on Job Scheduling Strategies for Parallel Processing (JSSPP '99)* IEEE Computer Society Press, Los Alamitos, California, 1999.
15. K. Czajkowski, S. Fitzgerald, I. Foster and C. Kesselman. *Grid Information Services for Distributed Resource Sharing, Proc. 10th IEEE International Symposium on High-Performance Distributed Computing* IEEE Press, 2001.
16. I. Foster, C. Kesselman, G. Tsudik and S. Tuecke. A Security Architecture for Computational Grids, In *Proc. of the 5th ACM Conference on Computer and Communication Security*, November 1998.
17. M. Humprey, F. Knabbe, A. Ferrari and A. Grimshaw. *Accountability and Control of Process Creation in the Legion Metasystem, Proc. of the 2000 Network and Distributed System Security Symposium DSS2000*, San Diego, California, February 2000.
18. K. Czajkowski, I. Foster, C. Kesselman, V. Sander and S. Tuecke. SNAP: A Protocol for negotiating service level agreements and coordinating resource management in distributed systems, *Lecture Notes in Computer Science*, 2537:153–183, 2002.
19. L. Pearlman, V. Welch, I. Foster, C. Kesselman and S. Tuecke. *A Community Authorization Service for Group Collaboration, Proc. IEEE 3rd International Workshop on Policies for Distributed Systems and Networks*, 2002.
20. R. Armstrong et al., *Toward a Common Component Architecture for High-Performance Scientific Computing, Proc. 8th IEEE International Symposium on High-Performance Distributed Computing*, IEEE Press, 1999.
21. M. van Steen, P. Homburg and A.S. Tanenbaum. Globe: A wide-area distributed system. *IEEE Concurrency*, January–March, 1999, pp. 70–78.

22. N. Karonis, B. Toonen and I. Foster, MPICH-G2: A grid-enabled implementation of the message passing interface, *Journal of Parallel and Distributed Computing (JPDC)*, Vol. 63, No. 5, pp. 551–563, May 2003.
23. F. Berman, G. Fox and A. Hey, *Grid Computing: Making the Global Infrastructure a Reality*, John Wiley and Sons, 2003.
24. M. Govindaraju et al., *Merging the CCA Component Model with the OGSI Framework, Proc. CCGrid2003, 3rd International Symposium on Cluster Computing and the Grid*, May 2003.
25. Scientific Portal Alliance Expedition, http://www.extreme.indiana.edu/alliance/.
26. Scientific Discovery through Advanced Computing, http://www.scidac.org/.

40

Improving Website Performance

Arun Iyengar

Erich Nahum

Anees Shaikh

Renu Tewari

CONTENTS

40.1 Introduction 40-1
40.2 Improving Performance at a Website 40-2
40.2.1 Load Balancing 40-2
40.2.2 Serving Dynamic Web Content 40-8
40.3 Server Performance Issues 40-9
40.3.1 Process-Based Servers 40-10
40.3.2 Thread-Based Servers 40-10
40.3.3 Event-Driven Servers 40-10
40.3.6 In-Kernel Servers 40-11
40.3.7 Server Performance Comparison 40-11
40.4 Web Server Workload Characterization 40-12
40.4.1 Request Methods 40-13
40.4.2 Response Codes 40-14
40.4.3 Document Popularity 40-14
40.4.4 File Sizes 40-15
40.4.5 Transfer Sizes 40-16
40.4.6 HTTP Version 40-18
40.4.7 Summary 40-19
Acknowledgment 40-19
References 40-19

40.1 Introduction

The World Wide Web has emerged as one of the most significant applications over the past decade. The infrastructure required to support Web traffic is significant, and demands continue to increase at a rapid rate. Highly accessed Websites may need to serve over 1,000,000 hits per min. Additional demands are created by the need to serve dynamic and personalized data.

This chapter presents an overview of techniques and components needed to support high-volume Web traffic. These include multiple servers at Websites that can be scaled to accommodate high request rates. Various load balancing techniques have been developed to efficiently route requests to multiple servers. Websites may also be dispersed or replicated across multiple geographic locations.

Web servers can use several different approaches for handling concurrent requests including processes, threads, event-driven architectures in which a single process is used with nonblocking I/O, and in-kernel servers. Each of these architectural choices has certain advantages and disadvantages. We discuss how these different approaches affect performance.

We also discuss Web server workload characterization and present properties of the workloads related to performance such as document sizes, popularities, and protocol versions used. Understanding these properties of a Website is critically important for optimizing performance.

1-58488-381-2/05/$0.00+$1.50
© 2005 by CRC Press LLC

40.2 Improving Performance at a Website

Highly accessed Websites may need to handle peak request rates of over 1,000,000 hits per min. Web serving lends itself well to concurrency because transactions from different clients can be handled in parallel. A single Web server can achieve parallelism by multithreading or multitasking between different requests, and additional parallelism and higher throughputs can be achieved by using multiple servers and load balancing requests among the servers.

Figure 40.1 shows an example of a scalable Website. Requests are distributed to multiple servers via a load balancer. The Web servers may access one or more databases or other back-end systems for creating content. The Web servers would typically store replicated content so that a request could be directed to any server in the cluster. One way to share static files across multiple servers is to use a distributed file system such as AFS or DFS [Kwan et al., 1995] for storage. Copies of files may be cached in one or more servers. This approach works fine if the number of Web servers is not too large and data do not change very frequently. For the large numbers of servers for which data updates are frequent, distributed file systems can be highly inefficient. Part of the reason for this is the strong consistency model imposed by distributed file systems. Shared file systems require copies of files to be consistent. In order to update a file in one server, other copies of the file may need to be invalidated before the update can take place. These invalidation messages add overhead and latency. At some Websites, the number of objects updated in temporal proximity to each other can be quite large. During periods of peak updates, the system might fail to perform adequately.

Another method of distributing content that avoids some of the problems of distributed file systems is to propagate updates to servers without requiring the strict consistency guarantees of distributed file systems. Using this approach, updates are propagated to servers without first invalidating all existing copies. This means that at the time an update is made, data may be inconsistent between servers for a little while. For many Websites, these inconsistencies are not a problem, and the performance benefits from relaxing the consistency requirements can be significant.

40.2.1 Load Balancing

40.2.1.1 Load Balancing via DNS

The load balancer in Figure 40.1 distributes requests among the servers. One method of load balancing requests to servers is via DNS servers. DNS servers provide clients with the IP address of one of the site's content delivery nodes. When a request is made to a Website such as http//www.research.ibm.com/

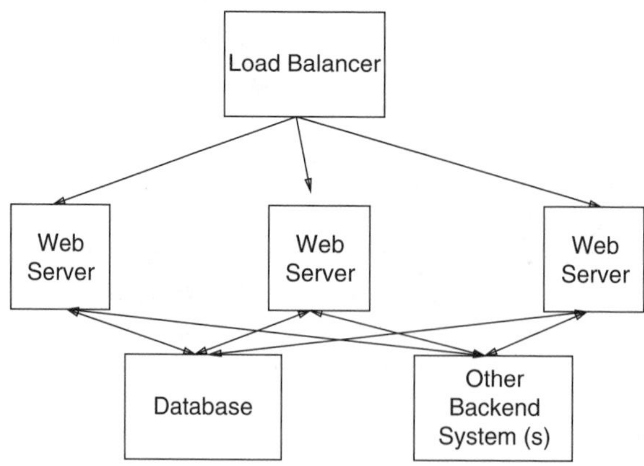

IGURE 40.1 Architecture of a scalable Website. Requests are directed from the load balancer to one of several Web servers. The Web servers may access one or more databases or other back-end systems for creating content.

compsci/, "www.research.ibm.com" must be translated to an IP address, and DNS servers perform this translation. A name associated with a Website can map to multiple IP addresses, each associated with a different Web server. DNS servers can select one of these servers using a policy such as *Round Robin* [Brisco, 1995].

There are other approaches for DNS load balances that offer some advantages over simple round robin [Cardellini et al., 1999b]. The DNS server can use information about the number of requests per unit time sent to a Website as well as geographic information. The Internet2 Distributed Storage Infrastructure Project proposed a DNS that would implement address resolution based on network proximity information, such as round-trip delays [Beck and Moore, 1998].

One of the problems with load balancing using DNS is that name-to-IP-address mappings resulting from a DNS lookup may be cached anywhere along the path between a client and a server. This can cause load imbalance because client requests can then bypass the DNS server entirely and go directly to a server [Dias et al., 1996]. Name-to-IP-address mappings have time-to-live attributes (TTL) associated with them that indicate when the mappings are no longer valid. Using small TTL values can limit load imbalances due to caching. One problem with this approach is that it can increase response times [Shaikh et al., 2001] and another is that not all entities caching name-to-IP-address mappings obey TTLs that are too short.

Adaptive TTL algorithms have been proposed in which the DNS assigns different TTL values for different clients [Cardellini et al., 1999a]. A request coming from a client with a high request rate would typically receive a name-to-IP-address mapping with a shorter lifetime than that assigned to a client with a low request rate. This prevents a proxy with many clients from directing requests to the same server for too long a period of time.

40.2.1.2 Load Balancing via Connection Routers

Another approach to load balancing is using a connection router (aka "dispatcher," "Web switch," or "content switch") in front of several back-end servers. Connection routers hide the IP addresses of the back-end servers. That way, IP addresses of individual servers will not be cached, eliminating the problem experienced with DNS load balancing. Connection routing can be used in combination with DNS routing for handling large numbers of requests. A DNS server can route requests to multiple connection routers. The DNS server provides coarse-grained load balancing, whereas the connection routers provide finer-grained load balancing. Connection routers also simplify the management of a Website because back-end servers can be added and removed transparently.

In such environments, a front-end connection router directs incoming client requests to one of the physical server machines, as shown in Figure 40.2. The physical servers often share one or more virtual IP addresses so that any server can respond to client requests. In other scenarios, the servers have only private addresses, so the connection router accepts all connections destined for the site virtual address. The request-routing decision can be based on a number of criteria, including server load, client request, or client identity.

Connection routers are typically required to perform several functions related to the routing decision:

- Monitor server load and distribute incoming requests to balance the load across servers
- Examine client requests to determine which server is appropriate to handle the request
- Identify the client to maintain session affinity with a particular server for e-business applications

In addition, many commercial connection routers provide functions important in a production data center environment. These include:

- Failover to a hot standby to improve availability
- Detection and avoidance of many common denial-of-service attacks
- SSL acceleration to improve the performance of secure applications
- Simplified configuration and management (e.g., Web browser-based configuration interface)

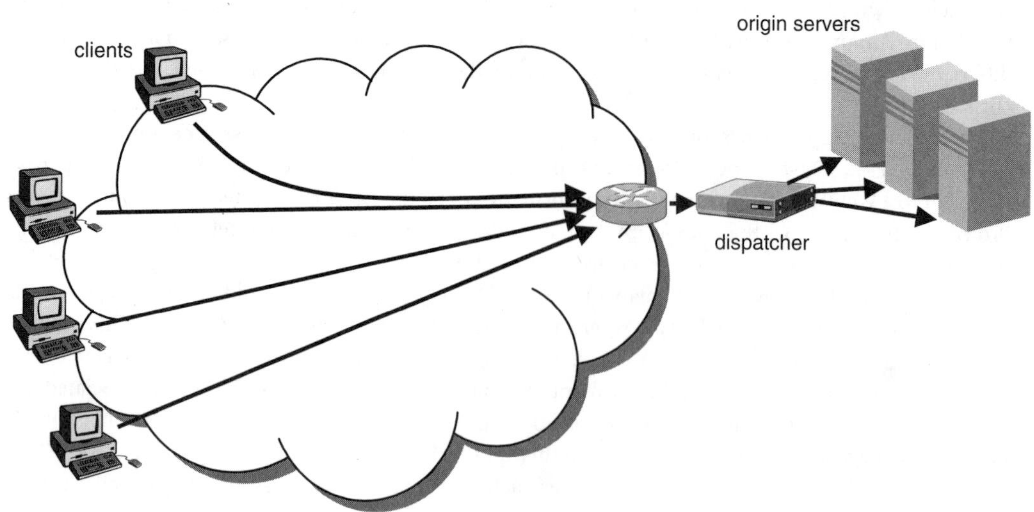

FIGURE 40.2 A server-side connection router (labeled *dispatcher*) directs incoming client Web requests to one of the physical servers in the cluster.

A variety of networking equipment and software vendors offer connection routers, including Cisco Systems [b, a], Nortel Networks [Alteon ACEdirector], IBM [Websphere edge server], Intel [Intel NetStructure], Foundry Networks [Foundry ServerIron], and F5 Networks [BIG-IP controller].

Request-routing may be done primarily in hardware, completely in software, or with a hardware switch and control software. For example, several of the vendors mentioned above offer dedicated hardware solutions consisting of multiple fast microprocessors, several Fast Ethernet and Gigabit Ethernet ports, and plenty of memory and storage. Others offer software-only solutions that can be installed on a variety of standard platforms.

IBM's Network Dispatcher [Hunt et al., 1998] is one example of a connection router that hides the IP address of back-end servers. Network Dispatcher uses *Weighted Round Robin* for load balancing requests. Using this algorithm, servers are assigned weights. Servers with a given weight have priority in receiving new connections over servers with a lesser weight. Consequently, servers with higher weights get more connections than those with lower weights, and servers with equal weights get an equal distribution of new connections.

With Network Dispatcher, requests from the back-end servers go directly back to the client, which reduces overhead at the connection router. By contrast, some connection routers function as proxies between the client and server, in which all responses from servers go through the connection router to clients.

Network Dispatcher has special features for handling client affinity to selected servers. These features are useful for handling requests encrypted using the Secure Sockets Layer protocol (SSL), which is commonly used for encryption on the Web. SSL can add significant overhead, however. When an SSL connection is made, a session key must be negotiated and exchanged. Session keys are expensive to generate; therefore, they have a lifetime (typically 100 sec), the period for which they exist after the initial connection is made. Subsequent SSL requests within the key lifetime reuse the key.

Network Dispatcher recognizes SSL requests by the port number (443), and allows certain ports to be designated as "sticky." Network Dispatcher keeps records of old connections on such ports for a designated affinity life span (e.g., 100 sec for SSL). If a request for a new connection from the same client on the same port arrives before the affinity life span for the previous connection expires, the new connection is sent to the same server that the old connection utilized.

Using this approach, SSL requests from the same client will go to the same server for the lifetime of a session key, obviating the need to negotiate new session keys for each SSL request. This can cause some load unbalance, particularly because the client address seen by Network Dispatcher may actually be a proxy representing several clients and not just the client corresponding to the SSL request. However, the reduction in overhead due to reduced session key generation is usually worth the load imbalance created. This is particularly true for sites that make gratuitous use of SSL. For example, some sites encrypt all of the image files associated with an HTML page and not just the HTML page itself.

40.2.1.3 Content-Based Routing

Connection routing is often done at layer 4 of the OSI model, where the connection router does not know the contents of the request. Another approach is to perform routing at layer 7. In layer 7 routing, also known as content-based routing, the router examines requests and makes its routing decisions based on the content of requests [Pai et al., 1998], which allows for more sophisticated routing techniques. For example, dynamic requests could be sent to one set of servers, whereas static requests could be sent to another set. Different quality of service policies could be assigned to different URLs, according to which the content-based router sends the request to an appropriate server based on the quality of service corresponding to the requested URL.

Content-based routing allows the servers at a Website to be asymmetrical. For example, information could be distributed at a Website so that frequently requested objects are stored on many or all servers, whereas infrequently requested objects are only stored on a few servers. This reduces the storage overhead of replicating all information on all servers. The content-based router can then use information on how objects are distributed to make correct routing decisions. The key problem with content-based routing is that the overhead incurred can be high. In order to examine the contents of a request, the router must terminate the connection with the client.

The use of layer-4 or layer-7 routers depends on the request-routing goal. Load balancing across replicated content servers, for example, typically does not require knowledge about the client request or identity, and thus is well suited to a layer-4 approach. Simple session affinity based on client IP address, or request direction to servers based on the application (e.g., port 80 HTTP traffic vs. port 110 POP3 traffic) are also easily accomplished by examining layer-3 or layer-4 headers of packets while in transit through the router. For example, the router may peek at the TCP header flags to determine when a SYN packet arrives from a client, indicating a new connection establishment. Then, once a SYN is identified, the source and destination port numbers and IP addresses may be used to direct the request to the right server. This decision is recorded in a table so that subsequent packets arriving with the same header fields are directed to the same server.

Layer-4 routers, due to their relative simplicity, are often implemented as specialized hardware as they need not perform any layer-4 protocol processing or maintain much per-connection state information. Although traffic from the clients must be routed via the router, the response traffic from the server — which accounts for the bulk of the data in HTTP transactions — can bypass the router, flowing directly back to the client. This is typically done by configuring each server to respond to traffic destined for the virtual IP address(es), using IP aliasing, for example.

Although layer-4 routers are usually deployed as front-end appliances, an alternative is to allow back-end servers to perform load-balancing themselves by redirecting connections to relatively underloaded machines [Bestavros et al., 1998]. However, even without such an optimization, hardware-based layer-4 routers are able to achieve very high scalability and performance.

Request-routing based on the URL (or other application-layer information), on the other hand, requires the router to terminate the incoming TCP connection and receive enough information to make a routing decision. In the case of Web traffic, for example, the router must accept the incoming TCP connection and then wait for the client to send an HTTP request in order to view application-layer information such as the requested URL or HTTP cookie. Once enough information to make a routing decision is received, the router can create a new connection to the appropriate server and forward the client request. The server response is then passed back to the client via the router on the client's original connection. Figure 40.3 outlines these steps.

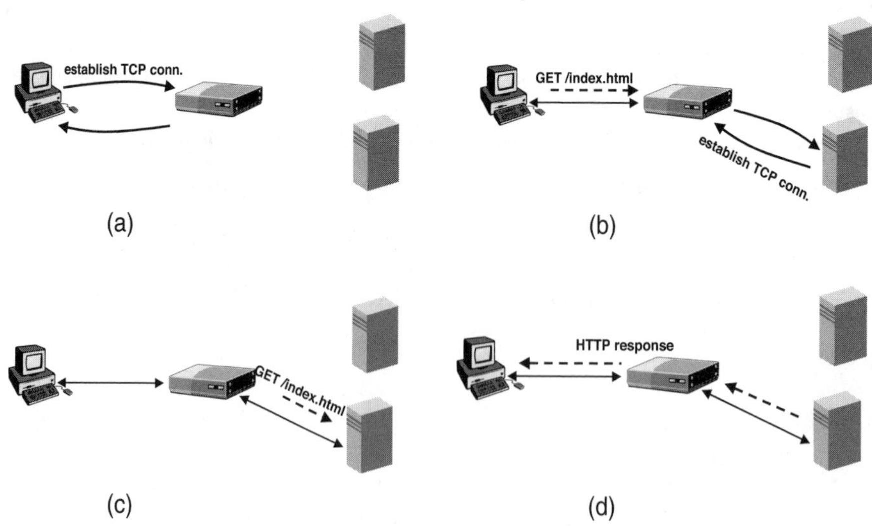

FIGURE 40.3 URL-based request routing. The connection router accepts the TCP connection transparently from the client (a). Next, in (b) the client sends a GET request on the established connection, prompting the router to open a new TCP connection to the appropriate server. In (c) the router forwards the client request to the server, and in (d) returns the response to the client.

In the simplest realization, a layer-7 router may be implemented as a software application-level gateway that transparently accepts incoming client connections (destined for port 80 to the server virtual IP address) and reads the requests. After deciding which server should handle the request, the application can forward it through a new or preestablished connection to the server. The router serves as a bridge between the two connections, copying data from one to the other. From a networking point of view, the router behaves much like a forward Web proxy installed at an enterprise site, though the forward proxy's primary function lies primarily in filtering and content caching rather than in request routing. In this approach, the router can quickly become a bottleneck, as it must perform connection termination and management for a large number of clients [Aron et al., 2000; Cohen et al., 1999]. This limits the overall scalability of the data center in terms of the number of clients it can support simultaneously.

Several techniques have been proposed to improve the performance and scalability of application-level gateways used in various contexts, including as HTTP proxies. TCP connection splicing is one such optimization technique, in which packets are forwarded from one connection to the other at the network layer, avoiding traversal of the transport layer and the user-kernel protection boundary [Maltz and Bhagwat, 1998; Spatscheck et al., 2000; Cohen et al., 1999]. TCP splicing mechanisms are usually implemented as kernel-level modifications to the operating system protocol stack, with an interface to allow applications to initiate the splice operation between two connections. Once the TCP splice is completed, data is relayed from one connection to the other without further intervention by the application. In Figure 40.4, (a) depicts the operation of TCP splicing. With splicing, care must be taken to ensure that TCP header fields such as sequence numbers, checksums, and options are correctly relayed. TCP splicing has been shown to improve the performance of application-layer gateways to the level of software IP routers. Variations in the kernel-based implementation include implementation in the kernel socket library (as opposed to the network layer) [Rosu and Rosu, 2002] and in a hardware switch [Apostolopoulos et al., 2000].

Though TCP splicing can improve the scalability of a layer-7 router, it is still limited by the fact that a centralized node must terminate incoming connections, examine application-layer information, and make request-routing decisions before initiating a splice. In addition, all traffic to and from the servers must pass through the router to allow the header translation to occur. To address these limitations, an alternate scheme using *connection handoff* was proposed [Pai et al., 1998; Aron et al., 2000; Song et al.,

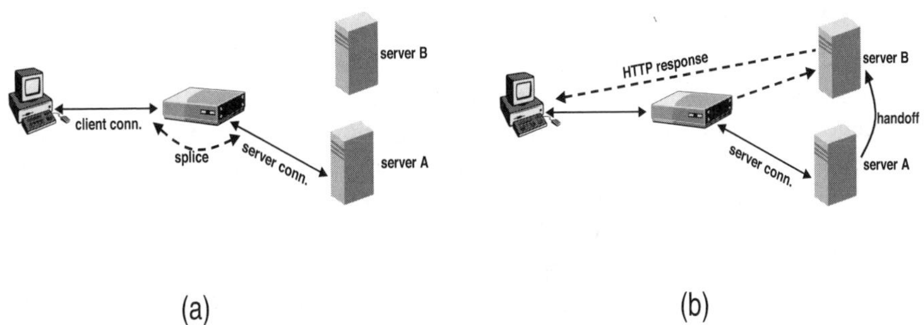

FIGURE 40.4 TCP splicing and TCP handoff. With TCP splicing (a), the connection router splices the two connections after determining that server B should handle the request. The response must be sent back through the router. In the TCP handoff approach (b), the router, a simpler layer-4 device, initially forwards the connection request to server A. After receiving the request, server A hands off the connection state to server B. The response traffic can then flow directly back to the client, bypassing the router. Client acknowledgements, however, still come through the router and must be forwarded to server B.

2002]. In this approach, each back-end server can function as a router, effectively distributing the content inspection and request-routing operations to multiple nodes. Client connections are initially routed to any one of the servers, perhaps using a fast hardware switch. If the initial server decides that another server is better suited to handle the request, it transfers the TCP connection state to the alternate server. Using the transferred state, the new server can resume the connection with the client without requiring that data pass through a front-end router. In Figure 40.4, (b) shows the operation of TCP connection handoff.

Although the connection handoff approach does remove the bottleneck of connection termination at a single front-end, its scalability and performance are ultimately limited by the number and overhead of TCP handoff operations. Furthermore, it still requires a special front-end layer-4 router, as incoming packets (e.g., TCP acknowledgments) must be forwarded to the appropriate server after the connection is handed off. Finally, TCP handoff requires kernel modifications to the server operating systems to support handoff. The splicing approach, on the other hand, is transparent to both servers and clients.

As Web application requirements evolve, there will be a need for more sophisticated connection routing, based on a variety of application information. This trend implies that the layer-4 approach of examining only transport-layer headers provides insufficient functionality. But layer-7 routers, although more sophisticated, suffer from the limitations on scalability and performance described above.

40.2.1.4 Client-Based Load Balancing

A number of client-based techniques have been proposed for load balancing. A few years ago, Netscape implemented a scheme for doing load balancing at the Netscape Website (before the company was purchased by AOL), in which the Netscape browser was configured to pick the appropriate server [Mosedale et al., 1997]. When a user accessed the Website www.netscape.com, the browser would randomly pick a number *i* between 1 and the number of servers and direct the request to www*i*.netscape.com.

Another client-based technique is to use the client's DNS [Fei et al., 1998; Rabinovich and Spatscheck, 2001]. When a client wishes to access a URL, it issues a query to its DNS to get the IP address of the site. The Website's DNS returns a list of IP addresses of the servers instead of a single IP address. The client DNS selects an appropriate server for the client. An alternative strategy is for the client to obtain the list of IP addresses from its DNS and itself do the selection. An advantage of the client itself making the selection is that the client can collect information about the performance of different servers at the site and make an intelligent choice based on this. The disadvantages of client-based techniques are that the Website loses control over how requests are routed, and such techniques generally require modifications to the client (or at least to the client's DNS server).

40.2.2 Serving Dynamic Web Content

Web servers satisfy two types of requests, static and dynamic. *Static requests* are for files that exist at the time a request is made. *Dynamic requests* are for content that has to be generated by a server program executed at request time. A key difference between satisfying static and dynamic requests is the processing overhead. The overhead of serving static pages is relatively low. A Web server running on a uniprocessor can typically serve several hundred static requests per sec. Of course, this number is dependent on the data being served; for large files, the throughput is lower.

The overhead for satisfying a dynamic request may be orders of magnitude greater than the overhead for satisfying a static request. Dynamic requests often involve extensive back-end processing. Many Websites make use of databases, and a dynamic request may invoke several database accesses. These database accesses can consume significant CPU cycles. The back-end software for creating dynamic pages may be complex. Although the functionality achieved by such software may not appear to be compute-intensive, such middleware systems are often not designed efficiently; commercial products for generating dynamic data can be highly inefficient.

One source of overhead in accessing databases is connecting to the database. Many database systems require a client to first establish a connection with a database before performing a transaction in which the client typically provides authentication information. Establishing a connection is often quite expensive. A naive implementation of a Website would establish a new connection for each database access. This approach could overload the database even with relatively low traffic levels.

A significantly more efficient approach is to maintain (one or more) long-running processes with open connections to the database. Accesses to the database are then made with one of these processes. That way, multiple accesses to the database can be made over a single connection.

Another source of overhead is the interface for invoking server programs in order to generate dynamic data. The traditional method for invoking server programs for Web requests is via the Common Gateway Interface (CGI). CGI forks off a new process to handle each dynamic request; this incurs significant overhead. There are a number of faster interfaces available for invoking server programs [Iyengar et al., 2000], which use one of two approaches. The first approach is for the Web server to provide an interface to allow a program for generating dynamic data to be invoked as part of the Web server process itself. IBM's GO Web server API (GWAPI) is an example of such an interface. The second approach is to establish long-running processes to which a Web server passes requests. Although this approach incurs some interprocess communication overhead, the overhead is considerably less than that incurred by CGI. FastCGI is an example of the second approach [Open Market].

In order to reduce the overhead for generating dynamic data, it is often feasible to generate data corresponding to a dynamic object once, store the object in a cache, and subsequently serve requests to the object from the cache instead of invoking the server program again [lyengar and Challenger, 1997]. Using this approach, dynamic data can be served at about the same rate as static data.

However, there are types of dynamic data that cannot be precomputed and served from a cache. For instance, dynamic requests that cause a side effect at the server, such as a database update, cannot be satisfied merely by returning a cached page. As an example, consider a Website that allows clients to purchase items using credit cards. At the point when a client commits to buying something, that information has to be recorded at the Website; the request cannot be solely serviced from a cache.

Personalized Web pages can also negatively affect the cacheability of dynamic pages. A personalized Web page contains content specific to a client, such as the client's name. Such a Web page cannot be used for another client. Therefore, caching the page is of limited utility because only a single client can use it. Each client would need a different version of the page.

One method that can reduce the overhead of generating dynamic pages and enable caching of parts of personalized pages is to define these pages as being composed of multiple fragments [Challenger et al., 2000]. In this approach, a complex Web page is constructed from several simpler fragments. A fragment may recursively embed other fragments. This is efficient because the overhead for assembling a Web page from simpler fragments is usually minor compared to the overhead for constructing the page from scratch, which can be quite high.

The fragment-based approach also makes it easier to design Websites. Common information that needs to be included on multiple Web pages can be created as a fragment. In order to change the information on all pages, only the fragment needs to be changed.

In order to use fragments to allow for partial caching of personalized pages, the personalized information on a Web page is encapsulated by (one or more) fragments that are not cacheable; however, the other fragments in the page are. When serving a request, a cache composes pages from its constituent fragments, many of which are locally available. Only personalized fragments have to be created by the server. As personalized fragments typically constitute a small fraction of the entire page, generating only them would require lower overhead than generating all of the fragments in the page.

Generating Web pages from fragments provides other benefits as well. Fragments can be constructed to represent entities that have similar lifetimes. When a particular fragment changes but the rest of the Web page stays the same, only the fragment needs to be invalidated or updated in the cache, not the entire page. Fragments can also reduce the amount of cache space taken up by multiple pages with common content. Suppose that a particular fragment contained in each of 2000 popular Web pages should be cached. Using the conventional approach, the cache would contain a separate version of the fragment for each page, resulting in as many as 2000 copies. By contrast, if the fragment-based method of page composition is used, only a single copy of the fragment needs to be maintained.

A key problem with caching dynamic content is maintaining consistent caches. It is advantageous for the cache to provide a mechanism (such as an API) allowing the server to explicitly invalidate or update cached objects so that they do not become obsolete. Web objects may be assigned expiration times that indicate when they should be considered obsolete. Such expiration times are generally not sufficient for allowing dynamic data to be cached properly because it is often not possible to predict accurately when a dynamic page will change.

40.3 Server Performance Issues

A central component of the response time seen by Web users is, of course, the performance of the origin server that provides the content. There is great interest, then, in understanding the performance of Web servers: How quickly can they respond to requests? How well do they scale with load? Are they capable of operating under overload, i.e., can they maintain some level of service even when the requested load far outstrips the capacity of the server?

A Web server is an unusual piece of software in that it must communicate with (potentially) thousands of remote clients simultaneously. The server thus must be able to deal with a large degree of *concurrency.* A server cannot simply respond to each client in a nonpreemptive, first-come-first-serve manner, for several reasons. Clients are typically located far away over the wide-area Internet, and thus connection lifetimes can last many seconds or even minutes. Particularly with HTTP 1.1, a client connection may be open but idle for some time before a new request is submitted. Thus, a server can have many concurrent connections open, and should be able to do work for one connection when another is quiescent. Another reason is that a client may request a file which is not resident in memory. While the server CPU waits for the disk to retrieve the file, it can work on responding to another client. For these and other reasons, a server must be able to multiplex the work it has to do through some form of concurrency.

A fundamental factor that affects the performance of a Web server is the *architectural model* that it uses to implement concurrency. Generally, Web servers can be implemented using one of four architectures: processes, threads, event-driven, and in-kernel. Each approach has its advantages and disadvantages, which are considered in more detail below. A central issue in this decision of which model to use is the sort of performance optimizations available under that model. Another issue is how well that model scales with the workload, i.e., how efficiently it can handle growing numbers of clients.

40.3.1 Process-Based Servers

The most common form of concurrency is perhaps provided by *processes*. The original NCSA server and the widely known Apache server [The Apache Project] use processes as the mechanism to handle large numbers of connections. In this model, a process is created for each new request, which can block when necessary, for example when waiting for data to become available on a socket or for file I/O to be available from the disk. The server handles concurrency by creating multiple processes.

Processes have two main advantages. First, they are consistent with a programmer's way of thinking, allowing the developer to proceed in a step-by-step fashion without worrying about managing concurrency. Second, they provide isolation and protection among different clients. If one process hangs or crashes, the other processes should be unaffected.

The main drawback to processes is performance. Processes are relatively heavyweight abstractions in most operating systems, and thus creating them, deleting them, and switching contexts between them is expensive. Apache, for example, tries to ameliorate these costs by preforking a number of processes and only destroying them if the load falls below a certain threshold. However, the costs are still significant, as each process requires memory to be allocated to it. As the number of processes grows, large amounts of memory are used. This puts pressure on the virtual memory system, which could use the memory for other purposes such as caching frequently-accessed data. In addition, sharing information (such as a cached file) across processes can be difficult.

40.3.2 Thread-Based Servers

Threads provide the next most common form of concurrency. Servers that use threads include JAWS [Hu et al., 1997] and Sun's Java Web Server [Sun Microsystems, b]. Threads are similar to processes but are considered more lightweight. Unlike processes, threads share the same address space and typically only provide a separate stack for each thread. Thus, creation costs and context-switching costs are usually much lower than for processes. In addition, sharing between threads is much easier. Threads also maintain the abstraction of an isolated environment, much like processes, although the analogy is not exact because programmers must worry more about issues like synchronization and locking to protect shared data structures.

Threads have several disadvantages as well. Because the address space is shared, threads are not protected from one another in the way processes are. Thus, a poorly programmed thread can crash the whole server. Threads also require proper operating system support; otherwise, when a thread blocks on something such as a file I/O, the whole address space will be suspended.

40.3.3 Event-Driven Servers

The third form of concurrency is known as the *event-driven* architecture. Servers that use this method include Flash [Pai et al., 1999] and Zeus [Zeus Inc., c]. With this architecture, a single process is used with *nonblocking* I/O. Nonblocking I/O is a way of doing asynchronous reads and writes on a socket or file descriptor. For example, instead of a process reading a file descriptor and blocking until data becomes available, an event-driven server will return immediately if there is no data. In turn, the operating system will let the server process know when a socket or file descriptor is ready for reading or writing through a *notification mechanism*. This notification mechanism can be an active one such as a signal handler, or a passive one requiring the process to ask the operating system, such as the `select()` system call. Through these mechanisms, the server process will necessarily respond to events and is typically guaranteed to never block.

Event-driven servers have several advantages. First, they are very fast. Zeus is frequently used by hardware vendors to generate high Web server numbers with the SPECWeb99 benchmark [SPEC, 1999]. Sharing is inherent, as there is only one process, and no locking or synchronization is needed. There are no context-switch costs or extra memory consumption, unlike the case with threads or processes. Maximizing concurrency is thus much easier than with the previous approaches.

Event-driven servers have downsides as well. As with threads, a failure can halt the whole server. Event-driven servers can tax operating system resource limits, such as the number of open file descriptors. Different operating systems have varying levels of support for asynchronous I/O, so a fully event-driven server may not be possible on a particular platform. Finally, event-driven servers require a different way of thinking on the part of the programmer, who must understand and account for the ways in which multiple requests can be in varying stages of progress simultaneously. In this approach, the degree of concurrency is fully exposed to the developer, with all the attendant advantages and disadvantages.

40.3.6 In-Kernel Servers

The fourth and final form of server architectures is the *in-kernel* approach. Servers that use this method include AFPA [Joubert et al., 2001] and Tux [Red Hat Inc., a]. All of the previous architectures place the Web server software in user space; in this approach, the HTTP server is in kernel space, tightly integrated with the host TCP/IP stack.

The in-kernel architecture has the advantage that it is extremely fast, as potentially expensive transitions to user space are completely avoided. Similarly, no data needs to be copied across the user-kernel boundary, which is another costly operation.

The disadvantages of in-kernel approaches are several. First, it is less robust and hence more vulnerable to programming errors; a server fault can crash the whole machine, not just the server! Development is much harder, as kernel programming is more difficult and much less portable than programming user-space applications. Kernel internals of Linux, FreeBSD, and Windows vary considerably, making deployment across platforms additional work. The socket and thread APIs, on the other hand, are relatively stable and portable across operating systems.

Dynamic content poses an even greater challenge for in-kernel servers because an arbitrary program may be invoked in response to a request for dynamic content. A full-featured in-kernel web server would need to have a PHP engine or Java runtime interpreter loaded along with the kernel! The way current in-kernel servers deal with this issue is to restrict their activities to the static-content component of Web serving, and pass dynamic content requests to a complete server in user space, such as Apache. For example, many entries in the SPECWeb99 site [SPEC, 1999] that use the Linux operating system follow this hybrid approach, with Tux serving static content in the kernel and Apache handling dynamic requests in user space.

40.3.7 Server Performance Comparison

Because we are concerned with performance, it is interesting to see how well the different server architectures perform. To compare them, we took an experimental test bed setup and evaluated the performance, using a synthetic workload generator [Nahum et al., 2001] to saturate the servers with requests for a range of web documents. The clients were eight 500 MHz PCs running FreeBSD, and the server was a 400 MHz PC running Linux 2.4.16. Each client had a 100 Mbps Ethernet connected to a Gigabit switch, and the server was connected to the switch using Gigabit Ethernet, Three servers were evaluated as representatives of their architecture: Apache as a process-based server, Flash as an event-driven server, and Tux as an in-kernel server.

Figure 40.5 shows the throughput of the three servers in HTTP operations/sec. As can be seen, Tux, the in-kernel server, is the fastest at 2193 ops/sec. However, Flash is only 10% slower at 2075 ops/sec, despite being implemented in user space. Apache, on the other hand, is significantly slower at 875 ops/sec. Figure 40.6 shows the response times for the three servers. Again, Tux is the fastest at 3 msec, Flash second at 5 msec, and Apache the slowest at 10 msec.

As multiple examples of each type of server architecture exist, there is clearly no consensus as to which is the best model. Instead, it may be that different approaches are better suited for different scenarios. For example, the in-kernel approach may be most appropriate for dedicated server appliances, or content distribution (CDN) nodes, whereas a back-end dynamic content server will rely on the full generality of

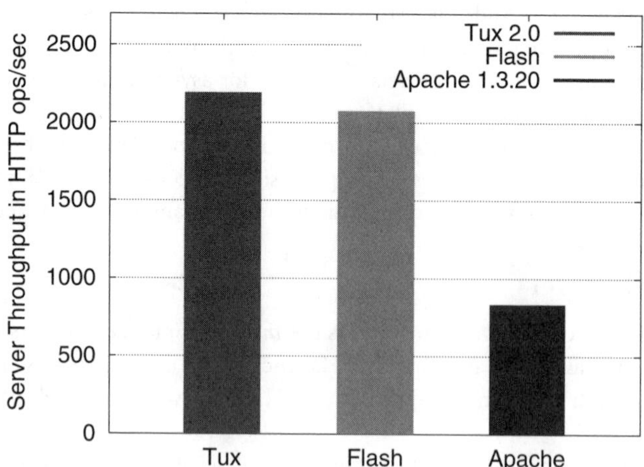

FIGURE 40.5 Server throughput.

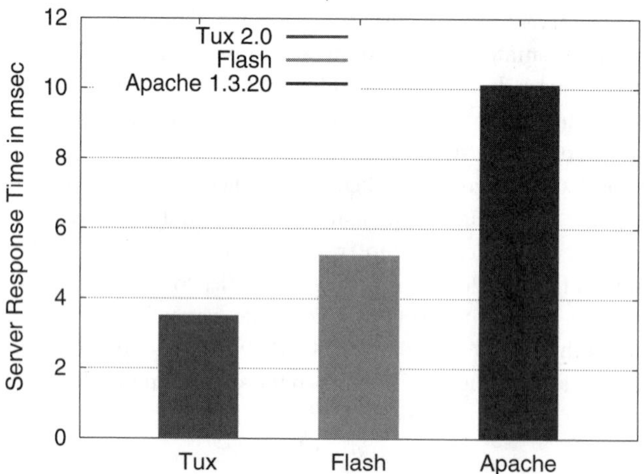

FIGURE 40.6 Server response time.

a process-based server like Apache. Still, website operators should be aware of how the choice of architecture will affect Web server performance.

40.4 Web Server Workload Characterization

Workload characterization is frequently necessary in order to better understand the performance of a Website. Typical workload questions include:

What do requests look like?
How popular are some documents vs. others?
How large are Web transfers?
What level of HTTP protocol deployment exists on the Web?

Over time, Web server workload characterization has answered some of these questions, and we provide an overview here. In this section, we describe various workload characteristics or performance metrics identified in the literature. To help illustrate the characteristics, we also present examples derived from several logs. Table 40.1 gives an overview of the logs used in the examples, several of which are taken

TABLE 40.1 Logs Used in Example Websites

Name	Chess 1997	Olympics 1998	IBM 1998	World Cup 1998	Department Store 2000	IBM 2001
Description	Kasparov–Deep Blue Event Site	Sporting Event Site	Corporate Presence	Sporting Event Site	Online Shopping	Corporate Presence
Period	2 weeks in May 1997	1 d in February 1998	1 d in June 1998	31 d in June 1998	12 d in June 2000	1 d in February 2001
Hits	1,586,667	11,485,600	5,800,000	1,111,970,278	13,169,361	12,445,739
Kbytes	14,171,711	54,697,108	10,515,507	3,952,832,722	43,960,527	28,804,852
Clients	256,382	86,021	80,921	2,240,639	254,215	319,698
URLs	2,293	15,788	30,465	89,997	11,298	42,874

from high-volume Websites that were managed by IBM. One log, of an online department store, is from a site hosted by but not designed or managed by IBM. We also include most of the 1998 World Cup logs [Arlitt and Jin, 2000], which are publicly available at the Internet Traffic Archive [Lawrence Berkeley National Laboratory]. Due to the size of these logs, we limit our analysis to the busiest 4 weeks of the trace, June 10th through July 10th (days 46 through 76 on the Website).

Because our analysis is based on Web logs, certain interesting characteristics cannot be examined. For example, persistent connections, pipelining, network round-trip times, and packet loss all have significant effects on both server performance and client-perceived response time. These characteristics are not captured in Apache Common Log format and typically require more detailed packet-level measurements using a tool such as tcpdump. These sort of network-level measurements are difficult to obtain due to privacy and confidentiality requirements.

An important caveat worth reiterating is that any one Website may not be representative of a particular application or workload. For example, the behavior of a very dynamic Website such as eBay, which hosts a great deal of rapidly changing content, is most likely very different from an online trading site like Schwab, which conducts most of its business using Secure Sockets Layer (SSL) encryption. Many example Websites given here were all run by IBM, and so may share certain traits not observed by previous researchers in the literature. As we will see, however, the characteristics of the IBM sites are consistent with those described in the literature.

Dynamic content [Amza et al., 2002; Cecchet et al., 2002; Challenger et al., 2000; Iyengar and Challenger, 1997] is becoming a central component of modern transaction-oriented Websites. Although dynamic content generation is clearly a very important issue, there is currently no consensus as to what constitutes a "representative" dynamic workload, and so we do not present any characteristics of dynamic content here.

40.4.1 Request Methods

The first trait we examine is how frequently different *request methods* appear in server workloads. Several methods were defined in the HTTP 1.0 standard [Berners-Lee et al., 1996] (e.g., HEAD, POST, and DELETE), and many others were added to the 1.1 specification [Fielding et al., 1997, 1999] (e.g., OPTIONS, TRACE, and CONNECT). GET requests are the primary method by which documents are retrieved; the method "means retrieve whatever information … is identified by the Request-URI" [Berners-Lee et al., 1996]. The HEAD method is similar to the GET method except that only metainformation about the URI is returned. The POST method is a request for the server to accept information from the client, and is typically used for filling out forms and invoking dynamic content-generation mechanisms. The literature has shown [Krishnamurthy and Rexford, 2001] that the vast majority of methods are GET requests, with a smaller but noticeable percentage being HEAD or POST methods. Table 40.2 shows the percentage of request methods seen in the various logs. Here, only those methods that appear as a nontrivial percentage are shown, in this case defined as greater than one hundredth of a percent. Although different logs have slightly varying breakdowns, they are consistent with the findings in the literature.

TABLE 40.2 HTTP Request Methods (Percent)

Request Method	Chess 1997	Olympics 1998	IBM 1998	World Cup 1998	Department Store 2000	IBM 2001
GET	92.18	99.37	99.91	99.75	99.42	97.54
HEAD	03.18	00.08	00.07	00.23	00.45	02.09
POST	00.00	00.02	00.01	00.01	00.11	00.22

TABLE 40.3 Server Response Codes (Percent)

Response Code	Chess 1997	Olympics 1998	IBM 1998	World Cup 1998	Department Store 2000	IBM 2001
200 OK	85.32	76.02	75.28	79.46	86.80	67.73
206 partial content	00.00	00.00	00.00	00.06	00.00	00.00
302 found	00.05	00.05	01.18	00.56	00.56	15.11
304 not modified	13.73	23.25	22.84	19.75	12.40	16.26
403 forbidden	00.01	00.02	00.01	00.00	00.02	00.01
404 not found	00.55	00.64	00.65	00.70	00.18	00.79

40.4.2 Response Codes

The next characteristic we study are the *response codes* generated by the server. Again, the HTTP specifications define a large number of responses, the generation of which depends on multiple factors such as whether or not a client is allowed access to a URL, whether or not the request is properly formed, etc. However, certain responses are much more frequent than others.

Table 40.3 shows those responses seen in the logs that occur with a nontrivial frequency, again defined as greater than one hundredth of a percent. We see that the majority of the responses are successful transfers, i.e., the 200 OK response code.

Perhaps the most interesting aspect of this data is, however, the relatively large fraction of 304 Not Modified responses. This code is typically in response to a client generating a GET request with the If-Modified-Since option, which provides the client's notion of the URL's last-modified time. This request is essentially a cache-validation option and asks the server to respond with the full document if the client's copy is out of date. Otherwise, the server should respond with the 304 code if the copy is OK. As can be seen, between 12 and 23% of responses are 304 codes, indicating that clients revalidating up-to-date content is a relatively frequent occurrence, albeit in different proportions at different Websites.

Other responses, such as 403 Forbidden or 404 Not Found, are not very frequent, on the order of a tenth of a percent, but appear occasionally. The IBM 2001 log is unusual in that roughly 15% of the responses use the 302 Found code, which is typically used as a temporary redirection facility.

40.4.3 Document Popularity

Numerous researchers [Almeida et al., 1996; Arlitt and Williamson, 1997; Crovella and Bestavros, 1997; Padmanabhan and Qui, 2000] have observed that, in origin Web servers, the relative probability with which a web page is accessed follows a Zipf-like distribution. That is,

$$p(r) \approx C/r^{\alpha}$$

where $p(r)$ is the probability of a request for a document with rank r, and C is a constant (depending on α and the number of documents) that ensures that the sum of the probabilities is one. Rank is defined by popularity; the most popular document has rank one, the second-most popular has rank two, etc. When α equals one, the distribution is a true Zipf; when α is another value, the distribution is considered "Zipf-like." Server logs tend to have α values of 1 or greater; proxy server logs have lower values ranging from 0.64 to 0.83 [Breslau et al., 1999].

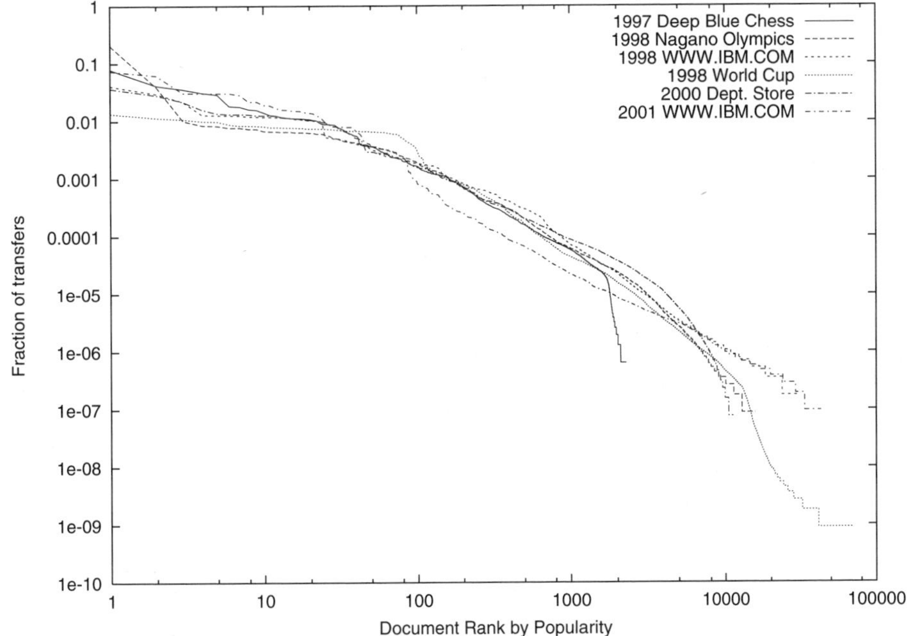

FIGURE 40.7 Document popularity.

Figure 40.7 shows the fraction of references based on document rank generated from the sample Web logs. Note that both the X and Y axes use a log scale. As can be seen, all the curves follow a Zipf-like distribution fairly closely, except towards the upper left of the graph and the lower right of the graph.

This Zipf property of document popularity is significant because it shows the effectiveness of document caching. For example, one can see that by simply caching the 100 most popular documents, assuming these documents are all cacheable, the vast majority of requests will find the document in the cache, avoiding an expensive disk I/O operation.

40.4.4 File Sizes

The next characteristic we examine is the range of sizes of the URLs stored on a Web server. File sizes give a picture of how much storage is required on a server and how much RAM might be needed to fully cache the data in memory. Which distribution best captures file size characteristics has been a topic of some controversy. There is consistent agreement that sizes range over multiple orders of magnitude and that the body of the distribution (i.e., that excluding the tail) is Log-Normal.

However, the shape of the tail of the distribution has been debated, with claims that it is Pareto [Crovella and Bestavros, 1997], Log-Normal [Downey, 2001], and even that the amount of data available is insufficient to statistically distinguish between the two [Gong et al., 2001]. Figure 40.8 shows the cumulative distribution function (CDF) of file sizes seen in the logs. Note that the X axis is in log scale. Table 40.4 presents the minimum, maximum, median, mean, and standard deviation of the file sizes. As can be seen, sizes range from 1 B to over 64 MB, varying across several orders of magnitude. In addition, the distributions show the rough "S" shape of the Log-Normal distribution.

As mentioned earlier, a metric of frequent interest in the research community is the "tail" of the distribution. Although the vast majority of files are small, most bytes transferred are found in large files. This is sometimes known as the "Elephants and Mice" phenomenon. To illustrate this property, we graphed the complement of the cumulative distribution function (CCDF) of the logs. These are shown in Figure 40.9. The Y values for this graph are essentially the complement of the corresponding Y values from Figure 40.8. Unlike Figure 40.8, however, note here that the Y-axis uses a log scale to better illustrate

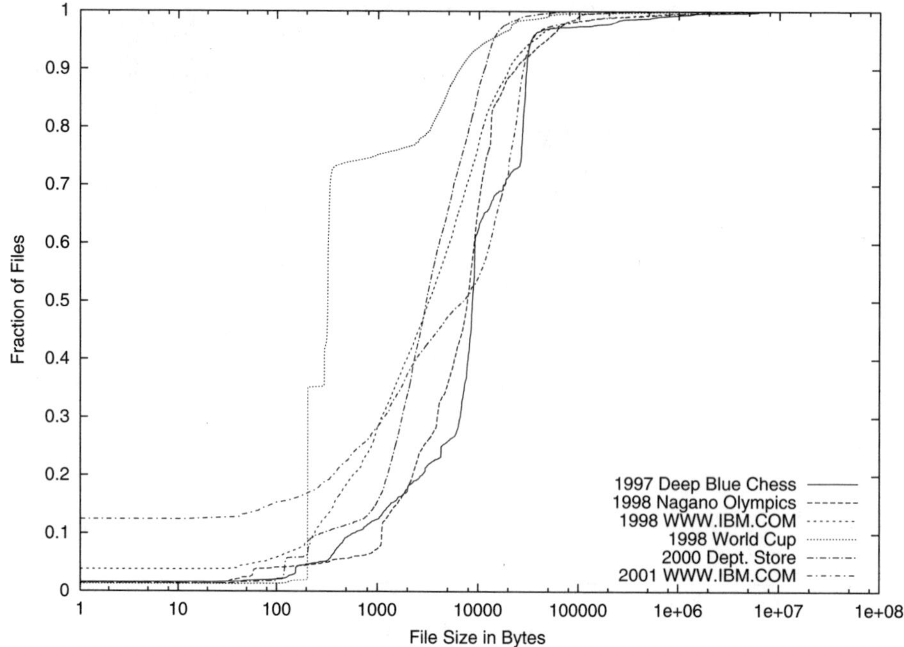

FIGURE 40.8 File size (CDF).

TABLE 40.4 File Size Statistics (Bytes)

Statistic	Chess 1997	Olympics 1998	IBM 1998	World Cup 1998	Department Store 2000	IBM 2001
Min	1	23	1	1	2	1
Median	8,697	7,757	3,244	328	3,061	7,049
Mean	45,012	12,851	20,114	6,028	4,983	29,662
Max	8,723,751	2,646,058	17,303,027	64,219,310	99,900	61,459,221
Standard deviation	384,175	44,618	193,892	253,481	6,115	394,088

the tail. We observe that all the logs have maximum file sizes in the range of 1 to 10 MB, with the exception of the Department Store log, which has no file size greater than 99,990 bytes.

40.4.5 Transfer Sizes

A metric related to Web file sizes is Web *transfer sizes,* or the size of the objects sent "over the wire." Transfer sizes are significant because they connote how much bandwidth is used by the Web server to respond to clients. In Apache Common Log Format, transfer size is based on the amount of *content* sent, and does not include the size from any HTTP headers or lower-layer bytes such as TCP or IP headers. Thus, here transfer size is based on the size of the content transmitted. The distribution of transfer sizes is thus influenced by the popularity of documents requested, as well as by the proportion of unusual responses such as 304 Not Modified and 404 Not Found.

Figure 40.10 shows the CDF of the object transfers from the logs. Table 40.5 presents the median, mean, and standard deviation of the transfer sizes. As can be seen, transfers tend to be small; for example, the median transfer size from the IBM 2001 log is only 344 bytes! An important trend is to note that a large fraction of transfers are for *zero* bytes, as much as 28 percent in the 1998 IBM log. The vast majority

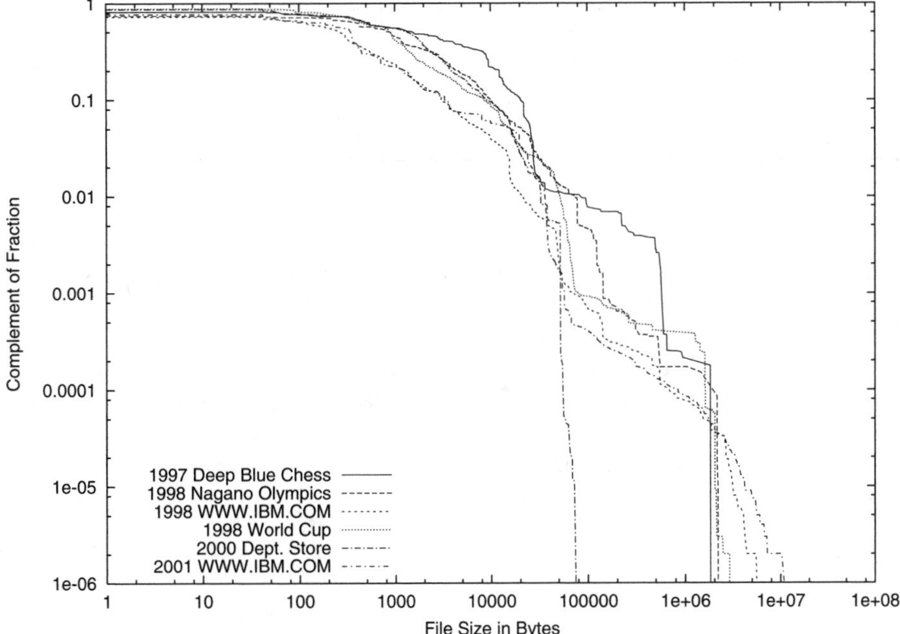

FIGURE 40.9 File size (CCDF).

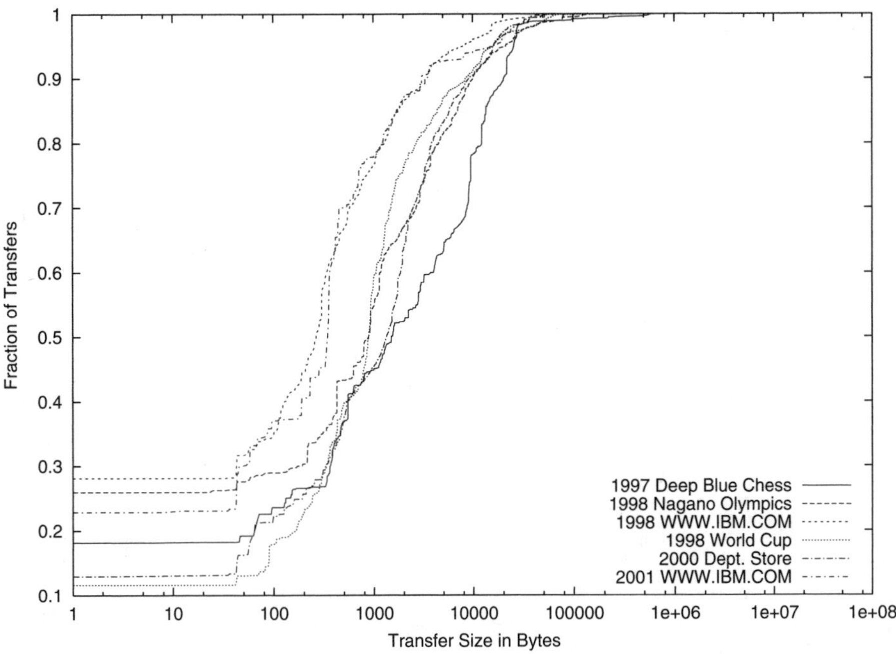

FIGURE 40.10 Transfer size (CDF).

TABLE 40.5 Transfer Size Statistics (Bytes)

Statistic	Chess 1997	Olympics 1998	IBM 1998	World Cup 1998	Department Store 2000	IBM 2001
Median	1,506	886	265	889	1,339	344
Mean	30,847	4,851	1,856	4,008	3,418	2,370
Standard Deviation	100,185	31,144	29,134	32,945	6,576	35,986

of these zero-byte transfers are the 304 Not Modified responses noted above in Section 40.4.2. When a conditional GET request with the If-Modified-Since option is successful, a 304 response is generated and no content is transferred. Other return codes, such as 403 Forbidden and 404 Not Found, also result in zero-byte transfers, but they are significantly less common. The exception is the IBM 2001 log, where roughly 15% of the 302 Found responses contribute to the fraction of zero-byte transfers.

Figure 40.11 shows the CCDF of the transfer sizes, in order to illustrate the "tail" of the distributions. Note again that the Y axis uses a log scale. The graph looks similar to Figure 40.9, perhaps because these transfers are so uncommon that weighting them by frequency does not change the shape of the graph, as it does with the bulk of the distribution in Figure 40.10.

40.4.6 HTTP Version

Another question we are interested in is the sort of HTTP protocol support being used by servers. HTTP 1.1 was first standardized in 1997 [Fielding et al., 1997]; the protocol has undergone some updating [Fielding et al., 1999; Krishnamurthy and Rexford, 2001] and in some ways is still being clarified [Krishnamurthy et al., 1999; Mogul, 2002]. The transition from 1.0 to 1.1 is a complex one, requiring support from browsers, servers, and proxy intermediaries (if any) as well.

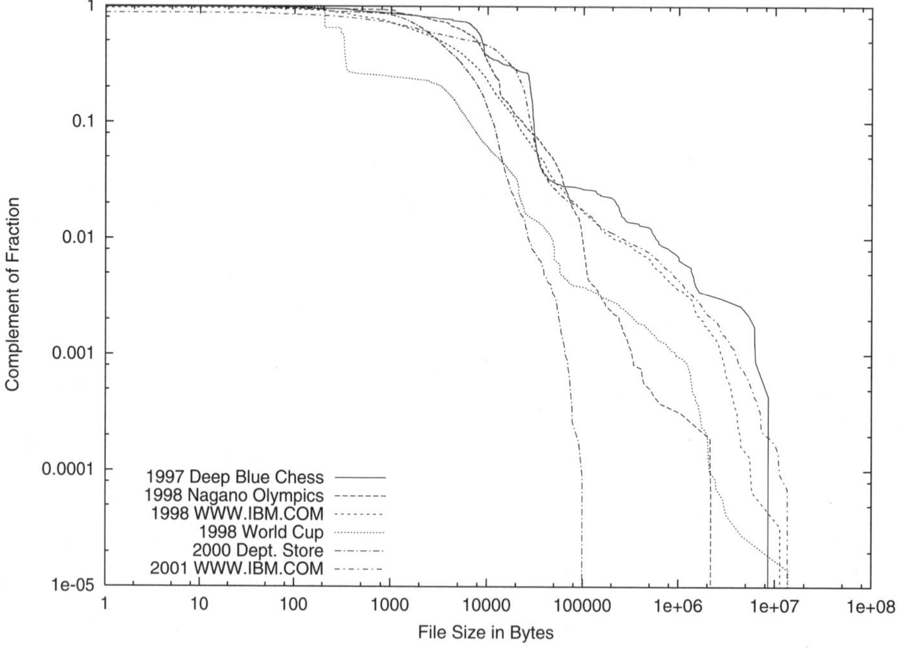

FIGURE 40.11 Transfer size (CCDF).

TABLE 40.6 HTTP Protocol Versions (Percent)

Protocol Version	Chess 1997	Olympics 1998	IBM 1998	World Cup 1998	Department Store 2000	IBM 2001
HTTP 1.0	95.30	78.56	77.22	78.62	51.1.3	51.08
HTTP 1.1	00.00	20.92	18.43	21.35	48.82	48.30
Unclear	04.70	00.05	04.34	00.02	00.05	00.06

Table 40.6 shows the HTTP protocol version that the server used in responding to requests. A clear trend is that over time, more requests are being serviced using version 1.1. In the most recent logs, from 2000 and 2001, HTTP 1.1 is used in just under half the responses.

Given the depth and complexity of the HTTP 1.1 protocol, the numbers above only scratch the surface of how servers utilize it. Many features have been added in version 1.1, including new mechanisms, headers, methods, and response codes. How these features are used in practice is still an open issue, and as mentioned earlier, server logs are insufficient to fully understand HTTP 1.1 behavior.

40.4.7 Summary

This work has presented some of the significant performance characteristics observed in Web server traffic. An interesting observation is that many of them have not fundamentally changed over time. Some performance characteristics, such as HTTP 1.1 support, have changed as expected, albeit more slowly than we anticipated. Still, certain characteristics seem to be invariants that hold true both across time and across different Websites.

Acknowledgment

Some of the material in this chapter appeared in the paper *Enhancing Web Performance* by the same authors in Proceedings of the 2002 IFIP World Computer Congress (Communication Systems: The State of the Art), edited by Lyman Chapin, published by Kluwer Academic Publishers, Boston, copyright 2002, International Federation for Information Processing.

References

Almeida, Virgilio, Azer Bestavros, Mark Crovella, and Adriana de Oliveira. Characterizing reference locality in the WWW. In *Proceedings of PDIS '96: The IEEE Conference on Parallel and Distributed Information Systems,* Miami Beach, FL, December 1996.

Alteon ACEdirector. http://www.nortelnetworks.com/products/01/acedir.

Amza, Cristiana, Emmanuel Cecchet, Anupam Chanda, Sameh Elnikety, Alan Cox, Romer Gil, Julie Marguerite, Karthick Rajamani, and Willy Zwaenepoel. Bottleneck characterization of dynamic Web site benchmarks. Technical Report TR02-388, Computer Science Department, Rice University, Houston, TX, February 2002.

Apostolopoulos, George, David Aubespin, Vinod Peris, Prashant Pradhan, and Debanjan Saha. Design, implementation, and performance of a content-based switch. In *Proceedings of IEEE INFOCOM,* March 2000.

Arlitt, Martin F. and Tai Jin. Workload characterization of the 1998 World Cup Web site. *IEEE Network,* 14(3): 30–37, May/June 2000.

Arlitt, Martin F. and Carey L. Williamson. Internet web servers: workload characterization and performance implications. *IEEE/ACM Transactions on Networking,* 5(5): 631–646, October 1997.

Aron, Mohit, Darren Sanders, Peter Druschel, and Willy Zwacnepoel. Scalable content-aware request distribution in cluster-based network servers. In *Proceedings of USENIX Annual Technical Conference,* San Diego, CA, June 2000.

Beck, Micah and Terry Moore. The Internet2 distributed storage infrastructure project: An architecture for Internet content channels. In *Proceedings of the 3rd International Web Caching Workshop,* 1998.

Berners-Lee, Tim, Roy Fielding, and Henrik Frystyk. Hypertext transfer protocol — HTTP/1.0. In *IETF RFC 1945,* May 1996.

Bestavros, Azer, Mark Crovella, Jun Liu, and David Martin. Distributed packet rewriting and its application to scalable server architectures. In *Proceedings of IEEE International Conference on Network Protocols,* Austin, TX, October 1998.

BIG-IP controller. http://www.f5.com/f5products/bigip/.

Breslau, Lee, Pei Cao, Li Fan, Graham Phillips, and Scott Shenker. Web caching and Zipf-like distributions: Evidence and implications. In *Proceedings of the Conference on Computer Communications (IEEE Infocom),* New York, March 1999.

Brisco, Thomas P. DNS support for load balancing. Number IETF RFC 1974, April 1995.

Cardellini, Valeria, Michele Colajanni, and Philip S. Yu. DNS dispatching algorithms with state estimators for scalable Web server clusters. *World Wide Web,* 2(2), July 1999a.

Cardellini, Valeria, Michele Colajanni, and Philip S. Yu. Dynamic load balancing on Web-server systems. *IEEE Internet Computing,* pp. 28–39, May/June 1999b.

Cecchet, Emmanuel, Anupam Chanda, Sameh Elnikety, Julie Marguerite, and Willy Zwaenepoel. A comparison of software architectures for e-business applications. Technical Report TR02-389, Computer Science Department, Rice University, Houston, TX, February 2002.

Challenger, Jim, Arun Iyengar, Karen Witting, Cameron Ferstat, and Paul Reed. A publishing system for efficiently creating dynamic Web content. In *Proceedings of IEEE INFOCOM,* March 2000.

Cisco CSS 1100. http://www.cisco.com/warp/public/cc/pd/si/11000/,a.

Cisco LocalDirector 400 series. http://www.cisco.com/warp/public/cc/pd/cxsr/400/,b.

Cohen, Ariel, Sampath Ragarajan, and Hamilton Slye. On the performance of TCP splicing for URL-aware redirection. In *Proceedings of USENIX Symposium on Internet Technologies and Systems,* Boulder, CO, October 1999.

Crovella, Mark and Azer Bestavros. Self-similarity in World Wide Web traffic: Evidence and possible causes. *IEEE/ACM Transactions on Networking,* 5(6): 835–846, November 1997.

Dias, Dan, William Kish, Rajat Mukherjee, and Renu Tewari. A scalable and highly available Web server. In *Proceedings of the 1996 IEEE Computer Conference (COMPCON),* February 1996.

Downey, Allen. The structural cause of file size distributions. In *Proceedings of the 9th International Symposium on Modeling, Analysis and Simulation of Computer and Telecommunication Systems (MASCOTS),* Cincinnati, OH, August 2001.

Fei, Zongming, Samrat Bhattacharjee, Ellen Zegura, and Mustapha Ammar. A novel server selection technique for improving the response time of a replicated service. In *Proceedings of IEEE INFOCOM,* 1998.

Fielding, Roy, Jim Gettys, Jeffrey Mogul, Henrik Frystyk, and Tim Berners-Lee. Hypertext transfer protocol — HTTP/1.1. In. *IETF RFC 2068,* January 1997.

Fielding, Roy, Jim Gettys, Jeffrey Mogul, Henrik Frystyk, Larry Masinter, Paul Leach, and Tim Berners-Lee. Hypertext transfer protocol — HTTP/1.1. In *IETF RFC 2616,* June 1999.

Foundry ServerIron. http://www.foundrynet.com/products/webswitches/serveriron.

Gong, Weibo, Yong Liu, Vishal Misra, and Don Towsley. On the tails of Web file size distributions. In *Proceedings of the 39th Allerton Conference on Communication, Control, and Computing,* Monticello, IL, October 2001.

Hu, James C., Irfan Pyarali, and Douglas C. Schmidt. Measuring the impact of event dispatching and concurrency models on Web server performance over high-speed networks. In *Proceedings of the 2nd Global Internet Conference (held as part of GLOBECOM '97),* Phoenix, AZ, November 1997.

Hunt, Guerney, German Goldszmidt, Richard King, and Rajat Mukherjee. Network dispatcher: A connection router for scalable Internet services. In *Proceedings of the 7th International World Wide Web Conference,* April 1998.

Intel NetStructure 7175 traffic director http://www.intel.com/network/idc/products/director_7175.htm

Iyengar, Arun and Jim Challenger. Improving Web server performance by caching dynamic data. In *Proceedings of the USENIX Symposium on Internet Technologies and Systems,* Monterey, CA, December 1997.

Iyengar, Arun, Jim Challenger, Daniel Dias, and Paul Dantzig. High-performance Web site design techniques. *IEEE Internet* Computing, 4(2), March/April 2000.

Joubert, Philippe, Robert King, Richard Neves, Mark Russinovich, and John Tracey. High-performance memory-based Web servers: Kernel and user-space performance. In *Proceedings of the USENIX Annual Technical Conference,* Boston, MA, June 2001.

Krishnamurthy, Balachander, Jeffrey C. Mogul, and David M. Kristol. Key differences between HTTP/1.0 and HTTP/1.1. In *Proceedings of WWW-8 Conference,* Toronto, Canada, May 1999.

Krishnamurthy, Balachander and Jennifer Rexford. *Web Protocols and Practice.* Addison-Wesley, Reading, MA, 2001.

Kwan, Thomas T., Robert E. McGrath, and Daniel A. Reed. NCSA's World Wide Web server: Design and performance. *IEEE Computer,* 28(11): 68–74, November 1995.

Lawrence Berkeley National Laboratory. The Internet traffic archive. http://ita.ee.lbl.gov/.

Maltz, David and Pravin Bhagwat. TCP splicing for application layer proxy performance. Technical Report RC 21139, IBM TJ Watson Research Center, 1998.

Mogul, Jeffrey C. Clarifying the fundamentals of HTTP. In *Proceedings of WWW 2002 Conference,* Honolulu, HI, May 2002.

Mosedale, Dan, William Foss, and Rob McCool. Lessons learned administering Netscape's Internet site. *IEEE Internet Computing,* 1(2): 28–35, March/April 1997.

Nahum, Erich M., Marcel Rosu, Srinivasan Seshan, and Jussara Almeida. The effects of wide-area conditions can WWW server performance. In *Proceedings of the ACM Sigmetrics Conference on Measurement and Modeling of Computer Systems,* Cambridge, MA, June 2001.

Open Market. FastCGI. http://www.fastcgi.com/.

Padmanabhan, Venkata N. and Lili Qui. The content and access dynamics of a busy Web site: Findings and implications. In *ACM SIGCOMM Symposium on Communications Architectures and Protocols,* pp. 111–123, 2000. URL: citeseer.nj.nec.com/padmanabhan00content.html.

Pai, Vivek, Peter Druschel, and Willy Zwaenepoel. Flash: An efficient and portable Web server. In *USENIX Annual Technical Conference,* Monterey, CA, June 1999.

Pai, Vivek S., Mohit Aron, Gaurav Banga, Michael Svendsen, Peter Druschel, Willy Zwaenepoel, and Erich M. Nahum. Locality-aware request distribution in cluster-based network servers. In *Proceedings of the Architectural Support for Programming Languages and Operating Systems,* pp. 205–216, 1998. URL: citeseer.nj.nec.com/article/pai98localityaware.html.

Rabinovich, Michael and Oliver Spatscheck. *Web Caching and Replication.* Addison-Wesley, Reading, MA, 2001.

Red Hat. The Tux WWW server. http://people.redhat.com/~mingo/TUX-patches/, a.

Rosu, Marcel-Catalin and Daniela Rosu. An evaluation of TCP splice benefits in Web proxy servers. In *Proceedings of the 11th International World Wide Web Conference* (WWW2002), Honolulu, HI, May 2002.

Shaikh, Anees, Renu Tewari, and Mukhesh Agrawal. On the effectiveness of DNS-based server selection. In *Proceedings of IEEE INFOCOM 2001,* 2001.

Song, Junehua, Arun Iyengar, Eric Levy, and Daniel Dias. Architecture of a Web server accelerator. *Computer Networks,* 38(1), 2002.

Spatscheck, Oliver, Jorgen S. Hansen, John H. Hartman, and Larry L. Peterson. Optimizing TCP forwarder performance. *IEEE/ACM Transactions on Networking,* 8(2): 146–157, April 2000.

Sun Microsystems. The Java Web server. http://wwws.sun.com/software/jwebserver/index.html, b.

The Apache Project. The Apache WWW server.http://httpd.apache.org.

The Standard Performance Evaluation Corporation. SPECWeb99. http://www.spec.org/osg/web99, 1999.

Websphere edge server. http://www.ibm.com/software/webservers/edgeserver/.

Zeus. The Zeus WWW server. http://www.zeus.co.uk, c.

41

Web Caching, Consistency, and Content Distribution

Arun Iyengar
Erich Nahum
Anees Shaikh
Renu Tewari

CONTENTS

41.1 Introduction 41-1
41.2 Practical Issues in the Design of Caches 41-2
41.3 Cache Consistency 41-3
41.3.1 Degrees of Consistency 41-4
41.3.2 Consistency Mechanisms 41-5
41.3.3 Invalidates and Updates 41-6
41.4 CDNs: Improved Web Performance through Distribution 41-7
41.4.1 CDN Architectural Elements 41-8
41.4.2 CDN Request-Routing 41-10
41.4.3 Request-Routing Metrics and Mechanisms 41-12
41.4.4 Consistency Management for CDNs 41-14
41.4.5 CDN Performance Studies 41-15
Acknowledgment 41-15
References 41-15

41.1 Introduction

Caching has been widely deployed to improve Web performance by reducing client-observed latency and network bandwidth usage, in addition to improving server scalability by reducing the load on the servers. Web caches can be deployed at various points in the network. Forward proxy caches are deployed close to the client at network entry points by ISPs to reduce the network bandwidth usage and improve client latency by caching frequently accessed data. Such caches can be either transparent to the client or manually configured. With transparent caching, the packets are intercepted by an intermediate router (layer-4 or layer-7 switch) and transparently routed to a cache, which in turn responds to the client directly [CiscoLocal]. Manual configuration requires the client to explicitly configure the browser to go via a proxy cache. In addition to forward proxies, caches can be deployed as a front-end to a server farm to reduce server load and increase server scalability. Such caches, called reverse proxies, are useful in eliminating the load of a hot-set from impacting the server performance. Typically, reverse proxies are in the same administrative domain as the server.

As with any caching system, Web caches need to use a cache replacement policy to decide what to keep in the cache and a consistency mechanism to keep it current. Various cache replacement algorithms, from LRU to Greedy-dual size, have been studied in the context of the Web to improve cache performance in terms of client response times and server throughput. For maintaining consistency, Web objects may have explicit expiration times associated with them, indicating when they become obsolete. The problem

1-58488-381-2/05/$0.00+$1.50
© 2005 by CRC Press LLC

with expiration times is that it is often not possible to tell in advance when Web data will become obsolete. Furthermore, expiration times are not sufficient for applications that have strong consistency requirements. Without expiration times, the proxy cache needs to always check the staleness of the data with the server using if-modified-since messages, thereby increasing client response times. Stale cached data and the inability in many cases to cache dynamic and personalized data limit the effectiveness of caching. Numerous proposals have been made to extend the support for consistency so that stronger requirements can be met [Li et al., 2000].

Simple proxy caching is limited by the space and processing capacity of a single caching server. To further improve performance, caching can be extended to include a group of cooperating caches deployed in the network either in a hierarchical or distributed manner. Hierarchical caches such as the NLANR Squid [Squid, 1997] cache consists of a single tree with parent–child relationship, whereas other organizations include meshes with hierarchical or centralized directories [Wolman et al., 1999]. A further extension of distributed caching is content distribution networks (CDNs), which supplement the client-side proxy caching to other points in the network controlled by the CDN service provider. A CDN is a shared network of servers or caches that delivers content to users on behalf of content providers by using various request routing techniques. The intent of a CDN is to serve content to a client from a CDN server so that response time is decreased compared to contacting the origin server directly. In doing so, CDNs also reduce the load on origin servers.

This chapter examines several issues related to cache management, consistency maintenance, and the overall architecture and techniques for routing requests in CDNs. We also provide insight into the performance improvements typically achieved by CDNs.

41.2 Practical Issues in the Design of Caches

Web caches can be implemented at the application level [Iyengar, 1999], kernel level [Joubert et al., 2001], or under an embedded operating system [Song et al., 2002]. Application-level caches are the easiest to design and have the potential for the most features. Kernel-level caches are harder to design but have the potential for better performance. Caches can also be designed for embedded operating systems that may be optimized for certain features such as communication. Such caches may offer performance comparable to kernel-level caches. A problem with using embedded operating systems is that as processor technology improves, it may not be feasible for the embedded operating system to keep up with new processors. This means that over time, the advantage achieved by a cache running under an embedded operating system may decrease.

HTTP provides a standard interface for applications to utilize caches. An HTTP interface alone is limiting, however, and does not provide adequate support for explicitly managing the contents of a cache. It is also not the most efficient interface and can be cumbersome for applications to use. It is, therefore, preferable for the cache to define an interface that an application program can use to explicitly add, delete, and update cached objects [Iyengar, 1999].

The number of transactions per unit time that a Web cache has to handle in order to achieve good performance is orders of magnitude less than that needed by a processor cache. Therefore, Web caches can employ more sophisticated consistency and replacement policies. Cache replacement policies are applied when a cache becomes full and it is necessary to determine what objects in the cache to keep. The least recently used algorithm (LRU) has been used for caching across a broad range of disciplines. In LRU, the object that was accessed the farthest in the past is selected for removal when the cache becomes full. LRU has the advantage that it is easy to implement. A doubly linked list is used to order objects by access times. Whenever an object is accessed, it is moved to the front of the list.

A number of cache replacement algorithms have been proposed that result in higher cache hit rates than LRU. One of the most commonly used such algorithms is the GreedyDual-Size algorithm [Cao and Irani, 1997]. The GreedyDual-Size algorithm associates a cost $C(o)$ with each object o. The cost would typically be associated with how expensive it is to fetch or create the object. It is preferable to cache more expensive objects because doing so results in greater savings in the event of a cache hit. GreedyDual-Size divides $C(o)$ by the size of o, $S(o)$, in order to arrive at an estimate $H(o)$ of the savings per unit of cache memory that would be achieved by caching the object.

When object o is first brought into the cache, $H(o)$ is set to $C(o)/S(o)$. When the cache becomes full and an object needs to be removed, the object with the lowest H value, H_{min}, is removed, and all objects reduce their H values by H_{min}. When an object is accessed, its H value is restored to $C(o)/S(o)$. That way, objects that are accessed frequently will on average have higher H values and are therefore less likely to be replaced.

A naive implementation would require n-1 subtractions every time an object is replaced (to update H values for the remaining cached objects), where n is the number of cached objects. This is inefficient. Instead, an inflation value, L, is maintained. When an object o is accessed, $H(o)$ is set to $C(o)/S(o) + L$. By adding L to compute the H value of an accessed object, it becomes unnecessary to reduce H values for all remaining objects when an object is replaced. L is initially set to 0. Whenever an object is replaced, L is updated to the H value of the replaced object.

The cost function C depends on the resources the cache is trying to minimize. If the objective is to maximize cache hit rates, then the cost function should be a constant for each object. If the objective is to minimize time consumed in fetching remote objects, then $C(o)$ could be the expected latency for fetching o. For a dynamic Web object, the CPU cycles consumed for creating the object may have the most significant effect on performance. The cost function for such an object could thus be proportional to the number of CPU cycles required for creating the object.

Caches can be implemented using both main memory and disk storage. Main memory offers better performance. In some cases, however, disk storage is essential. If the cache size exceeds the main memory size, it may be desirable to store colder objects on disk instead of deleting the objects to keep the cache within memory limits. Disk storage is also essential for persistence when a cache must be shot down and later restarted. If the cache is totally purged each time the machine is shut down, then performance is likely to be poor while the machine is being brought to a warm state after start-up. If, on the other hand, cached information is maintained on disk before the shutdown, the cache can be brought to a warm state right after the system is restarted. Disk storage is also important for fault tolerance. When a cache fails, if hot objects are maintained on disk, then the cache can be quickly brought to a warm state after the failure.

File systems and databases can be used for persistently storing cached data. A key problem with file systems and databases is that they can be inefficient. For Web caches, the rate at which objects are added to and deleted from caches can be high [Markatos et al., 1999]. If a file system is used and a different file is used to store each object, the overhead for creating and deleting files can be significant. Customized disk storage allocators can often achieve much better performance. Good performance for Web workloads has been achieved by maintaining multiple objects in a single file and efficiently managing the storage space within the file [Iyengar et al., 2001]. A portable disk storage allocator we built in Java achieves considerably better performance than both file systems and databases.

41.3 Cache Consistency

Caching has proven to be an effective and practical solution for improving the scalability and performance of Web servers. Static Web page caching has been applied both at client browsers and at intermediaries that include isolated proxy caches or multiple caches or servers within a CDN network. As with caching in any system, maintaining cache consistency is one of the main issues that a Web caching architecture needs to address. As more of the data on the Web is dynamically assembled, personalized, and constantly changing, the challenges of efficient consistency management become more pronounced. To prevent stale information from being transmitted to clients, an intermediary cache must ensure that the locally cached data is consistent with that stored on servers. The exact cache consistency mechanism and the degree of consistency employed by an intermediary depends on the nature of the cached data; not all types of data need the same level of consistency guarantees. Consider the following example: A Web server offers online auctions over the Internet. For each item being sold, the server maintains information such as its latest bid price (which changes every few minutes) as well as other information such as photographs and reviews for the item (all of which change less frequently). Consider an intermediary that caches this information. Clearly, the bid price returned by the intermediary cache should always be consistent with that at the server. In contrast, reviews of items need not always be up-to-date because a user may be willing to receive slightly stale information.

The above example shows that an intermediary cache will need to provide different degrees of consistency for different types of data. The degree of consistency selected also determines the mechanisms used to maintain it, and the overheads incurred by both the server and the intermediary.

41.3.1 Degrees of Consistency

In general, the degrees of consistency that an intermediary cache can support fall into the following four categories:

- *Strong consistency:* A cache consistency level that always returns the results of the latest (committed) write at the server is said to be strongly consistent. Due to the unbounded message delays in the Internet, no cache consistency mechanism can be strongly consistent in this idealized sense. Strong consistency is typically implemented using a two-phase message exchange along with timeouts to handle unbounded delays.
- *Delta consistency:* A consistency level that returns data that is never outdated by more than δ time units, where δ is a configurable parameter, with the last committed write at the server said to be delta consistent. In practice, the value of delta should be larger than t, which is the network delay between the server and the intermediary at that instant, i.e., $t < \delta < \infty$.
- *Weak consistency:* For this level of consistency, a read at the intermediary does not necessarily reflect the last committed write at the server but some correct previous value.
- *Mutual consistency:* A consistency guarantee in which a group of objects are mutually consistent with respect to each other. In this case, some objects in the group cannot be more current than the others. Mutual consistency can coexist with the other levels of consistency.

Strong consistency is useful for mirror sites that need to reflect the current state at the server. Some applications based on financial transactions may also require strong consistency. Certain types of applications can tolerate stale data as long as it is within some known time-bound. For such applications, delta consistency is recommended. Delta consistency assumes that there is a bounded communication delay between the server and the intermediary cache. Mutual consistency is useful when a certain set of objects at the intermediary (e.g., the fragments within a sports score page or within a financial page) need to be consistent with respect to each other. To maintain mutual consistency, the objects need to be automatically invalidated such that they all either reflect the new version or maintain the earlier stale version.

Most intermediaries deployed in the Internet today provide only weak consistency guarantees [Gwertzman and Seltzer, 1996; Squid, 1997]. Until recently, most objects stored on Web servers were relatively static and changed infrequently. Moreover, this data was accessed primarily by humans using browsers. Because humans can tolerate receiving stale data (and manually correct it using browser reloads), weak cache consistency mechanisms were adequate for this purpose. In contrast, many objects stored on Web servers today change frequently and some objects (such as news stories or stock quotes) are updated every few minutes [Barford et al., 1999]. Moreover, the Web is rapidly evolving from a predominantly read-only information system to a system where collaborative applications and program-driven agents frequently read as well as write data. Such applications are less tolerant of stale data than humans accessing information using browsers, These trends argue for augmenting the weak consistency mechanisms employed by today's proxies with those that provide strong consistency guarantees in order to make caching more effective. In the absence of such strong consistency guarantees, servers resort to marking data as uncacheable and thereby reduce the effectiveness of proxy caching.

41.3.2 Consistency Mechanisms

The mechanisms used by an intermediary and the server to provide the degrees of consistency described earlier fall into three categories: (1) *client-driven,* (2) *server-driven,* and (3) *explicit* mechanisms.

Server-driven mechanisms, referred to as *server-based invalidation,* can be used to provide strong or delta consistency guarantees [Yin et al., 1999b]. Server-based invalidation requires the server to notify proxies when the data changes. This approach substantially reduces the number of control messages exchanged between the server and the intermediary (because messages are sent only when an object is modified). However, it requires the server to maintain per-object state consisting of a list of all proxies that cache the object; the amount of state maintained can be significant especially at popular Web servers. Moreover, when an intermediary is unreachable due to network failures, the server must either delay write requests until it receives all the acknowledgments or a timeout occurs, or risk violating consistency guarantees. Several protocols have been proposed to provide delta and strong consistency using server-based invalidations. Web cache invalidation protocol (WCIP) is one such proposal for propagating server invalidations using application-level multicast while providing delta consistency [Li et al., 2000]. Web content distribution protocol (WCDP) is another proposal that supports multiple consistency levels using a request–response protocol that can be scaled to support distribution hierarchies [Tewari et al., 2002].

The client-driven approach, also referred to as *client polling,* requires that intermediaries poll the server on *every read* to determine if the data has changed [Yin et al., 1999b]. Frequent polling imposes a large message overhead and also increases the response time (because the intermediary must await the result of its poll before responding to a read request). The advantage, though, is that it does not require any state to be maintained at the server, nor does the server ever need to delay write requests (because the onus of maintaining consistency is on the intermediary).

Most existing proxies provide only weak consistency by (1) explicitly providing a server-specified lifetime of an object (referred to as the *time-to-live (TTL)* value), or (2) by *periodic polling* of the server to verify that the cached data is not stale [Cate, 1992; Gwertzman and Seltzer, 1996; Squid, 1997]. The TTL value is sent as part of the HTTP response in an Expires tag or using the Cache-Control headers. However, *a priori* knowledge of when an object will be modified is difficult in practice, and the degree of consistency is dependent on the clock skew between the server and the intermediaries. With periodic polling, the length of the period determines the extent of the object staleness. In either case, modifications to the object before its TTL expires or between two successive polls causes the intermediary to return stale data. Thus both mechanisms are heuristics and provide only weak consistency guarantees. Hybrid approaches where the server specifies a time-to-live value for each object, and the intermediary polls the server only when the TTL expires, also suffer from these drawbacks.

Server-based invalidation and client polling form two ends of a spectrum. Whereas the former minimizes the number of control messages exchanged but may require a significant amount of state to be maintained, the latter is stateless but can impose a large control message overhead. Figure 41.1 quantitatively compares these two approaches with respect to (1) the server overhead, (2) the network overhead, and (3) the client response time. Due to their large overheads, neither approach is appealing for Web environments. A strong consistency mechanism suitable for the Web must not only reduce client response time but also balance both network and server overheads.

One approach that provides strong consistency, while providing a smooth trade-off between the state space overhead and the number of control messages exchanged, is *leases* [Gray and Cheriton, 1989]. In this approach, the server grants a lease to each request from an intermediary. The lease duration denotes the interval of time during which the server agrees to notify the intermediary if the object is modified. After the expiration of the lease, the intermediary must send a message requesting renewal of the lease. The duration of the lease determines the server and network overhead. A smaller lease duration reduces the server state space overhead, but increases the number of control (lease renewal) messages exchanged and *vice versa.* In fact, an infinite lease duration reduces the approach to server-based invalidation, whereas a zero lease duration reduces it to client polling. Thus, the leases approach spans the entire spectrum between the two extremes of server-based invalidation and client polling.

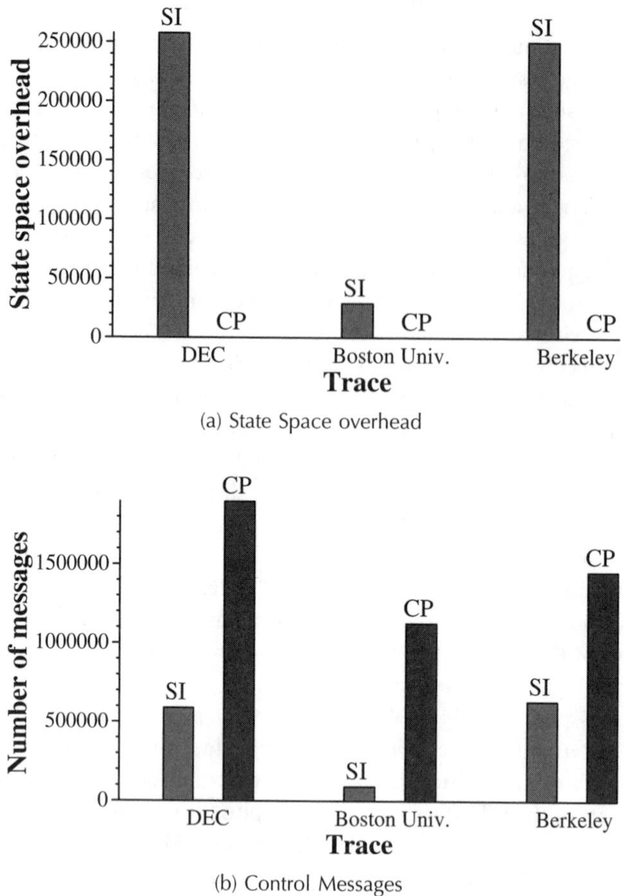

FIGURE 41.1 Efficacy of server-based invalidation and client polling for three different trace workloads (DEC, Berkeley, Boston University). The figure shows that server-based invalidation has the largest state space overhead; client polling has the highest control message overhead.

The concept of a lease was first proposed in the context of cache consistency in distributed file systems [Gray and Cheriton, 1989]. The use of leases for Web proxy caches was first alluded to in Liu and Cao [1997] and was subsequently investigated in detail in Yin et al. [1999b]. The latter effort focused on the design of *volume leases* — leases granted to a collection of objects so as to reduce (1) the lease renewal overhead and (2) the blocking overhead at the server due to unreachable proxies. Other efforts have focused on extending leases to hierarchical proxy cache architectures [Yin et al., 1999a; Yu et al., 1999]. The adaptive leases effort described analytical and quantitative results on how to select the optimal lease duration based on the server and message exchange overheads [Duvvuri et al., 2000].

A qualitative comparison of the overheads of the different consistency mechanisms is shown in Table 41.1. The message overheads of an invalidation-based or lease-based approach is smaller than that of polling, especially when reads dominate writes, as in the Web environment.

41.3.3 Invalidates and Updates

With server-driven consistency mechanisms, when an object is modified, the origin server notifies each "subscribing" intermediary. The notification consists of either an invalidate message or an updated (new) version of the object. Sending an invalidate message causes an intermediary to mark the object as invalid; a subsequent request requires the intermediary to fetch the object from the server (or from a designated

TABLE 41.1 Overheads of Different Consistency Mechanisms

Overheads	Polling	Periodic Polling	Invalidates	Leases	TTL
File Transfer	W'	$W' - \delta$	W'	W'	W'
Control Msgs.	$2R - W'$	$2R/t - (W' - \delta)$	$2W'$	$2W'$	W'
Staleness	0	t	0	0	0
Write delay	0	0	notify(all)	min(t, notify(all$_t$))	0
Server State	None	None	All	All$_t$	None

Note: t is the period in periodic polling or the lease duration in the leases approach. W' is the number of nonconsecutive writes. All consecutive writes with no interleaving reads are counted as a single write. R is the number of reads. The number of writes δ were not notified to the intermediary as only weak consistency was provided. "All" means all of the subscribers for server-driven invalidation. "All$_t$" means all of the servers within lease duration t.

site). Thus, each request after a cache invalidate incurs an additional delay due to this remote fetch. An invalidation adds to two control messages and a data transfer (an invalidation message, a read request on a miss, and a new data transfer) along with the extra latency. No such delay is incurred if the server sends out the new version of the object upon modification. In an update-based scenario, subsequent requests can be serviced using locally cached data. A drawback, however, is that sending updates incurs a larger network overhead (especially for large objects). This extra effort is wasted if the object is never subsequently requested at the intermediary. Consequently, cache invalidates are better suited for less popular objects, while updates can yield better performance for frequently requested small objects. Delta encoding techniques have been designed to reduce the size of the data transferred in an update by sending only the changes to the object [Krishnamurthy and Wills, 1997]. Note that delta encoding is not related to delta consistency. Updates, however, require better security guarantees and make strong consistency management more complex. Nevertheless, updates are useful for mirror sites where data need to be "pushed" to the replicas when they change. Updates are also useful for preloading caches with content that is expected to become popular in the near future.

A server can dynamically decide between invalidates and updates based on the characteristics of an object. One policy could be to send updates for objects whose popularity exceeds a threshold and to send invalidates for all other objects. A more complex policy is to take both popularity and *object* size into account. Because large objects impose a larger network transfer overhead, the server can use progressively larger thresholds for such objects (the larger an object, the more popular it needs to be before the server starts sending updates).

The choice between invalidation and updates also affects the implementation of a strong consistency mechanism. For invalidations only, with a strong consistency guarantee, the server needs to wait for all acknowledgments of the invalidation message (or a timeout) to commit the write at the server. With updates, on the other hand, the server updates are not immediately committed at the intermediary. Only after the server receives all the acknowledgments (or a timeout) and then sends a commit message to all the intermediaries is the new update version committed at the intermediary. Such two-phase message exchanges are expensive in practice and are not required for weaker consistency guarantees.

41.4 CDNs: Improved Web Performance through Distribution

End-to-end Web performance is influenced by numerous factors such as client and server network connectivity, network loss and delay, server load, HTTP protocol version, and name resolution delays. The content-serving architecture has a significant impact on some of these factors, as well as factors not related to performance such as cost, reliability, and ease of management. In a traditional content-serving architecture, all clients request content from a single location, as shown in Figure 41.2(a). In this architecture, scalability and performance are improved by adding servers, without the ability to address poor performance due to problems in the network. Moreover, this approach can be expensive because the site must be overprovisioned to handle unexpected surges in demand.

Some ISPs address performance bottlenecks in the network by deploying caching proxies near clients to reduce network traffic and improve client performance. Caching proxies are limited, however, because they operate based only on user demand for a very large and diverse set of content. Most proxy cache studies, for example, find that they achieve only a 20–40% hit rate [IRCache Project Daily Reports, 2002; Wolman et al., 1999].

Another way to address poor performance due to network congestion, or flash crowds at servers, is to distribute popular content to servers or caches located closer to the edges of the network, as shown in Figure 41.2(b). Such a distributed network of servers comprises a content distribution network (CDN). A CDN is simply a network of servers or caches that delivers content to users on behalf of content providers. The intent of a CDN is to serve content to a client from a CDN server such that the response-time performance is improved over contacting the origin server directly. CDN servers are typically shared, delivering content belonging to multiple Websites, though all servers may not be used for all sites. Because CDN servers receive requests only for hosted content, cache misses typically occur only for compulsory misses due to the initial request for some content.

CDNs have several advantages over traditional centralized content-serving architectures, including [Verma, 2002]:

- Improving client-perceived response time by bringing content closer to the network edge, and thus closer to end-users
- Off-loading work from origin servers by serving larger objects, such as images and multimedia, from multiple CDN servers
- Reducing content provider costs by reducing the need to invest in more powerful servers or more bandwidth as user population increases
- Improving site availability by replicating content in many distributed locations

Content distribution service providers (CDSPs) manage and operate the CDN, thus freeing content providers from the tasks of maintaining the servers themselves. Some network service providers offer a CDN service in addition to network access service (e.g., AT&T). Other CDSPs focus primarily on providing a variety of CDN services (e.g., Akamai and Speedera).

CDN servers may be configured in tree-like hierarchies [Yu et al., 1999] or clusters of cooperating proxies that employ content-based routing to exchange data [Gritter and Cheriton, 2001]. Commercial CDNs also vary significantly in their size and service offerings. CDN deployments range from a few tens of servers (or server clusters), to over 10,000 servers placed in hundreds of ISP networks. A large footprint allows a CDSP to reach the majority of clients with very low latency and path length.

Content providers use CDNs primarily for serving static content like images or large stored multimedia objects (e.g., movie trailers and audio clips). A recent study of CDN-served content found that 96% of the objects served were images [Krishnamurthy et al., 2001]. However, the remaining few objects accounted for 40–60% of the bytes served, indicating a small number of very large objects. Increasingly, CDSPs offer services to deliver streaming media and dynamic data such as localized content or targeted advertising.

41.4.1 CDN Architectural Elements

As illustrated in Figure 41.3(a), CDNs have three key architectural elements in addition to the CDN servers themselves: a distribution system, an accounting/billing system, and a request-routing system [Day et al., 2002]. The distribution system is responsible for moving content from origin servers into CDN servers and ensuring data consistency. Section 41.4.4 describes some techniques used to maintain consistency in CDNs. The accounting/billing system collects logs of client accesses and tracks CDN server usage for use primarily in administrative tasks. Finally, the request-routing system is responsible for directing client requests to appropriate CDN servers. The request-routing system may also interact with the distribution system to keep an up-to-date view of what content resides on which CDN servers.

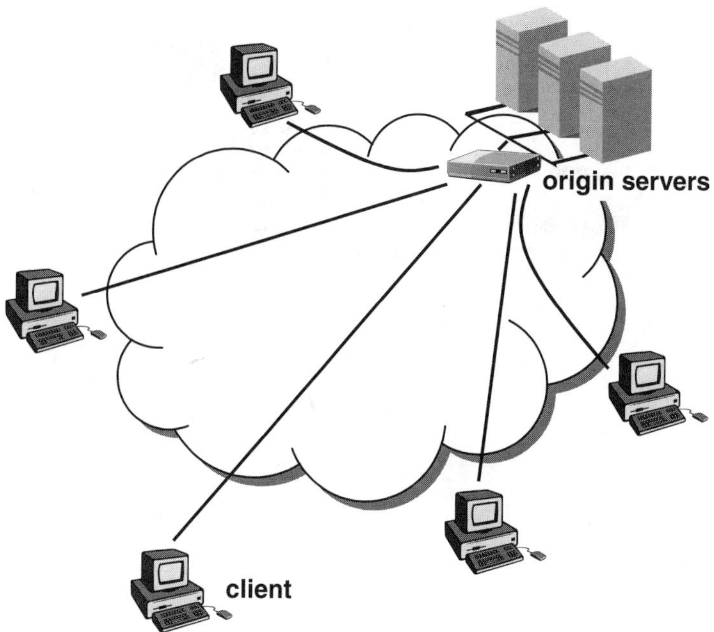

(a) Traditional centralized architecture

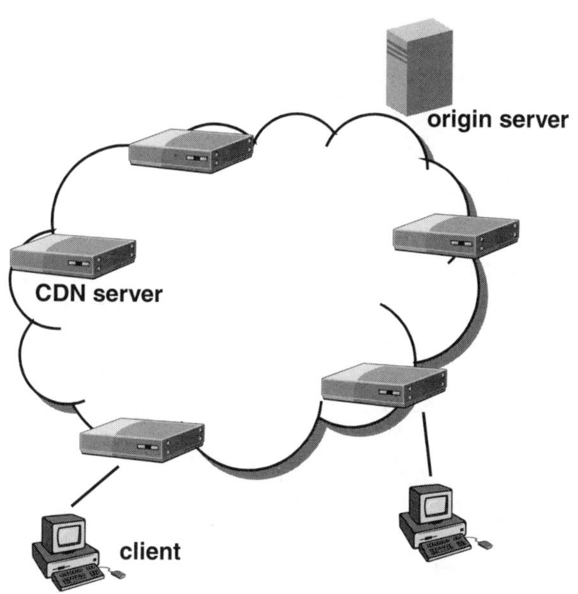

(b) Distibuted CDN architecture

FIGURE 41.2 Content-serving architectures.

The request-routing system operates as shown in Figure 41.3(b). Clients access content from the CDN servers by first contacting a request router (step 1). The request router makes a server selection decision and returns a server assignment to the client (step 2). Finally, the client retrieves content from the specified CDN server (step 3).

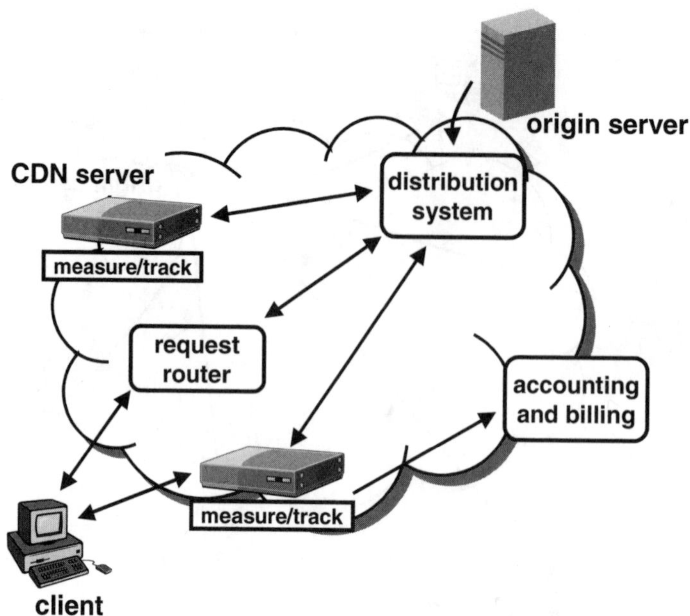

(a) CDN architectural elements

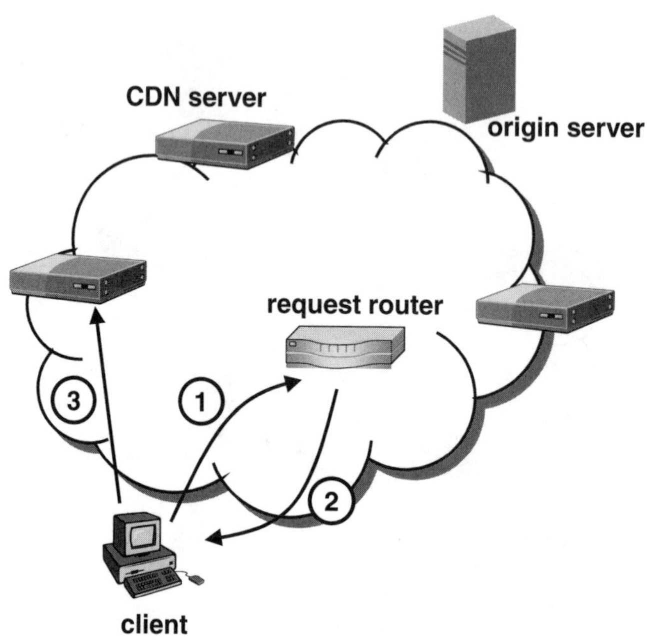

(b) CDN request-routing

FIGURE 41.3 CDN architecture and request-routing.

41.4.2 CDN Request-Routing

Clearly, the request-routing system has a direct impact on the performance of the CDN. A poor server selection decision can defeat the purpose of the CDN, namely, to improve client response time over accessing the origin server. Thus, CDNs typically rely on a combination of static and dynamic information when choosing the best server. Several criteria are used in the request-routing decision, including the

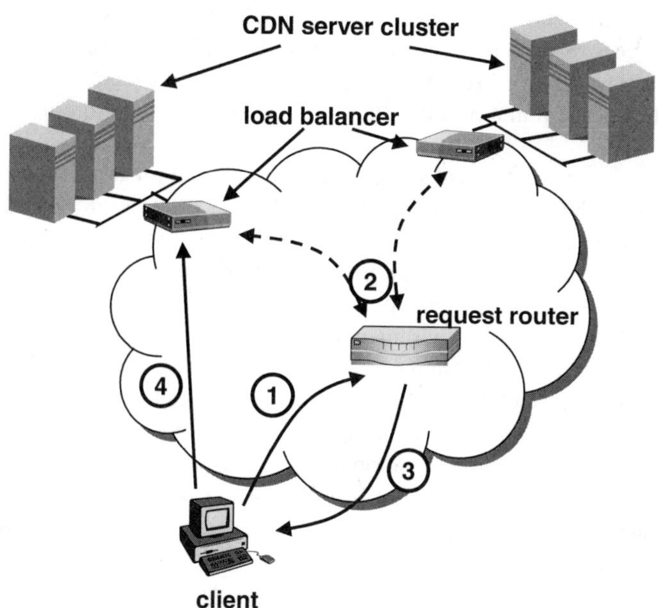

FIGURE 41.4 Interaction between request router and CDN servers.

content being requested, CDN server and network conditions, and client proximity to the candidate servers.

The most obvious request routing strategy is to direct the client to a CDN server that hosts the content being requested. This is complicated if the request router does not know what content is being requested, for example, if request-routing is done in the context of name resolution. In this case the request contains only a server hostname (e.g., www.service.com) as opposed to the full HTTP URL.

For good performance the client should be directed to a relatively unloaded CDN server. This requires that the request router actively monitor the state of CDN servers. If each CDN location consists of a cluster of servers and local load-balancer or connection router, it may be possible to query a server-side agent for server load information, as shown in Figure 41.4. After the client makes its request, the request router consults an agent at each CDN site load-balancer (step 2), and returns an appropriate answer back to the client.

As Web response time is heavily influenced by network conditions, it is important to choose a CDN server to which the client has good connectivity. Upon receiving a client request, the request router can determine which CDN server is closest to the client and then respond to the client appropriately.

A common strategy used in CDN request-routing is to choose a server "nearby" the client, where proximity is defined in terms of network topology, geographic distance, or network latency. Examples of proximity metrics include autonomous system (AS) hops or network hops. These metrics are relatively static compared with server load or network performance, and are also easier to measure.

Note that it is unlikely that any one of these metrics will be suitable in all cases. Most request routers use a combination of proximity and network or server load to make server selection decisions. For example, client proximity metrics can be used to assign a client to a "default" CDN server, which provides good performance most of the time. The selection can be temporarily changed if load monitoring indicates that the default server is overloaded.

Request-routing techniques fall into three main categories: transport-layer mechanisms, application-layer redirection, and DNS-based schemes [Barbir et al., 2002]. Transport-layer request routers use information in the transport-layer headers to determine which CDN server should serve the client. For example, the request router can examine the client IP address and port number in a TCP SYN packet and forward the packet to an appropriate CDN server. The target CDN server establishes the TCP connection and proceeds to serve the requested content. Forward traffic (including TCP acknowledgments) from the client to the target server continues to be sent to the request router and forwarded to the CDN server. The bulk of traffic (i.e., the requested content) will travel on the direct path from the CDN server to the client.

Application-layer request-routing has access to much more information about the content being requested. For example, the request-router can use HTTP headers like the URL, HTTP cookies, and Language. A simple implementation of an application-layer request router is a Web server that receives client requests and returns an HTTP redirect (e.g., status code 302) to the client indicating the appropriate CDN server. The flexibility afforded by this approach comes at the expense of added latency and overhead because it requires TCP connection establishment and HTTP header parsing.

With request-routing based on the Domain Name System (DNS), clients are directed to the nearest CDN server during the name resolution phase of Web access. Typically, the authoritative DNS server for the domain or subdomain is controlled by the CDSP. In this scheme, a specialized DNS server receives name resolution requests, determines the location of the client, and returns the address of a nearby CDN server or a referral to another nameserver. The answer may only be cached at the client side for a short time so that the request router can adapt quickly to changes in network or server load. This is achieved by setting the associated TTL field in the answer to a very small value (e.g., 20 sec).

DNS-based request routing may be implemented with either full- or partial-site content delivery [Krishnamurthy et al., 2001]. In full-site delivery, the content provider delegates authority for its domain to the CDSP or modifies its own DNS servers to return a referral (CNAME or NS record) to the CDSP's DNS servers. In this way, all requests for www.company.com, for example, are resolved to a CDN server that then delivers all of the content. With partial-site delivery, the content provider modifies its content so that links to specific objects have hostnames in a domain for which the CDSP is authoritative. For example, links to http://www.company.com/image.gif are changed to http://cdsp.net/company.com/image.gif. In this way, the client retrieves the base HTML page from the origin server, but retrieves embedded images from CDN servers to improve performance. This type of URL rewriting may also be done dynamically as the base page is retrieved, though this may increase client response time.

The appeal of DNS-based server selection lies in both its simplicity (it requires no change to existing protocols) and its generality (it works across any IP-based application regardless of the transport-layer protocol being used). This has led to adoption of DNS-based request routing as the *de facto* standard method by many CDSPs and equipment vendors. Using the DNS for request-routing does have some fundamental drawbacks, however, some of which have been recently studied and evaluated [Shaikh et al., 2001; Mao et al., 2002; Barbir et al., 2002].

One problem is that request-routing is done on the granularity of DNS domains rather than per-object, thus limiting the ability to make object-specific server selection decisions. A second problem is that requests usually come to the DNS server not from clients but from their local nameservers. Hence, the CDN server is chosen based on the local nameserver address instead of the client, which may lead to poor decisions if clients and their local nameservers are not proximal. Finally, as mentioned earlier, DNS request-routers return answers with small TTLs to facilitate fine-grained load balancing. This may actually increase Web access latency because clients must contact the DNS server more frequently to refresh the name-to-address mapping.

41.4.3 Request-Routing Metrics and Mechanisms

Request-routing systems use a number of metrics and techniques in deciding which CDN server is best suited for a given client. This section describes some specific metrics and techniques used in commercially available load-balancing and request-routing products. Note that these techniques are not necessarily limited to CDNs; they are applicable to load-balancing and request-routing in many replicated content-serving architectures.

41.4.3.1 Determining Server Availability and Load

Server availability is often the most critical criterion used in request-routing. Availability is usually determined using "health checks" initiated by the request-router. These probes may be implemented at layer-3 with ICMP (Internet Control Message Protocol) ping, or layer-4, for example, by checking that TCP connections can be established. In addition, the request-router is often configured to perform application-layer health checks, such as retrieving a specified file using HTTP or FTP, or interacting with an IMAP mail server or telnet server. Application-layer checks are important in detecting cases when a host machine may be operational but a mission-critical application is not, hence making the server unsuitable for handling client requests.

As described in Section 41.4.2, the request-router may consult a local load-balancing switch at each site to determine the relative load at candidate server sites. The local load balancer typically keeps track of statistics such as the number of active client connections, the aggregate packet and connection arrival rates, and number of available servers. Using agents that reside on the servers themselves, the local load balancer may also collect information such as per-server CPU load and memory usage. All or some of these statistics can be queried by the request-router to assess the relative load of each server or server cluster.

In most vendor solutions, the request-router is tightly integrated with an agent at the server-side load-balancer that reports statistics or an aggregate "score." This scheme usually requires that the request-router and load-balancer are from the same vendor because they often communicate using proprietary protocols. Limited support may also be available for communicating with heterogeneous local load-balancers or servers. This is often done using the Simple Network Management Protocol (SNMP) because most products and operating systems support SNMP queries of information such as packet arrival rate or number of active concurrent connections. The request-router may also use the responsiveness of application-layer health checks as an indication of the site or server load. These checks appear as normal client requests and thus do not require special protocols.

41.4.3.2 Determining Network Proximity and Performance

Because network performance plays an important role in overall end-to-end Web performance, the request-router tries to direct clients to the nearest server in terms of geographic or topological location, or network latency. In a typical DNS-based request-routing system, however, this is complicated by several factors. The network performance (e.g., delay, loss, throughput) may change dynamically and dramatically over time, requiring that the notion of "nearest" be updated regularly. Also, the client's actual location may be difficult to determine if the local nameserver that sends DNS requests on behalf of the client is not nearby the client. Finally, the network performance must be determined from the point of view of each server site rather than from the request-router.

One approach is for the request-router to ask candidate CDN servers to measure network latency to the client (or its nameserver) using ICMP echo (i.e., ping) and report the measured values. The request-router then responds to the client with the address of the CDN server reporting the lowest delay. Because these measurements are done online, this technique has the advantage of adapting the request-routing decision to the most current network. Measurement results are reported back to the request-router and can be cached for a short time to serve subsequent requests from the same or nearby clients. On the other hand, this technique can introduce additional latency for the client as the request-router waits for responses from the CDN servers.

In a slightly different approach, the request-router can forward the request to agents at several sites simultaneously, each of which then respond directly to the client. The client uses the response that reaches it first, thus automatically choosing the nearest site. For a fair "race," the request-router must know its one-way latency to each site and delay the forwarding accordingly to ensure that each site receives the forwarded request at the same time. This approach avoids actively probing the client nameserver from each server site, but it does require that each responding agent spoof the IP address of the request-router (to which the request was originally sent). Otherwise, the client may not accept the response.

Another alternative approach is to passively monitor client connections to the CDN servers to build a performance database that can be consulted by the request-router when making its decision. For example, the local load-balancer can capture and examine TCP packets to estimate the round-trip time between the site and a particular client. Using these estimates, the request-router can determine which site has the lowest delay to the same group of clients. This technique must address several issues, such as how to collect a sufficient number of samples at each CDN server and how to aggregate client performance statistics. It also requires tight integration between the request-router and the performance monitoring entity at each server. Note that all three of these approaches for determining client network proximity have been used in vendor products.

In addition to dynamic metrics such as network latency, request-routing systems often depend on more static notions of network proximity, based either on hopcount or geographic location. A hopcount-based metric may be implemented by simply using a UDP-based traceroute from each server site to the client nameserver, similar to the ICMP echo technique described above. If the request router has access to network routers at the server sites, it can consult interdomain routing tables at each site to find out the distance between the site and the client subnet in terms of AS-hops. This, however, requires a specialized agent or protocol on the network routers. Moreover, several studies have shown hopcount to be a poor predictor of network latency [Crovella and Carter, 1995; Obraczka and Silva, 2000].

Many request-routing systems attempt to direct clients to the geographically nearest site, often based on coarse notions of regions (e.g., U.S. East coast) or continents (e.g., Asia-Pacific clients). Determining geographic proximity based on IP addresses remains an active and open research topic, and though a number of heuristics have been developed, they are not always accurate [Moore et al., 2000; Padmanabhan and Subramanian, 2001]. Nevertheless, it is possible to use information published by regional Internet registries to obtain rough per-country address block allocations [IR Cache Project, 2003]. These can be used to determine, to some extent, the location of the client in order to direct it to the nearest site. Most request-routing products also offer the ability to manually specify IP addresses and their associated geographic regions. This is useful, for example, when the requests are anticipated from known clients (e.g., remote branch offices).

41.4.4 Consistency Management for CDNs

An important issue that must be addressed in a CDN is that of *consistency maintenance.* The problem of consistency maintenance in the context of a single proxy used several techniques such as TTL values, client-polling, server-based invalidation, adaptive refresh [Srinivasan et al., 1998], and leases [Yin et al., 2001]. In the simplest case, a CDN can employ these techniques at each individual CDN server or proxy; each proxy assumes responsibility for maintaining consistency of data stored in its cache and interacts with the server to do so independently of other proxies in the CDN. Because a typical CDN may consist of hundreds or thousands of proxies (e.g., Akamai has a footprint of more than 14,000 servers), requiring each proxy to maintain consistency independently of other proxies is not scalable from the perspective of the origin servers (because the server will need to individually interact with a large number of proxies). Further, consistency mechanisms designed from the perspective of a single proxy (or a small group of proxies) do not scale well to large CDNs. The leases approach, for instance, requires the origin server to maintain per-proxy state for each cached object. This state space can become excessive if proxies cache a large number of objects or some objects are cached by a large number of proxies within a CDN.

A cache consistency mechanism for hierarchical proxy caches was discussed in Yu et al. [1999]. The approach does not propose a new consistency mechanism; rather, it examines issues in instantiating existing approaches into a hierarchical proxy cache using mechanisms such as multicast. They argue for a fixed hierarchy (i.e., a fixed parent–child relationship between proxies). In addition to consistency, they also consider pushing of content from origin servers to proxies. Mechanisms for scaling leases are studied in Yin et al. [2001]. The approach assumes volume leases where each lease represents multiple *objects* cached by a stand-alone proxy. They examine issues such as delaying invalidations until lease renewals, and discuss prefetching and pushing lease renewals.

Another effort describes *cooperative consistency* along with a mechanism, called cooperative leases, to achieve it [Ninan et al., 2002]. Cooperative consistency enables proxies to cooperate with one another to reduce the overheads of consistency maintenance. By supporting delta consistency semantics and by using a single lease for multiple proxies, the cooperative leases mechanism allows the notion of leases to be applied in a scalable manner to CDNs. Another advantage of the approach is that it employs application-level multicast to propagate server notifications of modifications to objects, which reduces server overheads. Experimental results show that cooperative leases can reduce the number of server messages by a factor of 3.2 and the server state by 20% when compared to original leases, albeit at an increased proxy–proxy communication overhead.

Finally, numerous studies have focused on specific aspects of cache consistency for content distribution. For instance, piggybacking of invalidations [Krishnamurthy and Wills, 1997], the use of deltas for sending updates [Mogul et al., 1997], an application-level multicast framework for Internet distribution [Francis, 2000] and the efficacy of sending updates vs. invalidates [Fei, 2001].

41.4.5 CDN Performance Studies

Several research studies have recently tried to quantify the extent to which CDNs are able to improve response-time performance. An early study by Johnson et al. [2000] focused on the quality of the request-routing decision. The study compared two CDSPs that use DNS-based request-routing. The methodology was to measure the response time to download a single object from the CDN server assigned by the request-router and the time to download it from all other CDN servers that could be identified. The findings suggested that the server selection did not always choose the best CDN server, but it was effective in avoiding poorly performing servers, and certainly better than choosing a CDN server randomly. The scope of the study was limited, however, because only three client locations were considered, performance was compared for downloading only one small object, and there was no comparison with downloading from the origin server.

A study done in the context of developing the request mirroring Medusa Web proxy server evaluated the performance of one CDN (Akamai) by downloading the same objects from CDN servers and origin servers [Koletsou and Voelker, 2001]. The study was done only for a single-user workload, but showed significant performance improvement for those objects that were served by the CDN when compared with the origin server.

More recently, Krishnamurthy et al. studied the performance of a number of commercial CDNs from the vantage point of approximately 20 clients [Krishnamurthy et al., 2001]. The authors conclude that CDN servers generally offer much better performance than origin servers, though the gains were dependent on the level of caching and the HTTP protocol options. There were also significant differences in download times from different CDNs. The study finds that, for some CDNS, DNS-based request-routing significantly hampers performance due to multiple name lookups.

Acknowledgment

Some of the material in this chapter appeared in the paper *Enhancing Web Performance* by the same authors in *Proceedings of the 2002 IFIP World Computer Congress (Communication Systems: The State of the Art)*, edited by Lyman Chapin and published by Kluwer Academic Publishers, Boston, copyright 2002, International Federation for Information Processing.

References

Barbir, Abbie, Brad Cain, Fred Douglis, Mark Green, Markus Hofmann, Raj Nair, Doug Potter, and Oliver Spatscheck. Known CDN request-routing mechanisms. Internet Draft (draft-ietf cdi-known-request-routing-00.txt), February 2002.

Barford, Paul, Azer Bestavros, Adam Bradley, and Mark Crovella. Changes in Web client access patterns: Characteristics and caching implications. In *Proceedings of the World Wide Web Journal,* 1999.

Cao, Pei and Sandy Irani. Cost-aware WWW proxy caching algorithms. In *Proceedings of the USENIX Symposium on Internet Technologies and Systems,* December 1997.

Cate, Vincent. Alex: A global file system. In *Proceedings of the 1992 USENIX File System Workshop,* May 1992.

Cisco LocalDirector 400series. http://www.cisco.com/warp/public/cc/pd/cxsr/400/.

Crovella, Mark B. and Robert L. Carter. Dynamic server selection in the Internet. In *Proceedings of the IEEE Workshop on the Architecture and Implementation of High Performance Communication Subsystems (HPCS '95),* 1995.

Day, Mark, Brad Cain, Gary Tomlinson, and Phil Rzewski. A model for content internetworking (CDI). Internet Draft (draft-ietf-cdi-model-Ol Axt), February 2002.

Duvvuri, Venkata, Prashant Shenoy, and Renu Tewari. Adaptive leases: A strong consistency mechanism for the World Wide Web. In *Proceedings of IEEE Infocom,* Tel Aviv, March 2000.

Fei, Zongming. A novel approach to managing consistency in content distribution networks. In *Proceedings of the 6th Workshop on Web Caching and Content Distribution,* Boston, June 2001.

Francis, Paul. Yoid: Extending the Internet multicast architecture. In *Technical report, AT&T Center for Internet Research at ICSI (ACIRI),* April 2000.

Gray, Cary G. and David R. Cheriton. Leases: An efficient fault-tolerant mechanism for distributed file cache consistency. In *Proceedings of the 12th ACM Symposium on Operating Systems Principles,* 1989.

Gritter, Mark and David R. Cheriton. An architecture for content routing support in the Internet. In *Proceedings of the USENIX Symposium on Internet Technologies and Systems,* San Francisco, March 2001.

Gwertzman, James and Margo Seltzer. World-Wide Web cache consistency. In *Proceedings of the 1996 USENIX Technical Conference,* January 1996.

IP address services. http://www.iana.org/ipaddress/ip-addresses.htm, January 2003.

IR Cache Project Daily Reports. http://www.ircache.net/Statistics/Summaries/Root/, April 2002.

Iyengar, Arun. Design and performance of a general-purpose software cache. In *Proceedings of the 18th IEEE International Performance, Computing, and Communications Conference (IPCCC '99),* February 1999.

Iyengar, Arun, Shudong Jin, and Jim Challenger. Efficient algorithms for persistent storage allocation. In *Proceedings of the 18th IEEE Symposium on Mass Storage Systems,* April 2001.

Johnson, Kirk L., John F. Carr, Mark S. Day, and M. Frans Kaashoek. The measured performance of content distribution networks. In *International Web Caching and Content Delivery* Workshop *(WCW),* Lisbon, Portugal, May 2000. http://www.terena.nl/conf/wcw/Proceedings/S4/S4-1.pdf.

Joubert, Philippe, Robert King, Richard Neves, Mark Russinovich, and John Tracey. High-performance memory-based Web servers: Kernel and user-space performance. In *Proceedings of the USENIX Annual Technical Conference,* Boston, June 2001.

Koletsou, Mimika and Geoffrey M. Voelker. The Medusa proxy: A tool for exploring user-perceived Web performance. In *Proceedings of International Web Caching and Content Delivery Workshop (WCW),* Boston, June 2001, Elsevier, Amsterdam.

Krishnamurthy, Bala and Craig Wills. Study of piggyback cache validation for proxy caches in the WWW, In *Proceedings of the 1997 USENIX Symposium on Internet Technologies and Systems,* Monterey, CA, December 1997.

Krishnamurthy, Balachander, Craig Wills, and Yin Zhang, On the use and performance of content distribution networks. In *Proceedings of ACM SIGCOMM Internet Measurement Workshop,* November 2001.

Li, Dan, Pei Cao, and Mike Dahlin. WCIP: Web cache invalidation protocol. *In IETF Internet Draft,* November 2000.

Liu, Chengjie and Pei Cao. Maintaining strong cache consistency in the World-Wide Web. In *Proceedings of the 17th International Conference on Distributed Computing Systems,* May 1997.

Mao, Zhuoqing Morley, Charles D. Cranor, Fred Douglis, Michael Rabinovich, Oliver Spatscheck, and Jia Wang. A precise and efficient evaluation of the proximity between Web clients and their local DNS servers. In *Proceedings of USENIX Annual Technical Conference,* June 2002.

Markatos, Evangelos, Manolis Katevenis, Dionisis Pnevmatikatos, and Michael Flouris. Secondary storage management for Web proxies. In *Proceedings of the 2nd USENIX Symposium on Internet Technologies and Systems,* October 1999.

Mogul, Jeffrey C., Fred Douglis, Anya Feldmann, and Bala Krishnamurthy. Potential benefits of delta encoding and data compression for HTTP. In *Proceedings of the ACM SIGCOMM,* September 1997.

Moore, David, Ram Periakaruppan, and Jim Donohoe. Where in the world is netgeo.caida.org? In *Proceedings of the Internet Society Conference (INET), 2000.* http://www.caida.org/outreach/papers/2000/inet_netgeo/.

Ninan, Anoop, Purushottam Kulkarni, Prashant Shenoy, Krithi Ramamritham, and Renu Tewari. Cooperative leases: Scalable consistency maintenance in content distribution networks. In *Proceedings of the 11th International World Wide Web Conference (WWW2002),* May 2002.

Obraczka, Katia and Fabio Silva. Network latency metrics for server proximity. In *Proceedings of IEEE GLOBECOM,* pp. 421–427, 2000.

Padmanabhan, Venkata N. and Lakshminarayanan Subramanian. An investigation of geographic mapping techniques for Internet hosts. In *Proceedings of ACM SIGCOMM,* San Diego, CA, August 2001.

Shaikh, Anees, Renu Tewari, and Mukesh Agrawal. On the effectiveness of DNS-based server selection. In *Proceedings of IEEE INFOCOM,* Anchorage, AK, April 2001.

Song, Junehua, Arun Iyengar, Eric Levy, and Daniel Dias. Architecture of a Web server accelerator. *Computer Networks,* 38(1), 2002.

Squid Internet object cache users guide. http://squid.nlanr.net, 1997.

Srinivasan, Raghav, Chao Liang, and Krithi Ramamritham. Maintaining temporal coherency of virtual warehouses. In *Proceedings of the 19th IEEE Real-Time Systems Symposium (RTSS98),* Madrid, December 1998.

Tewari, Renu, Thirurnale Niranjan, and Srikanth Ramamurthy. WCDP: Web content distribution protocol. In *IETF Internet Draft,* March 2002.

Verma, Dinesh C. *Content Distribution Networks: An Engineering Approach.* John Wiley & Sons, New York, 2002.

Wolman, Alex, Geoffrey M. Voelker, Nitin Sharma, Neal Cardwell, Anna Karlin, and Henry M. Levy. On the scale and performance of cooperative Web proxy caching. In *Proceedings* of *ACM Symposium on Operating Systems Principles (SOSP),* pp. 16–31, December 1999.

Yin, Jian, Lorenzo Alvisi, Mike Dahlin, and Arun Iyengar. Engineering server-driven consistency for large-scale dynamic Web services. In *Proceedings of the 10th World Wide Web Conference,* Hong Kong, May 2001.

Yin, Jian, Lorenzo Alvisi, Mike Dahlin, and Calvin Lin. Hierarchical cache consistency in a WAN. In *Proceedings of the Usenix Symposium on Internet Technologies and Systems (USITS'99),* Boulder, CO, October 1999a.

Yin, Jian, Lorenzo Alvisi, Mike Dahlin, and Calvin Lin. Volume leases for consistency in large-scale systems. *IEEE Transactions on Knowledge and Data Engineering,* 11(4): 563–576, 1999b.

Haobo Yu, Lee Breslau, and Scott Shenker. A scalable Web cache consistency architecture. In *Proceedings of the ACM SIGCOMM,* Boston, September 1999.

42

Content Adaptation and Transcoding

Surendar Chandra

CONTENTS

42.1 Introduction 42-1
42.1.1 Client-Side Constraints 42-2
42.1.2 Server-Side Constraints 42-3
42.1.3 Differentiated Services to Manage Resources 42-3
42.2 Transcoding Techniques 42-4
42.2.1 Textual Content 42-4
42.2.2 Image Content 42-5
42.2.3 Streaming Media 42-6
42.2.4 Content Adaptation of Composite Web Objects 42-7
42.2.5 Quality-Aware Transcoding 42-7
42.3 Technologies That Utilize Transcoding Operation 42-7
42.3.1 Web Content Adaptation Service Architecture 42-7
42.3.2 Automatic Transcoding by Proxies and Web Servers 42-9
42.3.3 Content Producer and Consumer Involvement 42-9
42.3.4 Systems That Have Utilized Transcoding Technologies 42-11
42.4 Challenges in the Effective Use of Transcoding Technologies 42-11
References 42-12

This chapter describes content adaptation as a means to customize Web objects for constrained environments. We first describe a number of Internet scenarios with diverse resource constraints, and show that the rich Web content should be transcoded to more appropriate forms to be applicable in these various scenarios. Some popular transcoding mechanisms are discussed, paying particular attention to transcoding as a mechanism to adapt multimedia content. Finally, some of the research challenges involved in utilizing these transcoding technologies are described.

42.1 Introduction

An explosion in the number and variety of devices is dramatically changing the world of Internet computing. Recent mobile computing devices range in complexity from tablets to palmtops to fully functional jewelry with computational resources. The ubiquitous computing future espoused by Mark Weiser [1993] and others calls for these connected devices to eventually vanish into the periphery. The cooltown prototype (www.cooltown.org) implements this vision by mapping physical people, places, and things to their Web presence [Kindberg et al., 2000]. In this model, each object is represented and named by its corresponding URL. Many mobile applications will use the mobile device as a window into vast amounts of data that can be delivered via the Internet, particularly in the form of Web content. With the

1-58488-381-2/05/$0.00+$1.50
© 2005 by CRC Press LLC

increasing availability of the Internet on a wide range of these devices, the Web is emerging as the preferred data delivery mechanism for a number of application scenarios.

Such access dynamics expose a number of inadequacies in the Web infrastructure. In the next few sections, we outline some of the issues faced by clients because of local and network resource constraints as well as constraints at the servers themselves.

42.1.1 Client-Side Constraints

There is a rapid movement toward embedding powerful processors in a large variety of consumer electronic devices. These devices have constraints on the available battery capacity, network, displays and output mechanisms, processing power, and local storage. Such constraints severely restrict the quality of Web experience for the end users.

- **Networks:** Clients access the Internet using a number of networking technologies — from wired high speed LAN networks and broadband and dialup networks to wireless technologies. Predominant wireless technologies include IEEE 802.11 WiFi LANs [LAN/MAN Standards Committee of the IEEE Computer Society, 1999] and 2G and 3G cellular networks. The wide variability in the access latency, connection reliability, available bandwidth, and access cost for these network technologies make a single approach for data delivery inappropriate. Accesses from cellular modems using technologies such as GSM typically operate in the range up to 19.2 Kb/sec while variants of wireless LAN technologies such as IEEE 802.11 offer bandwidths of tens of Mb/sec. Mobile networks are also expensive, usually charging by the amount of useful data consumed. For example, in the U.S. access costs range in the order of \$.03 to \$.05 per kilobyte of cellular data transferred.
  - The problem of slow and expensive networks is compounded by the large size of multimedia objects that are becoming such a prevalent part of Web content. Studies [All Things Web, 1999] have shown that the average Web page is about 60 KB in size. Accessing 60 KB using Cellular Digital Packet Data (CDPD) cellular technologies would take about 3 min and cost around \$3. For many users, such access costs are prohibitive.
- **Processing power:** The compression mechanisms used in representing multimedia objects trade off the amount of compression with the client CPU power requirements to decode an object. In CPU resource-constrained mobile devices, the trade-off can translate to transferring more data (i.e., less compression) because of client CPU characteristics. The image may be uncompressed on the servers, and the uncompressed images may be transferred to the device; network capacity now becomes the bottleneck.
- **Displays:** In a typical mobile device, factors such as the screen size and color depth severely restrict the image that could be displayed to a user. Display technologies such as back-lit displays can consume significant amounts of energy but may allow images to be viewed at an apparently better quality than when viewed with reflective screens. Also, some devices are expected to completely lack display capabilities. Hence; the users may want to view variations of the objects that are better suited for their local display characteristics.
- **Battery capacity:** Mobile device system components such as wireless network interface card (WNIC) displays and processors consume significant amounts of energy. A necessary feature for mass acceptance of a mobile device is acceptable battery duration. Newer hardware improvements are reducing the power consumption of system components such as back-lit displays, CPUs, etc. However, WNICs operating at the same range continue to consume significant power. For example, a 2.4 GHz IEEE 802.11b Wavelan card [Stemm et al., 1996; Havinga, 2000; Feeney and Nilsson, 2001] consumes 177 mW while in *sleep* state, and about 1400 mW while in *idle* state as well as while receiving data. In comparison, a fully operational HP-Compaq iPAQ PDA only consumes about 1000 mW. Future trends in battery technologies alone (along with the continual pressure for further device miniaturization) do not promise dramatic improvements that will make this issue disappear. Hence, it is important to look at techniques to reduce the energy consumed by

the network interface. Techniques to reduce the amount of data transferred can be expected to offer significant energy savings.

In general, there is a huge mismatch between the rich multimedia content available on the World Wide Web and the characteristics of mobile devices that are used to access this Web content. Users would like to keep the access costs from becoming unaffordable.

42.1.2 Server-Side Constraints

The Web is emerging as the primary data dissemination mechanism for a variety of applications. The following important trends are evolving as the economic model for Websites.

The first trend is toward e-commerce and subscription-based services. Sites such as ESPN (espn.com) want to prioritize their consumers based on their subscription status, prior access history, and their current status (e.g., a customer with a purchasing history or one with a full shopping cart). To retain their paying customers, these sites need to maintain the quality of service for the preferred customers. In order to convince new users to subscribe, these sites need to provide quality teaser objects. Such teaser objects may include article headlines, abstracts, and image thumbnails.

The other trend is toward Web-hosting services (e.g., Yahoo!) that create and maintain Web pages on behalf of their customers. These services charge their hosted sites based on the size of the Website and the aggregate bandwidth consumed in a month. One of the major problems encountered by Web-hosting services is through *flash users.* Flash crowds (sometimes referred to as the *slashdot effect*) refers to the sudden upsurge in traffic to a Website because of breaking news events; for example, September 11 attacks on the U.S., Victoria's Secrets Webcast, etc. Flash crowds to another Website hosted by the same hosting service can degrade the performance for all the other sites hosted by the service. Current Web-hosting services use a *laissez faire* approach in managing their bandwidth. However, customers of these hosting services will demand performance guarantees, forcing the servers to provide differentiated services so that they can charge the sites not only for the aggregate bandwidth but also for guaranteed bandwidth during high-demand peak hours.

In general, the primary goal of a Web service is providing low latency access to its contents. However, during times of high demand, this goal is compromised by locally available network bandwidth. Even sites that utilize high capacity networks can be overwhelmed by flash crowds during periods of high demand. In the past, Websites have used *ad hoc* solutions to deal with peak loads. For example, when significant news stories break, news sites such as MSNBC and CNN export a lightweight version of the site with little or no graphics to conserve bandwidth. Caching at local Web proxies and content delivery networks (e.g., Akamai) are other traditional techniques to address bandwidth limitations; this is done by replicating and careful placement of objects. Sites have also utilized overprovisioning of resources to combat flash crowds. Gratuitous use of overprovisioning can lead to wasted resources and may prove too expensive for many applications. Also, much Web content is dynamically generated (e.g., maps, stock charts, etc.) or becomes uncacheable (e.g., sites selling access to multimedia contents such as images or movies). To fully realize the benefits of the Internet, one needs to solve the fundamental problems exacerbated by these Web trends.

42.1.3 Differentiated Services to Manage Resources

In such a scenario, differentiated service can allow the system to provide a more appropriate level of service for a user, based on the current operating environment. Differentiated service means that proxies can match object sizes with the network bandwidth available on the last hop to mobile clients. In this way, a network proxy server can provide different versions of the same object to different clients. Traditional human factors research [Nielsen, 1993] has shown that the response time for accessing a resource should be in the 1 to 10 sec range for information to be useful. If the response time is longer than this range, the users tend to lose interest and move on. The system server can choose variations of an object such that the objects are served at a uniform latency of less than 5 sec, based on the type of network link used in accessing the Web. Differentiated service can also allow the system to provide better service for certain customers, based on their status and the current network environment [Abdelzaher and Bhatti, 1999; Chandra et al., 2000a]. To summarize, differentiated service enables:

- Web services to customize Web objects for mobile users, depending on the client and network characteristics
- Web services to dynamically allocate the available bandwidth among different user classes
- Subscription services to provide different versions of contents to clients, based on customer status (subscriber vs. nonsubscriber)
- Web-hosting services to share their bandwidth for different classes of hosted clients
- E-commerce sites to allocate their bandwidth to customers who are making a sale
- Flexibility in redirecting unused preferred resources to nonpreferred customers

The key insight is that allocation of critical network resources should be done dynamically in response to client access patterns. The feasibility of such a differentiated service scheme depends on the availability of a range of variations for the content so that the server can choose the correct variation for the current network-operating environment. While the content provider can manually provide a number of different variations for use by the system, an automatic technique may be preferable to allow the system to dynamically adapt to variability in network performance and client characteristics.

42.2 Transcoding Techniques

One promising technique for providing differentiated quality of service is transcoding which can be used to serve variations of the same multimedia object at different sizes. Transcoding is defined as a transformation that is used to convert a multimedia object from one form to another, frequently trading off object fidelity for size. Transcoding operation is a lossy transformation; the user has to explicitly request the original to restore the lost quality. By their very nature, multimedia objects are amenable to soft access through a quality-vs.-size tradeoff. Transcoding allows Web services to transmit variations of the same multimedia object at different sizes, allowing some control over the amount of bandwidth consumed in transmitting a page to a particular client. By some estimates [Ortega et al., 1997], about 77% of the data bytes accessed across the Web are from multimedia objects such as images and audio and video clips. Of these, 67% of the data are transferred for images. Hence, it is important to focus our attention on multimedia data.

Transcoding being a lossy operation, information is usually lost and the transformed object becomes of lower fidelity. Transcoding operation is frequently applied to fit an object to a certain device (e.g., transcoding a color image to a gray-scale image can allow the original image to be viewed in an gray-scale monitor). We focus our attention on transcoding to provide differentiated service. In this case, the system intentionally chooses a lower fidelity version even though the original object can be displayed, albeit violating the service requirements. In the next section, we briefly describe some of the popular transcoding operations for typical Web contents available in textual, image, and streaming media formats.

42.2.1 Textual Content

Textual transcoding can be performed either to fit an object to the current environment or, more generally, to meet certain service requirements. For example, transcoding can be applied to richly annotated textual formats such as HTML [Figure 42.1(a)] to produce a simpler textual representation [Figure 42.1(b)]. Such transformations can preserve the original content while sacrificing presentation aesthetics. Automatic language translators (e.g., Babel Fish translations from altavista.com) can also transform the original language of the article.

On the other hand, one can also imagine transformations that generate textual abstract of a longer report. Such operations can be performed using automatic information processing techniques (see Chapter 14) or explicitly by the original content providers themselves. Such transformations are lossy;

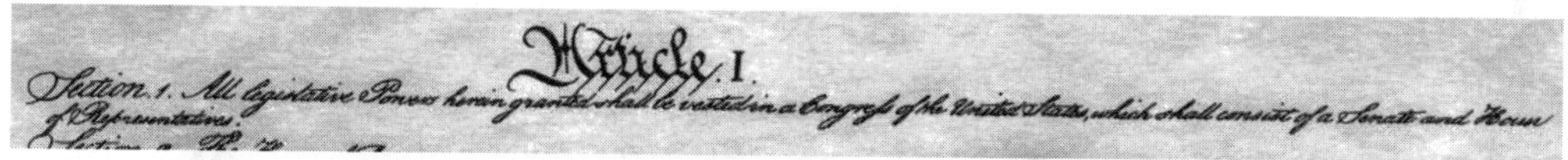

(a) Formatted text

All legislative Powers herein granted shall be vested in a Congress of the United States, which...

(b) ASCII text

FIGURE 42.1 Transcoding a formatted text to other variants (excerpts from the U.S. Constitution).

it is not possible to regenerate the original article from the transcoded version. The user is required to explicitly request the higher-quality original.

42.2.2 Image Content

Image content forms a rich part of Web technologies. Typical images tend to be large. The very nature of image compression and representation formats allow for easier and productive transcoding to various alternative formats. Most image compression and transcoding algorithms are lossy. Popular image transcoding operations include reducing the color depth, reducing the image geometry (i.e., thumbnailing), cropping the irrelevant parts of the image, and reducing the compression factor, as well as recompressing the object to an alternative format. For example, the various transcodings of the original image of the Earth rising above the lunar horizon [Figure 42.2(a)] to a lower quality image compression factor [Figure 42.2(b)], cropped image [Figure 42.2(c)] as well as to a thumbnail [Figure 42.2(d)] are illustrated in Figure 42.2. Progressive format can also allow clients to actively participate in the quality-choosing process by terminating the transfer once the required image fidelity is received. Note that transformations such as thumbnailing are easier to generate automatically than the correct cropping operation. The range of transcoded versions also depends on the image-encoding formats. For example, color JPEG images can be encoded as a 24-bit TrueColor image or an 8-bit pseudo color format.

However, not all transcodings that reduce the image fidelity lead to image file-size savings. Image-compression techniques exploit the image color distribution characteristics and human acuity models to achieve better compression ratios with least visual quality loss. For example, human eyes are less sensitive to chrominance (color) values than luminance (brightness) values. Hence, it is possible to drop more chrominance components without noticeable loss in image quality. Without proper precautions, transcoding operations can perturb such optimizations, leading simultaneously to reduced compression ratios (i.e., increased image size) as well as lower-quality images. For example, an analysis of typical Web images [Chandra et al., 2001] showed that 40% of GIF images that were reduced by a factor of 2 along the x and y axis transcoded to a size that is larger than the original image file size. GIF [Corn, 1987, 1989] uses a variation of the LZW compression algorithm to reduce the number of bits required to store frequently occurring color snap values. Pixels can be represented by 3 to 12 bits, depending on the occurrence frequency of a particular color value. A transcoding that reduces the spatial geometry tends to increase the number of unique colors in an image as original color values are replaced by a new average color value. Since it takes more bits to represent less frequent pixels, introducing less frequent color values with low occurrence frequency leads to an increase in the output image size. This is against our goal for transcoding an image to reduce its size. We will use an example to illustrate this problem. One of the GIF images (geometry 345 × 145, file size 2999 bits, and 7 unique colors) was transcoded to a GIF image of geometry 173 × 73. The new image was of 3005 bits and had 133 unique colors. The popular color (which will be represented by 3 bits) occurred 90.88% of the time in the original image, whereas it only occurred 85% of the time in the transcoded image. Naive transcoding approaches can provide disappointing results due to subtle interactions. Even if such transcoded output images are eventually discarded, valuable compute resources may still have been wasted. Note that it is not possible to always predict if a given transcoding can offer any space savings.

(a) Original JPEG image (56 KB)

(b) Lower Quality Factor JPEG image (16 KB)

(c) Cropped JPEG image (8 KB)

(d) Thumbnail JPEG image (8 KB).

FIGURE 42.2 Transcoding an image to other variants (pictures courtesy of NASA).

42.2.3 Streaming Media

As the number of clients with high bandwidth links to the last mile increases, streaming video and audio formats increasingly become popular. Video streams also carry an associated audio track that can be encoded at a different fidelity than the video track itself. By their very nature, streaming media tend to consume large amounts of data and also offer tremendous options for transcoding operations. Popular transcoding for streaming formats include clipping and highlighting, quality reduction (e.g., color depth and audio fidelity reduction), as well as transcoding to alternative formats (e.g., removing the video track completely). Streaming media are designed to operate under dynamic network conditions and, hence, allow transcoding decisions to be dynamic. The stream quality can be dynamically improved or reduced based on the current network conditions. By contrast, transcoding decisions for images cannot usually be undone half way through transmission.

42.2.4 Content Adaptation of Composite Web Objects

Typical Web documents can consist of a number of inlined images and other multimedia objects. It may be more appropriate to base transcoding decisions on the entire document group rather than on each individual component. Such analysis might expose the component intent, allowing for more appropriate decisions. For example, document bullet icons can be easily identified by their corresponding HTML tags and appropriately replaced with a textual icon, rather than transcoding each image to a lower-fidelity version.

42.2.5 Quality-Aware Transcoding

In general, for transcoding to provide the degree of control needed to deliver differentiated service, we need to understand its inherent trade-off characteristics: the information quality loss, the computational overhead required in computing the transcoding, and the potential benefits of reduced bandwidth requirements. Formal measurement of object quality loss can help in choosing the appropriate transformation from various transcodings (illustrated in Figure 42.2). Without such characterization, transcoding policies do not have the ability to measure information quality loss of a transcoded image; hence, it is impossible to ensure that preferred clients experience low degradation of quality. Without this characterization, systems that attempt to use transcoding to reduce bandwidth requirements can only transcode images to *ad hoc* Quality factor values, potentially leading to an increase in size for certain images. Systems have traditionally countered this by (unnecessarily) transcoding all images to a conservatively low Quality factor value.

To illustrate the benefits of such quantification for one specific case, Chandra and Ellis, [1999] characterized the information quality tradeoffs, the computational requirements, and the potential size reductions for transcodings that change the JPEG [Pennebaker and Mitchell, 1993] compression metric. It had been shown that the JPEG Quality factor parameter reflects a user's perception of image quality [Ford, 1997; Jacobson et al., 1997]. Chandra et al. analyzed typical Web images [Chandra et al., 2001] and developed techniques for measuring the initial Quality factor of a JPEG image, as well as for predicting the computational cost and potential space benefits achieved by the transcoding. Similar analysis of the multimedia objects can reduce unnecessary quality loss from a transcoding operation. Such results are useful in any system that uses transcoding to reduce access latencies and increase effective storage space, as well as reduce access costs.

42.3 Technologies That Utilize Transcoding Operation

In the last section, we described some of the transcoding operations for various Web object types. In this section, we describe various mechanisms on the Web server that utilize transcoding to customize the object.

42.3.1 Web Content Adaptation Service Architecture

A typical Web service architecture is illustrated in Figure 42.3. The system consists of a number of clients requesting Web objects using varying network conditions. Web proxies are strategically placed in the network infrastructure to provide caching, traffic control, and aggregation. Such proxies can be placed closer to the client in the Internet back bone and at the servers themselves as reverse proxies. For simplicity, we illustrated these proxies as a single entity. These proxy servers can be expected to perform the content

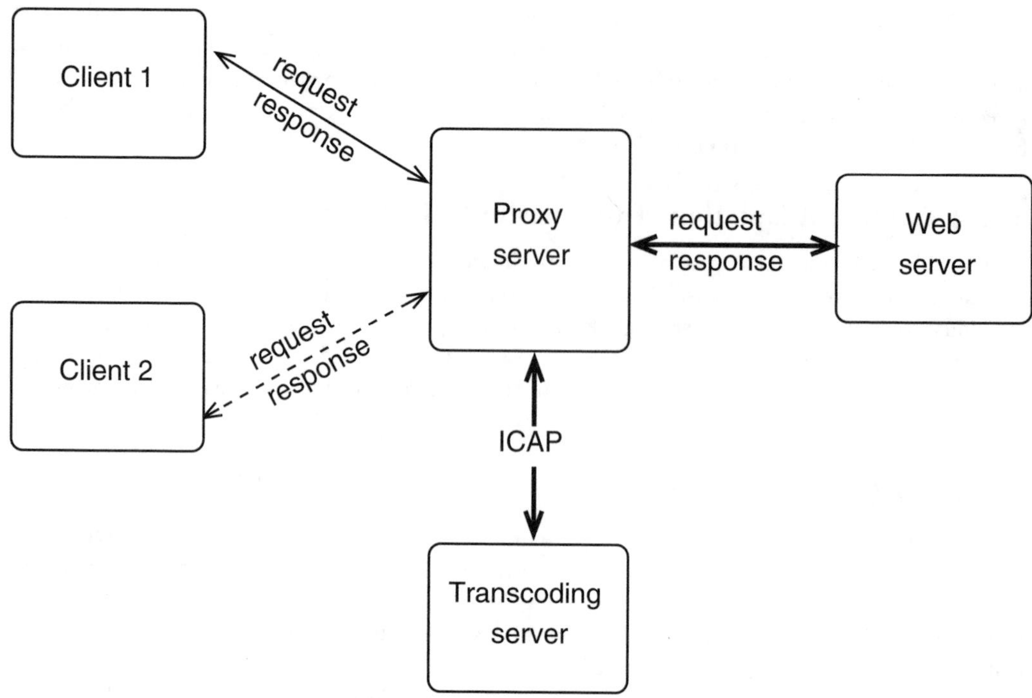

FIGURE 42.3 Web content adaptation architecture.

adaptation without modifying the origin servers themselves. These proxies can dynamically invoke transcoding operations using dedicated servers. The proxy servers can request transcoding services using the Internet Content Adaptation Protocol (ICAP) [Elson et al., 2000; Elson and Cerpa, 2003].

42.3.1.1 Internet Content Adaptation Protocol (ICAP)

ICAP is a request–response protocol designed to off-load transformation and processing for Web content to dedicated servers. ICAP allows ICAP clients to pass HTTP messages to ICAP servers for "adaptation." Typical Web clients interact with ICAP-capable surrogates (proxies) that in turn interact with one or more ICAP servers. ICAP provides a light-weight remote procedure call (RPC) functionality for Web services. Typical operations envisioned include language translation, virus checking, filtering, and transcoding. The server executes its transformation service on messages and sends back responses to the client, usually with modified messages. The adapted messages may be either HTTP requests (request modification mode) or HTTP responses (response modification mode). All ICAP transactions are similar to standard HTTP 1.1 request semantics. ICAP defines three methods of operation:

1. REQMOD: This mode is used by the ICAP-capable surrogate to request modification of an HTTP request from a Web client before forwarding to the origin server. The ICAP server can also directly provide a response to the original request (e.g., to indicate an error).
2. RESPMOD: This mode is used by the ICAP-capable surrogate to request modification of the HTTP response from the origin server before responding to the Web client. The request also encapsulates the original client request that caused the response from the origin server. On-demand transcoding would typically be implemented by an ICAP server during this mode of operation.
3. OPTIONS: This method allows the ICAP client to retrieve configuration information from the ICAP server.

42.3.2 Automatic Transcoding by Proxies and Web Servers

The simplest method of using transcoding is to automatically transcode objects to another form that is deemed useful to the end user. Typically, such transcodings are implemented in the Web proxy server which are closer to the end user. For example, America Online transcodes all JPEG images to a proprietary Johnson Grace ART format. Such transcodings are typically transparent to the end user. However, such transcoding operations run the risk of making inappropriate decisions. For example:

Lower quality object for end user: Automatic transcodings can make quality decisions that the user is not aware of; the user may actually want to pay higher cost to view higher quality objects because they are interested in them. Automatic transcoding is transparent and hides the fact that the images were transcoded.

Loss of content provider control: Even though transcodings can make Web objects useful and applicable in many scenarios, content providers may want to exercise greater control, potentially denying service to certain clients. For example, classical artists may not want users to listen to their music at a quality factor lower than some threshold (even if that means denying service to a potential listener).

Copyright issues: Automatic transcoders may also not be aware of local copyright restrictions. Translating and derived work on copy righted work can have legal implications in certain jurisdictions. Depending on the location of the transcoding proxies, their operations may violate local tenets.

Content provider and user interaction influencing the transcoding process is preferable.

42.3.3 Content Producer and Consumer Involvement

Involving the content provider and the consumer is attractive for a number of reasons. The content provider can provide *a priori* transcodings that are specifically customized to provide the objects in a format that is acceptable. Also, the clients can better specify their tolerances and capabilities of their devices in order to influence the object-matching process. The user would want to provide bounds on access latency and access cost. In general, we need a mechanism that addresses the following:

End users: Allow end users to specify their tolerances (i.e., formats, size expressed as cost, or latency tolerance).

Proxy servers: Negotiate variations of objects on behalf of a specific end user, while still serving other clients experiencing different requirements.

Content providers: Maximum flexibility in controlling the content variations deemed acceptable for a particular situation.

HTTP 1.1 [Fielding et al., 1999; Holtman and Mutz, 1998a, b] provides such mechanisms to allow the server and client to negotiate contents. There are two kinds of content negotiation which are possible in HTTP 1.1: server driven and agent driven. Both are orthogonal and, thus, may be used separately or in combination. For example, proxy caches can use one such hybrid mechanism, referred to as *transparent negotiation,* utilizing agent-driven negotiation for information provided by the origin server, in order to provide server-driven negotiation for subsequent client requests.

42.3.3.1 Server-Driven Negotiation

Server-driven negotiation is illustrated in Figure 42.4. The server selects the appropriate variation based on client-specified preferences. The clients can specify their preferences using additional HTTP headers with the original request. Clients can use headers such as Accept, Accept-Language, Accept-Encoding, etc. [Fielding et al., 1999] to specify their preferred encoding, language, etc. The efficacy of this scheme depends on the richness with which clients can specify their preference. However, adding such rich specification to every HTTP request can lead to inefficient network usage for all requests. Even if the browser specified all its capabilities, it still may not capture the user's intent for a specific object. Users may want to change their preferred capabilities, depending on the specific object. For example, users

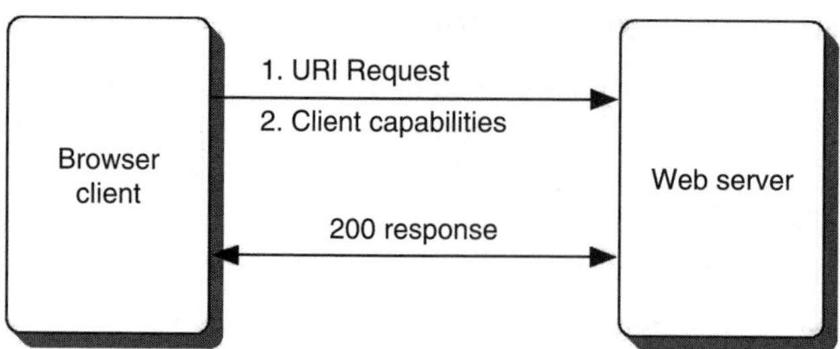

FIGURE 42.4 Server-driven negotiation.

may want to view a lower quality version but generate a higher quality hard copy printout. Also, there are privacy concerns in transmitting the exact capabilities of the user's browser to every Web server (servers can keep track of the applications installed in the client). Such policies can also burden the server with the decision process for each request.

If the server were not able to negotiate a preferred object (either because it does not have the requested object or is unwilling to serve the requested object), it could use a Vary header to provide alternatives.

In summary, server-driven negotiation requires clients to expose their capabilities and to leverage processing to power available on content servers to receive the appropriate content variation.

42.3.3.2 Agent-Driven Negotiation

With agent-driven negotiation (Figure 42.5), the selection of the best representation is performed by the client. The server responds with all the available variations to the initial request. The server can generate this list using static, user-provided variations or may generate these variations dynamically. For example, Apache web servers utilize ".Var" files to allow users to specify these variations. The client browser chooses the acceptable format and specifically requests a (nonnegotiated) URL. This selection can either happen automatically by the browser itself or via an explicit query to the end user.

In summary, agent-driven negotiation trades off the extra request to the server for the necessity to expose its entire capabilities to the server. The decision overhead is paid by each client, rather than by the origin server.

42.3.3.3 Transparent Negotiation

It is also possible to utilize a combination of server- and agent-driven negotiation in the form of transparent negotiation. Proxy caches can utilize information about the object variations available in the origin server to initiate server-driven negotiation with subsequent client requests. Transparent negotiation distributes the negotiation overhead with the proxy servers. However, transparent negotiation could

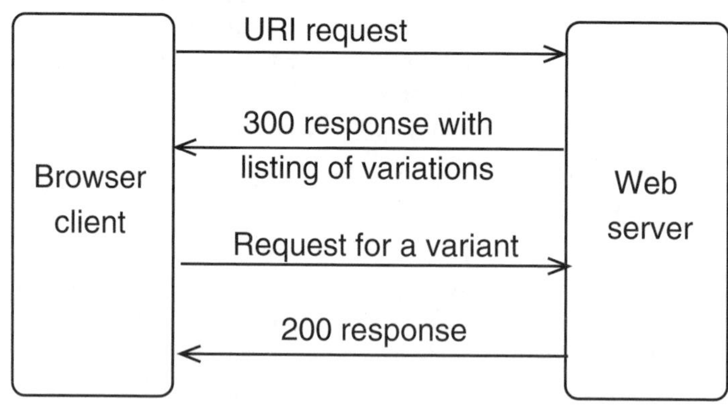

FIGURE 42.5 Agent-driven negotiation.

potentially miss out on dynamic variations that become available on the server since the original agent-driven negotiation.

42.3.4 Systems That Have Utilized Transcoding Technologies

A number of systems and research prototypes have used transcoding to fit images to the current operating environment.

The Mowgli project [Liljeberg et al., 1995, 1996] described content type specific lossless and lossy compression techniques to improve the Web experience over wireless mobile links. The GloMop [Fox and Brewer, 1996; Fox et al., 1996] project used transcoding to generate thumbnails on the fly to speed up image access from slow modems. Han et al. [1998] presented an analytical framework for determining whether to transcode an image and how much of it to transcode. Odyssey [Noble et al., 1997] manipulated the JPEG Compression metric as a distillation technique for a Web browser that adapts to changing network environments. The Soft Caching project [Ortega et al., 1997; Kangasharju et al., 1998] used progressive JPEG to produce lower quality images at the same spatial resolution. CaubWeb [Mazer et al., 1998] described a generic proxy service with content type specific application transducers that perform filtering operations such as transcoding. Some of those systems do not use an informed transcoding technique that can quantify the information loss from the chosen transcoding operation. Chandra et al. [2001] analyzed typical Web images to develop a quality-aware, informed transcoding mechanism for these JPEG images [Chandra and Ellis, 1999]. They utilized this technology in a proxy server to manage client network restrictions [Chandra et al., 1999] in a Web server to provide differentiated quality of service [Chandra et al., 2000a], and in a mobile content creation device (digital camera) to manage the available storage and battery capacity [Chandra et al., 2000b].

Commercial products such as WebExpress [Floyd et al., 1998] from IBM, QuickWeb technology from Intel, and Fastlane from Spectrum Information Technology have used various forms of compression and transcoding operations to improve Web access from slow networks. The WebQoS system from HP provides quality of service on the Web by using priority levels to determine admission priority and performance level. WebQoS uses parameters such as source IP address, destination IP address, URL, port number, hostname, and IP type-of-service to classify requests. The system uses these priorities in controlling the allocation of CPU and disk resources. Higher-priority requests are sent to servers running in separate ports that operate under different system priorities. The system uses priorities to delay or deny service to lower-priority clients. Though this policy leads to predictable service for the preferred clients, the lower-priority clients can be turned away.

42.4 Challenges in the Effective Use of TranscodingTechnologies

In this chapter, we described transcoding as a technique to adapt content to constrained environments. We summarize the important features discussed:

Transcoding is a lossy operation to generate variations of objects that are better suited to the current operating environment. Such variation allows Web access from constrained clients as well as busy servers.

If care is not taken to understand the transcoding characteristics, such operations can yield objects that may not convey the intended information. With the emergence of new representation formats (e.g., JPEG 2000), it is imperative that informed transcoding decisions are made for these newer formats.

Transcoding operations that are appropriate for the client's usage patterns, and that can also preserve the content provider's intent without violating the provider's rights and content control is a major challenge in enabling transcoding technologies.

References

Abdelzaher, Tarek F. and Nina Bhatti. Web content adaptation to improve server overload behavior. In *8th International World Wide Web Conference/Computer Networks,* Vol. 31, pp. 1563–1577, Toronto, Canada, May 1999.

All Things Web. Third State Of the Web Survey (SOWS III). http://www.pantos.org/atw/35654.html, May 1999.

Chandra, Surendar and Carla Schlatter Ellis. JPEG compression metric as a quality aware image transcoding. In *Proceedings of the 2nd USENIX Symposium on Internet Technologies and Systems USITS-99,* pp. 81–92, Boulder, CO, October 1999. USENIX Association.

Chandra, Surendar, Carla Schlatter Ellis, and Arvin Vahdat. Multimedia Web services for mobile clients using quality aware transcoding. In *Proceedings of the 2nd ACM International Workshop on Wireless and Mobile Multimedia (WoWMoM'99),* pp. 99–108, Seattle, WA, August 1999. ACM SIGMOBILE.

Chandra, Surendar, Carla Schlatter Ellis, and Amin Vahdat. Differentiated multimedia web services using quality aware transcoding. In *INFOCOM — 19th Annual Joint Conference Of The IEEE Computer and Communications Societies,* pp. 961–969, Tel Aviv, Israel, March 2000a. IEEE.

Chandra, Surendar, Carla Schlatter Ellis, and Amin Vahdat. Managing the storage and battery resources in an image capture device (digital camera) using dynamic transcoding. In *Proceedings of the 3rd ACM International Workshop on Wireless and Mobile Multimedia (WoWMoM'00),* pp. 73–82, Boston, August 2000b. AGM SIGMOBILE.

Chandra, Surendar, Ashish Gehani, Carla Schlatter Ellis, and Amin Vahdat. Transcoding characteristics of web images. In Martin Kienzle and Wu Chi Feng, Eds., *Multimedia Computing and Networking (MMCN'01),* Vol. 4312, pp. 135–149, San Jose, CA, January 2001. SPIE — The International Society of Optical Engineering.

Elson, Jeremy and Alberto Cerpa. *Internet Content Adaptation Protocol (ICAP).* University of California, Los Angeles, April 2003.

Elson, Jeremy, John Martin, Edward Sharp, John Schuster, Alberto Cerpa, Peter Danzig, Chuck Neerdaels, and Gary Tomlinson. ICAP — the Internet content adaptation protocol. Technical report, ICAP forum, http://www.i-cap.org, March 2000.

Feeney, Laura Marie and Martin Nilsson. Investigating the energy consumption of a wireless network interface in an *ad hoc* networking environment. In *Proceedings IEEE INFOCOM 2001,* Vol. 3, pp. 1548–1557, Anchorage, AK, April 2001.

Fielding, R., J. Gettys, J. Mogul, H. Frystyk, L. Masinter, P. Leach, and T. Berners-Lee. Hypertext Transfer Protocol — HTTP/1.1, June 1999.

Floyd, Rick, Barron Housel, and Carl Tait. Mobile Web access using eNetwork Web express. *IEEE Personal Communications,* 5(5), October 1998.

Ford, Adrian M. Relations between image quality and still image compression. Ph.D thesis, University of Westminster, London, May 1997.

Fox, Armando and Eric A. Brewer. Reducing www latency and bandwidth requirements via real-time distillation. In *Proceedings of the 5th International World Wide Web Conference,* pp. 1445–1456, Paris, May 1996.

Fox, Armando, Steven D. Gribble, Eric A. Brewer, and Elan Amir. Adapting to network and client variability via on-demand dynamic distillation. *ACM SIGPLAN Notices,* 31(9): 160–170, September 1996.

Graphics Interchange Format (GIF) — A standard defining a mechanism for the storage and transmission of raster-based graphics information. CompuServe, Columbus, OH, June 1987.

Graphics Interchange Format — Version. 89a. CompuServe, Columbus, OH, 1989.

Han, Richard, Pravin Bhagwat, Richard LaMaire, Todd Mummert, Veronique Perret, and Jim Rubas. Dynamic adaptation in an image transcoding proxy for mobile web browsing. *IEEE Personal Communications Magazine,* 5(6): 8–17, December 1998.

Havinga, Paul J. M. Mobile multimedia systems. Ph.D thesis, University of Twente, Enschede, February 2000.

Holtman, Koen and Andrew H. Mutz. HTTP Remote Variant Selection Algorithm — RVSA/1.0, March 1998a.

Holtman, Koen and Andrew H. Mutz. *Transparent Content Negotiation in HTTP,* March 1998b.

Jacobson, R. E., A. M. Ford, and G. G. Attridge. Evaluation of the effects of compression on the quality of images on a soft display. In *Proceedings of the SPIE: Human Vision and Electronic Imaging II,* Vol. 3016-14, San Jose, CA, February 1997.

Kangasharju, Jussi, Younggap Kwon, and Antonio Ortega. Design and implementation of a soft caching proxy. In *Proceedings of the 3rd International WWW Caching Workshop,* Vol. 30, pp. 2113–2121, Manchester, U.K., June 1998.

Kindberg, Tim, John Barton, Jeff Morgan, Gene Becker, Ilja Bedner, Debbie Caswell, Philippe Debaty, Gita Gopal, Marcos Frid, Venky Krishnan, Howard Morris, Celine Pering, John Schettino, Bill Serra, and Mirjana Spasojevic. People, places, things: Web presence for the real world. In *3rd IEEE Workshop on Mobile Computing Systems and Applications (WMCSA),* Monterey, CA, December 2000. IEEE.

LAN/MAN Standards Committee of the IEEE Computer Society. *Part II: Wireless LAN Medium Access Control (MAC) and Physical Layer (PHY) Specifications.* IEEE, New York, 1999.

Liljeberg, Mika, Timo Alanko, Markku Kojo, Heimo Laamanen, and Kimmo Raatikainen. Optimizing world-wide web for weakly connected mobile workstations: An indirect approach. In *Proceedings of the 2nd International Workshop on Services in Distributed and Networked Environments (SDNE'95),* Whistler, Canada, June 1995.

Liljeberg, Mika, Heikki Helin, Markku Kojo, and Kimmo Raatikainen. Mowgli www software: Improved usability of www in mobile wan environments. In *IEEE Global Internet 1996,* London, November 1996. IEEE Communications Society.

Mazer, Murray S., Charlie Brooks, John LoVerso, Louis Theran, Fredrick Hirsch, Stavros Macrakis, Steve Shapiro, and Dennis Rockwell. Distributed clients for enhanced usability, reliability, and adaptability in accessing the national information environment. Technical report, The Open Group Research Institute, Cambridge, MA, 1998.

Nielsen, Jakob. *Usability Engineering.* Academic Press, Boston, MA, 1993.

Noble, Brian D., M. Satyanarayanan, Dushyanth Narayanan, J. Eric Tilton, Jason Flinn, and Kevin R. Walker. Application-aware adaptation for mobility. In *Proceedings of the 16th ACM Symposium on Operating Systems and Principles,* pp. 276–287, Saint-Malo, France, October 1997.

Ortega, Antonio, Fabio Carignano, Serge Ayer, and Martin Vetterli. Soft Caching: Web cache management techniques for images. In *IEEE Signal Processing Society 1997 Workshop on Multimedia Signal Processing,* Princeton, NJ, June 1997.

Pennebaker, William B. and Joan L. Mitchell. *JPEG — Still Image Data Compression Standard.* Van Nostrand ReinHold, New York, 1993.

Stemm, Mark, Paul Gauthier, Daishi Harada, and Randy H. Katz. Reducing power consumption of network interfaces in hand-held devices. In *Proceedings of the 3rd International Workshop on Mobile Multimedia Communications (MoMuc-3),* Princeton, NJ, September 1996.

Weiser, Mark. Ubiquitous computing. *IEEE Computer Hot Topics,* October 1993.

Part 5

Engineering and Management

43

Software Engineering for Internet Applications

CONTENTS

43.1 Nature of Internet Application Development 43-1
43.2 Traditional Software Development Models 43-2
43.3 Agile Software Development Models 43-4
43.3.1 Empirical Process Control 43-5
43.3.2 Emergence 43-5
43.3.3 Self-Organization 43-6
43.4 Survey of Representative Agile Methodologies 43-7
43.4.1 Extreme Programming 43-7
43.4.2 Scrum 43-9
43.4.3 Crystal Methods 43-9
43.5 Agile Methods: Meeting the Challenges of Internet Application Development 43-10
43.5.1 Requirements Volatility 43-10
43.5.2 Collaboration 43-10
43.5.3 Time-to-Market Pressure 43-10
43.5.4 Security and Privacy 43-10
43.5.5 High Availability 43-11
43.6 Research Issues and Summary 43-11
References 43-12

Laurie Williams

43.1 Nature of Internet Application Development

The acceleration of change and intense competition brought on by the Internet revolution continues to thrive. The stronger companies have survived, but these same firms need to constantly innovate, or risk failure or lack of profitability. As a result, software applications for the Internet have compressed development schedules. Additionally, these applications must continuously evolve to meet new demands to outpace (or meet) competitive offerings. Often, Internet applications are updated on a daily basis. By moving a development file onto the server, the software could essentially be released daily to possibly millions of web users.

Controlling change is a challenge. Internet application development is complex in that it often involves consideration and implementation of attractively presented content, procedural processing, and business rules. This unification proves significantly more complex than traditional development that involved only procedural processing. Handling this added multifaceted complexity requires increased collaboration and teamwork. Security is paramount in Internet applications. Privacy concerns are heightened. Entire businesses could be brought down if their Internet application failed, making the availability of the application highly critical. In short, the Internet application development environment is complex, agitated, tumultuous, chaotic, competitive, and ruthless. Internet application development teams must

1-58488-381-2/05/$0.00+$1.50

© 2005 by CRC Press LLC

deal with requirements volatility, intense time-to-market pressure, high availability, security, and privacy. They must also work very collaboratively. Software engineering techniques must be adapted to respond to this environment.

Alan MacCormack of the Harvard Business School led an extensive, 2-year study [MacCormack et al., January 2001; MacCormack, Winter 2001] on necessary management practices for developing products in uncertain and dynamic environments — that is, products that need to be developed on "Internet time." MacCormack analyzed data from 29 projects, including some defining products such as Netscape's Navigator 3.0 and 4.0 browsers, Microsoft's Explorer 3.0 and 4.0 browsers, and Yahoo!'s My Yahoo! service. He found that a more flexible, evolutionary development process is associated with better-performing projects. Performance was measured based on a panel of experts' (14 industry observers) assessment of product quality. Quality was defined as a combination of product features, technical performance, and reliability. Because innovative firms developing software quickly cannot predict every design choice up front, their ability to affordably generate and respond to new information for as long as possible, can determine their profitability and competitiveness. However, the practices that allow organizations to be reactive often challenge the strongly held conventional software engineering belief in the importance of careful, thorough requirements elicitation and analysis, planning, and design early in the development process.

In this chapter, we will discuss the appropriateness of traditional software development life cycle models related to the development of Internet applications. We will then discuss the emerging agile software development model that appears to be better suited for the fast-paced, volatile environment of Internet application development. A survey of agile methodologies will be presented. Finally, we will assess how agile methods meet the challenges of Internet application development.

43.2 Traditional Software Development Models

In the early days of computing, most of the programming was done by scientists trying to solve specific, relatively small mathematical problems. The programming model that emerged from those days has been called the "code-and-fix model … [which] denotes a development process that is neither precisely formulated nor carefully controlled" [Ghezzi et al., 1991]. Ghezzi et al. [1991] describe the code-and-fix model as consisting of two steps: (1) write code and (2) fix code (to eliminate errors, enhance existing functionality, or add a new feature).

Through time, computers became cheaper and more common. More and more people started using them to solve larger and larger problems, still using the evolving original programming model.

The code-and-fix model is often still used today, particularly when development is rushed, as it is with Internet applications. Code-and-fix *feels* faster because no time is *wasted* on documentation, planning, or design. However, the model is not adequate to handle the complexities of complex or large-scale software development, and generally leads to unpredictable results. In a code-and-fix process, developers have an end goal in mind. Through coding and fixing, the development team may or may not come close to reaching their desired results. Many projects are declared failures and aborted midcourse. Additionally, the model does not lead to the most productive efforts due to the high potential of rework, and long testing and debugging cycles. Moreover, Ghezzi et al. [1991] attribute the failure of the code-and-fix model with the recognition of what is currently known as the *software crisis* [Gibbs, 1994]. The code-and-fix model can be summarized in the words of management thinker Gary Hamel [Hamel, 2002][1], "fire, fire, fire, fire, aim again, fire, fire, fire — there is no time for 'ready'."

As a result, the waterfall process model was developed to make the software development process more predictable and controllable. This process was first published in 1970 by Winston Royce [Royce, 1970]. Royce broke the software development process into five steps: requirements definition, system and software design, implementation and unit testing, integration and system testing, and finally operation

[1] Thanks to Jim Highsmith for first using this quote in a description of the code-and-fix model in a presentation at the XP Universe conference in 2001.

and maintenance. A key to the waterfall process is that each step had to be completed and inspected and deemed "perfect" because once a step was done, it was theoretically never revisited. All further developmental steps used the artifacts of previous steps without looking back. In contrast to the code-and-fix model, the waterfall model can be summarized, "ready, ready, ready, aim, aim, aim, fire."

Many organizations have graduated from code-and-fix to the waterfall model. However, current thinking says the "never look back" philosophy is not advisable when requirements are rapidly changing as with Internet applications. In fact, the waterfall model was deemed inappropriate in cases where requirements are uncertain (as early as 1976, by the software engineering thought leader Harlan Mills) because of the apparent lack of feedback loops to prior steps when new information has been learned [Mills, 1976; Larman and Basili, June 2003]. "There are dangers ... in the conduct of the stages in sequence and not in iteration — i.e., that development is done in an open loop, rather than a closed loop with user feedback between iterations." Due to the lack of feedback loops, the original planned result may not be the desired result of the customer when the product has considerable requirements volatility. The model may lead to the efficient delivery of the system to the customer as originally specified, but not the system that the customer really needs.

The shortcomings of the waterfall model motivated the incremental and iterative software development models. With incremental development, all requirements are analyzed before development begins. However, the set of requirements are then broken into increments of stand-alone functionality. The system as specified in the requirements is partitioned into small subsystems by functionality. New functionality is added with each new release. The first increment implements the core functionality and future releases enhance the core product. Each increment is a mini-waterfall cycle. Possibly, the development of successive increments can have some overlap in time (an increment can begin before its predecessor has completed). The prime motivation of incremental models is to improve cycle time because the project can be broken into pieces that can be developed with overlapping schedules. As the requirements are frozen early, incremental models remain nearly as unsatisfactory for Internet application development as the waterfall model. As with the waterfall model, there are no prescribed feedback loops to revisit the original set of requirements based on new information. Strict adherence to this kind of model would lead to a similar phenomenon as with the waterfall model, whereby the development team would likely not deliver the system currently desired by the customer.

Similar to incremental development, iterative development process models break a project into pieces. First, a basic core system is delivered and each subsequent release adds new functionality to the system. Again, each release has its own waterfall cycle. The key difference is that iterative development deals better with change, as the only complete requirements necessary are for the current iteration. Requirements for future releases remain preliminary, awaiting analysis at a later date. Traditional iterative models allow for limited changing technology or requirements with minimal impact on the project's momentum. The iterative software development model is a much better match for the volatile environment of Internet application development.

However, traditional iterative models still have some shortcomings. Traditional iterative and incremental models are similar in that each increment or release follows a mini-waterfall process. As a result, they can sometimes be viewed as document driven because many waterfall milestones are grounded in the delivery and inspection of complete documents. Products that utilize a full waterfall model might undergo development cycles in years. In comparison, both incremental and iterative models have cycles of multiple months. Both also require the development of a detailed plan for increments and releases (and the mini-waterfall milestones) before development begins. As a result, these models are often referred to as plan driven [Boehm, 2002]. Boehm reflects that plan-driven models may not be appropriate for the exploratory, fast-paced development common in Internet application development:

> Plan-driven methods work best when developers can determine the requirements in advance ... and when the requirements remain relatively stable, with change rates on the order of 1% per month. In the increasingly frequent situations in which the requirements change at a much higher rate than

this, the traditional emphasis on having complete, consistent, precise, testable, and traceable requirements will encounter difficult to insurmountable requirements-update problems.

Doubtlessly, typical Internet application development requirements change rates are significantly higher than 1% per month relative to the initial requirements statement. Therefore, alternatives to plan-driven models are needed for this type of development.

43.3 Agile Software Development Models

Beginning in the mid–late 1990s (which coincides with the beginning of the Internet age), many practitioners found the development of a complete initial requirements documentation frustrating and, perhaps, impossible. The industry and the technology move too fast. Customers have become increasingly unable to definitively state their needs up front. Additionally, they found the extensive documentation required to be bureaucratic and overly time consuming. They also felt the documentation made it more difficult to respond to change.

As a result, several consultants independently developed methodologies and practices to embrace and respond to the inevitable change they were experiencing. The consultants developed these methodologies independently, though there were fundamental similarities in their practices and philosophies. In February 2001, these consultants joined forces. They decided to classify their similar methodologies as *agile* — a term with a decade of use in flexible manufacturing practices [Agile, 1996]. Taking a historical view, these agile methodologies and practices are based on iterative enhancement, a technique introduced in 1975 [Basili and Turner, 1975]. Currently, agile methodologies are rapidly gaining in popularity. However, they are comprised of a mix of practices that are currently considered best practices as well as practices that are considered unproven (and therefore controversial). As a result, the validation of these practices and methodologies is an active research area. Testimonies of successes with agile methodologies can be found in the proceedings of conferences dedicated toward these methodologies [Beck, 2000; Marchesi and Succi, 2001; Marchesi et al., 2002; Wells and Williams, 2002].

The consultants formed the Agile Alliance[2] and wrote the Agile Manifesto [Agile, 1996; Beck et al., 2001; Fowler and Highsmith, August 2001]. The Agile Manifesto is listed and explained below:

We are uncovering better ways of developing software by doing it and helping others do it. Through this work we have come to value:

- *Individuals and interactions* over processes and tools
- *Working software* over comprehensive documentation
- *Customer collaboration* over contract negotiation
- *Responding to change* over following a plan

That is, while there is value in the items on the right, we value the items on the left more.

In the first phrase, the Manifesto authors express that they value *individuals and interactions* over processes and tools. The agilists feel that most often problems surface when someone should have spoken with someone else, but instead relied upon their own judgment or upon their interpretation of something in a document. As a result, the agile philosophy stresses daily face-to-face interaction between sponsors, developers, and business people. They feel this is much more important than "blindly" following a predetermined software development process or using a required tool.

In the second phrase, the Manifesto authors express that they value *working software* over comprehensive documentation. They feel the best way to clarify a requirement or to validate that an implementation meets the customer's desires is to create working software that the customer can try. It is better to work out these issues in an "executable" requirements document, e.g., working code, rather than by digesting

[2] http://agilealliance.com/home

or inspecting a standard requirements document or design. Documentation may be produced, but is intentionally minimized.

In the third phrase, the Manifesto authors express that they value *customer collaboration* over contract negotiation. The focus is to get feedback from the customer very often, if not every day. Customers clarify requirements and change their minds if they need to. The focus is on producing what the customers actually want, rather than what they originally said they wanted in the initial contract.

In the fourth phrase, the Manifesto authors express that they value a willingness to *respond to change* over following a plan. They appreciate that change will happen, and they are prepared to embrace rather than steer clear of change.

The methodologies originally embraced by the Agile Alliance were Adaptive Software Development (ASD) [Highsmith, 1999], Crystal Methods [Cockburn, 2001], Dynamic Systems Development Method (DSDM) [Stapleton, 1997], Extreme Programming (XP) [Beck, 2000], Feature Driven Development (FDD) [Coad et al., 1999], and Scrum [Schwaber and Beedle, 2002]. Since then, other methodologies have striven to be classified as agile. In the next section, we provide a representative overview of three of the original methodologies: XP, Scrum, and Crystal. All of these methodologies emphasize flexibility of the development process, the need to be able to respond to changing customer desires. Unaware of MacCormack's work, they had arrived at the same conclusions on the importance of flexibility and responsiveness. Advocates of agile methodologies profess that these methods are superior for dealing with change and, therefore, for providing customers with what they want, when they want it, and with acceptable defect rates.

Agile Alliance member Ken Schwaber [Schwaber, 2001] documented three overarching criteria for a methodology to be considered agile. These three criteria are now discussed.

43.3.1 Empirical Process Control

In many engineering fields, processes can be classified as defined or empirical (Ogunnaike and Ray, 1994). A defined process can be started and allowed to run to completion, with the same results most every time. (Schwaber and Beedle, 2002) An example of a defined process is the assembly of an automobile. Engineers can design a process to assemble the car by specifying an assembly order and actions for assembly-line workers, machines, and robots. If these predefined steps are followed, a high-quality car will most often be produced. We are beginning to realize that software development cannot be considered a defined process. There is simply too much change during the time that the team is developing the product to think that any set of predefined steps would lead to a desirable, predictable outcome. Requirements change, technology changes, people are added to or taken from the team, and so on. Software development, instead, can be classified as an empirical (or nonlinear) process. In an engineering context, empirical processes necessitate short "inspect-and-adapt" cycles and frequent, short feedback loops (Schwaber and Beedle, 2002). It is likely that these short inspect-and-adapt cycles of agile methodologies enable them to better handle the conflicting and unpredictable demands of the Internet application industry.

43.3.2 Emergence

> *The form of made things is always subject to change in response to their real or perceived shortcomings, their failures to function properly … This principle governs all invention, innovation, and ingenuity; it is what drives all inventors, innovators, and engineers.* — Henry Petroski [Petroski, 1994]

Petroski, a civil engineering professor, further states that the best way to determine how a product needs to evolve is to use it. When faced with changing requirements and technologies, agile methodologists do not believe that a software application can be fully specified up-front. Instead, the true requirements should be allowed to emerge over time. The agile methodologies welcome changing requirements, even late in the development cycle. A minimal timebox is dedicated toward collecting an initial set of requirements. A specified, agreed-upon subset of these requirements is completed in each short iteration. The focus is on producing working software, rather than documents. At the end of each iteration, the customer is shown the product and asked, "How do you like that? Now that you see what we've done, what would you like us to do for the next iteration?" In this manner, the requirements emerge, leading toward the delivery of what the customer really wants and providing them with the best competitive advantage possible in the face of constant change and turbulence.

43.3.3 Self-Organization

Agile methodologies give the entire development team the autonomy to self-organize themselves to determine the best way to get the job done. Team members are not constrained by predetermined roles or required to execute obsolete task plans. Managers of agile teams place a great deal of trust and confidence in the entire team. In the self-organization, the emphasis is on face-to-face conversations rather than on communicating through formal (or informal) documents. Software developers talk with software developers, business people talk with software developers, customers talk directly with either business people or software developers. Agile methodologies also advocate the use of retrospective meetings in which team members reflect on how to become more effective. The team then tunes and adjusts its behavior accordingly.

Through these values and philosophies, agile methodologists contend that they are better able to respond to volatile demands, such as are found in Internet applications. Because change is allowed, some rework is inevitable. The goal is to increase the possibility of delivering what the customer really wants, rather than their initial view of their desires. Agile methodologies can be summarized as having rapid "ready, aim, fire, ready, aim, fire" cycles.

Because of the inevitable rework in product development due to embracing change, the affordability and efficiency of agile methodologies is often questioned. Anecdotally, agile methodologies have not been shown to be more expensive or to take longer and have been shown to produce products of high quality. Consider the following scenario. First, a plan-driven development team plans to have two members spend 3 months developing a solid, well-specified requirements document. It is certain that a total of six person months will be spent on this requirements document. Instead, an agile team might have two people spend a week documenting the best set of requirements the customer can come up with at the time. In this case, it is certain that this team has spent a half of a person month developing this first version of the requirements "document." The agile team has just "banked" 5 months. They can later spend this time they just banked on revising requirements or the product based on current knowledge, including the customer's perception of the working software that has been developed thus far. The risk is that they will not have to spend more than the saved 5 months to rework the requirements and the product, and that the customer's current view of their desires is better than their initial view.

With plan-driven development practices, the majority of the expense is spent on "new development," whether it be the new development of a document, design, or code. Inevitably, there will be some rework during the development cycle, though the aim is to minimize rework expense. In such a project, development proceeds with a "do it right the first time" philosophy with the objective of "satisfying the original contract." This is an appropriate approach for projects without a significant degree of requirements variability.

Conversely, in an agile project, the expense incurred for new development and the expense incurred for rework has a much different profile, demonstrating a large increase in rework and a large decrease in new development. However, strong anecdotal evidence suggests that the additional rework does not exceed the expense that would have been incurred had extensive up-front requirements engineering, planning, and designing taken place. In an agile project, development proceeds with a "do it right the last time" philosophy with the objective of "delighting the customer." This approach has been demonstrated to be appropriate for projects with any significant degree of requirements variability.

43.4 Survey of Representative Agile Methodologies

In this section, we provide a brief summary of three representative agile methodologies: XP, Scrum, and Crystal.

43.4.1 Extreme Programming

Extreme Programming (XP) [Beck, 2000; Auer and Miller, 2001; Beck and Fowler, 2001; Jeffries et al., 2001] is currently considered to be the most popular agile methodology.

43.4.1.1 Requirements Definition

Requirements are gathered and scheduled using User Stories and the Planning Game [Beck and Fowler, 2001]. To begin, customers work with developers or requirements analysts to gather requirements. This is done very informally. Customers speak about a requirement and the analyst writes down their requirement on an index card (called a user story card) almost verbatim — in the customers' natural language. There is no intent to completely specify the customers' requirement on the index card. The card is technically a commitment for future conversation between the developer and the customers. When all the cards are done, the customers are asked to prioritize these stories.

The developers then work with the stories. They read the stories and ask questions of each customer (preferably in person) about what is desired. From this, the developers estimate how long the story will take to implement. To come up with the estimate, the developers look at how long other stories have taken them to complete. They also assess the risk associated with the story.

With each story card having a description, a priority, a resource estimate, and a risk assessment, the Planning Game is started, using these cards as the playing pieces. The game is played by laying all the cards out on the table and sliding them around to rearrange or reorganize them. The object of the game is to decide which stories get implemented in the next release, trying to maximize the number of high-priority and high-risk stories. With XP, a release is generally no more than 3 months long and each release must deliver functionality or value to the customer. Each release is broken up into three to four iterations. Each iteration is an internal development deliverable (that is, not a supported release delivered to customers). At the end of each iteration, the team assesses the stories they have been able to complete and makes a new plan for successful completion of the iteration (perhaps involving a new, mini Planning Game to reset the release plan based on actual results).

At the end of the Planning Game, the team knows which stories will be done in the next iteration. Developers sign up to implement the user stories. One developer is the owner of the story. All production work is done using pair programming [Williams et al., 2000], whereby two developers work together at one computer. Therefore, the story owner recruits partners to work with on the various subtasks of the story, based on the technical aspects of that particular subtask. While the story is being developed, many questions will arise because the developer only has an index card with a short description on it. With XP, it is very important to have an onsite customer representative to clarify these requirements at any time. XP espouses that many problems arise because requirements are not completely or accurately specified or because the developers simply do not understand the requirements. An onsite customer can clarify the requirements, which will prevent the developer from having to make as many assumptions on what the customer probably wanted.

Another important job of the onsite customer is to work with the developers to create acceptance test cases. They define what kinds of things they want to see the system do in order to feel assured that a user story has been properly implemented. In developing these test cases, the customer is also clarifying his or her requirements.

Once a release has been completed and delivered to customers, the planning game is replayed to determine which user stories will be implemented in the next release. At this time, the customers have ample opportunity to change their priorities and the stories.

43.4.1.2 System and Software Design

XP does not produce detailed design documents before commencing a development effort. Extreme Programmers call these large design documents "BDUF" (Big Design Up Front) and argue that the development of these documents is not cost effective. They believe that software engineers change their mind about how they will implement something once they try it and once they get feedback on completed work. By skipping the documentation step, the production of software assets that have value to customers are produced, not just documents. Some XP teams also create some bare-bones documentation after a release is completed that records what they did, not what they planned to do.

In traditional development, the system architecture is developed to provide a map of how all the pieces fit together. The closest that XP comes to architecture is the system metaphor. For example, the original XP project, the C3 project, was a payroll system. The metaphor that was used was an assembly line. As a paycheck went down the assembly line, it was given all the pieces of information it needed. When it got to the end of the assembly line, it had all of the necessary information. All the developers on the C3 project had the assembly-line mental model in their heads, which helped them to visualize how all the pieces fit together. "The metaphor gives the team a consistent picture they can use to describe the way the existing system works, where new parts fit, and what form they should take" [Auer and Miller, 2001].

In general, a pair starts to work on a story. They read the story and do whatever they need to do to feel comfortable that they can handle it. This may be to have an impromptu CRC card session [Bellin and Simone, 1997] or it may be to write down a quick UML class diagram [Jacobson et al., 1999] or flowchart on a whiteboard. They use any tools and technique that any other software developer might use to devise a design; however, they only use what they feel is necessary to tackle the task at hand. The artifacts of this process, the notes, cards, and diagrams are never officially archived for future use. Once the pair feels comfortable that they have a good view of their direction, they begin to code.

A important part of XP's design philosophy is "do the simplest thing that could possibly work." This philosophy is intended to reduce the tendencies software engineers may have to design and code functionality and structure to handle anticipated future needs that may never materialize. XP reminds developers to focus on the user story at hand.

Finally, XP strongly encourages refactoring [Fowler et al., 1999], which is the technique of improving code without changing its functionality. Martin Fowler has championed this technique. The first pass on the code is to do the simplest thing that could possibly work to get the user story to work. Then there is an explicit, deliberate time where the pair refactors code in order to improve the design of their working implementation.

43.4.1.3 Implementation and Unit Testing

As previously mentioned, a pair of programmers obtains a user story and tackles whatever is needed to ensure they have a reasonable plan of attack. This may entail CRC cards, a quick sketch of a class diagram, or even nothing at all. With their reasonable plan in hand, the pair begins the implementation. Prior to writing any functional code, the pair must first write automated unit test cases in a practice referred to as test-driven development (TDD) [Beck, 2003]. These tests show that the program works as intended and handles error conditions gracefully. Naturally, since the code has not been written, these test cases will initially fail. As more and more functionality is incrementally implemented, more and more test cases will pass. The task is not completed until the functionality for the user story is completed, and 100% of the new unit tests and 100% of all the unit test cases in the code base run successfully. Only then can the developer know that the new functionality was successfully implemented without breaking anything else in the project. This type of efficient, automated regression testing necessitates the use of a tool; xUnit[3] tools are most popular among Extreme Programmers.

There exist other notable implementation factors. First, the entire development team owns the entire code base (called "collective code ownership"). This means that any developer can change any line of

---

[3] see http://www.xprogramming.com/software.htm for several testing frameworks.

code in the entire code base in order to add functionality or to refactor. The developer does not need to ask permission to change any code. They just have to make sure that all the unit tests ever written on the code base still run at 100%.

Lastly, because of pair programming and collective code ownership, XP has commenting guidelines and follows a coding standard. This is necessary because everyone on the team needs to be able to quickly understand anyone else's code to enhance it or refactor it. The emphasis on the commenting is to have self-documenting or self-revealing code rather than too many comments (which might get out of date).

43.4.1.4 Integration and System Testing

A common trend in modern software development is to integrate code often. XP practices "Continuous Integration" whereby developers incorporate new functionality into the code base as soon as all the unit test cases pass. Ideally, this happens at least once a day for every pair.

System testing is conducted using the acceptance test cases developed and agreed upon by the customer. These tests measure the progress of system development (either in terms of the absolute number or percentage of test cases that pass) and make a statement of how much functionality has been properly implemented in the system. It is always good to automate these tests too so that it is easy to tell if any new functionality breaks any functional test cases that had previously passed and so that they are run often.

43.4.2 Scrum

XP is highly specified in that the individual development practices are specified. Conversely, Scrum is a managerial wrapper for existing practices. While operating under the agile philosophy, Scrum teams have the freedom to choose many of their individual software development practices. In the Scrum process [Highsmith, 2002; Schwaber and Beedle, 2002], a Sprint Planning meeting is held with the development team, management, and the Product Owner. The Product Owner is a representative of the customer or a contingent of customers. The Product Owner creates and prioritizes the Product Backlog, a list of all business and technology features to be incorporated into the product. In the planning meeting, the Product Owner expresses which features are desired for the next 30-day increment (called a Sprint). The development team figures out the tasks and resources required to deliver those features. Jointly, they determine a reasonable number of features to be included in the next Sprint. Once this set of features has been identified, no reprioritization takes place during the ensuing 30-d Sprint in which features are designed, implemented, and tested. At the end of a Sprint, a Post-Sprint meeting takes place to review progress, demonstrate features to the customer, and review the project from a technical perspective. Next a Sprint Planning meeting takes place to choose the features for the next Sprint.

43.4.3 Crystal Methods

The family of Crystal Methods [Cockburn, 2001] were developed by Alistair Cockburn in the early 1990s. All the Crystal Methods emphasize the importance of people in developing software. "[Crystal] focuses on people, interaction, community, skills, talents, and communication as first order effects on performance. Process remains important, but secondary" [Highsmith, 2002]. Consistent with the agile philosophy, Crystal emphasizes person-to-person communication.

To the extent that you can replace written documentation with face-to-face interactions, you can reduce the reliance on written work products and improve the likelihood of delivering the system. The more frequently you can deliver running, tested slices of the system, the more you can reduce the reliance on written "promissory" notes and improve the likelihood of delivering the system.

Crystal is a family of methods because Cockburn believes that there is no "one-size-fits-all" development process. As such, the different methods are assigned colors arranged in ascending opacity; the most Agile version is Crystal Clear, followed by Crystal Yellow, Crystal Orange, and Crystal Red. The version of crystal you use depends on the number of people involved and the criticality of the project, where the criteria for ascending criticality is based on whether a failure results in loss of (1) comfort, (2) discretionary money, (3) essential money, or (4) life. Based on the team size and the criticality, the corresponding Crystal methodology is identified; each methodology has a set of recommended practices, a core set of roles, work products, techniques, and notations.

43.5 Agile Methods: Meeting the Challenges of Internet Application Development

Agile methods have been shown to handle many of the complexities of Internet quite well. Research is underway to extend the capabilities of agile methods to better handle other factors of Internet application development, as discussed below.

43.5.1 Requirements Volatility

Requirements volatility is a significant challenge in Internet application development due to environmental factors. We once again assert that a prime motivation of agile methods is dealing with and embracing change in requirements and technology.

43.5.2 Collaboration

As said, Internet application development requires the unification of (1) attractively presented content, (2) procedural processing, and (3) business rules. This unification proves significantly more complex than more traditional development because these three items are usually handled by three different people with potentially specialized skills. Handling this multifaceted complexity requires increased collaboration and teamwork. Collaboration is also a strength of agile methods. These methods emphasize person-to-person (preferably face-to-face) interaction and the development of personal collaboration. For example, the procedural processing programmer and the Web designer would spend time together, potentially pair programming, to work on the development of an attractive interface with the necessary processing.

Most agile teams are relatively small (no more than 12 team members) and work collocated in one location, preferably in one room. However, larger and distributed agile teams have been known to be successful [Crocker, 2004].

43.5.3 Time-to-Market Pressure

Internet application must release software as quickly as possible and re-release often to stay competitive. Agile software development focuses on producing "working" software and has short release times. As a result, at least partial functionality can be released to customers early, and releases with increased functionality delivered often.

43.5.4 Security and Privacy

> *As e-commerce blossoms, and the Internet works its way into every nook and cranny of our lives, security and privacy come to play an essential role.* [Viega and McGraw, 2002]

While they are the most suitable model for Internet application development, the current agile methods may need to be adapted when security and privacy are an issue. There are two concerns with the ability of the methods to handle this aspect of Internet applications. First, generally agile methods de-emphasize the creation of inspectable documents and, instead, emphasize working software. A recommended practice for minimizing security risk is for specially trained security personnel to inspect various project artifacts, including requirements and design documents. Indeed, security experts [Viega and McGraw, 2002] express concern that XP will have a negative impact on software security due to the lack of artifacts for security personnel to review. When security is a concern, the artifacts necessary for security and privacy reviews must be produced. Producing these documents will likely add "weight" and time to the process, but the necessary trade-offs must be made using prudent judgment concerning the project risks. XP has a philosophy of "doing the simplest thing that could possibly work" [Beck, 2000]. This philosophy should be considered in adding necessary documentation back to an agile process.

Another security and privacy concern related to agile methods is that these methods are functional requirement-centric, and security and privacy are nonfunctional requirements. The users share and prioritized their desirable "stories" or features. Extra effort must be made on the part of the development team to (1) actively elicit the customer's security and privacy requirements and (2) communicate these requirements to the entire development team because the implications of these special nonfunctional requirements potentially impact each and every story or feature. Specific customer acceptance test cases need to be developed to verify that security and privacy requirements have been met. These test cases should validate that security cannot be violated and that systems respond to such attacks gracefully. These security and privacy requirements might also necessitate the need for more up-front architecture or high-level design than is often done with agile methods.

Pair programming could be used to spread privacy and security knowledge throughout the development team. In this scenario, a programmer could be specially trained on dealing with security and privacy in software development. This programmer could pair program with different team members each day. In doing this, the security and privacy programmers could intimately review the details of each programmer's work and could impart their specialized knowledge throughout the team. The integration of security and privacy practices within agile software development is an active research area.

43.5.5 High Availability

> *Agile methods work less well for critical, reliable, and safe systems. Agile methods fit applications that can be built quickly and don't require extensive quality assurance. Agile methods work less well for critical, reliable, and safe systems.* [Boehm, 2002].

When an Internet application is mission-critical to an organization, it needs to be highly available. Downtime on the part of an Internet-based company, such as Amazon, can have dire financial implications. Many agilists disagree with Boehm's statement above, though it should be taken as a cautionary message. Crystal Methods author Alistair Cockburn has stated[4] that large projects (with more than 20 people) with essential-money criticality need to adapt agile methods. Similar to security and privacy, availability and reliability must be important nonfunctional requirements elicited from the customer early in the cycle.

Extreme Programming has testing practices that are superior to the state-of-the-practice in many software organizations [George, 2002; George and Williams, 2003; Maximilien and Williams, 2003; Williams et al., 2003; George and Williams, 2003, in press]. Through the TDD practice, extensive automated unit and functional testing is done throughout the entire development cycle. This practice may be able to be composed with some Software Reliability Engineering practices [Musa, 1999] to lead to a system of high availability and reliability [Williams et al., 2002]. The integration of reliability practices and agile software development is an active research area.

43.6 Research Issues and Summary

Internet application development has several complicating factors that distinguish it from traditional software development. These complicating factors are (1) intense requirement and technology volatility; (2) the need for collaboration between team members with various specialties; (3) increased privacy and security concerns; (4) the need for highly available systems. In combination, these factors make traditional software development models problematic. Emerging agile methods have been shown to successfully address the first two concerns. Adaptations to current agile methods are suggested in this chapter to

[4] In keynote presentation of XP/Agile Universe 2002 conference in Chicago, Illinois.

address the last two factors. Active research in this area is also underway. Risk analysis can be used to determine if a product would best be developed by a plan-driven or by an agile software development methodology. Alternatively, risk analysis can be used for structuring a hybrid methodology that contains both agile and plan-driven practices [Boehm and Turner, June 2003].

References

Agile (1996). Agile Competition is Spreading to the World. http://www.ie.lehigh.edu/.

Auer, Ken and Roy Miller (2001). *XP Applied*. Addison-Wesley, Reading, MA.

Basili, Victor R. and Albert J. Turner (1975). Iterative enhancement: a practical technique for software development. *IEEE Transactions on Software Engineering*, 1(4).

Beck, Kent (2000). *Extreme Programming Explained: Embrace Change*. Addison-Wesley, Reading, MA.

Beck, Kent (2003). *Test Driven Development — by Example*. Addison Wesley, Reading, MA.

Beck, Kent, Mike Beedle et al. (2001). The Agile Manifesto. http://www.agileAlliance.org.

Beck, Kent and Martin Fowler (2001). *Planning Extreme Programming*. Addison-Wesley, Reading, MA.

Bellin, David and Susan S. Simone (1997). *The CRC Card Book*. Addison-Wesley, Reading, MA.

Boehm, Barry (2002). Get ready for agile methods, with care. *IEEE Computer*, 35(1): pp. 64–69.

Boehm, Barry and Richard Turner (June 2003). Using risk to balance agile and plan-driven methods. *IEEE Computer*, 36(6): pp. 57–66.

Coad, Peter, Jeff deLuca et al. (1999). *Java Modeling in Color with UML*. Prentice Hall, Upper Saddle River, NJ.

Cockburn, Alistair (2001). *Agile Software Development*. Addison-Wesley Longman, Reading, MA.

Crocker, Ron (2004). *Large-Scale Agile Software Development*. Addison-Wesley, Reading, MA.

Fowler, Martin, Kent Beck et al. (1999). *Refactoring: Improving the Design of Existing Code*. Addison-Wesley, Reading, MA.

Fowler, Martin and Jim Highsmith (August 2001). The agile manifesto. *Software Development*, pp. 28–32.

George, B. (2002). Analysis and Quantification of Test Driven Development Approach. M.S. Thesis, Computer Science. North Carolina State University, Raleigh, NC.

George, B. and L. Williams (2003a). An Initial Investigation of Test-Driven Development in Industry. *ACM Symposium on Applied Computing*, Melbourne, FL.

George, B. and L. Williams (2003b). Structured experiments of test-driven development. *Information and Software Technology (IST)*, 4 6(5): pp. 337–342.

Ghezzi, Carlo, Mehdi Jazayeri et al. (1991). *Fundamentals of Software Engineering*. Prentice Hall, Upper Saddle River, NJ.

Gibbs, W. Wayne (1994). Software's chronic crisis. *Scientific American*, pp. 86–95.

Hamel, Gary (2002). *Leading the Revolution: How to Thrive in Turbulent Times by Making Innovation a Way of Life*, Plume, New York.

Highsmith, James (1999). *Adaptive Software Development*. Dorset House, New York.

Highsmith, Jim (2002). *Agile Software Development Ecosystems*. Addison-Wesley, Reading, MA.

Jacobson, Ivar, Grady Booch et al. (1999). *The Unified Software Development Process*. Addison-Wesley, Reading, MA.

Jeffries, Ron, Ann Anderson et al. (2001). *Extreme Programming Installed*. Addison-Wesley, Reading, MA.

Larman, Craig and Victor Basili (June 2003). A history of iterative and incremental development. *IEEE Computer*, 36(6): pp. 47–56.

MacCormack, Alan (Winter 2001). How Internet companies build software. *MIT Sloan Management Review*, pp. 75–84.

MacCormack, Alan, Roberto Verganti et al. (January 2001). Developing products on "Internet Time": the anatomy of a flexible development process. *Management Science*, 47(1): pp. 133–150.

Marchesi, Michele and Giancarlo Succi, Eds. (2001). *Extreme Programming Examined*. In The XP Series. Addison-Wesley, Reading, MA.

Marchesi, Michele, Giancarlo Succi et al. Eds. (2002). *Extreme Programming Perspectives.* In The XP Series. Addison-Wesley, Reading, MA.

Maximilien, E. Michael and Laurie Williams (2003). Assessing Test-driven Development at IBM. International Conference of Software Engineering, Portland, OR.

Mills, Harlan D. (1976). Software development. *IEEE Transactions on Software Engineering,* 2(4): pp. 265–273.

Musa, John D. (1999). *Software Reliability Engineering.* McGraw-Hill, New York.

Ogunnaike, Babatunde A. and W. Harmon Ray (1994). *Process Dynamics, Modeling, and Conrol.* Oxford University Press, New York.

Petroski, Henry (1994). *The Evolution of Useful Things.* Vintage Books, New York.

Royce, Winston W. (1970). Managing the development of large software systems: concepts and techniques. IEEE WESTCON, Los Angeles, CA.

Schwaber, Ken (2001). Will the Real Agile Process Please Stand Up. *E-Project Management Advisory Service, Cutter Consortium* 2(8).

Schwaber, Ken and Mike Beedle (2002). *Agile Software Development with SCRUM.* Prentice Hall, Upper Saddle River, NJ.

Stapleton, Jennifer (1997). *DSDM: The Method in Practice.* Addison-Wesley Longman, Reading, MA.

Viega, John and Gary McGraw (2002). *Building Secure Software.* Addison-Wesley, Reading, MA.

Wells, Don and Laurie Williams, Eds. (2002). *Extreme Programming and Agile Methods — XP/Agile Universe 2002. Lecture Notes in Computer Science.* Springer-Verlag, Berlin.

Williams, Laurie, Robert Kessler et al. (2000). Strengthening the case for pair-programming. *IEEE Software,* 17: pp. 19–25.

Williams, Laurie, E. Michael Maximilien et al. (2003). Test-Driven Development as a Defect-Reduction Practice. IEEE International Symposium on Software Reliability Engineering, IEEE Computer Society, Denver, CO.

Williams, Laurie, Lili Wang et al. (2002). "Good Enough" Reliability for Extreme Programming. Fast Abstract, International Symposium on Software Reliability Engineering, Annapolis, MD.

44

Website Usability Engineering

CONTENTS

Abstract 44-1
44.1 Moving toward a Usable WWW 44-1
44.2 Website Considerations 44-2
44.2.1 Users 44-2
44.2.2 Computing Devices 44-3
44.2.3 Implementation Technology 44-3
44.2.4 In Summary 44-4
44.3 Usability Engineering Process 44-4
44.4 Discovery Phase 44-5
44.4.1 Assessing Users' Needs 44-5
44.4.2 Documenting Users' Needs 44-6
44.5 Design Exploration Phase 44-7
44.5.1 Information Design 44-7
44.5.2 Interaction Design 44-7
44.6 Design Refinement Phase 44-9
44.7 Production Phase 44-10
44.7.1 High-Fidelity Testing 44-10
44.7.2 Automated Assessment 44-11
44.8 Quality Assurance Phase 44-12
44.9 Maintenance Phase 44-12
44.10 Participatory Website Design 44-12
44.11 Summary and Research Issues 44-13
References 44-14

Melody Y. Ivory

Abstract

The World Wide Web enables broad dissemination of information and services, yet most sites have inadequate usability and accessibility. Creating an effective Website is a complex task that requires practitioners to follow a well-defined process. The Website usability engineering process and software tools that can support practitioners in following this process are described. Research directions that should produce better tools to support practitioners are proposed.

44.1 Moving toward a Usable WWW

The WWW has become the predominate means for communicating information on a broad scale. Unfortunately, despite the abundance of design recommendations, recipes, and guidelines for building effective sites, Website usability and accessibility continue to be pressing problems. The current state of

1-58488-381-2/05/$0.00+$1.50
© 2005 by CRC Press LLC

the WWW is attributable mostly to the fact that many people who build sites are trained in the technical aspects of Website creation (i.e., HTML coding), but they are not trained in usability engineering or ergonomic factors. Compounding this problem is the fact that many guidelines and texts offer prescriptive guidance that is often voluminous, vague, conflicting, or divorced from the context in which sites are being designed, which makes such guidance difficult to apply. Furthermore, many of the guidelines have not been empirically validated; hence, conforming to them may or may not improve Website quality.

Our intent in this chapter is to discuss the way in which effective Websites are designed and developed. Whether we are aware of it or not, we all follow a process when we design and develop Websites. However, we may not follow a process that will produce sites that are usable and accessible. A process that incorporates usability engineering at every stage will produce a high-quality site. As more sites are developed in this manner, we will move toward a usable WWW.

This chapter begins with a discussion of Website considerations of which every practitioner needs to be aware: specifically, user abilities, computing devices, and Website implementation technologies. We then describe the Website design and development process, and illustrate how usability engineering fits within it. Subsequent sections describe each process phase — discovery, design exploration, design refinement, production, quality assurance, and maintenance. An example class Website is used in our discussion of these phases, and we highlight the software tools that are available to support practitioners. Participatory design approaches that the practitioner can use to expedite the entire process are described. The chapter concludes with a summary and a discussion of promising research areas.

44.2 Website Considerations

Every Website development effort should produce a site that is universally usable [Shneiderman, 2002]. Universal usability refers to both a site's usability or "the extent to which a [site] can be used by specified users, to achieve specified goals, with effectiveness, efficiency and satisfaction, in a specified context of use" [International Organisation for Standardisation, 1998], and its accessibility or "usability of a [site] by people with the widest range of capabilities" [International Organisation for Standardisation, 2002]. There are three primary considerations while working toward accomplishing this goal: the users, computing devices, and implementation technology. Unfortunately, they represent moving targets. We discuss trends and key considerations for practitioners in this section.

44.2.1 Users

Early Website developers were privileged to cater to a small, well-defined user population — primarily technology-savvy researchers [Berners-Lee et al., 1992]. Now, there is a broad range of users online, people who differ with respect to age, education, computer expertise, and physical and cognitive abilities [World Wide Web Consortium, 2001]. Although there is frequent mention of the need to support blind users (e.g., by specifying alternative text for images), there is still a broader range of user abilities that needs to be supported [Ivory et al., 2003]. For example, at least 8% of Web users have an impairment and at least 9% are 55 or older[1] [GVU Center, 1998]; the elderly are more likely to have multiple impairments [Paciello and Paciello, 2000]. We also need to consider that Website use is a two-way exchange; users need to be able to get information from the site and to send information to it. There are many ways to access information on a site (e.g., via a screen reader or a textual browser), but support for sending information lags behind [Ivory et al., 2003]. We discuss some limitations below:

Mouse Use: Some users have little or no ability to use a mouse. So they need to use alternative input devices to send information to a site (e.g., a keyboard or a Braille device). Website use is even more complex for users who have severe mobility or other impairments that make it impossible for them to use a mouse and a keyboard [Ivory et al., 2003].

[1] The disability and age statistics are from a 1998 survey; current Web user demographics are difficult to find.

Keyboard Use: If users are unable to use a mouse due to physical impairments, they may not be able to use a keyboard either [Ivory et al., 2003]. Typically, users can employ speech input rather than a keyboard, Braille device, or mouse to send information to the site. However, speech recognition is still slow and error-prone, especially for users with speech impairments [Christian et al., 2000]. Users may also consider speech input to be awkward.

Vision: There has been substantial consideration to supporting users who have little or no eyesight. For example, there are tools to read Web page contents to users and automated tools to help practitioners ensure that their Web pages provide textual equivalents for images, sound, and other objects [Ivory et al., 2003]. Nonetheless, sending information to the site (e.g., completing forms and navigating among frames) remains a problem for users with vision impairments.

Hearing: If a user with a hearing impairment also suffers from a visual or other impairment, then he will experience similar difficulties in sending information to a Website. Multimedia poses a big challenge for users with hearing impairments. These users need textual equivalents for audio, video, and other multimedia elements. Furthermore, textual equivalents need to be synchronized with the original sources to facilitate comprehension.

Cognition: Some users have attention deficits and reading or other comprehension impairments that impede their ability to access and send information to a Website. For example, animated images may make it difficult for these users to concentrate on the information presented on a Web page. As another example, the level at which the content is written within a site may make it hard for some users to understand the text or to complete forms.

Some issues may not be easy to address at design time, and are perhaps more appropriately addressed by operating-system support or automated-transformation tools [Ivory et al., 2003]. Nonetheless, the practitioner needs to be aware of these varying user abilities. Practitioners can establish requirements for supporting users with diverse abilities early in the design process (see "Discovery Phase," Section 44.4), and make design and implementation decisions to ensure that the site satisfies the requirements.

44.2.2 Computing Devices

We cannot assume that users are accessing Websites via a desktop computer with a specific Web browser, an Internet connection of reasonable speed, and a 15-in. monitor, anymore. Users are accessing sites via a broad range of Web browsers and with different versions, including textual browsers and screen readers [Nielsen, 2000, Chapter 2]. Furthermore, they are using a broad range of low-bandwidth, small-screen mobile devices (e.g., cellular phones and PDAs), large-screen devices (e.g., MSN and AOL TV), and tablet and pad devices (e.g., tablet PCs) [Nielsen, 2001; Pearrow, 2002; Waterson et al., 2002]. The way in which users access and send information to sites with these devices is considerably different from how these tasks are accomplished with desktop computers. For example, users may only be able to access a small portion of a page at one time, and they may have to use a limited set of buttons to send information to the site. Practitioners need to consider these differences in computing devices; requirements can be established early in the design process, and used to inform design and implementation decisions.

44.2.3 Implementation Technology

Given the broad range of users and computing devices, practitioners need to carefully choose an implementation technology. For example, if we decide to use Macromedia Flash to develop a site, then users with certain screen readers or browsers will have only limited, if any, access to the site. Although Macromedia has made it possible for users to access Flash content under certain conditions[2] [Macromedia, 2002], such support does not exist for most non-HTML technologies.

[2] At publication time, users must use Flash Player 6, the Internet Explorer browser, and the Window-Eyes screen reader to access Macromedia Flash content.

44.2.4 In Summary

Website users today are not the same as those of yesterday. They have different abilities and use a range of computing devices. As practitioners, we need to consider these differences and choose implementation technologies that promote usability and accessibility. Following a usability engineering process is essential to accomplish this objective. By establishing requirements to address these issues early in the design process, and by using these requirements to inform design and implementation decisions, we can produce sites that better support today's Website users.

44.3 Usability Engineering Process

Creating a Website that is universally usable is a far more complex task than most new (and some experienced) developers realize. It may be straightforward to code HTML pages, but creating an effective site requires us to follow a well-defined process. To gain insight about the process of creating sites, Newman and Landay [2000] conducted an ethnographic study of professional Website designers. They discovered that designers go through several design iterations with paper sketches and create different representations of a site: site maps depicting the site's organization, storyboards depicting task-completion steps, and schematics illustrating page layouts. Other literature sources [e.g., De Troyer and Leune, 1998; Ford and Boyarski, 1997; Fuccella, 1997; Heller and Rivers, 1996; Nielsen, 2000; Sano, 1996; Shneiderman, 1997] describe similar Website design and development processes. The circles in Figure 44.1 summarize the phases that are often mentioned.

Site requirements are identified during the discovery phase; requirements need to address many issues, such as who the intended users are, what tasks they need to accomplish, what constraints need to be met, and the considerations discussed in the preceding section. Requirements and other insights from the discovery phase need to inform decisions throughout subsequent design and development phases.

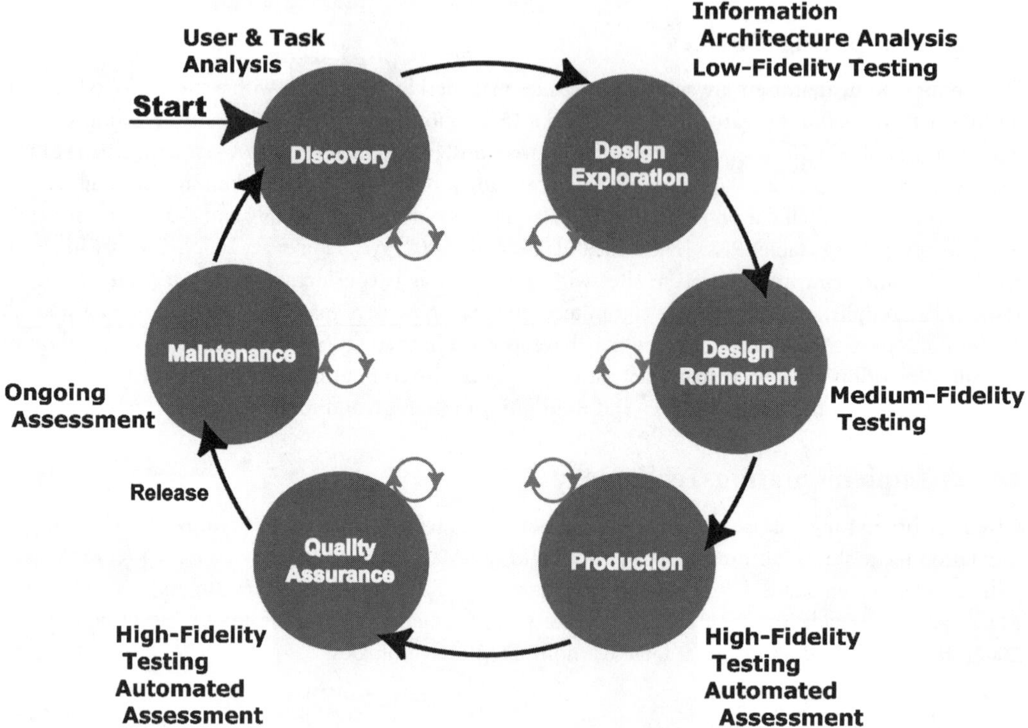

FIGURE 44.1 Website design and development process, which incorporates usability engineering.

Several design alternatives may be explored during the design exploration phase. However, one design idea is chosen and improved upon during the design refinement phase. Site implementation typically occurs during the production phase; design templates or style guides may be developed initially to guide the implementation effort. The implemented site is tested and repaired, if necessary, during the quality assurance phase; afterwards, it is released for broad use. Once the site is released, it undergoes continuous maintenance (e.g., updating content). Iteration may occur during each of these phases, as depicted in Figure 44.1.

Although we discussed the outcome of each phase, we did not discuss how these outcomes are produced; specifically, how a practitioner goes about developing a site that is universally usable. The annotations next to the phases in Figure 44.1 provide some insight; they demonstrate how usability engineering can be incorporated into the design and development process. The remainder of this chapter discusses this expanded process and emphasizes the usability engineering tasks that need to be completed and the tools that are available to support practitioners in completing them.

44.4 Discovery Phase

The discovery phase is the most important phase in the Website design and development process. Assuming that we begin this phase with a clear purpose for the site (e.g., to support motor vehicle or class registration), the discovery phase enables us to understand the intended users, the tasks they need to accomplish, the information they need to access, and the technology they will use. (It also enables us to discover the client's needs, if we are developing a site for a client.) These insights need to be represented as a list of tasks to be supported, usage scenarios, personas, work artifacts, etc., so that it is available to inform design and development decisions in subsequent phases.

44.4.1 Assessing Users' Needs

We discover information to guide Website development by conducting an initial competitive analysis and a detailed user and task analysis, rather than by making assumptions. A competitive analysis involves reviewing sites that offer functionality that is similar to the one being developed; this analysis may provide preliminary information about the types of users and tasks that need to be supported. The competitive analysis also involves reviewing the existing version of the site, if it is undergoing revision or being redesigned. For example, an analysis of user feedback or usage patterns captured in Web server logs may reveal ways in which the site can be improved.

Several literature sources discuss procedures for the user and task analysis in great detail [see Badre, 2002; Beyer and Holtzblatt, 1998; Brinck et al., 2001; Cato, 2001; van Duyne et al., 2002; Mayhew, 1999; Nielsen, 1993; Rosson and Carroll, 2002]; we summarize the key steps below:

1. **Identify intended users.** We need to determine for whom the site is being developed (e.g., vehicle owners or college students). We then need to identify a representative set of users or participants to include in our user and task analysis. As discussed in Section 44.2, we need to make sure that this set of users reflects the broad range of user abilities and computing technologies that need to be supported.
2. **Choose analysis methodology.** We need to choose one or more methodologies for conducting the analysis. Table 44.1 provides descriptions of inquiry methods, such as surveys, questionnaires, interviews, focus groups, and field observations that are often used. Each method has its own advantages, disadvantages, and requirements [see Fuccella and Pizzolato, 1998, for a discussion]. For instance, a questionnaire can be relatively easy to create and administer to a large number of users via the WWW, but it does not provide insight that can be gained only by observing users in their natural environments. Contextual inquiry is a systematic and complex inquiry method that entails observing users in their environments, but it is difficult to employ, and requires considerable time and resources [Beyer and Holtzblatt, 1998]. The Usability Evaluation Methods

TABLE 44.1 Inquiry Methods for Conducting a User and Task Analysis

Inquiry Method	Description
Questionnaires	User provides answers to specific questions
Surveys	Interviewer asks user specific questions
Interviews	One user participates in a discussion session
Focus groups	Multiple users participate in a discussion session
Field observation	Interviewer observes system use in user's environment
Contextual inquiry	Interviewer questions users in their environment

Note: Methods are ordered by increased complexity and relative cost. The ordering also reflects a potential decrease in the number of users who can be studied.

[Zhang, 1999b] and the Usability Methods Toolbox [Hom, 1998] are Web resources that describe a range of inquiry methods. In addition, there is an advisory tool, Ask Usability Advisor, that can recommend methods to use based on specified requirements [Zhang, 1999a].

3. **Employ analysis methodology.** We need to employ the selected methodology with the identified participants. The data collected depends on the method employed. For example, we may have responses to a questionnaire or notes from field observations. There are several Web tools for conducting questionnaires. For example, NetRaker's usability research tools enable practitioners to create custom HTML questionnaires via a template interface; it also provides a graphical summary of results even while studies are in progress [NetRaker, 2000]. Sinha [2000] compiled a comprehensive discussion and comparison of available tools.

44.4.2 Documenting Users' Needs

We need to analyze the collected data so that we can understand the site's intended users and their tasks (see Fuccella et al. [1998] for an in-depth example). Furthermore, we need to document this information so that we can use it to guide the remaining phases of the Website design and development process. A requirements or specification document, which details the site's functionality, constraints, and high-level goals, needs to be created. Personas (descriptions of users) and scenarios (descriptions of tasks) should also be created during this phase [Cooper, 1999]. Personas and scenarios should reflect the broad range of user abilities and computing devices previously mentioned. Figure 44.2 depicts an example persona for the class Website; an example scenario is below.

> Example scenario: Allen is frantically trying to prepare for his database class, and he has only an hour to do so. Allen loads the class Website. He checks the schedule to find out what is happening this week. He discovers that Lab Assignment #1 is due. Although he did the assignment earlier, he forgot to upload it. Allen submits this assignment for grading. He reviews the lecture materials and has a hard time understanding the normalization concept. Allen decides that he needs to get some extra help. He posts a message to see if any of his peers can respond quickly. Allen did not receive a timely response, so he decides to send a message to the instructor. The instructor promptly responds, and Allen is able to complete his preparation for class.

In addition to requirements, personas, and scenarios, there are various other documents that a practitioner could create like an affinity diagram (hierarchical diagram for representing findings across users)

Allen is a second year undergraduate student in the Informatics program. He is a good student, gets frustrated easily, and is not very comfortable with technology. Allen is taking three demanding classes this quarter, and he has an attention deficit disorder.

FIGURE 44.2 Example persona. The persona describes a student who might use the class Website. A picture is included to add realism to the persona.

or work models (flow diagrams for depicting interactions between user roles to accomplish tasks) [Beyer and Holtzblatt, 1998]. Regardless of the design documents produced, the objective is to be able to use these artifacts to inform design and development decisions. For example, a requirements document may suggest the tasks to support, and scenarios may suggest one or more ways to implement task completion sequences. User personas may suggest effective ways to organize content or provide functionality on the site or how to layout pages so that they can be used by users with diverse abilities. Several texts discuss how to create and use such artifacts systematically throughout the design process [Beyer and Holtzblatt, 1998; Mayhew, 1999; Rosson and Carroll, 2002].

44.5 Design Exploration Phase

Design activities are considered situations of problem solving in cognitive psychology [Malhotra et al., 1980] because the objective is to produce a design that fits a specific function — supporting the intended users and tasks identified during the discovery phase — while satisfying different requirements. There is typically more than one design that can satisfy requirements; thus, during the design exploration phase, we may develop several design alternatives for consideration. The focus is on both the information design and the interaction design in this phase.

44.5.1 Information Design

Information design entails determining how to organize information within the site and how to enable users to browse and search this information [Fleming, 1998; Rosenfeld and Morville, 1998]; this organization and structure is referred to as the *information architecture.* Category matching and card sorting are two analysis techniques for informing these decisions. For a category matching study, we analyze artifacts from the discovery phase to propose the site's major content areas or categories and to identify representative content items for each category. We then give users the list of categories and content items and ask them to match each content item to the category in which they would expect to find the item. By analyzing content assignments across multiple users, we can determine whether or not the proposed information organization is aligned with users' expectations. For example, if 80% of users assigned each content item to the proposed category, then we can conclude that the proposed information organization is likely to be usable. If not, then the results of the category matching study will reveal areas for improvement. Fuccella et al. [1998] provide guidance for conducting a category matching study, and the NIST Web Category Analysis Tool enables practitioners to conduct online studies [National Institute of Standards and Technology, 2002].

For a card sorting study, we identify representative content items and ask users to group them in a logical manner, rather than proposing categories. We then use statistical cluster analysis to analyze groupings across users [Martin, 1999]; this analysis will provide insight on an optimal information organization. Practitioners can use the IBM USort and EZCalc tools to conduct a card sorting study [Dong et al., 2001]. The USort program enables participants to sort virtual cards, instead of using physical cards (Figure 44.3). It also allows participants to specify names for each category. The EZCalc program performs cluster analyses on card sorting results. EZCalc generates tree diagrams that allow direct adjustment of the cluster thresholds (Figure 44.4); by adjusting cluster thresholds, practitioners can explore alternative information organizations. It may be beneficial to evaluate the derived information organization with a category matching study to verify that the category names are meaningful to users other than study participants.

44.5.2 Interaction Design

During the design exploration phase, interaction design involves exploring task flows and basic page layouts to support them, rather than choosing the specific images, colors, or fonts to be used [Newman and Landay, 2000]. Multiple design concepts may be generated, and practitioners can test these design

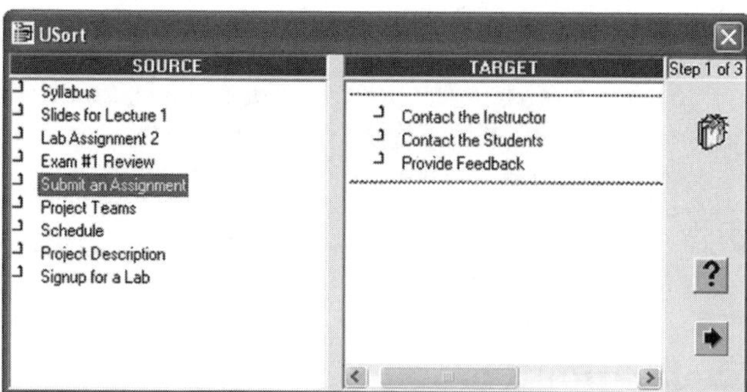

FIGURE 44.3 Example card sorting exercise. We used IBM USort to demonstrate how a student might group content for the class Website. Students can drag content items from the left side of the window to create groups on the right side of the window.

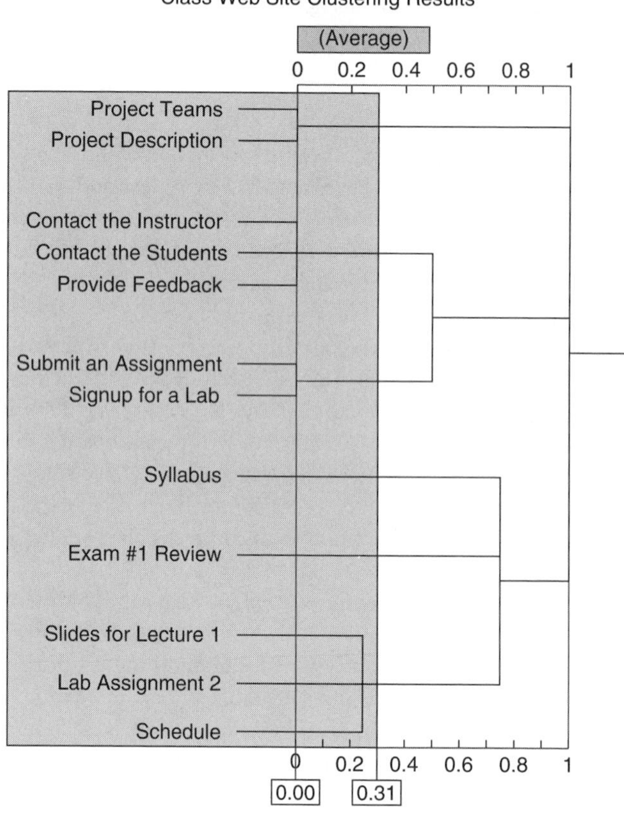

FIGURE 44.4 Example clustering results for the card sorting exercise. We used IBM EZCalc to do cluster analysis for two groupings. The clusters on the left show content items that the students grouped together most frequently. These groupings could be used for the main content categories in the class Website.

concepts with representative users. We refer to this level of testing as *low-fidelity testing* because it entails testing unfinished, rough prototypes (e.g., paper sketches or images; see Figure 44.5).[3] Test results will

[3] We discuss testing Websites at the low-, medium-, and high-fidelity levels; the levels are based on the degree to which sites or page layouts are developed. For instance, the tested site may consist of several paper sketches (low fidelity), a PowerPoint prototype (medium fidelity), or a functional HTML implementation (high fidelity).

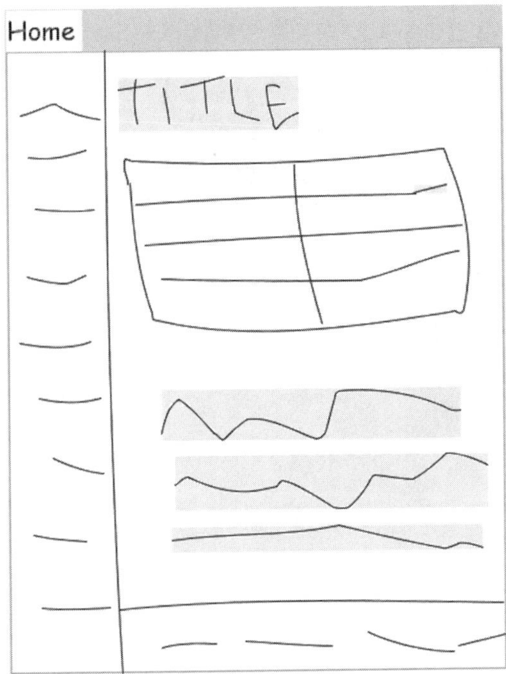

FIGURE 44.5 Example sketch of a Web page. The sketch depicts the home page for the class Website. Sketches similar to the one depicted can be used in low-fidelity testing. We used the DENIM tool to create this sketch.

enable practitioners to gauge user acceptance early in the design process. Practitioners can use tools, such as NetRaker and others, to conduct low-fidelity tests [Sinha, 2000].

Quinn [2002] and Farnum [2002] discuss low-fidelity testing in detail. Hong et al. [2001b] and Walker et al. [2002] have conducted empirical studies to document the effectivness of low-fidelity testing. These studies show that it is possible to identify the same types of usability issues with low-fidelity representations as with high-fidelity ones (i.e., HTML pages), The studies also show that participants tend to focus on low-level issues, such as colors and fonts, more so with high-fidelity representations than with low-fidelity ones.

44.6 Design Refinement Phase

After exploring and evaluating alternative information and interaction designs, we need to choose a design path and to continue developing it. For example, we may need to refine the labels used for content categories or merge design concepts from multiple page layouts. We also need to integrate the information and interaction designs; these merged representations are typically referred to as schematics [Newman and Landay, 2000]. Other useful design artifacts to develop include: site maps to demonstrate site structure and storyboards to demonstrate task completion steps. DENIM is a sketch-based tool for constructing site maps, storyboards, and schematics for a site [Lin et al., 1999]. Figure 44.6 depicts example design artifacts created with DENIM.

Similarly to the preceding phase, *medium-fidelity testing* should be conducted to inform design refinements. Testing can examine either a broad set of features with little depth (i.e., horizontal prototype) or a narrow set of features in depth (i.e., vertical prototype) [Nielsen, 1993]. Rather than evaluating design concepts, the testing should evaluate task flows that are appropriate for the prototype. Furthermore, the evaluated design should reflect the integrated information and interaction design. Test results should reveal potential usability issues that can be addressed before moving on to later design phases.

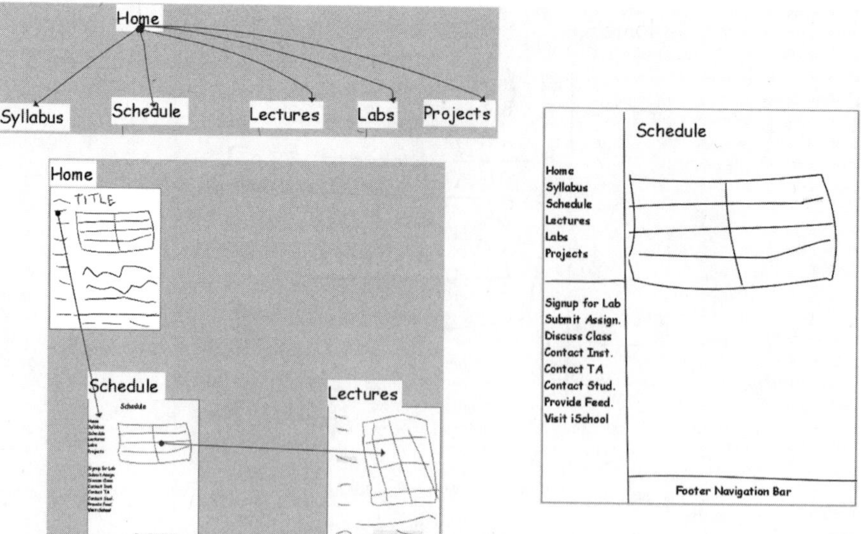

FIGURE 44.6 Example site map (top left), storyboard (bottom left), and schematic (right). The site map shows the major content areas for the class Website. The storyboard depicts a task wherein a student is using the site to prepare for class; the student starts at the home page, views the schedule to get an overview of activities for the week, and then explores the lecture materials. The schematic depicts the layout of the schedule page, We used the DENIM tool to create these design artifacts.

Practitioners can use tools such as NetRaker and others to conduct medium-fidelity tests [Sinha, 2000]. The DENIM tool has a run mode that enables participants to simulate Website use, making it possible to test Website representations. DENIM can also generate an HTML version of a design for testing.

44.7 Production Phase

Site development occurs during the production phase. We may begin this phase by translating the design artifacts (site maps, schematics, and storyboards) into a design style guide or Web page templates, that can promote consistency during site implementation. We may also produce mock-ups or a high-fidelity prototype for the major content areas or functional elements; see Figure 44.7 for an example. As we progress through this phase, the key usability engineering activities include testing the high-fidelity prototype or mock-ups and using automated assessment tools to identify and mitigate common usability, accessibility, coding, and other issues that they can detect.

44.7.1 High-Fidelity Testing

It would be beneficial to conduct *high-fidelity testing* with the mock-ups, the prototype, or the partially implemented site during this phase. Such testing should identify potential problems prior to the completion of site development, when it is easier to make changes. It is important to include users with a broad range of abilities and computing devices during this phase; Coyne and Nielsen [2001] provide guidance on how to conduct tests with impaired users. NetRaker and other online tools may be used to conduct tests. For example, NetRaker makes it possible to periodically capture the contents of participants' computer screens and to chat with them during the testing session. It also makes it possible to capture click stream data.

There are several publicly available tools for capturing click stream data during a usability test. WebQuilt uses proxy-based logging to capture usage data [Hong et al., 2001a]. The system automatically captures Web server requests using a special proxy server, logs requests, and subsequently routes requests

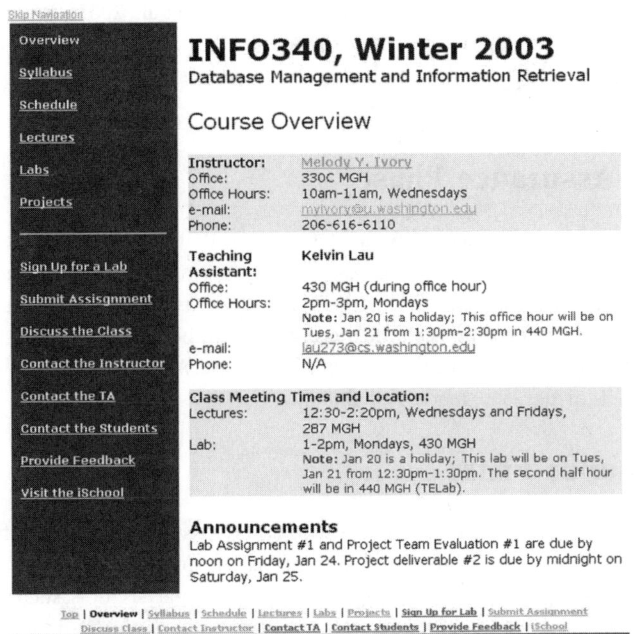

FIGURE 44.7 Example mock-up of a Web page. The mock-up is an HTML version of the home page for the class Website.

to the Web server. All of the links in Web pages are also redirected to the proxy server; this automated redirection eliminates the need for users to manually configure their browsers to route requests to the proxy. The system captures more accurate site usage details (e.g., use of the back button) than server-side logging. WebQuilt makes it possible to track participants individually because extra information like the participant's identifier can be encoded in the study URL. Although the system does not capture many client-side details, such as the use of page elements or window resizing, it does simplify instrumenting a site for logging because it is done automatically. The WebQuilt system also supports task-based analysis and visualization of captured usage data.

The NIST WebMetrics tool suite captures client-side usage data [Scholtz and Laskowski, 1998]. This suite includes WebVIP (Web Variable Instrumentor Program), a visual tool that enables the evaluator to specify the types of user events to log (e.g., link selection, use of back button, or window resizing). These choices are converted to JavaScripts which are then inserted into each page of the site after it is copied. This code automatically records the page identifier and a time stamp in an ASCII file every time a user selects a link. (This package also includes a visualization tool, VISVIP, for viewing logs collected with WebVIP [Cugini and Scholtz, 1999].) Using this client-side data, the evaluator can accurately measure time spent on tasks or particular pages as well as study use of the back button and click stream data. The tool cannot be used in sessions wherein a screen reader, textual browser, or other assistive technology that does not support scripts is employed; WebQuilt is better suited for these cases.

44.7.2 Automated Assessment

Once we have HTML versions of Web pages or the entire site, there is a broad range of automated evaluation tools available for use. Some tools, such as UsableNet's LIFT [UsableNet, 2002] and the 508 Accessibility Suite [Macromedia, 2001], are embedded within HTML authoring environments like Microsoft FrontPage and Macromedia Dreamweaver. Thus, it is easy to use these embedded tools throughout the implementation process. There are also over 30 stand-alone and Web-based evaluation tools (see [Ivory, 2001, Chapter 2] and [Ivory et al., 2003] for discussions). Automated evaluation tools may help practitioners to learn about effective design practices or to ensure that sites conform to these practices, but the tools should not be used as a substitute for testing [Ivory et al., 2003]. In general, the tools focus on low-level details (e.g., use of alternative text for images and sound files or use of adequate font sizes), and do not address high-level issues, such as whether the content is comprehensible or users can complete certain tasks efficiently.

44.8 Quality Assurance Phase

The quality assurance phase is concerned with making sure that all Website elements are functioning properly before the site is released. In most cases, usability testing is done only during this phase, if at all. High-fidelity testing on the fully implemented site can reveal usability problems and inform improvements. As discussed in the preceding section, there are a number of usability testing and automated evaluation tools that can support practitioners during this phase. After the site clears all quality assurance tests, it is then released for use (i.e., made available on-line).

44.9 Maintenance Phase

After the site is released, we need to continuously gauge users' satisfaction and identify aspects to enhance during future site revisions. Research study and questionnaire tools (see Section 44.2.4) make it possible to gather feedback from site users. It is also possible to use log file analysis tools, such as NetIQ WebTrends, to analyze server log files [NetIQ, 2002]. Such tools use statistical approaches to analyze site traffic (e.g., pages-per-visitor or visitors-per-page) and usage patterns (e.g., click streams and page-view durations). Statistical analysis is largely inconclusive for Web server logs because they provide only a partial trace of user behavior and timing estimates may be skewed by network latencies. Server log files are also missing valuable information about what tasks users want to accomplish. Nonetheless, statistical analysis techniques have been useful for improving usability and enable ongoing, cost-effective evaluation after a site is released.

44.10 Participatory Website Design

If we incorporated usability engineering into every phase of the Website development process, as proposed in this chapter, this process may be slowed down considerably. One way to streamline and expedite it is to employ a participatory design [Schuler and Namioka, 1993] or joint application design/development (JAD) [Wood and Silver, 1995] approach. Essentially, representatives from each group of stakeholders (users, developers, clients, designers, etc.) collaborate to produce solutions during each design phase. All stakeholders participate in a series of interactive working sessions or workshops wherein design aspects are explored and simultaneously evaluated. Remote meeting or group support tools, such as Meetingworks [Meetingworks, 2002] and GroupSystems [GroupSystems, 2002], are available to help facilitate the participatory site design process; see Leventhal [1995] for a discussion.

The participatory or JAD approach is in contrast to the typical design-prototype-evaluate approach, which involves exploring design ideas, creating a prototype to reflect those ideas, and then evaluating the prototype with users [van Duyne et al., 2002, Chapter 5]. The major disadvantage of the latter approach is that practitioners may not collect all the information that they need during the discovery phase; thus, some design decisions will be best guesses. Having representative users in design sessions will eliminate guessing; the users may also raise issues that practitioners may have overlooked. Practitioners need to understand that users should not function as designers in participatory design sessions. Their role should be to provide feedback on design proposals that are presented in a format that they can understand (i.e., paper sketches or HTML pages rather than requirements).

Participatory design approaches are most appropriate for complex sites, especially sites for which the user community or tasks are not well understood. For instance, it would not be appropriate to design the example course Website using this approach. Complex sites or sites to support a novel purpose may raise numerous issues that can only be addressed by having users closely involved in the design and development process. Depending on the breadth of the user community, it may be necessary to periodically replace user participants to get fresh perspectives on the site.

44.11 Summary and Research Issues

Website design and development is not as straightforward as many novice and some professional practitioners may think. What complicates the process is the fact that Website users have different abilities and use a range of computing devices. Thus, we need to consider these differences and choose implementation technologies that promote usability and accessibility.

We have outlined a process whereby practitioners can produce Websites that are universally usable. This process entails incorporating usability engineering into all phases of site design and development — discovery, design exploration, design refinement, production, quality assurance, and maintenance. We discussed the use of participatory design approaches as a way to expedite this process.

In addition to describing the design and development process, we emphasized the software tools that are available to assist practitioners during each phase. For example, there are questionnaire tools to facilitate the early design phases and automated evaluation tools to assist practitioners in the later phases. In general, existing design, evaluation, and authoring tools support practitioners during the later stages of design — production, quality assurance, and maintenance — when changes are more expensive and less likely to be made [Nielsen, 1993]. Furthermore, current support is inadequate and needs to be extended to better support practitioners. We propose several promising research directions to address these issues below.

Web design archive. During the design exploration and refinement stages, practitioners could use an archive of sites exhibiting good design practices (e.g., usability and accessibility) to identify pages and sites that satisfy one or more design criteria, such as effective navigation schemes within sites that provide information on health issues or good page layouts, color palettes, and site map designs. Currently, no such archive exists. Lin and Landay [2002] are developing a system called Damask to support practitioners in the early design stages, specifically for designing sites that are targeted at multiple devices. Damask is a sketch-based tool that will contain design patterns from van Duyne et al. [2002]. Some design patterns are represented as sketches that can be dragged into a site that is being designed. Damask is being incorporated into the DENIM tool.

Web design generator. Similarly to automated graphical interface generators [Balbo, 1995; Mackinlay, 1986; Sears, 1995], a site builder could be developed for informational as opposed to functional (i.e., applications) Websites. The site builder could simply require a practitioner to specify the site's content (in some raw format like an outline) and then generate an initial design. Ideally, the design tool would: (1) break content into logical units, (2) structure these units for readability, (3) determine an effective navigation structure, and (4) construct pages based on effective design practices.

Web design simulator. During design exploration and before sites are released, practitioners could use a simulation to provide objective measures for evaluating alternative designs and for improving the usability and accessibility of sites. The simulator could consider the effects of page and site design elements and employ user models representing a diverse set of users and devices, unlike existing simulators. Several researchers — Card and colleagues, Blackmon and colleagues, and Miller and Remington — attempt to simulate information-seeking behavior within an implemented site [Blackmon et al., 2002; Card et al., 2001; Chi et al., 2001; Miller and Remington, 2002]. These simulators model hypothetical users traversing the site from specified start pages, making use of information *scent* (i.e., common keywords between the user's goal and content on linked pages) to make navigation decisions. None of these approaches account for the effects of various Web page attributes, such as the amount of text or layout of links, on navigation behavior.

Universal accessibility checker. An automated tool that can evaluate early design representations and implemented sites, as well as modify them to conform to design guidelines is crucial for supporting practitioners. Ideally, this tool would evaluate designs based on multiple user models, as well as evaluate the design itself and the HTML code. WebTango researchers have developed a prototype to evaluate implemented sites and are working on tools to evaluate early designs [Ivory et al., 2001; Ivory, 2001; Ivory and Hearst, 2002a, b]. The WebTango approach entails deriving design guidelines (i.e., design patterns) from empirical data, which is in contrast to other guideline conformance tools like WatchFire Bobby and the W3C HTML Validator. The general approach involves: (1) identifying an exhaustive set of quantitative interface measures; (2) computing measures for a large sample of rated interfaces; (3) deriving statistical models from the measures and ratings; (4) using the models to predict ratings for new interfaces; and (5) validating model predictions. They have been able to develop predictive models with over 90% accuracy and have demonstrated that the models can improve designs.

Navigation pattern analyzer. Practitioners use log file analysis heavily for ongoing assessment of Websites; however, we are not aware of any research that analyzes navigation behavior to automatically identify patterns indicative of potential problems. This analysis is still largely a manual process, and practitioners could benefit from automated support. The previously mentioned simulation tools constitute a step in this direction. There has also been some work on mining usage patterns to determine ways to reorganize sites based on user preferences (for instance, see Büchner and Mulvenna [1998]; Chi et al. [2002]; Zaïane et al. [1998]).

Developing better training programs and tools to support practitioners is necessary to improve the overall quality of the Web. Existing and new tools should enable all practitioners — novices and professionals — to create usable and accessible sites.

References

Badre, Albert N. *Shaping Web Usability: Interaction Design in Context.* Boston: Addison-Wesley, 2002.

Balbo, Sandrine. Automatic evaluation of user interface usability: Dream or reality. In Sandrine Balbo, Ed., *Proceedings of the Queensland Computer-Human Interaction Symposium,* Queensland, Australia, 1995. Bond University.

Berners-Lee, T., R. Cailliau, J-F. Groff, and B. Pollermann. World-Wide Web: An information infrastructure for high-energy physics. In *Proceedings of the Workshop on Software Engineering, Artificial Intelligence and Expert Systems for High Energy and Nuclear Physics,* 1992.

Beyer, Hugh and Karen Holtzblatt. *Contextual Design: Defining Customer-Centered Systems.* Morgan Kaufmann Publishers, San Francisco, 1998.

Blackmon, Marilyn Hughes, Peter G. Polson, Muneo Kitajima, and Clayton Lewis. Cognitive walkthrough for the web. In *Proceedings of the Conference on Human Factors in Computing Systems,* Vol. 4 of *CHI Letters,* pp. 463–470, Minneapolis, MN, 2002.

Brinck, Tom, Darren Gergle, and Scott D. Wood. *Usability for the Web: Designing Websites That Work.* Morgan Kaufmann, San Francisco, 2001.

Büchner, Alex G. and Maurice D. Mulvenna. Discovering Internet marketing intelligence through online analytical web usage mining. *SIGMOD Record,* 27(4): 54–61, 1998.

Card, Stuart K., Peter Pirolli, Mija Van Der Wege, Julie B. Morrison, Robert W. Reeder, Pamela K. Schraedley, and Jenea Boshart. Information scent as a driver of web behavior graphs: Results of a protocol analysis method for web usability. In *Proceedings Conference on Human Factors in Computing Systems,* pp. 498–505, 2001.

Cato, John. *User-Centered Web Design.* Addison-Wesley, Boston, 2001.

Chi, Ed H., Peter Pirolli, Kim Chen, and James Pitkow. Using information scent to model user information needs and actions on the web. In *Proceedings of the Conference on Human Factors in Computing Systems,* Vol. 1, pp. 490–497, Seattle, WA, 2001. ACM Press, New York.

Chi, Ed H., Adam Rosien, and Jeffrey Heer. Lumberjack: intelligent discovery and analysis of web user traffic composition. In *WEBKDD,* Edmonton, Canada, 2002.

Christian, Kevin, Bill Kules, Ben Shneiderman, and Adel Youssef. A comparison of voice controlled and mouse controlled web browsing. In *4th Annual ACM Conference on Assistive Technologies,* pp. 72–79. ACM Press, New York, 2000.

Cooper, Alan. *The Inmates Are Running the Asylum.* Sams, Indianapolis, IN, 1999.

Coyne, Kara Pernice and Jakob Nielsen. How to conduct usability evaluations for accessibility: Methodology guidelines for testing websites and intranets with users who use assistive technology. Nielsen Norman Group Report, 2001.

Cugini, John and Jean Scholtz. VISVIP: 3D visualization of paths through Websites. In *Proceedings of the International Workshop on Web-Based Information Visualization,* pp. 259–263, Florence, Italy, 1999. Institute of Electrical and Electronics Engineers.

De Troyer, O. M. F. and C. J. Leune. WSDM: a user centered design method for Websites. *Computer Networks and ISDN Systems,* 30(1–7): 85–94, 1998.

Dong, Jianming, Shirley Martin, and Paul Waldo. A user input and analysis tool for information architecture. Available at http://www-3.ibm.com/ibm/easy/eou_ext.nsf/Publish/410/File/EZSortPaper.pdf, 2001.

Farnum, Chris. What an IA should know about prototypes for user testing. *Boxes and Arrows,* July 29, 2002. Available at http://www.boxesandarrows.com/archives/2002_07.php.

Fleming, Jennifer. *Web Navigation: Designing the User Experience.* O'Reilly, Sebastopol, CA, 1998.

Ford, Shannon and Dan Boyarski. Design@Carnegie Mellon: A web story. In Gerrit van der Veer, Austin Henderson, and Susan Coles, Eds., *Proceedings of the Conference on Designing Interactive Systems: Processes, Practices, Methods, and Techniques (DIS-97),* pp. 121–124, ACM Press, New York, 1997.

Fuccella, Jeanette. Using user centered design methods to create and design usable Websites. In *ACM 15th International Conference on Systems Documentation,* pp. 69–77, 1997.

Fuccella, Jeanette and Jack Pizzolato. Creating Website designs based on user expectations and feedback. *Internetworking, 1(1),* 1998. Available at http://www.internettg.org/newsletter/june98/web_design.html.

Fuccella, Jeanette, Jack Pizzolato, and Jack Franks. Website user centered design: Techniques for gathering requirements and tasks. *Internetworking,* 1(1), 1998. Available at http://www.internettg.org/newsletter/june98/user_requirements.html.

GroupSystems. GroupSystems products. Available at http://www.groupsystems.com/products.htm, 2002.

GVU Center. 10th WWW survey. Available at http://www.cc.gatech.edu/gvu/user_surveys/survey 1998-10/, 1998.

Heller, Hagan and David Rivers. So you wanna design for the web? *Interactions,* 3(2): 19–23, 1996.

Hom, James. The usability methods toolbox. Available at http://www.best.com/jthom/usability/usable.htm, 1998.

Hong, Jason I., Jeffrey Heer, Sarah Waterson, and James A. Landay. Webquilt: a proxy-based approach to remote web usability testing. *ACM Transactions on Information Systems,* 19: 263–285, 2001a.

Hong, Jason I., Francis C. Li, James Lin, and James A. Landay. End-user perceptions of formal and informal representations of Websites. In *Proceedings of the Conference on Human Factors in Computing Systems,* number Extended Abstracts, pp. 385–386, Seattle, WA, 2001b. ACM Press, New York.

International Organisation for Standardisation. *IS09241 Ergonomic Requirements for Office Work with Visual Display Terminals (VDTs), Part II: Guidance on Usability.* International Standard. Geneva, Switzerland, 1998.

International Organisation for Standardisation. *ISO16071 Ergonomics of Human-System Interaction — Guidance on Software Accessibility.* Technical Specification. Geneva, Switzerland, 2002.

Ivory, Melody Y. *An Empirical Foundation for Automated Web Interface Evaluation.* Ph.D. thesis, University of California, Berkeley, Computer Science Division, 2001.

Ivory, Melody Y. and Marti A. Hearst. Improving Website design. *IEEE Internet* Computing, 6(2): 56–63, 2002a.

Ivory, Melody Y. and Marti A. Hearst. Statistical profiles of highly-rated Website interfaces. In *Proceedings of the Conference on Human Factors in Computing Systems,* Vol. 4 of *CHI Letters,* pp. 367–374, Minneapolis, MN, 2002b.

Ivory, Melody Y., Jennifer Mankoff, and Audrey Le. Using automated tools to improve Website usage by users with diverse abilities. *IT&Society,* 1(3), 2003. Available at http://www.stanford.edu/group/siqss/itandsociety/v01i03/v01i03a11.pdf.

Ivory, Melody Y., Rashmi R. Sinha, and Marti A. Hearst. Empirically validated web page design metrics. In *Proceedings of the Conference on Human Factors in Computing Systems,* Vol. 1, pp. 53–60, Seattle, WA, 2001.

Leventhal, Naomi S. Using groupware tools to automate joint application development. *Journal of Systems Management,* 1995.

Lin, James and James A. Landay. Damask: A tool for early-stage design and prototyping of multi-device user interfaces. In *Proceedings of the 8th International Conference on Distributed Multimedia Systems,* pp. 573–580, San Francisco, CA, 2002.

Lin, James, Mark Newman, Jason I. Hong, and James A. Landay. DENIM: Finding a tighter fit between tools and practice for Website design. Submitted for publication, 1999.

Mackinlay, Jock. Automating the design of graphical presentations of relational information. *ACM Transactions on Graphics,* 5(2): 110–141, 1986.

Macromedia. Macromedia exchange — 508 accessibility suite extension detail page. Available at http://www.macromedia.com/cfusion/exchange/index.cfmloc=ensview=sn121viewName=Dreamweaver, 2001.

Macromedia. Accessibility: Macromedia Flash Player 6 overview. Available at http://www.macromedia.com/macromedia/accessibility/features/flash/player.html, 2002.

Malhotra, A., J. C. Thomas, J. M. Carroll, and L. A. Miller. Cognitive processes in design. *International Journal of Man-Machine Studies,* 12(2): 119–140, 1980.

Martin, Shirley. Cluster analysis for Website organization: Using cluster analysis to help meet users' expectations in site structure. *Internetworking,* 2(3), 1999. Available at http://www.internettg.org/newsletter/dec99/cluster_analysis.html.

Mayhew, Deborah J. *The Usability Engineering Lifecycle: A Practitioner's Handbook for User Interface Design.* Morgan Kaufmann, San Francisco, 1999.

Meetingworks. Meetingworks software. Available at http://www.meetingworks.com/html/meetingworks_software.html, 2002.

Miller, Craig S. and Roger W. Remington. Effects of structure and lable ambiguity on information navigation. In *Proceedings of the Conference on Human Factors in Computing Systems,* Extended Abstracts, pp. 630–631, Minneapolis, MN, 2002.

National Institute of Standards and Technology. WebCAT overview. Available at http://zing.ncsl.nist.gov/WebTools/WebCAT/, 2002.

NetIQ. Webtrends reporting center. Available at http://www.netiq.com/products/wrc/default.asp, 2002.

NetRaker. The NetRaker suite. Available at http://www.netraker.com/info/applications/index.asp, 2000.

Newman, Mark W. and James A. Landay. Sitemaps, storyboards, and specifications: A sketch of Website design practice. In *Proceedings of Designing Interactive Systems: DIS 2000,* pp. 263–274, New York, 2000.

Nielsen, Jakob. *Usability Engineering.* Academic Press, Boston, 1993.

Nielsen, Jakob. *Designing Web Usability: The Practice of Simplicity.* New Riders, Indianapolis, IN, 2000.

Nielsen, Jakob. Mobile devices will soon be useful. Available at http://www.useit.com/alertbox/20010916.html, 2001.

Paciello, Michael G. and Mike Paciello. *Web Accessibility for People With Disabilities.* CMP Books, Gilroy, CA, 2000.

Pearrow, Mark. *The Wireless Web Usability Handbook.* Charles River Media, Hingham, MA, 2002.

Quinn, Laura S. Defining feature sets through prototyping. *Boxes and Arrows,* Novemer 11, 2002. Available at http://www.boxesandarrows.com/archives/defining_feature_sets_through_prototyping.php.

Rosenfeld, Louis and Peter Morville. *Information Architecture for the World Wide Web.* O'Reilly, Sebastopol, CA, 1998.

Rosson, Mary Beth and John M. Carroll. *Usability Engineering: Scenario-Based Development* of *Human-Computer Interaction.* Morgan Kaufmann, San Francisco, 2002.

Sano, Darrell. *Designing Large-Scale Websites: A Visual Design Methodology.* John Wiley & Sons, New York, 1996.

Scholtz, Jean and Sharon Laskowski. Developing usability tools and techniques for designing and testing web sites. In *Proceedings of the 4th Conference on Human Factors and the Web,* Basking Ridge, NJ, 1998, Available at http://www.research.att.com/conf/hfweb/proceedings/scholtz/index.html.

Schuler, Douglas and Aki Namioka, Eds. *Participatory Design Principles and Practices.* Lawrence Erlbaum, Hillsdale, NJ, 1993.

Sears, Andrew. AIDE: A step toward metric-based interface development tools. In *Proceedings of the 8th ACM Symposium on User Interface Software and Technology,* pp. 101–110, Pittsburg, PA, 1995. New York: ACM Press.

Shneiderman, Ben. Designing information-abundant Websites: Issues and recommendations. *International Journal of Human-Computer Studies,* 47(1): 5–29, 1997.

Shneiderman, Ben. *Leonardo's Laptop: Human Needs and the New Computing Technologies.* Cambridge, MA: MIT Press, 2002.

Sinha, Rashmi. Survey software review. Available at http://sims.berkeley.edu/sinha/teaching/Infosys271_2000/surveyproject/surveysoftware.html, 2000.

UsableNet. UsableNet — products and services. Available at http://www.usablenet.com/products_services/products_services.html, 2002.

van Duyne, Douglas K., James A. Landay, and Jason I. Hong. *The Design of Sites: Patterns, Principles, and Processes for Crafting a Customer-Centered Web Experience.* Addison-Wesley, Boston, 2002.

Walker, Miriam, Leila Takayama, and James A. Landay. High-fidelity or low-fidelity, paper or computer? Choosing attributes when testing web prototypes. In *Proceedings of the Conference on Human Factors and Ergonomics Society 46th Annual Meeting,* Baltimore, MD, 2002. HFES.

Waterson, Sarah, James A. Landay, and Tara Matthews. In the lab and out in the wild: Remote web usability testing for moblie devices. In *Proceedings of the Conference on Human. Factors in Computing Systems,* Extended Abstracts, pp. 796–797, Minneapolis, MN, 2002.

Wood, Jane and Denise Silver. *Joint Application Development.* 2nd ed. John Wiley & Sons, New York, 1995.

World Wide Web Consortium. How people with disabilities use the web. W3C Working Draft, 2001. Available at http://www.w3.org/WAI/EO/Drafts/PWD-Use-Web/.

Zaïane, Osmar R., Man Xin, and Jiawei Han. Discovering web access patterns and trends by applying OLAP and data mining technology on web logs. In *Advances in Digital Libraries,* pp. 19–29, 1998.

Zhang, Zhijun. Ask usability advisor. Available at http://www.pages.drexel.edu/zwz22/Advisor.html, 1999a.

Zhang, Zhijun. Usability evaluation methods. Available at http://www.pages.drexel.edu/zwz22/Usability-Home.html, 1999b.

45

Distributed Storage

CONTENTS

Abstract 45-1
45.1 Introduction 45-1
45.2 Data Location 45-3
45.2.1 Cooperative Web Caching 45-3
45.2.2 Distributed File Systems 45-5
45.2.3 Distributed Hash Tables 45-6
45.3 Cache Coherence 45-6
45.4 Load Balancing 45-9
45.4.1 The Web 45-9
45.4.2 Peer-to-Peer Storage Systems 45-10
45.5 Array Designs 45-11
45.6 Weakly Connected Wide-Area Environments 45-12
45.7 Security 45-14
45.8 Conclusion 45-16
References 45-16

Sumeet Sobti

Peter N. Yianilos

Abstract

As distributed storage has emerged as a central feature of today's computing environment, complex issues and algorithmic challenges have appeared. This chapter identifies and discusses many of these, and provides a large set of references to the published literature and also to current Web-published online work in industry.

45.1 Introduction

Distributed storage is an integral part of our everyday computing landscape. For the casual, nonspecialist computer users, it usually takes the form of services available through the World Wide Web, and more recently, the peer-to-peer file-sharing systems. For users of shared computing resources at social institutions, such as academic departments and commercial enterprises, it often means access to a common file-storage facility. In general, wherever there is need to share data between multiple computers or users, one needs a solution that falls in the broad field of distributed storage.

This field comprises a large variety of applications and systems. Many of these are widely used and commercially supported, whereas many others have been developed only as research prototypes. The most prominent example of large-scale distributed storage is the World Wide Web that originated in the early 1990s, and that already includes millions of servers and users distributed all across the globe. It is a prime example of the client-server architecture, where clients access data residing at the servers using a stateless request–response protocol.

1-58488-381-2/05/$0.00+$1.50
© 2005 by CRC Press LLC

Due to its strong cultural and economic impact, Web-related technologies have been extensively researched and developed in both academic as well as commercial arenas. The issues of caching and load balancing have received considerable attention in the context of the Web. Many widely used content-distribution networks incorporate sophisticated, proprietary algorithms to manage Web caching and load balancing [Wang et al., 2002]. In fact, almost all Web-based and Web-related services need to address distributed storage issues, including Web indices and search engines, Web-based email services, Web-mirrors, data repositories, and e-commerce sites.

Another distributed storage application that has vast populations of users all over the world is peer-to-peer file-sharing. Examples of these systems include Napster, Gnutella, Kazaa, and many others. In the peer-to-peer model [Oram, 2001], several computers contribute CPU, storage, and network resources to build a symmetric system. The members do not have specialized roles as clients or servers; all members cooperatively serve and consume data.

Following the peer-to-peer model, many systems have emerged with the goal of building a publishing substrate that has properties such as censorship resistance, anonymity for producers and consumers of content, and deniability for participants hosting content. A seminal paper by Ross Anderson [1996] that proposed a storage medium with such properties has been the inspiration for many of these systems. Examples include Freenet [Clarke et al., 2000], FreeHaven [Dingledine et al., 2000] and Publius [Waldman et al., 2000].

A significant amount of effort in the area of peer-to-peer systems has gone into designing what are called "distributed hash tables" (DHTs). A DHT is a distributed data structure to maintain a mapping between names of data objects and names of machines or participants. The key properties that make DHTs attractive are very high scalability, fault-tolerance in the presence of multiple simultaneous failures, and very low space overhead in maintaining the mapping. Chord [Stoica et al., 2001], Pastry [Rowstron and Druschel, 2001a], and CAN [Ratnasamy et al., 2001] are a few example DHTs. These DHTs can then be used to build data archives with strong availability and survivability properties [Dabek et al., 2001; Rowstron and Druschel, 2001b].

Distributed file systems are yet another prevalent form of distributed storage. Such systems have been constructed for many diverse environments ranging from enterprise-scale systems to wide-area configurations. Within the enterprise, three forms of distributed storage have emerged. The file server is a computer system that exports file storage to users typically using the Network File System (NFS) or the Common Internet File System (CIFS) protocol. When a file server is packaged commercially as a turn-key appliance, the result is referred to as Network Attached Storage (NAS). The primary benefit of NAS is simpler administration and that the entire system may have been performance-optimized for the storage task. As the cost of high-speed local-area networking has fallen, another architecture has emerged that turns the traditional computer-storage picture inside out. Storage Area Networks (SAN) separate the physical storage devices (individual disk drives) from the processors, placing them at the other end of a high-speed network. These drives hold the *state* of an enterprise's data processing operation, and it is sometimes attractive to manage all of this information centrally in a large pool of storage. Because of the high bandwidth requirements of this architecture, fiber-optic interconnections are frequently employed.

File servers, NAS, and SAN are generally used within a trusted security domain, that is, an enterprise's LAN, sometimes extended by virtual private networks (VPN). They are, in effect, LAN extensions of the traditional single-computer processor–storage interconnection.

These technologies are widely used and continue to be the focus of intense and fast-moving development in industry. Most recently, the notion of *virtualizing* an enterprise's storage assets has emerged. This, as the name suggests, introduces a new management layer between the exporter and user of storage services. We will not discuss particular implementations of NAS, SAN, or other enterprise storage architectures any further. We will, however, cover many basic ideas that they may depend on.

The discussion in this chapter is organized around a set of design issues that are common to many distributed storage systems. Each of the following sections focuses on one particular issue and illustrates various approaches to addressing the issue using examples from a variety of systems. The examples include the Web, many peer-to-peer storage systems, and distributed file systems of various kinds. Section

45.2 concerns the issue of how to locate a data object in a system where many storage devices are used and not all of them store all the data objects. The major approaches to addressing this problem are illustrated through examples from Web-caching systems, distributed file systems, and peer-to-peer storage systems. Sections 45.3 and 45.4 discuss the caching and load-balancing issues in the context of these systems. In Section 45.5 we discuss various techniques used in disk arrays to achieve better performance and reliability. Section 45.6 discusses storage systems for weakly connected, wide-area environments. Security-related issues are discussed in Section 45.7. Section 45.8 presents our conclusion.

The chapter also includes an extensive bibliography that represents a snapshot of issues and systems in this field. Admittedly, the snapshot is somewhat biased due to our own interests in the field. In particular, we do not attempt to cover the rich literature on distributed databases and digital libraries. In our view, these fields address applications and interfaces that are qualitatively different from those in the "file-based" systems, which we focus on in this chapter.

45.2 Data Location

A typical distributed storage system consists of multiple storage devices that are used to store multiple data objects. Because not every device may store every data object, the following data location problem arises: given the name of a data object, how to locate a storage device that contains a copy of the data object. The solution adopted for the data location problem usually affects almost all other aspects of the system design.

Figure 45.1 illustrates four major approaches to data location. A natural approach is to maintain a centralized "data directory" for the system at a fixed place. For each data object, the directory records the set of devices that store a copy of that object. Thus, any data object can be located just by querying the directory once. The second approach is to hierarchically partition the directory into smaller pieces that can be stored and managed separately, possibly on different devices. This approach alleviates some of the drawbacks of the previous approach where the central directory can potentially become a performance bottleneck or a single point of failure.

The next two approaches do not maintain the data directory in any form, thus avoiding the time, space, and communication overheads associated with maintaining an accurate directory. In the hashing-based approach, the locations of a data object are obtained by hashing the name of the object using a deterministic hash function. In the search-based approach, data location requests are forwarded to all the storage devices, and a storage device that contains a copy of the requested data typically sends a reply back. There are, however, other limitations to these approaches. For example, in the hash-based approach, the system does not have sufficient flexibility to adapt the location of data objects to changing usage patterns or network conditions, whereas request forwarding imposes performance penalty in the search-based approach.

Variants and combinations of these approaches have been extensively used in all kinds of distributed storage systems. In the rest of this section, we present many such examples.

45.2.1 Cooperative Web Caching

Web objects are located by their URLs (Uniform Resource Locator). A URL is a text string specifying (1) the host name of a machine running a Web server, and (2) the name of a specific object on that machine. Typically, a client Web browser obtains data by connecting directly to the Web server. In many cases, client accesses go through a "proxy server" that fetches the Web objects on behalf of the client. Proxy servers are usually set up for groups of clients, e.g., for all users in an organization. Proxy servers often cache the Web objects they fetch, thus improving client access latency for objects that are found in the cache.

Many systems allow multiple proxy servers to share their local caches with each other. For such cooperative caching [Wolman et al., 1999], one often needs a mechanism to look up contents of multiple proxy caches for each client access. Several schemes have been developed for this purpose.

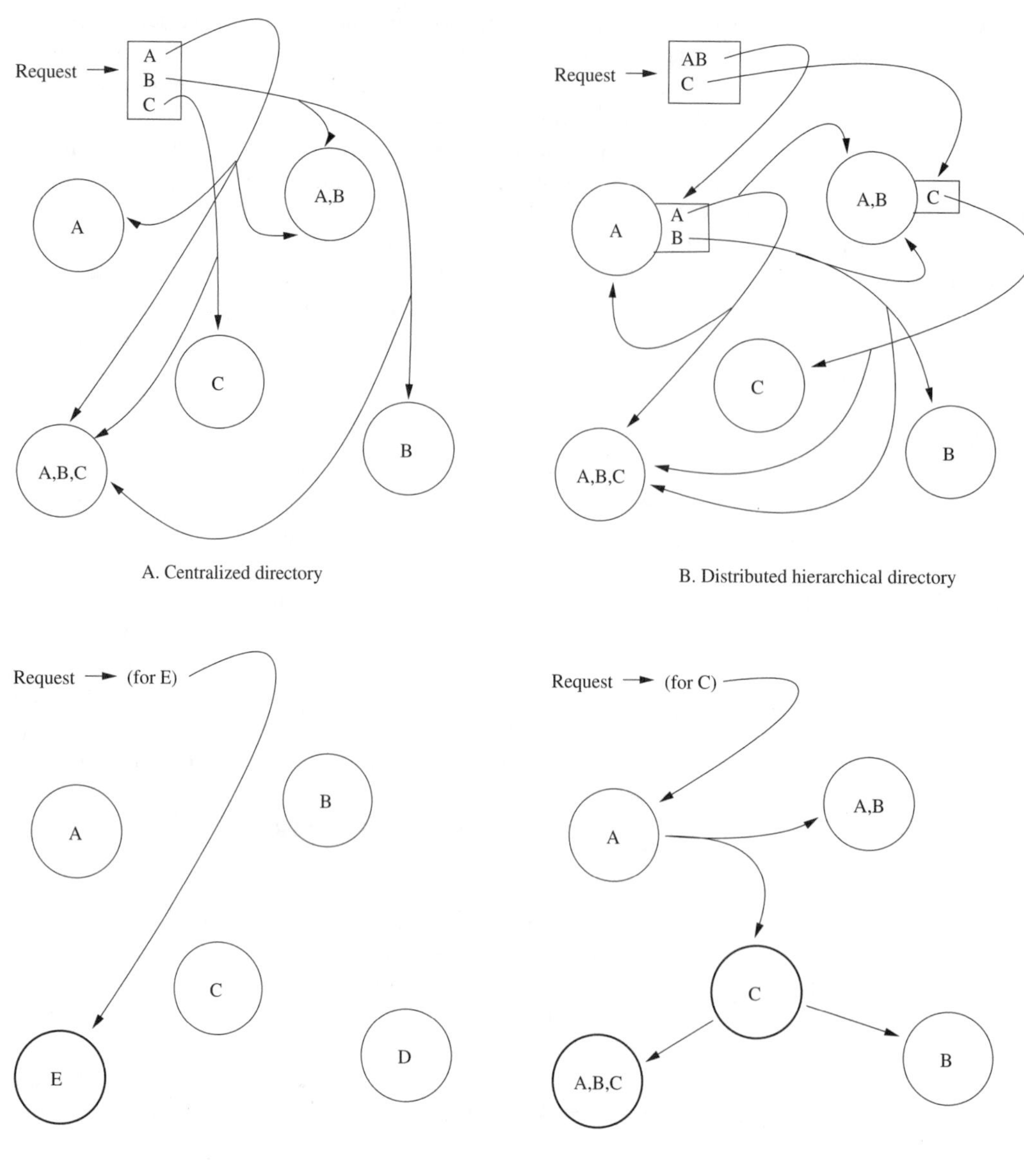

FIGURE 45.1 A conceptual depiction of the four major approaches to data location. Storage devices are represented by circles and data objects are denoted by letters A, B, C, etc. (A) For each data object, the central directory contains the names of devices that hold a copy of the object. (B) The directory can be hierarchically partitioned, and pieces of it can be distributed among multiple devices. (C) Each data object is mapped to a fixed set of devices using a deterministic hash function. Given the name of an object, a single hash computation leads to a device containing the object. No data directory is maintained. (D) Request for a data object is forwarded from one device to another until it reaches a device that contains the object. The path(s) along which a request is forwarded can be flexibly and dynamically determined. Again, no data directory is maintained.

The Harvest cache [Chankhunthod et al., 1996] organizes the set of proxy caches in the form of a hierarchy. When a proxy cache receives a request and does not have the object locally, it probes its immediate neighborhood for the object, including its siblings and parent in the hierarchy. If the object is not found in the immediate neighborhood, the request is passed up the hierarchy. Alternatively, a proxy cache can also obtain the object from the home Web server directly. Note that in this scheme, information about the contents of a given proxy cache is stored only on itself.

A directory-based alternative is used in many distributed Web-caching systems including Menaud et al. [1998]; Tewari et al. [1999]; and Gadde et al. [1998]. Here, a possibly distributed directory is maintained that contains information about the contents of all the participating caches. On each client request, the directory is looked up (via no or a few messages) to find a cache that holds the requested object. A protocol is implemented to keep the directory consistent as the contents of the caches change over time. Usually, this protocol works lazily in the background because an inconsistent directory in this application only poses a potential performance hit and is not a correctness problem. Summary cache [Fan et al., 2000] and cache digests [Rousskov and Wessels, 1998] use Bloom filters to keep the directory representation compact.

Hash-based schemes [Karger and Sherman, 1999; Valloppillil and Ross, 1998] partition the space of URLs using a hash function and assign each partition to a unique cache in the system. Object location proceeds by computing the hash of the object's URL and then directly requesting the corresponding cache. This eliminates the cost of extra look-up messages of the previous two alternatives. It also eliminates the space overhead of any directory-based scheme. This scheme, however, does not let the system control the placement of objects. This can lead to poor performance, especially when the pair-wise network distances between the caches are nonuniform.

45.2.2 Distributed File Systems

The oldest and perhaps the most popular abstraction for persistent storage is that of a file system. The file system abstraction typically consists of a hierarchy of *directories*, and data is stored in *files* that are located at arbitrary positions in the directory hierarchy. Popular examples of "local" disk-based file systems include flavors of the Unix file system [Ritchie and Thompson, 1974; McKusick et al., 1984; Rosenblum and Ousterhout, 1992] and various Windows file systems (FAT, FAT32, and NTFS). Key aspects in which these file systems differ from each other are application program interface (API), consistency semantics, reliability guarantees, and performance trade-offs.

Several network protocols have been designed that allow application programs to access file systems resources from a remote server. These include NFS [Sandberg et al., 1985], AFS [Howard et al., 1988], and CIFS. These protocols support APIs of many traditional local file systems, such as those mentioned above. Therefore, application programs are able to use them without any modifications.

Many cluster-based "distributed" file systems have been proposed that do not fit the traditional client-server model. Examples of such systems include xFS [Anderson et al., 1995] and Frangipani [Thekkath et al., 1997; Lee and Thekkath, 1996]. These systems distribute data at a fine granularity among multiple storage servers. In general, copies of a given data object may reside on any subset of these servers. Therefore, the system needs to keep track of the locations of these copies, both for being able to fetch those copies for satisfying reads and for being able to invalidate them as new writes occur. These systems use what can be called a *manager-based* approach for data location.

In these systems, a "manager" is responsible for tracking the current locations of the replicas of an object. Read or write requests are first sent to the manager, which in turn forwards the read requests or sends invalidation messages to the hosts that have replicas. To avoid bottlenecks at a single manager, the management of the whole data set is typically distributed among multiple managers. Each object is mapped to one manager that keeps full record of the copies.

Because each replica of each object is tracked accurately, the managers can become a scalability bottleneck if the number of nodes or objects becomes too large. Also, because each request has to go through a manager, managers can become performance bottlenecks in cases of highly nonuniform workloads. The manager-based approach has also been used in distributed memory systems [Feeley et al., 1995; Li and Hudak, 1989] and cluster applications such as scalable email services [Saito et al., 1999].

45.2.3 Distributed Hash Tables

Recently, there has been considerable interest in designing systems that consist of large number of independent but cooperating participants. Such systems were pioneered by file-sharing systems like Napster and Gnutella. More *recent* work includes PAST [Rowstron and Druschel, 2001b], CFS [Dabek et al., 2001], and OceanStore [Kubiatowicz et al., 2000; Rhea et al., 2003]. At the heart of these systems lies their particular solution to the object location problem that scales well with the number of nodes and objects in the system. These systems have come to be called peer-to-peer systems because all participating nodes in these systems are designed to be peers to each other, as opposed to having specialized roles like clients or servers.

Because these systems are intended to scale to millions of nodes with minimal centralized control or administration, solutions that work fine for other systems of more modest size do not work for these systems. For example, even maintaining a consistent and up-to-date list of the current participants in the system at a single node can be troublesome when system membership changes frequently. The common solution used by many of these systems is to use a "distributed hash table."

A distributed hash table (DHT) is an abstraction that allows one to map a set of objects to a set of nodes in a scalable and completely distributed manner. A typical DHT works as follows. All objects and nodes in the system are assigned unique names from a common name space. Unique names for objects can, for example, be obtained by hashing the contents of the object using a cryptographically strong hash function like MDs or SHA-1 and extracting some bits from the hash. Similarly, unique names for nodes can be obtained by hashing their IP addresses, for example. One advantage of using a cryptographically strong hash function for this purpose is that this results in an approximately uniform distribution of names in the name space. This leads to an approximately uniform assignment of objects to nodes because often an object is assigned to a set of nodes whose names are "close" to the object name (under some definition of "closeness").

If each node is allowed to locally maintain a complete and consistent database containing names and identities of all other nodes in the system, then object locations, when needed, can be determined by querying the database at any node. As we mentioned above, however, maintaining such a database is not feasible if the system is to scale to millions of nodes and if the membership in the system changes frequently. Therefore, a scheme is used where each node maintains information about only a few other nodes (called *neighbors*) in the system. When a node receives a query for an object, if it does not have the object, it forwards the query to a neighbor whose name is closer to the object name than the current node's name. Thus, in a sequence of hops, the query reaches a node whose name is sufficiently close to the object name, and who, therefore, contains a copy of the object. The neighbors of each node are carefully selected in a manner so that object location always succeeds in a reasonably small number of hops. In other words, the "neighborhood graph" is structured so as to have a small diameter. These schemes are also called *structured overlay networks* because of this property.

DHT implementations that fit this description include Chord [Stoica et al., 2001], Pastry [Rowstron and Druschel, 2001a], SCAN [Ratnasamy et al., 2001], Tapestry [Hildrum et al., 2002], and Kademlia [Maymounkov and Mazieres, 2002]. Technically, these implementations differ in their choice and organization of the name space, the closeness metric, and the structure of the neighborhood graph in the chosen name space. Many of these DHTs achieve a worst-case upperbound of $O(log\ n)$ logical hops per object lookup at the cost of $O(log\ n)$ storage overhead per node, where n is the total number of nodes in the system. Some of these also use information about the underlying physical topology in building the neighborhood graph, so that the logical hops are physically short hops, thus optimizing overall query latency.

45.3 Cache Coherence

Caching is probably the most widely used performance-optimization technique in almost all computer systems. From processors to disk-based systems to Internet-wide distributed services — wherever data is involved — caching is a primary technique employed to reduce data access latency. The basic idea is to keep a copy of the data in a place that is more efficiently accessible from where it is likely to be accessed in the future. For example, all modern processors have one or more levels of on-chip fast memory, which acts as a cache for the data stored in the main memory. Similarly, all disk-based file systems maintain a cache of disk blocks in the main memory.

Another prime example of caching as a means of reducing network latency of data access is the Web. Most Web browsers use a local on-disk cache where recently accessed Web objects are stored. Similarly, most Web proxy servers cache the Web objects that they fetch on behalf of their clients (see Section 45.2.1). Many content-delivery systems, such as Akamai (www.akamai.com) and Mirror Image (www.mirror-image.com), deploy Web caches close to the edges of the network (e.g., in *points of presence* of major ISPs), and direct client requests to caches that are close to them. (The techniques used by Akamai are described in more detail in Section 45.4.1.). This, in addition to reducing the access latency for the clients, serves as a means of reducing the load on the target Web servers during periods of high activity. Therefore, in the context of such distributed services, caching simultaneously serves three main objectives: reduction in data access latency, reduction in overall network traffic, and reduction in load on the server.

With caching comes the issue of *cache coherence* (or *cache consistency*). When there are multiple copies of a piece of data and one (or more) of them is updated/overwritten, then the copies are said to have become *inconsistent*. Inconsistency is resolved either by discarding all but one of the copies, or by refreshing all the copies with the latest version of the data, or by a suitable combination of the two. Cache coherence may or may not be a problem depending on the application involved and on the semantics or consistency guarantees provided by the application. For example, in a uniprocessor machine, because all reads and writes are performed by a single processor, the memory cache hierarchy does not present a severe cache coherence problem. In the case of a shared-memory multiprocessor machine, however, cache coherence becomes a significant problem and is often dealt with by sophisticated cache-coherence protocols [Patterson and Hennessy, 1996].

In distributed storage systems also, cache coherence may or may not be a major problem, depending on the particular application involved. In general, there is a trade-off between the application coherence semantics and the efficiency and complexity of implementation of the coherence mechanism. The HTTP protocol [Fielding et al., 1999], for example, enforces cache coherence using two simple mechanisms — the expiration mechanism and the validation mechanism. For each cacheable Web object, an expiration time is specified before which the object can be served from the cache without contacting the server. After the expiry time, the server can be sent a "validation" request for the cached object. In reply, the server either confirms the freshness of the cached object or sends a new version of the object. Similar expiration-time-based mechanisms are used in other systems as well, such as the Domain Name System (DNS).

In many peer-to-peer systems, such as those mentioned in Section 45.2.3, the name of a piece of data is often cryptographically tied to its content (see also Section 45.4.2). So, whenever the content of a block changes, its name changes. Therefore, in effect, the cache coherence problem is *defined away* for the most part. The upper layers of the system, however, are then responsible for maintaining coherence of names and for keeping track of the names of fresh versions of data. So far, most peer-to-peer systems allow read-only access to data where the coherence problem does not arise [Dabek et al., 2001; Rowstron and Druschel, 2001b]. The Ivy system [Muthitacharoen et al., 2002] allows write access. It allows a fixed number of possibly concurrent writers. Writes from each writer create a separate "operation log." Reads in the system are satisfied by traversing these logs backward and combining their content. To ensure that read operations always get the latest version of data, the reader must consult the latest tail of each log. To ensure that the reader always accesses the latest tail of each log, caching of the tail records is disabled. Therefore, some performance penalty is paid during each read access to ensure freshness.

Distributed file systems have traditionally attempted to offer stricter coherence guarantees to client processes running on different client machines. The goal is to try to achieve the same semantics in a distributed setting as those provided by a local disk-based file system, although this is not always possible.

In most distributed file systems, the permanent "home" of data is a disk at a server machine. When a client process tries to access a piece of data, there are many places other than the "home" where data can be found more quickly. These include memory at the client machine, disk at the client machine, memory at other client machines in the system, and memory at the server machine. All of these can potentially serve as a cache for the server disk. Most existing distributed file systems, such as NFS [Sandberg et al., 1985], AFS [Howard et al., 1988], Sprite [Nelson et al., 1988], LOCUS [Walker et al., 1983], and xFS [Anderson et al., 1995] use client memory and server memory as caches. AFS also uses client disk as a cache. xFS employs cooperative caching [Dahlin et al., 1994] among all client machines, so one client can potentially serve a data block to another client. Cooperative caching is also used in the Global Memory Service [Feeley et al., 1995].

Cache coherence semantics vary from system to system. Figure 45.2 shows some examples to illustrate how cache coherence is handled in four systems: NFS, AFS, Sprite, and LOCUS. The NFS server is designed to be completely "stateless" in the sense that it does not store any information about any of the clients. In particular, it does not keep track of whether a particular piece of data is being written at some client. As a consequence, an application on one client machine can potentially read stale data even when more fresh data has been written at another client machine. A reading client can potentially check with the server to see if its cached copy is up to date. This, however, is not sufficient because the data at the server can also be stale.

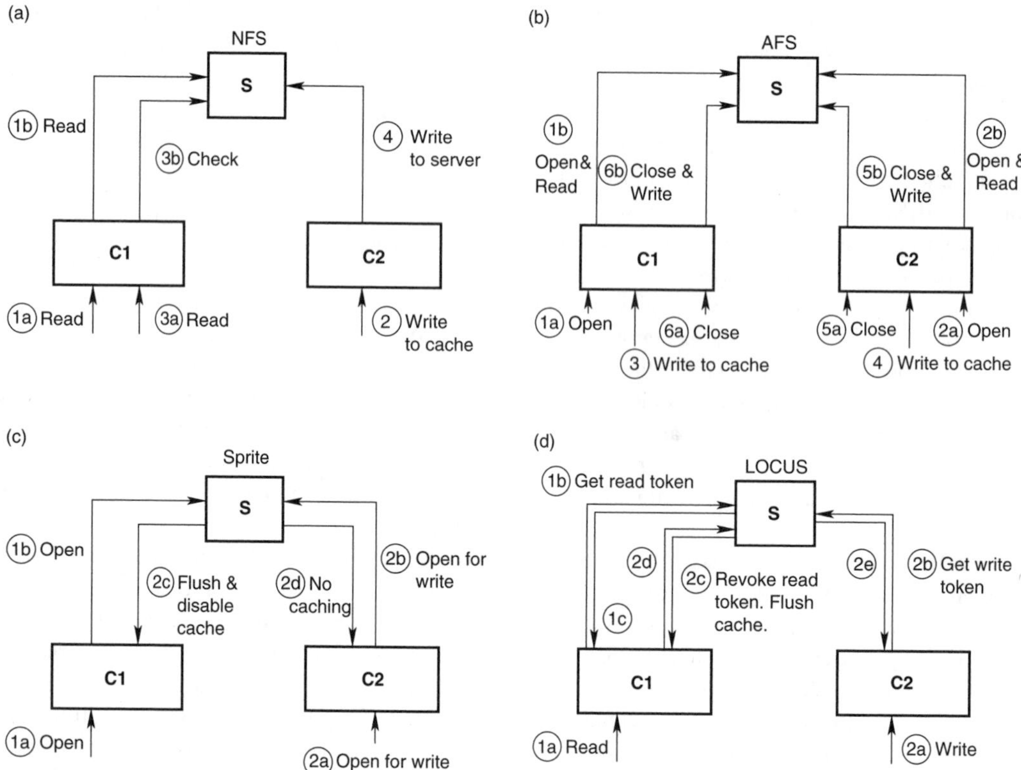

FIGURE 45.2 Cache coherence examples. S is the server and C_1 and C_2 are two clients that open the same file concurrently. (a) NFS. The operations occur in the sequence 1a, 1b, 2, 3a, 3b, 4. In Step 3, C_1 reads stale data because fresh data written by C_2 in Step 2 is absorbed by its local cache. The data at the server is also stale in Step 3. (b) AFS. C_1 and C_2 operate on their locally cached copies of the file. Updates made by one are not visible to the other. Note that updates made by C_2 are completely lost because C_1 closes the file after C_2 does. (c) Sprite. Caching is completely disabled during concurrent write sharing. (d) LOCUS. Data is always cacheable. Access to data is controlled by tokens. The local cache at a client is flushed when it loses its token.

AFS uses the local disk on the client as a cache. Conceptually, during a file open operation, the file is copied from the server to the local disk cache. Subsequent reads and writes are satisfied from the cache. Newly written data, if any, is sent to the server only when the file is closed. Therefore, data written by one client is visible to another client only if the former closes the file and flushes the new data to the server *before* the latter opens the file. As a natural optimization, the file copying during open is avoided if the client cache already has the server's version of the file. For each file, the server records the set of clients that have a cached copy of the file. When it receives a new version of the file from some client, all other clients holding that file are notified.

In Sprite, when two or more clients have opened a file and at least one of them is writing, then caching is completely disabled, and all the read and write operations are forced to go to the server. The implementation becomes somewhat simpler by disabling caching completely during concurrent write sharing. In LOCUS, data is always cacheable, although access to data is controlled by the use of read-and-write tokens. To perform a read-or-write operation, the client must receive an appropriate token from the server. When a client loses its token, possibly because some other client asked for a write token, its local cache is flushed. In both Sprite and LOCUS, per-file state is maintained at the server and a strong coherence guarantee is provided to the clients.

45.4 Load Balancing

As the number of clients or the total amount of data in a storage system grows to a size that is beyond the capabilities of a single server, multiple servers are deployed to share the load of servicing the clients. This brings up the problem of balancing the servicing load among multiple available servers. The goal is to avoid a situation where certain servers are heavily loaded, leading to poor quality or even complete denial of service to the clients they are serving, while certain other servers are sitting idle. Clearly, in such a situation, overall quality of service can be improved by offloading some responsibility from the overloaded servers to the underloaded ones. Often, it is desirable to perform this kind of load balancing in a manner that is completely transparent to the clients. In this section, we describe some load-balancing techniques used in the context of the Web and some peer-to-peer systems.

45.4.1 The Web

The Web is known to often experience the "hot spot" problem, where a large number of clients try to access data from the same Website in a very short window of time. This usually overloads the target server and the network, resulting in poor quality of service to the clients or even complete denial of service when the server or the network is unable to handle the load. The natural solution to such problems is to try to distribute the load among a set of servers that may or may not be physically located at the same site.

There are several approaches to achieving such distribution of requests in a client-unaware fashion. A simple *application-based* approach is to use the HTTP redirect feature [Andresen et al., 1996]: a Web server accepts client requests, and it either chooses to serve the client locally or it redirects the request to one of the other replica servers. The choice of the replica server may depend on the location of the client, the object being accessed, and the load and location of the replica servers. There are several drawbacks of such an application-based approach. The clients may perceive increased latency of access, as redirection may result in multiple connection setups and name lookups. Also, the redirecting server can itself become overloaded as it needs to communicate with each client that connects to it.

A different *naming-based* (or *DNS-based*) approach is to use the DNS to direct clients to different servers. The client (a Web browser, for example) uses DNS to translate the domain name of the Web server in the URL into an IP address to connect to. The DNS can be configured so that clients are returned different server IP addresses, depending on their location and the time at which the DNS query is performed [Katz et al., 1994]. Obviously, the current load and location of the available servers and the content being requested may also be considered in deciding which server a given client is directed to.

A prime example of a commercial content delivery system that uses the DNS-based approach is Akamai (www.akamai.com). The Web objects served by Akamai servers have URLs where the host name is of the form a248.e.akamai.net. The name is resolved by a two-level hierarchy of Akamai name servers. The upper-level name servers resolve e.akamai.net by returning the IP address of a lower-level Akamai name server that is close to the client in network distance. Typically, this server is in the same geographical region as the client. The lower-level name server then chooses an appropriate server in its geographical region that can serve the desired content, and returns its IP address. In a given region, not every server has all the content. Instead, content is partitioned among the set of available servers using a hashing scheme called "Consistent Hashing" [Karger and Sherman, 1999]. The first part of the host name ("a248" in the example above), in fact, identifies the hash bucket of the desired content. Consistent Hashing is shown to have nice load-balancing properties [Karger et al., 1997].

There are a number of shortcomings in the DNS-based approach. First, DNS query results (resource records) (RR) are cached by name servers so that future DNS queries from the client can be answered without much network communication. The time period for which these results are cached is determined by the Time-To-Live (TTL) field in the RR. A small value for TTL increases the client-perceived latency in the case of multiple sequential accesses. A large value for TTL, however, does not allow highly dynamic load balancing. Many DNS implementations do not allow TTL values that are less than 300 seconds [Brisco, 1995]. In such cases, the DNS-based approach can only achieve a relatively coarse-grained load balancing. Second, at DNS lookup time, DNS is presented with little information about the content being accessed (although it is possible to encode some limited information about the content in the host name, as the Akamai example described above does). Thus, it is hard to do fine-grained content-based load balancing via the DNS-based approach.

The *routing-based* approach to load balancing is to advertise a single address for the target Web server and then to use specialized routing techniques to route requests from different clients to different physical servers. An example of this approach is to use *Anycasting*. In this scheme, a single anycast address is advertised and multiple Web servers with different IT addresses are configured to correspond to the anycast address. Client requests can be routed to one of the servers using network-layer [Partridge et al., 1993] or application-layer techniques [Zegura et al., 2000].

Another form of routing-based load balancing is *server-side request redirection* [Schroeder et al., 2000; Pai et al., 1998]. A typical configuration of this form consists of a fast cluster of machines, each running an instance of the server. Requests enter the cluster through a single "front end." This front end is usually a smart switch which implements some load-balancing functionality, possibly in hardware. The main function of the front end is to choose a server and forward the client request to it in a completely client-unaware fashion. Because the communication between the client and the chosen server typically takes more than one messages to complete, the front end needs to maintain a mapping between the currently active client connections and the corresponding servers.

In practice, usually a combination of the techniques mentioned above is used. It must also be noted that many of these techniques are independent of the Web and are more widely applicable.

45.4.2 Peer-to-Peer Storage Systems

Two simple load-balancing ideas are used in almost all peer-to-peer storage systems, namely, *randomization* and *caching*. First, objects are assigned to nodes using bits from the output of a cryptographically strong hash function, as described in Section 45.2.3. This leads to an almost-random assignment of objects to nodes, thus achieving some level of load balancing. Second, as an object is accessed, the request passes through a sequence of nodes before reaching the object. Most systems also end up caching the object along the request-forwarding path. The result is that highly popular objects get widely cached throughout the network, and therefore, many later accesses are satisfied by the cached copies. This further contributes to distributing the load of serving a popular object among multiple nodes in a completely decentralized manner.

As an example, consider the CFS system [Dabek et al., 2001], which uses the Chord object location protocol [Stoica et al., 2001]. Chord, in turn, uses Consistent Hashing [Karger et al., 1997] to assign objects to nodes. The name space for objects and nodes is the set of all m-bit strings. The name space is organized in a circular fashion using the lexicographic ordering among the strings. An object is assigned to a node whose name is closest to the object name in the clockwise order on the circle. The *neighborhood* of a node (that is, the set of nodes whose names and identities are stored at this node) contains the two nodes that are closest to it on the circle in each direction. In addition, it has about $O(\log n)$ other nodes that are at geometrically increasing distance from it in the clockwise direction along the circle, where n is the total number of nodes in the system.

The object-to-node assignment described above does not always produce an optimal load balance. First, different nodes may have different capabilities, and different objects may have different sizes. So, an optimal assignment may not even distribute the objects uniformly among the nodes. Second, even in the case where all nodes have identical capabilities and all objects have the same size, if object and node names are chosen randomly, the load on the most heavily loaded node in the assignment is likely to be about $\log n$ times more than the average load. To alleviate this problem, the idea of "virtual nodes" is used. Each physical node is configured to take the responsibilities of a number of virtual nodes. The object location protocol works with virtual nodes. Therefore, each virtual node gets a unique name and stores information about $O(\log N)$ other virtual nodes, where N is the total number of virtual nodes in the system. Each physical node gets the responsibility for the objects that are assigned to its virtual nodes. Clearly, this scheme increases the storage overhead of the protocol at each physical node. The advantage, however, is a better load distribution among the physical nodes if each one of them is assigned virtual nodes in proportion to its capabilities. The load can also be dynamically adjusted by changing the number of virtual nodes or by moving virtual nodes from overloaded physical nodes to the underloaded ones [Rao et al., 2003]. Several other load-balancing techniques for peer-to-peer systems appear in Rowstron and Druschel [2001b] and Byers et al. [2003].

45.5 Array Designs

Disk arrays have been extensively researched and used since early 1980s. The basic idea in disk arrays is to use multiple independent disks to achieve better *performance* and *reliability*. Performance is mainly achieved through *striping* where data is distributed across multiple disks. This improves performance in two ways. First, requests involving large amounts of data are now able to use multiple disks in parallel, thereby improving the effective I/O bandwidth. Second, successive or concurrent requests are now potentially distributed across multiple disks, which results in less waiting in queues, thereby achieving higher overall request throughput. Reliability is improved by adding *redundancy* to data so that if one or more disks fail, lost data can be reconstructed. There are numerous schemes for striping data and adding redundancy, and their combinations give rise to a large number of array designs with different performance, reliability properties, and costs. A seminal paper by Patterson et al. [1988] on RAIDs (redundant arrays of inexpensive disks) introduces several such designs. Chen et al. [1994] present a comprehensive survey of the literature related to disk arrays.

A major practical problem relating to disk arrays, however, is that of configuration and management. Different array designs have vastly different cost, performance, and reliability characteristics. Therefore, given a set of disks, choosing which design to employ for a target workload is a nontrivial problem. The problem is only aggravated by the fact that workloads are usually difficult to characterize and they often change. So, even experienced administrators find it hard to tune and adapt the array designs to the workloads. As existing disks in an array fail and as new disks are added, data needs to be redistributed among disks, preferably while keeping it available. This administration task often becomes cumbersome and error prone. Many projects have attempted to automate the process of configuring and adapting array designs to workload characteristics, and to hardware removal and addition dynamically. The HP AutoRAID system [Wilkes et al., 1996] is one such effort. The system organizes the available disks in the array into two levels. The upper level uses *mirroring* to provide excellent performance to active (i.e., frequently written) data at a relatively higher storage cost, whereas the lower level uses *RAID-5* organization (see Chen et al., 1994) for the remaining data, providing significantly better storage cost with somewhat lower performance. Both levels provide redundancy to guard against single-disk failures. The system includes algorithms to dynamically change the organization of the two levels and to migrate data between the two levels. More recent work addressing the configuration problem includes Anderson et al. [2002a, b].

Disk areal density has been increasing at about 60% each year. This has resulted in significant improvements in disk I/O bandwidth. Array designs, as mentioned above, further improve the effective data rate. Disk access *latency*, however, is limited by the movement of mechanical components and has only been improving at about 10% per year. A disk array also presents opportunity for reducing the effective access latency seen by the applications. Techniques like striping, mirroring, rotational replication of data within a disk track [Ng, 1991] and eager-writing [Chao et al., 1992; Wang et al., 1999) are ways of trading off extra disk space for improved latency. Recent work from the MimdRAID project [Yu et al., 2000; Zhang et al., 2002] focuses on how to use these techniques in an optimal way to reduce data access latency in a disk array. A new storage technology based on microelectromechanical systems (MEMS) is soon expected to provide persistent storage alternatives with significantly better access latency than magnetic disks [Carley et al., 2000; Griffin et al., 2000]. Uysal et al. [2003] explore ways of incorporating such storage in disk arrays to improve their performance and cost–performance ratio.

45.6 Weakly Connected Wide-Area Environments

Many storage systems are designed to serve populations of mobile users who wish to access data at times and locations where the network connectivity may not be perfect. As an extreme example, consider a hypothetical user who likes to have access to all of his or her data at all times. The locations from where the user desires access include a well-connected office, DSL-connected home, and vacation house where only cellular-modem-based wireless connectivity is available. The user also likes to access his or her data when traveling. At the airport, the user may have an 802.11-based access to the Internet, whereas inside the airplane, no or only a poor-quality connection may be available at a very high price. As the example illustrates, the kind and quality of network connectivity may vary significantly over time and as the user moves. Ideally, the user would like to be oblivious to such changes in the environment and would like the system to transparently provide access to his or her data while making the best use of the available network resources.

Another key requirement from a usage viewpoint is the ability to share data among multiple collaborating users. In general, the users may be mobile or widely distributed over a large geographic region. In this section, we describe several systems that are designed to work in such highly heterogeneous and weakly connected wide-area environments. These systems use storage and network resources in complementary ways to provide transparent, continuous, and ubiquitous access to shared data. Caching and replication, used in conjunction with optimistic concurrency control, are prime techniques for enhancing availability of data and performance of access while avoiding delays associated with weak network links. We describe these in the following discussion.

Coda [Kistler and Satyanarayanan,1992] is a file system based on the client-server model, which allows clients to continue to operate without interruption even when the server temporarily becomes inaccessible. This mode of operation is called "disconnected operation." It is achieved as follows. When the server is accessible, selected data is fetched and cached on the client's local disk. In the disconnected state, reads are satisfied from the local disk cache, and update operations are logged. When the client reconnects to the server, the update log is replayed at the server. Later versions of Coda [Mummert et al., 1995] also include mechanisms to take advantage of a weak network link when one is available. A drawback of the system is that it does not allow clients to exchange data among themselves. Each client only exchanges data with the server. This may be overly restrictive and inefficient, for example, when collaborating users are located close to each other and are able to communicate with each other via a connection that is better than their connection to the remote server.

The Coda system employs optimistic concurrency control in the sense that multiple clients are allowed to update the same data concurrently. Conflicts are detected and resolved when the updates are replayed at the server. This approach works fairly well in practice because concurrent write-sharing, and hence conflicting updates, are rare in most file-system workloads. Most systems described in this section use optimistic concurrency control as opposed to pessimistic concurrency control where techniques like locking are used to make sure that conflicting updates do not happen. In a disconnected or partially connected environment, pessimism limits availability and performance, often unnecessarily.

Fluid Replication [Kim et al., 2002] is a Coda extension that introduces an intermediate level of storage between mobile clients and stationary servers, called "WayStations." The idea is to flush new data aggressively from the clients to WayStations near them, which provides a higher degree of data reliability without incurring communication costs across the wide area. Data movement from the WayStations to the servers is deferred, but knowledge of updates (i.e., only metadata) is exchanged between the WayStations and the servers periodically. This separation of data from metadata allows the system to use different mechanisms for ensuring reliability and visibility of data.

As mentioned above, the lack of client-to-client interactions can be overly restrictive in many situations. The Bayou system [Terry et al., 1995; Petersen et al., 1997] uses a peer-to-peer architecture to manage a set of mobile clients. Each device hosts a complete replica of a database and alternates between two distinct states of operation: disconnected and merging. In the disconnected state, the user of the device only sees local state stored on the device. In the merging state, a device communicates with any other device, and updates known to each are propagated to the other. After merging, newly propagated data is available for use on each device. Bayou includes an application-level toolkit for building distributed collaborative applications. Application-specific conflict detectors and resolvers can be specified, which are invoked when the system merges updates from two devices. The Bayou system, however, much like Coda, combines data and metadata into a single update log. A Bayou client is not allowed to read new data on a nearby device without replaying the new update log onto its own local copy of the database. This can lead to significant user-perceived latency, especially when the log contains a lot of data.

The PersonalRAID system [Sobti et al., 2002], designed to manage a distributed set of devices owned by a single user, completely separates data from metadata. When two devices are near each other, metadata is quickly exchanged and data on either device is immediately available for use in a coherent manner without waiting for any data propagation. An enhanced version of the PersonalRAID system is currently under development [Garg et al., 2002]. It includes support for data sharing among multiple users, and in addition, it does not mandate any kind of data or metadata propagation before allowing access to data on different devices in a coherent manner.

Ficus [Page et al., 1998] and Pangaea [Saito et al., 2002] are wide-area file systems that widely replicate data at a fine granularity of file and directories. These systems also allow independent concurrent updates to replicas and propagate these updates in a peer-to-peer manner. Conflicts are detected and resolved during update propagation. Pangaea includes mechanisms to track all replicas of a given object, and to aggressively and efficiently push updates to them in a network-aware fashion.

OceanStore [Kubiatowicz et al., 2000; Rhea et al., 2003] and Ivy [Muthitacharoen et al., 2002] are DHT based wide-area file systems. OceanStore attempts to provide strong consistency semantics for the stored objects. It uses primary-copy replication to reduce communication costs. Each object has *a primary replica* that serializes all updates to the object. The primary replica itself is implemented by a small community of servers that use a fault-tolerant Byzantine agreement protocol to commit the updates. Ivy, on the other hand, maintains a separate update log for each writer in the system, all of whom are allowed to proceed independently and concurrently. Reads are performed by consulting the tails of these logs and resolving conflicts among them, if any.

Experience from systems described above suggests that write-sharing and conflicting updates are rare, especially in file-system workloads [Reiher et al., 1994; Kistler and Satyanarayanan,1992]. Also, most of the conflicts are automatically resolvable. Therefore, optimistic concurrency control seems to be a good choice in favor of availability and performance in weakly connected environments.

45.7 Security

Many issues relating to storage systems properly lie under the umbrella of security. In this section we go beyond the conventional and somewhat narrow definition of security, giving instead a sketched perspective that includes more dimensions of the storage problem, and systems that range from single disk drives to self-organized worldwide publishing systems.

As storage has evolved from single disk drives attached to an isolated CPU, to its increasingly distributed form, these security-related issues have come to the forefront. What brings them together is that the conceptual, algorithmic, and mathematical ideas of cryptography, coding theory, and distributed algorithms often lead to principled solutions and techniques.

At the lowest level, the system must provide some assurance that the data retrieved are, in fact, the data that were deposited. This issue is important for any storage system, but it becomes critical when one moves to a distributed solution. These low-level issues have been effectively addressed for the simplest storage systems, i.e., a disk drive attached to a CPU. The *mean time before failure* (MTBF) of such systems is several years. But MTBF erodes roughly linearly as nodes are added to a distributed system. So low-level issues that are seldom a problem for single-node systems become critical for large distributed systems.

Defects in a disk's magnetic media can cause isolated data errors. The entire drive can also fail. Here, the *adversary* we are concerned with is *nonmalicious*, generally random in its choice of hardware system targets.

Specially designed *error correcting codes* (ECC) protect against media failure in today's media drives. They are generally based on classical Reed–Solomon codes, using the Berlekamp–Massey decoding algorithm [Berlekamp 1968; Massey 1969]. Binary exclusive or XOR, perhaps the simplest example of an *erasure code*, is the dominant method used to protect against the failure of entire drives [Patterson et al., 1988] and also against errors that a drive's ECC mechanism cannot correct.

Another source of nonmalicous attacks is *software bugs*. The operating system, or particular file system implementations, might scribble on the disk. So the data appear to be valid (based on low-level hardware error-correcting codes), but they are not.

Even DRAM can cause problems. Current operating systems maintain large RAM caches of disk blocks, and DRAM errors can lead to bad data in the storage system. DRAM with ECC is an effective precaution in large systems.

One way to add another layer of protection against data errors in a large system is to use strong hash functions or message authentication codes (MACs). A data block's content is hashed with its address and, in some cases, with a revision or time value. The reader can then receive a strong assurance that the block's data have not changed since the time they were written. Authentication approaches can also be implemented at the file or object level. It is much more difficult to protect against widespread systemic problems such as a bug in the operating system software.

If bad data blocks or objects are detected, or entire nodes fail, or if portions of the network fail, large systems must use redundancy of some form to keep their data secure and available. Algorithms here range from simple mirroring to distributed erasure coding schemes such as OceanStore [Kubiatowicz et al., 2000; Rhea et al., 2003] to Intermemory [Goldberg and Yianilos, 1998; Chen et al., 1999]. These approaches are also relevant in keeping data secure against natural disasters and other large-scale system failures. This illustrates that distributed storage can help deal with the important administrative issue of *backups* — a subject we do not otherwise discuss in this chapter.

Secure administration of distributed storage systems is also a challenge. Approaches range from simple centralized control to limited redundancy to fully *symmetric* schemes based on advanced algorithms from the field of distributed algorithms. Many distributed administration issues can be reduced to well-studied subproblems from this field, such as Byzantine agreement. See Lynch [1996] for an excellent discussion and bibliography.

Ongoing industry work aims to standardize many practical aspects of storage management and enterprise computing. At the center is the Common Information Model standard, the Distributed Management Task Force [DMTF 1999], the Storage Networking Industry Association [SNIA 2003], and other broader and related efforts. The issue of administration is also financially critical. In recent years, many analysts within the storage industry have demonstrated that administration is by far the largest component of a storage system's total cost of ownership — larger than the hardware and software that compose the system.

It is almost always important to keep data secure against unauthorized access. In centralized storage systems, physical security is often enough. That is, access to the data center is tightly controlled. When one moves to physically distributed designs, the situation is more complicated, and well-known algorithms from both public-key and symmetric-key cryptography provide the necessary foundation. But providing useful security of this kind in a distributed system is complicated as soon as one moves beyond the simple matter of encrypting each user's data so that only he or she can read it. Workgroup security must be supported. But as soon as more than one person has access rights, the matter of rights *revocation* should be considered, as well as other issues such as key *escrow*. Producing an effective working system is a complicated design and engineering task built on cryptographic foundations.

When it comes to intellectual property security issues, there is an interesting division of perspectives and algorithmic work.

On the one hand, applied cryptographers are striving to create systems that prevent unlicensed access to the intellectual property of others (such as digital music or movies). Here the individual seeking unauthorized access is the adversary, and the owner seeks to keep its content secure against misuse. The story of early peer-to-peer file-sharing systems such as Napster demonstrated not only the *power* of public distributed storage but also the dedication of intellectual property owners (in this case music publishers) to asserting their copyrights through the legal system.

At the same time, there is a community that views institutions and governments as the adversary and seeks to ensure the ability to use digital storage systems to freely *publish*. Here also ideas from cryptography have been applied [Clarke et al., 2000; Dingledine et al., 2000; Waldman et al., 2000]. A related notion is that of *anonymity* with respect to both reading from and writing to such systems. If such systems achieve widespread use, they will continue our society's debate regarding the legal liability of network providers and individuals who operate participating nodes.

Returning to the protection of intellectual property, it seems clear that the world is moving toward cryptographic standards to protect content. When implemented purely as software, special ideas and algorithms are required to make the result as *tamper resistant* as possible. Over the past several years security has started to move into the operating system and system hardware [Franklin, 2000]. All current approaches combine some form of cryptographic authentication based on public-key cryptography to create a *chain of trust* establishing a holder's rights.

The simplest situation involves an individual obtaining rights to use information on a particular machine. When license transfers are considered (both to other individuals and other machines), issues such as time restrictions on access, and other rules relating to acceptable use, make the situation much more complicated. Here, proprietary applied cryptographic algorithms developed in industry (Intertrust/Microsoft) are being used to address this issue.

We conclude our discussion of security by identifying what is, perhaps, the most insidious and unavoidable adversary, *the passage of time*. If a data owner would like to be secure against this adversary, that is, store it *archivally*, then widely distributed storage systems can help by ensuring that the bits are preserved even when individual storage nodes or network elements fail. We close this section by remarking that preserving the *bits* is not enough. To be truly secure against the passage of time, one must somehow ensure that their semantic value can be evaluated at some later time. The two broad algorithmic approaches to achieving this are: object standardization and emulation. The former seems clear. The latter involves preserving the semantics of data objects by preserving the correct operation of computer programs via emulation. We see this is an interesting and important area for future work.

45.8 Conclusion

Distributed storage has come a long way, both in terms of its impact on the world and as an area for academic study. Still, our society's demand for benefits of distributed storage seems to be accelerating, while there is increased sensitivity to the issues of intellectual property ownership, privacy, manageability, multiuser performance, and many others.

While simple single-user raw performance will always be an issue, we see these other dimensions of the storage problem as the new focus. We were forced, in effect, to accept the performance penalties associated with high-level programming languages so that we could correctly express more complex ideas. In the same way, we may be forced to place simple, raw performance in the *back seat* in order to deploy the worldwide distributed storage systems of immense proportions that we are surely heading toward. The result will be many opportunities for algorithmic invention, synthesis, and for the developments of much-needed standards.

The field of distributed storage is the result of tremendous improvements in storage and networking technology. Both have advanced, but architecture turns on ratios, and we are left to wonder what the future holds. If the cost to store a bit falls much faster than the cost to send one, then our offices and homes may be full of data. On the other hand, if transmission wins, they may be nearly empty. This is, of course, a simplification that ignores latency and other considerations, but may, nevertheless, represent a valid *macroeconomic* viewpoint. The authors' own computations suggest that over the past 25 years, storage is in the lead, but not by much. Progress has not been uniform during this period, so the long-term result is not yet clear.

References

Anderson, Eric, Michael Hobbs, Kimberly Keeton, Susan Spence, Mustafa Uysal, and Alistair Veitch. Hippodrome: Running circles around storage administration. In *Proceedings of the 1st Conference on File and Storage Technologies,* January 2002a. www.usenix.org/publications/library/proceedings/fast02/andersonHIP.html.

Anderson, Eric, Ram Swaminathan, Alistair Veitch, Guillermo A. Alvarez, and John Wilkes. Selecting RAID levels for disk arrays. In *Proceedings of the 1st Conference on File and Storage Technologies,* January 2002b. www.usenix.org/publications/library/proceedings/fastC2/andersonRAID.html.

Anderson, Ross J. The eternity service. In *Proceedings of the PRAGOCRYPT.* CTU Publishing House, Prague, 1996.

Anderson, Thomas E.F., Michael D. Dahlin, Jeanna M. Neefe, David A. Patterson, Drew S. Roselli, and Randolph Y. Wang. Serverless network file systems. In *Proceedings 15th ACM Symposium on Operating Systems Principles,* pp. 109–126, December 1995.

Andresen, D., T. Yang, V. Holmedahl, and O. Ibarra. Sweb: Towards a scalable World Wide Web server on multicomputers. In *Proceedings of the 10th IEEE International Symposium on Parallel Processing,* pp. 850-856,1996. citeseer.nj.nec.com/article/andresen96sweb.html.

Berlekamp, Elwyn. *Algebraic Coding Theory.* McGraw Hill, New York, 1968.

Brisco, T. RFC 1794: DNS support for load balancing, April 1995.

Byers, John, Jeffrey Considine, and Michael Mitzenmacher. Simple load balancing for distributed hash tables. In *Proceedings of the International Workshop on Peer-to-Peer Systems,* February 2003.

Carley, L. Richard, Gregory R. Ganger, and David F. Nagle. MEMS-based integrated-circuit mass-storage systems. *Communications of the ACM,* 43(11):72-80, November 2000.

Chankhunthod, Anawat, Peter B. Danzig, Chuck Neerdaels, Michael F. Schwartz, and Kurt J. Worrell. A hierarchical internet object cache. In *Proceedings of the USENIX Annual Technical Conference,* pp. 153-164,1996. citeseer.nj.nec.com/chankhunthod95hierarchical.html.

Chao, C., R. English, D. Jacobson, A. Stepanov, and J. Wilkes. Mime: A high-performance parallel storage device with strong recovery guarantees. In *Hewlett Packard Laboratories Report, HPL-CSP-92-9,* November 1992.

Chen, P., E. Lee, G. Gibson, R. Katz, and D. Patterson. RAID: High-performance, reliable secondary storage. *ACM Computing Surveys,* 26(2): 145–188, June 1994.

Chen, Yuan, Jan Edler, Andrew Goldberg, Allan Gottlieb, Sumeet Sobti, and Peter Yianilos. A prototype implementation of archival intermemory. In *Proceedings of the 4th ACM Conference on Digital libraries,* 1999.

Clarke, I., O. Sandberg, B. Wiley, and T. Hong. Freenet: A distributed anonymous information storage and retrieval system. In *Proceedings of the ICSI Workshop on Design Issues in Anonymity and Unobservability,* Berkeley, CA, 2000.

Dabek, Frank, M. Frans Kaashoek, David Karger, Robert Morris, and Ion Stoica. Wide-area cooperative storage with CFS. In *Proceedings of the 18th ACM Symposium on Operating Systems Principles,* October 2001. citeseer.nj.nec.com/dabek01widearea.html.

Dahlin, M., R. Wang, T. Anderson, and D. Patterson. Cooperative caching: Using remote client memory to improve file system performance. In *Proceedings of the 1st Symposium on Operating Systems Design and Implementation,* pp. 267–280, November 1994.

Dingledine, Roger, Michael J. Freedman, and David Molnar. The Free Haven project: Distributed anonymous storage service. In *Proceedings of the Workshop on Design Issues in Anonymity and Unobservability,* July 2000.

The Distributed Mangagement Task Force DMTF. Common Information Model (CIM) specification, v2.2. http://www.dmtf.org/standards/standard_wbem.php, June 1999.

Fan, Li, Pei Cao, Jussara Almeida, and Andrei Z. Broder. Summary cache: a scalable wide-area Web cache sharing protocol. *IEEE/ACM Transactions on Networking,* 8(3): 281–293, 2000. citeseer.nj.nec.com/fan00summary.html.

Feeley, M. J., W. E. Morgan, F. P. Pighin, A. R. Karlin, H. M. Levy, and C. A. Thekkath. Implementing global memory management in a workstation cluster. In *Proceedings of the 15th ACM Symposium on Operating Systems Principles,* pp. 201–212, December 1995.

Fielding, R., J. Gettys, J. Mogul, H. Frystyk, L. Masinter, P. Leach, and T. Berners-Lee. RFC 2616: Hypertext transfer protocol–http/1.1, June 1999.

Franklin Electronic Publishers. http://www.franklin.com, 2000.

Gadde, Syam, Jeff Chase, and Michael Rabinovich. A taste of crispy squid. In *Proceedings of the Workshop on Internet Server Performance,* 1998. citeseer.nj.nec.com/gadde98taste.html.

Garg, Nitin, Yilei Shao, Elisha Ziskind, Sumeet Sobti, Fengzhou Zheng, Junwen Lai, Arvind Krishnamurthy, and Randolph Wang. A peer-to-peer mobile storage system. Technical Report TR-664-02, Princeton University Computer Science, October 2002.

Goldberg, Andrew V. and Peter N. Yianilos. Towards an archival intermemory. In *Proceedings of the IEEE International Forum on Research and Technology Advances in Digital Libraries,* pp. 147–156. IEEE Computer Society, April 1998.

Griffin, J., S. Schlosser, G. Ganger, and D. Nagle. Modeling and performance of MEMS-based storage devices. In *Proceedings of the ACM Sigmetrics Conference on Measurement and Modeling of Computer Systems,* pp. 56–65, June 2000.

Hildrum, Kirsten, John D. Kubiatowicz, Satish Rao, and Ben Y. Zhao. Distributed object location in a dynamic network. In *Proceedings of the 14th ACM Symposium on Parallel Algorithms and Architectures,* pp. 41–52, August 2002. cite-seer.nj.nec.com/article/hildrum02distributed.html.

Howard, J., M. Kazar, S. Menees, D. Nichols, M. Satyanarayanan, R. Sidebotham, and M. West. Scale and performance in a distributed file system. ACM *Transactions on Computer Systems,* 6(1):51–81, February 1988.

Karger, David R., Eric Lehman, Frank Thomson Leighton, Rina Panigrahy, Matthew S. Levine, and Daniel Lewin. Consistent hashing and random trees: Distributed caching protocols for relieving hot spots on the World Wide Web. In *Proceedings of the ACM Symposium on Theory of Computing,* pp. 654–663, 1997. L citeseer.nj.nec.com/karger97consistent.html.

Karger, David and Alex Sherman. Web caching with consistent hashing. In *Proceedings of the 8th International World Wide Web Conference,* 1999.

Katz, E. D., M. Butler, and R. McGrath. A scalable HTTP server: the NCSA prototype. *Computer Networks and ISDN Systems,* 27: 155–164, 1994.

Kim, M., L. P. Cox, and B. D. Noble. Safety, visibility, and performance in a wide-area file system. In *Proceedings of the 1st Conference on File and Storage Technologies,* January 2002.

Kistler, J. and M. Satyanarayanan. Disconnected operation in the Coda file system. ACM *Transactions on Computer Systems,* 10(1): 3–25, February 1992.

Kubiatowicz, John, David Bindel, Yan Chen, Steven Czerwinski, Patrick Eaton, Dennis Geels, Ramakrishna Gummadi, Sean Rhea, Hakim Weatherspoon, Westley Weimer, Chris Wells, and Ben Zhao. Oceanstore: An architecture for global-scale persistent storage. In *Proceedings of the 9th International Conference on Architectural Support for Programming Languages and Operating Systems,* November 2000.

Lee, E. K. and C. E. Thekkath. Petal: Distributed virtual disks. In *Proceedings of the 7th International Conference on Architectural Support for Programming Languages and Operating Systems,* pp. 84–92, October 1996.

Li, K. and P. Hudak. Memory coherence in shared virtual memory systems. *ACM Transactions on Computer Systems,* 7(4): 321–359, November 1989.

Lynch, Nancy. *Distributed Algorithms.* Morgan Kaufmann, New York, 1996.

Massey, J. L. Shift register synthesis and BCH decoding. *IEEE Transactions on Information Theory,* 15: 122–127, 1969.

Maymounkov, P. and D. Mazieres. Kademlia: A peer-to-peer information system based on the XOR metric. In *Proceedings of the International Workshop on Peer-to-Peer Systems,* 2002. citeseer.nj.nec.com/article/maymounkov02kademlia.html.

McKusick, M., W. Joy, S. Leffler, and R. Fabry. A fast file system for UNIX. *ACM Transactions on Computer Systems,* 2(3): 181–197, August 1984.

Menaud, Jean-Marc, Valerie Issarny, and Michel Banatre. A new protocol for efficient cooperative transversal web caching. In *Proceedings of the International Symposium on Distributed Computing,* pp. 288–302, 1998. citeseer.nj.nec.com/277818.html.

Mummert, L. B., M. R. Ebling, and M. Satyanarayanan. Exploiting weak connectivity for mobile file access. In *Proceedings of the 15th ACM Symposium on Operating Systems Principles,* December 1995.

Muthitacharoen, Athicha, Robert Morris, Thomer M. Gil, and Benjie Chen. Ivy: A read/write peer-to-peer file system. In *Proceedings of the 5th Symposium on Operating Systems Design and Implementation,* 2002. citeseer.nj.nec.com/muthitacharoen02ivy.html.

Nelson, M., B. Welch, and J. Ousterhout. Caching in the sprite network file system. *ACM Transactions on Computer Systems,* 6(1), February 1988.

Ng, S. W. Improving disk performance via latency reduction. *IEEE Transactions on Computers,* 40(1): 22–30, January 1991.

Oram, Andy, Ed. *Peer-To-Peer: Harnessing the Power of Disruptive Technologies.* O'Reilly and Associates, Sebastopol, CA, March 2001.

Page, T. W., Jr., R. G. Guy, J. S. Heidemann, D. H. Ratner, P. L. Reiher, A. Goel, G. H. Kuenning, and G. Popek. Perspectives on optimistically replicated peer-to-peer filing. *Software — Practice and Experience,* 28(2): 155–180, 1998. cite-seer.nj.nec.com/page97perspectives.html.

Pai, Vivek S., Mohit Aron, Gaurav Banga, Michael Svendsen, Peter Druschel, Willy Zwaenepoel, and Erich M. Nahum. Locality-aware request distribution in cluster-based network servers. In *Proceedings of the Architectural Support for Programming Languages and Operating Systems,* pp. 205–216, 1998. citeseer.nj.nec.com/article/pai98localityaware.html.

Partridge, C., T. Mendez, and W. Milliken. RFC 1546: Host anycasting service. November 1993, ftp.internic.net/rfc/rfcIS46.txt, ftp://ftp.math.utah.edu/pub/rfc/rfc1546.txt.

Patterson, D., G. Gibson, and R. Katz. A case for redundant arrays of inexpensive disks (RAID). In *Proceedings of the International Conference on Management of Data,* pp. 109–116, June 1988.

Patterson, D. A. and J. L. Hennessy, *Computer architecture: A quantitative Approach.* Morgan Kaufmann, San Francisco, 1996.

Petersen, Karin, Mike J. Spreitzer, Douglas B. Terry, Marvin M. Theimer, and Alan J. Demers. Flexible update propagation for weakly consistent replication. In *Proceedings of the 16th ACM Symposium on Operating Systems Principles,* pp. 288–301, October 1997.

Rao, Ananth, Kartbik Lakshminarayanan, Sonesh Surana, Richard Karp, and Ion Stoica. Load balancing in structured p2p systems. In *Proceedings of the International Workshop on Peer-to-Peer Systems,* February 2003.

Ratnasamy, Sylvia, Paul Francis, Mark Handley, Richard Karp, and Scott Shenker. A scalable content-addressable network. In *Proceedings of the ACM SIGCOMM Conference,* August 2001.

Reiher, Peter, John Heidemann, David Ratner, Greg Skinner, and Gerald Popek. Resolving File Conflicts in the Ficus File System. In *Proceedings of the Summer USENIX Conference,* pp. 183–195, June 1994.

Rhea, Sean, Patrick Eaton, Dennis Geels, Hakim Weatherspoon, Ben Zhao, and John Kubiatowicz. POND: The Oceanstore prototype. In *Proceedings of the 2nd Conference on File and Storage Technologies,* March 2003.

Ritchie, D. M. and K. Thompson. The UNIX time-sharing system. *Communications of the ACM,* 17(7): 365–375, July 1974.

Rosenblum, Mendel and John K. Ousterhout. The design and implementation of a log-structured file system. *ACM Transactions on Computer Systems,* 10(1): 26–52, 1992. cite-seer.nj.nec.com/rosenblum91design.html.

Rousskov, Alex and Duane Wessels, Cache digests. *Computer Networks and ISDN Systems,* 30(22–23): 2155–2168,1998. citeseer.nj.nee.com/rousskav98cache.html.

Rowstron, A., and P. Druschel, Pastry: Scalable, distributed object location and routing for large-scale peer-to-peer systems. In *Proceedings of the IFIP/ACM International Conference on Distributed Systems Platforms (Middleware),* pp. 329–350, November 2001a.

Rowstron, Antony I. T. and Peter Druschel. Storage management and caching in PAST, a large-scale, persistent peer-to-peer storage utility. In *Proceedings of the Symposium on Operating Systems Principles,* pp. 188–201, 2001b. citeseer.nj.nec.com/rowstron01storage.html.

Saito, Yasushi, Brian N. Bershad, and Henry M. Levy. Manageability, availability, and performance in Porcupine: A highly scalable, cluster-based mail service. In *Proceedings of the Symposium on Operating Systems Principles,* pp. 1–15, 1999. URL cite-seer.nj.nec.com/article/saito99manageability.html.

Saito, Yasushi, Christos Karamanolis, Magnus Karlsson, and Mallik Mahalingam. Taming aggressive replication in the Pangaea wide-area file system. In *Proceedings of the 5th Symposium on Operating Systems Design and Implementation,* December 2002.

Sandberg, R., D. Goldberg, S. Kleiman, D. Walsh, and B. Lyon. Design and implementation of the Sun network filesystem. In *Proceedings of the Summer USFNIX Conference,* pp. 119–130, June 1985.

Schroeder, T., S. Goddard, and B. Ramamurthy. Scalable web server clustering technologies. *IEEE Network,* 14(3): 38–45, May/June 2000.

The Storage Networking Industry Association SNIA. The storage management initiative. http://www.snia.org/smi/home, 2003.

Sobti, Sumeet, Nitin Garg, Chi Zhang, Xiang Yu, Arvind Krishnamurthy, and Randolph Wang. PersonalRAID: Mobile storage for distributed and disconnected computers. In *Proceedings of the 1st Conference on File and Storage Technologies,* pp. 159–174, January 2002.

Stoica, Ion, Robert Morris, David Karger, Frans Kaashoek, and Hari Balakrishnan. Chord: A scalable peer-to-peer lookup service for Internet applications. In *Proceedings of the ACM SIGCOMM Conference,* pp. 149–160, 2001. citeseer.nj.nec.com/stoica01chord.html.

Terry, Douglas B., Marvin M. Theimer, Karin Petersen, Alan J. Demers, Mike J. Spreitzer, and Carl Hauser. Managing update conflicts in Bayou, a weakly connected replicated storage system. In *Proceedings of the 15th Symposium on Operating Systems Principles,* pp. 172–183, December 1995.

Tewari, Renu, Michael Dahlin, Harrick M. Vin, and Jonathan S. Kay. Design considerations for distributed caching on the Internet. In *Proceedings of the International Conference on Distributed Compacting Systems,* pp. 273–284, 1999. citeseer.nj.nec.com/tewari99design.html.

Thekkath, Chandramohan, Timothy Mann, and Edward Lee. Frangipani: A scalable distributed file system. In *Proceedings of the 16th ACM Symposium on Operating Systems Principles,* pp. 224–237, October 1997.

Uysal, Mustafa, Arif Merchant, and Guillermo A. Alvarez. Using MEMS-based storage in disk arrays. In *Proceedings of the 2nd Conference on File and Storage Technologies,* March 2003.

Valloppillil, Vinod and Keith W. Ross. Cache Array Routing Protocol v1.0, 1998. Available at: http://icp.ircache.net/carp.txt.

Waldman, Marc, Aviel D. Rubin, and Lorrie Faith Cranor. Publius: A robust, tamper-evident, censorship-resistant, web publishing system. In *Proceedings of the 9th USENIX Security Symposium,* pp. 59–72, August 2000.

Walker, B., G. Popek, R. English, C. Kline, and G. Thiel. The LOCUS distributed operating system. In *Proceedings of the 5th Symposium on Operating Systems Principles,* pp. 49–69, October 1983.

Wang, Limin, Vivek Pai, and Larry Peterson. The effectiveness of request redirection on CDN robustness. In *Proceedings of the Symposium on Operating Systems Design and Implementation,* December 2002.

Wang, R., T. Anderson, and D. Patterson. Virtual log-based file systems for a programmable disk. In *Proceedings of the Symposium on Operating Systems Design and Implementation,* 1999.

Wilkes, John, Richard A. Golding, Carl Staelin, and Tim Sullivan. The HP AutoRAID hierarchical storage system. *ACM Transactions on Computer Systems,* 14(1):108–136, February 1996.

Wolman, Alec, Geoffrey M. Voelker, Nitin Sharma, Neal Cardwell, Anna R. Karlin, and Henry M. Levy. On the scale and performance of cooperative web proxy caching. In *Proceedings of the Symposium on Operating Systems Principles,* pp. 16–31, 1999. cite-seer.nj.nec.com/wolman99scale.html.

Yu, X., B. Gum, Y. Chen, R. Y. Wang, K. Li, A. Krishnamurthy, and T. E. Anderson. Trading capacity for performance in a disk array. In *Proceedings of the 4th Symposium on Operating Systems Design and Implementation,* October 2000.

Zegura, Ellen W., Mostafa H. Ammar, Zongming Fei, and Samrat Bhattacharjee. Application-layer any-casting: a server selection architecture and use in a replicated Web service. *IEEE/ACM Transactions on Networking,* 8(4): 455–466, 2000. cite-seer.nj.nec.cam/411865.html.

Zhang, C., X. Yu, A. Krishnamurthy, and R. Y Wang. Configuring and scheduling an eager-writing disk array for a transaction processing workload. In *Proceedings of the 1st Conference on File and Storage Technologies,* January 2002.

46

System Management and Security Policy Specification[1]

CONTENTS

Abstract 46-1
Keywords 46-2
46.1 Introduction 46-2
46.2 Security Policy Specification 46-3
46.2.1 Role-Based Access Control 46-3
46.2.2 IBM's Trust Policy Language 46-5
46.2.3 Other Security Policy Specification Approaches 46-6
46.3 Management Policy Specification 46-7
46.3.1 Lucent's Policy Definition Language 46-7
46.3.2 CIM Policy Model 46-7
46.3.3 Other Approaches to Policy Specification 46-10
46.4 Ponder 46-11
46.4.1 Domains 46-12
46.4.2 Ponder Primitive Policies 46-12
46.4.3 Ponder Composite Policies 46-13
46.5 Research Issues 46-16
46.5.1 Conflict Analysis 46-16
46.5.2 Refinement 46-17
46.5.3 Multiple Levels of Policy 46-17
46.6 Conclusions 46-17
Acknowledgements 46-18
References 46-18

Morris Sloma
Emil Lupu

Abstract[1]

Policies are rules governing the choices in the behavior of a system. They are increasingly being used as a means of implementing flexible and adaptive systems for management of Internet services, networks, and security systems. There is also a need for a common specification of security policy for large-scale, multi-organizational systems where access control is implemented in a variety of heterogeneous components. In this paper we survey both security and management policy specification approaches, concentrating on practical systems in which the policy specification can be directly translated into an implementation.

[1]This chapter is a revised version of a paper that appeared in IEEE Network Special Issue on Policy Based Networking, Vol. 16, No. 2, March 2002, pp. 10–19.

1-58488-381-2/05/$0.00+$1.50
© 2005 by CRC Press LLC

Keywords

Policy-based management, security policy, security management, authorization policy, obligation policy, roles, adaptive systems.

46.1 Introduction

Modern network components such as routers and switches need to be programmable to support the adaptive quality of service (QoS) required by multimedia applications and mobile computing users. Portable intelligent communicators will always have limited battery life, processing, storage, and communication capability compared to desktop workstations and so will need to make use of local network services to provide a seamless, ubiquitous computing environment. This environment will require fast service creation and resource management through a combination of network-aware applications and application-aware networks. Adaptive networks must support rapid deployment of customized services tailored for potentially mobile, corporate, and individual users.

Policies provide a simple and flexible means of specifying network management strategy for adaptation to specific events such as component failures, application requests for new services, and dynamic changes in application requirements [Sloman and Lupu, 1999]. This provides a more constrained means for specifying adaptive behavior than the many mechanisms being promoted for programming network components. The alternative approaches to "programming behaviors" include code carrying IP packets which are executed by the routers traversed by the packets; scripts or interpreted code loaded via a management interface and mobile agents carrying both code and data which autonomously migrate around the network [Konstantinos, 1999]. All of these are very powerful mechanisms which can potentially destroy the network if the code contains malicious or inadvertent bugs.

Internet and e-commerce applications typically involve many different organizations — manufacturers, merchants, network service providers, banks, and customers. Specifying and analyzing the security for such environments to determine who can access what resources or information can be extremely difficult. The problem is that the access control is distributed across many heterogeneous components such as databases, operating systems, firewalls, and filtering routers controlled by the different interacting organizations. Even in the programmable network environment there is a need to define who is authorized and under what conditions they can inject code packets, scripts, mobile agents, or policy rules into network components and what resources or functions these "programs" can access. Security management needs similar adaptivity to that of network management to specify actions to take in response to simple security violations such as excessive login attempts but also for changing the behavior and applicable policies in firewalls or Web servers under denial of service or other network based attacks.

Policies are rules governing the choices in behavior of a system [Sloman, 1994]. *Obligation policies* are event triggered condition–action rules which can be used to define the conditions for reserving network resources, changing queuing strategy, or loading code onto a router. Some policies may be user or application specific, such as what information to filter when bandwidth or device capabilities are limited. *Authorization policies* are used to define what services or resources a subject (management agent, user, or role) can access. Policies are persistent so that a one-off command to perform an action is not a policy. Scripts and mobile agents are often based on powerful interpreted languages such as Java, and so can be used to introduce new functionality into network components. Policies define choices in behavior in terms of the conditions under which predefined operations or actions can be invoked rather than changing the functionality of the actual operations themselves.

The main motivation for the recent interest in policy-based services, networks, and security systems is to support dynamic adaptability of behavior by changing policy without recoding or stopping the system [Sloman, 1994]. This implies that it should be possible to dynamically update the policy rules interpreted by distributed entities to modify their behavior.

Large-scale systems may contain millions of users and resources. It is not practical to specify policies relating to individual entities; instead, it must be possible to specify policies relating to groups of entities and also to nested groups such as sections within departments and within sites in different countries in an international organization. It is also useful to group the policies pertaining to the rights and duties of a role or position within an organization such as a network operator, nurse in a ward, or mobile computing "visitor" in a hotel.

Policies are derived from business goals, service level agreements, or trust relationships within or between enterprises. The refinement of these abstract policies into policies relating to specific services and then into policies implementable by specific devices supporting the service is not easy and not amenable to automation.

This paper provides a survey of some of the work on policy specification for both security management and policy-driven network management. Rather than covering many papers superficially, we have concentrated on a few exemplary approaches in more detail, with emphasis on practical rather than theoretical approaches. In Section 2 we describe examples of security policy specification. Role Based Access Control provides the grouping of policies (permissions) related to an organizational position and the Trust Policy Language indicates the use of credential-based policies. In Section 3 we cover various approaches to management policy specification, namely the Policy Description Language for event-triggered policies from Lucent and the ongoing work in the IETF/DMTF on standardization of policy information models. In Section 4 we describe our Ponder policy specification language which combines many of the above concepts and can be used for both security and management policies. We do not discuss routing policies as these have been described in Stone et al. [2001]. WWW references to much of the work on policy and many of the papers described here can be found from Policies [Policies].

46.2 Security Policy Specification

46.2.1 Role-Based Access Control

Although Role-Based Access Control (RBAC) is not directly concerned with policy specification, it has been accepted as a security model which permits the specification and enforcement of organizational access control policies. The fundamental concept on which RBAC relies is that permissions are associated with roles rather than users, thus separating the assignment of users to roles from the assignment of permissions to roles. Users acquire access rights by virtue of their role memberships, and they can be dynamically assigned or removed from roles without changing the permissions associated with their role. For example, in Figure 46.1, the permissions related to the head of a department do not have to be changed when a new person is appointed to the role. Multiple users can be assigned to the same role, so there can be multiple people assigned to the lecturer or teaching assistant role in Figure 46.1, and multiple roles can be assigned to the same users.

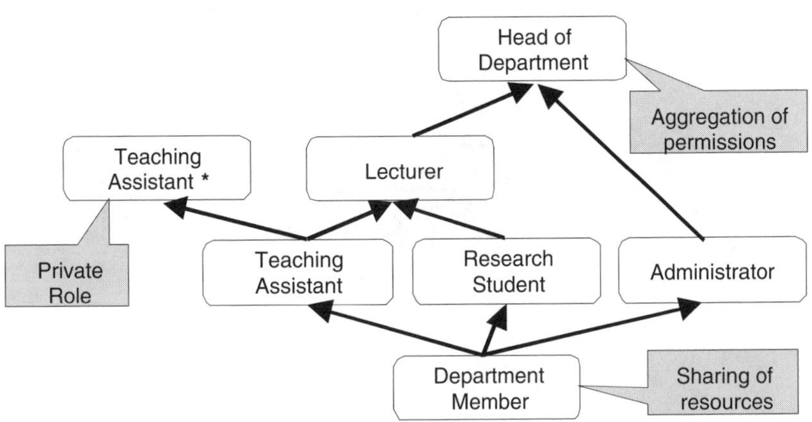

FIGURE 46.1 Role hierarchies.

Although the concept of role has existed in a fairly similar form for a long time, both in systems security and in Role Theory, the work presented by Sandhu et al. [1996] has prompted a renewed interest in this approach which is now adopted, to different degrees, in many commercial products and has been proposed as a standard [Ferraiolo et al., 2001].

The main goal of RBAC goes beyond the concept of role and aims to simplify permission management in large organizations. To achieve this, roles must be combined in a structured way and permissions must be "reused." The most popular approach relies on role inheritance where senior roles such as team leader or project supervisor inherit the permissions of junior roles such as employee or team member. However, other approaches such as assigning roles to other roles can also be found in the literature.

A possible role hierarchy for an academic institution is described in Figure 46.1 which illustrates how role inheritance provides the sharing of resources through common lower level roles and the aggregation of permissions through inheritance of permissions to the higher level. However, such role inheritance hierarchies are not without shortcomings as there are numerous exceptions to the rule that senior roles inherit all the permissions of junior roles. Most notably, access to private files and permissions granted by virtue of a competency are not inherited. For example, the head of a department does not usually inherit the access right of a system administrator. To accommodate such situations it is necessary to create private roles as shown in Figure 46.1, which group the permissions that are not inherited upwards in the hierarchy. Implementing an RBAC system with inheritance of permissions between roles considerably reduces the number of permissions in the system. However, in a distributed system it may also render access control checks, performed on each invocation, more complex since the inherited roles may be stored remotely and checking the inherited permissions may require several remote invocations. To avoid this increased complexity, a capability-based system may be more appropriate for RBAC since it shifts the responsibility for collecting the inherited permissions to the user (subject) system, and this is done prior to the access control check.

Several constraints may apply to an RBAC model across its associations, between users and roles, and roles and permissions or between roles themselves (inheritance) [Ahn and Sandhu, 2000]. Among these, separation of duty constraints, which identify mutually exclusive sets of permissions that a user is not allowed to hold, have been the subject of the most intensive work.

Figure 46.2 shows an overview of the RBAC model with the various relationships between user roles and permissions, and the constraints. The model presented in Sandhu et al. [1996] also introduces the concept of sessions. A session groups the permissions from a selected number of roles which the user may want active at a given moment in time, i.e., in a work session. The user can then use all access rights from the active roles in order to carry out his tasks.

The RBAC model presented above is the foundation from which many variations have developed. In particular, the team-based access control has developed as a means of simultaneously activating a set of related roles, e.g., the surgeon, nurses, and other personnel in an operating theatre.

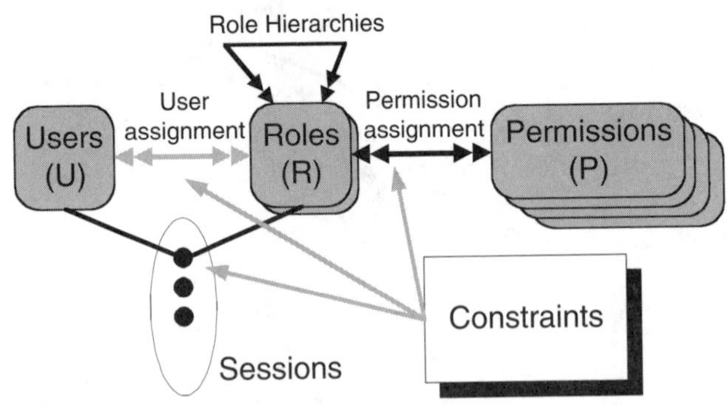

FIGURE 46.2 RBAC model. (From Sandhu, R. S., E. J. Coyne, H. L. Feinstein, and C. E. Youman. Role-based access control models. *IEEE Computer*, 29(2): 38–47, 1996.)

Although the concept of role is not new, the introduction of RBAC models has fostered a change from the traditional mandatory and discretionary access control models to new frameworks where the access control policy is neither rigidly embedded in the implementation nor left to the owner of each resource but can be implemented on the basis of clearly specified organizational policies. This is far more flexible than the simple access rules supported by most operating systems where owners of files or resources specify individuals or groups who can access their resources, and then these rules have to be changed as people leave or change access requirements as a result of reorganization of the enterprise. RBAC allows role hierarchies, user assignment to roles, and permission assignment to resources to be done independently by different administrators, not just resource owners.

46.2.2 IBM's Trust Policy Language

There are a number of groups which use the term "trust management" for frameworks that support sophisticated authorization policy specification and implementation using public key certificates as credentials to authenticate identities or membership of groups [Blaze et al., 1996, 1998]. This can be particularly useful for e-commerce and Internet applications. We will describe the IBM Trust Policy Language (TPL) and support system as an exemplar of this approach [Herzberg et al., 2000].

TPL uses XML to define policy rules which specify the criteria for a client to be assigned to a group which is similar to a role. Standard database or operating system authorization rules are then used to define what resources a group can access as shown in Figure 46.3. Thus, the authority that issues certificates for a role can be completely different from the one that defines access control policy for a resource.

The TPL policy consists of a list of required X509 certificates which permit a client to become a member of a group and a Boolean function relating the values of the fields in one or more certificates to define group membership criteria. Example groups include employees of an organization, registered customers for a merchant, or recognized cardiologists at a hospital. An example medical group membership criteria could be possession of 3 recommendation certificates from hospitals with recommendation level > 3.

A certificate identifies the issuer, the subject to whom they are issued (which could be a name or an anonymous public key), and a list of name-value pairs which may contain information such as age, sex, and rank of the holder. The certificate may also contain links to repositories for additional certificates for the issuer or subject as multiple certificates may be needed for a client to be assigned to a group.

The TPL policies shown in Figure 46.4 are for a *retailer* (the *self* group in the policy) to give discounts to preferred customers who are employees of a department of a partner company. Entities that have *partner* certificates, signed by the retailer, are placed in the group *partners*. The group *department* is defined as any user having a *partner* certificate signed by the partners group. Finally, the *customer* group consists of anyone who has an *employee* certificate signed by a member of the departments group who has a rank > 3.

TPL is available for download from IBM Alphaworks and is very flexible as a means for defining authorization policy for an Internet service, but the XML syntax is rather verbose and unreadable. Although a policy can have multiple rules, there is no inheritance or reuse between different policy

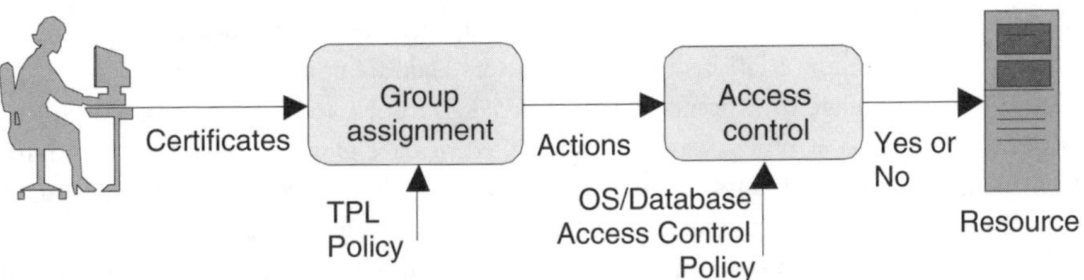

FIGURE 46.3 Separation of group assignment from access control decision.

```
<POLICY>

    <GROUP NAME="self">

    <GROUP NAME="partners">
        <RULE>
            <INCLUSION ID="partner" TYPE="partner" FROM "self"> </INCLUSION>
        </RULE>
    </GROUP>

    <GROUP NAME="departments">
        <RULE>
            <INCLUSION ID="partner" TYPE="partner" FROM="partners"> </INCLUSION>
        </RULE>
    </GROUP>

    <GROUP NAME="customers">
        <RULE>
            <INCLUSION ID="customer" TYPE="employee" FROM="departments"> </INCLUSION>
            <FUNCTION>
                <GT>
                    <FIELD ID="customer" NAME="rank"></FIELD>
                    <CONST>3/ >
                </GT>
            </FUNCTION>
        </RULE>
    </GROUP>

</POLICY>
```

FIGURE 46.4 TPL example.

specifications, and it is not suitable for specifying security *management* policy, such as the actions to take in response to security violations.

46.2.3 Other Security Policy Specification Approaches

There has been considerable interest in logic-based approaches to specifying authorization policy as exemplified by Jajodia et al. [1997, 2001]. They assume a strong mathematical background, which can make them difficult to use and understand, and they do not easily map onto implementation mechanisms. The ASL language [Jajodia et al., 1997] includes a form of metapolicies called *integrity rules* to specify application-dependent rules that limit the range of acceptable access control policies. Although it provides support for role-based access control, the language does not scale well to large systems because there is no way of grouping rules into structures for reusability. A separate rule must be specified for each action. There is no explicit specification of delegation or rights from a user to an agent to perform actions on the user's behalf and no means of specifying authorization rules for groups of objects that are not related by type. A recent paper describes extensions using predicates to evaluate hierarchical or other relationships between the elements of a system such as the membership of users in groups, inclusion relationships between objects, or supervision relationship between users [Jajodia et al., 2001].

In Ortalo [1998], a language to express security policies in information systems based on the logic of permissions and obligations, a type of modal logic called *deontic logic*, is described. Standard deontic logic centers on impersonal statements of the form "it is obliged that p" which do not necessarily identify to whom the obligation applies. Ortalo accepts the axiom $Pp = \neg O \neg p$ ("permitted p is equivalent to not p being not obliged") as a suitable definition of permission which leads to the fact that an obligation implies required permission. In our view, an obligation policy requires a relevant authorization policy (which may be specified independently of the obligation) to permit the actions defined in the obligation but an obligation policy does not automatically imply an authorization policy exists.

Others focus on the specification and implementation of access control policies for mobile agent systems. In most cases the studies focus on reconfigurable access control policies in the Java environment. They have included both the translation of higher level policies into Java security policies as well as different access control mechanisms within Java [Corradi et al., 2000; Hashii et al., 2000].

46.3 Management Policy Specification

46.3.1 Lucent's Policy Definition Language

The Policy Definition Language (PDL) was developed at Lucent Bell Labs for network management and is based on active database declarative event-condition-action rules, i.e., obligation policies [Lobo et al., 1999]. It is a very simple language with two main constructs — a policy rule corresponding to an obligation policy and a rule for triggering other events:

Policy Rule: *event* ***causes*** *action* $(t_1= v_1, \ldots .t_k=v_k)$ ***if*** *condition*

Policy Defined Event (pde): *event* ***triggers*** *pde* $(t_1= v_1, \ldots .t_k=v_k)$ ***if*** *condition*

where t_i = parameter (attribute) type and v_i = value. Every event has a timestamp and the URL of the source which generated it. Primitive events can be generated by the managed environment or by other PDEs. An Epoch is an application specific time window in which a set of events are considered to occur simultaneously, e.g., day, hour, second. Complex events include:

- e1 & e2 &... & en = conjunction of events in an epoch
- e1 | e2 |... | en = disjunction of events in an epoch
- !e = no occurrence of event in an epoch
- ^e = a sequence of zero or more occurrences of an event
- group (e1 & e2 &... & en) or group (e1 | e2 |... | en) = a single event grouping all instances of the complex event. If there are n instance of e1 and m instances of event e2 in an epoch, then there is only one occurrence of group (e1 & e2) instead of m x n occurrences.
- count (e, e.x > 4) counts the number of occurrences of event e for which parameter x > 4 is true
- Other aggregations, such as max., min. or ave. relating to event parameters.

An example is given in Figure 46.5 of the use of PDL to define a policy for a network component which rejects call requests when in overload mode.

PDL is implemented and used in Lucent switching products [Virmani et al., 2000]. It shows that event–condition–action rules are a very flexible approach to specifying management policy. It has also been extended to specify complex workflow tasks. However, it does not support any form of policy composition or reuse of specifications.

46.3.2 CIM Policy Model

The Distributed Management Task Force (DMTF) [DMTF, 1999], in collaboration with the Internet Engineering Task Force (IETF) Policy work group [IETF], are defining a policy information model as an extension to the Core Information Model (CIM), known as PCIM [Moore et al., 2001]. An information model is "an abstraction and representation of the entities in a managed environment — their properties, operation, and relationships." This is independent of any specific repository, application, protocol, or platform. The IETF are defining a mapping of the PCIM model to a directory schema so that an LDAP (Lightweight Directory Access Protocol) directory can be used as a repository. The CIM defines generic objects such as managed system elements, logical and physical elements, system services, and service access points. The Policy Model defines a policy rule and its component policy conditions and policy actions as shown in Figure 46.6. The assumption is that a policy rule is of the form if <condition set> then do <action list>. The condition set can be expressed in either disjunctive or conjunctive normal form. The action list can specify that actions are to be executed sequentially or that they can be executed in any order. Application specific policies can be formulated by defining specific subclasses derived from

```
Overload threshold: (time outs/calls made) > t
Normal threshold: (time outs/calls made) < n

Events:    normal, restricted : pde
           callMade, timeOut, powerOn: system events
           /* the above events have no attributes */

Actions:   restrictCalls, acceptAll

Policies:  powerOn triggers
           normal // a normal event is triggered when the component is switched on

           normal, ^(callMade | timeOut) triggers restricted if count (timeOut) > t*count(callMade)
           /* after a normal event, a sequence of callMade or timout events will trigger
           restrictCalls if the overload threshold is exceeded */

           restricted causes restrictCalls // sets overload mode to reject call requests

           restricted, ^(callMade | timeout) triggers normal if count (timeOut) < n*count(callMade)
           /* when in overload mode, a sequence of callMade or timeOut events will trigger
           normal if the normal threshold is exceeded */

           normal causes acceptAll

           /* Assumes only one callMade or timeOut event per epoch */
```

FIGURE 46.5 PDL example.

the CIM_PolicyCondition and CIM_PolicyAction classes. The CIM_VendorPolicyCondition and CIM_VendorPolicy action are included to allow vendor specific extensions to the Core Policy Model. The CIM_PolicyTimePeriodCondition class permits to specify sophisticated time validity period constraints for a policy rule in terms of times, masks for days in a week, days at beginning or end of month, months in year, and so on. Note that there is no explicit triggering event specified as part of the policy, although it is similar to an obligation policy. It is assumed that the agent interpreting the event will evaluate the policy when an implicit event occurs such as a new session is set up or possibly on every message.

Policy rules are always defined in some context as specified by the association to the CIM_System class. Furthermore, policy rules may be aggregated into nested policy groups to define the policies pertaining to an organisation, department or user. Conditions and actions may be specific to a rule or can optionally be stored separately in a policy repository (as shown by the PolicyConditionInPolicyReposiory and PolicyActionInPolicyRepository associations) and reused by multiple rules. The PolicyRoles attribute of a policy rule is a type of property that is used as a means of identifying elements to which a policy applies, e.g., a *gigabit Ethernet* role policy may apply to all elements which have a gigabit Ethernet interface. Policy rules are also assigned specific priority values in order to resolve conflicts, although it is not entirely clear how the conflicts are detected in the general case. At the time of writing the CIM model, including its PCIM part, is being continuously revised and updated, the latest information is available from http://www.dmtf.org.

The IETF has also defined extensions to PCIM called the QoS Policy Information Model (QPIM) which contains a set of abstractions specific to IntServ and DiffServ management [Snir et al., 2003]. For example they have abstract objects for classifiers, meters, shapers, droppers, and queues, which can be mapped onto the control elements provided by a specific DiffServ router. A simple DiffServ QPIM Policy is shown in Figure 46.7.

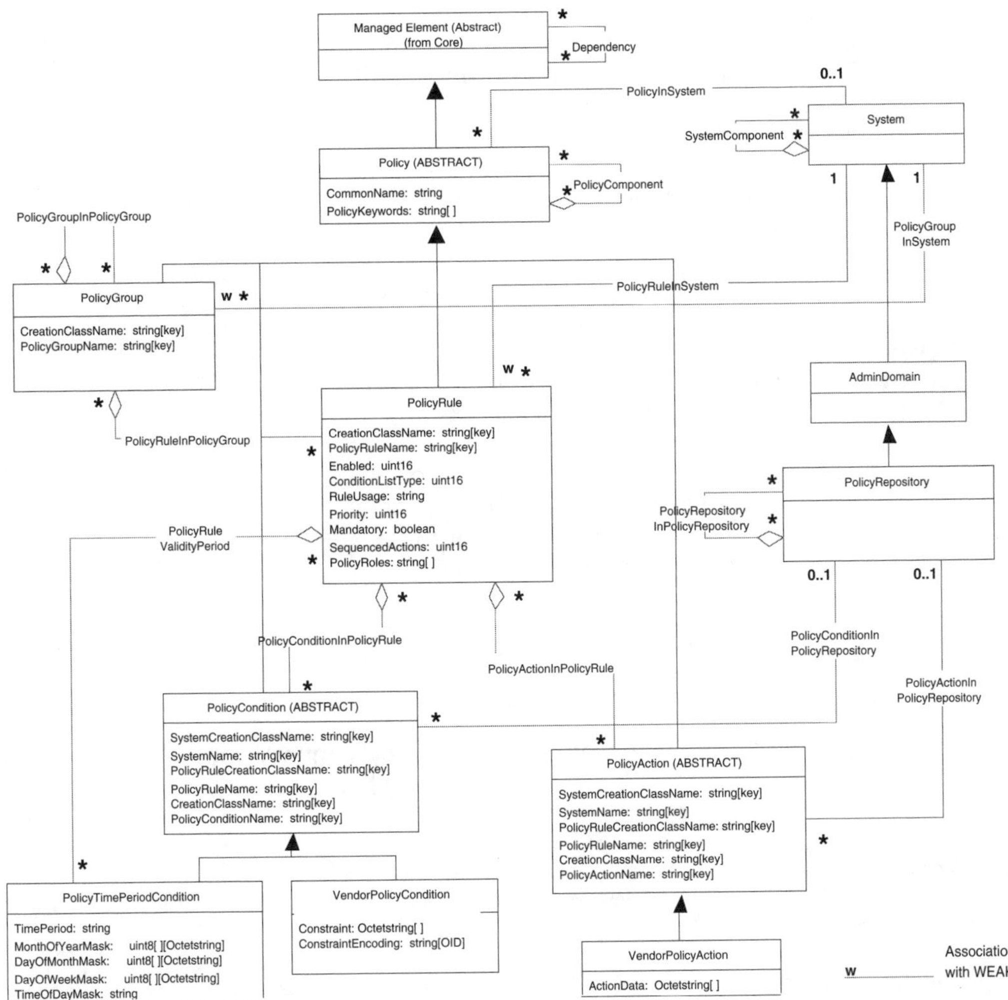

FIGURE 46.6 (PCIM) Policy Core Information Model. (From Strassner, J., Policy and the IETF — Theory and Practice, Invited presentation, *Policy 2001: Policies for Distributed Systems and Networks*, HP Labs, Bristol, U.K., January 2001, www-dse.doc.ic.ac.uk/events/policy-2001/.)

As can be seen from Figure 46.7, the LDAP schema representation of a very simple policy is extremely verbose and not really aimed at human interpretation. It is assumed this will be generated from a graphical tool of the form provided by a number of vendors or from a high level language such as Ponder, described in Section 46.4. PCIM does not distinguish between authorization and obligation policies. The emphasis in the working groups has been on QoS policies which can be considered obligation policies without event triggers. However, they can specify simple admission control policies by means of an action to accept or deny a message to give the effect of a positive or negative authorization policy.

It is assumed that policies are objects stored in an LDAP directory service. A policy consumer (policy decision point [PDP]) retrieves policies from the policy repository (e.g., LDAP server). A policy execution point (PEP), such as a router, requests policy decisions using the Common Open Policy Service Protocol (COPS). The PEP enforces the policy, for example, by permitting/forbidding requests or allocating packets from a connection to a particular queue. A PEP and PDP could be combined into a single component. Note that there is no explicit event to trigger an IETF policy. An implicit event such a packet arrival at a router will trigger the search for applicable policies. The policy condition will typically define a specific address or range of source or destination IP addresses or ports to which a policy applies so this information

```
If (SourcePort == MyWebServerPort) then Color DSCP=5

This is represented by the following LDAP Objects

Objectclass:qosPolicyRule
    Type: 1
    Direction: out
    Priority: 1

    Objectclass: qosColorPolicyAction
        DSCPValue: 5

    Objectclass: qosColorPolicyCondition
        Type: Integer OID
        Operator: "=="

        Objectclass: qosPolicyVariable
            Name: SourcePort
            Type: IntegerOID

        Objectclass: qosPolicyConstant
            Name: MyWebServerPort
            Type: IntegerOID

            Objectclass: qosPolicyNumberValue
                Type: IntegerOID
                PortValue: 80
```

FIGURE 46.7 QPIM example policy. (From Strassner, J., Policy and the IETF — Theory and Practice, Invited presentation, *Policy 2001: Policies for Distributed Systems and Networks*, HP Labs, Bristol, U.K., January 2001, www-dse.doc.ic.ac.uk/events/policy-2001/.)

taken from the packet can be used to find which policies apply. It is only practical for the PEP to query the PDP for a decision on comparatively infrequent packets such as an IntServ request related to a connection set up. If policy decisions are needed on every packet, such as for DiffServ, then these have to be preloaded into the PEP by the PDP using the COPS policy provisioning mode. Note that the use of COPS is not mandated by the IETF and other protocols such as SNMP, HTTP, or CORBA could be used for transferring policy information. There have been suggestions for extensions to the IETF approach to include explicit events for policies that deal with failures, overload situation, and so on, but so far nothing concrete has emerged.

46.3.3 Other Approaches to Policy Specification

There are a number of vendors producing policy-based management tools, usually with a graphical user interface for defining policies and allocating them to devices to which they apply. These generally follow the approach of the IETF policy work, which is still in progress, so there are not yet any standards to be compliant with. Initially, the main focus was on QoS management, but some of the vendors are extending this to include support for provisioning of virtual private networks (see [Policy] for links to vendor Web sites). Verma [2000] gives a detailed description of the architectures and algorithms needed to support an approach similar to that from the IETF for policy-based networking. It covers some of the issues of translating from higher level service level goals into policies for specific components in the network and how to distribute the policies to the relevant components. A description of an IETF-based language and architecture for implementing policy-based network resource management using CORBA is given in Flegkas et al. [2003].

Cfengine is a language-based administration system targeted primarily at configuration management of Unix and, to a lesser extent, Windows operating systems connected via a TCP/IP network [Burgess, 1995, 2001]. Cfengine grew out of the need to replace complex shell scripts used for the automation of administration tasks on Unix systems and allows the creation of single, central configuration files which describe how every host on the network should be configured. It uses the idea of classes to group hosts and dissect a distributed environment into overlapping sets. Host-classes are essentially labels which document the attributes of different systems. The following classes are meaningful in the context of a particular host: (1) the identity of the machine, including hostname, address, and network, (2) the operating system and architecture of the host, (3) an abstract user-defined group to which the host belongs, and (4) the result of any proposition about the system, including the time or date. Policies are specified for classes of hosts and define a sequence of actions regarding the configuration of a host. The following example demonstrates the use of the language for configuration management :

```
files:
    (linux|solaris).Hr12.OnTheHour.!exception_host::
        /etc/passwd mode=0644 action=fixall inform=true
```

The first line simply defines the name *files* for the action. The second line identifies the class of hosts for which the action is to be executed, followed by the actual command. The command line specifies that the cfengine agent, which is always the subject of the policy, must search for all password files with an invalid mode, fix them, and inform the administrator. The class membership expression specifies all hosts which are of type *linux* or *solaris*, during the time interval from 12:00 a.m. to 12:59 a.m., apart from a host labelled with the class *exception_host*. The second line identifies the target of the policy, i.e., all the hosts falling within the classification, the condition for execution of the policy, which is a time interval, and a trigger which specifies that the action must be executed *on the hour*. Policies are stored in a central repository, accessible to every host, and an active cfengine agent on each host executes the policies which apply only to that host.

Cfengine is a powerful and concise declarative scripting language, suitable for system administrators to automate common administrative tasks on Unix systems. However, it cannot be used to specify authorization policies and lacks support for object-oriented concepts such as inheritance and parameterized instantiation.

Minsky and colleagues' "law governed systems" specify permissions and prohibition as a set of rules which are similar to positive and negative authorisations. Their approach supports a common global set of constraints, similar to obligation policies, which are implemented by means of filters in every node which check that all interactions are consistent with a global law [Minsky and Pal, 1997].

46.4 Ponder

The Ponder language for specifying Management and Security policies [Damianou et al., 2001] evolved out of work on policy management at Imperial College over a period of about 10 years. Ponder is a declarative, object-oriented language that can be used to specify security policies that map onto various access control mechanisms for firewalls, operating systems, databases, and Java [Corradi, 2000]. It supports obligation policies that are event-triggered condition–action rules for policy-based management of networks and distributed systems. Ponder can also be used for security management activities such as registration of users or logging and auditing events for dealing with access to critical resources or security violations. Key concepts of the language include domains to group the objects to which policies apply, roles to group policies relating to a position in an organization [Lupu and Sloman, 1997], relationships to define interactions between roles and management structures to define a configuration of roles and relationships pertaining to an organizational unit such as a department or a section.

46.4.1 Domains

Domains provide a means of grouping objects that policies apply to and can be used to partition the objects in a large system according to geographical boundaries, object type, responsibility, and authority or for the convenience of human managers. Membership of a domain is explicit and not defined in terms of a predicate on object attributes. A domain does not encapsulate the objects it contains but merely holds references to objects. A domain is thus very similar in concept to a file system directory but may hold references to any type of objects, including a person. A domain that is a member of another domain, is called a *subdomain* of the parent domain. A subdomain is not a subset of the parent domain, in that an object included in a subdomain is not a *direct* member of the parent domain. It is an *indirect* member of a parent directory — similar to a file in a subdirectory. An object or subdomain may be a member of multiple parent domains, i.e., domains can overlap so an object or subdomain can have multiple parent domains. This is different from most directory systems. Domain hierarchies do not imply any form of inheritance or class hierarchy. Policies can have a subject and a target scope (see Section 46.4.2) defined in terms of domains that facilitate adding and removing objects from the domains to which policies apply without having to change the policies. Domains have been implemented as directories in an extended LDAP Service.

46.4.2 Ponder Primitive Policies

Authorization policies define what activities a member of the subject domain can perform on the set of objects in the target domain, where the subject domains could define users and the target domain defines the resources they are permitted to access. These are essentially access control policies to protect resources and services from unauthorized access. A positive authorization policy defines the actions that subjects are permitted to perform on target objects. A negative authorization policy specifies the actions that subjects are forbidden to perform on target objects.

The language provides reuse by supporting the definition of policy types to which any policy element can be passed as formal parameter. Multiple instances can then be created and tailored for the specific environment by passing actual parameters as shown by switchPolicyOps & routersPolicyOps in Figure 46.8.

Policies can also be declared directly without using a type as shown in the negative authorization policy in Figure 46.9, which also shows the use of a time based constraint to limit the applicability of the policy.

Ponder also supports a number of other basic policies for specifying security policy: *Information filtering* policy can be used to transform input or output parameters in an interaction. For example, a location service might permit access to detailed location information, such as the specific room in which a person is, only to users within the department. External users can only determine whether a person is at work or not. *Delegation* policy permits subjects to grant privileges that they possess (due to an existing authorization policy) to grantees to perform an action on their behalf, e.g., passing read rights to a printer spooler in order to print a file. *Refrain* policies act as restraints on the actions that subjects perform and

```
type auth+ PolicyOpsT (subject s, target <PolicyT> t) {
        action load(), remove(), enable(), disable() ; }

inst auth+ switchPolicyOps=PolicyOpsT(/NetworkAdmins, /Nregion/switches);
inst auth+ routersPolicyOps=PolicyOpsT(/QoSAdmins, /Nregion/routers);
```

The two policy instances created from a *PolicyOpsT* type allow members of /NetworkAdmins and /QoSAdmins (subjects) to load, remove, enable or disable objects of type PolicyT within the /Nregion/switches and /Nregion/routers domains (targets) respectively.

FIGURE 46.8 Example ponder authorization policies.

```
inst auth– /negativeAuth/testRouters {
    subject      /testEngineers/trainee ;
    action       performance_test() ;
    target       <routerT> /routers ;
    when         time.between ("0900", "1700")
}
```

Trainee test engineers are forbidden to perform performance tests on routers between the hours of 0900 and 1700. The policy is stored within the negativeAuth domain.

FIGURE 46.9 Direct policy declaration.

are similar to negative authorizations but are implemented by the subjects rather than the targets. For example, a Web browser (the subject) could enforce refrain policies to prevent children from accessing unsuitable target Web sites as they cannot be trusted to implement a negative authorization policy. See Damianou et al. [2001] for more details and examples of these policies.

Obligation policies are event-triggered, condition-action rules similar to Lucent's PDL, and define the activities subjects (human or automated manager components) must perform on objects in the target domain. Events can be simple, i.e., an internal timer event, or external events notified by monitoring service components, e.g., a temperature exceeding a threshold or a component failing. Composite events can be specified using event composition operators.

46.4.3 Ponder Composite Policies

Ponder composite policies facilitate policy management in large, complex enterprises. They provide the ability to group policies and structure them to reflect organizational structure, preserve the natural way system administrators operate, or simply provide reusability of common definitions. This simplifies the task of policy administrators.

Roles provide a semantic grouping of policies with a common subject, generally pertaining to a position within an organization such as department manager, project manager, analyst, or ward nurse. Specifying organizational policies for human managers in terms of manager positions rather than persons permits the assignment of a new person to the manager position without respecifying the policies referring to the duties and authorizations of that position. A role can also specify the policies that apply to an automated component acting as a subject in the system. Organizational positions can be represented as domains, and we consider a role to be the set of authorization, obligation, refrain, and delegation policies

```
inst oblig loginFailure {
    on                3*loginfail(userid) ;
    subject           s = /NRegion/SecAdmin ;
    target <userT>    t = /NRegion/users ^ {userid} ;
    do                t.disable() -> s.log(userid) ;
}
```

This policy is triggered by 3 consecutive loginfail events with the same userid. The NRegion security administrator (SecAdmin) disables the user with userid in the /NRegion/users domain and then logs the failed userid by means of a local operation performed in the SecAdmin object. The '->' operator is used to separate a sequence of actions in an obligation policy. Names are assigned to both the subject and the target. They can then be reused within the policy. In this example we use them to prefix the actions in order to indicate whether the action is on the interface of the target or local to the subject.

FIGURE 46.10 Example of ponder obligation policy.

with the *subject domain* of the role as their subject. A role is just a group of policies in which all the policies have the same subject, which is defined implicitly, as shown in Figure 46.11.

Managers acting in organizational positions (roles) interact with each other. A *relationship* groups the policies defining the rights and duties of roles towards each other. It can also include policies related to resources that are shared by the roles within the relationship. It thus provides an abstraction for defining policies that are not the roles themselves but are part of the interaction between the roles. The syntax of a relationship is very similar to that of a role but a relationship can include definitions of the roles participating in the relationship. However roles cannot have nested role definitions. Participating roles can also be defined as parameters within a relationship type definition as shown below.

```
type role ServiceEngineer (CallsDB callsDb) {
  inst oblig serviceComplaint {
          on              customerComplaint(mobileNo) ;
          do              t.checkSubscriberInfo(mobileNo) ->
                          t.checkPhoneCallList(mobileNo) ->
                          traceSignalReception(mobileNo);
          target   t = callsDb ; // calls register }

  inst oblig deactivateAccount { . . . }
  inst auth+ serviceActionsAuth { . . . }
  // other policies
}
```

The role type ServiceEngineer models a service engineer role in a mobile telecommunications service. A service engineer is responsible for responding to customer complaints and service requests. The role type is parameterised with the calls database, a database of subscribers in the system and their calls. The obligation policy serviceComplaint is triggered by a customerComplaint event with the mobile number of the customer given as an event attribute. On this event, the subject of the role must execute a sequence of actions on the calls-database in order check the information of the subscriber whose mobile-number was passed in through the complaint event, check the phone list and then trace the signal reception. Note that the obligation policy does not specify a subject as all policies within the role have the same implicit subject.

FIGURE 46.11 Example of role policy.

```
type rel ReportingT (ProjectManagerT pm, SecretaryT secr) {
    inst oblig reportWeekly {
          on              timer.day ("monday") ;
          subject         secr ;
          target          pm  ;
          do              mailReport() ;
  }
  //. . . other policies
}
```

The ReportingT relationship type is specified between a ProjectManager role type and a Secretary role type. The obligation policy reportWeekly specifies that the subject of the SecretaryT role must mail a report to the subject of the ProjectManagerT role every Monday. The use of roles in place of subjects and targets implicitly refers to the subject of the corresponding role.

FIGURE 46.12 Example of relationship type.

Many large organizations are structured into units such as branch offices, departments, and hospital wards that have similar configurations of roles and policies. Ponder supports the notion of *management structures* to define a configuration in terms of instances of roles, relationships, and nested management structures relating to organizational units. For example, a management structure *type* would be used to define a branch in a bank or a department in a university and then *instantiated* for particular branches or departments. A management structure is thus a composite policy containing the definition of roles, relationships, and other nested management structures, as well as instances of these composite policies.

Figure 46.13 shows a simple management structure for a software development company consisting of a project manager, software developers, and a project contact secretary. Figure 46.14 gives the definition of the structure.

Ponder allows specialization of policy types, through inheritance. When a type extends another, it inherits all of its elements, adds new elements, and overrides elements, with the same name. This is particularly useful for specialization of composite policies. For example, it would be possible to define a new type of mobile systems project manager from a project manager role, as defined in Figure 46.13, but with additional policies.

In Ponder a person can be assigned to multiple roles but rights from one role cannot be used to perform actions relating to another role. A person can also have policies that pertain to him as an

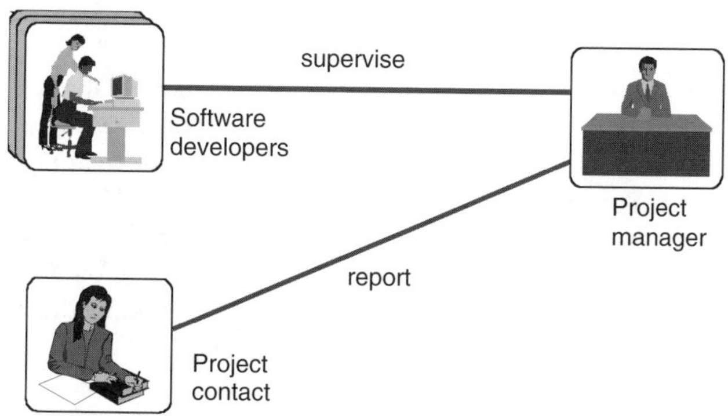

FIGURE 46.13 Simple management structure.

```
type mstruct BranchT (...) {
    inst  role  projectManager = ProjectManagerT(...);
          role  projectContact  = SecretaryT(...);
          role  softDeveloper   = SoftDeveloperT(...);

    inst  rel   supervise = SupervisionT (projectManager, softDeveloper);
          rel   report = ReportingT (projectContact, projectManager);
}
inst      mstruct branchA = BranchT(...);
          mstruct branchB = BranchT(...);
```

This declares instances of the 3 roles shown in Figure 46.13. Two relationships govern the interactions between these roles. A supervise relationship between the softDeveloper and the projectManager, and a reporting relationship between the ProjectContact and the projectManager. Two instances of the BranchT type are created for branches within the organisation that exhibit the same role-relationship requirements.

FIGURE 46.14 Software company management structure.

individual and have nothing to do with any roles. In RBAC, inheritance is based on policy instances and all policies are defined in terms of roles. This means RBAC requires a much more complicated role structure to separate the policies that are inherited from those that are private.

A compiler has been implemented for the Ponder language. Various backends have also been implemented to generate firewall rules, Windows access control templates, Java security policies [Corradi, 200], and Java obligation policy rules for interpretation by a policy agent. We also have a system to automatically disseminate policies to the relevant agents that will interpret them, i.e., to subjects for obligation and refrain policies and access control agents for authorization and filter policies [Dulay et al., 2001]. The application of Ponder for managing a DiffServ network is described in Lymberopoulos et al. [2003].

46.5 Research Issues

46.5.1 Conflict Analysis

There are a number of different conflicts that can arise from policies. For example, it might be possible that different QPIM policies apply for marking packets relating to a specific flow as one selects packets based on a range of IP addresses and another selects packets based on port numbers. The solution proposed in the IETF framework advocates assigning explicit priorities to policies but this is notoriously difficult to do in large systems where many different people are responsible for specifying policies. In Ponder "modality conflicts" arise when a positive policy permits an action and a negative policy prevents the action for the same subjects and targets [Lupu and Sloman, 1999]. Thus, a syntactic analysis for overlaps of subject, target, and action between the policies would identify all the potential situations in which modality conflicts occur. However, a conflict may not necessarily occur since constraints may limit the applicability of the policy to disjoint sets of circumstances, e.g., different times of day or different states of the target objects. In order to take into account these constraints it is necessary to employ formal reasoning techniques and model the system's behavior as well as the policies themselves. The approach adopted in Ponder is to translate policies to an Event Calculus representation. An Event Calculus representation of the system's behavior is then derived from a state transition model, and abduction is used to identify the conflicts. The advantage of using abduction is that in addition to detecting conflicts it also provides an explanation of the sequence of events that have led to the conflict occurring. Furthermore, abduction permits reasoning even when only partial specifications of the system's behavior are available. While modality conflicts do not depend on which particular subjects, actions, or targets the policies refer to, other conflicts can only be determined by understanding the actions being performed by the policies. For example, there will be a conflict between two policies that result in the same packet being placed on two different queues. Similarly, separation of duty conflicts arise from authorization policies that permit the same person to approve payments and sign checks. Generally, these conflicts are application-specific and to detect them it is necessary to specify the conditions that result in conflict. The approach is therefore to specify constraints on the set of policies (i.e., metapolicies) using a suitable notation and then analyze the policy set against these constraints to determine if there are any conflicts [Lupu and Sloman, 1999]. Furthermore, when conflicting policies are detected it is not obvious how to resolve the conflicts automatically. Explicit priority may work in some cases. In some situations, negative authorization policies should override positive ones, but in other situations the positive authorization is an exception to a more general negative authorization. In some situations more specific policies that apply to a department may override general policies applying to the whole organization. Metapolicies can be used to define application specific precedence relationships between conflicting policies.

Although some progress has been made in dealing with policy conflicts [Bandara et al., 2003; Lupu and Sloman, 1999; Verma, 2000], significant challenges remain to be addressed. In particular, it is difficult to account for the different levels of abstraction at which policy is specified. Conflicts between organizational goals will inevitably lead to conflicts between the policies derived from these goals. Some policies will trigger complex management procedures which require the execution of actions that may be specified as part of different policies. Thus, the task of ensuring the consistency of a policy specification is rendered much more complex.

46.5.2 Refinement

Both network and security policies are specified with a view to fulfilling organizational goals and service level objectives. The process of deriving a more concrete specification from a higher-level objective is termed *refinement*. Although the goal of automating refinement of management and security policies from higher-level objectives remains a worthy one, it is not practical for all but the most trivial scenarios. However, this does not preclude the partial automation of this process or providing tools that can assist human managers to refine high-level abstract policies into more concrete ones. This will require representing and taking into account domain specific knowledge as well as more general techniques and methodologies. Currently the most promising approach seems to be investigated in requirements engineering and relies on identifying, recognizing, and instantiating *refinement patterns* [Darimont et al., 1998]. A much simpler approach of integrating Service Level goals with policies is described in Bearden et al. [2001].

An essential requirement when refining a policy is to ensure that the goal achieved by that policy would still be achieved by the set of subpolicies into which it is refined. We are exploring the use of event calculus to represent policies and using abductive reasoning based on goal regression to derive a plan of action for achieving a specified end goal [Bandara et al., 2003]. The desired end goal will be determined from the postconditions of the operation specified in the base policy to be refined, and goal regression can be applied to derive the set of subject/operation tuples that will be used by the refined policy set. Because this procedure is based on a formal proof procedure, the derived set of subjects and operations will be correct and minimal. Using established patterns would help towards maintaining both these properties as patterns can be previously checked for completeness and for conflicts. Note that this does not ensure that conflicts will not occur with other policies in the system; thus, even when instantiating established patterns, it is necessary to perform conflict analysis.

46.5.3 Multiple Levels of Policy

Policies can be used to support adaptability at multiple levels in a network (1) within network-aware applications, (2) within application-aware networks, and (3) at the hardware level to support adaptability in the packet forwarding "fastpath" of network elements. Research is needed on defining interfaces for the exchange of policies between these levels. For example, an application specific policy may be more efficiently interpreted within a network component or an application may need to adapt its behavior as a result of adaptation within the network. However it is not easy to map the semantics of the policies between the different levels. The application may not be aware of what components exist within the network and so how can it specify policies to be interpreted by them?

An interesting variation of the above is to consider a policy feedback loop where the system is monitored to see whether it is performing according to high-level policies or to determine changes in the systems due to faults, new applications, or users appearing and, hence, to dynamically modify the lower level policies in order to adapt the behavior [Lymberopoulos et al., 2003].

46.6 Conclusions

Management and security are closely linked. Access control is essential to protect objects so that only authorized manager agents can perform management operations on them. Security needs to be managed to disseminate relevant policies to the agents that will implement them, specific actions are needed to deal with security violations, and flexible policies are needed for response to intrusions. Concepts such as roles are useful both for security and management for grouping the policies applying to a position in an organization.

In this paper we have described approaches to specifying both security and management policy and then discussed the Ponder framework which caters for both types of policy. Ponder is just a public domain tool from a university and there are no comparable industry-supported tools to allow easy specification, refinement, and analysis of policies. Often a single organization may have many different notations for specifying policy for slightly different application areas. These are some of the factors currently preventing widespread adoption. It would not be technically difficult to define a suitable standard policy notation but producing refinement and analysis tools is still a major research challenge.

There is considerable interest in policy-based systems from standardization groups, industry, and academia, particularly for the support of autonomic self-management. However most of the literature relates to proposed systems or small academic prototypes. There is no reported experience of deploying policy-based management in any large-scale system. Without substantial experience, it is difficult to determine what the performance issues that need to be addressed are and whether adaptive policies can lead to system instability. It is thus too early to judge whether the promised flexibility of policy-based adaptive systems will actually materialize.

Acknowledgements

We gratefully acknowledge the support of EPSRC for research grants GR/R 31409/01 (PolyNet). We also acknowledge the contribution of our colleagues, Arosha Bandara, Naranker Dulay, Nicodemos Damianou, and Leonidas Lymberopoulos to the ideas expressed in this paper.

References

Ahn, G. and Sandhu, R., Role-based authorization constraints specification. *ACM Transactions on Information and System Security* (TISSEC), 3(4): 207–226.

Bandara A. K., E. C. Lupu, and A. Russo, Using Event Calculus to Formalise Policy Specification and Analysis, *4th IEEE Workshop on Policies for Networks and Distributed Systems (Policy 2003)*, Lake Como, Italy, 2003.

Bearden M., S. Garg, and W. Lee, Integrating Goal specification in Policy-Based Management, *Policies for Distributed Systems and Networks (Policy2001)*, Springer, LNCS 1995, pp.153–217, HP Labs, Bristol, U.K., 29–31 January 2001.

Blaze M., Feigenbaum J., and Keromytis A. D., KeyNote: Trust Management for Public-Key Infrastructures, *Security Protocols International Workshop*, pp. 59–63, Cambridge, U.K., 1998, pp. 59–63. http://www.cis.upenn.edu/~angelos/Papers/keynote-position.ps.gz.

Blaze M., Feigenbaum J., and Lacy J., Decentralized Trust Management, *IEEE Conference on Security and Privacy*, pp. 164–173, Oakland, CA, 1996, http://www.crypto.com/papers/policymaker.pdf.

Burgess, M., A site configuration engine. *USENIX Computing systems*. 8(3), 1995.

Burgess, M., Recent Developments in CfEngine. *Unix NL Conference*, The Hague, 2001.

Corradi, A., R. Montanari, C. Stefanelli, E. Lupu, and M. Sloman, Flexible Access Control for Java Mobile Code, *16th Annual Computer Security Applications Conference (ACSAC2000)*, New Orleans, LA, December 2000.

Damianou, N., N. Dulay, E. Lupu, and M. Sloman, The Ponder Policy Specification Language, *Policies for Distributed Systems and Networks (Policy2001)*, Springer, LNCS 1995, pp. 18–38, HP Labs Bristol, U.K., 29–31 Jan 2001.

Darimont, R., E. Delor, P. Masonet, and A. van Lamsweerde, GRAIL/KAOS: An Environment for Goal-Driven Requirements Engineering, *Proceeding of the 20th International IEEE Conference on Software Engineering (ICSE'98)*, Vol. 2, pp. 58–62, Kyoto, Japan, April 1998.

DMTF: Distributed Management Task Force Inc., Common Information Model (CIM) Specification, version 2.2, www.dmtf.org/standards/standard_cim.php, June 14, 1999.

Dulay, N., E. Lupu, M. Sloman, and N. Damianou, A Policy Deployment Model for the Ponder Language, *Proceedings of the IEEE/IFIP International Symposium on Integrated Network Management, (IM'2001)*, pp. 529–544, Seattle, WA, May 2001, IEEE Press.

Ferraiolo, D., R. Sanhdu, S. Gavrila, D. R. Kuhn, and R. Chandramouli, Proposed NIST standard for RBAC. *ACM Transactions On Information & Systems*, 4(3): 224–274, August 2001.

Flegkas, P., Trimintzios, P., Pavlou, P., Liotta, A., Design and Implementation of a Policy-Based Resource Management Architecture, *IEEE/IFIP Integrated Management Symposium*, pp. 215–229, Colorado Springs, CO, March 2003.

Hashii, B., S. Malabarba, R. Pandey, and M. Bishop, Supporting reconfigurable security policies for mobile programs, *Computer Networks*, 33, 77–93, June 2000.

Herzberg, A., Y. Mass, J. Mihaeli, D. Naor, and Y. Ravid, Access Control Meets Public Key Infrastructure, or: Assigning Roles to Strangers, *IEEE Symposium on Security and Privacy*, pp. 2–14, Oakland, CA, 2000. www.hrl.il.ibm.com/TrustEstablishment/paper.asp.

IETF Internet Engineering Task Force, Policy Working Group www.ietf.org/html.charters/policy-charter.html.

Jajodia S., P. Samarati, and V. S. Subrahmanian, A Logical Language for Expressing Authorisations, *IEEE Symposium on Security and Privacy*, pp. 31–42, Oakland, CA, 1997.

Jajodia, S., Samarati, P., Sapino, M., Subrahmanian, V., Flexible support for multiple access control policies. *ACM Transactions on Database Systems*, 26(2): 214–260, June 2001.

Konstantinos P., Active networks: applications, security, safety, and architectures. *IEEE Communications Surveys and Tutorials*, 2(1), 1999. http://www.comsoc.org/livepubs/surveys/index.html.

Lobo, J., R. Bhatia, and S. Naqvi, A Policy Description Language, *Proceedings of AAAI*, pp. 291–298, Orlando, FL, MIT Press, July 1999.

Lupu, E. C. and M. Sloman, Towards a role-based framework for distributed systems management, *Journal of Network and Systems Management*, 5(1): 5–30, 1997.

Lupu, E. C. and M. Sloman, Conflicts in policy-based distributed systems management, *IEEE Transactions on Software Engineering*, 25(6): 852–869, November 1999.

Lymberopoulos, L., E. Lupu, and M. Sloman, An adaptive policy based framework for network services management, *Journal of Networks and Systems Management (JNSM)*, Special Issue on Policy Based Management of Networks and Services, 11: 3, September 2003.

Minsky, N. H. and Pal, P., Law-governed regularities in object systems — part 2: a concrete implementation, *Theory and Practice of Object Systems (TAPOS)*, 3(2), John Wiley & Sons, New York, 1997.

Moore, B., Ellesson, E., Strassner, J., and A. Westerinen, Policy Core Information Model — Version 1 Specification, RFC 3060, February 2001. www.ietf.org/rfc/rfc3060.txt.

Ortalo, R., A Flexible Method for Information System Security Policy Specification, *Proceedings of the 5th European Symposium on Research in Computer Security (ESORICS 98)*, pp. 67–84, Louvain-la-Neuve, Belgium, Springer-Verlag, 1998.

Policies http://www-dse.doc.ic.ac.uk/Research/policies/.

Policy 2001, Policies for Distributed Systems and Networks. M. Sloman, J. Lobo., and E. Lupu, Eds., *Lecture Notes in Distributed Systems Vol. 1995*, Springer, January 2001. http://link.springer.de/link/service/series/0558/tocs/t1995.htm.

Policy 2002, IEEE 3rd International Workshop on Policies for Distributed Systems and Networks, Monterey, CA, June 2002.

Sandhu, R. S., E. J. Coyne, H. L. Feinstein, and C. E. Youman. Role-based access control models. *IEEE Computer*, 29(2): 38–47, 1996.

Sloman, M., Policy driven management for distributed systems, *Journal of Network and Systems Management*, 2(4): 333–360, 1994.

Sloman, M. and E. Lupu, Policy Specification for Programmable Networks, *Proceedings of the 1st International Working Conference on Active Networks (IWAN'99)*, S. Covaci, Ed., pp. 73–84, Berlin, Springer LNCS, June 1999.

Snir, Y., Y. Ramberg, and J. Strassner, Policy QoS Information Model, May 2003. www.ietf.org/internet-drafts/draft-ietf-policy-qos-info-model-05.txt

Stone, G., B. Lundy, and G. Xie, Network policy languages: a survey and a new approach, *IEEE Network*, 15(1): 10–21, January 2001.

Strassner, J., Policy and the IETF — Theory and Practice, Invited presentation, *Policy 2001: Policies for Distributed Systems and Networks*, HP Labs, Bristol, U.K., January 2001, www-dse.doc.ic.ac.uk/events/policy-2001/.

Verma, D., *Policy Based Networking: Architecture and Algorithms*, New-Riders, Berkley, CA, 2000.

Virmani, A., J. Lobo, and M. Kohli, Netmon: Network Management for the SARAS Softswitch, *IEEE/IFIP Network Operations and Management Symposium, (NOMS2000)*, J. Hong and R. Weihmayer, Eds., pp. 803–816, Hawaii, May 2000.

47

Distributed Trust

CONTENTS

Abstract 47-1
47.1 Access Control and Trust Management 47-1
47.2 Technical Foundations 47-2
47.2.1 Authentication 47-2
47.2.2 Public Key Certificates 47-3
47.3 Distributed Trust Management 47-3
47.3.1 PolicyMaker 47-5
47.3.2 KeyNote 47-8
47.4 Applications of Trust Management Systems 47-10
47.4.1 Network-Layer Access Control 47-10
47.4.2 Distributed Firewalls and the STRONGMAN Architecture 47-11
47.4.3 Grid Computing and Transferable Micropayments 47-12
47.4.4 Micropayments: Microchecks and Fileteller 47-12
47.4.5 Active Networking 47-12
47.5 Other Trust-Based Systems 47-13
47.6 Closing Remarks 47-13
Acknowledgments 47-14
References 47-14

John Ioannidis

Angelos D. Keromytis

Abstract

This chapter explores the concept of trust management in access control. We introduce the concepts behind trust management and discuss two such systems. The first, PolicyMaker [Blaze et al., 1996], first introduced the concepts of trust management, which were further explored in the work on the KeyNote credential language [Blaze et al., 1999b]. We discuss some applications of trust management systems, as well as other related work. Our focus is on the concepts and design rather than the details of particular approaches or mechanisms. Our goal is to impart enough information to the readers to make informed decisions as to how best to use the power and expressiveness of trust management.

47.1 Access Control and Trust Management

Authorization and access control is the process by which a security enforcement point determines whether an entity should be allowed to perform a certain action. Authorization takes place after said entity has been authenticated. Furthermore, authorization occurs within the scope of an access control policy. In simpler terms, the first step in making an access control decision is determining who is making a request; the second step is determining, based on the result of the authentication as well as additional information (the access control policy), whether that request should be allowed.

One security mechanism often used in operating systems is the Access Control List (ACL). Briefly, an ACL is a list describing which access rights a principal has on an object (resource). For example, an entry

1-58488-381-2/05/$0.00+$1.50

© 2005 by CRC Press LLC

might read "User Foo Can Read File Bar." Such a list (or table) need not physically exist in one location but may be distributed throughout the system. The *Unix*™ filesystem "permissions" mechanism is essentially an ACL.

ACLs have been used in distributed systems because they are conceptually easy to grasp and because there is an extensive literature about them. However, there are a number of fundamental reasons that ACLs are inadequate for distributed-system security:

- *Authentication:* In an operating system, the identity of a principal is well known. This is not so in a distributed system, where some form of authentication has to be performed before the decision to grant access can be made.
- *Delegation:* Necessary for scalability of a distributed system. It enables *decentralization* of administrative tasks. Existing distributed-system security mechanisms usually delegate directly to a "certified entity." In such systems, policy (or authorizations) may only be specified at the last step in the delegation chain (the entity enforcing policy), most commonly in the form of an ACL. The implication is that high-level administrative authorities cannot directly specify overall security policy; rather, all they can do is "certify" lower-level authorities. This authorization structure leads easily to inconsistencies among locally specified subpolicies.
- *Expressibility and Extensibility:* A generic security mechanism must be able to handle new and diverse conditions and restrictions. The traditional ACL approach has not provided sufficient expressibility or extensibility. Thus, many security policy elements that are not directly expressible in ACL form must be hard-coded into applications. This means that changes in security policy often require reconfiguration, rebuilding, or even rewriting of applications.
- *Local trust policy:* The number of administrative entities in a distributed system can be quite large. Each of these entities may have a different trust model for different users and other entities. For example, system A may trust system B to authenticate its users correctly, but not system C; on the other hand, system B may trust system C. It follows that the security mechanism should not enforce uniform and implicit policies and trust relations.

The *trust-management approach* to distributed-system security was developed as an answer to the inadequacy of traditional authorization mechanisms. Trust-management engines avoid the need to resolve "identities" in an authorization decision. Instead, they express privileges and restrictions in a programming language. This allows for increased flexibility and expressibility, as well as standardization of modern, scalable security mechanisms. Further advantages of the trust-management approach include proofs that requested transactions comply with local policies and system architectures that encourage developers and administrators to consider an application's security policy carefully and specify it explicitly.

Section 47.1 provides some background material on the concepts of authentication and public key certificates, which form the basis on which trust management systems are built. Section 47.2.2 describes the trust-management approach to authorization and access control, focusing on the Policy-Maker and KeyNote systems. Section 47.3.2 describes various applications of trust management systems (especially KeyNote), while Section 47.4.5 briefly describes some related work.

47.2 Technical Foundations

47.2.1 Authentication

The term authentication in network security and cryptography is used for the purpose of "identifying" itself to indicate the process by which one entity convinces another that it has possession of some secret information. This identification does not necessarily correspond to a real-world entity; rather, it implies the continuity of a relationship (or, as stated in Schneier [2000], knowing who to trust and who not to trust).

In computer networks, strong authentication is achieved through cryptographic protocols and algorithms. In particular, public key cryptography forms the basis for many protocols and proposals for scalable network-based authentication. A discussion of the various constraints and goals in these protocols is beyond the scope of this chapter. The interested reader is referred to Schneier [1996].

47.2.2 Public Key Certificates

Public key certificates are statements made by a principal (an entity, such as a user or a process acting on behalf of a user, that can undertake an action in the system and is identified by a cryptographic public key) about another principal (also identified by a public key). Public key certificates are cryptographically signed, such that anyone can verify their integrity (the fact that they have not been modified since the signature was created). Public key certificates are utilized in authentication because of their natural ability to express delegation (more on this in Section 47.3.1).

A traditional public key certificate cryptographically binds an identity to a public key. In the case of the X.509 standard [CCITT, 1989], an identity is represented as a "distinguished name," e.g.,

```
/C=US/ST=PA/L=Philadelphia/O=University of Pennsylvania/
OU=Department of Computer and Information Science/CN=Jonathan M. Smith
```

In more recent public key certificate schemes [Ellison et al., 1999; Blaze et al., 1999b] the identity is the public key, and the binding is between the key and the permissions granted to it. Public key certificates also contain expiration and revocation information.

Revoking a public key certificate means notifying entities that might try to use it that the information contained in it is no longer valid, even though the certificate itself has not expired. Possible reasons for this include theft of the private key used to sign the certificate (in which case all certificates signed by that key need to be revoked) or discovery that the information contained in the certificate has become inaccurate. This happened when Verisign, a commercial Certification Authority (CA), mistakenly issued an X.509 certificate to an unknown person with the common name "Microsoft Corporation." There exist various revocation methods (Certificate Revocation Lists [CRLs], Delta-CRLs, Online Certificate Status Protocol [OCSP], refresher certificates), each with its own tradeoffs in terms of the amount of data that needs to be kept around and transmitted, any online availability requirements, and the window of vulnerability.

47.3 Distributed Trust Management

A traditional "system-security approach" to the processing of a signed request for action (such as access to a controlled resource) treats the task as a combination of *authentication* and *authorization*. The receiving system first determines *who* signed the request and then queries an internal database to decide *whether* the signer should be granted access to the resources needed to perform the requested action. It has been argued that this is the wrong approach for today's dynamic, Internet-worked world [Blaze et al., 1996, 1999a; Ellison, 1999; Ellison et al., 1999]. In a large, heterogeneous, distributed system, there is a huge set of people (and other entities) who may make requests, as well as a huge set of requests that may be made. These sets change often and cannot be known in advance. Even if the question "who signed this request?" could be answered reliably, it would not help in deciding whether or not to take the requested action if the requester is someone or something from whom the recipient is hearing for the first time.

The right question in a far-flung, rapidly changing network becomes "Is the key that signed this request *authorized* to take this action?" Traditional name-key mappings and precomputed access-control matrices are inadequate — the former because they do not convey any access control information, the latter because of the amount of state required: given N users, M objects to which access needs to be restricted, and X variables that need to be considered when making an access control decision. We would need access control lists of minimum size $N \times X$ associated with each object, for a total of $N \times M$ policy rules of size X in our system. As the conditions under which access is allowed or denied become more refined (and thus larger), these products increase. In typical systems, the number of users and objects (services) is large, whereas the number of variables is small; however, the combinations of variables in expressing access control policy can be nontrivial (and arbitrarily large, in the worst case). Furthermore, these rules have to be maintained, securely distributed, and stored across the entire network. Thus, one needs a more flexible, more "distributed" approach to authorization.

The *trust-management approach,* initiated by Blaze et al. [1996], frames the question as follows: "Does the set *C* of *credentials* prove that the *request r complies* with the local security policy P?" This difference is shown graphically in Figure 47.1.

Each entity that receives requests must have a policy that serves as the ultimate source of authority in the local environment. The entity's policy may directly authorize certain keys to take certain actions, but more typically it will *delegate* this responsibility to credential issuers that it trusts to have the required domain expertise as well as relationships with potential requesters. The *trust-management engine* is a separate system component that takes (*r, C, P*) as input, outputs a decision about whether compliance with the policy has been proven, and may also output some additional information about how to proceed if the required proof has not been achieved. Figure 47.2 shows an example of the interactions between an application and a trust-management system.

An essential part of the trust-management approach is the use of a *general-purpose, application-independent* algorithm for checking proofs of compliance. Why is this a good idea? Any product or service that requires some form of proof that requested transactions comply with policies, could use a special-purpose algorithm or language implemented from scratch. Such algorithms/languages could be made more expressive and tuned to the particular intricacies of the application. Compared to this, the trust-management approach offers two main advantages.

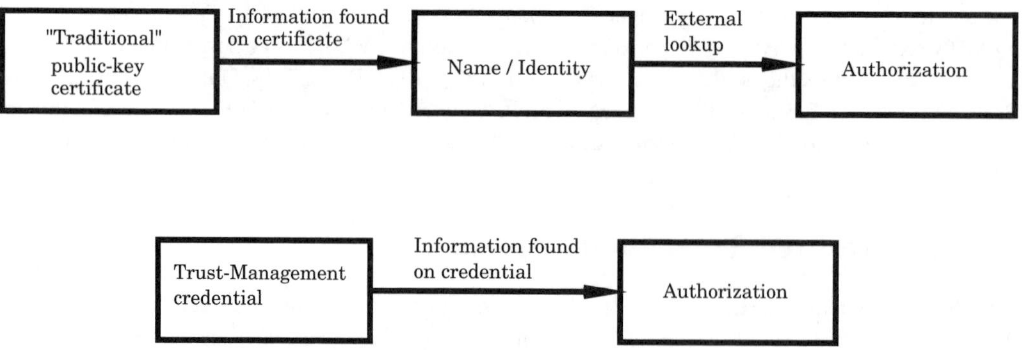

FIGURE 47.1 The difference between access control using traditional public-key certificates and trust management.

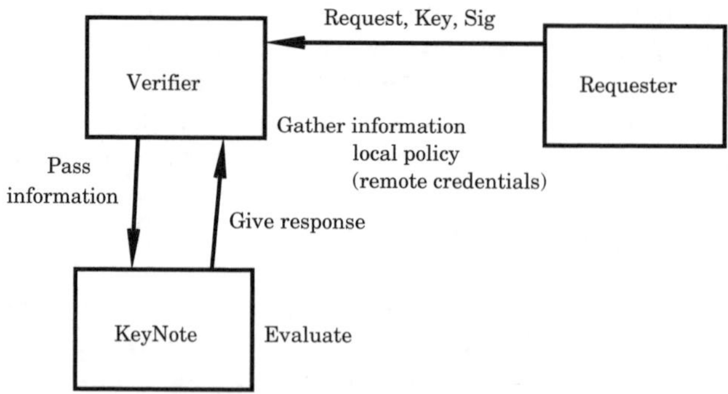

FIGURE 47.2 Interaction between an application and a trust-management system.

The first is simply one of engineering: it is preferable (in terms of simplicity and code reuse) to have a "standard" library or module, and a consistent API, that can be used in a variety of different applications.

The second, and perhaps most important gain, is in soundness and reliability of both the definition and the implementation of "proof of compliance." Developers who set out to implement a "simple," special-purpose compliance checker (in order to avoid what they think are the overly "complicated" syntax and semantics of a universal "metapolicy") discover that they have underestimated their application's need for proof and expressiveness. As they discover the full extent of their requirements, they may ultimately wind up implementing a system that is as general and expressive as the "complicated" one they set out to avoid. A general-purpose compliance checker can be explained, formalized, proven correct, and implemented in a standard package, and applications that use it can be assured that the answer returned for any given input (r, C, P) depends only on the input and not on any implicit policy decisions (or bugs) in the design or implementation of the compliance checker.

Basic questions that must be answered in the design of a trust-management engine include:

- How should "proof of compliance" be defined?
- Should policies and credentials be fully or only partially programmable? In which language or notation should they be expressed?
- How should responsibility be divided between the trust-management engine and the calling application? For example, which of these two components should perform the cryptographic signature verification? Should the application fetch all credentials needed for the compliance proof before the trust-management engine is invoked, or may the trust-management engine fetch additional credentials while it is constructing a proof?

At a high level of abstraction, trust-management systems have five components:

- A language for describing *actions*, which are operations with security consequences that are to be controlled by the system
- A mechanism for identifying *principals*, which are entities that can be authorized to perform actions
- A language for specifying application *policies*, which govern the actions that principals are authorized to perform
- A language for specifying *credentials*, which allow principals to delegate authorization to other principals
- A *compliance checker*, which provides a service to applications for determining how an action requested by principals should be handled, given a policy and a set of credentials

By design, trust management unifies the notions of security policy, credentials, access control, and authorization. An application that uses a trust-management system can simply ask the compliance checker whether a requested action should be allowed. Furthermore, policies and credentials are written in standard languages that are shared by all trust-managed applications; the security configuration mechanism for one application carries exactly the same syntactic and semantic structure as that of another, even when the semantics of the applications themselves are quite different.

47.3.1 PolicyMaker

PolicyMaker was the first example of a "trust-management engine." That is, it was the first tool for processing signed requests that embodied the trust-management principles articulated in Section 47.2.2. It addressed the authorization problem directly, rather than handling the problem indirectly via authentication and access control, and it provided an application-independent definition of "proof of compliance" for matching up requests, credentials, and policies. PolicyMaker was introduced in the original trust-management paper by Blaze et al. [1996], and its compliance-checking algorithm was later fleshed out in [Blaze et al., 1998]. A full description of the system can be found in Blaze et al. [1996, 1998], and experience using it in several applications is reported in Blaze et al. [1997] and Lacy et al. [1997].

PolicyMaker credentials and policies (collectively referred to as "assertions") are programmable: they are represented as pairs (*f*, *s*), where *s* is the source of authority, and *f* is a program describing the nature of the authority being granted as well as the party or parties to whom it is being granted. In a policy assertion, the source is always the keyword POLICY. For the PolicyMaker trust-management engine to be able to make a decision about a requested action, the input supplied to it by the calling application must contain one or more policy assertions; these force the "trust root," i.e., the ultimate source of authority for the decision about this request, as shown in Figure 47.3. In a credential assertion, the source of authority is the public key of the issuing entity. Credentials must be signed by their issuers, and these signatures must be verified before the credentials can be used.

PolicyMaker assertions can be written in any programming language that can be "safely" interpreted by a local environment that has to import credentials from diverse (and possibly untrusted) issuing authorities. A version of AWK without file UO operations and program execution time limits (to avoid denial of service attacks on the policy system) was developed for early experimental work on PolicyMaker (see [Blaze et al., 1996]) because AWK's pattern-matching constructs are a convenient way to express authorizations. For a credential assertion issued by a particular authority to be useful in a proof that a request complies with a policy, the recipient of the request must have an interpreter for the language in which the assertion is written (so that the program contained in the assertion can be executed). Thus, it would be desirable for assertion writers ultimately to converge on a small number of assertion languages so that receiving systems have to support only a small number of interpreters and so that carefully crafted credentials can be widely used. However, the question of which languages these will be was left open by the PolicyMaker project. A positive aspect of PolicyMaker's not insisting on a particular assertion language is that all of that work that has gone into designing, analyzing, and implementing the PolicyMaker

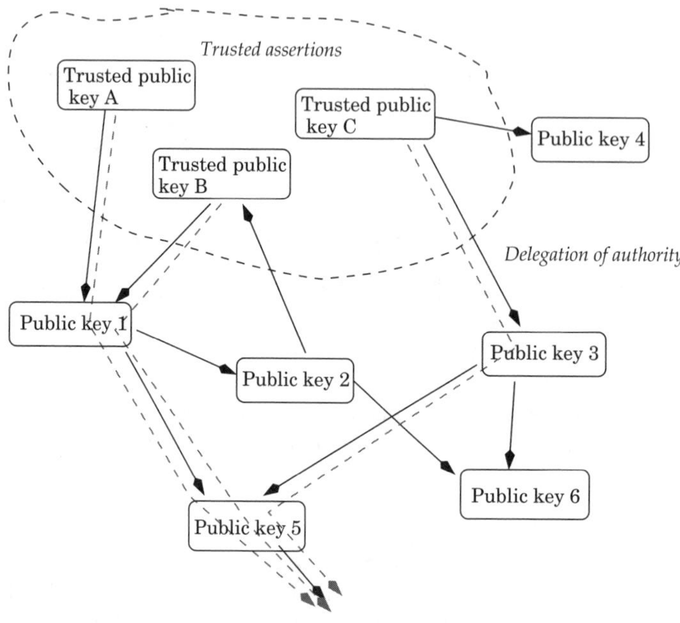

FIGURE 47.3 Delegation in PolicyMaker, starting from a set of trusted assertions. The dotted lines indicate a delegation path from a trusted assertion (public key) to the user making a request. If all the assertions along that path authorize the request, it will be granted.

compliance-checking algorithm will not have to be redone every time an assertion language is changed or a new language is introduced. The "proof of compliance" and "assertion-language design" problems are orthogonal in PolicyMaker and can be worked on independently.

One goal of the PolicyMaker project was to make the trust-management engine as small as possible and analyzable. Architectural boundaries were drawn so that a fair amount of responsibility was placed on the calling application rather than the trust-management engine. In particular, the calling application was made responsible for all cryptographic verification of signatures on credentials and requests. One pleasant consequence of this design decision is that the application developer's choice of signature scheme(s) can be made independently of his choice of whether or not to use PolicyMaker for compliance checking. Another important responsibility that was assigned to the calling application is credential gathering. The input (r, C, P) supplied to the trust-management module is treated as a claim that credential set C contains a proof that request r complies with Policy P. The trust-management module is not expected to be able to discover that C is missing just one credential needed to complete the proof and to go fetch that credential from, for example, the corporate database, the issuer's Website, the requester himself, or elsewhere. Later trust-management engines, including KeyNote [Blaze et al., 1999b] and REFEREE [Chu et al., 1997] divide responsibility between the calling application and the trust-management engine differently than the way PolicyMaker divides it.

The main technical contribution of the PolicyMaker project is a notion of "proof of compliance" that is fully specified and analyzed. We give an overview of PolicyMaker's approach to compliance checking here; a complete treatment of the compliance checker can be found in Blaze et al. [1998].

The PolicyMaker runtime system provides an environment in which the policy and credential assertions fed to it by the calling application can cooperate to produce a proof that the request complies with the policy (or can fail to produce such a proof). Among the requirements for this cooperation are a method of interassertion communication and a method for determining that assertions have collectively succeeded or failed to produce a proof.

Interassertion communication in PolicyMaker is done via a simple, append-only data structure on which all participating assertions record intermediate results. Specifically, PolicyMaker initializes the proof process by creating a "blackboard" containing only the request string r and the fact that no assertions have thus far approved the request or anything else. Then PolicyMaker runs the various assertions, possibly multiple tunes each. When assertion (f_i, s_i) is run, it reads the contents of the blackboard and then adds to the blackboard one or more *acceptance records* (i, S_i, R_{ij}). Here R_{ij} is an application-specific action that source s_i approves, based on the partial proof that has been constructed thus far. R_{ij} may be the input request r, or it may be some related action that this application uses for interassertion communication. Note that the meanings of the action strings R_{ij} are understood by the application-specific assertion programs f_i, but they are not understood by PolicyMaker. All PolicyMaker does is run the assertions and maintain the global blackboard, making sure that the assertions do not erase acceptance records previously written by other assertions, fill up the entire blackboard so that no other assertions can write, or exhibit any other noncooperative behavior. PolicyMaker never tries to interpret the action strings R_{ij}.

A proof of compliance is achieved if, after PolicyMaker has finished running assertions, the blackboard contains an acceptance record indicating that a policy assertion approves the request r. For example, consider the following two assertions, the first of which is a policy and the second a credential:

```
Source: POLICY
if (has_posted(KEY_ALICE) = = "accept") post("accept");
Source: KEY_ALICE
if (action_string("file") == "/home/angelos/html/index.html" &&
      action_string("operation.") == "read") post("accept");
Signature: ...
```

The policy assertion will approve a request if an assertion that has been signed by KEY_ALICE approves it. The second assertion will approve any "read" request to the given file. The function *action-string()* is used to read the action strings, *post()* is used to issue assertion acceptance records, and *has-posted()* reads acceptance records posted on the board.

Among the nontrivial decisions that PolicyMaker must make are (1) in what order assertions should be run, (2) how many times each assertion should be run, and (3) when an assertion should be discarded because it is behaving in a noncooperative fashion. Blaze et al. [1998] provide:

- A mathematically precise formulation of the PolicyMaker compliance-checking problem.
- Proof that the problem is undecidable in general and is NP-hard even in certain natural special cases.
- One special case of the problem that is solvable in polynomial-time, is useful in a wide variety of applications, and is implemented in the current version of PolicyMaker, as described below.

Although the most general version of the compliance-checking problem allows assertions to be arbitrary functions, the computationally tractable version that is analyzed in [Blaze et al., 1998] and implemented in PolicyMaker is guaranteed to be correct only when all assertions are monotonic. (Basically, if a monotonic assertion approves action *a* when given evidence set *E*, then it will also approve action *a*, when given an evidence set that contains *E* (see Blaze et al. [1998] for a formal definition.) In particular, correctness is guaranteed only for monotonic *policy* assertions, and this excludes certain types of policies that are used in practice, most notably those that make explicit use of "negative credentials" such as revocation lists. Although it is a limitation, the monotonicity requirement has certain advantages. One of them is that, although the compliance checker may not handle all potentially desirable policies, it is at least analyzable and provably correct on a well-defined class of policies. Furthermore, the requirements of many nonmonotonic policies can often be achieved by monotonic policies. For example, the effect of requiring that an entity not occur on a revocation list can also be achieved by requiring that it present a "certificate of non-revocation"; the choice between these two approaches involves trade-offs among the (system-wide) costs of the two kinds of credentials and the benefits of a standard compliance checker with provable properties. Finally, restriction to monotonic assertions encourages a conservative, prudent approach to security: In order to perform a potentially dangerous action, a user must present an adequate set of affirmative credentials; no potentially dangerous action is allowed "by default," simply because of the absence of negative credentials.

47.3.2 KeyNote

KeyNote [Blaze et al., 1999b] was designed according to the same principles as PolicyMaker, using credentials that directly authorize actions instead of dividing the authorization task into authentication and access control. Two additional design goals for KeyNote were standardization and ease of integration into applications. To address these goals, KeyNote assigns more responsibility to the trust-management engine than PolicyMaker does and less to the calling application; for example, cryptographic signature verification is done by the trust-management engine in KeyNote and by the application in PolicyMaker. KeyNote also requires that credentials and policies be written in a specific assertion language, designed to work smoothly with KeyNote's compliance checker. By fixing a specific assertion language that is flexible enough to handle the security policy needs of different applications, KeyNote goes further than PolicyMaker toward facilitating efficiency, interoperability, and widespread use of carefully written credentials and policies, at the cost of reduced expressibility and interaction between different policies (compared to PolicyMaker).

A calling application passes to a KeyNote evaluator a list of credentials, policies, and requester public keys, and an "Action Attribute Set." This last element consists of a list of attribute/value pairs, similar in some ways to the *Unix*™ shell environment (described in the *environ(7)* manual page of most Unix installations). The action attribute set is constructed by the calling application and contains all information deemed relevant to the request and necessary for the trust decision. The action-environment attributes and the assignment of their values must reflect the security requirements of the application accurately. Identifying the attributes to be included in the action attribute set is perhaps the most

```
KeyNote-Version : 2
Authorizer: "rsa-hex:1023abcd"
Licensees: "dsa-hex:986512a1" || "rsa-hex: 19abcd02"
Comment: Authorizer delegates read access to either of the
         Licensees
Conditions: (file = = "/etc/passwd" &&
               access = = "read") -> "true";
Signature: sig-rsa-md5-hex: "f00f5673"
```

FIGURE 47.4 Sample KeyNote assertion authorizing the two keys appearing in the Licensees field to read the file "/etc/passwd." No one else is authorized to read this file, based on this credential.

important task in integrating KeyNote into new applications. The result of the evaluation is an application-defined string (perhaps with some additional information) that is passed back to the application. In the simplest case, the result is of the form "authorized."

The KeyNote assertion format resembles that of e-mail headers. An example (with artificially short keys and signatures for readability) is given in Figure 47.4.

As in PolicyMaker, policies and credentials (collectively called assertions) have the same format. The only difference between policies and credentials is that a policy (that is, an assertion with the keyword POLICY in the *Authorizer* field) is locally trusted (by the compliance-checker) and thus needs no signature.

KeyNote assertions are structured so that the *Licensees* field specifies explicitly the principal or principals to which authority is delegated. Syntactically, the Licensees field is a formula in which the arguments are public keys and the operations are conjunction, disjunction, and threshold. Intuitively, this field specifies which combinations of keys must approve a request to satisfy the issuer of the assertion. Thus, for example, a request may be granted if the CEO, or all three vice-presidents together, or any 5 out of 10 members of the Board approve it. The full semantics of these expressions are specified in Blaze et al. [1999b].

The programs in KeyNote are encoded in the *Conditions* field and are essentially tests of the action attributes. These tests are string comparisons, numerical operations and comparisons, and pattern-matching operations.

The design choice of using a simple language for KeyNote assertions was based on the following reasons:

- AWK, one of the first assertion languages used by PolicyMaker, was criticized as too heavy-weight for most relevant applications. Because of AWK's complexity, the footprint of the interpreter is considerable, and this discourages application developers from integrating it into a trust-management component. The KeyNote assertion language is simple and has a small-size interpreter.
- In languages that permit loops and recursion (including AWK), it is difficult to enforce resource-usage restrictions, but applications that run trust-management assertions written by unknown sources often need to limit their memory and CPU usage.
- In retrospect, a language without loops, dynamic memory allocation, and certain other features seems sufficiently powerful and expressive [Blaze et al., 2001b,a]. The KeyNote assertion syntax is restricted so that resource usage is proportional to the program size. Similar concepts have been successfully used in other contexts [Hicks and Keromytis, 1999].
- Assertions should be both understandable by human readers and easy for a tool to generate from a high-level specification. Moreover, they should be easy to analyze automatically, so that automatic verification and consistency checks can done. This is currently an area of active research.
- One of the design goals is to use KeyNote as a means of exchanging policy and distributing access control information otherwise expressed in an application-native format. Thus, the language should be easy to map to a number of such formats (e.g., from a KeyNote assertion to packet-filtering rules).
- The language chosen was adequate for KeyNote's evaluation model.

This last point requires explanation.

In PolicyMaker, compliance proofs are constructed via repeated evaluation of assertions, along with an arbitrated "blackboard" for storage of intermediate results and interassertion communication.

In contrast, KeyNote uses an algorithm that attempts (recursively) to satisfy at least one policy assertion. Referring again to Figure 47.3, KeyNote treats keys as vertices in the graph, with (directed) edges representing assertions delegating authority. The prototype implementation uses a Depth First Search algorithm, starting from the set of trusted (POLICY) assertions and trying to construct a path to the key of the user making the request. An edge between two vertices in the graph exists only if:

- There exists an assertion where the *Authorizer* and the *Licensees* are the keys corresponding to the two vertices.
- The predicate encoded in the *Conditions* field of that KeyNote assertion authorizes the request.

Thus, satisfying an assertion entails satisfying both the *Conditions* field and the *Licensees* key expression. If no such graph exists (due to missing credentials or requests that were not accepted by an assertion's predicate), the request will be denied. Thus, as in PolicyMaker, it is the responsibility of the requester to provide all necessary material for the request to succeed.

Note that there is no explicit interassertion communication as in PolicyMaker; the *acceptance records* returned by program evaluation are used internally by the KeyNote evaluator and are never seen directly by other assertions. Because KeyNote's evaluation model is a subset of PolicyMaker's, the latter's compliance-checking guarantees are applicable to KeyNote. In PolicyMaker, the programs contained in the assertions and credentials can interact with each other by examining the values written on the blackboard and reacting accordingly. This, for example, allows for a negotiation of sorts: "I will approve, if you approve," "I will also approve, if you approve," "I approve," "I approve as well"…. In KeyNote, each assertion is evaluated exactly once and cannot directly examine the result of another assertion's evaluation. Whether the more restrictive nature of KeyNote allows for stronger guarantees to be made, e.g., a tighter space/time evaluation bound, is an open question requiring further research.

Ultimately, for a request to be approved, an assertion graph must be constructed between one or more policy assertions and one or more keys that signed the request. Because of the evaluation model, an assertion located somewhere in a delegation graph can effectively only refine (or pass on) the authorizations conferred on it by the previous assertions in the graph. (This principle also holds for PolicyMaker, although its evaluation model differs.) For more details on the evaluation model, see Blaze et al. [1999b].

It should be noted that PolicyMaker's restrictions regarding "negative credentials" also apply to KeyNote. Certificate revocation lists (CRLs) are not built into the KeyNote (or the PolicyMaker) system; these can be provided at a higher (or lower) level, perhaps even transparently to KeyNote. (Note that the decision to consult a CRL is [or should be] a matter of local policy.) The problem of credential discovery is also not explicitly addressed in KeyNote.

Finally, note that KeyNote, like other trust-management engines, does not directly *enforce* policy; it only provides "advice" to the applications that call it. KeyNote assumes that the application itself is trusted and that the policy assertions are correct. Nothing prevents an application from submitting misleading assertions to KeyNote or from ignoring KeyNote altogether.

47.4 Applications of Trust Management Systems

In this section we briefly describe the use of KeyNote in various systems. Although the ability to use KeyNote in such a wide range of applications validates its generality, we also identify several shortcomings.

47.4.1 Network-Layer Access Control

One of the first applications of KeyNote was providing access control services for the IPsec [Kent and Atkinson, 1998] architecture. The IPsec protocol suite, which provides network-layer security for the

Internet, has been standardized in the IETF and is beginning to make its way into commercial implementations of desktop, server, and router operating systems. IPsec does not itself address the problem of managing the policies governing the handling of traffic entering or leaving a node running the protocol. By itself, the IPsec protocol can protect packets from external tampering and eavesdropping, but does nothing to control which nodes are authorized for particular kinds of sessions or for exchanging particular kinds of traffic. In many configurations, especially when network-layer security is used to build firewalls and virtual private networks, such policies may necessarily be quite complex.

Blaze et al. [2001b, 2002] introduced a new policy management architecture for IPsec. A *compliance check* was added to the IPsec architecture that tests packet filters proposed when new security associations are created for conformance with the local security policy, based on credentials presented by the peer node. Security policies and credentials can be quite sophisticated (and specified in KeyNote), while still allowing very efficient packet-filtering for the actual IPsec traffic. The resulting implementation [Hallqvist and Keromytis, 2000] has been in use in the OpenBSD [de Raadt et al., 1999] operating system for several years. This system has formed the basis of several commercial VPN and SoHo firewall products.

47.4.2 Distributed Firewalls and the STRONGMAN Architecture

Conventional firewalls rely on topology restrictions and controlled network entry points to enforce traffic filtering. The fundamental limitation of the firewall approach to network security is that a firewall cannot filter traffic it does not see; by implication, everyone on the protected side has to be considered trusted. While this model has worked well for small-to-medium-size networks, networking trends such as increased connectivity, higher line speeds, extranets, and telecommuting threaten to make it obsolete. To address the shortcomings of traditional firewalls, the concept of a *distributed firewall* has been proposed [Bellovin, 1999]. In this scheme, security policy is still centrally defined, but enforcement is left up to the individual endpoints. Credentials distributed to every node express parts of the overall network policy. The use of KeyNote for access control at the network layer enabled us to develop a prototype distributed firewall [Ioannidis et al., 2000]. Under certain circumstances, the prototype exhibited better performance than the traditional firewall approach, as well as handled the increasing protocol complexity and the use of end-to-end encryption.

This functionality has been used in other projects where dynamic access control was necessary. In Keromytis et al. [2002], the ability to effectively control a large number of firewalls, any of which can be contacted by any of a large number of potential users, was allowed to build a distributed denial of service (DDoS) resistant architecture for allowing authorized users to contact sites that are under attack.

The distributed firewall concept was later generalized in the STRONGMAN architecture, which allowed coordinated and decentralized management of a large number of nodes and services throughout the network stack [Keromytis et al., 2003, Keromytis, 2001]. STRONGMAN offers three new approaches to scalability, applying the principle of local policy enforcement complying with global security policies. First is the use of a compliance checker to provide great local autonomy within the constraints of a global security policy. Second is a mechanism to compose policy rules into a coherent enforceable set, e.g., at the boundaries of two locally autonomous application domains. Third is the lazy instantiation of policies to reduce the amount of state that enforcement points need to maintain. STRONGMAN is capable of managing such diverse resources and protocols as fire walls, Web access control (discussed later), filesystem accesses, and process sandboxing. Work on STRONGMAN is continuing, focusing on the ease of management and correctness components of the system [Ioannidis et al., 2003].

Another use of KeyNote has been in Web access control, where it is used to mediate requests for pages or access to CGI scripts [Levine et al., 2003]. This is implemented as a module for the Apache Web server, *mod_keynote,* which performs the compliance checking functions on a per-request basis. This module has also been distributed with the OpenBSD operating system for several years, and the functionality has been folded into the STRONGMAN architecture.

47.4.3 Grid Computing and Transferable Micropayments

KeyNote is used to manage the authorization relationships in the Secure WebCom Metacomputer [Foley et al., 2001, 2002]. WebCom [Morrison et al., 1999] is a client/server-based system that may be used to schedule mobile application components for execution across a network. In Secure WebCom, KeyNote credentials are used to determine the authorization of X.509-authenticated SSL connections between WebCom masters and clients. Client credentials are used by WebCom masters to determine what operations the client is authorized to execute; WebCom master credentials are used by clients to determine if the master has the authorization to schedule the (trusted) mobile computation that the client is about to execute.

Systems that provide access to their resources can be paid using hash-chain-based micropayments [Foley and Quillinan, 2002]. KeyNote credentials are used to codify hash-chain micropayment contracts; determining whether a particular micropayment should be accepted amounts to a KeyNote compliance check that the micropayment is authorized. This scheme is generalized in [Foley, 2003] to support the efficient transfer of micropayment contracts whereby a transfer amounts to delegation of authorization for the contract. Characterizing a payment scheme as a trust management problem means that trust policies that are based on both monetary and conventional authorization concerns can be formulated.

47.4.4 Micropayments: Microchecks and Fileteller

One of the more esoteric uses of KeyNote has been as a micropayment scheme that requires neither online transactions nor trusted hardware for either the payer or payee. Each payer is periodically issued certified credentials that encode the type of transactions and circumstances under which payment can be guaranteed. A risk management strategy, taking into account the payer's history and other factors, can be used to generate these credentials in a way that limits the aggregated risk of uncollectible or fraudulent transactions to an acceptable level. Blaze et al. [2001a] showed a practical architecture for such a system that used KeyNote to encode the credentials and policies, and described a prototype implementation of the system in which vending machine purchases were made using off-the-shelf consumer PDAs.

Ioannidis et al. [2002] uses this micropayment architecture to build a credential-based network file storage system with provisions for paying for file storage and getting paid when others access files. Users get access to arbitrary amounts of storage anywhere in the network, and use a micropayments system to pay for both the initial creation of the file and any subsequent accesses. Wide-scale information sharing requires that a number of issues be addressed; these include distributed access, access control, payment, accounting, and delegation (so that information owners may allow others to access their stored content). Utilizing the same mechanism for both access control and payment results in an elegant and scalable architecture.

Ongoing work in this area is examining at distributed peer-to-peer filesystems and pay-per-use access to 802.11 networks.

47.4.5 Active Networking

Finally, STRONGMAN has been used in the context of active networks [Alexander et al., 1998a] to provide access control services to programmable elements [Alexander et al., 1998b, 2000, 2001]. An active network is a network infrastructure that is programmable on a per-user or even per-packet basis. Increasing the flexibility of such network infrastructures invites new security risks. Coping with these security risks represents the most fundamental contribution of active network research. The security concerns can be divided into those which affect the network as a whole and those which affect individual elements. It is clear that the element problems must be solved first, as the integrity of network-level solutions will be based on trust of the network elements. In the SANE architecture, KeyNote was used to limit the privileges of network users and their mobile code by specifying the operations such code was allowed to perform on any particular active node. KeyNote was used in a similar manner in the FLAME architecture [Anagnostakis et al., 2001, 2002a,b], and to provide an economy for resources in an active network [Anagnostakis et al., 2000].

47.5 Other Trust-Based Systems

The REFEREE system of Chu et al. [1997] is like PolicyMaker in that it supports full programmability of assertions (policies and credentials). However, it differs in several important ways. It allows the trust-management engine, while evaluating a request, to fetch additional credentials and to perform cryptographic signature-verification. (Recall that PolicyMaker places the responsibility for both of these functions on the calling application and insists that they be done before the evaluation of a request begins.) Furthermore, REFEREE's notion of "proof of compliance" is more complex than PolicyMaker's; for example, it allows nonmonotonic policies and credentials. The REFEREE proof system also supports a more complicated form of interassertion communication than PolicyMaker does. In particular, the REFEREE execution environment allows assertion programs to call each outer as subroutines and to pass different arguments to different subroutines, whereas the PolicyMaker execution environment requires each assertion program to write anything it wants to communicate on a global "blackboard" that can be seen by all other assertions.

REFEREE was designed with trust management for Web browsing in mind, but it is a general-purpose language and could be used in other applications. Some of the design choices in REFEREE were influenced by experience (reported in [Blaze et al., 1997]), using PolicyMaker for Web-page filtering based on PICS [Resnick and Miller, 1996] labels and users' viewing policies. It is unclear whether the cost of building and analyzing a more complex trust-management environment such as REFEREE is justified by the ability to construct more sophisticated proofs of compliance than those constructible in PolicyMaker. Assessing this tradeoff would require more experimentation with both systems, as well as a rigorous specification and analysis of the REFEREE proof system, similar to the one for PolicyMaker given in Blaze et al. [1998].

The Simple Public Key Infrastructure (SPKI) project of Ellison et al. [1999] and Ellison [1999] has proposed a standard format for authorization certificates. SPKI shares with our trust-management approach the belief that certificates can be used directly for authorization rather than simply for authentication. However, SPKI certificates are not fully programmable; they are data structures with the following five fields: "Issuer" (the source of authority), "Subject" (the entity being authorized to do something), and "Delegation" (a Boolean value specifying whether or not the subject is permitted to pass the authorization on to other entities), "Authorization" (a specification of the power that the issuer is conferring on the subject), and "Validity dates."

The SPKI documentation [Ellison, 1999] states that the processing of certificates and related objects to yield an authorization result is the province of the developer of the application or system. The processing plan presented in that document is an example that may be followed, but its primary purpose is to clarify the semantics of an SPKI certificate and the way it and various other kinds of certificate might be used to yield an authorization result.

Thus, strictly speaking, SPKI is not a trust-management engine, according to our use of the term, because compliance checking (referred to above as "processing of certificates and related objects") may be done in an application-dependent manner. If the processing plan presented in [Ellison, 1999] were universally adopted, then SPKI would be a trust-management engine. The resulting notion of "proof of compliance" would be considerably more restricted than PolicyMaker's; essentially, proofs would take the form of chains of certificates. On the other hand, SPKI has a standard way of handling certain types of nonmonotonic policies, because validity periods and simple CRLs are part of the proposal.

47.6 Closing Remarks

Trust management is a powerful approach to specifying and enforcing access control policies. The fundemental concepts behind trust management are the inherent constrained-delegation capability, *assertion monotonicity,* and a policy evaluation model that ensures safety and correctness. We briefly identified some uses of trust management systems in various applications, which should demonstrate the versatility and adaptability of the concepts and mechanisms presented.

The opportunity to use trust-management techniques exists in many projects that require some security component. Although the specific approaches we discussed may not be appropriate for any given application, the concepts are general enough and should be applicable in any context. The designer should carefully consider the system's needs and determine how best to use trust management. Doing so will allow them to easily manage fine-grained authorization and access control in a scalable and powerful way.

Acknowledgments

We would like to thank our collaborators in this work, Matt Blaze and Joan Feigenbaum.

References

Alexander, D. S., W. A. Arbaugh, M. Hicks, P. Kakkar, A. D. Keromytis, J. T. Moore, C. A. Gunter, S. M. Nettles, and J. M. Smith. The Switch Ware Active Network Architecture. *IEEE Network Magazine, special issue on Active and Programmable Networks,* 12(3): 29–36, 1998a.

Alexander, D. S., W. A. Arbaugh, A. D. Keromytis, S. Muir, and J. M. Smith. Secure quality of service handling (SQoSH). *IEEE Communications,* 38(4): 106–112, April 2000.

Alexander, D. S., W. A. Arbaugh, A. D. Keromytis, and J. M. Smith. A Secure Active Network Environment Architecture: Realization in Switch Ware. *IEEE Network Magazine, special issue on Active and Programmable Networks,* 12(3):37–45, 1998b.

Alexander, D. S., P. B. Menage, A. D. Keromytis, W. A. Arbaugh, K. G. Anagnostakis, and J. M. Smith. The Price of Safety in an Active Network. *Journal of Communications (JCN), special issue on programmable switches and routers,* 3(1): 4–18, March 2001.

Anagnostakis, K. G., M. B. Greenwald, S. Ioannidis, and S. Miltchev. Open Packet Monitoring on FLAME: Safety, Performance and Applications. In *Proceedings of the 4th International Working Conference on Active Networks (IWAN),* December 2002a.

Anagnostakis, K. G., M. W. Hicks, S. Ioannidis, A. D. Keromytis, and J. M. Smith. Scalable Resource Control in Active Networks. In *Proceedings of the 2nd International Working Conference on Active Networks (IWAN),* pp. 343–357, October 2000.

Anagnostakis, K. G., S. Ioannidis, S. Miltchev, J. Ioannidis, Michael B. Greenwald, and J. M. Smith. Efficient Packet Monitoring for Network Management. In *Proceedings of the IFIP/IEEE Network Operations and Management Symposium (NOMS) 2002,* April 2002b.

Anagnostakis, K. G., S. Ioannidis, S. Miltchev, and J. M. Smith. Practical Network Applications on a Lightweight Active Management Environment. In *Proceedings of the 3rd International Working Conference on Active Networks (IWAN),* October 2001.

Bellovin, S. M. Distributed Firewalls. *;login: magazine, special issue* on *security,* pp. 37–39, November 1999.

Blaze, M., J. Feigenbaum, J. Ioannidis, and A. Keromytis. The Role of Trust Management in Distributed Systems Security. In *Secure Internet Programming,* Vol. 1603 of *Lecture Notes in Computer Science,* pp. 185–210. Springer-Verlag, Berlin, 1999a.

Blaze, M., J. Feigenbaum, J. Ioannidis, and A. D. Keromytis. The KeyNote Trust Management System Version 2. Internet RFC 2704, September 1999b.

Blaze, M., J. Feigenbaum, and J. Lacy. Decentralized Trust Management. In *Proceedings of the 17th IEEE Symposium on Security and Privacy,* pp. 164–173, 1996.

Blaze, M., J. Feigenbaum, P. Resnick, and M. Strauss. Managing Trust in an Information Labeling System. In *European Transactions on Telecommunications, 8,* pp. 491–501, 1997.

Blaze, M., J. Feigenbaum, and M. Strauss. Compliance Checking in the PolicyMaker Trust-Management System. In *Proceedings of the Financial Cryptography Conference, Lecture Notes in Computer Science,* Vol. 1465, pp. 254–274. Springer-Verlag, Berlin, 1998.

Blaze, M., J. Ioannidis, and A. D. Keromytis. Offline Micropayments without Trusted Hardware. In *Proceedings of the 5th International Conference on Financial Cryptography,* pp. 21–40, February 2001a.

Blaze, M., J. Ioannidis, and A.D. Keromytis. Trust Management for IPsec. In *Proceedings of the Network and Distributed System Security Symposium (NDSS),* pp. 139–151, February 2001b.

Blaze, M., J. Ioannidis, and A.D. Keromytis. Trust Management for IPsec. *ACM Transactions on Information and System Security (TISSEC),* 32(4): 1–24, May 2002.

CCITT. *X.509: The Directory Authentication Framework.* International Telecommunications Union, Geneva, 1989.

Chu, Y.-H., J. Feigenbaum, B. LaMacchia, P. Resnick, and M. Strauss. REFEREE: Trust Management for Web Applications. In *World Wide Web Journal, 2,* pp. 127–139,1997.

de Raadt, T., N. Hallqvist, A. Grabowski, A. D. Keromytis, and N. Provos. Cryptography in OpenBSD: An Overview. In *Proceedings of the USENIX Annual Technical Conference, Freenix Track,* pp. 93–101, June 1999.

Ellison, C. SPKI Requirements. Request for Comments 2692, Internet Engineering Task Force, September 1999. URL ftp://ftp.isi.edu/in-notes/rfc2693.txt.

Ellison, C., B. Frantz, B. Lampson, R. Rivest, B. Thomas, and T. Ylonen. SPKI Certificate Theory. Request for Comments 2693, Internet Engineering Task Force, September 1999. URLftp://ftp.isi.edu/in-notes/rfc2693.txt,

Foley, S. N. Using Trust Management to Support Transferable Hash-Based Micropayments. In *Proceedings of the International Financial Cryptography Conference,* January 2003.

Foley, S. N. and T. B Quillinan. Using Trust Management to Support Micropayments. In *Proceedings of the Annual Conference on Information Technology and Telecommunications,* October 2002.

Foley, S. N., T. B. Quillinan, and J. P. Morrison. Secure Component Distribution Using WebCom. In *Proceedings of the 17th International Conference on Information Security (IFIP/SEC),* May 2002.

Foley, S. N., T. B. Quillinan, J. P. Morrison, D. A. Power, and J. J. Kennedy. Exploiting KeyNote in WebCom: Architecture Neutral Glue for Trust Management. In *5th Nordic Workshop on Secure IT Systems,* Oct 2001.

Hallqvist, Niklas and Angelos D. Keromytis. Implementing Internet Key Exchange (IKE). In *Proceedings of the Annual USENIX Technical Conference, Freenix Track,* pp. 201–214, June 2000.

Hicks, Michael and Angelos D. Keromytis. A Secure PLAN. In Stefan Covaci, Ed., *Proceedings of the 1st International Working Conference on Active Networks,* Vol. 1653 of *Lecture Notes in Computer Science,* pp. 307–314. Springer-Verlag, Berlin, June 1999. URL http://www.cis.upenn.edu/switchware/papers/iwan99.ps.

Ioannidis, John, Sotiris Ioannidis, Angelos Keromytis, and Vassilis Prevelakis. Fileteller: Paying and Getting Paid for File Storage. In *Proceedings of the 6th International Conference on Financial Cryptography,* March 2002.

Ioannidis, S., S. M. Bellovin, I. Ioannidis, A. D. Keromytis, and J. M. Smith. Design and Implementation of Virtual Private Services. In *Proceedings of the IEEE International Workshops on Enabling Technologies: Infrastructure for Collaborative Enterprises (WETICE), Workshop on Enterprise Security, Special Session on Trust Management in Collaborative Global Computing,* June 2003.

Ioannidis, S., A. D. Keromytis, S. M. Bellovin, and J. M. Smith. Implementing a Distributed Firewall. In *Proceedings of the ACM Conference on Computer and Communications Security (CCS),* pp. 190–199, November 2000.

Kent, S. and R. Atkinson. Security Architecture for the Internet Protocol. Request for Comments (Proposed Standard) 2401, Internet Engineering Task Force, November 1998. URL. ftp://ftp.isi.edu/in-notes/rfc2401.txt,

Keromytis, A. D. *STRONGMAN: A Scalable Solution To Trust Management In Networks.* Ph.D. thesis, University of Pennsylvania, Philadelphia, November 2001.

Keromytis, A. D., S. Ioannidis, M. B, Greenwald, and J. M. Smith. The STRONGMAN Architecture. In *Proceedings of DISCEX III,* April 2003.

Keromytis, Angelos D., Vishal Misra, and Daniel Rubenstein. SOS: Secure Overlay Services. In *Proceedings of ACM SIGCOMM,* pp. 61–72, August 2002.

Lacy, J., J. Snyder, and D. Maher. Music on the Internet and the Intellectual Property Protection Problem. In *Proceedings of the International Symposium on Industrial Electronics,* pp. SS77–83, 1997.

Levine, A., V. Prevelakis, J. Ioannidis, S. Ioannidis, and A. D. Keromytis. WebDAVA: An Administrator-Free Approach to Web Pile-Sharing. In *Proceedings of the IEEE International Workshops on Enabling Technologies: Infrastructure for Collaborative Enterprises (WETICE), Workshop on Distributed and Mobile Collaboration,* June 2003.

Morrison, J. P., D. A. Power, and J. J. Kennedy. WebCom: A Web Based Distributed Computation Platform. In *Proceedings of Distributed computing on the Web,* June 1999.

Resnick, P. and J. Miller. PICS: Internet Access Controls Without Censorship. *Communications of the ACM,* pp. 87–93, October 1996.

Schneier, B. *Applied Cryptography.* John Wiley & Sons, New York, 1996.

Schneier, B. *Secrets and Lies: Digital Security in a Networked World.* John Wiley & Sons, New York, 2000.

48

An Overview of Intrusion Detection Techniques

CONTENTS

Abstract 48-1
48.1 Introduction 48-1
48.2 Modeling and Analysis Approaches 48-4
48.2.1 Misuse Detection 48-4
48.2.2 Anomaly Detection 48-4
48.2.3 Alert Analysis 48-5
48.3 Network and System Issues 48-6
48.3.1 Deployment Strategies 48-6
48.3.2 Performance Optimization and Adaptation 48-7
48.4 Summary 48-8
48.5 To Learn More 48-8
References 48-9

Wenke Lee

Abstract

Intrusion detection is an essential component of the defense-in-depth network security mechanisms. In this chapter, we give an overview of the basic concepts, principles, and techniques of intrusion detection. There are two train modeling and analysis approaches. Misuse detection relies on patterns of known intrusions or vulnerabilities to detect attack instances. Anomaly detection uses normal profiles to detect deviations caused by intrusions. Intrusion alerts often require further analysis in order to recognize attack plan or trend. There are several implementation and deployment strategies for intrusion detection. A network-based intrusion detection system (IDS) monitors activities to a network. A host-based IDS only monitors activities on the host operating system. A network-node IDS monitors network activities associated with the host.

Despite more than two decades of research and development efforts, there are still serious limitations in intrusion detection technologies. These include high false alarm rate, the lack of early warning capabilities, and the lack of automated response and recovery mechanisms.

48.1 Introduction

As the Internet plays an increasingly important role in our society, e.g., the infrastructure for e-commerce and digital government, it is no longer just the playground of recreational hackers. The Internet has become the target of criminals and enemies who are devising and launching sophisticated intrusions

1-58488-381-2/05/$0.00+$1.50
© 2005 by CRC Press LLC

(i.e., attacks) with financial, political, and even military objectives. It is imperative that we provide the best protection possible for our network infrastructure.

Contrary to the myth that there can be a panacea in security, no single technology alone is the answer. Security is a process (or a chain) that is as secure as its weakest link; flaws in hardware, software, and networks, as well as human errors can all lead to security failure [Schneier, 2000]. For example, confidential data transmitted via an encrypted link can still be stolen from the end systems because of break-ins that exploit "weak passwords" or system software bugs. Experience has taught us that we need to deploy defense-in-depth network security mechanisms, which include these necessary technologies: security policy, vulnerability scanning and patching, access control and authentication, encryption, program wrappers, firewalls, *intrusion detection* (ID), and intrusion response and tolerance. Using each of these technologies alone is not sufficient to secure a network. However, each technology is valuable because it provides an additional layer of protection. For example, a company can use a virtual private network (VPN) over the Internet to authenticate and encrypt traffic between different sites. In addition, the company can deploy firewalls to block illegitimate access to its networks and intrusion detection systems (IDSs) to monitor its network activities to detect and respond to attacks that manage to bypass the firewalls. Further, the company can use replication and secret sharing techniques so that its critical services and data remain available even when an attack manages to bypass the firewalls and IDSs.

The primary assumptions of intrusion detection are that user and program activities are observable, for example via system auditing mechanisms, and, more important, normal and intrusion activities have distinct behavior. Intrusion detection, therefore, involves capturing audit data and reasoning about the evidence in the data to determine whether the system is under attack. Based on the type and source of audit data used, an intrusion detection system (IDS) can be network-based, host-based, or network-node-based. A network-based IDS normally runs at the gateway of a network and "captures" and examines network traffic to and from the network. A network-node-based IDS runs at a network node, for example, on its network interface to examine network packets to and from the node. A host-based IDS relies on operating system audit data to monitor and analyze the events generated by programs or users on the host. These implementation and deployment strategies have different strengths in handling traffic load, resisting evasion attempts, and analyzing distributed and coordinated attacks.

There are two major categories of analysis techniques in intrusion detection: *misuse detection* and *anomaly detection.* Misuse detection uses the "signatures" of known attacks, i.e., the patterns of attack behavior or effects, to identify a matched activity as an attack instance. By definition, misuse detection is not effective against *new* attacks, i.e., those that do not have known signatures. Anomaly detection uses established normal profiles, i.e., the expected behavior, to identify any unacceptable deviation as the result of an attack. Anomaly detection is intended for catching new attacks. However, new legitimate behavior can also be falsely identified as an attack, resulting in a false alarm.

Alerts of intrusions can be further analyzed and correlated to identify attack plans and trends. Reports of attacks (and plans) can trigger response actions (e.g., termination of the offending connections). Figure 48.1 depicts the layers of processing and analysis tasks in an IDS. At the lowest level, the IDS collects audit data. For example, the network interface receives network traffic data, and the data capturing and filtering unit then selects only a portion of the audit data, according to the IDS configuration policy. The data is then preprocessed (e.g., reassembled to connection data), and only the important events (e.g., a new connection is established) or suspicious events (e.g., an attempted connection to a closed port) are extracted. The event analysis engine then uses an intrusion detection algorithm(s) to piece the event data together and produces an alert when it believes that an intrusion is occurring. Alerts can trigger local response, and can be sent to a global correlater for analysis of distributed attacks and long-term attack plans (or scenarios).

The most commonly used IDS performance metrics are the true positive rate (or detection rate) and false alarm rate. Let I and $-I$ denote the intrusive or nonintrusive (or normal) behavior, and A and $-A$ denote the presence or absence of an intrusion alert from the IDS. The detection rate is $P(A|-I)$ (i.e., the probability that the IDS outputs an alert when an intrusion is present) and false alarm rate is $P(A|-I)$

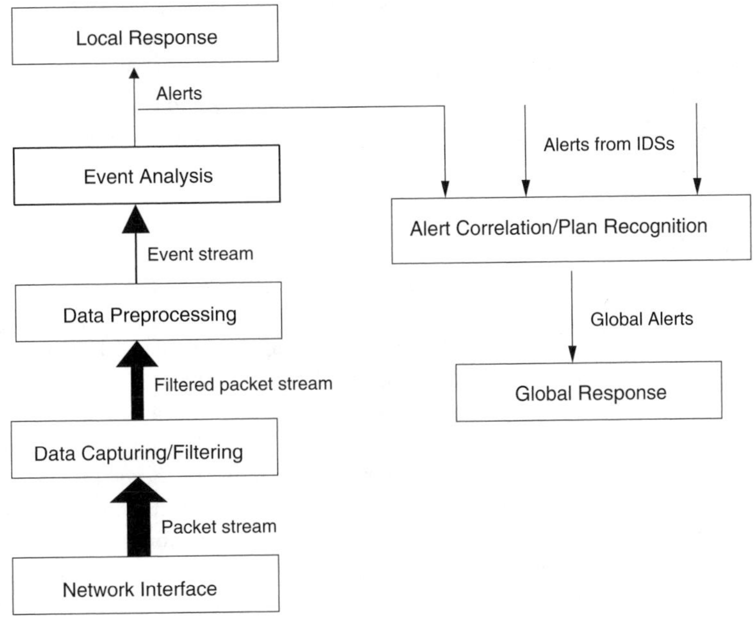

FIGURE 48.1 In an IDS, data is processed and analyzed at the appropriate semantic levels.

(i.e., the probability that the IDS outputs an alert when there is no intrusion). From a usability point of view, a critical performance measurement is the Bayesian detection rate [Axelsson, 2000] $P(A|I)$ (i.e., the probability that an intrusion is present when the IDS outputs an alert). It measures IDS credibility because it indicates how likely an intrusion is present when there is an alert. Using Bayes' theorem, it can be computed as [Axelsson, 2000]:

$$P(I \mid A) = \frac{P(I)P(A \mid I)}{P(I)P(A \mid I) + P(\neg I)P(A \mid \neg I)} \tag{48.1}$$

Although a higher detection rate and lower false alarm rate will lead to a higher Bayesian detection rate, as pointed out by Axelsson, if the base rate (the probability of intrusion data), P(I). is extremely low, say 2×10^{-5}, then even with a perfect detection rate, P(A|I) = 1, and a perhaps unattainably low false alarm rate, say P(A|–I) = 1×10^{-5}, the Bayesian detection rate is only 66% [2000].

Although some would claim that we have done enough research and that commercial products are effective, IDSs are really still in their infancy [Schneier, 2000]. In 1998 and 1999, DARPA conducted evaluations to survey the state-of-the-art in ID research. The results showed that the best research systems (some with both misuse and anomaly detection models) had detection rates (i.e., the percentages of attack incidents correctly identified) below 70% [Lippmann et al., 2000a, b]. Most of the missed intrusions were new attacks or *stealthy* attacks (i.e., clever variations of known attacks) that can lead to unauthorized user or root access to the mocked military network used in the evaluations. CMU/SEI recently conducted a comprehensive study of ID technologies [Allen et al., 2000]. The report finds that most commercial IDSs use only misuse detection techniques and thus are unable to detect new attacks. In addition, they have high false alarm rates and have difficulties dealing with switched network environments and high-speed network traffic.

The performance objectives of an IDS include broad detection coverage, economy in resource usage, and resilience to stress [Puketza et al., 1996]. An IDS must also resist attacks upon itself [Ptacek and Newsham, 1998; Paxson, 1999]. In the remainder of this paper, we discuss how these objectives influence the design and implementation of IDSs.

48.2 Modeling and Analysis Approaches

In order to provide broad detection coverage, an IDS needs to employ models that can accurately identify a wide range of attack behaviors. The two main categories of modeling and analysis techniques are misuse detection and anomaly detection. In addition, there are alert correlation algorithms for attack scenario analysis.

48.2.1 Misuse Detection

Misuse detection systems, e.g., IDIOT [Kumar and Spafford, 1995] and STAT [Ilgun et al., 1995], use patterns of well-known attacks or weak spots of the system to match and identify known intrusions. For example, a signature rule for the "guessing password attack" can be "there are more than 4 failed login attempts within 2 minutes." The main advantage of misuse detection is that it can accurately and efficiently detect instances of known attacks. For example, to detect the "Land" attack,[1] an IDS only needs to have a simple rule that checks whether a packet to a host has the same source and destination port. The main disadvantage is that it lacks the ability to detect the truly innovative (i.e., newly invented) attacks.

A misuse detection system needs to have a general yet concise representation of attack patterns. For example, in STAT, an attack pattern is represented as the path of state transitions from an initial legitimate system state to a compromised system state. The transitions represent actions or steps of the attack, and the intermediate states represent the results of the actions. Part(s) of two different attack patterns may be represented by the same state transition path(s). Another issue in misuse detection is the real-time pattern matching speed. This is particularly important when there are many attack patterns. For example, Snort [Roesch, 1999] has more than 2,000 rules (patterns). It organizes rules for each protocol in a structure where a linked list of rule headers each contains the common conditions (e.g., on IP addresses and ports) of a set of rules. A packet is first compared with the rule headers, and when there is match, the rules linked to the header will be checked one by one. Advanced string matching algorithms, e.g., Coit et al. [2001], can also speed up the pattern-matching process.

Misuse detection systems need to be updated whenever there is a newly discovered attack or vulnerability. Similar to the antivirus industry, several vendors have developed and deployed an infrastructure where installed IDSs can automatically receive and install updated rules from the vendors.

48.2.2 Anomaly Detection

Anomaly detection (sub)systems — for example, IDES [Lunt et al., 1992] — flag observed activities that deviate significantly from the established normal usage profiles as anomalies, i.e., possible intrusions. For example, the normal profile of a user may contain the averaged frequencies of some system commands used in his or her login sessions, If, for a session that is being monitored, the frequencies are significantly lower or higher, then an anomaly alarm will be raised. The main advantage of anomaly detection is that it does not require prior knowledge of intrusion and can thus detect new intrusions. The main disadvantage is that it may not be able to describe what the attack is and may have high false positive rate when a statistical approach is used.

The main issue in building an anomaly detection system is to control the false alarm rate to an "acceptable" level while detecting as many attacks as possible. There are several anomaly detection approaches.

Specification-based approaches, e.g., Ko et al. [1997] and Ko [2000], specify the intended (legitimate) behavior of a program (or system) and detect a violation (or misbehavior) as an anomaly (i.e., probably an intrusion). Static analysis approaches, e.g., Wagner and Dean [2001], use automated source code analysis techniques to build a specification (or model) of a program and detect a violation of the program

[1] The attacker crafts a packet with the same source and destination port and address, causing the receiving host running the vulnerable operating system to crash.

model as an anomaly. For example, the *callgraph* model characterizes the expected system call traces using static analysis of the program code [Wagner and Dean, 2001]. The control-flow graph is naturally transformed to a nondeterministic finite-state automaton (NDFA). The NDFA is then used to monitor the program execution in real-time. The operation of the NDFA is simulated on the observed system call trace. If all paths are blocked at some execution point of the program, there is an anomaly. Specification-based and static analysis approaches can produce models with no false alarms because all intended normal program behaviors are carefully studied and represented. Detection rates are not 100% because the models may include some possible intrusion paths. These approaches also do not scale well to complex programs.

Statistical approaches characterize the behavior of a program (or system) using some temporal and statistical measures. For example, in a network-based IDS, a traffic model can use the service request rate to a Web server to detect the anomaly caused by a Denial-of-Service (DoS) attack. As another example, in a host-based IDS, a system call model can record the length-k short sequences of consecutive system calls made by a program and detect an anomaly when there are a large percentage of unknown sequences during the program execution [Forrest et al., 1996]. Statistical approaches require extensive training (or learning), e.g., by exhaustively running the program under various (normal) conditions to account for its execution paths, so that the temporal and statistical measures capture as much of the normal behaviors as possible. With the help of data modeling and machine learning algorithms, the process of building a statistical anomaly model can be semiautomated [Lee and Stolfo, 2000]. Expert knowledge is still needed in selecting what measures to compute and in examining (validating) the output models. Statistical approaches are scalable to complex software but usually have false alarms.

48.2.3 Alert Analysis

In a large-scale network, the security devices, e.g., IDSs, often output a large amount of low-level or incomplete (fragmented) alert information because there is a large amount of network and system activities being monitored, and multiple IDSs can each report some aspects of the same (coordinated) security event. For example, a single distributed port-scan to a corporate network can result in many alerts from multiple IDSs. The intrusion response system and the security staff can be overwhelmed by the sheer volume of alert data and cannot adequately understand the security state and initiate appropriate response in a timely fashion. An intelligent adversary can exploit this situation by launching a large number of nuisance attacks with the purpose of masking his intended intrusions. Alert analysis is therefore a critical element in the intrusion detection and response process. The main analysis tasks include *clustering* and *correlation*. Clustering alerts of the same event (e.g., a security violation) to a single high-level alert can reduce the amount of alert data and hence also the number of false alarms. Correlating alerts of the related attack steps to identify an *attack scenario* can help forensic analysis, response, and recovery, and even the prediction of forthcoming attacks.

Alert fusion and clustering techniques are relatively straightforward. For example, each alert can have a number of attributes such as *timestamp, source IP, destination IP, port(s), user name, attack class,* and *sensor ID,* which are defined in the standard document "Intrusion Detection Message Exchange Format (IDMEF)" Group [2002] drafted by the IETF Intrusion Detection Working Group. In alert fusion, alerts that have the same attribute values except the timestamps and sensor IDs can be combined together. The timestamps can be slightly different, say, 2 sec apart. That is, alerts of the same attack that are output by different sensors within a short time window are combined. Alert clustering is then used to further group similar alerts together, based on the site-specific similarity measures on alert attributes [Valdes and Skinner, 2001] [Julisch and Dacier, 2002].

Alert correlation, on the other hand, is a much more challenging task. Although recently there have been several proposals (e.g., Debar and Wespi, 2001; Valdes and Skinner, 2001; Goldman et al., 2001; Porras et al., 2002; Ning et al., 2002), most of these proposed approaches have very limited capabilities because they rely on various forms of predefined knowledge of attack conditions and consequences, and cannot recognize a correlation when an attack is new (previously unknown) or the relationship between the attacks is new. In other words, these approaches in principle are similar to *misuse detection* techniques, which use the "signatures" of known attacks to perform pattern matching and cannot detect new attacks. For example, a popular approach is to use the prerequisites of intrusions to correlate alerts [Cuppens and Miége, 2002; Ning et al., 2002]. The assumption is that when an attacker launches a scenario, prior attack steps are preparing for later ones, and therefore the consequences of earlier attacks have a strong relationship with the prerequisites of later attacks. The correlation engine searches for alert pairs that have a consequences and prerequisites match and builds a correlation graph with such pairs. There are several limitations with this approach. First, a new attack cannot be paired with any other attack because its prerequisites and consequences are not yet defined. Second, even for known attacks, it is infeasible to predefine all possible prerequisites and consequences.

It is obvious that the number of possible correlations is very large, potentially a combinatorial of the number of (known and new) attacks. Therefore, in general, it is not feasible to know *a priori* and encode all possible matching conditions between attacks. To further complicate the matter, the more dangerous and intelligent adversaries will always invent new attacks and novel attack sequences. Clearly, any correlation approach that relies solely on predefined knowledge is bound to fail. Therefore, we need to develop new and significantly better alert correlation algorithms that can discover sophisticated and new attack sequences. Qin and Lee [2003] recently developed a statistical causality analysis approach for alert analysis and showed that this approach can discover new alert relationships as long as the alerts of the attacks can be statistically correlated.

48.3 Network and System Issues

As discussed in Section 48.1, the *performance objectives* of an IDS include good detection coverage, economy in resource usage, and resilience to stress [Puketza et al., 1996]. An IDS must resist attacks upon itself [Ptacek and Newsham, 1998; Paxson, 1999]. These objectives can be conflicting goals. For example, for broad coverage and high detection accuracy, an IDS needs to perform stateful analysis on a large quantity of audit data. This requires a large amount of resources (in both memory and detection time). A resource-intensive IDS is then vulnerable to stress and overload attacks. Recent studies (e.g., Lippmann et al., 2000a,b; Allen et al., 2000) showed that the current generation of intrusion detection systems (IDSs) are unable to detect new attacks, have high false alarm rates, and have difficulties dealing with high-speed and high-volume network traffic. These problems severely undermine the utility of IDSs. Worse, attackers can exploit these shortcomings to defeat the IDSs and accomplish their malicious goals [Ptacek and Newsham, 1998; Paxson, 1999; Shipley and Mueller, 2001]. For example, an attacker can launch overload or DoS attacks against an IDS to a point that it "drops" audit data, hence missing key evidence and failing to detect the intended attack (or its detection may be too late or slow to prevent the damage of the attack). These IDS performance problems can only get worse because as networking technologies continue to leap forward, both the speed and volume of network traffic, as well as the complexities of network services, will increase rapidly. Clearly, we need to carefully consider the performance tradeoffs in IDS implementation and deployment.

48.3.1 Deployment Strategies

A popular way to monitor the high-speed and high-volume traffic to a network is to run several network-based IDSs (NIDSs) and use load-balancing techniques to split the traffic among them in some meaningful way (e.g., Top Layer Networks and Internet Security Systems, 2000; Kruegel et al., 2002). A difficulty of this approach is that some distributed attacks may be missed because the evidence at any of the IDSs may be below the detection threshold. Another emerging approach is to use one or more specialized high-end hardware processors to implement an NIDS. Both of these approaches are expensive and do not scale well because more dedicated hardware will be needed. as the network bandwidth grows. A common and more serious problem with NIDSs is that they typically do not have sufficient knowledge of the network topology and which operating systems are running on the network hosts. As a consequence, an NIDS and an end-host could be seeing or interpreting connections differently. This weakness allows attackers to evade detection by sending attack traffic that looks harmless from the perspective of NIDS [Ptacek and Newsham, 1998; Paxson, 1999]. In addition, NIDSs generally do not have the necessary keys to examine end-to-end encrypted traffic, thus giving attackers another means to evade detection. A remedy to these problems is to use network-node-based IDSs (NNIDSs) that each monitor only the traffic to a host or to use a host-based IDS (HIDS) to monitor the operating system activities on the host. The NNIDSs (or HIDSs) can unambiguously check the traffic data (or operating system data) and have access to the key(s) to examine encrypted data. However, there are also problems with this approach. An NNIDS or HIDS typically runs as kernel- or application-level software, and as a result, its overhead can severely affect the performance of other applications running on the host. Furthermore, if attackers manage to compromise a host, they can also disable the NNIDS or HIDS so that all of their malicious activities will go undetected. These problems can be addressed by implementing an NNIDS on a (general purpose) network processor rather than on top of the host operating system.

Network processors will be widely available and affordable in the near future, and can be integrated into a network interface card (NIC). Having an NNIDS run on network processors not only allows analysis to be carried out close to the data source (without going through the data bus), thus achieving high-speed, but also frees the host processor from being a dedicated resource for intrusion detection. This makes it feasible to distribute these NNIDSs to nodes throughout the network. This deployment scheme can scale to large and complex networks because each NNIDS runs on an affordable NIC and unambiguously checks only the traffic to a node. There is also added security for the NNIDS itself. An attacker cannot disable the NNIDS even if he penetrates the host because the control flows to the network processors can be very restrictive.

There are several research issues with NNIDSs. The security policy that dictates network intrusion detection functions must now be managed and enforced in a distributed fashion. This problem is similar to managing distributed firewalls [Bellovin, 1999]. The NNIDSs also need to perform event-sharing and collaborative analysis to detect distributed attacks and share the work load when needed. This problem is not necessarily unique to NNIDSs because as discussed above, an NIDS, when implemented using load-balancing techniques, needs to deal with the same issue.

48.3.2 Performance Optimization and Adaptation

It is extremely difficult, if not impossible, for an IDS to be 100% accurate, especially when there are limited resources, e.g., CPU, memory, and response time. The optimal performance of an IDS should be determined by not only its tradeoffs between detection rate and false alarm rate, but also its cost metrics (e.g., damage cost of intrusion) and the probability of intrusion [Gaffney and Ulvila, 2001; Lee et al., 2002b]. That is, the expected value of an IDS, calculated as the prevention of damages by intrusions, should be considered. Accordingly, performance optimization and adaptation means that an IDS should continuously maximize its cost-benefits for the given (current) operational conditions. For example, if an IDS is forced to miss some intrusions (that can otherwise be detected using its "signature base"), for example, due to stress or overload attacks, it should still ensure that the best value (or minimum damage) is provided according to cost-analysis on the circumstances. As a simple example, if we regard buffer-overflow as more damaging than port-scan (and, for argument's sake, all other factors, i.e., attack probability and detection probability, are equal), then missing a port-scan is better than missing a buffer-overflow.

Lee et al. [2002a] studied the IDS performance optimization and adaptation problem. As with most optimization problems, IDS optimization seeks to maximize the "value" with bounded "cost." Let C^{β} denote the damage cost (e.g., data loss) of intrusion and C^{α} denote the false alarm cost (e.g., labor cost in investigation). The total value of an IDS [Lee et al., 2002a] is:

$$V = C^{\beta}P(I)P(A \mid I) + C^{\alpha}P(\quad I)P(A \mid - \qquad (48.2)$$

The first term is the loss (damage) prevented because of true detection, and the second term is the loss incurred because of false alarms. In terms of computational cost of an IDS, compute time is a reasonable choice because a real-time IDS needs to provide timely detection so that appropriate response actions can prevent or minimize damages. Depending on the specific application scenario, one can derive timing constraints for an IDS. For example, an IDS needs to finish analysis within packet inter-arrival time. The optimization problem can be stated as follows: include as many data analysis and intrusion detection tasks as possible so that the overall IDS value is maximized while the total compute time is within the timing constraints. This is known as the Knapsack problem (e.g., [Martello and Toth, 1990; Papadimitriou and Steiglitz, 1982]) in the optimization literature.

As traffic (and attack) conditions change, the configurations of the IDS also need to adapt in order to provide the (new) best value. That is, the Knapsack algorithm is invoked to solve the new optimization problem according to the current traffic conditions and timing constraints. Data analysis tasks and detection rules (modules) can be turned on and off as a result of the reconfiguration.

48.4 Summary

Intrusion detection is a very important network security mechanism. An IDS needs to provide broad and accurate detection coverage. Towards this end, misuse detection and anomaly detection approaches are needed to detect known and new attacks. Alert analysis algorithms are needed to process outputs from IDSs to recognize attack plans and trends. An IDS also needs to be economical in resource utilization, resilient to stress, and resistant to attacks upon itself. According to site-specific policies and the available computing and human resources, an intrusion detection solution can be any combination of network-based, host-based, and network-node-based. In addition, performance optimization and adaptation techniques are needed to consider the IDS performance tradeoffs.

There are still serious limitations in intrusion detection technologies. Various studies (e.g., Allen et al. [2000]) have shown that current IDSs generate many false alarms, cannot detect new attacks, and cannot handle high-speed and high-volume audit data streams. In order to address these problems, we must research and develop new approaches. For example, as discussed in Section 48.1, the base rate (the prior probability) in high-volume audit data stream is extremely low. Even if we can develop a detection algorithm with very low false alarm rate, the Bayesian detection rate is still low. Therefore, instead of focusing only on developing new detection algorithms, we need to also develop efficient filtering techniques, e.g., to remove normal data as much as possible in order to increase the base rate of the data stream before it is analyzed by detection algorithms.

The future for intrusion detection is both exciting and challenging. For example, taking advantages of networking technologies that continue to leap forward, attackers are developing large-scale and fast intrusions — for example, various Distributed Denial-of-Service attacks and fast worms. We need to develop Internet-scale distributed and cooperated intrusion detection infrastructure with a large number of IDSs (individual sensors) and alert aggregation and correlation centers. We also need to develop early sensing or warning capabilities with low false alarm rates to stop the fast spread of Internet worms.

48.5 To Learn More

Intrusion detection is a very active field for both research and development. This paper provides only an overview of the basic concepts, principles, and techniques. There is a lot more to learn. For researchers, there are several premier journals and conferences that report advances in intrusion detection research. These include the ACM Transactions on information and System Security, the Journal of Computer Security, the IEEE Symposium on Security and Privacy, the International Symposium on Recent Advances in Intrusion Detection, the ACM Conference on Computer and Communications Security, the USENIX Security Symposium, and the Network and Distributed System Security Symposium. For practitioners, there are numerous trade magazines and newsletters that report the latest products. In addition, the

Computer Security Division of the Information Technology Laboratory at the National Institute of Standards and Technology (NIST) has been publishing guidelines on security technologies, including intrusion detection (e.g., Bace and Mell [2001]; Mell et al. [2003]).

References

Allen, J., A. Christie, W. Fithen, J. McHugh, J. Pickel, and E. Stoner. State of the practice of intrusion detection technologies. Technical Report CMU/SEI-99-TR-028, CMU/SEI, 2000.

Axelsson, S. The base-rate fallacy and the difficulty of intrusion detection. *ACM Transactions on Information and System Security,* 3(3), 2000.

Bace, R. and P. Mell. Intrusion detection systems (ids). Technical Report NIST SP 800-31, National Institute of Standards and Technology, November 2001.

Bellovin, S. M. Distributed firewalls. *;login:,* November 1999.

Coit, C., S. Staniford, and J. McAlerney. Towards faster string matching for intrusion detection or exceeding the speed of snort. In *Proceedings of the 2001 DARPA Information Survivability Conference and Exposition (DISCEX II),* June 2001.

Cuppens, F. and A. Miége. Alert correlation in a cooperative intrusion detection framework. In *Proceedings of the IEEE Symposium on Research in Security and Privacy,* pp. 202–215, Oakland CA, May 2002.

Debar, H. and A. Wespi. Probabilistic alert correlation. In *4th International Symposium on Recent Advances in Intrusion Detection (RAID),* October 2001.

Forrest, S., S. A. Hofmeyr, A. Somayaji, and T. A. Longstaff. A sense of self for Unix processes. In *Proceedings of the 1996 IEEE Symposium on Security and Privacy,* pp. 120–128, Los Alamitos, CA, 1996. IEEE Computer Society Press.

Gaffney, J. E. and J. W. Ulvila. Evaluation of intrusion detectors: A decision theory approach. In *Proceedings of the 2001 IEEE Symposium on Security and Privacy,* May 2001.

Goldman, R. P., W. Heimerdinger, and S. A. Harp. Information modeling for intrusion report aggregation. In *DARPA Information Survivability Conference and Exposition (DISCEX 2001),* June 2001.

IETP Intrusion Detection Working Group. Intrusion detection message exchange format. http://www.ietf.org/internet-drafts/draft-ietf-idwg-idmef-xml-09.txt, 2002.

Ilgun, K., R. A. Kemmerer, and P. A. Porras. State transition analysis: A rule-based intrusion detection approach. *IEEE Transactions on Software Engineering,* 21(3): 181–199, March 1995.

Julisch, K., and M. Dacier. Mining intrusion detection alarms for actionable knowledge. In *The 8th ACM International Conference on Knowledge Discovery and Data Mining,* July 2002.

Ko, C., Logic induction of valid behavior specifications for intrusion detection. In *Proceedings of the 2000 IEEE Symposium on Security and Privacy,* May 2000.

Ko, C., M. Ruschitzka, and K. Levitt. Execution monitoring of security-critical programs in distributed systems: A specification-based approach. In *Proceedings of the 1997 IEEE Symposium on Security and Privacy,* May 1997.

Kruegel, C., F. Valour, G. Vigna, and R. A. Kemmerer. Stateful intrusion detection for high-speed networks. In *Proceedings of 2002 IEEE Symposium on Security and Privacy,* May 2002.

Kumar S., and E. H. Spafford. A software architecture to support misuse intrusion detection. In *Proceedings of the 18th National Information Security Conference,* pp. 194–204, 1995.

Lee, W., J. B. D. Cabrera, A. Thomas, N. Balwalli, S. Saluja, and Y. Zhang. Performance adaptation in real-time intrusion detection systems. In *Proceedings* of *the 5th International Symposium on Recent Advances in Intrusion Detection (RAID 2002),* October 2002a.

Lee, W., W. Fan, M. Miller, S. J. Stolfo, and E. Zadok. Toward cost-sensitive modeling for intrusion detection and response. *Journal of Computer Security,* 10(1, 2), 2002b.

Lee, W., and S. J. Stolfo. A framework for constructing features and models for intrusion detection systems. *ACM Transactions on Information and System Security,* 3(4), November 2000.

Lippmann, R., D. Fried, I. Graf, J. Haines, K. Kendall, D. McClung, D. Weber, S. Webster, D. Wyschogrod, R. Cunninghan, and M. Zissman. Evaluating intrusion detection systems: The 1998 DARPA off-line intrusion detection evaluation. In *Proceedings of the 2000 DARPA Information Survivability Conference and Exposition,* January 2000a.

Lippmann, R., J. Haines, D. Fried, J. Korba, and K. Das. Analysis and results of the 1999 DARPA off-line intrusion detection evaluation. In *Proceedings of the 3rd International Workshop on Recent Advances in Intrusion Detection (RAID 2000),* October 2000b.

Lunt, T., A. Tamaru, F. Gilham, R. Jagannathan, P. Neumann, H. Javitz, A. Valdes, and T. Garvey. A real-time intrusion detection expert system (IDES) — final technical report. Technical report, Computer Science Laboratory, SRI International, Menlo Park, CA, February 1992.

Martello, S. and P. Toth. *Knapsack Problems: Algorithms and Computer Implementations.* John Wiley & Sons, New York, 1990.

Mell, P., V. Hu, R. Lippmann, J. Haines, and M. Zissman. An overview of issues in testing intrusion detection systems. Technical Report NIST IR 7007, National Institute of Standards and Technology, June 2003.

Ning, P., Y. Cui, and D. S. Reeves. Constructing attack scenarios through correlation of intrusion alerts. In *9th ACM Conference on Computer and Communications Security,* November 2002.

Papadimitriou, C. H. and K. Steiglitz. *Combinatorial Optimization — Algorithms and Complexity.* Prentice-Hall, Upper Saddle River, NJ, 1982.

Paxson, V. Bro: A system for detecting network intruders in real-time. *Computer Networks,* 31 (23–24), December 1999.

Porras, Phillip A., Martin W. Fong, and Alfonso Valdes. A Mission-Impact-Based approach to INFOSEC alarm correlation. In *5th International Symposium on Recent Advances in Intrusion Detection (RAID),* October 2002.

Ptacek, T. H. and T. N. Newsham. Insertion, evasion, and denial of service: Eluding network intrusion detection. Technical report, Secure Networks Inc., January 1998. http://www.aciri.org/vern/Ptacek-Newsham-Evasion-98.ps.

Puketza, N., K. Zhang, M. Chung, B. Mukherjee, and R. Olsson. A methodology for testing intrusion detection systems. *IEEE Transactions on Software Engineering,* 22(10), October 1996.

Qin, X. and W. Lee. Statistical causality analysis of infosec alert data. In *Proceedings of the 6th International Symposium on Recent Advances in Intrusion Detection (RAID 2003),* September 2003.

Roesch, M. Snort — lightweight intrusion detection for networks. In *Proceedings of the USENIX LISA Conference,* November 1999. Snort is available at http://www.snort.org.

Schneier, B. *Secrets and Lies: Digital Security in a Networked World.* John Wiley & Sons, New York, 2000.

Shipley, G. and P. Mueller. Dragon claws its way to the top. In *Network Computing.* TechWeb, August 2001.

Top Layer Networks and Internet Security Systems. Gigabit Ethernet intrusion detection solutions: Internet security systems RealSecure network sensors and top layer networks AS3502 gigabit AppSwitch performance test results and configuration notes. White Paper, July 2000.

Valdes, A. and K. Skinner. Probabilistic alert correlation. In *Proceedings of the 4th International Symposium on Recent Advances in Intrusion Detection (RAID 2001),* October 2001.

Wagner, D. and D. Dean. Intrusion detection via static analysis. In *Proceedings of the 2001 IEEE Symposium on Security and Privacy,* May 2001.

49

Measuring the Internet

CONTENTS

Abstract ... 49-1
49.1 Introduction ... 49-1
49.2 Measurement Methodology ... 49-2
49.2.1 Metrics ... 49-3
49.2.2 Techniques ... 49-4
49.2.3 Active Measurements ... 49-4
49.2.4 Passive Measurements: Link Behavior, SNMP ... 49-5
49.2.5 Passive Measurements: Packet Traces ... 49-6
49.2.6 Passive Measurements: Traffic Characterization ... 49-7
49.2.7 Passive Measurements: Traffic Flows ... 49-7
49.2.8 Examples of Traffic Characterization ... 49-11
49.2.9 Measurement of Global Routing Infrastructure ... 49-11
49.2.10 General Considerations for Network Measurement Data ... 49-13
49.3 Measurement Research Topics ... 49-13
49.3.1 Packet Statistics ... 49-13
49.3.2 Mice, Elephants, Dragonflies, and Tortoises ... 49-14
49.3.3 Internet Topology ... 49-15
49.3.4 BGP Topology Data Analysis ... 49-16
49.4 Conclusion ... 49-17
49.4.1 IMRG: Priorities for Future Effort ... 49-17
References ... 49-18

Nevil Brownlee

kc claffy

Abstract

In this chapter we present a representative sample of operationally relevant measurement methodologies and describe measurement metrics and techniques, including active and passive measurements. We discuss off-line techniques such as collecting packet header traces, and contrast them with near-realtime analysis techniques. We categorize the many definitions of network traffic *flows* in common use, and describe tools available to work with each of them. We then highlight a few important topics in current measurement research, including packet statistics, traffic modeling, and analyzing Internet topology data. We conclude with some priorities for future measurement efforts.

49.1 Introduction

Most people using the Internet do so as part of their daily life without needing to understand the details of how it works. Most of us assume that it is based on well-understood engineering principles, similar to the telephone network. If that were true, network operators would have a set of parameters describing network performance, and they would monitor those parameters on their infrastructure, then use variations as indications of reduced network performance. We call this kind of measurement *network monitoring*.

1-58488-381-2/05/$0.00+$1.50
© 2005 by CRC Press LLC

However, the Internet has only existed for 30 years or so, and has grown relentlessly in the last 10 along many dimensions, so that today:

- The number of people using today's commodity Internet numbers hundreds of millions all over the world.
- The Internet now consists of thousands of interconnected independent networks called *autonomous systems* (AS). Some networks are *global* in scale, many more are *regional*, covering a country, and others serve only cities or targeted user communities.
- Link speeds have increased dramatically, allowing network providers to upgrade their links; as of early 2003, many global providers use 2.5 Gb/sec links for the busiest parts of their global backbones.
- Computer system speeds have increased, allowing users to work with gigabyte files, taking advantage of the ever-increasing speed of communications links.

Internet fundamentals such as its underlying TCP/IP fabric and architectural tenets such as the *end-to-end connectivity* principle provide some framework for Internet Service Providers (ISPs) to build their networks. Predictions of overwhelming growth in demand led ISPs and equipment vendors to focus on building up their networks as quickly as possible, regardless of how those networks were actually used. New services such as Network Address Translators (NATs) and firewalls were deployed rapidly, with related standards development lagging significantly behind.

As a consequence, the longer term need to develop a better understanding of how our own networks behave, how they work as a system, and the resulting macroscopic behavior of the global Internet has taken second place to building out capacity and keeping networks running from day to day. The Internet bubble deflation with its accompanying increased focus on matching capital expenditures with actual future demand has begun to change this situation; Internet measurement is now becoming a more widely supported activity, although, unfortunately, still far more as art than science.

"Internet measurement" is a wide-ranging subject covering many different activities. This chapter begins by discussing network measurement and monitoring, and reviewing measurements used by network operators, for example:

- To verify that the network is operating correctly, e.g., that the connection provided by an ISP is performing within the limits specified by its Service Level Agreement (SLA)
- To determine amounts of traffic being carried for capacity planning, usage charging, etc.
- To monitor network traffic to detect patterns of unusual behavior, e.g. to detect attacks on the network or to observe the rise in popularity of new applications

The second part of this chapter gives a brief overview of measurement-based Internet research, covering a few selected topics on the behavior of:

- Packets on links, i.e., *packet statistics*
- Aggregates of packets, i.e., *streams and flows*
- Global routing system behavior

The chapter concludes with some observations about future priorities in network measurement.

49.2 Measurement Methodology

To measure network behavior we need the ability to observe packets on wires, i.e., to determine packet sizes and arrival times. Packet and byte counters in network equipment yield overall packet and bit rates, simple observed metrics for the link being observed. We can also actively send probe packets through a network cloud and observe them at various points within that network. Other measurement approaches include passive packet capture and active peering agents that record data, e.g., routing data carried by the Border Gateway Protocol (BGP).

49.2.1 Metrics

When users notice that network performance is poor they can perform some simple tests. For example, they can use *ping* to determine whether a target host is responding, and measure its latency (round-trip time, RTT) from the probe source. Unfortunately results from such simple tests depend on tool implementation details and are inadequate for rigorous analysis.

In general terms, a *metric* is a quantity whose value can serve to quantify the behavior of the system being considered. For example, in a routing system, one could use hop count as a metric for selecting the link on which to forward a packet. For network performance measurement we especially need metrics that:

- Are well-defined and easily understood
- Are reproducible, i.e., produce the same results whenever their values are measured
- Are not influenced by implementation details of links along the network path being measured

Several international standards bodies have produced network metrics documents. These bodies include the Internet Engineering Task Force (IETF) and the International Telecommunication Union (ITU).

ITU-T is the Telecommunication Standardization Sector within the ITU. Its mission is to produce standards covering all systems and technologies comprising the emerging global information infrastructure. ITU-T has produced E.800, a set of recommended terms and definitions relating to network performance [E.800, 1994]. E.800 takes a top–down view of packet networks to arrive at a set of metrics covering quality of service for specified network paths.

The Internet Engineering Task Force (IETF) is responsible for standards development within the Internet. Within the IETF, the IP Performance Metrics (IPPM) Working Group developed a framework for measuring network performance [Paxson et al., 1998]. The framework presents terms for describing networks, together with a collection of background material covering issues such as clock synchronization, wire time, and techniques for generating and analyzing sample packets so as to minimize measurement bias.

IPPM's framework also provides a layout for later standards documents, each of which describes a particular network metric. For example, Almes et al. [1999] covers *one-way delay metrics,* i.e., measures of the time taken by packets to travel from a source to destination along a network path via a series of intermediate hops. Demichelis and Chimento [2002] cover *IP Packet Delay Variation,* i.e., measures of the variation in delay between successive packets traveling along a path, often informally called *jitter.*

IPPM metrics measure network performance directly, since they are based on observing well-specified test packets sent through the network. Although the various IPPM standards discuss implementation issues, the metrics themselves do not depend on any particular implementation.

The metrics discussed above are quantities measured directly on the network. For many purposes they are sufficient; for example, an ISP may specify the maximum round-trip delay time within the ISP's network. In other cases we may be more interested in how users perceive the service provided. For instance, a user making a Voice-Over-IP call via the Internet cares about the overall quality of the call which we cannot measure directly, rather than about easily measurable network metrics such as percentage packet loss and variation in one-way packet delay.

In general, the relationship between user-perceived service quality and network metrics is not well understood. However, there is considerable experience with evaluating voice calls [P.862, 2001]. To determine the quality of a system carrying human voices, it is common to get a large group of people to listen to the received voice after it is sent through the system under test. Each listener ranks the received voice on a scale from 1 (bad) to 5 (excellent). The arithmetic mean of the individual rankings is then taken to be the Mean Opinion Score (MOS) for the test. Although admittedly subjective, the MOS metric is widely used.

We would like to model the voice performance of an IP communications channel in order to derive an algorithm that would map network metrics such as one-way delay, delay variation, and packet loss patterns into an MOS. Such a model would allow us to monitor a voice system by computing MOS from network metrics in real time. Developing such a model remains a research topic.

49.2.2 Techniques

We can categorize network measurements along several dimensions, which we consider in some detail in this section:

- Passive vs. active measurement
- Observing all packets vs. sampling a subset of packets
- Collecting "trace" files for later analysis vs. near-realtime data reduction

Application area is another possible dimension for network measurement; for example, one can make use of measurement in areas such as:

- *Topology.* How is the network structured at various layers, e.g., physical (links and routers), logical (autonomous systems), etc.?
- *Routing.* Do routing tables correctly mirror the network so as to provide efficient paths to hosts and is the routing system stable as a whole?
- *Workload.* How much traffic is flowing within and between logical networks, which applications are contributing most to that traffic, and how does the traffic change over various time scales?
- *Performance. How* well are network resources being used and are users receiving the level of service they expect?

General considerations for network measurements are discussed in Section 49.2.10.

49.2.3 Active Measurements

Active measurements are the simplest for an end user to make since they require no special access to the network. Instead one generates probe packets and observes them as they pass through the network. Active probes are often made between pairs of hosts in a network, and can contribute to a reasonable picture of network behavior. Two widely used Unix probing tools are *ping* and *traceroute.*

With *ping*, ICMP (Internet Control Message Protocol) echo request packets are sent to a target host; the time taken is measured for corresponding ICMP echo reply packets to return, revealing whether the target host is *available*, i.e., responding to request packets and estimating the (two-way) delay for packets to reach that host and return.

The *traceroute* tool sets the "time to live (TTL)" field in an IP packet and sends it to a target host. That packet's TTL is decremented at each intermediate host as specified by the IP protocol [Postel, 1981b], and any host that decrements it to zero sends an ICMP Time Exceeded message back to the sending host [Postel, 1981a]. By using TTL values increasing from 1, *traceroute* produces a hop-by-hop listing of two-way delays for each node along its forward[1] path to its target host.

Apart from *traceroute's* widespread use as a network diagnostic tool, the *traceroute* algorithm can be used to gather two-way delay measurements along the forward path to a specified target host. Such data can be used to make inferences about path behavior over time — for example, to quantify long-term degradation of network performance. CAIDA's *skitter* [McRobb and claffy, 1998] is a tool that automatically performs *traceroutes* to every host in its destinations list. Researchers use the resulting *skitter* data to study Internet topology and routing dynamics. We discuss some results from this work in Section 49.2.9.

Beluga [Koga, 2002] is a tool that visualizes *traceroute*-like per-hop data, producing realtime plots of delay and packet loss along a path, together with some history of that path. *Beluga's* "history" plot also shows where delays have recently occurred along the path, and whether the path is deteriorating or improving over time. For example, in Figure 49.1 we see rather typical banding features, especially on the trans-Pacific link between hops 8 and 9 where there is an 85 msec delay.

[1] Note that *traceroute* can only produce information about its forward path.

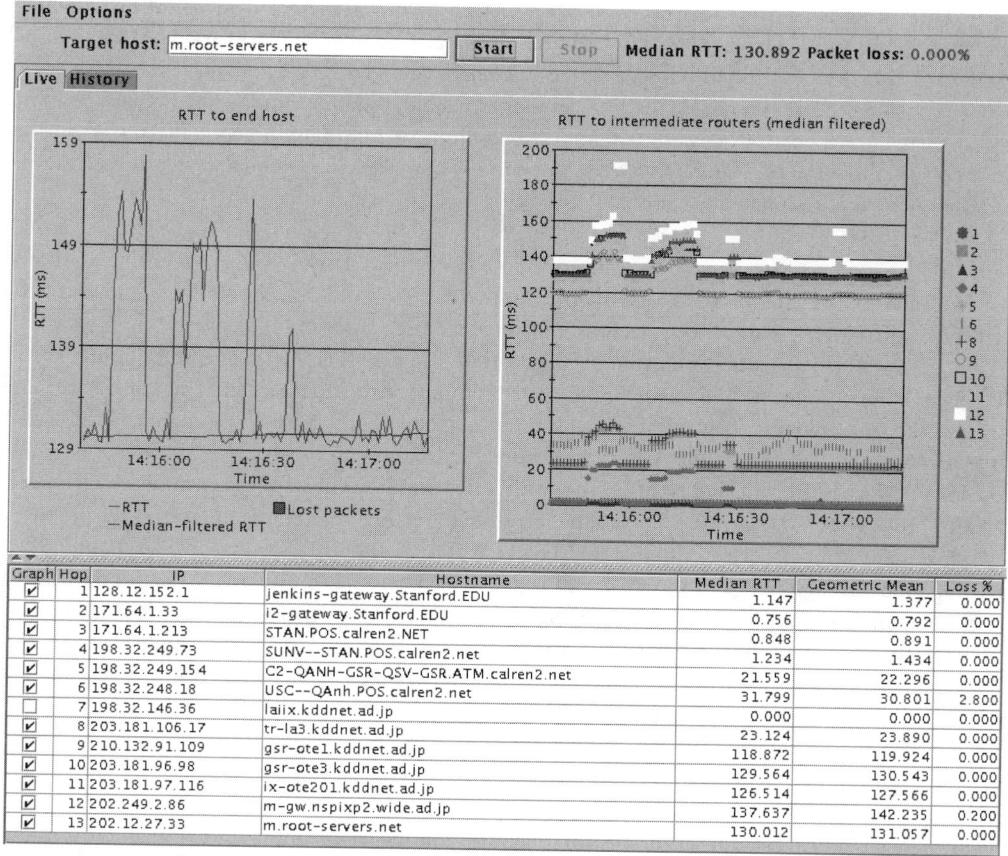

Graph	Hop	IP	Hostname	Median RTT	Geometric Mean	Loss %
✔	1	128.12.152.1	jenkins-gateway.Stanford.EDU	1.147	1.377	0.000
✔	2	171.64.1.33	i2-gateway.Stanford.EDU	0.756	0.792	0.000
✔	3	171.64.1.213	STAN.POS.calren2.NET	0.848	0.891	0.000
✔	4	198.32.249.73	SUNV--STAN.POS.calren2.net	1.234	1.434	0.000
✔	5	198.32.249.154	C2-QANH-GSR-QSV-GSR.ATM.calren2.net	21.559	22.296	0.000
✔	6	198.32.248.18	USC--QAnh.POS.calren2.net	31.799	30.801	2.800
☐	7	198.32.146.36	laiix.kddnet.ad.jp	0.000	0.000	0.000
✔	8	203.181.106.17	tr-la3.kddnet.ad.jp	23.124	23.890	0.000
✔	9	210.132.91.109	gsr-ote1.kddnet.ad.jp	118.872	119.924	0.000
✔	10	203.181.96.98	gsr-ote3.kddnet.ad.jp	129.564	130.543	0.000
✔	11	203.181.97.116	ix-ote201.kddnet.ad.jp	126.514	127.566	0.000
✔	12	202.249.2.86	m-gw.nspixp2.wide.ad.jp	137.637	142.235	0.200
✔	13	202.12.27.33	m.root-servers.net	130.012	131.057	0.000

FIGURE 49.1 Beluga display showing path behavior, Stanford to Tokyo (m-root name server), on August 16, 2002.

Furthermore, there are a few transient increases in delay that propagate to all hops (producing the step effect in the plots for hops 9 and above), most likely due to router queuing at hop 4.

The above measurements all rely on direct observation of the behavior of probe packets. Alternatively, observation of probe packet behavior can be used to infer the state of the network. For example, bandwidth estimation tools such as *pathrate* operate by sending a test sequence of packets with specified sizes and interpacket times and observing the dispersion the network produces in the test packets' interarrival times. Such *packet dispersion* techniques are discussed by Dovrolis et al. [2001].

49.2.4 Passive Measurements: Link Behavior, SNMP

Active measurements are unparalleled in convenience but since they inject test traffic they may negatively affect normal user traffic. Worse, some techniques, e.g. measuring available bandwidth by sending large test data files, can introduce artificially high loads on the network. To avoid such distortion we can use *passive* measurement techniques, where we simply observe packets crossing a network link. The easiest way to do this is indirectly, using SNMP.

Simple Network Management Protocol (SNMP) [Case et al., 2002] is the Internet standard system for managing network-attached devices. Indeed most network equipment such as hubs, switches, and routers come with built-in SNMP agents. An SNMP agent maintains a database of information specified in a Management Information Database (MIB) for the device. Hosts on the network can find out about the device by reading objects in its MIB, and if authorized can change the device's configuration by writing new values to some of those objects.

For network measurements one needs to decide which MIB objects will yield useful measurement data. Many MIBs are available, both standard and proprietary. One of the most useful MIBs, Internet Standard MIB-II [McCloghrie and Rose, 1991], is implemented on nearly every managed network device manufactured in the last 10 years. MIB-II has an interface table, with one entry for each interface on the device. Each interface entry has counters for various kinds of information about the traffic on that interface, including *ifInOctets* and *ifOutOetets*, the number of octets received on and transmitted out the interface, including framing characters.

To collect SNMP data from a network device one can easily write programs in languages such as Perl, using an open source API to interact with the device's SNMP agent. For everyday monitoring one could merely display SNMP counter values — *InOctets* and *OutOctets* — to plot the traffic rate to and from the device. Alternatively, a database could store the data for later analysis and display.

One widely used system for storing and displaying traffic data rates is MRTG, the Multi Router Traffic Grabber [Oetiker, 1996]. MRTG reads SNMP data from specified interfaces on a list of devices and produces Web page plots of their traffic rates for periods of hours, days, weeks, and months. MRTG data for a single interface is periodically consolidated into a fixed amount of disk space, allowing its users to store long periods of historical data. MRTG's compact storage allows a network operator to collect data for many routers on a large network, in exchange for decreasing time resolution as the data's age increases.

A more powerful data storage tool is Tobi Oetiker's Round Robin Database tool, RRDtool [1999]. Enjoying global recognition and use, RRDtool is a system to store and display time-series data in the same manner as MRTG, i.e., aggregating at progressively coarser time scales as it archives further back in time to maintain manageable archive size. RRDtool does not replace MRTG, it simply provides data management and display tools. These features nonetheless make it useful in many time series monitoring and analysis systems.[2]

49.2.5 Passive Measurements: Packet Traces

SNMP polling measurements are easy to perform, mostly because virtually all network devices have built-in SNMP agents so one does not have to deploy any special hardware to gather the data. However, they are limited in that most MIB objects are simple counters, providing aggregate information about the total traffic through an interface. A more detailed view of the network, including kinds of applications using the network and how bursty traffic is, requires capturing data about each packet, at least regarding:

- Its *timestamp,* the time it arrived, in microseconds
- The *packet header,* the first *n* bytes, where *n* is at least 40, i.e., large enough to carry IP and TCP header information for packets with no IP or TCP header options

We call a file of such data a *packet trace file.* Analyzing other aspects of application-level behavior may require software that gathers data even beyond the packet header, e.g., URLs for requested Web objects, specific host name of DNS queries, or peer-to-peer filenames.

The simplest way to gather a trace file is to use a utility such as *tcpdump,* a Unix utility that collects timestamped packet headers, writes them to a trace file, and reads (copies of) the headers via the Berkeley Packet Filter (BPF). One can configure BPF to ignore packets that do not meet specified filter criteria.

Trace files have several advantages, most importantly that they can be reused for different kinds of analysis. Trace files can be collected at regular intervals to provide data on long-term traffic trends for a single observation point. The biggest disadvantage of trace files is their enormous size, an hour's trace

---

[2] For more information and examples see [Oetiker, 1999]. Note that at the time of writing, several developers were working on new versions of MRTG, using RRDtool instead of MRTG's built-in data management system, and generating image files for Web pages "on the fly." Tobi Oetiker summarized these developments in [2003]: "MRTG-3 will be based entirety on RRDtool technology."

from an OC48 (2.5 Gb/sec) link carrying about 1.6 Gb/sec fills about 30 GB of disk. Processing such large data sets requires careful design and implementation of a storage and analysis system.

One way to reduce the size of trace data sets is to sample traffic, i.e., only record an average of one out of every *n* packets, assuming it will form a representative subset of the link's total traffic. One might use sampling because it is the only way to cope with a high bandwidth link, to reduce the amount of trace data one must retrieve over the network, or simply to keep the trace file down to a manageable size.

The IETF's Packet Sampling (PSAMP) Working Group is chartered to produce a standard method for collecting sampled trace data. Issues being discussed by the Working Group include:

- A standard set of sampling algorithms
- A standard way to specify which part of each packet's header bytes to capture
- A standard method for collecting sampled data from a remote PSAMP device

An alternative approach is to perform data reduction in near-realtime, most commonly used to aggregate packets into flows as discussed in Section 49.2.7. Other kinds of processing are also possible, for example, monitoring TCP streams to track the number of resent bytes. Realtime data reduction has the advantage of greatly reducing the space needed for data storage, but the tradeoff of only being able to analyze the data once. If a different kind of analysis is warranted, one must collect new data.

49.2.6 Passive Measurements: Traffic Characterization

One common use for trace data is for use in *traffic characterization* and associated reports on trends over time. Aspects of interest include:

- Total traffic rate
- Proportions of traffic by protocol (UDP TCP, etc.)
- Proportions of traffic by application, e.g., mail, Web, etc.
- Traffic matrix, showing traffic (in each direction) between pairs of hosts

Each transport packet header includes the IP header, in which the value of its IP protocol field indicates a packet's transport protocol, usually TCP or UDP. TCP/UDP port numbers indicate its application, and IP addresses provide source and destination addresses for use in constructing a traffic matrix. For many applications, especially older ones such as telnet, SNMP, and HTML, TCP/UDP port numbers are well-known and unique. Newer and increasingly popular applications such as peer-to-peer file sharing programs generally do not used fixed port numbers. They may use a fixed port number to find a server, perhaps to locate a copy of a particular file, but then select a pseudo-random port number for the TCP session that actually transfers that file.

Coping with applications that do not use fixed port numbers requires heuristics to classify packets by application. For example, if we observe a TCP session from host *X* to another host which we know is a KaZaA server, after which host *X* transfers a large file from host *Y*, we may assume that the transfer was a KaZaA file transfer. Unfortunately, such heuristics rely on knowledge of how the application works, which means that as new applications become widely used, new heuristics must be developed to classify their packets.

To characterize traffic we could simply write a dedicated program to read trace data and produce the reports we want, A better approach is to use a utility package such as CoralReef, designed to support trace data analysis. We discuss CoralReef in Section 49.2.8.1.

49.2.7 Passive Measurements: Traffic Flows

All Internet interactions are composed of packet exchanges between *end hosts* with common distinguishing features, i.e., *attributes.* For example, if host *A* sends a *file* to host *B,* all data packets will have source IP address A. Such sets of packets are normally referred to as traffic *flows.*

Each flow corresponds to a well-defined activity on the network, hence *flows* provide a natural aggregation unit of network traffic. We can use flows to get a good indication of the transactions taking place over the network. Such transactions may be user-initiated, e.g., Web page requests and downloads, or system-initiated, e.g., streaming media broadcasts.

Many different definitions of flows are in common use. We summarize the most common definitions in Figure 49.2, and describe them in the following subsections.

Figure 49.2 shows the various types of flows, arranged in terms of their endpoints. Flows in the center column may have one or two endpoints; a single-ended flow could represent all packets from a host or network. The three rows represent increasing generality when defining endpoints. The bottom row allows endpoints to be completely general; they either consist of *microflows* which have been aggregated into host or network flows, or have been defined in a general way (e.g., all packets from host *A* to network *X*). Flows that may be aggregations of *microflows* appear in the area inside the dotted line.

Often called *microflows,5-tuple* flows have the simplest definition. All packets in a 5-tuple flow have the same IP protocol, and the same source and destination IP addresses and port numbers. These flows are unidirectional, i.e., all their packets travel in the same direction.

To analyze network activities involving packets in two directions, one can simply use two unidirectional flows. Alternatively, some systems produce bidirectional flows directly; such flows correspond to the area inside the black line.

49.2.7.1 Claffy, Polyzos, and Braun (CPB) Flows

To analyze flows one typically reads packet headers (from either a trace file or a live network interface), building a table of 5-tuple flows. One examines each packet header to determine its 5-tuple, then uses that as a key into the flow table. Obviously the number of flows in the table increases as one works through the trace file; eventually the table becomes full. To avoid overflowing the flow table, one must decide which flows have terminated in order to recover their space. The simplest way to recognize a terminated flow is to use a *fixed timeout* interval; if no packets appear for at least the timeout interval we say that the flow has terminated.

In a seminal paper on Internet flow characterization, claffy et al. [1995] analyzed trace files with varying timeout intervals and observed that shorter timeouts tend to break a long running flow into a sequence of short flows, all with the same 5-tuple. Longer timeouts reduce this effect at the expense of increased flow table size. Since the number of flows detected in the trace file depends on the value of the timeout interval used in the analysis, claffy et al. proposed that one should specify the timeout value being used. *CPB flows* take their name from the seminal paper's authors, claffy, Polyzos, and Braun.

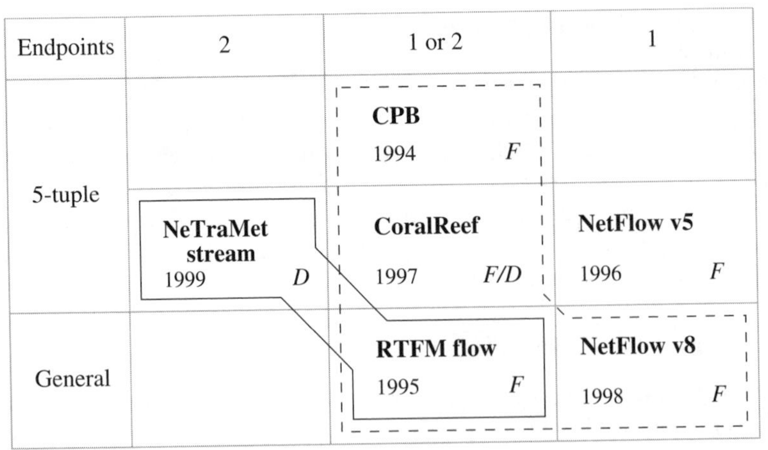

Endpoints
2
1 or 2
1
5-tuple
General
CPB
1994 F
NeTraMet stream
1999 D
CoralReef
1997 F/D
NetFlow v5
1996 F
RTFM flow
1995 F
NetFlow v8
1998 F
Timeout:
F = fixed
D = Dynamic
May be aggregated
Bidirectional

FIGURE 49.2 Taxonomy for commonly-used definitions of *flow.*

Claffy et al. used IP address and port pairs to specify flow endpoints and observed that such tuples can be aggregated into *hosts* and *networks.* They also remarked that although one mostly uses two-ended flows, single-ended flows, e.g., all packets from network *X,* may also be useful.

In summary, *CPB flows* are 5-tuple flows, analyzed using a specified *(fixed) flow timeout* which may be aggregated in various ways.

49.2.7.2 CoralReef Flows

CoralReef is a comprehensive software suite developed by CAIDA to collect and analyze passive measurement data, in real time or from trace files. The CoralReef API [Keys et al., 2001; Moore et al., 2001] provides a rich set of functions, including:

- Reading trace data in common formats such as *tcpdump and DAG data* [DAG], either from data files or from live network interfaces
- Stripping encapsulation layers from packet headers from a trace record so as to gain access to a packet's underlying IP header
- Building tables indexed by port number

Using CoralReef one can write simple scripts quickly, with the CoralReef API doing most of the work. In principle, CoralReef allows its users complete flexibility when defining flows and their level of aggregation. However, most CoralReef users begin by using one or more of the applications packaged with the CoralReef distribution [CoralReef]. For example, *crl_flow* produces summaries of traffic flows, using 5-tuples with either fixed or dynamic timeout and aggregation by mask length, i.e., by network block.

49.2.7.3 RTFM Flows, NeTraMet

All flow definitions above are low-level in that they aggregate packet data into flow data for every 5-tuple observed on a network link. Coarser aggregation, often needed to reduce the amount of data from a flow measurement system, requires more powerful flow specification heuristics.

In 1995 the IETF's Realtime Traffic Flow Measurement (RTFM) working group began to develop a system to enable realtime traffic data reduction and to minimize the size of captured measurements. The RTFM traffic measurement system [Brownlee et al., 1999] uses a general model of traffic flows in conjunction with a distributed asynchronous system for measuring those flows.

Three network entities comprise an RTFM system:

- *Meters* gather data from packets so as to produce flow data.
- *Meter readers* collect flow data from meters.
- *Managers* specify realtime data reduction by downloading configuration data (called *rulesets*) to meters, while specifying the frequency interval at which meter readers read flow data.

RTFM flows are arbitrary groupings of packets defined by attributes of their endpoints. Each endpoint definition may be as specific as a single protocol, IP address, and port number, or more general, such as a set of IP network prefixes, e.g., 192.168.3/24 and 192.168.4/24.

RTFM flows are bidirectional: an RTFM meter maintains two sets of packet and byte counters, one for each direction of the flow. Use of bidirectional flows roughly halves the number of flows to be read from a meter at each reading interval.

RTFM rulesets are written in SRL [Brownlee, 1999], a high-level (C-like) language, specifying:

- Which packets are of interest (other packets are ignored).
- Which direction of the flow is the "to" direction.
- What level of address detail is required for flows. If packets for a specified endpoint network are of interest, the meter can build flows for any address subset within that network.

NeTraMet [1993] is the first implementation of an RTFM system. A step-by-step introduction is given by Brownlee [2001]. NeTraMet is open source software and includes RTFM meters that run on Unix or

Linux systems, an SRL compiler, and combined Manager/Readers. NeTraMet comprises a toolkit suitable for constructing production flow measurement systems.

Routers are another potential source of flow data, which they export using systems such as Cisco's NetFlow (discussed in Section 49.2.7.5). One way to collect and analyze NetFlow data is to use NetflowMet, a variant of the NeTraMet meter. Whereas the NeTraMet meter reads packet header data directly from several network interfaces, NetFlowMet reads it via UDP from several NetFlow routers. This approach makes it possible to use SRL rulesets to analyze NetFlow data.

49.2.7.4 NeTraMet Streams, Flows, and Torrents

The collection of all RTFM flows constituting the total traffic on a network link is defined as a *torrent.* Within a torrent there are usually a few RTFM flows carrying large amounts of traffic and many others carrying only small amounts. Categorizing flows by their size (in bytes or packets) has inspired terminology such as *mice,* flows that are too short for the network to exercise any congestion control feedback over them, and *elephants* for flows so large that, although few in number, carry a high percentage of the torrent's traffic.

Upon examination of RTFM flows, one may discern *streams,* comprising bidirectional 5-tuples. Streams are individual IP sessions, e.g., TCP or UDP between ports on pairs of hosts [Brownlee and Murray, 2001]. An extension of the NeTraMet meter maintains queues of streams belonging to currently active flows.

Streams introduce a new dimension into NeTraMet, making it possible to observe distributions of stream properties such as their sizes (in bytes and packets) and lifetimes. NeTraMet can also maintain a queue of information about recent packets for each stream, making it possible to match pairs of packets, e.g., requests and responses within a DNS stream. Note that although NeTraMet uses streams to produce detailed information about RTFM flows, NeTraMet streams cannot be aggregated in arbitrary ways, as we indicate in Figure 49.2.

49.2.7.5 NetFlow Flows, IPFIX

In the mid-1990s Cisco Systems introduced NetFlow [1996] in their routers. When NetFlow is enabled for an interface, the router builds a flow table for *inbound* packets through that interface. The router sends information about timed-out flows via a UDP stream to a specified IP address and port number. Information about longer running flows can also be sent at specified intervals.

One collects NetFlow data by using either a proprietary NetFlow collector system, or an open source tool such as *cflowd* [McRobb, 1999] or *flow-tools* [Fullmer, 1999]. Such systems gather NetFlow data from routers and switches and write it to disk files. *NetFlow flows* use 5-tuples as the key to their flow table, but their data records are extended with other attributes, such as the value of their TOS byte (DiffServ Code Point), source and destination AS number, etc.

Because NetFlow is supported by Cisco routers and switches, it provides a convenient way for users to collect flow data. They do not need to install extra hardware, a relief for ISPs who often have limited rack space and power resources in network Points of Presence (PoPs). However, NetFlow users need to ensure that there is a secure network path from the router or switch to the host collecting the exported NetFlow data. Furthermore, such an export path must have sufficient capacity to avoid data loss during peaks in NetFlow's data rate, e.g., those caused by Denial-Of-Service attacks.

NetFlow has evolved steadily since its introduction, from version 1 to version 8.[3] Version 5, introduced about 1996, was the first version to provide AS numbers, rather than just IP addresses and port numbers. It is probably the most widely used version. In 1998 version 8 introduced the capability to aggregate the extended 5-tuples in various ways. One can retrieve only the aggregated records, thereby significantly reducing the amount of data a NetFlow collector must retrieve from a NetFlow router or switch.

In the last few years other equipment vendors have begun to implement flow measurement capabilities within their network devices. In 2001 the IETF chartered the IP Flow Information eXport Working Group

[3] Only version 1.5 and (less so) version 8 ever saw significant use in the field.

[IPFIX] to produce a standard definition of flows and a standard way to export them from network devices. An IPFIX standard will be a remarkable improvement in flow measurement technology.

49.2.8 Examples of Traffic Characterization

49.2.8.1 CoralReef

Claffy et al. [1998] used CoralReef to produce a set of interesting traffic analyses of traffic on the MCI backbone network in April 1998, together with some supporting background detail about the Coral system.

A current example of realtime traffic characterization implemented with CoralReef appears on CAIDA'S SD-NAP Web pages [SD-NAP]. SD-NAP (San Diego Network Access Point) is a neutral network traffic exchange facility for San Diego area ISPs to exchange Internet traffic, and also to provide a platform for traffic analysis by CAIDA researchers with the goal of promoting a robust, scalable global Internet infrastructure.

The SD-NAP Web page displays pie charts and tables of absolute and percentage bytes, packets, and flows, and they are dynamically updated every 5 min. Pie charts provide breakdowns not only by protocol and application but also by source and destination autonomous systems (ASes) and countries. Time series plots show breakdowns for the current hour, day, week, month and year.

For example, Figure 49.3 is a bar graph showing the daily average data rate for various common applications. Each shading pattern on the plot indicates an application, and the key lists the applications in the order they appear going up each day's bar. For example, Web traffic is the second shading on each bar, seldom accounting for more than about 30% of the day's total traffic. Note also the annual traffic rate variation, with a clear minimum in late December.

49.2.8.2 FlowScan

Dave Plonka's FlowScan [2000] takes NetFlow data files from *flow-tools*[4] and puts their flow data into an RRDtool database. FlowScan plots provide a useful network management tool, since they can make it easy to quickly observe changes in traffic characteristics, e.g., the breakdown of flows by source or destination network, application, or protocol mix, etc.

49.2.9 Measurement of Global Routing Infrastructure

So far we have focused on measurement based on direct observation of network traffic. Earlier we discussed active measurement of network links and paths, then we described passive measurements of packets and flows. However, the Internet can only forward packets to any destination IP address if every router maintains an accurate routing table. In this section we consider the global Internet routing system and the BGP protocol upon which it relies. Note that there are also a variety of important research and analysis activities in the area of intradomain routing measurement, i.e., within a single autonomous system. Due to space constraints we do not cover intradomain routing analysis. There are few published studies in this area, notably by Sprint ATL [Sridharan et al., 2003] and AT&T research labs [Shaikh and Greenberg, 2001].

The Internet consists of a large number of autonomous systems (ASes), each of which provides connectivity to its customer networks. That is, each AS maintains a list of address prefixes for networks to which it is directly connected. Border routers for ASes peer with other ASes, and pass routing information back and forth using BGP. In this way a border router can maintain a table of all the address prefixes reachable anywhere in the Internet — the global routing table. A router decides which AS it must forward a packet to by looking up a packet's address block in the routing table. BGP information is transported by long-running TCP connections between pairs of border routers, *i.e., BGP peerings,* over

[4] Flowscan can also use data files from *cflowd*, but *cflowd* is no longer being maintained, making *flow-tools* a better choice.

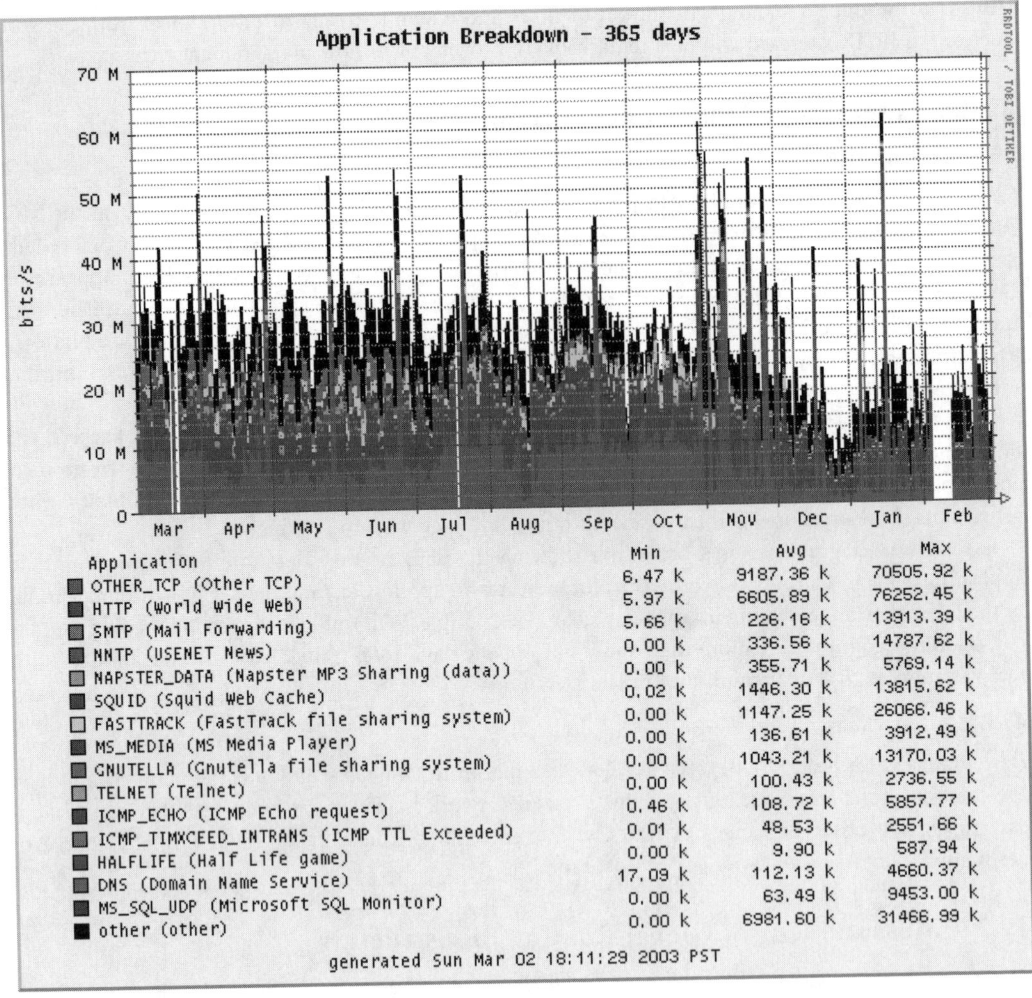

FIGURE 49.3 SD-NAP plot.

which realtime routing updates are exchanged between ASes. Routers store their accumulated routing data in tables. Changes in BGP tables over time usually indicate changes in connectivity as links fail and repair; they are also influenced by router configurations, and quirks in various BGP implementations. A BGP table differs from AS to AS, reflecting routing policies of the various ASes. BGP tables from many ASes, together with topology data from other sources such as *skitter*, can be used to build topology diagrams for the global Internet [AS Core].

There are two large-scale historical repositories of global routing tables: RouteViews at the University of Colorado and RIPE NCC's Routing Information System project, RIS.

RouteViews [Meyer, 2000] is a collaborative endeavor to archive global routing tables from the perspectives of many different backbones and locations around the Internet. The RouteViews router, route-views.oregon-ix.net, uses multihop BGP peering sessions with backbones at various locations. RouteViews uses AS 65534 in its peering sessions, and routes received from BGP peers are never propagated nor used to forward traffic.

As of February 2003, RIPE NCC's Routing Information Service [RIS] has nine BGP "route collector" hosts at various locations around the world, peering with routers in various participating networks. The route collectors send routing updates to a central database. Users retrieve routing table information from the database for any location and date/time via a Web interface. Alternatively, they can download files containing various types of raw BGP data.

49.2.10 General Considerations for Network Measurement Data

Collecting data simultaneously from several different measurement points implies that the realtime clocks of all the systems should be accurately synchronized. As a starting point, host clocks should be synchronized using NTP, the Network Time Protocol; NTP will usually maintain the time with an accuracy of a few milliseconds. Many studies over over the last decade have explored the three main approaches to achieving packet timestamp accuracy:

- Implement packet timestamps in kernel or interface card drivers. Working in kernel space gives more control of the timing, since one can minimize interrupt latency and reduce system overheads.
- Devise measurement algorithms that are less dependent on synchronized clocks or algorithms that track skew in a host's clock oscillator. This approach seems promising; recent papers such as Paxson [1998] and Elson et al.[2002] have reported achieved accuracies of better than one millisecond.
- Neither of the two preceding methods addresses the problem of synchronizing clocks at far separated hosts. The only reliable way to achieve precise synchronization is to use specialized hardware, e.g., GPS-synchronized network interface cards such as the DAG cards [ENDACE; DAG], which can achieve submicrosecond time resolution and also synchronize timestamps across DAG cards to better than a microsecond of UTC. Unfortunately, it is often difficult or expensive to get a GPS signal into a network PoP, which limits the usefulness of this approach. Current work in this area suggests that CDMA time signals may be a useful alternative to GPS for synchronizing host clocks [EndRun, 2002].

Before embarking on a network measurement project it is important to consider how the project's data will be collected, stored, analyzed, and distributed. In particular one must be aware of the requirements for security and privacy of network users, which implies keeping the data secure and available only to people directly involved in the project, as well as caution when publishing results to prevent leaking unintended details about network users. Publishing network measurement data for other researchers to use typically requires anonymization to prevent disclosure of IP addresses and port numbers. The usual approach for rendering a trace file for public consumption is to sanitize the IP addresses using tools such as *tcpdpriv* [Minshall, 1997], which rewrites IP addresses to random numbers with various levels of prefix-preserving semantics. Techniques for preserving user privacy as well as the file's address and port number structure, in particular at line speed, are the subject of ongoing research, for example, *Crypto-PAn* [Xu et al., 2001] provides an API for sanitizing IP addresses.[5]

Time synchronization and data privacy are only a few of the issues to be considered when planning Internet measurements. For a more wide-ranging view, see claffy and Braun [1995], Moore [2002] and Paxson [2001].

49.3 Measurement Research Topics

We have surveyed a representative sample of operationally relevant measurement methodologies and activities. We now highlight some of the more exciting or at least critically important topics in measurement research for the next several years.

49.3.1 Packet Statistics

Telephone networks have a long tradition of mathematical modeling, dating from about 1909. when Anger Erlang published his *Theory of Probabilities and Telephone Conversations.* Telephone traffic models normally assume a Poisson arrival process and are well handled by standard queuing theory.

For packet-based networks, work on measuring packet interarrival times began with the rise of Ethernet as a physical communication layer in the late 1980s. At that time there was concern that Ethernet's media

[5] As distinct from *tcpdpriv*, a tool that anonymizes IP addresses in *tcpdwnp* trace files.

access control would severely limit its achievable data rates. Early measurement work was aimed at understanding how Ethernet works in practice, especially the way it behaves as the load increases in an Ethernet collision domain.

Once Ethernet LANs had become well established, attention shifted to the question of how packets aggregated on links interconnecting separate LANs. In particular, Ethernet traffic was known to be bursty. Did the level of burstiness reduce as the degree of traffic aggregation increased?

In a landmark contribution, Leland et al. [1993] established that interarrival time distributions for packet network traffic for Ethernet LANs are *self-similar*. Their work was based on Ethernet packet traces with accurate timestamps, gathered on 10 Mb/sec Ethernets using a "custom-built Ethernet monitor." They divided their data collection time into 10 msec timeslots, and plotted the number of packets in a sequence of increasing timeslots, producing a characteristically bursty pattern. By successively aggregating their 10 msec timeslots into groups of ten, they produced a series of packets/timeslot plots for time scales of 100 msec, 1 sec, 10 sec, and 100 sec. All these plots look very much the same; there is no sign of any smoothing in the plots at the higher time scales, demonstrating underlying self-similarity in the data.

Self-similar packet behavior has several implications, including

- More complex models incorporating some notion of burstiness must be used instead of simple Poisson distributions.
- Older burstiness measures are clearly inappropriate, e.g., *peak-to-mean ratio* values depend on the time interval during which peaks and means are measured, and peaks measured over short intervals can easily be large.
- Because there is no limit to the size of a traffic burst; it is nearly impossible to allocate buffers sufficiently large so as to completely avoid queue overflows in network equipment.

Unfortunately, the revelation of self-similarity has yet to make significant operational difference. To the extent that it was affordable, most forwarding equipment already had buffers large enough to keep packet losses to an acceptably low level, and links could be run with utilization levels low enough to handle the bursts [Ferguson, 2003].

However, link speeds were and are inevitably increasing. In 1985, 10 Mb/sec Ethernet was standardized and became widely used by the early 1990s. In 1995,100 Mb/sec Ethernet was standardized, and Gigabit Ethernet in 2001. Today, 100 Mb/sec Ethernet is the norm and Gigabit Ethernet increasingly common. Backbone speeds have similarly increased as developments in fiber optics have allowed vendors to produce OC3 (155 Mb/sec), OC12 (622 Mb/sec), and OC48 (2.5 Gb/sec) routers and switches.

From about 1995 backbone operators began to observe that traffic seemed less bursty on higher-speed links. Whereas T1 (1.5 Mb/sec) links were usually run at about 50% utilization, T3 (45 Mb/sec) links could be run at 70%, suggesting that at higher levels of aggregation, the effects of self-similarity in individual Ethernet sessions became less significant. This notion was supported by a series of papers published by a group of Bell Labs researchers in 2002 [Cao et al., 2002]. Examining many packet traces they found that overall Internet traffic behavior tends towards Poisson as the number of streams being aggregated increases even though individual streams are self-similar. In short, *there are multiplexing gains* on backbone links carrying large numbers of streams.

49.3.2 Mice, Elephants, Dragonflies, and Tortoises

In Section 49.2.7.4 we defined *mice* as flows whose size is small, and *elephants* as large flows. Since the early 1990s, various researchers have noted that in any torrent most of the flows are mice, but that most of the traffic bytes are carried by elephants; this is commonly expressed by an "80/20 rule" statement such as "80% of the flows carry 20% of the data."

In areas such as accounting and network management, interest is in torrent-dominating flows. There are routing implications. For example, Mahajan et al. [2001] proposed RED-PD, a router congestion management scheme that recognizes high-bandwidth flows and controls their throughput when the router is congested, thereby allowing the mice a fairer share of link bandwidths.

Particularly relevant to operational accounting considerations, Estan and Varghese [2001] noted that when measuring flows, the primary problem is to create and accurately maintain a flow table quickly enough to produce reliable flow data while staying within a fixed amount of flow table memory. They suggested that at least for accounting purposes it would be sufficient to produce data only for elephants, and proposed a compelling *threshold accounting* scheme based on this notion.

More recent attention has turned to interactions among elephants and mice. For example, Joo et al. [1999] found that multiple elephants can synchronize with each other, which may cause routers to drop packets. They stated that "although elephants are responsible for a major proportion of the bytes on the network, the number of packets generated by mice can be sufficient to create losses from time to time." Joo et al. also examined the dynamics of packet drops and concluded that mice can break up synchronization effects, leading to more efficient use of network resources. This breakup effect may explain why best-effort datagram delivery has served the Internet so well as a lowest common denominator of network service.

As an alternative to classifying flows by size (number of bytes), i.e., as *elephants* or *mice,* one can also classify flows by their lifetime (in seconds). In a recent survey paper, Brownlee and claffy [2002] observed stream lifetimes at two different university Internet gateways, at the University of Auckland (UA)(155 Mb/sec, rate limited to 9 Mb/sec), and at the University of California–San Diego (622 Mb/sec). They found that stream size is independent of stream lifetime and classified streams by lifetime as:

- Very short *dragonflies,* lasting up to 2 sec
- Short, lasting up to 15 min
- Long running *tortoises,* lasting more than 15 min

For this data sample, 40% (UCSD) to 70% (UA) of the streams were dragonflies, most of which appeared to be Web traffic, and about 1.5% of UA and UCSD streams were tortoises. However, 5% (UA) to 50% (UCSD) of all bytes were carried by tortoises, and most of them appear to be non-Web traffic. Together, short streams and dragonflies can account for 70 to 98% of the streams in a torrent. Routers cannot establish enough information about dragonflies to forward them in an optimal way. Therefore, routers must be capable of handling large numbers of dragonflies while also optimizing forwarding performance for tortoises. Brownlee and claffy also noted that the overall character of Internet traffic is changing, with stream sizes and lifetimes both increasing.

49.3.3 Internet Topology

As the Internet continues to grow, so does the diversity of connectivity among nodes. The number of different paths among a given set of nodes depends upon unknown but crucial interconnection points beyond control of individual users and end customers. Internet topology research seeks insight into measures of infrastructural redundancy and robustness through analysis of Internet path data at various network scales, e.g., IP address, network prefix, and autonomous system.

Mapping macroscopic Internet topology is a daunting task, and all topology data has limitations. Analysis of Internet connectivity was pioneered by Vern Paxson in his Ph.D. thesis [1997] and in a follow-up study [Zhang et al., 2000], Paxson acquired data over several months via traceroutes among academic hosts. A smaller collection of data on Internet connectivity was gathered by Pannisot and Grad [1988].

Siamwalla et al. [1998] present heuristics found useful for discovery of logical Internet topology, including SNMP queries, DNS zone transfers, and broadcast pings. They correctly concluded that topology obtained by traceroutes from one source may be too sparsely sampled to be legitimately representative and that many sources are necessary to observe cross-links. Savage et al.[1999] collected and analyzed data among dozens of traceroute servers in the Detour project. These two studies focused on analyzing the stability and optimality of paths. Each of these studies dealt with fewer than 290,000 traceroutes.

Bill Cheswick and Hal Burch [2000] began a large-scale Internet mapping project in 1997 and made available on their Website traceroutes from a single source in New Jersey to about 100,000 selected destinations, including six best paths to each destination over approximately one year. Cheswick and

Burch also developed a novel algorithm for IP address level graph layout [Burch and Cheswick, 1999; Peacock maps].

Govindan and Tangmunuarunkit [2000] developed Mercator, an Internet topology discovery tool to build a router-level Internet map by intelligent probing from a single workstation. One strength of its design is its few *a priori* assumptions about Internet topology. They offer several valuable caveats of Internet topology acquisition. However, Mercator is considerably slower at processing probes than skitter and uses source routing to discover cross-links not captured by standard traceroute. This practice tends to generate more user and ISP complaints[6] and is less practical for large-scale longitudinal studies.

Radolavov et al. [2000] compare canonical graph models such as a grid or a tree, with the Mercator, AS, and Mbone graphs and with topology generators. They focus on the impact of topological properties on the performance of various flavors of multicast protocols.

CAIDA's skitter tool [McRobb and claffy, 1998] was explicitly designed and extensively tested to gather forward IP path data. A 2001 study by Broido and claffy [2001] has presented an analysis of the largest IP topology data set to date, including an algorithm for extracting the bidirectionally connected part of the graph; structural analysis of observed IP graphs in acyclic (downstream) and strongly connected (backbone) portion; measures of node importance such as sizes of neighborhoods, cones, and stub trees rooted at a node; and demonstration that Weibull distributions provide a good fit to a variety of topological object sizes.

The authors recognize that a reasonable framework for analyzing properties of Internet connectivity requires consideration of not only probed topology data, but also of the dynamics between topology data and BGP data gathered at the AS-level granularity. In the next section we discuss the importance of this latter type of different but equally important topological data.

49.3.4 BGP Topology Data Analysis

Several studies on Internet connectivity have used AS (autonomous system) data extracted from BGP routing tables [NLANR; Meyer, 2000; McCreary and Woodcock]. Compared to traceroute (forward IP) path data, BGP tables are easier to parse, process, and comprehend. It is understandable that researchers who do not collect their own topology data try to study Internet topology using BGP AS connectivity.

BGP data is useful for determining correspondence among IP addresses, prefixes, and ASes [AS Core], and in analyzing different routing policies in the Internet [Broido and claffy, 2001]. However, BGP connectivity does not qualify redundancy of different parts of the network. BGP tables only show the *selected* (best) routes, rather than all possible routes stored in the router. Nor does the BGP table show public and private exchange points within the infrastructure, or short-term AS path variation and AS load balancing. BGP data may also not be directly comparable to traversed path data due to the presence of *transit-only* ASes, i.e., ASes who do not announce global reachability of their networks but show up in forward AS paths [Hyun et al., 2003a, b]. In addition to engineering factors, BGP behavior reflects contractual business relationships among Internet service providers, specifying which companies agree to exchange traffic. It does not guarantee that this traffic will actually traverse listed administrative systems.

As such, using BGP data to obtain a topology map incurs significant distortion of network connectivity. In building graphs of topology core, graphs obtained by parsing even many dozen backbone BGP tables are extremely sparse. They represent some downstream (backbone to customers) connectivity, but no lateral connectivity. For example, extracting the largest component of bidirectionally connected nodes from RouteViews data [2000] yields less than 3% of all nodes, even when contributing routers number in dozens, carry full backbone tables, and are geographically and infrastructurally diverse. In contrast, for topology data gathered by active probing from many sites, the largest bidirectionally connected

[6] Internet providers often flag source routing as a security threat.

component comprises 8% of IP-level nodes and 35% of AS-level nodes. BGP data thus represents a relatively meager projection of Internet connectivity. It is thus imprudent to infer Internet properties from BGP data alone. In particular, Internet vulnerability, e.g., resilience to attacks, can only be approximated using current BGP instrumentation.

As the close relationship between the Internet routing system and the robustness of critical Internet infrastructure grows ever more obvious, a comprehensive understanding of both interdomain and intradomain routing is of increasing priority to both the research and operational communities. Practical routing analysis and modeling is an integral component of any Internet research agenda.

49.4 Conclusion

This chapter began by discussing practical aspects of network measurement. As well as describing techniques, the authors pointed out some of the available measurement tools and various aspects of experimental design and practice. We believe we have presented a fairly complete overview, sufficient for readers wanting to measure and monitor their own networks, but a single chapter can only serve as an introduction to this topic.

49.4.1 IMRG: Priorities for Future Effort

On a larger scale, although network measurement is now regarded as an important activity, it is still in its infancy, and we are far from "understanding the Internet." However, the research community is making progress; one sign of which is that the Internet Research Task Force (IRTF) has chartered an Internet Measurement Research Group [IMRG]. The IMRG's charter sets out some clear network measurement priorities, which we summarize as:

- Tackle the issue of sharing measurement data within the community. The RG could define a systematic way for storing measurements and any needed metadata that should be kept with the measurements, while meeting Acceptable Usage Policy (AUP) and privacy requirements.
- Provide a venue for assessing new measurement techniques and a forum for sharing preliminary findings in rough form to encourage further work and collaboration.
- Provide a venue for developing models based on network measurements, helping to better understand network dynamics and aiding researchers attempting to conduct useful simulations of the network.
- Foster communication between the research and operations communities. For example, operators could provide feedback to researchers as to what sorts of network properties/characteristics they would like to see measured and how well current techniques work.
- Tackle outstanding issues dealing with measurement infrastructures such as: scalability of meshes, security of measurement tools in the mesh, access control, resource control, and scheduling issues.

Two exogenous forces will also influence research and operational directions:

1. U.S. homeland security concerns, which will place further, although well overdue, emphasis on protection of critical infrastructure, a particular challenge in the current Internet which lacks any public systematic studies of outages or assessment of various causes of damage.
2. Global economic downturn, which will render it essential that any analyses incorporate actual operating and capital expenditure ratios (opex/capex) and tradeoffs where possible in order to capture and maintain attention from providers.

These outlined priorities form a sound, if intimidating, community research agenda for the coming decade, one that will require concerted and cooperative efforts from dozens of groups worldwide.

References

Almes, G. S. Kalidindi, and M. Zekauskas. A one-way delay metric for IPPM. RFC 2679, September 1999.

AS Core. AS Core poster. URL http://www.caida.org/analysis/topology/as_core_network/about.xml.

Broido, A. and k. claffy. Complexity of global routing policies. In *Proceedings of the Network-Related Data Management Workshop, Santa Barbara, CA*. ACM SIGMOD/PODS, May 2001. URL http://www.caida.org/outreach/papers/2001/CGR.

Brownlee, N. SRL: A language for describing traffic flows and specifying actions for flow groups. RFC 2723, October 1999.

Brownlee, N. Using NeTraMet for production traffic measurement. In *Proceedings of the IM2001*. IEEE, May 2001.

Brownlee, N. and k. claffy. Understanding Internet streams: dragonflies and tortoises. *IEEE Communications*, October 2002.

Brownlee, N, C. Mills, and G. Ruth. Traffic flow measurement: Architecture. RFC 2722, October 1999.

Brownlee, N. and M. Murray. Streams, flows and torrents. In *Proceedings of the PAM2001*. PAM, April 2001. URL http://www.caida.org/outreach/papers/2001/StreamsFlowsTorrents.

Burch, H. and B. Cheswick. Mapping the Internet. *IEEE Computer*, April 1999.

Cao, J., W. Cleveland, D. Lin, and D. Sun. Internet traffic tends *Toward* poisson and independent as the load increases. In C. Holmes, D. Dennison, M. Hansen, B. Yu, and B. Mallick, Eds., *Nonlinear Estimation and Classification*. Springer, New York, 2002. URL http://cm.bell-labs.com/cm/ms/departments/sia/Internet'Traffic/webpapers.html.

Case, J., R. Mundy, D. Partain, and B. Stewart. Introduction and applicability statements for Internet standard management framework. RFC 3410, December 2002.

Cheswick, B. and H. Burch. Internet mapping project, 2000. URL http://research.lumeta.com/ches/map.

claffy, k. and H. Braun. Post-nsfnet statistics collection. In *Proceedings of the INET'95*. Internet Society, June 1995, URL http://www.caida.org/outreach/papers/1995/pnsc/postns.pdf.

claffy, k., G. Miller, and K. Thompson. The nature of the beast: Recent traffic measurements from an Internet backbone. In *Proceedings of the INET 1998*, July 1998. URL http://www.caida.org/outreach/papers/1998/Inet98/Inet98.html.

claffy, k., G. Polyzos, and H. Braun. A parameterizable methodology for Internet traffic flow profiling. *IEEE Journal on Selected Areas in Communications*, 1995. URL http://www.caida.org/outreach/papers/1995/pmi.

CoralReef. CoralReef Web pages. URL http://www.caida.org/tools/measurement/coralreef.

DAG. Dag card Web pages. URL http://dag.cs.waikato.ac.nz.

Demichelis, C. and P. Chimento. IP packet delay variation metric for IP performance metrics (IPPM). RFC 3393, November 2002.

Dovrolis, C., P. Ramanathan, and D Moore. What do packet dispersion techniques measure? In *INFOCOM*, pp. 905-914,2001. URL citeseer.nj.nec.com/479183. html.

E.800. Terms and definitions related to quality of service and network performance including dependability. Technical report, ITU-T, August 1994.

Elson, J., L. Girod, and D. Estrin. Fine-grained network time synchronization using reference broadcasts, December 2002. URL http://Iecs.cs.ucla.edu/Publications/papers/broadcast-osdi.ps.

ENDACE. Endace Web pages. URL http://www.endace.com.

EndRun. GPS vs. CDMA Web page, 2002. URL, http://www.endruntechnologies.com/gps-cdma.htm.

Estan, C. and G. Varghese. New directions in traffic measurement and accounting. In *Proceedings of the IMW2001*. ACM SIGCOMM IMW, November 2001. URL http://www.icir.org/vern/imw-2001/program.html.

Ferguson, D. Queue size of routers. IRTF e2e mailing list archive, January 2003. URL http://www.postel.org/pipermail/end2end-interest/2003-january/002744.html, http://www.postel.org/pipermail/end2end-interest/2003-January/002752.html.

Fullmer, M. Flow-tools Web pages, 1999. URL http://www.splintered.net/sw/flow-tools.

Govindan, R and H. Tangmunuarunkit. Heuristics for Internet map discovery. In *Proceedings of the IEEE INFOCOM 2000*. IEEE INFOCOM, March 2000.

Hyun, Y., A. Broido, and k claffy. On third-party addresses in traceroute paths. In *Proceedings of the PAM2003*. PAM, March 2003a. URL http://www.caida.org/outreach/papers/2003/3rdparty.

Hyun, Y., A. Broido, and k. claffy. Traceroute and BGP AS path incongruities. Technical report, CAIDA, March 2003b. URL http://www.caida.org/outreach/papers/2003/ASP.

IMRG. IRTF's IMRG home page. URL http://www.irtf.org/charters/imrg.html.

IPFIX. IPFIX WG home page. URL http://net.doit.wisc.edu/ipfx.

Joo, Y., V Ribeiro, A. Feldmann, A. Gilbert, and W. Willinger. On the impact of variability on the buffer dynamics in IP networks. In *Proceedings of the 37th Annual Allerton Conference on Communication, Control and Computing*, September 1999. URL http://www.dsp.rice.edu/publications.

Keys, K., D. Moore, R. Koga, E. Lagache, M. Tesch, and k. claffy. The architecture of the CoralReef Internet traffic monitoring software suite. In *Proceedings of the PAM2001*. PAM, April 2001. URL http://www.caida.org/outreach/papers/2001/CoralArch.

Koga, R. BELUGA Web pages, 2002. URL http://www.caida.org/tools/measurement/beluga.

Leland, W., M. Taqqu, W. Willinger, and D. Wilson. On the self-similar nature of Ethernet traffic. In *Proceedings of the SIGCOMM 1993*. ACM SIGCOMM, 1993.

Mahajan, R., S. Floyd, and D. Wetherall. Controlling high-bandwidth flows at the congested router. In *Proceedings of the ICNP 2001*. IEEE Computer Society, November 2001. URL http://www.cs.washington.edu/homes/ratul/red-pd.

McCloghrie, K. and M. Rose. Management information base for network management of TCP/IP-based internets: MIB-II. RFC 1213, March 1991.

McCreary, S. and B. Woodcock. PCH RouteViews archive. URL http://www.pch.net/documents/data/routing-tables.

McRobb, D. cflowd Web pages, 1999. URL http://www.caida.org/tools/measurement/cflowd.

McRobb, D. and k. claffy. Skitter Web pages, 1998. URL http://www.caida.org/tools/measurement/skitter.

Meyer, D. RouteViews Web pages, 2000. URL http://antc.uoregon.edu/route-views.

Minshall, G., tcpdpriv: Program for eliminating confidential information from trace, 1997. URL http://ita.ee.lbl.gov/html/contrib/tcpdpriv.html.

Moore, D. Pitfalls and problems with Internet data, March 2002. URL http://www.caida.org/outreach/presentations/2002/ipam0203/ipam0203.pdf. Presented at the IPAM Large-Scale Communication Networks Conference.

Moore, D., K. Keys, R. Koga, E. Lagache, and k. claffy. CoralReef software suite as a tool for system and network administrators. In *Proceedings of the Usenix LISA 2001. Usenix LISA*, December 2001. URL http://www.caida.org/outreach/papers/2001/CoralApps.

Netflow. NetFlow services and applications (an introduction and overview). 1996. URL http://www.cisco.com/warp/public/cc/pd/iosw/loft/nef1ct/tech/napps_wp.htm.

NeTraMet. NeTraMet Web pages, 1993. URL http://www.auckland.ac.nz/net/NeTraMet.

NLANR. Nlanr routing tables. URL http://moat.nlanr.net/Routing/rawdata.

Oetiker, T. MRTG Web pages, 1996. URL http://people.ee.ethz.ch/oetiker/webtools/mrtg/mrtg.html.

Oetiker, T. RRDtool Web pages, 1999. URL http://www.caida.org/tools/utilities/rrdtool.

Oetiker, T. mrtg-rrd Web pages, 2003. URL http://people.ee.ethz.ch/ oetiker/webtools/mrtg/mrtg-rrd.htm.

P.862. Perceptual evaluation of speech quality (PESQ). Technical report, ITU-T, February 2001.

Panisot, J-J. and D. Grad. On routes and multicast trees in the Internet. ACM *SIGCOMM Computer Communications Review*, January 1988.

Paxson, V. *Measurements and* Analysis *of End-to-End Internet Dynamics*. Ph.D. thesis, University of California, Berkeley, 1997.

Paxson, V. On calibrating measurements of packet transit tunes. In *Measurement and Modeling of Computer Systems*, pp. 11–21, 1998. URL citeseer.nj.nec.com/paxson98calibrating.html.

Paxson, V. Some not-so-pretty admissions about dealing with Internet measurements, October 2001. URL http://www.icir.org/vern/talks/vp-nrdm01.ps.gz.

Paxson, V., G. Almes, J. Mahdavi, and M. Mathis. Framework for IP performance metrics. RFC 2330, May 1998.

Peacock Maps. Peacock maps, 2001. URL http://www.peacockmaps.com/index.html.

Plonka, D. Flowscan Web pages, 2000. URL http://www.caida.org/tools/utilities/flowscan.

Postel, J. Internet control message protocol (ICMP) specification. RFC 792, September 1981a.

Postel, J. Internet protocol (IP) specification. RFC 791, September 1981b.

Radolavov, P., H. Tangmunuarunkit, H. Yu, S. Govindan, Rand Shenker, and D. Estrin. On characterizing network topologies and analyzing their impact on protocol design. Technical Report 00-731, USC Computer Science Dept., May 2000. URL http://www.isi.edu/hongsuda/publication/USCTechOO_731.ps.gz.

RIS. RIPE routing information service, RIS. URL http://www.ripe.net/ris.

Savage, S., A. Collins, E. Hoffman, J. Snell, and T. Anderson. The end-to-end effects of Internet path selection. In *Proceedings of the SIGCOMM 1999*. ACM SIGCOMM, September 1999. URL http://www.cs.ucsd.edu/users/savage/papers/Sigcomn99.ps.

SD-NAP. SD-NAP workload analysis (CoralReef demonstration pages). URL http://www.caida.org/dynamic/analysis/workload/sdnap.

Shaikh, A. and A. Greenberg. Experience in black-box OSPF measurement, November 2001. URL http://www.research.att.com/~albert/black-box-measurement.html.

Siamwalla, R., R. Sharma, and S. Keshav. Discovering Internet topology, July 1998. URL http://www.cs.cornell.edu/skeshav/papers/discovery.pdf. Submittedto IEEE INFOCOM 1999.

Sridharan, A., R. Guérin, and C. Diot. Achieving near-optimal traffic engineering solutions for current OSPF/IS-IS networks. In *IEEE Infocom*, San Francisco, March 2003.

Xu, J., J. Fan, N. Ammar, and S. Moon. Cryptography-based prefix-preserving anonymization: Web site, 2001. URL http://www.cc.gatech.edu/computing/Telecomm/cryptopan.

Zhang, Y., V. Paxson, and S. Shenker. The stationarity of Internet path properties: routing, loss, and throughput. Technical report, ACIRI, May 2000.

50
What Is Architecture?

CONTENTS

50.1 Introduction 50-1
50.1.1 Architecture Roots 50-2
50.1.2 Value of Network Architecture 50-3
50.1.3 The Potential Pitfalls of Architecture 50-4
50.2 The Road from Architecture to Implementation 50-5
50.2.1 Infrastructure Hierarchy 50-6
50.2.2 Network Design 50-6
50.2.3 Implementation 50-7
50.3 Closed Architecture 50-8
50.3.1 Design Changes 50-9
50.3.2 Architecture Change 50-10
50.4 Open Architecture 50-11
50.4.1 A Clear Set of Benefits 50-12
50.5 Architecture Directions 50-14
50.6 Business Cases 50-15
50.6.1 An Example of the Impact of Open Architecture 50-16
50.6.2 The Mobile Wireless Challenge 50-16
50.6.3 ClosedCel: Closed Architecture 50-17
50.6.4 OpenCel: Open Architecture 50-17
50.6.5 The Voice-Over IP (VoIP) Challenge 50-18
50.6.6 Benefits 50-18
50.6.7 Maximize Application and Service Opportunities 50-19
50.6.8 Summary 50-19
50.7 Conclusion 50-20
References 50-20

Wayne Clark

John Waclawsky

50.1 Introduction

Network architecture can be found in two broad technology settings: producers of standards and consumers of standards. An example of a consumer of standards is an enterprise company (we include service providers in the category of an enterprise company). In both settings architecture is about technology planning. At an enterprise organization, architecture can be quickly created and driven by a single individual or a small group of individuals often referred to as architects. These architects attempt to address questions about the selection of products and services from the marketplace that will help an enterprise use technology to meet business needs (e.g., make money, save money, or obtain a competitive advantage). This is done against the backdrop of selecting *existing* technology with *well-known* economic, migration, complexity, ease of management, and use considerations.

An example of a producer of standards is a standards organization that is created and driven by a large number of individuals through slow consensus and compromise. Via architecture, they typically try to define what will be available in the marketplace and attempt to steer the usage and evolution of technology.

1-58488-381-2/05/$0.00+$1.50

© 2005 by CRC Press LLC

Unfortunately, standards organizations are typically working on things that do not exist, and evaluations cannot be done against any backdrop of reliable economic, complexity, migration, ease of management, and use considerations. Note that the difficulty of standards work varies by the technology level and the marketplace the standards address. For example, contrast physical link technologies with applications and systems management. Applications and systems management are more complex and are therefore different in terms of economics, complexity, interoperability issues, investment requirements, and deployment times.

Architecture deals with a large number of complex questions, such as what to make, what to buy, how to use it, when to buy, when to make, and when to deploy. This chapter attempts to describe a fundamental change in how technology is invented, deployed, and utilized around the world. The change is being brought about not directly by the Internet (although that is part of it) but by processes at work that are shaping the Internet. Due to the huge expense, ever-increasing complexity, and high cost of technology ownership, there is a very significant risk in deploying technology in rapidly changing environments. Much of the risk is due to the rapid obsolescence of technology, products, and standards. Today, standards organizations need to complete their work very quickly or run the real risk of becoming obsolete before any deployment is complete or even before the standards themselves are completed (witness OSI).

50.1.1 Architecture Roots

Architecture is a borrowed term that is often overused (and abused) in technology forums. *Webster's New World Dictionary* [Webster, 1978] defines architecture as "the science, art, or profession of designing and constructing buildings." However, outside the world of construction, the term *architecture* is a poorly understood concept.

Webster's architecture definition can be extended to computer networking by simply changing a single word: the science, art or profession of designing and constructing *networks*. While we can understand the concrete concept of a building and the process of building construction, many of us have trouble understanding the more abstract concepts of a computer and a network and, similarly, the process of constructing a computer network. And just like buildings, there are many different kinds of networks.

Even though architecture involves some well-defined activities, our first attempt at a definition uses the words *art* along with *science*. Unfortunately, for practical purposes, this definition is much too vague. But one thing the definition does indirectly tell us is that architecture is simply part of the process of building things. For example, when you are building networks, they are being built for a purpose and, when complete, are expected to have certain required characteristics (heterogeneity, robustness, economy, scalability, reliability, and capacity). The purpose of a network is usually described to a network architect by means of requirements documents that provide goals and usage information for the network that is to be built. Architects are typically individuals who use architecture and have extensive experience in building things to meet specific requirements. Designing architecture for either buildings or networks begins the process of turning requirements into engineering and then subsequently guides engineering into an implementation that ultimately satisfies the requirements.

The definition of architecture means many things to many people. In fact, numerous networking as well as nonnetworking architecture definitions can be found:

- "An architecture is a description of system structures" [Bass et al., 1998].
- "A network Architecture defines the messages and data formats as well as the protocols and other standards to which hardware and software must conform to meet desired network objectives" [Ralston and Reilly, 1983].
- "A set of principles and basic mechanisms that guide network engineering" [Braden, 2002].
- A typical (and traditional) vendor definition of architecture (e.g., IBM SNA's implicit definition) is a "blueprint for building interoperable products."

The meaning of architecture is in the eye of the beholder and we cannot find a well-accepted definition of the term or activity called *architecture*. But, we conclude that when you inspect the numerous theoretical

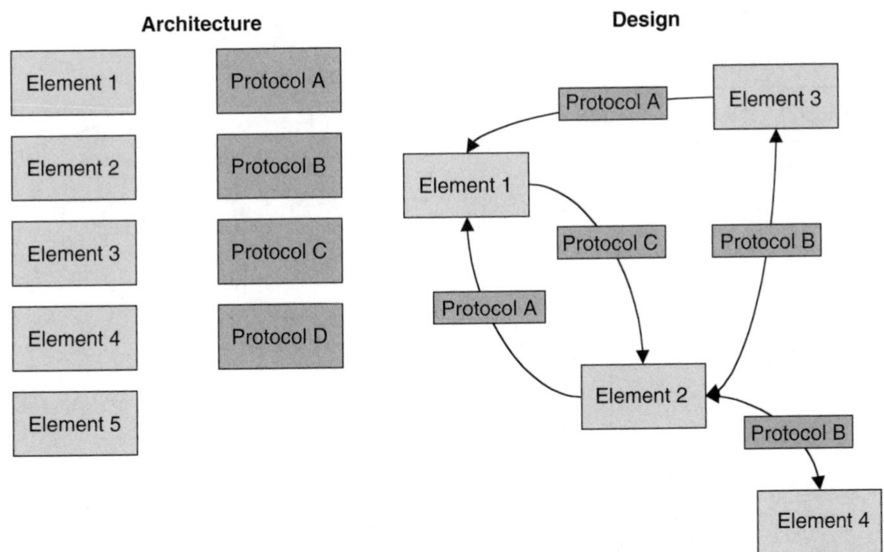

FIGURE 50.1 Architecture vs. design.

architecture definitions and discover their practical meanings to networking, all these definitions boil down to the fact that a *network architecture is just a list of elements*[1] *and protocols*[2*] employed in the design and construction of a network. An example architecture list is shown on the left side of Figure 50.1. An architect chooses this list as building blocks. These elements and protocols are used to develop a network design. In addition, usually being a fixed list, it often restricts implementation choices as the technology marketplace evolves. A simple example of a network design using the architecture elements is shown on the right side of Figure 50.1. We can conclude that an architecture is simply a finite list of elements and protocols that represent a pool of resources which are drawn upon by all designs.

Later on, we will discuss network design more deeply and also investigate the relationship between the flexibility of the architecture (i.e., open or closed) and the ability to change the list of elements and protocols.

50.1.2 Value of Network Architecture

Architecture is chiefly used to communicate future system behavior to stakeholders and specify the building blocks for satisfying requirements. A stakeholder is usually a person who pays for the effort and/or uses the end result. For example, a stakeholder could be the owner or a tenant of a future building, or a business owner or user of an anticipated network, or even an Internet user. Architecture and design drawings are frequent tools used to communicate attributes of the system to the stakeholders before the system is actually built. In fact, the communication of multiple attributes usually requires multiple drawings. And, unfortunately, architecture diagrams (usually multiple drawings) are often used incorrectly as design diagrams or vice versa.

This principle of multiple attributes of an architecture being communicated through multiple drawings is shown in Figure 50.2a, Figure 50.2b, and Figure 50.2c. To further illustrate the flexibility and value of architecture drawings, this set of diagrams describes the architecture for a computing system rather than a network. Figure 50.2a simply shows the major components or modules that comprise a computing system. No relationships are shown by Figure 50.2a. They are merely modules of the system. This figure best meets our definition of architecture.

[1] An element is a component where a particular networking function is located.

[2*] A protocol is a set of messages and procedures used for communication between elements.

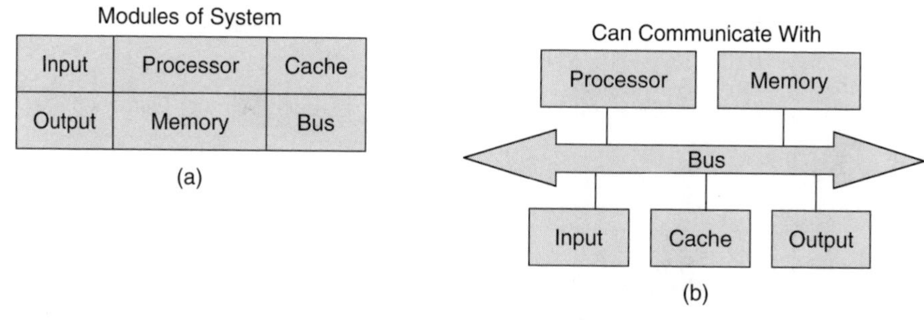

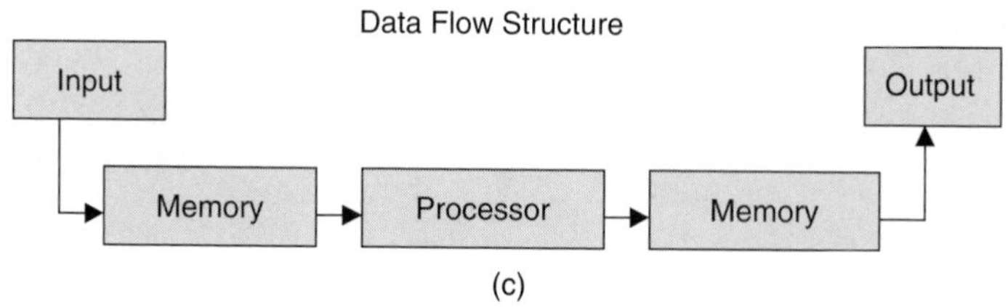

Figure 50.2 Attributes of an architecture being communicated through multiple drawings.

Figure 50.2b shows which modules are able to communicate with one another and gives some insight into the overall roles of the components. While the architecture building blocks are all shown, they don't have to be. The drawings become more of a design as dependencies are introduced. Finally, Figure 50.2c illustrates the path that data follows as it flows through the system. Notice this is more of a design drawing since not all architecture building blocks are shown. Any number of additional drawings could be developed to show other aspects of system behavior and/or dependencies. There is no clear demarcation point when architecture becomes a design. Generally when relationships are fixed or specified then drawings should be referred to as a design.

Network architecture has value to the customer/business enterprise and also has value for the vendor. When used by an individual company, architecture guides the deployment of a network for use as a tool supporting critical business processes. In the case of an enterprise network, architecture is the broadest term in what can be seen as an *infrastructure hierarchy* in organizations that deploy a computer network. When used in a standards body or industry forum, network architecture simply refers to an array of elements and protocols upon which the various market players in the industry will build products and applications. An open architecture allows companies, standards organizations, and the industry at large to fully exploit new and existing technology and provides the flexibility to build, evolve, and change their offerings based on marketplace demands. A closed architecture generally does the opposite and hinders the adoption of new technology. The concepts of an Infrastructure hierarchy and open and closed architectures will be discussed in more detail later.

50.1.3 The Potential Pitfalls of Architecture

As noted earlier, most architecture definitions can be condensed to a collection of elements and protocols. Two practical lessons that many implementers have learned about architecture:

1. *Leveraging new technologies lesson:* Once you have picked the elements and protocols for your architecture, you have effectively limited your flexibility. A very good example is the existing mobile

wireless systems, where wireless LANs were unforeseen. This means they are not deployable within the current mobile wireless systems. Anytime you limit flexibility you begin "gambling" that you have thought about everything. Experience tells us this is a poor bet.

2. *Architecture vs. design lesson:* Things get more interesting, and the bet grows larger when you progress from architecture to the design phase of a project. Once you have your list of elements and protocols, you continue further and draw protocol lines between elements. At this point you have a design because you have introduced dependencies between elements and protocols. Dependencies become stronger by fixing what elements are allowed to communicate. This further limits flexibility and possible future innovation. Often design pictures are presented as architecture but they are not! No matter what you call the picture, it is easy to see that once you "draw the lines" you are gambling more heavily that you can see the future clearly or at least long enough to get some return on investment (ROI) on the deployed network.

Architecture can be a very serious and expensive gamble for any company that uses architecture as a guide in a rapidly moving technology marketplace. It is a huge risk to build a system with a fixed set of elements and protocols. A fixed set of elements is a symptom of a closed architecture. A closed architecture does not allow for the possibility of very rapid and unexpected technology change, which has been the marketplace norm lately.

50.2 The Road from Architecture to Implementation

Network architecture is the usual starting point for the selection of components for an anticipated future network infrastructure(s). Architecture has two practical *situational meanings*. When used from an industry-wide perspective (e.g., in a standards body or industry forum), network architecture essentially refers to a collection of elements and protocols upon which the various vendors in the industry will build products. This perspective was shown on the left in Figure 50.1 and labeled architecture. The standards groups influence what is available in the marketplace tomorrow by increasing or constraining technology and/or connectivity choices. Unfortunately, dependencies between elements and protocols can be built into the standards (e.g., Box A won't work without Box J or Protocol P). When this occurs the design picture on the right of Figure 50.1 will often be referred to as architecture. The standards architect is assumed to be neutral. He is betting that he has considered all possible technology choices and made the best, cost effective selections for an industry.

The relationship between architectures and standards is a close one. When standards bodies heavily influence architectures, these architectures are often referred to as *de jure* standards and the terms *standard* and *architecture* are used almost as if they were interchangeable. Conversely, when a vendor defines architecture without input from other organizations and these architectures enjoy widespread implementation in the industry, they are often referred to as *de facto* "standards." Perhaps the most well known network example of a de facto standard was IBM's Systems Network Architecture (SNA).

Occasionally, vendors who produce architectures that are considered *de facto* standards do bring those technologies forward to standards organizations. If the standards bodies endorse these architectures, it is possible for *de facto* standards to make the transition to *de jure* standards. However, this transition is becoming increasingly rare since adoption of proprietary technology tends to undermine the openness of standards organizations.

When an individual company uses the term *architecture,* it refers to a very limited number of elements and protocols, which will populate the anticipated network infrastructure for the organization. The architect is betting on making a wise choice and hoping the resulting implementation will, in fact, meet business needs. His choices are limited to the standards organization he is following and what is available in the marketplace today (hopefully with an eye on tomorrow). In this sense, the process started by architecture results in a deployed network for use as a business tool.

50.2.1 Infrastructure Hierarchy

In the case of an individual company, network architecture is the broadest term in what can be seen as an *infrastructure hierarchy* in organizations that deploy a computer network. An example infrastructure hierarchy is shown in Figure 50.3 emanating from Architecture Ai. It involves the steps of creating *architecture* (Ai in this case), developing a *design*, and constructing an *implementation*. Each step becomes progressively more difficult, complex, and expensive with implementations of network infrastructures (e.g., mobile wireless), potentially reaching tens of billions of dollars.

Deployment mistakes are inevitable simply because new technologies and evolving business trends and directions will, over time, change the fundamental requirements that the architecture was supposed to satisfy. How do we deal with the fact that the technology and business worlds are constantly and rapidly changing? How does an architect hedge his bets?

The most effective approach to accommodate changes in technology and the marketplace is to embrace an open architecture. An open architecture allows companies to fully exploit new (and existing) technology and provides the flexibility to modify network infrastructures in support of changing business offerings based on marketplace demands. So what is an open architecture? Let's start with an example.

Because of the fuzzy definitions, the word *architecture* can be quite controversial. An interesting argument from many is that the Internet doesn't have an architecture; others say it does. Rather than debate this, we have decided to use the Internet as a manifestation of an "open architecture." By an open architecture we are referring to the perspective that the Internet is a simply a topology of interconnects using a set of agreed protocols. A key point is that Internet architecture elements are *not* defined. All that is defined is the protocols. Products compete to become elements in an Internet environment via implementing Internet protocols along with new or improved functions.

50.2.2 Network Design

What is network design? The term *network design* is often used in networking to refer to some high level view of a system that can be used as a roadmap for an eventual future deployment. We often use rough drawings to express a design. The drawings are usually populated by boxes, clouds, or regions showing the placement of function and often include arrows showing connectivity or expressing relationships. Relationships include possibilities such as: *is used by, communicates with, is a subsystem to,* and *provides*

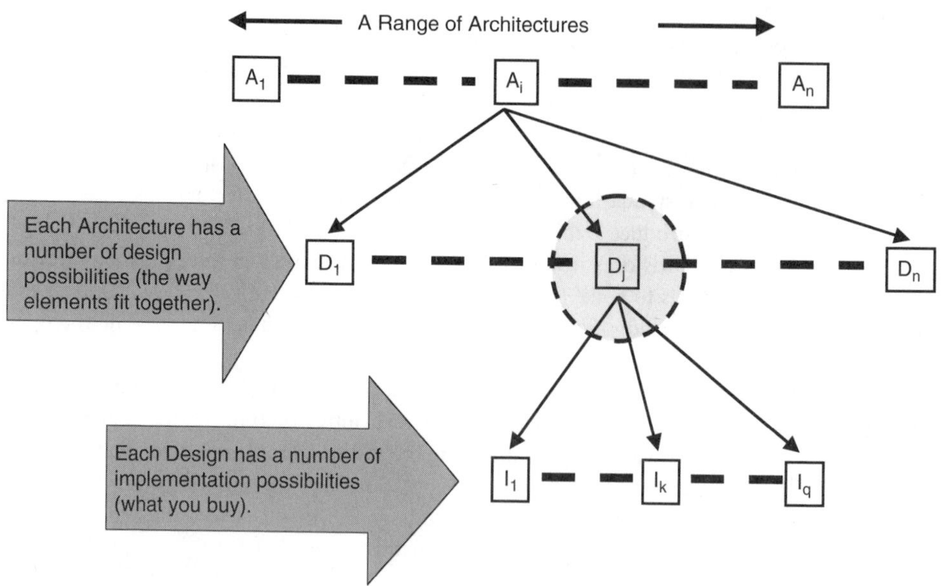

FIGURE 50.3 Domain of infrastructure hierarchies.

redundancy for. For example, many designs show two or more paths to a server location to prevent the loss of a single link from isolating a key server from the network. In this case, the design expresses the concept of redundancy improving availability.

One misunderstanding that happens frequently in the industry is that designs are presented as architecture. Design is much more limiting than architecture. When you draw communications lines between elements on a diagram, you begin the process of fixing network relationships (e.g., what elements are allowed to communicate) and therefore limiting flexibility and hindering innovation. Two extremes that affect flexibility are hierarchical vs. peer-to-peer network designs. In hierarchical network designs, elements of the architecture live within the narrow confines of a rigid hierarchy. Conversely, in peer-to-peer network designs, elements of the architecture operate with one another in a more equal fashion.

Design typically refers to the interconnection of networked elements. Specifying relationships between network elements is the demarcation point between architecture and design. Once you "draw the lines" you are now betting more heavily that you can see the future clearly. Unfortunately, this frequently has the unintended side-effect of limiting flexibility.

As Figure 50.2 and 50.3 illustrate, there can be many designs based upon a single architecture. In business terms, the major difference between design and architecture is that a design offers less flexibility to change business models or to offer different services in response to market demands, new technologies, or competitive threats. This is because a design specifies the interconnectivity of elements that often results in (or encourages) element interdependencies. If a company decides to change one element, the interdependency usually cascades to changes in other elements as well. This impedes the ability to scale the resulting infrastructure as service demand requires, reduces market place choice, and increases costs. As a result, changing from one design to another within a closed architecture tends to be a slow and expensive proposition.

As is shown in Figure 50.3, if an architecture is sufficiently broad in scope and flexible in intent, it is possible to derive a number of different designs from it. Similarly, if the network design is flexible, it is possible to derive numerous different implementations from a single design. These different implementations may vary not only in terms of the elements that are used but also in the protocols that are used to communicate between elements.

There are many examples of past architectures with rigid designs within the data networking industry. These architectures and designs have not offered their creators and adopters the flexibility to respond quickly to market changes. IBM's SNA and Digital Equipment Corporation's DECNET are two recent examples. Although they were both successful designs for their time, the lack of flexibility along with marketplace and competitive changes eventually rendered them both extinct.

50.2.3 Implementation

Implementation is the lowest but most expensive level in the infrastructure hierarchy and, consequently, is the most confining and resistant to change. As the name suggests, an implementation is usually associated with detailed specifications of the functions of particular elements in the design (e.g., disk and memory requirements, algorithm types, number of ports and speeds, equipment location, and brand names). There can be many implementations of a single design by simply varying specification parameters. For economic reasons, implementations usually either follow a vendor's roadmap or a standards organization roadmap. The vendor, alone or in conjunction with a standards organization, guides the customer's implementation choices, making it relatively easy to change an implementation and evolve along with the vendor's products. Implementation changes are dependent on the vendor's interpretation of the design and overall architecture. Effectively, the customer has bought into its equipment vendor's vision of the future.

What if the equipment vendor or standards organization vision of the future is faulty? For instance, what if a new technology comes along that was totally unforeseen and threatens existing business models? For an organization to make major changes, such as those that might be involved in offering a new service or changing business objectives, it may have to leave one vendor's roadmap or implementation in favor of another. This takes substantial time and provides an opportunity for competitors to gain market share at a customer's expense simply due to the lack of flexibility in the customer's network implementation.

50.3 Closed Architecture

A *closed architecture* is one in which *all* network elements and protocols are specified by a single vendor or industry forum. It is apparent that architecture decisions result in implementations that can run the gamut of being totally open or totally closed or possibly a hybrid somewhere in between with limited openness (at the edges for example). In the past, a single vendor such as IBM or DEC owned most closed network architectures. There is one notable exception, however. Telephony architectures are driven by a small consortium of telecommunications companies with the endorsement of governments (e.g., 3GPP and 3GPP2). Closed architectures lead to implementations that have numerous dependencies between network elements. These dependencies are an artifact of buying into the view of a single standards organization or vendor.

Closed architectures are popular because they provide a sense of control. This is especially true with people who have extensive past architecture experience with closed and proprietary systems. They have a very difficult time with flexibility and choice in a less structured and open setting. In a sense, closed architectures appeal to the structured nature of traditional enterprises and standards organizations. In essence, an open architecture asks individuals to embrace flexibility, evaluate choices, and enable the widest possible range of technology and deployment decisions. Also, people are naturally resistant to change. But history has taught us that changes in network implementations are inevitable.

In closed architectures, changes are anticipated and usually planned for by following an implementation roadmap. This is shown in Figure 50.4. By following the roadmap of a vendor and/or standards organization, they are making choices for a client and offering the client the ability to manage change under their control. Usually in doing so, the cost of change is typically low. History shows this low cost

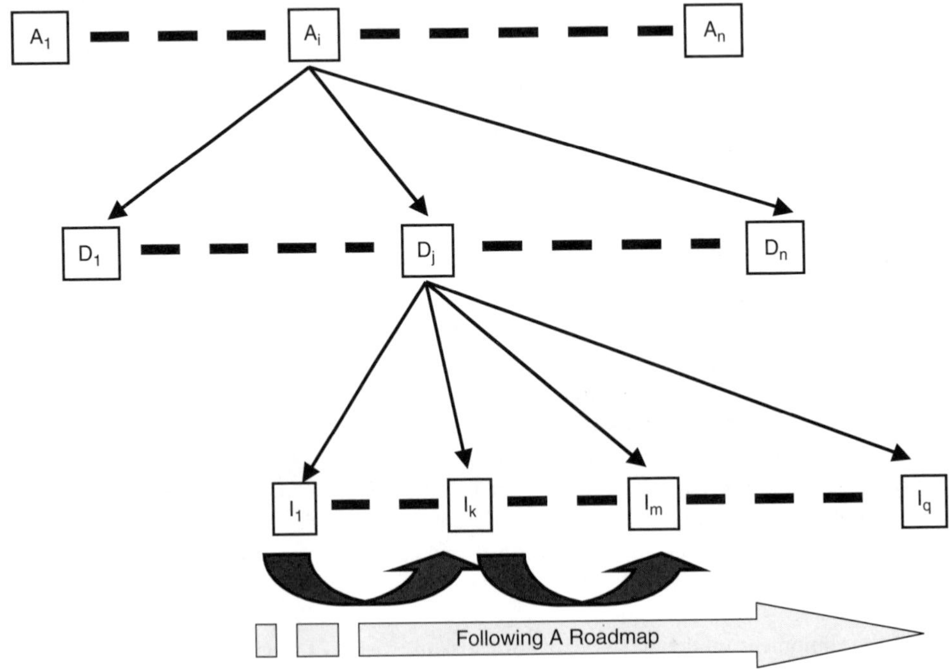

FIGURE 50.4 Typical closed architecture roadmap.

aspect is promised but never realized. The hidden danger is that by following a vendor or standards forum roadmap the customer is taking a gamble that they see the future clearly. Is this a good business bet?

There is also an important economic aspect related to the choice of an architecture. Closed architectures are almost always more confining and costly than open architectures. Closed architectures are typically used as barriers of entry to a particular market and often result in sole-sourced network implementations. Sole sourcing severely limits marketplace competition and inflates the cost of deployment and operation. This directly impedes innovation. In addition, a closed architecture results in a closed infrastructure that is usually built for the specific needs of a single application (e.g., telephony). Changes to meet new leading edge application requirements by vendors or standards organizations take time. Substantial time is necessary for evaluating the request and determining if their product or standards directions align with new application needs. In the mean time, companies can fall behind in the market waiting for their vendors or standards organizations to respond.

50.3.1 Design Changes

What drives the need for a change in design? It is usually a reaction to unexpected new products appearing in the market that are typically not planned for, or part of, an existing vendor or standards forum roadmap. Because of dependencies, the cost to move between designs is shown in Figure 50.5. It is much more than just an implementation change.

It is the natural reaction of incumbent vendors to resist making major changes to their implementation or product roadmaps. Since they are usually a sole source supplier in closed architectures, incumbent vendors typically offer partial solutions; i.e., solutions that will give some, but not all, of what is needed to deploy a new application or service over the existing network infrastructure. With this approach, a customer will often be forced to use vendor equipment in a manner that was not intended from the outset. In addition, adding functions that were not designed into the original system can result in increased system complexity and can cause unforeseen problems in other parts of the network. Closed architectures can be a significant problem because system functionality often depends on extensive proprietary communications between elements. Therefore, adding a new product to the system requires

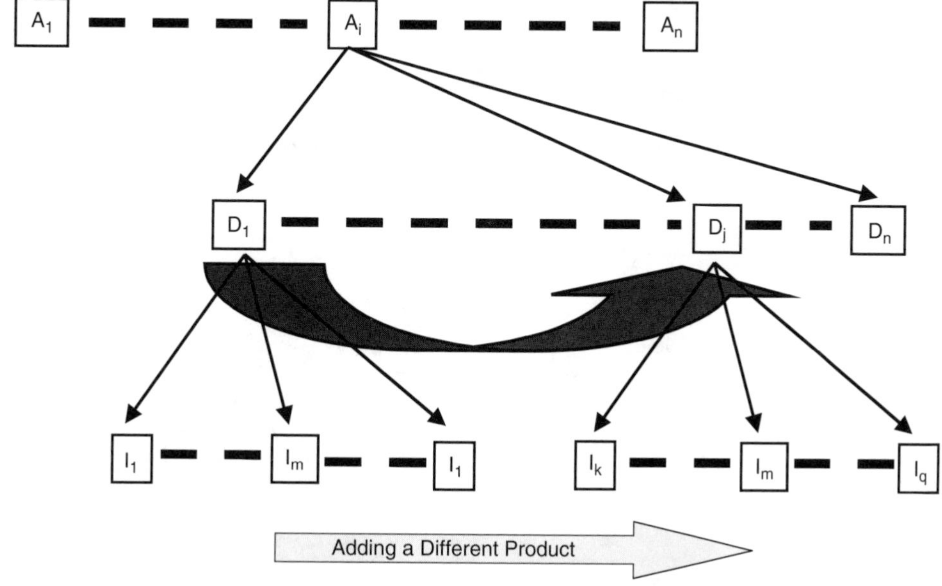

FIGURE 50.5 The cost to move within designs.

changing surrounding elements. This will also drive up the cost of operation and maintenance and impede future changes.

50.3.2 Architecture Change

What drives the need for a change in architecture? Most likely, it is unexpected competition enabled by new technology requiring new elements or protocols for the deployment of new services. With an infrastructure built on a closed architecture, a company is locked into its vendor's (or a standards organization) roadmap. Its flexibility to implement a new service quickly ahead of its competitors may be compromised. Therefore architecture change, shown in Figure 50.6, is a costly and expensive undertaking.

In a closed architecture, a business that is eager to exploit new technologies quickly will have many dependencies on their vendors' products. Will their vendor be able to deliver this partial solution in a timely fashion and at a reasonable cost? Will this solution offer enough functionality to deploy the application effectively? Does it require modifications to, or replacement of, existing network infrastructure elements? Will this (partial) solution impede a company from offering a more advanced application service in the future?

Closed architectures are attractive for the short-sighted hope of having a contained and controlled future. In reality, closed architectures are simply attempts to dictate the outcome of networking and control technology evolution and technology usage in a particular market. These attitudes by standards organizations are much more damaging to the marketplace due to the amplifying effect that standard organizations have on a number of vendors.

In summary, a closed architecture becomes an unnecessary and expensive gamble in a rapidly moving technology marketplace. It is a huge risk to build any constraints (either implicit or explicit dependencies) into a network using past closed architecture practices that assume a long ROI (e.g., telephony ROI is 10 years or more). It is naïve to think that a group of architects can predict and control the future. Network architecture neither controls application innovation above the network nor the technologies in the systems supporting the network. They will evolve on their own trajectories. With this in mind, any

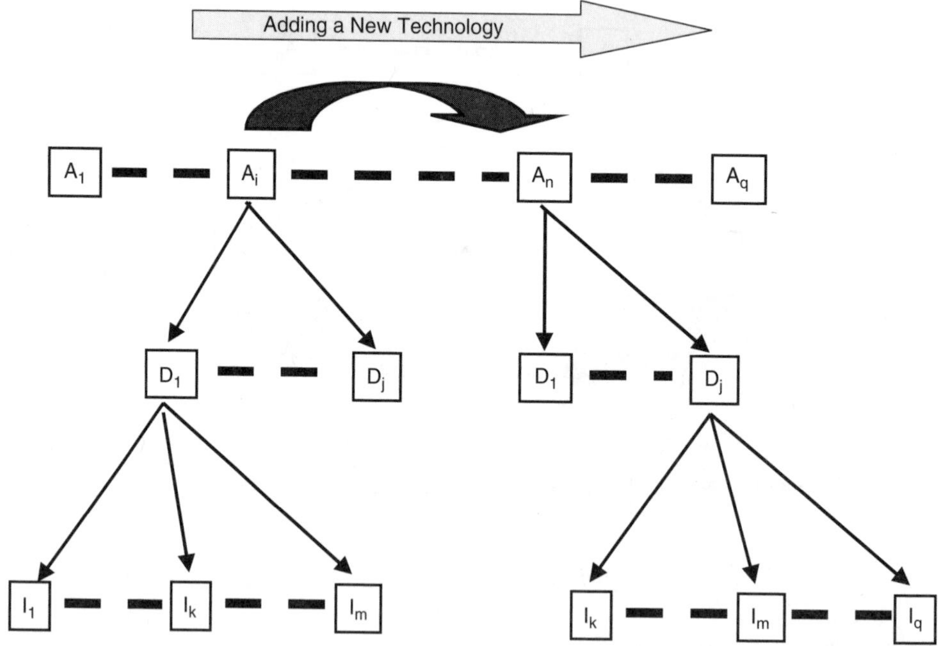

FIGURE 50.6 Cost to move between architectures.

network architecture needs to be flexible enough to allow for the possibility of very rapid and unexpected technology change, which is a marketplace certainty. A closed architecture is simply a bet that all contingencies were thought of.

50.4 Open Architecture

An open architecture in a practical sense is one in which there is no list of elements. Elements are simply not defined and the list of protocols is allowed to continually grow in response to changing market needs and technology innovation. Since elements are not defined; dependencies between them are difficult, if not impossible, to construct. Open architectures are usually the result of a collaborative effort by a variety of individuals, vendors, customers, and other organizations under the auspices of a standards body. These standards bodies promote equality of all members of the community and encourage innovation. A prime example of such a standards body is the IETF.

Open architectures also result in designs and implementations that do not have a dependency on any vendor roadmap or particular technology. They also are open in the regard that you are not making any bets about the direction that technology will take in the future. Figure 50.7 illustrates how openness affects the speed of change, facilitates the implementation of new ideas, improves time to market, and encourages competition. It is possible to move from one implementation to another without excessive constraints introduced by design dependencies or architecture limitations.

An open infrastructure is the result of an open architecture. The term *open infrastructure* simply means that companies have a choice in their equipment vendors. They can, in fact, easily mix and match equipment from various vendors. They are not tied to a specific design or a particular vendor's implementation. At a minimum the lack of dependencies is one key attribute of an open infrastructure that also promotes extreme flexibility known as "plug and play." From an economic standpoint, "open" means no bets are placed on which technology will win.

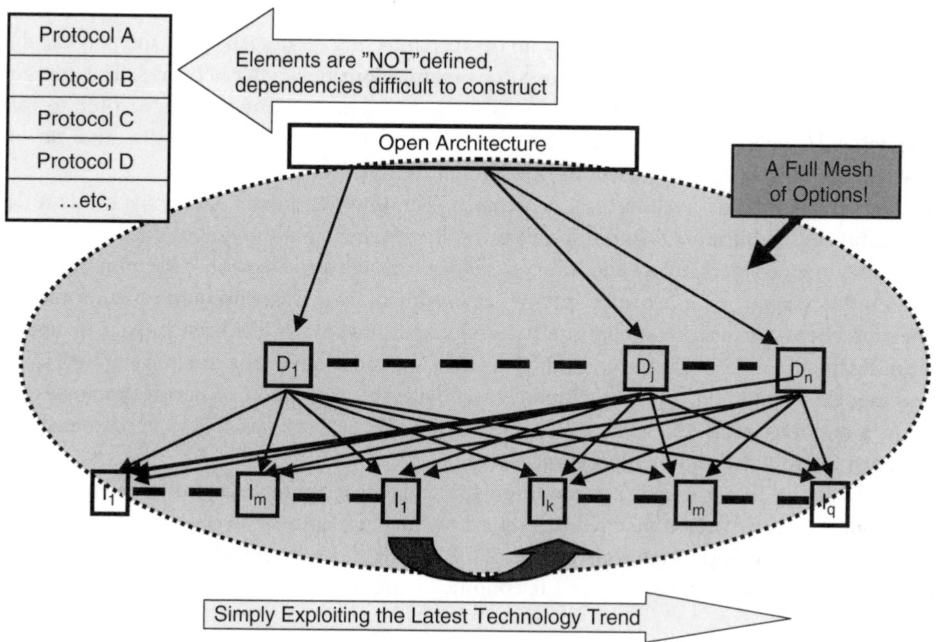

FIGURE 50.7 Open architecture flexibility.

An open architecture provides a *separation of concerns*[3] and encourages innovation. Conversely, closed architectures lock in too many otherwise unrelated decisions at once, meaning if you change one decision it results in cascading effects on others. Typically, an open architecture does not appeal to those who have a strong sense of control. By embracing open architectures, one admits that future marketplace requirements and technology direction are unknown. This means that one should anticipate surprises by maintaining a flexible plug and play posture.

Finally, an open architecture does not mean loss of control or loss of market opportunity for individual organizations. Open architectures result in infrastructures that are based upon open standards. In this environment, each vendor, regardless of size, has the opportunity to offer a competitive equipment solution.

A direct benefit of this opportunity is that an organization can evaluate all options and select the vendor with technology that provides an innovative, cost-effective, and timely solution. When a network is based upon open standards, new products and technology can easily integrate with minimal disruption. Since any new equipment does not require a major investment or overhaul of a network, new services can be rolled out gradually and then scaled as demand requires.

The Internet and intranets based upon open standards are great examples of open infrastructures. By referring to the Internet as *open infrastructure* we mean that the Internet is a simply a topology of interconnections using a set of protocols selected from open standards organizations. By using IETF protocols, the architecture of the Internet allows such diverse services as Web browsing, email, e-commerce, instant messaging, entertainment, video conferencing, and voice services just to name a few. And because the Internet is based on a loosely coupled set of functional elements communicating via IETF protocols, content and service providers within the Internet industry have the freedom to mix and match vendors, suppliers, and partners to innovate and create the most effective network infrastructure. With an open architecture, one selects the elements from the marketplace and interconnects them with the necessary open protocols.

50.4.1 A Clear Set of Benefits

Unlike closed and proprietary infrastructures, an open architecture creates true plug and play capabilities with an improved equipment selection process for organizations deploying networks. The competitive process of equipment selection provides impetus for innovation and fosters reduced time to market. Both capital and operational expenditures are minimized in an open environment because new functionality need not increase management or operational complexity.

The Internet has minimal architectural constraints. The Internet doesn't encourage element dependencies because the architecture does not specify any elements. It is simply an unbounded list of protocols that continues to grow. Marketplace adoption of protocols by network elements determines the protocol winners. Products compete for use in an Internet environment by implementing Internet protocols with their new or improved functions. The Internet architecture approach has been proven to result in a flexible infrastructure that is a fertile ground for innovation and ideal for the discovery of new ideas and business models. In addition, it works extremely well with the absorption of new technologies as they appear. As a recent example, the wireless LAN protocol (IEEE 802.11) was able to fit seamlessly within existing open Internet and intranet infrastructures.

Figure 50.8 shows the expected race condition that current telephony standards organizations face. The race condition is between a planned theoretical architecture approach versus an evolved, practical, and market-driven approach. Architects for the 3GPP standards are attempting to construct a planned future for mobile wireless networking and telephony. Figure 50.8 illustrates that this is being done in

[3] *Separation of concerns* is a concept that is at the core of software engineering. It refers to the ability to identify, encapsulate, and manipulate those parts of software that are relevant to a particular concern (concept, goal, purpose) [Dijkstra, 1976].

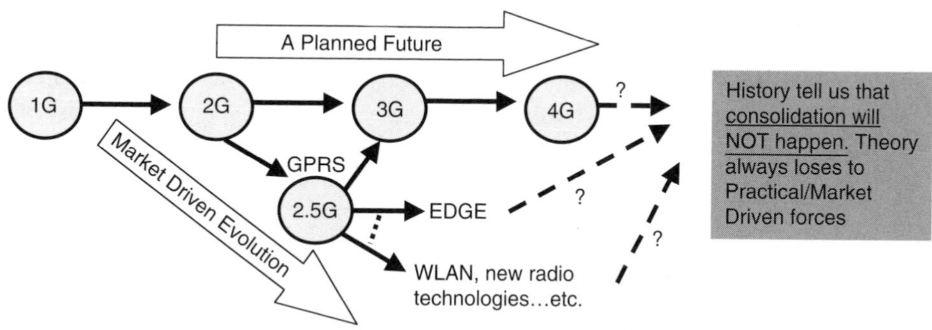

FIGURE 50.8 Planned vs. market driven future.

parallel with marketplace forces that are demanding the immediate deployment of newly developed technology, ideas, and business models.

In Figure 50.8, the telephony architect expects that consolidation will eventually happen and his vision of the future will be vindicated. But is this expectation realistic? If consolidation does not happen, then a lot of time, effort, and money are wasted. In fact, planned architecture from standards organizations could be a very bad approach. The marketplace is usually slowed by attempts to force consolidation. This negatively impacts the natural evolution of technology deployment and suppresses healthy responses to marketplace needs. Let's consider some recent history lessons to understand if consolidation is likely to happen.

50.4.1.1 ISO's Open System Interconnection (OSI)

In the 1980s, the International Standards Organization (ISO) proposed the comprehensive seven-layered OSI model for communications between distributed systems. The eventual goal of OSI was to replace the protocols used to interconnect distributed systems using TCP/IP and a number of *ad hoc* TCP-based applications.

Defining OSI took a considerable length of time. During that time, the popularity of TCP/IP increased as it was used to solve real-world problems, The IETF attempted to coordinate the peaceful coexistence of TCP/IP and OSI through elaborately crafted accommodation strategies. In the end, TCP/IP networks (such as the Internet) proliferated and the OSI effort was abandoned.

50.4.1.2 IBM's Systems Network Architecture (SNA)

IBM's proprietary Systems Network Architecture (SNA) dominated corporate enterprise networking for two decades from the mid-1970s through the mid-1990s. The early SNA architecture was purely hierarchical in nature, reflecting the dominant position occupied by the mainframe in these corporate networks.

As business processes changed and corporate functions became more distributed, the systems that were interconnected on the corporate network needed to take on more of a peer-to-peer relationship. IBM defined a major extension to hierarchical SNA called Advanced Peer-to-Peer Networking (APPN) to accommodate these distributed systems. Again, this variant of SNA was proprietary with the closed specification being completely defined by IBM.

By the time that IBM was ready to implement APPN on their computing platforms in the mid-1990s, the marketplace acceptance of TCP/IP had grown dramatically. Neither variant of SNA networking (hierarchical or peer-to-peer) could meet the price/performance ratio of IP networking. Even though SNA is still running in a few corporate networks, there are no new SNA networks being deployed and most of the existing networks have migrated to TCP/IP.

50.4.1.3 IPv4 to IPv6 Migration

Even IP itself is not impervious to marketplace realities as the present situation with IP Version 6 illustrates. The IPv6 effort was initiated to ostensibly handle the impending depletion of the IP address space. But as the formulation of IPv6 by the IETF took longer than anticipated, a more immediate solution for IPv4 was needed.

As a result, the practitioners in the networking industry significantly mitigated the IPv4 address space problem by implementing capabilities such as IP subnetting, IP supernetting (using Classless Inter-Domain Routing), and private IP address assignments (using Network Address Translation). Even though these solutions were defined under the auspices of the IETF, the organization's own grand scheme for adoption of IPv6 as the next generation of IP networking has been delayed.

50.5 Architecture Directions

Closed network architectures of the past were owned either by large companies such as IBM and Digital Equipment Corporation or by a consortium of companies such as the telephony vendors. Prime examples of these closed network architectures are IBM's SNA, Digital's DECnet, and telephony's landline and wireless architectures. In all cases, the active force behind the architecture was the vendor or a closed consortium of vendors. Telephony vendors continue to control the telephony architecture and select its elements. Even though multiple vendors are involved in a consortium, their collective vision of the future is still limited and rarely unified due to competition or being encumbered by individual vendor business goals.

However, network architecture is now taking on a new meaning because of the popularity of the Internet. The marketplace now largely determines what goes into an architecture. Also, the tempo of technology change has increased dramatically. There is no longer enough time to construct complex architectures and designs or to wait for others to develop them. Recent history has shown that new technologies and new classes of devices have been very disruptive in the marketplace in general and to closed architectures in particular.

Prime examples of recent disruptive network technology choices driven by the marketplace are:

- Ethernet instead of token ring and other layer 2 connection protocols
- IP instead of OSI, SNA, DECnet or IPX
- Routers and switches instead of front-end processors and cluster controllers
- Mobile wireless devices in addition to wireline devices
- Unregulated spectrum usage (e.g., 802.11) exploding and encroaching on the utility of licensed spectrum
- Packet switched voice over IP instead of circuit switched voice

Most closed architectures have a long cycle for technology deployment and return on investment (ROI). By defining these long, complex, and expensive ROI architectures, vendors and standards organizations are asserting that they know where the marketplace is going. To complicate matters even further, governments have dictated many aspects of closed telephony architectures such as their connectivity options and reliability requirements. However, history has not been kind to closed architectures. Some of them have passed away such as SNA, DECnet, and NetWare. Based upon the above discussion, we can predict that other closed architectures will eventually be eliminated as well.

Closed architectures typically result in tightly coupled infrastructures biased towards one or a few particular implementations. When elements are tightly coupled, groups of elements must be provisioned together because individual functional elements cannot work alone. This is shown by the monolith closed infrastructure on the left in Figure 50.9. These tightly coupled infrastructures increase operational complexity, introduce migration headaches, stifle innovation, reduce choice, increase costs, and impede the ability to scale the infrastructure.

Today's open architectures simply comprise an ever-growing list of protocols. Open architecture participants admit that it is impossible to see the future so they strive for flexibility and allow extensions needed for market needs and technical innovation. This flexibility gives you the power of plug-and-play and results in what is often referred to as an Internet-style infrastructure. Organizations such as the IETF, W3C, IEEE, and others recognize that the future is in open architectures.

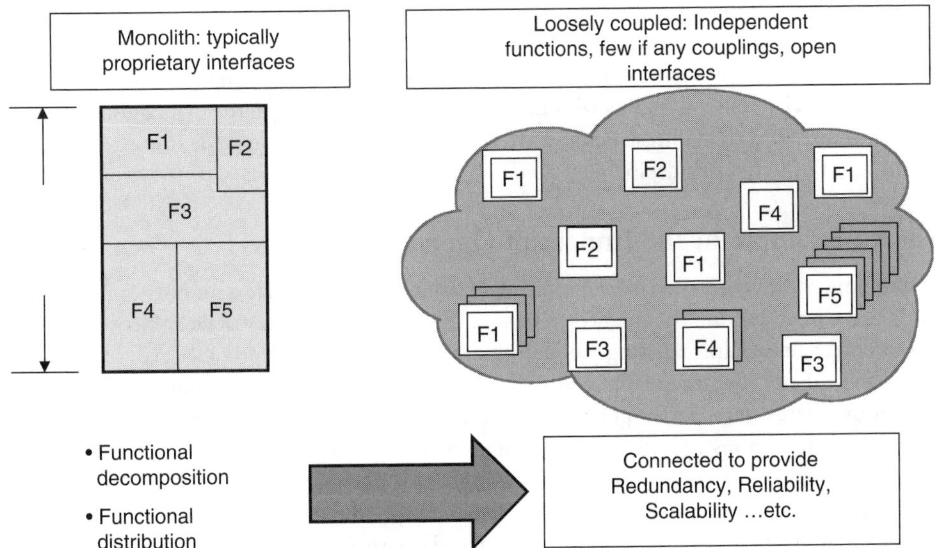

FIGURE 50.9 Tightly vs. loosely coupled architectural environments.

Even though some would argue that "open" implies chaos of choices and integration headaches, an open architecture really means that organizations have fewer technology restrictions. They are no longer tied to a specific design or a particular vendor or consortium roadmap. With an open architecture you are able to evaluate a variety of vendors and select the best solution from a competitive field. An open, Internet-style infrastructure encourages competition which, in turn, facilitates innovation and reduces equipment and implementation costs. In short, open does not mean a *loss* of control but *increased* control over business and technology decisions.

Today, the networking industry is moving from vertically integrated closed architecture environments to loosely coupled open architecture environments, which is shown on the right side of Figure 50.9. The active force for change is the marketplace. The definitions of the network protocols are becoming the most important aspect to networking. The role of architecture is greatly diminished in the plug-and-play environment, and the roles of system engineering, sound protocol design, and good project management are paramount in our future technology world.

The best strategy is one that strives to take advantage of technological innovation. History has taught us that technology innovation is a tremendous "wild card." It is unpredictable and often the catalyst for an avalanche of change, typically disrupting markets and destroying existing business models. Since no organization can control or even accurately predict the future, the best strategy is a flexible one. Because of this, open architectures are the ones that are most likely to survive a future of ongoing rapid technological innovation.

No one knows what the future will be, and no one can predict the countless applications and technological innovations that will exist. When the market creates a demand for a new service or when a new application creates huge revenue potential, an open architecture provides the flexibility to rapidly take advantage of any technology-driven opportunities.

50.6 Business Cases

In this section we will provide a business discussion and economic rationale, with examples of the enormous positive impact that an open architecture can have in a business environment and the industry in general. We will cover two emerging trends in the use of the Internet Protocol, which is central to open architectures. One is mobile wireless data communications, and the other is the transition to voice communications over Internet technology. Both trends are merely a continuation of the evolution of networking begun by the Internet.

Open Internet architectures have completely overwhelmed company-driven network architectures and now threaten the telephony monoliths with their cartel approach to standards that continue to result in closed architectures and rigid infrastructures. We will use the term "Internet-style Infrastructure" to refer to the end result of a network deployment guided by an open architecture.

50.6.1 An Example of the Impact of Open Architecture

The first example describes a current economic and business struggle where a number of mobile wireless standards are competing for dominance (all those involved are industry stakeholders in the mobile wireless Internet value chain, including mobile operators, ISPs, equipment providers, consultants, content providers, and others). Because of the enormous infrastructure expense, the mobile wireless industry approach to standards starts with a poor assumption. That assumption is that they have to carefully try to determine a future infrastructure that will enable them to harness the revenue opportunities that the mobile wireless market offers. Current organizations such as 3GPP and 3GPP2 are little more than misguided attempts trying to predict the future, control telephony, and steer technology evolution. While the 3G groups are working to bring about their particular visions of the future, the result is contention and delay in the availability of services for end users and missed revenue streams for all the industry stakeholders.

The solution for any contemplated networking environment, including the mobile wireless Internet, lies in reframing the question about predicting or controlling the future. Wireline or mobile wireless operators shouldn't be asking, "Which future infrastructure should we develop?" The question should be, "How can we best be prepared for *any* future?" The answer to that question is simply this: by starting with an open architecture that will enable industry stakeholders to rapidly accommodate inevitable technology changes.

The technological infrastructure that serves as the foundation for enterprise as well as mobile wireless Internet services will affect costs. It will play a key rôle in determining how the industry and individual businesses will realize both the profit and revenue opportunities that will inevitably exist in this market. Determining the infrastructure of the mobile wireless Internet in particular is a very important strategic business decision that affects the health of the entire mobile wireless industry. Let's consider a small mobile wireless business case study that we call: "*A Tale of Two Architectures*."

50.6.2 The Mobile Wireless Challenge

Look a few years into the future and assume that current efforts result in a new generation of mobile wireless infrastructure standards. Equipment vendors have manufactured products and sold them to mobile operators, who have invested billions of dollars on infrastructure and now hundreds of millions of dollars more to roll out their initial mobile data services. Data services are expanding rapidly as consumers and business users around the world access the Internet with an ever-increasing variety of wireless devices. Technological innovation continues its rapid acceleration. Mobile operators are evaluating dozens of new applications and services every year.

ClosedCel and OpenCel, two hypothetical mobile operators with global customer bases, have just been approached by WearAbouts, Inc., an applications developer who has a unique new mobile wireless application that could fundamentally change how people shop for clothes.

After evaluating the potential revenue figures and market research with WearAbouts, both ClosedCel and OpenCel decide that this mobile wireless data service is in line with their business strategies and has significant revenue potential from consumers, clothing manufacturers, and the retail industry alike. In fact, it has the attributes of a killer application. (A "killer application" is one that is used *frequently* by a *large* number of individuals. Voice is a great example of a killer application.)

As usual, both mobile operators quickly realize that application deployment depends heavily on new infrastructure technology. How will the process of modifying their infrastructure differ within a closed mobile wireless infrastructure vs. an open, Internet-style infrastructure? Within each infrastructure, how would each of the following be affected?

- Flexibility of choice
- Implementation costs
- Operations and maintenance costs
- Time-to-market
- Competitive advantage
- Revenue potential

50.6.3 ClosedCel: Closed Architecture

ClosedCel approaches its sole-source equipment vendor with a request for an equipment solution that will meet the needs of the WearAbout application. WearAbout's technology is leading edge, so none of ClosedCel's current equipment will support it. The equipment vendor must take the time to evaluate the request and determine if their product directions align with WearAbout's needs. The equipment vendor needs to evaluate if they can fit new development into their current implementation roadmap, or whether this is a major innovation to warrant a change in that roadmap. In either case, it is a time consuming and expensive proposition.

It is the natural reaction of the vendor to resist making major changes to its implementation roadmap. Since it is the sole source supplier for ClosedCel, the equipment vendor can offer a partial solution. This solution will give ClosedCel some, but not all, of what it needs to optimally deploy the WearAbouts application the way it had planned. With this solution, ClosedCel will be forced to use the vendor equipment in a manner that was not intended from the outset. ClosedCel is concerned that adding functions that were not considered by the original architecture will result in increased system complexity and cause unforeseen problems in other parts of its network. ClosedCel has come to realize that this can be a significant problem because its system's functionality depends on extensive proprietary communications between components. Therefore, adding a new component to the system requires changing surrounding components. And this, ClosedCel fears, will drive up the cost of operation and maintenance and impede future changes.

ClosedCel is eager to get to market quickly, but it has many questions. Will its equipment vendor be able to deliver this partial solution in a timely fashion and at a reasonable cost? Will this solution offer enough functionality to deploy the WearAbouts application effectively? Does it require modifications to, or replacement of, ClosedCel's existing network infrastructure elements? Will this partial solution impede ClosedCel from offering a more advanced WearAbouts service in the future?

With a closed infrastructure, ClosedCel is locked into its equipment vendor's roadmap. Its flexibility to implement a new service quickly ahead of its competitors is compromised. Also, its ability to deliver differentiated services and realize fresh revenue streams is in jeopardy. It will be many months before ClosedCel can offer its customers the WearAbouts service, if ever. In the meantime, WearAbouts is marketing aggressively to OpenCel, another wireless data service provider.

50.6.4 OpenCel: Open Architecture

Since its infrastructure is based on open standards, OpenCel has the option to investigate multiple, potential solutions from a variety of vendors. It approaches two established vendors whose equipment it has in its network and two newer, smaller vendors whose equipment it does not currently have. In this competitive environment, each vendor quickly offers an equipment solution with implementation costs and delivery schedules.

OpenCel evaluates all the options and perhaps selects one of the smaller vendors with new technology that provides an innovative, cost-effective, and timely solution. Because its network is based on open standards, OpenCel easily integrates the new equipment into its existing network via open protocols, and within a few weeks is test-marketing the WearAbouts service to its customers. Since the WearAbouts application does not require OpenCel to make a major investment or to overhaul of its network, OpenCel plans to roll out the WearAbouts service gradually and then scale the service as demand requires.

50.6.5 The Voice-Over IP (VoIP) Challenge

Historically, voice traffic has been carried over proprietary circuit-switched network infrastructures built from closed architectures. PBXs ruled the day and network designers were accustomed to proprietary designs consisting of a relatively few network elements from an even fewer number of vendors. When data transport was needed, it was carried over a telephony infrastructure. As the speed of networks improved in the 1990s and the convergence to the IP protocol began, telco standards bodies attempted to use traditional closed architecture methods to transport both IP data and voice over existing telco network infrastructures. This *data-over-voice* approach has been discredited as being too complex, too limiting, and too expensive.

The benefit of IP-based telephony was always clearer for enterprises. Corporate network administrators had always maintained two disjoint networks: a voice network and a data network. If voice traffic could be carried reliably over a data network, the separate voice network could be eliminated, resulting in a tremendous cost savings to the corporation. A *voice-over-data* approach based on an open architecture is credited as being a cost effective, flexible, and future-proof solution.

With the advent of the Internet and open, Internet-based services such as email, instant messaging, and on-line chat, most carriers and service providers saw their voice traffic decline. There was a parallel decline in revenue since much of their existing billing methods were tied to voice traffic. Service providers discovered that by transporting voice services over an increasingly data-centric open network, they could contain their costs while offering more advanced and desirable integrated voice and data services.

There were many challenges to carrying voice over an IP packet-based network:

- Interoperability: testing interoperability between various devices that support a set of standards that are themselves evolving,
- Quality: circuit-switched networks benefited from over 100 years of incremental quality improvements. The starting point for the quality of IP-based telephony was very high.
- Features: Similar in nature to the quality challenge, circuit switched voice had evolved over the years to address a number of advance features. IP-based telephony had to accommodate those features in order to be considered competitive.

Over the past decade, open industry consortia (such as the International Softswitch Consortium) and standards bodies (such as the IETF, IEEE, and W3C) have addressed all of these major challenges. Today's voice-over-IP vendors who base their solution on the results of these open architectures have IP packet-based solutions that not only achieve parity with its circuit switched predecessor but does so at a much better price/performance ratio and provides the ability to integrate new data-centric services in the future.

50.6.6 Benefits

Unlike closed architectures, an open architecture creates true plug-and-play capabilities and an improved equipment selection process (as we demonstrated for VoIP technology and the mobile operators). Open architectures improve purchasing power and reduce the cost of operation in several ways. First, it will create standard, open interfaces among network element components. This will greatly facilitate maintenance and repairs. Technology changes are easier to deploy and therefore less costly with an open architecture. By using commodity rather than proprietary technology, organizations that deploy networks will be able to "break open" boxes of equipment and buy only the functions they need when they need them, merely plugging them into their network. They will be able to buy the latest technologies and the best vendor equipment with flexible, competitive choices.

TABLE 50.1 How Open Architectures Compare

	Closed Architecture	Open Architecture
Flexibility of choice	Rely on a single source solution Locked into vendor's implementation roadmap Get partial or restricted solutions to new challenges	Compare multiple vendors' solutions Cost effectively mix-and-match equipment in network Get complete solutions to new challenges
Implementation costs	Expensive to implement new technologies and applications Changes in one part of network may impact other network components	Implement new technologies and applications with minimum expense Network absorbs changes easily with little impact to other components
Operations and maintenance costs	Managing added-on functionality (not integral to original design) increases operations and maintenance costs	New functionality is integrated easily without adding operational, management, or maintenance complexity
Time-to-market	Takes 6–18 months (and possibly many years in telephony) to implement new technologies and applications	Takes a few months or even weeks to implement new technologies and applications
Competitive advantage	Nimbler competitors can offer new services more quickly	Can be the first to offer new services
Revenue potential	Revenue streams are delayed	Revenue and profits are realized quickly

Second, an open architecture will save operational costs and management costs. While closed architecture solutions often have effective built-in management capabilities, the expansion of multivendor environments due to the Internet has lead to numerous composite management solutions from companies such as IBM (Tivoli), Computer Associates and Hewlett-Packard, to name a few. These open management solutions often exceed the capabilities of proprietary systems and result in lower operating costs in two ways:

- First, they provide a flexible framework that can easily accommodate new functionality.
- Second, they coordinate the expert management of individual components.

50.6.7 Maximize Application and Service Opportunities

An open architecture benefits both individual organizations and the industry as a whole. Opportunities will naturally arise from a common service and applications development environment. The community of developers for Internet open architecture solutions is quite large, estimated to be in the hundreds of thousands. Leveraging this huge source of innovation within an open environment creates virtually endless business and consumer services that result in more savings and greater revenue opportunities. In addition, an open architecture provides opportunities for a convergence of media and access technologies.

While proprietary designs claim to offer some of these advantages, they always fall short of this goal. A flexible open architecture will provide all of these benefits with reduced deployment costs, with an accelerated timeframe and within reasonable operating expenses. Table 50.1 summarizes the ways in which open and closed infrastructures will impact VoIP and the mobile wireless industry costs and revenue.

50.6.8 Summary

An open architecture represents the greatest business value for an individual organization. It is not a design, nor does it favor any implementation. From a business perspective, its goal is a flexible, dynamic, and integrated infrastructure in which developers can produce endless business and consumer services. An open architecture solution preserves the unique strengths of the Internet, reduces costs, and maximizes revenue opportunities for every organization in the networking value chain.

From a technical perspective, an open architecture is the necessary flexible base for any future network infrastructure. The Internet Protocol (IP) is the network glue that holds the open architecture vision together. Operating like electricity, IP accommodates an unbounded list of open protocols, both present and future. This common base will allow organizations to develop and deliver a wide range of services using a network infrastructure comprised of many vendors' products. The result will be significant cost savings, improved flexibility in adopting new technologies, easier accommodation of new business models, and the proliferation of worldwide services.

50.7 Conclusion

We have learned:

- **Architecture** is at the top of the infrastructure hierarchy. Architecture simply specifies the set of protocols and elements that are common to all of the designs and implementations. An open architecture containing a loosely coupled set of protocols offers the greatest degree of freedom and flexibility to exploit marketplace changes and new technologies.
- A **design** is less flexible than architecture. Designs require interconnections between elements and this reduces deployment flexibility. A rigid design (one that specifies elements tightly coupled by proprietary protocols) offers still less flexibility.
- An **implementation** is less flexible than design. In the infrastructure hierarchy, implementation is the vendor's interpretation of a design and offers the least freedom and flexibility.

Business history is filled with failed companies and whole industries that neglected to take a proactive approach to change. A recent example is that many landline telephone companies thought mobile networks would never catch on. Mobile wireless companies are now cannibalizing local and long distance wired markets. Neither IBM nor DEC understood the power of an open architecture and their network technology is now extinct. The technological infrastructure upon which to base a network is a strategic business decision that affects the ability to compete, deliver services, minimize costs and maximize revenues.

It is critical that open architectures guide your networking technology choices in order to maintain flexibility and freedom to manage the future, rather than a restrictive closed architecture that is based on a particular vision of the future since that vision is likely to be faulty. The only safe approach is to create a technological foundation that will support *any* future.

References

Bass, Len, Clements, Paul, and Kazman, Rick, *Software Architecture in Practice*, Addison-Wesley, Reading, MA, 1998.

Braden, Bob, Architectural Principles of the Internet, http://www.isi.edu/newarch/DOCUMENTS/rtb.IPAM.mar02.pdf, 2002.

Dijkstra, E.W., *A Discipline of Programming*, Prentice Hall, Englewood Cliffs, NJ, 1976.

Ralston, Anthony and Reilly, Edwin, *Encyclopedia of Computer Science, Engineering*, 1st ed., Van Nostrand Reinhold, New York, 1983.

Webster's New World Dictionary of the American Language, William Collins & World, New York, 1978.

51

Overlay Networks

CONTENTS

Abstract 51-1
51.1 Introduction 51-1
51.2 Background 51-2
51.2.1 Definitions 51-2
51.2.2 Issues 51-3
51.3 Tunneling 51-3
51.4 Complex Structures 51-3
51.5 Variations 51-5
51.5.1 Manual Overlays 51-5
51.5.2 Automated Overlays 51-5
51.5.3 VPN 51-6
51.5.4 Provider-Provisioned vs. End-to-End 51-6
51.5.5 Peer-to-Peer 51-6
51.6 Applications 51-7
51.6.1 VPNs/Security 51-8
51.6.2 Shared Use 51-8
51.6.3 Incremental Deployment of Services 51-8
51.7 Challenges 51-9
51.7.1 Security 51-9
51.7.2 Support 51-10
51.7.3 Deployment and Management 51-11
51.8 Recommendations 51-11
Acknowledgments 51-11
References 51-12

Joseph D. Touch

Abstract

Overlay networks are virtual infrastructure created on top of existing networks using tunnels. Overlays include the multicast backbone (MBone) and IPv6 backbone (6Bone), as well as virtual private networks (VPNs) and peer-to-peer file sharing networks. These virtual networks enable new protocols to be incrementally deployed, allow testbeds and private networks to share production infrastructure, and support new services using existing, underlying networks. They present unique challenges to applications and operating systems and require special attention to routing, forwarding, and tunneling. Overlay networks support unique opportunities to explore new network services and topologies independent from widescale ubiquitous upgrades.

51.1 Introduction

Overlay networks represent one of the few new paradigms in network technology in the past two decades. Most of the Internet's architecture, protocols, and services have remained constant, albeit optimized,

1-58488-381-2/05/$0.00+$1.50
© 2005 by CRC Press LLC

since the mid-1980s. For the first decade of TCP/IP, the introduction of new network-level services required modifying routers and connecting them with dedicated links. This created a barrier to new services such as content-based routing, secure internetworking, and multicast. Overlay networks overcome this barrier using the existing legacy infrastructure as a platform that enables flexibility, rather than impeding it.

Overlay networks include manually configured overlays, automated overlays, virtual private networks (VPNs [Scott et al., 1998]), and peer-to-peer networks (P2Ps [Oram, 2001]). Each of these kinds of overlay includes one fundamental component — the tunnel. Real networks are connected by links, which include direct physical connections between two network nodes. Tunnels create virtual links out of paths of other links, and so need not correlate to the physical layout of a network. This flexibility allows overlays to be deployed as virtual networks on top of existing networks, to deploy new services and protocols on a subset of the infrastructure without needing rewiring.

51.2 Background

The use of networks-on-networks goes at least as far back as TCP/IP, to the Cronus encapsulation tunnels used to "contain" network experiments in the early 1980s [MacGregor and Tappan, 1982]. The concept was reinvigorated by the development of multicast in the early 1990s [Eriksson, 1994]. Multicast was a new network-layer service based on modifications to the IP protocol and the addition of multicast routing protocols (e.g., Protocol Independent Multicast {PIM}, Distance-Vector Multicast Routing Protocol {DVMRP}, and Multicast Open Shortest Path First {MOSPF}). Like most new network protocols, multicast worked only if supported at every router along a network path. This meant that multicast would not be available until it was ubiquitous; there would be no incentive for early adopters because it would not work until thoroughly deployed.

Overlays were applied to multicast as a way to allow incremental deployment. Rather than requiring that every router on a path supported multicast, tunnels were used to connect multicast-capable components, hopping over nonmulticast routers. The nonmulticast Internet was thus used to enable multicast deployment, rather than acting as an impediment.

The basic concept of using tunnels to interconnect a subset of routers supporting a new capability allowed incremental development and deployment without requiring dedicated infrastructure. At that point, the Internet becomes a vehicle for its own development and research. Overlays emerged as a critical technology to support service development and deployment. Other playgrounds for developing new services, e.g., via programmable routers (Active Networks [Tennenhouse et al., 1997]) or even at the application layer (peer-to-peer [Oram, 2001]) ultimately depend on an overlay at some level.

51.2.1 Definitions

An overlay network is just a network overlaid on an existing network, i.e., a network composed of tunnels that rely on an existing multihop network for connectivity (Figure 51.1). This distinguishes an overlay, which is virtual, from the real network on which it relies, where the real network is itself a multihop interconnection of single-hop (at least as far as the network level is concerned) point-to-point or multipoint links.

An overlay network virtualizes all the components of a conventional network, including hosts, routers, links, forwarding, routing, addressing, and naming. Components of an overlay are generally a subset of components in the underlying infrastructure, where, for example, some infrastructure routers participate in a particular overlay and others enable connectivity that supports tunnels but do not otherwise participate in that overlay (see Figure 51.1).

By providing an alternate infrastructure on the base network, an overlay can present an alternate view of each of these components, for example, providing geographic routing (GeoNet [Finn and Touch, 2002]), URLs as hostnames, or as strictly ordered links even where they are not part of the underlying network infrastructure.

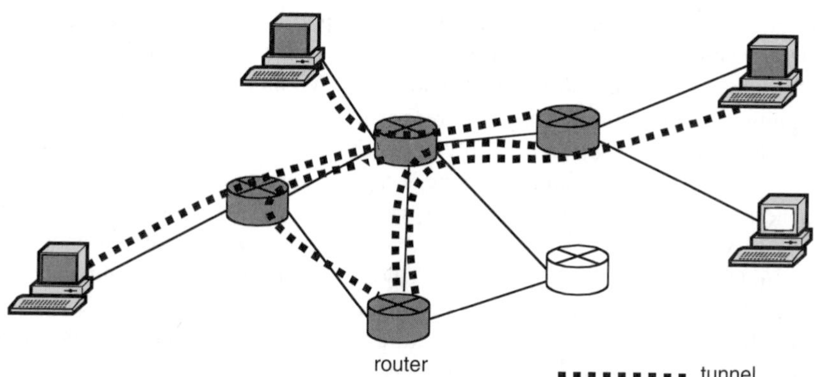

FIGURE 51.1 Overlay composed of tunnels.

51.2.2 Issues

There are several challenges to providing and using overlay networks. These include the efficiency and ubiquity of tunnel support, the need to support concurrent overlays, complex overlay hierarchies, revisitation and component reuse, and integration of overlays. The details of this section are not intended to provide a complete discussion of the issues but to lend insight into the complexities of supporting overlays in general.

51.3 Tunneling

Tunneling can be difficult to perform correctly and efficiently [Touch, 2001]. Tunnels can be achieved by a variety of mechanisms, the most common of which are (1) by adding options to existing headers and (2) by encapsulation, which adds additional headers. The first multicast overlay focused on header options, notably IP's Loose Source Routing (LSR) option [Eriksson, 1994]. This allowed the multicast address to be hidden from intermediate, nonmulticast routers, using the existing LSR support of legacy routers in between. Unfortunately, although the LSR option was expected in these legacy systems, it was often unused and thus untested; where it was implemented correctly, it often was handled in the "slow path," in unoptimized portions of software. The result was unreliable and slow processing at legacy routers. The lesson of the early MBone [Eriksson, 1994] was that when legacy components are traversed with tunnels, the tunnels should use only the most common legacy protocol functions.

The alternative to LSR header options is encapsulation, where a full IP header is added to the packet [Perkins, 1996]. The addition of a new header means that the entire packet — not just the header — is hidden inside the body of a conventional legacy packet. This allows not only new address semantics (e.g., multicast addresses), but also entirely new header formats (e.g., IPv6) to be utilized. Legacy routers in between process these encapsulated packets as they do all other packets, using the "fast path," optimized code. As a result, legacy routers exhibit no unexpected flaws.

Tunneling using encapsulation can be somewhat inefficient [Perkins, 1996]. For small packets, the addition of an extra header (e.g., 20 bytes for IP) can nearly double the packet size. For large packets, the extra 20 bytes can necessitate fragmentation of the encapsulated packet, which needs to be reassembled at the tunnel destination. Although these additional overheads can be reduced by compression, the effect on fragmentation can remain. This can have further effects on path maximum transmission unit (MTU) discovery, resulting in MTU black holes if not handled properly.

51.4 Complex Structures

Overlays can be used in more complex ways than just as single, simple, isolated topology on top of a base network infrastructure. Multiple overlays can coexist, concurrently sharing individual components.

Overlays can be stacked on other overlays. Individual components can be visited multiple times in a single overlay, emulating larger networks. In many ways, this advanced complexity of network virtualization parallels the impact of memory virtualization (i.e., VM) on computing.

Running multiple overlays concurrently requires particular attention to address separation and multihoming [Touch and Hotz, 1998]. Individual components can participate in multiple overlays at the same time only if their forwarding and demuxing is capable of keeping traffic of the different overlays separate. Applications need to be able to attach to specific overlays, or to be restricted from attaching except to a single overlay. Support for resource reservation is generally useful, but network interfaces, routing tables, and forwarding capacity must be strictly allocated to ensure partitioning of overlay traffic, as well as to avoid deadlock in provisioning and configuring attachments to different overlays.

Overlays may be deployed on the base infrastructure or layered or stacked recursively on top of other overlays, called recursive or stacked overlays [Campbell et al., 1999; Touch and Hotz, 2001]. Overlays on other overlays can be used to compartmentalize projects or to provide layers of abstraction. The former occurs when a consortium creates a wide-area overlay to allocate resources, as for a testbed. Individual projects would be deployed as overlays on that testbed overlay to allocate resources for experiments only from the testbed resources. The latter occurs when general experiments have component experiments, such as when a hypercube-based peer overlay topology includes small groups of rings. Each ring would be represented as a single node in the hypercube. The rings would support small, tightly-coupled groups of servers for scalable processing capability.

In principle, the networking provided by an overlay should enable overlays to be deployed on top of each other, but in practice there are a number of specific challenges [Touch and Hotz, 2001]. First, the addressing and routing of different layers must be kept distinct, whether using partitioned address spaces or separate VPN identifiers [Fox and Gleeson, 1999]. Second, the network service provided by an overlay must be complete, including overlay versions of infrastructure-layer services such as the DNS and address resolution (ARP), among others. Currently, network protocols are typically layered on link protocols, and a number of these glue protocols are provided (DHCP, ARP, BOOTP), some of which are not prepared to use network layer addresses as link addresses.

Revisitation requires particular attention to the different properties of virtual network and virtual link layers [Touch and Hotz, 2001; Guruprasad et al., 2003]. Specifically, it requires two layers of tunneling in order to differentiate between different "visits" to the same component. The link tunnels need to support the "strong host" model, in which incoming packets are filtered against the address of the incoming interfaces. Network tunnels need to support the "weak host" model in which incoming packets are matched to the address of any interface on the node. In this latter case, a node's interfaces must be partitioned into groups that correspond to the different instances on each overlay, e.g., as supported by Clonable Network Stacks, in order to keep traffic on different overlays separate.

For example, consider two overlays, X and Y, sharing the same nodes as a base network, A and a third overlay, Z, which also uses A but does not use the same nodes of A as either X or Y. Each node participating in X, Y, and A must maintain sufficient information to keep forwarding and routing entries of the different networks distinct. This is accomplished either by keeping different address spaces on these three networks, or by adding a separate VPN identifier (VPN-ID [Fox and Gleeson, 1999]), which effectively augments the address space and has the same effect. Addresses in X and Y must differ from each other as well as those of A, and addresses in Z must differ from those of A. Note, however, that addresses in Z need not be distinct from either X or Y; tunneling will ensure traffic separation, provided the addresses are not shared at common components.

This address separation issue is a key aspect of multihoming. A multihomed host has more than one network address, i.e., more than one interface (real, or virtual as for a tunnel). Applications in the host need to determine which interface outgoing traffic should use; this is often determined by the forwarding rules inside the host [Braden, 1989]. Routers are already multihomed, forwarding traffic between interfaces anyway. However, multihoming for routers usually means a router is a member of more than one network, as defined by how its routing protocols share those interfaces. This occurs when different groups of interfaces do not share traffic, such as when the router keeps traffic from two overlays separate from each other and from the base network.

It is useful to consider the parallelism between overlays and virtual memory (VM) [Touch et al., 2003b]. VM can support concurrent, separate address spaces for different processes; there can be multiple concurrent overlays using the same nodes. A VM can be run on another VM, e.g., to emulate one OS inside another (VMware on Windows is one example); an overlay can be run on top of another overlay. Finally, VM can present a process with a larger virtual address space than the physical memory provides; an overlay can "revisit" nodes to present a larger virtual topology than exists physically.

51.5 Variations

Overlays go by a variety of names, including virtual networks. Some networks can be considered special cases of overlays, e.g., VPNs and peer-to-peer networks. Other variants address whether an overlay is manually configured or automatic, or whether it exists in the core of the backbone of the network or end-to-end, out to the edges.

51.5.1 Manual Overlays

Manual overlays are typically used to support distributed testbeds for new protocols, and to interconnect early adopters or experiment participants. The MBone [Eriksson, 1994] interconnects islands of multicast; the 6Bone interconnects islands of IPv6 (vs. IPv4, the current IP in common use). The ABone [Braden and Ricciulli, 1999] supported experiments in Active Networks [Tennenhouse et al., 1997], in which packets are processed by routers, rather than just being forwarded. The Q-Bone supported experiments in quality-of-service (QoS).

The MBone [Eriksson, 1994] is an early and probably the most notable manually configured overlay. Software on a subset of routers, e.g., *mrouted* on Unix PC routers, is configured with the IP address of tunnel endpoints. The tunnels of the overlay enable multicast-capable routers to exchange packets with new formats across paths through routers that would not understand those formats. Other manual overlays used a variety of tunneling mechanisms, including UDP encapsulation for the ABone [Braden and Ricciulli, 1999] and IPv6 in IPv4 tunnels for the 6-Bone, or options that support new address ranges, e.g., the VPN identifier (VPN-ID) [Fox and Gleeson, 1999].

51.5.2 Automated Overlays

Automation enables overlays to be deployed and managed more efficiently. Graphical interfaces on automation provides "do-what-I-mean" (DWIM) configuration [Touch, 1998], and the automation itself ensures that overlays are deployed correctly and without interference with each other or the base network. Other forms of automation enable remote users to link into existing overlays or even private networks using tunnels, e.g., as with Remote Access Servers (RASs) for VPNs [Scott et al., 1998].

One of the more challenging aspects of overlays is management. Manual overlays are achieved by explicitly configuring remote components, usually via secure remote logins (e.g., SSH). However, when a component becomes disconnected, such as by a physical link failure or misconfiguration, it is impossible to access the remote node to correct or adjust it, or even to clean up its configuration to remove parameters of overlays no longer accessible. Automated management can ensure that only valid configurations are used, that backup configurations are tried when primary configurations fail, and that the vestiges of old overlays are removed when no longer pertinent. There are a variety of automated overlay systems, including those that self-configure the overlays themselves (peer-to-peer nets [Oram, 2001]), those enabling individual remote participants to join an overlay in progress (VPN RASs), and those providing DWIM interfaces and coordinating multiple overlays (X-Bone [Touch, 1998; Touch and Hotz, 2001]).

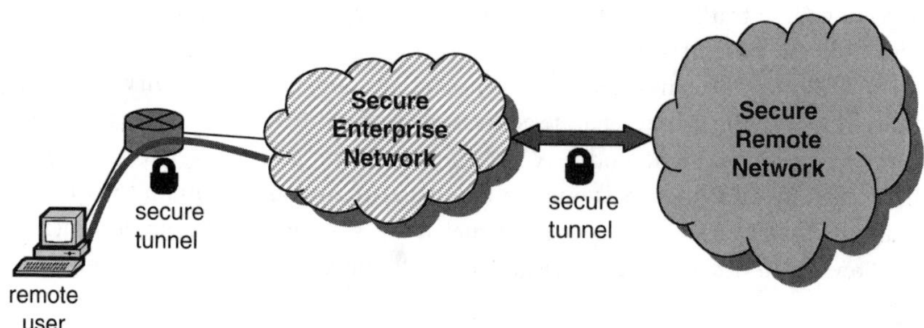

FIGURE 51.2 VPN using secure tunnels.

51.5.3 VPN

A VPN is a particular subset of overlay networks, notably whose links are secure [Scott et al., 1998]. VPNs often are deployed as partial networks, tethering single hosts back to a home network (left, Figure 51.2), or interconnecting two office networks over a public Internet (right, Figure 51.2). VPNs also tend to tie into existing naming at one of the original underlying networks rather than proving their own naming and addressing. In a sense, a VPN is a way of extending private, protected infrastructure over the public Internet.

VPNs are usually utilized differently from the more generic overlay networks on which they are based. Besides usually assuming an existing, secure network (center Figure 51.2), they also tend to limit participation. An end host (left, Figure 51.2) or other remote network (right, Figure 51.2) generally connects to only one VPN at a time. VPNs, besides using encrypted tunnels, also generally have strict boundary firewalls to ensure that external traffic does not leak into the VPN, and VPN traffic does not leak out onto the external network.

51.5.4 Provider-Provisioned vs. End-to-End

End-to-end overlays were shown in Figure 51.1, including tunnels that connect the end hosts as well as those connecting router hops in the middle. A Provider-Provisioned overlay or VPN (i.e., PPVPN [L3VPN], as they can be called) combines aspects of a VPN with that of an overlay. Edge hosts connect to RAS-like access points at the boundary of the VPN, as in a conventional VPN [Scott et al., 1998]. The VPN itself is composed of an overlay over a nonsecure public network, rather than being a private enterprise network (as with a conventional VPN). This combined structure is shown in Figure 51.3.

The access points of a PPVN may be split across the boundary of the overlay provider and that of the customer, and, as with many telephony technologies, there are a variety of names by which they are called, e.g., customer-premises access points, or provider-premises access points. However, regardless of where the box is placed or how its functionality is split, it basically acts as a RAS, connecting remote hosts via secure tunnels to an existing core infrastructure.

51.5.5 Peer-to-Peer

Peer-to-peer networks combine the virtualization of topology provided by an overlay network with application-based forwarding [Oram, 2001]. Rather than using existing network services, e.g., those based on IP addresses and nearest-hop metrics, peer networks allow application code at the routing points to determine where to forward packets, as shown in Figure 51.4. Routers in peer networks tend to be conventional end hosts, rather than core network routers as with network-level overlays and VPNs.

Peer networks network services tend to focus on distributed indexing for shared storage and retrieval. Recent examples include Napster, Gnutella, and Kazaa, used to share files among PC users. Other application-layer forwarding mechanisms have supported user-level versions of network services, such as multicast in Yallcast [Francis, 1999].

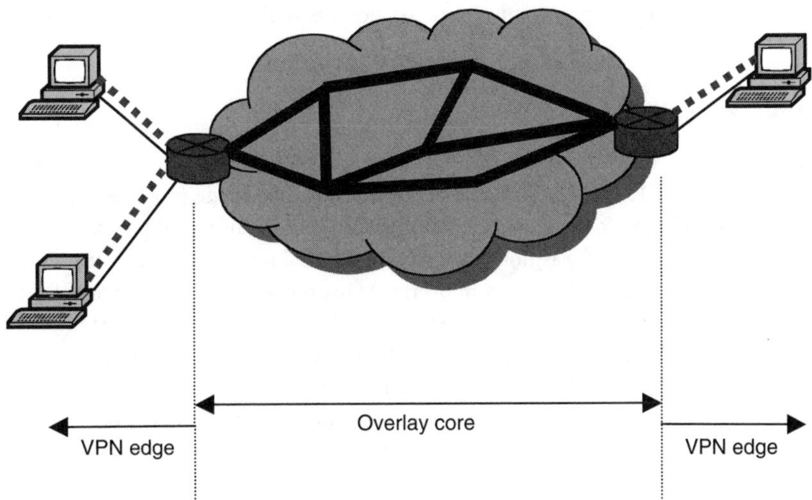

FIGURE 51.3 Provider-provisioned network; VPN at the edge, overlay inside the core.

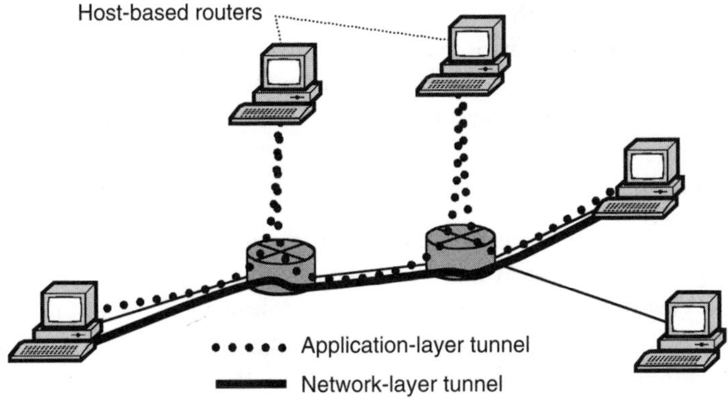

FIGURE 51.4 Application vs. network overlays.

By using applications for forwarding, combined with automated discovery and configuration protocols, peer networks are often self-configuring and self-managing. Some, such as Napster, used a central server to discover possible peers, whereas others, such as Gnutella, utilize a more distributed mechanism. Their self-organizing property is not without concern, especially because these systems are often used to share commercial software or music files illegally.

Peer networks demonstrate the power of overlays in that they can create instant virtual infrastructure even in the hands of casual users. However, they can be very inefficient, copying data many times between the OS and application space, compared to network-layer forwarding. They also recapitulate the design of the base network, often rediscovering issues in routing protocol stability because they need to reimplement the whole of network services at the application layer. Finally, because each peer system runs in its own application space, it can be difficult (if not impossible) to integrate multiple peer systems to create new services.

51.6 Applications

Overlays virtualize network infrastructure. Like their virtual memory counterparts, overlays allow different networks to operate concurrently and independently, keeping their resources isolated. Those properties enable incremental deployment of new services, as well as experiments to coexist with operational use.

51.6.1 VPNs/Security

As noted earlier, overlays can provide secure networking, or VPNs, using paths of a public, insecure infrastructure [Scott et al., 1998]. The links of the network can be made more secure, even more fault tolerant, for example, by using forward error correction or multiple dispersed paths. An overlay can also provide dynamic routing where it was not supported in the underlying base infrastructure.

The use of multiple dispersed paths allows an overlay to use more diverse routing than the base network on which it is built. An overlay can send packets over different tunnels, each terminated at different destination routers. Using different tunnels with different endpoints results in path diversity, which supports dynamic routing or allows copies of individual packets to be sent over the different paths. This diversity allows the overlay to tolerate packet losses in the underlying network, or even of loss of components in that network, without disrupting traffic at the overlay level. This diversity can be augmented by using diverse routing algorithms over different paths, or different tunneling protocols or encryption algorithms, as in DynaBone [Touch et al., 2003b].

Security can involve encrypting or authenticating packets on the overlay tunnels, as well as enforcing admission control at the boundary of the overlay. Although hop-by-hop tunnel security does not directly translate into end-to-end privacy, it provides additional protection even where end systems or applications do not support it. This is why VPNs and RAS servers are often used for remote access to enterprise email and file servers; conventional protocols (e.g., SMTP, POP, and IMAP for email; FTP for files) can be used even though they transmit clear-text (unencrypted) passwords.

51.6.2 Shared Use

Like other forms of virtualization, overlays enable concurrent use of network infrastructure. In the simplest form, a single overlay operates concurrently with as well as over the base infrastructure. However, overlays allow more powerful forms of concurrency, e.g., allowing multiple concurrent virtual networks sharing the base network. Unlike other forms of networking, *sharing* in this context means *isolation*; sharing a common infrastructure for different overlays means that the overlays must be separated from each other, as well as from the base network on which they are built.

Besides supporting parallel experiments and the so-called "ships in the night" (SITN) deployment of nonoverlapping protocols and services, concurrency enables more complex paradigms [Touch, 2001]. Just as virtual machines layer virtual memory on virtual memory (e.g., VMware), overlays can be deployed on top of other overlays. Layered overlays support hidden augmented capabilities such as fault tolerance, internal dynamic routing, forward error correction, or content-specific routing, completely hidden from network endpoints. They also support revisitation — emulating a larger network by visiting components more than one time for a single overlay.

Revisitation involves reusing network components inside a single overlay. Consider a group of ten routers; overlays using just those routers could be connected in a variety of ways, but would always include 10 or less routers in each overlay. A single overlay could revisit some of the nodes, either multiple times in a row (P, P, P, then Q, then R) or could alternate the visits (P, Q, P, R, P, R, Q). This reuse allows those original 10 routers to emulate a larger topology, such as one including a hundred, or even thousands, of routers.

51.6.3 Incremental Deployment of Services

Overlays were first used to isolate experimental protocols from the operational infrastructure without necessitating dedicated links. Those links are replaced with tunnels, interconnecting only components capable of handling new protocols, where the routers were configured to keep traffic on the overlay separate from operational traffic. This allows the development of new services among components that are not directly connected but rather leverage off the existing infrastructure to support distributed deployment. Isolation was first used to deploy IP separately inside the Cronus network [MacGregor and Tappan, 1982], but later used conversely to isolate multicast IP from IP via the MBone [Eriksson, 1994].

Just as overlays can be used to deploy experimental services, they can also be used to incrementally deploy new protocols, e.g., the MBone (multicast IP), the 6-Bone (IPv6 over IPv4), or GeoNet (geographic forwarding [Finn and Touch, 2002]). Such overlays can be deployed using path-constrained IP (e.g., Loose Source Route IP, as in the early MBone [Eriksson, 1994]), or using encapsulation-based tunnels, e.g., IP in IP at the network layer [Perkins, 1996] (later MBone, 6-Bone), or at the application layer in UDP (e.g., ABone [Braden and Ricciulli, 1999]) or TCP (e.g., peer nets [Oram, 2001]).

Besides deploying new services, overlays can deploy new versions of existing services with alternate configurations. In particular, alternative routing configurations can be used to provide enhanced path connectivity (Resilient Overlay Networks {RONs} [Andersen et al., 2001], Detour [Savage et al., 1999], Secure Overlay Services {SOS} [Keromytis et al., 2002]). Alternately, the topology of the tunnels themselves can provide an abstraction of the underlying network that itself is a capability, e.g., to connect stars of Web caches in a hierarchy, or to support topology-driven services such as distributed hash-table peer indexing.

One aspect of such new services involves resource reservation. Resources at overlay routers, such as queues, processing capacity, or transcoding services, can easily be reserved and assigned to a particular overlay. Resources on particular links can also be modified. Most link parameters can be limited (e.g., bandwidth under 1 Mbps), and some can be changed (e.g., increase delay by 10 ms). Other properties can be added (e.g., forward error correction {FEC}, or reducing jitter by adding playout buffers), typically properties added at tunnel endpoints. Properties of the tunnel path typically cannot be guaranteed or reserved (e.g., bandwidth at least 1 Mbps, delay under 10 ms) unless such a path can be reserved directly in the underlying infrastructure. It is particularly difficult to support QoS in an overlay when it does not already exist in the underlying network, but that does not undermine the value of other services that can be added.

51.7 Challenges

Although overlays have been in common use for over a decade, there are a number of open research areas. Because overlays operate on preexisting infrastructure, they afford a unique opportunity where communication exists prior to, and out-of-band from, the network being managed (the overlay). Automated overlay deployment, management, and monitoring can be provided in ways not feasible for the underlying infrastructure itself. Overlays virtualize network infrastructure, so many related issues of virtualization need to be considered in this new context. Finally, overlays make additional requirements on network components and software systems, and provide new opportunities to use networks as a tool rather than just a medium.

51.7.1 Security

Overlays, especially VPNs, necessitate attention to security. Security can be enforced internal to the overlay in its tunnels, at the boundary of the overlay, and between multiple concurrent overlays when they utilize the same network components.

VPNs, as well as some other overlays, use encrypted or authenticated tunnels (or both). Either method significantly increases the size of the tunneled packet, often resulting in substantially decreased tunnel MTUs or necessitating fragmentation. Fragmentation has substantial effects on tunnel efficiency, as noted earlier. Boundary protection requires that each node in an overlay strictly enforce traffic separation, preventing overlay traffic from leaking onto the base network or onto other overlays on the same nodes, and preventing the converse traffic from leaking into that overlay.

Implementing these security mechanisms presents unique challenges. The number of keys required to support per-tunnel security scales with the number of tunnels (key pair per tunnel). Enforcing boundary protection often requires end-to-end keys throughout the overlay, which requires N^2 keys, where N is the number of endpoints. One alternative would be group keys, which enable shared use of fewer keys but requires more complex key management, especially for revocation (i.e., when a node leaves an overlay). Fortunately, boundary protection can also be achieved by the use of firewall configuration at all nodes of an overlay, where firewalling can be strictly coordinated.

As noted above, security is complicated by the incremental addition and deletion of overlay members. It is also complicated by variations in overlay configuration, especially those that support dynamic routing. Dynamic routing may interfere with some tunneling mechanisms, notably those that combine security and tunneling in a single step (e.g., tunnel mode IPsec) [Touch, 2001].

51.7.2 Support

Overlays require support from the underlying OSs of hosts and host-based routers, or from the firmware of dedicated routers, as well as from applications and routing protocols [Touch, 2001]. End hosts need to be able to select outgoing overlays and localize traffic among the overlays they share. Routers need to partition forwarding and routing tables. Both kinds of components need to be able to support tunneling, sometimes tunnels in tunnels, as well as groups of addresses on the same tunnel. All components need to be able to dedicate other resources, such as bandwidth, memory, and processing capacity, on a per-overlay basis as well.

Routers need to be able to differentiate rules for different overlays, which usually implies partitioned forwarding tables and routing algorithms. Forwarding should be able to include ingress interface as context for forwarding lookups. The same address arriving on different interfaces should be routable to different outgoing interfaces to enable address reuse and avoid tunnel configuration conflicts. Routing algorithms must be able to distinguish groups of interfaces that participate in different instances of the same routing algorithm, but which do not share information. For example, different overlays using the same router may want to use OSPF; the OSPF algorithm should keep the information of these different overlays distinct, both to enforce traffic separation as well as to reduce the unnecessary overhead of the local computations.

All components must be able to support many concurrent tunnels, as well as tunnels on tunnels. Some systems routinely use three or more layers of tunnels, as well as hundreds of concurrent tunnels. Sometimes tunnels must be achieved by aliasing, where hundreds of addresses map to the same tunnel interface. Both routers and hosts need to support a variety of multihoming styles, e.g., both "weak" and "strong" in the same network stacks.

These capabilities require particular attention to the virtualization of resources inside a node. Each overlay requires separate virtual interfaces, and the address allocations must be coordinated to avoid overlap where overlays share nodes in common. Processes in a particular overlay need to be protected from processes in other overlays, as well as need to have limited access to only their own network interfaces — e.g., as provided by FreeBSD Jails and Linux VServers.

Many network services are deployed on end systems using wildcard addresses (for the Internet, e.g., *INADDR_ANY*). When a host or router is a member of more than one network concurrently (base infrastructure and overlay, or multiple overlays as well), such wildcards may not have the intended operation. In other cases, the use of wildcards is desired within the interfaces of a virtual router, e.g., for routing daemons, but there must be provisions for using a wildcard for each virtual router on a real router. This is further complicated by the layering of overlays on each other.

The issue of wildcards touches briefly on the general issue of naming and addressing, which affects host OSs as well. Overlays require a unique name and address space to distinguish the identities (IDs) of overlay endpoints from those of the base infrastructure. This can be accomplished by extending the ID space of the base infrastructure (e.g., the VPN ID [Fox and Gleeson, 1999]), or by using a reserved portion of the base infrastructure ID space (e.g., reserved IP addresses). Overlay IDs need not be globally unique, however; they can be reused. It is only required that reused overlay IDs do not utilize the same host or router components, i.e., that all overlays using virtual routers on a single router do not have overlapping ID spaces. Management of overlay ID space presents unique opportunities in this regard, but also presents substantial challenges, notably when overlays are merged or split, or when virtual components are relocated.

51.7.3 Deployment and Management

Overlays are deployed on existing infrastructure and so can utilize that existing network to discover and configure overlay components, as well as perform ongoing monitoring. These capabilities can rely on out-of-band communication, using the infrastructure level for command and control, rather than needing to rely solely on in-band mechanisms. For example, overlay components can be discovered using a network-based directory or multicast search and then configure tunnels separately, as is done in peer-to-peer networks [Oram, 2001].

The current state of overlay deployment and management, although advanced in comparison to managing the underlying infrastructure, has substantial room for growth. Support for some overlays is hand-coded into configuration files (*mrouted* for the MBone [Eriksson, 1994], *squid* for Web proxy caches), or discovered and configured by code embedded in individual applications (peer apps, other Web caching systems). Overlays provide the opportunity, given an appropriate API, to offload that effort from individual applications. This also allows integration of tuned network configurations in an overlay deployment service, rather than reinventing that capability in each application. Further, the use of network-layer overlays allows the use of network protocols, e.g., for dynamic routing or security management. This effort then need not be recapitulated (along with mistakes and the corresponding solutions already learned in network-level solutions) at the application layer.

51.8 Recommendations

There are a variety of overlay styles to meet a variety of uses and needs. VPNs are the oldest and most established overlay technology, and they provide security and enable remote users to securely access services of an enterprise network. VPNs work best when the remote user and enterprise are part of a single VPN; where multiple overlays are needed, more general network-based solutions can be more flexible. Generic overlays enable network experiments, testbeds, and the incremental deployment of new protocols and services, but are themselves more experimental in nature. Peer-to-peer networks are very useful in supporting distributed indexing services, especially where end host resources are plentiful, but can be inefficient and difficult to integrate with each other and other styles of overlays.

There are several generic recommendations to support overlay networks, regardless of what style of overlay is used. Network support should assume thousands of tunnels, thousands of aliases, and support forwarding partitions. Applications should support overlays by assuming that multiple instances can run on a single node, i.e., such that all configuration information is instance-specific.

Finally, when selecting an overlay technology, it may be as important to consider the effort in configuring, managing, and dismantling the overlay. Networks are notoriously difficult to configure and debug, and overlays compound that complexity. Fortunately, because they operate over existing networks, overlays provide a unique opportunity for automation — an opportunity that should be considered in any deployed solution.

Acknowledgments

Effort partially sponsored by the Defense Advanced Research Projects Agency (DARPA) and Air Force Research Laboratory, Air Force Material Command, USAF, under agreement number F30602-01-2-0529 entitled "DynaBone." The views and conclusions contained herein are those of the authors and should not be interpreted as necessarily representing the official policies or endorsements, either expressed or implied, of DARPA, the Air Force Research Laboratory, or the U.S. Government.

This material is also based upon work supported by the National Science Foundation under Grants No. NETFS: ANI-0129689 and STI-XTEND: ANI-0230789. Any opinions, findings, and conclusions or recommendations expressed in this material are those of the authors and do not necessarily reflect the views of the National Science Foundation.

References

Andersen, David G., Hari Balakrishnan, M. Frans Kaashoek, M. Robert Morris, Resilient Overlay Networks, Proceedings of the 18th ACM Symposium on Operating Systems Principles (SOSP), Oct. 2001, pp. 131–145.

Braden, Robert, Ed., Requirements for Internet Hosts — Application and Support, RFC-1123, Oct. 1989.

Braden, Robert, Livio Ricciulli, L., A Plan for a Scalable ABone — A Modest Proposal, July 1999. ftp://ftp.isi.edu/pub/braden/ActiveNets/ABone.whpaper.ps.

Campbell, Andrew T., Michael E. Kounavis, Daniel A. Villela, John B. Vicente, Hermann G. De Meer, Kazuho Miki, Kalai S. Kalaichelvan, Spawning Networks, *IEEE Network,* July/August 1999, pp. 16–29.

Delgrossi, Luca, Domenico Ferrari, A Virtual Network Service for Integrated-Services Internetworks, 7th International Workshop on Network and OS Support for Digital Audio and Video (NOSSDAV), May 1997.

Eriksson, Hans, MBone: The multicast backbone, *Communications of the ACM,* August 1994, pp. 54–60.

Finn, Gregory G., Joseph D. Touch, Network Construction and Routing in Geographic Overlays, ISI Technical Report ISI-TR-2002-564, July 2002.

Fox, Barbara, Bryan Gleeson, Virtual Private Networks Identifier, RFC-2685, September 1999.

Francis, Paul, Yallcast: Extending the Internet Multicast Architecture, Technical report, NTT Information Sharing Platform Laboratories, September 1999. http://www.yallcast.com/.

Guruprasad, Shashi, Leigh Stoller, Mike Hibler, Jay Lepreau, Scaling Network Emulation with Multiplexed Virtual Resources, SIGCOMM 2003 Poster Abstract, August 2003.

Keromytis, Angelos D., Vishal Misra, Dan Rubenstein, SOS: Secure Overlay Services, Proceedings of ACM SIGCOMM, Pittsburgh, PA, August 2002, pp. 61-72.

L3VPN WG in the IETF, http://www.ietf.org/html.charters/13vpn-charter.html.

MacGregor, William I., Daniel Tappan, The Cronus Virtual Local Network, RFC-824, August 1982.

Oram, Andy, Ed., *Peer-To-Peer: Harnessing the Power of Disruptive Technologies*, O'Reilly & Associates, Sebastopol, CA, 2001.

Perkins, Charles E., IP Encapsulation within IP, RFC-2003, October 1996.

Peterson, Larry, Tom Anderson, David Culler, Timothy Roscoe, A Blueprint for introducing disruptive technology into the internet, *ACM Computer Communications Review,* January 2003, pp. 59–64.

Savage, Stefan, Tom Anderson, Amit Aggarwal, David Becker, Neal Cardwell, Andy Collins, Eric Hoffman, John Snell, Amin Vahdat, Geoff Voelker, John Zahorjan, Detour: a case for informed internet routing and transport, *IEEE Micro*, 19(1): 50–59, January 1999.

Scott, Charlie, Paul Wolfe, Mike Erwin, *Virtual Private Networks*, O'Reilly & Associates, Sebastapol, CA, 1998.

Tennenhouse, David L., Jonathan M. Smith, W. David Sincoskie, David J. Wetherall, Gary J. Minden, A Survey of Active Network Research, *IEEE Communications Magazine*, January 1997, pp. 80–86.

Touch, Joe, Steve Hotz, The X-Bone, Proceedings of the Third Global Internet Mini-Conference at Globecom'98 Sydney, Australia November. 8–12, 1998, pp. 59–68 (pp. 44–52 of the mini-conference).

Touch, Joe, Dynamic internet overlay deployment and management using the X-Bone, *Computer Networks,* July 2001, pp. 117–135. An earlier version appeared in Proc. ICNP 2000, Osaka, pp. 59-68.

Touch, Joseph D., Gregory G. Finn, Yu-Shun Wang, Lars Eggert, DynaBone: Dynamic Defense Using Multi-layer Internet Overlays, Proceedings of the 3rd DARPA Information Survivability Conference and Exposition (DISCEX-III), Washington, D.C., USA, April 22–24, 2003, Vol. 2, pp. 271–276.

Touch, Joseph D., Yu-Shun Wang, Lars Eggert, Gregory G. Finn, Virtual Internet Architecture, ISI Technical Report ISI-TR-2003-570, March 2003b.

Yemini, Yechiam, Sushil da Silva, Towards Programmable Networks, IFIP/IEEE International Workshop on Distributed Systems: Operations and Management, L'Aquila, Italy, October 1996.

52

Network and Service Management

CONTENTS

Abstract 52-2
52.1 Introduction 52-2
52.2 Management Standards 52-2
52.3 Example of a Managed Network 52-3
52.4 SNMP 52-5
52.5 SNMPv1 52-7
52.5.1 Organization 52-7
52.5.2 Architecture 52-8
52.5.3 Information 52-10
52.5.4 Communication 52-13
52.6 SNMPv2 52-13
52.7 SNMPv3 52-14
52.8 RMON 52-15
52.9 Enterprise Network and Service Management 52-15
52.9.1 Fault Management 52-15
52.9.2 Configuration Management 52-16
52.9.3 Performance Management 52-16
52.9.4 Security Management 52-16
52.9.5 Account Management 52-17
52.9.6 OSS/Network Management System 52-17
52.10 Virtual Private Network 52-17
52.11 Broadband Access Networks 52-18
52.11.1 Cable Access Network Management 52-19
52.11.2 Digital Subscriber Line Management 52-19
52.11.3 Fixed Wireless 52-20
52.11.4 Wireless LAN 52-20
52.12 Future Trends and Challenges 52-20
52.13 QoS Management 52-21
52.14 Mobile and Enterprise Wireless Management 52-21
52.15 Customer Premises Network Management 52-21
52.16 Integrated Service Management 52-23
Acknowledgments 52-23
References 52-23
Further Information 52-28

Mani Subramanian

1-58488-381-2/05/$0.00+$1.50

© 2005 by CRC Press LLC

Abstract

Key concepts and challenges in managing the expanding enterprise Internet is the topic covered in this chapter. The history of Internet network management and management standards are discussed, followed by a detailed treatement of the three versions of simple network management protocol (SNMP). Five categories of OSI (Open System Interface) standard management application functions are addressed and compared to the traditional telecommunications operation. Management of virtual private networks deployed by enterprises have the added complexity of security and privacy. Broadband services are emerging at a rapid rate, and the management of broadband access networks is addressed in detail. The chapter ends with consideration of future trends in network and service management including end-to-end QoS and service level agreement (SLA), as well as operations support systems (OSSs), which enable service providers to network service in a cost-effective manner.

52.1 Introduction

The subject of network management has gone beyond just managing the network, with Internet now playing a significant role in the daily activities of common people, not just technical professionals. Internet is now reaching residences, small-office-home-offices (SOHO), and small and medium enterprises (SME) using applications such as email, Web-based services, and broadband services that carry voice, video, and data on the same medium. Such a proliferation of Internet has clouded the clear distinction between management of enterprise and customer premises networks, as well as between management of computer and telecommunications networks. This produces new challenges to the operation, administration, maintenance, and provisioning (OAMP) of networks.

This chapter addresses the key concepts and challenges in managing the expanding enterprise Internet. We will briefly discuss the history of Internet network management and the management standards in Section 52.2. We present in Section 52.3 an example of centralized network management that illustrates the power of remote network management. Section 52.4 reviews the three versions of simple network management protocol (SNMP). In Section 52.8, we discuss the management application functions — fault, configuration, account, performance, and security (FCAPS) — that enable service providers and enterprise network managers to perform OAMP. Section 52.10 addresses the special case of management of virtual private networks. Subscribers to and users of broadband services who are accustomed to the reliability and quality of service (QoS) in telephone network expect the same performance in the integrated Internet from telecommunications and computer communications service providers. Management of emerging broadband access networks is discussed in Section 52.11.

Section 52.12 addresses future trends in network and service management. QoS and service level agreement (SLA) play a significant role in the end-to-end network performance. The management issues associated with the mobile wireless networks add extra dimensions to network management. We will look at the protocols that have been developed but not deployed universally to accomplish end-to-end QoS with embedded active network management using dynamic service provisioning. The remote management of customer premises network is in embryonic stage. We cover the subject of operations support systems (OSSs), which are support systems that perform the functions of OAMP in a cost-effective manner.

For easy reading, the convention followed in this chapter is to indicate the commands and messages in sans serif and the managed objects in italics. There is an extensive bibliography given at the end of the chapter, most of which are not referenced in the text.

52.2 Management Standards

There are several management standards in use today. They are Open System Interface (OSI) model and Telecommunications Management Network (TMN), both developed by International Standards Organization (ISO); Internet model developed by Internet Engineering Task Force; and Web-based management. OSI model uses common management information protocol (CMIP) and common management information services (CMIS). TMN defines the functions of five layers of management, namely element, element management, network management, service management, and business management. Web-based management uses Web technology to implement different management protocols.

Simple Network Management Protocol (SNMP), the Internet model, is the one that is relevant to Internet network management. It is an application layer protocol primarily based on TCP/IP (transport control protocol/Internet protocol) protocol suite and IEEE Ethernet LAN protocol, IEEE 802.3. It has been adopted as the standard by the Internet Engineering Task Force (IETF), the organization that has the responsibility for developing Internet protocol development and standardization.

Although no formal architectural network model was developed for SNMP, it is beneficial to use the OSI network management model shown in Figure 52.1, which is ISO standard and is complete. The four components of the model are organization, information, communication, and functional model.

The organization model describes the components of a network management system, their functions, and their infrastructure. The information model deals with the structure and organization of management information. The former is concerned with the structure of management information (SMI) and specifies how the management information is structured. The latter defines the management information base (MIB), which deals with the relationship and storage of management objects. The communication model is concerned with the transactions between the components. The functional model deals with the user-oriented applications of network management.

There are five functional application areas defined in OSI, namely fault, configuration, accounting, performance, and security. These are referred to by the acronym FCAPS. OAMP, as defined by the telecommunications industry, could be correlated to FCAPS and the functions defined by TMN. Operations encompass fault, performance, and security, element, and network management. Administration addresses account, service, and business management. Maintenance is installation and maintenance of network. Provisioning defines configuration management of network and network elements.

52.3 Example of a Managed Network

Let us now illustrate some of the results of a network management system remotely monitoring a subnetwork using a commercial network management system (NMS) [110]. Figure 52.2 shows a managed LAN that was discovered by a network management system. We show here only a subnetwork of a larger network managed by the NMS. An NMS can automatically discover any component in the network as long as the component has a management agent, which is defined in the next section. The management agent could be as simple as a TCP/IP suite that responds to a ping by NMS. However, the agents in the modern network components are more sophisticated.

The managed subnetwork that we are discussing here is an Ethernet LAN that is shown below the backbone cloud in Figure 52.2. It consists of a router and two hubs and is connected to the backbone network. The LAN IP address is 172.16.46.1, and the two hub addresses have been configured as 172.16.46.2 and 172.16.46.3. The LAN IP address, 172.16.46.1, is the address assigned to the interface

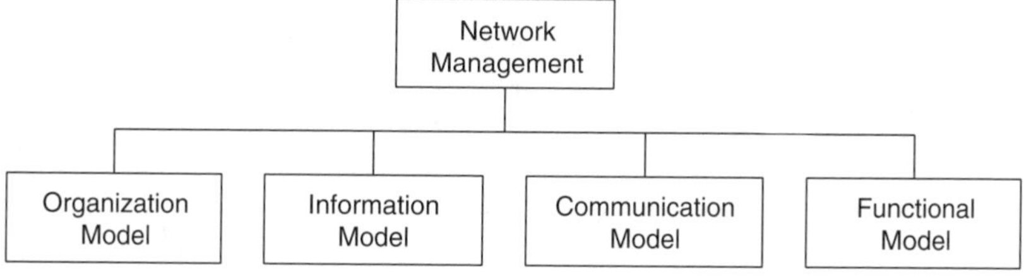

FIGURE 52.1 OSI network management model.

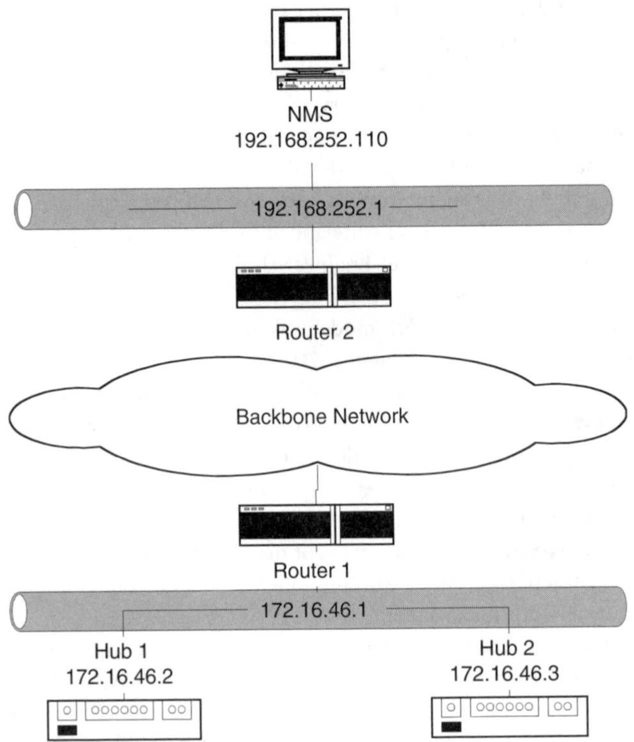

FIGURE 52.2 A managed LAN network.

card in the router. The interface cards in the router and the interface card in each of the hubs are connected by a cat5 cable forming the Ethernet LAN.

The network management system, whose IP address is 192.168.252.110, is physically and logically located remotely from the 172.16.46.1 LAN. It is configured on the LAN 192.168.252.1 and is connected to the backbone network. Information system managers establish conventions to designate a network and a subnetwork. A 0 in the fourth decimal position of an IP address designates a network, and a subnetwork is designated with a 1 in the fourth position of the dotted decimal notation. Thus, 172.16.46.1 is a LAN subnetwork in the network 172.16.46.0.

Once the network components have been discovered and mapped by the NMS, we can query and acquire information on system parameters and statistics on the network elements. Figure 52.3 presents the system information on the three network elements in the managed LAN gathered by the NMS sending specific queries asking for the system parameters.

Figure 52.3 a shows that the network element is designated by 172.16.46.2. No specific title or name has been assigned to it. System description indicates that it is a hub made by 3Com vendor, with its model and software version. It also gives the system object ID and how long the system has been up without failure. The format of the System ID refers to 3Com ID and will become clear in the next section. The *System Up Time* indicates that the system has been operating without failure for over 286 days. The number in parenthesis is in SNMP units of one hundredths of a second. Thus, the hub designated by the IP address 172.16.46.2 has been up for 2,475,380,437 hundredths of a second, or for 286 d, 12 h, 3 min, and 24.37 sec. System Description and System Object ID are factory set and the rest are user settable.

Figure 52.3b shows similar parameters for the second hub, 172.16.46.3, on the LAN. Figure 52.3c presents the system information sent by the router on the network to the NMS's queries. The system name for the router has been configured and hence the query received the response of the name, router1.gatech.edu. We could also obtain information on all protocol layers of these elements.

```
Name or IP Address:  172.16.46.2

System Name           :
System Description    :  3Com LinkBuilder FMS, SW version:3.02
System Contact        :
System Location       :
System Object ID      :  .iso.org.dod.internet.private.enterprises.43.1.8.5
System Up Time        :  (2475380437)  286 days, 12:03:24.37

                              (a) System Information on 172.16.46.2 Hub
```

```
Title:  System Information:  172.16.46.3
Name or IP Address:  172.16.46.3

System Name           :
System Description    :  3 Com LinkBuilder FMS, SW version: 3.12
System Contact        :
System Location       :
System Object ID      :  .iso.org.dod.internet.private.enterprises.43.1.8.5
System UpTime         :  (3146735182)  364 days, 4:55:51.82
                              (b) System Information on 172.16.46.3 Hub
```

```
Title:  System Information: router1.gatech.edu
Name or IP Address:  172.16.252.1

System Name           :  router1.gatech.edu
System Description    :  Cisco Internetwork Operating System Software
                      :  IOS (tm) 7000 Software  (C7000-JS-M), Version
                      : 11.2(6),RELEASE SOFTWARE (ge1)
                      :  Copyright  (c)  1986-1997 by Cisco Systems, Inc.
                      :  Compiled Tue  06-May-97 19:11 by kuong
System Contact
System Location       :
System Object ID      :  iso.org.dod.internet.private.enterprises.cisco.ciscoProducts.
                         cisco 7000
System Up Time        :  (315131795)  36 days, 11:21:57.95
                              (c) System Information on Router
```

FIGURE 52.3 System information acquired by NMS.

52.4 SNMP

SNMP management really began in the 1970s. Internet Control Message Protocol (ICMP) was developed to manage ARPANET (Advanced Research Project Agency NETwork). It is a mechanism to transfer control messages between nodes. A popular example of this is Packet Internet Groper (PING, a.k.a. ping), which is part of the TCP/IP suite now. The ping is a very simple command that is used to investigate the health of a node and the robustness of communication with it from the source node. It started as an early form of network-monitoring tool.

ARPANET, which started in 1969, developed into the Internet in the 1980s with the advent of UNIX and the popularization of client–server architecture. With the growth of the Internet, it became essential to have the capability to remotely monitor and configure gateways. Simple Gateway Monitoring Protocol (SGMP) was developed for this purpose as an interim solution. SNMP is a further enhancement of SGMP that was developed by IETF. Even the SNMP management, referred to as SNMPv1, was intended to be another interim solution, with the long-term solution being migration to the OSI standard CMIP/CMIS. However, due to the enormous simplicity of SNMP and its extensive deployment, it has become the *de facto* management standard for Internet. SNMPv2 was developed to make it independent of OSI standard as well as adding more features. SNMPv2 has only partially overcome some of the limitations of SNMP. The final version, SNMPv2C (community-based), was released in 1996 without one of its major enhancements on a security feature due to strong differences of opinion amongst the workgroups. SNMPv3 was then developed and released in 1998 with the focus on security features.

We describe the basic principles of SNMP, primarily using SNMPv1, and follow up with a discussion of the significant enhancements in SNMPv2 and SNMPv3. The specifications of SNMP are described in Internet Request for Documents (RFC). SNMPv3 defines a structure for organizing the documents, which is shown in Figure 52.4 [RFC 2271]. Two sets of documents are of general nature. One of them is the set of documents on roadmap, applicability statement, and coexistence and transition. They are placeholders for documents yet to be written.

The other set of documents, SNMP Frameworks, comprise the three versions of SNMP. An SNMP Framework represents the integration of a set of subsystems and models. The SNMP Frameworks document set is not explicitly shown in the pictorial presentation in [RFC 2271], as we have done here; [RFC 1901] in SNMPv2 and [RFC 2271] in SNMPv3 are SNMP framework documents.

The information module and MIBs cover SMI (Structure of Management Information), textual conventions, and conformance statements, as well as various MIBs. These are covered in STD 16 and STD 17 documents along with SMIv2 documents [RFCs 1902–1904].

Message Handling and PDU (Protocol Data Unit) Handling sets of documents address transport mappings, message processing and dispatching, protocol operations, applications, and access control. [RFCs 2273–2275] address these in SNMPv3.

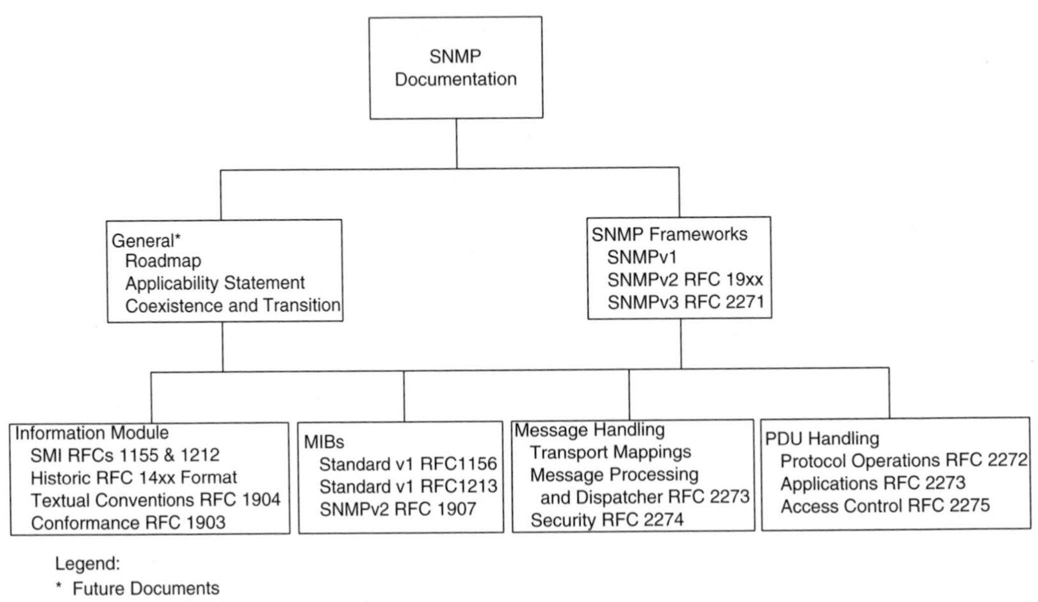

FIGURE 52.4 SNMP documentation (recommended in SNMPv3).

52.5 SNMPv1

52.5.1 Organization

The initial organization model of SNMP management is a simple two-tier model shown in Figure 52.5a. It consists of a management agent residing in each network element (managed object and manager) and a database in the manager. The agent residing in the manager manages the managed network comprising managed objects. The agent in the management object responds to any management system that communicates with it using SNMP protocol. Thus, multiple managers can interact with one agent.

In the two-tier model, the network manager receives raw data from agents and processes it. It is beneficial sometimes for the network manager to obtain preprocessed data. For example, we may want to get the temporal data of data traffic in a LAN. Instead of the network manager continuously monitoring the events remotely and calculating the statistics, an intermediate agent called RMON (Remote Monitoring) is inserted between the managed object and the network manager. This introduces a three-tier architecture as shown in Figure 52.5b. The network manager receives data from the managed objects as well as from the RMON agent about the managed objects. The three-tier architecture is also implemented in a hierarchical architecture.

The pure SNMP management system consists of SNMP agents and SNMP managers. However, an SNMP manager can manage a network element, which does not have an SNMP agent. This application occurs in many situations, such as with legacy systems management and telecommunications management networks. In these cases, they are part of an overall network that has to be managed on an integrated basis. As an example in a legacy case, we may want to manage outside plant and customer premises equipment for an HFC (hybrid fiber coax) access system in broadband services to home. There are amplifiers on the outside cable plant that do not have SNMP agents built into them. The outside cable plant uses some of the existing cable technology and has monitoring tools built into it, as for example transponders that measure the various amplifier parameters. The information from the amplifiers could be transmitted to a central (head end) location using telemetry facilities. We can have a proxy server at the central location that converts the data into a set that has SNMP compatible parameters and communicates with SNMP manager. This may be classified under the three-tier architecture.

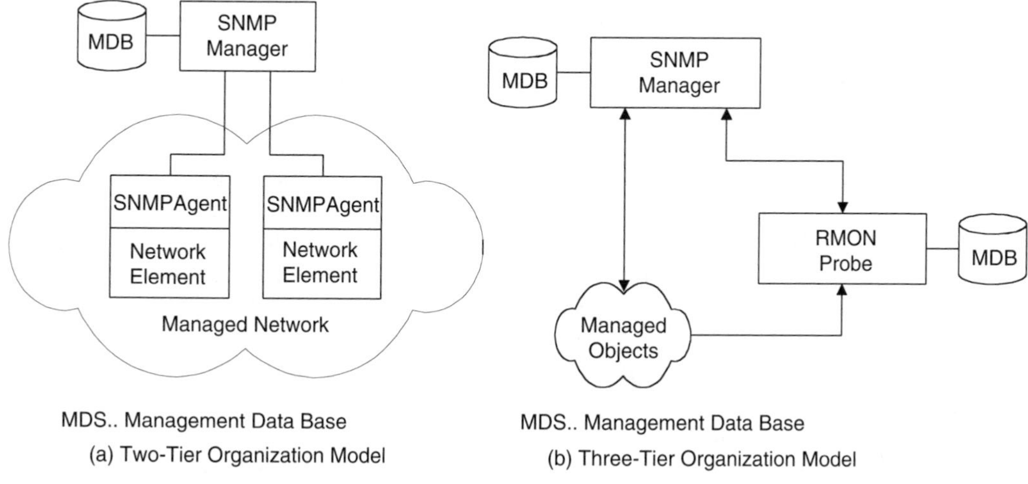

FIGURE 52.5 Organization model.

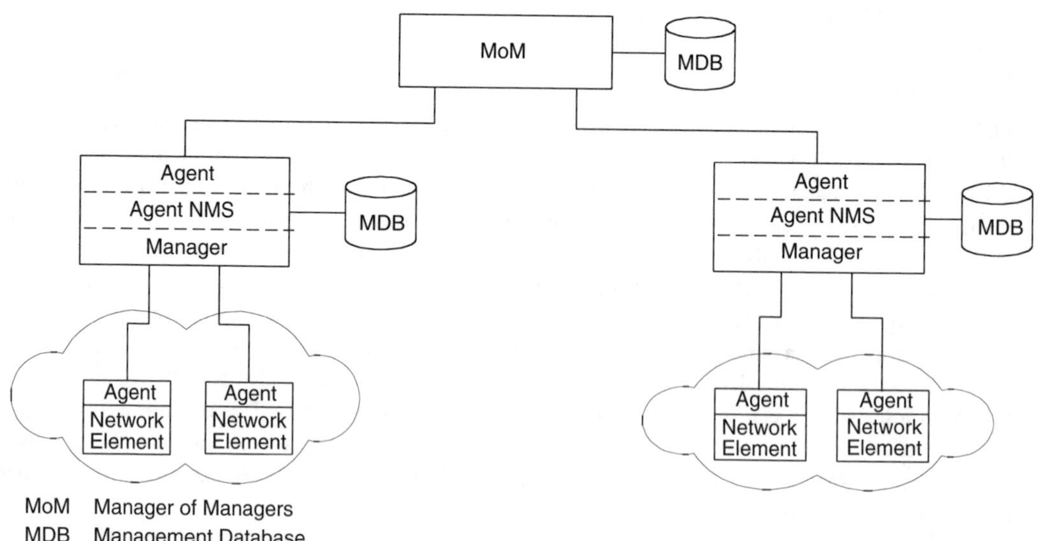

FIGURE 52.6 Network management organization model with MoM.

A network management system can behave as an agent as well as a manager. This is similar to client–server architecture, where a host can function as both server and client. Figure 52.6 shows the architecture of an NMS behaving as manager of managers (MoM) communicating with agent NMSs, each managing its own domain. A domain could be geographical, functional, or vendor-specific NMS.

52.5.2 Architecture

Figure 52.7 shows SNMPv2 network management architecture, which is a superset of SNMPv1. It portrays the data path between the manager application process and the agent application process via the four transport function protocols — UDP, IP, DLC (Data Link Control), and PHY (Physical). Internet is only concerned with the TCP/IP suite of protocols and does not address the layers above or below it. Thus, layers 1 (physical) and 2 (data link control) in the transport layers can be anything of users' choice. In practice, SNMP interfaces to the TCP/IP with UDP as the transport layer protocol and Ethernet in DLC layer.

Figure 52.7 also shows the local transfer of SNMP data packets, called protocol data unit (PDU) between SNMP layers and application PDU (APDU) between application layers.

As the name implies, the SNMP protocol has been intentionally designed to be simple and versatile; this surely has been accomplished as indicated by its success. The communication of management information among management entities is realized through exchange of just seven protocol messages, five in SNMPv1 and two more added in SNMPv2. Four of these (get-request, get-next-request, get-bulk-request, and set-request) are initiated by manager. Two messages, get-response and SNMPv2-trap, are generated by the agent. The inform-request is a manager-to-manager message. The message generation is called an *event*. In SNMP management scheme, the manager monitors the network by polling the agents as to their status and characteristics. However, efficiency is increased by agents generating unsolicited alarm messages, i.e., traps.

The get-request message is generated by the manager requesting the value of an object. The value of an object is a scalar variable. The system group parameters shown in Figure 52.3 are single instance values and could be obtained using get-request message. They could also be acquired using the get-request on the first parameter (system Description) and then get-next-request on the rest, one at a time.

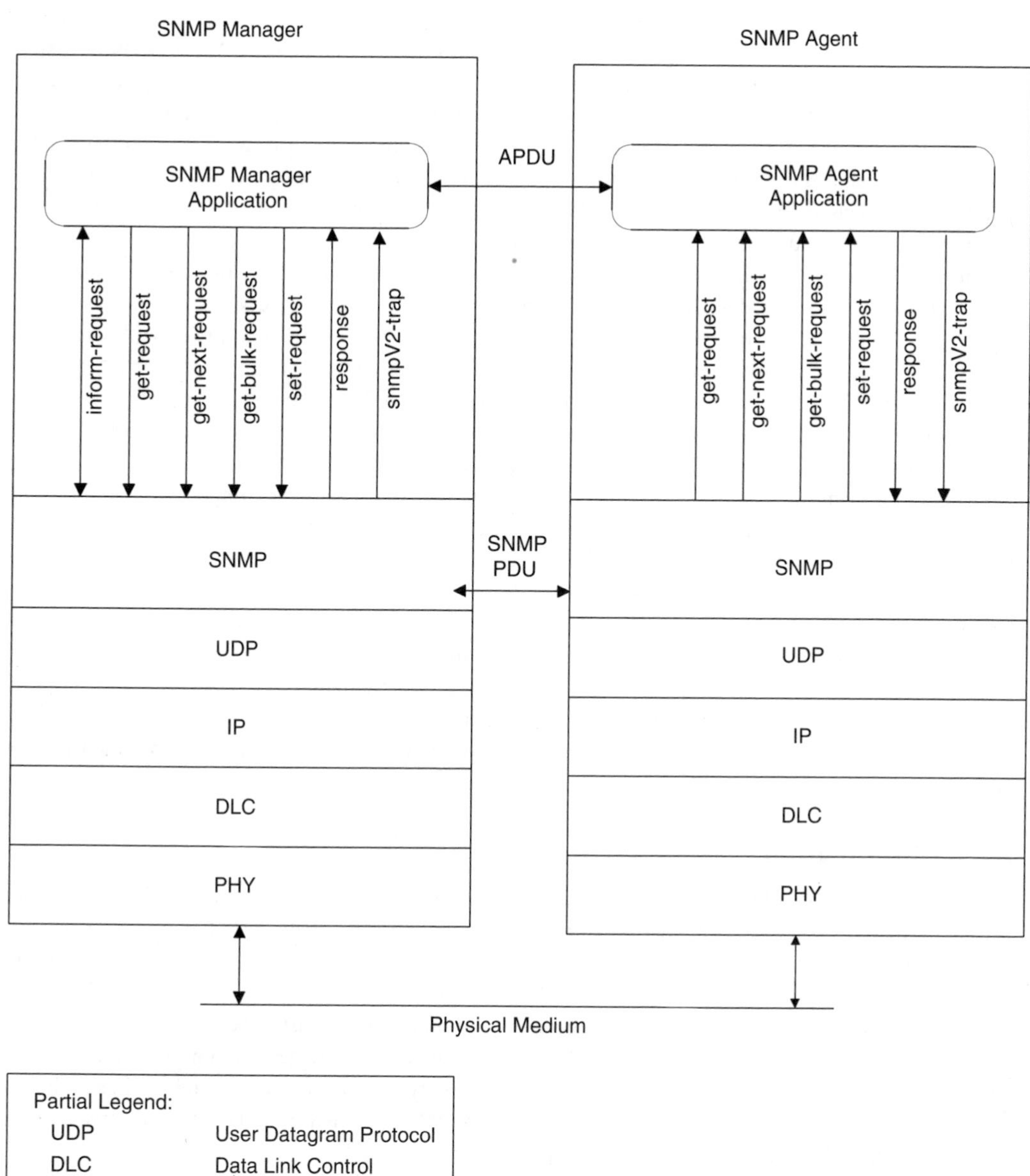

FIGURE 52.7 SNMPv2 network management architecture.

The get-next-request, simply called get-next, is very similar to get-request. In many situations, an object may have multiple values because of multiple instances of the object. For example, the routing table of a router has multiple values (instances) for each object. In such situations, the get-next-request obtains the value of the next instance of the object.

The get-bulk-request is used to transfer large amounts of data from the agent to the manager, especially if it includes tabular data such as in the routing table. The set-request is generated by the manager to initialize or reset the value of an object variable.

The get-response message is generated by the agent. It is generated on receipt of a get-request, get-next-request, get-bulk-request, or set-request message from a manager. The get-response process involves filling the value of the requested object with any success or error message associated with the response. An inform-request message by one manager to another is responded by a get-response message.

The other message that the agent generates is trap. A trap is an unsolicited message generated by an agent process without any message arriving from the manager. It occurs when it observes the occurrence of a preset parameter in the agent module. For example, a node can send traps when an interface link goes up or down. Or, if a network object has a threshold value set for a parameter, such as maximum number of packets queued up, a trap could be generated and transmitted by the agent application whenever the threshold is crossed in either direction.

The SNMP manager residing in the NMS has a database and polls the managed objects for management data. It contains two sets of data: one on the information about the objects, the Management Information Base (MIB), and a second on the values of the objects. These two are often confused with each other. MIB is a virtual data (information) base and is static. In fact, it needs to be there when an NMS discovers a new object in the network. It is compiled in the manager during the implementation. If the information about the managed object is not in the manager, it could still detect the object but would mark it as unidentifiable. This is because the discovery process involves broadcast ping command by NMS and responses to it from the network components. Thus, a newly added network component would respond if it has a TCP/IP stack that normally has built in ICMP (Internet Control Message Protocol). However, the response contains only the IP address. MIB needs to be implemented in both the manager and the agent to acquire the rest of the information, such as system group information shown in Figure 52.3.

The second database is dynamic and contains the measured values associated with the object. This is a true database. While MIB has a formalized structure, the database containing the actual values can be implemented using any database architecture chosen by the implementers.

It is worth noting in Figure 52.6 the SNMP manager has a database, which is the physical database. SNMP agent does not have a physical database. However, both have MIB, which are compiled into the software module and not shown in the figure.

52.5.3 Information

Figure 52.7 shows the information exchange between agent and manager. In a managed network, there are many managers and agents. For information to be exchanged intelligently between manager and agent processes, there has to be common understanding on both the syntax and semantics. The syntax used to describe management information is ASN.1, which is used to specify the SNMP SMI [RFC 1155]. The specifications of managed objects and the grouping of, and relationship between, managed objects are addressed in Management Information Base [RFC 1213]. There are generic objects that are defined by IETF and can be managed by any SNMP-compatible network management system. Objects that are defined by private vendors, if they conform to SMI and MIB specified by IETF RFC standards, can be managed by SNMP-compatible network management systems.

A managed object can be considered to be composed of an object type and an object instance. SMI is concerned only with the object type and not object instance. For example, Figure 52.3a and Figure 52.3b present data on two 3Com hubs. They are both identical hubs except for a minor software release difference. The object types associated with both hubs are represented by the identical object ID, iso.org.dod.internet.private.enterprises.43.1.8.5. Hub 1 with an IP address 172.16.46.2 and hub 2 with an IP address 172.16.46.3 are two instances of the object. Figure 52.8 shows managed object with multiple instances of an object type.

Object type, which is a data type, has a name, syntax, and an encoding scheme. The name is represented uniquely by a descriptor and object identifier. The syntax of an object type is defined using the abstract syntax notation ASN.1. Basic encoding rules (BER) have been adopted as the encoding scheme for transfer of data types between agent and manager processes, as well as between manager processes.

Every object type, i.e., every name, is uniquely identified by a DESCRIPTOR and an associated OBJECT IDENTIFIER. DESCRIPTOR and OBJECT IDENTIFIER are in uppercase since they are ASN.1 keywords.

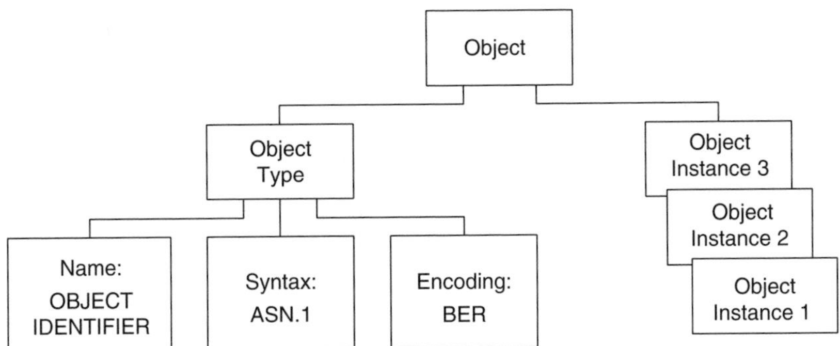

FIGURE 52.8 Managed object: type with multiple instances.

The DESCRIPTOR defining the name is mnemonic and is all in lowercase letters — or at least begins with lowercase letters, i.e., Internet object as *internet*. Since it is mnemonic and should be easily readable, uppercase letters are used as long as they are not the beginning letter. For example, the object IP address table is defined as *ipAddrTable*. OBJECT IDENTIFIER is a unique name and number in the MIB, which has a hierarchical tree structure. Thus, the managed object *internet* has its OBJECT IDENTIFIER 1.3.6.1, which indicates that it occupies the position in the MIB tree of node 1 (*iso*), then going down the tree node 3 (*org*), node 6 (*dod*), and then node 1 (*internet*).

Any object in Internet MIB will start with the prefix 1.3.6.1 or *internet*. For example, there are three management-related objects under the *internet* object, as shown in Figure 52.9. These three objects are defined as:

```
mgmt            OBJECT IDENTIFIER ::= {internet 2}
experimental    OBJECT IDENTIFIER ::= {internet 3}
private         OBJECT IDENTIFIER ::= {internet 4}
```

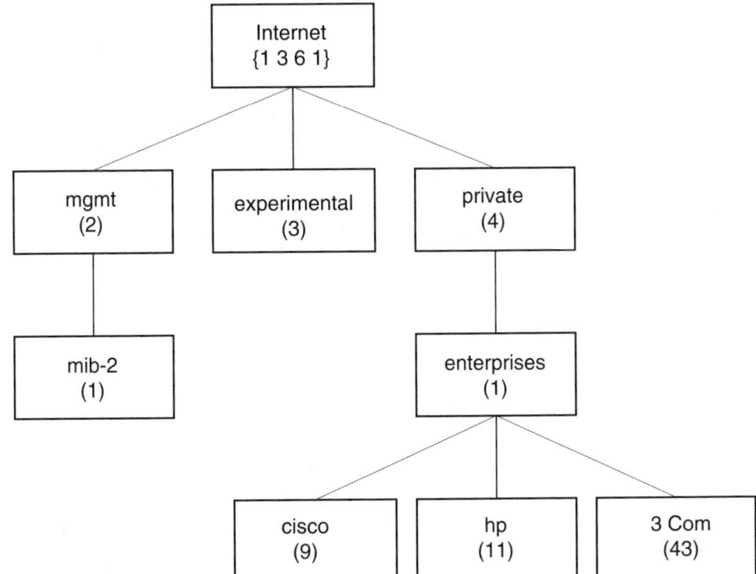

FIGURE 52.9 Management subnodes under internet node in SNMPv1.

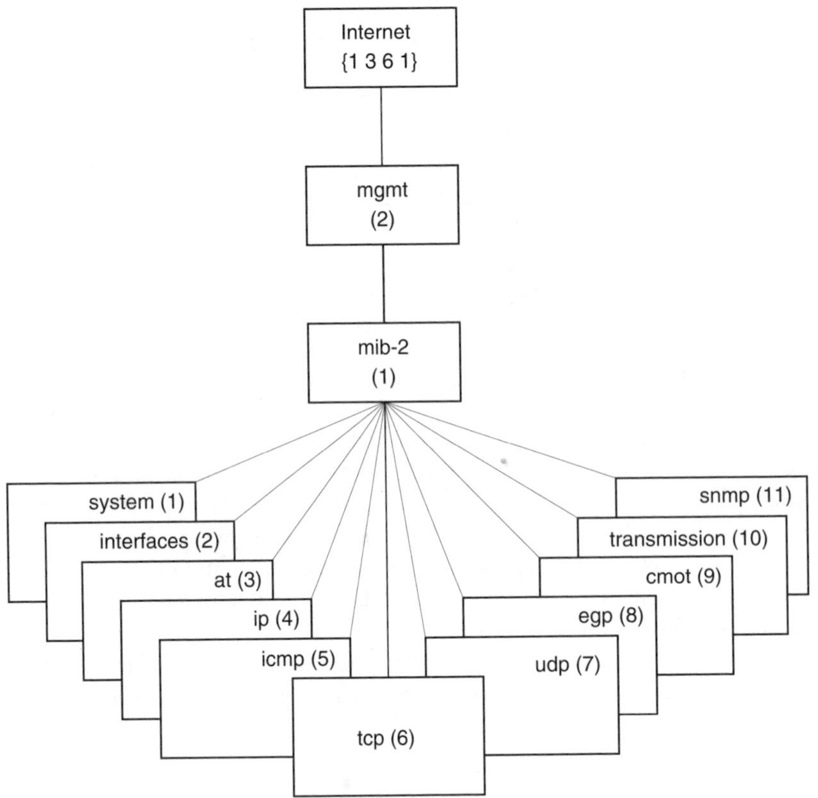

FIGURE 52.10 Internet MIB-II group.

The first line in this example states that the object, *mgmt,* is defined as node 2 under the object, *internet.* The node connected to {internet 2} through *mgmt* is *mib-2.* The *experimental* node was created to define objects under IETF experiments. The node *private* is a heavily used node, and commercial vendors can acquire a number under *enterprises.* Figure 52.9 shows an example of three commercial vendors, Cisco, HP, and 3Com, which are registered as nodes 9, 11, and 43, respectively, under *enterprises.* Nodes under any of these nodes are entirely left to the discretion of the vendors. Note in Figure 52.3, the 3Com hub has *43* following *enterprises* for system object ID and *cisco* following *enterprises* for the Cisco router.

MIB-II specified in RFC 1213 is the current standard, STD 17. It is a superset of MIB-I or simply MIB. Objects that are related are grouped into object groups.

There are eleven groups defined in MIB-II as specified in SNMPv1 and are shown as tree structure in Figure 52.10. Table 52.1 presents the name, object identification (OID), and a brief description of each group.

System group is the basic group in the Internet standard MIB. Its elements are probably one of the most accessed managed objects. After an NMS discovers all the components in a network or newly-added components in the network, it has to obtain information on the system it discovered, such as system name, object ID, etc. The NMS will initiate the get-request or get-next-request command on the objects in this group for this purpose. The data on the systems shown in Figure 52.3 were obtained by the NMS using this group. The group also has administrative information, such as contact person and physical location that helps a network manager in tracking troubles.

The Interfaces group contains layer-2 protocol information on interfaces. The Address Translation group consists of a table that converts network address to a physical address, as for example an IP to Ethernet address, for all the interfaces of the system. This table has been deprecated in SNMPv2 since the same information is available in the IP group. As can be observed from Table 52.1, each of the group

TABLE 52.1 MIB-II Groups in SNMPv1

Group	OID	Description (brief)
system	mib-2.1	System description and administrative information
interfaces	mib-2.2	Interfaces of the entity and associated information
at	mib-2.3	Address translation between IP and physical address
ip	mib-2.4	Information on IP protocol
icmp	mib-2.5	Information on ICMP protocol
tcp	mib-2.6	Information on TCP protocol
udp	mib-2.7	Information on UDP protocol
egp	mib-2.8	Information on EGP protocol
cmot	mib-2.9	Placeholder for OSI protocol
transmission	mib-2.10	Placeholder for transmission information
snmp	mib-2.11	Information on SNMP protocol

contains parameters associated with a protocol layer/protocol. We refer you to [108, 110] for detailed discussion of these groups.

52.5.4 Communication

SNMPv1 security implementation of authentication and authorization is based on exchange of SNMP information between members within a community. A pair of entities with the same common community name can communicate with each other. With the one-to-many, many-to-one, and many-to-many communication links between managers and agents, basic authentication scheme and access policy have been specified in SNMP. The authentication of requests is done using the community name. The agent authorizes the information accessed by a manager based on one of the four access modes: not-accessible, read-only, write-only, and read-write. The community names along with the access privilege constitute SNMPv1 access policy in SNMPv1 and SNMPv2. This has been enhanced greatly in SNMPv3.

52.6 SNMPv2

Several significant changes were introduced in SNMPv2. One of them was to have been to improve the security function that SNMPv1 lacked. Unfortunately, after significant effort, due to lack of consensus this was dropped from the final specifications, and SNMPv2 was released with the rest of the changes as SNMPv2C, the addition of C to indicate that the security is community based. There are significant differences between the two versions of SNMP, and unfortunately, version 2 is not backward compatible with version 1. [RFC 1908] presents implementation schemes that include proxy function for the coexistence of the two versions.

The basic components of network management in SNMPv2 are the same as version 1. They are agent and manager, both performing the same functions. The manager-to-manager communication, shown in Figure 52.7, is formalized in version 2 by adding an additional message. Thus, the organizational model in version 2 remains essentially the same.

Besides the manager-to-manager message, bulk data transfer message, get-bulk-request, was added as shown in Figure 52.7. This speeds up the get-next-request process and is especially useful to retrieve data from tables.

In SNMPv1, SMI is specified as STD 16, which is described in [RFC 1155 and 1212], along with [RFC 1215] describing traps. They have been consolidated and rewritten in [RFCs 1902–1904] for SMI in SNMPv2. [RFC 1902] deals with SMIv2, [RFC 1903] with textual conventions, and [RFC 1904] with conformances.

Many of the RFC specifications of SNMPv1 were formalized in SNMPv2 with the revision of SMI to SMIv2. The conformance and compliance statements issued were clearly defined so that the equipment vendors could exactly specify these. This helps the customer objectively compare the features of the various products. Compliance defines a minimum set of capabilities. Additional capabilities may be offered as options in the product by the vendors.

In SNMPv2, Internet node in MIB has two new sub-groups: security and snmpv2. There are significant changes to System and SNMP groups of version 1.

52.7 SNMPv3

Although SNMPv3 was developed primarily to enhance the security feature, it ended up addressing more than just security. It is a framework for all three versions of SNMP. It is designed to accommodate future development in SNMP management with minimum impact to existing management entities. A modular architecture was specified and documentation infrastructure shown in Figure 52.4 was developed [RFC 2271].

The design of the architecture integrated the SNMPv1 and SNMPv2 specifications with the newly proposed SNMPv3. This enables the continued usage of legacy SNMP entities along with SNMPv3 agents and manager. That is good news, as there are tens of thousands of SNMPv1 and SNMPv2 agents in the field. An SNMP engine is defined with explicit subsystems comprising dispatcher, message processing subsystem [RFC 2272], security subsystem [RFC 2274], and access control [RFC 2275], as shown in Figure 52.11. It manages all three versions of SNMP to coexist in a management entity. The application modules in the architecture define the message modules.

The primary feature in SNMPv3 is the improved security feature. The configuration can be set remotely with secured communication that protects against modification of information and masquerade (altering the source address) by using encryption schemes. It also tries to ensure against malicious modification of messages by reordering and time delaying of message streams, as well as protects against eavesdropping of messages.

The access policy used in SNMPv1 and SNMPv2 is continued and formalized in access control in SNMPv3, designated view-based access control model. The SNMP engine defined in the architecture

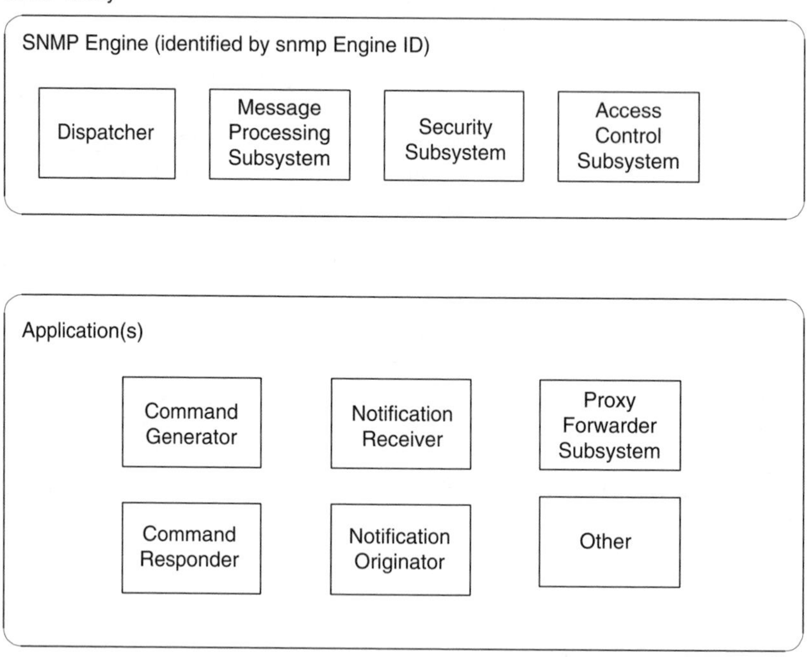

FIGURE 52.11 SNMPv3 architecture.

checks whether a specific type of access (read, write, create, notify) to a particular object (instance) is allowed.

52.8 RMON

We mentioned in Section 52.5.1 monitoring of data using remote monitoring (RMON) device (see Figure 52.5b). The use of RMON devices has several advantages. First, each RMON device monitors the local network segment and does the necessary analyses. It relays the necessary information in both solicited and unsolicited fashion to the network management system. For example, RMON could be locally polling the network elements in a segment. If it detects an abnormal condition such as heavy packet loss or excessive collisions, it would send an alarm. Because the polling is local, the information is more reliable. This example of local monitoring and reporting to a remote network management system significantly reduces SNMP traffic in the network.

There are more chances that the monitoring packets using ping, could get dropped in monitoring remotely. This could wrongly be interpreted by the network management system as the managed object being down. RMON pings locally and hence has less chance of losing packets, thus increasing the reliability of monitoring. Another advantage of local monitoring using RMON is that the individual segments can be monitored on a more continuous basis. This provides better statistics and greater ability for control. Thus, a fault could be diagnosed quicker by the RMON and reported to the NMS. In some situations, a failure could even be prevented by proactive management.

52.9 Enterprise Network and Service Management

Enterprise network and service management is concerned with the health of the network and quality of service provided by an enterprise. We define an enterprise as an organization, public or private, that provides a product or service. With the proliferation of broadband network, the domain of management has extended beyond LANs and WANs and into the broadband access networks that include wired and wireless devices and medium.

Network management applications are fault, configuration, account, performance, and security (FCAPS) management. Operations support systems including NMS perform these applications by collecting the data from managed data using SNMP and other management protocols.

52.9.1 Fault Management

Fault in a network is normally associated with failure of a network component and subsequent loss of connectivity. Fault management involves a five-step process. They are (1) fault detection, (2) fault location, (3) restoration of service, (4) identification of root cause of the problem, and (5) problem resolution. The fault should be detected as quickly as possible by the centralized network management system, preferably before or at about the same time the users notice it. Fault location involves identifying where the problem is located. We distinguish this from problem isolation, although in practice it could be the same. The reason for doing this is that it is important to restore service to the users as quickly as possible, using alternative means. The restoration of service takes a higher priority over diagnosing the problem and fixing it. However, it may not always be possible to do this.

Fault detection is accomplished using either polling scheme (network management system polling management agents periodically for status) or by generation of traps (management agents based on information from the network elements sending unsolicited alarms to the NMS). An application program generates the ping command periodically and waits for response. Connectivity is declared broken when a preset number of consecutive responses are not received. The frequency of the pinging and the preset number for failure detection may be optimized for balance between the traffic overhead and the rapidity with which failure is detected.

The alternative detection scheme is to use traps. For example, the generic trap messages *linkDown* and *egpNeighborLoss* in SNMPv1 can be set in the agents with capability to report the events to the network management system with the legitimate community name. One of the advantages of traps is that the failure detection is accomplished faster with less traffic overhead.

After having located where the fault is, the next step is to do fault isolation (i.e., determine the source of the problem). Identification of the root cause of the problem could be a complex process, and there are several techniques used in OSS to accomplish this [110]. After identifying the source of the problem, a trouble ticket can be generated to resolve the problem. In an automated network operations center, a ticket could be generated automatically in a trouble-tracking system by the NMS.

52.9.2 Configuration Management

Configuration management is normally used in the context of discovering network topology, mapping the network, and setting up the configuration parameters in management agents and management systems. However, network management in the broad sense also includes network provisioning, which includes planning and design of network.

Network management is based on knowledge of network topology. As network grows, shrinks, or changes are made, the network topology needs to be kept updated automatically. This is done by the discovery application in NMS. The discovery process needs to be constrained as to the scope of the network that it discovers. For example, arp command can discover any network component that responds with an IP address, which can then be mapped by the NMS. Auto discovery can be done using the broadcast ping on each segment and following up with further SNMP queries to gather further details on the system. The more efficient method is to look at the arp cache in the local router. The arp cache table is large and contains the addresses of all the recently communicated hosts and nodes. Using this table, subsequent arp queries could be sent to other routers. This process is continued until the information is obtained on all IP addresses defined by the scope of the auto discovery procedure. A map, showing network topology, is presented by the auto discovery procedure, after the addresses of the network entities have been discovered.

Network parameters of managed elements in the network could be set and monitored using SNMP commands. Network provisioning in packet-switched Internet is quite different from the circuit-switched telephone network. In a connectionless packet-switched circuit, each packet takes an independent path and the routers at various nodes switch each packet based on the load in the links. The links are provisioned based on the average and peak demands. Network provisioning is done for packet-switched network based on performance statistics and quality of service requirements.

52.9.3 Performance Management

Performance management is concerned with the performance behavior of the network and strongly influences the quality of service (QoS) provided. The status of the network is displayed by a network monitoring system that measures the traffic and performance statistics on the network. The network statistics include data on traffic volume, network availability, and network delay. The traffic data can be captured based on traffic volume in the various segments of the network. Performance monitoring tools such as network analyzers, can gather statistics of all protocol layers. We can analyze the various application-oriented traffic such as Web traffic, Internet mail, file transfers, etc. The statistics on applications could be used to make policy decisions on managing the applications. Performance data on availability and delay is useful for tuning the network to increase the reliability and to improve its response time. This becomes especially important if the network segments are carrying broadband traffic.

52.9.4 Security Management

Security management covers a broad range of security aspects. It involves physically securing the network, access to the network resources, and secured communication over the network. A security database is established and maintained by network operations center (NOC) for access to the network and network information. Any unauthorized access to the network resources generates an alarm caused by a trap to NMS at NOC. Firewalls are implemented to protect corporate networks and network resources being accessed by unauthorized personnel and programs, including virus programs. Secured communication is concerned with the tampering of information as it traverses the network. The content of the information should neither be accessed nor altered by unauthorized personnel. Cryptography plays a vital part in security management.

52.9.5 Account Management

Account management administers cost allocation of the usage of network. Metrics is established to measure the usage of resources and services provided. Traffic data gathered by performance management serves as input to this process. In the case of service providers, account management includes billing of subscribers. This area is getting a lot of attention recently with broadband services and video-on-demand service. Broadband services are offered with multiple tariffs to subscribers providing various classes of service.

52.9.6 OSS/Network Management System

A network management system (NMS), which can be considered as an operations support system (OSS), is an automated system tool that helps the networking personnel perform their functions efficiently. For SNMP-based NMS, this involves the implementation of the management functions FCAPS. The SNMP-based NMS manages those network components that have SNMP agent integrated in them. Non-SNMP components can be managed by an SNMP NMS using proxy server, which creates equivalent SNMP MIB objects for non-SNMP objects.

The implementation of NMS is being extended by many vendors using Web-based technology. With the rapid growth of Internet and World Wide Web, Web technology composed of Web server and Web browsers, has become universal in the enterprise environment. With the universality of Web technology, it appears logical to marry the two technologies — network or system management to gather data, and Web technology to display the information at multiple locations on Web browsers. We are no longer restricted to displaying the information on a centralized monitor associated with an NMS running on a proprietary platform. Further, we can go beyond just displaying the information on a Web browser. We can use the Web technology to both gather data and display it. This would take us to the realm of inserting active Web agents that monitor and control components on networks, systems, and applications and interact with them using Web interface. There are several technologies that are evolving in this area; refer to [110] for a detailed discussion on this subject.

52.10 Virtual Private Network

Global and national enterprises establish private networks as intranet for internal communication. Although many private networks are run and managed by the enterprises which own them, public carriers are providing the VPN service to enterprises and are referred to as virtual private networks (VPNs). There is a VPN consortium, referred to as VPNC, which is the international trade association for manufacturers in the VPN market.

There are three important VPN technologies: trusted VPNs, secure VPNs, and hybrid VPNs. Trusted VPNs use the public facilities of service provider. The privacy afforded by the communications provider assured the customer that no one else would use the same circuit. This allowed customers to have their own IP addressing and their own security policies. However, there is always the possibility that this network could be compromised. Technologies for trusted layer 2 VPNs include ATM circuits, frame relay circuits, and transport of layer 2 frames over MPLS.

VPN equipment vendors implemented protocols that would allow traffic to be encrypted at the edge of one network or at the originating computer, and tunneled through the Internet like any other data, then decrypted when it reached the corporate network or a receiving computer. Networks that are constructed using encryption are called secure VPNs. Secure VPNs use IPsec with encryption in either tunnel or transport modes. IPsec is described in many RFCs, including 2401, 2406, 2407, 2408, and 2409. Second implementation of security is using IPsec inside of L2TP as described in RFC 3193 and has significant deployment for client–server remote access secure VPNs. Both of these technologies are standardized by the IETF.

A secure VPN can be run as part of a trusted VPN, creating a third type of VPN, hybrid VPN. Management of all three types of VPNs is implemented in the same manner as enterprise network service management. The VPNC does not establish standards, but is supported by IETF standards. The MIBs for all the protocols and IPsec have already been established or in the process of development. Detailed list of protocols and IETF references are presented in the VPNC Web site.

52.11 Broadband Access Networks

Broadband service traffic is carried over the wide area network on asynchronous transfer mode/synchronous optical network (ATM/SONET) or Internet IP network. From edge routers, it is then transmitted to the residential and small-office-home-office/small and medium enterprise (SOHO/SME) customer premises over broadband access network such as hybrid fiber–coaxial (HFC) cable, digital subscriber line (DSL), or wireless–satellite, fixed and mobile medium. This is shown in Figure 52.12 [110]. The access network to the enterprise/business customer is either synchronous or asynchronous today and may be migrating to Internet connection in the future. The access point network could be Internet/LAN. Irre-

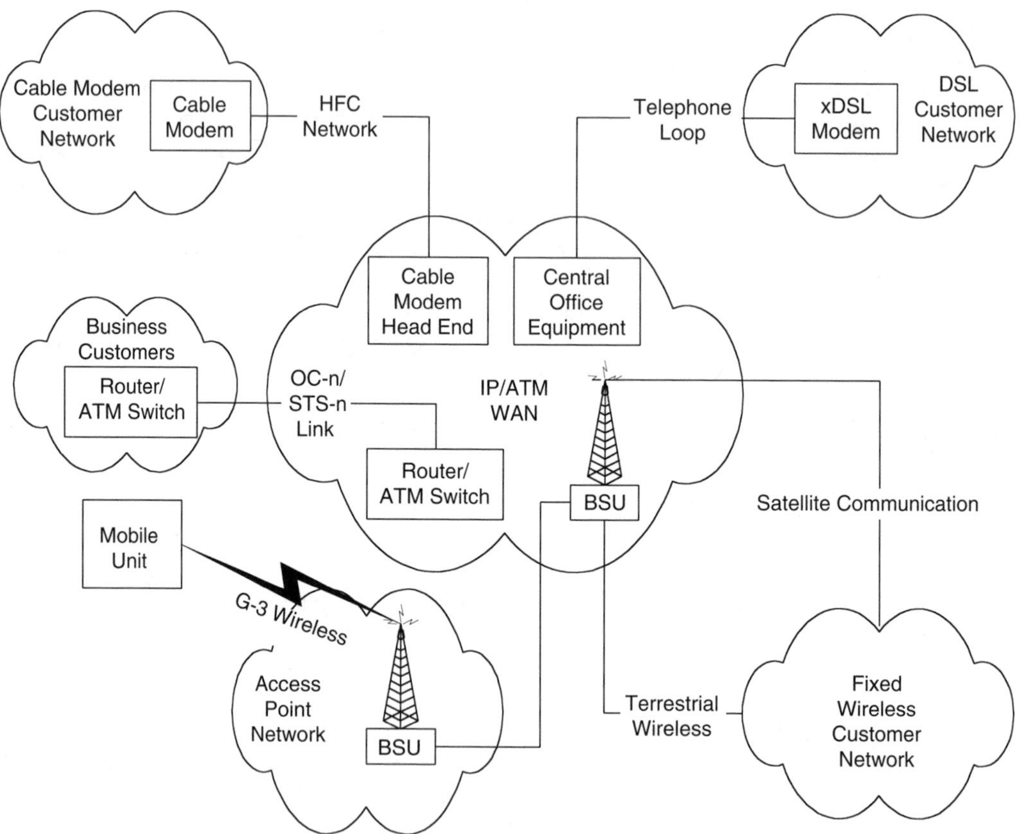

FIGURE 52.12 Broadband access networks.

spective of the transmission medium, they all carry IP traffic. Wwe discuss briefly in this section the network management of broadband access networks.

52.11.1 Cable Access Network Management

The cable or HFC technology is a two-way interactive multimedia communication system that uses fiber and coaxial cable facilities and cable modems. At the head end, signals from various sources, such as traditional satellite services, analog and digital services using WAN, and data from Internet Service Provider (ISP) services using private backbone network, are multiplexed and up-converted from electrical (radio frequency) to optical signal. The communication is one way on the optical fiber. There is a pair of optical fibers from the head end to a fiber node, each carrying one-way traffic in the opposite direction. The optical signal is down-converted to RF at the fiber node and traverses over the coaxial cable in a duplex mode. At the customer premises, there is a network interface unit (NIU) that is the demarcation point between the customer network and service provider network. The analog signal is split at NIU. The TV signal is directed to TV and the data to cable modem. The cable modem converts the analog signal to an Ethernet output feeding either a PC or LAN. Telephone signal is also transmitted along with video and data in some cable sites.

There are several components that need to be managed in HFC network. They are: (1) broadband LAN; (2) asymmetric bandwidth allocation to achieve two-way communication; (3) radio-frequency spread-spectrum technique to carry multiple signals over the HFC; and (4) radio-frequency spectrum allocation to carry multimedia services of telephony (voice), television (video), and computer communication (data). Besides managing data, the cable system management involves signal level and spectrum level management for the downstream and upstream signals at CMTS, cable modems, fiber nodes, and the amplifiers in coaxial cables. This makes it more complex than management of computer communications network in LAN and WAN.

A consortium of cable service providers (CableLabs) has developed data over cable interface specifications (DOCSIS) for the cable modem termination system (CMTS) and cable modems. IETF and CableLabs have developed SNMP MIBs for the cable access system components to be managed; refer to [110] for details on management of cable access systems.

52.11.2 Digital Subscriber Line Management

Amongst all the types of digital subscriber lines, Asymetric Digital Subscriber Loop (ADSL) is the technology that is being deployed now in the U.S. and the world. A simplified access network using ADSL consists of an ADSL modem (referred to as ADSL transmission unit, i.e., ATU) and a splitter at each end of the ADSL line. The data and video signal from the broadband network is converted to analog signal by a DSL access multiplexer (DSLAM) that has embedded ADSL modems and multiplexed and demultiplexed. The POTS (Plain Old Telephone Service) voice signal and the broadband signal are combined for each subscriber and transmitted over the ADSL line. The reverse process occurs at the customer premises.

The upstream and downstream signals are placed asymmetrically in the frequency spectrum. The POTS signal is always allocated the base band of 4 kHz, upstream 25 to 200 kHz and the downstream 200 to 1.1 MHz. The combined signal is modulated using either Discrete MultiTone (DMT) (standard adopted by DSL Forum) or Carrier Amplitude Phase (CAP) modulation. The former modulation scheme is more efficient, but more complex and costly. Both schemes are currently in use.

Those subscribers who have direct copper connection from the central office are served with ADSL from the DSLAM located in the central office. Many of the newer residential complexes have fiber cable to the neighborhood (FTTN) and twisted pair from FTTN to the residence. For these residences, ADSL service is provided from a mini-DSLAM located in the pedestal that terminates the fiber at the neighborhood.

DSL Forum, an industry consortium, has developed network management specifications that address the parameters, operations, and protocols associated with configuration, fault, and performance management. Security is addressed by other models such as SNMP security management.

The management function at the physical layer involves three entities: physical channel, fast channel, and interleaved channel. The fast and interleaved channels need to be managed separately. These two use the physical transmission medium, which also needs to be managed. Besides management of physical links and channel parameters, the parameters associated with the type of line coding, DMT or CAP, need to be monitored.

The various parameters that need to be managed for configuration of line and ADSL modems are associated with line type and coding, noise margins, rate adaptation parameters, interleave delay, and alarm thresholds. The link could be configured in one of five options: no separation of channels, fast, interleaved, either, or both. Because of the large number of ATUs associated with a DSLAM, management of configuration profiles is accomplished using specially developed MIB tables.

Fault management involves isolation of troubles associated with line and ADSL modems. After the automatic indication of faults, self tests are done by the ATUs at both ends. ADSL line status shows the current state of the line as to whether it is operational, or there is a loss of any of the parameters on frame, signal, power, or link. It also indicates initialization errors. Alarms are generated when the preset counter reading exceeds 15 min on loss of signal, frame, power, link, and error-seconds.

Each ATU's performance in terms of line attenuation, noise margin, total output power, current and previous data rate, along with the maximum attainable rate, channel data block length (on which CRC check is done), and interleave delay can be monitored. In addition, statistics are gathered for a 15-min interval and a 1-day interval on the error-seconds statistics. Two counters are maintained by each ATU for each error condition to measure these. The error statistics are maintained for loss of signal seconds, loss of frame seconds, loss of power seconds, loss of link seconds, errored seconds, transmit blocks, receive blocks, corrected blocks, and uncorrectable blocks.

52.11.3 Fixed Wireless

There is sparse deployment of broadband fixed wireless technologies, multichannel multipoint distribution service (MMDS) at 2.5 to 2.686 GHZ and local multipoint distribution service (LMDS) at 27.5 to 28.35 and 31.0 to 31.3 GHz channels. These are implemented using cable modem technology at both ends and replacing HFC medium with wireless. Thus, the network management considerations are similar to the ones discussed under cable access network.

52.11.4 Wireless LAN

Wireless LAN (WLAN) has been recently introduced as part of the broadband access network in the commercial market, such as in an enterprise environment, as "hot spots" in hotel and coffee shops, and as extension to DSL in city telephone booths. The network configuration comprises WLAN connected to wired network via an access point. The WLAN protocol that has become popular for this purpose is deviations of IEEE 802.11–802.11a/b/g, with data rates of 11, 54, and 54 mps, respectively. Management of some of these WLAN networks is accomplished using SNMP and traditional SNMP management system. The vendors have embedded private MIBs in the access points for managing them remotely. The private MIB database could be compiled in any standard SNMP network management system, and the wireless access points and any end devices connected to it could be managed along with the wired network components.

52.12 Future Trends and Challenges

Network management has been evolving as has networking technology. The demarcation between telecommunications and computer communications network has been made fuzzy. Enhancements in computer hardware, fiber, and Ethernet technologies have extended the range and bandwidth of networks at a rapid rate. Enterprise and nonenterprise networking are migrating towards one another; so are LAN and WAN. The need to access information is becoming ubiquitous. The wired enterprise network has become wired *and* mobile network and the provisioning of networks is no longer static but dynamic. With the widespread use of Internet, network management is no longer just an appendix to the network, but also part of the network. More and more of network management is dynamic and embedded into the network itself, i.e., evolution of active network management. We will now look at some of the ramifications of this.

52.13 QoS Management

As we noted earlier, quality of service (QoS) offered on the Internet strongly depends on the performance and configuration of a network. Internet has become global and today's Internet only provides best-effort service. The packet-switched traffic carrying broadband service is processed at each node as quickly as possible. There is no guarantee as to timeliness and jitter in QoS on an end-to-end basis. Requirements for voice, video, and data being different, it is essential to provide several classes of service as it is done in ATM network. Several service models and mechanisms are being studied by IETF to meet the demand for end-to-end QoS [120]. These are integrated services (IntServ)/Resource Reservation Protocol (RSVP), the differentiated services (diffServ), and multiprotocol label switching (MPLS). The resources are reserved ahead using RSVP in the intServ model. In diffServ, packets are encoded to create many packet classes. In MPLS, packets are assigned labels at the ingress node and subsequent classification, forwarding, and services are based on the label. It is beyond the scope of our presentation to go into details on these. However, this impacts network management in the following manner. First, these protocols require managed objects and, hence, MIBs to monitor them. The network needs to be dynamically provisioned and would require more sophisticated monitoring and alarm in a distributed architecture, yet centrally able to localize problems in the event of failure to meet end-to-end QoS as per SLA.

52.14 Mobile and Enterprise Wireless Management

Mobile wireless as part of the Internet is being deployed rapidly with 802.11 WLAN protocols and network of access points. Mobile IP has been specified by IETF [RFC 2002] to access mobile devices. Figure 52.13 shows the architecture of cell network that could use either 3-G or 802.11 wireless LAN for communication with a mobile node (MN) embedded with a mobile agent (MA). The mobile node, having mobile IP, registers with a foreign agent as it roams. It communicates through the foreign agent to its home agent in the home network. IETF has developed MIBs for mobile node, foreign agent, and home agent, which are shown in Table 52.2. When this is implemented by equipment vendors, service providers will be able to manage the integrated mobile and wireless network much better.

The management of mobility, resource allocation, security, and power adds complexity to the wireless enterprise network over that of wired network. Location tracking, which is part of mobility management, includes discovery of foreign agents by mobile units, broadcasting and advertising by foreign agents to locate mobile units, and solicitation by mobile units. The resource management includes scheduling and call admission control, load balancing between access points and mobile nodes, and power management to reduce interference between neighboring cells.

52.15 Customer Premises Network Management

With Internet reaching residences, SOHOs, and SMEs, networks are being established at the customer premises. With no technical knowledge being available for the users of customer premises network, it is essential for the service provider to manage the subscribers' networks remotely. There are efforts that

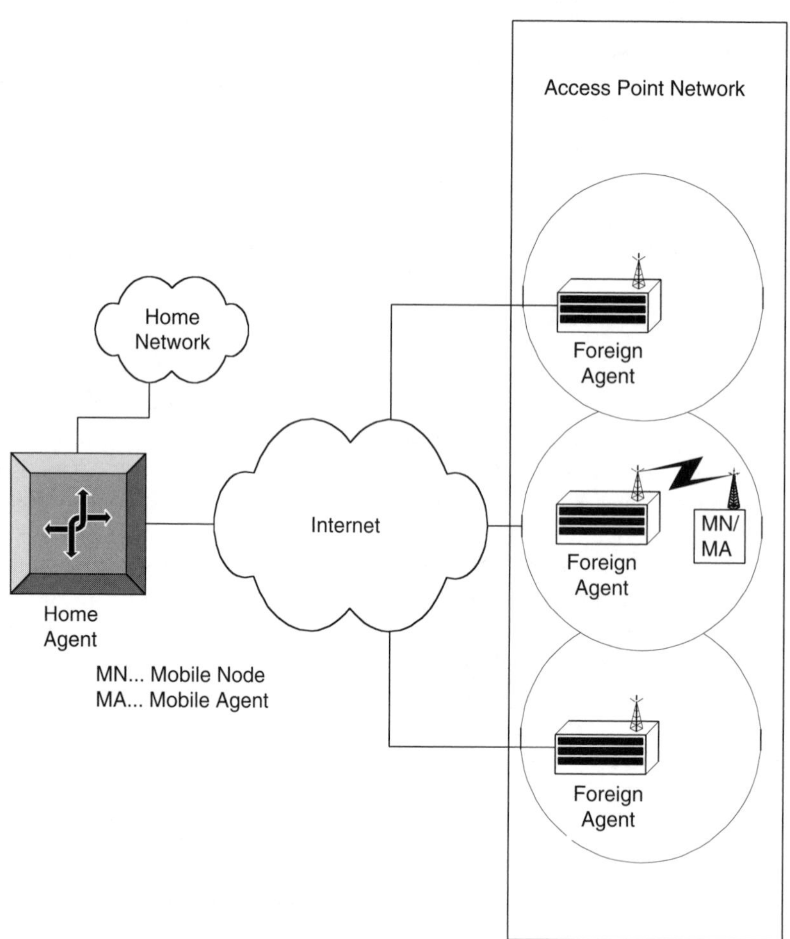

FIGURE 52.13 Access point network using mobile IP.

TABLE 52.2 Mobile IP MIB Groups

Groups	Mobile Node	Foreign Agent	Home Agent
mipSystemGroup	X	X	X
mipSecAssociationGroup	X	X	X
mipSecViolationGroup	X	X	X
mnSystemGroup	X		
mnDiscoveryGroup	X		
mnRegistrationGroup	X		
maAdvertisementGroup		X	X
faSystemGroup		X	
faAdvertisementGroup		X	
faRegistrationGroup		X	
haRegistrationGroup			X
haRegNodeCountersGroup			x

have started in many fronts in the design and management of customer premises networks. This area is still in the embryonic stage.

52.16 Integrated Service Management

Several of the operations support systems performing the functions of various network management applications are mostly stand-alone systems. However, with dynamic provisioning and service–on-demand such as video-on-demand, the OSSs performing the various functions need to communicate with each other in real time. For example, if a subscriber makes a request at 8 p.m. to watch a movie at 10 p.m. from a content server at the head end, the system should ensure that there is adequate bandwidth available for a video service at that time. Performance management OSS should be able to predict the availability of bandwidth at the requested time. This requires knowledge of the traffic pattern in predicting the usage, which would help service provisioning. Currently, it is done on best-effort basis. Further, service-provisioning system, i.e., configuration management OSS and the billing system, i.e., account management OSS need to be coordinated for implementing this feature for provisioning and billing. The current scheme is manual entry in both systems. See [111] for a detailed discussion on this.

Acknowledgments

The author wishes to acknowledge Addison-Wesley Publishers for use of material from the book *Network Management: Principles and Practice* by Mani Subramanian [2000].

References

1. Abe, George, *Residential Broadband*, Cisco Press, Indianapolis, IN,1997.
2. Adams, Elizabeth K. and Willetts, Keith J., *The Lean Communications Provider: Surviving the Shakeout through Service Management Excellence*, McGraw-Hill, New York, 1996.
3. Ahmed, Masuma and Vecchi, Mario P., Definitions of Managed Objects for HFC RF Spectrum Management Version 2.0, RF Spectrum Management MIB, April 21, 1995.
4. Air Interface for Fixed Broadband Wireless Access Systems, P802.16/D5 (C/LM) Standard for Local and Metropolitan Area Networks, IEEE 802.16.
5. Akyildiz, I. F., McNair, J., Ho, J. S. M., Uzunalioglu, H., and Wang, W., Mobility management in next generation wireless systems, *IEEE Proceedings Journal*, 87(8): 1347–1385, August 1999.
6. Azzam, Albert, *High-Speed Cable Modems*, McGraw-Hill, New York, 1997.
7. Biesecker, Keith, The Promise of Broadband Wireless, *IT Pro*, November/December 2000.
8. Black, Darryl P., *Managing Switched Local Area Networks — A Practical Guide*, Addison-Wesley Longman, Reading, MA, 1998.
9. Black, Uyless, *Network Management Standards*, McGraw-Hill, New York, 2nd ed., 1995.
10. Black, Uyless, *Emerging Communications Technologies*, Prentice Hall, Upper Saddle River, NJ, 2nd ed., 1997.
11. Black, Uyless, *Residential Broadband Networks*, Xdsl, Hfc, and Fixed Wireless Access, Prentice Hall, Upper Saddle River, NJ, 1997.
12. Cable Data Modems — A Primer for Non-Technical Readers, CableLabs, April 1996.
13. Cable Data Modem Performance Evaluation — A Primer for Nontechnical Readers, Cable Television Laboratories, November 1996.
14. Chadayammuri, Prabha G., A platform for building integrated telecommunications network management applications, *Hewlett-Packard Journal*, October 1996.
15. Chapman, D. Brent, Network (In)Security Through IP Filtering, USENIX Security Symposium III Proceedings, September 14–16, 1992.
16. CheetahNet Technical Summary, Superior Electronics Group, 1996.

17. CIM: Desktop Management Task Force, Common Information Model: Core Model White Paper, Version 2.0, 1999, http:///www.dmtf.org/.
18. Cisco/RMON: http://www.cisco.com/warp/public/cc/cisco/mkt/enm/cwsiman/tech/rmon2_wp.html.
19. Compaq DMI: Intelligent Manageability, Compaq White Paper, May 1998, http://www.compaq.com/im/dmi2.html.
20. Cong, M. Hamlen and Perkins, C., The Definitions of Managed Objects for IP Mobility Support using SMIv2, IETF RFC 2006, October 1996.
21. Cooper, Frederic J., et al., *Implementing Internet Security*, New Riders, Indianapolis, IN, 1995.
22. Cronk, R, Callahan, P., and Berstein, L., Rule-based expert systems for network management operations: an introduction, *IEEE Network Magazine*, September 1988.
23. DMI 2.0s, Desktop Management Interface Specification Version 2.0s, Desktop Management Task force, June 24, 1998, ftp://ftp.dmtf.org or http://www.dmtf.org
24. DMI/SNMP: Desktop Management Task Force, DMI to SNMP Mapping Standard, Version 1.0, November 1997, ftp://ftp.dmtf.org or http://www.dmtf.org.
25. Feldmeir, John, Network Traffic Management, *Unix Review*, November 1997.
26. Fratto, Mike, Mobile and wireless technology: wireless security, *Network Computing*, January 22, 2001.
27. Giaordano, Silvia, Salsano, Stefano, Van den Berghe, Steven, Ventre, Giorgio, and Giannakopolous, Dimitrios, Advanced QoS provisioning in IP networks: the European premium IP projects, *IEEE Communications Magazine*, January 2003.
28. Goralski, Walter, *ADSL and DSL Technologies*, McGraw-Hill, New York, 1998
29. Greggains, David, ADSL and high bandwidth over copper lines, *International Journal of Network Management*, 7, 277–287, 1997.
30. Hajela, Sujai, HP OEMF: alarm management in telecommunications networks, *Hewlett-Packard Journal*, October 1996.
31. Hegering, H. G. and Abeck, S., *Integrated Network and System Management*, Addison-Wesley, Wokingham, U.K., 1995.
32. Hegering, H. G. and Yemini, Y. Eds., *Integrated Network Management III*, Proceedings of the IFIP TC6/WG6.6, 3rd International Symposium on Integrated Network Management, San Francisco, CA, IFIP Transactions, Amsterdam, North Holland, 1993.
33. Heilbronner, Stephen and Wies, Rene, Managing PC networks, *IEEE Communications Magazine*, October 1997.
34. Hill, Ed, Simple System/Network Monitoring — Spong v1.1, http://strobe.wee.g.,uiowa.edu/~edhill/public/spong/
35. Hong, James Won-Ki, et al., Web-based intranet services and network management, *IEEE Communications Magazine*, October 1997.
36. HP WBM, HP Proactive Networking: The Networking Management Component, White Paper, January 1998.
37. Hyde, Douglas, Web-Based Management: The New Paradigm for Network Management, http://www.3com.com/nsc/500627, 1997.
38. IEEE 802.11 Management Information Base, http://standards.ieee.org/reading/ieee/std/lanman/MIB-D6.2.txt.
39. IEEE Communications, Special Issue on Wireless Broadband Communication Systems, *IEEE Communications Magazine*, January 1997.
40. JDMK WP, Java Dynamic Management Kit — A White Paper, http://java.sun.com/products/JavaManagement/, February 1998.
41. JMX WP, Java Management Extensions White Paper, http://java.sun.com/products/JavaManagement/, June 15, 1999.
42. Kliger, S., Yemini, S., Yemini, Y, Ohsie, D., and Stolfo, S., A Coding Approach to Event Correlations, Proceedings of the Fourth International Symposium on Integrated Network Management, 1995.

43. Lazar, A.; Saracco, R.; and Stadler, R. Eds., *Integrated Network Management V*, Chapman & Hall, London, 1997.
44. Leinwand, Allan and Conroy, Karen Fang, *Network Management: A Practical Perspective*, Addison-Wesley Longman, Reading, MA, 2nd ed., 1996.
45. Lewis, Lundy and Frey, Jim, Incorporating Business Process Management into Network and Systems Management, Proceedings of the Third International Symposium on Automated Decentralized Systems, Berlin, Germany, April 9–11, 1997.
46. Lewis, Lundy, A Fuzzy Logic Representation of Knowledge for Detecting/Correcting Network Performance Deficiencies, In *Network Management and Control — Volume 2*, Ivan T. Frisch, Manu Malek, and Shivendra S. Panwar, Eds., Plenum Press, New York and London, 1994.
47. Lewis, Lundy, A Case-Based Reasoning Approach to the Management of Faults in Communication Networks, IEEE Infocom '93 Proceedings, Volume 3, San Francisco, CA, March 28–April 1, 1993.
48. Lewis, Lundy, *Managing Computer Networks: A Case-Based Reasoning Approach*, Artech House, Norwood, MA, 1995.
49. Lewis, Lundy, Implementing policy in enterprise networks, *IEEE Communications Magazine*, January 1996.
50. Lewis, Lundy, *Service Level Management for Enterprise Networks*, Artech House, Norwood, MA, 1999.
51. MacGuire, Sean, Big Brother: A Web-based Systems and Network Monitoring and Notification System, http://www.maclawran.ca/bb/bb-info.html/.
52. Maxwell, Kim and Maxwell, Kimberly, *Residential Broadband, An Insider's Guide to the Battle for the Last Mile*, John Wiley & Sons, New York, 1998.
53. Miller, Mark A., *Managing Internetworks with SNMP*, M&T Books, New York, 1995.
54. NIST: Keeping Your Site Comfortably Secure: An Introduction to Internet Firewalls, http://csrc.ncsl.nist.gov/nistpubs/800-10.ps, December 1994.
55. Oetiker, Tobias and Rand, Dave, Multi Router Traffic Grapher, http://ee-staff.ethz.ch/~oetiker/webtools/mrtg/mrtg.html.
56. Perkins, C., IP Mobility support, IETF RFC 2002, October 1996.
57. S. Baudet, C. Besset-Bathias, P. Frene, and N. Giroux, QoS implementation in UMTS networks, Alcatel Telecommunications Review, 1st Quarter 2001.
58. Perkins, Charles E., Mobile Networking Through Mobile IP, Internet Computing on Line, www.computer.org/internet/v2n1/perkins.htm.
59. Perkins, David and McGinnis, Evan, *Understanding SNMP MIBs*, Prentice Hall, Upper Saddle River, NJ, 1997.
60. Perry, Ed and Ramanathan, Srinivas, Network management for residential broadband interactive data services, *IEEE Communications Magazine*, November 1996.
61. RFC 1155: M. Rose and K. McCloghrie, Structure and Identification of Management Information for TCP/IP-based Internets, Request for Comments 1155, May 1990.
62. RFC 1157: J. Case, M. Fedor, M. Schoffstall and J. Davin, A Simple Network Management Protocol, Request for Comments 1157, May 1990.
63. RFC 1212: M. Rose and K. McCloghrie, Concise MIB Definitions, Request for Comments 1212, March 1991.
64. RFC 1213: M. Rose, Management Information Base for Network Management of TC/IP-based Internets: MIB-II, Request for Comments 1213, March 1991.
65. RFC 1215: M. Rose, A Convention for Defining Traps for Use with the SNMP, Request for Comments 1215, March 1991.
66. RFC 1244: P. Holbrook and J. Reynolds, Site Security Handbook, Request for Comments 1244, July 1991.
67. RFC 1284: J. Cook, Definitions of Managed Objects for the Ethernet-like Interface Types, Request for Comments 1284, December 1991.

68. RFC 1285: J. Case, FDDI Management Information Base, Request for Comments 1285, January 1992.
69. RFC 1354: F. Baker, IP Forwarding Table MIB, Request for Comments 1354, July 1992.
70. RFC 1398: F. Kastenholz, Definitions of Managed Objects for the Ethernet-like Interface Types, Request for Comments 1354, January 1993.
71. RFC 1421: J. Linn, Privacy Enhancement for Internet Electronic Mail: Part I: Message Encryption and Authentication Procedures, Request for Comments 1421, February 1993.
72. RFC 1422: S. Kent, Privacy Enhancement for Internet Electronic Mail: Part II: Certificate-Based Key Management, Request for Comments 1422, February 1993.
73. RFC 1423: D. Balenson, Privacy Enhancement for Internet Electronic Mail: Part III: Algorithms, Modes, and Identifiers, Request for Comments 1423, February 1993.
74. RFC 1424: B. Kaliski, Privacy Enhancement for Internet Electronic Mail: Part IV: Key Certification and Related Services, Request for Comments 1424, February 1993.
75. RFC 1445:J. Glavin and K. McCloghrie, Administrative Model for version 2 of the Simple Network Management Protocol (SNMPv2), Request for Comments 1445, April 1993.
76. RFC 1446: J. Glavin and K. McCloghrie, Security Protocol for version 2 of the Simple Network Management Protocol (SNMPv2), Request for Comments 1446, April 1993.
77. RFC 1470: FYI on a Network Management Tool Catalog: Tools for Monitoring and Debugging TCP/IP Internets and Interconnected Devices, R. Enger and J. Reynolds, Eds., Request for Comments 1470, June 1993.
78. RFC 1513: S. Waldbusser, Token Ring Extensions to the Remote Network Monitoring MIB, Request for Comments 1513, September 1993.
79. RFC 1748: K. McCloghrie and E. Decker, IEEE 802.5 Token Ring MIB using SMIv2, Request for Comments 1748, December 1994.
80. RFC 1757: S. Waldbusser, Remote Network Monitoring Management Information Base, Request for Comments 1757, February 1995.
81. RFC 1901: SNMPv2 Working Group, J. Case, K. McCloghrie, M. Rose, and S. Waldbusser, Introduction to Community-based SNMPv2, Request for Comments 1901, January 1996.
82. RFC 1902: SNMPv2 Working Group, J. Case, K. McCloghrie, M. Rose, and S. Waldbusser, Structure of Management Information for Version w of the Simple Network Management Protocol (SNMPv2), Request for Comments 1902, January 1996.
83. RFC 1903: SNMPv2 Working Group, J. Case, K. McCloghrie, M. Rose, and S. Waldbusser, Textual Conventions for Version 2 of the Simple Network Management Protocol (SNMPv2), Request for Comments 1903, January 1996.
84. RFC 1904: SNMPv2 Working Group, J. Case, K. McCloghrie, M. Rose, and S. Waldbusser, Conformance Statements for Version 2 of the Simple Network Management Protocol (SNMPv2), Request for Comments 1904, January 1996.
85. RFC 1905: SNMPv2 Working Group, J. Case, K. McCloghrie, M. Rose, and S. Waldbusser, Protocol Operations for Version 2 of the Simple Network Management Protocol (SNMPv2), Request for Comments 1905, January 1996.
86. RFC 1906: SNMPv2 Working Group, J. Case, K. McCloghrie, M. Rose, and S. Waldbusser, Transport Mappings for Version 2 of the Simple Network Management Protocol (SNMPv2), Request for Comments 1906, January 1996.
87. RFC 1907: SNMPv2 Working Group, J. Case, K. McCloghrie, M. Rose, and S. Waldbusser, Management Information Base for Version 2 of the Simple Network Management Protocol (SNMPv2), Request for Comments 1907, January 1996.
88. RFC 1908: SNMPv2 Working Group, J. Case, K. McCloghrie, M. Rose, and S. Waldbusser, Coexistence between Version 1 and Version 2 of the Simple Network Management Protocol (SNMPv2), Request for Comments 1908, January 1996.
89. RFC 2021: S. Waldbusser, Remote Network Monitoring Management Information Base Version 2, Request for Comments 2021, January 1997.

90. RFC 2063: N. Brownlee, C. Mills, and G. Ruth, Traffic Flow Measurement: Architecture, Request for Comments 2063, January 1997.
91. RFC 2064: N. Brownlee, Traffic Flow Measurement: Meter MIB, Request for Comments 2064, January 1997.
92 RFC 2074: A. Bierman, Remote Network Monitoring MIB Protocol Identifiers, Request for Comments 2074, January 1997.
93. RFC 2123: N. Brownlee, Traffic Flow Measurement: Experiences with NeTraMet, Request for Comments 2123, March 1997.
94. RFC 2196: B. Fraser, Site Security Management, Request for Comments 2196, September 1997.
95. RFC 2213: F. Baker, J. Krawczyk, and A. Sastry, Integrated Services Management Information Base using SMIv2, Request for Comments September 1997.
96. RFC 2271: D. Harrington, R. Presuhn, and B. Wijnen, An Architecture for Describing SNMP Management Frameworks, Request for Comments 2271, January 1998.
97. RFC 2272: J. Case, D. Harrington, R. Presuhn, and B. Wijnen, Message Processing and Dispatching for the Simple Network Management Protocol (SNMP), Request for Comments 2272, January 1998.
98. RFC 2273: D. Levi, P. Meyer, and B. Stewart, SNMPv3 Applications, Request for Comments 2273, January 1998.
99. RFC 2274: U. Blumenthal, and B. Wijnen, User-based Security Model (USM) for version 3 of the Simple Network Management Protocol (SNMPv3), Request for Comments 2274, January 1998.
100. RFC 2275: B. Wijnen, R. Presuhn, and K. McCloghrie, View-based Access Control Model (VACM) for the Simple Network Management Protocol (SNMP), Request for Comments 2275, January 1998.
101. RFC 2358: J. Flick and J. Johnson, Definitions of Managed Objects for the Ethernet-like Interface Types, Request for Comments 2358, June 1998.
102. RFC 854: J. Postel and J. K. Reynolds, Telnet Protocol Specifications, Request for Comments 854, May 1, 1983.
103. Rivest, R. L., Shamir, A., and Adelman, L., A method for obtaining digital signatures and public-key cryptosystems, *Communications of the ACM*, February 1978.
104. Rose, M. T., *The Simple Book: An Introduction to Network Management*, Prentice Hall, Upper Saddle River NJ, 1996.
105. Sethi, A. S., Raynaud, Y., and F. Faure-Vincent, Eds., *Integrated Network Management IV*, Chapman & Hall, London, 1995.
106. Sloman, Morris, Ed., *Network and Distributed Systems Management*, Addison-Wesley, Workingham, U.K., 1994.
107. Sloman, Morris, Mazumdar, Subrata, and Lupu, Emil, *Integrated Network Management VI*, IEEE, IEEE Communications Society, and IFIP, Piscataway, NJ, 1999.
108. Stallings, William, *SNMP, SNMPv2 SNMPv3, and RMON 1 and 2*, Addison-Wesley, Reading, MA, 1998.
109. Strutt, Colin and Sylor, Mark W., Digital Equipment Corporation's Enterprise Management Architecture, in *Network and Distributed Systems Management*, Morris Sloman, Ed., Addison-Wesley, Workingham, U.K., 1994.
110. Subramanian, Mani, *Network Management: Principles and Practice*, Addison-Wesley, Reading, MA, 2000.
111. Subramanian, Mani and Lewis, Lundy, QoS and Bandwidth Management in Broadband Cable Access Network, Computer Networks Special Issue on Management of Services, 2003.
112. M. Subramanian, A Software-Based Elastic Broadband Cable Access Network, Proceedings Manual of Cable-Tec Expo 2002, San Antonio, TX, June 2002.
113. Sun Enterprise: Solstice Enterprise Manager 2.1 — A Technical White Paper, Copyright 1994–1998. Sun Microsystems, Palo Alto, CA.

114. Sun Site and Domain: Solstice Site Manager and Solstice Domain Manager 2.3 — A Technical White Paper, Copyright 1994–1998. Sun Microsystems, Palo Alto, CA.
115. Thompson, J. Patrick, Web-based enterprise management architecture, *IEEE Communications Magazine*, March 1998.
116. Tivoli: A series of product and technical documents at http://www.tivoli.com/.
117. Varshney, Upkar and Vetter, Ron, Emerging mobile and wireless networks, *Communications of the ACM*, June 2000.
118. Wack, John P. and Carnahan, Lisa J., Keeping Your Site Comfortably Secure: An Introduction to Internet Firewalls, NIST Special Publication 800-10, U.S. Department of Commerce, National Institute of Standards and Technology, Gaithersburg, MD, 1994.
119. WHI, Concepts and Terminology Important to Understanding WMI and CIM, WinHEC 99 White Paper, Windows® Hardware Engineering Conference: Advancing the Platform, 1999.
120. Wu, Chwan-Hwa "John" and Irwin, J. David, *Emerging Multimedia Computer Communication Technologies*, Prentice Hall, Upper Saddle River, NJ, 1998.
121. Xiao, Xipeng and Ni, Lionel M., Internet QoS, a big picture, *IEEE Network*, March/April 1999.
122. Yemini, S. A., Kliger, S., Mozes, E., Yemini, Y., and Olsie, D., High speed and robust event correlation, *IEEE Communications Magazine*, May 1996.
123. ZDNet, Task Masters: Network Monitoring Tools, ZDNet UK, 1998, http://www.microsite.co.uk/tivoli/tme.
124. Zeltserman, Dave and Puoplo, Gerard, *Building Network Management Tools with Tcl/Tk*, Prentice Hall, Upper Saddle River, NJ, 1998.

Further Information

Further information on the subject could be obtained from the following sources.

Journals:
IEEE Communications Magazine
IEEE Network
IEEE Wireless Communications
Journal of Network and Systems Management
Technical Association Specifications and Reports:
IETF RFCs
CableLabs publications
DSL Forum publications
Popular Magazines:
Network Computing
Network Magazine
Telecommunications Magazine

Conferences:
IEEE/IFIP International Symposium on Integrated Network Management
IEEE/IFIP Network Operations and Management

Part 6

Systemic Matters

53

Web Structure

CONTENTS

Abstract.......... 53-1
53.1 Introduction 53-1
53.2 Small-World Networks.......... 53-2
53.2.1 Properties of Small-World Networks 53-3
53.2.2 Web as a Small-World Network.......... 53-4
53.3 Power-Law Distributions 53-4
53.3.1 Copying Over.......... 53-5
53.3.2 Preferential Attachment.......... 53-5
53.3.3 Winners Don't Take All.......... 53-6
53.3.4 Local Actions.......... 53-6
53.4 Link Structure Heuristics 53-6
53.4.1 PageRank.......... 53-6
53.4.2 HITS 53-8
53.4.3 Topic-Sensitive PageRank.......... 53-10
53.4.4 Problems 53-10
53.5 Community Structures 53-11
53.5.1 Maximum Flows.......... 53-11
53.5.2 Bipartite Cores.......... 53-12
53.6 Directions.......... 53-13
References.......... 53-14

Pınar Yolum

Abstract

The way Web pages link to each other brings about a structure. This structure is dynamic, with pages and links being added and removed frequently. The dynamism of the structure gives rise to interesting properties that also hold for other complex systems. From a theoretical standpoint, understanding why or how these properties emerge is important. Models that can explain the structural properties of the Web are crucial to the study of their intrinsic properties in more depth. From a practical standpoint, the Web structure can be exploited to devise heuristics for Web applications, such as ranking query results from a Web search or identifying communities of Web pages.

53.1 Introduction

The Web consists of pages that are connected to each other through hyperlinks. The way the pages link to each other creates or rather induces a structure. The structure emerges based on each Web page's choice of which other pages to link to. Importantly, the structure is dynamic because the number of Web pages and the number of hyperlinks between the pages is constantly changing. The *Web* was not at its current state a year ago, and will be different a year later. New Websites will become online, and new hyperlinks will be added while some of the existing sites and hyperlinks disappear.

1-58488-381-2/05/$0.00+$1.50
© 2005 by CRC Press LLC

The most common methods of studying the structure of the Web view the Web as a directed graph where the nodes of the graph are Web pages and the edges of the graph are hyperlinks. This graph is usually built by crawling the Web. Hence, the graph captures a static snapshot of the Web. With a snapshot of the Web in hand, many graph-theoretic properties of the graph, such as the shortest distances between nodes or the number of links pointing to each node, can be computed.

The dynamic and emergent nature of the Web resemble many other complex systems, ranging from the way ants coordinate to the way cities settle [Johnson, 2002]. Viewing the Web as a complex system, understanding its global properties is theoretically valuable. The theoretical results established for other complex systems can be used to understand the Web and *vice versa*. More practically, structural properties of the Web can be exploited in Web applications.

While measures on a snapshot of the Web yield important results, they pertain to one instance of the Web. Naturally, the computed properties may vary from one snapshot to another. For understanding how and why these properties emerge, it is more important to model the Web in terms of dynamic properties, such as the number of Web pages or the number of links. To this end, there are several models of the Web that can generate graphs that exhibit the known properties of the Web. Developing accurate models is important for several reasons. One, these models can be used to predict other properties of the Web. Generating an up-to-date Web graph is generally expensive. Web crawlers need to traverse the Web to collect data to build the graph. Instead, if an accurate model of the Web exists, then that model can be used to generate a Web graph and compute other properties of interest. Two, the models can be used as predictors to foretell the future of the Web. This is important because Web applications can then also take into account the future Web structure.

From a practical perspective, properties of the Web structure can be used to develop heuristics or build algorithms. Among the most well known of these algorithms are those that use the Web link structure to rank Web pages that are returned for a search query. Effective methods for indexing or searching information on the Web are becoming more and more essential. Until recently, these methods were solely based on the content of the Web pages, benefiting only from traditional information retrieval techniques. However, the content of the Web pages can be augmented with information on how Web pages relate to each other through the hyperlinks they provide. Different social metrics, such as authoritativeness of a page, can be identified based on the link structure. These metrics can then be used to rank Web pages.

In what follows, we provide the state-of-the-art in the Web structure research. Section 53.2 describes some interesting properties of complex systems, including human societies, and explains how these properties relate to the Web structure. Section 53.3 discusses current models of the Web that predict existing in-degree distributions. Section 53.4 and Section 53.5 explain how the Web structure can be used to devise heuristics and discuss some leading approaches. Section 53.6 outlines some directions for further research.

53.2 Small-World Networks

The small-world phenomenon is seen frequently in everyday life: two unrelated people who just meet discover a chain of mutual acquaintances. Scientific studies of this phenomenon go back to a seminal experiment by Stanley Milgram [Milgram, 1967].

Milgram conducted an experiment to understand how tightly connected people in the U.S. are. He selected some residents of Kansas and gave each of them a package addressed to an individual living in Boston. Each person was asked to send the package to an acquaintance (and not simply mail the package to the final recipient). That is, each package moved among people who knew each other at least remotely. Then, Milgram calculated how many intermediate people helped to get each package to its final destination. Interestingly, the packages that arrived at the final recipients changed only six hands on average. In other words, two randomly chosen individuals were separated from each other by only six degrees, hence the famous motto of "six degrees of separation."

53.2.1 Properties of Small-World Networks

Informally, small-world networks can be described by three properties:

1. Each node in the network is connected to a small subset of the whole population. For example, in Milgram's experiments, each individual would have known a small fraction of the U.S. population.
2. Individuals who know each other usually have a circle of friends who also know one another.
3. Despite the first two properties, nodes in the network can reach each other through short paths.

This is the definitive conclusion of Milgram's experiments. That is, even though the U.S. population is quite large, and even though people usually know only those around themselves, the paths that get the packages into their destinations are quite short. If everyone knew everyone else in the population, it would not be hard to come up with short paths. Or, if people were not grouped together, who mostly know one another, again it would not be hard to find short paths. However, short paths exist even though each person mostly knows a few people around himself or herself.

Watts and Strogatz [1998] formalize these properties of small-world networks and show how such networks can be generated. Consider an undirected graph with n nodes, each having a degree of k. A graph is undirected if the edges do not have directions; if you can reach from node A to node B, then you can also reach node A from node B. The degree of node A denotes the number of edges that have A as an endpoint.

To satisfy the first property of small-world networks, k should be substantially smaller than n ($k \ll n$). Next, the nodes should be highly *clustered*. Clustering pertains to how mutually connected the neighbors of a node are. For example, if all k neighbors of a node are also neighbors with each other, the clustering of that node is high. The clustering of the network is the average clustering of all the nodes in the network. Finally, the network should have a short characteristic path length. A path length between two nodes is the smallest number of edges that need to be traversed to reach from one node to the other. The characteristic path length of the network is the average shortest path lengths of all node pairs in the network. By having short characteristic path lengths, small-world networks ensure that it is easy to reach from any node to any other node in the network.

It is interesting to conceptualize small-world networks in relation to random networks and regular networks. Random networks are graphs where the existence of an edge between two nodes is decided randomly. Hence, a node may have a higher degree than another node. Regular networks, on the other hand, are graphs where the nodes have an equal degree.

Small-world networks, like random networks, have short characteristic path lengths. However, unlike small-world networks, random networks have low clustering. In terms of its clustering properties, small-world networks are similar to regular networks: both have high clustering. However, small-world networks are different from regular networks in terms of their short characteristic path lengths, whereas regular networks have long characteristic path lengths.

Because small world networks are somewhat between random networks and regular networks, one way to generate a small-world network is by starting from a regular network and modifying the graph toward a random network [Watts and Strogatz, 1998]. The procedure starts with n nodes, laid out in a circle. Each node is connected to k of its closest nodes, yielding a ring structure. Ideally, n is substantially greater than k such that the graph is sparse. Being a regular network, this formed ring structure has high clustering, but also a large characteristic path length. In order to decrease the characteristic path length, we relocate some of the edges. Going in clockwise order, each node u is visited once. Let (u, v) be an edge. With probability p, this edge is replaced by an edge (u, v') where v' is picked from the set of nodes randomly. In essence, this procedure adds random edges that serve as shortcuts, which decrease the characteristic path length of the whole network. Notice that if the value of p is zero, then the network will remain a regular network. Conversely, if the value of p is one, then the regular network will transform into a random network.

53.2.2 Web as a Small-World Network

Because of their short characteristic path length, information flows fast in small-world networks. This is especially important in settings such as the Web where there is tremendous need to enable fast access to information. Hence, characterizing the Web as a small-world network is important for building search algorithms that can benefit from the shortcuts that exist in small-world networks.

One of the earlier works that study the small-world characteristics of the Web is by Adamic [1999]. Using Web crawls of 1998, Adamic studies the clustering and the characteristic path lengths of a subgraph of the Web. Here the Web sites (rather than Web pages) are taken to be the nodes of the graph. After pruning the sites that do not have any outgoing edges, the graph contains approximately 150,000 sites. To study the shortest paths, the graph needs to be connected in the first place. Hence, Adamic extracts the largest strongly connected component from the graph, in which any site can be reached from any other site by following a directed path. Among the 150,000 sites, approximately 60,000 fall into the largest strongly connected component. Adamic's results show that the characteristic path length of the strongly connected component is 4.228 and the clustering is considerably greater than that of a random network generated from the same number of nodes and the same number of edges. This confirms the hypothesis that the strongly connected component actually is a small-world network.

Albert et al. [1999] study the characteristic path length of the Web, which they term the *diameter.* Rather than using a crawl of the Web, they use the Web pages of the *nd.edu* domain as experimental and derive an equation that relates the diameter of the Web graph to the number of Web pages included in the graph. Their findings show that the diameter of the Web is linearly related to the logarithm of the number of nodes in the graph. That is, even if the number of nodes in the graph increase exponentially, the diameter increases little. For the nd.edu domain that contains 325,739 pages, the diameter is calculated as 11.6. Using the same equation to predict the diameter of the Web (with 8×10^8 pages) yields 18.59. That is, even though the number of pages is increased three orders of magnitude, the diameter increases by just six clicks. This also suggests that even though the Web keeps growing at high rates, the small-world characteristics of the Web will continue to hold.

53.3 Power-Law Distributions

Another interesting property of the Web is how the *in-degrees* and *out-degrees* of the nodes are distributed. The in-degree of a node is the number of edges pointing at the node, and the out-degree of a node is the number of edges that the node is pointing to. Initially, the Web was thought of as a random network, for example, like an Erdös–Rényi random network [Erdös and Rényi, 1960]. One way to generate a random network is as follows. At each time point, two nodes are picked randomly and an edge is added between the two nodes. This suggests that the in-degree and out-degree of the nodes are distributed uniformly, and so each node can be picked with equal probability.

However, measurements of both in-degree and out-degree distributions of the Web show a different distribution [Barabási et al., 2000; Adamic and Huberman, 2000; Broder et al., 2000]. Whereas a few nodes receive most of the in-degree, the remaining nodes all have low in-degree. Interestingly, this is quite similar to Pareto's law for income distributions, known as the law of vital few and trivial many. According to this law, 20% of the world population own 80% of world's wealth. There are variants of this *power-law* that hold for many other distributions, such as business firm sizes [Axtell, 2001], the city populations in the U.S. [Marsili and Zhang, 1998], frequency of word usage in natural languages [Zipf, 1949], as well as in-degree and out-degree of the Internet backbone [Faloutsos et al., 1999].

Equation 53.1 gives the generic form of the power-law distribution for the in-degree distributions on the Web, where i denotes a node, $P(i, k)$ denotes the probability that node i has in-degree k, κ is a constant, and α is the power-law exponent.

$$k = \kappa P(i, k)^{-\alpha} \tag{53.1}$$

The power-law distribution of in-degrees captures the linear relation between the logarithm of the in-degree k of the node and the logarithm of the probability that the node has in-degree k. The exponent α denotes the slope of the line and $log\ \kappa$ denotes the y-intercept. Taking logs of both sides, we have

$$\log k = \log \kappa + \alpha \log P(i, k) \tag{53.2}$$

In other words, the in-degree of a node i is proportional to $1/i^{\alpha}$. Recent studies estimate the exponent α to be 2.1 [Barabási et al., 2000]. That is, the probability of having a certain in-degree drops exponentially with an increase in the in-degree. While different measurements confirm the power-law distribution of degrees, the processes that create these distributions are still being studied. The following are some of the leading models developed to explain the power-law distribution of in-degrees on the Web.

53.3.1 Copying Over

Kleinberg et al. [1999] explain the emergence of power-law distributions through four stochastic processes. The first two processes describe how the pages are created and deleted. Both of these processes assume that at each discrete time point a Web page is created or deleted with some probability p. The next two processes are about how the edges between the nodes are created or deleted. For the creation of an edge, using some probability distribution, a node v is chosen. This is the node that will have new outgoing edges. Next, the number of edges that will be added to the graph is chosen. The new edges can be added in two different ways: (1) with some probability β, k nodes are chosen at random and edges between v and these k nodes are added, (2) with probability $1 - \beta$, a random vertex y is chosen and k out-edges of y are copied over to v (i.e., v ends up with out-edges pointing to same nodes as the original edges). These operations mimic the following: A user views k pages and adds links to these pages from his or her own Web page. Or, the user views a page y and copies over k links from this Web page to his or her own page.

53.3.2 Preferential Attachment

Barabási and Albert [1999] propose a different stochastic process for generating the power-law distributions on the Web based on the idea of *preferential attachment*. The underlying intuition is that the pages that already have high in-degree will be preferred (hence linked) more than the pages that have lower in-degree. This model starts with a small set of Websites. At each time point, a new site with k edges is added to the graph. Rather than picking k nodes randomly as in traditional random graphs, the k nodes are picked based on the in-degree of the nodes.

According to this model, a Website with high in-degree can easily acquire more in-degree because it will be preferred to be linked by new sites that are added to the graph. A new Website, on the other hand, will need a longer duration to gain in-degree and be preferred by newcomers. Thus, a few Websites will increase their in-degree at a faster rate, while many new Websites will start with a low in-degree and wait to be linked. This process yields a power-law distribution wherein those few Websites that increase their in-degree fast will share most of the in-degree among themselves, while many newcomers will not be pointed to by others. This suggests that older pages are more likely to be pointed to than newer pages because those pages can be preferred early on in the process. This model is sometimes criticized because the age of a page may not be indicative of the in-degree it will receive [Adamic and Huberman, 2000]. That is, in principle, a newer page with better content may receive more in-degree than an older page.

Huberman and Adamic describe the distribution of in-degrees on the Web using an earlier model they have developed for capturing the distribution of Web pages per Websites [Huberman and Adamic, 1999]. In this model, each site has a growth rate that determines how fast that Website will acquire new links. Based on the growth rate, the in-degree of a site increases by some fraction of its current in-degree. This way, even new sites (if they have sufficiently high growth rates) can substantially increase their in-degree.

53.3.3 Winners Don't Take All

Pennock et al. [2002] study subgraphs induced by Web pages that belong to different information categories, such as universities, corporate sites (as opposed to, say, Web portals), and so on. Whereas the overall Web graph follows a power-law distribution, all specific categories of the Web do not. In some categories the in-degrees are distributed more uniformly than a pure power-law would indicate. To account for this possible uniformity, Pennock et al. extend the Barabási and Albert model by factoring in the possible uniform attachment by some Web pages. Pure preferential attachment nodes with high in-degree increase their in-degree fast, whereas nodes with low in-degree continue to have low in-degree. The addition of a factor of uniform attachment suggests that nodes with low in-degree can also increase their in-degree as much as the uniform attachment probability allows.

This extended model of Pennock et al. can predict distributions of in-degrees for different categories by adjusting the probability of uniform attachment. For example, for some categories, the probability of uniform attachment is low, and hence the model works based on preferential attachment, yielding a power-law distribution. For some other categories (such as university Web pages), the probability of uniform attachment is higher, yielding a deviation from a power-law.

53.3.4 Local Actions

Notice that among the above models, Kleinberg et al.'s model is the only one that generates a power-law distribution based on the local actions of users. For copying over models, a user need not know the global structure of the network, but he or she can still visit a node at random and copy over some of the links. However, both pure preferential attachment as well as a combination of preferential and uniform attachments generate a graph based on the assumption that each node is aware of the whole network. Even though these centralized approaches can generate the current degree distributions, they do not actually capture the evolution of the Web based on individual actions.

53.4 Link Structure Heuristics

The link structure of the Web can be used to devise heuristics for searching the Web or more specifically for ranking Web pages that are returned for a query. Generally, a search query yields many matches, i.e., many Web pages are related to the query and can potentially be shown to the user. However, ordering the matches so that the user sees the more relevant matches before others is challenging.

Link-based heuristics generally assign values to Web pages based on the incoming and outgoing link structure around the Web page. Then, instead of returning the matches of a query in arbitrary order, the Web pages that have gained higher values are listed before the less-valued pages.

53.4.1 PageRank

PageRank algorithm uses the link structure of the Web to assign a PageRank value to Web pages [Brin and Page, 1998]. The PageRank of a Web page measures its *authoritativeness* based on who is pointing at the Web page. Informally, a Web page gains a high PageRank only if it is pointed to by Web pages with high PageRanks, i.e., if other authoritative pages view this page as authoritative.

Figure 53.1 shows an example graph where the nodes denote Web pages and the edges denote hyperlinks. If a hyperlink exists from a page *A* to page *B*, it is assumed that page *A* endorses page *B*. Initially, each page is assumed to be equally authoritative. At each iteration, each page distributes its PageRank to those pages that it endorses, i.e., the targets of its out-edges. If a page links to many other pages, those pages receive a small share of the page's PageRank. However, if a page links to, say, only one page, then that page receives all of the PageRank of the pointing page. This way, a page can gain high PageRank if authorities link to it, and if those authorities do not link to many other pages.

PageRank algorithm can be formalized as follows. Let K_i denote Web pages that point to page i, $C(i)$ denote the PageRank that a Web page i receives from the Web pages in K_i, and N_j denote the set of Web

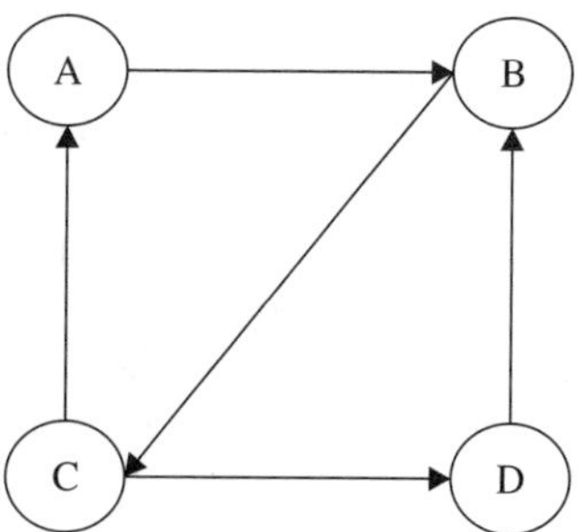

FIGURE 53.1 An example network.

pages that are pointed to by Web page *j*. At each iteration of the PageRank algorithm, a page *j* distributes its PageRank ($P(j)$) equally among the Web pages that it points to. Thus, a page *i* receives PageRank from all the pages that point at it as shown in Equation 53.3.

$$C(i) = \sum_{j \in K_i} \frac{P(j)}{|N_j|} \tag{53.3}$$

This equation captures the intuition of distributing PageRanks, but with a small caveat. Consider two Web pages pointing only to each other, and a third page that points to one of these two pages, as shown in Figure 53.2. This graph will have PageRank flowing in, but not flowing out. In other words, at each iteration the PageRanks of pages *B* and *C* will superficially increase.

In order to avoid PageRanks to be multiply accounted, Equation 53.3 is slightly modified so that a page can sometimes distribute some of its PageRank to a random Web page that is not necessarily one of its neighbors. A damping factor *d* is used to adjust the ratio of the PageRank a node receives from its neighbors compared to the PageRank the node receives from nodes other than its neighbors. For some reported experiments, *d* is taken to be 0.85 [Brin and Page, 1998]. The PageRank of a Web page $P(i)$ is then calculated using Equation 53.4.

$$P(i) = d \sum_{j \in K_i} \frac{P(j)}{|N_j|} + (1-d)(1/N) \tag{53.4}$$

The PageRank algorithm can be explained with a random walker pattern. When a user is viewing a Web page, he or she can either choose to follow one of the links from the page or move to a random page. If the user follows a link, he or she can choose any of the links with equal probability. Again, the option of moving to a random Web page prevents the surfer from getting stuck in a loop. This would infinitely increase the PageRank values of the pages in the loop, making those pages falsely authoritative. However, because the model allows randomly moving to a page outside the loop, the pages in loops receive PageRank from the pages that point at them and without any double counting. With this interpretation, the damping factor *d* corresponds to the probability of following a link from the given page. That is, with probability *d*, a user will follow the links from a page *i*. All the pages in the set N_j will share *i*'s PageRank equally. With probability (1 *Nd*), the user will jump to any one of the *N* Web pages randomly.

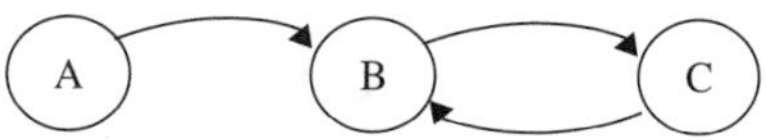

FIGURE 53.2 Nodes B and C form a loop.

One way to compute the PageRank of a Web page is by iterating Equation 53.4 on a Web graph until the PageRank values stabilize, i.e., until the difference between two successive calculations of PageRank values are below a small threshold. The second way of calculating the PageRank values is through matrix operations.

Let M be a transition matrix between the nodes in the graph, normalized so that the sum of each row equals 1. For example, Matrix 53.5 shows the transition matrix for the graph in Figure 53.1. The value 1 of (A, B) shows that node B is the only neighbor of A. The values ½ at (C, A) and (C, D) show that node C is neighbors with both nodes A and D. Thus, a random surfer can move from node C to either A or D with equal probability of ½.

$$\begin{array}{c} \\ a \\ b \\ c \\ d \end{array} \begin{array}{c} \begin{array}{cccc} a & b & c & d \end{array} \\ \begin{pmatrix} 0 & 1 & 0 & 0 \\ 0 & 0 & 1 & 0 \\ 1/2 & 0 & 0 & 1/2 \\ 0 & 1 & 0 & 0 \end{pmatrix} \end{array} \tag{53.5}$$

Let R be the random transition probabilities. For the PageRank calculations, R equals the uniform probability distribution; e.g., each page can be picked with equal probability of $1/N$. Equation 53.6 gives the characterization for the PageRank vector.

$$p = d(M^T)p + (1 - d)R \tag{53.6}$$

Then, the principal eigenvector p equals the probabilities that a random surfer will visit each page. For example, again for the graph in Figure 53.1 solving Equation 53.6 yields the following vector:

$$\begin{array}{c} a \\ b \\ c \\ d \end{array} \begin{pmatrix} 0.344 \\ 0.660 \\ 0.635 \\ 0.344 \end{pmatrix} \tag{53.7}$$

This vector denotes that the PageRank of nodes A is 0.344, the PageRank of node B is 0.660, and so on. The interpretation is that among the four nodes in Figure 53.1, node B is the most authoritative, while node C is the second authoritative, and nodes A and D are the least authoritative.

53.4.2 HITS

Hyperlink Induced Topic Search (HITS) is another Web-page ranking technique that uses Web structure [Kleinberg, 1999]. In addition to identifying some Web pages as authoritative, HITS identifies some pages as *hubs*. Each page in the graph is assigned a positive authority value and a positive hubness value.

A page is a good authority if it is being pointed to by many good hubs. For example, in Figure 53.3(a), if nodes A, B, and C are good hubs, then node D is a good authority. Similarly, a Web page is a good hub if it points to many good authorities. In Figure 53.3(b), if nodes B, C, and D are good authorities, then node A is a good hub. The authoritativeness and hubness values are calculated using Equations 53.8 and 53.9. As shown in Equation 53.8, the authoritativeness of a node i is the sum of the hubness values of the nodes that point at i.

$$a_i = \sum_{j \in K_i} h_j \tag{53.8}$$

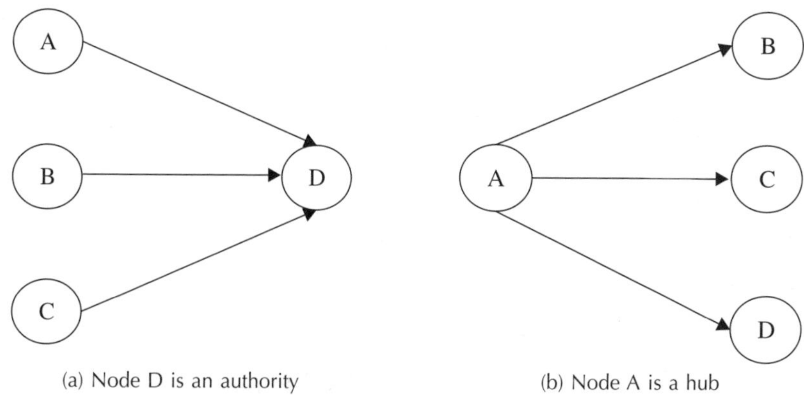

FIGURE 53.3 Hubs and authorities based on link structure.

As shown in Equation 53.9, the hubness value of a node i is the sum of the authoritativeness values of the nodes that i points to.

$$h_i = \sum_{j \in N_i} a_j \tag{53.9}$$

A Web page can increase its hubness value by pointing to a greater number of authoritative pages. On the other hand, the authoritativeness of a page is decided by other pages. A Web page has high authoritativeness if pages with high hubness values point to it.

Similar to PageRank calculations, the hubness and authoritativeness values can be calculated by iterating over Equations 53.8 and 53.9 until a fixed point is reached (i.e., successive iterations have negligible differences). The other method is to use eigenvectors. Let A be the adjacency matrix of a graph, such that if there is an edge between two nodes, the entry is 1, and if there is no edge, then the entry is 0. Equations 53.10 and 53.11 give the fix-point characterizations of authority and hubness vectors, respectively.

$$a^{(t+1)} = A^T h^{(t)} = (A^T A) a(t) \tag{53.10}$$

$$h^{(t+1)} = A a^{(t+1)} = (A A^T) h(t) \tag{53.11}$$

HITS runs on a subgraph of the Web. First, the query is run on a search engine and the top t ranked Web pages are retrieved. These pages make up the *root set* for the given query. The value for t is usually kept small (around 200). That is, the root set contains a small number of relevant pages about a given query. Intuitively, because the root set already contains pages relevant to the query, other pages that are pointed to by the root set as well as pages that point to the pages in the root set are likely to be relevant to the query as well. Hence, the root set is expanded into a *base set* by first adding all the Web pages that point to the pages in the root set. For the pages that are pointed to by the root set, a threshold value d is defined. If the number of pages that point to those in the root set is less than d, then all of these pages are added. Otherwise, a d number of pages among the pointers are chosen arbitrarily and added to the base set. The hubs and authorities are then computed on this base set. The results of a query are two separate lists of Web pages such that one list ranks the pages based on their authority values and the other list ranks the pages based on their hubness values. For example, again for the graph in Figure 53.1, solving equations 53.10 and 53.11 yield the following authority and hubness vectors, respectively.

$$a = \begin{pmatrix} 0 \\ 1 \\ 0 \\ 0 \end{pmatrix} h = \begin{pmatrix} 0 \\ 0 \\ 1 \\ 0 \end{pmatrix} \tag{53.12}$$

When the authoritativeness of the nodes are considered, then the only node that is more authoritative than the other nodes is node *B*. However, if the hubness of the nodes are considered, then node *C* is a better hub than all the other nodes in the graph.

53.4.3 Topic-Sensitive PageRank

As described above, the PageRank algorithm derives a PageRank vector from a Web graph. This PageRank vector is derived once and used for all search queries. This vector is generated independent of a given query. However, a Web page can be authoritative intuitively for a given query, but not authoritative for many other queries. That is, the context of the query is important to judge which documents would be more authoritative for a given query [Henzinger, 2001].

Haveliwala [2002] develops an algorithm called topic-sensitive PageRank to exploit this intuition. The algorithm starts with a predefined set of topics. For the reported experiments, the number of topics chosen is 16 and the topics correspond to the top-mast category titles of the Open Directory Project. Let c_i denote the *i*th category and the T_i denote the URLs listed under that category. The topic-sensitive PageRank builds the random probability vector *R* such that the probability of moving to a URL not listed under the category is zero, whereas the probability of moving to a URL under that category is uniformly distributed. For example, if a given query is about, say, arts, you may jump to a Web page on this topic but not to any other Web page related to a different topic. Notice that for the PageRank algorithm, the probability vector *R* was taken from a uniform distribution; each page could be chosen with the same probability of $1/N$. By using a different *R* for each topic, different topic-sensitive PageRank vectors are computed.

When a query comes in, the algorithm first decides how related the query is to each one of the predefined topics. To calculate this, for each domain, a term vector is built. This vector keeps the number of occurrence of each term in the category. The terms in the query are compared to the term vector of each category and a similarity probability ($P(c_i|q)$) is calculated for each domain. Then, each document *d* is ranked based on how authoritative it is for category *i* ($rank_{id}$).

$$s_{qd} = \sum_i P(c_i \mid q) \times rank_{id} \tag{53.13}$$

A document's topic-sensitive importance (s_{qd}) is based on these two values, as shown in Equation 53.13. If a document is authoritative in a category that the query belongs, then the document is ranked high in the search results for the query.

53.4.4 Problems

Even though ranking methods based on link analysis of the Web have been useful, some problems still remain.

53.4.4.1 Topic Drift

HITS algorithm uses a base set of URLs to build a search graph. The choice of URLs in the base set can influence the search results [Borodin et al., 2001]. If a base set has a subset of documents that all point at each other, the documents in the subset come out as authoritative, even if the documents are not related to the query. The obvious way of coping with this problem of topic drift is by choosing the base set uniformly and by eliminating some of the links among these pages, so that the links among the pages do not superficially decide on an authority up front. However, there is no obvious way to automate this process.

53.4.4.2 Stability

The stability of an algorithm shows how resistant the algorithm is to changes in the data. For example, on the Web, many new Websites become online and many others become offline on a regular basis. Intuitively, if a Website is authoritative, the addition of some new sites or the absence of a couple of existing sites should not affect that site's authoritativeness, Similarly, a Website that is not found to be authoritative should not become a top authority just because a few new sites become online.

Ng et al. [2001a] study the stability of HITS and PageRank algorithms. Their initial results show that in general, PageRank algorithm is more stable than the HITS algorithm. However, the stability of the HITS algorithm can improve with some minor changes. One improvement they suggest is to constrain the hubness values. In the original HITS algorithm, a node is a good hub if it points to authorities. But, a hub is not penalized for also pointing to nodes that are not authoritative. Ng et al. propose that a node should be viewed as a good hub if it is pointing only to authorities.

53.5 Community Structures

While the link-based heuristics have certainly improved the ranking of search results, there are still some shortcomings about searching the Web. Most importantly, because indexing is expensive, the crawlers cannot index the Web on frequent intervals. Thus, a page that is returned for a query may be offline, or its content may be different from when it was indexed. Further, when the crawlers index the Web, they can only index a small portion of it [Lawrence and Giles, 1999]. For this reason, it may be more viable for a Web crawler to find only a *community* of Web pages for a given query rather than crawling the Web offline for a general set of queries.

An interesting application of Web structure is the identification of communities on the Web. Even though different approaches adopt different definitions of communities, the most used definitions pertain to similar Web pages or Web pages that tightly link to one another.

53.5.1 Maximum Flows

Flake et al. [2002] use the frequency of ties among the Web pages as a defining characteristic of communities. This approach views a Web community as a set of Web pages that has more hyperlinks to the pages of the community than to the pages outside the community.

Given this definition, finding the Web communities can be expressed as a variant of the graph-partitioning problem, which is NP-complete. However, if some *seed* Web pages can be assumed to be part of a community, then the problem can be cast as a maximum-flow (or minimum-cut) problem.

A maximum-flow problem works on a directed graph, where two nodes are labeled as a source node and a sink node, respectively. All the edges are labeled with the maximum capacity that they allow to be transmitted from their end-points. A solution to the maximum-flow problem on such a graph gives the maximum information that can flow from the source node to the sink node, such that each edge carries at most the size of its maximum capacity. An easy way to calculate the maximum flow of a graph is by finding the graph's minimum cut. The minimum cut of a graph contains a set of edges such that when these edges are removed, the source and the sink nodes end up in different components. Figure 53.4 shows an example graph where node A denotes the virtual source and node M denotes the virtual sink.

To find the communities, the Web graph is cast into a maximum flow problem. The seed pages all together make up the source node of the maximum flow. A set of portal sites (such as Yahoo!, because of their high in-degree) are then connected into a virtual sink node. The reason for this is that the Websites with high in-degrees are pointed to by many other Websites of varying topics. Hence, they are not necessarily part of any focused community.

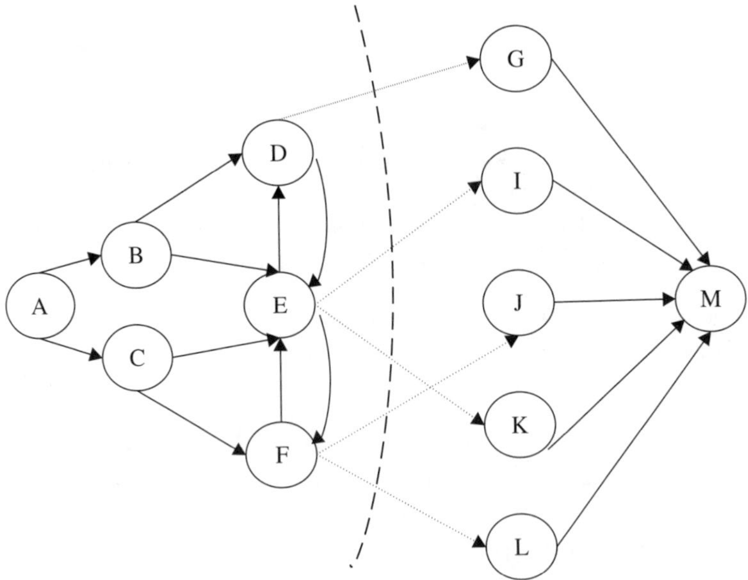

FIGURE 53.4 The dashed line separates the community from the rest of the network.

Finding a cut of the graph corresponds to dividing the graph into two components. The initial seed pages fall into one component, and the more general, less focused pages, such as portal sites, fall into the second component. The first component contains pages that are more related to the seed pages and thus form a focused community. In Figure 53.4, the dashed line separates the community (left side of the dashed line) from the rest of the network. Notice that the edges of the Web graph do not have any capacities. Instead, two threshold values $s^{\#}$ and $t^{\#}$ are defined. The former denotes the number of edges the source node should have within the community, whereas the latter denotes the number of edges the sink node should have with the Web pages outside the community. The requirement is that the cut size should be greater than both $s^{\#}$ and $t^{\#}$. The cut size being greater than $s^{\#}$ means that the community is big enough, i.e., it does not contain the seed pages alone. The cut size being greater than $t^{\#}$ means that the other component is kept big enough, so the community is still focused.

53.5.2 Bipartite Cores

Kumar et al. [1999] propose an alternative link-based definition of communities using a form of *co-citation*, as frequently used in the studies of bibliography sciences. The underlying idea is that if two pages link to same pages, then they are likely to be on the same topic, and thus belong to the same community.

To capture this intuition, Kumar et al. use *bipartite cores,* which are bipartite co-citation patterns. Figure 53.5 shows an example of a bipartite core. The characteristic property of a bipartite core is that

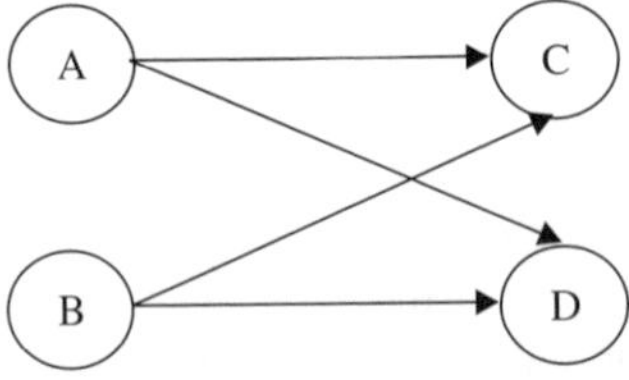

FIGURE 53.5 A (2, 2) bipartite graph.

the nodes can be partitioned into two sets such that every node in the first set links to every other node in the second set. For example, in Figure 53.5, *A* and *B* make up the first set and *C* and *D* make up the second set. Both *A* and *B* point to both *C* and *D*. The nodes that point to other nodes are called *fans* and the nodes that are pointed to by fans are called *centers*. For example, nodes *A* and *B* in Figure 53.5 would map to fans, and nodes *C* and *D* would map to centers.

Kumar et al. propose that any community structure should contain a bipartite core. A graph contains a bipartite core if any subgraph of the graph is a bipartite core. If all n fans point to a set of m centers, then they are likely to share a common topic and therefore be a community. Especially in the case of high n and m, the likelihood of being a community is assumed to be higher. The community is extended by adding nodes that are related to the bipartite core: (1) all the nodes that are pointed to by the fans are added, and (2) the nodes that point to at least two centers are added. For example, in the graph in Figure 53.6, the nodes *A* to *F* could denote a community where the nodes *A* and *B* are the fans, and the nodes *C* and *D* are the centers. These nodes constitute the bipartite core (as also shown in Figure 53.5). The node *E* is added because it points to two centers (*C* and *D*) and the node *F* is added because it is pointed to by a fan (*B*). After a community is extracted from the Web graph, the authorities and hubs for that community can be computed by running the HITS algorithm on the community.

53.6 Directions

Even though Web structure has been studied only in the last several years, the initial results show promising ideas for application and for further research. Link-based ranking heuristics, such as PageRank, are being used by search engines. Considerable effort is being made to evaluate, augment, or develop other link-based heuristics [Amento et al., 2000; Bharat and Mihaila, 2002; Ng et al., 2001b; Borodin et al., 2001].

Whereas most of these heuristics use the structure as aggregated in central servers, the more fruitful applications will come about as the structural properties begin to be exploited locally. For example, in Section 53.2, we discussed some experiments that show the Web is a small-world network. These studies are important because they show that short paths exist on the Web. However, this knowledge does not show how to use these paths. More specifically, an individual navigates the Web based on local information only, without knowing the global structure of the Web. Even though shortcuts do exist, a user, in principle, may never discover them [Kleinberg, 2000; Menczer, 2002]. Hence, the users, or perhaps their intelligent assistants, need mechanisms to discover these shortcuts, and adapt as appropriate, to exploit structural

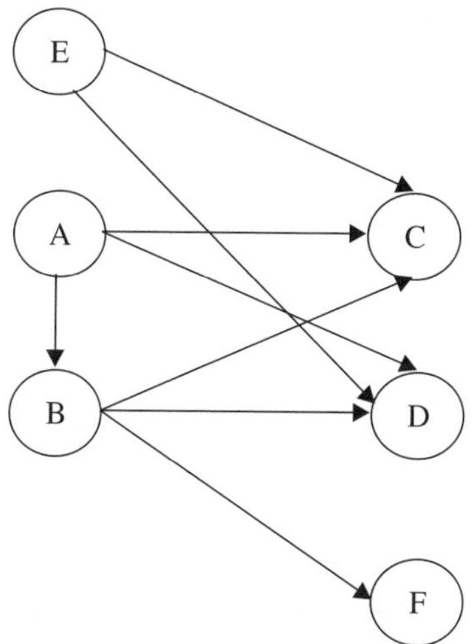

FIGURE 53.6 Nodes *A* to *F* form a bipartite community.

properties of their network. Once these shortcuts are discovered, the next step will be to identify the ones that are trustworthy.

Trust is an essential element of human dealings. Thus, it is not surprising to expect it to play a crucial role for the interactions on the Web. A bigger challenge then is to design mechanisms to identify trustworthy parties to interact with on the Web [Yolum and Singh, 2003]. Intuitively, trust plays a crucial role in the formation of communities as well. Current approaches identify Web communities centrally to serve information to the community members. However, with current approaches, members of the communities generally cannot use their communities as humans do to find information. More expressive representation of communities should include the levels of trust members have for each other. Most of the current approaches are geared towards understanding the structure of the static Web. However, increasingly more information resides in the *deep Web.* That is, information is generated as responses to individual queries based on individual information needs. Hence, most information is not accessible through traditional Web crawling techniques [Raghavan and Garcia-Molina, 2001]. For example, a Website may expose information or provide hyperlinks based on the credentials of the querying user or Website. The structure induced by such subtle interactions can be substantially different from that of the static Web as studied today [Singh, 2002]. Current models are strictly stochastic, not taking into account any factors other than some probabilistic distributions. However, to understand the structure of the deep Web, we need better models that can take into account more intricate information, such as why individual Websites prefer to link to one Website over another, the grounds on which a Web page copies over links to a new page, and so on.

References

Adamic, Lada. The small world Web. In *Proceedings of the 3rd European Conference on Research and Advanced Technology for Digital Libraries*, vol. 1696 of *Lecture Notes in Computer Science*, pp. 443–452. Springer-Verlag, Berlin, 1999.

Adamic, Lada A. and Bernardo A. Huberman. Power-law distribution of the World Wide Web. *Science*, 287(5461): 2115, March 24, 2000.

Albert, Réka, Hawoong Jeong, and Albert-László Barabási. Diameter of the World-Wide Web. *Nature*, 401: 130–131, September 1999.

Amento, Brian, Loren Terveen, and Will Hill. Does "authority" mean quality? Predicting expert quality ratings of Web documents. In *Proceedings of Annual ACM Conference on Research and Development in Information Retrieval (SIGIR)*, pp. 296–303, Athens, Greece, July 2000.

Axtell, Robert L. Zipf distribution of U.S. firm sizes. *Science*, 293: 1818–1820, September 7, 2001.

Barabási, Albert-László and Réka Albert. Emergence of scaling in random networks. *Science*, 286: 509-512, October 1999.

Barabási, Albert-László Réka, Albert, and Hawoong Jeong. Scale-free characteristics of random networks: The topology of the World Wide Web. *Physica A*, 281: 69–77, June 15, 2000.

Bharat, Krishna and George A. Mihaila. When experts agree: using non-affiliated experts to rank popular topics. *ACM Transactions on Information Systems*, 20(1): 47–58, January 2002.

Borodin, Allan, Gareth O. Roberts, Jeffrey S. Rosenthal, and Panayiotis Tsaparas. Finding authorities and hubs from link structures on the World Wide Web. In *Proceedings of the 10th International World Wide Web Conference (WWW10)*, pp. 415–429. ACM Press, New York, 2001.

Brin, Sergey and Lawrence Page. The anatomy of a large-scale hypertextual Web search engine. *Computer Networks and ISDN Systems*, 30(1-7): 107–117, 1998.

Broder, Andrei, Ravi Kumar, Farzin Maghou, Prabhakar Raghavan, Sridhar Rajagopalan, Raymie Stata, Andrew Tomkins, and Janet Wiener. Graph structure in the Web. *Computer Networks*, 33 (1–6): 309–320, 2000.

Erdós, Paul and Alfred Rényi. On the evolution of random graphs. *Publications of the Mathematical Institute of the Hungarian Academy of Sciences*, 5: 17–61, 1960.

Faloutsos, Michalis, Petros Faloutsos, and Christos Faloutsos. On power law relationships of the Internet topology. In *Proceedings of the ACM SIGCOMM Conference on Applications, Technologies, Architectures, and Protocols for Computer Communication*, pp. 251–262, 1999.

Flake, Gary William, Steve Lawrence, C. Lee Giles, and Frans M. Coetzee. Self-organization and identification of Web communities. *IEEE Computer,* 35(3): 66–70, March 2002.

Haveliwala, Taber H. Topic-sensitive PageRank. In *Proceedings of the 11th International World Wide Web Conference (WWW11)*, pp. 517–526. ACM Press, New York, May 2002.

Henzinger, Monika R. Hyperlink analysis for the Web. *IEEE Internet Computing,* 5(1): 45–50, January 2001.

Huberman, Bernardo A. and Lada A. Adamic. Internet: Growth dynamics of the World-Wide Web. *Nature,* 401: 131, September 1999.

Johnson, Steven. *Emergence: The Connected Lives of Ants, Brains, Cities, and Software.* Touchstone, New York, 2002.

Kleinberg, Jon M. Authoritative sources in a hyperlinked environment. *Journal of the ACM,* 46(5): 604–632,1999.

Kleinberg, Jon M. Navigation in a small world. *Nature,* 406: 845, August 2000.

Kleinberg, Jon, Ravi Kumar, Prabhakar Raghavan, Sridhar Rajagopalan, and Andrew Tomkins. The Web as a graph: Measurements, models, and methods. *Proceedings of the 5th International Conference on Combinatorics and Computing,* 1627: 1–18, 1999.

Kumar, Ravi, Prabhakar Raghavan, Sridhar Rajagopalan, and Andrew Tomkins. Extracting large-scale knowledge bases from the Web. In *Proceedings of the 25th Very Large Databases Conference,* pp. 639–650. Morgan Kaufmann, San Francisco, CA, 1999.

Lawrence, Steve and C. Lee Giles. Accessibility of information on the Web. *Nature,* 400: 107–109, July 8, 1999.

Marsili, Matteo and Yi-Cheng Zhang. Interacting individuals leading to Zipf's law. *Physical Review Letters,* 80(12): 2741–2735, 1998.

Menczer, Filippo. Growing and navigating the small world Web by local content. *Proceedings of the National Academy of Sciences,* 99(22): 14014–14019, October 2002.

Milgram, Stanley. The small world problem. *Psychology Today,* 2: 60–67, 1967.

Ng, Andrew Y., Alice X. Zheng, and Michael I. Jordan. Link analysis, eigenvectors and stability. In *Proceedings of the International Joint Conference on Artificial Intelligence (IJCAI),* pp. 903–910, 2001a.

Ng, Andrew Y., Alice X. Zheng, and Michael I. Jordan. Stable algorithms for link analysis. In *Proceedings of the 24th Annual ACM Conference on Research and Development in Information Retrieval (SIGIR),* September 2001b.

Pennock, David, Gary Flake, Steve Lawrence, Eric Glover, and C. Lee Giles. Winners don't take all: Characterizing the competition for links on the Web. *Proceedings of the National Academy of Sciences,* 99(8): 5207–5211, April 2002.

Raghavan, Sriram and Hector Garcia-Molina. Crawling the hidden Web. In *Proceedings of the 27th International Conference on Very Large Databases (VLDB),* pp. 129–138. Morgan Kaufmann, San Francisco, CA, 2001.

Singh, Munindar P. Deep Web structure. *IEEE Internet Computing,* 6(5): 4–5, November 2002. Instance of the column *Being Interactive.*

Watts, Duncan J. and Steven H. Strogatz. Collective dynamics of "small-world" networks. *Nature,* 393: 440–442, June 1998.

Yolum, Pınar and Munindar P. Singh. Emergent properties of referral systems. In *Proceedings of the 2nd International Joint Conference on Autonomous Agents and MultiAgent Systems (AAMAS),* pp. 592–599. ACM Press, New York, July 2003.

Zipf, G. K. *Human Behavior and the Principle of Least-Effort.* Addison-Wesley, Reading, MA, 1949.

54

The Internet Policy and Governance Ecosystem

CONTENTS

54.1 Major Historical Policy and Governance Developments..... 54-2
- 54.1.1 Meta Internet Ecosystem Transitions 54-2
- 54.1.2 Centers of Authority..... 54-4
- 54.1.3 Centers of Authority — the NIC 54-4
- 54.1.4 Centers of Authority — The NOC 54-5
- 54.1.5 Centers of Authority — The Research and Development Framework 54-5
- 54.1.6 International Politics of Control..... 54-6

54.2 Definitions 54-7
- 54.2.1 Protocols 54-8
- 54.2.2 Network Boundaries and Variables 54-8
- 54.2.3 Legal Constructs 54-8

54.3 Business Sector 54-9
- 54.3.1 Hardware and Software Vendors 54-9
- 54.3.2 Large Commercial Users 54-10
- 54.3.3 Major Service Providers 54-10

54.4 User Sector 54-11
- 54.4.1 Developers 54-11
- 54.4.2 End User and SOHOs 54-12
- 54.4.3 Advocacy and Academic Groups 54-12

54.5 Government Sector 54-12
- 54.5.1 Regulatory Constructs and Requirements for Internet Service Provisioning 54-13
- 54.5.2 Law..... 54-14

54.6 Standards and Administrative Sector 54-14
- 54.6.1 Legacy Standards and Administrative Forums 54-15
- 54.6.2 The Universe of Internet Standards and Administrative Forums 54-15

54.7 Emerging Trends 54-16
- 54.7.1 Security..... 54-16
- 54.7.2 Diversity 54-16
- 54.7.3 Assimilation 54-16

References..... 54-16

Appendix..... 54-18

Anthony M. Rutkowski

Other chapters in this book deal with many different facets of Internet computing. Policy and governance topics thread through nearly all of them. In this chapter, these topics are dealt with comprehensively as an ecosystem of controls on behavior that are assumed by or imposed upon myriad parties in four sectors that comprise or enable the Internet today: (1) a business sector consisting of vendors of Internet products, including large service providers; (2) a user sector consisting of major corporations or institutions, plus

1-58488-381-2/05/$0.00+$1.50

© 2005 by CRC Press LLC

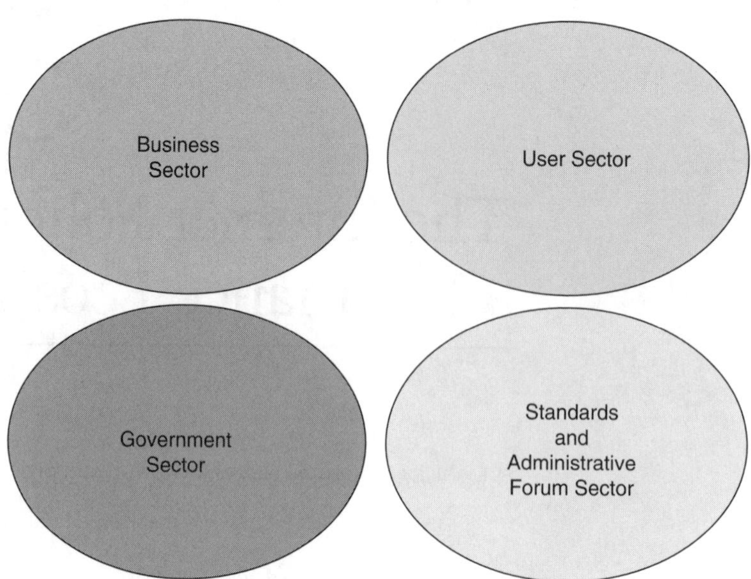

FIGURE 54.1 The Internet policy and governance ecosystem.

individuals or small offices; (3) a government sector; and (4) a standards and administrative forum sector. Wrapped around this ecosystem are important basics such as history, definitions, and emerging trends, as well as extensive references to additional information sources.

Such an ecosystem approach is necessary because of one simple fact — what is known as the Internet is not a network at all in the traditional sense. Rather, the Internet is a means for achieving autonomous resource sharing based on information systems, accomplished by largely independent cooperative actions among the parties constituting the four ecosystem sectors. A better term is perhaps "Internetworking," rather than Internet, and the constituent agglomerations exist because parties make available computer and transmission resources. Where we are dealing with topics like policy and governance, a common understanding of these essential basic elements is critical.

In large measure, this chapter will only focus on the generic Internet ecosystem (Figure 54.1). It will not treat two other prominent Internet related domains that include (1) the enormous number of application and syntax level arenas such as the World Wide Web or Internet Telephony, and (2) underlying transport media such as wireline, wireless, satellite, or cable.

First, it is important to examine the historical context within which the Internet came into existence and evolved at a rapid pace over the last three decades of the 20th century that has produced what we have today.

54.1 Major Historical Policy and Governance Developments

Historically, three somewhat separate sets of developments substantially shaped the Internet policy and governance environment. The first development revolves around the constituents that formed the Internet ecosystem over four distinct periods of time. The second development involves the centers of administrative authority within those periods. The third development represents a larger global "war" between two contending factions over the development, deployment, and control of information networking technology that subsequently just went away.

54.1.1 Meta Internet Ecosystem Transitions

From the point the Internet was first conceptualized as a host-to-host protocol network by Bob Kahn and his team of research developers in the early 1970s, four relatively distinct historical periods have ensued. The first is an initial DOD Advance Research Projects Agency (DARPA) period that presages the

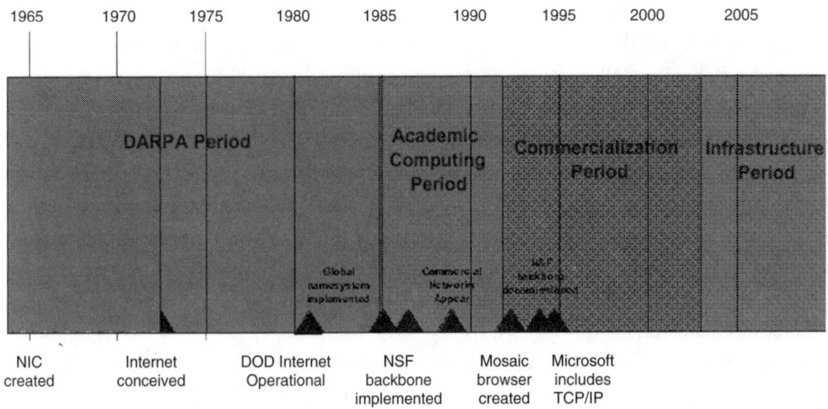

FIGURE 54.2 MetaHistory of the Internet ecosystem.

Internet, then nurtures, adopts, and scales it through the mid-80s. Although the DARPA program office played the dominant role in this period, as the Internet grew and evolved and become more important to DOD, other research centers and offices began to play important roles (Figure 54.2).

By 1982, as the DoD adopted TCP/IP as a protocol of choice in tactical, logistic, and messaging systems, including a mobile packet radio network and the ARPANET, the Defense Communications Agency (now the Defense Information Systems Agency) begins to play a significant role. The ARPANET was at that time a packet-switched technology-based network that had been developed in the 1960s also by DARPA and became an operational backbone for DOD operations and included multiple satellite facilities.

By the mid-1980s, an increasingly large number of parties external to DOD begin to assume important ecosystem roles in small commercial business user communities and large academic computing, and U.S. Federal networking and university Computer Science Network communities oriented around the National Science Foundation (NSF), Department of Energy, NASA, and equivalent institutions in other countries.

A particularly catalytic development was NSF's obtaining about $1.2 billion from the U.S. Congress over the late 1980s and early 1990s to fund the construction of a national TCP/IP backbone, international connectivity, and an enormous amount of applications research among centers across the U.S. The expenditure of this amount of money as national policy decision at such a critical point in the development of networking technology was in retrospect quite an extraordinary move. Particularly sage was the allocation of funds to largely generic applications development, in contrast with decisions made in other countries to allocate similar sums of money explicitly tied to specified communications protocols or standards.

The Academic Period begins to diminish in the early 1990s, as the Internet infrastructure becomes increasingly privatized and a large commercial and consumer marketplace begins to dominate the Internet's management and evolution. This last transition, however, was subject to its own considerable controversy as many academic community actors fought the transfer of "their" technologies and applications to a larger commercial universe encompassing the general public. Ultimately, however, it was large commercial players — especially Microsoft — whose commitment to Internet technology resulted in the scaling of the Internet to encompass the hundreds of millions of users today.

The Commercial Period itself evolved in new directions as it grew. After scaling as a social, economic, and even political phenomenon during the 1990s, followed by several years of a "bursting bubble" descendent phase, the Internet at the time of this publication seems to be finding a niche among synergistic technologies and products, even as it has been thoroughly assimilated by commercial business and an increasingly large portion of modern society.

Indeed, it is this very assimilation that is now giving rise to an Infrastructure Phase. This new phase is marked by an increased focus on security in terms of technologies, operation, and public policy and law. Not only beneficial developments have been manifested through the Internet, but increasingly, the

Internet has been a home to large-scale fraud, identity theft, destructive software agents, and myriad other criminal activities. A hallmark characteristic of this new and long-term steady-state phase of the Internet is security. Behavior will continue to be autonomous, but it will not be anonymous.

None of these transition points is very distinct. For example, commercialization of the Internet as a technology and corporate infrastructure began in the mid-1980s with the creation of such early pioneer companies as Sun Microsystems and Cisco Systems, who marketed their products to corporate IT managers at Interop trade shows. Similarly, the emergence of a consumer mass market could be mapped by the appearance of Internet-related articles in the major newspapers that ultimately led to the commitment of Microsoft Corporation to bundling TCP/IP in the next major release of its operating system. At any point in time, hundreds of events were in play, collectively pushing the envelope of change from day to day.

Like human genetic code, today's Internet policy and governance ecosystem reflect these major historical periods, which continue to shape an ongoing evolution. Not only the norms, but in many cases the roles, if not the powers of institutional parties, are traceable to earlier historical periods.

54.1.2 Centers of Authority

One of the frequently overlooked historical innovations of the Internet's development and evolution is the use of competency centers as sources of authority, many of which persist today. Ecosystems based on autonomously shared resources require an unusual degree of acceptance by the participants, in contrast to dictated power centers of highly regulated traditional infrastructures. In the Internet ecosystem, centers of authority solved this "buy-in" requirement rather nicely by relying on multiple self-initiative among principal actors in the community (Figure 54.3).

54.1.3 Centers of Authority — the NIC

In the Internet's earliest years, when the infrastructure and activities were largely under the control of government research program offices or academic institutions, various competence centers of authority began to emerge. One of the first was the Network Information Center (NIC). The initial DARPA period is traceable back to the assumption by the agency of a packet network research role in the early 1960s. The creation of an NIC is generally credited to computer networking pioneer Doug Engelbart at Menlo Park, California, and was run by the Stanford Research Institute (SRI). If the DARPA Program Office was the ultimate source of power during this early period, the NIC on a day-to-day basis played a key role in the Internet's development and coordination over the first two periods of its development.

During the late 1980s, the NIC began to be broken up into many pieces worldwide based on geographical or governmental jurisdiction, as well as to be increasingly privatized. Yet, even as these developments occurred at regional and national levels, the idea of the NIC competency center was replicated hundreds of times.

The NSF and Commercial Periods also witnessed significant NIC internationalization, beginning with coordination roles under the U.K. Internet pioneer Peter Kirstein at University College London. This was rapidly followed by NICs appearing in multiple countries and the emergence of world regional NICs — the *Reseaux Internet Protocol Europeen* Network Coordination Center (RIPE-NCC) in Amsterdam in the late 1980s, and the Asia-Pacific NIC (AP-NIC) in the early 1990s.

The original primary NIC at Menlo Park was transferred to the Defense Information Systems Agency and became known as the DISA-NIC. In the early 1990s, most of these functions were then transferred to the NSF and renamed InterNIC. The NIC contractors also shifted from SRI to Government Systems, Inc. (GSI) to Network Solutions, Inc. (NSI) (now a part of VeriSign, Inc.).

Today, the NIC as a center of authority is completely distributed among hundreds of cooperating institutions worldwide (See Section 54.6).

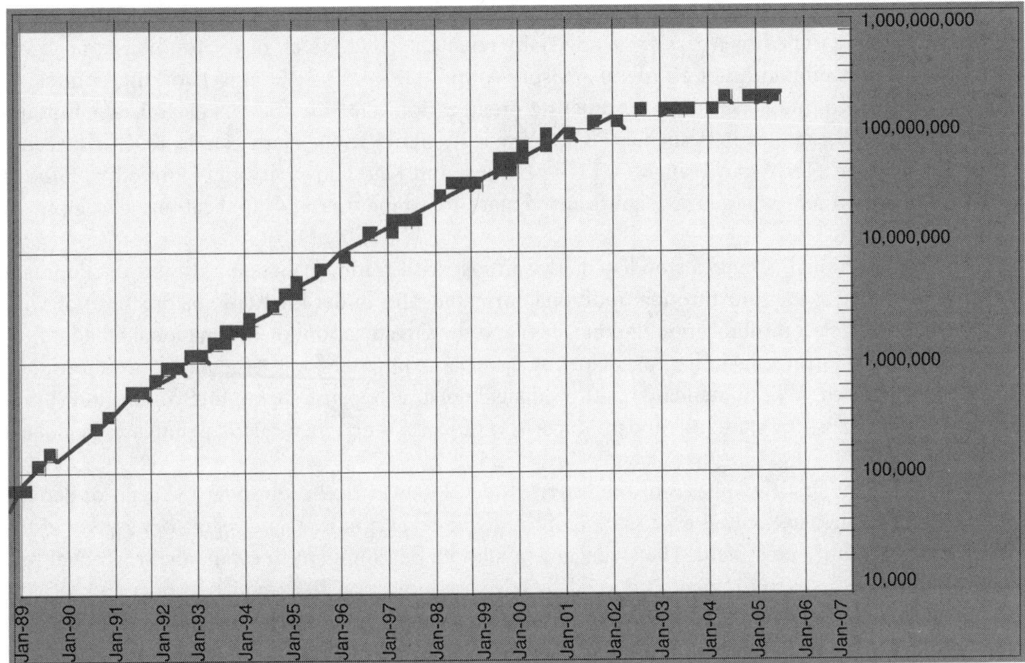

FIGURE 54.3 Publicly reachable Internet hosts (Mark Lottor data).

54.1.4 Centers of Authority — The NOC

A second early DARPA management innovation that ensued at about the same time as the ARAPA NIC was the creation of a Network Operations Center (NOC) operated by Bolt, Beranek, and Newman, Inc. (BBN) near Cambridge, Massachusetts. This responsibility during the NSF Period was largely transferred to Merit Network, Inc. for the domestic U.S. Internet infrastructure and Sprint Corp. for the international infrastructure. The underlying infrastructure itself was provided through an MCI–IBM joint venture known as Advanced Network and Services (ANS) which focused on domestic U.S. networks through multiple regional U.S. networks. Internationally, Sprint Corp. provided equivalent capabilities.

The NOC functions eventually transitioned in the mid-1990s to individual Internet Service Providers (ISPs), coordinated through a combination of bilateral arrangements and multilateral forums that included the Commercial Internet Exchange (CIX) (now known as the U.S. Internet Service Provider Association), and three global regional groups (North American Operators Group or NANOG, the *Reseaux IP Européen* or RIPE group, and the AP Networking Group or APNG). Internationally, this includes the Coordinating Committee for Intercontinental Research Networking (CCIRN).

54.1.5 Centers of Authority — The Research and Development Framework

The third management innovation involved standards making and applications development processes. During the early 1970s, Keith Uncapher started a DOD information systems thinktank in Marina del Rey, California — the Information Sciences Institute (ISI) under the University of Southern California. As initial Internet-focused Internet standards activity began to emerge during the 1970s, ISI — chiefly through the efforts of one of its graduate students, Jon Postel, who operated an Internet Assigned Numbers Authority (IANA) function for DOD — began to play an important standards coordination role that was shared with the Internet Engineering Task Force (IETF) Secretariat in the mid-1980s as the IETF began to emerge as an initial standards development body. The IETF Secretariat was run by the Corporation for National Research Initiatives (CNRI) after it was started in the mid-1980s by Bob Kahn. The secretariat remains at CNRI today.

During the DARPA and NSF periods, these standards and applications processes blossomed with significant funding to nearly every major university research center. Much of the funding was coordinated through a combination of a Federal advisory committee — the Federal Networking Council — and a university computer science coordinating organization. Outside the U.S., significant funding also occurred at research centers such as UCL in the U.K., SURF in the Netherlands, KTH in Sweden, UNI-C in Denmark, INRIA in France, CERN in Europe, and Keio University and University of Tokyo in Japan, all of which emerged as significant centers for standards and applications development activities.

These nontraditional activities stood in stark contrast with traditional standards and development activities occurring at the time through traditional formal forums under international organizations like the International Telecommunication Union (ITU) and the Organization for International Standardization (ISO). These forums, including participating agencies, companies, and academic institutions, had developed their own suite of standards and products known as Open Systems Interconnection (OSI). For most of the 1980s, OSI standards and products were officially sanctioned, and in many cases mandated by law for use.

The fact that the first two phases of the Internet's development occurred under the aegis of defense and scientific research agencies is especially significant with respect to legal and regulatory aspects of the policy and governance ecosystem. The arrangement allowed development to escape the traditional regulatory treatment and requirements imposed by telecommunication law upon networks and services made available to the public. Additionally, the sponsoring agencies assumed the civil liability and policing responsibilities. These roles began to diminish significantly as the Internet's commercial phase began in the mid-1990s. Vestiges of that transition are still underway.

54.1.6 International Politics of Control

The development of the Internet occurred over the years against a backdrop of several major developments that for lack of a better term are cast as international politics, although in many cases these developments had national counterparts. These principally include attempted control over the Internet's (1) ability to exist, (2) standards, and (3) administration of identifiers. Internationally, these controls were principally manifested by the International Telecommunication Union (ITU), which is a United Nations specialized agency of government telecommunication ministries that also serves as an umbrella for legacy telecommunication providers.

Under the international telecommunication regime of treaties affected by the ITU, all telecommunication and information network services and facilities were supposed to exist only under strict rules and standards established by ITU bodies and enforced by national governments worldwide. Although provision was made for large agencies and companies acquiring dedicated private circuit capacity for the purposes of building their own networks, neither the capacity nor the resulting services were to be made externally available. It was simply an international cartel for the purposes of controlling the marketplace for all public telecommunication services.

The notion of an Internet was inimical to this long-standing regime. What ensued was a succession of tactics that first sought to ban the existence of Internets, followed by a coordinated effort to erect economic impediments through costly leased-line tariffs, followed by official dismissiveness of the existence of a massively growing Internet infrastructure and marketplace, followed by attempted control over key administrative functions like identifier administration.

The international telecommunications cartel began to crumble in the early 1980s with a series of actions taken by the Federal Communications Commission in the U.S., beginning with the Computer II decision that established a policy of complete regulatory forbearance toward Internet-like networks. This action in turn induced similar actions like the Open Network Policy (ONP) of the Commission of the European Union, followed by initiatives within the General Agreement on Tariffs and Trade (GATT, now the World Trade Organization) and ultimately in the ITU itself at a 1988 conference that adopted a treaty provision explicitly allowing for an Internet to exist under international law.

Slowly, over the 1990s, as the Internet public marketplace and infrastructure began to scale to the point where it could no longer be ignored, most legacy telecommunication providers began to find ways to cooperate with the emerging array of Internet Service Providers. First the provisioning barriers fell, then the economic impediments of line and access costs began to moderate. Even today, however, in many locales worldwide, the artificial high metered costs of a local access line connection represent a continuing impediment.

A significant component of the global attempt to impede Internet developments began in the late 1970s in the form of a rigorous standards regime that existed on paper in parallel with the Internet's development. This Open Systems Interconnection (OSI) regime took the form of treaty provisions, national law and regulation, services and provisioning controls, and funding strictures. It dominated the formal telecommunications and information networking environment and institutions over nearly a 20-year period, consuming billions of dollars, millions of meeting hours, and whole forests of paper devoted to standards development and regulations. As the Internet and its TCP/IP protocol suite continued to grow in the early 1990s, the frictions and rhetoric grew to the point where the situation was referred to as the "TCP/IP vs. OSI wars." Ultimately OSI completely disappeared circa 1996 as if it had never existed. It did represent, however, an example of the limits of government and tradition industry to dictate market-product specifications in the face of evolutions in technology coupled with the innovations and large-scale public demand.

The next chapter in this history of the international politics of control took the form in 1996 of abortive attempts by the ITU and its constituents to assume power over the Internet's identifier administration provided by the NICs. This initial foray was an abortive one where the ITU General Secretariat attempted in 1996–1997 to craft an international agreement that ceded NIC authority to the ITU through a rump International Ad-Hoc Committee (IAHC).

After intervention by the U.S. Department of State which squelched the initiative, the matter was raised formally in an ITU treaty-making conference in 1998 with a majority of the ITU's constituents crafting an ITU Resolution that called for continuing discussion of the matter. A subsequent conference in 2002 readopted the resolution with minor modifications, and the dialogue continues. During these forays, the U.S. government conducted a policy-making proceeding in the 1997–1998 timeframe that led to a switch of IANA coordinating functions from the Institute for Information Sciences to another nonprofit organization known as the Internet Corporation for Internet Names and Numbers (ICANN). The NICs were essentially unaffected worldwide except for a few of the largest domain-name registration activities that were voluntarily segmented by the provider, Network Solutions, to allow sales opportunities for other providers.

Given the reality that Internet users and providers are unlikely to accept an ITU-dictated regime, coupled with the impracticability of enforcement and the continuing opposition of the U.S. on fundamental policy grounds, the ITU-based international politics of control seems likely to continue indefinitely.

A majority of its national administration members through the ITU do have the power to mandate a treaty-based result that asserts control over Internet names and numbers. At the time of publication of this book, the ITU World Summit on the Information Society (WSIS) is emerging as a venue for advancing such a result. As national regulatory authorities worldwide move toward regulatory frameworks for Internet services, it is inevitable that significant new arrangements will emerge where countries cooperate on some Internet matters thru the ITU. This seems likely to include the use of Internet addresses and some DNS domains as a stable permanent arrangement acceptable to most if not all countries.

54.2 Definitions

In any policy and governance ecosystem or regime, definitions play a key threshold role. This is particularly critical with respect to the Internet because the construct is purely virtual. There is no physical facilities basis for the Internet. It is constituted solely by protocols for sharing virtual information resources.

The threshold challenge is to define the Internet for policy and governance purposes. The challenge is magnified by the reality of the Internet as an abstraction for a chaotic ensemble of millions of networks encompassing hundreds of millions of host computers supporting billions of processes and service capabilities, all of which are autonomously shared in ways that are constantly changing.

54.2.1 Protocols

Generally, the Internet is defined solely as the use of the Internet Protocol, i.e., RFC 791, to exchange datagrams within a core architecture. RFC 791 specifies the Standard Internet Protocol, which "is designed for use in interconnected systems of packet-switched computer communication networks ... and provides for transmitting blocks of data called datagrams from sources to destinations, where sources and destinations are hosts identified by fixed length addresses." Although other Internet protocols exist, the almost universal practice over the past two decades is to confine the term "Internet" to the concatenation of networks using the RFC 791 specification.

54.2.2 Network Boundaries and Variables

At network boundaries or within networks under common management, however, the definition becomes more difficult to apply. The use of proxy servers and firewall gateways allow well-defined constraints on the use of the Internet Protocol to reach connected host computers. The use of further packet encapsulation is capable of creating myriad virtual Internets within the Internet. Terms like "intranet" or "extranet" have been invented to market these creations.

Entirely different network protocols can be used on one side of a gateway or where the gateways are dedicated to specific applications, and encompass entirely distinct, independent networks. One of the most extensive networks involves voice telephony and the existing Public Switched Telephone Network (PSTN). The Internet has long encompassed a larger "matrix" of multiple commercial, academic, and personal user networks such as America On-Line, Bitnet, CSnet, UUCP networks, and Fidonets, as well as gateways to the OSI world's X.400 messaging system, and assorted proprietary messaging networks such as Microsoft Mail, MCI Mail, Sprint Mail. The key requirement is the existence of a connecting gateway to the core Internet concatenation.

54.2.3 Legal Constructs

One of the first definitions developed and widely adopted within legal constructs was that of the Federal Networking Council (FNC) written in 1995 for use within the U.S. government.

> "Internet" refers to the global information system that —
>
> (i) is logically linked together by a globally unique address space based on the Internet Protocol (IP) or its subsequent extensions/follow-ons;
>
> (ii) is able to support communications using the Transmission Control Protocol/Internet Protocol (TCP/IP) suite or its subsequent extensions/follow-ons, and/or other IP-compatible protocols; and
>
> (iii) provides, uses or makes accessible, either publicly or privately, high level services layered on the communications and related infrastructure described herein.

As of 2002, in key proceedings of the Federal Communications Commission dealing with the exercise of regulatory authority over the provisioning of Internet access services, this FNC definition was recited as authoritative as an Internet definition.

Recently, a relatively simple definition was prepared by the T1 Committee of the telephony-oriented Alliance for Telecommunication and Information Standards (ATIS) as American National Standard T1.523-2001

Internet [the]: 1. A worldwide interconnection of individual networks (a) with an agreement on how to talk to each other, and (b) operated by government, industry, academia, and private parties.

The most widely used definition in U.S. domestic legislation and regulations follow the lead of Title 47, Secion 230 of the U.S. Code:

2. The international computer network of both federal and nonfederal interoperable packet switched data networks.

The quandary for regulatory authorities is that the Internet inherently consists of resource sharing among large numbers of "private" resources to create a single common aggregation. Private resources in this context refers to those resources manifested by privately owned computers or networks, and are not subject to national or international obligations to provide to the public as telecommunication facilities or services at network or application layers.

The private vs. public distinction has over the past 150 years formed a fundamental distinction in governing electronic networks. The Internet has come into existence and evolved over the past 30 years as a "private user network" either by virtue of government agency sponsorship or subsequent corporate implementations. This single common aggregation constituting the Internet occurred not by design or regulatory mandate, but rather through the choice of the many participating parties to share those resources for perceived common benefit.

The architectures of this resource sharing are also highly variable through the countless application and network-level gateways that implement locally administered rules for traversing the gateways. This does not imply anything, however, about definitive laws applying to Internet usage and behavior. The Internet, whatever its definitional construct, stands distinct and transparent with respect to individual or institutional behaviors and actions manifested using Internet resources. A rather significant constellation of policy and law applies, as covered in Section 54.5 below.

54.3 Business Sector

Providers of Internet hardware and software products, as well as Internet services, have long played the most significant role in the governance ecosystem since the mid-1980s when TCP/IP began to emerge as the internetworking protocol of choice. It includes companies and other kinds of organizations such as government agencies who procure intranet/extranet infrastructure for their own use.

This rather significant role of the Internet business frequently gets subordinated by other ecosystem sectors that depend on public self-promotion. In the final analysis, however, it is the individual and collective business decisions of vendors that substantially govern the Internet and implement the provisions of the other sectors.

The largest vendors also have significant resources that can be deployed to create their own independent development and standards communities that are extraordinarily valuable, bringing about rapid innovation and widespread deployment of new technology. This kind of entrepreneurial "just do it" behavior stands in striking contrast to traditional legacy practices in the telecommunications industry that rely on formal, hierarchical international, regional, and national standards bodies and development activity, with decade-long cycles. The formality and rigidity can be exacerbated and cycles stretched out over even longer timeframes through overlays of formal government-sponsored R&D activity frequently endemic in Europe and Asia.

54.3.1 Hardware and Software Vendors

Although there are many vendor-specific Internet development forums in existence today, the most prominent include large hardware and software vendors who have chosen to create their own communities, including devoting large grants to independent developer institutions such as Microsoft, Cisco, Sun Microsystems, and IBM. In some cases, such as Sun Microsystems with Java, the development activity was largely spun off as an independent group. This is not to say that other industry vendors are not significant and highly influential in the governance realm; only that the largest ones, who have also chosen to create an extensive community penumbra, emerge as the most prominent.

Because the Internet is fundamentally a software-based construct, it is not surprising that the vendors who control most of the operating systems extant on the hundreds of millions of Internet host computers emerge at the top of the governance ecosystem. "Code as Law" has even given rise to book-length treatises. However, these vendors are not alone. Constant changes in technology, agile competitors producing compelling new applications, marketplace conditions, constraints imposed by other suppliers in the Internet food chain such as telecom operators, and government agencies — all constrain the power of even the largest actors.

Prominent collective industry groups that have emerged to represent this sub-sector in the U.S. include the Information Technology Association of America (ITAA) and the Software Publishers Association.

54.3.2 Large Commercial Users

From the earliest years of the Internet and indeed the X.25 data network universe preceding it, the interests and role of large commercial users have been paramount in policy and governance. In this context, "commercial user" includes corporations, government agencies, and institutions, especially educational ones. One has only to look at the allocations of Class A blocks of Internet addresses to get a listing of commercial users who early on expressed their interests in the form of resource allocations.

The first large conferences devoted to the Internet were, not surprisingly, the Interop trade shows and seminars begun in 1986 to provide a means for the commercial user sector to meet, discuss current policy and governance developments, view new products, and express their common interests and needs to vendors. The phenomenon has continued over the years and dispersed worldwide. The number of commercial trade shows and seminars focused on the Internet today has blossomed to such an extent that it is difficult to discover all of them.

Large commercial users have also played significant roles within advisory bodies, as well as formal regulatory, legislative, and judicial forums, resulting in shaping some of the most fundamental Internet policies and governance regimes at domestic national and international levels. This has occurred both through individual corporate and institutional initiative, as well as collectively through common user organizations. Especially notable over many years have been ADEPSO, CBEMA, INTUG, the International Chamber of Commerce, and EDUCOM (now EDUCAUSE).

54.3.3 Major Service Providers

The Internet was largely ignored by service providers until it began to scale significantly as a business opportunity in the late 1980s. The first entrants as stand-alone Internet providers were UUNET and Performance Systems International. At about the same time, MCI obtained part of an NSF award to construct a national backbone (NSFnet), followed a few years later by Sprint garnering a similar award for international connectivity (International Connections Manager).

The late 1980s saw the emergence of mixes of educational and specialized Internet-related provider organizations appropriate to the times. These included creatures like FARNET and USENET, as well as comparable regional organizations like RARE (Réseaux Associés pour la Recherche Européenne), EARN (European Academic and Research Network, which subsequently joined with RARE to form TERENA in 1994), RIPE (which also emerged as an administrative organization), and a plethora of national-level bodies.

In the early 90s, the Commercial Internet Exchange (CIX) organization was formed among the then existing providers to play a major policy and governance role, including supporting a traffic exchange mechanism. The CIX subsequently evolved into the U.S. Internet Service Provider Association (US ISPA). *Boardwatch* magazine also emerged as an Internet service provider advocacy organization through its semiannual conferences of ISPs and policy-making initiatives that were institutionalized in the U.S. Internet Industry Association.

As the Internet Service Provider business grew and merged to a significant extent with mainline telecommunications provisioning, the boundaries between telecom, online (especially America On-Line), and Internet provisioning are substantially blurred. This has been reflected in turn in the associated ISP bodies in most countries, regions, and states. Hybrid organizations such as the Cellular Telecommunications and Internet Association (CTIA) in the U.S. are exemplary of the evolution within legacy industry organizations.

Like other business sector users, providers — individually and collectively through their industry bodies — constitute critical components of the policy and governance ecosystem because of the scaling, deployment, development, and economics of the Internet through advocacy and decisions taken within their organizations and in the marketplace.

Some significant business sector organizations dealing with governance and policy span broad interests. One of the less visible but nonetheless influential is the Internet Law and Policy Forum (ILPF). The ILPF consists principally of representatives from the general counsel or government relations offices of many significant providers of Internet products and services, and has been influential in harmonizing transnational law that affects the Internet.

54.4 User Sector

The Internet by definition is an edge network consisting of host applications and processes reachable by a combination of unique host addresses and TCP/UDP ports. Individual users and local system administrators have the ultimate ability to govern the Internet with respect to the user domain.

During the 1970s and 1980s and up through the mid-1990s, the collective power of users was especially strong because of the ability of most users until that time to set up their own Internet services and applications. As the Internet became a mass-market phenomenon — first with Microsoft bundling TCP/IP into the Windows operating system, and then with AOL connecting its infrastructure to the Internet via a gateway — the effective policy and governance power of end users began to decline. A prominent exception is developers.

54.4.1 Developers

The developer community, that is, individuals and groups that actually write "running code," has always been one of the principal strengths and forces within the Internet environment. Even those operating on the "dark side" as hackers of various sorts significantly shape the Internet's ecosystem.

For many years, the developer community existed largely with the university computer science community and the National Labs, and then, over time, migrated into existing companies or crafted new startups. The university Internet developer community was significantly well funded especially between 1985 and 1995 through the National Science Foundation that expended more than $500 million to create a renaissance for application development, which was enhanced through additional funding through DARPA.

Scores of new mass market applications, some successful and many unsuccessful, reshaped the Internet environment and led to new policy and governance mechanisms and developments. These included almost everything identified with the Internet today: e-mail, World Wide Web, network caching, search engines, file sharing, Internet domain names, Voiceover IP, and dialup access. All emerged from developer communities and institutions. Some subsequently evolved into continuing research institutions such as the Cooperative Association for Internet Data Analysis (CAIDA) spearheaded by kim (kc) claffy.

Perhaps the most significant developer forum is also a standards body — the Internet Engineering Task Force. In the Web development environment, the World Wide Web Consortium (W3C) is a forum led by Web developer Tim Berners-Lee, and enhanced through a companion staff developer team as well as an International World Wide Web Conference.

The U.S. was not alone in these endeavors. Almost every large country and region has maintained well-funded Internet development initiatives as a manifestation of national policy. The Commission of the European Union's Information Society programme is among the largest.

As noted above in the context of the business sector, major hardware and software vendors began to create their own large, active Internet user communities as the market opportunities grew. All of these communities coexist and in complex ways through scores of forums, large and small, many in the form of Internet-based virtual organizations.

54.4.2 End User and SOHOs

End users and Small Offices/Home Offices exercise broad power to make macro decisions affecting Internet governance and policy through their marketplace choices and through political pressure placed on officials in government or administrative positions. Their procurement choices also represent an enormous embedded economic base of capital investment. The Internet itself is an effective tool in rapidly organizing end users and reaching decision makers.

In the early 1990s, several small end-user advocacy organizations emerged. The Internet Society was formed primarily as an organization to promote common interests of the educational user community. The Society expanded its scope, created numerous national chapters, and subsequently asserted Intellectual Property ownership of the IETF standards and represented the IETF's interests in other standards bodies.

54.4.3 Advocacy and Academic Groups

A significant number of small advocacy organizations across the political spectrum have also emerged to play policy and governance-shaping roles. Some of the more prominent Internet libertarian groups include the U.S.-oriented Electronic Frontier Foundation (EFF) and Center for Democracy and Technology (CDT), Computer Professionals for Social Responsibility (CPSR), the Foundation for Information Policy Research in the U.K., and, internationally, the Soros Foundation Open Society Institute, the Global Internet Liberty Campaign (GILC), and the Global Internet Policy Initiative (GIPI). Others concerned with Internet content include ProtectKids.com.

A large number of prominent academic groups are involved in Internet policy and governance. Some of the more prominent in the U.S. include Harvard's Berkman Center, Chicago Law School's the Stanford Center for Internet and Society, Chicago-Kent College of Law, and the Georgia Tech Information Security Center (GTISC).

54.5 Government Sector

Public bodies have the ability to substantially shape the behavior of other governance ecosystem sectors, frequently with substantial interaction through public consultative proceedings or funding decisions. Government policy is manifested both through funding decisions such as discussed earlier under other sectors, or governance actions that are both direct (i.e., specific legal and regulatory provisions that apply to Internet use) as well as indirect (i.e., generic provisions that apply to all networking or other kinds of uses).

In almost all government systems, the governance and policy-making activities are effected through legislative, executive, judicial, or independent agency bodies that can exist at national as well as local levels. Additionally, national governments may establish bilateral agreements between themselves, or multilateral agreements among any number of nations through global and regional intergovernmental organizations, typically in the form of treaty instruments (Figure 54.4).

A large and rapidly growing body of law and policy applies both generally and explicitly to Internet operation and user conduct that may be regulatory in nature, or which establish civil and criminal causes of action. Where multiple law and policy of different jurisdictions concurrently applies to Internet

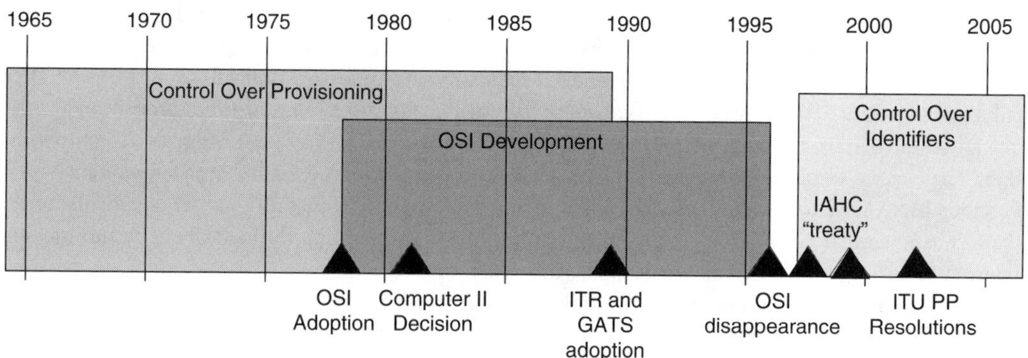

FIGURE 54.4 Historical periods of international control attempts.

architecture, service, or user behavior, instances of Conflict of Law occur, and for which there are some generally accepted guides for weighing competing claims and interests in crafting an equitable and just result.

54.5.1 Regulatory Constructs and Requirements for Internet Service Provisioning

The most enduring and significant regulatory construct applicable to the provisioning of network-based communication services is that between "public" and "private." Public services in some countries as the U.S., are often referred to as Common Carrier services. Since 1850, public network services have generally been subject to domestic and international government oversight, while private networks and service have not. The Internet is for regulatory purposes an amalgamation of private networks.

Until the early 1990s, the Internet operated under a difficult regulatory bifurcation where it was unregulated in the U.S., impeded in most other countries, and banned internationally. The ban, instituted by an ITU provision that prohibited international leased line capacity to be made available to third parties, was circumvented through government ownership of the network infrastructure.

The international ban stood in marked contrast to the actions of the Federal Communications Commission that, in the U.S., decided in the 1982 Computer II Decision to "forbear" indefinitely from exercising regulatory authority over "enhanced services" such as those provided via the Internet. The decision led significantly to the Internet's rapid growth and innovation, as anyone with the incentive and a modest investment could become an Internet provider. This FCC stance, however, left a vacuum that was partially filled in 1997 by a much more regulatory oriented Executive Branch agency — the U.S. Department of Commerce — in its imposing a classic legacy common-carrier regulatory regime on the provisioning of Internet domain name services. In actions related to national security, the Department also assumed control over the administration of many Internet addresses and the operation of root DNS servers, functions formerly controlled by the U.S. DOD and the NSF.

By 2002, as the Internet emerged as necessary infrastructure, as it began to support legacy telephone services, and an assortment of criminal and terrorist behaviors emerged, the FCC began to propose regulatory requirements for Internet service providers. In early 2004, the FCC instituted multiple new Internet-related rulemaking proceedings, including an omnibus rulemaking of profound importance to the Internet — establishing the basic direction for Internet governance for the future. At almost the same time, the European Commission instituted a very similar proceeding that is likely to have a similar result. The FCC exercised explicit jurisdiction over all "IP-enabled Services" — replacing them with a new regulatory framework. The new framework includes several fundamental steps: 1) bringing the 25-year-old basic-enhanced Computer II dichotomy to an end, 2) the preemption of most state regulation of IP-enabled services, and 3) the application of twelve kinds of regulatory mandates to defined classes of Internet services.

The Internet's new regulatory framework will revolve around requirements that include: public safety (E911) needs, disability assistance, law enforcement support, competition (Computer III/number portability/1996 Act requirements), fraud prevention, reliability and reporting obligations, restoration after failures, call prioritization during emergencies, privacy and data protection, consumer protection against unwanted intrusions, universal service and other contributory obligations, and intercarrier compensation. The precise initial requirements will be decided during 2004 and evolve over subsequent years. Previous IANA functions seem likely to transition to a combination of institutional arrangements under the FCC and similar national regulatory agencies worldwide cooperating through the ITU, and include Internet Service Provider organizations like ARIN, RIPE, and AP-NIC.

In most countries outside the U.S., the public provisioning of Internet services was not allowed until the early 1990s. This was followed by a period of several years where economic disincentives were instituted to impede Internet use — typically to promote higher-priced and officially favored public telecommunication service alternatives. The mechanisms included costly leased-line tariffs, high metered charges for dialup line use, restrictions on modem use, and prohibitions on specific services like Voiceover IP. In most countries, these impediments have largely disappeared except for the last, which is still prevalent in most of the world. Essentially, every country also imposes Lawful Access and Interception regulatory requirements upon the Internet service providers in the same fashion as for any telecommunication service.

54.5.2 Law

The diverse systems of law have always applied transparently to the conduct of Internet service providers and users. The laws pertaining to crime, fraud, contract, libel, contracts, intellectual property, and the like do not distinguish among kinds of media used, and judicial decisions over the years sought to adapt existing provisions to cases in controversy occurring via the Internet. Communication networks, however, have always posed occasional difficult questions of jurisdiction over the conduct or actors, and the characteristics of the Internet exacerbate jurisdictional issues.

During the 1990s, specific Internet-related law began to emerge to deal with specific issues or difficulties posed. These included enabling law such as the recognition of digital signatures for the purposes of providing assent, the acceptance of forms of digital documents as being sufficient in legal actions, and e-mail as being sufficient in providing notice. The law also began to deal with Internet cybercrime and other unique new developments in the form of malicious harm to computer systems, stalking, protection of minors, fraudulent communications, gambling, consumer protection, data protection, privacy protection, content regulation, intellectual property protection (e.g., copyright and trademark) fraud, identity theft, terrorism, unsolicited e-mail (SPAM), and taxation of online sales.

There is now a large and rapidly growing body of Internet law emerging in almost every legal jurisdiction throughout the world. Some international harmonization of this law was effected in 2002 in the form of the Convention on Cybercrime — a broad treaty instrument among 30 signatory countries that will likely come into force in 2004 and expand to include other nations. The Convention establishes a model for other areas of international harmonization and cooperation with respect to Internet law.

54.6 Standards and Administrative Sector

A variety of standards bodies and forums have developed technical and operational specifications among providers and users — occasionally with public body involvement. In some instances, there is some type of administrative body associated with the forum that implements the registration and notification requirements associated with some standards.

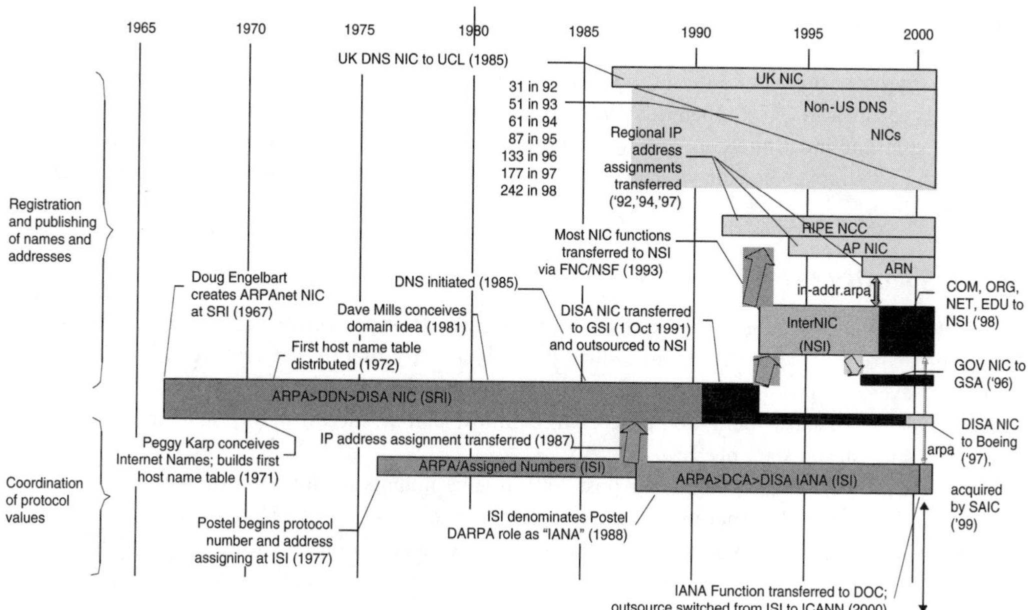

FIGURE 54.5 **EVOLUTION** of Internet NICs.

54.6.1 Legacy Standards and Administrative Forums

During the 1970s and 1980s, Internet standards were the province of the DARPA-sponsored committees that produced the specifications in the form of Requests for Comment (RFC). This activity and the standards were formalized by the U.S. Department of Defense in 1982 and published by DARPA and the Defense Communications Agency (now DISA). The standards development activity became institutionalized in the form IETF, maintained through an IETF Secretariat under the aegis of the Corporation for National Research Initiatives (CNRI). The IETF itself has become associated with the Internet Society. This configuration remains today, and the authoritative standards are published by the IETF Secretariat on its Website. The IETF work is managed through an Internet Engineering Steering Group (IESG) and an Internet Architecture Board (IAB) — also supported by the IETF Secretariat.

During the 1970s, the USC Information Sciences Institute (ISI) in Marina del Rey, CA, in cooperation with the Menlo Park, CA, NIC, began to provide some of the administrative functions necessary to implement the Internet standards. The ISI activity subsequently became institutionalized in the late 1980s as the Internet Assigned Numbers Authority (IANA). The evolution of the IP address and DNS components of this function are depicted in Figure 54.5. There were scores of other functions, however, that remain with the IANA, which is maintained as an outsourced contractor activity by the U.S. Department of Commerce's National Institute of Standards and Technology (NIST).

54.6.2 The Universe of Internet Standards and Administrative Forums

As the Internet grew, so did the standards and administrative forums of various kinds. There are now more than 100 different bodies and forums of various kinds that are far too numerous to describe here. Table 54A.1 — Internet Standards Forums — lists most of them (See Appendix).

Some of these forums operate essentially independently of each other. Many serve specialized technologies, applications, or constituencies.

54.7 Emerging Trends

Like all ecosystems, those for Internet policy and governance continue to evolve to accommodate the needs of its constituents. The inherently autonomous, self-organizing characteristics of the Internet will no doubt continue indefinitely to stress governmental attempts to encourage beneficial actor conduct and punish undesirable behavior, which is what policy and governance mechanisms are meant to accomplish.

54.7.1 Security

The most obvious emerging trends revolve around two kinds of protective and security-related needs. One is proactive, involving actions to reduce the vulnerability of Internet resources, including users subject to adverse behavior. The other is reactive, involving a need to identify bad Internet actors and to acquire evidence for subsequent legal proceedings. Almost all new, successful infrastructure technologies have these same steady-state needs.

These needs have grown dramatically post 2001 as governments worldwide have witnessed dramatic increases in malevolent Internet use. The needs seem unlikely to abate. An almost certain result will be to impose user authentication requirements and the maintenance of usage records. Accountability cannot otherwise exist. At the same time, encryption as a means of both protecting sensitive information and verifying content will expand.

54.7.2 Diversity

The Internet, because of its growing ubiquity, seems destined to support an increasing diversity of uses, both in terms of an expanding number of transport options and as increasing numbers of users and services. This "hourglass effect" of the Internet protocol becomes ever more attractive as a universal glue between transport options and applications, especially with expanded address options supported by IP version 6. On the other hand, single infrastructures create their own vulnerabilities, and because of increasing concerns regarding security and survivability, the all-encompassing expansion of the Internet is likely self-limiting.

54.7.3 Assimilation

Just as all of the precedent technologies before it, the Internet has moved into a mass-market assimilation phase where its identity has substantially merged into a common infrastructure together with a vast array of "always on" access devices, networks, and services. The price of success, however, is the adaptation and adoption of the infrastructure and the emergence of vulnerabilities as it becomes a vehicle for unintentional or intentional harm with profound adverse consequences for people, commerce, and society. The vulnerabilities exist for any significant infrastructure, whether communications, power, or transport.

Going forward, the challenges faced with this larger infrastructure will be not be those of innovation and growth alone, but include ever more prominently the imposition of policies and requirements that lessen infrastructure vulnerabilities.

References

Where possible, readers are urged to access source documents rather than secondary material.

1. Bootstrap Institute, Interview 4, 1987 *Interviews with Douglas Engelbart*, http://www.sul.stanford.edu/depts/hasrg/histsci/ssvoral/engelbart/engfmst4-ntb.html.
2. *Request for Comments Repository*, http://www.ietf.org/rfc.html.
3. J. McQuillan, V. Cerf, *A Practical View of Computer Communications Protocols*, IEEE Computer Society, 1978.

4. FCC, *Computer II Final Order*, 77 FCC2d 384 (1980).
5. *DoD Policy on Standardization of Host-to-Host Protocols for Data Communications Networks*, The Under Secretary of Defense, Washington, D.C., 23 March 1982.
6. *Internet Protocol Implementation Guide*, Network Information Center, SRI International, Menlo Park, CA, August 1982.
7. R. E. Kahn, A. Vezza, and A. Roth, Electronic Mail and Message Systems: Technical and Policy Perspectives, AFIPS, Arlington, VA, 1981.
8. *Internet Protocol Transition Workbook*, Network Information Center, SRI International, Menlo Park, CA, March 1982.
9. Mark Lottor, *Internet Domain Survey*, Network Wizards, http://www.nw.com/. Lottor has since 1982 engaged in Internet metrics research, data collection, and analysis.
10. *DDN Protocol Handbook*, DDN Network Information Center, SRI International, Menlo Park,CA, 1985.
11. Towards a Dynamic Economy — Green Paper on the Development of the Common Market for Telecommunications Services and Equipment. Communication by the Commission. COM (87) 290 final (June 30, 1987).
12. OECD, Committee for Information, Computer and Communications Policy, Value-Added Services: Implications for Telecommunications Policy, Paris 1987.
13. National Science Foundation, Network Information Services Manager(s) for NSFnet and the NREN, NSF 92-24 (1992).
14. Internet Engineering Task Force Secretariat, www.ietf.org.
15. RIPE NCC, www.ripe.net.
16. Cooperative Association for Internet Data Analysis, www.caida.org.
17. Internet International *Ad Hoc* Committee, www.iahc.org.
18. Commission of the European Union, Information Society, http://europa.eu.int/information_society. See also, EU Law + Policy Overview, The Internet, The Information Society and Electronic Commerce, http://www.eurunion.org/legislat/interweb.htm.
19. U.S. Department. of Congress, National Telecommunications and Information Administration, Management of Internet Names and Addresses, www.ntia.doc.gov/ntiahome/domainname/.
20. FCC, Notice of Proposed Rulemaking, Appropriate Framework for Broadband Access to the Internet over Wireline Facilities, CC Docket No. 02-33, FCC 02-42, 15 February 2002.
21. FCC, Notice of Proposed Rulemaking, In the Matter of IP-Enabled Services, Docket No. 04-36, Document No. FCC 04-28, 10 March 2004.
22. EC Information Society Directorate, IP Voice and Associated Convergent Services, Final Report for the European Commission, 28 January 2004.
23. FCC, Comment Sought on CALEA Petition for Rulemaking, RM-10865, DA No. 04-700, 12 March 2004.
24. FCC, Notice of Proposed Rulemaking, in the Matter of Rules and Regulations Implementing the Controlling of the Assault of Non-Solicited Pornography and Marketing Act of 2003, CG Docket No. 04-53, Doc. FCC 05-52, 19 March 2004.

Appendix

54A.1 Internet standards and Administrative Forums

Name	Acronym	URL	Type	Focus
3RD Generation Partnership Project	3GPP	www.3gpp.org	standards	telecom
Accredited Standards Committee (ASC) X12		www.x12.org	standards	data exchange
Aim, Inc.		www.aimglobal.org	standards	identifiers
Alliance for Telecommunications Industry Solutions	ATIS	www.atis.org	standards	telecom
American Library Association		www.ala.org	standards	library
American National Standards Institute	ANSI	www.ansi.org	standards	diverse
American Registry for Internet Numbers	ARIN	www.arin.net	operations	internet
American Society for Information Science and Technology	ASIS	www.asis.org	standards	general
ANSI X9		www.x9.org	standards	financial
Asia Pacific Networking Group	APNG	www.apng.org	operations	internet
Asia-Pacific Telecommunity Standardization Program	ASTAP	www.aptsec.org/astap/	standards	telecom
Association for Information and Image Management International	AIIM	www.aiim.org	standards	imaging
Bluetooth Consortium		www.bluetooth.com	standards	wireless
Cable Labs		www.cablelabs.org	standards	telecom
Computer Emergency Response Team	CERT	www.cert.org	operations	security
Content Reference Forum	CRF	www.crforum.org	standards	digital content
Critical Infrastructure Assurance Office	CIAO	www.ciao.gov	government	security
Cross Industry Working Team	XIWT	www.xiwt.org	standards	internet
Data Interchange Standards Association	DISA	www.disa.org	standards	application
Department of Justice	DOJ	www.doj.gov	government	security
Digital Library Federation	DLF	www.diglib.org	standards	library
Digital Video Broadcasting Consortium	DVB	www.dvb.org	standards	broadcasting
Directory Services Markup Language Initiative Group	DSML	www.dsml.org	standards	directory

54A.1 Internet standards and Administrative Forums (continued)

Name	Acronym	URL	Type	Focus
Distributed Management Task Force	DMTF	www.dmtf.org	standards	management
DOI Foundation		www.doi.org	standards	application
ebXML		www.ebxml.org	standards	application
EC Diffuse Project		www.diffuse.org/fora.html	standards	reference
Electronic Payments Forum	EPF	www.epf.org	standards	financial
Electronics Industry Data Exchange Association	EIDX	www.eidx.org	standards	data exchange
Enterprise Computer Telephony Forum	ECTF	www.ectf.org	standards	telecom
ENUM Forum	ENUM	www.enum-forum.org	standards	telecom
EPCGlobal		www.epcglobalinc.com	standards	identity/RFID
European Commission	EC	europa.eu.int/comm/index_en.htm	government	telecom
European Committee for Electrotechnical Standardization	CENELEC	www.cenelec.org	standards	general
European Committee for Standardization	CEN	www.cenorm.be	standards	general
European Computer Manufacturers Association	ECMA	www.ecma.ch	standards	telecom
European Forum for Implementers of Library Automation	EFILA	www.efila.dk	standards	classification
European Telecommunications Standards Institute	ETSI	www.etsi.org	standards	telecom
European Umbrella Organisation for Geographic Information	EUROGI	www.eurogi.org	standards	location
Federal Communications Commission	FCC	www.fcc.gov	government	telecom
Federal Trade Commission	FTC	www.ftc.gov	government	diverse
FidoNet Technical Standards Committee	FSTC	www.ftsc.org	standards	network
Financial Information eXchange (FIX) protocol		www.fixprotocol.org	standards	financial
Financial products Markup Language Group		www.fpml.org	standards	financial
financial services industry		www.x9.org	standards	financial
Financial Services Technology Consortium	FSTC	www.fstc.org	standards	financial

554A.1 Internet standards and Administrative Forums (continued)

Name	Acronym	URL	Type	Focus
Forum for metadata schema implementers		www.schemas-forum.org	standards	application
Forum of Incident Response and Security Teams	FIRST	www.first.org	operations	security
Global Billing Association		www.globalbilling.org	standards	
Global Standards Collaboration	GSC	www.gsc.etsi.org	standards	telecom
Group on Electronic Document Interchange	GEDI	lib.ua.ac.be/MAN/T02/ t51.html	standards	classification
GSM Association	GSM	www.gsmworld.com	standards	telecom
IEEE Standards Association		standards.ieee.org	standards	diverse
IMAP Consortium		www.impa.org	standards	application
Information and Communications Technologies Board	ICTSB	www.ict.etsi.org	standards	authentication
Infraguard Alliance		www.infraguard.net	government	security
Institutute of Electrical and Electronic Engineers	IEE802.11	www.ieee.org	standards	
Interactive Financial eXchange (IFX) Forum		www.ifxforum.org	standards	financial
International Confederation of Societies of Authors and Composers	CISAC	www.cisac.org	standards	classification
International Digital Enterprise Alliance	IDEA	www.idealliance.org/	standards	metadata
International Federation for Information Processing	IFIP	www.ifip.or.at	standards	application
International Federation of Library Associations	IFLA	www.ifla.org	standards	classification
International Imaging Industry Association		www.i3a.org	standards	imaging
International Multimedia Telecommunications Forum	IMTC	www.imtc.org	standards	telecom
International Organization for Standardization	ISO	www.iso.ch	standards	diverse
International Telecommunication Union	ITU	www.itu.org	standards	telecom
International Telecommunication Union	ITU	www.itu.int	government	telecom
International Telecommunications Advisory Committee	ITAC	www.state.gov/www/issues/ economic/cip/itac.html	standards	telecom

54A.1 Internet standards and Administrative Forums (continued)

Name	Acronym	URL	Type	Focus
International Webcasting Association	IWA	www.iwa.org	standards	broadcasting
Internet Architecture Board	IAB	www.iab.org	standards	internet
Internet Corporation for Names and Numbers	ICANN	www.icann.org		
Internet Engineering Task Force	IETF	www.ietf.org	standards	network
Internet Mail Consortium	IMC	www.imc.org	standards	application
Internet Security Alliance	ISA	www.isalliance.org	operations	security
IPDR (Internet Protocol Detail Record) Organization, Inc	IPDR	www.ipdr.org	standards	telecom
IPV6 Forum		www,ipv6.org	standards	internet
ISO/TC211		www.isotc211.org	standards	location
Java APIs for Integrated Networks	JAIN	jcp.org/jsr/detail/035.jsp	standards	telecom
Java Community		java.sun.com	standards	application
Liberty Alliance		www.projectliberty.net	standards	
Library of Congress		www.loc.gov/standards/	standards	classification
Localisation Industry Standard Association	LISA	www.lisa.org	standards	application
Mobile Games Interoperability Forum	MGIF	www.mgif.org	standards	games
Mobile Payment Forum		www.mobilepaymentforum.org	standards	financial
Mobile Wireless Internet Forum	MWIF	www.mwif.org	standards	wireless
Multiservice Switch Forum	MSF	www.msforum.org	standards	telecom
National Association of Regulatory and Utility Commissioners	NARUC	www.naruc.org	government	telecom
National Automated Clearing House Association	NACHA	www.nacha.org		
National Committee for Information Technology Standards	NCITS	www.ncits.org	standards	security
National Communications System	NCS	www.ncs.gov/ncs/html/NCSProjects.html	standards	telecom
National Emergency Number Association	NENA	www.nena.org	standards	telecom
National Exchange Carriers Association	NECA	www.neca.org	government	telecom

54A.1 Internet standards and Administrative Forums (continued)

Name	Acronym	URL	Type	Focus
National Genealogical Society		www.ngsgenealogy.org/comstandards.htm	standards	application
National Information Assurance Partnership	NIAP	niap.nist.gov	standards	security
National Information Standards Organization	NISO	www.niso.org	standards	security
National Infrastructure Protection Center	NIPC	www.nipc.gov	government	security
National Institute for Standards and Technology	NIST	www.nist.gov	government	security
National Security Agency	NSA	www.nsa.org	government	security
National Standards System Network	NSSN	www.nssn.org/developer.html	standards	reference
National Telecommunications and Information Administration	NTIA	www.ntia.doc.gov	government	telecom
Network Applications Consortium	NAC	www.netapps.org	standards	application
Network Reliability & Interoperability Council	NRIC		operations	telecom
NIMA Geospatial and Imagery Standards Management Committee	NIMA GSMC ISMC	http://164.214.2.51/	standards	location
NIST Computer Security Resource Center	CSRC	csrc.nist.gov	standards	security
North American Numbering Council	NANC	www.fcc.gov/ccb/Nanc/	operations	telecom
North American Operators Group	NANOG	www.nanog.org	operations	internet
Object Management Group	OMG	www.omg.org	standards	general
Online Computer Library Center	Dublin Core	www.oclc.org	standards	metadata
Ontology.org		www.ontology.org	standards	metadata
Open Applications Group	OAGI	www.openapplications.org	standards	application
Open Archives Forum	OAF	edoc.hu-berlin.de/oaf	standards	archive
Open Bioinformatics Foundation		www.open-bio.org	standards	application
Open Directory Project		www.dmoz.org	standards	directory
Open GIS Consortium	OGC	www.opengis.org	standards	location
Open H323 Forum		www.openh323.org	standards	multimedia
Open LS		www.openls.org	standards	location
Open Mobile Alliance		www.openmobilealliance.org	standards	wireless

54A.1 Internet standards and Administrative Forums (continued)

Name	Acronym	URL	Type	Focus
Open Services Gateway Initiative	OSGi	www.osgi.org	standards	application
Organization for Economic Cooperation and Development	OECD	www.oecd.gov	government	political
Organization for the Advancement of Structured Information Standards	OASIS	www.oasis-open.org	standards	
PKIForum	PKI Forum	www.pkiforum.org	standards	security
Presence and Availability Management Forum	PAM Forum	www.pamforum.org	standards	wireless
Project MESA		www.projectmesa.org	standards	wireless
Reseau IP EuropeenRéseaux IP Européens	RIPE	www.ripe.net	operations	internet
Security Industry Association	SIA	www.siaonline.org	standards	security
SIP Forum		www.sipforum.com/	standards	telecom
Smart Card Alliance	SCA	www.smartcardalliance.org/	standards	identifiers
Society of Motion Picture and Television Engineers	SMPTE	www.smpte.org	standards	imaging
Softswitch Consortium		www.softswitch.org/	standards	telecom
Speech Application Language Tags	SALT	www.saltforum.org	standards	application
SyncML Initiative, Ltd	SyncML	www.syncml.org	standards	wireless
Telecommunications Industry Association	TIA	www.tiaonline.org	standards	standards
TeleManagment Forum		www.tmforum.org	standards	telecom
The Alliance for Technology Access	ATA	www.ataccess.org	standards	handicaped
The Electronic Payments Association	NACHA	www.nacha.org	standards	financial
The European Forum for Electronic Business	EEMA	www.eema.org	standards	financial
The Open Group		www.opengroup.org	standards	general
The PARLAY Group	PARLAY	www.parlay.org	standards	telecom
The Portable Application Standards Committee		www.pasc.org	standards	application
TruSecure		www.trusecure.com	standards	security
Trusted Computing Group	TCG	www.trustedcomputinggrou p.org	standards	security
UMTS Forum	UMTS	www.umts-forum.org	standards	wireless

54A.1 Internet standards and Administrative Forums (continued)

Name	Acronym	URL	Type	Focus
Unicode Consortium		www.unicode.org	standards	identifiers
Uniform Code Council	EAN-UCC	www.uc-council.org/	standards	identifiers
Universal Description, Discovery and Integration Community	UDDI	www.uddi.org	standards	application
Universal Plug and Play Forum	UPnP	www.upnp.org	standards	network
Universal Wireless Communications Consortium	UWC	www.uwcc.org	standards	wireless
Value Added Services Alliance	VASA	www.vasaforum.org	standards	telecom
Voice XML Initiative		www.voicexml.org	standards	wireless
Web Services Interoperability Organization	WS-I	www.ws-i.org	standards	
Web3d		www.web3d.org	standards	games
WiFi Alliance		www.wirelessethernet.org	standards	wireless
Wireless LAN Association	WLANA	www.wlana.org	standards	wireless
Wireless Location Industry Association	WLIA	www.sliaonline.com	standards	location
World Wide Web Consortium	W3C	www.w3.org	standards	
XML Forum		www.xml.org	standards	application
XML/EDI Group		www.xmledi-group.org	standards	data exchange

55

Human Implications of Technology

CONTENTS

Abstract 55-1
55.1 Overview 55-1
55.2 Technological Determinism 55-2
55.3 Social Determinism 55-3
55.4 Values in Design 55-4
55.5 Human Implications of the Security Debates 55-5
55.6 Conceptual Approaches to Building Trustworthy Systems 55-6
55.6.1 Trusted Computing Base 55-6
55.6.2 Next Generation Secure Computing Platform 55-7
55.6.3 Human-Centered Trusted Systems Design 55-9
55.6.4 Identity Examples 55-13
55.6.5 Data Protection vs. Privacy 55-14
55.7 Network Protocols as Social Systems 55-15
55.8 Open vs. Closed Code 55-16
55.9 Conclusions 55-17
Additional Resources 55-17
References 55-18

L. Jean Camp

Ka-Ping Yee

Abstract

The relationship between technology and society is characterized by feedback. Technological determinist and social determinist perspectives offer informative but narrow insights into both sides of the process. Innovative individuals can generate new technologies that alter the lives, practices, and even ways of thinking in a society. Societies alter technologies as they adopt them, often yielding results far from the hopes or fears of the original designers. Design for values, also called value-sensitive design, consists of methods that explicitly address the human element. As the example of usability in security illustrates, a designer who is cognizant of human implications of a design can produce a more effective technological solution.

55.1 Overview

The human implications of a technology, especially for communications and information technology, begin in the design stage. Conversely, human implications of technology are most often considered after the widespread adoption of the technology. Automobiles cause pollution; televisions may cause violence in children.

Social values can be embedded at any stage in the development process: invention, adoption, diffusion, and iterative improvement. A hammer wielded by Habitat for Humanity and a hammer wielded in a

1-58488-381-2/05/$0.00+$1.50
© 2005 by CRC Press LLC

violent assault cannot be said to have the same human value at the moment of use, yet the design value of increasing the efficacy of human force applies in both situations.

In the case of the hammer, the laws of physics limit the designer. Increasingly the only limit to a designer in the virtual world is one of imagination. Thus, designs in search engines, browsers, and even network protocols are created from a previously inconceivably vast range of alternatives.

How important are the choices of the designer? There are two basic schools, one which privileges technical design as the driver of the human condition, and one which argues that technologies are the embodiment of social forces beyond the designers' control. After introducing these boundary cases, and identifying the emerging middle, this chapter focuses on a series of studies of particular technical designs. The most considered case is that of security and trust. The chapter closes with a pointer to the significant debates on open or closed code, and the potential of network protocols themselves to embody value-laden choices. The final word is one of caution to the designer that the single reliable principle of responsible design is full disclosure, as any obfuscation implicitly assumes omnipotence and correctness about the potential human implications of the designers' technical choice.

55.2 Technological Determinism

Technological determinism argues that the technologically possible will inevitably be developed and the characteristics of the newly developed technologies will alter society as the technology is adopted [Winner, 1986] [Eisenstein, 1987]. Some find optimism from such a viewpoint, arguing that technology will free us from the human condition [Negroponte, 1995; Pool, 1983]. Others find such a scenario to be the source of nightmares, arguing that information and communications technologies (ICT) have "laid waste the theories on which schools, families, political parties, religion, nationhood itself" and have created a moral crisis in liberal democracy [Postman, 1996].

Marx has been identified as perhaps the most famous technological determinist in his descriptions of the manner in which the industrial revolution led to the mass exploitation of human labor. Yet technological determinism is not aligned exclusively with any particular political viewpoint.

Technological determinism may be overarching, as with Marx and Winner. In this case, both observed that large complex technologies require large complex organizational systems for their management. The state was required by the structure of large capitalist institutions, which were required by the technologies of the factory and the railroad. Even government was a function of the organization of capital through the state, as Engels noted, "the proletariat needs the state, not in the interests of freedom but in order to hold down its adversaries, and as soon as it becomes possible to speak of freedom the state as such ceases to exist" [Tucker, 1978]. Thus when means and methods of production changed hands the state would no longer be needed. Urban industrialization created that moment in time, at which workers (previously peasants) would be empowered to organize and rise up, overthrowing their oppressors and thereby removing the burden of oppression and government simultaneously.

An echo of this determinist came from the libertarian viewpoint with the publication of the Declaration of Cyberspace. At the peak of the technologically deterministic embrace of the then-emerging digital network, this declaration argued that past technologies had created an invasive oppressive state. Once again the new technologies created under the old state would assure its downfall.

"You claim there are problems among us that you need to solve. You use this claim as an excuse to invade our precincts. Many of these problems do not exist. Where there are real conflicts, where there are wrongs, we will identify them and address them by our means. We are forming our own Social Contract. This governance will arise according to the conditions of our world, not yours. Our world is different." [Barlow, 1996]

Both of these are reductionist political arguments that the state and the society were a function of previous technologies. The changes in technologies would therefore yield radical changes — in both cases, the destruction of the state.

Technological determinism is reductionist. The essential approach is to consider two points in time that differ in the technology available. For example, the stirrup enabled the creation of larger towns by making the care of far-flung fields feasible. Larger towns created certain class and social practices. Therefore determinism would say that the stirrup created social and class realignments and sharpened class distinctions.

In the case of technological determinism of communications the argument is more subtle. Essentially the ICT concept of technological determinism is that media is an extension of the senses in the same way transport is an extension of human locomotion. Thus, communications technologies frame the personal and cultural perspective for each participant, as expressed most famously, in "The medium is the message." [McLuhan, 1962]

In the case of communications and determinism, Marshall McLuhan framed the discourse in terms of the tribal, literate, print, and electronic ages. For McLuhan each technology — the phonetic alphabet, the printing press, and the telegraph — created a new world view. Tribal society was based on stories and magic. Phonetic societies were based on myth and the preservation of cultural heritage. Print societies were based on rational construction and the scientific method. Text reaches for the logical mind; radio and television call out to the most primitive emotional responses. The new hypertext society will be something entirely different, both more reasoned and less rational.

A related debate is geographical determinism [Diamond, 1999] versus cultural determinism [Landes, 1999]. In this argument the distance from the equator and the availability of natural resources enabled technological development, with the role of culture in innovation being hotly contested. It is agreed that technological development then determined the relative power of the nations of the world.

In McLuhan's view the media technology determined our modern world, with a powerful rational scientific North and a colonized South. In the view of the geographical determinist the geography determined the technology, and technology then determined social outcomes.

55.3 Social Determinism

A competing thesis holds that technology is a product of society. This second view is called *social construction* [Bijker et al., 2001]. Technologies are physical representations of political interests. Social implications are embedded by the stakeholders including inventors and governments, on the basis of their own social values. Some proponents of this view hold that users are the only critical stakeholders. This implies that adoption is innovation, and thus the users define technology [Fischer, 1992].

As technical determinists look at two points in time and explain all changes as resulting from technology as a social driver, social constructionists look at two points in technological evolution and use social differences as the sole technical driver.

One example of how society drives innovation is in telephony. Both automated switching and low-end telephones were invented for identifiable, explicit social reasons. Telephones were initially envisioned as business technology and social uses were discouraged, by company policy and advertising campaigns. As long as patents in the U.S. protected the technology there was no rural market as the technology was also assumed by the Bell Company to be inherently urban. When the patents expired, farmers used their wire fences for telephony and provided part-time low-cost service. The existence of the service, not the quality of the service, was the critical variable. Once intellectual property protections were removed, families adopted phones and social uses of phones came to dominate. Thus, telephones were developed that were cheap, less reliable, and of lower quality as opposed to the high-end systems developed by the Bell Company. The design specifications of the telephones were socially determined, but the overall function was technically determined.

In the case of automatic switching, the overall function was socially determined. The goal of an automatic switch was to remove the human switchboard operator. An undertaker, believing that the telephone operator was connecting the newly bereaved to her brother (the competing town undertaker), invented automated switching [Pierce and Noll, 1990]. His design goal was socially determined yet the implementation and specifications for the switch were technical.

Social determinism reflects the obvious fact that identical technologies find different responses in different societies. The printing press created a scientific revolution in Western Europe, but coincided with a decline of science and education in Persian and Arab regions. The printing press had had little effect on the practice of science and literacy in China by 1600, despite the fact that paper and the movable type press had been invented there centuries earlier.

Yet as Barlow illustrates, the extreme of technological determinism, social determinism, also produces extremes. Robert Fogel received a Nobel Prize in economics for "The Argument for Wagons and Canals" in which he declared that the railroad was of no importance as canals would have provided the same benefit. The sheer impossibility of building a canal over the Sierra Mountains or across Death Valley were not elements of his thesis, which assumed that canals were feasible within 40 mi of any river — including the Rio Grande and various *arroyos*. A second thesis of Fogel's was that internal combustion would have developed more quickly without the railroad. This thesis ignores the contributions of the railroad engine in the technological innovation and the investments of railroads in combustion innovation. Most important, this does not acknowledge the dynamics of the emerging engineering profession, as the education and creation of human capital created by the railroad industry were important in internal combustion innovations. Such innovations are less likely to have arisen from those educated to dig canals. This finding requires a reductionist perspective that eliminates all technical distinction from the consideration. In fact, an appeal of this paper was the innovation of treating very different technologies as replaceable "black boxes" while maintaining a mathematically consistent model for their comparison. As is the case with technical determinism, social determinism is reductionist, ignoring technical fundamentals rather than the ubiquitous social environs.

55.4 Values in Design

Clearly, the social implications of technology cannot be understood by ignoring either the technologies or the social environment. Inclusive perspectives applicable to information and communications technologies have emerged in human-centered design and design for values. These perspectives recognize that technologies can have biases and attempt to address the values in an ethical manner in the design process. Further, the perspectives explicitly address technologies as biased by the fundamentals of the technology itself as well as through the adoption process.

Social determinism either argues for designers as unimportant cogs to controlling destiny or trivial sideshows to the economic and social questions. Technical determinists view designers as oblivious to their power or omniscient. The mad scientist is the poster child of dystopian technological determinism.

The design for values or human-centered design schools conceives of designers as active participants yet acknowledges the limits of technological determinism [Friedman and Nissenbaum, 2001]. From this design-for-values perspective, designers offer technological systems and products that bundle functions and values. If the functions are sufficiently useful, the system may be adopted despite undesirable values. If the functions and the values align with the needs of the potential adopters, then widespread adoption is inevitable. While this description sounds clear, determining the values embedded in a system is not trivial.

The examination of values in communications technology is enhanced by past research in the classification of specific values [Spinello, 1996]. In privacy there are definitions of *private* based on system purpose and use (for example the American Code of Fair Information Practice [Federal Trade Commission, 2000]) or only on system use (as with the European Directive on Data Protection [European Union, 1995]). In both cases, there are guidelines intended for users of technology that can be useful for designers of technology. Complicating the lack of examination of inherent and emergent (but theoretically economically predictable) biases is the reality that once adopted, technical standards are difficult to replace.

Biases in communications technologies result from the omission of variables in the design stage (e.g., packet-based networks are survivable and, incidentally, quality of service is difficult).

Some decisions which may exist in reality as values-determinant are framed by economics only. For example, backward compatibility for nth generation wireless systems is a determinant of cost and therefore of accessibility. Backwards compatibility enables the adoption of obsolete technology for regions that have less capital. In this case a choice on the basis of expense and compatibility can reasonably be said to result in a system which values the marginal first-world consumer more or less against the infrastructure needs of the third-world consumer. Yet the decision would be made based on the expectations of migration of users of earlier generation technology. Such economic biases are inherent in engineering and design decisions.

In other cases values are embedded through the technical assumptions of the designers, a case well made in a study of universal service and QoS. Similarly privacy risks created in ecommerce are arguably based on the assumption of middle-class designers about the existence of identity-linked accounts. Identity-linked accounts are not available to much of the population. Those without banks or credit cards obviously cannot use identity-linked ecommerce mechanisms. An awareness of privacy would have resulted in technologies usable by those in the cash economy. The plethora of ecash design illustrates that such assumptions can be avoided.

Often design for computer security requires developing mechanisms to allow and refuse trust and thus it may be necessary to embed social assumptions. The issue of human/computer interaction further complicates the design for values system. Any interface must make some assumption about the nature of simplification, and the suitable metaphors for interface (e.g., why does a button make sense rather than a switch or path?). Yet choosing against a simplifying interface is itself a values-laden choice. (See the section on human-computer interaction and security, for example.)

Even when technologists set out to design for a specific value, sometimes the result is not always as intended [Herkert, 1999]. For example, the Platform for Privacy Preferences has been described by Computer Scientists for Social Responsibility as a mechanism for ensuring that customer data are freely available to merchants, while its designers assert that the goal was customer empowerment. Similarly PICS has been described as a technology for human autonomy [Resnick, 1997] and as "the devil" [Lessig, 1997] for its ability to enhance the capabilities of censorship regimes worldwide. In both of these cases (not incidentally developed by the World Wide Web Consortium) the disagreement about values is a result of assumptions of the relative power of all the participants in an interaction — commercial or political.

If the designers' assumptions of fundamental bargaining equality are correct then these are indeed "technologies of freedom" [Pool, 1983]. On the other hand, the critics of these technologies are evaluating the implementation of these technologies in a world marked by differences in autonomy, ranging from those seeking Amnesty International to the clients of the Savoy.

There is no single rule to avoid unwanted implications for values in design, and no single requirement that will embed values into a specific design. However, the following examples are presented in order to provide insights on how values are embedded in specific designs.

55.5 Human Implications of the Security Debates

As computer security is inherently the control over the human uses of information, it is a particularly rich area for considering the human implications of technology. The technologically determinant, socially constructed, and design for values models of technological development all have strong parallels in the causes of failures in computer security. Privacy losses result from inherent characteristics of the technology, elements of specific product design, and implementation environments, as well as the interaction of these three.

Security failures are defined as coding errors, implementation errors, user errors, or so-called human engineering [Landwehr et al., 1994].

Security failures are defined as either coding errors or emergent errors, where coding errors are further delineated into either logical flaws in the high level code or simple buffer overruns. The logical, human error and environment errors correspond loosely to technical determinant (accidental), social determinant, and iterative embedding of security values in the code.

Coding errors, which can be further delineated into either logical flaws in the high level code or simple buffer overruns, are technologically determinant causes of security failures. The flaw is embedded into the implementation of the technology.

Implementation faults result from unforeseen interactions between multiple programs. This corresponds to the evolutionary perspective, as flaws emerge during adaptation of multiple systems over time. Adoption and use of software in unanticipated technical environments, as opposed to assumptions about the human environment, is the distinguishing factor between this case and the previous one.

User errors are security vulnerabilities that result from the individual failing to interact in the manner prescribed by the system; for example, users selecting weak passwords. These flaws could be addressed by design for values or human-centered design. Assumptions about how people should be (e.g., valuable sources of entropy) as opposed to how they are (organic creatures of habit) are a core cause of these errors. Alternatively organizational assumptions can create security failures, as models of control of information flow create a systemic requirement to undermine security. Another cause is a lack of knowledge about how people will react to an interface. For example, the SSL lock icon informs the user that there is confidentiality during the connection. Yet the interface has no mechanism to distinguish between confidentiality on the connection and security on the server.

Finally, human engineering means that the attacker obtains the trust of the authorized user and convinces that person to use his or her authorization unwisely. This is a case of a socially determined security flaw. As long as people have autonomy, people will err. The obvious corollary is that people should be removed from the security loop, and security should be made an unalterable default. It follows that users must be managed and prevented from harming themselves and others on the network. In this case the options of users must be decreased and mechanisms of automated user control are required. Yet this corollary ignores the fallible human involved in the design of security and fails to consider issues of autonomy.

Thus, enabling users to be security managers requires educating users to make choices based on valid information. Designs should inform the user and be informed by principles of human-computer interaction.

This section begins with a historical approach to the development of trustworthy systems beginning with the classic approach of building a secure foundation and ending with the recognition that people interact through interfaces, not in raw streams of bits. Both perspectives are presented, with systems that implement both ideas included in the examples.

Computer security is the creation of a trustworthy system. A secure system is trustworthy in the narrow sense that no data are altered, accessed, or without authorization. Yet such a definition of trustworthy requires perfect authorization policies and practice.

Trustworthy protocols can provide certainty in the face of network failures, memory losses, and electronic adversaries. An untrusted electronic commerce system cannot distinguish a failure in a human to comply with implicit assumptions from an attack; in either case, transactions are prevented or unauthorized access is allowed. When such failures can be used for profit then certainly such attacks will occur.

Trust and security are interdependent. A trusted system that is not compatible with normal human behavior can be subverted using human engineering. Thus, the system was indeed trusted, but it was not secure. Trust requires security to provide authentication, integrity, and irrefutability. Yet trust is not security; nor does security guarantee trust.

Ideal trustworthy systems inherently recognize the human element in system design. Yet traditional secure systems focus entirely upon the security of a system without considering its human and social context [Computer Science and Telecommunications Board, 1999].

Currently there is an active debate expressed in the legal and technical communities about the fundamental nature of a trustworthy system [Anderson, 2003; Camp, 2003a; Clark and Blumenthal, 2000]. This debate can be summed up as follows: Trusted by whom?

55.6 Conceptual Approaches to Building Trustworthy Systems

55.6.1 Trusted Computing Base

The fundamental concept of secure computing is the creation of a secure core and the logical construction of provably secure assertions built upon that secure base. The base can be a secure kernel that prevents unauthorized hardware access or secure special-purpose hardware.

The concept of the trusted computing base (TCB) was formalized with the development of the Trusted Computer System Evaluation Criteria (TCSEC) by the Department of Defense [Department of Defense, 1985]. The trusted computing base model sets a series of standards for creating machines, and grades machines from A1 to C2 according to the design and production of the machine. (There is a D rating, which means that the system meets none of the requirements. No company has applied to be certified at the D level.) Each grade, with C2 being the lowest, has increasingly high requirements for security beginning with the existence of discretionary access control, meaning that a user can set constraints on the files. The C level also requires auditing and authentication. The higher grade, B, requires that users be able to set security constraints on their own resources and be unable to alter the security constraints on documents owned by others. This is called mandatory access control.

The TCB model as implement in the Trusted Computer System Evaluation Criteria is ideal for special-purpose hardware systems. The TCB model becomes decreasingly applicable as the computing device becomes increasingly general purpose. A general purpose machine, by definition, can be altered to implement different functions for different purposes. The TCSEC model addresses this by making functionality and implementation distinct. Yet logging requirements, concepts of document ownership, and the appropriate level of hardening in the software change over time and with altered requirements.

While multiple systems have sought and obtained C level certification under the TCSEC, the certification is not widely used in the commercial sector. Military emphasis on information control differs from civilian priorities in three fundamental ways. First, in military systems it is better that information be destroyed than exposed. In civilian systems the reverse is true: bank records, medical records, and other personally identifiable information are private but critical. Better a public medical decision than a flawed diagnosis based on faulty information.

Second, the military is not sensitive to security costs. Security is the reason for the existence of the military; for others it is an added cost to doing business.

Third, the Department of Defense is unique in its interactions with its employees and members, and with its individual participants is uniquely tightly aligned. There is no issue of trust between a soldier and commander. If the Department determines its policies then the computer can implement those policies. Civilians, businesses, families, and volunteer organizations obviously have very different organizational dynamics.

One goal of the TCB is to allow a centralized authority to regulate information system use and access. Thus, the Trusted Computing Based may be trusted by someone other than the user to report upon the user. In a defense context, given the critical nature of the information, users are often under surveillance to prevent information leakage or espionage. In contrast, a home user may want to be secure against the Internet Service Provider as well as remote malicious users.

Microsoft's Next Generation Secure Computing Platform is built on the trusted computing base paradigm.

55.6.2 Next Generation Secure Computing Platform

Formerly known as Palladium, then the Trusted Computing Base, then the Next Generation Secure Computing Platform, the topic of this section is now called Trusted Computing (TC). Regardless of the name, this is a hotly contested security design grounded in the work of Microsoft. TC requires a secure coprocessor that does not rely on the larger operating system for its calculations. In theory, TC can even prevent the operating system from booting if the lowest level system initiation (the BIOS or basic input/output system which loads the operating system) is determined by TC to be insecure. TC must include at least storage for previously calculated values and the capacity for fast cryptographic operations.

A primary function of TC is to enable owners of digital content to control that content in digital form [Anderson, 2003]. TC binds a particular data object to a particular bit of software. Owners of high value commodity content have observed the widespread sharing of audio content enabled by overlay networks, increased bandwidth, and effective compression. By implementing players in "trusted" mode, video and audio players can prevent copying, and enforce arbitrary licensing requirements by allowing only trusted players. (In this case the players are trusted by the content owners, not the computer user.) This is not limited to multimedia formats, as TC can be used for arbitrary document management. For example, the Adobe eBook had encryption to prevent the copying of the book from one device to another, prohibit audio output of text books, and to enforce expiration dates on the content [Adobe, 2002]. The Advanced eBook Processor enabled reading as well as copying, with copying inherently preventing deletion at the end of the licensed term of use. Yet, with TC an Adobe eBook could only be played with an Adobe eBook player, so the existence of the Advanced eBook Processor would not threaten the Abode licensing terms.

In physical markets, encryption has been used to bind future purchases to past purchases, in particular to require consumers of a durable good to purchase related supplies or services from the manufacturer. Encryption protects ink cartridges for printers to prevent the use of third-party printer cartridge use. Patents, trade secrets and encryption link video games to consoles. Encryption connects batteries to cellular phones, again to prevent third-party manufacturers from providing components. Encryption enables automobile manufacturers to prevent mechanics from understanding diagnostic codes and, if all is working smoothly, from turning off warning lights after an evaluation. [Prickler, 2002]

Given the history of Microsoft and its use of tying, the power of TC for enforcing consumer choices is worth consideration. It is because of the potential to limit the actions of users that the Free Software Foundation refers to TC as "treacherous computing" [Stallman, 2002]. In 1999, the United States Department of Justice found Microsoft guilty of distorting competition by binding its Explorer browser to its ubiquitous Windows operating system, and in 2004, the European Union ruled that Microsoft abused its market power and broke competition laws by tying Windows Media Player to the operating system. The attestation feature of TC, which enables a remote party to determine whether a user is using software of the remote party's choice, could be used to coerce users into using Explorer and Windows Media Player even if they were not bundled with the operating system.

Microsoft also holds a monopoly position in desktop publishing, spreadsheet, and presentation software, with proprietary interests in the corresponding Word document (.doc), Excel spreadsheet (.xls), and PowerPoint presentation (.ppt) file formats. Currently Microsoft is facing competition from open code, including StarOffice and GNU/Linux. The encrypted storage feature of TC could enforce Microsoft lock-in. That is, TC could refuse to decrypt documents to any application other than one that could attest to being a legal copy of a Microsoft Office application running on Windows. Were competitors to reverse-engineer the encryption to enable users to read their own documents using other applications or on non-Windows operating systems, their actions would be felonious under the Digital Millennium Copyright Act.

Initially, Palladium explicitly included the ability to disable the hardware and software components, so that the machine could run in "untrusted" mode. The recent integration of document control features in the MS Office suite requires that TC be enabled for any manipulation of the MSOffice documents (reading, saving, altering).

TC centralizes power and trust. The centralization of trust is a technical decision. Using IBM implementations of TC it is possible to load Linux. IBM has made the driver code for the TCPA compatible chip (which they distinguish from Palladium) available over the Internet with an open license, and offer the product with Linux.

TC makes it possible to remove final (or root) authority from the machine owner or to allow the owner to control her own machine more effectively. Thus, the design leverages and concentrates power in a particular market and legal environment.

The design for values perspective argues the TC is valuable only if it provides root access and final authority to the end user. Yet TC is built in order to facilitate removal of owner control. TC offers two-party authorization — an operator who is assumed to have physical access to the machine and an owner with remote access. The remote owner is specifically enabled in its ability to limit the software run by the owner. A design goal of TC is that the operator cannot reject alterations made by the owner, in that the typical operator cannot return the machine to a previous state after an owner's update.

In short, TC is designed to remove control from the end user (or operator) and place that control with a remote owner. If the operator is also the owner, TC has the potential to increase user autonomy by increasing system security. If the owner is in opposition to the operator then the operator has lost autonomy by virtue of the increased security. The machine is more secure and less trustworthy from the perspective of the owner.

55.6.3 Human-Centered Trusted Systems Design

Technical systems, as explained above, embody assumptions about human responses [Camp et al., 2001]. That humans are a poor source of randomness is well documented, and the problems of "social engineering" are well known [Anderson, 2002]. Yet the consideration of human behavior has not been included in classic axiomatic tests [Aslam et al., 1996; Anderson, 1994].

For example, designers of secure systems often make assumptions about the moral trust of humans, which is a psychological state, and strategic trust of machines [Shneiderman, 2000][Friedman et al., 2000]. Yet user differentiation between technical failures and purposeful human acts of malfeasance has never been tested. Despite the fact that the current software engineering process fails to create trustworthy software [Viega et al., 2001] much work on trust-based systems assumes only purposeful betrayals or simply declares that the user should differentiate [Friedman et al., 2000].

The inclusion of human factors as a key concern in the design of secure systems is a significant move forward in human-centered design. Human-centered design attempts to empower users to make rational trust decisions by offering information in an effective manner. "Psychological acceptability" was recognized a security design principle over a quarter century ago [Saltzer and Schroeder, 1975], and users and user behavior are commonly cited as the "weak link" in computer security. Passwords and other forms of authentication are among the more obvious ways that security features appear as part of the human-computer interface. But the relevance of computer-human interaction to computer security extends far beyond the authentication problem, because the expectations of humans are an essential part of the definition of security. For example, Garfinkel and Spafford suggested the definition: "A computer is secure if you can depend on it and its software to behave as you expect" [1996]. Since goals and expectations vary from situation to situation and change over time in the real world, a practical approach to computer security should also take into account how those expectations are expressed, interpreted, and upheld.

Although both the security and usability communities each have a long history of research extending back to the 1960s, only more recently have there been formal investigations into the interaction between these two sets of concerns. Some usability studies of security systems were conducted as early as 1989 [Karat, 1989; Mosteller and Ballas, 1989]. However, with the advent of home networking in the late 1990s, the study of computer-mediated trust has significantly expanded.

55.6.3.1 General Challenges

It is fairly well known that usability problems can render security systems ineffective or even motivate users to compromise security measures. While HCI principles and studies can help to inform the design of usable security systems, merely applying established HCI techniques to design more powerful, convenient, or lucid interfaces is not sufficient to solve the problem; the challenge of usable security is uniquely difficult. There are at least six special characteristics of the usable security problem that differentiate it from the problem of usability in general [Whitten and Tygar, 1999; Sasse, 2003]:

1. *The barn door property.* Once access has been inadvertently allowed, even for a short time, there is no way to be sure that an attacker has not already abused that access.
2. *The weakest link property.* When designing user interfaces, in most contexts a deficiency in one area of an interface does not compromise the entire interface. However, a security context is less forgiving. The security of a networked computer is only as strong as its weakest component, so special care needs to be taken to avoid dangerous mistakes.
3. *The unmotivated user property.* Security is usually a secondary goal, not the primary purpose for which people use their computers. This can lead users to ignore security concerns or even subvert them when security tasks appear to interfere with the achievement of their primary goal.

4. *The abstraction property.* Security policies are systems of abstract rules, which may be alien and unintuitive to typical computer users. The consequences of making a small change to a policy may be far-reaching and non obvious.
5. *The lack of feedback property.* Clear and informative user feedback is necessary in order to prevent dangerous errors, but security configurations are usually complex and difficult to summarize.
6. *The conflicting interest property.* Security, by its very nature, deals with conflicting interests, such as the interests of the user against the interests of an attacker or the interests of a company against the interests of its own employees.

HCI research typically aims to optimize interfaces to meet the needs of a single user or a set of cooperating users, and is ill-equipped to handle the possibility of active adversaries. Because computer security involves human beings, their motivations, and conflicts among different groups of people, security is a complex socio-technical system.

55.6.3.2 Authentication

Since user authentication is a very commonly encountered task and a highly visible part of computer security, much of the attention in usable security research has been devoted to this problem. The most common authentication technique, of course, is the password. Yet password authentication mechanisms fail to acknowledge even well-known HCI constraints and design principles [Sasse et al., 2001]. Cognition research has established that human memory decays over time, that nonmeaningful items are more difficult to recall than meaningful items, that unaided recall is more difficult than cued recall, and that similar items in memory compete and interfere with each other during retrieval. Password authentication requires perfect unaided recall of nonmeaningful items. Furthermore, many users have a proliferation of passwords for various systems or have periodically changing passwords, which forces them to select the correct password from a set of several remembered passwords. Consequently, people often forget their passwords and rely on secondary mechanisms to deal with forgotten passwords.

One solution is to provide a way to recover a forgotten password or to reset the password to a randomly generated string. The user authorizes recovery or reset by demonstrating knowledge of a previously registered secret or by calling a helpdesk. There are many design choices to make when providing a challenge-based recovery mechanism [Just, 2003]. Another common user response to the problem of forgetting passwords is to write down passwords or to choose simpler passwords that are easier to remember, thereby weakening the security of the system. One study of 14,000 UNIX passwords found that nearly 25% of all the passwords were found by trying variations on usernames, personal information, and words from a dictionary of less than 63,000 carefully selected words [Klein, 1990].

In response to all the shortcomings of user-selected string passwords, several password alternatives have been proposed. Jermyn et al. [1999] have examined the possibility of using hand-drawn designs as passwords. Others have looked at recognition-based techniques, in which users are presented with a set of options and are asked to select the correct one, rather than performing unaided recall. Brostoff and Sasse [2000] studied the effectiveness of images of human faces in this manner, and Dhamija and Perrig [2000] studied the use of abstract computer-generated images. In contexts where it is feasible to use additional hardware, other solutions are possible. Users can carry smart cards that generate password tokens or that produce responses to challenges from a server. Paul et al. [2003] describe a technique called "visual cryptography" in which the user overlays a uniquely coded transparency over the computer screen to decrypt a graphical challenge, thereby proving possession of the transparency.

There is considerable interest in using biometrics for user authentication. A variety of measurable features can be used, such as fingerprints, voiceprints, hand geometry, faces, or iris scans. Each of the various methods has its own advantages and disadvantages. Biometrics offer the potential for users to authenticate without having to remember any secrets or carry tokens. However, biometrics have the fundamental drawback that they cannot be reissued. A biometric is a password that can never be changed. Once compromised, a biometric is compromised forever. Biometrics raise significant concerns about the creation of centralized databases of biometric information, as a biometric (unless hashed) creates a universal identifier. Biometrics also have value implications in that biometric systems most often fail for minorities [Woodward et al., 2003]. Biometrics present class issues as well; for example, biometric records for recipients of government aid are already stored in the clear in California. The storage of raw biometric data makes compromise trivial and thus security uncertain.

55.6.3.3 User Perceptions and Trust

User perceptions of security systems are crucial to their success in two different ways. First, the perceived level of reliability or trustworthiness of a system can affect the decision of whether to use the system at all; second, the perceived level of security or risk associated with various choices can affect the user's choice of actions. Studies [Cheskin, 1999; Turner et al., 2001] have provided considerable evidence that user perception of security on e-commerce Web sites is primarily a function of visual presentation, brand reputation, and third-party recommendations. Although sufficiently knowledgeable experts could obtain technical information about a site's security, for ordinary consumers, "feelings about a site's security were for the most part not influenced by the site's visible use of security technology" [Turner et al., 2001].

With regard to the question of establishing trust, however, perceived security is not the whole story. Fogg conducted a large study of over 1400 people to find out what factors contributed to a Web site's credibility [2001]. The most significant factors were those related to "real-world feel" (conveying the real-world nature of the organization, such as by providing a physical address and showing employee photographs), "ease of use," and "expertise" (displaying credentials and listing references). There is always a response to presenting photographs of people on Web sites, but the effect is not always positive [Riegelsberger, 2003].

In order to make properly informed decisions, users must be aware of the potential risks and benefits of their choices. It is clear that much more work is needed in this area. For example, a recent study showed that many users, even those from a high-technology community, had an inaccurate understanding of the meaning of a secure connection in their Web browser and frequently evaluated connections as secure when they were not or *vice versa* [Friedman et al., 2002].

A recent ethnographic study [Dourish et al., 2003] investigated users' mental models of computer security. The study revealed that users tend to perceive unsolicited e-mail, unauthorized access, and computer viruses as aspects of the same problem, and envision security as a barrier for keeping out these unwanted things. The study participants blended privacy concerns into the discussion, perceiving and handling marketers as threats in much the same way as hackers. However, there seemed to be an "overwhelming sense of futility" in people's encounters with technology. The perception that there will always be cleverer adversaries and new threats leads people to talk of security in terms of perpetual vigilance.

In order to make properly designed systems, designers must be aware of human practices with respect to trust. Studies of trust in philosophy and social science argue that humans trust readily and in fact have a need to trust. Further, humans implement trust by aggregating rather than differentiating. That is, humans sort entities into trustworthy and untrustworthy groups. Thus, when people become users of computers, they may aggregate all computers into the class of computers, and thus become increasingly trusting rather than increasingly differentiating over time [Sproull and Kiesler, 1992]. Finally, when using computers humans may or may not differentiate between malicious behavior and technical incompetence. Spam is clearly malicious, whereas privacy violations may be a result of an inability to secure a Web site or a database. Competence in Web design may indicate a general technical competence, thus mitigating concerns about competence. However, competence in Web design may also indicate efficacy in obtaining user's trust for malicious purposes. Only the ability to discern the technical actions and understand the implications can provide users with the ability to manage trust effectively on the network.

55.6.3.4 Interaction Design

A number of studies have shown the potential for problems in interaction design to seriously undermine security mechanisms. Whitten and Tygar [1999] demonstrated that design problems with PGP made it very difficult for even technically knowledgeable users to safely use e-mail encryption; a study by Good and Krekelberg [2003] identifies problems in the user interface for KaZaA that can lead to users unknowingly exposing sensitive files on their computer. Carl Ellison has suggested that each mouse click required to use encryption will cut the base of users in half. [Ellison, 2002]

Results of such studies and personal experiences with security systems have led researchers to propose a variety of recommendations for interaction design in secure systems. Yee [2002] has proposed ten principles for user interaction design in secure systems. At a higher level are recommendations to apply established HCI techniques to the design process itself. Karat [1989] described the benefits of applying rapid prototyping techniques to enable an iterative process involving several rounds of field tests and design improvements. Zurko and Simon [1996] suggest applying user-centered design to security — that is, beginning with user needs as a primary motivator when defining the security model, interface, or features of a system. Grinter and Smetters [2003] suggested beginning the design process with a user-centered threat model and a determination of the user's security-related expectations. Techniques such as contextual design [Wixon et al., 1990] and discount usability testing [Nielsen, 1989] are also applicable to interaction design for secure systems.

Some have suggested that the best way to prevent users from making incorrect security decisions is to avoid involving users in security at all. Others argue that only the user really knows what they want to do, and knowledge of the user's primary goal is essential for determining the correct security action. It is clear that forcing users to perform security tasks irrelevant to their main purpose is likely to bring about the perception that security interferes with real work. Yee [2002] has suggested the *principle of the path of least resistance,* which recommends that the most natural way to perform a task should also be the safest way. Sasse [2003] highlighted the importance of designing security as an integral part of the system to support the user's particular work activity. Grinter and Smetters [2003] have proposed a design principle called *implicit security,* in which the system infers the security-related operations necessary to accomplish the user's primary task in a safe fashion. The perspectives are similar, and all recognize the necessity of harmonizing security and usability goals rather than pitting them against each other.

In order for users to manage a computer system effectively, there must be a communication channel between the user and the system that is safe in both directions. The channel should protect against masquerading by attackers pretending to be authorized users, and protect users from being fooled by attackers spoofing messages from the system. This design goal was identified by Saltzer and Schroeder as the *trusted path* [1975]. Managing complexity is a key challenge for user interfaces in secure systems. Grinter and Smetters [2003] and Yee [2002] have identified the need to make the security state visible to the user so that the user can be adequately informed about the risks, benefits, and consequences of decisions. However, a literal representation of all security relationships would be overwhelming. Cranor [2003] applied three strategies to deal with this problem in designing a policy configuration interface: (1) reducing the level of detail in the policy specification, (2) replacing jargon with less formal wording, and (3) providing the option to use prepackaged bundles of settings. Whitten and Tygar [2003] has suggested a technique called *safe staging,* in which users are guided through a sequence of stages of system use to increase understanding and avoid errors. Earlier stages offer simpler, more conservative policy options for maintaining security; then as the user moves to later stages, she gains progressively greater flexibility to manipulate security policy while receiving guidance on potential new risks and benefits.

Whitten and Tygar's usability study of PGP [1999] showed strong evidence that designing a user interface according to traditional interface design goals alone was not sufficient to achieve usable security. Whitten recommends that usable security applications should not only be easy to use but also should teach users about security and grow in sophistication as the user demonstrates increased understanding. Ackerman and Cranor [1999] proposed "critics" — intelligent agents that offer advice or warnings to users, but do not take action on the user's behalf. Such agents could inform users of nonobvious consequences of their actions or warn when a user's decision seems unusual compared to decisions previously made in similar situations.

55.6.4 Identity Examples

Systems for identity management are inherently social systems. These systems implement controls that place risk, control data flows, and implement authentication. The strong link between control and computer security, and between identity and privacy, make these ideal examples for considering social implications of technology.

55.6.4.1 PKI

Public signatures create bindings between identifiable cryptographic keys and specific (signed) documents. Public key infrastructures serve to link knowledge of a particular key to particular attribute. Usually that attribute is a name, but there are significant problems with using a name as a unique identifier. [Ellison and Camp, 2003]

The phrase "public key infrastructure" has come to refer to a hierarchy of authority. There is a single root or a set of root keys. The root keys are used to sign documents (usually quite short, called *certificates*) that attest to the binding between a key and an attribute. Since that attribute is so often identity, the remainder of the section assumes it is indeed identity.

Thus each binding between a key and identity is based on a cryptography verification from some other higher-level key. At the base is a key that is self-certified. Standard PKI implements a hierarchy with the assumption of a single point from which all authority and trust emanates. The owner of the root key is a certificate authority.

The current public key infrastructure market and browser implementation (with default acceptable roots) create a concentration of trust. Matt Blaze [2003] argues that the SSL protects you from any institution that refuses to give Verisign money. The cryptographer Blaze arguably has an accurate assessment, as the purchaser of the cryptographic verification determines Verisign's level of investigation into the true identity of the certificate holder.

Public key infrastructures centralize trust by creating a single entity that signs and validates others. The centralization of trust is further implemented by the selection of a set of keys which are trusted by default by a market that is a duopoly or monopoly.

55.6.4.2 PGP

Confidentiality in communications was the primary design goal of Pretty Good Privacy. PGP was conceived as a privacy enhancing technology as opposed to a technology for data protection. [Garfinkel, 1999] This fundamental distinction in design arguably explains the distinct trust models in the two technologies.

PGP allows any person to assert an identity and public key binding. It is then the responsibility of the user to prove the value of that binding to another. In order to prove the binding the user presents the key and the associated identity claim to others. Other individuals who are willing to assert that the key/identity binding is correct sign the binding with their own public keys. This creates a network of signatures, in which any user may or may not have a trusted connection.

PGP utilizes social networks. If PKI can be said to model an authoritarian approach, PGP is libertarian. PKI has roots that are privileged by default. Each PGP user selects parties that are trusted not only for their assertions about their own binding of key and identity but also for their judgment in verifying the linkage of others. Those who have trusted attestations are called introducers. These introducers serve as linking points between the social networks formed by the original user and other social networks.

PGP has monotonically increasing trust. When an introducer is selected, that introducer remains trusted over an infinite period of time unless manually removed. If an introducer is manually removed, all the introduced parties remain trusted, as the paths to a trusted entity are not recorded after introduction.

PGP promotes trust as an increasing number of introducers know another entity. PGP does not decrease trust if one introducer declares a lack of knowledge, regardless of the properties of the social network.

PGP was designed to enable secure email. Secure email provides both integrity of the content and authentication of the sender. PGP enables confidential email by providing the endpoints with the capacity to encrypt.

PGP places final trust in the hands of users, and allows users to implement their own social network by creating introducers. The same underlying cryptography is used by PGP and PKI but the values choices embedded are distinct.

55.6.5 Data Protection vs. Privacy

The focus on human implications of design have focused on trust and trusted systems. However, privacy as well as security is an element of trust.

Data surveillance, privacy violations, or abuse of data (depending on the jurisdiction and action) can be both ubiquitous and transparent to the user. Computers may transmit information without the users' knowledge; collection, compilation and analysis of data is tremendously simplified by the use of networked information systems. Because of these facts, the balance between consumers and citizens who use services and those that offer digital services cannot be maintained by simply moving services on-line.

A consideration of social implications of technology should include the dominant privacy technologies. The two most widely used (and implemented) privacy enhancing technologies are the anonymizer and P3P. The anonymizer implements privacy while P3P implements a data protection regime.

Data protection and privacy have more commonalities than differences. The differences have philosophical and as well as technical design implications.

Technical decisions determine the party most capable of preventing a loss of security; policy decisions can motivate those most capable. Current policy does not reflect the technical reality. End users are least technically capable and most legally responsible for data loss. For the vast majority of users on the Internet there are programs beyond their understanding and protocols foreign to their experience. Users of the Internet know that their information is sometimes transmitted across the globe. Yet there is no way for any but the most technically savvy consumers to determine the data leakage that results from Internet use.

There is a comprehensive and developed argument for data protection. The privacy argument for data protection is essentially that when the data are protected privacy is inherently addressed. One argument against privacy is that it lacks a comprehensive, consistent underlying theory. There are competing theories [Camp, 2003b] [Trublow, 1991] [Kennedy 1995], yet the existence of multiple, complete, but disjoint theories does illustrate the point that there is limited agreement. Data protection regimes address problems of privacy via prohibition of data reuse and constraints on compilation. By focusing on practical data standards, data protection sidesteps difficult questions of autonomy and democracy that are inherent in privacy. Data protection standards constrain the use of personally identifiable information in different dimensions according to the context of the compilation of the information and the information itself.

Unlike data protection, privacy has a direct technical implementation: anonymity.

For both privacy and data protection, one significant technical requirement is enabling users to make informed choices. Given the role of security technology in enhancing privacy, human-computer design for security can enhance both privacy and data protection.

Anonymity provides privacy protection by preventing data from being personally identifiable. Anonymity provides the strongest protection possible for privacy, and anonymity provides uniform protection for privacy across all circumstances. Anonymity is of limited in value in situations where there is a need to link repeated transaction. In such cases pseudonyms are needed. Pseudonyms can have no link to other roles or true identity; for example a pseudonym may be a name used in an imagined community such as a role-playing game. Pseudonyms allow for personalization without privacy violations. Repeated use of a pseudonym in multiple transactions with the same entities leaks little information. Use of a pseudonym in multiple contexts (for example, with multiple companies) causes the pseudonym to converge with the identity of the user.

Privacy enhancing technologies include technologies for obscuring the source and destination of a message (onion routing) and preventing information leakage while browsing (the anonymizer).

Onion routing encrypts messages per hop, suing an overlay network of routers with public keys. At each router, the message provides the address of the next router and a message encrypted with the public key of the next router. Thus, each router knows the source of the message and the next hop, but not the original source nor the final destination. However, the router records could be combined to trace a message across the network. Yet even with the combined records of the routers, the confidentiality of the message would remain.

The anonymizer is a widely used privacy-enhancing proxy. The anonymizer implements privacy by functioning as an intermediary so that direct connections between the server and the browser are prevented. The anonymizer detects Web bugs. Web bugs are 1×1 invisible images embedded into pages to allow entities other than the serving page to track usage. Since Web bugs are placed by a server other than the one perceived by the user, Web bugs allow for placement of cookies from the originating server. This subverts user attempts to limit the use of third-party cookies. Note that browsers have a setting that allows users to reject cookies from any server other than one providing the page — so-called third party cookies. Web bugs enable placement of third party cookies. The anonymizer also limits Java script and prevents the use of ftp calls to obtain user email addresses. The anonymizer cannot be used in conjunction with purchasing; it is limited to browsing.

In contrast, data protection encourages protection based on policy and practice.

The Platform for Privacy Preferences is a technology that implements the data protection approach. [Cranor and Reagle 1998]. The Platform for Privacy preferences was designed to create a technical solution to the problem of data sharing. Ironically, of all privacy-enhancing technologies, it depends on regulatory enforcement of contracts as opposed to offering technical guarantees. [Hochheiser, 2003]

The Platform for Privacy Preferences includes a schema (or language) for expressing privacy preferences. P3P allows the user to select a set of possible privacy policies by selecting a number <1,10> on a sliding scale. P3P also has a mechanism for a service provider to express its privacy practices. If there is agreement between the server policy and user preference then user data are transmitted to the server. Otherwise, no data are sent.

P3P also includes profiles where the user enters data. The inclusion of profiles has been a significant source of criticism as it removes from the user the ability to provide incorrect data. By including profiles, P3P removed from the user a common defense against data re-use (obfuscation, lying, or misdirection) by automating the transmission of correct data. P3P, therefore, had an enforcement role with respect to user data; either the user was consistently dishonest or consistently honest. P3P had no corresponding enforcement mechanism for the server. The server attests to its own privacy policy and the protocol assumes the server implements its own policy. The most recent version of P3P removes profiles; however the MS Explorer implementation still maintains profiles.

55.7 Network Protocols as Social Systems

Network protocols for reservation of system resources in order to assure quality of service represent a technical area of research. However, even quality of service designs have significant impact on the economics, and therefore the social results, of such systems. There is no more clear example of politics in design than the change in the design of Cisco routers to simplify so-called "digital wiretaps." Wiretaps refer to aural surveillance of a particular set of twisted pair lines coming from local switching office [IEEE, 1997]. This simple model was often abused by federal authorities in the U.S., and use of aural surveillance against dissidents, activists, and criminals has been widespread across the globe [Diffie and Landau, 1997]. "Digital telephony" is a phrase used by law enforcement to map the concept of wiretaps to the idea of data surveillance of an individual on the network. If the observation of digital activity is conceived as a simple digital mapping, then the risks can be argued as the same. Yet the ease of compilation and correlation of digital information argues that there is not a direct mapping.

Cisco implemented a feature for automatically duplicating traffic from one IP address to a distinct location — a feature desired by law-abiding consumers of the product and by law enforcement. However, Cisco included no controls or reporting on this feature. Therefore Cisco altered the balance of power between those under surveillance and those implementing surveillance by lowering the work factor for the latter. Additionally, the invisibility of surveillance at the router level further empowers those who implement surveillance. The ability to distinguish the flow from one machine across the network and to duplicate that flow for purposes of surveillance is now hard-wired into the routers. Yet the oversight necessary to prevent abuse of that feature is not an element of the design; not even one that must be disabled. An alternative configuration would require selection of a default email address to which notifications of all active taps would be sent. The email could be set according to the jurisdiction of the purchaser; thus implementing oversight.

A more subtle question is the interaction of quality of service mechanisms and universal service. The experience of the telephone (described in a previous section) illustrates how high quality service may limit the range of available service. Universal service may require a high degree of service sharing and certainly requires an easy to understand pricing method. Ubiquitous service reservation, and the resulting premium pricing, can undermine the potential of best effort service to provide always on connections at a flat price. [Camp and Tsang, 2002]

55.8 Open vs. Closed Code

There exists a strong debate on how the availability of code alters society as digital systems are adopted. The initiator of this dialogue was Richard Stallman, who foresaw the corporate closing of code [Stallman, 1984]. Other technical pioneers contributed both to the code base and the theory of free and open code [Oram, 1999; Raymond, 1999]. Legal pioneers [Branscomb, 1984] clearly saw the problem of closing information by the failure of property regimes to match the economic reality of digital networked information. By 2000 [Lessig, 1999] there was widespread concern about the social implications of the market for code. Code can be distributed in a number of forms that range from completely open to completely closed. Code is governed by licenses as well as law; yet the law is sufficiently slow to adapt that graduations of and innovations in openness are provided by licenses.

Computer code exists along a continuum. At one end is source code. Source code is optimized for human readability and malleability. Source code is high level code, meaning that it is several degrees removed from the physical or hardware level. An example of inherently human readable code is markup languages. There are technical means to prohibit the trivially easy reading and viewing of a document source, and methods for writing increasingly obtuse source code are proliferating. For example, popular Web-authoring documents use unnecessary Java calls to confuse the reading. At the most extreme is converting html-based Web pages to Shockwave formats which cannot be read. Markup languages and scripting languages such as JavaScript and CGI scripts are designed to be readable. Such scripts are read (thus the name), and then the listed commands are played in the order received.

Between the highest level and the physical addresses required by the machine, there is assembly language. Assembly is the original coding language. Grace Hopper (who found the original computer bug, a moth in the machine at Harvard in 1945) implemented programs in assembly. Assembly requires that humans translate program into the binary language that machines understand. For example, adding two numbers in assembly takes many lines of code. The computer must be instructed to read the first number, bit by bit, and store it in an appropriate location. Then the computer must read the second number. Then the numbers must be placed at the input to the arithmetic logic unit, then added, and the result placed in an output register. Grace Hopper invented the breakthrough of the assembler, the forerunner to the compiler.

The earliest code was all binary, of course, and thus clearly the most basic binary codes can be read. In these early binary the commands were implemented by women who physically linked nodes to create the binary "1" of the commands. For each mathematician creating a code there existed a large machine and a score of women to implement the commands by connecting two relays, thus programming a "1."

Current programs are vastly more complex than those implemented in hand-wired binary. The breaking of the Enigma code was a vast enterprise. (Alan Turing was honored for this achievement with a statue in the U.K.) Today, the same endeavor is an advanced undergraduate homework assignment. Thus, machine (sometimes called binary) code for today's programs is unreadable.

It is the ability to read code that makes it open or closed. Code that can be read can be evaluated by the user or the representative of the user. Code that is closed and cannot be read requires trust in the producer of the code. Returning to social forces that influence technology, this is particularly problematic. A user wants a secure machine. The producer of commercial code has an incentive to create code as fast as possible, meaning that security is not *a priori*ty. The user wants control over his or her personal information. The producer of commercial code may want information about the user, particularly to enforce intellectual property rights [Anderson, 2003] or to implement price discrimination [Odlyzko, 2003].

The ability to read code grants the practical ability to modify it. Of course, the existence of closed code does not prevent modifications. This can be seen most clearly in the modifications of the Content Scrambling System. The decryption of the Content Scrambling System enabled users to watch digital video disks from any region on any operating system. CSS implements the marketing plans, specifically regional price discrimination that is the traditional and profitable practice of the owners of mass-produced high-value video content.

Open code encourages innovation by the highly distributed end users by optimizing opportunities for innovation. Closed code encourages innovation by increasing the rewards to the fewer centralized innovators. Thus open and closed code implement different visions of innovation in society.

Open code offers transparency. Closed code is not transparent. If code is law, then the ability to view and understand law is the essence of freedom [Lessig, 1999; Stallman, 1984; Syme and Camp, 2001]. The inability to examine law is a mark of a totalitarian state [Solzhenitsyn, 1975].

55.9 Conclusions

Can the assumption of values be prevented by the application of superior engineering? Or are assumptions about values and humans an integral part of the problem-solving process? Arguably both cases exist in communications and information technologies. These two cases cannot be systematically and cleanly distinguished so that the designer can know when the guidance of philosophy or social sciences is most needed. When is design political? One argument is that politics inevitably intrudes when there are assumptions about trust and power embedded into the design. Yet trust assumptions may be as subtle as reservation of router resources or as obvious as closed code.

Technologies embed changes in power relationships by increasing the leverage of applied force. Yet the social implications of amplification of one voice or one force cannot be consistently or reliably predicted.

Designers who turn to the study of computer ethics find the field not yet mature. There are some who study ethics that argue that computers create no new ethical problems, but rather create new instantiations of previous ethical problems [Johnson, 2001]. Others argue that digital networked information creates new classes or cases of ethical conundrums [Walter 1996] [Moor, 1985]. Ethicists from all perspectives worked on the relevant professional codes. Thus, the professional can be guided by the ACM/IEEE- CS Software Engineering Code of Ethics and Professional Practice.

Integrity and transparency are the highest calling. No engineer should implement undocumented features, and all designers should document their technical choices.

The most risk-averse principle of a design scientist may be "Do no harm"; however, following that principle may result in inaction. Arguably inaction in the beginning of a technological revolution is the least ethical choice of all, as it denies society the opportunity to make any choices, however technically framed.

Additional Resources

The IEEE Society on Social Implications of Technology:
http://radburn.rutgers.edu/andrews/projects/ssit/default.htm
ACM SIGCAS: Special Interest Group on Computers and Society:

http://www.acm.org/sigcas/
An extended bibliography on technology and society, developed by a reference librarian:
http://www.an.psu.edu/library/guides/sts151s/stsbib.html
A listing of technology and society groups, and electronic civil liberties organizations:
http://www.ljean.org/eciv.html

References

Ackerman, Michael and Lorrie Cranor. Privacy Critics: UI Components to Safeguard Users' Privacy. *Proceedings of CHI 1999.*

Adobe Corporation. Adobe eBook FAQ, 2002. http://www.adobe.com/support/ebookrdrfaq.html.

Alderman, Ellen and Caroline Kennedy. *The Right to Privacy.* Alfred A Knopf, New York, 1995.

Anderson, Ross. Why cryptosystems fail. *Communications of the ACM*, 37(11): 32–40, November 1994.

Anderson, Ross. Cryptography and Competition Policy — Issues with Trusted Computing. 2nd Annual Workshop on Economics and Information Security (May 29–30, 2003, Robert H. Smith School of Business, University of Maryland).

Anderson, Ross. *Security Engineering: A Guide to Building Dependable Distributed Systems.* John Wiley & Sons, New York, 2002.

Aslam, Taimur, Ivan Krsul, and Eugene Spafford. A Taxonomy of Security Vulnerabilities. *Proceedings of the 19th National Information Systems Security Conference* (October 6, 1996, Baltimore, MD), 551–560.

Barlow, John. A Declaration of Independence of Cyberspace. http://www.eff.org/~barlow/Declaration-Final.html, 1996 (last viewed September, 2003).

Bijker, Wiebe, Thomas P. Hughes, and Trevor Pinch *The Social Construction of Technological Systems.* MIT Press, Cambridge, MA, 2001.

Blaze, Matt. Quotes, August 31, 2003. http://world.std.com/~cme/html/quotes.html (last viewed September, 2003).

Branscomb, Anne W. *Who Owns Information?* HarperCollins, New York, 1994.

Brostoff, Saacha and M. Angela Sasse. Are Passfaces More Usable than Passwords? A Field Trial Investigation. *Proceedings of HCI 2000* (September 5–8, Sunderland, U.K.), pp. 405–424. Springer-Verlag, 2000.

Camp, L. Jean, First principles for copyright for DRM design. *IEEE Internet Computing*, 7(3): 59–65, 2003a.

Camp, L. Jean. Design for Trust. In *Trust, Reputation, and Security: Theories and Practice.* Rino Falcone, Ed., Springer-Verlag, New York, 2003b.

Camp, L. Jean, Cathleen McGrath, and Helen Nissenbaum. Trust: A Collision of Paradigms. *Proceedings of Financial Cryptography 2001*, 91–105. Springer-Verlag, 2001.

Camp, L. Jean and Rose Tsang. Universal service in a ubiquitous digital network. *Journal of Ethics and Information Technology*, 2(4): 211–221, 2001.

Cheskin and Studio Archetype/Sapient. eCommerce Trust Study. January 1999.

Clark, David and Marjory Blumenthal. Rethinking the design of the Internet: The end to end arguments vs. the brave new world. Telecommunications Policy Research Conference, Washington, D.C., September 2000.

Computer Science and Telecommunications Board. *Trust in Cyberspace.* National Academy Press, Washington, D.C., 1999.

Cranor, Lorrie. Designing a Privacy Preference Specification Interface: A Case Study. *CHI 2003 Workshop on HCI and Security.* April 2003.

Cranor, Lorrie and Joseph Reagle. Designing a Social Protocol: Lessons Learned from the Platform for Privacy Preferences. In *Telephony, the Internet, and the Media.* Jeffrey K. MacKie-Mason and David Waterman, Eds., Lawrence Erlbaum, Hillsdale, NJ, 1998.

Department of Defense. *Department of Defense Trusted Computer System Evaluation Criteria*. National Computer Security Center, 1985.

Dhamija, Rachna and Adrian Perrig. Déjà Vu: A User Study Using Images for Authentication. *Proceedings of the 9th USENIX Security Symposium*, August 2000.

Diamond, Jared. *Guns, Germs, and Steel: The Fates of Human Societies*, W. W. Norton & Company, New York, 1999.

Diffie, Whit and Susan Landau. *Privacy on the Line*. MIT Press, Cambridge, MA, 1997.

Dourish, Paul, Jessica Delgado de la Flor, and Melissa Joseph. Security as a Practical Problem: Some Preliminary Observations of Everyday Mental Models. *CHI 2003 Workshop on HCI and Security*, April 2003.

Eisenstein, Elizabeth L. *The Printing Press as an Agent of Change*. Cambridge University Press, Cambridge, U.K., 1979.

Ellison, Carl. Improvements on Conventional PKI Wisdom. *1st Annual PKI Research Workshop*, Dartmouth, NH, April 2002.

Ellison, Carl and L. Jean Camp. Implications with Identity in PKI. http://www.ksg.harvard.edu/digital center/conference/references.htm (last viewed September 2003).

European Union. Directive 95/46/EC of the European Parliament and of the Council of 24 October 1995, *Official Journal of the European Communities*, L. 281: 31, 23 November 1995.

Evans, Nathanael, Avi Rubin, and Dan Wallach. Authentication for Remote Voting. *CHI 2003 Workshop on HCI and Security*. April 2003.

Federal Trade Commission. *Privacy Online: Fair Information Practices in the Electronic Marketplace*. Federal Trade Commission Report to Congress, 2000.

Fischer, Charles. *America Calling: A Social History of the Telephone to 1940*. University of California Press, Berkeley, CA, 1992.

Fogg, B.J., Nicholas Fang, Jyoti Paul, Akshay Rangnekar, John Shon, Preeti Swani, and Marissa Treinen. What Makes A Web Site Credible? A Report on a Large Quantitative Study. *Proceedings of ACM CHI 2001 Conference on Human Factors in Computing Systems*, pp. 61–68. ACM Press, New York, 2001.

Friedman, Batya, Ed. *Human Values and the Design of Computer Technology*. CSLI Publications, Stanford, CA, 2001.

Friedman, Batya, David Hurley, Daniel C. Howe, Edward Felten, and Helen Nissenbaum. Users' Conceptions of Web Security: A Comparative Study. *Extended Abstracts of the ACM CHI 2002 Conference on Human Factors in Computing Systems*, pp. 746–747. ACM Press, New York, 2002.

Friedman, Batya, Peter H. Kahn, Jr., and Daniel C. Howe. Trust online. *Communications of the ACM*, 43(12): 34–40, December 2000.

Friedman, Batya and Lynette Millett. Reasoning About Computers as Moral Agents. *Human Values and the Design of Computer Technology*. B. Friedman, Ed., CSLI Publications, Stanford, CA, 2001.

Garfinkel, Simson. *Pretty Good Privacy*. O'Reilly, Sebastapol, CA, 1999.

Garfinkel, Simson and Gene Spafford. *Practical UNIX and Internet Security*, 2nd ed. O'Reilly, Sebastapol, CA, 1996.

Good, Nathaniel and Aaron Krekelberg. Usability and Privacy: A Study of Kazaa P2P File-Sharing. *Proceedings of the ACM CHI 2003 Conference on Human Factors in Computing Systems*, pp. 137–144. ACM Press, New York, 2003.

Grinter, Rebecca E. and Diane Smetters. Three Challenges for Embedding Security into Applications. *CHI 2003 Workshop on HCI and Security*. April 2003.

Herkert, Joseph R., Ed. *Social, Ethical, and Policy Implications of Engineering: Selected Readings*. IEEE Wiley, New York, 1999.

Hochheiser, Harry. Privacy, policy, and pragmatics: An examination of P3P's Role in the Discussion of Privacy Policy. Draft, 2003.

IEEE United States Activities Board. Position Statement on Encryption Policy. In *The Electronic Privacy Papers*, 543. B. Schneier and D. Banisar, Eds. John Wiley & Sons, New York, 1997.

Jermyn, Ian, Alain Mayer, Fabian Monrose, Michael K. Reiter, and Aviel D. Rubin. The Design and Analysis of Graphical Passwords. *Proceedings of the 8th USENIX Security Symposium*, August 1999.

Johnson, Deborah. *Computer Ethics,* 3rd ed. Prentice Hall, Upper Saddle River, NJ, 2001.

Just, Mike. Designing Secure Yet Usable Credential Recovery Systems With Challenge Questions. *CHI 2003 Workshop on HCI and Security.* April 2003.

Karat, Clare-Marie. Iterative Usability Testing of a Security Application. *Proceedings of the Human Factors Society 33rd Annual Meeting,* pp. 273–277, 1989.

Klein, Daniel V. Foiling the Cracker — A Survey of, and Improvements to, Password Security. *Proceedings of the 2nd USENIX Workshop on Security,* pp. 5–14, 1990.

Landes, David S. *The Wealth and Poverty of Nations: Why Some Are So Rich and Some So Poor.* W. W. Norton & Company, New York, 1999.

Landwehr, Carl E., A. R. Bull, J. P. McDermott, and W. S. Choi. A taxonomy of computer program security flaws, with examples. *ACM Computing Surveys,* 26(3): 211–254, September 1994.

Lessig, Larry. Tyranny in the Infrastructure. *Wired,* 5(7), 1997.

Lessig, Larry. *Code and Other Laws of Cyberspace.* Basic Books, New York, 1999.

Maner, Walter. Unique ethical problems in information technology. *Science and Engineering Ethics,* 2(2): 137–154, February 1996.

McLuhan, Marshall. *The Gutenberg Galaxy: The Making of Typographic Man.* Toronto: University of Toronto Press, Toronto, 1962.

Moor, James H. What is computer ethics? *Metaphilosophy,* 16(4): 266–275, October 1985.

Mosteller, William and James Ballas. Usability Analysis of Messages from a Security System, *Proceedings of the Human Factors Society 33rd Annual Meeting,* 1989.

Negroponte, Nicholas. *Being Digital — The Road Map for Survival on the Information Superhighway.* Alfred A. Knopf, New York, 1995.

Nielsen, Jakob. Usability engineering at a discount. *Designing and Using Human-Computer Interfaces and Knowledge Based Systems,* G. Salvendy and M. J. Smith, Eds., Elsevier Science, Amsterdam, 1989, pp. 394–401.

Odlyzko, Andrew M. Privacy, Economics, and Price Discrimination on the Internet. *Proceedings of ICEC '03.* ACM Press, New York, 2003, in press.

Oram, Andy. *Open Sources: Voices from the Revolution.* O'Reilly, Sebastapol, CA, 1999.

Pierce, John and Michael Noll. *Signals: The Science of Telecommunications.* Scientific American Press, New York, 1990.

Pool, Ithiel De Sola. *Technologies of Freedom.* Harvard University Press, Cambridge, MA, 1983.

Postman, Neil. *Technopoly: The Surrender of Culture to Technology.* Vintage Books, New York. 1996.

Prickler, Nedra, Mechanics Struggle with Diagnostics. *AP Wire,* 24 June 2002.

Raymond, Eric. *The Cathedral and the Bazaar.* O'Reilly, Sebastopol, CA, 1999.

Resnick, Paul. A Response to "Is PICS the Devil?" *Wired,* 5(7), July 1997.

Riegelsberger, Jens, M. Angela Sasse, and John McCarthy. Shiny Happy People Building Trust? Photos on e-Commerce Websites and Consumer Trust. *Proceedings of the ACM CHI 2003 Conference on Human Factors in Computing Systems,* April 5–10, Ft. Lauderdale, FL, pp. 121–128. ACM Press, New York, 2003.

Saltzer, Jerome H. and Michael D. Schroeder. The protection of information in computer systems. *Proceedings of the IEEE,* 63 (9): 1278–1308, 1975.

Sasse, M. Angela. Computer Security: Anatomy of a Usability Disaster, and a Plan for Recovery. CHI 2003 Workshop on HCI and Security. April 2003.

Sasse, M. Angela, Sacha Brostoff, and Dirk Weirich. Transforming the weakest link — a human/computer interaction approach to usable and effective security. *BT Technology Journal,* 19(3): 122–131, July 2001.

Adi Shamir. How to share a secret. *Communications of the ACM,* 22(11): 612–613, July 1979.

Ben Shneiderman. Designing trust into online experiences. *Communications of the ACM,* 43(12): 57–59, December 2000.

Solzhenitsyn, Alexander. The Law Becomes A Man. In *Gulag Archipelago*. Little, Brown and Company, New York, 1975 (English translation).

Sproull, Lee and Sara Kiesler, *Connections*. MIT Press, Cambridge, MA, 1992.

Spinello, Richard A., Ed. *Case Studies in Information and Computer Ethics*, Prentice Hall, Upper Saddle River, NJ, 1996.

Stallman, Richard. The GNU Manifesto. http://www.fsf.org/gnu/manifesto.html, 1984 (last viewed September 2003).

Stallman, Richard. Can You Trust Your Computer? http://www.gnu.org/philosophy/no-word-attachments.html, posted October 2002 (last viewed September 2003).

Syme, Serena and L. Jean Camp. Open land and UCITA land. *ACM Computers and Society*, 32(3): 86–101.

Trublow, George. *Privacy Law and Practice*. Times Mirror Books, Los Angeles, CA, 1991.

Tucker, Robert C., Ed. Marx and Engels to Babel, Liebknecht, Branke, and Others. In *Marx-Engels Reader*. W. W. Norton, New York, 1978, pp. 549–555.

Turner, Carl, Merrill Zavod, and William Yurcik. Factors That Affect The Perception of Security and Privacy of E-Commerce Web Sites. *Proceedings of the 4th International Conference on Electronic Commerce Research*, pp. 628–636, November 2001, Dallas, TX.

Viega, John, Tadayoshi Kohno, and Bruce Potter. Trust (and mistrust) in secure applications. *Communications of the ACM* 44(2): 31–36, February 2001.

Whitten, A. and J. Douglas Tygar. Why Johnny Can't Encrypt: A Usability Evaluation of PGP 5.0. *Proceedings of 8th USENIX Security Symposium*, 1999.

Whitten, Alam and J. Douglas Tygar. Safe Staging for Computer Security. *CHI 2003 Workshop on HCI and Security*. April 2003.

Woodward, John D., Katherine W. Webb, Elaine M. Newton et al. Appendix A, Biometrics: A Technical Primer. *Army Biometric Applications: Identifying and Addressing Sociocultural Concerns*, RAND/MR-1237-A, S RAND, 2001.

Winner, Langdon, *The Whale and the Reactor: A Search for Limits in an Age of High Technology*. Chicago University Press, Chicago, IL, 1986.

Wixon, Dennis, Karen Holtzblatt, and Stephen Knox. Contextual Design: An Emergent View of System Design. *Proceedings of the ACM CHI 1990 Conference on Human Factors in Computing Systems*, pp. 329–336, 1990.

Yee, Ka-Ping. User Interaction Design for Secure Systems. *Proceedings of the 4th International Conference on Information and Communications Security*, December 2002, Singapore.

Zurko, Mary Ellen and Richard T. Simon. User-Centered Security. *Proceedings of the UCLA Conference on New Security Paradigms*, Lake Arrowhead, CA, pp. 27–33, September 17–20, 1996.

56

The Geographical Diffusion of the Internet in the United States

CONTENTS

Abstract 56-1
56.1 Introduction 56-2
56.2 Brief History of the Internet 56-2
56.3 The Diffusion Process 56-3
56.3.1 Standard Diffusion Analysis 56-3
56.3.2 Demand for Business Purposes 56-5
56.3.3 Supply by Private Firms 56-6
56.3.4 Supply by Regulated Telephone Firms 56-6
56.4 Mapping the Internet's Dispersion 56-8
56.4.1 Backbone 56-8
56.4.2 Domain Name Registrations 56-8
56.4.3 Hosts, Internet Service Providers, and Points of Presence ... 56-9
56.4.4 Content and E-Commerce 56-10
56.5 Diffusion of Advanced Internet Access 56-10
56.5.1 Provision and Adoption 56-10
56.5.2 Rural vs. Urban Divides 56-11
56.6 Overview 56-13
56.6.1 What Happened during the First Wave of Diffusion? ... 56-13
56.6.2 Open Questions 56-13
References 56-14

Shane Greenstein

Jeff Prince

Abstract[1]

This chapter analyzes the rapid diffusion of the Internet across the U.S. over the past decade for both households and companies. The analysis explains why dialup connection has reached the saturation point while high-speed connection is far from it. Specifically, we see a geographic digital divide for high-speed access. We put the Internet's diffusion into the context of general diffusion theory where we consider costs and benefits on the demand and supply side. We also discuss several pictures of the Internet's current physical presence using some of the main techniques for Internet measurement to date. Through this analysis we draw general lessons about how other innovative aspects of the Internet diffuse.

[1] Northwestern University, Department of Management and Strategy, Kellogg School of Management, and Department of Economics, respectively. We thank the Kellogg School of Management for financial support. All errors contained here are our responsibility.

1-58488-381-2/05/$0.00+$1.50
© 2005 by CRC Press LLC

56.1 Introduction

The National Science Foundation (NSF) began to commercialize the Internet in 1992. Within a few years there was an explosion of commercial investment in Internet infrastructure in the U.S. By September 2001, 53.9 million homes (50.5%) in the U.S. had Internet connections (National Telecommunications and Information Administration [NTIA, 2002]).

The diffusion of the Internet has thus far proceeded in two waves. To be connected to the Internet is to have access at any speed. Yet, there is a clear difference between low-speed/dialup connection and high-speed/hardwire connection. In the early 1990s, those with dialup connection were considered on the frontier, but by the turn of the millennium, dialup connection had clearly become a nonfrontier technology, with the new frontier consisting of high-speed connections, mainly through xDSL and cable.

As with any new technology, the diffusion of the Internet and its related technologies follows predictable regularities. It always takes time to move a frontier technology from a small cadre of enthusiastic first users to a larger majority of potential users. This process displays systemic patterns and can be analyzed. Furthermore, the patterns found through analysis of the early diffusion of the Internet are general. These patterns provide insight about the processes shaping the diffusion of other advanced technologies today, such Wi-Fi, Bluetooth, XML, and supply-chain standards.

Accordingly, this essay has two goals: to tell a specific story and to communicate general lessons. We provide a survey of the literature concerning the diffusion of the Internet in the U.S. This is a specific story told about a specific technology in a particular time period. Throughout the essay we use this story to understand broader questions about the workings of diffusion processes. We also discuss why these broader processes are likely to continue or not in the future.

For the specific story we focus on a few key questions:

1. How has the Internet diffused over geographic space? How and why does location matter for adoption and provision of frontier and nonfrontier Internet technology?
2. Do we see differences in Internet access and use between rural and urban areas, as well as within urban areas?
3. How long will it take the commercial Internet to cover most geographic areas, and, as a result, realize the promise of reducing the importance of distance?

The remainder of this chapter will place the diffusion of the Internet in the context of general diffusion theory, analyze the costs and benefits for providers and adopters, and discuss various measurements of Internet presence in order to address these questions.

56.2 Brief History of the Internet

The Internet is a malleable technology whose form is not fixed across time and location. To create value, the Internet must be embedded in investments at firms that employ a suite of communication technologies, Transmission Control Protocol/Internet Protocol (TCP/IP) protocols, and standards for networking between computers. Often, organizational processes also must change to take advantage of the new capabilities.

What became the Internet began in the late 1960s as a research project of the Advanced Research Projects Administration of the U.S. Defense Department, or ARPANET. From these origins sprang the building blocks of a new communications network. By the mid-1980s, the entire Internet used TCP/IP packet-switching technology to connect universities and defense contractors.

Management for large parts of the Internet was transferred to the National Science Foundation (NSF) in the mid-1980s. Through NSFNET, the NSF was able to provide connection to its supercomputer centers and a high-speed backbone from which to develop the Internet. Since use of NSFNET was limited to only academic and research locations, Alternet, PSInet, and SprintLink developed their own private backbones for corporations looking to connect their systems with TCP/IP (Kahn, 1995).

By the early 1990s the NSF developed a plan to transfer ownership of the Internet out of government hands and into the private sector. When NSFNET was shut down in 1995, only for-profit organizations were left running the commercial backbone. Thus, with the Internet virtually completely privatized, its diffusion path within the U.S. was dependent on economic market forces (Greenstein, 2003).

56.3 The Diffusion Process

General diffusion theory is an effective guide for analyzing the geographical diffusion of the Internet. According to general diffusion theory, new products are adopted over time, not just suddenly upon their entrance into the market. In addition, the rate of adoption of a new technology is jointly determined by consumers' willingness to pay for the new product and suppliers' profitability from entering the new market. We consider each of these factors in turn.

56.3.1 Standard Diffusion Analysis

At the outset we begin our analysis with simple definitions. Any entity (household, individual, or firm) is considered connected to the Internet if it has the capability of communicating with other entities (information in and information out) via the physical structure of the Internet. We will defer discussion about connections coming at different speeds (56 k dial up vs. broadband) and from different types of suppliers (AOL vs. a telephone company).

With regard to consumers, it is the heterogeneity of adopters that generally explains differences in the timing of adoption (Rogers, 1995). In this case, a good deal of heterogeneity is the direct result of another technology's diffusion — that of personal computers (PCs). The Internet is a "nested innovation" in that heterogeneity among its potential adopters depends heavily on the diffusion process of PCs (Jimeniz and Greenstein, 1998). Then, on top of this nesting, within the class of PC users, there are also differences in their willingness to experiment and the intensity of their use.[2]

The following five attributes of a new technology are widely considered as the most influential for adoption speed across different types of users: relative advantage, compatibility, complexity, trialability, and observability. Any increase in the relative advantage over the previous technology, the compatibility of the new technology with the needs of potential adopters, the ability of adopters to experiment with the new technology, or the ability of users to observe the new technology will speed up the diffusion process. Similarly, any decrease in technological complexity will also speed up the diffusion process (Rogers, 1995).

The Internet has relative advantages along many dimensions. It provides written communication faster than postal mail, allows for purchases online without driving to the store, and dramatically increases the speed of information gathering. The Internet is also easy to try (perhaps on a friend's PC or at work), easy to observe, and compatible with many consumer needs (information gathering, fast communication); and its complexity has been decreasing consistently. All of these attributes have contributed toward increasing Americans' propensity to adopt.

The above attributes hold across the U.S., but the degree to which they hold is not geographically uniform. Specifically, we see differences between rural and urban areas. For example, people living in rural areas might find greater relative advantage because their next-best communication is not as effective as that of their urban counterparts. Also, they might find the Internet more difficult to try or observe and possibly more complex if they have less exposure or experience with PCs.

Beyond differences in the levels of the five major attributes, there are differences in types of adopters across regions. Generic diffusion theory points to five categories of adopters: innovators, early adopters, early majority, late majority, and laggards. When a technology first enters on the frontier, the group of innovators adopts first, and over time, the technology moves down the hierarchy. If these groups are not evenly dispersed geographically, there will be an uneven rate of adoption across regions of the country.

[2] For more on the diffusion of PCs, see Goolsbee and Klenow (1999), United States Department of Agriculture (2000), or NTIA (1995, 1997, 1998, 2002).

36.3.1.1 Cost–Benefit Framework

Within general diffusion theory there can be much dispute as to why the adoption of a new technology is actualized in a specific way. Here, we apply a general-purpose technology (GPT) framework to our study of the diffusion of the Internet. According to the GPT framework, some consumers use a new technology in its generic form, but for the technology to spread it must be customized for different subsets of users. This customization is why there is a delayed pattern of adoption.

Bresnahan and Trajtenberg (1995) define a GPT as "a capability whose adaptation to a variety of circumstances raises the marginal returns to inventive activity in each of these circumstances." General-purpose technologies are often associated with high fixed costs and low marginal costs to use. The Invention of the Internet follows this pattern, in the sense that the technology was largely invented and refined by the early 1990s (Bresnahan and Greenstein, 2001).

The GPT framework further predicts that additional benefit from the technology comes from "co-invention" of additional applications. Co-invention costs are the costs affiliated with customizing a technology to particular needs in specific locations at a point in time. These costs can be quite low or high, depending on the idiosyncrasy and complexity of the applications, as well as on economies of scale within locations.

Provision of the Internet in a region involves high fixed costs of operating switches, lines, and servers. We expect to see firms wishing to minimize fixed costs or exploit economies of scale by serving large markets. Also, the cost of "last mile" connection (e.g., xDSL or cable) in rural areas is far greater due to their longer distance from the backbone. This basic prediction frames much of the research on the diffusion of the Internet. There will necessarily be a margin between those who adopt and those who do not. What factors are correlated with the observed margin? We can divide these factors into those associated with raising or lowering the costs of supply or the intensity of demand.

56.3.1.2 Demand by Households

Households will pay for Internet connection when the benefits outweigh the costs. Internet literature points to several household characteristics that strongly correlate with Internet usage, namely, income, employment status, education, age, and location. We address each of these characteristics in turn.

According to the NTIA (2002) study, as of 2001, approximately 56.5% of American homes owned a PC, with Internet participation rates at 50.5% in 2001. By 2001, Internet usage correlated with higher household income, employment status, and educational attainment. With regard to age, the highest participation rates were among teenagers, while Americans in their prime working ages (20–50 years of age) were also well connected (about 70%) (NTIA, 2002). Although there did not appear to be a gender gap in Internet usage, there did appear to be a significant gap in usage between two widely defined racial groups: (1) whites, Asian Americans, Pacific Islanders (approximately 70%), and (2) Blacks and Hispanics (less than 40%) (NTIA, 2002). Much of this disparity in Internet usage can be attributed to observable differences in education and income. For example, at the highest levels of income and education there are no significant differences in adoption and use across ethnicities.

A great deal of literature points to a digital divide between rural and urban areas, contending that rural residents are less connected to the Internet than urban ones. Some argue that rural citizens are less prone to using computers and digital networks because of exacerbating propensities arising from lower income, and lower levels of education and technological skills (on average) compared to those living in the city. The evidence for this hypothesis is mixed, however, with many rural farms using the Internet at high rates (U.S. Department of Agriculture, 2000). In addition, over the 2-year span from 1998 to 2000, Internet access went up from 27.5 to 42.3% in urban areas, 24.5 to 37.7% in central cities, and 22.2 to 38.9% in rural areas. Thus, there was at least a narrowing of the gap in participation rates between rural and urban areas, and there certainly was no evidence of the gap widening on any front.[3]

[3] For the full historical trend, see also NTIA (1995, 1997, 1998).

Furthermore, when we divide American geography into three sections — rural, inner city urban, and urban (not inner city) — we see lower participation in the first two categories, with inner city participation also being low potentially due to a greater percentage of citizens with lower income and education levels. With the higher concentration of Blacks and Hispanics in the inner city, there then arises the correlation between education and income and socioeconomics. As we previously stated, ethnicity is not the cause of lower adoption rates; instead, lower education and income levels, which in turn are caused by socio-economic factors, create lower adoption rates in the inner cities. (Strover, 2001).

It has been argued that the benefits of adoption are greater for rural areas because rural residents can use the Internet to compensate for their distance from other activities. Adopting the Internet improves their retail choices, information sources, education options, and job availability more than those of urban residents (Hindman, 2000). However, these benefits may or may not be translated into actual demand.

56.3.2 Demand for Business Purposes

Business adoption of the Internet came in a variety of forms. Implementation for minimal applications, such as email, was rather straightforward by the late 1990s. It involved a PC, a modem, a contract with an Internet Service Provider, and some appropriate software. Investment in the use of the Internet for an application module in a suite of Enterprise Resource Planning software, for example, was anything but routine during the latter half of the 1990s. Such an implementation included technical challenges beyond the Internet's core technologies, such as security, privacy, and dynamic communication between browsers and servers. Organizational procedures usually also changed.

Businesses adopt different aspects of Internet technology when anticipated benefits outweigh the costs. In the standard framework for analyzing diffusion, the decision to adopt or reject the Internet falls under three categories: (1) optional, where the decision is made by the individual, (2) collective, where it is made by consensus among members, or (3) authoritative, where it is made by a few people with authority (Rogers, 1995). For businesses, the decision process falls under one of the latter two categories.

Either a consensus of members of the organization or top-level management will assess whether adopting the Internet is expected to improve the overall profitability of the company and then proceed accordingly. Again, as in individual adoption decisions, the five key attributes of the new technology will again be important. So there is every reason to expect basic Internet use in business to be as common as that found in households.

A further motivating factor will shape business adoption: competitive pressure. As Porter (2000) argues, there are two types of competitive motives behind Internet adoption. First, the level of "table stakes" may vary by region or industry. That is, there may be a minimal level of investment necessary just to be in business. Second, there may be investments in the Internet that confer competitive advantage vis-à-vis rivals. Once again, these will vary by location, industry, and even the strategic positioning of a firm (e.g., price leader, high service provider) within those competitive communities. The key insight is that such comparative factors shape competitive pressure.

Several recent studies look empirically at determining factors for Internet adoption by firms and the possible existence of a digital divide among them. Premkumar and Roberts (1999) test the former by measuring the relevance of 10 information technology attributes for the adoption rate of small rural businesses. The 10 attributes are: relative advantage, compatibility, complexity, cost-effectiveness, top-management support, information technology expertise, size, competitive pressure, vertical linkages, and external support. They find that relative advantage, cost-effectiveness, top-management support, competitive pressure, and vertical linkages were significant determinants for Internet adoption decisions.

Forman (2002) examines the early adoption of Internet technologies at 20,000 commercial establishments from a few select industries. He concentrates on a few industries with a history of adoption of frontier Internet technology and studies the microeconomic processes shaping adoption. He finds that rural establishments were as likely as their urban counterparts to participate in the Internet and to employ advanced Internet technologies in their computing facilities for purposes of enhancement of these facilities. He attributes this to the higher benefits received by remote establishments, which otherwise had no access to private fixed lines for transferring data. Forman, Goldfarb, and Greenstein (2002, 2003) measure national Internet adoption rates for medium and large establishments from all industries.[4] They distinguish between two purposes for adopting, one simple and the other complex. The first purpose, labeled *participation*, relates to activities such as email and Web browsing. This represents minimal use of the Internet for basic communications. The second purpose, labeled *enhancement*, relates to investment in frontier Internet technologies linked to computing facilities. These latter applications are often known as *e-commerce* and involve complementary changes to internal business computing processes. The economic costs and benefits of these activities are also quite distinct; yet, casual analysis in the trade press tends to blur the lines between the two.

Forman, Goldfarb, and Greenstein examine business establishments with 100 or more employees in the last quarter of 2000. They show that adoption of the Internet for purposes of participation is near saturation in most industries. With only a few exceptional laggard industries, the Internet is everywhere in medium to large businesses establishments. Their findings for enhancement contrast sharply with their findings for participation. There is a strong urban bias to the adoption of advanced Internet applications. The study concludes, however, that location, per se, does not handicap adoption decisions. Rather, the industries that "lead" in advanced use of the Internet tend to disproportionately locate in urban areas.

They conclude that a large determinant of the location of the Internet in e-commerce was the preexisting distribution of industrial establishments across cities and regions. This conclusion highlights that some industries are more information intensive than others, and, accordingly, make more intensive use of new developments in information technologies, such as the Internet, in the production of final goods and services. Heavy Internet technology users have historically been banking and finance, utilities, electronic equipment, insurance, motor vehicles, petroleum refining, petroleum pipeline transport, printing and publishing, pulp and paper, railroads, steel, telephone communications, and tires (Cortada, 1996).

56.3.3 Supply by Private Firms

Since the Internet's privatization in 1995, private incentives have driven the supply side of Internet access. Internet Service Providers (ISPs) are divided into four classes: (1) transit backbone ISPs, (2) downstream ISPs, (3) online service providers (e.g., AOL), and (4) firms that specialize in Website hosting. Provision incentives are profit based, and for a technology with significant economies of scale, profits will likely be higher in markets with high sales quantity. Thus, we see high numbers of ISPs in regions with high population concentrations (Downes and Greenstein, 1998, 2002).

The ISPs also decide on the services they provide (e.g., value-added services) and the price at which they provide them. Greenstein (2000a, 2000b) highlights two types of activities other than basic access in which ISPs partake — high-bandwidth applications and services that are complementary to basic access. He notes that differences in firm choices are due to "different demand conditions, different quality of local infrastructure, different labor markets for talent," or differing qualities of inherited firm assets.

Geography plays a role in these differences and can explain much of the variation in quality of access. We expect the quality of local infrastructure to be higher in urban areas. The quality of ISP service will be higher there as well. Additionally, rural ISPs often have less incentive to improve due to lack of competition, that is, they are the only provider in their area and thus have little incentive to enhance their service.

56.3.4 Supply by Regulated Telephone Firms

Every city in the U.S. has at least one incumbent local telephone provider. The deregulation of local telephony has been proceeding in many parts of the U.S. since the AT&T divestiture in the early 1980s.

---

[4] See, also, Atrostic and Nguyen (2002), who look at establishments in manufacturing. To the extent that they examine adoption, their study emphasizes how the size of establishments shapes the motives to adopt networking for productivity purposes.

This movement is an attempt to increase the number of potential providers of local voice services beyond this monopoly incumbent, and in so doing, to increase the competitiveness of markets for a variety of voice and data services. This form of deregulation became linked to the growth of broadband because the rules affecting telephony shaped the price of providing broadband. Deregulation had an impact on the Internet's deployment because it altered the organization of the supply of local data services, primarily in urban areas.[5]

Prior to the commercialization of the Internet, decades of debate in telephony had already clarified many regulatory rules for interconnection with the public switch network, thereby eliminating some potential local delays in implementing this technology on a small scale. By treating ISPs as an enhanced service and not as competitive telephone companies, the Federal Communication Commission (FCC) did not pass on access charges to them, which effectively made it cheaper and administratively easier to be an ISP (Oxman, 1999; Cannon 2001). [6]

The new competitor for the deregulated network is called a Competitive Local Exchange Company (CLEC). No matter how it is deployed, CLECs have something in common: each offers phone service and related data carrier services that interconnect with the network resources offered by the incumbent provider (e.g., lines, central switches, local switches). In spite of such commonalities, there are many claims in the contemporary press and in CLEC marketing literature that these differences produce value for end users. In particular, CLECs and incumbent phone companies offer competing versions of (sometimes comparable) DSL services and networking services.

Something akin to CLECs existed prior to the 1996 Telecommunications Act, the watershed federal bill for furthering deregulation across the country. These firms focused on providing high-bandwidth data services to business. After its passage, however, CLECs grew even more. And they quickly became substantial players in local networks, accounting for over $20 billion a year in revenue in 2000.[7] More to the point, CLECs became the center of focus of the deregulatory movement. Many CLECs grew rapidly and often took the lead in providing solutions to issues about providing the last mile of broadband, particularly to businesses and targeted households. In addition, many CLECs already were providing direct line (e.g., T-1) services to businesses (as was the incumbent local phone company).

The incumbent delivered services over the switch and so did CLECs. In recognition of the mixed incentives of incumbents, regulators set rules for governing the conduct of the transactions. As directed by the 1996 Telecommunications Act, this included setting the prices for renting elements of the incumbent's network, such as the loops that carried the DSL line.[8]

For our purposes here, the key question is: Did the change in regulations shape the geographic diffusion of Internet access across the U.S.? The answer is almost certainly, yes, at least in the short run. The answer, however, is more ambiguous in the long run. By the end of the millennium the largest cities in the U.S. had dozens of potential and actual competitive suppliers of local telephone service that interconnected with the local incumbent. By the end of 2000, over 500 cities in the U.S. had experience with at least a few competitive suppliers of local telephony, many of them focused on providing related Internet and networking services to local businesses, in addition to telephone service (New Paradigm Research Group, 2000).

This opportunity extended to virtually all cities with a population of more than 250,000, as well as to many cities with a population under 100,000. Very few rural cities, however, had this opportunity except in the few states that encouraged it. So, at the outset, if there were any effects at all, the entry of CLECs only moderately increased broadband supply in just the urban locations.

---

[5] For a comprehensive review of the literature, see Woroch (2001).

[6] The FCC's decision was made many years earlier for many reasons and extended to ISPs in the mid-1990s, with little notice at the time, because most insiders did not anticipate the extent of the growth that would arise. As ISPs grew in geographic coverage and revenues threatened to become competitive voice carriers, these interconnection regulations came under more scrutiny (Werbach, 1997; Kende, 2000; Weinberg, 1999).

[7] See Crandall and Alleman (2002).

[8] For review of the determinants of pricing within states, see Rosston and Wimmer (2001).

Due to the uneven availability of the Internet in some locations, local public government authorities also have intervened to speed deployment. Local governments act as an agent for underserved demanders by motivating broadband deployment in some neighborhoods through select subsidies or granting of right-of-ways (Strover and Berquist, 2001). There also often is help for public libraries, where the presence of a federal subsidy enables even the poorest rural libraries to have Internet access at subsidized rates (Bertot and McClure, 2000).

56.4 Mapping the Internet's Dispersion

A number of alternative methods have been devised for measuring the Internet's presence or its adoption in a location. None is clearly superior, as they are all valid ways of measuring the diffusion of the technology across geographic regions.

56.4.1 Backbone

The commercial Internet comprises hubs, routers, high-speed switches, points of presence (POPs), and high-speed high-capacity pipe that transmit data. These pipes and supporting equipment are sometimes called *backbone* for short. Backbone comprises mostly fiberoptic lines of various speeds and capacity. However, no vendor can point to a specific piece of fiber and call it "backbone." This label is a fiction, but a convenient one. Every major vendor has a network with lines that go from one point to another, but it is too much trouble to refer to it as "transmission capacity devoted primarily to carrying traffic from many sources to many sources."

One common theme in almost every article addressing the Internet's backbone is the following: a handful of cities in the U.S. dominate in backbone capacity, and, by extension, dominate first use of new Internet technology. Specifically, San Francisco/Silicon Valley, Washington, D.C., Chicago, New York, Dallas, Los Angeles, and Atlanta contain the vast majority of backbone capacity (Moss and Townsend, 2000a, 2000b). As of 1997, these seven cities accounted for 64.6% of total capacity, and the gap between this group and the rest remained even during the intense deployment of new networks and capacity between 1997 and 1999. By 1999, even though network capacity quintupled over the previous 2 years, the top seven still accounted for 58.8% of total capacity.

The distribution of backbone capacity does not perfectly mimic population distribution because metropolitan regions such as Seattle, Austin, and Boston have a disproportionately large number of connections (relative to their populations), whereas larger cities such as Philadelphia and Detroit have disproportionately fewer connections (Townsend, 2001a, 2001b). In addition, the largest metropolitan areas are well served by the backbone, whereas areas such as the rural South have few connections (Warf, 2001).

56.4.2 Domain Name Registrations

Domain names are used to help map intuitive names (such as www.northwestern.edu) to the numeric addresses computers use to find each other on the network (Townsend, 2001). This address system was established in the mid-1990s and diffused rapidly along with the commercial Internet.

The leaders in total domain names are New York and Los Angeles; however, Chicago — normally considered along with New York and Los Angeles as a global city — only ranks a distant fifth, far behind the two leaders. Furthermore, when ranking metropolitan areas according to domain names per 1000 persons, of these three cities, only Los Angeles ranks among the top twenty (17th). The full ranking of domain name density indicates that medium-sized metropolitan areas dominate, whereas global cities remain competitive and small metropolitan areas show very low levels of Internet activity (Townsend, 2001a).

Moss and Townsend (1998) look at the growth rate for domain name registrations between 1994 and 1997. They distinguish between *global information centers* and *global cities* and find that global information centers such as Manhattan and San Francisco grew at a pace six times the national average. In contrast, global cities such as New York, Los Angeles, and Chicago grew only at approximately one to two times the national average.

Kolko (2000) examines domain names in the context of questioning whether the Internet enhances the economic centrality of major cities in comparison to geographically isolated cities.[9] He argues, provocatively, that reducing the "tyranny of distance" between cities does not necessarily lead to proportional economic activity between them. That is, a reduction of communications costs between locations has ambiguous predictions about the location of economic activity in the periphery or the center. Lower costs can reduce the costs of economic activity in isolated locations, but it can also enhance the benefits of locating coordinative activity in the central location. As with other researchers, Kolko presumes that coordinative activity is easier in a central city where face-to-face communications take place.

Kolko (2000) documents a heavy concentration of domain name registrations in a few major cities. He also documents extraordinary per capita registrations in isolated medium-sized cities. He argues that the evidence supports the hypothesis that the Internet is a complement, not a substitute for face-to-face communications in central cities. He also argues that the evidence supports the hypothesis that lowering communication costs helps business in remote cities of sufficient size (i.e., medium-sized, but not too small).

56.4.3 Hosts, Internet Service Providers, and Points of Presence

Measurements of host sites, ISPs, and POPs also have been used to measure the Internet's diffusion. Indeed, the growth of the Internet can be directly followed in the successive years of *Boardwatch Magazine (Boardwatch Magazine)*. The earliest advertisements for ISPs in *Boardwatch Magazine* appear in late 1993, growing slowly until mid-1995, at which point *the magazine* began to organize their presentation of pricing and basic offerings. There was an explosion of entry in 1995, with thousands being present for the next few years. Growth only diminished after 2001.

Internet hosts are defined as computers connected to the Internet on a full-time basis. Host-site counting may be a suspect measurement technique due to its inability to differentiate between various types of equipment and to the common practice by firms of not physically housing Internet-accessible information at their physical location. Nevertheless, we do see results similar to those found with other measurement techniques, because, as of 1999, five states (California, Texas, Virginia, New York, and Massachusetts) contain half of all Internet hosts in the U.S. (Warf, 2001).

For ISPs, Downes and Greenstein (2002) analyze their presence throughout the U.S. Their results show that, while low entry into a county is largely a rural phenomenon, more than 92% of the U.S. population had access by a short local phone call to seven or more ISPs as of 1997.

Strover, Oden, and Inagaki (2002) look directly at ISP presence in areas that have traditionally been underserved by communications technologies (e.g., the Appalachian region). They examine areas in the states of Iowa, Texas, Louisiana, and West Virginia. They determine the availability and nature of Internet services from ISPs for each county and find that rural areas suffer significant disadvantages for Internet service (see, also, Strover, 2001).

Measurements of POPs help to identify "urban and economic factors spurring telecommunication infrastructure growth and investment" (Grubesic and O'Kelly, 2002). The POPs are locations where Internet service providers maintain telecommunication equipment for network access. Specifically, this is often a switch or router that allows Internet traffic to enter or proceed on commercial Internet backbones. Through POP measurement, Grubesic and O'Kelly derive similar results to those concerning the backbone, namely the top seven cities; Chicago, New York, Washington, D.C., Los Angeles, Dallas, Atlanta, and San Francisco provide the most POPs. Furthermore, cities such as Boston and Seattle are emerging Internet leaders.

Grubesic and O'Kelly (2002) use POPs to measure which metropolitan areas are growing the fastest. Their data indicates that areas such as Milwaukee, Tucson, Nashville, and Portland saw major surges in

[9] See, also, Kolko (2002).

POPs at the end of the 1990s. They provide several explanations for these surges: (1) proximity to major telecommunication centers (Tucson and Milwaukee), (2) intermediation between larger cities with high Internet activity (Portland), and (3) centralized location (Nashville).

56.4.4 Content and E-Commerce

Zook (2000, 2001) proposes two additional methods for measuring the presence of the Internet. The first measures the Internet by content production across the U.S. Zook defines the content business as "enterprises involved in the creation, organization, and dissemination of informational products to a global marketplace where a significant portion of the business is conducted via the Internet." He finds the location of each firm with a dot-com Internet address and plots it: San Francisco, New York, and Los Angeles are the leading centers for Internet content in the U.S. with regard to absolute size and degree of specialization.[10]

The second method looks at the locations of the dominant firms in e-commerce. Again, Zook (2000, 2001) finds the top Internet companies based on electronically generated sales and other means and their location. His analysis shows San Francisco, New York, and Los Angeles as dominant in e-commerce with Boston and Seattle, beating out the remainder of the top seven. When measured on a scale relative to the number of Fortune 1000 companies located in the region, his results indicate greater activity on the coasts (especially the West coast) with many Midwestern cities such as Detroit, Omaha, Cincinnati, and Pittsburgh lagging.

56.5 Diffusion of Advanced Internet Access

Internet connection generally comes in two forms: (1) dialup (technology now off the frontier) and (2) broadband (the new frontier technology).

56.5.1 Provision and Adoption

While dialup connection has moved past the frontier stage and is approaching the saturation point in the U.S., broadband access is still on the frontier and far from ubiquitous.[11] However, as the volume and complexity of traffic on the Internet increases dramatically each year, the value of universal "always-on" broadband service is constantly increasing. Furthermore, broadband access will enable providers to offer a wider range of bundled communications services (e.g., telephone, email, Internet video, etc.) as well as promote more competition between physical infrastructure providers already in place.

The diffusion of Internet access has been very much supply-driven in the sense that supply-side issues are the main determinants of Internet adoption. As ISPs face higher fixed costs for broadband due to lack of preexisting infrastructure, the spread of broadband service has been much slower and less evenly distributed than that of dialup service. In particular, ISPs find that highly populated areas are more profitable due to economies of scale and lower last-mile expenses.

As of September 2001, 19.1% of Internet users possessed high-speed connection; the dominant types of broadband access were cable modems and xDSL. The national broadband penetration of 19.1% can be partitioned into central city, urban, and rural rates of 22, 21.2, and 12.2%, respectively. We note that the rate of 22% for central cities is likely biased upward due to the presence of universities in the centers of many cities. Consistent with the supply-side issues, the FCC estimates that high-speed subscribers were present in 97% of the most densely populated zip codes by the end of 2000, whereas they were present in only 45% of the zip codes with the lowest population density (NTIA, 2002).

---

[10] Degree of specialization is measured by relating the number of .com domains in a region relative to the total number of firms in a region to the number of .com domains in the U.S. relative to the total number of firms in the U.S. (Zook, 2000).

[11] Broadband is defined by the FCC as "the capability of supporting at least 200 kbps in the consumer's connection to the network," both upstream and downstream (Grubesic and Murray, 2002).

Augereau and Greenstein (2001) analyze the evolution of broadband provision and adoption by looking at the determinants for upgrade decisions for ISPs. Although their analysis only looks at upgrades from dialup service to 56K modem or ISDN service occurring by 1997, it addresses issues related to the provision of high-speed service and warrants mention here. In their model, they look for firm-specific factors and location-specific factors that affect firms' choices to offer more advanced Internet services. Their main finding is that "the ISPs with the highest propensity to upgrade are the firms with more capital equipment and the firms with propitious locations." The most expansive ISPs locate in urban areas. They further argue that this could lead to inequality in the quality of supply between ISPs in high-density and low-density areas.

Grubesic and Murray (2002) look at differences in xDSL access for different regions in Columbus, Ohio. They point out that xDSL access can be inhibited for some consumers due to the infrastructure and distance requirements. The maximum coverage radius for xDSL is approximately 18,000 ft from a central switching office (CO), which is a large, expensive building. Furthermore, the radius is closer to 12,000 ft for high-quality, low-interruption service. Therefore, those living outside this radius from all the COs already built before xDSL was available will more likely suffer from lack of service. As a counterintuitive result, such affluent areas as Franklin County in Ohio might lack high-speed access, which is contrary to the usual notion of there being a socioeconomic digital divide (Grubesic and Murray, 2002). However, this does give more insight into a reason why many rural residents (those living in places with more dispersed population) might also lack high-speed access.

Lehr and Gillett (2000) compile a database consisting of communities in the U.S. where cable modem service is offered and link it to county-level demographic data. They find that broadband access is not universal. Only 43% of the population lives in counties with available cable modem service.[12] Broadband access is typically available in counties with large populations, high per capita income, and high population density; and there is a notable difference in strategy for cable operators with some being more aggressive than others.

In a very data-intensive study, Gabel and Kwan (2001) examined deployment of DSL services at central switches throughout the country and provided a thorough census of upgrade activity at switches. They examined providers' choice to deploy advanced technology to make broadband services available to different segments of the population. The crucial factors that affect the decision to offer service are listed as (1) cost of supplying the service, (2) potential size of the market, (3) cost of reaching the Internet backbone, and (4) regulations imposed on Regional Bell Operating Companies.[13] They found that advanced telecommunications service is not being deployed in low-income and rural areas.

In summary, the spread of broadband service has been much slower and much less evenly distributed than that of dialup. This is not a surprise once their basic economics are analyzed. The broadband ISPs find highly dense areas more profitable due to economies of scale in distribution and lower expenses in build-out. Moreover, the build-out and retrofit activities for broadband are much more involved and expensive than what was required for the build-out of the dialup networks. So, within urban areas, there is uneven availability. Thus, even before considering the impact of geographic dispersion in demand, the issues over the cost of supply guarantee that the diffusion process will take longer than dialup ever did. Until a low-cost wireless solution for providing high-bandwidth applications emerges, these economic factors are unlikely to change.

56.5.2 Rural vs. Urban Divides

To date, there has been a significant amount of research concerning the digital divide. Many researchers emphasize this divide along socioeconomic lines, such as wealth or race. Many others focus on a geo-

[12] They point out that this population is actually closer to 27% (as was stated by Kinetic Strategies), but explain that their data is not fine enough to show this measurement.

[13] Data was obtained concerning wire centers; also data on DSL and cable modem service availability was collected via Websites and calling service providers. They supplemented it with U.S. Census data.

graphical divide either contrasting rural vs. urban or rural vs. urban (center city) vs. urban (noncentral city).

We can make several key observations concerning a geographical divide. First, the divide for basic Internet services is generally nonexistent. Due to the preexisting infrastructure from telephone service, the cost of provision is relatively low; thus, we see over 92% of households just a local call away from Internet connection. Furthermore, as of 2001, 52.9% of rural residents were using the Internet, not far below the national average of 57.4% (NTIA, 2002).

Businesses participate at high rates, over 90%. While we do see lower basic participation rates in rural areas, this essentially is due to the type of industries we find there (i.e., industries deriving less relative benefit from Internet connection). Thus, in this particular case, we see that it is not necessarily availability of Internet access but largely the private incentives of the adopters (commercial businesses) determining the adoption rate.

Augereau and Greenstein (2001) warn of the possibility of the divide in availability worsening as large firms in large cities continue to upgrade their services rapidly while smaller firms in smaller cities move forward more slowly. As basic service has almost entirely saturated the country, the real issue of concern is the evolution of *quality* of service geographically, as well as value per dollar. Several authors warn that we may be headed down a road of bifurcation where large urban areas get better service at a faster pace while smaller cities and urban areas fall behind. Greenstein (2000a) suggests that urban areas get more new services due to two factors: "(1) increased exposure to national ISPs, who expand their services more often; and (2) the local firms in urban areas possess features that lead them to offer services with propensities similar to the national firms."

By a different line of argument, Strover (2001) arrives at a comparatively pessimistic assessment, one shared by many observers.[14] She points out that the cost structure for ISPs is unfavorable because of their dependence on commercial telecommunications infrastructure providers, which are reluctant to invest in rural areas due to the high costs necessary to reach what often are relatively few customers. Furthermore, a lack of competition in rural areas among telephone service providers serves to exacerbate the low incentives.

Even though we have suggested that high levels of participation by firms are largely determined by industry and not geography, the fact that industries heavily reliant on advanced telecommunications are concentrated in major cities provides an incentive for ISPs to place their main focus there. Furthermore, the fact that the economics of small cities is governed more by the private sector than government initiatives makes small cities less prone to initiating plans to develop telecommunications (Alles et al., 1994).

Many studies place a much greater emphasis on other variables along which they find the divide is much more pronounced. Hindman (2000) suggests that there is no strong evidence of a widening gap between urban and rural residents' use of information technologies, but that such predictors as income, education, and age have become even more powerful in predicting usage over the years (specifically from 1995 to 1998).

Forman, Goldfarb, and Greenstein (2002) find that, as of December 2000, 12.6% of establishments engage in some form of Internet enhancement activities. Furthermore, they find much higher enhancement adoption rates in large cities (consolidated metropolitan statistical areas) as the top 10 ranged from Denver at 18.3% to Portland at 15.1%. In addition, enhancement adoption rates in large urban counties (metropolitan statistical areas) is 14.7%, while that of small counties is only 9.9% on average.

However, they also find that the industries of "management of companies and enterprises" (NAICS 55) and "media, telecommunications, and data processing" (NAICS 51) had enhancement adoption rates of 27.9 and 26.8%, respectively — rates far exceeding all other industries. This strongly points to the idea that geographical differences may largely be explained by the preexisting geographical distribution of industries.

---

[14] See also Garcia (1996), Parker (2000), Hindman (2000).

56.6 Overview

The geographic diffusion of Internet infrastructure, such as the equipment to enable high-speed Internet access, initially appeared to be difficult to deploy and use. Technically, difficult technologies favor urban areas, where there are thicker labor markets for specialized engineering talent. Similarly, close proximity to thick technical labor markets facilitates the development of complementary service markets for maintenance and engineering services.

Labor markets for technical talent are relevant to the diffusion of new technologies in the Internet. As with many high-tech services, areas with complementary technical and knowledge resources are favored during the early use of technology. This process will favor growth in a few locations, such as Silicon Valley, the Boston area, or Manhattan — for a time, at least, particularly when technologies are young. But will it persist? This depends on how fast the technology matures into something standardized that can be operated at low cost in areas with thin supply of technical talent.

56.6.1 What Happened during the First Wave of Diffusion?

It was unclear at the outset which of several potential maturation processes would occur after commercialization. If advancing Internet infrastructure stayed exotic and difficult to use, then its geographic distribution would depend on the location of the users most willing to pay for infrastructure. If advancing Internet infrastructure became less exotic to a greater number of users and vendors, then commercial maturation would produce geographic dispersion over time, away from the areas of early experimentation. Similarly, as advanced technology becomes more standardized, it is also more easily serviced in outlying areas, again contributing to its geographic dispersion.

As it turned out, the first wave of the diffusion of the Internet (from 1995 to 2000) did not follow the most pessimistic predictions. The Internet did not disproportionately diffuse to urban areas with their complementary technical and knowledge resources. The location of experiments was necessarily temporary, an artifact of the lack of maturity of the applications. As this service matured — as it became more reliable and declined in price so that wider distribution became economically feasible — the geographic areas that were early leaders in technology lost their comparative lead or ceased to be leaders. As such, basic ISP technology diffused widely and comparatively rapidly after commercialization.

Open questions remain as the next wave proceeds. There is little experience with uncoordinated commercial forces developing a high-speed communication network with end-to-end architecture. This applies to the many facets that make up advanced telecommunications services for packet switching, such as switching using frame relay or Asynchronous Transfer Mode, as well as Synchronous Optical Network equipment or Optical Carrier services of various numerical levels (Noam, 2001).

To be sure, the spread of broadband service has been much slower and much less evenly distributed than that of dialup. This is not a surprise once their basic economics are analyzed. The broadband ISPs find highly dense areas more profitable due to economies of scale in distribution and lower expenses in build-out. Moreover, the build-out and retrofit activities for broadband are much more involved and expensive than what was required for the build-out of the dialup networks. So, within urban areas, there is uneven availability. This situation is unlikely to change until a low-cost wireless solution for providing high-bandwidth applications emerges.

56.6.2 Open Questions

Within this survey we have drawn a picture of where the Internet is today and discussed the main forces behind how it got there. While the geographical divide has all but vanished for basic connection, it still exists for advanced connection, and we are yet to determine whether this divide will improve or worsen.

Many interesting questions for this field still remain. The bilateral relationship between geography and the Internet still has many properties to be explained. Will Internet connection via satellite emerge as the connection of choice, and if so, how much would this dampen the argument that location matters?

Will another fixed wireless solution emerge for delivery of high-speed data services, and will it exhibit low enough economies of scale to spread to suburban areas?

As the majority of American homes become hardwired, how drastic will the effect be on local media, such as local newspapers (see. e.g., Chyi and Sylvie, 2001)? If individuals can access any radio station in the country at any time, can all these stations possibly stay in existence?

We can also ask how the spread of the Internet will affect the diffusion of other new products. In other words, are inventions spreading faster now than in the past because everything is connected? For example, use of some peer-to-peer technologies, such as ICQ and Napster, spread very fast worldwide. These were nested within the broader use of the Internet at the time. Was their speed of adoption exceptional, a by-product of the early state of the commercial Internet, or something we should expect to see frequently?

There are related questions about the spread of new technologies supporting improvements in the delivery of Internet services. Will the diffusion of IPv6 occur quickly because its use is nested within the structure of existing facilities? Will various versions of XML spread quickly or slowly due to the interrelatedness of all points on the Internet? What about standards supporting IP telephony? Will 802.11b (aka Wi-Fi) diffuse to multiple locations because it is such a small-scale technology, or will its small scale interfere with a coordinated diffusion?

As we speculate about future technologies, two overriding lessons from the past shape our thinking: First, when uncertainty is irreducible, it is better to rely on private incentives to develop mass-market services at a local level. Once the technology was commercialized, private firms tailored it in multiple locations in ways that nobody foresaw. Indeed, the eventual shape, speed, growth, and use of the commercial Internet was not foreseen within government circles (at NSF), despite (comparatively) good intentions and benign motives on the part of government overseers, and despite advice from the best technical experts in the world. If markets result in a desirable outcome in this set of circumstances in the U.S., then markets are likely to result in better outcomes most of the time.

Second, Internet infrastructure grew because it is malleable, not because it was technically perfect. It is better thought of as a cheap retrofit on top of the existing communications infrastructure. No single solution was right for every situation, but a TCP/IP solution could be found in most places. The U.S. telephone system provided fertile ground because backbone used existing infrastructure when possible but interconnected with new lines when built. CLECs and commercial ISPs provided Internet access to homes when local telephone companies moved slowly, but incumbent telephone companies and cable companies proved to be agile providers in some situations.

Ultimately, however, it is as fun to speculate as to watch the uncertain future unfold. The answers to these questions will become more approachable and important as the Internet continually improves and spreads across the U.S. at an extraordinary pace.

References

Alles, P., A. Esparza, and S. Lucas. 1994. Telecommunications and the Large City–Small City Divide: Evidence from Indiana cities. *Professional Geographer* 46: 307–16.

Atrostic, Barbara K. and Sang V. Nguyen. 2002. Computer Networks and U.S. Manufacturing Plant Productivity: New Evidence from the CNUS Data. Working Paper #02–01, Center for Economic Studies, U.S. Census Bureau, Washington, D.C.

Augereau, A. and S. Greenstein. 2001. The need for speed in emerging communications markets: Upgrades to advanced technology at Internet service providers. *International Journal of Industrial Organization*, 19: 1085–1102.

Bertot, John and Charles McClure. 2000. *Public Libraries and the Internet, 2000,* Reports prepared for National Commission on Libraries and Information Science, Washington, D.C., http://www.nclis.gov/statsurv/2000plo.pdf.

Boardwatch Magazine, Various years, Directory of Internet Service Providers, Littleton, CO.

Bresnahan, T. and Shane Greenstein. 2001. The economic contribution of information technology: Towards comparative and user studies." *Evolutionary Economics*, 11: 95–118.

Bresnahan, T. and Manuel Trajtenberg. 1995. General purpose technologies: Engines of growth? *Journal of Econometrics.* 65 (1): 83–108.

Cannon, Robert. 2001. Where Internet Service Providers and Telephone Companies Compete: A Guide to the Computer Inquiries, Enhanced Service Providers, and Information Service Providers, In *Communications Policy in Transition: The Internet and Beyond,* Benjamin Compaine and Shane Greenstein, Eds., Cambridge, MA: MIT Press.

Chyi, H. I. and Sylvie, G. 2001. The medium is global, the content is not: The role of geography in online newspaper markets. *Journal of Media Economics* 14(4): 231–48.

Cortada, James W. 1996. *Information Technology as Business History: Issues in the History and Management of Computers.* Westport, CT: Greenwood Press.

Crandall, Robert W. and James H. Alleman. 2002. *Broadband: Should We Regulate High-Speed Internet Access?* AEI–Brookings Joint Center for Regulatory Studies, Washington, D.C.

Downes, Tom and Shane Greenstein. 1998. Do commercial ISPs provide universal access? in *Competition, Regulation and Convergence: Current Trends in Telecommunications Policy Research,* Sharon Gillett and Ingo Vogelsang, Eds., pp. 195–212. Hillsdale, NJ: Lawrence Erlbaum Associates.

Downes, Tom and Shane Greenstein. 2002. Universal access and local Internet markets in the U.S. *Research Policy,* 31: 1035–1052.

Forman, Chris, 2002. The Corporate Digital Divide: Determinants of Internet Adoption. Working Paper, Graduate School of Industrial Administration, Carnegie Mellon University. Available at http://www.andrew.cmu.edu/~cforman/research/corp_divide.pdf.

Forman, Chris, A. Goldfarb, and S. Greenstein. 2003a. Digital Dispersion: An Industrial and Geographic Census of Commercial Internet Use. Working Paper, NBER, Cambridge, MA.

Forman, Chris, A. Goldfarb, and S. Greenstein. 2003b. The geographic dispersion of commercial Internet use. In *Rethinking Rights and Regulations: Institutional Responses to New Communication Technologies,* Steve Wildman and Lorrie Cranor, Eds., in press. Cambridge, MA: MIT Press.

Gabel, D. and F. Kwan. 2001. Accessibility of Broadband Communication Services by Various Segments of the American Population. In *Communications Policy in Transition: The Internet and Beyond,* Benjamin Compaine and Shane Greenstein, Eds., Cambridge, MA: MIT Press.

Garcia, D. L. 1996. Who? What? Where? A look at Internet deployment in rural America, *Rural Telecommunications* November/December: 25–29.

Goolsbee, Austan and Peter Klenow. 1999. Evidence on Learning and Network Externalities in the Diffusion of Home Computers. Working Paper # 7329, NBER, Cambridge, MA.

Gorman, S. P. and E. J. Malecki. 2000, The networks of the Internet: an analysis of provider networks in the USA. *Telecommunications Policy,* 24(2): 113–34.

Greenstein, Shane, 2000a. Building and delivering the virtual world: Commercializing services for Internet access. *The Journal of Industrial Economics,* 48(4): 391–411.

Greenstein, Shane, 2000b. Empirical Evidence on Commercial Internet Access Providers' Propensity to Offer new Services, In *the Internet Upheaval, Raising Questions, and Seeking Answers in Communications Policy,* Benjamin Compaine and Ingo Vogelsang, Eds., Cambridge, MA: MIT Press.

Greenstein, Shane, 2003, The economic geography of Internet infrastructure in the United States, In *Handbook of Telecommunications Economics,* Vol. 2, Martin Cave, Sumit Majumdar and Ingo Vogelsang, Eds., Amsterdam: Elsevier.

Grubesic, T. H. and A. T. Murray. 2002. Constructing the divide: Spatial disparities in broadband access. *Papers in Regional Science,* 81(2): 197–221.

Grubesic, T. H. and M. E. O'Kelly. 2002. Using points of presence to measure accessibility to the commercial Internet. *Professional Geographer,* 54(2): 259–78.

Hindman, D. B. 2000. The rural-urban digital divide. *Journalism and Mass Communication Quarterly,* 77(3): 549–60.

Jimeniz, Ed and Shane Greenstein. 1998. The emerging Internet retailing market as a nested diffusion process. *International Journal of Innovation Management,* 2(3).

Kahn, Robert. 1995. The role of government in the evolution of the Internet, In *Revolution in the U.S. Information Infrastructure,* by National Academy of Engineering, Washington, D.C.: National Academy Press.

Kende, Michael. 2000. The Digital Handshake: Connecting Internet Backbones, Working Paper No. 32., Federal Communications Commission, Office of Planning and Policy, Washington D.C.

Kolko, Jed. 2000. The Death of Cities? The Death of Distance? Evidence from the Geography of Commercial Internet Usage, In *The Internet Upheaval: Raising Questions, Seeking Answers in Communications Policy,* Ingo Vogelsang and Benjamin Compaine, Eds., Cambridge, MA: MIT Press.

Kolko, Jed. 2002. Silicon Mountains, Silicon Molehills, Geographic Concentration and Convergence of Internet Industries in the U.S., *Economics of Information and Policy.*

Lehr, William and Sharon Gillett. 2000. Availability of Broadband Internet Access: Empirical Evidence. Workshop on Advanced Communication Access Technologies, Harvard Information Infrastructure Project, Kennedy School of Government, Harvard University, Available at www.ksg.harvard.edu/iip/access/program.html/.

Moss, M. L. and A.M. Townsend. 1998. The Role of the Real City in Cyberspace: Understanding Regional Variations in Internet Accessibility and Utilization. Paper presented at the Project Varenius Meeting on Measuring and Representing Accessibility in the Information Age, Pacific Grove, CA.

Moss, M. L. and A.M. Townsend. 2000a. The Internet backbone and the American metropolis. *Information Society,* 16(1): 35–47.

Moss, M. L. and A.M. Townsend. 2000b. *The Role of the Real City* in *Cyberspace: Measuring and Representing Regional variations in Internet Accessibility, Information, Place, and Cyberspace,* Donald Janelle and David Hodge, Eds., Berlin: Springer-Verlag.

Mowery, D.C. and T.S. Simcoe, 2002, The Origins and Evolution of the Internet, in *Technological Innovation and Economic Performance,* R. Nelson, B. Steil, and D. Victor, Eds., Princeton, NJ: Princeton University Press.

NTIA, National Telecommunications and Information Administration. 1995. Falling Through the Net: A Survey of the "Have Nots" in Rural and Urban America. http://www.ntia.doc.gov/reports.html.

NTIA, National Telecommunications and Information Administration. 1997. Falling Through the Net: Defining the Digital Divide. http://www.ntia.doc.gov/reports.html.

NTIA, National Telecommunications and Information Administration. 1998. Falling Through the Net II: New Data on the Digital Divide. http://www.ntia.doc.gov/reports.html.

NTIA, National Telecommunications and Information Administration. 2002. A Nation Online: How Americans Are Expanding Their Use of the Internet. http://www.ntia.doc.gov/reports.html.

New Paradigm Resources Group. 2000. *CLEC Report,* Chicago, IL.

Noam, Eli. 2001. *Interconnecting the Network of Networks.* Cambridge, MA: MIT Press.

Oxman, Jason, 1999, The FCC and the Unregulation of the Internet, Working paper 31, Federal Communications Commission, Office of Planning and Policy, Washington, D.C.

Parker, E. B. 2000. Closing the digital divide in rural America. *Telecommunications Policy,* 24(4): 281–90.

Porter, Michael. 2000. Strategy and the Internet. *Harvard Business Review,* 79(March), 63–78.

Premkumar, G. and M. Roberts. 1999., Adoption of new information technologies in rural small businesses. *Omega–International Journal of Management Science,* 27(4): 467–484.

Rogers, Everett M. 1995. *Diffusion of Innovations.* New York: Free Press.

Rosston, Greg., and Brad Wimmer. 2001. "From C to Shining C": Competition and Cross-Subsidy in Communications, In *Communications Policy in Transition: The Internet and Beyond,* Benjamin Compaine and Shane Greenstein, Eds., Cambridge, MA: MIT Press.

Strover, Sharon. 2001. Rural Internet connectivity. *Telecommunications Policy,* 25(5): 331–47.

Strover, Sharon and Lon Berquist. 2001. Ping telecommunications infrastructure: State and Local Policy Collisions, In *Communications Policy in Transition: The Internet and Beyond* Benjamin Compaine and Shane Greenstein, Eds., Cambridge, MA: MIT Press.

Strover, Sharon, Michael Oden, and Nobuya Inagaki. 2002. Telecommunications and rural economies: Findings from the Appalachian region, In *Communication Policy and Information Technology: Promises, Problems, Prospects,* Lorrie Faith Cranor and Shane Greenstein, Eds., Cambridge, MA: MIT Press.

Townsend, A. M. 2001a. Network cities and the global structure of the Internet. *American Behavioral Scientist,* 44(10): 1697–1716.

Townsend, A. M. 2001b, The Internet and the rise of the new network cities, 1969–1999. *Environment and Planning B-Planning and Design* 28(1): 39–58.

U.S. Department of Agriculture. 2000. *Advanced Telecommunications in Rural America: The Challenge of Bringing Broadband Communications to All of America.* http://www.ntia.doc.gov/reports/ruralbb42600.pdf.

Warf, B. 2001. Segue ways into cyberspace: Multiple geographies of the digital divide. *Environment and Planning B-Planning and Design* 28(1): 3–19.

Weinberg, Jonathan. 1999. The Internet and telecommunications services: Access charges, universal service mechanisms, and other flotsam of the regulatory system. *Yale Journal of Regulation,* 16(2) .

Werbach, Kevin. 1997. A Digital Tornado: The Internet and Telecommunications Policy. Working Paper 29, Federal Communication Commission, Office of Planning and Policy, Washington, D.C.

Woroch, Glenn. 2001. Local Network Competition, In *Handbook of Telecommunications Economics,* Martin Cave, Sumit Majumdar, and Ingo Vogelsang, Eds., Amsterdam: Elsevier.

Zook, M. A. 2000. The Web of production: The economic geography of commercial Internet content production in the United States. *Environment and Planning A* 32(3): 411–426.

Zook, M. A. 2001. Old hierarchies or new networks of centrality? The global geography of the Internet content market. *American Behavioral Scientist,* 44(10): 1679–1696.

57

Intellectual Property, Liability, and Contract

CONTENTS

Abstract 57-1
57.1 Cyberlaw 57-1
57.2 Copyright 57-2
57.2.1 Basics of Copyright 57-2
57.2.2 Registration of Copyright 57-2
57.2.3 Copying in a Digital Medium 57-3
57.2.4 Websites 57-3
57.2.5 Databases 57-4
57.2.6 Software 57-4
57.2.7 Open Source Software 57-5
57.2.8 Digital Rights Management Systems 57-5
57.3 Trademarks and Service Marks 57-6
57.3.1 Domains 57-6
57.3.2 Cybersquatting — Reverse Domain Name Hijacking 57-7
57.4 Patents 57-7
57.4.1 Software 57-8
57.4.2 Business Methods 57-9
57.5 Liability 57-9
57.5.1 Linking 57-9
57.5.2 Providers 57-12
57.6 E-Commerce 57-12
57.6.1 E-Contract 57-12
57.6.2 Taxes 57-13
References 57-13

Jacqueline Schwerzmann

Abstract

Most principles of law apply identically to the physical world as well as to the digital one. But some new rules emerged with the rise of digital technology and the Internet: copyright protection of new types of work like software, databases or Websites, patenting of business methods, and new forms of trademark infringement like cybersquatting or the liability for linking. New forms of contract, concluded digitally without the physical presence of the parties, are challenging the legal community. This chapter provides a legal overview of special cyberlaw problems.

57.1 Cyberlaw

In the 1990s, with the increasing awareness of the sociological and economic impact of the Internet and the World Wide Web, legal scholars around the world created the terms *cyberlaw* and *Internet law*, implying

1-58488-381-2/05/$0.00+$1.50
© 2005 by CRC Press LLC

the dawn of a new category of law. Now, *cyberlaw* summarizes a loose collection of special rules regarding digital media, and most principles of law remain the same; a contract is still based on offer and acceptance, whether concluded electronically, e.g., via e-mail or physically.

What changed the manner of doing business more dramatically is the absence of national borders within the environment of the Internet. Internet-based transactions raise many international legal questions addressing potential jurisdiction conflicts. At the forefront of Internet business law is the use of forum clauses (contract clauses that determine the court in which legal questions have to be presented) and the choice of law (the selection of national legal rules that are applicable). These two clauses are essential parts of every contract in e-commerce. (For short overviews of Internet law, see Spindler and Boerner [2002] and Isenberg [2002]).

57.2 Copyright

57.2.1 Basics of Copyright

Almost every creative work, whether published or unpublished, digital or nondigital, is protected by copyright. These protected *works of authorship* include: literary works; musical works; dramatic works; pantomimes and choreographic works; pictorial, graphic and sculptural works; motion pictures and other audiovisual works; sound recordings and architectural works. It is essential to understand that the product does not have to be a piece of art; everyday work is also protected as intellectual property.

Copyright protection exists automatically upon creation, without any formalities and generally lasts until 70 years after the death of the creator in Europe and the U.S. Recently the *Sonny Bono Copyright Term Extension Act* extended the term of copyright for works created on or after January 1, 1978, from 50 to 70 years after the creator's death (17 U.S.C. 302). This preference of creators and producers caused heavy criticism in the scholarly community (see, e.g., http://cylber.law.harvard.edu/openlaw/eldred-vashcroft/). But in its decision on *Eldred v. Ashcroft*, the U.S. Supreme Court ruled that Congress was within its rights when it passed legislation to extend copyright terms, shooting down an argument by independent online booksellers and other groups who believed repeated extensions are unconstitutional.

To be covered by copyright, a work must be original and in a concrete "medium of expression" or a tangible form (e.g., a book or a CD; courts have even recognized that a computer program is protected when it exists in the RAM of a computer). A simple procedure or method cannot be protected as concrete work; however, patent registration might be possible.

The copyright owner is usually the creator of a work, or the employer or contracting body in the case of work for hire of an employee or a contractor. In Europe the creator is always the initial copyright owner but can pass the right to make use of the work to the employer; this is usually part of labor contracts.

A creator has several rights over reproduction, distribution, etc., which can be transferred to others. If the rights are transferred exclusively to somebody else, a written and signed agreement is mandatory. Single rights can be licensed separately and partial transfers to others are possible that only last for a certain period of time, for a specific geographical region, or for a part of the work.

Copyrights are automatically generated only within the borders of the country in which the creator is native or works or in which the work is first published. There is no "international copyright" which will automatically protect an author's work throughout the entire world, in contrast to international patents. However, multilateral or bilateral international contracts guarantee a similar level of copyright protection in most of the industrial countries and facilitate registration up until a country's specific national laws require special formalities for complete protection.

57.2.2 Registration of Copyright

Copyrights can be registered in the U.S. upon creation on a voluntary basis. Later, registration is necessary for works of U.S. origin before an infringement suit may be filed in court, and if registration is made within 3 months after publication of the work or prior to an infringement of the work, pursuit of statutory damages and attorney's fees in court is an option to the copyright owner. Otherwise, only the award of actual damages and profits is available to the copyright owner. For documentation purposes, the U.S. Copyright Act establishes a mandatory deposit requirement for works published in the U.S. within 3 months: two copies or phonorecords for the Library of Congress. In contrast, copyright registration is not available in European countries.

It is generally wise to publish a work with a *copyright note,* even if it is not required by law. The public knows to whom the rights belong and how long they will be valid. In case of infringement, an infringer cannot argue that he did not realize that the work was protected. A copyright notice has to contain "the letter C in a circle," "the letter P in a circle" for pbonorecords or the word "Copyright," or the abbreviation "Copr.," the year of first publication and the name of the copyright owner. A copyright notice is also not mandatory in European countries, but can be used to prove the bad faith of an infringer.

57.2.3 Copying in a Digital Medium

What is copying in a digital medium? Printing or storing of a digital document on a hard drive is a form of copying and not allowed without permission. Caching is another specific form of digital copying. The *Digital Millennium Copyright Act (DMCA),* which is an amendment to the U.S. Copyright Act as a response to new legal issues regarding digital media, allows *proxy caching* by creating an exception to specifically exempt this technically necessary process. However, there are elaborate provisions in the DMCA requiring that the system providers comply with "rules concerning the refreshing, reloading, or other updating of the material when specified by the person making the material available online in accordance with a generally accepted industry standard data communications protocol for the system or network."

Whether *client caching* is legal copying as well is not clearly addressed in the DMCA, although most legal scholars agree that client caching is allowed. Several court decisions support this interpretation and suggest that *client caching,* at least, is fair use.

The "fair use" doctrine allows certain limited exemptions to the strict copyright interdiction for personal, noncommercial use. In the *Betamax case* (*Sony Corp. of America v. Universal City Studios,* 464 U.S. 417, 455 [1984]), the Supreme Court held that "time-shifting" of copyrighted television programs with VCRs constitutes fair use under the Copyright Act and therefore is not an infringement of copyright. In *Recording Industry Association of America v. Diamond Multimedia Systems* (Court of Appeals (9th Cir.), Case No. 98-56727, June 1999) involving the use of portable digital music recorders for downloading (possibly illegal) MP3 files from the Internet, the U.S. Court of Appeals for the Ninth Circuit observed: "The Rio [the name of the player was Rio] merely makes copies in order to render portable, or 'space-shift,' those files that already reside on a user's hard drive ... Such copying is paradigmatic noncommercial personal use entirely consistent with the purposes of the Act."

57.2.4 Websites

Websites are normally multimedia works which are copyright protected as a whole or the different parts separately (texts, pictures, tables, and graphics can be trademarks as well). The use of copyrighted material belonging to a third party for creating a Webpage demands the permission of the copyright owner. It can be difficult to find the address or even the name of a copyright owner in order to get a permission (license). However if the work is registered, the Copyright Office may have this information.

A Website, created partly with licensed material, still owes its origin as a whole to the new developer, but not the licensed material. Alteration and reuse of third-party material without permission is copyright infringement, although "fair use" exemptions may allow use for purposes such as criticism, comment, news reporting, teaching, scholarship, or research.

Fair use is a complex legal issue with no adequate general definition. The courts usually seek to verify the noncommercial purpose and character of the use, the nature of the copyrighted work, the amount of the portion used in relation to the copyrighted work as a whole so that no more than was necessary was utilized, and the effect of the use on the potential market for the copyrighted work. The extent of the U.S. fair use exemption is determined by the economic effects and possible damages incurred by unauthorized use. (For further information about fair use, see Patry [1996]).

57.2.5 Databases

Databases are protected by copyright if the compilation has the quality of an original work of authorship. A collection of hyperlinks, such as the results of a search engine, could be protected if they are original enough, with a minimum of creativity of the collector as a necessary precondition. A database must therefore be original in its selection, coordination, and arrangement of data to have copyright protection as the mere collection of the data ("sweat of the brow") is not sufficiently original for protection. The same principles apply to Europe. This is why commercial database owners often protect their work through additional contracts to ensure financial compensations if the data are used against their will. Even if the database is not protected by copyright law, users can be liable for breaching the contract, not for copyright infringement.

An example is *ProCD, Inc. v. Zeidenberg* (86 F.3d 1447, 7th Cir. Wis. 1996). ProCD sold a CD ROM telephone database which was not original enough to be copyright-protected and defendant–user Zeidenberg was liable for making portions of the CD's data available over the Internet by breaking a shrink-wrap license and not for infringing on the copyright (see Section 57.6.1).

57.2.6 Software

Software is protected by copyright law (all kinds of formats: source code, object code or microcode). If software is simply copied without authorization, copyright infringement is obvious. More difficult is determining if the development of new software may present copyright infringement when it is, in some way, modeled after existing software to which the developer has been exposed.

In *Computer Associates International v. Altai, Inc.* (982 F2d 693, 1992) the U.S. Court of Appeal for the Second Circuit developed a three-part test for determining whether a new software infringes the copyright of an existing computer program. The test is called a *abstraction/filtration/comparison test* and is widely but not uniformly accepted by the courts and not yet confirmed by a Supreme Court decision.

1. *Abstraction:* A computer program is broken down into its structural parts. The idea is to retrace and map each of the designer's steps — in the opposite order in which they were taken during the program's creation.
2. *Filtration:* This level should separate protected expression from nonprotected material. This process entails examining the structural components of a given program at each level of abstraction to determine whether their particular inclusion at that level was "idea" or was dictated by considerations of efficiency, so as to be necessarily incidental to that idea; required by factors external to the program itself; or taken from the public domain and hence is nonprotectable expression:

Efficiency means that when there is essentially only one way to express an idea because no other way is as efficient, the idea and its expression are inseparable and copyright is no bar to copying that expression. The more efficient a set of modules are, the more closely they approximate the idea or process embodied in that particular aspect of the program's structure and should not be protected by copyright.

External factors can be mechanical specifications of the computer on which a particular program is intended to run; compatibility requirements of other programs with which a program is designed to operate in conjunction; computer manufacturers' design standards; demands of the industry being serviced; and widely accepted programming practices within the computer industry.

Public domain can be an expression that is, if not standard, then commonplace in the computer software industry.

3. *Comparison:* Comparing the remaining elements with those of the allegedly infringing software, there have to be substantial similarities. The inquiry focuses on whether the defendant copied any aspect of the protected expression, as well as an assessment of the copied portion's relative importance with respect to the plaintiff's overall program.

57.2.7 Open Source Software

Open Source Software is software that carries its source code with it or requires that this source code be kept open for others; it comes with a special open license. The Open Source Definition (OSD) specifies the license rules under which the Open Source Software shall be distributed. The license (often the GNU General Public License, but there are several others that have to be approved by the Open Source Initiative) has to guarantee free (but not necessarily cost-free) redistribution and openness of the source code and has further to allow modifications and derived works. The license must not discriminate against any person or group of persons and not restrict anyone from making use of the program in a specific field of endeavor. It also must be technology-neutral. The rights attached to the program must apply to all to whom the program is redistributed.

Open Source Software is copyright protected as is every proprietary software and must not be in the public domain, where no copyright protection applies. But the license agreements are more open than normally used and allow wider reuse rights, A developer still enjoys copyright protection and can earn money with its development, but copying, changing, and redistributing of the work has to remain allowed for third parties.

Linux is the most successful example of an open-source platform. In March 2003, the SCO Group, which succeeded Novell Networks as inheritor of the intellectual property for the Unix operating system, sued IBM, one of the most popular Linux distributors, for $3 billion, alleging that IBM had used parts of Unix or AIX (IBM's derivative Unix) in their contribution to the development of Linux (see CNET News.com [2003a]). IBM licensed Unix from initial Unix owner AT&T in the 1980s and was permitted to build on that Unix technology, but SCO argues that IBM violated its contract by transferring some of those modifications to Linux. The SCO Group claims that their Unix System 5 code is showing up directly, line-by-line, inside of Linux (see CNET News.com [2003b]). SCO alleges IBM of misappropriation of trade secrets, unfair competition, breach of contract, and tortious interference with SCO's business. Further, SCO sent out letters to 1,500 of the largest companies around the world to let them know that by using Linux they violate intellectual property rights of SCO. A court decision has not yet been made.

57.2.8 Digital Rights Management Systems

As a result of the ease with which digital material can be duplicated, copyright owners try to protect their content from being copied and distributed legally and also technically. Technical solutions are often called *digital rights management systems.* In a legal context, the term covers a lot of different technical measures, such as watermarking, copy-protection (hardware- and software-based), and more complex e-commerce systems which implement elements of duplicate protection, use control, or online payment to allow the safe distribution of digital media like e-books and e-newspapers.

International treaties (World Intellectual Property Organization's (WIPO) Copyright and Phonogram Treaty) and the Digital Millennium Copyright Act (DMCA) in the U.S., protect copyright systems and the integrity of copyright management information against circumvention and alteration. A similar Act to the DMCA exists in the European Union (Directive 2001/29/EC on the harmonization of certain aspects of copyright and related rights in the information society), and several countries have adopted these circumvention rules. It is forbidden to circumvent accessor copyright technology or to distribute tools and technologies used for circumvention. Copyright management information (identification of the work; name of the author/copyright owner/performer of the work; terms and conditions for the use of the work) is protected by law against removal or alteration. Civil remedies and criminal charges carrying up to $1 million in fine or imprisonment of 10 years apply in case of an infringement. There are also several exceptions which enable reverse engineering, to achieve interoperability of computer programs, and for encryption research and security testing.

The rules of the Digital Millennium Copyright Act concerning the protection of copyright mechanisms are controversial. Critics argue that they shift the copyright balance towards the copyright owner and are against consumer interests because fair use exemptions can be eliminated, security or encryption research will be difficult, and a cartelization of media publishers could develop. Several cases in which the DMCA applied are public already (for further information, see http://www.eff.org/IP/DMCA/). Following are two:

In September 2000, the *Secure Digital Music Initiative (SDMI),* a multiindustrial conglomerate, encouraged encryption specialists to crack certain watermarking technologies intended to protect digital music. Princeton Professor Edward Felten and a team of researchers at Princeton, Rice, and Xerox successfully removed the watermarks, and they were kept from publishing their results by SDMI representatives, as allowed by the DMCA.

In December 2002 a jury in San Jose acquitted the Russian software company *ElcomSoft* of all charges under the DMCA. ElcomSoft distributed over the Internet a software program called Advanced e-Book Processor, which allowed converting Adobe's protected e-book format into ordinary pdf files. The company faced charges under the DMCA for designing and marketing software that could be used to crack e-book copyright protections. ElcomSoft was acquitted only because the jury believed the company did not violate the DMCA's circumvention rules intentionally. One year prior to this litigation ElcomSoft's Russian programmer Dmitry Sklyarov was jailed for several weeks while attending a conference in the U.S., also under reference to the DMCA.

In *Lexmark International, Inc. v. Static Control Components, Inc. (SCC),* the printer company sued SCC for violation of the DMCA because SCC produced Smartek microchips, which allowed third party toner cartridges to work in several printers of Lexmark, circumventing Lexmarks software programs. Lexmark's complaint alleged that the Smartek microchips were sold by Static Control to defeat Lexmark's technological controls which guarantee that only Lexmark's printer cartridges can be used. The U.S. District Court in the Eastern District of Kentucky ordered SCC to cease production and distribution of the Smartek microchips.

57.3 Trademarks and Service Marks

Trademarks or Service Marks are used to identify goods or services and to distinguish them from those made or sold by others. Trademarks can be words, phrases, sounds, symbols or designs, or their combination, as well as the color of an item or a product or container shape (for example the shape of a Coca-Cola bottle). Trademark rights in the U.S. secure the legitimate use of the mark, and registration is not required. This is different from most European countries, where registration is mandatory. However registration is common and has several advantages such as: a legal presumption of the registrant's ownership of the mark; the ability to litigate the use of a mark in federal court; and the use of the U.S. registration as a basis to obtain registration in foreign countries or an international registration. A mark which is registered with federal government should be marked with the symbol R in a circle. Unregistered trademarks should be marked with a "tm," while unregistered service marks should be marked with a "sm." Trademarks are protected against infringement and dilution. Only famous marks are protected against dilution, for example if cited in connection with inferior products which weaken the distinctive quality of the famous trademark.

57.3.1 Domains

A domain name can be registered in the U.S. as a trademark, but only if it functions as a source identifier. The domain should not merely be an information used to access a Website. Printed on an article for sale, the domain should indicate to the purchaser the name of the product or company and not the address where information is found.

57.3.2 Cybersquatting — Reverse Domain Name Hijacking

Cybersquatting is the practice of registering and using a well-known name or trademark as a domain to keep it away from its owner or to extort a substantial profit from the owner in exchange for the return of the name. This sort of "business" prospered in the initial phase of e-commerce and has now become difficult due to the following legal actions available to an individual or a trademark owner:

- The *Anticybersquatting Consumer Protection Act (ACPA)* of 1999, an amendment to the U.S. trademark laws, protects trademark owners and living individuals against unauthorized domain name use by allowing them to take over or to enforce the cancellation of domain names that are confusingly similar or identical to their names or valid trademarks, and make the cybersquatter liable for damages. To be successful however, it must proven that the cybersquatter acted in bad faith.
- As an alternative to pursuing a domain name dispute through the courts, it is possible to use the administrative domain name dispute policies that have been developed by the organizations that assign domain names, This administrative procedure is often faster and cheaper. The Internet Corporation for Assigned Names and Numbers (ICANN) created a *Uniform Domain Name Dispute Resolution Policy (UDRP)*. The World Intellectual Property Organization (WIPO) in Geneva is the leading ICANN-accredited domain name dispute resolution service provider. As of the end of 2001, some 60% of all the cases filed under the UDRP were filed with the WIPO. Generally, the WIPO is known to favor the complainant in its decisions.

Reverse domain name hijacking in contrast refers to the bad faith attempt of a trademark owner to deprive a registered domain name holder of the domain by using the dispute resolution. In such cases a complaint in a dispute will not be successful. Trademark owners, though, have to be careful not to get accused — in reverse — of domain name hijacking. This can be the case if confusion between a registered domain and the trademark is very unlikely, if a complaint was brought despite knowledge that the domain name holder has a right or legitimate interest in the domain name or that the name was registered in good faith.

57.4 Patents

A patent is an exclusive right given by law to enable inventors to control use of their inventions for a limited period of time. Under U.S, law, the inventor must submit a patent application to the U.S. Patent and Trademark Office (USPTO), where it will be reviewed by an examiner to determine if the invention is patentable. U.S. patents are valid only within the territory of the U.S. The Patent Cooperation Treaty (PCT) allows International Patents for selected countries or all member countries. Europe has the single Community Patent for the E.U. It is a form of a monopoly right to the inventor and excludes others from making, using, and selling the protected invention. There are three kinds of patents protecting different kinds of innovations and they all have different requirements:

- The most common type of patent is a *utility patent*. It covers innovations which have to be useful, new, and nonobvious. The innovation itself has to be a process, a machine, an article of manufacture, or a composition of matter, or a combination of these factors. Prohibited is the patenting of abstract ideas, laws of nature, or natural phenomena, while applications of such discoveries are patentable. The innovation has to have a practical application to limit patent protection to inventions that possess a certain level of real world value. Furthermore, the innovation must be novel and different from existing patents or standard technology, with no public disclosures of the invention having been made before. If two persons are working independently and come to a result at roughly the same time, the first inventor gets patent protection, not the person who first filed an application. That is in contrast to European countries. "Nonobvious" means an innovation not previously recognized by people in that special field. U.S. patents generally expire 20 years after patent application. There are many Internet-related inventions protected by utility patents like data compression or encryption techniques and communications protocols, etc. IBM is the most prolific patent generator in information technology, topping the list of corporate patent awards for the last 10 years. The company filed 3,288 patents in 2002, bringing its total over the past 10 years to more than 22,000.

- *Design patents* protect ornamental designs. A design patent protects only the appearance of an article. If a design is utilitarian in nature as well as ornamental, a design patent will not protect the design. Such combination inventions (both ornamental and utilitarian) can only be protected by a utility patent. A design patent has a term of 14 years from the date of issuance. For design patents, the law requires that the design is novel, nonobvious, and nonfunctional. Design patents are used in the Internet industry to protect the look of hardware (monitor, mouse, modem, etc.), while icons can be protected either by design patents or as a trademark, such as the Apple or Windows logos.
- *Plant patents* protect new varieties of asexually reproducing plants.

57.4.1 Software

In contrast to Europe, where the E.U. is currently discussing a revision of the practice, software is generally patentable in the U.S. where the Supreme Court's 1981 opinion in *Diamond v. Diehr* (450 U.S. 175 [1981]) opened the way for patent protection for computer software — in addition to copyright protection as discussed above.

Computer programs can be claimed as part of a manufacture or machine — and are legally part of that invention. A computer program can also be patentable by itself, if used in a computerized process where the computer executes the instructions set forth in the program. In that case, the software has to be on a computer readable medium needed to realize the computer program's functionality. The interaction between the medium and the machine makes it a process and the software patentable. The innovation still has to have practical application and it cannot be a simple mathematical operation. For example, a computer process that simply calculates a mathematical algorithm that models noise is not patentable. However, a process for digitally filtering noise employing the mathematical algorithm can be patented.

The number of software patents issued in recent years is immense. Examples are patents on graphical user interface software (e.g., Apple's Multiple Theme Engine patent), audio software and file formats (e.g., Fraunhofer MP3 patents), Internet search engines (e.g., Google's patent for a method of determining the relevance of Web pages in relation to search queries), or Web standards (e.g., Microsoft's Style Sheet patent).

One of the reasons for extensive software patenting is the common practice of cross-licensing. Among companies with patent portfolios, it is common defense strategy to cross-license one or more of its own patents if accused of infringing a patent belonging to another company. Instead of paying damages, suits are settled by granting mutual licenses which offer advantages to both sides. Cross-licenses between two parties give each company the right to use the patent of its adversary with or without additional compensation.

Software is protected by copyright or patent law; choosing which course to pursue is sometimes difficult. Proving that third party software infringes copyright can be difficult. On the other hand protecting software through patents is expensive and time-consuming. In general patent protection is broader than copyright protection: e.g., if someone develops by chance a computer program similar to yours, copyright law does not forbid this — but patent rules do. Consequently, patent protection is the first choice, especially for financially strong companies.

57.4.2 Business Methods

Processes involving business methods are patentable in the U.S. but not in Europe. This U.S. practice has been in place since July 1998, when a federal court upheld a patent for a method of calculating the net asset value of mutual funds (*State Street Bank & Trust Co. v. Signal Financial Group, Inc.* 149 F.3d 1368 [Fed. Cir. 1998] cert denied 119 S. Ct. 851, 1999). One of the most famous patents in this field is Amazon's 1-Click Technology by Hartman et al. [1999], which allows customers to skip several steps in the checkout process by presetting credit card and shipping information. A long-running patent infringement suit between Barnes & Noble and Amazon ended with a settlement, the details of which were not disclosed.

Examples of patented business methods include: *Priceline.com's* patent for reverse auctions [Walker et al., 1998]; *Open Market's* patents related to secure online credit-card payments or online shopping cards [Payne et al., 1998]; *NetZero's* patent covering the use of pop-up advertising windows [Itakura et al., 2000], and a business model for recovery of missing goods, persons, or fugitive or disbursements of unclaimed goods using the Internet [Frankel et al., 2002].

In the initial phase of electronic commerce development, the U.S. Patent and Trademark Office (USPTO) granted a wide range of patents, which often affected the fundamental technologies behind e-commerce. In the following years, lawsuits between patent owners and several e-commerce companies that breached patents became numerous. USPTO now is more cautious in registering business methods, but there are still critics of this form of patent.

57.5 Liability

57.5.1 Linking

Links are functional core elements of the World Wide Web and in most cases connect one document to another. They consist of an HTML-command and the address of the linked document. URI and links are considered facts and are not copyrightable. Linking is normally legal. Judge Harry L. Hupp decided for the Central District of California in *Ticketmaster Corp. v Tickets.com* (99-7654) that "Hypertext linking does not itself involve a violation of the Copyright Act ... since no copying is involved."

But the linking of documents and the way links are used can nevertheless be legally problematic, despite the technical ethos of "free linking." In certain constellations linking can violate rules of unfair competition, intellectual property, or trademark. Since 1996, when the first linking cases were filed, a number of court decisions have been made, but clear legal guidelines are still not available (for an overview of court decisions see: http://www.linksandlaw.com/linkingcases.htm and Sableman [2001] or Sableman [2002]).

57.5.1.1 Framing and Inlining

Framing and inlining are likely to be copyright infringement, if the source is integrated into the new Website without permission of the copyright owner and if no fair use excuse applies.

- In the Dilbert case, programmer Dan Wallach used inline-links to display United Media's Dilbert cartoon on his Website, where he created more and differently arranged comics. He received a cease-and-desist letter from the company and removed the links, as he had infringed the right of the original author by making derivative works.
- Several major news organizations, including CNN, the *Washington Post*, and the *Wall Street Journal*, sued *TotalNews* for copyright and trademark infringement for its practice of framing their content within its own site. TotalNews retained its own frames — complete with advertisements — and put the other site or story within its main frame. The case was settled, and TotalNews ceased to frame the external content and a linking agreement guaranteed that TotalNews would link to Websites only via hyperlinks consisting of the names of the linked sites in plain text and in a way that would not lead to the confusion of the consumer.

57.5.1.2 Deep Linking

Deep Linking is still a controversial subject. A lot of commercial sites do not want to be deep linked because they fear financial losses if pages with advertisements are bypassed or less hits on their Web pages are recorded. Some claim trademark infringement if the link description uses their trademark logos or protected words. Other claims are unfair competition (if consumers are confused or trade practices are unfair) or a violation of trespass rules if spiders crawl servers excessively for linking information. Contract breaches are also claimed if the terms of service do not allow deep links. In Europe the protection of databases has to be taken into consideration as there are court decisions denying the right to deep link as an infringement of the database copyright.

Therefore, it depends on the concrete circumstances whether deep links are legal or not. Noncommercial sites are probably more often allowed to deep link to other sites than commercial sites. Commercial sites have to make it obvious that the user will enter a new Website, and third party trademark signs cannot be used. Deep links to competitors is an especially delicate issue. In controversial linking situations it is wise to reach a linking agreement, although the practice of commercial sites including a linking license in their terms-of-use policy generally does not bind a user. A lot of legal conflicts about deep linking in the U.S. have been settled out of court, which is why strict rules about legal or illegal deep links are still missing.

- *Shetland Times Ltd. v. Shetland News:* The *Shetland Times* newspaper and online site filed a copyright lawsuit against the digitally produced Shetland News for linking to the *Times*' online headlines. The court decided that headlines are copyright protected and that using the headlines of the *Shetland Times* as a description for the links was copyright infringement on the part of the Shetland News. The case was settled before a final decision; the Shetland News was granted permission to link to the *Times*' headlines, but must label individual articles as "A *Shetland Times* Story." Near such stories, the Shetland News also promised to feature a button with the *Times*' masthead logo that links to the newspaper's home page.
- The results of a search engine query consists of links. Until now, no major search engine has ever been sued in connection with deep linking. In the case *Kelly v. Arriba Soft Corp.*, the U.S. Court of Appeals for the 9th Circuit decided that the display of copyrighted thumbnail pictures as a search result of an image search engine is not a copyright infringement, as long as the original image URL was linked from the thumbnails. The display of the thumbnails qualified as fair use.
- *Tickets.com* was an online provider of entertainment, sports, and travel tickets and made available to its customer information about events as well as hypertext links to ticketsellers for tickets not available at Tickets.com. Tickets.com prefaced the link with the statement "These tickets are sold by another ticketing company." Ticketmaster sued Tickets.com claiming copyright infringement (because the information about the events was extracted from Ticketmaster's Website) and unfair competition among several other issues. The court decided in favor of Tickets.com, arguing that the information Tickets.com copied was only factual (date, location, etc.) and not copyrightable. Furthermore, the deep linking to Ticketmaster's Website, where customers could order tickets, was not unfair competition because the whole situation was made transparent to customers. Ticketmaster advertisement banners were displayed on the event page.
- In *StepStone v. OFIR* (two online recruitment companies) OFIR listed both its "own" vacancies and those gleaned from other online recruitment agencies on its Website such that this was not obvious to visitors at the first glance. The Cologne County Court in Germany dismissed OFIR's contention and upheld StepStone's arguments, agreeing that the list of job vacancies did constitute a database which warranted protection and that the bypassing of StepStone's welcome page did undermine the company's advertising revenue.
- In *Danish Newspaper Publishers Association (DDF) v. Newsbooster,* a news search service providing users with relevant headlines with deep links to articles, was recognized as copyright infringement by a Danish court. The content of electronic newspapers was protected as a database. Newsbooster had its own commercial interest in the news search service and charged an annual subscription

fee from its subscribers. The court also decided that Newsbooster used unfair marketing practices. Meanwhile the Danish search company has begun to offer its Danish clients a version of its Newsbooster service that operates in a similar fashion to the decentralized file-sharing networks like Kazaa and Gnutella.

- *Dallas Morning News:* Belo, the parent corporation of the *Dallas Morning News*, sent a letter to the Website BarkingDogs.org demanding that it stop deep linking to specific news articles within the paper's site, rather than from the homepage. Belo's lawyers justified this demand by citing a loss of advertising revenue, consumer confusion, and reference to its term of service which allowed only linking to the homepage. In the end, Belo did not pursue their claims.
- Intensive-searching Websites for obtaining information and collecting descriptions can be illegal. In *eBay v. Bidder's Edge*, eBay sued Bidder's Edge for crawling the eBay Website for auction listing information. The court concluded that by continuing to crawl eBay's site after being warned to stop, Bidder's Edge was trespassing on eBay's Web servers. The "Trespass Theory," which indicates the forbidden interference with property similar to physical trespassing on land, is also used in connection with spamming or unsolicited emails. A substantial amount of eBay's server capacity was blocked by the Bidder's Edge robotics.

57.5.1.3 Search Engines

Mark Nutritionals (MNI) as the trademark owner of Body Solutions filed suits against the search engines AltaVista, FindWhat, Kanoodle, and Ouverture for their practice of selling keywords (called "paid-placement"). If a user typed the plaintiff's trademark "Body Solutions" into the search bar, the Websites of advertisers who bought this keyword were shown more prominently than the plaintiff's Web page. MNI claimed trademark infringement and unfair competition.

57.5.1.4 Metatags

Metatags are dangerous, if competing business's trademarks are used. This often results in liability under the U.S. Lanham Act, which protects trademarks. Adding third-party trademarks as metatags can also be a violation of competition law. In *SNA v. Array* (51 F. Supp. 2d 554, 555, E.D.Pa. 1999) the court involved ruled that the defendants had intentionally used the plaintiff's trademark "to lure Internet users to their site instead of the official site." The defendants repeatedly inserted the word "Seawind" as a metatag, which was a trademark of the plaintiff. The trademark owner manufactured aircraft kits under the name of Seawind and the defendant produced turbine engines.

57.5.1.5 Links to Illegal Content

Linking to illegal or prohibited content can be illegal and can cause derivative liability. Linking can be interpreted as a contributory copyright infringement, if copyright infringement is involved in the Website to which the link points. Linking can also be illegal to Websites or information which breach the rules of the Digital Millennium Copyright Act and are therefore illegal (for example, circumvention technology). Until now, the courts decided only cases where the dissemination of illegal content was pursued in an active way by promoting the banned content. Whether links posted for strictly informational purposes are illegal as well is doubtful. Unanswered also is the question as to where the limits are to contributory infringement: How many link levels still create liability? The popular search engine Google changed its policy and removes offending links when a third party informs Google about potential infringements. A notice will inform users that the link has been removed. Google feared liability under the DMCA (see 1.5.2). Other link issues have included:

- *2600.com*. This was one of several Websites that began posting DeCSS near the end of 1999 and links to sites where the code was available to download. DeCSS is a program that allows the copying of DVDs protected by the CSS (Content Scramble System) technology. Several members of the Motion Picture Association filed a lawsuit and the U.S. Court of Appeals for the Second Circuit in Manhattan ruled in favor of the Motion Picture Association of America. The injunction barred Eric Corley and his company, 2600 Enterprises, Inc., from posting the software code designed to

crack DVD-movie copy protection on their Website and from knowingly linking their Website to any other site at which the DeCSS software was posted. DeCSS was seen as circumvention technology, illegal under the Digital Millennium Copyright Act. The higher court upheld a linking test to determine if those responsible for the link (1) know at the relevant time that the offending material is on the linked-to site, (2) know that it is circumvention technology that may not lawfully be offered, and (3) create or maintain the link for the purpose of disseminating that technology.
- In *Bernstein v. J.C. Penney, Inc.* the department store J.C. Penney and cosmetics company Elizabeth Arden were sued by photographer Gary Bernstein because of an unauthorized reproduction of one of his photographs. The picture was three clicks away from the Website of the defendants. The case did not reach a final decision but the district court denied the motion for a preliminary injunction.

57.5.2 Providers

The Digital Millennium Copyright Act rules when online service providers are responsible for the content on their networks.

1. Service providers are not liable for "transitory digital network communications" that they simply forward "through an automatic technical process without selection of the material by the service provider" from a customer to its intended destination.
2. If a user posts information on the system or network run by the service provider, without the service provider's knowledge that the information is infringing, then the service provider is generally not liable. If notified of an infringement, the service provider must respond "expeditiously to remove, or disable access to, the material that is claimed to be infringing." However, a service provider has first to designate an agent from the U.S. Copyright Office to receive notification.
3. Service providers are not liable for information location tools like hyperlinks, directories, and search engines that connect customers to infringing material, if they don't know about the infringing nature of the material and do not receive a financial benefit from the link. Nevertheless they have to take down or block access to the material upon receiving notice of a claimed infringement.
4. If a copyright owner requests a subpoena from the appropriate court for identification of an alleged infringer, a service provider is required to reveal the identity of any of its subscribers accused of violating the copyright.

57.6 E-Commerce

57.6.1 E-Contract

The purpose of commercial contract law is to facilitate and support commerce. Therefore, concluding a contract has to be as easy as possible. Most contracts in the physical world are valid without any formalities (e.g., written form, signature), just by offer and acceptance. Internet business benefits from this principle: Digital contracts can be automated acceptance-offer procedures but still valid.

- *Click-wrap agreements,* with the click of the mouse on the agreement button, are valid, whether a consumer buys merchandise, information, or software or just says yes to the terms and conditions of using a Website or mailservice. Even if a purchaser has not read the terms and conditions, they will usually get part of the contract. But terms of agreement have to be available to the consumer before shopping, and they have to be understandable and reasonable. To be on the safe side, conditions should be displayed to the consumer during the shopping procedure and should get accepted with a formal click.
- *Browse-wrap agreements* are agreements on a Website that a viewer may read, but no affirmative action, like clicking on a button, is necessary . The agreements can be in a link at the bottom of a Web page or even displayed, without enforcing a consumer to accept by clicking on an icon. At

least one court has denied the validity of such browse-wrap agreements: in *Specht v. Netscape* a software license agreement asked the users to "review and agree to the agreement," but without doing so they still could download the software. The agreement was held as not enforceable.

- *Shrink-wrap agreements* are not digital but printed agreements inside a product-box and apply mainly to software purchases. The term "shrink-wrap" refers to the act of opening a box that is sealed by plastic or cellophane. By breaking the cellophane the consumer should be bound to the terms and agreements inside. Shrink-wrap agreements are controversial because consumers normally see the agreement in detail not until the purchase is over, even if they get informed of the existence of the later agreement terms outside of the box. Some critics argue that consumers never really accept terms this way. U.S. courts generally rule that shrink-wrap agreements are enforceable. The most famous case is *ProCD, Inc. v. Zeidenberg* concerning the purchase of a CD ROM telephone database (see Section 57.2.5). The proposed federal UCITA (Uniform Computer Transactions Act) aims to make such adhesion contracts enforceable. In Europe shrink-wrap agreements are mostly seen as invalid.

57.6.2 Taxes

In November 2001, U.S. Congress extended a ban on Internet taxes in the U.S. for three years (Internet Tax Freedom Act). The idea is to give e-commerce a good chance to grow. But the Internet is not entirely tax-free. States can charge sales tax on e-commerce transactions between companies and consumers within their borders (Seattle-based Amazon.com's Washington state customers, for instance, must pay Washington's sales tax). And existing laws enacted previously by several states and local governments are still valid. Further, some companies collect taxes when they have a physical store in a state. The problem is to find a tax system that is enough simplified to be used in every day transactions. Amazon and other companies note that there are thousands of tax jurisdictions nationwide, each with its own definition of taxable goods and its own rate schedule.

In the E.U. online sales have been taxed since July 1 2003, when a new E.U. directive went into effect requiring all Internet companies to account for value added tax, or VAT, on digital sales. The levy adds 15% to 25% on select Internet transactions such as software and music downloads, monthly subscriptions to an Internet service provider and on any product purchased through an online auction. Digital goods have to be taxed in the country where they are supplied. For example, an E.U. company that sells an MP3 file or an e-book to a U.S.-based customer is required to collect the E.U.'s value-added tax (VAT). The directive requires U.S. companies to collect the VAT on digital goods sold in E.U.

References

CNET News.com. SCO sues Big Blue over Unix, Linux, March 2003a. http://news.com.com/2100-1016_3-991464.html?tag=rn.

CNET News.com. Why SCO decided to take IBM to court, June 2003b. http://news.com.com/2008-1082_3-1017308.html?tag=rn.

Frankel, Fred et al. Business model for recovery of missing goods, persons, or fugitive or disbursements of unclaimed goods using the internet, September 2002. U.S. Patent 6,157,946.

Hartman, Peri et al. Method and system for placing a purchase order via a communications network, September 1999. U.S. Patent 5,960,411.

Isenberg, Doug. *The GigaLaw Guide to Internet Law.* Random House, Washington, D.C., 2002.

Itakura, Yuichiro et al. Communication system capable of providing user with picture meeting characteristics of user and terminal equipment and information providing device used for the same, December 2000. U.S. Patent 6,157,946.

Patry, William. *The Fair Use Privilege in Copyright Law.* BNA Books, 2nd ed., Washington, D.C., 1996.

Payne, Andrew C. et al. Network sales system, February 1998. U.S. Patent 5,715,314.

Sableman, Mark. Link law revisited: Internet linking law at five years. *Berkeley Technology Law Journal*, 3(16): 1237, 2001.

Sableman, Mark. Link law revisited: Internet linking law at five years. July 2002 Supplement, http://www.thompsoncoburn.com/pubs/MS005.pdf.

Spindler, Gerald and Fritjof Boerner, Eds. *E-Commerce Law in Europe and the USA*. Springer, Berlin, 2002.

Walker, Jay S. et al. Method and apparatus for a cryptographically assisted commercial network system designed to facilitate buyer-driven conditional purchase offers, August 1998. U.S. Patent 5,794,207.

Index

A

A2A, *see* Application-to-application
AAA, *see* Authentication, authorization, and accounting
ABone, 51-5
Abstractions, 33-2
Abstract Process, 27-15
Abstract Syntax Notation One (ASN.1), 23-1, 24-7, 34-23
Acceptable ranking, 12-12
Access control
 mechanism, 26-2
 model, 26-3, 26-11
 role-based, 33-14, 46-3
Accessibility and user interface, 2-9
ACD, *see* Automatic call distribution
Acknowledgments, 27-9
ACPA, *see* Anticybersquatting Consumer Protection Act
Acquaintance spam, 7-15
Acronym extraction, 14-11
Active Bat, 37-8
Active Directory, 22-18, 22-20, 34-3
Active measurements, 49-4
Active Networks, 51-2, 51-5
Activity Lifecycle Service (ALS), 31-9
Actor
 creation, 38-6, 38-7
 language, 38-6
 model, 38-4
Adaptation effect, 1-2, 1-3
Adaptive annotation, 1-5
Adaptive content selection, 1-2, 1-3
Adaptive Hypermedia and Adaptive Web, 1-1 to 1-14
 Adaptive Hypermedia, 1-2 to 1-6
  adaptive navigation support, 1-3 to 1-6
  what can be adapted, 1-3
 Adaptive Web, 1-7 to 1-11
  Adaptive Hypermedia and Mobile Web, 1-7
  Adaptive Hypermedia and Semantic Web, 1-8 to 1-11
  open corpus Adaptive Hypermedia, 1-8
Adaptive navigation support, 1-2, 1-3
Adaptive ordering, 1-3
Adaptive presentation, 1-2, 1-3
Adaptive Server Enterprise (ASE), 25-18
Adaptive System, 1-1, 1-2, 46-1
Adaptive Web, 1-7
Address prefixing, 7-12
Ad hoc On Demand Distance Vector (AODV), 36-7
Admission control, 51-8
Advanced Peer-to-Peer Networking (APPN), 50-13
Adventure games, 11-2
Advertising games, 11-5
Agents, 16-1 to 16-14
 anatomy of agent, 16-2 to 16-7
  agent architecture, 16-3
  communication, 16-6 to 16-7
  decision-making behavior, 16-4 to 16-6
  effectors, 16-7
  example agent, 16-7
  mobility, 16-7
  perception, 16-4
  sensors, 16-3 to 16-4
 anthropomorphic, 16-10
 autonomous, 16-1, 16-2, 16-10
 definition of intelligent agents, 16-2
 -driven content negotiation, 42-10
 further information, 16-12
 glossary, 16-12 to 16-13
 intelligent agents on Internet, 16-8 to 16-9
  agents behind Websites, 16-9
  bots, 16-8 to 16-9
 multiagent teams, 16-8
 -oriented approaches, 33-14
 research issues related to agents and Internet, 16-10 to 16-11
  autonomic computing, 16-11
  human–computer interfaces, 16-10
  privacy, 16-10 to 16-11
  security, 16-11
Aggregate usage profiles, 15-3
AI, *see* Artificial intelligence
Alias, 51-10, 51-11
ALS, *see* Activity Lifecycle Service
alt.sewdish.chef.bork.bork.bork, 7-17
Amos II, 28-6, 28-9
Annotations, 5-16, 5-18
 multimedia, 5-2
 threaded, 5-16
 variable-granularity, 5-16
Anonymity, identification vs., 35-19
Anonymization, 49-13
ANSI, 6-5, 7-12
Anticybersquatting Consumer Protection Act (ACPA), 57-7
AODV, see *Ad hoc* On Demand Distance Vector
Apache Common Log format, 40-13
API, *see* Application Programming Interface
Applets, 38-16

Application
-to-application (A2A), 8-4
domain, 33-4
failures, 32-7
integration, 8-3
server, 6-8
setting, 33-5
sharing, 5-5
Application Programming Interface (API), 23-9, 39-4
evolution, 38-6
query-based database, 28-1
APPN, *see* Advanced Peer-to-Peer Networking
Architectural model, 40-9
Architecture, 50-1 to 50-20
architecture directions, 50-14 to 50-15
architecture roots, 50-2 to 50-2
business cases, 50-15 to 50-19
benefits, 50-18 to 50-19
ClosedCel, 50-17
example of impact of open architecture, 50-16
maximizing application and service opportunities, 50-19
mobile wireless challenge, 50-16 to 50-17
OpenCel, 50-17 to 50-18
voice-over IP challenge, 50-18
closed architecture, 50-8 to 50-11
architecture change, 50-10 to 50-11
design changes, 50-9 to 50-10
open architecture, 50-11 to 50-14
potential pitfalls of architecture, 50-4 to 50-5
road from architecture to implementation, 50-5 to 50-8
implementation, 50-7 to 50-8
infrastructure hierarchy, 50-6
network design, 50-6 to 50-7
value of network architecture, 50-3 to 50-4
Archives, 4-7
ARHP, *see* Association Rule Hypergraph partitioning
Arjuna, 31-8
ARPANET, 56-2
Artificial intelligence (AI), 11-6, 16-2
in games, 11-10
researchers, 11-1
ASE, *see* Adaptive Server Enterprise
Asheron's call, 11-5
AS hops, *see* Autonomous system hops
ASN.1, *see* Abstract Syntax Notation One
Association Rule Hypergraph partitioning (ARHP), 15-19, 15-27
Association rules, 15-3, 15-14, 15-20
Assumptions, 31-5
Asynchronous communication, 38-2
Asynchronous Transfer Mode (ATM), 36-3, 56-13
AT, *see* Atomic Transaction
Atari, 11-6
ATM, *see* Asynchronous Transfer Mode
Atomic process, 31-12
Atomic Transaction (AT), 27-10
Attribute types, 34-12
Audio clips, 41-8
Authentication, 51-8, 51-9
Authentication, authorization, and accounting (AAA), 6-2
Authoring policies, 26-10
Authorization rules, 26-3
Authorship ascription, 14-8
Automated discovery, 31-5, 31-6, 31-14
Automated query expansion, 12-2
Automatic call distribution (ACD), 6-15
Automatic Web Service Composition, 31-4
Automatic Web Service Discovery, 31-4, 31-8
Automatic Web Service Execution, 31-5
Automation, islands of, 32-2
Autonomous system (AS) hops, 41-11
Autonomy, 35-19
Availability
forecasting, 5-21
presence and, 5-22
Avatar, 5-8, 5-9

B

BA, *see* Business Activity
Backbone, 56-8
Bag of words model, 14-5, 14-7
Basic Encoding Rules (BER), 34-23
B2B
e-commerce, 33-9
integration, *see* Business-to-business integration
BDI agents, *see* Belief-Desire-Intention agents
Belief-Desire-Intention (BDI) agents, 16-3
Beluga, 49-5
Beowulf clusters, 39-11
BER, *see* Basic Encoding Rules
Berkeley DB, 34-7
BGP, *see* Border Gateway Protocol
Bill of Material (BOM), 8-6, 8-13
Binary Run-time Environment for Wireless (BREW), 11-12
Binding, 27-11, 27-12, 28-15, 34-2
Biotechnology, 31-14
Bipartite cores, 53-12
Black hole, 51-3
Blacklist, 7-16
Blogging, *see* Web logging
Bluetooth, 36-3, 56-2
Bluetooth Service Discovery Protocol, 36-7
Boardwatch Magazine, 56-9
BODs, *see* Business Object Documents
BOM, *see* Bill of Material
Border Gateway Protocol (BGP), 49-2, 49-16
Born digital, 33-11
Bots, 16-1
BPEL4WS, *see* Business Process Execution Language for Web Services
BPMI, 33-9
BPR, *see* Business process reengineering
BPSS, *see* Business Process Specification Schema
BREW, *see* Binary Run-time Environment for Wireless
Broadband access, 56-10

Broadvision, 15-2
Browser reloads, 41-4
Browsing policies, 26-10
Bulletin board, 5-1, 5-14, 5-15
Business
-to-business (B2B) integration, 8-4
logics mediation, 31-8
methods, patentable, 57-9
models, 33-7
performance, *see* E-Learning technology
Business Activity (BA), 27-10
Business Content Layer, 8-2
Business Internet Consortium, 8-2
Business Object Documents (BODs), 8-4
Business process, 8-2, 8-3, 8-4, 32-1 to 32-15
data management, 32-7
failure and exception handling, 32-7 to 32-9
need for business process automation, 32-2 to 32-4
reengineering (BPR), 33-7
resource assignment, 32-6 to 32-7
scheduling, 32-5 to 32-6
Web services and business processes, 32-10 to 32-13
Web services, 32-10 to 32-11
Web services and business process orchestration, 32-11 to 32-13
WfMS architectures, 32-9 to 32-10
workflow management systems overview, 32-4
Business Process Execution Language for Web Services (BPEL4WS), 27-5, 27-15, 32-12, 39-12
Business Process Layer, 8-2
Business Process Specification Schema (BPSS), 8-4, 8-5
Business Registry, 27-14
Butterfly.net, 11-10

C

CA, *see* Certificate Authority
Cache, 35-11
consistency, 41-3
replacement algorithms, 41-1
Call
routing, 6-18
setup, 6-12
Call Processing Language (CPL), 6-19
Call for Proposal (CFP), 36-11
CAN, 35-11, 35-15
Canonical XML, 23-7
Capability Maturity Model Integrated (CMMI), 33-3
Carrier route, 8-14
Cart, 8-13
Cascading Style Sheets (CSS), 24-2
CASE tools, 33-5
Catalog, 8-13
CCA, *see* Common Component Architecture
CCM, *see* Credential Checking Module
CDM, *see* Common data model
CDN, *see* Content distribution network
CDSPs, *see* Content distribution service providers
Central mediator, 28-2
Central switching office (CO), 56-11
Certificate Authority (CA), 39-8
Cfengine, 46-11
CFP, *see* Call for Proposal
CGI, *see* Common Gateway Interface
Chaining, 34-6, 34-7, 34-16
Chain letter, 7-15
Change management, 33-13
Character Large Object (CLOB), 25-10
Character set, 7-4
Chat, 5-12, 7-9
CIM, *see* Core Information Model
Classification, P2P systems, 35-4
Classroom 2000 project, 37-9
CLEC, *see* Competitive Local Exchange Company
Client
caching, 57-3
polling, 41-5
Client-server
architecture, 22-3
systems, 35-2
CLOB, *see* Character Large Object
Clonable Network Stacks, 51-4
Clustering, 15-3, 35-6, 35-11, 53-3
approaches, 15-17
document, 14-7, 14-16
page, 13-17
CMMI, *see* Capability Maturity Model Integrated
CMS, *see* Content Management System
CO, *see* Central switching office
Collaboration, natural, 5-2, 5-9
Collaboration Protocol Agreement (CPA), 32-12
Collaboration Protocol Profile and Agreement, 8-5
Collaboration Protocol Profiles (CPP), 32-12
Collaborative applications, 5-1 to 5-26
augmenting natural collaboration, 5-9 to 5-23
anonymity, 5-9
automatic redirection of message and per-device presence and availability forecasting, 5-21 to 5-23
chat history, 5-12 to 5-13
control of presence information, 5-10
disruptions caused by messages, 5-20
divergent views and concurrent input, 5-12
meeting browsing, 5-10 to 5-12
multitasking, 5-10
notifications, 5-19 to 5-20
prioritizing messages, 5-20 to 5-21
robust annotations, 5-18 to 5-19
scripted collaboration, 5-13
threaded articles discussion and annotations, 5-16
threaded chat, 5-14
threaded e-mail, 5-15 to 5-16
variable-granularity annotations to changing documents, 5-16 to 5-18
dual goals, 5-2
mimicking natural collaboration, 5-2 to 5-9
graphical chat, 5-8 to 5-9
horizontal time line, 5-6 to 5-8
multipoint lecture, 5-4
overview + speaker, 5-3 to 5-4
single audio/video stream transmission, 5-2 to 5-3

slides video vs. application sharing, 5-5 to 5-6
state-of-the-art chat, 5-6
supporting large number of users, 5-8
vertical time line, 5-8
video-production-based lecture, 5-4 to 5-5
Comfort noise, 6-9
Commercial off-the-shelf (COTS) systems, 33-10 to 33-11, 39-2
Common Component Architecture (CCA), 39-14
Common data model (CDM), 28-2
Common Gateway Interface (CGI), 40-8
Common Object Request Broker Architecture (CORBA), 38-2, 38-16
Common Open Policy Service Protocol (COPS), 46-9
Common Warehouse Metamodel (CWM), 22-10
Communication
agent, 16-2, 16-4, 16-6
model, 38-16
Compatibility, 31-8
Compensation, 31-7
Competitive Local Exchange Company (CLEC), 56-7, 56-14
Complementarity, 31-8
Composable mediators, 28-5
Composite process, 31-12
CompuServe, 11-4
Computational grids, 38-13, 39-4
Computer-supported cooperative work (CSCW), 26-5, 37-6
Computing device, 36-1
Concept Index, 12-11
Concurrent execution, 31-7
Concurrent input and output, 31-7
Connection
handoff, 40-6
routing, 40-3
Constrained broadcast, 35-3
Constrained flooding, 35-3
Consumption, 8-8, 8-13
Content
-based routing, 40-5
classification, 1-9
delivery network, 42-3
distribution service providers (CDSPs), 41-8
Content adaptation and transcoding, 42-1 to 42-13
challenges in effective use of transcoding technologies, 42-11
client-side constraints, 42-2 to 42-3
differentiated services to manage resources, 42-3 to 42-4
server-side constraints, 42-3
technologies utilizing transcoding operation, 42-7 to 42-11
automatic transcoding by proxies and Web servers, 42-9
content producer and consumer involvement, 42-9 to 42-11
systems utilizing transcoding technologies, 42-11
Web content adaptation service architecture, 42-7 to 42-8
transcoding techniques, 42-4 to 42-7
content adaptation of composite Web objects, 42-7
image content, 42-5
quality-aware transcoding, 42-7
streaming media, 42-6 to 42-7
textual content, 42-4 to 42-5
Content distribution networks (CDN), 41-2, 41-7
architectural elements, 41-8
consistency maintenance, 41-14
performance studies, 41-15
request-routing, 41-10
server, 41-11, 41-15
Content Management System (CMS), 9-1, 9-2, 9-3, 9-7
Context service, 31-9
Contextual Design, 33-6
Contiguous sequential pattern (CSP), 15-15, 15-20
Control flow, 31-7
Controlled replication, 35-18
Controlled vocabulary, 12-3
Convergence protocol, 34-4
Conversational agents, 10-1 to 10-16
applications, 10-2 to 10-3
enabling technologies, 10-8 to 10-14
enterprise integration technologies, 10-12 to 10-14
natural language processing technologies, 10-8 to 10-12
technical challenges, 10-3 to 10-8
enterprise delivery requirements, 10-7 to 10-8
natural language requirements, 10-3 to 10-7
Conversational QoS, 10-7
Cooperation, 31-8
Coordinate matching, 14-18
Coordination services, 38-8
COPS, *see* Common Open Policy Service Protocol
Copyright
basics, 57-2
management information, 57-5
note, 57-3
registration, 57-2
CoralReef, 49-9, 49-11
CORBA, *see* Common Object Request Broker Architecture
Core Components, 8-5
Core Information Model (CIM), 46-7
Core Representation Layer, 8-2
Cost and Usage Report, 8-6
COTS, *see* Commercial off-the-shelf systems
CPA, *see* Collaboration Protocol Agreement
CPL, *see* Call Processing Language
CPP, *see* Collaboration Protocol Profiles
CPU
cycles, 41-3
load, 41-13
Crawler
attacks on, 13-6
crawl order
breadth-first search, 13-4
depth-first search, 13-4
databases, 13-2, 13-3, 13-12
indexer, 13-2, 13-7
link extraction, 13-3
traps, 13-6

URL list, 13-3, 13-4
Cray X1, 39-11
Credential Checking Module (CCM), 2-13
Credentials, 26-9
Credit reference, 8-14
Critical Angle, 34-7
CRM, *see* Customer relationship management
Cronus, 51-2, 51-9
Cross-impacts, 33-6
Cryptographic mechanisms, 26-2
CSCW, *see* Computer-supported cooperative work
CSP, *see* Contiguous sequential pattern
CSS, *see* Cascading Style Sheets
Currency conversion, 8-14
Customer relationship management (CRM), 10-3, 33-9
CWM, *see* Common Warehouse Metamodel
Cybersquatting, 57-1, 57-7

D

DAC, *see* Discretionary access control
DADs, *see* Data Access Definitions
DAML, *see* DARPA Agent Markup Language
DAML-S, *see* DAML for Services
DAML for Services (DAML-S), 31-11, 31-12
Dark Age of Camelot, 11-5
DARPA, *see* Defense Advanced Research Projects Agency
DARPA Agent Markup Language (DAML), 31-3, 31-11, 36-8
Data
  access layer, 35-4
  completeness, 26-2, 26-14
  content, 15-5
  document, 25-2
  flow, 10-4, 31-7
  incidental, 25-2
  integration, 2-6, 25-2
  management, 35-7
  mediation, 31-6
  mining, 14-1, 14-2
  model mappings, 28-4
  replication, 35-18, 36-9
  sources
    distribution of, 9-3
    heterogeneous, 9-3
  structure, 15-5
  usage, 15-5, 15-6
  warehousing, 28-3, 33-8
Data Access Definitions (DADs), 25-13 to 25-14
Database Management System (DBMS), 24-7, 26-4
Databases, copyright protection of, 57-4
DBMS, *see* Database Management System
DCMI, *see* Dublin Core Metadata Initiative
DCOM, *see* Distributed Component Object Model
DDDS, *see* Dynamic Delegation Discovery System
Dead reckoning, 11-11
Deep Linking, 57-10
Deep Web, 13-14, 53-14
De facto standards, 50-5
Defense Advanced Research Projects Agency (DARPA), 6-18, 31-3, 31-11
De jure standards, 50-5
Delivery assurances, 27-8
Delphi, 11-4
Delta consistency, 41-4, 41-5
Denial-of-service (DOS) attacks, 35-18, 35-19
Design
  patents, 57-8
  reasoning, 33-6
Destination-Sequenced Distance Vector (DSDV), 36-7
DG, *see* Digital government
DHT, *see* Distributed hash table
Dialog(s)
  mixed-initiative, 10-11
  model, 10-10
Dialog Manager, 10-6
Dictionary, 8-3, 8-4
dig, *see* Domain information groper
Digital government (DG), 2-1 to 2-18
  applications, 2-4 to 2-6
    electronic voting, 2-4
    geographic information systems, 2-5
    government portals, 2-5
    social and welfare services, 2-5 to 2-6
    tax filing, 2-4 to 2-5
  brief history of digital government, 2-3 to 2-4
  case study, 2-9 to 2-15
    implementation, 2-14
    ontological organization of government databases, 2-10
    preserving privacy in WebDG, 2-13 to 2-14
    WebDG scenario tour, 2-14 to 2-15
    Web services support for digital government, 2-10 to 2-13
  issues in building e-government infrastructures, 2-6 to 2-9
    accessibility and user interface, 2-9
    data integration, 2-6
    interoperability of government services, 2-6 to 2-7
    privacy, 2-8
    scalability, 2-6
    security, 2-7 to 2-8
    trust, 2-9
Digital libraries (DLs), 4-1 to 4-13
  architecture, 4-6 to 4-7
  inception, 4-8 to 4-9
    digital library initiative, 4-8
    global DL trends, 4-9
    networked digital libraries, 4-8
  interfaces, 4-5 to 4-6
  personalization and privacy, 4-10
  theoretical foundation, 4-3 to 4-5
Digital Millennium Copyright Act (DMCA), 57-3, 57-5, 57-6, 57-12
Digital rights management (DRM), 21-1 to 21-16
  legal issues, 21-13 to 21-15
  overview, 21-2 to 21-4
  systems, 57-5
  tools, 21-4 to 21-13
    protection mechanisms, 21-5 to 21-13
    software cracking techniques and tools, 21-5
Digital signatures, 26-2

Direct guidance, 1-3
Directory information tree (DIT), 34-15
Discovery, 31-2, 31-3, 36-7
Discretionary access control (DAC), 26-4
Disintermediation, 33-10
Disk storage, fault tolerance and, 41-3
Distinguished Name (DN), 34-9
Distributed Component Object Model (DCOM), 22-6, 38-2
Distributed hash table (DHT), 35-14
Distributed indexing, 51-6, 51-11
Distributed storage, 45-1 to 45-20
  array designs, 45-11 to 45-12
  cache coherence, 45-6 to 45-9
  data location, 45-3 to 45-6
    cooperative Web caching, 45-3 to 45-5
    distributed file systems, 45-5
    distributed hash tables, 45-6
  load balancing, 45-9 to 45-11
    peer-to-peer storage systems, 45-10 to 45-11
    Web, 45-9 to 45-10
  security, 45-14 to 45-15
  weakly connected wide-area environments, 45-12 to 45-13
DIT, *see* Directory information tree
DLs, *see* Digital libraries
DM, *see* Document Management
DMCA, *see* Digital Millennium Copyright Act
DN, *see* Distinguished Name
DNS, *see* Domain Name System
Document
  clustering, 14-7, 14-16
  data, 25-2
  frequency, 14-18
  retrieval, 14-4
  similarity metrics, 14-6
  type definition (DTD), 23-2, 23-11, 25-15, 26-7
Document Management (DM), 9-4
Document Object Model (DOM), 23-8, 23-9, 24-4
Document Schema Definition Languages (DSDL), 23-12, 23-15
DOM, *see* Document Object Model
Domain(s), 46-12, 57-6
  information groper (dig), 7-4
  knowledge, 31-4
  name registration, 56-8
Domain Name System (DNS), 7-4, 7-8, 34-2, 38-8
  load balancing via, 40-2
  lookup, 40-3
  query, 7-4
  request-routing based on, 41-12
DOS attacks, *see* Denial-of-service attacks
Do-what-I-mean (DWIM), 51-5
Dragonflies and tortoises, 49-14
DRM, *see* Digital rights management
DSDL, *see* Document Schema Definition Languages
DSDV, *see* Destination-Sequenced Distance Vector
DSL, 56-7, 56-11
DTD, *see* Document type definition
Dublin Core Metadata Initiative (DCMI), 1-9
DWIM, *see* Do-what-I-mean
DynaBone, 51-8
Dynamic Delegation Discovery System (DDDS), 6-17
Dynamic pages, 13-14
Dynamic requests, 40-8
Dynamic service
  binding, 31-7
  invocation mediation, 31-8

E

EAI, *see* Enterprise Application Integration
Earth Simulator, 39-11
ebXML, *see* Electronic business XML
ECA rules, *see* Event-Condition-Action rules
ECDM, *see* Edutella Common Data Model
ECM, *see* Semantic enterprise content management
ECMAScript, 23-10
E-commerce, 33-8, 56-6, 56-10
E-contract, 57-12
EDI, *see* Electronic Data Interchange
Edutella Common Data Model (ECDM), 1-11
Effector, agent, 16-3, 16-7
E-Government, 2-2
EJB, *see* Enterprise Java Bean
E-Learning technology, 3-1 to 3-18
  e-learning and business performance, 3-2
  e-learning standards, 3-7 to 3-10
    SCORM specification, 3-8 to 3-10
    standards organizations, 3-7 to 3-8
  evolution of learning technologies, 3-3 to 3-4
  improving business performance using e-learning technology, 3-10 to 3-16
    advancements in infrastructure technology, 3-14 to 3-16
    delivering business knowledge, 3-10 to 3-11
    improving business processes, 3-12 to 3-13
    lifelong learning, 3-13 to 3-14
  Web-based e-learning environments, 3-4 to 3-7
    creation and delivery of e-learning, 3-6 to 3-7
    learning theories and instructional design, 3-4 to 3-5
    types of e-learning environments, 3-5
Electronic business XML (ebXML), 8-5, 8-17, 31-4, 33-9
Electronic Data Interchange (EDI), 2-3
Electronic mail, *see* E-mail
Electronic voting, 2-4
Elephants and Mice phenomenon, 40-15
E-mail, 7-1, 7-2
  address, reuse of, 6-14
  @ (at) symbol, 7-3
  body, 7-4
  chat and, 5-14
  forwarding, 7-2
  header, 7-4
  naming, 7-3
  point-to-point communication, 7-7
  remote access, 7-3
  resolving of name, 7-3
  response, out-of-office, 5-21
  threaded, 5-15
EMS, *see* Enhanced Messaging Service

Encapsulation, UDP, 51-5, 51-9
Encryption, SSL, 40-13
Engineering
change, 8-14
information, 8-10, 8-14
Enhanced Messaging Service (EMS), 7-12, 7-21
Enterprise
architecture, 22-2, 22-3, 22-17
manufacturing, 8-2
object, 8-2
resource planning (ERP) systems, 8-4, 8-7, 33-4, 33-8
Enterprise Application Integration (EAI), 22-3, 22-1, 27-3, 31-2, 31-14, 32-4, 33-8
Enterprise architectures, 22-1 to 22-23
comparisons of J2EE and .NET, 22-20 to 22-21
from client–server to *n*-tier architectures, 22-3 to 22-6
client–server architecture, 22-3 to 22-4
messaging systems, 22-5
n-tier architectures, 22-5 to 22-6
remote procedure calls, 22-4 to 22-5
global architecture, 22-21 to 22-28
J2EE architecture, 22-12 to 22-16
application architectures, 22-16
container model, 22-13 to 22-14
EJB container, 22-14 to 22-15
Java Database Connectivity, 22-16
Java Message Service, 22-15
Java Naming and Directory Interface, 22-15 to 22-16
layered approach, 22-12 to 22-13
Web container, 22-14
.NET architecture, 22-16 to 22-20
Active Directory, 22-20
application architectures, 22-18
basic principles, 22-17
Microsoft Message Queue, 22-20
Microsoft Transaction Server and language runtime, 22-18 to 22-20
new keys to interoperability, 22-6 to 22-12
Meta Data Registries, 22-9 to 22-10
OMG model-driven architecture, 22-10 to 2 2-12
SOAP, 22-7
Web services, 22-7 to 22-9
XML, 22-6 to 22-7
Enterprise Content Management (ECM), *see* Semantic enterprise content management
Enterprise Java Bean (EJB), 22-13, 22-14,
Container, 22-13, 22-14
Server, 22-17, 22-18
session bean, 22-15
Entity extraction, 14-10
Entity-Relationships (ER) modeling, 33-3
Epic Games, 11-2
ER modeling, *see* Entity-Relationships modeling
ERP systems, *see* Enterprise resource planning systems
Error
data, 31-7
recovery mechanism, 26-2
Esperonto, 31-3
Ethernet, 39-3, 50-14
Ethnographic studies, 33-6
Event-Condition-Action (ECA) rules, 32-5, 32-8
Event-driven servers, 40-10
Everquest, 11-5, 11-7
Exception handling, 31-7
Exchange rate, 8-14
Executable process, 27-15
Execution, 31-2
Extended Boolean retrieval model, 12-8
Extended request, 34-26
Extensible Markup Language (XML), 8-3, 11-8, 12-15, 22-6, 23-1, 26-2, 33-7, 56-2, 56-14
address-book, 25-7
Canonical, 23-7
document(s)
authentication techniques, 26-12
building blocks of, 26-6
secure structure of, 26-14
features, relational database system, 25-13
Inclusions, 23-2
Infoset, *see* XML Information Set
Namespaces, 23-2, 23-5
photo metadata in, 25-3
Schema, 23-2, 23-8, 23-9, 23-11
Signature, 26-12, 26-13
tree, encoding of, 25-10
Extensible Messaging and Presence Protocol (XMPP), 7-9
Extensible Stylesheet Language (XSL), 23-7, 24-2, 24-3
Externalization service, 38-3

F

Factory Planning, 8-4
Factory Scheduler, 8-7
Failure, 31-7
Failure resilience, 35-4, 35-6
Family and Social Services Administration (FSSA), 2-9
Fast Ethernet, 40-4
FastTrack, 35-10
Fault(s)
signal occurrence, 27-8 to 27-9
tolerance, disk storage and, 41-3
FCC, *see* Federal Communications Commission
FDMA, *see* Frequency Division Multiple Access
FEC, *see* Forward error correction
Federal Communications Commission (FCC), 56-7, 56-10
File sharing, 35-2, 51-1
File Transfer Protocol (FTP), 24-7
Filtering policies, 26-15
Finite state automaton (FSA), 10-10
Flash, 11-5
Flash crowd, 42-3
Flat P2P systems, 35-8
FlowScan, 49-11
Forecast, 8-14
Forking, 6-15
Formatting objects, *see* XSL Formatting Objects
Forward error correction (FEC), 51-9
Foundry Networks, 40-4

Framing, 57-9
FreeBSD, 40-11
Freenet, 35-11, 35-12
Free riding, 35-6
Frequency Division Multiple Access (FDMA), 36-3
Frequent Sequence Trie (FST), 15-24, 15-25
FSA, *see* Finite state automaton
FSSA, *see* Family and Social Services Administration
FST, *see* Frequent Sequence Trie
FTP, *see* File Transfer Protocol
Fujitsu, 31-8
Fully decentralized system, 35-3
Fuzzy set retrieval model, 12-7

G

Game genre, 11-2
  action games, 11-2
  adventure games, 11-2
  arcade simulations, 11-2
  casual games, 11-3
  fighting games, 11-3
  first-person shooters, 11-2
  Groove Alliance, 11-5
  real-time strategy games, 11-2
  role-playing games, 11-2
  simulation games, 11-2
  sports games, 11-3
  strategy games, 11-2
Games, Internet-based, 11-1 to 11-14
  background and history, 11-2 to 11-5
    genre, 11-2 to 11-3
    short history of online games, 11-3 to 11-5
  further information, 11-12
  games, gameplay, and Internet, 11-5 to 11-7
  games and mobile devices, 11-11 to 11-12
  implementation issues, 11-7 to 11-11
    consistency, 11-10 to 11-11
    system architecture, 11-10
GATE, *see* General Architecture for Text Engineering
General Architecture for Text Engineering (GATE), 14-19
General purpose technology (GPT), 56-4
General Switched Telephone Network (GSTN), 6-2
Generation, WebDG, 2-12, 2-13
Geographical information systems (GIS), 33-13
Geographic Information Systems, 2-5
Gigabit Ethernet, 40-4
GIIS, *see* Grid Index Information Service
GIS, *see* Geographical information systems
Global data management, 36-10
Global information centers, 56-8
Global routing infrastructure, 49-11
Global System for Mobile Communications (GSM), 36-2
Globus Metacomputing Directory Service, 39-9
Gnutella, 35-6, 35-8, 35-14, 36-4, 51-6
Goal
  reference, 31-7
  repositories, 31-6
Google, 12-13, 13-11, 35-1, 57-8
Government portals, 2-5
GPS location service, 37-5
GPT, *see* General purpose technology
Graphical User Interface (GUI), 2-14
  -based windowing systems, 37-7
  environment, OntoEdit, 9-14
GreedyDual-Size algorithm, 41-2
Grid
  environments, 39-11
  frameworks, *see* Metacomputing and grid frameworks
  service handles (GSH), 39-12, 39-13
Grid Index Information Service (GIIS), 39-7, 39-9
Grouping service, 38-3
GSH, *see* Grid service handles
GSM, *see* Global System for Mobile Communications
GSM-MAP, 7-12
GSTN, *see* General Switched Telephone Network
GUI, *see* Graphical User Interface

H

Habitat, 11-4
HAL, *see* High-level Actor Language
HCI, *see* Human–computer interaction
HDLC, *see* High-level Data Link Control
Heterogeneous data sources, 9-3
Hidden Markov models (HMMs), 14-18
Hierarchical P2P systems, 35-8, 35-10
High-level Actor Language (HAL), 38-15
High-level Data Link Control (HDLC), 36-3
High Performance Knowledge Bases (HPKB), 31-3
HITS, *see* Hyperlink Induced Topic Search
HLR, *see* Home Location Register
HMMs, *see* Hidden Markov models
Hoax, 7-15
Home Location Register (HLR), 7-11
Hosted IP PBX, 6-5
Howes, Tim, 34-6
HPKB, *see* High Performance Knowledge Bases
HTML, *see* Hypertext Markup Language
HTTP, *see* Hypertext Transport Protocol
Hubs and authorities, 14-18
Human–computer interaction (HCI), 36-2
Human resources, 8-3
  management, 8-7
  procurement of, 8-15
Hyperlink Induced Topic Search (HITS), 53-8
Hypertext Markup Language (HTML), 12-13, 12-2, 14-15, 33-7
Hypertext Transport Protocol (HTTP), 10-12, 24-7, 33-7
  Basic Authentication, 27-7
  header, 42-9
  protocol support, 40-18
  proxies, 40-6
  servers, 36-5
  URL, 41-11

I

IAD, *see* Integrated access device
IANA, *see* Internet Assigned Numbers Authority

IBM
Aglets, 38-16
GO Web server API, 40-8
MQ Series Workflow, 32-4
SecureWay, 34-7
Systems Network Architecture, 50-13
Websphere edge server, 40-4
ICAP, *see* Internet Content Adaptation Protocol
ICE, *see* Intelligent Concept Extraction
ICMP, *see* Internet Control Message Protocol
ICQ (I seek you), 7-7, 56-14
IDEF, 33-3
Identification, anonymity vs., 35-19
IDL, *see* Interface definition languages
IETF, *see* Internet Engineering Task Force
IM, *see* Instant messaging
IMAP, *see* Internet Message Access Protocol
i-mode, 7-11
Incidental data, 25-2
Indexing
exhaustivity, 12-3
forward index, 13-8
inverted index, 13-9
lexicon, 13-7
manual, 12-3
phrase, 12-5
single-term, 12-3
stop words, 13-7, 13-8
term frequency, 13-9
Index replication, 35-18
Informal games, 11-5
Information
access technology, 37-6
discovery of, 3-16
extraction, 14-4, 14-9, 14-12
managed, 4-1
presence, 5-10
sharing, 4-2
uncontrolled, 4-1
Information modeling, Web, 30-1 to 30-23
definition of metadata, 30-3 to 30-6
means for modeling information, 30-4 to 30-6
usage in various applications, 30-3 to 30-4
metadata expressions, 30-6 to 30-14
InfoHarness system, 30-7 to 30-10
metadata-based logical semantic Webs, 30-10 to 30-12
modeling languages and markup standards, 30-13 to 30-14
ontology, 30-14 to 30-20
controlled vocabulary for digital media, 30-16
expanding terminological commitments across multiple ontologies, 30-20
medical vocabularies and terminologies, 30-17 to 30-20
ontology-guided metadata extraction, 30-17
terminological commitments, 30-14 to 30-16
Information retrieval (IR), 12-1 to 12-23, 14-6
indexing documents, 12-2 to 12-6
multiterm or phrase indexing, 12-5 to 12-6
single-term indexing, 12-3 to 12-5
IR products and resources, 12-18
language modeling approach, 12-8 to 12-9
metasearch engines, 12-16 to 12-18
component techniques, 12-16 to 12-18
software component architecture, 12-16
multimedia and markup documents, 12-15 to 12-16
MPEG-7, 12-15
XML, 12-15 to 12-16
query expansion and relevance feedback techniques, 12-9 to 12-12
automated query expansion and concept-based retrieval models, 12-9 to 12-11
relevance feedback techniques, 12-11 to 12-12
research direction, 12-19 to 12-20
retrieval models, 12-6 to 12-8
with ranking of output, 12-7 to 12-8
without ranking of output, 12-7
retrieval models for Web documents, 12-12 to 12-15
HITS algorithm, 12-14
link analysis based page ranking algorithm, 12-14
topic-sensitive PageRank, 12-15
Web graph, 12-13 to 12-14
Information security, 26-1 to 26-18
access control for Web documents, 26-7 to 26-12
reference access control model for protection of XML documents, 26-9 to 26-12
requirements for Web data, 26-7 to 26-9
authentication techniques for XML documents, 26-12 to 26-14
signature policies, 26-13 to 26-14
XML signature, 26-12 to 26-13
basic concepts, 26-2 to 26-7
access control mechanisms, 26-2 to 26-6
brief introduction to XML, 26-6 to 26-7
data completeness and filtering, 26-14 to 26-16
data completeness, 26-14 to 26-15
filtering, 26-15 to 26-16
Information systems, 33-1 to 33-19
challenges in Internet Age, 33-8 to 33-12
automation, 33-11 to 33-12
nonfunctional qualities, 33-9 to 33-10
products and processes, 33-8 to 33-9
social structures, 33-10 to 33-11
conceptual abstractions, 33-12 to 33-14
how and why, 33-13 to 33-14
what and when, 33-12 to 33-13
where, 33-13
who, 33-14
information systems before advent of Internet, 33-2 to 33-5
automation, 33-4 to 33-5
nonfunctional qualities, 33-3 to 33-4
processes, 33-2 to 33-3
products, 33-3
social structures, 33-4
Internet computing news, 33-7 to 33-8
world as seen by, 33-5 to 33-7
automation, 33-7
nonfunctional qualities, 33-5 to 33-6
processes and products, 33-5
social structures, 33-6

In-kernel servers, 40-11
Inline links, 57-9
Innosoft International, 34-8
Input and output data, 31-7
Inspection, 8-13
Instant messaging, 5-1, 5-24, 7-1, 7-2, 7-7
channel, 7-8
chat room, 7-9
naming, 7-10
presence, 7-9
term-talk, 7-7
Integrated access device (IAD), 6-7, 6-8, 6-20
Integration objective, 8-2
Intellectual capital, 33-4
Intellectual property, liability, and contract, 57-1 to 57-14
copyright, 57-2 to 57-6
basics, 57-2
copying in digital medium, 57-3
databases, 57-4
digital rights management systems, 57-5 to 57-6
open source software, 57-5
registration, 57-2 to 57-3
software, 57-4 to 57-5
Websites, 57-3 to 57-4
cyberlaw, 57-1 to 57-2
e-commerce, 57-12 to 57-13
e-contract, 57-12 to 57-13
taxes, 57-13
liability, 57-9 to 57-12
linking, 57-9 to 57-12
providers, 57-12
patents, 57-7 to 57-9
business methods, 57-9
software, 57-8
trademarks and service marks, 57-6 to 57-7
cybersquatting, 57-7
domains, 57-6 to 57-7
Intelligent Concept Extraction (ICE), 13-10
Intel Net-Structure, 40-4
Interface definition languages (IDL), 32-11
Inter-gatekeeper communications, 6-13
International Telecommunications Union (ITU), 34-4
Internet
access, advanced, 56-10
banking, 56-6
-based architectures, 22-1
computing, 2-1
diffusion theory, 56-1, 56-2, 56-3
dispersion, mapping of, 56-8
filtering systems, 26-15
hosts, definition of, 56-9
law, 57-1
mapping, 49-15
nested innovation, 56-3
protocol (IP), 6-2
Centrex, 6-5
Flow Information Export (IPFIX), 49-10 to 49-11
packet delay variation, 49-3
relay chat (IRC), 7-7
routing system, 49-11, 49-17
Internet, geographical diffusion of in United States, 56-1 to 56-17
brief history of Internet, 56-2 to 56-3
diffusion of advanced Internet access, 56-10 to 56-12
provision and adoption, 56-10 to 56-11
rural vs. urban divides, 56-11 to 56-12
diffusion process, 56-3 to 56-8
demand for business purposes, 56-5 to 56-6
standard diffusion analysis, 56-3 to 56-5
supply by private firms, 56-6
supply by regulated telephone firms, 56-6 to 56-8
mapping of Internet's dispersion, 56-8 to 56-10
backbone, 56-8
content and e-commerce, 56-10
domain name registrations, 56-8 to 56-9
hosts, Internet service providers, and points of presence, 56-9 to 56-10
overview, 56-13 to 56-14
first wave of diffusion, 56-13
open questions, 56-13 to 56-14
Internet Assigned Numbers Authority (IANA), 34-10
Internet Content Adaptation Protocol (ICAP), 42-8
-capable surrogate, 42-8
request mode of operation, 42-8
response mode of operation, 42-8
servers, 42-8
Internet Control Message Protocol (ICMP), 41-13, 41-14
Internet Engineering Task Force (IETF), 6-5, 37-4, 39-8, 49-3, 50-11, 50-13
Internet measurement, 49-1 to 49-20
methodology, 49-2 to 49-13
active measurements, 49-4 to 49-5
examples of traffic characterization, 49-11
general considerations for network measurement data, 49-13
link behavior, 49-5 to 49-6
measurement of global routing infrastructure, 49-11 to 49-12
metrics, 49-3
packet traces, 49-6 to 49-7
techniques, 49-4
traffic characterization, 49-7
traffic flows, 49-7 to 49-11
research topics, 49-13 to 49-17
BGP topology data analysis, 49-16 to 49-17
Internet topology, 49-15 to 49-16
mice, elephants, dragonflies, and tortoises, 49-14 to 49-15
packet statistics, 49-13 to 49-14
Internet Message Access Protocol (IMAP), 7-3
Internet messaging, 7-1 to 7-18
comparison, 7-12 to 7-14
current Internet solutions, 7-2 to 7-11
electronic mail, 7-2 to 7-5
instant messaging, 7-7 to 7-10
network news, 7-5 to 7-7
Web logging, 7-10 to 7-11
outlook, 7-17 to 7-18
telecom messaging, 7-11 to 7-12
naming, 7-11
principal operation, 7-11

Short Message Service, 7-11 to 7-12
unified messaging, 7-16 to 7-17
unsolicited messaging, 7-14 to 7-16
protection mechanisms, 7-15 to 7-16
spam, 7-15
spreading viruses, 7-14 to 7-15
Internet policy and governance ecosystem, 54-1 to 54-24
business sector, 54-9 to 54-11
hardware and software vendors, 54-9 to 54-10
large commercial users, 54-10
major service providers, 54-10 to 54-11
definitions, 54-7 to 54-9
legal constructs, 54-8 to 54-9
network boundaries and variables, 54-8
protocols, 54-8
emerging trends, 54-16
assimilation, 54-16
diversity, 54-16
security, 54-16
government sector, 54-12 to 54-14
law, 54-14
regulatory constructs and requirements for Internet service provisioning, 54-13 to 54-14
Internet standards and administrative forums, 54-18 to 54-24
major historical policy and governance developments, 54-2 to 54-7
centers of authority, 54-4
international politics of control, 54-6 to 54-7
meta Internet ecosystem transitions, 54-2 to 54-4
NIC, 54-4
NOC, 54-5
research and development framework, 54-5 to 54-6
standards and administrative sector, 54-14 to 54-15
legacy standards and administrative forums, 54-15
universe of Internet standards and administrative forums, 54-15
user sector, 54-11 to 54-12
advocacy and academic groups, 54-12
developers, 54-11 to 54-12
end user and SOHOs, 54-12
Internet Service Provider (ISP), 7-3, 56-6
cost structure, 56-12
upgrade decisions, 56-11
Internet telephony, 6-1 to 6-28
architecture, 6-5 to 6-7
brief history, 6-18 to 6-19
core protocols, 6-10 to 6-18
call routing, 6-18
call setup and control, 6-12 to 6-17
device control, 6-11 to 6-12
media transport, 6-10 to 6-11
telephone number mapping, 6-17
discovery, 49-16
glossary, 6-20 to 6-21
killer application, 6-20
media encoding, 6-8 to 6-10
audio, 6-8 to 6-10
video, 6-10
motivation, 6-3 to 6-4
efficiency, 6-3
functionality, 6-3 to 6-4
integration, 6-4
overview of components, 6-7 to 6-8
service creation, 6-19
standardization, 6-4 to 6-5
Internet Telephony administrative domains (ITADs), 6-18
Interoperability, 8-4, 31-8
definition of, 22-3
government services, 2-6, 2-7
information systems, 22-2
new keys to, 22-6
stack, 8-5
Interrelation, 31-8
Intranets, 33-7
Intrusion detection techniques, 48-1 to 48-10
modeling and analysis approaches, 48-4 to 48-6
alert analysis, 48-5 to 48-6
anomaly detection, 48-4 to 48-5
misuse detection, 48-4
network and system issues, 48-6 to 48-8
deployment strategies, 48-6 to 48-7
performance optimization and adaptation, 48-7 to 48-8
Inventory, 8-14
Invoice, 8-13
IONA, 31-8
IP, *see* Internet protocol
IPFIX, *see* IP Flow Information Export
IPv4, 50-13
IPv6, 50-13
IR, *see* Information retrieval
IRC, *see* Internet relay chat
IS-41, 7-12
Islands of automation, 32-2
ISODE, *see* ISO Development Environment
ISO Development Environment (ISODE), 34-4
ISP, *see* Internet Service Provider
ITADs, *see* Internet Telephony administrative domains
IT architecture, 22-2, 22-3
Item Description, 8-6, 8-8
ITU, *see* International Telecommunications Union

J

JAD, *see* Joint Application Development
Java
Database Connectivity (JDBC), 22-16
Message Service (JMS), 22-5, 22-15
2 Micro Edition (J2ME), 11-12
Naming and Directory Interface (JNDI), 22-15
PI for XML Parsing (JAXP), 23-9
RMI, 38-2
serializable objects, 38-10
Server Pages (JSPs), 22-14
JAXP, *see* Java API for XML Parsing
JDBC, *see* Java Database Connectivity
JDOM, 23-9, 23-10
J2EE

architecture, 22-12
Container Model, 22-13
Jitter, 49-3
J2ME, *see* Java 2 Micro Edition
JMS, *see* Java Message Service
JNDI, *see* Java Naming and Directory Interface
Joint Application Development (JAD), 33-3
JPEG Quality factor parameter, 42-7
JSPs, *see* Java Server Pages
Junk-mail, 7-15
JXTA, 35-10

K

Kazaa, 35-10, 35-19, 51-6
Keyphrase
assignment, 14-9
extraction, 14-9
identification, 14-8
Kille, Steve, 34-4, 34-7
KM, *see* Knowledge management
k-Nearest-Neighbor (*k*NN) classification, 15-2, 15-8, 15-20
*k*NN classification, see *k*-Nearest-Neighbor classification
Knowledge
base, 9-10
discovery, 9-16
engineering, 14-6, 14-17
management, 3-2, 3-10, 31-2, 31-14, 33-14
structuring, 33-12
Web, 31-3
work, 33-11

L

Labor Resources, 8-7, 8-8, 8-10
Lancaster Guide project, 37-9
Language
identification, 14-4, 14-8
modeling, 12-2, 12-8
LCMS, *see* Learning Content Management Systems
LDAP, *see* Lightweight Directory Access Protocol
LDAP Interchange Format (LDIF), 34-13
LDIF, *see* LDAP Interchange Format
Learning, 16-3, 16-4, 16-6
objects metadata, 1-9
theory, 3-9
Learning Content Management Systems (LCMS), 3-7
Learning Management Systems (LMS), 3-7
Learning Objects Metadata Standard (LOM), 1-9
Least recently used algorithm (LRU), 41-2
Legacy telephony, 6-2
Life-cycle service, 38-3
Lifelong learning, *see* E-Learning technology
Lightweight Directory Access Protocol (LDAP), 34-1 to 34-32
directory services, 34-14 to 34-23
basic directory services, 34-15 to 34-18
high availability directory services, 34-18
LDAP proxy server, 34-19 to 34-20
master–slave replication, 34-18 to 34-19
multimaster replication, 34-21 to 34-22
replication standardization, 34-22 to 34-23
evolution of, 34-4 to 34-9
first- and second-generation directory services, 34-6 to 34-8
next-generation directory services, 34-8 to 34-9
past, present, and future generations of LDAP directories, 34-4 to 34-6
naming model, 34-9 to 34-12
Internet domain-based naming, 34-10 to 34-11
naming entries within organization, 34-11 to 34-12
X.500, 34-9 to 34-10
X.500 early alternatives, 34-10
X.500 limitations, 34-10
protocol and C language client API, 34-23 to 34-30
abandoning of operation, 34-28 to 34-29
add, delete, modify, and modifyDN operations, 34-29 to 34-30
bind, 34-25 to 34-26
compare request, 34-29
extended request, 34-26
general result handling, 34-24 to 34-25
protocol exchange, 34-32 to 34-24
searching, 34-26 to 34-27
search responses, 34-28
unbind, 34-26
schema model, 34-12 to 34-14
attribute-type definitions, 34-12
object-class definitions, 34-12 to 34-13
object classes for entries representing people, 34-13 to 34-14
other typical object classes, 34-14
Linear interpolation smoothing, 12-9
Link
analysis, 14-3
annotation, 1-11
generation, 1-3, 1-5
hiding, 1-5
Linux, 40-11, 51-10
LMS, *see* Learning Management Systems
Load balancing
client-based, 40-7
DNS, 40-2
Lobby Services, 11-6
LOM, *see* Learning Objects Metadata Standard
Longest repeating subsequences (LRS), 15-16
Loosely coupled P2P systems, 35-8
Loose source routing (LSR), 51-3
LRS, *see* Longest repeating subsequences
LRU, *see* Least recently used algorithm
LSR, *see* Loose source routing
LucasFilm, 11-4

M

Machine
-human understanding, 31-3
learning, 14-3
processability, 31-3
Mail
exchange records (MX), 7-4

servers, 7-2, 7-4
user agent (MUA), 7-2, 7-3
Mailbox, 7-2
Mail Transfer Agent, 6-15
Management policy, 46-1, 46-3, 46-7
Mandatory access control, 26-5
MANET, *see* Mobile *ad hoc* networks
Manual indexing, 12-3
Manufacturing enterprise systems interoperability, 8-1 to 8-19
advanced developments in support of, 8-12 to 8-17
disclaimer, 8-17
ebXML, 8-15 to 8-16
National Institute of Standards and Technology, 8-16 to 8-17
OAG, 8-15
RosettaNet, 8-16
semantic Web activity, 8-17
interoperable information systems, 8-6 to 8-12
classification framework, 8-10 to 8-11
description framework, 8-6 to 8-7
example inter-enterprise scenarios of integration, 8-11 to 8-12
manufacturing enterprise information systems, 8-7
OAG semantic integration standards in manufacturing sector, 8-8
overview of approaches, 8-2 to 8-6
general concepts, 8-2 to 8-3
selected approaches, 8-3 to 8-6
Manufacturing Execution, 8-7
Manufacturing Order, 8-7, 8-8
MAP, *see* Mobile Application Part
Market mechanisms, 33-9
Markov chains, 15-16
Markov models, 15-16
Master Schedule, 8-7
Matchmaking, WebDG, 2-12
Matilda's Smart House project, 37-11, 37-12
Maximum transmission unit (MTU), 51-3
MCUs, *see* Multipoint control units
MDA, *see* Message delivery agent
MDS, *see* Metacomputing Directory Service
Mean Opinion Score (MOS), 49-3
Media
encoding, 6-8
encryption, 6-4
gateway controller (MGC), 6-11
Mediators for querying heterogeneous data, 28-1 to 28-18
Amos II approach to composable mediation, 28-6 to 28-16
composed functional mediation, 28-11 to 28-14
functional data model of Amos II, 28-7 to 28-11
implementing wrappers, 28-15 to 28-16
mediator architectures, 28-2 to 28-6
composable mediators, 28-5 to 28-6
reconciliation, 28-4 to 28-5
wrappers, 28-3 to 28-4
Medium Access Control, 36-3
Message(s)
content, 8-2, 8-4
delivery agent (MDA), 7-2
exchange protocol, 31-7
flooding, 35-9
protocols mediation, 31-8
submit agent (MSA), 7-2
transfer agent (MTA), 7-2
unsolicited, 7-14
Message Authentication Code, 7-17
Message-oriented middleware (MOM), 32-10
Message Passing Interface (MPI), 39-2, 39-4
Messaging
i-mode, 7-11
layer, 8-2, 8-3
solutions, 8-2, 8-4
systems, 22-5, 22-7
taxonomy, 7-12, 7-13
telecom, 7-11
Meta-actors, 38-11, 38-13
Metacomputing Directory Service (MDS), 39-9
Metacomputing and grid frameworks, 39-1 to 39-16
applicability issues, 39-10 to 39-11
clusters and standards, 39-11
grid environments, 39-11
performance issues, 39-11
simplicity and flexibility, 39-10 to 39-11
computational grids, 39-4 to 39-10
definitions, 39-5 to 39-7
example grid infrastructures, 39-8 to 39-9
programming grids, 39-10
current trends, 39-11 to 39-14
components and portals, 39-14
OGSA, OGSI, and GTK3, 39-12 to 39-14
historical evolution of network computing, 39-2 to 39-4
MPI and network computing, 39-4
Metadata, 4-2, 9-3, 13-13, 14-4
annotation, 9-15
explicit, 27-3
extraction, 9-14
resources, 9-6
semantic, 9-7
standards, 9-4
structural, 9-7
syntactic, 9-7
Meta Data Registries, 22-6
Meta Object Facility (MOF), 22-10
Metasearch, 12-2, 12-16, 13-14
Metatags, 57-11
Metcalfe's Law, 33-7
MGC, *see* Media gateway controller
Mice and elephants, 49-10, 49-14
Microsoft
digital entertainment lifestyle, 11-11
SQL Server, 25-16
X-Box, 11-5
Middleware, *see* Worldwide computing middleware
Migration service, 38-8
MIME, *see* Multipurpose Internet Mail Extensions
Mining query language, 15-17
MIT SpectrumWare, 37-3
Mixed-initiative dialogs, 10-11
MLS/DBMSs, *see* Multilevel secure database management systems

MMS, *see* Multimedia Messaging Service
Mobile *ad hoc* networks (MANET), 36-4
Mobile agents, 16-2
Mobile Application Part (MAP), 7-12
Mobile computing, 36-1 to 36-16
application architectures, 36-4 to 36-5
application framework, 36-5 to 36-12
communications layer, 36-6 to 36-7
data management layer, 36-9 to 36-10
discovery layer, 36-7 to 36-8
location management layer, 36-8
security plane, 36-12
service management layer, 36-11 to 36-12
system management plane, 36-12
wired-network computing vs., 36-3 to 36-4
Mobile games, 11-12
Mobile handheld guides, 1-7
Mobile People Architecture, 7-16
Model(s)
access control, 26-3, 26-11
actor, 38-4
architectural, 40-9
bag of words, 14-5, 14-7
common data, 28-2
communication, 38-16
discretionary access control, 26-4
driven architecture, 22-6, 22-10
hidden Markov, 14-18
mandatory access control, 26-5
mappings, 28-4
Markov, 15-16
native programming, 39-10
peer-to-peer interaction, 27-3
policy-neutral, 26-6
resource, 32-6
retrieval, 12-6, 12-10
role-based access control, 26-5
Service Oriented Computing, 27-3
supplier–consumer, 34-18
Two-Level Actor Machine, 38-17
universal naming, 38-7
Web, 53-1, 53-2
Modeling techniques, 33-3
MOF, *see* Meta Object Facility
MOM, *see* Message-oriented middleware
MOS, *see* Mean Opinion Score
Movie trailers, 41-8
Moving Pictures Expert Group (MPEG), 12-15
MP3, 6-9
MPEG, *see* Moving Pictures Expert Group
MPEG-7, 12-15
MPI, *see* Message Passing Interface
MRTG, *see* Multi Router Traffic Grabber
MSA, *see* Message submit agent
MSO, *see* Multisystem operator
MTA, *see* Message transfer agent
MTU, *see* Maximum transmission unit
MUA, *see* Mail user agent
MUDs, *see* Multi-User Dungeons
Multiagent systems for Internet applications, 17-1 to 17-19
agent implementations of Web services, 17-6
benefits of approach based on multiagent systems, 17-3
brief history of multiagent systems, 17-3 to 17-4
building Web-service agents, 17-6 to 17-17
agent communication languages, 17-14 to 17-15
agent types, 17-6 to 17-14
cooperation, 17-17
knowledge and ontologies for agents, 17-15 to 17-16
reasoning systems, 17-16 to 17-17
composing cooperative Web services, 17-17
infrastructure and context for Web-based agents, 17-4 to 17-5
directory services, 17-5
semantic Web, 17-4
standards and protocols, 17-4 to 17-5
Multihoming, 51-4
Multilevel secure database management systems (MLS/DBMSs), 26-5
Multimedia content description interface, 12-15
Multimedia Messaging Service (MMS), 7-12
Multipoint control units (MCUs), 6-13
Multipurpose Internet Mail Extensions (MIME), 7-4
alternative subtype, 7-5
header, 7-5
mixed subtype, 7-5
multipart, 7-5
parallel subtype, 7-5
Multi Router Traffic Grabber (MRTG), 49-6
Multistrategy filtering systems, 26-16
Multisystem operator (MSO), 6-7
Multi-User Dungeons (MUDs), 11-4
MX, *see* Mail exchange records
Myrinet, 39-11

N

NADF, *see* North American Directory Forum
Naming service, 38-3
Napster, 35-5, 35-17, 35-19, 51-7, 56-14
National Institute of Standards and Technologies (NIST), 8-16
National Science Foundation (NSF), 31-6, 56-2
National Telecommunications and Information Administration (NTIA), 56-2, 56-4
NATs, *see* Network address translators
Natural language
components, conversational agent, 10-5
processing (NLP), 10-3, 10-8, 14-3
software, 10-2
Navigational pattern mining, 15-3
Negative feedback, 12-11
Neighborhood, 35-3
.NET architecture, 22-18
NetFlow, 49-11
Netnews (network news), 7-5
articles, 7-7
at @ symbol, 7-7
center, 7-5
cross posting, 7-6
header, 7-6
hierarchical naming, 7-7

naming, 7-6
Net Perceptions, 15-2
Netscape, 34-7, 40-7
Network(s)
address translators (NATs), 6-20
-based concurrent computing, 39-1
COTS, 39-2
effect, 33-7
Freenet, 35-11
latency, 10-7, 11-7
maintenance, 35-7
management, 46-2, 46-7
metrics, 49-3
news, *see* Netnews
overlay, 35-5
peer-to-peer, 13-15, 51-2
small world, 53-2
technology, 37-4
Network Dispatcher, 40-4
Networked organization/enterprise, 22-2, 22-3
Network File Systems, 34-3
Networking layer, 35-4
Network News Transfer Protocol (NNTP), 7-6
Network and service management, 52-1 to 52-28
broadband access networks, 52-18 to 52-20
cable access network management, 52-19
digital subscriber line management, 52-19 to 52-20
fixed wireless, 52-20
wireless LAN, 52-20
customer premises network management, 52-21 to 52-23
enterprise network and service management, 52-15 to 52-17
account management, 52-17
configuration management, 52-16
fault management, 52-15 to 52-16
OSS/network management system, 52-17
performance management, 52-16
security management, 52-16 to 52-17
example of managed network, 52-3 to 52-4
future trends, 52-20 to 52-21
integrated service management, 52-23
management standards, 52-2 to 52-3
mobile and enterprise wireless management, 52-21
QoS management, 52-21
RMON, 52-15
SNMP, 52-5 to 52-6
SNMPv1, 52-7 to 52-13
architecture, 52-8 to 52-10
communication, 52-13
information, 52-10 to 52-13
organization, 52-7 to 52-8
SNMPv2, 52-13 to 52-14
SNMPv3, 52-14 to 52-15
virtual private network, 52-17 to 52-18
Network Time Protocol (NTP), 49-13
Network Voice Control Protocol (NVCP), 6-18
Network Voice Protocol (NVP), 6-18
News feed, 7-6
Newsgroup, 7-6
News Transfer Agent (NTA), 7-6
News User Agent (NUA), 7-6
n-grams, 14-8
Nintendo Gamecube, 11-5
NIST, *see* National Institute of Standards and Technologies
NLP, *see* Natural language processing
NNTP, *see* Network News Transfer Protocol
Node connectivity, power-law distribution of, 35-6
Nonexhaustive indexing, 12-3
Nonfunctional parameters, 31-7
Nortel Alteon ACEdirector, 40-4
North American Directory Forum (NADF), 34-10
NOS directories, 34-3
Notifications, 5-19
NSF, *see* National Science Foundation
NSFNET, 56-2
NTA, *see* News Transfer Agent
NTIA, *see* National Telecommunications and Information Administration
n-tier architecture, 22-3, 22-5
NTP, *see* Network Time Protocol
NUA, *see* News User Agent
NVCP, *see* Network Voice Control Protocol
NVP, *see* Network Voice Protocol

O

OAG, *see* Open Applications Group
OAG Integration Specification (OAGIS), 8-4, 8-5, 8-17, 33-9
OAGIS, *see* OAG Integration Specification
Object
classes, 34-12, 34-13, 34-14
conceptual, 8-2
enterprise, 8-2
identifiers (OIDs), 28-7, 28-10
Objective terms, 12-2
Object Management Group (OMG), 8-7, 22-6, 22-10, 22-11
Object-oriented (OO) abstractions, 28-2
Object-oriented modeling, 33-13
Obsolescence, 33-10
OGSA, *see* Open Grid Service Architecture
OGSI, *see* Open Grid Services Infrastructure
OIDs, *see* Object identifiers
OLAP, *see* Online Analytical Processing
OMG, *see* Object Management Group
One-way delay, 49-3
Online Analytical Processing (OLAP), 15-13
Online marketplaces, 19-1 to 19-13
auctions, 19-4 to 19-9
complex, 19-8 to 19-9
configuration and market design, 19-6 to 19-8
types, 19-5 to 19-6
definition of online marketplace, 19-1 to 19-2
establishment of marketplace, 19-9 to 19-11
achieving critical mass, 19-10 to 19-11
technical issues, 19-9 to 19-10
future of, 19-11 to 19
market services, 19-2 to 19-4
discovery services, 19-3 to 19-4

transaction services, 19-4
Online reputation mechanisms, 20-1 to 20-18
ancient concept in new setting, 20-3 to 20-5
example, 20-5 to 20-8
new opportunities and challenges, 20-12 to 20-15
coping with easy name changes, 20-14
eliciting sufficient and honest feedback, 20-12 to 20-13
exploiting information processing capabilities of feedback mediators, 20-13
exploring alternative architectures, 20-14 to 20-15
understanding impact of scalability, 20-12
reputation in game theory and economics, 20-8 to 20-11
basic concepts, 20-8 to 20-9
reputation dynamics, 20-9 to 20-11
Ontology(ies), 1-8, 2-10, 31-3, 31-6
competency, 3-16
description languages, 9-11
modeling, 9-13
OntoWeb, 31-3
OO abstractions, *see* Object-oriented abstractions
Open Applications Group (OAG), 8-4
Open Corpus Adaptive Hypermedia, 1-8
Open Development Process, 8-15
Open Directory Project, 12-13
Open Grid Service Architecture (OGSA), 39-12
Open Grid Services Infrastructure (OGSI), 39-12
Open Service Gateway initiative (OSGi), 37-12
Open source, 33-11
Open Source Definition (OSD), 57-5
Open Source Software, copyright protection of, 57-5
Open standards, 33-7
Open Systems Interconnection (OSI), 23-1, 24-7, 34-4, 36-3, 50-13
Operational support system (OSS), 6-8
Oracle, 25-17, 31-8
Organization, naming entries within, 34-11
OSD, *see* Open Source Definition
OSGi, *see* Open Service Gateway initiative
OSI, *see* Open Systems Interconnection
OSS, *see* Operational support system
Outsourcing, 33-10
Overlay networks, 35-5, 51-1 to 51-13
applications, 51-7 to 51-9
incremental deployment of services, 51-8 to 51-9
shared use, 51-8
VPNs/security, 51-8
background, 51-2 to 51-3
definitions, 51-2
issues, 51-3
challenges, 51-9 to 51-11
deployment and management, 51-11
security, 51-9 to 51-10
support, 51-10 to 51-11
complex structures, 51-3 to 51-5
recommendations, 51-11
tunneling, 51-3
variations, 51-5 to 51-7
automated overlays, 51-5
manual overlays, 51-5
peer-to-peer, 51-6 to 51-7
provider-provisioned vs. end-to-end, 51-6
VPN, 51-6
OWL, *see* Web Ontology Language
Owner replication, 35-18

P

Packet
delay variation, 49-3
encoding rules, 6-13
loss concealment (PLC), 6-9
sampling, 49-7
statistics, 49-2, 49-13
PACT, *see* Profile Aggregation Based on Clustering Transactions
PageRank, 12-14, 13-11, 53-10, 53-11, 53-13
Page ranking
global ranking, 13-11
hubs and authorities, 13-10
local ranking, 13-10
Pageview-feature matrix, 15-20
Paging, 7-11
Paid-placement, 57-11
Pairing procedure, 36-12
PAPI, *see* Personal and Private Information
Parallel Virtual Machine (PVM), 39-2
Participatory Design, 33-6
Partner Interface Processes (PIPs), 8-5
Partner role, 8-13
Part-of-speech (POS) tagging, 10-9
Party, 8-13
Patent Cooperation Treaty (PCT), 57-7
Patents, 57-7
design, 57-8
plant, 57-8
Path
-decoupled signaling, 6-15
replication, 35-18
Pathnodes, 11-11
PC, *see* Personal computer
PCIM, 46-7
PCT, *see* Patent Cooperation Treaty
PDAs, 36-2, 37-3
PDF, *see* Portable Document Format
PDL, *see* Policy Definition Language
PDUs, *see* Protocol data units
Peak-to-mean ratio, 49-14
Peer actors, 38-7
Peer-to-peer infrastructures, 1-9
Peer-to-peer interaction model, 27-3
Peer-to-peer networks (P2Ps), 13-15, 51-2
Peer-to-peer systems, 35-1 to 35-22
comparative evaluation, 35-16 to 35-20
functional and qualitative properties, 35-17 to 35-20
performance, 35-16 to 35-17
fundamental concepts, 35-3 to 35-7
classification of P2P systems, 35-4 to 35-5
emergent phenomena in P2P systems, 35-5 to 35-7

principles of P2P architecture, 35-3 to 35-4
resource location, 35-7 to 35-16
hierarchical P2P systems, 35-10
properties and categories, 35-7 to 35-8
structured P2P systems, 35-10 to 35-16
unstructured P2P systems, 35-8 to 35-9
PEP, *see* Policy execution point
Periodic polling, 41-5
Persistence service, 38-3, 38-7
Persistent worlds, 11-4
Personal computer (PC), 56-3
Personal information managers (PIM), 1-7
Personalization, concepts and practice of, 18-1 to 18-14
discussion, 18-11 to 18-14
advice, 18-11 to 18-12
futures, 18-13 to 18-14
metrics, 18-12 to 18-13
key applications and historical development, 18-3 to 18-8
desktop applications such as e-mail, 18-3 to 18-4
knowledge management, 18-6 to 18-7
mobile applications, 18-7 to 18-8
Web applications such as e-commerce, 18-4 to 18-6
key concepts, 18-8 to 18-11
individual vs. collaborative, 18-9
representation and reasoning, 18-9 to 18-11
motivation, 18-1 to 18-3
Personal and Private Information (PAPI), 1-10
Pervasive computing, 37-1 to 37-14
technologies, 37-2 to 37-7
device technology, 37-3
environment technology, 37-4 to 37-5
information access technology, 37-6 to 37-7
network technology, 37-4
software technology, 37-5 to 37-6
user interface technology, 37-7
ubiquitous computing systems, 37-7 to 37-12
Active Bat, 37-8 to 37-9
Classroom 2000, 37-9
Lancaster Guide system, 37-9 to 37-11
Matilda's Smart House, 37-11 to 37-12
vision of, 37-1 to 37-2
P-Grid, 35-14, 35-20
Phrase indexing, 12-5
Piconet, 37-4
PIM, *see* Personal information managers
Ping, 49-3
PIPs, *see* Partner Interface Processes
PKI, *see* Public key infrastructures
Plain old telephone service (POTS), 6-2
Planning
agent, 16-3
schedule, 8-14
Plant and equipment resources, 8-7
Plant patents, 57-8
PLATO, 11-4
PLC, *see* Packet loss concealment
Points of presence (POPs), 56-8, 56-9
Policy
execution point (PEP), 46-9
-neutral models, 26-6
propagation, 26-8
Policy Definition Language (PDL), 46-7
Polling, 41-5
Polymorphic virus, 7-16
Ponder, 46-3, 46-11
Ponzi scheme, 7-15
POPs, *see* Points of presence
Portable computers, phenomenal growth of, 36-1
Portable Document Format (PDF), 24-3
Portal Expedition, 39-14
portType, 27-11
Positive feedback, 12-11
POS tagging, *see* Part-of-speech tagging
Postconditions, 31-6
Post Office Protocol, 7-3
Post Schema Validation Infoset (PSVI), 23-9, 23-13
POTS, *see* Plain old telephone service
Power-law distribution, 35-6, 53-6
PPM, *see* Privacy Profile Manager
P2Ps, *see* Peer-to-peer networks
PPVPN, *see* Provider-provisioned VPN
Preconditions, 31-6
Preferred mail server, 7-4
Prefix routing, 35-12, 35-13
Presence
availability and, 5-22
information, 5-10, 7-9
Preserving privacy, WebDG, 2-13
PRF, *see* Pseudo-Relevance Feedback
Price list, 8-13
Privacy, 2-8, 2-13, 33-11, 33-14
Privacy Profile Manager (PPM), 2-13
Probabilistic retrieval models, 12-8
Problem domain, 33-5
Process
-based servers, 40-10
control, 8-7, 31-11
mediation, 31-6
ontology, 31-11
redesign, 33-9
specifications, 8-7
tracking, 32-3
Procurement, 8-3, 8-10
Prodigy, 11-4
Product
availability, 8-13
data management, 8-4
release, 8-14
specification, 8-6
Profile Aggregation Based on Clustering Transactions (PACT), 15-22,15-27
Programming grids, 39-10
Project charter, 3-6
Promotion, 34-19
Proprietary technologies, 33-7
Protection, boundary
Protocol data units (PDUs), 34-23
Provider-provisioned VPN (PPVPN), 51-6
Proxy
caching, 57-3

server, 42-7, 42-8
types, 28-8
Pseudo-code, 14-15
Pseudo-Relevance Feedback (PRF), 12-9
PSTN, *see* Public switched telephone network
PSVI, *see* Post Schema Validation Infoset
Public key certificates, 34-10
Public key infrastructures (PKI), 35-19
Public switched telephone network (PSTN), 6-2
Publish-and-subscribe, 38-3
Purchase request, 8-7, 8-10
PVM, *see* Parallel Virtual Machine
Pyramid game, 7-15

Q

Q-Bone, 51-5
QoS, *see* Quality of service
QPIM, 46-8
Quality-aware transcoding, 42-7
Quality of service (QoS), 27-4, 27-7, 27-13, 51-5
Query(ies), 1-9, 1-10
expansion, 12-9
service, 38-3
splitting, 12-12
Quipu, 34-6
Quote, 8-13

R

Random replication, 35-18
Random walker, 35-9
Rapid Knowledge Formation (RKF), 31-3
RAS, *see* Remote access server
RBAC, *see* Role-Based Access Control
RDBMS, *see* Relational Database Management System
RDF, *see* Resource Description Framework
Realtime Traffic Flow Measurement (RTFM), 49-9
Real-Time Transport Protocol (RTP), 6-10
Reasoning, 16-2, 16-3, 16-4
Reference resolution, 10-5
Refinement, 46-17
Reflection, 38-5, 38-11
Regular Language description for XML (RELAX), 23-14
Relational Database Management System (RDBMS), 24-7, 24-8
Relational databases, semistructured data in, 25-1 to 25-19
relational schemas for semistructured data, 25-4 to 25-13
graph representation, 25-10 to 25-12
representing ancestors of TGEs, 25-5 to 25-8
representing deep structure, 25-4 to 25-5
semistructured components, 25-9 to 25-10
storing unsparsed XML, 25-12 to 25-13
using tuple-generating elements, 25-4
varying components, 25-8 to 25-9
running example, 25-3 to 25-4
sources of semistructured data, 25-2
using XML features of relational database systems, 25-13 to 25-19
IBM DB2, 25-13 to 25-16
Microsoft SQL Server, 25-16 to 25-17
Oracle XSU, 25-17 to 25-18
Sybase, 25-18 to 25-19
RELAX, *see* Regular Language description for XML
RELAX NG, 23-11, 23-14
Relevance
feedback techniques, 12-11
ranking, 14-5
Remote access server (RAS), 51-5
Remote Message Sending Protocol (RMSP), 38-10
Remote operation service element (ROSE), 34-5, 34-22
Remote Procedure Call (RPC), 22-3, 22-4, 27-6, 38-2
-based query language, 34-3
implementation of, 31-2
Replication, 34-2, 34-7, 34-18, 35-18, 38-8
Repository technologies, 33-3
Reputation management, 35-19
Request methods, 40-13
Request for quote, 8-13
Requirements engineering, 33-14
Residential gateway (RG), 6-8
Resilient Overlay Networks (RONs), 51-9
Resource
-abstraction, 39-7
broker, 32-6, 32-9
model, 32-6
reservation, 51-4
sharing, 35-2
virtualization, 39-5
Resource Description Framework (RDF), 1-9, 3-14, 9-3, 9-5
Resource Specification Language (RSL), 39-7
Response codes, 40-14
Response Generator, 10-6
Retrieval
engine, 13-2, 13-7
models, 12-6
status value (RSV), 12-7
Return on investment (ROI), 50-5
Reverse domain name hijacking, 57-7
Reverse proxy, 41-1, 42-7
Reward structures, 33-6
RG, *see* Residential gateway
RIS, *see* Routing Information Service
RKF, *see* Rapid Knowledge Formation
RMSP, *see* Remote Message Sending Protocol
RNIF, *see* RosettaNet Implementation Framework
Robot exclusion (or robots.txt), 13-4, 13-5
Robust annotations, 5-18
ROI, *see* Return on investment
Role-Based Access Control (RBAC), 26-5, 46-3
Role-playing games (RPGs), 11-2
RONs, *see* Resilient Overlay Networks
ROSE, *see* Remote operation service element
Rose, Marshall, 34-4
Rosettanet, 33-9
RosettaNet, 8-5
RosettaNet Implementation Framework (RNIF), 8-4
Round Robin, 40-3
RouteViews, 49-12

Routing
content-based, 40-5
dynamic, 51-8
information, 8-14
protocol
Ad hoc On Demand Distance Vector, 36-7
Destination-Sequenced Distance Vector, 36-7
proxy, 34-6
request-, 41-10, 41-12
table, 35-11
Routing Information Service (RIS), 49-12
RPC, *see* Remote Procedure Call
RPGs, *see* Role-playing games
RSL, *see* Resource Specification Language
RSV, *see* Retrieval status value
RTFM, *see* Realtime Traffic Flow Measurement
RTP, *see* Real-Time Transport Protocol
Rule(s)
association, 15-14
-based tree, 12-10
Event-Condition-Action, 32-5, 32-8
learning, 14-14
translator, 28-16

S

SADT, 33-3
Sales, 8-13, 10-3
SALSA, *see* Simple Actor Language System and Architecture
SAML, *see* Security Assertions Markup Language
SASL, *see* Simple Authentication and Security Layer
SAX, *see* Simple API for XML
Scalability, e-government, 2-6
Schema mapping rules, 28-4
Schematron, 8-5, 8-15, 23-8, 23-16
Scope notes, 12-3
SCORM, *see* Sharable Content Object Reference Model
SDD, *see* Service data description
Seamless game world, 11-9
Search
latency, 35-16
mechanism, 35-12
predicates, 35-17
Search engines, 57-11
crawler, 13-2
future directions, 13-2
indexer, 13-7
metasearch, 13-14
page ranking, 13-11
retrieval engine, 13-12
user interface, 13-13
Search for Extra Terrestrial Intelligence (SETI) project, 38-14
Secure Overlay Services (SOSs), 51-9
Secure Sockets Layer (SSL), 39-8, 39-9, 40-4, 40-13
Security
breaches, categories of, 26-1
digital government, 2-7
management, 46-2, 46-3, 46-6
policy, *see* System management and security policy specification
Security Assertions Markup Language (SAML), 27-7
Sega Dreamcast, 11-5
SEKT, 31-3
Selection, WebDG, 2-12, 2-13
Self-organization, 35-3, 35-4, 35-5
Self similarity, 49-14
Semantic associations, 9-8
Semantic definition, 8-8
Semantic enterprise content management, 9-1 to 9-21
applying semantics in ECM, 9-13 to 9-18
knowledge discovery, 9-16 to 9-18
semantic metadata annotation, 9-15
semantic metadata extraction, 9-14
semantic querying, 9-15 to 9-16
toolkits, 9-13 to 9-14
core components, 9-6 to 9-13
classification, 9-6
metadata, 9-6 to 9-9
ontologies, 9-9 to 9-13
primary challenges, 9-3 to 9-4
data size and relevance factor, 9-4
distribution of data sources, 9-3 to 9-4
heterogeneous data sources, 9-3
rise of semantics, 9-4 to 9-6
enabling interoperability, 9-4 to 9-5
semantic Web, 9-5 to 9-6
Semantic metadata, 9-7
Semantic querying, 9-15
Semantic Web, 1-8, 2-3, 3-1, 3-14, 8-15, 9-5, 13-6, 33-9
Semantic Web Services, 31-2, 31-3, 31-4
Sensor, agent, 16-3
Sequential patterns (SPs), 15-15, 15-20
Servent, 35-3
Server(s)
availability, 41-13
-based invalidation, 41-5
-driven content negotiation, 42-9
event-driven, 40-10
in-kernel, 40-11
performance, 40-9
process-based, 40-10
proxy, 42-7, 42-8
thread-based, 40-10
workload, 40-12
Service
composition, 2-7, 2-12
data description (SDD), 39-13
layer, 35-4
level agreements (SLA), 32-13
management, *see* Network and service management
matching mechanism, 36-7
model, 31-11
profile, 31-11
transaction management, 36-11
Service Discovery Protocol, 36-7
Service Grounding, 31-11, 31-12
Service Marks, 57-6
Service Oriented Computing (SOC), 27-2, 27-3
Service Provisioning Markup Language (SPML), 34-9

Session Description Protocol, 6-11, 6-14
Session Initiation Protocol (SIP), 6-14
  request methods, 6-16
  user agents, 6-15
SETI project, *see* Search for Extra Terrestrial Intelligence project
SGML, *see* Standard Generalized Markup Language
Sharable Content Object Reference Model (SCORM), 3-7
Shard, 11-8
Shipping information, 8-14
Ships in the night (SITN), 51-8
Shockwave, 11-5
Shopbots, 33-9
Shopping cart, 8-13
Short message service (SMS), 7-1, 7-11
  address prefixing, 7-12
  center, 7-12
  cross messaging, 7-12
  keywording, 7-12
  naming, 7-12
Signaling
  conversion, 6-8
  path-decoupled, 6-15
Signaling System number Seven (SS7), 7-12
Signature policies, 26-12, 26-13
Sign duty, 26-14
Significance filtering, 15-8
Simple Actor Language System and Architecture (SALSA), 38-6, 38-9, 38-17
Simple API for XML (SAX), 23-9, 23-10, 24-4
Simple Authentication and Security Layer (SASL), 34-25
Simple Mail Transfer Protocol (SMTP), 7-2, 24-7
Simple Object Access Protocol (SOAP), 22-6, 22-7, 24-7, 31-4, 31-12, 34-8
  -over-HTTP, 27-8, 27-12
  remote procedure calls using, 27-6
Simple process, 31-12
Single-term indexing, 12-3
SIP, *see* Session Initiation Protocol
SITN, *see* Ships in the night
Skewed data distributions, 35-15
Skitter, 49-4, 49-12
SLA, *see* Service level agreements
Slashdot effect, 42-3
Small-world graphs, 35-6
Small world networks, 53-2
SMS, *see* Short message service
SMTP, *see* Simple Mail Transfer Protocol
SMTP Mail Transfer Agent, 6-15
SNA, *see* Systems Network Architecture
SOAP, *see* Simple Object Access Protocol
SOC, *see* Service Oriented Computing
Social agents, 16-3
Social dynamics, 33-10
Social relationships, 33-6
Social Structures, 33-4
Social and Welfare Services, 2-5
Softswitch, 6-7
Soft Systems Methodology, 33-6
Software
  agents, 33-10
  copyright protection of, 57-4
  crisis, 33-4
  process improvement, 33-4
Software engineering for Internet applications, 43-1 to 43-13
  agile methods, 43-10 to 43-11
    collaboration, 43-10
    high availability, 43-11
    requirements volatility, 43-10
    security and privacy, 43-10 to 43-11
    time-to-market pressure, 43-10
  agile software development models, 43-4 to 43-6
    emergence, 43-5 to 43-6
    empirical process control, 43-5
    self-organization, 43-6
  nature of Internet application development, 43-1 to 43-2
  research issues, 43-11 to 43-12
  survey of representative agile methodologies, 43-7 to 43-10
    crystal methods, 43-9 to 43-10
    Extreme Programming, 43-7 to 43-9
    Scrum, 43-9
  traditional software development models, 43-2 to 43-4
SONET, 6-4
Sony Playstation 11-5
SOS, *see* Secure Overlay Services
Spam, 7-15, 13-6
Specification, WebDG, 2-12
Spider, *see* Crawler
Split-and-merge service, 38-8
SPML, *see* Service Provisioning Markup Language
SPs, *see* Sequential patterns
SQL, *see* Structured Query Language
Squid (web proxy cache), 51-11
SS7, see, Signaling System number Seven
SSL, *see* Secure Sockets Layer
SSRC, *see* Synchronization source identifier
Standard(s)
  content, 8-2
  de facto, 50-5
  de jure, 50-5
  e-learning, 3-7
  horizontal, 8-3
  vertical, 8-3
Standard Generalized Markup Language (SGML), 23-2
Standardization
  digital library, 4-2, 4-7
  e-government, 2-8
  Internet telephony, 6-4
  user profile, 1-10
Standardized metadata, 1-9
Standards for Technology in Automotive Retail (STAR), 8-5
STAR, *see* Standards for Technology in Automotive Retail
Stopword list, 12-3
Strong decoupling, 31-6
Strong mediation, 31-6
Structural metadata, 9-7
Structured P2P systems, 35-8, 35-10
Structured Query Language (SQL), 24-8, 25-8

Subnetting, 50-14
Subsystem
attentional, 16-4
behavioral, 16-3
perceptual, 16-4
Sun Microsystems, 31-8
Supercomputers, 39-11
Supernetting, 50-14
Super-peer architectures, 35-10
Supervised learning, 14-7
Supply chain
integration, 8-3
management, 33-9
partner, 8-2
planning, 8-3
SWIFT wire transfer, 32-2
SWWS, 31-3
Sybase Adaptive Server Enterprise, 25-18
Symmetric P2P systems, 35-2
Symmetry of roles, system architecture, 35-3
Synchronization source identifier (SSRC), 6-10
Synchronizing clocks, 49-13
Syntactic analysis, 14-3
Syntactic metadata, 9-7
System
development life cycle, 33-2
management plane, 36-12
System management and security policy specification, 46-1 to 46-20
management policy specification, 46-7 to 46-11
CIM policy model, 46-7 to 46-10
Lucent Policy Definition Language, 46-7
other approaches, 46-10 to 46-11
Ponder, 46-11 to 46-16
composite policies, 46-13 to 46-16
domains, 46-12
primitive policies, 46-12 to 46-13
research issues, 46-46 to 46-17
conflict analysis, 46-16 to 46-17
multiple levels of policy, 46-17
refinement, 46-17
security policy specification, 46-3 to 46-7
IBM Trust Policy Language, 46-5 to 46-6
other approaches, 46-6 to 46-7
role-based access control, 46-3 to 46-5
Systems Network Architecture (SNA), 50-5, 50-13

T

Tablet PCs, 36-2
Targeted Immutable Short Message (TISM), 7-16
TASI, *see* Time-Assigned Speech Interpolation
Tax filing, 2-4
Taxonomy for messaging, 7-12, 7-13
address dimension, 7-13
audience dimension, 7-13
direction dimension, 7-13
time dimension, 7-13
TCP
connection establishment, 41-12
splicing, 40-7
Tcpdump, 49-6
TCP/IP, *see* Transmission Control Protocol/Internet Protocol
TDMA, *see* Time Division Multiple Access
Teams, multiagent, 16-8
Teaser games, 11-5
Technology, human implications of, 55-1 to 55-21
conceptual approaches to building trustworthy systems, 55-6 to 55-15
data protection vs. privacy, 55-14 to 55-15
human-centered trusted systems design, 55-9 to 55-12
identity examples, 55-13 to 55-14
next generation secure computing platform, 55-7 to 55-9
trusted computing base, 55-6 to 55-7
human implications of security debates, 55-5 to 55-6
network protocols as social systems, 55-15 to 55-16
open vs. closed code, 55-16 to 55-17
overview, 55-1 to 55-2
social determinism, 55-3 to 55-4
technological determinism, 55-2 to 55-3
values in design, 55-4 to 55-5
Telecom messaging, 7-2, 7-11
in-band data, 7-11
naming, 7-11
out-of-band data, 7-11
paging, 7-11
Telecommunications
infrastructure providers, 56-12
service, deployment of advanced, 56-11
Telecommunications Act (1996), 56-7
Tele-Game, 11-5
Telephone number mapping, 6-17
Telephony, 50-8, 50-9, 50-14, 56-14
Telephony Routing over IP protocol (TRIP), 6-18
Telepresence, 5-2
Term
discrimination value, 12-4
frequency, 14-18
grouping, 12-6
set, document, 12-3
specificity, 12-3
-talk, 7-7
Text
categorization, 14-6, 14-9
summarization, 14-4, 14-5
Text mining, 14-1 to 14-22
data mining and, 14-2 to 14-3
human text mining, 14-16 to 14-17
mining of plain text, 14-4 to 14-14
assessing document similarity, 14-6 to 14-7
extracting information for human consumption, 14-4 to 14-6
extracting structured information, 14-9 to 14-14
language identification, 14-8 to 14-9
mining of structured text, 14-14 to 14-16
natural language processing and, 14-3
techniques and tools, 14-17 to 14-19
token identification, 14-17 to 14-19
training vs. knowledge engineering, 14-17

Text REtrieval Conference (TREC), 12-18
TGE, *see* Tuple-generating element
Thick client, 22-4
Thread-based servers, 40-10
Threaded articles, 5-16
Threaded chat, 5-14
Threaded e-mail, 5-14
3-tier architecture, 22-5, 22-6
Threshold accounting, 49-15
Tightly coupled P2P systems, 35-8
Tim Berners-Lee, 31-2
Time-Assigned Speech Interpolation (TASI), 6-18
Time Division Multiple Access (TDMA), 36-3
Time-to-market, 33-10
Timeshared virtual machine, 39-11
Timestamp accuracy, 49-13
TISM, *see* Targeted Immutable Short Message
TLAM model, *see* Two-Level Actor Machine model
tModel, 27-14
Token identification, 14-17
Token ring, 50-14
Topological routing, 35-15
Traceroute, 49-4
Trademarks, 57-6
Trading network integration, 8-3
Traffic characterization, 49-7, 49-11
Traffic light metaphor, 1-5
Training, knowledge engineering vs., 14-17
Transactional service, 38-3
Transaction–pageview matrix, 15-11
Transcoding, *see* Content adaptation and transcoding
Transfer sizes, 40-16, 40-17
Transformation API for XML (TrAX), 24-1
Translator rules, 28-16
Transmission Control Protocol/Internet Protocol (TCP/IP), 33-7, 50-13, 56-2, 56-14
Transparent content negotiation, 42-9
Transport and encoding, 27-4, 27-5
Transport service, 38-7
TrAX, *see* Transformation API for XML
TREC, *see* Text REtrieval Conference
Tree Regular Expressions for XML (TREX), 23-14
TREX, *see* Tree Regular Expressions for XML
TRIP, *see* Telephony Routing over IP protocol
Trojan, 7-15
Trust, 2-9, 53-14
Trust, distributed, 47-1 to 47-16
  access control and trust management, 47-1 to 47-2
  applications of trust management systems, 47-10 to 47-13
    active networking, 47-12 to 47-13
    distributed firewalls and STRONGMAN architecture, 47-11
    grid computing and transferable micropayments, 47-12
    micropayments, 47-12
    network-layer access control, 47-10 to 47-11
  distributed trust management, 47-3 to 47-10
    KeyNote, 47-8 to 47-10
    PolicyMaker, 47-5 to 47-8
  other trust-based systems, 47-13
  technical foundations, 47-2 to 47-3
    authentication, 47-2 to 47-3
    public key certificates, 47-3
Trust Policy Language, 46-3
Tunneling, two-layer, 51-4
Tuple-generating element (TGE), 25-4, 25-5
Turbine Games, 11-10
Two-Level Actor Machine (TLAM) model, 38-17

U

UAL, *see* Universal Actor Locator
UANP, *see* Universal Actor Naming Protocol
Ubiquitous computing, 36-2, 37-1, 37-7
UBR, *see* UDDI Business Registry
UCC, *see* Uniform Code Council
UCE, *see* Unsolicited Commercial E-mail
UDDI, *see* Universal Description, Discovery and Integration
UDDI Business Registry (UBR), 27-14
UDFs, *see* User-defined functions
UDP encapsulation, 51-5, 51-9
UDRP, *see* Uniform Domain Name Dispute Resolution Policy
UIs, *see* User interfaces
Ultima Online, 11-4
UM, *see* Unified Messaging
UML, *see* Unified Modeling Language
Unbind request, 34-26
UN/CEFACT, 8-4, 8-15, 33-9
Unified Messaging (UM), 7-16
Unified Modeling Language (UML), 22-10
Uniform Code Council (UCC), 8-5
Uniform Domain Name Dispute Resolution Policy (UDRP), 57-7
Uniform Resource Identifiers (URI), 13-2, 15-7, 38-8
Universal Actor Locator (UAL), 38-8
Universal Actor Naming Protocol (UANP), 38-9
Universal Description, Discovery and Integration (UDDI), 27-1, 22-8, 31-4, 33-10, 34-8, 36-7
Universal naming model, 38-7
Universal service fund (USF), 6-20
Unreal Championship, 11-2
Unsolicited Commercial E-mail (UCE), 7-15
Unsolicited messages, 7-14
  acquaintance spam, 7-15
  coercion, 7-15
  hoax, 7-15
  impersonation, 7-15
  malware, 7-15
  protection mechanisms, 7-15
  self-replication, 7-15
  signatures of, 7-16
  spam, 7-15
  spreading viruses, 7-14
  statistical comparison, 7-16
  Trojan, 7-15
  viruses, 7-15

worms, 7-15
Unstructured P2P systems, 35-8
URI, *see* Uniform Resource Identifiers
Use Case Diagrams, 3-6
USENET news, 7-5
User
-abstraction, 39-7
-defined functions (UDFs), 28-5
interfaces (UIs), 2-9
layer, 35-4
-oriented clustering, 12-12
profiles, 1-10
virtualization, 39-5
USF, *see* Universal service fund
U.S. Patent and Trademark Office (USPTO), 57-9
USPTO, *see* U.S. Patent and Trademark Office
Utility patents, 57-7

V

VAD, *see* Voice activity detection
Var files, 42-10
Vector space retrieval model, 12-7
Videoconferencing, 5-1, 5-5
Virtual enterprise, 33-8
Virtual machines, 38-4
Virtual memory (VM), 51-5
Virtual private network (VPN), 51-1, 51-2
definition of, 51-6
identifier (VPN ID), 51-4, 51-5
provider-provisioned, 51-6
security, 51-8
VM, *see* Virtual memory
VMware, 51-5
VoATM, *see* Voice-over-ATM
VoDSL, *see* Voice-over-DSL
Voice activity detection (VAD), 6-9
Voice-over-ATM (VoATM), 6-2
Voice-over-DSL (VoDSL), 6-2
Voice-over-IP (VoIP), 50-18
VoIP, *see* Voice-over-IP
Volume leases, 41-6
VPN, *see* Virtual private network
VPN ID, *see* Virtual private network identifier
Vservers, 51-10

W

Wahl, Mark, 34-7
WAIS, 34-2
WAP, *see* Wireless Application Protocol
W3C, World Wide Web Consortium
WCDP, *see* Web content distribution protocol
WCIP, *see* Web cache invalidation protocol
Web
-based self-service applications, 10-2
cache invalidation protocol (WCIP), 41-5
communities, 53-11, 53-14
content adaptation service architecture, 42-7
content distribution protocol (WCDP), 41-5
graph, 12-13
logging (blogging), 7-10 to 7-11
proxy cache, 5-11
servers, 40-1
service(s), 2-2, 3-11, 3-14, 22-7
transfer sizes, 40-16, 40-17
usage data, sequential patterns in, 15-15
Web caching, consistency, and content distribution, 41-1 to 41-17
cache consistency, 41-3 to 41-7
consistency mechanisms, 41-5 to 41-6
degrees of consistency, 41-4
invalidates and updates, 41-6 to 41-7
CDNs, 41-7 to 41-15
architectural elements, 41-8 to 41-10
consistency management, 41-14 to 41-15
performance studies, 41-15
request-routing, 41-10 to 41-12
request-routing metrics and mechanisms, 41-12 to 41-14
practical issues in design of caches, 41-2 to 41-3
Web crawling and search, 13-1 to 13-19
essential concepts and well-known approaches, 13-2 to 13-14
crawler, 13-2 to 13-7
databases, 13-12
improving search engines, 13-13 to 13-14
indexer, 13-7 to 13-9
relevance ranking, 13-9 to 13-11
retrieval engine, 13-12 to 13-13
research activities and future directions, 13-14 to 13-17
clustering and categorization of pages, 13-17
detecting duplicated pages, 13-16 to 13-17
searching dynamic pages, 13-14 to 13-15
semantic Web, 13-16
spam deterrence, 13-17
utilizing peer-to-peer networks, 13-15 to 13-16
WebDG, 2-2, 2-4, 2-9, 2-13
Web Ontology Language (OWL), 31-3, 31-11, 36-8
Web semantics, introduction to, 29-1 to 29-13
background and rationale, 29-3 to 29-4
creating information on Web, 29-3 to 29-4
sharing information over Web, 29-4
understanding information on Web, 29-3
discussion, 29-10 to 29-12
domain-specific ontologies, 29-11
methodologies and tools, 29-11 to 29-12
semantic Web services and processes, 29-11
historical remarks, 29-2 to 29-3
key ontology languages, 29-6 to 29-10
OWL, 29-8 to 29-10
RDF and RDF schema, 29-6 to 29-8
ontologies, 29-4 to 29-6
Web Service Composite Application Framework (WS-CAF), 27-11, 31-8
Web Service Modeling Framework (WSMF), 31-6
Web Service name, 31-7
Web services, semantic aspects of, 31-1 to 31-17

epistemological ontologies for describing services, 31-11 to 31-14
DAML-S elements, 31-11 to 31-13
DAML-S and OWL-S, 31-11
limitations, 31-14
relevant frameworks, 31-5 to 31-10
frameworks comparison, 31-10
WS-CAF, 31-8 to 31-10
WSMF, 31-6 to 31-8
semantic Web, 31-2 to 31-3
semantic Web services, 31-3 to 31-5
Web services, understanding, 27-1 to 27-20
composition, 27-15 to 27-17
description, 27-11 to 27-14
framework for defining quality of service, 27-13
functional definition of Web services, 27-11 to 27-12
service discovery, 27-13 to 27-14
quality of service, 27-7 to 27-11
coordination, 27-9 to 27-11
reliability, 27-8 to 27-9
security, 27-7 to 27-8
service oriented computing, 27-3 to 27-4
transport and encoding, 27-5 to 27-6
understanding Web services stack, 27-4 to 27-5
Web Services Context (WS-CTX), 31-8, 31-9
Web Services Description Language (WSDL), 22-8, 24-7, 27-2, 27-4, 27-11
Web Services Flow Language (WSFL), 32-12
Web Services Policy Framework (WS-Policy), 27-13
Web Services Transaction Management (WS-TXM), 31-8
Website performance, improving, 40-1 to 40-21
improving performance at Website, 40-2 to 40-9
load balancing, 40-2 to 40-7
serving dynamic Web content, 40-8 to 40-9
server performance issues, 40-9 to 40-12
event-driven servers, 40-10 to 40-11
in-kernel servers, 40-11
process-based servers, 40-10
server performance comparison, 40-11 to 40-12
thread-based servers, 40-10
Web server workload characterization, 40-12 to 40-19
document popularity, 40-14 to 40-15
file sizes, 40-15 to 40-16
HTTP version, 40-18 to 40-19
request methods, 40-13 to 40-14
response codes, 40-14
transfer sizes, 40-16 to 40-18
Website usability engineering, 44-1 to 44-17
design exploration phase, 44-7 to 44-9
information design, 44-7
interaction design, 44-7 to 44-9
design refinement phase, 44-9 to 44-10
discovery phase, 44-5 to 44-7
assessing users' needs, 44-5 to 44-6
documenting users' needs, 44-6 to 44-7
maintenance phase, 44-12
moving toward usable WWW, 44-1 to 44-2
participatory Website design, 44-12 to 44-13
production phase, 44-10 to 44-12
automated assessment, 44-11 to 44-12
high-fidelity testing, 44-10 to 44-11
quality assurance phase, 44-12
research issues, 44-13 to 44-14
usability engineering process, 44-4 to 44-5
Website considerations, 44-2 to 44-4
computing devices, 44-3
implementation technology, 44-3
summary, 44-4
users, 44-2 to 44-3
Web structure, 53-1 to 53-15
community structures, 53-11 to 53-13
bipartite cores, 53-12 to 53-13
maximum flows, 53-11 to 53-12
link structure heuristics, 53-6 to 53-11
HITS, 53-8 to 53-10
PageRank, 53-6 to 53-8
problems, 53-10 to 53-11
topic-sensitive PageRank, 53-10
power-law distributions, 53-4 to 53-6
copying over, 53-5
local actions, 53-6
preferential attachment, 53-5
winners don't take all, 53-6
small-world networks, 53-2 to 53-4
properties, 53-3
Web as, 53-4
Web usage mining and personalization, 15-1 to 15-31
background, 15-1 to 15-3
data preparation and modeling, 15-3 to 15-12
data integration from multiple sources, 15-10 to 15-12
postprocessing of user transactions data, 15-8 to 15-10
sources and types of data, 15-5 to 15-6
usage data preparation, 15-6 to 15-8
outlook, 15-26 to 15-28
pattern discovery from Web usage data, 15-12 to 15-20
data-mining tasks for Web usage data, 15-13 to 15-20
levels and types of analysis, 15-12 to 15-13
using discovered patterns for personalization, 15-20 to 15-26
association rules, 15-22 to 15-23
clustering, 15-21 to 15-22
*k*NN-based approach, 15-20 to 15-21
sequential patterns, 15-23 to 15-26
Web Utilization Miner (WUM), 15-17
WfMS, *see* Workflow Management Systems
Whiteboard, 5-12, 5-24
Whois, 34-2
Whois^{++}, 34-2
Wide area networks, high-speed, 39-4
Wideband codecs, 6-9
Wi-Fi, 56-2, 56-14

Wild Tangent, 11-5
WIMP, *see* Windows, Icons, Menus, Pointing devices
Windows, 40-11
Windows, Icons, Menus, Pointing devices (WIMP), 37-7
Wired computing, 36-3
Wired-network computing, 36-3
Wireless Application Protocol (WAP), 36-4
Wireless LAN, 50-4, 50-12
Wireless Transaction Protocol (WTP), 36-4
Workflow
  analyzer, 32-10
  definition manager, 32-9
  designer, 32-9
  engine, 32-9
  variables, 32-7
Workflow Management Coalition, 32-5
Workflow Management Systems (WfMS), 32-4, 33-8
Workplace democracy, 33-6
Work in process, 8-7, 8-10, 8-14
Works of authorship, protected, 57-2
World-Wide Computer (WWC), 38-1, 38-4, 38-7
Worldwide computing middleware, 38-1 to 38-21
  further information, 38-18
  glossary, 38-18
  middleware, 38-1 to 38-4
    adaptability and reflection, 38-4
    asynchronous communication, 38-2 to 38-3
    higher-level services, 38-3
    virtual machines, 38-4
  related work, 38-13 to 38-17
    adaptive and reflective middleware, 38-17
    languages for distributed and mobile computation, 38-15 to 38-16
    naming middleware, 38-16
    remote communication and migration middleware, 38-16
    worldwide computing, 38-13 to 38-15
  research issues, 38-17
  worldwide computing, 38-4 to 38-13
    actor model, 38-4 to 38-5
    language and middleware infrastructure, 38-5 to 38-6
    middleware services, 38-7 to 38-8
    reflection, 38-11 to 38-13
    remote communication and mobility, 38-9 to 38-11
    universal actor model and implementation, 38-6
    universal naming, 38-8 to 38-9
World Wide Web Consortium (W3C), 8-3, 8-5, 23-2
Wrapper(s), 28-2, 28-3
  implementing of, 28-15
  induction, 14-14
  query optimization methods, 28-4
WS-CAF, *see* Web Service Composite Application Framework
WS-Coordination, 27-9
WS-CTX, *see* Web Services Context
WSDL, *see* Web Services Description Language
WSFL, *see* Web Services Flow Language
WS-Inspection, 27-5, 27-13
WSMF, *see* Web Service Modeling Framework
WS-Policy, *see* Web Services Policy Framework
WS-Reliable-Messaging, 27-8
WS-Security, 27-7
WS-Transaction, 27-9
WS-TXM, *see* Web Services Transaction Management
WTP, *see* Wireless Transaction Protocol
WUM, *see* Web Utilization Miner
WWC, *see* World-Wide Computer

X

XER, *see* XML Encoding Rules
XLink, *see* XML Linking Language
XMI, *see* XML Metadata Interchange
XML, *see* Extensible Markup Language
XML core technologies, 23-1 to 23-18
  core standards, 23-2 to 23-6
    XML, 23-2 to 23-5
    XML namespaces, 23-5 to 23-6
  data models, 23-6 to 23-11
    XML Application Programming Interface, 23-9 to 23-11
    XML Information Set, 23-6 to 23-7
    XML Path Language, 23-7 to 23-9
  schema languages, 23-11 to 23-17
    document schema definition languages, 23-15 to 23-17
    RELAX NG, 23-14 to 23-15
    XML Schema, 23-12 to 23-14
XML Encoding Rules (XER), 24-7
XML Information Set (XML Infoset), 23-6
XML Linking Language (XLink), 23-5, 24-2
XML Metadata Interchange (XMI), 22-10
XML Path Language (XPath), 23-7, 24-4
XML Query Language (XQuery), 24-6, 24-9
XML SQL Utility (XSU), 25-17
XML technologies, advanced, 24-1 to 24-10
  style sheet languages, 24-1 to 24-4
    Cascading Style Sheets, 24-2
    Extensible Stylesheet Language, 24-3 to 24-4
  XML and databases, 24-7 to 24-9
    native XML databases, 24-8 to 24-9
    XML and relational databases, 24-8
  XML processing, 24-4 to 24-7
    distributed programming, 24-7
    processing pipelines, 24-6 to 24-7
    programming with XML, 24-4
    transforming XML, 24-4 to 24-6
XMPP, *see* Extensible Messaging and Presence Protocol
XPath, *see* XML Path Language
XQuery, *see* XML Query Language
XSL, *see* Extensible Stylesheet Language

XSL-FO, *see* XSL Formatting Objects
XSL Formatting Objects (XSL-FO), 24-3
XSLT, *see* XSL Transformations
XSL Transformations (XSLT), 23-7, 23-9, 24-3, 24-4
XSU, *see* XML SQL Utility

Y

Yahoo!, 35-1
Yallcast, 51-6
Yeong, Wengyik, 34-6

Z

Zeus, 40-10
Zona, 11-10
Zones, game, 11-8
Zork, 11-4